ANALOG INTERFACE AND DSP SOURCEBOOK

ANALOG INTERFACE AND DSP SOURCEBOOK

ALAN CLEMENTS

McGRAW-HILL BOOK COMPANY

London · New York · St Louis · San Francisco · Auckland · Bogotá
Caracas · Hamburg · Lisbon · Madrid · Mexico · Milan · Montreal
New Delhi · Panama · Paris · San Juan · São Paulo · Singapore
Sydney · Tokyo · Toronto

Published by McGRAW-HILL Book Company Europe
Shoppenhangers Road, Maidenhead, Berkshire, SL6 2QL, England
Telephone 0628 23432 Fax 0628 770224

British Library Cataloguing in Publication Data
Clements, Alan
Analog Interface and DSP Sourcebook
I. Title
621.382

ISBN 0–07–707694–X

Library of Congress Cataloging-in-Publication Data
Clements, Alan,
The analog interface and DSP sourcebook / Alan Clements.
p. cm.
Includes bibliographical references and index.
ISBN 0–07–707694–X
1. Analog-to-digital converters. 2. Signal processing—Digital techniques.
I. Title.
TK7887.6.C56 1993
621.3815—dc20 92–36743
CIP

12345 CL 96543

Printed and bound in Great Britain by Clays Ltd, St. Ives, plc.

Contents

Chapter 2 TRANDUCERS AND REFERENCE SOURCES

Chapter 3 DIGITAL TO ANALOG CONVERTERS

Chapter 4 ANALOG TO DIGITAL CONVERTERS

Chapter 5 MULTIPLEXERS AND SAMPLE AND HOLD CIRCUITS

Chapter 6 VOLTAGE TO FREQUENCY AND FREQUENCY TO VOLTAGE CONVERTERS

Chapter 7 AN INTRODUCTION TO DIGITAL SIGNAL PROCESSING

Chapter 8 DIGITAL SIGNAL PROCESSORS

Chapter 9 DIGITAL SIGNAL PROCESSOR APPLICATIONS

Preface

The Analog Interface and DSP Sourcebook offers a selection of the best tutorial material written by semiconductor manufacturers on both the analog interface and digital signal processor (DSP). These two topics complement each other, as it is necessary first to capture analog signals before a DSP can operate on them. This book has been compiled because of the rapid growth of interest in DSP. Although scientists and engineers have been interested in DSP for some time, they have been unable to carry out real-time DSP because of the high cost of suitable hardware. Until recently, the so-called microprocessor revolution was limited to conventional CPUs, memory devices and peripherals. Today, many of the major semiconductor manufacturers have designed dedicated DSP devices that can be applied economically to a wide range of applications.

Engineers and academics employ two sources of information when they design either analog or digital systems: primary sources and secondary sources. Secondary sources are the textbooks written by academics and engineers. The advantage of using a secondary source of information is that the writer has done a lot of work for you. He or she has collated a large amount of primary information, digested it and summarized it in a readable form. The disadvantage is that the writer has decided what should and should not be included and has interpreted the source material in his or her own way (which may not always be correct).

Primary sources of information are produced by semiconductor manufacturers to describe and define their products. This book is composed of primary information sources and can be used as a reference when either designing or studying the analog interface and the DSP subsystem.

We have attempted to provide more than a catalogue of device types. This book encompasses a range of important topics. We cover both practical topics (e.g., designing with operational amplifiers) and theoretical topics (e.g., correlation and DSP). We look at some of the important design issues relevant to anyone involved in analog/digital systems design and include bread and butter topics often neglected in other texts. We have attempted to describe a representative range of devices currently available.

By taking some of the best papers, articles and application notes from leading semiconductor manufacturers, the Analog Interface and DSP Sourcebook is a comprehensive and authoritative source book with a strongly tutorial nature. It is not simply a design cookbook, as that would be too specific and would rapidly become out of date. Engineers need to know the underlying design principles of any circuit or system. Since the material is supplied by major component manufacturers, it is highly practical. Another use of this book is to give the student an idea of how components are actually used in practice.

The first part of the book includes three fundamental topics often neglected in basic texts on microprocessor systems design: the analog interface, the digital to analog converter, the analog to digital converter, and an introduction to analog systems.

In the second part we look at the DSP that is replacing the conventional microprocessor in high-performance signal processing systems. We begin with some tutorials that introduce some of the background material. Then we look at the details of four typical digital signal processors. Finally, we provide a series of tutorials on the applications of digital signal processors.

Acknowledgements

Data sheets, application notes and technical briefs have been reproduced in this book with the kind permission of the following companies:

Advanced Micro Devices (UK) Limited
Analog Devices Limited
Burr-Brown International Limited
Electronic Engineering
GEC Plessey Semiconductors
INMOS Limited
Linear Technology (UK) Limited
Maxim (UK) Limited
Motorola Limited
National Semiconductor (UK) Limited
SenSym Inc
SGS-Thomson Microelectronics Group
Texas Instruments Limited
TRW LSI Products Inc

Chapter 1 Operational amplifiers

Analog signals come from a variety of sources ranging from microphones to temperature sensors. However, many of these signals have a very low amplitude and must be amplified before they can be presented to an analog to digital converter for sampling. This first chapter examines the operational amplifiers that perform this function.

It is tempting to say that an operational amplifier is just a simple building block designed to amplify signals. While this is true to a first approximation, it is not the whole story. When dealing with low level signals, second-order or even third-order imperfections in the op-amp have to be taken into account – along with the effects of noise and even the circuit layout.

This chapter covers several topics from the world of the op-amp. We look at the basic op-amp itself and the high-precision instrumentation amplifier (which is itself constructed from op-amps). Another important type of analog interface is the isolation amplifier. This device performs the same basic function as the instrumentation amplifier, but is designed to prevent any d.c. path between the input signal and the output signal. This feature is particularly necessary in bio-medical applications such as patient monitoring.

We cover the subject of noise in some detail as it is one of the most important topics to the designer of analog interfaces. Associated with the problems of noise are the construction, layout and shielding of input circuits.

We often need to process a signal in a non-linear fashion. For example, the voltage from a temperature sensor may not be a linear function of temperature. Linearizing such a voltage can be done either after the signal has been digitized or before it as part of input processing. Consequently, we take a look at a simple circuit that provides an output voltage that is proportional to the logarithm of the input voltage.

Orientation
Operational Amplifiers

The amplifiers listed in this catalog are intended to provide cost-effective solutions to the bulk of op-amp requirements in precision measurement and control, as well as to more-general requirements in electronic circuits. The technical data included here* cover the properties of some 36 op-amp families, comprising about 100 distinct types. Some are general purpose, others provide near-optimum performance for specific classes of applications.

They differ in a variety of ways, for example, circuit technology, circuit architecture, input properties, output properties, operating temperature range and in terms of the many performance specifications.

BACKGROUND

The operational amplifier is today the most-widely used analog subassembly. It is safe to say that its *basic* properties and applications are sufficiently understood by most circuit designers and builders. However, the basis for choice, the subtleties of using op amps in circuits for best results (especially in precision measurement and control), and the varieties of possible applications are less clearly understood by op amp users, in varying degrees.

In these few pages, we shall address the question of making a proper choice of op amp type for an application, in relation to the extensive array of device properties presented in the data sheets that follow.

For those users requiring basic tutorial material, and detailed information on getting the most out of op amps, we have provided on page 4–16 a bibliography that should make available up to 99% of information needed now and then, with "fanout" to the vast body of literature that – with some redundancy – will provide the remainder. It should come as no surprise to successful users of Analog Devices op amps that a number of the references are to the applications sections of data sheets included in this catalog.

SELECTION PRINCIPLES

In selecting the right device for a specific application, you should have clearly in mind your design objectives and a firm understanding of what published specifications mean. Beyond this, you should detail the significant variables that are pertinent to your application. The purpose of this section is to put these many decision factors into perspective to help you make the most meaningful buying decisions.

To properly choose an operational amplifier for any given set of requirements, the designer must have:

1. *A complete definition of the design objectives.* Signal levels, accuracy desired, bandwidth requirements, circuit impedance, environmental conditions and other factors must be well defined before selection can be effectively undertaken.

2. *Firm understanding of what the manufacturer means by the numbers published for the parameters.* Frequently, any two manufacturers may have comparable published specifications, which may have been arrived at using differing measurement techniques. This creates a pitfall in op amp selection. To avoid these difficulties, the designer must know what the published specifications mean and how these parameters are measured and then must be able to translate these published specifications in terms meaningful to the design requirements.

There are three fundamental aspects to the rational selection of an operational amplifier for a given application: (1) establishing the circuit architecture, (2) defining the performance levels, and (3) choosing the amplifier(s).

1. To obtain a circuit building block to implement a defined functional job, the principal choices are either to purchase a committed functional device or to design a circuit employing op amps to perform the function. For example, to obtain a difference between two voltages, one may either purchase an instrumentation or isolation amplifier, or design a suitable subtraction circuit using op amps. If a committed functional building block, with appropriate specs and price, is not available, the circuit designer must start by developing schematic diagrams of circuits that will perform the function simply using "ideal" operational amplifiers. Many commonly used circuits can be found in textbooks, "cookbooks", and linear circuit books, as well as in application notes and data sheets.

2. Recognizing that the choice of an op amp depends on both the overall circuit requirements and the characteristics of available op amps, the designer should interpret the desired overall performance in terms of the parameters of op amps, and establish acceptable ranges of parameters, and their variation with time, temperature, supply voltage, etc. Examples of the key parameters are the input offset voltage, input bias and offset currents, and the high-frequency performance and transient behavior of the op-amp block (and its effect on the closed-loop circuit) for large and small signals. It will be helpful to develop an application checklist, which includes such considerations as the character of the input signals and their impedance, the output load, the desired accuracy – static and dynamic – and the environmental conditions.

3. The designer must then relate acceptable performance of the op-amp building block to the specifications and prices of available devices from preferred suppliers, bearing in mind a firm understanding of the way in which manufacturers define their specifications, and how definitions can differ in a way that may be misleading. A set of definitions used by Analog Devices follows the next section.

APPLICATION CHECKLIST

By way of an application checklist, the designer will need to account for the following:

Character of the application: The character of the application (inverter, follower, differential amplifier, etc.) will often influence the choice of amplifier.

*In addition to the products listed here, which are recommended for new designs, a number of older products are still available; data sheets are available upon request.

Accurate description of the input signal: It is extremely important that the input signal be thoroughly characterized. Is the input a voltage source or current source? Range of amplitude? Source impedance? Time/frequency characteristics?

Environmental conditions: What is the maximum range of temperature, time, and supply voltage over which the circuits must operate (to the required accuracy) without readjustment?

Accuracy desired: The accuracy requirement determines the extent to which the foregoing considerations are critical, and ultimately points the way to a device (or series of devices) which are acceptable. Accuracy must, of course, be defined in terms meaningful to the application with regard to bandwidth, DC offset, and other parameters.

SELECTION PROCESS

In general, the objective of amplifier selection should be to choose the least expensive device which will meet the physical, electrical, and environmental requirements imposed by the application. This suggests that a "General Purpose" amplifier will be the best choice in all applications where the desired performance requirements can be met. Where this is not possible, it is generally because of limitations encountered in two areas – bandwidth requirements, and/or offset and drift parameters.

To make it easier to relate bandwidth requirements with the drift and offset characteristics, a capsule view of bandwidth considerations precedes the DC discussions below. The reader is then returned to an expanded discussion of gain-bandwidth considerations.

Gain Bandwidth Considerations, A Capsule View

Although all selection criteria must be met simultaneously, determination of the bandwidth requirements is a logical starting point because:

A) If DC information is not of interest, a suitable blocking capacitor can be connected at the amplifier input and all of the "drift" specifications may usually be ignored, and

B) Where high frequency (>10MHz) characteristics are of primary importance, the choice will be limited to those amplifiers designated "Wide Bandwidth/Fast Settling."

Where DC information is required and where frequency requirements are relatively modest (full power response below 100kHz, unity gain of less than 1.5MHz) other criteria will probably influence the final choice. It is important, however, to choose an amplifier with which an adequate value of loop gain is assured (at the maximum frequency of interest) to obtain the desired accuracy. Loop gain is the excess of open loop gain over closed loop gain, and is responsible for the diminishing error due to fluctuations in the open loop gain due to time, temperature, etc. For example, if the closed-loop gain is 1000, the open-loop gain must be at least 100,000 to yield an error of no more than 1%, and 1,000,000 to yield an error no greater than 0.1%. Where undistorted response is required, the specifications for full linear response and slewing rate should be chosen such that they are not exceeded at the highest frequency of operation.

Offset and Drift Considerations

In the majority of op-amp applications, final selection is determined by the DC offset and drift characteristics. To undertake amplifier selection in these cases, it is necessary to translate the requirements listed above as follows. (It is assumed that bandwidth requirements and temperature range have been established at this point.)

1. *What input impedance must the circuit present to the signal source?* This depends primarily on the source impedance, R_s, and the amount of loading error which is acceptable. Most amplifier circuits are designed around either the inverting or noninverting circuit of Figure 1. The choice is often made between the two to accommodate the impedance requirement. Input impedance for the inverting circuit is approximately equal to the summing impedance, R_i and the upper limit on the magnitude of R_i is determined by the allowable drift error because of input bias current as discussed below. The noninverting circuit offers inherently higher input impedance than the inverting circuit (due to "bootstrapping" feedback) and in this case input impedance is approximately equal to the common mode impedance of the amplifier R_{cm}.

2. *How much drift error can be tolerated?* The question is related to the input signal level, e_s, and the required accuracy. For example, to amplify or otherwise manipulate a DC input signal of one volt with an accuracy of 0.1%, the offset drift error, V_d, must be one millivolt or less. (This assumes that other sources of error such as input loading, noise and gain error have already been allowed for.) By the same reasoning, the allowable drift error for a 1 volt signal and 0.01% accuracy would be 100μV.

When this has been defined, the allowable limits of offset voltage (e_{os}), bias current (i_b), and difference current can be calculated by the equations of Figure 1.

Figure 1 gives the equations which relate offset voltage (e_{os}), bias current (i_b), difference current (i_d) and the external circuit impedances to the drift error, V_d, for both the inverting and the noninverting circuits. From these equations it can be seen how the input impedance requirements of the foregoing paragraphs are related to the drift error.

For example, in the case of the inverting circuit, an offset error voltage, i_bR_i, is generated by the bias current flowing through the summing impedance. This error increases for increasing R_i. Since R_i also sets the input impedance, there is a conflict between high input impedance and low offset errors. Likewise, for a given offset error, higher values for R_i can be used with an amplifier which has lower bias current.

$$e_o = -\frac{R_f}{R_i}\left[e_s + e_{os}\left(\frac{R_f + R_i}{R_f}\right) + I_b R_i\right] \quad \text{For } R_C = 0 \text{ and } R_s << R_i$$

Signal Input Drift Error = V_d

$$e_o = -\frac{R_f}{R_i}\left[e_s + e_{os}\frac{R_f + R_i}{R_f} + i_d R_i\right] \quad \text{For } R_C = R_i R_f/(R_i + R_f) \text{ and } R_s << R_i$$

Signal Input Drift Error = V_d

Input Impedance $R_{IN} \approx R_i$

$$\% \text{ Drift Error} = \frac{100V_d}{e_s}$$

Figure 1A. Inverting Configuration

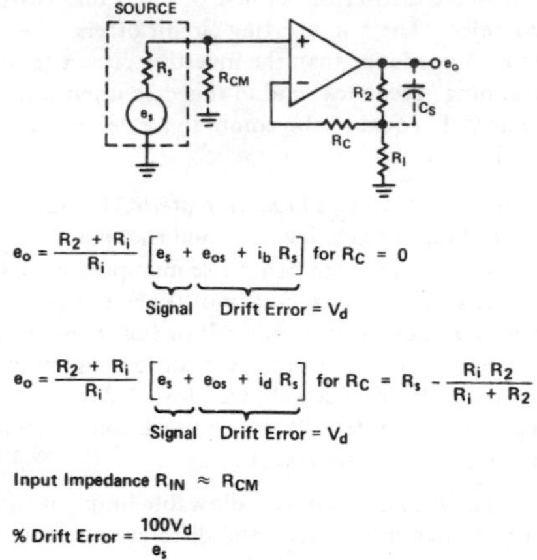

SOURCE
$e_o = \frac{R_2 + R_i}{R_i}\left[e_s + e_{os} + i_b R_s\right]$ for $R_C = 0$
Signal Drift Error = V_d
$e_o = \frac{R_2 + R_i}{R_i}\left[e_s + e_{os} + i_d R_s\right]$ for $R_C = R_s - \frac{R_i R_2}{R_i + R_2}$
Signal Drift Error = V_d
Input Impedance $R_{IN} \approx R_{CM}$
% Drift Error = $\frac{100V_d}{e_s}$

Figure 1B. Noninverting Configuration

Where it will otherwise function properly, the noninverting circuit generally makes a better choice for high input impedance circuits. Also, for the same source and input impedance requirement, a given amplifier will generate lower offset errors for the noninverting circuit than for the inverting circuit. This is so because the bias current flows only through R_s for the noninverter and this will always be less than the input impedance, R_i, of the inverter. Input impedance of the noninverter (approximately R_{CM}) is typically 10^7 ohms even for the least expensive bipolar amplifiers and up to 10^{11} ohms for FET types.

Unfortunately, however, the noninverting configuration cannot always be used since it is not convenient to use for many circuit functions such as integration or summation. A further limitation occurs in high accuracy applications, where common mode errors may rule out this circuit configuration.

Initial offsets can usually be zeroed at room temperature so that only the maximum temperature excursion (ΔT) from +25°C need be considered. For example, over the range of -25°C to +85°C, the maximum temperature excursion (ΔT) from +25°C would be 60°C. As a practical matter, offset errors due to supply voltage and time drift can generally be neglected since errors due to temperature drift are usually much greater.

Current Amplifier Considerations

Before leaving the subject of offset errors, we shall discuss briefly the current amplifier configuration which is shown in Figure 2A. The obvious approach to measuring current is to develop a voltage drop across a load resistor, R_f, and to measure this potential with a high impedance amplifier as shown in Figure 2B.

This approach has several disadvantages as compared to the circuit of Figure 2A. First the noninverting amplifier introduces common mode errors which do not occur for Figure 2A. Second, an ideal current meter would have zero impedance whereas, R_f in Figure 2B may become very large since this resistor determines the sensitivity of the measurement. Third, the changes of input impedance, R_{cm},

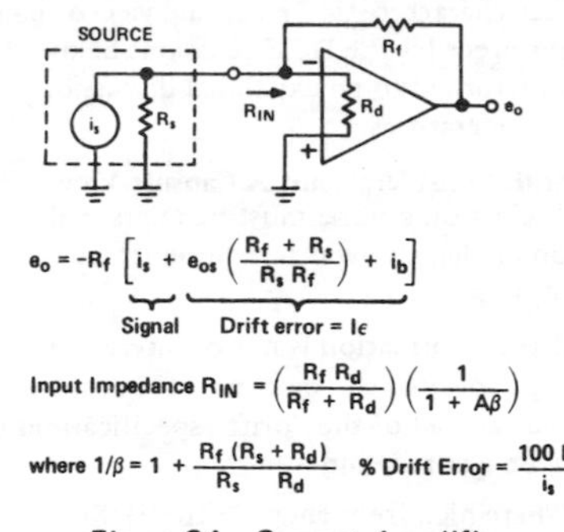

SOURCE
$e_o = -R_f\left[i_s + e_{os}\left(\frac{R_f + R_s}{R_s R_f}\right) + i_b\right]$
Signal Drift error = $I\epsilon$
Input Impedance $R_{IN} = \left(\frac{R_f R_d}{R_f + R_d}\right)\left(\frac{1}{1 + A\beta}\right)$
where $1/\beta = 1 + \frac{R_f (R_s + R_d)}{R_s R_d}$ % Drift Error = $\frac{100 I\epsilon}{i_s}$

Figure 2A. Current Amplifier

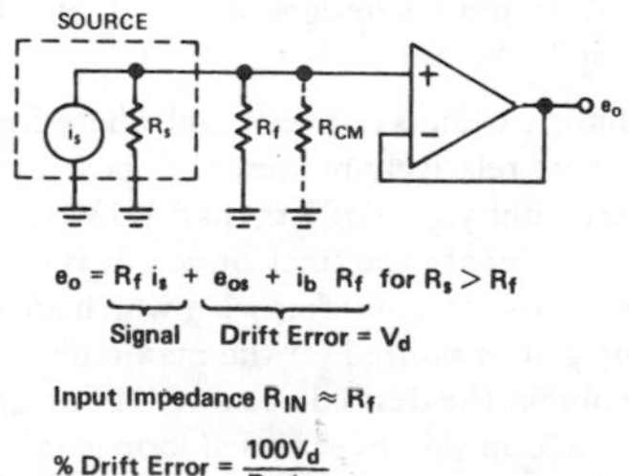

SOURCE
$e_o = R_f i_s + e_{os} + i_b R_f$ for $R_s > R_f$
Signal Drift Error = V_d
Input Impedance $R_{IN} \approx R_f$
% Drift Error = $\frac{100V_d}{R_f i_s}$

Figure 2B. Voltage Amplifier with Sampling Resistor

for the noninverting amplifier with temperature will cause variable loading on R_f and hence a change in sensitivity.

The current amplifier of Figure 2A circumvents all of these difficulties and approaches an ideal current meter; that is, there is essentially no voltage drop across the measuring circuit, since with enough open loop gain, A, the input impedance R_{IN} becomes very small.

In selecting a current amplifier, the most important consideration is current noise, and bias current drift. Measuring accuracy is largely the ratio of current noise and drift to signal current, i_s. To obtain the drift of error current I_ϵ referred to the input, use the following expression.

$$\Delta I_\epsilon = \left[\frac{\Delta e_{os}}{\Delta T}\left(\frac{R_f + R_s}{R_f R_s}\right) + \frac{\Delta i_B}{\Delta T}\right]\Delta T$$

Now, to make a proper selection you must pick an amplifier with an error current, I_ϵ, over the operating temperature which is small compared to the signal current, i_s. Do not overlook current noise which may be more important than current drift in many applications.

Gain Bandwidth Considerations, Expanded Discussion

From the previous discussion, it is apparent that most general purpose operational amplifiers will usually give adequate performance for the DC and audio frequency range applications. However, to obtain unity gain bandwidth above 2MHz, full power response above 20kHz and slewing rate above 6V/μsec, in general, requires special design techniques. All amplifiers with wideband, fast response characteristics have been listed in the wide bandwidth group to simplify the selection for higher frequency applications.

One factor often overlooked is that stray capacitance and impedance levels of the external feedback circuit can be the major limitation in high frequency applications. For example, in Figure 1A, if R_f were one megohm, and stray capacitance, C_S, were one picofarad then the closed loop bandwidth would be limited to 160kHz ($1/(2\pi R_F C_S)$) regardless of how fast the amplifier is. Moreover, output slewing rate will be limited by how fast C_S can be charged which in turn is related to signal level, e_s, and input impedance, R_i, by $de_o/dt = -e_s/R_iC_s$. For these reasons it is usually not possible to obtain both fast response and high input impedance for an inverting circuit since both R_1 and R_f must be large to obtain high input impedance.

Another advantage of the noninverting circuit (Figure 1B) is that input impedance, being determined by potentiometric feedback, does not depend on the impedance levels for R_1 and R_2. Therefore, a low impedance can be used for R_2 so that stray capacitance of C_s will not limit the circuit's bandwidth. In this case the minimum value for R_2 is constrained only by the output current rating of the amplifier. Again the trade-off between the frequency response and input impedance of the inverting and noninverting circuits must be evaluated in light of the common mode rejection error introduced by the noninverter.

For greater emphasis wideband applications can be separated into two categories – steady state and transient. Since the amplifier requirements for the two are somewhat different, these categories will be discussed separately.

A. Steady State Applications

Steady state applications involve amplifying or otherwise manipulating *continuous* sinusoidal, complex or random waveforms. In these applications the significant issues in choosing an amplifier are as follows:

1. *Is DC coupling required?* If DC information is of no consequence, then the offset drift errors are not usually important and a capacitor can be used if necessary to block the output DC offset. Your only concern here is that DC offset at the output does not become so large, as might be the case with a high gain stage, that the output is saturated or the dynamic swing for AC signals is limited. One way to circumvent the latter problem is to use feedback to limit the gain at DC as shown in Figure 3. The gain of these circuits can be small at DC but large at high frequencies.

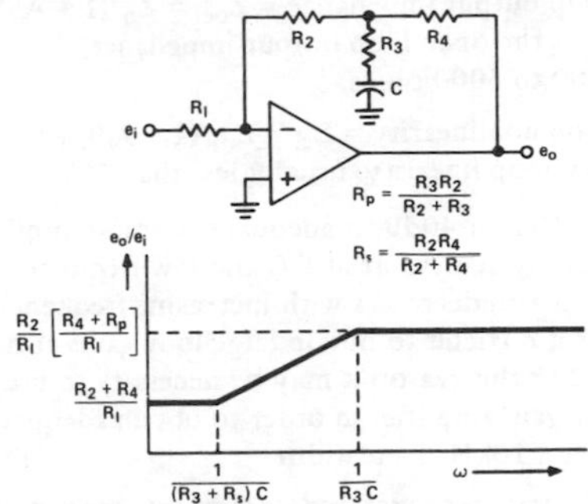

Figure 3. DC Feedback Minimizes Output Offset for AC Applications

2. *What closed loop gain and bandwidth are required?* Closed loop gain, G, is dictated by the application. To a first approximation the intersection of the open and closed loop gain curves in Figure 4 gives the closed loop bandwidth, f_{c1}(-3dB). For high gain, wideband requirements, it may be necessary, or more economical, to use two amplifiers in cascade each at lower gain.

3. *What loop gain is required or alternatively what gain stability, output impedance and/or linearity are necessary?* The available loop gain at a particular frequency or over a range of frequencies is very often more important than closed loop bandwidth in selecting an amplifier. Loop gain as illustrated in Figure 4, is defined as the difference, in dB, or as the ratio, arithmetically, of the open to closed loop gain ($A\beta$ = A/G). You will find in most of the equations defining the closed loop characteristic of a feedback amplifier that the loop gain ($A\beta$) is the determining factor in performance. Some of the more notable examples of this point are as follows:

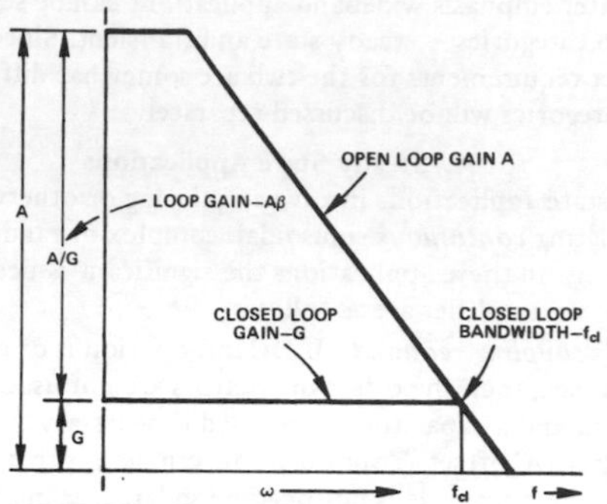

Figure 4. Closed Loop Bandwidth and Loop Gain

a. Closed loop gain stability = $\Delta G/G$
$\Delta G/G = (\Delta A/A)\ [1/(1 + A\beta)]$ where $\Delta A/A$ is the open loop gain stability, usually about 1%/°C.

b. Closed loop output impedance = $Z_{ocl} = Z_o/(1 + A\beta)$, where Z_o is the open loop output impedance, usually 200 to 5000 ohms.

c. Closed loop nonlinearity = $L_{cl} = L_{ol}/(1 + A\beta)$, where L_{ol} is the open loop linearity, usually less than 5%.

Loop gain of 100, or 40dB, is adequate for most applications and this is readily achievable at DC and low frequencies. But note that loop gain decreases with increasing frequency which makes it difficult to obtain large loop gains at high frequencies. For this reason it may be necessary to use a 10MHz unity gain amplifier in order to obtain adequate feedback over a 10kHz bandwidth.

4. *What full power response and/or slew rate are required?* You should examine your expected output waveform and select an amplifier whose slewing rate exceeds the maximum rate of change of output signal. For a sinusoidal waveform with a peak voltage output equal to the rated amplifier output the frequency should not exceed f_p, the full power response of the amplifier. As the output signal voltage is reduced below the rated output voltage, the usable maximum frequency can be extended proportionately. If you do not observe these restrictions you will get distortion and unexpected DC offsets at the output of the amplifier.

For some monolithic amplifier designs available today their frequency response is not a simple 6dB roll-off; the response may be shaped with external RC components for improved performance. Using feedforward or phase lag compensation networks, gain-bandwidth product and/or full power response may be shaped to meet varying design requirements. Most internally compensated op amps offer a stable 6dB per octave roll-off with specified unity gain-bandwidth and slew rate thereby limiting maximum speed and response to those published specifications.

B. Transient Applications

In applications such as A/D and D/A converters and pulse amplifiers, the *transient response* of the wideband amplifier is generally more important than the *gain bandwidth* characteristic described above. Slewing rate, overload recovery and settling time are the specifications which determine the transient response.

When applying the high frequency amplifier, it is important to understand how amplifier performance is affected by component selection as well as impedance levels used around the amplifier.

Settling Time

Settling time is defined as the time elapsed from the application of a perfect step input to the time when the amplifier output has entered and remained within a specified error band symmetrical about the final value (Figure 5). Settling time therefore includes the time required for the amplifier to slew from the initial value, recover from slew rate limited overload, and settle to a given error in the linear range.

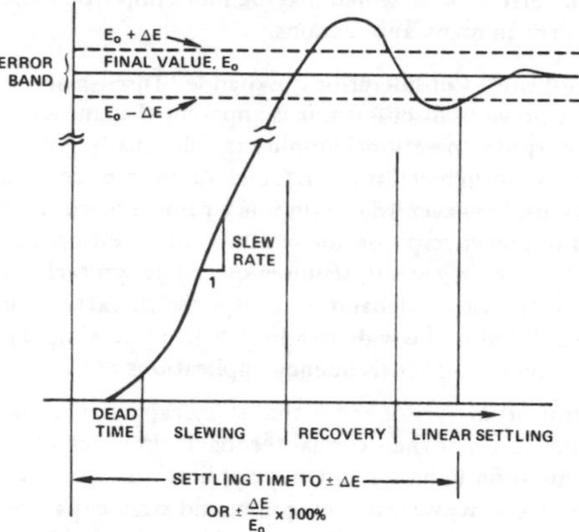

Figure 5. Typical Settling Time Characteristics

The time and frequency response of a linear, bilateral network or amplifier are related by well known mathematics. For example, the step response for a well behaved, ideally linear, 6dB/octave amplifier with a closed loop bandwidth of ω_{cl} is shown in Figure 6.

However, since settling time is determined by a combination of amplifier characteristics (both linear and nonlinear) and because it is a closed loop parameter, it cannot be readily predicted from the open loop specifications such as slew rate, small signal bandwidth, etc.

Analog Devices specifies settling time for the condition of unity gain, relatively low impedance levels, and no capacitive loading. A full-scale step input is used to determine settling time and the step is generally unipolar – i.e.: from zero to plus or minus full scale. The settling time indicated is generally the longest time resulting from a step of either polarity and is given as a percentage of the full scale step transition.

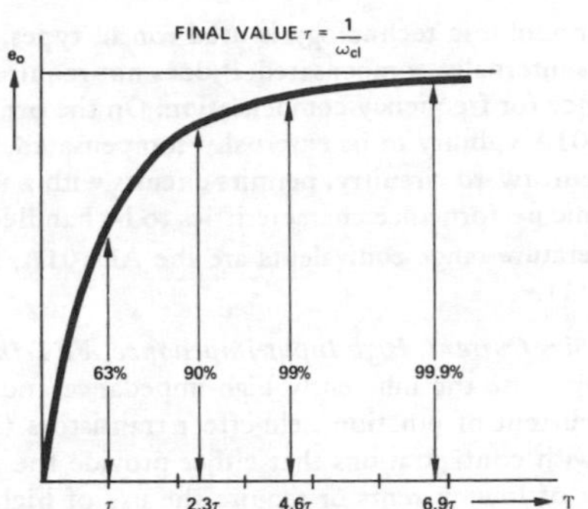

Figure 6. Step Response for Linear 6dB/Octave Amplifier

Settling time is a nonlinear function. It varies with the input signal level and it is greatly affected by impedances external to the amplifier.

ERRORS DUE TO NOISE

A major criterion in the selection of an amplifier for low level signals is the amplifier input noise, since this is usually the limiting factor on system resolution. In the general case, amplifier noise can be characterized by a voltage source in series with the summing junction and a current source in parallel with the summing junction. Whenever high source impedance is encountered, current noise flowing through the source impedance will appear as an additional voltage noise, combining with the amplifier voltage noise. The sum of these noise sources will then be amplified along with the desired signal. For this reason, selection of a particular amplifier must consider both the amplifier noise performance as well as the source impedance.

Consideration must also be given to noise sources other than the amplifier whenever determining total system noise. RF noise may be fed into an amplifier through any connecting wire, including power supply and output leads. Adequate shielding and low-pass filters on all incoming leads will usually prevent noise pick-up.

Thermal noise is generated in any conductor or resistor as a result of thermal agitation of the electrons. This noise voltage source, sometimes referred to as "Johnson Noise", is generated in the resistive component of any impedance and has a value:

$e_n = \sqrt{4KTBR}$

where e_n = the rms value of the noise voltage
K = Boltzman's Constant (1.38×10^{23} joules/°K)
T = absolute temperature of the resistance, °K
B = the bandwidth in which the noise is measured

Since noise is related to the bandwidth over which the measurement is made, no noise specification is meaningful unless the bandwidth for the specification is given. Although the Thermal Noise equation may appear unwieldy for practical noise calculations, all that is required to enable rapid approximations is to apply a few simple rules of thumb.

Rules of Thumb

(1) Remember that a 100kΩ resistor generates 40nV rms in a 1Hz bandwidth. The noise voltages generated by other values of resistances in other bandwidths can be calculated by remem bering that the noise is proportional to the square root of the resistance and the bandwidth; i.e.

$$e_n\ (\text{rms}) = (40\text{nV}/\sqrt{\text{Hz}})\left(\sqrt{\frac{R}{100\text{k}\Omega}\ (\text{BW})}\right)$$

(2) To convert the rms noise to a p-p value, a conversion factor of 6.6μV p-p/μV rms is applied for less than 0.1% probability of noise peaks exceeding calculated limits.

(3) The total rms noise contribution due to several noise sources is determined by the square root of the sum of the squares:

$$e_t = \sqrt{e_a^2 + e_b^2 + e_c^2 + \ldots\ e_n^2}$$

If any noise source is less than a third of another, it may be neglected. The resulting error will be approximately 5%.

(4) Restricting the bandwidth of a system to the minimum usable and using the lowest impedances possible are ways to reduce noise.

DESIGN EXAMPLE

Figure 7A illustrates a typical circuit with noise calculations shown for each noise source. The total of the noise sources is obtained by adding each of the individual sources in a RMS fashion.

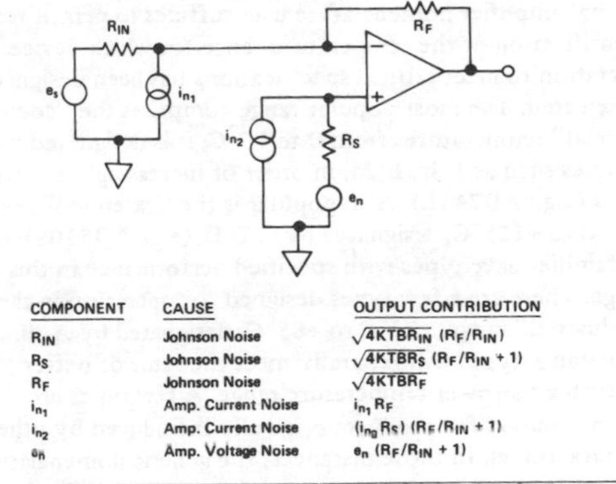

COMPONENT	CAUSE	OUTPUT CONTRIBUTION
R_{IN}	Johnson Noise	$\sqrt{4KTBR_{IN}}\ (R_F/R_{IN})$
R_S	Johnson Noise	$\sqrt{4KTBR_S}\ (R_F/R_{IN} + 1)$
R_F	Johnson Noise	$\sqrt{4KTBR_F}$
i_{n_1}	Amp. Current Noise	$i_{n_1} R_F$
i_{n_2}	Amp. Current Noise	$(i_{n_2} R_S)\ (R_F/R_{IN} + 1)$
e_n	Amp. Voltage Noise	$e_n\ (R_F/R_{IN} + 1)$

$$\text{TOTAL NOISE} = \sqrt{(e_{R_{IN}}\ G)^2 + [e_{R_S}\ (G+1)]^2 + e^2_{R_F} + (i_{n_1}\ R_F)^2 + [(i_{n_2}\ R_S)\ (G+1)]^2 + [e_n\ (G+1)]^2}$$

Figure 7A. Noise Components

Figure 7B illustrates how the Rules of Thumb may be applied in a practical case to approximate the total output noise. In this example, AD504, or a low noise type amplifier is being used with a 50kΩ source impedance. The two major noise sources, in addition to the AD504M input voltage noise of 0.6μV p-p, are the Johnson noise (58μV p-p) and current noise (2.5μV p-p).

GAIN = 100
BW = 0.01 TO 10Hz
R_S = 50kΩ
R_F = 10kΩ
R_{IN} = 100Ω

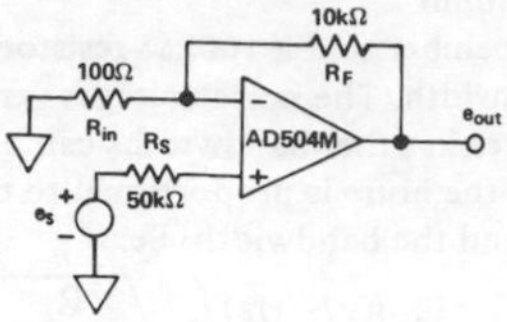

10kΩ R_F 100Ω R_{in} R_S 50kΩ AD504M e_s e_{out}

1) RESISTOR NOISE: $R_F \to 13nV/\sqrt{Hz}$
$R_{IN} \to (1.3nV/\sqrt{Hz})\ 100$
$R_S \to (28nV/\sqrt{Hz})\ 101 = 2.8\mu V/\sqrt{Hz}$

TOTAL RESISTOR NOISE IN 10Hz BW =
$(2.8\mu V/\sqrt{Hz})\ (\sqrt{10Hz})\ 6.6\mu V$ p-p/μV rms = 58μV p-p

2) AMPLIFIER CURRENT NOISE: (50pA p-p) (50k) (101) = 252μV
(50pA p-p) (10k) = 0.5μV

3) AMPLIFIER VOLTAGE NOISE: (0.6μV p-p) (101) = 60.6μV p-p

TOTAL OUTPUT NOISE = $\sqrt{(252)^2 + (60.6)^2 + (58)^2} \approx 265\mu V$ p-p

Figure 7B. Design Example

HOW THE OPERATIONAL AMPLIFIERS ARE CLASSIFIED

To assist the designer in distinguishing among the many types available from Analog Devices, we have provided a Selection Guide, in which amplifiers are grouped in terms of common properties which have been optimized in order to satisfy the needs of specific classes of applications. Once the choice has been narrowed to the manageable number of types in any group, distinctions can be drawn in terms of other requirements or considerations.

Temperature Range and Nomenclature. Analog Devices operational-amplifier nomenclature uses suffixes to permit ready identification of the temperature range for which device operation to meet critical specifications has been designed or selected. The most popular range comprises the "commercial" temperatures from 0 to 70°C; it is designated by suffixes such as J, K, L, M, in order of increasingly tighter specs (e.g., AD741L). Also popular is the "extended" range, -55°C to +125°C, designated by S, T, U, (e.g., AD510S); not all families have types with specified performance in this range. There are a few types designed for operation in the "industrial" range, -25°C to +85°C, designated by A, B. Wide-range types will generally meet the same or better specs in a narrower temperature range. A few types are second-sources for products originally introduced by other manufacturers. In those instances, the generic nomenclature is used (AD741C) or enlarged upon, if superior selections are offered (e.g., AD301AL).

1. *General-Purpose ICs.* Amplifiers in this group include our lowest-cost devices. They are best-suited for general purpose designs with moderate drift requirements, down to 5μV/°C max (AD301AL), and gain-bandwidth to 8MHz (AD301A). Typical applications include summing, inverting, impedance buffering (followers), and active filtering. They are also useful for developing nonlinear transfer functions, with appropriate external circuitry.

Bipolar monolithic technology is used for all types. The AD741 is internally compensated; it does not require external capacitance for frequency compensation. On the other hand, the AD301A's ability to be externally compensated, by either lag or feedforward circuitry, permits circuits with a wide range of dynamic performance characteristics to be handled. Extended-temperature-range equivalents are the AD101A, AD201A, and AD741.

4

2. *Low Bias-Current, High Input-Impedance, FET-Input ICs.* These types use the inherently high impedance and low leakage current of junction field-effect transistors (FET's) to deal with configurations that either provide the measurement of low currents or require the use of high-resistance circuitry.

Typical applications range from general-purpose high-impedance circuitry to integrators, current-to-voltage converters, and log-function generation, to measurements with high-impedance transducers, such as photomultipliers, flame detectors, pH cells, and radiation detectors.

The performance range is from the 75fA (75×10^{-15} A) maximum bias current of the AD515L electrometer to the 100pA max of the general purpose, lowest-cost AD611. The AD542 is a low-cost, laser-wafer-trimmed (LWT) monolithic implanted FET input amplifier with low offset and drift. The AD544 is similar, but has higher speed. Low bias current does not necessarily imply large voltage offsets; the AD515K combines a 150fA (0.15pA) max bias current with 1.0mV max offset and 15μV/°C max voltage drift; comparable figures for the AD547L are 25pA, 0.25mV and 1μV/°C.

The types of amplifiers in this group either are completely monolithic or employ matched FET's and a special bipolar amplifier chip designed to accommodate the input FET's electrically. In nearly all the IC's, thin-film resistors are deposited on the chip at critical circuit locations to ensure stability; low offsets and drift are achieved by laser-trimming of circuit balance. All FET-input op amps from Analog Devices are manufactured to meet their published bias-current specifications *after full warmup* (some manufacturers specify *initial* current, which is lower than warmed-up bias current). Our published max bias-current specification applies to either input (some manufacturers call "bias current" the *average* of the two input currents). Bias current of junction FET's approximately doubles for every 10°C increase of temperature.

3. *FET-Input Dual ICs.* The AD642, AD644, and AD647 are a single-chip pair of trimmed implanted-FET-input (TRIFET) op amps similar to the AD542, AD644, and AD547 with low warmed-up bias current (35pA max – K, L, S), low offset voltage (0.5mV max – L), low offset-voltage drift (2.5μV/°C max – L), and excellent V_{OS} matching (0.25mV max – L). Besides applications calling for more than one FET-input op amp at low cost per function, the AD647 is especially useful

in applications calling for matched duals, such as log-ratio amplifiers, FET-input instrumentation amplifiers, and buffering of differential signals. The AD644, a wideband version, was designed for fast DAC amplifiers, sample and hold, filters and wideband instrument amplifiers.

4. *Electrometers.* This class comprises the lowest bias-current devices, the AD515. The AD515L, with its 75fA input bias current, 1mV max offset, and 25μV/°C offset tempco, has differential inputs, and can be used in voltage measurements at high impedance, as a follower, or in current measurements, as an inverter, or even differentially.

5. *High-Accuracy Low-Drift Differential-Input ICs.* "Chopperless" low-drift designs with differential inputs, optimized for voltage offset and drift, dc open-loop gain, and CMR, should be considered for high-accuracy instrumentation, low-level transducer bridge circuits, precision voltage comparators, and for impedance buffer designs.

Performance of internally compensated premium amplifiers in this group ranges from the ADOP-07A's 25μV max offset voltage and 0.6μV/°C drift, and the AD517L's 50μV max offset voltage and 1.3μV/°C drift, combined with 1nA max bias current (1.5nA max over the temperature range), and CMR of 110dB min, to the low-cost AD741L's maximum offset of 0.5mV and max offset tempco of 5μV/°C, with 100nA max bias current over the temperature range, and CMR of 90dB min.

The ADOP-07 is a superior second source to other OP-07 families; for example, ADOP-07AH has minimum gain of 3×10^6 V/V compared to 3×10^5 V/V.

Among *uncompensated* op amps, the premium range is from the AD OP-27 with 25μV maximum offset voltage, 0.6μV/°C max drift, 40nA max bias current over the temperature range, and 114dB CMR, to the low-cost AD301AL, with max offset of 0.5mV, max drift of 5μV/°C, max bias current of 45nA over the temperature range, and minimum CMR of 90dB. For applications in which low noise is essential, the AD OP-27 has 100%-tested guaranteed maximum voltage noise of 0.18μV p-p, for the frequency range 0.1 to 10Hz, and maximum spot noise of 5.5 and 3.8V/$\sqrt{Hz}$ and 4.0 and 0.6pA/$\sqrt{Hz}$, at 10Hz, and 1000Hz, respectively.

The AD741J/K/L and the AD301AL are selected from production lots of the generic AD741 and AD101A types. The AD504, AD510, and AD517 are thermally balanced for low drift and high gain (independent of output loading), with inputs that are bootstrapped for high CMR and protected against overloads to prevent bias-current degradation due to reverse breakdown. Thin-film resistors, deposited on the chip, are another key to the stability of these amplifiers. The AD510 and the AD517 employ super-beta input transistors to achieve low bias current, and they are laser-trimmed at the wafer-probe stage (LWT) to achieve their excellent offset-voltage specifications at low cost. Since the bias currents are always of one polarity, they can be nulled at a given temperature with simple circuitry; and the change over the temperature range will be considerably less than for low-cost FET-input amplifiers having comparable specifications.

Extended-temperature-range equivalents are AD504S, AD510S, AD714S, and AD517S.

6. *Wide Bandwidth, Fast-Settling ICs.* High-speed op amps are characterized by high slewing rates, fast settling time, and wide bandwidth. Fast settling time is especially important in applications with rapidly changing or switched analog data, in buffers, d/a converters, and multiplexer circuitry; wide small-signal bandwidth is important in preamplification and in handling low-level wideband ac signals; high slewing rate is associated with fast settling time and is also important in handling ac signals having large magnitudes with minimal distortion, since the large-signal bandwidth is closely related to the slewing rate.

The products in this category with outstanding specifications are models HOS-050, AD3554 and AD380. Settling of the hybrid HOS-050 is to within 0.01% in 300ns in the inverting connection. Model AD3554 max slewing rate is 1000V/μs inverting, and small-signal unity-gain bandwidth is 70MHz; full-power bandwidth is 16MHz, min. In addition, all of these devices will deliver ±100mA of output current at ±10V, an important factor in video and line-driver circuitry, and in driving capacitive loads. For example, the current required to sustain 500V/μs in a 100pF load is $I = C\,dV/dt = 50mA$. AD380 is optimized for settling time: 250ns maximum to 0.01%, inverting or noninverting, with output of ±50mA at ±10V.

There are three families of monolithic ICs listed in this category, with slewing rates ranging from 25V/μs min to 100V/μs min. The AD509S is the fastest slewing (100V/μs min) and settling (500ns min to 0.1% and 2.5μs min to 0.01%). The AD507K is the best all-around performer, with small-signal bandwidth of 35MHz, slewing rate of 25V/μs min, and typical settling to 0.1% within 900ns, in addition to open-loop dc gain of 10^5 min, drift of 15μV/°C max, and bias current of 15nA max. The AD518J is the lowest in cost, yet it slews at 50V/μs min, and typically settles to within 0.1% in 800ns, with single-capacitor compensation.

Extended-temperature-range equivalents are models AD507S, AD509S and AD518S.

DEFINITIONS OF SPECIFICATIONS

Absolute Maximum Differential Voltage

Under most operating conditions, feedback maintains the error voltage between inputs to nearly zero volts. However, in some applications, such as voltage comparators, the voltage between the inputs can become large. This specification defines the maximum voltage which can be applied between inputs without causing permanent damage to the amplifier.

Common-Mode Rejection

An ideal operational amplifier responds only to the difference voltage between inputs ($e^+ - e^-$) and produces no output for a *common-mode voltage*, that is, when both inputs are at the same potential. However, due to slightly different gains between the plus and minus inputs, or variations in offset voltage as a function of common-mode level, common-mode input voltages are not eliminated at the output. If the output error voltage, due to a known magnitude of common-mode voltage, is referred to the input (dividing by the closed-loop gain), it reflects the equivalent *common-mode error voltage* (CME) between the inputs. Common-mode rejection ratio (CMRR) is defined as the ratio of common-mode voltage to the resulting common-mode error voltage. Common-mode rejection is often expressed logarithmically: CMR (in dB) = $20 \log_{10}$ (CMRR).

The precise specification of CMR is complicated by the fact that the common-mode voltage error can be a highly nonlinear function of common-mode voltage and also varies with temperature. As a consequence, CMR data published by Analog Devices are average figures, assuming an end-point measurement over the common-mode range specified. The incremental CMR about small values of common-mode voltage may be greater than the average CMR specified (on the other hand, the incremental CMR may be less in the neighborhood of large CMV). Published CMR specifications for op amps pertain to very low-frequency voltages, unless specified otherwise; CMR decreased with increasing frequency.

Common-Mode Voltage, Maximum

For differential-input amplifiers, the voltage at both inputs can swing about ground (power-supply common) level. *Common-mode voltage* is defined as any voltage (above or below ground) that could be observed at both inputs. The maximum common-mode voltage is defined as that voltage which will produce less than a specified value of common-mode error. This establishes the maximum input voltage for the voltage-follower connection.

Drift vs. Supply

Offset voltage, bias current, and difference current vary as supply voltage is varied. Usually, dc errors due to this effect are negligible compared to drift with temperature. No inference may be drawn from this low-frequency specification concerning the effects of rapid variation of voltage at the supply terminals.

Drift vs. Temperature

Offset voltage, bias current, and difference current all change, or "drift", from their initial values with temperature. This is by far the most important source of error in most precision applications. The temperature coefficients (tempcos) of those parameters are all defined as the average slope over a specified temperature range. Drift can be a nonlinear function of temperature (though it is often quite linear over limited temperature ranges); the slopes generally are greater at the extremes of temperature than around normal ambient (+25°C), which generally means that for small temperature excursions in the vicinity of +25°C, the specification is conservative.

Analog Devices precision operational amplifiers are specified by three- (or more-) point measurements, at 25°C and at the high and low extremes of the range (T_H, T_L), with the amplifier adjusted to zero at room temperature. The sum of the magnitudes of the drifts in the two ranges must be less than the specified drift rate (μV/°C or nA/°C) multiplied by the total temperature range (modified "butterfly"), or, in some cases, the magnitude of the drifts in both ranges must be less

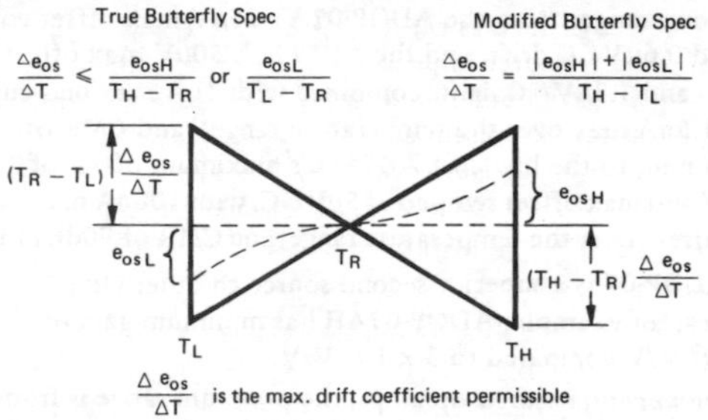

than the specified drift rate multiplied by the respective temperature ranges ("true butterfly").

The lowest-cost second-source IC amplifiers are specified only in terms of the maximum value of the parameter (e.g., offset voltage) over temperature in the specified range.

Drift vs. Time

Offset voltage, bias current, and difference current change with time as components age. It is important to realize that drift with time is random, and rarely – if ever – accumulates linearly for healthy devices. For example, voltage drift for a chopper-stabilized amplifier might be quoted at 1μV/day, whereas cumulative drift over 30 days might not exceed 5μV, or 15μV in a year (e.g., model 235). A convenient rule of thumb for extrapolation is to divide the drift for a stated interval by the square root of its ratio to any other interval of interest.

Full-Power Response

The large-signal and small-signal response characteristics of operational amplifiers differ substantially. An amplifier's output will not respond to large signal changes as fast as the small-signal bandwidth characteristics would predict, primarily because of slew-rate limiting in the output stages. Full-power response is specified in two ways: full linear response and full peak response. Full linear response is specified in terms of the maximum frequency, at unity closed-loop gain, for which a sinusoidal input signal will produce full output at rated load without exceeding a pre-determined distortion level. There is no industry-wide accepted value for the distortion level which determines the full-linear-response limitation, but we use 3% as a maximum acceptable limit for modules.

In many applications, the distortion caused by exceeding the full linear response can be comfortably ignored, but a more-serious effect (often overlooked) is an effect equivalent to a dc offset voltage that can be generated when full linear response is exceeded, due to rectification of the asymmetrical feedback waveform or overloading of the input stage by large distortion signals at the summing junction.

Another frequency response that is often if interest is the maximum frequency at which full output swing may be obtained, irrespective of distortion. This is termed "full peak response" and can often be found in a plot of output voltage swing vs. frequency.

Initial Bias Current
Bias current is defined as the current required at either input from an infinite source impedance to drive the output to zero (assuming zero common-mode voltage). For differential amplifiers, bias current is present at both the negative and the positive input. All Analog Devices specifications pertain to the *larger* of the two, *not the average.* For single-ended amplifiers (i.e., chopper types), bias current refers to the current at the input terminal.

Analog Devices specifies initial bias current, I_b, as the bias current at either input, specified at +25°C ambient with the input junctions *at normal operating temperature* (some manufacturers specify initial bias current at power turn-on. Such specifications may be misleading. For example, in FET-input amplifiers, bias current is doubled for each 10°C increase; since junction temperatures may warm up to 20°C or more above ambient, the "initial bias current" spec used by some manufacturers may be met only during a brief interval after the power is burned on, and I_b may be quadrupled under ordinary operation conditions.)

Initial Difference Current
Difference current is defined as the difference between the bias currents at the two inputs. The input circuitry of differential amplifiers is generally symmetrical, so that bias currents at both inputs tend to be equal and tend to track with changes in temperature and supply voltage. Therefore, difference current is often about 0.1 times the bias current at either input, assuming that initial bias current has not been compensated at the input terminals. For amplifiers in which bias currents track, it is often possible to reduce voltage errors due to bias current and its variations by the use of equal resistance loads at both inputs.

Input Impedance
Differential input impedance is defined as the impedance between the two input terminals at +25°C, assuming that the error voltage is nulled or very near zero volts. To a first approximation, dynamic impedance can be represented by a capacitor in parallel with a resistor.

Common-mode impedance, expressed as a resistance in parallel with a capacitance, is defined as the impedance between each input and power-supply common, specified at +25°C. For most circuits, common-mode impedance on the negative input has little significance, except for the capacitance which it adds at the summing junction (one exception is electrometer circuitry). However, common-mode impedance on the plus input sets the upper limit on closed-loop input impedance for the non-inverting configuration. Common-mode impedance is a nonlinear function of both temperature and common-mode voltage. For FET-input amplifiers, common-mode resistance is reduced by a factor of two for each 10° of temperature rise. As a function of common-mode voltage, the resistive component is defined as the average resistance for a common-mode change from zero to the maximum common-mode voltage. Incremental resistance may be less than the specified average value, especially at full-scale for some FET-input amplifiers.

Input Offset Voltage
Offset voltage is defined as the voltage required at the input from zero source impedance to drive the output to zero; its magnitude is measured by closing the loop (using low values of resistance) to establish a large fixed gain, measuring the amplified error at the output, and dividing the measured value by the gain.

The initial offset voltage is specified at +25°C and rated supply voltage. In most amplifiers, provisions are made to adjust initial offset to zero with an external trim potentiometer.

Input Noise
Input voltage- and current-noise characteristics can be specified and analyzed in much the same way as offset-voltage and bias-current characteristics. In fact, long-term drift can be considered as noise which occurs at very low frequencies. The primary difference is that, when evaluating noise performance, bandwidth must be considered. Also rms noise from different sources is summed by root-sum-of-squares, rather than linear, addition. Depending on the amplifier design, noise may have differing characteristics as a function of frequency, being dominated by "1/f noise", resistor noise, or junction noise, at various frequencies.

For this reason, several noise specifications are given. Low-frequency noise in the band 0.01 to 1Hz (or 0.1 to 10Hz) is specified as peak-to-peak, with a 3.3σ uncertainty, signifying that 99.9% of the observed peak-to-peak excursions will fall within the specified limits. Wideband noise is specified as rms. For some types, spectral-density plots or "spot noise", at specific frequencies, in $\mu V/\sqrt{Hz}$ or $pA/\sqrt{Hz}$, are provided.

Open-Loop Gain
Open-loop gain is defined as the ratio of a change of output voltage to the voltage applied between the amplifier inputs to produce the change. Gain is specified at dc. In many applications, the frequency dependence of gain is important; for this reason, the typical open-loop gain as a function of frequency is published for each amplifier type. See also *unity gain small-signal response.*

4

Overload Recovery
Overload recovery is defined as the time required for the output voltage to recover to the rated output voltage from a saturated condition caused by a 50% overdrive. Published specifications apply for low impedances and contain the assumption that overload recovery is not degraded by stray capacitance in the feedback network.

Rated Output
Rated output *voltage* is the minimum peak output voltage which can be obtained at rated current or a specified value of resistive load before clipping or out-of-spec nonlinearity occurs. Rated output *current* is the minimum guaranteed value of current supplied at the rated output voltage (or other specified voltage). Load impedances less than the specified (or implied) value can be used, but the maximum output voltage will decrease, distortion may increase, and the open-loop gain will be reduced. (All models are short-circuit protected to ground, and many are safe against shorts to the supplies.)

Settling Time
Settling time is defined as the time elapsed from the application of a perfect step input to the time when the amplifier output has entered and remained within a specified error band symmetrical about the final value. Settling time, therefore, includes the time required: for the signal to propagate through the amplifier, for the amplifier to slew from the initial value, recover from slew-rate-limited overload (if it occurs), and settle to a given error in the linear range. It may also include a "long tail" due to the time required to reach thermal equilibrium, or the settling time of compensation circuits. Settling time is usually specified for the condition of unity gain, relatively low impedance levels, and no (or a specified value of) capacitive loading, and any specified compensation. A full-scale unipolar step input is used, and both polarities are tested.

Although settling time can generally be grossly inferred from the other amplifier specifications (an amplifier that has extra-wide small-signal bandwidth, extra-fast slewing, and excellent full-power response may reasonably – but not always – be expected to have fast settling), the settling time cannot usually be rationally predicted from the other dynamic specifications.

Slewing Rate
The slewing rate of an amplifier, usually in volts per microsecond (V/μs), defines the maximum rate of change of output voltage for a large input step change.

Unity-Gain Small-Signal Response
Unity-gain small-signal response is the frequency at which the open-loop gain falls to 1V/V, or 0dB under a specified compensation condition. "Small signal" indicates that, in general, it is not possible to obtain large output voltage swing at high frequencies because of distortion due to slew-rate limiting or signal rectification. For amplifiers with symmetrical response for signals applied to either input, the dynamic behavior will be consistent for both inverting and non-inverting configurations. However, if feedforward compensation is used, fast response will be available only on the negative input, restricting fast applications of the device to the inverting mode.

A BRIEF BIBLIOGRAPHY ON OP AMPS

BOOKS (Not available from Analog Devices except where noted)

IC Op-Amp Cookbook by Walter Jung, Howard Sams & Co., Second Edition, 1980, down-to-earth and practical paperback

Linear Integrated Circuit Applications by George B. Clayton, The Macmillan Press Ltd., London, 1975

Modern Operational Circuit Design, by J. I. Smith, John Wiley & Sons, Inc., 1971

Nonlinear Circuits Handbook, edited by D. H. Sheingold. 1976. $5.95. Analog Devices, Box 796, Norwood, MA 02062

Operational Amplifiers and Linear IC's, by R. F. Coughlin and F. F. Driscoll, Prentice-Hall, Second Edition, 1982. Practical textbook

Operational Amplifiers, Theory and Practice, by J. K. Roberge, J. Wiley & Sons, 1975. Authoritative book on op amp principles and circuitry; contains extensive material on compensation to optimize dynamic performance

Transducer Interfacing Handbook, edited by D. H. Sheingold. 1980. $14.50. Analog Devices, Box 796, Norwood, MA 02062

ARTICLES AND APPLICATION NOTES (Available Upon Request; ask for specific issue of Analog Dialogue)

"Analog Signal Handling for High Speed and Accuracy" by A. P. Brokaw, ANALOG DIALOGUE 11-2

"Current Inverter with Wide Dynamic Range" by Barrie Gilbert, ANALOG DIALOGUE 9-1, 1975

"How to Select Operational Amplifiers", Application Note Section 20 of Volume I

"An IC-Amplifier User's Guide to Decoupling, Grounding, and Making Things Go Right for a Change," by A. P. Brokaw, Application Note Section 20 of Volume I

"Laser-Trimming on the Wafer, A Powerful New Tool for IC's" by R. Wagner, ANALOG DIALOGUE 9-3, 1975

"Simple Rules for Choosing Resistance Values in Adder-Subtractor Circuits" by D. Sheingold, ANALOG DIALOGUE 10-1, 1976

"Specifying and Measuring a Low-Noise FET-Input IC Op Amp" by Bill Maxwell, ANALOG DIALOGUE 8-2, 1974

"How to Test Operational Amplifier Parameters", Application Note Section 20 of Volume I

USEFUL TUTORIAL MATERIAL IN DATA SHEETS

Electrometer Circuitry, see AD515

High-Speed Amplifiers, see AD518 and Models 50/51

Low-Drift Differential Op Amp Performance, see AD504

Low-Level Applications of Chopper-Stabilized Amplifiers:
Inverting, see Models 234, 235
Non-Inverting, see Model 261

AN-241

Working with High Impedance Op Amps

National Semiconductor
Application Note 241

Robert J. Widlar
Apartado Postal 541
Puerto Vallarta, Jalisco
Mexico

Abstract. *New developments have dramatically reduced the error currents of IC op amps, especially at high temperatures. The basic techniques used to obtain this peformance are briefly described. Some of the problems associated with working at the high impedance levels that take advantage of these low error currents are discussed along with their solutions. The areas involved are printed-circuit board leakage, cable leakage and noise generation, semiconductor-switch leakages, large-value resistors and capacitor limitations.*

introduction

A new, low cost op amp reduces dc error terms to where the amplifier may no longer be the limiting factor in many practical circuits. FET bias currents are equalled at room temperature; but unlike FETs, the bias current is relatively stable even over a −55°C to 125°C temperature range. Offset voltage and drift are low because bipolar inputs and on-wafer trimming are used. The 100 μV offset voltage and 25 pA bias current are expected to advance the state of the art for high impedance sensors and signal conditioners.

bias currents

There has been a continual effort to reduce the bias current of IC op amps ever since the μA709 was introduced in 1965. The LM101A, announced in 1968, dropped this current by an order of magnitude through improved processing that gave better transistor current gain at low operating currents. In 1969, super-gain transistors (see appendix) were applied in the LM108 to beat FET performance when temperatures above 85°C were involved.

In 1974 FETs were integrated with bipolar devices to give the first FET op amp produced in volume, the LF155. These devices were faster than general purpose bipolar op amps and had lower bias current below 70°C. But FETs exhibit higher offset voltage and drift than bipolars. Long-term stability is also about an order of magnitude worse. Typically, this drift is 100 μV/year, but a small percentage could be as bad as 1 mV. Laser trimming and other process improvements have lowered initial offset but have not eliminated the drift problem.

The new IC is an extension of super-gain bipolar techniques. As can be seen from *Figure 1*, it provides low bias currents over a −55°C to 125°C temperature range. The offset current is so low as to be lost in the noise. This level of performance has previously been unavailable for either low-cost industrial designs or high reliability military/space applications.

This low bias current has not been obtained at the expense of offset voltage or drift. Typical offset voltage is under a millivolt and provision is made for on-water trimming to get it below 100 μV. The low drift exhibited in *Figure 2* indicates that the circuit is inherently balanced for exceptionally low drift, typically 1 μV/°C below 100°C.

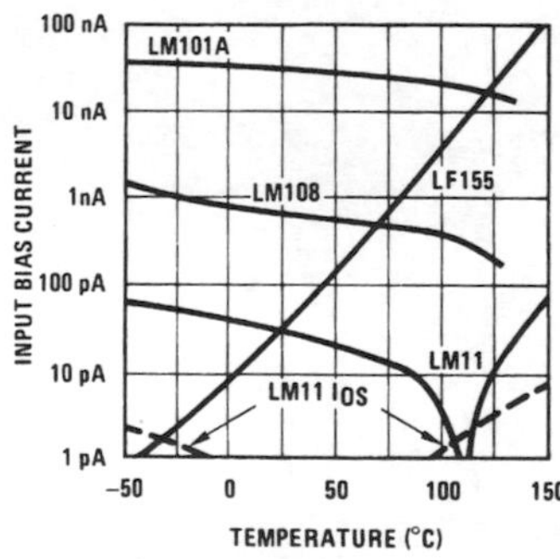

TL/H/7478–1

Figure 1. Comparison of typical bias currents for various types of IC op amps. New bipolar device not only has lower bias current over practical temperature ranges but also lower drift. Offset current is unusually low with the new design.

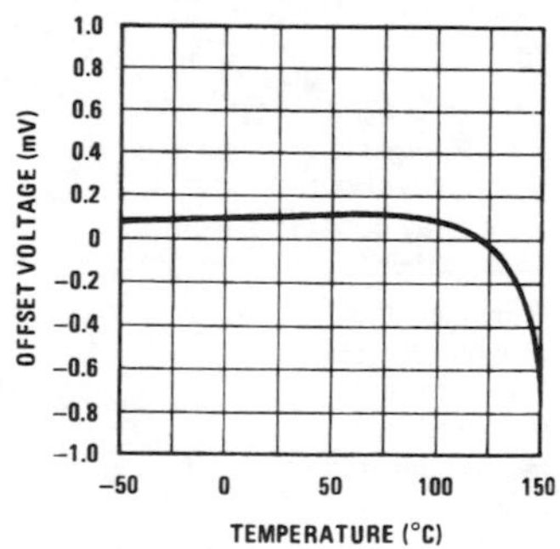

TL/H/7478–2

Figure 2. Bipolar transistors have inherently low offset voltage and drift. The low drift of the LM11 over a wide temperature range shows that there are no design problems degrading performance.

the new op amp

The LM11 is, in essence, a refinement of the LM108. A modified Darlington input stage has been added to reduce bias currents. With a standard Darlington, one transistor is biased with the base current of the other. This degrades dc amplifier performance because base current is noisy, subject to wide variation and generally unpredictable.

Supplying a bleed current greater than the base current, as shown in *Figure 3*, removes this objection. The 60 nA provided is considerably in excess of the 1 nA base current. The bleed current is made to vary as absolute temperature to maintain constant impedance at the emitters of Q1 and Q2. This stabilizes frequency response and also reduces the thermal variation of bias current. Parasitic capacitances of the current generator have been bootstrapped so that the 0.3 V/μs slew rate of the basic amplifier is unaffected.

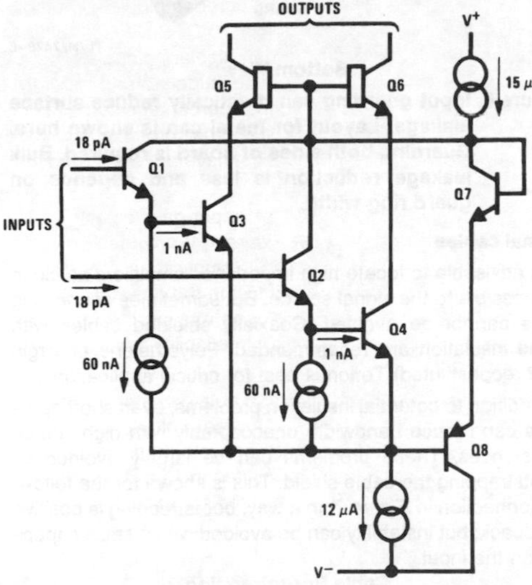

TL/H/7478–3

Figure 3. Modifying Darlington with bleed current reduces offset voltage, drift and noise. Unique circuitry provides well-controlled current with minimal stray capacitance so that speed of the basic amplifier is unaffected.

Results to date suggest that the base currents of this modified Darlington input are better matched than the simple differential amplifier. In fact, offset current is so low as to be unmeasurable on production test systems. Therefore, guaranteed limits are determined by the test equipment rather than the IC.

noise

Operating transistors at very low currents does increase noise. Thus, the LM11 is about a factor of four noisier than the LM108. But the low frequency noise, plotted in *Figure 4*, is still slightly less than that of FET amplifiers. Long-term measurements indicate that the offset voltage shift is under 10 μV.

In contrast to the noise voltage, low frequency noise current is subject to greater unit-to-unit variation. Generally, it is below 1 pA, peak-to-peak, about the same magnitude as the offset current.

With the LM11, both voltage and current related dc errors have been reduced to the point where overall circuit performance could well be noise limited, particularly in limited temperature range applications.

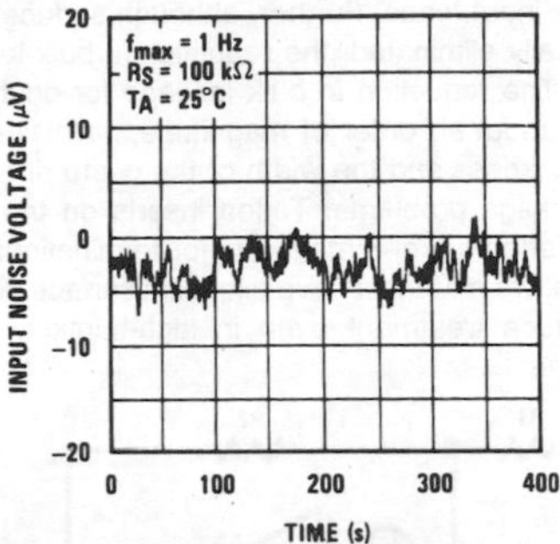

TL/H/7478–4

Figure 4. Lower operating currents increase noise, but low frequency noise is still slightly lower than IC FET amplifiers. Long-term stability is much improved.

reliability

The reliability of the LM11 is not expected to be substantially different than the LM108, which has been used extensively in military and space applications. The only significant difference is the input stage. The low current nodes introduced here might possibly be a problem were they not bootstrapped, biased and guarded to be virtually unaffected by both bulk and surface leakages. This opinion is substantiated by preliminary life-test data.

This IC could, in fact, be expected to improve reliability when used to replace discrete or hybrid amplifiers that use selected components and have been trimmed and tweaked to give the required performance.

From an equipment standpoint, reliability analysis of insulating materials, surface contamination, cleaning procedures, surface coating and potting are at least as important as the IC and other components. These factors become more important as impedance levels are raised. But this should not discourage designers. If poor insulation and contamination cause a problem when impedance levels are raised by an order of magnitude, it is best found out and fixed.

Even so, it may not be advisable to take advantage of the full potential of the LM11 in all cases, especially when hostile environments are involved. For example, there should be no great difficulty in finding an LM11 with offset current less than 5 pA over a −55°C to 125°C temperature range. But anyone designing high-reliability equipment that is going to be in trouble if combined leakages are greater than 10 pA at 125°C had best know what he is about.

electrical guarding

The effects of board leakage can be minimized using an old trick known as guarding. Here the input circuitry is surrounded by a conductive trace that is connected to a low impedance point at the same potential as the inputs. The electrical connection of the guard for the basic op amp configurations is shown in *Figure 5*. The guard absorbs the leakage from other points on the board, drastically reducing that reaching the input circuitry.

To be completely effective, there should be a guard ring on both sides of the printed-circuit board. It is still recommended for single-sided boards, but what happens on the unguarded side is difficult to analyze unless Teflon inserts are used on the input leads. Further, although surface leakage can be virtually eliminated, the reduction in bulk leakage is much less. The reduction in bulk leakage for double-sided guarding is about an order of magnitude, but this depends on board thickness and the width of the guard ring. If there are bulk leakage problems, Teflon inserts on the through holes and Teflon or kel-F standoffs for terminations can be used. These two materials have excellent surface properties without surface treatment even in high-humidity environments.

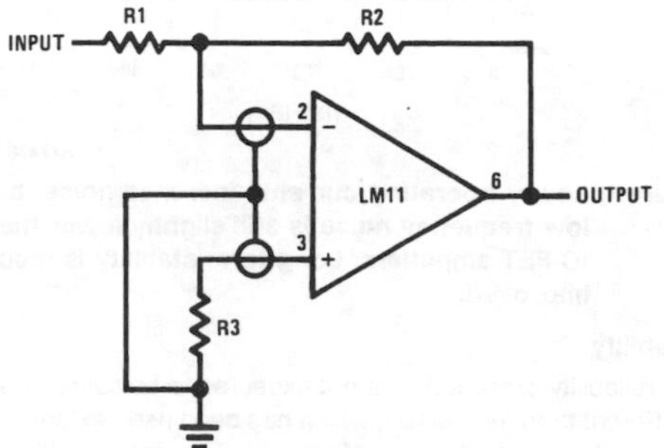

a. inverting amplifier

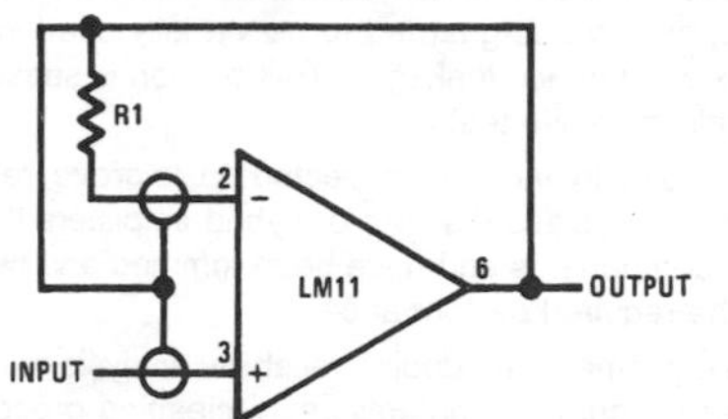

b. follower

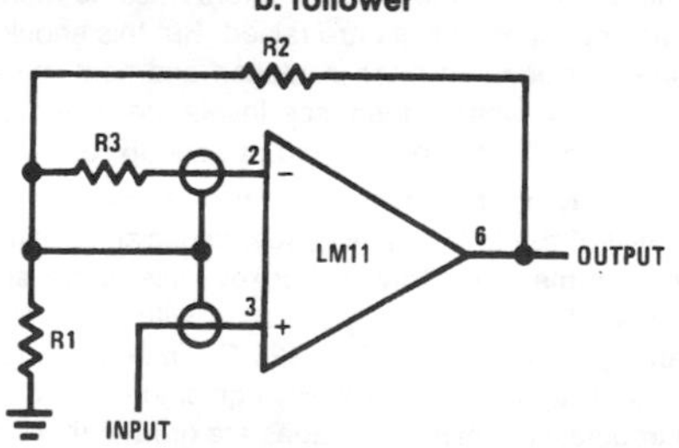

c. non-inverting amplifier

Figure 5. Input guarding for various op amp connections. The guard should be connected to a point at the same potential as the inputs with a low enough impedance to absorb board leakage without introducing excessive offset.

An example of a guarded layout for the metal-can package is shown in *Figure 6*. Ceramic and plastic dual-in-line packages are available for critical applications with guard pins adjacent to the inputs both to facilitate board layout and to reduce package leakage. These guard pins are not internally connected.

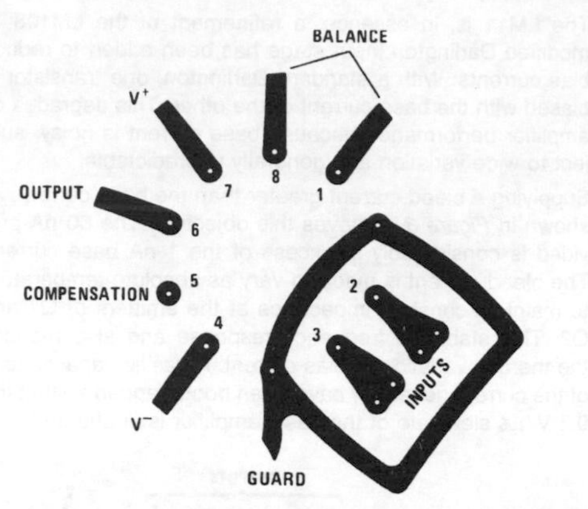

Bottom View

Figure 6. Input guarding can drastically reduce surface leakage. Layout for metal can is shown here. Guarding both sides of board is required. Bulk leakage reduction is less and depends on guard ring width.

signal cables

It is advisable to locate high impedance amplifiers as close as possible to the signal source. But sometimes connecting lines cannot be avoided. Coaxially shielded cables with good insulation are recommended. Polyethelene or virgin (not reconstituted) Teflon is best for critical applications.

In addition to potential insulation problems, even short cable runs can reduce bandwidth unacceptably with high source resistances. These problems can be largely avoided by bootstrapping the cable shield. This is shown for the follower connection in *Figure 7*. In a way, bootstrapping is positive feedback; but instability can be avoided with a small capacitor on the input.

Cable Bootstrapping

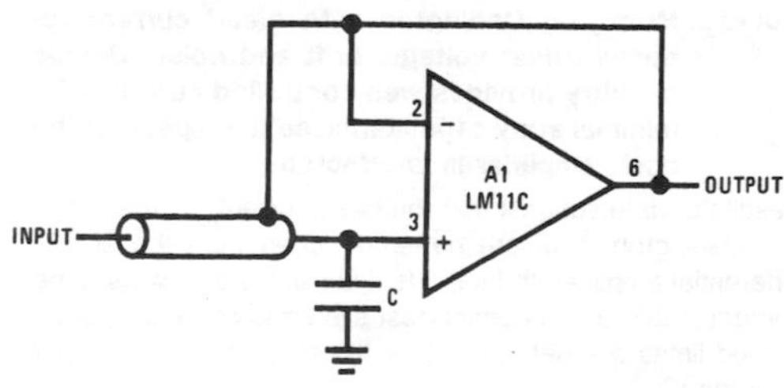

Figure 7. Bootstrapping input shield for a follower reduces cable capacitance, leakage and spurious voltages from cable flexing. Instability can be avoided with small capacitor on input.

With the summing amplifier, the cable shield is simply grounded, with the summing node at virtual ground. A small feedback capacitor may be required to insure stability with the added cable capacitance. This is shown in *Figure 8*.

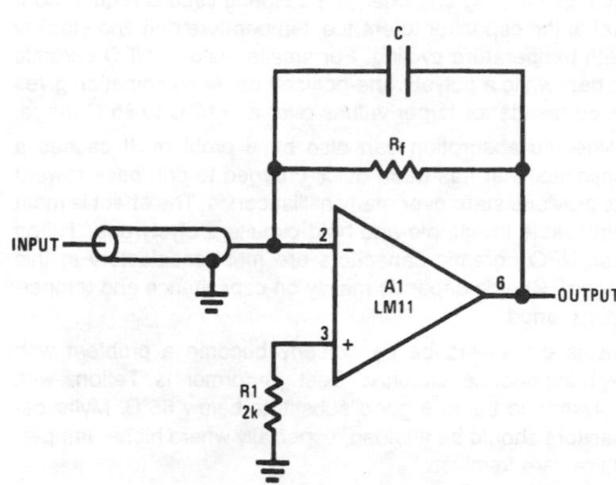

Figure 8. With summing amplifier, summing node is at virtual ground so input shield is best grounded. Small feedback capacitor insures stability.

An inverting amplifier with gain may require a separate follower to drive the cable shield if the influence of the capacitance, between shield and ground, on the feedback network cannot be accounted for.

High impedance circuits are also prone to mechanical noise (microphonics) generated by variable stray capacitances. A capacitance variation will generate a noise voltage given by

$$e_n = \frac{\Delta C}{C} V,$$

where V is the dc bias on the capacitor. Therefore, the wiring and components connected to sensitive nodes should be mechanically rigid.

This is also a problem with flexible cables, in that bending the cable can cause a capacitance change. Bootstrapping the shield nearly eliminates dc bias on the cable, minimizing the voltage generated. Another problem is electrostatic charge created by friction. Graphite lubricated Teflon cable will reduce this.

switch leakage

Semiconductor switches with leakage currents as low as the bias current of the LM11 are not generally available when operation much above 50°C is involved. The sample-and-hold circuit in *Figure 9* shows a way around this problem. It is arranged so that switch leakage does not reach the storage capacitor.

Isolating leakage current requires that two switches be connected in series. The leakage of the first, Q1, is absorbed by R1 so that the second, Q2, only has the offset voltage of the op amp across its junctions. This can be expected to reduce leakage by at least two orders of magnitude. Adjusting the op amp offset to zero at the maximum operating temperature will give the ultimate leakage reduction, but this is not usually required with the LM11.

MOS switches with gate-protection diodes are preferred in production situations as they are less sensitive to damage from static charges in handling. If used, D1 and R2 should be included to remove bias from the protection diode during hold. This may not be required in all cases but is advised since leakage from the protection diode depends on the internal geometry of the switch, something the designer does not normally control.

A junction FET could be used for Q1 but not Q2 because there is no equivalent to the enhancement mode MOSFET. The gate of a JFET must be reverse biased to turn it off, and leakage on its output cannot be avoided.

high-value resistors

Using op amps at very high impedance levels can require unusually large resistor values. Standard precision resistors are available up to 10 MΩ. Resistors up to 1 GΩ can be obtained at a significant cost premium. Larger values are quite expensive, physically large and require careful handling to avoid contamination. Accuracy is also a problem. There are techniques for raising effective resistor values in op amp circuits. In theory, performance is degraded; in practice, this may not be the case.

With a buffer amplifier, it is sometimes desirable to put a resistor to ground on the input to keep the output under control when the signal source is disconnected. Otherwise it will saturate. Since this resistor should not load the source, very large values can be required in high-impedance circuits.

Figure 10 shows a voltage follower with a 1 GΩ input resistance built using standard resistor values. With the input disconnected, the input offset voltage is multiplied by the same factor as R2; but the added error is small because the offset voltage of the LM11 is so low. When the input is connected to a source less than 1 GΩ, this error is reduced. For an ac-coupled input, a second 10 MΩ resistor could be connected in series with the inverting input to virtually eliminate bias current error; bypassing it would give minimal noise.

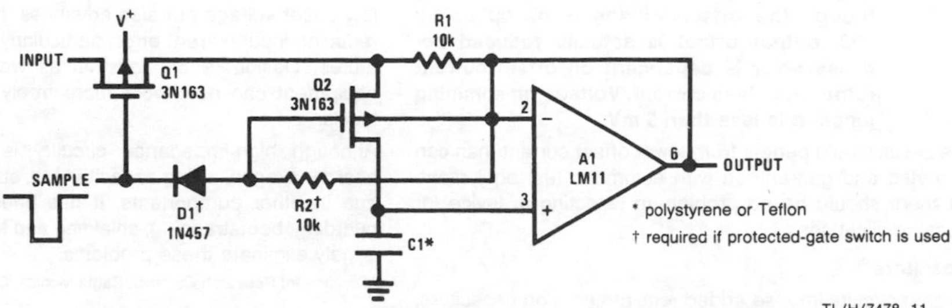

Figure 9. Switch leakage in this sample and hold does not reach storage capacitor. If Q2 has an internal gate-protection diode, D1 and R2 must be included to remove bias from its junction during hold.

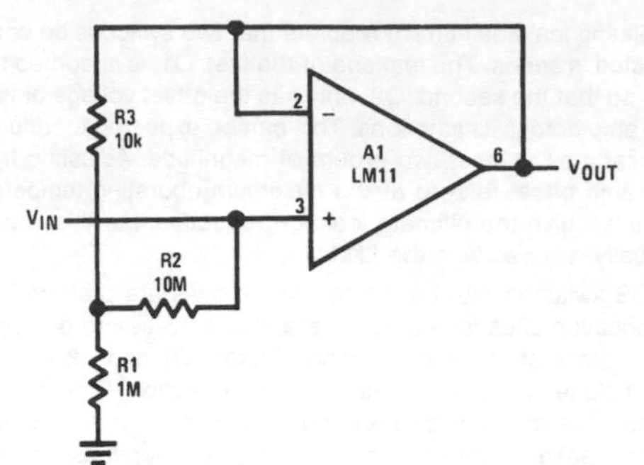

TL/H/7478-12

Figure 10. Follower input resistance is 1 GΩ. With the input open, offset voltage is multiplied by 100, but the added error is not great because the op amp offset is low.

The voltage-to-current converter in *Figure 11* uses a similar method to obtain the equivalent of a 10 GΩ feedback resistor. Output offset is reduced because the error can be made dependent on offset current rather than bias current. This would not be practical with large value resistors because of cost, particularly for matched resistors, and because the summing node would be offset several hundred millivolts from ground. In *Figure 11*, this offset is limited to several millivolts. In addition, the output can be nulled with the usual balance potentiometer. Further, gain trimming is easily done.

Resistance Multiplication

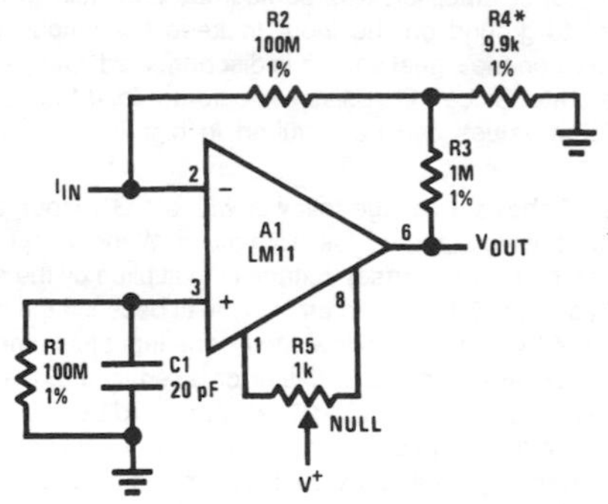

TL/H/7478-13

Figure 11. Equivalent feedback resistance is 10 GΩ, but only standard resistors are used. Even though the offset voltage is multiplied by 100, output offset is actually reduced because error is dependent on offset current rather than bias current. Voltage on summing junction is less than 5 mV.

This circuit would benefit from lower offset current than can be tested and guaranteed with automatic test equipment. But there should be no problem in selecting a device for critical applications.

capacitors

Op amp circuits impose added requirements on capacitors, and this is compounded with high-impedance circuitry. Frequency shaping and charge measuring circuits require control of the capacitor tolerance, temperature drift and stability with temperature cycling. For smaller values, NPO ceramic is best while a polystyrene-polycarbonate combination gives good results for larger values over a −10°C to 85°C range.

Dielectric absorption can also be a problem. It causes a capacitor that has been quick-charged to drift back toward its previous state over many milliseconds. The effect is most noticeable in sample-and-hold circuits. Polystyrene, Teflon and NPO ceramic capacitors are most satisfactory in this regard. Choice depends mainly on capacitance and temperature range.

Insulation resistance can clearly become a problem with high-impedance circuitry. Best performer is Teflon, with polystyrene being a good substitute below 85°C. Mylar capacitors should be avoided, especially where higher temperatures are involved.

Temperature changes can also alter the terminal voltage of a capacitor. Because thermal time constants are long, this is only a problem when holding intervals are several minutes or so. The effect is reported to be as high as 10 mV/°C, but Teflon capacitors that hold it to 0.5 mV/°C are available*.

An op amp with lower bias current can ease capacitor problems, primarily by reducing size. This is obvious with a sample-and-hold because the capacitor value is determined by the hold interval and the amplifier bias current. The circuit in *Figure 12* is another example. An RC time constant of more than a quarter hour is obtained with standard component values. Even when such long time constants are not required, reducing capacitor size to where NPO ceramics can be used is a great aid in precision work.

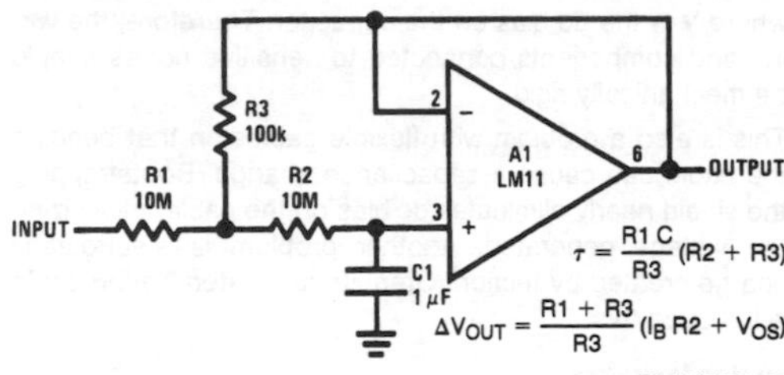

TL/H/7478-14

Figure 12. This circuit multiplies RC time constant to 1000 seconds and provides low output impedance. Cost is lowered because of reduced resistor and capacitor values.

conclusions

A low cost IC op amp has been described that not only has low offset voltage but also advances the state of the art in reducing input current error, particularly at elevated temperatures. Designers of industrial as well as military/space equipment can now work more freely at high impedance levels.

Although high-impedance circuitry is more sensitive to board leakages, wiring capacitances, stray pick-up and leakage in other components, it has been shown how input guarding, bootstrapping, shielding and leakage isolation can largely eliminate these problems.

*Component Research Co., Inc., Santa Monica, California.

acknowledgment

The author would like to acknowledge the assistance of the staff at National Semiconductor in implementing this design and sorting out the application problems. Discussions with Bob Dobkin, Bob Pease, Carl Nelson and Mineo Yamatake have been most helpful.

appendix

super-gain techniques

Super-gain transistors are not new, having been developed for the LM102/LM110 voltage followers in 1967 and later used on the LM108 general-purpose op amp. They are similar to regular transistors, except that they are diffused for high current gains (2,000–10,000) at the expense of breakdown voltage. A curve-tracer display of a typical device is shown in *Figure A1*. In an IC, super-gain transistors can be made simultaneously with standard transistors by including a second, light base predeposition that is diffused less deeply.

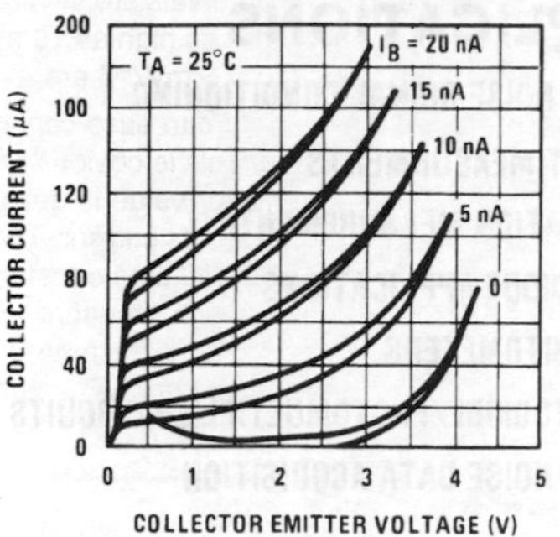

TL/H/7478–15

Figure A1. curve tracer display of a super-gain transistor

Super-gain transistors can be connected in cascode with regular transistors to form a composite device with both high gain and high breakdown. The simplified schematic of the LM108 input stage in *Figure A2* shows how it is done. A common base pair, Q3 and Q4, is bootstrapped to the input transistors, Q1 and Q2, so that the latter are operated at nearly zero collector-base voltage, no matter what the input common-mode. The regular NPN transistors are distinguished by drawing them with wider base regions.

Operating the input transistors at very low collector-base voltage has the added advantage of drastically reducing collector-base leakage. In this configuration bipolar transistors are affected little by the leakage currents that limit performance of FET amplifiers.

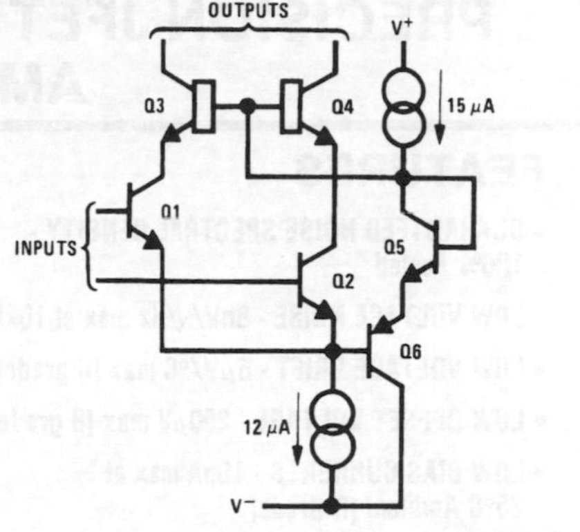

TL/H/7478–16

Figure A2. A bootstrapped input stage

OPA101
OPA102

Low Noise - Wideband PRECISION JFET INPUT OPERATIONAL AMPLIFIER

FEATURES

- GUARANTEED NOISE SPECTRAL DENSITY - 100% Tested
- LOW VOLTAGE NOISE - 8nV/$\sqrt{Hz}$ max at 10kHz
- LOW VOLTAGE DRIFT - 5μV/°C max (B grade)
- LOW OFFSET VOLTAGE - 250μV max (B grade)
- LOW BIAS CURRENTS - 10pA max at 25°C Ambient (B Grade)
- HIGH SPEED - 10V/μsec min (OPA102)
- GAIN BANDWIDTH PRODUCT - 40MHz (OPA102)

APPLICATIONS

- LOW NOISE SIGNAL CONDITIONING
- LIGHT MEASURMENTS
- RADIATION MEASUREMENTS
- PIN DIODE APPLICATIONS
- DENSITOMETERS
- PHOTODIODE/PHOTOMULTIPLIER CIRCUITS
- LOW NOISE DATA ACQUISITION

DESCRIPTION

The OPA101 and OPA102 are the first FET operational amplifiers available with noise characteristics (voltage spectral density) guaranteed and 100% tested.

The amplifiers have a complementary set of specifications permitting low errors in signal conditioning applications; low noise, low bias current, high open-loop gain, high common-mode rejection, low offset voltage, low offset voltage drift, etc.

In addition, the amplifiers have moderately high speed. The OPA101 is compensated for unity gain stability and has a slew rate of 5V/μsec, min. The OPA102 is compensated for gains of 3V/V and above and has a slew rate of 10V/μsec, min.

Each unit is laser-trimmed for low offset voltage and low offset voltage drift versus temperature. Bias currents are specified with the units fully warmed up at +25°C ambient temperature.

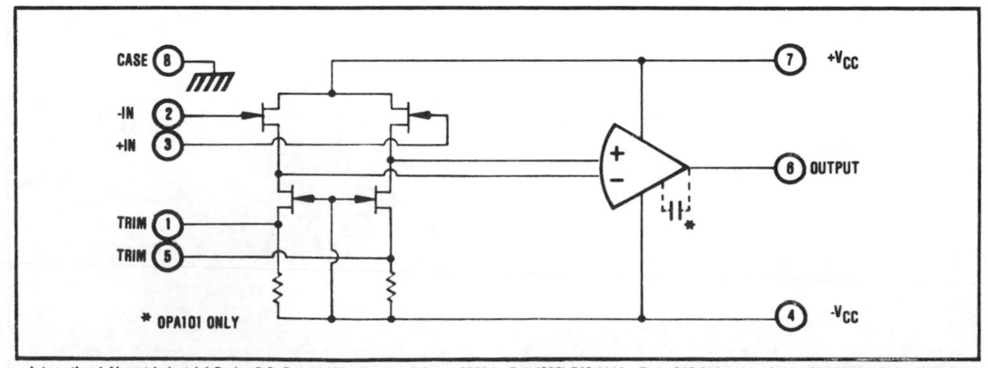

International Airport Industrial Park - P.O. Box 11400 - Tucson, Arizona 85734 - Tel. (602) 746-1111 - Twx: 910-952-1111 - Cable: BBRCORP - Telex: 66-6491

PDS-434B

SPECIFICATIONS

ELECTRICAL

Specifications at T_A = +25°C and $\pm V_{CC}$ = ±15VDC unless otherwise noted.

MODEL		OPA101/102AM			OPA101/102BM			
PARAMETER	**CONDITION**	**MIN**	**TYP**	**MAX**	**MIN**	**TYP**	**MAX**	**UNITS**
INPUT NOISE								
Voltage Noise Density	f_o = 1Hz(1)		100	200		80	100	nV/$\sqrt{Hz}$
	f_o = 10Hz		32	60		25	30	nV/$\sqrt{Hz}$
	f_o = 100Hz		14	30		11	15	nV/$\sqrt{Hz}$
	f_o = 1kHz		9	15		8	12	nV/$\sqrt{Hz}$
	f_o = 10kHz		7	8		7	8	nV/$\sqrt{Hz}$
	f_o = 100kHz		6.5	8		6.5	8	nV/$\sqrt{Hz}$
f_c; 1/f Corner Frequency			125			100		Hz
Voltage Noise	f_B = 0.1Hz to 10Hz(1)		1.3	2.6		1.0	1.3	μV, p-p
	f_B = 10Hz to 10kHz		1.0	1.2		0.8	1.0	μV, rms
	f_B = 10Hz to 100kHz		2.1	2.6		2.1	2.6	μV, rms
Current Noise Density	f_o = 0.1Hz thru 10kHz		2.0			1.4		fA/$\sqrt{Hz}$
Current Noise	f_B = 0.1Hz to 10Hz		38			26		fA, p-p
	f_B = 10Hz to 10kHz		200			140		fA, rms
DYNAMIC RESPONSE								
Bandwidth, Unity Gain	Small Signal							
OPA101			10			•		MHz
OPA102			Note 2			•		
Gain-Bandwidth Product	A_{CL} = 100							
OPA101			20			•		MHz
OPA102			40			•		MHz
Full Power Bandwidth	V_o = 20V, p-p; R_L = 1kΩ							
OPA101		80	100		•	•		kHz
OPA102		160	210		•	•		kHz
Slew Rate	V_o = ±10V; R_L = 1kΩ							
OPA101	A_{CL} = -1	5	6.5		•	•		V/μsec
OPA102	A_{CL} = -3	10	14		•	•		V/μsec
Settling Time (OPA101)	V_o = ±5V; A_{CL} = -1; R_L = 1kΩ							
ε = 1%			2			•		μsec
ε = 0.1%			2.5			•		μsec
ε = 0.01%			10			•		μsec
Settling Time (OPA102)	V_o = ±5V; A_{CL} = -3; R_L = 1kΩ							
ε = 1%			1			•		μsec
ε = 0.1%			1.5			•		μsec
ε = 0.01%			8			•		μsec
Small-Signal Overshoot	R_L = 1kΩ; C_L = 100pF							
OPA101	A_{CL} = +1		15			•		%
OPA102	A_{CL} = +3		20			•		%
Rise Time	10% to 90%, Small Signal							
OPA101			40			•		nsec
OPA102			30			•		nsec
Phase Margin	R_L = 1kΩ							
OPA101	A_{CL} = +1		60			•		Degrees
OPA102	A_{CL} = +3		45			•		Degrees
Overload Recovery(3)								
OPA101	A_{CL} = -1, 50% overdrive		1			•		μsec
OPA102	A_{CL} = -3, 50% overdrive		0.8			•		μsec
OPEN-LOOP GAIN, DC								
Full Load	V_o = ±10V; R_L = 1kΩ	94	105		•	•		dB
No Load	V_o = ±10V; R_L ≥ 10kΩ	96	108		•	•		dB
RATED OUTPUT								
Voltage	I_o = ±12mA	±12	±13		•	•		V
Current	V_o = ±12V	±12	±30		•	•		mA
Output Resistance	Open-Loop, f = DC		500			•		Ω
Short-Circuit Current			±45			•		mA
Capacitive Load Range	Phase Margin ≥ 25°							
OPA101	A_{CL} = +1		500			•		pF
OPA102	A_{CL} = +3		300			•		pF
INPUT OFFSET VOLTAGE								
Initial Offset	T_A = +25°C		±100	±500		±50	±250	μV
vs Temperature	-25°C ≤ T_A ≤ +85°C		±6	±10		±3	±5	μV/°C
vs Supply Voltage	±5VDC ≤ | V_{CC} | ≤ ±20VDC		±10	±50		•	•	μV/V
vs Time			±10			•		μV/mo.
Adjustment Range	Circuit in "Connection Diagram"		±1			•		mV
INPUT BIAS CURRENT								
Initial Bias	T_A = +25°C		-12	-15		-6	-10	pA
vs Temperature			Note 4			•		
vs Supply Voltage			Note 5			•		

ELECTRICAL (CONT)

MODEL		OPA101/102AM			OPA101/102BM			
PARAMETER	CONDITION	MIN	TYP	MAX	MIN	TYP	MAX	UNITS
INPUT DIFFERENCE CURRENT								
Initial Difference	$T_A = +25°C$		±3	±6		±1.5	±4	pA
vs Temperature			Note 4			*		
vs Supply Voltage			Note 5			*		
INPUT IMPEDANCE								
Differential								
Resistance			10^{12}			*		Ω
Capacitance			1			*		pF
Common-mode								
Resistance			10^{13}			*		Ω
Capacitance			3			*		pF
INPUT VOLTAGE RANGE								
Common-mode Voltage Range	Linear Operation		$\pm(\lvert V_{CC}\rvert - 3)$			*		V
Common-mode Rejection	$f_o = DC$, $V_{CM} = \pm10V$	80	105		*	*		dB
POWER SUPPLY								
Rated Voltage			±15			*		VDC
Voltage Range	Derated Performance	±5		±20	*			VDC
Current, Quiescent			5.8	8		*	*	mA
TEMPERATURE RANGE								
Specification		-25		+85	*		*	°C
Operating	Derated Performance	-55		+125	*		*	°C
Storage		-65		+150	*		*	°C

NOTES: *Specifications same as for OPA101/102AM.

1. Parameter is untested and is not guaranteed. This specification is established to a 90% confidence level.
2. Minimum stable gain for the OPA102 is 3V/V.
3. Time required for output to return from saturation to linear operation following the removal of an input overdrive signal.
4. Doubles approximately every 8.5°C.
5. See Typical Performance Curves.

MECHANICAL SPECIFICATIONS

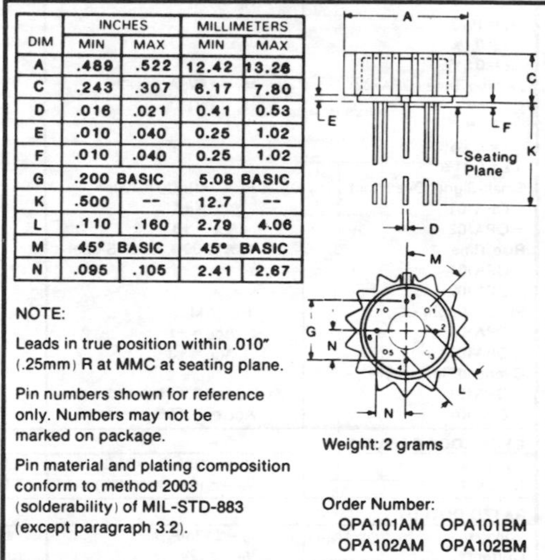

DIM	INCHES MIN	INCHES MAX	MILLIMETERS MIN	MILLIMETERS MAX
A	.489	.522	12.42	13.26
C	.243	.307	6.17	7.80
D	.016	.021	0.41	0.53
E	.010	.040	0.25	1.02
F	.010	.040	0.25	1.02
G	.200 BASIC		5.08 BASIC	
K	.500	--	12.7	--
L	.110	.160	2.79	4.06
M	45° BASIC		45° BASIC	
N	.095	.105	2.41	2.67

NOTE:

Leads in true position within .010" (.25mm) R at MMC at seating plane.

Pin numbers shown for reference only. Numbers may not be marked on package.

Pin material and plating composition conform to method 2003 (solderability) of MIL-STD-883 (except paragraph 3.2).

Weight: 2 grams

Order Number:
OPA101AM OPA101BM
OPA102AM OPA102BM

ABSOLUTE MAXIMUM RATINGS

Supply	±20VDC
Internal Power Dissipation(1)	750mW
Differential Input Voltage(2)	±20VDC
Input Voltage, Either Input(2)	±20VDC
Storage Temperature Range	-65°C to +150°C
Operating Temperature Range	-55°C to +125°C
Lead Temperature (soldering, 10 seconds)	+300°C
Output Short-Circuit Duration(3)	60 seconds
Junction Temperature	+175°C

NOTES:

1. Package must be derated according to the details in the Application Information section.
2. For supply voltages less than ±20VDC, the absolute maximum input is equal to the supply voltage.
3. Short-circuit may be to ground only. See discussion of Thermal Model in the Application Information section.

CONNECTION DIAGRAM

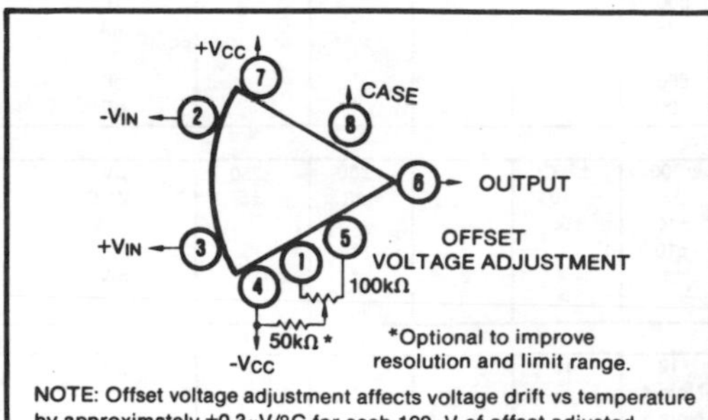

NOTE: Offset voltage adjustment affects voltage drift vs temperature by approximately ±0.3μV/°C for each 100μV of offset adjusted.

PIN CONFIGURATION

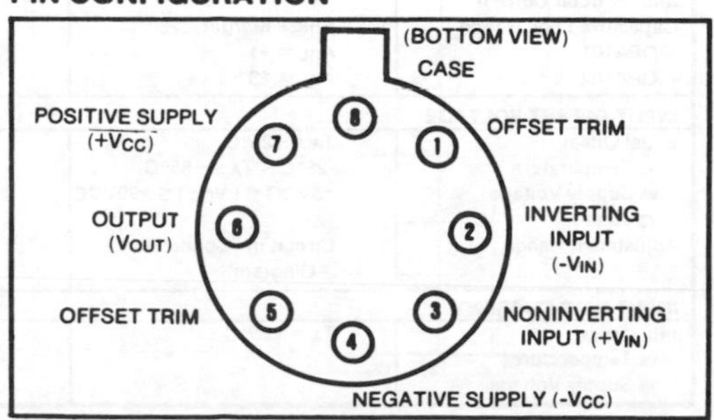

OPA101/102

2

OPERATIONAL AMPLIFIERS

TYPICAL PERFORMANCE CURVES

(T_A = +25°C, $\pm V_{CC}$ = ±15VDC, unless otherwise noted. Performance curves apply to both OPA101 and OPA102 unless otherwise noted.)

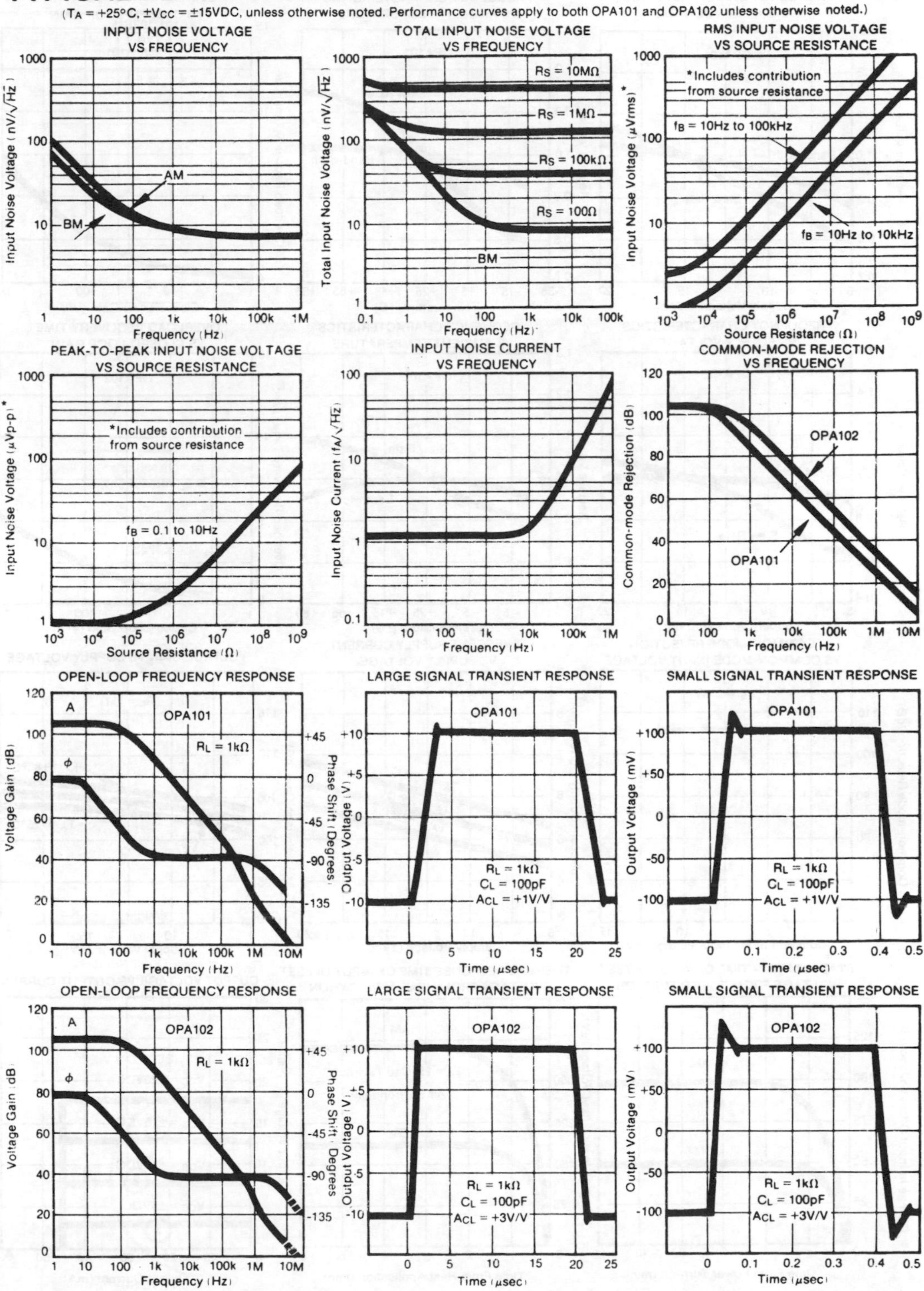

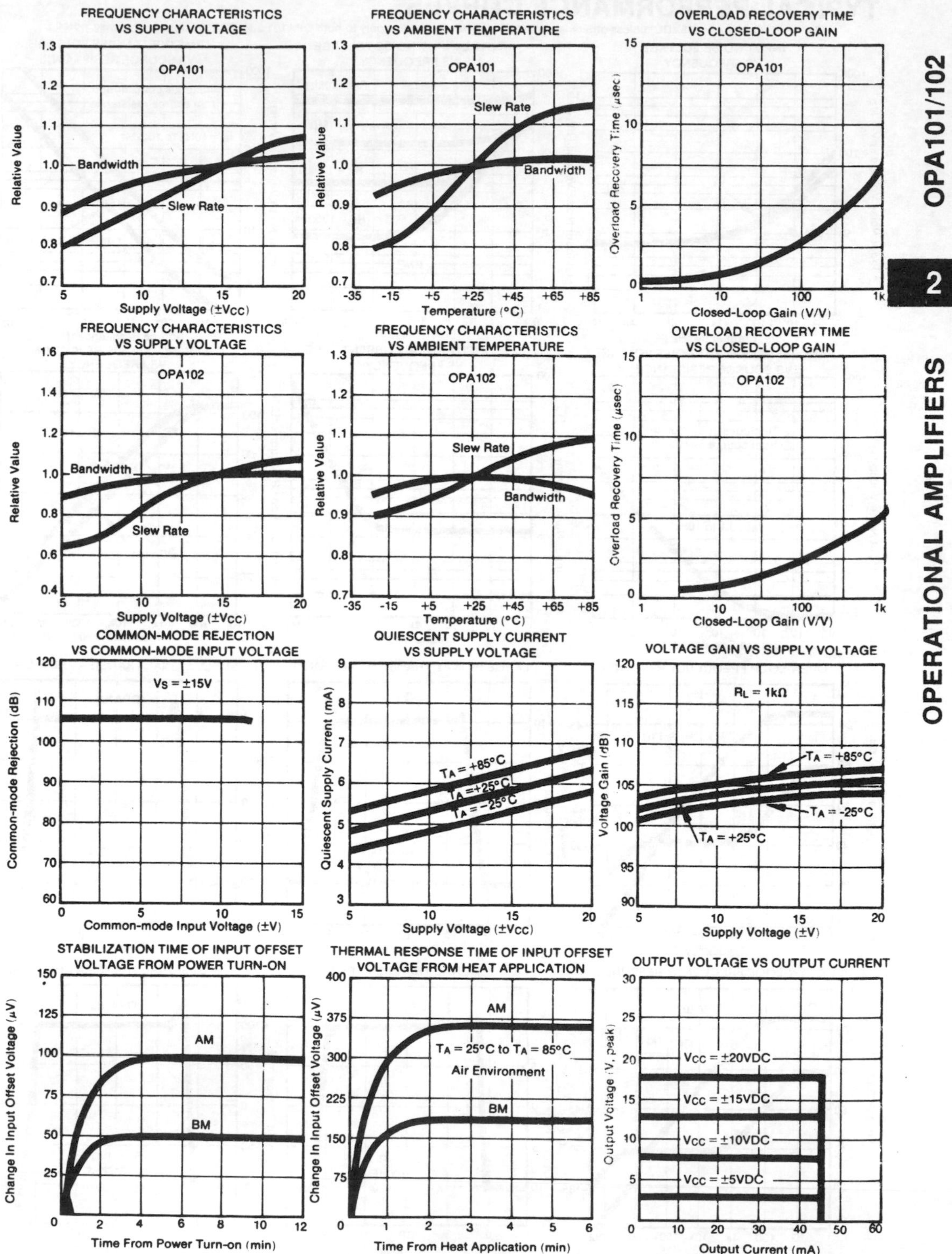
FREQUENCY CHARACTERISTICS VS SUPPLY VOLTAGE
OPA101
Bandwidth
Slew Rate
Relative Value
Supply Voltage (±VCC)
FREQUENCY CHARACTERISTICS VS AMBIENT TEMPERATURE
OPA101
Slew Rate
Bandwidth
Relative Value
Temperature (°C)
OVERLOAD RECOVERY TIME VS CLOSED-LOOP GAIN
OPA101
Overload Recovery Time (μsec)
Closed-Loop Gain (V/V)
FREQUENCY CHARACTERISTICS VS SUPPLY VOLTAGE
OPA102
Bandwidth
Slew Rate
Relative Value
Supply Voltage (±VCC)
FREQUENCY CHARACTERISTICS VS AMBIENT TEMPERATURE
OPA102
Slew Rate
Bandwidth
Relative Value
Temperature (°C)
OVERLOAD RECOVERY TIME VS CLOSED-LOOP GAIN
OPA102
Overload Recovery Time (μsec)
Closed-Loop Gain (V/V)
COMMON-MODE REJECTION VS COMMON-MODE INPUT VOLTAGE
VS = ±15V
Common-mode Rejection (dB)
Common-mode Input Voltage (±V)
QUIESCENT SUPPLY CURRENT VS SUPPLY VOLTAGE
TA = +85°C
TA = +25°C
TA = −25°C
Quiescent Supply Current (mA)
Supply Voltage (±VCC)
VOLTAGE GAIN VS SUPPLY VOLTAGE
RL = 1kΩ
TA = +85°C
TA = −25°C
TA = +25°C
Voltage Gain (dB)
Supply Voltage (±V)
STABILIZATION TIME OF INPUT OFFSET VOLTAGE FROM POWER TURN-ON
AM
BM
Change In Input Offset Voltage (μV)
Time From Power Turn-on (min)
THERMAL RESPONSE TIME OF INPUT OFFSET VOLTAGE FROM HEAT APPLICATION
AM
TA = 25°C to TA = 85°C
Air Environment
BM
Change In Input Offset Voltage (μV)
Time From Heat Application (min)
OUTPUT VOLTAGE VS OUTPUT CURRENT
VCC = ±20VDC
VCC = ±15VDC
VCC = ±10VDC
VCC = ±5VDC
Output Voltage (V, peak)
Output Current (mA)
OPA101/102
2
OPERATIONAL AMPLIFIERS

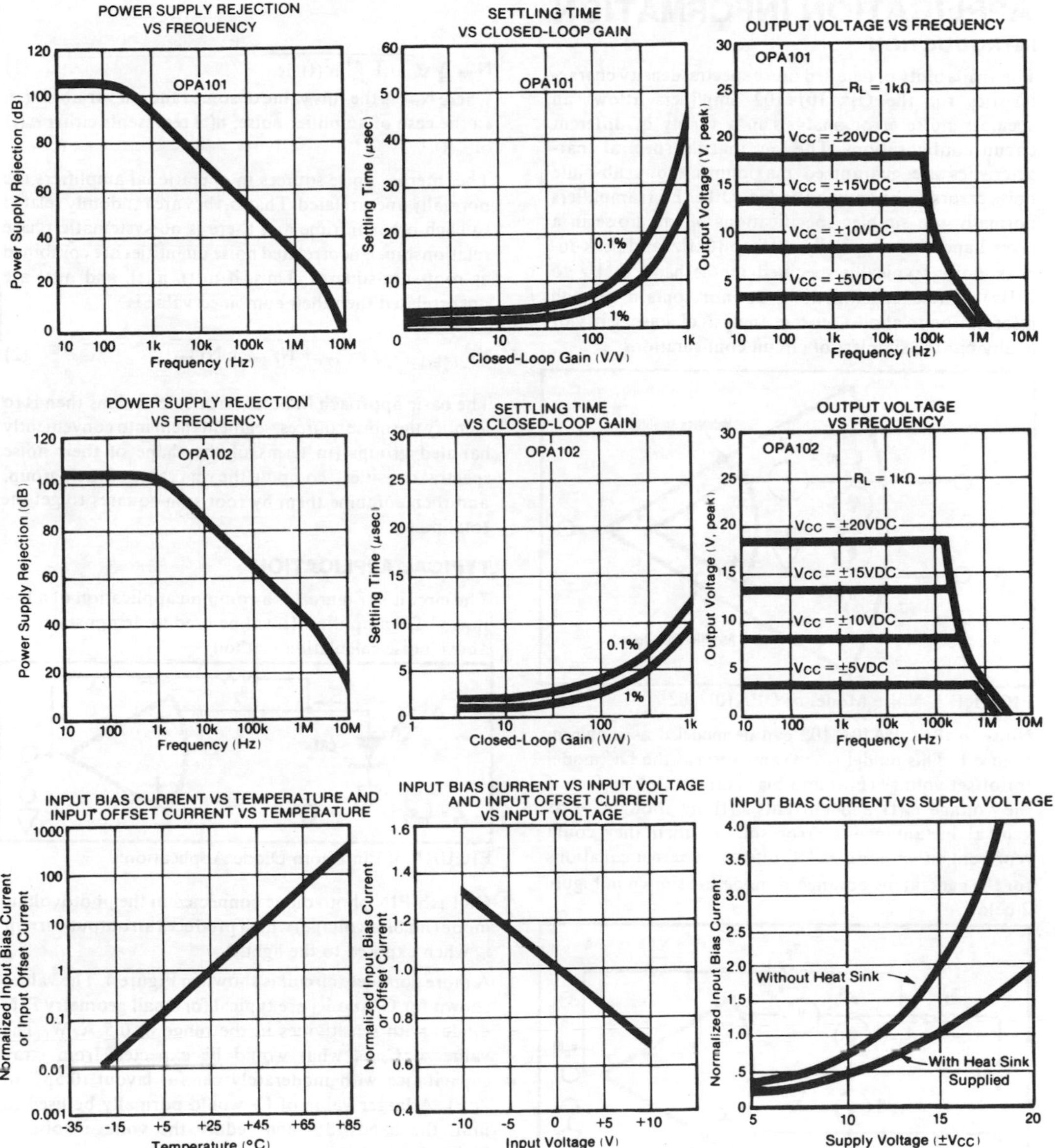

POWER SUPPLY REJECTION VS FREQUENCY
OPA101
Power Supply Rejection (dB)
Frequency (Hz)
SETTLING TIME VS CLOSED-LOOP GAIN
OPA101
Settling Time (μsec)
0.1%
1%
Closed-Loop Gain (V/V)
OUTPUT VOLTAGE VS FREQUENCY
OPA101
RL = 1kΩ
Vcc = ±20VDC
Vcc = ±15VDC
Vcc = ±10VDC
Vcc = ±5VDC
Output Voltage (V, peak)
Frequency (Hz)
POWER SUPPLY REJECTION VS FREQUENCY
OPA102
Power Supply Rejection (dB)
Frequency (Hz)
SETTLING TIME VS CLOSED-LOOP GAIN
OPA102
Settling Time (μsec)
0.1%
1%
Closed-Loop Gain (V/V)
OUTPUT VOLTAGE VS FREQUENCY
OPA102
RL = 1kΩ
Vcc = ±20VDC
Vcc = ±15VDC
Vcc = ±10VDC
Vcc = ±5VDC
Output Voltage (V, peak)
Frequency (Hz)
INPUT BIAS CURRENT VS TEMPERATURE AND INPUT OFFSET CURRENT VS TEMPERATURE
Normalized Input Bias Current or Input Offset Current
Temperature (°C)
INPUT BIAS CURRENT VS INPUT VOLTAGE AND INPUT OFFSET CURRENT VS INPUT VOLTAGE
Normalized Input Bias Current or Offset Current
Input Voltage (V)
INPUT BIAS CURRENT VS SUPPLY VOLTAGE
Without Heat Sink
With Heat Sink Supplied
Normalized Input Bias Current
Supply Voltage (±Vcc)

APPLICATION INFORMATION

INTRODUCTION

The availability of detailed noise spectral density characteristics for the OPA101/102 amplifiers allows an accurate noise error analysis in a variety of different circuit configurations. The fact that the spectral characteristics are guaranteed maximums allows absolute noise errors to be truly bounded. Other FET amplifiers normally use simpler specifications of rms noise in a given bandwidth (typically 10Hz to 10kHz) and peak-to-peak noise (typically specified in the band 0.1Hz to 10Hz). These specifications do not contain enough information to allow accurate analysis of noise behavior in any but the simplest of circuit configurations.

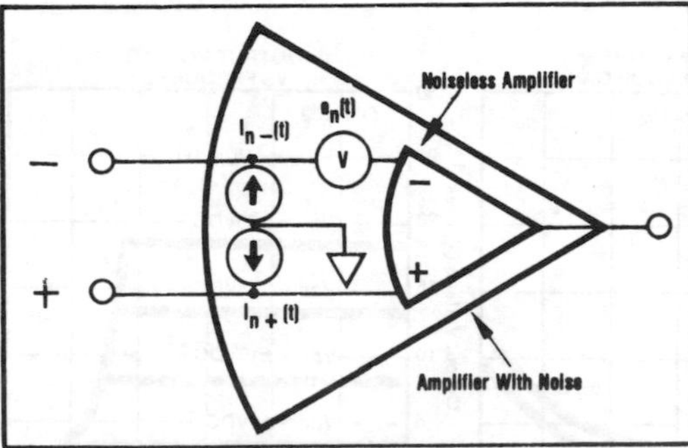

FIGURE 1. Noise Model of OPA101/102.

Noise in the OPA101/102 can be modeled as shown in Figure 1. This model is the same form as the DC model for offset voltage (E_{OS}) and bias currents (I_B). In fact, if the voltage $e_n(t)$ and currents $i_n(t)$ are thought of as general instantaneous error sources, then they could represent either noise or DC offsets. The error equations for the general instantaneous model are shown in Figure 2 below.

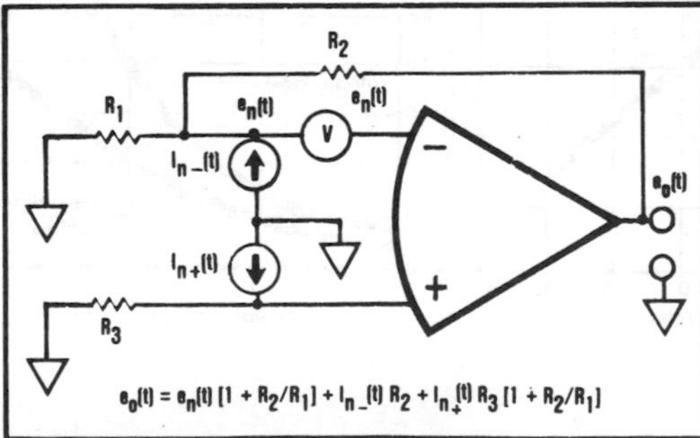

FIGURE 2. Circuit With Error Sources.

If the instantaneous terms represent DC errors (i.e., offset voltage and bias currents) the equation is a useful tool to compute actual errors. It is not, however, useful in the same <u>direct</u> way to compute noise errors. The basic problem is that noise cannot be predicted as a function of time. It is a random variable and must be described in probabilistic terms. It is normally described by some type of average - most commonly the rms value.

$$N_{rms} \triangleq \sqrt{1/T \int_0^T n^2(t)\,dt} \qquad (1)$$

where N_{rms} is the rms value of some random variable $n(t)$. In the case of amplifier noise, $n(t)$ represents either $e_n(t)$ or $i_n(t)$.

The internal noise sources in operational amplifiers are normally uncorrelated. That is, they are randomly related to each other in time and there is no systematic phase relationship. Uncorrelated noise quantities are combined as root-sum-squares. Thus, if $n_1(t)$, $n_2(t)$, and $n_3(t)$ are uncorrelated then their combined value is

$$N_{TOTAL_{rms}} = \sqrt{N_1{}^2{}_{rms} + N_2{}^2{}_{rms} + N_3{}^2{}_{rms}} \qquad (2)$$

The basic approach in noise error calculations then is to identify the noise sources, segment them into conveniently handled groups (in terms of the shape of their noise spectral densities), compute the rms value of each group, and then combine them by root-sum-squares to get the total noise.

TYPICAL APPLICATION

The circuit in Figure 3 is a common application of a low noise FET amplifier. It will be used to demonstrate the above noise calculation method.

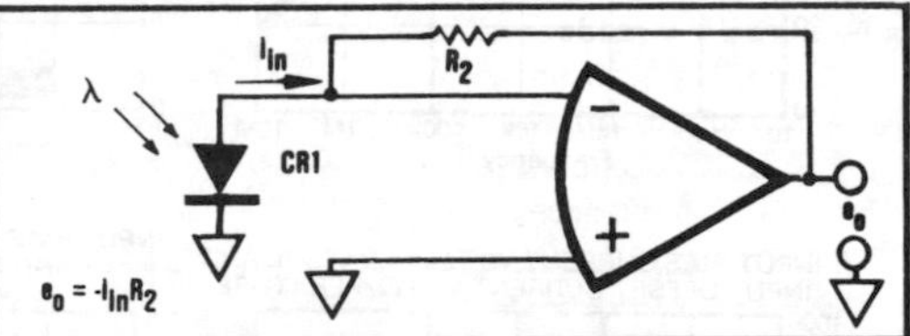

FIGURE 3. Pin Photo Diode Application.

CR1 is a PIN photo diode connected in the photovoltaic mode (no bias voltage) which produces an output current i_{in} when exposed to the light, λ.

A more complete circuit is shown in Figure 4. The values shown for C_1 and R_1 are typical for small geometry PIN diodes with sensitivities in the range of 0.5 A/W. The value of C_2 is what would be expected from stray capacitance with moderately careful layout (0.5pF to 2pF). A larger value of C_2 would normally be used to limit the bandwidth and reduce the voltage noise at higher frequencies.

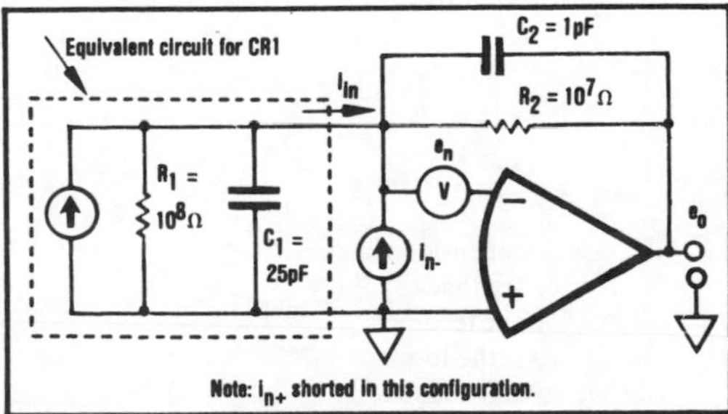

FIGURE 4. Noise Model of Photo Diode Application.

In Figure 4, e_n and i_n represent the amplifier's voltage and current spectral densities, $e_n(\omega)$ and $i_n(\omega)$ respectively. These are shown in Figure 5.

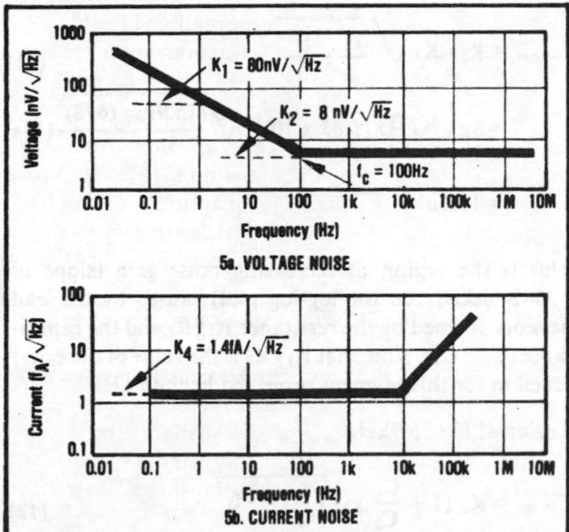

FIGURE 5. Noise Voltage and Current Spectral Density.

Figure 6 shows the desired "gain" of the circuit (transimpedance of $e_o/i_{in} = Z_2(s)$). It has a single-pole rolloff at $f_2 = 1/(2\pi R_2C_2) = \omega_2/2\pi$. Output noise is minimized if f_2 is made smaller. Normally R_2 is chosen for the desired DC transimpedance based on the full scale input current (i_{in} full scale) and maximum output (e_o max). Then C_2 is chosen to make f_2 as small as possible consistent with the necessary signal frequency response.

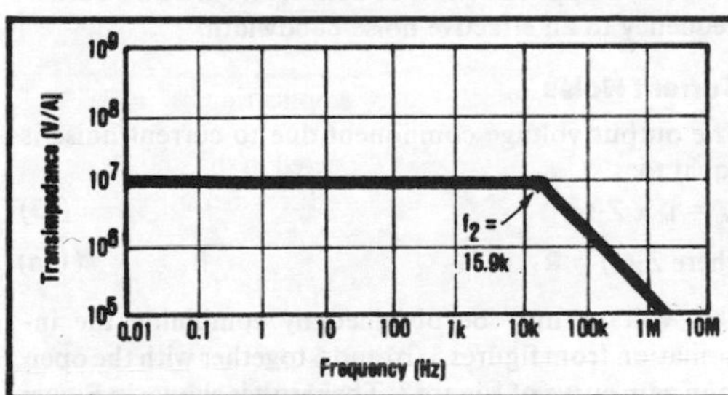

FIGURE 6. Transimpedance.

Voltage Noise

Figure 7 shows the noise voltage gain for the circuit in Figure 4. It is derived from the equation

$$e_o = e_n \left[\frac{A}{1 + A\beta} \right] = e_n \frac{1}{\beta} \left[\frac{1}{1 + \frac{1}{A\beta}} \right] \quad (3)$$

where:

A $= A(\omega)$ is the open-loop gain

β $= \beta(\omega)$ is the feedback factor. It is the amount of output voltage feedback to the input of the op amp.

$A\beta = A(\omega)\,\beta(\omega)$ is the loop gain. It is the amount of the output voltage feedback to the input and then amplified and returned to the output.

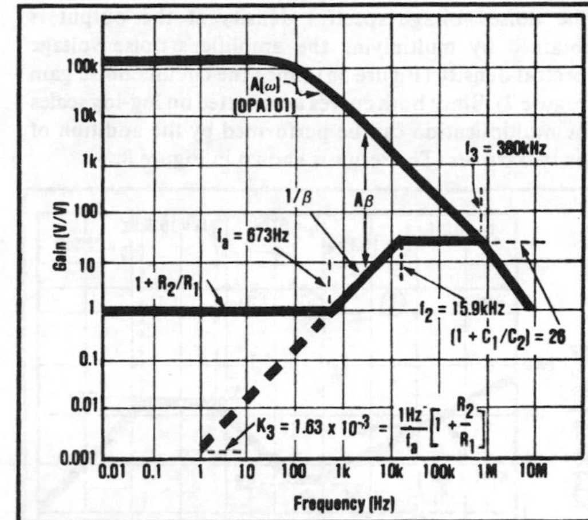

FIGURE 7. Noise Voltage Gain.

Note that for large loop gain ($A\beta >> 1$)

$$e_o \approx e_n \frac{1}{\beta} \quad (4)$$

For the circuit in Figure 4 it can be shown that

$$\frac{1}{\beta} = 1 + \frac{R_2(R_1C_1s + 1)}{R_1(R_2C_2s + 1)} \ . \quad (5)$$

This may be rearranged to

$$\frac{1}{\beta} = \frac{R_2 + R_1}{R_1} \left[\frac{\tau_a s + 1}{\tau_2 s + 1} \right] \quad (5a)$$

$$\text{where } \tau_a = (R_1 \parallel R_2)(C_1 \parallel C_2) \quad (5b)$$

$$= \left[\frac{R_1R_2}{R_1 + R_2}\right](C_1 + C_2)$$

$$\text{and} \quad \tau_2 = R_2C_2 \ . \quad (5c)$$

$$\text{Then, } f_a = \frac{1}{2\pi\tau_a} \text{ and } f_2 = \frac{1}{2\pi\tau_2} \ . \quad (5d)$$

For very low frequencies ($f << f_a$), s approaches zero and equation 5 becomes

$$\frac{1}{\beta} = 1 + \frac{R_2}{R_1} \ . \quad (6)$$

For very high frequencies ($f >> f_2$), s approaches infinity and equation 5 becomes

$$\frac{1}{\beta} = 1 + \frac{C_1}{C_2} \ . \quad (7)$$

The noise voltage spectral density at the output is obtained by multiplying the amplifier's noise voltage spectral density (Figure 5a) times the circuits noise gain (Figure 7). Since both curves are plotted on log-log scales the multiplication can be performed by the addition of the two curves. The result is shown in Figure 8.

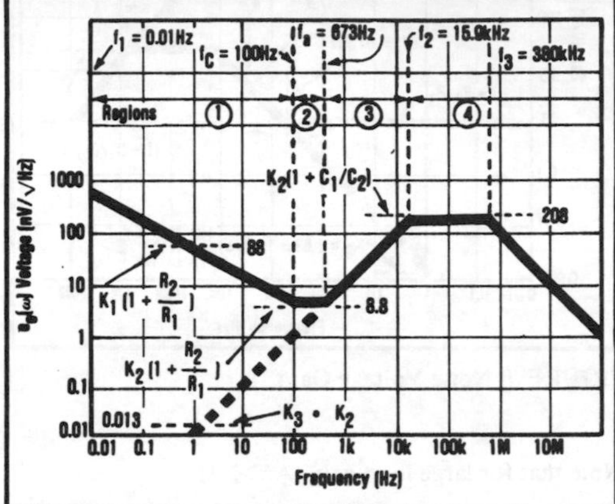

FIGURE 8. Output Noise Voltage Spectral Density.

The total rms noise at the amplifier's output due to the amplifier's internal voltage noise is derived from the $e_o(\omega)$ function in Figure 8 with the following expression:

$$E_{o\ rms} = \sqrt{\int_{-\infty}^{+\infty} e_o^2(\omega)\, d\omega} \tag{8}$$

It is both convenient and informative to calculate the rms noise using a piecewise approach (region-by-region) for each of the four regions indicated in Figure 8.

Region 1; $f_1 = 0.01$Hz to $f_c = 100$Hz

$$E_{n1\ rms} = K_1 \left(1 + \frac{R_2}{R_1}\right) \sqrt{\ln(f_c/f_1)} \tag{9}$$

$$= 80\text{nV}/\sqrt{\text{Hz}} \left(1 + \frac{10^7}{10^8}\right) \sqrt{\ln \frac{100}{0.01}} \tag{9a}$$

$$= 0.267\mu\text{V}$$

This region has the characteristic of 1/f or "pink" noise (slope of -10dB per decade on the log-log plot of $e_n(\omega)$). The selection of 0.01 Hz is somewhat arbitrary but it can be shown that for this example there would be only negligible additional contribution by extending f_1 several decades lower. Note that $K_1(1 + R_2/R_1)$ is the value of e_o at f = 1Hz.

Region 2; $f_c = 100$Hz to $f_a = 673$Hz

$$E_{n2\ rms} = K_2 \left(1 + \frac{R_2}{R_1}\right) \sqrt{f_a - f_c} \tag{10}$$

$$= 8\text{nV}/\sqrt{\text{Hz}} \left(1 + \frac{10^7}{10^8}\right) \sqrt{673 - 100} \tag{10a}$$

$$= 0.21\mu\text{V}$$

This is a region of "white" noise which leads to the form of equation (10).

Region 3; $f_a = 673$Hz to $f_2 = 15.9$kHz

$$E_{n3\ rms} = K_2 \bullet K_3 \sqrt{\frac{f_2^3}{3} - \frac{f_a^3}{3}} \tag{11}$$

$$= 8\text{nV}/\sqrt{\text{Hz}} (1.63 \times 10^{-3}) \sqrt{\frac{(15.9\text{k})^3}{3} - \frac{(673)^3}{3}} \tag{11a}$$

$$= 15.1\mu\text{V}$$

This is the region of increasing noise gain (slope of +20dB/decade on the log-log plot) caused by the lead network formed by the resistance $R_1 \| R_2$ and the capacitance $(C_1 + C_2)$. Note that $K_3 \bullet K_2$ is the value of the $e_o(\omega)$ function for this segment projected back to 1Hz.

Region 4; $f > 15.9$kHz

$$E_{n4\ rms} = K_2 \left(1 + \frac{C_1}{C_2}\right) \sqrt{\left[\frac{\pi}{2}\right] f_3 - f_2} \tag{12}$$

$$= 8\text{nV}/\sqrt{\text{Hz}} \left(1 + \frac{25}{1}\right) \sqrt{\left[\frac{\pi}{2}\right] 380\text{k} - 15.9\text{k}} \tag{12a}$$

$$= 158.5\mu\text{V}$$

This is a region of white noise with a single order rolloff at $f_3 = 380$kHz caused by the intersection of the $1/\beta$ curve and the open-loop gain curve. The value of 380kHz is obtained from observing the intersection point of Figure 7. The $\pi/2$ applied to f_3 is to convert from a 3dB corner frequency to an effective noise bandwidth.

Current Noise

The output voltage component due to current noise is equal to:

$$E_{ni} = i_n \times Z_2(s) \tag{13}$$

$$\text{where } Z_2(s) = R_2 \ \| \ X_{C_2} \tag{13a}$$

This voltage may be obtained by combining the information from figures 5 (b) and 6 together with the open loop gain curve of Figure 7. The result is shown in Figure 9 below.

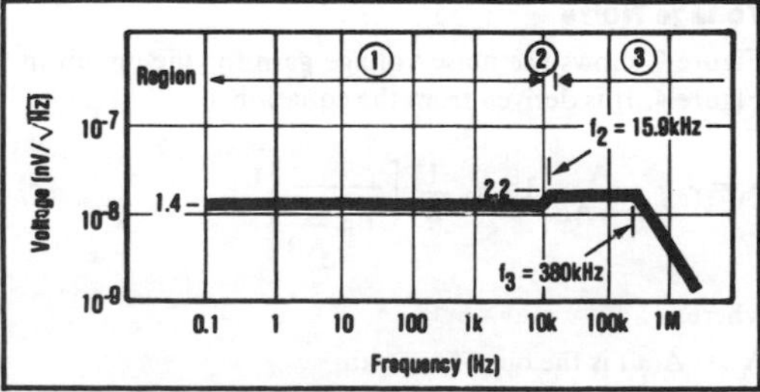

FIGURE 9. Output Voltage Due to Noise Current.

Using the same techniques that were used for the voltage noise:

Region 1; 0.1Hz to 10kHz

$$E_{ni1} = 1.4 \times 10^{-8} \sqrt{10k - 0.1} \quad (14)$$

$$= 1.4\mu V$$

Region 2; 10kHz to 15.9kHz

$$E_{ni2} = 1.4 \times 10^{-12} \sqrt{\frac{(15.9k)^3}{3} - \frac{(10k)^3}{3}} \quad (14a)$$

$$= 1.4\mu V$$

Region 3; f > 15.9kHz

$$E_{ni3} = 2.2 \times 10^{-8} \sqrt{\frac{\pi}{2} 380k - 15.9k} \quad (14b)$$

$$= 16.8\mu V$$

$$E_{ni\ total} = 10^{-6} \sqrt{(1.4)^2 + (1.4)^2 + (16.8)^2} \quad (14c)$$

$$= 16.9\mu V_{rms}$$

Resistor Noise

For a complete noise analysis of the circuit in Figure 4, the noise of the feedback resistor, R_2, must also be included. The thermal noise of the resistor is given by:

$$E_{R\ rms} = \sqrt{4kTRB} \quad (15)$$

K = Boltzmann's constant = 1.38×10^{-23} Joules/°Kelvin
T = Absolute temperature (degrees Kelvin)
R = Resistance (ohms)
B = Effective noise bandwidth (Hz) (ideal filter assumed)

At 25°C this becomes

E_R rms ≈ $0.13 \sqrt{RB}$
E_R rms in μV
R in MΩ
B in Hz

For the circuit in Figure 4

$R_2 = 10^7\Omega = 10M\Omega$

$$B = \frac{\pi}{2}(f_2) = \frac{\pi}{2}\ 15.9k$$

Then

$$E_R\ rms = (411nV/\sqrt{Hz})\sqrt{B}$$

$$= (411nV/\sqrt{Hz})\sqrt{\frac{\pi}{2}\ 15.9kHz}$$

$$= 64.9\mu V\ rms$$

Total Noise

The total noise may now be computed from

$$E_{n\ total} = \sqrt{E_{n1}^2 + E_{n2}^2 + E_{n3}^2 + E_{n4}^2 + E_{nR}^2 + E_{ni}^2} \quad (16)$$

$$= \sqrt{0.267^2 + 0.21^2 + 15.1^2 + 158.5^2 + 64.9^2 + 16.9^2} \quad (16a)$$

$$= \sqrt{0.07 + 0.04 + 228 + 25122 + 4212 + 286} \quad (16b)$$

$$= 173\mu V\ rms$$

Conclusions

Examination of the results in equation (16b) together with the curves in Figure 8 leads to some interesting conclusions. In this example 84% of the noise comes from E_{n4}. From Figure 8 it is seen that this is the area beyond the pole formed by R_2 and C_2.

The E_{n4} contribution could be reduced several ways. The most common method is to increase C_2. This reduces f_2 and the value of K2(1 + C1/C2) (see Figure 8). It also reduces the signal bandwidth (see Figure 6) and the final value of C_2 is normally a compromise between noise gain and necessary signal bandwidth.

It should be noted that increasing C_2 will also affect f_a since f_a is determined by $(C_1 + C_2)$ (see equation (5b)). Normally C_2 is larger than C_1 and f_2 will change more than for a given change in C_2.

The other means of reducing the noise in region 4 involves changing amplifier parameters. For example, the use of a slower amplifier would move the open-loop gain curve to the left and decrease f_3. Of course, reducing the value of K_2, the noise floor, would also reduce the noise in this region.

The second largest component is the resistor noise E_{nR} (14% of the total noise). A lower resistor value decreases resistor noise as a function of $\sqrt{R}$, but it also lowers the desired signal gain as a direct function of R. Thus, lowering R reduces the signal-to-noise ratio at the output which shows that the feedback resistor should be as large as possible. The noise contribution due to R_2 can be decreased by raising the value of C_2 (lowering f_2) but this reduces signal bandwidth.

It is interesting to note that the current noise of the amplifier accounted for only 1% of the total E_n. This is different than would be expected when comparing the current and voltage spectral densities with the size of the feedback resistor. For example, if we define a characteristic value of resistance as

$$R_{characteristic} = \frac{e_n(\omega)}{i_n(\omega)} \text{ at } f = 10kHz \quad (17)$$

$$= \frac{8nV/\sqrt{Hz}}{1.4fA/\sqrt{Hz}}$$

$$= 5.7M\Omega$$

Thus, in simple transimpedance circuits with feedback resistors greater than the characteristic value, the amplifier's current noise would cause more output noise than the amplifier's voltage noise. Based on this and the 10MΩ feedback resistor in the example, the amplifier noise current would be expected to have a higher contribution than the noise voltage. The reason it does not in the example of Figure 4 is that the noise voltage has high gain at higher frequencies (Figure 7) and the noise current does not (Figure 6).

The fourth largest component of total noise comes from E_{n3} (0.8%). Decreasing C_1 will also lower the term $K_2(1 + C_1/C_2)$. In this case, f_2 will stay fixed and f_a will move to the right (i.e., the +20dB/decade slope segment will move to the right). This can have a significant reduction on noise without lowering the signal bandwidth. This points out the importance of maintaining low capacitance at the amplifier's input in low noise applications.

Shielding and Guarding

The low noise, low bias current and high input impedance of the OPA101/102 are well suited to a number of precision applications. In order to fully benefit from the outstanding specifications of this unit, careful layout, shielding, and guarding are required. Careless signal wiring or printed circuit board layout can easily degrade circuit performance several orders of magnitude below the capability of the OPA101/102.

As in any situation where high impedances are involved, careful shielding is required to reduce "hum" pickup in input leads. If large feedback resistors are used, they should also be shielded along with the external input circuitry. The metal case of the OPA101/102 is connected to pin 8 and is not connected to any internal amplifier circuitry. Thus it is possible to use the case as a shield to reduce noise pickup.

Unless care is used, leakage currents across printed circuit boards can easily exceed the bias current of the OPA101/102. To avoid leakage problems, it is recommended that a Teflon IC socket be used or that at least the signal input lead of the amplifier be wired to a Teflon standoff. If this is not done and instead the OPA101/102 is to be soldered directly into a printed circuit board, utmost care must be used in planning the board layout. A "guard" pattern should completely surround the two amplifier input leads and should be connected to a low impedance point which is at the signal input potential (see Figure 10). The amplifier case, pin 8, should also be connected to the guard. This insures that the entire amplifier circuitry is fully surrounded by the guard potential. This minimizes the voltage placed across any leakage paths and thus reduces leakage currents. In addition, noise pickup is also reduced.

Figures 11, 12, and 13 show typical applications using the guard and case shielding.

Cleanliness is also a prime concern in low bias current circuits. It is recommended that after installation is complete the assembly be washed with a low residue solvent such as TMC Freon followed by rinsing with deionized water. The use of some form of high dielectric conformal coating such as a good two-part urathane should be considered if the assembly will be used in air environment which could deposit contaminants on the low current circuitry.

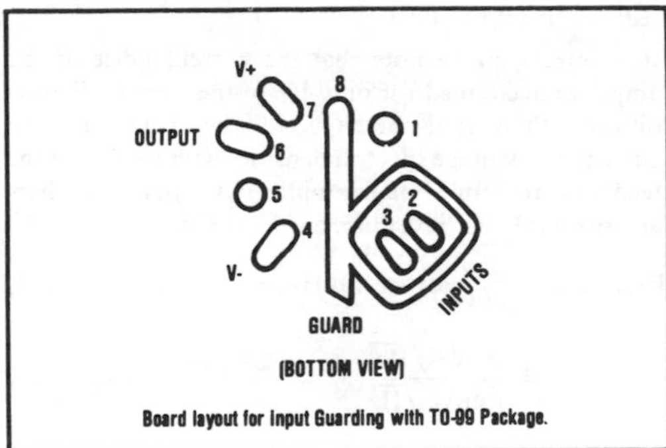

FIGURE 10. Connection of Case Guard and Input Guard.

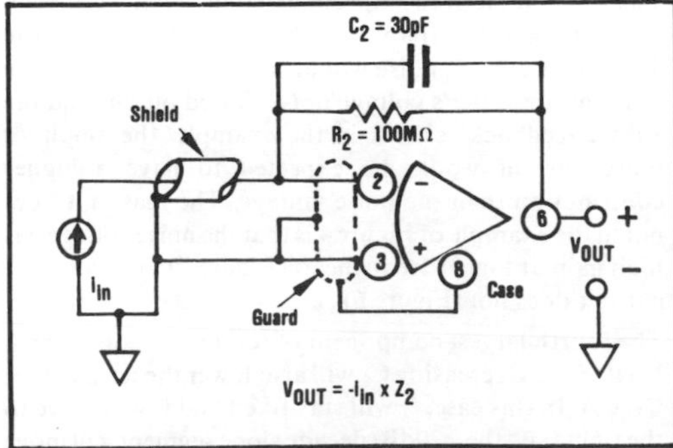

FIGURE 11. Ultra-Low Current to Voltage Converter.

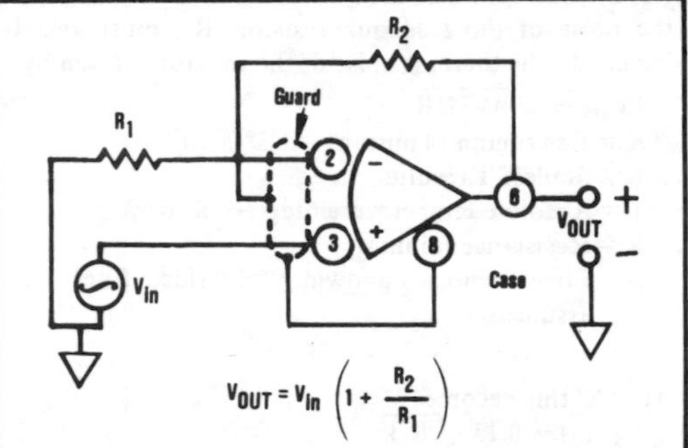

FIGURE 12. Ultra-High Input Impedance Noninverting Circuit.

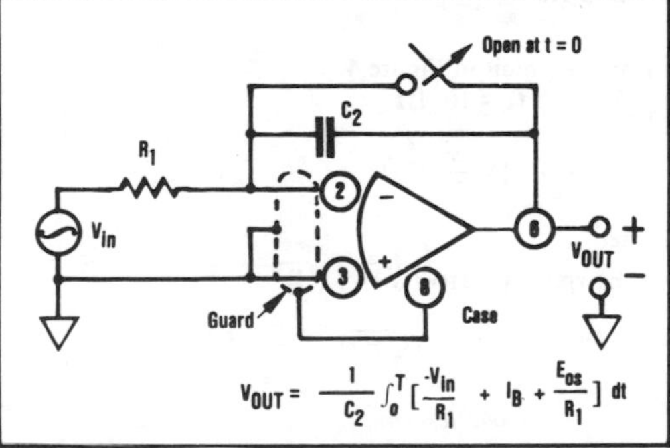

$$V_{OUT} = \frac{1}{C_2}\int_0^T \left[\frac{-V_{in}}{R_1} + I_B + \frac{E_{os}}{R_1}\right] dt$$

FIGURE 13. Low Drift Integrator.

Thermal Model

Figure 14 is the thermal model for the OPA101/102 where:

T_J = Junction temperature (output load)
T_J^* = Junction temperature (no load)
T_C = Case temperature
T_A = Ambient temperature
θ_{CA} = Thermal resistance, case-to-ambient

θ_{HS} = Effective thermal resistance of the heat sink
P_{DQ} = Quiescent power dissipation
$|+V_{CC}|\ I_{+QUIESCENT} + |-V_{CC}|\ I_{-QUIESCENT}$
P_{DX} = Power dissipation in the output transistor
$= (V_{OUT} - V_{CC})\ I_{OUT}$

(In a complementary output stage only one output transistor is conducting current at a time.)

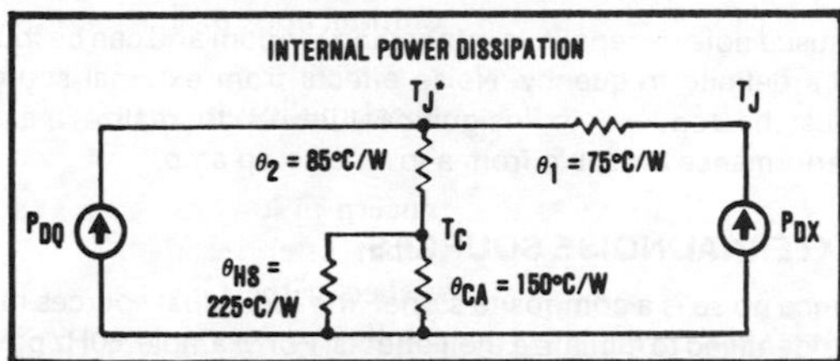

FIGURE 14. OPA101/102 Thermal Model

This model is obviously not the simple one-power source model used with most linear integrated circuits. It is, however, a more accurate model for multichip hybrid integrated circuits where the quiescent power is dissipated in the input stage and the internal power dissipation due to the load is dissipated in a somewhat physically separated output stage.

The model in Figure 14 must be used in conjunction with the OPA101/102's absolute maximum ratings of internal power dissipation and junction temperature to determine the derated power dissipation capability of the package.

As an example of how to use this model, consider this problem: Determine the output transistor junction temperature when the output has its maximum load resistance and is operated at the worst-case output voltage conditions. Assume $V_{CC} = \pm 15VDC$ and $T_A = 25°C$.

Maximum P_{DX} occurs where $V_{OUT} = 1/2V_{CC}$. Then

$$P_{DX\ max} = \frac{(V_{CC})^2}{4R_{load}} \quad (18)$$

$$T_j = T_A + P_{DQ}\left[\theta_2 + (\theta_{HS} \parallel \theta_{CA})\right] + P_{DX}\left[\theta_1 + \theta_2 + (\theta_{HS} \parallel \theta_{CA})\right] \quad (19)$$

where $(\theta_{HS} \parallel \theta_{CA}) = \dfrac{\theta_{HS}\theta_{CA}}{\theta_{HS} + \theta_{CA}} = 90°C/W$

Substituting appropriate values yields

$$T_j = 25° + (30V \times 8mA)[85°C/W + 90°C/W]$$

$$+ \frac{(15V)^2}{4 \times 1k\Omega}[75°C/W + 85°C/W + 90°C/W]$$

$$= 25°C + 42°C + 14°C = T_A + 56°C$$

$$= 81°C$$

The conclusion is that under a worst-case output voltage condition and with a 1kΩ load the junction temperature rise is 56°C above ambient. Thus, under these conditions, the device could be operated in an ambient up to 119°C without exceeding the 175°C junction temperature rating.

A similar analysis for conditions of the output short-circuited to ground where

$$P_{DX\ SS} = V_{CC}\ I_{(output\ limit)} \quad (20)$$

shows that the maximum junction temperature rating of 175°C is exceeded. Thus, the output should not be shorted to ground for sustained periods of time.

HEAT SINK

The heat sink used on the OPA101/102 should not be removed. It has the effect of reducing the package thermal resistance from 150°C/W to about 90°C per watt. Removing the heat sink would naturally increase the junction temperature of the amplifier which would in turn raise the input bias current. The change in thermal resistance also affects the noise performance. Removing the heat sink would increase the noise in the 1/f region.

AN-15

MINIMIZATION OF NOISE IN OPERATIONAL AMPLIFIER APPLICATIONS

Precision Monolithics Inc.

APPLICATION NOTE 15

INTRODUCTION

Since operational amplifier specifications such as Input Offset Voltage and Input Bias Current have improved tremendously in the past few years, noise is becoming an increasingly important error consideration. To take advantage of today's high performance op amps, an understanding of the noise mechanisms affecting op amps is required. This paper examines noise contributions, both internal and external to an op amp, and provides practical methods for minimizing their effects.

BASIC NOISE PROPERTIES

Noise, for purposes of this discussion, is defined as any signal appearing in an op amp's output that could not have been predicted by DC and AC input error analysis. Noise can be random or repetitive, internally or externally generated, current or voltage type, narrowband or wideband, high frequency or low frequency; whatever its nature, it can be minimized.

The first step in minimizing noise is source identification in terms of bandwidth and location in the frequency spectrum; some of the more common sources are shown in Figure 1, an 11-decade frequency spectrum chart. Some preliminary observations can be made: noise is present from DC to VHF from sources which may be identified in terms of bandwidth and frequency. Noise source bandwidths overlap, making noise a composite quantity at any given frequency. Most externally caused noise is repetitive rather than random and can be found at a definite frequency. Noise effects from external sources must be reduced to insignificant levels to realize the full performance available from a low noise op amp.

EXTERNAL NOISE SOURCES

Since noise is a composite signal, the individual sources must be identified to minimize their effects. For example, 60Hz power line pickup is a common interference noise appearing at an op amp's output as a 16ms sine wave. In this and most other situations, the basic tool for external noise source frequency characterization is the oscilloscope sweep rate setting. Recognizing the oscilloscope's potential in this area, Tektronix® manufactures an oscilloscope vertical amplifier with variable upper and lower −3dB points, which allows quick noise source frequency identification. Another basic identification tool is the simple low pass filter as shown in Figure 2, where the bandpass is calculated by:

$$(1) \quad f_O \cong \frac{1}{2\pi RC}$$

With such a filter, measurement bandpass can be changed from 10Hz to 100kHz (C = 4.7μF to 470pF), attenuating higher frequency components while passing frequencies of interest. Once identified, noise from an external source may be minimized by the methods outlined in Table 1—the external noise source chart.

Tektronix® is a registered trademark of Tektronix, Inc., Oregon.

FIGURE 1: Frequency Spectrum of Noise Sources Affecting Operational Amplifier Performance

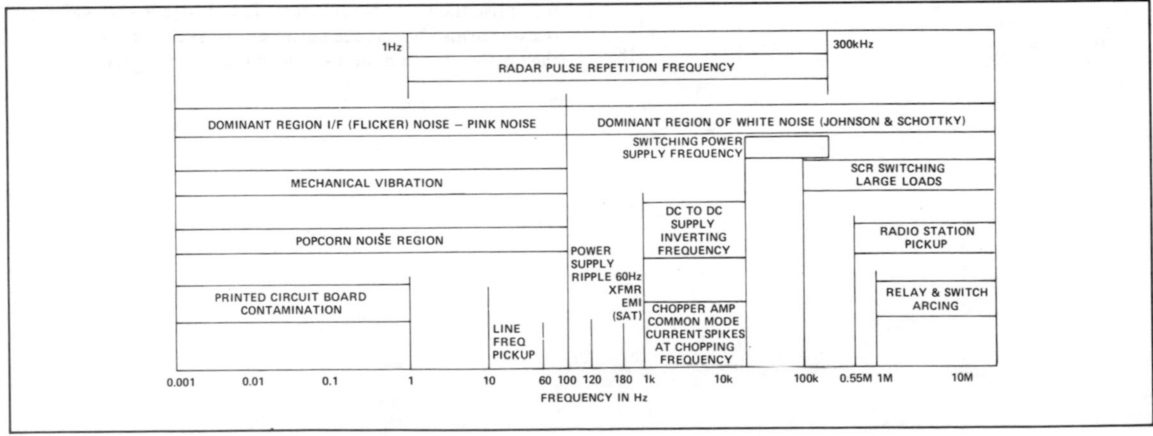

TABLE 1: External Noise Source Chart

Source	Nature	Causes	Minimization Methods
60Hz	Repetitive Interference	Powerlines physically close to op amp inputs. Poor CMRR at 60Hz. Power transformer primary-to-secondary capacitive coupling.	Reorientation of power wiring. Shielded transformers. Single point grounding. Battery power.
120Hz Ripple	Repetitive	Full wave rectifier ripple on op amp's supply terminals. Inadequate ripple consideration. Poor PSRR at 120Hz.	Thorough design to minimize ripple. RC decoupling at the op amp. Battery power.
180Hz	Repetitive EMI	180Hz radiated from saturated 60Hz transformers.	Physical reorientation of components. Shielding. Battery power.
Radio Stations	Standard AM Broadcast Through FM	Antenna action anyplace in system.	Shielding. Output filtering. Limited circuit bandwidth.
Relay and Switch Arcing	High Frequency Burst At Switching Rate	Proximity to amplifier inputs, power lines, compensation terminals, or nulling terminals.	Filtering of HF components. Shielding. Avoidance of ground loops. Arc suppressors at switching source.
Printed Circuit Board Contamination	Random Low Frequency	Dirty boards or sockets.	Thorough cleaning at time of soldering followed by a bakeout and humidity sealant.
Radar Transmitters	High Frequency Gated At Radar Pulse Repetition Rate	Radar transmitters from long range surface search to short range navigational—especially near airports.	Shielding. Output filtering of frequencies ≫ PRR.
Mechanical Vibration	Random < 100Hz	Loose connections, intermittent contact in mobile equipment.	Attention to connectors and cable conditions. Shock mounting in severe environments.
Chopper Frequency Noise	Common Mode Input Current At Chopping Frequency	Abnormally high noise chopper amplifier in system.	Balanced source resistors. Use bipolar input op amps instead. Use premium low noise chopper.
Switching Power Supply	Repetitive High Frequency Glitches In Supply And Ground	Improper ground return. Radiated noise from switching circuit.	Analog ground return to AC return. Shield power supply. Liberal power supply bypass at the op amp.

FIGURE 2: Noise Frequency Analysis RC Low Pass Filter

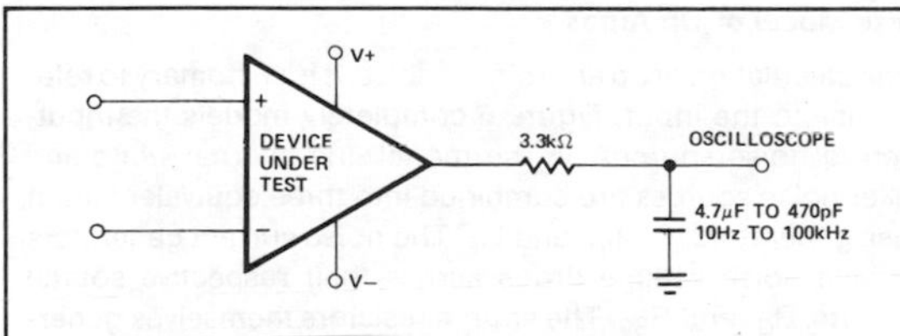

Power Supply Ripple

Power supply ripple at 120Hz is not usually thought of as a noise, but it should be. In an actual op amp application, it is quite possible to have a 120Hz noise component that is equal in magnitude to all other noise sources combined, and, for this reason, it deserves a special discussion.

To be negligible, 120Hz ripple noise should be between 10nV and 100nV referred to the input of an op amp. Achieving these low levels requires consideration of three factors: the op amp's 120Hz power supply rejection ratio (PSRR), the regulator's ripple rejection ratio, and finally, the regulator's input capacitor size.

PSRR at 120Hz for a given op amp may be found in the manufacturer's data sheet curves of PSRR versus frequency as shown in Figure 3. For the amplifier shown, 120Hz PSRR is

FIGURE 3: PSRR vs Frequency (OP-77)

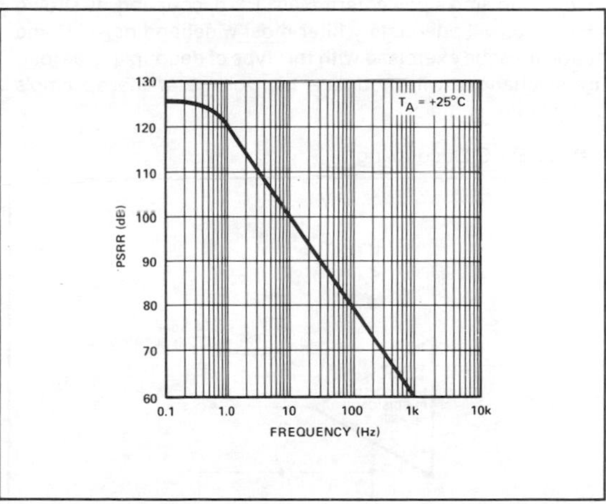

about 76dB, and to attain a goal of 100nV referred to the input, ripple at the power terminals must be less than 0.6mV. Today's IC regulators provide about 60dB of ripple rejection; in this case the regulator input capacitor must be made large enough to limit input ripple to 0.6V.

AN-15

Externally-compensated low noise op amps can provide improved 120Hz PSRR in high closed-loop gain configurations. The PSRR versus frequency curves of such an op amp are shown in Figure 4. When compensated for a closed-loop gain of 1000, 120Hz PSRR is 115dB. PSRR is still excellent at much higher frequencies allowing low ripple-noise operation in exceptionally severe environments.

FIGURE 4: PSRR vs Frequency (OP-06)

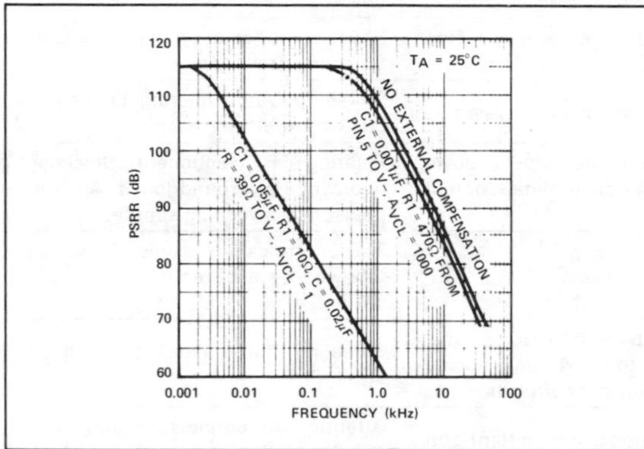

Power Supply Bypassing

Usually, 120Hz ripple is not the only power supply associated noise. Series regulator output typically contain at least 150μV of noise in the 100Hz to 100kHz range; switching types contain even more. Unpredictable amounts of induced noise can also be present on power leads from many sources. Since high frequency PSRR decreases at 20dB/decade, these higher frequency supply noise components must not be allowed to reach the op amp's power terminals. RC decoupling, as shown in Figure 5, will adequately filter most wideband noise. Some caution must be exercised with this type of decoupling, as load current changes will modulate the voltage at the op amp's supply pins.

FIGURE 5: RC Decoupling

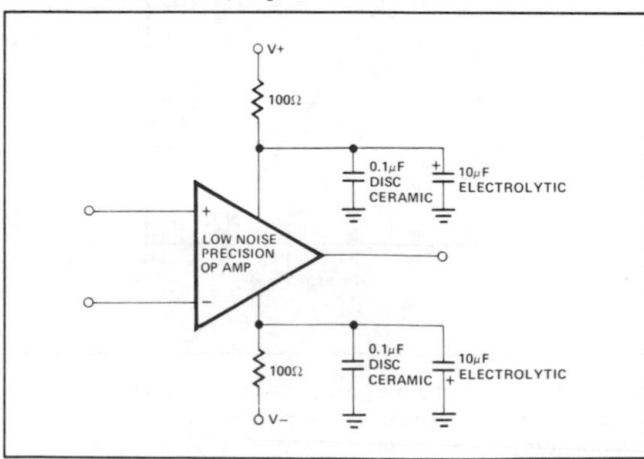

Power Supply Regulation

Any change in power supply voltage will have a resultant effect referred to an op amp's inputs. For the op amp of Figure 3, PSRR at DC is 126dB (0.5μV/V) which may be considered as a potential low frequency noise source. Power supplies for low noise op amp applications should, therefore, be both low in ripple and well-regulated. Inadequate supply regulation is often mistaken to be low frequency op amp noise.

When noise from external sources has been effectively minimized, further improvements in low noise performance are obtained by specifying the right op amp and through careful selection and application of the associated components.

OPERATIONAL AMPLIFIER INTERNAL NOISE

Most completely specified low-noise op amp data sheets specify current and voltage noises in a 1Hz bandwidth centered on 10Hz, 100Hz, and 1kHz, as well as low frequency noise over a range of 0.1Hz to 10Hz. To minimize total noise, a knowledge of the derivation of these specifications is useful. In this section, the reader is provided with an explanation of basic op amp-associated random noise mechanisms and introduced to a simplified method for calculating total input-referred noise in typical applications.

Op amp-associated noise currents and voltages are random in nature. They are aperiodic and uncorrelated to each other; and typically have Gaussian amplitude distributions, with the highest noise amplitudes having the lowest probability. There is a statistical relationship between the peak-to-peak value of random noise and its rms value. Where the amplitude distribution is Gaussian, the rms value may be multiplied by six to yield a peak-to-peak value that will not be exceeded 99.73% of the time (this is a handy rule-of-thumb for noise calculations).

Noise Model of Op Amps

In the calculation of op amp circuit noise, it is customary to refer all noise to the input. Figure 6 completely models the input-referred noise sources. In the model, the internal white and flicker noise sources are combined into three equivalent input noise generators, E_n, I_{n1}, and I_{n2}. The noise current generators produce noise voltage drops across their respective source resistors, R_{S1} and R_{S2}. The source resistors themselves generate thermal noise voltages, E_{t1} and E_{t2}. Total rms input-referred voltage noise, over a given bandwidth, is the square root of the sum of the squares of the five noise voltage generators over that bandwidth.

FIGURE 6: Op Amp Noise Model

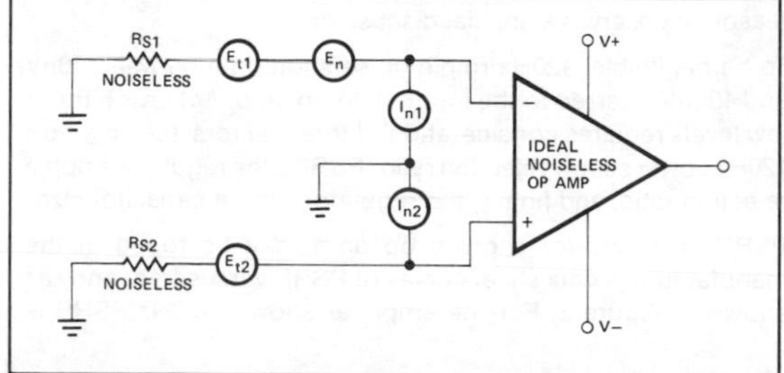

(2) $E_n(f_H, f_L) = \sqrt{E_n^2 + (I_{n1} \cdot R_{S1})^2 + (I_{n2} \cdot R_{S2})^2 + E_{t1}^2 + E_{t2}^2}$

Equation 2 describes, in total, all noise sources of an op amp circuit. It will be used throughout this application note.

Minimization of total noise requires an understanding of the mechanisms involved in each of the five generators. First, the white noise mechanisms, thermal and shot, are discussed, followed by other low frequency noise mechanisms, flicker and popcorn.

Noise Mechanisms of Op Amps

The two basic types of op amp-associated noises are white noise and flicker noise (1/f). White noise contains equal amounts of power in each hertz of bandwidth. Flicker noise is different in that it contains equal amounts of power in each decade of bandwidth. This is best illustrated by spectral noise density plots such as in Figures 7 and 8. Above a certain corner frequency, white noise dominates; below that frequency, flicker (1/f) noise is dominant. Low noise corner frequencies in conjunction with a low white noise magnitude distinguish low noise op amps from general purpose devices.

FIGURE 7: OP-77 Noise Voltage

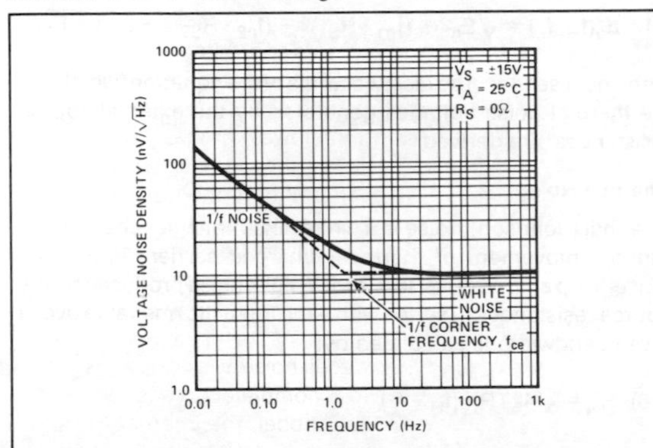

FIGURE 8: OP-77 Noise Current

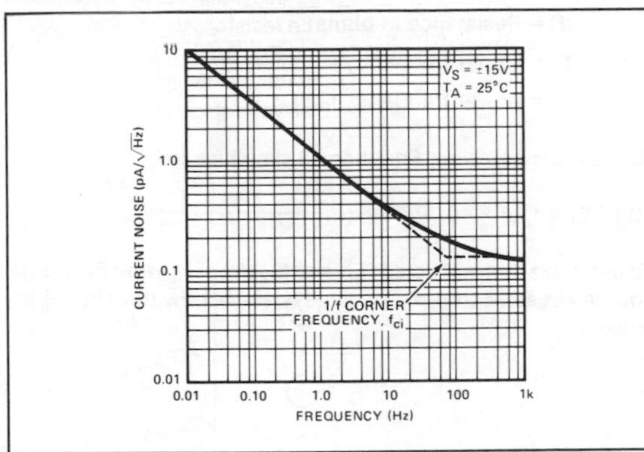

Mathematically, noise spectral density may be expressed as:

(3a) $e_n^2 = \frac{E_n^2}{\Delta f}$ (3b) $i_n^2 = \frac{I_n^2}{\Delta f}$

Where: e_n, i_n = Spectral noise density of voltage and current, respectively

E_n, I_n = Total rms voltage and current noise in a frequency band, respectively

Δf = Bandwidth of 1Hz

From Equation 3, the total rms noise in a frequency band from f_L to f_H is then,

(4a) $E_n^2 = \int_{f_L}^{f_H} e_n^2\, df$ (4b) $I_n^2 = \int_{f_L}^{f_H} i_n^2\, df$

Where: f_H = Upper frequency limit of interest

f_L = Lower frequency limit of interest

Equation 4 means that three things must be known to evaluate total voltage noise (E_n) or current noise (I_n): f_H, f_L, and a knowledge of noise behavior over frequency.

White Noise

White noise contains many frequency components and is so named in analogy to white light which is made up of many colors. The important point to remember is that white noise has equal noise power in each hertz of bandwidth. In other words, the noise spectral density of white noise is **constant** with varying frequency. Thus, Equation 4 may be rewritten to describe white noise over a frequency band.

(5a) $E_{nW} = e_{nW}\sqrt{f_H - f_L}$ (5b) $I_{nW} = i_{nW}\sqrt{f_H - f_L}$

When $f_H \geq 10 f_L$, the white noise expressions may be reduced to:

(6a) $E_{nW} = e_{nW}\sqrt{f_H}$ (6b) $I_{nW} = i_{nW}\sqrt{f_H}$

Flicker Noise

Unlike white noise, flicker (1/f) noise is not constant with respect to frequency, but has a power spectral density that is inversely proportional (K_e, K_i) to the frequency of interest as described in Equation 7.

(7a) $e_{nF}^2(f) = \frac{K_e^2}{f}$ (7b) $i_{nF}^2(f) = \frac{K_i^2}{f}$

or,

(8a) $e_{nF}(f) = \frac{K_e}{\sqrt{f}}$ (8b) $i_{nF}(f) = \frac{K_i}{\sqrt{f}}$

Where: K_e, K_i are constants of proportionality.

The constants of proportionality depend on a number of parameters internal to the amplifier. It will be shown later that the constants will drop out mathematically.

In order to calculate total voltage and current noise, the concept of corner frequency is useful. Referring to the graphs of e_n or i_n versus frequency as in Figures 7 and 8, we can see that it is a composite of a zero-slope line (white noise) summed with a line

AN-15

of slope −1/2 (1/f noise, or flicker noise). The projected intersection of these lines occurs where the two noise powers are equal, at a frequency called the *corner frequency*. Therefore, it follows that at the corner frequency, f_{ce} or f_{ci},

(9a) $e_{nW}^2 = e_{nF}^2(f_{ce}) = \frac{K_e^2}{f_{ce}}$ (9b) $i_{nW}^2 = i_{nF}^2(f_{ci}) = \frac{K_i^2}{f_{ci}}$

rearranging,

(10a) $K_e^2 = e_{nW}^2 \cdot f_{ce}$ (10b) $K_i^2 = i_{nW}^2 \cdot f_{ci}$

substituting in Equation 7,

(11a) $e_{nF}^2(f) = e_{nW}^2 \cdot \frac{f_{ce}}{f}$ (11b) $i_{nF}^2(f) = i_{nW}^2 \cdot \frac{f_{ci}}{f}$

or,

(12a) $e_{nF}(f) = e_{nW}\sqrt{\frac{f_{ce}}{f}}$ (12b) $i_{nF}(f) = i_{nW}\sqrt{\frac{f_{ci}}{f}}$

We can find the rms flicker noise in a band as follows:

(13a) $$E_{nF}^2 = \int_{f_L}^{f_H} e_{nF}^2(f)\,df = e_{nW}^2 \cdot f_{ce} \cdot \ln\left(\frac{f_H}{f_L}\right)$$

(13b) $$I_{nF}^2 = \int_{f_L}^{f_H} i_{nF}^2(f)\,df = i_{nW}^2 \cdot f_{ci} \cdot \ln\left(\frac{f_H}{f_L}\right)$$

Typical bipolar op amp corner frequencies for voltage noise are in the range of 1 to 20Hz; and for current noise, 10 to 1,000Hz. In comparison, FET input op amps have voltage noise corner frequencies in the range of 100Hz to 500Hz. Still higher are CMOS op amps whose corner frequencies are typically on the order of 1kHz.

Now that we have the mathematical expressions describing white noise and flicker noise, we can sum (by root-sum-square method) the two components to yield a total spectral density expression.

(14a) $e_n^2 = e_{nW}^2 + e_{nF}^2(f)$ (14b) $i_n^2 = i_{nW}^2 + i_{nF}^2(f)$

substituting from Equation 11,

(15a) $e_n = e_{nW}\sqrt{1 + \frac{f_{ce}}{f}}$ (15b) $i_n = i_{nW}\sqrt{1 + \frac{f_{ci}}{f}}$

Equation 15 is an expression frequently used to describe noise (voltage and current) curves seen in op amp data sheets.

The rms noise in a band is then:

(16) $$E_n(f_H, f_L) = e_{nW}\sqrt{f_{ce} \cdot \ln\left(\frac{f_H}{f_L}\right) + (f_H - f_L)}$$

(17) $$I_n(f_H, f_L) = i_{nW}\sqrt{f_{ci} \cdot \ln\left(\frac{f_H}{f_L}\right) + (f_H - f_L)}$$

Where: e_{nW} = White noise voltage spectral density
i_{nW} = White noise current spectral density
f_{ce} = Voltage noise corner frequency
f_{ci} = Current noise corner frequency
f_H = Upper frequency limit of interest
f_L = Lower frequency limit of interest

The two most important internally-generated noise minimization rules are derived from Equation 16 and 17: a) limit the circuit bandwidth, and b) use operational amplifiers with low white noise specifications in conjunction with low corner frequencies. So far we have derived the noise voltage (E_n) and noise current (I_n) components (Equations 16 and 17) for the first three terms of Equation 2, which is reproduced below.

(2) $$E_n(f_H, f_L) = \sqrt{E_n^2 + (I_{n1} \cdot R_{S1})^2 + (I_{n2} \cdot R_{S2})^2 + E_{t1}^2 + E_{t2}^2}$$

In the next section, the last two terms of the equation, which are the thermal noise voltages generated by the external source resistances, are derived.

Thermal Noise

Thermal (Johnson) noise is a white noise voltage generated by random movement of thermally-charged carriers in a resistance; in op amp circuits, this is the type of noise produced by the source resistances in series with each input. Its rms value over a given bandwidth is calculated by:

(18) $$E_t = \sqrt{4kTR \cdot (f_H - f_L)}$$

Where: k = Boltzmann's constant = 1.38×10^{-23} joules/K
T = Absolute temperature, kelvin
R = Resistance in ohms
f_H = Upper frequency limit in hertz
f_L = Lower frequency limit in hertz

At room temperature, Equation 18 simplifies to:

(19) $$E_t = 1.28 \times 10^{-10}\sqrt{R \cdot (f_H - f_L)}$$

To minimize thermal noise (E_{t1} and E_{t2}) from R_{S1} and R_{S2}, large source resistors and excessive system bandwidth should be avoided.

Thermal noise is also generated inside the op amp, principally from r_{bb}', the base-spreading resistances in the input stage transistors. These noises are included in E_n, the total equivalent input voltage noise generator.

All the component noise sources of Equation 2 have now been derived. Total noise of an op amp circuit may be easily calculated using the equation. In the next sections, examples using several precision op amps will be calculated to illustrate the noise minimization techniques as well as to contrast the different noise performance of these devices.

TOTAL NOISE CALCULATION

With data sheet curves and specifications, and a knowledge of source resistance values, total input-referred noise may be calculated for a given application. To illustrate the method, noise information from the Precision Monolithics OP-77A and OP-27A data sheets are reproduced in Figure 9. The first step is to determine the current and voltage noise corner frequencies so that the E_n and I_n terms of Equation 2 may be calculated using Equations 16 and 17.

Corner Frequency Determination

In the input spot noise versus frequency curves of Figure 9, it may be seen that voltage noise ($R_S = 0$) begins to rise at about 3Hz. Lines projected from the horizontal (white noise) portion and the sloped (flicker noise) portion intersect at 2Hz, the voltage noise corner frequency (f_{ce}). In the center curve, excluding thermal noise from the source resistance, current noise multiplied by 200kΩ is plotted as a voltage noise. Lines projected from the horizontal portion and sloped portions intersect at 80Hz, the current noise corner frequency (f_{ci}).

FIGURE 9A: OP-77 Input Spot Noise Voltage vs Frequency

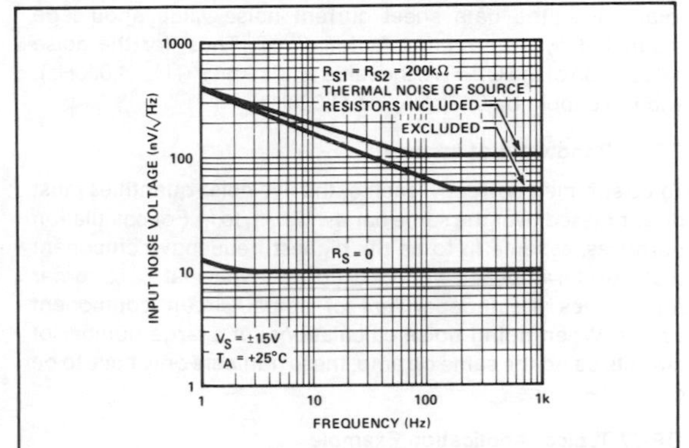

FIGURE 9B: OP-77/OP-27 Ultra-Low Offset Voltage Op Amps

ELECTRICAL CHARACTERISTICS at $V_S = \pm15V$ and $T_A = 25°C$, unless otherwise noted.

			OP-77A			OP-27A			
PARAMETER	**SYMBOL**	**CONDITIONS**	**MIN**	**TYP**	**MAX**	**MIN**	**TYP**	**MAX**	**UNITS**
Input Noise Voltage	$e_{np\text{-}p}$	0.1Hz to 10Hz	—	0.35	0.6	—	0.08	0.18	$\mu V_{p\text{-}p}$
Input Noise Voltage Density	e_n	f_O = 10Hz f_O = 100Hz f_O = 1000Hz	— — —	10.3 10.0 9.6	18.0 13.0 11.0	— — —	3.5 3.1 3.0	5.5 4.5 3.8	nV/$\sqrt{Hz}$
Input Noise Current	$i_{np\text{-}p}$	0.1Hz to 10Hz	—	14	30	—	—	—	$pA_{p\text{-}p}$
Input Noise Current Density	i_n	f_O = 10Hz f_O = 100Hz f_O = 1000Hz	— — —	0.32 0.14 0.12	0.80 0.23 0.17	— — —	1.7 1.0 0.4	4.0 2.3 0.6	pA/$\sqrt{Hz}$
Input Offset Voltage	V_{OS}		—	10	25	—	10	25	μV
Input Offset Voltage Drift	TCV_{OS}	$-55°C \leq T_A \leq +125°C$	—	0.1	0.3	—	0.2	0.6	μV/°C
Long Term Input Offset Voltage Stability	V_{OS}/Time		—	0.2	1.0	—	0.2	1.0	μV/Mo
Input Offset Current	I_{OS}		—	0.3	1.5	—	7	35	nA
Input Bias Current	I_B		—	±1.2	±2.0	—	±10	±40	nA

INPUT NOISE VOLTAGE ($e_{np\text{-}p}$)
The peak-to-peak noise voltage in a specified frequency band.

INPUT NOISE VOLTAGE DENSITY(e_n)
The rms noise voltage in a 1Hz band surrounding a specified value of frequency.

INPUT NOISE CURRENT ($i_{np\text{-}p}$)
The peak-to-peak noise current in a specified frequency band.

INPUT NOISE CURRENT DENSITY (i_n)
The rms noise current in a 1Hz band surrounding a specified value of frequency.

Equations 16 and 17 also require e_{nW} and i_{nW} for calculation of E_n and I_n. To find e_{nW} and i_{nW}, use the data sheet specifications a decade or more above the respective corner frequencies; in the case of the OP-77A, e_{nW} is $9.6V/\sqrt{Hz}$ (1,000Hz), and i_{nW} is $0.12pA/\sqrt{Hz}$ (1,000Hz). At this time, it should be noted that the noise current, $0.12pA/\sqrt{Hz}$, is a value that has been incorrectly derived from the standardized, commonly-used test method on virtually ALL commercially available op amps. The value is off by a factor of $\sqrt{2}$. Therefore, in order to calculate the correct total noise, the data sheet current noise value should be multiplied by a correction factor of $\sqrt{2}$. Thus, for the noise calculation of the OP-77A, the value e_{nW} is $9.6nV/\sqrt{Hz}$ (1,000Hz), and i_{nW} should be $0.17pA/\sqrt{Hz}$ (1,000Hz).

OP-77 Bandwidth of Interest

To be summed correctly, each of the five noise quantities must be expressed over the same bandwidth, f_H to f_L. For calculation purposes, assume f_H to be the highest frequency component that must be amplified without distortion. Note that e_n, i_n, corner frequencies are independent of actual circuit component values. When doing noise calculations for a large number of circuits using the same op amp, these numbers only have to be calculated once.

OP-77 Typical Application Example

Figure 10A shows a typical ×10 gain stage with a 10kΩ source resistance. In Figure 10B, the circuit is redrawn to show five noise voltage sources. To evaluate total input-referred noise, the values of each of the five sources must be determined.

FIGURE 10A: Noise Analysis Circuit

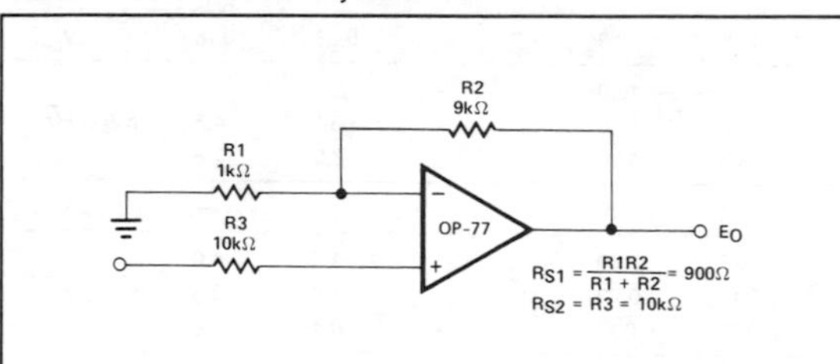

FIGURE 10B: Noise Analysis Equivalent Circuit

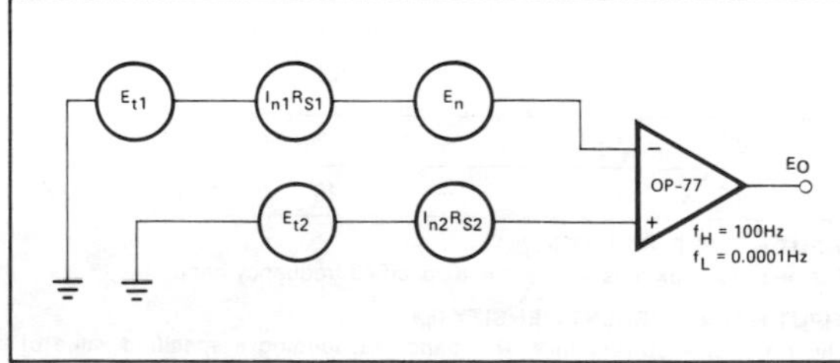

Using Equation 19: $E_t = 1.28 \times 10^{-10} \sqrt{R \cdot (f_H - f_L)}$

$E_{t1} = 1.28 \times 10^{-10} \sqrt{(900\Omega)\ (100Hz)} = 0.04\mu Vrms$

$E_{t2} = 1.28 \times 10^{-10} \sqrt{(10k\Omega)\ (100Hz)} = 0.128\mu Vrms$

Next, calculate I_n using Equation 17:

$$I_n = i_n \sqrt{f_{ci} \cdot \ln\left(\frac{f_H}{f_L}\right) + f_H - f_L}$$

$$= 0.17pA \sqrt{80 \cdot \ln\left(\frac{100Hz}{0.0001Hz}\right) + 100 - 0.0001}$$

$$= 5.9pArms$$

and:

$I_{n1} \cdot R_{S1} = 5.9pA \cdot (900\Omega) = 0.0053\mu Vrms$

$I_{n2} \cdot R_{S2} = 5.9pA \cdot (10k\Omega) = 0.059\mu Vrms$

Finally, E_n from Equation 16:

$$E_n = e_n \sqrt{f_{ce} \cdot \ln\left(\frac{f_H}{f_L}\right) + f_H - f_L}$$

$$= 9.6nV \sqrt{2 \cdot \ln\left(\frac{100Hz}{0.0001Hz}\right) + 100 - 0.0001}$$

$$= 0.108\mu Vrms$$

Substituting in Equation 2:

$$E_n(f_H - f_L) = \sqrt{E_n{}^2 + I_{n1}{}^2\, R_{S1}{}^2 + I_{n2}{}^2\, R_{S2}{}^2 + E_{t1}{}^2 + E_{t2}{}^2}$$

$$= \sqrt{\begin{matrix}(0.108\mu V)^2 + (0.0053\mu V)^2 + (0.059\mu V)^2 + \\ (0.04\mu V)^2 + (0.128\mu V)^2\end{matrix}}$$

$$= 0.18\mu Vrms$$

Total input-referred noise = $1.08\mu V$ peak-to-peak (0.0001Hz to 100Hz).

Notice that of the five terms in the equation, the first and the last terms dominate. Since the first term is the total rms noise voltage inherent of the amplifier, nothing can be done by the system designer to lower its noise other than to choose a device having inherently low noise characteristics. As can be seen in Equation 16, two key parameters determine the total rms noise of an amplifier—low white noise density and low noise corner frequency.

Notice that the thermal noise voltage (last term) of Equation 2 is determined by the 10kΩ value selected for R3. Had the value been reduced to 1kΩ, the thermal noise voltage would have been $0.04\mu Vrms$ instead of $0.128\mu Vrms$. As a result, total rms noise voltage would have become $0.122\mu V$, a remarkable 32% reduction in total noise.

Indeed, low noise design requires the system designer not only to choose an amplifier with low noise characteristics, but also to pay close attention in selecting appropriately low source resistances in the input circuit.

741 Calculation Example

The preceding calculation determined total noise in a given bandwidth using a low noise op amp. To place this level of performance into perspective, a calculation using the industry-standard 741 op amp in the circuit of Figure 10 is useful. Once again the starting point is corner frequency determination, using the data sheet curves of Figure 11: $f_{ce} = 200Hz$; $f_{ci} = 2kHz$; $e_n = 20nV/\sqrt{Hz}$; $i_n = (\sqrt{2}) \cdot (0.5pA/\sqrt{Hz}) = 0.71pA/\sqrt{Hz}$.

Using these corner frequencies and noise magnitudes, E_n and I_n are calculated to be 1.07μVrms and 118pArms respectively. Multiplying this noise current by the source resistance gives terms 2 and 3 of Equation 2 as shown below:

(2) $E_n(f_H, f_L) = \sqrt{E_n{}^2 + I_{n1}{}^2 R_{S1}{}^2 + I_{n2}{}^2 R_{S2}{}^2 + E_{t1}{}^2 + E_{t2}{}^2}$

substituting in the equation:

$$E_n(f_H, f_L) = \sqrt{(1.07\mu V)^2 + (0.106\mu V)^2 + (1.18\mu V)^2 + (0.04\mu V)^2 + (0.128\mu V)^2}$$

$$= 1.6\mu Vrms$$

Total input-referred noise = 9.6μV peak-to-peak (0.0001Hz to 100Hz). This is more than 8 times that of the low noise OP-77 example. Notice further in this example, the third term of the equation becomes an additional dominant term. It is due to a higher noise current flow in the 10kΩ source resistance.

The calculation examples illustrate four rules for minimizing noise in operational amplifier applications:

Rule 1. Use an op amp with low noise characteristics.

Rule 2. Use an op amp with low noise corner frequencies.

Rule 3. Keep source resistances as low as practical.

Rule 4. Limit circuit bandwidth to signal bandwidth.

FIGURE 11A: Input Noise Voltage as a Function of Frequency

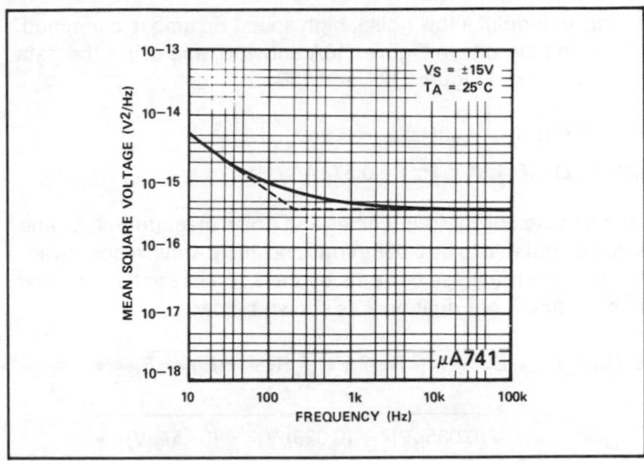

FIGURE 11B: Input Noise Current as a Function of Frequency

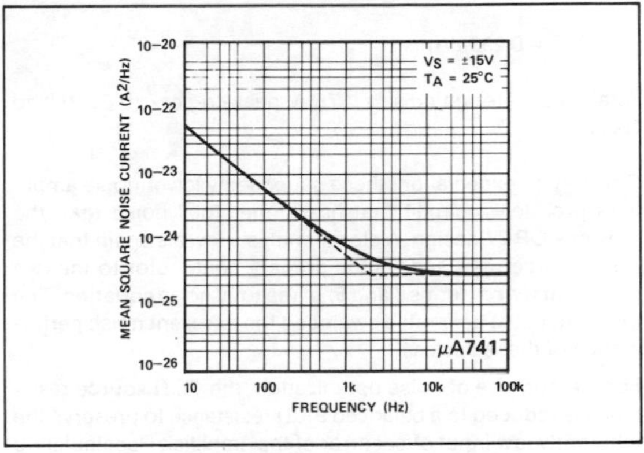

FIGURE 12: OP-27, OP-37, and OP-227 Noise Voltage and Current as a Function of Frequency

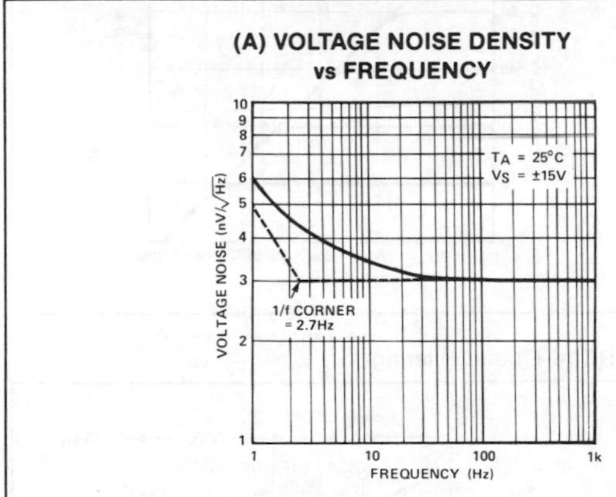

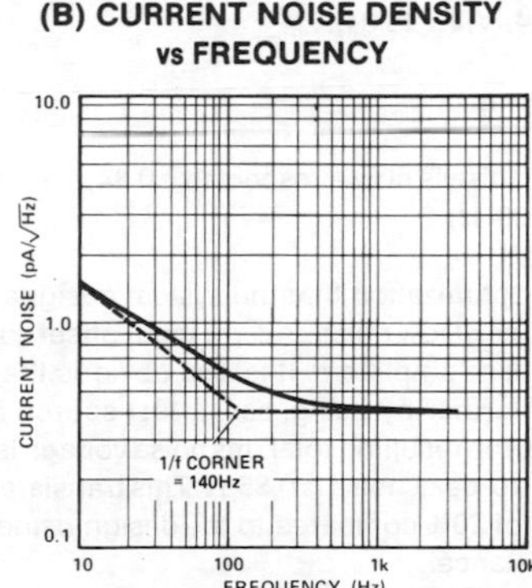

AN-15

OP-27/OP-227/OP-37 Noise Optimization Design

In this example, a low noise, high speed op amp is examined. Using the circuits in Figures 10A and 10B, and using the data sheet curves of Figures 12A and 12B:

$f_{ce} = 2.7\text{Hz};\ f_{ci} = 140\text{Hz};\ e_n = 3.0\text{nV}/\sqrt{\text{Hz}};$

$i_n = (\sqrt{2}) \cdot (0.4\text{pA}/\sqrt{\text{Hz}}) = 0.57\text{pA}/\sqrt{\text{Hz}}$

Using these corner frequencies and noise magnitudes, E_n and I_n are calculated to be 0.035μVrms and 25.7pArms, respectively. Multiplying the noise currents by the source resistances yield terms 2 and 3 of Equation 2 as shown below:

$$E_n(f_H, f_L) = \sqrt{E_n^2 + I_{n1}^2 R_{S1}^2 + I_{n2}^2 R_{S2}^2 + E_{t1}^2 + E_{t2}^2}$$

$$= \sqrt{(0.035\mu V)^2 + (0.023\mu V)^2 + (0.257\mu V)^2 + (0.04\mu V)^2 + (0.128\mu V)^2}$$

$$= 0.293\mu \text{Vrms}$$

Total input-referred noise = 1.76μV peak-to-peak (0.0001Hz to 100Hz).

Contrary to expectation, these supposedly lower noise amplifiers produce a circuit that has higher total noise than the previous OP-77 design. A closer analysis reveals again that the 10kΩ source resistance is the primary contributor to the two dominant terms (terms 3 and 5) of the total noise equation. The resulting noise generated swamped the excellent noise performance of these devices.

For the purpose of noise optimization, the 10kΩ source resistance is reduced to a balanced 910Ω resistance to preserve the inherently low input offset error of the amplifier. Recalculating Equation 2,

$$E_n(f_H, f_L) = \sqrt{(0.035\mu V)^2 + (0.023\mu V)^2 + (0.023\mu V)^2 + (0.04\mu V)^2 + (0.04\mu V)^2}$$

$$= 0.074\mu \text{Vrms}$$

Total input-referred noise is now a respectable 0.44μV peak-to-peak (0.0001Hz to 100Hz).

It is clear from this optimization that the system designer can achieve both a balance of low noise and low input offset voltage performance with these amplifiers. It is also obvious that one can optimize noise further by using, say, a 10Ω source resistance; in which case, the resulting total rms noise voltage is now 0.058μV, and a peak-to-peak noise is 0.35μV. This translates to a net noise reduction of 20% compared to the design using 1kΩ balance source resistance.

LIMITING BANDWIDTH TO MINIMIZE NOISE

Effective circuit bandwidth must not be much greater than signal bandwidth or amplification of undesirable high frequency noise components will occur. Throughout the preceding calculations, an assumption of "bandwidth-of-interest" was made, while in actual application the amplifier's bandwidth must be considered.

In Figure 13, the OP-77 frequency response curves show a rolloff of 20dB/decade; integration of the area under the curve will show the effective circuit noise bandwidth to be 1.57 times the 3dB bandwidth. In most closed-loop gain configurations, the amplifier's bandwidth may be greater than required, and output filtering, such as in Figure 14, could be used. As an alternate to output filtering, an integrating capacitor may be connected across the feedback resistor. Bandwidth may also be limited in some applications by overcompensating an externally-compensated low noise op amp, such as the OP-06.

FIGURE 13A: OP-77 Open-Loop Frequency Response

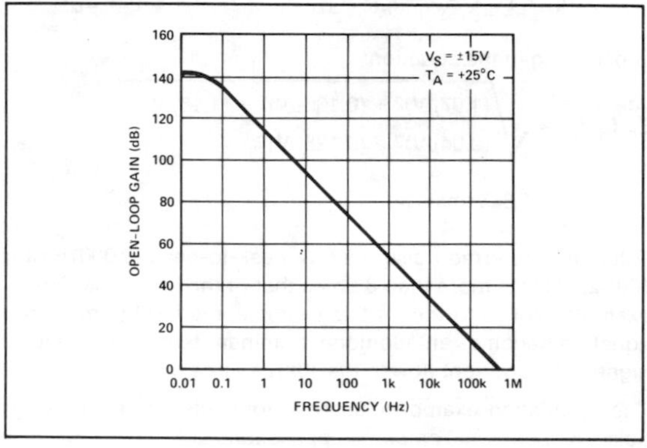

FIGURE 13B: OP-77 Closed-Loop Response for Various Gain Configurations

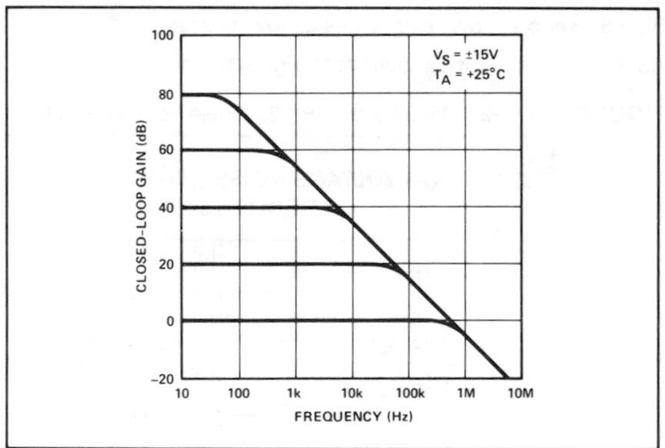

FIGURE 14: Output Filtering

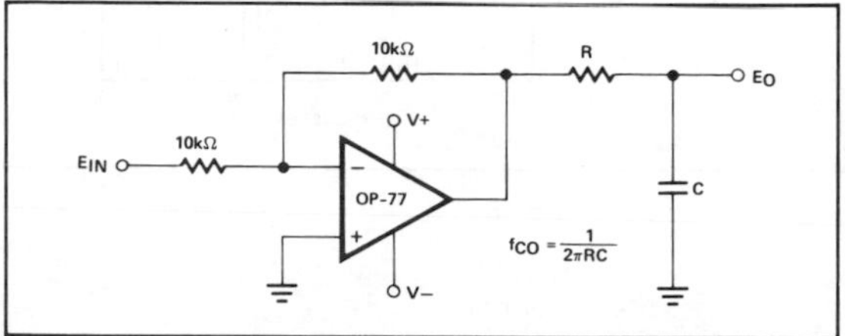

MISCELLANEOUS NOISE MINIMIZATION METHODS

Certain other noise mechanisms merit consideration: use metal film resistors; carbon resistors exhibit "excess noise," with both 1/f and white noise content being related to DC applied voltage. The use of balanced source resistors, while sometimes good for DC error purposes, will increase noise; the balancing resistor is not required for op amps such as the OP-07 and OP-77, since $I_{OS} \cong I_B$. Keep noise in its proper perspective; minimize it without introducing additional DC errors. Use low noise op amps with overall DC specifications that will satisfy the application.

OTHER NOISES

Shot noise (Schottky noise) is a white noise current associated with the fact that current flow is actually a movement of discrete charged particles (electrons) across a potential barrier, such as a PN junction of a transistor or diode. Shot noise is a component of i_n, and indirectly, e_n. In Figure 6, I_{n1} and I_{n2}, above the 1/f frequency, are shot noise currents which are related to the amplifier's DC input bias currents:

$$(20)\quad I_{sh} = \sqrt{2qI_{DC}\,(f_H - f_L)}$$

Where: I_{sh} = rms shot noise value in amps

q = Charge of an electron = 1.602×10^{-19} C

I_{DC} = DC bias current in amps

f_H = Upper frequency limit in hertz

f_L = Lower frequency limit in hertz

At room temperature Equation 20 simplifies to:

$$(21)\quad I_{sh} = 5.66 \times 10^{-10}\sqrt{I_{DC}\,(f_H - f_L)}$$

Shot noise currents also flow in the input-stage emitter dynamic resistances (r_e), producing input noise voltages. These voltages, along with the r_{bb}' thermal noise, make up the white noise portion of E_n; the total equivalent input noise voltage generator.

Shot noise can also be generated from external sources such as PIN photodiodes, zener diodes, and other semiconductor junction devices. Noise current from these sources may be calculated using Equation 20 or 21.

In limited bandwidth, very low frequency applications, *flicker* (1/f) noise is the most critical noise source. An op amp designer minimizes flicker noise by keeping current noise components in the input and second stages from contributing to input voltage noise. Equation 22 illustrates this relationship:

$$(22)\quad \frac{i_{n\ \text{second stage}}}{g_{m\ \text{first stage}}} = e_{n\ \text{input}}$$

Another critical factor is corner frequency. For minimum noise, the current and voltage noise corner frequencies must be low; this is crucial. As shown in Figure 15, low-noise corner frequencies distinguish low-noise op amps from ordinary industry-standard 741 types.

The photograph in Figure 16, taken using the test circuit of Figure 17, illustrates the flicker noise performance of the OP-77. This device demonstrates proper attention to low noise circuit design and wafer processing and achieves a remarkable 0.35μV peak-to-peak input voltage noise in the 0.1Hz to 10Hz bandwidth.

FIGURE 15: Noise Voltage Comparison

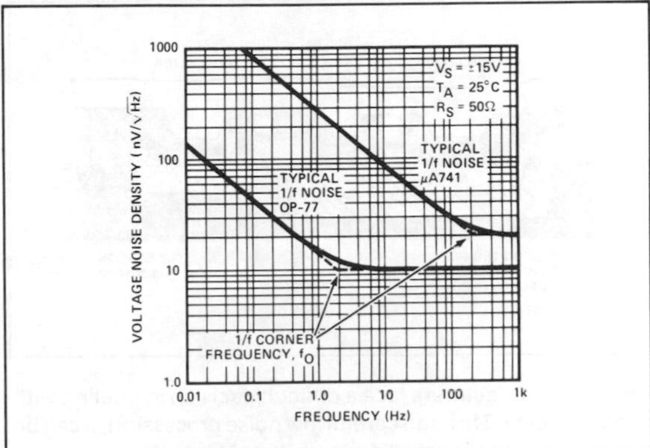

FIGURE 16: OP-77 Low Frequency Noise

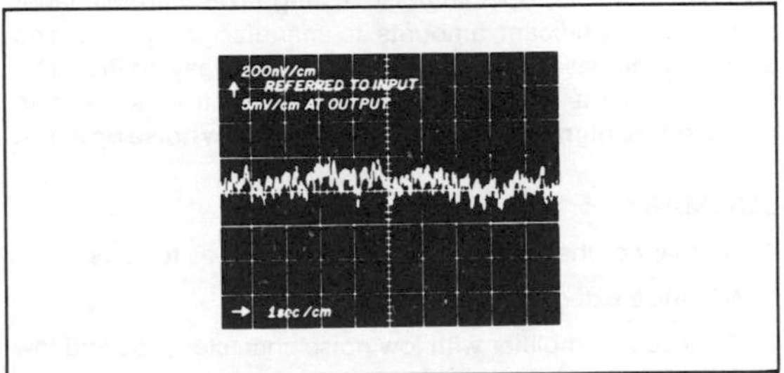

FIGURE 17: Low Frequency Noise Test Circuit

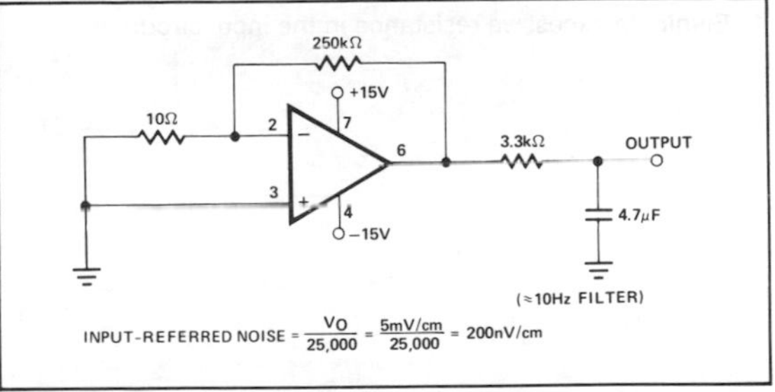

Popcorn noise (burst noise) is a momentary change in input bias current usually occurring below 100Hz, and is caused by imperfect semiconductor surface conditions incurred during wafer processing. Precision Monolithics minimizes this problem through careful surface treatment, general cleanliness, and a special three-step process known as "Triple Passivation."

To begin the process, a specially-treated thermal silicon dioxide layer is grown. This protects the junctions and also attracts any residual ionic impurities to the top surface of the oxide, where they are held fixed. Next, a layer of silicon nitride is applied to prevent the entry of any potential contamination or impurities.

AN-15

The third step is the thick glass overcoat which leaves only the bonding pads exposed. A cutaway view of a finished device is shown in Figure 18.

FIGURE 18: Triple Passivated Integrated Circuit Process

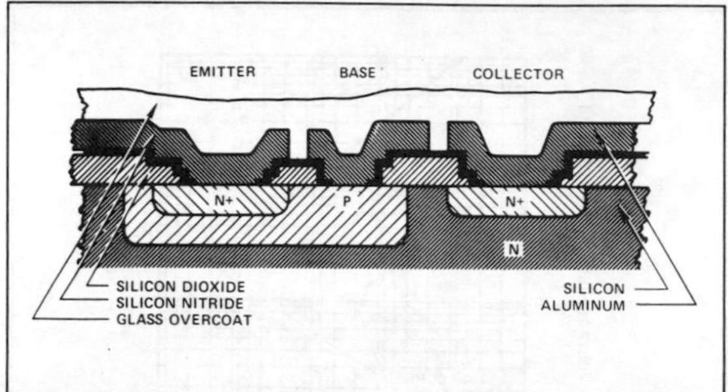

Op amp manufacturers face a difficult decision in dealing with popcorn noise. Through careful low noise processing, it can be eliminated from almost all devices; alternatively, the processing may be relaxed, and finished devices must be individually tested for this parameter. Special noise testing takes valuable labor time, adds significant amounts to manufacturing cost, and ultimately increases the price a customer has to pay. At Precision Monolithics, the low noise process alternative is used to manufacture high volumes of cost-effective low noise op amps.

SUMMARY

A summary of the major points to consider is as follows:

1. Minimize externally-generated noise.
2. Choose an amplifier with low noise characteristics and low 1/f noise corner frequencies.
3. Limit the circuit bandwidth to signal bandwidth.
4. Eliminate excessive resistance in the input circuit.

CONCLUSION

Recent improvements in IC op amp DC specifications have made noise an important error consideration. From data sheet information and source resistance values, total input-referred noise over a given bandwidth can be easily calculated. Total noise can be minimized by a thorough understanding of the various noise-generation mechanisms.

NOISE BIBLIOGRAPHY

1. "The OP-07 Ultra-Low Offset Voltage Op Amp—A Bipolar Op Amp That Challenges Choppers, Eliminates Nulling," D. Soderquist and G. Erdi, Precision Monolithics Inc., Application Note AN-13.
2. *Low Noise Electronic Design*, C.D. Motchenbacher and F.C. Fitchen, John Wiley and Sons, 1973.
3. "Low Frequency Noise Predicts When a Transistor Will Fail," A. Van der Ziel and H. Tong, *Electronics*, 39, November 28, 1966.
4. "Bistable Noise in Operational Amplifiers," S. Hsu, *IEEE Solid-State Circuit SC-6*, December 1971.
5. "Low-Noise Transistor Amplifiers," E. Chenette, *Solid State Design*, 5, February, 1964.
6. *Electrical Noise*, A. Bennett, McGraw-Hill, 1960.
7. *Noise*, A. Van der Ziel, Prentice-Hall, 1954.

SECTION 1—BASIC IN-AMP THEORY

Introduction

The fact that some op-amps are dubbed instrumentation amplifiers (or in-amps) by their suppliers does not make them an in-amp, even though they may be used in instrumentation. Likewise, an isolation amplifier is also not an instrumentation amplifier. This application note will explain what an in-amp actually is, how it operates, and how and where to use it.

What Is an Instrumentation Amplifier?

An instrumentation amplifier is a closed-loop gain block which has a differential input and an output which is singled-ended with respect to a reference terminal. Most commonly, the impedances of the two input terminals are balanced and have high values, typically $10^9\ \Omega$ or greater. As with op-amps, output impedance is very low, nominally only a few milliohms. Unlike an op-amp, which has its closed-loop gain determined by external resistors connected between its inverting input and its output, an in-amp employs an internal feedback resistor network which is isolated from its signal input terminals. With the input signal applied across the two differential inputs, gain is either preset internally or is user-set by an internal (via pins) or external gain resistor, which is also isolated from the signal inputs. Figure 1 contrasts the differences between op-amp and in-amp input characteristics.

Common-mode rejection, the property of cancelling out any signals which are common (the same potential on both inputs) while amplifying any signals which are differential (a potential difference between the inputs), is the most important function an instrumentation amplifier provides. Common-mode gain (A_{CM}) is the ratio of change in output voltage to a change in common-mode input voltage. This is the net gain (or attenuation) from input to output for voltages common to both inputs. For example, an in-amp with a common-mode gain of 1/1,000 and a 10 volt common-mode voltage at its inputs will exhibit a 10 mV output change. The differential or "normal mode" gain (A_D) is the gain between input and output for voltages applied differentially (or across) the two inputs. The common-mode rejection ratio (CMRR) is simply the ratio of

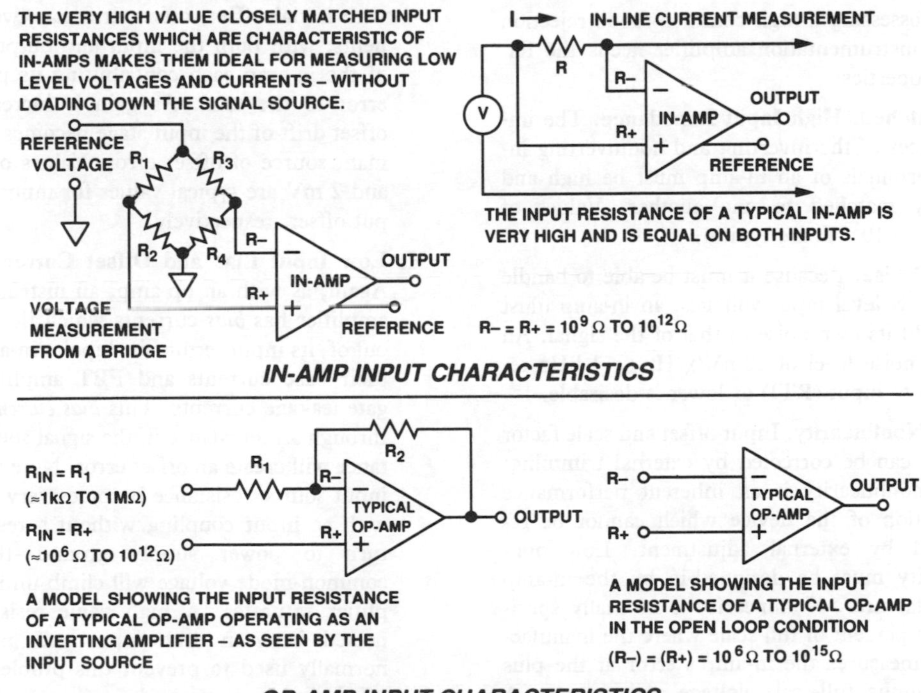
THE VERY HIGH VALUE CLOSELY MATCHED INPUT RESISTANCES WHICH ARE CHARACTERISTIC OF IN-AMPS MAKES THEM IDEAL FOR MEASURING LOW LEVEL VOLTAGES AND CURRENTS – WITHOUT LOADING DOWN THE SIGNAL SOURCE.
REFERENCE VOLTAGE
R_1
R_3
R_2
R_4
R–
R+
IN-AMP
OUTPUT
REFERENCE
VOLTAGE MEASUREMENT FROM A BRIDGE
I
IN-LINE CURRENT MEASUREMENT
R
V
OUTPUT
THE INPUT RESISTANCE OF A TYPICAL IN-AMP IS VERY HIGH AND IS EQUAL ON BOTH INPUTS.
$R- = R+ = 10^9\ \Omega$ TO $10^{12}\ \Omega$
IN-AMP INPUT CHARACTERISTICS
$R_{IN} = R_1$
(≈1kΩ TO 1MΩ)
$R_{IN} = R+$
($\approx 10^6\ \Omega$ TO $10^{12}\ \Omega$)
TYPICAL OP-AMP
OUTPUT
A MODEL SHOWING THE INPUT RESISTANCE OF A TYPICAL OP-AMP OPERATING AS AN INVERTING AMPLIFIER – AS SEEN BY THE INPUT SOURCE
A MODEL SHOWING THE INPUT RESISTANCE OF A TYPICAL OP-AMP IN THE OPEN LOOP CONDITION
$(R-) = (R+) = 10^6\ \Omega$ TO $10^{15}\ \Omega$
OP-AMP INPUT CHARACTERISTICS

Figure 1. Op-Amp vs. In-Amp Input Characteristics

the differential gain, A_D, to the common-mode gain (A_{CM}).

Common-mode rejection is usually specified for a full-range common-mode voltage (CMV) change at a given frequency, and a specified imbalance of source impedance (e.g., 1 kΩ source unbalance, at 60 Hz). The term CMR is a logarithmic expression of the common-mode rejection ratio (CMRR).

That is: $CMR = 20 \text{ Log}_{10} \text{ CMRR}$.

In order to be effective, an in-amp needs to be able to amplify microvolt-level signals while simultaneously rejecting volts of common-mode at its inputs.

This requires that instrumentation amplifiers have very high common-mode rejection—typical values of CMR are 70 dB to over 100 dB, with CMR usually improving at higher gains. While it is true that operational amplifiers, connected as subtractors, also provide common-mode rejection, the user must provide closely matched external resistors. On the other hand, monolithic in-amps with their pre-trimmed resistor networks, are far easier to apply.

What Other Properties Define a High Quality In-Amp?

Besides possessing a high common-mode rejection ratio, an instrumentation amplifier needs the following properties:

A. **A Matched, High Input Impedance.** The impedances of the inverting and noninverting input terminals of an in-amp must be high and closely matched to one another. Values of 10^9 Ω to 10^{12} Ω are typical.

B. **Low Noise.** Because it must be able to handle very low level input voltages, an in-amp must not add its own noise to that of the signal. An input noise level of 10 nV/$\sqrt{\text{Hz}}$ @ 1 kHz referred to input (RTI) or lower is desirable.

C. **Low Nonlinearity.** Input offset and scale factor errors can be corrected by external trimming; but, nonlinearity is an inherent performance limitation of the device which cannot be removed by external adjustment. Low nonlinearity must be designed-in by the in-amp manufacturer. Nonlinearity is normally specified in percent of full scale where the manufacturer measures the in-amp's error at the plus and minus full-scale voltage and at zero. A nonlinearity error of 0.01% is typical for a high quality in-amp; some even have levels as low as 0.0001%.

D. **Adequate Bandwidth.** An instrumentation amplifier must provide sufficient bandwidth needed for the particular application. Since typical unity gain small-signal bandwidths fall between 500 kHz and 4 MHz, performance at low gains is easily achieved, but at higher gains, bandwidth becomes much more of an issue.

F. **Simple Gain Selection.** Gain selection should be quick and easy to apply. Gain selection via a single external resistor is one common method. Many in-amps provide a choice of internally preset gains (often pin selectable) which are stable over temperature.

G. **Low Offset Voltage and Offset Voltage Drift.** As with an operational amplifier, an in-amp must have a low offset voltage. Since an instrumentation amplifier consists of two independent sections: an input stage and an output amplifier, total output offset will equal the sum of the gains, times the input offset, plus the output offset. Although the initial offset voltage may usually be nulled with external trimming, offset voltage drift cannot be adjusted out. As with initial offset, offset drift has two components, with both the input and output section of the in-amp each contributing its portion of error to the total. As gain is increased, the offset drift of the input stage becomes the dominant source of offset error. Values of 100 μV and 2 mV are typical values for input and output offset, respectively.

H. **Low Input Bias and Offset Current Errors.** Again, as with an op-amp, an instrumentation amplifier has *bias* currents which flow into, or out of, its input terminals: bipolar in-amps with their base currents and FET amplifiers with gate leakage currents. This *bias current* flowing through an imbalance in the signal source resistance will create an offset error. Note that if the input source resistance becomes very large—as with ac input coupling without a resistive return to power supply ground—the input common-mode voltage will climb until the amplifier saturates. A high value resistor, connected between each input and ground, is normally used to prevent this problem. Input *offset current* errors are defined as the mismatch between the bias currents flowing in the two inputs. Typical values of input bias current for a bipolar in-amp range from 1 nA to 0.5 μA; for a FET input device, values of 50 pA are typical at room temperature.

Where Is an Instrumentation Amplifier Used?

Data Acquisition. In-amps find their primary use amplifying signals from low output level transducers in noisy environments. The amplification of pressure or temperature transducer signals is a common in-amp application. Common bridge applications include strain and weight measurement using "load cells" and temperature measurement using resistive temperature detectors or "RTD's."

Medical Instrumentation. In-amps are also widely used in medical equipment such as ECG & EEG monitors and blood pressure monitors.

Audio Applications. Again, because of their high common-mode rejection, instrumentation amplifiers are sometimes used for audio (as microphone preamps, etc.) to extract a weak signal from a noisy environment and to minimize offsets and noise due to ground loops.

High Speed Signal Conditioning. As the speed and accuracy of modern video data acquisition systems have increased, there has developed a growing need for high bandwidth instrumentation amplifiers—particularly in the field of CCD imaging equipment where offset correction and input buffering are required. Here double-correlated sampling techniques are often used for offset correction of the CCD image. Two sample-and-hold amplifiers monitor the pixel and reference levels, and a dc corrected output is provided by feeding their signals into an instrumentation amplifier.

In-Amps: an External View

Figure 2, shows a functional block diagram of an instrumentation amplifier.

Since an ideal instrumentation amplifier detects only the difference in voltage between its inputs, any common-mode signals (potentials which are equal on both inputs), such as noise or voltage drops in ground lines, are rejected at the input stage without being amplified.

Normally, either a single resistor or resistor pair is used to program the in-amp for the desired gain. The manufacturer will then provide a transfer function or gain equation that allows the user to calculate the required values of resistance for a given gain.

The output of an instrumentation amplifier has its own reference terminal which allows it to drive ground referenced loads, such as those usually found in instruments. Figure 2 shows the input and output commons being returned to the same point where they are connected to power supply "ground." This "star" ground connection is a very effective means for minimizing ground loops in the circuit; however some residual "common-mode"

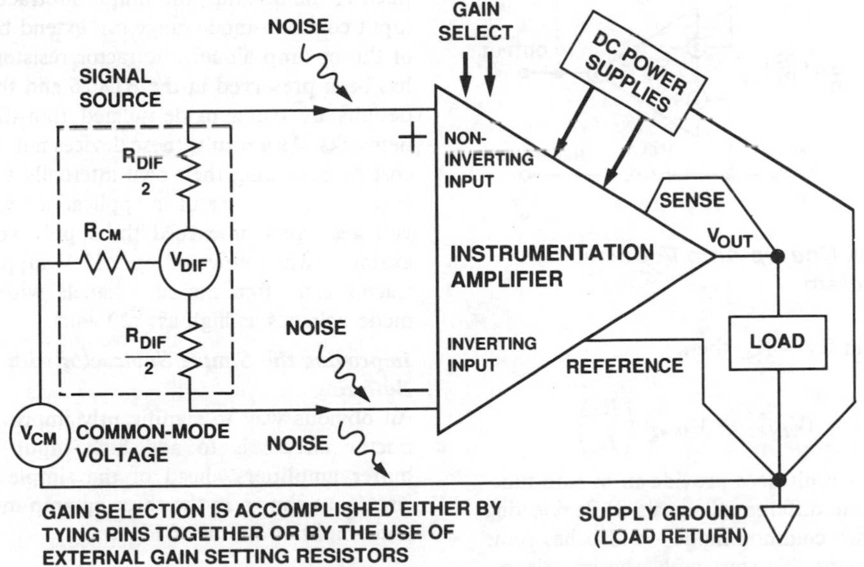

Figure 2. Basic Instrumentation Amplifier

ground currents will still remain. These currents flowing through R_{CM} will develop a common-mode voltage error, V_{CM}. The in-amp, by virtue of its high common-mode rejection, will amplify the differential signal while rejecting V_{CM} and any common-mode noise.

Of course, power must be supplied to the IA—as with op amps, this is normally a dual supply voltage which will operate the in-amp over a specified range. Alternatively, only a single ground referenced supply may be required, for some in-amps.

An instrumentation amplifier may be assembled using one or more operational amplifiers or it may be of monolithic design. Both approaches have their virtues and limitations. In general, op-amp in-amps offer wide flexibility at low cost and sometimes can provide performance unattainable from monolithic designs, such as in very high bandwidth applications. In contrast, monolithic designs provide the complete in-amp function, fully specified and, in many cases, factory trimmed to high accuracy. Op-amp designs will be discussed first.

Inside an Instrumentation Amplifier

A Simple Op-Amp Subtractor Provides an In-Amp Function. The most simple (and very useful) method of implementing a differential gain block is shown in Figure 3.

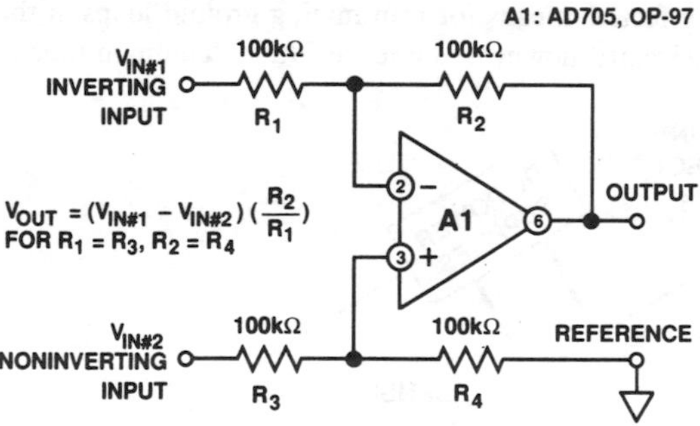

Figure 3. A One Op-Amp IA Circuit Functional Diagram

If $R_1 = R_3$ and $R_2 = R_4$, then:

$$V_{OUT} = (V_{IN\#1} - V_{IN\#2})\left(\frac{R_2}{R_1}\right)$$

Although this circuit does provide an in-amp function—amplifying differential signals while rejecting those which are common-mode—it also has some serious limitations. To start with, the impedances of the inverting and noninverting inputs are relatively low and unequal. This requires that the sources driving this amplifier be of very low impedance. While, in this example, the inverting input has a 100 kΩ impedance (because R_1 connects directly to the amplifier's summing junction, a "virtual ground"), the impedance of the noninverting input is twice that, equaling 200 kΩ. Additionally, signals on the noninverting input will drive the inverting input through 100 kΩ at a gain of 0.5.

Furthermore, this circuit requires a very close ratio match between resistors pairs R_1/R_2 and R_3/R_4; otherwise, the gain from each input will be different—directly affecting common-mode rejection. For example, at a gain of 1—with all resistors of equal value—a 0.1% mismatch in just one of the resistors will degrade the CMR to a level of 66 dB (1 part in 2000).

In spite of these problems, this type of "bare bones" IA circuit is useful as a building block within higher performance in-amps. It is also very practical as a stand-alone functional circuit in video and other high speed uses, or in low frequency, high CMV applications, where the input resistors also provide input protection for the amplifier. Monolithic in-amps such as the Analog Devices' AD626 and AMP-03 benefit from their internal pre-trimmed resistor networks; both employ variations of the simple subtractor in their design.

Because the circuit divides down the voltage applied to the op-amp, the simple subtractor circuit's input common-mode range can extend beyond that of the op-amp alone. Subtractor resistor matching has been preserved in the AD626 and the AMP-03 designs by using oxide-isolated thin-film resistor networks. As a result, these devices not only reduce cost by providing their own internally trimmed resistors, they are useful in applications where input voltages equal or exceed the supply voltage. For example, when powered by ±15 V supplies, a subtractor can often measure signals with common-mode voltages as high as ±20 volts.

Improving the Simple Subtractor with Input Buffering

An obvious way to significantly improve the subtractor circuit is to add high input impedance buffer amplifiers ahead of the simple subtractor circuit, as shown in the three op-amp instrumentation amplifier circuit of Figure 4.

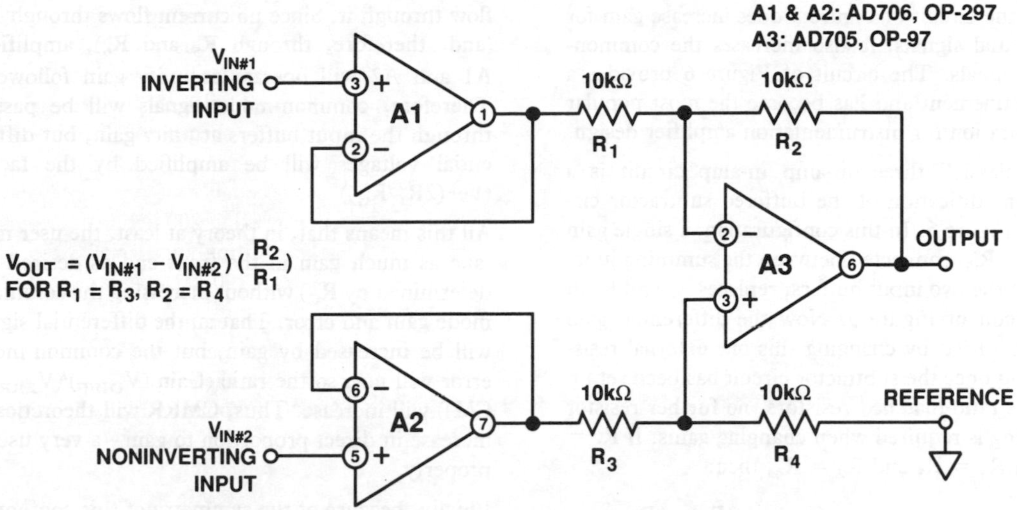

Figure 4. A Subtractor Circuit with Input Buffering

This circuit now provides matched, high impedance inputs so that the impedances of the input sources will have a minimal effect on the circuit's common-mode rejection. The use of a dual op-amp for the two input buffer amplifiers is preferred since they will better track each other over temperature. Although the resistance values are different, this circuit has the same transfer function as the circuit of Figure 3.

Figure 5 shows a further improvement: now the input buffers are operating with gain, which provides a circuit with more flexibility.

If $R_5 = R_8$ and $R_6 = R_7$ and, as before, $R_1 = R_3$ and $R_2 = R_4$, then:

$$V_{OUT} = (V_{IN\#1} - V_{IN\#2})\left(1 + \frac{R_5}{R_6}\right)\left(\frac{R_2}{R_1}\right)$$

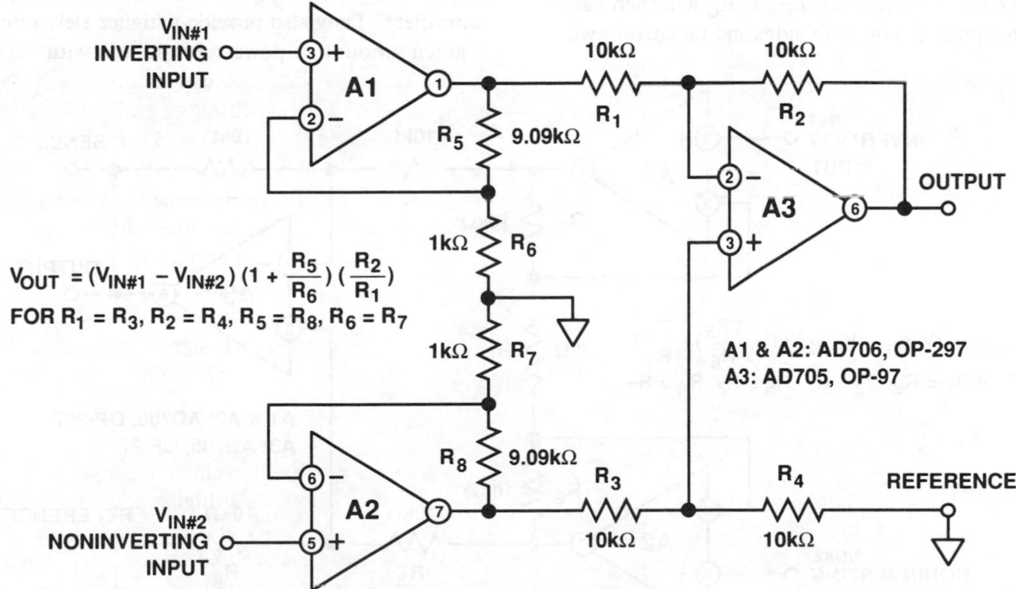

Figure 5. A Buffered Subtractor Circuit Which Has the Buffer Amplifiers Operating with Gain

While the circuit of Figure 5 does increase gain for differential signals, it also increases the common-mode signals. The circuit of Figure 6 provides a final refinement and has become the most popular configuration for instrumentation amplifier design.

The "classic" three op-amp in-amp circuit is a clever modification of the buffered subtractor circuit of Figure 5. In this configuration, a single gain resistor, R_G, connected between the summing junctions of the two input buffers, replaces R_6 and R_7 in the circuit of Figure 5. Now the differential gain may be varied by changing this one external resistor. And once the subtractor circuit has been set up with its ratio-matched resistors, no further resistor matching is required when changing gains. If $R_5 = R_6$ and $R_1 = R_3$ and $R_2 = R_4$, then:

$$V_{OUT} = (V_{IN\#1} - V_{IN\#2})\left(\frac{2R_5}{R_G}\right)\left(\frac{R_2}{R_1}\right)$$

As with the previous circuit, op-amps A1 and A2 of Figure 6 buffer the input voltage. Because resistor R_G is connected between the summing junctions of these two amplifiers, the full input voltage will appear across R_G when a differential voltage is applied to the inputs of the in-amp. Since the voltage across R_G equals V_{IN}, the current through R_G will equal: (V_{IN}/R_G). Amplifiers A1 and A2 will, therefore, operate with gain and amplify the input signal. Note, however, that if a common-mode voltage is applied to the amplifier inputs, R_G will then have the same potential on both sides and no current will flow through it. Since no current flows through R_G (and, therefore, through R_5 and R_6), amplifiers A1 and A2 will operate as unity gain followers. Therefore, common-mode signals will be passed through the input buffers at unity gain, but differential voltages will be amplified by the factor $(1 + (2R_F/R_G))$.

All this means that, in theory at least, the user may take as much gain in the front end as desired (as determined by R_G) without increasing the common-mode gain and error. That is, the differential signal will be increased by gain, but the common-mode error will not, so the ratio (Gain $(V_{DIFF})/(V_{ERROR}$ CM)) will increase. Thus, CMRR will theoretically increase in direct proportion to gain—a very useful property.

Finally, because of the symmetry of this configuration, common-mode errors in the input amplifiers, if they track, tend to be canceled out by the output stage subtractor. These features explain the popularity of this configuration.

Three Op-Amp In-Amps—Design Considerations

Three op-amp instrumentation amplifiers may be constructed using either FET or bipolar input operational amplifiers. FET input op-amps have very low bias currents and are generally well-suited for use with very high ($>10^6\ \Omega$) source impedances. FET amplifiers usually have lower CMR, higher offset voltage, and higher offset drift than bipolar amplifiers. They also provide a higher slew rate for a given amount of power. Amplifiers with bipolar

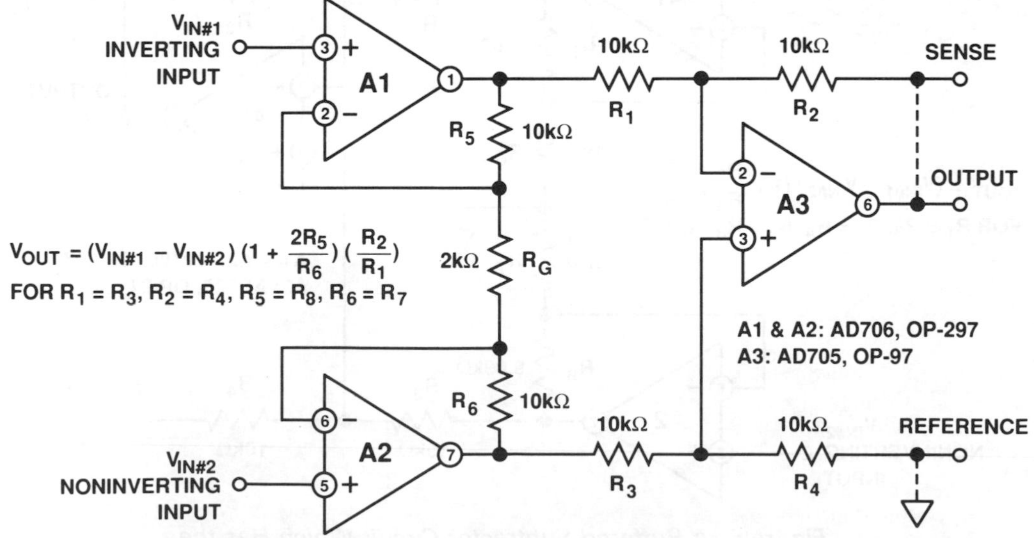

Figure 6. The "Classic" Three Op-Amp In-Amp Circuit

input stages will tend to achieve both higher CMR and lower input offset voltage drift than FET input amplifiers. Superbeta bipolar input stages combine many of the benefits of both FET and bipolar processes, with even lower I_B drift than FET devices.

A common (but frequently overlooked) pitfall for the unwary designer using a three op-amp in-amp design is the reduction of common-mode voltage range which occurs when the in-amp is operating at high gain. Figure 7 is a schematic of a three op-amp in-amp which is operating at a gain of 1000.

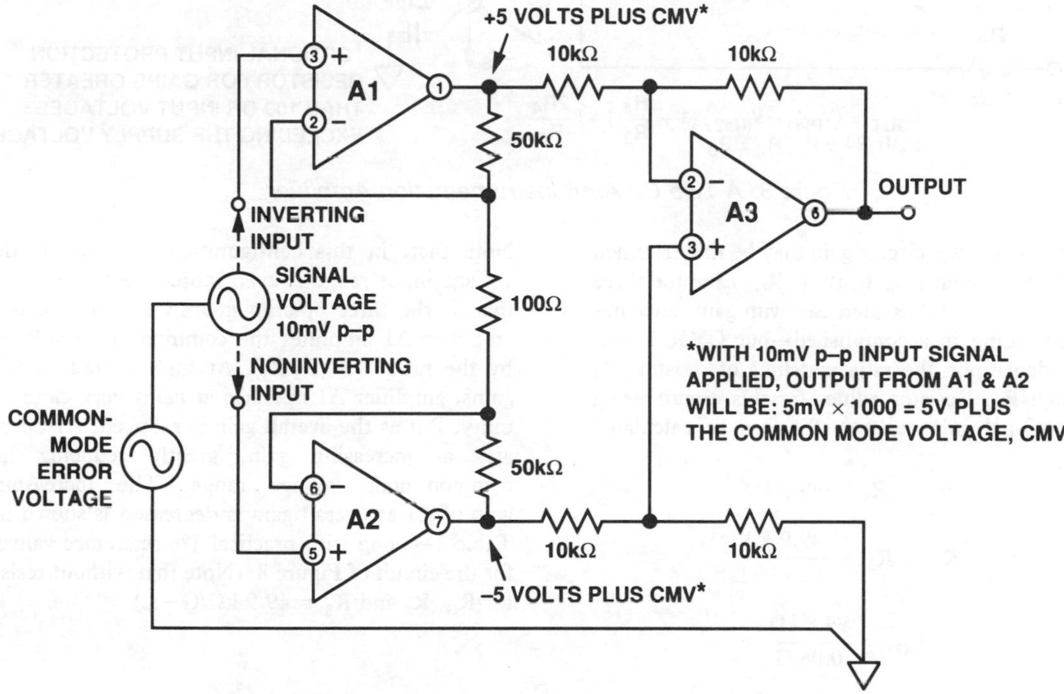

Figure 7. A Three Op Amp IA Showing Reduced CMV Range

In this example, the input amplifiers, A1 and A2, are operating at a gain of 1000, while the output amplifier is providing unity gain. This means that the voltage at the output of each input amplifier will equal one half the peak to peak input voltage times 1000, plus any common-mode voltage that is present on the inputs (the common-mode voltage will pass through at unity gain regardless of the differential gain). Therefore, if a 10 mV differential signal is applied to the amplifier inputs, then amplifier A1's output will equal +5 volts plus the common-mode voltage and A2's output will be −5 volts plus the common-mode voltage. If the amplifiers are operating from ±15 volt supplies, they will usually have 7 volts or so of headroom left, thus permitting an 8 volt common-mode voltage—but not the full 12 volts of CMV which typically would be available at unity gain. Higher gains, or lower supply voltages, will further reduce the common-mode voltage range.

The Basic Two Op-Amp Instrumentation Amplifier

The two op-amp in-amp approach provides a special class of in-amps: those which are very useful for a few limited applications, but which have serious drawbacks that make them inappropriate for general use.

Figure 8 shows just such a circuit; it has the obvious advantage of requiring only two operational amplifiers, rather than three, with subsequent savings in cost and power consumption. The transfer function of this circuit (without R_G) is:

$$V_{OUT} = (V_{IN\#1} - V_{IN\#2})\left(1 + \frac{R_4}{R_3}\right)$$

$$for\ R_1 = R_4\ and\ R_2 = R_3$$

Input resistance is high, thus permitting the signal source to have an unbalanced output impedance.

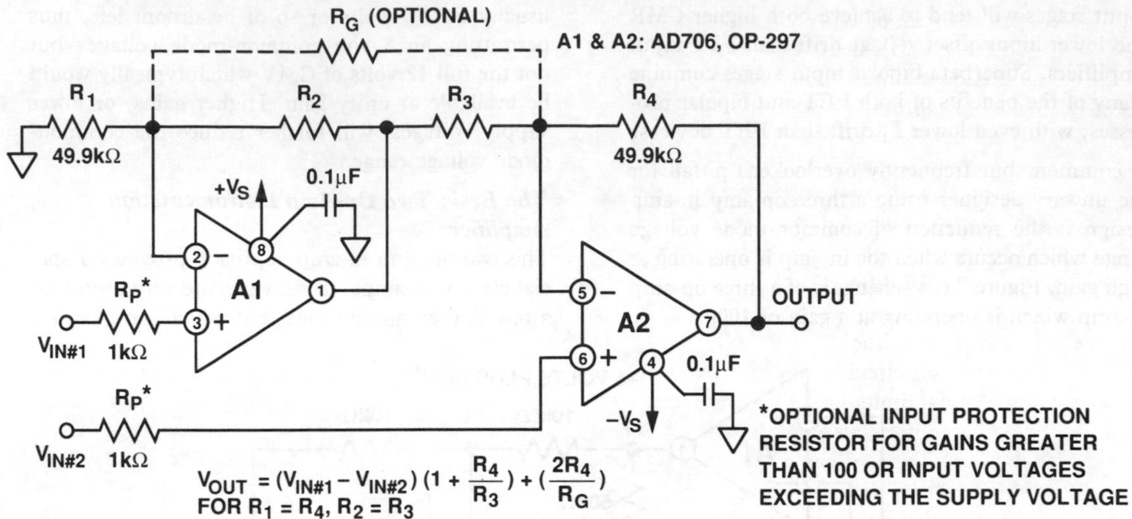

Figure 8. A Two Op-Amp Instrumentation Amplifier

Furthermore, the circuit gain may be fine trimmed using an optional trim resistor, R_G. Like the three op-amp circuit, CMR increases with gain, once initial trimming is accomplished—but CMR is still dependent upon the ratio matching of resistors R_1 through R_4. Resistor values for this circuit using the optional gain resistor, R_G, can be calculated using:

$$R_1 = R_4 = 49.9\ k\Omega$$

$$R_2 = R_3 = \frac{49.9\ k\Omega}{0.9\ G-1}$$

$$R_G = \frac{99.8\ k\Omega}{0.06\ G}$$

where G = Desired Circuit Gain

Note that, in this configuration, common-mode voltage input range is even more affected by gain than is the three op-amp in-amp circuit, because amplifier A1 amplifies the common-mode voltage by the ratio $1 + (R_2/R_1)$. At high overall circuit gains, amplifier A1 operates at gains very close to unity. But as the overall gain is reduced, A1 operates at increasing gain, greatly reducing its common-mode voltage range. The increasing gain of A1 as overall gain is decreased is shown in Table 1—along with practical 1% resistance values for the circuit of Figure 8. (Note that without resistor R_G, R_2 and R_3 = 49.9 kΩ/G−1.)

Table 1. Operating Gains of Amplifiers A1 and A2 and Practical 1% Resistor Values for the Circuit of Figure 8

Circuit Gain	Gain of A1	Gain of A2	R_2, R_3	R_1, R_4
1.10	11.00	1.10	499 kΩ	49.9 kΩ
1.33	4.01	1.33	150 kΩ	49.9 kΩ
1.50	3.00	1.50	100 kΩ	49.9 kΩ
2.00	2.00	2.00	49.9 kΩ	49.9 kΩ
10.1	1.11	10.10	5.49 kΩ	49.9 kΩ
101.0	1.01	101.0	499 Ω	49.9 kΩ
1001	1.001	1001	49.9 Ω	49.9 kΩ

For example, at a gain of 100, the gain of A1 would be 1.01 (i.e., 1 + 0.01), therefore, leaving A1 with 99% of its maximum input voltage range. But, at an overall gain of three (all five resistors equal 49.9 kΩ) amplifier A1 is now operating at a gain of 2 (1 + 1), which will reduce its maximum common-mode range by 50%. These large reductions in input common-mode range at low gains can easily lead to saturation of A1, thus leaving no "headroom" to amplify the differential signal of interest. Also note that the required value of R_2 and R_3 approaches infinity as the circuit gain approaches unity; therefore, this circuit cannot be operated at a gain of one. A final limitation of this circuit is that since the two amplifiers are operating at different closed-loop gains (and thus at different bandwidths), there will be generally poor ac common-mode rejection without the use of an ac CMR trim capacitor (connected between Pin 2 of A1 and ground).

A Two Op-Amp IA Circuit with Post Filtering

The circuit of Figure 9 combines a basic two op-amp in-amp circuit with a four-pole post filter to improve performance. Benefits include: a 30 pA input bias current, a 0.2 μV/°C offset voltage drift and only 1.5 mA quiescent current for the entire circuit. Providing a low cost, high gain preamp for transducer applications, it will operate at gains of two or greater. Figure 9 gives component values for a total circuit gain of 10.

The 1 Hz, 4-pole active filter provides high dc precision at low cost while requiring a minimum number of components. The low levels of current noise, input offset and input bias currents in the quad op-amp (either an AD704 or OP-497) allow the use of 1 MΩ resistors without sacrificing the 1 μV/°C drift of the op-amp. Thus lower capacitor values may be used, reducing cost and space. Furthermore, since the input bias current of these op-amps is as low as their input offset currents over most of the MIL temperature range, there rarely is a need to use the normal balancing resistor (along with its noise-reducing bypass capacitor). Note, however, that adding the optional balancing resistor will enhance performance at temperatures above 100°C.

Monolithic Instrumentation Amplifiers
Advantages Over Op-Amp In-Amps

To satisfy the demand for in-amps which would be easier to apply, monolithic IC instrumentation amplifiers were developed. These circuits often incorporate the same design approaches previously used, while providing laser-trimmed resistors and other benefits of monolithic IC technology. Since both active and passive components are now within the same die they can be closely matched—this will insure that the device provides a high CMR. In addition, these components will stay matched over temperature, assuring excellent performance over a wide temperature range. IC technologies such as laser wafer trimming allow monolithic integrated circuits to be "tuned-up" to very high accuracy and provide low cost, high volume manufacturing.

1 Hz, 4-Pole Low-Pass Filter Recommened Component Values

Desired Low Pass Response	Section 1 Freq (Hz)	Section 1 Q	Section 2 Freq (Hz)	Section 2 Q	C1 (μF)	C2 (μF)	C3 (μF)	C4 (μF)
Bessel	1.43	0.522	1.60	0.806	0.116	0.107	0.160	0.0616
Butterworth	1.00	0.541	1.00	1.31	0.172	0.147	0.416	0.0609
0.1 dB Chebychev	0.648	0.619	0.948	2.18	0.304	0.198	0.733	0.0385
0.2 dB Chebychev	0.603	0.646	0.941	2.44	0.341	0.204	0.823	0.0347
0.5 dB Chebychev	0.540	0.705	0.932	2.94	0.416	0.209	1.00	0.0290
1.0 dB Chebychev	0.492	0.785	0.925	3.56	0.508	0.206	1.23	0.0242

Specified values are for a −3 dB point of 1.0 Hz. For other frequencies simply scale capacitors C1 through C4 directly; i.e., for 3 Hz Bessel response, C1 = 0.0387 μF, C2 = 0.0357 μF, C3 = 0.0533 μF, C4 = 0.0205 μF.

In-Amp Section–Recommended Component Values

Circuit Gain	R_1 & R_3	R_G	−3 dB BW
10	6.34 kΩ	166 kΩ	50 kHz
100	526 Ω	16.6 kΩ	5 kHz
1,000	56.2 Ω	1.66 kΩ	0.5 kHz

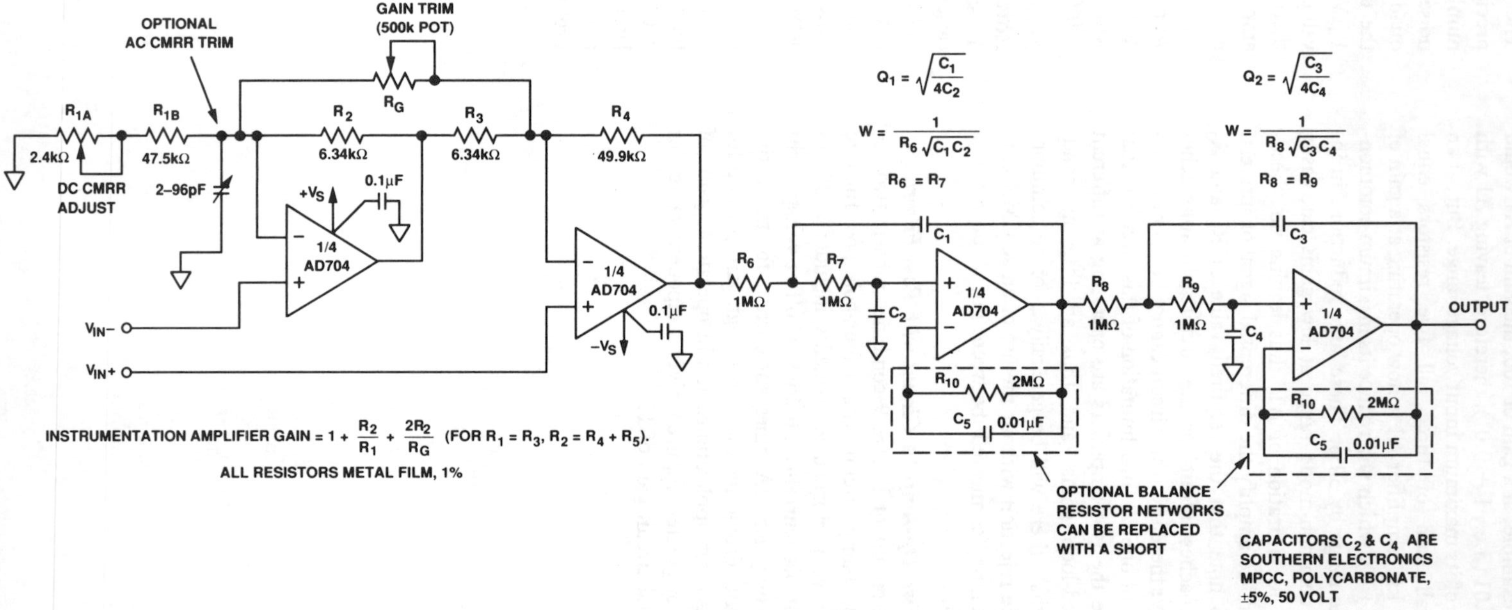

Figure 9. A Two Op-Amp In-Amp With Post Filtering

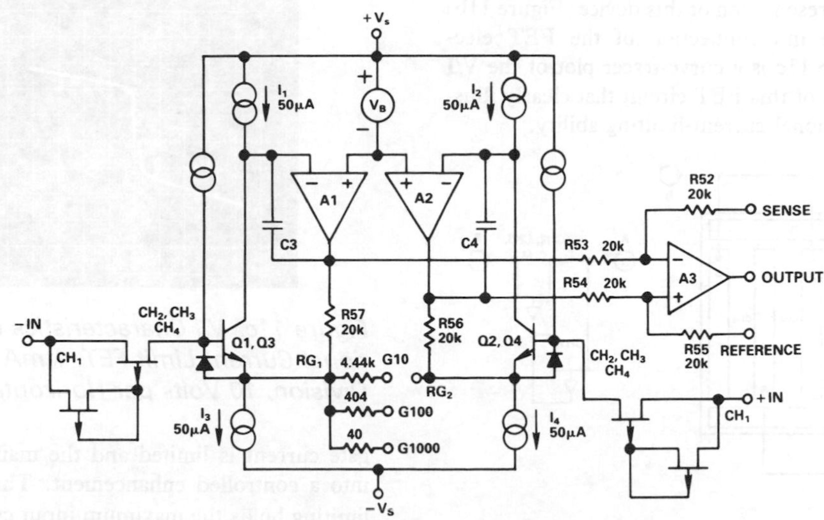

Figure 10. A Simplified Schematic of the AD524 Monolithic Instrumentation Amplifier

Monolithic In-Amp Design—The Inside Story

The AD524 is a good example of current monolithic instrumentation amplifier design. It is based on the classic three op-amp in-amp circuit previously described. The AD524 simplified schematic is shown in Figure 10. Note that the AD620 and AMP-02 both use a similar circuit architecture and offer many of the same performance benefits.

In these designs, the desired gain is selected by varying the value of external resistor R_G. Feedback forces the collector currents of Q1, Q2, Q3, and Q4 to be constant which impresses the input voltage across R_G. As R_G is reduced, thus increasing the programmed gain, the transconductance of the input preamp increases until it equals the transconductance of the input transistors.

The AD524 achieves a very high open-loop gain of 3×10^8 while operating at a programmed closed-loop gain of 1000. This reduces gain related errors to 30 ppm. It also features a high gain/bandwidth product of 25 MHz, determined by capacitors C3 and C4 and the input transconductance. Additionally, the AD524 is a very low noise in-amp—with only 7 nV/$\sqrt{\text{Hz}}$ noise @ at gain of 1000.

The AD524 has its own internal input protection. As interface amplifiers for data acquisition systems, instrumentation amplifiers are often subjected to input overload, i.e., voltage levels in excess of the full scale for the selected gain range. At low gains, (10 or less) the gain resistor acts as a current-limiting element in series with the inputs. At high gains, the lower value for R_G will not adequately protect the inputs from excessive currents. Standard practice would be to place a limiting resistor in series with each input. But limiting input current below 5 mA, with the full differential overload of 36 volts applied, requires over 7 kΩ of resistance. This added resistance will increase noise by 10 nV$\sqrt{\text{Hz}}$. Unfortunately, using a normal FET device will allow the input to go negative with respect to the drain, causing the gate-drain junction to be forward biased (Figure 11a). The FET then acts as a large-area diode, with forward current increasing exponentially with applied voltage.

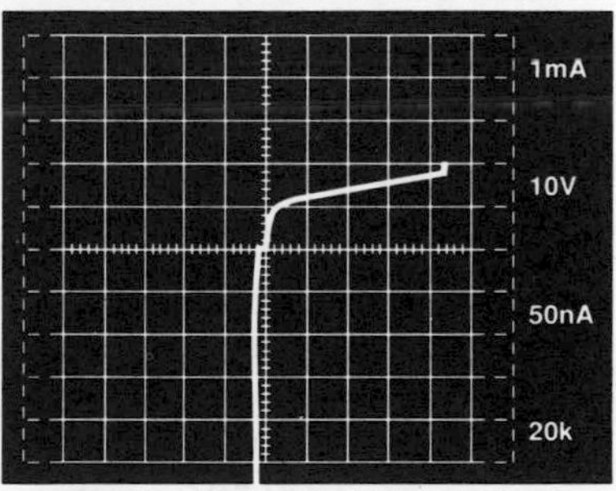

Figure 11a. V/I Characteristics of FET with Gate Shorted to Source. 1 mA per Vertical Division, 10 Volts per Horizontal Division.

Therefore, to overcome this problem, a special series protection FET was used in the AD524 to provide both input protection and low noise. The schematic representation of this device (Figure 11b) illustrates the interconnection of the FET electrodes. Figure 11c is a curve-tracer plot of the V/I characteristics of this FET circuit that clearly illustrates bidirectional current-limiting ability.

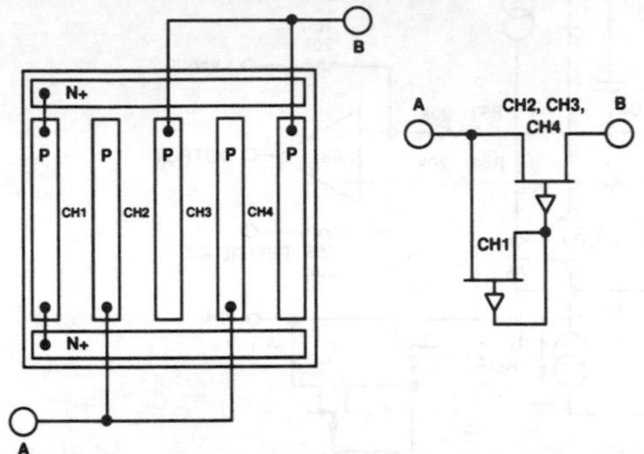

Figure 11b. Bidirectional Current Limit FET and Its Schematic Representation

This protects the in-amp from both positive and negative input overloads. Under nonoverload conditions, the three channels, CH2, CH3, CH4, act as an ≈1 kΩ resistance in series with the input. During an overload in the positive direction, the fourth channel, CH1, acts as a small resistance (≈3 kΩ) in series with the gate, which draws only the leakage current, and the FET limits I_{DSS}. When the FET enhances under a negative overload, the gate current must go through the small FET formed by CH1; and when this FET goes into saturation, the

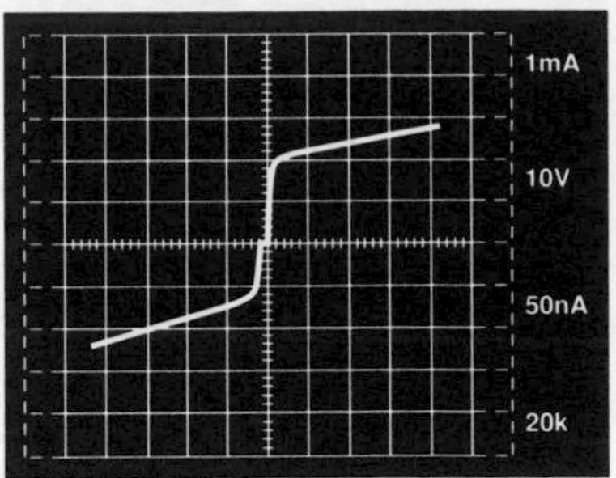

Figure 11c. V/I Characteristics of Bidirectional Current Limit FET. 1 mA per Vertical Division, 10 Volts per Horizontal Division.

gate current is limited and the main FET will go into a controlled enhancement. This bidirectional limiting holds the maximum input current to 3 mA at the maximum overload voltage of 36 volts .

The AMP-01, another monolithic in-amp, incorporates a unique current feedback architecture into its design, achieving both a very high common-mode rejection and a relatively constant bandwidth over a wide range of gains. A simplified schematic of the AMP-01 is shown in Figure 12.

Amplifier gain may be set between 0.1 to 10,000 and is determined by the ratio of two resistors, R_{SCALE} (R_S) and R_{GAIN} (R_G) according to the expression:

$$Gain = \frac{20\,R_{SCALE}}{R_{GAIN}}$$

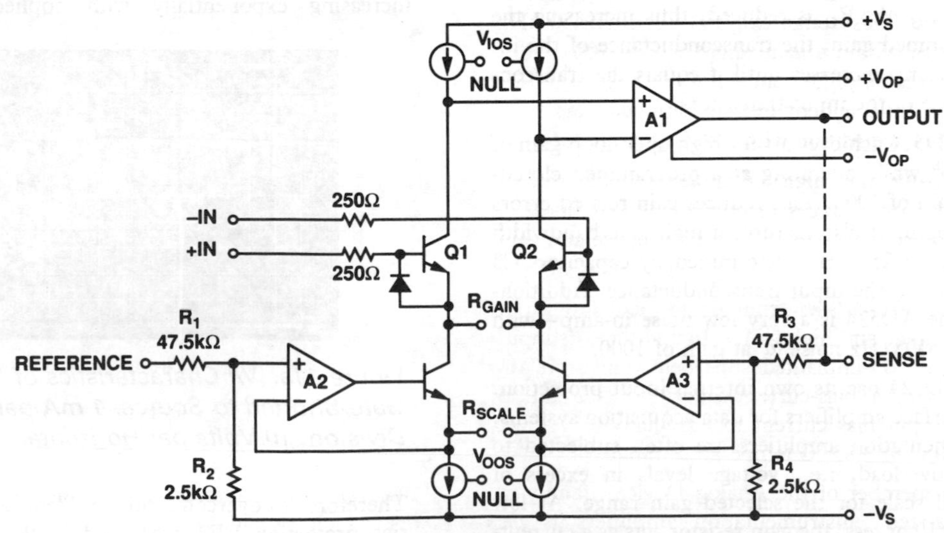

Figure 12. A Simplified Schematic of the AMP-01 Monolithic In-Amp

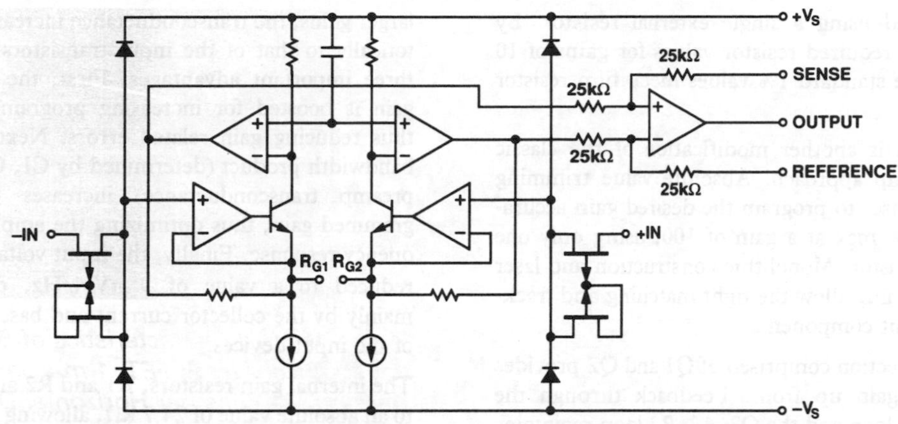

Figure 13a. A Simplified Schematic of the AMP-02 Monolithic In-Amp

The overall amplifier gain temperature coefficient is better than 15 ppm/°C when the temperature coefficient of resistors R_S and R_G is matched to 5 ppm/°C or better.

Referring to the simplified schematic, the current feedback works as follows. As voltage is applied across the inputs of the AMP-01, that same voltage develops across resistor R_{GAIN} which steers current from one leg of the differential stage to the other. This causes the output amplifier to swing, feeding signal back via the sense pin, which converts a fraction of that signal into a current that steers the current flowing in R_{GAIN} into R_{SCALE}—but in the reverse direction, to equalize the imbalance. This current feedback scheme results in a common-mode rejection better than 120 dB over a wide temperature range. This design exhibits a constant 50 kHz bandwidth within a gain range of 1 to 100. Additionally, the amplifier can settle in 15 µs to 0.01% accuracy, independent of gain thus allowing high speed data acquisition with a high throughput rate.

Additionally, the (high) combined gains of the input and output stages result in excellent linearity and gain accuracy—to 16-bit performance at a gain of 1000.

The AMP-01 is laser trimmed to yield low input offset voltage and high CMR. In addition, its input stage uses ion-implanted superbeta transistors together with a bias-current-cancellation circuit to reduce input bias current to less than 10 nA over the −40°C to +85 °C temperature range.

A third member of the Analog Devices' family of "rugged-ized" instrumentation amplifiers is *the AMP-02*. Similar to the AD524, it also employs a series input FET protection scheme, as shown in Figure 13a. This protection circuit limits the input current to ±4 mA and prevents damage to the inputs with differential overloads as high as 60 V.

Figure 13b shows the input current limiting action as the input overload voltage increases. Note that the protection is still active with, or without, power applied. In addition to current limiting, this design uses a pair of internal diodes to clamp the inputs to the power supply rails, when either input is exposed to a voltage higher than that of the supply. This effectively prevents damage due to input stage breakdown.

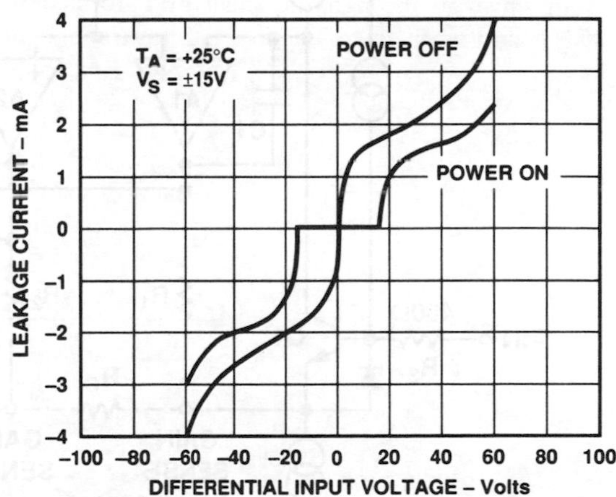

Figure 13b. The Overload Characteristics of the AMP-02

The AD620 is a monolithic instrumentation amplifier which provides a low cost, low power in-amp function in an 8-pin SOIC package. Gain is programmed using a single external resistor. By design, the required resistor values for gains of 10 and 100 are standard 1% values metal film resistor values.

The design is another modification of the classic three op-amp approach. Absolute value trimming allows the user to program the desired gain accuracy—to 0.5% max at a gain of 100, using only one external resistor. Monolithic construction and laser wafer trimming allow the tight matching and tracking of circuit components.

A preamp section comprised of Q1 and Q2 provides additional gain up front. Feedback through the Q1–A1–R1 loop and the Q2–A2–R2 loop maintains a constant collector current through the input devices Q1, Q2, thereby impressing the input voltage across the external gain setting resistor R_G. This creates a differential gain from the inputs to the A1/A2 outputs given by $G = (R1 + R2)/R_G + 1$. The unity gain subtractor A3 removes any common-mode signal, yielding a single-ended output referred to the REF pin potential.

The value of R_G also determines the transconductance of the preamp stage. As R_G is reduced for larger gains, the transconductance increases asymptotically to that of the input transistors. This has three important advantages: First, the open-loop gain is boosted for increasing programmed gain, thus reducing gain related errors. Next, the gain bandwidth product (determined by C1, C2 and the preamp transconductance) increases with programmed gain, thus optimizing the amplifier's frequency response. Finally, the input voltage noise is reduced to a value of 9 $nV/\sqrt{Hz}$, determined mainly by the collector current and base resistance of the input devices.

The internal gain resistors, R1 and R2 are trimmed to an absolute value of 24.7 kΩ, allowing the gain to be programmed accurately with a single external resistor. The gain equation is then:

$$G = \frac{49.4\ k\Omega}{R_G} + 1$$

so that:

$$R_G = \frac{49.4\ k\Omega}{G - 1}$$

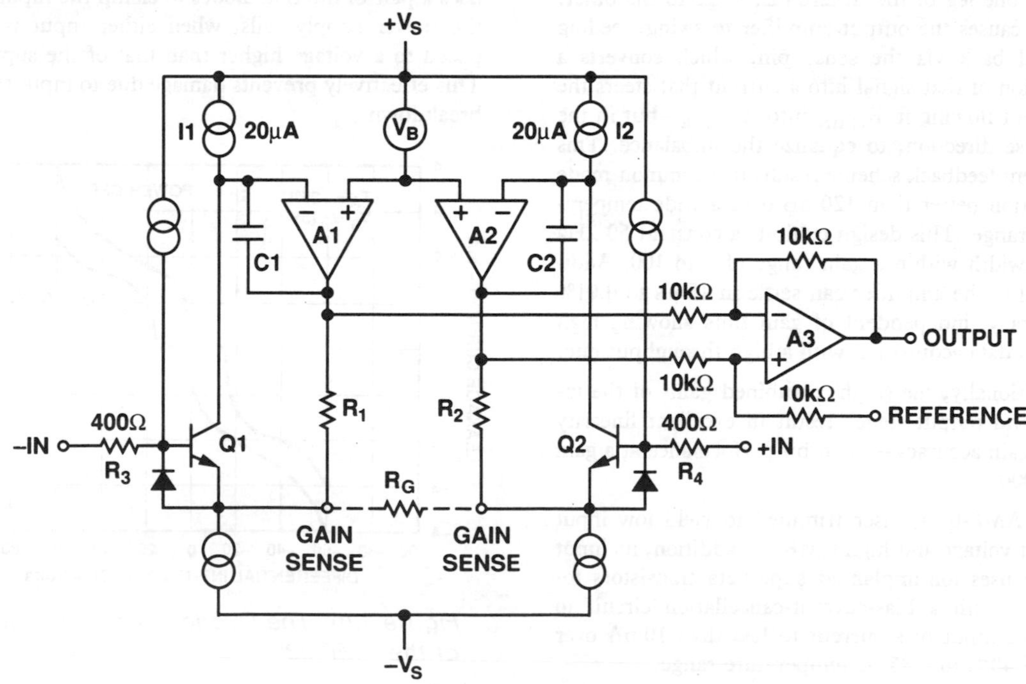

Figure 14a. A Simplified Schematic of the AD620

Monolithic In-Amps Optimized for High Performance

The AD624, is a monolithic in-amp similar to the AD524, but optimized for even higher accuracy and lower noise. Figure 14b is a simplified schematic of the AD624.

The gain equation for this circuit is:

$$Gain = \frac{40{,}000\ \Omega}{R_G} + 1$$

Note that the chief differences between the AD524 and AD624 circuits are in their input sections and in their internal scaling resistors, allowing the AD624 to provide different preset gains than those of the AD524. For details concerning the entire line of monolithic in-amps produced by Analog Devices, refer to Appendix B.

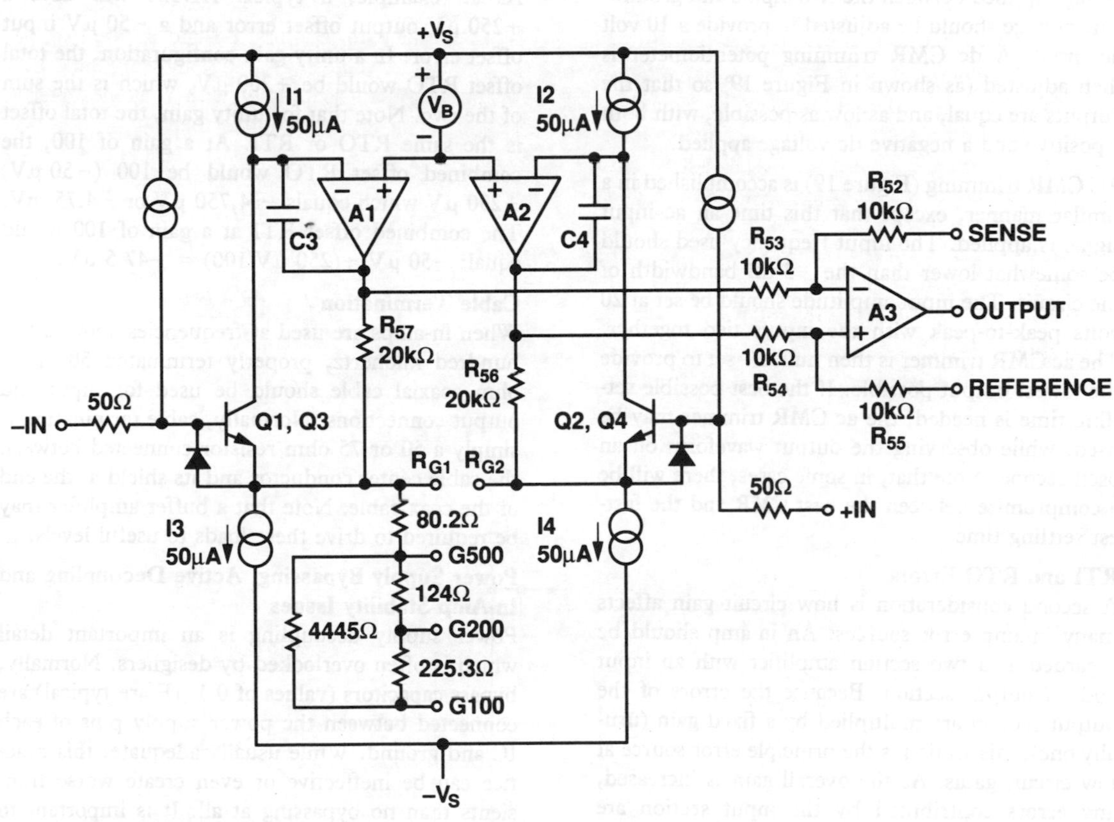

Figure 14b. AD624 Simplified Schematic

SECTION II—DESIGN CONSIDERATIONS FOR INSTRUMENTATION AMPLIFIERS

External CMR and Settling Time Adjustments
One of the virtues of the three amplifier in-amp design is that it may be readily "tuned-up" for best performance by external trimming. The dc CMR should always be trimmed first, since it affects CMRR at all frequencies. The $+V_{IN}$ and $-V_{IN}$ terminals should be tied together and a dc input voltage applied between the two inputs and ground. The voltage should be adjusted to provide a 10 volt dc input. A dc CMR trimming potentiometer is then adjusted (as shown in Figure 19) so that the outputs are equal, and as low as possible, with both a positive and a negative dc voltage applied.

AC CMR trimming (Figure 19) is accomplished in a similar manner, except that this time an ac input signal is applied. The input frequency used should be somewhat lower than the −3 dB bandwidth of the circuit. The input amplitude should be set at 20 volts peak-to-peak with the inputs tied together. The ac CMR trimmer is then nulled—set to provide the lowest output possible. If the best possible settling time is needed, the ac CMR trimmer may be used, while observing the output waveform on an oscilloscope. Note that, in some cases, there will be a compromise between the best CMR and the fastest settling time.

RTI and RTO Errors
A second consideration is how circuit gain affects many in-amp error sources. An in-amp should be regarded as a two section amplifier with an input and an output section. Because the errors of the output section are multiplied by a fixed gain (usually one), this section is the principle error source at low circuit gains. As the overall gain is increased, any errors contributed by the input section are multiplied directly and thereby become dominant. All input-related specifications are classified as "referred to input" (RTI) errors while all output-related specifications are considered "referred to output" (RTO) errors.

By separating these errors, it is possible to evaluate the total error, independent of the selected gain setting. For a given gain, an in-amp's input and output errors can be combined to provide a total error specification which is either referred to the input (RTI) or to the output (RTO) by using the following formulas:

Total Error, RTI =
Input Error + (Output Error/Gain)

Total Error, RTO =
(Gain × Input Error) + Output Error

As an example, a typical AD524 will have a +250 μV output offset error and a −50 μV input offset error. In a unity gain configuration, the total offset RTO would be +200 μV, which is the sum of the two. Note that for unity gain, the total offset is the same RTO or RTI. At a gain of 100, the combined offset RTO would be 100 (−50 μV) +250 μV which equals − 4,750 μV or −4.75 mV. The combined offset RTI at a gain of 100 would equal: −50 μV +(250 μV/100) = −47.5 μV.

Cable Termination
When in-amps are used at frequencies above a few hundred kilohertz, properly terminated 50 or 75 ohm coaxial cable should be used for input and output connections. Normally, cable termination is simply a 50 or 75 ohm resistor connected between the cable center conductor and its shield at the end of the coax cable. Note that a buffer amplifier may be required to drive these loads to useful levels.

Power Supply Bypassing, Active Decoupling and In-Amp Stability Issues
Power supply decoupling is an important detail which is often overlooked by designers. Normally, bypass capacitors (values of 0.1 μF are typical) are connected between the power supply pins of each IC and ground. While usually adequate, this practice can be ineffective or even create worse transients than no bypassing at all. It is important to consider where the circuit's currents originate, where they will return, and by what path. Then, once that has been established, bypass these currents around ground and other signal paths. In general, most monolithic in-amps have their integrators referenced to the negative supply (such as the AD524, AD624, AD625, and AD620) and should be decoupled with respect to the output reference terminal. This means that, for each chip, a bypass capacitor should be connected between

each power supply pin and the point on the board where the in-amp's reference terminal is connected. For a much more comprehensive discussion of these issues refer to the application note: "An I.C. Amplifier Users' Guide to Decoupling, Grounding, and Making Things Go Right for a Change," by Paul Brokaw, available free from Analog Devices.

INA120

Precision INSTRUMENTATION AMPLIFIER

FEATURES

- LOW OFFSET VOLTAGE: 25μV max
- LOW OFFSET VOLTAGE DRIFT: 0.25μV/°C max
- PIN-STRAPPED GAINS: 1, 10, 100, 1000
- LOW GAIN DRIFT: 30ppm/°C max at G = 100
- HIGH COMMON-MODE REJECTION: 106dB at 60Hz, G = 100

APPLICATIONS

- BRIDGE AMPLIFIER
- THERMOCOUPLE AMPLIFIER
- RTD SENSOR AMPLIFIER
- MEDICAL INSTRUMENTATION
- DATA ACQUISITION SYSTEM
- SWITCHED-GAIN AMPLIFIER

DESCRIPTION

The INA120 is a precision instrumentation amplifier ideal for accurate signal acquisition. It combines precision, protected-input operational amplifiers, laser-trimmed gain-setting resistors, and a high common-mode rejection difference amplifier on a single chip.

Simple pin-strapped connections set precise gains of 1, 10, 100 or 1000. External resistors can be used to set any gain from one to 5000. Gains can be digitally selected with an external multiplexer. Gain-sense connections on the INA120 maintain accuracy when using multiplexer or gain-switching circuitry. Low power dissipation and careful on-chip thermal management reduce warm-up drift and assure excellent long-term stability.

The INA120 is available in both plastic and ceramic 18-pin DIP packages, specified for the industrial temperature range.

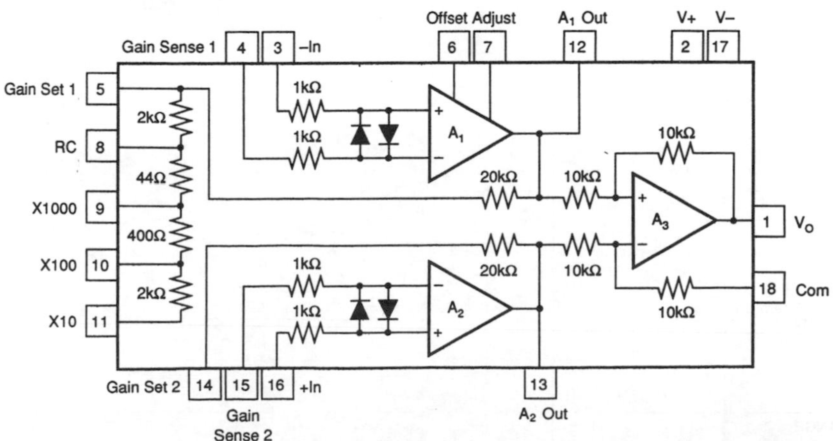

International Airport Industrial Park • Mailing Address: PO Box 11400 • Tucson, AZ 85734 • Street Address: 6730 S. Tucson Blvd. • Tucson, AZ 85706
Tel: (602) 746-1111 • Twx: 910-952-1111 • Cable: BBRCORP • Telex: 066-6491 • FAX: (602) 889-1510 • Immediate Product Info: (800) 548-6132

PDS-1071
Printed in U.S.A. July, 1990

SPECIFICATIONS

ELECTRICAL

At T_A = +25°C and V_S = ±15V unless otherwise specified.

		INA120CG			INA120BG/BP			INA120AP			
PARAMETER	**CONDITIONS**	**MIN**	**TYP**	**MAX**	**MIN**	**TYP**	**MAX**	**MIN**	**TYP**	**MAX**	**UNITS**
GAIN											
Range of Gain		1		1000	1		1000	1		1000	V/V
Gain Equation			$1 + (2R_F/R_G)$			$1 + (2R_F/R_G)$			$1 + (2R_F/R_G)$		V/V
Gain Error	G = 1		0.01	0.05		0.01	0.05		0.02	0.1	%
	G = 10		0.05	0.1		0.05	0.2		0.1	0.2	%
	G = 100		0.1	0.2		0.1	0.3		0.2	0.5	%
	G = 1000		0.3	0.5		0.3	1		0.5	1	%
Gain Temp Coefficient	G = 1		4	10		4	20		6	20	ppm/°C
	G = 10		4	10		4	20		8	40	ppm/°C
	G = 100		6	30		6	40		10	60	ppm/°C
	G = 1000		22	50		22	50		40	100	ppm/°C
Nonlinearity	G = 1		0.001	0.005		0.001	0.01		0.001	0.01	% of FS
	G = 10		0.002	0.005		0.002	0.01		0.002	0.01	% of FS
	G = 100		0.004	0.01		0.004	0.02		0.004	0.02	% of FS
	G = 1000		0.008	0.05		0.008	0.1		0.008	0.1	% of FS
OFFSET VOLTAGE											
Initial Offset			(10+ 300/G)	(25+ 600/G)		(50+ 300/G)	(100+ 1000/G)		(50+ 600/G)	(200+ 2000/G)	µV
vs Temperature				(.25 + 10/G)			(1 + 20/G)		(2 + 20/G)		µV/°C
vs Power Supply	V_S = ±6V to ±18V		(1 + 20/G)	(10 + 150/G)		(1 + 20/G)	(20 + 250/G)		(1 + 20/G)	(40 + 300/G)	µV/V
INPUT BIAS CURRENT											
Initial Bias Current			±7	±20		±7	±20		±20	±50	nA
vs Temperature			±0.2			±0.2			±0.2		nA/°C
Initial Offset Current			±5	±10		±5	±20		±10	±50	nA
vs Temperature			±0.2			±0.2			±0.2		nA/°C
Impedance: Differential			$10^{10} \parallel 3$			$10^{10} \parallel 3$			$10^{10} \parallel 3$		Ω ∥ pF
Common-Mode			$10^{10} \parallel 3$			$10^{10} \parallel 3$			$10^{10} \parallel 3$		Ω ∥ pF
INPUT VOLTAGE RANGE											
Range, Linear Response		±10	±12.5		±10	±12.5		±10	±12.5		V
CMRR (DC, 1kΩ Source Imbalance)	G = 1	80	90		74	90		70	85		dB
	G = 10	96	106		90	106		86	95		dB
	G = 100	106	110		106	110		100	105		dB
	G = 1000	106	110		106	110		100	105		dB
NOISE											
Input Voltage Noise											
f_B = 0.1Hz to 10Hz	G = 1000		0.7			0.7			0.7		µV p-p
Density; f = 10Hz	G = 1000		14			14			14		nV/√Hz
f = 100Hz			11			11			11		nV/√Hz
f = 1000Hz			10			10			10		nV/√Hz
Input Current Noise											
f_B = 0.1Hz to 10Hz			50			50			50		pAp-p
Density; f = 10Hz			1.8			1.8			1.8		pA/√Hz
f = 1kHz			0.4			0.4			0.4		pA/√Hz
Output Voltage Noise											
f_B = 0.1Hz to 10Hz			8			8			8		µVp-p
DYNAMIC RESPONSE											
Small Signal Bandwidth (–3dB)	G = 1		2			2			2		MHz
	G = 10		200			200			200		kHz
	G = 100		20			20			20		kHz
	G = 1000		2			2			2		kHz
Slew Rate		0.4	0.6		0.4	0.6		0.4	0.6		V/µs
Settling Time to 0.01%	G = 1		24			24			24		µs
	G = 10		30			30			30		µs
	G = 100		50			50			50		µs
	G = 1000		200			200			200		µs
Full Power Bandwidth, $G < 200$	V_O = ±10V, R_L = 2kΩ		9			9			9		kHz
Overload Recovery	50% Overdrive		2			2			2		µs
OUTPUT											
Voltage, R_L = 2kΩ	Over Temperature	±10.5	±12.8		±10.5	±12.8		±10.5	±12.8		V
Current	Over Temperature	5	15		5	15		5	15		mA
Short-Circuit Current			24			24			24		mA
Capacitive Load, Stable Operation			4000			4000			4000		pF
POWER SUPPLY											
Rated Voltage			±15			±15			±15		V
Voltage Range		±6		±18	±6		±18	±6		±18	V
Supply Current	V_O = 0V		±2.7	±4		±2.7	±4		±2.7	±4	mA
TEMPERATURE RANGE											
Specification		–25		+85	–25		+85	–25		+85	°C
Operation BP,AP					–40		+85	–40		+85	°C
Operation CG,BG		–55		+125	–55		+125				°C
Storage		See Absolute Maximum Table.									

PIN CONFIGURATION

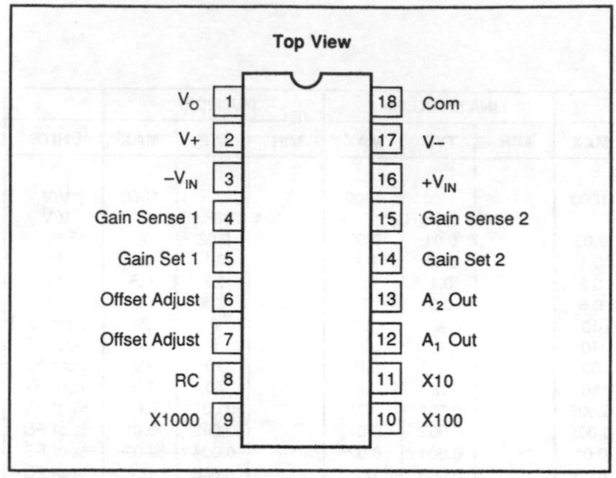

ABSOLUTE MAXIMUM RATINGS

Supply Voltage .... ±18V
Input Voltage Range .... (V+) +2 to (V–) –2V
Differential Input Voltage .... Total V_s +4V
Operating Temperature
  Ceramic G Package .... –65°C to +150°C
  Plastic P Package .... –40°C to +125°C
Storage Temperature
  Ceramic G Package .... –65°C to +150°C
  Plastic P Package .... –40°C to +125°C
Junction Temperature
  Ceramic G Package .... +175°C
  Plastic P Package .... +125°C
Lead Temperature (soldering, 10s) .... +300°C

ORDERING INFORMATION

MODEL	PACKAGE	TEMPERATURE RANGE
INA120AP	Plastic DIP	–25°C to +85°C
INA120BP	Plastic DIP	–25°C to +85°C
INA120BG	Ceramic DIP	–25°C to +85°C
INA120CG	Ceramic DIP	–25°C to +85°C

MECHANICAL

P Package — 18-Pin Plastic DIP

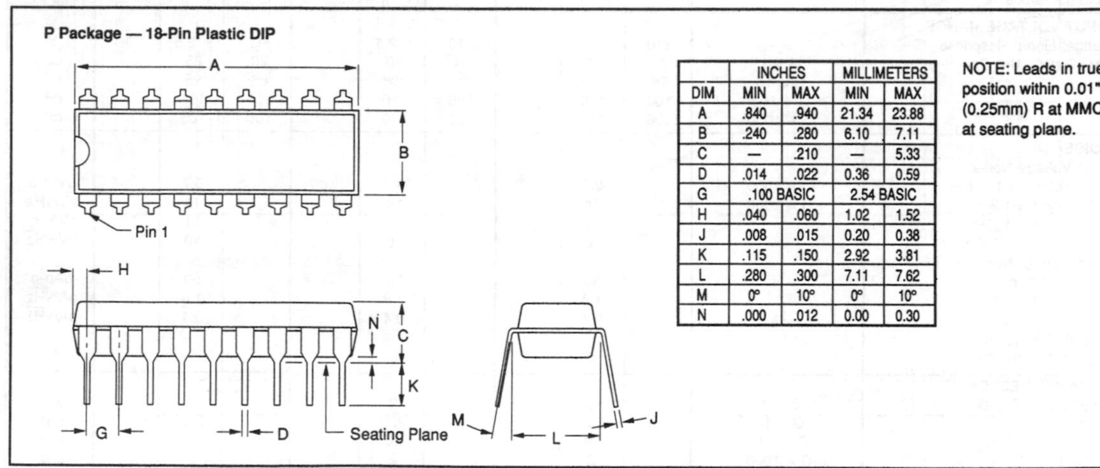

DIM	INCHES MIN	INCHES MAX	MILLIMETERS MIN	MILLIMETERS MAX
A	.840	.940	21.34	23.88
B	.240	.280	6.10	7.11
C	—	.210	—	5.33
D	.014	.022	0.36	0.59
G	.100 BASIC		2.54 BASIC	
H	.040	.060	1.02	1.52
J	.008	.015	0.20	0.38
K	.115	.150	2.92	3.81
L	.280	.300	7.11	7.62
M	0°	10°	0°	10°
N	.000	.012	0.00	0.30

NOTE: Leads in true position within 0.01" (0.25mm) R at MMC at seating plane.

G Package — 18-Pin Ceramic DIP

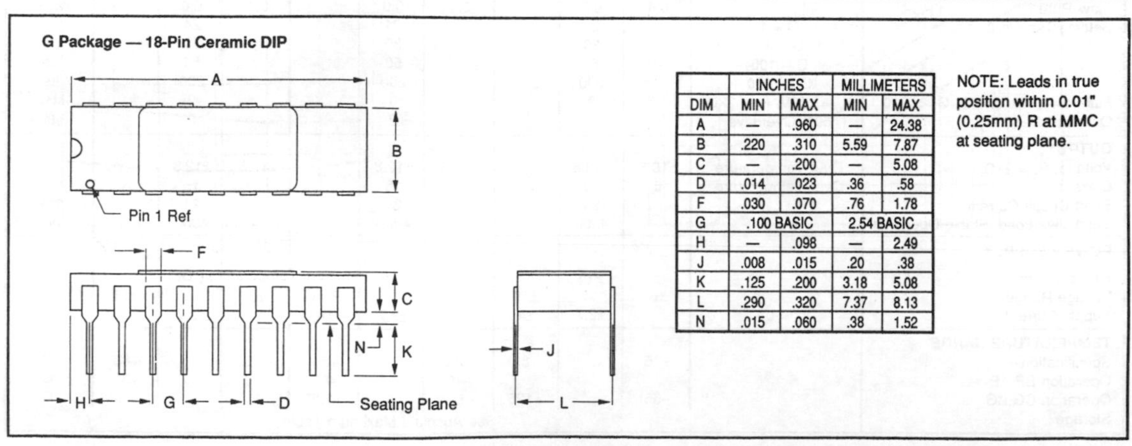

DIM	INCHES MIN	INCHES MAX	MILLIMETERS MIN	MILLIMETERS MAX
A	—	.960	—	24.38
B	.220	.310	5.59	7.87
C	—	.200	—	5.08
D	.014	.023	.36	.58
F	.030	.070	.76	1.78
G	.100 BASIC		2.54 BASIC	
H	—	.098	—	2.49
J	.008	.015	.20	.38
K	.125	.200	3.18	5.08
L	.290	.320	7.37	8.13
N	.015	.060	.38	1.52

NOTE: Leads in true position within 0.01" (0.25mm) R at MMC at seating plane.

TYPICAL PERFORMANCE CURVES

T_A = +25°C, V_S = ±15V unless otherwise noted.

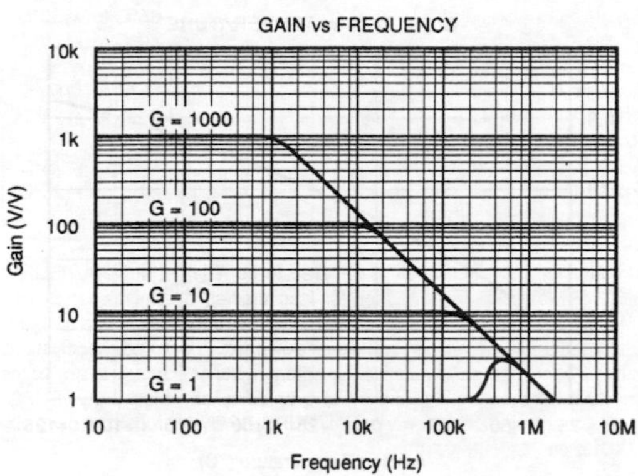

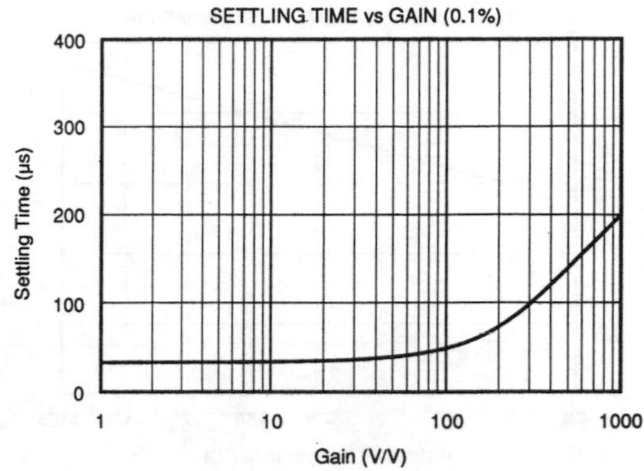

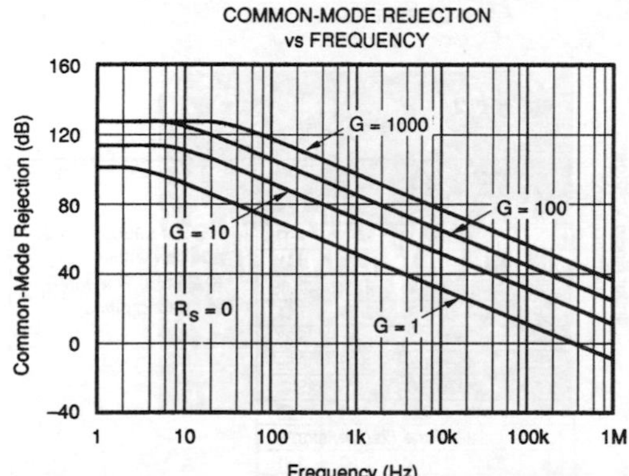

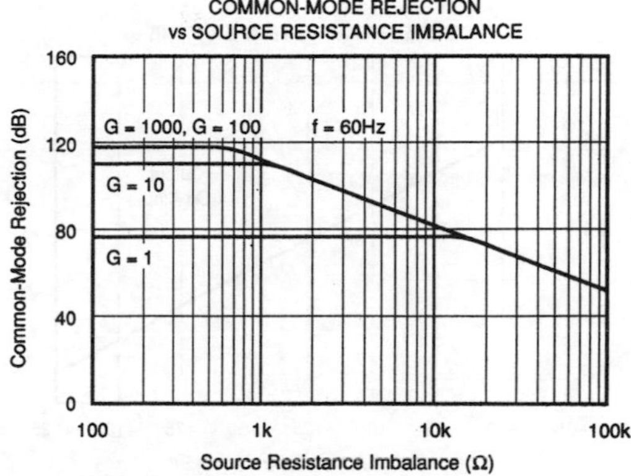

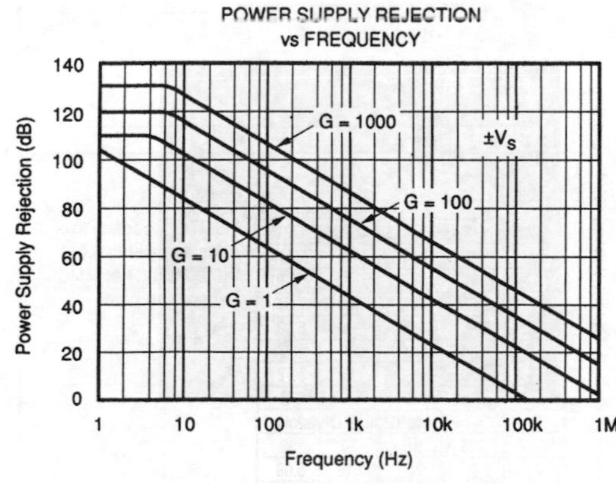

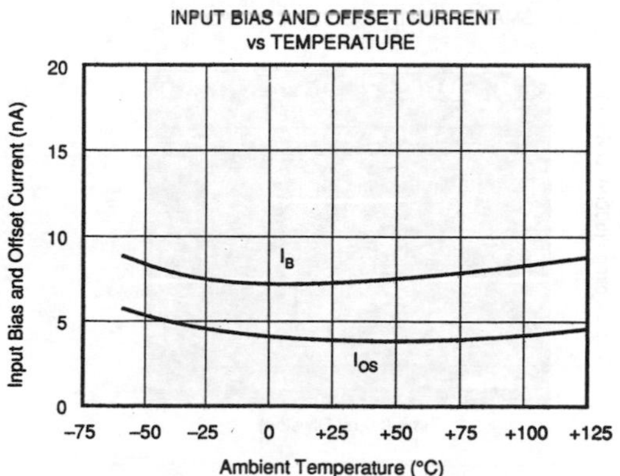

TYPICAL PERFORMANCE CURVES (CONT)

$T_A = +25°C$, $V_S = \pm 15V$ unless otherwise noted.

QUIESCENT CURRENT vs TEMPERATURE

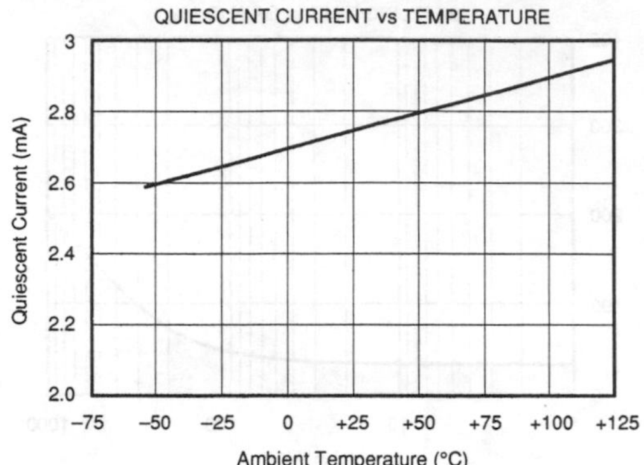

SLEW RATE vs TEMPERATURE

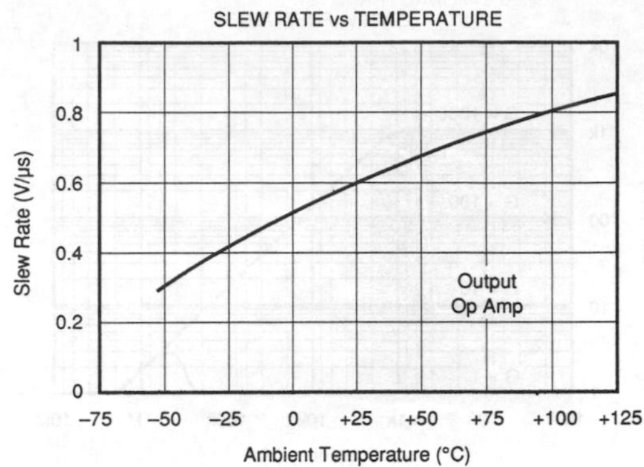

CURRENT LIMIT vs TEMPERATURE

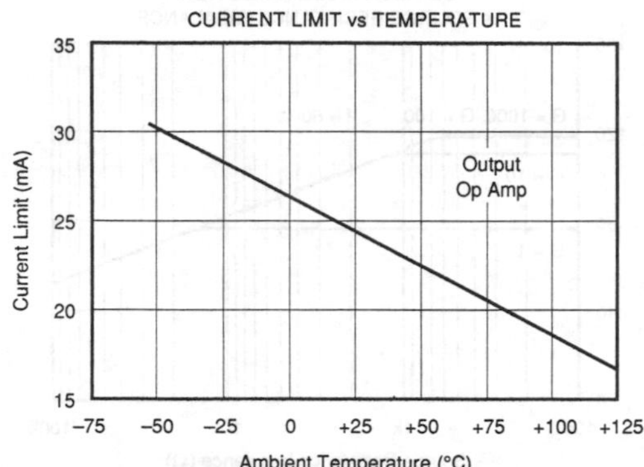

INPUT-REFERRED NOISE
G = 1000

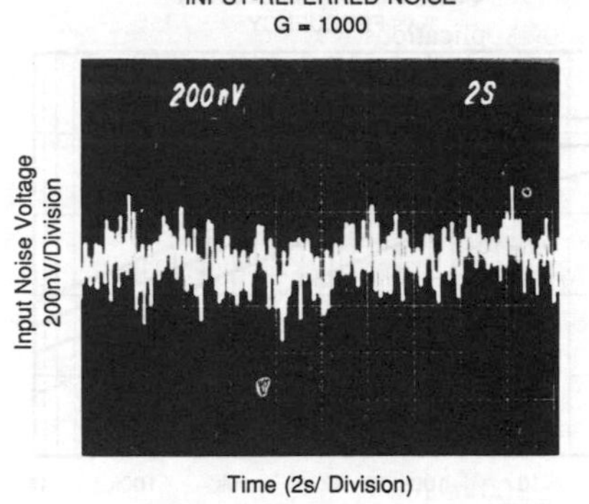

Input Noise Voltage 200nV/Division

Time (2s/ Division)

SMALL-SIGNAL TRANSIENT RESPONSE
G = 1

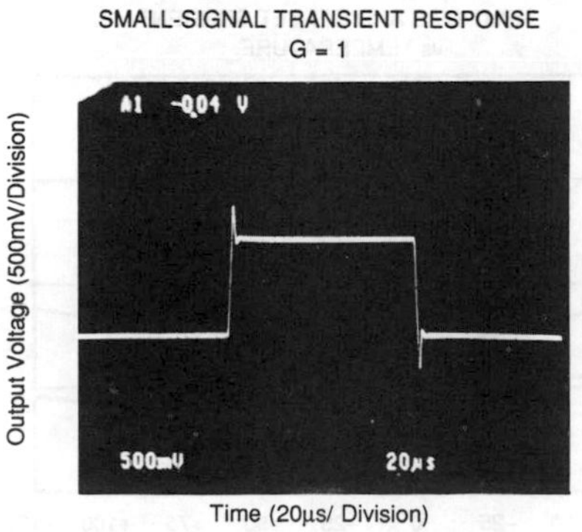

Output Voltage (500mV/Division)

Time (20µs/ Division)

LARGE-SIGNAL TRANSIENT RESPONSE G = 1

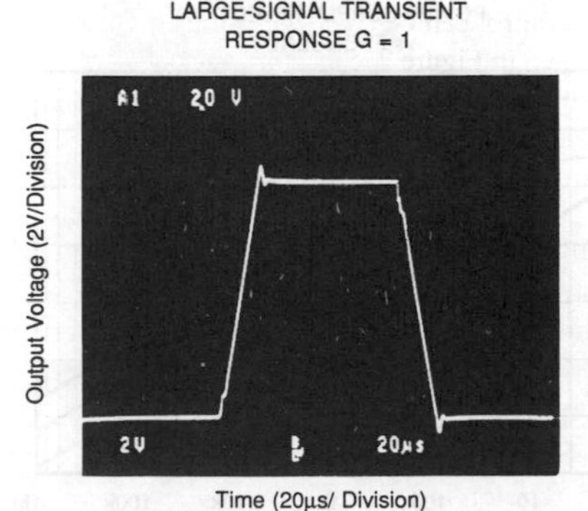

Output Voltage (2V/Division)

Time (20µs/ Division)

TYPICAL PERFORMANCE CURVES (CONT)

T_A = +25°C, V_S = ±15V unless otherwise noted.

SMALL-SIGNAL TRANSIENT RESPONSE
G = 100

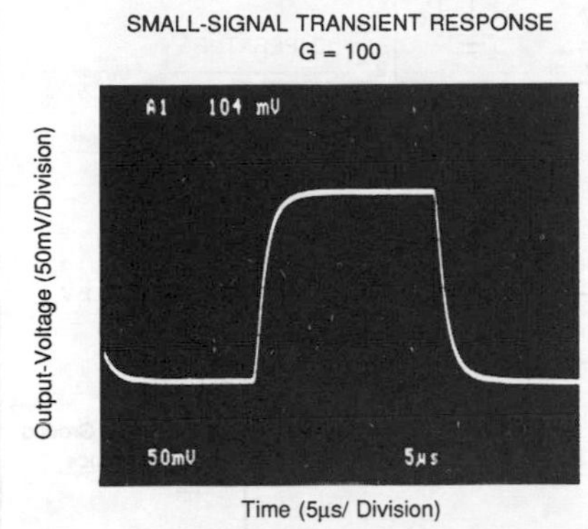

Time (5µs/ Division)

LARGE-SIGNAL TRANSIENT
RESPONSE G = 100

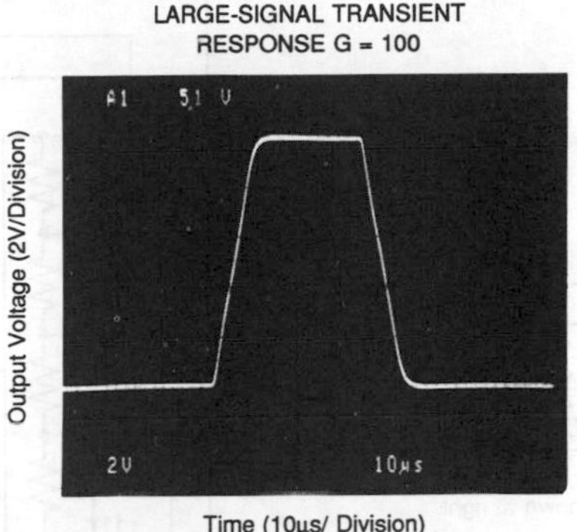

Time (10µs/ Division)

APPLICATION INFORMATION

Figure 1 shows the basic connections required for operation of the INA120. Applications with noisy or high impedance power supply lines may require decoupling capacitors close to the device pins as shown. The differential input voltage is applied to pins 16 and 3.

The output is referred to the output common reference terminal, pin 18. This terminal must have a low-impedance connection to ground. A resistance of 1Ω or greater in series with the common terminal could degrade common-mode rejection beyond the specified value.

SETTING THE GAIN

Gains of 1, 10, 100 or 1000 can be configured by interconnecting the gain-set pins as shown in the table of Figure 1. These pin-strapped gains provide best gain accuracy and drift because they are determined by the ratios of accurately trimmed and matched on-chip resistors.

Digital gain control can be achieved using an analog multiplexer as shown in Figure 2. Since the switches are in series with the high impedance gain-sense connections, pins 4 and 15, their series resistance does not significantly affect gain error or drift. Gain error at G = 1 is slightly higher than with direct pin connections shown in Figure 1. The gain is selected with a two-bit address, A_0 and A_1. The Multiplexer Enable control is directly connected to V+ since a logic "low" on this line would cause the input amplifiers to run open-loop.

Other gains may be set by connecting an external resistor, R_G, as shown in Figure 3a. Gain accuracy using an external gain-setting resistor is a function of R_G and the internal 20kΩ resistors. The internal resistors are typically within ±0.2% of nominal value and their drift under ±80ppm/°C. Inaccuracy and drift of R_G will contribute additional gain error and drift.

Figure 3b shows an external gain-setting resistor connected in parallel with internal resistors. By forming a portion of the effective R_G with internal resistors, gain accuracy and drift can be somewhat improved.

Connections available on the INA120 allow all input stage gain-setting resistors to be provided externally. A custom precision resistor network could be connected to provide the highest accuracy and lowest gain drift for non-standard gains. Impedance of this external network should be made close to that of the internal network for best performance.

OFFSET TRIMMING

Many applications require no external offset voltage trimming. Figure 4 shows optional circuits for trimming offset voltage. Since the INA120 has two amplification stages, the offset voltage is comprised of two components—the input stage offset and output stage offset.

The input stage offset is equal to the combined offset of op amps A_1 and A_2. This input stage offset dominates at high gain. When used in gains of 100 to 1000, it is often sufficient to adjust the input stage offset with a potentiometer connected to pins 6 and 7 as shown. Connect both inputs to ground and adjust for 0V at the output, pin 1. Do not use pins 6 and 7 to trim offset voltage at G = 1 or to correct for offset in devices following the INA120 since this can cause excessive offset voltage drift.

At G = 1, offset is dominated by the output stage. Output stage offset can be trimmed by applying a correction voltage at the output reference terminal, pin 18. Low impedance must be maintained at this node to preserve the high CMR of the INA120. This is achieved by buffering the trim voltage with an op amp as shown.

At intermediate gains it may be necessary to provide both input stage and output stage offset adjustments. Again, ground both inputs. Connect a jumper between pins 9 and 11 (temporarily connects the INA120 in high gain) and adjust R_1 for 0V at the output, pin 1. Then disconnect the jumper and adjust the output offset control for 0V output.

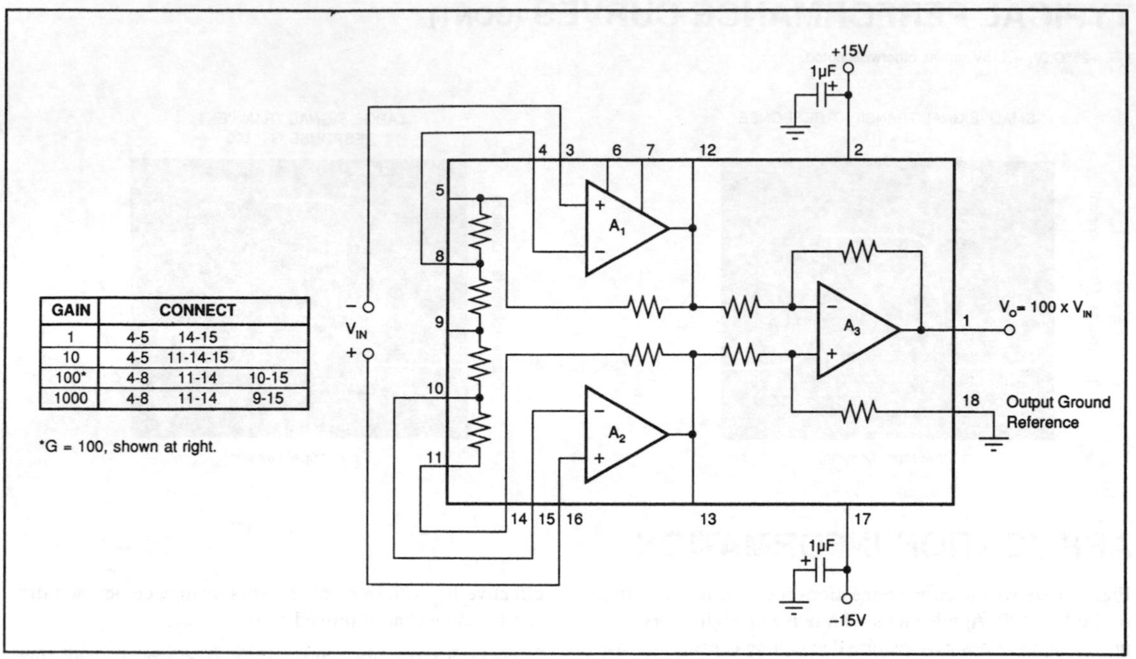

GAIN	CONNECT		
1	4-5	14-15	
10	4-5	11-14-15	
100*	4-8	11-14	10-15
1000	4-8	11-14	9-15

*G = 100, shown at right.

FIGURE 1. Basic Connection.

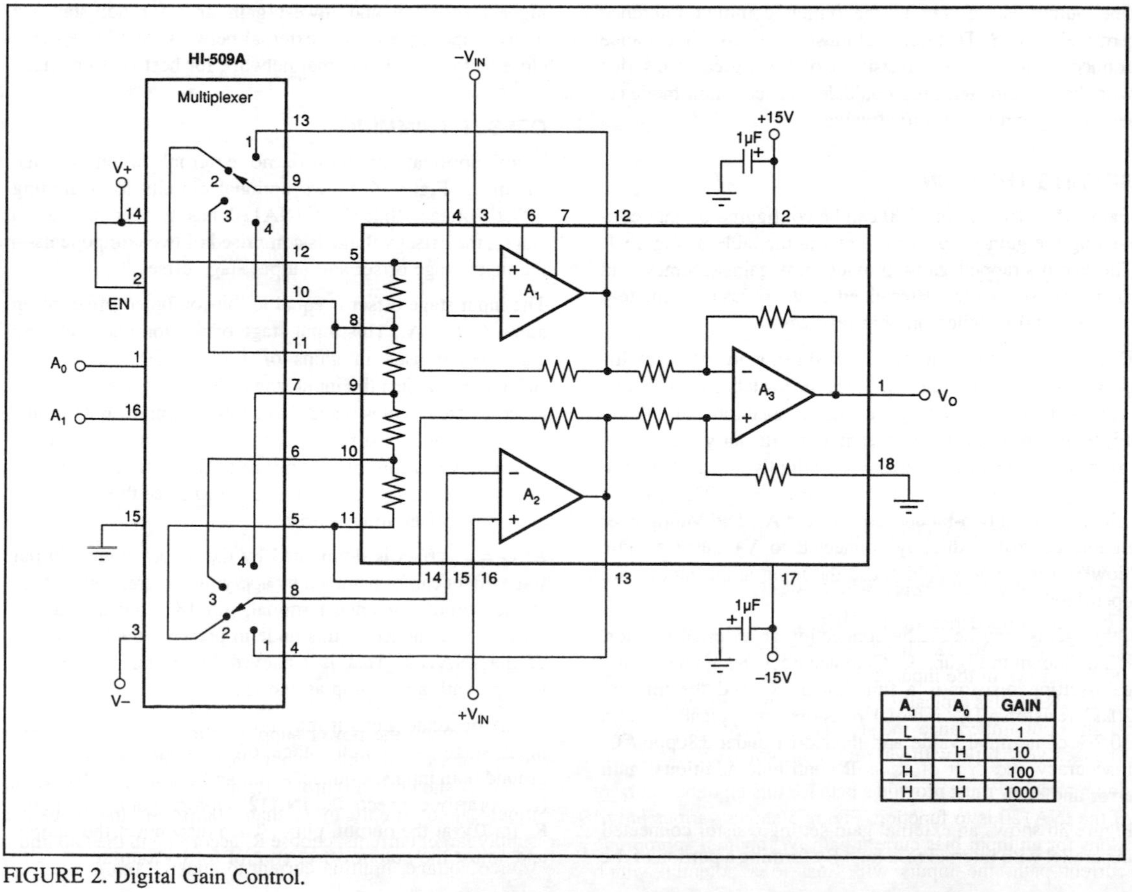

A_1	A_0	GAIN
L	L	1
L	H	10
H	L	100
H	H	1000

FIGURE 2. Digital Gain Control.

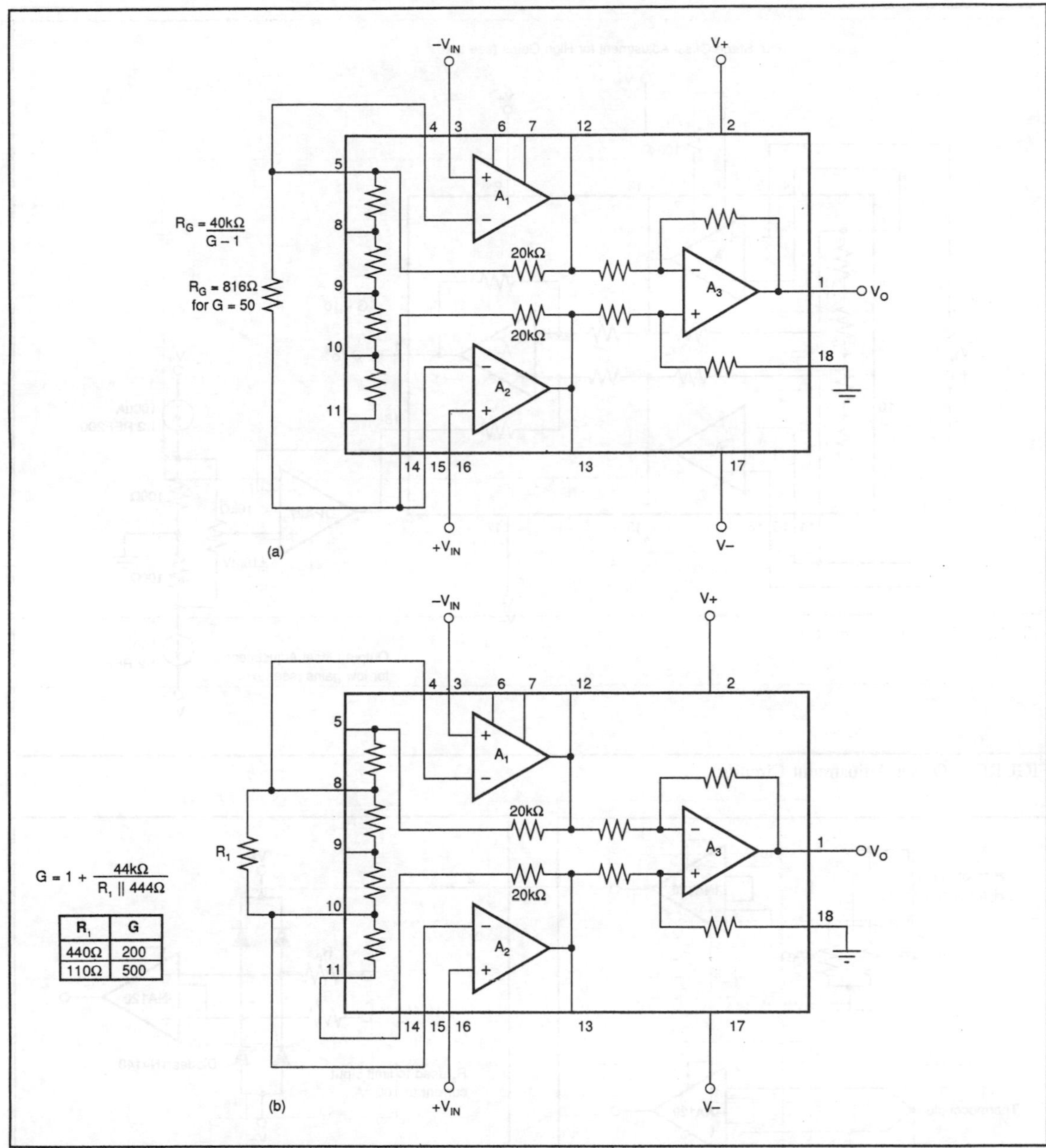

R_1	G
440Ω	200
110Ω	500

FIGURE 3. External Gain-Setting Resistors.

INPUT BIAS CURRENT RETURN PATH

The input impedance of the INA120 is extremely high—approximately $10^{10}\Omega$. This does not mean, however, that no current flows in the input terminals. The input bias current of the INA120 is typically ±10nA (it can be either polarity). High input impedance means that this input bias current changes very little with varying input voltage.

Input circuitry must provide a path for this input bias current if the INA120 is to function. Figure 5 shows various provisions for an input bias current path. Without an appropriate current path, the inputs will float to a potential which exceeds the common-mode range of the INA120 and the input amplifiers will saturate.

INPUT PROTECTION

The inputs of the INA120 are protected for input voltages up to 2V beyond the power supply voltages. If the input can exceed these conditions, input clamp diodes should be provided as shown in Figure 6. R_S may not be required if the input cannot supply more than 100mA. If the input can supply larger currents, choose R_S according to the maximum source voltage, limiting current to under 100mA.

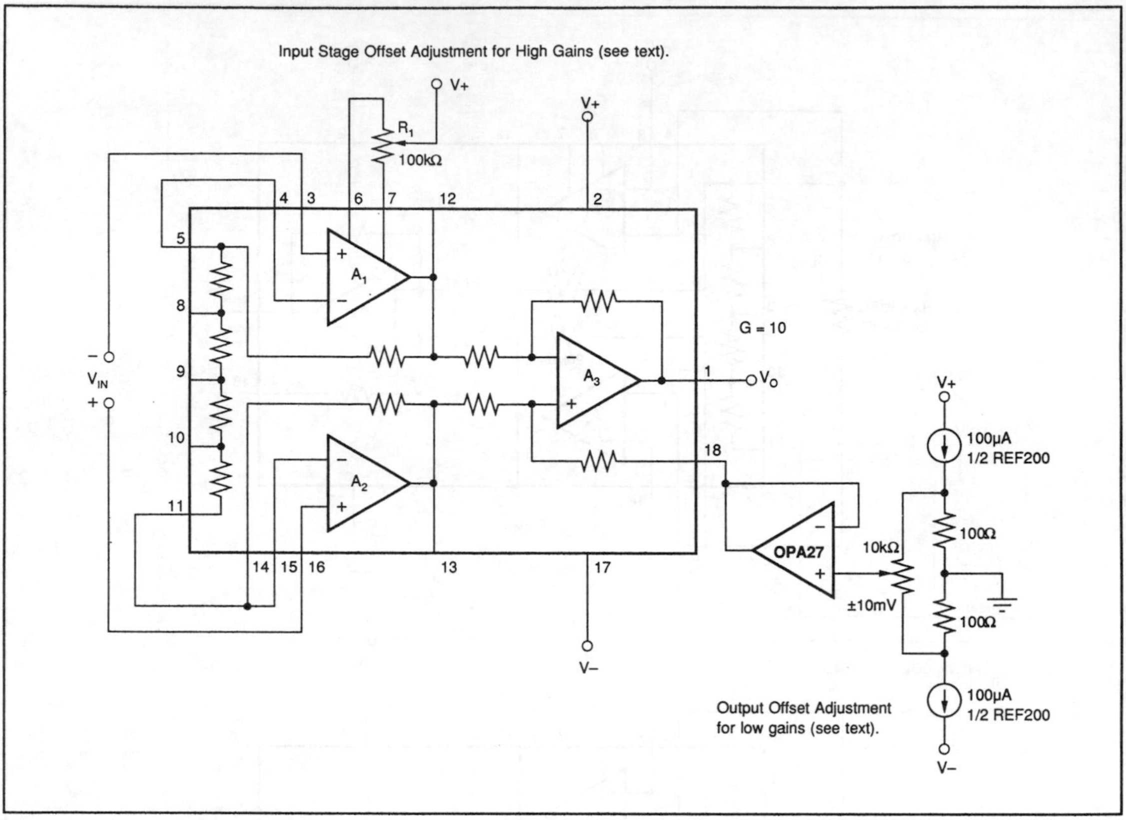

Input Stage Offset Adjustment for High Gains (see text).

G = 10

Output Offset Adjustment for low gains (see text).

FIGURE 4. Offset Adjustment Circuits.

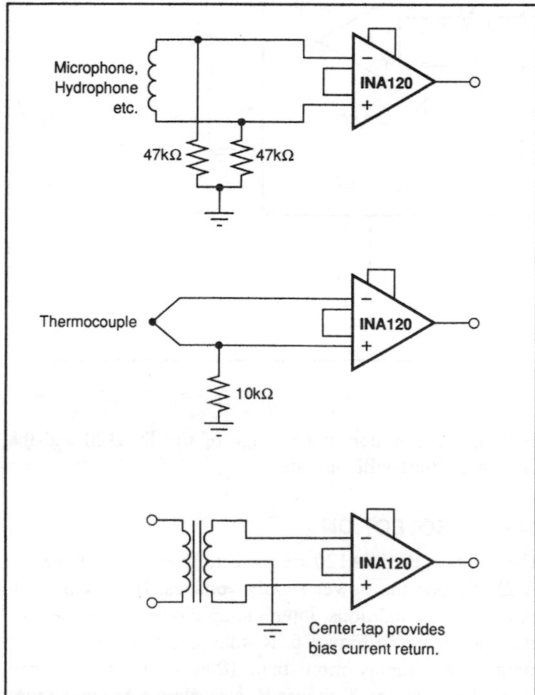

FIGURE 5. Providing an Input Bias Current Path.

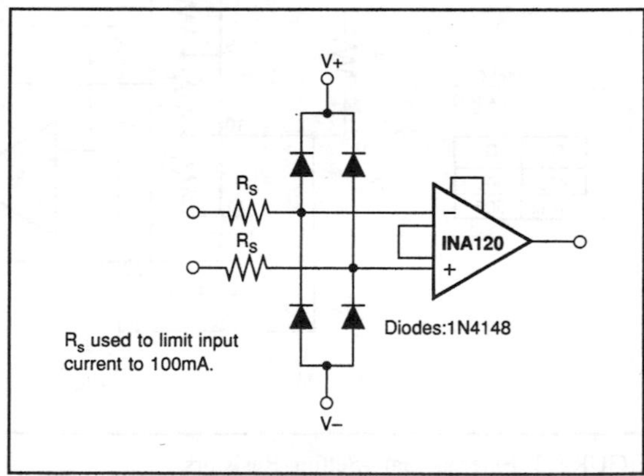

FIGURE 6. Input Protection Circuit.

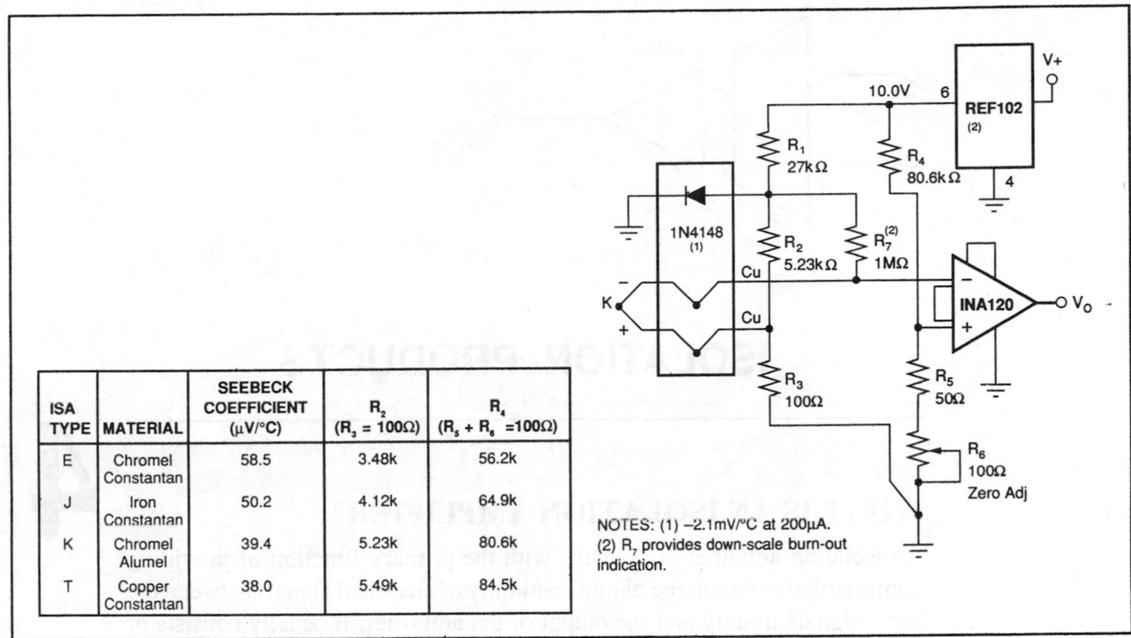

ISA TYPE	MATERIAL	SEEBECK COEFFICIENT (μV/°C)	R_2 (R_3 = 100Ω)	R_4 ($R_5 + R_6$ =100Ω)
E	Chromel Constantan	58.5	3.48k	56.2k
J	Iron Constantan	50.2	4.12k	64.9k
K	Chromel Alumel	39.4	5.23k	80.6k
T	Copper Constantan	38.0	5.49k	84.5k

FIGURE 7. Thermocouple Amplifier With Cold Junction Compensation.

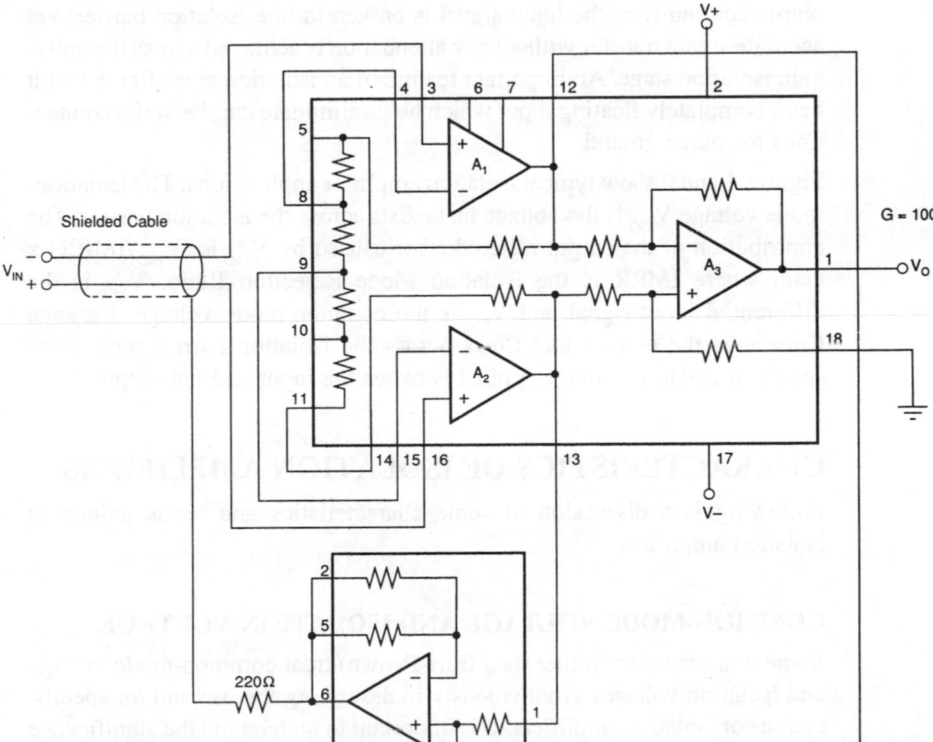

FIGURE 8. Guard Drive Circuit.

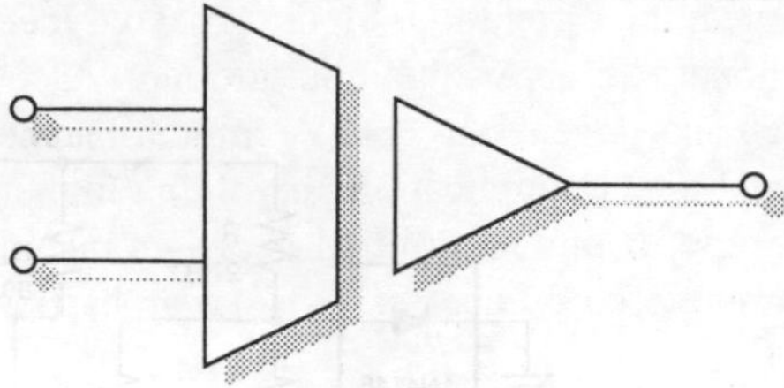

ISOLATION PRODUCTS

4

WHAT IS AN ISOLATION AMPLIFIER?

An isolation amplifier is a device with the primary function of providing ohmic isolation (break the ohmic continuity of electrical signal) between the input signal/circuitry and the output of the amplifier. It usually consists of an input operational amplifier or instrumentation amplifier followed by a unity-gain isolation stage. The sole purpose of the unity-gain isolation stage is to completely isolate the input from the output of the device. Ideally, the ohmic continuity of the input signal is broken (at the isolation barrier) yet accurate signal transfer without any attenuation is achieved across the unity-gain isolation stage. An important feature of an isolation amplifier is that it has a completely floating input which helps eliminate cumbersome connections to source ground.

Figures 1 and 2 show typical isolation amplifier applications. The isolation-mode voltage V_{ISO} is the voltage that exists across the isolation barrier. The contribution of the output referred error caused by V_{ISO} is (V_{ISO}/IMRR) x Gain where IMRR is the Isolation Mode Rejection Ratio. V_{SIG} is the differential input signal and V_{CM} is the common-mode voltage. Leakage Current is the current that flows across the isolation barrier with some specified isolation voltage applied between the input and the output.

CHARACTERISTICS OF ISOLATION AMPLIFIERS

Following is a discussion of some characteristics and terms unique to isolation amplifiers.

COMMON-MODE VOLTAGE AND ISOLATION VOLTAGE

Some manufacturers (other than Burr-Brown) treat common-mode voltage and isolation voltages synonymously in describing the use and /or specifications of isolation amplifiers. It is important to understand the significance of these terms and the difference between them.

When the input common is grounded, the differential input signal V_D (see Figure 1) can be floated by the amount V_{CM} above the input ground. V_{CM} is the common-mode voltage (CMV) and is generally ±10V, limited by the CMV rating of the input stage amplifier. In applications involving higher systems common-mode voltages, input common terminal is not grounded and the common-mode voltages are referenced across the isolation barrier to the output common terminal.

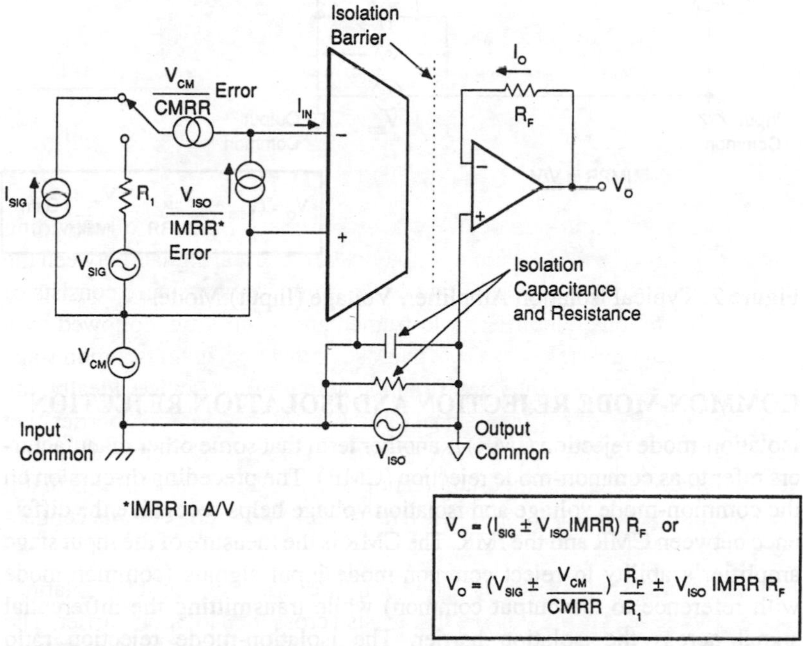

Figure 1. Typical Isolation Amplifier, Current (Input) Mode.

The isolation voltage V_{ISO} shown in Figure 1 is the potential difference between the input common and the output common terminals. The isolation voltage rating describes the amount of voltage that the isolation barrier can withstand without breakdown. This feature of the isolation amplifier allows two distinct ground connections to be made when necessary. It allows the isolation amplifier to be used in applications involving very-high common-mode voltages and in applications of breaking ground loops.

Many applications involve a large "system common-mode voltage." In such applications, the isolation amplifier's input common terminal is not connected to any ground but the output common terminal is connected to the system ground. In such a case, the term V_{CM} shown in Figures 1 and 2 becomes negligible and V_{ISO} determines the safe limit for the system common-mode voltage. In this manner, the isolation amplifier can accommodate common-mode voltages of 2000V or more.

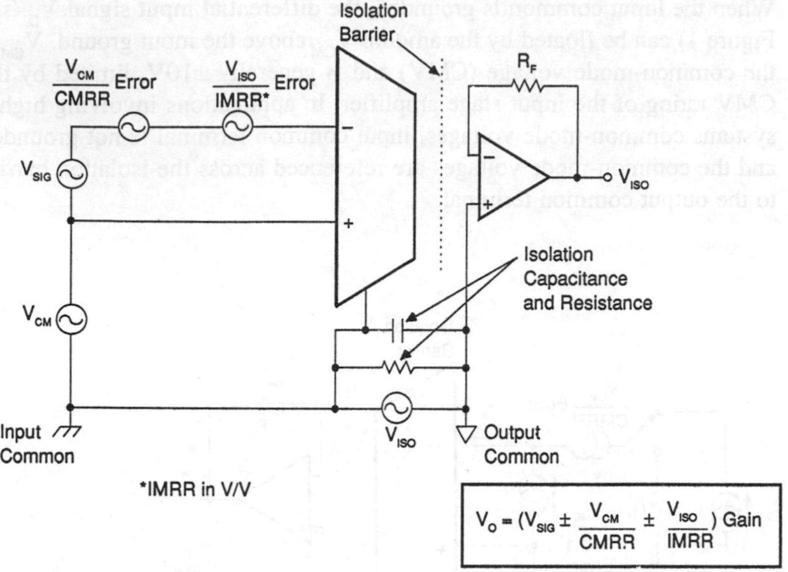

Figure 2. Typical Isolation Amplifier, Voltage (Input) Mode.

COMMON-MODE REJECTION AND ISOLATION REJECTION

Isolation-mode rejection (IMR) is another term that some other manufacturers refer to as common-mode rejection (CMR). The preceding discussion on the common-mode voltage and isolation voltage helps recognize the difference between CMR and the IMR. The CMR is the measure of the input stage amplifier's ability to reject common-mode input signals (common-mode with reference to the output common) while transmitting the differential signal across the isolation barrier. The isolation-mode rejection ratio (IMRR) is defined by the equations shown in Figures 1 and 2. Thus, understanding the IMR capability of isolation amplifiers allows their meaningful use in applications requiring very high common-mode rejection ratios such as 100dB to 140dB.

ISOLATION VOLTAGE RATINGS, TEST VOLTAGE

It is important to understand the significance of the continuous derated isolation voltage specification and its relationship to the actual test voltage applied to the unit. Since a "continuous" test is impractical in a product manufacturing situation (implies infinite test duration) it is generally accepted practice to perform a production test at a higher voltage (higher than the continuous rating) for some shorter length of time.

The important consideration is then "what is the relationship between actual test conditions and the continuous derated minimum specification?" There are several rules of thumb used throughout the industry to establish this relationship. For most isolation amplifiers, Burr-Brown has chosen a very

conservative one: $V_{TEST} = (2 \times V_{CONTINOUS\ RATING}) + 1000V$. This relationship is appropriate for conditions where the system transient voltages are not well defined.* Where the real voltages are well defined or where the isolation voltage is not continuous the user may chose to use a less conservative derating to establish a specification from the test voltage.

Beginning with the introduction of the ISO120 and ISO121, new introductions are being tested for partial discharge. To accommodate poorly defined transients, the part under test is exposed to a voltage 1.6 times the continuous rated voltage and must display a partial discharge level of ≤5pC in a 100% production test. This method is described in detail in the ISO120 data sheet.

APPLICATIONS OF ISOLATION AMPLIFIERS

When one or more of the following conditions/requirements are present in an application, an isolation amplifier would generally be the right choice as a signal conditioning device:

- When ohmic isolation between the signal source and the output is a requirement (isolation impedance between the input and the output is >10MΩ).
- When common-mode noise and voltage rejection requirements are >100dB).
- When is is necessary to process signals in the presence of, or riding on, high common-mode voltages (CMV >> 10V).

In general, most applications can be broadly categorized into the following four types:

- Amplifying and measuring low level signals in the presence of high common-mode voltages.
- Breaking ground loops and/or eliminating source ground connections. The isolation amplifier provides full floating input, eliminating the need for connections to source ground, and thus allows two-wire hook-up to the signal sources.
- Providing an interface between medical patient monitoring equipment and the transducer/devices that may be in physical contact with the patients. Such applications require high isolation voltage levels and very-low leakage currents.
- Providing isolation protection to electronic instruments/equipment. Large common-mode voltages occasionally cause hazardous electronic faults. Low leakage currents and high isolation voltage capability of isolation amplifiers help protect instruments against damage caused by such faults.

---

*Reference National Electrical Manufacturers Association (NEMA) Standards Parts ICS 1-109 and ICS 1-111.

Isolation amplifier performance requirements vary significantly, depending on the type of requirement. In applications where bandwidth and speed of response are more important than gain accuracy and linearity, the optically or capacitatively coupled amplifiers will be the best choice. For applications where gain accuracy and linearity are key parameters, Burr-Brown's family of transformer or capacitatively coupled amplifiers are the suitable choice.

ISOLATION AMPLIFIERS SELECTION GUIDES

The following Selection Guides show parameters for the high grade. Refer to the Product Data Sheet for a full selection of grades. Models shown in **boldface** are new products introduced since publication of the previous *Burr-Brown IC Data Book.*

TRANSFORMER-COUPLED AMPLIFIERS

Descrip	Model	Isolation Voltage (V) Cont Peak	Isolation Voltage (V) Pulse/ Test, Peak	Isolation Mode Rejection, typ DC (dB)	Isolation Mode Rejection, typ 60Hz (dB)	Leakage Current at Test Voltage (μA)	Iso Impedance (Ω)	Iso Impedance (pF)	Gain Nonlinearity (%)	Gain Nonlinearity max (%)	Voltage Drift typ (±μV/°C) max	Bias Current max	±3dB Freq (kHz)	Ext Iso Power Req	Temp[1]	Pg
High Isolation Voltage	3656G	±3500	±8000	160	125	0.5	10^{12}	6	±0.05	±0.03	5+ (1000/G_1)	100nA	30	No	Ind	4-108

NOTES: All packages are DIPs. (1) Ind = –25°C to +85°C.

OPTICALLY COUPLED AMPLIFIERS

Descrip	Model	Isolation Voltage (V) Cont Peak	Isolation Voltage (V) Pulse/ Test, Peak	Isolation Mode Rejection, typ DC (dB)	Isolation Mode Rejection, typ 60Hz (dB)	Leakage Current at Test Voltage (μA)	Iso Impedance (Ω)	Iso Impedance (pF)	Gain Nonlinearity max (%)	Gain Nonlinearity typ (%)	Voltage Drift (±μV/°C) max	Bias Current max	±3dB Freq (kHz)	Ext Iso Power Req	Temp[1]	Pg
Balanced Current Input	3650G	±2000	±5000	140	120	0.25[2]	10^{12}	1.8	±0.05	±0.02	5	10nA	15	Yes[3]	Ind	4-100
Balanced FET Input	3652G	±2000	±5000	140	120	0.25[2]	10^{12}	1.8	±0.1	±0.05	25	50nA	15	Yes	Ind	4-100
Low Drift Wide BW	ISO100P	750	2500	146[3]	108[3]	0.3	10^{12}	2.5	0.07	0.02	4[3]	10nA	60	Yes	Ind	4-8

NOTES: All packages are DIPs. (1) Ind = –25°C to +85°C. (2) At 240V/60Hz. (3) R_{IN} = 10k, Gain = 100.

CAPACITOR COUPLED, HERMETICALLY SEALED AMPLIFIERS

Boldface = NEW

Descrip	Model	Isolation Voltage (V) Cont Peak	Isolation Voltage (V) Pulse/ Test, Peak	Isolation Mode Rejection, typ DC (dB)	Isolation Mode Rejection, typ 60Hz (dB)	Leakage Current at Test Voltage (μA)	Iso Impedance (Ω)	Iso Impedance (pF)	Gain Nonlinearity max (%)	Gain Nonlinearity typ (%)	Voltage Drift (±μV/°C) max	Bias Current max	±3dB Freq (kHz)	Ext Isol Power Req	Temp[1]	Pg
1500VAC Isolation	**ISO102B**	**±2121**	**±4000**	**160**	**120**	**1.0**	**10^{14}**	**6**	**±0.025**	**±0.02**	**±250**	**100μA**	**70**	**Yes**	**Ind**	**4-20**
	ISO120B	**±2121**	**±3535[2]**	**160**	**115**	**0.5**	**10^{14}**	**2**	**±0.01**	**±0.005**	**±150**	**50μA**	**60**	**Yes**	**Ind**	**4-44**
	ISO122	**±2121**	**±3394**		**140**	**0.5**	**10^{14}**	**2**	**±0.01**	**±0.005**	**±200**	**50μA**	**50**	**Yes**	**Com**	**4-57**
3500VAC Isolation	**ISO106B**	**±4950**	**±8000**	**160**	**130**	**1.0**	**10^{14}**	**6**	**±0.012**	**±0.007**	**±250**	**100μA**	**70**	**Yes**	**Ind**	**4-20**
	ISO121B	**±4950**	**±5600[2]**	**160**	**115**	**0.5**	**10^{14}**	**2**	**±0.01**	**±0.005**	**±150**	**50μA**	**60**	**Yes**	**Ind**	**4-44**

NOTES: All packages are DIPs. (1) Ind = –25°C to +85°C. Com = 0°C to +70°C. (2) Partial discharge voltage.

ISO122P

Precision Lowest Cost ISOLATION AMPLIFIER

FEATURES

- 100% TESTED FOR HIGH-VOLTAGE BREAKDOWN
- RATED 1500Vrms
- HIGH IMR: 140dB at 60Hz
- BIPOLAR OPERATION: $V_O = \pm 10V$
- SINGLE-WIDE 16-PIN PLASTIC DIP
- EASE OF USE: Fixed Unity Gain Configuration
- 0.020% max NONLINEARITY
- ±4.5V to ±18V SUPPLY RANGE

APPLICATIONS

- INDUSTRIAL PROCESS CONTROL: Transducer Isolator, Isolator for Thermocouples, RTDs, Pressure Bridges, and Flow Meters, 4mA to 20mA Loop Isolation
- GROUND LOOP ELIMINATION
- MOTOR AND SCR CONTROL
- POWER MONITORING
- PC-BASED DATA ACQUISITION
- TEST EQUIPMENT
- VENDING MACHINES

DESCRIPTION

The ISO122P is a precision isolation amplifier incorporating a novel duty cycle modulation-demodulation technique. The signal is transmitted digitally across a 2pF differential capacitive barrier. With digital modulation the barrier characteristics do not affect signal integrity, resulting in excellent reliability and good high frequency transient immunity across the barrier. Both barrier capacitors are imbedded in the plastic body of the package.

The ISO122P is easy to use. No external components are required for operation. The key specifications are 0.020% max nonlinearity, 50kHz signal bandwidth, and 200μV/°C V_{OS} drift. A power supply range of ±4.5V to ±18V and quiescent currents of ±4.5mA on V_{S1} and ±5.0mA on V_{S2} make these amplifiers ideal for a wide range of applications.

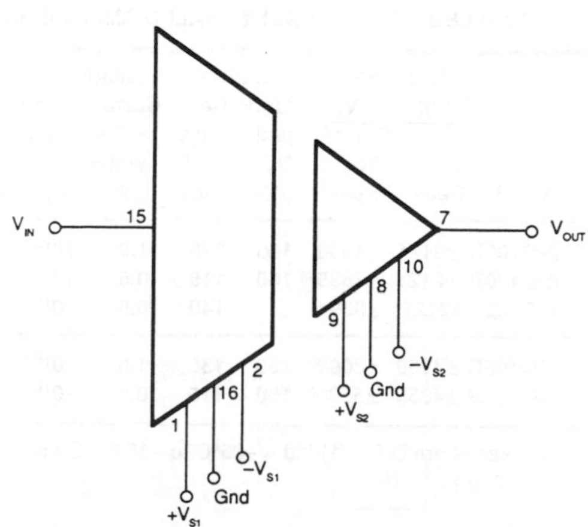

International Airport Industrial Park • Mailing Address: PO Box 11400 • Tucson, AZ 85734 • Street Address: 6730 S. Tucson Blvd. • Tucson, AZ 85706
Tel: (602) 746-1111 • Twx: 910-952-1111 • Cable: BBRCORP • Telex: 066-6491 • FAX: (602) 889-1510 • Immediate Product Info: (800) 548-6132

PDS-857D

Printed in U.S.A. August, 1991

SPECIFICATIONS

At $T_A = +25°C$, $V_{S1} = V_{S2} = \pm15V$, and $R_L = 2k\Omega$ unless otherwise noted.

PARAMETER	CONDITIONS	MIN	TYP	MAX	UNITS
ISOLATION					
Voltage Rated Continuous AC 60Hz		1500			VAC
100% Test [1]	1s, 5pc PD	2400			VAC
Isolation Mode Rejection	60Hz		140		dB
Barrier Impedance			10^{14} || 2		Ω || pF
Leakage Current at 60Hz	V_{ISO} = 240Vrms		0.18	0.5	μArms
GAIN	$V_O = \pm10V$				
Nominal Gain			1		V/V
Gain Error			±0.05	±0.50	%FSR
Gain vs Temperature			±10		ppm/°C
Nonlinearity P			±0.008	±0.020	%FSR
JP			±0.020	±0.050	%FSR
INPUT OFFSET VOLTAGE					
Initial Offset			±5	±50	mV
vs Temperature			±200		μV/°C
vs Supply			±2		mV/V
Noise			4		μV/√Hz
INPUT					
Voltage Range[2]		±10	±12.5		V
Resistance			200		kΩ
OUTPUT					
Voltage Range		±10	±12.5		V
Current Drive		±5	±20		mA
Capacitive Load Drive			0.1		μF
Ripple Voltage[3]			20		mVp-p
FREQUENCY RESPONSE					
Small Signal Bandwidth			50		kHz
Slew Rate			2		V/μs
Settling Time	$V_O = \pm10V$				
0.1%			50		μs
0.01%			350		μs
Overload Recover Time			150		μs
POWER SUPPLIES					
Rated Voltage			±15		V
Voltage Range		±4.5		±18	V
Quiescent Current: V_{S1}			±4.5	±7.0	mA
V_{S2}			±5.0	±7.0	mA
TEMPERATURE RANGE					
Specification		0		70	°C
Operating		–25		85	°C
Storage		–25		85	°C
θ_{JA}			100		°C/W

NOTES: (1) Tested at 1.6 X rated, fail on 5pC partial discharge. (2) Input Range is ±10V independent of input $\pm V_{S1}$. (3) Ripple frequency is at carrier frequency (500kHz).

CONNECTION DIAGRAM

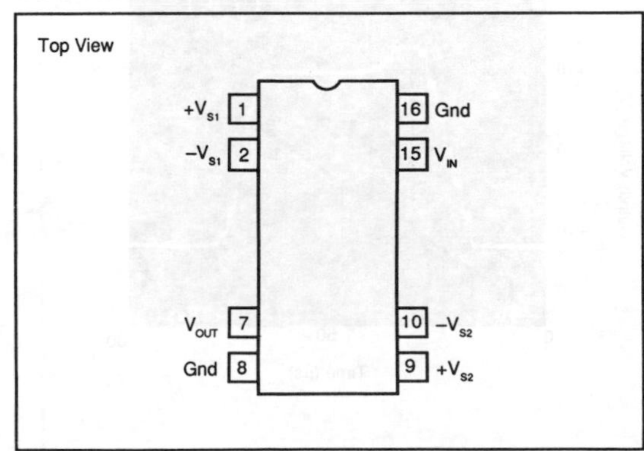

ABSOLUTE MAXIMUM RATINGS

Supply Voltage	±18V
V_{IN}	±100V
Continuous Isolation Voltage	1500Vrms
Junction Temperature	+150°C
Storage Temperature	+85°C
Lead Temperature (soldering, 10s)	+300°C
Output Short to Common	Continuous

MECHANICAL

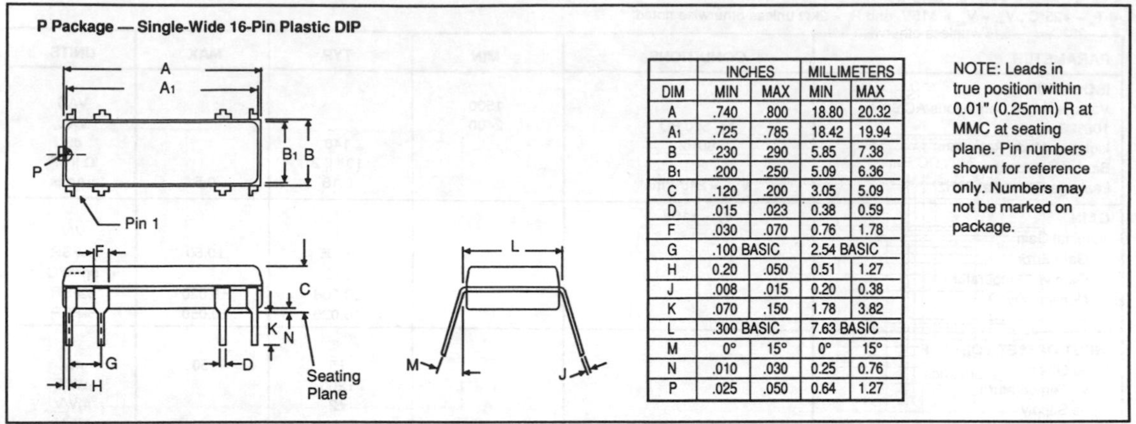

DIM	INCHES MIN	INCHES MAX	MILLIMETERS MIN	MILLIMETERS MAX
A	.740	.800	18.80	20.32
A1	.725	.785	18.42	19.94
B	.230	.290	5.85	7.38
B1	.200	.250	5.09	6.36
C	.120	.200	3.05	5.09
D	.015	.023	0.38	0.59
F	.030	.070	0.76	1.78
G	.100 BASIC		2.54 BASIC	
H	0.20	.050	0.51	1.27
J	.008	.015	0.20	0.38
K	.070	.150	1.78	3.82
L	.300 BASIC		7.63 BASIC	
M	0°	15°	0°	15°
N	.010	.030	0.25	0.76
P	.025	.050	0.64	1.27

NOTE: Leads in true position within 0.01" (0.25mm) R at MMC at seating plane. Pin numbers shown for reference only. Numbers may not be marked on package.

TYPICAL PERFORMANCE CURVES

T_A = +25°C, V_S = ±15V unless otherwise noted.

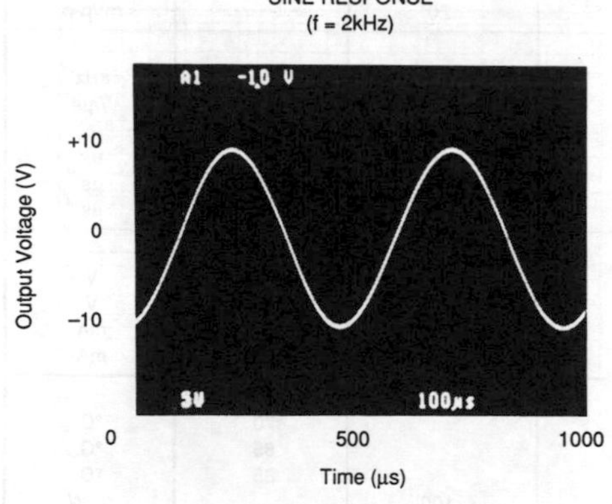

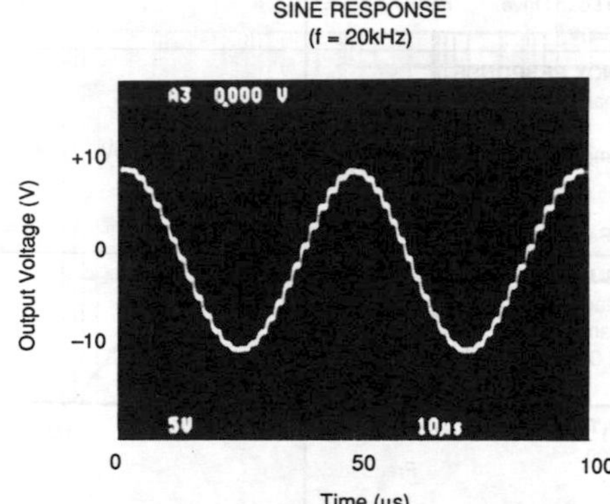

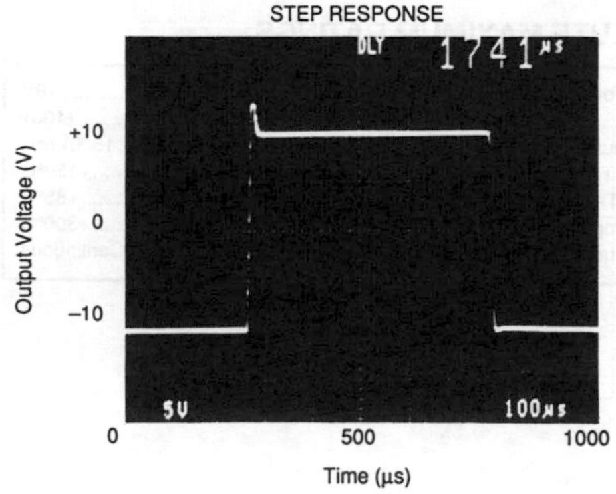

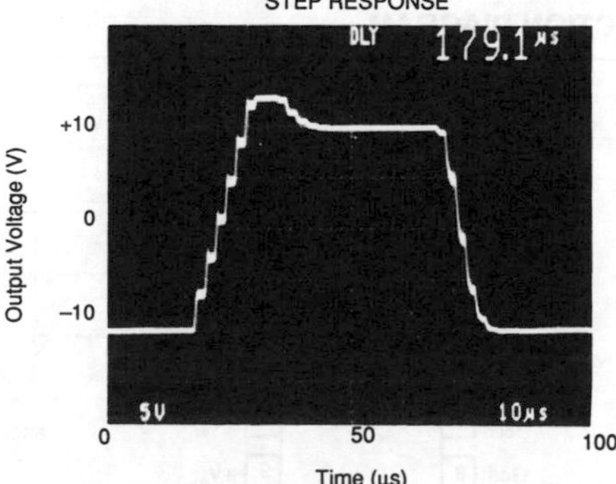

TYPICAL PERFORMANCE CURVES

$T_A = +25°C$, $V_S = ±15V$ unless otherwise noted.

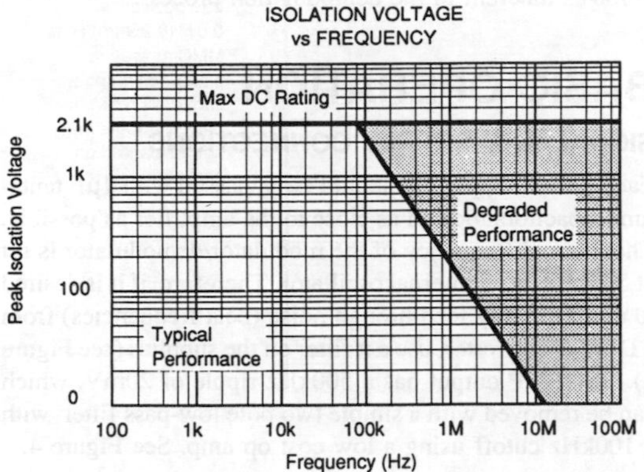

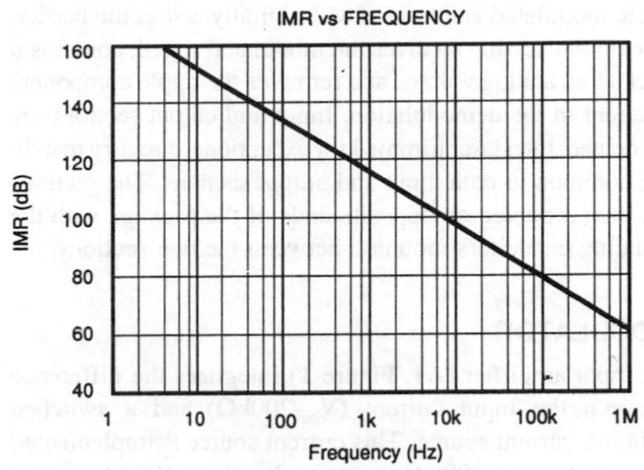

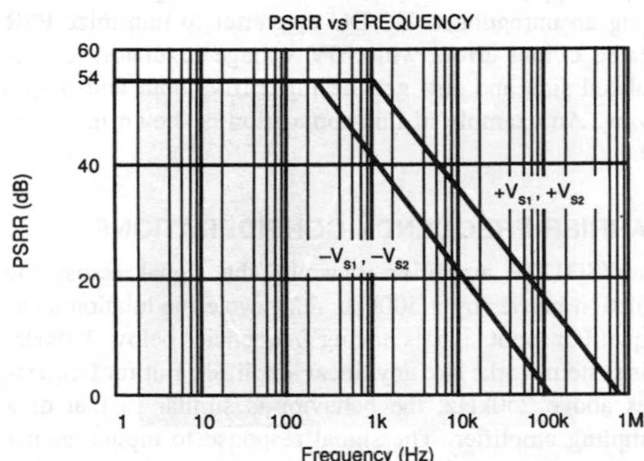

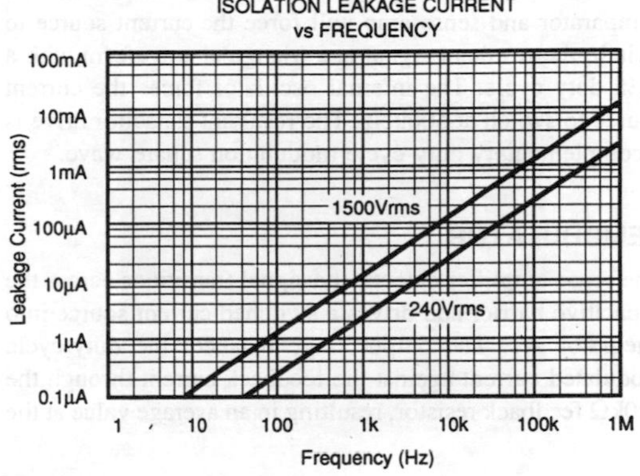

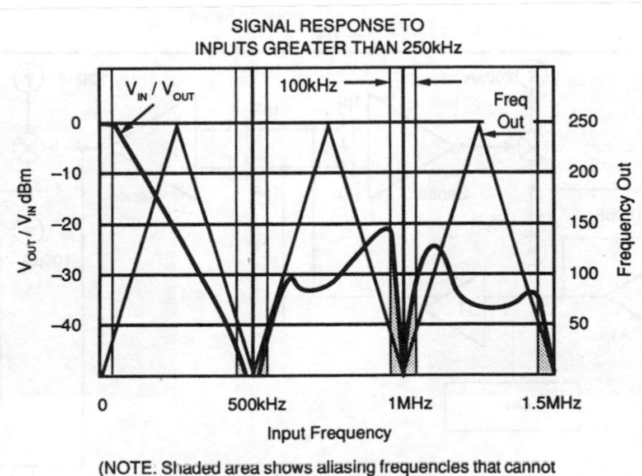

(NOTE. Shaded area shows aliasing frequencies that cannot be removed by a low-pass filter at the output.)

THEORY OF OPERATION

The ISO122P isolation amplifier uses an input and an output section galvanically isolated by matched 1pF isolating capacitors built into the plastic package. The input is duty-cycle modulated and transmitted digitally across the barrier. The output section receives the modulated signal, converts it back to an analog voltage and removes the ripple component inherent in the demodulation. Input and output sections are fabricated, then laser trimmed for exceptional circuitry matching common to both input and output sections. The sections are then mounted on opposite ends of the package with the isolating capacitors mounted between the two sections.

MODULATOR

An input amplifier (A1, Figure 1) integrates the difference between the input current (V_{IN}/200kΩ) and a switched ±100µA current source. This current source is implemented by a switchable 200µA source and a fixed 100µA current sink. To understand the basic operation of the modulator, assume that V_{IN} = 0.0V. The integrator will ramp in one direction until the comparator threshold is exceeded. The comparator and sense amp will force the current source to switch; the resultant signal is a triangular waveform with a 50% duty cycle. The internal oscillator forces the current source to switch at 500kHz. The resultant capacitor drive is a complementary duty-cycle modulation square wave.

DEMODULATOR

The sense amplifier detects the signal transitions across the capacitive barrier and drives a switched current source into integrator A2. The output stage balances the duty-cycle modulated current against the feedback current through the 200kΩ feedback resistor, resulting in an average value at the V_{OUT} pin equal to V_{IN}. The sample and hold amplifiers in the output feedback loop serve to remove undesired ripple voltages inherent in the demodulation process.

BASIC OPERATION

SIGNAL AND SUPPLY CONNECTIONS

Each power supply pin should be bypassed with 1µF tantalum capacitors located as close to the amplifier as possible. The internal frequency of the modulator/demodulator is set at 500kHz by an internal oscillator. Therefore, if it is desired to minimize any feedthrough noise (beat frequencies) from a DC/DC converter, use a π filter on the supplies (see Figure 4). ISO122P output has a 500kHz ripple of 20mV, which can be removed with a simple two pole low-pass filter with a 100kHz cutoff using a low cost op amp. See Figure 4.

The input to the modulator is a current (set by the 200kΩ integrator input resistor) that makes it possible to have an input voltage greater than the input supplies, as long as the output supply is at least ±15V. It is therefore possible when using an unregulated DC/DC converter to minimize PSR related output errors with ±5V voltage regulators on the isolated side and still get the full ±10V input and output swing. An example of this application is shown in Figure 10.

CARRIER FREQUENCY CONSIDERATIONS

The ISO122P amplifier transmits the signal across the isolation barrier by a 500kHz duty cycle modulation technique. For input signals having frequencies below 250kHz, this system works like any linear amplifier. But for frequencies above 250kHz, the behavior is similar to that of a sampling amplifier. The signal response to inputs greater

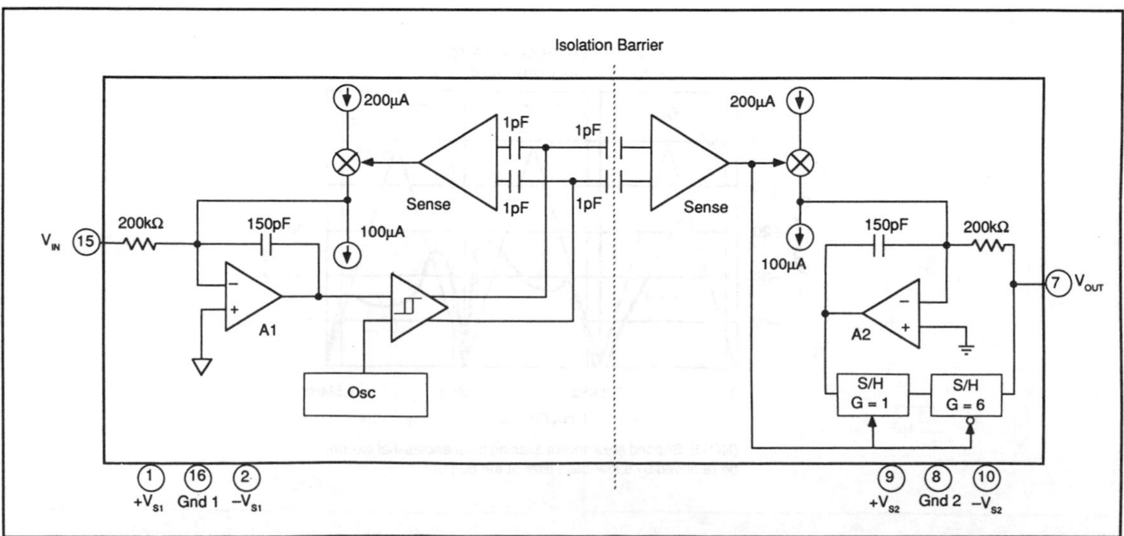

FIGURE 1. Block Diagram.

than 250kHz performance curve shows this behavior graphically; at input frequencies above 250kHz the device generates an output signal component of reduced magnitude at a frequency below 250kHz. This is the aliasing effect of sampling at frequencies less than 2 times the signal frequency (the Nyquist frequency). Note that at the carrier frequency and its harmonics, both the frequency and amplitude of the aliasing go to zero.

ISOLATION MODE VOLTAGE INDUCED ERRORS

IMV can induce errors at the output as indicated by the plots of IMV vs Frequency. It should be noted that if the IMV frequency exceeds 250kHz, the output also will display spurious outputs (aliasing), in a manner similar to that for $V_{IN} > 250$kHz and the amplifier response will be identical to that shown in the Signal Response to Inputs Greater Than 250kHz performance curve. This occurs because IMV-induced errors behave like input-referred error signals. To predict the total error, divide the isolation voltage by the IMR shown in the IMR vs Frequency curve and compute the amplifier response to this input-referred error signal from the data given in the Signal Response to Inputs Greater than 250kHz performance curve. For example, if a 800kHz 1000Vrms IMR is present, then a total of [(–60dB) + (–30dB)] x (1000V) = 32mV error signal at 200kHz plus a 1V, 800kHz error signal will be present at the output.

HIGH IMV dV/dt ERRORS

As the IMV frequency increases and the dV/dt exceeds 1000V/µs, the sense amp may start to false trigger, and the output will display spurious errors. The common mode current being sent across the barrier by the high slew rate is the cause of the false triggering of the sense amplifier. Lowering the power supply voltages below ±15V may decrease the dV/dt to 500V/µs for typical performance.

HIGH VOLTAGE TESTING

Burr-Brown Corporation has adopted a partial discharge test criterion that conforms to the German VDE0884 Optocoupler Standards. This method requires the measurement of minute current pulses (<5pC) while applying 2400Vrms, 60Hz high voltage stress across every ISO122 isolation barrier. No partial discharge may be initiated to pass this test. This criterion confirms transient overvoltage (1.6 x 1500Vrms) protection without damage to the ISO122. Lifetest results verify the absence of failure under continuous rated voltage and maximum temperature.

This new test method represents the "state of the art" for non-destructive high voltage reliability testing. It is based on the effects of non-uniform fields that exist in heterogeneous dielectric material during barrier degradation. In the case of void non-uniformities, electric field stress begins to ionize the void region before bridging the entire high voltage barrier. The transient conduction of charge during and after the ionization can be detected externally as a burst of 0.01-0.1µs current pulses that repeat on each AC voltage cycle. The minimum AC barrier voltage that initiates partial discharge is defined as the "inception voltage." Decreasing the barrier voltage to a lower level is required before partial discharge ceases and is defined as the "extinction voltage." We have characterized and developed the package insulation processes to yield an inception voltage in excess of 2400Vrms so that transient overvoltages below this level will not damage the ISO122. The extinction voltage is above 1500Vrms so that even overvoltage induced partial discharge will cease once the barrier voltage is reduced to the 1500Vrms (rated) level. Older high voltage test methods relied on applying a large enough overvoltage (above rating) to break down marginal parts, but not so high as to damage good ones. Our new partial discharge testing gives us more confidence in barrier reliability than breakdown/no breakdown criteria.

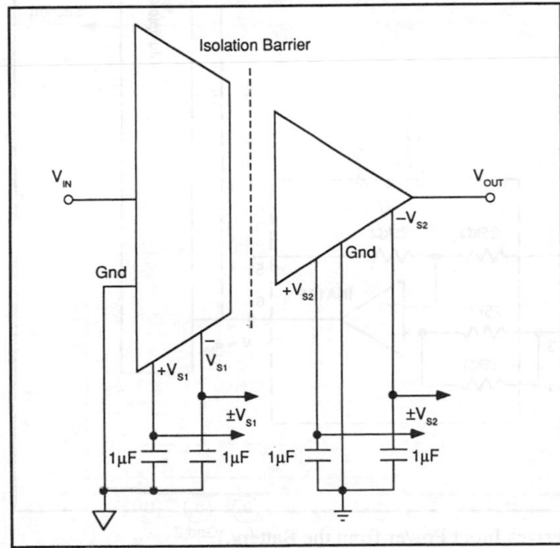

FIGURE 2. Basic Signal and Power Connections.

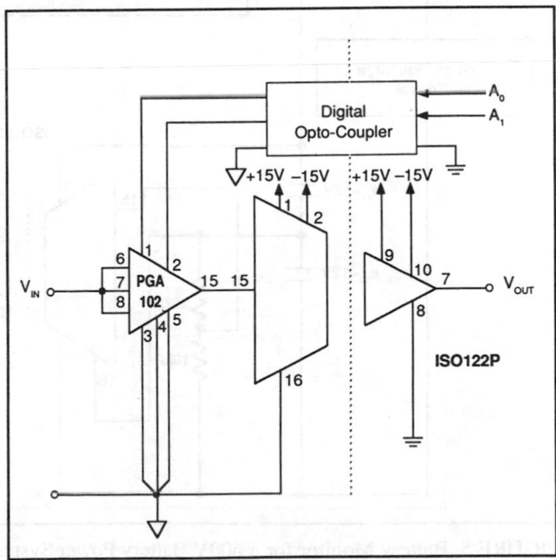

FIGURE 3. Programmable-Gain Isolation Channel with Gains of 1, 10, and 100.

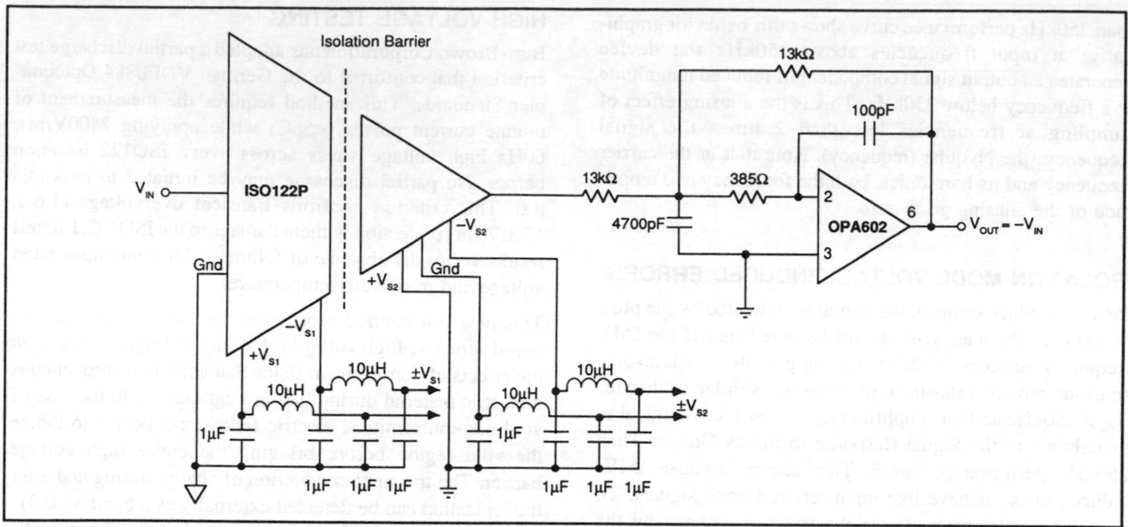

FIGURE 4. Optional π Filter to Minimize Power Supply Feedthrough Noise; Output Filter to Remove 500kHz Carrier Ripple.

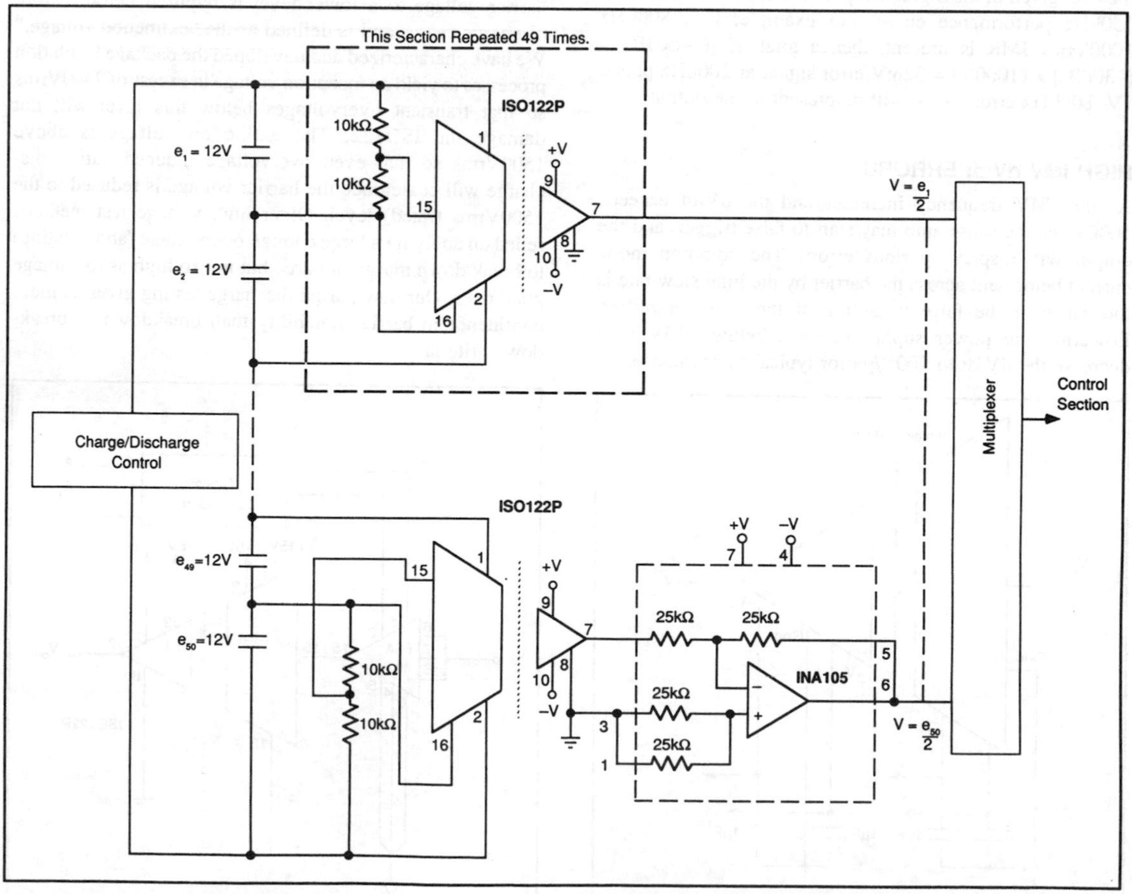

FIGURE 5. Battery Monitor for a 600V Battery Power System. (Derives Input Power from the Battery.)

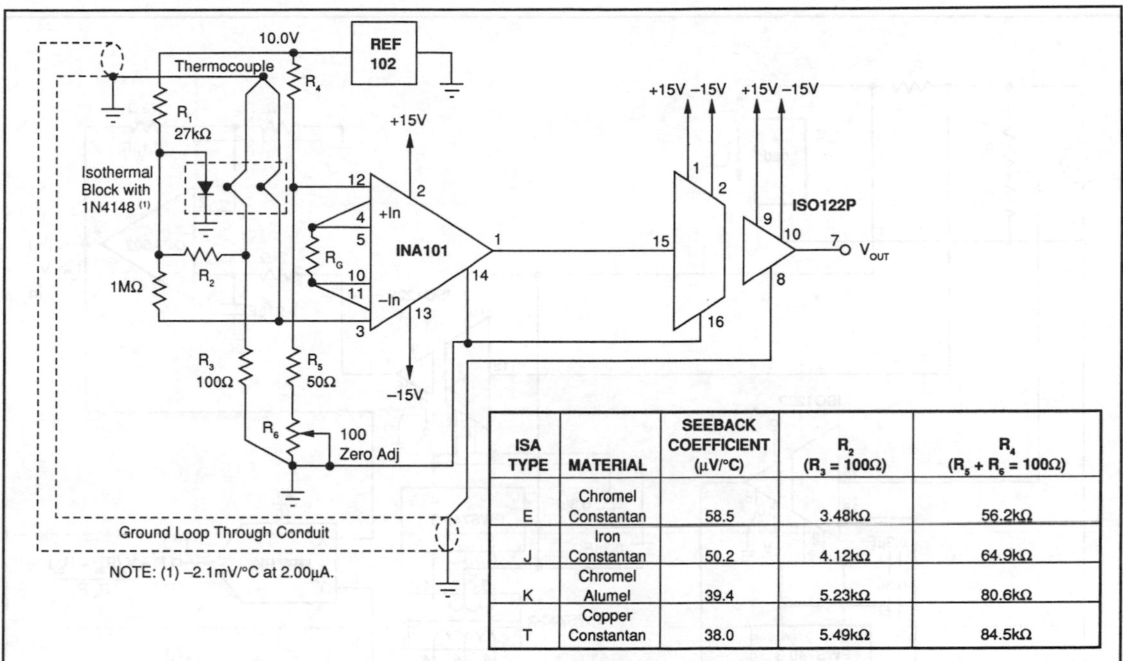

ISA TYPE	MATERIAL	SEEBACK COEFFICIENT (µV/°C)	R_2 (R_3 = 100Ω)	R_4 (R_5 + R_6 = 100Ω)
E	Chromel Constantan	58.5	3.48kΩ	56.2kΩ
J	Iron Constantan	50.2	4.12kΩ	64.9kΩ
K	Chromel Alumel	39.4	5.23kΩ	80.6kΩ
T	Copper Constantan	38.0	5.49kΩ	84.5kΩ

FIGURE 6. Thermocouple Amplifier with Ground Loop Elimination, Cold Junction Compensation, and Up-scale Burn-out.

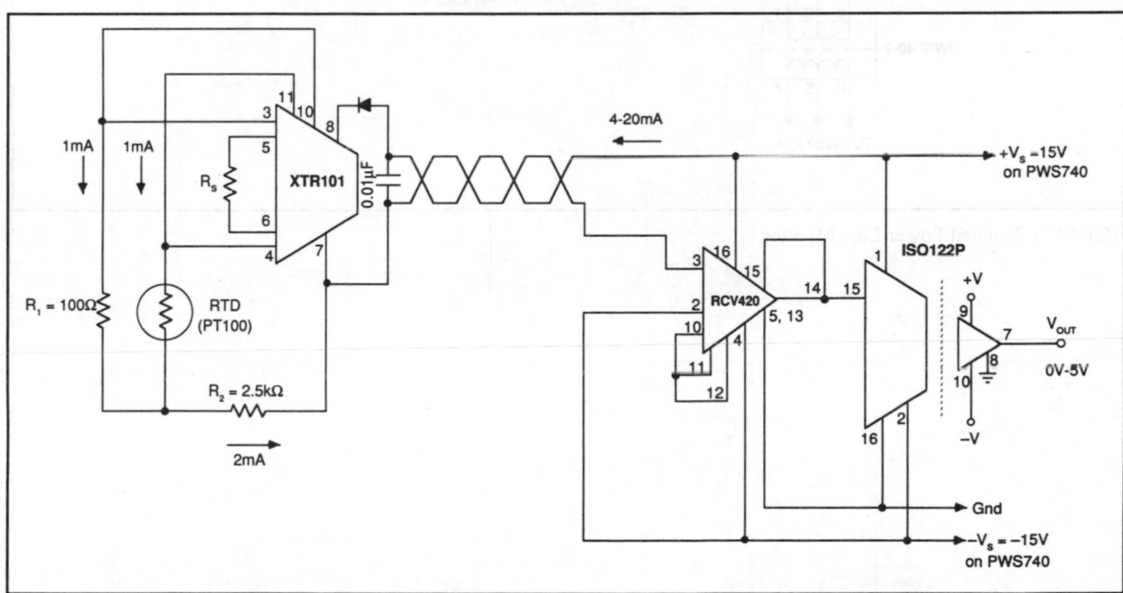

FIGURE 7. Isolated 4-20mA Instrument Loop. (RTD shown.)

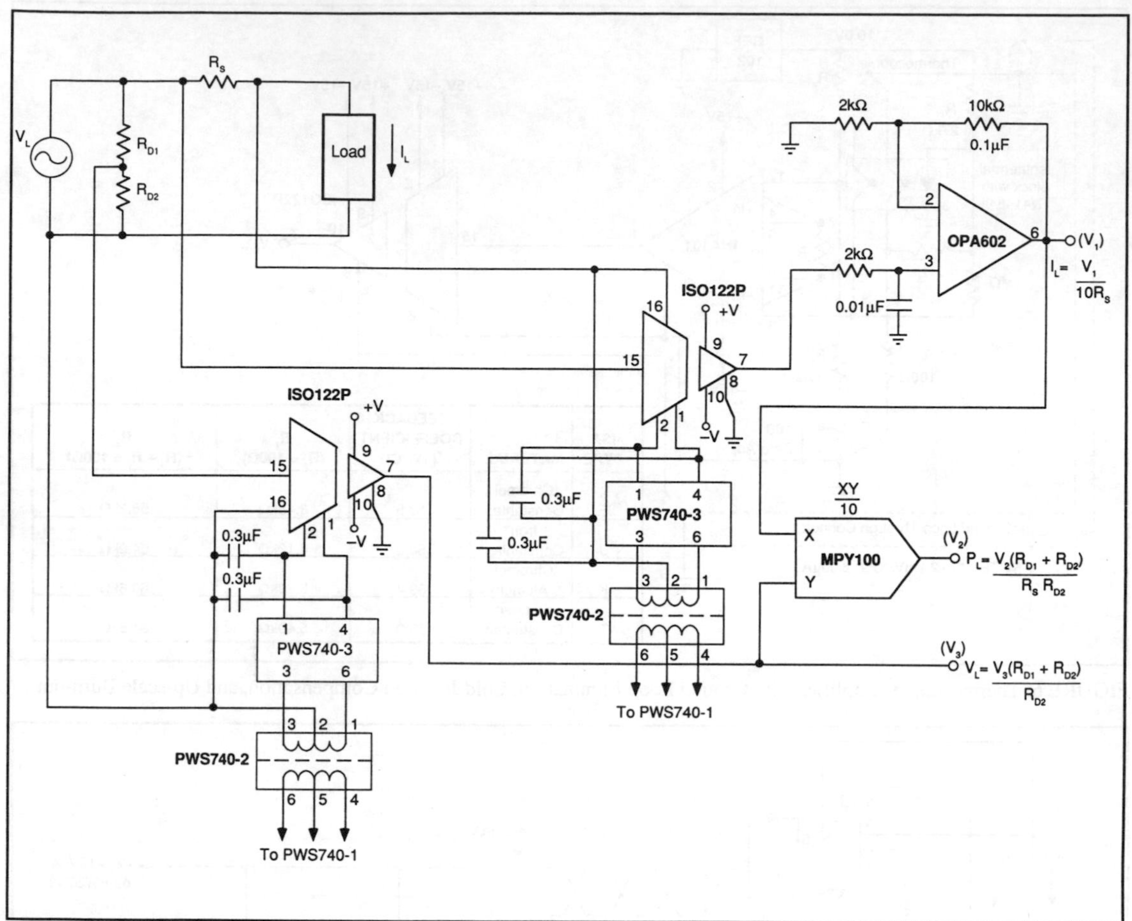

FIGURE 8. Isolated Power Line Monitor.

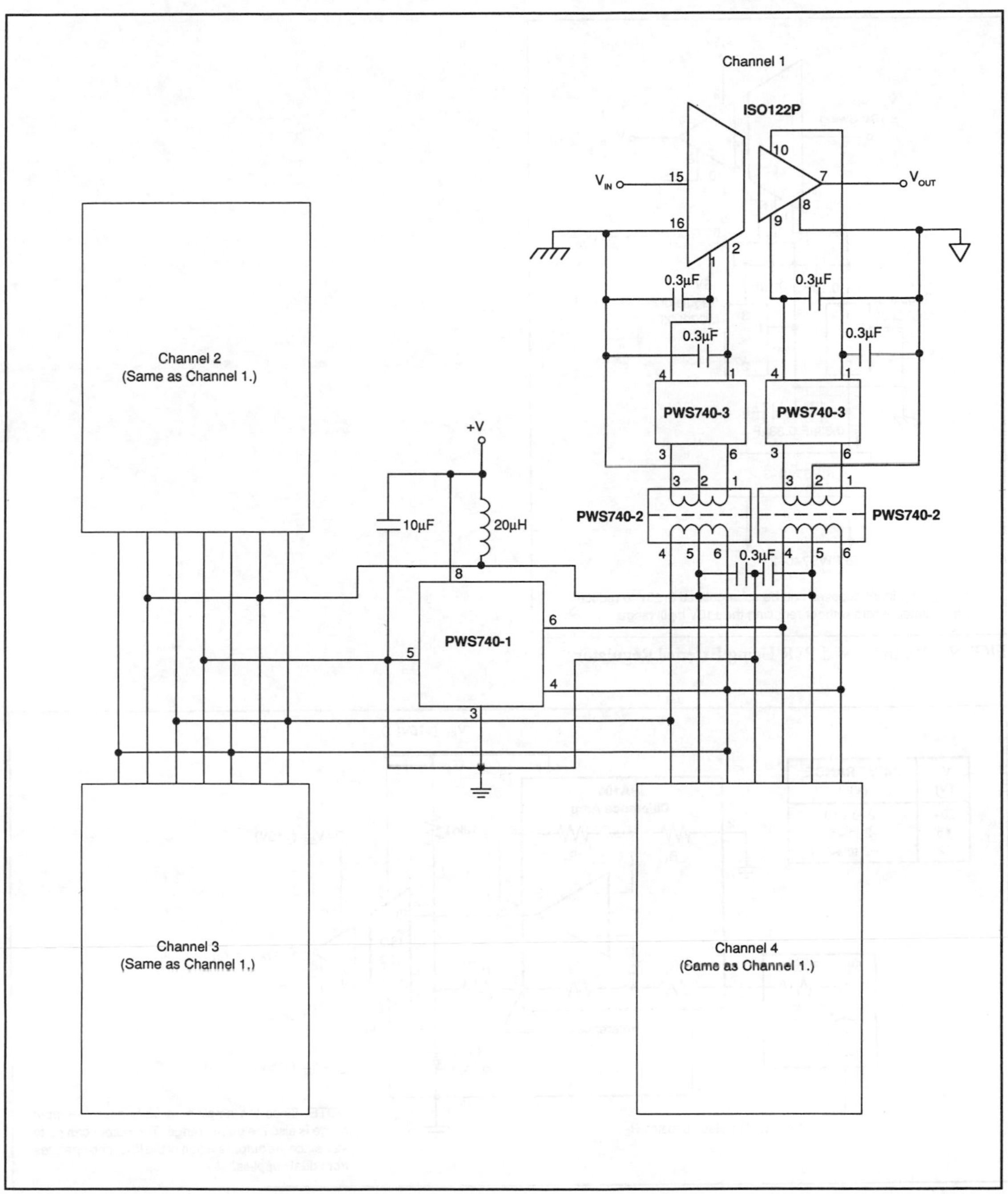

FIGURE 9. Three-Port, Low-Cost, Four-Channel Isolated, Data Acquisition System.

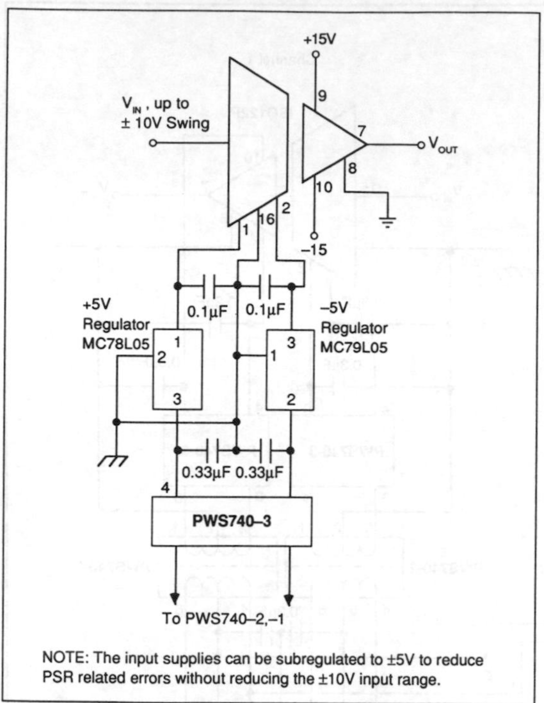

FIGURE 10. Improved PSR Using External Regulator.

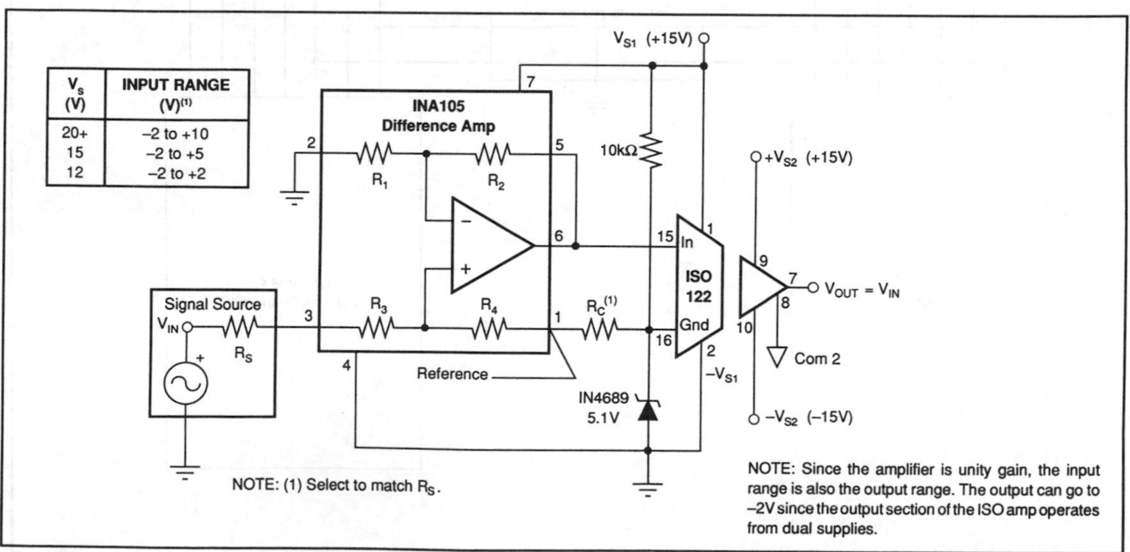

V_S (V)	INPUT RANGE (V)[1]
20+	–2 to +10
15	–2 to +5
12	–2 to +2

FIGURE 11. Single Supply Operation of the ISO122 Isolation Amplifier.

AN-30

Log Converters

National Semiconductor
Application Note 30

One of the most predictable non-linear elements commonly available is the bipolar transistor. The relationship between collector current and emitter base voltage is precisely logarithmic from currents below one picoamp to currents above one milliamp. Using a matched pair of transistors and integrated circuit operational amplifiers, it is relatively easy to construct a linear to logarithmic converter with a dynamic range in excess of five decades.

The circuit in *Figure 1* generates a logarithmic output voltage for a linear input current. Transistor Q_1 is used as the non-linear feedback element around an LM108 operational amplifier. Negative feedback is applied to the emitter of Q_1 through divider, R_1 and R_2, and the emitter base junction of Q_2. This forces the collector current of Q_1 to be exactly equal to the current through the input resistor. Transistor Q_2 is used as the feedback element of an LM101A operational amplifier. Negative feedback forces the collector current of Q_2 to equal the current through R_3. For the values shown, this current is 10 μA. Since the collector current of Q_2 remains constant, the emitter base voltage also remains constant. Therefore, only the V_{BE} of Q_1 varies with a change of input current. However, the output voltage is a function of the difference in emitter base voltages of Q_1 and Q_2:

$$E_{OUT} = \frac{R_1 + R_2}{R_2}(V_{BE_2} - V_{BE_1}). \quad (1)$$

For matched transistors operating at different collector currents, the emitter base differential is given by

$$\Delta V_{BE} = \frac{kT}{q} \log_e \frac{I_{C_1}}{I_{C_2}}, \quad (2)$$

where k is Boltzmann's constant, T is temperature in degrees Kelvin and q is the charge of an electron. Combining these two equations and writing the expression for the output voltage gives

$$E_{OUT} = \frac{-kT}{q}\left[\frac{R_1 + R_2}{R_2}\right] \log_e \left[\frac{E_{IN} R_3}{E_{REF} R_{IN}}\right] \quad (3)$$

for $E_{IN} \geq 0$. This shows that the output is proportional to the logarithm of the input voltage. The coefficient of the log term is directly proportional to absolute temperature. Without compensation, the scale factor will also vary directly with temperature. However, by making R_2 directly proportional to temperature, constant gain is obtained. The temperature compensation is typically 1% over a temperature range of −25°C to 100°C for the resistor specified. For limited temperature range applications, such as 0°C to 50°C, a 430Ω sensistor in series with a 570Ω resistor may be substituted for the 1k resistor, also with 1% accuracy. The divider, R_1 and R_2, sets the gain while the current through R_3 sets the zero. With the values given, the scale factor is 1V/decade and

$$E_{OUT} = -\left[\log_{10}\left|\frac{E_{IN}}{R_{IN}}\right| + 5\right] \quad (4)$$

where the absolute value sign indicates that the dimensions of the quantity inside are to be ignored.

Log generator circuits are not limited to inverting operation. In fact, a feature of this circuit is the ease with which non-inverting operation is obtained. Supplying the input signal to A_2 and the reference current to A_1 results in a log output that is not inverted from the input. To achieve the same 100 dB dynamic range in the non-inverting configuration, an LM108 should be used for A_2, and an LM101A for A_1. Since the LM108 cannot use feedforward compensation, it is frequency compensated with the standard 30 pF capacitor.

The only other change is the addition of a clamp diode connected from the emitter of Q_1 to ground. This prevents damage to the logging transistors if the input signal should go negative.

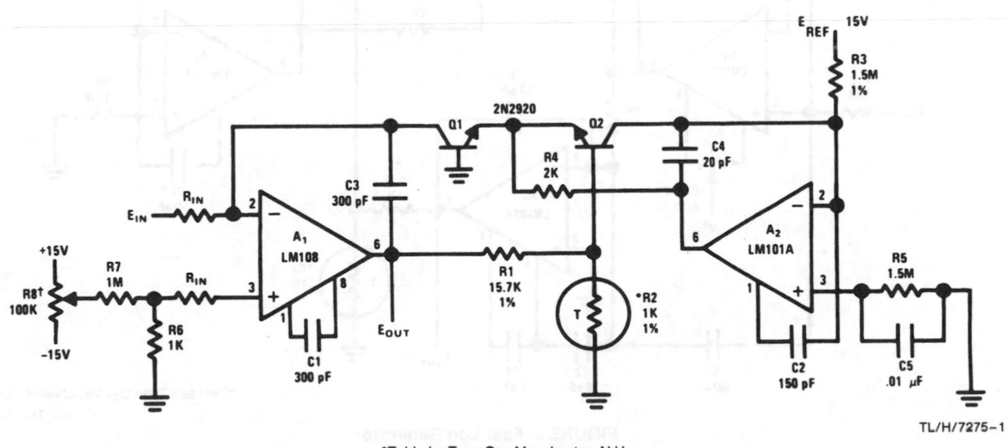

*Tel Labs Type Q_{81} Manchester, N.H.

†Offset Voltage Adjust

FIGURE 1. Log Generator with 100 dB Dynamic Range

The log output is accurate to 1% for any current between 10 nA and 1 mA. This is equivalent to about 3% referred to the input. At currents over 500 μA the transistors used deviate from log characteristics due to resistance in the emitter, while at low currents, the offset current of the LM108 is the major source of error. These errors occur at the ends of the dynamic range, and from 40 nA to 400 μA the log converter is 1% accurate referred to the input. Both of the transistors are used in the grounded base connection, rather than the diode connection, to eliminate errors due to base current. Unfortunately, the grounded base connection increases the loop gain. More frequency compensation is necessary to prevent oscillation, and the log converter is necessarily slow. It may take 1 to 5 ms for the output to settle to 1% of its final value. This is especially true at low currents.

The circuit shown in *Figure 2* is two orders of magnitude faster than the previous circuit and has a dynamic range of 80 dB. Operation is the same as the circuit in *Figure 1*, except the configuration optimizes speed rather than dynamic range. Transistor Q_1 is diode connected to allow the use of feedforward compensation[1] on an LM101A operational amplifier. This compensation extends the bandwidth to 10 MHz and increases the slew rate. To prevent errors due to the finite h_{FE} of Q_1 and the bias current of the LM101A, an LM102 voltage follower buffers the base current and input current. Although the log circuit will operate without the LM102, accuracy will degrade at low input currents. Amplifier A_2 is also compensated for maximum bandwidth. As with the previous log converter, R_1 and R_2 control the sensitivity; and R_3 controls the zero crossing of the transfer function. With the values shown the scale factor is 1V/decade and

$$E_{OUT} = -\left[\log_{10}\left|\frac{E_{IN}}{R_{IN}}\right| + 4\right] \quad (5)$$

from less than 100 nA to 1 mA.

Anti-log or exponential generation is simply a matter of rearranging the circuitry. *Figure 3* shows the circuitry of the log converter connected to generate an exponential output from a linear input. Amplifier A_1 in conjunction with transistor Q_1 drives the emitter of Q_2 in proportion to the input voltage. The collector current of Q_2 varies exponentially with the emitter-base voltage. This current is converted to a voltage by amplifier A_2. With the values given

$$E_{OUT} = 10^{-[E_{IN}]}. \quad (6)$$

Many non-linear functions such as $X^{1/2}$, X^2, X^3, 1/X, XY, and X/Y are easily generated with the use of logs. Multiplication becomes addition, division becomes subtraction and powers become gain coefficients of log terms. *Figure 4* shows a circuit whose output is the cube of the input. Actually, any power function is available from this circuit by changing the values of R_9 and R_{10} in accordance with the expression:

$$E_{OUT} = E_{IN}^{\frac{16.7 R_9}{R_9 + R_{10}}}. \quad (7)$$

Note that when log and anti-log circuits are used to perform an operation with a linear output, no temperature compensating resistors at all are needed. If the log and anti-log transistors are at the same temperature, gain changes with temperature cancel. It is a good idea to use a heat sink which couples the two transistors to minimize thermal gradients. A 1°C temperature difference between the log and anti-log transistors results in a 0.3% error. Also, in the log converters, a 1°C difference between the log transistors and the compensating resistor results in a 0.3% error.

Either of the circuits in *Figures 1* or *2* may be used as dividers or reciprocal generators. Equation 3 shows the outputs of the log generators are actually the ratio of two currents:

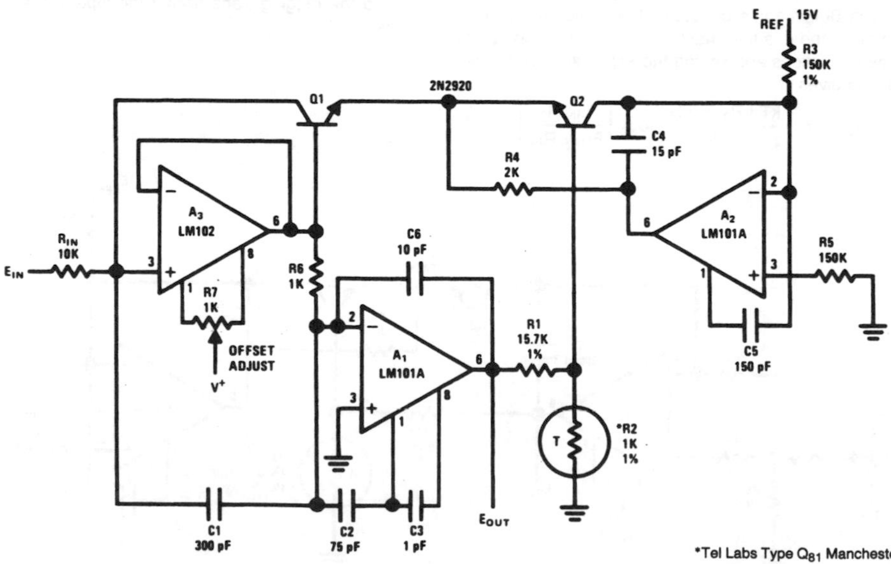

*Tel Labs Type Q_{81} Manchester, N.H.

TL/H/7275–2

FIGURE 2. Fast Log Generator

the input current and the current through R_3. When used as a log generator, the current through R_3 was held constant by connecting R_3 to a fixed voltage. Hence, the output was just the log of the input. If R_3 is driven by an input voltage, rather than the 15V reference, the output of the log generator is the log ratio of the input current to the current through R_3. The anti-log of this voltage is the quotient. Of course, if the divisor is constant, the output is the reciprocal.

A complete one quadrant multiplier/divider is shown in *Figure 5.* It is basically the log generator shown in *Figure 1* driving the anti-log generator shown in *Figure 3.* The log generator output from A_1 drives the base of Q_3 with a voltage proportional to the log of E_1/E_2. Transistor Q_3 adds a voltage proportional to the log of E_3 and drives the anti-log transistor, Q_4. The collector current of Q_4 is converted to an output voltage by A_4 and R_7, with the scale factor set by R_7 at $E_1 E_3/10E_2$.

Measurement of transistor current gains over a wide range of operating currents is an application particularly suited to log multiplier/dividers. Using the circuit in *Figure 5*, PNP current gains can be measured at currents from 0.4 μA to 1 mA. The collector current is the input signal to A_1, the base current is the input signal to A_2, and a fixed voltage to R_5 sets the scale factor. Since A_2 holds the base at ground, a single resistor from the emitter to the positive supply is all that is needed to establish the operating current. The output is proportional to collector current divided by base current, or h_{FE}.

In addition to their application in performing functional operations, log generators can provide a significant increase in

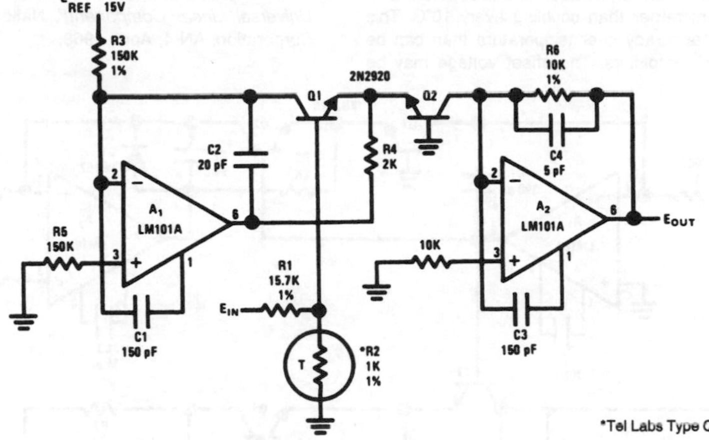

FIGURE 3. Anti-Log Generator

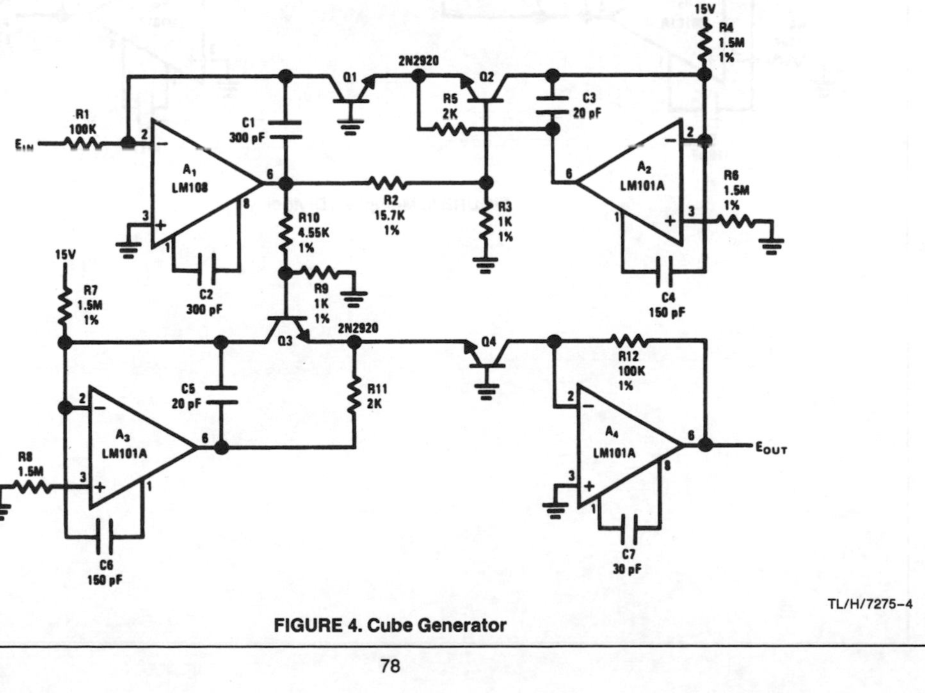

TL/H/7275-4

FIGURE 4. Cube Generator

the dynamic range of signal processing systems. Also, unlike a linear system, there is no loss in accuracy or resolution when the input signal is small compared to full scale. Over most of the dynamic range, the accuracy is a percent-of-signal rather than a percent-of-full-scale. For example, using log generators, a simple meter can display signals with 100 dB dynamic range or an oscilloscope can display a 10 mV and 10V pulse simultaneously. Obviously, without the log generator, the low level signals are completely lost.

To achieve wide dynamic range with high accuracy, the input operational amplifier necessarily must have low offset voltage, bias current and offset current. The LM108 has a maximum bias current of *3 nA* and offset current of *400 pA* over a −55°C to 125°C temperature range. By using equal source resistors, only the offset current of the LM108 causes an error. The offset current of the LM108 is as low as many FET amplifiers. Further, it has a low and constant temperature coefficient rather than doubling every 10°C. This results in greater accuracy over temperature than can be achieved with FET amplifiers. The offset voltage may be zeroed, if necessary, to improve accuracy with low input voltages.

The log converters are low level circuits and some care should be taken during construction. The input leads should be as short as possible and the input circuitry guarded against leakage currents. Solder residues can easily conduct leakage currents, therefore circuit boards should be cleaned before use. High quality glass or mica capacitors should be used on the inputs to minimize leakage currents. Also, when the +15V supply is used as a reference, it must be well regulated.

REFERENCES

1. R. C. Dobkin, *"Feedforward Compensation Speeds Op Amp"*, National Semiconductor Corporation, Linear Brief 2, April, 1969.
2. R. J. Widlar, *"Monolithic Operational Amplifiers—The Universal Linear Component"*, National Semiconductor Corporation, AN-4, April, 1968.

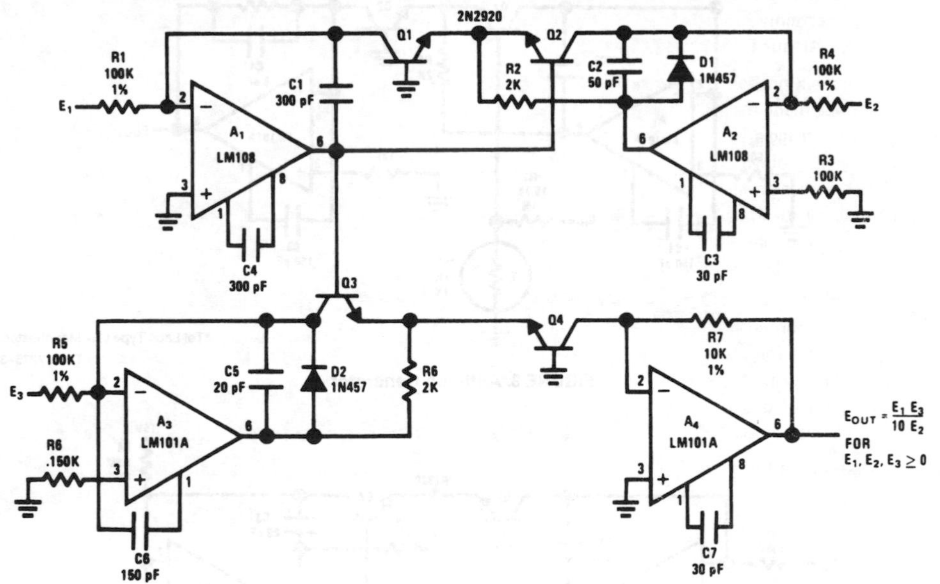

FIGURE 5. Multiplier/Divider

APPLICATION NOTE

An I.C. Amplifier Users' Guide To Decoupling, Grounding, And Making Things Go Right For A Change

by Paul Brokaw

"There once was a breathy baboon
Who always breathed down a bassoon,
For he said "It appears
that in billions of years
I shall certainly hit on a tune"
(Sir Arthur Eddington)

This quotation seemed a proper note with which to begin on a subject which has made monkeys of most of us at one time or another. The struggle to find a suitable configuration for system power, ground, and signal returns too frequently degenerates into a frustrating glitch hunt. While a strictly experimental approach can be used to solve simple problems, a little forethought can often prevent serious problems and provide a plan of attack if some judicious tinkering is later required.

The subject is so fragmented that a completely general treatment is too difficult for me to tackle. Therefore, I'd like to state one general principle and then look a bit more narrowly at the subject of decoupling and grounding as it relates to integrated circuit amplifiers.

. . . Principle: Think—where the currents will flow.

I suppose this seems pretty obvious, but all of us tend to think of the currents we're interested in as flowing "out" of some place and "through" some other place but often neglect to worry how the current will find its way back to its source. One tends to act as if all "ground" or "supply voltage" points are equivalent and neglect (for as long as possible) the fact that they are parts of a network of conductors through which currents flow and develop finite voltages.

In order to do some advance planning it's important to consider where the currents originate and to where they will return and to determine the effects of the resulting voltage drops. This in turn requires some minimum amount of understanding of what goes on inside the circuits being decoupled and grounded. This information may be lacking or difficult to interpret when integrated circuits are part of the design.

Operational amplifiers are one of the most widely used linear I.C.'s, and fortunately most of them fall into a few classes, so far as the problems of power and grounding are concerned. Although the configuration of a system may pose formidable problems of decoupling and signal returns, some basic methods to handle many of these problems can be developed from a look at op-amps.

OP AMPS HAVE FOUR TERMINALS:

A casual look through almost any operational amplifier text might leave the reader with the impression that an ideal op-amp has three terminals: a pair of differential inputs and an output as shown in Figure 1. A quick review of fundamentals, however, shows that this can't be the case. If the amplifier has an output voltage it must be measured with respect to some point . . . a point to which the amplifier has a reference. Since the ideal op-amp has infinite common mode rejection, the inputs are ruled out as that reference so that there must be a fourth amplifier terminal. Another way of looking at it is that if the amplifier is to supply output current to a load, that current must get into the amplifier somewhere. Ideally, no input current flows, so again the conclusion is that a fourth terminal is required.

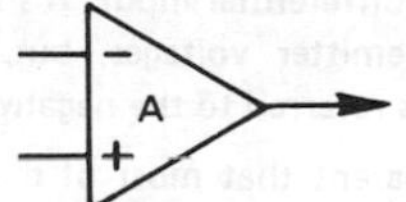

Figure 1. Conventional "Three Terminal" Op Amp

A common practice is to say, or indicate in a diagram, that this fourth terminal is "ground." Well, without getting into a discussion of what "ground" may be we can observe that most integrated circuit op-amps (and a lot of the modular ones as well) don't have a "ground" terminal. With these circuits the fourth terminal is one or both of the power supply terminals. There's a temptation here to lump together both supply voltages with the ubiquitous ground. And, to the extent that the supply lines really do present a low impedance at all frequencies within the amplifier bandwidth, this is probably reasonable. When the impedance requirement isn't satisfied, however, the door is left open to a variety of problems including noise, poor transient response, and oscillation.

20

DIFFERENTIAL TO SINGLE-ENDED CONVERSION: One fundamental requirement of a simple op-amp is that an applied signal which is fully differential at the input must be converted to a single-ended output. Single ended, that is, with respect to the often neglected fourth terminal. To see how this can lead to difficulties, take a look at Figure 2.

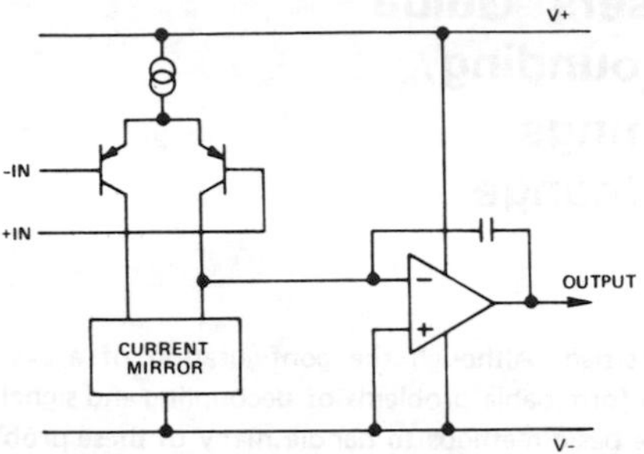

Figure 2. Simplified "Real" Op Amp

The signal flow illustrated by Figure 2 is used in several popular integrated circuit families. Details vary, but, the basic signal path is the same as the 101, 741, 748, 777, 4136, 503, 515, and other integrated circuit amplifiers. The circuit first transforms a differential input voltage into a differential current. This input stage function is represented by PNP transistors in Figure 2. The current is then converted from differential to single-ended form by a current mirror which is connected to the negative supply rail. The output from the current mirror drives a voltage amplifier and power output stage which is connected as an integrator. The integrator controls the open-loop frequency response, and its capacitor may be added externally, as in the 101, or may be self-contained, as in the 741. Most descriptions of this simplified model don't emphasize that the integrator has, of course, a differential input. It's biased positive by a couple of base emitter voltages, but, the non-inverting integrator input is referred to the negative supply.

It should be apparent that most of the voltage difference between the amplifier output and the negative supply appears across the compensation capacitor. If the negative supply voltage is changed abruptly the integrator amplifier will *force* the output to follow the change. When the entire amplifier is in a closed loop configuration the resulting error signal at its input will tend to restore the output, but, the recovery will be limited by the slew rate of the amplifier. As a result, an amplifier of this type may have outstanding low frequency power supply rejection, but, the negative supply rejection is fundamentally limited at high frequencies. Since it is the feedback signal to the input that causes the output to be restored, the negative supply rejection will approach zero for signals at frequencies above the *closed loop* bandwidth. This means that high-speed, high-level circuits can "talk to" low-level circuits through the common impedance of the negative supply line.

Note that the problem with these amplifiers is associated with the negative supply terminal. Positive supply rejection may also deteriorate with increasing frequency, but, the effect is less severe. Typically, small transients on the positive supply have only a minor effect on the signal output. The difference between these sensitivities can result in an apparent asymmetry in the amplifier transient response. If the amplifier is driven to produce a positive voltage swing across its rated load it will draw a current pulse from the positive supply. The pulse may result in a supply voltage transient, but, the positive supply rejection will minimize the effect on the amplifier output signal. In the opposite case, a negative output signal will extract a current from the negative supply. If this pulse results in a "glitch" on the buss, the poor negative supply rejection will result in a similar "glitch" at the amplifier output. While a positive pulse test may give the amplifier transient response, a negative pulse test may actually give you a pretty good look at your negative supply line transient response, instead of the amplifier response!

Remember that the impulse response of the power supply itself is not what is likely to appear at the amplifier. Thirty or forty centimeters of wire can act like a high Q inductor to add a high-frequency component to the normally overdamped supply response. A decoupling capacitor near the amplifier won't always cure the problem either, since the supply must be decoupled *to* somewhere. If the decoupled current flows through a long path, it can still produce an undesirable glitch.

Figure 3 illustrates three possible configurations for negative supply decoupling. In 3a the dotted line shows the negative signal current path through the decoupling and along the ground line. If the load "ground" and decoupled "ground" actually join at the power supply the "glitch" on the ground lines is similar to the "glitch" on the negative supply buss. Depending upon how the feedback and signal sources are "grounded" the effective disturbance *caused* by the decoupling capacitor may be larger than the disturbance which it was intended to prevent. Figure 3b shows how the decoupling capacitor can be used to minimize disturbance of V− and ground busses. The high-frequency component of the load current is confined to a loop which doesn't include any part of the ground path. If the capacitor is of sufficient size and quality, it will minimize the glitch on the negative supply without disturbing input or output signal paths. When the load situation is more complex, as in 3c, a little more thought is required. If the amplifier is driving a load that goes to a virtual ground, the actual load current does not return to ground. Rather, it must be supplied by the amplifier creating the virtual ground as shown in the figure. In this case, decoupling the negative supply of the first amplifier to the positive supply of the second amplifier closes the fast signal current loop without disturbing ground or signal paths. Of course, it's still important to provide a low impedance path from "ground" to V− for the second amplifier to avoid disturbing the input reference.

The key to understanding decoupling circuits is to note where the actual load and signal currents will flow. The key to optimizing the circuit is to bypass these currents around ground and other signal paths. Note, that as in figure 3a, "single point grounding" may be an oversimplified solution to a complex problem.

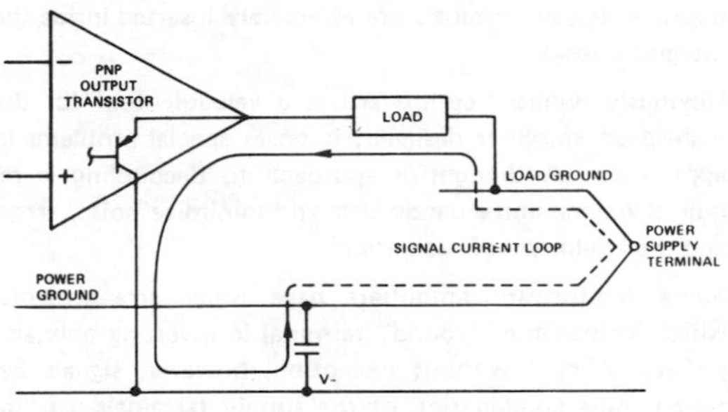

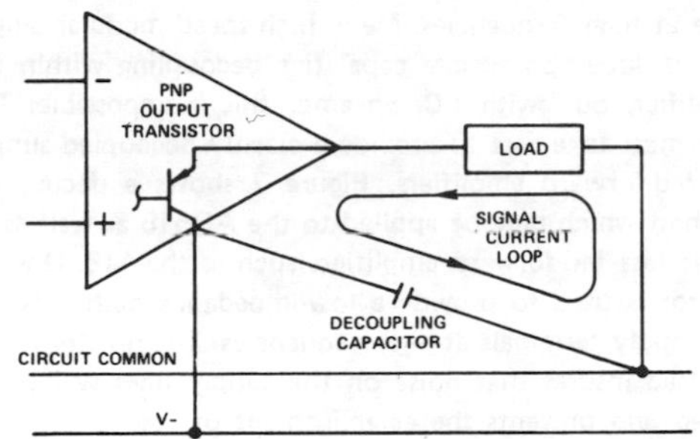

Figure 3a. Decoupling for Negative Supply Ineffective

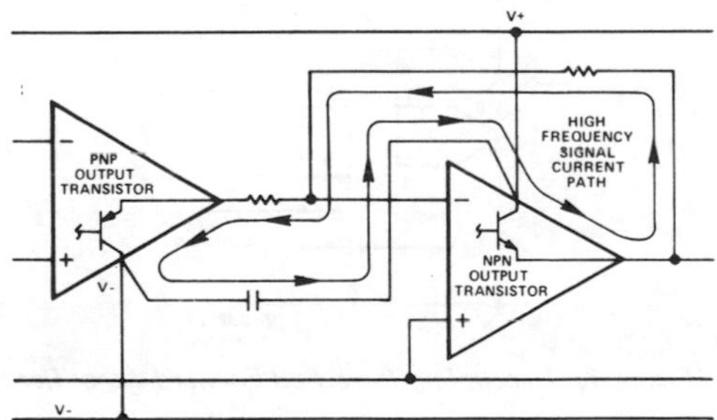

Figure 3b. Decoupling Negative Supply Optimized for "Grounded" Load

Figure 3c. Decoupling Negative Supply Optimized for "Virtual Ground" Load

Figure 3b and 3c have been simplified for illustrative purposes. When an entire circuit is considered conflicts frequently arise. For example, several amplifiers may be powered from the same supply, and an individual decoupling capacitor is required for each. In a gross sense the decoupling capacitors are all paralleled. In fact, however, the inductance of the interconnecting power and ground lines convert this harmless-looking arrangement into a complex L-C network that often rings like the "Avon Lady". In circuits handling fast signal wavefronts, decoupling networks paralleled by more than a few centimeters of wire generally mean trouble. Figure 4 shows how small resistors can be added to lower the Q of the undesired resonant circuits. The resistors can generally be tolerated since they convert a bad high-frequency jingle to a small damped signal at the op amp supply terminal. The residual has larger *low frequency* components, but, these can be handled by the op-amp supply rejection.

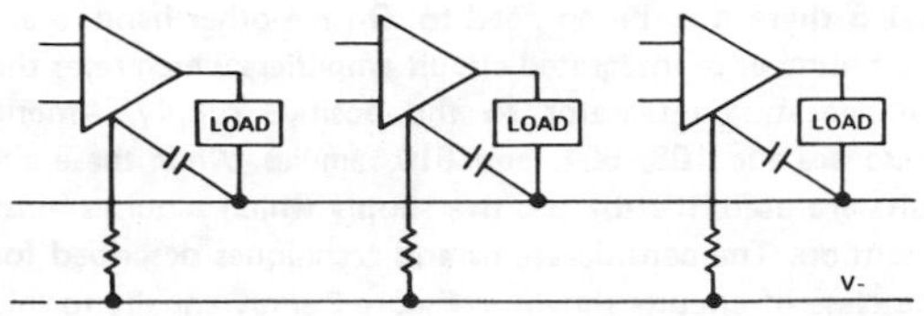

Figure 4. Damping Parallel Decoupling Resonances

FREQUENCY STABILITY:

There's a temptation to forget about decoupling the negative supply when the system is intended to handle only low-frequency signals. Granted that decoupling may not be required to handle low-frequency signals, but it may still be required for frequency stability of the op-amps.

Figure 5 is a more-detailed version of Figure 2 showing the output stage of the I.C. separated from the integrator (since this is the usual arrangement) and showing the negative power supply and wiring impedance lumped together as a single constant. The amplifier is connected as a unity gain follower. This makes a closed-loop path from the amplifier output through the differential input to the integrator input. There is a second feedback path from the collector of the output PNP transistor back to the other integrator input. The net input to the integrator is the difference of the signals through these two paths. At low frequencies this is a net, negative feedback. The high-frequency feedback depends upon both the load reactance and the reactance of the V– supply.

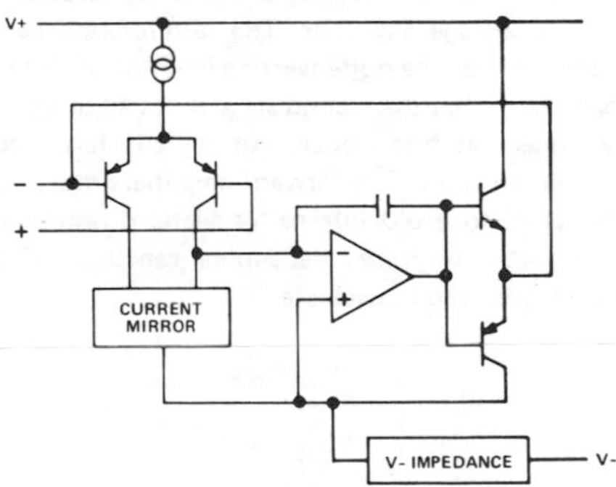

Figure 5. Instability Can Result from Neglecting Decoupling

When the supply lead reactance is inductive, it tends to destabilize the integrator. This situation is aggravated by a capacitive load on the amplifier. Although it's difficult to predict under exactly what circumstances the circuit will become unstable, it's generally wise to decouple the negative supply if there is any substantial lead inductance in the V– lead *or* in the common return to the load and amplifier input signal source. If the decoupling is to be effective, of course, it must be with respect to the *actual* signal returns, rather than to some vague "ground" connection.

POSITIVE SUPPLY DECOUPLING:

Up to this point we haven't considered decoupling the positive supply line, and with amplifiers typified by Figures 2 and 5 there may be no need to. On the other hand, there are a number of integrated circuit amplifiers which refer the compensating integrator to the positive supply. Among these are the 108, 504, and 510 families. When these circuits are used, it's the positive supply which requires most attention. The considerations and techniques described for the class of circuits shown in Figure 2 apply equally to this second class, but, should be applied to the positive supply rather than the negative.

FEED-FORWARD:

A technique which is most frequently used to improve bandwidth is called feed-forward. Generally, feed-forward is used to bypass an amplifier or level translator stage which has poor high frequency response. Figure 6 illustrates how this may be done. Each of the amplifiers shown is really a subcircuit, usually a single stage, in the overall amplifier. In the illustration, the input stage converts the differential input to a single-ended signal. The signal drives an intermediate stage (which in practice often includes level translator circuitry) which has low-frequency gain, but, limited bandwidth. The output of this stage drives an integrator-amplifier and output stage. The overall compensation capacitor feeds back to the input of the second stage and includes it in the integrator loop. The compromises necessary to obtain gain and level translation in the intermediate stage often limit its bandwidth and slow down the available integrator response. A feed-forward capacitor permits high-frequency signals to bypass this stage. As a result, the overall amplifier combines the low-frequency gain available from 3 stages with the improved frequency response available from a 2-stage amplifier. The feed-forward capacitor also feeds back to the non-inverting input of the intermediate stage. Note that the second stage is not an integrator, as it may appear at first glance, but actually has a positive feedback connection. Fed-forward amplifiers must be carefully designed to avoid internal oscillations resulting from this connection. Improper decoupling can upset this plan and permit this loop to oscillate.

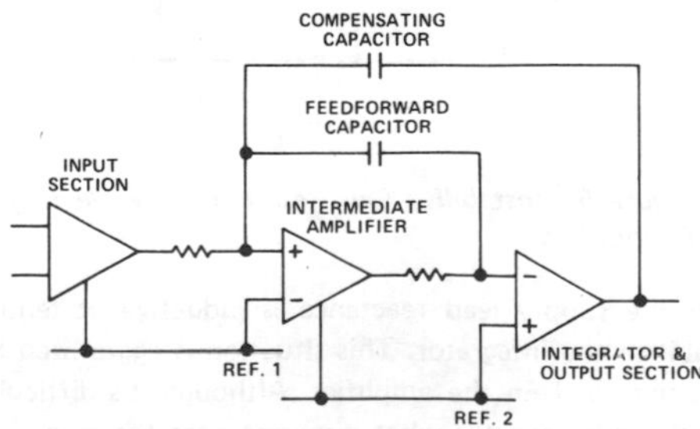

Figure 6. Fast Fed-Forward Amplifier

Note that the internal input stages are shown as being referred to separated reference points. Ideally, these will be the same reference so far as signals are concerned, although they may differ in bias level. In practice this may not be the case. Examples of fed-forward amplifiers are the AD518, the AD118, and the OP-05. In these amplifiers, signal Reference 1 is the positive supply, while signal Reference 2 is the negative supply. Signals appearing between the positive and negative supply terminals are effectively inserted inside the integrator loop!

Obviously, while feed-forward is a valuable tool for the high-speed amplifier designer, it poses special problems in application. A thoughtful approach to decoupling is required to maximize bandwidth and minimize noise, error, and the likelihood of oscillation.

Some fed-forward amplifiers have other arrangements, which include the "ground" terminal in inverting only amplifiers. Almost without exception, however, signals between some combination of the supply terminals get "inside" the amplifier. It is vital to proper operation that the involved supply terminals present a common low impedance at high frequencies. Many high-speed modular amplifiers include appropriate capacitive decoupling within the amplifier, but, with I.C. op amps this is impossible. The user must take care to provide a cleanly decoupled supply for fed-forward amplifiers. Figure 7 shows a decoupling method which may be applied to the AD518 as well as to other fast fed-forward amplifiers such as the 118. One capacitor is used to provide a low-impedance path between the supply terminals at high frequencies. The resistor in the V+ lead insures that noise on the supply lines will be rejected *and* prevents the establishment of resonances with other decoupling circuits. The second capacitor decouples the low side of the integrator to the load.

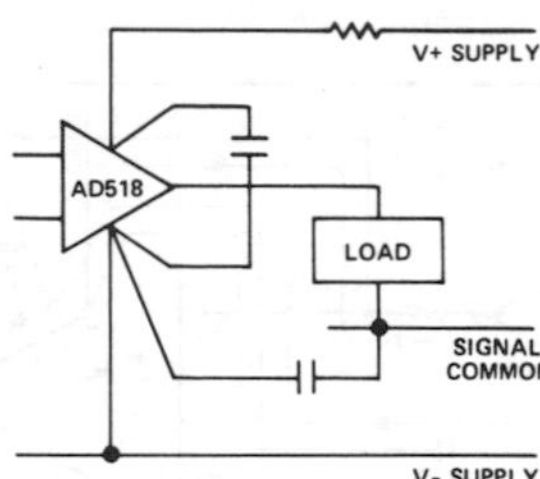

Figure 7. Decoupling for a Fed-Forward Amplifier

Alternatives include a resistor in both supply leads and/or decoupling from V+ to the load. In principle, the positive and negative supply should be tied in a "tight knot" with the signal return. To the extent that this cannot be done, there is a slight advantage to favoring the negative supply due to the high frequency limitations of PNP transistors used in junction-isolated I.C.'s.

OTHER COMPENSATION:

While most integrated circuit amplifiers use one of the three compensation schemes already described, a significant fraction use some other plan. The 725 type amplifiers combine a V– referred integrator with a network which the manufacturers recommend to be connected from signal ground to the integrator input. This makes the circuit extremely liable to pick up noise between V– and ground. In many circumstances it may be wiser to connect the external compensation to the negative supply, rather than to signal ground.

One more class of amplifiers is typified by the Analog Devices AD507 and AD509. In these circuits, a single capacitor may be used to induce a dominant pole of response without resorting to an integrator connection. The high-frequency response of the amplifier will appear with respect to the "ground" end of the compensation capacitor. In these amplifiers a small internal capacitance is connected between V+ and the compensation point. Unity gain compensation can be added in parallel and the pin-out is arranged to make this simple. The free end of the compensation capacitor can also be connected either to V– or signal common. It is extremely important that the signal common and the compensation connect directly or through a low-impedance decoupling.

Although the main signal path of these amplifiers can be compensated in a variety of ways, some care is required to insure the stability of internal structures. It's always wise to use extra care in decoupling wideband amplifiers to avoid problems with the output stage and other subcircuits which are similar to the main integrator problem illustrated by Figure 5. An effective compensation and decoupling circuit for the AD509 is shown in Figure 8. This arrangement is similar to Figure 7, and one of these two circuits is likely to be suitable for many types of wideband amplifier. Depending upon the power distribution, a small (10Ω to 50Ω) resistor may be appropriate in both of the supply leads to reduce power lead resonance and interference both to and from circuits sharing the power supply.

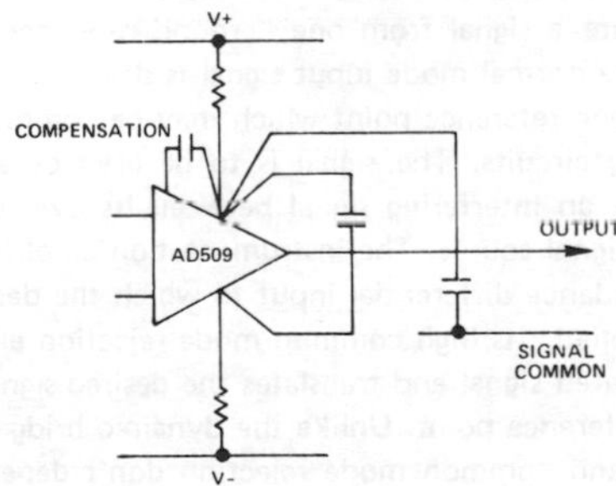

Figure 8. Decoupling a Wideband Amplifier

GROUNDING ERRORS:

Ground in most electronic equipment is not an actual connection to earth ground, but a common connection to which signals and power are referred. It is frequently immaterial to the function of the equipment whether or not the point actually connects to earth ground. I myself prefer some distinguishing name or names for these common points to emphasize that they must be *made* common. The term "ground" too often seems to be associated with a sort of cure-all concept, like snake oil, money or motherhood. If you're one of those who regards ground with the same sort of irrational reverence that you hold for your mother, remember that while you can always trust your mother, you should *never* trust your "ground." Examine and think about it.

It's important to have a look at the currents which flow in the ground circuit. Allowing these currents to share a path with a low-level signal may result in trouble. Figure 9 illustrates how careless grounding can degrade the performance of a simple amplifier. The amplifier drives a load which is represented by the load resistor. The load current comes from the power supply and is controlled by the amplifier as it amplifies the input signal. This current must return to the supply by some path; suppose that points A and B are alternative power supply "ground" connections. Assuming that the figure represents the proper topology or ordering of connections along the "ground" bus, connecting the supply at A will cause the load current to share a segment of wire with the input signal connection. Fifteen centimeters of number 22 wire in this path will present about 8 milliohms of resistance to the load current. With a 2k load, a 10-volt output signal will result in about 40 microvolts between the points marked "ΔV." This signal acts in series with the non-inverting input and can result in significant errors. For example, the typical gain of an AD510 amplifier is 8 million so that only 1¼μV of input signal is required to produce a 10 volt output. The 40μV ground error signal will result in a 32 times increase in the circuit gain error! This degradation could easily be the most serious error in a high-gain precision application. Moreover, the error represents positive feedback so that the circuit will latch up or oscillate for large closed-loop gains with R_f/R_i greater than about 250k.

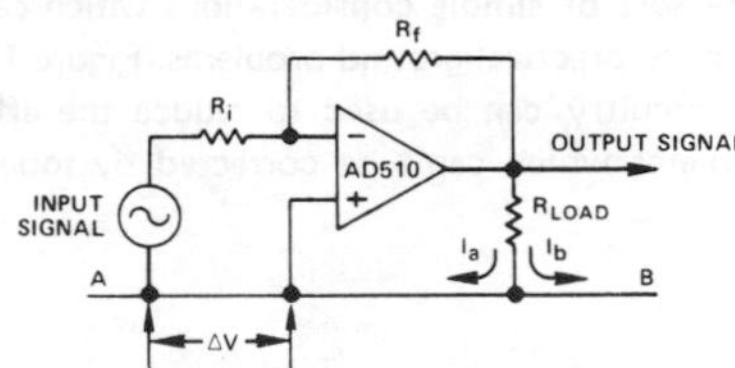

Figure 9. Proper Choice of Power Connections Minimizes Problems

Reconnecting the power supply to point B will correct the problem by eliminating the common impedance feedback connection. In a real system, the problem may be more complex. The input signal source, which is represented as floating in Figure 9, may also produce a current which must return to the power supply. With the supply at point B, any current which flows in additional loads (other than R_i) may interfere with the operation of the amplifier shown. Figure 10 illustrates how amplifiers can be cascaded and still drive auxiliary loads without common impedance coupling. The

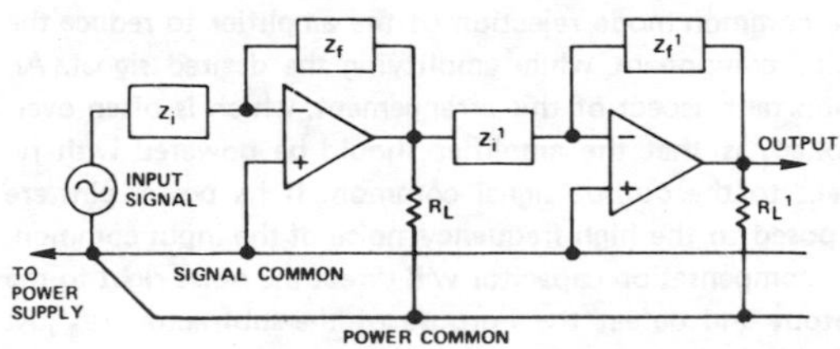

Figure 10. Minimizing Common Impedence Coupling

20

output currents flow through the auxiliary loads and back to the power supply through power common. The currents in the input and feedback resistors are supplied from the power supply by way of the amplifiers as previously illustrated in Figure 3c. The only current flowing in signal common is the amplifier's input current, and its effect is generally negligibly small.

Having given an example of a simple "grounding error" and its solution, I will now get back on my soap box and say that grounding errors result from neglect based on the assumption that a ground, is a ground, is a ground. *Some* impedance will be present in any interconnection path, and its effect should be considered in the overall design of a system. Quantitative approaches are quite useful in specialized applications. In fast TTL and ECL logic circuitry the characteristic impedance of interconnections is controlled so that proper terminations can reduce problems. In RF circuitry the unavoidable impedances are taken into account and incorporated into the design of the circuit. With op-amp circuitry, however, impedance levels do not lend themselves to transmission line theory, and the power and ground impedances are difficult to control or analyze. The most expedient procedure, short of difficult and restrictive quantitative analysis, seems to be to arrange the unavoidable impedances so as to minimize their effects and arrange the circuitry to overcome the effects. Figures 9 and 10 illustrate the sort of simple considerations which can substantially reduce practical ground problems. Figure 11 illustrates how circuitry can be used to reduce the effect of ground problems which can't be corrected by topological tricks.

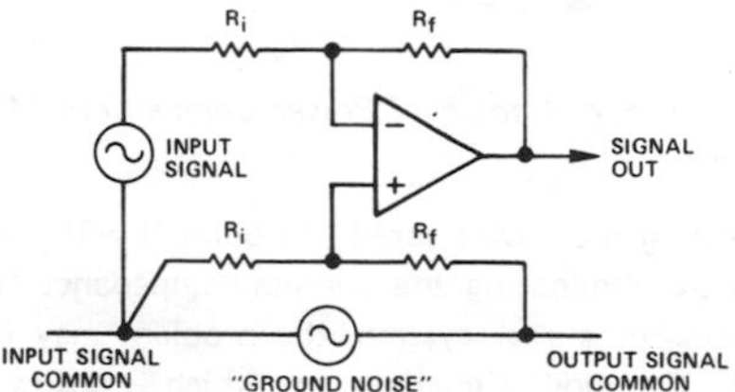

Figure 11. Subtractor Amplifier Rejects Common Mode Noise

GETTING AROUND THE PROBLEM:

In Figure 11 a subtractor circuit is used to amplify a normal mode input signal and reject a ground noise signal which is common to both sides of the input signal. This scheme uses the common-mode rejection of the amplifier to reduce the noise component while amplifying the desired signal. An important aspect of this arrangement, which is often overlooked, is that the amplifier should be powered with respect to the *output* signal common. If its power pins are exposed to the high-frequency noise of the input common, the compensation capacitor will direct the noise right to the output and defeat the purpose of the subtractor. It's just this kind of effect which makes it important to use care in grounding and decoupling. A subtractor or dynamic bridge, like Figure 11, will be ineffective in correcting a grounding problem if the amplifier itself is carelessly decoupled. In general, an op-amp should be decoupled to the point which is the reference for measuring or using its output signal. In "single-ended" systems it should also be decoupled to the input signal return as well. When it is impossible to satisfy both these requirements at once, there's a high probability of either a noise or oscillation problem or both. Frequently the difficulty can be resolved with a subtractor, like Figure 11, where a network like the single-ended feedback network (which needn't be all resistive) joins the input and output signal reference points and provides a "clean" reference point for the non-inverting input of the amplifier.

A problem with the subtractor is that it uses a balanced bridge to reject the common mode signal between the input and output reference points. The arms of the network must be carefully balanced, since to the extent they don't match, the unwanted signal will be amplified. Although even a poorly matched network will probably eliminate oscillation problems, noise rejection will suffer in direct proportion to any mismatches. An easier way to reject large "ground noise" signals is to use a true instrumentation amplifier.

INSTRUMENTATION AMPLIFIERS:

A true instrumentation amplifier has a very visible "fourth terminal." The output signal is developed with respect to a well defined reference point which is usually a "free" terminal that may be tied to the output signal common. The instrumentation amplifier also differs from an op amp in that the gain is fixed and well defined, but there is no feedback network coupling input and output circuits. Figure 12 shows how an instrumentation amplifier can be used to translate a signal from one "ground reference" to another. The normal mode input signal is developed with respect to one reference point which may be common to its generating circuits. The signal is to be used by a system which has an interfering signal between its own common and the signal source. The instrumentation amplifier has a high impedance differential input to which the desired signal is applied. Its high common mode rejection eliminates the unwanted signal and translates the desired signal to the output reference point. Unlike the dynamic bridge circuit, the gain and common mode rejection don't depend on a network connecting the input and output circuits. The gain is set, in Figure 12, by the ratio of a pair of resistors which are connected inside the amplifier. The amplifier has a very high input impedance, so that gain and common mode rejection are not greatly affected by variations or unbalance in source impedance.

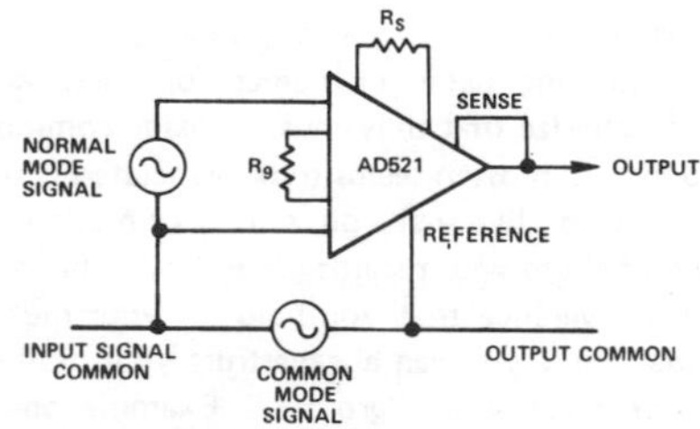

Figure 12. Applying an In-Amp

Since instrumentation amplifiers have a reference or "ground" terminal, they have the potential to be free of the power supply sensitivities of op amps. In practice, however, most instrumentation amplifiers have internal frequency compensation which is referred to the power supply. In the case of the AD521, the compensation integrator is referred to the negative supply terminal. The decoupling of this terminal is particularly important, and it should be decoupled with respect to the output reference terminal, or actually to the point to which this terminal refers. The AD520 instrumentation amplifier, on the other hand, has an internal integrator which is referred to the positive supply terminal. For best results both the V+ and V– terminals should be decoupled to the output reference point.

THE "OTHER" INPUT:

Most I.C. op-amps and in-amps include offset voltage nulling terminals. These terminals generally have a small voltage on them and by loading the terminals with a potentiometer the amplifier offset voltage can be adjusted. While their impedance level is much lower than the normal input, the null terminals can act as another differential input to the amplifier. Although the null terminals aren't generally looked at as inputs, most amplifiers are quite sensitive to signals applied here. For example, in 741 family amplifiers the output voltage gain from the null terminals is greater than the gain from the normal input!

An illustration of the type of problems that can arise with the "other" input is shown in Figure 13. The figure is an op-amp circuit with some of the offset null detail shown.

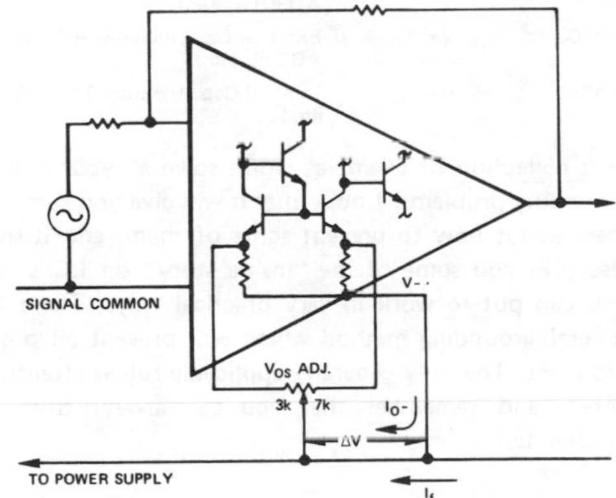

Figure 13. Details of V_{OS} Nulling – the "Other" Input

As it's drawn, the V_{OS} null pot wiper connects to a point along a V– "clothesline" which carries both the return current from the amplifier and currents from other circuits back to the power supply. These currents will develop a small voltage, ΔV, along the conductor between the amplifier V– terminal and the null pot wiper. If the null pot is set on center, the equal halves will form a balanced bridge with the resistors inside the amplifier. The effect of the voltage generated along the wire is balanced at the V_{OS} terminals and will have little effect on the amplifier output. On the other hand, if the null pot is unbalanced, to correct an amplifier offset, the bridge will no longer balance. In this case voltages developed along the "clothesline" will result in a difference voltage at the V_{OS} terminals. For instance, suppose that a 10k null pot balances out the op amp offset when it is set with 3k and 7k branches as shown in the figure. In a 741 the internal resistors are about 1k so that the difference signal at the V_{OS} terminals will be about 1/8 ΔV. The gain from these terminals is about twice the gain from the normal input, so that the disturbance acts as if it were an input signal of about 1/4 ΔV. Using the same assumptions as in the discussion of Figure 9. the current I_O– will result in a 10 microvolt input error signal. In this case, however, the error will appear *only* when the amplifier load current comes from the negative supply. When the load is driven positive the error will disappear. As a result, the V_{OS} input signal will result in distortion rather than a simple gain error!

An additional problem is created by I_f, a current returning to the power supply from other circuits. The current from other circuits is not generally related to the op amp signal, and the voltage developed by it will manifest itself as noise. This signal at the null terminals can easily be the dominant noise in the system. A few milliamps of V– current through a few centimeters of wire can result in interference which is orders of magnitude larger than the inherent input noise of the amplifier. The remedy is to make the connection from the null pot wiper direct to the V– pin of the amplifier, as shown in Figure 14. Some amplifiers such as the AD504 and AD510 refer to the null offset terminals to V+. Obviously, the pot wiper should go to the V+ terminal of this type of amplifier. It's important to connect the line directly to the op amp terminal so as to minimize the common impedance shared by the op amp current and the null pot connection.

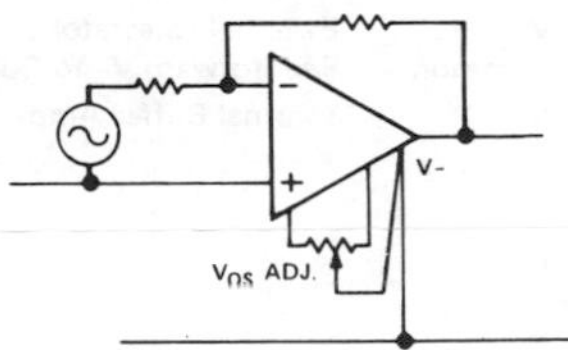

Figure 14. Connecting the Null Pot for Trouble Free Operation

The considerations for op-amp null pots also apply to the similar trimmers on almost all types of integrated circuits. For example, the AD521 In-Amp null terminals exhibit a gain of about 30 to the output. Although this is much less than in the case of most op-amps, it still warrants care in controlling the null pot wiper return. Table I lists the integrated circuits manufactured by Analog Devices, including some popular second-source families, and indicates how internal conversions from differential to single ended are referred. That is, the signals are made to appear with respect to the terminal(s) listed.

	Internal Integrator Referred to:	Comments
AD502	V-	
AD503	V-	
AD504	V+	External Cap
AD506	V-	
AD507	—	External Cap to Signal Common or V+
AD509	—	External Cap to Signal Common or V+
AD510	V+	
AD511	V-	
AD512	V-	
	V-, -in	External Caps, Optional Feedforward to -in
AD514	V-	
AD515	V-	
AD517	V+	
AD518	V+, V-	Internal Feedforward Cap V+ to V- and Integrator to Output
AD520	V+, V-	Internal Integrator Refers to V+, Internal Input Stage Cap Refers to V-, External Output Caps Refer to V+ and Common
AD521	V-	Output Amplifier Integrator Refers to V-
AD522	V+, V-	Input Amplifier Refers to V+ Output Amplifier Refers to V-
AD523	V-	
AD528	V+, V-	Internal Feedforward Cap V+ to V- and Integrator to Output
AD530	V+	Multiplier Output Amplifier Integrator Refers to V+
AD531	V+	Multiplier Output Amplifier Integrator Refers to V+
AD532	V+	Multiplier Output Amplifier Integrator Refers to V+
AD533	V+	Multiplier Output Amplifier Integrator Refers to V+
AD534	V-	Output Amplifier
AD535	V-	Output Amplifier
AD536A	V-, V+, Common	External Integrator to V+, Internal Feedforward V- to Common
AD537	V-	Internal Buffer Amp

	Internal Integrator Referred To:	Comments
AD540	V-	
AD542	V-	
AD544	V-	
AD545	V-	
AD559	V-, Common	DAC Control Loop Integrator Referred Between V- and Common
AD561	V-, Common	DAC Control Loop Integrator and Ref. Amp Refer to Common Ref. Bias Amp Refers to V-
AD562	V-	DAC Control Loop Integrator Referred to V-. Reference Input Common to Control Loop Isolated from DAC Output Common
AD563	V-	DAC Control Loop Integrator Referred to V-. Reference Input Common to Control Loop Isolated from DAC Output Common
AD565	V-	DAC Control Loop Integrator Referred to V-. Reference Input Common to Control Loop Isolated from DAC Output Common
AD566	V-	DAC Control Loop Integrator Referred to V-, Reference Input Common to Control Loop Isolated from DAC Output Common
AD580	V-	
AD581	V-	
AD582	V-	
AD584	V-	
AD101A	V-	External Cap (Includes AD201A, AD301A, etc.)
AD108	V+	External Cap (Includes AD208, AD308, etc.)
AD741	V-	Internal Cap (Includes 741J, K, L, etc.)

This collection of examples won't solve all your potential grounding problems. I hope that it will give you some good ideas about how to prevent some of them, and it should also give you some of the "inside story" on I.C.'s which you can put to work in very practical ways. There is no general grounding method which will prevent all possible problems. The only generally applicable rule is attention to detail, and remember that you can always trust your mother, but. . .

Table I.

SHIELDING AND GUARDING

How to Exclude Interference-Type Noise What to Do and Why to Do It—A Rational Approach

by Alan Rich

This is the second of two articles dealing with interference noise. In the last issue of *Analog Dialogue* (Vol. 16, No. 3, pp. 16-19), we discussed the nature of interference, described the relationship between sources, coupling channels, and receivers, and considered means of combatting interference in systems by reducing or eliminating one of those three elements.

One of the means of reducing noise coupling is *shielding*, Our purpose in this article is to describe the correct uses of shielding to reduce noise. The major topics we will discuss include noise due to capacitive coupling, noise due to magnetic coupling, and driven shields and guards. A set of guidelines will be included, with do's and don'ts.

From the outset, it should be noted that shielding problems are always rational and do not involve the occult; but they are not always straightforward. Each problem must be analyzed carefully. It is important first to identify the noise source, the receiver, and the coupling medium. Improper shielding and grounding, based on faulty identification of any of these elements, may only make matters worse or create a new problem.

You can think of shielding as serving two purposes. First, shielding can be used to confine noise to a small region; this will prevent noise from extending its reach and getting into a nearby critical circuit. However, the problem with such shields is that noise captured by the shield can still cause problems if the return path the noise takes is not carefully planned and implemented by understanding of the ground system and making the connections correctly.

Second, if noise is present in a system, shields can be placed around critical circuits to prevent the noise from getting into sensitive portions of the circuits. These shields can consist of metal boxes around circuit regions or cables with shields around the center conductors. Again, where and how the shields are connected is important.

CAPACITIVELY COUPLED NOISE

If the noise results from an electric field, a shield works because a charge, Q_2, resulting from an external potential, V_1, cannot exist on the interior of a closed conducting surface (Figure 1).

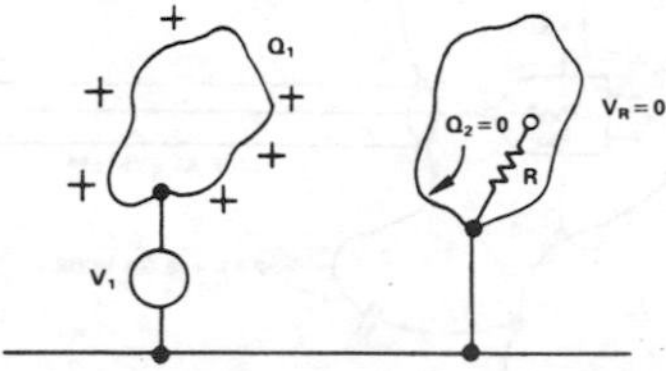

Figure 1. Charge Q_1 cannot create charge inside a closed metal shell.

Coupling by mutual, or stray, capacitance can be modeled by the circuit of Figure 2. Here, V_n is a noise source (switching transistor, TTL gate, etc.), C_s is the stray capacitance, Z is the impedance of a receiver (for example, a bypass resistor connected between the input of a high-gain amplifier and ground), and V_{no} is the output noise developed across Z.

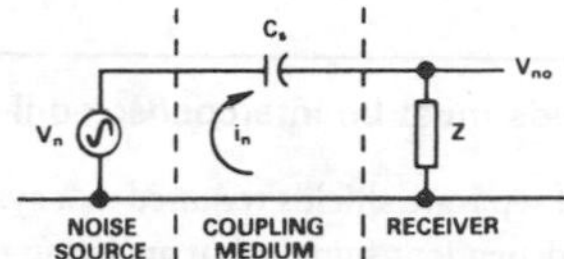

Figure 2. Equivalent circuit of capacitive coupling between a source and a nearby impedance.

A noise current, $i_n = V_n/(Z + Z_{Cs})$, will result, producing a noise voltage, $V_{no} = V_n/(1 + Z_{Cs}/Z)$. For example, if $C_s = 2.5$ pF, $Z = 10k\Omega$ (resistive), and $V_n = 100$mV at 1.3 MHz, the output noise will be 20 mV (0.2% of 10V, i.e., 8 LSBs of 12 bits).

It is important to recognize the effect that very small amounts of stray capacitance will have on sensitive circuits. This becomes increasingly critical as systems are being designed to combine circuits operating at lower power (implying higher impedance levels), higher speed (implying lower nodal stray capacitance, faster edges, and higher frequencies), and higher resolution (much less output noise permitted).

When a shield is added, the change to the situation of Figure 2 is exemplified by the circuit model of Figure 3. With the assumption that the shield has zero impedance, the noise current in loop A-B-D-A will be V_n/Z_{Cs1}, but the noise current in loop D-B-C-D will be zero, since there is no driving source in that loop. And, since no current flows, there will be no voltage developed across Z. The sensitive circuit has thus been shielded from the noise source, V_n.

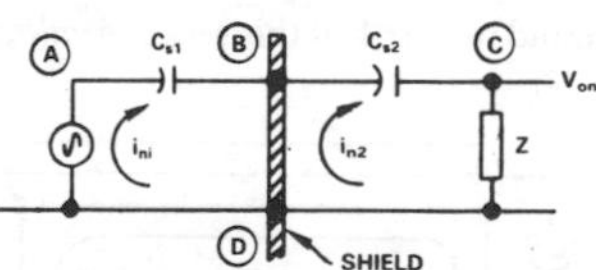

Figure 3. Equivalent circuit of the situation of Figure 2, with a shield interposed between the source and the impedance.

Guidelines for Applying Electrostatic Shields

- An electrostatic shield, to be effective, should be connected to the reference potential of any circuitry contained within the shield. If the signal is earthed or grounded (i.e., connected to a metal chassis or frame, and/or to earth), the shield must be earthed or grounded. But grounding the shield is useless if the signal is not grounded.
- The shield conductor of a shielded cable should be connected to the reference potential at the signal-reference node (Figure 4).
- If the shield is split into sections, as might occur if connectors are used, the shield for each segment must be tied to those for the

Reprinted from Analog Dialogue 17-1, 1983

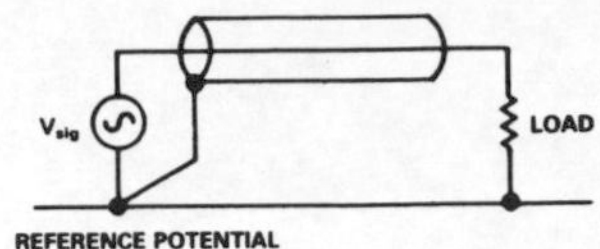

Figure 4. Grounding a cable shield.

adjoining segments, and ultimately connected (only) to the signal-reference node (Figure 5).

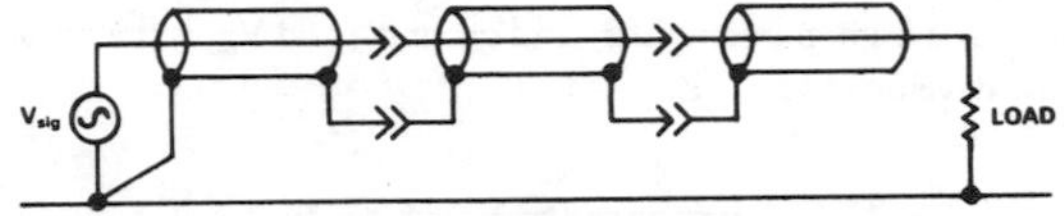

Figure 5. Shields must be interconnected if interrupted.

•The number of separate shields required in a system is equal to the number of independent signals that are being measured. Each signal should have its own shield, with no connections to other shields in the system, unless they share a common reference potential (signal "ground"). If there is more than one signal ground (Figure 6), each shield should be connected to its own reference potential.

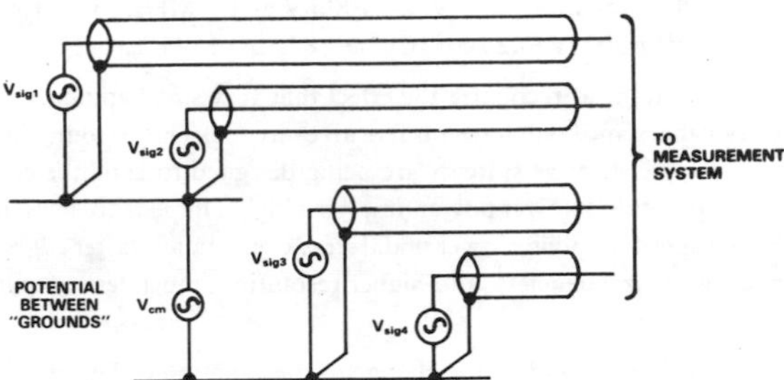

Figure 6. Each signal should have its own shield connected to its own reference potential.

•*Don't connect both ends of the shield to "ground"*. The potential difference between the two "grounds" will cause a shield current to flow (Figure 7). The shield current will induce a noise voltage into the center conductor via magnetic coupling. An example of this can be found in Part 1 of this series, *Analog Dialogue* 16-3, page 18, Figure 10.

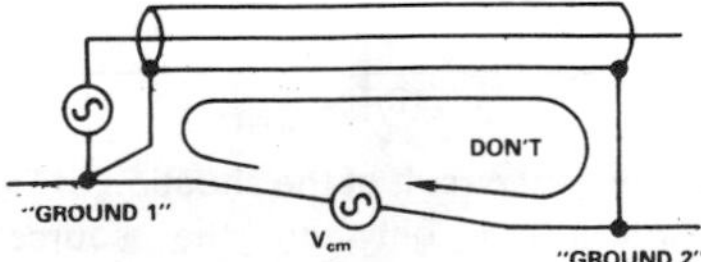

Figure 7. Don't connect the shield to ground at more than one point.

•*Don't allow shield current to exist* (except as noted later in this article). The shield current will induce a voltage in the center conductor.

•*Don't allow the shield to be at a voltage with respect to the reference potential* (except in the case of a guard shield, to be described). The shield voltage will couple capacitively to the center conductor (or conductors in a multiple-conductor shield). With a noise voltage, V_s, on the shield, the situation is as shown in Figure 8.

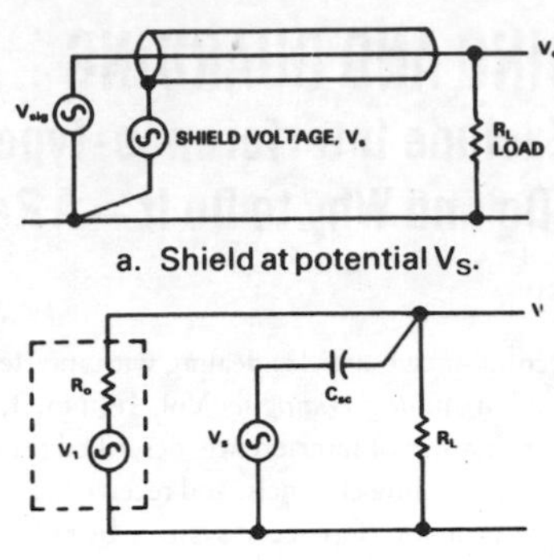

Figure 8. Don't permit the shield to be at a potential with respect to the signal.

The fraction of V_s appearing at the output will be

$$V_o = \frac{V_s}{\sqrt{1 + \frac{1}{(2\pi f R_{eq} C_{sc})^2}}}, \tag{1}$$

where V_1 is the open-circuit signal voltage, R_o is the signal's source impedance, C_{sc} is the cable's shield-to-conductor capacitance, and R_{eq} is the equivalent parallel resistance of R_o and R_L. For example, if V_s = 1V at 1.5MHz, C_{sc} = 200pF (10 feet of cable), R_o = 1000 ohms, and R_L = 10kΩ, the output noise voltage will be 0.86 volts. This is an often-ignored guideline; serious noise problems can be *created* by inadvertently applying undesired potentials to the shield.

•*Know by careful study how the noise current that has been captured by the shield returns to "ground."* An improperly returned shield can cause shield voltages, can couple into other circuits, or couple into other shields. The shield return should be as short as possible to minimize inductance.

Here is an example that illustrates the problems that can arise in relation to these last two guidelines: Consider the improperly configured shield system shown in Figure 9, in which a precision voltage source, V1, and a digital logic gate share a common shield connection. This situation can occur in a large system where analog and digital signals are cabled together.

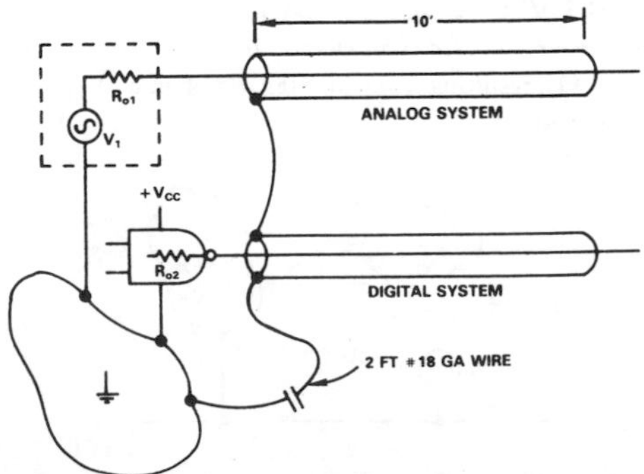

Figure 9. A situation that generates transient shield voltages.

A step voltage change in the output of the logic circuit couples capacitively to its shield, creating a current in the common 2-foot shield return. This, in turn, develops a shield voltage common to both the analog and digital shields. An equivalent circuit is shown in Figure 10, in which V(t) is a 5-volt step from a TTL logic gate, R_{o2} is the 13-ohm output impedance of the logic gate, C_{ws} is the 470-pF capacitance from the shield to the center conductor of the shielded cable, and R_s and L_s are the 0.1-ohm resistance and 1-microhenry inductance of the 2-foot wire connecting the shield to the system ground.

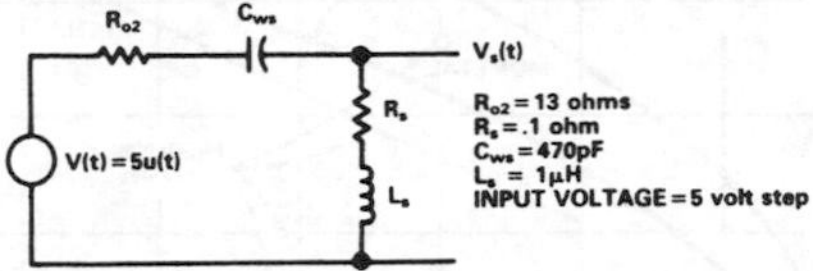

Figure 10. Equivalent circuit for generating shield voltage.

The shield voltage, $V_s(t)$, can be solved for by conventional circuit-analysis techniques, or simulated by actually building and carefully making measurements on a circuit with the given parameters. For the purpose of demonstration, the calculated response waveform, illustrated in Figure 11, with a 5-volt initial spike, resonant frequency of 7.3 MHz, and damping time constant of 0.15μs, is sufficient to illustrate the nature of the voltage that appears on the shield and is capacitively coupled to the analog input. If the voltage is looked at with a wideband oscilloscope, it will look like a noise "spike." We can see that this transient will couple a fast damped waveform of significant peak amplitude to the analog system input.

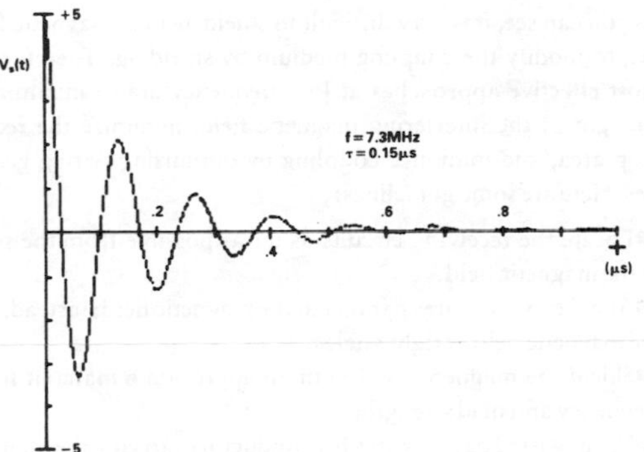

Figure 11. Computed response of circuit of Figure 10.

Even in a purely digital system, noise glitches can be caused to appear in apparently remote portions of a system having the kind of situation shown. This can often explain some otherwise inexplicable system bugs.

In quite a few cases, the proper choice of shield connection among the many possibilities may not be immediately obvious, and the guidelines may not provide us with a clear choice. There is no alternative but to analyze the various possibilities and choose the approach for which the lowest noise may be calculated.

For example, consider the case illustrated in Figure 12, in which the measurement system and the source have differing ground potentials. Should we connect the shield to A: the low side at the measurement-system input, B: ground at the system input, C: ground at the signal source, or D: the low side at the source?

A is a poor choice, since noise current is allowed to flow in a signal

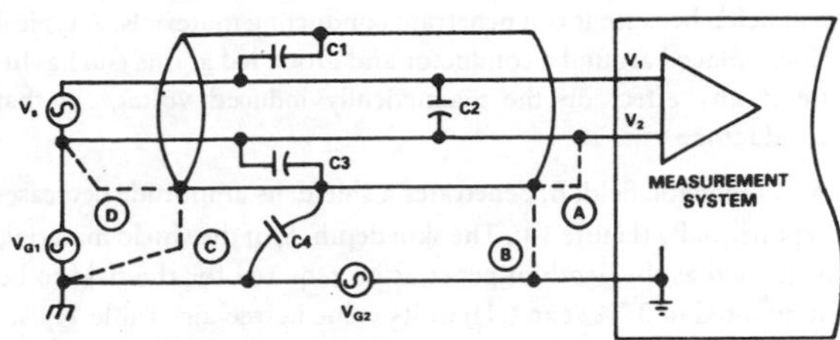

Figure 12. Possible grounds where system and source have differing ground potentials.

conductor. The path of the noise current due to V_{G1}, as it returns through C4, is shown in Figure 13a.

B is also a poor choice, since the two noise sources in series, V_{G1} and V_{G2}, produce a component across the two signal wires, developed by the source impedance in parallel with C_2, in series with C_1, as shown in Figure 13b.

C is poor, too, since V_{G1} produces a voltage across the two signal wires, by the same mechanism as (B), as Figure 13c shows.

D is the best choice, under the given assumptions, as can be seen in Figure 13d. It also tends to confirm the grounding guideline to connect the shield at the signal's reference potential.

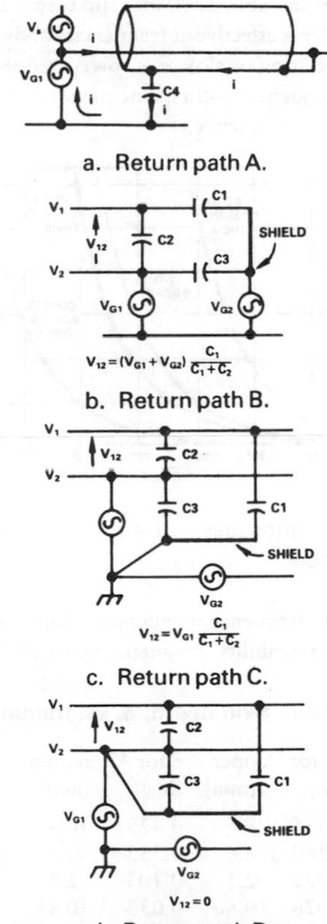

Figure 13. Equivalent circuits.

NOISE RESULTING FROM A MAGNETIC FIELD

Noise in the form of a magnetic field induces voltage in a conductor or circuit; it is much more difficult to shield against than electric fields because it can penetrate conducting materials. A typical shield placed around a conductor and grounded at one end has little if any effect on the magnetically induced voltage in that conductor.

As a magnetic field, B, penetrates a shield, its amplitude decreases exponentially (Figure 14). The skin depth, δ, of the shield material, is defined as the depth of penetration required for the field to be attenuated to 37% (exp (-1)) of its value in free air. Table 1[1] lists typical values of δ for several materials at various frequencies. You can see that any of the materials will be more effective as a shield at high frequency, because δ decreases with frequency, and that steel provides at least an order of magnitude more effective shielding at any frequency than copper or aluminum.

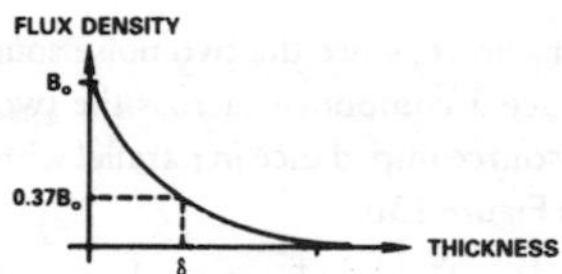

Figure 14. Magnetic field in a shield as a function of penetration depth.

Figure 15 compares absorption loss as a function of frequency for two thicknesses of copper and steel. ⅛-inch steel becomes quite effective for frequencies above 200 Hz, and even a 20-mil (0.5 mm) thickness of copper is effective at frequencies above 1 MHz. However, all show a glaring weakness at lower frequencies, including 50-60-Hz line frequencies—the principal source of magnetically coupled noise at low frequency.

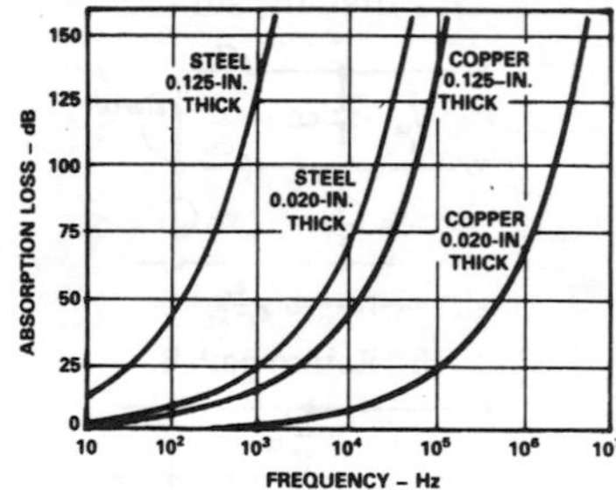

Figure 15. Absorption loss vs. frequency for two thicknesses of copper and steel.

For improved low-frequency magnetic shielding, a shield consisting of a high-permeability magnetic material (e.g., Mumetal)

Table 1. Skin depth, δ, vs. frequency

Frequency	δ for Copper (in.)	(mm)	δ for Aluminum (in.)	(mm)	δ for Steel (in.)	(mm)
60Hz	0.335	8.5	0.429	10.9	0.034	0.86
100Hz	0.260	6.6	0.333	8.5	0.026	0.66
1kHz	0.082	2.1	0.105	2.7	0.008	0.2
10kHz	0.026	0.66	0.033	0.84	0.003	0.08
100kHz	0.008	0.2	0.011	0.3	0.0008	0.02
1MHz	0.003	0.08	0.003	0.08	0.0003	0.008

[1]Table 1 and Figures 15 and 16 are from Ott, H.W., *Noise Reduction Techniques in Electronic Systems* (New York: John Wiley & Sons, © 1976).

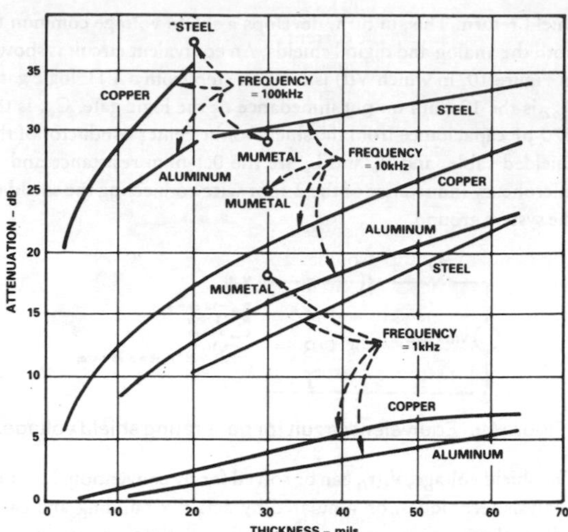

Figure 16. Shielding attenuation of Mumetal and other materials at several frequencies.

should be considered. Figure 16 compares a 30-mil thickness of Mumetal with various materials at several frequencies. It shows that, below 1 kHz, Mumetal is more effective than any of the other materials, while at 100kHz it is the least effective. However, Mumetal is not especially easy to apply, and if it is saturated by an excessively strong field, it will no longer provide an advantage.

As you can see, it is very difficult to shield against magnetic fields, i.e., to modify the coupling medium by shielding. Therefore, the most effective approaches at low frequency are to minimize the strength of the interfering magnetic field, minimize the receiver loop area, and minimize coupling by optimizing wiring geometries. Here are some guidelines:

- Locate the receiving circuits as far as possible from the source of the magnetic field.
- Avoid running wires parallel to the magnetic field; instead, cross the magnetic field at right angles.
- Shield the magnetic field with an appropriate material for the frequency and field strength.
- Use a twisted pair of wires for conductors carrying the high-level current that is the source of the magnetic field. If the currents in the two wires are equal and opposite, the net field in any direction

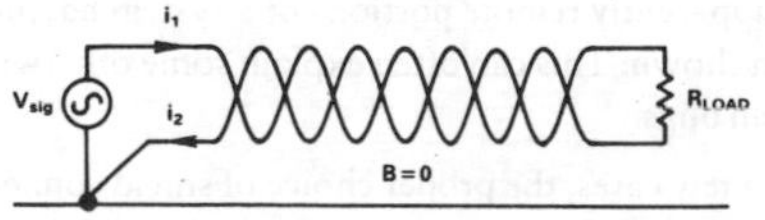

a. Correct connection with balanced currents.

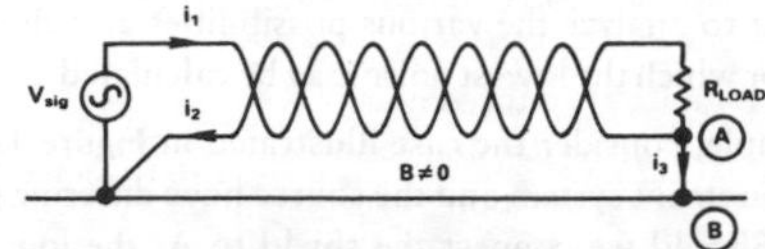

b. Incorrect connection forming ground loop.

Figure 17. Connections to a twisted pair.

over each cycle of twist will be zero (Figure 17a). For this arrangement to work, none of the current can be shared with another conductor, for example, a ground plane. Figure 17b shows what can happen if a ground loop is formed; if part of the current flows through the ground plane (depending on the ratio of conductor resistance to ground resistance), it will form a loop with the twisted pair, generating a field determined by i_3 ($= i_1 - i_2$).

The ground connection between A and B need not be as simple as a short circuit to cause trouble. Any stray unbalanced capacitance or resistance from R_{load} circuits to the ground plane will also unbalance the currents and produce a net current through the wires and the ground plane, producing a ground loop and a related magnetic field. For this reason, it is also good practice to run the twisted pair close to the ground plane to tend to balance the capacitances from each side to ground, as well as to minimize loop area.

•Use a shielded cable with the high-level source circuit's return current carried in the shield (Figure 18). If the shield current, I_2 is equal and opposite to that in the center conductor, the center-conductor field and the shield field will cancel, producing a zero net field. In this case, which seems to violate the "no shield current" rule for receiver circuits, the concentric cable is not used to shield the center lead; instead, the geometry produces cancellation.

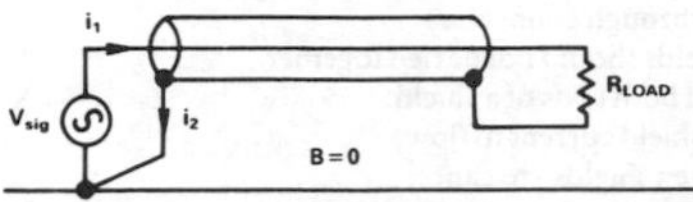

Figure 18. Use of shield for return current to noisy source.

This scheme can be usefully employed in an ATE system where ac curate measurements must be performed on devices with high power-supply currents that may be noisy. For example, Figure 19 shows the application of this technique to the connections for the high-current logic supply for an a/d converter under test—at the end of a test cable.

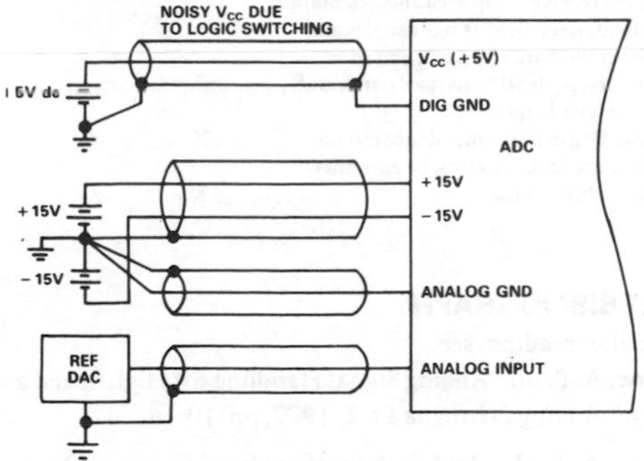

Figure 19. Application of circuit of Figure 18 in a test system.

•Since magnetically induced noise depends on the area of the receiver loop, the induced voltage due to magnetic coupling can be reduced by reducing the loop's area. What is the receiver loop? In the example shown in Figure 20, the signal source and its load are connected by a pair of conductors of length L and separation D. The circuit (assuming it has a rectangular configuration) forms a loop with area D · L.

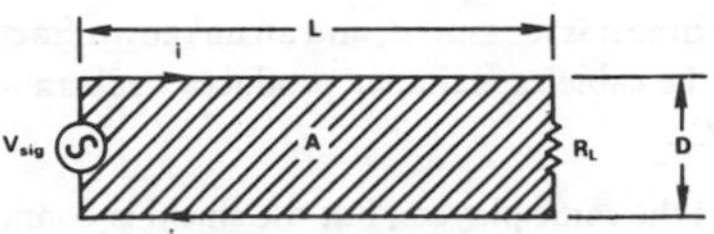

Figure 20. Area of a loop that receives magnetically coupled noise.

The voltage induced in series with the loop is proportional to the area and the cosine of its angle to the field. Thus, to minimize noise, the loop should be oriented at right angles to the field, and its area should be minimized.

The area can be reduced by decreasing the length of and/or decreasing the distance between the conductors. This is easily accomplished with a twisted pair, or at least a tightly cabled pair, of conductors. It is good practice to pair conductors so that the circuit wire and its return path will always be together. To do this, the designer must be certain of the actual path that the return current takes in getting back to the signal source. Quite often, the current returns by a path not intended in the original design layout.

If wires are moved (for example, by a technician troubleshooting some other problem), the loop area and orientation to the field may change, so that yesterday's acceptable noise level may be transformed to tomorrow's disastrous noise level. Which may lead to a service call . . . and another repetition of the cycle. The bottom line: Know the loop area and orientation, do what must be done to minimize noise—*and permanently secure the wiring*!

DRIVEN SHIELDS AND GUARDING

We have discussed the role of a current-driven shield carrying an equal and opposite current to reduce generated noise by reducing the magnetic field around a conductor.

Guarding is similar, in that it involves driving a shield, at low impedance, with a potential essentially equal to the common-mode voltage on the signal wire contained within the shield. Guarding has many useful purposes: It reduces common-mode capacitance, improves common-mode rejection, and eliminates leakage currents in high-impedance measurement circuits.

Figure 21 shows an example of an op amp with negligible bias current connected as a high-impedance non-inverting amplifier with gain. The purpose of the cable is to shield the high input-impedance signal conductor from capacitively coupled noise and to minimize leakage currents. The signal comes from a 10-megohm source, and the cable is assumed to have 1000 megohms of leakage resistance (which may change as a function of temperature, humidity, etc.) from conductor to shield. If connected as shown, the equivalent input circuit is an attenuator which loses 1% of the

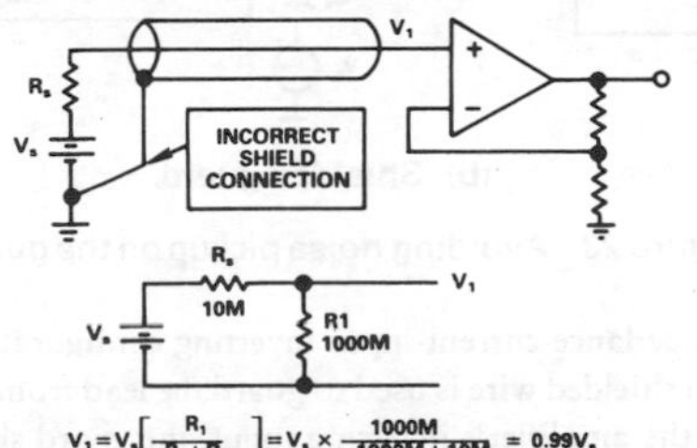

Figure 21. Op amp connected as high-impedance non-inverting amplifier with gain, with shielded input lead.

20

signal at the time it is measured, and an unknown fraction at other times. Also, the cable capacitance produces a substantial lag time constant, R_sC_c.

Figure 22 has the same players, but the shield is connected to the tap of the gain divider (usually at low impedance). Being connected to the inverting input of the op amp, it should be at the same potential as the amplifier's non-inverting input. Since there is no voltage across the cable's leakage resistance, there is no current through it and its resistance value doesn't matter; V_1 must therefore be equal to V_s, since bias current was assumed negligible.

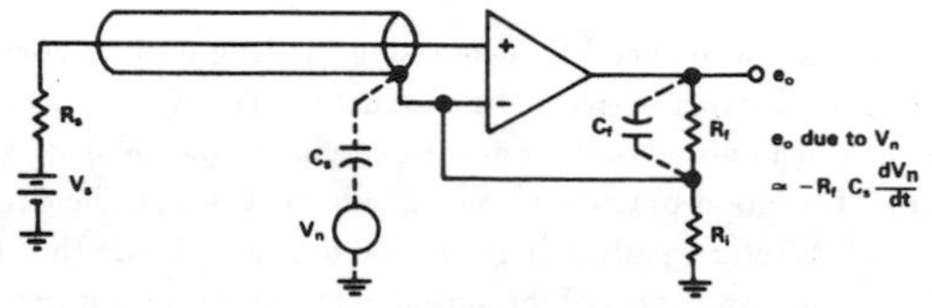

Figure 22. Same as Figure 21, but cable shield connected as a guard.

Also, there is no voltage across the cable capacitance, hence no charging or discharging of the cable; thus the lag time constant depends mainly on circuit strays and the amplifier's input capacitance. For stability, capacitance should be connected between the output and the negative input, such that $C_fR_F = C_sR_i$, where C_s is sum of the stray capacitance between shield and ground and the input capacitance.

There must be no noise voltage applied to the guard. In noisy systems, as Figure 22 shows, capacitively coupled noise will be differentiated, emphasizing the higher-frequency components. This can be avoided (Figure 23) by either using a buffer follower with fast response and low output impedance to drive the guard (a) or a second shield, around the guard, grounded to the signal common (b).

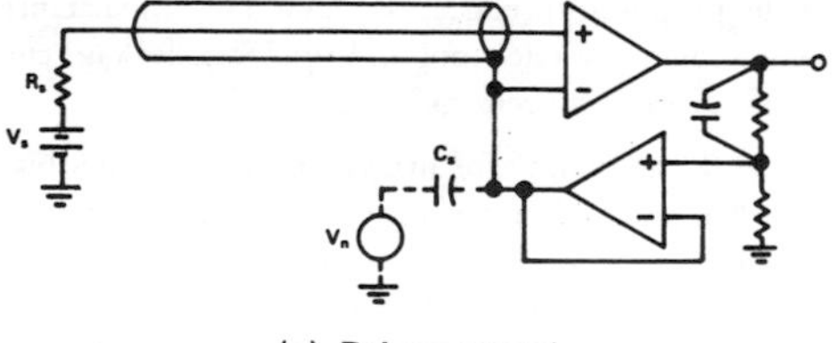

(a) Driven guard.

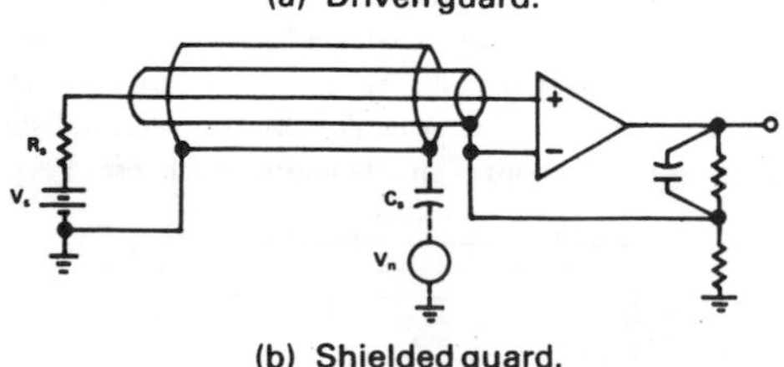

(b) Shielded guard.

Figure 23. Avoiding noise pickup on the guard.

In high-impedance current-input inverting configurations, where a length of shielded wire is used to guard the lead from the current source to the amplifier's inverting input, the guard should either be driven by a buffer at the same potential as the non-inverting input (and connected nowhere else), or be tied directly to the non-inverting input, with a second outer shield connected to the signal's reference point.

SUMMARY

Table 2 summarizes the important points made in this article. All are important to maintaining a high-integrity shield system. However, we cannot emphasize too strongly the two subjects that are most-often ignored: appearance of noise voltage on signal shields and proper disposition of shield noise currents. *Noise voltage must not exist on the shield*; shield-to-conductor capacitance will couple the noise directly to the center conductor. *If shield currents are not returned properly, they can show up in a remote part of the system* and perhaps cause trouble in a location totally unrelated to the shielding problem that was "solved." ■

Table 2. Applicability of shielding considerations

Consideration	Universal	Electric	Magnetic
Know the noise source, coupling medium, and receiver.	X	X	X
Different shielding techniques are required for different noise sources, coupling channels, and receivers.	X	X	X
In most situations, conventional circuit analysis using lumped elements can be used.	X	X	X
Connect the shield at the signal-source end only.		X	
Carry shields through connectors.		X	
Individual shields should not be tied together.		X	
Do not ground both ends of a shield.		X	
Do not allow shield current to flow, except for driven shields - to cancel magnetic fields		X	X
Do not allow voltage on a shield, except for guarding.		X	
Know *exactly* where noise current from the shield will flow.		X	
Use short connections to return noise current from the shield.		X	
Electrostatic shields have little effect in reducing noise resulting from magnetic fields.			X
Reduce magnetic fields by physical separation proper orientation, twisted pairs, and/or driven shields.			X
Know the receiver loop area and orientation to the field. Keep field at right angles and reduce the loop area by using paired conductors, preferably twisted pairs, and minimize wire lengths.			X
Use guarding in high-impedance circuits	X	X	
In high-impedance circuits, be extremely careful of shield noise	X	X	

BRIEF BIBLIOGRAPHY

For further reading, see:

Brokaw, A. Paul. "Analog Signal Handling for High Speed and Accuracy," *Analog Dialogue* 11-2, 1977, pp. 10-16.

Brokaw, A. Paul. "An I.C. Amplifier Users' Guide to Decoupling, Grounding, and Making Things Go Right for a Change," *Analog Devices Data-Acquisition Databook 1982*, Volume I, Pages 21-13 to 21-20.

Morrison, Ralph. *Grounding and Shielding Techniques in Instrumentation* Second Edition. (New York: John Wiley & Sons, 1977).

Ott, Henry W. *Noise Reduction Technique in Electronic Systems.* (New York: John Wiley & Sons, 1976).

Chapter 2 Transducers and reference sources

Although this book is concerned primarily with the processing of analog signals as opposed to their generation, we have included articles on two types of transducer – the temperature transducer and the pressure transducer. Both of these devices are fabricated using semiconductor technology. Furthermore, we have selected the temperature transducer because thermal effects are very important in analog systems. Moreover, the internal circuitry of this type of transducer is of interest to the analog systems designer. We have selected the pressure transducer because its interface illustrates some of the considerations we introduced in Chapter 1. The application notes in this chapter include full details of the pressure transducers described.

In addition to transducers we cover the voltage reference sources that are required by all analog to digital and digital to analog systems. We have chosen one reference that employs the emitter-base voltage of a transistor and another that employs the Zener effect.

Two-Terminal IC Temperature Transducer

AD590*

FEATURES
Linear Current Output: 1μA/K
Wide Range: -55°C to +150°C
Probe Compatible Ceramic Sensor Package
Two-Terminal Device: Voltage In/Current Out
Laser Trimmed to ±0.5°C Calibration Accuracy (AD590M)
Excellent Linearity: ±0.3°C Over Full Range (AD590M)
Wide Power Supply Range: +4V to +30V
Sensor Isolation from Case

AD590 FUNCTIONAL BLOCK DIAGRAM

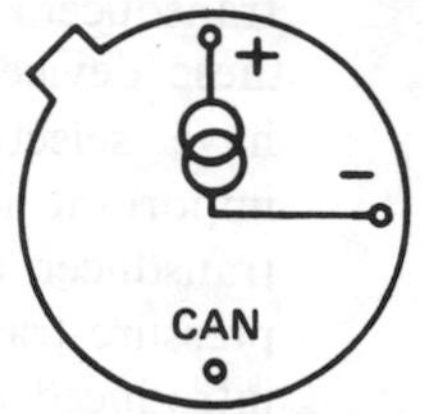

TO-52
BOTTOM VIEW

8

PRODUCT DESCRIPTION

The AD590 is a two-terminal integrated circuit temperature transducer which produces an output current proportional to absolute temperature. For supply voltages between +4V and +30V the device acts as a high impedance, constant current regulator passing 1μA/K. Laser trimming of the chip's thin film resistors is used to calibrate the device to 298.2μA output at 298.2K (+25°C).

The AD590 should be used in any temperature sensing application below +150°C in which conventional electrical temperature sensors are currently employed. The inherent low cost of a monolithic integrated circuit combined with the elimination of support circuitry makes the AD590 an attractive alternative for many temperature measurement situations. Linearization circuitry, precision voltage amplifiers, resistance measuring circuitry and cold junction compensation are not needed in applying the AD590.

In addition to temperature measurement, applications include temperature compensation or correction of discrete components, biasing proportional to absolute temperature, flow rate measurement, level detection of fluids and anemometry. The AD590 is available in chip form making it suitable for hybrid circuits and fast temperature measurements in protected environments.

The AD590 is particularly useful in remote sensing applications. The device is insensitive to voltage drops over long lines due to its high impedance current output. Any well-insulated twisted pair is sufficient for operation hundreds of feet from the receiving circuitry. The output characteristics also make the AD590 easy to multiplex: the current can be switched by a CMOS multiplexer or the supply voltage can be switched by a logic gate output.

*Covered by Patent No. 4,123,698

PRODUCT HIGHLIGHTS

1. The AD590 is a calibrated two terminal temperature sensor requiring only a dc voltage supply (+4V to +30V). Costly transmitters, filters, lead wire compensation and linearization circuits are all unnecessary in applying the device.
2. State-of-the-art laser trimming at the wafer level in conjunction with extensive final testing insures that AD590 units are easily interchangeable.
3. Superior interference rejection results from the output being a current rather than a voltage. In addition, power requirements are low (1.5mW's @ 5V @ +25°C). These features make the AD590 easy to apply as a remote sensor.
4. The high output impedance (>10MΩ) provides excellent rejection of supply voltage drift and ripple. For instance, changing the power supply from 5V to 10V results in only a 1μA maximum current change, or 1°C equivalent error.
5. The AD590 is electrically durable: it will withstand a forward voltage up to 44V and a reverse voltage of 20V. Hence, supply irregularities or pin reversal will not damage the device.

SPECIFICATIONS (@ +25°C and V_S = 5V unless otherwise noted)

Model	AD590I Min	AD590I Typ	AD590I Max	AD590J Min	AD590J Typ	AD590J Max	AD590K Min	AD590K Typ	AD590K Max	Units
ABSOLUTE MAXIMUM RATINGS										
Forward Voltage (E + to E −)			+44			+44			+44	Volts
Reverse Voltage (E + to E −)			−20			−20			20	Volts
Breakdown Voltage (Case to E + or E −)			±200			±200			±200	Volts
Rated Performance Temperature Range[1]	−55		+150	−55		+150	−55		+150	°C
Storage Temperature Range[1]	−65		+155	−65		+155	−65		+155	°C
Lead Temperature (Soldering, 10 sec)			+300			+300			+300	°C
POWER SUPPLY										
Operating Voltage Range	+4		+30	+4		+30	+4		+30	Volts
OUTPUT										
Nominal Current Output @ +25°C (298.2K)		298.2			298.2			298.2		μA
Nominal Temperature Coefficient		1			1			1		μA/K
Calibration Error @ +25°C			±10			±5.0			±2.5	°C
Absolute Error (over rated performance temperature range)										
Without External Calibration Adjustment			±20			±10			±5.5	°C
With +25°C Calibration Error Set to Zero			±5.8			±3.0			±2.0	°C
Nonlinearity			±3.0			±1.5			±0.8	°C
Repeatability[2]			±0.1			±0.1			±0.1	°C
Long Term Drift[3]			±0.1			±0.1			±0.1	°C
Current Noise		40			40			40		pA/$\sqrt{Hz}$
Power Supply Rejection										
$+4V \leq V_S \leq +5V$		0.5			0.5			0.5		μA/V
$+5V \leq V_S \leq +15V$		0.2			0.2			0.2		μA/V
$+15V \leq V_S \leq +30V$		0.1			0.1			0.1		μA/V
Case Isolation to Either Lead		10^{10}			10^{10}			10^{10}		Ω
Effective Shunt Capacitance		100			100			100		pF
Electrical Turn-On Time		20			20			20		μs
Reverse Bias Leakage Current[4] (Reverse Voltage = 10V)		10			10			10		pA
PACKAGE OPTION[5]										
"H" Package: TO-52		AD590IH			AD590JH			AD590KH		
"F" Package: Flat Pack (F2A)		AD590IF			AD590JF			AD590KF		

NOTES

[1]The AD590 has been used at −100°C and +200°C for short periods of measurement with no physical damage to the device. However, the absolute errors specified apply to only the rated performance temperature range.

[2]Maximum deviation between +25°C readings after temperature cycling between −55°C and +150°C; guaranteed not tested.

[3]Conditions: constant +5V, constant +125°C; guaranteed, not tested.

[4]Leakage current doubles every 10°C.

[5]See Section 19 for package outline information.

Specifications subject to change without notice.

Specifications shown in boldface are tested on all production units at final electrical test. Results from those tests are used to calculate outgoing quality levels. All min and max specifications are guaranteed, although only those shown in boldface are tested on all production units.

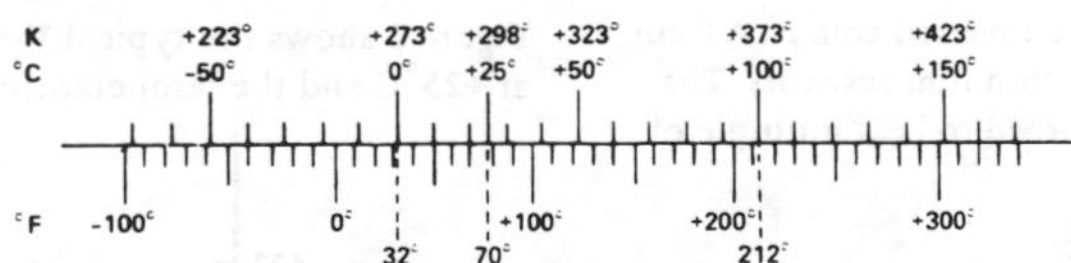

TEMPERATURE SCALE CONVERSION EQUATIONS

$$^\circ C = \frac{5}{9}(^\circ F - 32) \qquad K = {}^\circ C + 273.15$$

$$^\circ F = \frac{9}{5}\,^\circ C + 32 \qquad ^\circ R = {}^\circ F + 459.7$$

Model	AD590L Min	AD590L Typ	AD590L Max	AD590M Min	AD590M Typ	AD590M Max	Units
ABSOLUTE MAXIMUM RATINGS							
Forward Voltage (E+ to E−)			+44			+44	Volts
Reverse Voltage (E+ to E−)			−20			−20	Volts
Breakdown Voltage (Case to E+ or E−)			±200			±200	Volts
Rated Performance Temperature Range[1]	−55		+150	−55		+150	°C
Storage Temperature Range[1]	−65		+155	−65		+155	°C
Lead Temperature (Soldering, 10 sec)			+300			+300	°C
POWER SUPPLY							
Operating Voltage Range	+4		+30	+4		+30	Volts
OUTPUT							
Nominal Current Output @ +25°C (298.2K)		298.2			298.2		μA
Nominal Temperature Coefficient		1			1		μA/K
Calibration Error @ +25°C			**±1.0**			**±0.5**	°C
Absolute Error (over rated performance temperature range)							
Without External Calibration Adjustment			**±3.0**			**±1.7**	°C
With +25°C Calibration Error Set to Zero			**±1.6**			**±1.0**	°C
Nonlinearity			**±0.4**			**±0.3**	°C
Repeatability[2]			±0.1			±0.1	°C
Long Term Drift[3]			±0.1			±0.1	°C
Current Noise		40			40		$pA/\sqrt{Hz}$
Power Supply Rejection							
$+4V \le V_S \le +5V$		0.5			0.5		μA/V
$+5V \le V_S \le +15V$		0.2			0.2		μA/V
$+15V \le V_S \le +30V$		0.1			0.1		μA/V
Case Isolation to Either Lead		10^{10}			10^{10}		Ω
Effective Shunt Capacitance		100			100		pF
Electrical Turn-On Time		20			20		μs
Reverse Bias Leakage Current[4] (Reverse Voltage = 10V)		10			10		pA
PACKAGE OPTION[5]							
"H" Package: TO-52		AD590LH			AD590MH		
"F" Package: Flat Pack (F2A)		AD590LF			AD590MF		

CIRCUIT DESCRIPTION[1]

The AD590 uses a fundamental property of the silicon transistors from which it is made to realize its temperature proportional characteristic: if two identical transistors are operated at a constant ratio of collector current densities, r, then the difference in their base-emitter voltages will be (kT/q)(In r). Since both k, Boltzman's constant and q, the charge of an electron, are constant, the resulting voltage is directly proportional to absolute temperature (PTAT).

In the AD590, this PTAT voltage is converted to a PTAT current by low temperature coefficient thin film resistors. The total current of the device is then forced to be a multiple of this PTAT current. Referring to Figure 1, the schematic diagram of the AD590, Q8 and Q11 are the transistors that produce the PTAT voltage. R5 and R6 convert the voltage to current. Q10, whose collector current tracks the collector currents in Q9 and Q11, supplies all the bias and substrate leakage current for the rest of the circuit, forcing the total current to be PTAT. R5 and R6 are laser trimmed on the wafer to calibrate the device at +25°C.

Figure 2 shows the typical V–I characteristic of the circuit at +25°C and the temperature extremes.

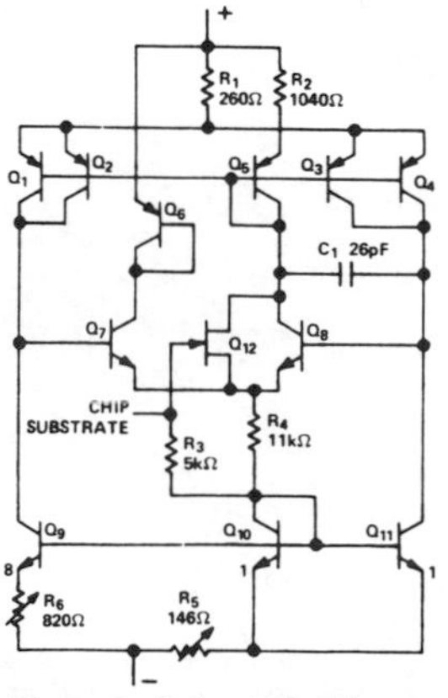

Figure 1. Schematic Diagram

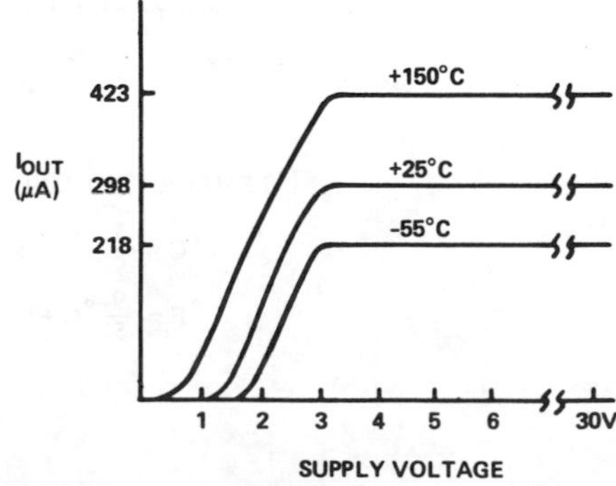

Figure 2. V–I Plot

[1] For a more detailed circuit description see M.P. Timko, "A Two-Terminal IC Temperature Transducer," IEEE J. Solid State Circuits, Vol. SC-11, p. 784-788, Dec. 1976.

EXPLANATION OF TEMPERATURE SENSOR SPECIFICATIONS

The way in which the AD590 is specified makes it easy to apply in a wide variety of different applications. It is important to understand the meaning of the various specifications and the effects of supply voltage and thermal environment on accuracy.

The AD590 is basically a PTAT (proportional to absolute temperature)[1] current regulator. That is, the output current is equal to a scale factor times the temperature of the sensor in degrees Kelvin. This scale factor is trimmed to 1μA/K at the factory, by adjusting the indicated temperature (i.e. the output current) to agree with the actual temperature. This is done with 5V across the device at a temperature within a few degrees of 25°C (298.2K). The device is then packaged and tested for accuracy over temperature.

CALIBRATION ERROR

At final factory test the difference between the indicated temperature and the actual temperature is called the calibration error. Since this is a scale factor error, its contribution to the total error of the device is PTAT. For example, the effect of the 1°C specified maximum error of the AD590L varies from 0.73°C at -55°C to 1.42°C at 150°C. Figure 3 shows how an exaggerated calibration error would vary from the ideal over temperature.

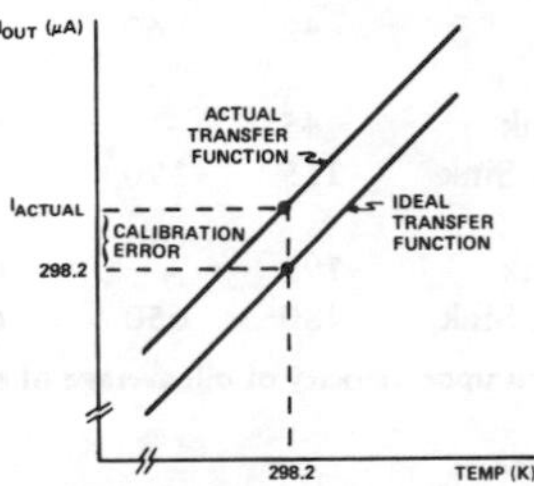

Figure 3. Calibration Error vs. Temperature

The calibration error is a primary contributor to maximum total error in all AD590 grades. However, since it is a scale factor error, it is particularly easy to trim. Figure 4 shows the most elementary way of accomplishing this. To trim this circuit the temperature of the AD590 is measured by a reference temperature sensor and R is trimmed so that V_T = 1mV/K at that temperature. Note that when this error is trimmed out at one temperature, its effect is zero over the entire temperature range. In most applications there is a current to voltage conversion resistor (or, as with a current input ADC, a reference) that can be trimmed for scale factor adjustment.

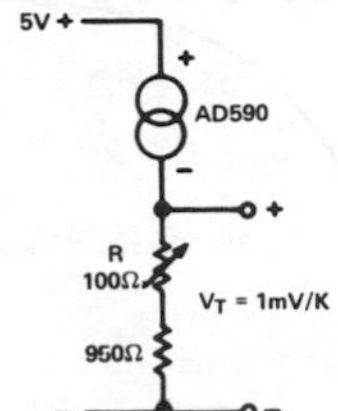

Figure 4. One Temperature Trim

[1] T(°C) = T(K) -273.2; Zero on the Kelvin scale is "absolute zero"; there is no lower temperature.

ERROR VERSUS TEMPERATURE: WITH CALIBRATION ERROR TRIMMED OUT

Each AD590 is also tested for error over the temperature range with the calibration error trimmed out. This specification could also be called the "variance from PTAT" since it is the maximum difference between the actual current over temperature and a PTAT multiplication of the actual current at 25°C. This error consists of a slope error and some curvature, mostly at the temperature extremes. Figure 5 shows a typical AD590K temperature curve before and after calibration error trimming.

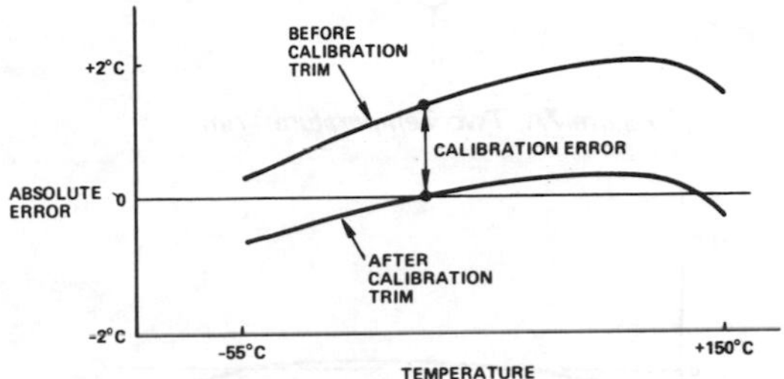

Figure 5. Effect of Scale Factor Trim on Accuracy

ERROR VERSUS TEMPERATURE: NO USER TRIMS

Using the AD590 by simply measuring the current, the total error is the "variance from PTAT" described above plus the effect of the calibration error over temperature. For example the AD590L maximum total error varies from 2.33°C at -55°C to 3.02°C at 150°C. For simplicity, only the larger figure is shown on the specification page.

NONLINEARITY

Nonlinearity as it applies to the AD590 is the maximum deviation of current over temperature from a best-fit straight line. The nonlinearity of the AD590 over the -55°C to +150°C range is superior to all conventional electrical temperature sensors such as thermocouples, RTD's and thermistors. Figure 6 shows the nonlinearity of the typical AD590K from Figure 5.

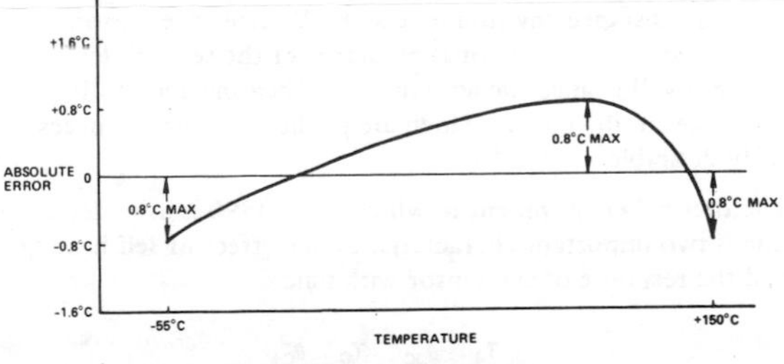

Figure 6. Nonlinearity

Figure 7A shows a circuit in which the nonlinearity is the major contributor to error over temperature. The circuit is trimmed by adjusting R_1 for a 0V output with the AD590 at 0°C. R_2 is then adjusted for 10V out with the sensor at 100°C. Other pairs of temperatures may be used with this procedure as long as they are measured accurately by a reference sensor. Note that for +15V output (150°C) the V+ of the op amp must be greater than 17V. Also note that V- should be at least -4V: if V- is ground there is no voltage applied across the device.

Understanding the AD590 Specifications

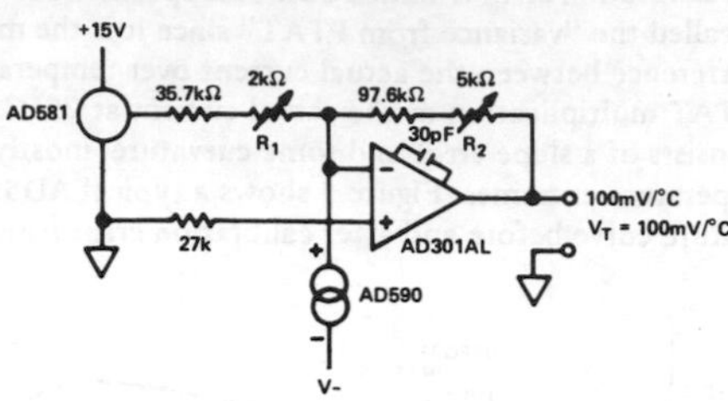

Figure 7A. Two Temperature Trim

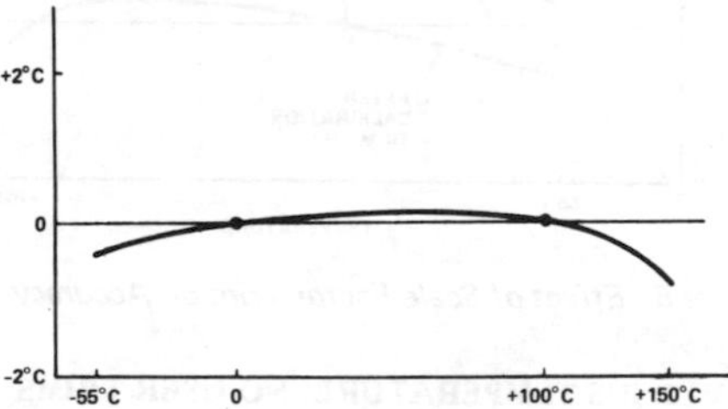

Figure 7B. Typical Two-Trim Accuracy

VOLTAGE AND THERMAL ENVIRONMENT EFFECTS

The power supply rejection specifications show the maximum expected change in output current versus input voltage changes. The insensitivity of the output to input voltage allows the use of unregulated supplies. It also means that hundreds of ohms of resistance (such as a CMOS multiplexer) can be tolerated in series with the device.

It is important to note that using a supply voltage other than 5V does not change the PTAT nature of the AD590. In other words, this change is equivalent to a calibration error and can be removed by the scale factor trim (see previous page).

The AD590 specifications are guaranteed for use in a low thermal resistance environment with 5V across the sensor. Large changes in the thermal resistance of the sensor's environment will change the amount of self-heating and result in changes in the output which are predictable but not necessarily desirable.

The thermal environment in which the AD590 is used determines two important characteristics: the effect of self heating and the response of the sensor with time.

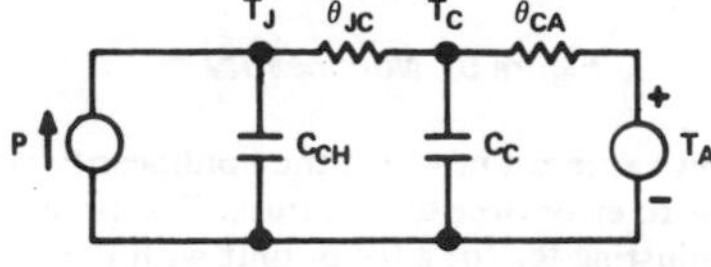

Figure 8. Thermal Circuit Model

Figure 8 is a model of the AD590 which demonstrates these characteristics. As an example, for the TO-52 package, θ_{JC} is the thermal resistance between the chip and the case, about 26°C/watt. θ_{CA} is the thermal resistance between the case and its surroundings and is determined by the characteristics of the thermal connection. Power source P represents the power dissipated on the chip. The rise of the junction temperature, T_J, above the ambient temperature T_A is:

$$T_J - T_A = P\,(\theta_{JC} + \theta_{CA}). \qquad \text{Eq. 1}$$

Table I gives the sum of θ_{JC} and θ_{CA} for several common thermal media for both the "H" and "F" packages. The heatsink used was a common clip-on. Using Equation 1, the temperature rise of an AD590 "H" package in a stirred bath at +25°C, when driven with a 5V supply, will be 0.06°C. However, for the same conditions in still air the temperature rise is 0.72°C. For a given supply voltage, the temperature rise varies with the current and is PTAT. Therefore, if an application circuit is trimmed with the sensor in the same thermal environment in which it will be used, the scale factor trim compensates for this effect over the entire temperature range.

8

MEDIUM	$\theta_{JC} + \theta_{CA}$ (°C/watt) H	F	τ (sec)(Note 3) H	F
Aluminum Block	30	10	0.6	0.1
Stirred Oil[1]	42	60	1.4	0.6
Moving Air[2]				
With Heat Sink	45	–	5.0	–
Without Heat Sink	115	190	13.5	10.0
Still Air				
With Heat Sink	191	–	108	–
Without Heat Sink	480	650	60	30

[1] Note: τ is dependent upon velocity of oil; average of several velocities listed above.

[2] Air velocity ≅ 9ft/sec.

[3] The time constant is defined as the time required to reach 63.2% of an instantaneous temperature change.

Table I. Thermal Resistances

The time response of the AD590 to a step change in temperature is determined by the thermal resistances and the thermal capacities of the chip, C_{CH}, and the case, C_C. C_{CH} is about 0.04 watt-sec/°C for the AD590. C_C varies with the measured medium since it includes anything that is in direct thermal contact with the case. In most cases, the single time constant exponential curve of Figure 9 is sufficient to describe the time response, T(t). Table I shows the effective time constant, τ, for several media.

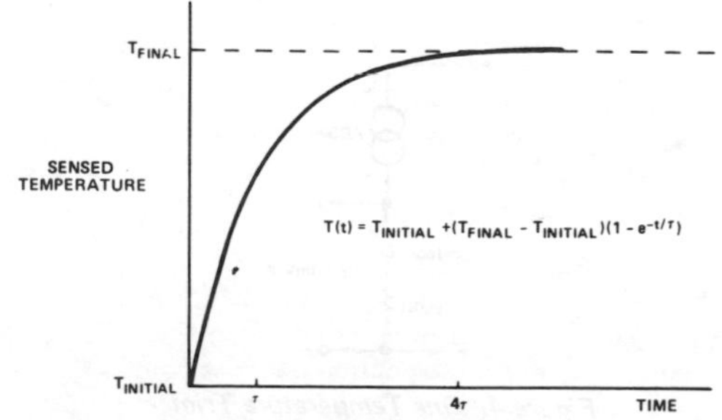

Figure 9. Time Response Curve

GENERAL APPLICATIONS

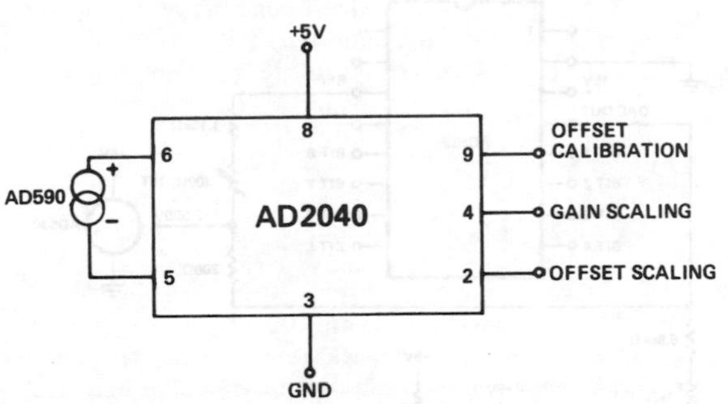

Figure 10. Variable Scale Display

Figure 10 demonstrates the use of a low-cost Digital Panel Meter for the display of temperature on either the Kelvin, Celsius or Fahrenheit scales. For Kelvin temperature Pins 9, 4 and 2 are grounded; and for Fahrenheit temperature Pins 4 and 2 are left open.

The above configuration yields a 3 digit display with 1°C or 1°F resolution, in addition to an absolute accuracy of ±2.0°C over the -55°C to +125°C temperature range if a one-temperature calibration is performed on an AD590K, L, or M.

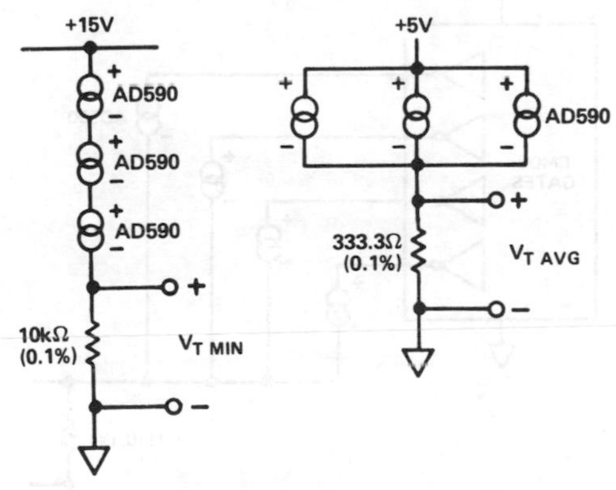

Figure 11. Series & Parallel Connection

Connecting several AD590 units in series as shown in Figure 11 allows the minimum of all the sensed temperatures to be indicated. In contrast, using the sensors in parallel yields the average of the sensed temperatures.

The circuit of Figure 12 demonstrates one method by which differential temperature measurements can be made. R_1 and R_2 can be used to trim the output of the op amp to indicate

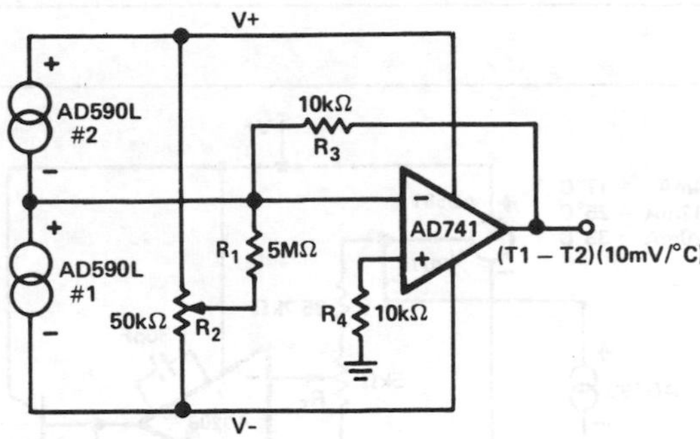

Figure 12. Differential Measurements

a desired temperature difference. For example, the inherent offset between the two devices can be trimmed in. If V+ and V- are radically different, then the difference in internal dissipation will cause a differential internal temperature rise. This effect can be used to measure the ambient thermal resistance seen by the sensors in applications such as fluid level detectors or anemometry.

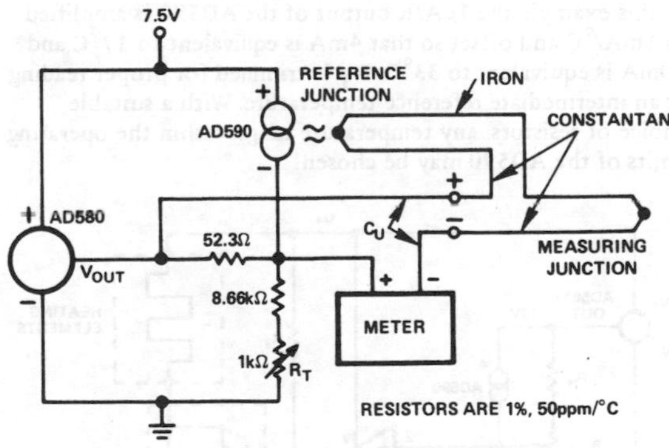

Figure 13. Cold Junction Compensation Circuit for Type J Thermocouple

Figure 13 is an example of a cold junction compensation circuit for a Type J Thermocouple using the AD590 to monitor the reference junction temperature. This circuit replaces an ice-bath as the thermocouple reference for ambient temperatures between +15°C and +35°C. The circuit is calibrated by adjusting R_T for a proper meter reading with the measuring junction at a known reference temperature and the circuit near +25°C. Using components with the T.C.'s as specified in Figure 13, compensation accuracy will be within ±0.5°C for circuit temperatures between +15°C and +35°C. Other thermocouple types can be accommodated with different resistor values. Note that the T.C.'s of the voltage reference and the resistors are the primary contributors to error.

Applying the AD590

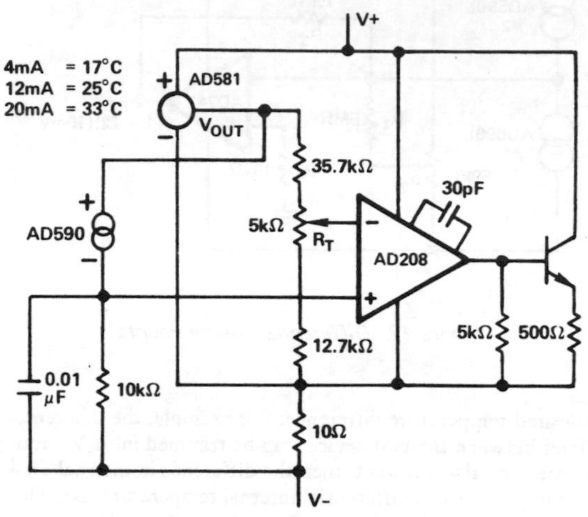

Figure 14. 4 to 20mA Current Transmitter

Figure 14 is an example of a current transmitter designed to be used with 40V, 1kΩ systems; it uses its full current range of 4mA to 20mA for a narrow span of measured temperatures. In this example the 1μA/K output of the AD590 is amplified to 1mA/°C and offset so that 4mA is equivalent to 17°C and 20mA is equivalent to 33°C. R_T is trimmed for proper reading at an intermediate reference temperature. With a suitable choice of resistors, any temperature range within the operating limits of the AD590 may be chosen.

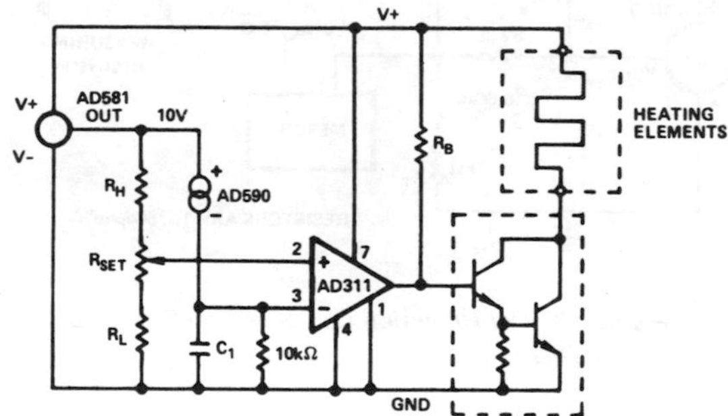

Figure 15. Simple Temperature Control Circuit

Figure 15 is an example of a variable temperature control circuit (thermostat) using the AD590. R_H and R_L are selected to set the high and low limits for R_{SET}. R_{SET} could be a simple pot, a calibrated multi-turn pot or a switched resistive divider. Powering the AD590 from the 10V reference isolates the AD590 from supply variations while maintaining a reasonable voltage (~7V) across it. Capacitor C_1 is often needed to filter extraneous noise from remote sensors. R_B is determined by the β of the power transistor and the current requirements of the load.

Figure 16 shows how the AD590 can be configured with an 8-bit DAC to produce a digitally controlled set point. This

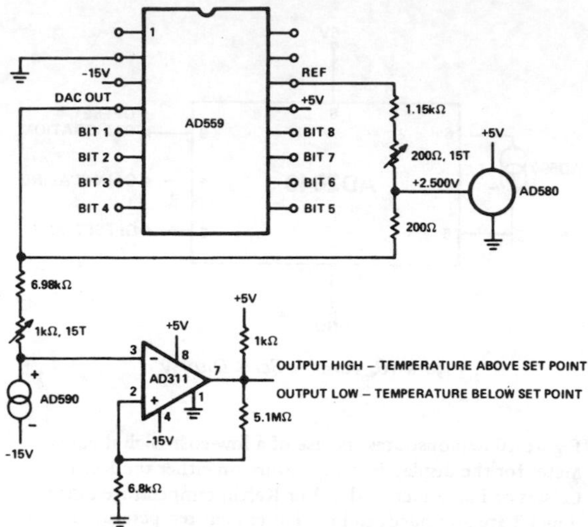

Figure 16. DAC Set Point

particular circuit operates from 0 (all inputs high) to +51°C (all inputs low) in 0.2°C steps. The comparator is shown with 1°C hysteresis which is usually necessary to guard-band for extraneous noise; omitting the 5.1MΩ resistor results in no hysteresis.

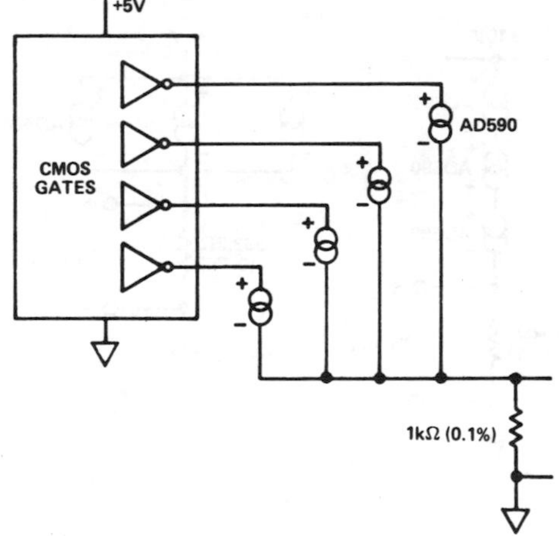

Figure 17. AD590 Driven from CMOS Logic

The voltage compliance and the reverse blocking characteristic of the AD590 allows it to be powered directly from +5V CMOS logic. This permits easy multiplexing, switching or pulsing for minimum internal heat dissipation. In Figure 17 any AD590 connected to a logic high will pass a signal current through the current measuring circuitry while those connected to a logic zero will pass insignificant current. The outputs used to drive the AD590's may be employed for other purposes, but the additional capacitance due to the AD590 should be taken into account.

8

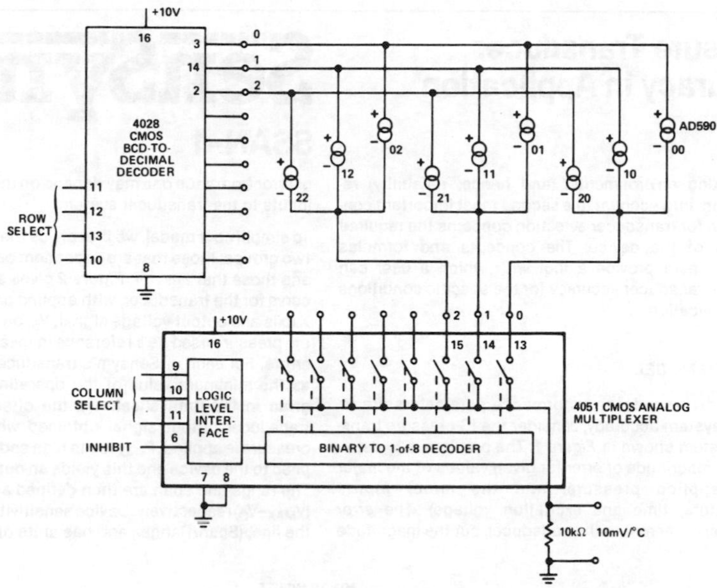

Figure 18. Matrix Multiplexer

CMOS Analog Multiplexers can also be used to switch AD590 current. Due to the AD590's current mode, the resistance of such switches is unimportant as long as 4V is maintained across the transducer. Figure 18 shows a circuit which combines the principal demonstrated in Figure 17 with an 8 channel CMOS Multiplexer. The resulting circuit can select one of eighty sensors over only 18 wires with a 7 bit binary word. The inhibit input on the multiplexer turns all sensors off for minimum dissipation while idling.

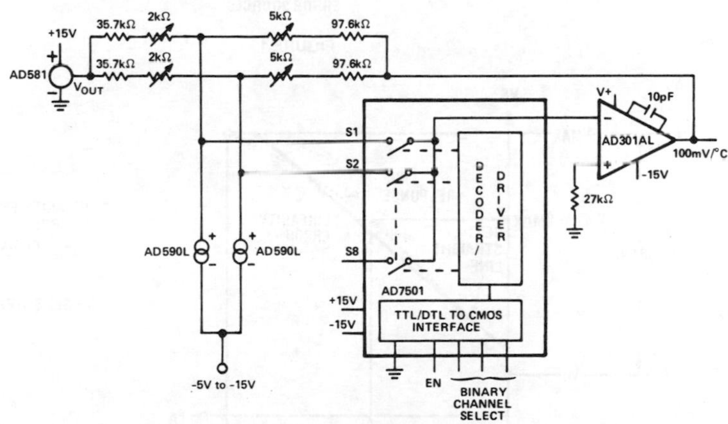

Figure 19. 8-Channel Multiplexer

Figure 19 demonstrates a method of multiplexing the AD590 in the two-trim mode (Figure 7). Additional AD590's and their associated resistors can be added to multiplex up to 8 channels of ±0.5°C absolute accuracy over the temperature range of –55°C to +125°C. The high temperature restriction of +125°C is due to the output range of the op amps; output to +150°C can be achieved by using a +20V supply for the op amp.

Pressure Transducer Accuracy in Application

SenSym

SSAN-1

After taking environmental (and hence, reliability) requirements into account, the second most important consideration for transducer selection concerns the required accuracy of the device. The concepts and formulas presented here provide a tool with which a user can calculate transducer accuracy for the specific conditions of his application.

A SYSTEM MODEL

To see how transducer performance parameters are related to system accuracy, consider the IC pressure transducer system shown in *Figure 1*. The problem is to determine the magnitude of error for given values of the *major input* (applied pressure) and the *minor inputs* (temperature, time and excitation voltage). The *error sources* are inherent to the transducer, but the magnitude of error from each one may depend on the major and minor inputs to the transducer system.

To simplify our model, we first divide the error sources into two groups: those that are dependent on applied pressure and those that are not. *Figure 2* gives a typical response curve for the transducer, with applied pressure, P_A, on the X-axis and output voltage signal, V_S, on the Y-axis. P_{REF} is the pressure used as a reference in measuring transducer errors. For each of Sensym's transducers, this is defined as the minimum value of the operating pressure range given in the data sheet. V_O, the offset voltage, is the transducer output signal obtained when the reference pressure is applied. P_{MAX} is the high endpoint pressure applied to the device and this yields an output voltage, V_{MAX}. The range and span are then defined as ($P_{MAX}-P_{REF}$) and ($V_{MAX}-V_O$) respectively. Device sensitivity, S, is the slope of the line, (Span/Range), and has units of volts/psi.

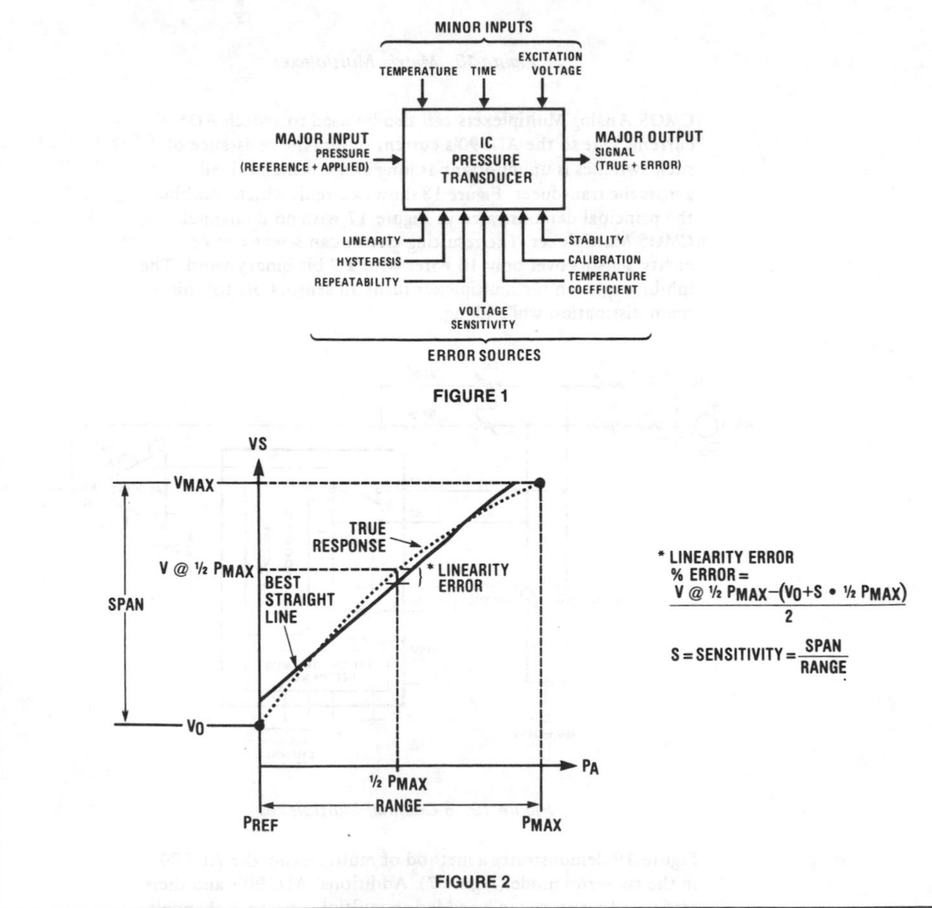

FIGURE 1

* LINEARITY ERROR

$$\%\ ERROR = \frac{V\ @\ \tfrac{1}{2} P_{MAX} - (V_O + S \cdot \tfrac{1}{2} P_{MAX})}{2}$$

$$S = SENSITIVITY = \frac{SPAN}{RANGE}$$

FIGURE 2

Because Sensym's transducers are inherently linear, the output signal can be given by:

$$V_S = V_O + S \bullet (P_A - P_{REF}) = V_O + \text{Span} \bullet \frac{(P_A - P_{REF})}{\text{Range}}$$

As a further result, the error in output signal, ΔV_S, can be expressed as:

$$\Delta V_S = \Delta V_O + \Delta \text{Span} \bullet \frac{(P_A - P_{REF})}{\text{Range}}$$

This equation shows that ΔV_O, the offset error, is independent of applied pressure while

$$\Delta \text{Span} \bullet \frac{(P_A - P_{REF})}{\text{Range}},$$

the span error, is proportional to the applied pressure range, $(P_A - P_{REF})$.

Offset errors, being independent of the major input variable (applied pressure), are equivalent to system *common-mode* errors, as shown in *Figure 3.* Because the offset error is the same regardless of pressure, it has the effect of translating the response line up or down, while the slope or sensitivity remains constant.

Span errors, being proportional to applied pressure, are equivalent to system *normal-mode* errors, as shown in *Figure 4.* Because the span error increases linearly with applied pressure, it has the effect of rotating the response line around the offset-reference pressure point.

While independent, the *offset* and *span* error groups both contain errors that are dependent on the minor input variables, as shown in Table I. These coefficients are used to specify the errors in Sensym's pressure transducers and to calculate overall accuracy.

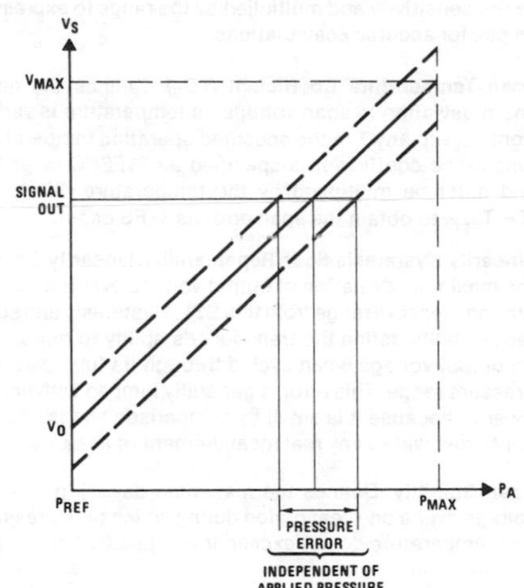

FIGURE 3

TABLE I. OFFSET AND SPAN ERRORS

OFFSET (Common-Mode)	SPAN (Normal-Mode)
Calibration	Calibration
Repeatability	Linearity-Hysteresis-Repeatability
Stability	Stability
Temperature Coefficient	Temperature Coefficient
Excitation Voltage Coefficient	Excitation Voltage Coefficient

SYSTEM ACCURACY

With the errors divided into 2 groups of independent coefficients, we can now compute both the worst-case error and the most probable error for any IC pressure transducer system.

Worst-Case Error: The worst-case overall error ϵ_{WC} is obtained by simple addition of all applicable errors:

$$\epsilon_{WC} = \sum_{1}^{n} \epsilon_j$$

where ϵ_j is the error resulting from the jth error coefficient and n is the number of error terms included in the calculation.

Most Probable Error: The most probable error ϵ_{MP} is obtained by computing the square root of the sum of the squares:

$$\epsilon_{MP} = \sqrt{\sum_{1}^{n} \epsilon_j^2}$$

We can now select the applicable error coefficients, calculate the error terms, ϵ_j, from the specifications given for any individual pressure transducer, and plug into the appropriate formula above to get system accuracy.

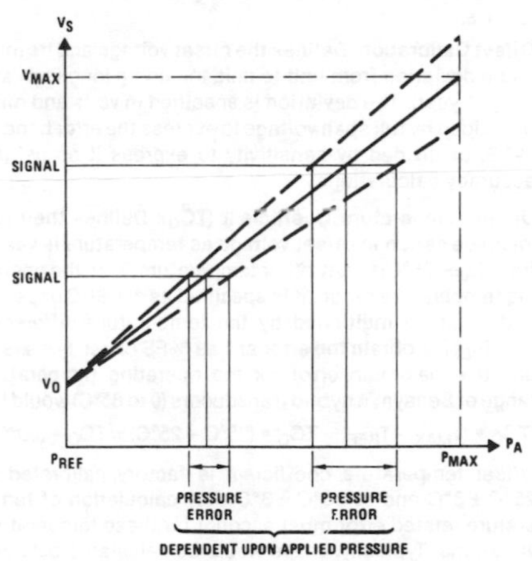

FIGURE 4

ACCURACY SPECIFICATIONS

By convention, system accuracy is expressed in the dimensions of the major input variable, in this case, psi. However, transducer accuracy is typically expressed as percent of full span (%FS). Fortunately, the transposition from one dimension to the other is analytically simple. An error coefficient expressed as %FS is changed to psi by multiplying by the range, $(P_{MAX} - P_{REF})$, of the device under consideration and dividing by 100. So the user can perform accuracy calculations in either dimension, both %FS and psi error values are given in the data sheets for linearity, hysteresis, repeatability, stability, and temperature coefficient.

Three additional points should be noted relative to the data sheet accuracy specifications. Offset calibration error is given directly as volts, since this is the parameter most users actually measure. To convert from volts to %FS, divide offset calibration error by the span voltage (typically 10V for signal-conditioned devices) and multiply by 100. If psi is desired, divide offset calibration error by the device sensitivity (mV/psi) and multiply by 1000 (mV/V). In a similar fashion, sensitivity calibration error is given directly as mV/psi. To convert to %FS, divide by the device sensitivity and multiply by 100. (Because these are linear devices, the ratio of sensitivity error to sensitivity is the same as the ratio of span error to span.) To get psi, divide sensitivity calibration by the sensitivity and multiply by the range.

Finally, supply voltage coefficient (voltage regulation error) is given directly as percentage of supply voltage change. Conversion to psi from %FS would use the same technique discussed in the first paragraph of this page.

OFFSET SPECIFICATIONS

The offset characteristics are measured at reference temperature with reference pressure applied. Although measured at the reference pressure, offset errors given in the data sheet, ΔV_O, are the same regardless of pressure and should be used in the accuracy formulas without any modification for user pressure range. They are defined as follows:

Offset Calibration: Defines the offset voltage and its maximum deviation from unit to unit, including long-term stability (1 year). The deviation is specified in volts and must be divided by full span voltage to express the error band as %FS, or divided by sensitivity to express it as psi, for accuracy calculations.

Offset Temperature Coefficient (TC_O): Defines the maximum deviation in offset voltage as temperature is varied from T_{REF} (25°C) to any other temperature, T, in the operating temperature range. It is specified as %FS/°C or psi/°C and must be multiplied by the temperature difference $|T - T_{REF}|$ to obtain the error at T as %FS or psi. For example, the maximum error for the operating temperature range of Sensym's hybrid transducers (0 to 85°C) would be:

$$TC_O \bullet (T_{MAX} - T_{REF}) = TC_O \bullet (85°C - 25°C) = TC_O \bullet (60°C).$$

Offset temperature coefficient is factory calibrated at 25°C ± 3°C and at 80°C ± 3°C. Any calculation of temperature related error must account for these temperature variations. Typically, errors would be calculated between two points that are at least 15°C apart.

Offset Repeatability: Defines the maximum deviation in offset voltage when applied pressure is cycled through its full range.

Offset Stability: Defines the maximum deviation in offset voltage over a one year period, during which time the pressure and temperature do not exceed their specified maximum ratings.

SPAN SPECIFICATIONS

Full span corresponds to the entire operating pressure range, (P_{REF} to P_{MAX}), specified on the data sheet for each device type. This yields a span voltage, measured at the reference temperature, equal to $(V_{MAX} - V_O)$. If an application utilizes a transducer's full operating pressure range, then the span error values given in the data sheet, ΔSpan, can be plugged directly into the error formulas to determine system accuracy. However, if only part of the range is used, the data sheet span errors must be reduced proportionally, since they are a linear function of applied pressure. This is accomplished by taking the actual pressure range used in the application, dividing by the range of the device, and then multiplying each of the data sheet span errors by this ratio (which is a number between 0 and 1). Note that although application span error,

$$\Delta Span \bullet \frac{(P_A - P_{REF})}{Range},$$

is defined to include P_{REF}, this is not a user requirement. Span error is simply

$$\Delta Span \bullet \frac{User\ Range}{Range}$$

for any user range. The data sheet span errors, ΔSpan, are specified as follows:

Sensitivity Calibration: Sensitivity is defined as span divided by the range, $(V_{MAX} - V_O)/(P_{MAX} - P_{REF})$. Sensitivity calibration defines the maximum deviation of sensitivity from unit to unit, including long term span stability (1 year). The deviation is specified as mV/psi and must be divided by the sensitivity to express the error as %FS, or divided by the sensitivity and multiplied by the range to express it as psi, for accuracy calculations.

Span Temperature Coefficient (TC_S): Defines the maximum deviation in span voltage as temperature is varied from T_{REF} to any T in the specified operating temperature range. The coefficient is specified as %FS/°C or psi/°C and must be multiplied by the temperature difference $|T - T_{REF}|$ to obtain the span error as %FS or psi.

Linearity-Hysteresis-Span Repeatability: Linearity defines the maximum deviation of output voltage over the full operating pressure range from this BSL. Hysteresis and span repeatability define the transducer's ability to reproduce an output voltage when cycled through its full operating pressure range. This error is generally lumped with linearity error because it is small by comparison and is usually contained within any real measurement of linearity.

Span Stability: Defines the maximum deviation in span voltage over a one year period during which time pressure and temperature do not exceed their specified maximum ratings.

SYSTEM ACCURACY CALCULATIONS

Voltage Regulation

In the example calculations that follow, we will assume that user excitation voltage is sufficiently regulated so as to make the voltage regulation error, ϵ_{VR}, negligible ($\leq$0.1%FS). ϵ_{VR} is an output signal change due solely to a change in excitation voltage. The percent regulation required to satisfy this condition is derived as follows. For signal-conditioned devices, ϵ_{VR} is given by:

$$\epsilon_{VR} = 0.5\% \bullet \Delta V_e$$

where 0.5% is the specified transducer output voltage change to excitation voltage change and ΔV_e is the excitation voltage deviation [from nominal (15V)]. To keep the regulation error below 0.1%FS, the required external power supply regulation is given by:

$$\frac{\Delta V_e}{V_e} = \left(\frac{1}{0.5\%}\right) \bullet \left(\frac{\epsilon_{VR}}{V_e}\right) = 200 \bullet \left(\frac{\epsilon_{VR}}{Span}\right) \bullet \left(\frac{Span}{V_e}\right)$$

Since $\frac{\epsilon_{VR}}{Span} = 0.1\%$,

$$\frac{\Delta V_e}{V_e} = 20\% \bullet \left(\frac{Span}{V_e}\right)$$

For Span = 10V and V_e = 15V,

$$\frac{\Delta V_e}{V_e} = 20\% \bullet \left(\frac{2}{3}\right) = \pm 13\% \text{ Regulation}$$

which holds for any signal-conditioned pressure transducer.

For monolithics, which do not have any internal regulation or signal conditioning, the formulation is not quite as neat since output characteristics vary from device to device. However, as an approximation, it can be assumed that output signal changes are roughly proportional to changes in excitation voltage. Therefore, to ensure $\epsilon_{VR} \leq 0.1\%$FS would require a power supply with $\leq$0.1% regulation.

If the signal conditioning or monolithic regulation requirements calculated above are met, then regulation error can be eliminated from essentially all error calculations, with the possible exception of ultra-high accuracy applications. However, the greater the deviation from these requirements, the more necessary it becomes to include ϵ_{VR} in the accuracy calculations. Since both ΔV_O and ΔSpan are affected, ϵ_{VR} would be included in both segments of the error calculation.

Interchangeable vs. Calibrated Accuracy

Interchangeable Accuracy: In calculating overall accuracy, the first question is whether each pressure transducer will be field calibrated upon installation or replacement. If you're going to just plug it in with no adjustments, you'll need the *interchangeable accuracy*, which allows for unit-to-unit calibration errors. In this case, you include Sensym's calibration errors but exclude stability error (the specified calibration error includes both calibration and stability errors). ϵ_I, the overall error allowing for direct exchange of transducers of the same type, includes calibration, TC, linearity, hysteresis, and repeatability errors.

Calibrated Accuracy: If you intend to calibrate each device upon installation, you will want to use the *calibrated accuracy*, ϵ_C, which holds only for one specific transducer. The calibrated overall accuracy excludes Sensym's calibration errors, but includes all other applicable specified errors including stability, TC, linearity, hysteresis, and repeatability.

Example Calculations

The LX1604D is chosen to show how error calculations would be performed for a typical pressure transducer under various conditions. Analogous procedures apply to any Sensym IC pressure transducer and can be extended for use in evaluating errors in a complete pressure system.

Table II is a reproduction of the applicable LX1604D data on page 5-5. The LX1604D operating pressure range is −15 psid to +15 psid. Therefore, $P_{REF} = -15$ psid, $P_{MAX} = +15$ psid, and Range $= (P_{MAX} - P_{REF}) = 30$ psid. V_O (at −15 psid) = 2.5V (from the offset calibration column) and $V_{MAX} = V_O + S \bullet (P_{MAX} - P_{REF}) = 2.5 + (0.333)(30) = 12.5$V (where the sensitivity value is obtained from the sensitivity calibration column and converted into V/psi). Therefore, Span $= V_{MAX} - V_O = 10$V.

To be consistent with Table I, the data divides the error terms into two categories: those for offset, ΔV_O, and those for span, ΔSpan. Table II further identifies each component of error as follows: ΔV_{O1} is the offset calibration error, ΔV_{O2} is the offset temperature coefficient error, ΔV_{O3} is the offset repeatability error, and ΔV_{O4} is the offset stability error. Likewise, sensitivity (and thus span) calibration error is $\Delta Span_1$, $\Delta Span_2$ is the span temperature coefficient error, $\Delta Span_3$ is the combined linearity, hysteresis, and repeatability error, and $\Delta Span_4$ is the span stability error. As mentioned previously, where appropriate, both %FS and psi errors are included.

The following calculations are performed using %FS error values. However, completely analogous results would be obtained using psi errors (psi results for each calculation are included for reference purposes).

Maximum Error—Case I

The maximum possible error would occur for the case where the full temperature and pressure ranges are used. Under these temperature conditions, each temperature coefficient is converted to %FS by multiplying the data sheet errors by $(T_{MAX} - T_{REF}) = (85°C - 25°C) = 60°C$. Then:

$$\Delta V_{O2} = 0.03 \times 60 = 1.8\%FS$$

and

$$\Delta Span_2 = 0.03 \times 60 = 1.8\%FS$$

Since the full pressure range is being used, it is not necessary to decrease any of the span errors proportionally. The %FS table values would be plugged directly into the accuracy formulas.

The only conversion remaining is to change calibration errors in the data to %FS. Offset calibration is converted by dividing by span voltage (10V) and multiplying by 100 while sensitivity calibration is converted by dividing by sensitivity (333 mV/psi) and multiplying by 100.

$$\Delta V_{O1} = \frac{100(0.35)}{10} = 3.5\%FS$$

and

$$\Delta Span_1 = \frac{100(6.7)}{333} = 2\%FS$$

Having all %FS values, it is now possible to calculate ϵ_I, interchangeable overall error, and ϵ_c, calibrated overall error, worst-case and most probable error values.

TABLE II. LX1604D PRESSURE TRANSDUCER SPECIFICATIONS

Offset Characteristics						
Offset Calibration V ± ΔV_{O1}	Temperature Coefficient ΔV_{O2}		Repeatability ΔV_{O3}		Stability ΔV_{O4}	
	± %FS/°C	± psi/°C	± %FS	± psi	± %FS	± psi
2.5 ± 0.35	0.03	0.009	0.4	0.12	1.7	0.5
Span Characteristics						
Sensitivity Calibration mV/psi $\Delta Span_1$	Temperature Coefficient $\Delta Span_2$		Linearity Hysteresis & Repeatability $\Delta Span_3$		Stability $\Delta Span_4$	
	± %FS/°C	± psi/°C	± %FS	± psi	± %FS	± psi
333 ± 6.7	0.03	0.009	0.67	0.20	0.3	0.1

For interchangeable overall error, offset stability, ΔV_{O4}, and span stability, $\Delta Span_4$, are eliminated. The remaining offset and span errors are plugged into the ϵ_{WC} and ϵ_{MP} formulas to yield

Worst-case: ϵ_{WCI}

$$= \Delta V_{O1} + \Delta V_{O2} + \Delta V_{O3} + \Delta Span_1 + \Delta Span_2 + \Delta Span_3$$

$$= \underbrace{(3.5 + 1.8 + 0.4)}_{\text{Offset}} + \underbrace{(2 + 1.8 + 0.67)}_{\text{Span}} = \pm 10.17\%FS$$

Most probable: ϵ_{MPI}

$$= \sqrt{\Delta V_{O1}^2 + \Delta V_{O2}^2 + \Delta V_{O3}^2 + \Delta Span_1^2 + \Delta Span_2^2 + \Delta Span_3^2}$$

$$= \sqrt{\underbrace{3.5^2 + 1.8^2 + 0.4^2}_{\text{Offset}} + \underbrace{2^2 + 1.8^2 + 0.67^2}_{\text{Span}}} = \pm 4.83\%FS$$

This corresponds to ± 3.05 psid and ± 1.45 psid respectively.

For calibrated overall error, both calibration errors, ΔV_{O1} and $\Delta Span_1$, are eliminated from the above calculation and the two stability errors, ΔV_{O4} and $\Delta Span_4$, are inserted.

Worst-case: ϵ_{WCC}

$$= \Delta V_{O2} + \Delta V_{O3} + \Delta V_{O4} + \Delta Span_2 + \Delta Span_3 + \Delta Span_4$$

$$= \underbrace{(1.8 + 0.4 + 1.7)}_{\text{Offset}} + \underbrace{(1.8 + 0.67 + 0.3)}_{\text{Span}} = \pm 6.67\%FS$$

Most probable: ϵ_{MPC}

$$= \sqrt{\Delta V_{O2}^2 + \Delta V_{O3}^2 + \Delta V_{O4}^2 + \Delta Span_2^2 + \Delta Span_3^2 + \Delta Span_4^2}$$

$$= \sqrt{\underbrace{1.8^2 + 0.4^2 + 1.7^2}_{\text{Offset}} + \underbrace{1.8^2 + 0.67^2 + 0.3^2}_{\text{Span}}} = \pm 3.17\%FS$$

This corresponds to ± 2.00 psid and ± 0.95 psid respectively.

Reducing Temperature Errors—Case II

Since the temperature coefficients are two of the main error components, a reduced temperature range can greatly reduce overall error. For 80% effective temperature compensation (reducing effective range from 60°C to 12°C), the offset and span temperature coefficients would be

$$\Delta V_{O2} = 0.03 \times 12 = 0.36\%FS$$

and

$$\Delta Span_2 = 0.03 \times 12 = 0.36\%FS$$

The interchangeable overall errors would then be reduced to

Worst-case: ϵ_{WCI}

$$= (3.5 + \underline{0.36} + 0.4) + (2 + \underline{0.36} + 0.67) = \pm 7.29\%FS$$

(Reduced TC Errors)

Most probable: ϵ_{MPI}

$$= \sqrt{3.5^2 + 0.36^2 + 0.4^2 + 2^2 + 0.36^2 + 0.67^2}$$

$$= \pm 4.14\%FS$$

This corresponds to ± 2.19 psid and ± 1.24 psid respectively, reduced from ± 3.05 psid and ± 1.45 psid in Case I.

A corresponding improvement is achieved for the calibrated accuracy:

Worst-case: ϵ_{WCC}

$$= (\underline{0.36} + 0.4 + 1.7) + (\underline{0.36} + 0.67 + 0.3) = \pm 3.79\%FS$$

(Reduced TC Errors)

Most probable: ϵ_{MPC}

$$= \sqrt{0.36^2 + 0.4^2 + 1.7^2 + 0.36^2 + 0.67^2 + 0.3^2}$$

$$= \pm 1.96\%FS$$

This corresponds to ± 1.14 psid and ± 0.59 psid respectively, reduced from ± 2.00 psid and ± 0.95 psid by 80% effective temperature compensation.

Reduced Pressure Range—Case III

When the full specified pressure range of a particular device is not being used, all span errors should be reduced by the ratio R, where R is defined as:

$$R = \frac{\text{User Range}}{\text{Device Specified Range}}$$

Assume, for example, that the user application is for +5 psid to + 15 psid. Then R = 10 psid/30 psid = 0.333, and each application span error would be

$$\Delta Span_1 = 2 \times 0.333 = \pm 0.67\%FS$$

$$\Delta Span_2 = 0.03 \times 0.333 = \pm 0.01\%FS/°C$$

$$\Delta Span_3 = 0.67 \times 0.333 = \pm 0.22\%FS$$

$$\Delta Span_4 = 0.3 \times 0.333 = \pm 0.1\%FS$$

If Case II conditions above are maintained, then the interchangeable overall errors would now be reduced to

Worst-case: ϵ_{WCI}

$$= (3.5 + 0.36 + 0.4) + \underbrace{(0.67 + 0.12 + 0.22)}_{\text{Reduced Span Errors}} = \pm 5.27\%FS$$

Most probable: ϵ_{MPI}

$$= \sqrt{3.5^2 + 0.36^2 + 0.4^2 + 0.67^2 + 0.12^2 + 0.22^2}$$

$$= \pm 3.61\%FS$$

This corresponds to ± 1.58 psid and ± 1.08 psid respectively, reduced from ± 2.19 psid and ± 1.24 psid in Case II.

NOTE: The reduced pressure range decreases span error values only. Offset errors remain unchanged!

For calibrated overall error

Worst-case: ϵ_{WCC}

$$= (0.36 + 0.4 + 1.7) + \underbrace{(0.12 + 0.22 + 0.1)}_{\text{Reduced Span Errors}} = \pm 2.90\%FS$$

Most probable: ϵ_{MPC}

$$= \sqrt{0.36^2 + 0.4^2 + 1.7^2 + 0.12^2 + 0.22^2 + 0.1^2}$$

$$= \pm 1.80\%FS$$

This corresponds to ± 0.87 psid and ± 0.54 psid, respectively, reduced from ± 1.14 psid and ± 0.59 psid in Case II, by using 1/3 of the LX1604D pressure range.

Auto-Reference Compensation—Case IV

A powerful, easy-to-use, and generally applicable technique, auto-referencing, can often eliminate all offset errors by period sampling of the offset voltage at reference pressure. With this technique, (see Section 6), only the span errors apply. Again, using the LX1604D specifications and assuming all Case III conditions hold, the interchangeable accuracy is:

Worst-case: ϵ_{WCI}

$$= \underbrace{(0.67 + 0.12 + 0.22)}_{\text{Span Errors}} = \pm 1.01\%FS$$

Most probable: ϵ_{MPI}

$$= \sqrt{0.67^2 + 0.12^2 + 0.22^2} = \pm 0.72\%FS$$

This corresponds to ± 0.30 psid and ± 0.22 psid, respectively, reduced from ± 1.58 psid and ± 1.08 psid in Case III.

Calibrated overall error would be

Worst-case: ϵ_{WCC}

$$= \underbrace{(0.12 + 0.22 + 0.1)}_{\text{Span Errors}} = \pm 0.44\%FS$$

Most probable: ϵ_{MPC}

$$= \sqrt{0.12^2 + 0.22^2 + 0.1^2} = \pm 0.27\%FS$$

This corresponds to ± 0.13 psid and ± 0.08 psid, respectively, reduced from ± 0.87 psid and ± 0.54 psid in Case III.

Auto-Reference + Temperature Control—Case V

For very high accuracy applications, both auto-referencing and complete temperature range reduction may prove valuable. In these cases, the additional temperature compensation may take the form of a temperature-controlled chamber designed to hold temperature within a few degrees of T_{REF} (which may be shifted to a higher temperature to allow use of an oven). In such a case, the only errors included are linearity, hysteresis, span repeatability and either span calibration or span stability. For interchangeable accuracy, span calibration error is included:

Worst-case: ϵ_{WCI}

$$= \underbrace{(0.67 + 0.22)}_{\text{Span Errors without TC}} = \pm 0.89\%FS$$

Most probable: ϵ_{MPI}

$$= \sqrt{0.67^2 + 0.22^2} = \pm 0.71\%FS$$

This corresponds to ± 0.27 psid and ± 0.21 psid respectively.

For calibrated accuracy, span stability error is including:

$$\text{Worst-case: } \epsilon_{WCC} = \underbrace{(0.22 + 0.1)}_{\text{Span Errors without TC}} = \pm 0.32\%FS$$

$$\text{Most probable: } \epsilon_{MPC} = \sqrt{0.22^2 + 0.1^2} = \pm 0.24\%FS$$

This corresponds to ± 0.10 psid and ± 0.07 psid respectively.

Periodic Span Calibration—Case VI

With overall error down to a fraction of a psi, resulting from auto-referencing and temperature control, the periodic recalibration of span may become worthwhile. Since the span stability error is a slowly aging variation of span voltage, a periodic recalibration may well reduce this error by an order of magnitude. This procedure eliminates calibration error, so only calibrated accuracy applies.

$$\text{Worst-case: } \epsilon_{WCC} = (0.22 + \underbrace{0.01}_{\text{Reduced Stability Error}}) = \pm 0.23\%FS$$

$$\text{Most probable: } \epsilon_{MPC} = \sqrt{0.22^2 + 0.01^2} = \pm 0.22\%FS$$

This corresponds to ± 0.07 psid in both cases.

Linearity Compensation—Case VII

For ultra-high accuracy applications, the remaining error, linearity-hysteresis-span repeatability, must be reckoned with. The hysteresis and repeatability components of this coefficient are so small as to approach the noise in the operational amplifier included in the signal-conditioned IC pressure transducer. This noise is about 0.4%FS for a 1kHz bandwidth and may require narrow bank filter techniques if ultra-high accuracy is to be achieved. We do know, however, that the linearity error is a large fraction of the remaining error, perhaps as high as 90%, and that it can be successfully compensated via curve-fitting techniques to reduce overall calibrated error to about ±0.1%FS, worst-case.

SUMMING IT UP

The foregoing offers the reader a quantitative technique for separating transducer errors and evaluating their contribution to system accuracy, as well as describing several methods for system optimization. It is hoped that this will encourage the transducer user to take a closer look at system requirements as they relate to each of the parameters and optimization techniques discussed, thereby allowing him to make optimum accuracy/cost design tradeoffs. Too often, transducer application specifications are unnecessarily tight, simply because an analysis similar to the one performed above has not been done. The result of this over-specification is unnecessary cost.

SCX C Series
O to 1 psi to 0 to 150 psi
Low Cost Compensated Pressure Sensors

FEATURES
- Low Cost
- Temperature Compensation
- Calibrated Zero and Span
- Small Size
- Low Noise
- High Impedance for Low Power Applications

APPLICATIONS
- Medical Equipment
- Computer Peripherals
- Pneumatic Controls
- HVAC

EQUIVALENT CIRCUIT

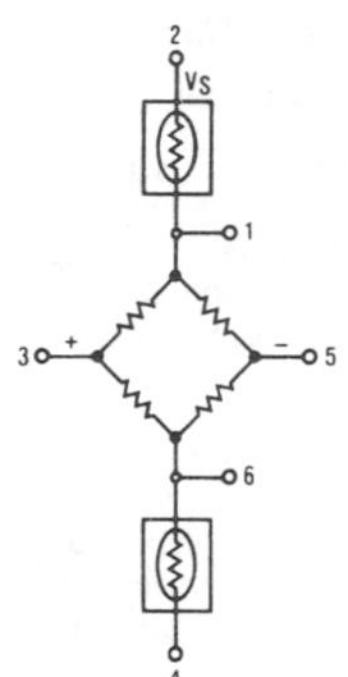

ELECTRICAL CONNECTION

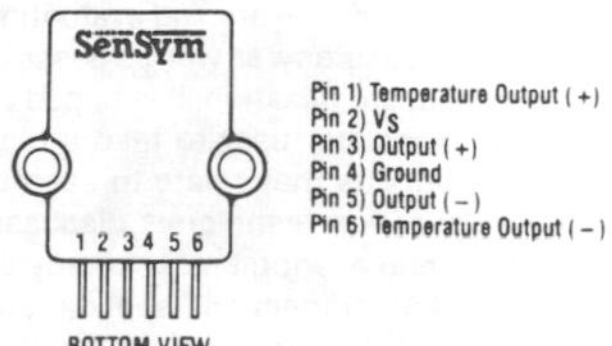

Note: The polarity indicated is for pressure applied to port B. (For Absolute devices, pressure is applied to port A and the output polarity is reversed.)

GENERAL DESCRIPTION

The SCX C series sensors will provide a very cost effective solution for pressure applications that require operation over a wide temperature range. These internally calibrated and temperature compensated sensors give an accurate and stable output over a 0°C to 70°C temperature range. This series is intended for use with non-corrosive, non-ionic working fluids such as air, dry gases, and the like.

Devices are available to measure absolute, differential, and gage pressures from 1psi (SCX01) up to 150psi (SCX150). The Absolute (A) devices have an internal vacuum reference and an output voltage proportional to absolute pressure. The Differential (D) devices allow application of pressure to either side of the pressure sensing diaphragm and can be used for gage or differential pressure measurements.

The SCX devices feature an integrated circuit sensor element and laser trimmed thick film ceramic housed in a compact nylon case. This package provides excellent corrosion resistance and provides isolation to external package stresses. The package has convenient mounting holes and pressure ports for ease of use with standard plastic tubing for pressure connection.

All SCX devices are calibrated for span to within ±5% and provide an offset (zero pressure output) of ±1 millivolt maximum. These parts were designed for low cost applications where the user can typically provide fine adjustment of zero and span in external circuitry. For higher accuracies, refer to the standard SCX series datasheet. If the application requires extended temperature range operation, beyond 0 to 70°C, two pins which provide an output voltage proportional to temperature are available for use with external circuitry.

The output of the bridge is ratiometric to the supply voltage and operation from any D.C. supply voltage up to +30V is acceptable.

Because these devices have very low noise and 100 microsecond response time they are an excellent choice for medical equipment, computer peripherals, and pneumatic control applications.

For further technical information on the SCX series, please contact your local Sensym office or the factory.

PRESSURE SENSOR CHARACTERISTICS — SCX C Series

STANDARD PRESSURE RANGES

PART NUMBER	OPERATING PRESSURE	PROOF PRESSURE*	FULL-SCALE SPAN
SCX01DN	0-1 psid	20 psid	18 mV
SCX05DN	0-5 psid	20 psid	60 mV
SCX15AN	0-15 pisa	30 psia	90 mV
SCX15DN	0-15 psid	30 psid	90 mV
SCX30AN	0-30 pisa	60 psia	90 mV
SCX30DN	0-30 psid	60 psid	90 mV
SCX100AN	0-100 pisa	150 psia	100 mV
SCX100DN	0-100 psid	150 psid	100 mV
SCX150DN	0-150 psid	200 psid	60 mV

*Maximum pressure above which causes permanent sensor failure.

Maximum Ratings (For All Devices)

Supply Voltage, V_S	$+30 V_{DC}$
Common-mode Pressure	50psig
Lead Temperature (Soldering, 10 seconds)	300°C

Environmental Specifications (For All Devices)

Temperature Range	
Compensated	0 to 70°C
Operating	−40°C to +85°C
Storage	−55°C to +125°C
Humidity Limits	0 to 100% RH

SCX01DNC PERFORMANCE CHARACTERISTICS (Note 1)

CHARACTERISTIC	MIN	TYP	MAX	UNIT
Operating Pressure Range		—	1	psid
Sensitivity	—	18	—	mV/psi
Full-scale Span (Note 2)	17.00	18.00	19.00	mV
Zero Pressure Offset	−1.0	0	+1.0	mV
Combined Linearity and Hysteresis (Note 3)	—	0.2	1.0	%FSO
Temperature Effect on Span (0-70°C) (Note 4)	—	0.4	2.0	%FSO
Temperature Effect on Offset (0-70°C) (Note 4)	—	±0.20	±1.0	mV
Repeatability (Note 5)	—	0.2	0.5	%FSO
Input Impedance (Note 6)	—	4.0	—	kΩ
Output Impedance (Note 7)	—	4.0	—	kΩ
Common-mode Voltage (Note 8)	5.7	6.0	6.3	V_{DC}
Response Time (Note 9)	—	100	—	• μsec
Long Term Stability of Offset and Span (Note 10)	—	0.1	—	%FSO

Specification Notes: (For All Devices)

Note 1: Reference Conditions: Unless otherwise noted: Supply Voltage, $V_S = 12 V_{DC}$, $T_A = 25°C$, Common-mode Line Pressure = 0 psig, Pressure Applied to Port B. For absolute devices only, pressure is applied to Port A and the output polarity is reserved.

Note 2: Span is the algebraic difference between the output voltage at full-scale pressure and the output at zero pressure. Span is ratiometric to the supply voltage.

Note 3: See Definition of Terms.
Hysteresis — the maximum output difference at any point within the operating pressure range for increasing and decreasing pressure.

Note 4: Maximum error band of the offset voltage and the error band of the span, relative to the 25°C reading.

Note 5: Maximum difference in output at any pressure with the operating pressure range and temperature within 0°C to + 70°C after:
a) 1,000 temperature cycles, 0°C to + 70°C
b) 1.5 million pressure cycles, 0 psi to full-scale span.

PRESSURE SENSOR CHARACTERISTICS (Cont.) SCX C Series

SCX05DNC PERFORMANCE CHARACTERISTICS (Note 1)

CHARACTERISTIC	MIN	TYP	MAX	UNIT
Operating Pressure Range		—	5	psid
Sensitivity	—	12.0	—	mV/psi
Full-scale Span (Note 2)	57.5	60.0	62.5	mV
Zero Pressure Offset	−1.0	0	+1.0	mV
Combined Linearity and Hysteresis (Note 3)	—	0.1	1.0	%FSO
Temperature Effect on Span (0–70°C) (Note 4)	—	0.4	2.0	%FSO
Temperature Effect on Offset (0–70°C) (Note 4)	—	±0.20	±1.0	mV
Repeatability (Note 5)	—	0.2	0.5	%FSO
Input Impedance (Note 6)	—	4.0	—	kΩ
Output Impedance (Note 7)	—	4.0	—	kΩ
Common-mode Voltage (Note 8)	5.7	6.0	6.3	V_{DC}
Response Time (Note 9)	—	100	—	μsec
Long Term Stability of Offset and Span (Note 10)	—	0.1	—	%FSO

SCX15C PERFORMANCE CHARACTERISTICS (Note 1)

CHARACTERISTIC	MIN	TYP	MAX	UNIT
Operating Pressure Range		—	15	psi
Sensitivity	—	6.0	—	mV/psi
Full-scale Span (Note 2)	85.0	90.0	95.0	mV
Zero Pressure Offset	−1.0	0	+1.0	mV
Combined Linearity and Hysteresis (Note 3)	—	0.1	1.0	%FSO
Temperature Effect on Span (0–70°C) (Note 4)	—	0.4	2.0	%FSO
Temperature Effect on Offset (0–70°C) (Note 4)	—	±0.20	±1.0	mV
Repeatability (Note 5)	—	0.2	0.5	%FSO
Input Impedance (Note 6)	—	4.0	—	kΩ
Output Impedance (Note 7)	—	4.0	—	kΩ
Common-mode Voltage (Note 8)	5.7	6.0	6.3	V_{DC}
Response Time (Note 9)	—	100	—	μsec
Long Term Stability of Offset and Span (Note 10)	—	0.1	—	%FSO

Specification Notes: (For All Devices)

Note 6: Input impedance is the impedance between pins 2 and 4.

Note 7: Output impedance is the impedance between pins 3 and 5.

Note 8: This is the common-mode voltage of the output arms (Pins 3 and 5) for $V_S = 12\ V_{DC}$.

Note 9: Response time for a 0 psi to full-scale span pressure step change, 10% to 90% rise time.

Note 10: Long term stability over a one year period.

PRESSURE SENSOR CHARACTERISTICS (Cont.) SCX C Series

SCX30C PERFORMANCE CHARACTERISTICS (Note 1)

CHARACTERISTIC	MIN	TYP	MAX	UNIT
Operating Pressure Range		—	30	psi
Sensitivity	—	3.0	—	mV/psi
Full-scale Span (Note 2)	85.0	90.0	95.0	mV
Zero Pressure Offset	−1.0	0	+1.0	mV
Combined Linearity and Hysteresis (Note 3)	—	±0.2	±1.0	%FSO
Temperature Effect on Span (0–70°C) (Note 4)	—	±0.4	±2.0	%FSO
Temperature Effect on Offset (0–70°C) (Note 4)	—	±0.2	±1.0	mV
Repeatability (Note 5)	—	±0.2	±0.5	%FSO
Input Impedance (Note 6)	—	4.0	—	kΩ
Output Impedance (Note 7)	—	4.0	—	kΩ
Common-mode Voltage (Note 8)	5.7	6.0	6.3	V_{DC}
Response Time (Note 9)	—	100	—	μsec
Long Term Stability of Offset and Span (Note 10)	—	±0.1	—	%FSO

SCX100C PERFORMANCE CHARACTERISTICS (Note 1)

CHARACTERISTIC	MIN	TYP	MAX	UNIT
Operating Pressure Range		—	100	psi
Sensitivity	—	1.0	—	mV/psi
Full-scale Span (Note 2)	95.0	100.0	105.0	mV
Zero Pressure Offset	−1.0	0	+1.0	mV
Combined Linearity and Hysteresis (Note 3)	—	±0.2	±1.0	%FSO
Temperature Effect on Span (0–70°C) (Note 4)	—	±0.4	±2.0	%FSO
Temperature Effect on Offset (0–70°C) (Note 4)	—	±0.20	±1.0	mV
Repeatability (Note 5)	—	±0.2	±0.5	%FSO
Input Impedance (Note 6)	—	4.0	—	kΩ
Output Impedance (Note 7)	—	4.0	—	kΩ
Common-mode Voltage (Note 8)	5.7	6.0	6.3	V_{DC}
Response Time (Note 9)	—	100	—	μsec
Long Term Stability of Offset and Span (Note 10)	—	±0.1	—	%FSO

SCX150C PERFORMANCE CHARACTERISTICS (Note 1)

CHARACTERISTIC	MIN.	TYP	MAX.	UNIT
Operating Pressure Range	---	---	150	psi
Sensitivity	---	0.6	---	mV/psi
Full-Scale Span (Note 2)	85.0	90.0	95.0	mV
Zero Pressure Offset	-1.0	0	+1.0	µV
Combined Linearity and Hysteresis (Note 3)	---	0.1	1.0	% FSO
Temperature Effect on Span (0-70ºC) (Note 4)	---	0.4	2.0	%FSO
Temperature Effect on Offset (0-70ºC) Note 4)	---	+/-0.20	+/-1.0	µV
Repeatability (Note 5)	---	0.2	0.5	%FSO
Input Impedance (Note 6)	---	4.0	---	kΩ
Output Impedance (Note 7)	---	4.0	---	kΩ
Common-mode Voltage (Note 8)	5.7	6.0	6.3	V_{DC}
Response Time (Note 9)	---	100	---	µsec
Long Term Stability of Offset and Span (Note 10)	---	0.1	---	%FSO

TYPICAL PERFORMANCE CHARACTERISTICS

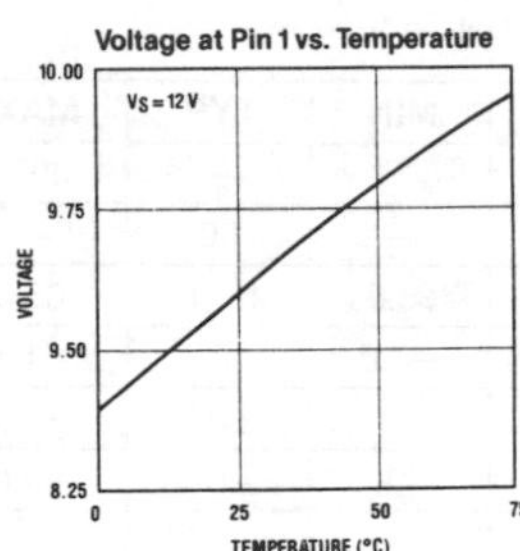

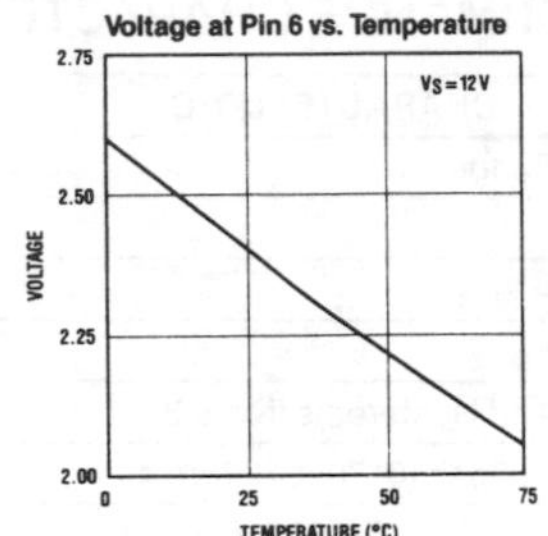

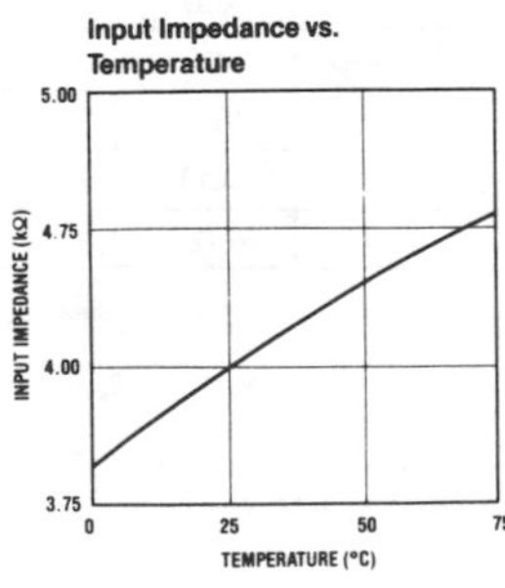

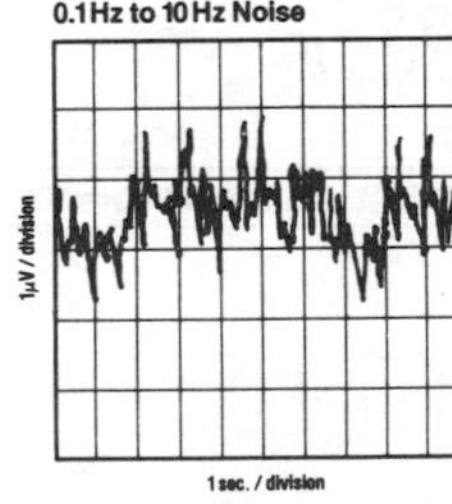

GENERAL DISCUSSION

The SCX series devices give a voltage output which is directly proportional to applied pressure. The devices will give an increasing positive going output when increasing pressure is applied to pressure port P_B of the device. If the input pressure connections are reversed, the output will increase with decreases in pressure. The devices are ratiometric to the supply voltage and changes in the supply voltage will cause proportional changes in the offset voltage and full-scale span. Since for absolute device pressure is applied to port P_A, output polarity will be reversed.

User Calibration

The SCX devices are fully calibrated for offset and span and should therefore require little user adjustment in most applications. For precise span and offset adjustments, refer to the applications section herein or contact the Sensym factory.

Vacuum Reference (Absolute Devices)

Absolute sensors have a hermetically sealed vacuum reference chamber. The offset voltage on these units is therefore measured at vacuum, 0 psia. Since all pressure is measured relative to a vacuum reference, all changes in barometric pressure or changes in altitude will cause changes in the device output.

Media Compatibility

SCX devices are compatible with most non-aqueous fuels, oils, refrigerants, hydraulic fluids, and non-corrosive gases. Because the circuitry is coated with a protective silicon gel, many otherwise corrosive environments can be compatible with the sensors. As shown in the physical construction diagram below, fluids must generally be compatible with silicon gel, plastic, aluminum, RTV, silicon, and glass for use with Port B. For questions concerning media compatibility, contact the factory.

MECHANICAL AND MOUNTING CONSIDERATIONS

The SCX nylon housing is designed for convenient pressure connection and easy PC board mounting. To mount the device horizontally to a PC board, the leads can be bent downward and the package attached to the board using either tie wraps or mounting screws. For pressure attachment, tygon or silicon tubing is recommended.

All versions of the SCX sensors have two (2) tubes available for pressure connection. For absolute devices, only port P_A is active. Applying pressure through the other port will result in pressure dead ending into the backside of the silicon sensor and the device will not give an output signal with pressure.

For gage applications, pressure should be applied to port P_B. Port P_A is then the vent port which is left open to the atmosphere. For differential pressure applications, to get proper output signal polarity, port P_B should be used as the high pressure port and P_A should be used as the low pressure port.

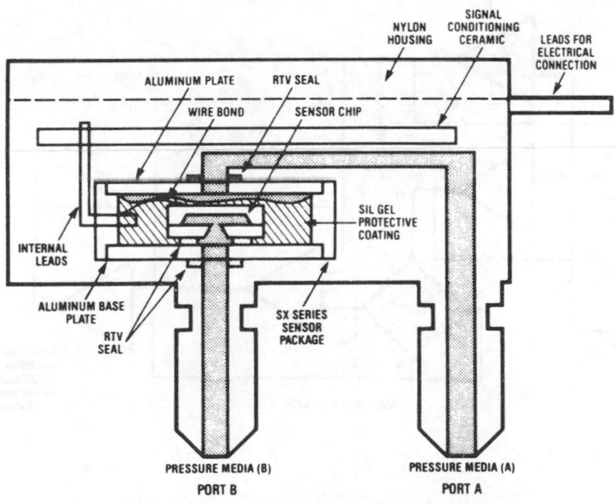

Physical Construction (Cutaway Diagram)

(Not Drawn to Scale)

APPLICATION INFORMATION

Shown here is a popular circuit which gives a high level 2-5 V_{DC} output for a 0–10 inch W.C. pressure input. Additional applications circuits are shown in the standard SCX series datasheet and SCX-EB evaluation board literature. For further applications information or assistance, please contact your nearest Sensym sales office or the Sensym factory.

Low Pressure Application

For sensing pressure below 1 psi, the circuit shown in Figure A uses the SCX01DN to provide a 2 to 5 V output for a 0 to 10 inch of water column input pressure. This output signal is compatible with many A/D converters and hence can be used to interface to a microprocessor system. This low-cost circuit is easily adaptable to lower full-scale pressure down to 5 inches of water column.

Circuit Description

The LM10 is used to provide a voltage reference for the excitation voltage (V_E), and for the voltage node V_{REF}. With this configuration, V_E and V_{REF} are not affected by noise or voltage variations in the 12 V power supply. R_3 is used to adjust V_{REF} to set the initial offset voltage at the output, V_{OUT}.

The pressure signal, V_{IN}, is amplified by amplifiers B_1, and B_2. (See Sensym Application Note SSAN-17 for details on this amplifier) R_2 is used to adjust the signal gain of the circuit. The output equation is given below.

$$V_{OUT} = V_{IN}\left[2\left(1+\frac{R}{R_1}\right)\right] + V_{REF}$$

For the best circuit performance, a careful selection of components is necessary. Use wirebound pots to insure low temperature coefficients and low long-term drift. A five-element resistor array (10 kΩ) SIP should be used for the resistors in the amplifier stage in order to obtain closely matched values and temperature coefficients. All other resistors should be 1% metal film. Amplifiers B_1, and B_2 should have low offset voltages and low noise. Signal lines should be as short as possible and the power supply should be capacitively bypassed on the PC board.

Adjustment Procedure

1. With zero-pressure applied, adjust the offset adjust R_3, until V_{OUT} = 2.000 V.
2. Apply full-scale pressure (10 in. W.C.) to port B, and adjust the full-scale adjust R_2, so that V_{OUT} = 5.000 V.
3. Repeat procedure if necessary.

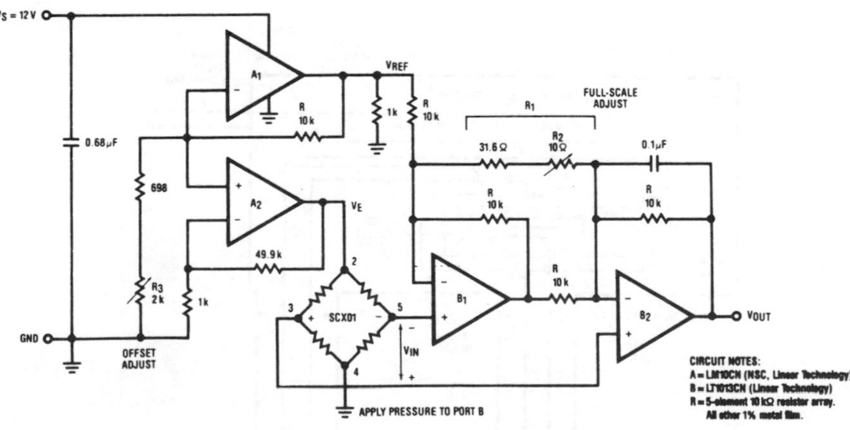

Figure A. Low Pressure Circuits Provide a 2 to 5 V Output for a 0–10 in. W.C. Pressure Input.

BP01
Blood Pressure Sensor

FEATURES

- Precise, Stable, Output
- Temperature Compensated
- Small Size
- Low Noise
- High Impedance for Low Power Portable Applications
- Manufactured in Accordance with GMP for Medical Devices

APPLICATIONS

- Non-Invasive Blood Pressure Monitoring
- Medical Pressure Monitoring

EQUIVALENT CIRCUIT

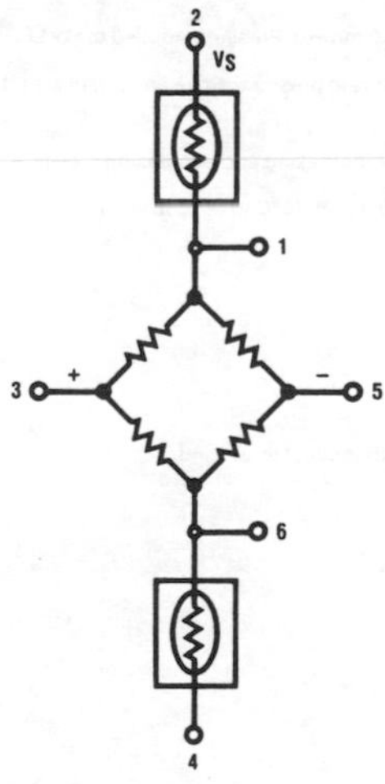

GENERAL DESCRIPTION

The BP01 sensor was specifically designed for use in medical applications; primarily non-invasive blood pressure monitoring. They provide excellent performance within the pressure range of 0 mmHg to 300 mmHg and over the temperature range from 10°C to 50°C.

The BP01 consists of a highly linear, low noise semiconductor pressure sensor in combination with a precision thick film ceramic, housed in a compact nylon case. This package offers small size and excellent isolation to external package stresses. It also gives convenient mounting holes and pressure ports for ease of use with standard Tygon* tubing.

The BP01 is internally calibrated and temperature compensated to achieve a high degree of accuracy, stability, and repeatability. Span is calibrated to within ±1% and zero pressure offset is set to ±300 microvolts maximum. As a result, many applications will not require additional trimming networks in the external signal conditioning circuitry.

The output of the BP01 is ratiometric to the supply voltage and operation from any DC supply voltage up to +30 V is acceptable. Because the BP01 is a relatively high impedance circuit (4 kΩ), it is also an ideal choice for portable or battery operated equipment. Given its low noise and fast response time, the BP01 provides an excellent choice for many other medical applications.

ELECTRICAL CONNECTION

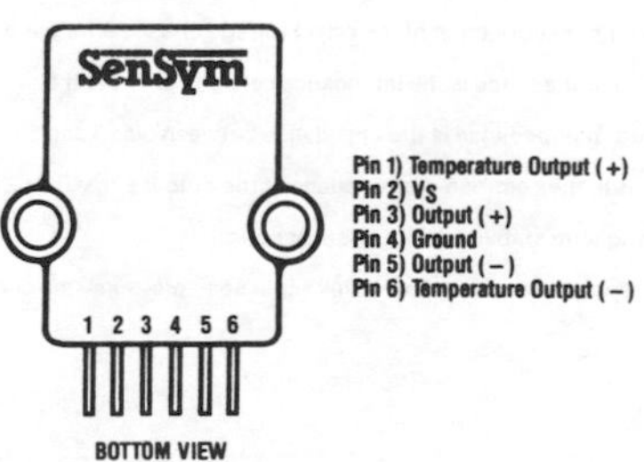

Note: The polarity indicated is for pressure applied to port B.

*Tygon is a trademark of the Norton company.

PRESSURE SENSOR CHARACTERISTICS BP01

Maximum Ratings

Supply Voltage, V_S +30 V_{DC}
Common-mode Pressure 2500 mmHg
Lead Temperature
(Soldering, 10 seconds) 300°C
Proof Pressure 1500 mmHg

Environmental Specifications

Temperature Range
Compensated +10 to +50°C
Operating 0 to +70°C
Storage −40 to +85°C
Humidity Limits 0 to 100% RH

BP01 PERFORMANCE CHARACTERISTICS (Note 1)

Characteristic	Min.	Typ.	Max.	Unit
Operating Pressure Range	−300	—	300	mmHg
Sensitivity	—	10	—	uV/v/mmHg
Full-scale Span (Note 2) AT +300 mmHg	14.85	15.0	15.15	mV
Zero Pressure Offset	−6	0	6	mmHg
Linearity 0–200 mmHg	—	—	0.1	%FSO
Linearity 0–300 mmHg	—	—	0.2	%FSO
Hysteresis (Note 3)	—	—	0.05	%FSO
Repeatability (Note 4)	—	—	0.25	%FSO
Temperature Effect on Span (10°C–50°C) (Note 5)	—	—	0.02	%FSO/°C
Temperature Effect on Offset (10°C–50°C) (Note 5)	—	—	0.2	mmHg/°C
Noise (Peak-to-Peak) 0.01 Hz to 10 Hz	—	0.04	—	mmHg
Input Impedance (Note 6)	—	4.0	—	kΩ
Output Impedance (Note 7)	—	4.0	—	kΩ
Common-mode Voltage (Note 8)	2.4	2.5	2.6	V
Long Term Stability of Offset and Span (Note 9)	—	0.1	—	%FSO
Response Time (Note 10)	—	1.0	—	msec

Specification Notes: (For All Devices)

Note 1: Unless otherwise noted: Supply Voltage, $V_S = 5.0 V_{DC}$, $T_A = 25°C$, Common-mode Line Pressure = 0 mmHg, Pressure Applied to Port B.

Note 2: Span is the algebraic difference between the output voltage at full-scale pressure and the output at zero pressure. Span and offset are ratiometric to the supply voltage.

Note 3: Hysteresis — the maximum output difference at any point within the operating pressure range for increasing and decreasing pressure.

Note 4: Maximum difference in output at any pressure with the operating pressure range and temperature within 10°C to 50°C after:
a) 1,000 temperature cycles, 10°C to 50°C
b) 1.5 million pressure cycles, 0 psi to full-scale span.

Note 5: Maximum error band of the offset voltage and the error band of the span, relative to the 25°C reading.

Note 6: Input impedance is the impedance between pins 2 and 4.

Note 7: Output impedance is the impedance between pins 3 and 5.

Note 8: This is the common-mode voltage of the output arms (Pins 3 and 5) for $V_S = 5.0 V_{DC}$. Measured with respect to Ground.

Note 9: Long term stability over a one year period.

Note 10: Response time for a 0 psi to full-scale span pressure step change, 10% to 90% rise time.

GENERAL DISCUSSION

BP01

Output Characteristics

The BP01 provides a voltage output which is directly proportional to applied pressure. The devices will give a positive going output when increasing pressure is applied to pressure Port B of the device (See Figure I). If the input pressure connections are reversed, the output will increase with decreases in pressure. The devices are ratiometric to the supply voltage and changes in the supply voltage will cause proportional changes in the offset voltage and full-scale span. For medical applications that require true differential pressure, Port B should be used as the high pressure port and Port A should be used as the low pressure port.

User Calibration

The BP01 is fully calibrated for offset and span and should therefore require little if any user adjustment in most applications. For precise span and offset adjustments, refer to the applications section herein or contact the Sensym factory.

MECHANICAL AND MOUNTING CONSIDERATIONS

The BP01 nylon housing is designed for convenient pressure connection and easy PC board mounting. The simplest PC board mounting technique is to use "xmas tree" clips as fasteners (Sensym Part #SCXCLIP), this technique is illustrated in Figure II. The xmas tree clip secures the BP01 to the board with minimum package stress. For electrical connections, we recommend a standard 6 pin right angle connector (Sensym Part # SCXCNCT). In blood pressure applications flexible Tygon or silicon tubing is suggested for pressure connection as it is easy to work with and quite pliable so package stresses are minimized. To secure the tubing properly to either port Sensym offers a standard easy to use hose clamp (Sensym Part #SCXSNP1).

For further information regarding mounting, refer to Sensym's application note SSAN-25.

Media Compatibility

The sensor is only recommended for use with air media for non-invasive blood pressure measurement.

The BP01 is not sterilized, therefore it is prohibited to bring blood or saline solution into direct contact with the sensor.

APPLICATION INFORMATION

Introduction

The following application information gives circuit ideas for some of the more common applications of the BP01. Please note however that Sensym cannot be an expert in any given application and although we provide application assistance on Sensym products both verbally and through our literature, it is still the total responsibility of the customer to determine the suitability of any product used in their application. For components suggested in any application information, it is also the responsibility of the customer to check with the individual component manufacturers for suitability of these products in the given application.

For additional application information, refer to Sensym's SCX series datasheet or contact the factory directly.

General

The circuit shown in Figure A provides a 0.5V to 3.5V output for a 0 to 300 mmHg input pressure. The circuit is easily calibrated and is not affected by changes in the supply voltage.

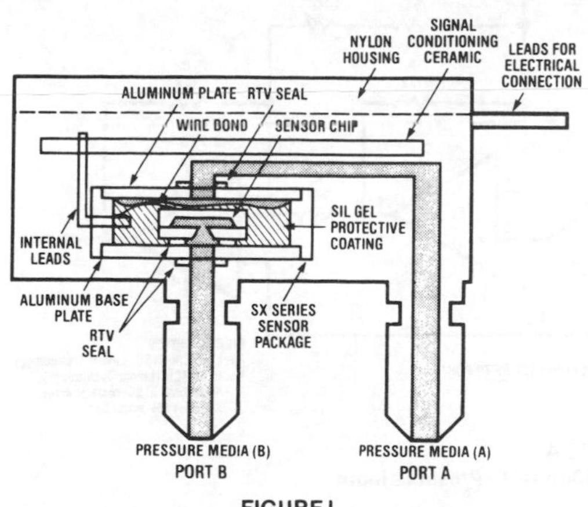

FIGURE I
SCX Physical Construction Diagram (Cutaway View)

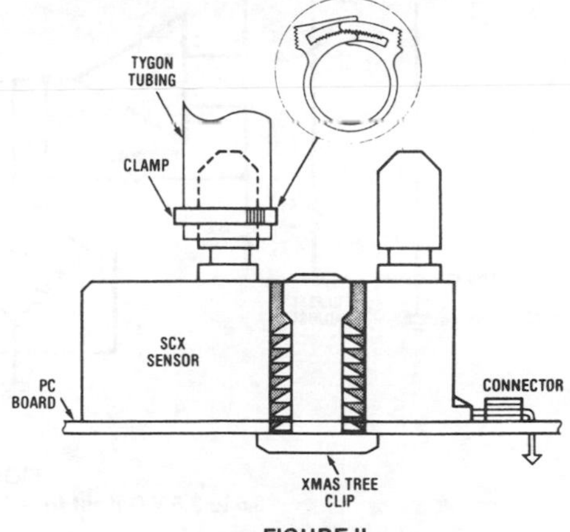

FIGURE II
Recommended PC Board Mounting Technique for the SCX Sensor

APPLICATION INFORMATION (Cont.)

Circuit Description

The LM10 (A_1) has a precision 200 mV reference present at Pin 8 which is amplified by A_2 to provide an excitation (V_E) of 4.2 volts to the top of the BP01 sensor. By providing a buffered reference voltage to the sensor, the circuit becomes insensitive to power supply variations and will operate from 5 volts to 20 volts without performance degradation. The same 200 mV reference is also amplified to provide voltage V_{REF}, which is adjusted by means of R_3 to set the initial 0.5 volt output voltage when there is no pressure applied.

From the BP01 specifications, it is noted that at 300 mmHg applied pressure, the output of the BP01 will nominally be 15 mV when using a 5 volt supply. Because the output is ratiometric to the supply voltage, the expected output for a supply of 4.2 volts will be 12.6 mV. The required gain can now be calculated. Since the output span required is 3 volts (3.5 volts minus 0.5 volts), and the input to the amplifier will be 12.6 mV, the gain required from amplifier configuration B_1 and B_2 is 238 V/V.

The output voltage equation (Eq. 1) and the gain equation derived (Eq. 3) are given below and are now used to solve for the unknown resistance, R_T.

$$V_{OUT} = V_{IN} \left| 2\left(1+\frac{R}{R_T}\right) \right| + V_0 \qquad (1)$$

or rewriting

$$V_{OUT} = V_{IN} \quad A_V + V_0 \qquad (2)$$

$$\text{where } A_V = \frac{\Delta\, V_{OUT}}{\Delta\, V_{IN}} = 2\left(1+\frac{R}{R_T}\right) \qquad (3)$$

and V_0 = initial output V_{OUT} for zero pressure applied.

With R = 10k, R_T is found to be 85Ω. To allow for component tolerances in the gain adjustment, R_T consists of a 1% metal film 75Ω resistor in series with a 20Ω multiturn cermet trim pot.

Adjustment Procedure

1. With zero-pressure applied, adjust the offset adjust R_3 until V_{OUT} = 0.500 V.
2. Apply full-scale pressure (300 mmHg) to Port B, and adjust R_2 until V_{OUT} = 3.500 V.
3. Repeat procedure if necessary.

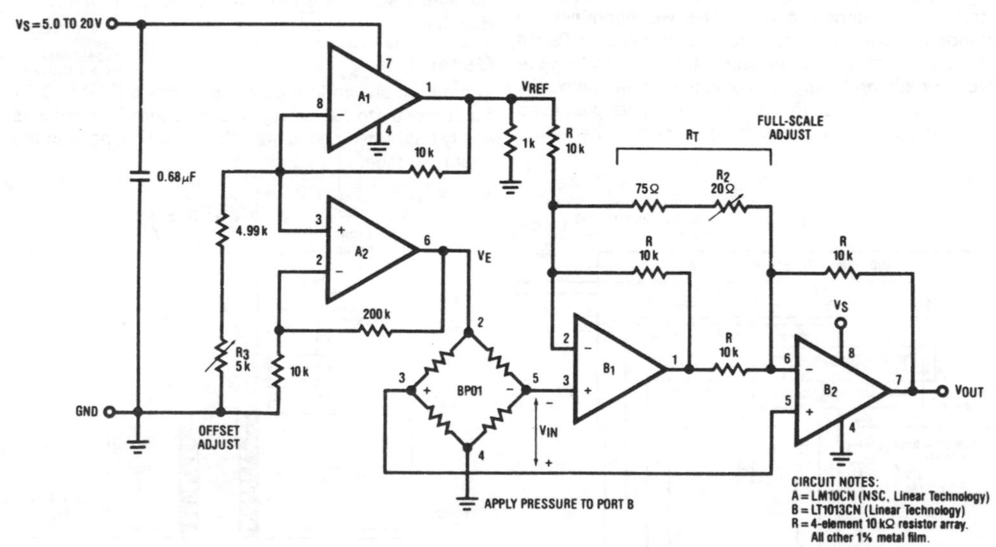

FIGURE A
0.5 to 3.5 V Output for a 0–300 mmHg Pressure Input

APPLICATION INFORMATION (Cont.)

BP01

0 V to 5 V Output

The circuit shown in Figure B is similar to that shown in Figure A, except, it provides a true 0 to 5 V output, for a 0 to 300 mmHg pressure input.

Circuit Description

In order to achieve an output voltage that will truely swing to zero volts (or even slightly below ground), a negative supply must be used. The circuit shown below makes use of a popular power supply inverter, the LTC1044, to achieve a negative supply voltage of −1.23 V (−V) using a single supply of 12 V and ground. The LTC1044 can be used in this application because the BP01 and the circuitry shown requires less than 5 mA of current.

The negative 1.23 V (−V) is used to power the amplifiers as well as a negative voltage for the zero adjust divider network. In this manner, the offset pot (R_0) adjusts V_{REF} to compensate for offset errors in the op amp circuitry and sensor, and enables the output to be set to 0.00 volts with zero pressure applied.

The 2.5 V reference, Z1, is amplified by A_1 to provide a 10 V excitation voltage (V_E) to the BP01. Since the output voltage of the BP01 is ratiometric to its supply, the sensor output will now be 30 mV for a 300 mmHg full scale pressure applied.

This 30 mV output voltage is now the differential input (V_{IN}) to A_3 and A_4 of the instrumentation amplifier stage. The output span in this example is 5 volts, therefore the required gain is 167 V/V. Using the gain equation given in (Eq. 3) R_T is found to be 1215 Ω.

Again, to allow for component tolerances, R_T is the combined series resistance of 1.1 k 1% metal film resistor (R_S) and a 200 Ω multiturn cermet trim pot (R_G).

Adjustment Procedure

1. With zero-pressure voltage applied, adjust the output using R_0 until V_{OUT} = 0.00 V.
2. Apply full-scale pressure (300 mmHg) to Port B, and adjust R_G until V_{OUT} = 5.0 V.
3. Repeat procedure if necessary.

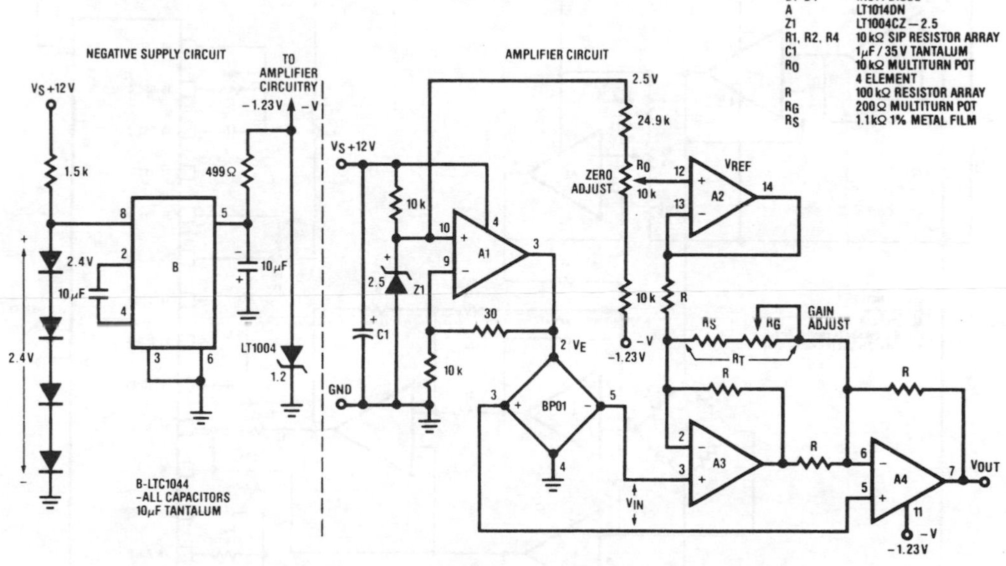

FIGURE B

True 0 to 5 V Output for 0-300 mmHg Pressure Input using a Single +12 V Supply and Ground

PORTABLE PRESSURE METER — BP01

General

The circuit shown in Figure C is a 0 to 200mmHg portable pressure meter. The 3½ digit display will read 199.9mmHg full-scale. This circuit operates from a single 9V battery and draws approximately 4.5mA supply current, thus providing a typical battery life in excess of 100 hours. The minimum battery voltage is approximately 6.5V.

This meter provides a resolution of 0.1mmHg. The same circuit can also be used for other pressure ranges simply by changing the sensor and gain.

Circuit Description

The LM10 provides a 200mV reference on pin 8 (A_1) which is amplified ($A_V = 1 + R_2/R_3$) and buffered to provide a stable +5V supply for the circuit. The resistor divider network comprised of two 100k resistors and pot R_4 provide the offset adjustment for the circuit. Since the BP01 provides an output which is ratiometric to its supply voltage, using the data sheet electrical specifications, the expected output voltage will be 10mV at 200mmHg when operating from a 5 volt supply. For the component values shown, a full-scale input voltage of 200mV (V_0) is required to the ICL7106 in order to display a full scale output of 1999, thus the gain required for the instrumentation amplifier is 20V/V. Again, from the gain equation (Eq. 3) with $R_1 = 10k$, R_T is found to be 1.1k. In this circuit, a fixed 1% metal film resistor has been used for R_T. The fine tuning of the full scale span adjustment is accomplished by using Pot R_8, which sets the reference voltage of the A/D converter.

Component Selection

The value of the components R_6–R_{10} and C_1–C_5 have been optimized for 200mV full-scale. (See the Intersil ICL7106 Datasheet). R_4 and R_8 should be 15 turn cermet pots, R_6–R_{10} are metal film 1% resistors. C_3 the integrating capacitor should be polypropylene, the reference and auto zero capacitors can be polystyrene or mylar, the clock capacitor, C_4, is mica.

Adjustment Procedure

Apply 165mmHg and adjust R_8 until the display reads 165. Apply 0mmHg and adjust R_4 until the display reads 000.0. Repeat if required.

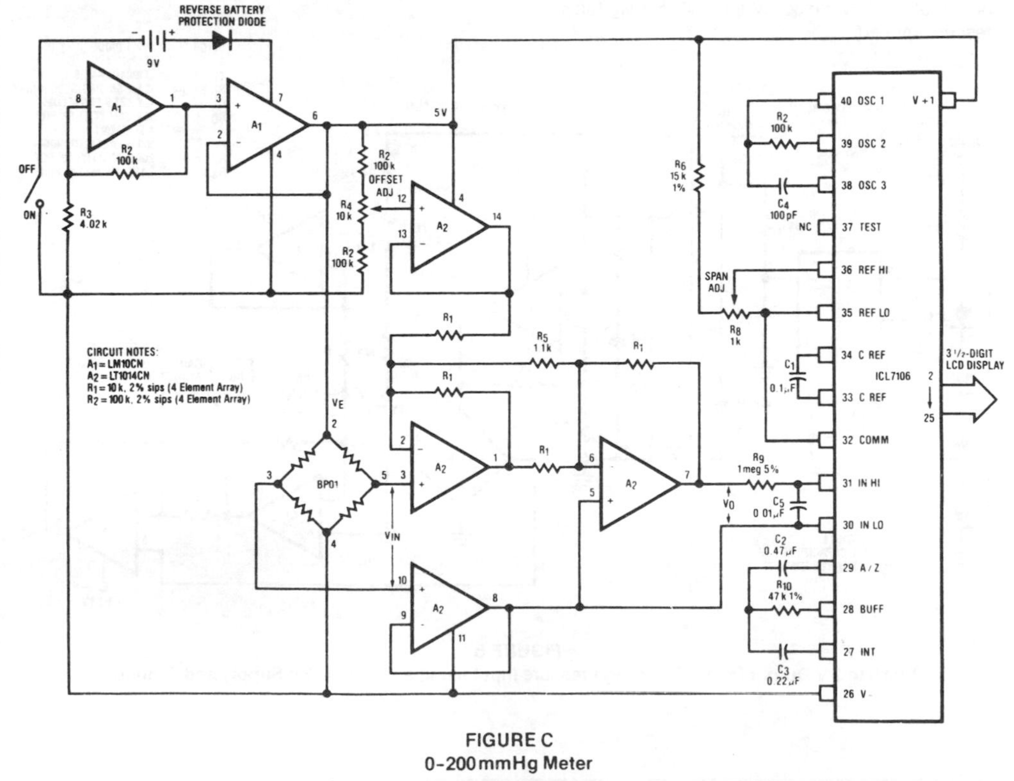

FIGURE C
0-200mmHg Meter

PARALLEL A/D CONVERSION

BP01

General

The BP01 sensor can be easily interfaced to a microprocessor bus. Using an A/D converter, for a 0 to 300 mmHg input, the circuit in Figure D will provide an eight-bit parallel output which is proportional to applied pressure. The circuit allows for easy calibration and uses a single 5 V supply.

Circuit Description

The LM10 provides both the reference voltage V_{REF} and the sensor excitation voltage V_E which are independent of supply voltage changes above the 5 V supply. V_{REF} is amplified by B_3 to provide the A/D converter with 1.5 volt reference voltage. With a 1.5 volt reference, a full-scale output of all "ONES" will occur when the input signal to the A/D is 3 volts.

The sensor excitation voltage V_E of 4.2 V will result in a differential sensor output of 12.6 mV when an input pressure of 300 mmHg is applied. This output signal is amplified by B_1 and B_2 to provide the converter with $V_{IN}(+)$. The gain is adjusted using pot R_G. The initial offset voltage is adjusted by R_3, which varies $V_{IN}(-)$ on the converter. The A/D converter, as shown, is a free-running configuration and the binary output is updated continuously.* The only requirement is that the WR and INTR must be momentarily grounded or taken to a logic low after power-up to ensure proper operation.

Adjustment Procedure

1. With no pressure applied, adjust the offset pot R_3 until all bits are zero except the LSB, which should be switching between one and zero.
2. Apply full-scale pressure (1 psig) to Port B, and adjust the full-scale pot R_G until all bits are ones except the LSB which should be flickering between one and zero.
3. Repeat procedure If necessary.

*For timing specifications and bus interface, see the ADC0804 Datasheet from National Semiconductor.

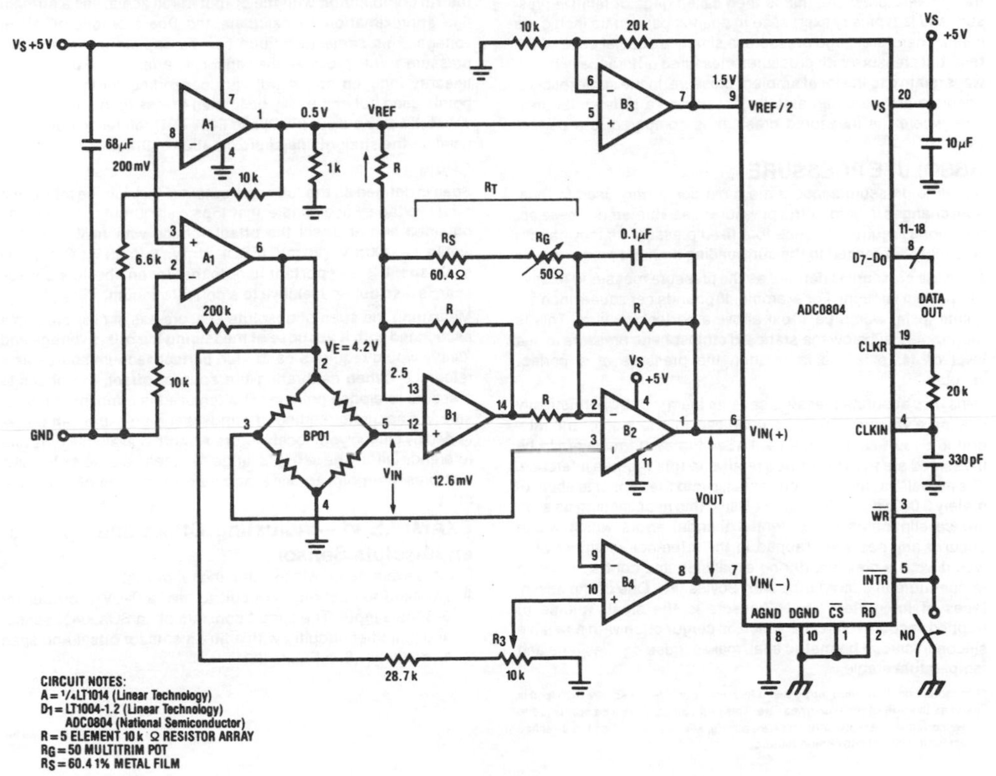

FIGURE D

A Parallel A/D Conversion Circuit for 0-300 mmHg Applications

Understanding Absolute Pressure Sensors

SSAN-24

INTRODUCTION

Most people are typically accustomed to dealing in gage pressure, that is, pressure relative to the normal atmospheric pressure which surrounds us. As such, "absolute" pressure and absolute pressure sensors which measure pressure relative to a perfect vacuum can be somewhat confusing. Also, because zero absolute pressure (a perfect vacuum) is impossible to achieve, it is much harder to measure and calibrate absolute pressure sensors. This application note will discuss what absolute pressure is, how it is best measured and how to calibrate absolute pressure sensors.

DIFFERENTIAL (GAGE) PRESSURE

It is often easier to understand absolute pressure if we have a clear understanding of differential and gage pressure which we are generally more familiar with.

Differential pressure is the pressure difference measured between two pressure sources. This is usually expressed in pounds per square inch differential (psid). When one source is the ambient pressure, this is then called gage or relative pressure and is typically expressed in pounds per square inch gage (psig). Therefore, gage pressure is simply a special case of differential pressure with pressures measured differentially but always relative to the local ambient pressure. In the same respect, absolute pressure can also be considered a differential pressure where the measured pressure is compared to a perfect vacuum.

ABSOLUTE PRESSURE

Absolute pressure sensors are most commonly used to measure changes in barometric pressure or as altimeters. These applications require reference to a fixed pressure as they cannot be simply referenced to the surrounding ambient pressure.

Absolute pressure is defined as the pressure measured relative to a perfect vacuum. For example, 10 pounds per square inch absolute (psia) would be 10 psi above a perfect vacuum. This is roughly 4.7 psi below the standard atmospheric pressure at sea level of 14.7 psia. 0 psia is then the pressure of a perfect vacuum.*

Sensym's absolute pressure sensors are made by hermetically sealing a vacuum reference chamber on one side of the integrated circuit sensing element. (See Figure I) Pressures to be measured are then measured relative to this vacuum reference. The actual "vacuum" which is sealed into the sensor is approximately 0.0005 psia (25 millitorr). Using this near vacuum as a reference eliminates any potential thermal errors which would occur if any gas was trapped in the reference chamber as it would exert a pressure during expansion and contraction with temperature in accordance with Boyles law. One of the advantages of integrated circuit sensors is the small volume of trapped vacuum reference which, in conjunction with a reliable silicon-to-silicon hermetic seal, makes these devices time and temperature stable.

*Footnote: For illustration purposes in this note, pounds per square inch (psi) is used as the unit of pressure measure. This unit can obviously be converted to other common pressure units such as mmHg, kPa, bar, etc.. (See the attached chart for individual conversion factors.)

CALIBRATING ABSOLUTE SENSORS

To use any sensor in an absolute application, we must be able to accurately calibrate the device for offset and span. This requires understanding offset and span in terms of absolute pressure.

OFFSET VOLTAGE

The Offset voltage is defined as the sensors output at *zero differential* pressure. For gage sensors, this is the output with ambient pressure (0 psig) applied to the sensor. As such, offset voltages are relatively easy to measure for gage sensors. However, *for an absolute device, the offset voltage is the output voltage of the sensor with a perfect vacuum (0 psia) applied to the sensor.* This means that with normal atmospheric pressure applied to the absolute sensor, there will be an output voltage which corresponds to approximately +14.7 psia at sea level.

Because a perfect vacuum is impossible or at least impractical to obtain, measuring the actual offset voltage for absolute sensors is not possible. At Sensym, we draw a vacuum to 1 psia and then in combination with the output at full scale, use a straight line approximation to calculate the 0 psia output or offset voltage. This same technique can be applied using any two pressure points provided the sensor is perfectly linear. The nonlinearity induced errors will vary depending on the pressure points used but can easily be limited to less than ±0.1% if the 10% full scale output (FSO) and 90% FSO reference points are used for the straight line approximation of 0 psia.

SPAN

Span is defined as the full scale output (FSO) voltage minus the offset voltage. For example, if at 15 psia an output of 101 mV was obtained and at 0 psia the offset voltage was 1 mV, the span would be 100 mV (101 mV (FSO) − 1 mV (offset) = 100 mV full scale span). It is important to note that for an absolute sensor, span is also defined relative to a perfect vacuum.

Measuring the span of absolute sensors has similar problems associated with it as those of measuring the offset voltage and ideally would require a calibrated perfect vacuum source as a reference. When calibrating the span or offset, the absolute pressure reference point most often used is atmospheric pressure. An accurate reading of atmospheric pressure can be obtained by calling your local airport. Any other available pressure reference within the sensors range can then be used as the second pressure point to allow accurate calibration of span and offset.

EXAMPLE #1 — Adjusting Offset and Span for an Absolute Sensor

For this example we will assume the following:

1) We need to calibrate a circuit to give a 0-5 V_{DC} output for 0-30 psia input. The circuit consists of an SCX30AN sensor and amplifier circuitry with adjustments for offset and span (sensitivity). (See Figure II).

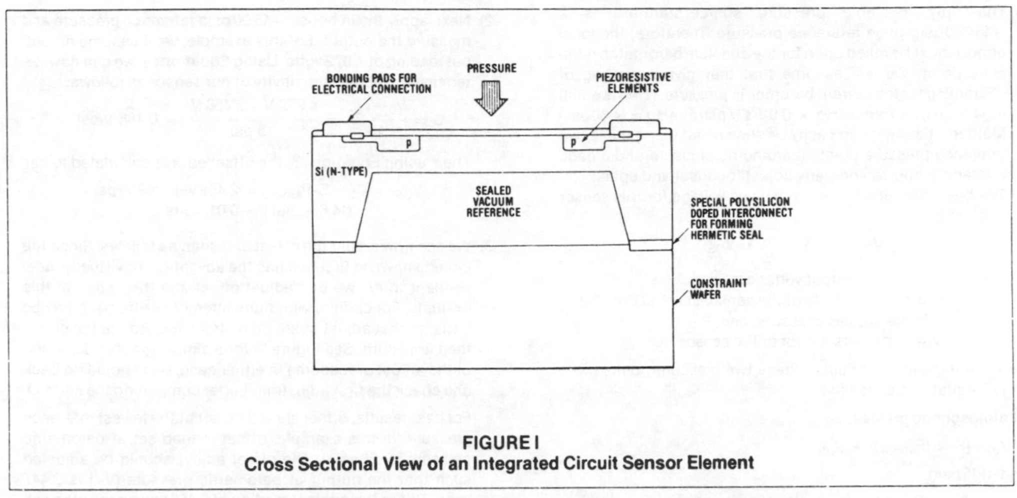

FIGURE I
Cross Sectional View of an Integrated Circuit Sensor Element

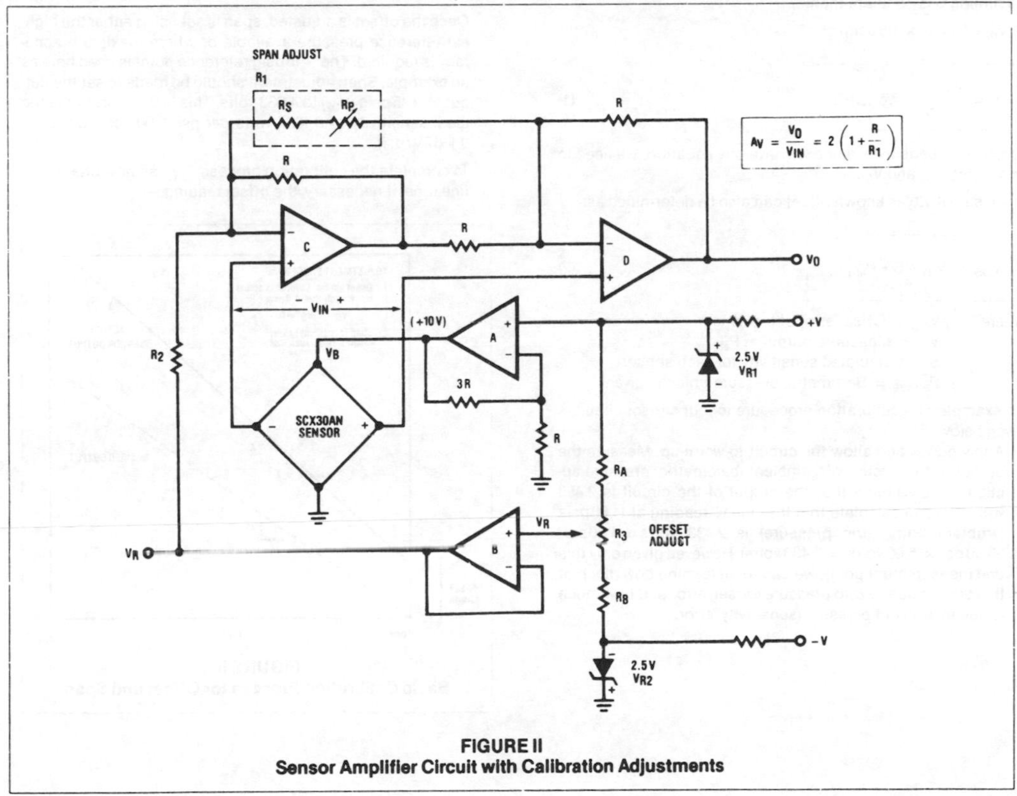

$$A_V = \frac{V_O}{V_{IN}} = 2\left(1 + \frac{R}{R_1}\right)$$

FIGURE II
Sensor Amplifier Circuit with Calibration Adjustments

2) The cnly available pressure source/standard is a +15.000psig gage reference pressure. Therefore, the local airport must be relied upon for the absolute barometric reference point. We will assume that they give a reading of 755mmHg for the current barometric pressure. (This is equal to 14.60psi, as 755mmHg × 0.019337psi/mmHg = 14.60psi.) Mathematically, it can easily be shown that using these two reference pressure points (barometric pressure and a gage reference), one can properly adjust for offset and span.
The basic output equation at a given voltage for this sensor is given by:

$$V_{OUT} = (S \times P) + V_{OS}$$

where: V_{OUT} is the output voltage in volts
S is the sensitivity of the sensor in volts/PSI
P is the applied pressure, and
V_{OS} is the offset error of the sensor.

Determining sensitivity using these two pressure points gives us two equations as follows:

At atmospheric pressure,

$$V_{01} = [S \times (P_{atmos})] + V_{OS}$$

and at 15psig,

$$V_{02} = [S \times (P_{atmos} + 15psig)] + V_{OS}$$

Assuming S is constant then:

$$V_{02} - V_{01} = S \times (15psig) \text{ or,}$$

$$S = \frac{V_{02} - V_{01}}{15} \text{ volts/PSI} \qquad (1)$$

To solve for sensitivity since we have one equation, we need to know both V_{02} and V_{01}.

Once sensitivity is known offset can also be determined as:

$$V_{OS} = V_{01} - S \cdot P_{atmos} \qquad (2)$$

where: V_{OS} = Offset error of the sensor
V_{01} = Measured output at P_{atmos}
S = Calculated sensitivity for each sensor
P_{atmos} = Barometric pressure which is given

An example of a calibration procedure for our sensor circuit is given below:

1) Apply power and allow the circuit to warm-up. Measure the output of the circuit with ambient (barometric) pressure applied. Let's assume that the output of the circuit is 2.458 volts. We can calculate that the "ideal" reading at 14.60psia (ambient barometric pressure) is 2.433 volts (14.60psi/30.00psi × 5.00volts = 2.433volts). However, given only this one measurement point, we cannot determine how much of this error is due to zero pressure (offset) error and how much is due to ambient pressure (sensitivity) error.

2) Next, apply the in-house +15.00psig reference pressure and measure the output. For this example, we'll assume an output reading of 4.972 volts. Using Equation (1), we can now determine the actual sensitivity of our sensor as follows:

$$S = \frac{V_{02} - V_{01}}{15} = \frac{4.972\text{ V} - 2.458\text{ V}}{15\text{ psi}} = 0.168\text{ V/psi}$$

Then, using Equation (2), the offset error is calculated to be:

$$V_{OS} = V_{01} - S \cdot P_{atmos} = 2.458\text{V} - [0.168\text{V/psi}\ (14.60\text{psia})] = 0.011\text{ volts}$$

3) We can now adjust the offset and span as follows: Since the circuit shown in Figure II has the advantage of virtually independent trims, we can adjust offset and then span in this example. For circuits with more interactive effects it may be better to first adjust span, then offset. (i.e. Rotate the curve, then level shift. See Figure III for a simple graphic depiction of the adjust procedure.) In either case, one should go back and check the first adjustment after completing the second.

For best results, offset should be set at the lowest reference pressure. In this example, offset is then set at barometric pressure (14.60psia). The offset adjust should be adjusted such that the output at barometric pressure (V_{01}) is 2.447 volts. This is the original reading of 2.458 volts minus the calculated offset error of 0.011 volts.

Once the offset is adjusted, span is set using either the highest reference pressure available or where maximum accuracy is required. The +15psig reference point is used here as an example. Span adjustment should be made to set the output at +15psig (V_{02}) to 4.933 volts. This is 29.60psi times the ideal sensitivity of 0.1667 volts per psi (5.00 volts/30psi = 0.1667V/psi).

To complete the calibration process, we can now check and fine tune, if necessary, the offset reading.

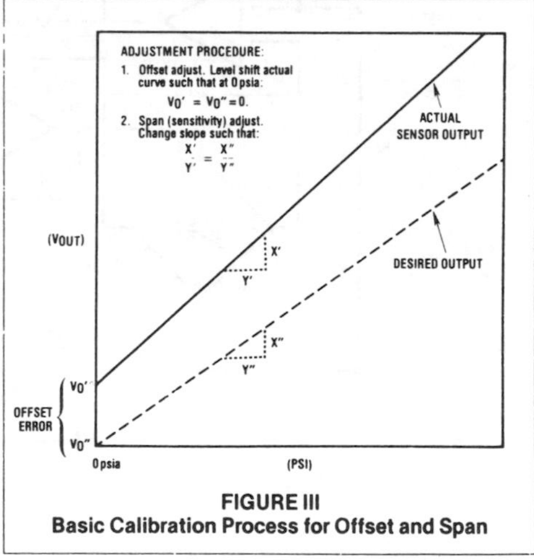
ADJUSTMENT PROCEDURE:
1. Offset adjust. Level shift actual curve such that at 0psia:
V0' = V0" = 0.
2. Span (sensitivity) adjust. Change slope such that:
X'/Y' = X"/Y"
ACTUAL SENSOR OUTPUT
DESIRED OUTPUT
(VOUT)
X'
Y'
X"
Y"
V0'
OFFSET ERROR
V0"
0psia
(PSI)

FIGURE III
Basic Calibration Process for Offset and Span

CONCLUSION

Although absolute pressure sensors are perhaps not as well understood or easily calibrated as gage pressure devices, by using barometric pressure as a reference in conjunction with any other reference point, gage or absolute, absolute pressure devices can be accurately calibrated.

PRESSURE UNIT CONVERSION CONSTANTS

(Most Commonly Used — Per International Conventions)

	PSI(1)	In. H_2O(2)	In. Hg(3)	kPa	millibar	cm H_2O(4)	mm Hg(5)
PSI(1)	1.000	27.680	2.036	6.8947	68.947	70.308	51.715
In. H_2O(2)	3.6127×10^{-2}	1.000	7.3554×10^{-2}	0.2491	2.491	2.5400	1.8683
In. Hg(3)	0.4912	13.596	1.000	3.3864	33.864	34.532	25.400
kPa	0.14504	4.0147	0.2953	1.000	10.000	10.1973	7.5006
millibar	0.01450	0.40147	0.02953	0.100	1.000	1.01973	0.75006
cm H_2O(4)	1.4223×10^{-2}	0.3937	2.8958×10^{-2}	0.09806	0.9806	1.000	0.7355
mm Hg(5)	1.9337×10^{-2}	0.53525	3.9370×10^{-2}	0.13332	1.3332	1.3595	1.000

Notes: 1. PSI — pounds per square inch 2. at 39°F 3. at 32°F 4. at 4°C 5. at 0°C 6. 1 Torr = 1 mmHg

1.2V Reference

National Semiconductor
Application Note 56

INTRODUCTION

Temperature compensated zener diodes are the most easily used voltage reference. However, the lowest voltage temperature-compensated zener is 6.2V. This makes it inconvenient to obtain a zero temperature-coefficient reference when the operating supply voltage is 6V or lower. With the availability of the LM113, this problem no longer exists.

The LM113 is a 1.2V temperature compensated shunt regulator diode. The reference is synthesized using transistors and resistors rather than a breakdown mechanism. It provides extremely tight regulation over a wide range of operating currents in addition to unusually low breakdown voltage and low temperature coefficient.

DESIGN CONCEPTS

The reference in the LM113 is developed from the highly-predictable emitter-base voltage of integrated transistors. In its simplest form, the voltage is equal to the energy-band-gap voltage of the semiconductor material. For silicon, this is 1.205V. Further, the output voltage is well determined in a production environment.

A simplified version of this reference[1] is shown in *Figure 1*. In this circuit, Q_1 is operated at a relatively high current density. The current density of Q_2 is about ten times lower, and the emitter-base voltage differential (ΔV_{BE}) between the two devices appears across R_3. If the transistors have high current gains, the voltage across R_2 will also be proportional to ΔV_{BE}. Q_3 is a gain stage that will regulate the output at a voltage equal to its emitter base voltage plus the drop across R_2. The emitter base voltage of Q_3 has a negative temperature coefficient while the ΔV_{BE} component across R_2 has a positive temperature coefficient. It will be shown that the output voltage will be temperature compensated when the sum of the two voltages is equal to the energy-band-gap voltage.

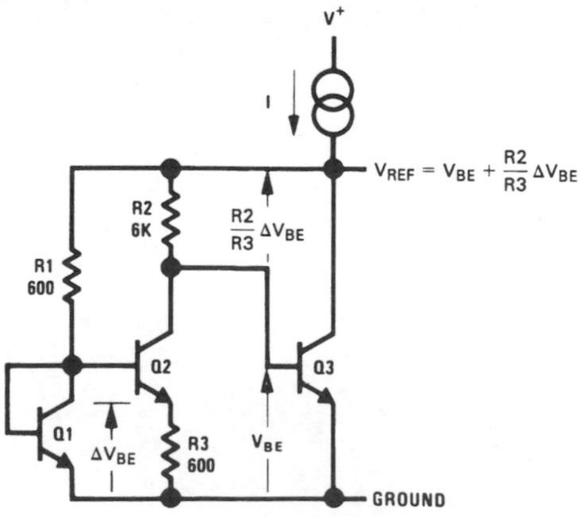

FIGURE 1. The Low Voltage Reference in One of Its Simpler Forms

Conditions for temperature compensation can be derived starting with the equation for the emitter-base voltage of a transistor which is[2]

$$V_{BE} = V_{g0}\left(1 - \frac{T}{T_0}\right) + V_{BE0}\left(\frac{T}{T_0}\right) + \frac{nkT}{q}\log_e\frac{T_0}{T} + \frac{kT}{q}\log_e\frac{I_C}{I_{C0}}, \quad (1)$$

where V_{g0} is the extrapolated energy-band-gap voltage for the semiconductor material at absolute zero, q is the charge of an electron, n is a constant which depends on how the transistor is made (approximately 1.5 for double-diffused, NPN transistors), k is Boltzmann's constant, T is absolute temperature, I_C is collector current and V_{BE0} is the emitter-base voltage at T_0 and I_{C0}.

The emitter-base voltage differential between two transistors operated at different current densities is given by

$$\Delta V_{BE} = \frac{kT}{q}\log_e\frac{J_1}{J_2} \quad (2)$$

where J is current density.

Referring to Equation (1), the last two terms are quite small and are made even smaller by making I_C vary as absolute temperature. At any rate, they can be ignored for now because they are of the same order as errors caused by non-theoretical behavior of the transistors that must be determined empirically.

If the reference is composed of V_{BE} plus a voltage proportional to ΔV_{BE}, the output voltage is obtained by adding (1) in its simplified form to (2):

$$V_{ref} = V_{g0}\left(1 - \frac{T}{T_0}\right) + V_{BE0}\left(\frac{T}{T_0}\right) + \frac{kT}{q}\log_e\frac{J_1}{J_2}. \quad (3)$$

Differentiating with respect to temperature yields

$$\frac{\partial V_{ref}}{\partial T} = -\frac{V_{g0}}{T_0} + \frac{V_{BE0}}{T_0} + \frac{k}{q}\log_e\frac{J_1}{J_2}. \quad (4)$$

For zero temperature drift, this quantity should equal zero, giving

$$V_{g0} = V_{BE0} + \frac{kT_0}{q}\log_e\frac{J_1}{J_2}. \quad (5)$$

The first term on the right is the initial emitter-base voltage while the second is the component proportional to emitter-base voltage differential. Hence, if the sum of the two are equal to the energy-band-gap voltage of the semiconductor, the reference will be temperature-compensated.

Figure 2 shows the actual circuit of the LM113. Q_1 and Q_2 provide the ΔV_{BE} term and Q_4 provides the V_{BE} term as in the simplified circuit. The additional transistors are used to decrease the dynamic resistance, improving the regulation of the reference against current changes. Q_3 in conjunction with current inverter, Q_5 and Q_6, provide a current source load for Q_4 to achieve high gain.

Q_7 and Q_9 buffer Q_4 against changes in operating current and give the reference a very low output resistance. Q_8 sets the minimum operating current of Q_7 and absorbs any leak-

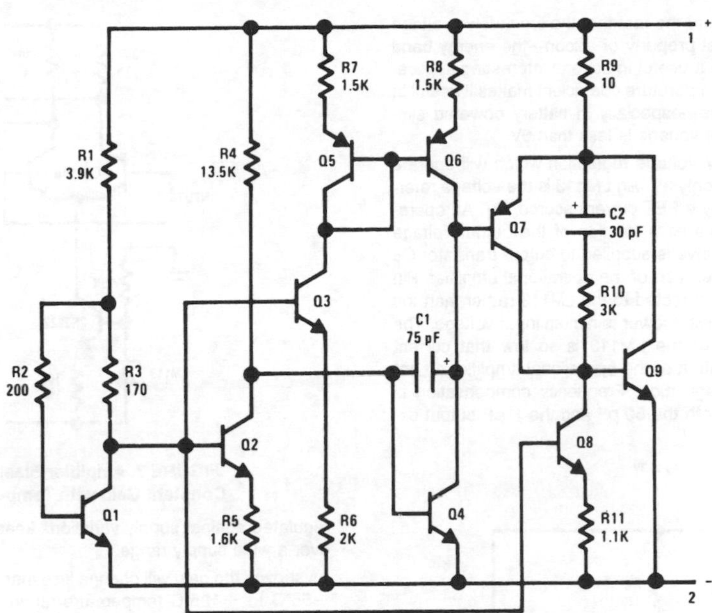

FIGURE 2. Schematic of the LM113

age from Q_9. Capacitors C_1, C_2 and resistors R_9 and R_{10} frequency compensate the regulator diode.

PERFORMANCE

The most important features of the regulator diode are its good temperature stability and low dynamic resistance. *Figure 3* shows the typical change in output voltage over a −55°C to +125°C temperature range. The reference voltage changes less than 0.5% with temperature, and the temperature coefficient is relatively independent of operating current.

Figure 4 shows the output voltage change with operating current. From 0.5 mA to 20 mA there is only 6 mV of change. A good portion of the output change is due to the resistance of the aluminum bonding wires and the Kovar leads on the package. At currents below about 0.3 mA the diode no longer regulates. This is because there is insufficient current to bias the internal transistors into their active region. *Figure 5* illustrates the breakdown characteristic of the diode.

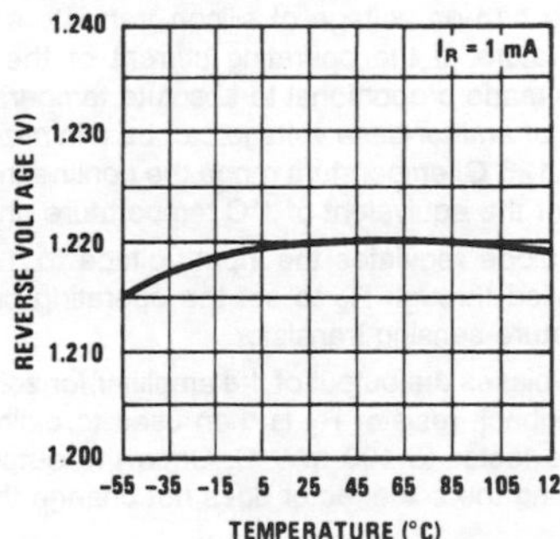

FIGURE 3. Output Voltage Change with Temperature

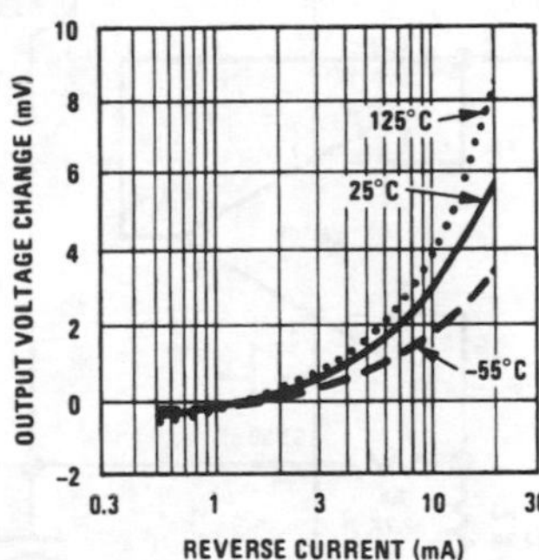

FIGURE 4. Output Voltage Change with Current

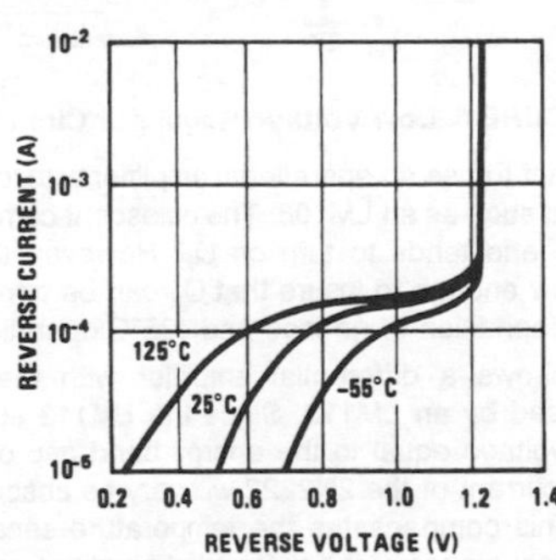

FIGURE 5. Reverse Breakdown Characteristics

APPLICATIONS

The applications for zener diodes are so numerous that no attempt to delineate them will be made. However, the low breakdown voltage and the fact that the breakdown voltage is equal to a physical property of silicon—the energy band gap voltage—makes it useful in several interesting applications. Also the low temperature coefficient makes it useful in regulator applications—especially in battery powered systems where the input voltage is less than 6V.

Figure 6 shows a 2V voltage regulation which will operate on input voltages of only 3V. An LM113 is the voltage reference and is driven by a FET current source, Q_1. An operational amplifier compares a fraction of the output voltage with the reference. Drive is supplied to output transistor Q_2 through the V^+ power lead of the operational amplifier. Pin 6 of the op amp is connected to the LM113 rather than the output since this allows a lower minimum input voltage. The dynamic resistance of the LM113 is so low that current changes from the output of the operational amplifier do not appreciably affect regulation. Frequency compensation is accomplished with both the 50 pF and the 1 μF output capacitor.

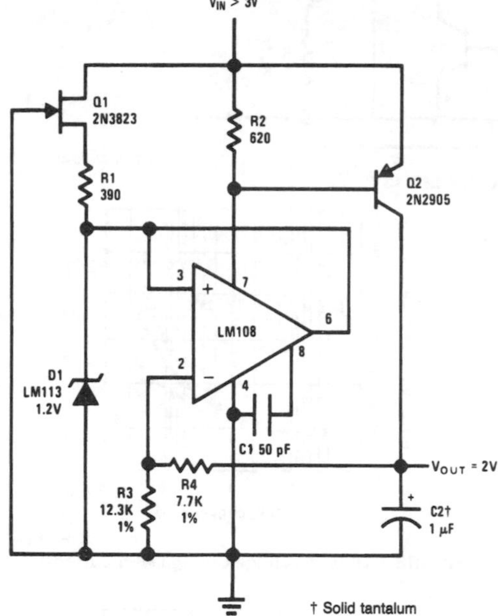

FIGURE 6. Low Voltage Regulator Circuit

It is important to use an operational amplifier with low quiescent current such as an LM108. The quiescent current flows through R_2 and tends to turn on Q_2. However, the value shown is low enough to insure that Q_2 can be turned off at worst case condition of no load and 125°C operation.

Figure 7 shows a differential amplifier with the current source biased by an LM113. Since the LM113 supplies a reference voltage equal to the energy band gap of silicon, the output current of the 2N2222 will vary as absolute temperature. This compensates the temperature sensitivity of the transconductance of the differential amplifier making the gain temperature stable. Further, the operating current is

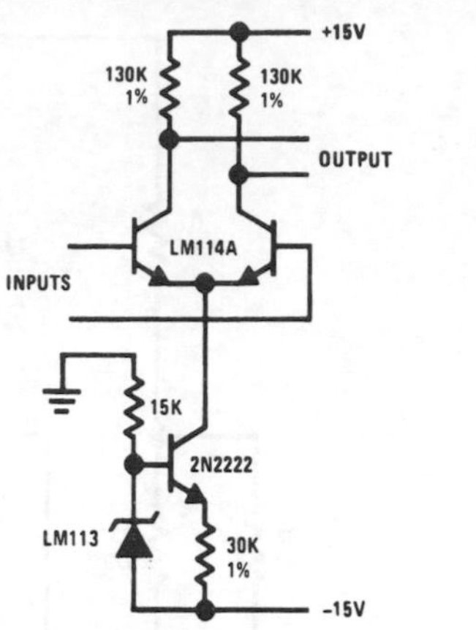

FIGURE 7. Amplifier Biasing for Constant Gain with Temperature

regulated against supply variations keeping the gain stable over a wide supply range.

As shown, the gain will change less than two per cent over a −55°C to +125°C temperature range. Using the LM114A monolithic transistor and low drift metal film resistors, the amplifier will have less than 2 μV/°C voltage drift. Even lower drift may be obtained by unbalancing the collector load resistors to null out the initial offset. Drift under nulled condition will be typically less than 0.5 μV/°C.

The differential amplifier may be used as a pre-amplifier for a low-cost operational amplifier such as an LM101A to improve its voltage drift characteristics. Since the gain of the operational amplifier is increased by a factor of 100, the frequency compensation capacitor must also be increased from 30 pF to 3000 pF for unity gain operation. To realize low voltage drift, case must be taken to minimize thermoelectric potentials due to temperature gradients. For example, the thermoelectric potential of some resistors may be more than 30 μV/°C, so a 1°C temperature gradient across the resistor on a circuit board will cause much larger errors than the amplifier drift alone. Wirewound resistors such as Evenohm are a good choice for low thermoelectric potential.

Figure 8 illustrates an electronic thermometer using an inexpensive silicon transistor as the temperature sensor. It can provide better than 1°C accuracy over a 100°C range. The emitter-base turn-on voltage of silicon transistors is linear with temperature. If the operating current of the sensing transistor is made proportional to absolute temperature the nonlinearily of emitter-base voltage can be minimized. Over a −55°C to 125°C temperature range the nonlinearily is less than 2 mV or the equivalent of 1°C temperature change.

An LM113 diode regulates the input voltage to 1.2V. The 1.2V is applied through R_2 to set the operating current of the temperature-sensing transistor.

Resistor R_4 biases the output of the amplifier for zero output at 0°C. Feedback resistor R_5 is then used to calibrate the output scale factor to 100 mV/°C. Once the output is zeroed, adjusting the scale factor does not change the zero.

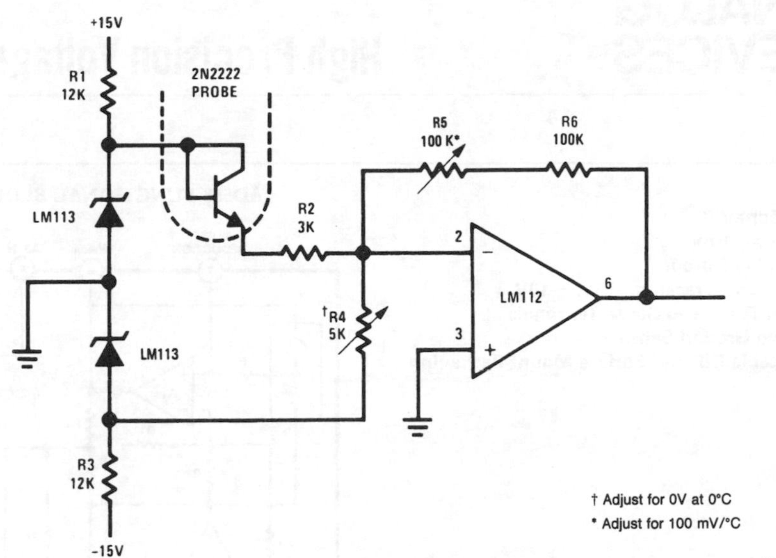

FIGURE 8. Electronic Thermometer

CONCLUSION

A new two terminal low voltage shunt regulator has been described. It is electrically equivalent to a temperature-stable 1.2V breakdown diode. Over a −55°C to 125°C temperature range and operating currents of 0.5 mA to 20 mA the LM113 has one hundred times better reverse characteristics than breakdown diode. Addiitionally, wideband noise and long term stability are good since no breakdown mechanism is involved.

The low temperature coefficient and low regulation voltage make it especially suitable for a low voltage regulator or battery operated equipment. Circuit design is eased by the fact that the output voltage and temperature coefficient are largely independent of operating current. Since the reference voltage is equal to the extrapolated energy-band-gap of silicon, the device is useful in many temperature compensation and temperature measurement applications.

REFERENCES

1. R.J. Widlar, *"On Card Regulator for Logic Circuits,"* National Semiconductor AN-42, February, 1971.

2. J.S. Brugler, *"Silicon Transistor Biasing for Linear Collector Current Temperature Dependence,"* IEEE Journal of Solid State Circuits, pp. 57–58, June, 1967.

High Precision Voltage Reference

AD588*

FEATURES
Low Drift – 1.5ppm/°C
Low Initial Error – 1mV
Pin-Programmable Output
+10V, +5V, ±5V Tracking, –5V, –10V
Flexible Output Force and Sense Terminals
High Impedance Ground Sense
Machine-Insertable DIP and Surface Mount Packaging

AD588 FUNCTIONAL BLOCK DIAGRAM

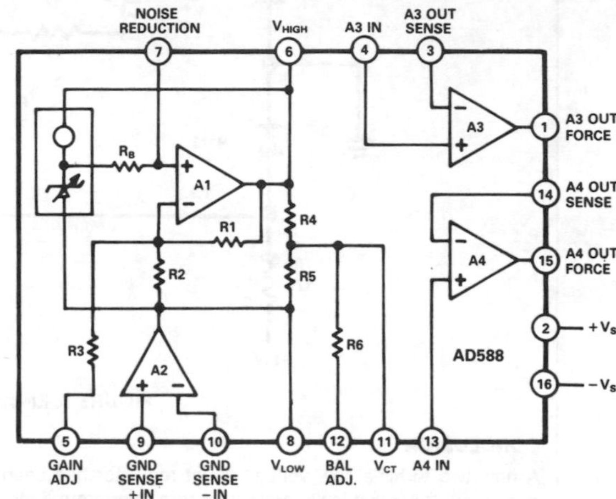

PRODUCT DESCRIPTION

The AD588 represents a major advance in the state-of-the-art in monolithic voltage references. Low initial error and low temperature drift give the AD588 absolute accuracy performance previously not available in monolithic form. The AD588 uses a proprietary ion-implanted buried Zener diode, and laser-wafer-drift-trimming of high stability thin-film resistors to provide outstanding performance at low cost.

The AD588 includes the basic reference cell and three additional amplifiers which provide pin-programmable output ranges. The amplifiers are laser-trimmed for low offset and low drift to maintain the accuracy of the reference. The amplifiers are configured to allow Kelvin connections to the load and/or boosters for driving long lines or high-current loads, delivering the full accuracy of the AD588 where it is required in the application circuit.

The low initial error allows the AD588 to be used as a system reference in precision measurement applications requiring 12-bit absolute accuracy. In such systems, the AD588 can provide a known voltage for system calibration in software and the low drift allows compensation for the drift of other components in a system. Manual system calibration and the cost of periodic recalibration can therefore be eliminated. Furthermore, the mechanical instability of a trimming potentiometer and the potential for improper calibration can be eliminated by using the AD588 in conjunction autocalibration software.

The AD588 is available in six versions. AD588AD and BD grades are packaged in a 16-pin side-brazed ceramic DIP and are specified for the –25°C to +85°C industrial temperature range. The ceramic AD588SD and TD grades are specified for the full military/aerospace temperature range. For surface mount applications, the AD588AE, SE and TE grades will also be available in 20-pin LCC packages.

PRODUCT HIGHLIGHTS

1. The AD588 offers 12-bit absolute accuracy without any user adjustments. Optional fine-trim connections are provided for applications requiring higher precision. The fine-trimming does not alter the operating conditions of the Zener or the buffer amplifiers and thus does not increase the temperature drift.
2. Output noise of the AD588 is very low – typically 6μV p-p. A pin is provided for additional noise filtering using an external capacitor.
3. A precision ±5V tracking mode with Kelvin output connections is available with no external components. Tracking error is less than one millivolt and a fine-trim is available for applications requiring exact symmetry between the +5V and –5V outputs.
4. Pin strapping capability allows configuration of a wide variety of outputs: ±5V, +5V & +10V, –5V & –10V dual outputs or +5V, –5V, +10V, –10V single outputs.
5. Extensive temperature testing at –55°C, –25°C, 0, +25°C, +50°C, +70°C, +85°C and +125°C ensures that the specified temperature coefficient is truly representative of device performance.

*Covered by Patent Number 4,644,253

8

SPECIFICATIONS (typical @ +25°C, +10V output, V_S = ±15V unless otherwise noted[1])

	AD588SD/SE			AD588AD/TD/AE/TE			AD588BD			
	Min	Typ	Max	Min	Typ	Max	Min	Typ	Max	Units
OUTPUT VOLTAGE ERROR										
+10V, −10V Outputs	**−5**		**+5**	**−3**		**+3**	**−1**		**+1**	mV
+5V, −5V Outputs	**−5**		**+5**	**−3**		**+3**	**−1**		**+1**	mV
±5V TRACKING MODE										
Symmetry Error	**−1.5**		**+1.5**	**−1.5**		**+1.5**	**−0.75**		**+0.75**	mV
OUTPUT VOLTAGE DRIFT										
0 to +70°C (A, B, C)					±2		**−1.5**		**+1.5**	ppm/°C
−25°C to +85°C (A, B)				**−3**		**+3**	**−3**		**+3**	ppm/°C
−55°C to +125°C (S, T)	**−6**		**+6**	**−4**		**+4**				ppm/°C
GAIN ADJ AND BAL ADJ[2]										
Trim Range		±4			±4			±4		mV
Input Resistance		150			150			150		kΩ
LINE REGULATION										
T_{min} to T_{max}[3]			**±200**			**±200**			**±200**	μV/V
LOAD REGULATION										
T_{min} to T_{max}										
+10V Output, $0<I_{OUT}<10mA$			**±50**			**±50**			**±50**	μV/mA
−10V Output, $-10<I_{OUT}<0mA$			**±50**			**±50**			**±50**	μV/mA
SUPPLY CURRENT										
T_{min} to T_{max}		6	**10**		6	**10**		6	**10**	mA
Power Dissipation		180	**300**		180	**300**		180	**300**	mW
OUTPUT NOISE (Any Output)										
0.1 to 10Hz		6			6			6		μV p-p
Spectral Density, 100Hz		100			100			100		$nV/\sqrt{Hz}$
LONG-TERM STABILITY (@ +25°C)	15				15			15		ppm/1000hr
BUFFER AMPLIFIERS										
Offset Voltage		100			100			100		μV
Offset Voltage Drift		1			1			1		μV/°C
Bias Current		20			20			20		nA
Open Loop Gain		110			110			110		dB
Output Current A3, A4	−10		+10	−10		+10	−10		+10	mA
Common Mode Rejection (A3, A4)										
V_{CM} = 1V p-p		100			100			100		dB
Short-Circuit Current		50			50			50		mA
TEMPERATURE RANGE										
Specified Performance										
A, B Grades				−25		+85	−25		+85	°C
S, T Grades	−55		+125	−55		+125				°C

NOTES

[1]Output — Configuration
+10V — Figure 2a
−10V — Figure 2c
+5V, −5V, ±5V — Figure 2b

Specifications tested using +10V configuration unless otherwise indicated.

[2]Gain and balance adjustments guaranteed capable of trimming output voltage error and symmetry error to zero.

[3]Test Conditions:
+10V Output $-V_S = -15V$, $13.5V \leq +V_S \leq 18V$
−10V Output $-18V \leq -V_S \leq -13.5V$, $+V_S = 15V$
±5V Output $+V_S = +18V$, $-V_S = -18V$
$+V_S = +10.8V$, $-V_S = -10.8V$

Specifications subject to change without notice.

Specifications shown in boldface are tested on all production units at final electrical test. Results from those tests are used to calculate outgoing quality levels. All min and max specifications are guaranteed, although only those shown in boldface are tested on all production units.

ORDERING GUIDE

Part Number	Initial Error	Temperature Coefficient	Temperature Range °C	Package Option[1]
AD588AD	3mV	3ppm/°C	−25 to +85	Ceramic (D-16)
AD588AE	3mV	3ppm/°C	−25 to +85	LCC (E-20A)
AD588BD	1mV	1.5ppm/°C	−25 to +85*	Ceramic (D-16)
AD588SD	5mV	6ppm/°C	−55 to +125	Ceramic (D-16)
AD588SE	5mV	6ppm/°C	−55 to +125	LCC (E-20A)
AD588TD	3mV	4ppm/°C	−55 to +125	Ceramic (D-16)
AD588TE	3mV	4ppm/°C	−55 to +125	LCC (E-20A)

*Temperature Coefficient specified from 0 to +70°C.
[1]See Section 14 for package outline information.

Applying the AD588

ABSOLUTE MAXIMUM RATINGS*

$+V_S$ to $-V_S$ 36V
Power Dissipation (+25°C)
D Package 600mW
Storage Temperature . . . . −65°C to +150°C
Lead Temperature (Soldering, 10sec) . . . . 300°C
Package Thermal Resistance
D (θ_{JA}/θ_{JC}) . . . . 90/25°C/W
Output Protection: All outputs safe if shorted to ground

*Stresses above those listed under "Absolute Maximum Ratings" may cause permanent damage to the device. This is a stress rating only and functional operation of the device at these or any other conditions above those indicated in the operational sections of this specification is not implied. Exposure to absolute maximum rating conditions for extended periods may affect device reliability.

PIN CONFIGURATIONS

DIP

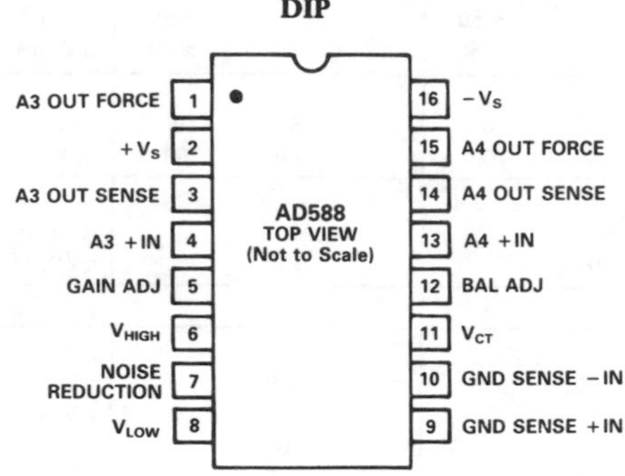

LCC

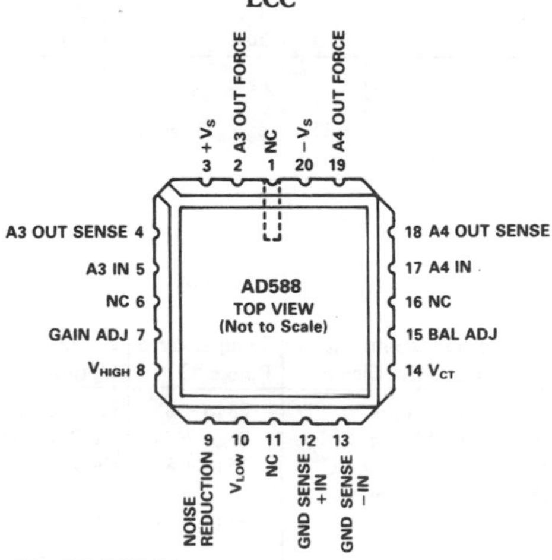

NC = NO CONNECT

THEORY OF OPERATION

The AD588 consists of a buried Zener diode reference, amplifiers used to provide pin programmable output ranges, and associated thin-film resistors as shown in the block diagram of Figure 1. The temperature compensation circuitry provides the device with a temperature coefficient of 1.5ppm/°C or less.

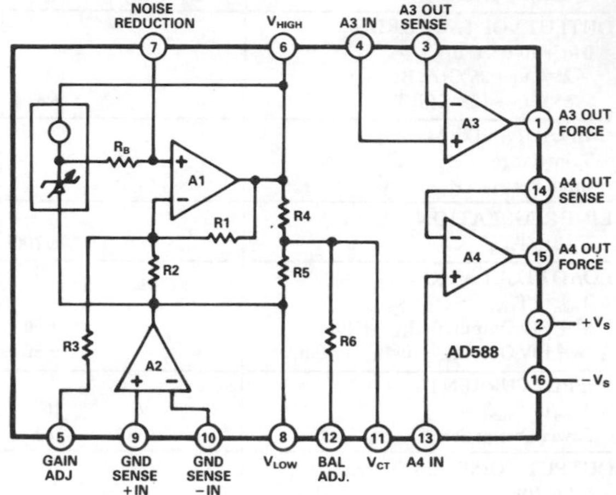

Figure 1. AD588 Functional Block Diagram

Amplifier A1 performs several functions. A1 primarily acts to amplify the Zener voltage from 6.5V to the required 10V output. In addition, A1 also provides for external adjustment of the 10V output through pin 5, the GAIN ADJUST. Using the bias compensation resistor between the Zener output and the non-inverting input to A1, a capacitor can be added at the NOISE REDUCTION pin (pin 7) to form a low pass filter and reduce the noise contribution of the Zener to the circuit. Two matched 10kΩ nominal thin film resistors (R4 & R5) divide the 10V output in half. Pin V_{CT} (pin 11) provides access to the center of the voltage span and pin 12 (BALANCE ADJUST) can be used for fine adjustment of this division.

Ground sensing for the circuit is provided by amplifier A2. The noninverting input (pin 9) senses the system ground which will be transferred to the point on the circuit where the inverting input (pin 10) is connected. This may be pin 6, 8 or 11. The output of A2 drives pin 8 to the appropriate voltage. Thus, if pin 10 is connected to pin 8, the V_{LOW} pin will be the same voltage as the system ground. Alternatively, if pin 10 is connected to the V_{CT} pin, it will be ground and pin 6 and pin 8 will be +5V and −5V respectively.

Amplifiers A3 and A4 are internally compensated and are used to buffer the voltages at pins 6, 8 and 11 as well as to provide a full Kelvin output. Thus, the AD588 has a full Kelvin capability by providing the means to sense a system ground and provide forced and sensed outputs referenced to that ground.

8

Applying the AD588

APPLYING THE AD588

The AD588 can be configured to provide +10V and −10V reference outputs as shown in Figures 2a and 2c respectively. It can also be used to provide +5V, −5V or a ±5V tracking reference as shown in Figure 2b. Table I details the appropriate pin connections for each output range. In each case, pin 9 is connected to system ground and power is applied to pins 2 and 16.

The architecture of the AD588 provides ground sense and uncommitted output buffer amplifiers which offer the user a great deal of functional flexibility. The AD588 is specified and tested in the configurations shown in Figure 2. The user may choose to take advantage of the many other configuration options available with the AD588. However, performance in these configurations is not guaranteed to meet the extremely stringent data sheet specifications.

As indicated in Table I, a +5V buffered output can be provided using amplifier A4 in the +10V configuration (Figure 2a). A −5V buffered output can be provided using amplifier A3 in the −10V configuration (Figure 2c). Specifications are not guaranteed for the +5V or −5V outputs in these configurations. Performance will be similar to that specified for the +10V or −10V outputs.

As indicated in Table I, unbuffered outputs are available at pins 6, 8 and 11. Loading of these unbuffered outputs will impair circuit performance.

Amplifiers A3 and A4 can be used interchangeably. However, the AD588 is tested (and the specifications are guaranteed) with the amplifiers connected as indicated in Figure 2 and Table I. When either A3 or A4 is unused, its output force and sense pins should be connected and the input tied to ground.

Two outputs of the same voltage may be obtained by connecting both A3 and A4 to the appropriate unbuffered output on pin 6, 8 or 11. Performance in these dual output configurations will typically meet data sheet specifications.

CALIBRATION

Generally, the AD588 will meet the requirements of a precision system without additional adjustment. Initial output voltage error of 1mV and output noise specs of 10μV p-p allow for accuracies of 12-16 bits. However, in applications where an even greater level of accuracy is required, additional calibration may be called for. Provision for trimming has been made through the use of the GAIN ADJUST and BALANCE ADJUST pins (pins 5 and 12 respectively).

The AD588 provides a precision 10V span with a center tap (V_{CT}) which is used with the buffer and ground sense amplifiers to achieve the voltage output configurations in Table I. GAIN ADJUST and BALANCE ADJUST can be used in any of these configurations to trim the magnitude of the span voltage and the position of the center tap within the span. The GAIN ADJUST should be performed first. Although the trims are not interactive within the device, the GAIN trim will move the BALANCE trim point as it changes the magnitude of the span.

Figure 2b. shows GAIN and BALANCE trims in a +5V and −5V tracking configuration. A 100kΩ 20-turn potentiometer is used for each trim. The potentiometer for GAIN trim is connected between pins 6 (V_{HIGH}) and 8 (V_{LOW}) with the wiper connected to pin 5 (GAIN ADJ). The potentiometer is adjusted to produce exactly 10V between pins 1 and 15, the amplifier outputs. The BALANCE potentiometer, also connected between pins 6 and 8 with the wiper to pin 12 (BAL ADJ), is then adjusted to center the span from +5V to −5V.

Trimming in other configurations works in exactly the same manner. When producing +10V and +5V, GAIN ADJ is used to trim +10V and BAL ADJ is used to trim +5V. In the −10V and −5V configuration, GAIN ADJ is again used to trim the magnitude of the span, −10V, while BAL ADJ is used to trim the center tap, −5V.

Range	Connect Pin 10 to Pin:	Unbuffered[1] Output on Pins −10V	−5V	0V	+5V	+10V	Buffered Output Connections	Buffered Output on Pins −10V	−5V	0V	+5V	+10V
+10V	8	–	–	8	11	6	11-13 & 14-15	–	–	–	15	–
							6-4 & 3-1	–	–	–	–	1
−5V or +5V	11	–	8	11	6	–	8-13 & 14-15	–	15	–	–	–
							6-4 & 3-1	–	–	–	1	–
−10V	6	8	11	6	–	–	8-13 & 14-15	15	–	–	–	–
							11-4 & 3-1	–	1	–	–	–
+5V	11	–	–	–	6	–	6-4 & 3-1	–	–	–	1	–
−5V		–	8	–	–	–	8-13 & 14-15	–	15	–	–	–

[1] "Unbuffered" outputs should not be loaded.

Table I. AD588 Connections

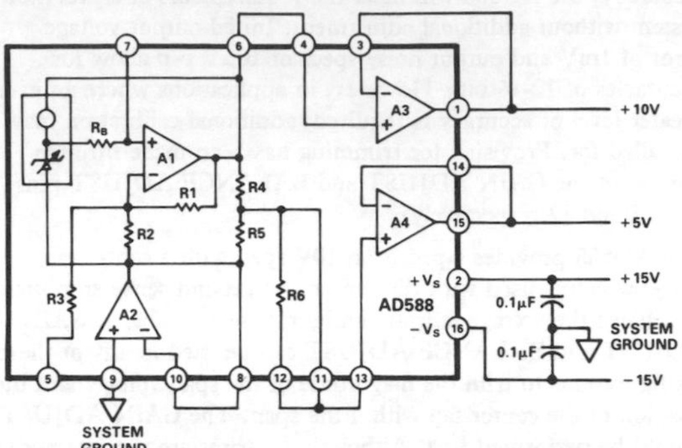

Figure 2a. +10V Output

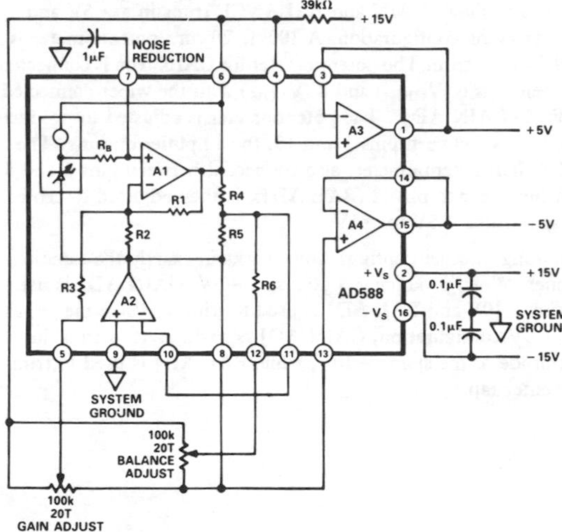

Figure 2b. +5V and −5V Outputs

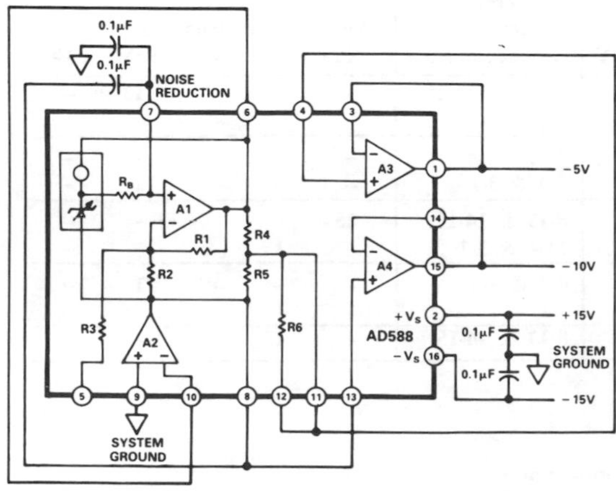

Figure 2c. −10V Output

In single output configurations, GAIN ADJ is used to trim outputs utilizing the full span (+10V or −10V) while BAL ADJ is used to trim outputs using half the span (+5V or −5V).

Input impedance on both the GAIN ADJUST and BALANCE ADJUST pins is approximately 150kΩ. The GAIN ADJUST trim network effectively attenuates the 10V across the trim potentiometer by a factor of about 1500 to provide a trim range of −3.5mV to +7.5mV with a resolution of approximately 550µV/turn (20 turn potentiometer). The BALANCE ADJUST trim network attenuates the trim voltage by a factor of about 1400, providing a trim range of ±4.5mV with resolution of 450µV/turn.

Trimming the AD588 introduces no additional errors over temperature so precision potentiometers are not required.

For single output voltage ranges, or in cases when BALANCE ADJUST is not required, pin 12 should be connected to pin 11. If GAIN ADJUST is not required, pin 5 should be left floating.

NOISE PERFORMANCE AND REDUCTION

The noise generated by the AD588 is typically less than 6µV p-p over the 0.1Hz to 10Hz band. Noise in a 1MHz bandwidth is approximately 600µV p-p. The dominant source of this noise is the buried Zener which contributes approximately $100nV/\sqrt{Hz}$. In comparison, the op amp's contribution is negligible. Figure 3 shows the 0.1Hz to 10Hz noise of a typical AD588.

8

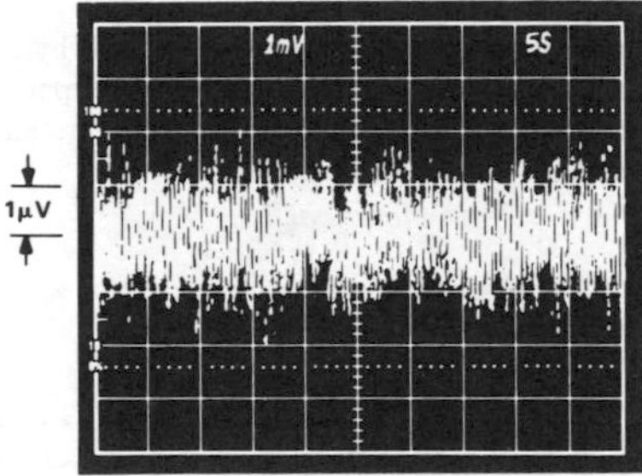

Figure 3. 0.1Hz to 10Hz Noise

If further noise reduction is desired, an optional capacitor may be added between the NOISE REDUCTION pin and ground as shown in Figure 2b. This will form a low pass filter with the 4kΩ R_B on the output of the Zener cell. A 1µF capacitor will have a 3dB point at 40Hz and will reduce the high frequency (to 1MHz) noise to about 200µV p-p. Figure 4 shows the 1MHz noise of a typical AD588 both with and without a 1µF capacitor.

Note that a second capacitor is needed in order to implement the NOISE REDUCTION feature when using the AD588 in the −10V mode (Figure 2c.). The NOISE REDUCTION capacitor is limited to 0.1µF maximum in this mode.

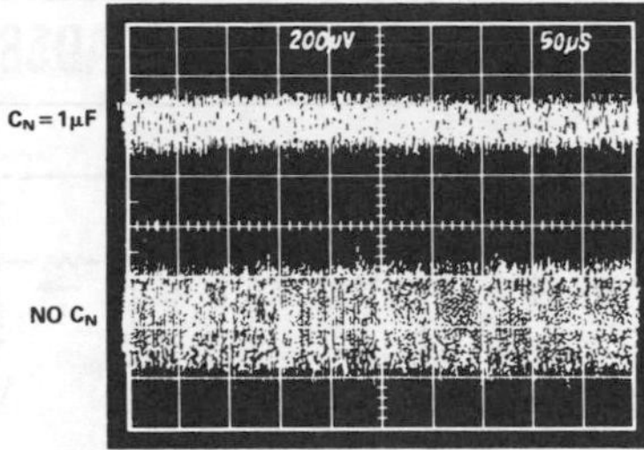

Figure 4. Effect of 1µF Noise Reduction Capacitor on Broadband Noise

TURN-ON TIME

Upon application of power (cold start), the time required for the output voltage to reach its final value within a specified error band is the turn-on settling time. Two components normally associated with this are: time for active circuits to settle and time for thermal gradients on the chip to stabilize. Figure 5 shows the turn-on characteristics of the AD588. It shows the settling to be about 600µs. Note the absence of any thermal tails when the horizontal scale is expanded to 2ms/cm in Figure 5b.

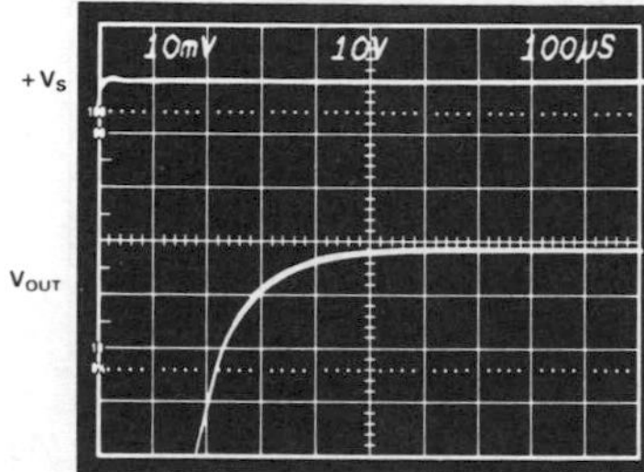

a. Electrical Turn-On

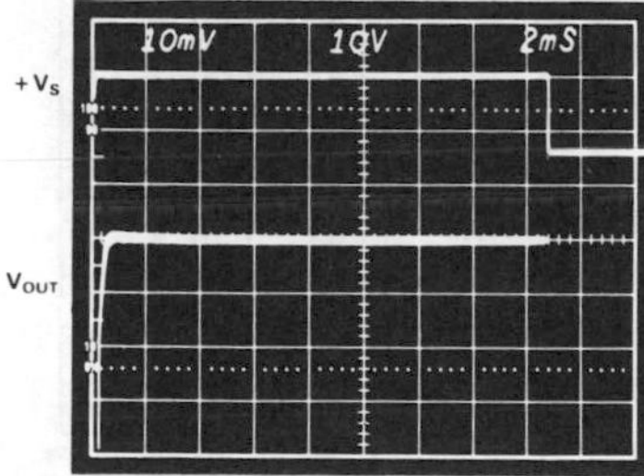

b. Extended Time Scale

Figure 5. Turn-On Characteristics

Output turn-on time is modified when an external noise reduction capacitor is used. When present, this capacitor presents an additional load to the internal Zener diode's current source, resulting in a somewhat longer turn-on time. In the case of a 1µF capacitor, the initial turn-on time is approximately 60ms (see Figure 6).

Note: If the NOISE REDUCTION feature is used in the ±5V configuration, a 39kΩ resistor between pins 6 and 2 is required for proper startup.

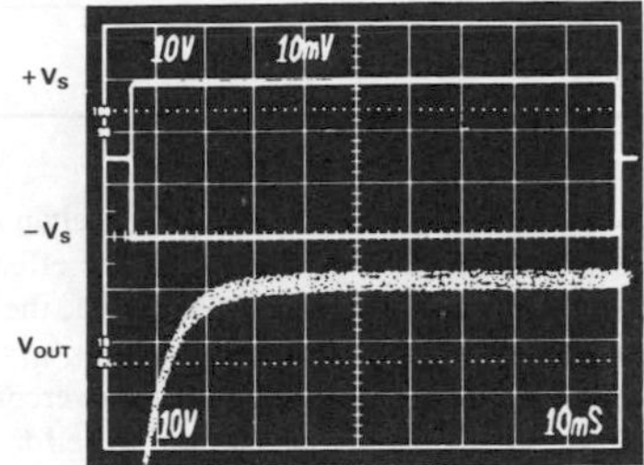

Figure 6. Turn-on with 1µF C_N

TEMPERATURE PERFORMANCE

The AD588 is designed for precision reference applications where temperature performance is critical. Extensive temperature testing ensures that the device's high level of performance is maintained over the operating temperature range.

Figure 7 shows typical output voltage drift for the AD588BD and illustrates the test methodology. The box in Figure 7 is bounded on the sides by the operating temperature extremes and on top and bottom by the maximum and minimum output voltages measured over the operating temperature range. The slope of the diagonal drawn from the lower left corner of the box determines the performance grade of the device.

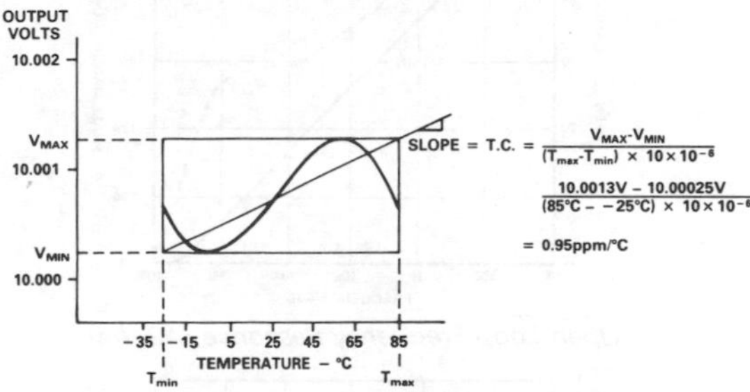

Figure 7. Typical AD588BD Temperature Drift

Each AD588A and B grade unit is tested at −25°C, 0°C, +25°C, +50°C, +70°C and +85°C. Each AD588S and T grade unit is tested at −55°C, −25°C, 0°C, +25°C, +50°C, +70°C and +125°C. This approach ensures that the variations of output voltage that occur as the temperature changes within the specified range will be contained within a box whose diagonal has a slope equal to the maximum specified drift. The position of the box on the vertical scale will change from device to device as initial error and the shape of the curve vary. Maximum height of the box for the appropriate temperature range is shown in Figure 8. Duplication of these results requires a combination of high accuracy and stable temperature control in a test system. Evaluation of the AD588 will produce a curve similar to that in Figure 7, but output readings may vary depending on the test methods and equipment utilized.

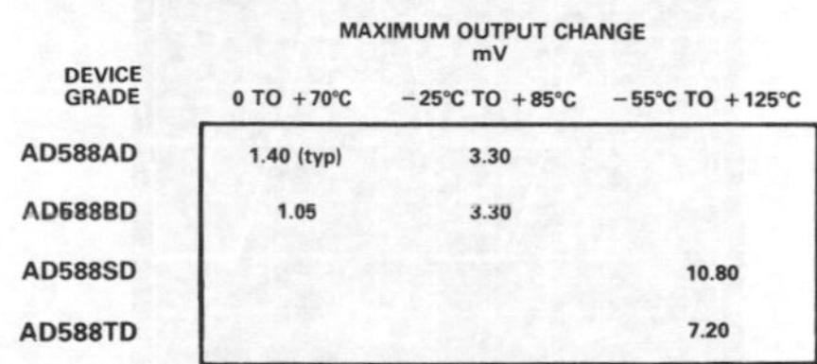

DEVICE GRADE	MAXIMUM OUTPUT CHANGE mV 0 TO +70°C	−25°C TO +85°C	−55°C TO +125°C
AD588AD	1.40 (typ)	3.30	
AD588BD	1.05	3.30	
AD588SD			10.80
AD588TD			7.20

Figure 8. Maximum Output Change – mV

KELVIN CONNECTIONS

Force and sense connections, also referred to as Kelvin connections, offer a convenient method of eliminating the effects of voltage drops in circuit wires. As seen in Figure 9a, the load current and wire resistance produce an error ($V_{ERROR} = R \times I_L$) at the load. The Kelvin connection of Figure 9b overcomes the problem by including the wire resistance within the forcing loop of the amplifier and sensing the load voltage. The amplifier corrects for any errors in the load voltage. In the circuit shown, the output of the amplifier would actually be at 10 volts + V_{ERROR} and the voltage at the load would be the desired 10 volts.

The AD588 has three amplifiers which can be used to implement Kelvin connections. Amplifier A2 is dedicated to the ground force-sense function while uncommitted amplifiers A3 and A4 are free for other force-sense chores.

In some single-output applications, one amplifier may be unused.

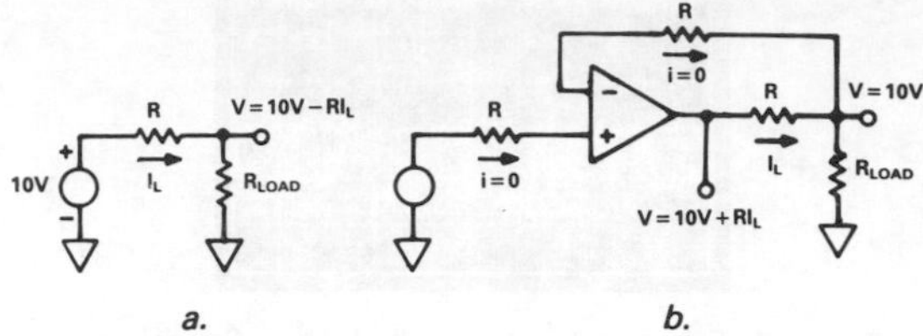

Figure 9. Advantage of Kelvin Connection

In such cases, the unused amplifier should be connected as a unity-gain follower (force + sense pin tied together) and the input should be connected to ground.

An unused amplifier section may be used for other circuit functions as well. The curves on this page show the typical performance of A3 and A4.

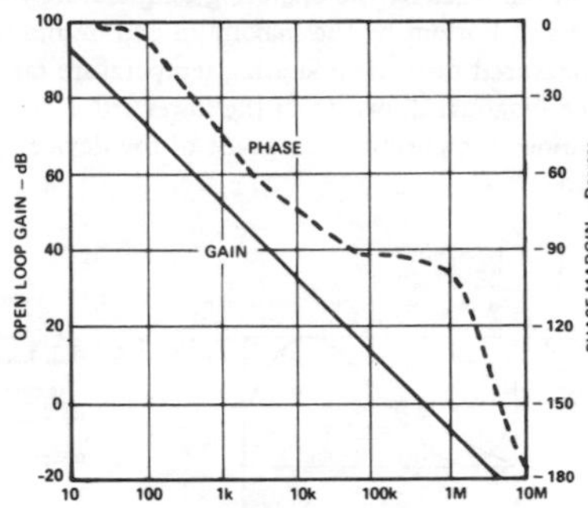

Open Loop Frequency Response (A3, A4)

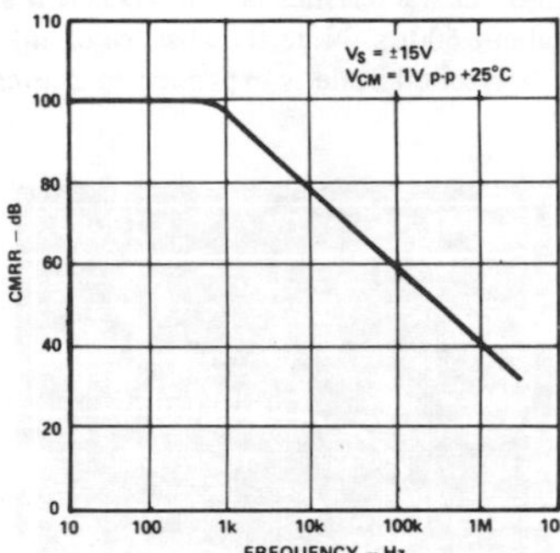

Common Mode Rejection vs. Frequency (A3, A4)

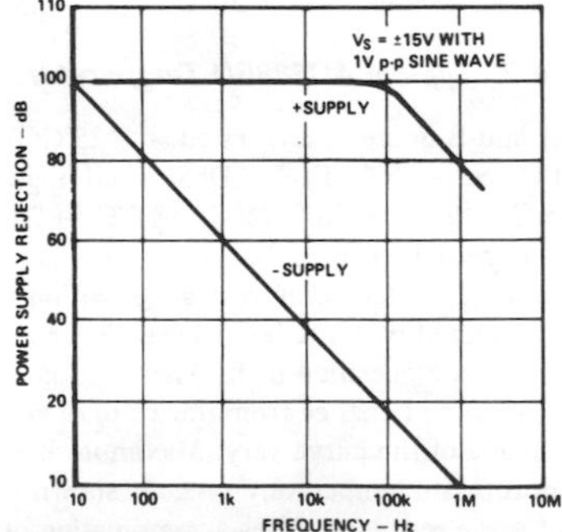

Power Supply Rejection vs. Frequency (A3, A4)

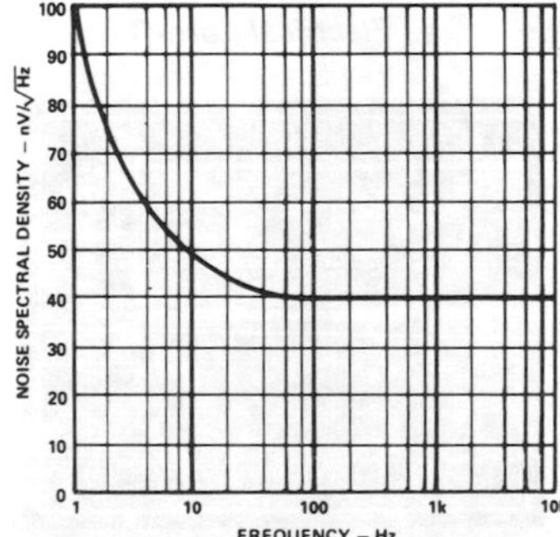

Input Noise Voltage Spectral Density

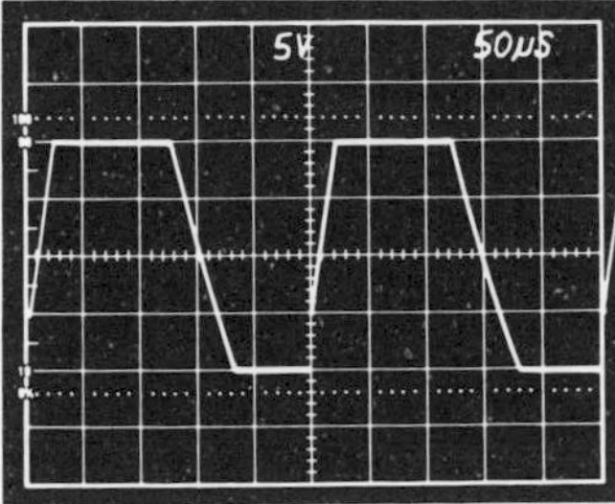

Unity Gain Follower Pulse Response (Large Signal)

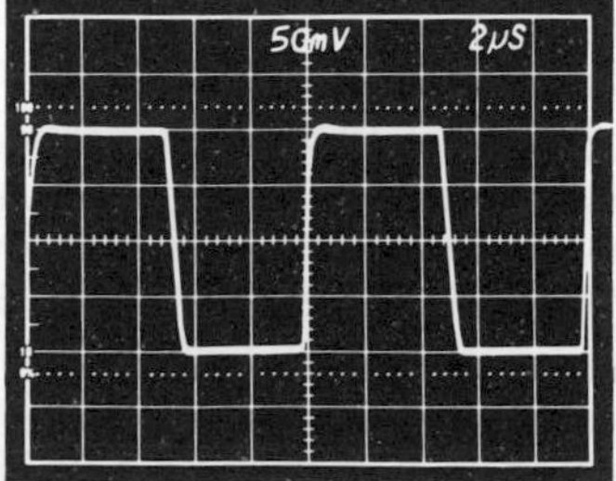

Unity Gain Follower Pulse Response (Small Signal)

8

DYNAMIC PERFORMANCE

The output buffer amplifiers (A3 and A4) are designed to provide the AD588 with static and dynamic load regulation superior to less complete references.

Many A/D and D/A converters present transient current loads to the reference, and poor reference response can degrade the converter's performance.

Figure 10 displays the characteristics of the AD588 output amplifier driving a 0 to 10mA load.

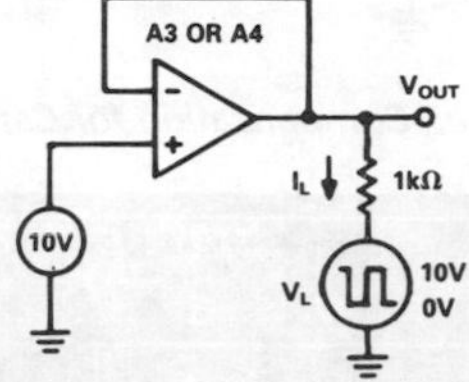

Figure 10a. Transient Load Test Circuit

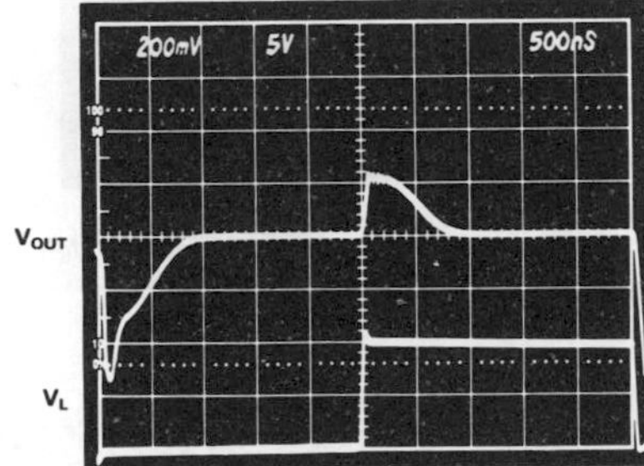

Figure 10b. Large-Scale Transient Response

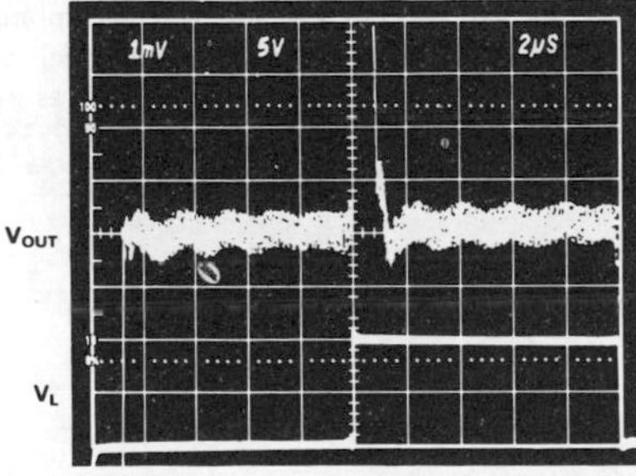

Figure 10c. Fine Scale Settling for Transient Load

Figure 11 displays the output amplifier characteristics driving a 5mA to 10mA load, a common situation found when the reference is shared among multiple converters or is used to provide a bipolar offset current.

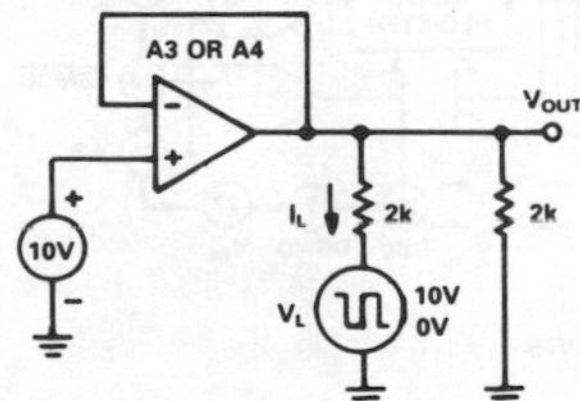

Figure 11a. Transient and Constant Load Test Circuit

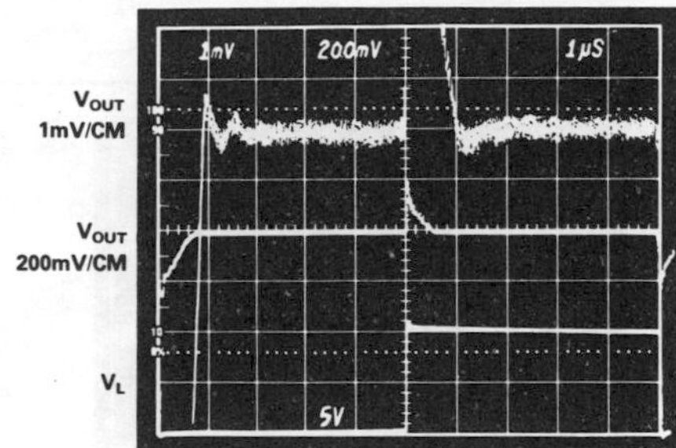

Figure 11b. Transient Response 5-10mA Load

In some applications, a varying load may be both resistive and capacitive in nature, or be connected to the AD588 by a long capacitive cable.

Figure 12 displays the output amplifier characteristics driving a 1,000pF, 0-to-10mA load.

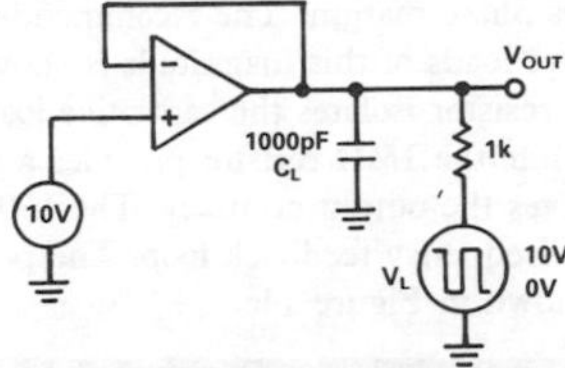

Figure 12a. Capacitive Load Transient Response Test Circuit

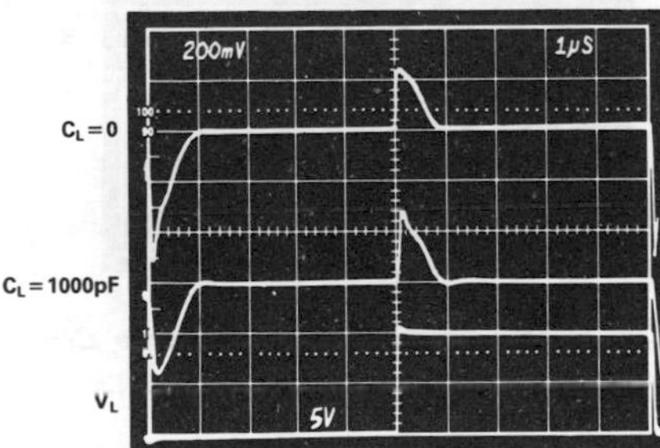

Figure 12b. Output Response with Capacitive Load

Figure 13 displays the crosstalk between output amplifiers. The top trace shows the output of A4, dc-coupled and offset by 10 volts, while the output of A3 is subjected to a 0-to-10mA load current step. The transient at A4 settles in about 1μs, and the load-induced offset is about 100μV.

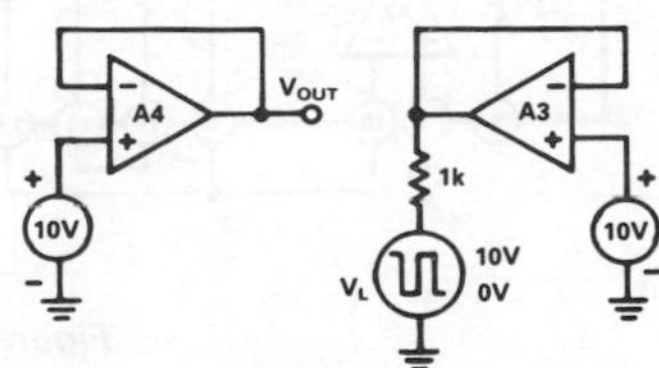

Figure 13a. Load Crosstalk Test Circuit

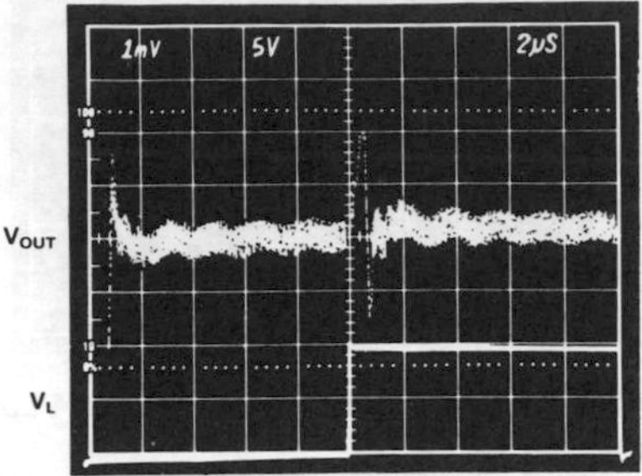

Figure 13b. Load Crosstalk

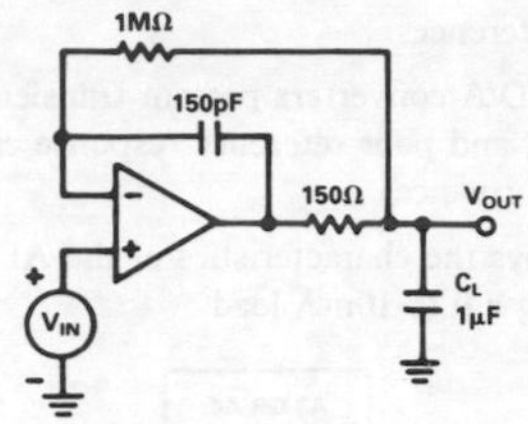

Figure 14b. Compensation for Capacitive Loads

Attempts to drive a large capacitive load (in excess of 1,000pF) may result in ringing or oscillation, as shown in the step response photo (Figure 14a). This is due to the additional pole formed by the load capacitance and the output impedance of the amplifier, which consumes phase margin. The recommended method of driving capacitive loads of this magnitude is shown in Figure 14b. The 150Ω resistor isolates the capacitive load from the output stage, while the 1MΩ resistor provides a dc feedback path and preserves the output accuracy. The 150pF capacitor provides a high-frequency feedback loop. The performance of this circuit is shown in Figure 14c.

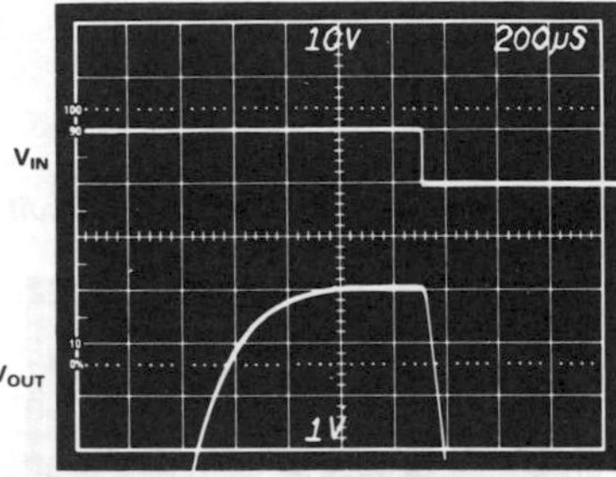

Figure 14c. Output Amplifier Step Response Using Figure 14b Compensation

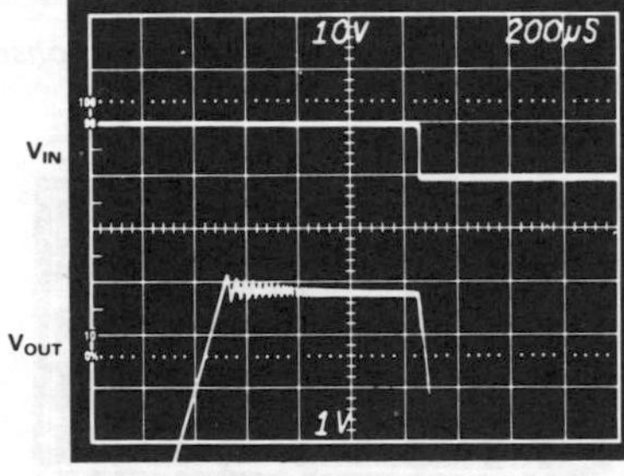

Figure 14a. Output Amplifier Step Response, $C_L = 1\mu F$

USING THE AD588 WITH CONVERTERS

The AD588 is an ideal reference for a wide variety of A/D and D/A converters. Several representative examples follow.

14-Bit Digital-to-Analog Converter – AD7535

High resolution CMOS D/A converters require a reference voltage of high precision to maintain rated accuracy. The combination of the AD588 and AD7535 takes advantage of the initial accuracy, drift and full Kelvin output capability of the AD588 as well as the resolution, monotonicity and accuracy of the AD7535 to produce a subsystem with outstanding characteristics.

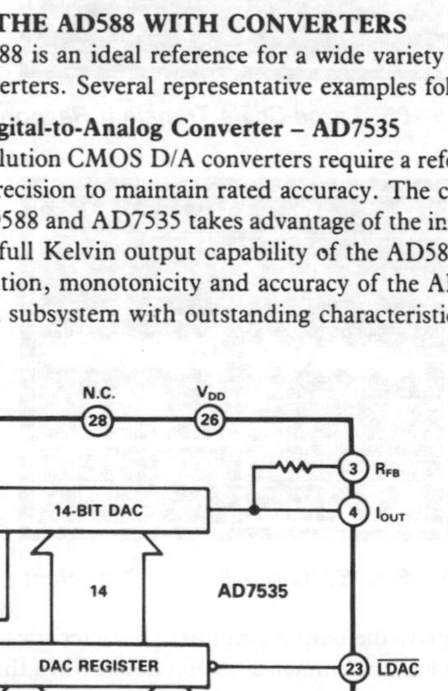

Figure 15. AD588/AD7535 Connections

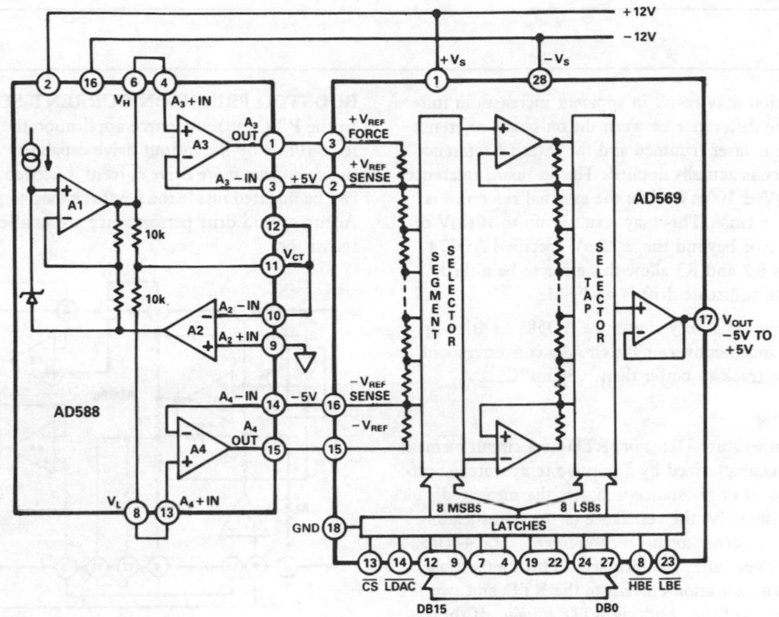

Figure 16. High-Accuracy ±5V Tracking Reference for AD569

16-Bit Digital-to-Analog Converter – AD569

Another application which fully utilizes the capabilities of the AD588 is supplying a reference for the AD569, as shown in Figure 16. Amplifier A2 senses system common and forces V_{CT} to assume this value, producing +5V and −5V at pins 6 and 8 respectively. Amplifiers A3 and A4 buffer these voltages out to the appropriate reference force-sense pins of the AD569. The full Kelvin scheme eliminates the effect of the circuit traces or wires and the wire bonds of the AD588 and AD569 themselves, which would otherwise degrade system performance.

SUBSTITUTING FOR INTERNAL REFERENCES

Many converters include built-in references. Unfortunately, such references are the major source of drift in these converters. By using a more stable external reference like the AD588, drift performance can be improved dramatically.

12-Bit Analog-to-Digital Conveter – AD574A

The AD574A is specified for gain drift from 10ppm/°C to 50ppm/°C, (depending on grade) using the on-chip reference. The reference contributes typically 75% of this drift. Therefore, the total drift using an AD588 to supply the reference can be improved by a factor of 3 to 4.

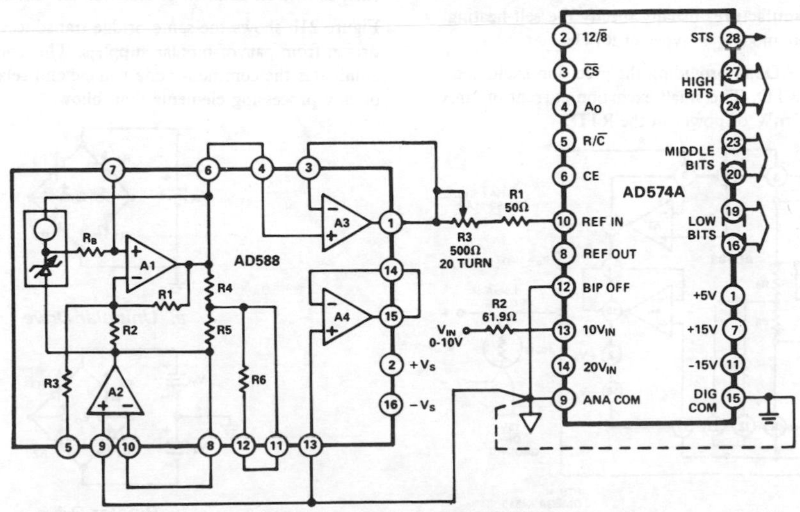

Figure 17. AD588/AD574A Connections

Using this combination may result in apparent increases in full-scale error due to the difference between the on-board reference by which the device is laser trimmed and the external reference with which the device is actually applied. The on-board reference is specified to be 10V ±100mV while the external reference is specified to be 10V ±1mV. This may result in up to 101mV of apparent full-scale error beyond the ±25mV specified AD574 gain error. Resistors R2 and R3 allow this error to be nulled. Their contribution to full-scale drift is negligible.

The high output drive capability allows the AD588 to drive up to 6 converters in a multi-converter system. All converters will have gain errors that track to better than ±5ppm/°C.

RTD EXCITATION

The Resistance Temperature Detector (RTD) is a circuit element whose resistance is characterized by a positive temperature coefficient. A measurement of resistance indicates the measured temperature. Unfortunately, the resistance of the wires leading to the RTD often adds error to this measurement. The 4-wire ohms measurement overcomes this problem. This method uses two wires to bring an excitation current to the RTD and two additional wires to tap off the resulting RTD voltage. If these additional two wires go to a high input impedance measurement circuit, the effect of their resistance is negligible. Therefore, they transmit the true RTD voltage.

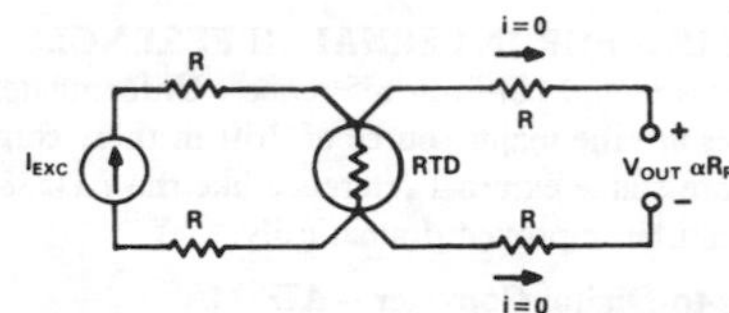

Figure 18. 4-Wire Ohms Measurement

A practical consideration when using the 4-wire ohms technique with an RTD is the self-heating effect that the excitation current has on the temperature of the RTD. The designer must choose the smallest practical excitation current that still gives the desired resolution. RTD manufactures usually specify the self-heating effect of each of their models or types of RTDs.

Figure 19 shows an AD588 providing the precision excitation current for a 100Ω RTD. The small excitation current of 1mA dissipates a mere 0.1mW of power in the RTD.

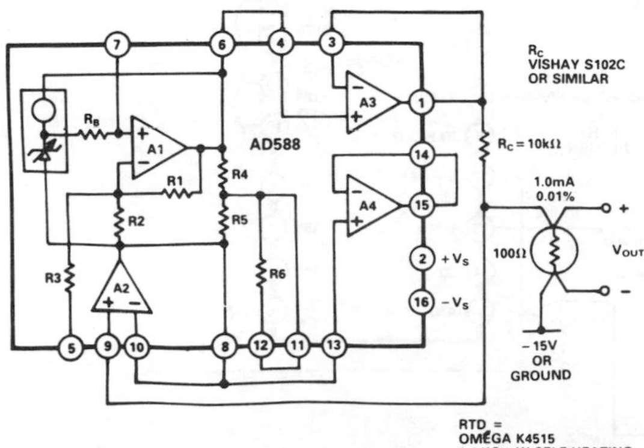

Figure 19. Precision Current Source for RTD

BOOSTED PRECISION CURRENT SOURCE

In the RTD current-source application the load current is limited to ±10mA by the output drive capability of amplifier A3. In the event that more drive current is needed, a series pass transistor can be inserted inside the feedback loop to provide higher current. Accuracy and drift performance are unaffected by the pass transistor.

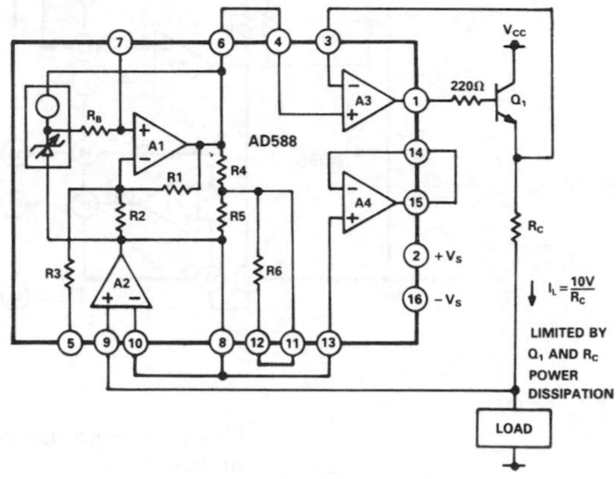

Figure 20. Boosted Precision Current Source

BRIDGE DRIVER CIRCUITS

The Wheatstone bridge is a common transducer. In its simplest form, a bridge consists of 4 two terminal elements connected to form a quadrilateral, a source of excitation connected along one of the diagonals and a detector comprising the other diagonal. Figure 21a shows a simple bridge driven from a unipolar excitation supply. E_O, a differential voltage, is proportional to the deviation of the element from the initial bridge values. Unfortunately, this bridge output voltage is riding on a common-mode voltage equal to approximately $V_{IN}/2$. Further processing of this signal may necessarily be limited to high common-mode rejection techniques such as instrumentation or isolation amplifiers.

Figure 21b shows the same bridge transducer, but this time it is driven from pair of bipolar supplies. This configuration ideally eliminates the common-mode voltage and relaxes the restrictions on any processing elements that follow.

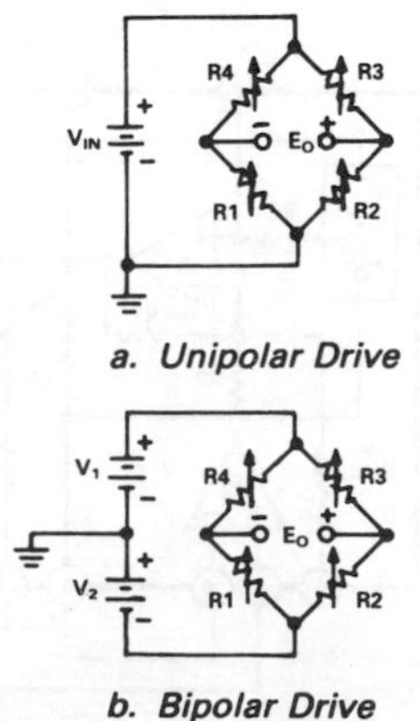

Figure 21. Bridge Transducer Excitation

8

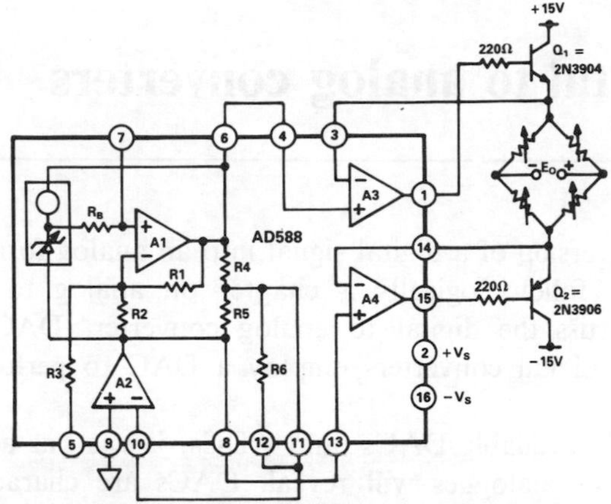

Figure 22. Bipolar Bridge Drive

As shown in Figure 22, the AD588 is an excellent choice for the control element in a bipolar bridge driver scheme. Transistors Q1 and Q2 serve as series pass elements to boost the current drive capability to the 28mA required by a typical 350Ω bridge. A differential gain stage may still be required if the bridge balance is not perfect. Such gain stages can be expensive.

Additional common-mode voltage reduction is realized by using the circuit illustrated in Figure 23. A1, the ground sense amplifier, servo's the supplies on the bridge to maintain a virtual ground at one center tap. The voltage which appears on the opposite center tap is now single-ended (referred to ground) and can be amplified by a less expensive circuit.

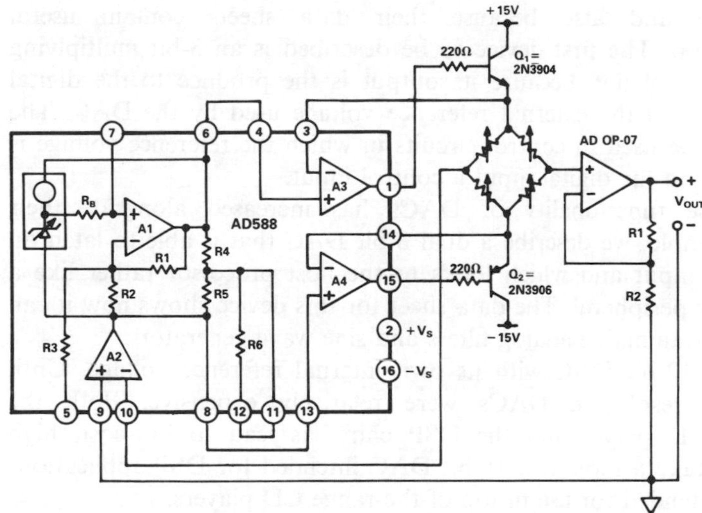

Figure 23. Floating Bipolar Bridge Drive with Minimum CMV

Chapter 3 Digital to analog converters

This chapter describes the conversion of a digital signal into an analog form. You might expect this chapter to follow logically a chapter on analog to digital conversion. However, we discuss the digital to analog converter (DAC) first, because many real analog to digital converters employ a DAC to perform the conversion process indirectly.

The number of commercially available DACs (and ADCs) is vast, as a glance at some of the manufacturers' catalogues will reveal. DACs are characterized according to precision, type, speed, construction, cost, and so on. In this chapter we have attempted to introduce a range of typical devices.

Before we look at actual examples of DACs, two articles provide an overview of some of the basic principles involved in digital to analog conversion. In particular, the error characteristics of DACs are described because the nature of the errors introduced in the conversion process is one principal factor that influences the engineer's choice of a certain device.

The devices described in this chapter have been chosen to provide a representative sample and also because their data sheets contain useful applications information. The first device to be described is an 8-bit multiplying DAC. It is called 'multiplying' because its output is the produce to the digital input to be converted and the external reference voltage used by the DAC. The multiplying DAC can be used in control circuits in which the reference voltage is a time-varying signal and the digital input a control input.

Over the years the functionality of DACs has increased alongside their performance. For example, we describe a dual 8-bit DAC that is able to latch on to the digital control input and which looks to the host processor rather like a typical microprocessor peripheral. The data sheet for this device shows how it can be used to design programmable analog filters and sine wave generators.

We then look at a 12-bit DAC with its own internal reference voltage. Until fairly recently, high resolution DACs were relatively expensive. Both the introduction of the CD player and the DSP chip has lead to low-cost, high precision DACs. We take a look at a 16-bit DAC intended for DSP applications and an 18-bit DAC intended for use in top-of-the-range CD players.

Some designers require high resolution, others require high speed. We therefore include a data sheet for an 8-bit video DAC. This is followed by the data sheet and application note for an 80MHz 6-bit triple DAC intended for applications in high-speed graphics processors.

The chapter ends with an introduction to the companding DAC. This device is used to process telephone signals by compressing their dynamic range. Essentially, unlike conventional linear DACs, it has a non-linear digital to analog transfer function.

Digital to Analogue Converters

1. A digital to analogue converter (DAC) is a device which converts a digital data input into a corresponding analogue output. This output usually takes the form of a voltage or current.

1.1 Ideal output characteristics

If a unipolar voltage output and normal binary coding are assumed, then the ideal transfer function of a linear DAC may be written as:

$V_{out} = V_{FS} \ (B_1.2^{-1} + B_2.2^{-2} + B_3.2^{-3} + \ldots. + B_n.2^{-n})$

where B_1 is the most significant bit input (MSB) and B_n is the least significant bit input (LSB). Bits 1 to n can each assume a value of '1' or '0'. The number of bit inputs a DAC possesses is known as the **resolution** of the converter.

The smallest increment of output voltage is that contributed by the LSB and is equal to $V_{FS}.2^{-n}$.

The terms 'MSB', 'LSB' etc., are frequently used interchangeably to describe either the digital input or the corresponding analogue output.

The maximum output from a DAC is known as full-scale output (V_{FSO}).

It occurs when all inputs are '1' and is equal to $V_{FS}\left(\frac{(2^n - 1)}{2^n}\right)$. For example the maximum output of a 3-bit DAC is ⅞ V_{FS}.

The transfer function graph of an ideal 3-bit DAC is shown in Fig. 1. For each of the 8 input codes there exists a discrete analogue output level,

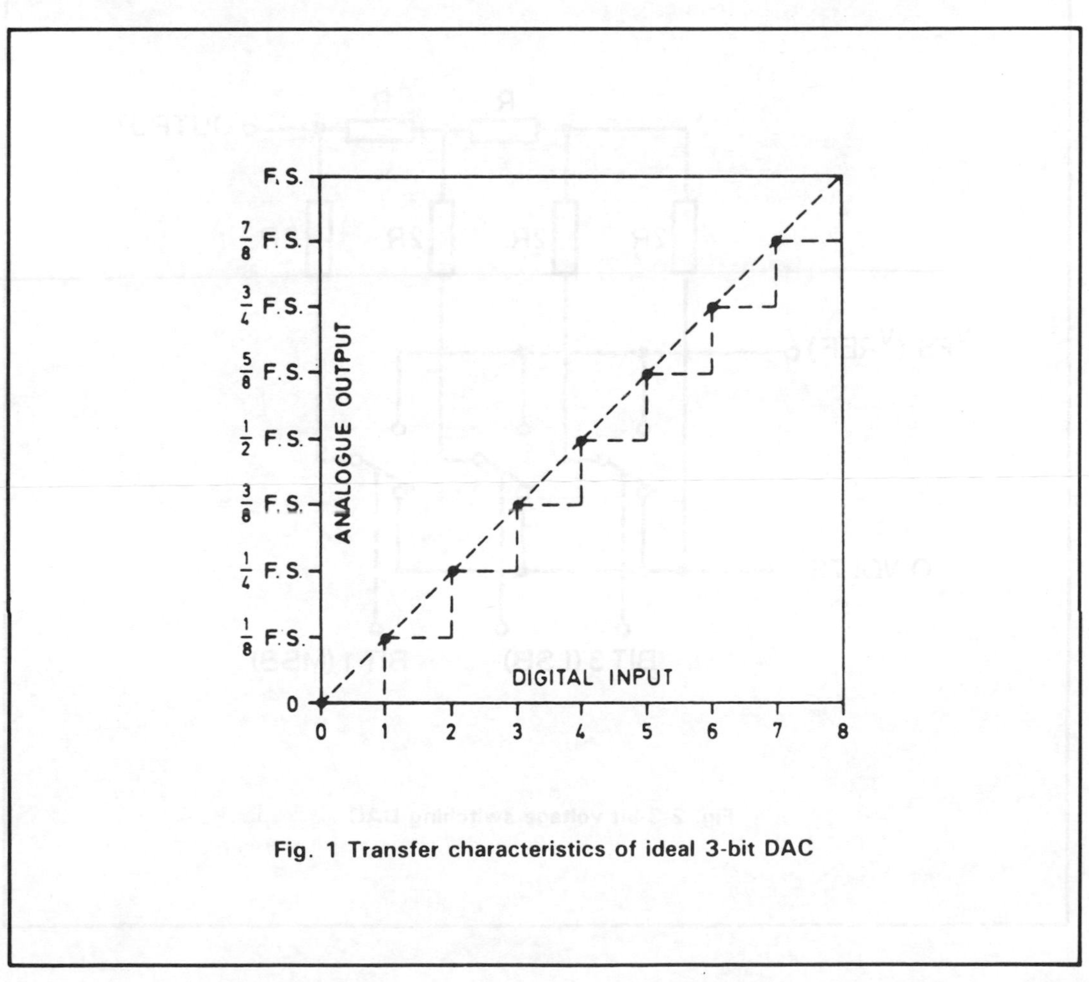

Fig. 1 Transfer characteristics of ideal 3-bit DAC

represented by a point on the graph. It should be emphasised that the transfer characteristic is not a continuous function and it is, therefore, not strictly correct to join the points with a continuous line, since this would imply that non-integral input codes and corresponding levels existed. However, a straight line is often drawn between zero and full-scale to represent the 'ideal' transfer function on which all the points should lie.

Similarly, if the input code of a DAC is incremented using, say a binary counter and clock generator, then the analogue output will be a staircase waveform. DAC transfer functions are frequently drawn as a staircase, since this is a convenient way of illustrating various errors that may occur in a DAC. However, such a graph is, strictly speaking, a plot of analogue output v time rather than output v input code.

1.2 Practical DAC circuits

Fig.2 shows an example of a 3-bit DAC circuit based on a voltage-switching R-2R ladder network, a technique widely used in Plessey converters.

Each 2R element is connected either to 0V or V_{FS} (V_{REF}) by transistor switches. Binary weighted voltages are produced at the output of the R-2R ladder, the value being proportional to the digital input number.

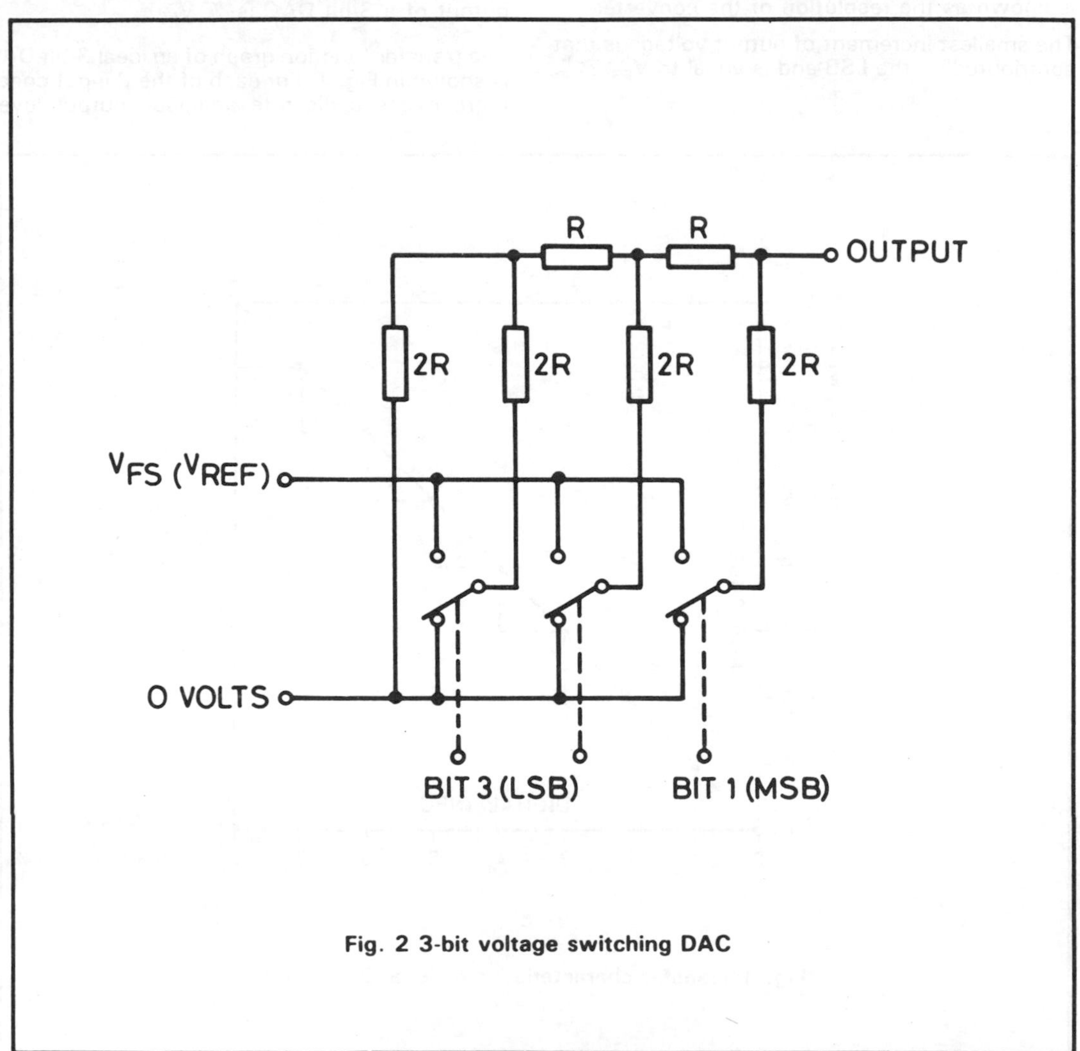

Fig. 2 3-bit voltage switching DAC

For example, it is fairly easy to see that if bit 1 is '1' and bit 2 and 3 are '0' then an output of $V_{FS}/2$ is produced. This is because the resistance of the ladder looking from the output through the first R is 2R, which forms a 2:1 attenuator with the 2R in series with the MSB switch. Output voltages for other input codes can similarly by calculated, and it can be seen that the ladder may be extended to any number of bits.

1.3 D-A parameters and definitions

1.3.1 Converter errors

The ideal DAC assumes that all the resistors are perfectly matched and that the switches have zero resistance. In a practical converter this will not be the case and various errors will occur in the output.

1.3.2 Monotonicity

When the input code of a DAC is increased in 1 LSB steps the analogue output of the DAC should also increase, staircase fashion. If the output always increases in this manner then the DAC is said to be monotonic, i.e. the output is a single-valued function of the input. If, due to errors in the bit weighting, the output of the DAC decreases at any step, as shown in Fig. 3, then the DAC is said to be non-monotonic.

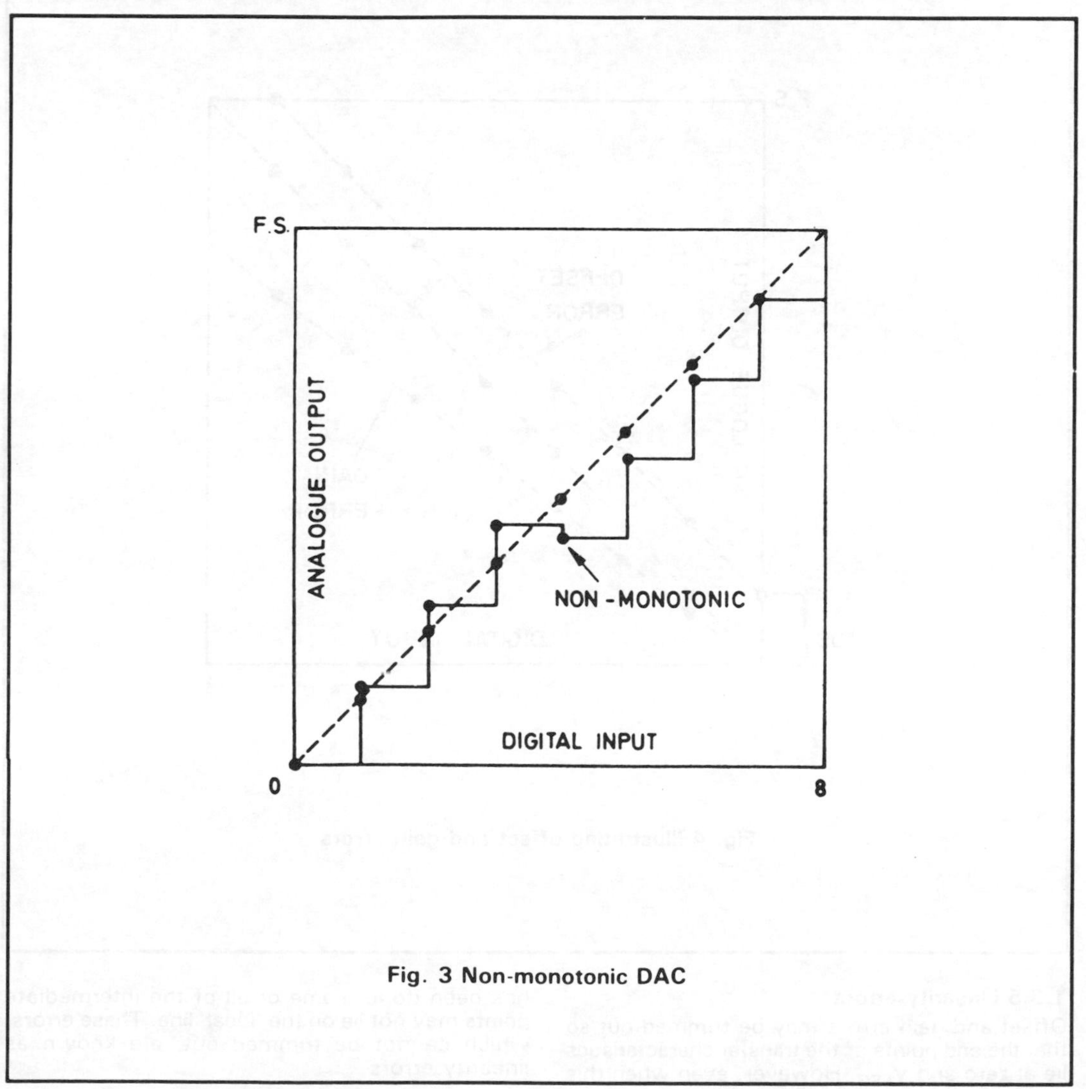

Fig. 3 Non-monotonic DAC

1.3.3 Offset (zero error)

Assuming unipolar operation and normal binary coding, when the input code is zero then the DAC outputs should also be zero. However, due to package lead resistances and offset voltages in the switches this will not be the case, and a small output offset may exist. This has the effect of shifting the transfer function so that it no longer passes through zero, as shown in Fig. 4.

1.3.4 Gain error

If the reference voltage of a DAC is exactly the nominal value then the transfer characteristics of the converter should follow the ideal straight line. However, due to imperfections in the converter the transfer function may diverge from this line, as shown in Fig. 4. This error is known as gain error and is the difference between the slope of the actual transfer characteristic and the slope of the ideal transfer characteristic.

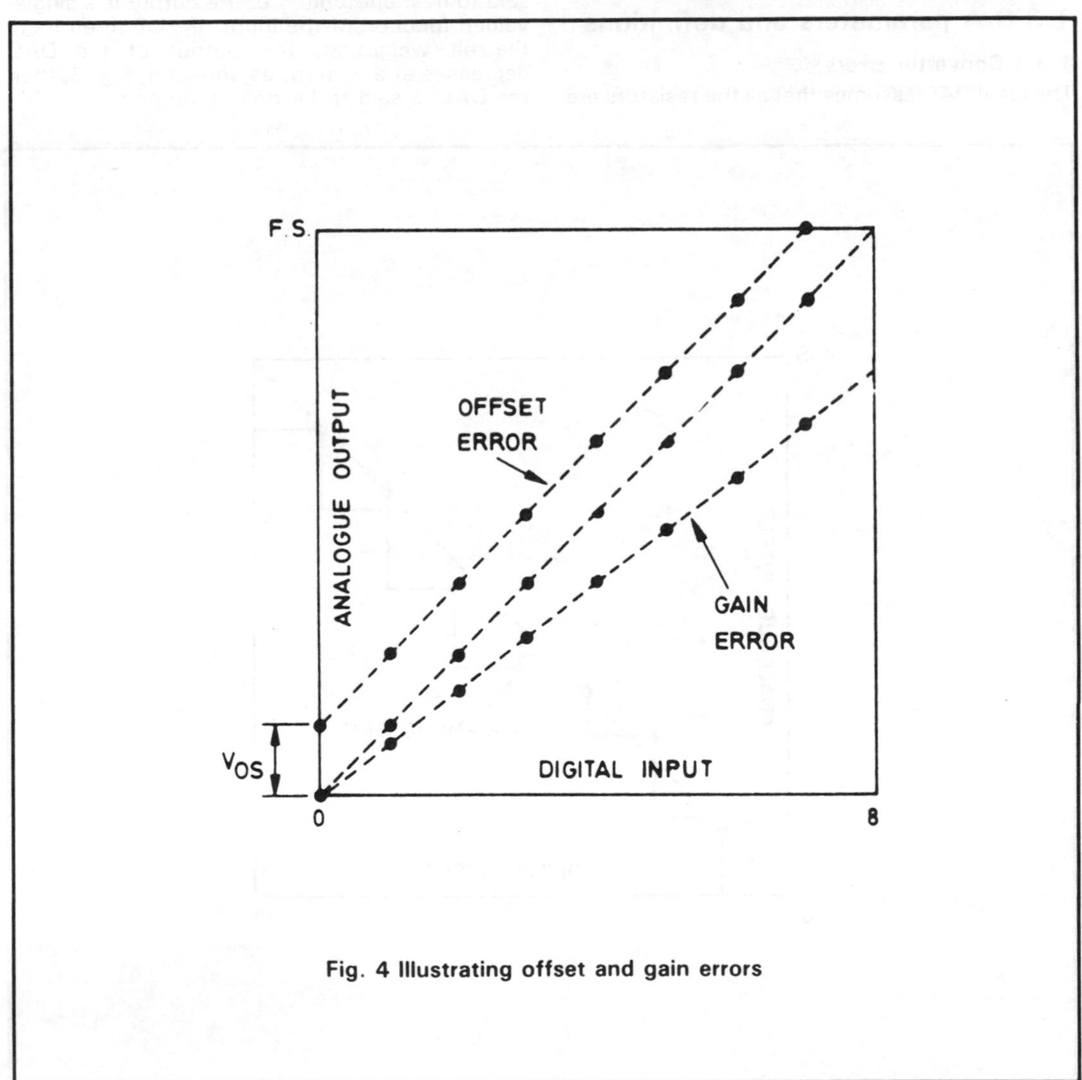

Fig. 4 Illustrating offset and gain errors

1.3.5 Linearity errors

Offset and gain errors may be trimmed out so that the end points of the transfer characteristics lie at zero and V_{FSO}. However, even when this has been done, some or all of the intermediate points may not lie on the 'ideal' line. These errors, which cannot be trimmed out, are known as linearity errors.

1.3.6 Non-linearity (linearity error)

This is the maximum amount, given either as a percentage of full-scale or a fraction of an LSB, by which any point on the transfer characteristic deviates from the ideal straight line passing through zero and V_{FSO}. Non-linearity is illustrated in Fig. 5. A linearity error within the range ± ½LSB assures monotonic operation. Note however that the converse is not true and a DAC may still be monotonic with large linearity errors, which is also shown by Fig. 5.

1.3.7 Differential non-linearity

This is the maximum difference, specified as a fraction of an LSB, between the actual and ideal size of any one LSB analogue increment. This can be seen as an error in the step height of a DAC staircase. A positive value of differential non-linearity means that the step height is larger than nominal, whilst a negative value means that it is smaller than nominal. If it is more negative than −1LSB then the DAC is non-monotonic. However, positive differential non-linearity may assume any value and a DAC can still be monotonic, as shown in Fig. 5.

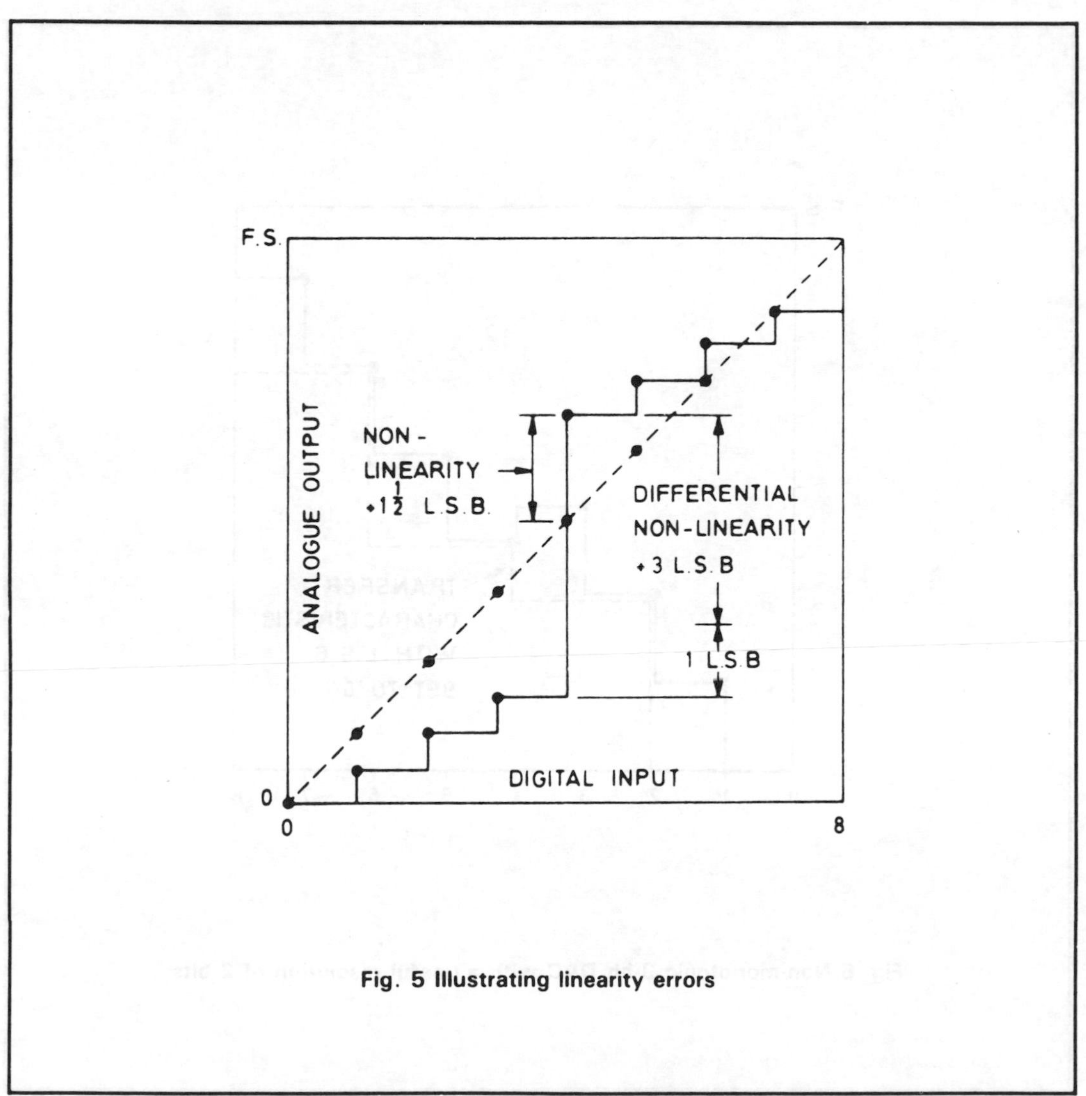

Fig. 5 Illustrating linearity errors

1.3.8 Resolution

As stated earlier, the resolution of a DAC is simply the number of bit inputs that a DAC possesses, which indicates the smallest analogue increment that the converter can produce as a fraction of V_{FS}. e.g. 8 bits = 1 part in 2^8(256). Resolution implies nothing about the accuracy of a DAC, which is defined by linearity and other errors.

1.3.9 Useful resolution

If an n bit DAC has a differential non-linearity of say −1.5LSB then it is non-monotonic. However, if the LSB input is made permanently '0' then the DAC becomes an n − 1 bit device with an LSB equal to twice the original LSB. The differential non-linearity error thus becomes −0.75(new)LSB and the device is monotonic at a resolution of n − 1 bits. This is illustrated in Fig. 6, which shows the transfer characteristic of a 3-bit DAC that has a useful resolution of 2 bits.

Due to manufacturing tolerances a proportion of n-bit converters will have only n - 1 or n - 2 bit useful resolution. In applications not requiring n - bit useful resolution these reduced resoution versions offer a significant price advantage. The useful resolution of Plessey DACs is guaranteed over their full operating temperature range.

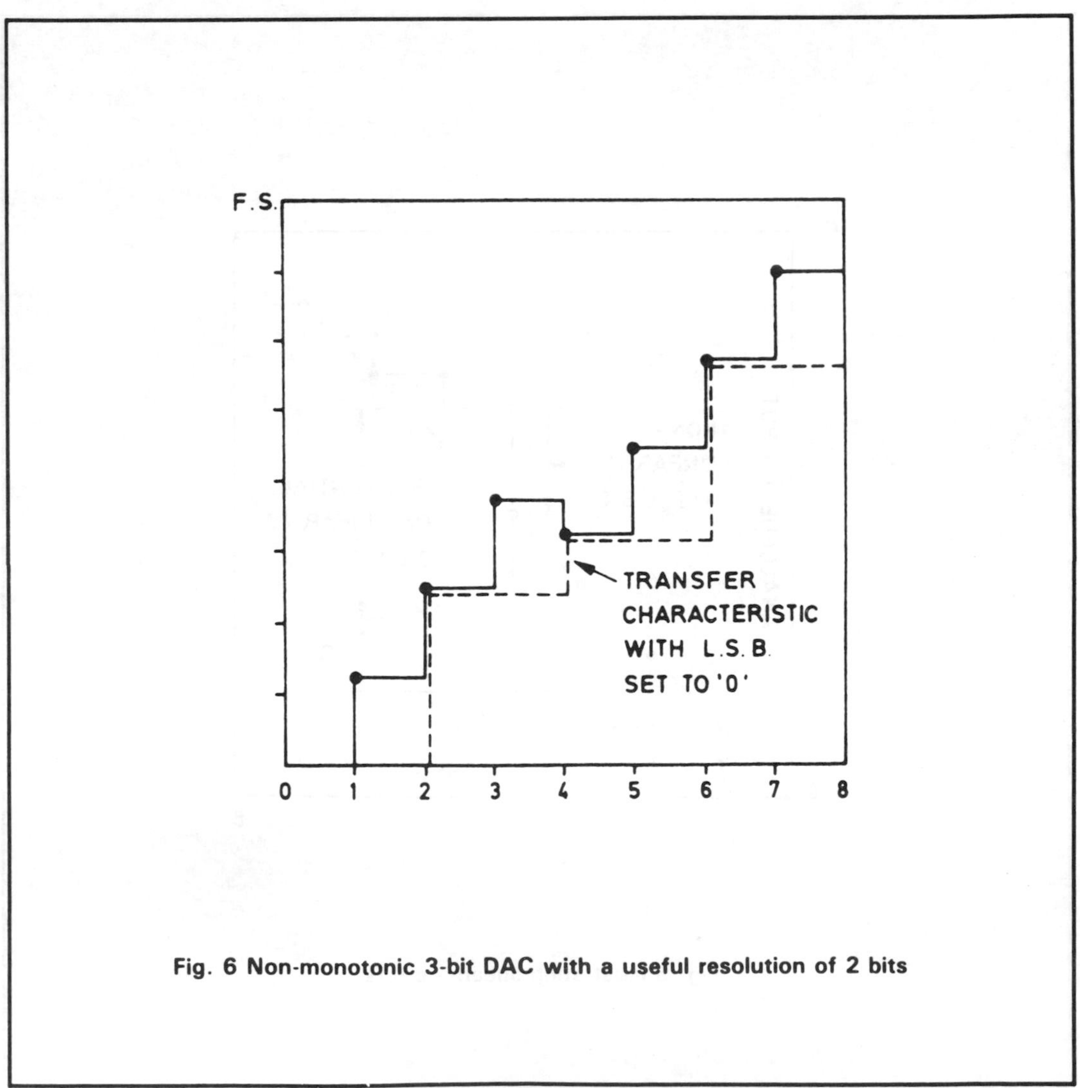

Fig. 6 Non-monotonic 3-bit DAC with a useful resolution of 2 bits

1.3.10 Settling time

Settling time is the time taken after a transition of the input code for the output of a DAC to settle to within ± ½LSB of its final value. This varies depending on which bits are being changed. It may be specified for a change of 1LSB which generally gives the most optimistic (fastest) figure. More conservative figures are given by the most major transition (where the MSB changes in one direction and all other bits change in the opposite direction, e.g. 01111111 to 10000000 or vice versa) or by a change from all bits off to all bits on (00000000 to 11111111) or vice versa.

1.4 Bipolar operation

The discussion so far has been concerned only with DACs producing a single polarity (usually positive) output voltage. In some applications a bipolar (both positive and negative) output range may be required.

This can be achieved by adding a negative offset of $\frac{V_{REF}}{2}$ to the analogue output, as shown in Fig. 7. For all input codes where the MSB is '0' the output voltage is then negative, and for output codes where the MSB is '1' the output voltage is positive. Where the input coding is normally binary but the output voltage is offset by $\frac{-V_{REF}}{2}$ then the input code is referred to as offset binary.

The transfer function of a 3-bit DAC with offset binary coding is shown in Fig. 8.

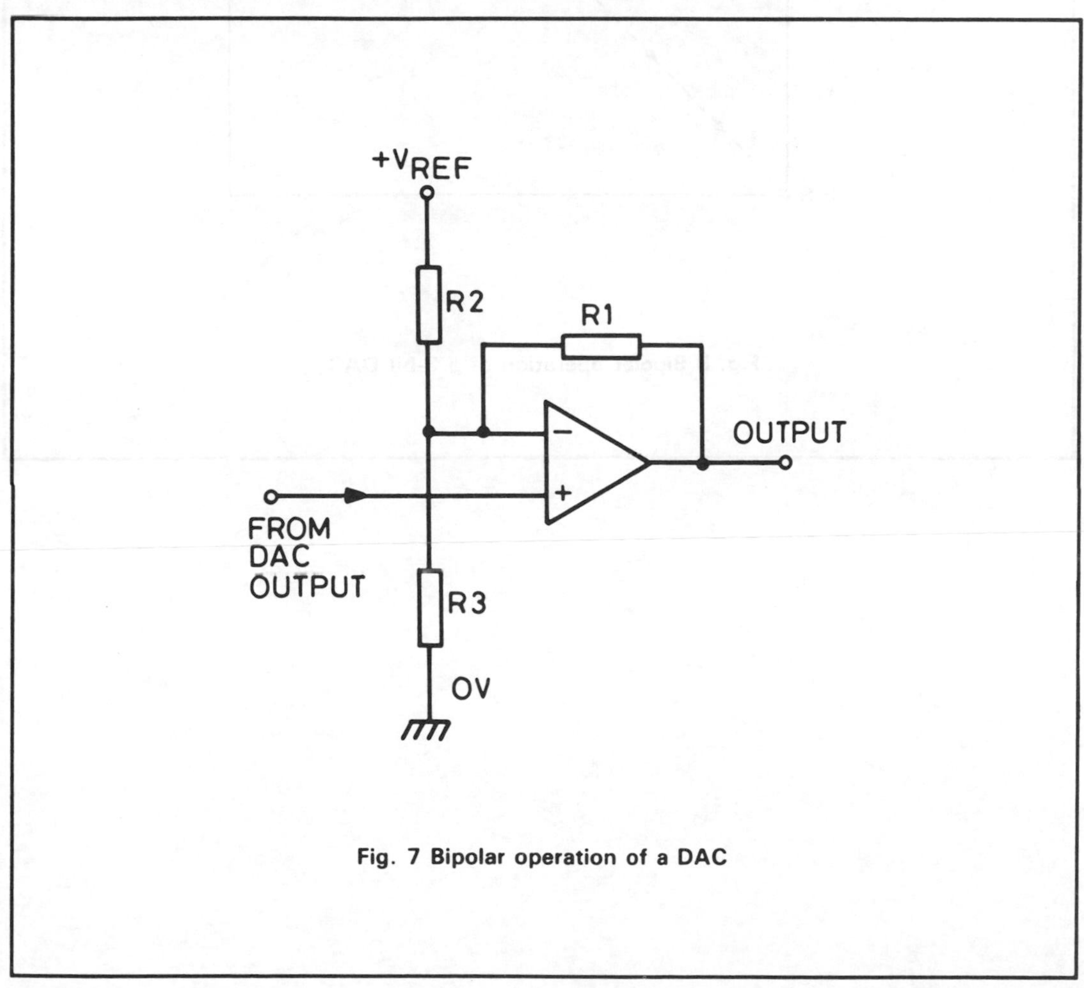

Fig. 7 Bipolar operation of a DAC

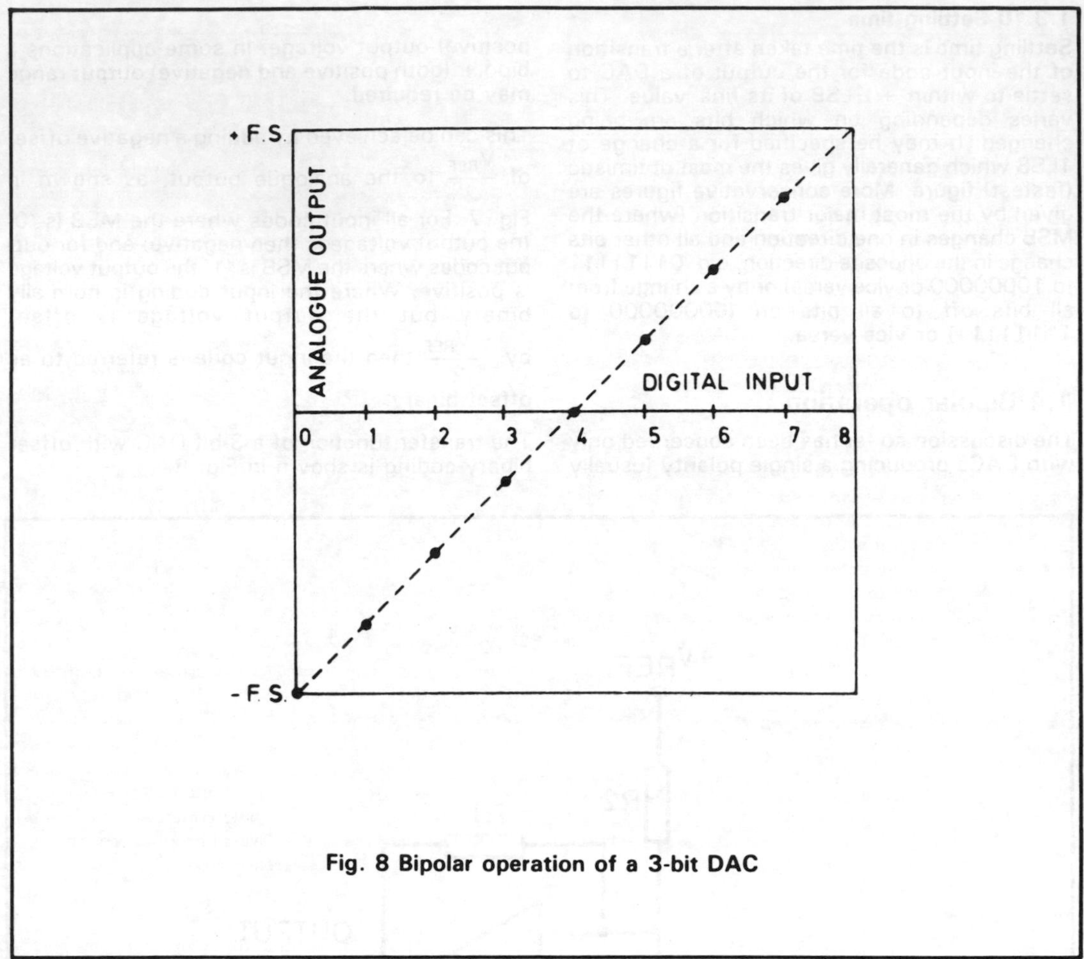

Fig. 8 Bipolar operation of a 3-bit DAC

Orientation
Digital-to-Analog Converters

FACTORS IN CHOOSING A D/A CONVERTER

In the current issue of this two-volume catalog, there are listed some 62 different families of digital-to-analog converters (DACs). If one were to consider all the variations, there would be more than 260 types to choose among. The reason for so many different types is the number of degrees of freedom in selection–technological, functional, and performance. Complete information on converters may be found in the 250-page book, ANALOG-DIGITAL CONVERSION NOTES, published by Analog Devices and available for $5.95 from P.O. Box 796, Norwood MA 02062.

FUNCTIONAL CHARACTERISTICS

The basic structure of all conventional D/A converters involves a network of precision resistors, a set of switches, and some form of level-shifting to adapt the switch drives to the specified logic levels. In addition, the device may contain output-conditioning circuitry, an output amplifier, a reference amplifier, an on-board reference, on-board buffer-registers (single- or dual-rank), configuration conditioning, and even high-voltage isolation.

Basic DAC

This form, which supplies a current, and consequently a small voltage across its internal impedance or an external low-impedance load, is used principally for high speed, for example, the 10ns HDS-0810E. Basic current-output DACs, such as AD566A, are inherently fast, but additional elements (such as an output op amp), furnished by the user to meet overall system specs, slow down the conversion. Some popular CMOS IC devices, such as the AD7523 and the AD7533, are quite simple (and correspondingly low in cost), but they usually require a buffering op amp.

While the basic DAC function is almost always linear, there are exceptions. For example, the AD7111 LOGDAC, which has linear two-quadrant analog response, has a digitally controlled exponential gain function, i.e., 0.375dB per bit; thus its gain at the input code 10000000 (binary 128) is –48dB (48 × 0.375), and the analog output swing for 10V p-p input is 0.04V p-p V_{IN} to $\exp - \left[\frac{0.375N}{20}\right]$.

Output Conditioning

The analog quantity that is the "output" of a DAC, representing the input digital data, may be a "gain" (multiplying DAC), a current, and/or a voltage. In order to obtain a substantial voltage output at low impedance, an op amp is required. It is generally provided on-board in modular and hybrid DACs (and in the monolithic AD558), but there are many ICs and other types that permit the user to choose an external op amp that will meet the particular needs of the application in stability, speed, and cost.

Almost all types of DACs provide one or more feedback resistors: they are matched to, and thermally track, resistances in the network, so that an external op amp, if used, will not require an external feedback resistor that might introduce tracking errors. If more than one feedback resistor is provided, a choice of analog output voltage ranges becomes available, e.g., 0–5V full-scale or 0–10V full-scale. If bipolar output-voltage ranges are specified, a bipolar-offset resistor is provided to subtract a half-scale value from the current flowing through the op amp summing point; it is usually derived from the DAC's reference (or analog) input to avoid additional tracking error. Multiplying DACs use an internal or external op amp for bipolar offset.

In order to avoid difficulties, the user must pay especial attention to the specified output polarity, its relationship to the reference (if external) and to the input digital code. This can be especially tricky if the output is bipolar and the input requires a complementary (negative-true) digital coding. Another such case is where a current-output DAC, specified for a particular output-voltage polarity when used with an inverting op amp, is used in a mode that develops an output voltage passively (without the op amp) across an external resistive load. In addition to polarity, in this case, the user should be aware of the output-compliance constraint and the specified resistive component of output impedance.

9

Reference Input

The reference may be specified as external or internal, fixed or variable, single-polarity or bipolar. If internal, it may be permanently connected (as in the AD561) or optionally connectible (as in the AD565A). If the DAC is a 4-quadrant multiplying type, the reference (or "analog input") is external, variable, and bipolar (e.g., AD7533, 7541, etc.) The user should check a converter's specifications to determine whether the full-scale accuracy specifications are overall or subdivided into a converter-gain spec and a reference spec.

Digital Data

There are a number of ways in which converters differ in regard to the input data: First, the *coding* must be appropriate (binary, offset-binary, two's-complement, BCD, arbitrary, etc.), and its sense should be understood (positive-true, negative-true). The *resolution* (number of bits) must be sufficient; in addition, the specifications must be checked to ascertain that the 2^n distinct binary input codes will not only be accepted, but that also they will (if necessary) correspond to 2^n output values in a monotonic progression at any temperature in the operating range, with sufficient accuracy. The *data levels* accepted by the converter must be checked (TTL, ECL, low-voltage CMOS, high-voltage CMOS), as must the input loading imposed by the converter, and the supply conditions under which the converter will respond to the data. Check the data notation (is the MSB Bit 1 or Bit (n–1)?)–misinterpretation can lead to connecting the data bits in backward order.

If *buffer registers* are desired, the converter should have an appropriate buffer configuration (for example, the AD558 and AD7226 have a set of TTL buffers', the AD667 and AD7548 have two ranks of buffering).

Controls

If the DAC has external digital controls–for example, register strobes– their drive levels, digital sense (true or false), loading, and timing must be considered. The function and use of configuration controls (where present), such as serial/parallel, short-cycle, or chip-select decoding should be understood, and the appropriate ways of disabling them when not needed should be employed.

Power Supplies
Appropriate power supplies should be made available, considering the logic levels and analog output signals to be employed in the system. The appropriate degree of power-supply stability to meet the accuracy specs should be employed. Any recommended external protection circuitry (e.g., Schottky diodes, to ensure that V_{CC} is never more than 0.4V above V_{DD} in the AD7522) should be planned for. In many cases separate analog and digital grounds are required; ground wiring should follow best practice to minimize digital interference with high-accuracy analog signals, while ensuring that a connection between the grounds can always exist at one point, even if the "mecca" point is inadvertently unplugged from the system.

SPECIFICATIONS AND TERMS

Definitions of the performance specifications, and related information, are provided on the next few pages, in alphabetical order.

Accuracy, Absolute
Error of a D/A converter is the difference between the actual analog output and the output that is expected when a given digital code is applied to the converter. Sources of error include gain (calibration) error, zero error, linearity errors, and noise. Error is usually commensurate with resolution, i.e., less than $2^{-(n+1)}$, or "½ LSB" of full scale. However, accuracy may be much better than resolution in some applications; for example, a 4-bit reference supply having only 16 discrete digitally chosen levels would have a resolution of 1/16, but it might have an accuracy to within 0.01% of each ideal value.

Absolute-accuracy measurements should be made under a set of standard conditions with sources and meters traceable to an internationally accepted standard.

Accuracy, Relative
Relative accuracy error, expressed in %, ppm, or fractions of 1 LSB, is the deviation of the analog value at any code (relative to the full analog range of the device transfer characteristics) from its theoretical value (relative to the same range), after the full-scale range (FSR) has been calibrated. Since the discrete analog output values corresponding to the digital input values ideally lie on a straight line, the relative-accuracy error of a linear DAC can be interpreted as a measure of nonlinearity (see *Linearity*).

Compliance-Voltage Range
For a current-output DAC, the maximum range of (output) terminal voltage for which the device will provide the specified current-output characteristics.

Common-Mode Rejection (CMR)
The ability of an amplifier to reject the effect of voltage applied to both input terminals simultaneously. Usually a logarithmic expression representing a "common-mode rejection ratio" e.g., 1,000,000:1 (CMRR) or 120dB (CMR). A CMRR of 10^6:1 means that a 1V common-mode voltage passes through the device as though it were a differential input signal of 1 microvolt.

Common-Mode Voltage
An undesirable signal picked up in a circuit by both wires making up the circuit, with reference to an arbitrary "ground." Amplifiers differ in their ability to amplify a desired signal accurately in the presence of a common-mode voltage.

Deglitcher
As the input code to a DAC is increased or decreased by small changes, it passes through what is known as major and minor transitions. The most major transition is at half-scale, when the DAC switches around the MSB, and all switches change state, i.e., 01111111 to 10000000. If, at major transitions, the switches are faster (or slower) to switch off than on, this means that, for a short time, the D/A will give a zero (or full-scale) output, and then return to the required 1 LSB above the previous reading. Such large transient spikes which differ widely in amplitude and are extremely difficult to filter out, are commonly known as "glitches", hence, a deglitcher is a device which removes these glitches or reduces them to a set of small, uniform pulses. It normally consists of a fast sample-hold circuit, which holds the output constant until the switches reach equilibrium. Glitch energy is smallest in fast-switching DACs driven by fast logic gates that have little time-skew between 0-1 and 1-0 transitions.

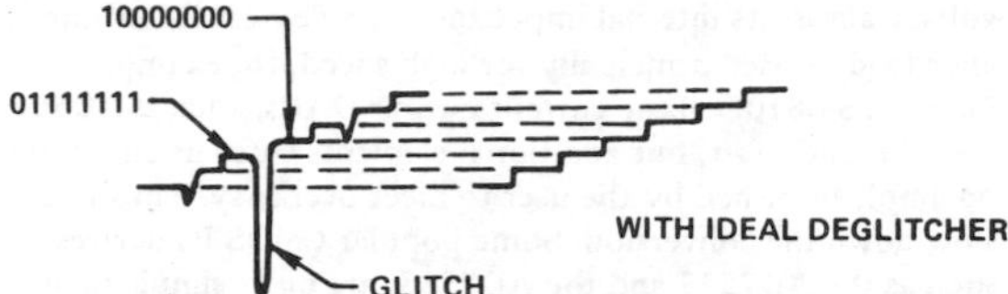

Feedthrough
Undesirable signal coupling around switches or other devices that are supposed to be turned off or provide isolation, e,g,. *feedthrough error* in a multiplying DAC. It is variously specified in %, ppm, fractions of 1 LSB, or fractions of 1 volt, with a given set of inputs, at a specified frequency.

Four-Quadrant
In a multiplying DAC, "four quadrant" refers to the fact that both the reference signal and the number represented by the digital input may be of either positive or negative polarity. A four-quadrant multiplier is expected to obey the rules of multiplication for algebraic sign.

Gain
The "gain" of a converter is that analog scale-factor setting that provides the nominal conversion relationship, e.g., 10V span for a full-scale code change, in a fixed-reference converter. For fixed-reference converters where the use of the internal reference is optional, the converter gain and the reference may be specified separately. Gain- and zero-adjustment are discussed under *Zero.*

Least-Significant Bit (LSB)

In a system in which a numerical magnitude is represented by a series of binary (i.e., two-valued) digits, the LSB is that bit that carries the smallest value, or weight. For example, in the natural-binary number 1101 (decimal 13, or $2^3 + 2^2 + 0 + 2^0$), the rightmost digit is the LSB. Its analog weight, relative to full scale, is 2^{-n}, where n is the number of binary digits. It represents the smallest analog change that can be resolved by an n-bit converter.

Linearity

Linearity error of a converter (also, *integral nonlinearity*, see *Linearity, Differential*), expressed in % or ppm of full-scale range, or (sub)multiples of 1 LSB, is a deviation of the analog values, in a plot of the measured conversion relationship, from a straight line. The straight line can be either a "best straight line", determined empirically by manipulation of the gain and/or offset to equalize maximum positive and negative deviations of the actual transfer characteristics from this straight line; or it can be a straight line passing through the end points of the transfer characteristic after they have been calibrated (sometimes referred to as "end-point" linearity). End-point linearity error is similar to *relative-accuracy* error.

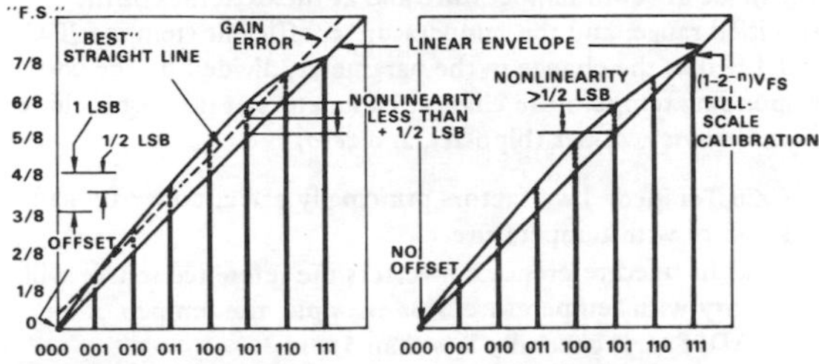

a. ½LSB Nonlinearity Achieved By Arbitrary Location of "Best Straight Line".

b. Nonlinearity Reference is Straight Line Through End Points. Nonlinearity >½LSB for Curve of a.

Comparison of Linearity Criteria for 3-Bit D/A Converter. Straight Line Through End Points is Easier to Measure, Gives More-Conservative Specification.

For multiplying D/A converters, the *analog* linearity error, at a specified digital code, is defined in the same way as for multipliers, i.e., by deviation from a "best straight line" through the plot of the analog output-input response.

Linearity, Differential

Any two adjacent digital codes should result in measured output values that are exactly 1 LSB apart (2^{-n} of full scale for an n-bit converter). Any deviation of the measured "step" from the ideal difference is called *differential nonlinearity*, expressed in (sub)multiples of 1 LSB. It is an important specification, because a differential linearity error greater than 1 LSB can lead to non-monotonic response in a D/A converter and missed codes in an A/D converter (see *Differential Linearity* in the A/D converter section for an illustration).

Monotonic

A DAC is said to be monotonic if the output either increases or remains constant as the digital input increases, with the result that the output will always be a single-valued function of the input. The specification "monotonic" (over a given temperature range) is sometimes substituted for a *differential nonlinearity* specification, since differential nonlinearity less than 1 LSB is a sufficient condition for monotonic behavior.

Most-Significant Bit (MSB)

In a system in which a numerical magnitude is represented by a series of binary (i.e., two-valued) digits, the MSB is that digit (or bit) that carries the largest value of weight. For example, in the natural-binary number 1101 (decimal 13, or $2^3 + 2^2 + 0 + 2^0$), the leftmost "1" is the MSB, with a weight of 2^{n-1}, or 8 LSBs. Its analog weight, relative to a DAC's full-scale span, is ½. In bipolar DACs, the MSB indicates the polarity of the number represented by the rest of the bits.

Multiplying DAC

A multiplying DAC differs from a fixed-reference DAC in being designed to operate with varying (or ac) reference signals. The output signal of such a DAC is proportional to the product of the "reference" (i.e., analog input) voltage and the fractional equivalent of the digital input number (see also *four-quadrant*).

Noise, Peak and rms

Internally generated random noise is not a major factor in D/A converters, except at extreme resolutions (e.g., DAC1138) and dynamic ranges (AD7111). Random noise is characterized by rms specifications for a given bandwidth, or as a spectral density (current or voltage per root hertz); if the distribution is Gaussian, the probability of peak-to-peak values exceeding 7x the rms value is less than 0.1%.

Of much greater importance in DACs is interference in the form of high-amplitude low-energy (hence low-rms) spikes appearing at the DAC's output, caused by coupling of digital signals in a surprising variety of ways; they include coupling via stray capacitance, via power supplies, via inadequate ground systems, via feedthrough, and by glitch-generation. Their presence underscores the necessity for maximum application of the designer's art, including layout, shielding, guarding, grounding, bypassing, and deglitching.

Offset

For almost all bipolar converters (e.g., ±10-volt output), instead of actually generating negative currents to correspond to negative numbers, a unipolar DAC is used, and the output is offset by half full scale (1 MSB). For best results, this offset voltage or current is derived from the same reference supply that determines the gain of the converter.

This makes the zero point of the converter independent of thermal drift of the reference, because the ½ scale offset cancels the weight of the MSB at zero, independently of the amplitude of both.

Power-Supply Sensitivity
The sensitivity of a converter to changes in the power-supply voltages is normally expressed in terms of percent-of-full-scale change in analog output value (or fractions of 1 LSB) for a 1% dc change in the power supply, e.g., 0.05%/%ΔV_S). Power supply sensitivity may also be expressed in relation to a specified dc shift of supply voltage. A converter may be considered "good" if the change in reading at full scale does not exceed ±½ LSB for a 3% change in power supply. Even better specs are necessary for converters designed for battery operation.

Quantizing Uncertainty (or "Error")
The analog continuum is partitioned into 2^n discrete ranges for n-bit processing. All analog values within a given range of output (of a DAC) are represented by the same digital code, usually assigned to the nominal midrange value. For applications in which an analog continuum is to be restored, there is an inherent quantization uncertainty of ±½ LSB, due to limited resolution, in addition to the actual conversion errors. For applications in which discrete output levels are desired (e.g., digitally controlled power supplies or digitally controlled gains), this consideration is not relevant.

Resolution
An n-bit binary converter should be able to provide 2^n distinct and different analog output values corresponding to the set of n-bit binary words. A converter that satisfies this criterion is said to have a *resolution* of n bits. The smallest output change that can be resolved by a linear DAC is 2^{-n} of the full-scale span. However, a nonlinear device, such as the AD7111 LOGDAC has a logarithmic gain resolution of 0.375/88.5dB = 1:256dB, which corresponds to a gain increment of 4.25%/step, or 26,600:1.

Settling Time
The time required, following a prescribed data change, for the output of a DAC to reach and remain within a given fraction (usually ±½ LSB) of the final value. Typical prescribed changes are full-scale, 1 MSB, and 1 LSB at a major carry. Settling time of current-output DACs is quite fast. The major share of settling time of a voltage-output DAC is usually contributed by the settling time of the output op-amp circuit.

Slew Rate (or Slewing Rate)
Slew rate of a device or circuit is a limitation in the rate of change of output voltage, usually imposed by some basic circuit consideration, such as limited current to charge a capacitor. Amplifiers with slew rate of a few V/µs are common, and moderate in cost. Slew rates greater than about 75 volts/µs are usually seen only in more sophisticated (and expensive) devices. The output slewing speed of a voltage-output D/A converter is usually limited by the slew rate of the amplifier used at its output (if one is used).

Stability
Stability of a converter usually applies to the insensitivity of its characteristics to time, temperature, etc. All measurements of stability are difficult and time consuming, but stability vs. temperature is sufficiently critical in most applications to warrant universal inclusion of temperature coefficients in tables of specifications (see "Temperature Coefficient").

Staircase
A voltage or current, increasing in equal increments as a function of time and having the appearance of a staircase (in a time plot), generated by applying a pulse train to a counter, and the output of the counter to the input of a DAC.

A very simple A/D converter can be built by comparing a staircase from a DAC with the unknown analog input. When the DAC output exceeds the analog input by a fraction of 1 LSB, the count is stopped, and the code corresponding to the count is the digital output.

Switching Time
In a DAC, the switching time is the time it takes for the switch to change from one state to the other ("delay time" plus "rise time" from 10%-90%), but does not include settling time, e.g. to <½ LSB.

Temperature Coefficients
In general, temperature instabilities are expressed as %/°C, ppm/°C, as fractions of 1 LSB/°C, or as a change in a parameter over a specified temperature range. Measurements are usually made at room temperature and at the extremes of the specified range, and the temperature coefficient (tempco, T.C.) is defined as the change in the parameter, divided by the corresponding temperature change. Parameters of interest include *gain, linearity, offset* (bipolar), and *zero.*

Gain Tempco: Two factors principally affect converter gain stability with temperature.

a) In fixed-reference converters the reference source will vary with temperature. For example, the tempco of an AD581L is generally less than 5ppm/°C

b) The reference circuitry and switches may add another 3ppm/°C in good 12-bit converters (e.g. AD566K/T). High-resolution converters require much better tempcos for accuracy commensurate with the resolution.

Linearity Tempco: Sensitivity of linearity ("integral" and/or differential linearity) to temperature (in % FSR/°C or ppm FSR/°C) over the specified range. Monotonic behavior is achieved if the differential nonlinearity is less than 1 LSB at any temperature in the range of interest. The *differential nonlinearity temperature coefficient* may be expressed as a ratio, as a maximum change over a temperature range, and/or implied by a statement that the device is monotonic over the specified temperature range.

Offset Tempco: The temperature coefficient of the all-DAC-switches-off (minus full scale) point of a bipolar converter (in % FSR/°C or ppm FSR/°C) depends on three major factors:

a) The tempco of the reference source
b) The voltage zero-stability of the output amplifier
c) The tracking capability of the bipolar-offset resistors and the gain resistors

Unipolar Zero Tempco (in % FSR/°C or ppm FSR/°C): The temperature stability of a unipolar fixed-reference DAC is principally affected by current leakage (current-output DAC), and offset voltage and bias current of the output op-amp (voltage-output DAC).

Zero- and Gain-Adjustment Principles

The output of a unipolar DAC is set to zero volts in the all-bits-off condition. The gain is set for F.S. $(1 - 2^{-n})$ with all bits on. The "zero" of an offset-binary bipolar DAC is set to -F.S. with all bits off, and the gain is set for +F.S. $(1 - 2^{-(n-1)})$ with all bits on. The data sheet instructions should be followed.

CMOS Dual 8-Bit Buffered Multiplying DAC

AD7528

2

FEATURES

On-Chip Latches for Both DACs
+5V to +15V Operation
DACs Matched to 1%
Four Quadrant Multiplication
TTL/CMOS Compatible
Latch Free (Protection Schottkys not Required)

APPLICATIONS

Digital Control of:
- **Gain/Attenuation**
- **Filter Parameters**
- **Stereo Audio Circuits**
- **X–Y Graphics**

AD7528 FUNCTIONAL BLOCK DIAGRAM

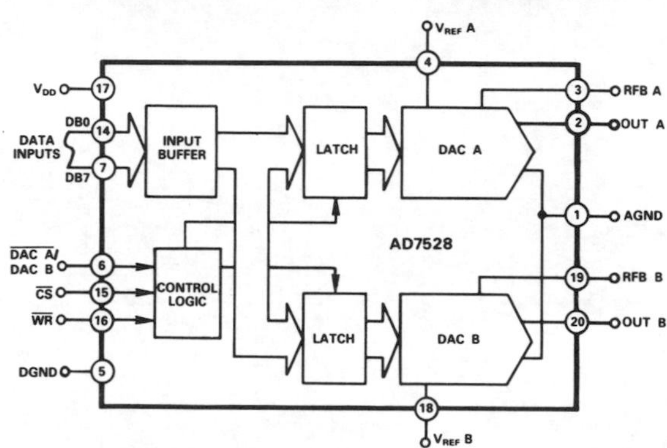

GENERAL DESCRIPTION

The AD7528 is a monolithic dual 8-bit digital/analog converter featuring excellent DAC-to-DAC matching. It is available in skinny 0.3″ wide 20-pin DIPs and in 20-terminal surface mount packages.

Separate on-chip latches are provided for each DAC to allow easy microprocessor interface.

Data is transferred into either of the two DAC data latches via a common 8-bit TTL/CMOS compatible input port. Control input $\overline{\text{DAC A}}$/DAC B determines which DAC is to be loaded. The AD7528's load cycle is similar to the write cycle of a random access memory and the device is bus compatible with most 8-bit microprocessors, including 6800, 8080, 8085, Z80.

The device operates from a +5V to +15V power supply, dissipating only 20mW of power.

Both DACs offer excellent four quadrant multiplication characteristics with a separate reference input and feedback resistor for each DAC.

PRODUCT HIGHLIGHTS

1. DAC to DAC matching: since both of the AD7528 DACs are fabricated at the same time on the same chip, precise matching and tracking between DAC A and DAC B is inherent. The AD7528's matched CMOS DACs make a whole new range of applications circuits possible, particularly in the audio, graphics and process control areas.
2. Small package size: combining the inputs to the on-chip DAC latches into a common data bus and adding a $\overline{\text{DAC A}}$/DAC B select line has allowed the AD7528 to be packaged in either a small 20-pin 0.3″ wide DIP or in 20-terminal surface mount packages.

PIN CONFIGURATIONS

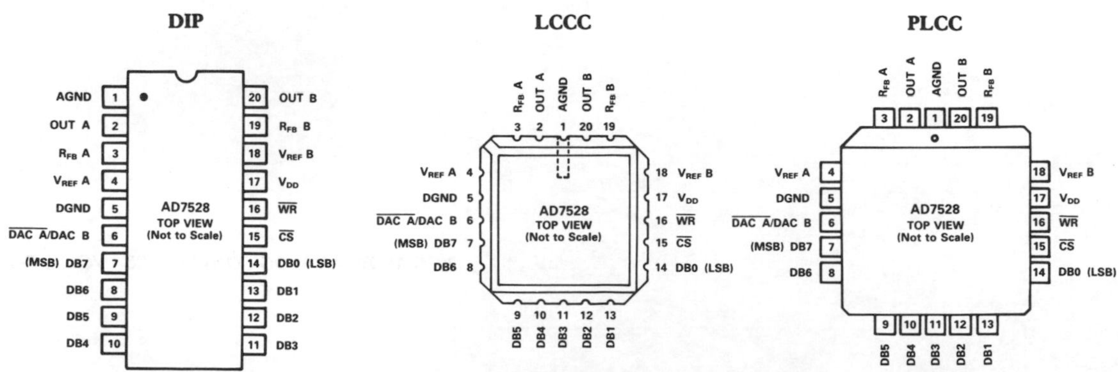

SPECIFICATIONS (V_{REF} A = V_{REF} B = +10V; OUT A = OUT B = 0V unless otherwise specified)

Parameter	Version[1]	V_{DD} = +5V T_A = +25°C	V_{DD} = +5V T_{min}, T_{max}	V_{DD} = +15V T_A = +25°C	V_{DD} = +15V T_{min}, T_{max}	Units	Test Conditions/Comments
STATIC PERFORMANCE[2]							
Resolution	All	8	8	8	8	Bits	
Relative Accuracy	J, A, S	±1	±1	±1	±1	LSB max	This is an Endpoint Linearity Specification
	K, B, T	±1/2	±1/2	±1/2	±1/2	LSB max	
	L, C, U	±1/2	±1/2	±1/2	±1/2	LSB max	
Differential Nonlinearity	All	±1	±1	±1	±1	LSB max	All Grades Guaranteed Monotonic Over Full Operating Temperature Range
Gain Error	J, A, S	±4	±6	±4	±5	LSB max	Measured Using Internal RFB A and RFB B.
	K, B, T	±2	±4	±2	±3	LSB max	Both DAC Latches Loaded with 11111111.
	L, C, U	±1	±3	±1	±1	LSB max	Gain Error is Adjustable Using Circuits of Figures 1 and 2.
Gain Temperature Coefficient[4] ΔGain/ΔTemperature	All	±0.007	±0.007	±0.0035	±0.0035	%/°C max	
Output Leakage Current							
OUT A (Pin 2)	All	±50	±400	±50	±200	nA max	DAC Latches Loaded with 00000000
OUT B (Pin 20)	All	±50	±400	±50	±200	nA max	
Input Resistance (V_{REF} A, V_{REF} B)	All	8	8	8	8	kΩ min	Input Resistance TC = −300ppm/°C, Typical
		15	15	15	15	kΩ max	Input Resistance is 11kΩ
V_{REF} A/V_{REF} B Input Resistance Match	All	±1	±1	±1	±1	% max	
DIGITAL INPUTS[3]							
Input High Voltage V_{IH}	All	2.4	2.4	13.5	13.5	V min	
Input Low Voltage V_{IL}	All	0.8	0.8	1.5	1.5	V max	
Input Current I_{IN}	All	±1	±10	±1	±10	μA max	V_{IN} = 0 or V_{DD}
Input Capacitance							
DB0–DB7	All	10	10	10	10	pF max	
$\overline{WR}$, $\overline{CS}$, $\overline{DAC A}$/DAC B	All	15	15	15	15	pF max	
SWITCHING CHARACTERISTICS[4]							See Timing Diagram
Chip Select to Write Set Up Time t_{CS}	All	200	230	60	80	ns min	
Chip Select to Write Hold Time t_{CH}	All	20	30	10	15	ns min	
DAC Select to Write Set Up Time t_{AS}	All	200	230	60	80	ns min	
DAC Select to Write Hold Time t_{AH}	All	20	30	10	15	ns min	
Data Valid to Write Set Up Time t_{DS}	All	110	130	30	40	ns min	
Data Valid to Write Hold Time t_{DH}	All	0	0	0	0	ns min	
Write Pulse Width t_{WR}	All	180	200	60	80	ns min	
POWER SUPPLY							
I_{DD}	All	2	2	2	2	mA max	All Digital Inputs V_{IL} or V_{IH}
	All	100	500	100	500	μA max	All Digital Inputs 0V or V_{DD}

AC PERFORMANCE CHARACTERISTICS[5] (Measured Using Recommended P.C. Board Layout and AD644 as Output Amplifiers)

Parameter	Version[1]	V_{DD} = +5V T_A = +25°C	V_{DD} = +5V T_{min}, T_{max}	V_{DD} = +15V T_A = +25°C	V_{DD} = +15V T_{min}, T_{max}	Units	Test Conditions/Comments
DC SUPPLY REJECTION (ΔGAIN/ΔV_{DD})	All	0.02	0.04	0.01	0.02	% per % max	ΔV_{DD} = ±5%
CURRENT SETTLING TIME[2]	All	350	400	180	200	ns max	To 1/2LSB. Out A/Out B load = 100Ω. $\overline{WR}$ = $\overline{CS}$ = 0V. DB0–DB7 = 0V to V_{DD} or V_{DD} to 0V
PROPAGATION DELAY (From Digital Input to 90% of Final Analog Output Current)	All	220	270	80	100	ns max	V_{REF} A = V_{REF} B = +10V. OUT A, OUT B Load = 100Ω C_{EXT} = 13pF. $\overline{WR}$, $\overline{CS}$ = 0V DB0–DB7 = 0V to V_{DD} or V_{DD} to 0V
DIGITAL TO ANALOG GLITCH IMPULSE	All	160	–	440	–	nV sec typ	For Code Transition 00000000 to 11111111
OUTPUT CAPACITANCE							
C_{OUT} A	All	50	50	50	50	pF max	DAC Latches Loaded with 00000000
C_{OUT} B		50	50	50	50	pF max	
C_{OUT} A		120	120	120	120	pF max	DAC Latches Loaded with 11111111
C_{OUT} B		120	120	120	120	pF max	
AC FEEDTHROUGH							
V_{REF} A to OUT A	All	−70	−65	−70	−65	dB max	V_{REF} A, V_{REF} B = 20V p-p Sine Wave @ 100kHz
V_{REF} B to OUT B		−70	−65	−70	−65	dB max	
CHANNEL TO CHANNEL ISOLATION							Both DAC Latches Loaded with 11111111.
V_{REF} A to OUT B	All	−77	–	−77	–	dB typ	V_{REF} A = 20V p-p Sine Wave @ 100kHz V_{REF} B = 0V.
V_{REF} B to OUT A		−77	–	−77	–	dB typ	V_{REF} B = 20V p-p Sine Wave @ 100kHz V_{REF} A = 0V.
DIGITAL CROSSTALK	All	30	–	60	–	nV sec typ	Measured for Code Transition 00000000 to 11111111
HARMONIC DISTORTION	All	−85	–	−85	–	dB typ	V_{IN} = 6V rms @ 1kHz

NOTES

[1]Temperature Ranges are J, K, L Versions; −40°C to +85°C
A, B, C Versions; −40°C to +85°C
S, T, U Versions; −55°C to +125°C

[2]Specification applies to both DACs in AD7528.

[3]Logic inputs are MOS Gates. Typical input current (+25°C) is less than 1nA.

[4]Guaranteed by design but not production tested.

[5]These characteristics are for design guidance only and are not subject to test.

Specifications subject to change without notice.

INTERFACE LOGIC INFORMATION

DAC Selection:

Both DAC latches share a common 8-bit input port. The control input $\overline{DAC\ A}$/DAC B selects which DAC can accept data from the input port.

Mode Selection:

Inputs $\overline{CS}$ and $\overline{WR}$ control the operating mode of the selected DAC. See Mode Selection Table below.

Write Mode:

When $\overline{CS}$ and $\overline{WR}$ are both low the selected DAC is in the write mode. The input data latches of the selected DAC are transparent and its analog output responds to activity on DB0–DB7.

Hold Mode:

The selected DAC latch retains the data which was present on DB0–DB7 just prior to $\overline{CS}$ or $\overline{WR}$ assuming a high state. Both analog outputs remain at the values corresponding to the data in their respective latches.

$\overline{DAC\ A}$/DAC B	$\overline{CS}$	$\overline{WR}$	DAC A	DAC B
L	L	L	WRITE	HOLD
H	L	L	HOLD	WRITE
X	H	X	HOLD	HOLD
X	X	H	HOLD	HOLD

L = Low State H = High State X = Don't Care

Mode Selection Table

CAUTION:

1. ESD sensitive device. The digital control inputs are diode protected; however, permanent damage may occur on unconnected devices subjected to high energy electrostatic fields. Unused devices must be stored in conductive foam or shunts.
2. Do not insert this device into powered sockets. Remove power before insertion or removal.

WRITE CYCLE TIMING DIAGRAM

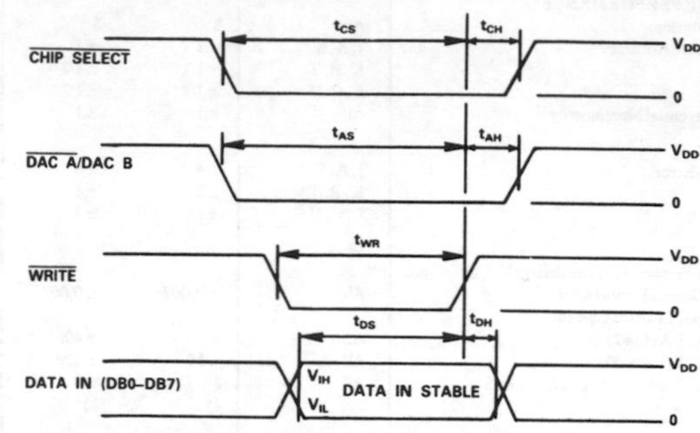

NOTES:
1. ALL INPUT SIGNAL RISE AND FALL TIMES MEASURED FROM 10% TO 90% OF V_{DD}.
V_{DD} = +5V, $t_r = t_f$ = 20ns;
V_{DD} = +15V, $t_r = t_f$ = 40ns.
2. TIMING MEASUREMENT REFERENCE LEVEL IS $\frac{V_{IH} + V_{IL}}{2}$

ABSOLUTE MAXIMUM RATINGS

(T_A = +25°C unless otherwise noted)

- V_{DD} to AGND . . . . . . 0V, +17V
- V_{DD} to DGND . . . . . . 0V, +17V
- AGND to DGND . . . . . . V_{DD} +0.3V
- DGND to AGND . . . . . . V_{DD} +0.3V
- Digital Input Voltage to DGND . . . . −0.3V, V_{DD} +0.3V
- V_{PIN2}, V_{PIN20} to AGND . . . . −0.3V, V_{DD} +0.3V
- V_{REF} A, V_{REF} B to AGND . . . . . . ±25V
- V_{RFB} A, V_{RFB} B to AGND . . . . . . ±25V
- Power Dissipation (Any Package) to +75°C 450mW
  - Derates above +75°C by . . . . . . 6mW/°C
- Operating Temperature Range
  - Commercial (J, K, L) Grades . . . . . . −40°C to +85°C
  - Industrial (A, B, C) Grades . . . . . . −40°C to +85°C
  - Extended (S, T, U) Grades . . . . . . −55°C to +125°C
- Storage Temperature . . . . . . −65°C to +150°C
- Lead Temperature (Soldering, 10 secs.) . . . . . . +300°C

ORDERING INFORMATION[1]

Relative Accuracy	Gain Error T_A = +25°C	Temperature Range and Package Options[2, 3]		
		−40°C to +85°C	−40°C to +85°C	−55°C to +125°C
		Plastic DIP (N-20)	**Hermetic (Q-20)**	**Hermetic (Q-20)**
±1LSB	±4LSB	AD7528JN	AD7528AQ	AD7528SQ
±1/2LSB	±2LSB	AD7528KN	AD7528BQ	AD7528TQ
±1/2LSB	±1LSB	AD7528LN	AD7528CQ	AD7528UQ
		PLCC[4] (P-20A)		**LCCC[5] (E-20A)**
±1LSB	±4LSB	AD7528JP		AD7528SE
±1/2LSB	±2LSB	AD7528KP		AD7528TE
±1/2LSB	±1LSB	AD7528LP		AD7528UE

NOTES

[1]To order MIL-STD-883, Class B processed parts, add/883B to part number. Contact your local sales office for military data sheet. For U.S. Standard Military Drawing (SMD), see DESC drawing #5962-87701.

[2]See Section 14 for package outline information.

[3]Also available in SOIC package (AD7528KR, AD7528LR).

[4]PLCC: Plastic Leaded Chip Carrier.

[5]LCCC: Leadless Ceramic Chip Carrier.

Applying the AD7528

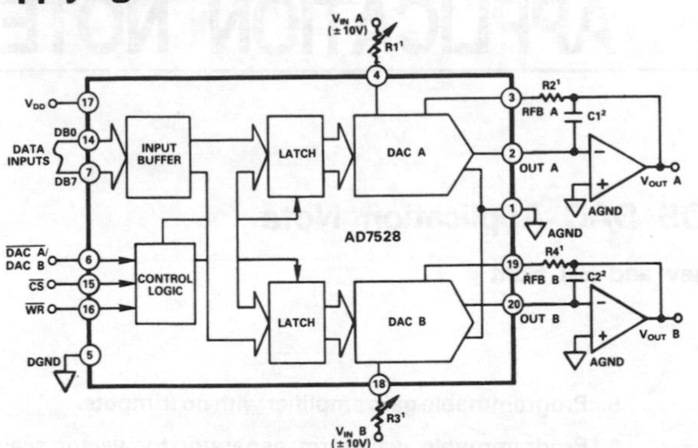

NOTES:
¹R1, R2 AND R3, R4 USED ONLY IF GAIN ADJUSTMENT IS REQUIRED. SEE TABLE 3 FOR RECOMMENDED VALUES.
²C1, C2 PHASE COMPENSATION (10pF–15pF) IS REQUIRED WHEN USING HIGH SPEED AMPLIFIERS TO PREVENT RINGING OR OSCILLATION.

Figure 1. Dual DAC Unipolar Binary Operation (2 Quadrant Multiplication). See Table I.

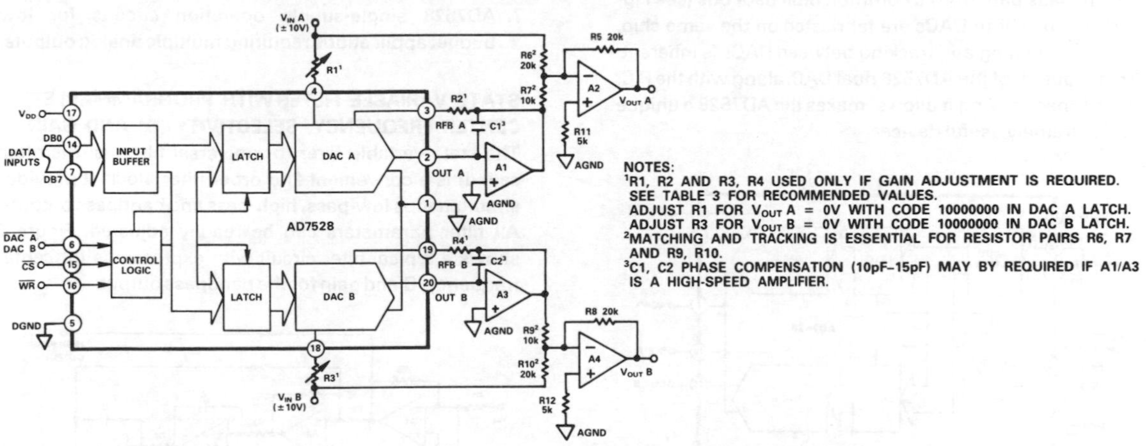

NOTES:
¹R1, R2 AND R3, R4 USED ONLY IF GAIN ADJUSTMENT IS REQUIRED. SEE TABLE 3 FOR RECOMMENDED VALUES. ADJUST R1 FOR V_{OUT} A = 0V WITH CODE 10000000 IN DAC A LATCH. ADJUST R3 FOR V_{OUT} B = 0V WITH CODE 10000000 IN DAC B LATCH.
²MATCHING AND TRACKING IS ESSENTIAL FOR RESISTOR PAIRS R6, R7 AND R9, R10.
³C1, C2 PHASE COMPENSATION (10pF–15pF) MAY BY REQUIRED IF A1/A3 IS A HIGH-SPEED AMPLIFIER.

Figure 2. Dual DAC Bipolar Operation (4 Quadrant Multiplication). See Table II.

DAC Latch Contents MSB LSB	Analog Output (DAC A or DAC B)
11111111	$-V_{IN}\left(\frac{255}{256}\right)$
10000001	$-V_{IN}\left(\frac{129}{256}\right)$
10000000	$-V_{IN}\left(\frac{128}{256}\right) = -\frac{V_{IN}}{2}$
01111111	$-V_{IN}\left(\frac{127}{256}\right)$
00000001	$-V_{IN}\left(\frac{1}{256}\right)$
00000000	$-V_{IN}\left(\frac{0}{256}\right) = 0$

Note: $1LSB = (2^{-8})(V_{IN}) = \frac{1}{256}(V_{IN})$

Table I. Unipolar Binary Code Table

DAC Latch Contents MSB LSB	Analog Output (DAC A or DAC B)
11111111	$+V_{IN}\left(\frac{127}{128}\right)$
10000001	$+V_{IN}\left(\frac{1}{128}\right)$
10000000	0
01111111	$-V_{IN}\left(\frac{1}{128}\right)$
00000001	$-V_{IN}\left(\frac{127}{128}\right)$
00000000	$-V_{IN}\left(\frac{128}{128}\right)$

Note: $1LSB = (2^{-7})(V_{IN}) = \frac{1}{128}(V_{IN})$

Table II. Bipolar (Offset Binary) Code Table

Trim Resistor	J/A/S	K/B/T	L/C/U
R1;R3	1k	500	200
R2;R4	330	150	82

Table III. Recommended Trim Resistor Values vs. Grade

ANALOG DEVICES

APPLICATION NOTE

AD7528 Dual 8-Bit CMOS DAC Application Note

By Paul Toomey and Bill Hunt

INTRODUCTION

The AD7528 is a monolithic dual 8-bit CMOS DAC packaged in a 20-pin DIP. Each DAC has its own 8-bit data latch which loads data from a common 8-bit data bus (see Figure 1). Since both DACs are fabricated on the same chip, precise matching and tracking between DACs is inherent. This property of the AD7528 dual DAC, along with the P.C. board space saving it allows, makes the AD7528 a unique and extremely useful device.

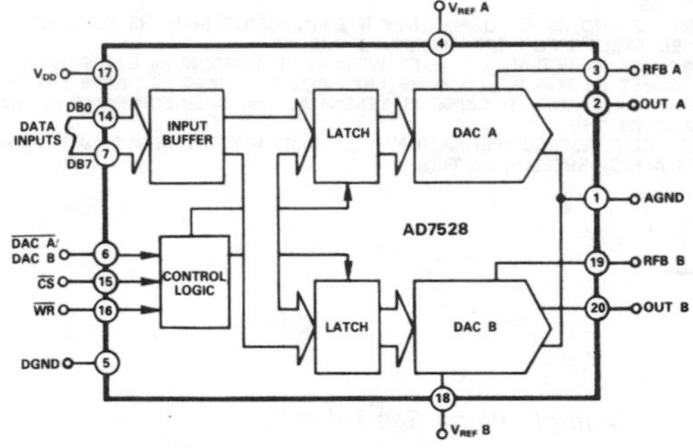

Figure 1. AD7528 Functional Diagram

This note discusses the AD7528 applications circuits listed below. Several of these circuits rely on the DAC to DAC matching provided by the AD7528. All of the circuits benefit from the high packing density the AD7528 allows, especially when used with dual and quad op-amps such as the AD644 or TL074. Not discussed in this note are basic details of AD7528 operation, consult the data sheet for this information.

AD7528 APPLICATIONS DISCUSSED IN THIS NOTE

1. State-variable filter (S.V.F.) with programmable center frequency, selectivity and gain.
2. Programmable sine wave oscillator with linear control.
3. Function fitting sine wave synthesizer with amplitude control facility and programmable phase shift.
4. Programmable voltage/current source, unipolar and bipolar circuits.
5. Programmable gain amplifier with no trimpots.
6. Programmable waveform generator for vector scan CRT displays.
7. AD7528 single-supply operation circuits for low budget applications requiring multiple analog outputs.

STATE VARIABLE FILTER WITH PROGRAMMABLE CENTER FREQUENCY, SELECTIVITY (Q) AND GAIN

The state variable filter (or universal filter as it is often called) is a convenient 2nd order filter block. It provides simultaneous low-pass, high-pass and bandpass outputs. All filter parameters can be readily adjusted. Figure 2 shows a typical filter circuit with expressions for center frequency, Q and gain for the bandpass output.

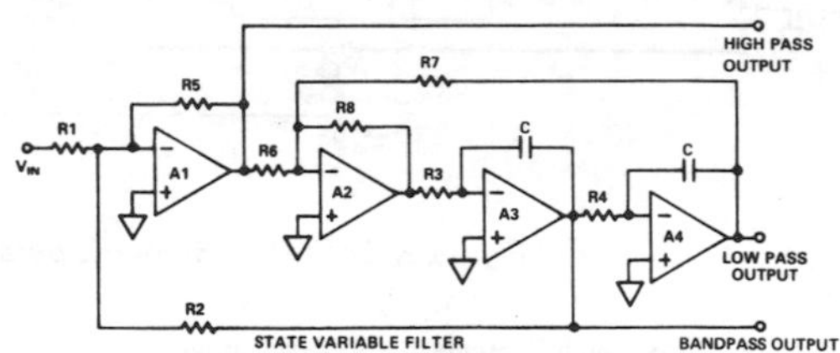

Figure 2. State Variable Filter

BANDPASS TRANSFER FUNCTION

$$\frac{V_{OUT}}{V_{IN}}(f) = \frac{A_O}{1 + jQ\left[\frac{f}{f_o} - \frac{f_o}{f}\right]}$$

$$f_o = \frac{1}{2\pi R3\,C} \cdot \sqrt{\frac{R8}{R7}} \quad (\text{For } R3 = R4)$$

$$Q = \frac{R6}{R8} \cdot \frac{R2}{R5} \cdot \sqrt{\frac{R8}{R7}}$$

$$A_O = -\frac{R2}{R1}$$

Where f = frequency of V_{IN}

A_O = gain at $f = f_o$

Q = circuit Q factor, i.e., $\frac{f_o}{\text{3dB Bandwidth}}$

f_o = resonant frequency.

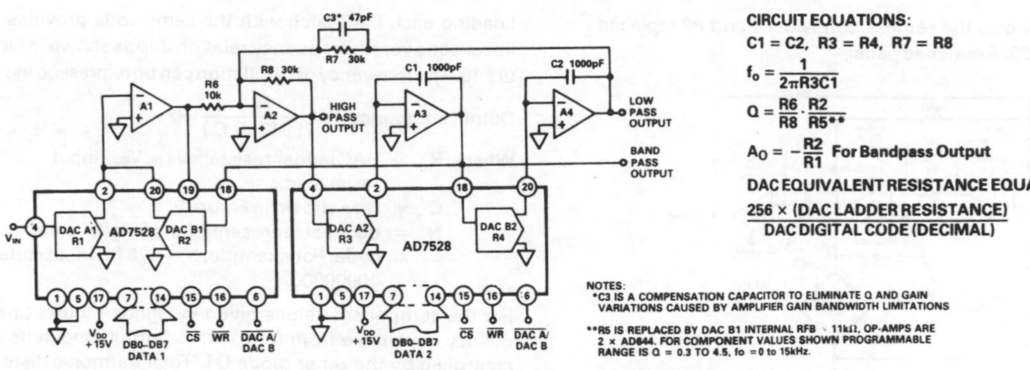

Figure 3. Digitally Controlled State Variable Filter

Introducing the DACs as Control Elements:

By replacing R1, R2 and R3, R4 with matched DAC pairs the filter parameters can be made programmable as shown in Figure 3. DAC A1 and DAC B1 control filter gain and Q, while DAC A2 and DAC B2 control center frequency (f_o). For the component values shown the programmable Q range is from 0.3 to 4.5 and is independent of f_o (see Figure 4). Center frequency (f_o) is programmable from 0 to 15kHz (see Figure 5) and is independent of Q.

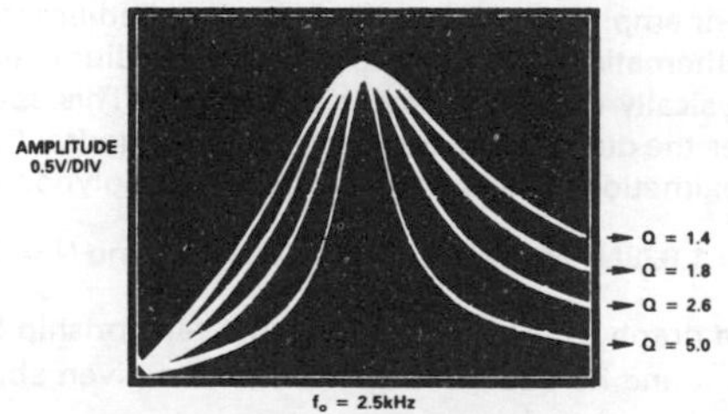

Filter 4. Filter Q Variation

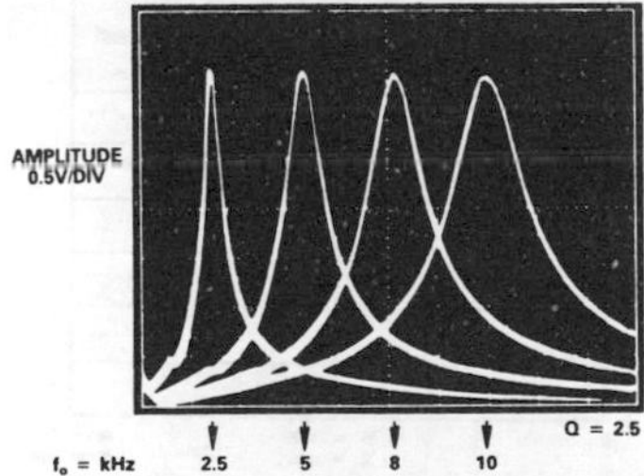

Figure 5. Filter f_o Variation

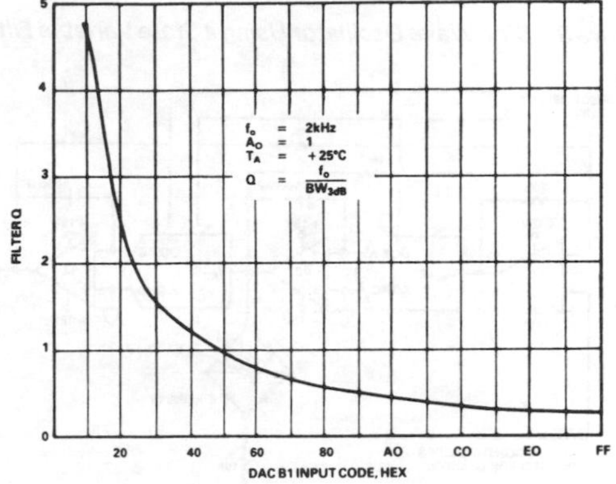

Figure 6. Filter Q Variation with Code

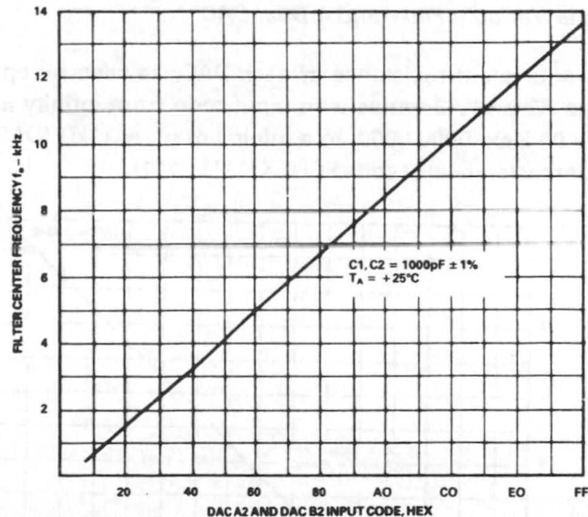

Figure 7. Filter f_o Variation with Code

Programming

The graph in Figure 6 shows how the circuit Q varies with DAC B1 code and Figure 7 shows how the center frequency varies with DAC 2 (A and B) code for the component values given in Figure 3. Gain variation alone is accomplished by changing DAC A1 code. Unity gain occurs when the data in DAC A1 and DAC B1 latches is identical. Since the AD7528's logic inputs are TTL or CMOS compatible, the DACs are readily interfaced to most microprocessors, (see data sheet for hookups) thus providing an ideal microprocessor-to-filter interface.

PROGRAMMABLE SINE WAVE OSCILLATOR WITH LINEAR CONTROL

Frequency control of many oscillator circuits can be accomplished using two ganged potentiometers. However, the two potentiometers must track precisely over their full temperature range if a linear response is required. Figure 8 shows a high performance sine-wave oscillator realized using a state-variable filter. The frequency of oscillation is set by ganged potentiometers P1 and P2.

Figure 9 shows the same circuit with P1 and P2 replaced by the AD7528 matched pairs.

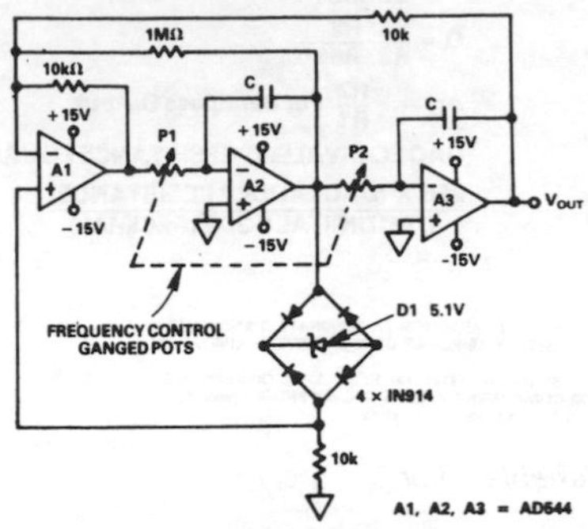

Figure 8. Sine Wave Oscillator Using a State Variable Filter

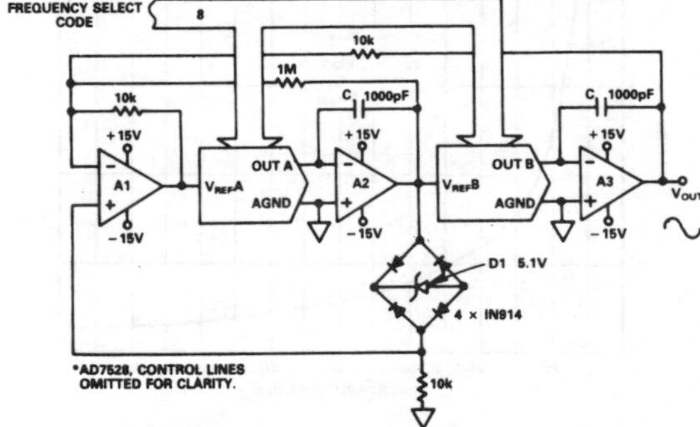

Figure 9. Programmable Sine Wave Oscillator Using a State Variable Filter and a Dual DAC

The equivalent resistance of each DAC, as seen by op-amps A2 and A3 varies with input code from infinity at code 00 Hex (0000 0000) to a minimum of ≈ 11kΩ (DAC ladder resistance) at code FF HEX (1111 1111).

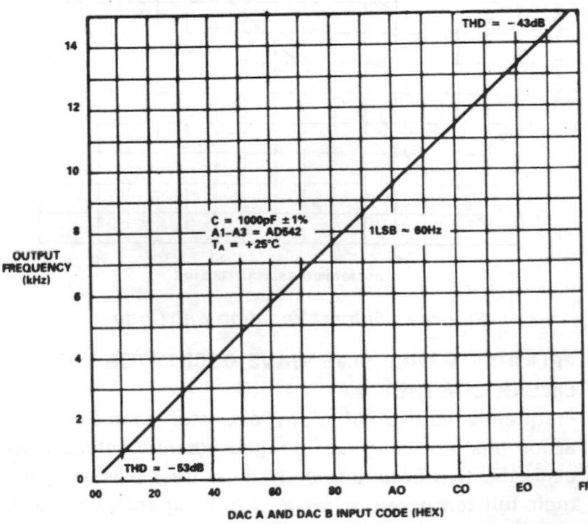

Figure 10. Frequency vs. DAC Code for Programmable Sine Wave Oscillator (Figure 9)

Loading each DAC latch with the same code provides a linear code versus frequency relationship as shown in Figure 10. The frequency of oscillation can be expressed as:

$$\text{Output Frequency} = \frac{N}{256\,(2\pi R\,C)}\ \text{Hz}$$

Where R = DAC ladder resistance i.e. V_{REF} input resistance.
C = is as shown in Figure 9.
N = decimal representation of digital input code. For example, N = 128 for input code 10000000.

For the component values given in Figure 9, output frequency is variable from 0 to 15kHz. Output amplitude is controlled by the zener diode D1. Total harmonic distortion for the circuit shown is −53dB at low frequencies (1kHz) and −43dB at higher frequencies (14kHz). Note that a cosine output is also available at the output of op-amp A2.

FUNCTION FITTING SINE WAVE SYNTHESIZER

In this application the multiplying capabilities of the two CMOS DACs are used to synthesize a sine wave based on a function fitting technique. This allows very low frequency, highly stable sine waves to be generated.

Function Fitting:

Function fitting is a technique for translating a mathematical or empirical relationship from one medium (such as a mathematical formula) to another medium (usually a physically realizable device or system). This application uses the dual DAC to implement a one quadrant sin X approximation in the form of the quadratic polynomial.

$$Y = 1.828N - 0.828\,N^2 \quad \text{where } 0 \le N \le 1 \text{ and } N = \frac{2}{\pi}X$$

The graph of Figure 11 shows the relationship between sin X and its quadratic approximation given above. The

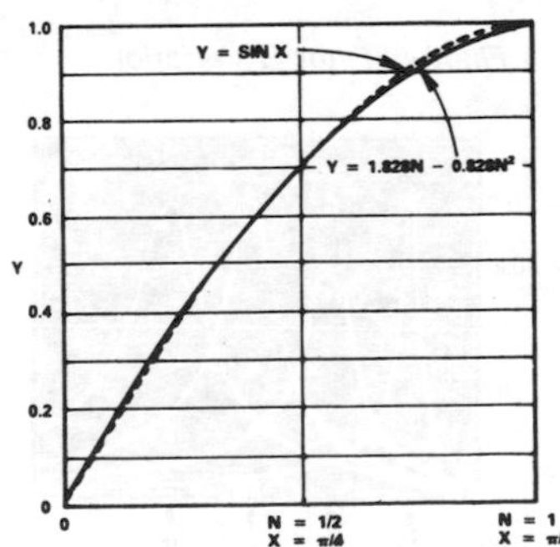

Figure 11. Relationship Between Sin X and its Quadratic Approximation

circuit of Figure 12 implements the function by ramping N up and down using an up/down counter, and switching the circuit output polarity. This generates sin X in four stages (see Figure 13).

Circuit Operation: (Figure 12)

An input clock drives the up/down counter in real time. The counter is connected so that it counts up and down continuously, providing an output pulse at "borrow" every time it reaches the all zeros count.

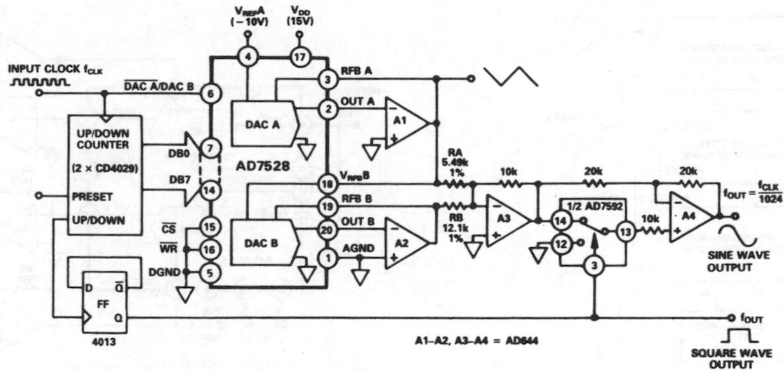

Figure 12. Function Fitting Sine Wave Generator

DAC A produces a triangle waveform consisting of two ramps of opposite slope, each generated in 256 steps at op-amp A1 output. This is the N variable.

DAC B is driven with the same digital word as DAC A. Its reference input is driven by op-amp A1, thus DAC B multiplies the digital version of N by the analog version of N to produce an output from op-amp A2 of $-N^2$ (negative sign is due to inversion through A2).

Since the N and $-N^2$ signals are of opposite polarity, the $Y = 1.828N - 0.828N^2$ expression is implemented by summing N and $-N^2$ signals in the correct ratios determined by RA and RB.

The analog switch, in conjunction with op-amp A4, changes the circuit's output polarity at one-half the triangle wave frequency, thus producing both positive and negative halves of the sine wave.

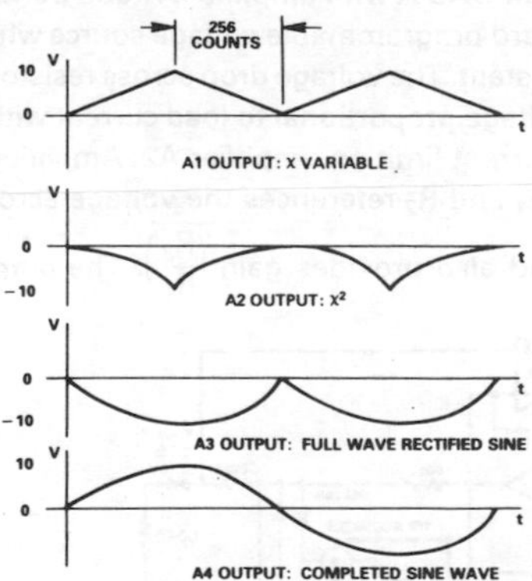

Figure 13. Sine Wave Synthesis Using Function Fitting

Distortion:

Distortion in the output sine wave is a function of the quadratic approximation fit to the sine curve. Errors in the values of R_A and R_B will, therefore, contribute directly to distortion. If R_A is made adjustable over a small range, it can be trimmed to minimize distortion. Distortion was measured for the circuit of Figure 12 at −33dB and was constant over sine-wave frequency 0–2.5kHz.

The circuit of Figure 12 generates constant amplitude sine waves in the 0–2.5kHz frequency range. Output frequency is given by

$$f_{OUT} = \frac{f\,clk}{1024}$$

It is possible to obtain rapid frequency sweeping by varying the input clock rate. The counter up/down output provides a useful zero crossing pulse. By applying an ac signal to the DAC A reference input, the output sine wave can be amplitude-modulated as shown in Figure 14.

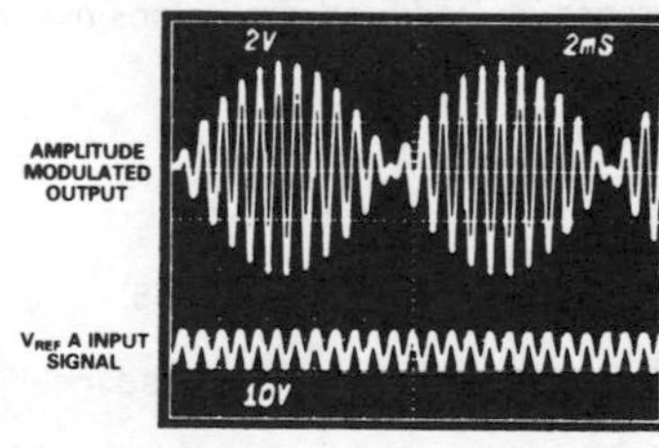

Figure 14. Amplitude Control Using DAC A Reference Input

Output amplitude is directly controlled by the voltage level on DAC A reference input. Ac or dc signals may be used within the range ±10 volts, 0 to 10kHz. Figure 14 shows the amplitude of a 1kHz sine wave being controlled by a 55Hz sine wave.

Programmable Phase

Two sine waves with 0 to 360°programmable phase relationship can be generated by running two of the circuits, shown in Figure 12, from the same input clock source. By allowing one circuit to start running from zero count a preset number of clock cycles before the other, a phase difference between their output sine waves is introduced (see Figure 15).

Since each complete cycle is generated by 1024 clock cycles, phase steps of 360/1024 degrees are programmable. Note also that the phase difference is independent of output frequency.

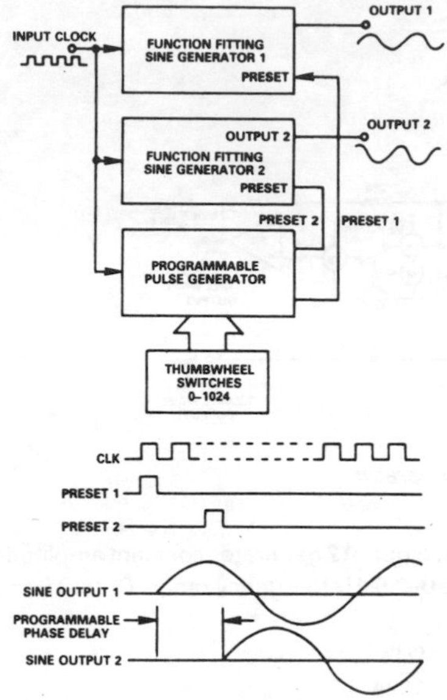

Figure 15. Programmable Phase Relationship

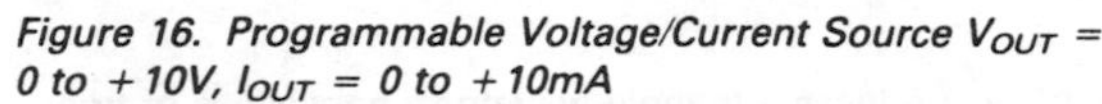

Figure 16. Programmable Voltage/Current Source V_{OUT} = 0 to + 10V, I_{OUT} = 0 to + 10mA

PROGRAMMABLE VOLTAGE/CURRENT SOURCE USING DUAL DAC AD7528

The circuit in Figure 16 is that of a unipolar V/I source. A negative reference is required for a positive output voltage.

The circuit provides;

(a) A programmable output voltage

$$V_{OUT} = +V_{REF}A \, \frac{N_A}{256}, \text{ for } N_A = 0 \text{ to } 255.$$

provided the current limit is not exceeded.

(b) In the voltage mode, a programmable load current limit given by:

$$I_{OUT(max)} = V_{REF}B \cdot \frac{R_1}{R_2} \cdot \frac{1}{R_{S2}} \cdot \frac{N_B}{256}$$

for $N_B = 0$ to 255

(c) A constant current feature by setting $N_A = 255$ i.e., maximum output voltage capability, and limiting the load resistance value R_L such that

$$I_{OUT(max)} \cdot R_L < \frac{255}{256} V_{REF}A$$

$$\text{with } I_{OUT\ (max)} = V_{REF}B \cdot \frac{R_1}{R_2} \cdot \frac{1}{R_{S2}} \cdot \frac{N_B}{256}$$

as in (b).

A useful feature of the circuit is the possibility of load current "readback" in the voltage mode (or load voltage readback in the current mode). By monitoring point X in Figure 16, as the current limit value is reduced, a state change will take place when current limit is attained. The set current limit value will correspond to the load current.

In the circuit DAC A with amplifier A1 and buffer A3 acts as a standard programmable voltage source when $V_{REF}A$ is held constant. The voltage drop across resistor R_{S2} provides a voltage proportional to load current with R_{S1} acting as a current limit on amplifier A2. Amplifier A4 with resistors R_1 and R_2 references the voltage across R_{S2} to ground and also provides gain $\left(\frac{R_2}{R_1}\right)$. The output of A4

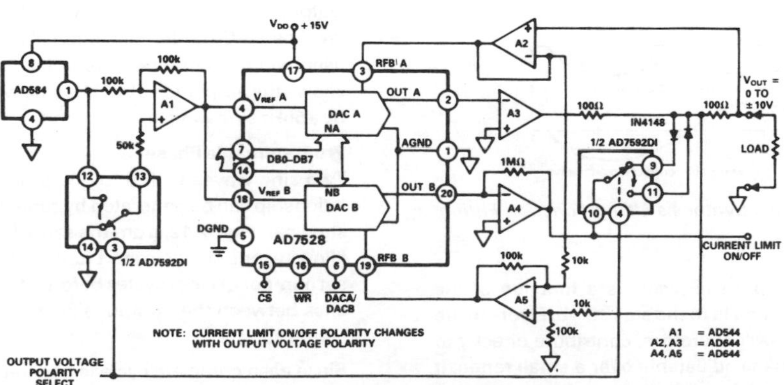

Figure 17. Programmable Voltage/Current Source with Bipolar Output

is compared with a proportion $\frac{N_B}{256}$ of $V_{REF}B$. (usually $V_{REF}A = V_{REF}B$) by amplifier A2. If $V_{OUT4} > V_{REF}B \frac{N_B}{256}$ then current limit is required and the output of A2 via diode D1 draws load current to maintain a constant load current.

If $V_{OUT4} < V_{REF}B \frac{N_B}{256}$ then the current limit is not required and amplifier A2 output is disconnected as diode D1 is reversed biased.

Figure 17 shows a similar circuit for bipolar operation, i.e., 0 to ±10V at 0 to ±10mA.

DUAL DAC PROGRAMMABLE GAIN AMPLIFIER WITH NO TRIMPOTS

A unique advantage of the matched DACs available in the AD7528 is utilized in the programmable gain/attenuation circuit shown in Figure 18. The equivalent resistance of each DAC from its reference input to its output is used to replace the input and feedback resistors in the standard inverting amplifier circuit. By loading DAC's A and B with suitable codes, programmable gain/attenuation over the range −48dB to +48dB can be achieved.

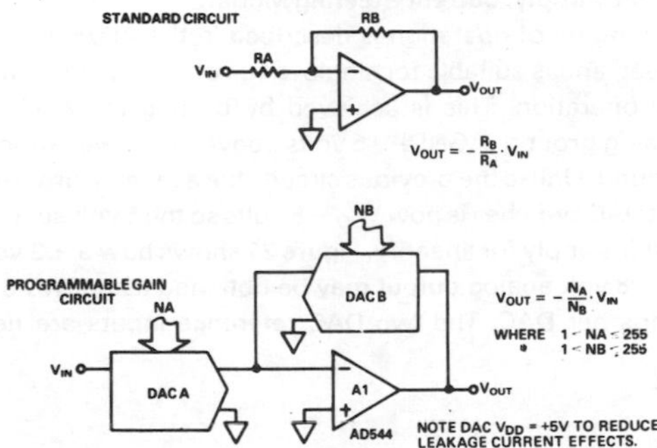

Figure 18. Dual DAC Programmable Gain Amplifier

In the circuit of Figure 18, the DAC equivalent resistances are given by:

$$R_{DAC}A = \frac{256\,R_{LD}A}{N_A} \quad \text{and} \quad R_{DAC}B = \frac{256R_{LD}B}{N_B}$$

Where: $R_{LD}A$ and $R_{LD}B$ are DAC A and DAC B R-2R ladder resistances respectively, N_A and N_B are the DAC codes in decimal (1–255).

The resultant gain expression for the circuit is

$$\frac{V_{OUT}}{V_{IN}} = -\frac{256\,R_{LD}B}{N_B} \cdot \frac{N_A}{256\,R_{LD}A}$$

This simplifies to:

$$\frac{V_{OUT}}{V_{IN}} = -\frac{R_{LD}B \cdot N_A}{R_{LD}A \cdot N_B}$$

But since DAC A and DAC B are a matched pair, $R_{LD}A = R_{LD}B$. This simplifies the expression even further to give:

$$\frac{V_{OUT}}{V_{IN}} = -\frac{N_A}{N_B} \qquad \begin{matrix} 1 \le N_A \le 255 \\ 1 \le N_B \le 255 \end{matrix}$$

Notice that DAC ladder resistance does not appear in the expression. Previous PGA circuits using DACs have always had to be trimmed to accommodate the DAC ladder resistance which usually has a wide tolerance (typically 8k to 20k). This circuit does not suffer from this problem since $R_{LD}A$ and $R_{LD}B$ are matched to better than 1%. Notice also that the circuit has a constant input resistance of $R_{LD}A$. The two unused feedback resistors, $R_{FB}A$ and $R_{FB}B$ are also precisely matched and could be used to provide other DAC code vs. gain relationships.

PROGRAMMABLE WAVEFORM GENERATOR FOR VECTOR SCAN CRT DISPLAYS

Figure 19 shows the dual DAC in a triangle/rectangle wave generator in which the period of each half cycle can be programmed. Such a circuit is useful for vector scan CRT displays to generate variable rate sweep signals (depending upon whether a long or short vector is to be drawn). DAC A determines the ramp rate for the positive going ramp of the triangle while DAC B determines the ramp rate for the negative going ramp. The integrator output voltage is sensed by comparators A4 and A5. When this voltage reaches +10V or −10V the comparators drive the R-S flip-flop G1 and G2 which selects the output of the appropriate DAC via the double-pole ganged analog switch SW1, SW2.

The switching arrangement shown has the advantage that high speed switches (such as CD 4016 or AD7519) can be used to change between the two DACs without introducing significant output glitches at the changeover.

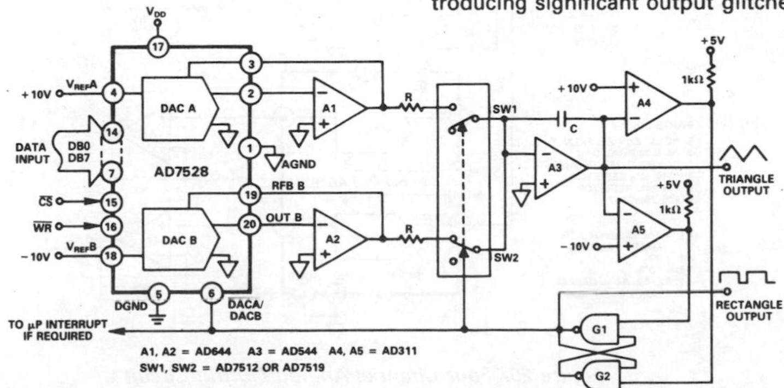

Figure 19. Digitally Programmable Waveform Generator

Furthermore, one DAC can be updated from the data bus and allowed to settle while the output of the other DAC is being used to generate the ramp signal. The output of flip-flop G1 and G2 automatically connects the "unused DAC" to the data bus for further data update if necessary. The output of the flip-flop can be used to drive the interrups of a microprocessor if required.

Selecting Waveform Parameters:
The period (t) of the waveforms generated by the circuit is given by:

$$t = 512\,RC\left[\frac{1}{N_A} + \frac{1}{N_B}\right]$$

where N_A and N_B are the DAC A, DAC B codes in decimal (1–255) respectively.

If DAC A and DAC B latches contain the same codes, the expression simplifies to:

$$t = \frac{1024\,RC}{N_A} \text{ i.e., output frequency } f = \frac{N_A}{1024\,RC}\text{ Hz}$$

The mark-to-space ratio of the rectangle wave output is dependent on the ratio of N_B to N_A

$$\text{Mark to Space Ratio} = \frac{N_A}{N_B}$$

A special case, exists when the code in either DAC is zero. In this case the circuit will stop oscillating as the integrator input voltage will be zero. If the all zeros condition can occur, it is advisable to connect a 10MΩ resistor from the V_{REF} terminal to the output terminal of each DAC, i.e, $V_{REF}A$ to OUT A and $V_{REF}B$ to OUT B. This provides sufficient bias current to keep the circuit oscillating, and does not affect frequency calculations significantly as the 10M resistor introduces only 1/4 LSB of additional error into each DAC output.

AD7528 SINGLE SUPPLY OPERATION
In low budget digital designs requiring analog outputs, the cost of adding an extra power supply rail for the DAC circuits can be a limiting factor. The AD7528 in the single supply configurations shown below provides an ideal cost effective solution for such applications (especially where multiple analog outputs are required).

Single Supply, Voltage Switching Mode:
In this mode, the normal DAC R-2R ladder is inverted. The reference voltage is applied to the DAC OUT A or OUT B terminal and the output voltage is taken from the DAC $V_{REF}A$ or $V_{REF}B$ terminal. For the DACs to retain their specified linearity, the reference voltage range must be restricted as follows:

For $V_{DD} = +15V$, $V_{REF}\max = +2.5V$

For $V_{DD} = +5V$, $V_{REF}\max = +0.5V$

Figure 20 shows a circuit for use with a +15 volt power supply giving four separate 0 to +10V outputs. The op-amps used have a Class A output configuration for small signal levels, thus allowing their outputs to go to zero volts for zero volts input. At higher signal levels, the outputs convert to Class B.

Single Supply, Current Steering Mode:
This mode of operation is described in the AD7528 data sheet, and is suitable for single +10 volt to +15 volt supply operation. This is achieved by biasing the AD7528 analog ground (AGND) +5 volts above the power supply ground. Unlike the previous circuits the available drive for the DAC switches is now $V_{DD} - 5$ volts so the 5 volt specifications apply for linearity. Figure 21 shows how a +2 volt to +8 volt analog output may be obtained using two op-amps per DAC. The two DAC reference inputs are tied

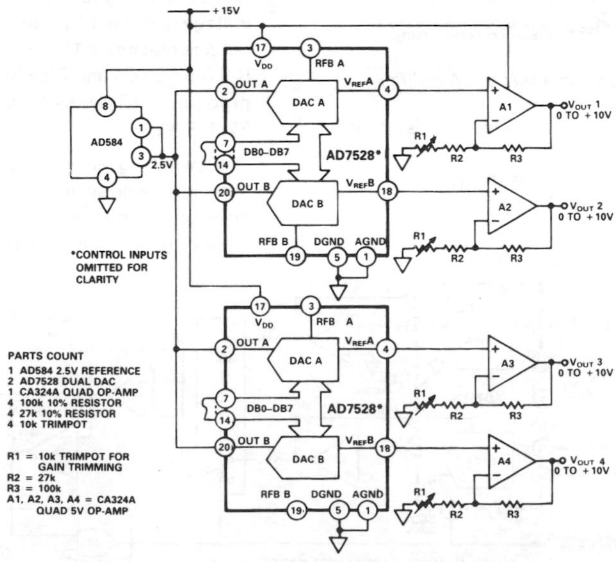

Figure 20. Four Channel Analog Output Circuit

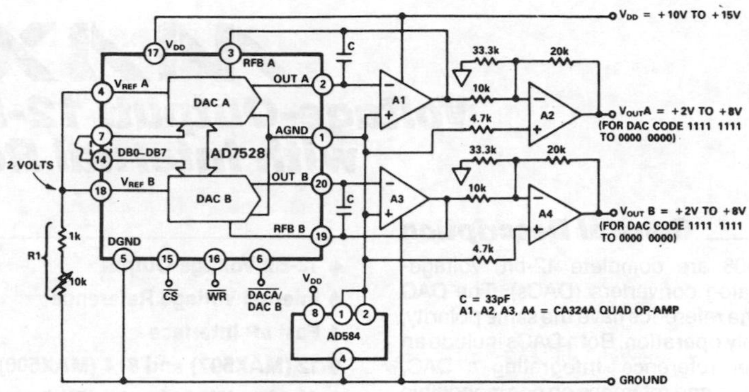

Figure 21. AD7528 Single Supply Operation with AGND Biased to +5 Volts

together and a reference input voltage is obtained without a buffer amplifier by making use of the constant and matched impedances of the DAC A and DAC B reference inputs. Current flows through the two DAC R-2R ladders into R1; R1 is adjusted until $V_{REF}A$ and $V_{REF}B$ inputs are at +2 volts. The adjustment is independent of either DAC code.

Each analog output channel has a +2 to +8 volt range for DAC codes 1111 1111 to 0000 0000.

Reference

Dan H. Sheingold, "Nonlinear Circuits Handbook," available from Analog Devices.

D. P. Burton, Application Note "Methods For Generating Complex Waveforms And Vectors Using Multiplying D/A Converters" Analog Devices Publication Number: E671–15–9/81.

Application Guide to CMOS Multiplying D/A Converters, Analog Devices Publication Number: G479–15–8/78.

19-4338; Rev 0; 10/91

MAXIM

Voltage-Output, 12-Bit DACs with Internal Reference

MAX507/MAX508

General Description

The MAX507/MAX508 are complete 12-bit, voltage-output digital-to-analog converters (DACs). The DAC output voltage and the reference have the same polarity, allowing single-supply operation. Both DACs include an internal buried-zener reference. Integrating a DAC, voltage-output amplifier, and reference on one monolithic device greatly enhances reliability over multi-chip circuits.

Double-buffered logic inputs interface easily to microprocessors (μPs). Data is transferred into the input register either from a 12-bit-wide data bus (MAX507) for 16-bit μPs, or in a right-justified (8+4)-bit format (MAX508) for 8- or 16-bit μPs. All logic signals are level triggered and are TTL and CMOS compatible. Interface timing specifications insure compatibility with all common μPs.

The DACs are specified and tested for both dual- and single-supply operation. Usable supplies range from single +12V to dual ±15V.

On-board gain-setting resistors allow three output-voltage ranges: 0V to +5V and 0V to +10V can be generated when using either single or dual supplies. With dual supplies, ±5V is also available. The output amplifier can drive a 2kΩ load to +10V.

Applications

Digital Offset and Gain Adjustment
Industrial Controls
Arbitrary Function Waveform Generators
Automatic Test Equipment
Automated Calibration
Machine and Motion Control

Features

- ♦ 12-Bit Voltage Output
- ♦ Internal Voltage Reference
- ♦ Fast μP Interface
- ♦ 12 (MAX507) and 8+4 (MAX508) Data-Bus Widths
- ♦ Single +12V to Dual ±15V Supply Operation
- ♦ 20- and 24-Pin DIP and Wide SO Packages

Ordering Information

PART	TEMP. RANGE	PIN-PACKAGE	ERROR (LSBs)
MAX507ACNG	0°C to +70°C	24 Narrow Plastic DIP	±1/2
MAX507BCNG	0°C to +70°C	24 Narrow Plastic DIP	±3/4
MAX507ACWG	0°C to +70°C	24 Wide SO	±1/2
MAX507BCWG	0°C to +70°C	24 Wide SO	±3/4
MAX507BC/D	0°C to +70°C	Dice*	±3/4
MAX507AENG	-40°C to +85°C	24 Narrow Plastic DIP	±1/2
MAX507BENG	-40°C to +85°C	24 Narrow Plastic DIP	±3/4
MAX507AEWG	-40°C to +85°C	24 Wide SO	±1/2
MAX507BEWG	-40°C to +85°C	24 Wide SO	±3/4
MAX507AMRG	-55°C to +125°C	24 Narrow CERDIP**	±1/2
MAX507BMRG	-55°C to +125°C	24 Narrow CERDIP**	±3/4

Ordering Information continued on page 12.

* *Contact factory for dice specifications.*
** *Contact factory for availability and processing to MIL-STD-883.*

Functional Diagram

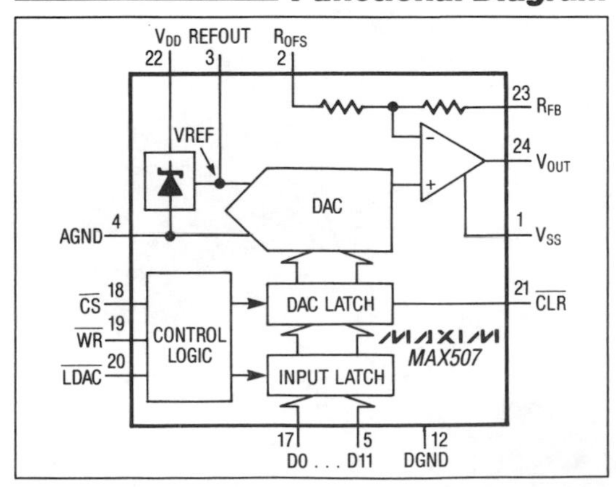

Pin Configurations

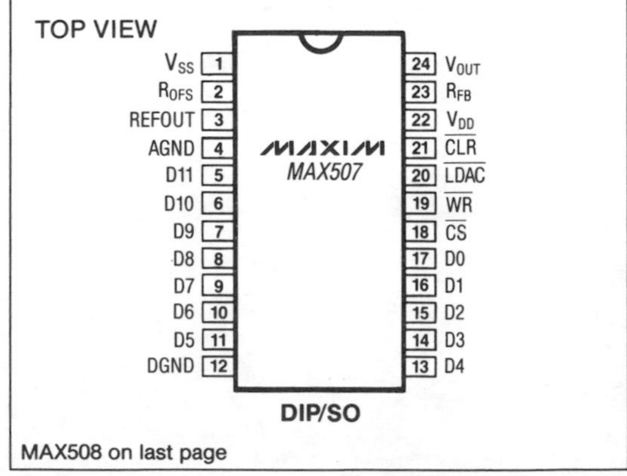

Voltage-Output, 12-Bit DACs with Internal Reference

ABSOLUTE MAXIMUM RATINGS

V_{DD} to AGND .... -0.3V, +17V
V_{DD} to DGND .... -0.3V, +17V
V_{DD} to V_{SS} .... -0.3V, +34V
AGND to DGND .... -0.3V, V_{DD}
Digital Input Voltage to GND .... -0.3V, V_{DD} +0.3V
V_{OUT} to AGND (Note 1) .... V_{SS}, V_{DD}
V_{OUT} to V_{SS} (Note 1) .... 0V, +34V
V_{OUT} to V_{DD} (Note 1) .... -34V, 0V
REFOUT to AGND (Note 1) .... -0.3V, V_{DD} +0.3V

Continuous Power Dissipation (any package)
to +75°C 450mW
derate above +75°C .... 6mW/°C
Operating Temperature Ranges:
MAX507_C__, MAX508_C__ .... 0°C to +70°C
MAX507_E__, MAX508_E__ .... -40°C to +85°C
MAX507_M__, MAX508_M__ .... -55°C to +125°C
Storage Temperature Range .... -65°C to +150°C
Lead Temperature (soldering, 10 sec) .... +300°C

Note 1: The output can be shorted to either supply rail if the package power dissipation is not exceeded. Typical short-circuit current to AGND is 25mA.

Stresses beyond those listed under "Absolute Maximum Ratings" may cause permanent damage to the device. These are stress ratings only, and functional operation of the device at these or any other conditions beyond those indicated in the operational sections of the specification is not implied. Exposure to absolute maximum rating conditions for extended periods may affect device reliability.

ELECTRICAL CHARACTERISTICS

Single Supply (V_{DD} = +11.4V to +15.75V, V_{SS} = AGND = DGND = 0V, R_L = 2kΩ, C_L = 100pF, REFOUT unloaded, all grades, T_A = T_{MIN} to T_{MAX}, unless otherwise noted.)

PARAMETER	SYMBOL	CONDITIONS		MIN	TYP	MAX	UNITS
STATIC PERFORMANCE							
Resolution	N			12			Bits
Relative Accuracy	INL	T_A = +25°C	MAX507/508A			±1/2	LSB
			MAX507/508B			±3/4	
		T_A = T_{MIN} to T_{MAX}	MAX507/508A			±3/4	
			MAX507/508B			±1	
Differential Nonlinearity	DNL					±1	LSB
Unipolar Offset Error		T_A = +25°C				±3	LSB
		T_A = T_{MIN} to T_{MAX}				±5	
DAC Gain Error						±2	LSB
Full-Scale Output Voltage Error		V_{DD} = +12V or +15V	T_A = +25°C			±0.2	%FSR
			T_A = T_{MIN} to T_{MAX}			±0.6	
Full-Scale Output Voltage Change		V_{DD} over full range	T_A = +25°C			±0.12	%FSR/V
			T_A = T_{MIN} to T_{MAX}			±0.2	
Full-Scale Tempco		MAX507/508_C/E				±30	ppm FSR/°C
		MAX507/508_M				±40	
Unipolar Offset Error Change		V_{DD} = +12V ± 5% or +15V ± 5%				±1	mV

Voltage-Output, 12-Bit DACs with Internal Reference

ELECTRICAL CHARACTERISTICS (continued)

Single Supply (V_{DD} = +11.4V to +15.75V, V_{SS} = AGND = DGND = 0V, R_L = 2kΩ, C_L = 100pF, REFOUT unloaded, all grades, T_A = T_{MIN} to T_{MAX}, unless otherwise noted.)

PARAMETER	SYMBOL	CONDITIONS		MIN	TYP	MAX	UNITS
REFERENCE							
Reference Output		V_{DD} = +12V or +15V	T_A = +25°C	4.99		5.01	V
Reference Voltage Change		V_{DD} = +12V ± 5% or +15V ± 5%	T_A = +25°C			2	mV/V
			T_A = T_{MIN} to T_{MAX}			6	
Reference Temperature Coefficient		MAX507/508_C/E			±30		ppm/°C
		MAX507/508_M			±40		
Reference Load Sensitivity		I_{LOAD} = 0µA to 100µA			±1		mV
ANALOG OUTPUT							
Ranges (Note 2)						0 to 5	V
						0 to 10	
Output Range Resistors				15		30	kΩ
DC Output Impedance					0.5		Ω
Short-Circuit Current					40		mA
DYNAMIC PERFORMANCE (Note 3)							
Voltage-Output Slew Rate				2			V/µs
V_{OUT} Settling Time		To ±1/2 LSB for full-scale change				5	µs
Digital Feedthrough					10		nV-s
Digtal-to-Analog Glitch Impulse		Major carry transition			30		nV-s
Output Load Resistance (Note 2)		V_{OUT} = 0V to +10V		2			kΩ
POWER SUPPLIES							
V_{DD} Range		For specified performance		11.4		15.75	V
I_{DD}		Outputs unloaded	T_A = +25°C			9	mA
			T_A = T_{MIN} to T_{MAX}			12	

Voltage-Output, 12-Bit DACs with Internal Reference

ELECTRICAL CHARACTERISTICS

Dual Supply (V_{DD} = +11.4V to +15.75V, V_{SS} = −11.4V to −15.75V, DGND = AGND = 0V, R_L = 2kΩ, C_L = 100pF, REFOUT unloaded, all grades, T_A = T_{MIN} to T_{MAX}, unless otherwise noted.)

PARAMETER	SYMBOL	CONDITIONS		MIN	TYP	MAX	UNITS
STATIC PERFORMANCE							
Resolution	N			12			Bits
Relative Accuracy	INL	T_A = +25°C	MAX507/508A			±1/2	LSB
			MAX507/508B			±3/4	
		T_A = T_{MIN} to T_{MAX}	MAX507/508A			±3/4	
			MAX507/508B			±1	
Differential Nonlinearity	DNL					±1	LSB
Bipolar Zero Offset Error	BZOE	MAX507/508A	T_A = +25°C			±2	LSB
			T_A = T_{MIN} to T_{MAX}			±4	
		MAX507/508B	T_A = +25°C			±3	
			T_A = T_{MIN} to T_{MAX}			±5	
DAC Gain Error						±2	LSB
Full-Scale Output Voltage Error		V_{DD} = +15V, V_{SS} = −15V	T_A = +25°C			±0.2	%FSR
			T_A = T_{MIN} to T_{MAX}			±0.6	
		V_{DD} = +12V, V_{SS} = −12V	T_A = +25°C			±0.2	
			T_A = T_{MIN} to T_{MAX}			±0.6	
Full-Scale Output Change with V_{DD}		V_{DD} = +12V ± 5% or +15V ± 5% V_{SS} = −12V or −15V	T_A = +25°C			±0.12	%FSR/V
			T_A = T_{MIN} to T_{MAX}			±0.2	
Full-Scale Output Change with V_{SS}	V_{SS}	V_{SS} = −12V ± 5% or −15V ± 5% V_{DD} = +12V or +5V				0.01	%FSR/V
Full-Scale Tempco		MAX507/508_C/E				±30	ppm FSR/°C
		MAX507/508_M				±40	
Bipolar Zero Offset Change		V_{DD} = +12V ± 5% or +15V ± 5% V_{SS} = −12V or −15V				±1	mV
		V_{SS} = −12V ± 5% or −15V ± 5% V_{DD} = +12V or +15V				±1	
REFERENCE							
Reference Output		V_{DD} = +12V or +15V	T_A = +25°C	4.99		5.01	V
Reference Output Change		V_{DD} over full range	T_A = +25°C			2	mV/V
			T_A = T_{MIN} to T_{MAX}			6	
Reference Temperature Coefficient		MAX507/508_C/E			±30		ppm/°C
		MAX507/508_M			±40		
Reference Load Sensitivity		I_{LOAD} = 0µA to 100µA				±1	mV

Voltage-Output, 12-Bit DACs with Internal Reference

ELECTRICAL CHARACTERISTICS (continued)

Dual Supply (V_{DD} = +11.4V to +15.75V, V_{SS} = −11.4V to −15.75V, DGND = AGND = 0V, R_L = 2kΩ, C_L = 100pF, REFOUT unloaded, all grades, T_A = T_{MIN} to T_{MAX}, unless otherwise noted.)

PARAMETER	SYMBOL	CONDITIONS		MIN	TYP	MAX	UNITS
ANALOG OUTPUT							
Ranges (Notes 2, 4)					0 to +5 or +10, −5 to +5		V
Output Range Resistors				15		30	kΩ
DC Output Impedance					0.5		Ω
Short-Circuit Current					40		mA
DYNAMIC PERFORMANCE (Note 3)							
Voltage-Output Slew Rate				2			V/μs
V_{OUT} Settling Time		to ±1/2 LSB				5	μs
Digital Feedthrough					10		nV-s
Digtal-to-Analog Glitch Impulse		Major carry transition			30		nV-s
Output Load Resistance		V_{OUT} = −5V to +10V		2			kΩ
POWER SUPPLIES							
V_{DD} Range		For specified performance		11.4		15.75	V
V_{SS} Range		For specified performance		−11.4		−15.75	V
I_{DD}		Outputs unloaded	T_A = +25°C			9	mA
			T_A = T_{MIN} to T_{MAX}			12	
I_{SS}		Outputs unloaded	T_A = +25°C			3	mA
			T_A = T_{MIN} to T_{MAX}			5	

Voltage-Output, 12-Bit DACs with Internal Reference

ELECTRICAL CHARACTERISTICS

Single or Dual Supply (V_{DD} = +11.4V to +15.75V, V_{SS} = 0V to −15.75V, DGND = AGND = 0V, REFOUT unloaded, R_L = 2kΩ, C_L = 100pF, all grades, T_A = T_{MIN} to T_{MAX}, unless otherwise noted.)

PARAMETER	SYMBOL	CONDITIONS		MIN	TYP	MAX	UNITS
DIGITAL INPUTS							
V_{INH}				2.4			V
V_{INL}						0.8	V
Input Current	I_{IN}	D0–D11	T_A = +25°C			±1	μA
			T_A = T_{MIN} to T_{MAX}			±10	
I_{INH}		CS, WR, LDAC, CLR	T_A = +25°C			±1	μA
			T_A = T_{MIN} to T_{MAX}			±10	
I_{INL}		$\overline{CS}$, $\overline{WR}$, $\overline{LDAC}$, $\overline{CLR}$	T_A = +25°C			±150	μA
			T_A = T_{MIN} to T_{MAX}			±200	
Digital Input Capacitance					8		pF

TIMING CHARACTERISTICS

(All grades, T_A = T_{MIN} to T_{MAX}, unless otherwise noted.)

PARAMETER	SYMBOL	CONDITIONS	MIN	TYP	MAX	UNITS
$\overline{CS}$ Pulse Width (Note 5)	t_1	T_A = +25°C	80			ns
		T_A = T_{MIN} to T_{MAX}	100			
$\overline{WR}$ Pulse Width	t_2	T_A = +25°C	80			ns
		T_A = T_{MIN} to T_{MAX}	100			
$\overline{CS}$ to $\overline{WR}$ Setup Time (Note 5)	t_3		0			ns
$\overline{CS}$ to $\overline{WR}$ Hold Time (Note 5)	t_4		0			ns
Data to $\overline{WR}$ Setup Time	t_5	T_A = +25°C	100			ns
		T_A = T_{MIN} to T_{MAX}	110			
Data to $\overline{WR}$ Hold Time	t_6		10			ns
$\overline{LDAC}$ Pulse Width	t_7	T_A = +25°C	80			ns
		T_A = T_{MIN} to T_{MAX}	100			
$\overline{CLR}$ Pulse Width (MAX507)	t_8	T_A = +25°C	80			ns
		T_A = T_{MIN} to T_{MAX}	100			

Note 2: V_{OUT} must be less than (V_{DD} − 2.5V).
Note 3: Dynamic performance is included for design guidance, not subject to test.
Note 4: The 0V to +5V or +10V ranges can be used with V_{SS} = −5V with no degradation.
Note 5: $\overline{CS}$ = $\overline{CSLSB}$ and $\overline{CSMSB}$ for MAX508.

Voltage-Output, 12-Bit DACs with Internal Reference

Detailed Description

Digital-to-Analog Converters

The MAX507/MAX508 are 12-bit, voltage-output DACs. The DAC output voltage has the same polarity as the reference, allowing single-supply operation.

The basic DAC circuit consists of a laser-trimmed, thin-film, R-2R resistor array with NMOS voltage switches (Figure 1).

Output-Buffer Amplifier

The output amplifier is noninverting and configurable for a gain of 1 or 2. Three output voltage ranges can be configured for: 0V to +5V, 0V to +10V, and -5V to +5V. The output amplifier can drive 2kΩ in parallel with 100pF connected to GND.

The MAX507/MAX508 can operate from a single supply with a 0V to +5V or a 0V to +10V output range by tying V_{SS} to 0V. However, the speed and current-sinking capability of the amplifier decreases as the output falls within 0.5V of V_{SS}. Speed and current-sinking capability can be maintained by including a negative supply. Table 1 lists the allowable single and dual supplies for each range.

The output amplifier's small-signal bandwidth is typically 2MHz. Output noise is approximately $25nV/\sqrt{Hz}$ at 1kHz, and output broadband noise is approximately $25\mu V_{RMS}$.

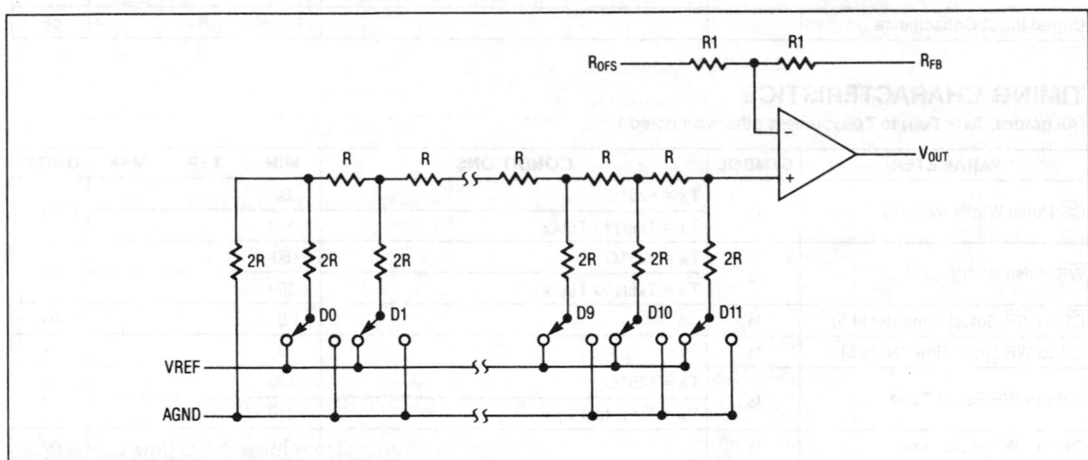

Figure 1. Simplified MAX507 DAC Circuit

Table 1. Output Voltage Range vs. Supply Voltage

Range	Single Supply V_{DD}	Dual Supply V_{DD}	Dual Supply V_{SS}
0V to +5V	+11.4V to +15.75V	+11.4V to +15.75V	-4.5V to -15.75V
0V to +10V	+14.25V to +15.75V	+14.25V to +15.75V	-4.5V to -15.75V
-5V to +5V		+11.4V to +15.75V	-11.4V to -15.75V

Voltage-Output, 12-Bit DACs with Internal Reference

Voltage Reference

The voltage at REFOUT is 5V ± 10mV at +25°C. The reference is internally connected to the DAC and is buffered to accommodate the DAC's variable impedance. This buffer is capable of driving the DAC, the R_{OFS} resistor, and up to 500μA of external current. MAX507/MAX508 specifications are determined with the internal reference. The reference should be decoupled at REFOUT with 10Ω in series with the recommended decoupling capacitors, 10μF in parallel with 0.1μF.

Digital Inputs and Interface Logic

All logic inputs are compatible with both TTL and 5V CMOS logic. Supply current is specified for TTL input levels, but is reduced by about 450μA when the data inputs are driven near DGND or V_{DD}. The control inputs ($\overline{\text{CLR}}$, $\overline{\text{LDAC}}$, $\overline{\text{WR}}$, $\overline{\text{CS}}$, $\overline{\text{CSMSB}}$, and $\overline{\text{CSLSB}}$) each draw 100μA from I_{DD} when low.

MAX507 Interface

Table 2 is the MAX507 truth table. The MAX507 accepts a 12-bit input word that can be latched or transferred directly to the DAC. $\overline{\text{CS}}$ and $\overline{\text{WR}}$ control the input latch, and $\overline{\text{LDAC}}$ transfers information from the input latch to the DAC latch.

Table 2. MAX507 Truth Table

$\overline{\text{CLR}}$	$\overline{\text{LDAC}}$	$\overline{\text{WR}}$	$\overline{\text{CS}}$	Function
1	0	0	0	Both latches transparent
1	1	1	X	Both latches latched
1	1	X	1	Both latches latched
1	1	0	0	Input latch transparent
1	1	↑	0	Input latch latched
1	0	1	1	DAC latch transparent
1	↑	1	1	DAC latch latched
0	X	X	X	DAC latch all 0s
↑	1	1	1	DAC latch latched with 0s; output at 0V or −5V
↑	0	0	0	Both latches transparent; output follows input data

1 = High State X = Don't Care
0 = Low State ↑ = Rising Edge

The input latch is transparent when $\overline{\text{CS}}$ and $\overline{\text{WR}}$ are low; the DAC latch is transparent when $\overline{\text{LDAC}}$ is low. Data is latched within the input latch on the rising edge of $\overline{\text{WR}}$ when $\overline{\text{CS}}$ is low. The rising edge of $\overline{\text{LDAC}}$ latches data into the DAC when $\overline{\text{CS}}$ and $\overline{\text{WR}}$ are low. After $\overline{\text{CS}}$ and $\overline{\text{WR}}$ are high, $\overline{\text{LDAC}}$ must be held low for t_7 or longer (Figure 2).

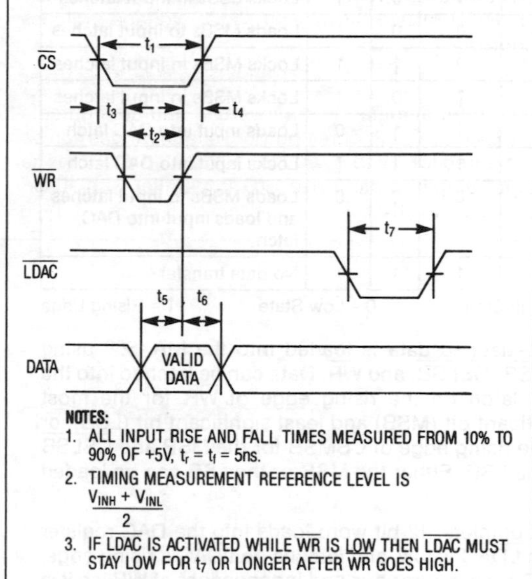

CS
t1
t3
t2
t4
WR
t7
LDAC
t5
t6
DATA
VALID DATA
NOTES:
1. ALL INPUT RISE AND FALL TIMES MEASURED FROM 10% TO 90% OF +5V, tr = tf = 5ns.
2. TIMING MEASUREMENT REFERENCE LEVEL IS (VINH + VINL)/2
3. IF LDAC IS ACTIVATED WHILE WR IS LOW THEN LDAC MUST STAY LOW FOR t7 OR LONGER AFTER WR GOES HIGH.

Figure 2. MAX507 Timing Diagram

The DAC latch is reset to zeros with $\overline{\text{CLR}}$ low. $\overline{\text{CLR}}$ acts as a zero override when the input latch and DAC latch are transparent. Then, a low-to-high $\overline{\text{CLR}}$ transition loads all zeros into the DAC latch, and the output remains low (0V to −5V).

MAX508 Interface

The MAX508's 8-bit-wide data bus interfaces with 8-bit μPs. The MAX508 contains an input latch and a DAC latch. The data held in the DAC latch determines the output of the DAC. Table 3 is the MAX508 truth table, Figure 3 shows the input control logic, and Figure 4 shows the write-cycle timing.

Table 3. MAX508 Truth Table

CSLSB	CSMSB	WR	LDAC	Function
0	1	0	1	Loads LSBs to input latches
0	1	↑	1	Locks LSBs in input latches
↑	1	0	1	Locks LSBs in input latches
1	0	0	1	Loads MSBs to input latches
1	0	↑	1	Locks MSBs in input latches
1	↑	0	1	Locks MSBs in input latches
1	1	1	0	Loads input into DAC latch
1	1	1	↑	Locks input into DAC latch
1	0	0	0	Loads MSBs to input latches and loads input into DAC latch
1	1	1	1	No data transfer

1 = High State   0 = Low State   ↑ = Rising Edge

Right-justified data is loaded into the MAX508 using CSMSB, CSLSB, and WR. Data can be latched into the input latch on the rising edge of WR for the most significant bit (MSB) and least significant bit (LSB), or on the rising edge of CSMSB for the MSB and CSLSB for the LSB. Either the MSB or the LSB can be loaded first.

The complete, 12-bit word loads into the DAC register when LDAC is low, and latches on LDAC's rising edge. LDAC is asynchronous and independent of WR, so it is ideal for simultaneously updating multiple MAX508 outputs. Because LDAC can occur during a write cycle, it must stay low for t_7 (or longer) after WR goes high to ensure correct data is latched to the output.

The MAX508 output can be updated in two write cycles by tying CSMSB and LDAC. In this automatic transfer mode, CSLSB and WR latch the lower 8 bits into the input latch; then CSMSB, WR, and LDAC load the upper 4 bits into the input latch and transfer the 12-bit word into the DAC latch. Alternatively, the MAX507 can be updated in two writes by tying CSLSB to LDAC if the upper 4 bits are input first, followed by the lower 8 bits.

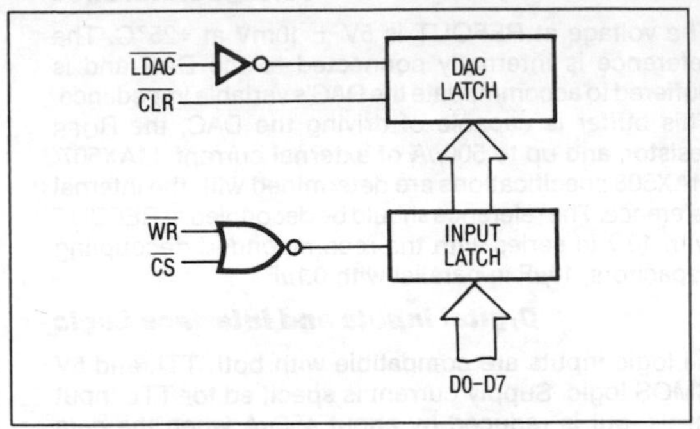

Figure 3a. MAX507 Input Control Logic

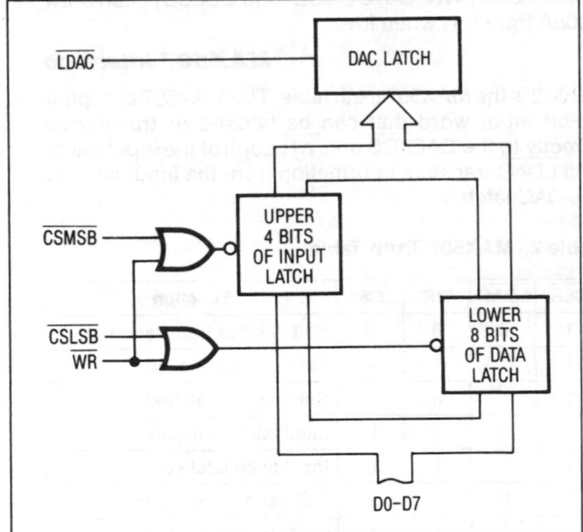

Figure 3b. MAX508 Input Control Logic

Voltage-Output, 12-Bit DACs with Internal Reference

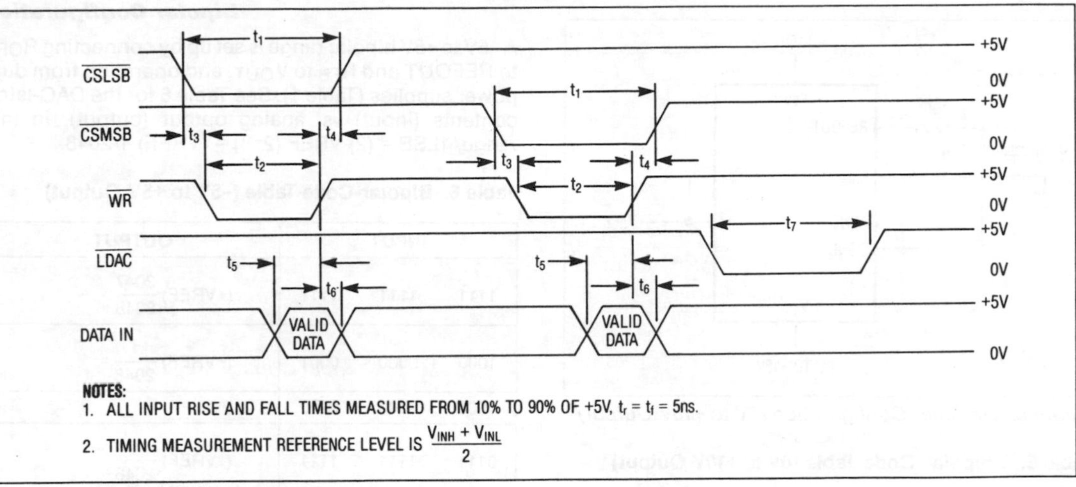

Figure 4. MAX508 Timing Diagram

Unipolar Configuration

The MAX507/MAX508 are set up for a 0V to +5V unipolar output range by connecting R_{OFS}, R_{FB}, and V_{OUT} (Figure 5). The converters operate from either a single or a dual supply in this configuration. See Table 4 for the DAC-latch contents (input) vs. analog output (output). In this range, 1LSB = VREF (2^{-12}).

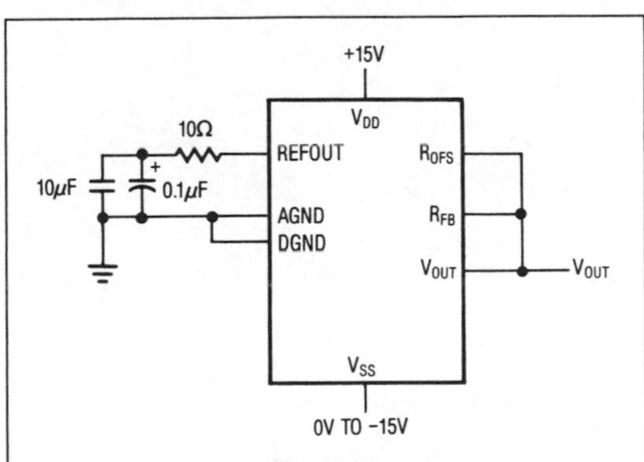

Figure 5. Unipolar Configuration (0V to +5V Output)

Table 4. Unipolar-Code Table (0V to +5V Output)

INPUT			OUTPUT
1111	1111	1111	(VREF) $\frac{4095}{4096}$
1000	0000	0001	(VREF) $\frac{2049}{4096}$
1000	0000	0000	(VREF) $\frac{2048}{4096}$ =+VREF/2
0111	1111	1111	(VREF) $\frac{2047}{4096}$
0000	0000	0001	(VREF) $\frac{1}{4096}$
0000	0000	0000	0V

A 0V to +10V unipolar output range is set up by connecting R_{OFS} to AGND and R_{FB} to V_{OUT} (Figure 6). See Table 5 for the DAC-latch contents (input) vs. analog output (output). The MAX507/MAX508 operate from either a single or a dual supply in this configuration. In this range, 1LSB = VREF (2^{-11}).

Voltage-Output, 12-Bit DACs with Internal Reference

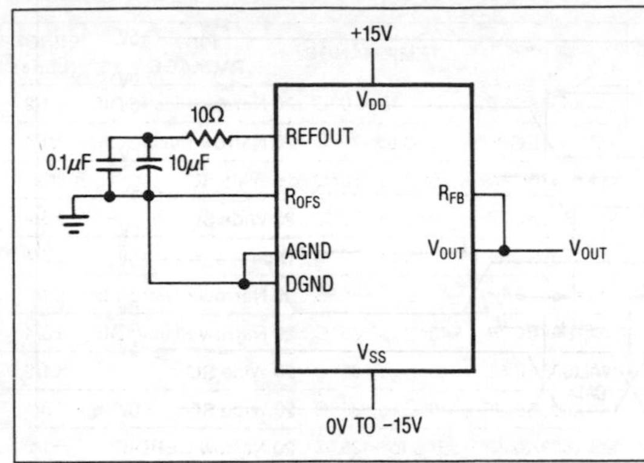

Figure 6. Unipolar Configuration (0V to +10V Output)

Table 5. Unipolar-Code Table (0V to +10V Output)

INPUT			OUTPUT
1111	1111	1111	$+2\ (VREF)\ \frac{4095}{4096}$
1000	0000	0001	$+2\ (VREF)\ \frac{2049}{4096}$
1000	0000	0000	$+2\ (VREF)\ \frac{2048}{4096} = +VREF$
0111	1111	1111	$+2\ (VREF)\ \frac{2047}{4096}$
0000	0000	0001	$+2\ (VREF)\ \frac{1}{4096}$
0000	0000	0000	0V

Bipolar Configuration

A -5V to +5V bipolar range is set up by connecting R_{OFS} to REFOUT and R_{FB} to V_{OUT}, and operating from dual power supplies (Table 1). See Table 6 for the DAC-latch contents (input) vs. analog output (output). In this range, 1LSB = (2) VREF (2^{-11}) = (VREF) 1/2048.

Table 6. Bipolar-Code Table (-5V to +5V Output)

INPUT			OUTPUT
1111	1111	1111	$(+VREF)\ \frac{2047}{2048}$
1000	0000	0001	$(+VREF)\ \frac{1}{2048}$
1000	0000	0000	0V
0111	1111	1111	$(-VREF)\ \frac{1}{2048}$
0000	0000	0001	$(-VREF)\ \frac{2047}{2048}$
0000	0000	0000	$(-VREF)\ \frac{2048}{2048} = -VREF$

Voltage-Output, 12-Bit DACs with Internal Reference

Pin Configurations (continued)

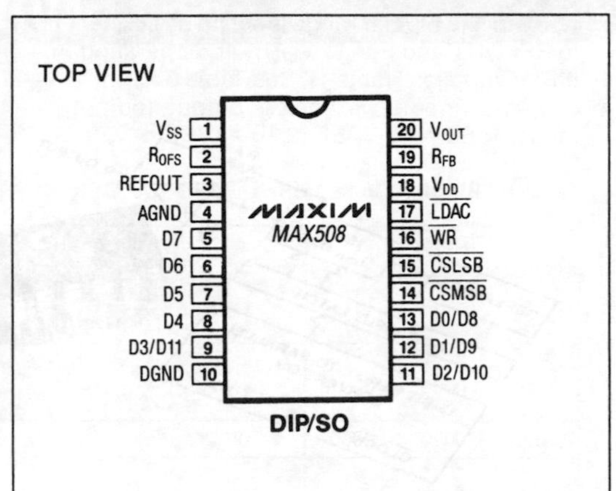

Ordering Information (continued)

PART	TEMP. RANGE	PIN-PACKAGE	ERROR (LSBs)
MAX508ACPP	0°C to +70°C	20 Narrow Plastic DIP	±1/2
MAX508BCPP	0°C to +70°C	20 Narrow Plastic DIP	±3/4
MAX508ACWP	0°C to +70°C	20 Wide SO	±1/2
MAX508BCWP	0°C to +70°C	20 Wide SO	±3/4
MAX508BC/D	0°C to +70°C	Dice*	±3/4
MAX508AEPP	-40°C to +85°C	20 Narrow Plastic DIP	±1/2
MAX508BEPP	-40°C to +85°C	20 Narrow Plastic DIP	±3/4
MAX508AEWP	-40°C to +85°C	20 Wide SO	±1/2
MAX508BEWP	-40°C to +85°C	20 Wide SO	±3/4
MAX508AMJP	-55°C to +125°C	20 Narrow CERDIP**	±1/2
MAX508BMJP	-55°C to +125°C	20 Narrow CERDIP**	±3/4

* *Contact factory for dice specifications.*
** *Contact factory for availability and processing to MIL-STD-883.*

16-Bit DSP DAC

AD766

FEATURES
0.0040% THD
Direct Interface to High Speed Digital Signal Processors
No External Components Required
Sample Rates up to 500 kSPS
Serial Input, 2s Complement
±3 V Output
Optional Trim Allows Superlinear Performance
±5 V to ±12 V Operation
16-Pin Plastic Package

APPLICATIONS
Digital Signal Processing
High Speed Modems
Speech Synthesis

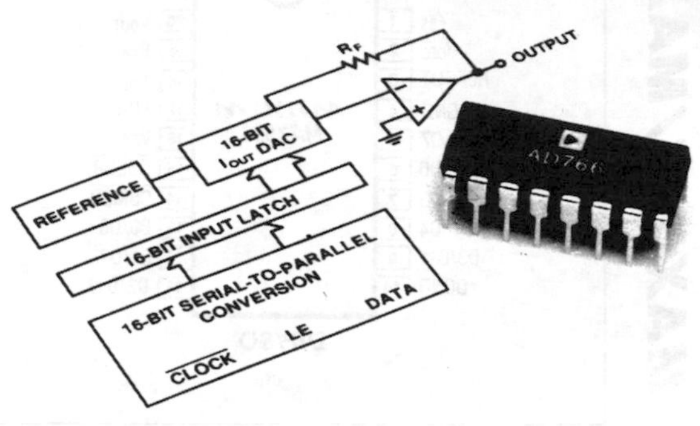

PRODUCT DESCRIPTION

The AD766 is a monolithic 16-bit DSP DAC. Each device provides a voltage output amplifier, 16-bit DAC, 16-bit serial-to-parallel input register and voltage reference. The digital portion of the AD766 is fabricated with CMOS logic elements that are provided by Analog Devices' BiMOS II process. The analog portion of the AD766 is fabricated with bipolar and MOS devices as well as thin film resistors.

The combination of circuit elements, as well as careful design and layout techniques, results in the AD766's excellent ac performance. Laser-trimming of the linearity error affords extremely low total harmonic distortion. An optional linearity trim pin is provided to allow residual differential linearity error at midscale to be eliminated. Output glitch is also small contributing to the overall high level of performance. The output amplifier achieves fast settling and high slew rates, providing a full ±3 V signal. The AD766 can be used to produce output signals up to 50 kHz. The output amplifier is also short circuit protected and can withstand indefinite shorts to ground.

The serial input interface consists of the clock, data and latch enable signals. The serial 2s complement data word is clocked into the DAC, MSB first, by the external data clock. The latch enable signal transfers the input word from the internal serial input register to the parallel DAC input register. This serial input port is directly compatible with the DSP56000, TMS320CXX, ADSP2101 and other popular DSPs, requiring no external components to achieve operation. The input clock can support a 12.5 MHz clock rate.

The AD766 operates with ±5 V to ±12 V power supplies. The digital supplies, $+V_L$ and $-V_L$, can be separated from the analog supplies, $+V_S$ and $-V_S$, for reduced digital crosstalk. Separate analog and digital ground pins are also provided.

Power dissipation is 120 mW typical with ±5 V supplies, 225 mW typical when +5 V/−12 V supplies are used, and 300 mW typical with ±12 V supplies.

The AD766 is available in a commercial version. The AD766JN is specified for operation over 0 to +70°C and is available in a 16-pin plastic DIP.

PRODUCT HIGHLIGHTS

1. Dynamic Specifications for DSP Users.
   The AD766 is specified for ac parameters particularly suited for DSP applications. Total harmonic distortion and signal-to-noise ratio are tested on every device.
2. Temperature Performance Specifications.
   The AD766 is specified and 100% tested over the full operating commercial temperature range.
3. Zero-Chip Microprocessor Interface.
   The AD766 is designed to achieve a zero-chip interface to the ADSP2101, TMS320CXX and DSP56000. The user no longer needs to worry about external "glue logic" to enable the DAC and μprocessor to communicate; three wires are required–clock, data input and latch enable.
4. Wide Dynamic Range of 96 dB.
5. Low Power Dissipation of 120 mW.
6. Space-Saving 16-Pin Plastic DIP Package.

One Technology Way, P.O. Box 9106, Norwood, MA 02062-9106, U.S.A.
Tel: 617/329-4700 Fax: 617/326-8703 Twx: 710/394-6577
Telex: 924491 Cable: ANALOG NORWOODMASS

SPECIFICATIONS (T_{min} to T_{max}, ±5 V supplies, F_S = 500 kSPS unless otherwise noted)

Parameter	Conditions	Min	Typ	Max	Units
RESOLUTION				**16**	Bits
DIGITAL INPUTS	V_{IH}	2.0		$+V_L$	V
	V_{IL}			0.8	V
	I_{IH}, $V_{IH} = V_L$			**1.0**	μA
	I_{IL}, V_{IL} = 0.4			**−10**	μA
Maximum Clock Input Frequency		12.5			MHz
ACCURACY					
Gain Error			±2.0		% of FSR
Gain Drift			±25		ppm of FSR/°C
Midscale Output Voltage Error			±30		mV
Bipolar Zero Drift			±4		ppm of FSR/°C
Differential Linearity Error			±0.001		% of FSR
TOTAL HARMONIC DISTORTION					
F_{OUT} = 1037 Hz	0 dB		−88	**−81**	dB
	−20 dB		−75	**−65**	dB
	−60 dB		−37	**−27**	dB
F_{OUT} = 49.07 kHz[1]	0 dB		−77	**−72**	dB
	−20 dB		−69	**−66**	dB
	−60 dB		−25	**−21**	dB
SIGNAL-TO-NOISE RATIO[2]					
Wideband					
20 Hz to 20 kHz			−102	**−94**	dB
20 kHz to 250 kHz			−83	**−79**	dB
SETTLING TIME (to ±0.0015% of FSR)					
Voltage Output	6 V Step		1.5		μs
	1 LSB Step		1.0		μs
	Slew Rate		9		V/μs
Current Output	1 mA Step 10 Ω to 100 Ω Load		350		ns
	1 kΩ Load		350		ns
MONOTONICITY			15		Bits
OUTPUT					
Voltage Output Configuration					
	Bipolar Range	±2.88	±3.0	±3.12	V
	Output Current		±8.0		mA
	Output Impedance		0.1		Ω
	Short Circuit Duration		Indefinite to Common		
Current Output Configuration					
	Bipolar Range	±0.7	±1.0	±1.3	mA
	Output Impedance (±30%)		1.7		kΩ
POWER SUPPLY (12.5 MHz Clock)					
Voltage	$+V_L$ and $+V_S$	4.75		13.2	V
	$-V_L$ and $-V_S$	−13.2		−4.75	V
Current; Case 1	+I, V_S and V_L = +5 V		12.0	**15.0**	mA
	−I, $-V_S$ and $-V_L$ = −5 V		−12.0	**−15.0**	mA
Case 2	+I, V_S and V_L = +12 V		10.5		mA
	−I, $-V_S$ and $-V_L$ = −12 V		14		mA
Case 3	+I, V_S and V_L = +5 V		12		mA
	−I, $-V_S$ and $-V_L$ = −12 V		14		mA
Power Dissipation	V_S and V_L = ±5 V		120	150	mW
	V_S and V_L = ±12 V		300		mW
	V_S and V_L = +5 V, $-V_S$ and $-V_L$ = −12 V		225		mW
TEMPERATURE RANGE					
Specified		0		+70	°C
Storage		−60		+100	°C
WARMUP TIME		1			min

NOTES
[1]Specified using external op amp, see Figure 7 for more details.
[2]Tested at full scale input.

All min and max specifications are guaranteed. Specifications in **boldface** are tested on all production units at final electrical test. Results from those tests are used to calculate outgoing quality levels.

Specifications subject to change without notice.

ABSOLUTE MAXIMUM RATINGS*

V_L to DGND . . . . 0 to 13.2 V
V_S to AGND . . . . 0 to 13.2 V
$-V_L$ to DGND . . . . −13.2 to 0 V
$-V_S$ to AGND . . . . −13.2 to 0 V
Digital Inputs to DGND . . . . −0.3 to V_L
AGND to DGND . . . . ±0.3 V
Short Circuit Protection . . . . Indefinite Short to Ground
Soldering . . . . +300°C, 10 sec
Storage Temperature (N) . . . . −60°C to +100°C

*Stresses greater than those listed under "Absolute Maximum Ratings" may cause permanent damage to the device. This is a stress rating only and functional operation of the device at these or any other conditions above those indicated in the operational section of this specification is not implied. Exposure to absolute maximum rating conditions for extended periods may affect device reliability.

CONNECTION DIAGRAM

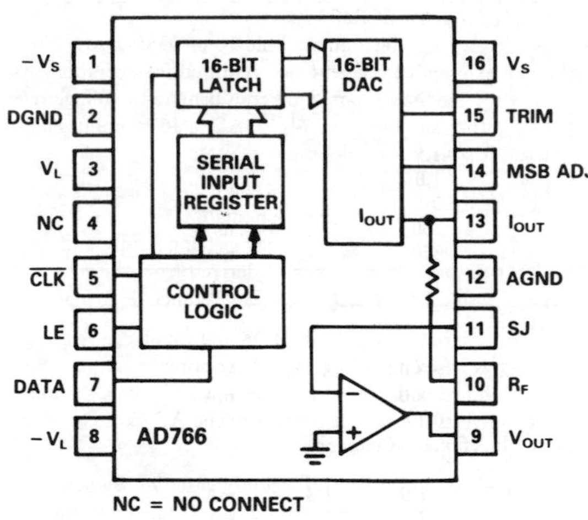

PIN DESIGNATIONS

Pin	Function	Description
1	$-V_S$	Analog Negative Power Supply
2	DGND	Digital Ground
3	V_L	Logic Positive Power Supply
4	NC	No Connection
5	$\overline{CLK}$	Clock Input
6	LE	Latch Enable Input
7	DATA	Serial Data Input
8	$-V_L$	Logic Negative Power Supply
9	V_{OUT}	Voltage Output
10	R_F	Feedback Resistor
11	SJ	Summing Junction
12	AGND	Analog Ground
13	I_{OUT}	Current Output
14	MSB ADJ	MSB Adjustment Terminal
15	TRIM	MSB Trimming Potentiometer Terminal
16	V_S	Analog Positive Power Supply

OUTLINE DIMENSIONS

Dimensions shown in inches and (mm).

16-Pin Plastic DIP

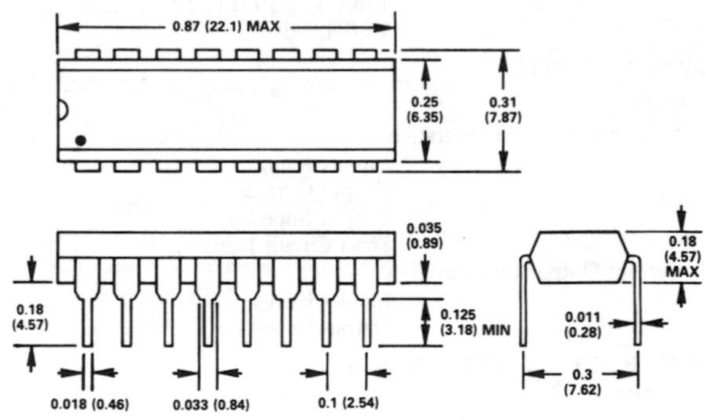

CAUTION

ESD (electrostatic discharge) sensitive device. The digital control inputs are diode protected; however, permanent damage may occur on unconnected devices subject to high energy electrostatic fields. Unused devices must be stored in conductive foam or shunts. The protective foam should be discharged to the destination socket before devices are removed.

Definition of Specifications

TOTAL HARMONIC DISTORTION

Total Harmonic Distortion (THD) is defined as the ratio of the square root of the sum of the squares of the values of the harmonics to the value of the fundamental input frequency. It is expressed in percent (%) or decibels (dB).

THD is a measure of the magnitude and distribution of integral linearity error and differential linearity error. The distribution of these errors may be different, depending on the amplitude of the output signal. Therefore, to be most useful, THD should be specified for both large and small signal amplitudes.

SETTLING TIME

Settling Time is the time required for the output to reach and remain within a specified error band about its final value, measured from the digital input transition. It is the primary measure of dynamic performance.

DYNAMIC RANGE

Dynamic Range is the specification that indicates the ratio of the smallest signal the converter can resolve to the largest signal it is able to produce. As a ratio, it is usually expressed in decibels (dB). The theoretical dynamic range of an n-bit converter is approximately $(6\times n)$ dB. In the case of the 16-bit AD766, that is 96 dB. The actual dynamic range of a converter is less than the theoretical value due to limitations imposed by noise and quantization and other errors.

BIPOLAR ZERO ERROR

Bipolar Zero Error or midscale error is the deviation of the actual analog output from the ideal output (0 V) when the 2s complement input code representing half scale (all 0s) is loaded in the input register.

DIFFERENTIAL LINEARITY ERROR

Differential Linearity Error is the measure of the variation in analog value, normalized to full scale, associated with a 1 LSB change in the digital input. Monotonic behavior requires that the differential linearity error not exceed 1 LSB in the negative direction.

MONOTONICITY

A D/A converter is monotonic if the output either increases or remains constant as the digital input increases.

SIGNAL-TO-NOISE RATIO

SNR is defined as the ratio of the square root of the sum of the squares for the values of all the nonfundamental, nonharmonic signals for a specified bandwidth. SNR is tested at full-scale input. The AD766 specifies SNR for 20 kHz and 250 kHz bandwidths.

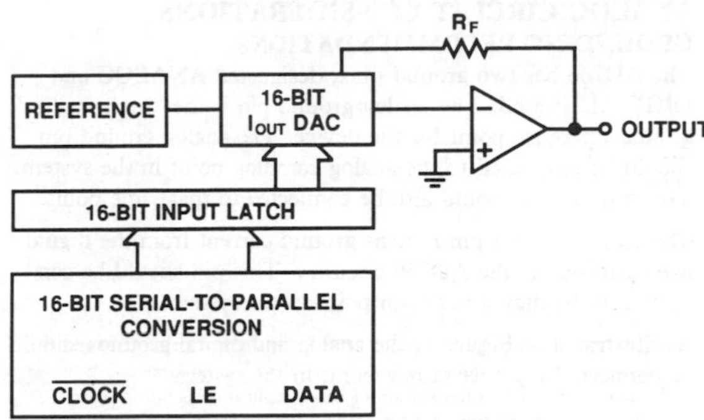

AD766 Block Diagram

FUNCTIONAL DESCRIPTION

The AD766 is a complete monolithic 16-bit DSP DAC. No additional external components are required for operation. As shown in the block diagram, each chip contains a voltage reference, an output amplifier, a 16-bit DAC, a 16-bit input latch and a 16-bit serial-to-parallel input register.

The voltage reference consists of a bandgap circuit and buffer amplifier. This combination of elements produces a reference voltage that is unaffected by changes in temperature and age. The DAC output voltage, which is derived from the reference voltage, is also unaffected by these environmental changes.

The output amplifier uses both MOS and bipolar devices to produce low offset, high slew rate and optimum settling time. When combined with the on-chip feedback resistor, the output op amp converts the output current of the AD766 to a voltage output.

The 16-bit D/A converter uses a combination of segmented decoding and R-2R architecture to achieve consistent linearity and differential linearity. The resistors which form the ladder structure are fabricated with silicon chromium thin film. Laser trimming of these resistors further reduces linearity error resulting in low output distortion.

The input register and serial-to-parallel converter are fabricated with CMOS logic gates. These gates result in fast switching speeds and low power consumption. This contributes to the overall low power dissipation of the AD766.

ANALOG CIRCUIT CONSIDERATIONS

GROUNDING RECOMMENDATIONS

The AD766 has two ground pins, designated ANALOG and DIGITAL ground. The analog ground pin is the "high quality" ground reference point for the device. The analog ground pin should be connected to the analog common point in the system. The output load should also be connected to that same point.

The digital ground pin returns ground current from the digital logic portions of the AD766 circuitry. This pin should be connected to the digital common point in the system.

As illustrated in Figure 1, the analog and digital grounds should be connected together at one point in the system.

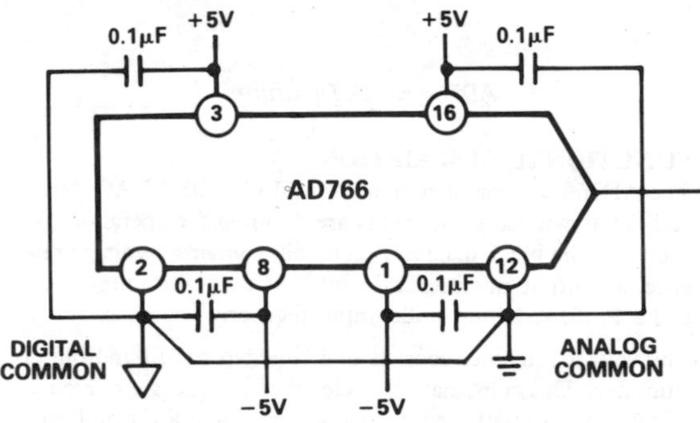

Figure 1. Recommended Circuit Schematic

POWER SUPPLIES AND DECOUPLING

The AD766 has four power supply input pins. $\pm V_S$ provide the supply voltages to operate the linear portions of the DAC including the voltage reference, output amplifier and control amplifier. The $\pm V_S$ supplies are designed to operate from ±5 V up to ±12 V.

The $\pm V_L$ supplies operate the digital portions of the chip including the input shift register and the input latching circuitry. The $\pm V_L$ supplies are also designed to operate from ±5 V up to ±12 V.

Decoupling capacitors should be used on all power supply pins. Furthermore, good engineering practice suggests that these capacitors be placed as close as possible to the package pins as well as the common points. The logic supplies, $\pm V_L$, should be decoupled to digital common; and the analog supplies, $\pm V_S$, should be decoupled to analog common.

The use of four separate power supplies will reduce feedthrough from the digital portion of the system to the linear portions of the system, thus contributing to good performance. However, four separate voltage supplies are not necessary for good circuit performance. For example, Figure 2 illustrates a system where only a single positive and a single negative supply are available. In this case, the positive logic and positive analog supplies may both be connected to the single positive supply. The negative logic and negative analog supplies may both be connected to the single negative supply. Performance would benefit from a measure of isolation between the supplies introduced by using simple low-pass filters in the individual power supply leads.

As with most linear circuits, changes in the power supplies will affect the output of the DAC. Analog Devices recommends that well regulated power supplies with less than 1% ripple be incorporated into the design of any system using these devices.

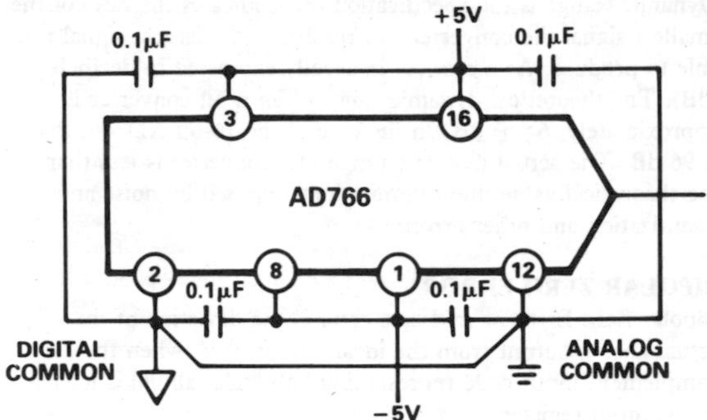

Figure 2. Alternate Recommended Schematic

TOTAL HARMONIC DISTORTION

The THD specification of a DSP DAC represents the amount of undesirable signal produced during reconstruction of a digital waveform. To account for the variety of operating conditions in signal processing applications, the DAC is tested at two output frequencies and over the full operating temperature ranges.

A block diagram of the test setup is shown in Figure 3. In this test setup, a digital data stream, representing a 0 dB, −20 dB or −60 dB sine wave is sent to the device under test. The frequencies used are 1037 Hz and 49.07 kHz. Input data is latched into the AD766 at 500 kSPS. The AD766 under test produces an analog output signal using the on-board op amp for 1 kHz and an external op amp for 50 kHz.

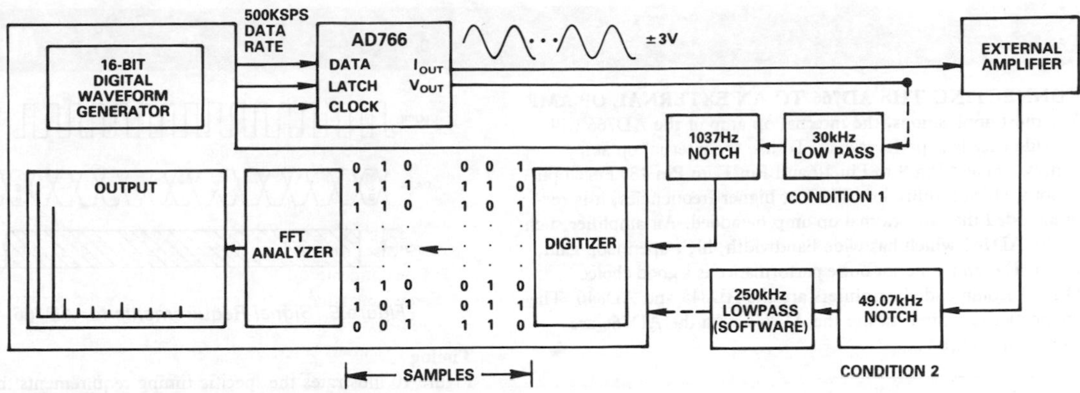

Figure 3. Block Diagram of Distortion Test Circuit

The automatic test equipment digitizes the output test waveform and then an FFT is performed on the results of the test. Based on the first 9 harmonics of the fundamental 1037 Hz and the first 3 harmonics of the 49.07 kHz output waves, the total harmonic distortion of the device is calculated. Neither a deglitcher nor an MSB trim is used during the THD test.

The circuit design, layout and manufacturing techniques employed in the production of the AD766 result in excellent THD performance. Figure 4 shows the typical unadjusted THD performance of the AD766 for various amplitudes of a 1 kHz and 50 kHz with external amplifier output signal. As can be seen, the AD766 offers excellent performance even at amplitudes as low as −60 dB. Figure 5 illustrates the typical THD versus frequency performance for a filtered AD766 output. At frequencies greater than approximately 30 kHz, depending on the low pass filter used, an improvement in THD of 3–4 dB over the performance shown in the figure can be expected. The graph in Figure 5 was generated using a 250 kHz low pass filter. Figure 6 illustrates the consistent THD performance of the AD766 over temperature.

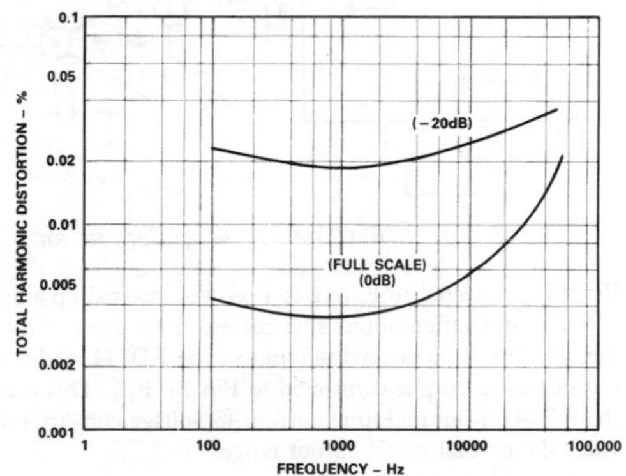

Figure 5. Typical THD vs. Frequency

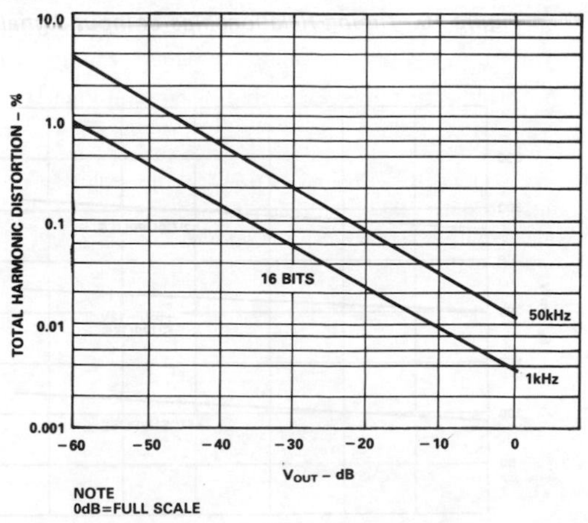

Figure 4. Typical Unadjusted THD

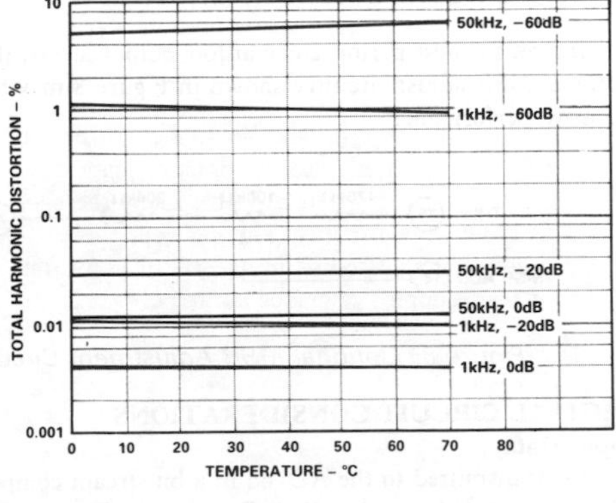

Figure 6. THD vs. Temperature

CONNECTING THE AD766 TO AN EXTERNAL OP AMP

For most applications, the internal op amp of the AD766 will provide excellent performance. To use the internal op amp, simply connect Pin 9 to Pin 10 and Pin 11 to Pin 13. For applications which require low THD at higher frequencies, it is recommended that an external op amp be added. An amplifier such as the AD744 which has wide bandwidth, high open loop gain, low THD, and superior noise performance is a good choice. Other recommended amplifiers are the AD845 and AD846. The connections required to use the AD744 with the AD766 are shown in Figure 7.

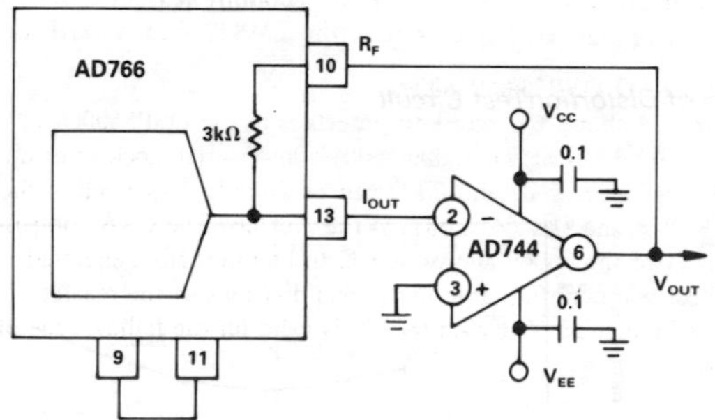

Figure 7. External Op Amp Connections

Pin 9 is connected to Pin 11 to keep the internal op amp in a closed-loop configuration. The current output pin of the AD766 directly drives the inverting input of the AD744 while the output of the op amp is connected to Pin 10 (R_F). This configures the AD744 as an inverting current-to-voltage converter and provides the normal ±3 V output range.

OPTIONAL MSB ADJUSTMENT

Use of an optional adjust circuit allows residual differential linearity error around midscale to be eliminated. This error is especially important when low amplitude signals are being reproduced. In those cases, as the signal amplitude decreases, the ratio of the midscale differential linearity error to the signal amplitude increases, thereby increasing THD.

Therefore, for best performance at low output levels, the optional MSB adjust circuitry shown in Figure 8 may be used to improve performance.

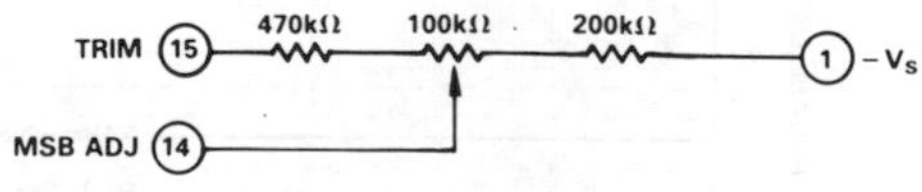

Figure 8. Optional MSB Adjustment Circuit

DIGITAL CIRCUIT CONSIDERATIONS

Input Data

Data is transmitted to the AD766 in a bit stream composed of 16-bit words with a serial, MSB first format. Three signals must be present to achieve proper operation: the data, clock and latch enable signals. Input data bits are clocked into the input register on the falling edge of the clock signal. The LSB is clocked in on the 16th clock pulse. When all data bits are loaded, a low-going latch enable pulse updates the DAC input. Figure 9 illustrates the general signal requirements for data transfer for the AD766.

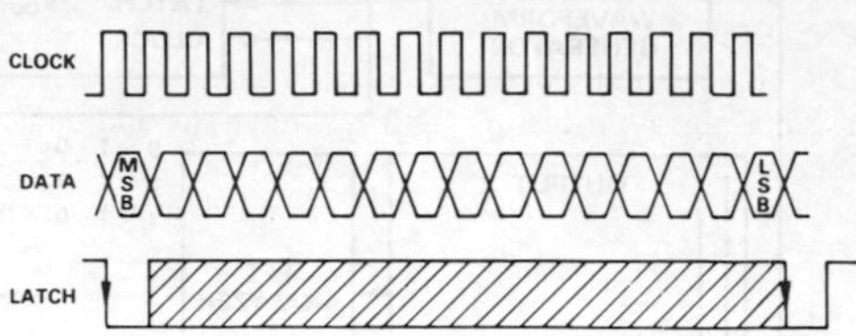

Figure 9. Signal Requirements for AD766

Timing

Figure 10 illustrates the specific timing requirements that must be met in order for the data transfer to be accomplished properly. The input pins of the AD766 are both TTL and 5 V CMOS compatible. The input requirements illustrated in Figures 9 and 10 are compatible with the data outputs provided by popular DSPs. The AD766 input clock can run at a 12.5 MHz rate. This clock rate will allow data transfer rates up to 500 kSPS. The application section of this datasheet contains additional guides for using the AD766 with various DSP engines from Texas Instruments, Motorola and Analog Devices.

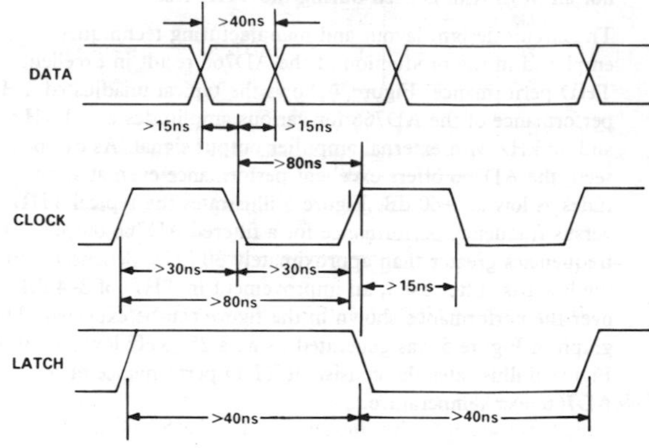

Figure 10. Timing Relationships of Input Signals

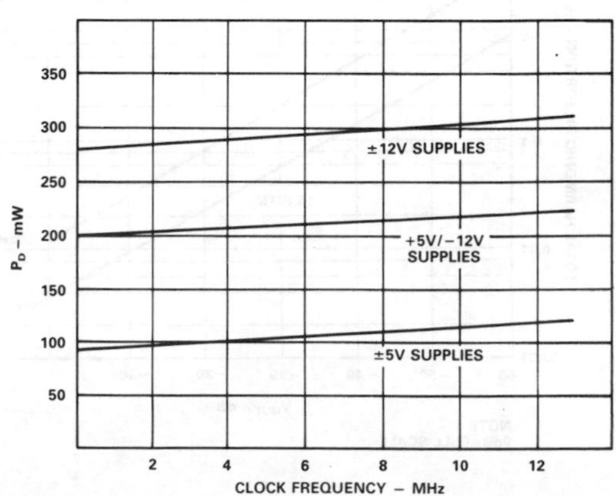

Figure 11. Power Dissipation vs. Clock Frequency

Applications

INTERFACING THE AD766 TO DIGITAL SIGNAL PROCESSORS

The AD766 is specifically designed to easily interface to several popular digital signal processors (DSP) without any additional logic. Such an interface reduces the possibility of interface problems and improves system reliability by minimizing component count.

AD766 TO ADSP-2101

The ADSP-2101 incorporates two complete serial ports which can be directly interfaced to the AD766 as shown in Figure 12. The SCLK, TFS and DT outputs of the ADSP2101 are connected directly to the $\overline{\text{CLK}}$, LE and DATA inputs of the AD766, respectively. SCLK is internally generated and can be programmed to operate from 94 Hz to 6.25 MHz. Data (DT) is valid on the falling edge of SCLK. After 16 bits have been transmitted, the falling edge of TFS updates the AD766's data latch. Using both serial ports of the ADSP-2101, two AD766's can be directly interfaced with no additional hardware.

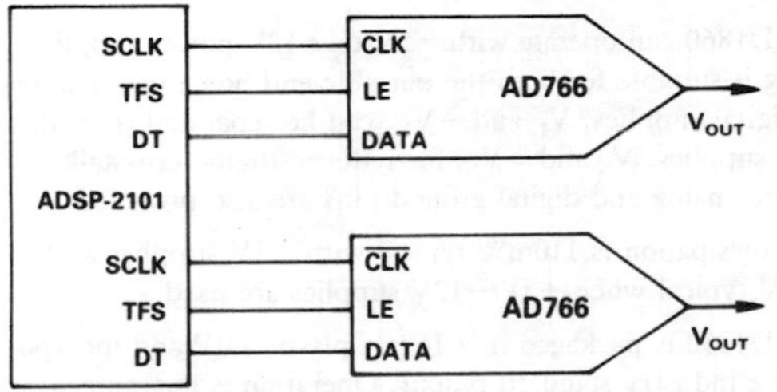

Figure 12. AD766 to ADSP-2101/ADSP-2102

AD766 TO TMS320C25

Figure 13 shows the zero-chip interface to the TMS320C25. The interface to other TMS320C2X processors is similar.

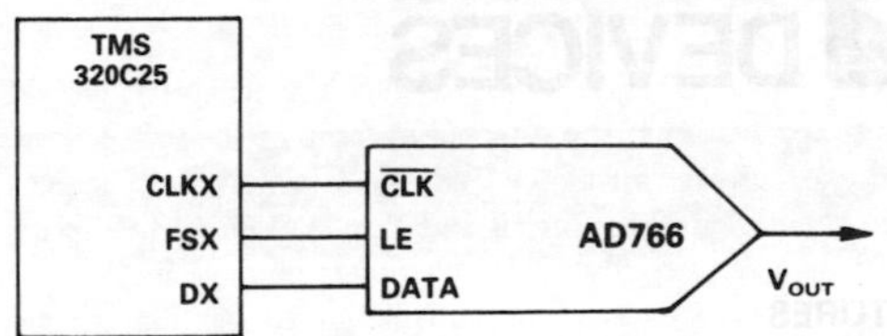

Figure 13. AD766 to TMS320C25

The CLKS, FSX and DX outputs of the TMS320C25 are connected to the $\overline{\text{CLK}}$, LE and DATA inputs of the AD766, respectively. Data (DX) is valid on the falling edge of CLKX. The maximum serial clock rate of the TMS320C25 is 5 MHz.

AD766 TO DSP56000/56001

Figure 14 shows the zero-chip interface to the DSP56000/56001. The SSI of the 56000/56001 allows serial clock rates up to fosc/4. SCK, SC2 and STD can be directly connected to the $\overline{\text{CLK}}$, LE and DATA inputs of the AD766. The CRA control register of the 56000 allows SCLK to be internally generated and software configurable to various divisions of the master clock frequency. The data (STD) is valid on the falling edge of SCK.

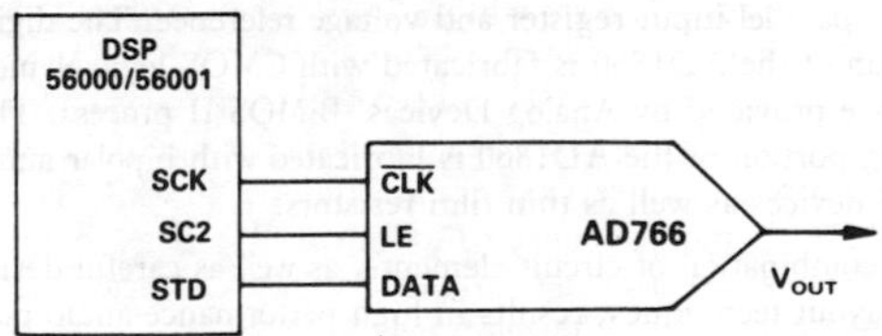

Figure 14. AD766 TO DSP56000/DSP56001

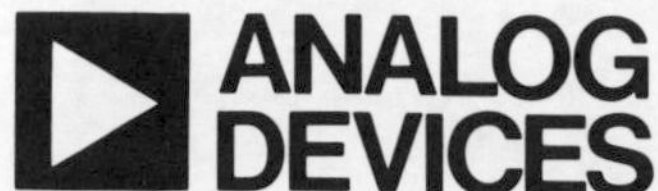

18-Bit PCM Audio DAC

AD1860

FEATURES
0.002% THD + Noise
Fast Settling Permits 8×Oversampling
±3V Output
Optional Trim Allows Superlinear Performance
±5V to ±12V Operation
16 Pin Plastic DIP Package
Industry Standard Pinout
2s Complement, Serial Input

APPLICATIONS
High End Compact Disc Players
Digital Audio Amplifiers
DAT Recorders and Players
Synthesizers and Keyboards

PRODUCT DESCRIPTION

The AD1860 is a monolithic 18-bit PCM Audio DAC. Each device provides a voltage output amplifier, 18-bit DAC, 18-bit serial to parallel input register and voltage reference. The digital portion of the AD1860 is fabricated with CMOS logic elements that are provided by Analog Devices' BiMOS II process. The analog portion of the AD1860 is fabricated with bipolar and MOS devices as well as thin film resistors.

This combination of circuit elements, as well as careful design and layout techniques, results in high performance audio playback. Laser trimming of the linearity error affords extremely low total harmonic distortion. An optional linearity trim pin is provided to allow residual differential linearity error at midscale to be eliminated. This feature is particularly valuable for low distortion reproductions of low amplitude signals. Output glitch is also small contributing to the overall high level of performance. The output amplifier achieves fast settling and high slew rates, providing a full ±3V signal at load currents up to 8mA. When used in current output mode, the AD1860 provides a ±1mA output signal. The output amplifier is short circuit protected and can withstand indefinite shorts to ground.

The serial input interface consists of the clock, data and latch enable pins. The serial 2s complement data word is clocked into the DAC, MSB first, by the external data clock. The latch enable signal transfers the input word from the internal serial input register to the parallel DAC input register. The input clock can support a 12.7MHz data rate. This serial input port is compatible with second generation digital filter chips used in consumer audio products. These filters operate at oversampling rates of 2×, 4× and 8×sampling frequencies.

The AD1860 can operate with ±5V to ±12V power supplies making it suitable for both the portable and home use markets. The digital supplies, V_L and $-V_L$, can be separated from the analog supplies, V_S and $-V_S$, for reduced digital crosstalk. Separate analog and digital ground pins are also provided.

Power dissipation is 110mW typical with ±5V supplies and is 225mW typical when +5V/−12V supplies are used.

The AD1860 is packaged in a 16-pin plastic DIP and incorporates the industry standard pinout. Operation is guaranteed over the temperature range of −25 to +70°C and over the voltage supply range of ±4.75 to ±13.2V.

PRODUCT HIGHLIGHTS

1. 18-bit resolution provides 108dB dynamic range.
2. No external components are required.
3. Operates with ±5V to ±12V supplies.
4. Space saving 16-pin plastic DIP package.
5. 110mW power dissipation.
6. 1.5μs settling time permits 2×, 4× and 8×oversampling.
7. ±3V or ±1mA output capability.
8. THD + Noise is 100% tested.

One Technology Way; P. O. Box 9106; Norwood, MA 02062-9106 U.S.A.
Tel: 617/329-4700 Twx: 710/394-6577
Telex: 924491 Cables: ANALOG NORWOODMASS

SPECIFICATIONS (T_A at +25°C and ±5V supplies unless otherwise noted)

	Min	Typ	Max	Units
RESOLUTION			**18**	Bits
DIGITAL INPUTS V_{IH}	2.0		$+V_L$	V
V_{IL}			0.8	V
I_{IH}, $V_{IH} = V_L$			**1.0**	μA
I_{IL}, $V_{IL} = 0.4$			**−10**	μA
Clock Input Frequency	12.5			MHz
ACCURACY				
Gain Error		±2.0		%
Midscale Output Voltage		±30		mV
Differential Linearity Error		±0.001		% of FSR
TOTAL HARMONIC DISTORTION + NOISE				
0dB, 990.5Hz AD1860N-K		0.002	**0.0025**	%
AD1860N-J		0.002	**0.004**	%
AD1860N		0.004	**0.008**	%
−20dB, 990.5Hz AD1860N-K		0.006	**0.020**	%
AD1860N-J		0.010	**0.020**	%
AD1860N		0.010	**0.040**	%
−60dB, 990.5Hz AD1860N-K		0.9	**2.0**	%
AD1860N-J		0.9	**2.0**	%
AD1860N		0.9	**4.0**	%
SIGNAL TO NOISE RATIO	**102**	108		dB
DRIFT (0 to +70°C)				
Total Drift		±25		ppm of FSR/°C
Bipolar Zero Drift		±4		ppm of FSR/°C
SETTLING TIME (to ±0.0015% of FSR)				
Voltage Output, 6V Step		1.5		μs
1LSB Step		1.0		μs
Slew Rate		9		V/μs
Current Output 1mA Step 10Ω to 100Ω Load		350		ns
1kΩ Load		350		ns
MONOTONICITY		15		Bits
OUTPUT				
Voltage Output Configuration				
Bipolar Range	**±2.88**	±3.0	**±3.12**	V
Output Current	±8			mA
Output Impedance		0.1		Ω
Short Circuit Duration		Indefinite to Common		
Current Output Configuration				
Bipolar Range (±30%)		±1.0		mA
Output Impedance (±30%)		1.7		kΩ
POWER SUPPLY				
Voltage V_L and V_S	4.75		13.2	V
Voltage $-V_L$ and $-V_S$	−13.2		−4.75	V
Current $+I$, V_L and $V_S = 5V$, 10MHz Clock		10.0	**13.0**	mA
$-I$, $-V_L$ and $-V_S = -5V$, 10MHz Clock		12.0	**−15.0**	mA
Current $+I$, V_L and $V_S = 12V$, 10MHz Clock		10.5		mA
$-I$, $-V_L$ and $-V_S = -12V$, 10MHz Clock		13.5		mA
Current $+I$, V_L and $+V_S = +5$, 10MHz Clock		10		mA
$-I$, $-V_L$ and $-V_S = -12V$, 10MHz Clock		14		mA
POWER DISSIPATION				
V_S and $V_L = \pm 5V$, 10MHz Clock		110		mW
V_S and $V_L = \pm 12V$, 10MHz Clock		300		mW
V_S and $V_L = +5V$, $-V_S$ and $-V_L = -12V$, 10MHz Clock		225		mW

	Min	Typ	Max	Units
TEMPERATURE RANGE				
Specification	0	+25	+70	°C
Operation	−25		+70	°C
Storage	−60		+100	°C
WARMUP TIME	1			min

Specifications subject to change without notice.

TYPICAL PERFORMANCE

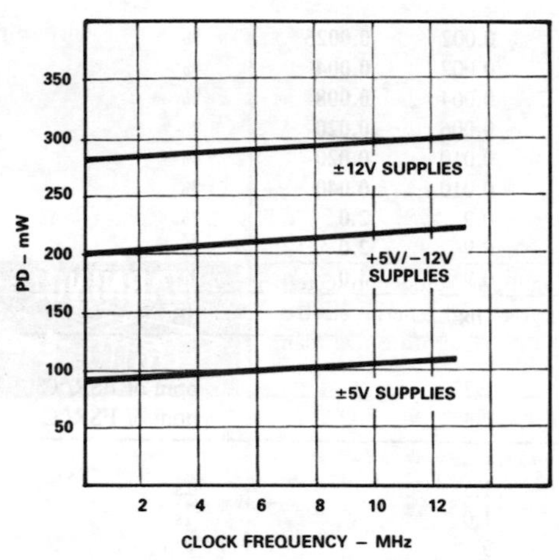

Power Dissipation vs. Clock Frequency

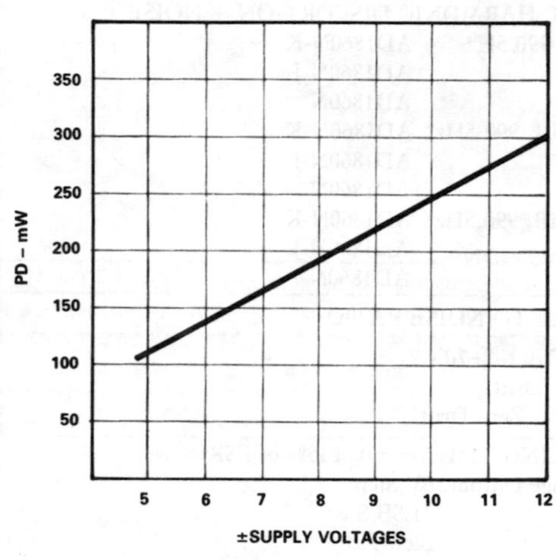

Power Dissipation vs. Supply Voltages

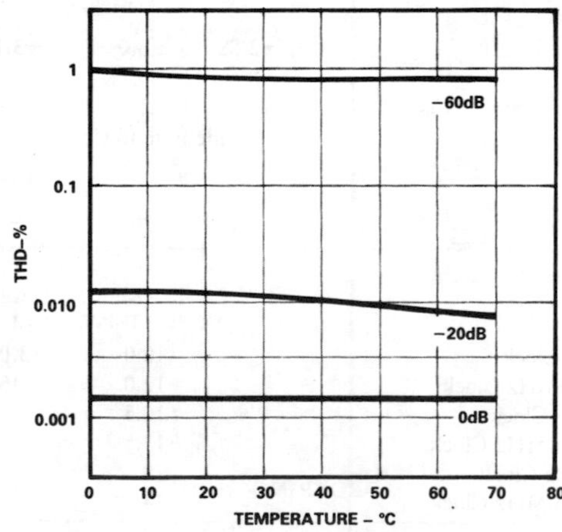

THD vs. Temperature

OUTLINE DIMENSIONS

Dimensions shown in inches and (mm).

N (Plastic) Package

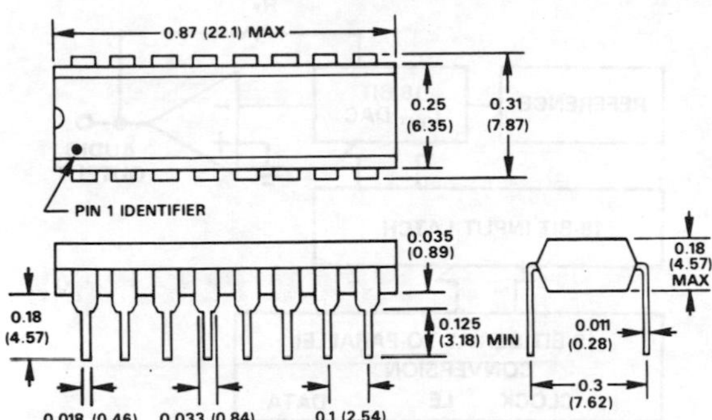

ABSOLUTE MAXIMUM RATINGS*

V_L to DGND . . . . 0 to 13.2V
V_S to AGND . . . . 0 to 13.2V
$-V_L$ to DGND . . . . −13.2 to 0V
$-V_S$ to AGND . . . . −13.2 to 0V
Digital Inputs to DGND . . . . −0.3 to V_L
AGND to DGND . . . . ± 0.3V
Short Circuit . . . . Indefinite Short to Ground
Soldering . . . . +300°C, 10sec
Storage Temperature . . . . −60°C to +100°C

Note
*Stresses greater than those listed under "Absolute Maximum Ratings" may cause permanent damage to the device. This is a stress rating only and functional operation of the device at these or any other conditions above those indicated in the operational section of this specification is not implied. Exposure to absolute maximum rating conditions for extended periods may affect device reliability.

CAUTION:
ESD (electrostatic discharge) sensitive device. The digital control inputs are diode protected; however, permanent damage may occur on unconnected devices subject to high energy electrostatic fields. Unused devices must be stored in conductive foam or shunts. The protective foam should be discharged to the destination socket before devices are inserted.

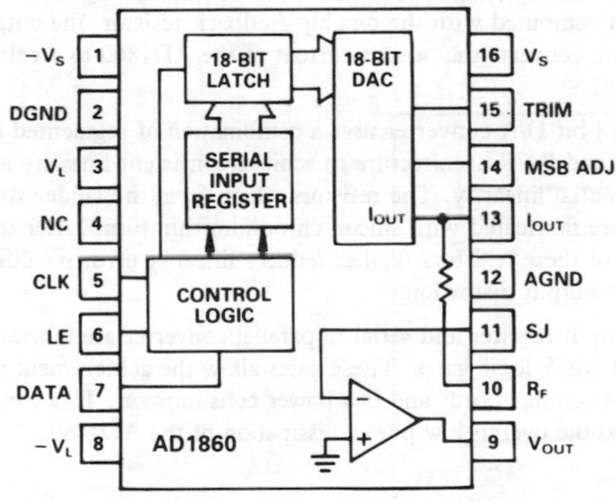

Functional Block Diagram

PIN ASSIGNMENTS		
1	$-V_S$	Analog Negative Power Supply
2	DGND	Logic Ground
3	V_L	Logic Positive Power Supply
4	NC	No Connection
5	CLK	Data Clock Input
6	LE	Latch Enable Input
7	DATA	Serial Data Input
8	$-V_L$	Logic Negative Power Supply
9	V_{OUT}	Voltage Output
10	R_F	Feedback Resistor
11	SJ	Summing Junction
12	AGND	Analog Ground
13	I_{OUT}	Current Output
14	MSB ADJ	MSB Adjustment Terminal
15	TRIM	MSB Trimming Potentiometer Terminal
16	V_S	Analog Positive Power Supply

ORDERING GUIDE

Model	THD @ FS
AD1860	0.008%
AD1860N-J	0.004%
AD1860N-K	0.0025%

TOTAL HARMONIC DISTORTION + NOISE

Total Harmonic Distortion plus Noise (THD+N) is defined as the ratio of the square root of the sum of the squares of the values of the harmonics and noise to the value of the fundamental input frequency. It is usually expressed in percent (%).

THD+N is a measure of the magnitude and distribution of linearity error, differential linearity error, quantization error and noise. The distribution of these errors may be different, depending on the amplitude of the output signal. Therefore, to be most useful, THD+N should be specified for both large and small signal amplitudes.

SETTLING TIME

Settling Time is the time required for the output to reach and remain within a specified error band about its final value, measured from the digital input transition. It is a primary measure of dynamic performance.

DYNAMIC RANGE

Dynamic Range is the specification that indicates the ratio of the smallest signal the converter can resolve to the largest signal it is able to produce. As a ratio, it is usually expressed in decibels (dBs). The theoretical dynamic range of an n-bit converter is ($6\times n$) dB. In the case of the 18-bit AD1860, that is 108dB. The actual dynamic range of a converter is less than the theoretical value due to limitations imposed by noise and other errors.

MIDSCALE ERROR

Midscale Error, or bipolar zero error, is the deviation of the actual analog output from the ideal output (0V) when the 2s complement input code representing half scale is loaded in the input register.

DIFFERENTIAL LINEARITY ERROR

Differential Linearity Error is the measure of the variation in analog value, normalized to full scale, associated with a 1LSB change in the digital input. Monotonic behavior requires that the differential linearity error not exceed 1LSB in the negative direction.

MONOTONICITY

A D/A converter is monotonic if the output either increases or remains constant as the digital input increases.

SIGNAL-TO-NOISE RATIO

The Signal-to-Noise Ratio is defined as the ratio of the amplitude of the output with no signal present to the amplitude of the output when a full-scale output is present.

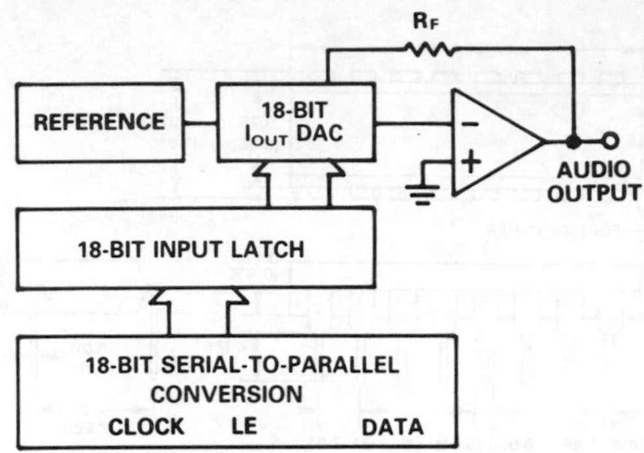

AD1860 Block Diagram

FUNCTIONAL DESCRIPTION

The AD1860 is a complete monolithic 18-bit PCM Audio DAC. No additional external components are required for operation. As shown in the block diagram, each chip contains a voltage reference, an output amplifier, an 18-bit DAC, an 18-bit input latch and an 18-bit serial to parallel input register.

The voltage reference consists of a bandgap circuit and buffer amplifier. This combination of elements produces a reference voltage that is unaffected by changes in temperature and age. The DAC output voltage, which is derived from the reference voltage, is also unaffected by these environmental changes.

The output amplifier uses both MOS and bipolar devices to produce low offset, high slew rate and optimum settling time. When combined with the on chip feedback resistor, the output op amp converts the output current of the AD1860 to a voltage output.

The 18-bit D/A converter uses a combination of segmented decoder and R-2R architecture to achieve consistent linearity and differential linearity. The resistors which form the ladder structure are fabricated with silicon chromium thin film. Laser trimming of these resistors further reduces linearity error resulting in low output distortion.

The input register and serial to parallel converter are fabricated with CMOS logic gates. These gates allow the achievement of fast switching speeds and low power consumption. This contributes to the overall low power dissipation of the AD1860.

Analog Circuit Considerations

GROUNDING RECOMMENDATIONS

The AD1860 has two pins, designated Analog and Digital ground. The analog ground pin is the "high quality" ground reference point for the device. The analog ground pin should be connected to the analog common point in the system. The output load should also be connected to that same point.

The digital ground pin returns ground current from the digital logic portions of the AD1860 circuitry. This pin should be connected to the digital common point in the system.

As illustrated in Figure 1, the analog and digital grounds should be connected together at one point in the system.

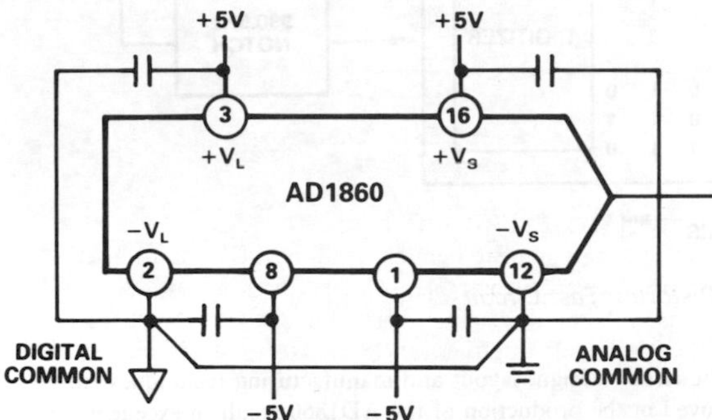

Figure 1. Recommended Circuit Schematic

POWER SUPPLIES AND DECOUPLING

The AD1860 has four power supply input pins. $\pm V_S$ provide the supply voltages to operate the linear portions of the DAC including the voltage reference, output amplifier and control amplifier. The $\pm V_S$ supplies are designed to operate from ±5V to ±12V.

The $\pm V_L$ supplies operate the digital portions of the chip including the input shift register and the input latching circuitry. The $\pm V_L$ supplies are also designed to be operated from ±5V to ±12 V subject only to the limitation that $-V_L$ may not be more negative than $-V_S$.

Decoupling capacitors should be used on all power supply pins. Furthermore, good engineering practice suggests that these capacitors be placed as close as possible to the package pins as well as the common points. The logic supplies, $\pm V_L$, should be decoupled to digital common; and the analog supplies, $\pm V_S$, should be decoupled to analog common.

The use of four separate power supplies will reduce feedthrough from the digital portion of the system to the linear portion of the system, thus contributing to good performance. However, four separate voltage supplies are not necessary for good circuit performance. For example, Figure 2 illustrates a system where only a single positive and a single negative supply are available.

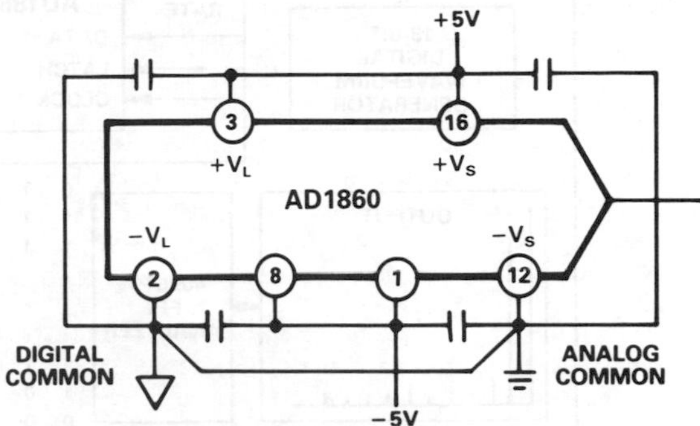

Figure 2. Typical Power Supply Sensitivity

Given that these two supplies are within the range of ±5V to ±12V, they may be used to power the AD1860. In this case, the positive logic and positive analog supplies may both be connected to the single positive supply. The negative logic and negative analog supplies may both be connected to the single negative supply. Performance would benefit from a measure of isolation between the supplies introduced by using simple low pass filters in the individual power supply leads.

As with most linear circuits, changes in the power supplies will affect the output of the DAC. Analog Devices recommends that well regulated power supplies with less than 1% ripple be incorporated into the design of any system using these devices.

TOTAL HARMONIC DISTORTION + NOISE

The THD figure of an audio DAC represents the amount of undesirable signal produced during reconstruction and playback of an audio waveform. The THD specification, therefore, provides a direct method to classify and choose an audio DAC for a desired level of performance.

By combining noise measurement with THD measurement, a THD+N specification is produced. This specification measures all undesirable signal produced by the DAC, including harmonic products of the test tone as well as noise.

Analog Devices tests and grades all AD1860s on the basis of THD+N performance. A block diagram of the test setup is shown in Figure 3. In this test setup, a digital data stream representing a 0dB, −20dB or −60dB sinewave is sent to the device under test. The frequency of this waveform in 990.5 Hz.

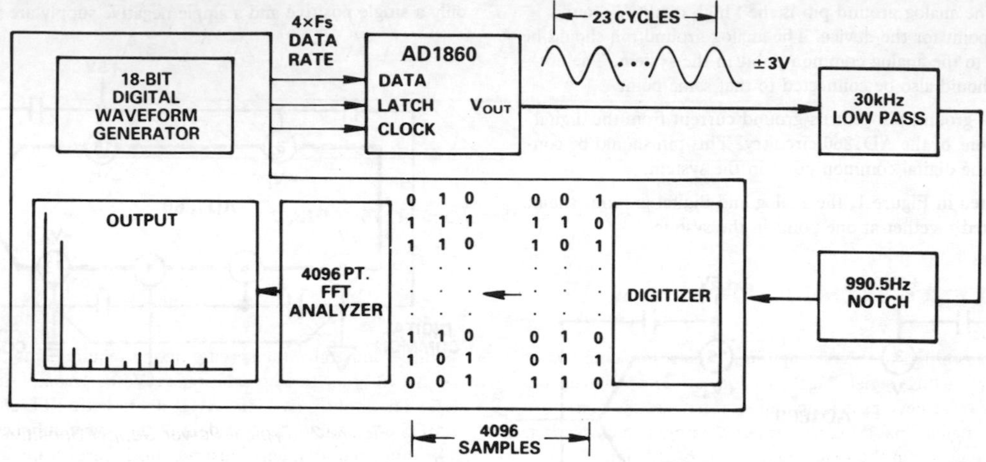

Figure 3. Block Diagram of DistortionTest Circuit

Input data is sent to the AD1860 at a 4×F_S rate (176.4kHz). The AD1860 under test produces an output signal with its onboard op amp. The automatic test equipment digitizes 4096 samples of the output test waveform, incorporating 23 complete cycles of the sinewave. A 4096 point FFT is performed on the results of the test. Based on the harmonics of the fundamental 990.5Hz test tone and the noise components, the total harmonic distortion + noise of the device is calculated. Neither a deglitcher nor an MSB trim is used during this test.

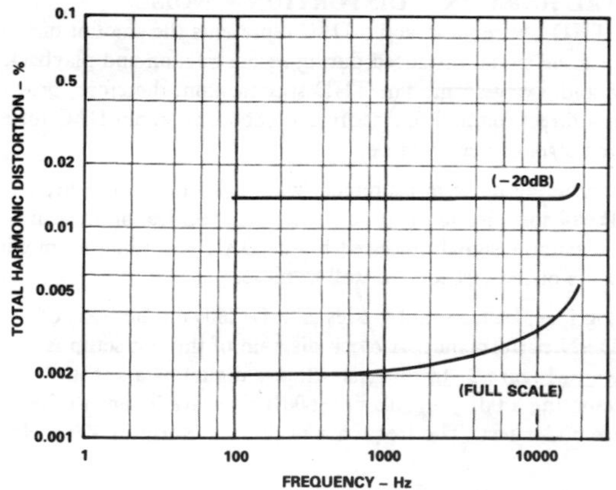

Figure 4. Typical THD vs Frequency

The circuit design, layout and manufacturing techniques employed in the production of the AD1860 result in excellent THD performance. Figure 4 shows the typical unadjusted THD performance of the AD1860 for various amplitudes and frequencies of output signals. As can be seen, the AD1860 offers excellent performance, even at low amplitudes.

OPTIONAL MSB ADJUSTMENT

Use of an optional adjust circuitry allows residual differential linearity error around midscale to be eliminated. This error is especially important when low amplitude signals are being reproduced. In those cases, as the signal amplitude decreases, the ratio of the midscale differential linearity error to the signal amplitude increases, thereby increasing THD.

Therefore, for best performance at low output levels, the optional MSB adjust circuitry shown in Figure 5 may be used to improve performance.

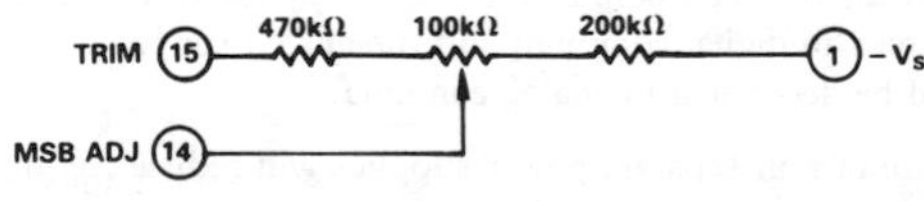

Figure 5. Optional THD Adjust Circuit

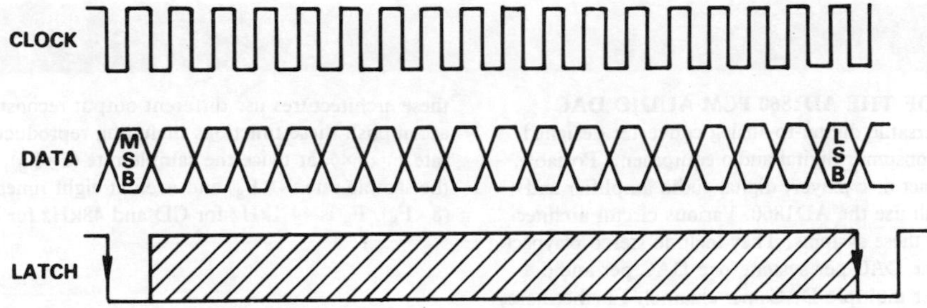

Figure 6. Signal Requirements for AD1860

DIGITAL CIRCUIT CONSIDERATIONS

Input Data

Data is transmitted to the AD1860 in a bit stream composed of 18-bit words with a serial, MSB first format. Three signals must be present to achieve proper operation. They are the Data, Clock and Latch Enable signals. Input data bits are clocked into the input register on the rising edge of the Clock signal. The LSB is clocked in on the 18th clock pulse. When all data bits are loaded, a low-going Latch Enable pulse updates the DAC input. Figure 6 illustrates the general signal requirements for data transfer for the AD1860.

Timing

Figure 7 illustrates the specific timing requirements that must be met in order for the data transfer to be accomplished properly. The input pins of the AD1860 are both TTL and 5V CMOS compatible, independent of the power supplies used. The input requirements illustrated in Figures 6 and 7 are compatible with the data outputs provided by popular DSP filter chips used in digital audio playback systems. The AD1860 input clock can run at a 12.7MHz rate. This clock rate will allow data transfer rates for 2×, 4× or 8×oversampling reconstruction. The application section of this datasheet contains additional guides for using the AD1860 with various DSP filter chips available from Sony, NPC and Yamaha.

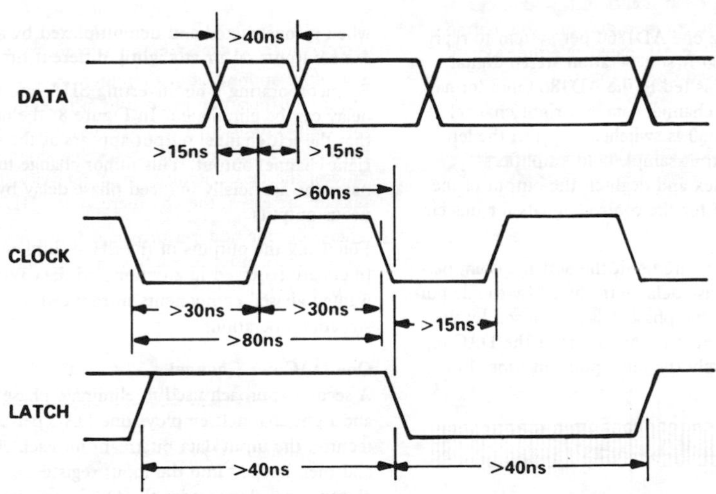

Figure 7. Timing Relationships of Input Signals

APPLICATIONS OF THE AD1860 PCM AUDIO DAC

The AD1860 is a versatile digital-to-analog converter designed for applications in consumer digital audio equipment. Portable, car and home compact disc player, digital audio amplifier and DAT schemes can all use the AD1860. Various circuit architectures are popular in these systems. They include stereo playback sections featuring one DAC per system, one DAC per audio channel (left/right) or multiple DACs per channel. Furthermore, these architectures use different output reconstruction rates to accomplish these functions including reproduction at the sample rate F_S (1×), at twice the sample rate (2×F_S), at four times the sample rate (4×F_S) and even at eight times the sample rate (8×F_S). F_S is 44.1kHz for CD and 48kHz for DAT applications.

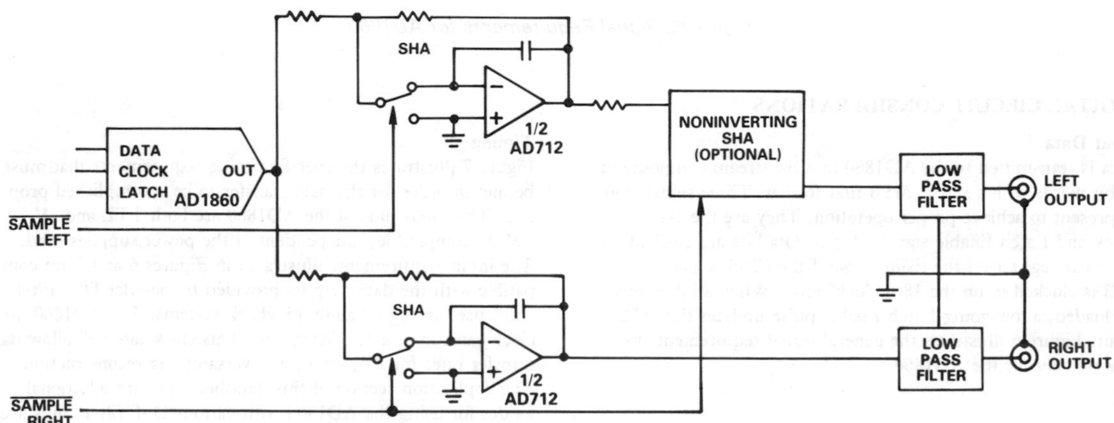

Figure 8. AD1860 in a One DAC per System Architecture

One DAC per System

Figure 8 shows a circuit using one AD1860 per system to reproduce both channels of a typical first generation stereo digital audio system. The input data is fed to the AD1860 in a format which alternates between left channel data and right channel data. The output of the AD1860 is switched between the left channel and right channel output sample/hold amplifiers (SHAs). The SHAs demultiplex and deglitch the output of the AD1860. The timing diagram for the control signals for this circuit are shown in Figure 9.

However, when only two SHAs are used, the actual system performance is limited by the phase delay introduced by the demultiplexed format. This undesirable phase delay is caused by the fact that the data words presented to the inputs of the DAC represent samples taken at precisely the same point in time. But when reconstructed and demultiplexed by a single DAC, these same outputs occur at slightly different times.

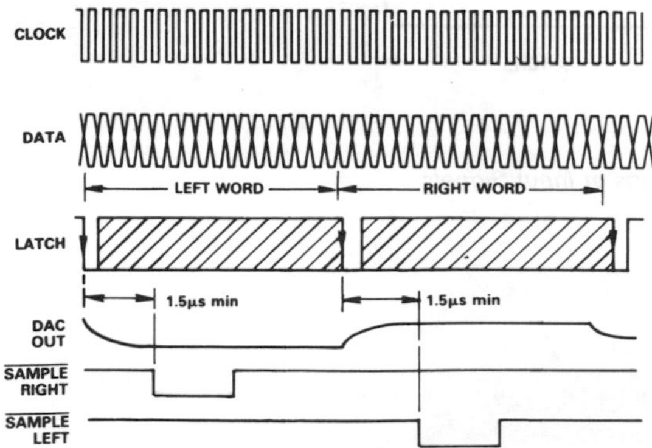

Figure 9. Control Signals for One DAC Circuit

By incorporating a noninverting SHA into the circuit, the phase delay can be eliminated. In Figure 8, the optional SHA ensures that the left channel output appears at the same time as the right channel output. This minor change to the circuit eliminates the artificially induced phase delay by restoring simultaneous outputs.

Following the outputs of the SHAs are low pass filters. These filters are required in any sampled data system to remove unwanted aliased components introduced by the sample and reconstruction operations.

One DAC per Channel

A second approach used to eliminate phase delay between left and right channels employs one DAC per channel. In this architecture, the input data bitstream for each channel is transmitted and then latched into the input register of each DAC. The logic circuitry which precedes the DACs has the task of demultiplexing the serial data into separate right and left channel data. Optional sample/hold amplifiers may be connected before the low pass filters to deglitch the DAC outputs. This "second generation" approach is illustrated in Figure 10.

Two DACs per Channel

Another architecture uses two DACs per channel. In this scheme each DAC reproduces one half of the output waveform. The advantage obtained with this structure is that midscale differential linearity crror no longer affects the zero crossing points of the waveforms. Its effects are shifted to the points where the output waveform crosses ±3/4 full scale. The result is that THD performance for low amplitude signals is greatly improved.

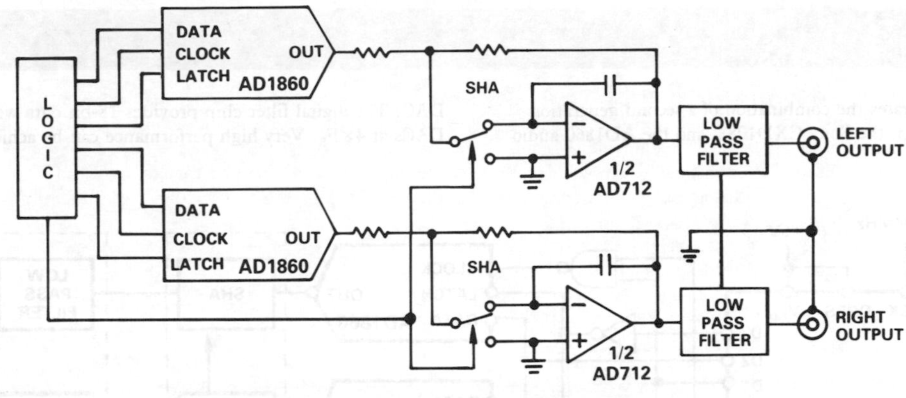

Figure 10. One DAC per Channel Architecture

DIGITAL FILTERING AND OVERSAMPLING

Oversampling is a term which refers to playback techniques in which the reconstruction frequency used is an integral (2 or more) multiple of the original quantized data rate. For example, in compact disc stereo digital audio playback units, the original quantized data sample rate is 44.1kHz. Popular oversampling rates are 2× or 4×F_S, yielding reconstruction rates of 88.2 and 176.4kHz, respectively.

Oversampling is used to ease the performance constraints of the low pass filters which follow the reconstruction DAC. In any signal reconstructed from sampled data, unwanted frequency components are introduced in the output spectrum; these components are centered at the reconstruction frequency. When a 44.1kHz reconstruction frequency is used, the actual frequency band of interest is 20Hz to 20kHz, and the band of unwanted "image" frequency components extends from 44.1kHz to approximately 24kHz. These unwanted components must be removed with a low-pass filter of very high order. First generation digital audio systems often used low-pass filters of 9, 11 and even 13 poles. Linear implementations of these filters are expensive, difficult to manufacture and can produce distortion due to varying group delay characteristics.

When a 2×reconstruction frequency (88.2kHz) is used, the lowest frequency components now extend down to approximately 68kHz. A 4×rate (176.4kHz) has unwanted components extending down to approximately 156kHz. The filter response needed to remove these frequency components can now be less steep. This means that a lower order filter may be used resulting in less distortion at lower cost. Linear filters with 3 or 5 poles are adequate to do the job and are quite common in digital audio products employing oversampling techniques.

Oversampling techniques require the serial input data stream to run at the same integral multiple of the original data rate. So, while the constraints on the output low-pass filter are eased, the constraints on the serial digital input port and the settling time of the output stage are not.

The actual oversampling operation takes place in the digital filter chip (DSP) which is located "upstream" from the DAC. The digital filter accepts data from the media and adds the additional reconstruction points according to the algorithm and coefficients stored in the filter chip. Since the digital filters actually interpolate these additional reconstruction points, they have earned the name "interpolation filters".

The AD1860 is compatible with popular digital filter chips used in digital audio products such as the Sony CXD1088 and Yamaha YM3414.

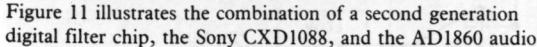

Figure 11 illustrates the combination of a second generation digital filter chip, the Sony CXD1088, and the AD1860 audio DAC. The digital filter chip provides 18-bit data words to the DACs at $4 \times F_S$. Very high performance can be achieved.

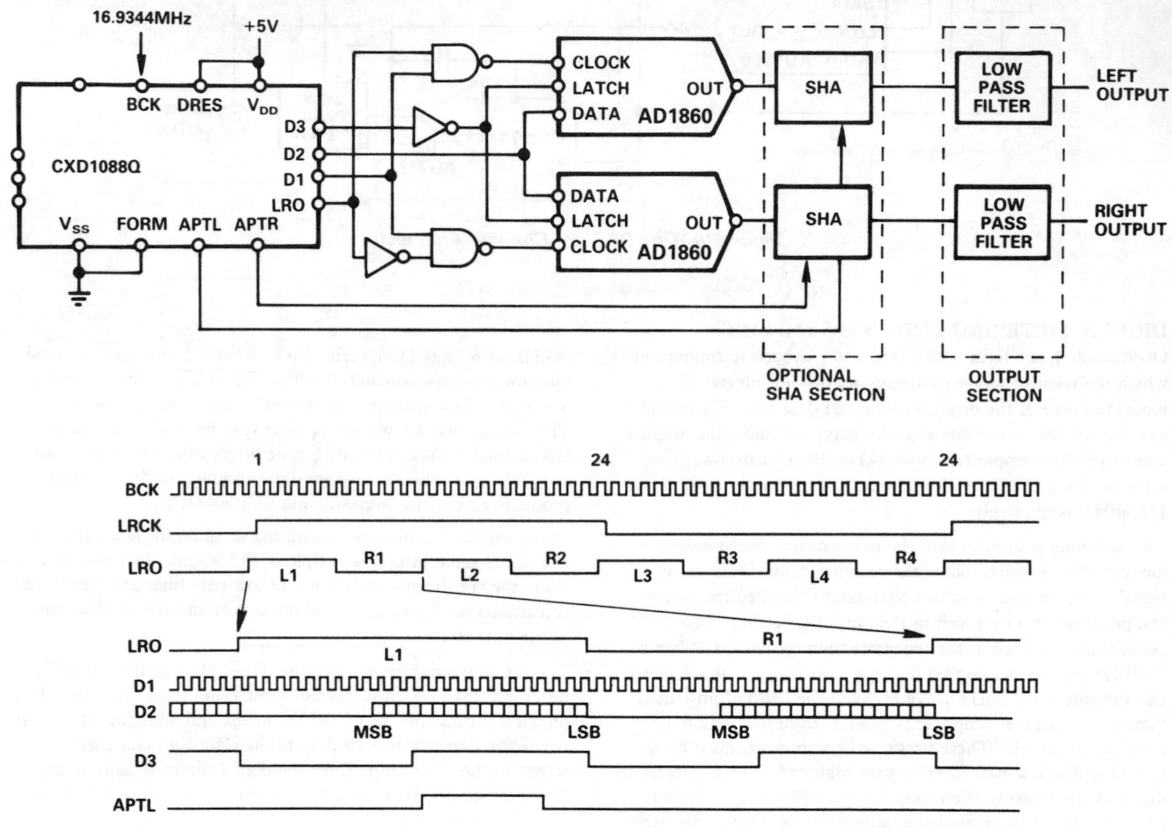

Figure 11. $4 \times F_S$ with the CXD1088Q

Figure 12 illustrates the combination of a Yamaha YM3414 digital filter chip and two AD1860 audio DACs. This combination of components results in $8 \times F_S$ oversampling reconstruction rates. This rate allows the use of lower order output low pass filters than would be required with lower oversampling rates, without sacrificing performance. In this high performance CD player application, the DAC input data is simultaneously transmitted to the input registers of the DACs through dedicated left and right channel output pins on the YM3414. Several logic gates are required to produce a latch enable pulse after the eighteenth data bit is transmitted. This is due to the fact that the YM3414 puts out a latch pulse after the sixteenth bit. This scheme recovers the last two data bits that would otherwise be "lost" and restores the dynamic range to the full 108dB afforded by an 18-bit audio DAC. As before, optional sample/hold signals are provided.

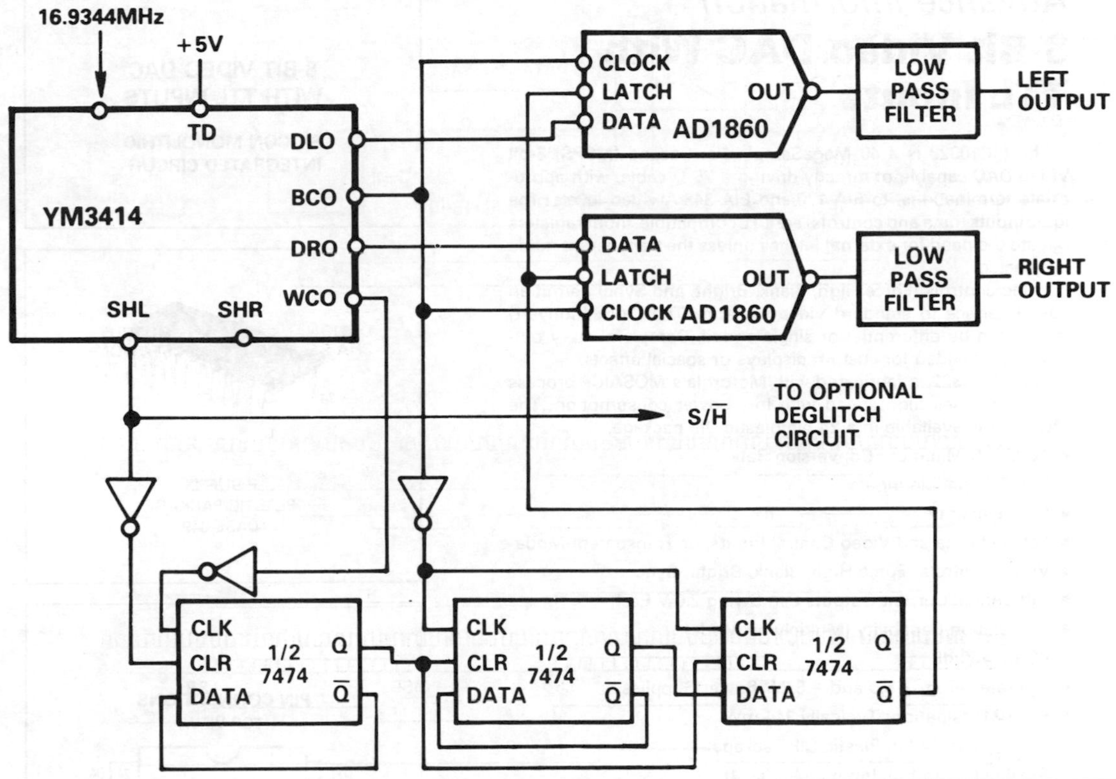

Figure 12. YM3414 and AD1860 Achieve $8 \times F_S$

Order this data sheet by MC10322/D

MOTOROLA SEMICONDUCTOR TECHNICAL DATA

MC10322

Advance Information

8-Bit Video DAC With TTL Inputs

The MC10322 is a 40 MegaSample Per Second (MSPS) 8-bit Video DAC capable of directly driving a 75 Ω cable, with appropriate terminations, to EIA-170 and EIA-343-A video levels. The logic inputs (data and controls) are TTL compatible. Input registers negate the need for external latches unless the transparent mode is selected.

Video controls (Force High, Blank, Bright, and Sync) permit an easy interface to standard video systems. The Clock (Convert) inputs can be differential or single-ended. Complementary outputs are provided for custom displays or special effects.

The MC10322 is fabricated with Motorola's MOSAIC® process which provides high speed with low power consumption. The MC10322 is available in a 24 pin plastic DIP package.

- 40 MSPS Minimum Conversion Rate
- TTL Compatible Inputs
- 8-Bit Linearity
- Latched Data and Video Control Inputs, or Transparent Mode
- Video Controls: Force High, Blank, Bright, Sync
- Differential Current Outputs Can Swing 2.0 V Each
- Modulation Capability (Multiplying Mode)
- PSRR >60 dB
- Operates From +5.0 and −5.2 V Power Supplies
- Power Dissipation: Typically 344 mW
- Available in 24 Pin Plastic DIP Package
- Available With ECL Inputs (MC10324)

8-BIT VIDEO DAC WITH TTL INPUTS

SILICON MONOLITHIC INTEGRATED CIRCUIT

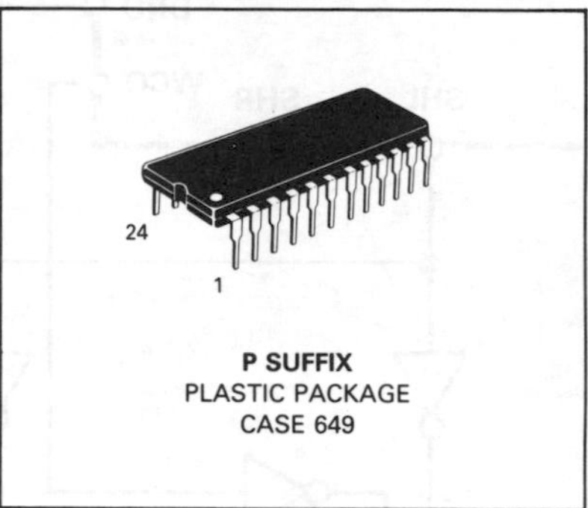

P SUFFIX
PLASTIC PACKAGE
CASE 649

SIMPLIFIED BLOCK DIAGRAM

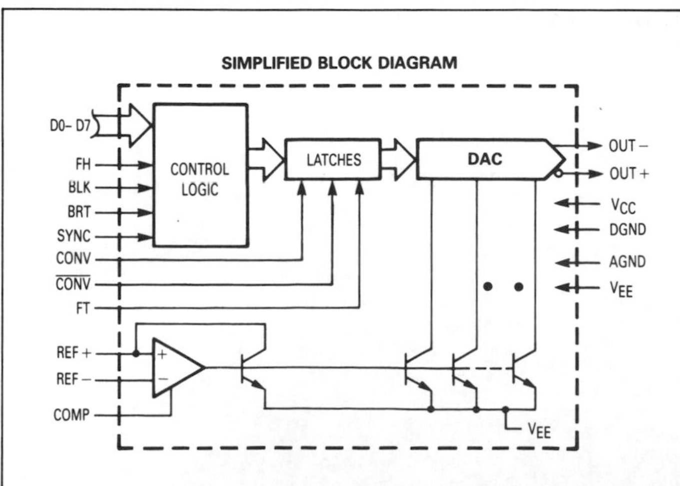

PIN CONNECTIONS
(TOP VIEW)

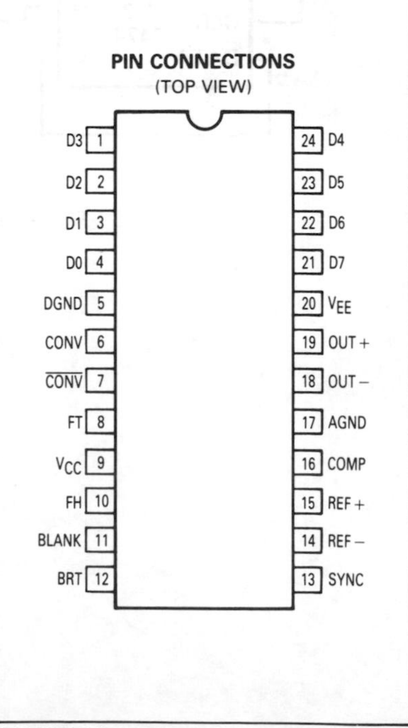

This document contains information on a new product. Specifications and information herein are subject to change without notice.

PIN DESCRIPTION

Symbol	Pin	Description
D0–D7	1–4, 21–24	Data inputs. D0 is the LSB and D7 is the MSB. Inputs are TTL compatible. Maximum update rate is typically 60 MHz. The eight bits control the Gray Scale amplitude only, and do not involve the Sync or Blank levels.
DGND	5	Connect to system digital ground. This pin must be within 100 mV of AGND (Pin 17).
CONV	6	Convert input. The rising edge latches data and controls if FT = 0. May be used single-ended (with Pin 7 at a fixed voltage), or differentially with Pin 7.
$\overline{\text{CONV}}$	7	$\overline{\text{Convert}}$ input. The falling edge latches data and controls if FT = 0. May be used single-ended (with Pin 6 at a fixed voltage), or differentially with Pin 6.
FT	8	Feedthrough. When high, the internal latches are transparent, and Pins 6 and 7 are unused. When low, data and controls are latched by Pins 6 and 7.
V_{CC}	9	Connect to +5.0 V, ±10%. This pin powers the digital portion of the IC.
FH	10	Force High. A logic high internally sets data inputs = 1, overriding external data inputs.
BLANK	11	A logic high overrides data inputs, sets the outputs to the video blanking level.
BRT	12	Bright. A logic high increases the Gray Scale output level by ≈11%, providing an enhanced display. Does not affect SYNC or BLANK.
SYNC	13	A logic high overrides all other inputs, and sets the output to video sync level.
REF−	14	Inverting Reference input. A high impedance input, normally set to a negative DC voltage in the range of −0.8 to −1.7 V. Can be used to modulate the output.
REF+	15	Noninverting Reference input. A virtual ground, current supplied to this pin (between 0.5 and 1.7 mA) sets the maximum output current.
COMP	16	Compensation. A capacitor between this pin and Pin 20 stabilizes the reference amplifier.
AGND	17	Connect to system analog ground. This pin must be within 100 mV of DGND (Pin 5).
OUT−	18	A high impedance current output. Video voltage levels are produced when connected to a 75 Ω cable with appropriate terminations. This output provides a "sync down" waveform. If unused, connect to Pin 17.
OUT+	19	Complementary output provides a "sync up" waveform. If unused, connect to Pin 17.
V_{EE}	20	Connect to −5.2 V, ±10%. This supply should be referenced to analog ground.

FIGURE 1 — TEST CIRCUIT

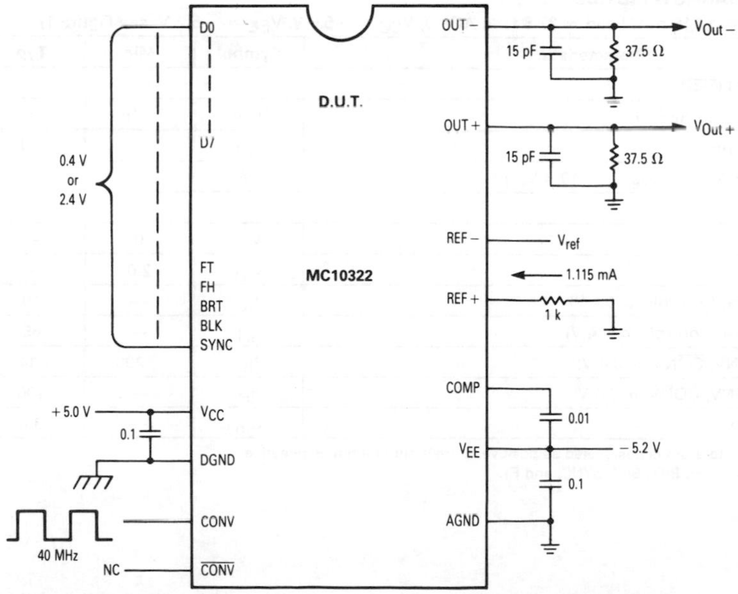

ABSOLUTE MAXIMUM RATINGS

Characteristic	Value	Units
V_{CC} (with respect to DGND and AGND)	+7.0, −0.5	Vdc
DGND (with respect to AGND)	−1.0, +0.5	Vdc
V_{EE} (with respect to AGND)	−7.0, +0.5	Vdc
Logic Input Voltage (with respect to DGND)	−0.5	Vdc
Logic Input Voltage (with respect to V_{CC})	+0.5	Vdc
Voltage at Reference Amp Inputs	+0.5, V_{EE}	Vdc
Current into REF+	+6.0, 0	mA
Voltage Applied to OUT+, OUT− (Normal Operation)	+0.5, −2.0	Vdc
Voltage Applied to OUT+, OUT− (V_{CC}, V_{EE} = 0)	+0.5, −1.2	Vdc
Junction Temperature	−65, +150	°C

Devices should not be operated at these values. The "Recommended Operating Conditions" table provides conditions for actual device operation.

RECOMMENDED OPERATING CONDITIONS

Characteristic	Symbol	Min	Typ	Max	Unit
Supply Voltage	V_{CC} — DGND DGND — AGND V_{EE} — AGND	4.5 −0.1 −5.72	5.0 0 −5.2	5.5 +0.1 −4.68	Vdc
Logic Input Voltage	V_{in}	DGND	—	V_{CC}	Vdc
Reference Current (for Video Standard Output) (for all other applications)	I_{ref}	— 0.5	1.115 —	— 1.7	mA
Voltage at REF−	V_{ref}	−1.7	—	−0.8	Vdc
Output Load Impedance	R_L	0	37.5	—	Ω
Output Compliance (with respect to AGND)	V_o	−1.7	—	+0.3	Vdc
Convert Frequency (FT = 0)	f_s	0	60	40	MHz
Data Update Frequency (FT = 1)	f_s	0	60	40	MHz
Convert, $\overline{\text{Convert}}$ Common Mode Range	V_{CM}	DGND + 1.3	—	V_{CC} − 2.0	Vdc
Operating Ambient Temperature	T_A	−40	—	+85	°C

All limits are not necessarily functional concurrently.

ELECTRICAL CHARACTERISTICS

(T_A = 25°C, I_{ref} = 1.115 mA, Load = 37.5 Ω to AGND, V_{CC} = +5.0 V, V_{EE} = −5.2 V, see Figure 1)

Characteristic	Symbol	Min	Typ	Max	Unit
REFERENCE AMPLIFIER					
Input Offset (REF+ to REF−)	V_{OS}	−15	±5.0	+15	mV
Bias Current into REF−	I_{BR}	—	1.4	5.0	μA
Bandwidth (C_C = 250 pF, V_{ref-} = 10 mVp-p)	BW	—	3.0	—	MHz
DIGITAL INPUTS					
Low Voltage	V_{IL}	0	—	0.8	Vdc
High Voltage	V_{IH}	2.0	—	V_{CC}	Vdc
Low Current (Data, Controls @ 0.4 V)	I_{IL}	—	10	25	μA
High Current (Data, Controls @ 2.4 V)	I_{IH}	—	65	110	μA
Low Current (CONV, $\overline{\text{CONV}}$ @ 0.4 V)	I_{IL}	−200	−144	—	μA
High Current (CONV, $\overline{\text{CONV}}$ @ 2.4 V)	I_{IH}	—	100	140	μA
Input Capacitance	C_{in}	—	3.0	—	pF

NOTES: 1. Current into a pin is designated as positive, current out of a pin as negative.
2. Controls = FH, BRT, BLK, SYNC, and FT.

(continued)

ELECTRICAL CHARACTERISTICS — continued

(T_A = 25°C, I_{ref} = 1.115 mA, Load = 37.5 Ω to AGND, V_{CC} = +5.0 V, V_{EE} = −5.2 V, see Figure 1)

Characteristic	Symbol	Min	Typ	Max	Unit
TRANSFER CHARACTERISTICS					
Resolution	Res	8.0	8.0	8.0	Bits
Integral Nonlinearity	INL	−1/2	0	+1/2	LSB
Differential Nonlinearity	DNL	−1/2	0	+1/2	LSB
Monotonicity	—		Guaranteed*		
Differential Gain	DG	—	1.0	—	%
Differential Phase	DP	—	0.5	—	Deg.
OUTPUTS					
Output Current at OUT− (Control inputs = 0 except as noted)					
Enhanced White (FH = BRT = 1)	I_{EH-}	0	18	100	µA
Normal White (FH = 1 or D0–D7 = 1)	I_{NW-}	1.75	1.94	2.13	mA
Normal Black (D0–D7 = 0) referred to Normal White	I_{NB-}	16.6	17.7	18.7	
Blank referred to Normal Black (BLK = 1)	I_{BLN-}	1.32	1.43	1.54	
Sync referred to Blank (SYNC = 1)	I_{SYNC-}	7.1	7.7	8.3	
Output Voltage at OUT− (Control inputs = 0 except as noted)					mV
Enhanced White (FH = BRT = 1)	V_{EH-}	0	−0.67	−3.75	
Normal White (FH = 1 or D0–D7 = 1)	V_{NW-}	−67.8	−73	−77.6	
Normal Black (D0–D7 = 0) referred to Normal White	V_{NB-}	−626	−663	−694	
Blank referred to Normal Black (BLK = 1)	V_{BLN-}	−50.6	−53.6	−56.6	
Sync referred to Blank (SYNC = 1)	V_{SYNC-}	−270	−288	−308	
Output Current at OUT+ (Control inputs = 0 except as noted)					mA
Enhanced White (FH = BRT = 1)	I_{EH+}	26.8	28.8	30.6	
Normal White (FH = 1 or D0–D7 = 1)	I_{NW+}	25	26.8	28.5	
Normal Black (D0–D7 = 0) referred to Normal White	I_{NB+}	−16.6	−17.7	−18.7	
Blank referred to Normal Black (BLK = 1)	I_{BLN+}	−1.32	−1.43	−1.54	
Sync referred to Blank (SYNC = 1)	I_{SYNC+}	−7.1	−7.7	−8.3	
Output Voltage at OUT+ (Control inputs = 0 except as noted)					mV
Enhanced White (FH = BRT = 1)	V_{EH+}	−1016	−1080	−1132	
Normal White (FH = 1 or D0–D7 = 1)	V_{NW+}	−949	−1005	−1057	
Normal Black (D0–D7 = 0) referred to Normal White	V_{NB+}	626	663	694	
Blank referred to Normal Black (BLK = 1)	V_{BLN+}	50.6	53.6	56.6	
Sync referred to Blank (SYNC = 1)	V_{SYNC+}	270	288	308	
Output Matching ($\lvert I_{NB+}\rvert - \lvert I_{NB-}\rvert$)	I_{FSER}	−50	0	+50	µA
Gain Error (Gray Scale at OUT−)	G_{ER}	−5.0	0	+5.0	%
Output Impedance (Gray Scale, −1.7 V < V_O < 0.3 V)	Z_O	25	100	—	kΩ
Output Capacitance	C_O	—	16	—	pF
Glitch Energy (Clocked Mode)					pV-sec
At Midscale Transition (D0–D7 = 127↔128)	E_{GM}	—	18	—	
Due to Clock Feedthrough (D0–D7 = constant)	E_{GC}	—	2.0	—	
Due to Data Feedthrough (Clock = constant)	E_{GD}	—	25	—	
Peak Glitch Current (Clocked Mode)					
At Midscale Transition (D0–D7 = 127↔128)	I_{GM}	—	0.2	—	mA
Due to Clock Feedthrough (D0–D7 = constant)	I_{GC}	—	55	—	µA
Due to Data Feedthrough (Clock = constant)	I_{GD}	—	0.5	—	mA
POWER SUPPLIES					
Supply Current					mA
(V_{CC} = +5.5 V)	I_{CC}	—	22	28	
(V_{EE} = −5.72 V, I_{ref} = 1.115 mA)	I_{EE}	−55	−45	—	
Power Dissipation	P_D	—	344	469	mW
Power Supply Sensitivity at Outputs					µA/V
(V_{EE} = −5.2 V, 4.5 < V_{CC} < 5.5 V)	PSSD	−25	1.0	+25	
(V_{CC} = +5.0 V, −5.72 < V_{EE} < −4.68 V)	PSSA	−100	20	+100	
V_{CC} Supply Rejection (V_{CC} = +5.0 V, V_{EE} = −5.2 V, f = 10 kHz)	PSRRD	—	85	—	dB
V_{EE} Supply Rejection (V_{CC} = +5.0 V, V_{EE} = −5.2 V, f = 10 kHz)	PSRRA	—	65	—	dB

*Guaranteed by linearity tests.

(continued)

TIMING CHARACTERISTICS
(T_A = 25°C, I_{ref} = 1.115 mA, Load = 37.5 Ω//15 pF, V_{CC} = +5.0 V, V_{EE} = −5.2 V, see Figures 2, 3)

Characteristic	Symbol	Min	Typ	Max	Unit
Maximum Conversion or Update Rate	F_S	40	60	—	MHz
Clocked Mode (FT = 0)					ns
Clock to Output Delay (Data, Controls)	t_{COD}	—	10	—	
Setup Time — Data to CONV rising edge	t_{SD}	—	8.0	—	
FH, SYNC to CONV rising edge	t_{SFS}	—	5.0	—	
BRT, BLK to CONV rising edge	t_{SB}	—	4.0	—	
Hold Time — All Inputs (after CONV rising edge)	t_H	—	0	—	
Minimum Clock Width (High or Low)	t_{PW}	—	6.0	—	
Transparent Mode (FT = 1)					ns
Data to Output Delay	t_{DO}	—	16	—	
SYNC to Output Delay	t_{SO}	—	13	—	
BLK to Output Delay	t_{BO}	—	14	—	
FH to Output Delay	t_{FO}	—	16	—	
BRT to Output Delay	t_{RO}	—	12	—	
Output Rise/Fall Time — 10 to 90% of Gray Scale	t_{RFG}	—	2.0	—	ns
Enhanced White to Sync Level	t_{RFF}	—	3.0	—	
Output Settling Time	t_{SET}	—	4.0	—	ns

FUNCTIONAL DESCRIPTION

Introduction

The MC10322 is an 8-bit DAC intended for video applications, employing TTL inputs for the data (natural binary code) and video controls, and outputs capable of directly driving a standard 50 Ω or 75 Ω monitor. Its use is not limited to video, however, any application requiring a high speed (typically 60 MHz) DAC, or a DAC with high output current capability (up to 44 mA) can use the MC10322. The input data and controls may be clocked into the internal registers, or the MC10322 may be used in the transparent mode eliminating the need for the clock.

The MC10322 may be used in the multiplying mode by varying the reference current along with the digital inputs producing the product of the two at the outputs. The reference current can be varied over a range of 0.5 to 1.7 mA. Standard power supplies are required: +5.0 V and −5.2 V, both ±10%. Power consumption is nominally 344 mW.

DAC Outputs

The outputs of the MC10322 are high impedance constant current sinks whose values depends on the reference current (I_{ref}), the binary value of the digital word at D0–D7, and the status of the video controls (SYNC, BLK, FH, and BRT). Complementary outputs are provided allowing an increased output swing (when used differentially), or for creation of special effects required by the application. For a given reference current, the sum of the two output currents is a constant equal to:

$$I_{O-} + I_{O+} = I_{ref} \times 25.86$$

The OUT− output provides a "sync down" waveform, while the OUT+ output provides a "sync up" waveform (see Table 1). Current flow is **into** each of the outputs. Each output's impedance is typically 100 kΩ over the compliance range of −1.7 to +0.3 V, and the output capacitance is typically 16 pF. An unused output cannot be left open — both outputs must be connected to a voltage within the compliance range. Both outputs should be equally loaded for best accuracy.

TABLE 1 — OUTPUT LEVELS

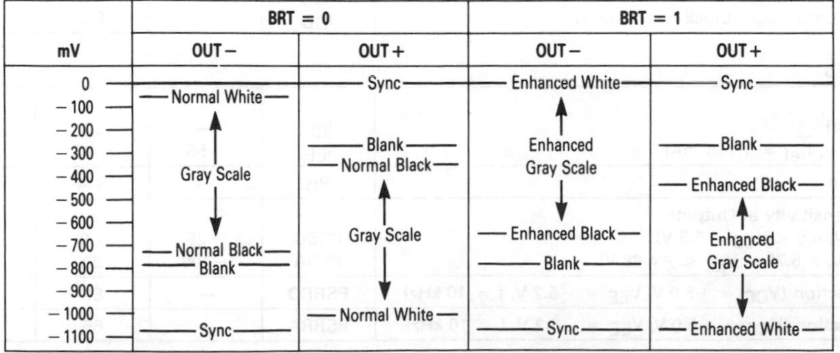

I_{ref} = 1.115 mA, R_L = 37.5 Ω

FIGURE 2 — TIMING DIAGRAM, CLOCKED MODE
(FT = 0)

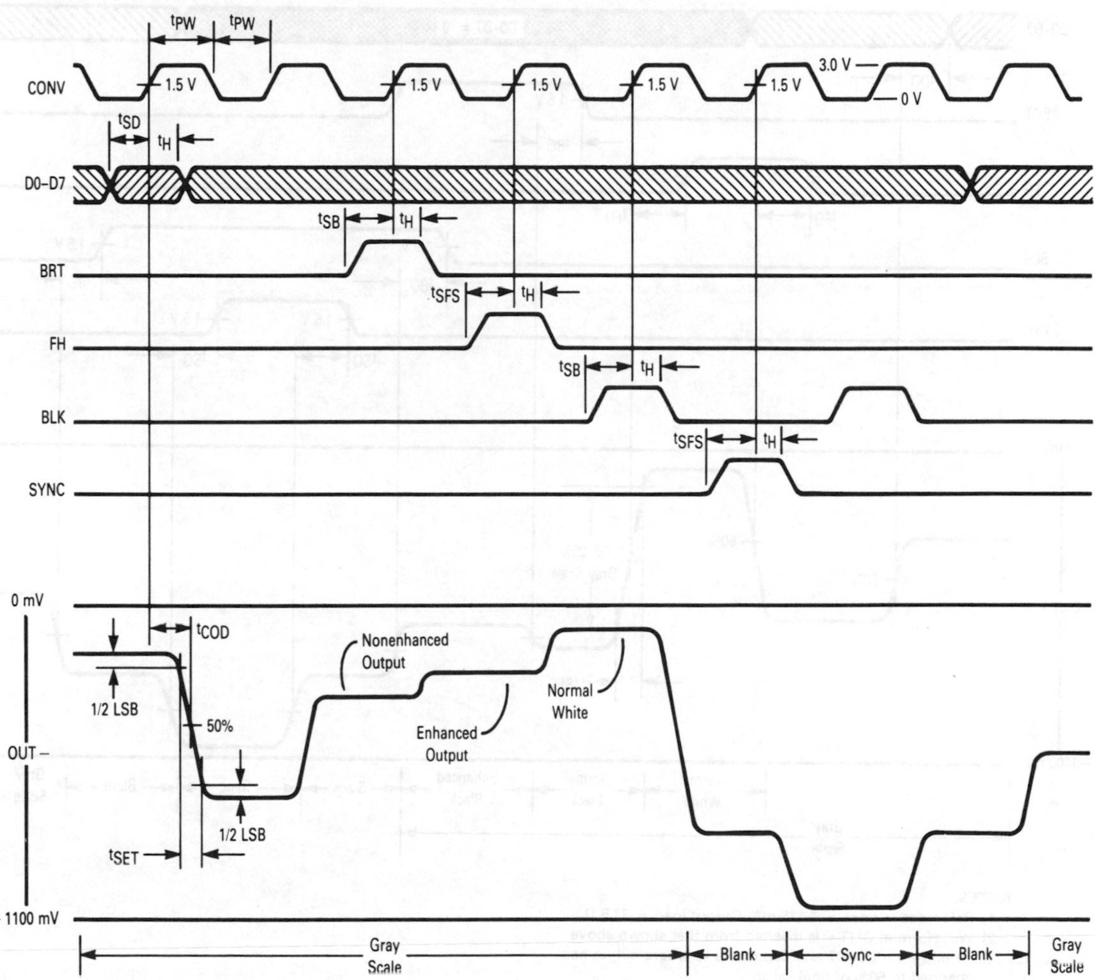

NOTES:

1) Single-ended clock used in production testing.
2) If differential clock is used, timing would be determined from the crossover point of the two clock signals.
3) If $\overline{\text{CONVERT}}$ is used in single-ended mode, timing would be measured from its falling edge.
4) Timing to output from data and controls is from CONVERT rising edge (threshold) to where the output has changed to 50% of final value.
5) Reference current = 1.115 mA. Output load = 37.5 Ω.
6) Waveform at OUT+ is inverted from that shown above.

FIGURE 3 — TIMING DIAGRAM, TRANSPARENT MODE (FT = 1)

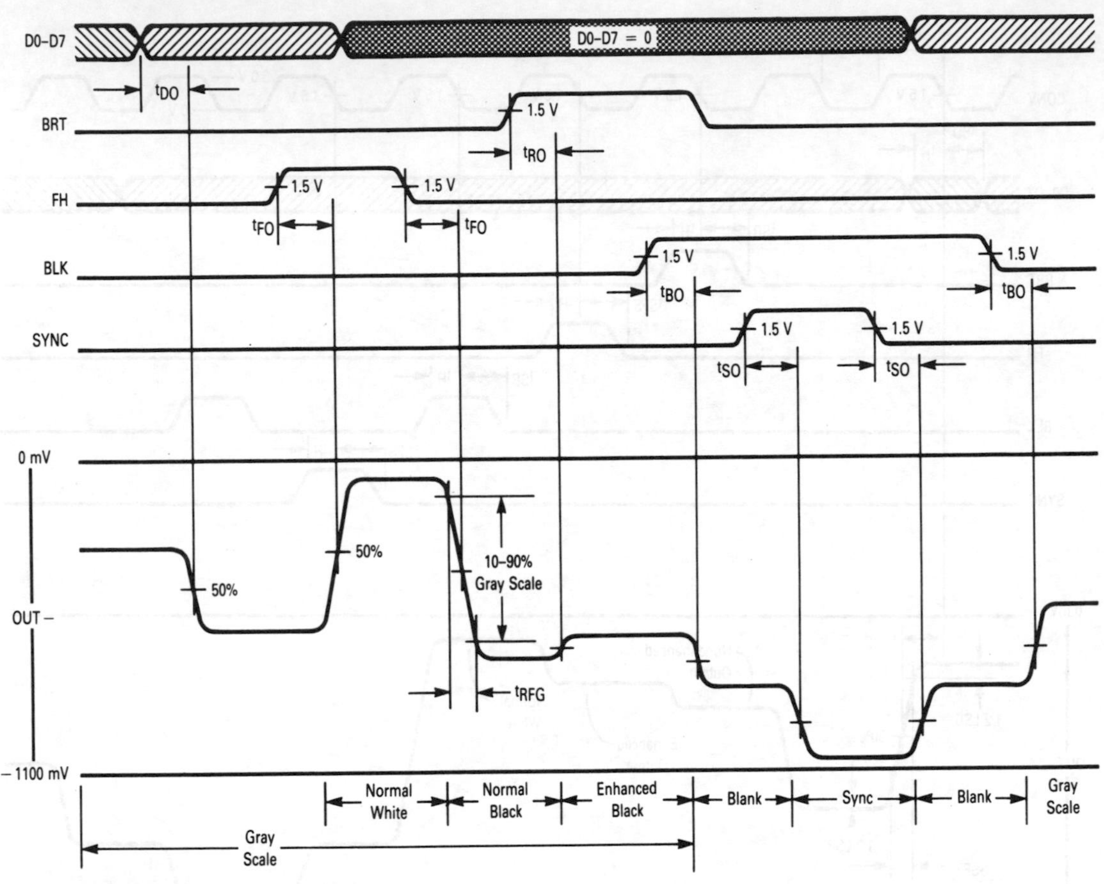

NOTES:

1) Reference current = 1.115 mA. Output load = 37.5 Ω.
2) Waveform at OUT+ is inverted from that shown above.
3) Timing from D0–D7 and Controls is to where output has changed to 50% of final value.

TRUTH TABLE

INPUTS					OUTPUTS				
Controls				Data	Out−		Out+		
Sync	Blank	Force High	Bright	D7–D0	(mA)	(mV)	(mA)	(mV)	Condition
1	X	X	X	X	28.8	−1080	0	0	Sync
0	1	X	X	X	21.1	−790	7.7	−289	Blank
0	0	0	0	000..	19.6	−736	9.1	−341	Normal Black
0	0	0	1	000..	17.7	−663	11.1	−416	Enhanced Black
0	0	1	0	X	1.94	−73	26.8	−1005	Normal White
0	0	0	0	111..	1.94	−73	26.8	−1005	Normal White
0	0	0	1	111..	0	0	28.8	−1080	Enhanced White
0	0	1	1	X	0	0	28.8	−1080	Enhanced White

NOTES: 1) Current flow is into the output pins.
2) Output voltage measured across a 37.5 Ω resistor to AGND.
3) Waveform at OUT+ is inverted from that at OUT−.
4) Reference Current = 1.115 mA.

DAC Gray Scale (D0–D7)

Within the Gray Scale (all 4 video controls = 0), the current at OUT− is controlled by the data inputs (D0–D7) according to the following equation:

$$I_{O-(GS)} = \frac{I_{ref} \times (255\text{-}A)}{16} + (I_{ref} \times 1.74)$$

The current at OUT+ in the Gray Scale is determined by:

$$I_{O+(GS)} = \frac{I_{ref} \times A}{16} + (I_{ref} \times 8.18)$$

For the test value of I_{ref} = 1.115 mA, I_{O-} varies from 19.6 mA to 1.94 mA as the digital inputs (A) vary from 0 to 255 (00_H to FF_H), and I_{O+} will vary from 9.12 mA to 26.8 mA. The data inputs are overridden by SYNC, BLK, or FH.

Figure 4 depicts a typical input stage configuration, and Figure 8 indicates the typical input current. The inputs' threshold is ≈ + 1.5 V, independent of V_{CC}. An open input is equivalent to a logic low, but good design practices dictate that inputs should never be left open. The inputs must be kept within the range of V_{CC} and DGND. If an input is taken more than 0.5 V above V_{CC} or below GND excessive currents will flow, and the DAC output waveform will be distorted.

FIGURE 4 — TYPICAL INPUT STAGE

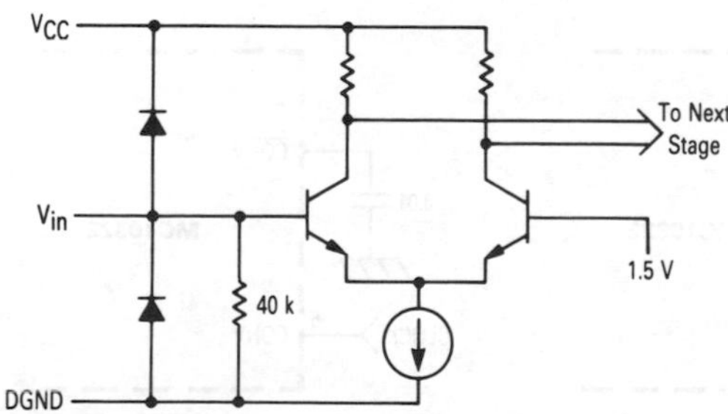

DAC — Video Controls

The four video controls (SYNC, BLK, FH, and BRT) are logic level inputs, TTL compatible, which permit setting the outputs to standard video levels. All four are active high. The Truth Table on page 7 indicates their priority.

The Force High input (FH) overrides the data inputs (D0–D7), setting the DAC inputs to all 1s (FF_H). In most applications, this is equivalent to the normal white level. FH can be used with the BRT input to create an enhanced white, but is overridden by SYNC or BLK.

The Bright input (BRT) shifts the Gray Scale by ≈11% in the high (white) direction. Typically this function is used to provide an enhanced, or brighter display so as to highlight certain portions of the screen. A highlighted cursor is a typical example.

The current change at each output is equal to:

$$\Delta I_{O(BRT)} = I_{ref} \times 1.74$$

The current at I_{O-} decreases in magnitude, while the current at I_{O+} increases, when BRT is asserted. BRT is ineffective when SYNC or BLK are asserted, but can be used at any point within the Gray Scale.

The Blank input (BLK) sets the output currents to the blanking level used during horizontal and vertical retrace.

The current at OUT− is:

$$I_{O-(BLK)} = I_{ref} \times 18.96$$

The current at OUT+ is:

$$I_{O+(BLK)} = I_{ref} \times 6.9$$

The BLK input will override the data inputs, FH and BRT, but is overridden by SYNC. Therefore, the BLK input may be left asserted during the sync time.

The Synchronizing input (SYNC) sets the output currents to the sync level used for normal horizontal and vertical picture synchronization.

The current at OUT− is:

$$I_{O-(SYNC)} = I_{ref} \times 25.86$$

The current at OUT+ will be leakage current only, typically <20 μA. The SYNC input will override all other control inputs as well as data inputs.

Figure 4 depicts a typical input stage configuration, and Figure 8 indicates the typical input current. The input's threshold is ≈ + 1.5 V, independent of V_{CC}. An open input is equivalent to a logic low, but good design practices dictate that inputs should never be left open. The inputs must be kept within the range of V_{CC} and DGND. If an input is taken more than 0.5 V above V_{CC} or below GND excessive currents will flow and the DAC output waveform will be distorted.

Feedthrough (FT) Input

With FT low, the internal registers are active, and the data and video controls are clocked through to the DAC on the rising edge of Pin 6 (CONV), or on the falling edge of Pin 7. In this mode, the data bits (D7–D0) which may appear asynchronously to the MC10322 are then presented synchronously to the DAC, reducing output glitches and noise. This mode is also useful for synchronizing control functions with other events. While hold times are typically 0 ns for all inputs, the setup times prior to the clock edge must be observed.

With FT high, the registers are transparent and the data and video controls feed through directly to the DAC. This mode may be used if the data is presented to the MC10322 from external latches, which ensure minimum skew among the data bits. In this mode Pins 6 and 7 are not used.

Figure 4 depicts the FT input stage configuration, and Figure 8 indicates the typical input current. The threshold is ≈ + 1.5 V, independent of V_{CC}. An open input is equivalent to a logic low, but good design practices dictate that inputs should never be left open. FT must be kept within the range of V_{CC} and DGND. If an input is taken more than 0.5 V above V_{CC} or below GND excessive currents will flow and the DAC output waveform will be distorted.

Convert Inputs

The Convert inputs (Pins 6 & 7) are used to clock in data and the video controls to the internal registers only if FT (Pin 8) is low. The input stage for these pins is shown in Figure 5. The pins are internally biased at ≈ +1.6 V with a nominal input impedance of 10 kΩ. The inputs may be driven from complementary TTL clock signals with the clocking action occurring on the rising edge of CONV and the fallinge edge of $\overline{\text{CONV}}$ as the signals cross each other in voltage.

A single-ended clock source may be used by connecting either Pin 6 or 7 to a fixed voltage to set the threshold and applying the clock signal to the other pin. If done this way, the fixed voltage must be within the range of + 1.3 V to $V_{CC} - 2.0$ V. Figure 6 shows three positive edge triggered examples. Interchanging Pins 6 and 7 provides negative edge triggered operation.

The input current required at each pin is shown in Figure 9, and is independent of the clocking mode used.

If FT is high, the Convert pins are nonfunctional, and **must** be connected to **different** voltages (e.g., V_{CC} and DGND). Leaving the pins open can result in high frequency oscillations or spurious noise.

CONV and $\overline{\text{CONV}}$ must be kept within the range of V_{CC} and DGND. If taken more than 0.5 V above V_{CC} or below GND excessive currents will flow, and the DAC output waveform will be distorted.

FIGURE 5 — CONVERT INPUT STAGE

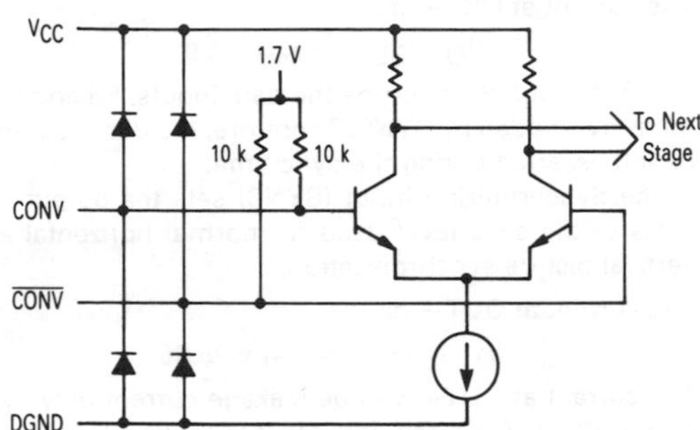

FIGURE 6 — SINGLE-ENDED CLOCK INPUT

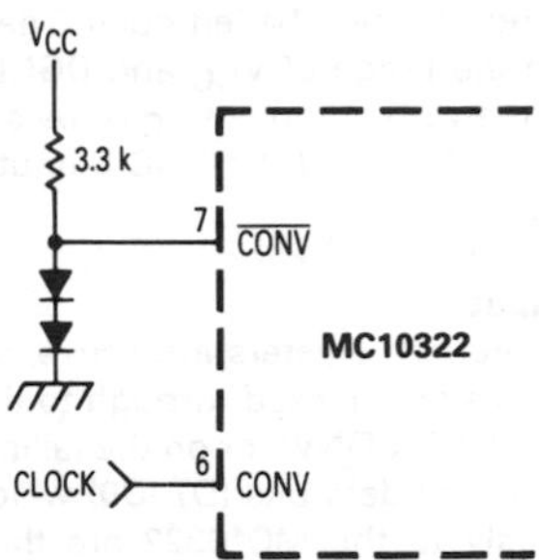

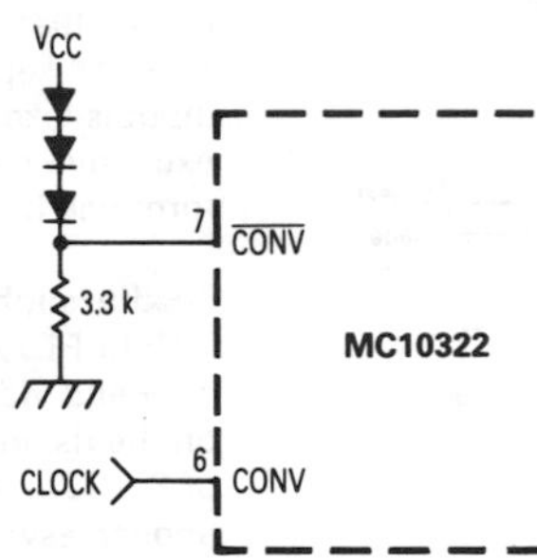

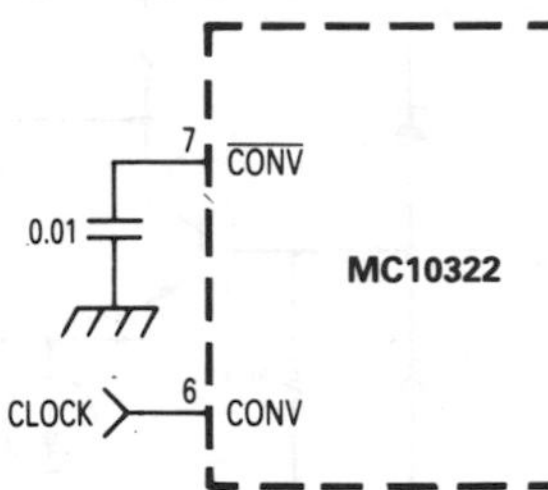

Reference Amplifier

The reference amplifier (Pins 14–16) is used to accept the externally supplied reference current for the DAC current switches (see Figure 7).

REF+ (Pin 15) is a low impedance (virtual ground) input into which the reference current flows (current cannot flow out of this pin). Due to the op amp's internal feedback, the voltage at REF+ is the same as that set at REF−, with a typical input offset of ±5.0 mV. The current into REF+ should be within the range of 0.5 mA to 1.7 mA to maintain 8-bit linearity and accuracy. A reference current of 1.115 mA is recommended to obtain EIA-170 and EIA-343-A voltage levels at the outputs if they are terminated with 37.5 Ω (double 75 Ω terminations).

REF− is a high impedance input (>10 MΩ) which must be set to a voltage within the range of −0.8 to −1.7 V. A nominal bias current of ≈1.4 μA will flow into this pin. In Figure 7, $I_{ref} = V_{ref}/R_{ref}$.

FIGURE 7 — REFERENCE AMPLIFIER

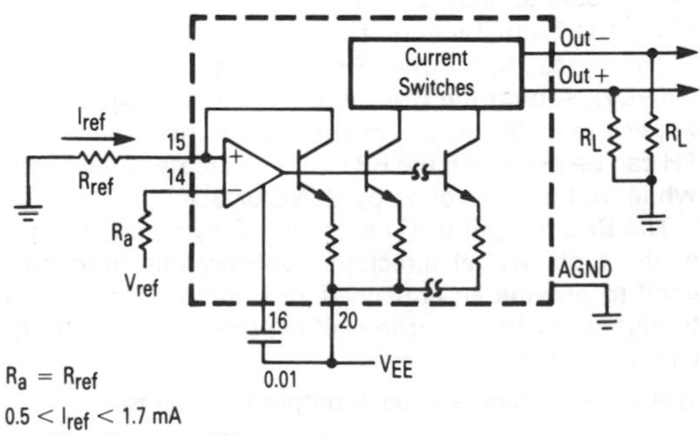

Power Supplies

The MC10322 requires both a +5.0 V and a −5.2 V supply, both ±10%. Nominal current requirements are 22 mA and 45 mA, respectively, (including a total output current of 29 mA). The supply current required at V_{EE} is dependent on the total output current (Pin 18 + Pin 19). The +5.0 V supply powers only the digital portion of the IC (control logic and latches), and should be referenced to Digital Ground (Pin 5). The −5.2 V supply powers the analog portion of the IC (reference amplifier and the DAC's current sources), and should be referenced to Analog Ground (Pin 17). See the Applications Section for additional information on power supplies, bypassing and PC board layout.

Timing

Figures 2 and 3 are the timing diagrams for the MC10322. Figure 2 is for the clocked mode where data and control inputs are latched into the input registers by the Convert (and/or $\overline{\text{Convert}}$) inputs. If the clock signal is single-ended, the data and control latching occurs on the rising edge of Convert, or the falling edge of $\overline{\text{Convert}}$. If a differential clock is used, latching occurs at the cross-over point of the two signals. The hold time for the data and controls is 0, but the setup times must be observed. The clock duty cycle is not important as long as the minimum pulse widths are observed.

Figure 3 is for the transparent (non-clocked) mode. The output responds to the application of new data or control inputs without the need for a clocking edge. The propagation delay to the output is different for each of the data and control signals. To prevent large glitches at the outputs, it is imperative that the data bits (D0–D7) arrive at the MC10322 simultaneously with minimum skew. If the synchronism of the 8-bits cannot be guaranteed, either an 8-bit latch should be used (F373 or F374 type), or the MC10322 should be used in the clocked mode.

FIGURE 8 — INPUT CURRENT, DATA AND CONTROLS

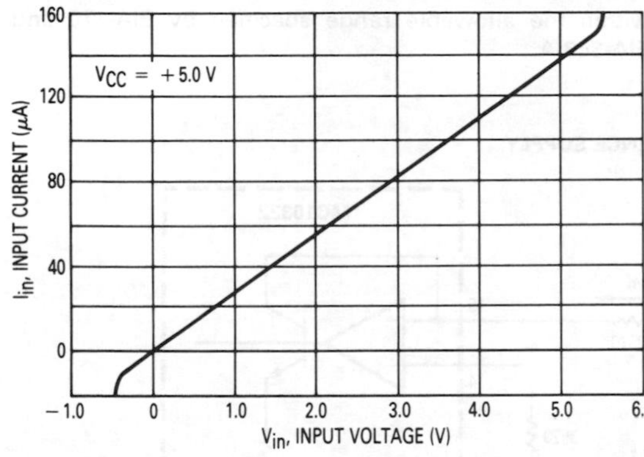

FIGURE 9 — INPUT CURRENT, CONVERT INPUTS

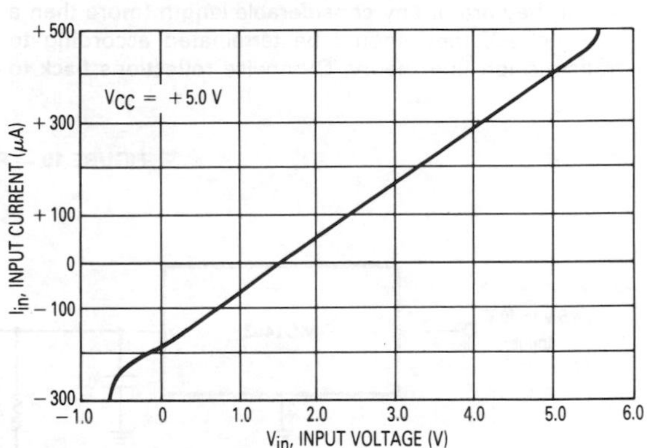

APPLICATIONS INFORMATION

Power Supplies, Grounding

The PC board layout and the quality of the power supplies and the ground system **at the IC** are very important in order to obtain proper operation. Noise from any source coming into the device on V_{CC}, V_{EE}, or ground can cause an incorrect output code due to interaction with the analog portion of the circuit. At the same time, noise generated within the MC10322 can cause incorrect operation if that noise does not have a clear path to AC ground.

Both the V_{CC} and V_{EE} power supplies must be decoupled to the appropriate ground **at the IC** (within 1″ max) with a 10 μF tantalum and a 0.1 μF ceramic. Tantalum capacitors are recommended since electrolytic capacitors simply have too much inductance at the frequencies of interest. The quality of the V_{CC} and V_{EE} supplies should then be checked at the IC with a high frequency scope. Noise spikes (whenever digital circuits are present) can easily exceed 400 mV peak, and if they get into the analog portion of the IC, the operation can be disrupted. Noise can be reduced by inserting resistors and/or inductors between the supplies and the IC.

If switching power supplies are used, there will usually be spikes of 0.5 V or greater at frequencies of 50 kHz to 1.0 MHz. These spikes are generally more difficult to reduce because of their greater energy content. In extreme cases three terminal regulators (MC78L05ACP, MC7905.2CT), with appropriate high frequency filtering should be used and dedicated to the MC10322.

The ripple content of the supplies should not allow their magnitude to exceed the values in the Recommended Operating Conditions.

The PC board tracks supplying V_{CC} and V_{EE} to the MC10322 should preferably not be at the tail end of the bus distribution after passing through a maze of digital circuitry. The MC10322 should be close to the power supply, or the connector where the supply voltages enter the board. If the V_{CC} and V_{EE} lines are supplying considerable current to other parts of the board, then it is preferable to have dedicated lines from the supply or connector directly to the MC10322.

The two ground pins (DGND and AGND) must eventually be connected together, usually near the power supply, although the specific board layout may dictate a different "best point." V_{CC} must be referenced to DGND, and V_{EE} must be referenced to AGND.

PC Board Layout

Due to the high frequencies involved, and in particular, the fast edges of the various digital signals, proper PC board layout is imperative. A solid ground plane is strongly recommended in order to have known transmission characteristics, and also to minimize coupling of the digital signals into the analog section. Use of wire wrapped boards should definitely be avoided.

Each PC track should be considered a transmission line, and if they are of any considerable length (more than a few inches), they should be terminated according to transmission line theory. Otherwise reflections back to the signal sources can occur, disrupting their operation. Additionally, the overshoots and undershoots which will occur at the MC10322's input pins can cause its operation to be disrupted, resulting in a noisy or incorrect output.

Additional information regarding the transmission characteristics of PC board tracks can be found in Motorola's MECL System Design Handbook (HB205).

Reference Circuits

Since the accuracy of the outputs are directly related to the accuracy and quality of the reference current and voltage, it is imperative that accurate and stable references be used at Pins 14 and 15. The voltage supply used for the digital circuitry should preferably **not** be used as a source for either the reference current or voltage due to the noise spikes and ripple present on the supply and its ground lines.

Figure 10 indicates a method for generating the reference signals from a positive supply. The MC1403 reference is a stable 2.5 V bandgap regulator (±1%), with a maximum temperature coefficient of 40 ppm/°C, and good ripple and high frequency noise rejection. In the figure, the circuit supplies −1.48 V to Pin 14, and a current of 1.113 mA to Pin 15. If the outputs of the MC10322 are terminated with 37.5 Ω, the voltage levels will be well within the allowable range specified by EIA-170 and EIA-343-A.

FIGURE 10 — REFERENCE SUPPLY

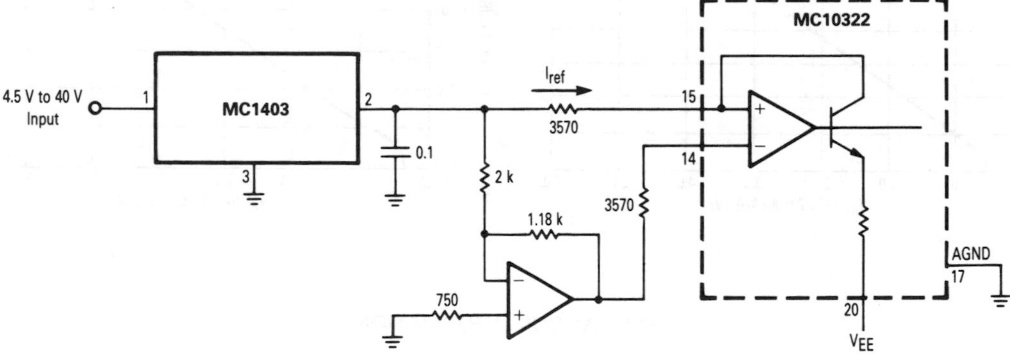

If the analog −5.2 V supply is fairly clean and free of digital noise the circuit of Figure 11 may be used. The TL431 is a stable 2.5 V bandgap reference (±1%) with an effective temperature coefficient of 50 ppm/°C. The 5k pot allows adjustment for precise output levels, or it may replaced with a precision resistor which provides the correct voltage at Pin 14.

FIGURE 11 — REFERENCE SUPPLY

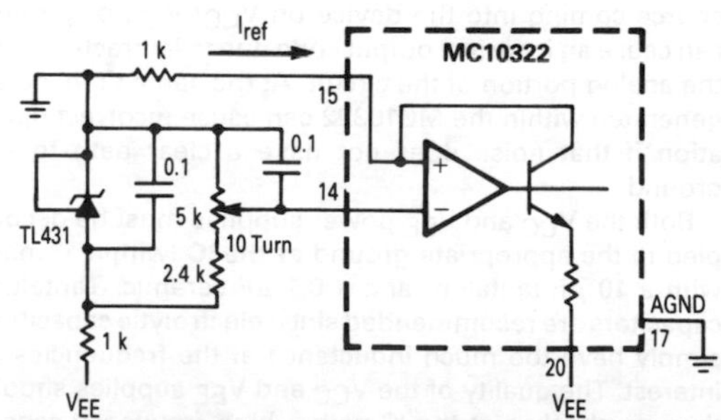

Figure 12 indicates another reference circuit using the LM385-1.2 reference diode. R_{ref} is chosen to provide the desired reference current to Pin 15 knowing that it is set at −1.235 V. The LM385 is a bandgap type reference with a ±2% initial accuracy. The 20 k resistor biases the diode at approximately 200 μA for minimum temperature variations. The 0.2 μF capacitor, with the 20 k resistor, filters out noise above ≈40 Hz.

FIGURE 12 — REFERENCE SUPPLY

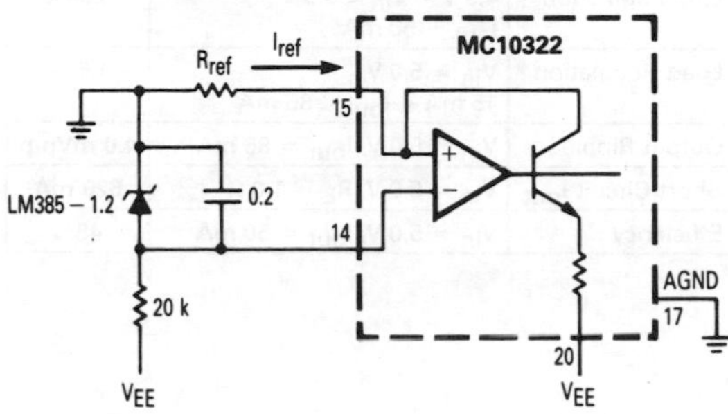

Digitally Modulating an Analog Signal

The MC10322 may be used to digitally modulate (or attenuate) an analog signal by applying the analog signal to the reference amplifier. Three methods of doing this are shown in Figures 13–15.

FIGURE 13 — APPLYING AN ANALOG SIGNAL DIRECTLY

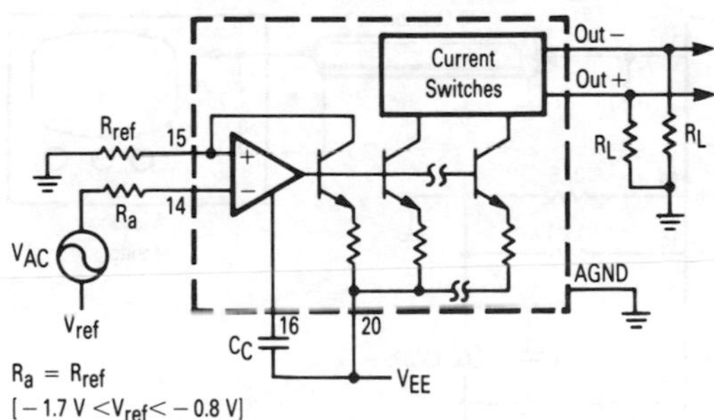

FIGURE 14 — CAPACITOR COUPLING THE AC VOLTAGE

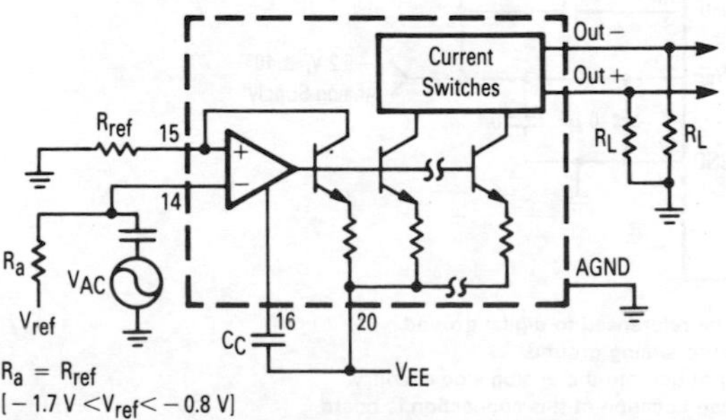

FIGURE 15 — APPLYING A MODULATING CURRENT

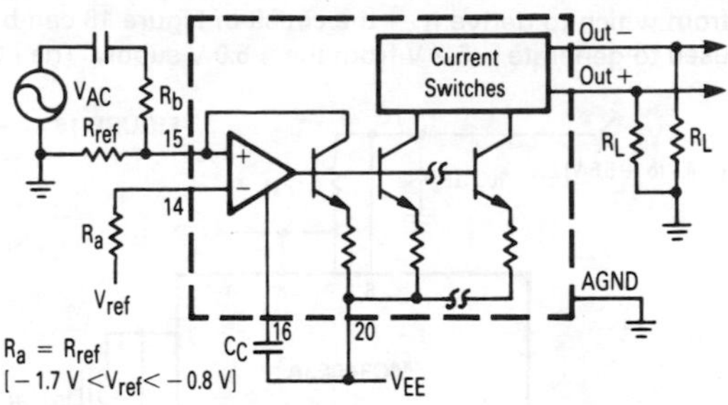

In all three examples the DC reference current is V_{ref}/R_{ref}. In Figure 13 the AC signal source is referenced to a negative voltage source (V_{ref}). In Figures 13 and 14 the AC reference current is equal to V_{AC} divided by R_{ref}. In Figure 15 the AC reference current is equal to V_{AC} divided by R_b. The AC signal at OUT− and OUT+ is determined by the following equations:

$$V_{O-}(AC) = \frac{I_{ref(AC)} \times (255\text{-}A) \times R_L}{16}$$

$$V_{O+}(AC) = \frac{I_{ref(AC)} \times A \times R_L}{16}$$

where "A" is the value of the digital word at D0–D7 (0 to 255).

When implementing any of the above schemes, or any other method of feeding an AC signal to the reference amplifier, the following operating limits must be observed:

1) The peak values of the reference current (AC + DC) must be within the range of 0.5 mA to 1.7 mA into Pin 15;
2) The peak values of the voltage at REF+ and REF− must be within the range of −0.8 V to −1.7 V;
3) The peak values of the voltage at OUT− and OUT+ must be within the range of −1.7 V to +0.3 V.

The maximum frequency which can be handled by the reference amplifier is dependent on the compensation capacitor (C_C) at Pin 16, and the signal amplitude according to the following equation:

$$f_{MAX} = \frac{1.59 \times 10^{-8}}{C_C \times I_{pk}}$$

where I_{pk} is the peak value of the AC reference current (1/2 of the peak-to-peak value). The small signal bandwidth of the reference amplifier is ≈3.0 MHz.

Components associated with the reference amplifier (Pins 14–16) should be physically close to the pins. The board layout should be neat, preventing unwanted stray capacitive coupling between the outputs and the reference amplifier. If C_C is smaller than 5000 pF a ground plane is strongly recommended. C_C should not be smaller than 250 pF.

Negative Voltage Regulator

In the case where a negative power supply is not available — neither the −5.2 V, nor a higher negative voltage from which to derive it — the circuit of Figure 16 can be used to generate −5.2 V from the +5.0 V supply. The PC board space required is small (≈2.0 in^2), and it can be located physically close to the MC10322. The MC34063A is a switching regulator, and in Figure 16 is configured in an inverting mode of operation. The regulator operating specifications are given in the figure.

FIGURE 16 — −5.2 VOLT REGULATOR

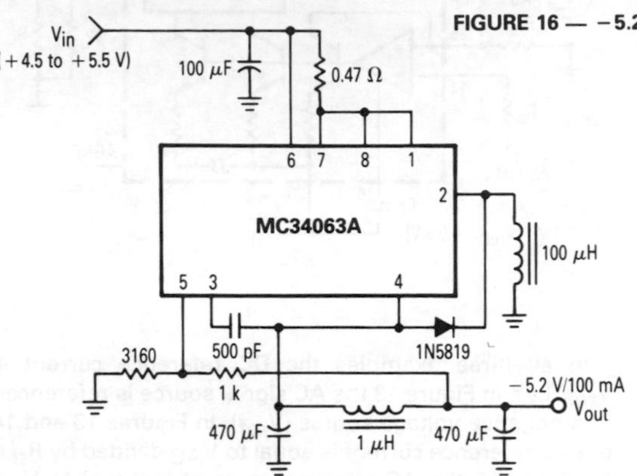

Line Regulation	4.5 V < V_{in} < 5.5 V, I_{out} = 50 mA	0.04%
Load Regulation	V_{in} = 5.0 V, 15 mA < I_{out} < 85 mA	1.5 %
Output Ripple	V_{in} = 5.0 V, I_{out} = 85 mA	4.0 mVp-p
Short Circuit I_{out}	V_{in} = 5.0 V, R_L = 1.0 Ω	620 mA
Efficiency	V_{in} = 5.0 V, I_{out} = 50 mA	48%

Typical Application Circuits

Figure 17 shows a typical video application circuit using the MC10322 in the clocked mode. The clock is single-ended, and the circuit updates the output on the rising edge of the clock. The OUT− pin feeds a standard 75 Ω monitor through a 75 Ω cable, which is terminated at both ends. The reference voltage is supplied by an LM385-1.2 regulator.

Figure 18 shows a circuit similar to that of Figure 17, except the MC10322 is used in the transparent mode. The source of the data bits must provide the 8-bits simultaneously, with minimum skew, to keep output glitches to a minimum. If latches, or other antiskew circuitry, are not available within the microprocessor circuitry, a set of 8-bit latches between it and the MC10322 is recommended, or the MC10322 should be used in the clocked mode.

FIGURE 17 — TYPICAL APPLICATION CIRCUIT, CLOCKED MODE

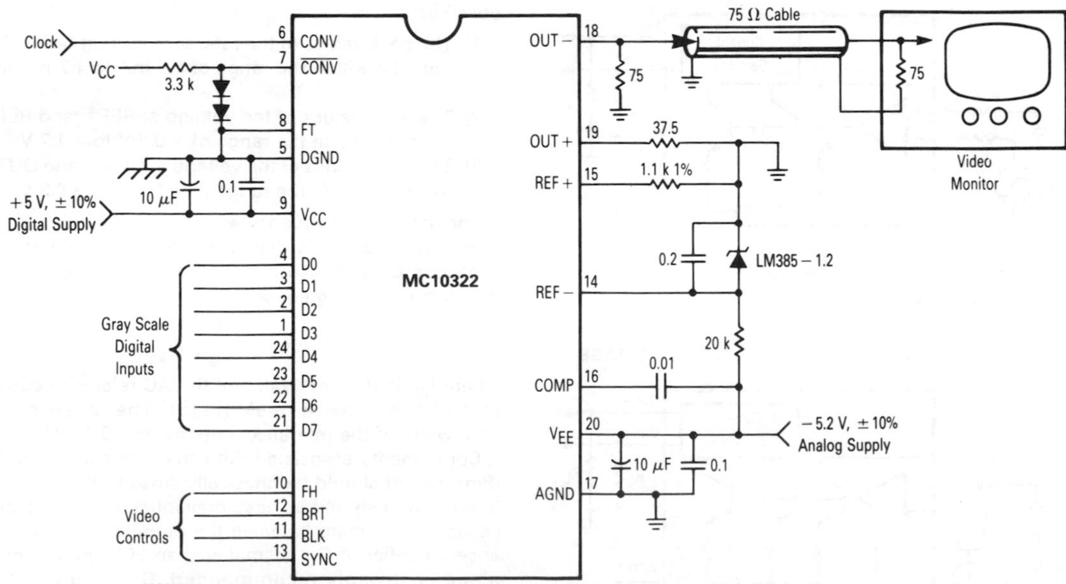

NOTES:

1) Gray Scale inputs, video controls, and clock are to be referenced to digital ground.
2) Outputs and reference circuitry are to be referenced to analog ground.
3) PC board layout to be such that digital noise does not get into the analog side circuitry.
4) Analog and digital grounds to be connected together. Location of this connection is board layout dependent, and is to be such that digital ground noise does not show up in the analog signals.

FIGURE 18 — TYPICAL APPLICATION CIRCUIT, TRANSPARENT MODE

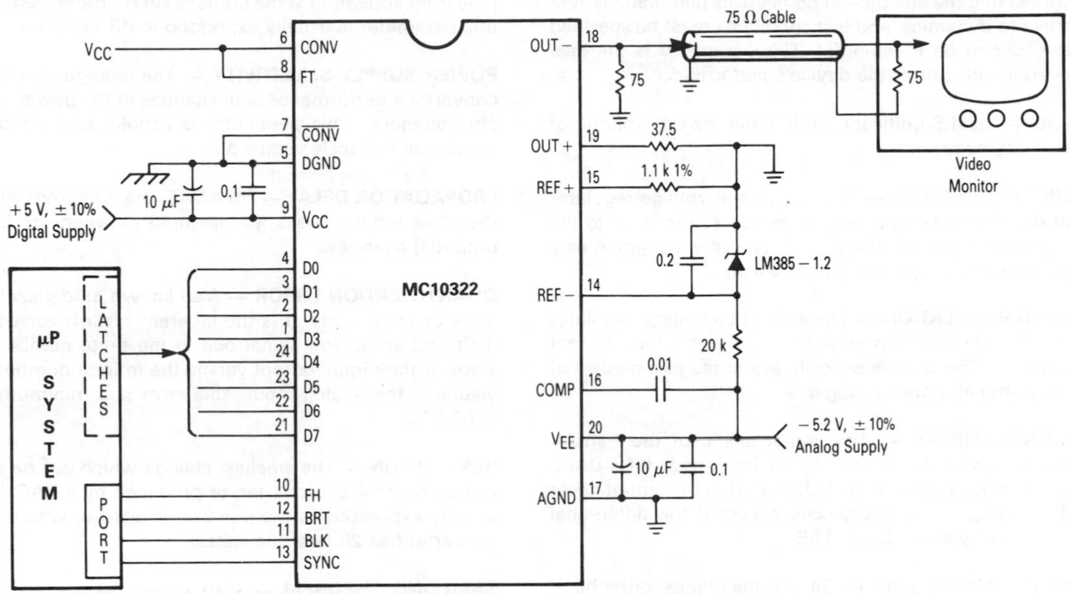

NOTES:

1) Gray Scale inputs and video controls are to be referenced to digital ground.
2) Outputs and reference circuitry are to be referenced to analog ground.
3) PC board layout to be such that digital noise does not get into the analog side circuitry.
4) Analog and digital grounds to be connected together. Location of this connection is board layout dependent, and is to be such that digital ground noise does not show up in the analog signals.

GLOSSARY

BANDGAP REFERENCE — A voltage reference circuit based on the predictable base-emitter voltage of a transistor. The silicon bandgap voltage of ≈1.2 V is the basis for generating other voltages which are stable with time and temperature.

BIPOLAR INPUT — A mode of operation whereby the analog input (of an A-D), or output (of a DAC), includes both negative and positive values. Examples are: −5.0 V to +5.0 V, −2.0 V to +8.0 V, etc.

DAC CURRENT GAIN — The internal gain the DAC applies to the reference current to determine the full scale output current. The actual maximum current out of a DAC is one LSB less than the full scale current.

DIFFERENTIAL GAIN — In video systems, differential gain is a component's change in gain as a function of luminance level. In a color picture, contrast will be affected if the differential gain is not zero.

DIFFERENTIAL NONLINEARITY — The maximum deviation in the actual step size (one transition level to another) from the ideal step size. The ideal step size is defined as the Full Scale Range divided by 2^n. This error must be within ±1 LSB for proper operation.

DIFFERENTIAL PHASE — In video systems, differential phase is the change in the phase modulation of the chrominance as a function of the luminance level. The hue in a color picture will be distorted if the differential phase is not zero.

FULL SCALE RANGE — The difference between the minimum and maximum end points of the analog input (of an A/D), or output (of a DAC), plus one LSB.

GAIN ERROR — The difference between the actual and expected gain (end point to end point), with respect to the reference, of a data converter. The gain error is usually expressed in LSBs, or percent.

GLITCH AREA — The energy content of a glitch, specifically in volt-seconds. It is the area under the curve of the glitch waveform. For a symmetrical glitch, the area and the energy can be zero.

GRAY CODE — Also known as *reflected binary code,* it is a digital code such that each code differs from adjacent codes by only one bit. Since more than one bit is never changed at each transition, race condition errors are eliminated.

INTEGRAL NONLINEARITY — The maximum error of an A/D or DAC, transfer function from the ideal straight line connecting the analog end points. This parameter is sensitive to dynamics, and test conditions must be specified in order to be meaningful. This parameter is the best overall indicator of the device's performance.

LSB — Least Significant Bit. It is the lowest order bit of a binary code.

LINE REGULATION — The ability of a voltage regulator to maintain a certain output voltage as the input to the regulator is varied. The error is typically expressed as a percent of the nominal output voltage.

LOAD REGULATION — The ability of a voltage regulator to maintain a certain output voltage as the load current is varied. The error is typically expressed as a percent of the nominal output voltage.

MONOTONICITY — The characteristic of the transfer function whereby increasing the input code (of a DAC), or the input signal (of an A/D), results in the output never decreasing. Nonmonotonicity occurs if the differential nonlinearity exceeds ±1 LSB.

MSB — Most Significant Bit. It is the highest order bit of a binary code.

NATURAL BINARY CODE — A binary code defined by:

$$N = A_n 2^n + \ldots + A_3 2^3 + A_2 2^2 + A_1 2^1 + A_0 2^0$$

where each "A" coefficient has a value of 1 or 0. Typically, all zeros correspond to a zero input voltage of an A/D, and all ones corresponds to the most positive input voltage.

NYQUIST THEORY — See Sampling Theorem.

OFFSET BINARY CODE — Applicable only to bipolar input (or output) data converters, it is the same as Natural Binary code, except that all zeros correspond to the most negative output signal (of a D/A), while all ones correspond to the most positive output.

OUTPUT COMPLIANCE — The maximum voltage range to which the DAC outputs can be subjected, and still meet all specifications.

POWER SUPPLY REJECTION RATIO — The ability of a device to reject noise and/or ripple on the power supply pins from appearing at the outputs. An AC measurement, this parameter is usually expressed in dB rejection.

POWER SUPPLY SENSITIVITY — The change in a data converter's performance with changes in the power supply voltage(s). This parameter is usually expressed in percent of full scale versus ΔV.

PROPAGATION DELAY — For a DAC, the time from when the clock input crosses its threshold to when the DAC output(s) changes.

QUANTITIZATION ERROR — Also known as *digitization error* or uncertainty. It is the inherent error involved in digitizing an analog signal due to the finite number of steps at the digital output versus the infinite number of values at the analog input. This error is a minimum of ±1/2 LSB.

RESOLUTION — The smallest change which can be discerned by an A/D converter, or produced by a DAC. It is usually expressed as the number of bits (n), where the converter has 2^n possible states.

SAMPLING THEOREM — Also known as the *Nyquist Theorem.* It states that the sampling frequency of an A/D must be no less than 2x the highest frequency (of interest) of the analog signal to be digitized in order to preserve the information of that analog signal.

SETTLING TIME — For a DAC, the time required for the output to change (and settle in) from an initial ±1/2 LSB error band to the final ±1/2 LSB error band.

TTL — Transistor-transistor logic.

TWO'S COMPLEMENT CODE — A binary code applicable to bipolar operation, in which the positive and negative codes of the same analog magnitude sum to all zeros, plus a carry. It is the same as *offset binary code,* with the MSB inverted.

UNIPOLAR INPUT — A mode of operation whereby the analog input range (of an A/D), or output range (of a D/A), includes values of a single polarity. Examples are: 0 to +10 V, 0 to −5.0 V, +2.0 V to +8.0 V, etc.

CMOS 80MHz Monolithic 256 × 24(18) Color Palette RAM-DAC

ADV478/ADV471

FEATURES
Personal System/2* Compatible
80MHz Pipelined Operation
Triple 8-Bit (6-Bit) D/A Converters
256 × 24(18) Color Palette RAM
15 × 24(18) Overlay Registers
RS-343A/RS-170 Compatible Outputs
Sync on All Three Channels
Programmable Pedestal (0 or 7.5 IRE)
External Voltage or Current Reference
Standard MPU Interface
+5V CMOS Monolithic Construction
44-Pin PLCC Package
Power Dissipation: 800mW

APPLICATIONS
High Resolution Color Graphics
CAE/CAD/CAM Applications
Image Processing
Instrumentation
Desktop Publishing

AVAILABLE CLOCK RATES
80MHz
50MHz
35MHz

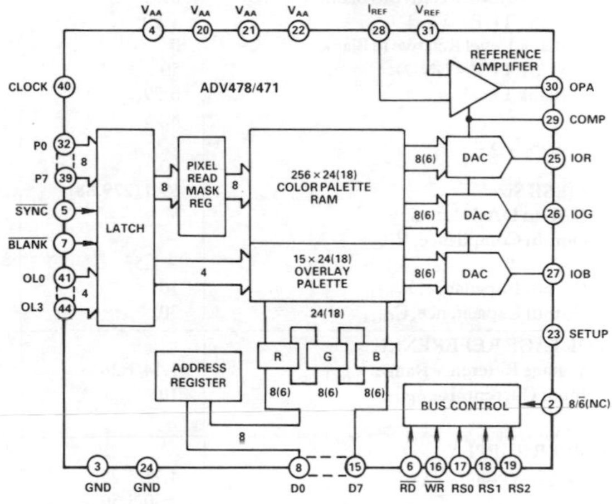

ADV478/ADV471 Functional Block Diagram

GENERAL DESCRIPTION

The ADV478 and ADV471 are pin compatible and software compatible RAM-DACs designed specifically for Personal System/2 compatible color graphics.

The ADV478 has a 256 × 24 color lookup table with triple 8-bit video D/A converters. It may be configured for either 6 bits or 8 bits per color operation. The ADV471 has a 256 × 18 color lookup table with triple 6-bit video D/A converters.

Options on both parts include a programmable pedestal (0 or 7.5 IRE) and use of an external voltage or current reference. Fifteen overlay registers provide for overlaying cursors, grids, menus, EGA emulation, etc. Also supported is a pixel read mask register and sync generation on all three channels.

The ADV478 and ADV471 generate RS-343A compatible video signals into a doubly terminated 75Ω load, and RS-170 compatible video signals into a singly terminated 75Ω load, without requiring external buffering. Differential and integral linearity errors are guaranteed to be a maximum of ± 1LSB for the ADV478 and ± 1/4LSB for the ADV471 over the full temperature range.

***Personal System/2 is a trademark of International Business Machines Corp.**

One Technology Way; P. O. Box 9106; Norwood, MA 02062-9106 U.S.A.
Tel: 617/329-4700 **Twx: 710/394-6577**
Telex: 924491 **Cables: ANALOG NORWOODMASS**

SPECIFICATIONS (V_{AA}[1] = +5V, SETUP = 8/$\overline{6}$ = V_{AA}, V_{REF} = +1.235V. R_{SET} = 148Ω (ADV478), R_{SET} = 140Ω (ADV471). All Specifications T_{min} to T_{max}[2] unless otherwise noted.)

Parameter	All Versions	Units	Test Conditions/Comments
STATIC PERFORMANCE			
Resolution (Each DAC)[3]	8 (6)	Bits	
Accuracy (Each DAC)[3]			
Integral Nonlinearity	±1 (1/4)	LSB max	
Differential Nonlinearity	±1 (1/4)	LSB max	Guaranteed Monotonic
Gray Scale Error	±5	% Gray Scale max	
Coding	Binary		
DIGITAL INPUTS			
Input High Voltage, V_{INH}	2	V min	
Input Low Voltage, V_{INL}	0.8	V max	
Input Current, I_{IN}	±1	μA max	V_{IN} = 0.4V or 2.4V
Input Capacitance, C_{IN}	7	pF max	
DIGITAL OUTPUTS			
Output High Voltage, V_{OH}	2.4	V min	I_{SOURCE} = 400μA
Output Low Voltage, V_{OL}	0.4	V max	I_{SINK} = 3.2mA
Floating-State Leakage Current	50	μA max	
Floating-State Output Capacitance	7	pF max	
ANALOG OUTPUTS			
Gray Scale Current Range	20	mA max	
Output Current			
White Level Relative to Blank	17.69	mA min	Typically 19.05mA
	20.40	mA max	
White Level Relative to Black	16.74	mA min	Typically 17.62mA
	18.50	mA max	
Black Level Relative to Blank	0.95	mA min	Typically 1.44mA
(SETUP = V_{AA})	1.90	mA max	
Black Level Relative to Blank	0	μA min	Typically 5μA
(SETUP = GND)	50	μA max	
Blank Level	6.29	mA min	Typically 7.62mA
	8.96	mA max	
Sync Level	0	μA min	Typically 5μA
	50	μA max	
LSB Size[3]	69.1 (279.68)	μA typ	8/$\overline{6}$ = Logical 1 for ADV478
DAC to DAC Matching	5	% max	Typically 2%
Output Compliance, V_{OC}	−1	V min	
	+1.5	V max	
Output Impedance, R_{OUT}	10	kΩ typ	
Output Capacitance, C_{OUT}	30	pF max	I_{OUT} = 0mA
VOLTAGE REFERENCE			
Voltage Reference Range, V_{REF}	1.14/1.26	V min/V max	
Input Current, I_{VREF}	10	μA typ	Tested in Voltage Reference Configuration with V_{REF} = 1.235V
POWER SUPPLY			
Supply Voltage, V_{AA}	4.75/5.25	V min/V max	80MHz Parts
	4.50/5.50	V min/V max	50MHz and 35MHz Parts
Supply Current, I_{AA}	200	mA max	Typically 160mA
Power Supply Rejection Ratio	0.5	%/% max	f = 1kHz, COMP = 0.1μF
Power Dissipation	1000	mW max	Typically 800mW, V_{AA} = 5V
DYNAMIC PERFORMANCE			
Clock and Data Feedthrough[4,5]	−30	dB typ	
Glitch Impulse[4,5]	75	pV secs typ	
DAC to DAC Crosstalk[6]	−23	dB typ	

NOTES
[1] ±5% for 80MHz parts; ±10% for 50MHz and 35MHz parts.
[2] Temperature Range (T_{min} to T_{max}); 0 to +70°C.
[3] Numbers in parentheses indicate ADV471 parameter value.
[4] Clock and data feedthrough is a function of the amount of overshoot and undershoot on the digital inputs. For this test, the digital inputs have a 1kΩ resistor to ground and are driven by 74HC logic. Glitch impulse includes clock and data feedthrough, −3dB test bandwidth = 2 × clock rate.
[5] TTL input values are 0 to 3 volts, with input rise/fall times ≤3ns, measured between the 10% and 90% points. Timing reference points at 50% for inputs and outputs. Analog output load ≤10pF, D0 – D7 output load ≤50pF. See timing notes in Figure 2.
[6] DAC to DAC crosstalk is measured by holding one DAC high while the other two are making low to high and high to low transitions.

Specifications subject to change without notice.

TIMING CHARACTERISTICS[1] (V_{AA}[2] = +5V, SETUP = 8/$\overline{6}$ = V_{AA}, V_{REF} = 1.235V. R_{SET} = 148Ω (ADV478), R_{SET} = 140Ω (ADV471). All Specifications T_{min} to T_{max}[3].)

Parameter	KP80 Version	KP50 Version	KP35 Version	Units	Conditions/Comments
f_{max}	80	50	35	MHz	Clock Rate
t_1	10	10	15	ns min	RS0 – RS2 Setup Time
t_2	10	10	15	ns min	RS0 – RS2 Hold Time
t_3	5	5	5	ns min	$\overline{RD}$ Asserted to Data Bus Driven
t_4	40	40	40	ns max	$\overline{RD}$ Asserted to Data Valid
t_5	20	20	20	ns max	$\overline{RD}$ Negated to Data Bus 3-Stated
t_6	10	10	15	ns min	Write Data Setup Time
t_7	10	10	15	ns min	Write Data Hold Time
t_8	50	50	50	ns min	$\overline{RD}$, $\overline{WR}$ Pulse Width Low
t_9	$4 \times t_{12}$	$4 \times t_{12}$	$4 \times t_{12}$	ns min	$\overline{RD}$, $\overline{WR}$ Pulse Width High
t_{10}	3	3	4	ns min	Pixel and Control Setup Time
t_{11}	3	3	4	ns min	Pixel and Control Hold Time
t_{12}	12.5	20	28	ns min	Clock Cycle Time
t_{13}	4	6	7	ns min	Clock Pulse Width High Time
t_{14}	4	6	9	ns min	Clock Pulse Width Low Time
t_{15}	30	30	30	ns max	Analog Output Delay
t_{16}	3	3	3	ns typ	Analog Output Rise/Fall Time
t_{17}[4]	13	20	28	ns typ	Analog Output Settling Time
t_{18}	2	2	2	ns max	Analog Output Skew
t_{PD}	$4 \times t_{12}$	$4 \times t_{12}$	$4 \times t_{12}$	ns min	Pipeline Delay

NOTES

[1]TTL input values are 0 to 3 volts, with input rise/fall times ≤3ns, measured between the 10% and 90% points. Timing reference points at 50% for inputs and outputs. Analog output load ≤10pF, D0 – D7 output load ≤50pF. See timing notes in Figure 2.

[2]±5% for 80MHz parts; ±10% for 50MHz and 35MHz parts.

[3]Temperature Range (T_{min} to T_{max}); 0 to +70°C.

[4]Settling time does not include clock and data feedthrough. For this test, the digital inputs have a 1kΩ resistor to ground and are driven by 74HC logic.

Specifications subject to change without notice

TIMING DIAGRAMS

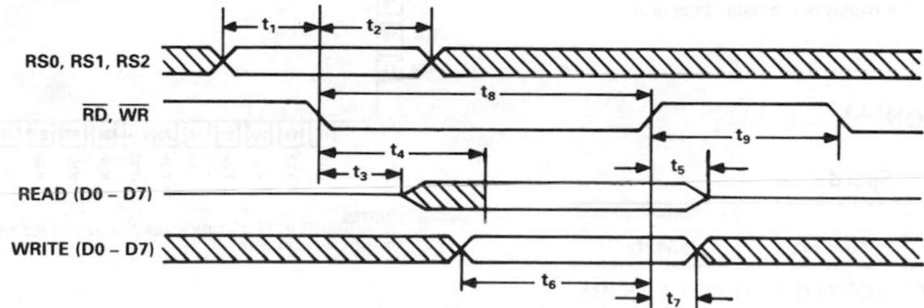

Figure 1. MPU Read/Write Timing

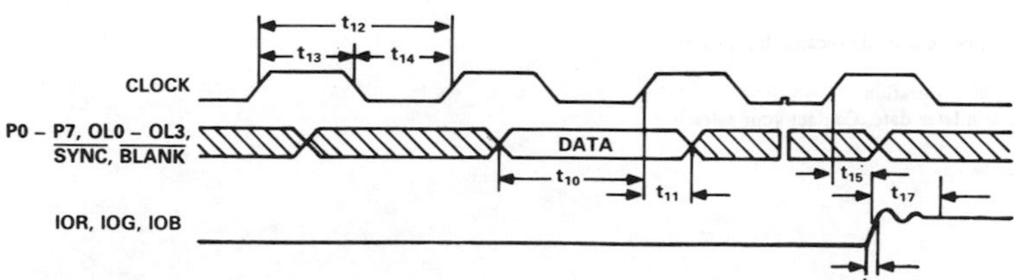

NOTES
1. OUTPUT DELAY (t_{15}) MEASURED FROM THE 50% POINT OF THE RISING EDGE OF CLOCK TO THE 50% POINT OF FULL SCALE TRANSISTION.
2. SETTLING TIME (t_{17}) MEASURED FROM THE 50% POINT OF FULL SCALE TRANSITION TO THE OUTPUT REMAINING WITHIN ±1LSB (ADV478) OR ±1/4LSB (ADV471).
3. OUTPUT RISE/FALL TIME (t_{16}) MEASURED BETWEEN THE 10% AND 90% POINTS OF FULL SCALE TRANSITION.

Figure 2. Video Input/Output Timing

RECOMMENDED OPERATING CONDITIONS

Parameter	Symbol	Min	Typ	Max	Units
Power Supply	V_{AA}				
80MHz Parts		4.75	5.00	5.25	Volts
50, 35MHz Parts		4.5	5.00	5.5	Volts
Ambient Operating Temperature	T_A	0		+70	°C
Output Load	R_L		37.5		Ω
Voltage Reference Configuration					
Reference Voltage	V_{REF}	1.14	1.235	1.26	Volts
Current Reference Configuration					
Reference Current	I_{REF}	−3		−10	mA

CAUTION:

ESD (Electro-Static Discharge) sensitive device. The digital control inputs are diode protected; however, permanent damage may occur on unconnected devices subject to high energy electrostatic fields. Unused devices must be stored in conductive foam or shunts. The protective foam should be discharged to the destination socket before devices are removed.

ABSOLUTE MAXIMUM RATINGS*

V_{AA} to GND . . . . +7V
Voltage on Any Digital Pin . . . GND −0.5V to V_{AA} +0.5V
Ambient Operating Temperature (T_A) . . . −55°C to +125°C
Storage Temperature (T_S) . . . . −65°C to +150°C
Junction Temperature (T_J) . . . . +175°C
Vapor Phase Soldering (2 minutes) . . . . TBD
IOR, IOB, IOG to GND[1] . . . . 0V to V_{AA}

NOTES
*Stresses above those listed under "Absolute Maximum Ratings" may cause permanent damage to the device. This is a stress rating only and functional operation of the device at these or any other conditions above those listed in the operational sections of this specification is not implied. Exposure to absolute maximum rating conditions for extended periods may affect device reliability.
[1]Analog output short circuit to any power supply or common can be of an indefinite duration.

ORDERING INFORMATION[1,2,3]

Color Palette RAM	Speed 80MHz	50MHz	35MHz
256 × 18	ADV471KP80	ADV471KP50	ADV471KP35
256 × 24	ADV478KP80	ADV478KP50	ADV478KP35

NOTES
[1]All devices are packaged in a 44-pin plastic leaded (J-lead) chip carrier, PLCC.
[2]All devices are specified for 0 to +70°C operation.
[3]A 66MHz version will be available at a later date. Contact your sales office for further information.

PLCC PIN CONFIGURATION

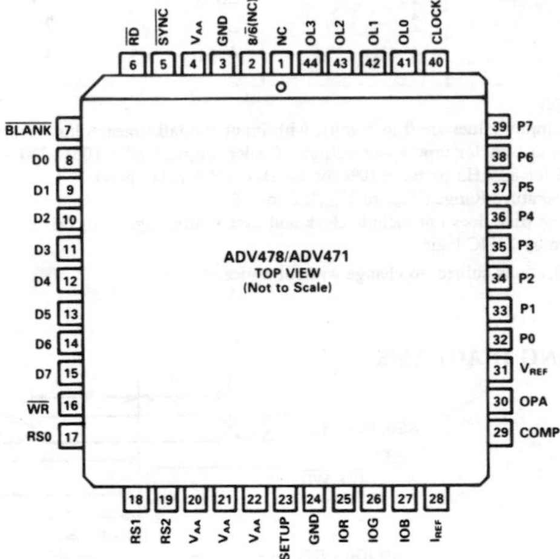

NOTES
1. NUMBERS IN PARENTHESIS INDICATE PIN NAMES FOR THE ADV471.
2. NC = NO CONNECT

PIN FUNCTION DESCRIPTION

Pin Mnemonic	Function
$\overline{BLANK}$	Composite blank control input (TTL compatible). A logic zero drives the analog outputs to the blanking level as illustrated in Tables IV and V. It is latched on the rising edge of CLOCK. When $\overline{BLANK}$ is a logical zero, the pixel and overlay inputs are ignored
SETUP	Setup control input. Used to specify either a 0 IRE (SETUP = GND) or 7.5 IRE (SETUP = V_{AA}) blanking pedestal.
$\overline{SYNC}$	Composite sync control input (TTL compatible). A logical zero on this input switches off a 40 IRE current source on the analog outputs (see Figures 3 and 4). $\overline{SYNC}$ does not override any other control or data input, as shown in Tables IV and V; therefore, it should be asserted only during the blanking interval. It is latched on the rising edge of CLOCK.
CLOCK	Clock input (TTL compatible). The rising edge of CLOCK latches the P0 – P7, OL0 – OL3, $\overline{SYNC}$, and $\overline{BLANK}$ inputs. It is typically the pixel clock rate of the video system. It is recommended that CLOCK be driven by a dedicated TTL buffer.
P0 – P7	Pixel select inputs (TTL compatible). These inputs specify, on a pixel basis, which one of the 256 entries in the color palette RAM is to be used to provide color information. They are latched on the rising edge of CLOCK. P0 is the LSB. Unused inputs should be connected to GND.
OL0 – OL3	Overlay select inputs (TTL compatible). These inputs specify which palette is to be used to provide color information, as illustrated in Table III. When accessing the overlay palette, the P0 – P7 inputs are ignored. They are latched on the rising edge of CLOCK. OL0 is the LSB. Unused inputs should be connected to GND.
IOR, IOG, IOB	Red, green, and blue current outputs. These high impedance current sources are capable of directly driving a doubly terminated 75Ω coaxial cable (Figures 5 and 6).
I_{REF}	Full-scale adjust control. Note that the IRE relationships in Figures 3 and 4 are maintained, regardless of the full-scale output current.

When using an external voltage reference (Figure 5), a resistor (R_{SET}) connected between this pin and GND controls the magnitude of the full-scale video signal. The relationship between R_{SET} and the full-scale output current on each output is:

$$R_{SET}\ (\Omega) = K * 1{,}000 * V_{REF}\ (V)/I_{OUT}\ (mA)$$

K is defined in the table below, along with corresponding R_{SET} values for doubly terminated 75Ω loads.

When using an external current reference (Figure 6), the relationship between I_{REF} and the full-scale output current on each output is:

$$I_{REF}\ (mA) = I_{OUT}\ (mA)/K$$

Mode	Pedestal	K	R_{SET} (Ω)
6-Bit	7.5 IRE	3.025	140
8-Bit	7.5 IRE	3.200	148
6-Bit	0 IRE	3.000	139
8-Bit	0 IRE	3.175	147

Pin Mnemonic	Function
COMP	Compensation pin. If an external voltage reference is used (Figure 5), this pin should be connected to OPA. If an external current reference is used, this pin should be connected to I_{REF}. A 0.1μF ceramic capacitor must always be used to bypass this pin to V_{AA}.
V_{REF}	Voltage reference input. If an external voltage reference is used (Figure 5), it must supply this input with a 1.2V (typical) reference. If an external current reference is used (Figure 6), this pin should be left floating, except for the bypass capacitor. A 0.1μF ceramic capacitor must always be used to decouple this input to V_{AA} as shown in Figures 5 and 6.
OPA	Reference amplifier output. If an external voltage reference is used (Figure 5), this pin must be connected to COMP. When using an external current reference (Figure 6), this pin should be left floating.
V_{AA}	Analog power. All V_{AA} pins must be connected to the Analog Power Plane.
GND	Analog ground. All GND pins must be connected to the Ground Plane.
$\overline{WR}$	Write control input (TTL compatible). D0 – D7 data is latched on the rising edge of $\overline{WR}$, and RS0 – RS2 are latched on the falling edge of $\overline{WR}$ during MPU write operations. See Figure 1.

PIN FUNCTION DESCRIPTION (Continued)

Pin Mnemonic	Function
$\overline{RD}$	Read control input (TTL compatible). To read data from the device, $\overline{RD}$ must be a logical zero. RS0 – RS2 are latched on the falling edge of $\overline{RD}$ during MPU read operations.
RS0, RS1, RS2	Register select inputs (TTL compatible). RS0 – RS2 specify the type of read or write operation being performed as illustrated in Tables I and II.
D0 – D7	Data bus (TTL compatible). Data is transferred into and out of the device over this 8-bit bidirectional data bus. D0 is the least significant bit.
$8/\overline{6}$	8-bit/6-bit select input (TTL compatible). This control input specifies whether the MPU is reading and writing 8-bits (logical one) or 6-bits (logical zero) of color information each cycle. For 8-bit operation, D7 is the most significant data bit during color read/write cycles. For 6-bit operation, D5 is the most significant data bit during color read/write cycles (D6 and D7 are ignored during color write cycles and are logical zero during color read cycles). This control input is implemented only on the ADV478.

TERMINOLOGY

Blanking Level
The level seperating the SYNC portion from the video portion of the waveform. Usually referred to as the front porch or back porch. At 0 IRE units, it is the level which will shut off the picture tube, resulting in the blackest possible picture.

Color Video (RGB)
This usually refers to the technique of combining the three primary colors of red, green and blue to produce color pictures within the usual spectrum. In RGB monitors, three DACs would be required, one for each color.

Composite SYNC Signal (SYNC)
The position of the composite video signal which synchronizes the scanning process.

Composite Video Signal
The video signal with or without setup, plus the composite SYNC signal.

Gray Scale
The discrete levels of video signal between reference black and reference white levels. An 8-bit DAC contains 256 different levels while a 6-bit DAC contains 64.

Raster Scan
The most basic method of sweeping a CRT one line at a time to generate and display images.

Reference Black Level
The maximum negative polarity amplitude of the video signal.

Reference White Level
The maximum positive polarity amplitude of the video signal.

Setup
The difference between the reference black level and the blanking level.

SYNC Level
The peak level of the composite SYNC signal.

Video Signal
That portion of the composite video signal which varies in gray scale levels between reference white and reference black. Also referred to as the picture signal, this is the portion which may be visually observed.

CIRCUIT DESCRIPTION

MPU Interface

As illustrated in the functional block diagram, the ADV478 and ADV471 support a standard MPU bus interface, allowing the MPU direct access to the color palette RAM and overlay color registers.

The RS0 – RS2 select inputs specify whether the MPU is accessing the address register, color palette RAM, overlay registers or read mask register, as shown in Table I. The 8-bit address register is used to address the color palette RAM and overlay registers, eliminating the requirement for external address multiplexers.

To write color data, the MPU writes to the address register (selecting RAM or overlay write mode) with the address of the color palette RAM location or overlay register to be modified. The MPU performs three successive write cycles (8 or 6 bits each of red, green and blue), using RS0 – RS2 to select either the color palette RAM or overlay registers. During the blue write cycle, the three bytes of color information are concatenated into a 24-bit word (18-bit word for the ADV471) and written to the location specified by the address register. The address register then increments to the next location which the MPU may modify by simply writing another sequence of red, green and blue data.

RS2	RS1	RS0	Addressed by MPU
0	0	0	Address Register (RAM Write Mode)
0	1	1	Address Register (RAM Read Mode)
0	0	1	Color Palette RAM
0	1	0	Pixel Read Mask Register
1	0	0	Address Register (Overlay Write Mode)
1	1	1	Address Register (Overlay Read Mode)
1	0	1	Overlay Registers
1	1	0	Reserved

Table I. Control Input Truth Table

To read color data, the MPU loads the address register (selecting RAM or overlay read mode) with the address of the color palette RAM location or overlay register to be read. The MPU performs three successive read cycles (8 or 6 bits each of red, green and blue), using RS0 – RS2 to select either the color palette RAM or overlay registers. Following the blue read cycle, the address register increments to the next location which the MPU may read by simply reading another sequence of red, green and blue data.

When accessing the color palette RAM, the address register resets to 00H following a blue read or write cycle to RAM location FFH. When accessing the overlay color registers, the address register increments following a blue read or write cycle. However, while accessing the overlay color registers, the four most significant bits of the address register (ADDR4 – 7) are ignored.

The MPU interface operates asynchronously to the pixel clock. Data transfers between the color palette RAM/overlay registers and the color registers (R, G and B in the block diagram) are synchronized by internal logic and occur in the period between MPU accesses. As only one pixel clock cycle is required to complete the transfer, the color palette RAM and overlay registers may be accessed at any time with no noticeable disturbance on the display screen.

To keep track of the red, green and blue read/write cycles, the address register has two additional bits (ADDRa, ADDRb) that count modulo three, as shown in Table II. They are reset to zero when the MPU writes to the address register and are not reset to zero when the MPU reads the address register. The MPU does not have access to these bits. The other eight bits of the address register, incremented following a blue read or write cycle (ADDR0 – 7), are accessible to the MPU and are used to address color palette RAM locations and overlay registers, as shown in Table II. ADDR0 is the LSB when the MPU is accessing the RAM or overlay registers. The MPU may read the address register at any time without modifying its contents or the existing read/write mode.

Figure 1 illustrates the MPU read/write timing.

	Value	RS2	RS1	RS0	Addressed By MPU
ADDRa,b (Counts Modulo 3)	00				Red Value
	01				Green Value
	10				Blue Value
ADDR0 – 7 (Counts Binary)	00H – FFH	0	0	1	Color Palette RAM
	XXXX 0000	1	0	1	Reserved
	XXXX 0001	1	0	1	Overlay Color 1
	XXXX 0010	1	0	1	Overlay Color 2
	•	•	•	•	•
	•	•	•	•	•
	XXXX 1111	1	0	1	Overlay Color 15

Table II. Address Register (ADDR) Operation

ADV478 Data Bus Interface

On the ADV478, the $8/\overline{6}$ control input is used to specify whether the MPU is reading and writing 8 bits ($8/\overline{6}$ = logical one) or 6 bits ($8/\overline{6}$ = logical zero) of color information each cycle.

For 8-bit operation, D0 is the LSB and D7 is the MSB of color data.

For 6-bit operation (and also when using the ADV471), color data is contained on the lower six bits of the data bus, with D0 being the LSB and D5 the MSB of color data. When writing color data, D6 and D7 are ignored. During color read cycles, D6 and D7 will be a logical zero.

ADV471 Data Bus Interface

Color data is contained on the lower six bits of the data bus, with D0 being the LSB and D5 the MSB of color data. When writing color data, D6 and D7 are ignored. During color read cycles, D6 and D7 will be a logical zero.

Frame Buffer Interface

The P0 – P7 and OL0 – OL3 inputs are used to address the color palette RAM and overlay registers, as shown in Table III.

OL0 – OL3	P0 – P7	Addressed by Frame Buffer
0H	00H	Color Palette RAM Location 00H
0H	01H	Color Palette RAM Location 01H
•	•	•
•	•	•
0H	FFH	Color Palette RAM Location FFH
1H	XXH	Overlay Color 1
2H	XXH	Overlay Color 2
•	•	•
•	•	•
FH	XXH	Overlay Color 15

Table III. Pixel and Overlay Control Truth Table (Pixel Read Mask Register = FFH)

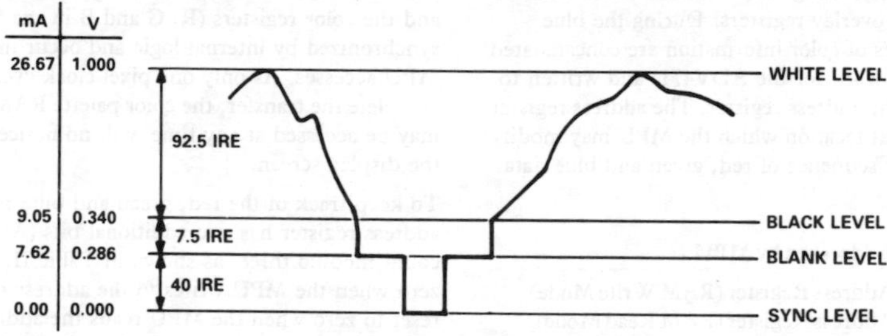

NOTES
1. CONNECTED WITH A 75Ω DOUBLY TERMINATED LOAD, SETUP = V_{AA}.
2. EXTERNAL VOLTAGE OR CURRENT REFERENCE ADJUSTED FOR 26.67mA FULL-SCALE OUTPUT.
3. RS-343A LEVELS AND TOLERANCES ASSUMED ON ALL LEVELS.

Figure 3. Composite Video Output Waveform (SETUP = V_{AA})

Description	I_{OUT} (mA)[1]	$\overline{SYNC}$	$\overline{BLANK}$	DAC Input Data
WHITE LEVEL	26.67	1	1	FFH
DATA	data + 9.05	1	1	data
DATA-SYNC	data + 1.44	0	1	data
BLACK LEVEL	9.05	1	1	00H
BLACK-SYNC	1.44	0	1	00H
BLANK LEVEL	7.62	1	0	xxH
SYNC LEVEL	0	0	0	xxH

NOTES
[1]Typical with full-scale IOG = 26.67mA, SETUP = V_{AA}.
External voltage or current reference adjusted for 26.67mA full-scale output.

Table IV. Video Output Truth Table (SETUP = V_{AA})

The contents of the pixel read mask register, which may be accessed by the MPU at any time, are bit-wise logically ANDed with the P0 – P7 inputs. Bit D0 of the pixel read mask register corresponds to pixel input P0. The addressed location provides 24 bits (18 bits for the ADV471) of color information to the three D/A converters.

The $\overline{\text{SYNC}}$ and $\overline{\text{BLANK}}$ inputs, also latched on the rising edge of CLOCK to maintain synchronization with the color data, add appropriately weighted currents to the analog outputs, producing the specific output levels required for video applications, as illustrated in Figures 3 and 4. Tables IV and V detail how the $\overline{\text{SYNC}}$ and $\overline{\text{BLANK}}$ inputs modify the output levels.

The SETUP input is used to specify whether a 0 IRE (SETUP = GND) or 7.5 IRE (SETUP = V_{AA}) blanking pedestal is to be used.

The analog outputs of the ADV478 and ADV471 are capable of directly driving a 37.5Ω load, such as a doubly terminated 75Ω coaxial cable.

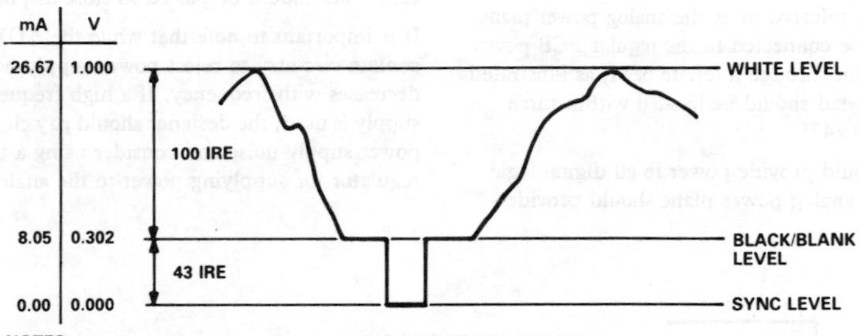

NOTES
1. CONNECTED WITH A 75Ω DOUBLY TERMINATED LOAD, SETUP = GND.
2. EXTERNAL VOLTAGE OR CURRENT REFERENCE ADJUSTED FOR 26.67mA FULL-SCALE OUTPUT.
3. RS-343A LEVELS AND TOLERANCES ASSUMED ON ALL LEVELS.

Figure 4. Composite Video Output Waveform (SETUP = GND)

Description	I_{OUT} (mA)[1]	$\overline{\text{SYNC}}$	$\overline{\text{BLANK}}$	DAC Input Data
WHITE LEVEL	26.67	1	1	FFH
DATA	data + 8.05	1	1	data
DATA-SYNC	data	0	1	data
BLACK LEVEL	8.05	1	1	00H
BLACK-SYNC	0	0	1	00H
BLANK LEVEL	8.05	1	0	xxH
SYNC LEVEL	0	0	0	xxH

NOTES
[1]Typical with full-scale IOG = 26.67mA, SETUP = GND
External voltage or current reference adjusted for 26.67mA full-scale output.

Table V. Video Output Truth Table (SETUP = GND)

PC BOARD LAYOUT CONSIDERATIONS

PC Board Considerations

The layout should be optimized for lowest noise on the ADV478/ADV471 power and ground lines by shielding the digital inputs and providing good decoupling. The lead length between groups of V_{AA} and GND pins should by minimized so as to minimize inductive ringing.

Ground Planes

The ground plane should encompass all ADV478/ADV471 ground pins, current/voltage reference circuitry, power supply bypass circuitry for the ADV478/ADV471, the analog output traces and all the digital signal traces leading up to the ADV478/ADV471.

Power Planes

The ADV478/ADV471 and any associated analog circuitry should have its own power plane, referred to as the analog power plane. This power plane should be connected to the regular PCB power plane (V_{CC}) at a single point through a ferrite bead, as illustrated in Figures 5 and 6. This bead should be located within three inches of the ADV478/ADV471.

The PCB power plane should provide power to all digital logic on the PC board, and the analog power plane should provide power to all ADV478/ADV471 power pins and current/voltage reference circuitry.

Plane-to-plane noise coupling can be reduced by ensuring that portions of the regular PCB power and ground planes do not overlay portions of the analog power plane, unless they can be arranged such that the plane-to-plane noise is common mode.

Supply Decoupling

For optimum performance, bypass capacitors should be installed using the shortest leads possible, consistent with reliable operation, to reduce the lead inductance.

Best performance is obtained with a 0.1μF ceramic capacitor decoupling each of the two groups of V_{AA} pins to GND. These capacitors should be placed as close as possible to the device.

It is important to note that while the ADV478 and ADV471 contain circuitry to reject power supply noise, this rejection decreases with frequency. If a high frequency switching power supply is used, the designer should pay close attention to reducing power supply noise and consider using a three terminal voltage regulator for supplying power to the analog power plane.

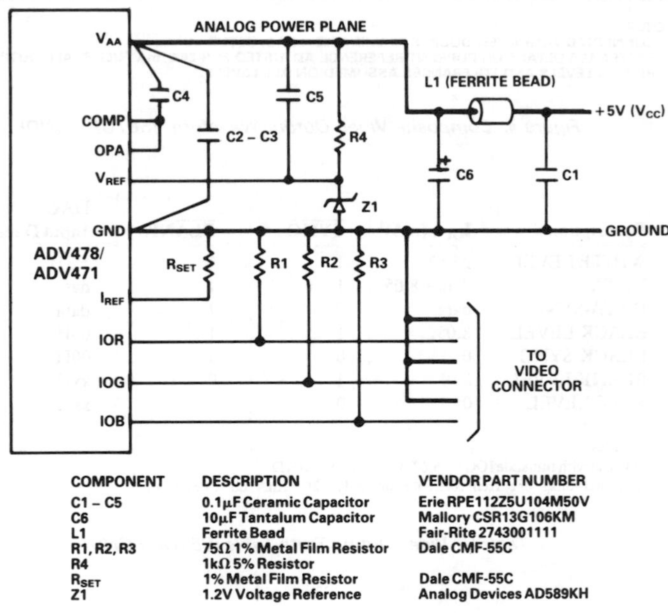

COMPONENT	DESCRIPTION	VENDOR PART NUMBER
C1 – C5	0.1μF Ceramic Capacitor	Erie RPE112Z5U104M50V
C6	10μF Tantalum Capacitor	Mallory CSR13G106KM
L1	Ferrite Bead	Fair-Rite 2743001111
R1, R2, R3	75Ω 1% Metal Film Resistor	Dale CMF-55C
R4	1kΩ 5% Resistor	
R_{SET}	1% Metal Film Resistor	Dale CMF-55C
Z1	1.2V Voltage Reference	Analog Devices AD589KH

Figure 5. Typical Connection Diagram and Component List (External Voltage Reference)

Digital Signal Interconnect
The digital inputs to the ADV478/ADV471 should be isolated as much as possible from the analog outputs and other analog circuitry. Also, these input signals should not overlay the analog power plane.

Due to the high clock rates involved, long clock lines to the ADV478/ADV471 should be avoided to reduce noise pickup.

Any active termination resistors for the digital inputs should be connected to the regular PCB power plane (V_{CC}), and not the analog power plane.

Analog Signal Interconnect
The ADV478/ADV471 should be located as close as possible to the output connectors to minimize noise pickup and reflections due to impedance mismatch.

The video output signals should overlay the ground plane, and not the analog power plane, to maximize the high frequency power supply rejection.

For maximum performance, the analog outputs should each have a 75Ω load resistor connected to GND. The connection between the current output and GND should be as close as possible to the ADV478/ADV471 to minimize reflections.

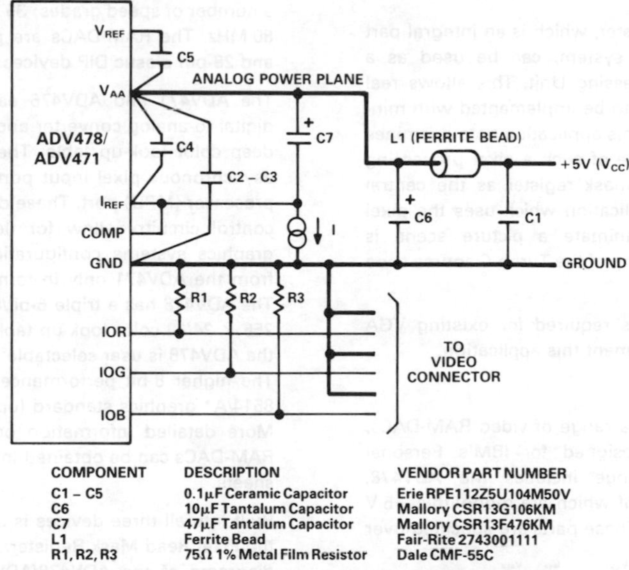

COMPONENT	DESCRIPTION	VENDOR PART NUMBER
C1 – C5	0.1μF Ceramic Capacitor	Erie RPE112Z5U104M50V
C6	10μF Tantalum Capacitor	Mallory CSR13G106KM
C7	47μF Tantalum Capacitor	Mallory CSR13F476KM
L1	Ferrite Bead	Fair-Rite 2743001111
R1, R2, R3	75Ω 1% Metal Film Resistor	Dale CMF-55C

Figure 6. Typical Connection Diagram and Component List (External Current Reference)

APPLICATION INFORMATION

External Voltage vs. Current Reference
The ADV478/ADV471 is designed to have excellent performance using either an external voltage or current reference. The voltage reference design (Figure 5) has the advantages of temperature compensation, simplicity, lower cost and provides excellent power supply rejection. The current reference design (Figure 6) requires more components to provide adequate power supply rejection and temperature compensation (two transistors, three resistors and additional capacitors).

RS-170 Video Generation
For generation of RS-170 compatible video, it is recommended that the DAC outputs be connected to a singly terminated 75Ω load. If the ADV478/ADV471 is not driving a large capacitive load, there will be negligible difference in video quality between doubly terminated 75Ω and singly terminated 75Ω loads.

If driving a large capacitive load (load RC> $1/(2\pi f_C)$), it is recommended that an output buffer (such as a MC1378 with an unloaded gain>2) be used to drive a doubly terminated 75Ω load.

APPLICATION NOTE

ONE TECHNOLOGY WAY • P.O. BOX 9106 • NORWOOD, MASSACHUSETTS 02062-9106 • 617/329-4700

Animation Using the Pixel Read Mask Register of the ADV47X Series of Video RAM-DACs

by Bill Slattery & Eamonn Gormley

INTRODUCTION

The Pixel Read Mask Register, which is an integral part of IBM's VGA* graphics system, can be used as a hardware-level Pixel Processing Unit. This allows real time motion or animation to be implemented with minimal software overhead. This application note examines the operation and structure of such a pixel processing unit with the pixel read mask register as the central controller. A practical application which uses the pixel read mask register to animate a picture scene is described. A complete listing of the Turbo-C source code is given in the appendix.

No additional hardware is required for existing VGA graphics systems to implement this application.

VIDEO RAM-DAC

Analog Devices produces a range of video RAM-DACs, which are specifically designed for IBM's Personal System/2* VGA. The range includes the ADV478, ADV471 and ADV476, all of which are monolithic +5 V CMOS video RAM-DACs. These parts are specified over a number of speed grades; 35 MHz, 50 MHz, 66 MHz and 80 MHz. The RAM-DACs are packaged as 44-pin PLCC and 28-pin plastic DIP devices.

The ADV471 and ADV476 each contain a triple 6-bit digital-to-analog converter and a 256 location by 18 bits deep color look-up table. The devices also include an asynchronous pixel input port and bidirectional microprocessor (MPU) port. These devices and the associated control circuitry allow for flexible interface to many graphics systems configurations. The ADV478 differs from the ADV471 only in terms of its color resolution. The ADV478 has a triple 6-bit/8-bit D/A converter with a 256 × 24/18 color look-up table. The color resolution of the ADV478 is user selectable between 6 bits and 8 bits. The higher 8-bit performance can be used with IBM's 8514/A* graphics standard (upgrade on standard VGA). More detailed information on these and other video RAM-DACs can be obtained in the relevant product data sheets.

Built into all three devices is an 8-bit register known as the Pixel Read Mask Register. Figures 1 and 2 are block diagrams of the ADV478/ADV471 and ADV476 which show the Pixel Read Mask Register.

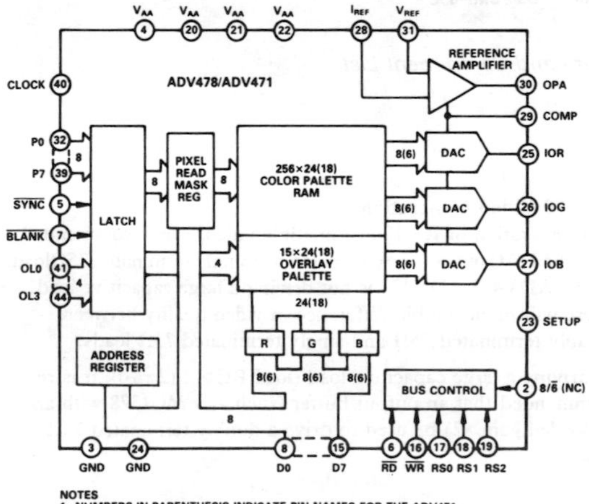

NOTES
1. NUMBERS IN PARENTHESIS INDICATE PIN NAMES FOR THE ADV471.
2. NC = NO CONNECT

Figure 1. ADV478/ADV471 Functional Block Diagram

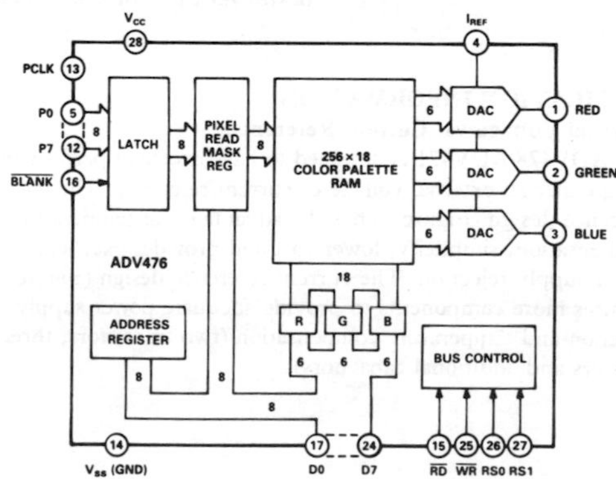

Figure 2. ADV476 Functional Block Diagram

*IBM, VGA, Personal System/2 and 8514/A are trademarks of International Business Machines Corp.

Some of the uses to which the Pixel Read Mask Register can be put include on-screen special effects such as real time animation, flashing objects and overlays.

PIXEL READ MASK REGISTER

The Pixel Read Mask Register is placed in the path of the pixel input stream of data as shown in Figure 3.

The input pixel data stream (P0–P7) is gated with the contents of the Pixel Read Mask Register. The operation is a bitwise logical ANDing of the pixel data. The contents of the Pixel Read Mask Register can be accessed and altered at any time by the MPU (D0–D7). Table I shows the relevant control signals. Under normal operating conditions, this register is loaded with all 1s, i.e., transparent mode.

In a VGA graphics system, the Pixel Read Mask Register is memory mapped and is accessible (read/write) by addressing memory location 36CH.

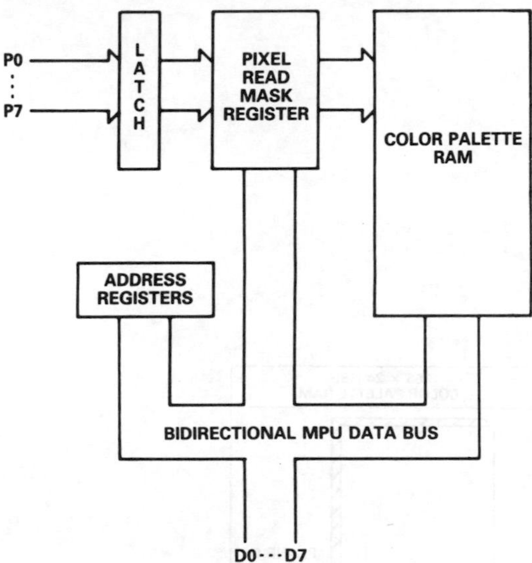

Figure 3. Video RAM-DAC Pixel & Data Ports Showing the Pixel Read Mask Register

RS2*	RS1	RS0	Addressed by MPU
0	0	0	Address Register (RAM Write Mode)
0	1	1	Address Register (RAM Read Mode)
0	0	1	Color Palette RAM
0	1	0	Pixel Read Mask Register

*RS2 is only present on ADV478/ADV471

Table I. Control Input Truth Table for Video RAM-DAC

Figure 4 shows the internal architecture of the pixel input port. The input word Pi, which corresponds to an on-screen pixel location, is ANDed with the contents of the Pixel Read Mask Register, Pm.

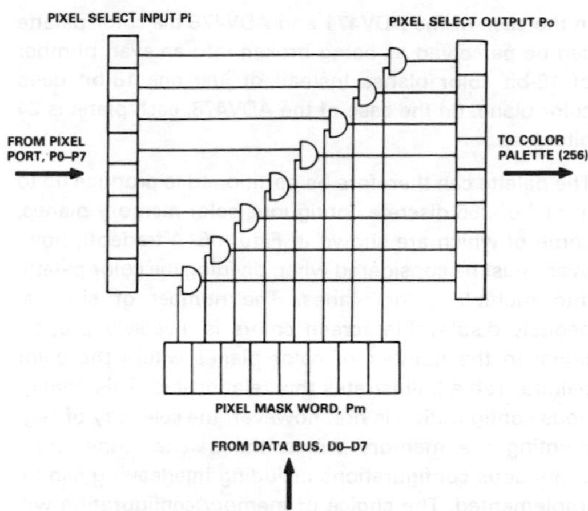

Figure 4. Internal Architecture of Pixel Input Port

The resulting output word, Po, determines which location in the color palette will be assigned to a particular on-screen pixel. Figure 5 shows the logical diagram for this masking operation.

Figure 5. Equivalent Logical Representation of Masking Operation

$$Po = Pi \cdot Pm \quad (1)$$

If Pm = 1, transparent mode, then

$$Po = Pi \quad (2)$$

The pixel stream of data, Po, which arrives at the color palette is a function of both the pixel input stream, Pi, from the Frame Buffer and the contents of the Pixel Read Mask Register, Pm. In the case of an animation application, which will be discussed later in this application note, the rate at which the pixel mask word is changed will determine the motion speed of the scene.

This pixel masking operation can be used to alter the displayed colors without changing the contents of either the video Frame Buffer or the Color Palette RAM. One interpretation of this operation is to consider the pixel input structure of the video RAM-DAC as an on board Pixel Processing Unit.

PIXEL PROCESSING UNIT

The Pixel Input Port (Pi), Pixel Read Mask Register (Pm) and Data Input Port (MPU) within the video RAM-DAC, are the hardware components of this Pixel Processing Unit (PPU). An associated software routine to control the operation is the final element in the complete PPU system. This interpretation enables the color palette to be configured as a multidimensional, paged memory address space, see Figure 6.

In the case of the ADV471 and ADV476 the color palette can be perceived as being broken into an even number of 18-bit color planes instead of just one 18-bit deep color plane. (In the case of the ADV478, each plane is 24 bits deep.)

The palette can therefore be partitioned to produce up to a total of 256 discrete contiguous color memory planes, some of which are shown in Figure 6. A tradeoff, however, must be considered when dividing the color palette into multiple color planes. The number of simultaneously displayable screen colors is inversely proportional to the number of color planes within the color palette. Table II illustrates this relationship. This contiguous configuration is not, however, the sole way of segmenting the memory within the palette. Other non-contiguous configurations including interleaving can be implemented. The choice of memory configuration will be determined by the particular application so as to make most efficient use of the available video memory (Image Frame Buffer and Color Palette RAM).

Number of Color Planes	Number of Simultaneously Displayable Colors
1	256
2	128
4	64
.	.
.	.
256	1

Table II. Simultaneously Displayable Screen Colors versus Number of Color Planes

To operate the PPU, two principal steps must be taken:

1. Load the image data in the correct paged configuration to both the frame buffer and color look-up table, e.g., four frame composite images to the frame buffer corresponding to four discrete planes of color to the palette RAM.
2. Generate the corresponding pixel mask words. These words are individually written to the Pixel Read Mask Register, Pm (at VGA memory location 36CH) and select which of the color planes is to be assigned to the incoming pixel data stream. In the case where four planes are implemented (see Figure 6), four pixel mask words are required.

The overall VGA System Block Diagram showing the PPU and an associated 8-page memory configuration of the color palette is shown in Figure 7.

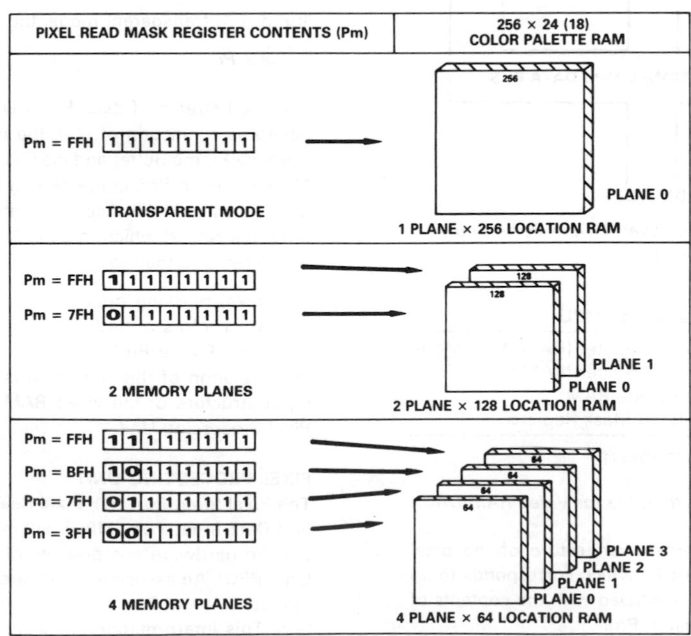

Figure 6. Some Color Palette RAM Configurations Showing Paged Memory Planes and Associated Pixel Read Mask Word (Pm)

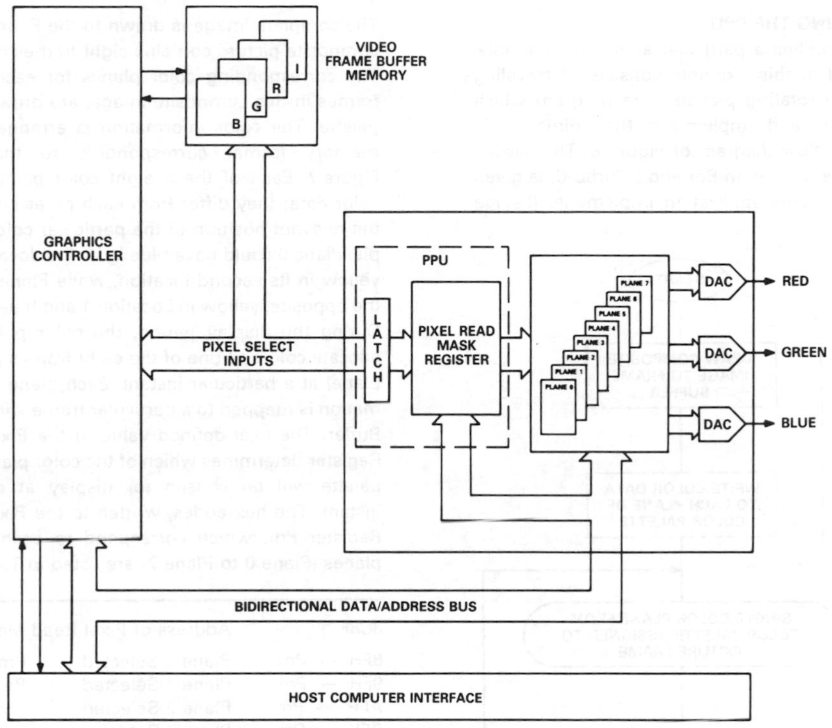

Figure 7. VGA System Block Diagram Showing Color Palette Broken into a Number of Color Planes

ANIMATION

Real time animation using the Pixel Processing Unit is based on the principle that rapidly changing the colors of a stationary object gives the illusion of motion. In other words, a number of similar images or frames, differing only by the relative position of the various colors, displayed in quick succession, can result in motion.

A simple example to explain the idea of animation is illustrated opposite. The animated image consists of three frames; each of the three frames is initially drawn as one composite picture (Frame Buffer image). The color palette contains three discrete, memory blocks or planes of color information, corresponding to three stages of animation. The animation effect in this example is "arm waving" of the cartoon character. By assigning the color planes one by one to the image in the Frame Buffer, the effect of animation can be perceived on the screen. The color plane assigned to the composite image is determined by the PPU which is controlled by the word in the Pixel Read Mask Register. Frame 1 is assigned Color Plane 0, this colors the down arm position in black, while the up and horizontal arm positions take on the background color. Frame number 2 is assigned Color Plane 1, this colors the horizontal arm position in black while the other two arm positions are assigned the back ground color.

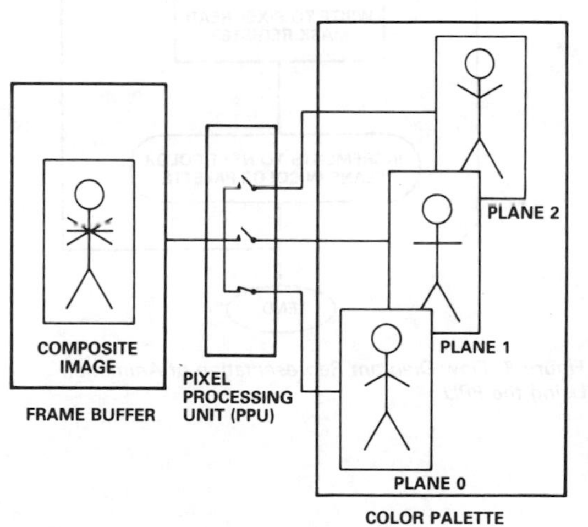

Finally, Frame 3 takes on Color Plane 2. This process is then repeated giving the illusion of motion. The rate at which each frame is selected determines the rapidity of the arm waving.

ANIMATION USING THE PPU

This section describes a particular animation example. The scene used in this example consists of traveling space ships and rotating planets. The program which draws the scene and implements the animation is described in the flow diagram of Figure 8. The associated source code, written in Borland's Turbo-C, is given in the Appendix. This application implements 8-stage animation.

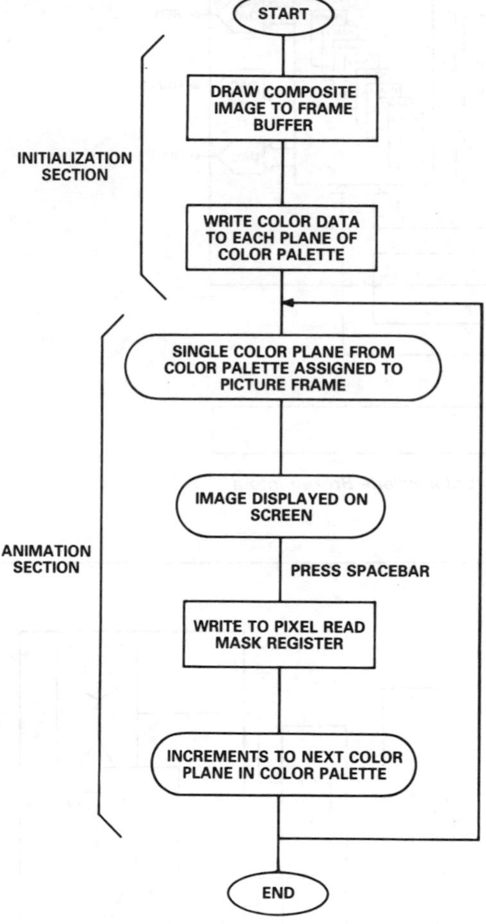

Figure 8. Flow Diagram Representation of Animation Using the PPU

The complete image is drawn to the Frame Buffer. This composite picture contains eight frames of information. The corresponding color planes for each of the eight frames in this composite image, are drawn to the color palette. The color information is arranged in a paged memory format corresponding to that shown in Figure 7. Each of these eight color planes has similar color data; they differ from each other only in terms of the relevant position of the particular colors. For example, Plane 0 could have blue in its first location and color yellow in its second location, while Plane 1 could have the opposite, yellow in Location 1 and blue in Location 2. During the display period, the color palette will only allocate colors to one of the eight frames (i.e., one color plane) at a particular instant. Each plane of color information is mapped to a particular frame within the Frame Buffer. The user-defined value of the Pixel Read Mask Register determines which of the color planes within the palette will be chosen for display at any particular instant. The hex codes, written to the Pixel Read Mask Register Pm, which correspond to each of the color planes (Plane 0 to Plane 7) are listed in Table III.

3C6F	: Address of Pixel Read Mask Register (Pm)	
8FH → Pm	: Plane 0 Selected	Pm = 1000 1111
9FH → Pm	: Plane 1 Selected	Pm = 1001 1111
AFH → Pm	: Plane 2 Selected	Pm = 1010 1111
BFH → Pm	: Plane 3 Selected	Pm = 1011 1111
CFH → Pm	: Plane 4 Selected	Pm = 1100 1111
DFH → Pm	: Plane 5 Selected	Pm = 1101 1111
EFH → Pm	: Plane 6 Selected	Pm = 1110 1111
FFH → Pm	: Plane 7 Selected	Pm = 1111 1111

Table III. Value Written to Pixel Read Mask Register and Associated Color Plane

Pressing the "Spacebar" increments the pixel read mask register corresponding to a jump of 16 locations in the color palette. As there are 16 colors in each color plane, a jump of 16 locations will select the corresponding color in the next highest plane. Continuously pressing the "Spacebar" cycles the incoming pixel stream of data through each of the eight color planes within the palette. This results in the apparent motion or animation of the image.

APPENDIX

C Program for ANIMATION EXAMPLE

Pixel Processing Using Video RAM-DACs "Rotating Planets & Spaceships"

```
#include <stdlib.h>  /* Turbo C include files */
#include <math.h>    /* these are available under most versions */
#include <dos.h>     /* of C for the IBM & compatibles */
#include <graphics.h>
void palette(int col,int red,int green,int blue);
void plot13(int x,int y,int col);                 /* function definitions */
void mode13();
void circle13(int x,int y,int r,double tilt);
void planets();
void stars();
void line13(int x1,int y1,int x2,int y2,int col);
void triangle();

main()
{
 int gd=0,gm=0,opt;
 union REGS reg;
 detectgraph(&gd,&gm);                     /* check for a VGA card */
 if (gd != 9){
  printf("This program cannot find a VGA card installed in this computer.\n");
  printf("A VGA card is necessary to run the tests.");
  exit(1); }

 planets();                                /* do demo */
 reg.h.ah = 0x00;
 reg.h.al = 0x03;
 int86(0x10,&reg,&reg);                    /* return to text mode when finished */
}

void mode13()                              /* set up mode hex 13 = decimal 19 */
{                                          /* this is a 256 color mode with */
 union REGS reg;                           /* 320 x 200 pixel resolution */
 reg.x.ax = 0x0013;
 int86(0x10,&reg,&reg);                    /* set mode 0x13 */
}

void palette(int col,int red,int green,int blue)
               /* assigns a physical color to a logical color */
{
  union REGS reg;
     reg.x.ax = 0x1010;
     reg.x.bx = col;
     reg.h.dh = red;
     reg.h.ch = green;
     reg.h.cl = blue;
     int86(0x10,&reg,&reg);      /* call bios routine to change palette */
}

void plot13(int x,int y,int col)   /* special plot routine for mode 0x13 */
{
     union REGS reg;
     if(x>=0 && y>=0 && x <320 && y<200){
        reg.x.dx = y;              /* set up registers */
        reg.x.cx = x;
        reg.h.ah = 0x0c;
        reg.h.al = col;
        int86(0x10,&reg,&reg);     /* call bios plot routine */
     }}

void circle13(int x,int y,int r,double tilt)
{                                    /* routine draws a single planet */
```

```
 int la,yy;
 double ang,oldx,oldy,newx,newy,sintl,costl,rcos,rsin;
 for(la = -r; la < r; la++)
 {                                    /* routine uses a fairly simple */
  yy = sqrt(r*r - la*la) + 1;         /* algorithm to draw a solid circle */
  line13(la+x,y-yy,la+x,y+yy,14); }

 costl = cos(tilt);                   /* set up some variables */
 sintl = sin(tilt);
 yy = 240;                            /* To draw lines of longitude: */
 for(la = r; la >= -r; la-=r/15)      /* draw portions of ellipses */
   {                                  /* and rotate them by tilt radians */
    oldx = x-r*sintl;
    oldy = y+r*costl;
    for(ang = -1.57;ang <1.57;ang+=.195) {
     newx = x+la*cos(ang)*costl+r*sin(ang)*sintl;
     newy = y-r*sin(ang)*costl+la*cos(ang)*sintl;
     line13(newx,newy,oldx,oldy,yy); /* line segment of ellipse */
     oldx = newx;                     /* store endpoints */
     oldy = newy;  }
    yy = (yy==247) ? 240 : ++yy;      /* incrememt color used */
   }
 for(ang=-1.57;ang<1.57;ang+=.39)     /* draw lines of latitude */
    {                                 /* ie sloped lines */
     rcos = r*cos(ang);
     rsin = r*sin(ang);
     line13(x+rcos*costl-rsin*sintl,y+rsin*costl+rcos*sintl
          ,x-rcos*costl-rsin*sintl,y+rsin*costl-rcos*sintl,15);
    }}
void planets()          /* routine to draw and animate the planets */
{
 int la,lb;

 mode13();
 palette(7,255,255,255);
 printf("       Pixel Read Mask Demo\n");
 printf("       ==========================\n");
 printf("  This program contains an animated     picture scene which ");
 printf("is initially drawn  on the screen and then ANIMATED ");
 printf("using   the Pixel Read Mask Register.\n");
 printf("\n  Press the spacebar to draw scene and  hold it down ");
 printf("when scene is ready for    animation. When finished, ");
 printf("press any     other key......");
 while(getch() != ' ');               /* wait for keypress */

 mode13();
 for (la=8;la<16;la++)                /* set up the palette for animation */
   {
    for (lb=0;lb<8;lb++)
    palette(la*16+lb,0,10,63);        /* set planet lines to blue */
    for (lb=8;lb<16;lb++)
    palette(la*16+lb,0,0,0);          /* stars are initially black */
   }
 for (la=128;la<256;la+=17)
    palette(la,63,63,63),             /* define one line on planet to white */
    palette(la+8,63,63,0);            /* and one star to yellow, per frame */

  palette(15,255,255,255);            /* set color 15 to pure white */
  palette(7,20,255,0);                /* color 7 to green */
  palette(14,0,10,63);                /* color 14 used for planet background */

  stars();                            /* draw stars in background */
  circle13(30,30,30,0.9);             /* draw the actual planets */
  circle13(280,35,35,4.0);
  circle13(130,100,70,-0.8);
```

```
   circle13(40,240,125,0.5);
   triangle();                              /* draw the spaceship thingy */
   gotoxy(30,21);printf("Space to");
   gotoxy(30,22);printf("animate.");        /* on screen instructions */
   gotoxy(30,24);printf("Other key");
   gotoxy(30,25);printf("to stop.");
    la=143;                                 /* 143 = %10001111 */
    do
     outportb(0x3c6,la),                    /* this part does the actual animation */
     la = (la<255) ? la+16 : 143;           /* loop through the palette */
    while((lb = getch()) == ' ');           /* while the spacebar is being pressed */
 }

void stars()                                /* routine to plot in the stars */
{
 int la,lb,lc,ld,le,col = 248;
 long q;
 srand(time(&q) % 37);                      /* set up random background */
 for (la=0;la<200;la+=5)  {
    lc = (rand()&0x7)-0x4;
    ld = la;
    le = (rand()&7)+3;
    for (lb=1;lb<320;lb+=le,ld=la+lc*lb/64)
       plot13(lb,ld,col),         /* plot the star */
       col = (col == 255) ? 248 : ++col;
   }}

void line13(int x1,int y1,int x2,int y2,int col)
{                                                     /* this routine draws a line in */
 int la,lb,lc;                                        /* graphics mode 13H */
 if (abs(x1-x2) > abs(y1-y2))  {                     /* line longer in x or y direction ? */
   lc = (x2-x1);lb = (x2 - x1 >=0) ? 1 : -1;
   for (la=x1;la!=x2;la+=lb)                          /* loop works out the points on */
   plot13(la,y1+(la-x1)*(y2-y1)/lc,col);             /* the line and plots them */
  }
  else  {
   lc = (y2-y1);lb = (y2 - y1 >=0) ? 1 : -1;
   for (la=y1;la!=y2;la+=lb)
   plot13(x1+(la-y1)*(x2-x1)/lc,la,col);
  }}

void triangle()                /* This routine draws a simple spacecraft-type */
{                              /* object for animation. */
 int la,lb=19,col=248;         /* starting size = 19, color = 248 */
 double tilt=0.5236;           /* starting tilt */

 for (la=200;lb>0;la-=lb,lb--,tilt += .3)
                               /* loop to draw 19 objects */
 {line13(200+la/2+lb*cos(tilt),la+lb*sin(tilt),
         200+la/2+lb*cos(tilt+2.0944),la+lb*sin(tilt+2.0944),col);
  line13(200+la/2+lb*cos(tilt+2.0944),la+lb*sin(tilt+2.0944),
         200+la/2+lb*cos(tilt+4.1888),la+lb*sin(tilt+4.1888),col);
  line13(200+la/2+lb*cos(tilt+4.1888),la+lb*sin(tilt+4.1888),200+la/2,la,col);
  line13(200+la/2,la,200+la/2+lb*cos(tilt),la+lb*sin(tilt),col);
  col = (col==255) ? 248 : ++col;        /* col = col + 1 until col = 255, when */
                                         /* col returns to zero */
 }}
```

AN-39
COMPANDING D/A CONVERTER

Precision Monolithics Inc.

APPLICATION NOTE 39

INTRODUCTION

A companding digital-to-analog converter (DAC) is the key component in PCM CODEC systems. (CODEC is an acronym for coder-decoder.) A CODEC performs the coding functions which consist of an analog-to-digital conversion (ADC) of the input analog (voice) signal and decoding, which consists of a digital-to-analog conversion (DAC) of the received digital input.

The DAC is used for both encoding and decoding; it is in a feedback loop to generate the ADC functions. Voice signals in telephony require a system with a very large dynamic range. The dynamic range (DR) of a CODEC is defined as the ratio of the largest resolvable signal to the smallest signal which can be encoded. The dynamic range of the CODEC is the same as that of the DAC used in either the decode mode or in the feedback loop of the successive approximation type ADC. The dynamic range of a DAC is simply the ratio of its output for a linear input of one least significant bit (LSB) to that of the largest, all "1s," input. This ratio is usually expressed in decibels using the equation:

$$DR = 20 \log_{10} \frac{I_{MAX.}}{I_{LSB}}$$

where for a current output DAC I_{MAX} is the output current for all "1s" input and I_{LSB} is the output current for one LSB input. Using this equation a linear bit DAC can be shown to resolve a ratio of 2^n:1 therefore:

$$DR = 20 \log_{10} \frac{2^n}{1} \approx 6^n$$

The wide dynamic range requirements of a telephone system require the equivalent dynamic range of a 12-bit system or 72dB. However, this system would not be satisfactory for telephone voice transmission because of its excessive bandwidth requirements. With present day T1 type transmission systems a 64kbits/sec data rate is required to transmit each voice channel. The use of the linear system would increase this bit rate to 96kbits/sec. This would provide more accuracy than is needed at the expense of excessive bandwidth.

For voice systems the most important criterion is the signal-to-noise ratio. In a PCM system noise is due almost entirely to quantizing distortion. Thus, a non-linear DAC has a non-linear transfer characteristic to compress the analog signal into a digital word and a complementary transfer characteristic to expand the digital words into analog signals with a wide dynamic range. For a telephone system a CODEC requires a fairly uniform signal-to-distortion ratio over its entire dynamic range. Achieving this uniform signal-to-distortion ratio over a wide dynamic range requires the use of non-uniform coding. A non-uniform CODEC is a coder-decoder pair whose input amplitude range is divided into steps of unequal widths, such that the width of the quantizing steps increase in proportion to the amplitude of the signal. To achieve uniform signal to distortion performance a logarithmic transfer function is required. The word compand, (compand is an acronym for compress — expand) was borrowed from analog systems to describe this non-uniform coding system where quantizing and coding is such that step size depends on the input amplitude.

COMPANDING PRINCIPLES

Companding requirements differ for different signal distributions. As mentioned above, voice signals require constant S/D performance over a wide dynamic range. In order to accomplish this the distortion must be proportional to the signal level. This feat is best achieved by the use of a logarithmic compression law. However, a truly logarithmic assignment of code words is not physically possible since this implies an infinite number of codes. Two methods for generating practical implementations of logarithmic transfer functions have been derived which have become industry standards. These methods are generally known by their transfer functions which are called μ-law and A-law respectively. Both of these transfer functions are normally implemented with eight-bit non-linear DACs to achieve a 72dB dynamic range. This is the equivalent dynamic range of a twelve-bit linear DAC. The μ-law and the A-law transfer functions are described by the following equations:

$$\mu\text{-law} \quad Y = \frac{\ln(1 + \mu|X|)}{\ln(1 + \mu)} \text{ sgn } X \quad \text{for } -1 \leq X \leq +1$$

$$\text{A-law} \quad Y = \frac{1 + \ln A|X|}{1 + \ln A} \text{ sgn } X \quad \text{for } 1/A \leq X \leq 1$$

$$Y = \frac{A|X|}{1 + \ln A} \text{ sgn } X \quad \text{for } 0 \leq X \leq 1/A$$

These laws have unique signal-to-distortion characteristics for each value of μ and A respectively. At present ATT has settled on a value of μ equal to 255 and CCITT specifications use a value of A equal to 87.6. Substituting these constants into the original equation above obtain:

$$\mu\text{-law} \quad Y = 0.18 \ln(1 + \mu|X|) \quad \text{sgn } X \text{ for } -1 \leq X \leq 1$$

$$\text{A-law} \quad Y = 0.18 \ln(1 + \ln|X|) \quad \text{sgn } X \text{ for } 1/A \leq |X| \leq 1$$

$$Y = 0.18\, A|X| \quad \text{sgn } X \text{ for } 0 \leq |X| = 1/A$$

The wideband (unfiltered) signal-to-distortion ratio over the useable dynamic range of voice transmissions is shown in Figure 1. This plot does not represent actual system performance; it is instead, a measure of the distortion which would be caused by an ideal quantizer.

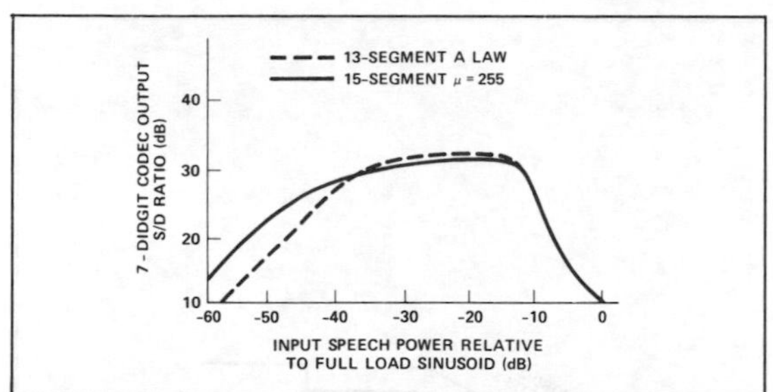

Figure 1. Input Speech Power Relative to Full Load Sinusoid (dB)

The practical implementation of the two transfer functions is accomplished by standardized piece-wise linear approximations. The transfer functions are implemented in chords or segments where the transfer function within any one chord is a linear staircase. Each chord has sixteen steps and the size of the step in each succeeding chord is double the size of the step in the preceeding chord. There are normally eight chords numbered zero through seven in both μ-law and A-law characteristics. For the A-law function the first two chords on either side of the origin have equal step sizes, whereas, for the μ-law function, the second chord after the origin has a step size which is double that of the first. For all remaining chords the steps double in size for each succeeding chord. This applies to both the μ-law and A-law functions. For the A-law function the four chords about the origin can be considered as a single segment so that the A-law characteristic is sometimes referred to as being a "13-segment" code. The A-law characteristic also differs from the μ-law characteristic in the manner in which the transfer function crosses the origin. The X-axis origin for the μ-law is at "mid-step" while the X-axis origin for the A-law is coincident with a "riser". This can be understood better from the "blow-ups" about the origin of Figures 2 and 3.

In order to obtain the best implementations of the transfer function, companded DACs are constructed such that encode and decode functions are offset by one-half step. With this technique the quantizing band for the encode DAC will be centered about the decode value. This can be seen in Figure 4, where the μ-law characteristics about the origin are shown. (The A-law characteristics would be identical except for the "mid-riser" phenomena at the origin.) As an example suppose that, for Figure 4, an analog input whose amplitude lies between levels 2 and 4 is being encoded. The best quantizing code to assign to this entire quantizing band is its mean value of 3. Thus the DAC used in the suc-

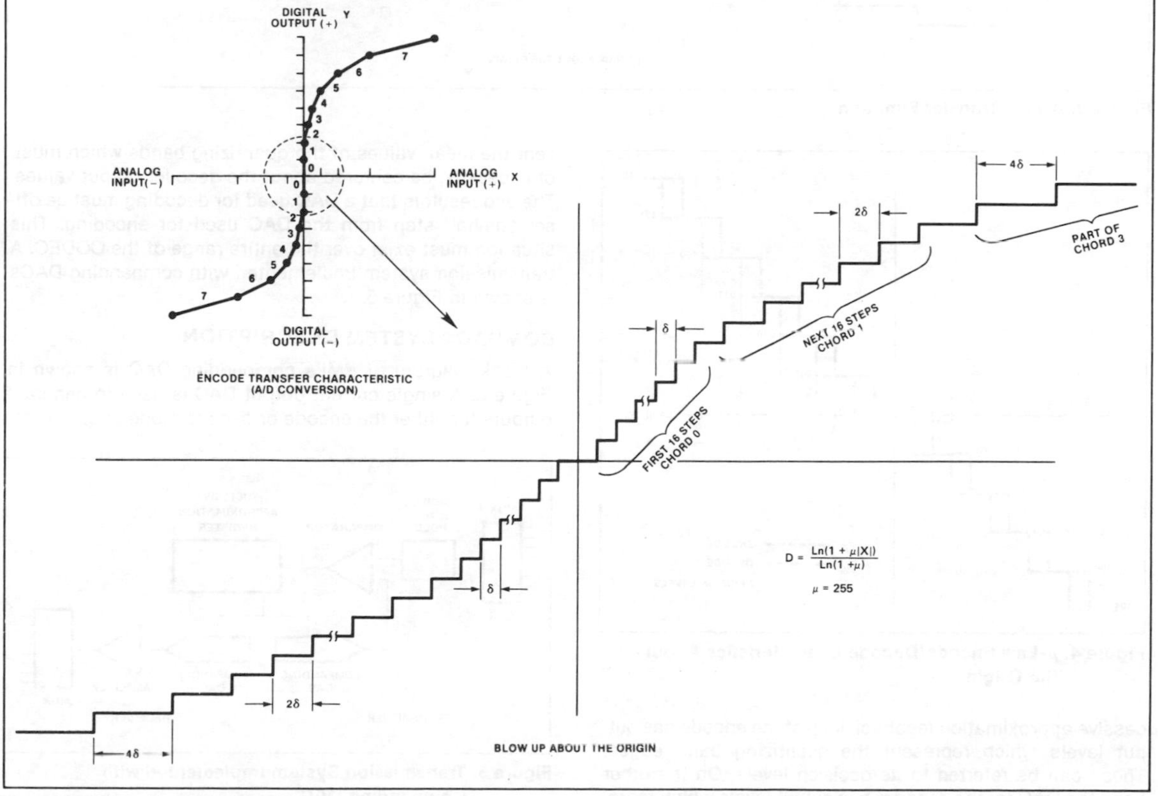

Figure 2. μ-Law Transfer Function

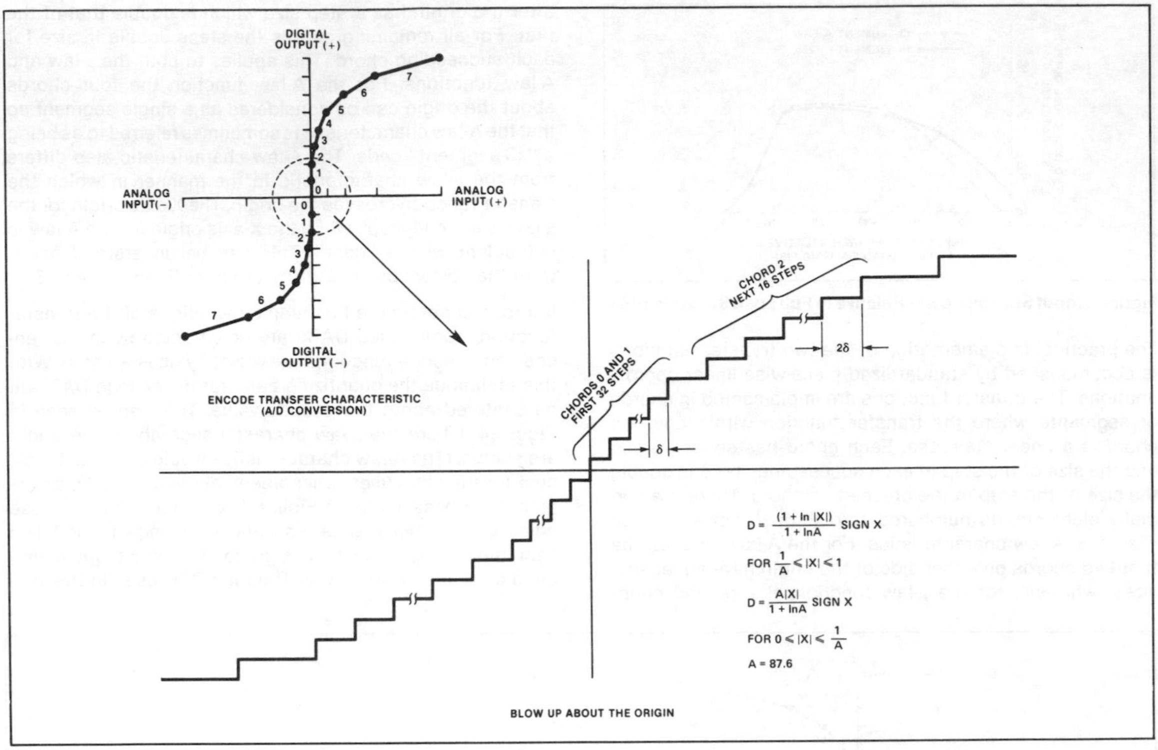

$$D = \frac{(1 + \ln |X|)}{1 + \ln A} \text{ SIGN } X$$

$$\text{FOR } \frac{1}{A} \leq |X| \leq 1$$

$$D = \frac{A|X|}{1 + \ln A} \text{ SIGN } X$$

$$\text{FOR } 0 \leq |X| \leq \frac{1}{A}$$

$$A = 87.6$$

Figure 3. A-Law Transfer Function

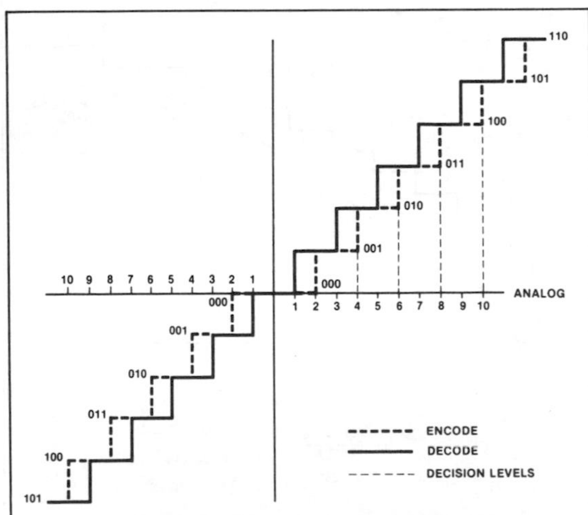

Figure 4. μ-Law Encode/Decode Characteristics About the Origin

cessive approximation feedback loop of the encode has output levels which represent the quantizing band edges. These can be referred to as decision levels. On the other hand the DAC for the decoder has output levels which represent the mean values of the quantizing bands which must, of necessity, be centered about the decoder output values. The end result is that a DAC used for decoding must be offset one-half step from the DAC used for encoding. This situation must exist over the entire range of the CODEC. A transmission system implemented with companding DACs is shown in Figure 5.

COMDAC® SYSTEM DESCRIPTION

A block diagram of PMI's companding DAC is shown in Figure 6. A single current output DAC is used to generate outputs for either the encode or decode mode of operation.

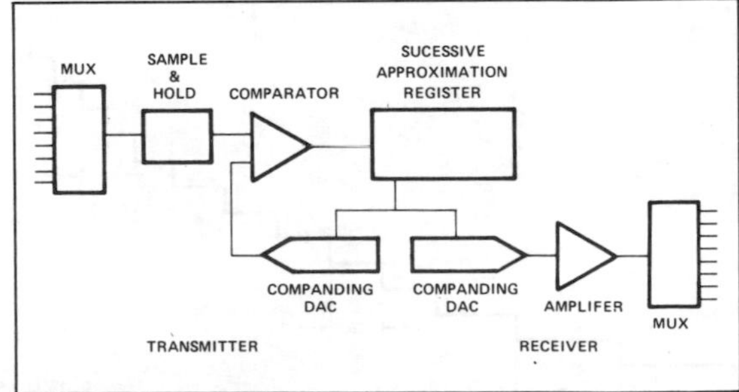

Figure 5. Transmission System Implemented with Companding DAC

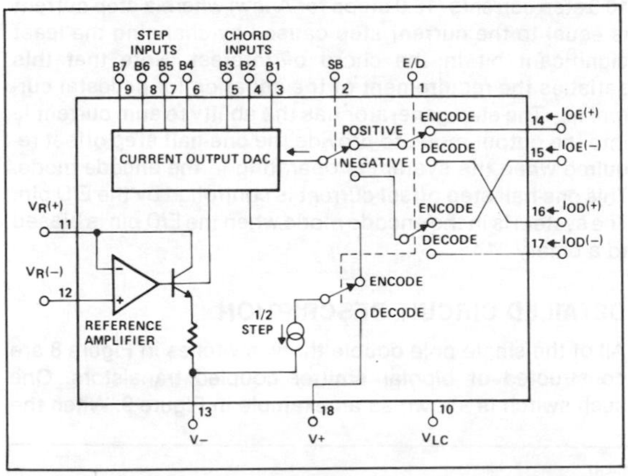

Figure 6. Equivalent Circuit and Pin Connection Diagram

Each companding DAC can be programmed to operate as either an encoder or a decoder by properly programming the E/D pin. The encode mode is offset one-half step from the decode mode by means of the current generator which is switched in during the encode mode. The reference amplifier establishes the current reference for the current output DAC. The sign bit pin (SB) controls the positive-negative switch which directs the output of the current output DAC.

This output will eventually end up at the positive (I_{OE+} or I_{OD+}) outputs or the negative (I_{OE-} or I_{OD-}) outputs depending on whether the SB pin is programmed to a binary "1" or a binary "0". The encode-decode switch E/D determines whether the DAC output shall be directed to the encode or decode terminations as shown in Figure 6. In addition, this same switch introduces the one-half step of offset current required during encode.

A better understanding of the COMDAC® circuitry is obtained by reviewing the previously discussed piece-wise linear approximation of the companding DAC transfer function to the desired μ-law or A-law transfer functions. Each chord or segment consists of 16 steps numbered from 0 to 15. The size of the steps double in size from one chord to the next as the number of the chord increases. The chords are numbered 0 to 7. In order to smooth out the characteristics during the transition from one chord to the next, the step current for step 0 of each chord is 1-1/2 times larger than the current of the highest step of the chord immediately preceding it. The succeeding 15 steps (steps 1 to 15) are then two times the size of the steps of this previous chord. These characteristics can be examined in Figure 7.

To implement the transfer function, the first chord (N = 0) uses 16 equal steps each of whose size, I_0 is 1/16 of chord current source I_{C0} for A-law, or 1/16.5 of current source I_{C0} for μ-law. The next chord, N = 1, must begin at $I_{C0} + 1.5I_0$ for both A-law or μ-law. Another way of saying this is that chord

AN-39

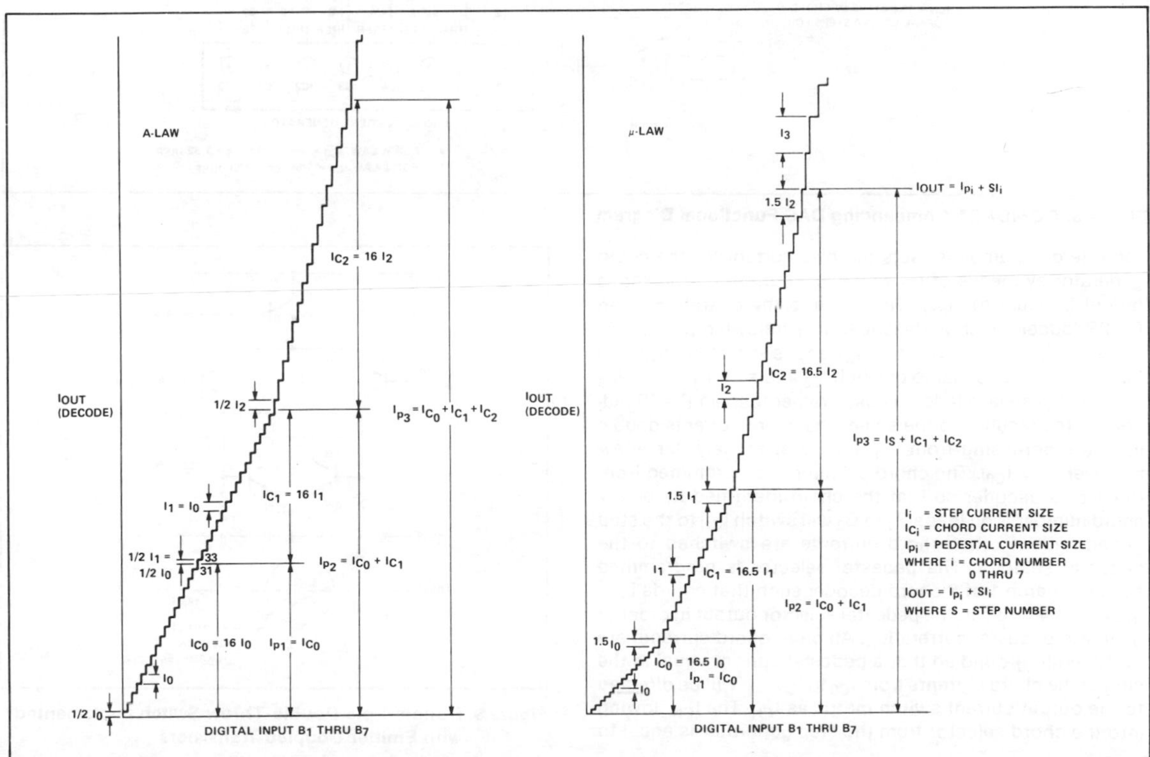

Figure 7. Construction of the Companding DAC Transfer Function

N = 1 begins 16.5 steps from the origin. In order to accomplish this a pedestal current must be directed toward the output whose magnitude is equal to $I_{C0} + 1.5I_0$. Chord C2 begins at $I_{C0} + 1.5I_0 + I_{C1} + 1.5I_1$ and ends at $I_{C0} + 1.5I_0 + I_{C1} + 1.5I_1 + I_{C2} + 1.5I_2$ and so forth. This process continues with pedestal currents for each chord number N described by the equation:

$$IPN = \sum_{i=0}^{N-1} (I_{Ci} + 1.5I_i) = 16.5 \sum_{i=0}^{N-1} I_i$$

note that $I_{PO} = 0$.

A functional diagram of a companding DAC which implements the proper transfer function discussed above is shown in Figure 8, which operates in the following manner:

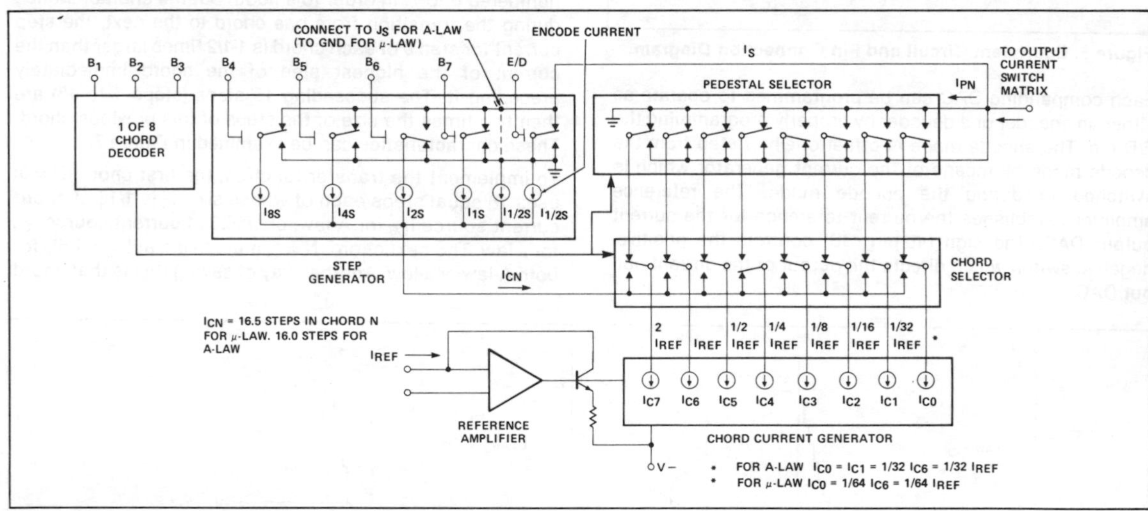

Figure 8. COMDAC® Companding DAC Functional Diagram

the reference amplifier sets the bias current for the chord generator by means of I_{C7} which is a current mirror whose output is equal to $2I_{REF}$. Next, due to the operation of an R − 2R ladder which is described in a following paragraph, I_{C6} is made equal to one-half I_{C7} and is therefore equal to I_{REF}. I_{C5} is made equal to one-half I_{C6} and so forth. From I_{C3} down to I_{C0} a slave ladder is used rather than an R − 2R ladder but the results are the same. The chord currents double in size progressing from I_{C0} to I_{C7} respectively (for A-law however $I_{C1} = I_{C0}$). The chord selector is programmed from the 1 of 8 decoder so that the chord identified by binary chord number N on leads B_1 to B_3 will switch I_{CN} to the step generator. All other chord currents are switched to the pedestal selector. The pedestal selector is programmed from the same 1 of 8 chord decoder such that chords I_0 to I_{N-1} are switched to the pedestal selector output in order to generate pedestal current I_{PN}. All other chord currents are switched to ground so that a pedestal current equal to the sum of the chord currents from I_{C0} to $I_{C(N-1)}$ will be directed to the output current switch matrix as I_{PN}. The I_{CN} flowing into the chord selector from the step generator is equal to 16.5 step currents (16.0 steps for A-law) where a step current is equal to the current step caused by changing the least significant bit in the chord of interest. Note that this satisfies the requirement of the equation for pedestal current I_{PN}. The step generator has the ability to sum current I_E into the output mode to provide the one-half step offset required when the system is operating in the encode mode. This one-half step offset current is controlled by the E/D pin. The system is in the encode mode when the E/D pin is biased to a binary "1".

DETAILED CIRCUIT DESCRIPTION

All of the single pole double throw switches in Figure 8 are constructed of bipolar emitter coupled transistors. One such switch is shown as an example in Figure 9. When the

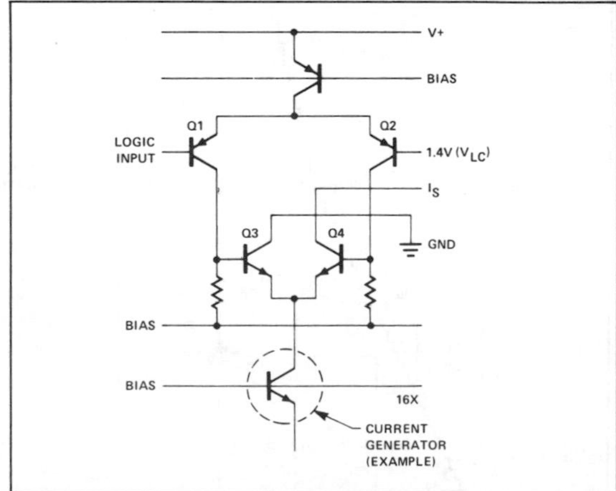

Figure 9. Double-Pole Double-Throw Switch Implemented with Emitter Coupled Transistors

logic input exceeds the logic level bias V_{LC} Q_1 is turned off and Q_2 is turned on. In turn Q_3 is turned off and Q_4 is turned on thus effectively switching the current generator, shown as an example, from the ground to I_S. Conversely, lowering the logic level input below V_{LC} will switch the current from I_S to ground. The V_{LC} Control permits the circuit to interface with a large range of logic levels.

The chord current generator circuit is shown in Figure 10. This circuit is the implementation of the chord current generator previously discussed. Q_0 is forced to operate at the reference input current I_{REF} and Q_1, with an emitter resistor one-half the size of the emitter resistor of Q_0, will then operate at $2I_{REF}$. Q_2 through Q_4 will operate at progressively smaller currents where each transistor operates at one-half the current of the transistor to its immediate left. To review this normal R−2R current-ladder function notice that Q_{4A} and Q_{4B} operate at equal currents and that the sum of their currents is equal to that of one transistor with an emitter resistor equal to R. When the series resistor R is added to the junction of the emitter resistors of Q_{4A} and Q_{4B} the current of Q_3 will be forced to equal the sum of the Q_{4A} and Q_{4B} currents. Thus Q_{4A} current equals one-half the Q_3 current. Now the current from Q_{4A}, Q_{4B} and Q_3 must all flow through the next series resistor R. This current is equal to twice that of Q_3; therefore it is easy to compute that the Q_2 current is twice that of the Q_3. The same reasoning may be used to proceed down the ladder to show that each transistor in the ladder sinks twice the current of the transistor on its immediate right. The slave ladder consisting of Q_5 through Q_{8A} and Q_{8B} continues to halve currents for each transistor proceeding to the right. However this part of the chord current generator uses scaled resistors instead of the R−2R ladder technique. Since Q_{4B} sinks constant current from the slave ladder, and since all the current must flow through the scaled emitter resistors, then the curent through each transistor must be inversely proportional to the size of its emitter resistor. By examination of the slave ladder it can be seen that each transistor proceeding to the right sinks one-half the current of the transistor to its immediate left. For the μ-law chord current generator Q_{8B} is simply diode connected such that the chord current for chord C_0 is roughly one-half the current of chord C_1. For the A-law chord current generator, however, the collectors of transistors Q_{8A} and Q_{8B} are tied together so that I_{C0} is exactly equal to I_{C1}. The currents flow to the chord current generator from an array of bipolar single pole double throw switches labeled "chord selector" in Figure 8. The actual switches are not shown in this paper.

The Step Current Generator is shown in Figure 11. Again the single pole double throw switches which connect the step generator to the output current matrix as shown in the companding DAC functional diagram are not represented. The step generator is connected to the chord selector which sinks I_{CN}. Ratioed emitters are used to divide the current. The largest emitter is 16 times the size of the smallest emitter and therefore sinks 16 times the current. The A-law step generator differs from the μ-law step generator in that each chord begins with a riser instead of a step. This also applies to the origin, therefore one-half step of current flows (decode mode) even when the binary input to the step generator is "0". Step switches controlled directly by the binary code connect the appropriate collectors of the step current generator transistors to the output current matrix. For both A-law and μ-law devices I_{CN} is one of the pedestal currents. The difference is that for the A-law device the pedestal current is equal to 16 steps whereas, for the μ-law, the pedestal current is equal to 16.5 steps.

NORMALIZED COMPANDING DAC OUTPUTS

It is convenient to generate tables of normalized values which correspond exactly to the CCITT (Consultative Committee for International Telephone and Telegraph) specifications. The following tables are normalized to the smallest DAC output which is equivalent to one-half step.

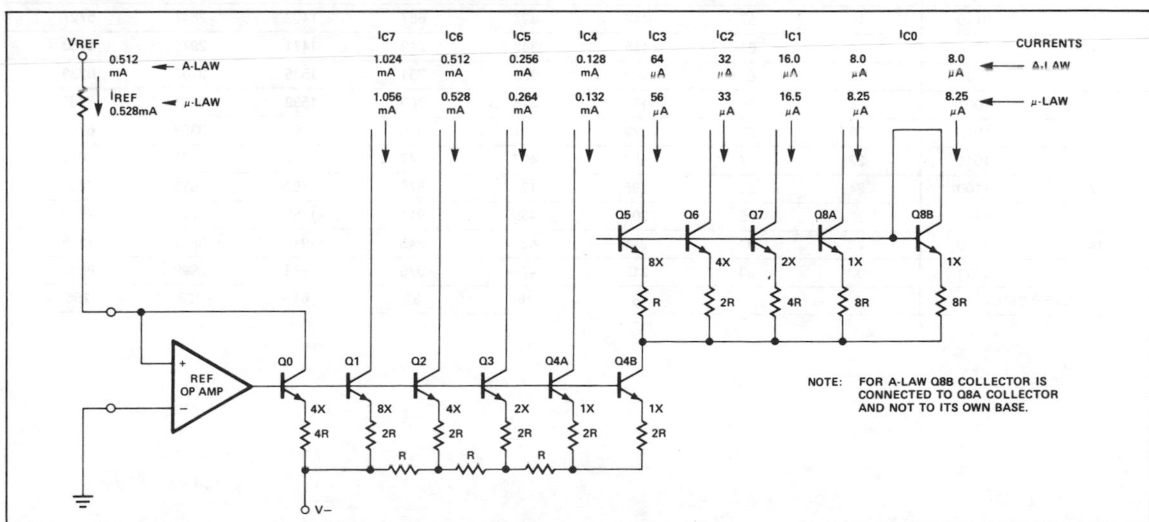

Figure 10. Chord Current Generator Diagram

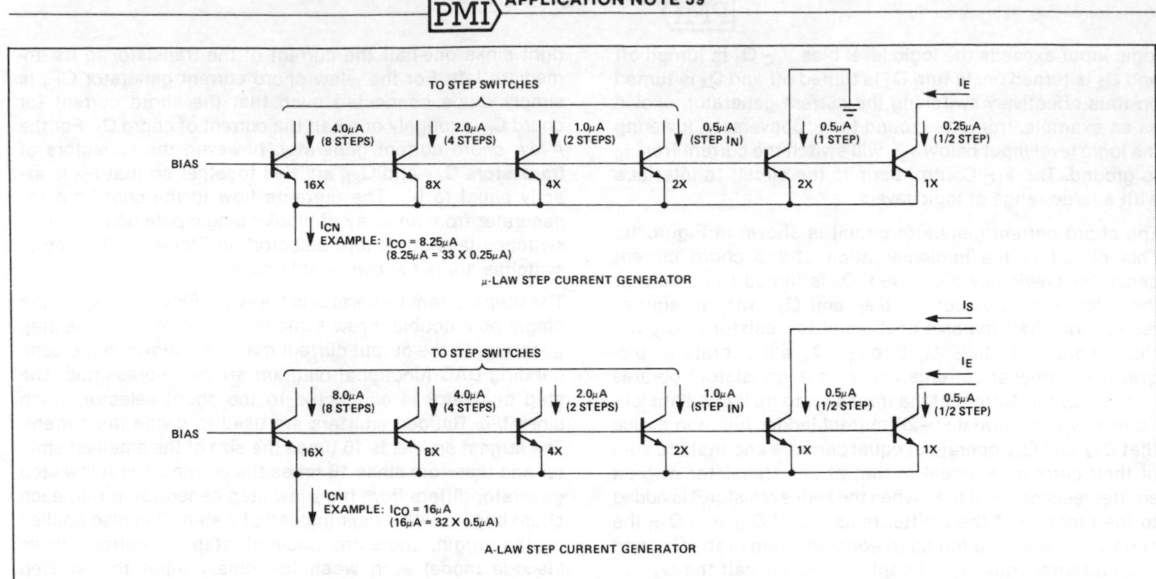

Figure 11. A-Law and μ-Law Step Current Generators

μ-Law Normalized Table

NORMALIZED DECODE OUTPUT (SIGN BIT EXCLUDED) $I_{C,S} = 2[2^C (S + 16.5) - 16.5]$

C = chord no. (0 through 7)
S = step no. (0 through 15)

STEP \ CHORD		0 000	1 001	2 010	3 011	4 100	5 101	6 110	7 111
0	0000	0	33	99	231	495	1023	2079	4191
1	0001	2	37	107	247	527	1087	2207	4447
2	0010	4	41	115	263	559	1151	2335	4703
3	0011	6	45	123	279	591	1215	2463	4959
4	0100	8	49	131	295	623	1279	2591	5215
5	0101	10	53	139	311	655	1343	2719	5471
6	0110	12	57	147	327	687	1407	2847	5727
7	0111	14	61	155	343	719	1471	2975	5983
8	1000	16	65	163	359	751	1535	3103	6239
9	1001	18	69	171	375	783	1599	3231	6495
10	1010	20	73	179	391	815	1663	3359	6751
11	1011	22	77	187	407	847	1727	3487	7007
12	1100	24	81	195	423	879	1791	3615	7263
13	1101	26	85	203	439	911	1855	3743	7519
14	1110	28	89	211	455	943	1919	3871	7775
15	1111	30	93	219	471	975	1983	3999	8031
STEP SIZE		2	4	8	16	32	64	128	256

μ-Law Normalized Table

NORMALIZED ENCODE LEVEL (SIGN BIT EXCLUDED) $I_{C,S} = 2[2^C (S + 17) - 16.5]$

C = chord no. (0 through 7)
S = step no. (0 through 15)

STEP \ CHORD		0 000	1 001	2 010	3 011	4 100	5 101	6 110	7 111
0	0000	1	35	103	239	511	1055	2143	4319
1	0001	3	39	111	255	543	1119	2271	4575
2	0010	5	43	119	271	575	1183	2399	4831
3	0011	7	47	127	287	607	1247	2527	5087
4	0100	9	51	135	303	639	1311	2655	5343
5	0101	11	55	143	319	671	1375	2783	5599
6	0110	13	59	151	335	703	1439	2911	5855
7	0111	15	63	159	351	735	1503	3039	6111
8	1000	17	67	167	367	767	1567	3167	6367
9	1001	19	71	175	383	799	1631	3295	6623
10	1010	21	75	183	399	831	1695	3423	6879
11	1011	23	79	191	415	863	1759	3551	7135
12	1100	25	83	199	431	895	1823	3679	7391
13	1101	27	87	207	447	927	1887	3807	7647
14	1110	29	91	215	463	959	1951	3935	7903
15	1111	31	95	223	479	991	2015	4063	8159
STEP SIZE		2	4	8	16	32	64	128	256

AN-39

A-Law Normalized Table

NORMALIZED DECODE OUTPUT (SIGN BIT EXCLUDED) $I_{CS} = 2^{N-1} (33 + 2S)$ For $N > 0$
$I_{CS} = 2S + 1$ For $N = 0$

STEP \ CHORD		0 000	1 001	2 010	3 011	4 100	5 101	6 110	7 111
0	0000	1	33	66	132	264	528	1056	2112
1	0001	3	35	70	140	280	560	1120	2240
2	0010	5	37	74	148	296	592	1184	2368
3	0011	7	39	78	156	312	624	1248	2496
4	0100	9	41	82	164	328	656	1312	2624
5	0101	11	43	86	172	344	688	1376	2752
6	0110	13	45	90	180	360	720	1440	2880
7	0111	15	47	04	188	376	752	1504	3008
8	1000	17	49	98	196	392	784	1568	3136
9	1001	19	51	102	204	408	816	1632	3264
10	1010	21	53	106	212	424	848	1696	3392
11	1011	23	55	110	220	440	880	1760	3520
12	1100	25	57	114	228	456	912	1824	3648
13	1101	27	59	118	236	472	944	1888	3776
14	1110	29	61	122	244	488	976	1952	3904
15	1111	31	63	126	252	504	1008	2016	4032
STEP SIZE		2	2	4	8	16	32	64	128

A-Law Normalized Table

$I_{CS} = 2^{N-1}(34 + 2S)$ For $N > 0$
$I_{CS} = 2S + 2$ For $S = 0$

NORMALIZED ENCODE DECISION LEVELS (SIGN BIT EXCLUDED)

STEP \ CHORD		0	1	2	3	4	5	6	7
		000	001	010	011	100	101	110	111
0	0000	2	34	68	136	272	544	1088	2176
1	0001	4	36	72	144	288	576	1152	2304
2	0010	6	38	76	152	304	608	1216	2432
3	0011	8	40	80	160	320	640	1280	2560
4	0100	10	42	84	168	336	672	1344	2688
5	0101	12	44	88	176	352	704	1408	2816
6	0110	14	46	92	184	368	736	1472	2944
7	0111	16	48	96	192	384	768	1536	3072
8	1000	18	50	100	200	400	800	1600	3200
9	1001	20	52	104	208	416	832	1664	3328
10	1010	22	54	108	216	432	864	1728	3456
11	1011	24	56	112	224	448	896	1792	3584
12	1100	26	58	116	232	464	928	1856	3712
13	1101	28	60	120	240	480	960	1920	3840
14	1110	30	62	124	248	496	992	1984	3968
15	1111	32	64	128	256	512	1024	2048	*4096
STEP SIZE		2	2	4	8	16	32	64	128

The numbers in these tables are directly proportional to the input reference current. However the exact relationship is somewhat complicated. A reference current of 528μA for the μ-law DAC will produce a step size of 0.5μA; thus, for the μ-law device driven by a reference current of 528μA, it is only necessary to multiply all the numbers in the normalized tables by one-half step or 0.25μA to obtain the output in μA. The table tabulated below corresponds to a 528μA reference.

μ-Law Current Output Table

IDEAL DECODE OUTPUT CURRENT IN MICROAMPS (SIGN BIT EXCLUDED)

STEP \ CHORD		0	1	2	3	4	5	6	7
		000	001	010	011	100	101	110	111
0	0000	0	8.25	24.75	57.75	123.75	255.75	519.75	1047.75
1	0001	0.5	9.25	26.75	61.75	131.75	271.75	551.75	1111.75
2	0010	1	10.25	28.75	65.75	139.75	287.75	583.75	1175.75
3	0011	1.5	11.25	30.75	69.75	147.75	303.75	615.75	1239.75
4	0100	2	12.25	32.75	73.75	155.75	319.75	647.75	1303.75
5	0101	2.5	13.25	34.75	77.75	163.75	335.75	679.75	1367.75
6	0110	3	14.25	36.75	81.75	171.75	351.75	711.75	1431.75
7	0111	3.5	15.25	38.75	85.75	179.75	367.75	743.75	1495.75
8	1000	4	16.25	40.75	89.75	187.75	383.75	775.75	1559.75
9	1001	4.5	17.25	42.75	93.75	195.75	399.75	807.75	1623.75
10	1010	5	18.25	44.75	97.75	203.75	415.75	839.75	1687.75
11	1011	5.5	19.25	46.75	101.75	211.75	431.75	871.75	1751.75
12	1100	6	20.25	48.75	105.75	219.75	447.75	903.75	1815.75
13	1101	6.5	21.25	50.75	109.75	227.75	463.75	935.75	1879.75
14	1110	7	22.25	52.75	113.75	235.75	479.75	967.75	1943.75
15	1111	7.5	23.25	54.75	117.75	243.75	495.75	999.75	2007.75
STEP SIZE		.50	1	2	4	8	16	32	64

*Virtual Decision Level

A similar exercise will yield a corresponding table for the A-law part. Multiplying all the numbers in the normalized A-law table, for instance, will produce a table of currents for a reference input of 512μA. A table based on 512μA reference current will have a step size of 1.0μA and is tabulated in the μ-law current output table.

A-Law Current Output Table

IDEAL DECODE OUTPUT CURRENT IN MICROAMPS (SIGN BIT EXCLUDED)

STEP \ CHORD		0	1	2	3	4	5	6	7
		000	001	010	011	100	101	110	111
0	0000	0.5	16.5	33	66	132	264	528	1056
1	0001	1.5	17.5	35	70	140	280	560	1120
2	0010	2.5	18.5	37	74	148	286	592	1184
3	0011	3.5	19.5	39	78	156	312	624	1248
4	0100	4.5	20.5	41	82	164	328	656	1312
5	0101	5.5	21.5	43	86	172	344	688	1376
6	0110	6.5	22.5	45	90	180	360	720	1440
7	0111	7.5	23.5	47	94	188	376	752	1504
8	1000	8.5	24.5	49	98	196	392	784	1568
9	1001	9.5	25.5	51	102	204	408	816	1632
10	1010	10.5	26.5	53	106	212	424	848	1696
11	1011	11.5	27.5	55	110	220	440	880	1760
12	1100	12.5	28.5	57	114	228	456	912	1824
13	1101	13.5	29.5	59	118	236	472	944	1888
14	1110	14.5	30.5	61	122	244	488	976	1952
15	1111	15.5	31.5	63	126	252	504	1008	2016
STEP SIZE		1	1	2	4	8	16	32	64

Reviewing the companding DAC functional diagram Figure 8 demonstrates the relationship between step size and I_{REF}. For a μ-law device I_{C0} equals 16.5 chord zero steps and for an A-law device I_{C0} equals 16 chord zero steps. I_{C6} is always equal to I_{REF} in either system. I_{C6} is then equal to 64 times I_{C0} for a μ-law system, and 32 times I_{C0} for an A-law system. The step size can then be related to I_{REF} by the following equations:

$$\text{step size} = I_{REF}/64 \times 16.5 = I_{REF}/1056 \ (\mu\text{-law})$$
$$\text{step size} = I_{REF}/32 \times 16 = I_{REF}/512 \ (\text{A-law})$$

Now for a reference current of 528μA the step size for a μ-law system is 528/1056 or 0.5μA. For a reference current of 512μA the step size for an A-law system is 512/512 or 1.0μA. These values concur with those used to generate the tables.

In the design of the PMI DAC-89 the biasing resistors were not scaled to exactly integer values. This was done deliberately to standardize somewhat on 528μA input reference current for both A-law and μ-law parts. The performance of the device is not affected, however the actual scaling is somewhat complicated and will not be discussed in this paper.

Finally if encode output tables were desired for current output they could be obtained by scaling to proper step size the normalized encode tables or adding one-half step to each value in the decode table, where the step size depends on the chord number.

DAC ACCURACY

Companding DACs must be manufactured to satisfy a unique set of parameters. The performance of a companded DAC used for telephony must satisfy the requirements of a communication system on an end-to-end basis. A voice channel is first encoded by one CODEC then decoded by a second CODEC such that the system performance can be measured on an audio-in-audio-out basis. The CODEC performance will be almost completely dominated by the Gain Tracking requirement.

GAIN TRACKING

Gain Tracking refers to the ability of a system to track its input power level. The test is normally made with a system such as that shown in Figure 12.

Gain Tracking is measured by monitoring the input and output levels in decibels. At an input level of −10Bm0 the output is recorded as the output reference level. For ideal Gain Tracking, any change (in dB) of the input level must be matched exactly by the same change in the output level.

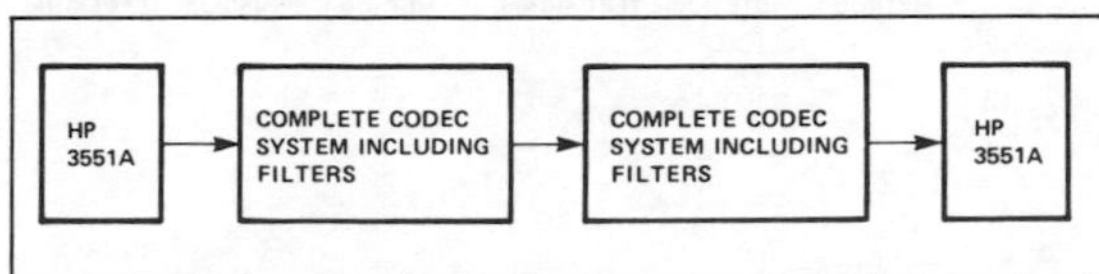

Figure 12. Gain Tracking or S/N Test

AN-39

This condition is monitored over all input power levels of interest. The extent to which these power level changes differ (again in dB) is a measure of Gain Tracking, also referred to as gain deviation. The ATT/D3 Gain Tracking specification is show in Figure 13.

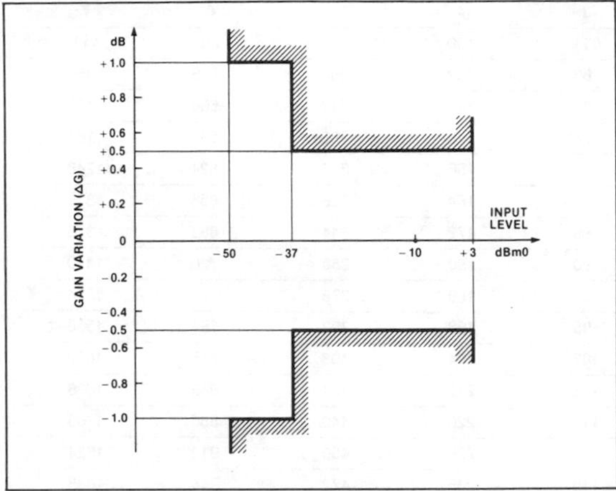

Figure 13. ATT/D3 Gain Tracking Specification

CCITT publishes two separate specifications for Gain Tracking. The apparatus used for making either of these tests is basically the same as that used in Figure 13 except that for the first part of the "method one" test the HP3551A would be replaced with a suitable white noise source at the input and an RMS reading voltmeter at the output. Gain Tracking masks equivalent to those found in CCITT publications are shown in Figure 14.

POWER LEVELS

For PCM channel performance measurements, power levels are characteristically expressed in dBm0. A reference level of 0dBm0 is established by referencing to a code in the **digital** transmission. The binary code pattern required to establish a reference level of 0dBm0 can be found in the CCITT publications. This pattern is reproduced in the PMI Telecommunications Handbook for the readers convenience. The constant repetition of these binary numbers at the normal sampling rate of 8kHz will produce a 1kHz sinusoid at a 0dBm0 reference level. Starting with this definition it can then be shown that a sinusoid whose peak value is just at the system saturation level (all "1s" PCM output) will have a power level of 3.14 and 3.17dBm0 for A-law and μ-law respectively.

SIGNAL-TO-DISTORTION MEASUREMENTS

Signal-to-Distortion is a measure of the total distortion a system will exhibit on an end-to-end basis. As with Gain Tracking this measurement is normally performed on an audio-to-audio basis. A typical setup for measuring Signal-to-Distortion is shown in Figure 15. A wideband (3kHz) filter may be substituted for the C-Message filter shown for some tests.

Figure 16 shows the ATT/D3 specification mask with the performance of a PMI demonstration COMDAC® based shared CODEC system superimposed. This method of measuring Signal-to-Distortion is applicable to either CCITT or ATT specifications.

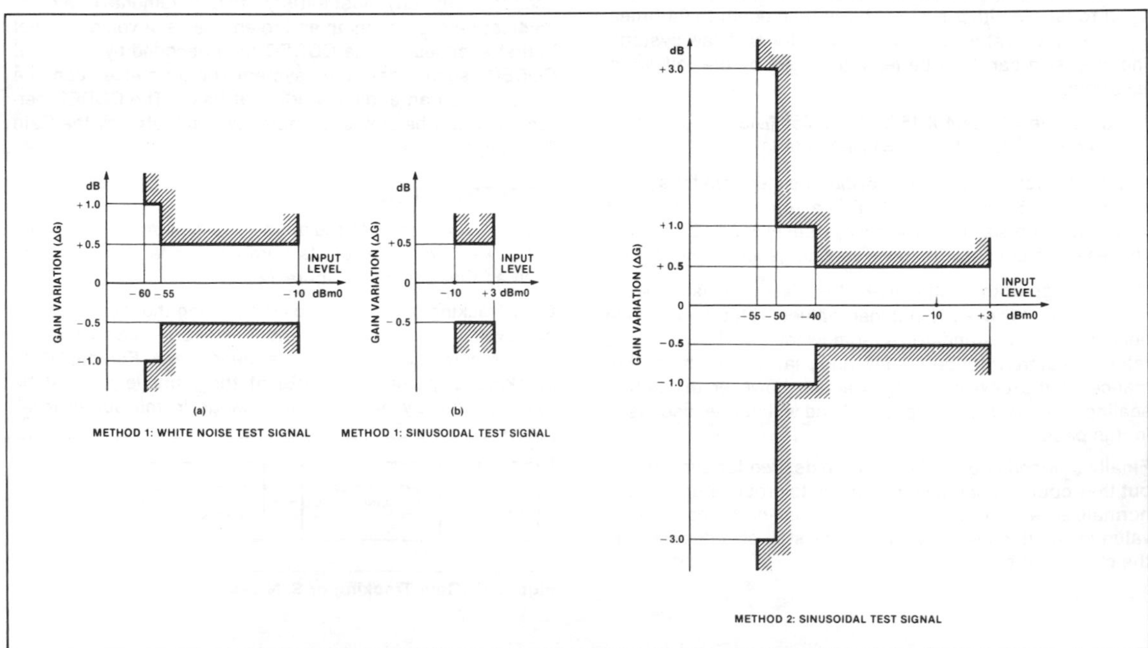

Figure 14. CCIT Gain Tracking Specification

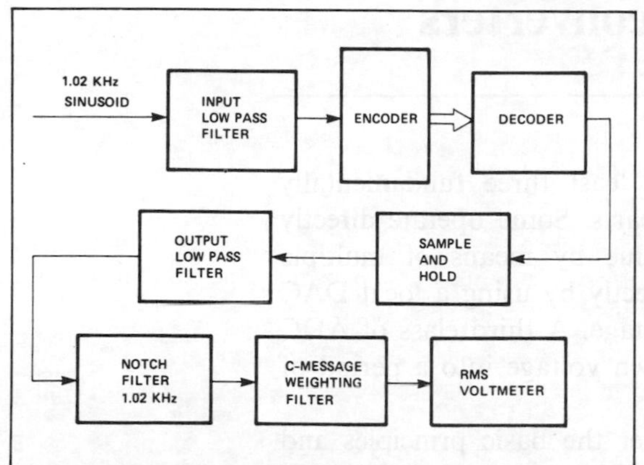

Figure 15. Signal-to-Distortion Test Setup

DAC ACCURACY VERSUS GAIN TRACKING AND SIGNAL-TO-DISTORTION RATIO

The analog portions of a PCM system usually make only a minor contribution to either Gain Tracking or Signal-to-Distortion errors. Thus, the major contribution to error is the inability of the companding DAC to accurately follow the encoding format. The process of quantizing and coding will cause some deviation from the ideal, however the errors made by the ideal CODEC system will be well within telephony specifications. To conform to the required Gain Tracking and Signal-to-Distortion specifications the DAC output currents must conform as closely as possible to the ideal transfer function as tabulated in the normalized tables. This corresponds to a specification of absolute error on the DAC output current with respect to its binary inputs. The DAC-86/89 companded DACs are guaranteed to plus or minus one-fourth step from ideal values in chord zero and to plus or minus one-half step elsewhere. This information can be transformed into tabular form by adding the allowable error to the DAC tables. Either the normalized tables or the current output tables can be used as a basis for this exercise.

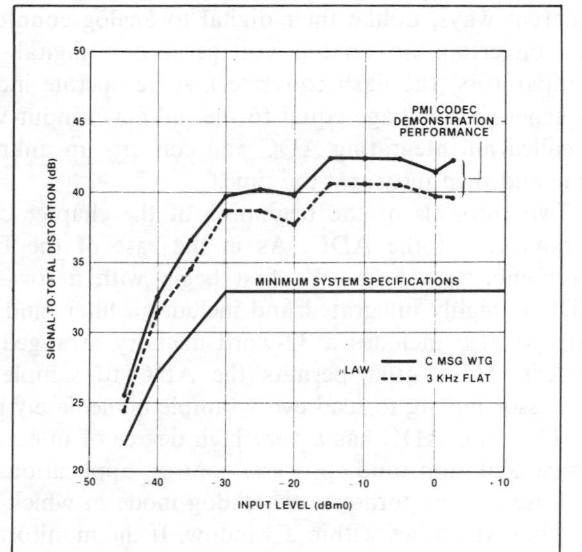

Figure 16. ATT/D3 Signal-to-Distortion Mask

AN-39

Chapter 4 Analog to digital converters

Analog to digital converters (ADCs) operate in at least three fundamentally different ways, unlike their digital to analog counterparts. Some operate directly by converting an analog voltage into a digital value by means of multiple comparators (the flash converter), some operate indirectly by using a local DAC to generate a voltage equal to the unknown input voltage. A third class of ADC is called an integrating ADC and converts an unknown voltage into a period of time and then measures the time.

Two tutorials at the beginning of the chapter cover the basic principles and parameters of the ADC. As in the case of the DAC, we look at a range of representative ADCs. We first begin with a low-cost 8-bit ADC. The second ADC is highly integrated and includes a filter, and a sample and hold circuit on chip. It also includes a 32-word memory arranged as a FIFO (first in first out buffer). This buffer permits the ADC to sample bursts of data without the processor having to read every sample immediately it becomes available.

The third ADC has a very high degree of integration. It is designed for use in instrumentation and process control applications. This device is interesting because it incorporates a watchdog mode in which one of the channels monitors a signal which lies within a window. If the monitored signal falls below the lower bound or above the upper bound, the device interrupts the host processor.

The fourth device is a 15-bit single-chip integrating ADC that can be used in a high-resolution data acquisition system. Data averaging is used to provide an effective resolution of 18-bits.

The final part of this chapter provides the data sheet and an application note for a 7-bit high-speed flash converter. The application note covers the design of an evaluation board for applications at video frequencies.

Analogue to Digital Converters

2. An analogue to digital converter (ADC) is a device which converts an analogue input into a corresponding digital output code.

2.1 Ideal output characteristics

Assuming a unipolar input voltage and binary coded output the transfer function of an ideal n-bit ADC is given by:

$V_{FS} \quad (B_1.2^{-1} + B_2.2^{-2} + \ldots. + B_n.2^{-n}) = V_{in}$
$\pm$ ½LSB.

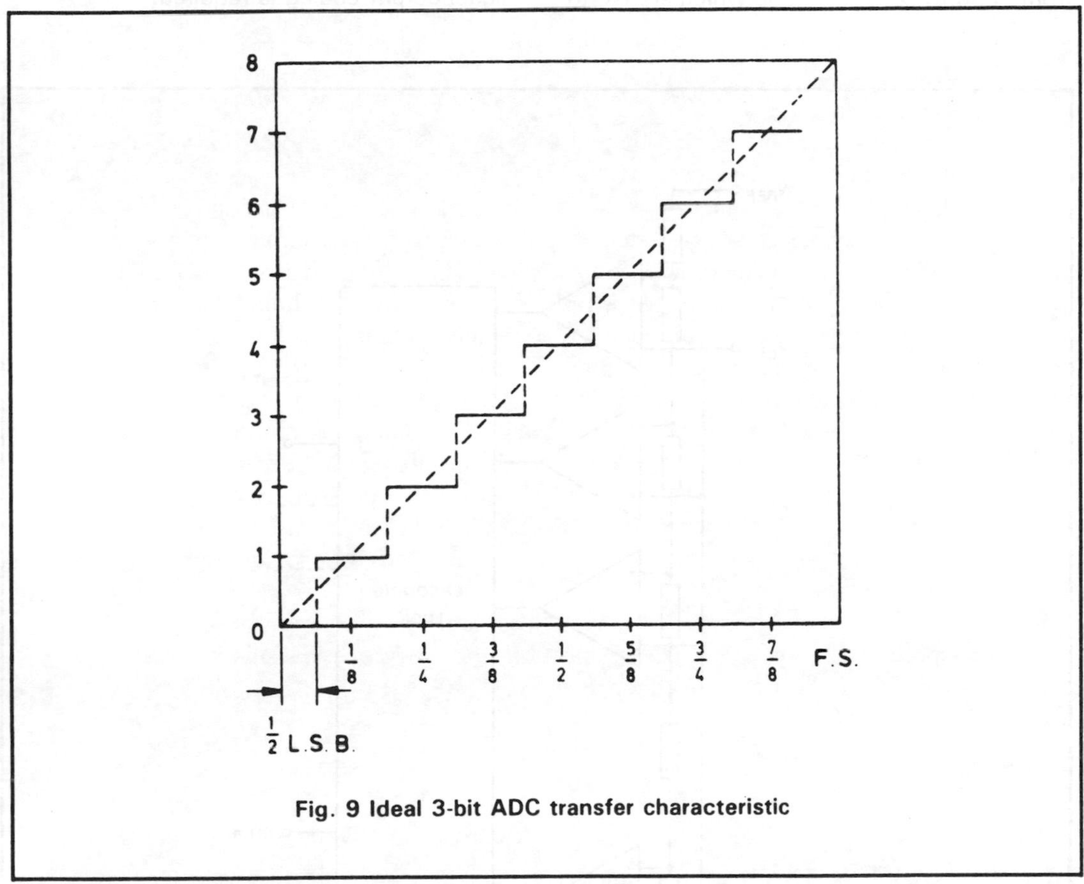

Fig. 9 Ideal 3-bit ADC transfer characteristic

The transfer function of an ideal 3-bit ADC is shown in Fig. 9. In this case there are 8 digital output codes corresponding to the 8 input codes of a DAC. However, unlike the analogue output of a DAC, the analogue input of an ADC can vary continuously, which means that each digital output code, with the exception of 0 and 7, exists over an analogue increment of 1LSB. The zero of an ADC is usually trimmed so that the transitions between codes occur $\pm$ ½LSB on either side of the nominal analogue input for a particular code. For example, the nominal input for output code 2 is ¼ V_{FS}. The transition from 1 to 2 occurs at $\frac{3}{16}$ V_{FS} and the transition from 2 to 3 occurs at $\frac{5}{16}$ V_{FS}.

As with a DAC, an 'ideal' straight line may be drawn through the transfer characteristic of an ADC.

2.2 Practical A-D conversion methods

There are many methods of performing an analogue to digital conversion; all of these methods are used in the current range of Plessey A-D converters.

2.2.1 Parallel (flash) conversion

In an n-bit parallel converter (Fig. 10) a resistor ladder is used to generate $2^n - 1$ voltage levels from 1LSB to $(2^n - 1) \times$ LSB which are fed to the reference inputs of $2^n - 1$ voltage comparators. The analogue input signal is fed to the second input of each comparator, and is thus compared simultaneously with each of the $2^n - 1$ voltage levels. At the point in the comparator chain where the reference voltage exceeds the input voltage the comparator outputs will change over from low to high. The comparator outputs are encoded into whatever digital output coding is required.

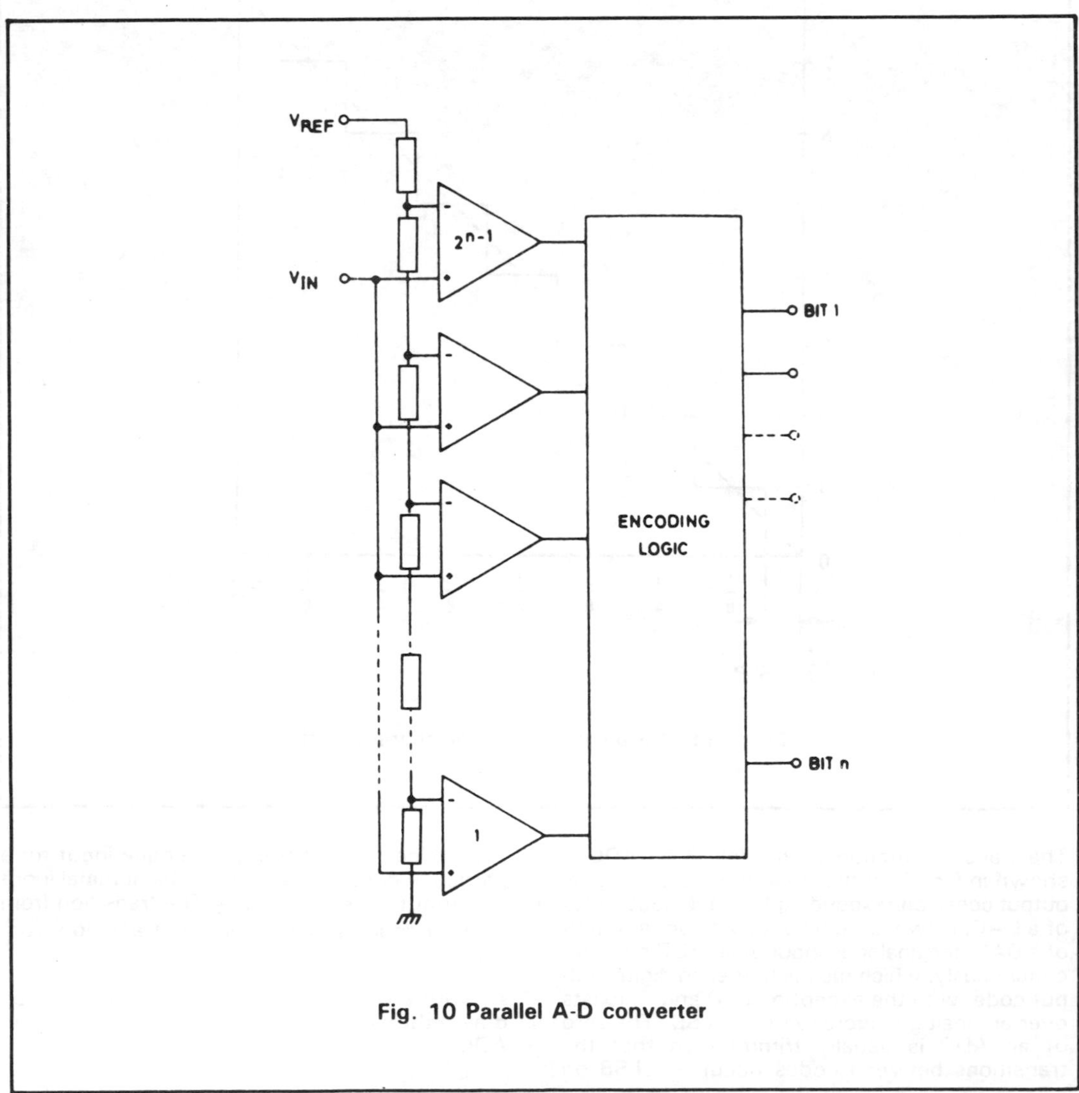

Fig. 10 Parallel A-D converter

Since the only delays involved in the conversion are the propagation delay of one comparator plus the logic propagation delays, parallel converters are very fast and may perform in excess of 10 million conversions per second. However, due to the large number of comparators required (63 for a 6-bit converter, 255 for an 8-bit converter) they are expensive to produce. Applications include digital video systems, digital storage oscilloscopes and radar data processing.

2.2.2 Staircase and comparator

In this type of ADC the input code of a DAC is incremented by a binary counter to give a staircase waveform, as shown in Fig. 11. This is compared with the analogue input and when the staircase exceeds the analogue voltage the comparator output changes state and stops the clock. The count reached by the binary counter is thus the ADC output code. This method of A-D conversion is relatively simple and cheap, but is also relatively slow, requiring $2^n - 1$ clock pulses for a full-scale conversion, where n is the number of bits. This conversion method is used in the ZN425 series of dual-purpose D-A/A-D converters.

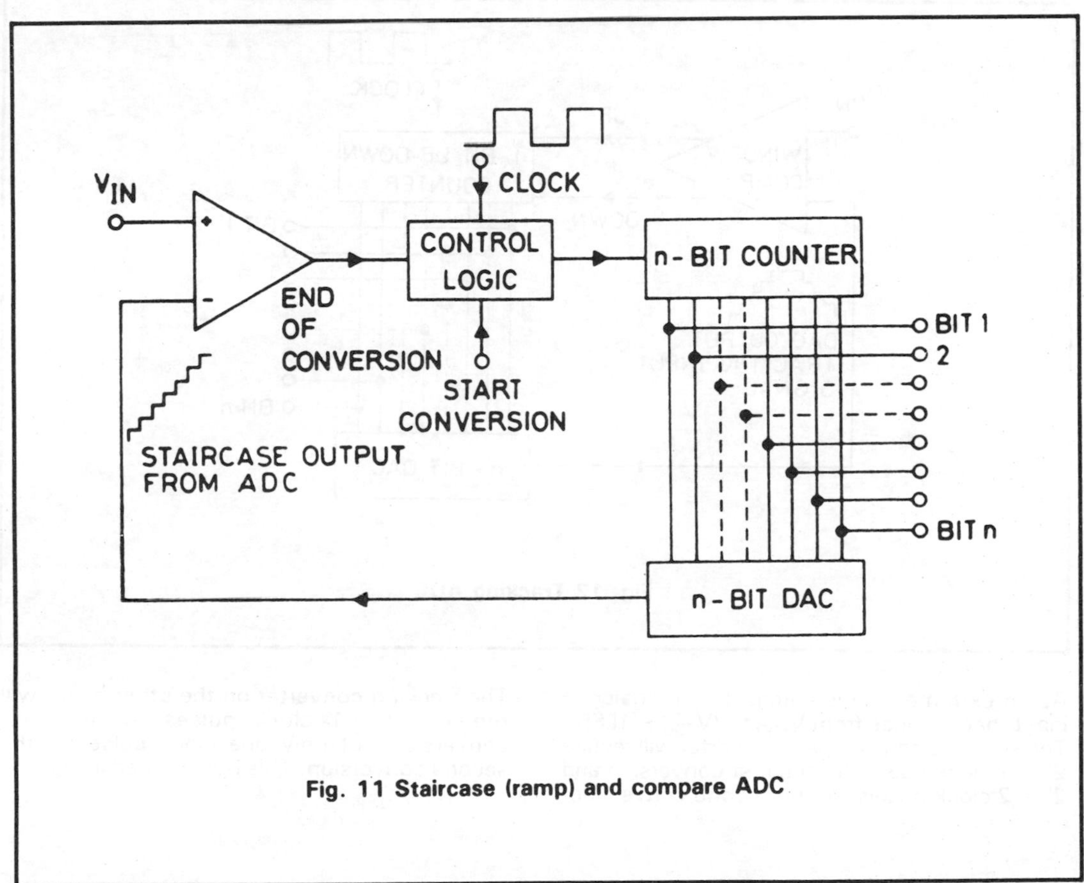

Fig. 11 Staircase (ramp) and compare ADC

2.2.3 Tracking converters

As its name implies, a tracking converter can follow changing analogue inputs. The principle operation is similar to that of the staircase and compare type of converter, but it uses an up/down counter and a window comparator, as shown in Fig. 12. When the DAC output is less than the analogue input the comparator instructs the counter to count up and the DAC output thus increases. If the DAC output is greater than the analogue input the comparator causes the counter to count down, thus decreasing the DAC output. When the DAC output is equal to the analogue input ± ½LSB, the input is within the 'window' of the comparator and the counter is stopped. This is illustrated in Fig. 13. A tracking converter has speed advantages over a staircase and compare type, since the counter of the latter type can only count up, and must therefore be reset between conversions. In the case of a tracking converter, once it has performed an initial conversion starting from zero, any subsequent conversions require only that number of clock pulses necessary to track any increase or decrease in input voltage.

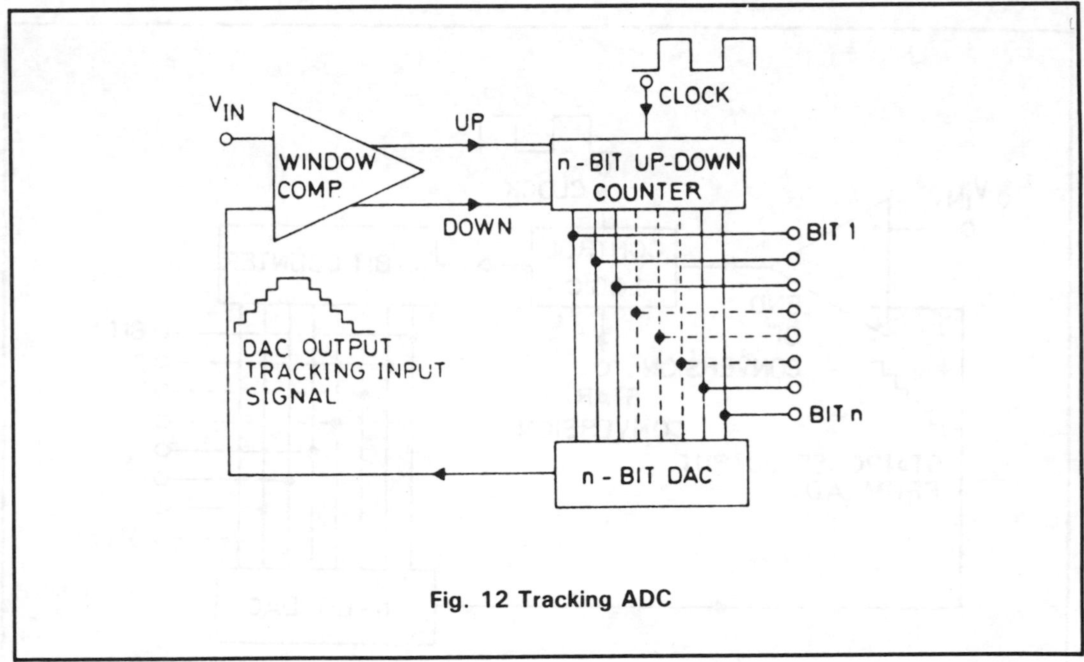

Fig. 12 Tracking ADC

As an extreme example consider an analogue input that changes from V_{FSO} to (V_{FSO} − 1LSB). The staircase and compare converter will require $2^n - 1$ clock pulses for the first conversion and $2^n - 2$ clock pulses for the second conversion. The tracking converter on the other hand, will require $2^n - 1$ clock pulses for the first conversion but only one clock pulse for the second conversion. This is illustrated in Fig. 14.

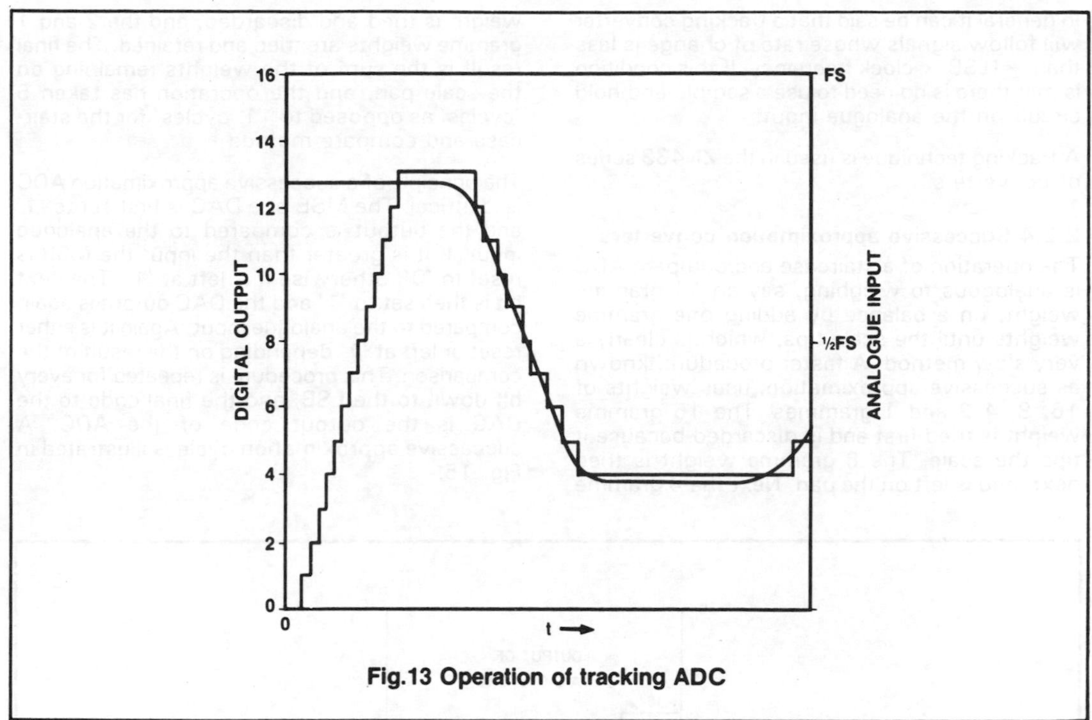

Fig.13 Operation of tracking ADC

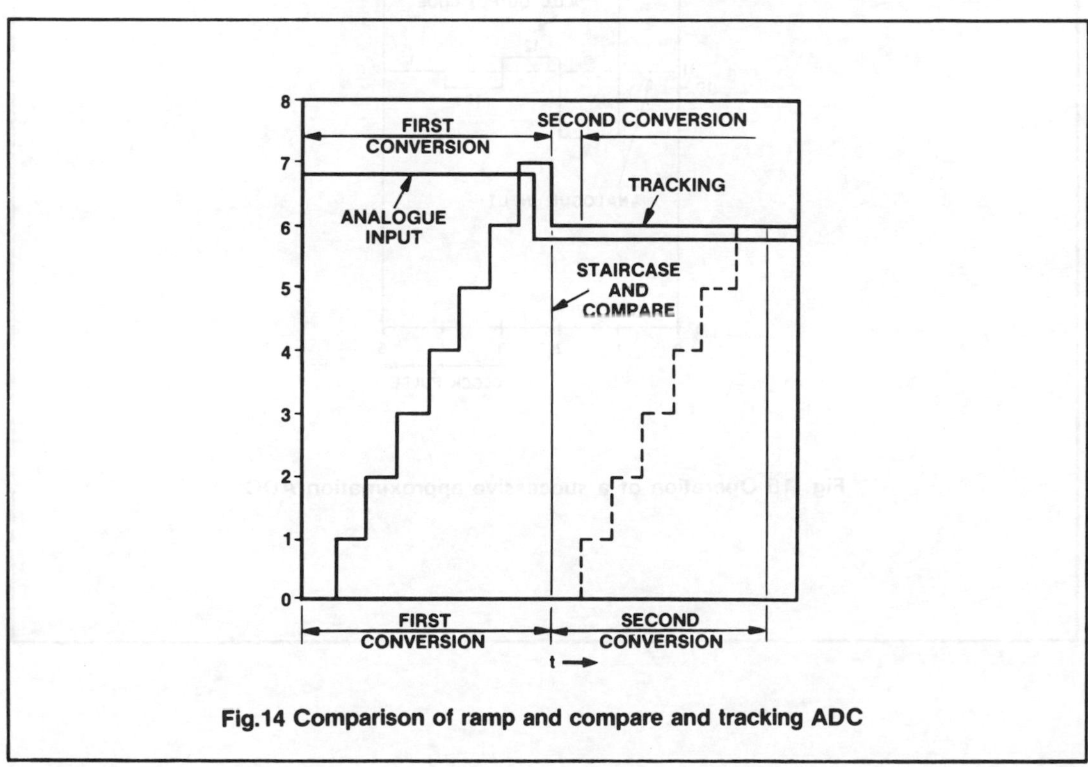

Fig.14 Comparison of ramp and compare and tracking ADC

In general it can be said that a tracking converter will follow signals whose rate of change is less than $\pm$ 1LSB × clock frequency. If this condition is met there is no need to use a sample-and-hold circuit on the analogue input.

A tracking technique is used in the ZN433 series of converters.

2.2.4 Successive approximation converters

The operation of a staircase and compare ADC is analogous to weighing, say an 11 gramme weight, on a balance by adding one gramme weights until the scale tips, which is clearly a very slow method. A faster procedure, known as successive approximation, uses weights of 16, 8, 4 2 and 1 grammes. The 16 gramme weight is tried first and is discarded because it tips the scale. The 8 gramme weight is tried next, and is left on the pan. Next the 4 gramme weight is tried and discarded, and the 2 and 1 gramme weights are tried and retained. The final result is the sum of the weights remaining on the scale pan, and the operation has taken 5 'cycles' as opposed to 11 'cycles' for the staircase and compare method.

The principle of a successive approximation ADC is identical. The MSB of a DAC is first set to '1' and the output is compared to the analogue input. If it is greater than the input the MSB is reset to '0', otherwise it is left at '1'. The next bit is then set to '1' and the DAC output is again compared to the analogue input. Again it is either reset or left at '1' depending on the result of the comparison. This procedure is repeated for every bit down to the LSB, and the final code to the DAC is the output code of the ADC. A successive approximation cycle is illustrated in Fig. 15.

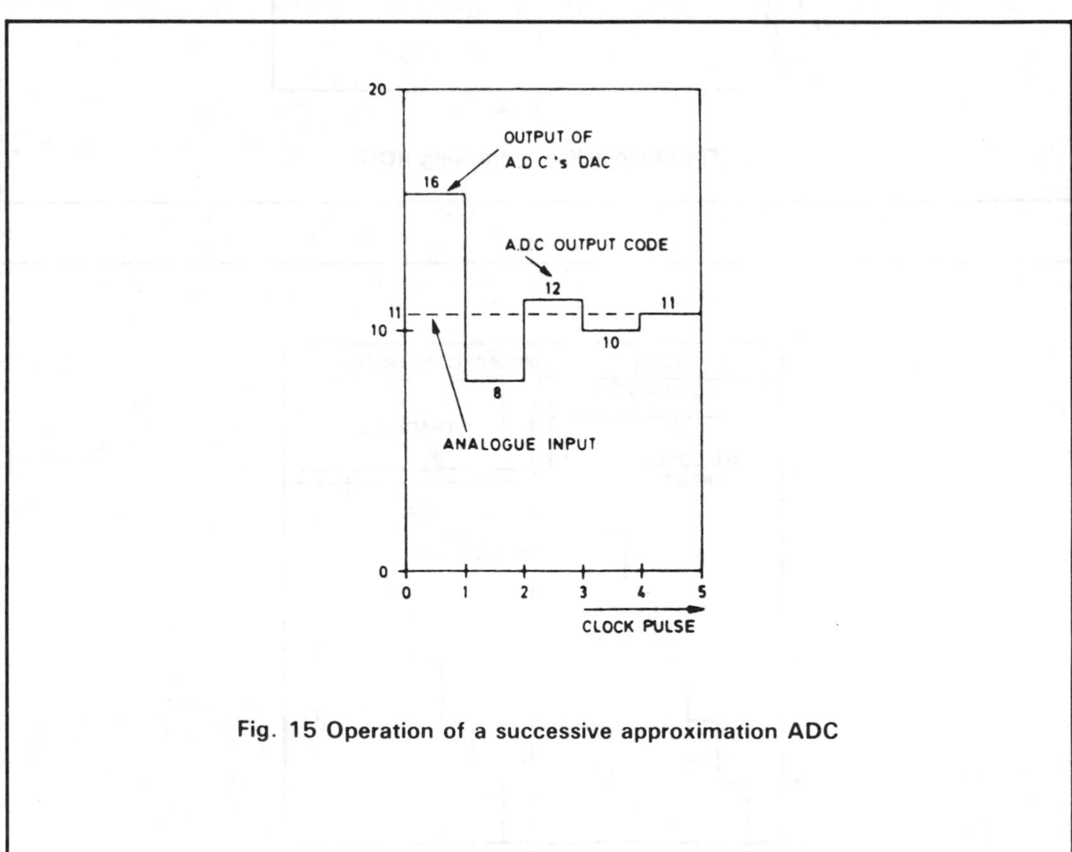

Fig. 15 Operation of a successive approximation ADC

The successive approximation technique is used in most of the Plessey data converters that operate below video speeds.

2.2.5 Dual-slope converters

Dual-slope integration is one of the slowest methods of A-D conversion, but it offers high resolution at a modest cost.

A block diagram of a dual-slope converter is shown in Fig. 16. It operates in the following manner: Switch S1 is closed by the control logic, S4 is opened and the input voltage is integrated for n clock periods, where n is usually the maximum count of the counter. At the end of this time the integrator output voltage, V_O, is $\frac{-V_{in}\,n\,T_c}{RC}$ where T_c is the clock period. This is shown in Fig. 17.

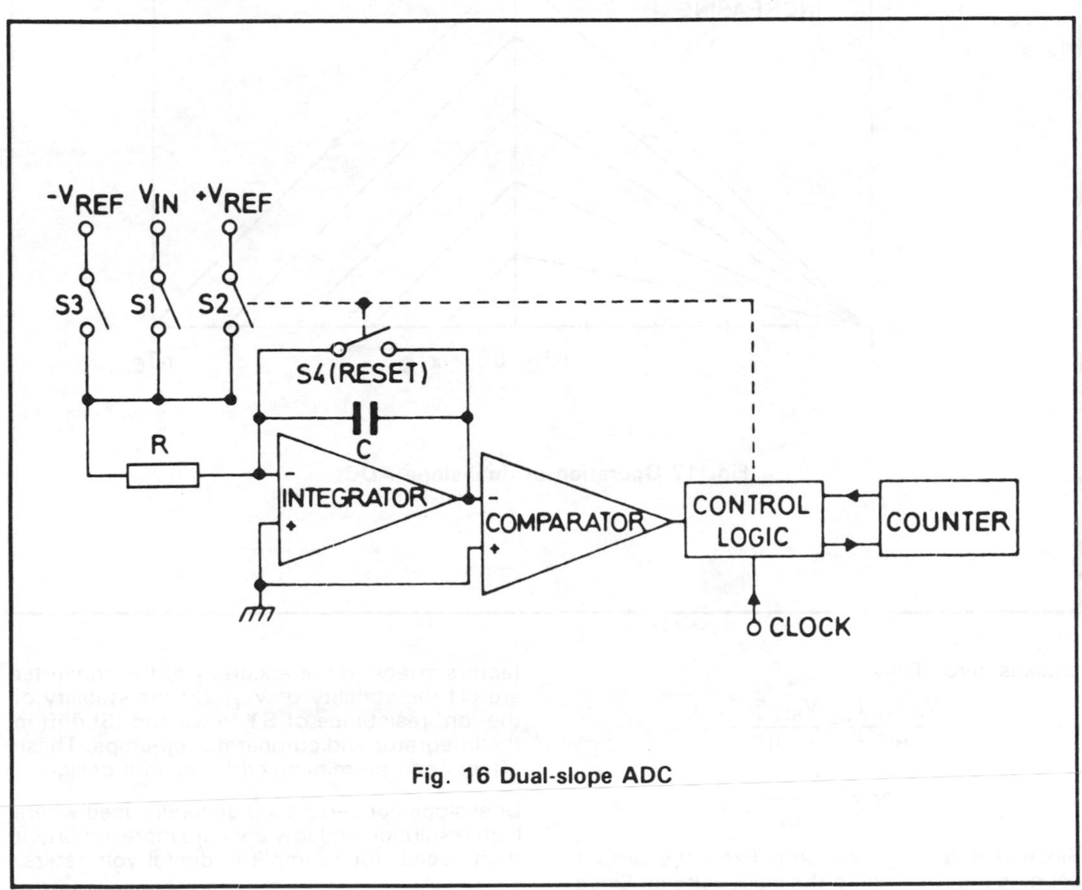

Fig. 16 Dual-slope ADC

During this period the polarity of the input signal is detected by the comparator. At the end of the integration period S1 is opened and, depending on the polarity of V_{in}, either S2 or S3 is closed to connect the integrator to a reference voltage of opposite polarity to V_{in}. The counter is now allowed to count from zero until the integrator output reaches 0V, when the comparator output changes state and the counter is stopped. Since the integration is over the same voltage range (V_O), $V_O = \frac{-V_{REF}\ X\ T_c}{RC}$, where X is the count reached by the time the integrator output

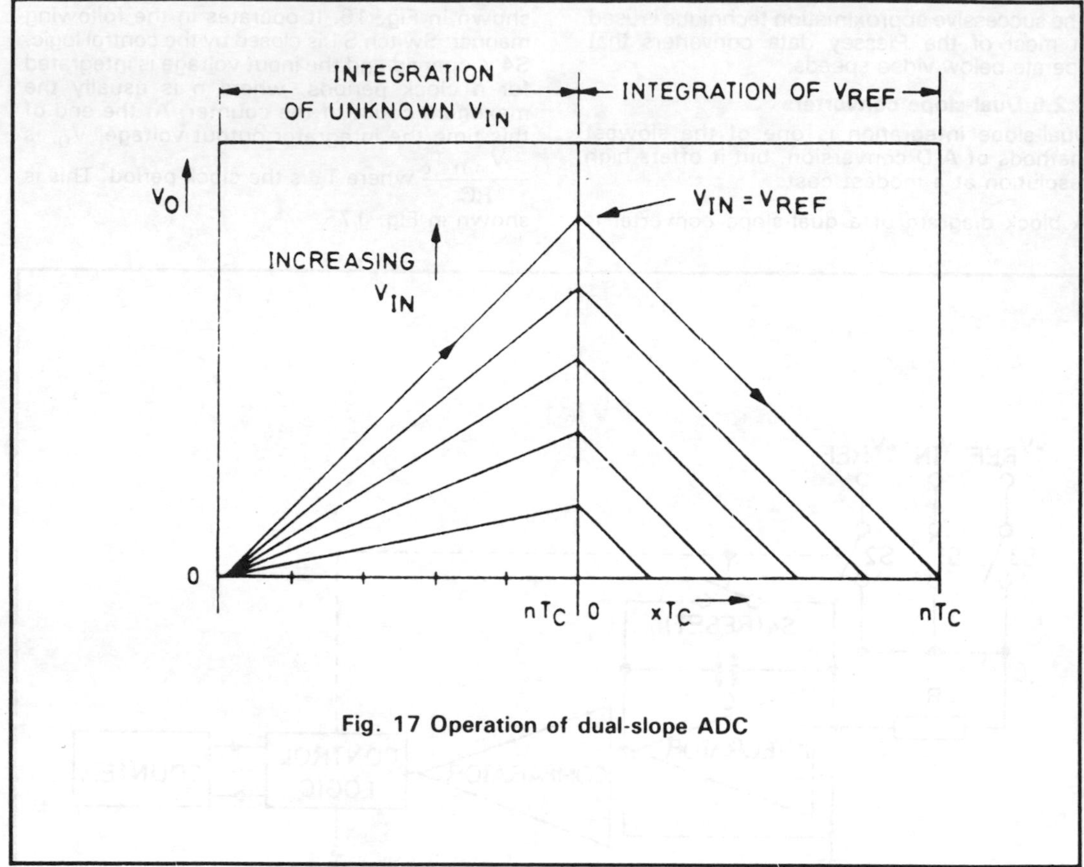

Fig. 17 Operation of dual-slope ADC

crosses zero. Thus

$$\frac{V_{in}.\ n\ T_c}{RC} = \frac{V_{REF}\ X\ T_c}{RC}$$

$$\text{or } X = \frac{V_{in}.\ n}{V_{REF}}$$

Since n and V_{REF} are both fixed the output count is proportional to the input voltage. Since both the first and second integrations occur under identical conditions the converter is unaffected by any long term variations in T_c, R or C, as demonstrated by the disappearance of these terms from the final equation. The only factors affecting the accuracy of the converter are (1) the stability of V_{REF} (2) the stability of the 'on' resistance of S1 to S3 and (3) drift in the integrator and comparator op-amps. These affects can be minimised by careful design.

Dual-slope converters are generally used where high resolution and low cost are more important than speed, for example in digital voltmeters.

The ZNA216 is a DVM logic sub-system containing the clock, counter and all control logic necessary for dual-slope converter or DVM.

2.3 A-D parameters and definitions

2.3.1. A-D converter errors

Like DACs, practical ADCs are subject to a number of error sources, and since most ADCs contain a reference DAC, many of these error sources are the same for both types of converter.

2.3.2 Quantising error (uncertainty)

Quantising error is an ADC specification that has no counterpart in DAC specifications. For each input code of a DAC there is a unique analogue output level, but for any ADC output code there is a 1LSB range of analogue input levels. It is thus not possible to tell from the output code the precise value of the analogue level, there being a quantising error or uncertainty of ± ½LSB. Since all ADCs have this inherent quantising error the parameter is frequently not quoted in specifications.

2.3.3 Missing codes

Missing codes are perhaps best explained by considering the operation of a staircase and compare type 3-bit ADC which has a non-monotonic DAC, as shown in Fig. 18. The reference DAC exhibits non-monotonicity at input code 4, i.e. step 4 of the staircase decreases. There is thus no way in which the counter can be stopped at this code. If the analogue input is less than the DAC output for code 3 then the comparator will stop the counter before 4 is reached. If the analogue input is greater than output 3 it must also be greater than output 4, so the comparator will not change state at code 4.

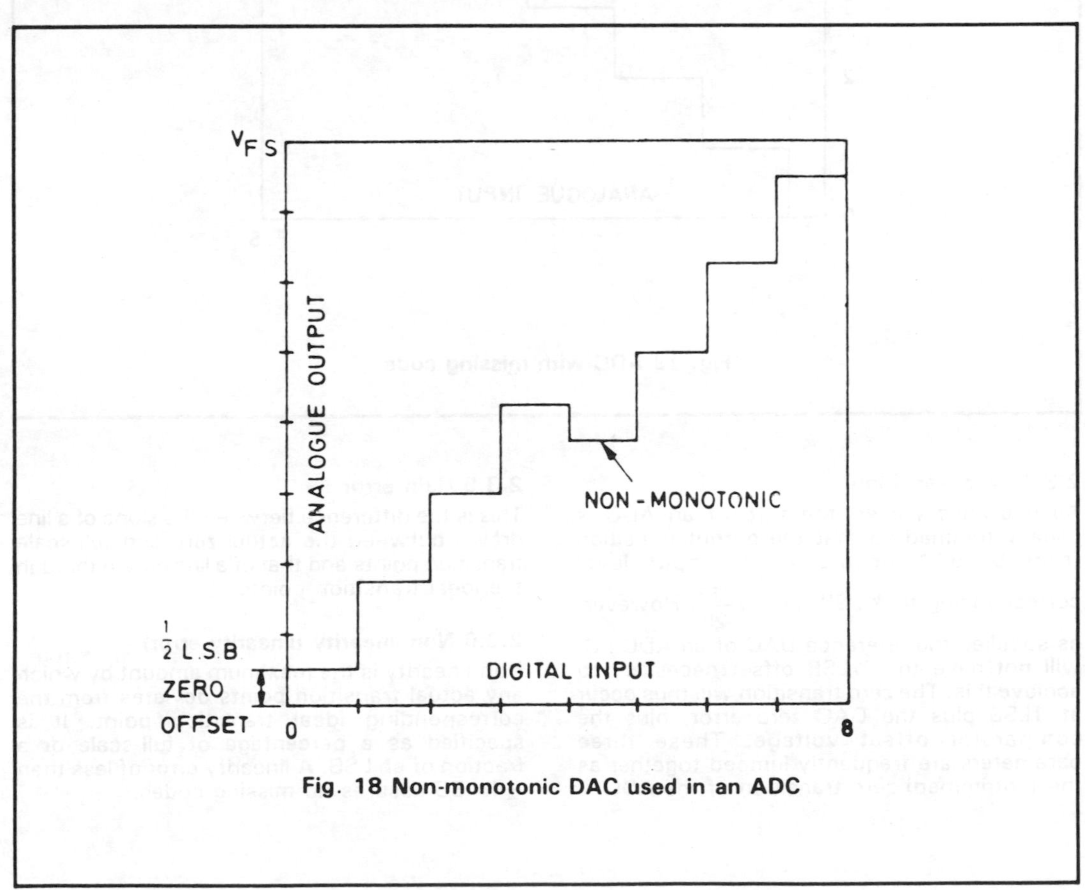

Fig. 18 Non-monotonic DAC used in an ADC

Output code 4 will thus never appear and is known as a 'missing' code. The transfer function of an ADC with a missing code is shown in Fig. 19.

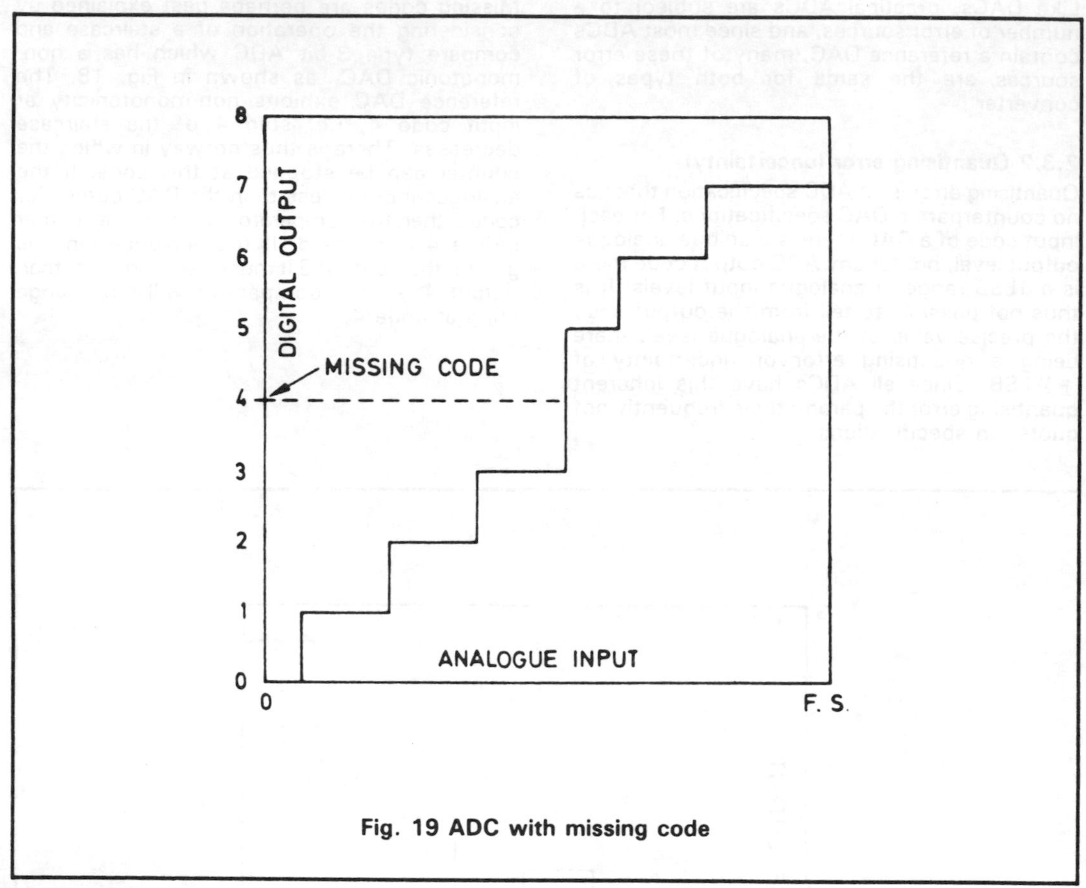

Fig. 19 ADC with missing code

2.3.4 Zero transition

As explained earlier, the zero of an ADC is usually trimmed so that the output transition from 0 to 1 occurs at an input level corresponding to ½LSB, i.e. $\frac{1}{2}\frac{V_{FS}}{2^n}$. However, as supplied the reference DAC of an ADC I.C. will not have the ½LSB offset necessary to achieve this. The zero transition will thus occur at 1LSB plus the DAC zero error, plus the comparator offset voltage. These three parameters are frequently lumped together as the (untrimmed) zero transition of the ADC.

2.3.5 Gain error

This is the difference between the slope of a line drawn between the actual zero and full-scale transition points and that of a line drawn through the ideal transition points.

2.3.6 Non-linearity (linearity error)

Non-linearity is the maximum amount by which any actual transition points deviates from the corresponding ideal transition point. It is specified as a percentage of full-scale or a fraction of an LSB. A linearity error of less than ± ½LBS assures no missing codes.

2.3.7 Differential non-linearity

This is the maximum difference between any 1LSB increment of the analogue input and the ideal size of an LSB increment $\frac{V_{FS}}{2^n}$. Differential non-linearity of less than 1LSB guarantees no missing codes.

2.3.8 Resolution

The resolution of an ADC is simply the number of bit outputs that the converter possesses. As with a DAC, resolution implies nothing about the accuracy of a device.

2.3.9 Useful resolution

Useful resolution is the resolution (number of bits) at which an ADC has no missing codes, which for Ferranti ADCs is guaranteed over the operating temperature range. As with DACs, an n-bit ADC may have a useful resolution less than n bits, for reasons previously explained.

2.3.10 Conversion time

The time taken for an ADC to perform a complete conversion is known as the conversion time. For successive approximation converters conversion time is fixed by the number of bits and the clock frequency. However, for other types, conversion time may vary with input voltage. For example, a ramp and compare ADC requires $2^n - 1$ clock pulses for a full-scale conversion but only one clock pulse for a one bit conversion. It is thus important to check exactly what is being specified.

2.4 Bipolar operation

As with a DAC, an ADC may be used for bipolar operation. Taking the ZN427 as an example the input is offset by $\frac{+V_{REF}}{2}$ so that the input voltage presented to the ADC is always positive, even with negative input voltages down to $\frac{-V_{REF}}{2}$. The principle of offsetting an ADC input is illustrated in Fig. 20, whilst the transfer function of a 3-bit bipolar ADC is shown in Fig. 21. In this case the **output** coding is known as offset binary.

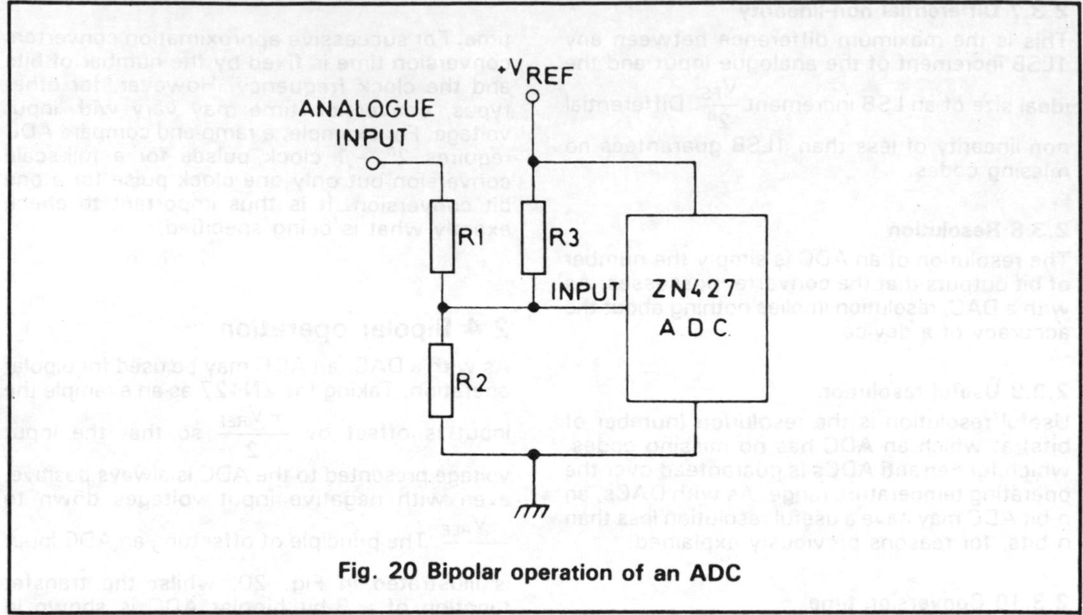

Fig. 20 Bipolar operation of an ADC

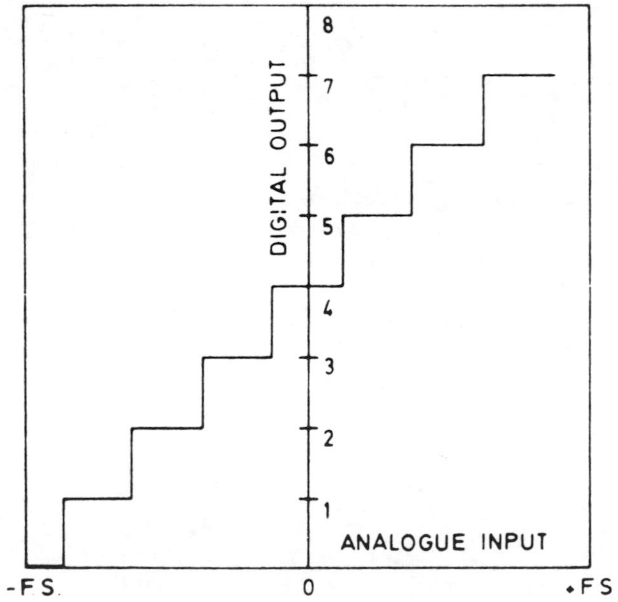

Fig. 21 Bipolar transfer characteristics of an ADC

Orientation Analog-to-Digital Converters

3

FACTORS IN CHOOSING AN A/D CONVERTER

In this catalog, there are listed approximately 50 different families of analog-to-digital converters (ADCs). If one were to consider all the variations, there would be considerably more than 100 different types to choose among. The reason for so many different types is the number of degrees of freedom in selection-technological, functional, performance and package. Complete information on converters may be found in the 700-page book, *Analog-Digital Conversion Handbook*, published by Prentice-Hall, Inc.

FUNCTIONAL CHARACTERISTICS

Block diagrams illustrating the various conversion techniques appear on individual data sheets.

The moderate-speed converters described in this catalog (<1MHz) employ two fundamental techniques – *successive approximations* for moderate-to-high resolution at moderate-to-high speed, and *integration* for high resolution at modest speeds. The AD574A and ADC1130/1131 are examples of the former, the AD1175 the latter.

Like a chemist's balance with binary weights (1/2, 1/4, 1/8, etc.), the *successive approximation* converter compares the unknown input with sums of accurately-known binary fractions of full scale starting with the largest (2^{-1}) and rejecting any that change the comparator's state ("tip the scale"). At the end of conversion (EOC), the output of the converter is a digital word representing the ratio of the input to full scale by a fractional-binary code.

Integrating types count pulses for a period proportional to the input. The charge balancing integrating converter (essentially a voltage-to-frequency converter) measures the input signal by balancing a proportional current against a train of precisely controlled reference pulses using an integrator (AD1170). During the integration phase, the input signal is measured; during the computation phase, the data from the first phase is processed and calibration factors applied. This type of converter can provide very high resolution and accuracy.

The video converters described here (AD9002, CAV-1205, etc.) employ two basic encoding techniques: simultaneous, or *flash* conversion, and serial-Gray-Code conversion. High resolution and high speed are obtained by *subranging* i.e., by performing an n-bit conversion in two steps; Analog Devices has perfected a form of subranging known as DCS – *digitally corrected subranging* – which permits accurate resolutions of 12 bits and more.*

In *flash* conversion, the analog signal is compared against $2^n - 1$ graded voltage levels using as many comparators, and the comparator output logic levels are processed by a priority encoder which converts the "thermometer" output to a binary (or Gray) code. Since the whole conversion occurs essentially simultaneously, it is the fastest means of conversion, but it requires many accurate comparators and large numbers of gates.

In *serial* analog-parallel-digital conversion, there are a number of cascaded stages, each having a gain of $+2$ for signals less than one-half the reference, and a gain of -2 for signals between one-half the reference and full scale. At each stage, a decision is made as to whether the signal is larger (1) or smaller (0) than one-half the reference; the stage's analog output becomes the input to the next stage. The complete time for one conversion is determined by the propagation delay of the analog signal through all stages; however, since the decision of each stage can be latched as soon as the stage has settled (and a new conversion can, in principle, be started as soon as the first bit has been latched), the rate at which conversions come out of the pipeline is considerably faster than the time for one sample to go through the conversion process. Though fast, this process is difficult to implement accurately for more than a few bits because of the compounding of gain (hence errors).

A *subranging* converter digitizes to a group of more-significant bits and stores them in a latch. A fast, very high-accuracy D/A converter converts them to an analog signal which is then subtracted from the input. The difference, or residue, is amplified and digitized, and (in DCS) the result is combined digitally in such a way as to correct for midscale conversion errors.

Whatever the technique, these A/D converters comprise several essential functions: an analog section, a digital data-generating section, data outputs and digital controls.

Analog Section

This section requires a reference, one or more high-gain comparators, and either a D/A converter (successive approximations) or a controllable integrator. The reference may be internal or external, fixed or variable, and of a specified polarity/sense in relation to the analog input. In ratiometric conversion, the reference is usually external and variable.

In successive approximation converters, the comparator is generally used in the *current-summing* mode; that is, the current output of the DAC is summed with the current developed in the DAC's "feedback resistor" by the input voltage (of opposite polarity), and the balancing action of the converter tends to bring the summing junction towards a voltage null (much like that of an op amp) at the end of conversion. The typical DAC feedback options, when applied in an ADC, provide input-scaling choices. When the bipolar-offset connection is jumpered to the summing point, input signals of both polarities can be handled. The current-switching action of the DAC, at the typically fast clock rates used in successive approximation converters, can disturb the output of the analog signal source, especially if it is a slow high-precision op amp. In such cases, buffering may be necessary.

Sample and Hold

When an ADC without a sample and hold is used, the analog input must not change by more than 1/2 LSB during the conversion. For some applications this constraint is not a concern, but it limits the bandwidth of the signal that can be applied to the converter. A sample-and-hold circuit must be used in front of the ADC if increased bandwidth is required. This sample and hold can be external, or an integral part of the converter (e.g., AD7579/AD7580).

***A considerable amount of useful information about the differences between video conversion and moderate-speed conversion can be found in the article "Very High Speed Data Acquisition," by Ed Graves in *Analog Dialogue* 13–2, available upon request.**

Digital Data-Generating Section
In successive-approximation types, this section consists of a discrete or integrated successive-approximations register (SAR), its controls and inputs from the comparator and clock (which is on-board, but in many cases permits external clock pulses, frequency adjustment and/or control). In integrating types, this section consists of the clock-pulse generator, the counter(s), the input from the comparator and the associated controls. Often, provisions are made for the pulse train to be jumpered to the counter externally so that the pulse train can be operated on externally, or can transmit its train of pulses to a remote counter. In a few types there are no on-board counters or registers; the pulse train, magnitude, overrange and control terminals are intended to communicate with external counters and registers.

Data Outputs
Factors to consider here include coding, resolution, overrange information, levels, format, validity and timing. *Coding* is usually binary including jumper-connected offset-binary and/or twos complement for bipolar input signals. For some types, BCD is available with sign-magnitude for bipolar inputs. Output coding specs should always be checked for digital polarity (positive- or negative-true) of both magnitude and sign information. The *resolution* (number of output bits) must be sufficient for the application; in addition, the specifications must be checked to ascertain that not only will all 2^n (binary) output codes be present (no missing codes), but they must all be present at any temperature in the operating range and related to the input with sufficient accuracy. Integrating types generally have no problems with missing codes (except sometimes at zero, with sign-magnitude coding); nevertheless, nonlinear integration can cause the conversion relationship to become nonlinear. Successive-approximation types have no way of determining *overrange*; they simply fill up. However, counter types roll over and put out a carry flag to signal overrange. Analog Devices offers ADCs, with 4- through 22-bit resolution, with a span of conversion times from milliseconds to nanoseconds.

The *data levels* available at the converter output must be checked (TTL, low-voltage CMOS, high-voltage CMOS, ECL), as must the load-driving capability and fanout, and the supply conditions under which appropriate output levels will be furnished. The available choice of output *formats* must also be as desired – parallel, serial, byte-serial, and/or pulse-train. If the converter is intended to communicate directly with an 8-bit data bus, the output should have three-state capability, and parallel outputs must be enabled in bytes of eight or fewer lines (AD573, AD574A). If the output is serial, it is usually NRZ (non-return-to-zero) and should be accompanied by a set of synchronized clock-pulses.

A *status* (or *busy* or EOC) output changes state to indicate when the data becomes *valid*. The exact nature of this transition should be specified – polarity, timing, levels, etc. For serial data, the exact relationship between the data and the synchronizing clock should be specified to indicate when each bit becomes valid, and for how long. In general, the *timing* of the whole conversion process must be clearly understood, especially if high speeds are necessary, either for conversion or for communication with a processor (or both). The timing diagrams on specification sheets are usually accompanied by adequate descriptions of the conversion process and specifications of the critical interface parameters.

Controls
The functions, action (levels or edges), polarity and timing of all control inputs and outputs should be clearly understood, as well as their loading characteristics and dependence on the supply. In addition to the essential *start-conversion-command* input and a *status* output, various control commands may be available, such as *clock inhibit, high- (low-) byte enable, status enable* and – for speeding up conversion at the cost of resolution in successive-approximation converters – *short-cycle*.

Many ADCs are designed to interface directly to the bus of a computer or microprocessor. These ADCs provide the necessary control and handshake lines, as well as data bit registers, to minimize and often eliminate the required interface circuitry. The bus timing should be studied with respect to the timing provided by the ADC interface, especially as the processor executes a data read cycle to the ADC to retrieve the conversion results. Systems with higher speed clocks require either shorter minimum write pulse widths (such as 50ns for the AD7579/AD7580) or the use of processor-wait states when the ADC is addressed.

STATIC AND DYNAMIC PERFORMANCE SPECIFICATIONS

All ADCs are specified using terms such as accuracy, linearity, offset, defined and explained below. These static, or "dc," parameters are necessary and sufficient for many applications; they may not be sufficient for others, such as those in digital signal processing, adaptive filtering or waveform generation. Dynamic ac specifications define how the ADC performs using parameters such as signal-to-noise ratio (SNR), intermodulation distortion (IMD) and total harmonic distortion (THD). These specifications characterize the performance of the ADC output in applications where the envelope of output changes and output timing errors are critical.

POWER SUPPLIES

Appropriate power supplies should be made available considering the logic levels and analog input signals to be employed in the system. The appropriate degree of power-supply stability to meet the accuracy specifications should be provided. Any recommended external protection circuitry should be planned for. In many cases, separate analog and digital grounds are required; ground wiring should follow best practice to minimize digital interference with high-accuracy analog signals while ensuring that a connection between grounds can always exist at one point, even if the "mecca" point is inadvertently unplugged from the system.

APPLICATION CHECKLIST

The designer will generally require specific information in the following categories before proceeding to the selection process:

- Accurate description of input and output
  1. Analog signal range and source or load impedance
  2. Digital code needed – binary, offset binary, twos complement, BCD, etc.
  3. Logic level system, i.e., TTL/DTL compatible

- What is the needed data throughput rate?
- What are the control interface details?
- What does the system error budget allow for the converter?
- What are environmental conditions – temperature range, time, supply voltage – over which the converter should operate to the desired accuracy?

For A/D converters, the following considerations are typical.

- What is the analog input voltage range, and to what resolution must the signal be measured?
- What is the requirement for linearity error (or relative accuracy error)?
- To what extent must the various sources of error be minimized as environmental temperature changes?
- How much time can be allowed in the system for each complete conversion? What aperture uncertainty and acquisition time are needed for the sample-hold?
- How stable is the system power supply? What errors will result from power supply terminal voltage variations of this order?
- Can the system tolerate missed codes under any conditions?
- What is the character of the input signal? Is it noisy, sampled, filtered, rapidly varying, slowly varying? What kind of pre processing is to be (or can be) done that will affect the choice (and cost) of the converter? Is aliasing a potential problem?

SPECIFICATIONS AND TERMS

Definitions of performance specifications and related information are to be found on the following pages in alphabetical order.*

Accuracy, Absolute

The error of an A/D converter at a given output code is the difference between the theoretical and the actual analog input voltages required to produce that code. Since the code can be produced by any analog voltage in a finite band (see Quantizing Uncertainty), the "input required to produce that code" is defined as the midpoint of the band of inputs that will produce the code. For example, if 5 volts (±1.2mV) will theoretically produce a 12-bit half-scale code of 100000000000, then a converter for which any voltage from 4.997V to 4.999V will produce that code will have absolute error of 1/2 (4.997 + 4.999) – 5 volts = –2mV.

Absolute error comprises gain error, zero error and nonlinearity, together with noise. Absolute-accuracy measurements should be made under a set of standard conditions with sources and meters traceable to an internationally accepted standard.

***For video converters, there are a number of additional application-oriented specifications pertaining to the device's use in a system (e.g., *noise power ratio, differential phase, differential gain, signal-to-noise ratio*). Some useful references for understanding such specifications can be found in the following publications available from Analog Devices, Computer Labs Division, 7910 Triad Center Drive, Greensboro, NC 27409.**

Kester, W.A., "PCM Signal Codecs for Video Applications," *SMPTE Journal*, Volume 88, November 1979, pp 770-778.

Pratt, W.J., "Test A/D Converters Digitally," *Electronic Design*, December 6, 1975.

Smith, B.F. and Pratt, W.J., "Understanding High-Speed A/D Converter Specifications," Computer Labs, 1974.

Accuracy, Relative

Relative accuracy error, expressed in %, ppm or fractions of an LSB, is the deviation of the analog value at any code (relative to the full analog range of the device transfer characteristic) from its theoretical value (relative to the same range) after the full-scale range (FSR) has been calibrated.

Since the discrete points on the theoretical transfer characteristic lie on a straight line, this deviation can also be interpreted as a measure of nonlinearity (see Linearity).

The "discrete points" of an A/D transfer characteristic are the midpoints of the quantization bands at each code (see Accuracy, Absolute).

Aperture Time

This is the interval between the application of the *hold* command to a sample/track-hold and the actual opening of the switch. The aperture time consists of a delay (which depends on the logic and the switching device – 5ns for HTS-0025) and an uncertainty (due to jitter – 20ps max rms for HTS-0025). When a sample-hold is used in an application where timing is critical, the timing of the hold command can be advanced to compensate for the known component of aperture delay. The jitter, however, imposes the ultimate limitation on timing accuracy. When a sample-hold is used with an ADC, the timing uncertainty of the conversion process is reduced by the ratio of aperture jitter to the conversion time, i.e., the maximum frequency which can be handled with less than 1LSB error due to timing is $2^{-n}/(\pi\ \tau_{au})$ instead of $2^{-n}/(\pi\ \tau_c)$, where τ_{au} is the aperture uncertainty and τ_c is the conversion time.

Common-Mode Rejection (CMR)

The ability of a device to reject the effect of voltage applied to both input terminals simultaneously. Usually expressed as the log of a "common-mode rejection ratio," e.g., 1,000,000:1 (CMRR) or 120dB (CMR). A CMRR of 1,000,000 to 1 means that a 1V common-mode voltage passes through the amplifier as though it were a differential signal of one microvolt at the input.

Conformance, Straight-Line

This indicates how closely the ADC transfer characteristic conforms to a reference straight line. This straight-line conformance is critical in DSP applications where deviations from a straight line are seen as distortion, while gain and offset errors are not as serious. The straight-line conformance error is measured from the center of each code to the best-fit straight line.

Conversion Time and Conversion Rate

The time required for a complete measurement by an ADC is called *conversion time*. For most converters (assuming no significant additional systemic delays), this is identical to the inverse of *conversion rate*. However, in some high-speed converters, because of pipelining, new conversions are initiated before the results of prior conversions have been determined; thus, for example, the CAV-1250 can provide 12-bit output data at a 3.85MHz word rate (260ns/conversion), even though the time for any one conversion, from start to finish, is two 280ns encode periods plus 195ns, or 755ns at 3.85MHz.

Dual-Slope Converter

An integrating analog-to-digital converter in which the unknown signal is converted to a proportional time interval, which is then measured digitally. This is done by integrating the unknown for a predetermined time. Then a reference input is switched to the integrator and integrates "down" from the level determined by the unknown until a "zero" level is reached. The time for the second integration process is proportional to the average of the unknown signal level over the predetermined integrating period. A digital time-interval meter (i.e., counter) is generally used as the output indicator.

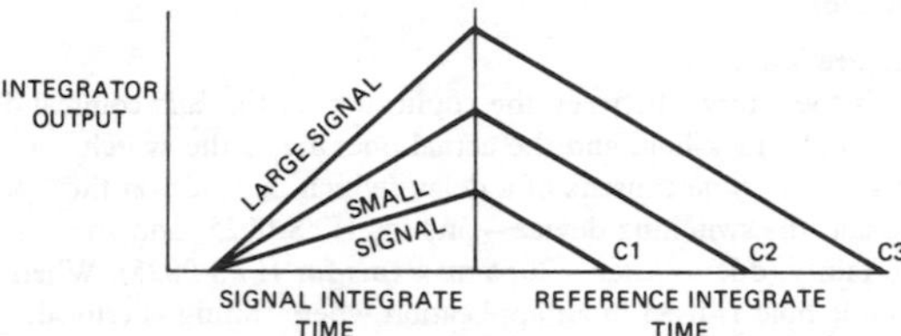

Feedthrough

Undesirable signal coupling around switches or other devices that are supposed to be turned off or provide isolation, e.g., *feedthrough error* in a multiplexer. It is variously specified in %, ppm, fractions of 1LSB, or fractions of 1 volt, with a given set of inputs at a specified frequency.

"Flash" Converter

A converter in which all the bit choices are made at the same time. It requires $2^n - 1$ voltage-divider taps and comparators, and a comparable amount of priority encoding logic. An extremely fast scheme, it requires large numbers of precision components. Flash converters are often used for partial conversions in *subranging converters*.

Full-Scale Error

The ideal difference between the first transition voltage and last transition voltage for an ADC is (F.S. −2LSB). Full-Scale Error is defined as the deviation between this ideal difference and the measured difference.

Gain Adjustment

The "gain" of a converter is that analog scale factor setting that provides the nominal conversion relationship, e.g., 10V full scale in a fixed-reference converter, or 100% of full scale in a ratiometric converter. Gain- and zero-adjustment principles are discussed under *zero*.

Harmonic Distortion (and Total Harmonic Distortion)

The ADC is driven by a spectrally pure, analog sine wave from a signal generator. The ADC outputs are analyzed via FFT and the ratio of the rms sum of the harmonics of the ADC output to the fundamental value is the THD. Usually, only the lower order harmonics are included, such as second through fifth:

$$\text{THD} = 20 \log \frac{(V_2^2 + V_3^2 + V_4^2 + V_5^2)^{1/2}}{V_1}$$

where V_1 is the rms amplitude of the fundamental and V_2, V_3, V_4 and V_5 are the rms amplitudes of the individual harmonics.

Intermodulation Distortion

The ADC is driven by an analog signal source producing two combined sine waves of frequencies f_a and f_b. As with any imperfectly linear device, distortion products (of order $m+n$) are produced at sum and difference frequencies of $mf_a \pm nf_b$ where $m, n = 0, 1, 2, 3 \ldots$ by the ADC. Intermodulation terms are those for which m or n is not equal to zero. The second order terms include (f_a+f_b), and $f_a-f_b)$, and the third order terms are $(2f_a+f_b)$, $(2f_a-f_b)$, (f_a+2f_b) and (f_a-2f_b). The ADC outputs are analyzed by FFT. IMD is defined as:

$$\text{IMD} = 20 \log \frac{\text{(rms sum of the sum and difference distortion products)}}{\text{rms amplitude of the fundamental}}$$

Least Significant Bit (LSB)

In a system in which a numerical magnitude is represented by a series of binary (i.e., two-valued) digits, the "least significant bit" is that digit (or "bit") that carries the smallest value or weight. For example, in the natural binary number 1101 (decimal 13, or $2^3 + 2^2 + 0 + 2^0$), the rightmost "1" is the LSB. Its analog weight, relative to full scale, is 2^{-n}, where n is the number of binary digits. It represents the smallest change that can be resolved by an n-bit converter.

Linearity Error

Linearity error of a converter, expressed in percent or parts-per-million of full-scale range, or fractions of a least-significant bit, is the deviation of the analog values from a straight line, in a plot of the measured conversion relationship. The straight line can be either a "best straight line," determined empirically by manipulation of the gain and/or offset to equalize maximum positive and negative deviations of the actual transfer characteristic from this straight line; or it can be a straight line passing through the end points of the transfer characteristic after they have been calibrated. Sometimes referred to as "end-point" nonlinearity, the latter is both a more conservative measure and is much easier to verify in actual practice. "End-point" nonlinearity is similar to relative accuracy error (see Accuracy, Relative). Linearity has two components – *differential* and *integral* nonlinearity.

Linearity, Differential and Integral

A digital output code should correspond to a quantum of analog input values exactly 1LSB in width (2^{-n} of full scale, for an n-bit converter). Any deviation of the measured "step" from the ideal width is called Differential Nonlinearity. It is an important specification, because a differential nonlinearity error greater than 1LSB can lead to nonmonotonic behavior of a D/A converter and missed codes in an A/D converter employing such a DAC. A flagrant example of differential nonlinearity is shown here.

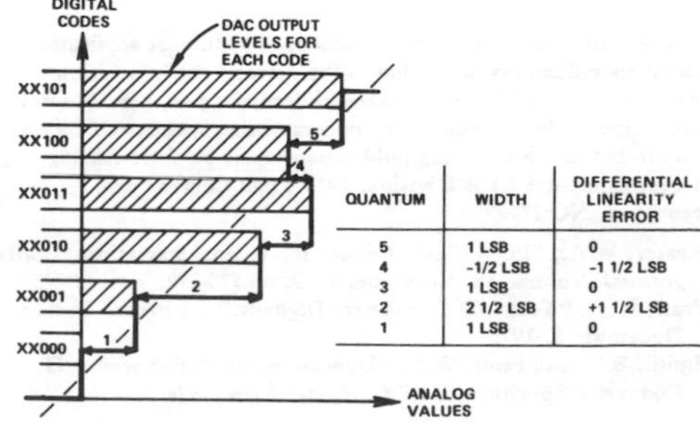

QUANTUM	WIDTH	DIFFERENTIAL LINEARITY ERROR
5	1 LSB	0
4	-1/2 LSB	-1 1/2 LSB
3	1 LSB	0
2	2 1/2 LSB	+1 1/2 LSB
1	1 LSB	0

In the illustration, the horizontal bars represent the measured DAC output values corresponding to six adjacent digital codes. The DAC is nonlinear in that the next least-significant bit (XX010) is 1 1/2LSB too large. Thus, instead of the five quanta, or steps, being all equal (= 1LSB), quantum 2 is 2 1/2LSB and quantum 4 is − 1/2LSB. The differential linearity error, the difference between the actual quantum width and the ideal 1LSB, is 1 1/2LSB for quantum 2 and − 1 1/2LSB for quantum 4.

When this DAC is used in successive-approximations conversion, it will lead to a missed code. Analog inputs slightly larger than the value of XX100 will be converted to XX100, and analog inputs slightly less than the value of XX100 will be converted to XX010. The code XX011 will not exist; it will be a *missed code.*

Often, instead of a maximum differential nonlinearity specification, there will be a simple specification of "no missed codes" which implies a differential nonlinearity less than 1LSB.

While differential nonlinearity deals with errors in step size, *integral nonlinearity* has to do with deviations of the overall shape of the conversion response. Even converters that are not subject to differential linearity errors (e.g., integrating types) have integral linearity (sometimes just "linearity") errors.

Power-Supply Sensitivity
The sensitivity of a converter to dc changes in power-supply voltages is normally expressed in terms of percentage change in analog input value (or fractions of the analog equivalent of 1LSB), corresponding to a given code for a 1% dc change in the power supply, e.g., 0.05%/%ΔV_S. Power-supply sensitivity may also be expressed in relation to a specified dc shift of the supply voltage. High-accuracy ADCs intended for battery operation require excellent rejection of large supply variations.

Quad-Slope Converter
This is an integrating analog-to-digital converter that goes through two cycles of *dual-slope* conversion, once with zero input and once with the analog input being measured. The errors determined during the first cycle are subtracted digitally from the result in the second cycle. The scheme results in an extremely accurate converter.

Quantizing Uncertainty (or "Error")
The analog continuum is partitioned into 2^n discrete ranges for n-bit conversion. All analog values within a given range are represented by the same digital code usually assigned to the nominal midrange value. There is, therefore, an inherent quantization uncertainty of ± 1/2LSB, in addition to the actual conversion errors. In integrating converters, this "error" is often expressed as "± 1 count."

Ratiometric Converter
The output of an A/D converter is a digital number proportional to the ratio of (some measure of) the input to a reference. Most requirements for conversions call for an absolute measurement, i.e., against a fixed reference. In some cases, where the measurement is affected by a changing reference voltage (e.g., the voltage applied to a bridge), it is advantageous to use that same reference as the reference for the conversion, to eliminate the effect of variation. Ratiometric conversion can also serve as a substitute for analog signal division (where the denominator changes but little during the conversion).

Signal-to-Noise Ratio
Signal-to-Noise Ratio (SNR) is measured signal to noise at the output of the ADC. The signal is the rms magnitude of the fundamental. Noise is the rms sum of all nonfundamental signals up to half the sampling frequency. SNR is dependent on the number of quantization levels used in the digitization process; the more levels, the smaller the quantization noise. The theoretical SNR for a sine-wave input is given by:

$$SNR = (6.02N + 1.76)\ dB,$$

where N is the number of bits in the ADC. Thus for an ideal 10-bit ADC, SNR = 62dB.

Slew Rate
Slew rate is the maximum allowable rate of change of input signal such that the digital sample values are not in error.

Stability
Stability of a converter usually applies to the insensitivity of its characteristics with time, temperature, etc. All measurements of stability are difficult and time consuming, but stability vs. temperature is sufficiently critical in most applications to warrant universal inclusion in tables of specifications (see "Temperature Coefficients").

Subranging Converters
In this type of converter, an extremely fast conversion produces the most significant portion of the output word. This portion is converted back to analog with a fast high-accuracy D/A converter and subtracted from the input. The resulting residue is converted to digital at high speed and combined with the results of the earlier conversion to form the output word. In *digitally corrected subranging* (DCS), the two bytes are combined in a manner that corrects for the error of the LSB of the most significant byte. For example, using 8-bit and 5-bit conversion, and this proprietary technique, a full-accuracy high-speed 12-bit converter can be built.

Successive Approximations
Successive approximations is a high-speed method of comparing an unknown against a group of weighted references. The operation of a successive approximations A/D converter is generally similar to the orderly weighing of an unknown quantity on a precision chemical balance using a set of weights such as: 1 gram, 1/2 gram, 1/4 gram, 1/8 gram, 1/16 gram, etc. The weights are tried in order, starting with the largest. Any weight that tips the scale is removed. At the end of the process, the sum of the weights remaining on the scale will be within 1LSB of the actual weight (± 1/2LSB, if the scale is properly biased – see *zero*).

Temperature Coefficients
In general, temperature instabilities are expressed in %/°C, ppm/°C, as fractions of 1LSB/°C, or as a change in a parameter over a specified temperature range. Measurements are usually made at room temperature and at the extremes of the specified range, and the temperature coefficient (tempco, T.C.) is defined as the change in the parameter divided by the corresponding temperature change. Parameters of interest include, *gain, linearity, offset* (bipolar) and *zero*. The last three are expressed in % or ppm of full-scale range per Celsius degree.

Gain Tempco: Two factors principally affect converter gain instability with temperature:

1. In fixed-reference converters, the reference source will vary with temperature. For example, the tempco of an AD581L is typically 5ppm/°C.
2. The ratiometric circuitry has a sensitivity to temperature.

Linearity Tempco: Sensitivity of linearity to temperature over the specified range. To avoid missed codes, it is sufficient that the differential nonlinearity error be less than 1LSB at any temperature in the range of interest. The ***differential nonlinearity temperature coefficient*** may be expressed as a ratio, as a maximum change over a specified temperature range, and/or implied by a statement that there are no missed codes when operating within a specified temperature range.

Offset Tempco: The temperature coefficient of the all-DAC-switches-off (minus full-scale) point, of a bipolar successive-approximations converter, is dependent on three variables:

1. The tempco of the reference source
2. The voltage stability of the input buffer and the comparator
3. The tracking capability of the bipolar-offset resistors and the gain resistors.

Unipolar Zero: The zero tempco of an ADC is dependent only on the zero stability of the integrator and/or the input buffer and the comparator. It may be expressed in μV/°C, or in percent or ppm of full-scale per degree C.

Zero- and Gain-Adjustment Principles
The zero adjustment of a unipolar ADC is set so that the transition from all-bits-off to LSB-on occurs at $1/2 \times 2^{-n}$ of nominal full scale. The gain is set for the final transition to all-bits-on to occur at F.S. $(1 - 3/2 \times 2^{-n})$. The "zero" of an offset-binary bipolar ADC is set so that the first transition occurs at $-$F.S. $(1-2^{-n})$ and the last transition at $+$F.S. $(1-3\times2^{-n})$. The data sheet instructions should be followed.

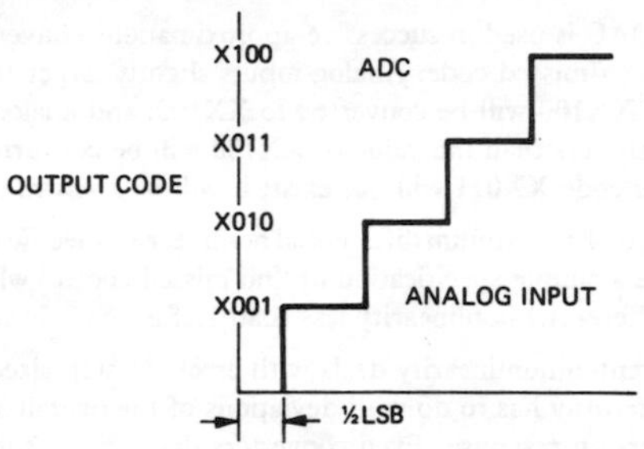

Zero Code Error
This is a measure of the difference between the ideal (0.5LSB) and the actual differential analog input level required to produce the first positive LSB code to transition (00 . . . 00 to 00 . . . 01).

Low Cost Signal Conditioning 8-Bit ADC

AD670

FEATURES

Complete 8-Bit Signal Conditioning A/D Converter Including Instrumentation Amp and Reference
Microprocessor Bus Interface
10µs Conversion Speed
Flexible Input Stage: Instrumentation Amp Front End Provides Differential Inputs and High Common-Mode Rejection
No User Trims Required
No Missing Codes Over Temperature
Single +5V Supply Operation
Convenient Input Ranges
20-Pin DIP or Surface-Mount Package
Low Cost Monolithic Construction

AD670 BLOCK DIAGRAM AND TERMINAL CONFIGURATION (ALL PACKAGES)

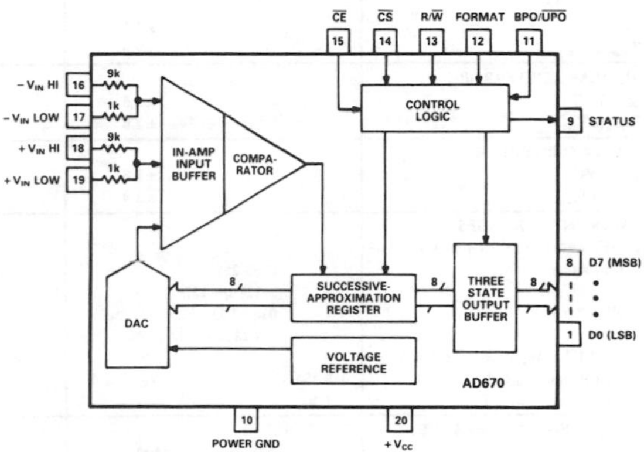

GENERAL DESCRIPTION

The AD670 is a complete 8-bit signal conditioning analog-to-digital converter. It consists of an instrumentation amplifier front end along with a DAC, comparator, successive approximation register (SAR), precision voltage reference, and a three-state output buffer on a single monolithic chip. No external components or user trims are required to interface, with full accuracy, an analog system to an 8-bit data bus. The AD670 will operate on the +5V system supply. The input stage provides differential inputs with excellent common-mode rejection and allows direct interface to a variety of transducers.

The device is configured with input scaling resistors to permit two input ranges: 0 to 255mV (1mV/LSB) and 0 to 2.55V (10mV/LSB). The AD670 can be configured for both unipolar and bipolar inputs over these ranges. The differential inputs and common-mode rejection of this front end are useful in applications such as conversion of transducer signals superimposed on common-mode voltages.

The AD670 incorporates advanced circuit design and proven processing technology. The successive approximation function is implemented with I^2L (integrated injection logic). Thin-film SiCr resistors provide the stability required to prevent missing codes over the entire operating temperature range while laser wafer trimming of the resistor ladder permits calibration of the device to within ± 1LSB. Thus, no user trims for gain or offset are required. Conversion time of the device is 10µs.

The AD670 is available in four package types and five grades. The J and K grades are specified over 0 to +70°C and come in 20-pin plastic DIP packages or 20-terminal PLCC packages. The A and B grades (−40°C to +85°C) and the S grade (−55°C to +125°C) come in 20-pin ceramic DIP packages.

The S grade is also available with optional processing to MIL-STD-883 in 20-pin ceramic DIP or 20-terminal LCC packages. The Analog Devices Military Products Databook should be consulted for details on these configurations.

PRODUCT HIGHLIGHTS

1. The AD670 is a complete 8-bit A/D including three-state outputs and microprocessor control for direct connection to 8-bit data buses. No external components are required to perform a conversion.
2. The flexible input stage features a differential instrumentation amp input with excellent common-mode rejection. This allows direct interface to a variety of transducers without preamplification.
3. No user trims are required for 8-bit accurate performance.
4. Operation from a single +5V supply allows the AD670 to run off of the microprocessor's supply.
5. Four convenient input ranges (two unipolar and two bipolar) are available through internal scaling resistors: 0 to 255mV (1mV/LSB) and 0 to 2.55V (10mV/LSB).
6. Software control of the output mode is provided. The user can easily select unipolar or bipolar inputs and binary or 2's complement output codes.

SPECIFICATIONS (@ V_{CC} = +5V and +25°C unless otherwise noted)

Model	AD670J			AD670K			
	Min	Typ	Max	Min	Typ	Max	Units
OPERATING TEMPERATURE RANGE	0		+70	0		+70	°C
RESOLUTION	8			8			Bit
CONVERSION TIME			10			10	μs
RELATIVE ACCURACY			±1/2			±1/4	LSB
T_{min} to T_{max}			±1/2			±1/2	LSB
DIFFERENTIAL LINEARITY ERROR							
T_{min} to T_{max}		GUARANTEED NO MISSING CODES ALL GRADES					
GAIN ACCURACY							
@ +25°C			±1.5			±0.75	LSB
T_{min} to T_{max}			±2.0			±1.0	LSB
UNIPOLAR ZERO ERROR							
@ +25°C			±1.5			±0.75	LSB
T_{min} to T_{max}			±2.0			±1.0	LSB
BIPOLAR ZERO ERROR							
@ +25°C			±1.5			±0.75	LSB
T_{min} to T_{max}			±2.0			±1.0	LSB
ANALOG INPUT RANGES							
DIFFERENTIAL ($-V_{IN}$ to $+V_{IN}$)							
Low Range		0 to +255			0 to +255		mV
		−128 to +127			−128 to +127		mV
High Range		0 to +2.55			0 to +2.55		V
		−1.28 to +1.27			−1.28 to +1.27		V
ABSOLUTE (Inputs to Power Gnd)							
Low Range T_{min} to T_{max}	−0.150		V_{CC} − 3.4	−0.150		V_{CC} − 3.4	V
High Range T_{min} to T_{max}	−1.50		V_{CC}	−1.50		V_{CC}	V
BIAS CURRENT (255mV RANGE)							
T_{min} to T_{max}		200	500		200	500	nA
OFFSET CURRENT (255mV RANGE)							
T_{min} to T_{max}		40	200		40	200	nA
2.55V RANGE INPUT RESISTANCE	8.0		12.0	8.0		12.0	kΩ
2.55V RANGE FULL SCALE MATCH + AND − INPUT		±1/2			±1/2		LSB
COMMON-MODE REJECTION RATIO (255mV RANGE)			1			1	LSB
COMMON-MODE REJECTION RATIO (2.55V RANGE)			1			1	LSB
POWER SUPPLY							
Operating Range	4.5		5.5	4.5		5.5	V
Current I_{CC}		30	45		30	45	mA
Rejection Ratio T_{min} to T_{max}			0.015			0.015	% of FS/%
DIGITAL OUTPUTS							
SINK CURRENT (V_{OUT} = 0.4V)							
T_{min} to T_{max}	1.6			1.6			mA
SOURCE CURRENT (V_{OUT} = 2.4V)							
T_{min} to T_{max}	0.5			0.5			mA
THREE-STATE LEAKAGE CURRENT			±40			±40	μA
OUTPUT CAPACITANCE		5			5		pF
DIGITAL INPUT VOLTAGE							
V_{INL}			0.8			0.8	V
V_{INH}	2.0			2.0			V
DIGITAL INPUT CURRENT							
($0 \leq V_{IN} \leq +5V$)							
I_{INL}	−100			−100			μA
I_{INH}			+100			+100	μA
INPUT CAPACITANCE		10			10		pF

NOTES

Specifications shown in boldface are tested on all production units at final electrical test. Results from those tests are used to calculate outgoing quality levels. All min and max specifications are guaranteed, although only those shown in boldface are tested on all production units.

Specifications subject to change without notice.

AD670

Model	AD670A Min	AD670A Typ	AD670A Max	AD670B Min	AD670B Typ	AD670B Max	AD670S Min	AD670S Typ	AD670S Max	Units
OPERATING TEMPERATURE RANGE	−40		+85	−40		+85	−55		+125	°C
RESOLUTION	8			8			8			Bit
CONVERSION TIME			**10**			**10**			**10**	μs
RELATIVE ACCURACY			**±1/2**			**±1/4**			**±1/2**	LSB
T_{min} to T_{max}			**±1/2**			**±1/2**			**±1**	LSB
DIFFERENTIAL LINEARITY ERROR										
T_{min} to T_{max}	**GUARANTEED NO MISSING CODES ALL GRADES**									
GAIN ACCURACY										
@ +25°C			**±1.5**			**±0.75**			**±1.5**	LSB
T_{min} to T_{max}			**±2.5**			**±1.5**			**±2.5**	LSB
UNIPOLAR ZERO ERROR										
@ +25°C			**±1.0**			**±0.5**			**±1.0**	LSB
T_{min} to T_{max}			**±2.0**			**±1.0**			**±2.0**	LSB
BIPOLAR ZERO ERROR										
@ +25°C			**±1.0**			**±0.5**			**±1.0**	LSB
T_{min} to T_{max}			**±2.0**			**±1.0**			**±2.0**	LSB
ANALOG INPUT RANGES										
DIFFERENTIAL ($-V_{IN}$ to $+V_{IN}$)										
Low Range		0 to +255			0 to +255			0 to +255		mV
		−128 to +127			−128 to +127			−128 to +127		mV
High Range		0 to +2.55			0 to +2.55			0 to +2.55		V
		−1.28 to +1.27			−1.28 to +1.27			−1.28 to +1.27		V
ABSOLUTE (Inputs to Power Gnd)										
Low Range T_{min} to T_{max}	**−0.150**		**V_{CC} − 3.5**	**−0.150**		**V_{CC} − 3.5**	**−0.150**		**V_{CC} − 3.5**	V
High Range T_{min} to T_{max}	−1.50		V_{CC}	−1.50		V_{CC}	−1.50		V_{CC}	V
BIAS CURRENT (255mV RANGE)										
T_{min} to T_{max}		200	**500**		200	**500**		200	**750**	nA
OFFSET CURRENT (255mV RANGE)										
T_{min} to T_{max}		40	**200**		40	**200**		40	**200**	nA
2.55V RANGE INPUT RESISTANCE	**8.0**		**12.0**	**8.0**		**12.0**	**8.0**		**12.0**	kΩ
2.55V RANGE FULL SCALE MATCH										
+ AND − INPUT		±1/2			±1/2			±1/2		LSB
COMMON-MODE REJECTION RATIO (255mV RANGE)			1			1			1	LSB
COMMON-MODE REJECTION RATIO (2.55V RANGE)			1			1			1	LSB
POWER SUPPLY										
Operating Range	4.5		5.5	4.5		5.5	4.75		5.5	V
Current I_{CC}		30	**45**		30	**45**		30	**45**	mA
Rejection Ratio T_{min} to T_{max}			**0.015**			**0.015**			**0.015**	% of FS/
DIGITAL OUTPUTS										
SINK CURRENT (V_{OUT} = 0.4V)										
T_{min} to T_{max}	**1.6**			**1.6**			**1.6**			mA
SOURCE CURRENT (V_{OUT} = 2.4V)										
T_{min} to T_{max}	**0.5**			**0.5**			**0.5**			mA
THREE-STATE LEAKAGE CURRENT			**±40**			**±40**			**±40**	μA
OUTPUT CAPACITANCE		5			5			5		pF
DIGITAL INPUT VOLTAGE										
V_{INL}			**0.8**			**0.8**			**0.7**	V
V_{INH}	**2.0**			**2.0**			**2.0**			V
DIGITAL INPUT CURRENT										
($0 \le V_{IN} \le +5V$)										
I_{INL}	**−100**			**−100**			**−100**			μA
I_{INH}			**+100**			**+100**			**+100**	μA
INPUT CAPACITANCE		10			10			10		pF

NOTES

Specifications shown in boldface are tested on all production units at final electrical test. Results from those tests are used to calculate outgoing quality levels. All min and max specifications are guaranteed, although only those shown in boldface are tested on all production units.

Specifications subject to change without notice.

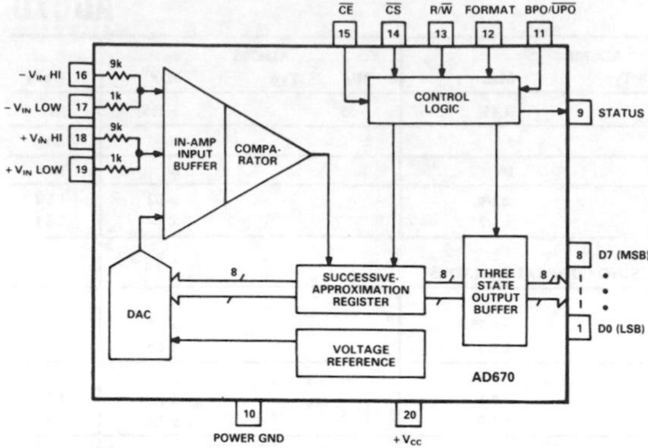

Figure 1. AD670 Block Diagram and Terminal Configuration (All Packages)

ABSOLUTE MAXIMUM RATINGS*

V_{CC} to Ground . . . . 0V to +7.5V
Digital Inputs (Pins 11-15) . . . . −0.5V to V_{CC} +0.5V
Digital Outputs (Pins 1-9) . Momentary Short to V_{CC} or Ground
Analog Inputs (Pins 16-19) . . . . −30V to +30V
Power Dissipation 450mW
Storage Temperature Range . . . . −65°C to +150°C
Lead Temperature (Soldering, 10sec) . . . . +300°C

*Stresses above those listed under "Absolute Maximum Ratings" may cause permanent damage to the device. This is a stress rating only and functional operation of the device at these or any other conditions above those indicated in the operational sections of this specification is not implied. Exposure to absolute maximum rating conditions for extended periods may affect device reliability.

AD670 ORDERING GUIDE

Model	Temperature Range	Relative Accuracy @ 25°C	Gain Accuracy @ 25°C	Package Options*
AD670JN	0 to +70°C	±1/2LSB	±1.5LSB	Plastic DIP (N-20)
AD670JP	0 to +70°C	±1/2LSB	±1.5LSB	PLCC (P-20A)
AD670KN	0 to +70°C	±1/4LSB	±0.75LSB	Plastic DIP (N-20)
AD670KP	0 to +70°C	±1/4LSB	±0.75LSB	PLCC (P-20A)
AD670AD	−40°C to +85°C	±1/2LSB	±1.5LSB	Ceramic DIP (D-20)
AD670BD	−40°C to +85°C	±1/4LSB	±0.75LSB	Ceramic DIP (D-20)
AD670SD	−55°C to +125°C	±1/2LSB	±1.5LSB	Ceramic DIP (D-20)

*Section 13 for package outline information.

CIRCUIT OPERATION/FUNCTIONAL DESCRIPTION

The AD670 is a functionally complete 8-bit signal conditioning A/D converter with microprocessor compatibility. The input section uses an instrumentation amplifier to accomplish the voltage to current conversion. This front end provides a high impedance, low bias current differential amplifier. The common-mode range allows the user to directly interface the device to a variety of transducers.

The A/D conversions are controlled by R/$\overline{W}$, $\overline{CS}$, and $\overline{CE}$. The R/$\overline{W}$ line directs the converter to read or start a conversion. A minimum write/start pulse of 300ns is required on either $\overline{CE}$ or $\overline{CS}$. The STATUS line goes high, indicating that a conversion is in process. The conversion thus begun, the internal 8-bit DAC is sequenced from MSB to LSB using a novel successive approximation technique. In conventional designs, the DAC is stepped through the bits by a clock. This can be thought of as a static design since the speed at which the DAC is sequenced is determined solely by the clock. No clock is used in the AD670. Instead, a "dynamic SAR" is created consisting of a string of inverters with taps along the delay line. Sections of the delay line between taps act as one shots. The pulses are used to set and reset the DAC's bits and strobe the comparator. When strobed, the comparator then determines whether the addition of each successively weighted bit current causes the DAC current sum to be greater or less than the input current. If the sum is less, the bit is turned off. After all bits are tested, the SAR holds an 8-bit code representing the input signal to within 1/2LSB accuracy. Ease of implementation and reduced dependence on process related variables make this an attractive approach to a successive approximation design.

The SAR provides an end-of-conversion signal to the control logic which then brings the STATUS line low. Data outputs remain in a high impedance state until R/$\overline{W}$ is brought high with $\overline{CE}$ and $\overline{CS}$ low and allows the converter to be read. Bringing $\overline{CE}$ or $\overline{CS}$ high during the valid data period ends the read cycle. The output buffers cannot be enabled during a conversion. Any convert start commands will be ignored until the conversion cycle is completed; once a conversion cycle has been started it cannot be stopped or restarted.

The AD670 provides the user with a great deal of flexibility by offering two input spans and formats and a choice of output codes. Input format and input range can each be selected. The BPO/$\overline{UPO}$ pin controls a switch which injects a bipolar offset current of a value equal to the MSB less 1/2LSB into the summing node of the comparator to offset the DAC output. Two precision 10 to 1 attenuators are included on board to provide input range selection of 0 to 2.55V or 0 to 255mV. Additional ranges of

-1.28 to 1.27V and -128 to 127mV are possible if the BPO/$\overline{UPO}$ switch is high when the conversion is started. Finally, output coding can be chosen using the FORMAT pin when the conversion is started. In the bipolar mode and with a logic 1 on FORMAT, the output is in two's complement; with a logic 0, the output is offset binary.

CONNECTING THE AD670

The AD670 has been designed for ease of use. All active components required to perform a complete A/D conversion are on board and are connected internally. In addition, all calibration trims are performed at the factory, assuring specified accuracy without user trims. There are, however, a number of options and connections that should be considered to obtain maximum flexibility from the part.

INPUT CONNECTIONS

Standard connections are shown in the figures that follow. An input range of 0 to 2.55V may be configured as shown in Figure 2a. This will provide a one LSB change for each 10mV of input change. The input range of 0 to 255mV is configured as shown in Figure 2b. In this case, each LSB represents 1mV of input change. When unipolar input signals are used, Pin 11, BPO/$\overline{UPO}$, should be grounded. Pin 11 selects the input format for either unipolar or bipolar signals. Figures 3a and 3b show the input connections for bipolar signals. Pin 11 should be tied to $+V_{CC}$ for bipolar inputs.

Although the instrumentation amplifier has a differential input, there must be a return path to ground for the bias currents. If it is not provided, these currents will charge stray capacitances and cause internal circuit nodes to drift uncontrollably causing the digital output to change. Such a return path is provided in Figures 2a and 3a (larger input ranges) since the 1k resistor leg is tied to ground. This is not the case for Figures 2b and 3b (the lower input ranges). When connecting the AD670 inputs to floating sources, such as transformers and ac-coupled sources, there must still be a dc path from each input to common. This can be accomplished by connecting a 10kΩ resistor from each input to ground.

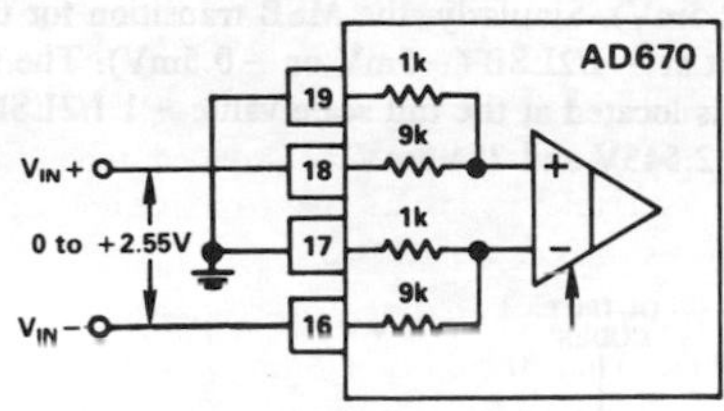

2a. 0 to 2.55V (10mV/LSB)

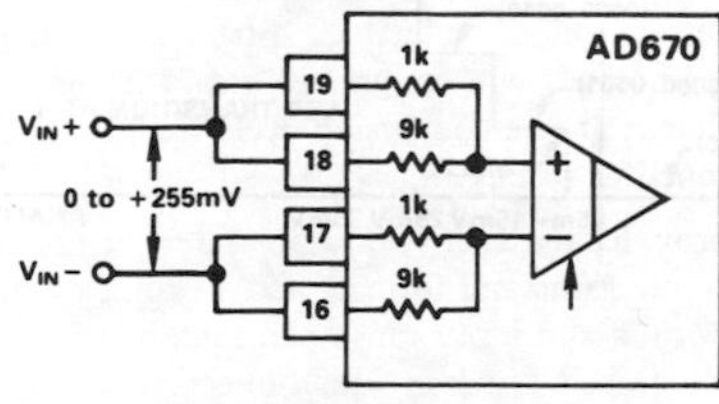

2b. 0 to 255mV (1mV/LSB)

NOTE: PIN 11, BPO/$\overline{UPO}$ SHOULD BE LOW WHEN CONVERSION IS STARTED.

Figure 2. Unipolar Input Connections

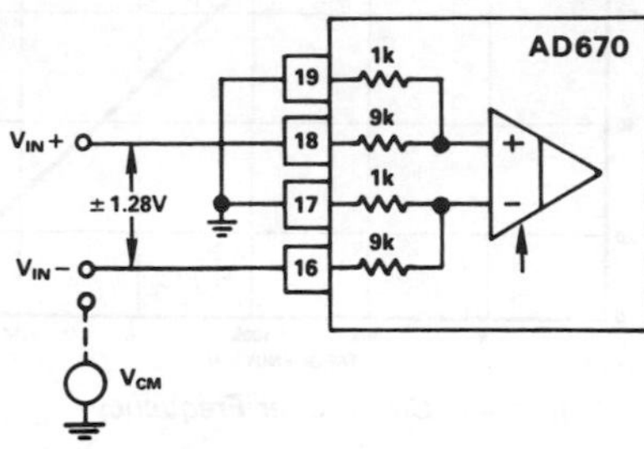

3a. ±1.28V Range

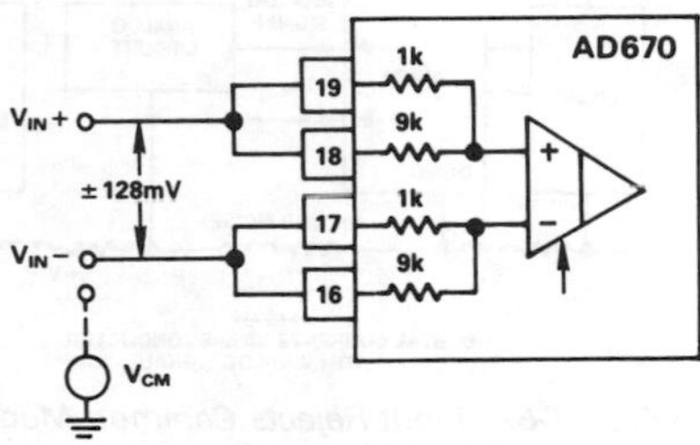

3b. ±128mV Range

NOTE: PIN 11, BPO/$\overline{UPO}$ SHOULD BE HIGH WHEN CONVERSION IS STARTED.

Figure 3. Bipolar Input Connections

Bipolar Operation

Through special design of the instrumentation amplifier, the AD670 accommodates input signal excursions below ground, even though it operates from a single 5V supply. To the user, this means that true bipolar input signals can be used without the need for any additional external components. Bipolar signals can be applied differentially across both inputs, or one of the inputs can be grounded and a bipolar signal applied to the other.

Common-Mode Performance

The AD670 is designed to reject dc and ac common-mode voltages. In some applications it is useful to apply a differential input signal V_{IN} in the presence of a dc common-mode voltage V_{CM}. The user must observe the absolute input signal limits listed in the specifications, which represent the maximum voltage $V_{IN} + V_{CM}$ that can be applied to either input without affecting proper operation. Exceeding these limits (within the range of absolute maximum ratings), however, will not cause permanent damage.

The excellent common-mode rejection of the AD670 is due to the instrumentation amplifier front end, which maintains the differential signal until it reaches the output of the comparator. In contrast to a standard operational amplifier, the instrumentation amplifier front end provides significantly improved CMRR over a wide frequency range (Figure 4a).

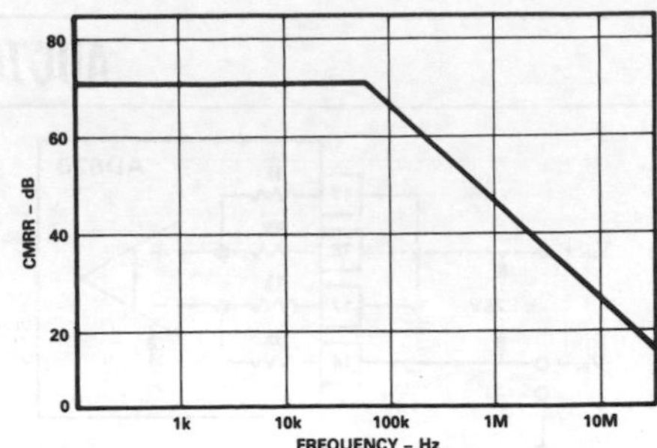

Figure 4a. CMRR over Frequency

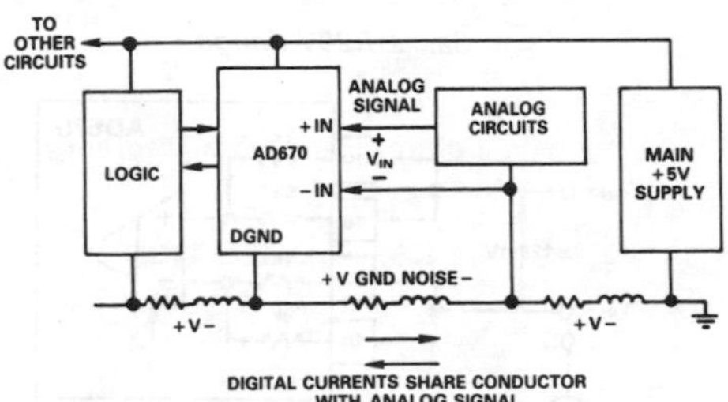

Figure 4b. AD670 Input Rejects Common-Mode Ground Noise

Good common-mode performance is useful in a number of situations. In bridge-type transducer applications, such performance facilitates the recovery of differential analog signals in the presence of a dc common-mode or a noisy electrical environment. High-frequency CMRR also becomes important when the analog signal is referred to a noisy, remote digital ground. In each case, the CMRR specification of the AD670 allows the integrity of the input signal to be preserved.

The AD670's common-mode voltage tolerance allows great flexibility in circuit layout. Most other A/D converters require the establishment of one point as the analog reference point. This is necessary in order to minimize the effects of parasitic voltages. The AD670, however, eliminates the need to make the analog ground reference point and A/D analog ground one and the same. Instead, a system such as that shown in Figure 4b is possible as a result of the AD670's common-mode performance. The resistors and inductors in the ground return represent unavoidable system parasitic impedances.

Input/Output Options

Data output coding (2's complement vs. straight binary) is selected using Pin 12, the FORMAT pin. The selection of input format (bipolar vs. unipolar) is controlled using Pin 11, BPO/$\overline{\text{UPO}}$. Prior to a write/convert, the state of FORMAT and BPO/$\overline{\text{UPO}}$ should be available to the converter. These lines may be tied to the data bus and may be changed with each conversion if desired. The configurations are shown in Table I. Output coding for representative signals in each of these configurations is shown in Figure 5.

An output signal, STATUS, indicates the status of the conversion. STATUS goes high at the beginning of the conversion and returns low when the conversion cycle has been completed.

BPO/$\overline{\text{UPO}}$	FORMAT	INPUT RANGE/OUTPUT FORMAT
0	0	Unipolar/Straight Binary
1	0	Bipolar/Offset Binary
0	1	Unipolar/2's Complement
1	1	Bipolar/2's Complement

Table I. AD670 Input Selection/Output Format Truth Table

$+V_{IN}$	$-V_{IN}$	DIFF V_{IN}	STRAIGHT BINARY (FORMAT = 0, BPO/$\overline{\text{UPO}}$ = 0)
0	0	0	0000 0000
128mV	0	128mV	1000 0000
255mV	0	255mV	1111 1111
255mV	255mV	0	0000 0000
128mV	127mV	1mV	0000 0001
128mV	–127mV	255mV	1111 1111

Figure 5a. Unipolar Output Codes (Low Range)

$+V_{IN}$	$-V_{IN}$	DIFF V_{IN}	OFFSET BINARY (FORMAT = 0, BPO/$\overline{\text{UPO}}$ = 1)	2's COMPLEMENT (FORMAT = 1, BPO/$\overline{\text{UPO}}$ = 1)
0	0	0	1000 0000	0000 0000
127mV	0	127mV	1111 1111	0111 1111
1.127V	1.000V	127mV	1111 1111	0111 1111
255mV	255mV	0	1000 0000	0000 0000
128mV	127mV	1mV	1000 0001	0000 0001
127mV	128mV	–1mV	0111 1111	1111 1111
127mV	255mV	–128mV	0000 0000	1000 0000
–128mV	0	–128mV	0000 0000	1000 0000

Figure 5b. Bipolar Output Codes (Low Range)

Calibration

Because of its precise factory calibration, the AD670 is intended to be operated without user trims for gain and offset; therefore, no provisions have been made for such user trims. Figures 6a, 6b, and 6c show the transfer curves at zero and full scale for the unipolar and bipolar modes. The code transitions are positioned so that the desired value is centered at that code. The first LSB transition for the unipolar mode occurs for an input of +1/2LSB (5mV or 0.5mV). Similarly, the MSB transition for the bipolar mode is set at –1/2LSB (–5mV or –0.5mV). The full scale transition is located at the full scale value –1 1/2LSB. These values are 2.545V and 254.5mV.

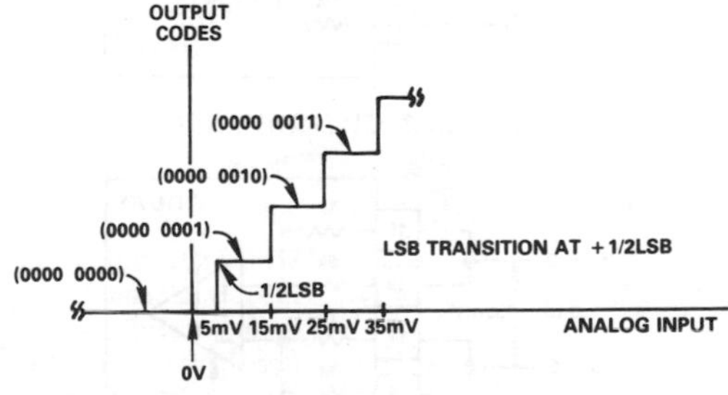

Figure 6a. Unipolar Transfer Curve

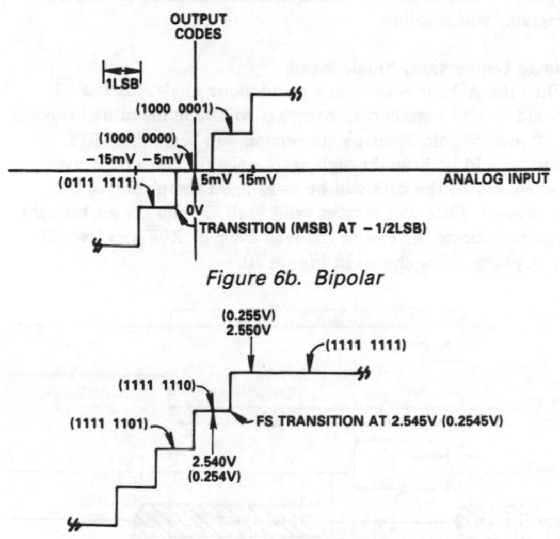

Figure 6b. Bipolar

Figure 6c. Full Scale (Unipolar)

Figure 6. Transfer Curves

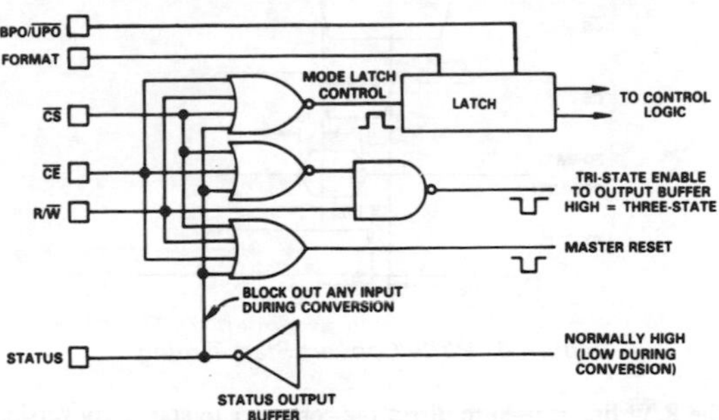

Figure 7. Control Logic Block Diagram

R/$\overline{\text{W}}$	$\overline{\text{CS}}$	$\overline{\text{CE}}$	OPERATION
0	0	0	WRITE/CONVERT
1	0	0	READ
X	X	1	NONE
X	1	X	NONE

Table II. AD670 Control Signal Truth Table

CONTROL AND TIMING OF THE AD670

Control Logic

The AD670 contains on-chip logic to provide conversion and data read operations from signals commonly available in microprocessor systems. Figure 7 shows the internal logic circuitry of the AD670. The control signals, $\overline{\text{CE}}$, $\overline{\text{CS}}$, and R/$\overline{\text{W}}$ control the operation of the converter. The read or write function is determined by R/$\overline{\text{W}}$ when both $\overline{\text{CS}}$ and $\overline{\text{CE}}$ are low as shown in Table II. If all three control inputs are held low longer than the conversion time, the device will continuously convert until one input, $\overline{\text{CE}}$, $\overline{\text{CS}}$, or R/$\overline{\text{W}}$ is brought high. The relative timing of these signals is discussed later in this section.

Timing

The AD670 is easily interfaced to a variety of microprocessors and other digital systems. The following discussion of the timing requirements of the AD670 control signals will provide the designer with useful insight into the operation of the device.

Write/Convert Start Cycle

Figure 8 shows a complete timing diagram for the write/convert start cycle. $\overline{\text{CS}}$ (chip select) and $\overline{\text{CE}}$ (chip enable) are active low and are interchangeable signals. Both $\overline{\text{CS}}$ and $\overline{\text{CE}}$ must be low for the converter to read or start a conversion. The minimum pulse width, t_W, on either $\overline{\text{CS}}$ or $\overline{\text{CE}}$ is 300ns to start a conversion.

Table III. AD670 TIMING SPECIFICATIONS

Boldface indicates parameters tested 100% unless otherwise noted. See Specifications page for explanation.

Symbol	Parameter	Min (@ +25°C)	Typ	Max	Units
WRITE/CONVERT START MODE					
t_W	Write/Start Pulse Width	300			ns
t_{DS}	Input Data Setup Time	200			ns
t_{DH}	Input Data Hold	10			ns
t_{RWC}	Read/Write Setup Before Control	0			ns
t_{DC}	Delay to Convert Start			700	ns
t_C	**Conversion Time**			**10**	µs
READ MODE					
t_R	Read Time	250			ns
t_{SD}	Delay from Status Low to Data Read			250	ns
t_{TD}	**Bus Access Time**		200	**250**	ns
t_{DH}	Data Hold Time	25			ns
t_{DT}	**Output Float Delay**			**150**	ns
t_{RT}	R/$\overline{\text{W}}$ before $\overline{\text{CE}}$ or $\overline{\text{CS}}$ low	0			ns

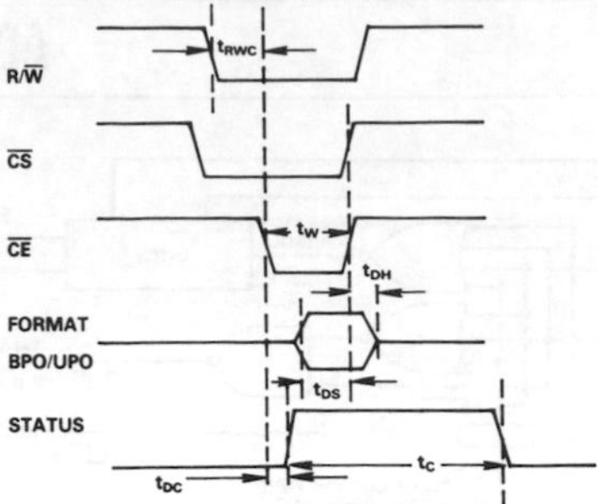

Figure 8. Write/Convert Start Timing

The $R/\overline{W}$ line is used to direct the converter to start a conversion ($R/\overline{W}$ low) or read data ($R/\overline{W}$ high). The relative sequencing of the three control signals ($R/\overline{W}$, $\overline{CE}$, $\overline{CS}$) is unimportant. However, when all three signals remain low for at least 300ns (t_W), STATUS will go high to signal that a conversion is taking place.

Once a conversion is started and the STATUS line goes high, convert start commands will be ignored until the conversion cycle is complete. The output data buffer cannot be enabled during a conversion.

Read Cycle

Figure 9 shows the timing for the data read operation. The data outputs are in a high impedance state until a read cycle is initiated. To begin the read cycle, $R/\overline{W}$ is brought high. During a read cycle, the minimum pulse length for $\overline{CE}$ and $\overline{CS}$ is a function of the length of time required for the output data to be valid. The data becomes valid and is available to the data bus in a maximum of 250ns. This delay between the high impedance state and valid data is the maximum bus access time or t_{TD}. Bringing $\overline{CE}$ or $\overline{CS}$ high during valid data ends the read cycle. The outputs remain valid for a minimum of 25ns (t_{DH}) and return to the high impedance state after a delay, t_{DT}, of 150ns maximum.

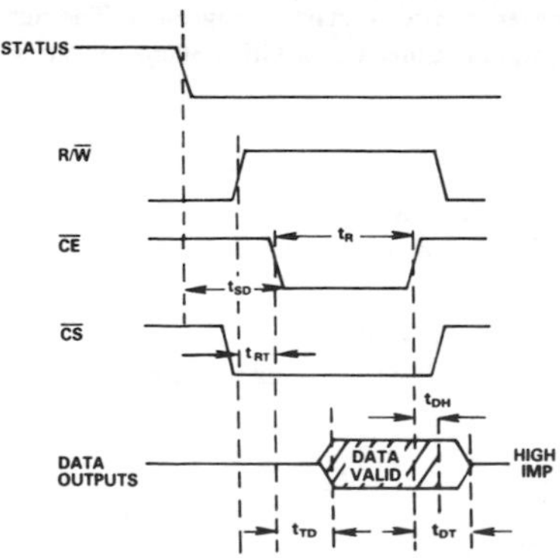

Figure 9. Read Cycle Timing

STAND-ALONE OPERATION

The AD670 can be used in a "stand-alone" mode, which is useful in systems with dedicated input ports available. Two typical conditions are described and illustrated by the timing diagrams which follow.

Single Conversion, Single Read

When the AD670 is used in a stand-alone mode, $\overline{CS}$ and $\overline{CE}$ should be tied together. Conversion will be initiated by bringing $R/\overline{W}$ low. Within 700ns, a conversion will begin. The $R/\overline{W}$ pulse should be brought high again once the conversion has started so that the data will be valid upon completion of the conversion. Data will remain valid until $\overline{CE}$ and $\overline{CS}$ are brought high to indicate the end of the read cycle or $R/\overline{W}$goes low. The timing diagram is shown in Figure 10.

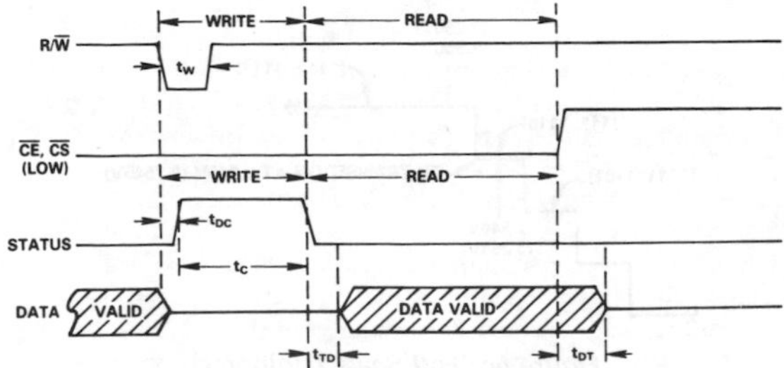

Figure 10. Stand-Alone Mode Single Conversion/ Single Read

Continuous Conversion, Single Read

A variety of applications may call for the A/D to be read after several conversions. In process control systems, this is often the case since a reading from a sensor may only need to be updated every few conversions. Figure 11 shows the timing relationships.

Once again, $\overline{CE}$ and $\overline{CS}$ should be tied together. Conversion will begin when the $R/\overline{W}$ signal is brought low. The device will convert repeatedly as indicated by the status line. A final conversion will take place once the $R/\overline{W}$ line has been brought high. The rising edge of $R/\overline{W}$ must occur while STATUS is high. $R/\overline{W}$ should not return high while STATUS is low since the circuit is in a reset state prior to the next conversion. Since the rising edge of $R/\overline{W}$ must occur while STATUS is high, $R/\overline{W}$'s length must be a minimum of 10.25μs ($t_C + t_{TD}$). Data becomes valid upon completion of the conversion and will remain so until the $\overline{CE}$ and $\overline{CS}$ lines are brought high indicating the end of the read cycle or $R/\overline{W}$ goes low initiating a new series of conversions.

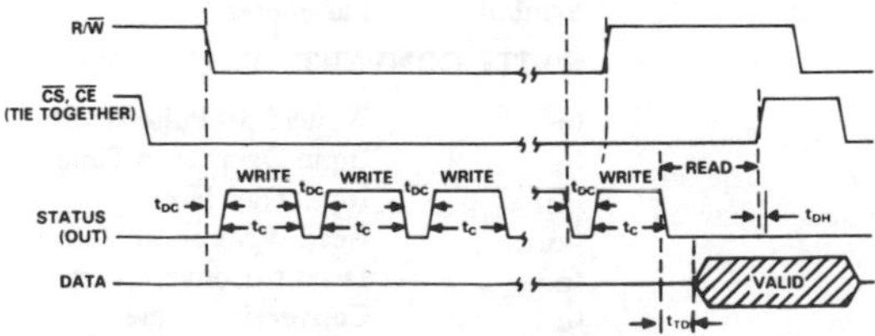

Figure 11. Stand-Alone Mode Continuous Conversion/ Single Read

APPLYING THE AD670

The AD670 has been designed for ease of use, system compatibility, and minimization of external components. Transducer interfaces generally require signal conditioning and preamplification before the signal can be converted. The AD670 will reduce and even eliminate this excess circuitry in many cases. To illustrate the flexibility and superior solution that the AD670 can bring to a transducer interface problem, the following discussions are offered.

Temperature Measurements

Temperature transducers are one of the most common sources of analog signals in data acquisition systems. These sensors require circuitry for excitation and preamplification/buffering. The instrumentation amplifier input of the AD670 eliminates the need for this signal conditioning. The output signals from temperature transducers are generally sufficiently slow that a sample/hold amplifier is not required. Figure 12 shows the AD590 IC temperature transducer interfaced to the AD670. The AD580 voltage reference is used to offset the input for 0°C calibration. The current output of the AD590 is converted into a voltage by R1. The high impedance unbuffered voltage is applied directly to the AD670 configured in the − 128mV to 127mV bipolar range. The digital output will have a resolution of 1°C.

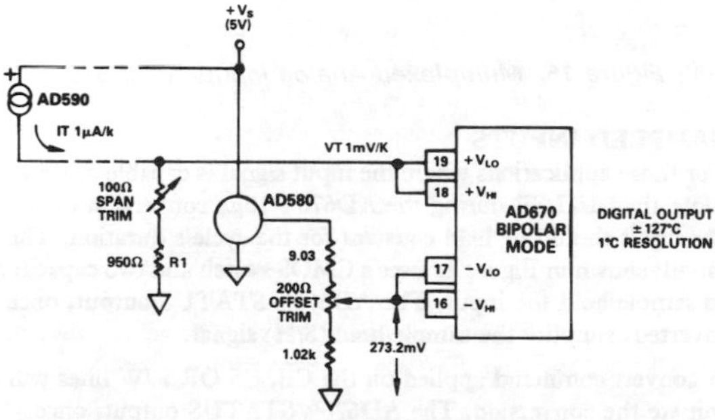

Figure 12. AD670 Temperature Transducer Interface

Platinum RTDs are also a popular, temperature transducer. Typical RTDs have a resistance of 100Ω at 0°C and change resistance 0.4Ω per °C. If a constant excitation current is caused to flow in the RTD, the change in voltage drop will be a measure of the change in temperature. Figure 13 shows such a method and the required connections to the AD670. The AD580 2.5V reference provides the accurate voltage for the excitation current and range offsetting for the RTD. The op-amp is configured to force a constant 2.5mA current through the RTD. The differential inputs of the AD670 measure the difference between a fixed offset voltage and the temperature dependent output of the op-amp which varies with the resistance of the RTD. The RTD change of approximately 0.4Ω/°C results in a 1mV/°C voltage change. With the AD670 in the 1mV/LSB range, temperatures from 0 to 255°C can be measured.

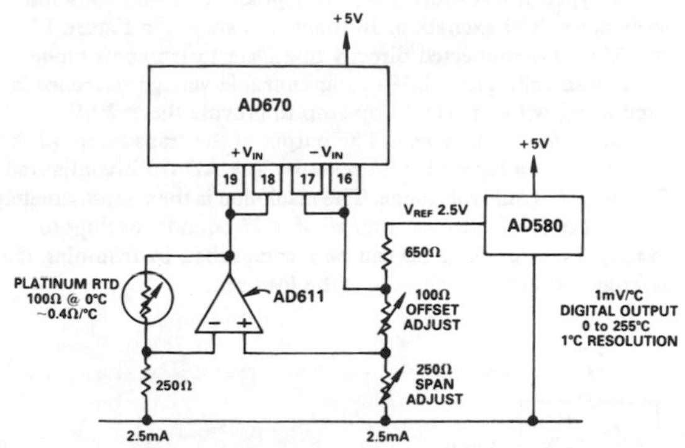

Figure 13. Low Cost RTD Interface

Differential temperature measurements can be made using an AD590 connected to each of the inputs as shown in Figure 14. This configuration will allow the user to measure the relative temperature difference between two points with a 1°C resolution. Although the internal 1k and 9k resistors on the inputs have ± 20% tolerance, trimming the AD590 is unnecessary as most differential temperature applications are concerned with the relative differences between the two. However, the user may see up to a 20% scale factor error in the differential temperature to digital output transfer curve.

This scale factor error can be eliminated through a software correction. Offset corrections can be made by adjusting for any difference that results when both sensors are held at the same temperature. A span adjustment can then be made by immersing one AD590 in an ice bath and one in boiling water and eliminating any deviation from 100°C. For a low cost version of this setup, the plastic AD592 can be substituted for the AD590.

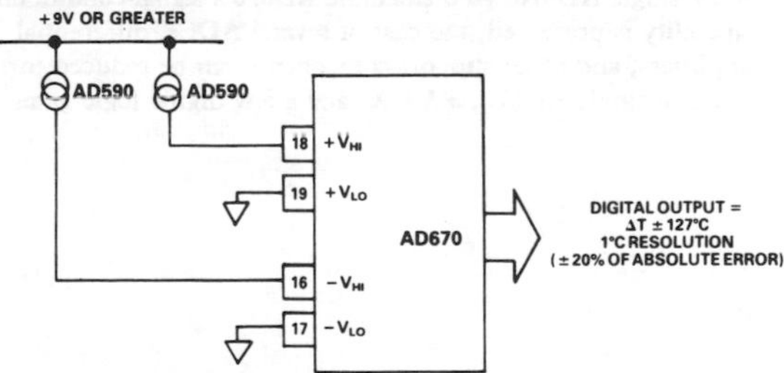

Figure 14. Differential Temperature Measurement Using the AD590

3

STRAIN GAUGE MEASUREMENTS

Many semiconductor-type strain gauges, pressure transducers, and load cells may also be connected directly to the AD670. These types of transducers typically produce 30 millivolts full-scale per volt of excitation. In the circuit shown in Figure 15, the AD670 is connected directly to a Data Instruments model JP-20 load cell. The AD584 programmable voltage reference is used along with an AD741 op-amp to provide the ±2.5V excitation for the load cell. The output of the transducer will be ±150mV for a force of ±20 pounds. The AD670 is configured for the ±128 millivolt range. The resolution is then approximately 2.1 ounces per LSB over a range of ±17 pounds. Scaling to exactly 2 ounces per LSB can be accomplished by trimming the reference voltage which excites the load cell.

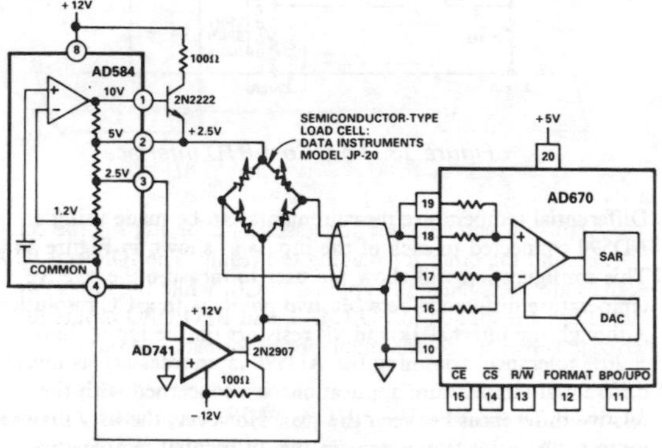

Figure 15. AD670 Load Cell Interface

MULTIPLEXED INPUTS

Most data acquisition systems require the measurement of several analog signals. Multiple A/D converters are often used to digitize these inputs, requiring additional preamplification and buffer stages per channel. Since these signals vary slowly, a differential MUX can multiplex inputs from several transducers into a single AD670. And since the AD670's signal-conditioning capability is preserved, the cost of several ADCs, differential amplifiers, and other support components can be reduced to that of a single AD670, a MUX, and a few digital logic gates.

An AD7502 dual 4-channel MUX appears in Figure 16 multiplexing four differential signals to the AD670. The AD7502's decoded address is gated with the microprocessor's write signal to provice a latching strobe at the flip-flops. A write cycle to the AD7502's address then latches the two LSBs of the data word thereby selecting the input channel for subsequent conversions.

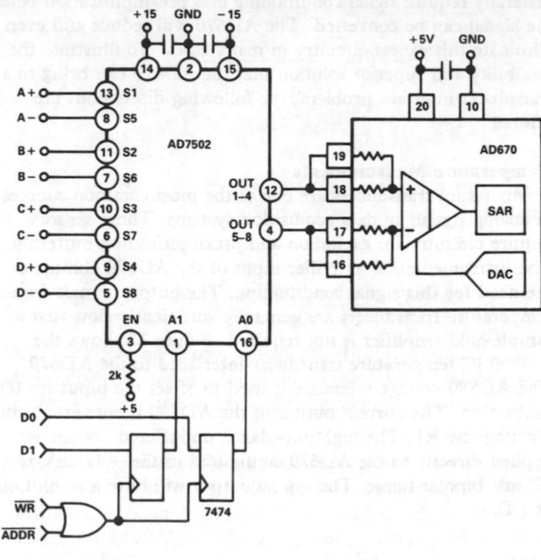

Figure 16. Multiplexed Analog Inputs to AD670

SAMPLED INPUTS

For those applications where the input signal is capable of slewing more than 1/2LSB during the AD670's 10μs conversion cycle, the input should be held constant for the cycle's duration. The circuit shown in Figure 17 uses a CMOS switch and two capacitors to sample/hold the input. The AD670's STATUS output, once inverted, supplies the sample/hold ($S/\overline{H}$) signal.

A convert command applied on the $\overline{CE}$, $\overline{CS}$ OR $R/\overline{W}$ lines will initiate the conversion. The AD670's STATUS output, once inverted, supplies the sample/hold signal to the CD4066. The CD4066 CMOS switch shown in Figure 17 was chosen for its fast transition times, low on-resistance and low cost. The control input's propagation delay for switch-closed to switch-open should remain less than 150ns to ensure that the sample-to-hold transition occurs before the first bit decision in the AD670.

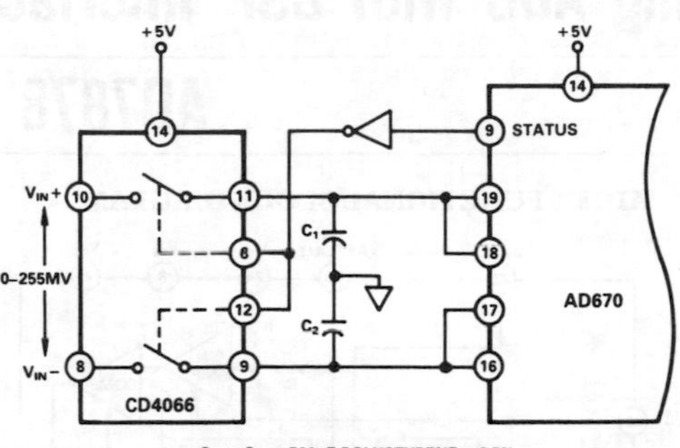

Figure 17. Low Cost Sample-and-Hold Circuit for AD670

Since settling to 1/2LSB at 8-bits of resolution requires 6.2 RC time constants, the 500pF hold capacitors and CD4066's 300Ω on-resistance yield an acquisition time of under 1μs, assuming a low impedance source.

This sample/hold approach makes use of the differential capabilities of the AD670. Because 500pF hold capacitors are used on both $V_{IN}+$ and $V_{IN}-$ inputs, the droop rate depends only on the offset current of the AD670, typically 20nA. With the matched 500pF capacitors, the droop rate is 40μV/μs. The input will then droop only 0.4mV (0.4LSB) during the AD670's 10μs conversion time. The differential approach also minimizes pedestal error since only the difference in charge injection between the two switches results in errors at the A/D.

The fast conversion time and differential and common-mode capabilities of the AD670 permit this simple sample-hold design to perform well with low sample-to-hold offset, droop rate of about 40μV/μs and acquisition time under 1μs. The effective aperture time of the AD670 is reduced by about 2 orders of magnitude with this circuit, allowing frequencies to be converted up to several kilohertz.

While no input anti-aliasing filter is shown, filtering will be necessary to prevent output errors if higher frequencies are present in the input signal. Many practical variations are possible with this circuit, including input MUX control, for digitizing a number of AC channels.

IBM PC INTERFACE

The AD670 appears in Figure 18 interfaced to the IBM PC. Since the device resides in I/O space, its address is decoded from only the lower ten address lines and must be gated with AEN (active low) to mask out internal (DMA) cycles which use the same I/O address space. This active low signal is applied to $\overline{CS}$. AO, meanwhile, is reserved for the R/$\overline{W}$ input. This places the AD670 in two adjacent addresses; one for starting the conversion and the other for reading the result. The $\overline{IOR}$ and $\overline{IOW}$ signals are then gated and applied to $\overline{CE}$, while the lower two data lines are applied to FORMAT and BPO/$\overline{UPO}$ inputs to provide software programmable input formats and output coding.

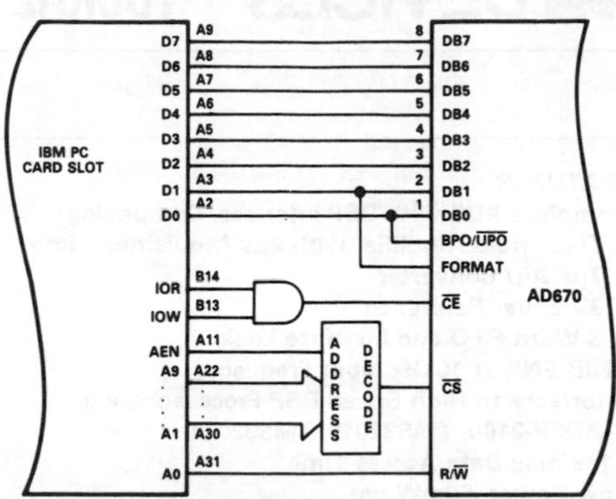

Figure 18. IBM PC Interface to AD670

In BASIC, a simple OUT ADDR, WORD command initiates a conversion. While the upper six bits of the data WORD are meaningless, the lower two bits define the analog input format and digital output coding according to Table IV. The data is available ten microseconds later (which is negligible in BASIC) and can be read using INP (ADDR+1). The 3-line subroutine in Figure 19, used in conjunction with the interface of Figure 18, converts an analog input within a bipolar range to an offset binary coded digital word.

NOTE: Due to the large number of options that may be installed in the PC, the I/O bus loading should be limited to one Schottky TTL load. Therefore, a buffer/driver should be used when interfacing more than two AD670's to the I/O bus.

DATA	INPUT FORMAT	OUTPUT CODING
0	Unipolar	Straight Binary
1	Bipolar	Offset Binary
2	Unipolar	2's Complement
3	Bipolar	2's Complement

Table IV.

```
10 OUT &H310,1                'INITIATE CONVERSION
20 ANALOGIN=INP (&H311)       'READ ANALOG INPUT
30 RETURN
```

Figure 19. Conversion Subroutine

LC²MOS Complete 12-Bit 100kHz Sampling ADC with DSP Interface

AD7878

FEATURES
Complete ADC with DSP Interface, Comprising:
- **Track/Hold Amplifier with 2μs Acquisition Time**
- **7μs A/D Converter**
- **3V Zener Reference**
- **8-Word FIFO and Interface Logic**

72dB SNR at 10kHz Input Frequency
Interfaces to High Speed DSP Processors, e.g.,
ADSP-2100, TMS32010, TMS32020
41ns max Data Access Time
Low Power, 60mW typ

APPLICATIONS
Digital Signal Processing
Speech Recognition and Synthesis
Spectrum Analysis
High Speed Modems
DSP Servo Control

AD7878 FUNCTIONAL BLOCK DIAGRAM

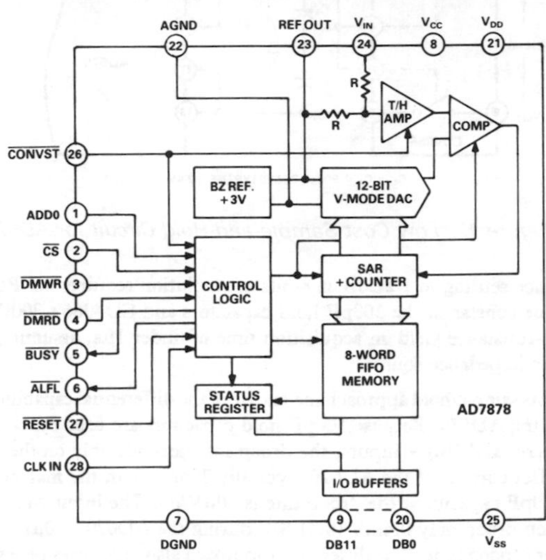

GENERAL DESCRIPTION

The AD7878 is a fast complete 12-bit A/D converter with a versatile DSP interface consisting of an 8-word, first-in, first-out (FIFO) memory and associated control logic.

The FIFO memory allows up to eight samples to be digitized before the microprocessor is required to service the A/D converter. The eight words can then be read out of the FIFO at maximum microprocessor speed. A fast data access time of 41ns allows direct interfacing to DSP processors and high speed 16-bit microprocessors.

An on-chip status/control register allows the user to program the effective length of the FIFO and contains the FIFO out of range, FIFO empty and FIFO word count information.

The analog input of the AD7878 has a bipolar range of ±3V. The AD7878 can convert full power signals up to 50kHz and is fully specified for dynamic parameters such as signal-to-noise ratio and harmonic distortion.

The AD7878 is fabricated in Linear Compatible CMOS (LC²MOS), an advanced, mixed technology process that combines precision bipolar circuits with low power CMOS logic. The part is available in four package styles, 28-pin plastic and hermetic dual-in-line package (DIP), leadless ceramic chip carrier (LCCC) or plastic leaded chip carrier (PLCC).

PRODUCT HIGHLIGHTS

1. Complete A/D Function with DSP Interface
   The AD7878 provides the complete function for digitizing ac signals to 12-bit accuracy. The part features an on-chip track/hold, on-chip reference and 12-bit A/D converter. The additional feature of an 8-word FIFO reduces the high software overheads associated with servicing interrupts in DSP processors.
2. Dynamic Specifications for DSP Users
   The AD7878 is fully specified and tested for ac parameters, including signal-to-noise ratio, harmonic distortion and intermodulation distortion. Key digital timing parameters are also tested and specified over the full operating temperature range.
3. Fast Microprocessor Interface
   Data access time of 41ns is the fastest ever achieved in a monolithic A/D converter and makes the AD7878 compatible with all modern 16-bit microprocessors and digital signal processors.

SPECIFICATIONS (V_{DD} = +5V ± 5%, V_{CC} = +5V ± 5%, V_{SS} = −5V ± 5%, AGND = DGND = 0V, f_{CLK} = 8MHz. All Specifications T_{min} to T_{max} unless otherwise noted.)

Parameter	J, A Versions[1]	K, L, B Versions	S Version	Units	Test Conditions/Comments
DYNAMIC PERFORMANCE[2]					
Signal-to-Noise Ratio (SNR)[3] @ 25°C	70	72	70	dB min	V_{IN} = 10kHz sine wave, f_{SAMPLE} = 100kHz
T_{min} to T_{max}	70	71	70	dB min	Typically 71.5dB for 0<V_{IN}<50kHz
Total Harmonic Distortion (THD)	−80	−80	−80	dB max	V_{IN} = 10kHz sine wave, f_{SAMPLE} = 100kHz Typically −86dB for 0<V_{IN}<50kHz
Peak Harmonic or Spurious Noise	−80	−80	−80	dB max	V_{IN} = 10kHz, f_{SAMPLE} = 100kHz Typically −86dB for 0<V_{IN}<50kHz
Intermodulation Distortion (IMD)					
Second Order Terms	−80	−80	−80	dB max	fa = 9kHz, fb = 9.5kHz, f_{SAMPLE} = 50kHz
Third Order Terms	−80	−80	−80	dB max	fa = 9kHz, fb = 9.5kHz, f_{SAMPLE} = 50kHz
Track/Hold Acquisition Time	2	2	2	μs max	See Throughput Rate section.
DC ACCURACY					
Resolution	12	12	12	Bits	
Minimum Resolution for which No Missing Codes are Guaranteed	12	12	12	Bits	
Relative Accuracy	±1/2	±1/4	±1/2	LSB typ	
Differential Nonlinearity	±1/2	±1/2	±1/2	LSB typ	
Bipolar Zero Error	±6	±6	±6	LSB max	
Positive Full Scale Error[4]	±6	±6	±6	LSB max	
Negative Full Scale Error[4]	±6	±6	±6	LSB max	
ANALOG INPUT					
Input Voltage Range	±3	±3	±3	Volts	
Input Current	±550	±550	±550	μA max	
REFERENCE OUTPUT[5]					
REF OUT	3	3	3	V nom	
REF OUT Error @ 25°C	±10	±10	±10	mV max	
T_{min} to T_{max}	±15	±15	±15	mV max	
Reference Load Sensitivity (ΔREF OUT/ΔI)	±1	±1	±1	mV max	Reference Load Current change (0–500μA). Reference Load should not be changed during conversion.
LOGIC INPUTS					
Input High Voltage, V_{INH}	+2.4	+2.4	+2.4	V min	V_{CC} = +5V ± 5%
Input Low Voltage, V_{INL}	+0.8	+0.8	+0.8	V max	V_{CC} = +5V ± 5%
Input Current, I_{IN}	±10	±10	±10	μA max	V_{IN} = 0 to V_{CC}
Input Capacitance, C_{IN}[6]	10	10	10	pF max	
LOGIC OUTPUTS					
Output High Voltage, V_{OH}	+2.7	+2.7	+2.7	V min	I_{SOURCE} = 40μA
Output Low Voltage, V_{OL}	+0.4	+0.4	+0.4	V max	I_{SINK} = 1.6mA
DB11–DB0					
Floating State Leakage Current	±10	±10	±10	±10	μA max
Floating State Output Capacitance[6]	15	15	15	15	pF max
CONVERSION TIME					
	7/7.125	7/7.125	7/7.125	μs min/μs max	Assuming no external Read/Write operations
	7/9.250	7/9.250	7/9.250	μs min/μs max	Assuming 17 external Read/Write operations See Internal Comparator Timing section
POWER REQUIREMENTS					
V_{DD}	+5	+5	+5	V nom	±5% for specified performance
V_{CC}	+5	+5	+5	V nom	±5% for specified performance
V_{SS}	−5	−5	−5	V nom	±5% for specified performance
I_{DD}	13	13	13	mA max	$\overline{CS}$ = $\overline{DMWR}$ = $\overline{DMRD}$ = 5V
I_{CC}	100	100	100	μA max	$\overline{CS}$ = $\overline{DMWR}$ = $\overline{DMRD}$ = 5V
I_{SS}	6	6	6	mA max	$\overline{CS}$ = $\overline{DMWR}$ = $\overline{DMRD}$ = 5V
Power Dissipation	95.5	95.5	95.5	mW max	Typically 60mW

NOTES
[1] Temperature range as follows:
J, K, L versions: 0 to +70°C
A, B versions: −25°C to +85°C
S version: −55°C to +125°C
[2] V_{IN} = ±3V. See Dynamic Specifications section.
[3] SNR calculation includes distortion and noise components.
[4] Measured with respect to the Internal Reference.
[5] For Capacitive Loads greater than 50pF a series resistor is required (see Internal Reference section).
[6] Sample tested @ 25°C to ensure compliance.
Specifications subject to change without notice.

TIMING CHARACTERISTICS[1] (V_{DD} = 5V ± 5%, V_{CC} = 5V ± 5%, V_{SS} = −5V ± 5%)

Parameter	Limit at T_{min}, T_{max} (L Grade)	Limit at T_{min}, T_{max} (J,K,A,B Grades)	Limit at T_{min}, T_{max} (S Grade)	Units	Conditions/Comments
t_1	65	65	75	ns max	CLK IN to $\overline{BUSY}$ Low Propagation Delay
t_2	65	65	75	ns max	CLK IN to $\overline{BUSY}$ High Propagation Delay
t_3	2 CLK IN cycles	2 CLK IN cycles	2 CLK IN cycles	min	$\overline{CONVST}$ Pulse Width
t_4	0	0	0	ns min	$\overline{CS}$ to $\overline{DMRD}$/REGISTER ENABLE Setup Time
t_5	0	0	0	ns min	$\overline{CS}$ to $\overline{DMRD}$/REGISTER ENABLE Hold Time
t_6	**45**	**60**	**60**	ns min	$\overline{DMRD}$ Pulse Width
	50	50	50	µs max	
t_7	**16**	**16**	**16**	ns min	ADD0 to $\overline{DMRD}$/REGISTER ENABLE Setup Time
t_8	0	0	0	ns min	ADD0 to $\overline{DMRD}$/REGISTER ENABLE Hold Time
t_9[2]	**41**	**57**	**57**	ns min	Data Access Time after $\overline{DMRD}$
t_{10}[3]	5	5	5	ns min	Bus Relinquish Time
	45	**45**	**45**	ns max	
t_{11}	**42**	**42**	**55**	ns min	REGISTER ENABLE Pulse Width
	50	50	50	µs max	
t_{12}	**20**	**20**	**30**	ns min	Data Valid to REGISTER ENABLE Setup Time
t_{13}	10	10	10	ns min	Data Hold Time after REGISTER ENABLE
t_{14}[2]	41	57	57	ns min	Data Access Time after $\overline{BUSY}$

NOTES

[1]Timing Specifications in **bold** print are 100% production tested. All other times are sample tested at +25°C to ensure compliance. All input signals are specified with tr = tf = 5ns (10% to 90% of 5V) and timed from a voltage level of 1.6V.

[2]t_9 and t_{14} are measured with the load circuits of Figure 1 and defined as the time required for an output to cross 0.8V or 2.4V.

[3]t_{10} is defined as the time required for the data lines to change 0.5V when loaded with the circuits of Figure 2.

Specifications subject to change without notice.

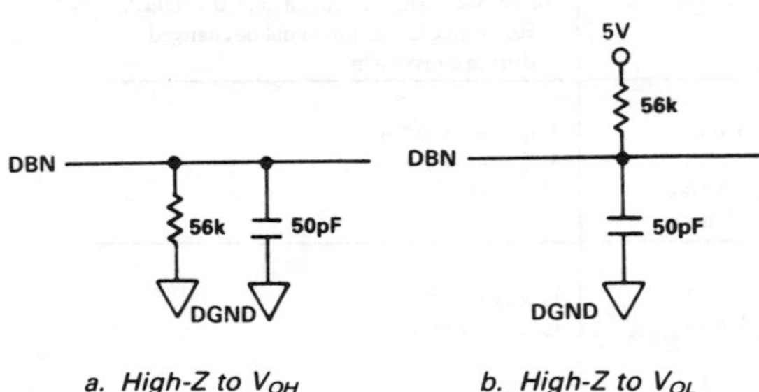

a. High-Z to V_{OH} b. High-Z to V_{OL}

Figure 1. Load Circuits for Access Time

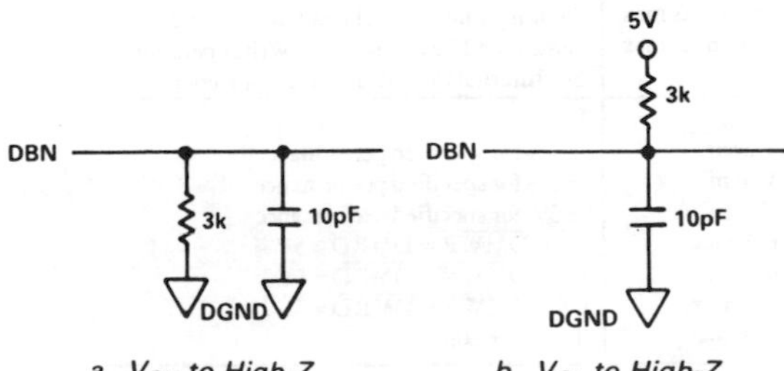

a. V_{OH} to High-Z b. V_{OL} to High-Z

Figure 2. Load Circuits for Output Float Delay

ABSOLUTE MAXIMUM RATINGS*

(T_A = +25°C unless otherwise stated)

V_{DD} to DGND . . . . −0.3V to +7V
V_{CC} to DGND . . . . −0.3V to +7V
V_{SS} to DGND . . . . +0.3V to −7V
V_{DD} to V_{CC} . . . . −0.3V to +0.3V
AGND to DGND . . . . −0.3V to V_{DD} +0.3V
V_{IN} to AGND . . . . −15V to +15V
REF OUT to AGND . . . . 0 to V_{DD}
Digital Inputs to DGND
CLK IN, $\overline{DMWR}$, $\overline{DMRD}$, $\overline{RESET}$,
$\overline{CS}$, $\overline{CONVST}$, ADD0 . . . . −0.3V to V_{DD} +0.3V
Digital Outputs to DGND
$\overline{ALFL}$, $\overline{BUSY}$ . . . . −0.3V to V_{DD} +0.3V
Data Pins
DB11 – DB0 . . . . −0.3V to V_{DD} +0.3V
Operating Temperature Range
J, K, L Versions . . . . 0 to +70°C
A, B Versions . . . . −25°C to +85°C
S Version . . . . −55°C to +125°C
Storage Temperature Range . . . . −65°C to +150°C
Lead Temperature (Soldering, 10secs) . . . . +300°C
Power Dissipation (Any Package) to +75°C 1000mW
Derates above +75°C by . . . . 10mW/°C

*Stresses above those listed under "Absolute Maximum Ratings" may cause permanent damage to the device. These are stress rating only and functional operation of the device at these or any other conditions above those indicated in the operational sections of this specification is not implied. Exposure to absolute maximum rating conditions for extended periods may affect device reliability.

CAUTION

ESD (electrostatic discharge) sensitive device. The digital control inputs are diode protected; however, permanent damage may occur on unconnected devices subject to high energy electrostatic fields. Unused devices must be stored in conductive foam or shunts. The protective foam should be discharged to the destination socket before devices are removed.

PIN FUNCTION DESCRIPTION

Pin Number	Pin Mnemonic	Function
1	ADD0	Address Input. This control input determines whether the word placed on the output data bus during a read operation is a data word from the FIFO RAM or the contents of the status/control register. A logic low accesses the data word from Location 0 of the FIFO while a logic high selects the contents of the register (see Status/Control Register section).
2	$\overline{CS}$	Chip Select. Active low logic input. The device is selected when this input is active.
3	$\overline{DMWR}$	Data Memory Write. Active low logic input. $\overline{DMWR}$ is used in conjunction with $\overline{CS}$ low and ADD0 high to write data to the status/control register. Corresponds to $\overline{DMWR}$ (ADSP-2100), R/$\overline{W}$ (MC68000, TMS32020), $\overline{WE}$ (TMS32010).
4	$\overline{DMRD}$	Data Memory READ. Active low logic input. $\overline{DMRD}$ is used in conjunction with $\overline{CS}$ low to enable the three-state output buffers. Corresponds directly to $\overline{DMRD}$ (ADSP-2100), $\overline{DEN}$ (TMS32010).
5	$\overline{BUSY}$	Active low logic output. This output goes low when the ADC receives a $\overline{CONVST}$ pulse and remains low until the track/hold has gone into its hold mode. The three-state drivers of the AD7878 can be disabled while the $\overline{BUSY}$ signal is low (see Extended READ/WRITE section). This is achieved by writing a logic 0 to DB5 ($\overline{DISO}$) of the status/control register. Writing a logic 1 to DB5 of the status/control register allows data to be accessed from the AD7878 while $\overline{BUSY}$ is low.
6	$\overline{ALFL}$	FIFO Almost Full. A logic low indicates that the word count (i.e., number of conversion results) in the FIFO memory has reached the programmed word count in the status/control register. $\overline{ALFL}$ is updated at the end of each conversion. The $\overline{ALFL}$ output is reset to a logic high when a word is read from the FIFO memory. It can also be set high by writing a logic 1 to DB7 ($\overline{ENAF}$) of the status/control register.
7	DGND	Digital Ground. Ground reference for digital circuitry.
8	V_{CC}	Digital supply voltage, +5V ± 5%. Positive supply voltage for digital circuitry.
9	DB11	Data Bit 11 (MSB). Three-state TTL output. Coding for the data words in FIFO RAM is 2s complement.
10-15	DB10-DB5	Data Bit 10 to Data Bit 5. Three-state TTL input/outputs.
16-19	DB4-DB1	Data Bit 4 to Data Bit 1. Three-state TTL outputs.
20	DB0	Data Bit 0 (LSB). Three-state TTL output.
21	V_{DD}	Analog positive supply voltage, +5V ± 5%.
22	AGND	Analog Ground. Ground reference for track/hold, reference and DAC.
23	REF OUT	Voltage Reference Output. The internal 3V analog reference is provided at this pin. The external load capability of the reference is 500μA.
24	V_{IN}	Analog Input. Analog input range is ± 3V.
25	V_{SS}	Analog negative supply voltage, − 5V ± 5%.
26	$\overline{CONVST}$	Convert Start. Logic input. A low to high transition on this input puts the track/hold into its hold mode and starts conversion. The $\overline{CONVST}$ input is asynchronous to CLK IN and independent of $\overline{CS}$, $\overline{DMWR}$ and $\overline{DMRD}$.
27	$\overline{RESET}$	Reset. Active low logic input. A logic low sets the words in FIFO memory to 1000 0000 0000 and resets the $\overline{ALFL}$ output and status/control register.
28	CLK IN	Clock Input. TTL-compatible logic input. Used as the clock source for the A/D converter. The mark space ratio of this clock can vary from 35/65 to 65/35.

PIN CONFIGURATIONS

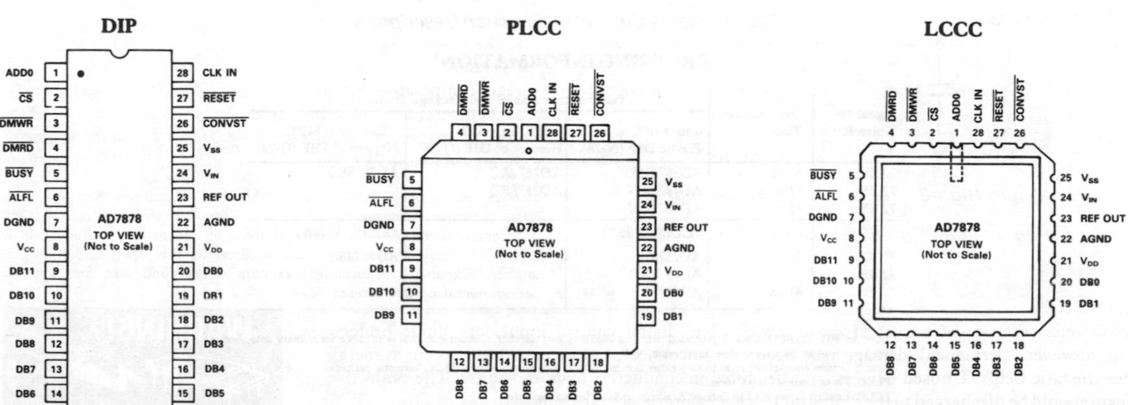

STATUS/CONTROL REGISTER

The status/control register serves the dual function of providing control and monitoring the status of the FIFO memory. This register is directly accessible through the data bus (DB11 – DB0) with a read or write operation while ADD0 is high. A write operation to the status/control register provides control for the $\overline{\text{ALFL}}$ output, bus interface and FIFO counter reset. This is normally done on power-up initialization. The FIFO memory address pointer is incremented after each conversion and compared with a preprogrammed count in the status/control register. When this preprogrammed count is reached, the $\overline{\text{ALFL}}$ output is asserted if the $\overline{\text{ENAF}}$ control bit is set to zero. This $\overline{\text{ALFL}}$ can be used to interrupt the microprocessor after any predetermined number of conversions (between 1 and 8). The status of the address pointer along with sample overrange and $\overline{\text{ALFL}}$ status can be accessed at any time by reading the status/control register. Note, reading the status/control register does not cause any internal data movement in the FIFO memory. Status information for a particular word should be read from the status register before the data word is read from the FIFO memory.

STATUS/CONTROL REGISTER FUNCTION DESCRIPTION

DB11 ($\overline{\text{ALFL}}$)

Almost Full Flag, Read only. This the same as Pin 6 ($\overline{\text{ALFL}}$ output) status. A logic low indicates that the word count in the FIFO memory has reached the preprogrammed count in bit locations DB10 – DB8. $\overline{\text{ALFL}}$ is updated at the end of conversion.

DB10 – DB8 (AFC2 – AFC0)

Almost Full Word Count, Read/Write. The count value determines the number of words in the FIFO memory which will cause $\overline{\text{ALFL}}$ to be set. When the FIFO word count equals the programmed count in these three bits, then both the $\overline{\text{ALFL}}$ output and DB11 of the status register are set to a logic low. For example, when a code of 011 is written to these bits, $\overline{\text{ALFL}}$ is set when Location 0 through Location 3 of the FIFO memory contains valid data. AFC2 is the most significant bit of the word count. The count value can be read back if required.

DB7 ($\overline{\text{ENAF}}$)

Enable Almost Full, Read/Write. Writing a 1 to this bit disables the $\overline{\text{ALFL}}$ output and status register bit DB11.

DB6 (FOVR/$\overline{\text{RESET}}$)

FIFO Overrun/RESET, Read/Write. Reading a 1 from this bit indicates that at least one sample has been discarded because the FIFO memory is full. When the FIFO is full (i.e., contains eight words) any further conversion results will be lost. Writing a 1 to this bit causes a system RESET as per the $\overline{\text{RESET}}$ input (Pin 27).

DB5 (FOOR/$\overline{\text{DISO}}$)

FIFO Out of RANGE/Disable Outputs, Read/Write. Reading a 1 from this bit indicates that at least one sample in the FIFO memory is out of range. Writing a 0 to this bit prevents the data bus from becoming active while $\overline{\text{BUSY}}$ is low regardless of the state of $\overline{\text{CS}}$ and $\overline{\text{DMRD}}$.

DB4 (FEMP)

FIFO Empty, Read Only. Reading a 1 indicates that there are no samples in the FIFO memory. When the FIFO is empty the internal ripple-down effects of the FIFO are disabled and further reads will continue to access the last valid data word in Location 0.

DB3 (SOOR)

Sample out of Range, Read Only. Reading a 1 indicates that the next sample to be read is out of range, i.e., the sample in Location 0 of the FIFO.

DB2 – DB0 (FCN2 – FCN0)

FIFO Word Count, Read Only. The value read from these bits indicates the number of samples in the FIFO memory. For example, reading 011 from these bits indicates that Location 0 through Location 3 contains valid data. Note, reading all 0s indicates that there is either one word or no word in the FIFO memory; in this case the FIFO Empty determines if there is no word in memory. FCN2 is the most significant bit.

BIT LOCATION	DB11	DB10	DB9	DB8	DB7	DB6	DB5	DB4	DB3	DB2	DB1	DB0
STATUS INFORMATION (READ)	$\overline{\text{ALFL}}$	AFC2	AFC1	AFC0	$\overline{\text{ENAF}}$	FOVR	FOOR	FEMP	SOOR	FCN2	FCN1	FCN0
CONTROL FUNCTION (WRITE)	X	AFC2	AFC1	AFC0	$\overline{\text{ENAF}}$	RESET	$\overline{\text{DISO}}$	X	X	X	X	X
RESET STATUS	1	0	0	0	0	0	0	1	0	0	0	0

X = DON'T CARE

Table I. Status/Control Bit Function Description

ORDERING INFORMATION[1]

Signal-to-Noise Ratio	Data Access Time	Temperature Range and Package Options[2]		
		0 to +70°C Plastic DIP (N-28)	−25°C to +85°C Hermetic[3] DIP (Q-28)	−55°C to +125°C Hermetic[3] DIP (Q-28)
70 dB	57 ns	AD7878JN	AD7878AQ	AD7878SQ
72 dB	57 ns	AD7878KN	AD7878BQ	
72 dB	41 ns	AD7878LN		
		PLCC[4] (P-28A)		LCCC[5] (E-28A)
70 dB	57 ns	AD7878JP		AD7878SE
72 dB	57 ns	AD7878KP		
72 dB	41 ns	AD7878LP		

NOTES

[1]To order MIL-STD-883, Class B processed parts, add/883B to part number. Contact our local sales office for military data sheet.

[2]See Section 14 for package outline information.

[3]Analog Devices reserves the right to ship either ceramic (D-28) packages or cerdip (Q-28) hermetic packages.

[4]PLCC: Plastic Leaded Chip Carrier.

[5]LCCC: Leadless Ceramic Chip Carrier. Available to 883B processing only.

INTERNAL FIFO MEMORY

The internal FIFO memory of the AD7878 consists of eight memory locations. Each word in memory contains 13 bits of information – 12 bits of data from the conversion result and one additional bit which contains information as to whether the 12-bit result is out of range or not. A block diagram of the AD7878 FIFO architecture is shown in Figure 3.

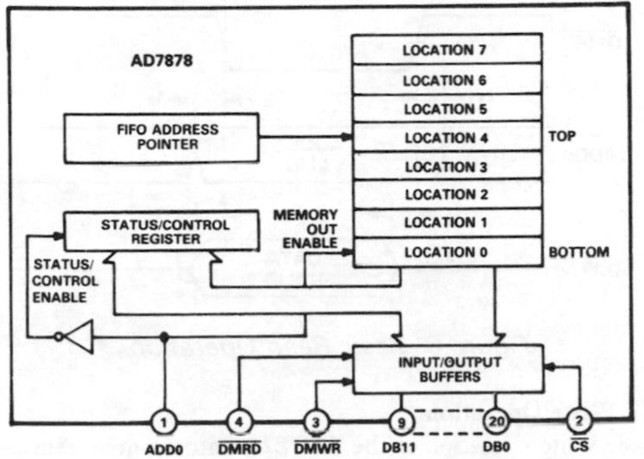

Figure 3. Internal FIFO Architecture

The conversion result is gathered in the successive approximation register (SAR) during conversion. At the end of conversion this result is transferred to the FIFO memory. The FIFO address pointer always points to the top of memory, i.e., the uppermost location which contains valid data. The pointer is incremented after each conversion. A read operation from the FIFO memory accesses data from the bottom of the FIFO, i.e., Location 0. On completion of the read operation each data word moves down one location and the address pointer is decremented by one. Therefore, each conversion result from the SAR enters at the top of memory, propagates down with successive reads until it reaches Location 0 from where it can be accessed by a microprocessor read operation.

The transfer of information from the SAR to the FIFO occurs in synchronization with the AD7878 input clock (CLK IN). The propagation of data words down the FIFO is also synchronous with this clock. As a result, a read operation to obtain data from the FIFO must also be synchronous with CLK IN to avoid Read/Write conflicts in the FIFO (i.e., reading from FIFO Location 0 while it is being updated). This requires that the microprocessor clock and the AD7878 CLK IN are derived from the same source.

INTERNAL COMPARATOR TIMING

The ADC clock, which is applied to CLK IN, controls the successive approximation A/D conversion process. This clock is internally divided by four to yield a bit trial cycle time of 500ns min (CLK IN = 8MHz clock). Each bit decision occurs 25ns after the rising edge of this divided clock. The bit decision is latched by the rising edge of an internal comparator strobe signal. There are 12 bit decisions, as in a normal successive approximation routine, and one extra decision which checks if the input sample is out of range. In a normal successive approximation A/D converter, reading data from the device during conversion can upset the conversion in progress. This is due to on-chip transients, generated by charging or discharging the data bus, concurrent with a bit decision. The scheme outlined below and shown in Figure 4 describes how the AD7878 overcomes this problem.

The internal comparator strobe on the AD7878 is gated with both $\overline{\text{DMRD}}$ and $\overline{\text{DMWR}}$ so that if a read or write operation occurs when a bit decision is about to be made, the bit decision point is deferred by one CLK IN cycle. In other words, if $\overline{\text{DMRD}}$ or $\overline{\text{DMWR}}$ goes low (with $\overline{\text{CS}}$ low) at any time during the CLK IN low-time immediately prior to the comparator strobing edge (t_{LOW} of Figure 4), the bit trial is suspended for a clock cycle. This makes sure that the bit decision is latched at a time when the AD7878 is not attempting to charge or discharge the data bus, thereby ensuring that no spurious transients occur internally near a bit decision point.

The decision point slippage mechanism is shown in Figure 4 for the MSB decision. Normally, the MSB decision occurs 25ns after the fourth rising CLK IN edge after $\overline{\text{CONVST}}$ goes high. However, in the timing diagram of Figure 4, $\overline{\text{CS}}$ and $\overline{\text{DMRD}}$ or $\overline{\text{DMWR}}$ are low in the time period t_{LOW} prior to the MSB decision point on the fourth rising edge. This causes the internal comparator strobe to be slipped to the fifth rising clock edge. The AD7878 will again check during a period t_{LOW} prior to this fifth rising clock edge; and if the $\overline{\text{CS}}$ and $\overline{\text{DMRD}}$ or $\overline{\text{DMWR}}$ are still low, the bit decision point will be slipped a further clock cycle.

The conversion time for the ADC normally consists of the 13 bit trials described above and one extra internal clock cycle during which data is written from the SAR to the FIFO. For an 8MHz input clock this results in a conversion time of 7μs. However, the software routine which services the AD7878 has the potential to read 16 times from the device during conversion – 8 reads from the FIFO and 8 reads from the status/control register. It also has the potential to write once to the status/control register. If these 17 (16 read plus 1 write) operations all occur during t_{LOW} time periods, it will cause the conversion time to slip by 17 CLK IN cycles. Therefore, if read or write operations can occur during t_{LOW} periods, it means that the conversion time for the ADC can vary from 7μs to 9.12μs (assuming 8MHz CLK IN). This calculation assumes that there is a slippage of one CLK IN cycle for each read or write operation.

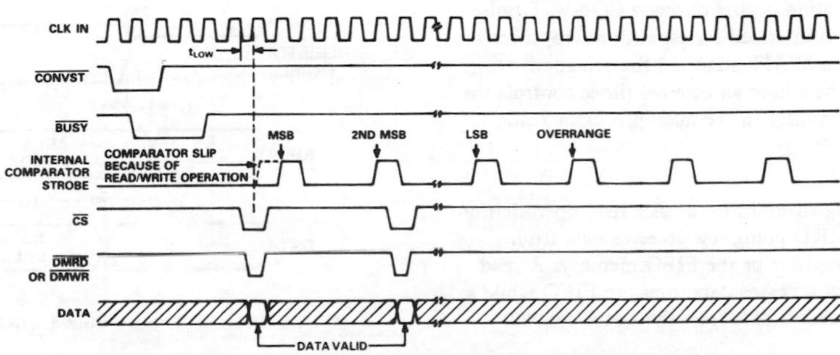

Figure 4. Operational Timing Diagram

INITIATING A CONVERSION

Conversion is initiated on the AD7878 by asserting the $\overline{\text{CONVST}}$ input. This $\overline{\text{CONVST}}$ input is an asynchronous input which is independent of either the ADC or DSP clocks. This is essential for applications where precise sampling in time is important. In these applications the signal sampling must occur at exactly equal intervals to minimize errors due to sampling uncertainty or jitter. In these cases the $\overline{\text{CONVST}}$ input is driven from a timer or some precise clock source. On receipt of a $\overline{\text{CONVST}}$ pulse, the AD7878 acknowledges by taking the $\overline{\text{BUSY}}$ output low. This $\overline{\text{BUSY}}$ output can be used to ensure no bus activity while the track/hold goes from track to hold mode (see Extended Read/Write section). The $\overline{\text{CONVST}}$ input must stay low for at least two CLK IN periods. The track/hold amplifier switches from the track to hold mode on the rising edge of $\overline{\text{CONVST}}$ and conversion is also initiated at this point. The $\overline{\text{BUSY}}$ output returns high after the $\overline{\text{CONVST}}$ input goes high and the ADC begins its successive approximation routine. Once conversion has been initiated another conversion start should not be attempted until the full conversion cycle has been completed. Figure 5 shows the timing diagram for the conversion start.

In applications where precise sampling is not critical, the $\overline{\text{CONVST}}$ pulse can be generated from a microprocessor $\overline{\text{WR}}$ or $\overline{\text{RD}}$ line gated with a decoded address (different to the AD7878 $\overline{\text{CS}}$ address). Note that the $\overline{\text{CONVST}}$ pulse width must be a minimum of two AD7878 CLK IN cycles.

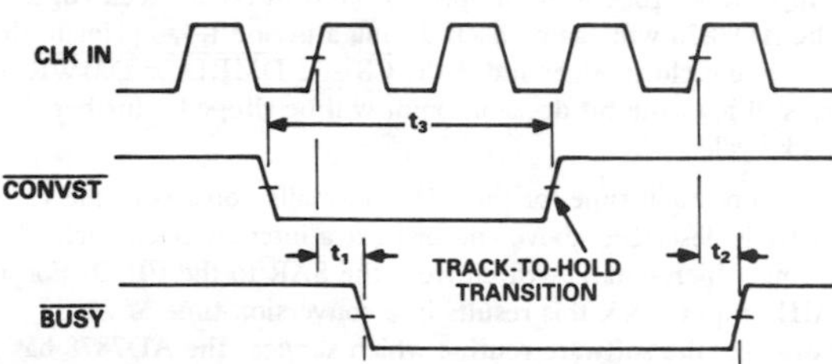

Figure 5. Conversion Start Timing Diagram

READ/WRITE OPERATIONS

The AD7878 read/write operations consist of reading from the FIFO memory and reading and writing from the status/control register. These operations are controlled by the $\overline{\text{CS}}$, $\overline{\text{DMRD}}$, $\overline{\text{DMWR}}$ and ADD0 logic inputs. A description of these operations is given in the following sections. In addition to the basic read/write operations there is an extended read/write operation. This can occur if a read/write operation occurs during a $\overline{\text{CONVST}}$ pulse. This extended read/write is intended for use with microprocessors which can be driven into a WAIT state and the scheme is recommended for applications where an external timer controls the $\overline{\text{CONVST}}$ input asynchronously to the microprocessor read/write operations.

Basic Read Operation

Figure 6 shows the timing diagram for a basic read operation on the AD7878. $\overline{\text{CS}}$ and $\overline{\text{DMRD}}$ going low accesses data from either the status/control register or the FIFO memory. A read operation with ADD0 low accesses data from the FIFO while a read with ADD0 high accesses data from the status/control register.

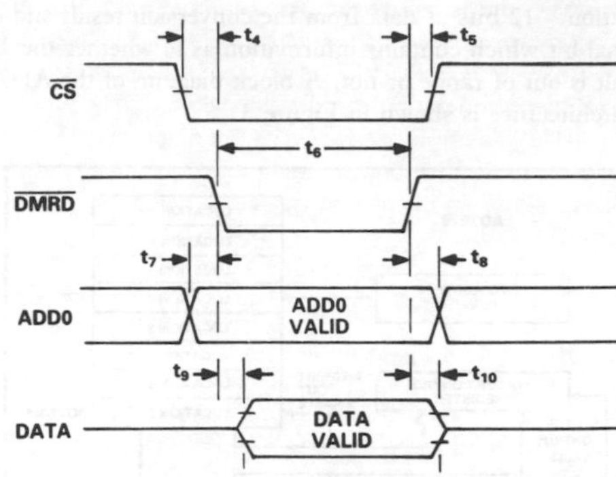

Figure 6. Basic Read Operation

Basic Write Operation

A basic write operation to the AD7878 status/control register consists of bringing $\overline{\text{CS}}$ and $\overline{\text{DMWR}}$ low with ADD0 high. Internally these signals are gated with CLK IN to provide an internal REGISTER ENABLE signal (see Figure 7). The pulse width of this REGISTER ENABLE signal is effectively the overlap between the CLK IN low time and the $\overline{\text{DMWR}}$ pulse. This may result in shorter write pulse widths, data setup times and data hold times than those given by the microprocessor. The timing on the AD7878 timing diagram of Figure 8 is therefore given with respect to the internal REGISTER ENABLE signal rather than the $\overline{\text{DMWR}}$ signal.

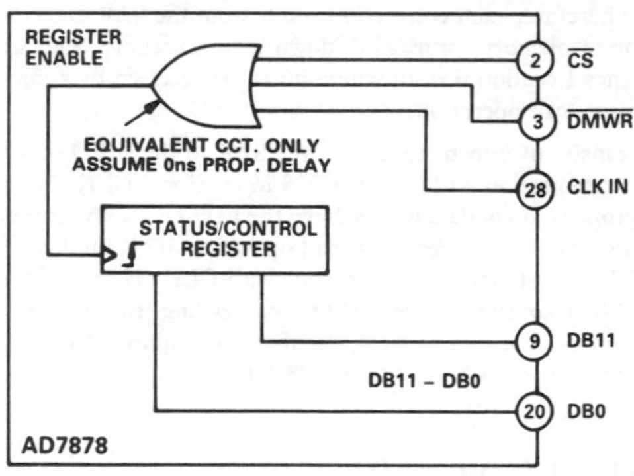

Figure 7. $\overline{\text{DMWR}}$ Internal Logic

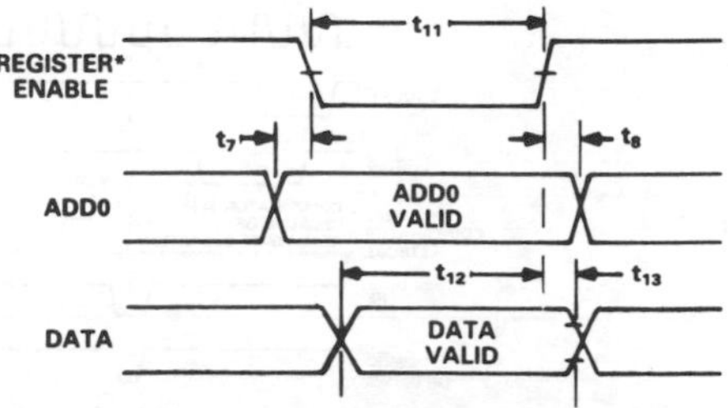

Figure 8. Basic Write Operation

3

Extended Read/Write Operation

As described earlier, a read/write operation to the AD7878 can cause spurious on-chip transients. Should these transients occur while the track/hold is going from track to hold mode it may result in an incorrect value of V_{IN} being held by the track/hold amplifier. Because the $\overline{CONVST}$ input has asynchronous capability, a read/write operation could occur while $\overline{CONVST}$ is low. The AD7878 allows the read/write operation to occur but has the facility to disable its three-state drivers so that there is no data bus activity and hence no transients while the track/hold goes from track to hold.

Writing a logic 0 to DB5 ($\overline{DISO}$) of the status/control register prevents the output latches from being enabled while the AD7878 $\overline{BUSY}$ signal is low. If a microprocessor read/write operation can occur during the $\overline{BUSY}$ low time, the $\overline{BUSY}$ should be gated with $\overline{CS}$ of the AD7878 and this gated signal used to stretch the instruction cycle using DMACK (ADSP-2100), READY (TMS32020) or $\overline{DTACK}$ (68000).

When $\overline{CONVST}$ goes low the AD7878 acknowledges by bringing $\overline{BUSY}$ low on the next rising edge of CLK IN. With a logic 0 in DB5, the AD7878 data bus cannot now be enabled. If a read/write operation now occurs, the $\overline{BUSY}$ and $\overline{CS}$ gated signal drives the microprocessor into a WAIT state, thereby extending the read/write operation. $\overline{BUSY}$ goes high on the second rising edge of CLK IN after $\overline{CONVST}$ goes high. The AD7878 data outputs are now enabled and the microprocessor is released from its WAIT state, allowing it to complete its read/write operation to the AD7878.

The microprocessor cycle time for the read/write operation is extended by the $\overline{CONVST}$ pulse width plus two CLK IN periods worst case. This is the maximum length of time for which $\overline{BUSY}$ can be low. Assuming a $\overline{CONVST}$ pulse width of two CLK IN periods and an 8MHz CLK IN, the instruction cycle is extended by 500ns maximum. Figure 9 shows the timing diagram for an extended read operation. In a similar manner, a write operation will be extended if it occurs during a $\overline{CONVST}$ pulse.

For processors which cannot be forced into a WAIT state, writing a logic 1 into DB5 of the status/control register allows the output latches to be enabled while $\overline{BUSY}$ is low. In this case $\overline{BUSY}$ still goes low as before, but it would not be used to stretch the read/write cycle and the instruction cycle continues as normal (see Figures 6 and 8).

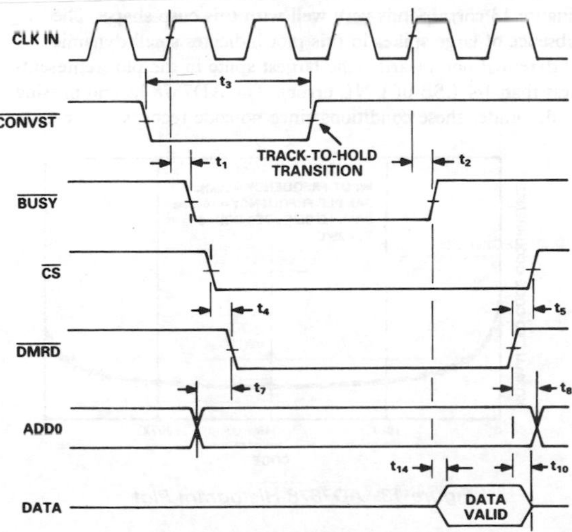

Figure 9. Extended Read Operation

AD7878 DYNAMIC SPECIFICATIONS

The AD7878 is specified and 100% tested for dynamic performance specifications rather than traditional dc specifications such as Integral and Differential Nonlinearity. These ac specifications provide information on the AD7878's effect on the spectral content of the input signal. Hence the parameters for which the AD7878 is specified include SNR, Harmonic Distortion, Intermodulation Distortion and Peak Harmonics. These terms are discussed in more detail in the following sections.

Signal-to-Noise Ratio (SNR)

SNR is the measured signal-to-noise ratio at the output of the ADC. The signal is the rms magnitude of the fundamental. Noise is the rms sum of all the nonfundamental signals (excluding dc) up to half the sampling frequency (fs/2). SNR is dependent upon the number of quantization levels used in the digitization process; the more levels, the smaller the quantization noise. The theoretical signal-to-noise ratio for a sine wave input is given by

$$SNR = (6.02N + 1.76)dB \; . \; . \; . \; . \; . \; (1)$$

where N is the number of bits. Thus for an ideal 12-bit converter, SNR = 74dB.

The output spectrum from the ADC is evaluated by applying a sine-wave signal of very low distortion to the V_{IN} input which is sampled at a 100kHz sampling rate. A Fast Fourier Transform (FFT) plot is generated from which the SNR data can be obtained. Figure 10 shows a typical 2048 point FFT plot of the AD7878KN with an input signal of 25kHz and a sampling frequency of 100kHz. The SNR obtained from this graph is 72.6dB. It should be noted that the harmonics are included in the SNR calculation.

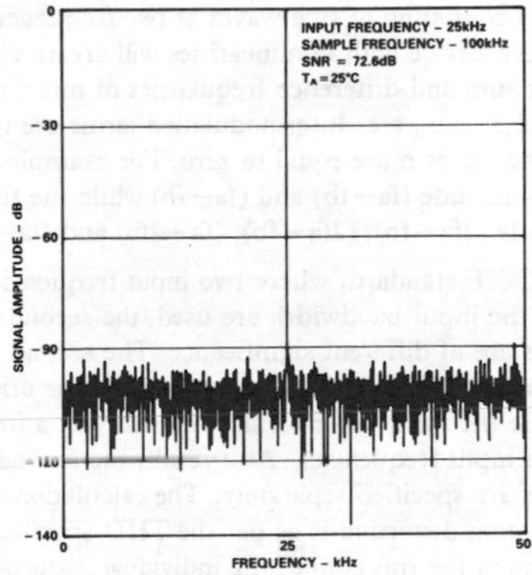

Figure 10. AD7878 FFT Plot

Effective Number of Bits

The formula given in (1) relates the SNR to the number of bits. Rewriting the formula, as in (2), it is possible to get a measure of performance expressed in effective number of bits (N). The effective number of bits for a device can be calculated directly from its measured SNR.

$$N = \frac{SNR - 1.76}{6.02} \; . \; . \; . \; (2)$$

Figure 11 shows a typical plot of effective number of bits versus frequency for an AD7878KN with a sampling frequency of 100kHz. The effective number of bits typically falls between 11.7 and 11.85 corresponding to SNR figures of 72.2 and 73.1dB.

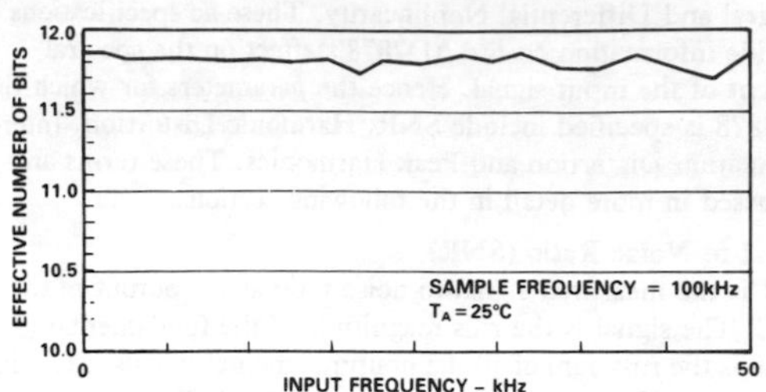

Figure 11. Effective Number of Bits vs. Frequency

Harmonic Distortion

Harmonic Distortion is the ratio of the rms sum of harmonics to the fundamental. For the AD7878, Total Harmonic Distortion (THD) is defined as:

$$\text{THD} = 20 \text{ Log } \frac{\sqrt{(V_2^2 + V_3^2 + V_4^2 + V_5^2 + V_6^2)}}{V_1}$$

where V_1 is the rms amplitude of the fundamental and V_2, V_3, V_4, V_5 and V_6 are the rms amplitudes of the second to the sixth harmonic. The THD is also derived from the FFT plot of the ADC output spectrum.

Intermodulation Distortion

With inputs consisting of sine waves at two frequencies, fa and fb, any active device with nonlinearities will create distortion products at sum and difference frequencies of mfa ± nfb where m,n = 0,1,2,3 . . . , etc. Intermodulation terms are those for which neither m or n are equal to zero. For example, the second order terms include (fa + fb) and (fa − fb) while the third order terms include (2fa + fb), (2fa − fb), (fa + 2fb) and (fa − 2fb).

Using the CCIF standard, where two input frequencies near the top end of the input bandwidth are used, the second and third order terms are of different significance. The second order terms are usually distanced in frequency from the original sine waves, while the third order terms are usually at a frequency close to the input frequencies. As a result, the second and third order terms are specified separately. The calculation of the intermodulation distortion is as per the THD specification where it is the ratio of the rms sum of the individual distortion products to the rms amplitude of the fundamental expressed in dBs.

Intermodulation distortion is calculated using an FFT algorithm but in this case the input consists of two equal amplitude, low distortion sine waves. Figure 12 shows a typical IMD plot for the AD7878.

Peak Harmonic or Spurious Noise

Peak harmonic or spurious noise is defined as the ratio of the rms value of the next largest component in the ADC output spectrum (up to fs/2 and excluding dc) to the rms value of the fundamental. Normally, the value of this specification will be determined by the largest harmonic in the spectrum, but for parts where the harmonics are buried in the noise floor the largest peak will be a noise peak.

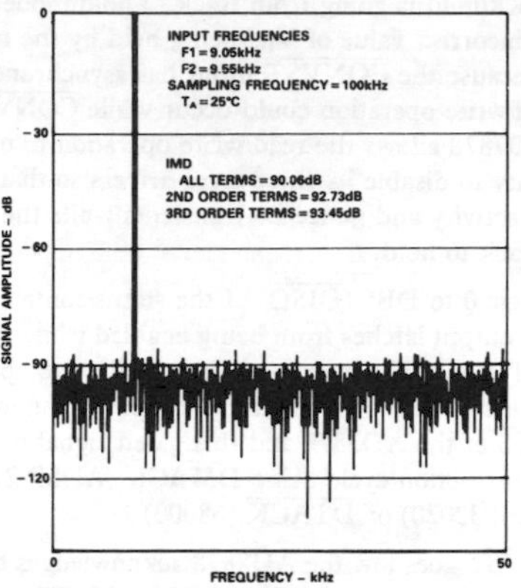

Figure 12. AD7878 IMD Plot

Histogram Plot

When a sine wave of a specified frequency is applied to the V_{IN} input of the AD7878 and several million samples are taken, it is possible to plot a histogram showing the frequency of occurrence of each of the 4096 ADC codes. If a particular step is wider than the ideal 1LSB width, then the code associated with that step will accumulate more counts than for the code for an ideal step. Likewise, a step narrower than ideal will have fewer counts. Missing codes are easily seen in the histogram plot because a missing code means zero counts for a particular code. Large spikes in the plot indicate large differential nonlinearity.

Figure 13 shows a histogram plot for the AD7878KN with a sampling frequency of 100kHz and an input frequency of 25kHz. For a sine-wave input, a perfect ADC would produce a cusp probability density function described by the equation:

$$p(V) = \frac{1}{\pi\sqrt{(A^2 - V^2)}}$$

where A is the peak amplitude of the sine wave and p(V) the probability of occurrence at a voltage V. The histogram plot of Figure 13 corresponds very well with this cusp shape. The absence of large spikes in this plot indicates small dynamic differential nonlinearity (the largest spike in the plot represents less than 1/4 LSB of DNL error). The AD7878 has no missing codes under these conditions since no code records zero counts.

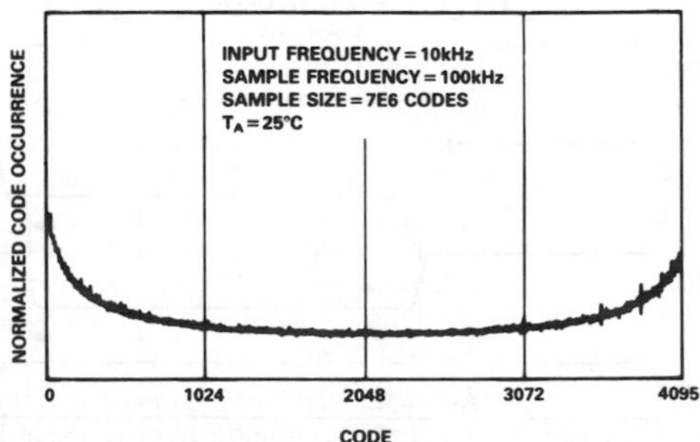

Figure 13. AD7878 Histogram Plot

3

CONVERSION TIMING

The track-and-hold on the AD7878 goes from track to hold mode on the rising edge of $\overline{CONVST}$ and the value of V_{IN} at this point is the value which will be converted. However, the conversion actually starts on the next rising edge of CLK IN after $\overline{CONVST}$ goes high. If $\overline{CONVST}$ goes high within approximately 30ns prior to a rising edge of CLK IN, that CLK IN edge will not be seen as the first CLK IN edge of the conversion process, and conversion will not actually start until one CLK IN cycle later. As a result, the conversion time (from $\overline{CONVST}$ to FIFO update) will vary by one clock cycle depending on the relationship between $\overline{CONVST}$ and CLK IN. A conversion cycle normally consists of 56 CLK IN cycles (assuming no read/write operations) which corresponds to a 7μs conversion time. If $\overline{CONVST}$ goes high within 30ns prior to a rising edge of CLK IN, the conversion time will consist of 57 CLK IN cycles, i.e., 7.125μs. This effect does not cause track/hold jitter.

INTERNAL REFERENCE

The AD7878 has an on-chip temperature compensated buried Zener reference (see Figure 14) which is factory trimmed to 3V ± 1%. Internally it provides both the DAC reference and the dc bias required for bipolar operation. The reference output is available (REF OUT) and is capable of providing up to 500μA to an external load.

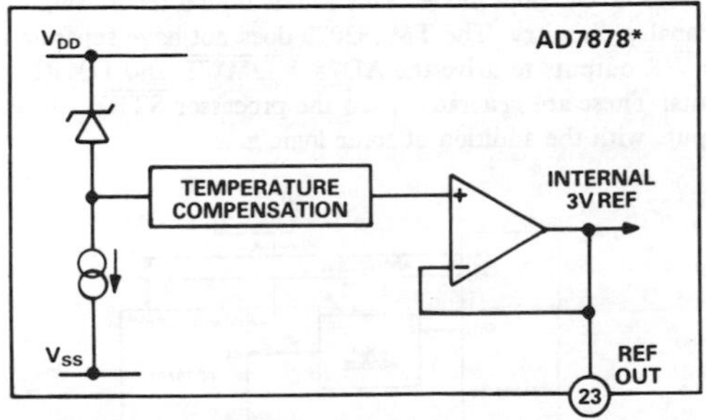

Figure 14. AD7878 Reference Circuit

The maximum recommended capacitance on REF OUT for normal operation is 50pF. If the reference is required for use external to the AD7878 it should be decoupled with a 200Ω resistor in series with a parallel combination of a 10μF tantalum capacitor and a 0.1μF ceramic capacitor. These decoupling components are required to remove voltage spikes caused by the AD7878's internal operation.

TRACK-AND-HOLD AMPLIFIER

The track-and-hold amplifier on the analog input of the AD7878 allows the ADC to accurately convert an input sine wave of 6V peak-peak amplitude to 12-bit accuracy. The input bandwidth of the track/hold amplifier is much greater than the Nyquist rate of the ADC even when operated at its minimum conversion time. The 0.1dB cutoff frequency occurs typically at 500kHz. The track/hold amplifier acquires an input signal to 12-bit accuracy in less than 2μs.

The operation of the track/hold amplifier is transparent to the user. The track/hold amplifier goes from its tracking mode to its hold mode at the start of conversion on the rising edge of $\overline{CONVST}$ and returns to track mode at the end of conversion.

ANALOG INPUT

Figure 15 shows the AD7878 analog input. The analog input range is ± 3V into an input resistance of typically 15kΩ. The designed code transitions occur midway between successive integer LSB values (i.e., 1/2LSB, 3/2LSBs, 5/2LSBs . . . . FS – 3/2LSBs). The output code is 2s complement binary with 1LSB = FS/4096 = 6V/4096 = 1.46mV. The ideal input/output transfer function is shown in Figure 16.

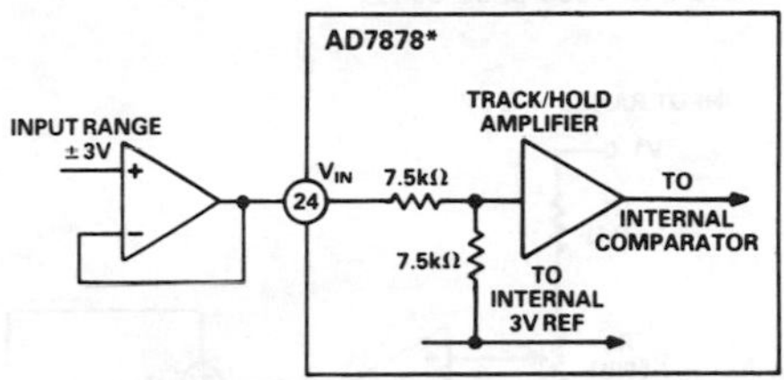

Figure 15. AD7878 Analog Input

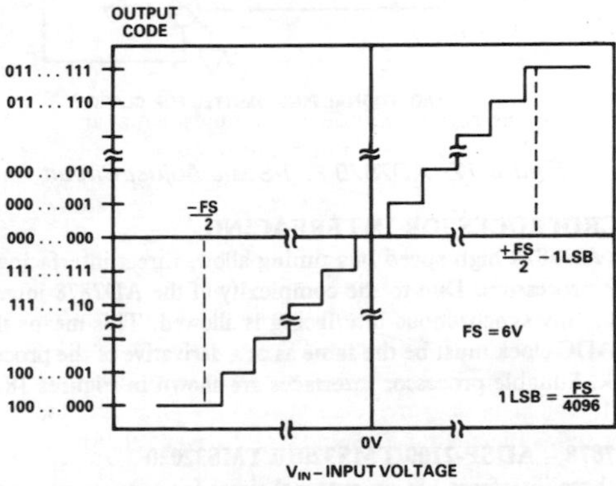

Figure 16. Input/Output Transfer Function

OFFSET AND FULL-SCALE ADJUSTMENT

In most Digital Signal Processing (DSP) applications offset and full-scale error have little or no effect on system performance. Offset error can always be eliminated in the analog domain by ac coupling. Full-scale error effect is linear and does not cause problems as long as the input signal is within the full dynamic range of the ADC. Some applications may require that the input signal span the full analog input dynamic range and accordingly offset and full-scale error will have to be adjusted to zero.

Where adjustment is required offset must be adjusted before full-scale error. This is achieved by trimming the offset of the op amp driving the analog input of the AD7878 while the input voltage is 1/2 LSB below ground. The trim procedure is as follows: apply a voltage of – 0.73mV (– 1/2 LSB) at V_1 and adjust the op amp offset voltage until the ADC output code flickers between 1111 1111 1111 and 0000 0000 0000.

Gain error can be adjusted at either the first code transition (ADC negative full scale) or the last code transition (ADC positive full scale). The trim procedures for both cases are as follows:

Positive Full-Scale Adjust

Apply a voltage of 2.9978V (FS/2 – 3/2LSBs) at V_1. Adjust R2 until the ADC output code flickers between 0111 1111 1110 and 0111 1111 1111.

Negative Full-Scale Adjust

Apply a voltage of −2.9993V (−FS/2 + 1/2 LSB) at V_1 and adjust R2 until the ADC output code flickers between 1000 0000 0000 and 1000 0000 0001.

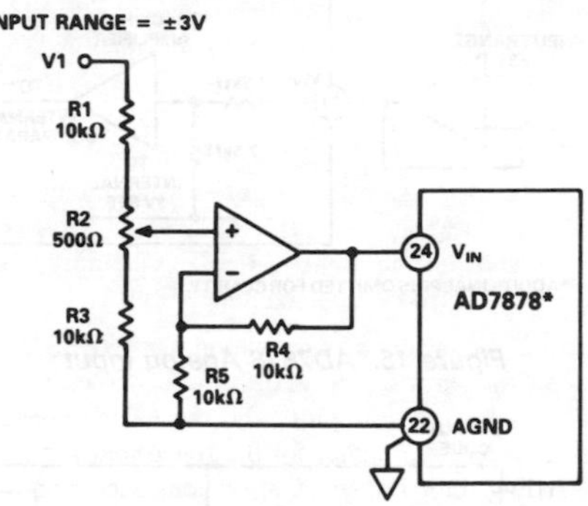

Figure 17. AD7878 Full-Scale Adjust Circuit

MICROPROCESSOR INTERFACING

The AD7878 high speed bus timing allows direct interfacing to DSP processors. Due to the complexity of the AD7878 internal logic, only synchronous interfacing is allowed. This means that the ADC clock must be the same as or a derivative of the processor clock. Suitable processor interfaces are shown in Figures 18 to 21.

AD7878 – ADSP-2100/TMS32010/TMS32020

All three interfaces use an external timer for conversion control. This allows the ADC to sample the analog input asynchronously to the microprocessor. The AD7878 $\overline{\text{ALFL}}$ output interrupts the processor when the FIFO preprogrammed word count is reached. The processor then reads the conversion results from the AD7878 internal FIFO memory.

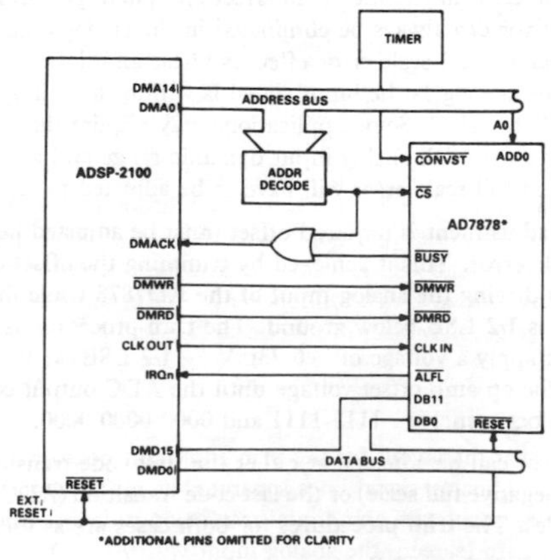

Figure 18. AD7878–ADSP-2100 Interface

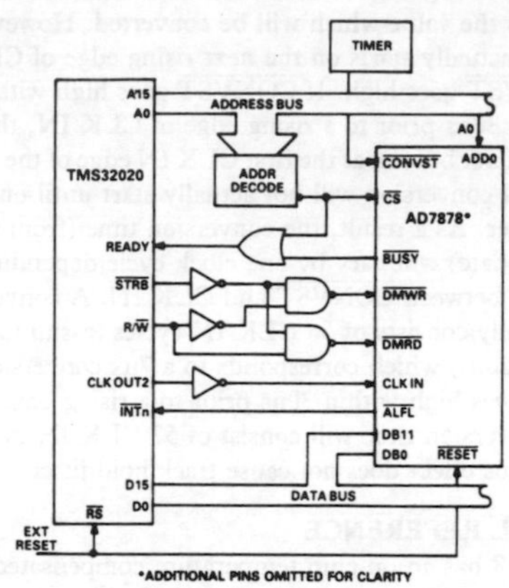

Figure 19. AD7878 – TMS32020 Interface

The interfaces to the ADSP-2100 and the TMS32020 gate the AD7878 $\overline{\text{CS}}$ and the $\overline{\text{BUSY}}$ to provide a signal which drives the processor into a wait state if a read/write operation to the ADC is attempted while the ADC track/hold amplifier is going from the track to the hold mode. This avoids digital feedthrough to the analog circuitry. The TMS32020 does not have separate $\overline{\text{RD}}$ and $\overline{\text{WR}}$ outputs to drive the AD7878 $\overline{\text{DMWR}}$ and $\overline{\text{DMRD}}$ inputs. These are generated from the processor $\overline{\text{STRB}}$ and R/$\overline{\text{W}}$ outputs with the addition of some logic gates.

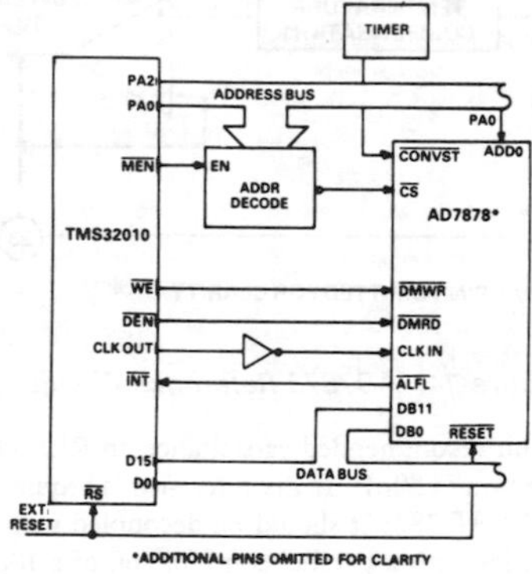

Figure 20. AD7878 – TMS32010 Interface

AD7878 – MC68000

This interface also uses an external timer for conversion control as described for the previous three interfaces. It is discussed separately because it needs extra logic due to the nature of its interrupts. The MC68000 has eight levels of external interrupt. When interrupting this processor one of these levels (0 to 7) has to be encoded onto the $\overline{\text{IPL2}}$ – $\overline{\text{IPL0}}$ inputs. This is achieved with a 74148 encoder in Figure 21, (interrupt Level 1 is taken for example purposes only). The MC68000 places this interrupt level on address bits A3 to A1 at the start of the interrupt service routine. Additional logic is used to decode this interrupt level on the address bus and the FC2 – FC0 outputs to generate a $\overline{\text{VPA}}$ signal for the MC68000. This results in an autovectored interrupt, the start address for the service routine must be loaded into the appropriate auto vector location during initialization. For further information on the 68000 interrupts consult the 68000 users manual.

The MC68000 $\overline{AS}$ and R/$\overline{W}$ outputs are used to generate separate $\overline{DMWR}$ and $\overline{DMRD}$ inputs for the AD7878. As with the previous three interfaces described earlier, WAIT states are inserted if a read/write operation is attempted while the track/hold amplifier is going from the track to the hold mode.

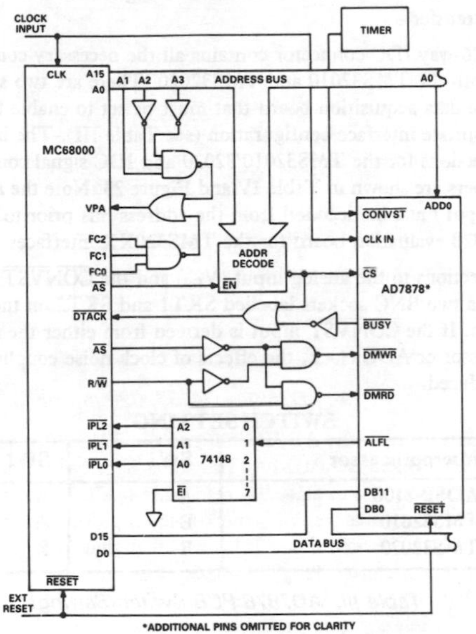

Figure 21. AD7878 – MC68000 Interface

Typical AD7878 Microprocessor Operating Sequence

After power up or reset the status/control register is initialized by writing to the AD7878. This enables the $\overline{ALFL}$ output if required for a microprocessor interrupt and sets the effective word length of the FIFO memory. The processor now executes the main body of the program while waiting for an ADC interrupt. This interrupt will occur when the preprogrammed number of samples are collected in the FIFO memory. The interrupt service routine first interrogates DB5(FOOR) of the status/control register to determine if any sample in the FIFO memory is out of range. If all data samples are valid then the program proceeds to read the FIFO memory. If, on the other hand, at least one sample is out of range then an overrange routine is called.

There are many actions which can be taken by the out of range routine, the selection of which is application dependent. One option is to ignore all the current samples residing in the FIFO memory, reinitialize the status/control register and return to the main body of the program. Another option is to check the individual out of range status of each word in the FIFO memory and discard the invalid ones. The underrange or overrange status of each word can also be determined and the analog input adjusted accordingly before returning to the main program.

Note there is no need to check the out of range status if the analog input is always assured to be within range.

THROUGHPUT RATE

The AD7878 has a maximum specified throughput rate (sample rate) of 100kHz. This is a worst case test condition and specifications apply for reduced sampling rates provided the Nyquist criterion is obeyed. The throughput rate must take into account ADC $\overline{CONVST}$ pulse width, ADC conversion time and the track/hold amplifier acquisition time. The time required for each of these tasks is shown in Table II for a selection of DSP processors. Since the ADC clock has to be synchronized to the microprocessor clock, the conversion time depends on the microprocessor used. In addition, time must be allowed for reading data from the AD7878. If this task is performed during the track/hold amplifier acquisition period then it does not impact on the overall throughput rate. However, if the read operations occur during a conversion, then they may stretch the conversion time and reduce the track/hold amplifier acquisition time. The track/hold amplifier requires a minimum of 2µs to operate to specification. The time required to read from the AD7878 depends on the number of FIFO memory locations to be read and the software organization.

As an example, consider an application using the ADSP-2100 and the AD7878 with a throughput rate of 100kHz. The time required for the $\overline{CONVST}$ pulse and the ADC conversion is 7.375µs. This leaves 2.625µs for the track/hold acquisition time and for reading the ADC (both operations occurring in parallel). The ADSP-2100, when operating from a 32MHz clock, has an instruction cycle of 125ns and an interrupt response time of 500ns. This allows adequate time to perform 16 read operations within the time budget allowed.

	$\overline{CONVST}$ Pulse Width	Conversion Time	T/H Acquisition Time
Number of Clock Cycles	2 min	57 max	Non-Applicable
ADSP-2100[1]	250ns min	7.125µs max	2µs min
TMS32010[2]	400ns min	11.14µs max	2µs min
TMS32020[2]	400ns min	11.14µs max	2µs min

NOTES
[1]ADSP-2100 Clock Freq. = 32MHz
[2]TMS320XX Clock Freq. = 20MHz

Table II. AD7878 Throughput Rate

APPLICATION HINTS

Good printed circuit board (PCB) layout is as important as the overall circuit design itself in achieving high speed A/D performance. The AD7878 is required to make bit decisions on an LSB size of 1.465mV. To achieve this, the designer has to be conscious of noise both in the ADC itself and in the preceding analog circuitry. Switching mode power supplies are not recommended as the switching spikes will feed through to the comparator causing noisy code transitions. Other causes of concern are ground loops and digital feedthrough from microprocessors. These are factors which influence any ADC, and a proper PCB layout which minimizes these effects is essential for best performance.

LAYOUT HINTS

Ensure that the layout for the printed circuit board has the digital and analog signal lines separated as much as possible. Take care not to run any digital track alongside an analog signal track. Guard (screen) the analog input with AGND.

Establish a single point analog ground (star ground) separate from the logic system ground at Pin 22 (AGND) or as close as possible to the AD7878 as shown in Figure 22. Connect all other grounds and Pin 7 (AD7878 DGND) to this single analog ground point. Do not connect any other digital grounds to this analog ground point. Low impedance analog and digital power supply common returns are essential to low noise operation of the ADC, so make the foil width for these tracks as wide as possible. The use of ground planes minimizes impedance paths and also guards the analog circuitry from digital noise. The circuit layout of Figures 25 and 26 have both analog and digital ground planes which are kept separated and only joined together at the AD7878 AGND pin.

NOISE

Keep the input signal leads to V_{IN} and signal return leads from AGND (Pin 22) as short as possible to minimize input noise coupling. In applications where this is not possible use a shielded cable between the source and the ADC. Reduce the ground circuit impedance as much as possible since any potential difference in grounds between the signal source and the ADC appears as an error voltage in series with the input signal.

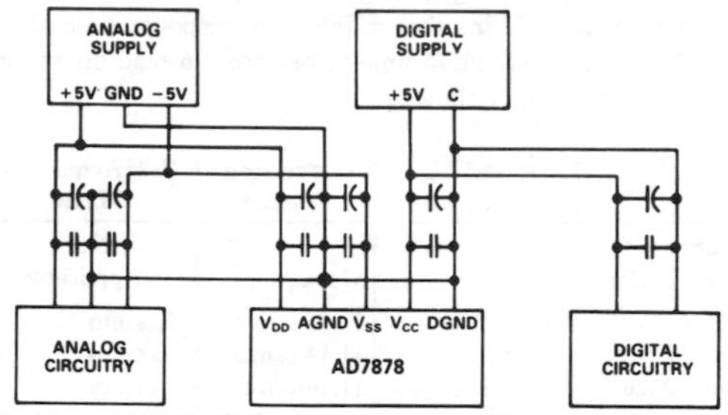

Figure 22. Power Supply Grounding Practice

DATA ACQUISITION BOARD

Figure 23 shows the AD7878 in a data acquisition circuit which will interface directly to either the ADSP-2100, TMS32010 or the TMS32020. The corresponding printed circuit board (PCB) layout and silkscreen are shown in Figures 24 to 26.

The only additional component required for a full data acquisition system is an antialiasing filter. There is a component grid provided near the analog input on the PCB which may be used for such a filter or any other conditioning circuitry. To facilitate this option, a wire link (labelled LK1 on the PCB) is required on the analog input track. This link connects the input signal to either the component grid or directly to the buffer amplifier driving the AD7878 analog input.

Microprocessor connections to the PCB can be made by either of two ways:

1. 96-contact (3 ROW) Eurocard connector.
2. 26-contact (2 ROW) IDC connector.

The 96-contact Eurocard connector is directly compatible with the ADSP-2100 Evaluation Board Prototype Expansion Connector. The expansion connector on the ADSP-2100 has eight decoded chip enable outputs labelled $\overline{ECE8}$ to $\overline{ECE1}$. $\overline{ECE6}$ is used to drive the AD7878 $\overline{CS}$ input on the data acquisition board. To avoid selecting onboard RAM sockets at the same time, LK6 on the ADSP-2100 board must be removed. In addition, the expansion connector on the ADSP-2100 has four interrupts labelled $\overline{EIRQ3}$ to $\overline{EIRQ0}$. The AD7878 $\overline{ALFL}$ output connects to $\overline{EIRQ0}$. The AD7878 and ADSP-2100 data lines are aligned for left justified data transfer.

The 26-way IDC connector contains all the necessary contacts for both the TMS32010 and TMS32020. There are two switches on the data acquisition board that must be set to enable the appropriate interface configuration (see Table III). The interface connections for the TMS32010/32020 and IDC signal contact numbers are shown in Table IV and Figure 23. Note the AD7878 $\overline{CS}$ input must be decoded from the address bus prior to the AD7878 evaluation board for the TMS320XX interfaces.

Connections to the analog input (V_{IN}) and the $\overline{CONVST}$ input are via two BNC sockets labelled SKT1 and SKT2 on the silk-screen. If the $\overline{CONVST}$ input is derived from either the micro-processor or ADC clock, the effects of clock noise coupling will be reduced.

SWITCH SETTING

Microprocessor	SW1	SW2
ADSP-2100	A	A
TMS32010	B	A
TMS32020	B	B

Table III. AD7878 PCB Switch Settings

POWER SUPPLY CONNECTIONS

The PCB requires two analog supplies and one 5V digital supply. Connections to the analog supplies are made directly to the PCB as shown on the silk screen in Figure 24. The connections are labelled V+ and V− and the range for both of these supplies is 12V to 15V. Connection to the 5V digital supply is made through either of the two microprocessor connectors. The +5V and −5V analog power supplies required by the AD7878 are generated from two voltage regulators on the V+ and V− power supply inputs (IC3 and IC4 in Figure 23).

COMPONENT LIST

IC1	AD711 Op Amp
IC2	AD7878 Analog-to-Digital Converter
IC3	MC78L05 5V Regulator
IC4	MC79L05 −5V Regulator
IC5*	74HC00 Quad NAND Gate
IC6*	74HC04 Hex Inverter
IC7	74HC02 Quad NOR Gate
SW1	Single Pole Double Throw
SW2	Double Pole Double Throw
LK1	Wire Link for Analog Input
C1, C3, C5, C7, C9 C11, C13, C15	10μF Capacitors
C2, C4, C6, C8, C10 C12, C14, C16	0.1μF Capacitors
R1*, R2*	10kΩ Resistors
SKT1, SKT2	BNC Sockets
SKT3	26-Contact (2 Row) IDC Connector
SKT4	96-Contact (3 Row) Eurocard Connector

*Not required for ADSP-2100 Interface

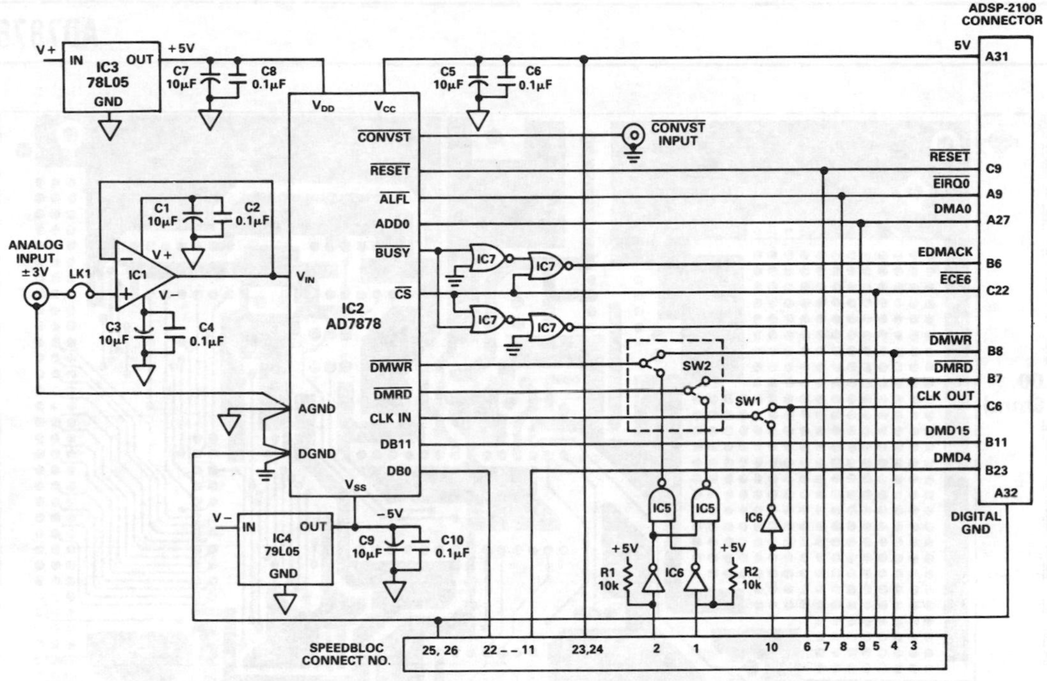

Figure 23. Data Acquisition Circuit Using the AD7878

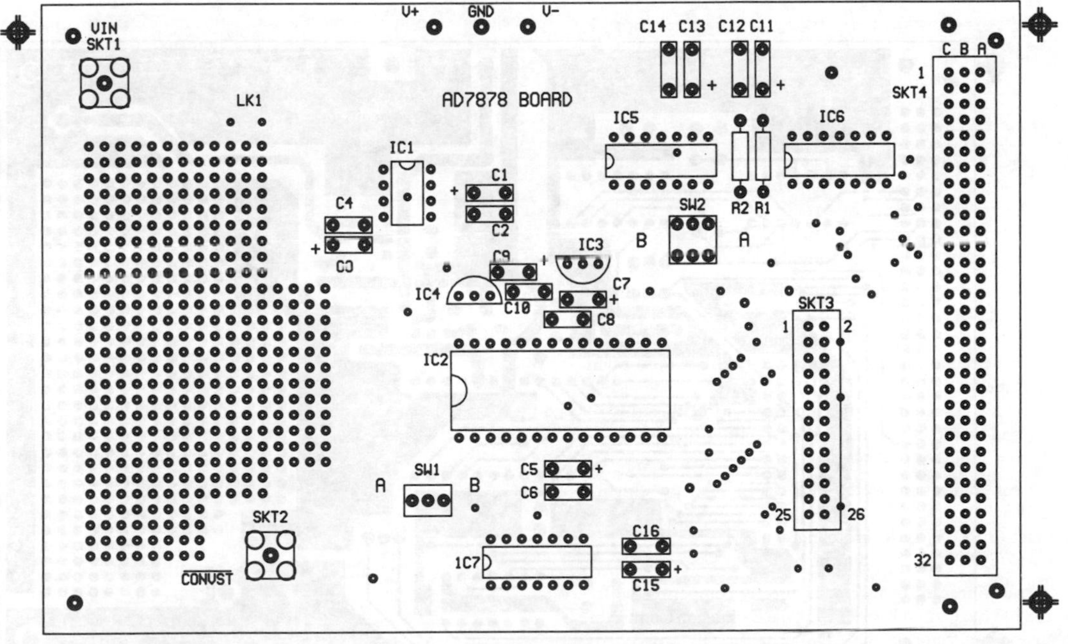

Figure 24. PCB Silkscreen for Figure 23

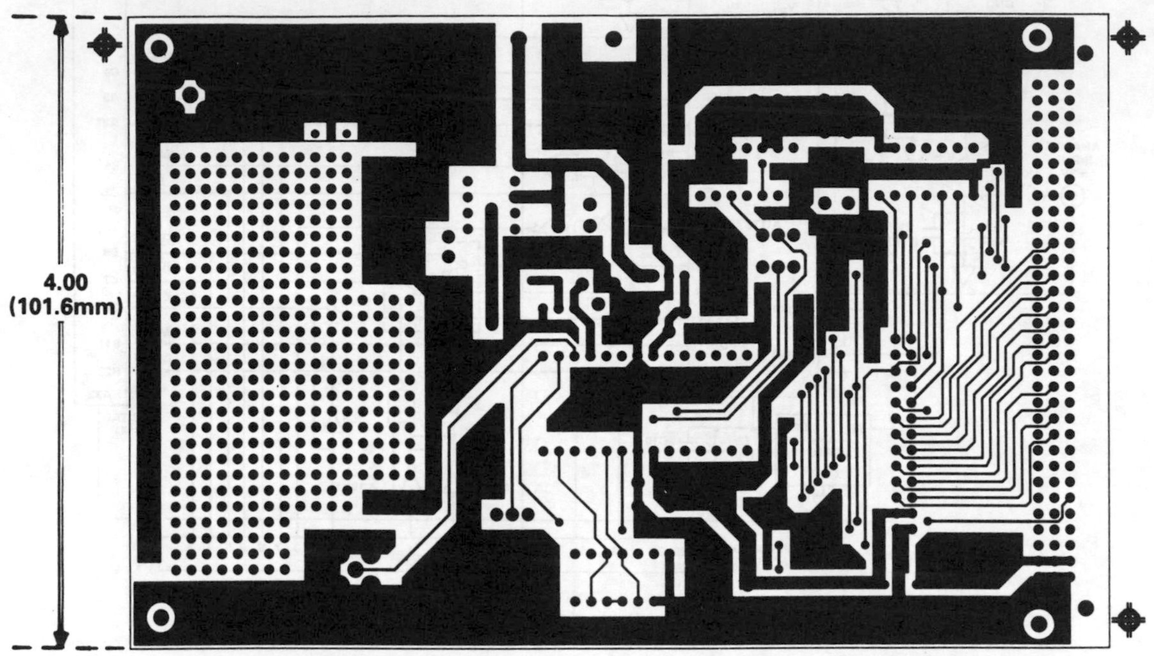

Figure 25. PCB Component Side Layout for Figure 23

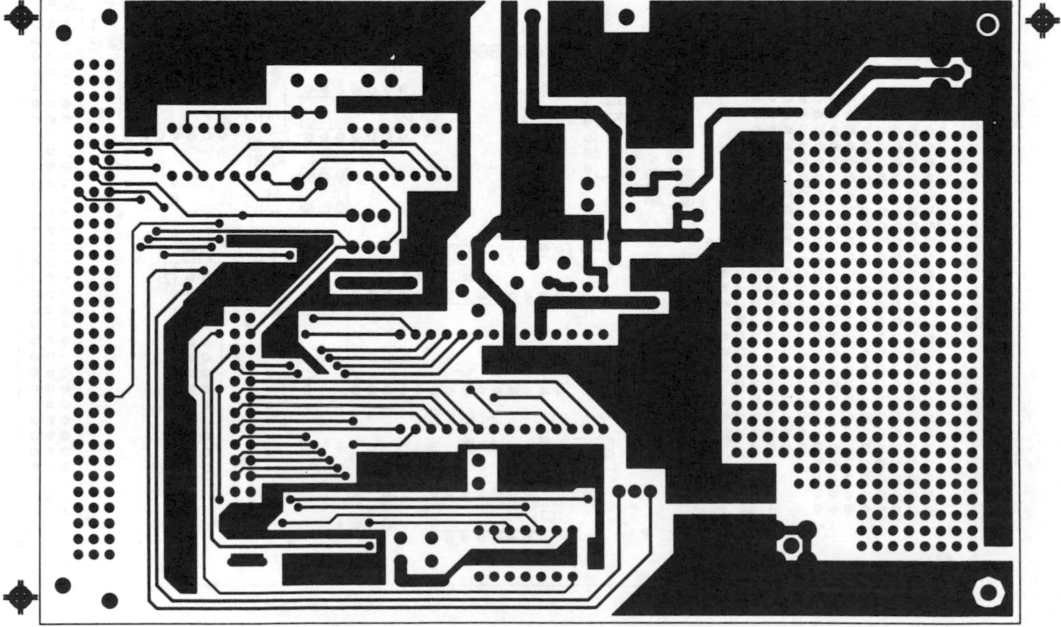

Figure 26. PCB Solder Side Layout for Figure 23

IDC Contact No.	Signal Connect Mnemonic	TMS32010 Signal	TMS32020 Signal
1	R/$\overline{W}$	–	R/$\overline{W}$
2	$\overline{STRB}$	–	$\overline{STRB}$
3	$\overline{DMRD}$	$\overline{DEN}$	–
4	$\overline{DMWR}$	$\overline{WE}$	–
5	$\overline{CS}$	$\overline{CS}$	$\overline{CS}$
6	READY	–	READY
7	$\overline{RESET}$	$\overline{RESET}$	$\overline{RESET}$
8	$\overline{ALFL}$	$\overline{INT}$	$\overline{INT}$
9	ADD0	PA0	A0
10	CLK	CLKOUT	CLKOUT2
11	DB10	D10	D10
12	DB11	D11	D11
13	DB8	D8	D8
14	DB9	D9	D9
15	DB6	D6	D6
16	DB7	D7	D7
17	DB4	D4	D4
18	DB5	D5	D5
19	DB2	D2	D2
20	DB3	D3	D3
21	DB0	D0	D0
22	DB1	D1	D1
23	5V	5V	5V
24	5V	5V	5V
25	GND	GND	GND
26	GND	GND	GND

Table IV. TMS32010/TMS32020 Interface Connections

National Semiconductor

October 1991

LM12458 12-Bit + Sign Data Acquisition System with Self-Calibration

General Description

The LM12458 is a highly integrated Data Acquisition System. Operating on just 5V, it combines a fully-differential self-calibrating (correcting linearity and zero errors) 13-bit (12-bit + sign) analog-to-digital converter (ADC) and sample-and-hold with extensive analog functions and digital functionality. Up to 32 consecutive conversions, using two's complement format, can be stored in an internal 32-word (16-bit wide) FIFO data buffer. An internal 8-word RAM can store the conversion sequence for up to eight acquisitions through the eight-input multiplexer. The LM12458 can also operate with 8-bit + sign resolution and in a supervisory "watchdog" mode that compares an input signal against two programmable limits.

Programmable acquisition times and conversion rates are possible through the use of internal clock-driven timers. The reference voltage input can be externally generated for absolute or ratiometric operation or can be derived using the internal 2.5V bandgap reference.

All registers, RAM, and FIFO are directly addressable through the high speed microprocessor interface to either an 8-bit or 16-bit databus. The LM12458 includes direct memory access (DMA) for high-speed conversion data transfer.

Key Specifications (f_{CLK} = 5.0 MHz)

- Resolution — 12-bit + sign or 8-bit + sign
- 13-bit conversion time — 8.8 μs (max)
- 9-bit conversion time — 4.2 μs (max)
- Through-put rate — 87k samples/s @ 13 bits (min)
- Comparison time ("watchdog" mode) — 2.2 μs/channel (max)
- Integral linearity error — ±1 LSB (max)
- Power dissipation — 30 mW (max)
- Stand-by mode — 50 μW (typ)

Features

- Three operating modes: 12-bit + sign, 8-bit + sign, and "watchdog"
- Fully differential inputs: −5V to 5V **differential** input voltage range, 0V to 5V **common-mode** input voltage range
- Built-in Sample-and-Hold and 2.5V bandgap reference
- Instruction RAM and event sequencer
- 8-channel multiplexer
- 32-word conversion FIFO
- Programmable acquisition times and conversion rates
- Self-calibration and diagnostic mode
- 8- or 16-bit wide databus microprocessor or DSP interface

Applications

- Data Logging
- Instrumentation
- Process Control
- Energy Management
- Robotics
- Signal Analysis

Connection Diagram

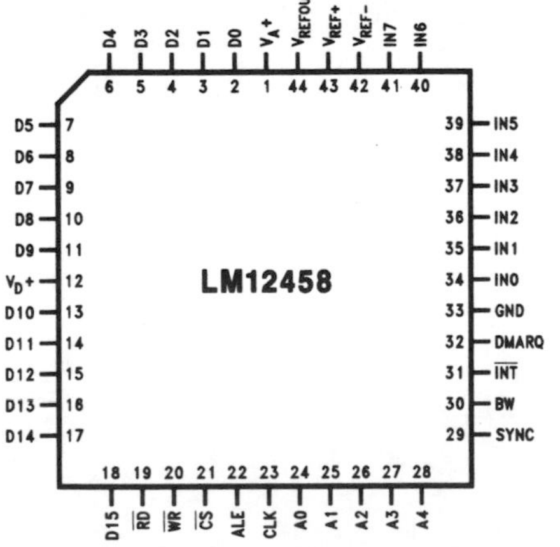

TL/H/11264-2

Order Number LM12458CIV
See NS Package Number V44A

Functional Diagram

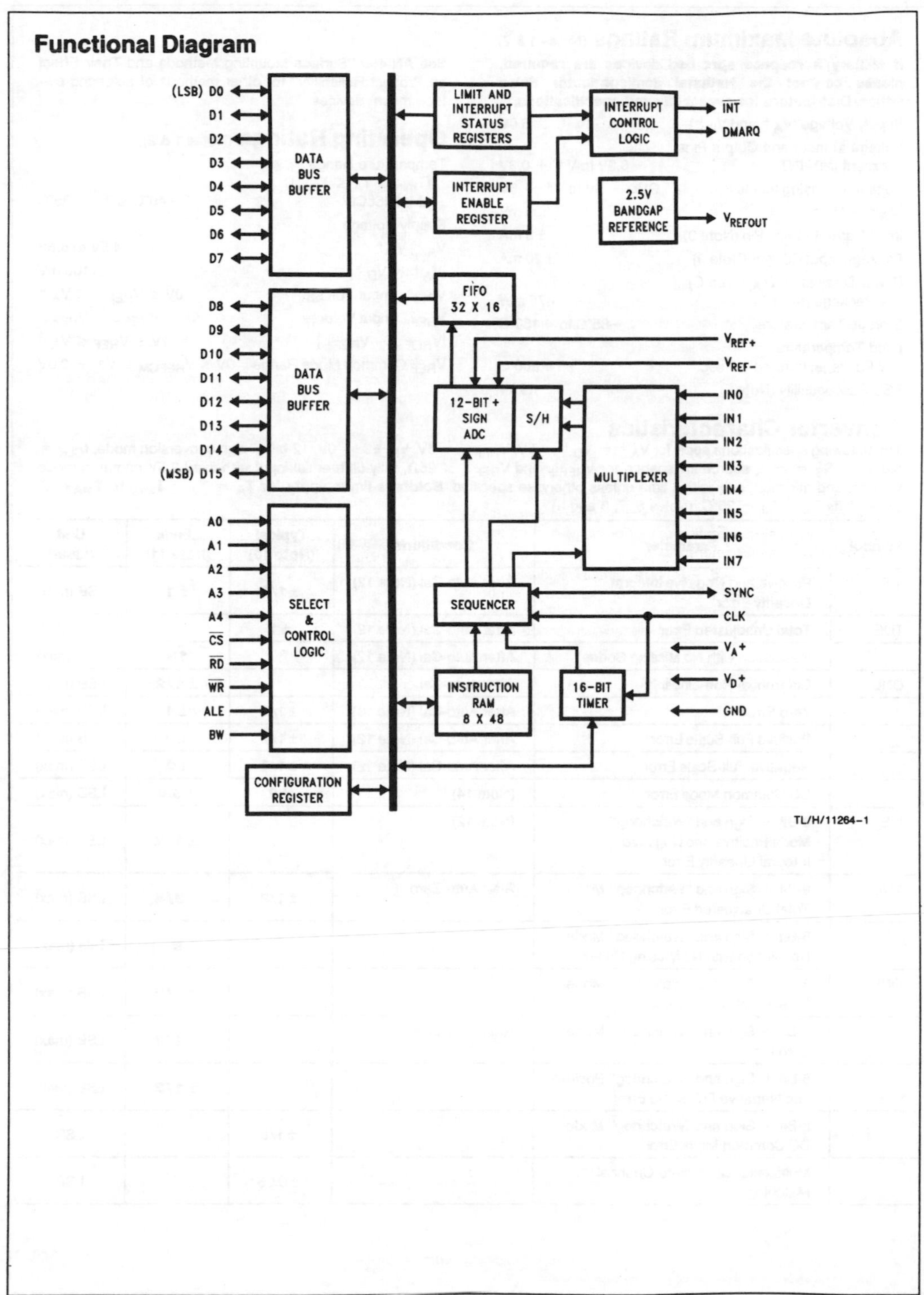

TL/H/11264–1

Absolute Maximum Ratings (Notes 1 & 2)

If Military/Aerospace specified devices are required, please contact the National Semiconductor Sales Office/Distributors for availability and specifications.

Supply Voltage (V_A+ and V_D+)	6.0V
Voltage at Input and Output Pins except IN0–IN7	−0.3V to V+ + 0.3V
Voltage at Analog Inputs IN0–IN7	GND − 5V to V+ + 5V
$\|V_A+ - V_D+\|$	300 mV
Input Current at Any Pin (Note 3)	±5 mA
Package Input Current (Note 3)	±20 mA
Power Dissipation (T_A = 25°C) V Package (Note 4)	875 mW
Storage Temperature	−65°C to +150°C
Lead Temperature V Package, Infrared, 15 sec.	+300°C
ESD Susceptibility (Note 5)	1.5 kV

See AN-450 "Surface Mounting Methods and Their Effect on Product Reliability" for other methods of soldering surface mount devices.

Operating Ratings (Notes 1 & 2)

Temperature Range ($T_{min} \leq T_A \leq T_{max}$) LM12458CIV	$-40°C \leq T_A \leq 85°C$
Supply Voltage	
V_A+, V_D+	4.5V to 5.5V
$\|V_A+ - V_D+\|$	≤100 mV
V_{REF+} Input Voltage	$0V \leq V_{REF+} \leq V_A+$
V_{REF-} Input Voltage	$0V \leq V_{REF-} \leq V_{REF+}$
$\|V_{REF+} - V_{REF-}\|$	$1V \leq V_{REF} \leq V_A+$
V_{REF} Common Mode Range	$0V \leq V_{REFCM} \leq V+ - 2.0V$

Converter Characteristics

The following specifications apply for V_A+ = V_D+ = 5V, V_{REF+} = 5V, V_{REF-} = 0V, 12-bit + sign conversion mode, f_{CLK} = 5.0 MHz, R_S = 25Ω, source impedance for V_{REF+} and V_{REF-} ≤ 25Ω, fully-differential input with fixed 2.5V common-mode voltage, and minimum acquisition time unless otherwise specified. **Boldface limits apply for $T_A = T_J = T_{MIN}$ to T_{MAX}**; all other limits $T_A = T_J$ = 25°C. (Notes 6, 7, 8 and 9)

Symbol	Parameter	Conditions	Typical (Note 10)	Limits (Note 11)	Unit (Limit)
ILE	Positive and Negative Integral Linearity Error	After Auto-Cal (Note 12)	±1/2	**±1**	LSB (max)
TUE	Total Unadjusted Error	After Auto-Cal (Note 12)	±1		LSB
	Resolution with No Missing Codes	After Auto-Cal (Note 12)		**13**	Bits (max)
DNL	Differential Non-Linearity	After Auto-Cal		**±1/2**	LSB (max)
	Zero Error	After Auto-Cal (Note 13)	±1/2	**±1**	LSB (max)
	Positive Full-Scale Error	After Auto-Cal (Note 12)	±1/2	**±2**	LSB (max)
	Negative Full-Scale Error	After Auto-Cal (Note 12)	±1/2	**±2**	LSB (max)
	DC Common Mode Error	(Note 14)	±2	**±3.5**	LSB (max)
ILE	8-Bit + Sign and "Watchdog" Mode Positive and Negative Integral Linearity Error	(Note 12)		**±1/2**	LSB (max)
TUE	8-Bit + Sign and "Watchdog" Mode Total Unadjusted Error	After Auto-Zero	±1/2	**±3/4**	LSB (max)
	8-Bit + Sign and "Watchdog" Mode Resolution with No Missing Codes			**9**	Bits (max)
DNL	8-Bit + Sign and "Watchdog" Mode Differential Non-Linearity			**±1/2**	LSB (max)
	8-Bit + Sign and "Watchdog" Mode Zero Error	After Auto-Zero		**±1/2**	LSB (max)
	8-Bit + Sign and "Watchdog" Positive and Negative Full-Scale Error			**±1/2**	LSB (max)
	8-Bit + Sign and "Watchdog" Mode DC Common Mode Error		±1/8		LSB
	Multiplexer Channel-to-Channel Matching		±0.05		LSB

Converter Characteristics (Continued)

The following specifications apply for $V_A+ = V_D+ = 5V$, $V_{REF+} = 5V$, $V_{REF-} = 0V$, 12-bit + sign conversion mode, f_{CLK} = 5.0 MHz, $R_S = 25\Omega$, source impedance for V_{REF+} and $V_{REF-} \leq 25\Omega$, fully-differential input with fixed 2.5V common-mode voltage, and minimum acquisition time unless otherwise specified. **Boldface limits apply for $T_A = T_J = T_{MIN}$ to T_{MAX};** all other limits $T_A = T_J = 25°C$. (Notes 6, 7, 8 and 9)

Symbol	Parameter	Conditions	Typical (Note 10)	Limits (Note 11)	Unit (Limit)
$\lvert V_{IN+} - V_{IN-} \rvert$	Differential Input Voltage Range			**GND** **V_A+**	V (min) V (max)
$\frac{V_{IN+} - V_{IN-}}{2}$	Common Mode Input Voltage Range			**GND** **V_A+**	V (min) V (max)
PSS	Power Supply Sensitivity (Note 15): Zero Error Full-Scale Error Linearity Error	$V_A+ = V_D+ = 5V \pm 10\%$ $V_{REF+} = 4.5V$, V_{REF-} = GND	±0.2 ±0.4 ±0.2	**±1.75** **±2**	LSB (max) LSB (max) LSB
C_{REF}	V_{REF+}/V_{REF-} Input Capacitance		85		pF
C_{IN}	Selected Multiplexer Channel Input Capacitance		75		pF

Converter AC Characteristics

The following specifications apply for $V_A+ = V_D+ = 5V$, $V_{REF+} = 5V$, $V_{REF-} = 0V$, 12-bit + sign conversion mode, f_{CLK} = 5.0 MHz, $R_S = 25\Omega$, source impedance for V_{REF+} and $V_{REF-} \leq 25\Omega$, fully-differential input with fixed 2.5V common-mode voltage, and minimum acquisition time unless otherwise specified. **Boldface limits apply for $T_A = T_J = T_{MIN}$ to T_{MAX};** all other limits $T_A = T_J = 25°C$. (Notes 6, 7, 8 and 9)

Symbol	Parameter	Conditions	Typical (Note 9)	Limits (Note 10)	Unit (Limit)
	Clock Duty Cycle		50	**40** **60**	% % (min) % (max)
t_C	Conversion Time	13-Bit Resolution, Sequencer State S5 *(Figure 7)*	44 (t_{CLK})	**44 (t_{CLK}) + 50 ns**	(max)
		9-Bit Resolution, Sequencer State S5 *(Figure 7)*	21 (t_{CLK})	**21 (t_{CLK}) + 50 ns**	(max)
t_A	Acquisition Time	Sequencer State S7 *(Figure 7)* Built-in minimum for 13-Bits	9 (t_{CLK})	**9 (t_{CLK}) + 50 ns**	(max)
		Built-in minimum for 9-Bits and "Watchdog" mode	2 (t_{CLK})	**2 (t_{CLK}) + 50 ns**	(max)
t_Z	Auto-Zero Time	Sequencer State S2 *(Figure 7)*	76 (t_{CLK})	**76 (t_{CLK}) + 50 ns**	(max)
t_{CAL}	Full Calibration Time	Sequencer State S2 *(Figure 7)*	4944 (t_{CLK})	**4944 (t_{CLK}) + 50 ns**	(max)
t_{WD}	"Watchdog" Mode Comparison Time	Sequencer States S6, S4, and S5 *(Figure 7)*	11 (t_{CLK})	**11 (t_{CLK}) + 50 ns**	(max)
BSNR	Bipolar Signal-to-Noise Ratio	$V_{IN} = \pm 5V$ f_{IN} = 1 kHz f_{IN} = 20 kHz f_{IN} = 40 kHz	77.5 75.2 74.7		dB dB dB
USNR	Unipolar Signal-to-Noise Ratio	$V_{IN} = 5 V_{p-p}$ f_{IN} = 1 kHz f_{IN} = 20 kHz f_{IN} = 40 kHz	69.8 69.2 66.6		dB dB dB
BSINAD	Bipolar Signal-to-Noise + Distortion Ratio	$V_{IN} = \pm 5V$ f_{IN} = 1 kHz f_{IN} = 20 kHz f_{IN} = 40 kHz	76.9 73.9 70.7		dB dB dB

Converter AC Characteristics (Continued)

The following specifications apply for $V_A+ = V_D+ = 5V$, $V_{REF+} = 5V$, $V_{REF-} = 0V$, 12-bit + sign conversion mode, $f_{CLK} = 5.0$ MHz, $R_S = 25\Omega$, source impedance for V_{REF+} and $V_{REF-} \leq 25\Omega$, fully-differential input with fixed 2.5V common-mode voltage, and minimum acquisition time unless otherwise specified. **Boldface limits apply for $T_A = T_J = T_{MIN}$ to T_{MAX}**; all other limits $T_A = T_J = 25°C$. (Notes 6, 7, 8 and 9)

Symbol	Parameter	Conditions	Typical (Note 9)	Limits (Note 10)	Unit (Limit)
USINAD	Unipolar Signal-to-Noise + Distortion Ratio	$V_{IN} = 5\ V_{p-p}$ $f_{IN} = 1$ kHz $f_{IN} = 20$ kHz $f_{IN} = 40$ kHz	 69.4 68.3 65.7		 dB dB dB
BTHD	Bipolar Total Harmonic Distortion	$V_{IN} = \pm 5V$ $f_{IN} = 1$ kHz $f_{IN} = 20$ kHz $f_{IN} = 40$ kHz	 −85.8 −79.9 −72.9		 dB dB dB
UTHD	Unipolar Total Harmonic Distortion	$V_{IN} = 5\ V_{p-p}$ $f_{IN} = 1$ kHz $f_{IN} = 20$ kHz $f_{IN} = 40$ kHz	 −80.3 −75.6 −72.8		 dB dB dB
	Bipolar Effective Number of Bits	$V_{IN} = \pm 5V$ $f_{IN} = 1$ kHz $f_{IN} = 20$ kHz $f_{IN} = 40$ kHz	 12.6 12.2 12.1		 Bits Bits Bits
	Unipolar Effective Number of Bits	$V_{IN} = 5\ V_{p-p}$ $f_{IN} = 1$ kHz $f_{IN} = 20$ kHz $f_{IN} = 40$ kHz	 11.3 11.2 10.8		 Bits Bits Bits
	Bipolar Spurious Free Dynamic Range	$V_{IN} = \pm 5V$ $f_{IN} = 1$ kHz $f_{IN} = 20$ kHz $f_{IN} = 40$ kHz	 87.2 78.9 72.8		 dB dB dB
t_{PU}	Power-Up Time		10		ms
t_{WU}	Wake-Up Time		10		ms

DC Characteristics

The following specifications apply for $V_A+ = V_D+ = 5V$, $V_{REF+} = 5V$, $V_{REF-} = 0V$, $f_{CLK} = 5.0$ MHz, and minimum acquisition time unless otherwise specified. **Boldface limits apply for $T_A = T_J = T_{MIN}$ to T_{MAX}**; all other limits $T_A = T_J = 25°C$. (Notes 6, 7 and 8)

Symbol	Parameter	Conditions	Typical (Note 9)	Limits (Note 10)	Unit (Limit)
I_D+	V_D+ Supply Current	$\overline{CS}$ = "1"	0.55	**1.0**	mA (max)
I_A+	V_A+ Supply Current	$\overline{CS}$ = "1"	3.1	**5.0**	mA (max)
I_{ST}	Stand-By Supply Current ($I_D+ + I_A+$)	Power-Down Mode Selected Clock Stopped 5 MHz Clock	 10 40		 μA (max) μA (max)
	Multiplexer ON-Channel Leakage Current	$V_A+ = 5.5V$ ON-Channel = 5.5V OFF-Channel = 0V ON-Channel = 0V OFF-Channel = 5.5V	 0.1 0.1	 **0.3** **0.3**	 μA (max) μA (max)
	Multiplexer OFF-Channel Leakage Current	$V_A+ = 5.5V$ ON-Channel = 5.5V OFF-Channel = 0V ON-Channel = 0V OFF-Channel = 5.5V	 0.1 0.1	 **0.3** **0.3**	 μA (max) μA (max)

Internal Reference Characteristics

The following specifications apply for $V_A+ = V_D+ = 5V$ unless otherwise specified. **Boldface limits apply for $T_A = T_J = T_{MIN}$ to T_{MAX}**; all other limits $T_A = T_J = 25°C$. (Notes 6 and 7)

Symbol	Parameter	Conditions	Typical (Note 9)	Limits (Note 10)	Unit (Limit)
V_{REFOUT}	Internal Reference Output Voltage		2.5	**2.5 ±4%**	V (max)
$\Delta V_{REF}/\Delta T$	Internal Reference Temperature Coefficient		40		ppm/°C
$\Delta_{REF}/\Delta I_L$	Internal Reference Load Regulation	Sourcing ($0 < I_L \leq +4$ mA) Sinking ($-1 \leq I_{IL} < 0$ mA)		**0.2** **1.2**	%/mA (max) %/mA (max)
ΔV_{REF}	Line Regulation	$4.5V \leq V_A+ \leq 5.5V$	3	**15**	mV (max)
I_{SC}	Internal Reference Short Circuit Current	$V_{REFOUT} = 0V$	13	**25**	mA (max)
$\Delta V_{REF}/\Delta t$	Long Term Stability		200		ppm/kHr
t_{SU}	Internal Reference Start-Up Time	$V_A+ = V_D+ = 0V \rightarrow 5V$ $C_L = 100\ \mu F$	10		ms

Digital Characteristics

The following specifications apply for $V_A+ = V_D+ = 5V$, $V_{REF+} = 5V$, $V_{REF-} = 0V$, $f_{CLK} = 5.0$ MHz. **Boldface limits apply for $T_A = T_J = T_{MIN}$ to T_{MAX}**; all other limits $T_A = T_J = 25°C$. (Notes 6, 7 and 8)

Symbol	Parameter	Conditions	Typical (Note 9)	Limits (Note 10)	Unit (Limit)
$V_{IN(1)}$	Logical "1" Input Voltage	$V_A+ = V_D+ = 5.5V$		**2.0**	V (min)
$V_{IN(0)}$	Logical "0" Input Voltage	$V_A+ = V_D+ = 4.5V$		**0.8**	V (max)
$I_{IN(1)}$	Logical "1" Input Current	$V_{IN} = 5V$	0.005	**1.0**	μA (max)
$I_{IN(0)}$	Logical "0" Input Current	$V_{IN} = 0V$	−0.005	**−1.0**	μA (max)
C_{IN}	D0–D15 Input Capacitance		6		pF
$V_{OUT(1)}$	Logical "1" Output Voltage	$V_A+ = V_D+ = 4.5V$ $I_{OUT} = -360\ \mu A$ $I_{OUT} = -10\ \mu A$		 **2.4** **4.25**	 V (min) V (min)
$V_{OUT(0)}$	Logical "0" Output Voltage	$V_A+ = V_D+ = 4.5V$ $I_{OUT} = 1.6$ mA		**0.4**	V (max)
I_{OUT}	TRI-STATE® Output Leakage Current	$V_{OUT} = 0V$ $V_{OUT} = 5V$	−0.01 0.01	**−3.0** **3.0**	μA (max) μA (max)

Digital Timing Characteristics

The following specifications apply for $V_A+ = V_D+ = 5V$, $t_r = t_f = 3$ ns, and $C_L = 100$ pF on data I/O, $\overline{INT}$ and DMARQ lines unless otherwise specified. **Boldface limits apply for $T_A = T_J = T_{MIN}$ to T_{MAX}**; all other limits $T_A = T_J = 25°C$. (Notes 6, 7 and 8)

Symbol (See *Figures 4a, 4b,* and *4c*)	Parameter	Conditions	Typical (Note 9)	Limits (Note 10)	Unit (Limit)
1, 3	$\overline{CS}$ or Address Valid to ALE Low Set-Up Time			**40**	ns (min)
2, 4	$\overline{CS}$ or Address Valid to ALE Low Hold Time			**20**	ns (min)
5	ALE Pulse Width			**45**	ns (min)
6	$\overline{RD}$ High to Next ALE High			**35**	ns (min)

Digital Timing Characteristics (Continued)

The following specifications apply for $V_A{}^+ = V_D{}^+ = 5V$, $t_r = t_f = 3$ ns, and $C_L = 100$ pF on data I/O, $\overline{INT}$ and DMARQ lines unless otherwise specified. **Boldface limits apply for $T_A = T_J = T_{MIN}$ to T_{MAX};** all other limits $T_A = T_J = 25°C$. (Notes 6, 7 and 8)

Symbol (See *Figures 4a, 4b,* and *4c*)	Parameter	Conditions	Typical (Note 9)	Limits (Note 10)	Unit (Limit)
7	ALE Low to $\overline{RD}$ Low			**20**	ns (min)
8	$\overline{RD}$ Pulse Width			**100**	ns (min)
9	$\overline{RD}$ High to Next $\overline{RD}$ or $\overline{WR}$ Low			**100**	ns (min)
10	ALE Low to $\overline{WR}$ Low			**20**	ns (min)
11	$\overline{WR}$ Pulse Width			**60**	ns (min)
12	$\overline{WR}$ High to Next ALE High			**75**	ns (min)
13	$\overline{WR}$ High to Next $\overline{RD}$ or $\overline{WR}$ Low			**140**	ns (min)
14	Data Valid to $\overline{WR}$ High Set-Up Time			**40**	ns (min)
15	Data Valid to $\overline{WR}$ High Hold Time			**30**	ns (min)
16	$\overline{RD}$ Low to Data Bus Out of TRI-STATE		40	**10** **70**	ns (min) ns (max)
17	$\overline{RD}$ High to TRI-STATE	$R_L = 1\ k\Omega$	30	**10** **110**	ns (min) ns (max)
18	$\overline{RD}$ Low to Data Valid (Access Time)		30	**10** **80**	ns (min) ns (max)
21	Address Valid or $\overline{CS}$ Low to $\overline{RD}$ High			**120**	ns (min)
22	Address Valid or $\overline{CS}$ Low to $\overline{WR}$ High			**80**	ns (min)
19, 20	Address Invalid or $\overline{CS}$ High from $\overline{RD}$ or $\overline{WR}$ High			**10**	ns (min)
23	$\overline{INT}$ High from $\overline{RD}$ Low		30	**10** **60**	ns (min) ns (max)
24	DMARQ Low from $\overline{RD}$ Low		30	**10** **60**	ns (min) ns (max)

Note 1: Absolute Maximum Ratings indicate limits beyond which damage to the device may occur. Operating Ratings indicate conditions for which the device is functional, but do not guarantee specific performance limits. For guaranteed specifications and test conditions, see the Electrical Characteristics. The guaranteed specifications apply only for the test conditions listed. Some performance characteristics may degrade when the device is not operated under the listed test conditions.

Note 2: All voltages are measured with respect to GND, unless otherwise specified.

Note 3: When the input voltage (V_{IN}) at any pin exceeds the power supply rails (V_{IN} < GND or $V_{IN} > (V_A{}^+$ or $V_D{}^+)$), the current at that pin should be limited to 5 mA. The 20 mA maximum package input current rating allows the voltage at any four pins, with an input current of 5 mA, to simultaneously exceed the power supply voltages.

Note 4: The maximum power dissipation must be derated at elevated temperatures and is dictated by T_{Jmax} (maximum junction temperature), Θ_{JA} (package junction to ambient thermal resistance), and T_A (ambient temperature). The maximum allowable power dissipation at any temperature is $PD_{max} = (T_{Jmax} - T_A)/\Theta_{JA}$ or the number given in the Absolute Maximum Ratings, whichever is lower. For this device, $T_{Jmax} = 150°C$, and the typical thermal resistance (Θ_{JA}) of the LM12458 in the V package, when board mounted, is 47°C/W.

Note 5: Human body model, 100 pF discharged through a 1.5 kΩ resistor.

Note 6: Two on-chip diodes are tied to each analog input through a series resistor, as shown below. Input voltage magnitude up to 5V above $V_A{}^+$ or 5V below GND will not damage the LM12458. However, errors in the A/D conversion can occur if these diodes are forward biased by more than 100 mV. As an example, if $V_A{}^+$ is 4.5 V_{DC}, full-scale input voltage must be ≤ 4.6 V_{DC} to ensure accurate conversions.

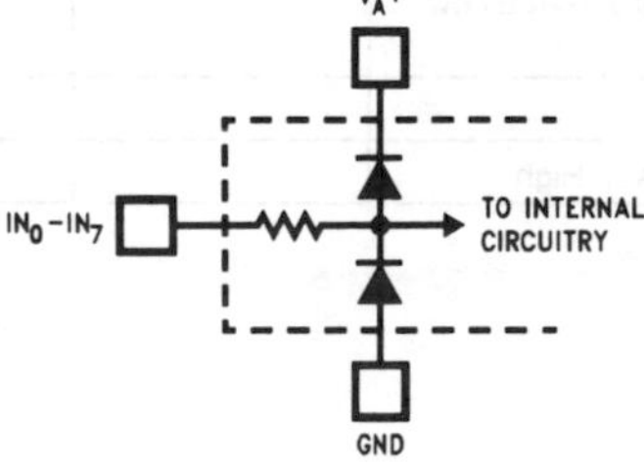

TL/H/11264–3

Digital Timing Characteristics (Continued)

Note 7: $V_A{}^+$ and $V_D{}^+$ must be connected together to the same power supply voltage and bypassed with separate capacitors at each V^+ pin to assure conversion/comparison accuracy.

Note 8: Accuracy is guaranteed when operating at f_{CLK} = 5 MHz.

Note 9: With the test condition for V_{REF} ($V_{REF+} - V_{REF-}$) given as +5V, the 12-bit LSB is 1.22 mV and the 8-bit/"Watchdog" LSB is 19.53 mV.

Note 10: Typicals are at T_A = 25°C and represent most likely parametric norm.

Note 11: Limits are guaranteed to National's AOQL (Average Output Quality Level).

Note 12: Positive integral linearity error is defined as the deviation of the analog value, expressed in LSBs, from the straight line that passes through positive full-scale and zero. For negative integral linearity error the straight line passes through negative full-scale and zero. (See *Figures 1b* and *1c*).

Note 13: Zero error is a measure of the deviation from the mid-scale voltage (a code of zero), expressed in LSB. It is the worst-case value of the code transitions between −1 to 0 and 0 to +1 (see *Figure 2*).

Note 14: The DC common-mode error is measured with both inputs shorted together and driven from 0V to 5V. The measured value is referred to the resulting output value when the inputs are driven with a 2.5V signal.

Note 15: Power Supply Sensitivity is measured after Auto-Zero and/or Auto-Calibration cycle has been completed with $V_A{}^+$ and $V_D{}^+$ at the specified extremes.

Electrical Characteristics (Continued)

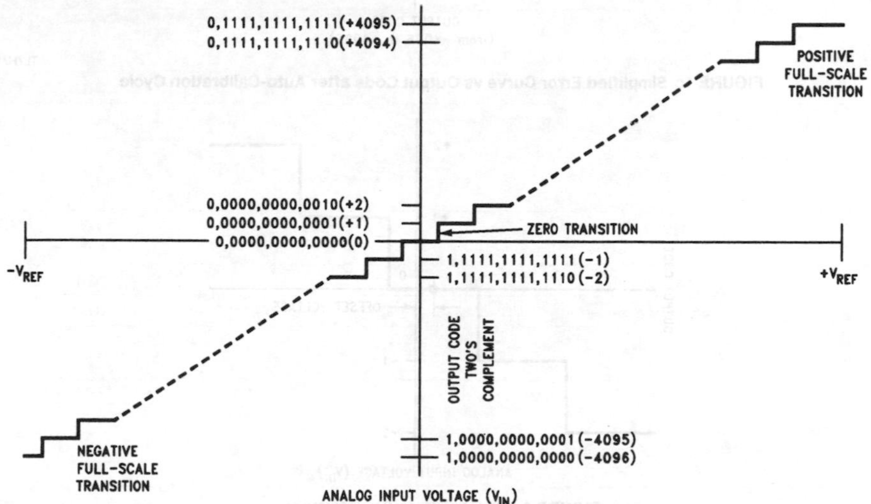

TL/H/11264–4

FIGURE 1a. Transfer Characteristic

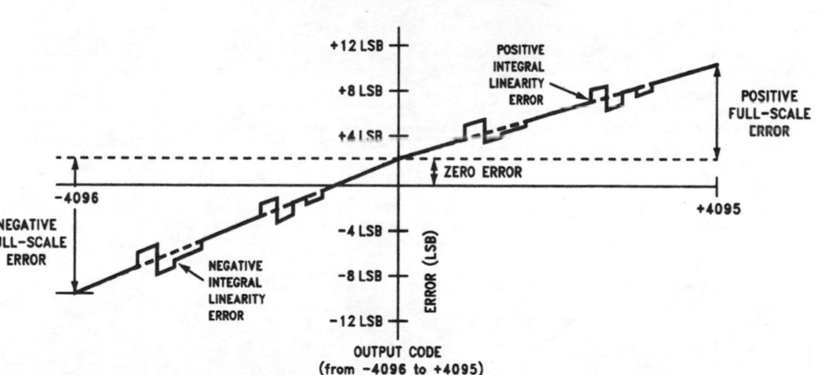

TL/H/11264–5

FIGURE 1b. Simplified Error Curve vs Output Code without Auto-Calibration or Auto-Zero Cycles

Electrical Characteristics (Continued)

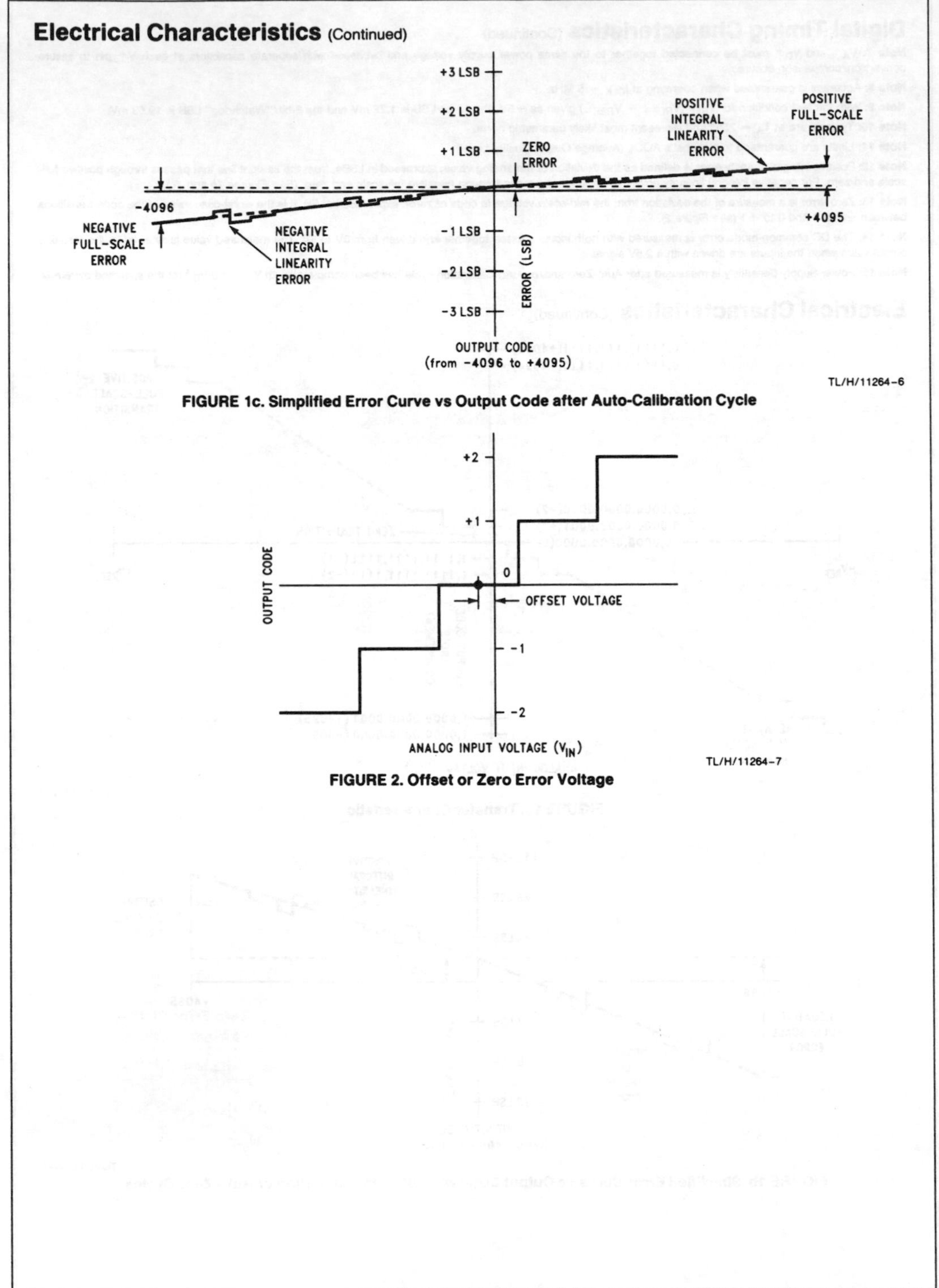

FIGURE 1c. Simplified Error Curve vs Output Code after Auto-Calibration Cycle

FIGURE 2. Offset or Zero Error Voltage

Typical Performance Characteristics

The following curves apply for 12-bit + sign mode after auto-calibration unless otherwise specified. The performance for 8-bit + sign and "watchdog" modes is equal to or better than shown. (Note 9)

Linearity Error Change vs Clock Frequency

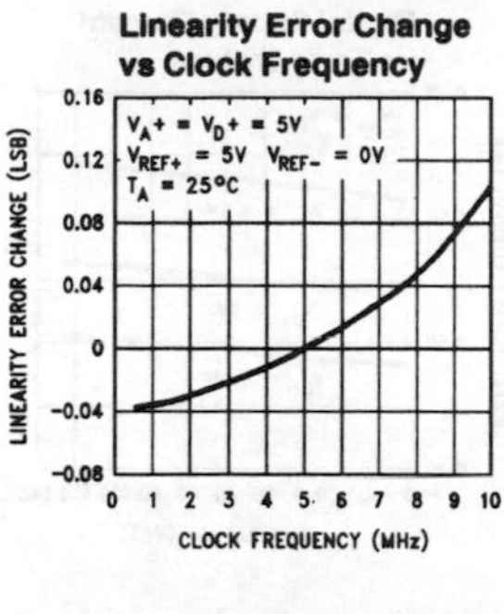

Linearity Error Change vs Temperature

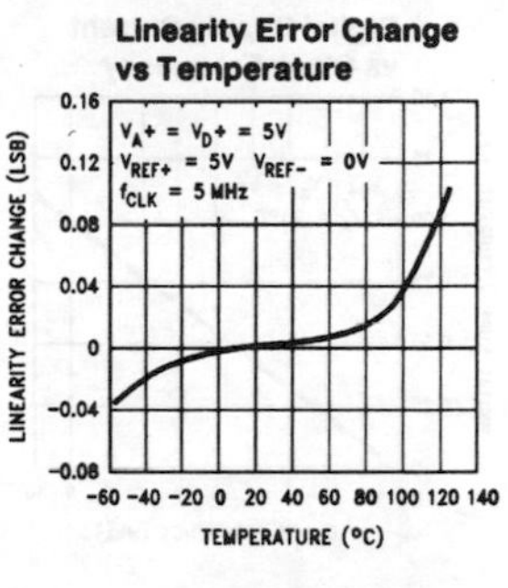

Linearity Error Change vs Reference Voltage

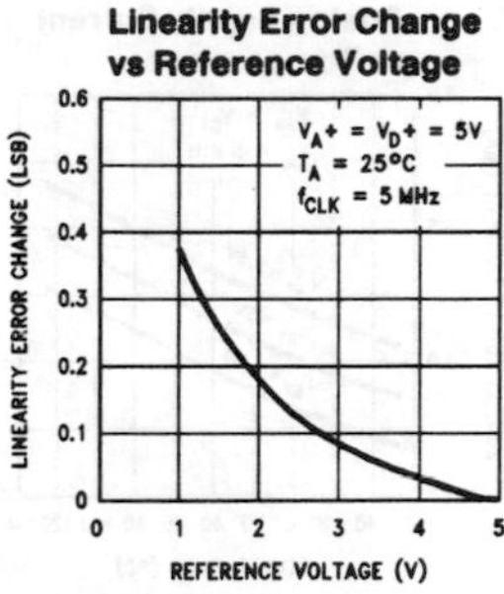

Linearity Error Change vs Supply Voltage

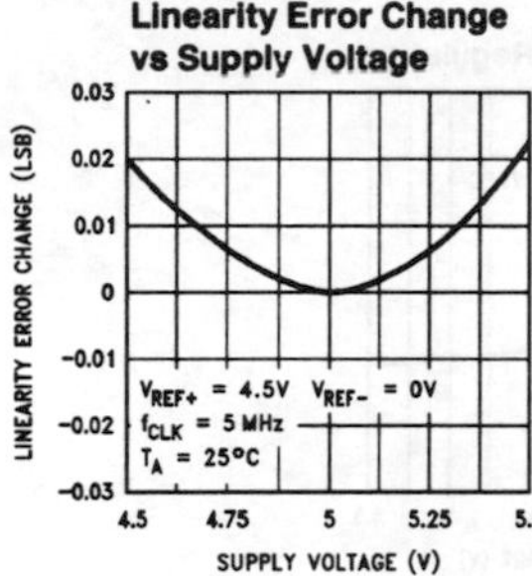

Full-Scale Error Change vs Clock Frequency

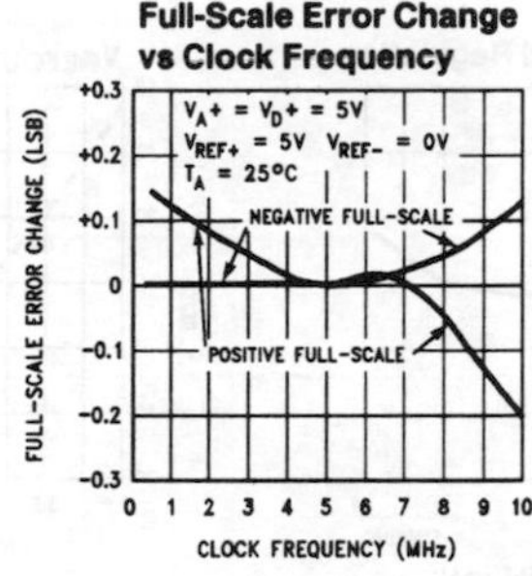

Full-Scale Error Change vs Temperature

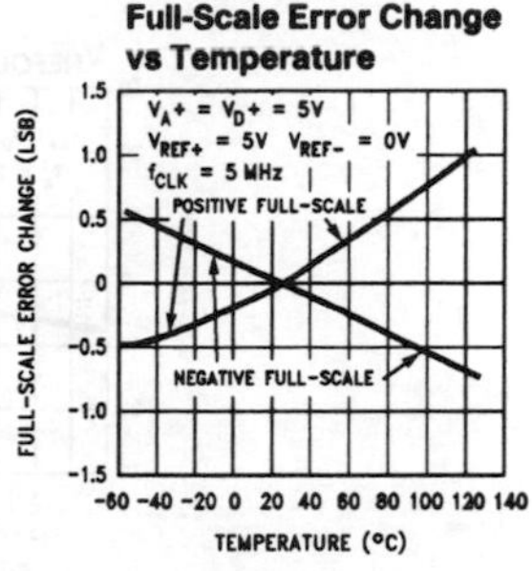

Full-Scale Error Change vs Reference Voltage

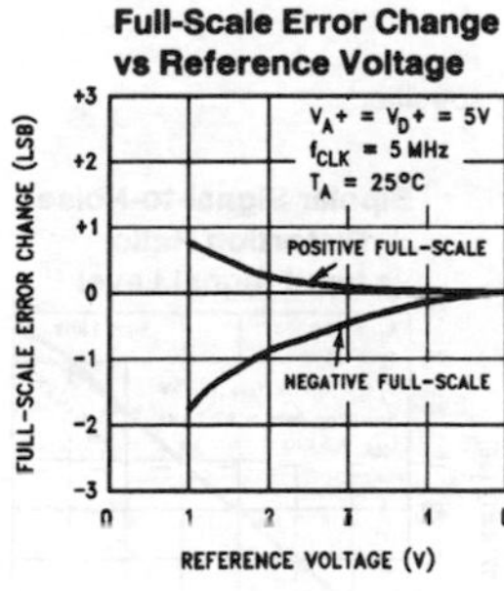

Full-Scale Error Change vs Supply Voltage

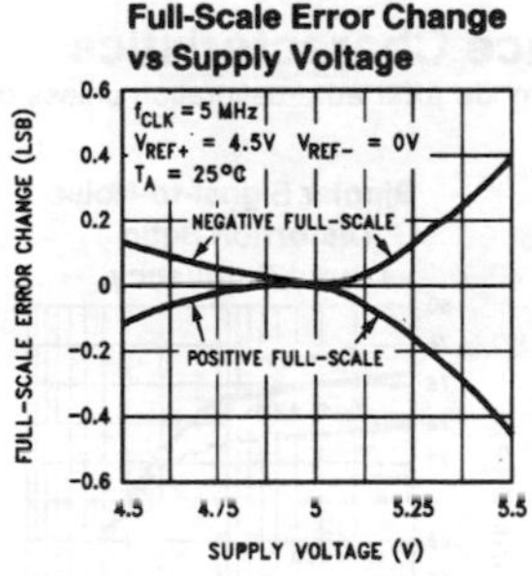

Zero Error Change vs Clock Frequency

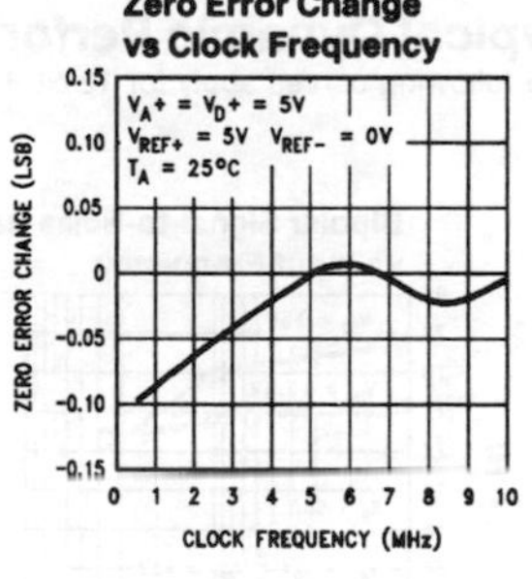

Zero Error Change vs Temperature

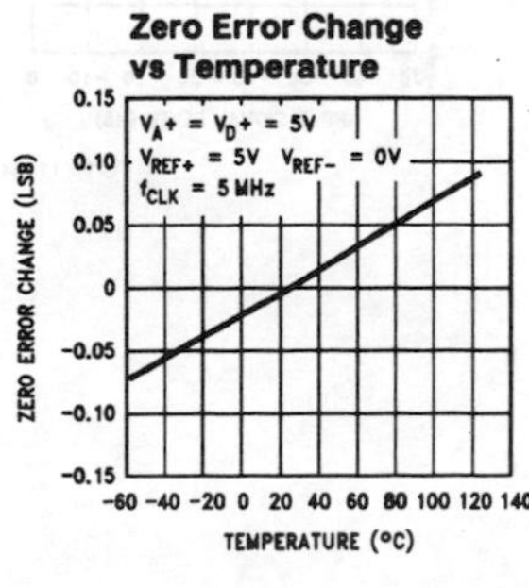

Zero Error Change vs Reference Voltage

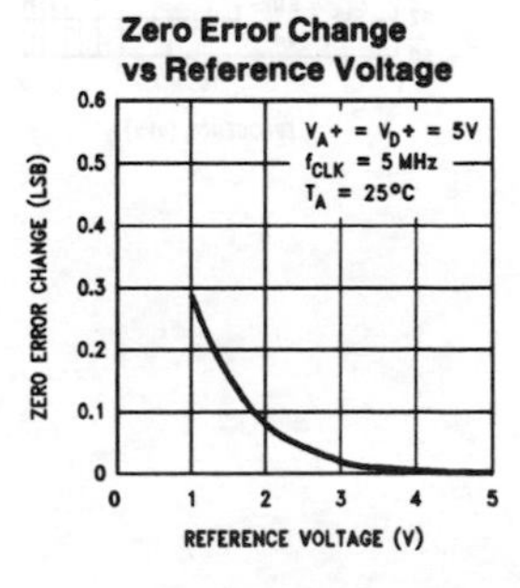

Zero Error Change vs Supply Voltage

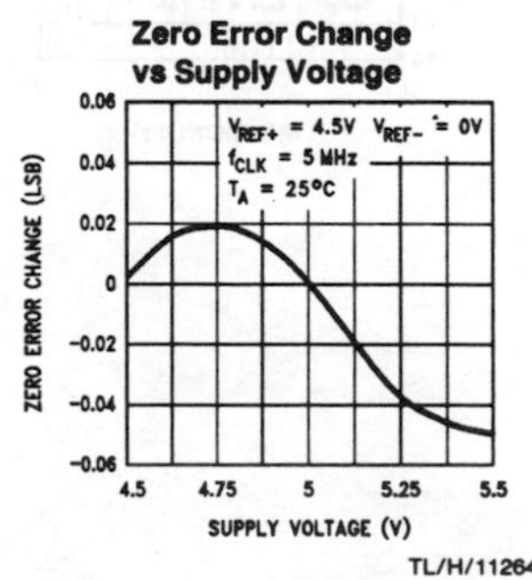

TL/H/11264–8

Typical Performance Characteristics (Continued)

The following curves apply for 12-bit + sign mode after auto-calibration unless otherwise specified. The performance for 8-bit + sign and "watchdog" modes is equal to or better than shown. (Note 9)

Analog Supply Current vs Temperature

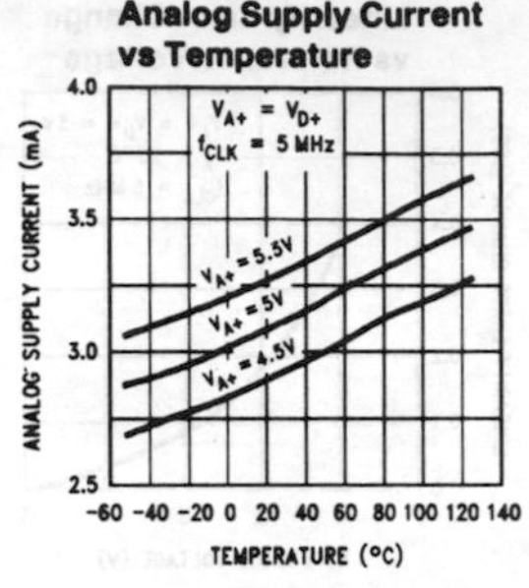

Digital Supply Current vs Clock Frequency

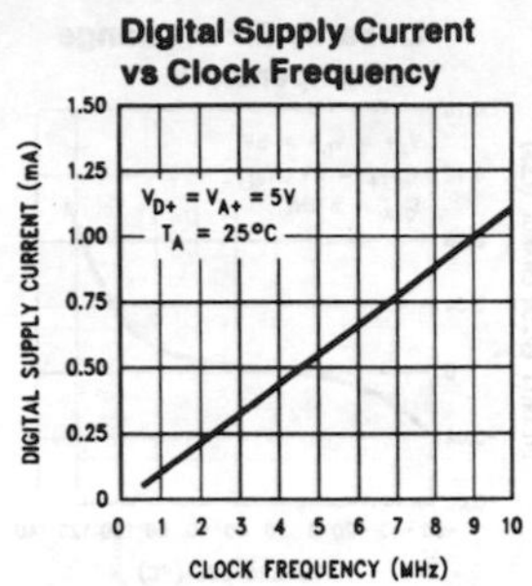

Digital Supply Current vs Temperature

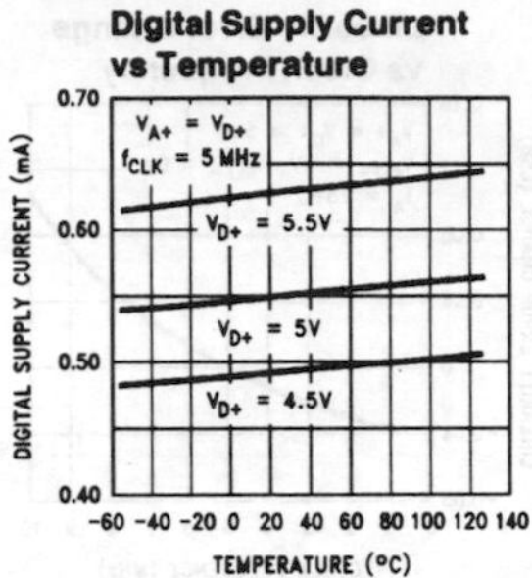

V_{REFOUT} Load Regulation

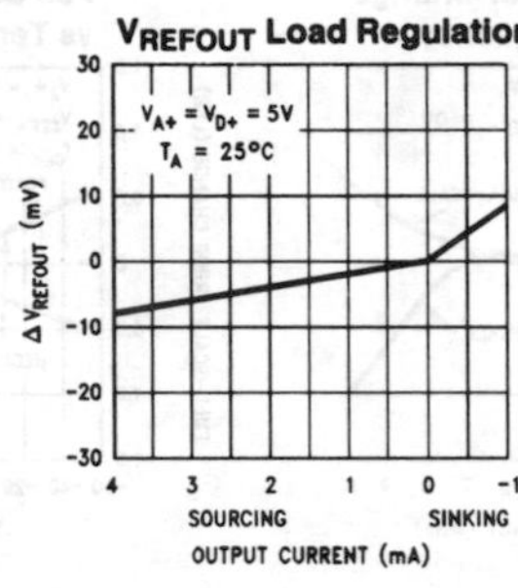

V_{REFOUT} Line Regulation

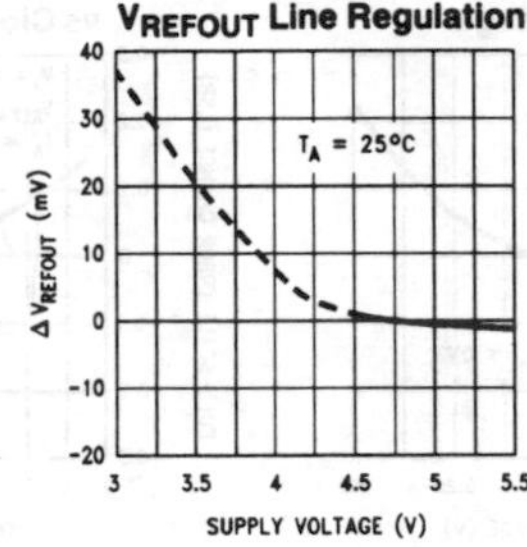

TL/H/11264–9

Typical Dynamic Performance Characteristics

The following curves apply for 12-bit + sign mode after auto-calibration unless otherwise specified.

Bipolar Signal-to-Noise Ratio vs Input Frequency

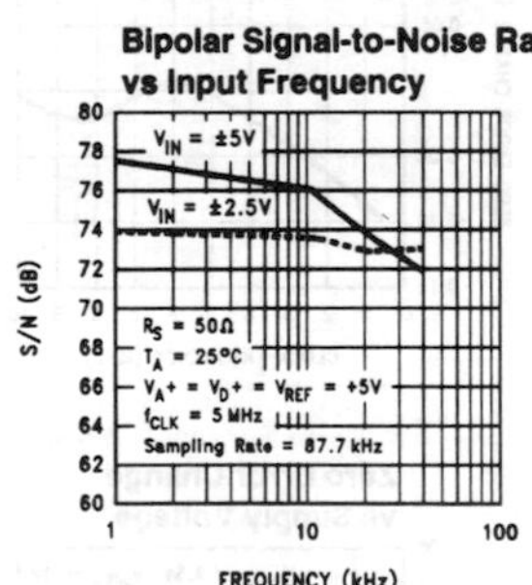

Bipolar Signal-to-Noise + Distortion Ratio vs Input Frequency

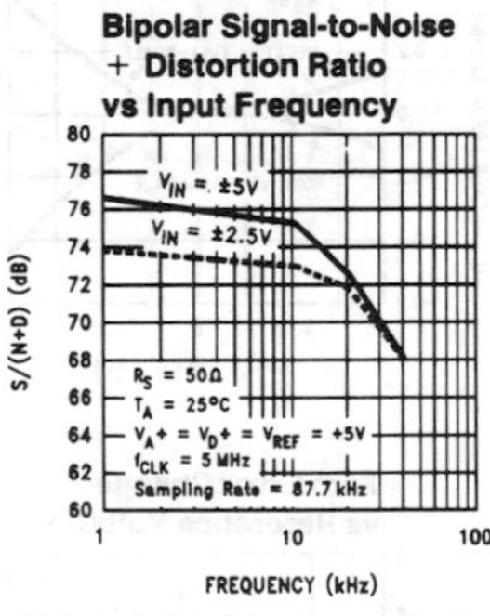

Bipolar Signal-to-Noise + Distortion Ratio vs Input Signal Level

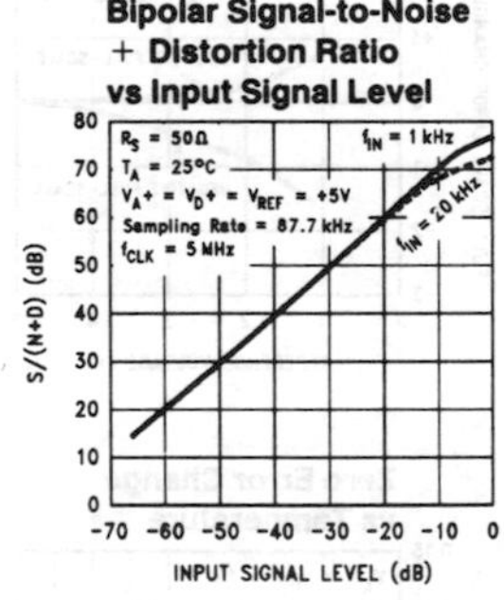

TL/H/11264–10

Typical Dynamic Performance Characteristics (Continued)

The following curves apply for 12-bit + sign mode after auto-calibration unless otherwise specified.

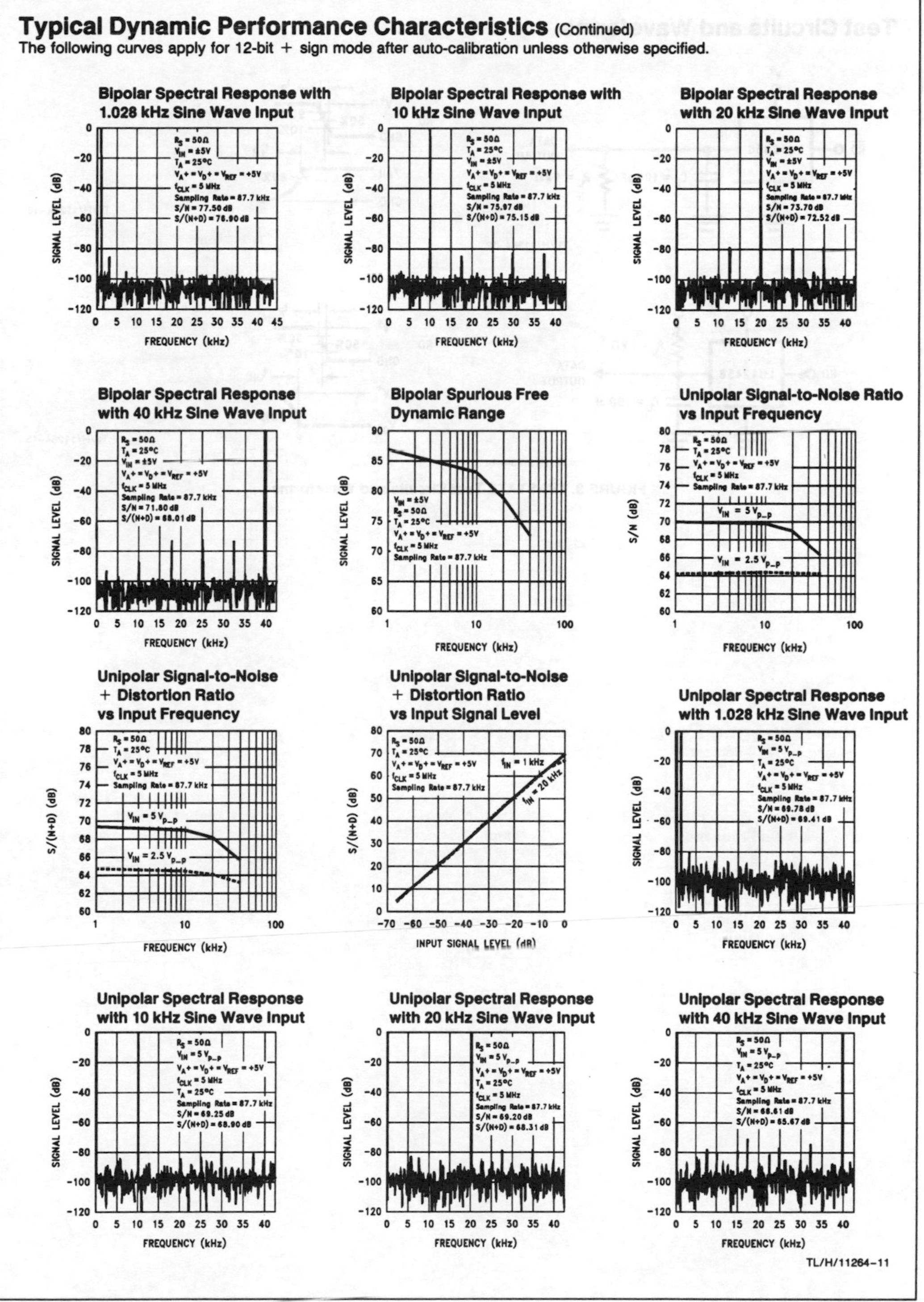

Test Circuits and Waveforms

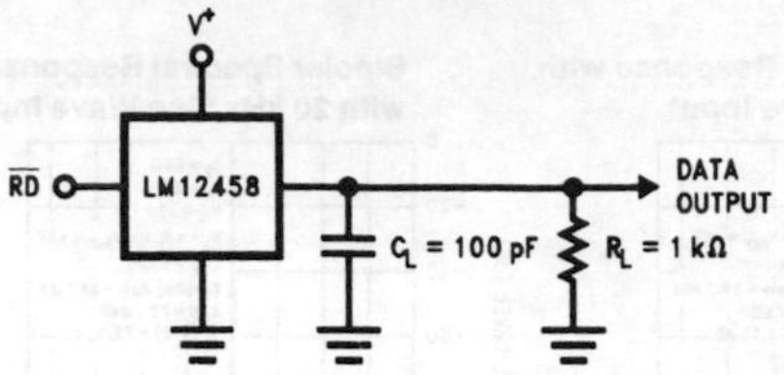

V+
RD
LM12458
DATA OUTPUT
C_L = 100 pF
R_L = 1 kΩ

TL/H/11264–12

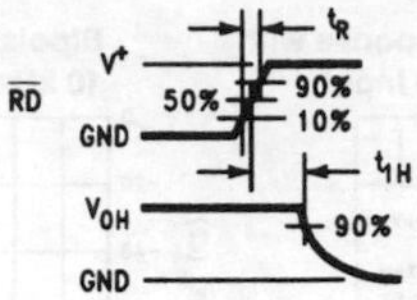

V+
RD
50%
90%
10%
GND
t_R
t_{1H}
V_{OH}
90%
GND

TL/H/11264–13

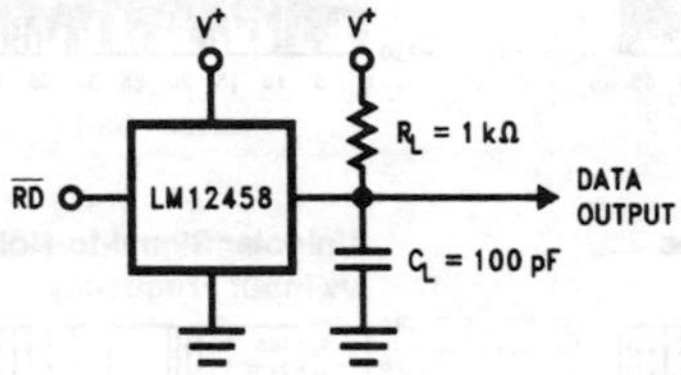

V+
V+
R_L = 1 kΩ
RD
LM12458
DATA OUTPUT
C_L = 100 pF

TL/H/11264–14

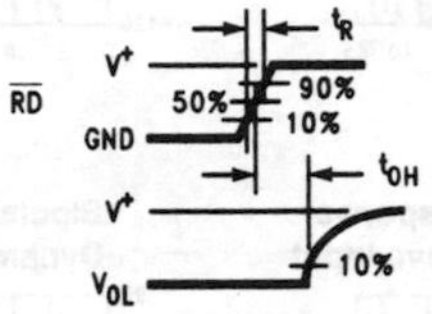

V+
RD
50%
90%
10%
GND
t_R
t_{OH}
V+
10%
V_{OL}

TL/H/11264–15

FIGURE 3. TRI-STATE Test Circuits and Waveforms

Timing Diagrams

$V_A{}^+ = V_D{}^+ = +5V$, $t_R = t_F = 3$ ns, $C_L = 100$ pF for the $\overline{INT}$, DMARQ, D0–D15 outputs.

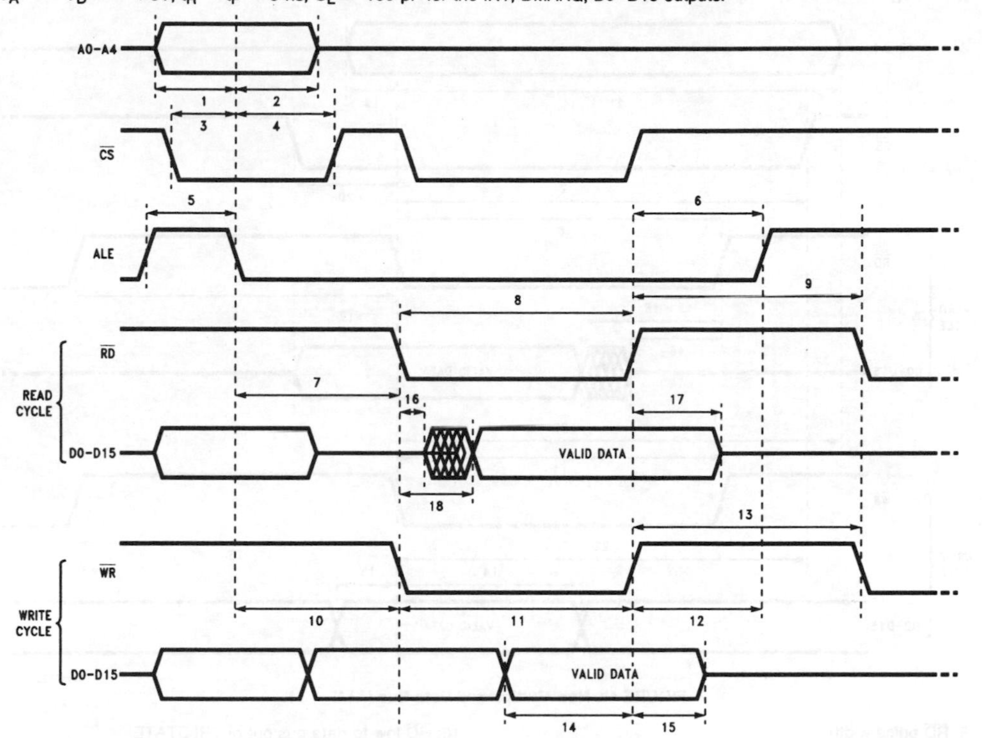

FIGURE 4a. Multiplexed Data Bus

1, 3: $\overline{CS}$ or Address valid to ALE low set-up time.
2, 4: $\overline{CS}$ or Address valid to ALE low hold time.
5: ALE pulse width
6: $\overline{RD}$ high to next ALE high
7: ALE low to $\overline{RD}$ low
8: $\overline{RD}$ pulse width
9: $\overline{RD}$ high to next $\overline{RD}$ or $\overline{WR}$ low
10: ALE low to $\overline{WR}$ low
11: $\overline{WR}$ pulse width
12: $\overline{WR}$ high to next ALE high
13: $\overline{WR}$ high to next $\overline{WR}$ or $\overline{RD}$ low
14: Data valid to $\overline{WR}$ high set-up time
15: Data valid to $\overline{WR}$ high hold time
16: $\overline{RD}$ low to data bus out of TRI-STATE
17: $\overline{RD}$ high to TRI-STATE
18: $\overline{RD}$ low to data valid (access time)

Timing Diagrams (Continued)

$V_A^+ = V_D^+ = +5V$, $t_R = t_F = 3$ ns, $C_L = 100$ pF for the $\overline{INT}$, DMARQ, D0–D15 outputs.

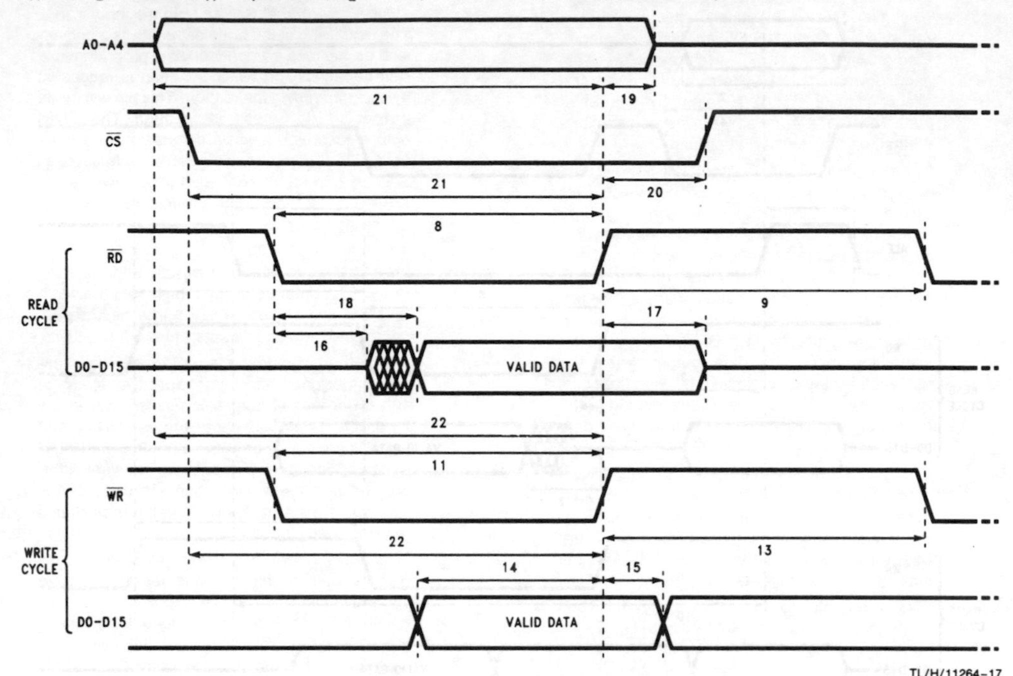

FIGURE 4b. Non-Multiplexed Data Bus (ALE = 1)

8: $\overline{RD}$ pulse width
9: $\overline{RD}$ high to next $\overline{RD}$ or $\overline{WR}$ low
11: $\overline{WR}$ pulse width
13: $\overline{WR}$ high to next $\overline{WR}$ or $\overline{RD}$ low
14: Data valid to $\overline{WR}$ high set-up time
15: Data valid to $\overline{WR}$ high hold time
16: $\overline{RD}$ low to data bus out of TRI-STATE
17: $\overline{RD}$ high to TRI-STATE
18: $\overline{RD}$ low to data valid (access time)
19, 20: Address invalid or $\overline{CS}$ high from $\overline{RD}$ or $\overline{WR}$ high (hold time)
21: $\overline{CS}$ low or address valid to $\overline{RD}$ high
22: $\overline{CS}$ low or address valid to $\overline{WR}$ high

$V_A^+ = V_D^+ = +5V$, $t_R = t_F = 3$ ns, $C_L = 100$ pF for the $\overline{INT}$, DMARQ, D0–D15 outputs.

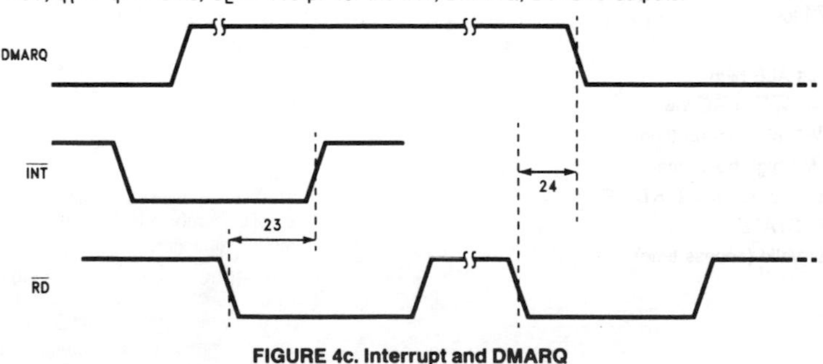

FIGURE 4c. Interrupt and DMARQ

23: $\overline{INT}$ high from $\overline{RD}$ low
24: DMARQ low from $\overline{RD}$ low

Pin Description

V_A+ / V_D+ — These are the analog and digital supply voltage pins. The LM12458's supply voltage operating range is +4.5V to +5.5V. Accuracy is guaranteed only if V_A+ and V_D+ are connected to the same power supply. Each pin should have a parallel combination of 10 μF (electrolytic or tantalum) and 0.1 μF (ceramic) bypass capacitors connected between it and ground.

D0–D15 — The internal data input/output TRI-STATE buffers are connected to these pins. These buffers are designed to drive capacitive loads of 100 pF or less. External buffers are necessary for driving higher load capacitances. These pins allows the user a means of instruction input and data output. With a logic **high** applied to the **BW** pin, data lines D8–D15 are placed in a high impedance state and data lines D0–D7 are used for instruction input and data output when the LM12458 is connected to an 8-bit wide data bus. A logic **low** on the **BW** pin allows the LM12458 to exchange information over a 16-bit wide data bus.

$\overline{RD}$ — This is the input for the active low READ bus control signal. The data input/output TRI-STATE buffers, as selected by the logic signal applied to the **BW** pin, are enabled when $\overline{RD}$ and $\overline{CS}$ are both low. This allows the LM12458 to transmit information onto the databus.

$\overline{WR}$ — This is the input for the active low WRITE bus control signal. The data input/output TRI-STATE buffers, as selected by the logic signal applied to the **BW** pin, are enabled when $\overline{WR}$ and $\overline{CS}$ are both low. This allows the LM12458 to receive information from the databus.

$\overline{CS}$ — This is the input for the active low Chip Select control signal. A logic low should be applied to this pin only during a READ or WRITE access to the LM12458. The internal clocking is halted and conversion stops while Chip Select is low. Conversion resumes when the Chip Select input signal returns high.

ALE — This is the Address Latch Enable input. It is used in systems containing a multiplexed databus. When ALE is asserted **high**, the LM12458 accepts information on the databus as a valid address. A high-to-low transition will latch the address data on A0–A4 and the logic state on the $\overline{CS}$ input. Any changes on A0–A4 and $\overline{CS}$ while ALE is low will not affect the LM12458. See *Figure 4a.*

CLK — This is the external clock input pin. The LM12458 operates with an input clock frequency in the range of 0.05 MHz to 10.0 MHz.

A0–A4 — These are the LM12458's address lines. They are used to access all internal registers, Conversion FIFO, and Instruction **RAM**.

SYNC — This is the synchronization input/output. When used as an output, it is designed to drive capacitive loads of 100 pF or less. External buffers are necessary for driving higher load capacitances. SYNC is an **input** if the Configuration register's "I/O Select" bit is **low**. A rising edge on this pin causes the internal S/H to hold the input signal. The next rising clock edge either starts a conversion or makes a comparison to a programmable limit depending on which function is requested by a programming instruction. This pin will be an **output** if "I/O Select" is set **high**. The SYNC output goes high when a conversion or a comparison is started and low when completed. (See Section 2.2). An internal reset after power is first applied to the LM12458 automatically sets this pin as an input.

BW — This is the Bus Width input pin. This input allows the LM12458 to interface directly with either an 8- or 16-bit databus. A logic high sets the width to 8 bits and places D8–D15 in a high impedance state. A logic low sets the width to 16 bits.

$\overline{INT}$ — This is the active low interrupt output. This output is designed to drive capacitive loads of 100 pF or less. External buffers are necessary for driving higher load capacitances. An interrupt signal is generated any time a non-masked interrupt condition takes place. There are eight different conditions that can cause an interrupt. Any interrupt is reset by reading the Interrupt Status register. (See Section 2.3.)

DMARQ — This is the active high Direct Memory Access Request output. This output is designed to drive capacitive loads of 100 pF or less. External buffers are necessary for driving higher load capacitances. It goes high whenever the number of conversion results in the conversion FIFO equals a programmable value stored in the Interrupt register. It returns to a logic low when the FIFO is empty.

GND — This is the LM12458 ground connection. It should be connected to a low resistance and inductance analog ground return that connects directly to the system power supply ground.

IN0–IN7 — These are the eight analog inputs. A given channel is selected through the instruction RAM. Any of the channels can be configured as an independent single-ended input. Any pair of channels, whether adjacent or non-adjacent, can operate as a fully differential pair.

V_{REF-} — This is the negative reference input. The LM12458 operates with $0V \le V_{REF-} \le V_{REF+}$. This pin should be bypassed to ground with a parallel combination of 10 μF and 0.1 μF (ceramic) capacitors.

V_{REF+} — This is the positive reference input. The LM12458 operates with $0V \le V_{REF+} \le V_A+$. This pin should be bypassed to ground with a parallel combination of 10 μF and 0.1 μF (ceramic) capacitors.

V_{REFOUT} — This is the internal 2.5V bandgap's output pin. when used, this pin should be bypassed to ground with a 100 μF capacitor. When not used, an external bypass capacitor is not needed. However, to ensure the reference's stability, keep stray capacitance below 50 pF.

Application Information

1.0 Functional Description

The LM12458 is a multi-functional Data Acquisition System that includes a fully differential 12-bit-plus-sign self-calibrating analog-to-digital converter (ADC) with a two's-complement output format, an 8-channel analog multiplexer, an internal 2.5V reference, a first-in-first-out (FIFO) register that can store 32 conversion results, and an Instruction RAM that can store as many as eight instructions to be sequentially executed. All of this circuitry operates on only a single +5V power supply.

The LM12458 has three modes of operation:

12-bit + sign with correction

8-bit + sign without correction

8-bit + sign comparison mode ("watchdog" mode)

The fully differential 12-bit-plus-sign ADC uses a charge re-distribution topology that includes calibration capabilities. Charge re-distribution ADCs use a capacitor ladder in place of a resistor ladder to form an internal DAC. The DAC is used by a successive approximation register to generate intermediate voltages between the voltages applied to V_{REF-} and V_{REF+}. These intermediate voltages are compared against the sampled analog input voltage as each bit is generated. The number of intermediate voltages and comparisons equals the ADC's resolution. The correction of each bit's accuracy is accomplished by calibrating the capacitor ladder used in the ADC.

Two different calibration modes are available; one compensates for offset voltage, or zero error, while the other corrects both offset error and the ADC's linearity error.

When correcting offset only, the offset error is measured once and a correction coefficient is created. During the full calibration, the offset error is measured eight times, averaged, and a correction coefficient is created. After completion of either calibration mode, the offset correction coefficient is stored in an internal offset correction register.

The LM12458's overall linearity correction is achieved by correcting the internal DAC's capacitor mismatch. Each capacitor is compared eight times against all remaining smaller value capacitors and any errors are averaged. A correction coefficient is then created and stored in one of the thirteen internal linearity correction registers. An internal state machine, using patterns stored in an internal 16 x 8-bit ROM, executes each calibration algorithm.

Once calibrated, an internal arithmetic logic unit (ALU) uses the offset correction coefficient and the 13 linearity correction coefficients to reduce the conversion's offset error and linearity error, in the background, during the 12-bit + sign conversion. The 8-bit + sign conversion and comparison modes use only the offset coefficient. The 8-bit + sign mode performs a conversion in less than half the time used by the 12-bit + sign conversion mode.

The LM12458's "watchdog" mode is used to monitor a single-ended or differential signal's amplitude. Each sampled signal has two limits. An interrupt can be generated if the input signal is above or below either of the two limits. This allows interrupts to be generated when analog voltage inputs are "inside the window" or, alternatively, "outside the window". After a "watchdog" mode interrupt, the processor can then request a conversion on the input signal and read the signal's magnitude.

The 8-channel analog input multiplexer can be configured for any combination of single-ended or fully differential operation. Each input is referenced to ground when a multiplexer channel operates in the single-ended mode. Fully differential analog input channels are formed by pairing any two channels together.

An internal 2.5V bandgap reference output is available at pin 44. This voltage can be used as the ADC reference for ratiometric conversion or as a virtual ground for front-end analog conditioning circuits. When using this built-in reference, the V_{REFOUT} pin should be bypassed to ground with a 100 μF capacitor. If not used, ensure that stray capacitance at the V_{REFOUT} pin is below 50 pF.

Microprocessor overhead is reduced through the use of the internal conversion FIFO. Thirty-two consecutive conversions can be completed and stored in the FIFO without any microprocessor intervention. The microprocessor can, at any time, interrogate the FIFO and retrieve its contents. It can also wait for the LM12458 to issue an interrupt when the FIFO is full or after any number (≤32) of conversions have been stored.

Conversion sequencing, internal timer interval, multiplexer configuration, and many other operations are programmed and set in the Instruction RAM.

A diagnostic mode is available that allows verification of the LM12458's operation. This mode internally connects the voltages present at the V_{REFOUT}, V_{REF+}, V_{REF-}, and GND pins to the internal V_{IN+} and V_{IN-} S/H inputs. This mode is activated by setting the Diagnostic bit (Bit 11) in the Configuration register to a "1". More information concerning this mode of operation can be found in Section 2.2.

2.0 Internal User-Programmable Registers

2.1 INSTRUCTION RAM

The instruction RAM holds up to eight sequentially executable instructions. Each 48-bit long instruction is divided into three 16-bit sections. READ and WRITE operations can be issued to each 16-bit section using the instruction's address and the 2-bit "RAM pointer" in the Configuration register. The eight instructions are located at addresses 0000 through 0111 (A4–A1, BW = 0) when using a 16-bit wide data bus or at addresses 00000 through 01111 (A4–A0, BW = 1) when using an 8-bit wide data bus. They can be accessed and programmed in random order.

Any Instruction RAM READ or WRITE can affect the sequencer's operation:

The Sequencer should be stopped by setting the RESET bit to a "1" or by resetting the START bit in the Configuration Register and waiting for the current instruction to finish execution before any Instruction RAM READ or WRITE is initiated.

A soft RESET should be issued by writing a "1" to the Configuration Register's RESET bit after any READ or WRITE to the Instruction RAM.

The three sections in the Instruction RAM are selected by the Configuration Register's 2-bit "RAM Pointer", bits D8 and D9. The first 16-bit Instruction RAM section is selected with the RAM Pointer equal to "00". This section provides multiplexer channel selection, as well as resolution, acquisition time, etc. The second 16-bit section holds "watchdog" limit #1, its sign, and an indicator that shows that an interrupt can be generated if the input signal is greater or less than the programmed limit. The third 16-bit section holds "watchdog" limit #2, its sign, and an indicator that shows that an interrupt can be generated if the input signal is greater or less than the programmed limit.

Instruction RAM "00"

Bit 0 is the LOOP bit. It indicates the last instruction to be executed in any instruction sequence when it is set to a "1". The next instruction to be executed will be instruction 0.

Bit 1 is the PAUSE bit. This controls the Sequencer's operation. The Sequencer will stop after reading the current instruction, but before executing it, and the START bit, in the Configuration register, is automatically reset to a "0" when the PAUSE bit is set ("1"). Setting the PAUSE also causes an interrupt to be issued. The Sequencer is restarted by placing a "1" in the Configuration register's Bit 0 (Start bit).

Bits 2–4 select which of the eight input channels ("000" to "111" for IN0–IN7) will be configured as non inverting inputs to the ADC. (See Table I.)

Bits 5–7 select which of the seven input channels ("001" to "111" for IN1 to IN7) will be configured as inverting inputs to the ADC. Fully differential operation is created by selecting two multiplexer channels, one operating in the non-inverting mode and the other operating in the inverting mode. A code of "000" selects ground as the inverting input for single ended operation. (See Table I.)

Bit 8 is the SYNC bit. Setting Bit 8 to "1" causes the Sequencer to suspend operation at the end of the internal S/H's acquisition cycle and before the ADC starts a conversion. The ADC will wait until a rising edge appears at the SYNC pin. When a rising edge appears, the S/H acquires the input signal magnitude and the ADC performs a conversion on the clock's next rising edge. When the SYNC pin is used as an input, the Configuration register's "I/O Select" bit (Bit 7) must be set to a "1". With SYNC configured as an input, it is possible to synchronize the start of a conversion to an external event. This is useful in applications such as digital signal processing (DSP) where the exact timing of conversions is important.

When the LM12458 is used in the "watchdog" mode with external synchronization, two rising edges on the SYNC input are required to initiate two comparisons. The first rising edge initiates the comparison of the selected analog input signal with Limit #1 (found in Instruction RAM "01") and the second rising edge initiates the comparison of the same analog input signal with Limit #2 (found in Instruction RAM "10").

Bit 9 is the TIMER bit. When Bit 9 is set to "1", the Sequencer will halt until the internal 16-bit Timer counts down to zero. During this time interval, no "watchdog" comparisons or analog-to-digital conversions will be performed.

Bit 10 selects the ADC conversion resolution. Setting Bit 10 to "1" selects 8-bit + sign and when reset to "0" selects 12-bit + sign.

Bit 11 is the "watchdog" comparison mode enable bit. When operating in the "watchdog" comparison mode, the selected analog input signal is compared with the programmable values stored in Limit #1 and Limit #2 (see Instruction RAM "01" and Instruction RAM "10"). Setting Bit 11 to "1" causes two comparisons of the selected analog input signal with the two stored limits. When Bit 11 is reset to "0", an 8-bit + sign or 12-bit + sign (depending on the state of Bit 10 of Instruction RAM "00") conversion of the input signal can take place.

2.0 Internal User-Programmable Registers (Continued)

A4	A3	A2	A1	Purpose	Type	D15	D14	D13	D12	D11	D10	D9	D8	D7	D6	D5	D4	D3	D2	D1	D0
0	0 to 1	0 to 1	0 to 1	Instruction RAM (RAM Pointer = 00)	R/W	Acquisition Time				Watch-dog	8/12	Timer	Sync	$V_{IN}-$			$V_{IN}+$			Pause	Loop
0	0 to 1	0 to 1	0 to 1	Instruction RAM (RAM Pointer = 01)	R/W	Don't Care						>/<	Sign	Limit #1							
0	0 to 1	0 to 1	0 to 1	Instruction RAM (RAM Pointer = 10)	R/W	Don't Care						>/<	Sign	Limit #2							
1	0	0	0	Configuration Register	R/W	Don't Care				DIAG	Test = 0	RAM Pointer		I/O Sel	Auto $Zero_{ec}$	Chan Mask	Stand-by	Full CAL	Auto-Zero	Reset	Start
1	0	0	1	Interrupt Enable Register	R/W	Number of Conversions in Conversion FIFO to Generate INT2					Sequencer Address to Generate INT1			INT7	INT6	INT5	INT4	INT3	INT2	INT1	INT0
1	0	1	0	Interrupt Status Register	R	Actual Number of Conversion Results in Conversion FIFO					Actual Sequencer Instruction Executed			INST7	INST6	INST5	INST4	INST3	INST2	INST1	INST0
1	0	1	1	Timer Register	R/W	Timer Preset High Byte								Timer Preset Low Byte							
1	1	0	0	Conversion FIFO	R	Address or Sign			Sign	Conversion Data: MSBs				Conversion Data: LSBs							
1	1	0	1	Limit Status Register	R	Limit #2: Status								Limit #1: Status							

FIGURE 5. LM12458 Memory Map for 16-Bit Wide Databus (BW = "0", Test Bit = "0" and A0 = Don't Care)

2.0 Internal User-Programmable Registers (Continued)

A4	A3	A2	A1	A0	Purpose	Type	D7	D6	D5	D4	D3	D2	D1	D0
0	0 to 1	0 to 1	0 to 1	0	Instruction RAM (RAM Pointer = 00)	R/W		V_{IN-}			V_{IN+}		Pause	Loop
0	0 to 1	0 to 1	0 to 1	1	Instruction RAM (RAM Pointer = 00)	R/W	Acquisition Time				Watch-dog	8/12	Timer	Sync
0	0 to 1	0 to 1	0 to 1	0	Instruction RAM (RAM Pointer = 01)	R/W	Comparison Limit #1							
0	0 to 1	0 to 1	0 to 1	1	Instruction RAM (RAM Pointer = 01)	R/W	Don't Care						>/<	Sign
0	0 to 1	0 to 1	0 to 1	0	Instruction RAM (RAM Pointer = 10)	R/W	Comparison Limit #2							
0	0 to 1	0 to 1	0 to 1	1	Instruction RAM (RAM Pointer = 10)	R/W	Don't Care						>/<	Sign
1	0	0	0	0	Configuration Register	R/W	I/O Sel	Auto $Zero_{ec}$	Chan Mask	Stand-by	Full Cal	Auto-Zero	Reset	Start
1	0	0	0	1	Configuration Register	R/W	Don't Care				DIAG	Test = 0	RAM Pointer	
1	0	0	1	0	Interrupt Enable Register	R/W	INT7	INT6	INT5	INT4	INT3	INT2	INT1	INT0
1	0	0	1	1	Interrupt Enable Register	R/W	Number of Conversions in Conversion FIFO to Generate INT2					Sequencer Address to Generate INT1		
1	0	1	0	0	Interrupt Status Register	R	INST7	INST6	INST5	INST4	INST3	INST2	INST1	INST0
1	0	1	0	1	Interrupt Status Register	R	Actual Number of Conversions Results in Conversion FIFO					Actual Sequencer Instruction Executed		
1	0	1	1	0	Timer Register	R/W	Timer Preset: Low Byte							
1	0	1	1	1	Timer Register	R/W	Timer Preset: High Byte							
1	1	0	0	0	Conversion FIFO	R	Conversion Data: LSBs							
1	1	0	0	1	Conversion FIFO	R	Address or Sign			Sign	Conversion Data: MSBs			
1	1	0	1	0	Limit Status Register	R	Limit #1 Status							
1	1	0	1	1	Limit Status Register	R	Limit #2 Status							

FIGURE 6. LM12458 Memory Map for 8-Bit Wide Databus (BW = "1" and Test Bit = "0")

2.0 Internal User-Programmable Registers (Continued)

Bits 12–15 are used to store the user programmable acquisition time. The Sequencer keeps the internal S/H in the acquisition mode for twice the number of clock cycles stored in Bits 12–15 (nine clock cycles, minimum for 12-bit + sign conversions and two cycles for 8-bit + sign conversions or "watchdog" comparisons and 39 clock cycles, maximum. The minimum delay compensates for the typical internal multiplexer series resistance of 2 kΩ). The necessary acquisition time is determined by the source impedance at the multiplexer input. If the source resistance (R_S) < 100Ω and the clock frequency is 5 MHz, the value stored in bits 12–15 (t_{ACQ} Factor) can be 0000. If R_S > 100Ω, the following equation determines the value stored in bits 12–15.

$$R_S \times f_{CLK} \times 0.45 = t_{ACQ}\ \text{Factor}$$

for 12-bits + sign

$$R_S \times f_{CLK} \times 0.36 = t_{ACQ}\ \text{Factor}$$

for 8-bits + sign and "watchdog"

R_S is in kΩ and f_{CLK} is in MHz. Round the result to the next higher integer value. If t_{ACQ} Factor is greater than 15, it is advisable to lower the source impedance through the use of an analog buffer between the signal source and the LM12458's multiplexer inputs.

Instruction RAM "01"

The second Instruction RAM section is selected by placing a "01" in Bits 8 and 9 of the Configuration register.

Bits 0–7 hold "watchdog" **limit #1**. When Bit 11 of Instruction RAM "00" is set to a "1", the LM12458 will perform a "watchdog" comparison of the sampled analog input signal with the limit #1 value first, followed by a comparison of the same sampled analog input signal with the value found in limit #2 (Instruction RAM "10").

Bit 8 holds limit #1's sign.

Bit 9's state determines the limit condition that generates a "watchdog" interrupt. A "1" causes a voltage greater than limit #1 to generate an interrupt, while a "0" causes a voltage less than limit #1 to generate an interrupt.

Bits 10–15 are not used.

Instruction RAM "10"

The third Instruction RAM section is selected by placing a "10" in Bits 8 and 9 of the Configuration register.

Bits 0–7 hold "watchdog" **limit #2**. When Bit 11 of Instruction RAM "00" is set to a "1", the LM12458 will perform a "watchdog" comparison of the sampled analog input signal with the limit #1 value first (Instruction RAM "01"), followed by a comparison of the same sampled analog input signal with the value found in limit #2.

Bit 8 holds limit #2's sign.

Bit 9's state determines the limit condition that generates a "watchdog" interrupt. A "1" causes a voltage greater than limit #2 to generate an interrupt, while a "0" causes a voltage less than limit #2 to generate an interrupt.

Bits 10–15 are not used.

2.2 CONFIGURATION REGISTER

The Configuration register, 1000 (A4–A1, BW = 0) or 1000x (A4–A0, BW = 1) is a 16-bit control register with read/write capability. It acts as the LM12458's "control panel" holding global information as well as start/stop, reset, self-calibration, and stand-by commands.

Bit 0 is the START/STOP bit. Reading Bit 0 returns an indication of the Sequencer's status. A "0" indicates that the Sequencer is stopped and waiting to execute the next instruction. A "1" shows that the Sequencer is running. Writing a "0" halts the Sequencer when the current instruction has finished execution. The next instruction to be executed is pointed to by the instruction pointer found in the status register. A "1" restarts the Sequencer with the instruction currently pointed to by the instruction pointer. (See Bits 8–10 in the Interrupt Status register.)

Bit 1 is the LM12458's system RESET bit. Writing a "1" to Bit 1 stops the Sequencer (resetting the Configuration register's START/STOP bit), resets the Instruction pointer to "000" (found in the Status register), clears the Conversion FIFO, and resets all interrupt flags. The RESET bit will return to "0" after two clock cycles unless it is forced high by writing a "1" into the Configuration register's Standby bit. A reset signal is internally generated when power is first applied to the part. No operation should be started until the RESET bit is "0".

Writing a "1" to **Bit 2** initiates an auto-zero offset voltage calibration. Unlike the eight-sample auto-zero calibration performed during the full calibration procedure, Bit 2 initiates a "short" auto-zero by sampling the offset once and creating a correction coefficient (full calibration averages eight samples of the converter offset voltage when creating a correction coefficient). If the Sequencer is running when Bit 2 is set to "1", an auto-zero starts immediately after the conclusion of the currently running instruction. Bit 2 is reset automatically to a "0" and an interrupt flag (Bit 3, in the Interrupt Status register) is set at the end of the auto-zero (76 clock cycles). After completion of an auto-zero calibration, the Sequencer fetches the next instruction as pointed to by the Instruction RAM's pointer and resumes execution. If the Sequencer is stopped, an auto-zero is performed immediately at the time requested.

Writing a "1" to **Bit 3** initiates a complete calibration process that includes a "long" auto-zero offset voltage correction (this calibration averages eight samples of the comparator offset voltage when creating a correction coefficient) followed by an ADC linearity calibration. This complete calibration is started after the currently running instruction is completed if the Sequencer is running when Bit 3 is set to "1". Bit 3 is reset automatically to a "0" and an interrupt flag (Bit 4, in the Interrupt Status register) will be generated at the end of the calibration procedure (4944 clock cycles). After completion of a full auto-zero and linearity calibration, the Sequencer fetches the next instruction as pointed to by the Instruction RAM's pointer and resumes execution. If the Sequencer is stopped, a full calibration is performed immediately at the time requested.

2.0 Internal User-Programmable Registers (Continued)

Bit 4 is the Standby bit. Writing a "1" to Bit 4 immediately places the LM12458 in Standby mode. Normal operation returns when Bit 4 is reset to a "0". The Standby command ("1") disconnects the external clock from the internal circuitry, decreases the LM12458's internal analog circuitry power supply current, and preserves all internal RAM contents. After writing a "0" to the Standby bit, the LM12458 returns to an operating state identical to that caused by exercising the RESET bit. A Standby completion interrupt is issued after a power-up completion delay that allows the analog circuitry to settle. The Sequencer should be restarted only after the Standby completion is issued. The Instruction RAM can still be accessed through read and write operations while the LM12458 is in Standby Mode.

Bit 5 is the Channel Address Mask. If Bit 5 is set to a "1", Bits 13–15 in the conversion FIFO will be equal to the sign bit (Bit 12) of the conversion data. Resetting Bit 5 to a "0" causes conversion data Bits 13 through 15 to hold the instruction pointer value of the instruction to which the conversion data belongs.

Bit 6 is used to select a "short" auto-zero correction for every conversion. The Sequencer automatically inserts an auto-zero before every conversion or "watchdog" comparison if Bit 6 is set to "1". No automatic correction will be performed if Bit 6 is reset to "0".

The LM12458's offset voltage, after calibration, has a typical drift of 0.1 LSB over a temperature range of −40°C to +85°C. This small drift is less than the variability of the change in offset that can occur when using the auto-zero correction with each conversion. This variability is the result of using only one sample of the offset voltage to create a correction value. This variability decreases when using the full calibration mode because eight samples of the offset voltage are taken, averaged, and used to create a correction value.

Bit 7 is used to program the SYNC pin (29) to operate as either an input or an output. The SYNC pin becomes an output when Bit 7 is a "1" and an input when Bit 7 is a "0". With SYNC programmed as an input, the rising edge of any logic signal applied to pin 29 will start a conversion or "watchdog" comparison. Programmed as an output, the logic level at pin 29 will go high at the start of a conversion or "watchdog" comparison and remain high until either have finished. See Instruction RAM "00", Bit 8.

Bits 8 and **9** form the RAM Pointer that is used to select each 48-bit instruction RAM's three sections during read or write actions. A "00" selects Instruction RAM section one, "01" selects section two, and "10" selects section three.

Bit 10 activates the Test mode that is used only during production testing. Leave this bit reset to "0".

Bit 11 is the Diagnostic bit. It can be activated by setting it to a "1" (the Test bit must be reset to a "0"). The Diagnostic mode, along with a correctly chosen instruction, allows verification that the LM12458's ADC is performing correctly. When activated, the inverting and non-inverting inputs are connected as shown in Table I. As an example, an instruction with "001" for both V_{IN+} and V_{IN-} while using the Diagnostic mode typically results in a full-scale output.

2.3 INTERRUPTS

The LM12458 has eight possible interrupts, all with the same priority. Any of these interrupts will cause a hardware interrupt to appear on the $\overline{INT}$ pin (31) if they are not masked (by the Interrupt register). The Status register is then read to determine which of the eight interrupts has been issued.

TABLE I. Diagnostic Mode Input Multiplexer Channel Configuration

Channel Selection Data	Normal Mode		Diagnostic Mode	
	V_{IN+}	V_{IN-}	V_{IN+}	V_{IN-}
000	IN0	GND	V_{REFOUT}	GND
001	IN1	IN1	V_{REF+}	V_{REF-}
010	IN2	IN2	IN2	IN2
011	IN3	IN3	IN3	IN3
100	IN4	IN4	IN4	IN4
101	IN5	IN5	IN5	IN5
110	IN6	IN6	IN6	IN6
111	IN7	IN7	IN7	IN7

The Interrupt Status register, 1010 (A4–A1, BW = 0) or 1010x (A4–A0, BW = 1) must be cleared by reading it after writing to the Interrupt Enable register. This removes any spurious interrupts on the INT pin generated during an Interrupt Enable register access.

Interrupt 0 is generated whenever the analog input voltage on a selected multiplexer channel crosses a limit while the LM12458 is operating in the "watchdog" comparison mode. Two sequential comparisons are made when the LM12458 is executing a "watchdog" instruction. Depending on the logic state of Bit 9 in the Instruction RAM's second and third sections, an interrupt will be generated either when the input signal's magnitude is greater than or less than the programmable limits. (See the Instruction RAM, Bit 9 description.) The Limit Status register will indicate which preprogrammed limit, #1 or #2 and which instruction was executing when the limit was crossed.

Interrupt 1 is generated when the Sequencer reaches the instruction counter value specified in the Interrupt register's bits 8–10. This flag appears before the instruction's execution.

Interrupt 2 is activated when the Conversion FIFO holds a number of conversions equal to the programmable value stored in the Interrupt Enable register's Bits 11–15.

The completion of the short, single-sampled auto-zero calibration causes the generation of **Interrupt 3**.

The completion of a full auto-zero and linearity self-calibration causes the generation of **Interrupt 4**.

Interrupt 5 is generated when the Sequencer encounters an instruction that has its Pause bit (Bit 1 in Instruction RAM "00") set to "1".

The LM12458 issues **Interrupt 6** whenever it senses that its power supply voltage is dropping below 4V (typ). This interrupt indicates the potential corruption of data returned by the LM12458.

Interrupt 7 is issued after a short delay (10 ms typ) while the LM12458 returns from Standby mode to active operation using the Configuration register's Bit 4. This short delay allows the internal analog circuitry to settle sufficiently, ensuring accurate conversion results.

2.0 Internal User-Programmable Registers (Continued)

2.4 INTERRUPT ENABLE REGISTER

The Interrupt Enable register at address location 1001 (A4–A1, BW = 0) or 1001x (A4–A0, BW = 1) has READ/WRITE capability. An individual interrupt's ability to produce an external interrupt at pin 31 ($\overline{INT}$) is accomplished by placing a "1" in the appropriate bit location. Any of the internal interrupt producing operations will set their corresponding bits in the Interrupt Status register regardless of the state of the associated bit in the Interrupt Enable register. See Section 2.3 for more information about each of the eight internal interrupts.

Bit 0 enables an external interrupt when an internal comparison limit interrupt has taken place.

Bit 1 enables an external interrupt when the Sequencer has reached the address stored in Bits 8–10 of the Interrupt Enable register.

Bit 2 enables an external interrupt when the Conversion FIFO's limit, stored in Bits 11–15 of the Interrupt Enable register, has been reached.

Bit 3 enables an external interrupt when the single-sampled auto-zero calibration has been completed.

Bit 4 enables an external interrupt when a full auto-zero and linearity self-calibration has been completed.

Bit 5 enables an external interrupt when a Pause interrupt has been generated.

Bit 6 enables an external interrupt when a low power supply condition (V_A+ < 4V) has generated an interrupt.

Bit 7 enables an external interrupt when the LM12458 returns from power-down to active mode.

Bits 8–10 form the storage location of the programmable Sequencer address. When the Sequencer reaches an address that is equal to the value stored in Bits 8–10, an internal interrupt is generated and appears in Bit 1 of the Interrupt Status register. If Bit 1 of the Interrupt Enable register is set to "1", an external interrupt will appear at pin 31 ($\overline{INT}$).

Bits 11–15 hold the number of conversions that must be stored in the Conversion FIFO in order to generate an internal interrupt. This internal interrupt appears in Bit 2 of the Interrupt Status register. If Bit 2 of the Interrupt Enable register is set to "1", an external interrupt will appear at pin 31 ($\overline{INT}$).

2.5 INTERRUPT STATUS REGISTER

This read-only register is located at address 1010 (A4–A1, BW = 0) or 1010x (A4–A0, BW = 1). The corresponding Interrupt Status flag will go high ("1") any time that an interrupt condition takes place, whether an interrupt is enabled or disabled in the Interrupt Enable register. Any of the active ("1") Interrupt Status flags are reset to "0" whenever this register is read or a device reset is issued (see Bit 1 in the Configuration Register).

Bit 0 is set to "1" when a comparison limit interrupt has taken place.

Bit 1 is set to "1" when the Sequencer has reached the address stored in Bits 8–10 of the Interrupt Enable register.

Bit 2 is set to "1" when the Conversion FIFO's limit, stored in Bits 11–15 of the Interrupt Enable register, has been reached.

Bit 3 is set to "1" when the single-sampled auto-zero has been completed.

Bit 4 is set to "1" when an auto-zero and full linearity self-calibration has been completed.

Bit 5 is set to "1" when a Pause interrupt has been generated.

Bit 6 is set to "1" when a low-supply voltage condition (V_A+ < 4V) has taken place.

Bit 7 is set to "1" when the LM12458 returns from power-down to active mode.

Bits 8–10 hold the Sequencer's actual address while it is running.

Bits 11–15 hold the actual number of conversions stored in the Conversion FIFO while the Sequencer is running.

2.6 LIMIT STATUS REGISTER

The read-only register is located at address 1101 (A4–A1, BW = 0) or 1101x (A4–A0, BW = 1). This register is used in tandem with the Limit #1 and Limit #2 registers in the Instruction RAM. Whenever a given instruction's input voltage exceeds the limit set in its corresponding Limit register (#1 or #2), a bit, corresponding to the instruction number, is set in the Limit Status register. Any of the active ("1") Limit Status flags are reset to "0" whenever this register is read or a device reset is issued (see Bit 1 in the Configuration register). This register holds the status of limits #1 and #2 for each of the eight instructions.

Bits 0–7 show the Limit #1 status. Each bit will be set high ("1") when the corresponding instruction's input voltage exceeds the threshold stored in the instruction's Limit #1 register. When, for example, instruction 3 is a "watchdog" operation (Bit 11 is set high) and the input to channel 3 meets the magnitude and/or polarity data stored in instruction 3's Limit #1 register, Bit 3 in the Limit Status register will be set to a "1".

Bits 8–15 show the Limit #2 status. Each bit will be set high ("1") when the corresponding instruction's input voltage exceeds the threshold stored in the instruction's Limit #2 register. When, for example, the input to instruction 6 meets the value stored in instruction 6's Limit #2 register, Bit 14 in the Limit Status register will be set to a "1".

2.7 TIMER

The LM12458 has an on-board 16-bit timer that includes a 5-bit pre-scaler. It uses the clock signal applied to pin 23 as its input. It can generate time intervals of 0 through 2^{21} clock cycles in steps of 2^5. This time interval can be used to delay the execution of instructions. It can also be used to slow the conversion rate when converting slowly changing signals. This can reduce the amount of redundant data stored in the FIFO and retrieved by the controller.

The user-defined timing value used by the Timer is stored in the 16-bit READ/WRITE Timer register at location 1011 (A4–A1, BW = 0) or 1011x (A4–A0, BW = 1) and is preloaded automatically. Bits 0–7 hold the preset value's low byte and Bits 8–15 hold the high byte. The Timer is activated by the Sequencer only if the current instruction's Bit 9 is set ("1"). If the equivalent decimal value "N" ($0 \leq N < 2^{16}$) is written inside the 16-bit Timer register and the Timer is enabled by setting an instruction's bit 9 to a "1", the Sequencer will delay the same instruction's execution by halting at state 3 (S3), as shown in *Figure 7*, for $32 \times N + 2$ clock cycles.

3.0 FIFO

The result of each conversion stored in an internal read-only FIFO (First-In, First-Out) register. It is located at 1100 (A4–A1, BW = 0) or 1100x (A4–A0, BW = 1). This register has 32 16-bit wide locations. Each location holds 13-bit data. Bits 0–3 hold the four LSB's in the 12 bits + sign mode or "1110" in the 8 bits + sign mode. Bits 4–11 hold the eight MSB's and Bit 12 holds the sign bit. Bits 13–15 can hold either the sign bit, extending the register's two's complement data format to a full sixteen bits or the instruction adress that generated the conversion and the resulting data. These modes are selected according to the logic state of the Configuration register's Bit 5.

The FIFO status should be read in the Status register (Bits 12–15) to determine the number of conversion results that are held in the FIFO before retrieving them. This will help prevent conversion data corruption that may take place if the number of reads are greater than the number of conversion results contained in the FIFO. Trying to read the FIFO when it is empty may corrupt new data being written into the FIFO. Writing more than 32 conversion data into the FIFO by the ADC results in loss of the first conversion data. Therefore, to prevent data loss, it is recommended that the LM12458's interrupt capability be used to inform the system controller that the FIFO is full.

The lower portion (A0 = 0) of the data word (Bits 0–7) should be read first followed by a read of the upper portion (A0 = 1) when using the 8-bit bus width (BW = 1). Reading the upper portion first causes the data to shift down, which results in loss of the lower byte.

Bits 0–12 hold 12-bit + sign conversion data. **Bits 0–3** will be 1110 when using 8-bit plus sign resolution.

Bits 13–15 hold either the instruction responsible for the associated conversion data or the sign bit. Either mode is selected with Bit 5 in the Configuration register.

4.0 Sequencer

The Sequencer uses a 3-bit counter (Instruction Pointer, or IP, in *Figure 7*) to retrieve the programmable conversion instructions stored in the Instruction RAM. The 3-bit counter is reset to 000 during chip reset or if the current executed instruction has its Loop bit (Bit 1 in any Instruction RAM "00") set high ("1"). It increments at the end of the currently executed instruction and points to the next instruction. It will continue to increment up to 111 unless an instruction's Loop bit is set. If this bit is set, the counter resets to "000" and execution begins again with the first instruction. If all instructions have their Loop bit reset to "0", the Sequencer will execute all eight instructions continuously. Therefore, it is important to realize that if less than eight instructions are programmed, the Loop bit on the last instruction must be set. Leaving this bit reset to "0" allows the Sequencer to execute "unprogrammed" instructions, the results of which may be unpredictable.

The Sequencer's Instruction Pointer value is readable at any time and is found in the Status register at Bits 8–11. The Sequencer can go through eight states during instruction execution:

State 0: The current instruction's first 16 bits are read from the Instruction RAM "00". This state is one clock cycle long.

State 1: Checks the state of the Calibration and Start bits. This is the "rest" state whenever the Sequencer is stopped using the reset, a Pause command, or the Start bit is reset low ("0"). When the Start bit is set to a "1", this state is one clock cycle long.

State 2: Perform calibration. If bit 2 or bit 6 of the Configuration register is set to a "1", state 2 is 76 clock cycles long. If the Configuration register's bit 3 is set to a "1", state 2 is 4944 clock cycles long.

State 3: Run the internal 16-bit Timer. The number of clock cycles for this state varies according to the value stored in the Timer register. The number of clock cycles is found by using the expression below

$$32T + 2$$

where $0 \le T < 2^{16}$.

State 7: Run the acquisition delay and read Limit #1's value if needed. The number of clock cycles for 12-bit + sign mode varies according to

$$9 + 2D$$

where $0 \le D \le 15$.

The number of clock cycles for 8-bit + sign or "watchdog" mode varies according to

$$2 + 2D$$

where $0 \le D \le 15$.

State 6: Perform first comparison. This state is 5 clock cycles long.

State 4: Read Limit #2. This state is 1 clock cycle long.

State 5: Perform a conversion or second comparison. This state takes 44 clock cycles when using the 12-bit + sign mode or 21 clock cycles when using the 8-bit + sign mode. The "watchdog" mode takes 5 clock cycles.

4.0 Sequencer (Continued)

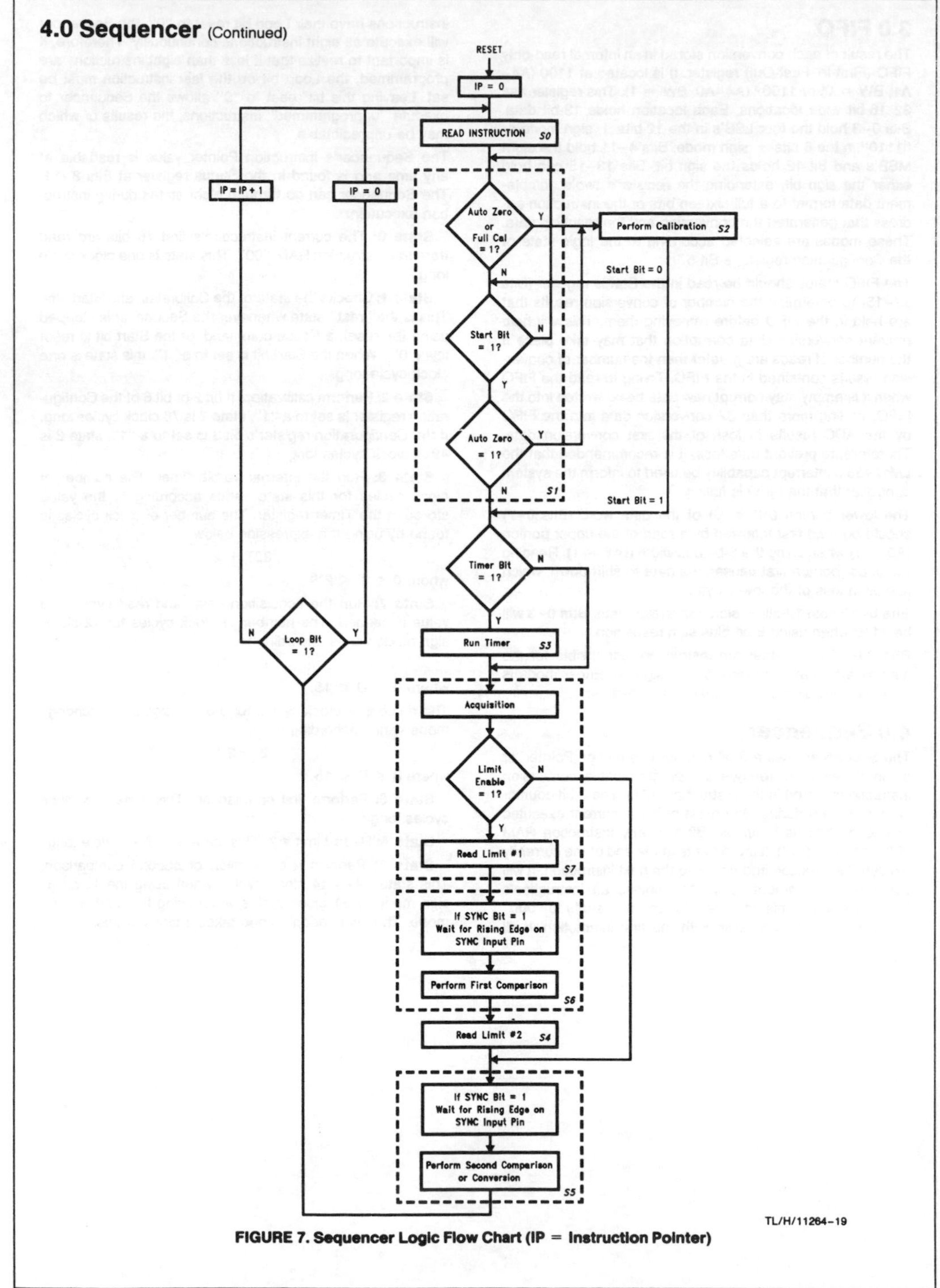

FIGURE 7. Sequencer Logic Flow Chart (IP = Instruction Pointer)

5.0 Analog Considerations

5.1 REFERENCE VOLTAGE

The difference in the voltages applied to the V_{REF+} and V_{REF-} defines the analog input voltage span (the difference between the voltages applied between two multiplexer inputs or the voltage applied to one of the multiplexer inputs and analog ground), over which 4095 positive and 4096 negative codes exist. The voltage sources driving V_{REF+} or V_{REF-} must have very low output impedance and noise. The circuit in *Figure 8* is an example of a very stable reference appropriate for use with the LM12458.

The ADC can be used in either ratiometric or absolute reference applications. In ratiometric systems, the analog input voltage is proportional to the voltage used for the ADC's reference voltage. When this voltage is the system power supply, the V_{REF+} pin is connected to V_A^+ and V_{REF-} is connected to GND. This technique relaxes the system reference stability requirements because the analog input voltage and the ADC reference voltage move together. This maintains the same output code for given input conditions.

For absolute accuracy, where the analog input voltage varies between very specific voltage limits, a time and temperature stable voltage source can be connected to the reference inputs. Typically, the reference voltage's magnitude will require an initial adjustment to null reference voltage induced full-scale errors.

When using the LM12458's internal 2.5V bandgap reference, a parallel combination of a 100 μF capacitor and a 0.1 μF capacitor connected to the V_{REFOUT} pin is recommended for low noise operation. When left unconnected, the reference remains stable without a bypass capacitor. However, ensure that stray capacitance at the V_{REFOUT} pin remains below 50 pF.

5.2 INPUT RANGE

The LM12458's fully differential ADC and reference voltage inputs generate a two's-complement output that is found by using the equation below.

$$\text{output code} = \frac{V_{IN+} - V_{IN-}}{V_{REF+} - V_{REF-}}(4096) - \tfrac{1}{2}$$

Round up to the next integer value if the result of the above equation is not a whole number. As an example, $V_{REF+} = 2.5V$, $V_{REF-} = 1V$, $V_{IN+} = 1.5V$ and $V_{IN-} =$ GND. The output code is positive full-scale, or 0,1111,1111,1111.

5.3 INPUT CURRENT

A charging current flows into or out of (depending on the input voltage polarity) the analog input pins, IN0–IN7 at the start of the analog input acquisition time (t_{ACQ}). This current's peak value will depend on the actual input voltage applied.

5.4 INPUT SOURCE RESISTANCE

For low impedance voltage sources (<100Ω), the input charging current will decay, before the end of the S/H's acquisition time, to a value that will not introduce any conversion errors. For higher source impedances, the S/H's acquisition time can be increased. As an example, operating with a 5 MHz clock frequency and maximum acquisition time, the LM12458's analog inputs can handle source impedance as high as 6.67 kΩ. Refer to Section 2.1, Instruction RAM "00", Bits 12–15 for further information.

5.5 INPUT BYPASS CAPACITANCE

External capacitors (0.01 μF–0.1 μF) can be connected between the analog input pins, IN0–IN7, and analog ground to filter any noise caused by inductive pickup associated with long input leads. It will not degrade the conversion accuracy.

5.6 NOISE

The leads to each of the analog multiplexer input pins should be kept as short as possible. This will minimize input noise and clock frequency coupling that can cause conversion errors. Input filtering can be used to reduce the effects of the noise sources.

5.7 POWER SUPPLIES

Noise spikes on the V_A^+ and V_D^+ supply lines can cause conversion errors; the comparator will respond to the noise. The ADC is especially sensitive to any power supply spikes that occur during the auto-zero or linearity correction. Low inductance tantalum capacitors of 10 μF or greater paralleled with 0.1 μF monolithic ceramic capacitors are recommended for supply bypassing. Separate bypass capacitors should be used for the V_A^+ and V_D^+ supplies and placed as close as possible to these pins.

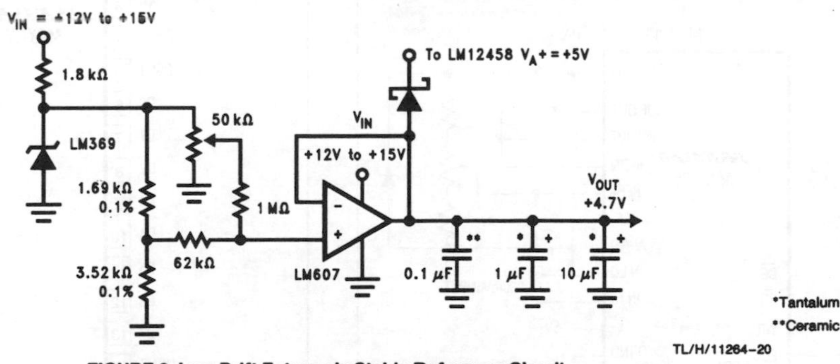

FIGURE 8. Low Drift Extremely Stable Reference Circuit

19-4131; Rev. 0; 9/91

MAXIM

15-Bit ADC with Parallel Interface

MAX135

General Description

The MAX135 is a CMOS 15-bit, binary-output analog-to-digital converter (ADC). Multi-slope integration provides low-noise and high-resolution conversions in less time than standard integrating ADCs: The MAX135 is tested at 16 conversions per second, but operates at up to 6 times that rate. The MAX135 uses Super LSBs with data averaging to achieve 18-bit resolution.

Supply current is 125µA maximum during normal operation and only 10µA maximum in sleep mode. Low conversion noise allows tested operation at only 300mV full scale (15µV per LSB). A simple 8-bit parallel data bus and three control lines easily interface to all common microprocessors, and twos-complement output coding simplifies bipolar measurements.

High resolution and compact size make the MAX135 ideal for data loggers, numerical control systems, weigh scales, data-acquisition systems, and panel meters. The MAX135 comes in 28-pin DIP and SO packages in both commercial and extended temperature grades.

Applications

- Data Acquisition
- Battery-Powered Instruments
- Control Applications
- Analog-Signal Measurement
  - Pressure, Flow, Temperature, Voltage, Current, Resistance, Weight

Functional Diagram

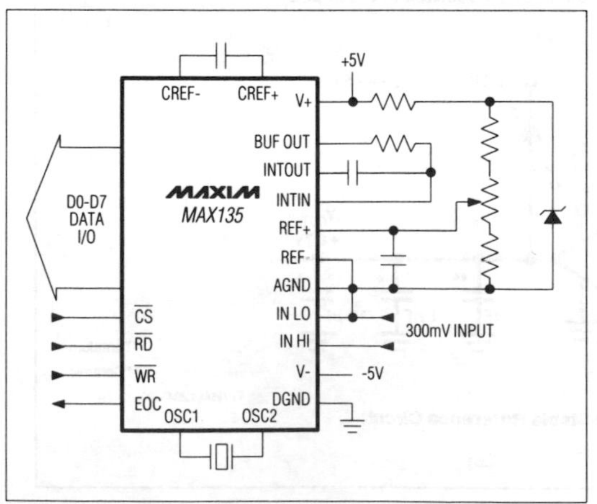

Features

- ◆ **15-Bit, Multi-Slope Integrating ADC**
- ◆ **15µV Resolution at 16 Conv/Sec**
- ◆ **Low Supply Current**
  - **- 125µA Max (Normal Operation)**
  - **- 10µA Max (Sleep-Mode Operation)**
- ◆ **±0.005% Accuracy at 16 Conv/Sec**
- ◆ **3 Super Bits for 18-Bit Resolution**
- ◆ **Low Noise - Operates at 300mV Full Scale**
- ◆ **Easy µP Interface - 8-Bit Parallel Data Bus**
- ◆ **±10pA Input Leakage Current**
- ◆ **Small 28-Pin DIP and SO Packages**

Ordering Information

PART	TEMP. RANGE	PIN-PACKAGE
MAX135CPI	0°C to +70°C	28 Plastic DIP
MAX135CWI	0°C to +70°C	28 Wide SO
MAX135C/D	0°C to +70°C	Dice*
MAX135EPI	-40°C to +85°C	28 Plastic DIP
MAX135EWI	-40°C to +85°C	28 Wide SO

** Contact factory for dice specifications.*

Pin Configuration

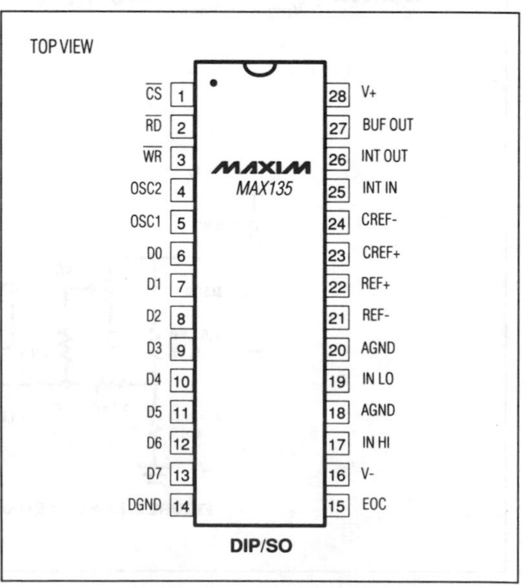

15-Bit ADC with Parallel Interface

ABSOLUTE MAXIMUM RATINGS

Supply Voltage
- V+ to DGND .... -0.3V < V+ < +6.0V
- V- to DGND .... +0.3V < V- < -9.0V
- V+ to V- .... +15V

Analog Input Voltage (any input) .... V+ to V-
Digital Input Voltage .... (DGND - 0.3V) to (V+ + 0.3V)
Continuous Power Dissipation (T_A = +70°C)
- 28-Pin Plastic DIP (derate 9.09mW/°C above +70°C) 786mW
- 28-Pin Wide SO (derate 12.50mW/°C above +70°C) . 688mW

Operating Temperature Ranges:
- MAX135C_ _ .... 0°C to +70°C
- MAX135E_ _ .... -40°C to +85°C

Storage Temperature Range .... -65°C to +160°C
Lead Temperature (soldering, 10 sec) .... +300°C

Stresses beyond those listed under "Absolute Maximum Ratings" may cause permanent damage to the device. These are stress ratings only, and functional operation of the device at these or any other conditions above those indicated in the operational sections of the specifications is not implied. Exposure to absolute maximum rating conditions for extended periods may affect device reliability.

ELECTRICAL CHARACTERISTICS

(V+ = 5V, V- = -5V, DGND = AGND = IN LO = REF- = 0V, REF+ = 545mV, RINT = 402kΩ, CINT = 0.0047µF, CREF = 0.1µF, f_{CLK} = 32,768Hz, 60Hz mode, T_A = T_{MIN} to T_{MAX}, unless otherwise noted.)

PARAMETER	SYMBOL	CONDITIONS		MIN	TYP	MAX	UNITS
Resolution		(Note 1)				20,000	LSB
Zero Error		IN HI = 0V	T_A = +25°C			±2	LSB
			T_A = T_{MIN} to T_{MAX}			±5	
Nonlinearity		(Notes 2, 3)	T_A = +25°C			±2	LSB
Rollover Error		(Note 4)	T_A = +25°C			±3	LSB
			T_A = T_{MIN} to T_{MAX}			±10	
Conversion Time				63			ms
Input Voltage Range		IN HI to IN LO				±300	mV
Leakage Current		IN HI, IN LO	T_A = +25°C			±10	pA
			T_A = T_{MIN} to T_{MAX}			±250	
Common-Mode Rejection		IN HI = IN LO	V_{CM} = ±500mV		3	10	LSB
			V_{CM} = ±3.5V		70		
Common-Mode Range		IN HI = IN LO			±3.5		V
Read Zero (50Hz/60Hz Range)						±2000	LSB
Noise (Zero-Reading Mode)		IN HI = IN LO	T_A = +25°C		1.5		LSB
Zero-Reading Drift		(Note 3)				0.1	LSB/°C
Scale Factor Temp. Coefficient		(Note 3)				5	ppm/°C
Positive Supply Rejection		FS change, V- = -5.0V, +4.5V ≤ V+ ≤ +5.5V				2	LSB
Negative Supply Rejection		FS change, V+ = 5.0V, -4.5V ≤ V- ≤ -5.5V				2	LSB
Positive Supply Current	I+				60	125	µA
Negative Supply Current	I-				-35	-65	
Digital Ground Supply Current					-25	-60	
Positive Sleep-Mode Current	I+				4	10	µA
Negative Sleep-Mode Current	I-				-4	-10	
Digital Ground Sleep-Mode Current					0	±2	

ELECTRICAL CHARACTERISTICS (continued)

(V+ = 5V, V- = -5V, DGND = AGND = IN LO = REF- = 0V, REF+ = 545mV, RINT = 402kΩ, CINT = 0.0047µF, CREF = 0.1µF, f_{CLK} = 32,768Hz, 60Hz mode, T_A = T_{MIN} to T_{MAX}, unless otherwise noted.)

PARAMETER	SYMBOL	CONDITIONS	MIN	TYP	MAX	UNITS
DIGITAL SECTION						
Output High	V_{OH}	D0-D7, I_{OUT} = -1mA	3.5	4.3		V
		DO-D7, I_{OUT} = -100µA	4.0	4.5		
		EOC, I_{OUT} = -100µA	4.0			
Output Low	V_{OL}	D0-D7, I_{OUT} = 1.6mA		0.2	0.4	V
		EOC, I_{OUT} = 100µA			0.4	
Input High	V_{IH}	Referred to DGND, $\overline{CS}$, $\overline{WR}$, $\overline{RD}$	2.4			V
Input Low	V_{IL}	Referred to DGND, $\overline{CS}$, $\overline{WR}$, $\overline{RD}$			0.8	V
Input Current	I_{IN}	$\overline{CS}$, $\overline{WR}$, $\overline{RD}$, D0-D7 when three-stated		±10	±500	nA
Input Capacitance	C_{IN}	$\overline{CS}$, $\overline{WR}$, $\overline{RD}$, D0-D7 when three-stated		5		pF

TIMING CHARACTERISTICS

(Test circuit of Figures 1 and 2, V+ = 5V, V- = -5V, DGND = AGND = 0V, T_A = +25°C, unless otherwise noted.) (Note 3)

PARAMETER	SYMBOL	CONDITIONS	MIN	TYP	MAX	UNITS
$\overline{CS}$ to $\overline{WR}$ Setup Time	t_1		0			ns
$\overline{WR}$ Data-Setup Time	t_2		200	<100		ns
$\overline{WR}$ Pulse Width	t_3		200	<100		ns
Data Hold after $\overline{WR}$	t_4		0			ns
$\overline{CS}$ to $\overline{RD}$ Setup Time	t_5		0			ns
$\overline{CS}$ to $\overline{RD}$ Hold Time	t_6		0			ns
$\overline{RD}$ to Data Valid	t_7		480	240		ns
Bus-Relinquish Time	t_8		380	190		ns
$\overline{WR}$ to $\overline{RD}$	t_9		300			ns
$\overline{RD}$ to $\overline{WR}$	t_{10}		200			ns
Delay between Write Operations	t_{11}		500	<250		ns

Note 1: 18-bit resolution achieved by averaging multiple conversions.
Note 2: Max deviation from best straight line fit, 1LSB = 15.63µV.
Note 3: Guaranteed by design, not tested.
Note 4: Difference in reading for equal positive and negative inputs near full scale.

15-Bit ADC with Parallel Interface

Typical Operating Characteristic

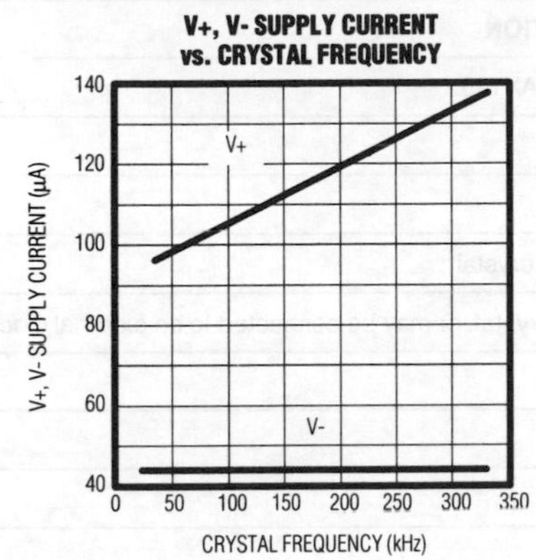

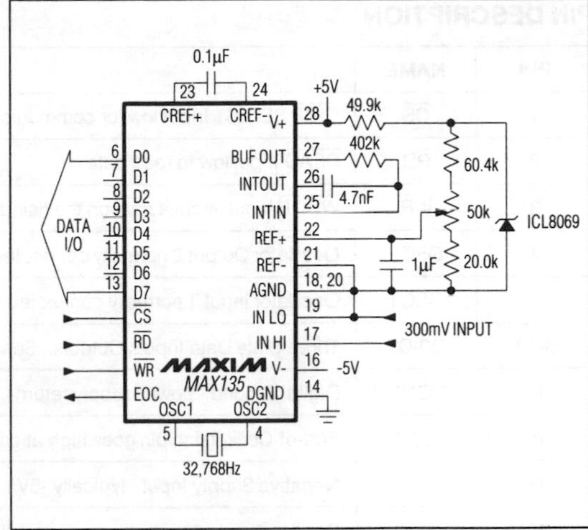

Figure 1. Test and Typical Application Circuit

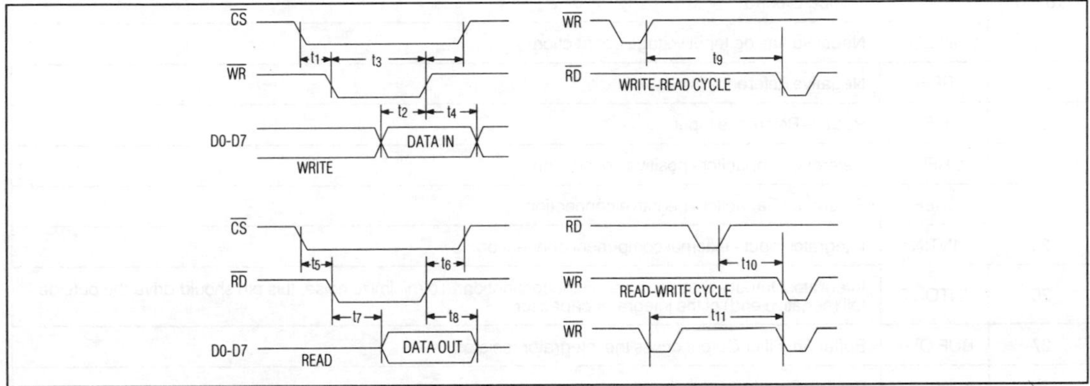

Figure 2. Parallel-Mode Timing

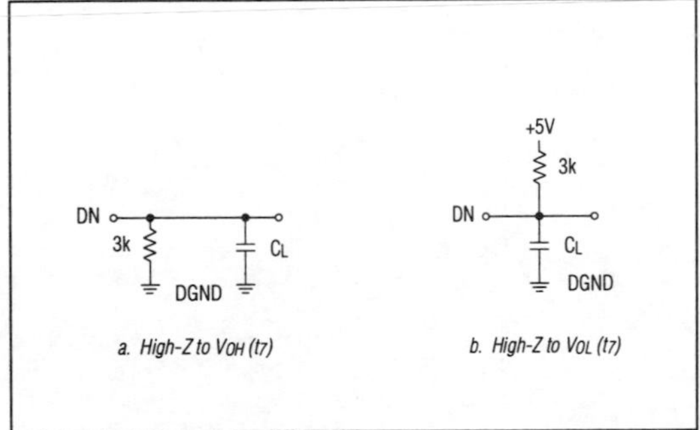

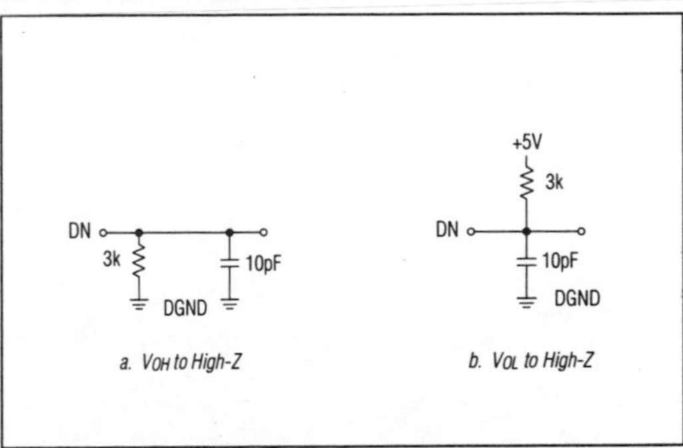

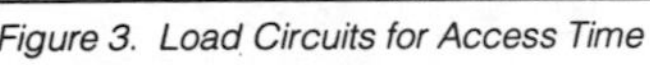

Figure 3. Load Circuits for Access Time

Figure 4. Load Circuits for Bus Relinquish Time

PIN DESCRIPTION

PIN	NAME	FUNCTION
1	$\overline{CS}$	$\overline{\text{CHIP SELECT}}$ Input low for communication with the MAX135.
2	$\overline{RD}$	$\overline{\text{READ}}$ Input low to read data.
3	$\overline{WR}$	$\overline{\text{WRITE}}$ Input latches data on the rising edge.
4	OSC2	Oscillator Output 2 normally connected to a 32,768Hz crystal.
5	OSC1	Oscillator Input 1 normally connected to a 32,768Hz crystal, or may be connected to an external clock.
6-13	D0-D7	Three-State Data Inputs/Outputs. See Tables 1 and 2.
14	DGND	Digital Ground - power-supply return
15	EOC	End-of-Conversion pin goes high at conversion end.
16	V-	Negative Supply Input - typically -5V
17	IN HI	Positive Analog Input voltage connection
18, 20	AGND	Analog Ground
19	IN LO	Negative Analog Input voltage connection
21	REF-	Negative Reference Input
22	REF+	Positive Reference Input
23	CREF+	Reference Capacitor - positive connection
24	CREF-	Reference Capacitor - negative connection
25	INTIN	Integrator Input - external component connection
26	INTOUT	Integrator Output - external component connection. To minimize noise, this pin should drive the outside foil (negative end) of the integrator capacitor.
27	BUF OUT	Buffer Amplifier Output drives the integrator resistor.
28	V+	Positive Supply, +5V

15-Bit ADC with Parallel Interface

Application Information

Figure 1 shows the basic MAX135 application circuit. The component values are selected for 16 conversions per second. Keep the analog lines and components away from the digital input/output (I/O) lines to prevent capacitive coupling to the analog circuitry.

Increasing Speed

Applications with greater conversion rates require different components; the *Components* section describes how to determine the correct values. Ground planes and shielding are essential for stable readings on faster conversion rates. When operating at crystal frequencies specified in Table 1 that are greater than 32,768Hz, use either the 50Hz or 60Hz mode; however, the 50Hz mode will improve performance due to a longer integration period. When operating in 50Hz mode, both the integrator capacitor and the reference voltage must be appropriately selected. For fast conversion rates where resolution is not important, use fewer digits as a quick solution. If maximum resolution is important, a trailing average of the data will provide highest resolution.

Increasing Resolution

For applications where resolution is important and speed is not, achieve additional resolution by using bits LSB/2, LSB/4, and LSB/8 found in the status register. When averaged, these bits can yield up to ±18-bit resolutions. For maximum resolution, use a trailing average of at least 100 readings.

Components

The MAX135 requires an integrator resistor (RINT) and capacitor (CINT), a reference capacitor (CREF), and a crystal. A 32,768Hz crystal frequency is used to test the MAX135. The crystal frequency, reference voltage, and integrator current dictate the values of RINT and CINT.

Integrator Resistor

The integrator resistor sets the maximum integrator output current for the integrate phase. A 402kΩ metal-film resistor is recommended for use with reference voltages between 345mV and 655mV. Best linearity is achieved when the integration current (IINT) does not exceed 2.5µA. For other reference voltages, select RINT as follows:

$$RINT = \frac{VREF}{2.5\mu A < IINT < 0.5\mu A} \qquad IINT = \frac{VREF}{RINT}$$

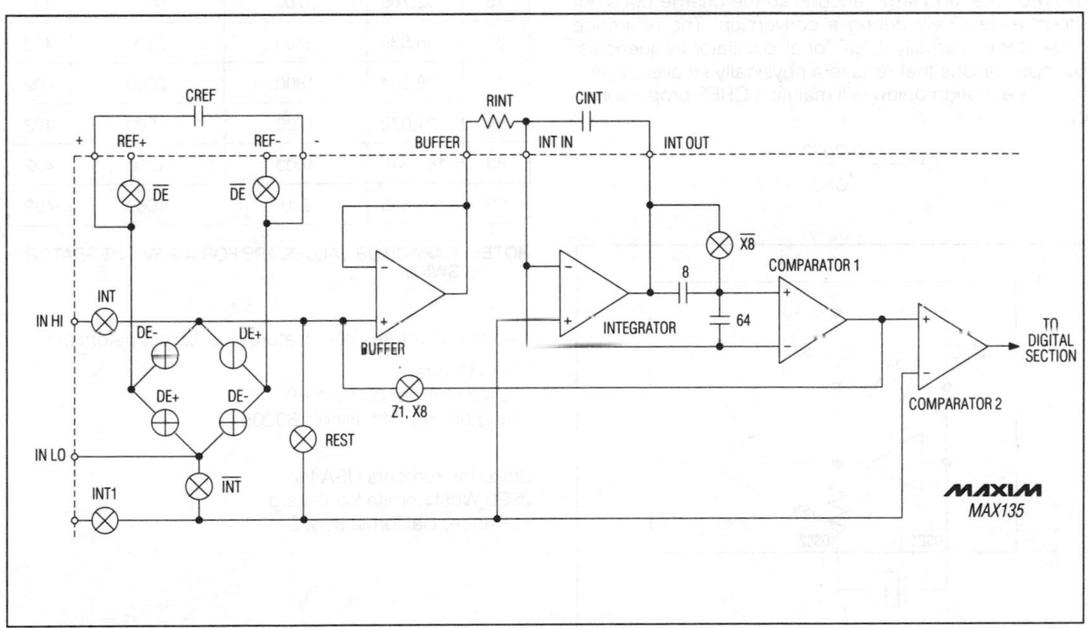

Figure 5. Analog Section Block Diagram

15-Bit ADC with Parallel Interface

Integrator Capacitor

The oscillator frequency, integrator resistor, and integrator capacitor set the maximum integrator output-voltage swing for full-scale reading. The voltage swing is about 3V for a 402kΩ integrator resistor and a 4.7nF integrator capacitor when the clock frequency is 32,768Hz. If different clock frequencies are used, select CINT using the following equations:

$$\text{tINT} = \left(\frac{1}{f_{OSC}}\right) \times \begin{pmatrix} 545 \text{ for } 60\text{Hz} \\ \text{or} \\ 655 \text{ for } 50\text{Hz} \end{pmatrix}$$

$$\text{CINT} = \frac{(V_{IN(FS)}/\text{RINT}) \times \text{tINT}}{3.5\text{V} > V_{SWING} > 1\text{V}}$$

The integrator capacitor's dielectric absorption directly affects integral nonlinearity. High-quality metal-film capacitors are recommended in the following order of preference: polypropylene, polystyrene, polycarbonate, and polyester (Mylar). A polyester capacitor will generate some integral nonlinearity.

Reference Capacitor

The reference capacitor value must be small enough to fully charge from a discharged state on power-up in the desired time, and large enough so the charge does not droop excessively during a conversion. The reference capacitor is normally 0.1μF for all oscillator frequencies. For applications that require a physically smaller capacitor, the equation below will maintain CREF proportionality.

$$\text{CREF} = \frac{0.0033}{f_{OSC}}$$

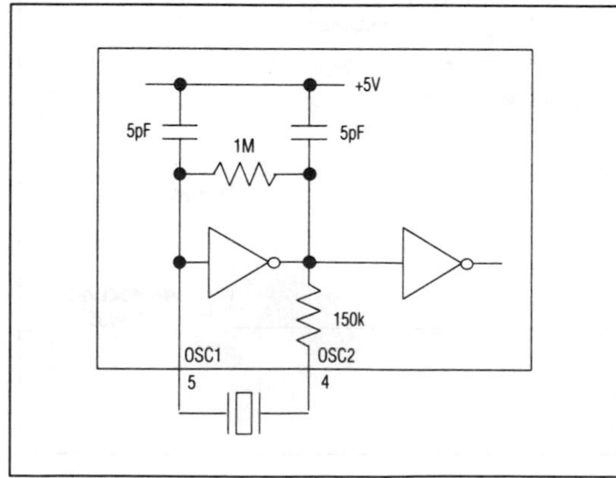

Figure 6. MAX135 Internal Oscillator Drive Circuitry

The reference capacitor must have low leakage, since it stores the reference voltage while floating during the deintegrate phase. Any leakage or charge loss during this phase changes the scale factor. Polypropylene, polystyrene, polycarbonate, and polyester metal-film capacitors are recommended, in this order, for low-leakage characteristics.

Crystal Frequency

The crystal frequency sets the conversion rate . Table 1 lists the crystal frequencies and integrator capacitors for 50Hz and 60Hz operation at particular conversion rates. These crystal frequencies complete a conversion in an integral number of line cycles. Figure 6 shows the internal oscillator drive circuitry used with external crystals.

Table 1. Crystal Frequencies and Integrator Capacitors for 50Hz to 60Hz Operation

Conv/ Sec	Hz	CINT/60Hz (pF)	CINT/50Hz (pF)	R kΩ
16	32,768	4700	6800	402
32	65,536	2700	3300	402
48	98,304	1800	2000	402
64	131,072	1200	1500	402
80	163,840	1000	1200	402
96	196,608	820	1000	402

NOTE: CAPACITOR VALUES ARE FOR A 2.5V INTEGRATOR SWING.

Two manufactures of miniature quartz resonators are:

Micro Crystal
702 West Algonquin Road
Arlington Heights, Illinois 60005

Seiko Instruments USA Inc.
2990 West Lomita Boulevard
Torrance, California 90505

15-Bit ADC with Parallel Interface

Reference Voltage Selection

For 300mV full scale, when the 60Hz mode is selected, the reference voltage is 523mV. If 50Hz is selected, the reference voltage is 629mV for 300mV full scale.

$$\text{Volts / LSB} = \frac{V_{IN(FS)}}{20{,}000}$$

$$VREF = \frac{(545 \text{ counts}) \times 64 \times V_{IN(FS)}}{20{,}000} \text{ 60Hz operation}$$

or

$$VREF = \frac{(655 \text{ counts}) \times 64 \times V_{IN(FS)}}{20{,}000} \text{ 50Hz operation}$$

Note: $V_{IN(FS)}$ = full-scale input voltage.

The ICL8069 is a 1.25V 50µA supply current reference, making it idealy suited for generating the MAX135's reference. Figure 7 shows how 1.25V can be divided for the desired reference voltage.

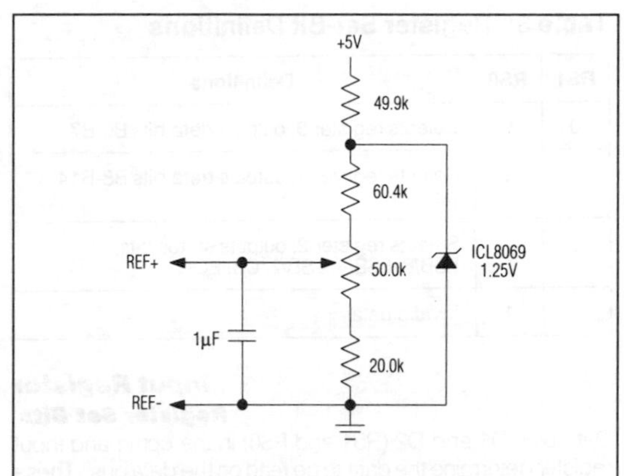

Figure 7. Dividing an ICL8069 to generate the MAX135's reference voltage.

Digital Interface

The MAX135 implements an 8-bit, bidirectional data bus with CHIP SELECT ($\overline{CS}$), READ ($\overline{RD}$), and WRITE ($\overline{WR}$). $\overline{CS}$ allows access to the I/O data bits and control lines. $\overline{WR}$ latches data into the command input register. $\overline{RD}$ sets the MAX135 data I/O lines to drive the data bus.

Initialize the ADC immediately after power-up to insure correct operation.

The MAX135 has four internal registers: the command input register, output register 0, output register 1, and the output status register. Table 2 defines their bit locations. Use $\overline{WR}$ to write to the command input register. Once the register bits RS0 and RS1 are set to the desired register address and the $\overline{WR}$ command is initiated, the designated register can be read. When $\overline{RD}$ is low, the designated register data is available on the data bus. When $\overline{RD}$ is high, the outputs go three-state and all I/O output lines return to input lines.

Table 2: MAX135 Register Map of Input and Output Data

		Data Bit							
		D7	**D6**	**D5**	**D4**	**D3**	**D2**	**D1**	**D0**
Command Input Register	1	Start Convert	50Hz Mode	Sleep Mode	Read Zero	Don't Care	RS0	RS1	Don't Care
	0	Returns to 0 at EOC	60Hz Mode	Run	Read IN HI	Don't Care	See Table 3		Don't Care
Output Register 0 RS1 = 0, RS0 = 0		B7	B6	B5	B4	B3	B2	B1	B0 LSB
Output Register 1 RS1 = 0, RS0 = 1	1	- Polarity	B14	B13	B12	B11	B10	B9	B8
	0	+ Polarity							
Output Status Register RS1 = 1, RS0 = 0	1	Collision	EOC	Integrating Input	Sleep	Always Low	Super LSBs		
	0	No Collision	Converting	Not Integrating	Run		LSB/2	LSB/4	LSB/8

15-Bit ADC with Parallel Interface

Table 3: Register Set-Bit Definitions

RS1	RS0	Definitions
0	0	Selects register 0; outputs data bits B0-B7
0	1	Selects register 1; outputs data bits B8-B14, polarity
1	0	Selects register 2; outputs status bits, LSB/8, LSB/4, LSB/2, CB, EOC
1	1	Invalid data

Input Register

Register Set Bits

Data pins D1 and D2 (RS1 and RS0) in the command input register determine the data to be read on the data bus. These bits select which register outputs data to the bus. Table 3 defines the bit values that determine the register in use.

Read-Zero Bit

The MAX135 performs a read-zero conversion on command - a calibration process that removes zero offset. The read-zero bit, when set to 1, internally shorts the inputs; when a start-conversion command is given, the zero error is converted. Subtract the results from the standard external measurement conversion when the read-zero conversion ends. If the read-zero bit is set to 0, the converter measures the voltage between IN HI and IN LO once a start bit is given.

An average of multiple read-zero measurements determines the most accurate read zero.

Sleep Bit

With the sleep bit set to 1 and 1 written to D5, the low-power sleep mode starts when EOC = 1. In sleep mode, the supply current is typically under 5µA, the oscillator shuts down, and data can be read. When sleep mode is released, the analog circuitry needs time to stabilize before the next conversion starts. Accomplish this by writing a separate instruction to emerge from sleep mode, and waiting at least one conversion cycle before writing a start instruction.

50Hz/60Hz

With a 32,768Hz crystal, the 50Hz/60Hz bit sets the integration period equal to one line cycle for 50Hz/60Hz environments. When D6 is set to 0, the integrator count is an integer multiple of 60Hz (32,768Hz/60Hz = 546 counts). When D6 is set to 1, the integrator count is an integer multiple of 50Hz (32,768Hz/50Hz = 655 counts). Achieve the highest AC rejection by adjusting the integration period for 50Hz or 60Hz.

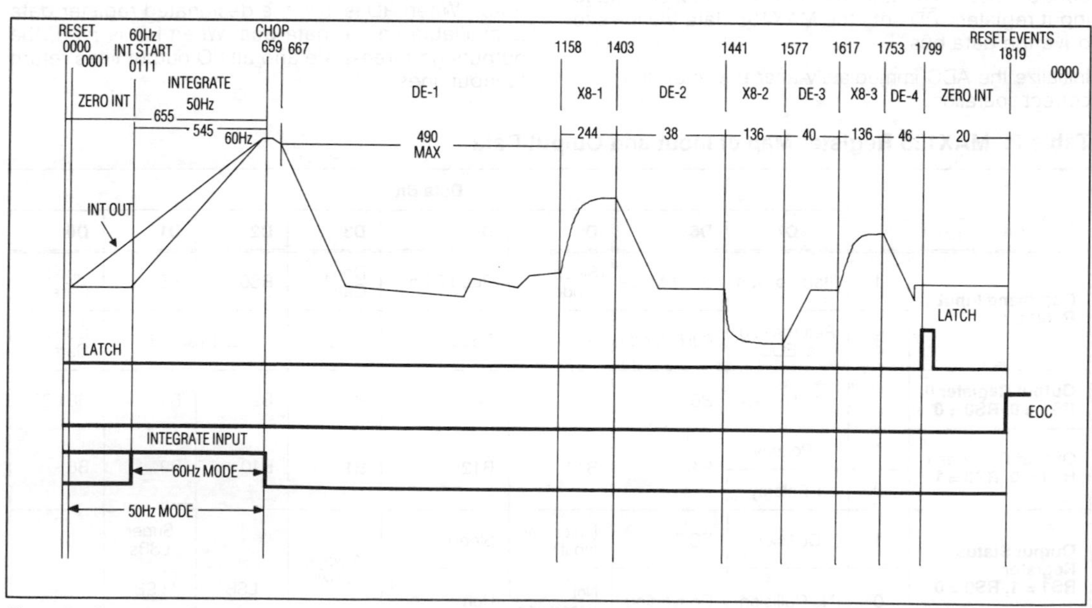

Figure 8. Conversion Timing

15-Bit ADC with Parallel Interface

Start Bit

The start bit initiates a conversion when set to 1 in the command input register. The MAX135 immediately starts a conversion, stops at conversion end, and waits for the next start-bit command. Send a start instruction to initiate each conversion.

To initiate a continuous data stream, write a separate start command for each conversion in one of three ways:

1. Wait longer than a known conversion time and write another start command.
2. Poll either the EOC status register bit or the EOC line to determine conversion end and start time for the next conversion. EOC becomes 1 at conversion end at count 0000 of the conversion counter (Figure 8).
3. Set the start bit to 1 before conversion end. The internal conversion counter is then checked for its count. If the count is 0000 (EOC = 1), a new conversion starts and the conversion counter is set to 0001. The start bit resets to 0 after 5 clock cycles. The MAX135 will not check the start bit again until the conversion counter returns to a 0000 count. This means a start command can be given any time after the 0005 internal conversion count; the next conversion starts when the counter returns to 0000.

Output Registers

Register 0

Register 0 contains the low-byte (bits B0-B7) conversion data. New data is available after EOC goes high. Access register 0 by setting RS0 and RS1 to 0. Output data is the sum of system offset (read zero) plus the results of the external input voltage measurement.

Register 1

Register 1 contains the high-byte (bits B8-B15) data. Data is in a twos-complement format, where the polarity bit is a 1 for negative polarity data. Access register 1 by setting control bits RS0 = 1 and RS1 = 0 when writing to the command input register.

Status Register

LSB/2, LSB/4, LSB/8 Bits (Super LSBs)

The LSB/2, LSB/4, and LSB/8 bits enhance resolution. At each conversion end they are updated and read back from the status register. When using these three bits, 18 bits of resolution are available. When using the 17 bits plus sign, average the readings to stabilize the result.

Integrate Bit

The integrate (INT) bit is set to 1 at the beginning of an integration and becomes 0 at the end. Poll INT to determine the earliest time the analog input can be changed without affecting the conversion.

End-of-Conversion Bit

The end-of-conversion (EOC) bit signals conversion status. If EOC is 1, the conversion is complete and the ADC waits in zero-integrate mode at count 0000 for the next start instruction. A conversion cycle has 1820 counts. EOC becomes 1 at count 0000 and 0 at count 0001.

Collision Bit

The collision bit warns the microprocessor that the register's data was changed during the read cycle. Once the status register is read, the collision bit resets to 0. Collisions do not occur if a conversion's read cycle is completed before the next conversion begins.

Analog Section

Sequence Counter and Results Counter

A binary sequence counter controls the sequencing or timing of the conversion phases. In integrate phase, both start and stop occur at preset counts. The deintegrate phases start at predetermined counts, and terminate when the comparator detects zero crossing at the integrator output.

The results counter accumulates counts during all deintegrate phases. It is an up/down binary counter, with the deintegrate polarity determining count direction. In the first deintegrate phase, the results counter counts by 64s. Since the second deintegrate phase deintegrates a residual voltage multiplied by 8, the results counter increments or decrements by 8s during this phase. It increments or decrements by 1s during the third deintegrate phase, and by 1/8s during the fourth deintegrate phase. The results counter's contents transfers to the results register at each conversion end.

Differential Reference Inputs

The reference inputs accept voltages anywhere within the converter's power-supply voltage range.

The main source of rollover error is common-mode voltage, which is caused by the reference capacitor losing or gaining charge to stray capacitance. A positive signal with a large common-mode voltage can cause the reference capacitor to gain charge (increase voltage). In contrast, the reference capacitor will lose charge (decrease voltage) when deintegrating a negative input signal. The rollover error is a direct result of the difference in reference for positive or negative input voltages. Use an optimum reference capacitor to hold rollover error under one-half count for worst-case conditions (see *Components* section). A common-mode voltage near or at AGND minimizes rollover error caused by these sources.

15-Bit ADC with Parallel Interface

Differential Input

Acceptable differential input voltages are dictated by the input amplifier's common-mode range (specifically from 1.5V below the positive supply to 1.5V above the negative supply). For optimum performance, the input voltage at IN HI and IN LO should not come within 2V of either the positive or negative supply. Do not saturate the integrator output, since the integrator also swings with the common-mode voltage.

Conversion Phases

For this section of an explanation of conversion phases refer to figures 5, and 8.

Integrate Phase

The MAX135 integrates the input signal by connecting the noninverting input of the integrator to IN LO and the buffer input to IN HI. The integration period is 545 counts for 60Hz mode and 655 counts for 50Hz mode.

Deintegrate Phase

The voltage polarity on the integrator capacitor at the end of integrate phase determines the polarity of the first deintegration phase. The first deintegrate phase ends when the comparator detects the integrator capacitor discharge. The MAX135 then goes into a rest phase, where both the buffer input and the integrator's noninverting input connect to AGND, integrating the system offset.

Near the end of the maximum allowable deintegration period, the integrator capacitor's voltage polarity is again sampled, resulting in either a positive or negative deintegrate cycle.

Rest Phase

A rest phase follows each deintegrate phase. Rest phase starts when the integrator crosses zero and ends when the maximum count for that deintegrate phase is reached.

Times-Eight Phase

When a zero crossing is detected at the end of the deintegrate phase , deintegration continues until the next clock cycle. This causes the integrator to overshoot zero crossing slightly, leaving a small residual voltage on the integrator capacitor. The times-eight (X8) phase inverts and multiplies this residual by a factor of 8.

Second Deintegrate Phase

The second deintegrate phase deintegrates the residual voltage on the integrator capacitor that has been through the X8 phase. Since the voltage across the integrator capacitor has been multiplied by 8, each clock cycle of deintegration corresponds to 1/8 of one clock cycle during the first deintegration.

Additional Times-Eight and Deintegrate Phases

At the end of the second and third deintegrate phases, the device X8 multiplies the residual voltage left on the integrator capacitor. A deintegration occurs after each of these X8 multiplications, resulting in a second, third, and fourth deintegrate phase. Each time the integrator capacitor's residual voltage is multiplied by 8, the following deintegrate phase has 8X finer resolution.

Zero-Integrate Phase

The zero-integrate phase zeros out the integrator to prepare for the next integration (Figure 8). This phase occurs at the beginning and end of each conversion. At power-up or in the hold mode prior to a conversion, the MAX135 continues to zero integrate until a conversion starts. When a conversion starts in 60Hz mode, another 111 clocks of zero integrate are completed before a conversion begins. In 50Hz mode, only one additional zero integrate is performed before the conversion starts. An additional 20 clocks of zero integrate occur at each conversion end.

Chip Topography

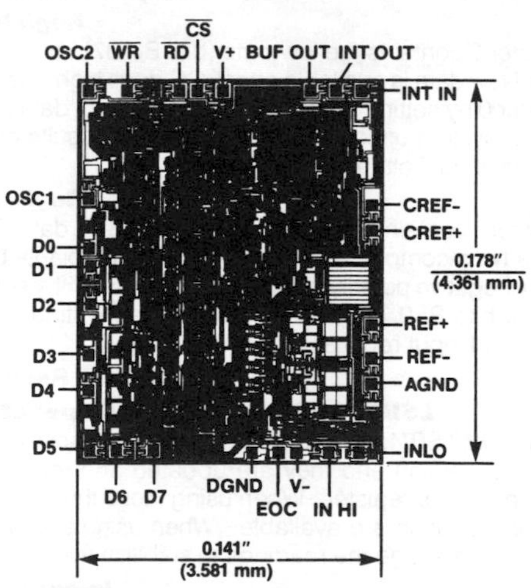

MOTOROLA
Semiconductors
Avenue Général-Eisenhower 31023 Toulouse CEDEX – FRANCE

MC10315L
MC10317L

Advance Information

HIGH SPEED 7-BIT ANALOG-TO-DIGITAL FLASH CONVERTER

SILICON MONOLITHIC INTEGRATED CIRCUIT

SEVEN-BIT PARALLEL HIGH SPEED A/D CONVERTER (WITH OVERRANGE)

The MC10315L/MC10317L is a 7-bit high speed parallel A/D converter which employs ECL processing. The device consists of 128 parallel latched comparators across a high quality input reference network. The 128 comparator outputs are then fed to a 128-to-7 encoder and latched to the outputs which are ECL compatible. An overrange bit is provided to allow overrange sensing, or to facilitate the connection of an MC10315L and MC10317L in parallel to produce an 8-bit A/D converter. The MC10315L and MC10317L are identical devices except for the method of overranging used, which simplifies the utilization of two 7-bit converters to produce an 8-bit conversion. (See ordering information and technical description.)

Applications include video display and radar signal processing, high speed instrumentation, and TV broadcast video encoding.

- 7-Bit Resolution/8-Bit Accuracy Plus Overrange
- Direct Interconnection for 8-Bit Conversion
- 15 MHz Sampling Rate
- Wide Range of Input Voltage: ±2.0 Volts
- Low Input Capacitance: ≤70 pF
- 1.2 Watt Power Dissipation
- No Sample and Hold Required for Video Bandwidth Signals
- Standard 24-Pin Package

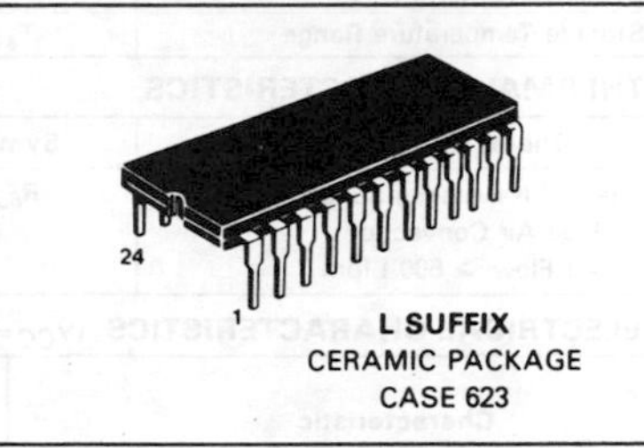

L SUFFIX
CERAMIC PACKAGE
CASE 623

MC10315L/MC10317L DEVICE/APPLICATION CONFIGURATION

(11) A_{in} · (9) CLK · (10,24) V_{CC}

(5) V_{RT} · (3) V_{RTac} · (6) V_{RM} · (8) V_{RBac} · (7) V_{RB}

128 Latched Comp. → 128 to 7 Encode → Output Latches

$\overline{OVR}$ (14) · OVR (15) · D6 (16) · D0 (22)

Gnd2 (1,13) · Gnd1 (23) · V_{EE} (2,12)

PIN DIAGRAM

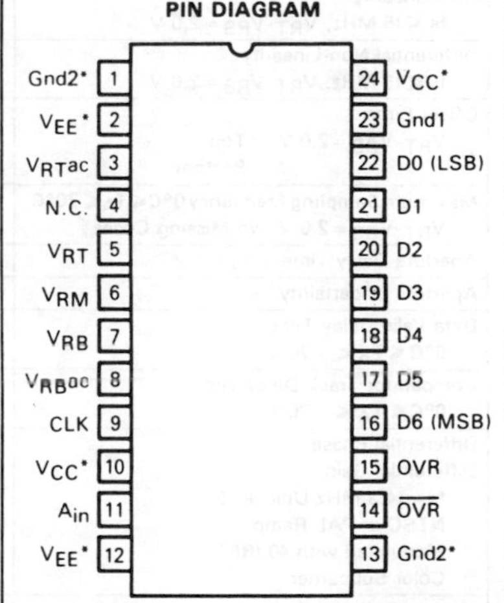

*V_{CC}, V_{EE} and Gnd2 are each available on two pins. Interconnections for the respective function are made on chip. To minimize I•R drops on chip and in the bonding wires, utilization of both pins for each function is recommended.

ORDERING INFORMATION**

Device	Overrange Function		
	Analog Input Condition	Logic Levels	
		OVR Bit	D0–D6 Bits
MC10315L	Overranged	High	High
MC10317L	Overranged	High	Low

**For information regarding an evaluation board, contact Linear Marketing.

ADI-654R1

This document contains information on a new product. Specifications and information herein are subject to change without notice.

MAXIMUM RATINGS (T_A = 25°C unless otherwise noted [Note 1])

Rating	Symbol	Value	Unit
Supply Voltage	V_{CC} V_{EE}	+7.0 −8.0	Vdc Vdc
Ground 1	Gnd1	−0.8, +3.0	Volts
Clock Input Voltage	V_{CLK}	0 to V_{EE}	Volts
Analog Inputs: A_{in}, V_{RT}, V_{RB} $\|V_{RT} - V_{RB}\|$		 ±2.5 2.5	Volts
Digital Output Source Current (per Output)	I_{source}	30	mA
Power Dissipation Free Air Convection Air Flow ≥ 500 Lfpm	$P_{D(max)}$	 2.8 4.0	W
Operating Temperature	T_A	0 to +70	°C
Junction Temperature	T_J	165	°C
Storage Temperature Range	T_{stg}	−65 to +150	°C

THERMAL CHARACTERISTICS

Characteristic	Symbol	Max	Unit
Thermal Resistance Junction to Air Free Air Convection Air Flow ≥ 500 Lfpm	$R_{\theta JA}$	 50 35	°C/W

ELECTRICAL CHARACTERISTICS (V_{CC} = +5.0 Vdc, V_{EE} = −5.2 Vdc, T_A = 25°C unless otherwise noted [Note 1])

Characteristic	Symbol	MC10315L/MC10317L Min	Typ	Max	Unit
Resolution 0°C ≤ T_A ≤ 70°C		–	–	7	Bits
Non-Linearity fs ≤ 15 MHz, V_{RT}-V_{RB} = 2.0 V	NL	–	±0.16	–	%
Differential Non-Linearity fs ≤ 15 MHz, V_{RT}-V_{RB} = 2.0 V	DNL	–	±0.10	–	%
Offset Error V_{RT}-V_{RB} = 2.0 V Top Bottom	 V_{OSRT} V_{OSRB}	 – –	 ±7.0 ±7.0	 – –	mV
Maximum Sampling Frequency 0°C ≤ T_A ≤ 70°C V_{RT}-V_{RB} = 2.0 V, No Missing Codes	fs	14.5	15	–	MHz
Aperture Delay Time	t_{ad}	–	3.0	–	ns
Aperture Uncertainty		–	80	–	ps
Data Valid Delay Time 0°C ≤ T_A ≤ +70°C	t_{vd}	–	–	43	ns
Comparator Track Delay Time 0°C ≤ T_A ≤ +70°C	t_{cd}	–	–	25	ns
Differential Phase Differential Gain fs = 14.3 MHz Unlocked NTSC or PAL Ramp Modulated with 40 IRE Color Subcarrier		– –	1.0 1.5	– –	Deg. %
Maximum Analog Input Slew Rate	SR	–	50	–	V/μs
Analog Input Bias Current V_{in} ≥ V_{RT}, 0°C ≤ T_A ≤ +70°C	I_{IB}	–	300	400	μA
Equivalent Analog Input Resistance V_{RT}-V_{RB} = 2.0 V, 0°C ≤ T_A ≤ +70°C	R_{in}	5.0	–	–	KΩ
Analog Input Capacitance V_{in} ≥ V_{RT}	C_{in}	–	70	–	pF
Reference Ladder Current V_{RT}-V_{RB} = 2.0 V	I_{ref}	24	31	47	mA
Reference Ladder Resistance (Total Resistance)	R_{ref}	42	64	83	Ω

ELECTRICAL CHARACTERISTICS (V_{CC} = +5.0 Vdc, V_{EE} = −5.2 Vdc, T_A = 25°C unless otherwise noted [Note 1]) continued

Characteristic	Symbol	MC10315L/MC10317L Min	Typ	Max	Unit
Reference Ladder Resistance Temperature Coefficient $0°C \leq T_A \leq +70°C$	T_{CR}	—	0.37	—	%/°C
Clock Input Logic Levels, $0°C \leq T_A \leq +70°C$					V
High Logic State	V_{IH}	-1.145	—	—	
Low Logic State Note 2	V_{IL}		—	-1.455	
Clock Input Current					μA
High Logic State	I_{IH}	—	150	—	
Low Logic State	I_{IL}	—	100	—	
Digital Output Logic Levels					V
High Logic State	V_{OH}	−1.020	—	—	
Low Logic State $0°C \leq T_A \leq +70°C$ Note 2	V_{OL}	—	—	−1.605	
Power Supply Current, $0°C \leq T_A \leq +70°C$					mA
$4.75\ V \leq V_{CC} \leq 5.25\ V$	I_{CC}	—	118	150	
$-4.94\ V \geq V_{EE} \geq -5.46\ V$	I_{EE}	—	-110	-140	

RECOMMENDED OPERATING CONDITIONS (Note 1)

Characteristic	Symbol	MC10315L/MC10317L Min	NOMINAL	Max	Unit
Power Supply Voltages	V_{CC}	4.75	5.0	5.25	Vdc
	V_{EE}	−5.46	−5.2	−4.94	
Ground 1	Gnd1	−0.3	0	+1.0	V
Reference Input, Top	V_{RT}	−1.0	0	+2.0	V
Reference Input, Bottom	V_{RB}	−2.0	0	+1.0	V
Reference Input Voltage Range (V_{RT}-V_{RB})	V_{RR}	1.0	—	2.0	V
Convert Clock Pulse Width, High	t_{pwH}	44	—	—	ns
Convert Clock Pulse Width, Low	t_{pwL}	25	—	—	
Digital Output Current	I_{OH}	—	10	—	mA
Operating Temperature Range	T_A	0	—	70	°C

Notes:

1. All voltage levels referenced to Ground 2 (Gnd2) unless otherwise noted.
2. MECL 10K logic levels are designed to meet the dc specifications after thermal equilibrium has been established with a transverse airflow greater than 500 Linear fpm and V_{EE} = −5.2 V ± 0.010 V. All outputs are specified driving 50Ω to −2.0 V.

FIGURE 1 — TIMING DIAGRAM

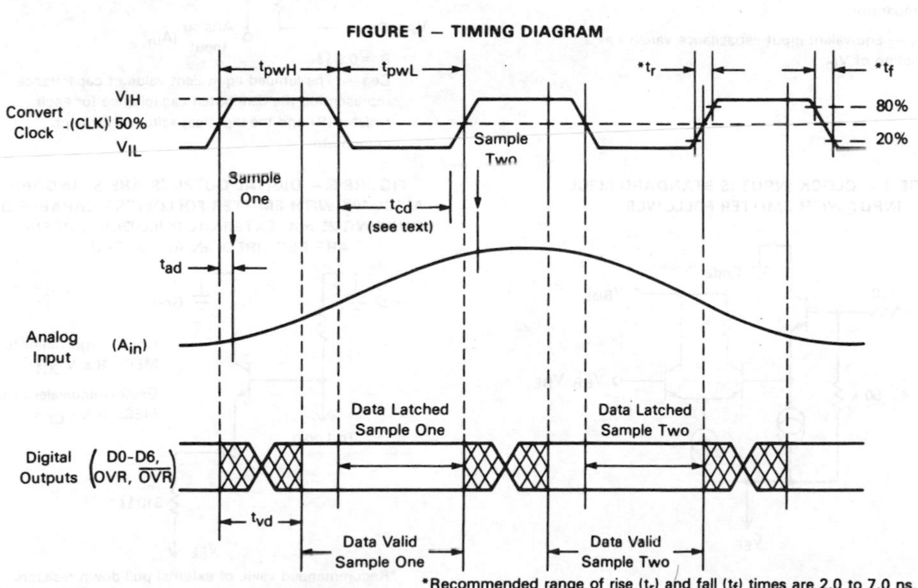

*Recommended range of rise (t_r) and fall (t_f) times are 2.0 to 7.0 ns.

FIGURE 2 — EQUIVALENT R_{in} AND C_{in} OF THE ANALOG INPUT

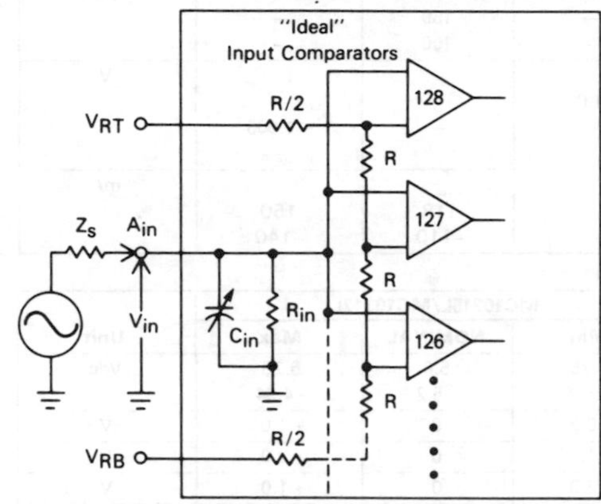

$$R_{in} \simeq \frac{|V_{RT}-V_{RB}|}{400\ \mu A}$$

$$^{*}C_{in} \simeq \frac{30pF}{|V_{RT}-V_{RB}|}\ |V_{in}-V_{RB}| + 40\ pF$$

*Valid for $V_{RT} \geq V_{in} \geq V_{RB}$

$R \simeq 0.5\,\Omega$

R_{in} — Effective input resistance representing the cumulative bias currents of the 128 input comparators.

C_{in} — Equivalent input capacitance variable as a function of V_{in}.

FIGURE 3 — EQUIVALENT CIRCUIT OF REFERENCE RESISTOR LADDER NETWORK

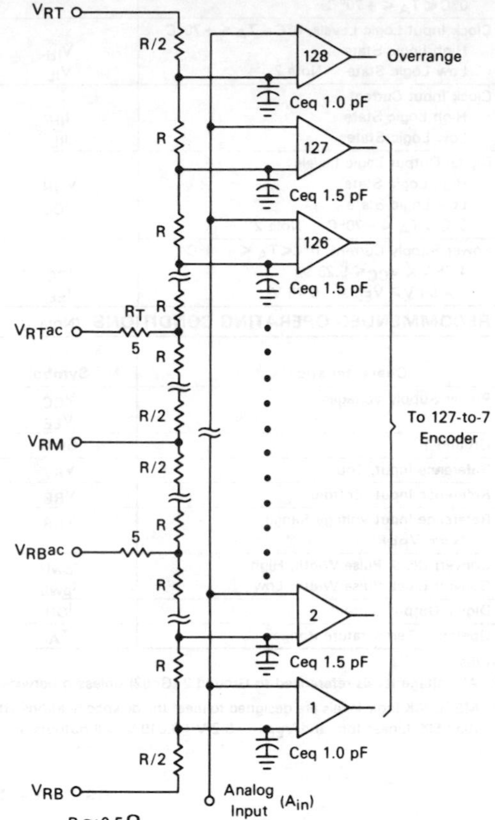

$R \simeq 0.5\,\Omega$

Ceq — The lumped equivalent value of capacitance representing the distributed capacitance for each resistor (R) and the input capacitance for each comparator.

FIGURE 4 — CLOCK INPUT IS STANDARD MECL INPUT WITH EMITTER FOLLOWER

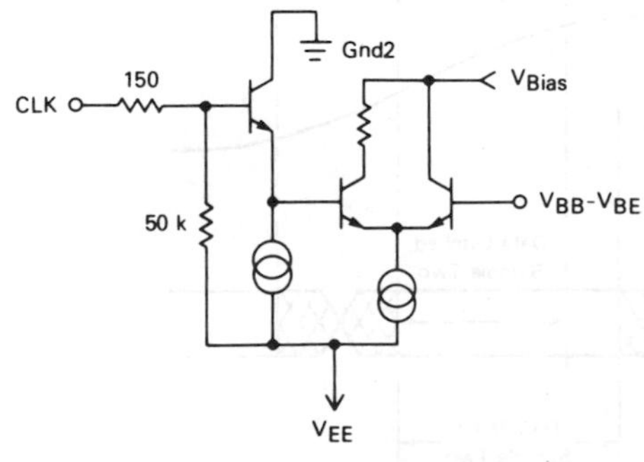

FIGURE 5 — DIGITAL OUTPUTS ARE STANDARD MECL 10K WITH EMITTER FOLLOWERS CAPABLE OF SOURCING 25 mA. EXTERNAL PULL-DOWN RESISTORS ARE REQUIRED ON ALL OUTPUTS.

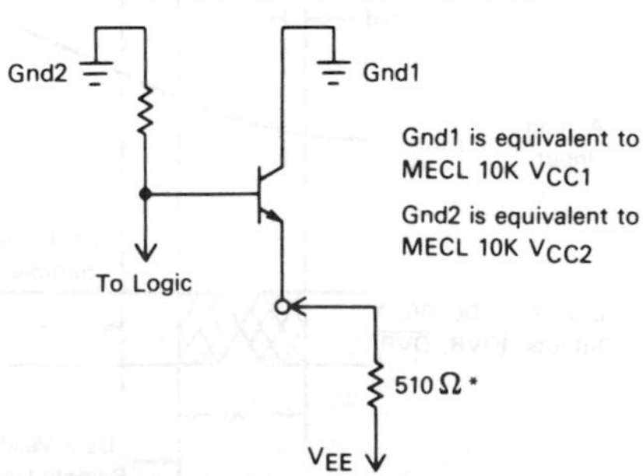

*Recommended value of external pull-down resistors for all outputs.

FIGURE 6 — OUTPUT CODING FOR THE MC10315L/MC10317L DEVICES*

Comparator Step	Analog Input Range (15.6 mV per LSB) −2.0 V to 0 V	0 V to 2.0 V	±1.0 V	MC10315L Data Bits (D0-D6)	MC10317L Data Bits (D0-D6)	Overrange Bit (OVR)	$\overline{\text{Overrange}}$ $\overline{\text{Bit}}$ (OVR)
000	−2.0000 V	+0.0000 V	−1.0000 V	0000000	0000000	0	1
001	−1.9922 V	+0.0078 V	−0.9922 V	0000001	0000001	0	
•	•	•	•	•	•		
•	•	•	•	•	•		
•	•	•	•	•	•		
063	−1.0234 V	+0.9766 V	−0.0234 V	0111111	0111111		
064	−1.0078 V	+0.9922 V	−0.0078 V	1000000	1000000		
065	−0.9922 V	+1.0078 V	+0.0078 V	1000001	1000001		
•	•	•	•	•	•		
•	•	•	•	•	•		
•	•	•	•	•	•		
126	−0.0391 V	+1.9609 V	+0.9609 V	1111110	1111110		
127	−0.0234 V	+1.9766 V	+0.9766 V	1111111	1111111		
128	−0.0078 V	+1.9922 V	+0.9922 V	1111111	0000000	1	0

*The MC10315L and MC10317L differ only in output coding at comparator step 128 where the device is overranged.

CIRCUIT DESCRIPTION

Conversion Timing

The MC10315L/MC10317L performs a conversion and outputs data within a single clock cycle. Referring to Figure 1 will indicate that the clock input is sensitive to the rising and falling clock edges. All significant operations are referenced to the edges. A rising clock edge holds the analog input by latching the (128) input comparators. The output latches are also released to toggle and update to the new digital value. The falling edge of the clock will latch the data outputs. Clock timing must be considered to ensure a valid conversion. With the rising edge of the clock, there will be an aperture delay (t_{ad}) which is the time from the threshold of the (50%) edge to the actual time the input comparators latch in the analog value of A_{in}. The data valid delay time (t_{vd}) is the time interval for valid data to appear at the outputs. t_{vd} is 43 ns from the rising clock edge. After this time, the clock can go low to latch the valid data at the outputs. The clock must remain low a minimum time before another rising edge in order for the input comparators to unlatch and begin to track. This comparator tracking delay time (t_{cd}) is 25 ns. After this minimum time, the conversion clock cycle is repeated, latching in a new analog input value.

The minimum recommended clock pulse width high time (t_{pwH}) is 44 ns and pulse width low time (t_{pwL}) is 25 ns for maximum recommended sample frequency (fs). Rise (t_r) and fall (t_f) time of the clock edges should be in the range from 2.0 to 7.0 ns to minimize the chance of clocking errors or uncertainty.

Analog Input (A_{in})

The dc current drive required by the analog input (A_{in}) is a function of the input voltage (V_{in}) and is directly attributable to the accumulation of input bias currents for each of the 128 comparators. When $V_{in} < V_{RB}$, the dc current is zero and when $V_{in} > V_{RT}$ the current is a maximum of 400 μA. Looking at this current as a function of V_{in} on a large signal basis, it will appear as a straight line approximation.

This input current loading on a driving source impedance can produce a dc gain error. Cancellation of this error is accomplished by utilizing an adjustable voltage reference at V_{RT}. If V_{RT} is tied to a fixed reference or grounded, the driving amplifiers offset can be adjusted. However, a zero error will now occur which can be cancelled by adjusting V_{RB}. Another method of reducing dc gain error due to analog input current is to use a driving amplifier with sufficiently low output impedance (Z_s). This can be determined by:

$$Z_s < \frac{\text{maximum gain error (V)}}{400\ \mu A}$$

i.e., with a 1.0 volt analog input range (V_{RT}-V_{RB} = 1.0 V), a 1/2 LSB of gain error = 3.9 mV

$$Z_s < \frac{3.9\text{ mV}}{400\ \mu A} < 9.76\ \Omega$$

Analog Input (A_{in}) (Continued)

The input capacitance (C_{in}) is also a function of input voltage (V_{in}). For $V_{in} < V_{RB}$, $C_{in} \simeq 40$ pF; $V_{in} > V_{RT}$, $C_{in} \simeq 70$ pF. The input capacitance on a large signal basis over the analog input range is a linear function. C_{in} can limit the analog input bandwidth if the driving source impedance is too great. This can introduce an ac gain error if the corner frequency f_c is not sufficiently extended from maximum input frequency (f_{in}).

For example, to keep the ac gain error to within 1/2 LSB of 7-bits, the corner frequency (f_c) of the effective single pole, low-pass filter created by the driving source impedance (Z_s) and the input capacitance (C_{in}) should be: $fc \geq 11.3\ f_{in}$

- C_{in} - analog input capacitance.
- f_{in} - maximum input frequency of A_{in}
- f_c - corner frequency determined by C_{in} and Z_s
- n - number of bits
- Z_s - driving source impedance of the analog input

For single Pole Filter: $$\frac{fc}{f_{in}} \geq \frac{1}{\sqrt{\dfrac{1}{\left(\dfrac{2^{n+1}-1}{2^{n+1}}\right)^2}-1}}$$

If measures have already been taken to keep dc gain error to within 3.9 mV (1/2 LSB for 1.0 V full scale) by providing a low Z_s as described earlier, the calculated $Z_s \leq 9.76\ \Omega$ will sufficiently extend the corner frequency of the input pole to $\simeq 233$ MHz.

Figure 2 illustrates the equivalent analog input in terms of an effective variable C_{in} and R_{in}.

Reference Inputs

As shown in Figure 3, a resistive (divider) ladder comprised of 128 matched resistors with a nominal value of 0.5 Ω each, provides a reference voltage to each of the 128 comparator inputs. Recommended range of reference voltage applied across the resistive ladder (V_{RT} to V_{RB}) is 1.0 volt to 2.0 volts. V_{RT} must be kept more positive than V_{RB}. V_{RT} must not exceed +2.5 volts above Gnd2 and V_{RB} must not become more negative than −2.5 volts below Gnd2. With 2.0 volts across the reference ladder (V_{RT}-V_{RB} = 2.0 V), the ladder network has a common-mode range capability about Gnd2, permitting analog input (A_{in}) ranging options such as ± 1.0 volt, 0 to − 2.0 volts and 0 to 2.0 volts. A minimum of 1.0 volt should be maintained across the ladder network to ensure linearity to 7-bits. Less than 1.0 volt will degrade linearity due to comparator offsets becoming a significant factor.

Additional taps on the reference ladder are pinned out, providing access to the middle (V_{RM}), 1/4 (V_{RB}ac) and 3/4 (V_{RT}ac) scale points. V_{RM} can be left open, but if ladder linearity adjustment is required, an appropriate reference voltage can be applied. The V_{RB}ac and V_{RT}ac pins are intended for ac bypassing if ladder noise presents a problem. Reference voltages can be applied to these pins if tighter ladder linearity is desired. If the reference ladder voltage is to be varied dynamically such as in an AGC application, ac bypassing of any of the reference taps would likely yield undesirable results.

Calibration is accomplished by adjusting V_{RB} and V_{RT} to set the first and 127th comparator thresholds to the desired voltages. If a 0 to −1.0 V input (A_{in}) range is desired, continuously strobe the convertor with −0.9961 V on the analog input, adjust V_{RB} for output toggling between codes 0000000 and 0000001. Then apply A_{in} = −0.0117 V and adjust V_{RT} for toggling between 1111110 and 1111111 (thresholds 126th and 127th). Rather than adjusting V_{RT}, it may be more convenient to connect V_{RT} to Gnd2 and adjust the driving amplifier offset control. V_{RB} can again be used as a gain adjust point to cancel the effects of using the offset control technique.

Application Information

8-bits of resolution and accuracy can be obtained by stacking two 7-bit converters and wire ORing the data outputs. Shown in Figure 7 is an MC10315L and MC10317L in an 8-bit A/D configuration. The circuit is quite straightforward with the analog input (A_{in}) for each converter tied together, forming a common input. The analog input range is negative unipolar with V_{RT} of the MC10315L grounded (Gnd2) or referenced very near ground. V_{RB} of the MC10315L is connected to V_{RT} of the MC10317L and referenced to $V_{REF}/2$ to ensure this node is midscale. Unit to unit variations in Reference Ladder Resistance of each device can shift this point if a reference is not used, causing linearity errors. Care should be taken when interconnecting V_{RB} and V_{RT} of the MC10315L and MC10317L respectively. Reference ladder current flowing through resistance of printed circuit board runs, sockets and even device pins and bonding wires can establish significant IR drops of several millivolts causing differential non-linearity errors at midscale. A negative reference of -2.000 V is applied to V_{RB} of the MC10317L. The remaining pins V_{RT}ac, V_{RB}ac and V_{RM} for both devices can be left open or be connected to additional external references if linearity improvements are required. V_{RT}ac and V_{RB}ac can also be used as ac decoupling points for the resistor ladders to reduce any transients which may exist due to current noise.

The clock (CLK) inputs are driven by a common clock. Depending on the input frequency to be encoded, it may be necessary to skew the rising clock edges to one of the devices to compensate for a slight difference in aperture delay time (t_{ad}) which may occur between the two devices.

Application Information (Continued)

The digital outputs, D0 through D6 are wire ORed. The overrange (OVR) bit of the MC10317L becomes the MSB for the 8-bit word in Binary coding.

The MC10315L and MC10317L differ only in the method of overranging (see Figure 6, output coding truth table). When the MC10317L input (A_{in}) is overranged, the overrange (OVR) bit goes high, all other bits (D0-D6) go low. This enables direct wire ORing of additional A/D outputs to expand to ≥8-bits. When the MC10315L is overranged, OVR goes high and the data bits (D0-D6) remain high. This device provides a true termination of a digital word when the system becomes overranged. Generally the MC10315L will be used in a 7-bit, stand-alone converter scheme, or as the upper scale A/D when stacking two or more devices to expand to ≥7-bits.

Pull-down resistors are required at the digital outputs. A recommended value of 510 Ω will provide proper output fall times in most applications and also hold down device power dissipation. The outputs are capable of sustaining MECL levels when terminated with a 50 Ω (to −2.0 V) characteristic load impedance to minimize reflections. Design rules for MECL 10K should be followed when using these devices.

Care must be taken in PC board ground layout to prevent digital ground currents from flowing through the analog ground. Separate grounds are provided on the MC10315L/MC10317L to help isolate the digital noise from the analog section of a system. Gnd1 is internally connected to only the collectors of the output emitter followers as shown in Figure 5. This provides a separate path for current transients of the switched output loads. All other internal circuitry is referenced to Gnd2.

Low and high frequency power supply bypassing should be provided physically close to the device, with V_{CC} and V_{EE} bypassed to Gnd2.

FIGURE 7 — CIRCUIT CONFIGURATION UTILIZING A MC10315L AND MC10317L A/D TO PERFORM A HIGH SPEED, 8-BIT CONVERSION

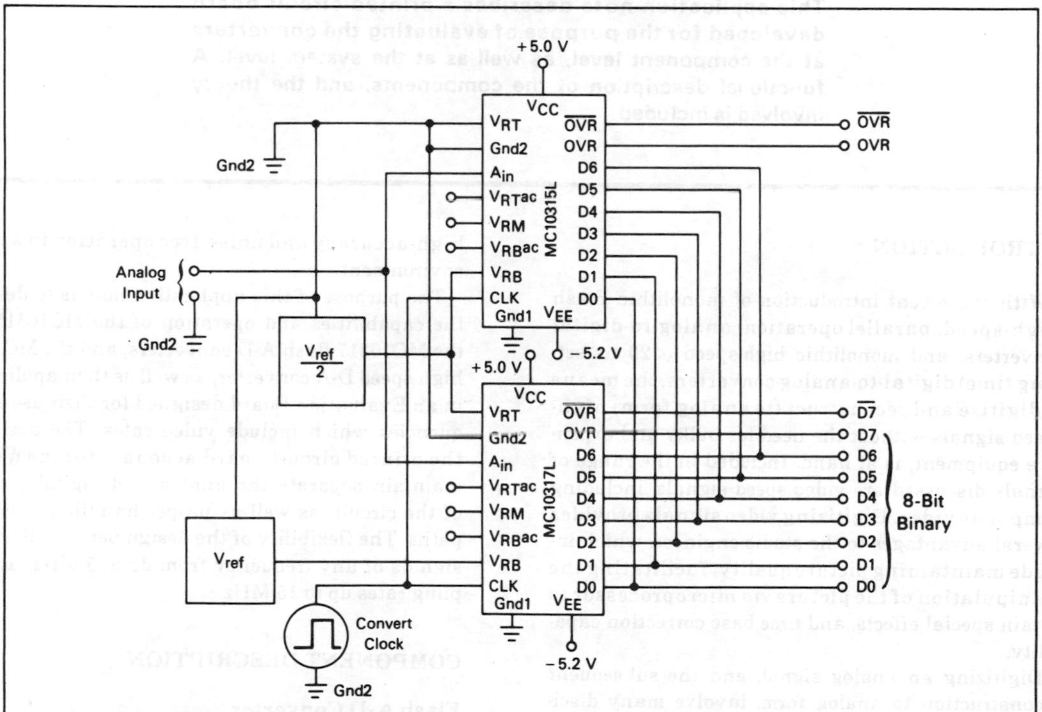

Note:
Required pull-down resistors at the digital outputs are not shown. All digital outputs should be treated according to MECL 10K design rules.

Ⓜ **MOTOROLA**
Semiconductor Products Inc.

AN-848
Application Note

AN EVALUATION SYSTEM FOR HIGH-SPEED A-D AND D-A CONVERTERS

The recent availability of low cost, high-speed monolithic analog-to-digital and digital-to-analog converters have opened new opportunities which until now have been prohibitively expensive or impractical. One of the major areas of application is in digitizing television signals at nearly all phases of the television system.

This application note describes a printed circuit board developed for the purpose of evaluating the converters at the component level, as well as at the system level. A functional description of the components, and the theory involved is included.

INTRODUCTION

With the recent introduction of monolithic Flash (high-speed, parallel operation) analog-to-digital converters, and monolithic high-speed (<20 ns settling time) digital-to-analog converters, the means to digitize and reconstruct (to analog form) high-speed signals without the need for bulky and expensive equipment, is at hand. Included in the range of signals discussed are video speed signals, including composite video. Digitizing video signals provides several advantages to the studio engineer, which include maintaining picture quality, facilitating the manipulation of the picture via microprocessor to obtain special effects, and time base correction capability.

Digitizing an analog signal, and the subsequent reconstruction to analog form, involve many disciplines of mathematics and electronics. A thorough study involves the Nyquist Sampling Theorem, Fourier analysis, and the study of beat frequencies and harmonics. Techniques for handling high-speed signals are necessary, along with the need to maintain high-accuracy and noise-free operation in a noisy environment.

The purpose of this application note is to describe the capabilities and operation of the MC10315 and the MC10317 flash A-D converters, and the MC10318 high-speed D-A converter, as well as their application in an Evaluation Board designed for their use at frequencies which include video rates. The design of the printed circuit board accounts for the need to maintain separate the analog and digital portions of the circuit, as well as proper handling of critical paths. The flexibility of the design permits digitizing signals of any frequency from dc to 5 MHz, at sampling rates up to 15 MHz.

COMPONENT DESCRIPTION

Flash A-D Converter

The MC10315 and MC10317 are 7-bit flash A-D converters, whose operation differ only in the overrange function. In the MC10315, the outputs remain high (Logic "1") when an overrange condition is reached, whereas in the MC10317 the outputs switch to low

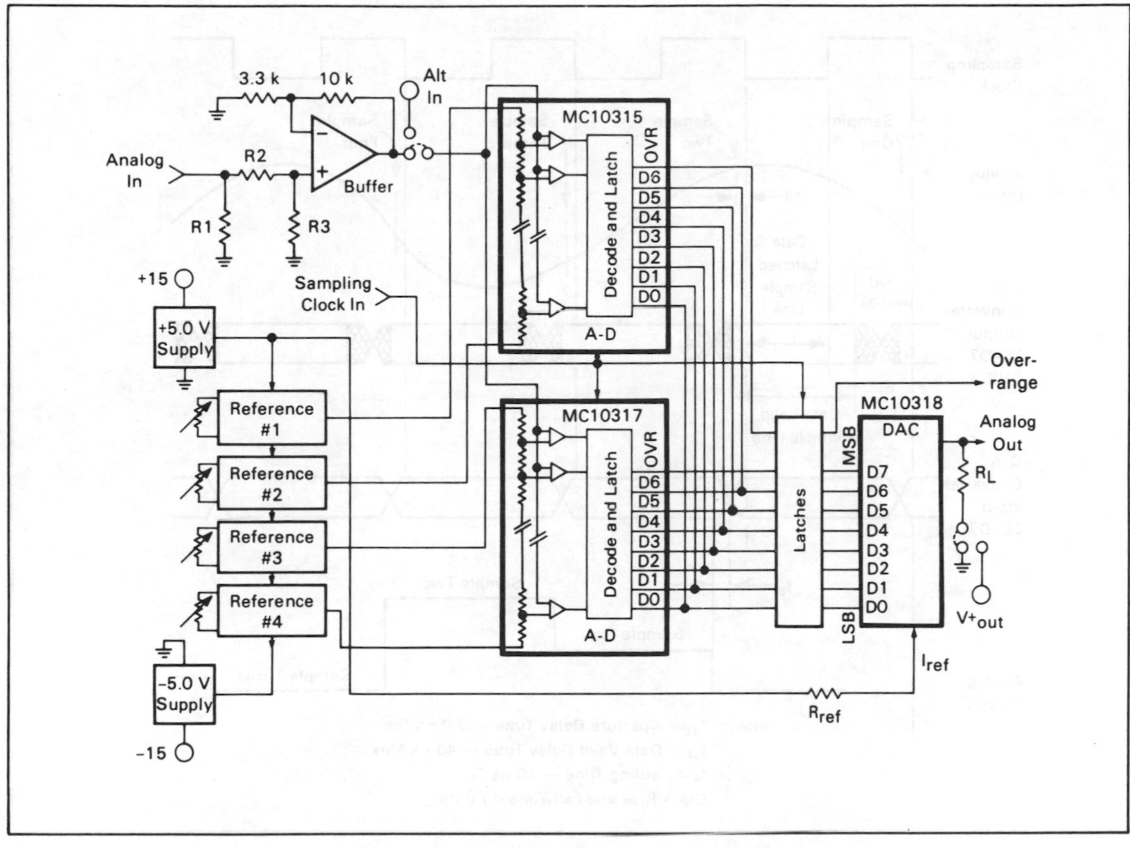

FIGURE 1 — Evaluation System Block Diagram

(Logic "0") on reaching an overrange condition. This difference is necessary in order to be able to parallel two devices into an 8-bit configuration.

The digital output of a flash converter is determined by comparing the analog input to a voltage on a resistor divider reference network. The digital outputs are Logic "0" when the analog input is within 1/2 LSB of the most negative reference voltage (V_{RB}), or less. The outputs are at a logic "1" when the input is within 1-1/2 LSB to 1/2 LSB below the most positive reference voltage (V_{RT}). In between these values are the remaining 126 possible binary codes. When the input exceeds V_{RT} - 1/2 LSB, the overange output is high. The reference span (V_{RT}-V_{RB}) must be less than 2 volts, but preferably greater than 1 volt, within an absolute voltage range of +2 V to -2 V.

The Flash converter requires a clock input. When the clock is high, the 128 input comparators are latched (effectively a sample-and-hold function), allowing the subsequent logic to decode their outputs into a 7-bit binary code. During this time, the output latches are transparent, and the output data is not stable or valid for the first (approximate) 43 ns. When the clock is low the output latches are latched, and the comparators are allowed to resume tracking the input signal. (See Figure 2.) Minimum high time is 44 ns, and minimum low time is 25 ns.

The outputs and clock signal are ECL compatible, the outputs therefore requiring pull-down resistors. The analog input impedance is 5 kΩ minimum in parallel with 70 pF, and the input signal may be bipolar or unipolar.

High-Speed D-A Converter

The MC10318 is a high-speed, 8-bit D-A converter, whose output settles, typically, in 10 ns. The inputs are ECL compatible, and no clock is required. The output (I_O) varies with the digital input in 256 steps from 0 to 51 mA when a reference current of 3.2 mA is supplied to the device. $\overline{I_O}$ is the complementary output, varying in opposite phase to I_O. Both outputs are current sources.

The voltage at either output may fall within the range of +2.5 V to -1.3 V by selection of appropriate load resistor and pullup voltage. Since the output current flows **into** the device, increasing the output current causes the output voltage to swing negative.

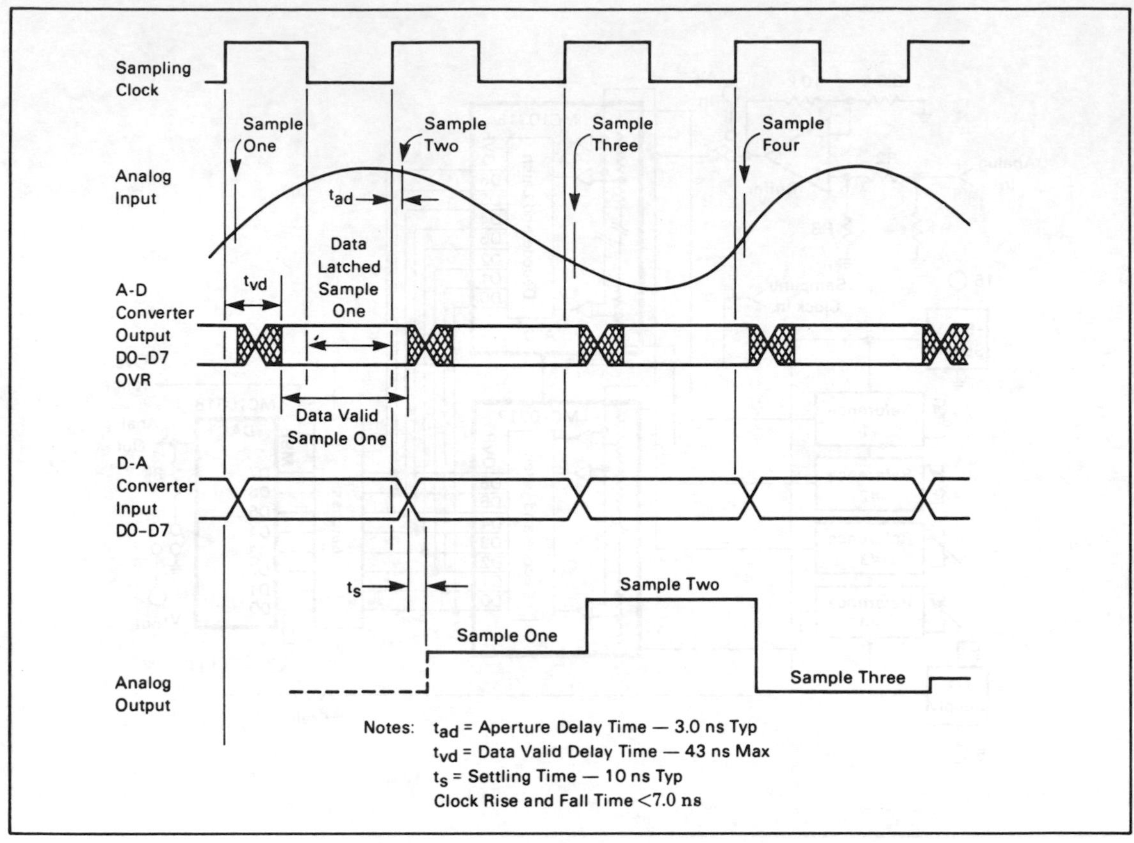

FIGURE 2 — A-D and D-A Conversion Timing

EVALUATION BOARD

Block Diagram

See Figure 1. The board has provisions for the following:

- One MC10315 A-D **and** one MC10317 A-D, to allow digitizing to 8-bit resolution. Either device may be deleted for 7-bit operation;
- One MC10318 D-A;
- Latches to transfer the data from the A-D to the D-A;
- Buffer/amplifier for the analog input signal;
- Adjustable references for the A-D converters;
- Reference current for the D-A converter;
- Various input voltage ranges and input impedances;
- Various output voltage ranges and output impedances.

By installing appropriate components, the board may be configured to produce any of the following functions:

- 8-bit A-D and D-A;
- 7-bit A-D and D-A;
- 8-bit A-D only;
- 7-bit A-D only;
- 8-bit D-A only.

Input Section

The input amplifier permits matching the input impedance to the source impedance, as well as amplifying low-level signals, where that is necessary. Selection of input resistors R1, R2, and R3 should be made from Table 1. Where the signal amplitude does not match exactly with a value in the table, the next higher range should be selected. The A-D references (described later) can then be adjusted to match the signal level.

The amplifier has been configured for stable operation, at a gain of +4, while driving two A-D converters (5 K ∥ 70 pF each). The –3 dB point is approximately 10 MHz.

A dc blocking capacitor is provided at the Analog In input, but may be bypassed by the user in the case of low frequency or dc signals.

The ALT IN connector provides a means of applying the input signal directly to the A-D converters, bypassing the amplifier. If the signal has proper amplitude (1–2 volts p-p for 7-bits, 2–4 volts for 8-bits), and has sufficient drive capability, this input can be used. The jumper adjacent to the connector must be relocated per Figure 13.

TABLE 1 — INPUT SELECTION

P-P	$R_{IN} = 50\ \Omega$			$R_{IN} = 75\ \Omega$			$R_{IN} = 600\ \Omega$			$R_{IN} = 1000\ \Omega$		
V_{IN}	R1	R2	R3	R1	R2	R3	R1	R2	R3	R1	R2	R3
1	51	0	∞	75	0	∞	1200	0	1200	1000	0	∞
2	51	1300	1300	100	150	150	1000	750	750	2000	1000	1000
4	51	3000	1000	91	300	100	1200	910	300	2000	1500	510
5	51	3000	750	150	120	30	1000	1200	300	3000	1200	300
10	51	2700	300	100	270	30	750	2700	300	1500	2700	300

TABLE 2 — REFERENCE VOLTAGE

JUMPER	VOLTAGE RANGE
None	−0.625 to +0.625
Odd No. Only	+0.5 to +2.5
Even No. Only	−2.5 to −0.5

TABLE 3 — OUTPUT VOLTAGE

OUT V+	R_L	ANALOG OUT
0 V	20	−1.0 V to 0 V
+2.5 V	50	0 V to +2.5 V
+2.5 V	76	−1.3 V to +2.5 V
+1.0 V	40	−1.0 V to +1.0 V

FIGURE 3 — Evaluation Board Schematic

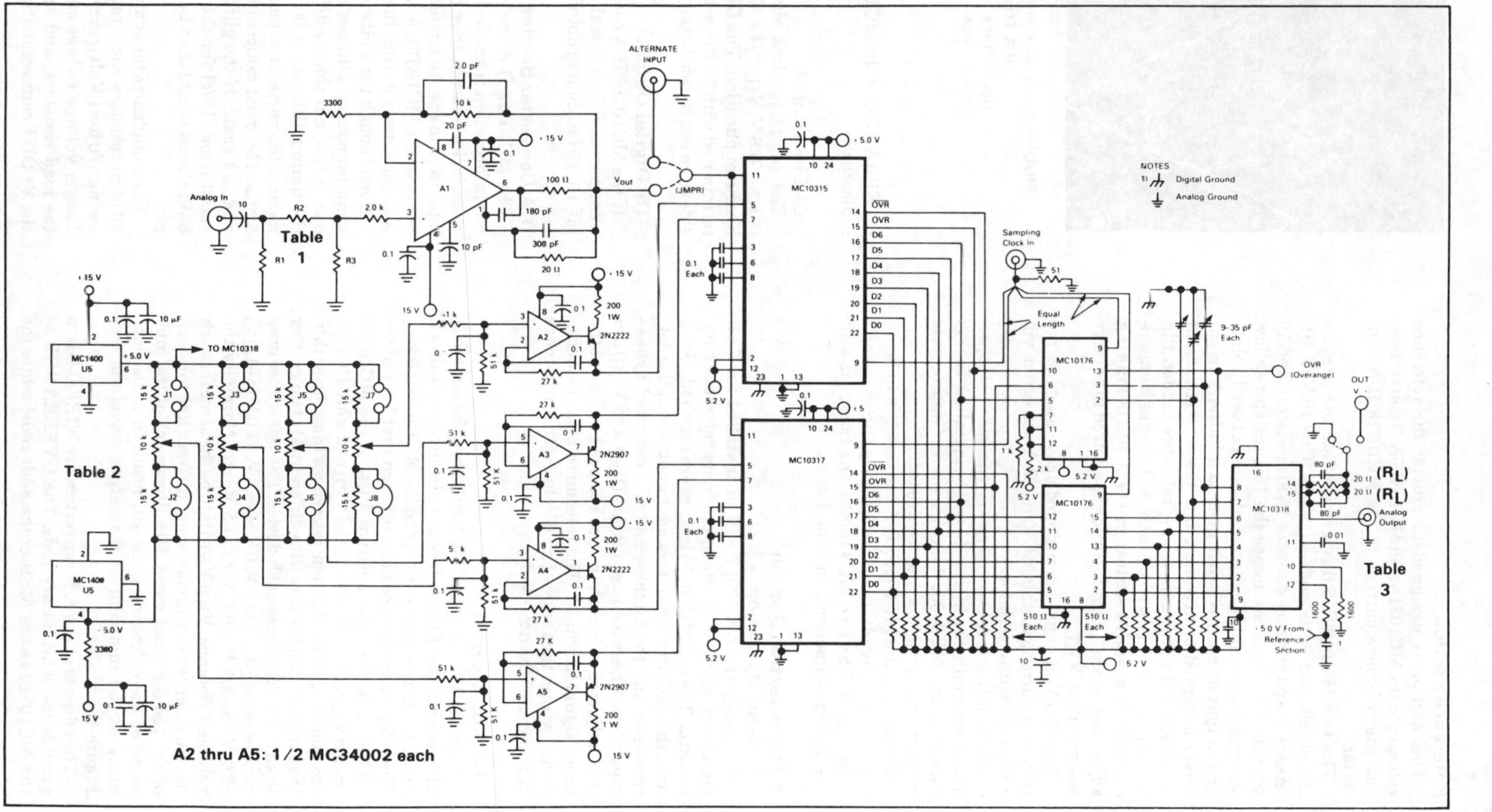

Reference Section

This section is designed to provide the reference voltages to the MC10315 and MC10317 A-D converters, and the reference current to the MC10318 D-A converter.

The two MC1400U5 devices provide the precise and stable +5.0 Volts and -5.0 Volts. This 10 Volt difference is spanned by 4 sets of 15 kΩ resistors and 10 kΩ pots. The jumpers across the 15 k resistors allow shifting the available range of adjustment per Table 2. The output of each trimpot goes to a buffer/current driver designed to accommodate the 32 mA current required by the A-D converters' reference ladder. Pin 5 (V_{RT}) or Pin 7 (V_{RB}) of each A-D should be monitored with a high resolution DVM while making adjustments. V_{RB} should **never** be allowed to be more positive than V_{RT}.

The +5.0 Volt supply mentioned above supplies the reference current to the MC10318 D-A converter via a current setting resistor.

Where only one A-D converter is required (7-bit operation), normally the MC10317 will be installed. In this case the upper MC34002 dual op-amp, and appropriate trimpots and resistors may be deleted.

Where only the MC10318 D-A converter is to be installed, only the MC1400U5 supplying the +5.0 Volts is required. The two MC34002's, the trimpots and associated components may be deleted.

A-D Convertion Section

A) 7-Bit Operation

Normally the MC10317 will be installed for 7-bit digitizing, along with the appropriate reference components. The two MC10176 latches are required, along with the pull-down resistors on the outputs of the A-D converter and the latches, and all associated bypass capacitors. The reference voltages (V_{RT} and V_{RB}) are to be adjusted to match the peak-to-peak values of the input signal at Pin 11. The references must be maintained within the range of ±2.0 Volts, with a difference of no greater than 2.0 Volts.

B) 8-Bit Operation

Both the MC10315 and MC10317 are to be installed, along with the four reference circuits, pull-down resistors, bypass capacitors, and the MC10176 latches. The references are to be adjusted such that V_{RT} of the MC10315 is the most positive, V_{RB} of the MC10315 equals V_{RT} of the MC10317, and V_{RB} of the MC10317 is the most negative. All four references must be within the range of ±2.0 Volts, with a differential of no greater than 2.0 Volts across each device. The references may thus be set at +2.0 V to 0 V, 0 V to -2.0 V, +1.0 V to -1.0 V, +2.0 V to -2.0 V, +0.5 V to -1.5 V, etc. Normally the voltage difference (V_{RT}-V_{RB}) is to be the same for each device in order to provide equal differential for each of the 256 digitized steps. Misadjustment of the midpoint of the reference span will result in a discontinuity at the midpoint of the output waveform (see Figure 4).

The outputs of the A-D converters are WIRED-OR to provide the 8-bit binary code. The OVERRANGE of the MC10317 is the MSB of the 8-bit code, while the

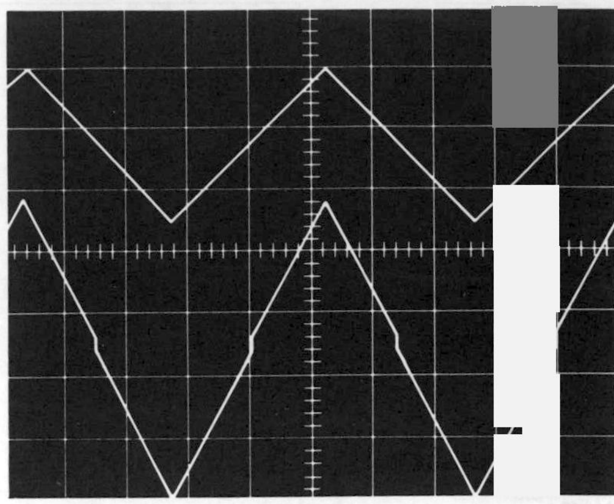

FIGURE 4 — Discontinuity due to misadjustment of A-D references
Upper Trace = Analog Input
Lower Trace = Analog Output

OVERRANGE of the MC10315 remains an overrange indicator.

C) Clock Input

The sampling clock signal is to be at ECL levels (V_{IH} = -0.8 V, V_{IL} = -1.8 V) at a frequency determined by the application. The CLK input BNC connector is terminated with 51 ohms, and the traces leading to the 4 devices are (intentionally) equal length.

D) Digital Output

If the MC10318 D-A converter is not to be included, a connector strip is installed in pin positions 1-8 of the MC10318 location to obtain the digital output signals.

D-A Converter Section

The MC10318 D-A converter receives its digital input information from the latches, after the rising edge of each clock signal. (See Figure 2.) The purpose of the latches is to ensure simultaneous transmission of the 8-bits to the MC10318, and to block the outputs of the A-D converters during the time they are invalid.

Even though the 8 bits are received by the MC10318 simultaneously, glitches can still occur on its output due to internal timing differences. Compensation for the timing differences is achieved, in part, by adjustment of the variable capacitors connected to the upper 3 bits. The most noticeable glitches occur at 1/4, 1/2, and 3/4 scale. If the glitches cannot be reduced to a satisfactory level for the application, a sample-and-hold circuit would then be required at the Analog Output.

The complementary current outputs are loaded with 20 ohm resistors each, resulting in an output voltage swing of 0 to -1 Volt (current flows **into** the MC10318). Larger voltage swings can be obtained by changing the load resistors, and applying a pullup voltage to the V+ OUT connector. (See Table 3.) The jumper near the ANALOG OUT connector must be located as per Figure 13. The output voltage is limited to a range of +2.5 V to -1.3 V.

If only the D-A function is to be evaluated (A-D Converters not installed), a connector strip is installed in pin positions 15-22 of the MC10317 location. The digital input (ECL level) can then be applied to the board at this connector.

THEORY

Sampling Theory

The process of digitizing an analog signal involves the concept of sampling (ideally instantaneously) the signal at various points with respect to time. The value measured at each sample is then converted into a corresponding digital word (comprised of several bits). While the analog signal represents a continuous, unbroken stream of information, the digital words are available only at a fixed, finite rate (the sampling rate) and each word represents a fixed value for a period of time. In order to reconstruct the information contained in the digital words, a number of the words must be "read" (converted to an analog value), and the discrete steps must (in some cases) be "averaged out" (by filtering or other means). See Figures 7 and 8. Obviously the higher the sample rate, compared to the frequency of the analog signal, the more information that will be available in a given period of time. See Figures 5 and 7. The Nyquist Sampling Theorem states that the sampling rate must be equal to or greater than twice the frequency of the highest frequency component of the signal of interest. Sampling at a rate less than the Nyquist limit generally will not permit truthful reproduction of the original analog signal.

The number of bits resolution largely determines the accuracy to which the original analog signal can be reconstructed. In an 8-bit system, there are 256 discrete steps, and therefore each digital code represents a range of 0.39% of the total available signal. As a result, each sampling and digitizing of an analog signal is accompanied by an uncertainty of ±0.195% (±0.39% for a 7-bit system, ±0.097% for a 9-bit system). The uncertainty shows up as a possible error in the amplitude of the reconstructed signal compared to the original signal.

The timing differences between the sampling clock signal and the analog signal causes another type of uncertainty (or error). In the cases where the two signals are not synchronized (the usual mode of operation), the reconstructed analog output will appear to have a phase jitter in an amount equal to the period of the sampling clock. This effect is easiest seen by applying a square wave to the analog input, and observing the jitter at the analog output (see Figure 9). The main effect of the timing uncertainty is the generation of beat frequencies, and harmonics. The greater the difference between the sampling frequency and the signal frequency, the easier it is to contend with the additional unwanted frequencies. But where the sampling rate is a small multiple (2-10) of the signal frequency, the beat frequencies become a very significant part of the output signal (see Figure 6).

In the cases where the sampling rate is locked to the input signal frequency, beat frequencies and harmonics will still be generated, but the timing uncertainty (apparent phase jitter) can be eliminated. For example, a sine wave sampled at exactly twice its frequency will result in a constant amplitude square wave when reconstructed. Recapturing the original sine wave will obviously require filtering at the fundamental frequency. (In the special case where sampling occurs at the zero-crossing point, the output will be a dc signal).

In the processing, or storing of the digital words, prior to reconstruction to analog form, the transmission rates of the words, or bits, becomes a significant factor. If the MC10317 is made to sample at its maximum frequency (≈15 MHz), it will be outputting data at the rate of 15 Megawords per second. If the data is to be transmitted serially, a transmission rate of 105 Megabits per second results (120 Megabits per second for an 8-bit system). A bandwidth of half the bit rate is required, resulting in bandwidth requirements much higher than that for the original analog signal. This fact is one of the major costs to be paid for the benefits of a digital signal.

Digital Video Signals

The highest frequency signal of interest to be digitized in a video signal is the 3.58 MHz sine wave used for the color reference and the subsequent color information. Discussions have been conducted as to whether the signal should be sampled at 3X (10.7 MHz) or at 4X (14.3 MHz). Currently the trend seems to support sampling at 4X, and some manufacturers of VTR's have used 4X to date.

Considerable discussion has been conducted as to the minimum number of bits resolution to which a video signal should be digitized. A general agreement is that 6 bits (64 quantizing levels) is the minimum number of bits which can maintain an acceptable picture. While 7 bits is being considered for many applications, 8 bits digitizing appears to be preferred in order to allow for the cumulative differential phase and gain errors resulting from multiple processing of the digital signals.

As mentioned in the previous section (Sampling Theory), digitizing a signal involves uncertainties, which show up as errors at the output. In video signals, the results are differential phase and gain errors, and a luminance shift. Per reference 1, an ideal 8-bit digitizing system will produce maximum errors of ±4.3% differential gain, ±2.5° differential phase and ±0.33 IRE unit luminance shift. These figures are based on a 10 IRE signal (20 IRE units p-p) within an available field of 166 IRE units, sampled at 3X. The luminance shift results from digitizing the dc component of the video signal with an uncertainty of ±1/2-LSB, and is independent of the color information. The maximum gain and phase errors occur only under certain sampling conditions, and in fact do not occur simultaneously. When instantaneous gain error is at a maximum, phase error is zero, and vice-versa. Since the sampling signal is asynchronous to the video information, a distribution of each error will result,

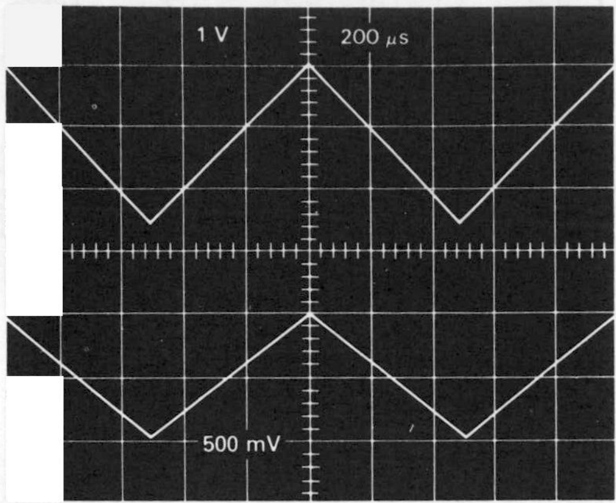

FIGURE 5 — 1.0 kHz Triangle Sampled at 10 MHz

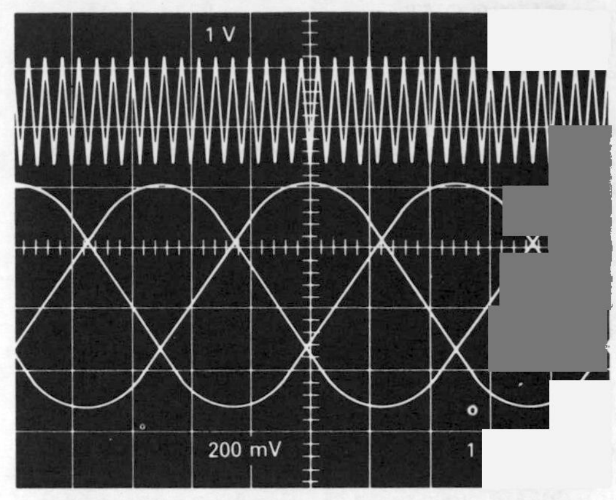

FIGURE 6 — 3.58 MHz Sine Wave Sampled at 10.7 MHz

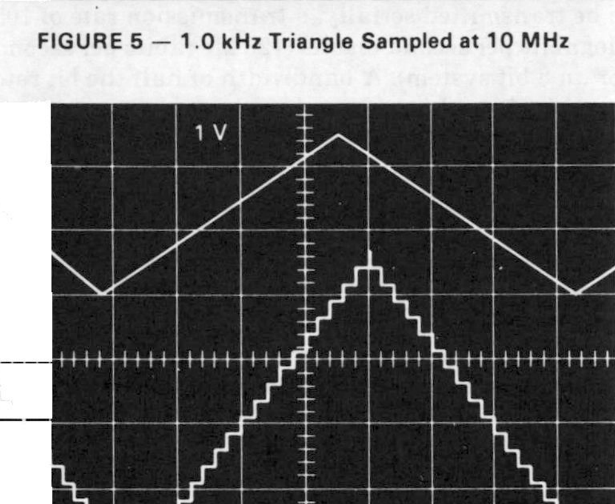

FIGURE 7 — 270 kHz Triangle Sampled at 10 MHz

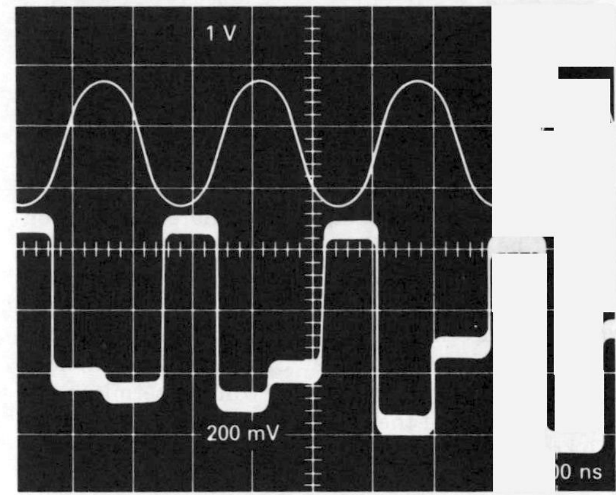

FIGURE 8 — 3.58 MHz Sine Wave Sampled at 10.7 MHz

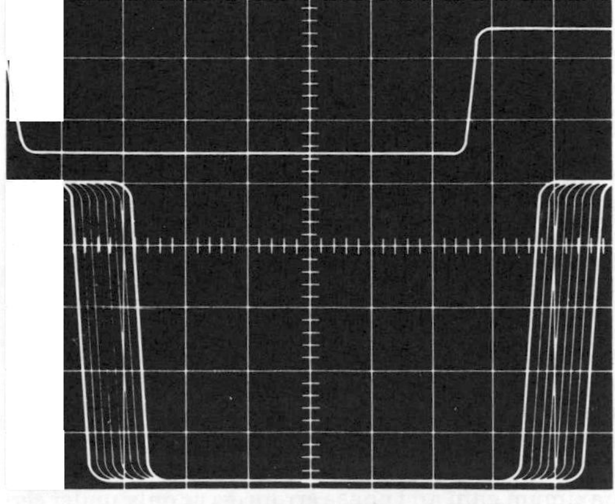

FIGURE 9 — Horizontal Scale = 100 ns/div
Sample Rate = 10 MHz

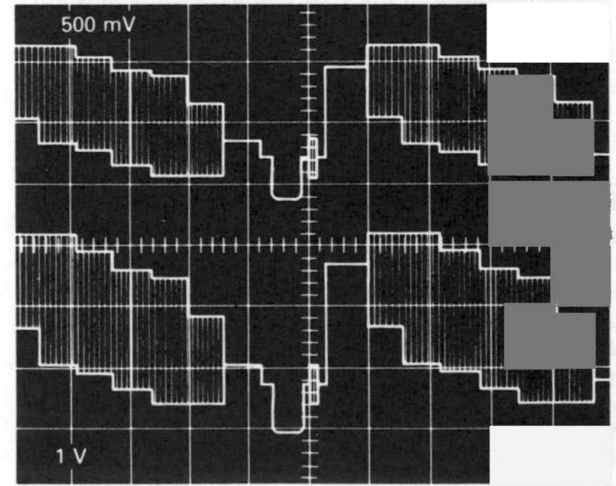

FIGURE 10 — Full Field Color Bars @ 10.7 MHz

Note: Upper Trace = Analog Input, Lower Trace = Analog Output (Each Figure)

which explains the "jittery" display on a vectorscope. Additionally, the errors decrease proportionately as the amplitude of the signal is increased.

If the composite video signal is sampled at 4X (14.3 MHz), then the sync signal will have an uncertainty (or jitter) of ±34.9 ns, which is ±0.055% of a horizontal scan period.

The Evaluation Board described in this article has been tested for differential phase and differential gain, using a standard video test signal (40 IRE subcarrier on a 100 IRE ramp, sampled at 14.3 MHz), with results of ±1% gain error and ±1° phase error. The signal was obtained from a Tektronix 147A video test generator applied to the ALT IN connector. The output (of the MC10318) was configured for 75 ohms output impedance, and applied to a Tektronix 520A Vectorscope.

Tests conducted with the Evaluation Board in a video system (video camera and a TV monitor) showed no visible degradation of picture quality (at 8-bits resolution). The board provides an easy means of testing picture quality at a reduced number of bits, or for conducting any test on a digitized video signal.

See Figure 10 for an example of a video test pattern.

Note: In order for the output of the Evaluation Board to comply with RS-170, make the following changes at the MC10318 DAC:

- Replace the 20 Ω resistor at Pin 15 with 75 Ω;
- Replace the 20 Ω resistor at Pin 14 with 37.5 Ω;
- Replace the 1600 Ω resistors at Pins 10 and 12 with 3000 Ω each;
- Replace the 80 pF capacitors at Pins 14 and 15 with 10 pF each;
- The jumper near the Analog Out connector must be in the Normal position (Figure 13).

REFERENCES

1. Felix, Michael O., *Differential Phase and Gain Measurements in Digitized Video Signals*, SMPTE J., 85: 76-79, Feb. 1976.
2. *The Near-Term Future for Digital Television*, 120th SMPTE Technical Conference, Nov. 1978.
3. *Interconnection Techniques for Motorola's MECL 10,000 Series Emitter Coupled Logic*, AN-556, Motorola, Inc., 1974.
4. *MECL System Design Handbook*, Third Edition, Motorola, Inc., 1980.
5. MC10315L/MC10317L Data Sheet, Motorola, Inc., 1981.
6. MC10318L Data Sheet, Motorola, Inc., 1982.

PARTS LIST

	8-Bit	7-Bit	8-Bit	7-Bit	8-Bit
Device	A-D and D-A		A-D	A-D	D-A
Semiconductors					
MC10315	1	—	1	—	—
MC10317	1	1	1	1	—
MC10318	1	1	—	—	1
MC34002	2	1	2	1	—
MC1400U5	2	2	2	2	1
MC10176	2	2	2	2	2
2N2222	2	1	2	—	—
2N2907	2	1	2	—	—
LH0024C	1	1	1	1	—
Resistors					
51K, 1/4 W	8	4	8	4	—
15K, 1/4 W	8	4	8	4	—
3.3K, 1/4 W	2	2	2	2	—
2K, 1/4 W	2	2	2	2	1
10K, 1/4 W	1	1	1	1	—
27K, 1/4 W	4	2	4	2	—
510, 1/4 W	20	17	20	17	16
20, 1/4 W	3	3	1	1	2
1600, 1/4 W	2	2	—	—	2
100, 1/4 W	1	1	1	1	—
51, 1/4 W	1	1	1	1	1
200, 1 W	4	2	4	2	—
1K, 1/4 W	1	1	1	1	1
Trimpots					
10K, 20 Turn (Rectangular)	4	2	4	2	—
Capacitors					
0.1 μF, 50 V	41	26	37	22	7
10 μF, 50 V Tant.	8	8	7	7	3
300 pF	1	1	1	1	—
80 pF	2	2	—	—	2
10 pF	1	1	1	1	—
20 pF	1	1	1	1	—
22 μF, 50 V Tant.	1	1	—	—	1
180 pF	1	1	1	1	—
9-35 pF Trimmer	3	3	—	—	3
0.01 μF, 50 V	1	1	—	—	1
2 pF	1	1	1	1	—
Misc.					
PC Board	1	1	1	1	1
Banana Jacks	8	8	7	6	5
BNC Connector	4	4	3	3	2
8-Pin Connector (SAMTEC TS120GD11)	—	—	1	1	1

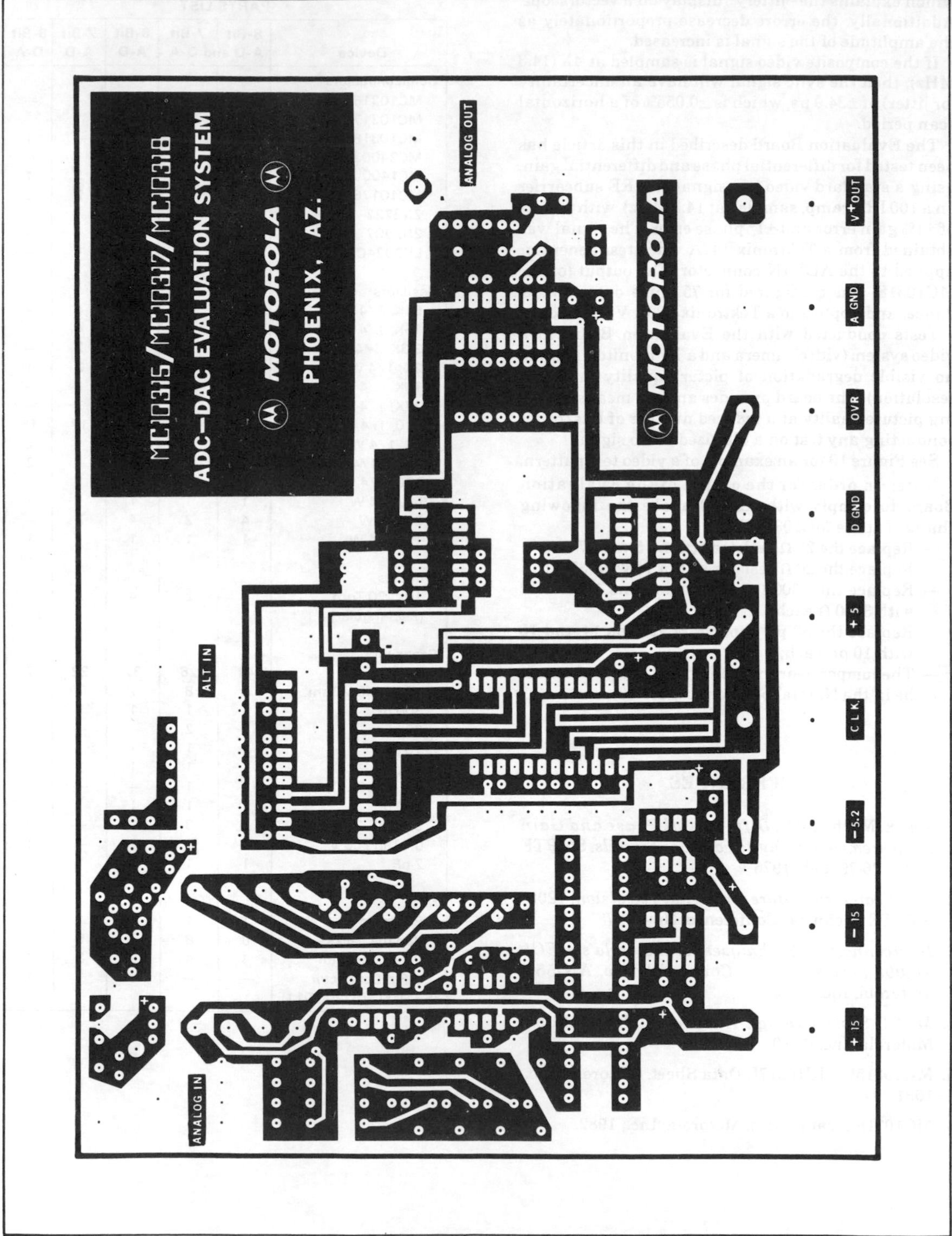

FIGURE 11 — Component Side Artwork (Top)
(Overall Size = 6.0″ × 8.00″)

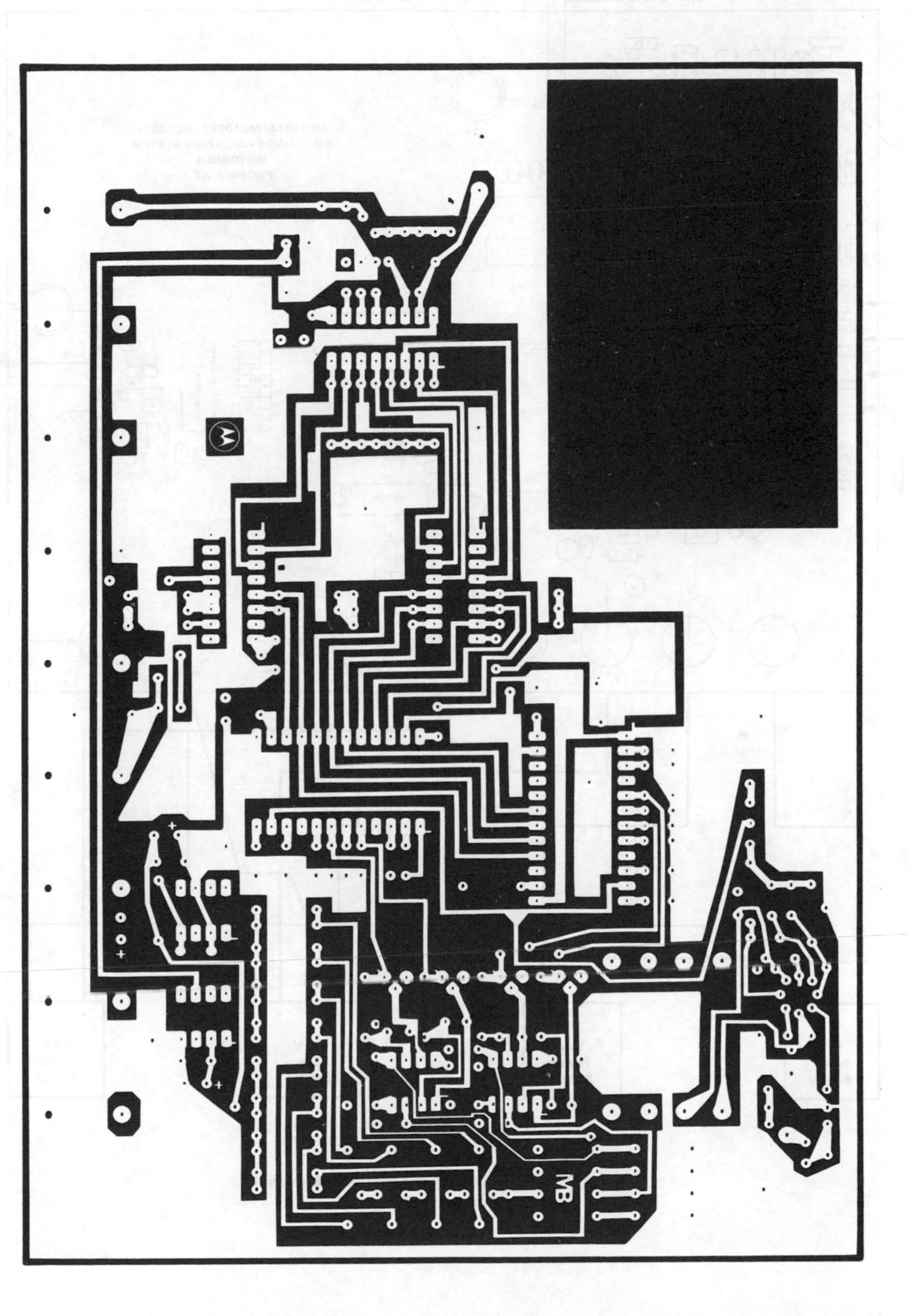

FIGURE 12 — Solder Side Artwork (Bottom)

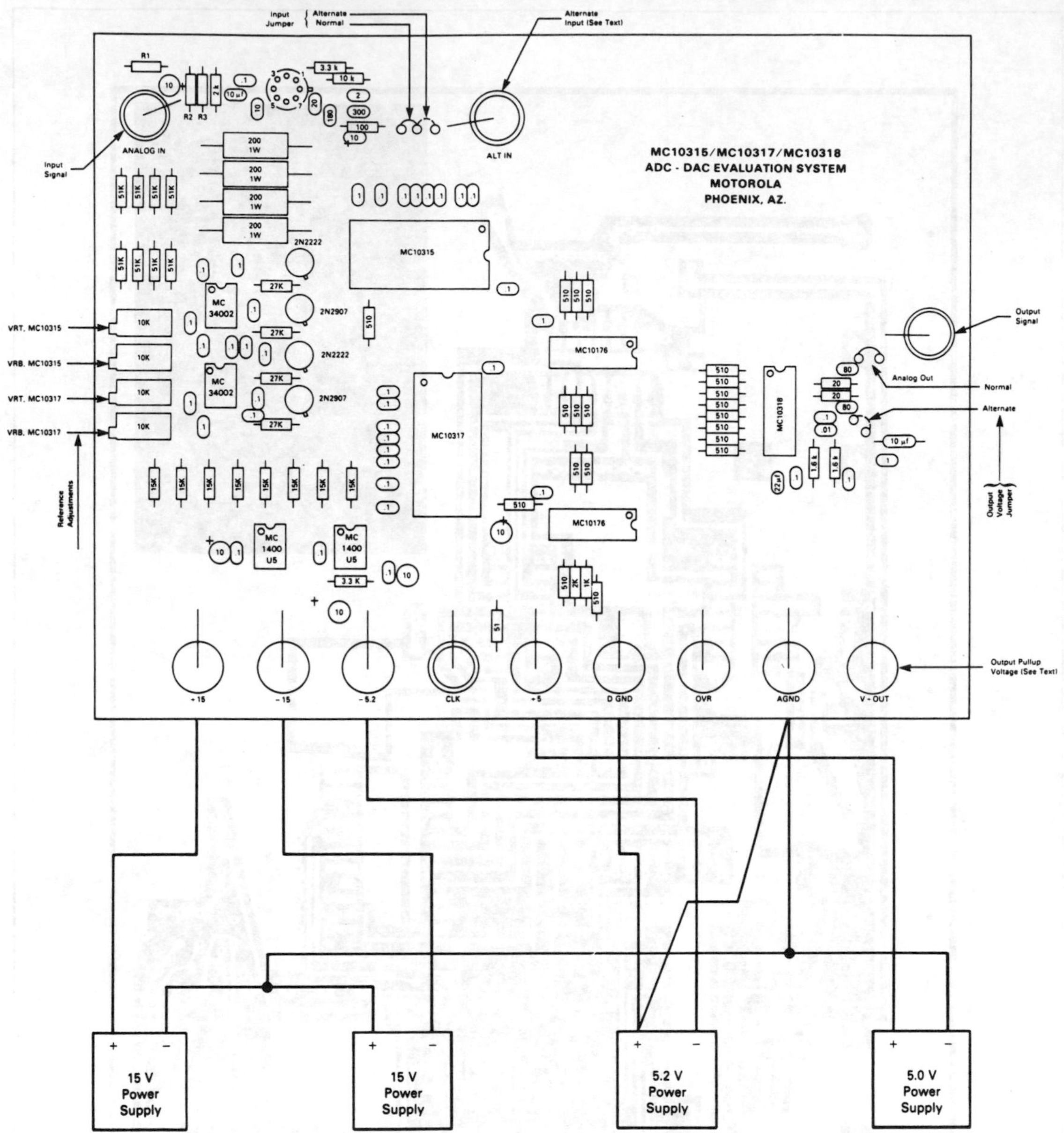

FIGURE 13 — Component Location and External Connections

Chapter 5 Multiplexers and sample and hold circuits

Apart from the op-amp and the instrumentation amplifier, analog interfaces often contain at least two other types of circuit. One is the multiplexer that electronically switches an analog input from one of several channels to the input of a single analog to digital converter. The multiplexer is a relatively low-cost circuit that avoids the expense of n ADCs in a system with an n-channel input.

We look at the data sheets of a typical single-ended 16-channel analog multiplexer. This device can also function as an 8-channel differential multiplexer. We provide an application note that analyses the effect of one of the principal limitations of the multiplexer – crosstalk between channels.

The largest part of this chapter is devoted to the sample and hold circuit. Analog to digital converters take a finite time to convert an analog input into a digital value. If the analog input changes while it is being digitized, the output of the ADC might be in error. The sample and hold circuit is able to take a sample of an analog signal and hold it constant while it is being digitized.

The characteristics and performance of sample and hold circuits are highly interrelated. For example, there is a trade-off between the time it takes to sample a signal and the rate at which the signal decays once it has been sampled. In this chapter we look at the principles of the sample and hold circuit and describe typical devices.

The final part of this chapter looks at a complete data acquisition system that includes a multiplexer, an instrumentation amplifier, a sample and hold circuit, and a 12-bit ADC, all on a hermetically sealed 1 inch square leadless chip carrier. A device like this can be interfaced to a processor to produce a computer-controlled system monitor in, for example, a process control environment.

HI-506A
HI-507A

MILITARY & DIE VERSIONS AVAILABLE

Single-Ended 16-Channel/Differential 8-Channel CMOS ANALOG MULTIPLEXERS

FEATURES

- ANALOG OVERVOLTAGE PROTECTION: 70Vp-p
- NO CHANNEL INTERACTION DURING OVERVOLTAGE
- ESD RESISTANT
- BREAK-BEFORE-MAKE SWITCHING
- ANALOG SIGNAL RANGE: ±15V
- STANDBY POWER: 7.5mW typ
- TRUE SECOND SOURCE

DESCRIPTION

The HI-506A is a 16-channel single-ended analog multiplexer and the HI-507A is an 8-channel differential multiplexer.

The HI-506A and HI-507A multiplexers have input overvoltage protection. Analog input voltages may exceed either power supply voltage without damaging the device or disturbing the signal path of other channels. The protection circuitry assures that signal fidelity is maintained even under fault conditions that would destroy other multiplexers. Analog inputs can withstand 70Vp-p signal levels and standard ESD tests. Signal sources are protected from short circuits should multiplexer power loss occur; each input presents a 1kΩ resistance under this condition. Digital inputs can also sustain continuous faults up to 4V greater than either supply voltage.

These features make the HI-506A and HI-507A ideal for use in systems where the analog signals orginate from external equipment or separately powered sources.

The HI-506A and HI-507A are fabricated with Burr-Brown's dielectrically isolated CMOS technology. The multiplexers are available in a hermetic ceramic DIP or plastic DIP. Commercial (0°C to +75°C) and military (−55°C to +125°C) versions are available.

FUNCTIONAL DIAGRAMS

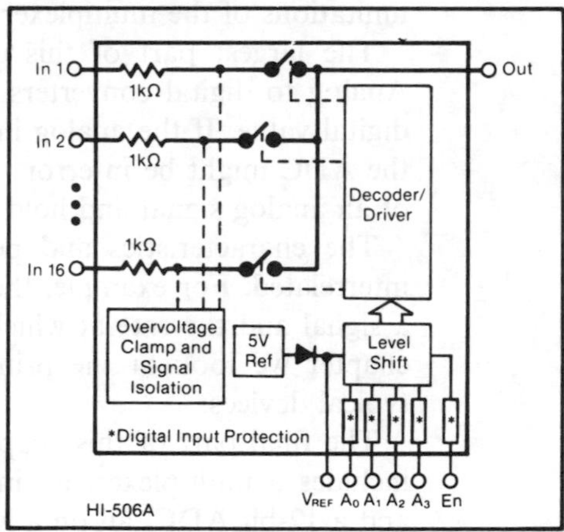

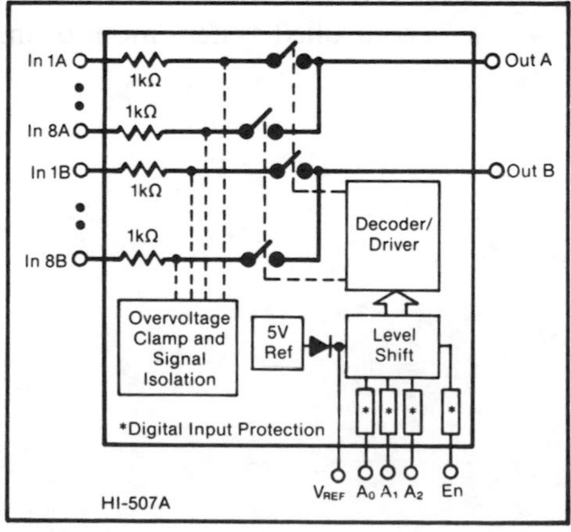

International Airport Industrial Park · P.O. Box 11400 · Tucson, Arizona 85734 · Tel. (602) 746-1111 · Twx: 910-952-1111 · Cable: BBRCORP · Telex: 66-6491

PDS-774A

SPECIFICATIONS

ELECTRICAL

Supplies = +15V, −15V; V_{REF} (Pin 13) = Open; V_{AH} (Logic Level High) = +4.0V; V_{AL} (Logic Level Low) = +0.8V unless otherwise specified.

PARAMETER	TEMP	HI-506A-2/HI-507A-2			HI-506A-5/HI-507A-5			UNITS
		MIN	TYP	MAX	MIN	TYP	MAX	
ANALOG CHANNEL CHARACTERISTICS								
V_S, Analog Signal Range	Full	−15		+15	−15		+15	V
R_{ON}, On Resistance[1]	+25°C		1.2	1.5		1.5	1.8	kΩ
	Full		1.5	1.8		1.8	2.0	kΩ
I_S (OFF), Off Input Leakage Current	+25°C		0.03			0.03		nA
	Full			50			50	nA
I_D (OFF), Off Output Leakage Current	+25°C		0.1			0.1		nA
HI-506A	Full			300			300	nA
HI-507A	Full			200			200	nA
I_D (OFF) with Input Overvoltage Applied[2]	+25°C		4.0			4.0		nA
	Full			2.0				μA
I_D (ON), On Channel Leakage Current	+25°C		0.1			0.1		nA
HI-506A	Full			300			300	nA
HI-507A	Full			200			200	nA
I_{DIFF} Differential Off Output Leakage Current (HI-507A Only)	Full			50			50	nA
DIGITAL INPUT CHARACTERISTICS								
V_{AL}, Input Low Threshold TTL Drive	Full			0.8			0.8	V
V_{AH}, Input High Threshold[3]	Full	4.0			4.0			V
V_{AL} MOS Drive[4]	+25°C			0.8			0.8	V
V_{AH}	+25°C	6.0			6.0			V
I_A, Input Leakage Current (High or Low)[5]	Full			1.0			1.0	μA
SWITCHING CHARACTERISTICS								
t_A, Access Time	+25°C		0.5			0.5		μs
	Full			1.0			1.0	μs
t_{OPEN}, Break-Before-Make Delay	+25°C	25	80		25	80		ns
t_{ON} (EN), Enable Delay (ON)	+25°C		300	500		300		ns
	Full			1000			1000	ns
t_{OFF} (EN), Enable Delay (OFF)	+25°C		300	500		300		ns
	Full			1000			1000	ns
Settling Time (0.1%)	+25°C		1.2			1.2		μs
(0.01%)	+25°C		3.5			3.5		μs
"OFF Isolation"[6]	+25°C	50	68		50	68		dB
C_S (OFF), Channel Input Capacitance	+25°C		5			5		pF
C_D (OFF), Channel Output Capacitance: HI-506A	+25°C		50			50		pF
HI-507A	+25°C		25			25		pF
C_A, Digital Input Capacitance	+25°C		5			5		pF
C_{DS} (OFF), Input to Output Capacitance	+25°C		0.1			0.1		pF
POWER REQUIREMENTS								
P_D, Power Dissipation	Full		7.5			7.5		mW
I+, Current Pin 1[7]	Full		0.5	2.0		0.5	2.0	mA
I−, Current Pin 27[7]	Full		0.02	1.0		0.02	1.0	mA

NOTES: (1) V_{OUT} = ±10V, I_{OUT} = −100μA. (2) Analog overvoltage = ±33V. (3) To drive from DTL/TTL circuits. 1kΩ pull-up resistors to +5.0V supply are recommended. (4) V_{REF} = +10V. (5) Digital input leakage is primarily due to the clamp diodes. Typical leakage is less than 1nA at 25°C. (6) V_{EN} = 0.8V, R_L = 1kΩ, C_L = 15pF, V_S = 7Vrms, f = 100kHz. Worst-case isolation occurs on channel 4 due to proximity of the output pins. (7) V_{EN}, V_A = 0V or 4.0V.

TRUTH TABLES

HI-506A

A_3	A_2	A_1	A_0	EN	"ON" CHANNEL
X	X	X	X	L	None
L	L	L	L	H	1
L	L	L	H	H	2
L	L	H	L	H	3
L	L	H	H	H	4
L	H	L	L	H	5
L	H	L	H	H	6
L	H	H	L	H	7
L	H	H	H	H	8
H	L	L	L	H	9
H	L	L	H	H	10
H	L	H	L	H	11
H	L	H	H	H	12
H	H	L	L	H	13
H	H	L	H	H	14
H	H	H	L	H	15
H	H	H	H	H	16

HI-507A

A_2	A_1	A_0	EN	"ON" CHANNEL PAIR
X	X	X	L	None
L	L	L	H	1
L	L	H	H	2
L	H	L	H	3
L	H	H	H	4
H	L	L	H	5
H	L	H	H	6
H	H	L	H	7
H	H	H	H	8

ABSOLUTE MAXIMUM RATINGS[1]

Voltage between supply pins 44V
V_{REF} to ground, V+ to ground 22V
V− to ground 25V
Digital input overvoltage:
V_{EN}, V_A: $V_{SUPPLY}(+)$ .... +4V
$V_{SUPPLY}(-)$ .... −4V
or 20mA, whichever occurs first.
Analog input overvoltage:
V_S: $V_{SUPPLY}(+)$ .... +20V
$V_{SUPPLY}(+)$ .... −20V
Continuous current, S or D 20mA
Peak current, S or D
(pulsed at 1ms, 10% duty cycle max) 40mA
Power dissipation* 2.0W
Operating temperature range:
HI-506A/507A-2.......... −55°C to +125°C
HI-506A/507A-5.......... 0°C to +75°C
Storage temperature range .... −65°C to +150°C
*Derate 20.0mW/°C above T_A = +75°C

NOTES: 1. Absolute maximum ratings are limiting values, applied individually, beyond which the serviceability of the circuit may be impaired. Functional operation under any of these conditions is not necessarily implied.

PIN CONFIGURATIONS

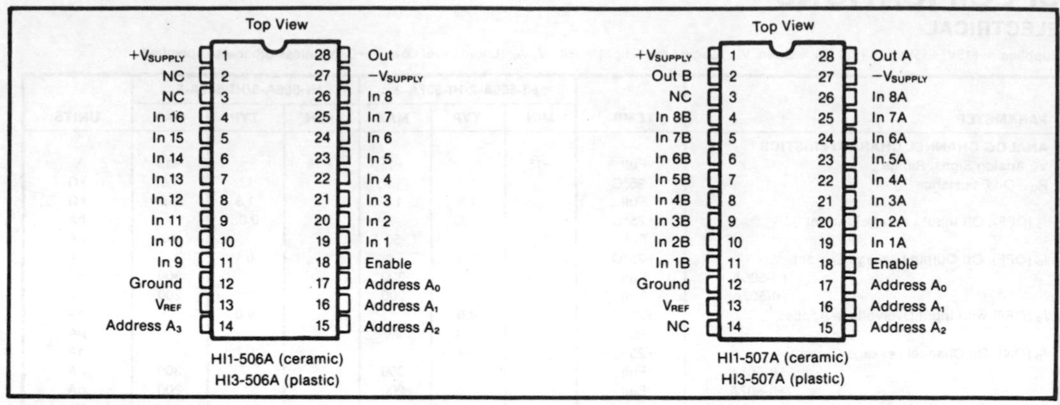

MECHANICALS

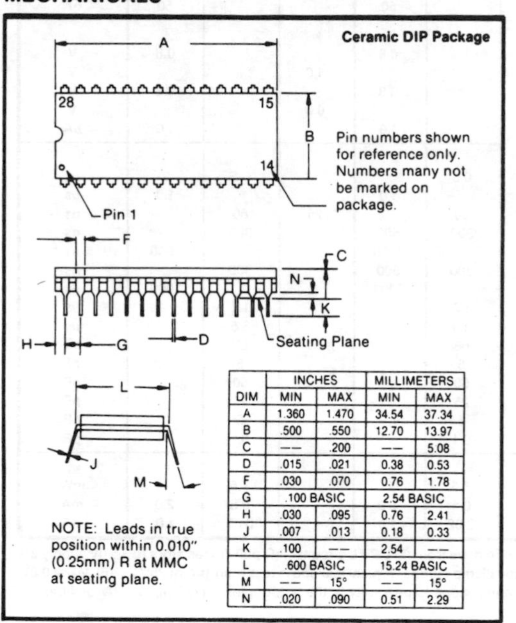

DIM	INCHES MIN	INCHES MAX	MILLIMETERS MIN	MILLIMETERS MAX
A	1.360	1.470	34.54	37.34
B	.500	.550	12.70	13.97
C	—	.200	—	5.08
D	.015	.021	0.38	0.53
F	.030	.070	0.76	1.78
G	.100 BASIC		2.54 BASIC	
H	.030	.095	0.76	2.41
J	.007	.013	0.18	0.33
K	.100	—	2.54	—
L	.600 BASIC		15.24 BASIC	
M	—	15°	—	15°
N	.020	.090	0.51	2.29

NOTE: Leads in true position within 0.010" (0.25mm) R at MMC at seating plane.

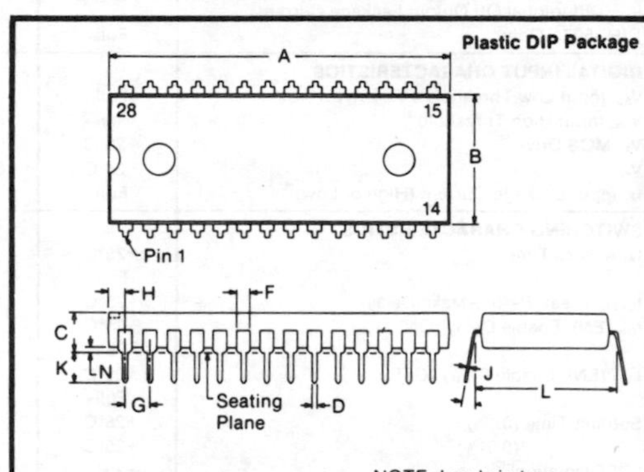

DIM	INCHES MIN	INCHES MAX	MILLIMETERS MIN	MILLIMETERS MAX
A	1.350	1.450	34.29	36.83
B	.520	.575	13.21	14.61
C	.169	.224	4.29	5.70
D	.015	.023	0.38	0.58
F	.043	.062	1.09	1.57
G	.100 BASIC		2.54 BASIC	
H	.030	.090	0.76	2.29
J	.008	.015	0.20	0.38
K	.100	.150	2.54	3.81
L	.600 BASIC		15.24 BASIC	
M	0°	15°	0°	15°
N	.018	.040	0.46	1.02

NOTE: Leads in true position within 0.010" (0.25mm) R at MMC at seating plane.

Pin numbers are shown for reference only. numbers may not be marked on package.

CASE: Plastic

MATING CONNECTOR: 2803MC

WEIGHT: 4.3 grams (0.15oz.)

ORDERING INFORMATION

Model	Package	Temperature Range	Description
HI3-0506A-5	28-Pin Plastic DIP	0°C to +75°C	16-Channel Single-Ended
HI1-0506A-5	28-Pin Ceramic DIP	0°C to +75°C	16-Channel Single-Ended
HI1-0506A-2	28-Pin Ceramic DIP	−55°C to +125°C	16-Channel Single-Ended
HI3-0507A-5	28-Pin Plastic DIP	0°C to +75°C	8-Channel Differential
HI1-0507A-5	28-Pin Ceramic DIP	0°C to +75°C	8-Channel Differential
HI1-0507A-2	28-Pin Ceramic DIP	−55°C to +125°C	8-Channel Differential

BURN-IN SCREENING OPTION
See text for details.

Model	Package	Temperature Range	Burn-In Temp. (160 Hours)[1]
HI3-0506A-5-BI	28-Pin Plastic DIP	0°C to +75°C	+85°C
HI1-0506A-5-BI	28-Pin Ceramic DIP	0°C to +75°C	+125°C
HI1-0506A-2-BI	28-Pin Ceramic DIP	−55°C to +125°C	+125°C
HI3-0507A-5-BI	28-Pin Plastic DIP	0°C to +75°C	+85°C
HI1-0507A-5-BI	28-Pin Ceramic DIP	0°C to +75°C	+125°C
HI1-0507A-2-BI	28-Pin Ceramic DIP	−55°C to +125°C	+125°C

NOTE: (1) Or equivalent combination of time and temperature.

DISCUSSION OF PERFORMANCE

DC CHARACTERISTICS

The static or DC transfer accuracy of transmitting the multiplexer input voltage to the output depends on the channel ON resistance (R_{ON}), the load impedance, the source impedance, the load bias current and the multiplexer leakage current.

Single-Ended Multiplexer Static Accuracy

The major contributors to static transfer accuracy for single-ended multiplexers are:

Source resistance loading error
Multiplexer ON resistance error
DC offset error caused by both load bias current and multiplexer leakage current.

Resistive Loading Errors

The source and load impedances will determine the input resistive loading errors. To minimize these errors:

- Keep loading impedance as high as possible. This minimizes the resisitive loading effects of the source resistance and multiplexer ON resistance. As a guideline, load impedances of $10^8\Omega$ or greater will keep resistive loading errors to 0.002% or less for 1000Ω source impedances. A $10^6\Omega$ load impedance will increase source loading error to 0.2% or more.
- Use sources with impedances as low as possible. A 1000Ω source resistance will present less than 0.001% loading error and 10kΩ source resistance will increase source loading error to 0.01% with a 10^8 load impedance.

Input resistive loading errors are determined by the following relationship: (see Figure 1)

Source and Multiplexer Resistive Loading Error

$$\epsilon_{(R_S + R_{ON})} = \frac{R_S + R_{ON}}{R_S + R_{ON} + R_L} \times 100\%$$

where R_S = source resistance
R_L = load resistance
R_{ON} = multiplexer ON resistance

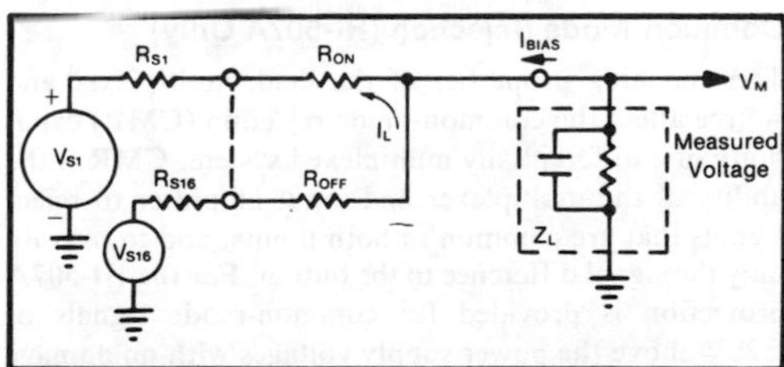

FIGURE 1. HI-506A Static Accuracy Equivalent Circuit.

Input Offset Voltage

Bias current generates an input OFFSET voltage as a result of the IR drop across the multiplexer ON resistance and source resistance. A load bias current of 10nA will generate an offset voltage of 20μV if a 1kΩ source is used. In general, for the HI-506A, the OFFSET voltage at the output is determined by:

$$V_{OFFSET} = (I_B + I_L)(R_{ON} + R_S)$$

where I_B = Bias current of device multiplexer is driving
I_L = Multiplexer leakage current
R_{ON} = Multiplexer ON resistance
R_S = Source resistance

Differential Multiplexer Static Accuracy

Static accuracy errors in a differential multiplexer are difficult to control, especially when it is used for multiplexing low-level signals with full-scale ranges of 10mV to 100mV.

The matching properties of the multiplexer, source and output load play a very important part in determining the transfer accuracy of the multiplexer. The source impedance unbalance, common-mode impedance, load bias current mismatch, load differenital impedance mismatch, and common-mode impedance of the load all contribute errors to the multiplexer. The multiplexer ON resistance mismatch, leakage current mismatch and ON resistance also contribute to differenital errors.

Referring to Figure 2, the effects of these errors can be minimized by following the general guidelines described in this section, especially for low-level multiplexing applications.

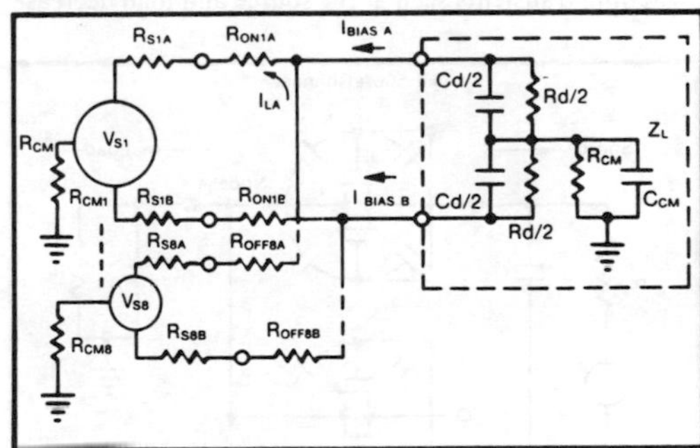

FIGURE 2. HI-507A Static Accuracy Equivalent Circuit.

Load (Output Device) Characteristics

- Use devices with very low bias current. Generally, FET input amplifiers should be used for low-level signals less than 50mV FSR. Low bias current bipolar input amplifiers are acceptable for signal ranges higher than 50mV FSR. Bias current matching will determine the input offset.
- The system DC common-mode rejection (CMR) can never be better than the combined CMR of the multiplexer and driven load. System CMR will be less than the device which has the lower CMR figure.
- Load impedances, differential and common-mode, should be $10^{10}\Omega$ or higher.

Source Characteristics

- The source impedance unbalance will produce offset, common-mode and channel-to-channel gain-scatter errors. Use sources which do not have large impedance unbalances if at all possible.
- Keep source impedances as low as possible to minimize resitive loading errors.
- Minimize ground loops. If signal lines are shielded, ground all shields to a common point at the system analog common.

If the HI-507A is used for multiplexing high-level signals of 1V to 10V full-scale ranges, the foregoing precautions should still be taken, but the parameters are not as critical as for low-level signal applications.

DYNAMIC CHARACTERISTICS

Settling Time

The gate-to-source and gate-to-drain capacitance of the CMOS FET switches, the RC time constants of the source and the load determine the settling time of the multiplexer.

Governed by the charge transfer relation $i = C\,(dV/dt)$, the charge currents transferred to both load and source by the analog switches are determined by the amplitude and rise time of the signal driving the CMOS FET switches and the gate-to-drain and gate-to-source junction capacitances as shown in Figures 3 and 4. Using this relationship, one can see that the amplitude of the switching transients seen at the source and load decrease

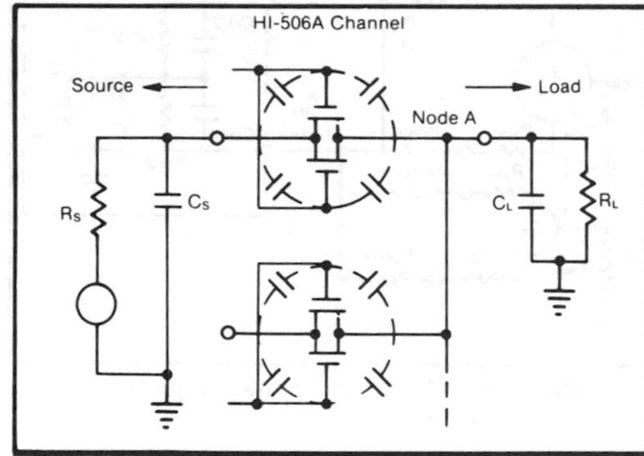

FIGURE 3. Settling Time Effects—HI-506A.

proportionally as the capacitance of the load and source increase. The tradeoff for reduced switching transient amplitude is increased settling time. In effect, the amplitude of the transients seen at the source and load are:

$$dV_L = (i/C)\,dt$$

where $i = C\,(dV/dt)$ of the CMOS FET switches

C = load or source capacitance

The source must then redistribute this charge, and the effect of source resistance on settling time is shown in the Typical Performance Curves. This graph shows the settling time for a 20V step change on the input. The

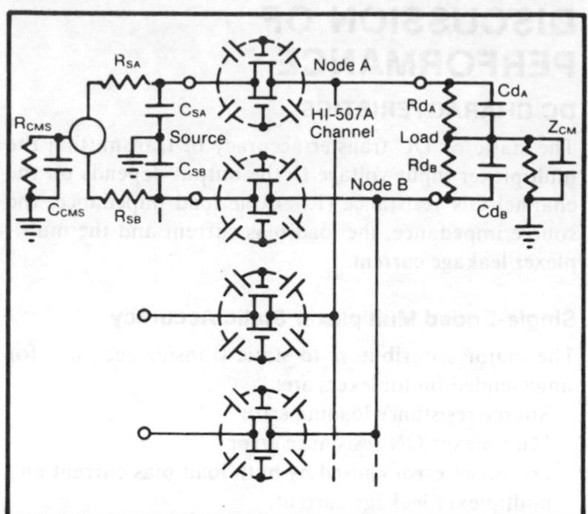

FIGURE 4. Settling and Common-Mode Effects HI-507A.

settling time for smaller step changes on the input will be less than that shown in the curve.

Switching Time

This is the time required for the CMOS FET to turn ON after a new digital code has been applied to the Channel Address inputs. It is measured from the 50 percent point of the address input signal to the 90 percent point of the analog signal seen at the output for a 10V signal change between channels.

Crosstalk

Crosstalk is the amount of signal feedthrough from the seven (HI-507A) or 15 (HI-506A) OFF channels appearing at the multiplexer output. Crosstalk is caused by the voltage divider effect of the OFF channel, OFF resistance and junction capacitances in series with the R_{ON} and R_S impedances of the ON channel. Crosstalk is measured with a 20Vp-p 1000Hz sine wave applied to all OFF channels. The crosstalk for these multiplexers is shown in the Typical Performance Curves.

Common-Mode Rejection (HI-507A Only)

The matching properties of the load, multiplexer and source affect the common-mode rejection (CMR) capability of a differentially multiplexed system. CMR is the ability of the multiplexer and input amplifier to reject signals that are common to both inputs, and to pass on only the signal difference to the output. For the HI-507A protection is provided for common-mode signals of ±20V above the power supply voltages with no damage to the analog switches.

The CMR of the HI-507A and Burr-Brown's INA110 Instrumentation Amplifier (G = 100) is 110dB at DC to 10Hz with a 6dB/octave rolloff to 70dB at 1000Hz. This measurement of CMR is shown in the Typical Performance Curves and is made with a Burr-Brown INA110 Instrumentation Amplifier connected for gains of 500, 100, and 10.

Factors which will degrade multiplexer and system DC CMR are:

- Amplifier bias current and differential impedance mismatch
- Load impedance mismatch
- Multiplexer impedance and leakage current mismatch
- Load and source common-mode impedance

AC CMR rolloff is determined by the amount of common-mode capacitances (absolute and mismatch) from each signal line to ground. Larger capacitances will limit CMR at higher frequencies; thus, if good CMR is desired at higher frequencies, the common-mode capacitances and unbalance of signal ines and multiplexer to amplifier wiring must be minimized. Use twisted-shielded pair signal lines wherever possible.

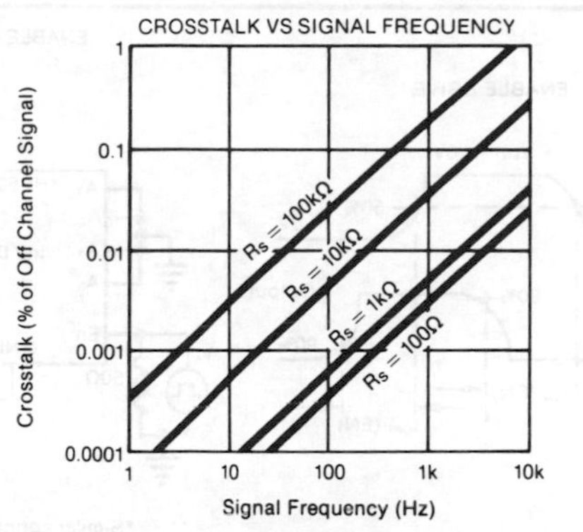

TYPICAL DYNAMIC PERFORMANCE CURVES

Typical at +25°C unless otherwise noted.

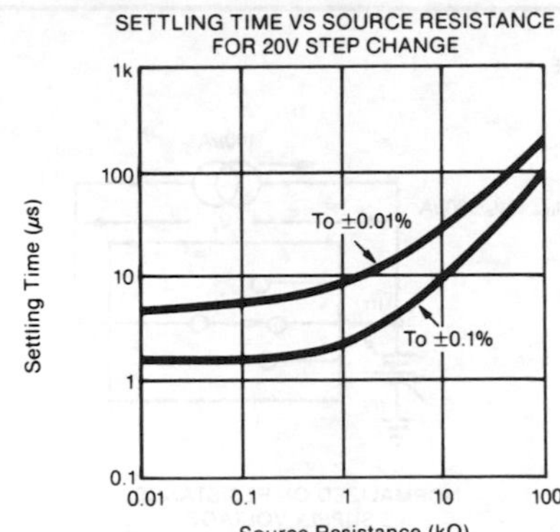

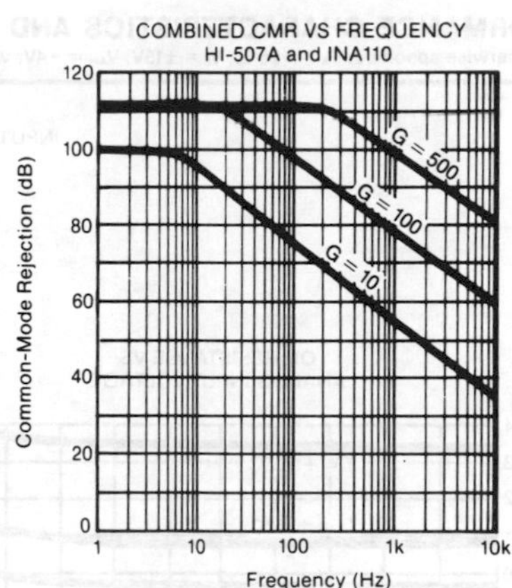

SWITCHING WAVEFORMS

Typical at +25°C unless otherwise noted.

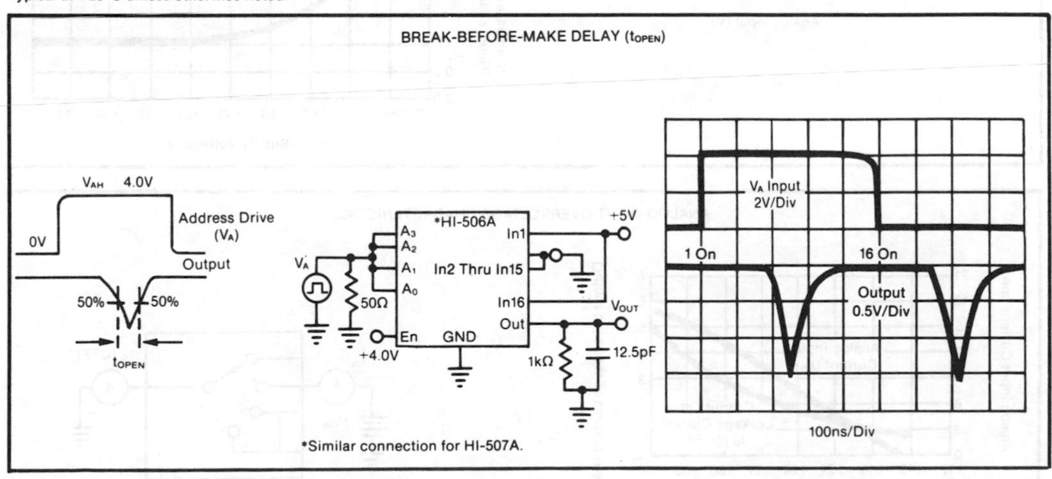

SWITCHING WAVEFORMS (CONT)

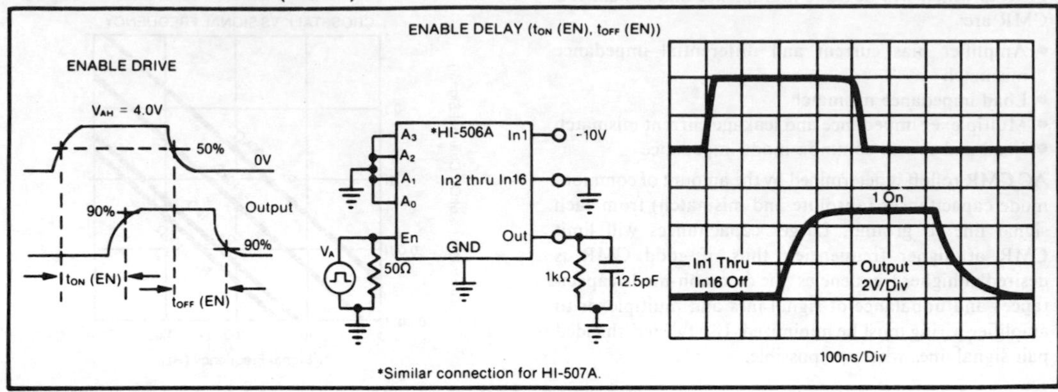

*Similar connection for HI-507A.

PERFORMANCE CHARACTERISTICS AND TEST CIRCUITS

Unless otherwise specified: $T_A = +25°C$, $V_S = \pm15V$, $V_{AH} = +4V$, $V_{AL} = 0.8V$ and V_{REF} = Open.

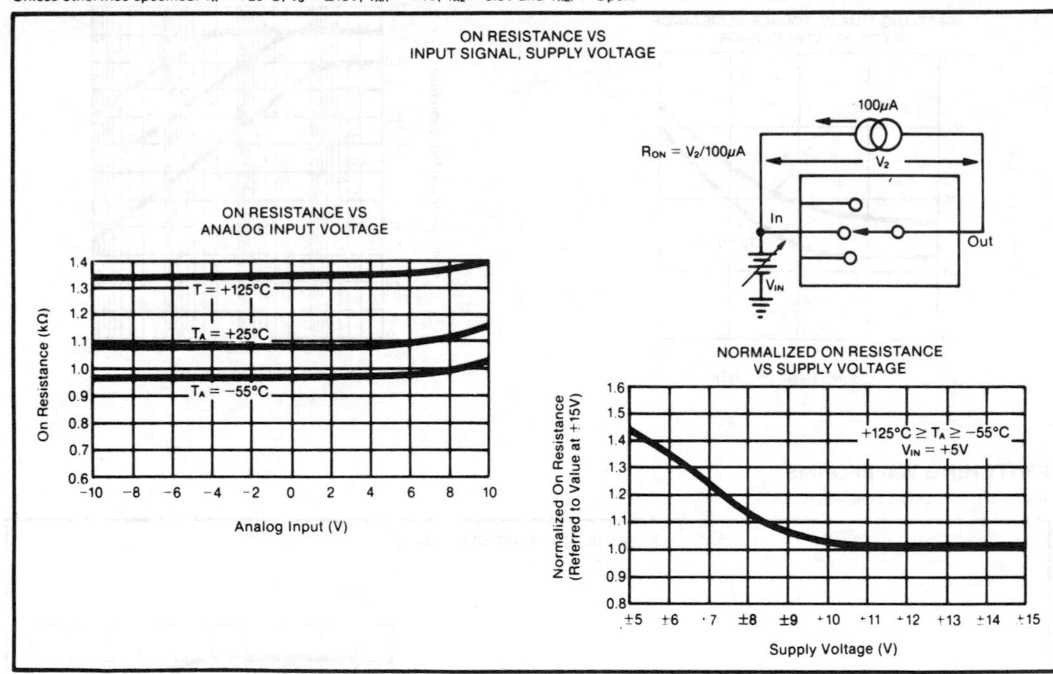

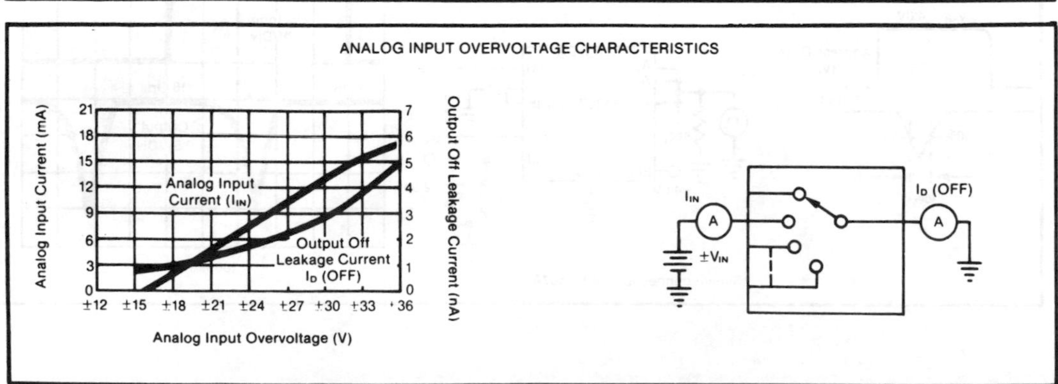

HI-506A/507A

7

ANALOG MULTIPLEXERS

PERFORMANCE CHARACTERISTICS AND TEST CIRCUITS (CONT)

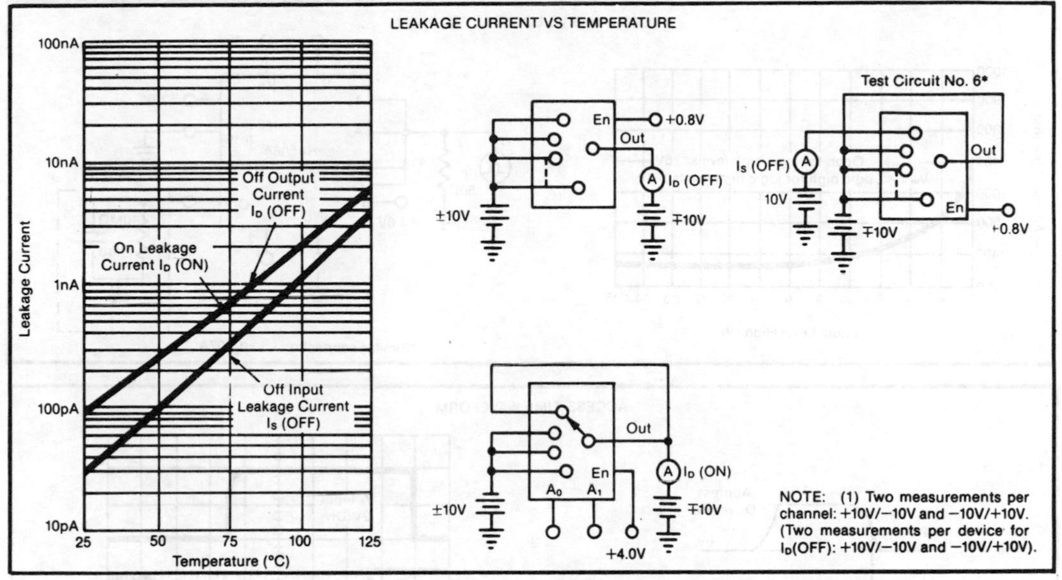

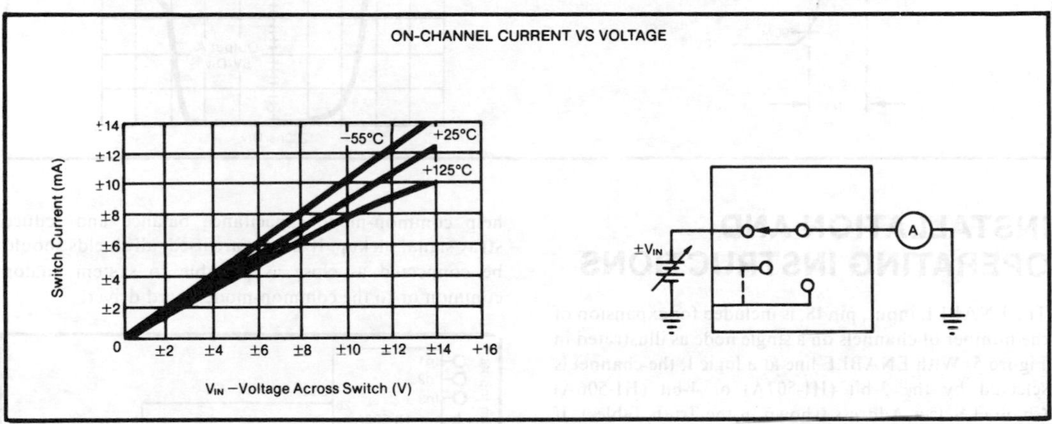

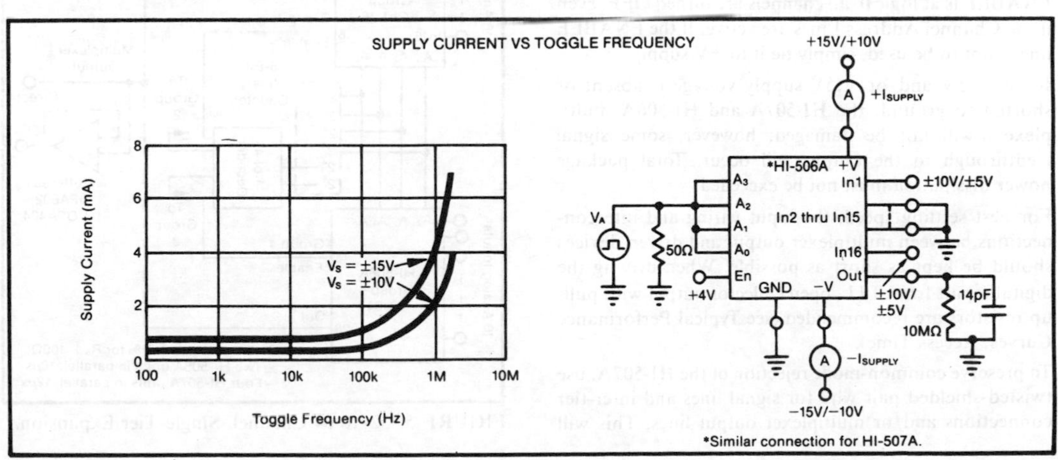

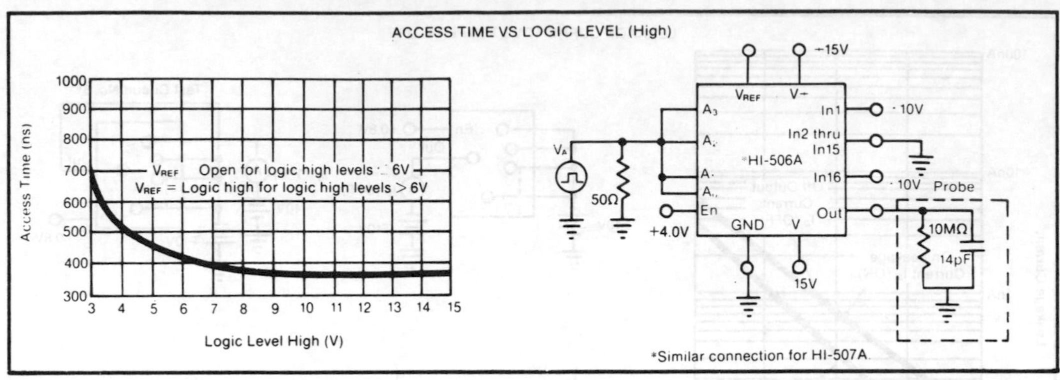

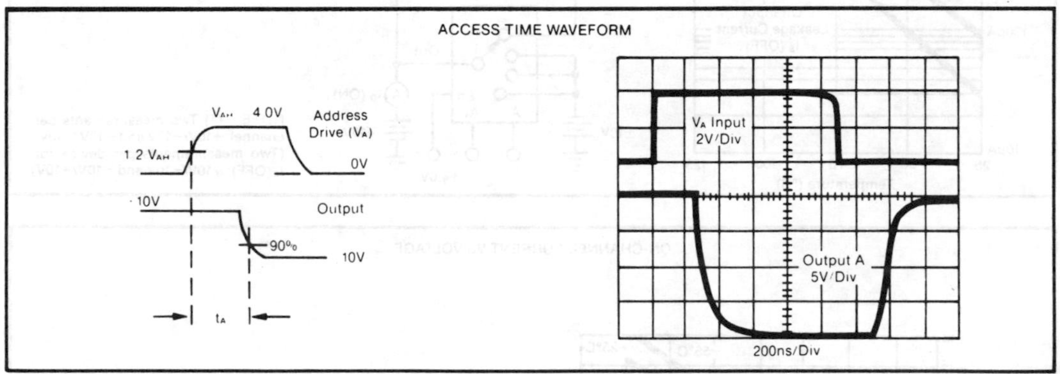

INSTALLATION AND OPERATING INSTRUCTIONS

The ENABLE input, pin 18, is included for expansion of the number of channels on a single node as illustrated in Figure 5. With ENABLE line at a logic 1, the channel is selected by the 3-bit (HI-507A) or 4-bit (HI-506A) Channel Select Address (shown in the Truth Tables). If ENABLE is at logic 0, all channels are turned OFF, even if the Channel Address Lines are active. If the ENABLE line is not to be used, simply tie it to +V supply.

If the +15V and/or −15V supply voltage is absent or shorted to ground, the HI-507A and HI-506A multiplexers will not be damaged; however, some signal feedthrough to the output will occur. Total package power dissipation must not be exceeded.

For best settling speed, the input wiring and interconnections between multiplexer output and driven devices should be kept as short as possible. When driving the digital inputs from TTL, open collector output with pull-up resistors are recommended (see Typical Performance Curves, Access Time).

To preserve common-mode rejection of the HI-507A, use twisted-shielded pair wire for signal lines and inter-tier connections and/or multiplexer output lines. This will help common-mode capacitance balance and reduce stray signal pickup. If shields are used, all shields should be connected as close as possible to system analog common or to the common-mode guard driver.

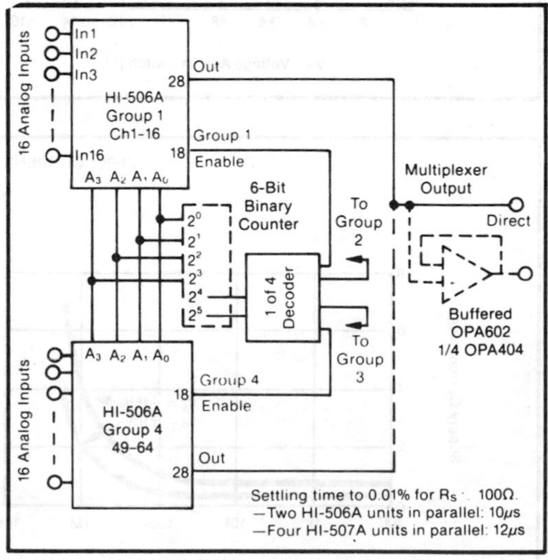

FIGURE 5. 32- to 64-Channel, Single-Tier Expansion.

CHANNEL EXPANSION

Single-Ended Multiplexer (HI-506A)

Up to 64 channels (four multiplexers) can be connected to a single node, or up to 256 channels using 17 HI-506A multiplexers on a two-tiered structure as shown in Figures 5 and 6.

Differential Multiplexer (HI-507A)

Single or multitiered configurations can be used to expand multiplexer channel capacity up to 64 channels using a 64 × 1 or an 8 × 8 configuration.

Single-Node Expansion

The 64 × 1 configuration is simply eight (HI-507A) units tied to a single node. Programming is accomplished with a 6-bit counter, using the 3LSBs of the counter to control Channel Address inputs A_0, A_1 and A_2 and the 3MSBs of the counter to drive an 8-of-1 decoder. The 8-of-1 decoder then is used to drive the ENABLE inputs (pin 18) of the HI-507A multiplexers.

Two-Tier Expansion

Using an 8 × 8 two-tier structure for expansion to 64 channels, the programming is simplified. The 6-bit counter output does not require an 8-of-1 decoder. The 3LSBs of the counter drive the A_0, A_1 and A_2 inputs of the eight first-tier multiplexers and the 3MSBs of the counter are applied to the A_0, A_1 and A_2 inputs of the second-tier multiplexer.

Single vs Multitiered Channel Expansion

In addition to reducing programming complexity, two-tier configuration offers the added advantages over single-node expansion of reduced OFF channel current leakage (reduced OFFSET), better CMR, and a more reliable configuration if a channel should fail ON in the single-node configuration, data cannot be taken from any channel, whereas only one channel group is failed (8 or 16) in the multitiered configuration.

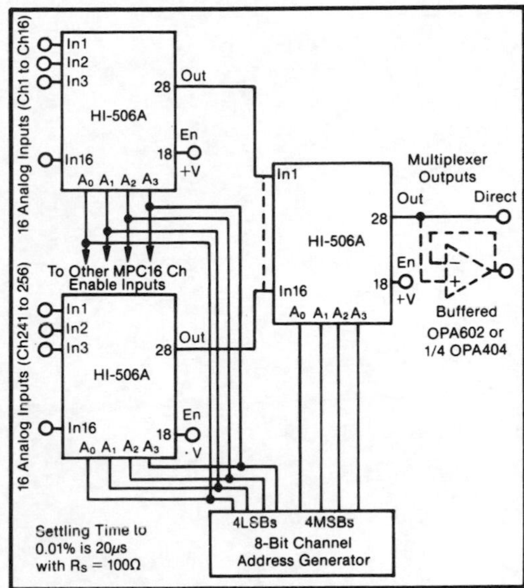

FIGURE 6. Channel Expansion Up to 256 Channels Using 16 × 16 Two-Tiered Expansion.

BURN-IN SCREENING

Burn-in screening is an option available for both plastic and ceramic package CMOS HI-050XA analog multiplexers. Burn-in duration is 160 hours at the temperature (or equivalent combination of time and temperature) indicated below:

Plastic "-BI" models: +85°C
Ceramic "-BI" models: +125°C

All units are 100% electrically tested after burn-in is completed. To order burn-in, add "-BI" to the base model number.

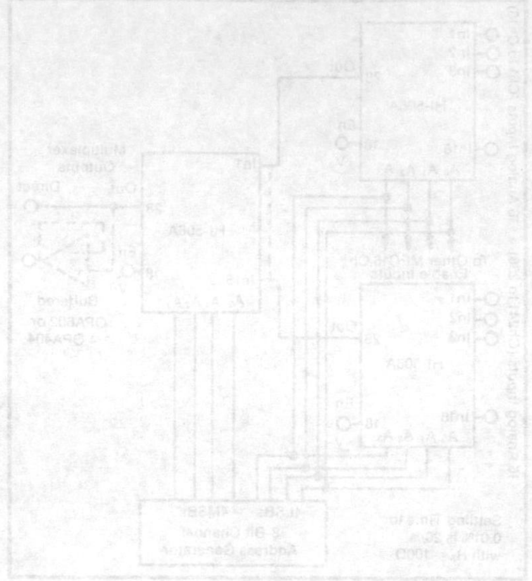

AN-35

UNDERSTANDING CROSSTALK IN ANALOG MULTIPLEXERS

Precision Monolithics Inc.

APPLICATION NOTE 35

INTRODUCTION

One of the most troublesome errors in analog multiplexers is crosstalk. Various schemes have been devised to reduce its effects. One designer will terminate the multiplexer in a 10kΩ resistive impedance. Another will short the multiplexer node to ground between address changes with an analog switch. A third engineer will terminate the multiplexer node in 1MΩ because he doesn't want to live with the attenuation which comes about with any lower impedance. What is confounding about these three situations is that the solution is correct in each case. THE CORRECT SOLUTION IS DICTATED BY THE APPLICATION.

To understand why the solution is application dependent, it is necessary to dig rather deeply into what crosstalk really is. When this is done, crosstalk is found to have not one, but three components in a multiplexer. To differentiate the components one from the other, it is convenient to give them names:

1. Static crosstalk (CT)
2. Dynamic crosstalk (DCT)
3. Adjacent Channel crosstalk (ACCT)

This application note explains the three crosstalk components qualitatively and quantitatively. The qualitative discussion tells what component(s) should be considered in various applications. The quantitative discussion uses both theoretical and empirical information to arrive at conclusions about what performance should be expected.

STATIC CROSSTALK (CT)

To introduce the concept of crosstalk, Figure 1 is helpful. A basic analog switch may be constructed with a FET (JFET or CMOS) and a suitable driver which switches it OFF and ON, as shown in Figure 1a. The equivalent circuit, as shown in Figure 1b, models the analog switch such that when the ideal switch (SW) is closed, the switch has an ON resistance R_{ON}. When SW is open, the OFF impedance is determined by C_{EQ}. A two-channel multiplexer circuit, made up of two analog switches connected as shown in Figure 1c, shows how signals from one channel can be coupled into the other channel. Theoretically, V_{OUT} consists of e_1 modified by the resistor divider formed by R_{ON1} and R_L (assumes reactance of C_L is $\gg R_L$). However, the capacitance of switch number two (C_{EQ2}) does couple some portion of e_2 into V_{OUT}. This is the simplest example of crosstalk.

The model which explains static crosstalk is relatively simple and may be derived from the OFF isolation model. Figure 2a shows the OFF isolation model as capacitive coupling from the input to the output of an OFF switch. This condition may be duplicated in Figure 1c by opening SW_1 and setting $e_2 = 0$. Coupling from input to output is accomplished through C_{EQ},

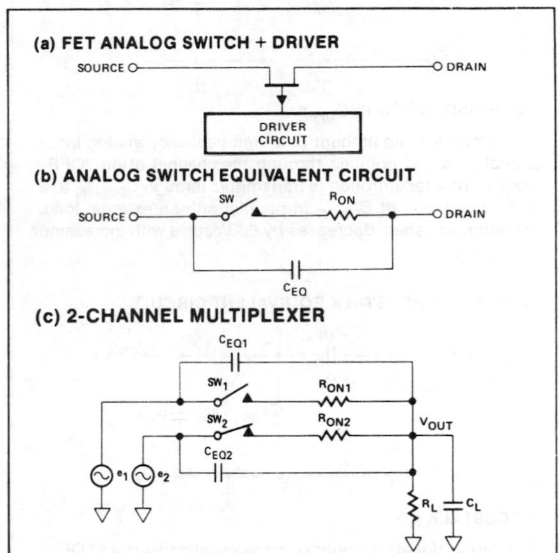

Figure 1. Essentials of an Analog Multiplexer

and this parameter may be computed from measurements of V_{IN}, V_{OUT}, and frequency. In the case of static crosstalk, C_{EQ} is shown coupling into a parallel combination of R_{ON} with R_L and C_L (Figure 2b). The two channel multiplexer shown in Figure 1c reduces to the circuit in Figure 2b, where $e_1 = 0$, $e_2 = V_{IN}$, and C_{EQ} is the coupling capacitance from e_2 to V_{OUT}.

Since R_L is generally 10kΩ or more, and typical analog switches are less than 1kΩ, static crosstalk is much smaller than OFF isolation. The crosstalk and OFF isolation numbers quoted on analog multiplexer data sheets are derived from the models shown in Figure 2. Unfortunately the one component of crosstalk specified is the least troublesome of the three. However the crosstalk figures on data sheets will alert the designer to those devices which absolutely will not satisfy his requirements.

There are applications where the static crosstalk specification given on data sheets is adequate. When the multiplexer is being used as a one-of-many switch, and is not being cycled through all channels on an automatic basis, then the static crosstalk component will give accurate prediction of the actual performance. Examples of such applications are:

1. Audio/Video Selector Switch
2. Programmable Gain Amplifier
3. Programmable Power Supply

AN-35

(a) OFF ISOLATION EQUIVALENT CIRCUIT

"OFF" ISOLATION (ISO_{OFF})

The proportionate amount of a high frequency analog input signal which is coupled through the channel of an "OFF" device. This feedthrough is transmitted through $C_{DS(OFF)}$ to a load comprised of $C_{D(OFF)}$ in parallel with an external load. Isolation generally decreases by 6dB/octave with increasing frequency.

(b) STATIC CROSSTALK EQUIVALENT CIRCUIT

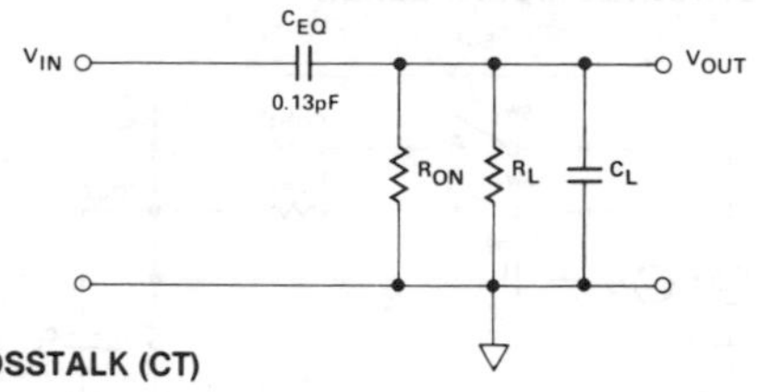

CROSSTALK (CT)

The proportionate amount of cross-coupling from an "OFF" analog input channel to the output of another "ON" output channel.

Figure 2. Model for Static Crosstalk

DYNAMIC CROSSTALK (DCT)

The dynamic crosstalk model can be derived from Figure 3. The switch SW_1 represents one condition on the multiplexer node (SW_1 is open). Actually SW_1 is continually switching between OFF and ON. This is represented in Figure 3b. In order to reduce crosstalk, multiplexers are designed to have break-before-make switching so that no two channels are addressed at the same time. The finite open time of SW_1 (shown in Figure 3b) represents the break-before-make action. There are two "open" conditions on the multiplexer node per cycle of the clock; thus the equivalent nodal resistance (R_{EQ}) may be computed as given in Figure 3b. Table I shows some typical values of static and dynamic crosstalk. Static crosstalk values are given in lines 1 and 12. There is a change in crosstalk as the clock frequency (f_{CLK}) is varied. Starting at line 4 notice the variation in crosstalk as R_L is varied from 10kΩ to 100kΩ while f_{CLK} remains constant at 100kHz. While Table I yields some theoretical values which give insight into the operation of dynamic crosstalk, a working multiplexer will have different values of f_{CLK} with respect to the maximum value of f_{SIG}. The real world situation will be analyzed in a later section of this paper.

Examples of multiplexer applications which are dynamic in nature are:

1. Industrial Process Control
2. Telephony
3. Data Acquisition Systems
4. Telemetry

Each one of the above applications are a form of Time Division Multiplexing. In other words, these are sampled-data

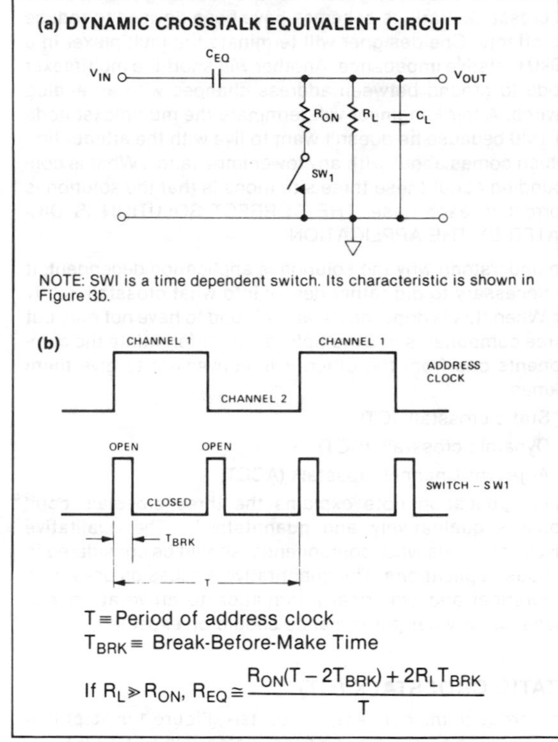

Figure 3. Model for Dynamic Crosstalk

Table 1. Computed Values of Static and Dynamic Crosstalk

LINE NO.	f_{SIG} Hz	f_{CLK} Hz	T μsec	T_{BRK} μsec	R_{ON} OHMS	R_L OHMS	R_{EQ} OHMS	C_{EQ} pF	CROSS-TALK dB
1	10K	0	—	0.80	300	10K	291	0.30	105
2	10K	20K	50	0.80	300	10K	602	0.30	99
3	10K	40K	25	0.80	300	10K	913	0.30	95
4	10K	100K	10	0.80	300	10K	1845	0.30	89
5	10K	100K	10	0.80	300	20K	3448	0.30	84
6	10K	100K	10	0.80	300	40K	6650	0.30	78
7	10K	100K	10	0.80	300	100K	16.25K	0.30	70
8	20K	50K	20	0.80	300	10K	1068	0.30	88
9	20K	50K	20	0.80	300	20K	1872	0.30	83
10	20K	50K	20	0.80	300	40K	3474	0.30	78
11	20K	50K	20	0.80	300	100K	8275	0.30	70
12	20K	0	—	0.80	300	100K	291	0.30	99

systems where each channel is being continuously sampled and the information for a given channel is contained in a given time slot. In these applications, the static crosstalk is almost meaningless, since the wrong choice of R_L (or f_{CLK}) can be disastrous.

ADJACENT CHANNEL CROSSTALK (ACCT)

Adjacent channel crosstalk is the most confusing component of crosstalk. In addition to its confusing nature, in some cases, it is the most dominant component. While both static and dynamic crosstalk are capacitive in nature, i.e., they vary with frequency at 6dB/octave, the adjacent channel crosstalk is **invariant with frequency**. In other words, it is possible to have crosstalk **when multiplexing DC signals** such as the outputs of thermocouples, pressure transducers, etc. The parameters which must be dealt with are R_L, C_L, R_{ON}, and f_{CLK}. In addition, the break-before-make time ($= T_{BRK}$) of the multiplexer is of importance. Before diving into the details of this component of crosstalk, it will be helpful to define what is meant by ACCT.

The term "adjacent" refers to time only. In other words, channel two is adjacent to channel one if channel two **immediately** follows channel one in time slots. Since the channel following is the "adjacent" channel, then channel one is not adjacent to channel two, but rather the other way around. Figure 4 illustrates the concept of adjacent channels. Assuming the multiplexer had, say, 1V on channel one, 2V on channel two, etc., then the output would look like the curve labeled "channel addressed." What is important about the waveforms in Figure 4 is the way the adjacent channel (in time) is shown. Note that while channel two is adjacent to channel one, channel one is itself adjacent to channel eight.

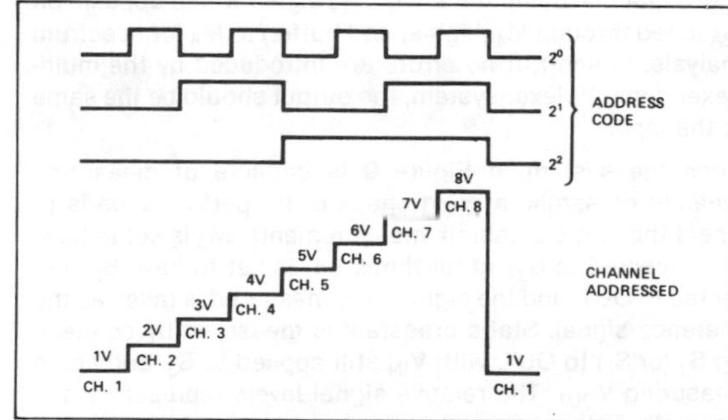

Figure 4. Adjacent Channel Concept

The fact that information is "carried forward" from one channel to the next (in time) suggests a storage mechanism as causing ACCT. Thus the multiplexer nodal capacitance becomes the prime suspect. Figure 5 illustrates how information is carried forward from one channel to the next as the addresses are changed. The address code is shown in Figure 5a, while Figure 5b shows the theoretical multiplexer output. Note that the even numbered channels have zero volt on them, while the odd channels have their channel number in volts. This arrangement best illustrates how the

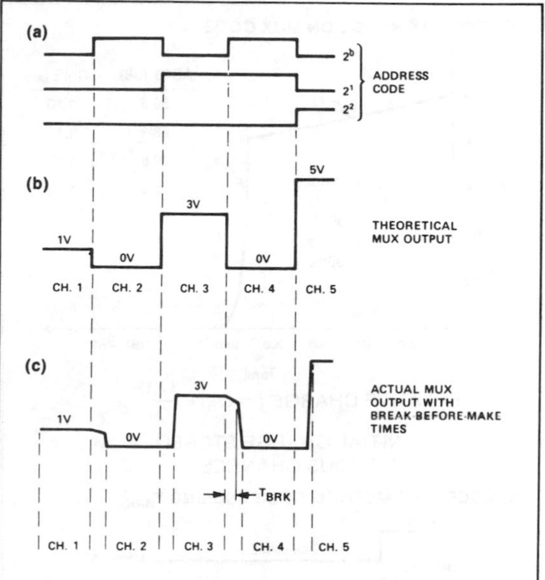

Figure 5. Adjacent Channel Crosstalk

information is transferred to the adjacent channel (as shown in Figure 5c). While the theoretical MUX output switches from channel three (3 volts) to channel four (0 volt) at the moment of the address change, note the delay in the actual MUX output caused by T_{BRK}. During this time the MUX node discharges along an RC curve determined by the load capacitance (C_L), and the load resistance (R_L). When the break-before-make time (T_{BRK}) is over, channel four is turned ON and the RC product is suddenly reduced to $R_{ON}C_L$. A curve which details how this all takes place is shown in Figure 6. Before leaving Figure 5, the arrangement suggests a method of avoiding adjacent channel crosstalk. In other words, the alternate grounding of channels prevents channel one signals from reaching channel three... channel three from reaching channel five, etc.

The curve in Figure 6a shows a typical nodal discharge for a set of real world conditions. The curve is normalized and T_{BRK} is chosen to be 900nsec. An accepted method of measuring T_{BRK} is from the 50% point of the channel which has been turned OFF to the 50% point of the channel which is being turned ON. This concept is illustrated in Figure 6b. In this case (Figure 6a) T_{BRK} is measured from the moment of the address change. While this is not totally correct, the agreement between theoretical and actual results is good enough to justify the simpler model which is derived. Since most designers are interested in crosstalk which is less than the resolution of the discharge curve, the ACCT vs. time graph gives crosstalk down to 90dB. In other words, the ACCT is down 90dB in less than 1.25μsec.

Adjacent channel crosstalk is a problem in every application where dynamic crosstalk must be considered; however there are techniques to minimize its effects. A popular way to diminish adjacent channel crosstalk is to short the

AN-35

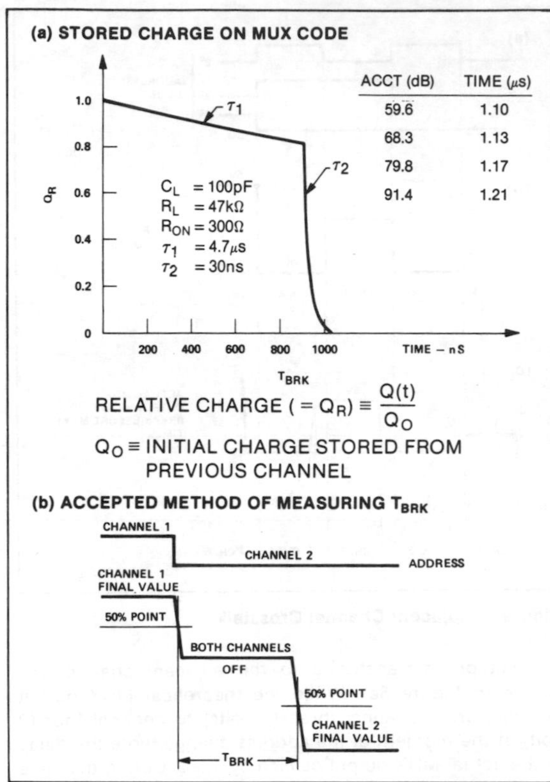

Figure 6. Stored Charge Decay and Definition of T_{BRK}

multiplexer node to ground between address changes. Tnis requires an additional analog switch which should be fast and have low R_{ON}. An alternative approach to reducing adjacent channel crosstalk is to ground every other channel in a multiplexer. This technique was illustrated in Figure 5.

MEASUREMENT OF STATIC CROSSTALK

Figures 7 and 8 give the element values for a typical PMI JFET MUX-08 on channel three. In the case shown, the OFF isolation was first measured and found to be 75dB. With R_L and f_{SIG} known, then C_{EQ} was calculated. Once C_{EQ} is known, then R_{EQ} may be calculated from the static crosstalk measurement made in Figure 8. R_{EQ} is the parallel combination of R_L and R_{ON}; thus it is possible to compute R_{ON} and this value is also shown in Figure 8. The measurements thus far are relatively simple and only require a voltmeter which is capable of measuring signals which are 100dB below the reference signal. On the other hand, the measurement of dynamic crosstalk is a bit more involved, and requires a more complex system.

MEASUREMENT OF DYNAMIC CROSSTALK

The crosstalk measuring system shown in Figure 9 is to be used for measuring dynamic crosstalk. The signal from M_5 is fed into M_1 where it is multiplexed onto the OUT terminal.

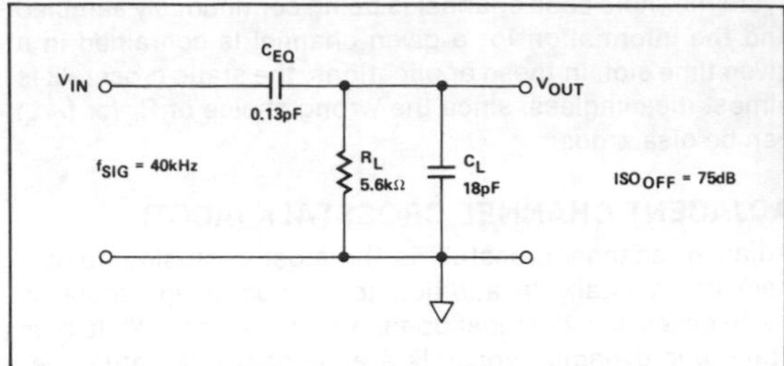

Figure 7. Typical OFF Isolation Element Values

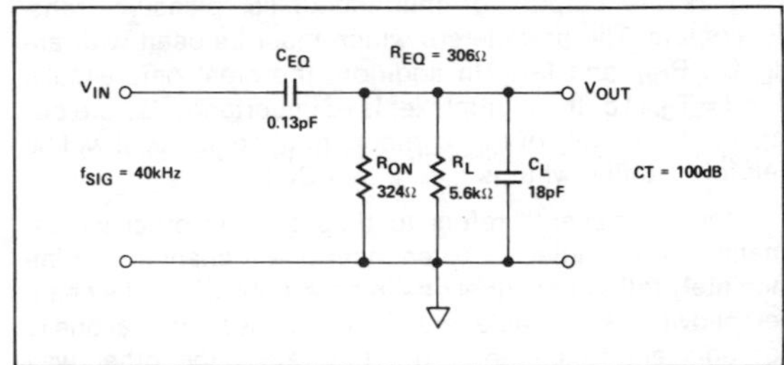

Figure 8. Typical Static Crosstalk Element Values

M_1 contains the multiplexer under test and a decoding circuit. The decoding circuit allows the selection of any two channels to be used as a two channel multiplexer. M_2 is a high-speed buffer used for driving the IN terminal of M_3. M_3 contains a multiplexer operated in a demultiplexer mode, along with decoding circuitry to allow several combinations of two channel demultiplexing. The signal which appears on S_{3A} is fed through M_4 (high-speed buffer) to M_8 for spectrum analysis. In short, if no errors are introduced by the multiplexer-demultiplexer system, the output should be the same as the input.

Since the system in Figure 9 is capable of measuring dynamic crosstalk, a good check of its performance is to repeat the static crosstalk measurements. M_3 is set to have IN connected to S_{3A} at all times. M_1 is set to have S_3 connected to OUT, and the signal thus measured is taken as the reference signal. Static crosstalk is measured by connecting S_1 (or S_8) to OUT, with V_{IN} still applied to S_3, and again measuring V_{OUT}. The relative signal levels represent static crosstalk. This measuring technique was used to verify the accuracy of the system.

The measurement of dynamic crosstalk leaves M_3 exactly as in the static case. With V_{IN} connected to S_3, M_1 is switched between S_1 and S_8. The signal frequency (f_{SIG}) was 40kHz and f_{CLK} was 100kHz (see Figure 10). From the crosstalk measured, the equivalent resistance (R_{EQ}) is computed to be 1150Ω (see Figure 10a). To verify the validity of this measurement, R_{EQ} was calculated using the formula in Figure 10c (T_{BRK} was measured separately). Since there is very good agreement between these two independently derived values, both the measurement technique and the dynamic crosstalk model are valid.

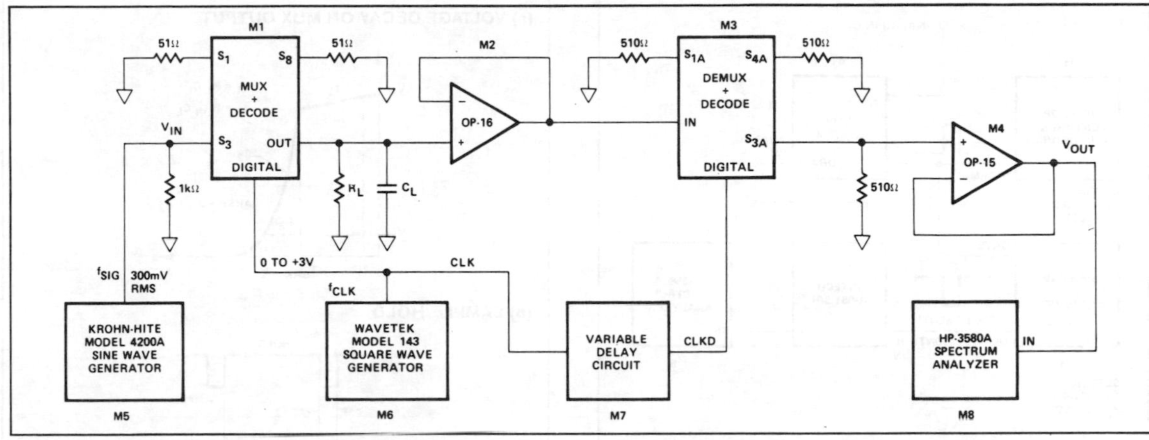

Figure 9. Dynamic Crosstalk Measuring System

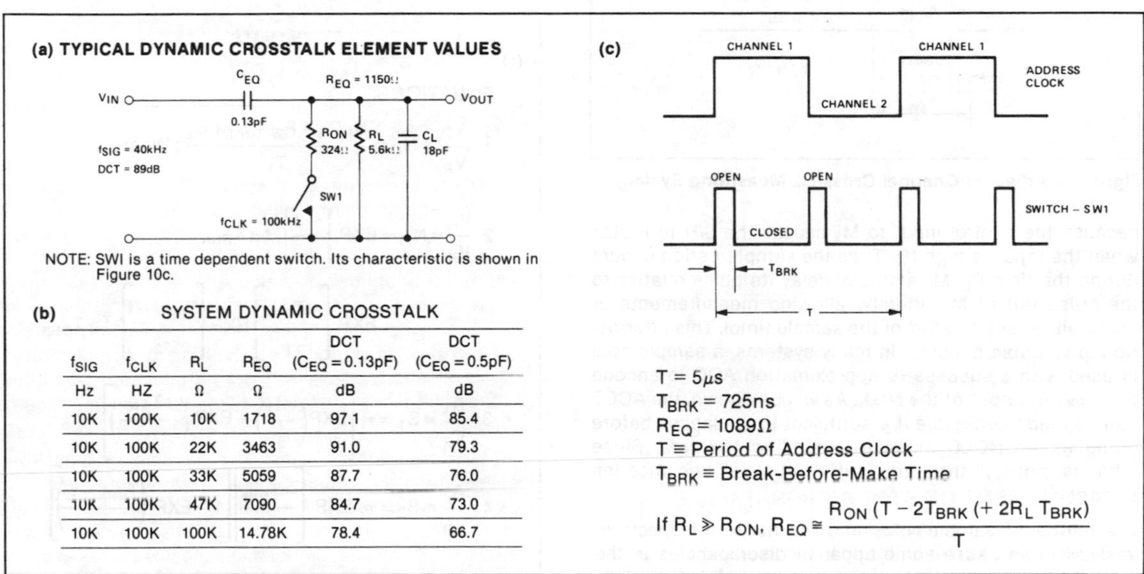

(b) SYSTEM DYNAMIC CROSSTALK

f_{SIG}	f_{CLK}	R_L	R_{EQ}	DCT (C_{EQ} = 0.13pF)	DCT (C_{EQ} = 0.5pF)
Hz	HZ	Ω	Ω	dB	dB
10K	100K	10K	1718	97.1	85.4
10K	100K	22K	3463	91.0	79.3
10K	100K	33K	5059	87.7	76.0
10K	100K	47K	7090	84.7	73.0
10K	100K	100K	14.78K	78.4	66.7

$T = 5\mu s$
$T_{BRK} = 725ns$
$R_{EQ} = 1089\Omega$
$T \equiv$ Period of Address Clock
$T_{BRK} \equiv$ Break-Before-Make Time

$$\text{If } R_L \gg R_{ON},\ R_{EQ} \cong \frac{R_{ON}\,(T - 2T_{BRK}\ (+\ 2R_L\ T_{BRK})}{T}$$

Figure 10. Computed Dynamic Crosstalk for Actual Multiplexer

The numbers shown in Figure 10 apply to the measurement system, but are unlikely in a real multiplexer. To satisfy sampling theory limitations, f_{SIG} must be less than one-half the sampling frequency. Assuming $f_{CLK} = 200kHz$ then each channel in a multiplexer is addressed for 5µsec. This means that it takes 40µsec to sample all channels of an eight channel multiplexer. In other words, **each channel** is sampled at a 25kHz rate. Thus the maximum value of f_{SIG} would be 12.5kHz. Figure 10b gives values of dynamic crosstalk (DCT) which would be experienced if the values of R_{ON} and T_{BRK} shown in Figures 10a and 10b were used. The first DCT column lists the values for a C_{EQ} of 0.13pF (measured value of channel three). The second DCT column shows the performance for $C_{EQ} = 0.5pF$. The purpose for the second column is to point out how critical minimizing stray capacitance is to good crosstalk performance.

MEASUREMENT OF ADJACENT CHANNEL CROSSTALK

The system shown in Figure 11 was used to measure adjacent channel crosstalk (ACCT). M_1 drives the address lines of the MUX system and the gating input of M_4. By setting the period of M_4 (T_2) to 10µsec, the pulse rate out of M_4 is controlled by the pulse rate of M_1 (40µsec) coming into the gate input of M_4. The output of M_4 is in the complement mode

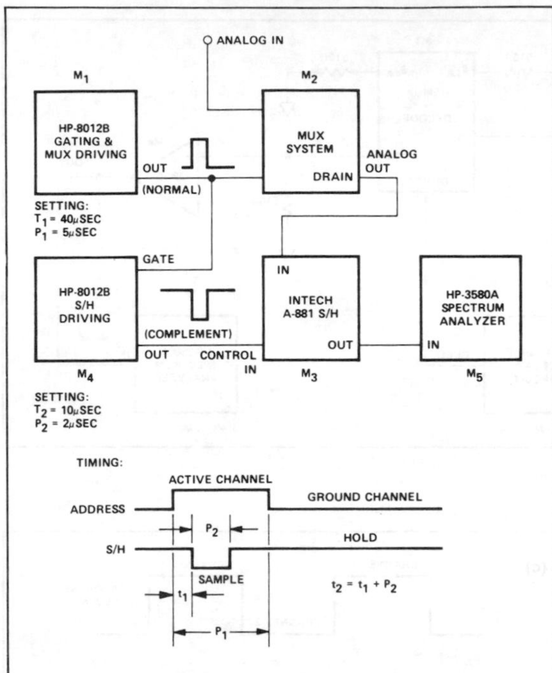

Figure 11. Adjacent Channel Crosstalk Measuring System

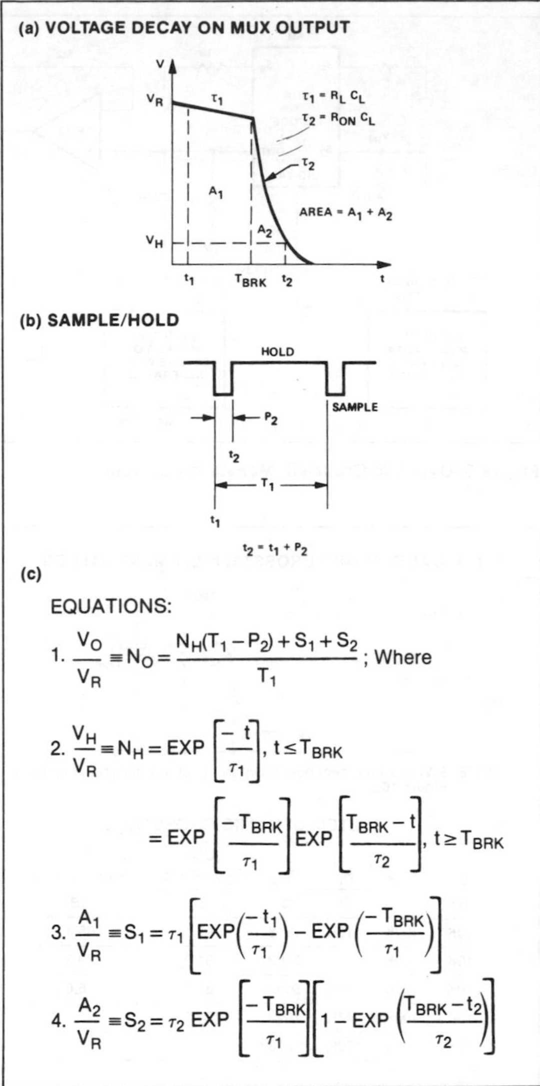

1. $$\frac{V_O}{V_R} \equiv N_O = \frac{N_H(T_1 - P_2) + S_1 + S_2}{T_1}\text{; Where}$$

2. $$\frac{V_H}{V_R} \equiv N_H = \text{EXP}\left[\frac{-t}{\tau_1}\right],\ t \le T_{BRK}$$

$$= \text{EXP}\left[\frac{-T_{BRK}}{\tau_1}\right]\text{EXP}\left[\frac{T_{BRK} - t}{\tau_2}\right],\ t \ge T_{BRK}$$

3. $$\frac{A_1}{V_R} \equiv S_1 = \tau_1\left[\text{EXP}\left(\frac{-t_1}{\tau_1}\right) - \text{EXP}\left(\frac{-T_{BRK}}{\tau_1}\right)\right]$$

4. $$\frac{A_2}{V_R} \equiv S_2 = \tau_2\,\text{EXP}\left[\frac{-T_{BRK}}{\tau_1}\right]\left[1 - \text{EXP}\left(\frac{T_{BRK} - t_2}{\tau_2}\right)\right]$$

Figure 12. Predicting the Measurement System Response

because the control input to M_3 causes the S/H to HOLD when the input is high (1). Thus the sample period occurs during the time P_2. M_4 also can delay its pulse relative to the pulse out of M_1, thereby allowing measurements of crosstalk versus t_1 (start of the sample time). This information is valuable because in many systems, a sample/hold is used with a successive approximation ADC to encode the analog output of the MUX. As will be shown, the ACCT can be made negligible if a sufficient time elapses before going to the HOLD mode for encoding the data. Since "time is money," the term "sufficient time" becomes important.

The nature of sample/holds and the nature of spectrum analyzers can cause some apparent discrepancies in the data observed by this measurement system. It is important to note the spectrum analyzer "sees" the average of **everything** that is presented to its input terminals. While it is true the sample/hold holds the last value it "saw," the spectrum analyzer also looks at the signal present during the sample/hold's sample time. Thus the equation which expresses the signal level present as a function of time must also account for the true averaging of the spectrum analyzer. Figure 12 shows the equations (12c) and the definitions of the terms used in the equations (12a and 12b). The term N_O is the **relative** signal level which the spectrum analyzer measures. If the model of the signal decay shown in Figure 12a is the correct one to explain the ACCT, then the computed value of N_O should correspond to the measured values. As will be shown in Figure 14, the agreement does in fact justify the model; however it was necessary to choose the measurement conditions **very carefully.**

In order to get good correlation between lab data and theoretical predictions, it was necessary to use fairly long time constants ($R_L = 22k\Omega$ and $C_L = 1000pF$). With $R_L = 22k\Omega$ and $C_L = 50pF$ ($R_{ON} = 300\Omega$), the theoretical plot of ACCT (as measured on the spectrum analyzer) vs. t_1 is shown in Figure 13. Note that the data is plotted between 900nsec and 1025nsec. The curve shows that a **10nsec error in t_1 can cause a 6dB error** in reading on the spectrum analyzer. The results shown in Figure 14 confirm the necessity of using large capacitances to obtain predictable results. The theoretical curve tracks the actual data well in both cases; however the 1000pF curve is better than the 300pF curve. Notice that there is good agreement both at DC and at 4kHz.

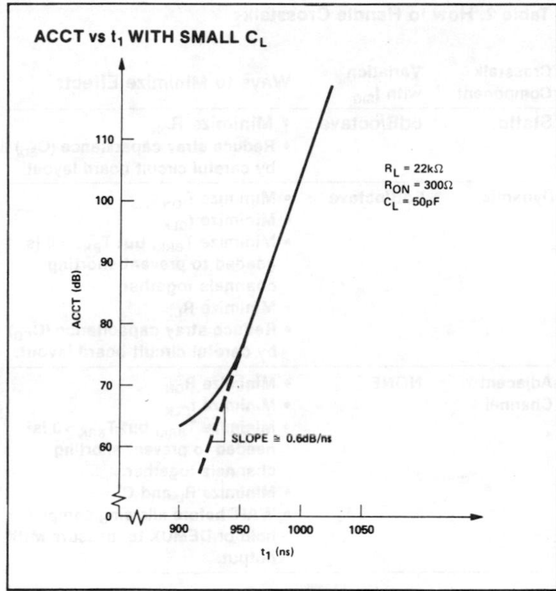

Figure 13. Measurement Errors Due To Small C_L

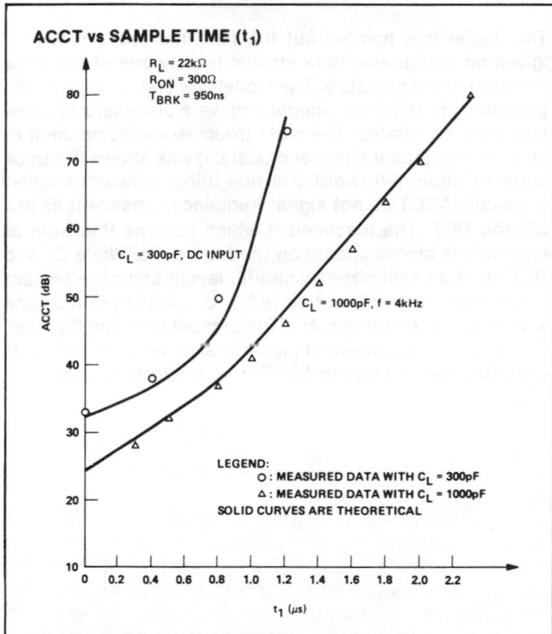

Figure 14. Agreement Between Measured and Computed ACCT

PREDICTING AND CONTROLLING ADJACENT CHANNEL CROSSTALK

The equations in Figure 12c can be used to predict how much adjacent channel crosstalk one might expect in an actual system. An all analog system will follow the MUX with a

A. Multiplexer-Demultiplexer System:

$N_H \equiv 0$ Therefore

1. $N_O = \dfrac{S_1 + S_2}{T_1}$, Where $T_1 = \dfrac{1}{f_{CLK}}$ x (No. of Channels)

2. $S_1 = \tau_1 \left[\text{EXP}\left(\dfrac{-t_1}{\tau_1}\right) - \text{EXP}\left(\dfrac{-T_{BRK}}{\tau_1}\right) \right]$

3. $S_2 = \tau_2 \ \text{EXP}\left[\dfrac{T_{BRK}}{\tau_1}\right]\left[1 - \text{EXP}\left(\dfrac{T_{BRK} - t_2}{\tau_2}\right)\right]$

Where $t_1 = T_D$ (Break-Before-Make Time of DEMUX)

$$t_2 = \frac{1}{f_{CLK}} - T_D$$

B. Multiplexer — Sample/Hold System

$S_1 = S_2 = P_2 \equiv 0$

4. $N_O = N_H = \text{EXP}\left[\dfrac{-t}{\tau_1}\right] \quad t \leq T_{BRK}$

$= \text{EXP}\left[\dfrac{-T_{BRK}}{\tau_1}\right] \text{EXP}\left[\dfrac{T_{BRK} - t}{\tau_2}\right], \quad t \geq T_{BRK}$

Where: $t = t_H$ (Hold Command for Sample/Hold as measured from Address Change Time)

Figure 15. Predicting Adjacent Channel Crosstalk

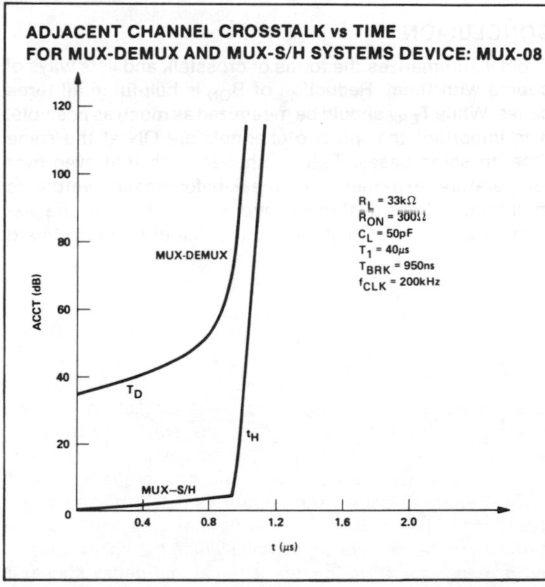

Figure 16. Computed ACCT vs Time for MUX-DEMUX and MUX-S/H Systems

AN-35

demultiplexer, which will have its own break-before-make delay. An analog to digital system will have a sample/hold amplifier in front of the A/D converter. Since the equations which apply to these situations are different, they will be discussed separately. Figure 15 summarizes the conditions and the equations which apply to them.

Since there is no held voltage, then $N_H = 0$ in the multiplexer-demultiplexer system. This reduces N_O to the simple form shown in equation (1). S_1 and S_2 follow in equations (2) and (3). Since $t_1 = T_D$ (break-before-make time of the DEMUX), that time will have a significant effect on ACCT. The MUX-sample/hold system imposes the condition $S_1 = S_2 = P_2 = 0$; thus $N_O = N_H$. It will be instructive to compare the levels of ACCT in these two systems versus their appropriate times.

Figure 16 looks at a "typical" system which will give approximately one percent transmission error (33kΩ R_L and 300Ω R_{ON}), and has 50pF C_L. The value of C_L is somewhat on the high side (20pF being typical for MUX-08 connected to a buffer amp), but it does give a conservative value for analysis. What Figure 16 shows is rather startling. The adjacent channel crosstalk, while inherent in the multiplexer itself, can be eliminated in **both** systems by the proper timing. In the case of the sample/hold it is only necessary to delay the hold command for approximately 1.2μsec to have the ACCT vanish completely. This is no problem, since most sample/holds need at least 2μsec to accurately acquire the signal (this is particularly true of monolithic devices). The plot for the MUX-DEMUX system relates to T_D, which is **not adjustable** for a given DEMUX. What is possible is to add some delay to the address change for the DEMUX. In this way, the DEMUX will not "look" at the MUX output until the charge from the previous channel has had a chance to dissipate.

CONCLUSION

Table II summarizes the forms of crosstalk and lists ways of coping with them. Reduction of R_{ON} is helpful in all three cases. While T_{BRK} should be minimized as much as possible, it is important that no two channels are ON at the same time. In some cases, T_{BRK} is chosen such that even over temperature extremes, the break-before-make feature is maintained. Since all three components of crosstalk are present in a dynamic multiplexer, the "careful circuit board layout" is important even though it is not listed in the ACCT section.

Table 2. How to Handle Crosstalk

Crosstalk Component	Variation with f_{SIG}	Ways to Minimize Effects
Static	**6dB/octave**	• **Minimize R_{ON}** • **Reduce stray capacitance (C_{EQ}) by careful circuit board layout.**
Dynamic	**6dB/octave**	• **Minimize R_{ON}** • **Minimize f_{CLK}** • **Minimize T_{BRK}, but $T_{BRK} > 0$ is needed to prevent shorting channels together.** • **Minimize R_L** • **Reduce stray capacitance (C_{EQ}) by careful circuit board layout.**
Adjacent Channel	**NONE**	• **Minimize R_{ON}** • **Minimize f_{CLK}** • **Minimize T_{BRK}, but $T_{BRK} > 0$ is needed to prevent shorting channels together.** • **Minimize R_L and C_L** • **WAIT before allowing sample/hold or DEMUX to measure MUX output.**

This paper has pointed out the fact that static crosstalk (given on multiplexer data sheets) is only **one** of the **three** components of crosstalk. The models for static and dynamic crosstalk are relatively simple and were discussed to show how they are related. The most troublesome component of crosstalk (adjacent channel crosstalk) was shown not to be quite so straight-forward. For one thing, adjacent channel crosstalk (ACCT) is **not signal frequency dependent** as are CT and DCT. The mechanism which governs this form of crosstalk is stored charge on the MUX node. While CT and DCT must be minimized by careful layout and once present in the multiplexer cannot be reduced, such is not the case with ACCT. Even though ACCT is present in the multiplexer, the proper timing of demultiplexer or sample/hold commands can effectively eliminate ACCT from the total system.

Monolithic sample and hold amplifiers

D. Bowers looks at the recent advances in sample/hold amplifiers

Sample/hold amplifiers are devices capable of acquisition and temporary storage of analogue information, and in principle are by no means new devices. Recent advances in high speed data acquisition systems have brought these devices to the forefront, and availability in monolithic form has meant that designers now have a more cost effective approach than discrete or hybrid designs can normally offer.

When the complexity and conflicting requirements of sample/hold amplifiers are examined, it is perhaps not surprising that monolithic technology has been slow in taking over, but new generations of single chip designs are replacing discretes and hybrids at an ever increasing rate.

Before examining the design problems, some general theory is appropriate.

Basic sample/hold designs

Many types of sample/hold device exist, for instance first order and polygonal hold devices perform extrapolation or interpolation between discrete samples. Also some high speed designs take short, fixed samples via pulsed switches and are incapable of following the input continuously. All presently available monolithic types though, function as analogue equivalents of the "transparent" latch so the terms "track and hold" and "sample and hold" will be considered equivalent for the purpose of this description.

The author is with Bourns (Trimpot)

The universal storage element for analogue information is the capacitor and from this the general circuit of Fig. 1 evolves. In monolithic designs this "hold" capacitor, as it is known, is external to the device, partly due to the difficulty of integrating sufficiently large capacitors, and partly to allow performance optimisation discussed later. When the switch S1 closes, Ch charges to the input voltage, and this charge remains when the switch opens. The buffer amplifiers ensure that the external load does not discharge the capacitor and also ensure that the charging current is supplied from the power supplies, rather than from the input. In high speed systems the switch must, of course, be a solid state type and this represents one of the first performance limitations in an S/H amplifier.

Static errors are those of the buffers themselves and are the same as for operational amplifiers: Vos, TCVos, CMRR, PSRR, etc. As with operational amplifiers, the most significant error is usually initial offset (Vos) and to minimise this the closed loop arrangement of Fig. 2 is often used. Since the offset of A2 is now within the feedback loop of A1, only the latter's Vos remains as an error term which is easily trimmed externally in the same manner as nulling operational amplifiers. This closed loop arrangement is particularly advantageous considering that a FET input buffer, with relatively high Vos, is usually used for A2 in order to keep the capacitor discharge current (usually referred to as the "droop" current) to a minimum. Another variation is the integrating mode design (Fig. 3). The output amplifier now acts as an integrator with the input amplifier acting as a transconductance drive to the capacitor. The advantage of this configuration is that the switch and input amplifier operate close to virtual ground, disadvantages are that the output amplifier must also supply the charging current, and that an open loop design is precluded. Other configurations including the use of multiple switches are possible. Overall settling time of a closed loop design will be longer than a comparable open loop design, and frequently resistance has to be placed in series with the capacitor to ensure loop stability, which places limits on the charge time. This brings us to a consideration of dynamic (sampling) errors which are somewhat more complex.

Hold to sample errors

Fig. 4 illustrates the dynamic performance at the instant of sampling. When the switch S1 closes, the capacitor does not charge instantaneously but at a rate dependent on the

Fig. 1: Basic open loop sample/hold amplifier.

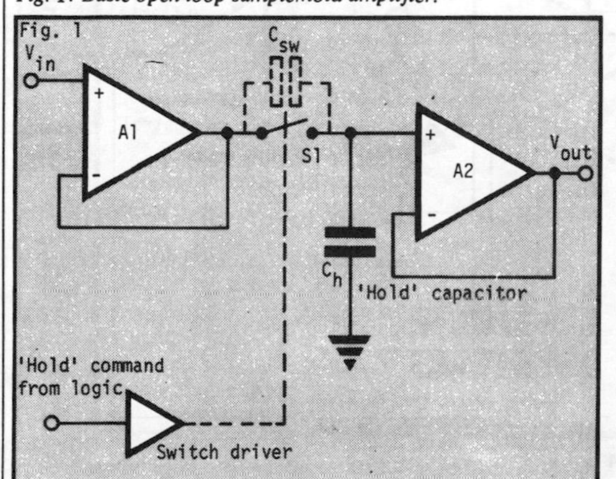

Fig. 2: Closed loop version of Fig. 1.

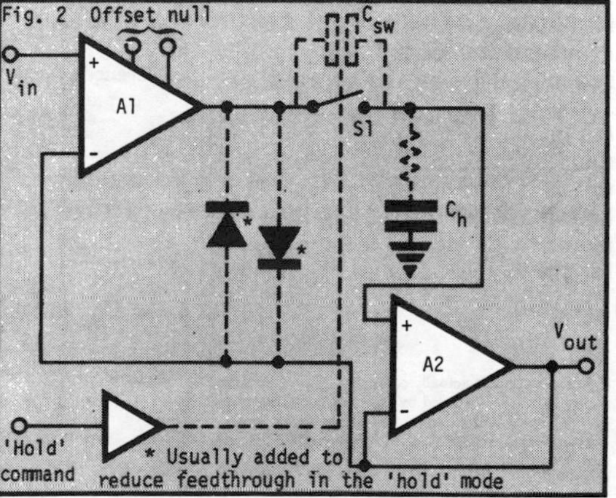

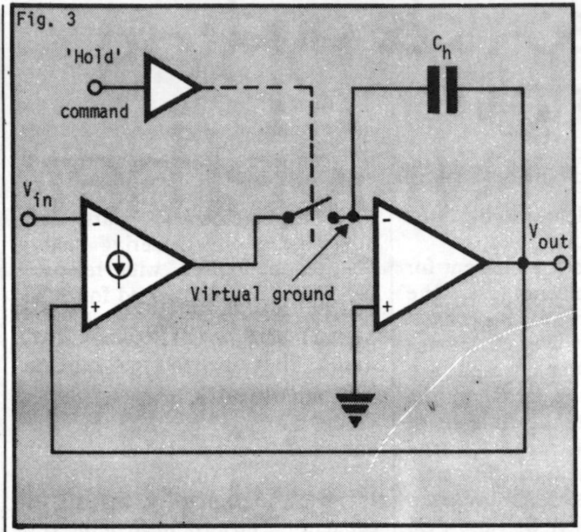

Fig. 3: Integrating mode design.

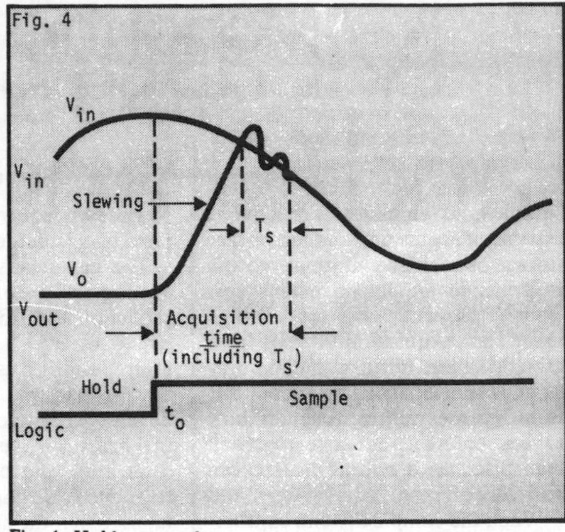

Fig. 4: Hold to sample errors.

capacitor size, switch resistance and available output current from the input buffer. Thus the overall charging curve can be quite complex and the diagram shows a rather idealised function. Acquisition time is the time taken to acquire an input signal (usually a 10V step) to a specified accuracy (usually 0.1%) with a specified capacitor size, and may or may not include the system settling time T_s which is inevitable with such a high capacitive load. Clearly, acquisition time will be inversely proportional to capacitor size and this point should be borne in mind when evaluating data sheets. With realistic capacitor sizes, monolithic designs are capable of acquisition times in the 1-10 microsecond region.

Sample to hold errors

The act of opening switch S1 places the amplifier in the "hold" mode (Fig. 5). S1 cannot open instantaneously when the "hold" command is received, but typically suffers a delay of the order of 50-250 ns known as the aperture time during which the output will continue to track the input.* Following this is a system settling time (T_{hs}) known as the hold settling time and is usually defined as the time taken for the output to settle within one

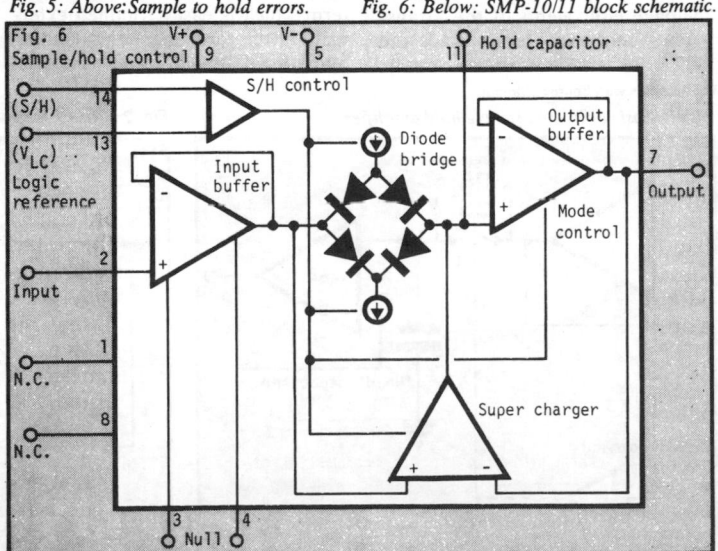

* Aperture time is actually composed of several terms, including effects imposed by the frequency limitations of the input buffer. In a very high speed system, aperture time can be partially compensated for in which case aperture uncertainty time becomes the dominant factor.

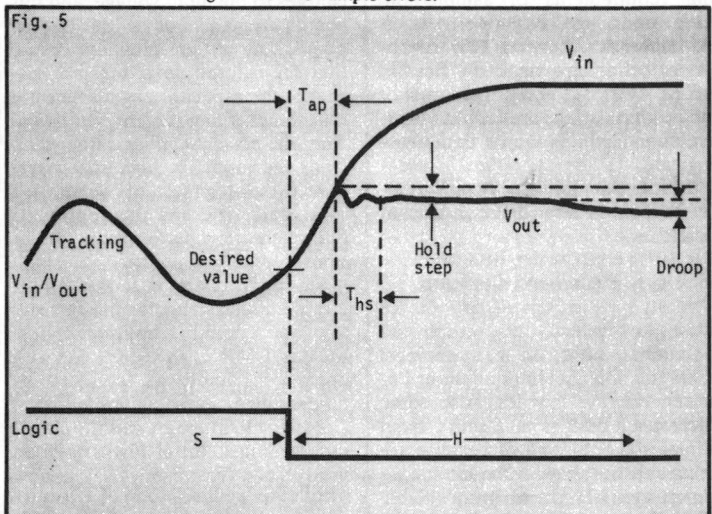

Fig. 5: Above: Sample to hold errors.

Fig. 6: Below: SMP-10/11 block schematic.

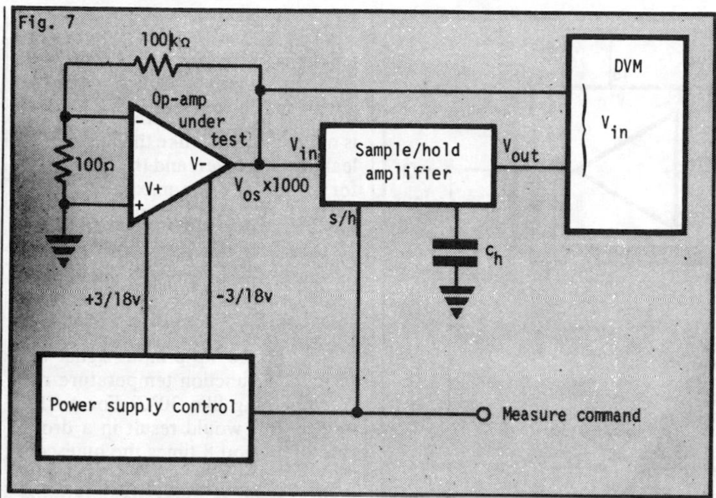

Fig. 7: Measuring op-amp PSRR with an S/H amplifier.

acquisition and droop rate. With the capacitor as an external component, this trade-off can be optimised for a particular application.

The errors during sampling are mostly those of the buffers themselves, but the capacitor loading on the input buffer limits the maximum slew rate available. Note also that if significant charging current remains when the "hold" command is applied, any resistance in series with the capacitor will produce an extra DC error between sample and hold.

millivolt of its final value. (This figure is usually in the 0.5-10 μs range.) The hold step occurs because of feedthrough of the sample pulse via the switch, and this causes extra charge (typically 5-100 pc) to be transferred to the "hold" capacitor resulting a small DC shift. Since errors are usually more important during hold than during sampling, the DC errors of the buffers and this charge transfer error are often combined in a parameter called "zero scale" error. Note that the hold step will be inversely proportional to capacitor size, so that zero scale error is only valid for a given value of "hold" capacitor. Additionally, since both gain error and hold step depend on input voltage**, zero scale error exhibits a dependence on analogue signal.

During hold, the charge on the capacitor will alter because of capacitor leakage, internal leakages and any current drawn by the output buffer. The early part of this change will be almost linear and, even though polarity is undefined, is known as the droop rate (droop rate is equal to droop current divided by the value of the "hold" capacitor). A large hold capacitor will result in a low droop rate, and vice-versa, and there is thus a trade-off between

** Hold step can be fairly constant in integrating mode designs.

Practical sample/hold amplifiers

The configurations so far described represent a highly simplified view of available amplifiers, and many manufacturers have developed variations on the basic design to suit their own particular processes. It is true to say, however, that FET technology has a large role to play in the realisation of monolithic sample/hold amplifiers. FET's can provide both the technology used for switching and the high input impedance required of the output buffer. Even with modern fabrication techniques, FET input buffers have relatively high Vos so the commonest configuration is the closed-loop configuration.

One device using JFET technology is the AD582, which features a droop current of 50 pA maximum at 25°C ambient. This device is an example of the closed loop integrat-

Fig. 8: 8 Bit DAC with S/H buffer.

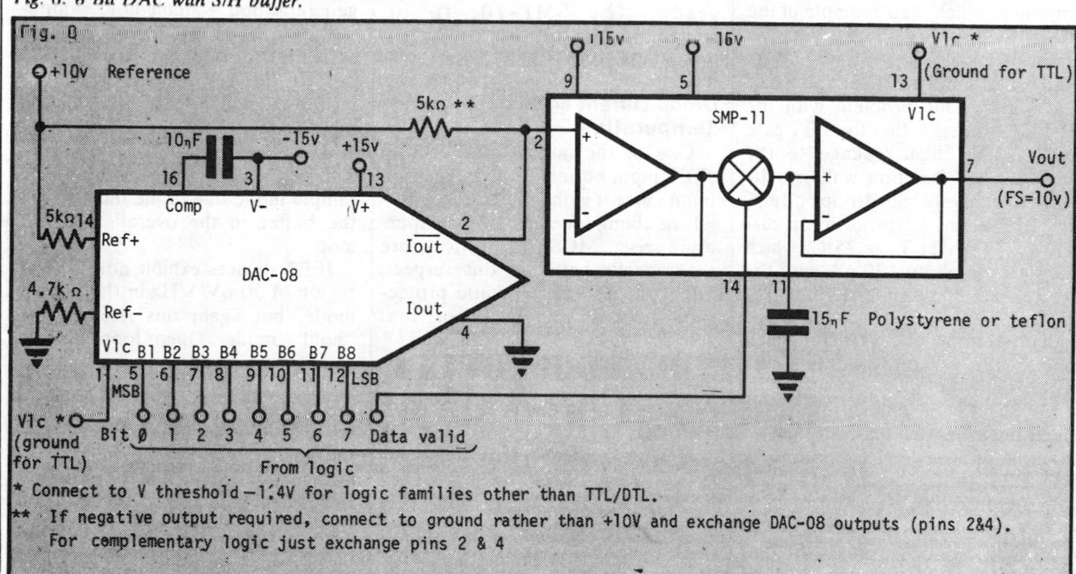

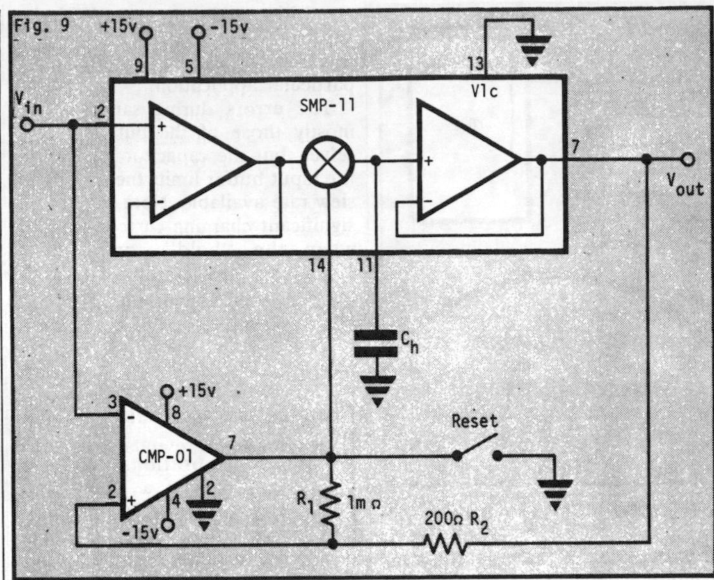

Fig. 9: Positive peak detector

ing mode configuration, but has the loop closed externally enabling the input amplifier to provide voltage gain in a similar manner to operational amplifiers.

Acquisiton time is 6 μs with C_H = 100 pF, and offset voltage is 6 mV maximum for the "K" grade.

A common process for integrating well matched JFETs on the same chip as bipolar transistors is the BI-FET process and naturally the advantages of this technique have been used to produce sample/hold devices.

The LF398 is a popular BI-FET amplifier, which is an example of the closed loop unity gain configuration featuring an acquisition time of 4 μs (C_H = 1 nF) and an offset voltage of 7 mV max. One problem with the BI-FET process is that the FET gate has a rather high leakage to the substrate and isolation walls resulting in a relatively high droop current. The LF398 has a typical droop current of 30 pA at T_j = 25°C, which translates to about 120 pA with the device warmed up (non-sampling) at 25°C ambient. This device features a high input impedance ($10^{10}\Omega$) which increases its versatility within a data acquisition system.

MOSFETs of course can have much lower leakage current, and an example of a device incorporating a MOSFET buffer is the HA2425, with a typical droop current of 5 pA at 25°C ambient. This device has an externally closed loop configuration, and features an acquisition time of 5 μS (C_H = 1 nF) and an offset voltage of 6 mV max.

An example of the unity gain open-loop configuration is the SMP-11 from Precision Monolithics (Fig. 6), a completely FET-less design. The diode bridge switch ensures a low charge transfer (5 pC) and the "Supercharger" (a modified transconductance amplifier) gives up to 50 mA of charge current, allowing a 3.5 μs acquisiton time (C_H = 5 nF). The dual mode superbeta output buffer gives a droop current of 300 pA at 25°C ambient, and zero scale error is 1 mV max. for the "E" grade. The SMP-10, to be announced later this year, will have a droop current five times lower[2].

Droop current against temperature

One of the major drawbacks of JFET input buffers is the increase in input current with temperature, usually doubling for every 10°C temperature rise. MOSFET devices are theoretically better in this respect, although the internal static protection circuits usually mean that roughly the same law is followed.

The result is that BI-FET droop currents can be 10 nA or so at T_A = 70°C, and up to 200 nA at 125°C. Diffused JFET designs can be somewhat better, typically leaking about 1 nA at 70°C (25 nA at 125°C). MOSFETs are capable of holding droop current below 1 nA over the commercial temperature range, although leakage can be up to 10 nA at 125°C.

In contrast, bipolar devices are at their worst at low temperatures due to the decrease in transistor beta with fall in temperature. This effect is not exponential like the increase in leakage, however, and the SMP-11A for example, has a maximum droop current of 4 nA from −55°C to 125°C. Allowance must also be made for junction temperature rise during sampling, which can be significant where high speed data is concerned. With a 10 nF hold capacitor, a 100 kHz 20 Vp-p input signal will produce a junction temperature rise of something like 30°C. For a FET device, this would result in a droop rate of around 8 times the quiescent figure, whereas this self-heating would tend to improve a bipolar device.

Other parameters

Other parameters may be of importance in some applications, input characteristics for example can be important in data acquisition systems where a sample/hold amplifier often directly follows an analogue multiplexer. One specification not so far mentioned is analogue feedthrough, which is the rejection of input signal in the "hold" mode. Closed loop designs frequently have clamp diodes to prevent the input buffer saturating during "hold", which would greatly degrade this parameter. Most monolithic devices offer a feedthrough rejection of around 90 dB at 1 kHz.

Noise can also be an important factor in some applications, and as a generalisation sample/hold amplifiers are relatively noisy devices. MOSFET devices are traditionally poor in this respect, and MOSFET buffered sample/holds frequently have noise in the region of 100 μV RMS in the hold mode. This figure will be considerably lower in the sample mode due to the inclusion of the buffer in the overall feedback loop.

JFET devices exhibit noise in the region of 50 nV/√Hz in the sample mode, but again this rises in the "hold" mode. Open loop devices would be expected to have noise more or less unchanged between modes, although the use of a dual mode output amplifier (such as in the SMP-11) will cause an increase in the "hold" mode due to the reduction of input stage currents within the buffer. The SMP-11 has a broadband noise of 15 μV over a 10 kHz bandwidth during "hold", falling by about 30% in the sample mode.

The "hold" capacitor

The choice of dielectric for the "hold" capacitor is important with all sample/hold amplifiers. Some dielectrics, notably ceramic and mylar, exhibit high dielectric polarisation — causing spurious acquisition values and high droop rates. Polystyrene is the recommended dielectric for industrial applications though polycarbonate and polypropylene also work well. Although polycarbonate will work up to 125°C, teflon capacitors are the usual choice for critical military applications, due to the extremely low dielectric absorption available with some types. Interestingly, the MOS capacitors used for compensation with many monolithic IC's also have good dielectric polarisation characteristics, however, the severe limitation on capacitance (≃200 pF) seems likely to keep the "hold" capacitor as an external component. Another factor is the insulation resistance, usually expressed in megohm microfarads. Suitable high quality capacitors often have insulation resistance greater than 5×10^5M Ω μF, which would contribute only 1% droop in 1½ hours.

Design precautions

Implementation of a sample/hold amplifier is not as straightforward as many other linear devices, since there are both logic and linear signal paths, and very high charging transients. As a guide, the following points should be considered.

(i) For good noise immunity, the sample/hold amplifier should be selected to ensure compatibility with the logic systems in use. Some devices (such as the AD582 and SMP-11) have logic threshold controls which facilitate their use with several logic families, and the control voltage for this input should be derived from the logic supply and decoupled to digital ground.

(ii) The capacitor should be connected to analogue rather than digital ground since any potential appearing at capacitor ground will pass directly to the output. Also note that this should be a low impedance ground, since large transient currents (up to 50 mA) can flow during sampling.

(iii) Any stray capacitance between the hold capacitor and the digital inputs can greatly increase the apparent magnitude of the "hold" step. Use of a guard ring around the capacitor input connected to the amplifier output can help to alleviate this, as well as reducing droop effects due to PCB surface leakage.

Applications of sample/hold amplifiers

One of the major uses of sample/hold amplifiers is to hold data steady during an A/D conversion, which even with successive approximation types can often take 20 μs or more. The sample/hold amplifier enables input signals with speeds approaching the sampling frequency to be sampled at precise intervals, and also enables optimum speed to be obtained from analogue multiplexer systems. Other advantages obtained with the sample/hold amplifier include much less susceptibility to noise, and the ability to capture fast signal peaks, as well as providing the necessary buffering between multiplexer and A/D converter.

Sample/hold amplifiers have equally important applications within the general field of instrumentation, a typical example being the "delta" test of Fig. 7, which shows a method for direct evaluation of PSRR of an operational amplifier. With the operational amplifier power supplies at ±3 V, resistors R1 and R2 cause amplification of the operational amplifier offset voltage, which is stored in the sample/hold amplifier by a latch pulse. Increasing the supplies to ±18 V now allows the DVM to read the change in offset, which with the values shown equals the PSRR × 30,000.

Another application is analogue deglitching/latching in data distribution systems. Fig. 8 shows an eight-bit glitch free voltage output D/A converter with a settling time of about 12 μs, which is capable of holding the output within ½LSB for 2 s after the digital input has disappeared. Sample/hold amplifiers also have applications as peak detectors, Fig. 9 showing a positive peak follower configuration. The comparator places the sample/hold in the sampling mode whenever the input voltage exceeds the previous maximum value (R1 and R2 add hysteresis to reduce threshold noise). The 50 mA charging current ensures that large values of C_H can be used without drastic degradation of response time: 1 μF, for instance would result in a droop rate around 350 μV/s. With extra complexity this principle can be extended to provide polarity programmability[3]

References

1. Kress, D. "Monolithic Sample/Hold Amplifiers: IC Flexibility Comes to the Sample/Hold Scene". Electronic Products magazine March 1977.
2. Erdi, G. & Henneuse, P. "A Precision FET-less Sample and Hold with High Charge to Droop Current Ratio". IEEE Journal of Solid State Circuits. December 1978, Volume SC-13, Number 6.
3. Soderquist, D. Precision Monolithics' Application Note AN-27. "A Polarity Programmable Peak Detector".

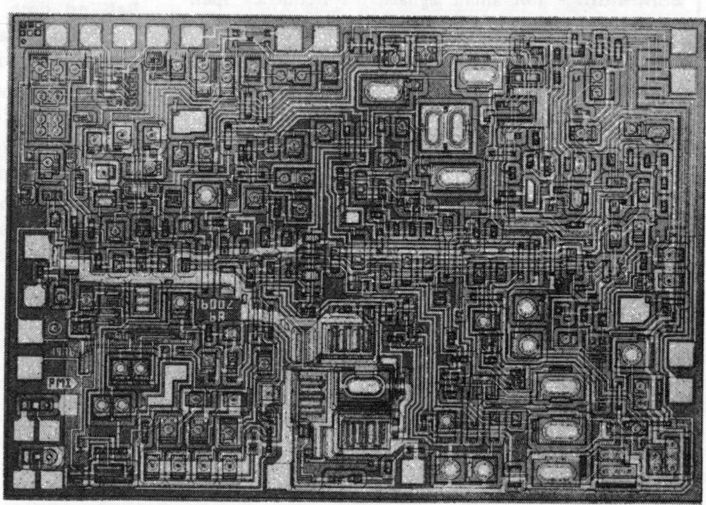

Pic A: sample/hold chip from Precision Monolithics uses no FETs. The probing pads at top left are used for zener-zap trimming of zero-scale error

TO SIDESTEP SAMPLE/HOLD PITFALLS, RECOGNIZE SUBTLE DESIGN ERRORS

By knowing the key parameters and practices that govern important sample/hold functions, you can obtain optimum performance from these deceptively simple devices.

You can avoid the potential design traps that lurk in sample/hold applications by understanding and applying proven sampling rules and definitions. Widely used for voltage storage in analog-signal-processing and data conversion systems, sample/hold devices provide seemingly simple operation that often misleads designers.

In practice, functional intricacies hide error sources that can degrade sample/hold performance. Unfortunately, when manufacturers describe these errors, further complications can arise because most vendors use their own nonstandard terminology. (For some widely accepted definitions, see below.)

To help clarify a muddled situation, therefore, this application note presents some sample/hold design-verification guidelines. By carefully defining the basic sample/hold types - their primary specifications and their chief time- and frequency-response considerations - these guidelines arm you with the design information needed for proper sample/hold application (see Table 1).

A GLOSSARY OF SAMPLE/HOLD TERMS

Acquisition time - Time after the sample-to-track command activates for the hold capacitor to charge to a full-scale voltage change and settle within a specified error band around the final voltage value.

Aperture delay - Elapsed time from activation of the sample-to-hold command to the opening of the switch in Hold mode.

Aperture time - Time for a switch to go from sample to Hold mode, measured from the 50% point of mode-control transition to when the output stops sampling the input.

Aperture uncertainty time - Variation in the time required for a switch to open after sample-to-hold transition occurs, or the time variation in aperture delay.

Bandwidth - For small signals, the frequency span between the points at which a sample/hold's gain goes down 3dB from its DC value (the frequency span between the points at which the output signal's amplitude equals 0.707 times the input signal). This parameter serves as a gauge of amplifier performance in Sample mode.

Charge injection - Offset error voltage on the hold capacitor when charge transfers from the capacitor to the gate-drive circuit via capacitive coupling at switch turn-off.

Droop rate - Hold capacitor's voltage-output decay or drift in Hold mode, arising from switch leakage current, hold-capacitor value and op amp bias current.

Feedthrough - Amount of input signal that appears at a sample/hold's otuput in Hold mode.

Gain accuracy - Expressed by the deviation in gain from its nominal value.

Monotonicity - In ADCs, describes an output that increases continuously with increasing input.

Quantizing error - Inherent uncertainty in digitizing an analog voltage to the nearest digital code word.

Sample/hold - Amplifier circuit that acquires analog input voltages during very short sampling times and stores them on a hold capacitor for a specified time.

Settling time - Time taken after a sample-to-hold transition for a sample/hold's output to assume final voltage value with a specified accuracy.

Spectrum response - Charcteristic that traces a sample/hold's amplitude - versus - frequency values.

Slew rate - Fastest rate, usually measured in volts per second, at which a sample/hold's output can change.

Zero-order hold - Filter that reconstructs an analog signal from a train of impulses by producing the first term of a power series approximation of the input.

TABLE I. Characteristics of Diode-Bridge and FET Sample/Hold.

DIODE BRIDGE TYPE	FET TYPE
Lower charge injection error.	Greater DC accuracy; has no offset and doesn't need feeback resistors.
Less generation of spikes; lower drive voltages are needed for bridge switching.	Faster in higher voltage applications; offers larger slewing current.
Shorter aperture delay; Diodes switch faster than FET's.	Smaller droop; Exhibits less inherent leakage.

FOR HIGH SPEED, USE DIODE-BRIDGE UNITS

High performance wideband and sample/holds come in two versions - junction FET and diode bridge types (see Figure 1). Diode bridge devices more readily accommodate applications that call for short RC time constants. Emphasing speed rather than accuracy, these applications generally involve 6-bit to 10-bit data acquisition systems with sampling rates of 1MHz to 50MHz.

Because diode bridge devices require up to 5mA bias, their slew rate becomes the limiting factor in large capacitor and large input signal applications. When the hold capacitor exceeds 15pF, for example, the diode bridge's slew rate starts to decrease. (In such cases, consider the FET-type sample/hold.)

An important diode bridge application centers on deglitching a fast DAC (see Figure 2). Diode bridges' straightforward interfacing capability allows them to serve well here, and the relatively low voltage switching levels involved lead to very fast operation. In display applications, especially, DAC deglitchers must deliver low noise signals with minimal spiking at high update rates (3MHz to 20MHz). The deglitchers operate by placing the sample/hold in Hold mode before updating the DAC.

CALL ON FET DEVICES FOR BOTH SPEED AND ACCURACY

For more demanding applications, FET-type sample/holds yield superior performance in high speed, high accuracy, 10-bit to 13-bit data acquisition systems. Accuracy and speed result from the use of two FET switches, both of which help cancel the charge-injection error caused by the gate's drive waveform, permitting a low hold capacitor value. Additional FET-type sample/hold advantages include higher input impedance than diode bridge types and the elimination of offset, bias, and feedback resistors. And because FET sample/holds exhibit much lower leakage than diode bridge units - as well as a balanced configuration - the FET type's output buffer amplifier contributes virtually no leakage current.

In high voltage applications, where slewing time can cause long operational delays, high speed FET sample/holds have 25mA to 40mA available for slewing the hold capacitor, compared with 5mA for the diode bridge types. And although a diode bridge's lower charge-injection error eliminates the need for a large hold capacitor, this

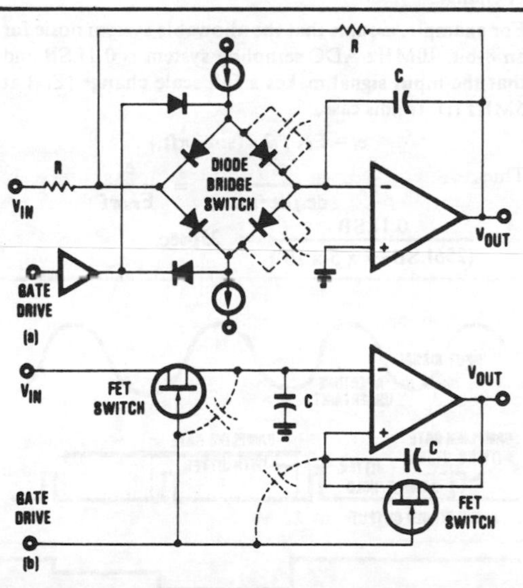

In a basic sample/hold circuit, an analog switch closes and charges a capacitor to the input voltage (Sample mode). When the switch opens, the capacitor passes the stored voltage to the output amplifier (Hold mode). For fast small-signal switching at reasonable accuracy, consider a diode bridge type sample/hold (a). To obtain high accuracy as well as fast large-signal switching, use an FET-type sample/hold (b).

FIGURE 1. Diode-Bridge and Juntion-FET Sample/Hold Circuits.

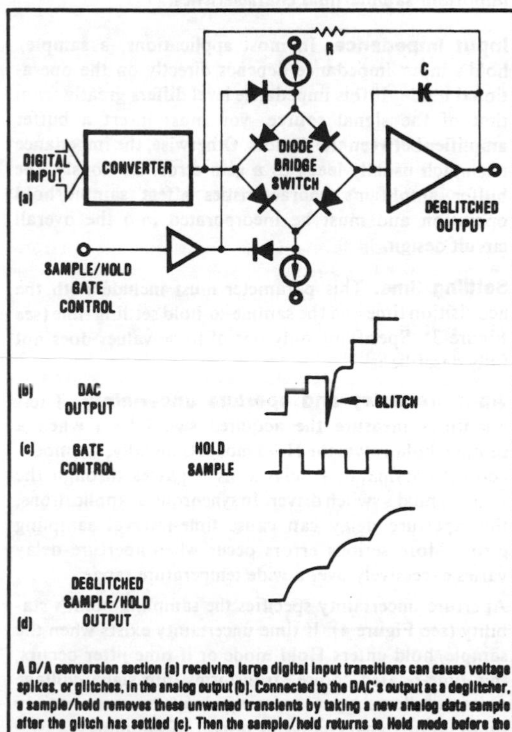

A D/A conversion section (a) receiving large digital input transitions can cause voltage spikes, or glitches, in the analog output (b). Connected to the DAC's output as a deglitcher, a sample/hold removes these unwanted transients by taking a new analog data sample after the glitch has settled (c). Then the sample/hold returns to Hold mode before the DAC's output changes again, resulting in a relatively smooth monotonic transition between both samples (d).

FIGURE 2. Diode-Bridge Sample/Hold Used for Deglitching a Fast DAC.

advantage still doesn't overcome its lower slewing-current limitation.

Further, FET devices' single time constant - based on feedback resistance and hold capacitor - is much shorter than that of the diode-bridge types. When an FET unit's time constant equals or becomes less than the device's buffer-amplifier time constant, the sample/hold's settling time tends to depend upon the amplifier's settling characteristics. Generally, this relationship applies for a hold capacitor of about 10pF and a feedback resistor of about 300Ω. But when the time constant produced by these components becomes much larger than the amplifier's time constant, the sample/hold's settling characteristics do not depend on the amplifier. In this case, the FET-type sample/hold provides faster settling than a diode bridge device.

CLEARING UP SOME FOGGY SPECS

Many sample/hold problems emerge not from lack of knowledge about circuit design, but from difficult-to-understand nonstandard specifications. Complex parameter descriptions frequently result in designer confusion and misunderstanding. Further compounding the problem, parameter nomenclatures and definitions differ markedly among vendors. To help remedy this problem, some detailed explanations can help clarify the more important sample/hold characteristics.

Input impedance. In most applications, a sample/hold's input impedance depends directly on the operational mode. If this impedance level differs greatly from that of the signal source, you must insert a buffer amplifier between the devices. Otherwise, the impedance mismatch usually leads to a gain error. Obviously, the buffer amplifier's characteristics affect sample/hold operation and must be incorporated into the overall circuit design.

Settling time. This parameter must include both the acquisition time and the sample-to-hold settling time (see Figure 3). Specifying only one of these values does not fully describe it.

Aperture delay and aperture uncertainty. These quantities measure the acquired signal level when a sample/hold enters the Hold mode. Generally, the mode-control command is delayed as it passes through the sample/hold's switch driver. In synchronous applications, this aperture delay can cause time-interval sampling error. More serious errors occur when aperture delay varies excessively over a wide temperature range.

Aperture uncertainty specifies the sampling point's stability (see Figure 4). If time uncertainty exists when the sample/hold enters Hold mode or if time jitter occurs, this time-related "noise" transforms into signal-voltage noise as follows: signal noise equals aperture uncertainty times input-signal change at instant of sampling.

To pinpoint aperture uncertainty time (T_A), use the equation:

$$T_A = \frac{e_n}{de_s/dt}$$

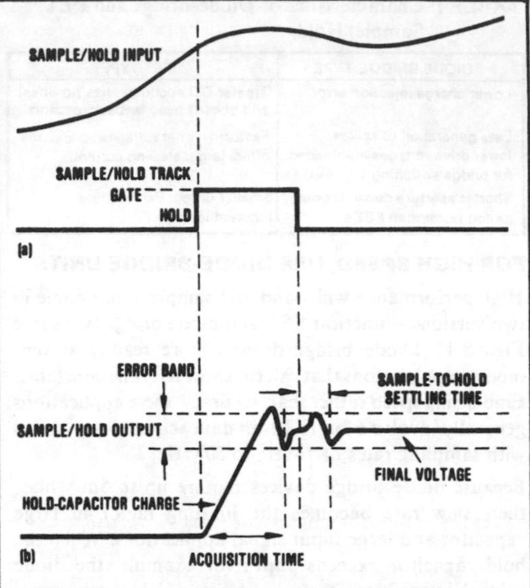

Considered the most important specification for sample/holds, acquisition time begins at the hold-to-sample gate transition (a). It measures how long the hold capacitor takes to charge to a full-scale voltage change and remain within a specified error band (b). A correct definition of this parameter must include the settling time incurred after the sample-to-hold transition.

FIGURE 3. Sample/Hold Response Time.

where e_n = allowable signal noise and de_s/dt = signal rate of change.

For example, assume that the allowable system nosie for an 8-bit, 10MHz ADC sampling system is 0.1LSB and that the input signal makes a full scale change (E_{FS}) at 5MHz (f). In this case,

$$e_s = E_{FS} / 2 \quad (\sin 2\pi ft.)$$

Thus,

$$T_A = \frac{e_n}{de_s/dt\,(\max)} = \frac{e_n}{E_{FS}\pi f}$$

$$= \frac{0.1LSB}{(256LSB)(\pi \times 5 \times 10^6)} = 25psec.$$

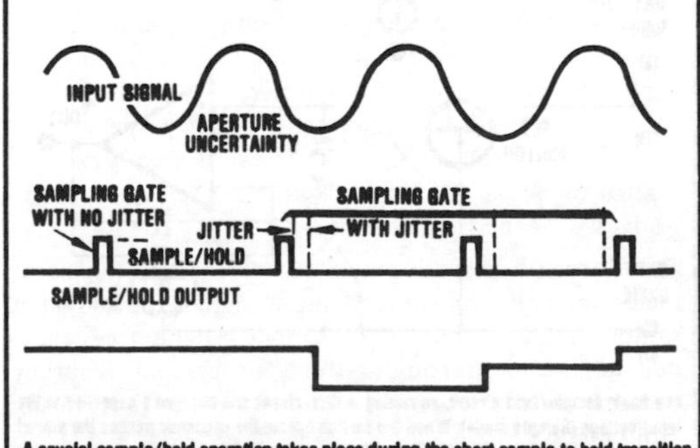

A crucial sample/hold operation takes place during the short sample-to-hold transition (aperture time) that triggers open the control switch. A related term, aperture delay time defines the elapsed time between hold-command initiation and switch opening. Yet another related term, aperture-uncertainty jitter describes the time variation in the opening of the switch or in aperture delay.

FIGURE 4. Aperture Time Uncertainty.

Change-transfer or injection error. This error usually occurs when a sample/hold's gate drive couples onto the hold capacitor via charge-injection capacitance. The injected charge (Q) depends on the relationship.

$$Q = C_C V_G$$

where C_C = coupling capacitance and V_G = voltage arising from gate-drive capacitance. With respect to the output, the induced error becomes:

$$Q = V_I C_H$$

where V_I = voltage (error) arising from charge injection and C_H = hold capacitance. Eliminating Q from both equations yields:

$$V_I = \frac{C_C}{C_H} V_G$$

Furthermore, because the output responds to the *differential* charge transfer, the voltage-error equation changes to

$$V_I = \frac{\Delta C_D}{C_H} V_G$$

where C_D = the difference between the two capacitances in Figure 1.

Typically, this voltage behaves like a DC offset error - the output voltage differs from the input voltage by a constant value. In practice, diode bridge sample/holds can maintain a charge-transfer error of 0.005pC over a ±50°C span, whereas high-performance FET sample/holds achieve about 0.25pC over the same temperature range.

Sample/hold spiking. In many display applications, unwanted spikes appear on the CRT's involved. Vendors commonly dimension this error in volts x seconds (V x T), because the spikes' amplitudes depend on either the measurement circuit's or the sample/hold drive circuit's bandwidth.

If you model the spike as a narrow pulse fed into a sample RC circuit, the voltage output equals

$$e_o = \frac{VT}{RC} \epsilon^{-t/RC}$$

In effect, the low-pass RC circuit lowers the spikes' amplitudes but spreads out their duration.

In conjunction with a charge-injection error, sample/hold spiking can cause output signal noise. To overcome this problem, use a filter that effectively cancels positive spiking with negative spiking. Further improvement results when the output signal's bandwidth greatly exceeds that of the spike. Of the two sample/hold types, diode bridges yield the smallest spike.

Feedthrough. Usually stated as a percentage, this parameter indicates how much input signal appears at a sample/hold's output in Hold mode. Highly frequency-dependent, feedthrough affects processing accuracy because the output signal should stay constant during Hold mode.

Droop. Vendors usually specify the hold capacitor's voltage decay (droop) during Hold mode as a leakage current when referring to an external capacitor, and in volts per second when the sample/hold contains an integral capacitor. In analog-to-digital applications, the amound of tolerable droop depends on the ADC's conversion time and accuracy.

DATA-SYSTEM MODEL AIDS SPECTRAL ANALYSIS

Another major group of sample/hold design problems stem from noise effects. Unlike a linear process, such as amplification, the sample/hold sampling process generates output noise at frequencies that intermix with the input signal frequencies. Obviously, this noise interferes with the signal flow.

To gain insight into a sample/hold's spectral response, consider a representative data sampling system comprising (in series from input to output) a sample/hold device, an ADC, a digital processor, and a DAC. Theoretically, the sample/hold's output spectral response resembles that of the DAC. In practice, however, the DAC's output contains quantizing noise and, possibly, spiking noise. To simplify system operation without markedly affecting spectral response, assume that the ADC directly drives the DAC.

DUAL-SWITCH SAMPLE/HOLD SIMPLIFIES SAMPLING

As an aid in developing the data-sampling system's spectral properties, consider a sample/hold model employing two switches (see Figure 5). This model allows partitioning of the sampling process into multiplier and sample/hold portions, thus separating the former's noise analysis from the latter's bandwidth analysis.

Recall that when you multiply signal frequency f_1 by signal frequency f_2, new frequencies $f_1 + f_2$ and $f_1 - f_2$ result. On a spectrum analyzer, therefore, the multiplier's output appears as vertical lines corresponding to the new frequencies.

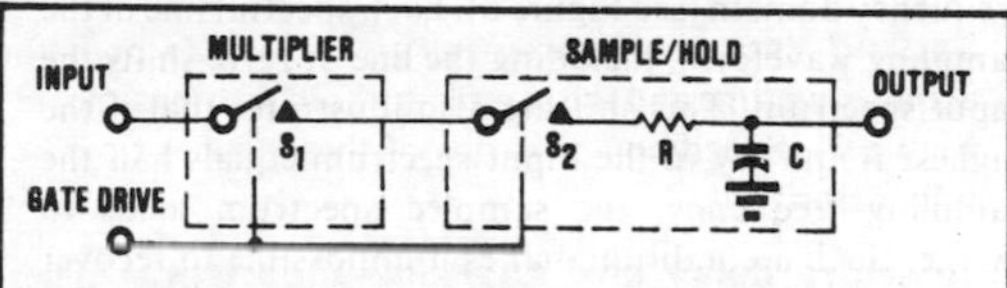

As a vehicle for explaining the sample/hold's spectral noise response, consider a circuit model that contains two control switches, representing a multiplier stage followed by a typical sample/hold stage. This separated-stage approach allows division of the noise analysis into workable segments.

FIGURE 5. Simplified Model of Sample/Hold.

Furthermore, assume that the gate drive signal consists of impulses. This assumption focuses the discussion of the spectral components' amplitudes at the sample/hold's output. As Figure 6 shows, the frequency spectrum at the DAC's output is independent of the gate width; therefore, the foregoing impulse-input assumption is valid for this analysis. Note also in Figure 6 that so long as the sample/hold's output value remains the same for both impulse and gate-sampling methods, the ADC makes the same conversion, provided the gate stays closed long enough to acquire the signal. With impulse sampling, you can thus mathematically describe the gate drive's waveform by:

$$f(t) = \sum_{0}^{\infty} \sigma(t - nT)$$

where $n = 0, 1, 2...$, and where $(t-nT)$ is an impulse that occurs when $t = nT$. T is the sampling interval.

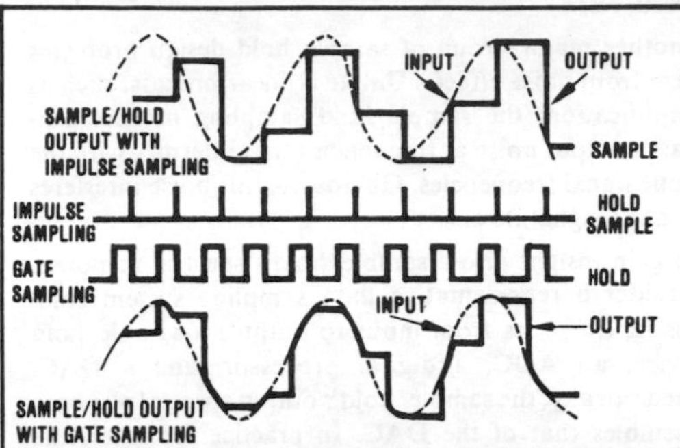

Sample/holds employ two principal sampling methods - In impulse sampling, extremely narrow gate pulses place the sample/hold in sample mode for very short time intervals and then quickly return the device to Hold mode. In gate sampling, relatively wide gate pulses keep the sample/hold in a Sample mode for longer time intervals before returning it to Hold mode. Note that the sample/hold's output value stays the same for both methods.

FIGURE 6. Sample/Hold Output as a Function of Gate Width.

USE IMPULSE RESPONSE FOR BANDWIDTH CHECK

Remember that the spectrum of an impulse train in the time domain can be represented as a series of lines in the frequency domain separated by the sampling frequency (see Figure 7). Note that as the pulse widths get narrower, the spectral lines' amplitudes become uniform. Thus, impulse sampling becomes equivalent to multiplying an input frequency f_1 by nf_2, where $n = 0, \pm1, \pm2$, etc.

If you use a complex input waveform containing many frequencies, the input spectrum periodically repeats in the frequency domain (see Figure 8). Each spectral line of the sampling waveform, including the line at zero, shifts the input spectrum. This shifting also illustrates that if the highest frequency in the input spectrum equals half the sampling frequency, the sampled spectrum tends to merge. Such a condition makes it impossible to recover the original signal by filtering (the classic Nyquist sampling approach).

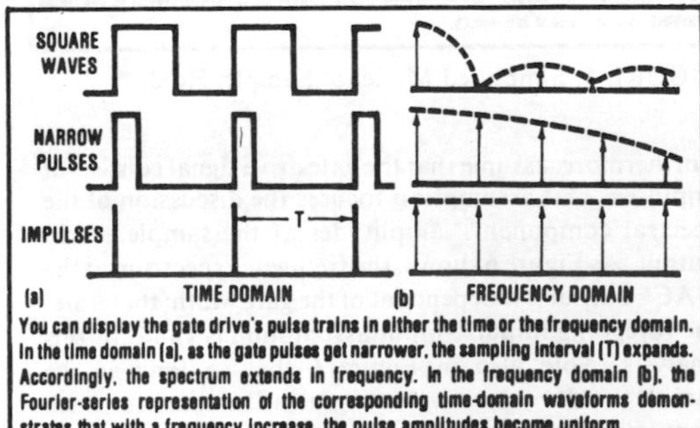

You can display the gate drive's pulse trains in either the time or the frequency domain. In the time domain (a), as the gate pulses get narrower, the sampling interval (T) expands. Accordingly, the spectrum extends in frequency. In the frequency domain (b), the Fourier-series representation of the corresponding time-domain waveforms demonstrates that with a frequency increase, the pulse amplitudes become uniform.

FIGURE 7. Time and Frequency Domain Descriptions of Gating Wave Forms.

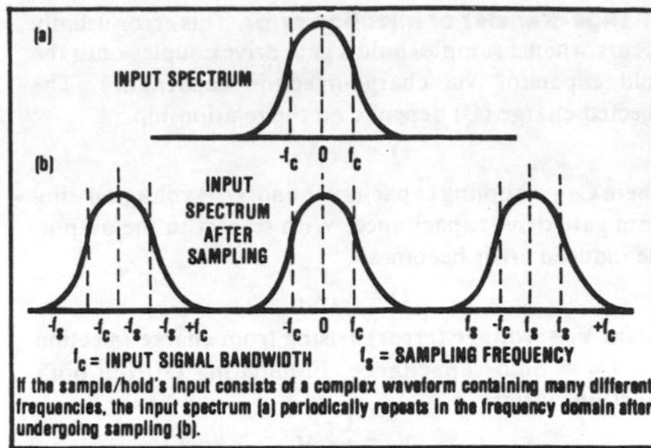

If the sample/hold's input consists of a complex waveform containing many different frequencies, the input spectrum (a) periodically repeats in the frequency domain after undergoing sampling (b).

FIGURE 8. Spectral Response Due to Sampling.

You can now compute the sample/hold's bandwidth by evaluating its impulse response. Figure 9 shows that a sample/hold's response to impulse inputs is a pulse whose width equals the sampling time. Taking the Fourier transform of the time-domain response thus yields the sample/hold's frequency characteristic:

$$\frac{F(f)}{F(0)} = \int_0^T f(t)\,\epsilon^{j2\pi ft}\,dt = \frac{f(T)\,\epsilon^{j2\pi ft} - 1}{j2\pi fT}$$

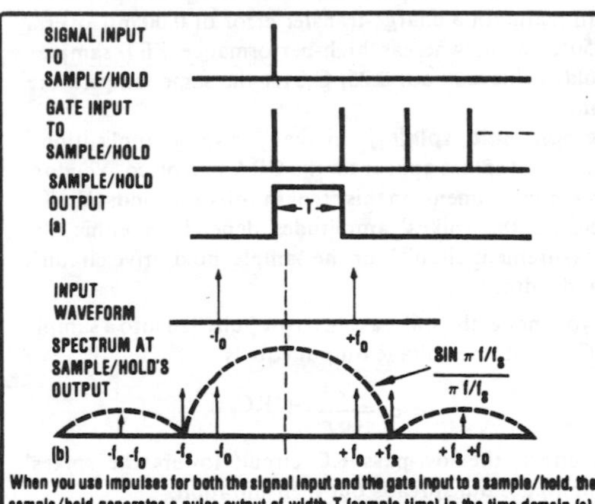

When you use impulses for both the signal input and the gate input to a sample/hold, the sample/hold generates a pulse output of width T (sample time) in the time domain (a). Taking Fourier transforms of this impulse response produces the sample/hold's frequency response (b).

FIGURE 9. Impulse and Spectral Response to Sample/Hold.

After some manipulation, this expression equals:

$$\frac{F(f)}{F(0)} = \epsilon^{j\pi f/f_s}\,\frac{\sin \pi f/f_s}{\pi f/f_s}$$

A sampled signal becomes harmonically distributed over the frequency domain, modified by this equation's sine term, where f is the input frequency and $1/f_s$ is the sampling period. With an applied input sine wave, the sample/hold's bandwidth drops 3dB down at $f/f_s = 0.443$.

AN-294

Special Sample and Hold Techniques

National Semiconductor
Application Note 294

Although standard devices (e.g., the LF398) fill most sample and hold requirements, situations often arise which call for special capabilities. Extended hold times, rapid acquisition and reduced hold step are areas which require special circuit techniques to achieve good results. The most common requirement is for extended hold time. The circuit of *Figure 1* addresses this issue.

EXTENDED HOLD TIME SAMPLE AND HOLD

In this circuit, extended hole time is achieved by "stacking" two sample and hold circuits in a chain. In addition, rapid acquisition time is retained by use of a feed-forward path. When a sample command is applied to the circuit (trace A, *Figure 2*), A1 acquires the input very rapidly because its 0.002 μF hold capacitor can charge very quickly. The sample command is also used to trigger the DM74C221 one-shot (trace B, *Figure 2*), which turns on the FET switch, S1. In this fashion, A1's output is fed immediately to the A3 output buffer. During the time the one-shot is high, A2 acquires the value of A1's output. When the one-shot drops low, S1 opens, disconnecting A1's output from A3's input. At this point A2's output is allowed to bias A3's input and the circuit output does not change from A1's initial sampled value. Trace C details what happens when S1 opens. A small glitch, due to charge transfer through the FET, appears but the steady state output value does not change. This circuit will acquire a 10V step in 10 μs to 0.01% with a droop rate of just 30 μV/second.

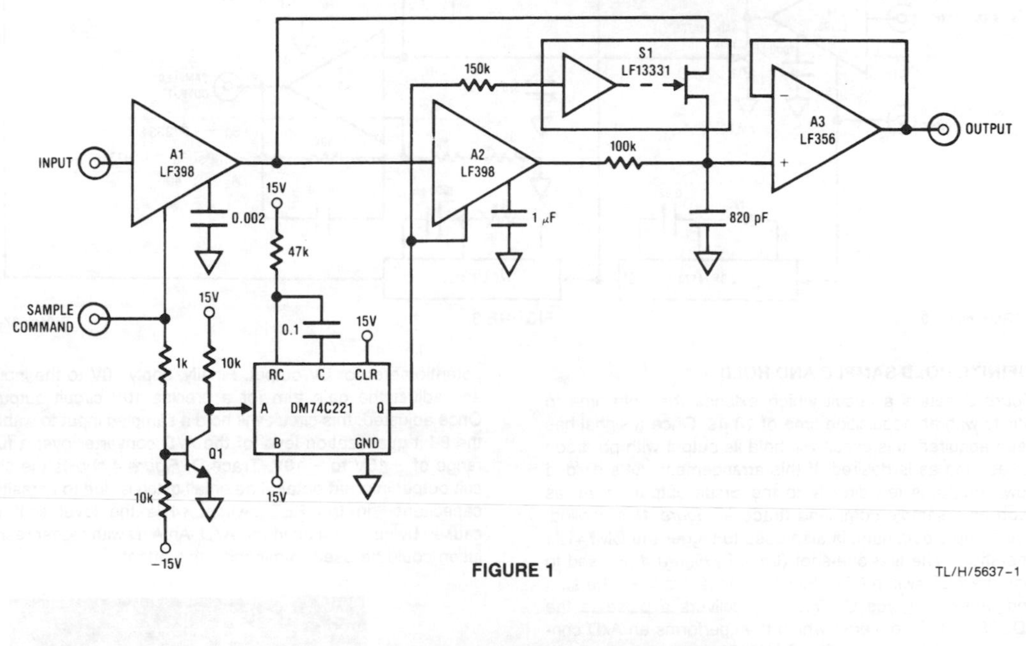

FIGURE 1

TL/H/5637–1

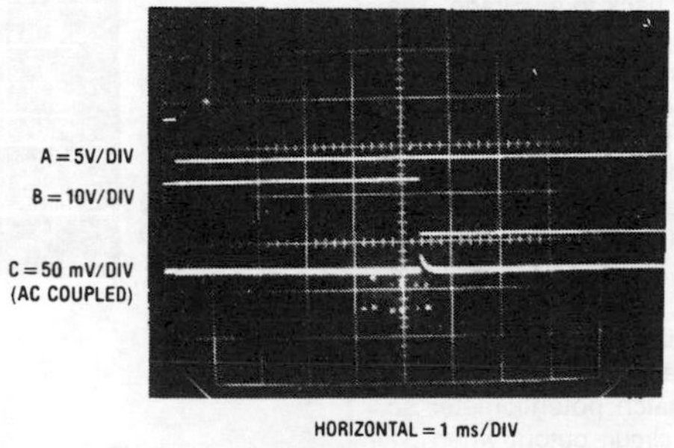

TL/H/5637–2

FIGURE 2

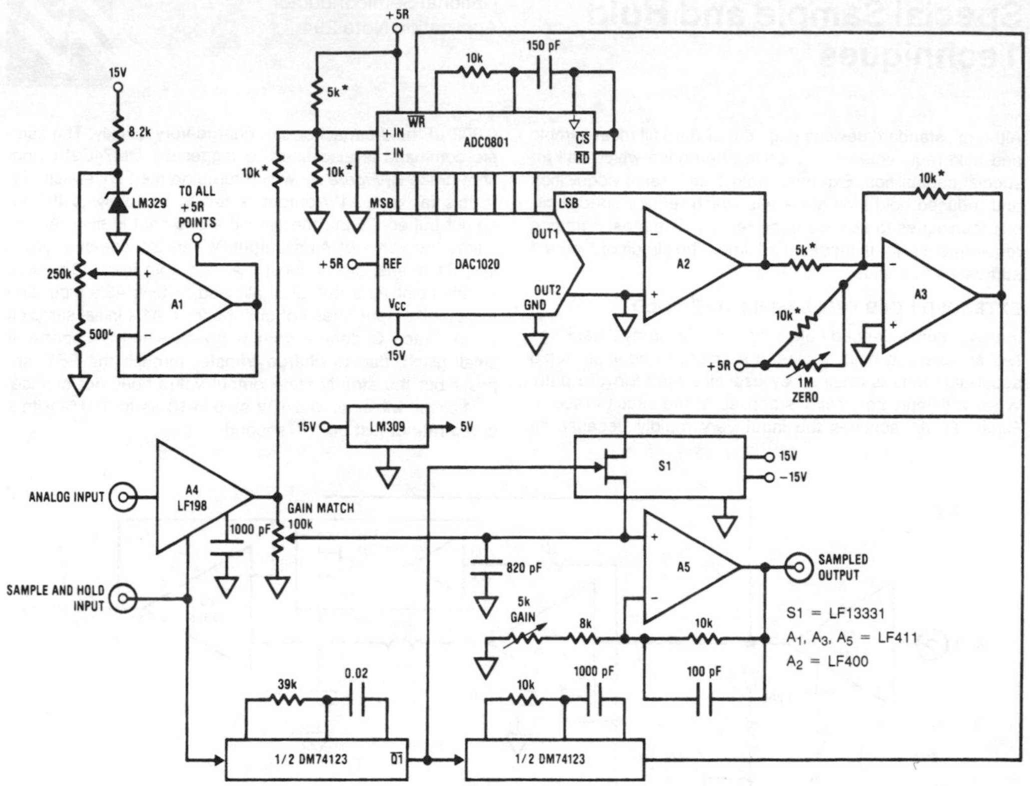

*Ratio match 0.1%

S1 = LF13331

A_1, A_3, A_5 = LF411

A_2 = LF400

FIGURE 3

TL/H/5637-3

INFINITE HOLD SAMPLE AND HOLD

Figure 3 details a circuit which extends the hold time to infinity with an acquisition time of 10 µs. Once a signal has been acquired, this circuit will hold its output with no droop for as long as is desired. If this arrangement, A4's divided down output is fed directly to the circuit output via A5 as soon as a sample command (trace A, *Figure 4)* is applied. The sample command is also used to trigger the DM74123 one-shots. The first one-shot (trace B, *Figure 4)* is used to bias the FET switch OFF during the time it is low. The second one-shot (trace C, *Figure 4)* delivers a pulse to the ADC0801 A/D converter which then performs an A/D conversion on A4's output. The DAC1020, in combination with A2 and A3, converts the A/D output back to a voltage. The A/D/A process requires about 100 µs. When the one-shot (trace B) times out, its output goes high, closing the FET switch. This action effectively connects A3's output to A5 while disconnecting A4's output. In this manner, the circuit output will remain at the DC level that was originally determined by A4's sampling action. Because the sampled value is stored digitally, no droop error can occur. The precision resistors noted in the circuit provide offsetting capability for the unipolar A/D output so that a −10V to +10V input range can be accommodated. To calibrate this circuit, apply 10V to the input and drive the sample command input with a pulse generator. Adjust the gain match potentiometer so that minimum "hop" occurs at the circuit output when S1 closes. Next, ground the input and adjust the zero potentiometer for 0V output. Finally, apply 10V to the input and adjust the gain trim for a precise 10V circuit output. Once adjusted, this circuit will hold a sampled input to within the 8-bit quantization level of the A/D converter over a full range of +10V to −10V. Trace D, *Figure 4* shows the circuit output in great detail. The small glitch is due to parasitic capacitance in the FET switch, while the level shift is caused by quantization in the A/D. An A/D with higher resolution could be used to minimize this effect.

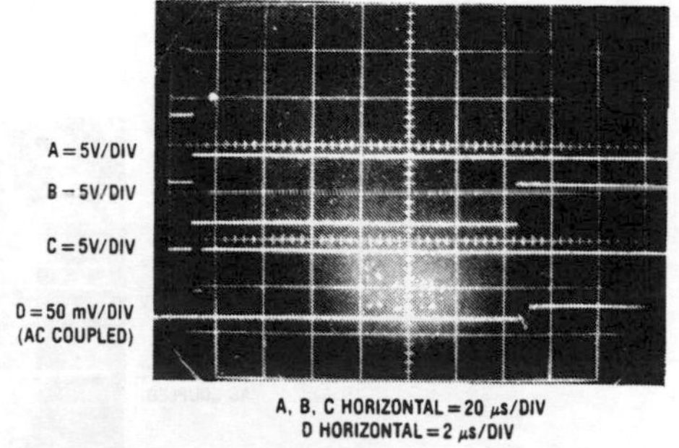

A, B, C HORIZONTAL = 20 µs/DIV
D HORIZONTAL = 2 µs/DIV

TL/H/5637-4

FIGURE 4

HIGH SPEED SAMPLE AND HOLD

Another requirement encountered in sample and hold work is high speed. Although conventional sample and hold circuits can be built for very fast acquisition times, they are difficult and expensive. If the input waveform is repetitive, the circuit of *Figure 5* can be employed. In this circuit a very fast comparator and a digital latch are placed in front of a differential integrator. Feedback is used to close a loop around all these elements. Each time an input pulse is applied, the DM7475 latch is opened for 100 ns. If the summing junction error at the LM361 is positive, A1 will pull current out of the junction. If the error is negative, the inverse will occur. After some number of input pulses, A1's output will settle at a DC level which is equivalent to the value of the level sampled during the 100 ns window. Note that the delay time of one-shot A is variable, allowing the sample pulse from one-shot B to be placed at any desired point on the input waveform. *Figure 6a* shows the circuit waveforms. Trace A is the circuit input. After the variable delay provided by one-shot A, one-shot B generates the sample pulse (trace B). In this case the delay has been adjusted so that sampling occurs at the mid-point of the input waveform, although any point may be sampled by adjusting the delay appropriately.

Figure 6b shows the circuit at work sampling a 1 MHz sine wave input. The optional comparator (C2) shown in dashed lines is used to convert the sine wave input into a TTL compatible signal for the DM74123 one-shot. Trace A is the sine wave input while trace B represents the output of C2. Trace C is the delay generated by one-shot A and trace D is the sample width window out of one-shot B. Note that this pulse can be positioned at any point on the high speed sine wave with the resultant voltage level appearing at A1's output.

REDUCED HOLD STEP SAMPLE AND HOLD

Another area where special techniques may offer improvement is minimization of hold step. When a standard sample and hold switches from sample to hold, a large amplitude high speed spike may occur. This is called hold step and is usually due to capacitive feedthrough in the FET switches

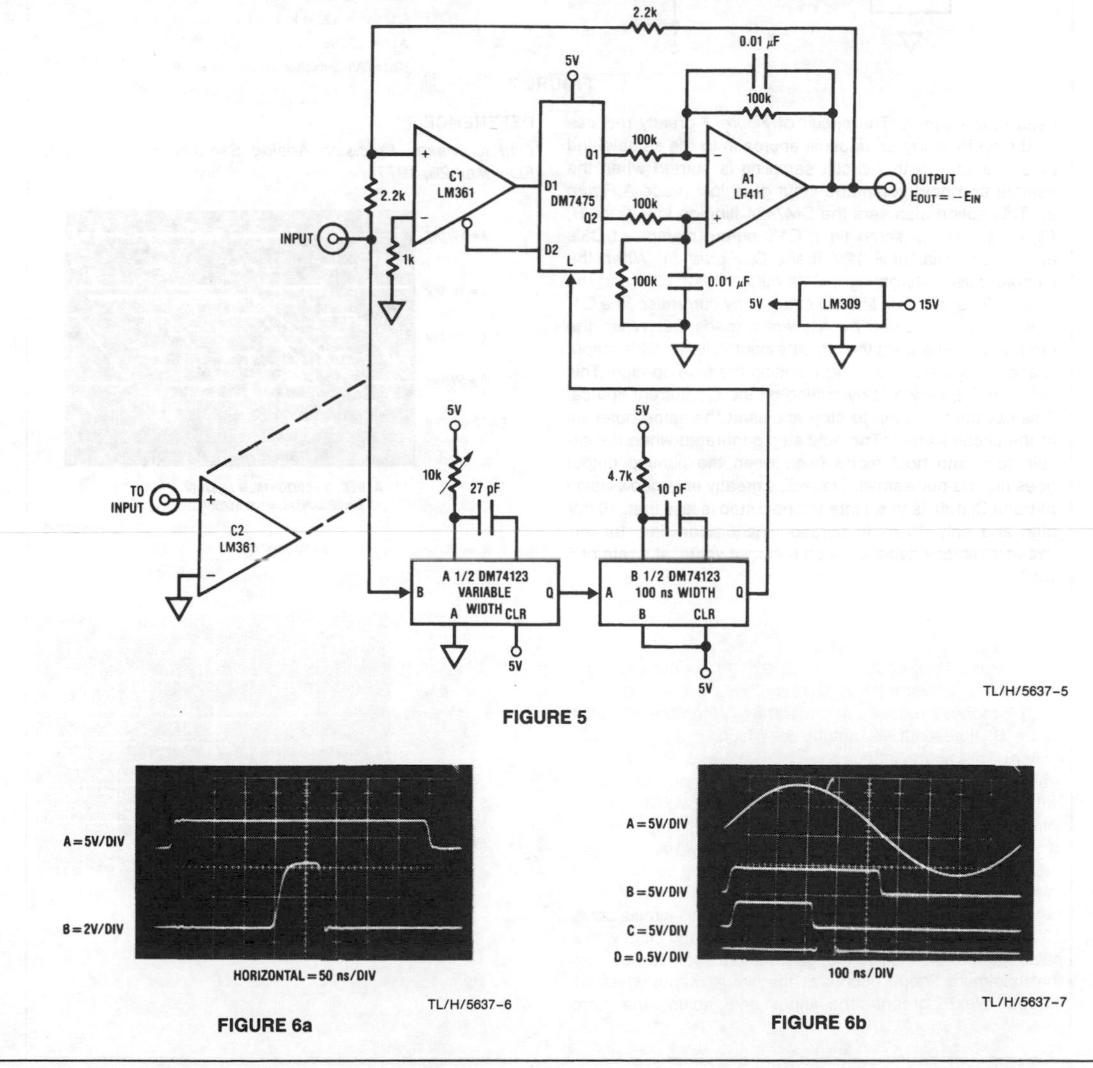

FIGURE 5

FIGURE 6a

FIGURE 6b

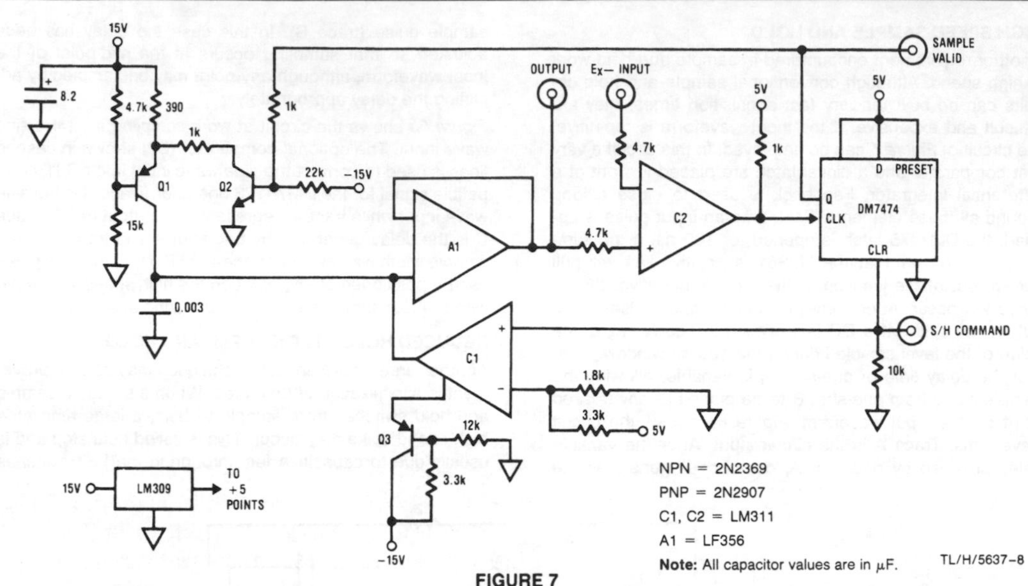

FIGURE 7

used in the circuit. The circuit of *Figure 7* greatly reduces hold step by using an unusual approach to the sample and hold function. In this circuit sampling is started when the sample and hold command input goes low (trace A, *Figure 8*). This action also sets the DM7474 flip-flop low (trace B, *Figure 8*). At the same time, C1's output clamps at Q3's emitter potential of −12V (trace C, *Figure 8*). When the sample pulse returns high, C1's output floats high and the 0.003 µF capacitor is linearly charged by current source Q1. This ramp is followed by A1, which feeds C2. When the ramp potential equals the circuit's input voltage, C2's output (trace D, *Figure 8*) goes high, setting the flip-flop high. This turns on Q2, very quickly cutting off the Q1 current source. This causes the ramp to stop and sit at the same potential at the circuit's input. The hold step generated when the circuit goes into hold mode (e.g., when the flip-flop output goes high) is quite small. Trace E, a greatly enlarged version of trace C, details this. Note the hold step is less than 10 mV high and only 30 ns in duration. Acquisition time for this circuit is directly dependent on the input value, at a rate of 5 µs/V.

REFERENCE

One IC Makes Precision Analog Sampler, S. Dendinger; EDN May 20, 1977.

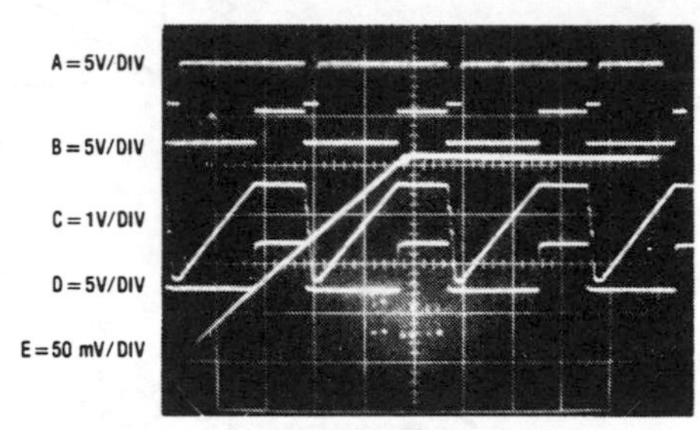

FIGURE 8

LF198A/LF398A
LF198/LF398
Precision Sample and Hold Amplifier

FEATURES

- *Guaranteed* 6µs Max. Acquisition Time
- *Guaranteed* 0.005% Max. Gain Error
- *Guaranteed* 1mV Max. Offset Voltage
- *Guaranteed* 1mV Max. Hold Step
- Very Low Feedthrough 86dB Min.
- High Input Impedance under All Conditions
- Logic Inputs Compatible with All Logic Families

APPLICATIONS

- 12-Bit Data Acquisition Systems
- Ramp Generators
- Analog Switches
- Staircase Generators
- Sample and Difference Circuits

DESCRIPTION

The LF198 is a precision sample and hold amplifier which uses a combination of bipolar and junction FET transistors to provide precision, high speed, and long hold times. A typical offset voltage of 1mV and gain error of 0.002% allow this sample and hold amplifier to be used in 12-bit systems. Dynamic performance can be optimized by proper selection of the external hold capacitor. Acquisition times can be as low as 4µs for small capacitors while hold step and droop errors can be held below 0.1mV and 30µV/sec respectively when using larger capacitors.

The LF198 is fixed at unity gain with $10^{10}\Omega$ input impedance independent of sample/hold mode. The logic inputs are high impedance differential to allow easy interfacing to any logic family without ground loop problems. A separate offset adjust pin can be used to zero the offset voltage in either the sample or hold mode. Additionally, the hold capacitor can be driven with an external signal to provide precision level shifting or ''differencing'' operation. The device will operate over a wide supply voltage range from ±5V to ±18V with very little change in performance, and key parameters are specified over this full supply range.

The LF198A version offers tightened electrical specifications for key parameters.

9

Basic Sample and Hold

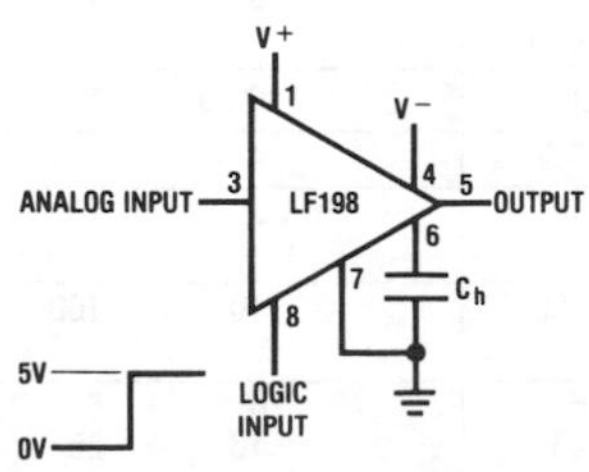

Acquisition Time

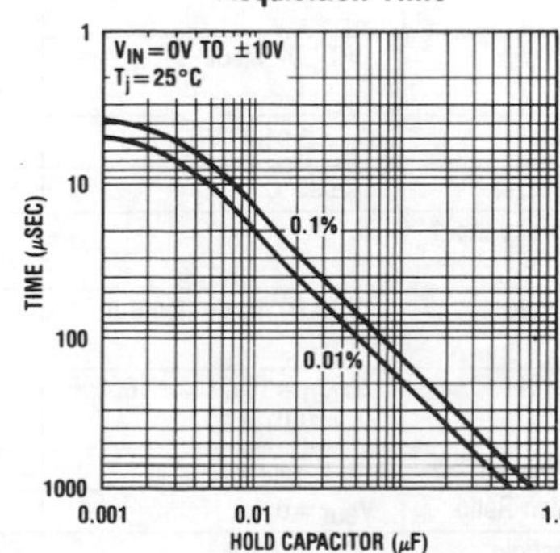

LF198A/LF398A
LF198/LF398

ABSOLUTE MAXIMUM RATINGS

Input Voltage .... Equal to Supply Voltage
Logic to Logic Reference Differential
  Voltage (Note 2) .... +30V, −30V
Output Short Circuit Duration .... Indefinite
Hold Capacitor Short Circuit Duration .... 10 sec
Lead Temperature (Soldering, 10 seconds) ...... 300°C
Supply Voltage .... ±18V
Power Dissipation (Package Limitation)
  (Note 1) 500mW
Operating Temperature Range
  LF198/LF198A .... −55°C to 125°C
  LF398/LF398A .... 0°C to 70°C
Storage Temperature Range .... −65°C to 150°C

PACKAGE/ORDER INFORMATION

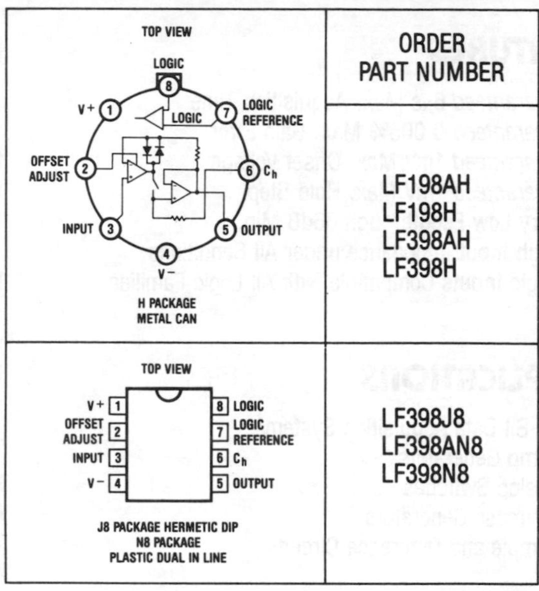

Package	ORDER PART NUMBER
H PACKAGE METAL CAN	LF198AH LF198H LF398AH LF398H
J8 PACKAGE HERMETIC DIP / N8 PACKAGE PLASTIC DUAL IN LINE	LF398J8 LF398AN8 LF398N8

ELECTRICAL CHARACTERISTICS (Note 3)

PARAMETER	CONDITIONS		LF198A MIN	LF198A TYP	LF198A MAX	LF398A MIN	LF398A TYP	LF398A MAX	UNITS
Input Offset Voltage (Note 6)				0.5	1		1	2	mV
		●			2			3	mV
Input Bias Current (Note 6)				5	25		10	25	nA
		●			75			50	nA
Input Impedance				10^{10}			10^{10}		Ω
Gain Error	$R_L = 10k$			0.001	0.005		0.001	0.005	%
		●			0.01			0.01	%
Feedthrough Attenuation Ratio at 1kHz	$C_h = 0.01\mu F$		86	96		86	96		dB
Output Impedance	"HOLD" Mode			0.5	1		0.5	1	Ω
		●			4			6	Ω
"HOLD" Step (Note 4)	$C_h = 0.01\mu F$, $V_{OUT} = 0$			0.25	1		0.25	1	mV
Supply Current (Note 6)	$T_j \geq 25°C$			4.5	5.5		4.5	6.5	mA
Logic and Logic Reference Input Current				2	10		2	10	μA
Leakage Current into Hold Capacitor (Note 6)	"HOLD" Mode (Note 5)			10	100		10	100	pA
Acquisition Time to 0.1%	$\Delta V_{OUT} = 10V$, $C_h = 1000pF$			4	6		4	6	μs
	$C_h = 0.01\mu F$			16	25		16	25	μs
Hold Capacitor Charging Current	$V_{IN} - V_{OUT} = 2V$			5			5		mA
Supply Voltage Rejection Ratio	$V_{OUT} = 0$		90	110		90	110		dB
Differential Logic Threshold			0.8	1.4	2.4	0.8	1.4	2.4	V

ELECTRICAL CHARACTERISTICS (Note 3)

PARAMETER	CONDITIONS		LF198 MIN	LF198 TYP	LF198 MAX	LF398 MIN	LF398 TYP	LF398 MAX	UNITS
Input Offset Voltage (Note 6)				1	3		2	7	mV
		●			5			10	mV
Input Bias Current (Note 6)				5	25		10	50	nA
		●			75			100	nA
Input Impedance				10^{10}			10^{10}		Ω
Gain Error	$R_L = 10k$			0.002	0.005		0.004	0.01	%
		●			0.02			0.02	%
Feedthrough Attenuation Ratio at 1kHz	$C_h = 0.01\mu F$		86	96		80	96		dB
Output Impedance	"HOLD" Mode			0.5	2		0.5	4	Ω
		●			4			6	Ω
"HOLD" Step (Note 4)	$C_h = 0.01\mu F$, $V_{OUT} = 0$			0.5	2.0		0.5	2.5	mV
Supply Current (Note 6)	$T_j \geq 25°C$			4.5	5.5		4.5	6.5	mA
Logic and Logic Reference Inout Current				2	10		2	10	μA
Leakage Current into Hold Capacitor (Note 6)	"HOLD" Mode (Note 5)			30	100		30	200	pA
Acquisition Time to 0.1%	$\Delta V_{OUT} = 10V$, $C_h = 1000pF$			4			4		μs
	$C_h = 0.01\mu F$			16			16		μs
Hold Capacitor Charging Current	$V_{IN} - V_{OUT} = 2V$			5			5		mA
Supply Voltage Rejection Ratio	$V_{OUT} = 0$		80	110		80	110		dB
Differential Logic Threshold			0.8	1.4	2.4	0.8	1.4	2.4	V

The ● denotes the specifications which apply over the full operating temperature range.

Note 1: T_j max for the LF198/LF198A is 150°C; T_j max for the LF398/LF398A is 100°C.

Note 2: The logic inputs are protected to ±30V differential as long as the voltage on both pins does not exceed the supply voltage. For proper operation, however, both logic and logic reference pins must be at least 2V below the positive supply and one of these pins must be at least 3V above the negative supply.

Note 3: Unless otherwise noted, $V_S = \pm 15V$, $T_j = 25°C$, $-11.5V \leq V_{IN} \leq +11.5V$, $C_h = 0.01\mu F$, $R_L = 10k\Omega$ and unit is in "sample" mode. Logic reference = 0V and logic voltage = 2.5V.

Note 4: The hold step is sensitive to stray capacitance coupling between input logic signals and the hold capacitor. 1pF, for instance, will create an additional 0.5mV step with a 5V logic swing and a 0.01μF hold capacitor. Magnitude of the hold step is inversely proportional to hold capacitor value.

Note 5: Leakage current is measured at a *junction* temperature of 25°C. The effects of junction temperature rise due to power dissipation or elevated ambient can be calculated by doubling the 25°C value for each 11°C increase in chip temperature. Leakage is guaranteed over full input signal range.

Note 6: These parameters are guaranteed over a supply voltage range of ±5V to ±18V.

FUNCTIONAL DIAGRAM

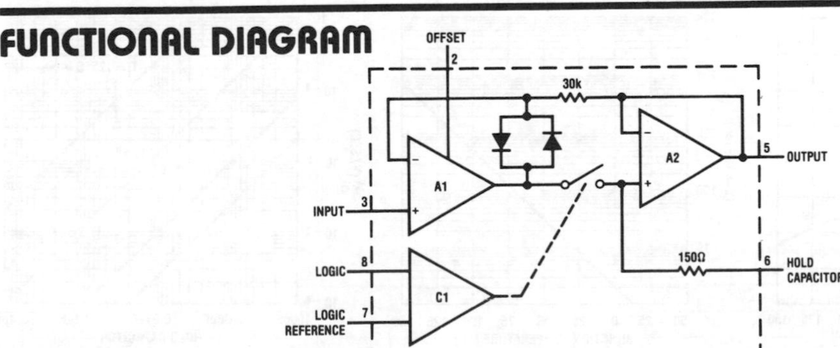

9

TYPICAL PERFORMANCE CHARACTERISTICS

Aperture Time*

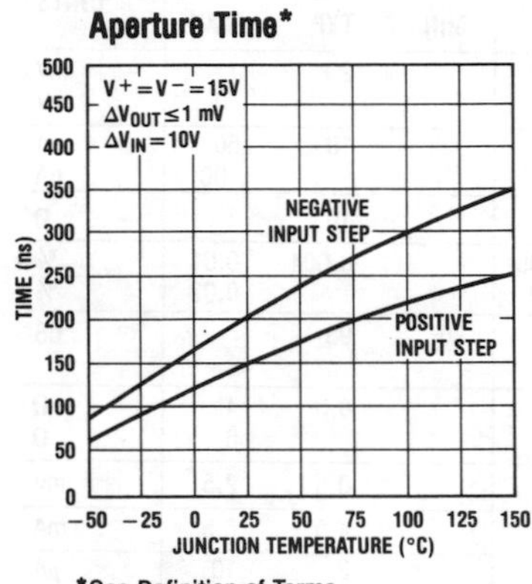

*See Definition of Terms

Dynamic Sampling Error*

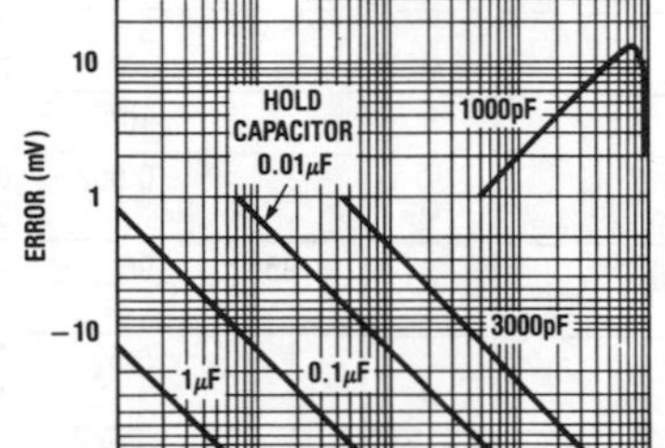

*See Definition of Terms

"Hold" Settling Time*

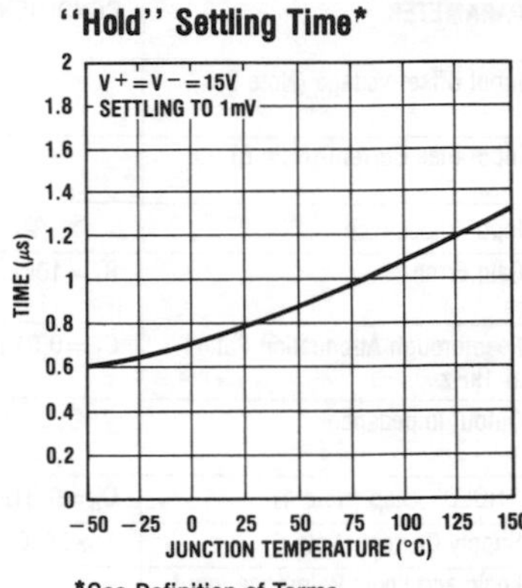

*See Definition of Terms

Hold Step

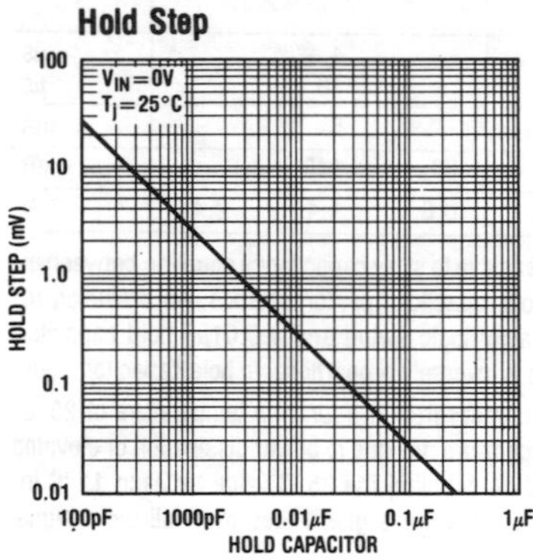

Hold Step vs Input Voltage

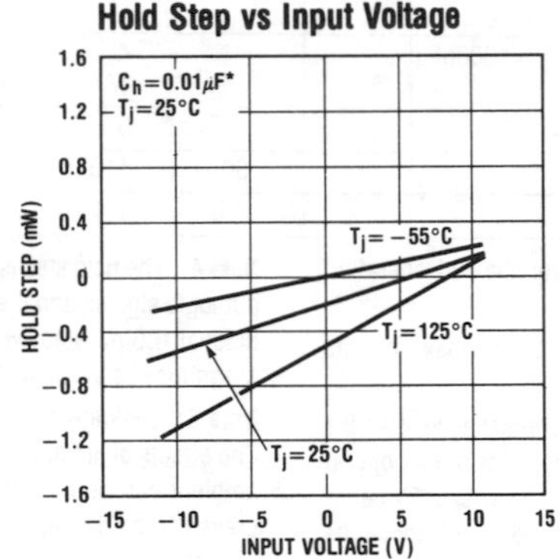

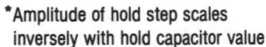

*Amplitude of hold step scales inversely with hold capacitor value

Hold Step vs Logic Slew Rate

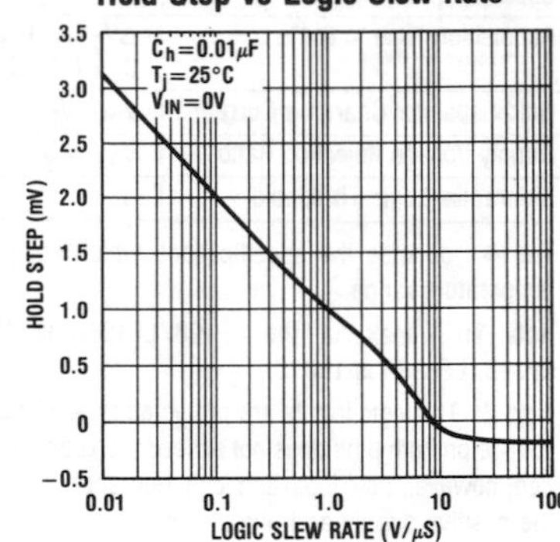

Input Bias Current

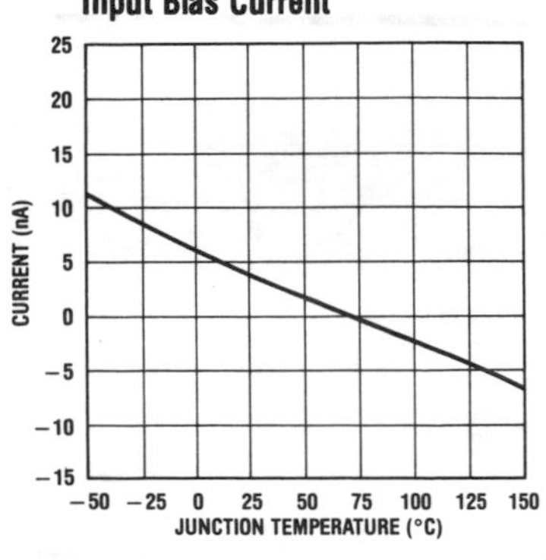

Capacitor Pin Leakage

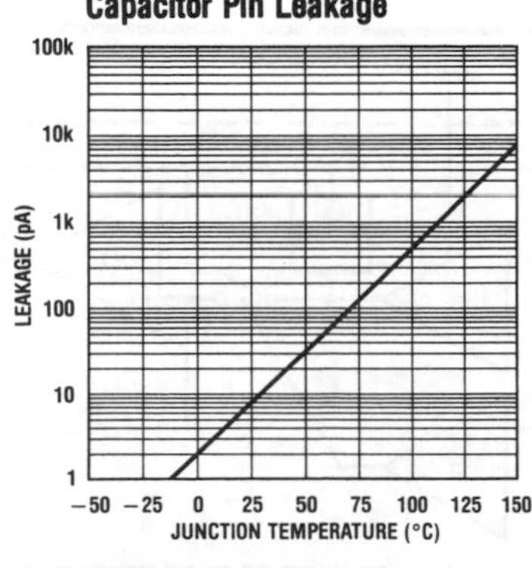

Output Droop Rate

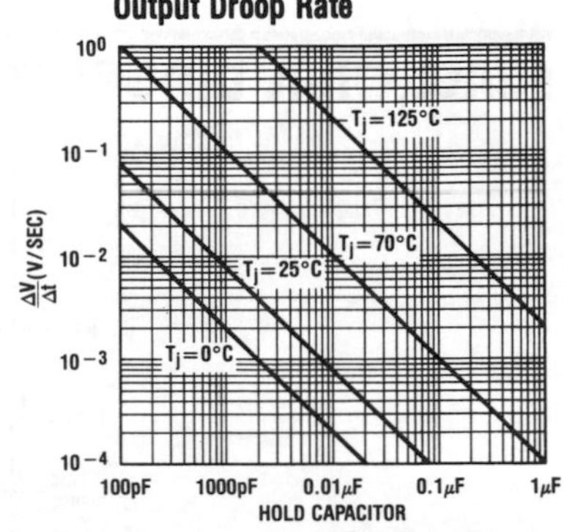

TYPICAL PERFORMANCE CHARACTERISTICS

Output Short Circuit Current

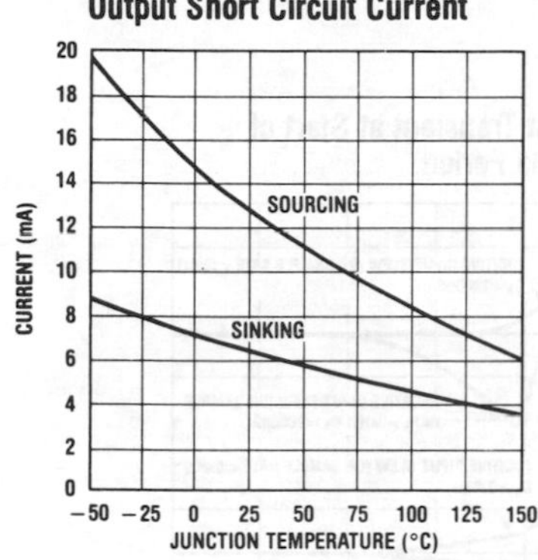

Output Noise

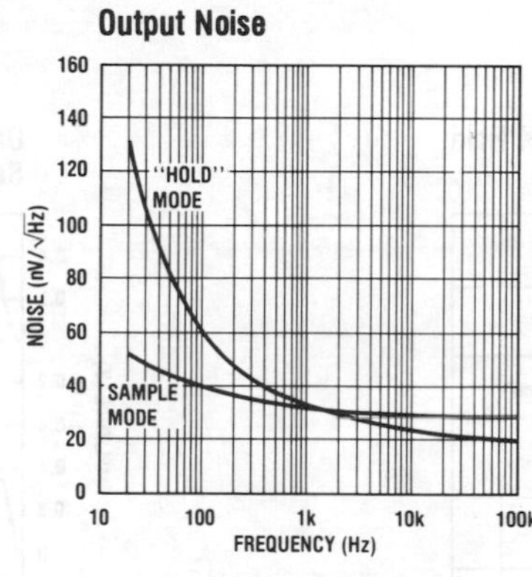

Gain Error

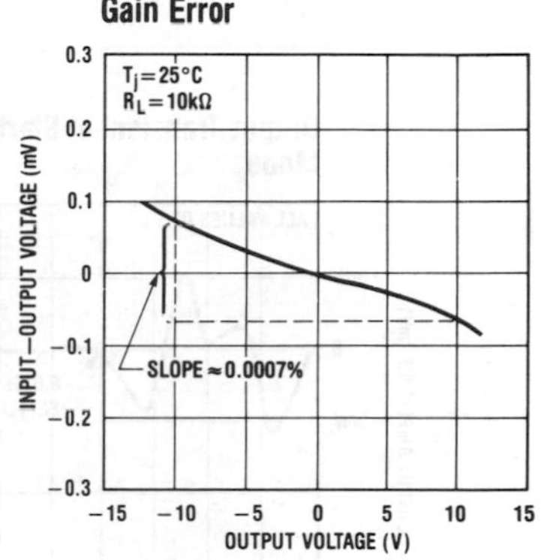

Phase and Gain (Input to Output, Small Signal)

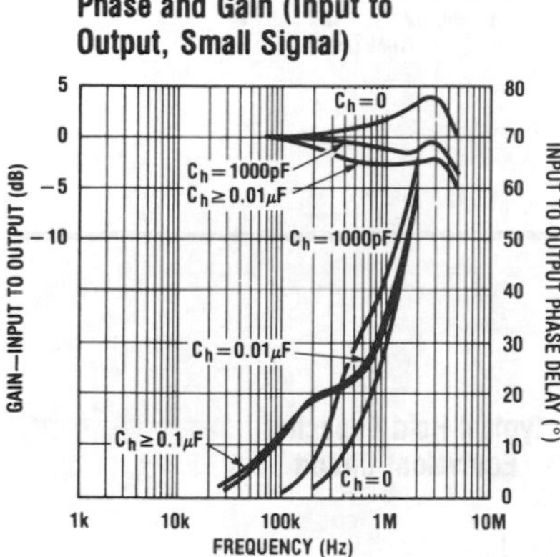

Feedthrough Rejection Ratio

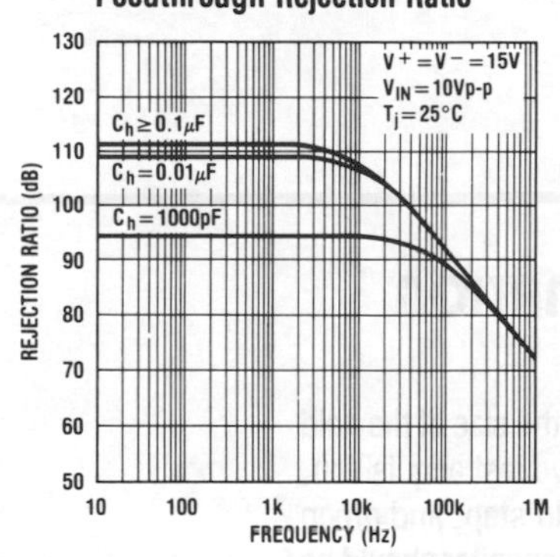

Power Supply Rejection

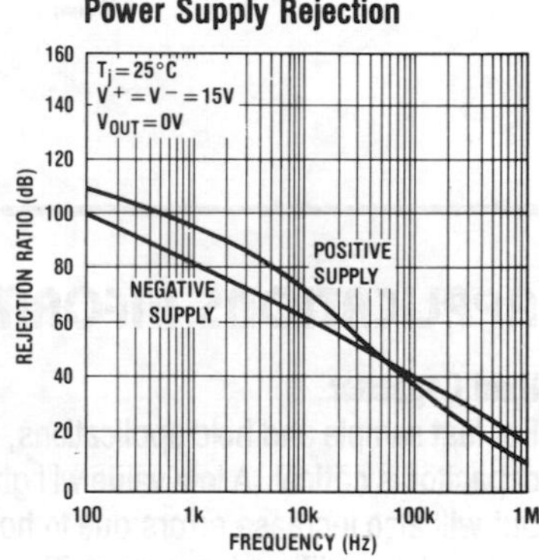

Capacitor Dielectric Absorption

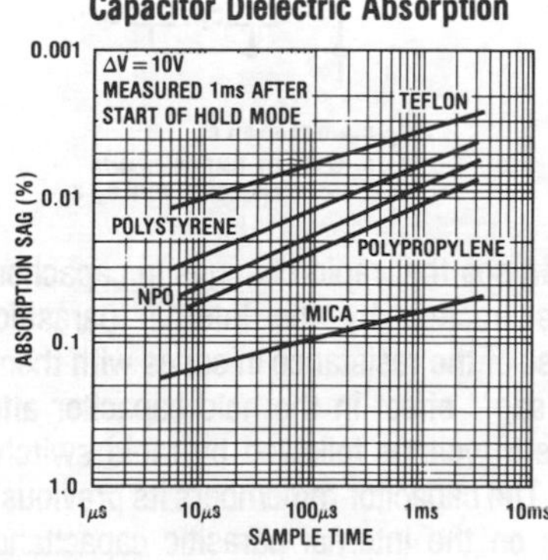

Capacitor Dielectric Absorption

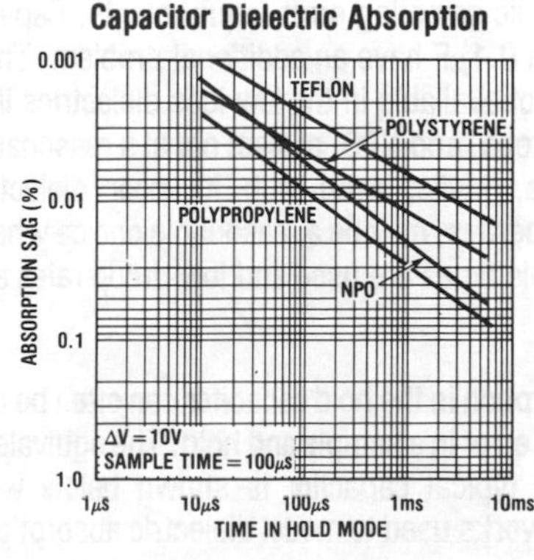

TYPICAL PERFORMANCE CHARACTERISTICS

Output Transient at Start of Hold Mode

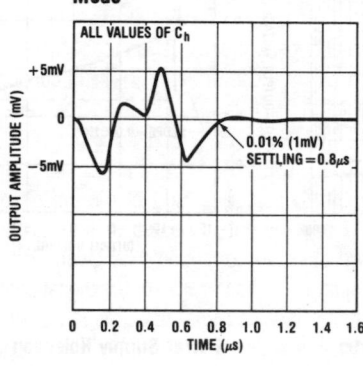

Output Transient at Start of Sample Period

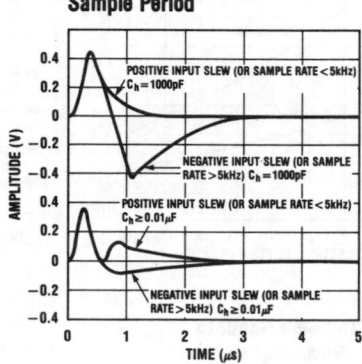

APPLICATIONS INFORMATION

Hold Capacitor

For fast sample and hold applications, the size of the hold capacitor is critical. A low value will give fast acquisition, but will also increase errors due to hold step, and droop caused by amplifier bias current. The capacitor should be made as large as possible, consistent with acquisition time and dynamic sampling error requirements. Capacitors larger than 0.1μF have an additional problem. They are generally not available in the low loss dielectrics like Teflon, Polystyrene, and NPO, at least not at a reasonable price and size. Mylar, even with its poor dielectric absorption properties, may be a reasonable choice where very long sample times are used and low droop rates are needed.

Dielectric absorption in the hold capacitor can often be the major source of error in a sample and hold. The equivalent "circuit" of a typical capacitor is shown below with parallel RC networks used to model dielectric absorption.

Typical Hold Capacitor Equivalent Circuit

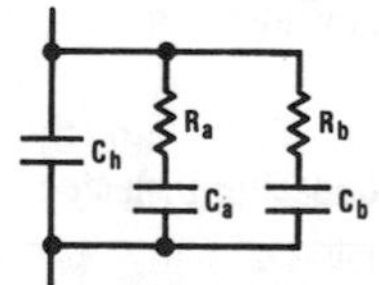

C_a, $C_b \approx$ 0.01 TO 0.1 C_h
R_a, R_b GENERATE TIME CONSTANTS OF 0.1-50 MILLISECONDS WITH C_a, C_b

One can see that rapid changes in capacitor voltage will not be tracked by the internal parasitic capacitors because of the resistance in series with them. This leads to a "sag" effect in the hold capacitor after a sudden change in voltage followed by rapid switch to the hold mode. The capacitor remembers its previous state via the charge on the internal parasitic capacitance and sags

APPLICATIONS INFORMATION

back slightly toward the previous voltage. The magnitude of the sag depends on the voltage change and the time spent sampling the new voltage. Several time constants are typically evident in the sag, although some capacitors tend to exhibit a single time constant, while others show a sag that indicates a blending of many time constants. The curves labeled CAPACITOR DIELECTRIC ABSORPTION show the amount of sag found after a 10V step with sample time at the new voltage and hold time at the new voltage as variables. It is obvious that sag problems are minimized by long sample times and short hold times. This is often in conflict with basic sampling requirements, but one point should be made: if at all possible, keep the sample and hold amplifier in the "tracking", or sampling, mode as much as possible to maximize the time the hold capacitor spends near the voltage at which it will eventually "hold".

The best capacitor for sample and hold applications is Teflon. It is clearly superior with regard to dielectric absorption and operates over the full $-55°C$ to $125°C$ temperature range. If size or price becomes a problem, the second choice for full temperature range operation is "NPO", or "COG" ceramic units. Some care must be used here—not all NPO capacitors use the low dielectric constant ceramic necessary for low dielectric absorption. For lower temperatures ($\leq 70°C$), Polystyrene has traditionally been the best hold capacitor. The best units are cylindrical and fairly large—there seems to be a strong correlation between small size and poorer dielectric performance. Polypropylene has nearly the same absorption properties as polystyrene and offers 85°C operation. It also tends to be smaller. Again, stay with cylindrically wrapped units. Other standard dielectrics such as mica, glass, mylar, and ordinary ceramic are much worse with regard to dielectric absorption. Mylar is sometimes used for large values when the ratio of sample to hold time is large and extremely low droop is required.

Dynamic Sampling Error

A significant sampling error can occur in any sample and hold if the input is moving when the unit is put into the hold mode. The two major causes for this error are digital delay in switch opening and analog delay across the hold capacitor. The switch opening delay is obvious and leads to a "held" output error of $(dv/dt) \times (T_d)$, where dv/dt is the slew rate of the input signal and T_d is switch delay. In the case of the LF198, T_d is approximately 150ns, giving a 4.5mV error when sampling the zero crossing of a 5V (peak) sine wave at 1kHz ($dv/dt = A \bullet 2\pi f = 5 \bullet 2\pi \bullet 10^3$). The analog delay is the difference between input signal and capacitor voltage. It is determined by the RC product of the hold capacitor and the effective series resistance, which in the case of the LF198 is about 150Ω. This analog delay with a 0.01µF hold capacitor is $R \bullet C = 150 \times 10^{-8} = 1.5\mu s$, or about ten times the delay of the switch. The sign of the analog delay is negative—the held output is related in time to the input voltage *before* the hold command was given. The overall dynamic sampling error is the sum of the digital and analog errors. The curve labeled *Dynamic Sampling Error* will be helpful in estimating these errors as a function of input slew rate and hold capacitor size.

Dynamic sampling error can be reduced by a factor of ten or more by inserting a delay in the logic input so that the "hold" command is delayed by an amount equal to the RC time constant of the LF198 and external hold capacitor. For a 0.01µF hold capacitor and the 150Ω resistor internal to the LF198, this is 1.5µs. A simple RC network can be used in front of the logic input for delays up to $\approx 1\mu s$. Longer delays require the addition of a logic gate to speed up the rise time of the delayed signal. See LOGIC RISE TIME in this section for further details.

9

Hold Step

Hold step is the small voltage step (after settling) seen at the output of a sample and hold amplifier when it is switched from the sample mode to the hold mode with a steady DC input. Hold step is typically the result of, or can be modeled as, a fixed quantity of charge transferred to the hold capacitor as a result of the internal switching that occurs during the hold command. In the case of the LF198, that charge is about 5 picocoulombs, giving a hold step of 0.5mV for a 0.01µF hold capacitor and 5mV for a 1000pF hold capacitor. ($V = Q/C$.) Hold step is reasonably independent of logic amplitude if care is taken to minimize the stray capacitance between the logic input

APPLICATIONS INFORMATION

and the hold capacitor. With thoughtful layout, including the guarding technique shown below, stray capacitance should be under 0.3pF, limiting charge variations to less than 0.3 picocoulombs per volt.

Guarding Technique

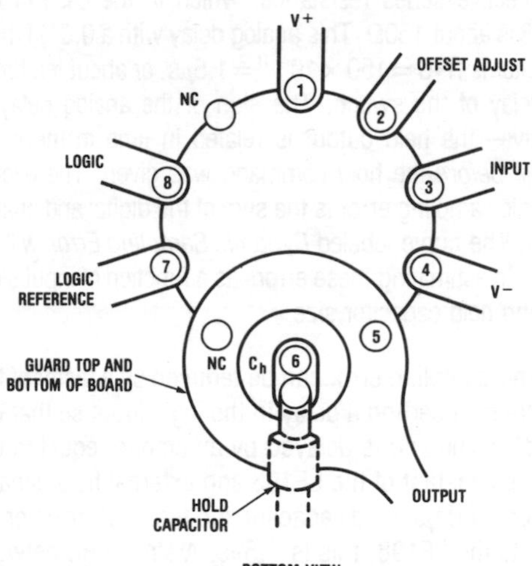

Use 10-pin layout. Guard around C_h is tied to output.

Hold step varies slightly with analog input voltage (see curves). A typical unit will change at 0.4 picocoulombs per volt. This manifests itself as a gain error when the amplifier is switched to the hold mode. With a 0.01μF capacitor, the resulting gain error will be (0.4 PC/V)/0.01μF = 0.004%. This gain error is in the opposite direction of DC (sample mode) gain error. At high values of hold capacitor, DC gain error will dominate and gain will be slightly below unity (0.002%). For low value capacitors (<0.01μF), hold step induced gain error will dominate and hold mode gain will be slightly above unity. Zeroing out hold step does not change the variation of hold step with regard to input voltage.

Offset Zeroing

A sample and hold amplifier has two distinct offset voltages. The first is just the DC offset of the amplifier while in the sample or "tracking" mode. It is identical to the input offset of any operational amplifier. The second offset voltage is the sum of the DC offset plus a dynamic term called hold step. Hold step is a change in output voltage when the amplifier is switched from sample mode to hold mode, with the input held steady. This second offset is often called hold mode offset. It can be less than or much greater than the DC offset, depending on the magnitude and sign of hold step.

A fairly accurate model for hold step is a fixed charge injected into the hold capacitor by the switch turn-off circuitry. The magnitude of the charge is reasonably independent of logic input amplitude. The resulting change in hold capacitor voltage is Q/C_h. The charge, Q, is typically 5 picocoulombs, giving a 0.5mV hold step with a 0.01μF hold capacitor. Since most sample and hold amplifiers are "used," i.e., have their outputs read by an A to D converter, etc., during the hold mode, hold mode offset is arguably much more important than sample mode DC offset.

DC offset adjustment is accomplished with a 1k low TC cermet potentiometer tied to V^+ with 0.6mA flowing through it and the wiper tied to pin 2. This allows pin 2 to be moved ±300mV around its nominal voltage (0.3V below V^+). Offset adjustment range is ±9mV, and the adjustment procedure nominally improves offset drift when the DC offset is reduced to zero. This offset method *can* be used to zero out hold mode offset, but at the expense of some induced offset drift. Each millivolt of hold step offset that is corrected by this method introduces 3.3μV/°C drift. For 0.002μF or larger hold capacitors where hold step is a few millivolts or less, this is a practical solution to hold mode offset. In precision wide temperature range applications, or where C_h is less than 0.002μF, a separate hold mode zeroing method should be used. The circuit shown in the application section using a logic inverter and a 5pF capacitor is recommended (DC AND AC ZEROING).

APPLICATIONS INFORMATION

Logic Fall Time

Hold step is independent of logic input fall time only for fall times faster than 10V/μs. For instance, as logic fall time changes from 10V/μs to 1V/μs, hold step with a 0.01μF hold capacitor will typically increase from 0.25mV to 1.0mV. See the curve labeled HOLD STEP vs LOGIC SLEW RATE for further data points. If logic slew rate is not constant, use the value at the threshold point (1.5V with respect to logic reference). An RC network will have a discharge slew rate of V_L/RC, where V_L is the logic threshold of the LF198. The delay generated by the network will be $RC \cdot \ln(V^+/V_L)$, where V^+ is logic amplitude. For a 1μs delay, with 5V logic, an RC time constant of 0.8μs is needed. This has a slew rate of 2V/μs at threshold, which will slightly degrade hold step. It is obvious that an RC delay network significantly longer then 1μs will have a large effect on hold step. If longer delays are required, they should be followed by several inverter stages or a Schmitt trigger to increase slew rate.

Adding Delay to Logic Input

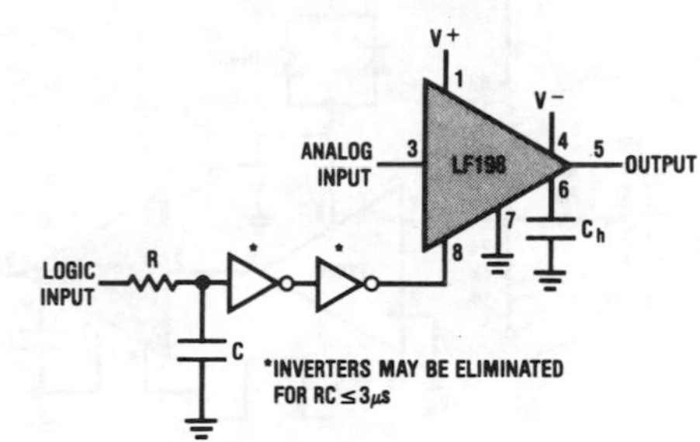

LOGIC INPUT CONFIGURATIONS*

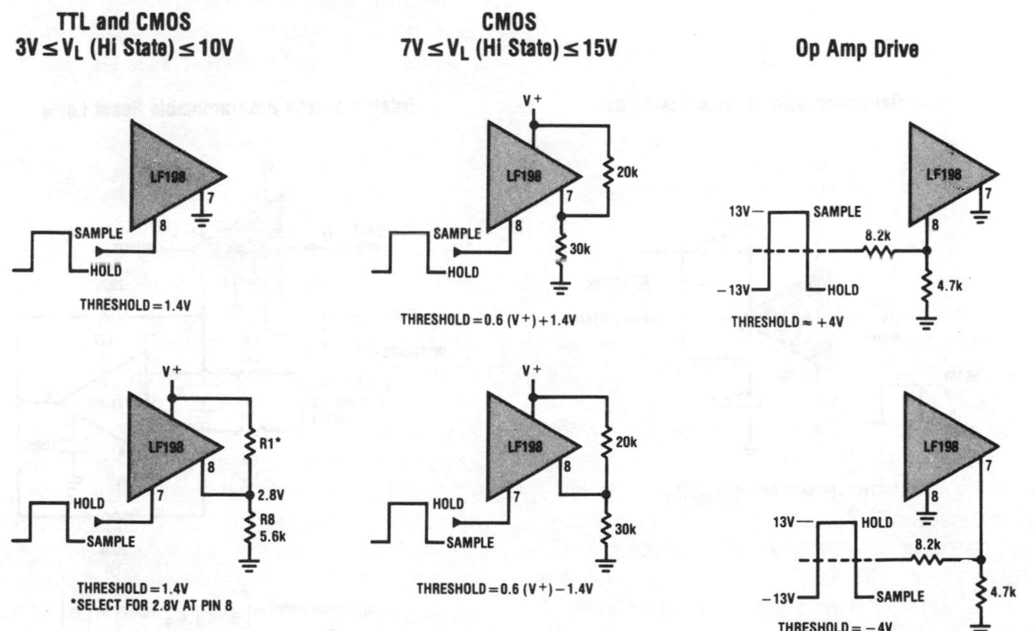

*The logic input signal high state must be at least 2V below the positive supply voltage of the LF198.

TYPICAL APPLICATIONS

X1000 Sample and Hold

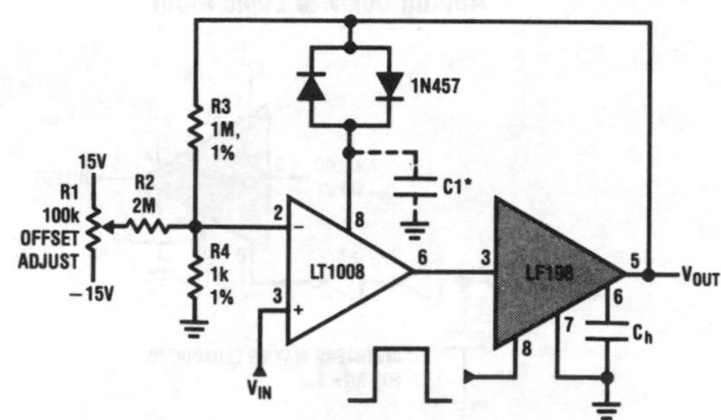

Sample and Difference Circuit (Output Follows Input in Hold Mode and Resets to V_B in Sample Mode)

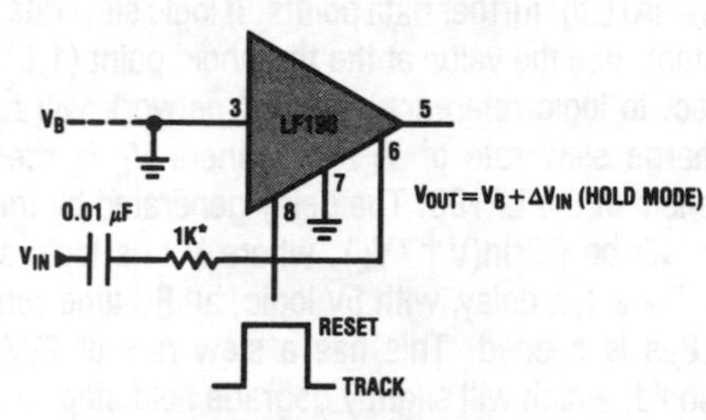

Ramp Generator with Variable Reset Level

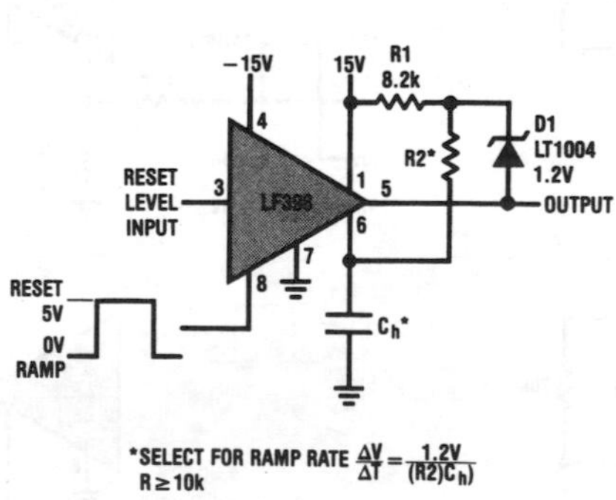

Integrator with Programmable Reset Level

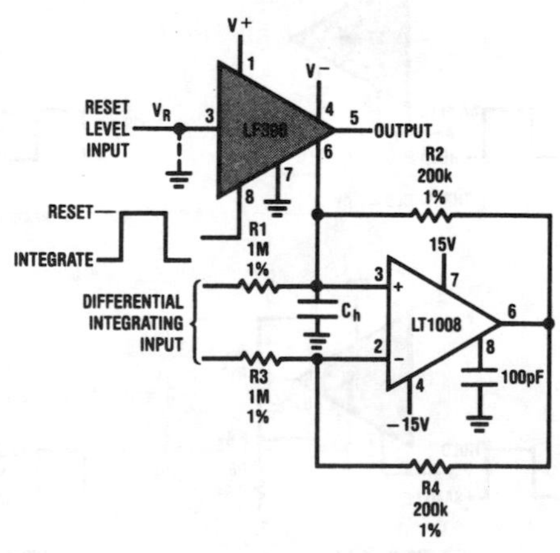

TYPICAL APPLICATIONS

Output Holds at Average of Sampled Input

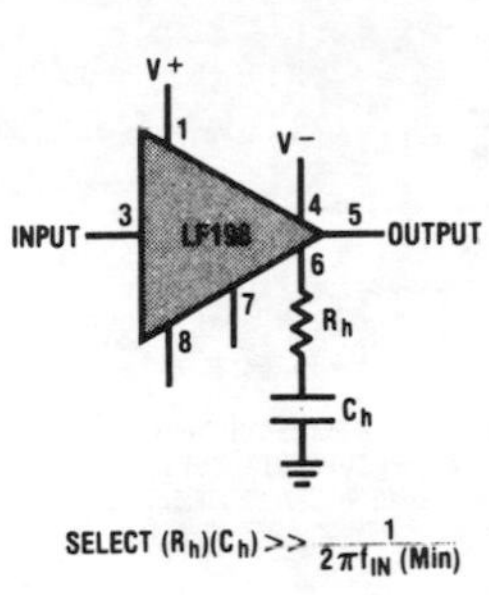

Fast Acquisition, Low Droop Sample and Hold

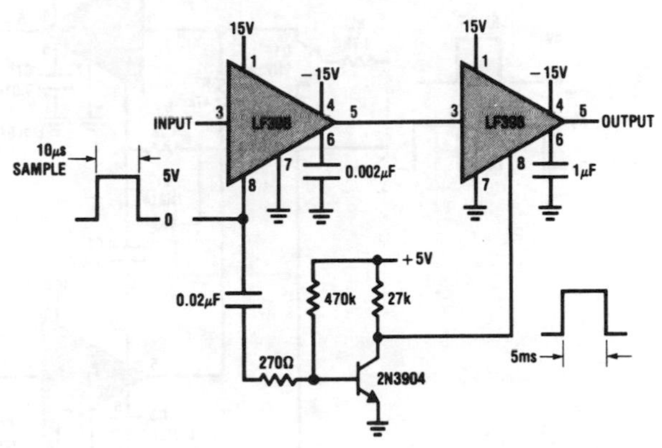

DC and AC Zeroing

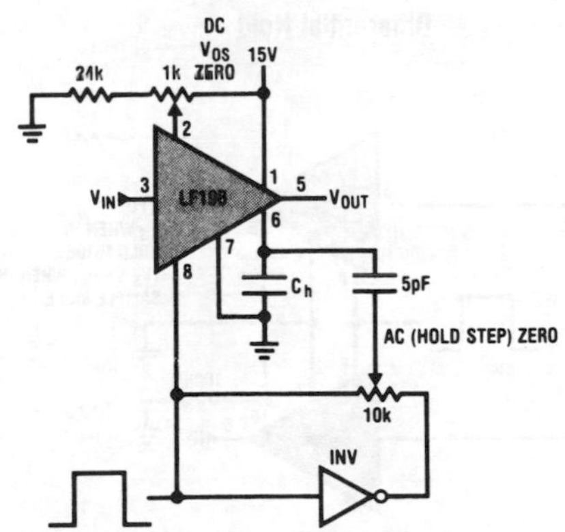

2-Channel Switch

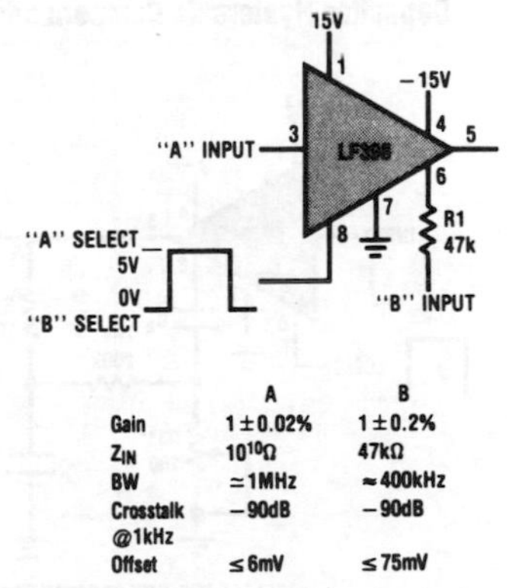

	A	B
Gain	1 ± 0.02%	1 ± 0.2%
Z_{IN}	$10^{10}\Omega$	47kΩ
BW	≅ 1MHz	≈ 400kHz
Crosstalk @1kHz	−90dB	−90dB
Offset	≤ 6mV	≤ 75mV

TYPICAL APPLICATIONS

Staircase Generator

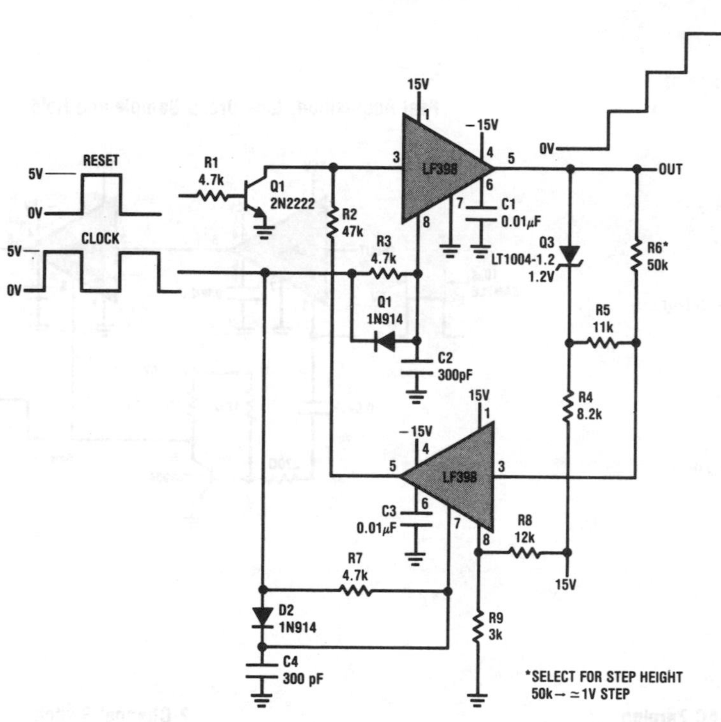

Capacitor Hysteresis Compensation

Differential Hold

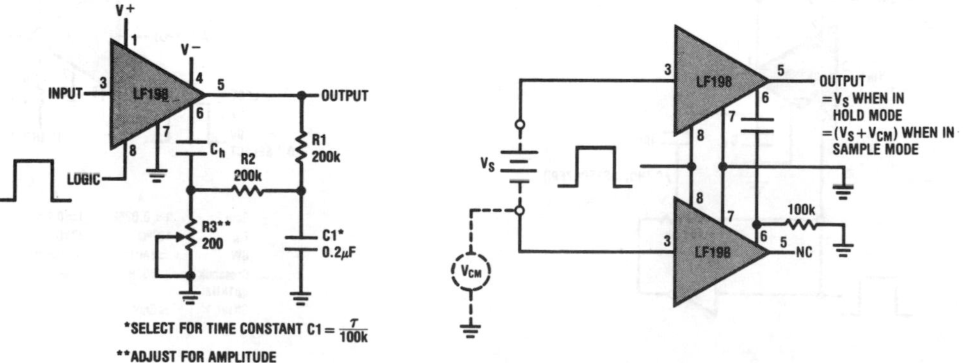

*SELECT FOR TIME CONSTANT $C1 = \frac{\tau}{100k}$

**ADJUST FOR AMPLITUDE

TYPICAL APPLICATIONS

Isolated Temperature Sensor

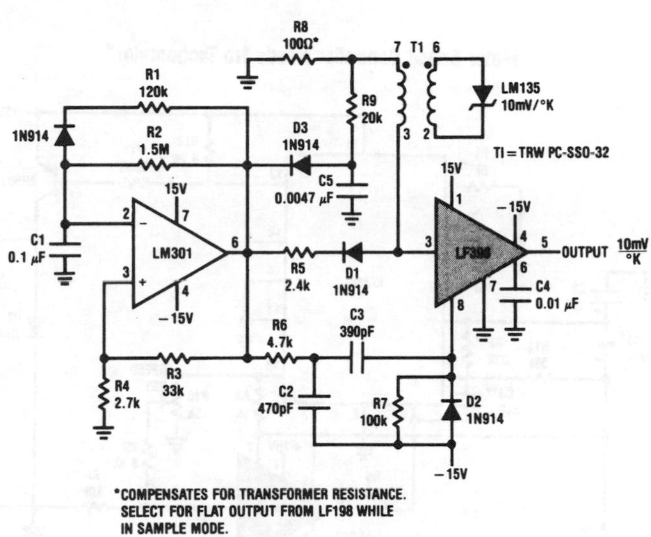

*COMPENSATES FOR TRANSFORMER RESISTANCE. SELECT FOR FLAT OUTPUT FROM LF198 WHILE IN SAMPLE MODE.

Pulse Width to Voltage Converter

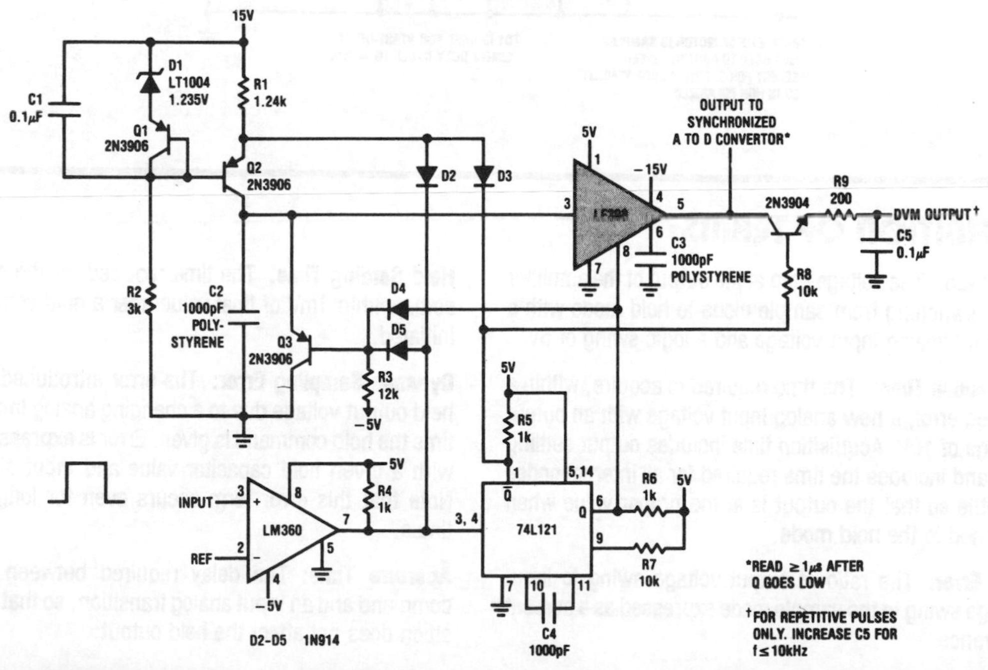

*READ ≥1μs AFTER Q GOES LOW

†FOR REPETITIVE PULSES ONLY. INCREASE C5 FOR f≤10kHz

TYPICAL APPLICATIONS

Motor Speed Controller Needs No Tachometer*

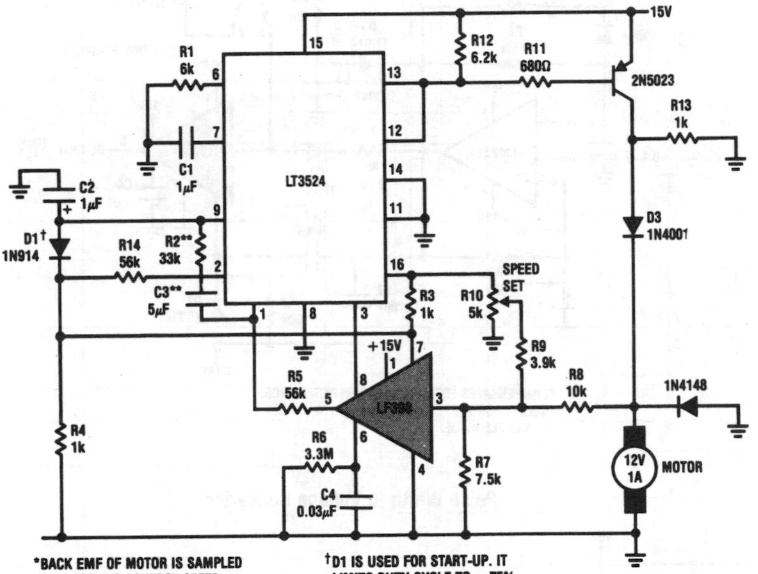

*BACK EMF OF MOTOR IS SAMPLED AND USED TO CONTROL SPEED.
**SELECT FOR OPTIMUM LOOP STABILITY. C3 IS NON POLARIZED

†D1 IS USED FOR START-UP. IT LIMITS DUTY CYCLE TO ≈75%

DEFINITION OF TERMS

Hold Step: The voltage step at the output of the amplifier when switching from sample mode to hold mode with a constant analog input voltage and a logic swing of 5V.

Acquisition Time: The time required to acquire, within a defined error, a new analog input voltage with an output change of 10V. Acquisition time includes output settling time and includes the time required for all internal nodes to settle so that the output is at the proper value when switched to the hold mode.

Gain Error: The ratio of output voltage swing to input voltage swing in the sample mode expressed as a percent difference.

Hold Settling Time: The time required for the output to settle within 1mV of final value after a hold command is initiated.

Dynamic Sampling Error: The error introduced into the held output voltage due to a changing analog input at the time the hold command is given. Error is expressed in mV with a given hold capacitor value and input slew rate. Note that this error term occurs even for long sample times.

Aperture Time: The delay required between "Hold" command and an input analog transition, so that the transition does not affect the held output.

SCHEMATIC DIAGRAM

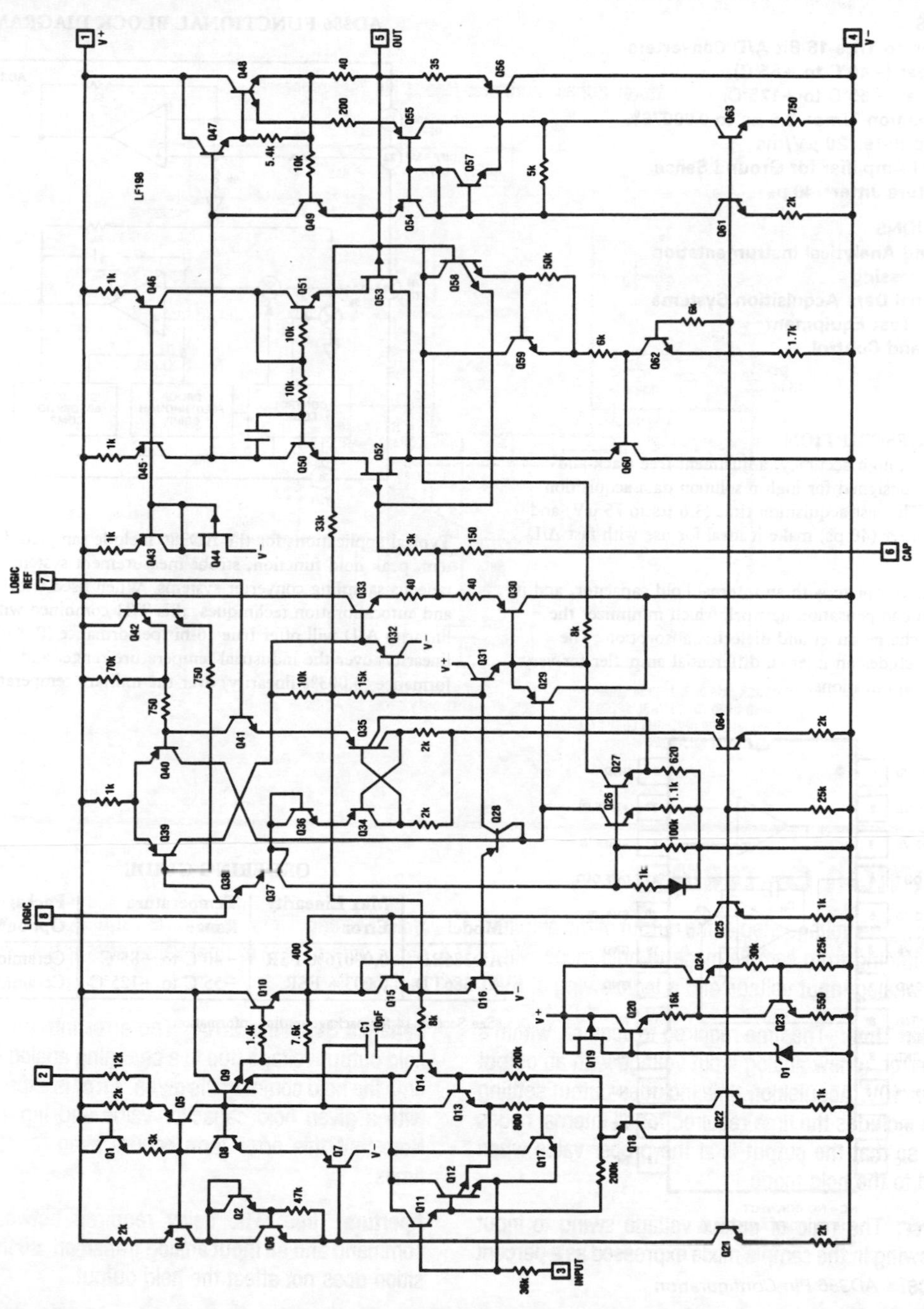

9

True 16-Bit Track-and-Hold Amplifier

AD386

FEATURES
Companion to True 16-Bit A/D Converters
16-Bit Linear (−40°C to +85°C)
14-Bit Linear (−55°C to +125°C)
Fast Acquisition Time: 3.6 μs to 0.00076%
Low Droop Rate: 20 μV/ms
Differential Amplifier for Ground Sense
Low Aperture Jitter: 40 ps

APPLICATIONS
Medical and Analytical Instrumentation
Signal Processing
Multichannel Data Acquisition Systems
Automatic Test Equipment
Guidance and Control
Sonar

AD386 FUNCTIONAL BLOCK DIAGRAM

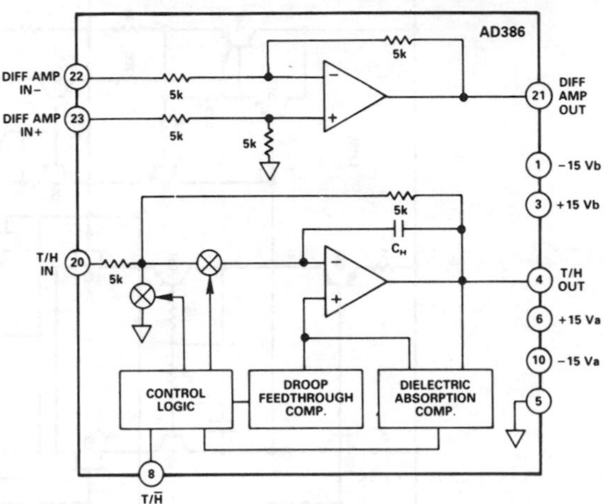

PRODUCT DESCRIPTION

The AD386 is a high accuracy, adjustment free track-and-hold amplifier designed for high resolution data acquisition applications. The fast acquisition time (3.6 μs to 75 μV) and low aperture jitter (40 ps) make it ideal for use with fast A/D converters.

The AD386 is complete with an internal hold capacitor, and it incorporates a compensation network which minimizes the track-to-hold charge offset and dielectric absorption. The AD386 also includes an internal differential amplifier for very high accuracy applications.

Typical applications for the AD386 include sampled data system, peak hold function, strobe measurement system and simultaneous sampling converter systems. When used with autozero and autocalibration techniques, this T/H combined with a high linearity A/D will offer true 16-bit performance (0.00076% linearity) over the industrial temperature range, and 14-bit performance (0.003% linearity) over the military temperature range.

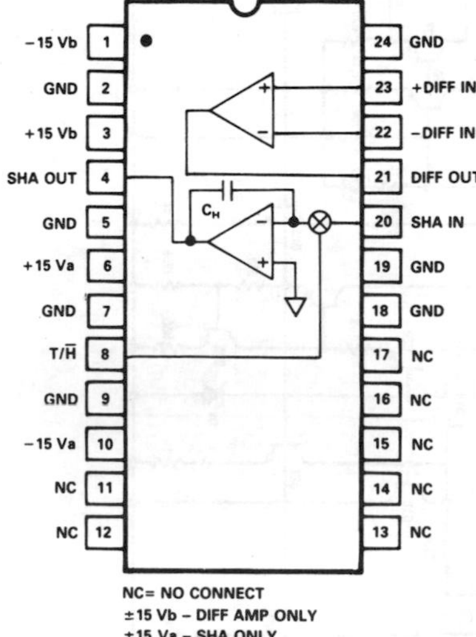

AD386 Pin Configuration

ORDERING GUIDE

Model	Max Linearity Error	Temperature Range	Package Option*
AD386BD	0.00076% FSR	−40°C to +85°C	Ceramic (DH-24B)
AD386TD	0.003% FSR	−55°C to +125°C	Ceramic (DH24B)

*See Section 14 for package outline information.

SPECIFICATIONS (@ +25°C unless otherwise noted, V_S = ±15 V ±10%)

Model	Conditions	AD386BD Min	AD386BD Typ	AD386BD Max	AD386TD Min	AD386TD Typ	AD386TD Max	Units
DIFFERENTIAL AMPLIFIER								
INPUT CHARACTERISTICS								
Input Range		±10			±10			V
Common-Mode Range		±10			±10			V
Input Resistance[1]								
Signal			5			5		kΩ
Ground Sense			10			10		kΩ
Offset[2]			0.6	**2.0**		0.6	**2.0**	mV
Offset Drift	T_{min} to T_{max}		10	30		10	30	μV/°C
CMRR	V_{CM} = ±10	**80**	90		**80**	90		dB
PSRR[3]		**76**	85		**76**	85		dB
TRANSFER CHARACTERISTICS								
Gain			−1			−1		V/V
Gain Error				**0.02**			**0.02**	%
Gain Error Drift	T_{min} to T_{max}		1	5		1	5	ppm/°C
Gain Linearity			0.0002	**0.00076**		0.0002	**0.003**	%
Gain Linearity Drift	T_{min} to T_{max}		0.01	0.05		0.01	0.05	ppm/°C
Noise (ENBW = 1.8 MHz)			32	45		32	45	μV rms
DYNAMIC CHARACTERISTICS								
Small Signal Bandwidth			6			6		MHz
Slew Rate			65			65		V/μs
Settling Time[4]								
10 V Step to 1/2 LSB16			2.0	3.0				μs
10 V Step to 1/2 LSB14			0.8	1.5		0.8	1.5	μs
20 V Step to 1/2 LSB16			2.0	**3.0**				μs
20 V Step to 1/2 LSB16	T_{min} to T_{max}		2.0	3.0				μ s
20 V Step to 1/2 LSB14			0.8	**1.5**		0.8	**1.5**	μs
20 V Step to 1/2 LSB14	T_{min} to T_{max}		0.8	1.5		0.8	1.5	μs
OUTPUT								
Voltage	R_{LOAD}>3.5 kΩ, T_{min} to T_{max}	±10			±10			V
Current	Short Circuit		15			15		mA
POWER SUPPLY								
Rated Performance			±15			±15		V
Operating Range		±5		±18	±5		±18	V
Quiescent Current			4.2	**5.0**		4.2	**5.0**	mA
TRACK-AND-HOLD								
INPUT CHARACTERISTICS								
Input Range		±10			±10			V
Input Resistance[1]			5			5		kΩ
Offset[2]			0.6	**2.0**		0.6	**2.0**	mV
Offset Drift	T_{min} to T_{max}		10	30		10	30	μV/°C
TRANSFER CHARACTERISTICS								
Gain			−1			−1		V/V
Gain Error				**0.02**			**0.02**	%
Gain Error Drift	T_{min} to T_{max}		1	5		1	5	ppm/°C
Gain Linearity			0.0002	**0.00076**		0.0002	**0.003**	%
Gain Linearity Drift	T_{min} to T_{max}		0.01	0.05		0.01	0.05	ppm/°C
PSRR[3]		**76**	85		**76**	85		dB
DYNAMIC CHARACTERISTICS								
Small Signal Bandwidth			2			2		MHz
Slew Rate			15			15		V/μs
TRACK-TO-HOLD SWITCHING								
Pedestal + Offset			0.5	**1.5**		0.5	**1.5**	mV
Pedestal + Offset	T_{min} to T_{max}			5.0			7.5	mV
Pedestal Linearity	T_{min} to T_{max}		0.0004	0.00076		0.0004	0.003	%
Aperture Delay			12			12		ns
Aperture Jitter			40			40		ps
Transient Settling[4]								
to 1/2 LSB16	T_{min} to T_{max}		600	**800**				ns
to 1/2 LSB14	T_{min} to T_{max}		400	**500**		400	**500**	ns

Model	Conditions	AD386BD Min	AD386BD Typ	AD386BD Max	AD386TD Min	AD386TD Typ	AD386TD Max	Units
HOLD MODE								
Droop Rate			20	**100**		20	**100**	mV/s
Droop Rate	T_{max}		0.2	1.0		3.6	18	V/s
Feedthrough[5]			−99	**−94**		−99	**−94**	dB
Noise (ENBW = 1.7 MHz)			32	50		32	50	µV rms
PSRR[3]		**60**	66		**60**	66		dB
Dielectric Absorption[6]			7	**10**		7	**10**	ppm
HOLD-TO-TRACK DYNAMICS								
Acquisition Time[4]								
10 V Step to 1/2 LSB16			3.6	4.1				µs
10 V Step to 1/2 LSB14			3.1	3.6		3.1	3.6	µs
20 V Step to 1/2 LSB16			3.6	**4.1**				µs
20 V Step to 1/2 LSB16	T_{min} to T_{max}		4.0	4.5				µs
20 V Step to 1/2 LSB14			3.1	**3.6**		3.1	**3.6**	µs
20 V Step to 1/2 LSB14	T_{min} to T_{max}		3.5	4.0		4.0	4.5	µs
DIGITAL INPUTS								
V_{IH}	T_{min} to T_{max}	**2.4**			**2.4**			V
V_{IL}	T_{min} to T_{max}			**0.8**			**0.8**	V
I_{IH}	T_{min} to T_{max}	**−10**		**+10**	**−10**		**+10**	µA
I_{IL}	T_{min} to T_{max}	**−10**		**+10**	**−10**		**+10**	µA
OUTPUT								
Voltage	R_{LOAD}>3.5 kΩ, T_{min} to T_{max}	±10			±10			V
Current	Short Circuit		15			15		mA
POWER SUPPLY								
Rated Performance			±15			± 15		V
Operating Range		± 8		±18	± 8		±18	V
Quiescent Current								
Positive Supply			8.0	**12.0**		8.0	**12.0**	mA
Negative Supply		**−6.0**	−5.4		**−6.0**	−5.4		mA
SYSTEM								
Gain Linearity	T_{min} to T_{max}		0.0003	0.00076		0.0003	0.003	%
Acquisition Time[4, 7]								
20 V Step to 1/2 LSB16			4.1	5.1				µs
20 V Step to 1/2 LSB16	T_{min} to T_{max}		4.5	5.4				µs
20 V Step to 1/2 LSB14			3.2	3.9		3.2	3.9	µs
20 V Step to 1/2 LSB14	T_{min} to T_{max}		3.6	4.3		4.1	4.8	µs
Power Dissipation			312	435		312	435	mW
TEMPERATURE RANGE								
Operating		−40		+85	−55		+125	°C
Storage		−60		+150	−60		+150	°C

NOTES
[1]Typical resistance tolerance is ±25%.
[2]After 5 minute warmup at +25°C.
[3]Test conditions: $+V_S$ = +15 V, $-V_S$ = −16 V to −14 V and $+V_S$ = +14 V to +16 V, $-V_S$ = −15 V.
[4]R_{LOAD} = 5 kΩ, C_{LOAD} = 10 pF, settling measured to 1/2 LSB at output.
[5] Measured at 1 kHz.
[6]Dielectric Absorption represents the magnitude of long-term settling artifacts for hold times up to 80 µs as a fraction of the difference in voltages between two successive held samples.
[7]Specifications also apply for 10 V step.
Specifications subject to change without notice.
Specifications in **bold** are 100% production tested.

ABSOLUTE MAXIMUM RATINGS[1]

Supply Voltage . . . . ±18 V
Internal Power Dissipation . . . . 800 mW
Input Voltage[2] . . . . ±18 V
T/$\overline{H}$ Input Voltage . . . . −0.5 V, + 16 V
Output Short Circuit Duration . . . . Indefinite
Storage Temperature Range . . . . −65°C to +150°C
Operating Temperature Range
AD386B . . . . −40°C to +85°C
AD386T . . . . −55°C to +125°C
Lead Temperature Range (Soldering 60 sec) . . . . +300°C

NOTES
[1]Stresses above those listed under "Absolute Maximum Ratings" may cause permanent damage to the device. This is a stress rating only, and functional operation of the device at these or any other conditions above those indicated in the operational section of this specification is not implied. Exposure to absolute maximum rating conditions for extended periods may affect device reliability.
[2]For supply voltages less than ±18 V, the absolute maximum input voltage is equal to the supply voltage.

Typical Performance Characteristics

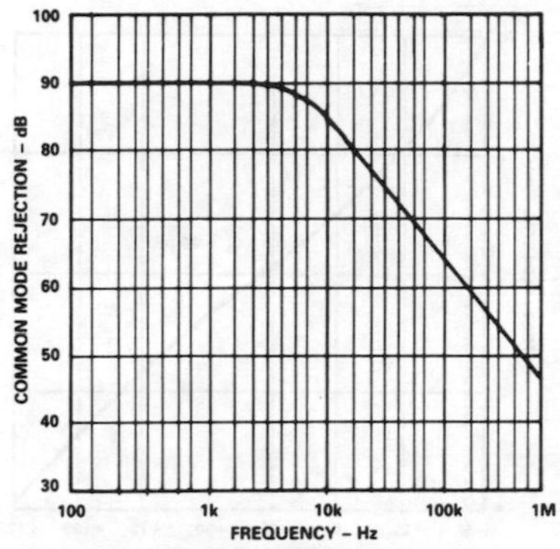

Figure 1. Differential Amplifier Common Mode Rejection vs. Frequency

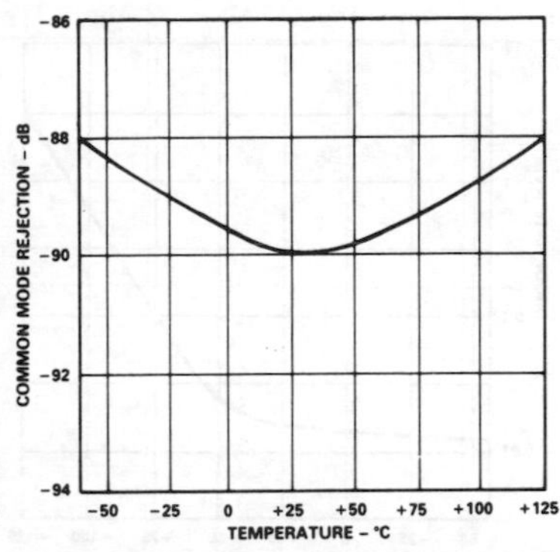

Figure 2. Differential Amplifier Common Mode Rejection vs. Temperature (100 Hz)

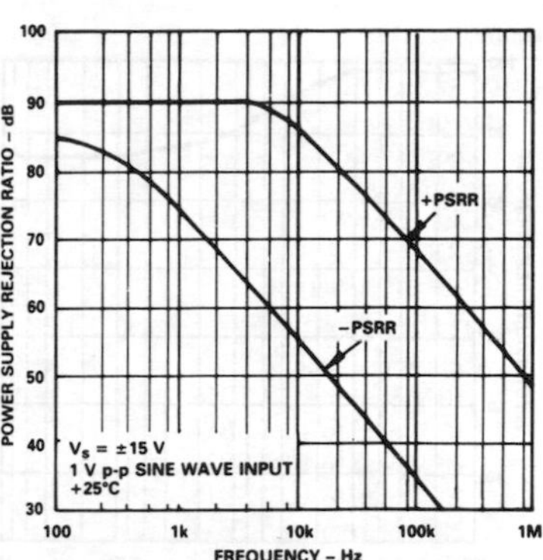

Figure 3. Differential Amplifier Power Supply Rejection vs. Frequency

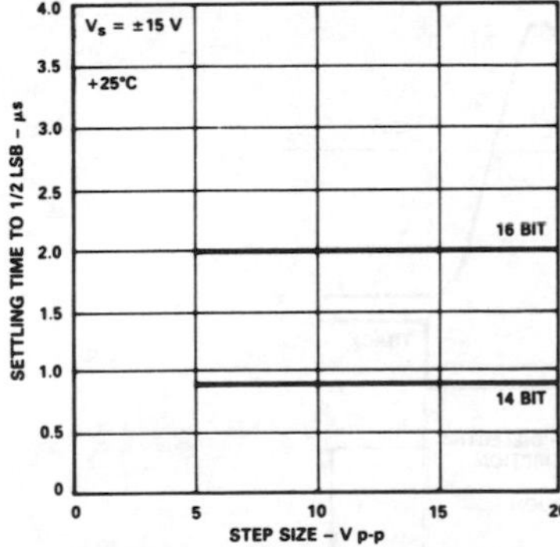

Figure 4. Differential Amplifier Settling Time vs. Step Size

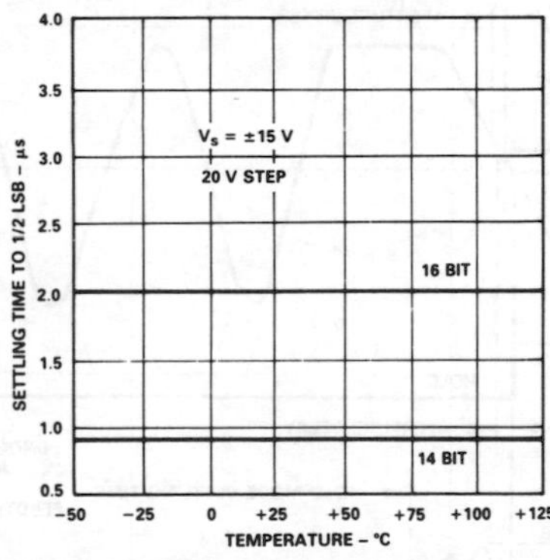

Figure 5. Differential Amplifier Settling Time vs. Temperature

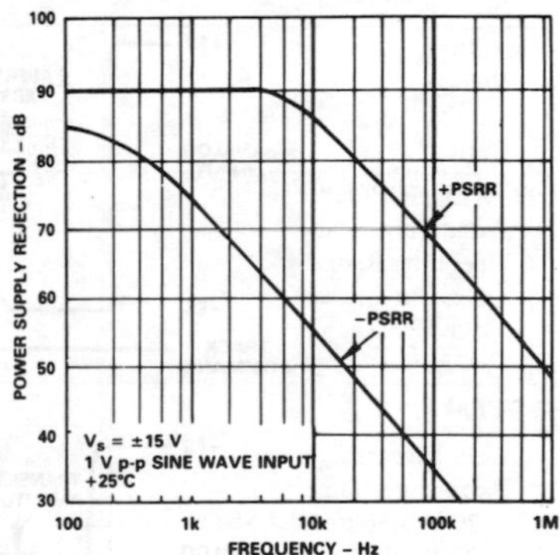

Figure 6. T/H Power Supply Rejection vs. Frequency, Track Mode

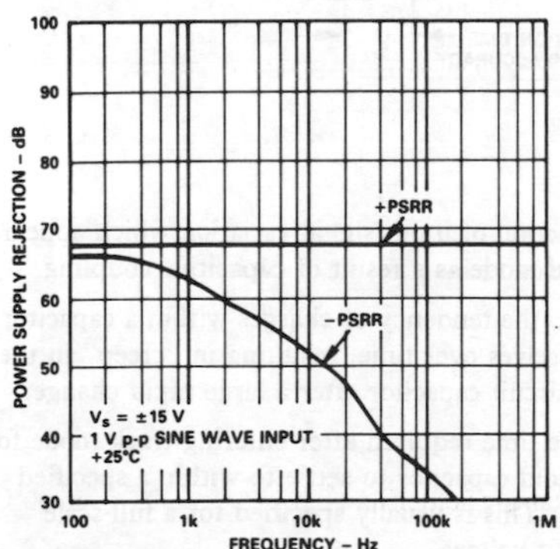

Figure 7. T/H Power Supply Rejection vs. Frequency, Hold Mode

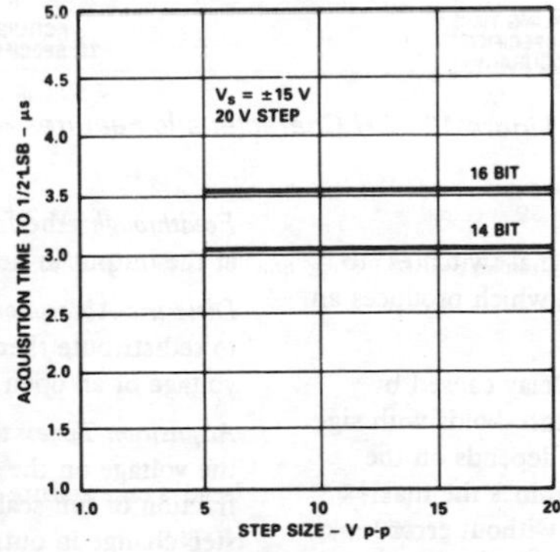

Figure 8. T/H Acquisition Time vs. Step Size

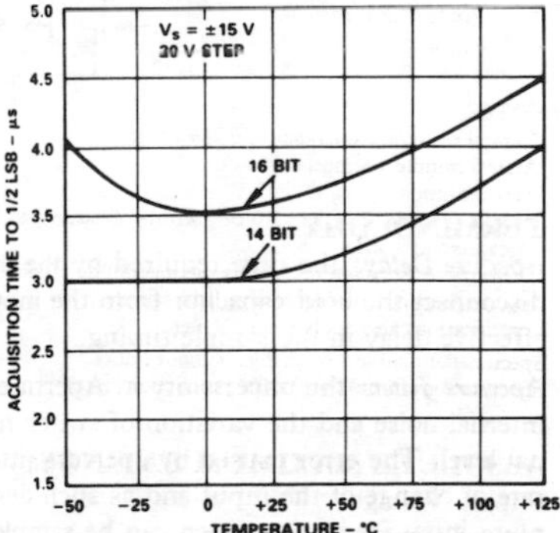

Figure 9. T/H Acquisition Time vs. Temperature

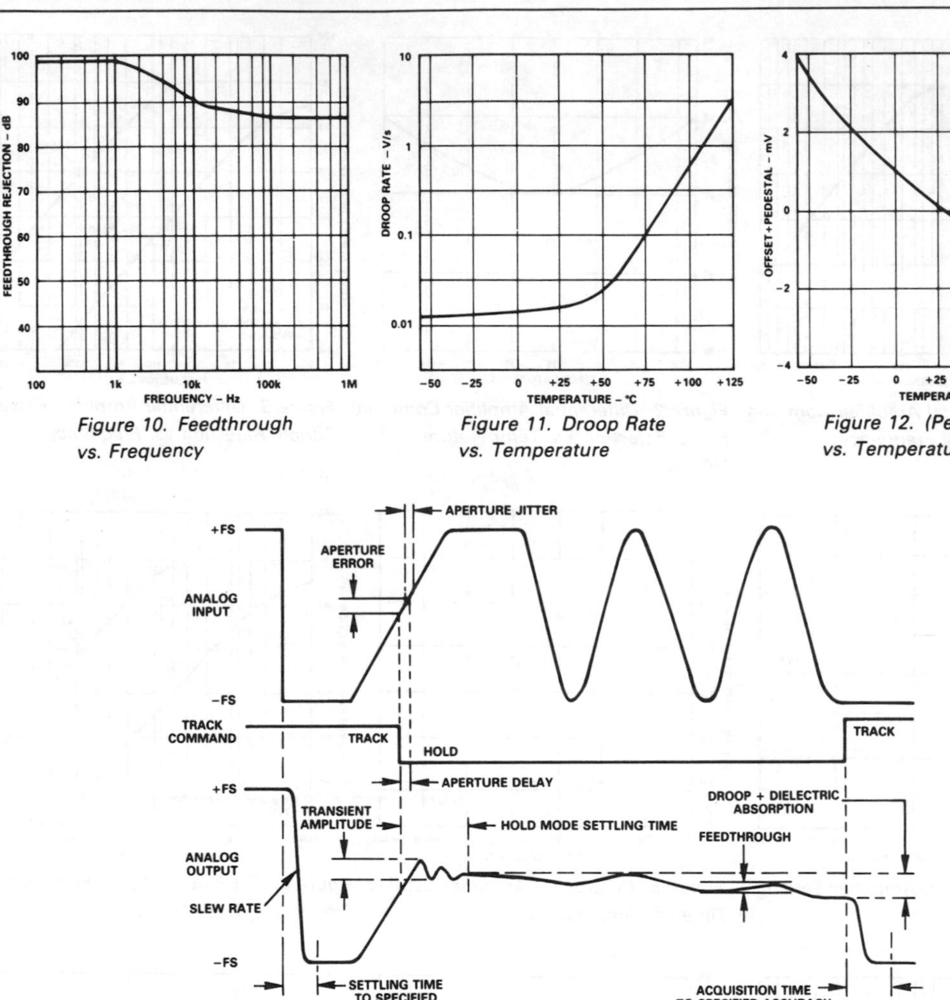

Figure 10. Feedthrough vs. Frequency

Figure 11. Droop Rate vs. Temperature

Figure 12. (Pedestal+Offset) vs. Temperature

Figure 13. T/H Characteristic Features

TERMINOLOGY

Aperture Delay: the time required by the internal switch(es) to disconnect the hold capacitor from the input, which produces an effective delay in the sample timing.

Aperture Jitter: the uncertainty in Aperture Delay caused by internal noise and the variation of switching thresholds with signal level. The error caused by aperture jitter depends on the rate of change of the input and as such determines the maximum input frequency which can be sampled without error.

Pedestal: a step change in the output voltage which occurs when switching from track mode to hold mode.

Hold Mode Settling Time: the time required for the pedestal to reach its final value to within a specified fraction of full scale.

Droop: the change in the held output voltage resulting from leakage currents.

Feedthrough: the fraction of input signal variation which appears at the output in hold mode as a result of capacitive coupling.

Dielectric Absorption: the tendency of charges within a capacitor to redistribute themselves over time, resulting in "creep" in the voltage of an open circuit capacitor after a large rapid change.

Acquisition Time: the time required after entering track mode for the voltage on the hold capacitor to settle to within a specified fraction of full scale. This is usually specified for a full-scale step change in output voltage.

Settling Time: the time required in track mode for the output to reach its final value within a specified fraction of full scale following a step change in the input voltage.

Nonlinearity: the degree to which a plot of output versus input deviates from the straight line defined by the end points. It is usually specified as a percentage of full scale.

THEORY OF OPERATION

The architecture of the AD386 differs from that usually encountered in inverting Track-and-Hold (T/H) circuits. The hold capacitor in a conventional T/H (Figure 14) is always connected from the amplifier's output to its inverting input. In track mode switch A is open and switch B is closed. Since the summing junction is a virtual ground, the voltage across the capacitor follows the input. The switches change state in hold mode which disconnects the capacitor from the input and holds the output voltage constant. The clamping action of switch A reduces the variations across switch B, improving feedthrough performance.

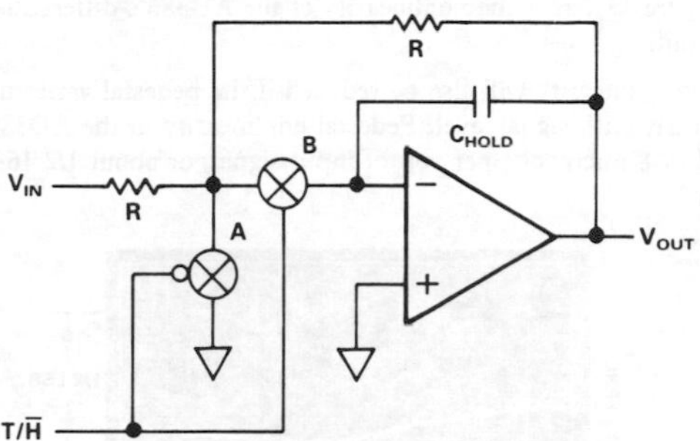

Figure 14. Conventional Inverting Integrator T/H

This circuit forces several tradeoffs. The hold capacitor's charging current is limited by the input resistor. Either the resistor or the capacitor, or both, must be made small to obtain fast acquisition times. A small resistor creates greater demands on the circuit which drives the T/H, while a small capacitor leads to increased pedestal and droop. In addition, the parallel combination of the feedback resistor and the hold capacitor acts as a low pass filter and constrains both bandwidth and acquisition time.

The AD386 uses a four-switch flyback architecture which removes the hold capacitor from the feedback loop during track mode (Figure 15). Switches A and C are open in track mode while switches B and D are closed. This maximizes bandwidth and provides minimum acquisition time because the charging

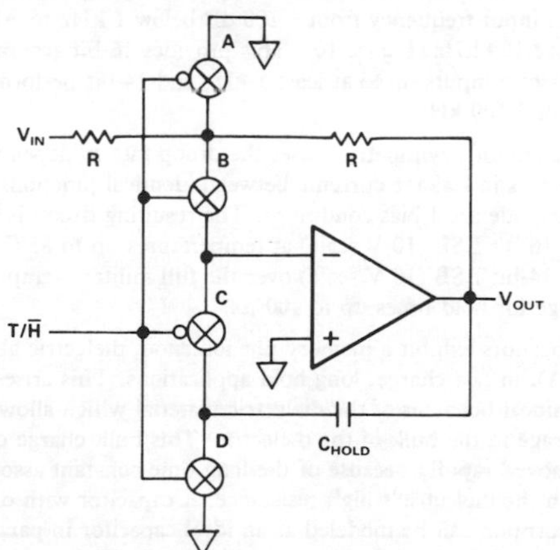

Figure 15. Four-Switch Inverting Flyback T/H

current delivered to the hold capacitor is limited only by the amplifier's output capability. The hold capacitor can be made larger, subject to amplifier stability, since it no longer appears in parallel with the feedback resistor. This helps to reduce droop and pedestal. Switches A and C close in hold mode while switches B and D open, which connects the hold capacitor to the amplifier's inverting input.

Additional switches and capacitors, not shown in the figure, provide first order cancellation of amplifier and switch leakage currents, switching charge injection, and switch feedthrough. Finally, a small amount of positive feedback is used to reduce dielectric absorption effects.

TRACK-AND-HOLD ERROR CONTRIBUTIONS IN SAMPLED-DATA SYSTEM

Any track-and-hold amplifier imposes performance limits on the system in which it is used. Some of these limits can be derived from the theory of sampled-data systems, some are intrinsic to the T/H, and some depend on details of the system design. Many subtle effects come into play as system resolution increases to 14 or 16 bits, and these can contribute significant errors. Understanding T/H error sources is critical to maintaining signal integrity in a high resolution data acquisition system.

FREQUENCY LIMITATIONS

Three factors set fundamental limits on system performance when digitizing high frequency signals. These are: T/H amplifier bandwidth, aperture uncertainty, and the maximum update rate of the T/H and A/D combination. The track mode bandwidth of the T/H must be significantly greater than the bandwidth of the signals being digitized to prevent the introduction of amplitude and phase errors. The 2 MHz small signal bandwidth of the AD386 attenuates a 35 kHz signal by 0.001 dB and shifts its phase by 1.0 degrees.

There are two different aperture related error terms. The first is aperture delay time, the delay between the HOLD command and the complete opening of internal switches in the T/H. This time amounts to a negative phase delay applied to the input signal because the T/H output can actually continue to track the input for a brief time after the HOLD command. Aperture delay time can be "tuned out" by advancing the assertion of HOLD.

Aperture jitter, the random variations in aperture delay time, causes errors which are directly related to the rate of change of the input signal and which cannot be eliminated by circuit adjustments.

A simple calculation provides the frequency at which aperture jitter produces an error of 1/2 LSB when the input is a full-scale sinusoid. The general result for an N-bit A/D converter is

$$F_{max} = \frac{V_{FS}}{V_{PP}} \times \frac{1}{2^{N+1} \times \pi \times Aperture\ Jitter}$$

where V_{FS} is the A/D converter's input range and V_{PP} is the peak-to-peak value of the input sinusoid. The worst case (minimum) value of F_{max} occurs when V_{PP} is equal to V_{FS}. If the T/H has an aperture jitter of 100 ps and is used with a 16-bit linear A/D, the maximum input frequency is 24.3 kHz.

The same T/H, when used with a 14-bit linear A/D, permits the processing of signals up to 97.1 kHz before aperture jitter-errors become observable. Figure 16 shows these errors as a function of frequency, assuming a full scale input sinusoid, for several values of aperture jitter.

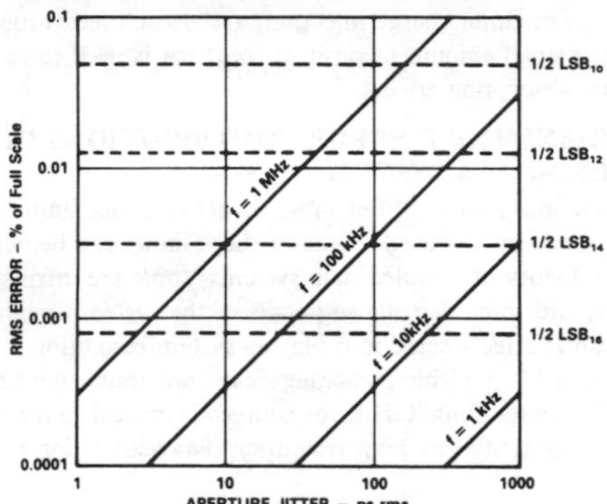

Figure 16. T/H Error vs. Aperture Jitter and Input Frequency

Aperture jitter is often expressed as an rms number. "Peak-to-peak" aperture jitter is usually defined as 6 times this rms value. This comes from probability theory, where 99.7% of the measurements of a random variable will be within 3 standard deviations of the variable's average value. Aperture jitter arises from broadband electrical noise, which is very nearly an ideal random process with a standard deviation equal to its rms value, so multiplication by 6 gives a good approximation to the noise's peak-to-peak value.

A second limit on the input frequency is imposed by the finite time required for signal acquisition and conversion. It is possible to reconstruct any uniformly sampled signal without loss of information provided the sampling rate is at least twice the bandwidth of the input signal; this is the Nyquist criterion, a fundamental result in sampling theory. This limits input frequency to

$$F_{max} = \frac{1}{2 \times (t_{ACQ} + t_{CONV} + t_{AP})}$$

where t_{ACQ} is the T/H acquisition time, T_{CONV} is the time required for the A/D conversion, and T_{AP} is the aperture delay of the T/H. The last term is usually very small and can be ignored. A system composed of a 3.6 μs T/H and a 10 μs A/D can be used successfully to digitize signals with frequency components up to 36.76 kHz. This limit is independent of input signal amplitude. Throughput rates and input frequency ranges for the AD386 in combination with various A/D converters are shown in Table I.

A/D	Conversion Time	Minimum Throughput
ADADC71	50 μs max	18.7 kHz
AD1376/78	17 μs max	48.8 kHz
AD1377	10 μs max	73.5 kHz

Table I. Throughput for AD386 with Various A/D Converters

NONLINEARITIES

Two phenomena directly affect the fidelity of a T/H's transfer function and can degrade system linearity. One of these error sources is track mode nonlinearity. It arises primarily from gain nonlinearity in the T/H's internal amplifier(s). Mismatches in the temperature coefficients of internal resistors may also contribute, but usually do so only for very low frequency signals. The AD386's track mode nonlinearity is about 1/6 16-bit LSB (Figure 17), as is the nonlinearity of the AD386's differential amplifier.

System linearity will also be reduced if the pedestal varies nonlinearly with signal level. Pedestal nonlinearity in the AD386 is below 8 microvolts per volt of input signal, or about 1/2 16-bit LSB.

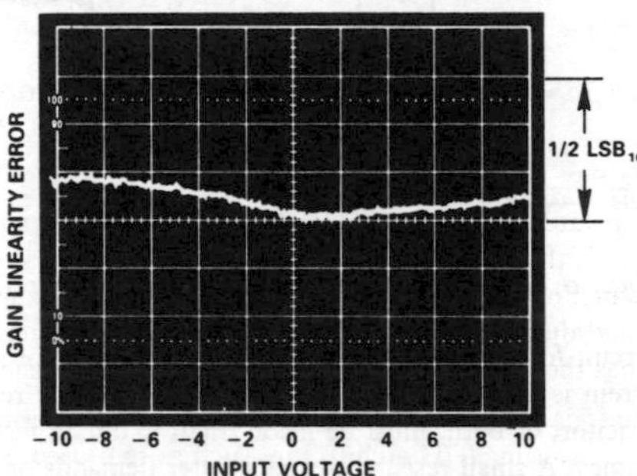

Figure 17. AD386 Track Mode Nonlinearity

FEEDTHROUGH, DROOP, AND DIELECTRIC ABSORPTION

Errors resulting from signal feedthrough and droop must be less than 1/2 LSB in order for the system's linearity to be maintained. The AD386 uses a symmetrical, compensated architecture to minimize both these effects. Feedthrough varies slightly with input frequency from −100 dB below 1 kHz to −86 dB above 100 kHz (Figure 10). This provides 16-bit accuracy for full-scale inputs up to at least 5 kHz and 14-bit performance to beyond 100 kHz.

The circuit's symmetry causes the droop rate to depend on differences in leakage currents between identical junctions under nearly identical bias conditions. The resulting droop is less than 1/2 16-bit LSB (10 V scale) at temperatures up to 85°C and 1/2 14-bit LSB (10 V scale) over the full military temperature range for hold times up to 100 μs.

Capacitors exhibit a memory phenomenon, dielectric absorption (DA), in fast charge, long hold applications. This arises from nonideal behavior of the dielectric material which allows charge storage in the bulk of the dielectric. This bulk charge cannot be removed rapidly because of the long time constant associated with the dielectric's high resistance. A capacitor with dielectric absorption can be modeled as an ideal capacitor in parallel with a series R-C circuit as shown in Figure 18. When such a capacitor is used as the hold capacitor in a T/H the held voltage will tend to creep back towards the voltage held for the previous conversion cycle. The degree and time constant of this behavior depends on the capacitor's dielectric material, as well as on the charge and hold time of the circuit.

6

Dielectric absorption will cause a variable "offset" if a T/H is used to sample multiple channels with widely varying signals. This causes an apparently nonlinear pedestal because the difference between the currently measured voltage and the previously measured voltage determines the magnitude of the DA error. The AD386 uses a high quality hold capacitor with low intrinsic DA. Residual DA errors are further reduced by laser trimming a compensation network during the manufacturing process. The trimming is performed under typical system timing conditions of 5 μs track, 45 μs hold. The post-trim dielectric absorption error is less than 1/2 16-bit LSB for full-scale changes between samples and hold times between 10 μs and 100 μs.

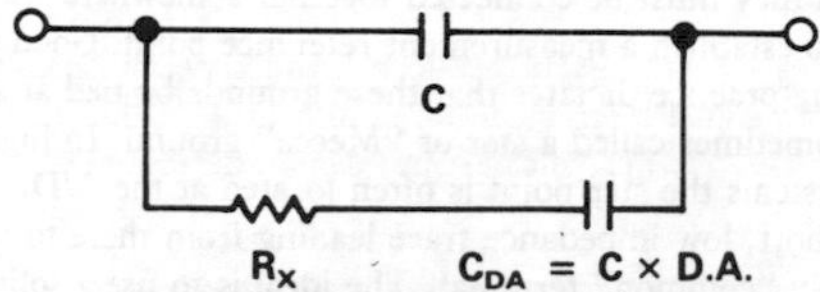

Figure 18. Capacitor Model with Dielectric Absorption

NOISE

Noise generated in a T/H adds to the held signal and causes variations in the output code of an A/D. This noise has two components, one which arises during track mode and another contributed during hold mode. The rms sum of these terms determines the noise performance of the T/H in the system.

Track noise is the noise which gets sampled when entering hold mode. An inverting T/H architecture such as that used in the AD386 has a noise gain of 2. This noise is low pass filtered in the R-C network comprised of the hold capacitor and the switch on resistance (see Figure 19a). The rms value of the track noise is

$$<e_{nT}> \; = \; (op\ amp\ noise) \times (noise\ gain) \times (ENBW)^{1/2}$$

Op amp noise is the rms sum of the amplifier's broadband voltage noise and the thermal noise contributions of the input and feedback resistors, about 17 nV/$\sqrt{\text{Hz}}$. Other noise sources, including amplifier current noise and switch thermal noise, are negligible. ENBW, the equivalent noise bandwidth, is

$$ENBW = \frac{\pi}{2} \times \frac{BW1 \times BW2}{BW1 + BW2}$$

where BW1 is the small signal bandwidth of the T/H in track mode (2 MHz for the AD386) and BW2 is the corner frequency of the R_{SWITCH}–C_{HOLD} combination (2.7 MHz). The resulting track noise in the AD386 is at most 46 μV rms.

Noise gain is reduced to 1 in hold mode, and input and feedback resistor thermal noise makes no contribution (Figure 19b). The equivalent noise bandwidth now depends on the T/H's small signal bandwidth and the characteristics of the A/D converter used in the system. This is because the signal at the input

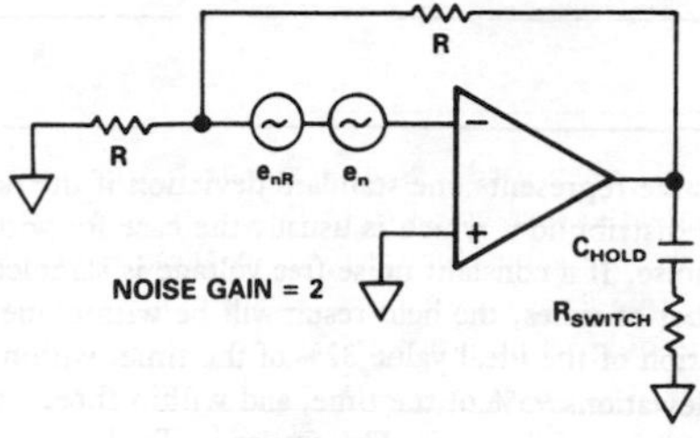

a. Track Mode

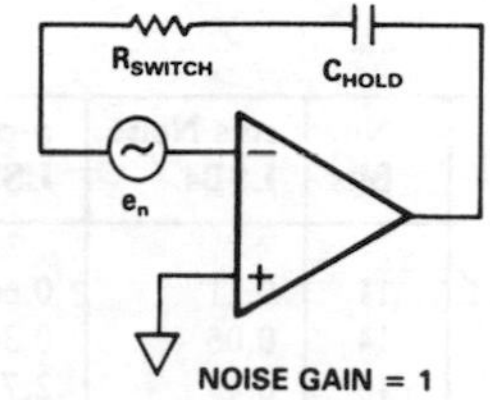

b. Hold Mode

Figure 19. Dominant AD386 Noise Sources

of the comparator in a successive approximation A/D converter is filtered by the converter's input resistance and the summing junction capacitance. ENBW is calculated as before, but now BW1 is the T/H's small signal bandwidth in hold mode (4 MHz for the AD386), and BW2 is the bandwidth of the A/D's input R-C. BW2 is about 700 kHz in the AD ADC71 and AD1376 and roughly 1.7 MHz in the AD1377 and AD1378 (assuming a 10 V span). The respective values of ENBW are 940 kHz and 1.9 MHz. The hold noise contribution of the AD386 is about 16 μV rms when used with the AD ADC71 or AD1376 and 22 μV rms when used with the AD1377 or AD1378; this noise is 30% less for a 20 V span and 40% greater for a 5 V span because changes in the A/D's input resistance cause changes in BW2.

The total noise is the rms sum of these two results:

$$<e_n> \; = \; [<e_{nT}^2> \; + \; <e_{nH}^2>]^{1/2}$$

This yields 49 μV rms and 51 μV rms for the two cases. Track noise dominates in both instances.

When the AD386's differential amplifier is used, its noise contribution will be band limited and sampled by the T/H. The equivalent bandwidth for this noise is also 1.8 MHz and the contribution to the track noise is 46 μV rms. The total track noise is the rms sum of 46 μV and 46 μV, or 65 μV rms, and the overall noise for the complete AD386 used with any of the above A/D converters is at most 70 μV rms.

The rms value represents one standard deviation if the noise has a Gaussian distribution, which is usually the case for wideband electrical noise. If a constant noise-free voltage is sampled a large number of times, the held result will be within one standard deviation of the ideal value 32% of the time, within two standard deviations 95% of the time, and within three standard deviations 99.7% of the time. The entries in Table II were calculated using three standard deviations as the definition of the peak-to-peak noise.

Span	No. Bits	rms Noise LSBs	p-p Noise LSBs
10 V	14	0.11	0.66
20 V	14	0.06	0.36
10 V	16	0.45	2.7
20 V	16	0.23	1.4

Table II. AD386 Noise Contribution as a Function of A/D Span and Resolution

POWER SUPPLY REJECTION

Variations on the power supply lines, both dc and ac, can lead to unwanted changes in the voltage acquired by a T/H. Power supply variations in track mode cause the output voltage, and hence the voltage across the hold capacitor, to vary. PSRR decreases with increasing frequency, making well regulated, low noise linear power supplies and proper bypassing essential in a high resolution data acquisition system.

Equally important, but usually forgotten or omitted, is hold PSRR. This is frequently much worse than track PSRR because parasitic capacitances which are not significant in track mode couple into the extremely high impedance nodes which exist in a T/H during hold mode. This specification is essential to the system designer, as hold mode PSRR often determines the performance required from the system's power supplies. The power supply rejection of the AD386 is specified and characterized in both track and hold modes.

Pedestal arises from the transfer of charge from the internal switching circuitry to the hold capacitor during the transition from track mode to hold mode. Pedestal in some T/H circuits is extremely sensitive to changes in the high and low levels of the external control signal. The AD386 uses an internal +5 V supply and logic buffers to prevent this behavior.

GROUNDING

All voltage measurements in a data acquisition system are eventually referenced to ground. Variations in the "ground" potential through the system resulting from resistive drops of power supply and signal return currents as well as from interference from external sources may add to the signal being digitized and produce false results. The grounding scheme in a high resolution system cannot be left to chance and must be planned as carefully as any other aspect of the system's design. Proper grounding and the reduction of externally induced ground noise are discussed at length in the following Applications section.

Applications

GROUNDING, DECOUPLING, AND LAYOUT CONSIDERATIONS

Many data acquisition systems have two or more ground pins which are not connected together within the device(s). These "grounds" may be referred to as Logic Power Return, Digital Return, Analog Ground, Analog Power Return, Signal Ground, etc., and they must be connected together somewhere within the system to establish a measurement reference point. Good grounding practice dictates that these grounds be tied at a single point, sometimes called a star or "Mecca" ground. In high resolution systems the star point is often located at the A/D, with a single, short, low impedance trace leading from there to the analog supply "common" terminal. The ideal is to use a solid analog ground plane beneath the T/H and A/D as the star point.

Because circuit traces have resistance and inductance, currents in the various ground runs can create voltage differences of hundreds of millivolts between "ground" in different parts of the system. Power supply and signal ground traces should be separate to prevent summing power supply return currents with analog signal currents, which would lead to measurement errors. It is also important to avoid closed circuit loops in system ground connections. A loop can act as a very effective antenna, coupling voltages created by stray magnetic fields into the measurement system.

Each of the AD386's power supply terminals should be capacitively bypassed to the ground plane as closely as possible to the device. This is best done using 0.01 μF to 0.1 μF ceramic capacitors. High frequency supply noise rejection may be further improved by placing small (4.7 Ω to 10 Ω) carbon composition resistors in series with the supply leads. These resistors, in combination with the ceramic capacitors, act as local low pass filters and prevent crosstalk between system components. The bypassing scheme should also include solid Tantalum capacitors of 1 μF to 10 μF from each supply to ground in the critical areas of the board. Proper grounding and bypassing techniques are shown in Figure 20.

All AD386 ground pins (Pins 2, 5, 7, 9, 18, 19, and 24) should be connected to the analog ground plane.

WARNING: Improper bypassing can result in poor settling performance or high frequency oscillations.

The metal cover of the AD386 is internally grounded to provide additional shielding. Do not make any external connection to the cover.

DIFFERENTIAL AMPLIFIER

Many high resolution applications require the ability to sense ground at the signal source. This is especially true in systems with physical or thermal constraints that make it necessary to locate the T/H and A/D at some distance from the transducer. Under these conditions stray electromagnetic fields may cause "ground" at the signal source to be at a different potential from "ground" at the A/D despite the designer's best efforts. This will give rise to measurement errors because the potential difference will appear to be added to the true signal. The AD386's differential amplifier may be used to eliminate this type of ground noise as shown in Figure 21.

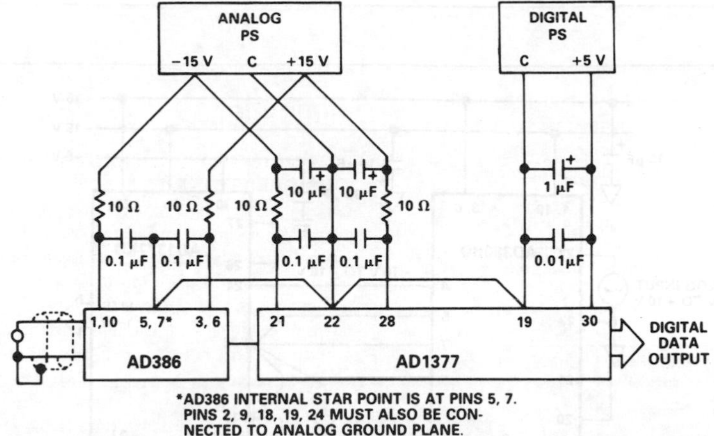

Figure 20. Proper Grounding and Supply Bypassing Techniques for a High Resolution Data Acquisition System

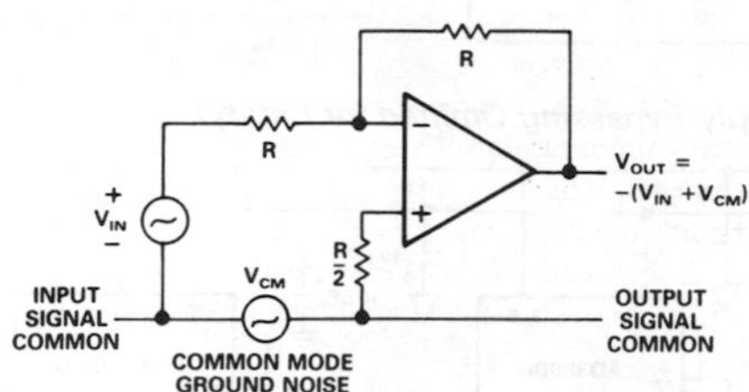

a. Without Differential Amplifier

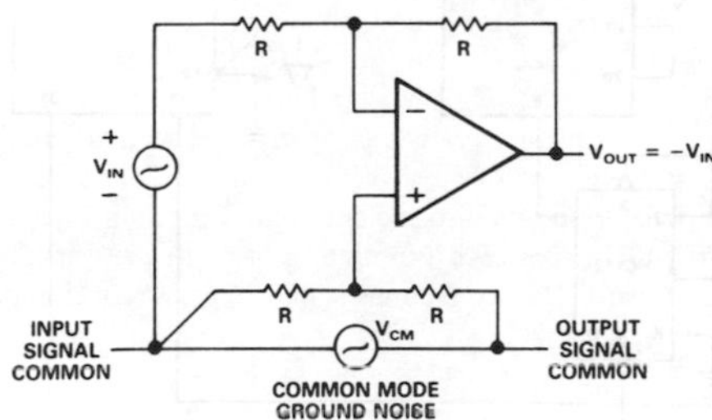

b. With Differential Amplifier

Figure 21. Effects of Common Mode Noise

In extremely noisy environments it may be necessary to connect the differential amplifier to the signal source with shielded twisted pair cable. The shield should be connected to ground at the transducer and should be left floating at the AD386. This shielding technique is shown in Figure 22. The cable presents a capacitive load, and the signal source must be capable of driving this load without ringing or oscillations. The differential amplifier's noninverting input should be connected to Pin 24 if ground sensing is not required.

Another use of the differential amplifier is to restore signal polarity. Like most high resolution T/H amplifiers, the T/H in the AD386 operates in the inverting mode. The differential amplifier may be used to provide a second inversion so that the T/H output has the same polarity as the sensor output.

The differential amplifier also provides a low dynamic source impedance to the T/H section. This absorbs transients produced when the T/H switches from hold mode to track mode, providing optimal settling performance.

The T/H and differential amplifier have independent power supply connections. This permits a reduction in system power dissipation when the differential amplifier function is not needed.

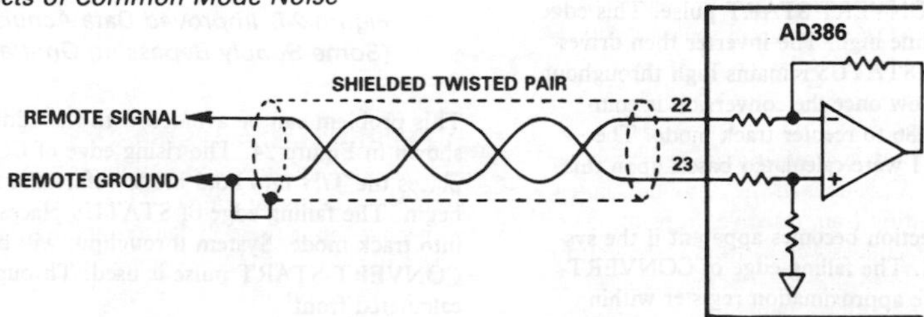

Figure 22. Remote Ground Sensing in a Noisy Environment

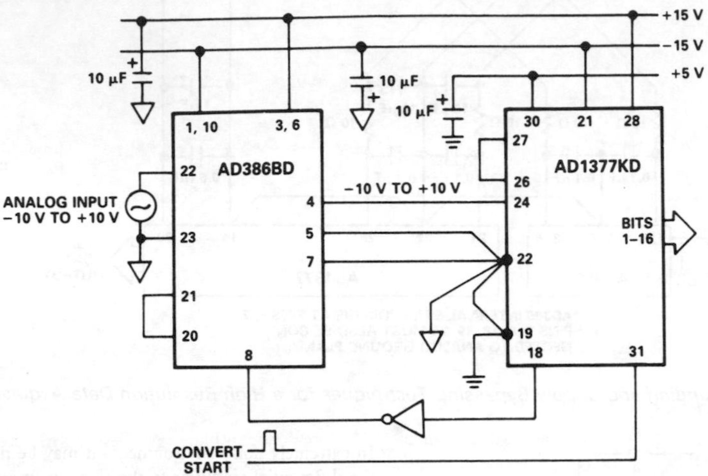

Figure 23. Basic Data Acquisition System (Some Supply Bypassing Omitted for Clarity)

GAIN AND OFFSET ADJUSTMENT

The usual practice in the design of data acquisition systems is to incorporate a single system level trim for offsets and a second for gain errors, rather than to trim each element in the signal processing chain. Traditionally these trims involve potentiometers or fixed resistors. The trims should be designed so that nulling static errors does not introduce new errors such as noise, increased thermal drift, or nonlinearity.

The offset, drift, and gain errors of the AD386 are laser trimmed during manufacture and no external adjustment capabilities are provided. This prevents the introduction of noise through offset adjust terminals and preserves the excellent gain linearity and drift performance. Most A/Ds provide for nulling gain and offset errors with a range sufficient to include the contributions of the AD386. Of course, it is also possible to include calibration routines in the system's software to eliminate mechanical adjustments.

HIGH RESOLUTION DATA ACQUISITION SYSTEM

The essential details of a high resolution data acquisition system using the AD386 are shown in Figure 23. Conversion is initiated by the falling edge of the CONVERT START pulse. This edge drives the A/D's STATUS line high. The inverter then drives the AD386 into hold mode. STATUS remains high throughout the conversion and returns low once the conversion is completed. This allows the AD386 to reenter track mode. The throughputs given in Table I were calculated based upon this circuit configuration.

One drawback of this connection becomes apparent if the system's grounding is marginal. The falling edge of CONVERT-START resets the successive approximation register within the A/D, causing transient currents in both the analog and digital return paths. These transients vary depending on the input signal and the prior conversion result. The same edge also drives the T/H into hold mode. The exact timing relationship of these two events depends upon differences in propagation delays. The T/H's held value may be affected if the A/D reset transient begins before the T/H has fully entered hold mode. The end result is system nonlinearity.

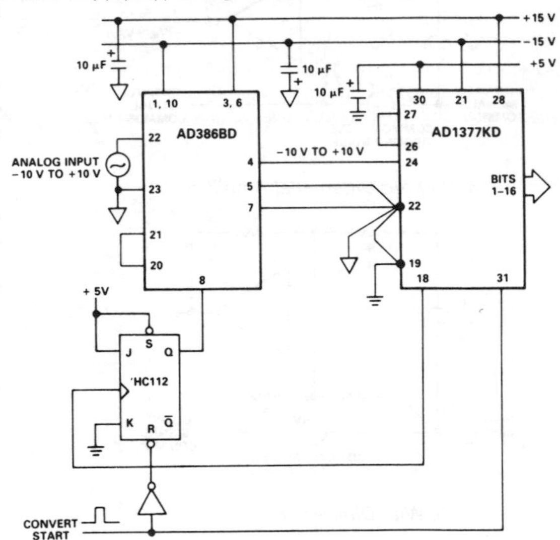

Figure 24. Improved Data Acquisition System (Some Supply Bypassing Omitted for Clarity)

This problem can be avoided with the addition of a flip flop as shown in Figure 24. The rising edge of CONVERT START places the T/H into hold mode before the A/D reset transients begin. The falling edge of STATUS places the AD386 back into track mode. System throughput will be reduced if a long CONVERT START pulse is used. Throughput can be calculated from

$$Throughput = \frac{1}{T_{ACQ} + T_{CONV} + T_{CS}}$$

where T_{ACQ} is the T/H acquisition time, T_{CONV} is the time required for the A/D conversion, and T_{CS} is the duration of CONVERT START. No significant T/H droop error will be introduced provided the width of CONVERT START is small compared with the A/D's conversion time.

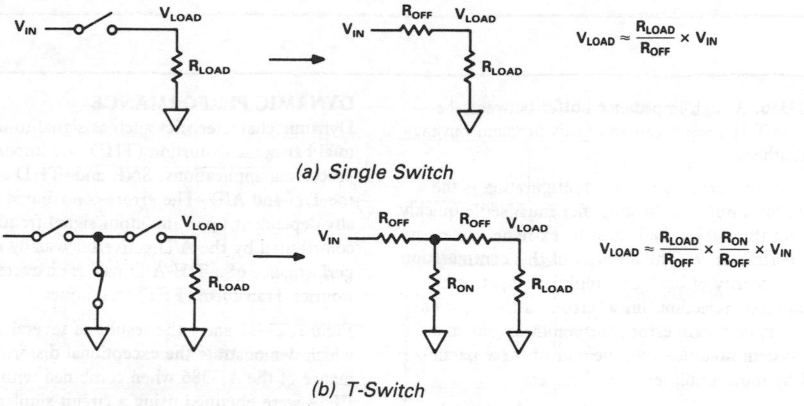

$$V_{LOAD} \approx \frac{R_{LOAD}}{R_{OFF}} \times V_{IN}$$

(a) Single Switch

$$V_{LOAD} \approx \frac{R_{LOAD}}{R_{OFF}} \times \frac{R_{ON}}{R_{OFF}} \times V_{IN}$$

(b) T-Switch

Figure 25. Single and "T" Analog Switches (Shown in OFF Position)

MULTICHANNEL SYSTEMS

The design of multiplexed data acquisition systems which maintain 14- or 16-bit signal fidelity is an extremely demanding task. One of the first difficulties encountered is the lack of adequate analog switches. The specified feedthrough performance of most switches and multiplexers is seldom better than −80 dB. This is an order of magnitude too high for a 16-bit system with its 8 parts-per-million sensitivity. A "T" switch configuration can be used to reduce feedthrough as shown in Figure 25. The improvement in "off" isolation relative to a single switch is substantial.

A few monolithic video T-switch ICs are now available and provide the necessary isolation in the dc-50 kHz frequency range. Unfortunately, these devices have voltage limitations which restrict their utility. It will usually be necessary to design a multiplexer using analog multiplexer and switch ICs. Figure 26 shows a simple 4-channel single-ended T-switch multiplexer and includes a high performance buffer (see below).

The on-resistance of analog switches and multiplexers is a nonlinear function of signal voltage. This will produce severe nonlinearity in a system in which a multiplexer supplies signals

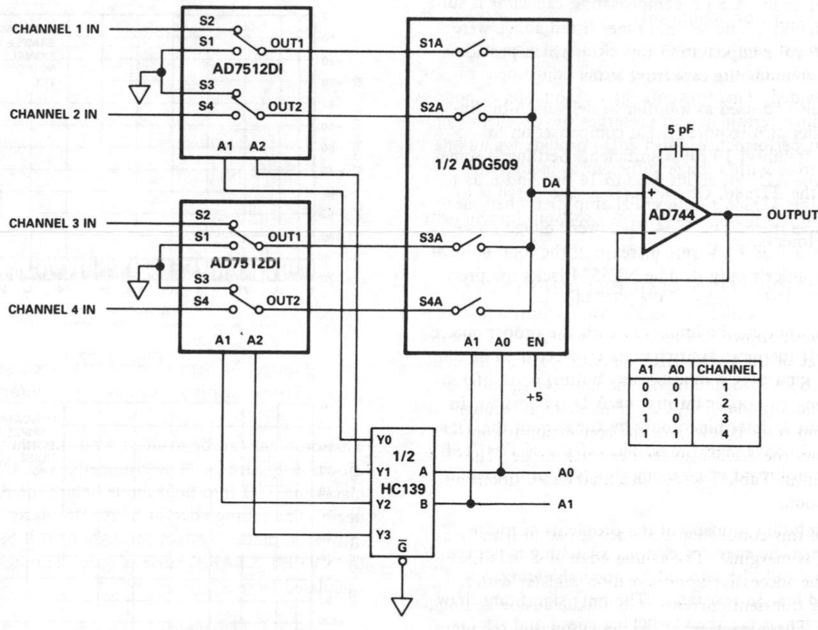

Figure 26. Four-Channel T-Switch Multiplexer (Power Supply Connections Not Shown)

directly to an AD386. A high-impedance buffer between the multiplexer and the T/H's input can solve this problem but may introduce several others.

An op amp in the noninverting gain-of-1 configuration is the obvious candidate for a buffer. The amplifier must settle quickly to maximize system throughput and must be extremely linear to maintain system performance. The linearity of this configuration depends upon the linearity of both the amplifier's open loop gain and common-mode rejection (linear errors in these parameters result only in system gain error, but nonlinear gain and CMRR produce system nonlinearity). Neither of these parameters is specified by most amplifier manufacturers.

A buffer may also increase system noise. Applications which require ground- sensing will require two buffers, resulting in 40% more noise than a one-buffer system.

Finally, a buffer will add its own offset to the signal being measured. Software calibration of the error and its drift is possible using a permanently grounded multiplexer channel.

The AD744 is a nearly ideal buffer for multiplexed systems. This amplifier provides offsets as low as 250 μV and an offset drift of 3 μV/°C while maintaining 16-bit linearity over the −40°C to +85°C temperature range. Typical settling times at room temperature are 2.3 μs (14 bits) and 3.5 μs (16 bits) for the AD744 combined with the AD386's differential amplifier. The increase in noise at the differential amplifier's output will be about 6 μV rms in a one-buffer system and roughly 12 μV rms in a two buffer system (recall that a 16-bit LSB in a 20 volt system is 305 μV). The AD744 is not unity-gain stable, and compensation is required. A 5 pF compensating capacitor is sufficient to ensure stability. The settling times listed above were measured using a 9 pF compensation capacitor which provides greater stability with moderate capacitive loads.

The NE5534 can also be used as a buffer to deliver 16-bit linearity. This amplifier also requires slight compensation to achieve unity-gain stability; 10 pF is sufficient. Settling is somewhat slower than the AD744, about 5 μs to 14 bits and 6 μs to 16 bits, including the AD386's differential amplifier when measured at room temperature. The 5534 has lower voltage noise and will cause only a 1 or 2 μV rms increase in the total noise at the differential amplifier's output. The NE5534 lacks the precision offset and drift performance of the AD744.

Multiplexed throughput can be improved with the proper choice of system timing. If the new input channel is selected while the AD386 is in Hold mode, then multiplexer, buffer, and differential amplifier settling can occur during the A/D conversion. In this case throughput is determined only by the sum of the T/H acquisition and A/D conversion times. The effects of T/H feedthrough must be considered when using this type of overlap in system timing.

There is another solution to many of the problems of multiplexed systems when the speed of channel switching is not critical: relays. Relays should be selected for good shielding, low thermal EMF, and low on-resistance. The only significant drawback of this approach, other than switching speed and size, is power dissipation. In all other respects relays offer a near-perfect solution to the problems of high resolution system design discussed above.

DYNAMIC PERFORMANCE

Dynamic characteristics such as signal-to-noise ratio (SNR) and total harmonic distortion (THD) are important in many signal processing applications. SNR and THD are affected by both the T/H and A/D. The errors contributed by the T/H are generally dependent upon the input signal frequency, while those contributed by the A/D converter usually are not. The dynamic performance of a T/H-A/D pair is characterized using Fast Fourier Transform (FFT) techniques.

Figures 27–31 show the results of several 1024-point FFTs which demonstrate the exceptional distortion and noise performance of the AD386 when combined with the AD1377. These FFTs were obtained using a circuit similar to that of Figure 24. The input signal was processed by both the differential amplifier and T/H sections of the AD386 and was sampled at an 83.333 kHz rate. The AD1377's clock was adjusted to yield an 8.0 μs conversion time, which provided 4.0 μs for the AD386 to acquire each new sample. The vertical scale for these figures is based on a full-scale input referenced as 0 dB. The system was configured for a 10 volt span.

Figures 27 and 28 illustrate the system's low frequency noise and distortion performance. The input frequency is 1.546 kHz. When the input is −0.3 dB, nearly full scale, the largest harmonic component is −102.8 dB (Figure 27). Total harmonic distortion, the rms sum of the second through fifth harmonics, is −99.9 dB. The signal to noise ratio is 89.9 dB. The ultimate noise floor can be determined using a lower level input. Reducing the input level about 20 dB, as in Figure 28, decreases the

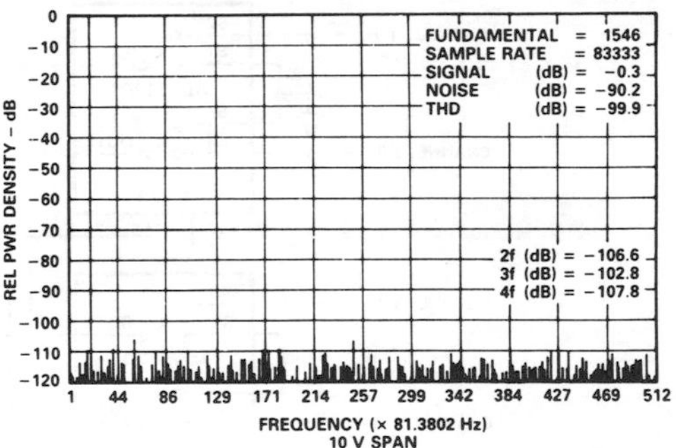

Figure 27.

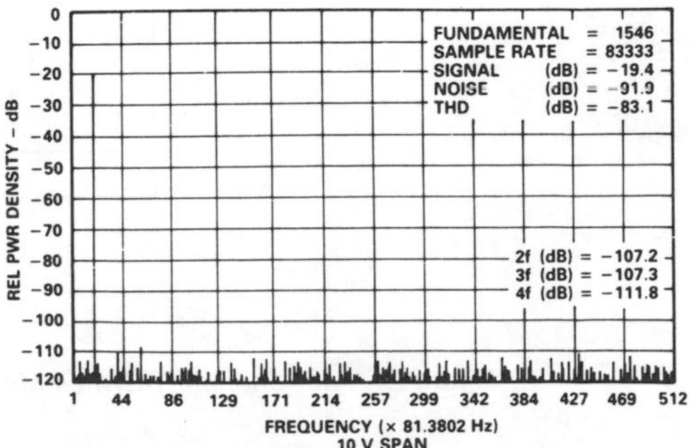

Figure 28.

6

noise floor by 1.8 dB to −91.9 dB. This corresponds to a total AD386 noise contribution of about 45 μV rms. The FFT noise floor would improve about 2 dB with the system configured for a 20 volt span because the effect of noise contributed by the AD386 is reduced as a result of the increased LSB size.

System performance just beyond the high end of the audio band is shown in Figure 29. Here the input is a −0.3 dB sinusoid at 21.24 kHz. The only significant harmonic component, the second harmonic, is −91.9 dB with respect to the fundamental, and THD is −91.1 dB. The noise floor is 0.5 dB greater than in Figure 27. The additional noise is contributed by higher-order harmonics; the second through fifth harmonics have been excluded from the noise floor calculations, but higher harmonics are considered to be "noise". These harmonics arise from the AD386's aperture jitter. The additional noise is consistent with an rms jitter of 40 ps.

In Figures 30 and 31, −0.3 dB and −20.1 dB inputs at 40.61 kHz show system performance near the Nyquist frequency. Even at this high frequency a full-scale input produces THD of only −84.6 dB, dominated by the second harmonic at −85.1 dB (Figure 31). In Figure 31 the harmonics have been eliminated by reducing the input level by a factor of 10.

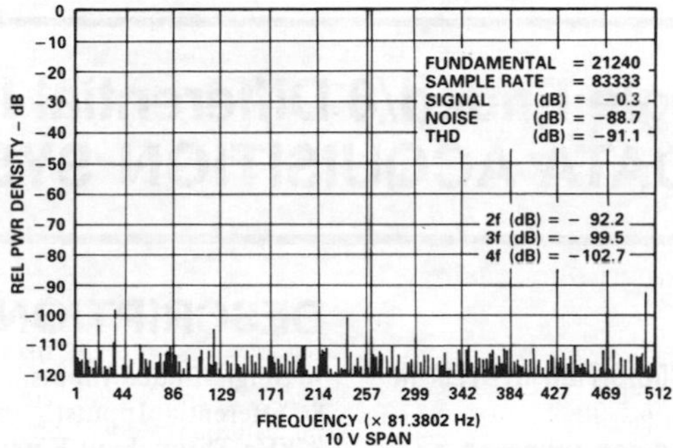

Figure 29.

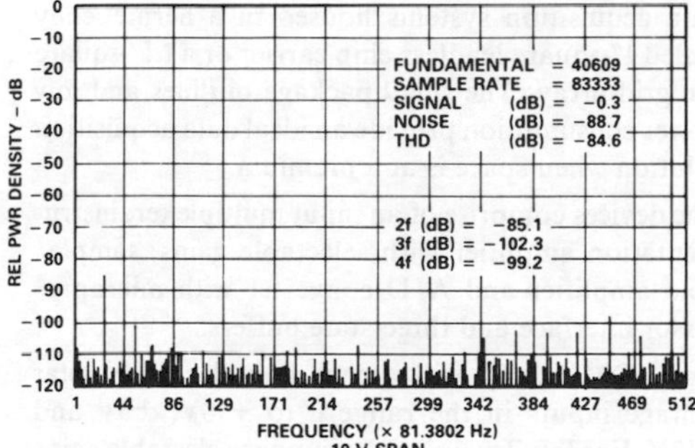

Figure 30.

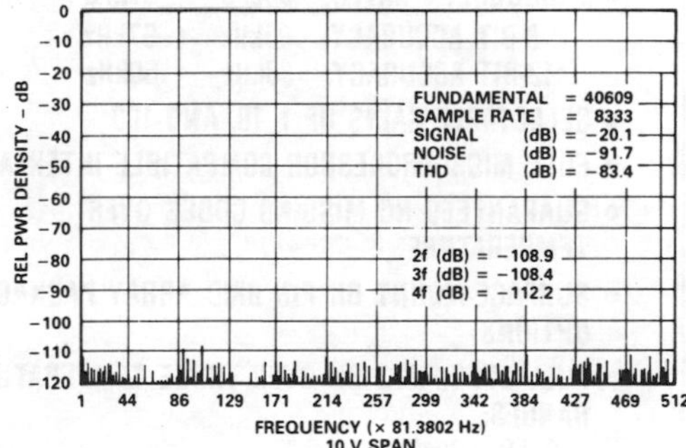

Figure 31.

SDM862 SDM863 SDM872 SDM873

16 Single Ended/8 Differential Input 12-BIT DATA ACQUISITION SYSTEMS

FEATURES

- COMPLETE 12-BIT DATA ACQUISITION SYSTEM IN A MINIATURE PACKAGE
- INPUT RANGES SELECTABLE FOR UNIPOLAR OR BIPOLAR OPERATION
- THROUGHPUT RATES:

	862/3	872/3
8-BIT ACCURACY:	45kHz	67kHz
12-BIT ACCURACY:	33kHz	50kHz

- SELECTABLE GAINS OF 1, 10, AND 100
- FULL MICROPROCESSOR COMPATIBLE INTERFACE
- GUARANTEED NO MISSING CODES OVER TEMPERATURE
- SURFACE-MOUNT OR PIN GRID ARRAY PACKAGE OPTIONS
- FULL SPECIFICATION OVER THREE TEMPERATURE RANGES:
  0 TO +70°C
  −25 TO +85°C
  −55 TO +125°C

DESCRIPTION

16 Single-Ended Inputs:	SDM862	SDM872
8 Differential Inputs:	SDM863	SDM873
33kHz Throughput Rate:	SDM862	SDM863
50kHz Throughput Rate:	SDM872	SDM873

The SDM components are complete, pin-compatible, data acquisition systems housed in a hermetically sealed 1″-square leadless chip carrier or a 1.1″-square pin grid array. The small package outlines and low power consumption provide an ideal data acquisition solution when space is at a premium.

The devices comprise of an input multiplexer, instrumentation amplifier with selectable gains, sample/hold amplifier and A/D converter with microprocessor interface and three-state buffers.

The SDM family will accept unipolar or bipolar voltage inputs in the range 0 to +10V, ±5V and ±10V. For low-level signals, jumper-selectable gains of 10 or 100 can be applied. The number of input channels can be expanded by the addition of multiplexers. System integration is simplified by the microprocessor interface and the facility of the sample/hold amplifier being controlled directly by the A/D converter.

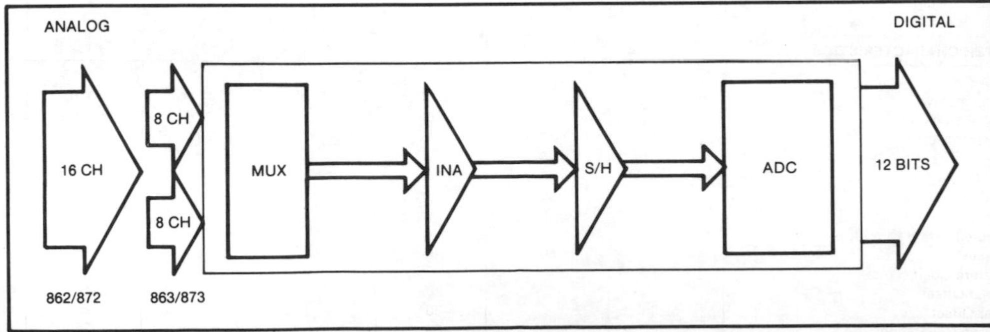

International Airport Industrial Park • P.O. Box 11400 • Tucson, Arizona 85734 • Tel.: (602) 746-1111 • Twx: 910-952-1111 • Cable: BBRCORP • Telex: 66-6491

PDS-686B

Printed in U.S.A. June, 1988

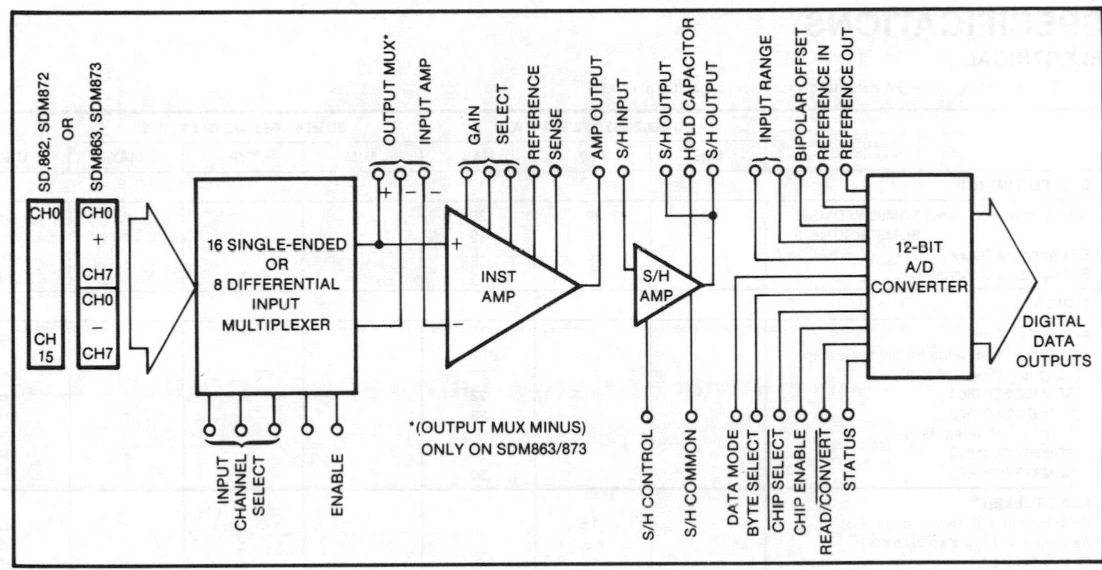

SPECIFICATIONS

ELECTRICAL

At 25°C, V_{CC} = ±15V, V_{DD} = 5V, external sample/hold capacitor of 4700pF. All grades are burned-in at +125°C for 48 hours min.

	SDM862/863/872/873 J, A, R			SDM862/863/872/873 K, B, S			
	MIN	TYP	MAX	MIN	TYP	MAX	UNITS
RESOLUTION			12			*	BITS
INPUT							
ANALOG							
Voltage Ranges: Bipolar			±5, ±10				V
Unipolar			0-10				V
Input Impedance: On Channel		10^{10}			*		Ω
Off Channel		10^{10}			*		Ω
Input Capacitance: On Channel		20			*		pF
Off Channel		20			*		pF
CMRR (20VDC to 1kHz)	80	85		*	*		dB
Crosstalk (20Vp-p, 1kHz)[1]		–85	–80		*	*	dB
Feedthrough (at 1kHz)[1]		–85	–80		*	*	dB
Offset (channel to channel) G = 1 [2]		30	100		*	*	μV
Input Bias Current/Channel		1	5		*	*	nA
Input Voltage Range [3]	+10	+11		*	*		V
	–10	–15		*	*		V
DIGITAL							
MUX Input Channel Select: Logic '1' (2V)		5	30		*	*	μA
Logic '0' (0.8V)		5	30		*	*	μA
S/H Command: Logic '1' (2V)		0.2			*		nA
Logic '0' (0.8V)		5	30		*	*	μA
ADC Section: Logic '1' (2.4V)			10			*	μA
Logic '0' (0.8V)			10			*	μA
TRANSFER CHARACTERISTICS							
ACCURACY							
Integral Linearity [4]			±0.024			±0.012	%FSR
Differential Linearity [4]			±0.024			±0.012	%FSR
Gain Error [5]: G = 1		0.7			*		%
: G = 100		0.9			*		%
Unipolar Offset Error [5]		16			*		mV
Bipolar Offset Error [5]		50			*		mV
Noise Error							
(Measured at S/H Output) G = 1		0.5	1		*	*	mVp-p
Droop Rate		50	500		*	*	μV/ms
Temperature Coefficients:							
Unipolar Offset			20			15	ppm of FSR/°C
Bipolar Offset			30			25	ppm of FSR/°C
Full-scale Calibration			60			35	ppm of FSR/°C

SPECIFICATIONS

ELECTRICAL

At 25°C, V_{CC} = ±15V, V_{DD} = 5V, external sample/hold capacitor of 4700pF.

	SDM862/863/872/873 J, A, R			SDM862/863/872/873 K, B, S			
	MIN	TYP	MAX	MIN	TYP	MAX	UNITS
SYSTEM TIMINGS							
ADC Conversion Time: SDM862/SDM863	15	20	25	*	*	*	μs
SDM872/SDM873	9	12	15	*	*	*	μs
S/H Aperture Delay		50			*		ns
S/H Aperture Uncertainty		2			*		ns
TIMING							
Acquisition Time (to 0.01% of final value for full scale step)		5			*		μs
Throughput (Serial Mode)							
SDM862/SDM863			22			*	kHz
SDM872/SDM873			28			*	kHz
(Overlap Mode):							
SDM862/SDM863			33			*	kHz
SDM872/SDM873			50			*	kHz
MULTIPLEXER[6]							
Switching time (between channels)		+1.5			*		μS
Settling time (10V step to 0.02%)		2.5			*		μS
Enable time 'ON'		1	2		*	*	μS
'OFF'		0.25	0.5		*	*	μS
INSTRUMENTATION AMPLIFIER[6]							
Settling time (20V step to 0.01%)							
G = 1		5	12.5		*	*	μS
G = 10		3	7.5		*	*	μS
G = 100		4	7.5		*	*	μS
Slew rate	12	17			*		V/μS
S/H AMPLIFIER[6]							
Acquisition time (10V step to 0.01%)		5			*		μS
Aperture delay		50			*		nS
Hold mode settling time		1.5			*		μS
Slew rate		10			*		V/μS
OUTPUT							
DIGITAL DATA							
Output Codes: Unipolar			Unipolar Straight Binary (USB)				
Bipolar			Bipolar Offset Binary (BOB)				
Logic Levels: Logic 0 (sink = 1.6mA)			+0.4			*	V
Logic 1 (source = 500μA)	+2.4			*			V
Leakage (Data Bits Only), High-Z State	−5	0.1	+5	*	*	*	μA
POWER SUPPLY REQUIREMENTS							
Rated Voltage: Analog (±V_{CC})	14.25	15	15.75	*	*	*	VDC
Digital (VDD)	4.75	5	5.25	*	*	*	VDC
Supply Drain: +15V		28	40		*	*	mA
−15V		36	45		*	*	mA
+5V		8	15		*	*	mA
Power Dissipation		1	1.4		*	*	W
TEMPERATURE RANGE							
Operating Temperature Range							
JH, KH/JL, KL	0		70	*		*	°C
AH, BH/AL, BL	−25		+85	*		*	°C
RH, SH/RL, SL	−55		+125	*		*	°C
Storage Temperature Range	−65		+150	*		*	°C

* Specification same as SDM862/863/872/873J, A, R grades.

NOTES: (1) Measured at the sample and hold output. (2) Measured with all input channels grounded. (3) The range of voltage on any input with respect to common over which accuracy and leakage current is guaranteed. (4) Applicable over full operating temperature range. NO MISSING CODES GUARANTEED OVER TEMPERATURE RANGE. (5) Adjustable to zero using external potentiometer or select-on-test resistor. (6) Specifications are at +25°C and measured at 50% level of transition.

DIGITAL TIMING

SYMBOL	PARAMETER	MIN	TYP	MAX	UNITS
CONVERT MODE					
tdsc	Status delay from CE		100	200	nS
thec	CE Pulse width	50	30		nS
tssc	$\overline{CS}$ to CE setup	50	20		nS
thsc	$\overline{CS}$ low during CE high	50	20		nS
tsrc	R/$\overline{C}$ to CE setup	50	0		nS
thrc	R/$\overline{C}$ low during CE high	50	20		nS
tsac	Byte select to CE setup	50	0		nS
thac	Byte selected valid during CE high	50	20		nS
tc 86X	Conversion time: 12 bit cycle	15	20	25	µS
	8 bit cycle	10	13	17	µS
tc 87X	Conversion time: 12 bit cycle	9	12	15	µS
	8 bit cycle	6	8	10	µS
READ MODE					
tdd	Access time from CE		75	150	nS
thd	Data valid after CE low	25	35		nS
thl	Output float delay		100	150	nS
tssr	$\overline{CS}$ to CE setup	50	0		nS
tsrr	R/$\overline{C}$ to CE setup	0	0		nS
tsar	Byte select to CE setup	50	25		nS
thsr	$\overline{CS}$ valid after CE low	0	0		nS
thrr	R/$\overline{C}$ high after CE low	0	0		nS
thar	Byte select valid after CE low	50	25		nS
ths 86X	Status delay after data valid	300	500	1000	nS
ths 87X	Status delay after data valid	100	300	600	nS

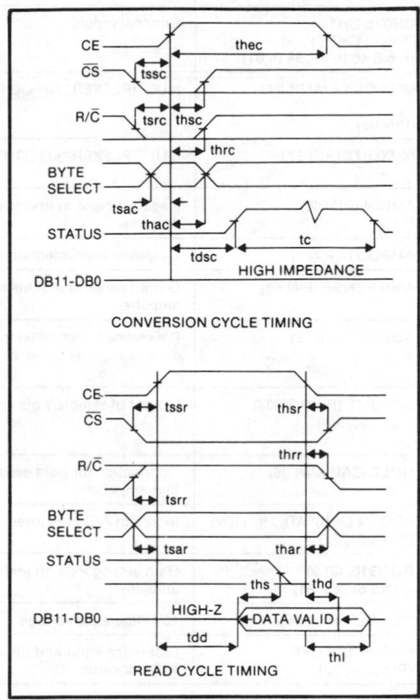

PIN CONFIGURATIONS

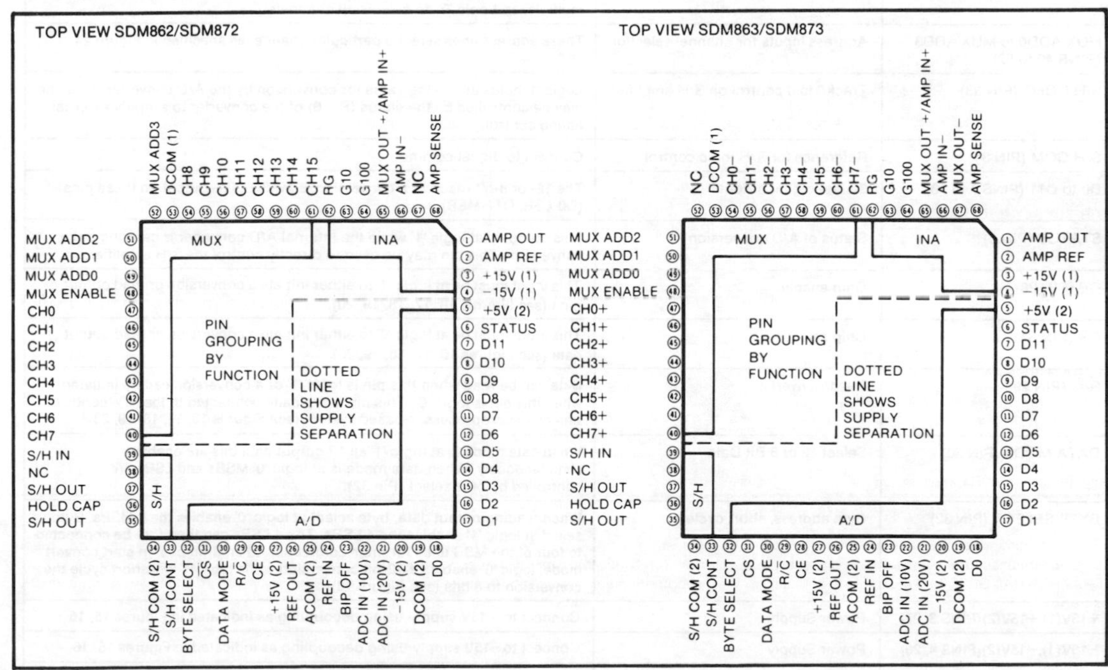

ABSOLUTE MAXIMUM RATINGS

+VCC TO ACOM	−0.5V TO +16V	ANALOG INPUT SIGNAL RANGE	+ VCC + 20V TO − VCC − 20V
− VCC TO ACOM	+0.5V TO −16V	DIGITAL INPUT SIGNAL	− 0.5V TO + VDD
+ VDD TO DCOM	−0.5V TO +5.5V	ACOM TO DCOM	± 1V

PIN DESIGNATION	DEFINITION	COMMENTS SDM8X2 = SDM862 or SDM872
CH0 to CH15 CH0 to CH7 (+, −) (PINS 40 to 47, 54 to 61)	Channel Inputs	Analog Inputs (Total 16) for single-ended and differential operation. Unused inputs must be connected to analog common.
MUX OUT+/AMP IN+ (PIN 65)	MULTIPLEXER "HI" OUTPUT	On the SDM8X2 this is the multiplexer output. On the SDM8X3 it is the output of the positive selected inputs. It is connected internally to the positive input of the instrumentation amplifier.
MUXOUT (PIN 67)	MULTIPLEXER "LO" OUTPUT	This pin is used on the SDM8X3 only. It should be connected to the negative input of the instrumentation amplifier.
AMPIN (PIN 66)	Negative Input of Instrumentation Amplifier	On the SDM8X2 this should be connected to analog common. On the SDM8X3 it should be connected to Muxout—(Pin 67).
AMPOUT (PIN 1)	Output of instrumentation amplifier	This pin should be connected to the input of the S/H amplifier (Pin 39).
AMP SENSE (PIN 68)	Output sense line of instrumentation amplifier.	This pin will normally be connected direct to AMP OUT (Pin 1).
AMP REF (PIN 2)	Reference for amplifier output	This pin will normally be connected to analog common. Care should be taken to minimize tracking and contact resistance to analog common to optimize system accuracy.
S/H OUT (PINS 35/37)	Output of sample/hold amplifier	Two pins are provided to facilitate a guard ring around the hold capacitor pin. These pins should be connected to either ADC in (20V) or ADC in (10V) depending on the desired range.
HOLD CAP (PIN 36)	Connection for hold capacitor on S/H amplifier	The tracking to the hold capacitor should be as short as possible and a guard ring employed using Pins 35 and 37.
ADC IN (20V); ADC IN (10V) (PINS 21, 22)	Inputs to A/D converter	Connect to S/H amplifier output. Use appropriate Pin for desired range.
RG, G10, G100 (PINS 62, 63, 64)	Gain setting Pins on instrumentation amplifier	For Gain = 1, no connections. For Gain = 10, connect G10 to RG. For Gain = 100, connect G100 to RG.
REF OUT (PIN 26)	10V Reference voltage	This is the reference voltage for the A/D converter.
REF IN, BIP OFF (PINS 24, 23)	Reference input and offset input to A/D converter	Connect trim potentiometers (or select-on-test resistors) to these pins for unipolar or bipolar operation as shown in Figures 12, 13.
S/H IN (PIN 39)	Input to sample/hold amplifier	Connect to amp out (Pin 1).
MUX ENABLE (PIN 48)	Multiplex enable/disable	Logic '1' on this pin will enable a selected channel on the internal multiplexer. Logic '0' de-selects all channels.
MUX ADD0 to MUX ADD3 (PINS 49 to 52)	Address inputs for channel selection	These address lines select a particular channel as specified in Figure 24.
S/H CONT (PIN 33)	Track/Hold control on S/H amplifier	Logic '1' holds an analog value for conversion by the A/D converter. This line may be controlled by the status (Pin 6) of the converter to simplify external timing control.
S/H COM (PIN 34)	Reference for S/H logic control	Connect to digital common
D0 to D11 (PINS 7 to 18)	3-state digital outputs	The 12- or 8-bit result of a conversion is available as output on these pins (D0-LSB, D11-MSB).
STATUS (PIN 6)	Status of A/D conversion	This output is at logic '1' while the internal A/D converter is carrying out a conversion. This pin may be used to directly control the S/H amplifier.
CE (PIN 28)	Chip enable	This input must be at logic '1' to either initiate a conversion or read output data (see Figures 10, 17, 18, 19, 20).
$\overline{CS}$ (PIN 31)	Chip select	This input must be at logic '0' to either initiate a conversion or read output data (see Figures 10, 17, 18, 19, 20).
R/$\overline{C}$ (PIN 29)	Read/convert	Data can be read when this pin is logic '1' or a conversion can be initiated when this pin is logic '0'. This pin is typically connected to the R/$\overline{W}$control line of a microprocessor-based system (see Figures 10, 17, 18, 19, 20).
DATA MODE (PIN 30)	Select 12 or 8 Bit Data	When data mode is at logic '1' all 12 output data bits are enabled simultaneously. When data mode is at logic '0' MSBs and LSBs are controlled by byte select (Pin 32).
BYTE SELECT (PIN 32)	Byte address, short cycle	When reading output data, byte select at logic '0' enables the 8 MSBs. Byte select at logic '1' enables the 4 LSBs. The 4 LSBs can therefore be connected to four of the MSB lines for inter-connection to an 8-bit bus. In start convert mode, logic '0' enables a 12-bit conversion while logic '1' will short cycle the conversion to 8 bits (see Figure 10).
+15V(1),+15V(2)(PINS 3, 27)	Power Supply	Connect to +15V supply using decoupling as indicated in Figures 15, 16.
−15V(1),−15V(2)(PINS 4, 20)	Power Supply	Connect to −15V supply using decoupling as indicated in Figures 15, 16.
ACOM(2) (PIN 25)	Analog common	Analog common connection. Note that a common (including digital common) should be connected together at one point close to the device.
DCOM (1) (PIN 53)	Reference for Mux logic control.	Connect to digital common.
+5V (PIN 5)	Logic power supply	Connect to +5V digital supply line with decoupling as in Figures 15, 16.
DCOM(2) (PIN 19)	Reference for A/D converter control lines	Connect to S/H common at one point close to device.
NC (PIN 38)	No internal connection	

SYSTEM DESCRIPTION

The SDM comprises four circuit elements—an input-protected multiplexer, an instrumentation amplifier, a sample/hold amplifier, and an analog-to-digital converter.

INSTALLATION

MULTIPLEXER

The SDM family has a choice of input multiplexers (MUX).

SDM862 and SDM872: 16 single-ended inputs
SDM863 and SDM873: 8 differential inputs

The select inputs are designed for use with TTL and CMOS logic levels and do not require pull-up resistors to ensure break-before-make operation.

On all models, the analog inputs may be expanded using the enable control. See Figure 1. When the enable is at a logic "0," the internal MUX is disabled, allowing additional multiplexers to be connected in parallel. The limiting factor for the number of additional multiplexers is the cumulative effect of leakage current flowing in the signal source impedance, causing offset errors.

Differential inputs will generally eliminate the noise associated wtih common system grounds, but care must be taken to ensure that neither of the differential inputs exceed the maximum input range. Otherwise, signal distortion will result. A return path for the input bias currents must always be provided. This prevents the charging of stray capacitances in applications using floating sources, such as transformers and thermocouples. Multiplexer inputs are protected from overvoltage, as indicated in the electrical specifications, and should be current limited to 25mA. To avoid signal distortion on the selected channel, MUX inputs that are not selected should have their input voltages limited to between $-V_{CC}$ and $+V_{CC}-4V$, as voltages outside of these values can turn on the non-selected channel. A graph of this characteristic is shown in Figure 2 with a possible circuit solution where it is known that the input voltages will exceed the above values.

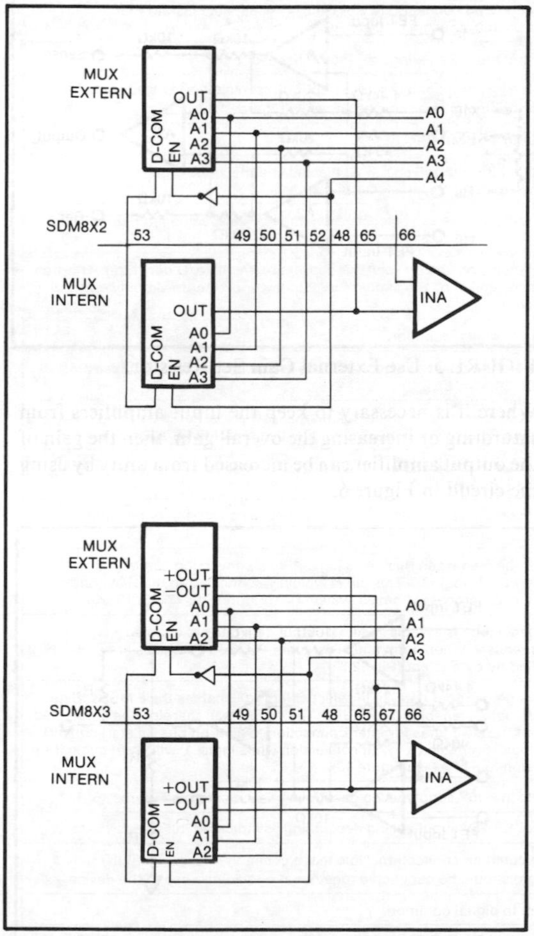

FIGURE 1. External Multiplexer Connections for Differential and Single-Ended Operation.

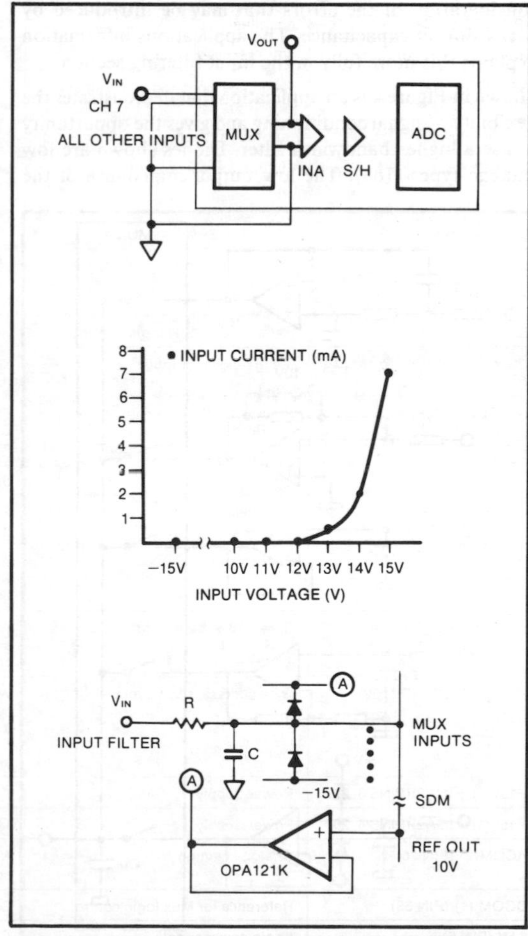

FIGURE 2. MUX Inputs With Limited Input Voltages and Possible Circuit Solution for Non-limited Cases.

Where high-speed operation is required and channels require rapid sampling, then it is important to buffer the inputs against the effect of current sharing between the MUX output capacitance and the input filter capacitance. See Figure 3.

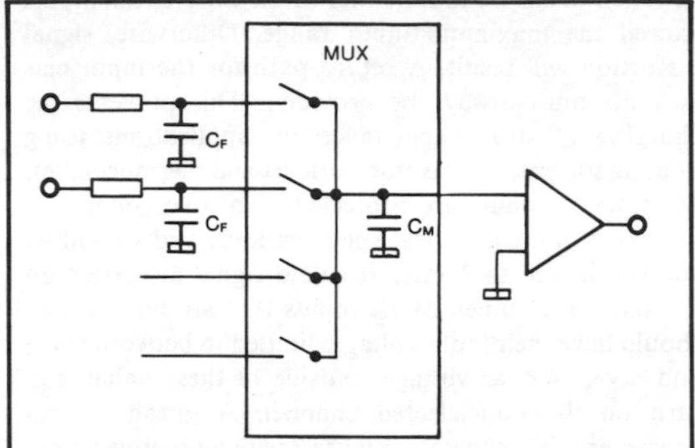

FIGURE 3. Filter and MUX Capacitance.

All data acquisition systems using a MUX require consideration of the errors that may be introduced by MUX output capacitance. The applications information explains this more fully in the input filtering section.

Shown in Figure 4 is an application that demonstrates the flexibility of signal conditioning and gives the opportunity to use a higher bandwidth filter. Diodes shown are low leakage types (1na). The low output impedance of the

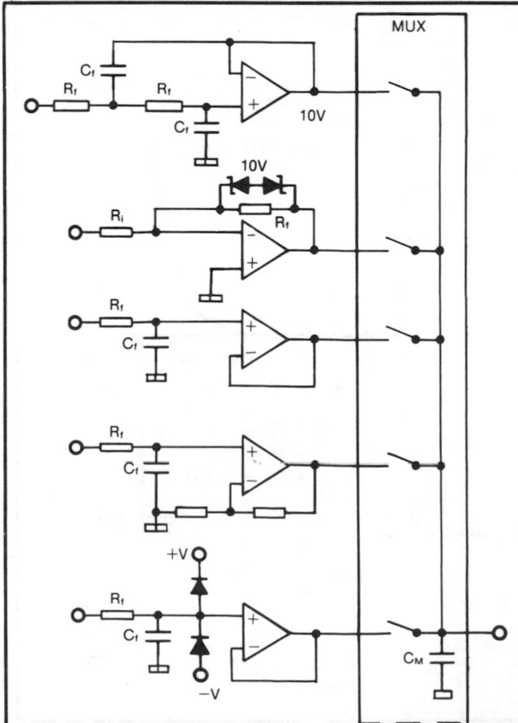

FIGURE 4. Example Application Illustrating Flexible Signal Conditioning.

amplifiers reduces the time taken to charge MUX capacitance C_M.

INSTRUMENT AMPLIFIER

The instrument amplifier (INA) presents a very high input impedance to the signal source, eliminating gain errors introduced by voltage divider action between the source output impedance and SDM input impedance. Where the differential models are used, the INA performs the differential to single-ended conversion required to drive the sample/hold amplifier. Gains may be set by using external jumpers, to values of 1 (no jumper), 10 and 100. For gains other than these presets, the following formula may be used to find an external resistor value to add in series with the G = 10 or G = 100 jumpers.

$$R_{ext} = \frac{40\ K\Omega}{G-1} - Ri \quad \text{Where } Ri = 4444\Omega,\ G = 10 \text{ input.} \quad 404\Omega,\ G = 100 \text{ input.}$$

It should be noted that the internal gain set resistors have a ±20% tolerance and ±20ppm/°C drift.

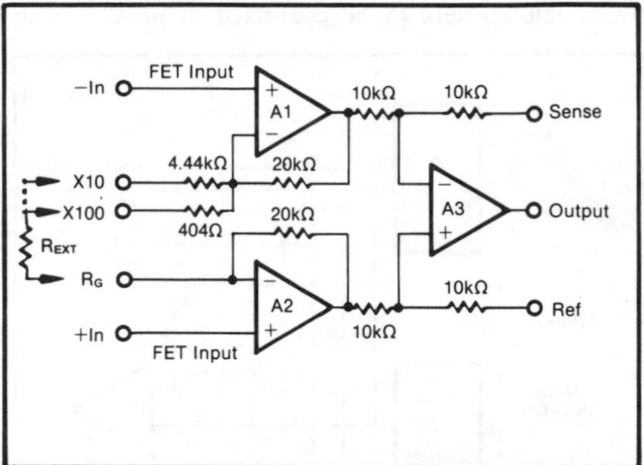

FIGURE 5. Use External Gain Set Resistor.

Where it is necessary to keep the input amplifiers from saturating or increasing the overall gain, then the gain of the output amplifier can be increased from unity by using the circuit in Figure 6.

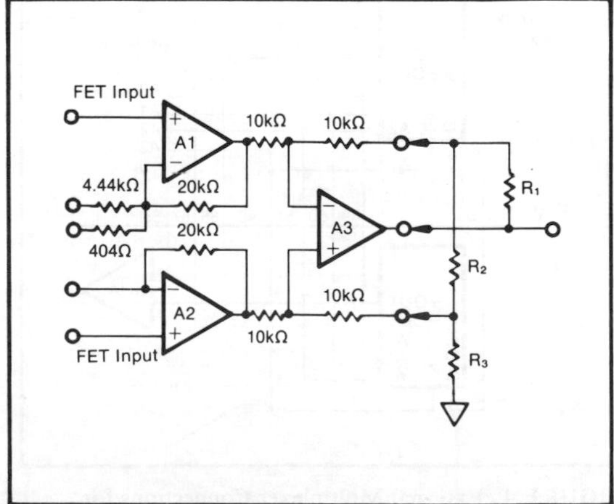

FIGURE 6. Increasing Output Amplifier Gain.

The values of the resistors in Figure 6 are in the following table.

O/P gain	R_1 & R_3 ohms	R_2 ohms
2	1200	2740
5	1000	511
10	1500	340

Matching of R_1 and R_3 is required to maintain high common mode rejection (CMR), R_2 sets the gain and may be varied without effect on CMR.

To ensure that the effects of temperature are minimized when altering the gain with external components, it is very important to use low tempco resistors. When connecting the output sense, ensure that series resistance is minimized because resistance present will degrade CMR.

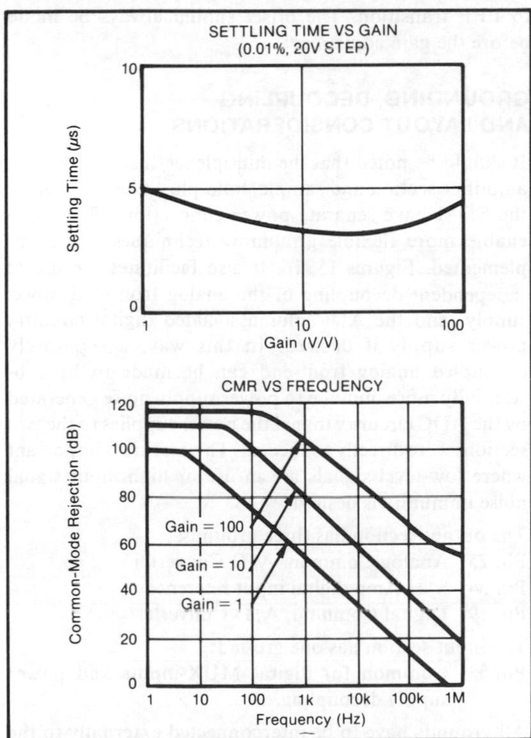

FIGURE 7. Typical INA Settling Time and CMR.

Some applications may require programmable gains. This may be realized with Figure 8.

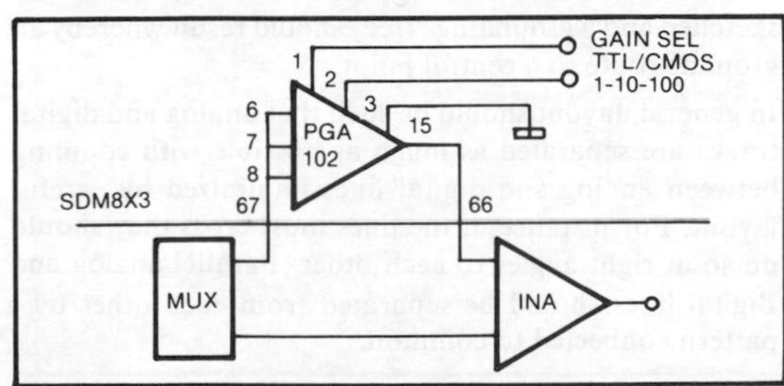

FIGURE 8. Setting Programmable Gains.

SAMPLE/HOLD AMPLIFIER

The Sample/Hold amplifier (S/H) is used to track the incoming signal and "hold" the required instantaneous value so that it does not change while the ADC is carrying out its conversion. Timing for the S/H may be derived from the STATUS output of the ADC, with care being taken to comply with the SDM timing considerations.

Capacitors with high insulation resistance and low dielectric absorption such as Teflon™, polystyrene or polypropylene should be used as storage elements. (Polystyrene should not be used above +80°C.) Teflon™ is recommended for high temperature operation. Care should be taken in the printed circuit layout to minimize stray capacitance and leakage currents from the capacitor to minimize charge offset and droop errors. The use of a guard ring driven by the S/H output around the pin connecting to the hold capacitor is recommended. (Refer to the application board layout for an example of this.)

The value of the external hold capacitor determines the droop rate, charge offset and acquisition time of the S/H, Figure 9. Droop rate for the SDM is specified with a hold capacitor value of 4700pf. There is a trade-off between acquisition time and droop rate, as the hold capacitor is increased in value it takes longer to charge, and hence there is a corresponding increase in acquisition time and reduction in droop rate. The droop rate is determined by the amount of leakage present in the SDM, board leakage and the dielectric absorption of the hold capacitance. The hold capacitor is also a compensation element for the S/H and should not be reduced below 2nf for good stability. The offset error in sample mode is not affected by the hold capacitor. However, during the transition to hold mode there is approximately 5pC of charge injected into the hold capacitor, causing an offset error that has been nulled for use with a 5nf hold capacitor. Any other value for the hold capacitor will cause a minor but fixed hold mode offset to be introduced, and is proportional to the change in value from 5nf. Therefore the SDM should be offset nulled with the S/H in hold mode.

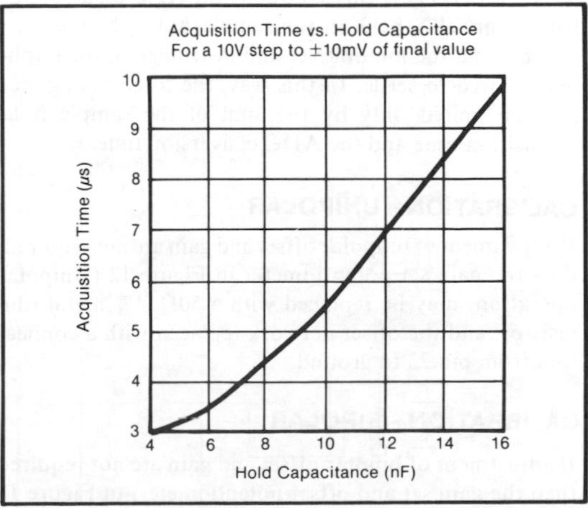

FIGURE 9. Acquisition Time vs. Hold Capacitance for a 10V Step Settling to ±10mV of Final Value.

ANALOG-TO-DIGITAL CONVERTER

This circuit element converts the analog voltage presented by the sample/hold amplifier to a digital number in binary format under control of the digital signals detailed in Figure 10. The converter can convert unipolar and bipolar signals in the range 10V and 20V. It can be calibrated to remove gain and offset errors from the entire system. The converter contains its own clock, voltage reference, and microprocessor interface with 3-state outputs. The converter will normally be used to digitize signals to 12-bit resolution, but it can be short-cycled to provide 8-bit resolution at higher speed. The digital output is compatible with 8- or 16-bit data buses, the data format being selected by control signals as detailed in Figure 10.

CE	$\overline{CS}$	$R/\overline{C}$	DATA MODE	BYTE SELECT	OPERATION
0	X	X	X	X	None
X	1	X	X	X	None
↑	0	0	X	0	Initiate 12-bit conversion
↑	0	0	X	1	Initiate 8-bit conversion
1	↓	0	X	0	Initiate 12-bit conversion
1	↓	0	X	1	Initiate 8-bit conversion
1	0	↓	X	0	Initiate 12-bit conversion
1	0	↓	X	1	Initiate 8-bit conversion
1	0	1	1	X	Enable 12-bit output
1	0	1	0	0	Enable 8 MSBs only
1	0	1	0	1	Enable 4 LSBs plus 4 trailing zeros

FIGURE 10. Control Input Truth Table.

OPERATING INSTRUCTIONS

OPERATING MODES

The SDM can operate in one of two modes, namely serial and overlap, as shown in Figure 11. In serial mode, control of the device is such that a multiplexer channel X is first selected, time is then allowed for the instrumentation amplifier to settle, the sample/hold amplifier is set to HOLD mode and finally a conversion is carried out. This procedure is then repeated for channel Y. Faster throughput can be obtained using overlap mode. While a conversion is being carried out by the ADC on a voltage from channel X held on the sample/hold, channel Y is selected and the multiplexer and instrumentation amplifier allowed to settle. In this way, the total throughput time is limited only by the sum of the sample/hold acquisition time and the ADC conversion time.

CALIBRATION - UNIPOLAR

If adjustment of unipolar offset and gain are not required, then the gain set potentiometer in Figure 12 (Unipolar operation) may be replaced with a 50Ω, 1% metal film resistor, and the offset network replaced with a connection from pin 23 to ground.

CALIBRATION - BIPOLAR

If adjustment of bipolar offset and gain are not required then the gain set and offset potentiometers in Figure 13 (Bipolar operation) may both be replaced with 50Ω, 1% metal film resistors.

CALIBRATION - GENERAL

The input voltage ranges of the ADC are 0–10V, ±5V and ±10V. Calibration in all ranges is achieved by adjusting the offset and gain potentiometers (indicated in Figures 12 and 13) such that the 000 to 001 code transition takes place at +1/2LSB from full-scale negative (−FS) and the FFE to FFF transition takes place at −3/2LSB from full-scale positive (+FS). The procedure is therefore to select the required range from Figure 14, apply the specified (−FS+1/2LSB) voltage to any selected input channel and adjust the offset potentiometer for the 000 to 001 transition. The (+FS−3/2LSB) voltage should then be applied to the same channel and the gain potentiometer adjusted for the FFE to FFF transition. The offset should always be made before the gain adjustment.

GROUNDING, DECOUPLING AND LAYOUT CONSIDERATIONS

It should be noted that the multiplexer/instrumentation amplifier section and sample/hold plus ADC section of the SDM have separate power connections. This is to enable more flexible grounding techniques to be implemented, Figures 15, 16. It also facilitates the use of independent decoupling of the analog front-end power supply, and the ADC plus associated digital circuitry power supply if desired. In this way, a separately decoupled analog front-end can be made to be substantially more immune to power supply noise generated by the ADC circuitry than if the power supplies to the two sections were directly connected. This feature is important where low-level signals are in use or high input signal noise immunity is desired.

The output section has three grounds:

Pin 25 Analog Common, A/D Converter
Pin 34 S/H Amp digital input reference
Pin 19 Digital Common, A/D Converter

The input section has one ground:

Pin 53 Common for digital MUX-inputs and power supply decoupling.

All grounds have to be interconnected externally to the SDM, and it is recommended that all grounds are connected via one track to a single point as close as possible to the SDM. To check that the grounding structure is correct, the ground tracking should be sketched and a grounding "tree" should result whereby all grounds route to a central point.

In general, layout should be such that analog and digital tracks are separated as much as possible with coupling between analog and digital lines minimized by careful layout. For instance, if the lines must cross they should do so at right angles to each other. Parallel analog and digital lines should be separated from each other by a pattern connected to common.

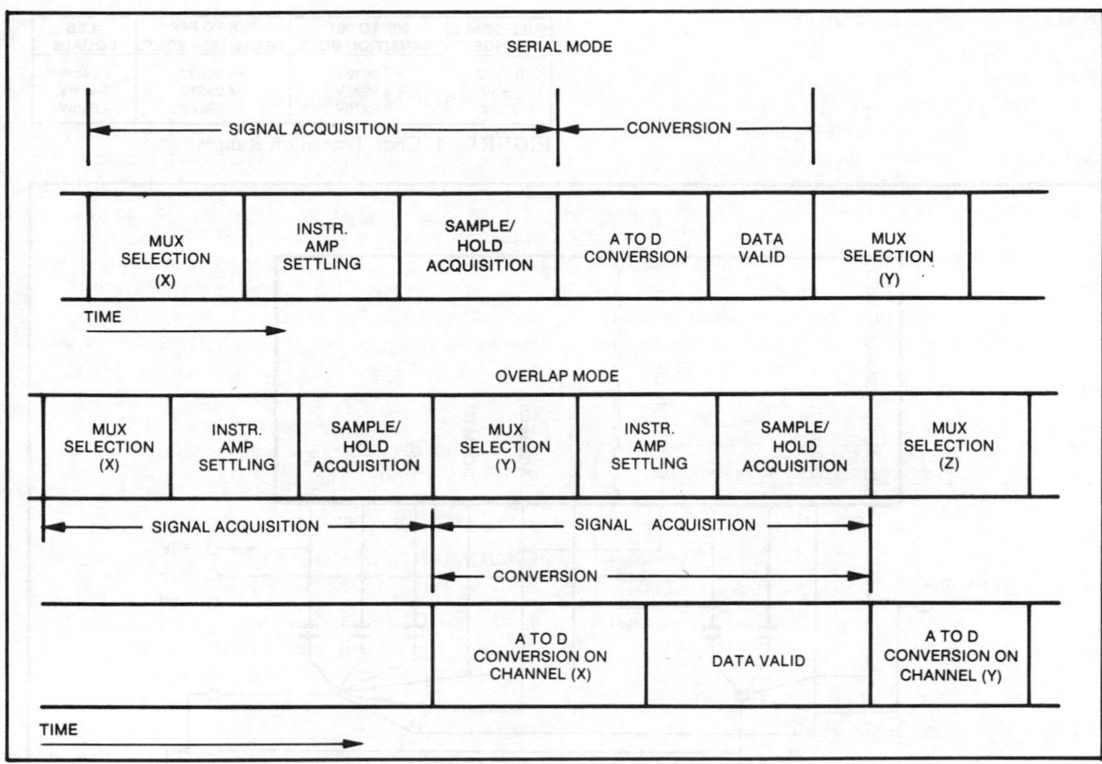

FIGURE 11. Serial and Overlap Modes of Operation.

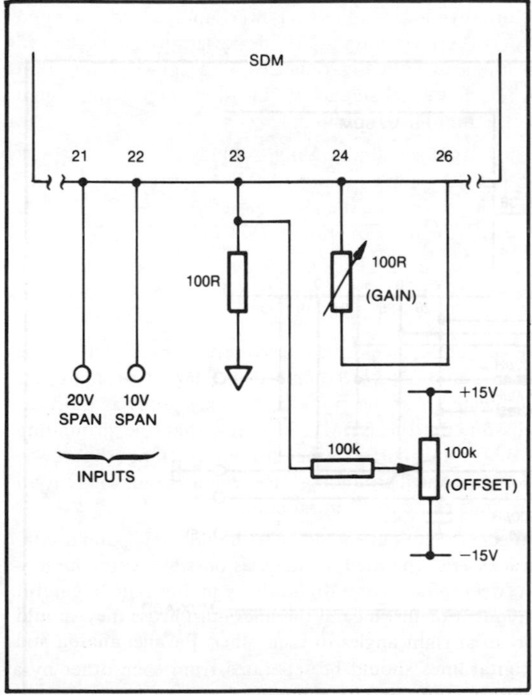

FIGURE 12. Unipolar Calibration.

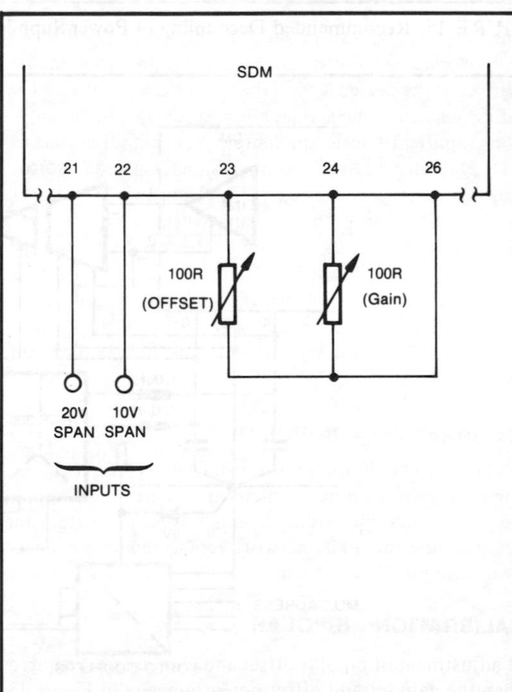

FIGURE 13. Bipolar Calibration.

FULL-SCALE RANGE	000 TO 001 TRANSITION VOLT.	FFE TO FFF TRANSITION VOLT.	1LSB EQUALS
0–10V	+0.0012V	+9.9963V	2.44mV
±5V	−4.9988V	+4.9963V	2.44mV
±10V	−9.9976V	+9.9927V	4.88mV

FIGURE 14. Code Transition Ranges.

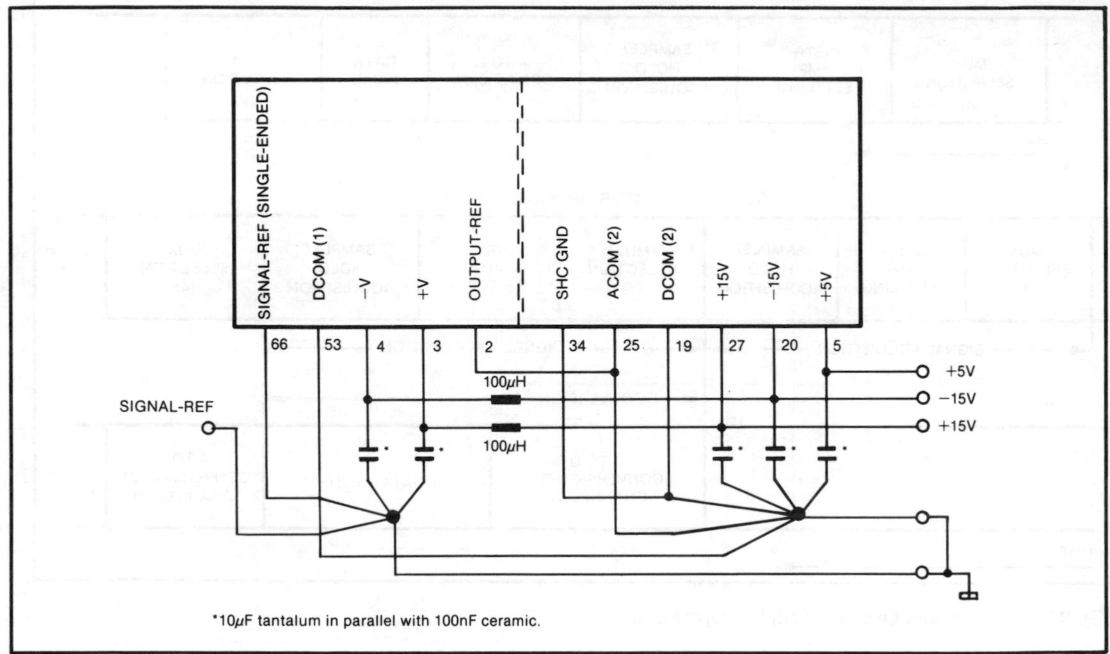

FIGURE 15. Recommended Decoupling of Power Supplies.

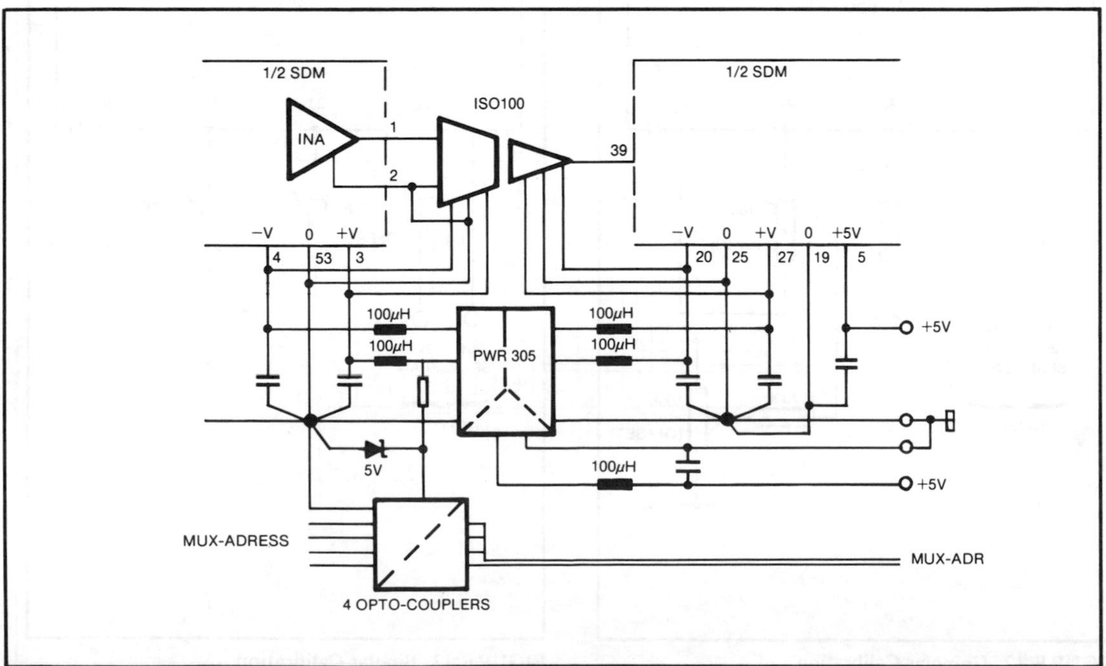

FIGURE 16. Galvanic Isolation Between Analog and Digital Signals.

CONTROLLING THE SDM

The Burr-Brown SDM family can be easily interfaced to most microprocessor systems, as shown in Figures 17-20. The microprocessor may control each conversion, or the converter may operate in a stand-alone mode controlled only by the R/$\overline{C}$ input.

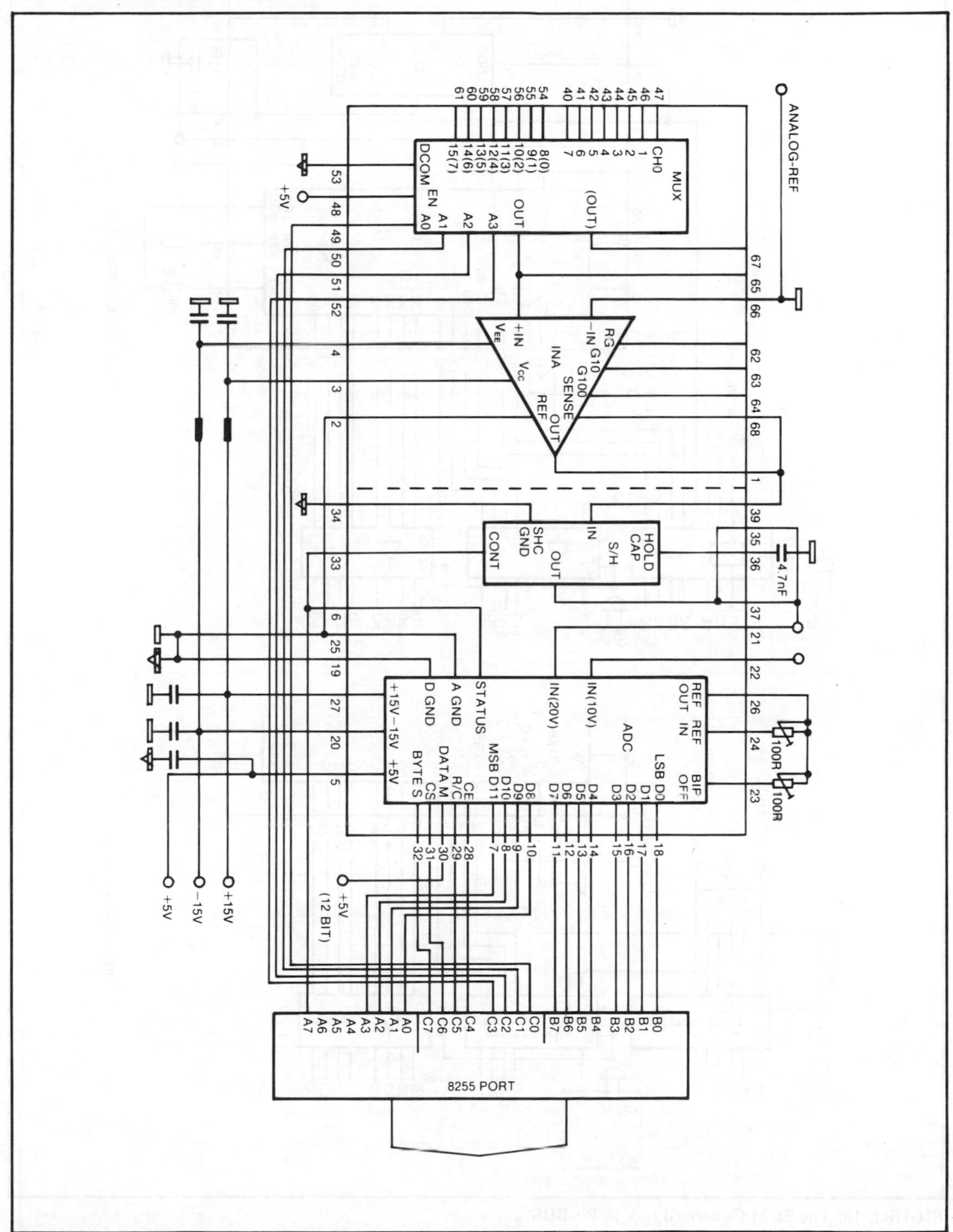

FIGURE 17. The SDM Connected to an Input/Output Port.

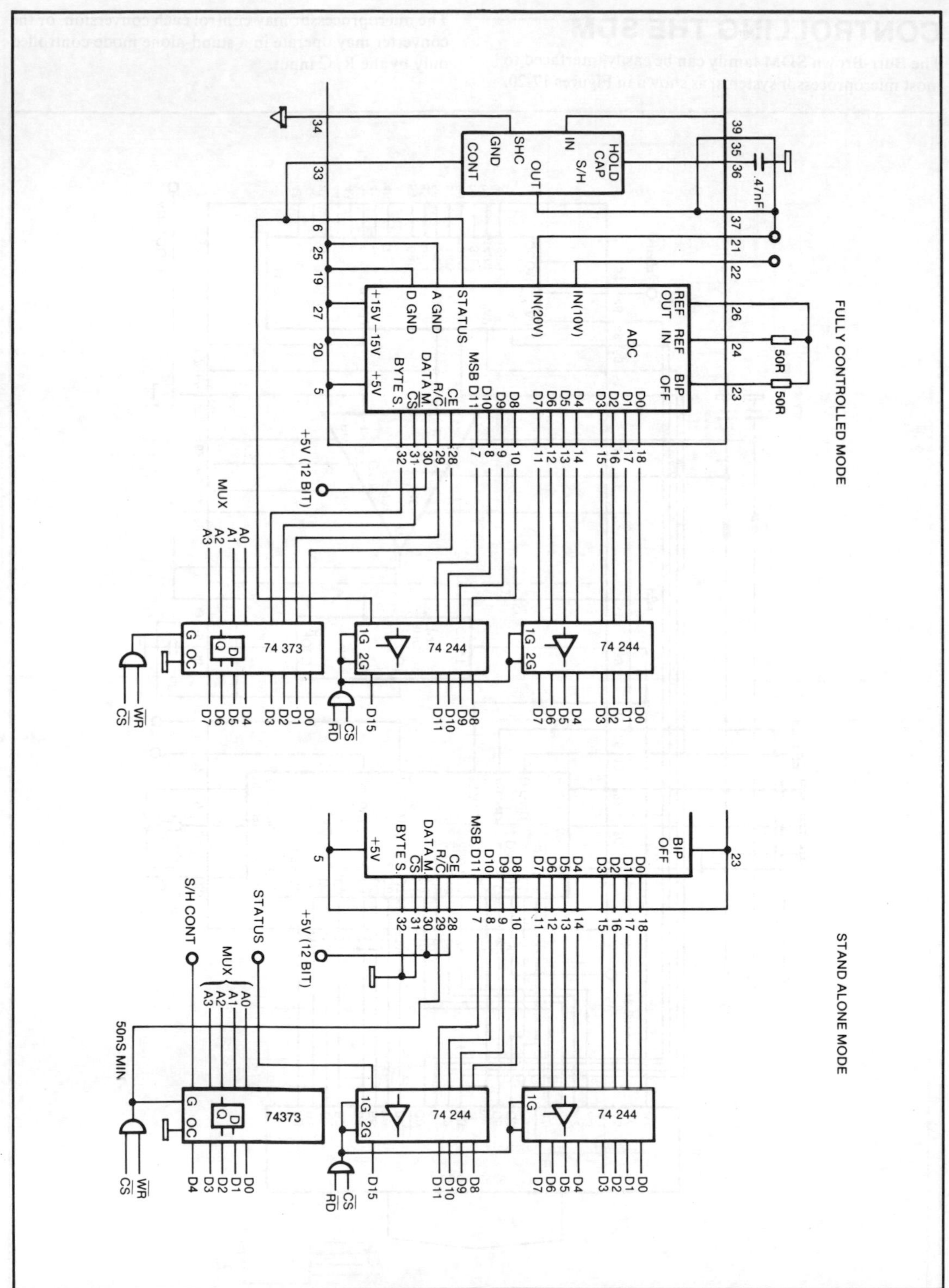

FIGURE 18. The SDM Connected to a 16-Bit-BUS.

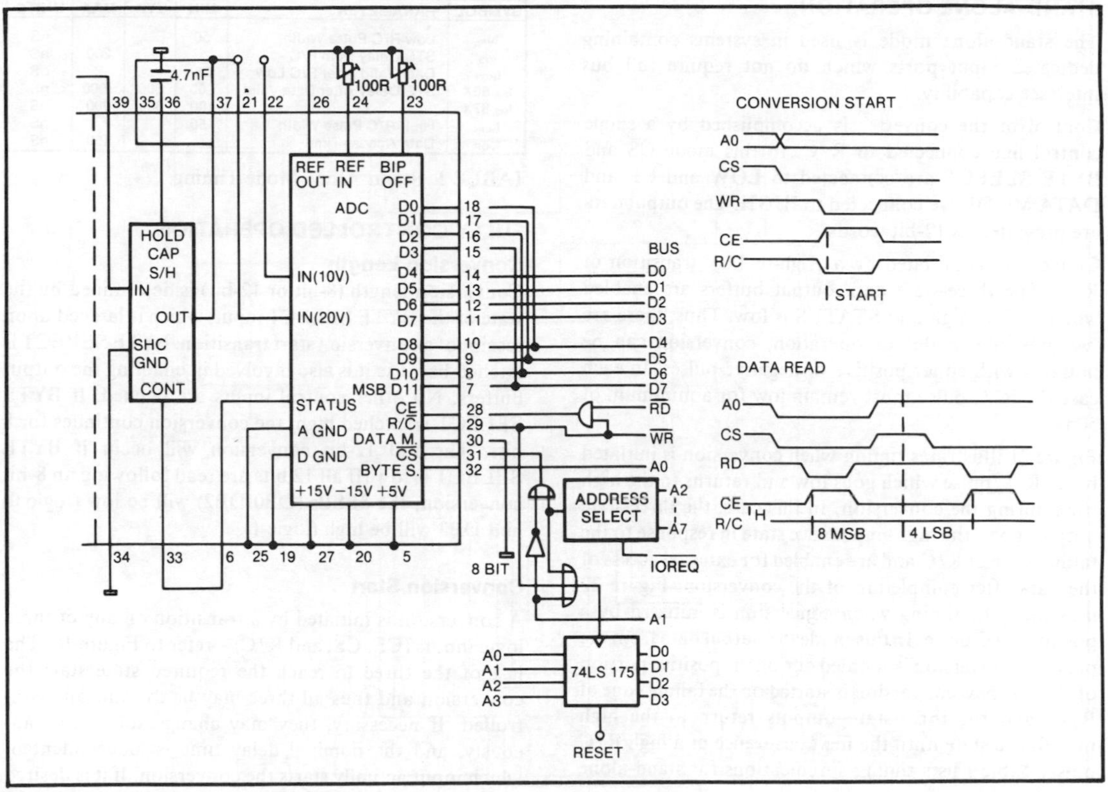

FIGURE 19. SDM on the Z80 BUS.

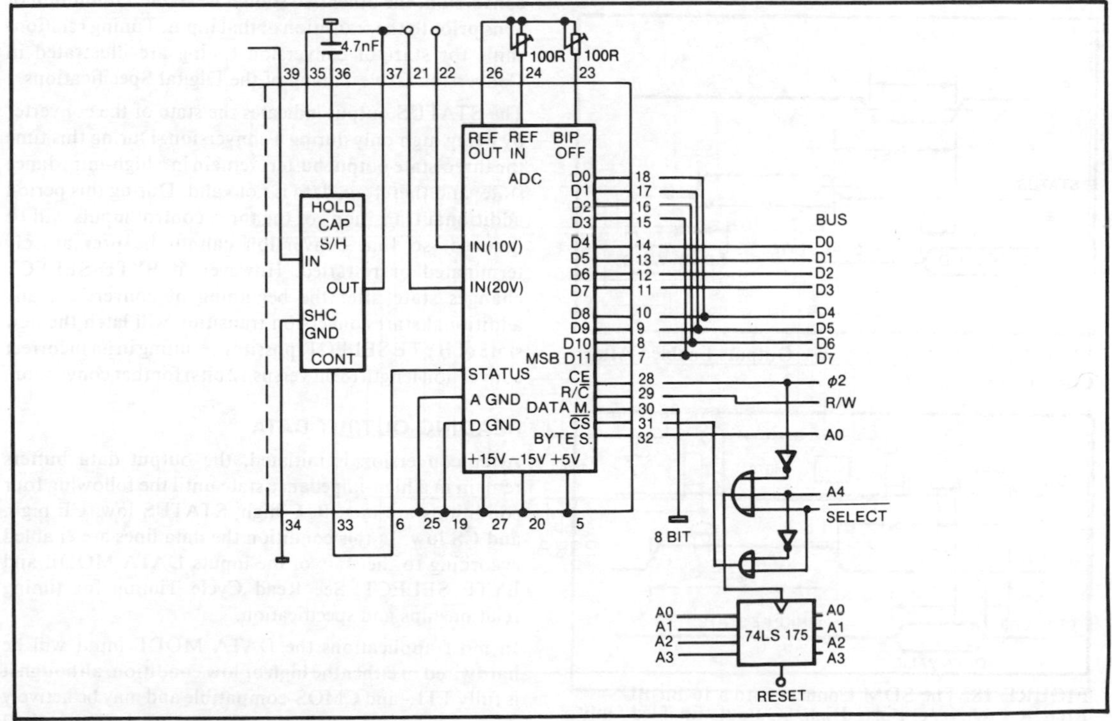

FIGURE 20. SDM on the 6502 BUS.

STAND-ALONE OPERATION

The stand-alone mode is used in systems containing dedicated input ports which do not require full bus interface capability.

Control of the converter is accomplished by a single control line connected to R/$\overline{C}$. In this mode $\overline{CS}$ and BYTE SELECT are connected to LOW and CE and DATA MODE are connected to HIGH. The output data are presented as 12-bit words.

Conversion is initiated by a High-to-Low transition of R/$\overline{C}$. The three-state data output buffers are enabled when R/$\overline{C}$ is high and STATUS is low. Thus, there are two possible modes of operation; conversion can be initiated with either positive or negative pulses. In each case the R/$\overline{C}$ pulse must remain low for a minimum of 50ns.

Figure 21 illustrates timing when conversion is initiated by an R/$\overline{C}$ pulse which goes low and returns to the high state during the conversion. In this case, the three-state outputs go to the high-impedance state in response to the falling edge of R/$\overline{C}$ and are enabled for external access of the data after completion of the conversion. Figure 22 illustrates the timing when conversion is initiated by a positive R/$\overline{C}$ pulse. In this mode the output data from the previous conversion is enabled during the positive portion of R/$\overline{C}$. A new conversion is started on the falling edge of R/$\overline{C}$, and the three-state outputs return to the high impedance state until the next occurence of a high R/$\overline{C}$ pulse. Table I lists timing specifications for stand-alone operation.

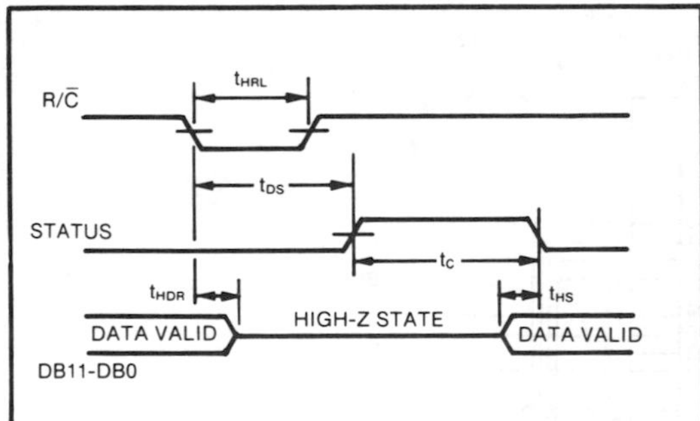

FIGURE 21. R/$\overline{C}$ Pulse Low—Outputs Enabled After Conversion.

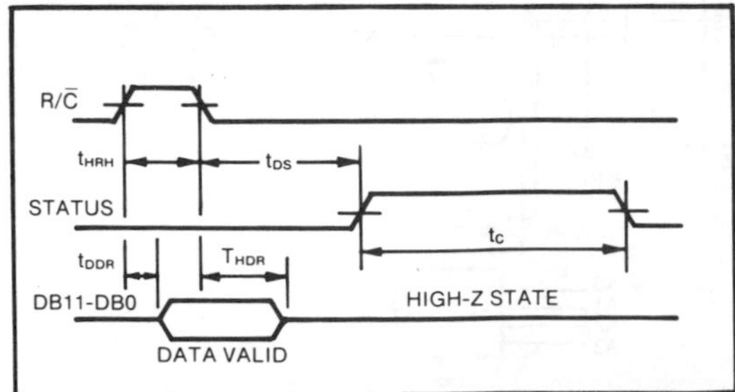

FIGURE 22. R/$\overline{C}$ Pulse High—Outputs Enabled Only Where R/$\overline{C}$ is High.

SYMBOL	PARAMETER	MIN	TYP	MAX	UNITS
t_{HRL}	Low R/C Pulse Width	50			nS
t_{DS}	STS Delay from R/C			200	nS
t_{HDR}	Data Valid After R/C Low	25			nS
t_{HS} 86X	STS Delay After Data Valid	300	500	1000	nS
t_{HS} 87X		100	300	600	nS
t_{HRH}	High R/C Pulse Width	150			nS
t_{DDR}	Data Access Time			150	nS

TABLE I. Stand-Alone Mode Timing.

FULLY CONTROLLED OPERATION

Conversion Length

Conversion length (8-bit or 12-bit) is determined by the state of the BYTE SELECT input, which is latched upon receipt of a conversion start transition. BYTE SELECT is latched because it is also involved in enabling the output buffers. No other control inputs are latched. If BYTE SELECT is latched high, the conversion continues for 8 bits. The full 12-bit conversion will occur if BYTE SELECT is low. If all 12 bits are read following an 8-bit conversion, the 3LSBs (DB0-DB2) will be low (logic 0) and DB3 will be high (logic 1).

Conversion Start

A conversion is initiated by a transition on any of three logic inputs (CE, $\overline{CS}$, and R/$\overline{C}$)—refer to Figure 10. The last of the three to reach the required state start the conversion and thus all three may be dynamically controlled. If necessary, they may change state simultaneously, and the nominal delay time is independent of which input actually starts the conversion. If it is desired that a particular input establish the actual start of conversion, the other two should be stable a minimum of 50ns prior to the transition of that input. Timing relationships for start of conversion timing are illustrated in Conversion Cycle Timing of the Digital Specifications.

The STATUS output indicates the state of the converter by being high only during a conversion. During this time the three-state output buffers remain in a high-impedance state, and therefore, data is not valid. During this period additional transitions of the three control inputs will be ignored, so that conversion cannot be prematurely terminated or restarted. However, if BYTE SELECT changes state after the beginning of conversion, any additional start conversion transition will latch the new state of BYTE SELECT, possibly resulting in an incorrect conversion length (8 bit versus 12 bits) for that conversion.

READING OUTPUT DATA

After conversion is initiated, the output data buffers remain in a high-impedance state until the following four conditions are met: R/$\overline{C}$ high, STATUS low, CE high, and $\overline{CS}$ low. In this condition the data lines are enabled according to the state of the inputs DATA MODE and BYTE SELECT. See Read Cycle Timing for timing relationships and specification.

In most applications the DATA MODE input will be hardwired in either the high or low condition, although it is fully TTL- and CMOS-compatible and may be actively driven if desired. When DATA MODE is high, all 12 outputs lines (DB0-DB11) are enabled simultaneously for full data word transfer to a 12-bit or 16-bit bus and the state of the BYTE SELECT is ignored.

When DATA MODE is low, the data is presented in the form of two 8-bit bytes, with selection of each byte by the state of BYTE SELECT during the read cycle.

The BYTE SELECT input is usually driven by the least significant bit of the address bus, allowing storage of the output data word in two consecutive memory locations.

When BYTE SELECT is low, the byte addressed contains the 8MSBs. When BYTE SELECT is high, the byte addressed contains the 4LSBs from the conversion followed by four zeros that have been forced by the control logic. The left-justified formats of the two 8-bit bytes are shown in Figure 23. The design of the SDM guarantees that the BYTE SELECT input may be toggled at any time without damage to the output buffers occuring.

In the majority of applications, the read operation will be attempted only after the conversion is complete and the status output has gone low. In those situations requiring the fastest possible access to the data, the read may be started as much as (t_{DD} max + t_{HS} max) before STATUS goes low. Refer to Read Cycle Timing for these timing relationships.

	Word 1								Word 2							
Processor	DB7	DB6	DB5	DB4	DB3	DB2	DB1	DB0	DB7	DB6	DB5	DB4	DB3	DB2	DB1	DB0
SDM	DB11	DB10	DB9	DB8	DB7	DB6	DB5	DB4	DB3	DB2	DB1	DB0	0	0	0	0

FIGURE 23. 12-Bit Data Format for 8-Bit Systems (connected as Figures 19 and 20).

APPLICATIONS INFORMATION

For the engineer who wishes to evaluate the SDM family, Burr-Brown has designed printed circuit boards on a single 'Eurocard' (shown here for LCC only). These boards enable the design engineer to experiment with various accuracy improvement techniques which are described below. Special consideration has been given to the grounding and circuit layout techniques required when dealing with 12-bit analog signals.

The printed circuit board has been designed so that the solutions to several of the problems likely to be encountered by the user can be examined.

It should not be thought that every user is required to adopt all of the techniques used on the circuit board. In many applications very few external components will be required. However, in following the application guidelines illustrated by the circuitry and accompanying notes, the designer will be able to select and adapt the solutions most suited to their own particular application or problem area.

Provisions for the following are made on the LCC PC board:

- 68 pin LCC socket (Burr-Brown Part No. MC 0068).
- 8 differential or 16 single-ended inputs.
- Input filtering with overvoltage protection for each channel.
- Socket for quad D-type flip-flop 74175 (MUX address latches).
- 7 additional I.C. sockets for easy interfacing to various BUS systems (connection by wire wrap techniques).
- 2 voltage regulators (15 volts).
- LC power supply decoupling.

The Layout pays particular attention to the requirements when operating with precision analog signals. This requires strict separation of the analog and digital areas. Analog and digital commons are totally separated and connected together only at the commons of the supply voltage. All common lines are low resistance and low inductance.

SUPPLY VOLTAGES

In order to avoid coupling between the external supply voltage 15 volt supplies, 2 voltage regulators (78M15, 79L15) are provided on the PC board. The unregulated supply voltage may vary from ±17 volts to ±25 volts.

The MUX/INA section and SHC/ADC section of the SDM have separate supply lines which can be inductively decoupled. This is recommended in order to suppress the high frequency noise which comes from the ADC during conversion.

The power supply rejection of the instrumentation amplifier reduces with increasing frequency. If high frequency noise on the supplies is not decoupled it will be injected into the signal path and cause errors. This effect can be

SDM862/872						SDM863/873				
MUX ADD3	MUX ADD2	MUX ADD1	MUX ADD0	MUX Enable	Channel Selected	MUX ADD2	MUX ADD1	MUX ADD0	MUX Enable	Channel Pair Selected
X	X	X	X	L	NONE	X	X	X	L	NONE
L	L	L	L	H	0	L	L	L	H	0
L	L	L	H	H	1	L	L	H	H	1
L	L	H	L	H	2	L	H	L	H	2
L	L	H	H	H	3	L	H	H	H	3
L	H	L	L	H	4	H	L	L	H	4
L	H	L	H	H	5	H	L	H	H	5
L	H	H	L	H	6	H	H	L	H	6
L	H	H	H	H	7	H	H	H	H	7
H	L	L	L	H	8			—		
H	L	L	H	H	9			—		
H	L	H	L	H	10			—		
H	L	H	H	H	11			—		
H	H	L	L	H	12			—		
H	H	L	H	H	13			—		
H	H	H	L	H	14			—		
H	H	H	H	H	15			—		

FIGURE 24. Channel Select Truth Table.

particularly pronounced when using the 'overlap' mode since the instrumentation amplifier is settling to a new analog value while the ADC is still carrying out the previous conversion.

The digital supply voltage is +5 volts and is also LC-filtered.

All supply lines are bypassed with a $10\mu F$ tantalum and a 100nF ceramic capacitor situated **as close as possible** to the package.

If the voltage regulators for the ±15 volts are not used, small inductors for decoupling of the supply voltages are recommended. If inductors are not fitted a dynamic ground loop will be created from supply lines via bypass capacitors to analog common.

INPUT PROTECTION

The multiplexer is protected up to an input voltage which can exceed the supply voltage by a maximum of 20 volts. This means, that with ±15 volts supply voltage, the input voltage can be ±35 volts without damage. This is also the case when the supply voltages are switched off (0 volts). The maximum input voltage can then be ±20 volts. For higher overvoltage protection a series resistor has to be used. The current via the multiplexer should be limited to a maximum of 1mA. For example, a 10kΩ series resistor would give an additional 10 volts overprotection.

For much higher overvoltages (e.g. 100 volts), high value series resistors cannot be used as offset errors would result. In practice, a combination of series resistors and diodes is used. The diodes are connected to ±15 volts and will conduct whenever the input voltage exceeds the ±15 volts supply voltage. The diodes are selected by signal source impedance, as well as filter resistance, as the diode leakage current across the series resistor can cause offset and linearity errors. In this circuit, IN4148 together with 10kΩ are used.

INPUT FILTER

Processor noise can be induced in the analog ground. Input filtering is therefore recommended for analog data aquisition. Such high frequency noise signals can cause dynamic overload of the instrumentation amplifier resulting in non-linear behavior. This leads directly to digitizing errors.

The design of the filter takes into account the characteristics of the SDM and of the signal source.

The following points have to be considered:

— The stray capacitance, output capacitance of the multiplexer and input capacitance of the instrument amplifier (60-80pf) has to be discharged in order to minimize errors caused by 'charge sharing.'

— The series resistor limits the current in the protection diodes, but it also has to be selected for the required filter time constant.

— The noise rejection of the filter has to be >80db in order to satisfy a 12-bit A/D conversion.

As well as considering the above, different calculations have to be carried out for single and differential input signals.

Single-Ended Measurement

R_f limits the maximum input current through the protection diodes. In this case, R_f has been chosen as 10kΩ and together with the capacitor C_g, forms the input filter time constant ($C_g = 0.47\mu F$). The time constant must be chosen according to the requirements of the input signal bandwidth and noise rejection. The multiplexer capacitance (C_m) is discharged mainly by C_g. This means C_g has to be sufficiently large compared with C_m or charged via R_f prior to re-sampling of the signal.

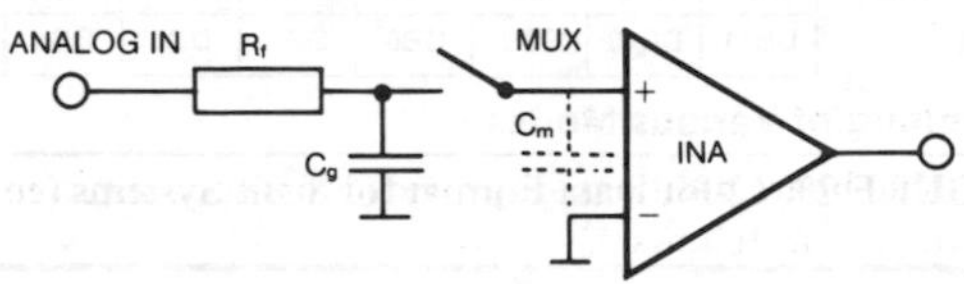

Differential Measurement

Capacitor C_f is used for limiting the input signal frequency. The bandwidth is calculated as follows:

$$F_g = \frac{1}{4_\pi R_f C_f} \quad \text{IF } C_f >> C_g$$

When selecting the value of C_f, it should be noted that C_m has to be discharged when switching the multiplexer channels. This means that the voltage error of C_f (induced by 'charge sharing' with C_m) has to be smaller than 1LSB. Therefore, C_f should have a minimum value of a $0.47\mu F$. The resistors R_f, together with the source impedance have to be sufficiently small in order to recharge C_f prior to signal sampling. This prevents errors in the signal value caused by the charge stored on C_m by the previously selected channel.

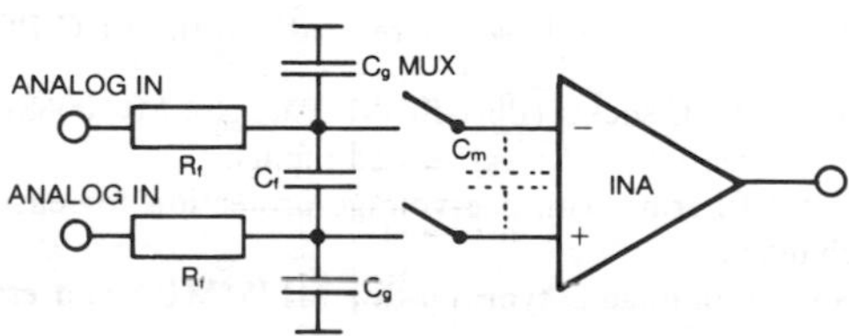

The 2 capacitors C_g form together with R_f a common-mode filter. This filter greatly improves accuracy in a noisy environment (decrease of common-mode rejection of instrumentation amplifier with increasing frequency). For good filter operation, both time constants R_f. C_g should match each other within 2%. Additional errors will be induced by a mismatch.

Selected values are: $C_f = 0.47\mu F$, $C_g = 10nF$, $R_f = 10k\Omega$. The filter reduces the signal slew rate so that the instrumentation amplifier can follow the voltage variation of the signal with the noise component eliminated.

In general, all measurements which require more than a gain of 10 should be done in differential mode. Single ended measurements should be limited to applications where current sources are measured via shunts or where signal voltages in the range of some volts are available.

Bus-Interface

As the outputs of the SDM are BUS compatible, only a few I.C.s are necessary to interface to various BUS systems. For such interfacing, 4 off 14-pin and 3 off 16-pin I.C. sockets are provided. Wiring is by wire wrap to the BUS connector.

Setting of Various Modes

Circuit board positions are provided for the connection of 'jumpers' as follows:

J1, J2—ADC analog input voltage settings.
J3—Set for differential (SDM8X3) or single ended (SMD8X2) operation.
J4—Instrumentation amplifier gain settings.

(a) 16 input channels, single ended:
—Use SDM8X2
—Consider single-ended filtering
—Connect J3 (pin 66) to common

(b) Differential inputs
—Use SDM8X3
—Consider differential filtering
—Connect J3 (pin 66) to pin 67

(c) Analog input

±10 volts	Connect J1 to pin 21
	Connect J2 to pot P2 (100Ω)
±5 volts	Connect J1 to pin 22
	Connect J2 to pot P2 (100Ω)
0 to +10 volts:	Connect J1 to pin 22
	Connect J2 to junction of R_1/R_2

(d) Gain of instrumentation amplifier

G = 1	Jumper J4 open
G = 10	Jumper J4 to pin 63
G = 100	Jumper J4 to pin 64

Other gains: use additional resistor between pin 62 and pin 63

Gain equation: $R_g = \frac{40k\Omega}{G-1}$ 4.444kΩ

Low tempco is recommended in order to minimize gain drift.

INPUT FILTER AND PROTECTION CIRCUITRY

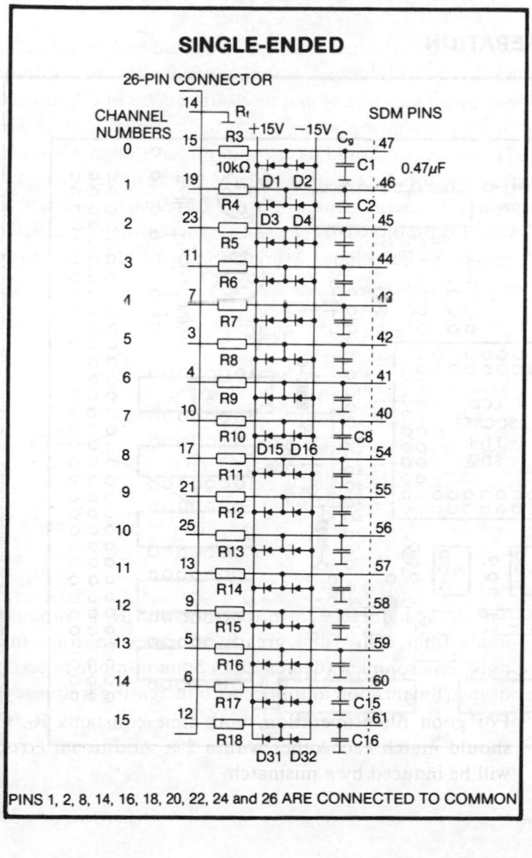

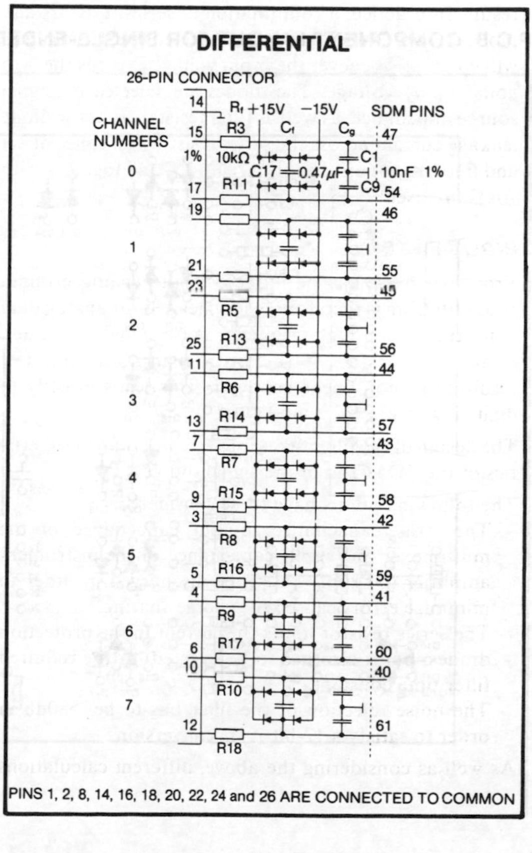

SINGLE-ENDED

PINS 1, 2, 8, 14, 16, 18, 20, 22, 24 and 26 ARE CONNECTED TO COMMON

DIFFERENTIAL

PINS 1, 2, 8, 14, 16, 18, 20, 22, 24 and 26 ARE CONNECTED TO COMMON

P.C.B. COMPONENT LAYOUT FOR DIFFERENTIAL OPERATION

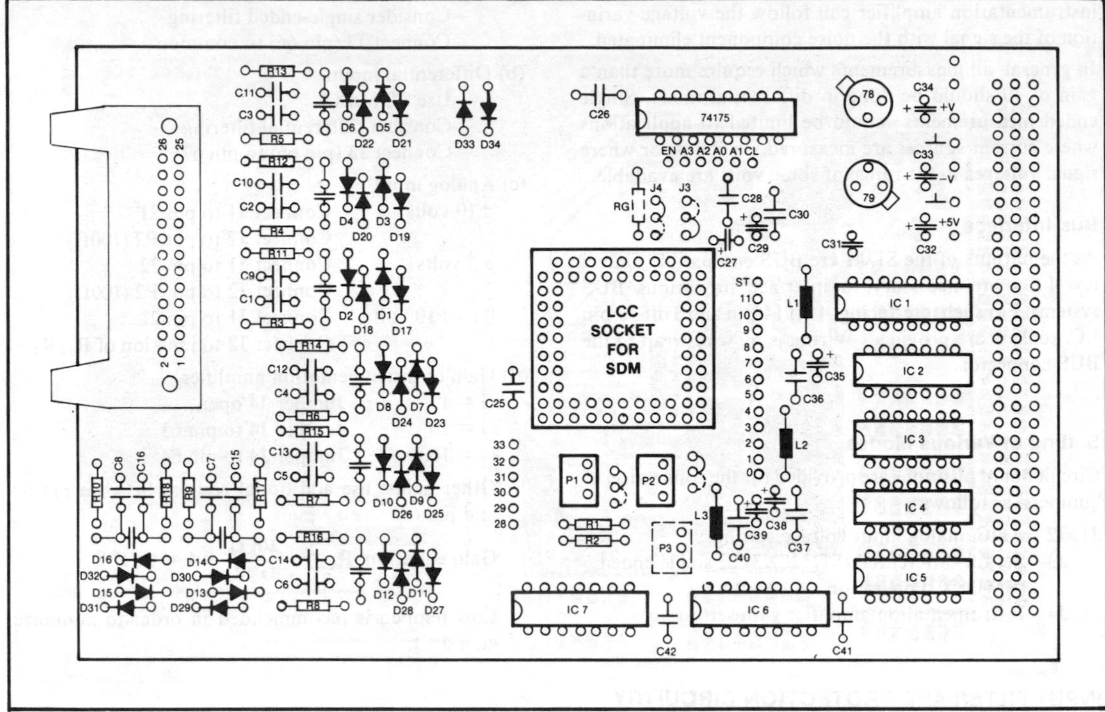

P.C.B. COMPONENT LAYOUT FOR SINGLE-ENDED OPERATION

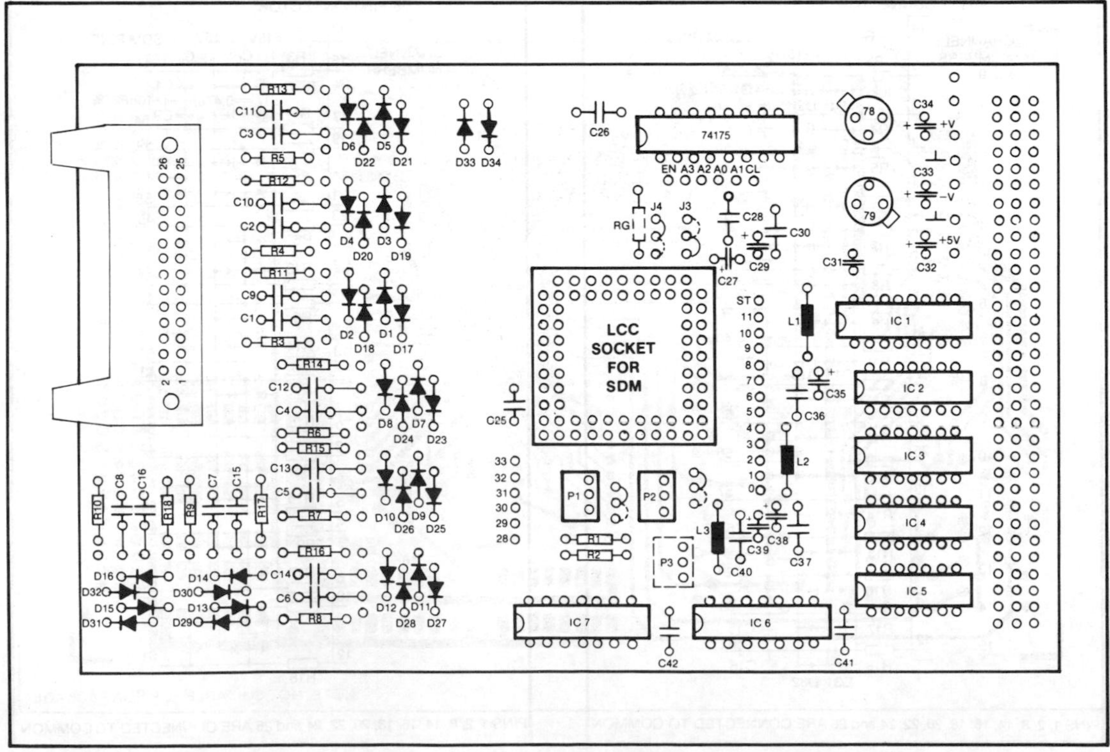

P.C.B. LAYOUT

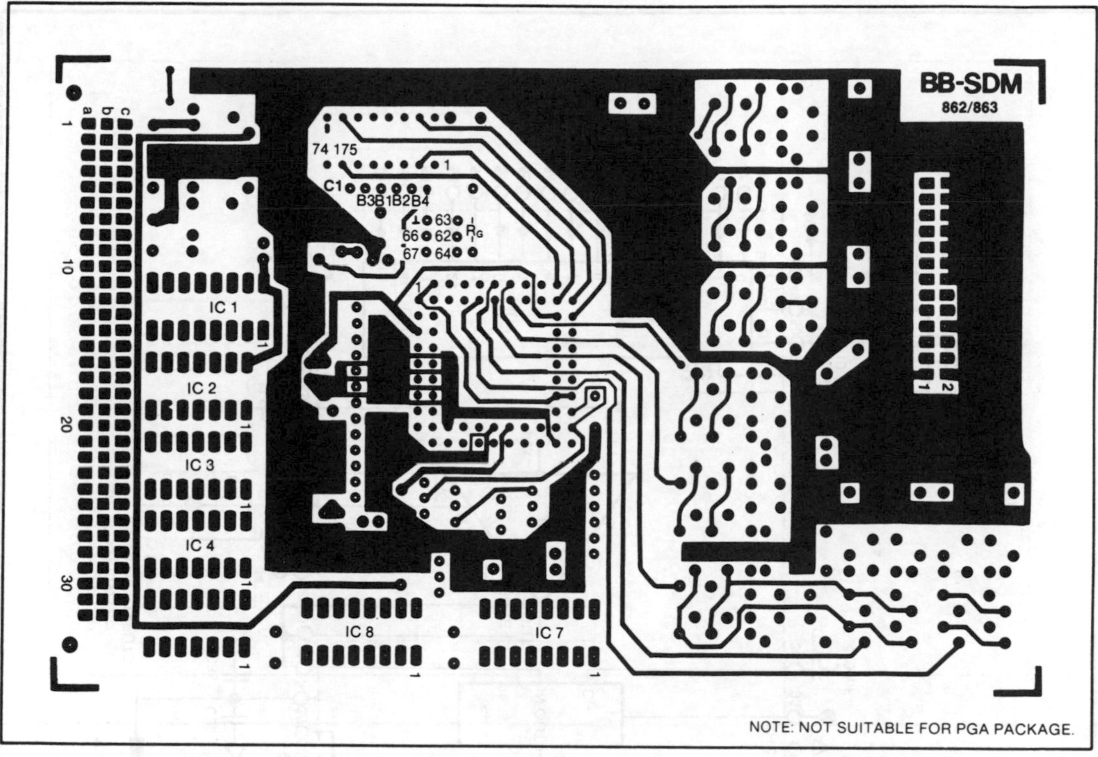
BB-SDM
862/863
74 175
C1
B3B1B2B4
63
66 62
67 64
RG
IC 1
IC 2
IC 3
IC 4
IC 8
IC 7
NOTE: NOT SUITABLE FOR PGA PACKAGE.

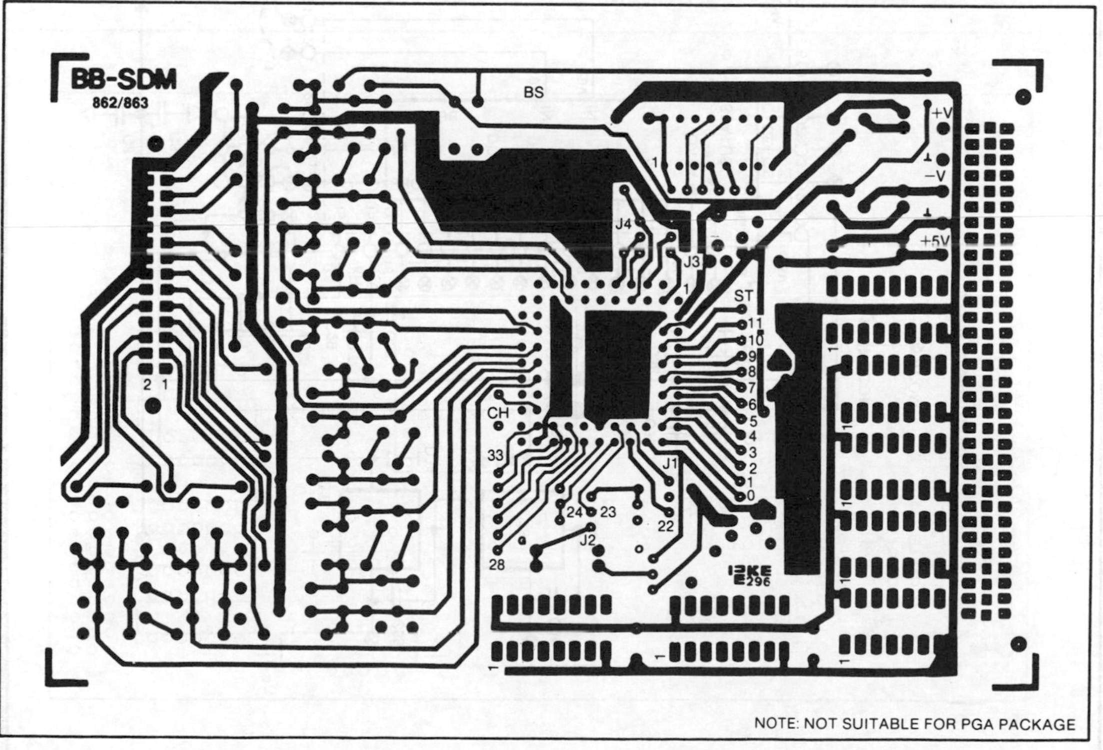
BB-SDM
862/863
BS
+V
−V
+5V
J4
J3
ST
CH
J1
J2
NOTE: NOT SUITABLE FOR PGA PACKAGE

CIRCUIT DIAGRAM—SDM P.C. BOARD

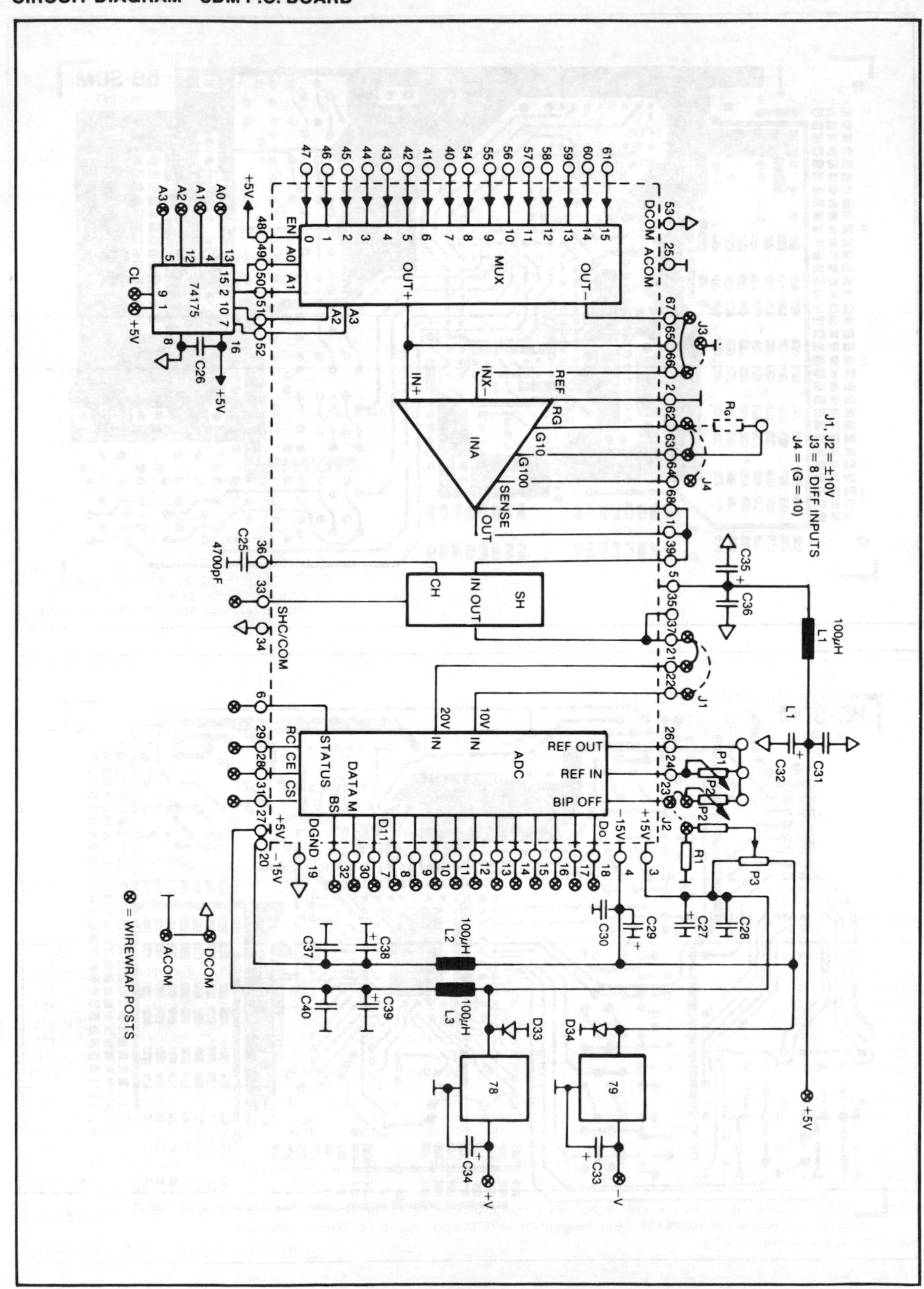

P.G.A. MECHANICAL OUTLINE

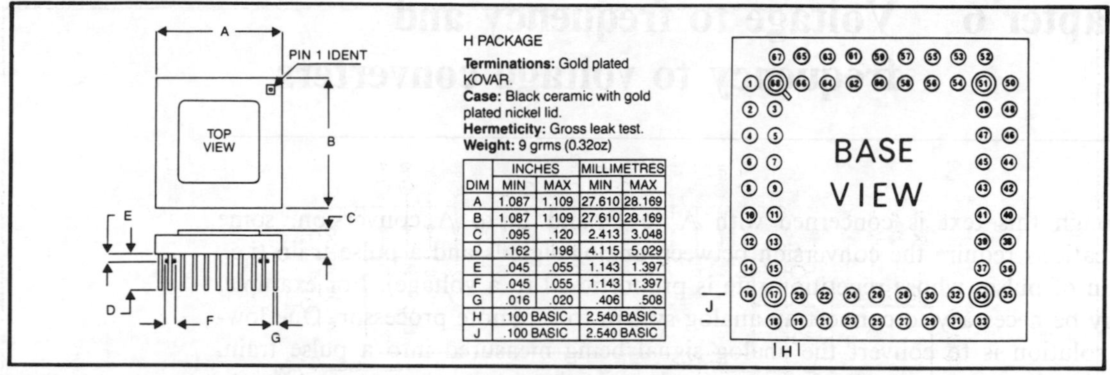

H PACKAGE

Terminations: Gold plated KOVAR.
Case: Black ceramic with gold plated nickel lid.
Hermeticity: Gross leak test.
Weight: 9 grms (0.32oz)

DIM	INCHES MIN	INCHES MAX	MILLIMETRES MIN	MILLIMETRES MAX
A	1.087	1.109	27.610	28.169
B	1.087	1.109	27.610	28.169
C	.095	.120	2.413	3.048
D	.162	.198	4.115	5.029
E	.045	.055	1.143	1.397
F	.045	.055	1.143	1.397
G	.016	.020	.406	.508
H	.100 BASIC		2.540 BASIC	
J	.100 BASIC		2.540 BASIC	

L.C.C. MECHANICAL OUTLINE

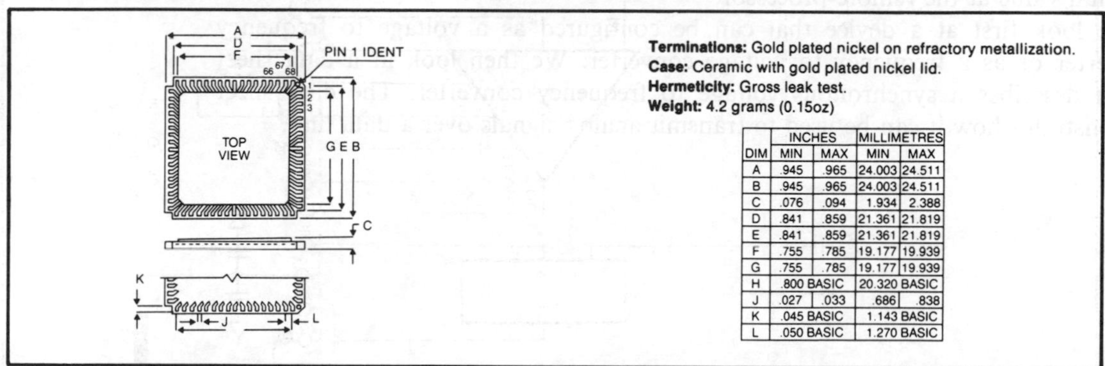

Terminations: Gold plated nickel on refractory metallization.
Case: Ceramic with gold plated nickel lid.
Hermeticity: Gross leak test.
Weight: 4.2 grams (0.15oz)

DIM	INCHES MIN	INCHES MAX	MILLIMETRES MIN	MILLIMETRES MAX
A	.945	.965	24.003	24.511
B	.945	.965	24.003	24.511
C	.076	.094	1.934	2.388
D	.841	.859	21.361	21.819
E	.841	.859	21.361	21.819
F	.755	.785	19.177	19.939
G	.755	.785	19.177	19.939
H	.800 BASIC		20.320 BASIC	
J	.027	.033	.686	.838
K	.045 BASIC		1.143 BASIC	
L	.050 BASIC		1.270 BASIC	

P.C.B. COMPONENTS PARTS LIST

R1	100kΩ For 0–10 Volts settling	C26	10nF Ceramic	P3	**100kΩ 0–10 volts range only**
R2	100Ω For 0–10 Volts settling	C27, C29, C35, C32, C38, C39	10μF Tantalum (Decoupling)	L1 . . . L3	100μH (Decoupling)
R3 . . . R18	10kΩ 1%	C28, C30, C31, C36, C37, C40	100nF Ceramic (Decoupling)	D1 . . . D32	1N4148 (Input Protection Diodes)
C1 . . . C16	0.47μF—Single ended input mode; 10nF 1%—Differential input mode	C33, C34	0.33μF Tantalum	D33, D34	1N4007
C17 . . . C24	0.47μF—Differential input mode	P1	100Ω	78	MC78M15CG
C25	4.700pF (Polypropylene, Polystyrene or Teflon™)	P2	100Ω **±5 volts, ±10 volts range only**	79	MC79L15CG
				74175	74LS175
				LCC Socket	MC0068

UNLESS OTHERWISE MARKED—RESISTORS ARE 1/4W, 5%, CAPACITORS ARE 10%

ORDERING INFORMATION[(1)]

Model	Input	LCC, PGA Pkg.	Accuracy [% FSR]	Throughput	Temp. Range [°C]	Model	Input	LCC, PGA Pkg.	Accuracy [% FSR]	Throughput	Temp. Range [°C]
SDM862J[(2)]	16SE	L, H	±0.024	33kHz	0 to +70	SDM863J	8DIF	L, H	±0.024	33kHz	0 to +70
SDM862K	16SE	L, H	±0.012	33kHz	0 to +70	SDM863K	8DIF	L, H	±0.012	33kHz	0 to +70
SDM862A	16SE	L, H	±0.024	33kHz	−25 to +85	SDM863A	8DIF	L, H	±0.024	33kHz	−25 to +85
SDM862B	16SE	L, H	±0.012	33kHz	−25 to +85	SDM863B	8DIF	L, H	±0.012	33kHz	−25 to +85
SDM862R	16SE	L, H	±0.024	33kHz	−55 to +125	SDM863R	8DIF	L, H	±0.024	33kHz	−55 to +125
SDM862S	16SE	L, H	±0.012	33kHz	−55 to +125	SDM863S	8DIF	L, H	±0.012	33kHz	−55 to +125
SDM872J	16SE	L, H	±0.024	50kHz	0 to +70	SDM873J	8DIF	L, H	±0.024	50kHz	0 to +70
SDM872K	16SE	L, H	±0.012	50kHz	0 to +70	SDM873K	8DIF	L, H	±0.012	50kHz	0 to +70
SDM872A	16SE	L, H	±0.024	50kHz	−25 to +85	SDM873A	8DIF	L, H	±0.024	50kHz	−25 to +85
SDM872B	16SE	L, H	±0.012	50kHz	−25 to +85	SDM873B	8DIF	L, H	±0.012	50kHz	−25 to +85
SDM872R	16SE	L, H	±0.024	50kHz	−55 to +125	SDM873R	8DIF	L, H	±0.024	50kHz	−55 to +125
SDM872S	16SE	L, H	±0.012	50kHz	−55 to +125	SDM873S	8DIF	L, H	±0.012	50kHz	−55 to +125

NOTES: (1) LCC Evaluation Board Part Number: PC862/863-1. PGA Evaluation Board Part Number: PC862/863-2. (2) 16 single-ended inputs, LCC package, with accuracy of 0.024% FSR, Temp Range of 0°C to 70°C and throughput of 33kHz = SDM862JL.

Teflon™ E.I. du Pont de Nemours & Co.

Chapter 6 Voltage to frequency and frequency to voltage converters

Although this text is concerned with A to D and D to A conversion, some applications require the conversion between analog signals and a pulse train (i.e., a train of pulses whose repetition rate is proportional to a voltage). For example, it may be necessary to connect an analog system to a remote processor. One low-cost solution is to convert the analog signal being measured into a pulse train. This signal can be transmitted over a simple twisted pair and then converted into a digital value at the remote processor.

We look first at a device that can be configured as a voltage to frequency converter or as a frequency to voltage converter. We then look at a data sheet which describes a synchronous voltage to frequency converter. The data sheet demonstrates how it can be used to transmit analog signals over a data link.

Voltage-to-Frequency and Frequency-to-Voltage Converter

AD650

FEATURES
V/F Conversion to 1MHz
Reliable Monolithic Construction
Very Low Nonlinearity
0.002% typ at 10kHz
0.005% typ at 100kHz
0.07% typ at 1MHz
Input Offset Trimmable to Zero
CMOS or TTL Compatible
Unipolar, Bipolar, or Differential V/F
V/F or F/V Conversion
Available in Surface Mount

AD650 PIN CONFIGURATION

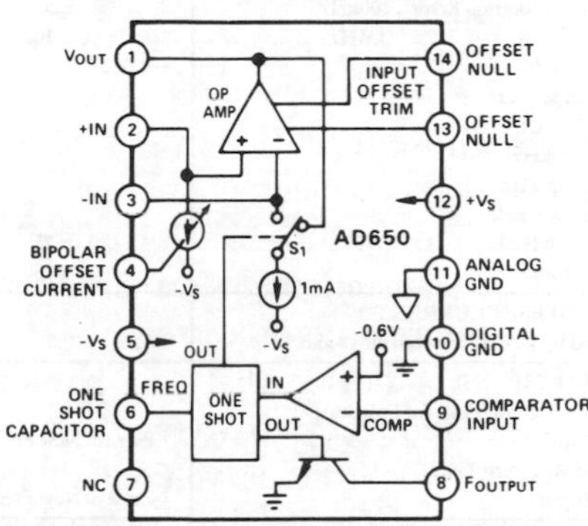

PRODUCT DESCRIPTION

The AD650 V/F/V (voltage-to-frequency or frequency-to-voltage converter) provides a combination of high frequency operation and low nonlinearity previously unavailable in monolithic form. The inherent monotonicity of the V/F transfer function makes the AD650 useful as a high-resolution analog-to-digital converter. A flexible input configuration allows a wide variety of input voltage and current formats to be used, and an open-collector output with separate digital ground allows simple interfacing to either standard logic families or opto-couplers.

The linearity error of the AD650 is typically 20ppm (0.002% of full scale) and 50ppm (0.005%) maximum at 10kHz full scale. This corresponds to approximately 14-bit linearity in an analog-to-digital converter circuit. Higher full-scale frequencies or longer count intervals can be used for higher resolution conversions. The AD650 has a useful dynamic range of six decades allowing extremely high resolution measurements. Even at 1MHz full scale, linearity is guaranteed less than 1000ppm (0.1%) on the AD650KN, KP, BD and SD grades.

In addition to analog-to-digital conversion, the AD650 can be used in isolated analog signal transmission applications, phased-locked-loop circuits, and precision stepper motor speed controllers. In the F/V mode, the AD650 can be used in precision tachometer and FM demodulator circuits.

The input signal range and full-scale output frequency are user-programmable with two external capacitors and one resistor. Input offset voltage can be trimmed to zero with an external potentiometer.

The AD650JN and AD650KN are offered in a plastic 14-pin DIP package. The AD650JP and AD650KP are available in a 20-pin plastic leaded chip carrier (PLCC). Both plastic packaged versions of the AD650 are specified for the commerical (0 to +70°C) temperature range. For industrial temperature range (−25°C to +85°C) applications, the AD650AD and AD650BD are offered in a ceramic package. The AD650SD is specified for the full −55°C to +125°C extended temperature range.

PRODUCT HIGHLIGHTS

1. In addition to very high linearity, the AD650 can operate at full scale output frequency up to 1MHz. The combination of these two features makes the AD650 an inexpensive solution for applications requiring high resolution monotonic A/D conversion.
2. The AD650 has a very versatile architecture that can be configured to accommodate bipolar, unipolar, or differential input voltages, or unipolar input currents.
3. TTL or CMOS compatibility is achieved using an open collector frequency output. The pullup resistor can be connected to voltages up to +30V, or +15V or +5V for conventional CMOS or TTL logic levels.
4. The same components used for V/F conversion can also be used for F/V conversion by adding a simple logic biasing network and reconfiguring the AD650.
5. The AD650 provides separate analog and digital grounds. This feature allows prevention of ground loops in real-world applications.

4

SPECIFICATIONS (@ +25°C with V_S = ±15V unless otherwise noted)

Model	AD650J/AD650A Min	Typ	Max	AD650K/AD650B Min	Typ	Max	AD650S Min	Typ	Max	Units
DYNAMIC PERFORMANCE										
Full Scale Frequency Range			1			1			1	MHz
Nonlinearity[1] f_{max} = 10kHz		0.002	0.005		0.002	0.005		0.002	0.005	%
100kHz		0.005	**0.02**		0.005	**0.02**		0.005	**0.02**	%
500kHz		0.02	0.05		0.02	0.05		0.02	0.05	%
1MHz		0.1			0.05	**0.1**		0.05	**0.1**	%
Full Scale Calibration Error[2], 100kHz		±5			±5			±5		%
1MHz		±10			±10			±5		%
vs. Supply[3]	−0.002		**+0.002**	−0.002		**+0.002**	−0.002		**+0.002**	% of FSR/V
vs. Temperature										
A, B, and S Grades										
at 10kHz			±75			±75			**±75**	ppm/°C
at 100kHz			±150			±150			**±150**	ppm/°C
J and K Grades										
at 10kHz		±75			±75					ppm/°C
at 100kHz		±150			±150					ppm/°C
BIPOLAR OFFSET CURRENT										
Activated by 1.24kΩ between pins 4 and 5	0.45	0.5	0.55	0.45	0.5	0.55	0.45	0.5	0.55	mA
DYNAMIC RESPONSE										
Maximum Settling Time for Full Scale										
Step Input	1 Pulse of New Frequency Plus 1μs			1 Pulse of New Frequency Plus 1μs			1 Pulse of New Frequency Plus 1μs			
Overload Recovery Time										
Step Input	1 Pulse of New Frequency Plus 1μs			1 Pulse of New Frequency Plus 1μs			1 Pulse of New Frequency Plus 1μs			
ANALOG INPUT AMPLIFIER (V/F Conversion)										
Current Input Range (Figure 1)	0		+0.6	0		+0.6	0		+0.6	mA
Voltage Input Range (Figure 5)	−10		0	−10		0	−10		0	V
Differential Impedance		2MΩ‖10pF			2MΩ‖10pF			2MΩ‖10pF		
Common Mode Impedance		1000MΩ‖10pF			1000MΩ‖10pF			1000MΩ‖10pF		
Input Bias Current										
Noninverting Input		40	**100**		40	**100**		40	**100**	nA
Inverting Input		±8	**±20**		±8	**±20**		±8	**±20**	nA
Input Offset Voltage										
(Trimmable to Zero)			**±4**			**±4**			**±4**	mV
vs. Temperature (T_{min} to T_{max})		±30			±30			±30		μV/°C
Safe Input Voltage		±V_S			±V_S			±V_S		C
COMPARATOR (F/V Conversion)										
Logic "0" Level	−V_S		−1	−V_S		−1	−V_S		+1	V
Logic "1" Level	0		+V_S	0		+V_S	0		+V_S	V
Pulse Width Range[4]	0.1		(0.3 × t_{OS})	0.1		(0.3 × t_{OS})	0.1		(0.3 × t_{OS})	μs
Input Impedance		250			250			250		kΩ
OPEN COLLECTOR OUTPUT (V/F Conversion)										
Output Voltage in Logic "0"										
I_{SINK} ≤ 8mA, T_{min} to T_{max}			0.4			0.4			0.4	V
Output Leakage Current in Logic "1"			**100**			**100**			**100**	nA
Voltage Range[5]	0		+36	0		+36	0		+36	V
AMPLIFIER OUTPUT (F/V Conversion)										
Voltage Range (1500Ω min load resistance)	0		+10	0		+10	0		+10	V
Source Current (750Ω max load resistance)	10			10			10			mA
Capacitive Load (Without Oscillation)			100			100			100	pF
POWER SUPPLY										
Voltage, Rated Performance	±9		**±18**	±9		**±18**	±9		**±18**	V
Quiescent Current			**8**			**8**			**8**	mA
TEMPERATURE RANGE										
Rated Performance – N Package	0		+70	0		+70				°C
D Package	−25		+85	−25		+85	−55		+125	°C
Storage – N Package	−25		+85	−25		+85				°C
D Package	−65		+150	−65		+150	−65		+150	°C
PACKAGE OPTIONS[6]										
PLCC (P-20A)	AD650JP			AD650KP						
Plastic DIP (N-14)	AD650JN			AD650KN						
Ceramic DIP (D-14)	AD650AD			AD650BD			AD650SD			

NOTES

[1]Nonlinearity is defined as deviation from a straight line from zero to full scale, expressed as a fraction of full scale.

[2]Full scale calibration error adjustable to zero.

[3]Measured at full scale output frequency of 10kHz.

[4]Refer to F/V conversion section of the text.

[5]Referred to digital ground.

[6]See Section 14 for package outline information.

Specifications subject to change without notice.

Specifications shown in boldface are tested on all production units at final electrical test. Results from those tests are used to calculate outgoing quality levels. All min and max specifications are guaranteed, although only those shown in boldface are tested on all production units.

Unipolar Operation – AD650

ORDERING GUIDE

Part Number	Gain Tempco ppm/°C 100kHz	1MHz Linearity	Specified Temperature Range °C	Package
AD650JN	150 typ	0.1% typ	0 to +70	Plastic DIP
AD650KN	150 typ	0.1% max	0 to +70	Plastic DIP
AD650JP	150 typ	0.1% typ	0 to +70	PLCC
AD650KP	150 typ	0.1% max	0 to +70	PLCC
AD650AD	150 max	0.1% typ	−25 to +85	Ceramic
AD650BD	150 max	0.1% max	−25 to +85	Ceramic
AD650SD	150 max	0.1% max	−55 to +125	Ceramic

ABSOLUTE MAXIMUM RATINGS

Total Supply Voltage $+V_S$ to $-V_S$	36V
Storage Temperature Ceramic	−55°C to +165°C
Plastic	−25°C to +125°C
Differential Input Voltage (Pins 2 & 3)	±10V
Maximum Input Voltage	$\pm V_S$
Open Collector Output Voltage Above Digital GND	36V
Current	50mA
Amplifier Short Ckt to Ground	Indefinite
Comparator Input Voltage (Pin 9)	$\pm V_S$

CIRCUIT OPERATION

UNIPOLAR CONFIGURATION

The AD650 is a *charge balance* voltage-to-frequency converter. In the connection diagram shown in Figure 1, or the block diagram of Figure 2a, the input signal is converted into an equivalent current by the input resistance R_{IN}. This current is *exactly* balanced by an internal feedback current delivered in short, timed bursts from the switched 1mA internal current source. These bursts of current may be thought of as precisely defined packets of charge. The required number of charge packets, each producing one pulse of the output transistor, depends upon the amplitude of the input signal. Since the number of charge packets delivered per unit time is dependent on the input signal amplitude, a linear voltage-to-frequency transformation will be accomplished. The frequency output is furnished via an open collector transistor.

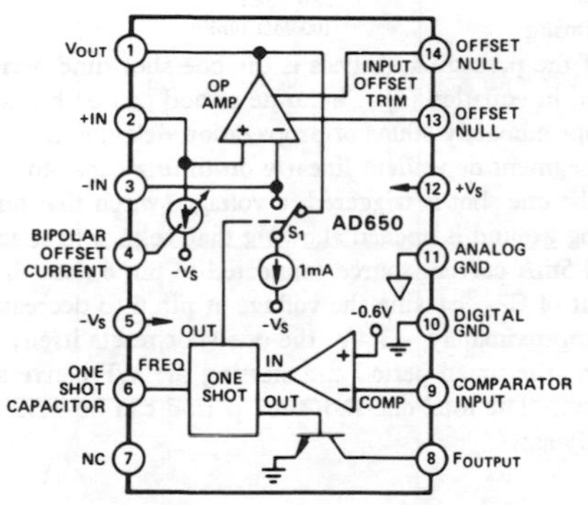

AD650 Pin Configuration

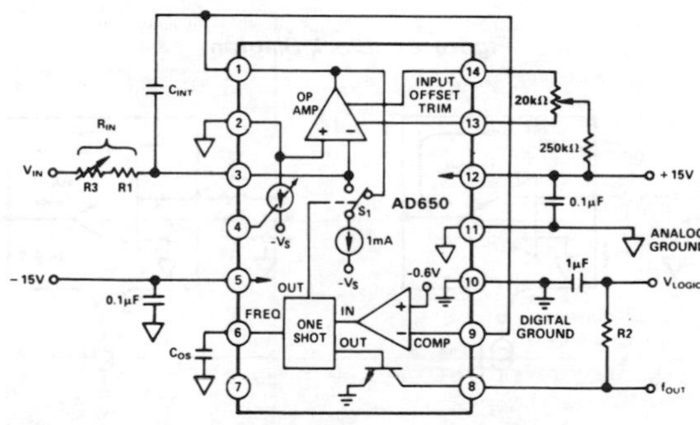

Figure 1. Connection Diagram for V/F Conversion, Positive Input Voltage

A more rigorous analysis demonstrates how the charge balance voltage-to-frequency conversion takes place.

A block diagram of the device arranged as a V to F converter is shown in Figure 2a. The unit is comprised of an input integrator, a current source and steering switch, a comparator and a one-shot. When the output of the one-shot is low, the current steering switch S_1 diverts all the current to the output of the op amp; this is called the Integration Period. When the one-shot has been triggered and its output is high, the switch S_1 diverts all the current to the summing junction of the op amp; this is called the Reset Period. The two different states are shown in Figure 2 along with the various branch currents. It should be noted that the output current from the op amp is the same for either state, thus minimizing transients.

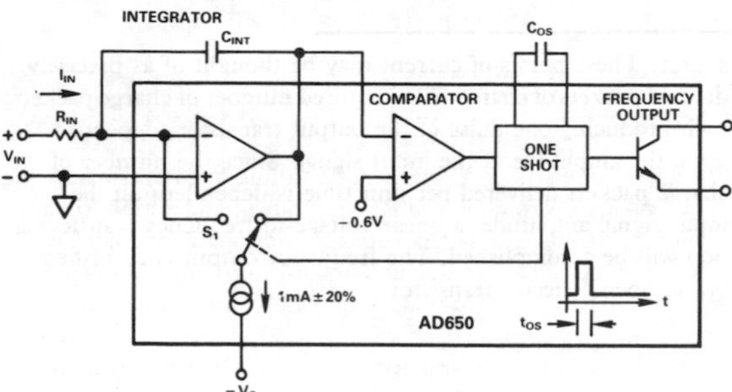

Figure 2a. Block Diagram

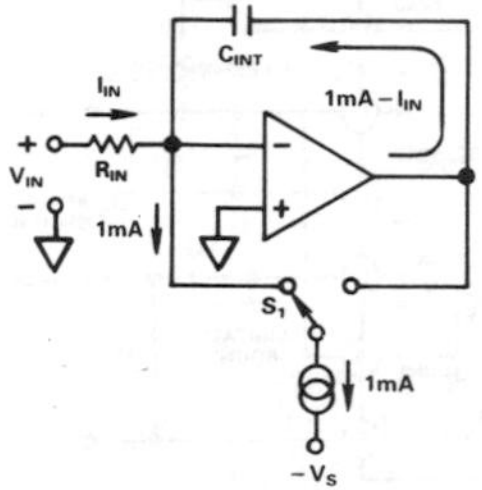

Figure 2b. Reset Mode

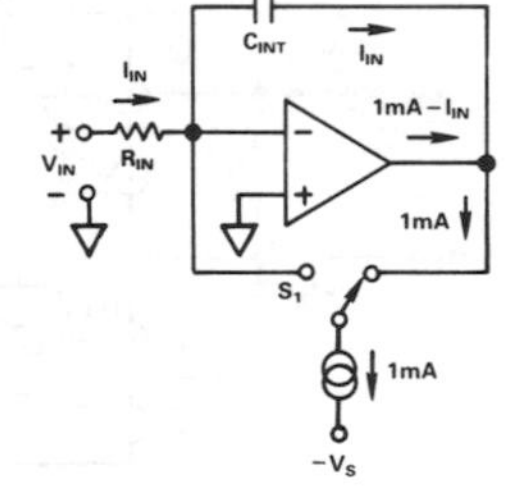

Figure 2c. Integrate Mode

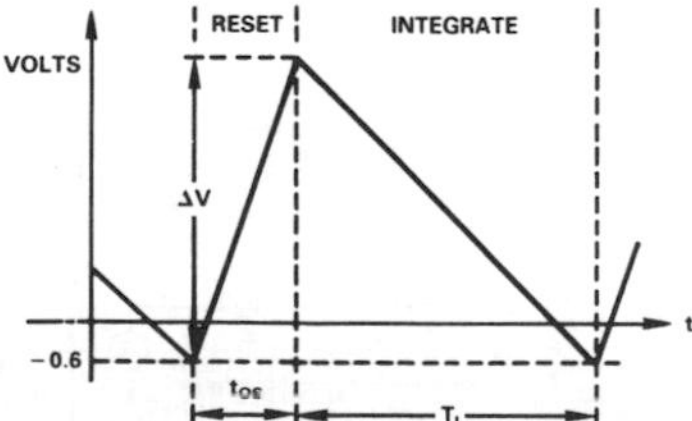

Figure 2d. Voltage Across C_{INT}

The positive input voltage develops a current ($I_{IN} = V_{IN}/R_{IN}$) which charges the integrator capacitor C_{INT}. As charge builds up on C_{INT}, the output voltage of the integrator ramps downward towards ground. When the integrator output voltage (pin 1) crosses the comparator threshold (−0.6 volt) the comparator triggers the one shot, whose time period, t_{OS} is determined by the one shot capacitor C_{OS}.

Specifically, the one shot time period is:

$$t_{OS} = C_{OS} \times 6.8 \times 10^3\ sec/F + 3.0 \times 10^{-7} sec \tag{1}$$

The Reset Period is initiated as soon as the integrator output voltage crosses the comparator threshold, and the integrator ramps upward by an amount:

$$\Delta V = t_{OS} \cdot \frac{dV}{dt} = \frac{t_{OS}}{C_{INT}}\left(1mA - I_{IN}\right) \tag{2}$$

After the Reset Period has ended, the device starts another Integration Period, as shown in Figure 2, and starts ramping downward again. The amount of time required to reach the comparator threshold is given as:

$$T_I = \frac{\Delta V}{\frac{dV}{dt}} = \frac{t_{OS}/C_{INT}(1mA - I_{IN})}{I_{IN}/C_{INT}} = t_{OS}\left(\frac{1mA}{I_{IN}} - 1\right) \tag{3}$$

The output frequency is now given as:

$$f_{OUT} = \frac{1}{t_{OS} + T_I} = \frac{I_{IN}}{t_{OS} \times 1mA} = 0.15\,\frac{F \bullet Hz}{A}\,\frac{V_{IN}/R_{IN}}{C_{OS} + 4.4 \times 10^{-11}F} \tag{4}$$

Note that C_{INT}, the integration capacitor has no effect on the transfer relation, but merely determines the amplitude of the sawtooth signal out of the integrator.

One Shot Timing

A key part of the preceding analysis is the one shot time period that was given in equation (1). This time period can be broken down into approximately 300ns of propagation delay, and a second time segment dependent linearly on timing capacitor C_{OS}. When the one shot is triggered, a voltage switch that holds pin 6 at analog ground is opened allowing that voltage to change. An internal 0.5mA current source connected to pin 6 then draws its current out of C_{OS}, causing the voltage at pin 6 to decrease linearly. At approximately −3.4V, the one shot resets itself, thereby ending the timed period and starting the V/F conversion cycle over again. The total one shot time period can be written mathematically as:

$$t_{OS} = \frac{\Delta V\ C_{OS}}{I_{DISCHARGE}} + T_{GATE\ DELAY} \tag{5}$$

substituting actual values quoted above,

$$t_{OS} = \frac{-3.4V \times C_{OS}}{-0.5 \times 10^{-3}A} + 300 \times 10^{-9} sec \tag{6}$$

This simplifies into the timed period equation given above.

COMPONENT SELECTION

Only four component values must be selected by the user. These are input resistance R_{IN}, timing capacitor C_{OS}, logic resistor R_2, and integration capacitor C_{INT}. The first two determine the input voltage and full scale frequency, while the last two are determined by other circuit considerations.

Of the four components to be selected, R_2 is the easiest to define. As a pull up resistor, it should be chosen to limit the current through the output transistor to 8mA if a TTL maximum V_{OL} of 0.4V is desired. For example, if a 5V logic supply is used, R_2 should be no smaller than 5V/8mA or 625Ω. A larger value can be used if desired.

R_{IN} and C_{OS} are the only two parameters available to set the full scale frequency to accommodate the given signal range. The

Bipolar Operation – AD650

"swing" variable that is affected by the choice of R_{IN} and C_{OS} is nonlinearity. The selection guide of Figure 3 shows this quite graphically. In general, larger values of C_{OS} and lower full scale input currents (higher values of R_{IN}) provide better linearity. In Figure 3, the implications of four different choices of R_{IN} are shown. Although the selection guide is set up for a unipolar configuration with a zero to 10V input signal range, the results can be extended to other configurations and input signal ranges. For a full scale frequency of 100kHz (corresponding to 10V input), you can see that among the available choices, $R_{IN} = 20k$ and $C_{OS} = 620pF$ gives the lowest nonlinearity, 0.0038%. Also, if you wish to use the highest frequency that will give the 20ppm minimum nonlinearity, it is approximately 33kHz (40.2kΩ and 1000pF).

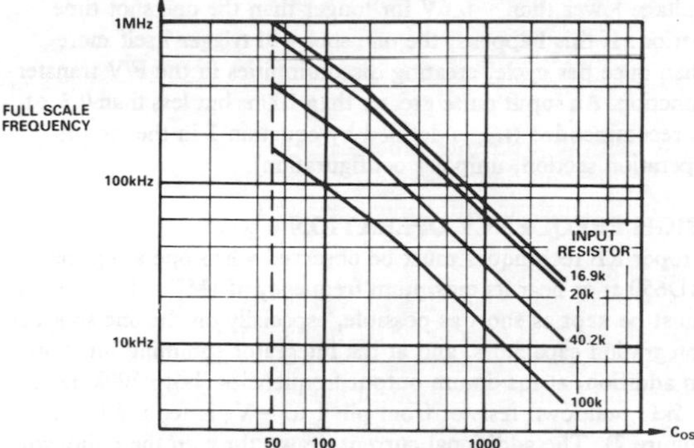

Figure 3a. Full Scale Frequency vs. C_{OS}

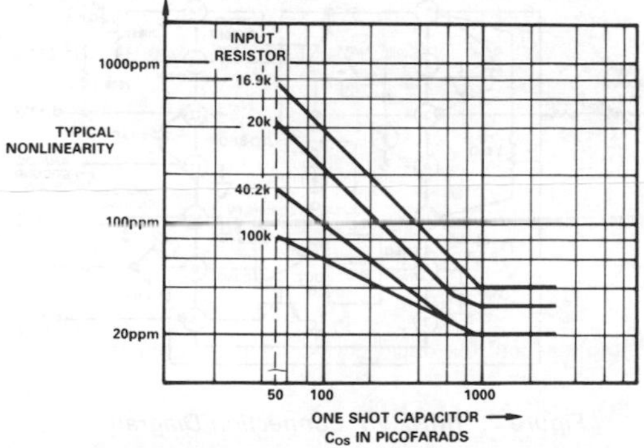

Figure 3b. Typical Nonlinearity vs. C_{OS}

For input signal spans other than 10V, the input resistance must be scaled proportionately. For example, if 100kΩ is called out for a 0–10V span, 10k would be used with a 0–1V span, or 200kΩ with a ±10V bipolar connection.

The last component to be selected is the integration capacitor C_{INT}. In almost all cases, the best value for C_{INT} can be calculated using the equation:

$$C_{INT} = \frac{10^{-4} F/sec}{f_{MAX}} \quad (1000pF \text{ minimum}) \qquad (7)$$

When the proper value for C_{INT} is used, the charge balance architecture of the AD650 provides continuous integration of the input signal, hence large amounts of noise and interference can be rejected. If the output frequency is measured by counting pulses during a constant gate period, the integration provides infinite normal mode rejection for frequencies corresponding to the gate period and its harmonics. However, if the integrator stage becomes saturated by an excessively large noise pulse, the continuous integration of the signal will be interrupted, allowing the noise to appear at the output. If the approximate amount of noise that will appear on C_{INT} is known (V_{NOISE}), the value of C_{INT} can be checked using the following inequality:

$$C_{INT} > \frac{t_{OS} \times 1 \times 10^{-3} A}{+V_S - 3V - V_{NOISE}} \qquad (8)$$

For example, consider an application calling for a maximum frequency of 75kHz, a 0–1 volt signal range, and supply voltages of only ±9 volts. The component selection guide of Figure 3 is used to select 2.0kΩ for R_{IN} and 1000pF for C_{OS}. This results in a one shot time period of approximately 7μs. Substituting 75kHz into equation 7 yields a value of 1300pF for C_{INT}. When the input signal is near zero, 1mA flows through the integration capacitor to the switched current sink during the reset phase, causing the voltage across C_{INT} to increase by approximately 5.5 volts. Since the integrator output stage requires approximately 3 volts head room for proper operation, only 0.5 volt margin remains for integrating extraneous noise on the signal line. A negative noise pulse at this time might saturate the integrator, causing an error in signal integration. Increasing C_{INT} to 1500 or 2000pF will provide much more noise margin, thereby eliminating this potential trouble spot.

BIPOLAR V/F

Figure 4 shows how the internal bipolar current sink is used to provide a half-scale offset for a ±5V signal range, while providing a 100kHz maximum output frequency. The nominally 0.5mA (±10%) offset current sink is enabled when a 1.24kΩ resistor is connected between pins 4 and 5. Thus, with the grounded 10kΩ nominal resistance shown, a −5V offset is developed at pin 2. Since pin 3 must also be at −5V, the current through R_{IN} is 10V/40kΩ = +0.25mA at $V_{IN} = +5V$, and 0mA at $V_{IN} = -5V$.

Components are selected using the same guidelines outlined for the unipolar configuration with one alteration. The voltage

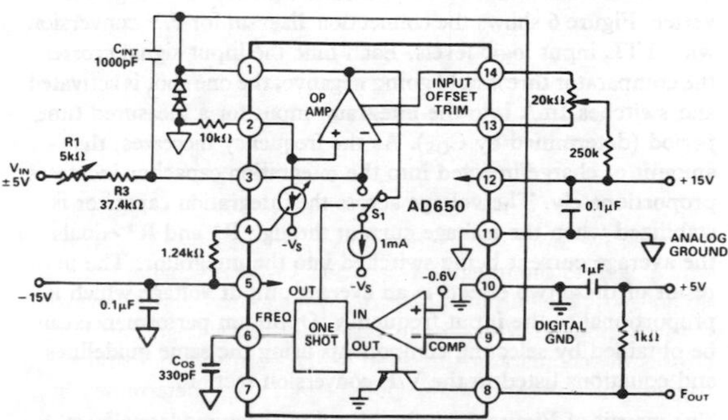

Figure 4. Connections for ±5V Bipolar V/F with 0 to 100kHz TTL Output

across the total signal range must be equated to the maximum input voltage in the unipolar configuration. In other words, the value of the input resistor R_{IN} is determined by the input voltage span, not the maximum input voltage. A diode from pin 1 to ground is also recommended. This is discussed further under "Other Circuit Conditions".

As in the unipolar circuit, R_{IN} and C_{OS} must have low temperature coefficients to minimize the overall gain drift. The 1.24kΩ resistor used to activate the 0.5mA offset current should also have a low temperature coefficient. The bipolar offset current has a temperature coefficient of approximately −200ppm/°C.

UNIPOLAR V/F, NEGATIVE INPUT VOLTAGE

Figure 5 shows the connection diagram for V/F conversion of negative input voltages. In this configuration full scale output frequency occurs at negative full scale input, and zero output frequency corresponds with zero input voltage.

A very high impedance signal source may be used since it only drives the noninverting integrator input. Typical input impedance at this terminal is 1GΩ or higher. For V/F conversion of positive input signals using the connection diagram of Figure 1, the signal generator must be able to source the integration current to drive the AD650. For the negative V/F conversion circuit of Figure 5, the integration current is drawn from ground through R1 and R3, and the active input is high impedance.

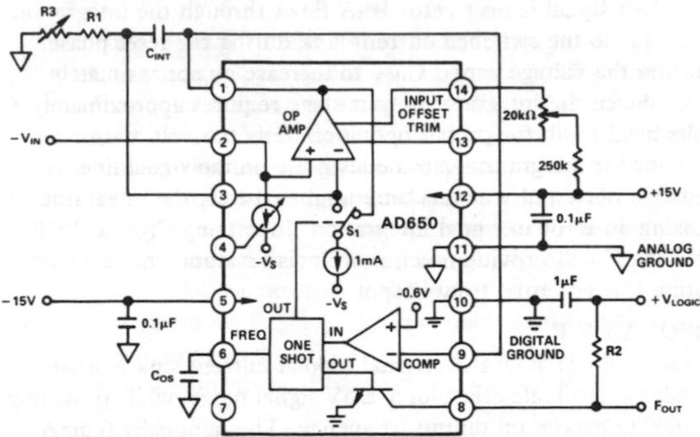

Figure 5. Connection Diagram for V/F Conversion, Negative Input Voltage

Circuit operation for negative input voltages is very similar to positive input unipolar conversion described in a previous section. For best operating results use component equations listed in that section.

F/V CONVERSION

The AD650 also makes a very linear frequency-to-voltage converter. Figure 6 shows the connection diagram for F/V conversion with TTL input logic levels. Each time the input signal crosses the comparator threshhold going negative, the one shot is activated and switches 1mA into the integrator input for a measured time period (determined by C_{OS}). As the frequency increases, the amount of charge injected into the integration capacitor increases proportionately. The voltage across the integration capacitor is stabilized when the leakage current through R1 and R3 equals the average current being switched into the integrator. The net result of these two effects is an average output voltage which is proportional to the input frequency. Optimum performance can be obtained by selecting components using the same guidelines and equations listed in the V/F conversion section.

The circuit of Figure 6 can be biased to accommodate almost any input signal waveform. With a TTL input, the 1000pF coupling capacitor and 2.2kΩ resistor creates a clean negative

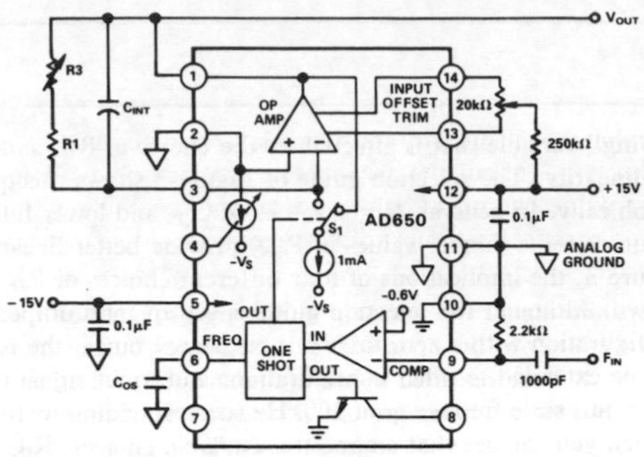

Figure 6. Connection Diagram for F/V Conversion

spike that triggers the one shot on negative going edges. For input signals with slower edges, a larger capacitor and/or resistor may be used as long as the comparator is never exposed to a voltage lower than −0.6V for longer than the one shot time period. If this happens, the one shot will trigger itself more than once per cycle, creating discontinuities in the F/V transfer function. An input pulse greater than 100ns but less than $0.3 \times t_{OS}$ is recommended (t_{OS} is defined by equation 1 in the circuit operation section, unipolar configuration).

HIGH FREQUENCY OPERATION

Proper RF techniques must be observed when operating the AD650 at or near its maximum frequency of 1MHz. Lead lengths must be kept as short as possible, especially on the one shot and integration capacitors, and at the integrator summing junction. In addition, at maximum output frequencies above 500kHz, a 3.6kΩ pulldown resistor from pin 1 to $-V_S$ is required (see Figure 7). The additional current drawn through the pulldown resistor reduces the op amp's output impedance and improves its transient response.

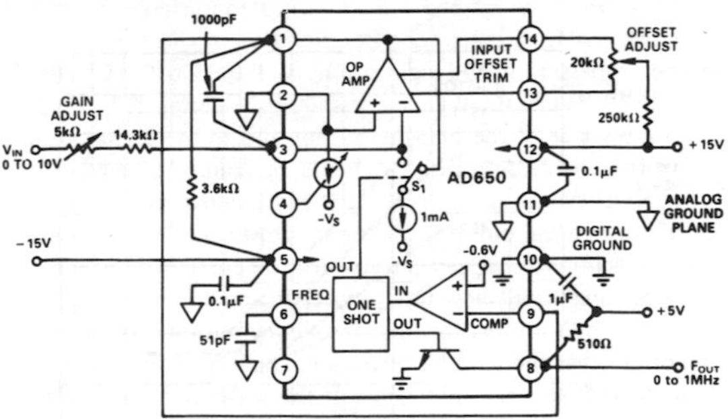

Figure 7. 1MHz V/F Connection Diagram

DECOUPLING AND GROUNDING

It is good engineering practice to use bypass capacitors on the supply-voltage pins and to insert small-valued resistors (10 to 100Ω) in the supply lines to provide a measure of decoupling between the various circuits in a system. Ceramic capacitors of 0.1μF to 1.0μF should be applied between the supply-voltage pins and analog signal ground for proper bypassing on the AD650.

In addition, a larger board level decoupling capacitor of 1μF to 10μF should be located relatively close to the AD650 on each power supply line. Such precautions are imperative in high resolution data acquisition applications where one expects to

Circuit Considerations – AD650

exploit the full linearity and dynamic range of the AD650. Although some types of circuits may operate satisfactorily with power supply decoupling at only one location on each circuit board, such practice is strongly discouraged in high accuracy analog design.

Separate digital and analog grounds are provided on the AD650. The emitter of the open collector frequency output transistor is the only node returned to the digital ground. All other signals are referred to analog ground. The purpose of the two separate grounds is to allow isolation between the high precision analog signals and the digital section of the circuitry. As much as several hundred millivolts of noise can be tolerated on the digital ground without affecting the accuracy of the VFC. Such ground noise is inevitable when switching the large currents associated with the frequency output signal.

At 1MHz full scale, it is necessary to use a pull-up resistor of about 500Ω in order to get the rise time fast enough to provide well defined output pulses. This means that from a 5 volt logic supply, for example, the open collector output will draw 10mA. This much current being switched will surely cause ringing on long ground runs due to the self inductance of the wires. For instance, #20 gauge wire has an inductance of about 20nH per inch; a current of 10mA being switched in 50ns at the end of 12 inches of 20 gauge wire will produce a voltage spike of 50mV. The separate digital ground of the AD650 will easily handle these types of switching transients.

A problem will remain from interference caused by radiation of electro-magnetic energy from these fast transients. Typically, a voltage spike is produced by inductive switching transients; these spikes can capacitively couple into other sections of the circuit. Another problem is ringing of ground lines and power supply lines due to the distributed capacitance and inductance of the wires. Such ringing can also couple interference into sensitive analog circuits. The best solution to these problems is proper bypassing of the logic supply at the AD650 package. A 1μF to 10μF tantalum capacitor should be connected directly to the supply side of the pull-up resistor and to the digital ground – pin 10. The pull-up resistor should be connected directly to the frequency output – pin 8. The lead lengths on the bypass capacitor and the pull up resistor should be as short as possible. The capacitor will supply (or absorb) the current transients, and large ac signals will flow in a physically small loop through the capacitor, pull up resistor, and frequency output transistor. It is important that the loop be physically small for two reasons: first, there is less self-inductance if the wires are short, and second, the loop will not radiate RFI efficiently.

The digital ground (pin 10) should be separately connected to the power supply ground. Note that the leads to the digital power supply are only carrying dc current and cannot radiate RFI. There may also be a dc ground drop due to the difference in currents returned on the analog and digital grounds. This will not cause any problem. In fact, the AD650 will tolerate as much as 0.25 volt dc potential difference between the analog and digital grounds. These features greatly ease power distribution and ground management in large systems. Proper technique for grounding requires separate digital and analog ground returns to the power supply. Also, the signal ground must be referred directly to analog ground (pin 11) at the package. All of the signal grounds should be tied directly to pin 11, especially the one-shot capacitor. More information on proper grounding and reduction of interference can be found in reference 1.

TEMPERATURE COEFFICIENTS

The drift specifications of the AD650 do not include temperature effects of any of the supporting resistors or capacitors. The drift of the input resistors R1 and R3 and the timing capacitor C_{OS} directly affect the overall temperature stability. In the application of Figure 2, a 10ppm/°C input resistor used with a 100ppm/°C capacitor may result in a maximum overall circuit gain drift of:

150ppm/°C (AD650A) + 100ppm/°C (C_{OS}) + 10ppm/°C (R_{IN}) = 260ppm/°C

In bipolar configuration, the drift of the 1.24kΩ resistor used to activate the internal bipolar offset current source will directly affect the value of this current. This resistor should be matched to the resistor connected to the op amp noninverting input (pin 2), see Figure 4. That is, the temperature coefficients of these two resistors should be equal. If this is the case, then the effects of the temperature coefficients of the resistors cancel each other, and the drift of the offset voltage developed at the op amp non-inverting input will be determined solely by the AD650. Under these conditions the TC of the bipolar offset voltage is typically – 200ppm/°C and is a maximum of – 300ppm/°C. The offset voltage always decreases in magnitude as temperature is increased.

Other circuit components do not directly influence the accuracy of the VFC over temperature changes as long as their actual values are not so different from the nominal value as to preclude operation. This includes the integration capacitor, C_{INT}. A change in the capacitance value of C_{INT} simply results in a different rate of voltage change across the capacitor. During the Integration Phase (refer to Figure 2), the rate of voltage change across C_{INT} has the opposite effect that it does during the Reset Phase. The result is that the conversion accuracy is unchanged by either drift or tolerance of C_{INT}. The net effect of a change in the integrator capacitor is simply to change the peak to peak amplitude of the sawtooth waveform at the output of the integrator.

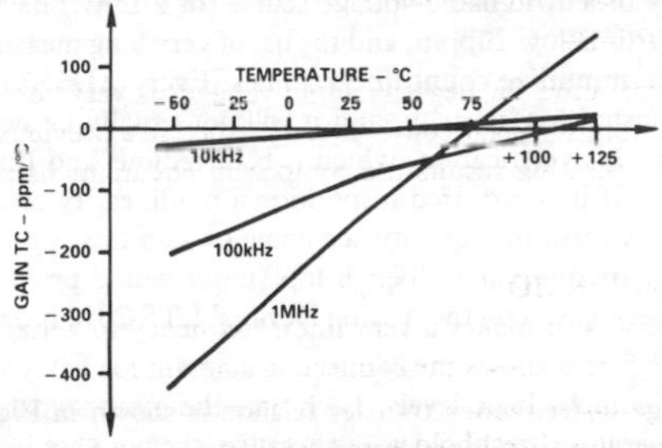

Figure 8. Gain TC vs. Temperature

The gain temperature coefficient of the AD650 is not a constant value. Rather the gain TC is a function of both the full scale frequency and the ambient temperature. At a low full scale frequency, the gain TC is determined primarily by the stability of the internal reference–a buried zener reference. This low speed gain TC can be quite good; at 10kHz full scale, the gain TC near 25°C is typically 0 ± 50ppm/°C. Although the gain TC changes with ambient temperature (tending to be more positive

[1]"Noise Reduction Techniques in Electronic Systems", by H. W. OTT, (John Wiley, 1976).

4

at higher temperatures), the drift remains within a ±75ppm/°C window over the entire military temperature range. At full scale frequencies higher than 10kHz dynamic errors become much more important than the static drift of the dc reference. At a full scale frequency of 100kHz and above, these timing errors dominate the gain TC. For example, at 100kHz full scale frequency (R_{IN}=40k and C_{OS}=330pF) the gain TC near room temperature is typically −80 ±50ppm/°C, but at an ambient temperature near +125°C, the gain TC tends to be more positive and is typically +15 ±50ppm/°C. This information is presented in a graphical form in Figure 8. The gain TC always tends to become more positive at higher temperatures. Therefore it is possible to adjust the gain TC of the AD650 by using a one-shot capacitor with an appropriate TC to cancel the drift of the circuit. For example, consider the 100kHz full scale frequency. An average drift of −100ppm/°C means that as temperature is increased, the circuit will produce a lower frequency in reponse to a given input voltage. This means that the one-shot capacitor must decrease in value as temperature increases in order to compensate the gain TC of the AD650; that is, the capacitor must have a TC of −100ppm/°C. Now consider the 1MHz full scale frequency. It is not possible to achieve very much improvement in performance unless the expected ambient temperature range is known. For example, in a constant low temperature application such as gathering data in an Arctic climate (approximately −20°C), a C_{OS} with a drift of −310ppm/°C is called for in order to compensate the gain drift of the AD650. However, if that circuit should see an ambient temperature of +75°C, the C_{OS} cap would change the gain TC from approximately 0ppm to +310ppm/°C.

The temperature effects of the components described above are the same when the AD650 is configured for negative or bipolar input voltages, and for F/V conversion as well.

NONLINEARITY SPECIFICATION

The linearity error of the AD650 is specified by the end point method. That is, the error is expressed in terms of the deviation from the ideal voltage to frequency transfer relation after calibrating the converter at full scale and "zero". The nonlinearity will vary with the choice of one-shot capacitor and input resistor (see Figure 3). Verification of the linearity specification requires the availability of a switchable voltage source (or a DAC) having a linearity error below 20ppm, and the use of very long measurement intervals to minimize count uncertainties. Every AD650 is automatically tested for linearity, and it will not usually be necessary to perform this verification, which is both tedious and time consuming. If it is required to perform a nonlinearity test either as part of an incoming quality screening or as a final product evaluation, an automated "bench-top" tester would prove useful. Such a system based on the Analog Devices' LTS-2010 is described in Reference 2.

The voltage-to-frequency transfer relation is shown in Figure 9 with the nonlinearity exaggerated for clarity. The first step in determining nonlinearity is to connect the end points of the operating range (typically at 10mV and 10V) with a straight line. This straight line is then the ideal relationship which is desired from the circuit. The second step is to find the difference between this line and the actual response of the circuit at a few points between the end points – typically ten intermediate points will suffice. The difference between the actual and the ideal response is a frequency error measured in hertz. Finally, these frequency errors are normalized to the full scale frequency and expressed either as parts per million of full-scale (ppm) or parts

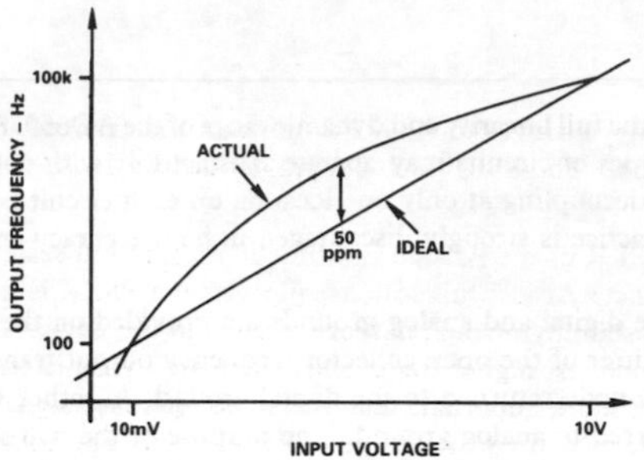

Figure 9a. Exaggerated Nonlinearity at 100kHz Full Scale

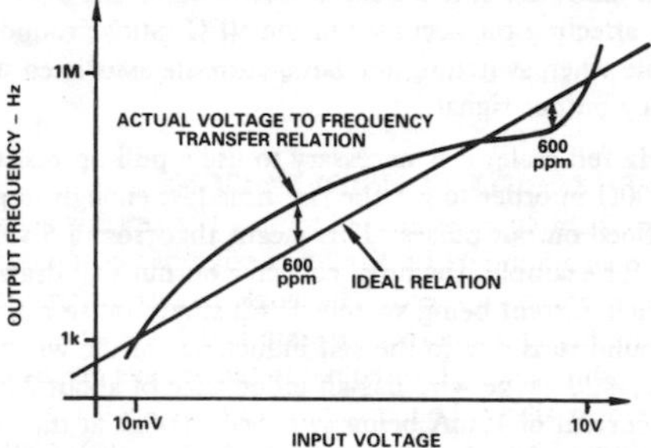

Figure 9b. Exaggerated Nonlinearity at 1MHz Full Scale

per hundred of full scale (%). For example, on a 100kHz full scale, if the maximum frequency error is 5Hz, the nonlinearity would be specified as 50ppm or 0.005%. Typically on the 100kHz scale, the nonlinearity is positive and the maximum value occurs at about midscale (Figure 9a). At higher full scale frequencies, (500kHz to 1MHz), the nonlinearity becomes "S" shaped and the maximum value may be either positive or negative. Typically, on the 1MHz scale (R_{IN}=16.9k, C_{OS}=51pF) the nonlinearity is positive below about 2/3 scale and is negative above this point. This is shown graphically in Figure 9b.

PSRR

The power supply rejection ratio is a specification of the change in gain of the AD650 as the power supply voltage is changed. The PSRR is expressed in units of parts-per-million change of the gain per percent change of the power supply – ppm/%. For example, consider a VFC with a 10 volt input applied and an output frequency of exactly 100kHz when the power supply potential is ±15 volts. Changing the power supply to ±12.5 volts is a 5 volt change out of 30 volts, or 16.7%. If the output

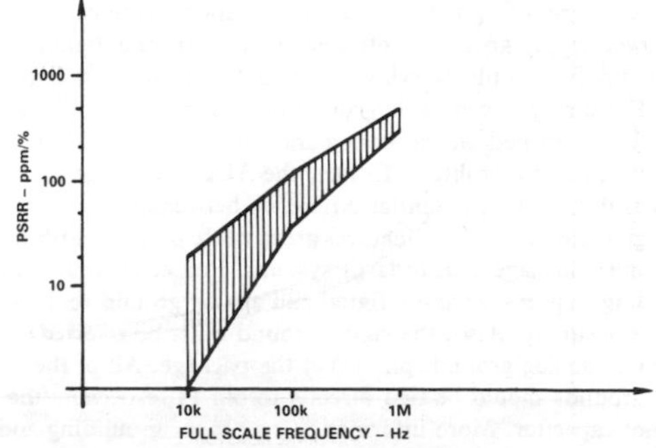

Figure 10. PSRR vs. Full Scale Frequency

[2]"V-F Converters Demand Accurate Linearity Testing", by L. DeVito, (Electronic Design, March 4, 1982)

Applications — AD650

frequency changes to 99.9kHz, the gain has changed 0.1% or 1000ppm. The PSRR is 1000ppm divided by 16.7% which equals 60ppm/%.

The PSRR of the AD650 is a function of the full scale operating frequency. At low full scale frequencies the PSRR is determined by the stability of the reference circuits in the device and can be very good. At higher frequencies there are dynamic errors which become more important than the static reference signals, and consequently the PSRR is not quite as good. The values of PSRR are typically 0 ±20ppm/% at 10kHz full scale frequency (R_{IN}=40k, C_{OS}=3300pF). At 100kHz (R_{IN}=40k, C_{OS}=330pF) the PSRR is typically +80 ±40ppm/%, and at 1MHz (R_{IN}=16.9kΩ, C_{OS}=51pF) the PSRR is +350 ±50ppm/%. This information is summarized graphically in Figure 10.

OTHER CIRCUIT CONSIDERATIONS

The input amplifier connected to pins 1, 2 and 3 is not a standard operational amplifier. Rather, the design has been optimized for simplicity and high speed. The single largest difference between this amplifier and a normal op amp is the lack of an integrator (or level shift) stage. Consequently the voltage on the output (pin 1) must always be more positive than 2 volts below the inputs (pins 2 and 3). For example, in the F to V conversion mode, see Figure 6, the noninverting input of the op amp (pin 2) is grounded, which means that the output (pin 1) will not be able to go below −2 volts. Normal operation of the circuit as shown in the figure will never call for a negative voltage at the output but one may imagine an arrangement calling for a bipolar output voltage (say ±10 volts) by connecting an extra resistor from pin 3 to a positive voltage. This will not work.

Care should be taken under conditions where a high positive input voltage exists at or before power up. These situations can cause a latch up at the integrator output (pin 1). This is a non-destructive latch and, as such, normal operation can be restored by cycling the power supply. Latch up can be prevented by connecting two diodes (e.g., 1N914 or 1N4148) as shown in Figure 4 thereby preventing pin 1 from swinging below pin 2.

A second major difference is that the output will only sink 1mA to the negative supply. There is no pulldown stage at the output other than the 1mA current source used for the V to F conversion. The op amp will source a great deal of current from the positive supply, and it is internally protected by current limiting. The output of the op amp may be driven to within 3 volts of the positive supply when it is not sourcing external current. When sourcing 10mA the output voltage may be driven to within 6 volts of the positive supply.

A third difference between this op amp and a normal device is that the inverting input, pin 3, is bias current compensated and the noninverting input is not bias current compensated. The bias current at the inverting input is nominally zero, but may be as much as 20nA in either direction. The noninverting input typically has a bias current of 40nA that always flows into the node (an npn input transistor). Therefore, it is not possible to match input voltage drops due to bias currents by matching input resistors.

The op amp has provisions for trimming the input offset voltage. A potentiometer of 20kΩ is connected to pins 13 and 14 and the wiper is connected to the positive supply through a 250kΩ resistor. A potential of about 0.6 volt is established across the 250kΩ resistor, and the 3μA current is injected into the null pins. It is also possible to null the op amp offset voltage by using only one of the null pins and use a bipolar current either into or out of the null pin. The amount of current required will be very small – typically less than 3μA. This technique is shown in the applications section of this data sheet: the auto-zero circuit uses this technique.

The bipolar offset current is activated by connecting a 1.24kΩ resistor between pin 4 and the negative supply. The resultant current delivered to the op amp noninverting input is nominally 0.5mA and has a tolerance of ±10%. This current is then used to provide an offset voltage when pin 2 is tied to ground through a resistor. The 0.5mA which appears at pin 2 is also flowing through the 1.24kΩ resistor and this current may be measured by observing the voltage across the 1.24kΩ resistor. An external resistor is used to activate the bipolar offset current source to provide the lowest tolerance and temperature drift of the resultant offset voltage. It is possible to use other values of resistance between pin 4 and $-V_S$ to obtain a bipolar offset current different than 0.5mA. Figure 11 is a graph of the relationship between the bipolar offset current and the value of the resistor used to activate the source.

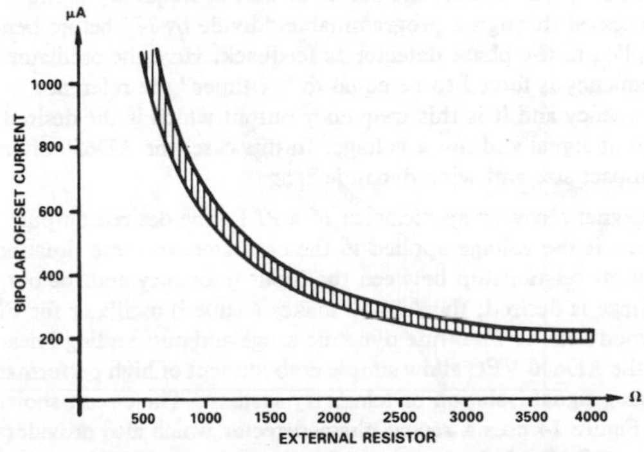

Figure 11. Bipolar Offset Current vs. External Resistor

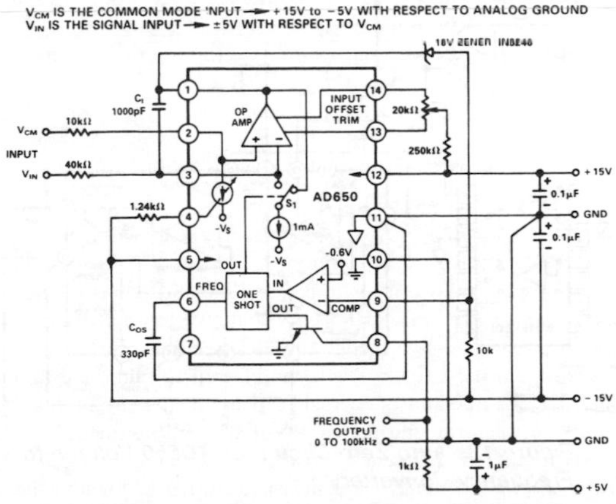

Figure 12. AD650 Differential Input

APPLICATIONS

DIFFERENTIAL VOLTAGE-TO-FREQUENCY CONVERSION

The circuit of Figure 12 accepts a true floating differential input signal. The common mode input, V_{CM}, may be in the range +15 to −5 volts with respect to analog ground. The signal input, V_{IN}, may be ±5 volts with respect to the common mode input. Both inputs are low impedance: the source which drives the common mode input must supply the 0.5mA drawn by the bipolar offset current source and the source which drives the signal input must supply the integration current.

If less common mode voltage range is required, a lower voltage zener may be used. For example, if a 5 volt zener is used, the V_{CM} input may be in the range +10 to −5 volt. If the zener is not used at all, the common mode range will be ±5 volts with repect to analog ground. If no zener is used, the 10k pulldown resistor is not needed and the integrator output (pin 1) is connected directly to the comparator input (pin 9).

AUTO ZERO CIRCUIT

In order to exploit the full dynamic range of the AD650 VFC, very small input voltages will need to be converted. For example, a six decade dynamic range based on a full scale of 10 volts will require accurate measurement of signals down to 10µV. In these situations a well-controlled input offset voltage is imperative. A constant offset voltage will not affect dynamic range but simply shift all of the frequency readings by a few hertz. However, if the offset should change, then it will not be possible to distinguish between a small change in a small input voltage and a drift of the offset voltage. Hence, the useable dynamic range is less. The circuit shown in Figure 13 provides automatic adjustment of the op amp offset voltage. The circuit uses an AD582 sample and hold amplifier to control the offset and the input voltage to the VFC is switched between ground and the signal to be measured via an AD7512DI analog switch. The offset of the AD650 is adjusted by injecting a current into or drawing a current out of pin 13. Note that only one of the offset null pins is used. During the "VFC Norm" mode, the SHA is in the hold mode and the hold capacitor is very large, 0.1µF, to hold the AD650 offset constant for a long period of time.

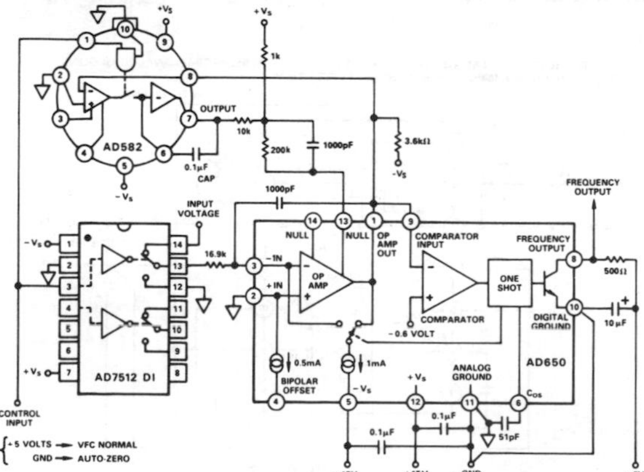

Figure 13. Auto-Zero Circuit for AD650 Voltage-to-Frequency Converter

When the circuit is in the "Auto Zero" mode the SHA is in sample mode and behaves like an op amp. The circuit is a variation of the classical two amplifier servo loop, where the output of the Device Under Test (DUT) – here the DUT is the AD650 op amp – is forced to ground by the feedback action of the control amplifier – the SHA. Since the input of the VFC circuit is connected to ground during the auto zero mode, the input current which can flow is determined by the offset voltage of the AD650 op amp. Since the output of the integrator stage is forced to ground it is known that the voltage is not changing (it is equal to ground potential). Hence if the output of the integrator is constant, its input current must be zero, so the offset voltage has been forced to be zero. Note that the output of the DUT could have been forced to any convenient voltage other than ground. All that is required is that the output voltage be known to be constant. Note also that the effect of the bias current at the inverting input of the AD650 op amp is also nulled in this circuit. The 1000pF capacitor shunting the 200kΩ resistor is compensation for the two amplifier servo loop. Two integrators in a loop requires a single zero for compensation. Note that the 3.6kΩ resistor from pin 1 of the AD650 to the negative supply is *not* part of the auto-zero circuit, but rather it is required for VFC operation at 1MHz.

PHASE LOCKED LOOP F/V CONVERSION

Although the F/V conversion technique shown in Figure 6 is quite accurate and uses only a few extra components, it is very limited in terms of signal frequency response and carrier feed-through. If the carrier (or input) frequency changes instantaneously, the output cannot change very rapidly due to the integrator time constant formed by C_{INT} and R_{IN}. While it is possible to decrease the integrator time constant to provide faster settling of the F to V output voltage, the carrier feedthrough will then be larger. For signal frequency response in excess of 2kHz, a phase locked F/V conversion technique such as the one shown in Figure 14 is recommended.

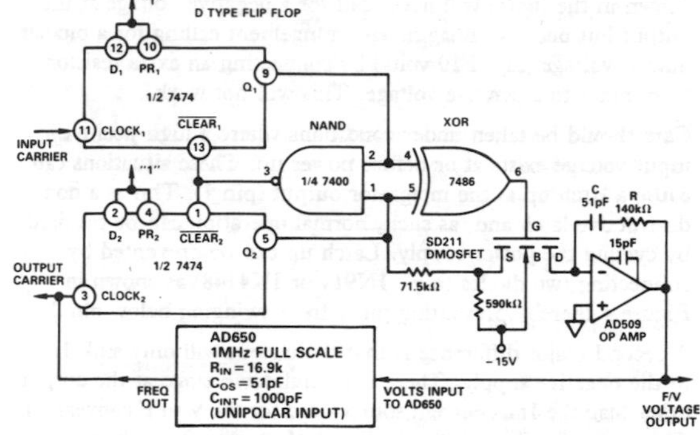

Figure 14. Phase Locked Loop F/V Conversion

In a phase locked loop circuit, the oscillator is driven to a frequency and phase equal to an input reference signal. In applications such as a synthesizer, the oscillator output frequency is first processed through a programmable "divide by N" before being applied to the phase detector as feedback. Here the oscillator frequency is forced to be equal to "N times" the reference frequency and it is this frequency output which is the desired output signal and not a voltage. In this case, the AD650 offers compact size and wide dynamic range.

In signal recovery applications of a PLL, the desired output signal is the voltage applied to the oscillator. In these situations a linear relationship between the input frequency and the output voltage is desired; the AD650 makes a superb oscillator for FM demodulation. The wide dynamic range and outstanding linearity of the AD650 VFC allow simple embodiment of high performance analog signal isolation or telemetry systems. The circuit shown in Figure 14 uses a digital phase detector which also provides proper feedback in the event of unequal frequencies. Such phase-frequency detectors (PFD's) are available in integrated form.

Applications – AD650

For a full discussion of phase lock loop circuits see Reference 3. An analysis of this circuit must begin at the 7474 dual D flip flop. When the input carrier matches the output carrier in both phase and frequency, the Q outputs of the flip flops will rise at exactly the same time. With two zero's, then two one's on the inputs of the exclusive or (XOR) gate, the output will remain low keeping the DMOS FET switched off. Also, the NAND gate will go low resetting the flip-flops to zero. Throughout the entire cycle just described, the DMOS integrator gate remained off, allowing the voltage at the integrator output to remain unchanged from the previous cycle. However, if the input carrier leads the output carrier by a few degrees, the XOR gate will be turned on for the small time span that the two signals are mismatched. Since Q_2 will be low during the mismatch time, a negative current will be fed into the integrator, causing its output voltage to rise. This in turn will increase the frequency of the AD650 slightly, driving the system towards synchronization. In a similar manner, if the input carrier lags the output carrier, the integrator will be forced down slightly to synchronize the two signals.

Using a mathematical approach, the $\pm 25\mu A$ pulses from the phase detector are incorporated into the phase detector gain, K_d.

$$K_d = \frac{25\mu A}{2\pi} = 4 \times 10^{-6} \quad \textit{amperes/radian} \tag{9}$$

Also, the V/F converter is configured to produce 1MHz in response to a 10 volt input, so its gain K_o is:

$$K_o = \frac{2\pi \times 1 \times 10^6 Hz}{10V} = 6.3 \times 10^5 \ \frac{radians}{volt \cdot sec} \tag{10}$$

The dynamics of the phase relationship between the input and output signals can be characterized as a second order system with natural frequency ω_n:

$$\omega_n = \sqrt{\frac{K_o K_d}{C}} \tag{11}$$

and damping factor

$$\zeta = \frac{R\sqrt{C\,K_o K_d}}{2} \tag{12}$$

For the values shown in Figure 14, these relations simplify to a natural frequency of 35kHz with a damping factor of 0.8. For those desiring a simple approach to determining component values for other PLL frequencies and VFC full scale voltage, the following cookbook steps can be used:

1. Determine K_o (in units of radians per volt second) from the maximum input carrier frequency F_{max} (in hertz) and the maximum output voltage V_{max}.

$$K_o = \frac{2\pi \times F_{max}}{V_{max}} \tag{13}$$

2. Calculate a value for C based upon the desired loop bandwidth, f_n. Note that this is the desired frequency range of the output signal. The loop bandwidth (f_n) is *not* the maximum carrier frequency (f_{max}): the signal may be very narrow even though it is transmitted over a 1MHz carrier.

$$C = \frac{K_o}{{f_n}^2} \cdot 1 \times 10^{-7} \frac{V \cdot F}{Rad \cdot sec} \qquad \begin{array}{l} \text{C units FARADS} \\ f_n \text{ units HERTZ} \\ K_o \text{ units RAD/VOLT·SEC} \end{array} \tag{14}$$

3. Calculate R to yield a damping factor of approximately 0.8 using this equation:

$$R = \frac{f_n}{K_o} \cdot 2.5 \times 10^6 \frac{Rad \cdot \Omega}{V} \qquad \begin{array}{l} \text{R units OHMS} \\ f_n \text{ units HERTZ} \\ K_o \text{ units RAD/VOLT·SEC} \end{array} \tag{15}$$

If in actual operation the PLL overshoots or hunts excessively before reaching a final value, the damping factor may be raised by increasing the value of R. Conversely, if the PLL is overdamped, a smaller value of R should be used.

PLL PERFORMANCE

The performance of the PLL circuit is demonstrated by the system shown in Figure 15; an analog signal is converted into a frequency, and then this frequency is converted back into an analog voltage by the PLL.

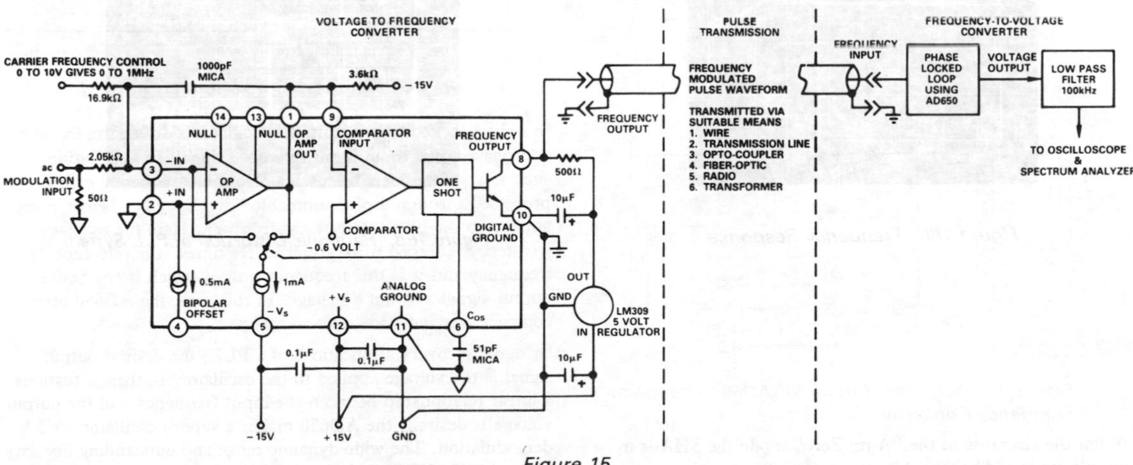

Figure 15.

[3]"Phase lock Techniques", by F.M. Gardner, 2nd Edition, 1979, John Wiley and Sons.

The source of the frequency input signal used to drive the PLL is an AD650 with two separate inputs: one for dc to set the carrier frequency, and one for ac to establish a modulation. Note how the summing junction input to the AD650 allows such flexibility. The output frequency is then relayed to the PLL via a jumper cable. The signal at this point is a 5 volt digital pulse train and as such may be transmitted in any fashion suitable to the application at hand. For example, galvanic isolation is achieved with a simple transformer or opto-isolator; extremely high voltage isolation or transmission through severe RF environments can be accomplished with a fiber-optic link; telemetry can be accomplished with a radio link. The actual method of conveying the pulses is not crucial to the system performance. The PLL is the circuit shown in Figure 14, and the filter shown on the output signal is simply to attenuate carrier feedthrough to allow easy interpretation of the signal with an oscilloscope and spectrum analyzer.

The step response of the system is shown in Figure 16a. The signal output is swinging between 5 volts and 10 volts, for an input step of 500kHz to 1MHz. Note that the AD650 is actually overshooting to 1.1MHz and the response remains well controlled. Note the slight irregularity during the transition: this is caused by cycleslipping during the slew where feedback is lost temporarily and the PLL actually loses phase lock. The frequency response of the system when driven with sinewave excitation is shown in Figure 16b. Here the output level is set to 2 volts peak to peak, and the carrier is 800kHz. Note that the −3dB bandwidth is about 70kHz, which is consistent with a damping factor of 0.8 and a natural frequency of 35kHz[4]. When an unmodulated carrier is applied to the PLL, the noise that appears at the output determines the dynamic range of the system. The spectrum of the noise at the output of the PLL is shown in Figure 16c. By comparing this with Figure 16b, the dynamic range of the system is seen to be 80dB. The harmonic distortion of the system is shown in Figure 16d. The output is a 2V p-p sinewave at 5kHz, and the amplitude of the first harmonic is seen to be 48dB below the fundamental. The harmonic distortion can be improved to the level of 60dB by reducing the amplitude of the modulation, but this is at the expense of dynamic range since the intensity of the noise floor remains constant.

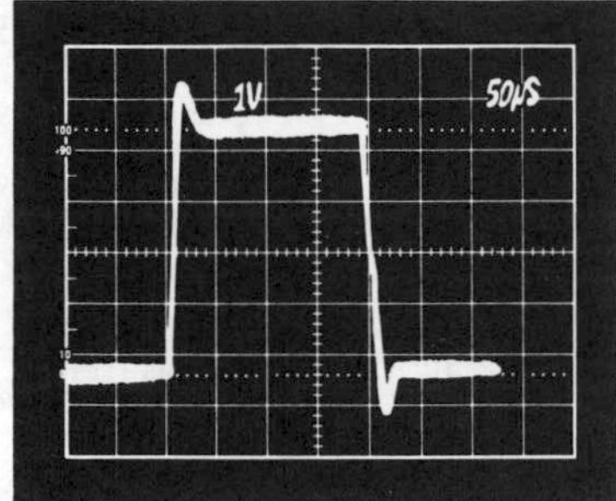

Figure 16a. Step Response

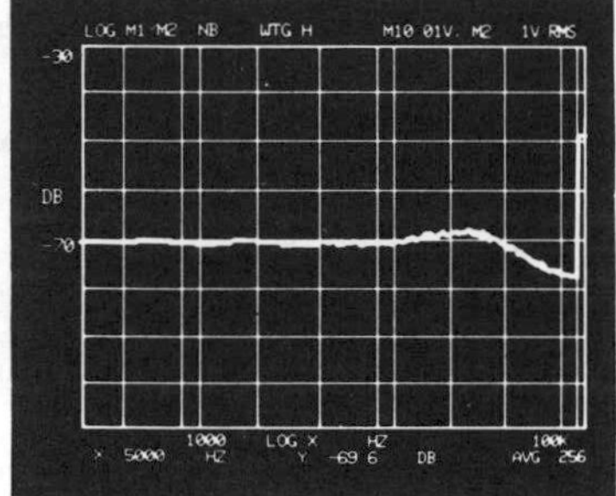

Figure 16c. Noise Output from PLL

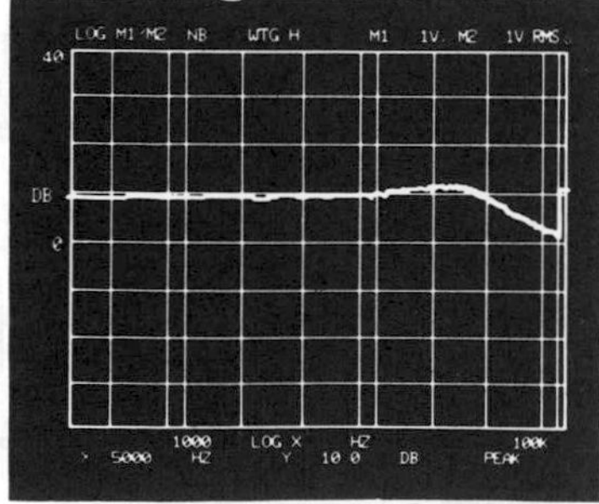

Figure 16b. Frequency Response

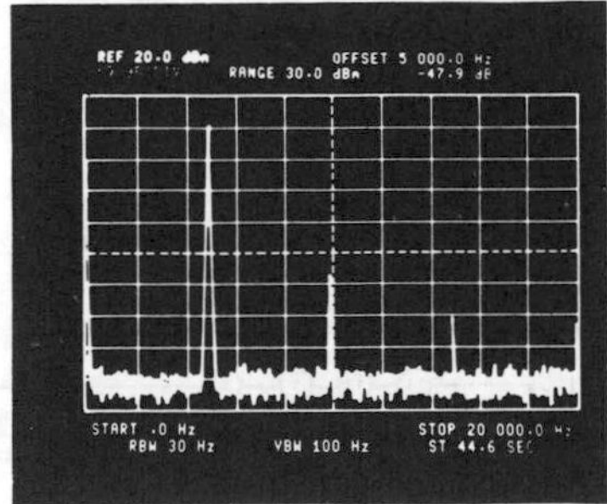

Figure 16d. Harmonic Distortion of PLL System

[4]See page 13 of reference 3.

VFC100

AVAILABLE IN DIE FORM

Synchronized VOLTAGE-TO-FREQUENCY CONVERTER

FEATURES

- FULL-SCALE FREQUENCY SET BY SYSTEM CLOCK, NO CRITICAL EXTERNAL COMPONENTS REQUIRED
- PRECISION 10V FULL-SCALE INPUT, 0.5% MAX GAIN ERROR
- ACCURATE 5V REFERENCE VOLTAGE
- EXCELLENT LINEARITY, 0.02% MAX AT 100kHz FS
  0.1% MAX AT 1MHz FS
- VERY-LOW GAIN DRIFT, 50ppm/°C

APPLICATIONS

- A/D CONVERSION
- PROCESS CONTROL
- DATA ACQUISITION
- VOLTAGE ISOLATION

DESCRIPTION

The VFC100 voltage-to-frequency converter is an important advance in VFCs. The well-proven charge balance technique is used, however, the critical reset integration period is derived from an external clock frequency. The external clock accurately sets an output full-scale frequency, eliminating error and drift from the external timing components required for other VFCs. A precision input resistor is provided which accurately sets a 10V full-scale input voltage. In many applications the required accuracy can be achieved without external adjustment.

The open collector active-low output provides fast fall time on the important leading edge of output pulses, and interfaces easily with TTL and CMOS circuitry. An output one-shot circuit is particularly useful to provide optimum output pulse widths for optical couplers and transformers to achieve voltage isolation. An accurate 5V reference is also provided which is useful for applications such as offsetting for bipolar input voltages, exciting bridges and sensors, and autocalibration schemes.

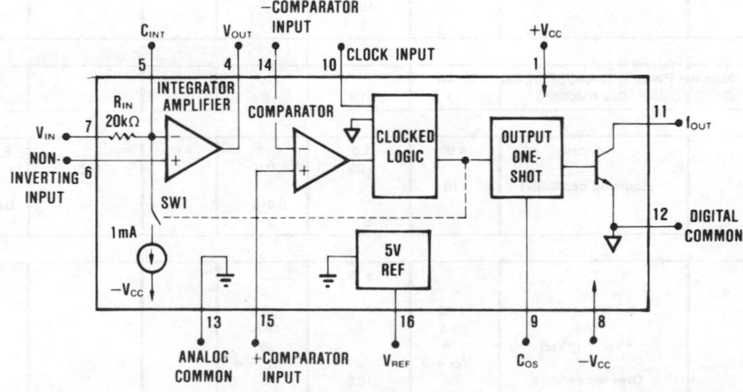

International Airport Industrial Park · P.O. Box 11400 · Tucson, Arizona 85734 · Tel. (602) 746-1111 · Twx: 910-952-1111 · Cable: BBRCORP · Telex: 66-6491

PDS-547D

SPECIFICATIONS

ELECTRICAL

At T_A = +25°C and ±15VDC supplies unless otherwise noted.

PARAMETER	CONDITIONS	VFC100AG/SG MIN	VFC100AG/SG TYP	VFC100AG/SG MAX	VFC100BG MIN	VFC100BG TYP	VFC100BG MAX	UNITS
TRANSFER FUNCTION								
Voltage-to-Frequency Mode	$f_{OUT} = f_{CLOCK} \times (V_{IN}/20V)$							
Gain Error[1]	FSR = 100kHz		±0.5	±1		±0.2	±0.5	% of FSR
Linearity Error	FSR = 100kHz, over temp.		±0.01	±0.025		•	±0.02	% of FSR
	FSR = 500kHz, C_{OS} = 60pF		±0.015			•	±0.05	% of FSR
	FSR = 1MHz, C_{OS} = 60pF		±0.025			•	±0.1	% of FSR
Gain Drift[2]	FSR = 100kHz		±70	±100		±30	±50	ppm of FSR/°C
Referred to Internal V_{REF}			10	±25		10	±15	ppm of FSR/°C
Offset Referred to Input			±1	±3		±1	±2	mV
Offset Drift			±12	±100		±6.5	±25	μV/°C
Power Supply Rejection	Full supply range			0.01			•	%/V
Response Time	to Step Input Change	One period of new output frequency plus one clock period						
Current-to-Frequency Mode	$f_{OUT} = f_{CLOCK} \times (I_{IN}/1mA)$							
Gain Error			±0.5	±1		±0.2	±0.5	% of FSR
Gain Drift[2]			±120	±200		±80	±140	ppm of FSR/°C
Frequency-to-Voltage Mode[3]	$V_{OUT} = 20V \times (f_{IN}/f_{CLOCK})$							
Gain Accuracy[1]	FSR = 100kHz		±0.5	±1		±0.2	±0.5	%
Linearity	FSR = 100kHz		±0.01	±0.025		•	±0.02	%
Input Resistor (R_{IN})								
Resistance		19.8	20	20.2	•	•	•	kΩ
Temperature Coefficient (T_C)[2]			±50	±100		•	•	ppm/°C
INTEGRATOR OP AMP								
V_{OS}[1]			±150	±1000		•	•	μV
V_{OS} Drift			±5			•		μV/°C
I_B			±50	±100		±25	±50	nA
I_{OS}			100	200		50	100	nA
A_{OL}	Z_{LOAD} = 5KΩ/10000pF	100	120		•	•		dB
CMRR		80	105		•	•		dBV
CM Range		−7.5		+0.1	•		•	V
V_{OUT} Range	Z_{LOAD} = 5kΩ/10000pF	−0.2		+12		•	•	V
Bandwidth			14			•		MHz
COMPARATOR INPUTS								
Input Current (operating)	$-11V < V_{COMPARATOR} < +V_{CC} - 2V$			5			•	μA
CLOCK INPUT (referenced to digital common)								
Frequency (maximum operating)			4.0			•		MHz
Threshold Voltage			1.4			•		V
	Over temperature	0.8		2.0	•		•	V
Voltage Range (operating)		$-V_{CC}$ + 2V		$+V_{CC}$	•		•	V
Input Current	$-V_{CC} < V_{CLOCK} < +V_{CC}$		0.5	5		•	•	μA
Rise Time				2			•	μsec
OPEN COLLECTOR OUTPUT (referenced to digital common)								
V_{OL}	I_{OUT} = 10mA			0.4			•	V
I_{OL}				15			•	mA
I_{OH} (off leakage)	V_{OH} = 30V		.01	10		•	•	μA
Delay Time, positive clock edge to output pulse			300			•		nsec
Fall Time			100			•		nsec
Output Capacitance			5			•		pF
OUTPUT ONE-SHOT								
Pulse Width Out	Nominal PW_{OUT} = (5nsec/pF) × C_{OS} − 90nsec; C_{OS} = 300pF	1	1.4	2	•	•	•	μsec
REFERENCE VOLTAGE								
Accuracy	No load	4.90	5.0	5.10	4.95	•	5.05	V
Drift[2]			±60	±150		±40	±100	ppm/°C
Current Output	(Sourcing capability)	10				•		mA
Power Supply Rejection				0.015			0.015	%/V
Output Impedance			0.5	2		•	•	Ω
POWER SUPPLY								
Rated Voltage			±15			•		V
Operating Voltage Range (see Figure 9)	$+V_{CC}$	+7.5		+28.5	•		•	V
	$-V_{CC}$	−7.5		−28.5	•		•	V
Total Supply	$+V_{CC} - (-V_{CC})$	15		36	•		•	V
Digital Common		$-V_{CC}$ + 2		$+V_{CC}$ − 4	•		•	V
Quiescent Current: $+I_{CC}$	Over temperature		10.6	15		•	•	mA
$-I_{CC}$			9.6	15		•	•	mA

ELECTRICAL (CONT)

At T_A = +25°C and ±15VDC supplies unless otherwise noted.

PARAMETER	CONDITIONS	VFC100AG/SG MIN	VFC100AG/SG TYP	VFC100AG/SG MAX	VFC100BG MIN	VFC100BG TYP	VFC100BG MAX	UNITS
TEMPERATURE RANGE								
Specification	AG/BG	−25		+85	*		*	°C
	SG	−55		+125				°C
Storage	AG/BG/SG	−65		+150	*		*	°C
θ Junction—ambient			150			*		°C/W
θ Junction—case			100			*		°C/W

*Specification same as AG grade.

NOTES: (1) Offset and gain error can be trimmed to zero. See text. (2) Specified by the box method: (Max. − Min.) ÷ (Avg. × ΔT). (3) Refer to detailed timing diagram in Figure 16 for frequency input signal timing requirements.

MECHANICAL

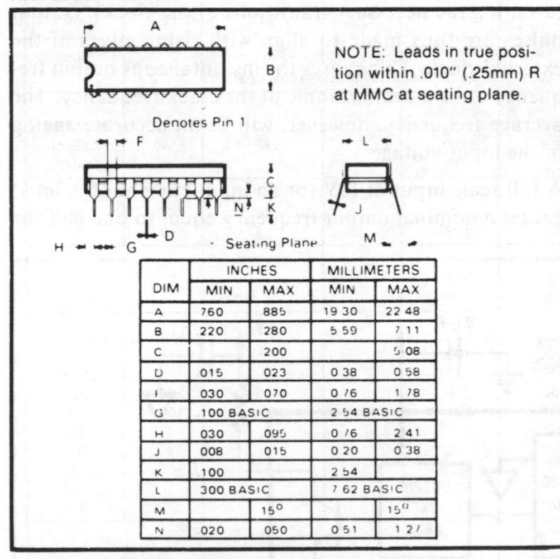

DIM	INCHES MIN	INCHES MAX	MILLIMETERS MIN	MILLIMETERS MAX
A	760	885	19 30	22 48
B	220	280	5 59	7 11
C		200		5 08
D	.015	023	0 38	0 58
F	030	070	0 76	1 78
G	.100 BASIC		2 54 BASIC	
H	030	095	0 76	2 41
J	008	015	0 20	0 38
K	.100		2 54	
L	300 BASIC		7 62 BASIC	
M		15°		15°
N	.020	050	0 51	1 27

ABSOLUTE MAXIMUM RATINGS

Power Supply Voltage ($+V_{CC}$ to $-V_{CC}$)	36V
$+V_{CC}$ to Analog Common	28V
$-V_{CC}$ to Analog Common	28V
Integrator Out Short-Circuit-to-Ground	Indefinite
Integrator Differential Input	±10V
Integrator Common-Mode Input	$-V_{CC}$ +5V to +2V
V_{IN} (pin 7)	$\pm V_{CC}$
Clock Input	$\pm V_{CC}$
V_{REF} Out Short-Circuit-to-Ground	Indefinite
Pin 9 (C_{OS})	0 to $+V_{CC}$
f_{OUT} (referred to digital common)	−0.5V to 36V
Digital Common	$\pm V_{CC}$
Storage Temperature Range	−65°C to +150°C
Lead Temperature (soldering 10sec)	300°C

ORDERING INFORMATION

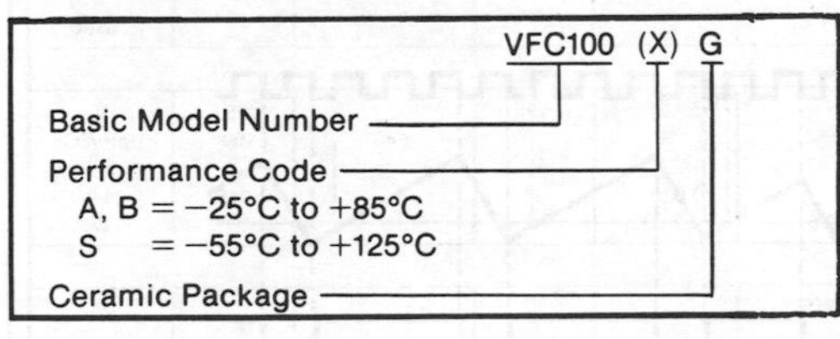

VFC100 (X) G
Basic Model Number
Performance Code
A, B = −25°C to +85°C
S = −55°C to +125°C
Ceramic Package

PIN CONFIGURATION

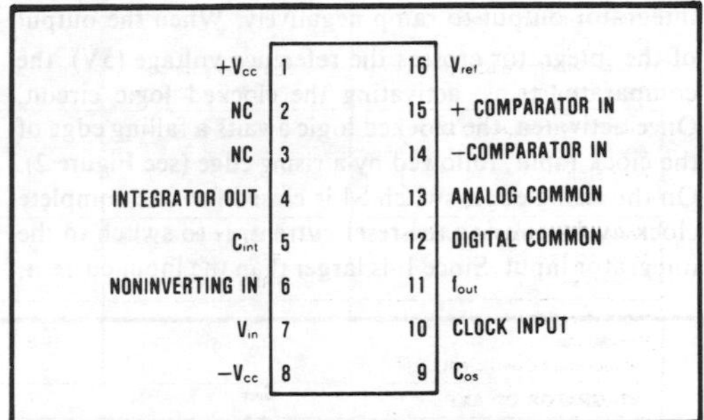

TYPICAL PERFORMANCE CURVES

At +25°C, $\pm V_{CC}$ = 15VDC, and in circuit of Figure 1 unless otherwise specified.

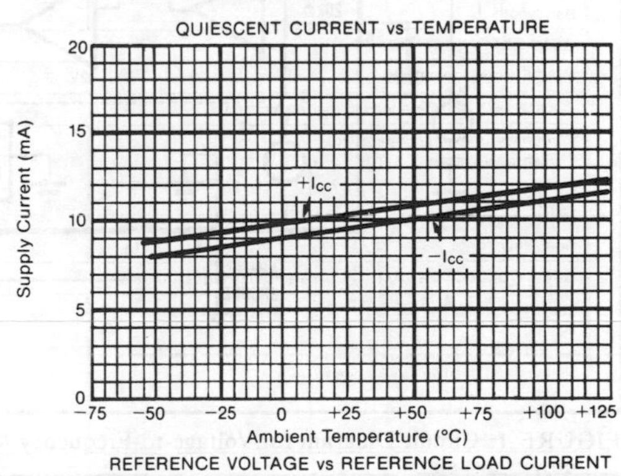

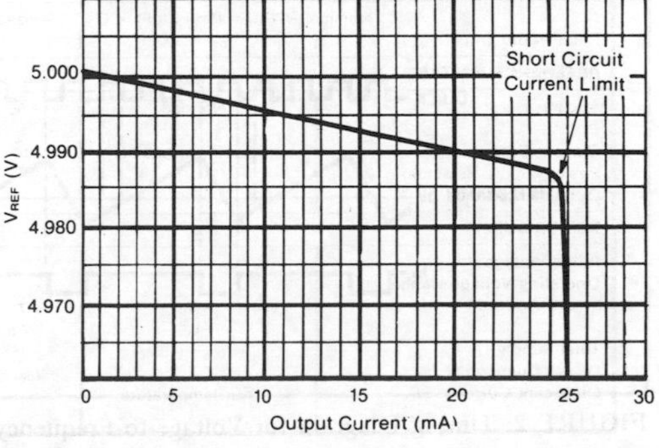

THEORY OF OPERATION

The VFC100 monolithic voltage-to-frequency converter provides a digital pulse train output with an average frequency proportional to the analog input voltage. The output is an active low pulse of constant duration, with a repetition rate determined by the input voltage. Falling edges of the output pulses are synchronized with rising edges of the clock input.

Operation is similar to a conventional charge balance VFC. An input operational amplifier (Figure 1) is configured as an integrator so that a positive input voltage causes an input current to flow in R_{IN}. This forces the integrator output to ramp negatively. When the output of the integrator crosses the reference voltage (5V), the comparator trips, activating the clocked logic circuit. Once activated, the clocked logic awaits a falling edge of the clock input, followed by a rising edge (see Figure 2). On the rising edge, switch S1 is closed for one complete clock cycle, causing the reset current, I_1 to switch to the integrator input. Since I_1 is larger than the input current, I_{IN}, the output of the integrator ramps positively during the one clock cycle reset period. The clocked logic circuitry also generates a VFC output pulse during the reset period.

Unlike conventional VFC circuits, the VFC100 accurately derives its reset period from an external clock frequency. This eliminates the critical timing capacitor required by other VFC circuits. One period (from rising edge to rising edge) of the clock input determines the integrator reset period.

When the negative-going integration of the input signal crosses the comparator threshold, integration of the input signal will continue until the reset period can start (awaiting the necessary transitions of the clock). Output pulses are thus made to align with rising edges of the external clock. This causes the instantaneous output frequency to be a subharmonic of the clock frequency. The average frequency, however, will be an accurate analog of the input voltage.

A full scale input of 10V (or an input current of 0.5mA) causes a nominal output frequency equal to one half the

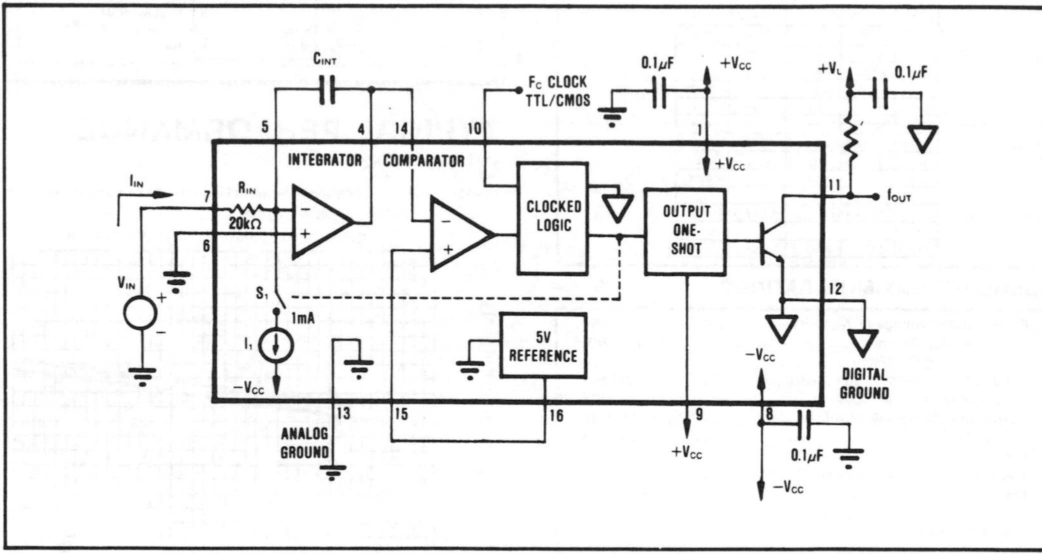

FIGURE 1. Circuit Diagram for Voltage-to-Frequency Mode.

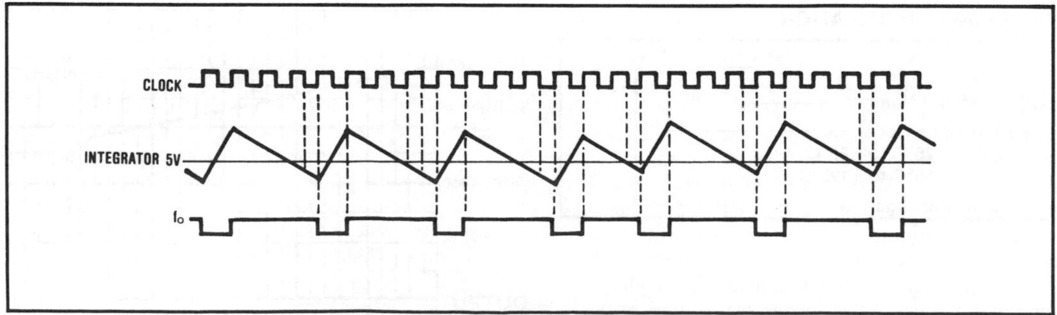

FIGURE 2. Timing Diagram for Voltage-to-Frequency Mode.

VOLTAGE-TO-FREQUENCY CONVERTERS

10

VFC100

clock frequency. The transfer function is

$$f_{OUT} = (V_{IN}/20V)\, f_{CLOCK}$$

Figure 3 shows the transfer function graphically. Note that inputs above 10V (or 0.5mA) do not cause an increase in the output frequency. This is an easily detectable indication of an overrange input. In the overrange condition, the integrator amplifier will ramp to its negative output swing limit. When the input signal returns to within the linear range, the integrator amplifier will recover and begin ramping upward during the reset period.

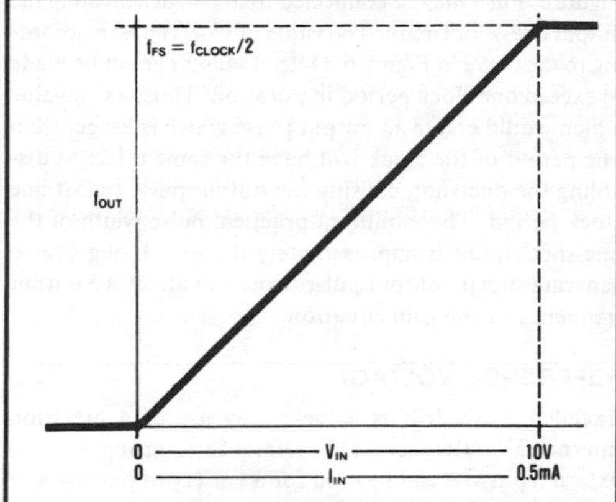

FIGURE 3. Transfer Function for Voltage-to-Frequency Mode.

INSTALLATION AND OPERATING INSTRUCTIONS

The integrator capacitor C_{INT} (see Figure 1) affects the magnitude of the integrator voltage waveform. Its absolute accuracy is not critical since it does not affect the transfer function. This allows a wide range of capacitance to produce excellent results. Figure 4 facilitates choosing an appropriate standard value to assure that the integrator waveform voltage is within acceptable limits. Good dielectric absorption properties are required to achieve best linearity. Mylar®, polycarbonate, mica, polystyrene, Teflon® and glass types are appropriate choices. The choice in a given application will depend on the particular value and size considerations. Ceramic capacitors vary considerably from type to type and some produce significant nonlinearities. Polarized capacitors should not be used.

Deviation from the nominal recommended +1V to −0.75V integrator voltage (as controlled by the integrator capacitor value) is permissible and will have a negligible effect on VFC operation. Certain situations may make deviations from the suggested integrator swing highly desirable. Smaller integrator voltages, for instance, allow more "headroom" for averaging noisy input signals. The VFC is a fully integrating input converter, able to reject large levels of interfering noise. This ability is limited only by the output voltage swing range of the integrator amplifier. By setting a small integrator voltage swing using a large C_{INT} value, larger levels of noise can be integrated without output saturation and loss of accuracy. For instance, with a 50kHz full-scale output and $C_{INT} = 0.1\mu F$, the circuit in Figure 1 can accurately average an input through the full 0 to 10V input range with 1V p-p superimposed 60Hz noise.

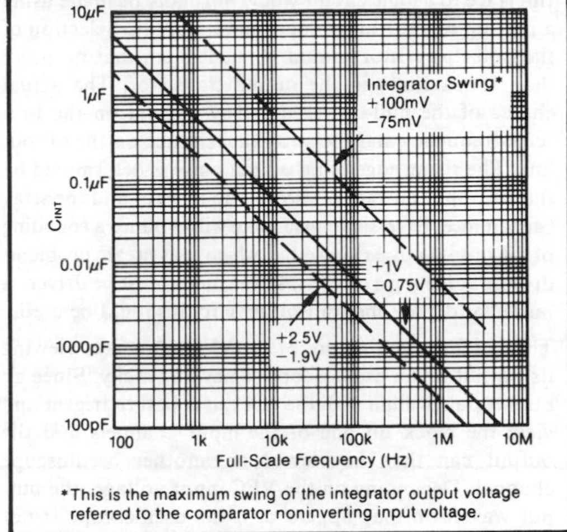

FIGURE 4. Integrator Capacitor Selection Graph.

The integrator output voltage should not be allowed to exceed +12V or −0.2V, otherwise saturation of the operational amplifier could cause inaccuracies. Operation with positive power supplies less than +15V will limit the output swing of the integrator operational amplifier. Smaller integrator voltage waveforms may be required to avoid output saturation of the integrator amplifier. See "Power Supply Considerations" for information on low voltage operation.

The maximum integrator voltage swing requirement is nearly symmetrical about the comparator threshold voltage (see Figure 12). One-third greater swing is required above the threshold than below it. Maximum demand on positive integrator swing occurs at low scale, while the negative swing is greatest just below full scale.

CLOCK INPUT

The clock input is TTL- and CMOS-compatible. Its input threshold is approximately 1.4V (two diode voltage drops) referenced to digital ground (pin 12). The clock "high" input may be standard TLL or may be as high as $+V_{CC} - 2V$. A CMOS clock should be powered from a voltage source at least 2V below the VFC100's $+V_{CC}$ to prevent overdriving the clock input. Alternatively, a resistive voltage divider may be used to limit the clock voltage swing to $+V_{CC} - 2V$ maximum. The clock input has a high input impedance, so no special drivers are required. Rise time in the transition region from 0.5V to 2V must be less than 2µsec for proper operation.

OUTPUT

The frequency output is an open collector current-sink transistor. Output pulses are active low such that the output transistor is on only during the reset integration period (see Shortened Output Pulses). This minimizes power dissipation over the full frequency range and provides the fastest logic edge at the beginning of the output pulse where it is most desirable.

Interface to a logic circuit would normally be made using a pull-up resistor to the logic power supply. Selection of the pull-up resistor should be made such that no more than 15mA flows in the output transistor. The actual choice of the pull-up resistor may depend on the full-scale frequency and the stray capacitance on the output line. The rising edge of an output pulse is determined by the RC time constant of the pull-up resistor and the stray capacitance. Excessive capacitance will produce a rounding of the output pulse rising edge, which may create problems driving some logic circuits. If long lines must be driven, a buffer or digital line transmitter circuit should be used.

The synchronized nature of the VFC100 makes viewing its output on an oscilloscope somewhat tricky. Since all output pulses align with the clock, it is best to trigger and view the clock on one of the input channels and the output can then be viewed on another oscilloscope channel. Depending on the VFC input voltage, the output waveform may appear as if the oscilloscope is not properly triggered. The output might best be visualized by imagining a constant output frequency which is locked to a submultiple of the clock frequency with occasional extra pulses or missing pulses to create the necessary average frequency. It is these extra or missing pulses that make the output waveform appear as if the oscilloscope is not properly triggered. This is normal. Experimentation with the input voltage and oscilloscope triggering will generally allow a stable view of the output and provides an understanding of its nature.

SHORTENED OUTPUT PULSES

In normal operation, the negative output pulse duration is equal to one period of the clock input. Shorter output pulses may be useful in driving optical couplers or transformers for voltage isolation or noise rejection. This can be accomplished by connecting capacitor C_{OS} as shown in Figure 5. Pin 9 may be connected to $+V_{CC}$, deactivating the output one-shot circuit. The value of C_{OS} is chosen according to the curve in Figure 6. Output pulses cannot be made to exceed one clock period in duration. Thus, a C_{OS} value which would create an output pulse which is longer than one period of the clock will have the same effect as disabling the one-shot, causing the output pulse to last one clock period. The minimum practical pulse width of the one-shot circuit is approximately 100nsec. Using C_{OS} to generate shorter output pulses does not affect the output frequency or the gain equation.

REFERENCE VOLTAGE

Excellent gain drift is achieved by use of a precision internal 5V reference. This reference is brought to an external pin and can be used for a variety of purposes. It is used to offset the noninverting comparator input in voltage-to-frequency mode (although a precise voltage is not requried for this function). It is very useful in many other applications such as offsetting the input to handle bipolar input signals. It can source up to 10mA and sink

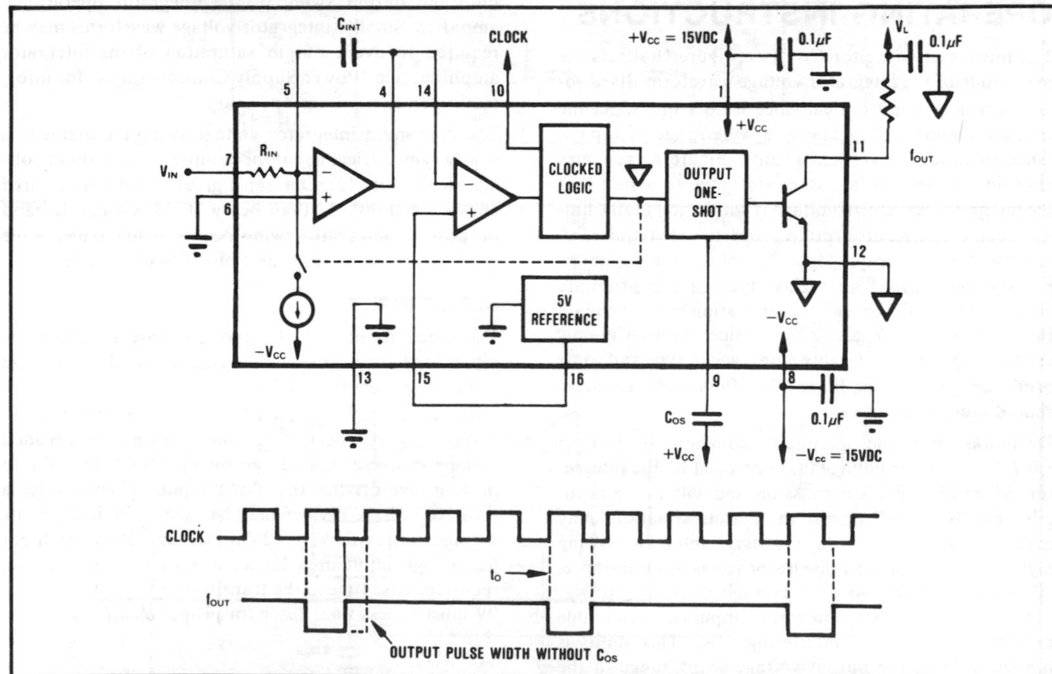

FIGURE 5. Circuit and Timing Diagram for Shortened Output Pulses.

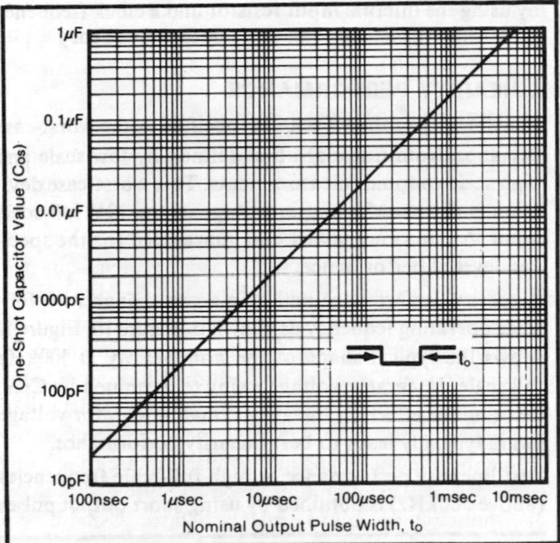

FIGURE 6. Output One-Shot Capacitor Selection Graph.

100μA. Heavy loading of the reference will change the gain of the VFC as well as affecting the external reference voltage. For instance, a 10mA load interacting with a 0.5Ω typical output impedance will change the VFC gain equation and reference voltage by 0.1%.

Figure 7 shows the reference used to offset the VFC transfer function to convert a −5V to +5V input to zero to 500kHz output. The circuit in Figure 8 uses the reference to excite a 300Ω bridge transducer. R_1 provides the majority of the current to the bridge while the V_{REF} output supplies the balance and accurately controls the bridge voltage. The VFC gain is inversely proportional to the reference voltage, V_{REF}. Since the bridge gain is directly proportional to its excitation voltage, the two equal and opposite effects cancel the effect of reference voltage drift on gain.

The reference output amplifier is specifically designed for excellent transient response to provide precision in a noisy environment. Although not required for normal operation, a 0.05μF bypasss capacitor from the reference

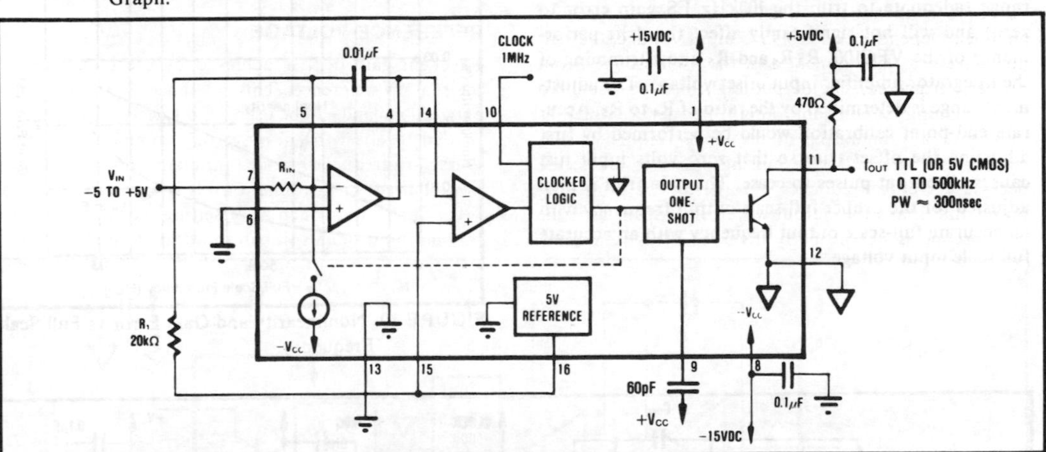

FIGURE 7. Circuit Diagram for Bipolar Input Voltages.

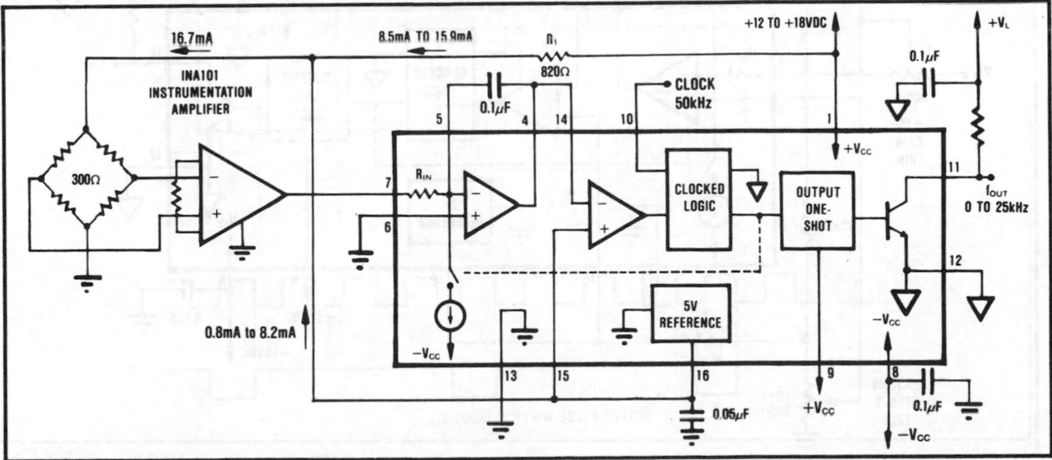

FIGURE 8. Circuit Diagram for Bridge Excitation Using V_{REF}.

output to analog ground (pin 13) may improve the rejection of digital noise from external circuitry.

OTHER INPUT VOLTAGE RANGES

The internal input resistor, $R_{IN} = 20k\Omega$, sets a full-scale input of 10V. Other input ranges can be created by using an external gain set resistor connected to pin 5. Since the excellent temperature drifts of the VFC100 are achieved by careful matching of internal temperature coefficients, use of an external gain set resistor will generally degrade this drift. Using an external resistor to set the gain, the resulting gain drift would be equal to the sum of the external resistor drift and the specified current gain drift of the VFC100. Different voltage input ranges are best implemented by using the internal input resistor, R_{IN}, in series or parallel with a high quality external resistor, thus maintaining as much of the precision temperature tracking as possible.

For best drift performance, the adjustment range of a fine gain trim should be made as narrow as practical. R_1 and R_2 in Figure 9 allow gain adjustment over a ±1% range (adequate to trim the 100kHz FS gain error to zero) and will not significantly affect the drift performance of the VFC100. R_3, R_4, and R_5 allow trimming of the integrator amplifier input offset voltage. The adjustment range is determined by the ratio of R_4 to R_5. Accurate end-point calibration would be performed by first adjusting the offset trim so that zero volts input just causes all output pulses to cease. The gain trim is then adjusted for the proper full-scale output frequency with an accurate full-scale output frequency with an accurate full scale input voltage.

by using the internal input resistor and a clock frequency of 10 times the desired full-scale output frequency.

LINEARITY PERFORMANCE

The linearity of the VFC100 is specified as the worst-case deviation from a straight line defined by low scale and high scale endpoint measurements. This worst-case deviation is expressed as a percentage of the 10V full-scale input. All units are tested and guaranteed for the specified level of performance.

Linearity performance and gain error change with full-scale operating fequency as shown in Figure 10. Figure 11 shows the typical shape of the nonlinearity at 100kHz full scale. Integrator voltage swing (determined by C_{INT}) has a minor effect on linearity. Small integrator voltage swing typically leads to best linearity performance.

Best linearity performance at high full-scale frequencies (above 500kHz) is obtained by using short output pulses

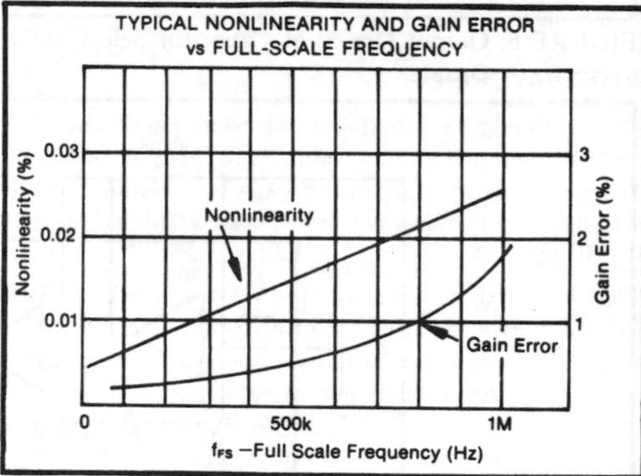

FIGURE 10. Nonlinearity and Gain Error vs Full Scale Frequency.

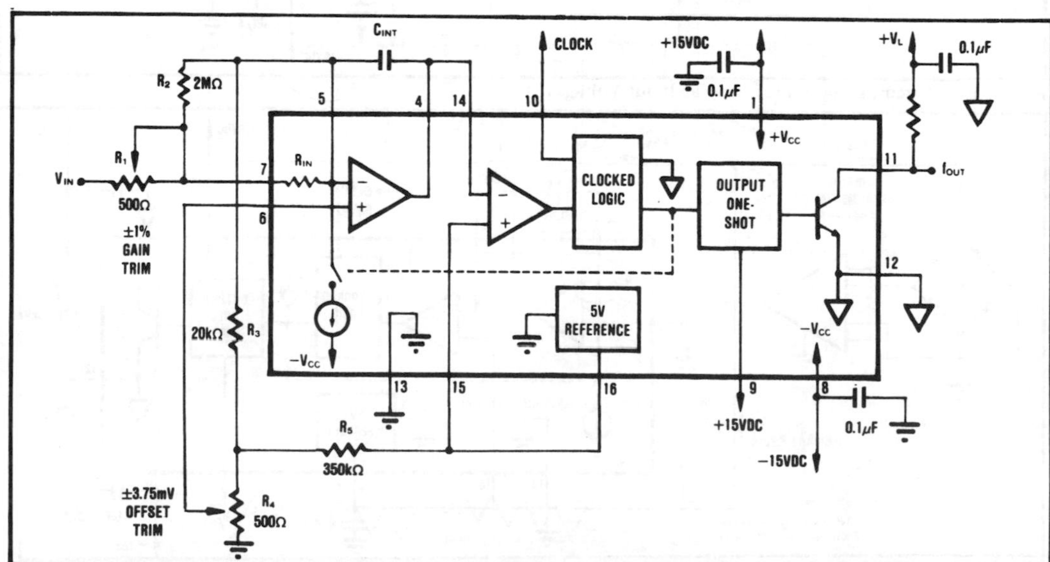

FIGURE 9. Circuit Diagram for Fine Offset and Gain Trim.

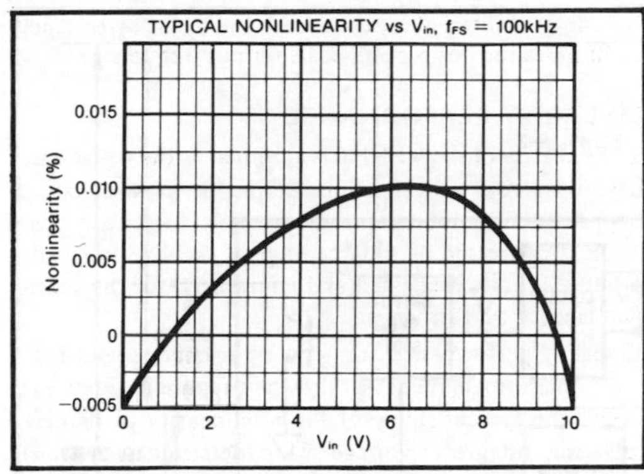

FIGURE 11. Typical Nonlinearity vs V_{IN}.

with a one-shot capacitor of 60pF. As with any high-frequency circuit, careful attention to good power supply bypassing techniques (see "Power Supplies and Grounding") is also required.

TEMPERATURE DRIFT

Conventional VFC circuits are affected significantly by external component temperature drift. Drift of the external input resistor and timing capacitor required with these devices may easily exceed the specified drift of the VFC itself.

When used with its internal input resistor, the gain drift of the complete VFC100 circuit is totally determined by the performance of the VFC100. Gain drift is specified at a full scale output frequency of 100kHz. Conventional VFC circuits usually specify drift at 10kHz and degrade significantly at higher operating frequency. The VFC-100's gain drift remains excellent at higher operating frequency, typically remaining within specification at f_{FS} = 1MHz.

Drift of the external clock frequency directly affects the output frequency, but by using a common clock for the VFC and counting circuitry this drift can be cancelled (see Counting the Output).

POWER SUPPLIES AND GROUNDING

Separate analog and digital grounds are provided on the VFC100 and it is important to separate these grounds to attain greatest accuracy. Logic sink current flowing in the f_{OUT} pin is returned to the digital ground. If this "noisy" current were allowed to flow in analog ground, errors could be created. Although analog and digital grounds may eventually be connected together at a common point in the circuitry, separate circuit connections to this common point can reduce the error voltages created by varying currents flowing through the ground return impedance. The +5V V_{REF} pin is referenced to analog ground.

The power supplies should be well bypassed using capacitors with low impedance at high frequency. A value of 0.1μF is adequate for most circuit layouts.

The VFC100 is specified for a nominal supply voltage of ±15V. Supply voltages ranging from ±7.5V to ±18V may be used. Either supply can be up to 28V as long as the total of both does not exceed 36V. Steps must be taken, however, to assure that the integrator output does not exceed its linear range. Although the integrator output is capable of 12V output swing with 15V power supplies, with 7.5V supplies, output swing will be limited to approximately 4.5V. In this case, the comparator input

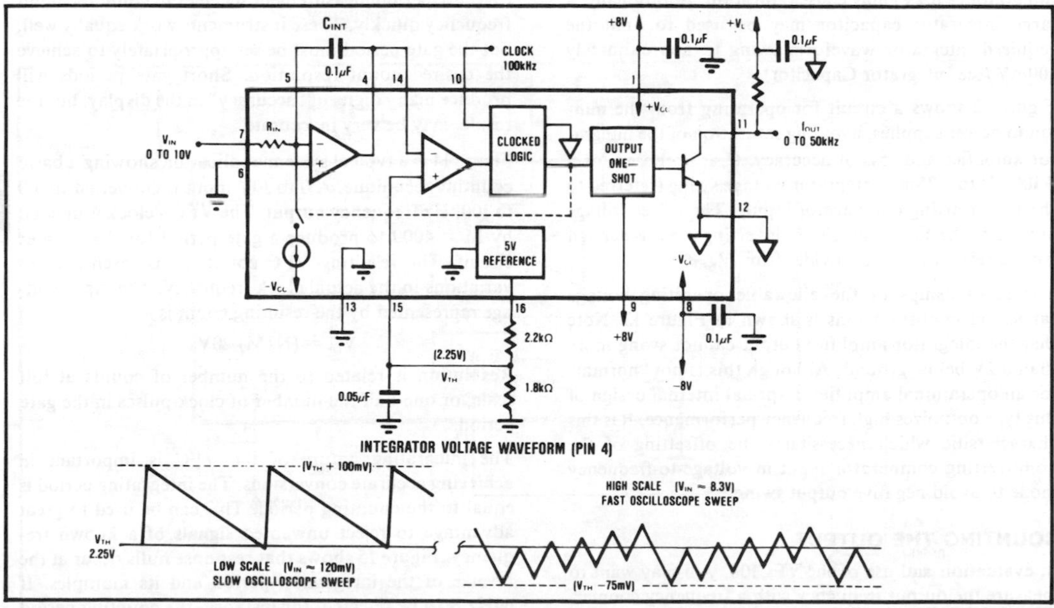

FIGURE 12. Circuit Diagram and Integrator Voltage Waveform for Low Power Supply Voltage Operation.

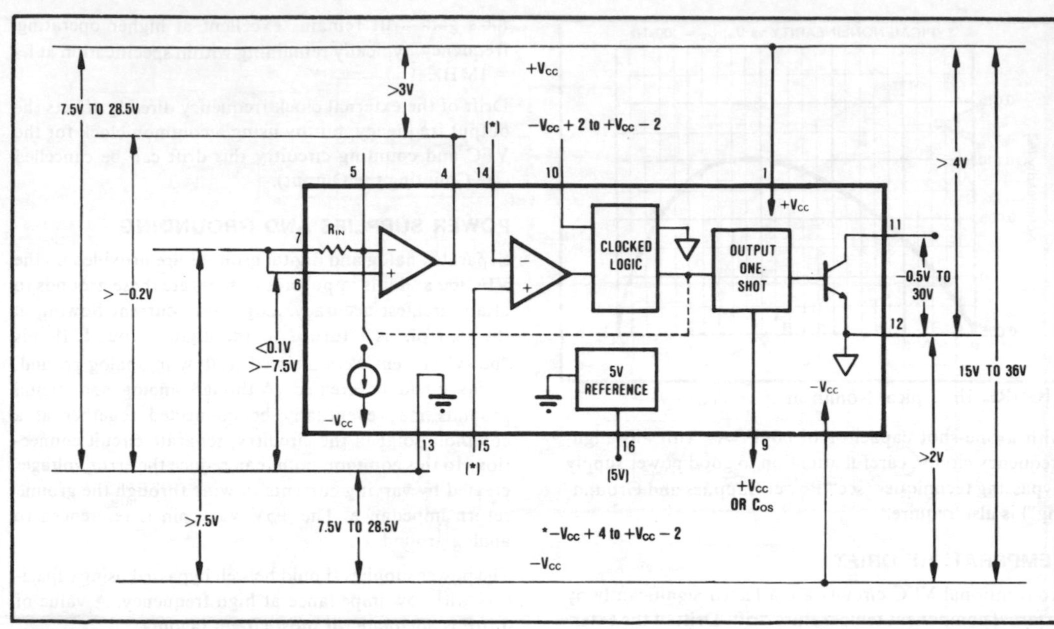

FIGURE 13. Relationships of Allowable Voltages.

cannot be offset by directly connecting to the 5V reference output pin. The comparator input must be connected to a lower voltage point (approximately 2V). This allows the integrator output to operate around a lower voltage point, assuring linear operation. This threshold voltage does not affect the accuracy or drift of the VFC as long as it is not noisy. It should not be made too small, however, or the negative output limitation of the integrator (−0.2V) may cause saturation. Additionally, a large integrator capacitor may be used to limit the required integrator waveform swing to approximately 100mV (see Integrator Capacitor).

Figure 12 shows a circuit for operating from the minimum power supplies, avoiding saturation of the integrator amplifier and loss of accuracy. C_{INT} is chosen for a +100mV to −75mV integrator voltage swing (referred to the noninverting comparator input). The offset voltage applied to the comparator's noninverting input is derived from a resistive voltage divider from V_{REF}.

The relationships of the allowable operating voltage ranges on important pins is shown in Figure 13. Note that the integrator amplifier output cannot swing more than 0.2V below ground. Although this is not "normal" for an operational amplifier, a special internal design of this type optimizes high frequency performance. It is this charactersitic which necessitates the offsetting of the noninverting comparator input in voltage-to-frequency mode to avoid negative output swing.

COUNTING THE OUTPUT

In evaluation and use of the VFC100, you may want to measure the output frequency with a frequency counter. Since synchronization of the VFC100 causes it to await a clock edge for any given output pulse, the output frequency is essentially quantized. The quantized steps are equal to one clock period of the counting gate period. The quantizing error can be made arbitrarily small by counting with long gate times. For instance, a one second counter gate period and a 100kHz full-scale frequency has a one part in 100,000 resolution. Many of the more sophisticated laboratory frequency counters, however, use period measurement schemes to count the input frequency quickly. These instruments work equally well, but the gate period must be set appropriately to achieve the desired count resolution. Short gate periods will produce many digits of "accuracy" in the display, but the results may be very inaccurate.

Figure 14 is a typical system application showing a basic counting technique. A 0 to 10V input is converted to a 0 to 100kHz frequency output. The VFC's clock is divided by M = 4000 to produce a gate period for the counter circuit. The resulting VFC count, N, is insensitive to variations in the actual clock frequency. The input voltage represented by the resulting count is

$$V_{IN} = (N/M)\ 20V$$

Resolution is related to the number of counts at full scale, or one-half the number of clock pulses in the gate period.

The integrating nature of the VFC is important in achieving accurate conversions. The integrating period is equal to the counting period. This can be used to great advantage to reject unwanted signals of a known frequency. Figure 15 shows that response nulls occur at the inverse of the integration period and its multiples. If 60Hz is to be rejected, for instance, the counting period

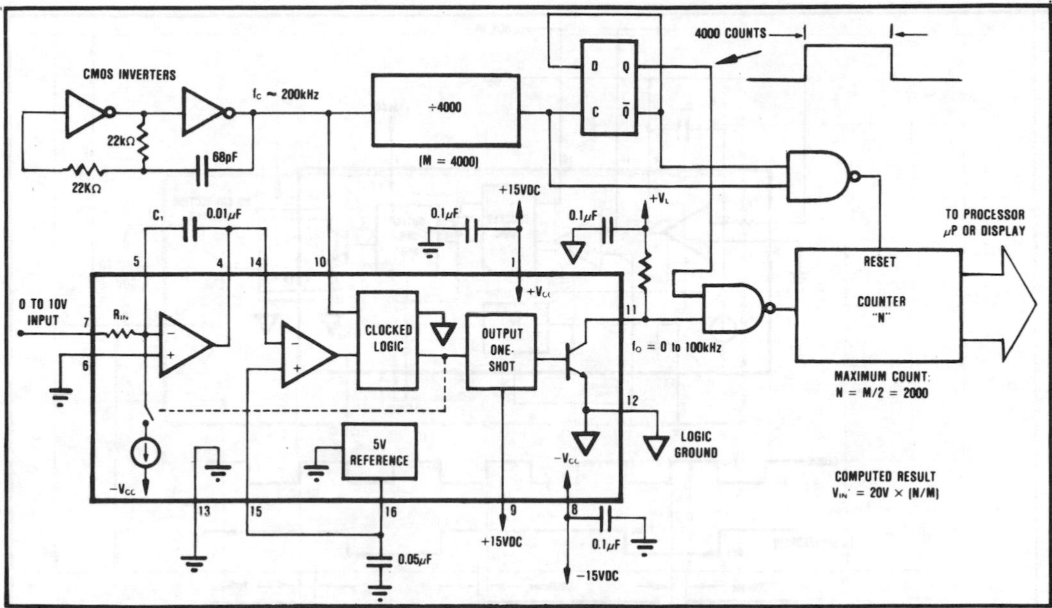

FIGURE 14. Diagram of a Voltage-to-Frequency Converter and Counter System.

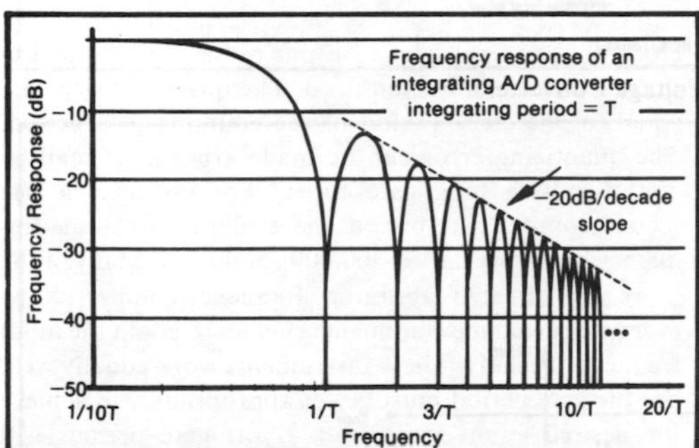

FIGURE 15. Frequency Response of an Integrating Analog-to-Digital Converter.

should be made equal to, or a multiple of 1/60 of a second.

FREQUENCY-TO-VOLTAGE MODE

The VFC100 can also function as a frequency-to-voltage converter by applying an input frequency to the comparator input as shown in Figure 16. The input resistor, R_{IN}, is connected as a feedback resistor. The voltage at the integrator amp output is proportional to the ratio of the input frequency to the clock frequency. The transfer function is

$$V_{OUT} = (f_{IN}/f_{CLOCK})\ 20V$$

This transfer function is complementary to the voltage-to-frequency mode transfer function, making voltage-to-frequency-to-voltage conversions simple and accurate.

Direct coupling of the input frequency to the comparator is easily accomplished by driving both comparators with complementary frequency input signals. Alternatively, one of the comparator inputs can be biased at half the logic voltage (using V_{REF} and a voltage divider) and the other input driven directly.

The proper timing of the input frequency waveform is shown in Figure 16. The input pulse should go low for one clock cycle, centered around a falling edge of the clock. The minimum acceptable input pulse width must fall no later than 200nsec before a negative clock edge and rise no sooner than 200nsec after the falling clock edge. An input pulse which remains low for more than one falling edge of the clock will produce incorrect output voltages. Positive (active high) input pulses can be accepted by reversing the connections to pins 14 and 15. Figure 17 shows a digital conditioning circuit which will accept any input duty cycle and provide the proper pulse width to the comparator. Each rising edge at this circuit's input generates the required negative pulse at the inverting comparator input. The noninverting comparator is driven by a complementary signal.

The integrator amplifier output is designed to drive up to 10,000pF and 5kΩ loads in frequency-to-voltage mode. This allows driving long lines in a large system.

Ripple voltage in the voltage output is unavoidable and is inversely proportional to the value of the integrator capacitor. Figure 18 shows the output ripple and settling time as a function of the C_{INT} value.

The ripple frequency is equal to the input frequency. Its magnitude can be reduced by using a large integrator capacitor value, but at the sacrifice of slow settling time at the voltage output in response to an input frequency change. The settling time constant is equal to $R_{IN} \times C_{INT}$. A better compromise between output ripple and settling time can be achieved by using a moderately low integra-

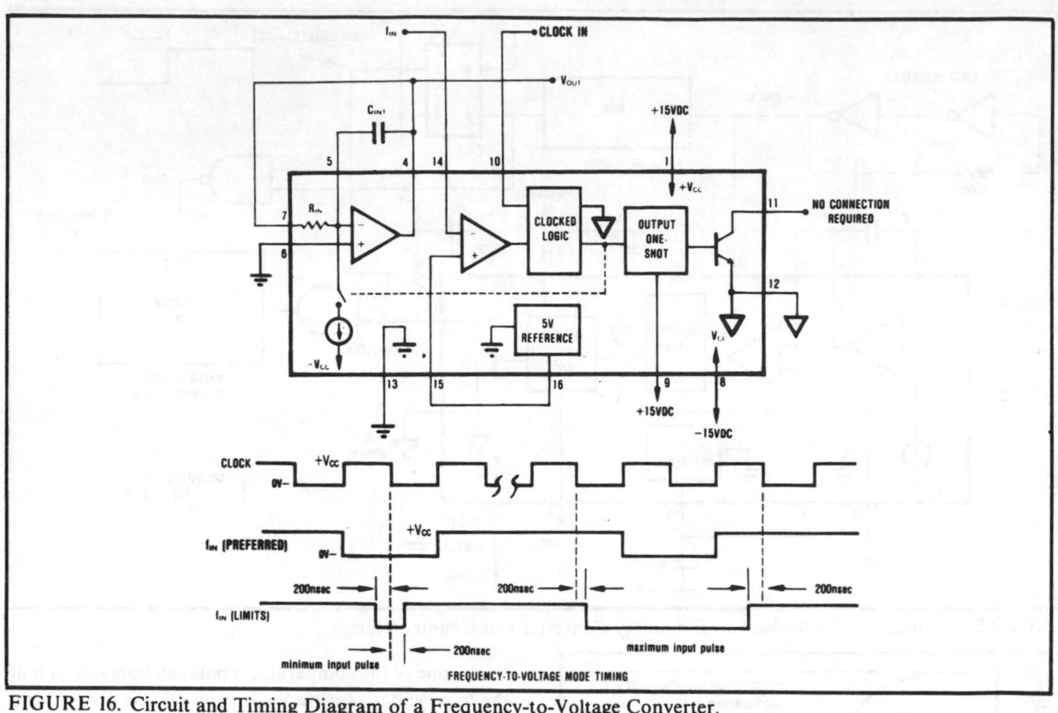

FIGURE 16. Circuit and Timing Diagram of a Frequency-to-Voltage Converter.

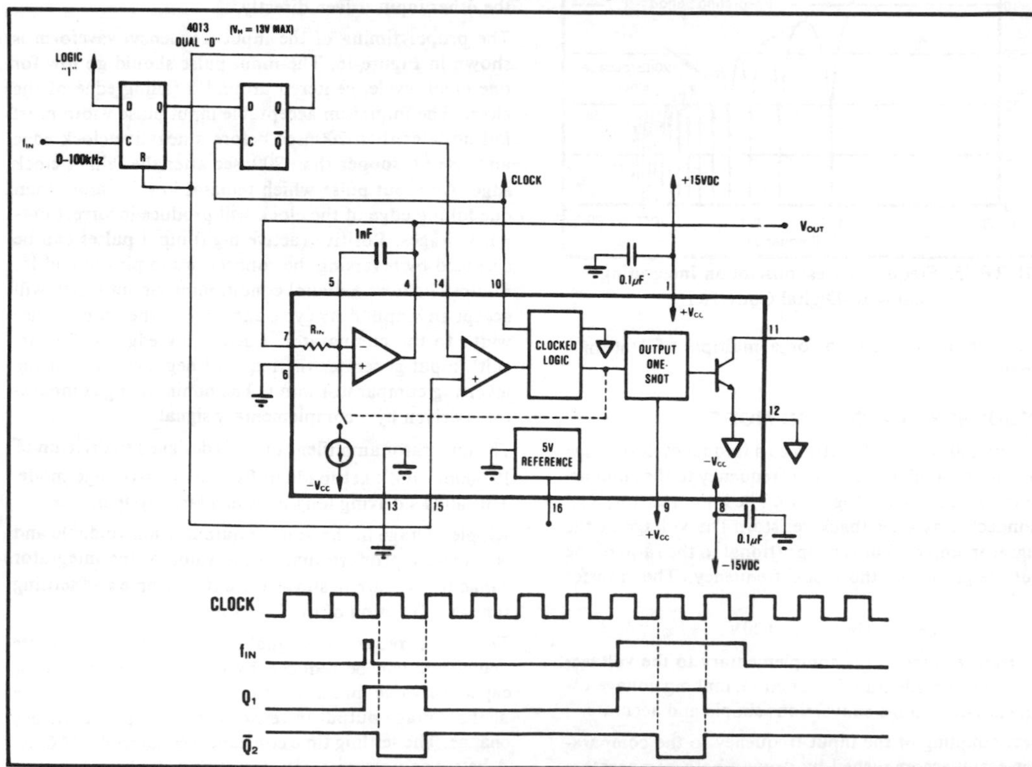

FIGURE 17. Digital Timing Input Conditioning Circuit for Frequency-to-Voltage Operation.

VOLTAGE-TO-FREQUENCY CONVERTERS

10

VFC100

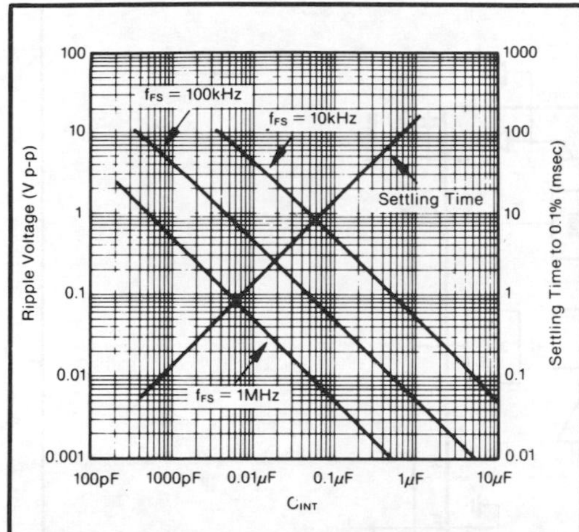

FIGURE 18. Frequency-to-Voltage Mode Output Ripple and Settling Time vs Integrator Capacitance.

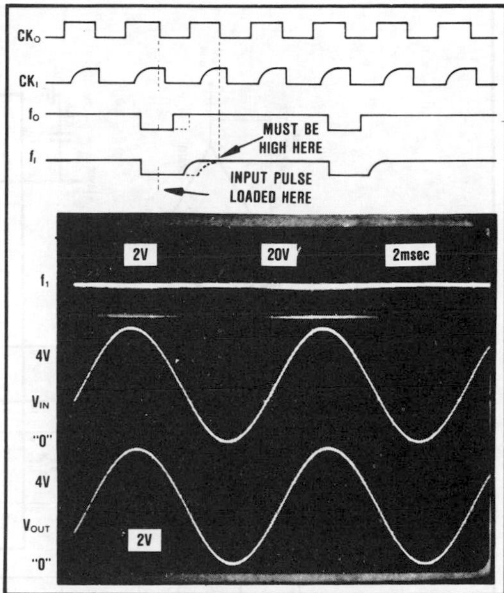

FIGURE 19. Timing Diagram and Oscilloscope Photo of Isolated Voltage-to-Frequency/ Frequency-to-Voltage System.

tor capacitor value and adding a low-pass filter on the analog output. The cutoff frequency of the filter should be made below the lowest expected input frequency to the frequency-to-voltage converter.

The system in Figure 20 makes use of both voltage-to-frequency and frequency-to-voltage mode to send a signal across an optically-isolated barrier. This technique is useful not only for providing safety in the presence of high voltages, but for creating high noise rejection in electrically noisy environments. The use of a common clock frequency causes the two devices to have complementary transfer functions, which minimizes errors.

Optical coupling is facilitated by use of the output one-shot feature. The output pulse is shortened (see Shortened Output Pulses) to allow for the relatively slow turn-off time of the LED. The timing diagram in Figure 19 shows how the accumulated delay of both optical couplers could produce too long an input pulse for the frequency-to-voltage converter, VFC_2 of Figure 20.

An output filter is used to reduce the ripple in the output of VFC_2. In order to most effectively filter the output, both input and output VFCs are offset. By connecting R_1 to V_{REF}, an accurate offset is created in the voltage-to-frequency function. Zero volts input now creates a 10kHz output. This offset is subtracted in the frequency-to-voltage conversion on the output side, by V_{REF} and R_5.

MORE PULSE POSITION RESOLUTION

Since output pulses must always align with clock edges, the instantaneous output frequency is quantized and appears to have phase jitter. This effect can be greatly reduced by using a high speed clock so that available clock edges come more frequently. This would also create a high full-scale frequency, but the technique shown in Figure 21 offers an alternative. A high speed clock is used to produce high resolution of the output pulse position, but a low full-scale frequency can be programmed.

When an output pulse is generated, the next rising edge of the high frequency clock is delayed for a programmable number of clock counts. Since the integrator reset period (which sets the full-scale range) is determined by the time from rising edge to rising edge at the VFC's clock input once the comparator is tripped, the effective clock frequency is $f_{CLOCK}/16$. The circuit shown can be programmed for any N from 2 to 16. Since an output pulse must propagate through the VFC before the next rising edge of the clock arrives, maximum clock frequency is limited by the delay time shown in the timing diagram.

With output pulses now able to align with greater resolution, the output has lower phase jitter. Using this technique, the output is suitable for ratiometric (period measurement) type counting. This counting technique achieves the maximum possible resolution for short gate periods (see Burr-Brown Application Note AN-130).

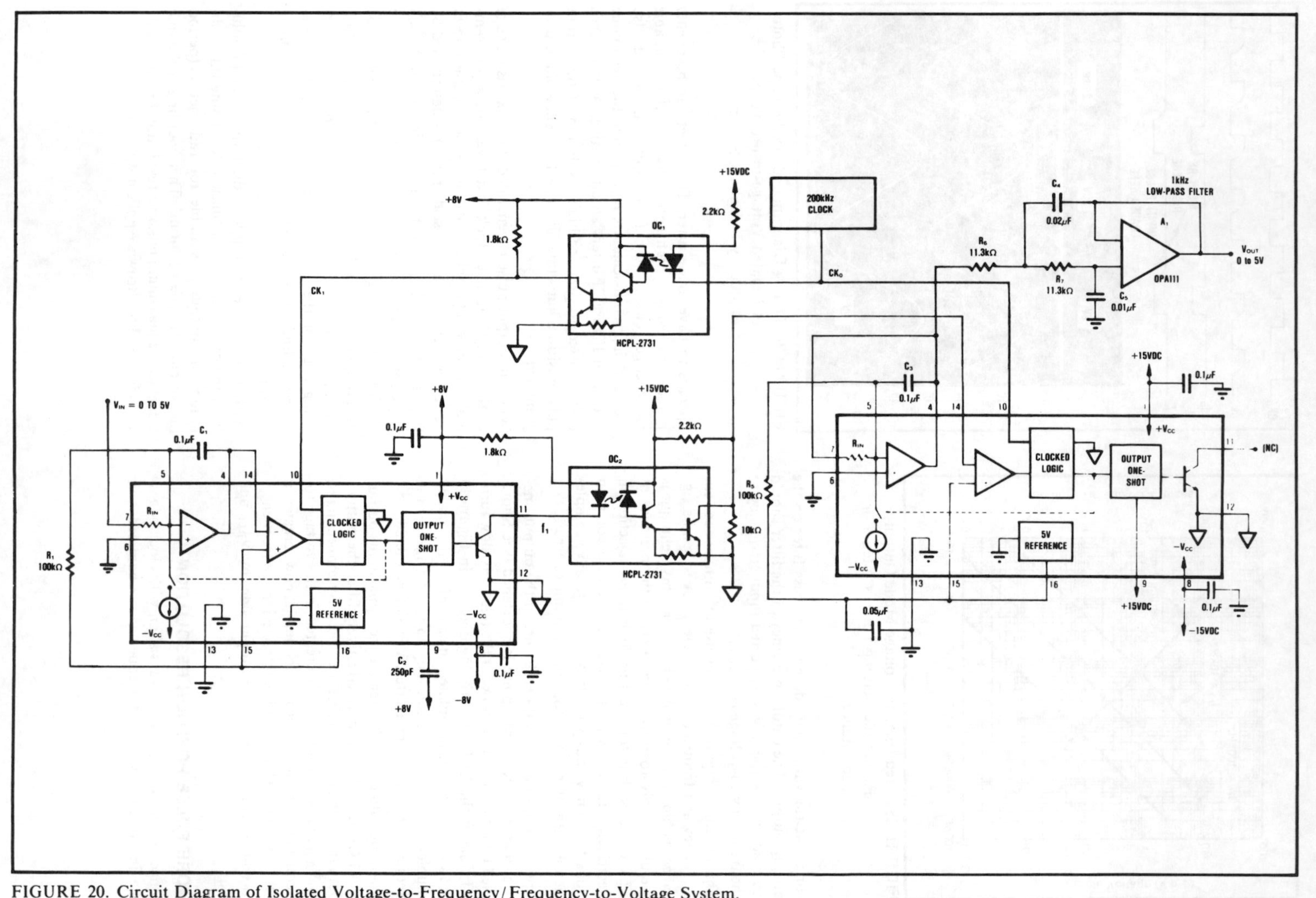

FIGURE 20. Circuit Diagram of Isolated Voltage-to-Frequency/Frequency-to-Voltage System.

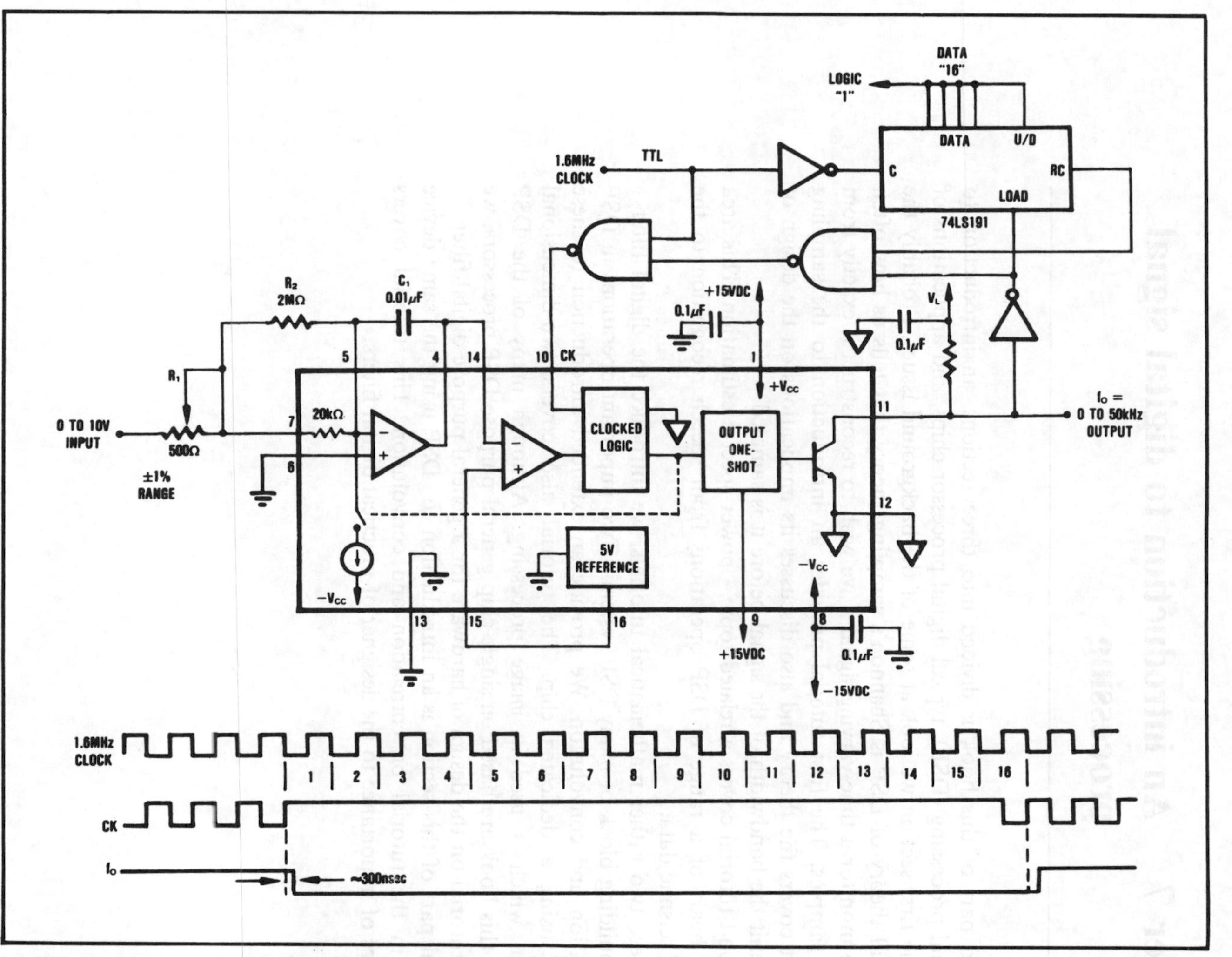

FIGURE 21. Circuit Diagram for Increased Pulse Position Resolution.

Chapter 7 An introduction to digital signal processing

The second part of this book is divided into three sections: an introduction to digital signal processing (DSP), typical digital processor chips, and applications of DSP. In the first section we look at some of the background issues. Probably the fundamental theory of DSP is Shannon's sampling theory that tells us how often we have to monitor a time-varying signal if we wish to reconstruct it exactly from its digital samples. The first tutorial provides an introduction to the sampling theorem. It covers the theory and also discusses its implications on the design of filters to limit the bandwidth of the signal before it is sampled.

The second tutorial covers a related topic – power spectra estimation. This area lies at the heart of a range of DSP operations from speech recognition to the analysis of seismic data.

After these two rather mathematical tutorials, we introduce the digital filter – the basic building block of many DSP systems. Two important operations in DSP are correlation and convolution. We provide an extended introduction to these operations using a dedicated chip. The tutorial also covers two-dimensional convolution which is used in image processing. Although many of the DSP systems in this book are based on single-chip general purpose DSP processors, we provide a tutorial on the design of hardware for a general purpose digital filter.

The final part of this section is an introduction to DSP using the same device described in the tutorial on correlation and convolution. This tutorial covers many topics of importance to the designer of practical digital filters.

An Introduction to the Sampling Theorem

National Semiconductor
Application Note 236
Carson Chen
January 1980

An Introduction to the Sampling Theorem

With rapid advancement in data acquisition technology (i.e. analog-to-digital and digital-to-analog converters) and the explosive introduction of micro-computers, selected complex linear and nonlinear functions currently implemented with analog circuitry are being alternately implemented with sample data systems.

Though more costly than their analog counterpart, these sampled data systems feature programmability. Additionally, many of the algorithms employed are a result of developments made in the area of signal processing and are in some cases capable of functions unrealizable by current analog techniques.

With increased usage a proportional demand has evolved to understand the theoretical basis required in interfacing these sampled data-systems to the analog world.

This article attempts to address the demand by presenting the concepts of aliasing and the sampling theorem in a manner, hopefully, easily understood by those making their first attempt at signal processing. Additionally discussed are some of the unobvious hardware effects that one might encounter when applying the sampling theorem.

With this . . . let us begin.

I. An Intuitive Development

The introduction of the sampling theorem by C.E. Shannon in 1949 places restrictions on the frequency content of the time function signal, f(t), and can be simply stated as follows:

> In order to recover the signal function f(t) exactly, it is necessary to sample f(t) at a rate greater than twice its highest frequency component.

Practically speaking for example, to sample an analog signal having a maximum frequency of 2Kc requires sampling at greater than 4Kc to preserve and recover the waveform exactly.

The consequences of sampling a signal at a rate below its highest frequency component results in a phenomenon known as *aliasing*. This concept results in a frequency mistakenly taking on the identity of an entirely different frequency when recovered. In an attempt to clarify this, envision the ideal sampler of Figure 1(a), with a sample period of T shown in (b), sampling the waveform f(t) as pictured in (c). The sampled data points of f'(t) are shown in (d) and can be defined as the sample set of the continuous function f(t). Note in Figure 1(e) that another frequency component, a'(t), can be found that has the

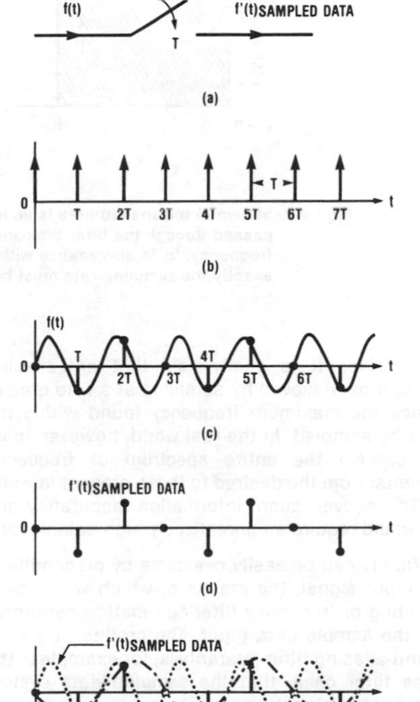

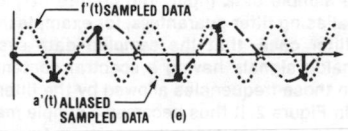

Figure 1. When sampling, many signals may be found to have the same set of data points. These are called aliases of each other.

same sample set of data points as f'(t) in (d). Because of this it is difficult to determine which frequency a'(t), is truly being observed. This effect is similar to that observed in western movies when watching the spoked wheels of a rapidly moving stagecoach rotate backwards at a slow rate. The effect is a result of each individual frame of film resembling a discrete strobed sampling operation flashing at a rate slightly faster than that of the rotating wheel. Each observed sample point or frame catches the spoked wheel slightly displaced from its previous position giving the effective appearance of a wheel rotating backwards. Again, aliasing is evidenced and in this example it becomes difficult to determine which is the true rotational frequency being observed.

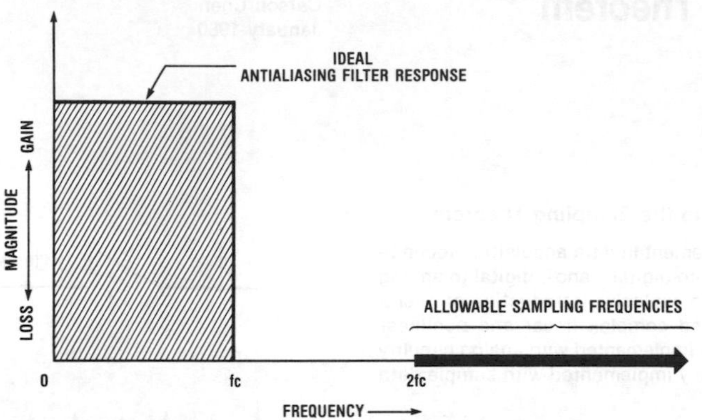

Figure 2. Shown in the shaded area is an ideal, low pass, anti-aliasing filter response. Signals passed through the filter are bandlimited to frequencies no greater than the cutoff frequency, fc. In accordance with the sampling theorem, to recover the bandlimited signal exactly the sampling rate must be chosen to be greater than 2fc.

On the surface it is easily said that anti-aliasing designs can be achieved by sampling at a rate greater than twice the maximum frequency found within the signal to be sampled. In the real world, however, most signals contain the entire spectrum of frequency components; from the desired to those present in white noise. To recover such information accurately the system would require an unrealizably high sample rate.

This difficulty can be easily overcome by preconditioning the input signal, the means of which would be a band-limiting or frequency filtering function performed prior to the sample data input. The prefilter, typically called anti-aliasing filter guarantees, for example in the low pass filter case, that the sampled data system receives analog signals having a spectral content no greater than those frequencies allowed by the filter. As illustrated in Figure 2, it thus becomes a simple matter to sample at greater than twice the maximum frequency content of a given signal.

A parallel analogy of band-limiting can be made to the world of perception when considering the spectrum of white light. It can be realized that the study of violet light wavelengths generated from a white light source would be vastly simplified if inital band-limiting were performed through the use of a prism or white light filter.

II. The Sampling Theorem

To solidify some of the intuitive thoughts presented in the previous section, the sampling theorem will be presented applying the rigor of mathematics supported by an illustrative proof. This should hopefully leave the reader with a comfortable understanding of the sampling theorem.

Theorem: If the Fourier transform $F(\omega)$ of a signal function f(t) is zero for all frequencies above $|\omega| \geqslant \omega_c$, then f(t) can be uniquely determined from its sampled values

$$f_n = f(nT) \tag{1}$$

These values are a sequence of equidistant sample points spaced $\frac{1}{2fc} = \frac{Tc}{2} = T$ apart. f(t) is thus given by

$$f(t) = \sum_{n=-\infty}^{\infty} f(nT) \frac{\sin \omega_c (t - nT)}{\omega_c (t - nT)} \tag{2}$$

Proof: Using the inverse Fourier transform formula:

$$f(t) = \frac{1}{2\pi} \int_{-\infty}^{\infty} F(\omega)\varepsilon^{j\omega t} \, d\omega \tag{3}$$

the band limited function, f(t), takes the form, Figure 3a,

$$f(t) = \frac{1}{2\pi} \int_{-\omega_c}^{\omega_c} F(\omega)\varepsilon^{j\omega t} \, d\omega \tag{4}$$

$f_n = f(n\frac{\pi}{\omega_c})$ is then given as

$$f_n = \frac{1}{2\pi} \int_{-\omega_c}^{\omega_c} F(\omega)\varepsilon^{j\omega \frac{n\pi}{\omega_c}} \, d\omega \tag{5}$$

See Figure 3c and e.

Expressing $F(\omega)$ as a Fourier series in the interval $-\omega_c \leqslant \omega \leqslant \omega_c$ we have

$$F(\omega) = \sum_{n=-\infty}^{\infty} C_n \varepsilon^{-j\omega \frac{n\pi}{\omega_c}} \tag{6}$$

Where,

$$C_n = \frac{1}{2\omega_c} \int_{-\omega_c}^{\omega_c} F(\omega)\varepsilon^{j\omega \frac{n\pi}{\omega_c}} d\omega \tag{7}$$

Further manipulating eq. (7)

$$C_n = \frac{2\pi}{2\omega_c} \frac{1}{2\pi} \int_{-\omega_c}^{\omega_c} F(\omega)\varepsilon^{j\omega \frac{n\pi}{\omega_c}} d\omega \tag{8}$$

C_n can be written as

$$C_n = \frac{\pi}{\omega_c} f_n \tag{9}$$

Substituting eq. (9) into eq. (6) gives the periodic Fourier Transform

$$F(\omega) = \sum_{n=-\infty}^{\infty} \frac{\pi}{\omega_c} f_n \varepsilon^{-j\omega \frac{n\pi}{\omega_c}} \tag{10}$$

of Figure 3f. Using Poisson's sum formula[1] $F(\omega)$ can be stated more clearly as

$$F(\omega) = \sum_{n=-\infty}^{\infty} F(\omega - 2\omega_{cn}) \tag{11}$$

Interestingly for the interval $-\omega_c \leqslant \omega \leqslant \omega_c$ the periodic function $F_p(\omega)$ and Figure 3f. equals $F(\omega)$ and Figure 3b. respectively. Analogously if $F_p(\omega)$ were multiplied by a rectangular pulse defined,

$$H(\omega) = 1 \quad -\omega_c \leqslant \omega \leqslant \omega \tag{12}$$

and

$$H(\omega) = 0 \quad |\omega| \geqslant \omega_c \tag{13}$$

then as pictured in Figures 4b, d, and f,

$$F(\omega) = H(\omega) \cdot F(\omega) = H(\omega) \sum_{n=-\infty}^{\infty} \frac{\pi}{\omega_c} f_n \varepsilon^{-j\omega \frac{n\pi}{\omega_c}} \tag{14}$$

Solving for f(t) the inverse Fourier transform eq (3) is applied to eq (14)

$$f(t) = \frac{1}{2\pi} \int_{-\omega_c}^{\omega_c} F(\omega)\varepsilon^{j\omega t} d\omega \tag{3}$$

$$= \frac{1}{2\pi} \int_{-\omega_c}^{\omega_c} \left[H(\omega) \sum_{n=-\infty}^{\infty} \frac{\pi}{\omega_c} f_n \, \varepsilon^{-j\omega \frac{n\pi}{\omega_c}} \right] \varepsilon^{j\omega t} d\omega$$

$$= \sum_{n=-\infty}^{\infty} fn \frac{1}{2\omega_c} \int_{-\omega_c}^{\omega_c} \varepsilon^{j\omega\left(t - \frac{n\pi}{\omega}\right)} d\omega$$

[1] Poisson's sum formula

$$\frac{1}{T} \sum_{n=-\infty}^{\infty} F(\omega - n\omega_s) = \sum_{n=-\infty}^{\infty} f(nT)\varepsilon^{-j\omega nT}$$

where $T = \frac{1}{fs}$ and fs is the sampling frequency

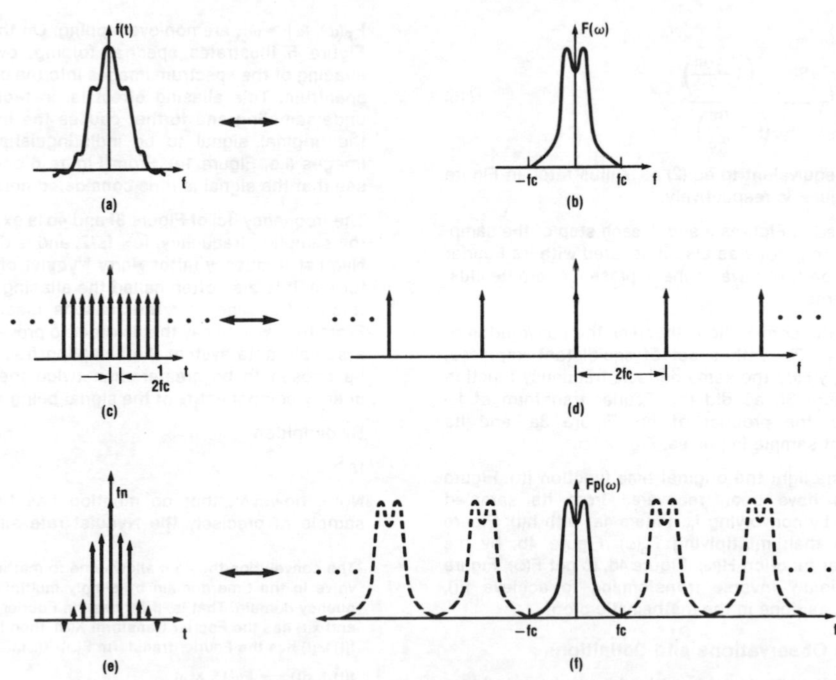

Figure 3. Fourier transform of a sampled signal.

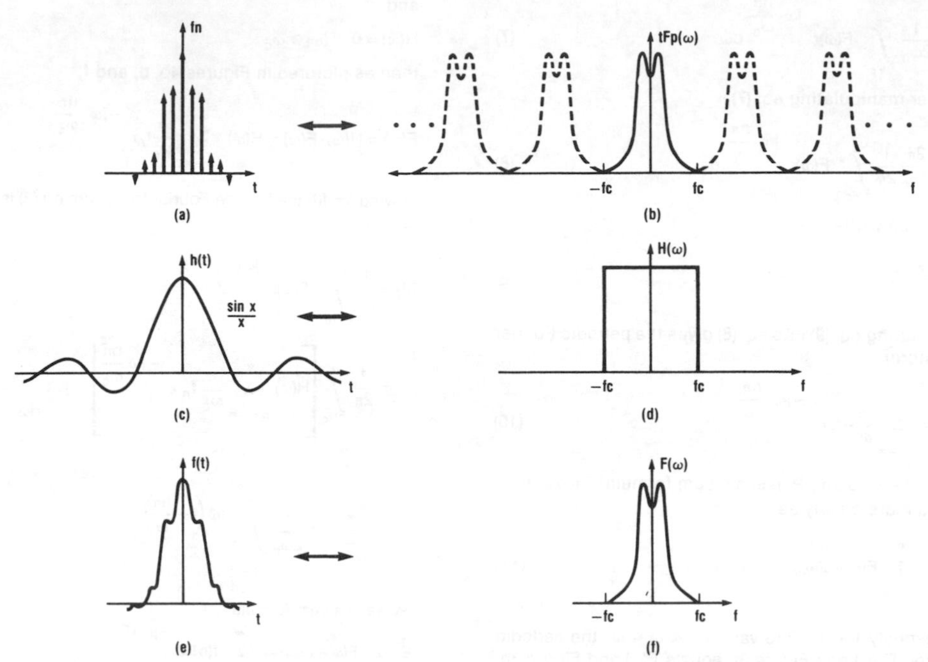

Figure 4. Recovery of a signal f(t) from sampled data information.

giving

$$f(t) = \sum_{n=-\infty}^{\infty} f_n \frac{\operatorname{Sin} \omega_c\left(t - \frac{n\pi}{\omega_c}\right)}{\omega_c\left(t - \frac{n\pi}{\omega_c}\right)} \qquad (15)$$

Eq (15) is equivalent to eq (2) as is illustrated in Figure 4e and Figure 3a respectively.

As observed in Figures 3 and 4, each step of the sampling theorem proof was also illustrated with its Fourier transform pair. This was done to present alternate illustrative proofs.

Recalling the convolution[2] theorem, the convolution of $F(\omega)$, Figure 3b, with a set of equidistant impulses, Figure 3d, yields the same periodic frequency function $F_p(\omega)$, Figure 3f, as did the Fourier transform of f_n, Figure 3e, the product of f(t), Figure 3a, and its equidistant sample impulses, Figure 3c.

In the same light the original time function f(t), Figure 4e, could have been recovered from its sampled waveform by convolving f_n, Figure 4a, with h(t), Figure 4c, rather than multiplying $F_p(\omega)$, Figure 4b, by the rectangular function $H(\omega)$, Figure 4d, to get $F(\omega)$, Figure 4f, and finally inverse transforming to achieve f(t), Figure 4e, as done in the mathematic proof.

III. Some Observations and Definitions

If Figures 3f or 4b are re-examined it can be noted that the original spectrum $F_P(\omega)$, $|\omega| \leqslant \omega_C$, and its images $F_P(\omega)$, $|\omega| \geqslant \omega_c$, are non-overlapping. On the other hand Figure 5 illustrates spectral folding, overlapping or aliasing of the spectrum images into the original signal spectrum. This aliasing effect is, in fact, a result of undersampling and further causes the information of the original signal to be indistinguishable from its images (i.e. Figure 1e). From Figure 6 one can readily see that the signal is thus considered non-recoverable.

The frequency |fc| of Figure 3f and 4b is exactly one half the sampling frequency, fc = fs/2, and is defined as the Nyquist frequency (after Harry Nyquist of Bell Laboratories). It is also often called the aliasing frequency or folding frequency for the reasons discussed above. From this we can say that in order to prevent aliasing in a sample data system the sampling frequency should be chosen to be greater than twice the highest frequency component f_n of the signal being sampled.

By definition

$$fs \geqslant 2fn \qquad (16)$$

Note, however, that no mention has been made to sample at precisely the Nyquist rate since in actual

[2]The convolution theorem allows one to mathematically convolve in the time domain by simply multiplying in the frequency domain. That is, if f(t) has the Fourier transform $F(\omega)$ and x(t) has the Fourier transform $X(\omega)$, then the convolution $f(t)*x(t)$ has the Fourier transform $F(\omega)\cdot X(\omega)$.

$$f(t) * x(t) \longleftrightarrow F(\omega) \cdot X(\omega)$$

$$f(t) \cdot x(t) \longleftrightarrow F(\omega) * X(\omega)$$

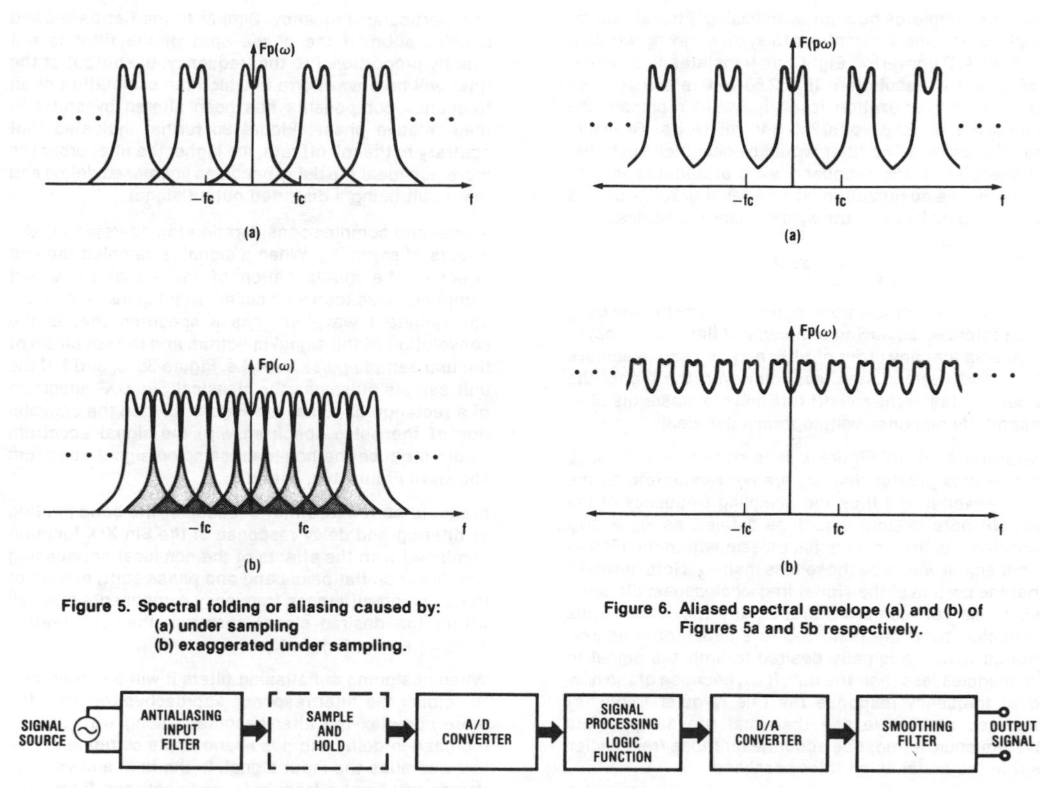

Figure 5. Spectral folding or aliasing caused by: (a) under sampling (b) exaggerated under sampling.

Figure 6. Aliased spectral envelope (a) and (b) of Figures 5a and 5b respectively.

Figure 7. Generalized single channel sample data system.

practice it is impossible to sample at fs = 2fn unless one can guarantee there are absolutely no signal components above fn. This can only be achieved by filtering the signal prior to sampling with a filter having infinite rolloff . . . a physical impossibility, see Figure 2.

IV. The Sampling Theorem and Its Hardware Implications

Though there are numerous sophisticated techniques of implementation, it is appropriate to re-emphasize that the intent of this article is to give the first time user a basic and fundamental approach toward the design of a sample data system. The method with which to achieve this goal will be to introduce a few of the common perils encountered when implementing such a system. We begin by considering the generalized block diagram of Figure 7.

As shown in Figure 7, prior to any signal processing manipulation the analog input signal must be preconditioned to prevent aliasing and thereafter digitized to logic signals usable by the logic function block. The antialiasing and digitizing functions are performed by an input filter and analog-to-digital converter respectively. Once digitized the signal can then be altered or processed and upon completion, reconstructed back to a continuous analog signal via a digital-to-analog converter followed by a smoothing filter.

To this point no mention has been made concerning the sample and hold circuit block depicted in Figure 7. In general the analog-to-digital converter can operate as a stand alone unit. In many high speed operations however, the converter speed is insufficient and thus requires the assistance of a sample and hold circuit. This will be discussed in detail further in the article.

A. The Antialiasing Input Filter

As indicated earlier in the text, the antialiasing filter should band-limit the input signal's spectrum to frequencies no greater than the Nyquist frequency. In the real world however, filters are non-ideal and have typical attenuation or band-limiting and phase characteristics as shown in Figure 8.[3] It must also be realized that true band-limiting of a specific frequency spectrum is not possible. In the sample data system band-limiting is achieved by attenuating those frequencies greater than the Nyquist frequency to a level undetectable or invisible to the system analog-to-digital (A/D) converter. This level would typically be less than the rms quantization[4] noise level defined by the specific converter being used.

[3] In order not to disrupt the flow of the discussion a list of filter terms has been presented in Appendix A.

[4] For an explanation of quantization refer to section IV. B. of this article.

As an example of how an antialiasing filter would be applied, assume a sample data system having within it an 8-bit A/D converter. Eight bits translates to $2^n = 2^8 = 256$ levels of resolution. If a 2.56 volt reference were used each quantization level, q, would represent the equivalent of 2.56 volts/256 = 10 millivolts. Realizing this the antialiasing filter would be designed such that frequencies in the stopband were attenuated to less than the rms quantization noise level of $q/2\sqrt{3}$ and thus appearing invisible to the system. More specifically

$$-20 \log_{10} \frac{\text{V full scale}}{V_{q/2\sqrt{3}}} \cong -59\text{dB} = A_{MIN}$$

It can be seen, for example in the Butterworth filter case (characterized as having a maximally flat pass-band) of Figure 9a that any order of filter may be used to achieve the −59dB attenuation level, however, the higher the order, the faster the roll off rate and the closer the filter magnitude response will approach the ideal.

Referring back to Figure 8 it is observed that those frequencies greater than ω_a are not recognized by the A/D converter and thus the sampling frequency of the sample data system would be defined as $\omega_s \geqslant 2\omega_a$. Additionally, the frequencies present within the filtered input signal would be those less than ω_a. Note however, that the portion of the signal frequencies least distorted are those between $\omega = o$ and ω_p and those within the transition band are distorted to a substantial degree, though it was originally desired to limit the signal to frequencies less than the cutoff ω_p, because of the non-ideal frequency response the true Nyquist frequency occurred at ω_a. We see then that the sample data system could at most be accurate for those frequencies within the antialiasing filter passband.

From the above example, the design of an antialiasing filter appears to be quite straight formward. Recall however, that all waveforms are composed of the sums and differences of various frequency components and as a result, if the reponse of the filter passband were not flat for the desired signal frequency spectra, the recovered signal would be an inaccurate summation of all frequency components altered by their relative attenuations in the pass-band.

Additionally the antialiasing filter design should not neglect the effects of delay. As illustrated in Figure 8 and 9b, delay time corresonds to a specific phase shift at a particular frequency. Similar to the flat pass-band consideration, if the phase shift of the filter is not exactly proportional to the frequency, the output of the filter will be a waveform in which the summation of all frequency components has been altered by shifts in their relative phase. Figure 9b further indicates that contrary to the roll off rate, the higher the filter order the more non-ideal the delay becomes (increased delay) and the result being a distorted output signal.

A final and complex consideration to understand is the effects of sampling. When a signal is sampled the end effect is the multiplication of the signal by a unit sampling pulse train as recalled from Figure 3a, c and e. The resultant waveform has a spectrum that is the convolution of the signal spectrum and the spectrum of the unit sample pulse train, i.e. Figure 3b, d, and f. If the unit sample pulse has the classical Sin X/X[5] spectrum of a rectangular pulse, see Figure 13, then the convolution of the pulse spectrum with the signal spectrum would produce the non-ideal sampled signal spectrum shown in Figure 10a, b, and c.

It should be realized that because of the band-limiting or filtering and delay response of the Sin X/X function combined with the effects of the non-ideal antialiasing filter (i.e. non-flat pass-band and phase shift) certain of the sum and difference frequency components may fall within the desired signal spectrum thereby creating aliasing errors, Figure 10c.

When designing antialiasing filters it will be found that the closer the filter response approaches the ideal the more complex the filter becomes. Along with this an increase in delay and pass-band ripple combine to distort and alias the input signal. In the final analysis the design will involve trade offs made between filter complexity, sampling speed and thus system bandwidth.

B. The Analog-to-Digital Converter

Following the antialiasing filter is the A/D converter which performs the operations of quantizing and coding the input signal in some finite amount of time. Figure 11 shows the quantization process of converting a continuous analog input signal into a set of discrete output levels. A quantization, q, is thus defined as the smallest step used in the digital representation of $f_q(n)$ where f(n)

[5]This will be explained more clearly in Section IV. of this article.

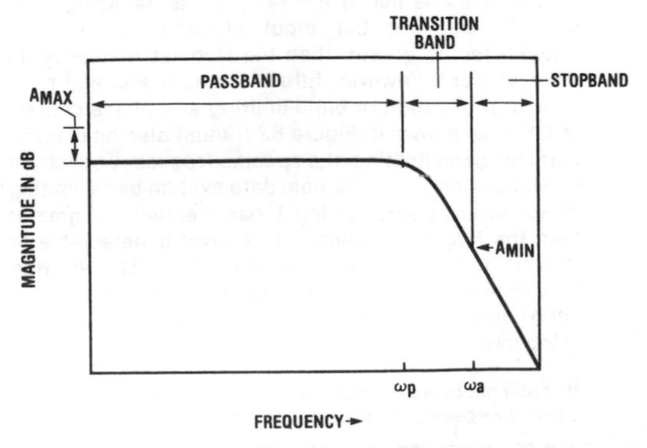

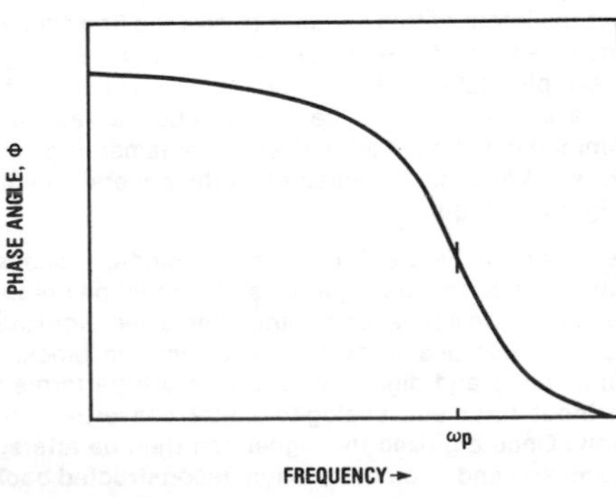

Figure 8. Typical filter magnitude and phase versus frequency response.

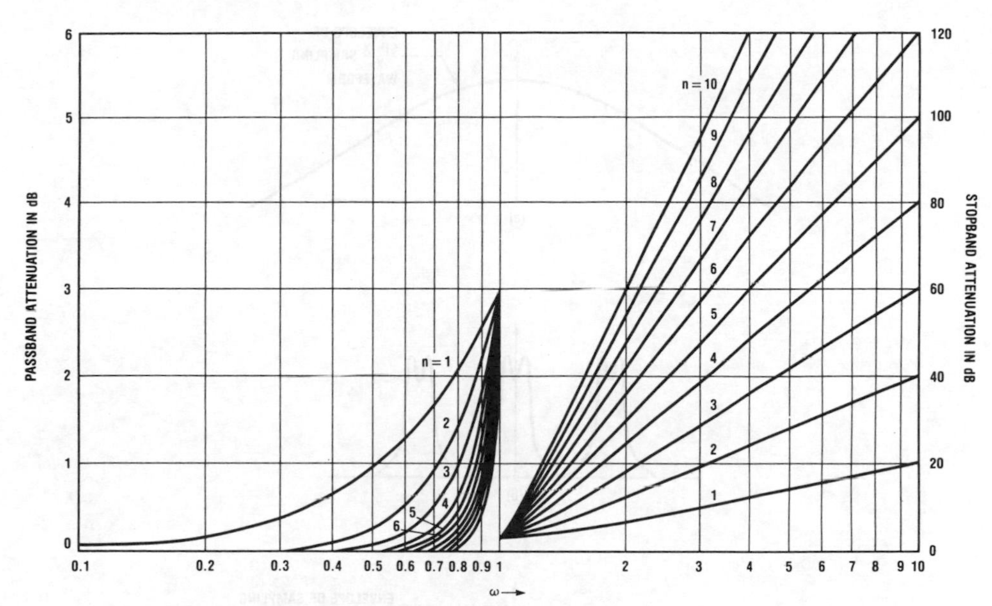

a) Attenuation characteristics of a normalized Butterworth filter as a function of degree n.

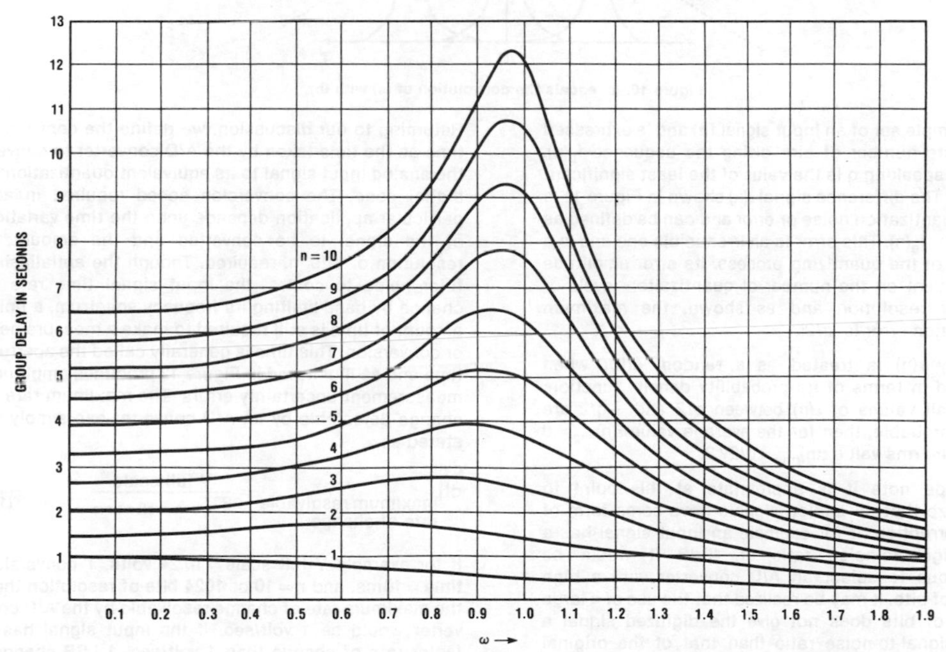

b) Group delay performances of normalized Butterworth lowpass filters as a function of degree n.

Figure 9.

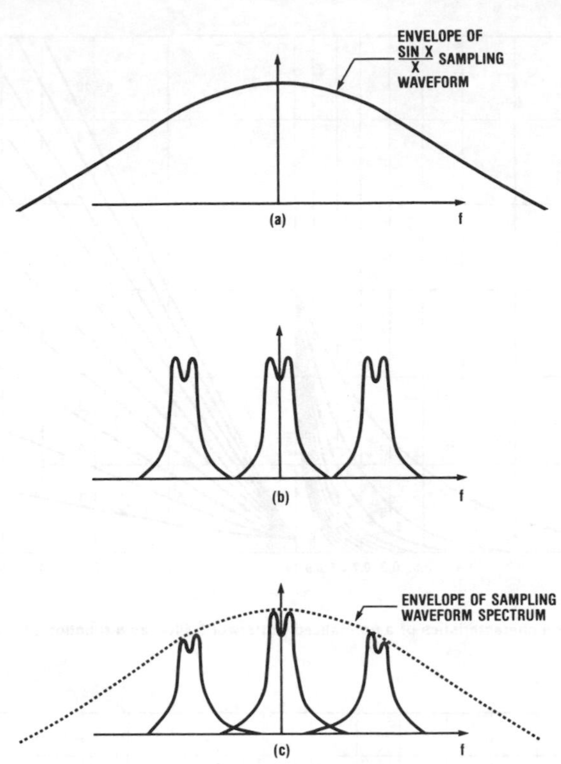

Figure 10. (c) equals the convolution of (a) with (b).

is the sample set of an input signal f(t) and is expressed by a finite number of bits giving the sequence $f_q(n)$. Digitally speaking q is the value of the least significant code bit. The difference signal $\varepsilon(n)$ shown in Figure 11 is called quantization noise or error and can be defined as $\varepsilon(n) = f(n) - f_q(n)$. This error is an irreducible one and is a function of the quantizing process. Its error amplitude is dependent on the number of quantizaiton levels or quantizer resolution and as shown, the maximum quantization error is $|q/2|$.

Generally $\varepsilon(n)$ is treated as a random error when described in terms of its probability density function, that is, all values of $\varepsilon(n)$ between q/2 and $-q/2$ are equally probable, then for the average value $\varepsilon(n)_{avg} = 0$ and for the rms value $\varepsilon(n)_{rms} = q/2\sqrt{3}$.

As a side note it is appropriate at this point to emphasize that all analog signals have some form of noise corruption. If for example an input signal has a finite signal-to-noise ratio of 40dB it would be superfluous to select an A/D converter with a high number of bits. It may be realized that the use of a large number of bits does not give the digitized signal a higher signal-to-noise ratio than that of the original analog input signal. As a supportive argument one may say that though the quantization steps q are very small with respect to the peak input signal the lower order bits of the A/D converter merely provide a more accurate representation of the noise inherent in the analog input signal.

Returning to our discussion, we define the conversion time as the time taken by the A/D converter to convert the analog input signal to its equivalent quantization or digital code. The conversion speed required in any particular application depends upon the time variation of the signal to be converted and the amount of resolution or bits, n, required. Though the antialiasing filter helps to control the input signal time rate of change by band-limiting its frequecy spectrum, a finite amount of time is still required to make a measurement or conversion. This time is generally called the aperture time and as illustrated in Figure 12 produces amplitude measurement uncertainty errors. The maximum rate of change detectable by an A/D converter can simply be stated as

$$\left.\frac{dv}{dt}\right|_{\substack{\text{maximum resolvable}\\ \text{rate of change}}} = \frac{\text{V full scale}}{2^n T_{\text{conversion time}}} \qquad (17)$$

If for example V full scale = 10.24 volts, T conversion time = 10ms, and n = 10 or 1024 bits of resolution then the maximum rate of change resolvable by the A/D converter would be 1 volt/sec. If the input signal has a faster rate of change than 1 volt/sec, 1 LSB changes cannot be resolved within the sampling period.

In many instances a sample-and-hold circuit may be used to reduce the amplitude uncertainty error by measuring the input signal with the smaller aperture time than the conversion time aperture of the A/D

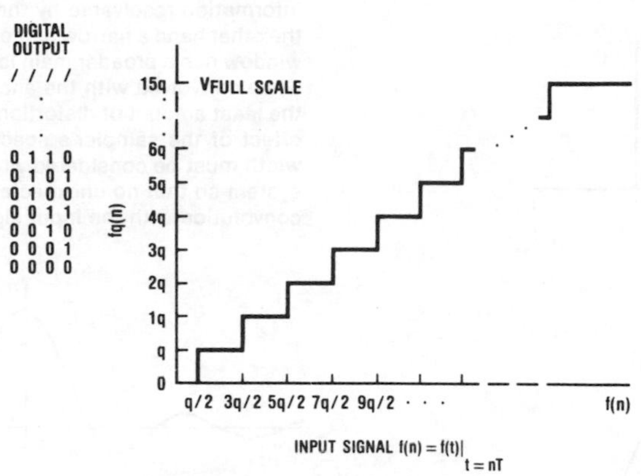

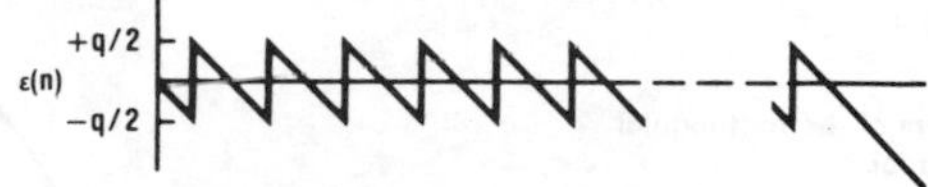

Figure 11. Quantization characteristics.

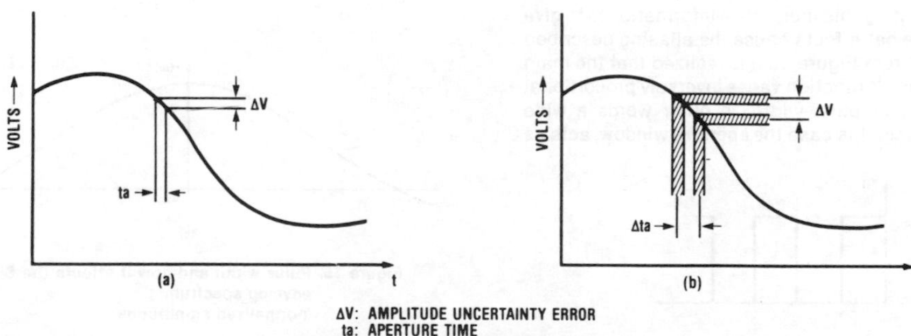

Figure 12. Amplitude uncertainty as a function of (a) a nonvarying aperture and (b) aperture time uncertainty.

converter. In this case the maximum rate of change resolvable by the sample-and-hold would be

$$\left.\frac{dv}{dt}\right|_{\substack{\text{maximum resolvable} \\ \text{rate of change}}} = \frac{\text{V full scale}}{\text{t aperture}} \qquad (18)$$

Note also that the actual calculated rate of change may be limited by the slew rate specification of the sample-and-hold in the track mode. Additionally it is very important to clarify that this does not imply violating the sampling theorem in lieu of the increased ability to more accurately sample signals having a fast time rate of change.

An ideal sample-and-hold effectively takes a sample in zero time and with perfect accuracy holds the value of the sample indefinitely. This type of sampler is also known as a zero order hold circuit and its effect on a sample data system warrants some discussion.

It is appropriate to recall the earlier discussion that the spectrum of a sampled signal is one in which the resultant spectrum is the product obtain by convolving the input signal spectrum with the SinX/X spectrum of the sampling waveform. Figure 13 illustrates the frequency spectrum plotted from the Fourier transform

$$F(\omega) = AT \frac{\sin \frac{\omega T}{2}}{\frac{\omega T}{2}} \qquad (19)$$

of a rectangular pulse. The SinX/X form occurs frequently in modern communication theory and is commonly called the sampling function.

The magnitude and phase of a typical zero order hold sampler spectrum

$$H(\omega) = A\left[\tau \frac{\sin \omega\tau}{\omega\tau} + j\frac{1}{\omega}(\cos \omega - 1)\right] \qquad (20)$$

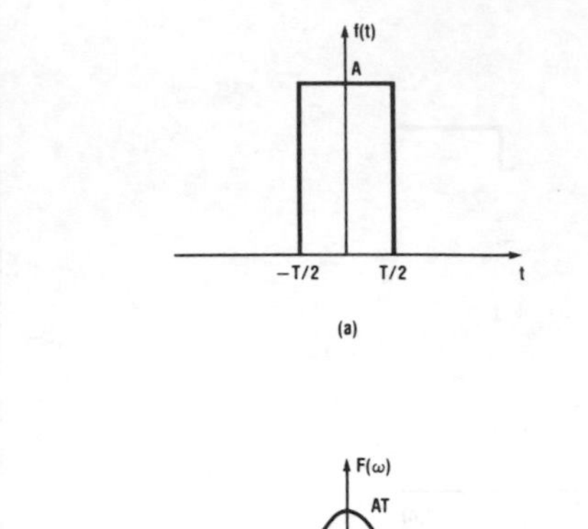

Figure 13. The Fourier transform of the rectangular pulse (a) is shown in (b).

is shown in Figure 14 and Figure 15 illustrates the spectrums of various sampler pulse-widths. The purpose of presenting this illustrative information is to give insight as to what effects cause the aliasing described in Figure 10. From Figure 15 it is realized that the main lobe of the SinX/X function varies inversely proportional with the sampler pulse-width. In other words a wide pulse-width, or in this case the aperture window, acts as

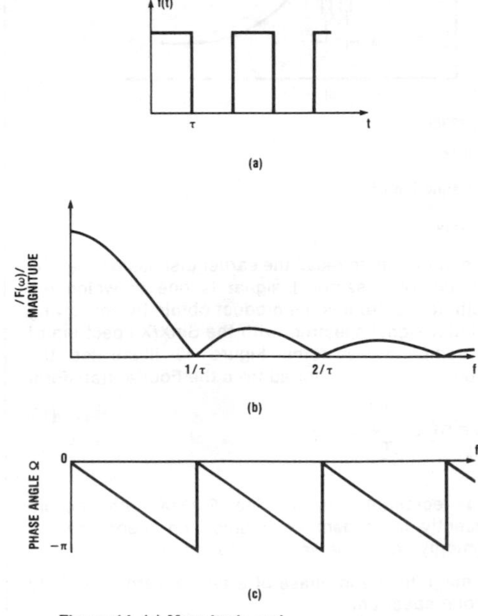

Figure 14. (a) Magnitude and (b) phase frequency response of the sampling pulse (a).

a low pass filtering function and limits the amount of information resolvable by the sample data system. On the other hand a narrow sampler pulse-width or aperture window has a broader main lobe or band-width and thus when convolved with the analog input signal produces the least amount of distortion. Understandably then the effect of the sampler's spectral phase and main lobe width must be considered when developing a sampling system so that no unexpected aliasing occurs from its convolution with the input signal spectrum.

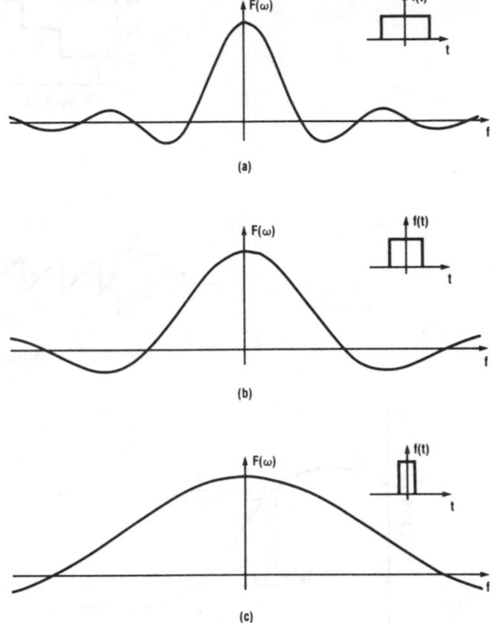

Figure 15. Pulse width and how it effects the SinX/X envelop spectrum (normalized amplitudes).

C. The Digital-to-Analog Converter and Smoothing Filter

After a signal has been digitally conditioned by the signal processing unit of Figure 7, a D/A converter is used to convert the sampled binary information back in to an analog signal. The conversion is called a zero order hold type where each output sample level is a function of its binary weight value and is held until the next sample arrives, see Figure 16. As a result of the D/A converter step function response it is apparent that a large amount of undesirable high frequency energy is present. To eliminate this the D/A converter is usually followed by a smoothing filter, having a cutoff frequency no greater than half the sampling frequency. As its name suggests the filter output produces a smoothed version of the D/A converter output which in fact is a convolved function. More simply said, the spectrum of the resulting signal is the product of a step function SinX/X spectrum and the band-limited analog filter spectrum. Analogous to the input sampling problem, the smoothed output may have aliasing effects resulting from the phase and attenuation relations of the signal recovery system (defined as the D/A converter and smoothing filter combination).

As a final note, the attenuation due to the D/A converter SinX/X spectrum shape may in some cases be compensated for in the signal processing unit by pre-processing using a digital filter with an inverse response X/SinX prior to D/A conversion. This allows an overall flat magnitude signal response to be smoothed by the final filter.

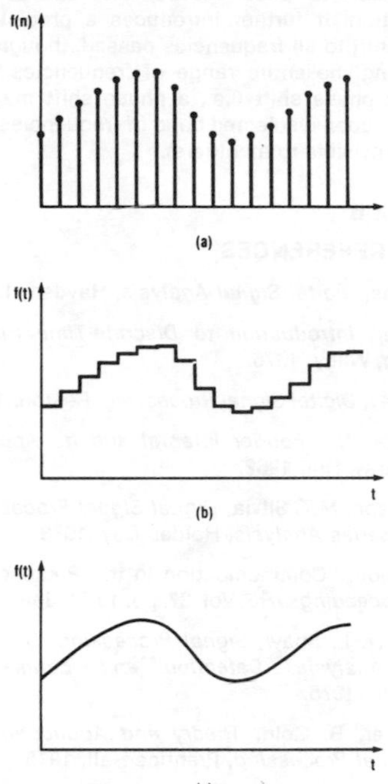

Figure 16. (a) Processed signal data points (b) output of D/A converter (c) output of smoothing filter.

V. A Final Note

This article began by presenting an intuitive development of the sampling theorem supported by a mathematical and illustrative proof. Following the theoretical development were a few of the unobvious and troublesome results that develop when trying to put the sampling theorem into practice. The purpose of presenting these thought provoking perils was to perhaps give the beginning designer some insight or guidelines for consideration when developing a sample data system's interface.

VI. Acknowledgements

The author wishes to thank James Moyer and Barry Siegel for their encouragement and the time they allocated for the writing of this article.

APPENDIX A

Basic Filter Concepts

A filter is a network used for separating signal waves on the basis of their frequency and is usually composed of passive, reactive and active elements such as resistors, capacitors, inductors, and amplifiers, or combinations thereof.

There are basically five types of filters used to pass or reject such signals and they are defined as follows:

1. A low-pass filter allows a specific band of frequencies to pass. This band of frequencies, called the *passband*, range from zero frequency or DC to a certain *cutoff frequency*, ω_c*, and in addition has a maximum attenuation or ripple level of A_{MAX} within the passband. See Figure 1.

*Recall that the radian frequency $\omega = 2\pi f$.

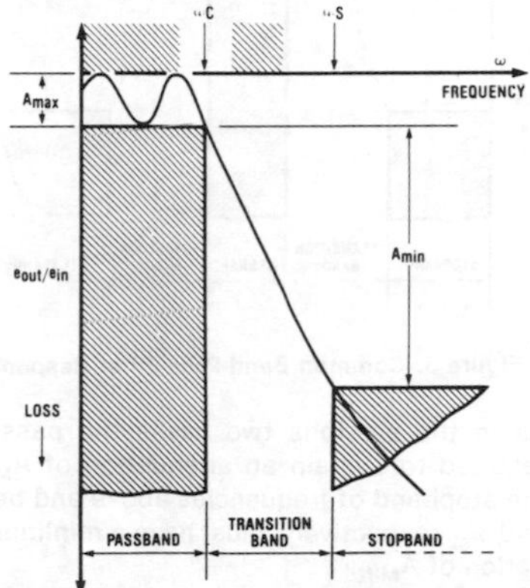

Figure 1. Common Low Pass Filter Response

Frequencies beyond the ω_c may have an attenuation greater than A_{MAX} but at a specific frequency ω_s defined as the *stopband frequency*, a minimum attenuation of A_{MIN} must prevail. The band of frequencies higher than ω_s and maintaining attenuation greater than or equal to A_{MIN} is called the *stopband*. The transition region or *transition band* is that band of frequencies between ω_c and ω_s.

2. A high-pass filter allows frequecies above the passband frequency, ω_c, to pass and rejects frequencies below this point. A_{MAX} must be maintained in the passband and frequencies equal to and below the stopband frequency, ω_s, must have a minimum attenuation of A_{MIN}. See Figure 2.

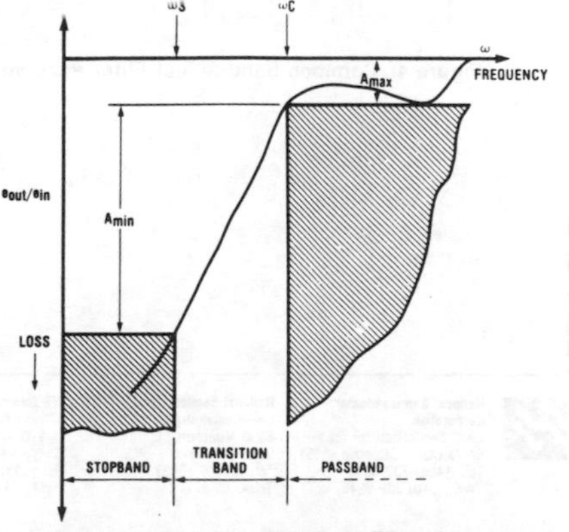

Figure 2. Common High Pass Filter Response

3. A bandpass filter performs the function of passing a specific band of frequecies while rejecting those frequencies above and below the band's respective upper ω_{c2} and lower, ω_{c1} cutoff frequency limits. See Figure 3.

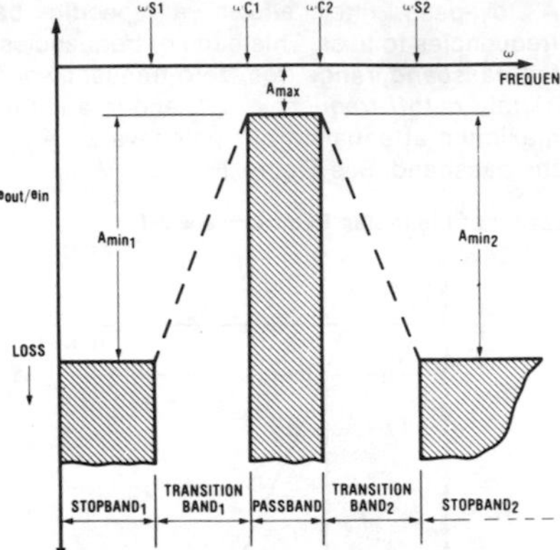

Figure 3. Common Band-Pass Filter Response

As in the previous two cases the passband is required to sustain an attenuation of A_{MAX}, and the stopband of frequencies above and below ω_{s2} and ω_{s2} respectively, must have a minimum attenuation of A_{MIN}.

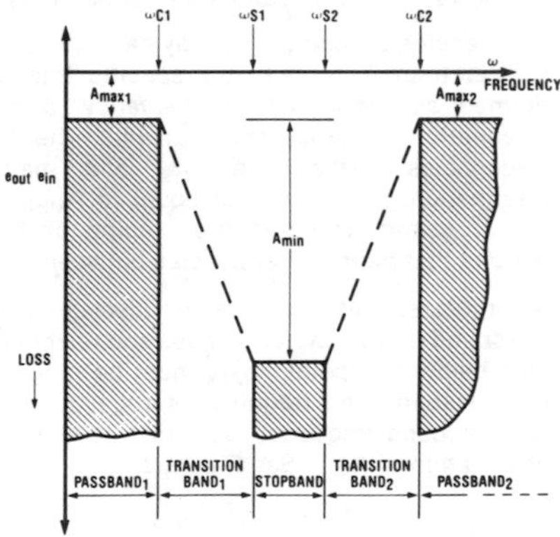

Figure 4. Common Band-Reject Filter Response

4. A band-reject filter or notch filter allows all but a specific band of frequencies to pass. As shown in Figure 4, those frequecies between ω_{s1} and ω_{s2} are filtered out and those frequecies above and below ω_{c2} and ω_{c1} respectively are passed. The attenuation requirements of the stopband A_{MIN} and passband A_{MAX} must still hold.

5. An all-pass or phase shift filter allows all frequencies to pass without any appreciable attenuation. It further introduces a predictable phase shift to all frequencies passed, though not restricting the entire range of frequencies to a specific phase shift (i.e., a phase shift may be imposed upon a selected band of frequencies and appear invisible to all others).

APPENDIX B

ARTICLE REFERENCES

S.D. Stearns, *Digital Signal Analysis*, Hayden, 1975.

S.A. Tretter, *Introduction to Discrete-Time Signal Processing*, Wiley, 1976.

W.D. Stanley, *Digital Signal Processing*, Reston, 1975.

A. Papoulis, *The Fourier Integral and its Applications*, McGraw-Hill, 1962.

E.A. Robinson, M.T. Silvia, *Digital Signal Processing and Time Series Analysis*, Holden-Day, 1978.

C.E. Shannon, "Communication in the Presence of Noise," *Proceedings IRE*, Vol. 37, pp. 10-21, Jan. 1949.

M. Schwartz, L. Shaw, *Signal Processing: Discrete Spectral Analysis, Detection and Estimation*, McGraw-Hill, 1975.

L.R. Rabiner, B. Gold, *Theory and Application of Digital Signal Processing*, Prentice-Hall, 1975.

W.H. Hayt, J.E. Kemmerly, *Engineering Circuit Analysis*, McGraw-Hill, 3rd edition, 1978.

E.O. Brigham, *The Fast Fourier Transform*, Prentice-Hall, 1974.

J. Sherwin, *Specifiying A/D and D/A converters*, National Semiconductor Corp., Application Note AN-156.

Analog-Digital Conversion Notes, Analog Devices Inc., 1974.

A.I. Zverev, *Handbook of Filter Synthesis*, Wiley, 1967.

National Semiconductor Corporation
2900 Semiconductor Drive
Santa Clara, California 95051
Tel.: (408) 737-5000
TWX: (910) 339-9240

National Semiconductor GmbH
Elsenheimerstrasse 61/II
8000 München 21
West Germany
Tel.: 089/9 15027
Telex: 05-22772

NS International Inc., Japan
Miyake Building
1-9 Yotsuya, Shinjuku-ku 160
Tokyo, Japan
Tel.: (03) 355-3711
TWX: 232-2015 NSCJ-J

National Semiconductor (Hong Kong) Ltd.
8th Floor,
Cheung Kong Electronic Bldg.
4 Hing Yip Street
Kwun Tong
Kowloon, Hong Kong
Tel.: 3-411241-8
Telex: 73866 NSEHK HX
Cable: NATSEMI

NS Electronics Do Brasil
Avda Brigadeiro Faria Lima 844
11 Andar Conjunto 1104
Jardim Paulistano
Sao Paulo, Brasil
Telex:
1121008 CABINE SAO PAULO

NS Electronics Pty. Ltd.
Cnr. Stud Rd. & Mtn. Highway
Bayswater, Victoria 3153
Australia
Tel.: 03-729-6333
Telex: 32096

Power Spectra Estimation

National Semiconductor
Application Note 255

1.0 INTRODUCTION

Perhaps one of the more important application areas of digital signal processing (DSP) is the power spectral estimation of periodic and random signals. Speech recognition problems use spectrum analysis as a preliminary measurement to perform speech bandwidth reduction and further acoustic processing. Sonar systems use sophisticated spectrum analysis to locate submarines and surface vessels. Spectral measurements in radar are used to obtain target location and velocity information. The vast variety of measurements spectrum analysis encompasses is perhaps limitless and it will thus be the intent of this article to provide a brief and fundamental introduction to the concepts of power spectral estimation.

Since the estimation of power spectra is statistically based and covers a variety of digital signal processing concepts, this article attempts to provide sufficient background through its contents and appendices to allow the discussion to flow void of discontinuities. For those familiar with the preliminary background and seeking a quick introduction into spectral estimation, skipping to Sections 6.0 through 11.0 should suffice to fill their need. Finally, engineers seeking a more rigorous development and newer techniques of measuring power spectra should consult the excellent references listed in Appendix D and current technical society publications.

As a brief summary and quick lookup, refer to the Table of Contents of this article.

TABLE OF CONTENTS

2.0 WHAT IS A SPECTRUM?

A spectrum is a relationship typically represented by a plot of the magnitude or relative value of some parameter against frequency. Every physical phenomenon, whether it be an electromagnetic, thermal, mechanical, hydraulic or any other system, has a unique spectrum associated with it.

In electronics, the phenomena are dealt with in terms of signals, represented as fixed or varying electrical quantities of voltage, current and power. These quantities are typically described in the time domain and for every function of time, f(t), an equivalent frequency domain function $F(\omega)$ can be found that specifically describes the frequency-component content (frequency spectrum) required to generate f(t). A study of relationships between the time domain and its corresponding frequency domain representation is the subject of Fourier analysis and Fourier transforms.

The *forward Fourier transform*, time to frequency domain, of the function x(t) is defined

$$F[x(t)] = \int_{-\infty}^{\infty} x(t)\epsilon^{-j\omega t}\,dt = X(\omega) \qquad (1)$$

and the *inverse Fourier transform*, frequency to time domain, of $X(\omega)$ is

$$F^{-1}[X(\omega)] = \frac{1}{2\pi}\int_{-\infty}^{\infty} X(\omega)\epsilon^{j\omega t}\,d\omega = x(t) \qquad (2)$$

(For an in-depth study of the Fourier integral see reference 19.) Though these expressions are in themselves self-explanatory, a short illustrative example will be presented to aid in relating the two domains.

If an arbitrary time function representation of a periodic electrical signal, f(t), were plotted versus time as shown in *Figure 1*, its Fourier transform would indicate a spectral content consisting of a DC component, a fundamental frequency component ω_o, a fifth harmonic component $5\omega_o$ and a ninth harmonic component $9\omega_o$ (see *Figure 2*). It is illustratively seen in *Figure 3* that the superposition of these frequency components, in fact, yields the original time function f(t).

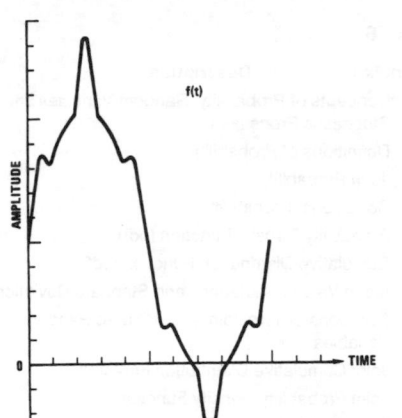

TL/H/8712-1

FIGURE 1. An Electrical Signal f(t)

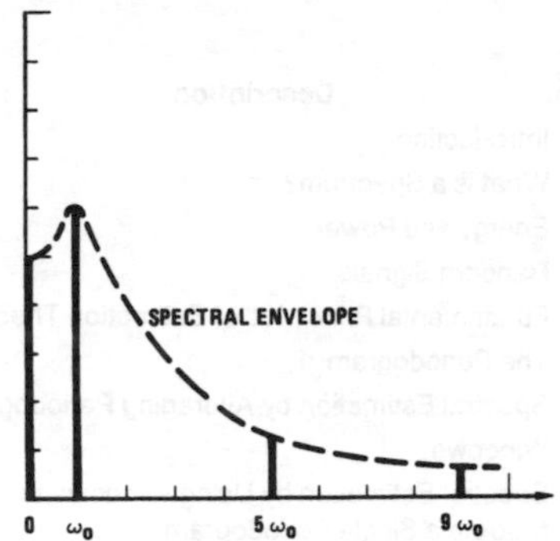

TL/H/8712-2

FIGURE 2. Spectral Composition or Spectrum $F(\omega)$ or f(t)

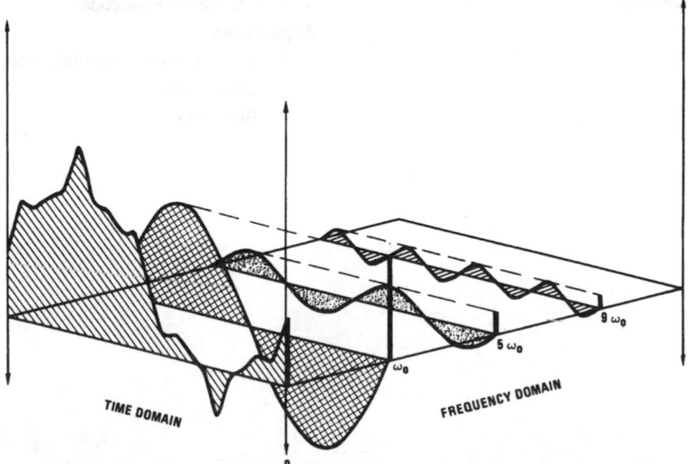

TL/H/8712-3

FIGURE 3. Combined Time Domain and Frequency Domain Plots

3.0 ENERGY AND POWER

In the previous section, time and frequency domain signal functions were related through the use of Fourier transforms. Again, the same relationship will be made in this section but the emphasis will pertain to signal power and energy.

Parseval's theorem relates the representation of energy, ω(t), in the time domain to the frequency domain by the statement

$$\omega(t) = \int_{-\infty}^{\infty} f_1(t)f_2(t)\,dt = \int_{-\infty}^{\infty} F_1(f)F_2(f)\,df \quad (3)$$

where f(t) is an arbitrary signal varying as a function of time and F(t) its equivalent Fourier transform representation in the frequency domain.

The proof of this is simply

$$\int_{-\infty}^{\infty} f_1(t)f_2(t)\,dt = \int_{-\infty}^{\infty} f_1(t)f_2(t)\,dt \quad (4a)$$

Letting $F_1(f)$ be the Fourier transform of $f_1(t)$

$$\int_{-\infty}^{\infty} f_1(t)f_2(t)\,dt = \int_{-\infty}^{\infty}\left[\int_{-\infty}^{\infty} F_1(f)\epsilon^{j2\pi ft}\,df\right]f_2(t)\,dt \quad (4b)$$

$$= \int_{-\infty}^{\infty}\left[F_1(f)\int_{-\infty}^{\infty} \epsilon^{j2\pi ft}\,df\right]f_2(t)\,dt \quad (4c)$$

Rearranging the integrand gives

$$\int_{-\infty}^{\infty} f_1(t)f_2(t)\,dt = \int_{-\infty}^{\infty} F_1(f)\left[\int_{-\infty}^{\infty} f_2(t)\epsilon^{j2\pi ft}\,dt\right]df \quad (4d)$$

and the factor in the brackets is seen to be $F_2(-f)$ (where $F_2(-f) = F_2^*(f)$ the conjugate of $F_2(f)$ so that

$$\int_{-\infty}^{\infty} f_1(t)f_2(t)\,dt = \int_{-\infty}^{\infty} F_1(f)F_2(-f)\,df \quad (4e)$$

A corollary to this theorem is the condition $f_1(t) = f_2(t)$ then $F(-f) = F^*(f)$, the complex conjugate of F(f), and

$$\omega(t) = \int_{-\infty}^{\infty} f^2(t)\,dt = \int_{-\infty}^{\infty} F(f)F^*(f)\,df \quad (5a)$$

$$= \int_{-\infty}^{\infty} |F(f)|^2\,df \quad (5b)$$

This simply says that the total energy† in a signal f(t) is equal to the area under the square of the magnitude of its Fourier transform. $|F(f)|^2$ is typically called the *energy density, spectral density,* or *power spectral density* function and $|F(f)|^2df$ describes the density of signal energy contained in the differential frequency band from f to f + dF.

In many electrical engineering applications, the instantaneous signal power is desired and is generally assumed to be equal to the square of the signal amplitudes i.e., $f^2(t)$.

†Recall the energy storage elements

Inductor $\quad v = L\dfrac{di}{dt}$

$$\omega(t) = \int_0^T vidt = \int_0^T L\frac{di}{dt}idt = \int_0^I Lidt = \frac{1}{2}LI^2$$

Capacitor $\quad i = c\dfrac{dv}{dt}$

$$\omega(t) = \int_0^T vidt = \int_0^T vc\frac{dv}{dt}dt = \int_0^V cvdv = \frac{1}{2}cv^2$$

This is only true, however, assuming that the signal in the system is impressed across a 1Ω resistor. It is realized, for example, that if f(t) is a voltage (current) signal applied across a system resistance R, its true instantaneous power would be expressed as $[f(t)]^2/R$ (or for current $[f(t)]^2R$) thus, $[f(t)]^2$ is the true power only if R = 1Ω.

So for the general case, power is always proportional to the square of the signal amplitude varied by a proportionality constant R, the impedance level in a circuit. In practice, however, it is undesirable to carry extra constants in the computation and customarily for signal analysis, one assumes signal measurement across a 1Ω resistor. This allows units of power to be expressed as the square of the signal amplitudes and the units of energy measured as volts²-second (amperes²-second).

For periodic signals, equation (5) can be used to define the average power, P_{avg}, over a time interval t_2 to t_1 by integrating $[f(t)]^2$ from t_1 to t_2 and then obtaining the average after dividing the result by $t_2 - t_1$ or

$$P_{avg} = \frac{1}{t_2 - t_1}\int_{t_1}^{t_2} f^2(t)\,dt \quad (6a)$$

$$= \frac{1}{T}\int_0^T f^2(t)\,dt \quad (6b)$$

where T is the period of the signal.

Having established the definitions of this section, energy can now be expressed in terms of power, P(t),

$$\omega(t) = \int_{-\infty}^{\infty} [f(t)]^2\,dt \quad (7a)$$

$$= \int_{-\infty}^{\infty} P(t)\,dt \quad (7b)$$

with power being the time rate of change of energy

$$P(t) = \frac{d\omega(t)}{dt} \quad (8)$$

As a final clarifying note, again, $|F(f)|^2$ and P(t), as used in equations (7b) and (8), are commonly called throughout the technical literature, *energy density, spectral density,* or *power spectral density* functions, PSD. Further, PSD may be interpreted as the average power associated with a bandwidth of one hertz centered at f hertz.

4.0 RANDOM SIGNALS

It was made apparent in previous sections that the use of Fourier transforms for analysis of linear systems is widespread and frequently leads to a saving in labor.

In view of using frequency domain methods for system analysis, it is natural to ask if the same methods are still applicable when considering a random signal system input. As will be seen shortly, with some modification, they *will* still be useful and the modified methods offer essentially the same advantages in dealing with random signals as with nonrandom signals.

It is appropriate to ask if the Fourier transform may be used for the analysis of any random sample function. Without proof, two reasons can be discussed which make the transform equations (1) and (2) invalid.

Firstly, $X(\omega)$ is a random variable since, for any fixed ω, each sample would be represented by a different value of the ensemble of possible sample functions. Hence, it is not a frequency representation of the process but only of one member of the process. It might still be possible, however, to use this function by finding its mean or expected value over the ensemble except that the second reason netages this approach. The second reason for not using the $X(\omega)$ of equations (1) and (2) is that, for stationary processes, it almost never exists. As a matter of fact, one of the conditions for a time function to be Fourier transformable is that it be integrable so that,

$$\int_{-\infty}^{\infty} |x(t)|\, dt < \infty \tag{9}$$

A sample from a stationary random process can never satisfy this condition (with the exception of generalized functions inclusive of impulses and so forth) by the argument that if a signal has nonzero power, then it has infinite energy and if it has finite energy then it has zero power (average power). Shortly, it will be seen that the class of functions having no Fourier integral, due to equation (9), but whose average power is finite can be described by statistical means.

Assuming $x(t)$ to be a sample function from a stochastic process, a truncated version of the function $x_T(t)$ is defined as

$$x_T(t) = \begin{cases} x(t) & |t| \le T \\ 0 & |t| > T \end{cases} \tag{10}$$

and

$$x(t) = \lim_{T \to \infty} x_T(t) \tag{11}$$

This truncated function is defined so that the Fourier transform of $x_T(t)$ can be taken. If $x(t)$ is a power signal, then recall that the transform of such a signal is not defined

$$\int_{-\infty}^{\infty} |x(t)|\, dt \text{ not less than } \infty \tag{12}$$

but that

$$\int_{-\infty}^{\infty} |x_T(t)|\, dt < \infty \tag{13}$$

The Fourier transform pair of the truncated function $x_T(t)$ can thus be taken using equations (1) and (2). Since $x(t)$ is a power signal, there must be a power spectral density function associated with it and the total area under this density must be the average power despite the fact that $x(t)$ is non-Fourier transformable.

Restating equation (5) using the truncated function $x_T(t)$

$$\int_{-\infty}^{\infty} x_T^2(t)\, dt = \int_{-\infty}^{\infty} |X_T(f)|^2\, df \tag{14}$$

and dividing both sides by 2T

$$\frac{1}{2T}\int_{-\infty}^{\infty} x_T^2(t)\, dt = \frac{1}{2T}\int_{-\infty}^{\infty} |X_T(f)|^2\, df \tag{15}$$

gives the left side of equation (15) the physical significance of being proportional to the average power of the sample function in the time interval $-T$ to T. This assumes $x_T(t)$ is a voltage (current) associated with a resistance. More precisely, it is the square of the *effective value* of $x_T(t)$ and for an ergodic process approaches the *mean-square* value of the process as T approaches infinity.

At this particular point, however, the limit as T approaches infinity cannot be taken since $X_T(f)$ is non-existent in the limit. Recall, though, that $X_T(f)$ is a random variable with respect to the ensemble of sample functions from which $x(t)$ was taken. The limit of the *expected value* of

$$\frac{1}{2T}|X_T(f)|^2$$

can be reasonably assumed to exist since its integral, equation (15), is always positive and certainly does exist. If the expectations $E\{\ \}$, of both sides of equation (15) are taken

$$E\left\{\frac{1}{2T}\int_{-\infty}^{\infty} x_T^2(t)\, dt\right\} = E\left\{\frac{1}{2T}\int_{-\infty}^{\infty} |x_T(f)|^2\, df\right\} \tag{16}$$

then interchanging the integration and expectation and at the same time taking the limit as $T \to \infty$

$$\lim_{T \to \infty} \frac{1}{2T}\int_{-\infty}^{\infty} \overline{x^2}(t)\, dt = \lim_{T \to \infty} \frac{1}{2T}\int_{-\infty}^{\infty} E\{|x_T(f)|^2\}\, df \tag{17}$$

results in

$$< \overline{x^2}(t) > = \int_{-\infty}^{\infty} \lim_{T \to \infty} \frac{E\{|x_T(f)|^2\}}{2T}\, df \tag{18}$$

where $\overline{x^2}(t)$ is defined as the mean-square value ($\overline{\ }$ denotes ensemble averaging and < > denotes time averaging).

For stationary processes, the time average of the mean-square value is equal to the mean-square value so that equation (18) can be restated as

$$\overline{x^2}(t) = \int_{-\infty}^{\infty} \lim_{T \to \infty} \frac{E\{|X_T(f)|^2\}}{2T}\, df \tag{19}$$

The integrand of the right side of equation (19), similar to equation (5b), is called the *energy density* spectrum or *power spectral density* function of a random process and will be designated by S(f) where

$$S(f) = \lim_{T \to \infty} \frac{E\{X_T(f)|^2\}}{2T} \tag{20}$$

Recall that letting $T \to \infty$ is not possible before taking the expectation.

Similar to the argument in Section III, the physical interpretation of power spectral density can be thought of in terms of average power. If $x(t)$ is a voltage (current) associated with a 1Ω resistance, $\overline{x^2}(t)$ is just the average power dissipated in that resistor and S(f) can be interpreted as the average power associated with a bandwidth of one hertz centered at f hertz.

S(f) has the units volts^2-second and its integral, equation (19), leads to the mean square value hence,

$$\overline{x^2}(t) = \int_{-\infty}^{\infty} S(f)\, df \tag{21}$$

Having made the tie between the power spectral density function S(f) and statistics, an important theorem in power spectral estimation can now be developed.

Using equation (20) and recalling that $X_T(f)$ is the Fourier transform of $x_T(t)$, assuming a nonstationary process,

$$S(f) = \lim_{T \to \infty} \frac{E\{|X_T(f)|^2\}}{2T} \tag{22}$$

$$S(f) = \lim_{T \to \infty} \frac{1}{2T} E\left\{\int_{-\infty}^{\infty} x_T(t_1)\epsilon^{j\omega t_1}\, dt_1 \int_{-\infty}^{\infty} x_T(t_2)\epsilon^{j\omega t_2}\, dt_2\right\} \tag{23}$$

Note that $|X_T(f)|^2 = X_T(f)X_T(-f)$ and that in order to distinguish the variables of integration when equation (23) is remanipulated the subscripts of t_1 and t_2 have been introduced. So, (see Appendix B)

$$S(f) = \lim_{T \to \infty} \left\{\frac{1}{2T} E\left[\int_{-\infty}^{\infty} dt_2 \int_{-\infty}^{\infty} \epsilon^{-j\omega(t_2 - t_1)} x_T(t_1)x_T(t_2)\, dt_1\right]\right\} \tag{24}$$

$$= \lim_{T \to \infty} \left\{\frac{1}{2T}\int_{-\infty}^{\infty} dt_2 \int_{-\infty}^{\infty} E[x_T(t_1)x_T(t_2)]\, \epsilon^{-j\omega(t_2 - t_1)}\, dt_1\right\} \tag{25}$$

Finally, the expectation $E[x_T(t_1)x_T(t_2)]$ is recognized as the autocorrelation, $R_{xx}(t_1, t_2)$ (Appendix A.14) function of the truncated process where

$$E[x_T(t_1)x_T(t_2)] = R_{xx}(t_1, t_2) \qquad |t_1|, |t_2| \le T$$
$$= 0 \qquad \text{everywhere else.}$$

Substituting

$$t_2 - t_1 = \tau \tag{26}$$

$$dt_2 = d\tau \tag{27}$$

equation (25) next becomes

$$S(f) = \lim_{T \to \infty} \frac{1}{2T}\int_{-\infty}^{\infty} d\tau \int_{-T}^{T} R_{xx}(t_1, t_1 + \tau)\epsilon^{-j\omega\tau}\, dt \tag{28}$$

or

$$S(f) = \left\{\left[\int_{-\infty}^{\infty} \lim_{T \to \infty} \frac{1}{2T} \int_{-T}^{T} R_{xx}(t_1, t_1 + \tau)\, dt_1\right]\epsilon^{-j\omega t}\right\} d\tau \tag{29}$$

We see then that the special density is the Fourier transform of the time average of the autocorrelation function. The relationship of equation (29) is valid for a nonstationary process.

For the stationary process, the autocorrelation function is independent of time and therefore

$$\langle R_{xx}(t_1, t_1, +\tau)\rangle = R_{xx}(\tau) \tag{30}$$

From this it follows that the spectral density of a stationary random process is just the Fourier transform of the autocorrelation function where

$$S(f) = \int_{-\infty}^{\infty} R_{xx}(\tau)\epsilon^{-j\omega t}\, d\tau \tag{31}$$

and

$$R_{xx}(\tau) = \int_{-\infty}^{\infty} S(f)\epsilon^{j\omega t}\, df \tag{32}$$

are described as the Wiener-Khintchine theorem.

The Wiener-Khintchine relation is of fundamental importance in analyzing random signals since it provides a link between the time domain [correlation function, $R_{xx}(\tau)$] and the frequency domain [spectral density, S(f)]. Note that the uniqueness is in fact the Fourier transformability. It follows, then, for a stationary random process that the autocorrelation function is the inverse transform of the spectral density function. For the nonstationary process, however, the autocorrelation function cannot be recovered from the spectral density. Only the time average of the correlation is recoverable, equation (29).

Having completed this section, the path has now been paved for a discussion on the techniques used in power spectral estimation.

5.0 FUNDAMENTAL PRINCIPLES OF ESTIMATION THEORY

When characterizing or modeling a random variable, estimates of its statistical parameters must be made since the data taken is a finite sample sequence of the random process.

Assume a random sequence x(n) from the set of random variables $\{x_n\}$ to be a realization of an ergodic random process (random for which ensemble or probability averages, E[ ], are equivalent to time averages, < >) where for all n,

$$m = E[x_n] = \int_{-\infty}^{\infty} x f_x(x)\, dx \tag{33}$$

Assuming further that the estimate of the desired averages of the random variables $\{x_n\}$ from a finite segment of the sequence, x(n) for $0 \le n \le N-1$, to be

$$m = \langle x_n \rangle = \lim_{N \to \infty} \frac{1}{2N+1} \sum_{n=-N}^{N} x_n \tag{34}$$

then for each sample sequence

$$m = \langle x(n) \rangle = \lim_{N \to \infty} \frac{1}{2N+1} \sum_{n=-N}^{N} x(n) \tag{35}$$

Since equation (35) is a precise representation of the mean value in the limit as N approaches infinity then

$$\hat{m} = \frac{1}{N} \sum_{n=0}^{N-1} x(n) \tag{36}$$

may be regarded as a fairly accurate estimate of m for sufficiently large N. The caret (∧) placed above a parameter is representative of an estimate. The area of statistics that pertains to the above situations is called *estimation theory*. Having discussed an elementary statistical estimate, a few common properties used to characterize the estimator will next be defined. These properties are *bias, variance* and *consistency*.

Bias

The difference between the mean or expected value $E[\hat{\eta}]$ of an estimate $\hat{\eta}$ and its true value η is called the *bias*.

$$\text{bias} = B_{\hat{\eta}} = \eta - E[\hat{\eta}] \tag{37}$$

Thus, if the mean of an estimate is equal to the true value, it is considered to be *unbiased* and having a *bias* value equal to zero.

Variance

The variance of an estimator effectively measures the width of the probability density (Appendix A.4) and is defined as

$$\sigma^2_{\hat{\eta}} E = \left[(\hat{\eta} - E[\hat{\eta}])^2 \right] \tag{38}$$

A good estimator should have a small variance in addition to having a small bias suggesting that the probability density function is concentrated about its mean value. The mean square error associated with this estimator is defined as

$$\text{mean square error} = E\left[(\hat{\eta} - \eta)^2 \right] = \sigma^2_{\hat{\eta}} + B^2_{\hat{\eta}} \tag{39}$$

Consistency

If the bias and variance both tend to zero as the limit tends to infinity or the number of observations become large, the estimator is said to be consistent. Thus,

$$\lim_{N \to \infty} \sigma^2_{\hat{\eta}} = 0 \tag{40}$$

and

$$\lim_{N \to \infty} B_{\hat{\eta}} = 0 \tag{41}$$

This implies that the estimator converges in probability to the true value of the quantity being estimated as N becomes infinite.

In using estimates the mean value estimate of m_x, for a Gaussian random process is the *sample mean* defined as

$$\hat{m}_x = \frac{1}{N} \sum_{i=0}^{N-1} x_i \tag{42}$$

the mean square value is computed as

$$E[\hat{m}^2_x] = \frac{1}{N^2} \sum_{i=0}^{N-1} \sum_{j=0}^{N-1} E[x_i x_j] \tag{43}$$

$$= \frac{1}{N^2} \left[\sum_{i=0}^{N-1} E[x_i^2] + \sum_{i=0}^{N-1} \sum_{\substack{j=0 \\ i \neq j}}^{N-1} E[x_i] \bullet E[x_j] \right] \tag{44}$$

$$= \frac{1}{N_2} \left[N(E[x_n^2]) + \left(\sum_{i=0}^{N-1} E[x_i] \right) \left(\sum_{\substack{j=0 \\ i \neq j}}^{N-1} E[x_j] \right) \right] \tag{45}$$

$$= \frac{1}{N} E[x_n^2] + (m_x)^2 \frac{N-1}{N} \tag{46}$$

thus allowing the variance to be stated as

$$\sigma^2_{\hat{m}_x} = E[(\hat{m}_x)^2] - \{E[\hat{m}_x]\}^2 \tag{47}$$

$$= \frac{1}{N} E[x_n^2] + (m_x^2) \frac{N-1}{N} - \{E[\hat{m}_x]\}^2 \tag{48}$$

$$= \frac{1}{N} E[x_n^2] + (m_x^2) \frac{N-1}{N} - m_x^2 \tag{49}$$

$$= \frac{1}{N} (E[x_n^2] - m_x^2) \tag{50}$$

$$= \frac{\sigma_x^2}{N} \tag{51}$$

This says that as the number of observations N increase, the variance of the sample mean decreases, and since the bias is zero, the sample mean is a consistent estimator.

If the variance is to be estimated and the mean value is a known then

$$\hat{\sigma}^2_x = \frac{1}{N} \sum_{i=0}^{N-1} (x_i - m_x)^2 \tag{52}$$

this estimator is consistent.

If, further, the mean and the variance are to be estimated then the sample variance is defined as

$$\hat{\sigma}^2_x = \frac{1}{N} \sum_{i=0}^{N-1} (x_i - \hat{m}_x)^2 \tag{53}$$

again $\hat{m}_x$ is the sample mean.

The only difference between the two cases is that equation (52) uses the true value of the mean, whereas equation (53) uses the estimate of the mean. Since equation (53) uses an estimator the bias can be examined by computing the expected value of $\hat{\sigma}^2_x$ therefore,

$$E[\hat{\sigma}^2_x] = \frac{1}{N} \sum_{i=0}^{N-1} (E[x_i] - E[\hat{m}_x])^2 \tag{54}$$

$$= \frac{1}{N} \sum_{i=0}^{N-1} \left\{ E[x_i^2] - 2E[x_i \hat{m}_x] + E[\hat{m}^2_x] \right\} \tag{55}$$

$$= \frac{1}{N} \sum_{i=0}^{N-1} E[x_i^2] - \frac{2}{N^2} \sum_{i=0}^{N-1} \left(\sum_{j=0}^{N-1} E[x_i x_j] \right) + \frac{1}{N^2} \sum_{i=0}^{N-1} \sum_{j=0}^{N-1} E[x_i x_j] \tag{56}$$

$$= \frac{1}{N}\sum_{i=0}^{N-1} E[x_i^2] - \frac{2}{N^2}\left(\sum_{i=0}^{N-1} E[x_i^2] + \sum_{i=0}^{N-1}\sum_{\substack{j=0\\ i\neq j}}^{N-1} E[X_i]\bullet E[x_j]\right) + \frac{1}{N^2}\left(\sum_{i=0}^{N-1} E[x_i^2] + \sum_{i=0}^{N-1}\sum_{j=0}^{N-1} E[x_i]\bullet E[x_j]\right) \quad (57)$$

$$= \frac{1}{N}\left(N\bullet E[x_i^2]\right) - \frac{2}{N^2}\left[N\bullet E[x_i^2] + N(N-1)m_x^2\right] + \frac{1}{N^2}\left[N\bullet E[x_i^2] + N(N-1)m_x^2\right] \quad (58)$$

$$= \frac{1}{N}\left(N\bullet E[x_i^2]\right) - \frac{2N}{N^2}\left(E[x_i^2]\right) - \frac{2N(N-1)}{N^2}m_x^2 + \frac{N}{N^2}E[x_i^2] + \frac{N(N-1)}{N^2}m_x^2 \quad (59)$$

$$= \frac{1}{N}\left(N\bullet E[x_i^2]\right) - \frac{2}{N}(E[x_i^2]) - \frac{2(N-1)}{N}m_x^2 + \frac{1}{N}E[x_i^2] + \frac{(N-1)}{N}m_x^2 \quad (60)$$

$$= \frac{1}{N}(N - E[x_i^2]) - \frac{1}{N}(E[x_i^2]) - \frac{(N-1)}{N}m_x^2 \quad (61)$$

$$= \frac{(N-1)}{N}(E[x_i^2]) - \frac{(N-1)}{N}m_x^2 \quad (62)$$

$$= \frac{(N-1)}{N}\sigma_x^2 \quad (63)$$

It is apparent from equation (63) that the mean value of the sample variance does not equal the variance and is thus biased. Note, however, for large N the mean of the sample variance asymptotically approaches the variance and the estimate virtually becomes unbiased. Next, to check for consistency, we will proceed to determine the variance of the estimate sample variance. For ease of understanding, assume that the process is zero mean, then letting

$$\psi = \hat{\sigma}_x^2 = \frac{1}{N}\sum_{i=0}^{N-1} x_i^2 \quad (64)$$

so that,

$$E[\psi^2] = \frac{1}{N^2}\sum_{i=1}^{N}\sum_{k=1}^{N} E[x_i^2 x_k^2] \quad (65)$$

$$= \frac{1}{N^2}\left[NE[x_n^4] + N(N-1)\,(E[x_n^2])^2\right] \quad (66)$$

$$= \frac{1}{N}\left[E[x_n^4] + (N-1)(E[x_n^2])^2\right] \quad (67)$$

the expected value

$$E[\psi] = E[x_n^2] \quad (68)$$

so finally

$$\text{var}[\hat{\sigma}_x^2] = E[\psi^2] - (E[\psi])^2 \quad (69)$$

$$= \frac{1}{N}\left[E[x_n^4] - (E[x_n^2])^2\right] \quad (70)$$

Re-examining equations (63) and (70) as N becomes large clearly indicates that the sample variance is a consistent estimate. Qualitatively speaking, the accuracy of this estimate depends upon the number of samples considered in the estimate. Note also that the procedures employed above typify the style of analysis used to characterize estimators.

6.0 THE PERIODOGRAM

The first method defines the sampled version of the Wiener-Khintchine relation, equations (31) and (32), where the power spectral density estimate $S_{N_{xx}}(f)$ is the discrete Fourier transform of the autocorrelation estimate $R_{N_{xx}}(k)$ or

$$S_{N_{xx}}(f) = \sum_{k=-\infty}^{\infty} R_{N_{xx}}(k)\,\epsilon^{-j\omega kT} \quad (71)$$

This assumes that x(n) is a discrete time random process with an autocorrelation function $R_{N_{xx}}(k)$.

For a finite sequence of data

$$x(n) = \begin{cases} x_n & \text{for } n = 0, 1, \ldots, N-1 \\ 0 & \text{elsewhere} \end{cases} \quad (72)$$

called a rectangular data window, the sample autocorrelation function (sampled form of equation A.14-9)

$$R_{N_{xx}}(k) = \frac{1}{N}\sum_{n=-\infty}^{\infty} x(n)x(n+k) \quad (73)$$

can be substituted into equation (71) to give the spectral density estimate

$$S_{N_{xx}}(f) = \frac{1}{N}|X_N(f)|^2 \quad (74)$$

called the periodogram.

$$\left(\text{Note: } \frac{|X_N(f)|^2}{N} = \frac{X_N(f)X_{N^*}(f)}{N} = \frac{X_N{}^2(f)_{real} + X_N{}^2(f)_{imag}}{N} = F\left[R_{N_{xx}}(k)\right] = F\left[E[x(n)x(n+k)]\right]\right).$$

Hence,

$$S_{N_{xx}}(f) = \sum_{k=-\infty}^{\infty} R_{N_{xx}}(k)\,\epsilon^{-j\omega kT} \quad (75)$$

$$= \sum_{k=-\infty}^{\infty}\left[\frac{1}{N}\sum_{n=-\infty}^{\infty} x(n)x(n+k)\right]\epsilon^{-j\omega kT} \quad (76)$$

so letting $1 = \epsilon^{j\omega nT} \epsilon^{-j\omega nT}$

$$S_{N_{XX}}(f) = \frac{1}{N}\left(\sum_{n=-\infty}^{\infty} x(n)\epsilon^{j\omega nT} \sum_{k=-\infty}^{\infty} x(n+k)\,\epsilon^{-j\omega(n+k)T}\right) \tag{77}$$

and allowing the variable m = n + k

$$S_{N_{XX}}(f) = \frac{1}{N} X_N(f) X_N{}^*(f) = \frac{1}{N}|X_N(f)|^2 \tag{78}$$

in which the Fourier transform of the signal is

$$X_N(f) = \sum_{n=-\infty}^{\infty} x(n)\,\epsilon^{-j\omega nT} \tag{79}$$

The current discussion leads one to believe that the periodogram is an excellent estimator of the true power spectral density S(f) as N becomes large. This conclusion is false and shortly it will be verified that the periodogram is, in fact, a poor estimate of S(f). To do this, both the expected value and variance of $S_{N_{XX}}(f)$ will be checked for consistency as N becomes large. As a side comment it is generally faster to determine the power spectral density, $S_{N_{XX}}(f)$, of the random process using equation (74) and then inverse Fourier transforming to find $R_{N_{XX}}(k)$ than to obtain $R_{N_{XX}}(k)$ directly. Further, since the periodogram is seen to be the magnitude squared of the Fourier transformed data divided by N, the power spectral density of the random process is unrelated to the angle of the complex Fourier transform $X_N(f)$ of a typical realization.

Prior to checking for the consistency of $S_{N_{XX}}(f)$, the sample autocorrelation must initially be found consistent. Proceeding, since the sample autocorrelation estimate

$$R_{N_{XX}}(k) = \frac{x(0)x(k)+x(1)x(|k|+1)+\ldots+x(N-1-|k|)x(N-1)}{N} \tag{80}$$

$$= \frac{1}{N}\sum_{n=0}^{N-1-|k|} x(n)x(n+|k|) \tag{81}$$

$$k = 0, \pm 1, \pm 2, \ldots, \pm N-1$$

which averages together all possible products of samples separated by a lag of k, then, the mean value of the estimate is related to the true autocorrelation function by

$$E[R_{N_{XX}}(k)] = \left(\frac{1}{N}\sum_{n=0}^{N-1-|k|} E[x(n)x(n+|k|)]\right) = \frac{N-|k|}{N}R(k) \tag{82}$$

where the true autocorrelation function R(k) is defined as (the sample equivalent of equation A.14-8)

$$R(k) = E[x(n)x(n+k)] \tag{83}$$

From equation (82) it is observed that $R_{N_{XX}}(k)$ is a biased estimator. It is also considered to be asymptotically unbiased since the term

$$\frac{N-|k|}{N}$$

approaches 1 as N becomes large. From these observations $R_{N_{XX}}(k)$ can be classified as a good estimator of R(k).

In addition to having a small bias, a good estimator should also have a small variance. The variance of the sample autocorrelation function can thus be computed as

$$\text{var}[R_{N_{XX}}(k)] = E[R^2_{N_{XX}}(k)] - E^2[R_{N_{XX}}(k)] \tag{84}$$

Examining the $E[R_{N_{XX}}(k)]$ term of equation (84), substituting the estimate of equation (81) and replacing n with m, it follows that

$$E[R^2_{N_{XX}}(k)] = E\left\{\left[\frac{1}{N}\sum_{n=0}^{N-1-|k|} x(n)x(n+|k|)\right]\left[\frac{1}{N}\sum_{m=0}^{N-1-|k|} x(m)x(m+|k|)\right]\right\} \tag{85}$$

$$= \frac{1}{N^2}\left(\sum_{n=0}^{N-1-|k|}\sum_{m=0}^{N-1-|k|} E[x(n)x(n+|k|)x(m)x(m+|k|)]\right) \tag{86}$$

If the statistical assumption that x(n) is a zero-mean Gaussian process, then the zero-mean, jointly Gaussian, random variables symbolized as X_1, X_2, X_3 and X_4 of equation (86) can be described as [Ref. (30)].

$$E[X_1X_2X_3X_4] = E[X_1X_2]\,E[X_3X_4] + E[X_1X_3]\,E[X_2X_4] + E[X_1X_4]\,E[X_2X_3] \tag{87}$$

$$= \Big[E[x(n)\,x(n+|k|)]\,E[x(m)\,x(m+|k|)] + E[x(n)x(m)]E[x(n+|k|)x(m+|k|)] + E[x(n)\,x(m+|k|)]\,E[x(n+|k|)\,x(m)]\Big] \tag{88}$$

Using equation (88), equation (84) becomes

$$\text{Var}[R_{N_{XX}}(k)] = \left\{\frac{1}{N^2}\sum_{n=0}^{N-1-|k|}\sum_{m=0}^{N-1-|k|} R_{N_{XX}}(k)R_{N_{XX}}(k) + R_{N_{XX}}(n-m)\,R_{N_{XX}}(n-m) + R_{N_{XX}}(n-m-|k|)\,R_{N_{XX}}(n-m+|k|)\right\} - \left[\frac{1}{N}\sum_{n=0} R_{N_{XX}}(k)\right]^2 \tag{89}$$

$$\text{Var}[R_{N_{XX}}(k)] = \frac{1}{N^2}\sum_{n=0}^{N-1-|k|}\sum_{m=0}^{N-1-|k|}\left\{R^2_{N_{XX}}(n-m) + R_{N_{XX}}(n-m-k)R_{N_{XX}}(n-m+|k|)\right\} \tag{90}$$

where the lag term $n - m$ was obtained from the lag difference between $\tau = n - m = (n + k) - (m + k)$ in the second term of equation (88). The lag term $n - k + m$ and $n - k - m$ was obtained by referencing the third term in equation (88) to n, therefore for

$$E[x(n)\,x(m + |k|)] \tag{91}$$

the lag term $\tau = n - (m + |k|)$ so

$$E[x(n)\,x(m + |k|)] = R_{N_{xx}}(n - m + |k|) \tag{92}$$

and for

$$E[x(n + |k|)\,x(m)] \tag{93}$$

first let $n - m$ then add $|k|$ so $\tau = n - m + |k|$ and

$$E[x(n + |k|)\,x(m)] = R_{N_{xx}}(n - m + |k|) \tag{94}$$

Recall that a sufficient condition for an estimator to be consistent is that its bias and variance both converge to zero as N becomes infinite. This essentially says that an estimator is consistent if it converges in probability to the true value of the quantity being estimated as N approaches infinity.

Re-examining equation (90), the variance of $R_{N_{xx}}(k)$, and equation (82), the expected value of $R_{N_{xx}}(k)$, it is found that both equations tend toward zero for large N and therefore $R_{N_{xx}}(k)$ is considered as a consistent estimator of R(k) for fixed finite k in practical applications.

Having established that the autocorrelation estimate is consistent, we return to the question of the periodogram consistency.

At first glance, it might seem obvious that $S_{N_{xx}}(f)$ should inherit the asymptotically unbiased and consistent properties of $R_{N_{xx}}(k)$, of which it is a Fourier transform. Unfortunately, however, it will be shown that $S_{N_{xx}}(f)$ does not possess these nice statistical properites.

Going back to the power spectral density of equation (71).

$$S_{N_{xx}}(f) = \sum_{k=\infty}^{\infty} R_{N_{xx}}(k)\,\epsilon^{-j\omega kT}$$

and determining its expected value

$$E[S_{N_{xx}}(f)] = \sum_{k=-\infty}^{\infty} E[R_{N_{xx}}(k)]\,\epsilon^{-j\omega kT} \tag{95}$$

the substitution of equation (82) into equation (95) yields the mean value estimate

$$E[S_{N_{xx}}(f)] = \sum_{K=-N}^{N} R(k)\left(1 - \frac{|k|}{N}\right)\epsilon^{-j\omega kT} \tag{96}$$

the $\left(1 - \frac{|k|}{N}\right)$

term of equation (96) can be interpreted as a(k), the triangular window resulting from the autocorrelation of finite-sequence rectangular-data window $\omega(k)$ of equation (72). Thus,

$$a(k) = \begin{cases} 1 - \dfrac{|k|}{N} & |k| < N - 1 \quad \text{(97a)} \\ 0 & \text{elsewhere} \quad \text{(97b)} \end{cases}$$

and the expected value of the periodogram can be written as the finite sum

$$E[S_{N_{xx}}(f)] = \sum_{k=-\infty}^{\infty} R_{N_{xx}}(k)\,a(k)\,\epsilon^{-j\omega kT} \tag{98}$$

Note from equation (98) that the periodogram mean is the discrete Fourier transform of a product of the true autocorrelation function and a triangular window function. This frequency function can be expressed entirely in the frequency domain by the convolution integral. From equation (98), then, the convolution expression for the mean power spectral density is thus,

$$E[S_{N_{xx}}(f)] = \int_{-1/2}^{1/2} S(\eta)\,A(f - \eta)\,d\eta \tag{99}$$

where the general frequency expression for the transformed triangular window function A(f) is

$$A(f) = \frac{1}{N}\left[\frac{\sin(2\pi f)\frac{N}{2}}{\sin\frac{(2\pi f)}{2}}\right]^2 \tag{100}$$

Re-examining equation (98) or (96) it can be said that equation (71) or (74) gives an asymptotically unbiased estimate of S(f) with the distorting effects of a(k) vanishing as N tends toward infinity. At this point equation (98) still appears as a good estimate of the power spectral density function. For the variance var $[S_{N_{xx}}(f)]$ however, it can be shown [Ref. (10)] that if the data sequence x(n) comes from a Gaussian process then the variance of $S_{N_{xx}}(f)$ approaches the square of the true spectrum, $S^2(f)$, at each frequency f. Hence, the variance is not small for increasing N,

$$\lim_{N \to \infty} \operatorname{var}[S_{N_{xx}}(f)] = S^2(f) \tag{101}$$

More clearly, if the ratio of mean to standard deviation is used as a kind of signal-to-noise ratio, i.e.

$$\frac{E[S_{N_{xx}}(f)]}{\{\operatorname{var}[S_{N_{xx}}(f)]\}^{1/2}} \cong \frac{S(f)}{S(f)} = 1 \tag{102}$$

it can be seen that the true signal spectrum is only as large as the noise or uncertainty in $S_{N_{xx}}(f)$ for increasing N. In addition, the variance of equation (101), which also is approximately applicable for non-Gaussian sequences, indicates that calculations using different sets of N samples from the same x(n) process will yield vastly different values of $S_{N_{xx}}(f)$ even when N becomes large. Unfortunately, since the variance of $S_{N_{xx}}(f)$ does not decrease to zero as N approaches infinity, the periodogram is thus an inconsistent estimate of the power spectral density and cannot be used for spectrum analysis in its present form.

7.0 SPECTRAL ESTIMATION BY AVERAGING PERIODOGRAMS

It was shown in the last section that the periodogram was not a consistent estimate of the power spectral density function. A technique introduced by Bartlett, however, allows the use of the periodogram and, in fact, produces a consistent spectral estimation by averaging periodograms. In short, Bartlett's approach reduces the variance of the estimates by averaging together several independent periodograms. If, for example $X_1, X_2, X_3, \ldots, X_L$ are uncorrelated random variables having an expected value E[x] and a variance σ^2, then the arithmetic mean

$$\frac{X_1 + X_2 + X_3 + \ldots + X_L}{L} \tag{103}$$

has the expected value E[x] and a variance of σ^2/L. This fact suggests that a spectral estimator can have its variance reduced by a factor of L over the periodogram. The procedure requires the observed process, an N point data sequence, to be split up into L nonoverlapping M point sections and then averaging the periodograms of each individual section.

To be more specific, dividing an N point data sequence x(n), $0 \leq n \leq N-1$, into L segments of M samples each the segments $X_M^{\ell}(n)$ are formed. Thus,

$$x_M^{\ell}(f) = x(n + \ell M - M) \quad \begin{cases} 0 \leq n \leq M-1 \\ 1 \leq \ell \leq L \end{cases} \tag{104}$$

where the superscript ℓ specifies the segment or interval of data being observed, the subscript M represents the number of data points or samples per segment and depending upon the choice of L and M, we have $N \geq LM$. For the computation of L periodograms

$$S_M^{\ell}(f) = \left| \frac{1}{M} \sum_{n=0}^{M-1} x_M^{\ell}(n)\, \epsilon^{-j\omega nT} \right|^2 \quad 1 \leq \ell \leq L \tag{105}$$

If the autocorrelation function $R_{N_{xx}}(m)$ becomes negligible for m large relative to M, $m > M$, then it can be said that the periodograms of the separate sections are virtually independent of one another. The corresponding *averaged periodogram estimator* $\hat{S}_M^{\ell}(f)$ computed from L individual periodograms of length M is thus defined

$$\hat{S}_M^{\ell}(f) = \frac{1}{L} \sum_{\ell=1}^{L} S_M^{\ell}(f) \tag{106}$$

Since the L subsidiary estimates are identically distributed periodograms, the averaged spectral estimate will have the same mean or expected value as any of the subsidiary estimates so

$$E[\hat{S}_M^{\ell}(f)] = \frac{1}{L} \sum_{\ell=1}^{L} E[S_M^{\ell}(f)] \tag{107}$$

$$= E[S_M^{\ell}(f)] \tag{108}$$

From this, the expected value of the Bartlett estimate can be said to be the convolution of the true spectral density with the Fourier transform of the triangular window function corresponding to the M sample periodogram where $M \leq N/L$ equations (98) or (99) we see that

$$E[\hat{S}_M^{\ell}(f)] = E[S_M^{\ell}(f)] = \frac{1}{M} \int_{-1/2}^{1/2} S(\eta)\, A(f-\eta)\, d\eta \tag{109}$$

where A(f) is the Fourier transformed triangular window function of equation (100). Though the averaged estimate is no different than before, its variance, however, is smaller. Recall that the variance of the average of L identical independent random variables is 1/L of the individual variances, equation (51). Thus, for L statistically independent periodograms, the variance of the averaged estimate is

$$\text{var}[\hat{S}_M^{\ell}(f)] = \frac{1}{L}\, \text{var}[S_{N_{xx}}(f)] \cong \frac{1}{L}\, [S(f)]^2 \tag{110}$$

So, again, the averaging of L periodograms results in approximately a factor of L reduction in power spectral density estimation variance. Since the variance of equation (110) tends to zero as L approaches infinity and through equation (98) and (99) $\hat{S}_M^{\ell}(f)$ is asymptotically unbiased, $\hat{S}_M^{\ell}(f)$ can be said to be a consistent estimate of the true spectrum.

A few notes are next in order. First, the L fold variance reduction or $(L)^{1/2}$ signal-to-noise ratio improvement of equation (102) is not precisely accurate since there is some dependence between the subsidiary periodograms. The adjacent samples will correlated unless the process being analyzed is white.

However, as indicated in equation (110), such a dependence will be small when there are many sample intervals per periodogram so that the reduced variance is still a good approximation. Secondly, the bias of $\hat{S}_M^{\ell}(f)$, equation (106), is greater than $\hat{S}_M^{\ell}(f)$, equation (105), since the main lobe of the spectral window is larger for the former. For this situation, then, the bias can be thought of as effecting spectral resolution. It is seen that increasing the number of periodograms for a fixed record length N decreases not only the variance but, the samples per periodograms M decrease also. This decreases the spectral resolution. Thus when using the Bartlett procedure the actual choice of M and N will typically be selected from prior knowledge of a signal or data sequence under consideration. The tradeoff, however, will be between the spectral resolution of bias and the variance of the estimate.

8.0 WINDOWS

Prior to looking at other techniques of spectrum estimation, we find that to this point the subject of spectral windows has been brought up several times during the course of our discussion. No elaboration, however, has been spent explaining their spectral effects and meaning. It is thus appropriate at this juncture to diverge for a short while to develop a fundamental understanding of windows and their spectral implications prior to the discussion of Sections 9 and 10 (for an in depth study of windows see *Windows, Harmonic Analysis and the Discrete Fourier Transform*; Frederic J. Harris; submitted to IEEE Proceedings, August 1976).

In most applications it is desirable to taper a finite length data sequence at each end to enhance certain characteristics of the spectral estimate. The process of terminating a sequence after a finite number of terms can be thought of as multiplying an infinite length, i.e., impulse response sequence by a finite width window function. In other words, the window function determines how much of the original impulse sequence can be observed through this window, see *Figures 4a, 4b,* and *4c*. This tapering by mulitiplying the sequence by a data window is thus analogous to multiplying the correlation function by a lag window. In addition, since multiplication in the time domain is equivalent to convolution in the frequency domain then it is also analogous to convolving the Fourier transform of a finite-length-sequence with the Fourier transform of the window function, *Figures 4d, 4e,* and *4f*. Note also that the Fourier transform of the rectangular window function exhibits significant oscillations and poor high frequency convergence, *Figure 4e*. Thus, when convolving this spectrum with a desired amplitude function, poor convergence of the resulting amplitude response may occur. This calls for investigating the use of other possible window functions that minimize some of the difficulties encountered with rectangular function.

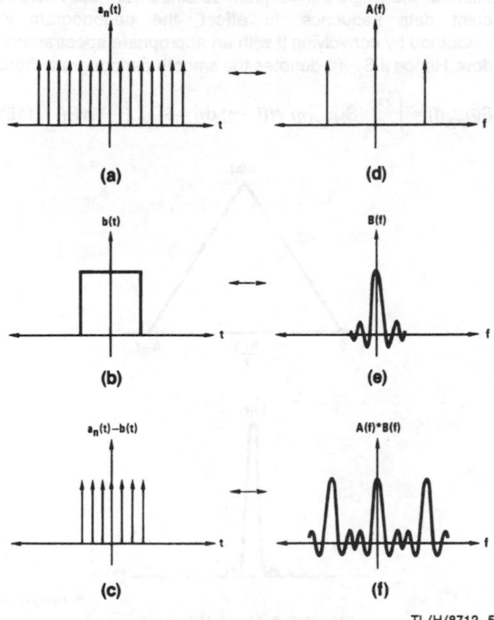

FIGURE 4

In order for the spectrum of a window function to have minimal effects upon the desired amplitude response, resulting from convolving two functions, it is necessary that the window spectrum approximate an impulse function. This implies that as much of its energy as possible should be concentrated at the center of the spectrum. Clearly, an ideal impulse spectrum is not feasible since this requires an infinitely long window.

In general terms, the spectrum of a window function typically consists of a main lobe, representing the center of the spectrum, and various side lobes, located on either side of the main lobe (see *Figures 6* thru *9*). It is desired that the window function satisfy two criteria; (1) that the main lobe should be as narrow as possible and (2) relative to the main lobe, the maximum side lobe level should be as small as possible. Unfortunately, however, both conditions cannot be simultaneously optimized so that, in practice, usable window functions represent a suitable compromise between the two criteria. A window function in which minimization of the main lobe width is the primary objective, fields a finer frequency resolution but suffers from some oscillations, i.e., the spectrum passband and substantial ripple in the spectrum stopband. Coversely, a window function in which minimization of the side lobe level is of primary concern tends to have a smoother amplitude response and very low ripple in the stopband but, yields a much poorer frequency resolution. Examining *Figure 5* assume a hypothetical impulse response, *Figure 5a*, whose spectrum is *Figure 5b*. Multiplying the impulse response by the rectangular window, *Figure 4b*, yields the windowed impulse response, *Figure 5c*, implying the convolution of the window spectrum, *Figure 4e*, with the impulse response spectrum, *Figure 5b*. The result of this convolution is seen in *Figure 5d* and is a distorted version of the ideal spectrum, *Figure 5b*, having passband oscillations and stopband ripple. Selecting another window, i.e., *Figure 9* with more desirable spectral characteristics, we see the appropriately modified windowed data, *Figure 5e*, results in a very good approximation of *Figure 5b*.

This characteristically provides a smoother passband and lower stopband ripple level but sacrifices the sharpness of the roll-off rate inherent in the use of a rectangular window (compare *Figures 5d* and *5f*). Concluding this brief discussion, a few common window functions will next be considered.

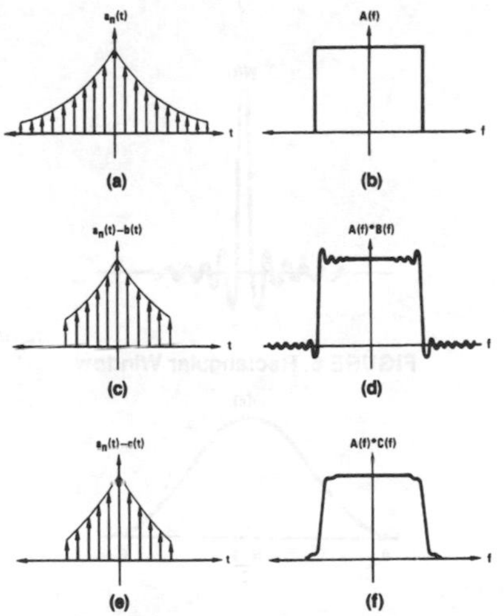

FIGURE 5. (a)(b) Unmodified Data Sequence (c)(d) Rectangular Windowed Data Sequence (e)(f) Hamming Windowed Data Sequence

Rectangular window: *Figure 6*

$$w(n) = 1 \qquad |n| < \frac{N-1}{2} \qquad (111)$$
$$= 0 \qquad \text{otherwise}$$

Bartlett or triangular window: *Figure 7*

$$w(n) = 1 - \frac{2|n|}{N} \qquad |n| < \frac{N-1}{2} \qquad (112)$$
$$= 0 \qquad \text{otherwise}$$

Hann window: *Figure 8*

$$w(n) = 0.5 + 0.5 \cos\left(\frac{2\pi n}{N}\right) \qquad |n| < \frac{N-1}{2} \qquad (113)$$
$$= 0 \qquad \text{otherwise}$$

Hamming window: *Figure 9*

$$w(n) = 0.54 + 0.46 \cos\left(\frac{2\pi n}{N}\right) \quad |n| < \frac{N-1}{2} \qquad (114)$$
$$= 0 \qquad \text{otherwise}$$

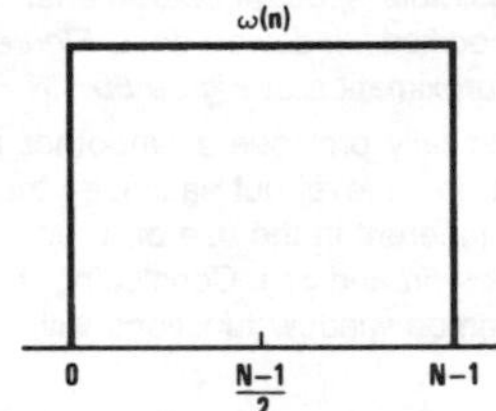

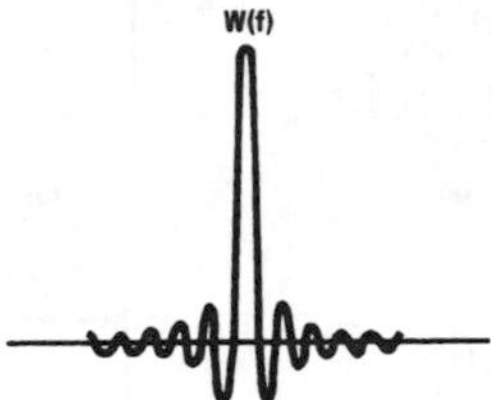

TL/H/8712-6

FIGURE 6. Rectangular Window

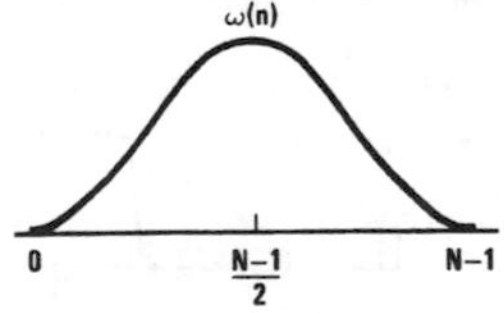

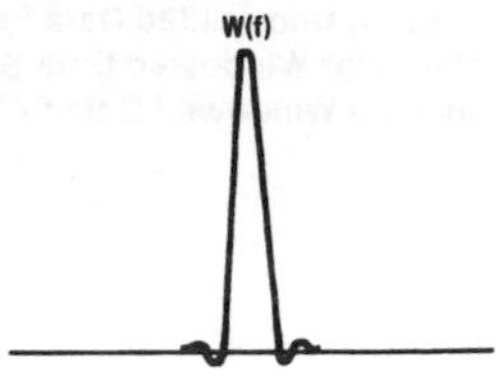

TL/H/8712-7

FIGURE 7. Bartlett or Triangular Window

Again the reference previously cited should provide a more detailed window selection. Nevertheless, the final window choice will be a compromise between spectral resolution and passband (stopband) distortion.

9.0 SPECTRAL ESTIMATION BY USING WINDOWS TO SMOOTH A SINGLE PERIODOGRAM

It was seen in a previous section that the variance of a power spectral density estimate based on an N point data sequence could be reduced by chopping the data into shorter segments and then averaging the periodograms of the individual segments. This reduced variance was acquired at the expense of increased bias and decreased spectral resolution. We will cover in this section an alternate way of computing a reduced variance estimate using a smoothing operation on the single periodogram obtained from the entire N point data sequence. In effect, the periodogram is smoothed by convolving it with an appropriate spectral window. Hence if $S_{XX}(f)$ denotes the smooth periodogram then,

$$S_{W_{XX}}(f) = \int_{-1/2}^{1/2} S_{N_{XX}}(\eta)\, W(f-\eta)\, d\eta = S_{N_{XX}}(\eta) * W(\eta) \qquad (115)$$

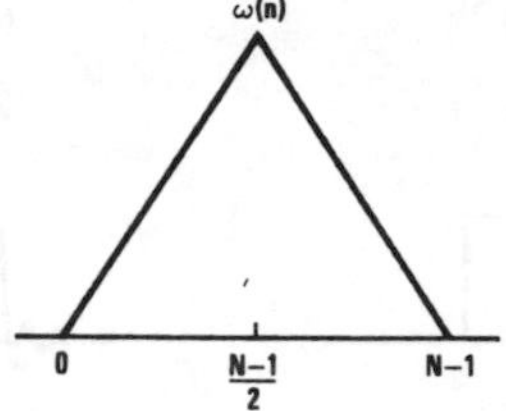

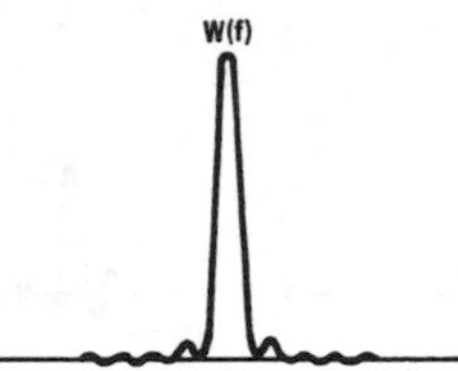

TL/H/8712-8

FIGURE 8. Hann Window

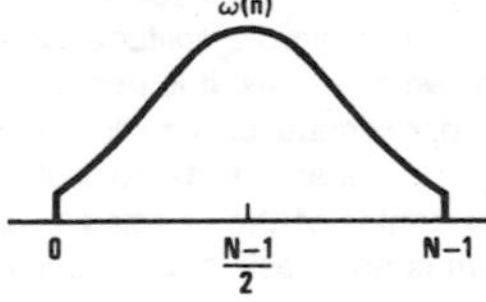

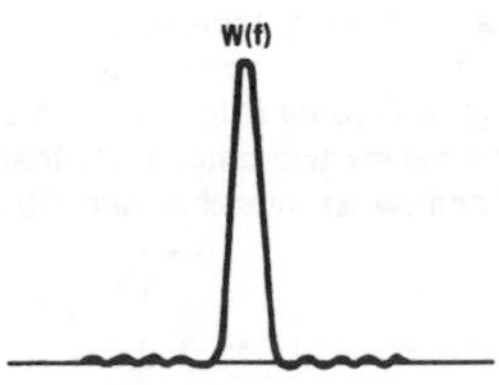

TL/H/8712-9

FIGURE 9. Hamming Window

where $W(f - \eta)$ is the spectral window and * denotes convolution. Since the periodogram is equivalent to the Fourier transform of the autocorrelation function $R_{N_{xx}}(k)$ then, using the frequency convolution theorem

$$F\{x(t)\,y(t)\} = X(f) * Y(\eta - f) \quad (116)$$

where F { } denotes a Fourier transform, $S_{XX}(f)$ is the Fourier transform of the product of $R_{N_{xx}}(k)$ and the inverse Fourier transform of W(f). Therefore for a finite duration window sequence of length 2K − 1,

$$S_{W_{xx}}(f) = \sum_{k=-(K-1)}^{K-1} R_{N_{xx}}(k)\, w(k)\, \epsilon^{-j\omega kT} \quad (117)$$

where

$$w(k) = \int_{-1/2}^{1/2} W(f)\, \epsilon^{j\omega kT}\, df \quad (118)$$

References (10)(16)(21) proceed further with a development to show that the smoothed single windowed periodogram is a consistent estimate of the power spectral density function. The highlights of the development, however, show that a smoothed or windowed spectral estimate, $S_{W_{xx}}(f)$, can be made to have a smaller variance than that of the straight periodogram, $S_{N_{xx}}(f)$, by β the variance ratio relationship

$$\beta = \frac{\text{var}\,[S_{W_{xx}}(f)]}{\text{var}\,[S_{N_{xx}}(f)]} = \frac{1}{N} \sum_{m=-(M-1)}^{M-1} w^2(m) = \frac{1}{N}\int_{-1/2}^{1/2} W^2(f)\, df \quad (119)$$

where N is the record length and 2M − 1 is the total window width. Note that the energy of the window function can be found in equation (119) as

$$E_\omega = \sum_{m=-(M-1)}^{M-1} w^2(m) = \int_{-1/2}^{1/2} W^2(f)\, df \quad (120)$$

Continuing from equation (119), it is seen that a satisfactory estimate of the periodogram requires the variance of $S_{W_{xx}}(f)$ to be small compared to $S^2_{N_{xx}}$ so that

$$\beta \ll 1 \quad (121)$$

Therefore, it is the adjusting of the length and shape of the window that allows the variance of $S_{W_{xx}}(f)$ to be reduced over the periodogram.

Smoothing is like a low pass filtering effect, and so, causes a reduction in frequency resolution. Further, the width of the window main lobe, defined as the symmetric interval between the first positive and negative frequencies at which W(f) = 0, in effect determines the bandwidth of the smoothed spectrum. Examining Table I for the following defined windows;

Rectangular window

$$w(m) = 1 \qquad |m| \le M - 1 \quad (122)$$
$$= 0 \qquad \text{otherwise}$$

Bartlett or triangular window

$$w(m) = 1 - \frac{|m|}{M} \qquad |m| \le M - 1 \quad (123)$$
$$= 0 \qquad \text{otherwise}$$

Hann window

$$w(m) = 0.5 + 0.5\cos\left(\frac{\pi m}{M-1}\right) \qquad |m| \le M - 1 \quad (124)$$
$$= 0 \qquad \text{otherwise}$$

Hamming window

$$w(m) = 0.54 + 0.46\cos\left(\frac{\pi m}{M-1}\right) \qquad |m| \le M - 1 \quad (125)$$
$$= 0 \qquad \text{otherwise}$$

We see once again, as in the averaging technique of spectral estimation (Section 7), the smoothed periodogram technique of this discussion also makes a trade-off between variance and resolution for a fixed N. A small variance requires a small M while high resolution requires a large M.

TABLE I

Window Function	Width of Main Lobe* (approximate)	Variance Ratio* (approximate)
Rectangular	$\frac{2\pi}{M}$	$\frac{2M}{N}$
Bartlett	$\frac{4\pi}{M}$	$\frac{2M}{3N}$
Hann	$\frac{3\pi}{M}$	$2M\left[\frac{\frac{(0.5)^2 + (0.5)^2}{2}}{N}\right]$
Hamming	$\frac{3\pi}{M}$	$2M\left[\frac{\frac{(0.54)^2 + (0.46)^2}{2}}{N}\right]$

*Assumes $M \gg 1$

10.0 SPECTRAL ESTIMATION BY AVERAGING MODIFIED PERIODOGRAMS

Welch [Ref. (36)(37)] suggests a method for measuring power spectra which is effectively a modified form of Bartlett's method covered in Section 7. This method, however, applies the window w(n) directly to the data segments before computing their individual periodograms. If the data sequence is sectioned into

$$L \le \frac{N}{M}$$

segments of M samples each as defined in equation (104), the L modified or windowed periodogram can be defined as

$$I^{\ell}_M(f) = \frac{1}{UM}\left|\sum_{n=0}^{M-1} x^{\ell}_M(n)\, w(n)\, \epsilon^{-j\omega nT}\right|^2 \quad 1 \le \ell \le L \quad (126)$$

where U, the energy in the window is

$$U = \frac{1}{M} \sum_{n=0}^{M-1} w^2(n) \qquad (127)$$

Note the similarity between equations (126) and (105) and that equation (126) is equal to equation (105) modified by functions of the data window w(n).

The spectral estimate $\hat{I}_M^{\ell}$ is defined as

$$\hat{I}_M^{\ell}(f) = \frac{1}{L} \sum_{\ell=1}^{L} I_M^{\ell}(f) \qquad (128)$$

and its expected value is given by

$$E[I_M^{\ell}(f)] = \int_{-1/2}^{1/2} S_{N_{xx}}(\eta)\, W(f-\eta)\, d\eta = S_{N_{xx}}(\eta) * W(\eta) \qquad (129)$$

where

$$W(f) = \frac{1}{UM} \left| \sum_{n=0}^{M-1} w(n)\, \epsilon^{-j\omega nT} \right|^2 \qquad (130)$$

The normalizing factor U is required so that the spectral estimate $\hat{I}_m^{\ell}(f)$, of the modified periodogram, $\hat{I}_m^{\ell}(f)$, will be asymptotically unbiased [Ref. (34)]. If the intervals of x(n) were to be nonoverlapping, then Welch [Ref. (37)] indicates that

$$\text{var}\,[\hat{I}_m^{\ell}(f)] \cong \frac{1}{L}\,\text{var}\,[S_{N_{xx}}(f)] \cong \frac{1}{L}\,[S(f)]^2 \qquad (131)$$

which is similar to that of equation (110). Also considered is a case where the data segments overlap. As the overlap increases the correlation between the individual periodograms also increase. We see further that the number of M point data segments that can be formed increases. These two effects counteract each other when considering the variance $I_m^{\ell}(f)$. With a good data window, however, the increased number of segments has the stronger effect until the overlap becomes too large. Welch has suggested that a 50 percent overlap is a reasonable choice for reducing the variance when N if fixed and cannot be made arbitrarily large. Thus, along with windowing the data segments prior to computing their periodograms, achieves a variance reduction over the Bartlett technique and at the same time smooths the spectrum at the cost of reducing its resolution. This trade-off between variance and spectral resolution or bias is an inherent property of spectrum estimators.

11.0 PROCEDURES FOR POWER SPECTRAL DENSITY ESTIMATES

Smoothed single periodograms [Ref. (21)(27)(32)]

A. 1. Determine the sample autocorrelation $R_{N_{xx}}(k)$ of the data sequence x(n)

2. Multiply $R_{N_{xx}}(k)$ by an appropriate window function w(n)

3. Compute the Fourier transform of the product

$$R_{N_{XX}}(k)\, w(n) \longleftrightarrow S_{W_{xx}}(f) \qquad (71)$$

B. 1. Compute the Fourier transform of the data sequence x(n)

$$x(n) \longleftrightarrow X(f)$$

2. Multiply X(f) by its conjugate to obtain the power spectral density $S_{N_{xx}}(f)$

$$S_{N_{xx}}(f) = \frac{1}{N}|X(f)|^2 \qquad (74)$$

3. Convolve $S_{N_{xx}}(f)$ with an appropriate window function W(f)

$$S_{W_{xx}}(f) = S_{N_{xx}}(f) * W(f) \qquad (115)$$

C. 1. Compute the Fourier transform of the data sequence x(n)

$$x(n) \longleftrightarrow X(f)$$

2. Multiply X(f) by its conjugate to obtain the power spectral density $S_{N_{xx}}(f)$

$$S_{N_{xx}}(f) = \frac{1}{N}|X(f)|^2 \qquad (74)$$

3. Inverse Fourier transform $S_{N_{xx}}(f)$ to get $R_{N_{xx}}(k)$

4. Multiply $R_{N_{xx}}(k)$ by an appropriate window function w(n)

5. Compute the Fourier transform of the product to obtain $S_{W_{xx}}(f)$

$$S_{W_{xx}}(f) \longleftrightarrow R_{N_{xx}}(k) \bullet w(n) \qquad (117)$$

Averaging periodograms [Ref. (32)(37)(38)]

A. 1. Divide the data sequence x(n) into $L \leq N/M$ segments, $x_M^{\ell}(n)$

2. Multiply a segment by an appropriate window

3. Take the Fourier transform of the product

4. Multiply procedure 3 by its conjugate to obtain the spectral density of the segment

5. Repeat procedures 2 through 4 for each segment so that the average of these periodogram estimates produce the power spectral density estimate, equation (128)

12.0 RESOLUTION

In analog bandpass filters, resolution is defined by the filter bandwidth, Δf_{BW}, measured at the passband half power points. Similarly, in the measurement of power spectral density functions it is important to be aware of the resolution capabilities of the sampled data system. For such a system resolution is defined as

$$\Delta f_{BW} = \frac{1}{NT} \qquad (132)$$

and for;

Correlation resolution

$$\Delta f_{BW} = \frac{1}{\tau_{max}} \qquad (133)$$

$\tau_{max} = mT$, where m is the maximum value allowed to produce τ_{max}, the maximum lag term in the correlation computation

Fourier transform (FFT) resolution

$$\Delta f_{BW} = \frac{1}{P_L} = \frac{m}{NT} = \frac{1}{LT}$$ where P is the record length, N, the number of data points and m, the samples within each L segment,

$$L = \frac{N}{M}. \quad (134)$$

Note that the above Δf_{BW}'s can be substantially poorer depending upon the choice of the data window. A loss in degrees of freedom (Section 13) and statistical accuracy occurs when a data sequence is windowed. That is, data at each end of a record length are given less weight than the data at the middle for anything other than the rectangular window. This loss in degrees of freedom shows up as a loss in resolution because the main lobe of the spectral window is widened in the frequency domain.

Finally, the limits of a sampled data system can be described in terms of the maximum frequency range and minimum frequency resolution. The maximum frequency range is the Nyquist or folding frequency, f_c,

$$f_c = \frac{f_s}{2} = \frac{1}{2T_s} \quad (135)$$

where f_s is the sampling frequency and T_s the sampling period. And, secondly, the resolution limit can be described in terms of a (Δf_{BW}) NT product where

$$\Delta f_{BW} \geq \frac{1}{NT} \quad (136)$$

or

$$(\Delta f_{BW})\, NT \geq 1 \quad (137)$$

13.0 CHI-SQUARE DISTRIBUTIONS

Statistical error is the uncertainty in power spectral density measurements due to the amount of data gathered, the probabilistic nature of the data and the method used in deriving the desired parameter. Under reasonable conditions the spectral density estimates approximately follow a chi-square, χ_n^2, distribution. χ_n^2 is defined as the sum of the squares of χ_n, $1 \leq n \leq N$, independent Gaussian variables each with a zero mean and unit variance such that

$$\chi_N^2 = \sum_{n=1}^{N} \chi_n^2 \quad (138)$$

The number n is called the *degrees of freedom* and the χ_n^2 probability density function is

$$f(\chi_n^2) = \frac{1}{2^{n/2}\Gamma\left(\frac{n}{2}\right)}\left[(\chi_n^2)^{\frac{n-2}{2}}\right]\epsilon^{\frac{-\chi_n^2}{2}} \quad (139)$$

where $\Gamma\left(\frac{n}{2}\right)$ is the statistical gamma function (Ref. (14)].

Figure 10 shows the probability density function for several n values and it is important to note that as n becomes large the chi-square distribution approximates a Gaussian distribution. We use this χ_n^2 distribution in our analysis to discuss the variability of power spectral densities. If x_n has a zero mean and N samples of it are used to compute the power spectral density estimate S(f) then, the probability that the true spectral density, S(f), lies between the limits

$$A \leq S(f) \leq B \quad (140)$$

is

$$P = (1 - \alpha) = \text{probability} \quad (141)$$

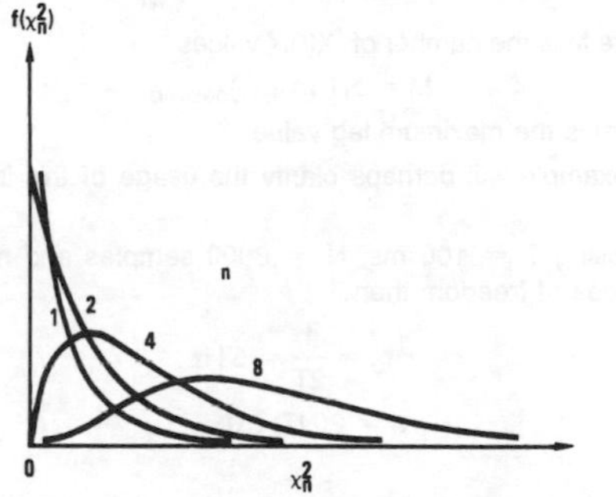

TL/H/8712-10

FIGURE 10

The lower A and upper B limits are defined as

$$A = \frac{n\hat{S}(f)}{\chi^2_{n;\frac{\alpha}{2}}} \quad (142)$$

and

$$B = \frac{n\hat{S}(f)}{\chi^2_{n;1-\frac{\alpha}{2}}} \quad (143)$$

respectively. $\chi^2_{n;\alpha}$ is defined by

$$\chi^2_{n;\alpha} = [\nu \text{ so that } \int_{\nu}^{\infty} f(\chi_n^2)\, d\chi_n^2] = \alpha \quad (144)$$

see *Figure 11* and the interval A to B is referred to as a confidence interval. From Otnes and Enrochson [Ref. (35) pg. 217] the degrees of freedom can be described as

$$n = 2(\Delta f_{BW})\, NT = 2(\Delta f_{BW})\, P_L \quad (145)$$

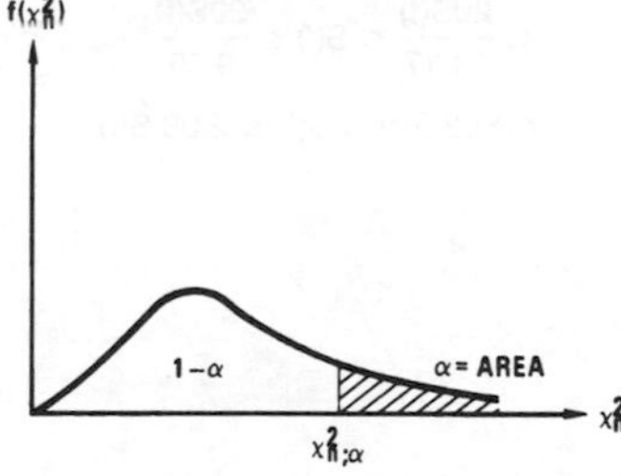

TL/H/8712-11

FIGURE 11

and that for large n i.e., $n \geq 30$ the χ_n^2 distribution approaches a Gaussian distribution so that the standard deviation or standard error, ϵ_o, can be given by

$$\epsilon_o = \frac{1}{\sqrt{\Delta f_{BW} NT}} \qquad (146)$$

The degrees of freedom and associated standard error for the correlation and Fourier transform are as follows:

$$\text{correlation: } n = \frac{2N}{m} \qquad \epsilon_o = \sqrt{\frac{m}{N}} \qquad (147)$$

$$\text{FFT: } n = 2M \qquad \epsilon_o = \sqrt{\frac{1}{M}} \qquad (148)$$

where M is the number of $|X(f)|^2$ values

$$M = NT\,(\Delta f_{BW})_{desired} \qquad (149)$$

and m is the maximum lag value.

An example will perhaps clarify the usage of this information.

Choosing T = 100 ms, N = 8000 samples and n = 20 degrees of freedom then

$$f_c = \frac{1}{2T} = 5\text{ Hz}$$

$$n = 2(NT)\,(\Delta f_{BW})$$

so

$$\Delta f_{BW} = \frac{n}{2NT} = 0.0125\text{ Hz}$$

If it is so desired to have a 95% confidence level of the spectral density estimate then

$$P = (1 - \alpha)$$

$$0.95 = 1 - \alpha$$

$$\alpha = 1 - 0.95 = 0.05$$

the limits

$$B = \frac{n\hat{S}(f)}{\chi^2_{n;\,1-\alpha/2}} = \frac{20\hat{S}(f)}{\chi^2_{20;\,0.975}}$$

$$A = \frac{n\hat{S}(f)}{\chi^2_{n;\,\alpha/2}} = \frac{20\hat{S}(f)}{\chi^2_{20;\,0.025}}$$

yield from Table II

$$\chi^2_{20;\,0.975} = 9.59$$

$$\chi^2_{20;\,0.025} = 34.17$$

so that

$$\frac{20\hat{S}(f)}{34.17} \leq S(f) \leq \frac{20\hat{S}(f)}{9.59}$$

$$0.5853\,\hat{S}(f) \leq S(f) \leq 2.08\,\hat{S}(f)$$

There is thus a 95% confidence level that the true spectral density function S(f) lies within the interval $0.5853\,\hat{S}(f) \leq S(f) \leq 2.08\,\hat{S}(f)$.

As a second example using equation (148) let T = 1 ms, N = 4000 and it is desired to have (Δf_{BW}) desired = 10 Hz.

Then,

$$NT = 4$$

$$f_c = \frac{1}{2T} = 500\text{ Hz}$$

$$\epsilon_o = \sqrt{\frac{1}{M}} = \sqrt{\frac{1}{NT\,(\Delta f_{BW})_{desired}}} = 0.158$$

$$N = 2M = 2NT\,(\Delta f_{BW})_{desired} = 80$$

If it is again desired to have a 95% confidence level of the spectral density estimate then

$$\alpha = 1 - p = 0.05$$

$$\chi^2_{80;\,0.975} = 5.75$$

$$\chi^2_{80;\,0.025} = 106.63$$

and we thus have a 95% confidence level that the true spectral density S(f) lies within the limits

$$0.75\,\hat{S}(f) \leq S(f) \leq 1.39\,\hat{S}(f)$$

It is important to note that the above examples assume Gaussian and white data. In practical situations the data is typically colored or correlated and effectively results in reducing number of degrees of freedom. It is best, then, to use the white noise confidence levels as guidelines when planning power spectral density estimates.

14.0 CONCLUSION

This article attempted to introduce to the reader a conceptual overview of power spectral estimation. In doing so a wide variety of subjects were covered and it is hoped that this approach left the reader with a sufficient base of "tools" to indulge in the mounds of technical literature available on the subject.

15.0 ACKNOWLEDGEMENTS

The author wishes to thank James Moyer and Barry Siegel for their support and encouragement in the writing of this article.

TABLE II. Percentage Points of the Chi-Square Distribution

n	α 0.995	0.990	0.975	0.950	0.050	0.025	0.010	0.005
1	0.000039	0.00016	0.00098	0.0039	3.84	5.02	6.63	7.88
2	0.0100	0.0201	0.0506	0.1030	5.99	7.38	9.21	10.60
3	0.0717	0.115	0.216	0.352	7.81	9.35	11.34	12.84
4	0.207	0.297	0.484	0.711	9.49	11.14	13.28	14.86
5	0.412	0.554	0.831	1.150	11.07	12.83	15.09	16.75
6	0.68	0.87	1.24	1.64	12.59	14.45	16.81	18.55
7	0.99	1.24	1.69	2.17	14.07	16.01	18.48	20.28
8	1.34	1.65	2.18	2.73	15.51	17.53	20.09	21.96
9	1.73	2.09	2.70	3.33	16.92	19.02	21.67	23.59
10	2.16	2.56	3.25	3.94	18.31	20.48	23.21	25.19
11	2.60	3.05	3.82	4.57	19.68	21.92	24.72	26.76
12	3.07	3.57	4.40	5.23	21.03	23.34	26.22	28.30
13	3.57	4.11	5.01	5.89	22.36	24.74	27.69	29.82
14	4.07	4.66	5.63	6.57	23.68	26.12	29.14	31.32
15	4.60	5.23	6.26	7.26	25.00	27.49	30.58	32.80
16	5.14	5.81	6.91	7.96	26.30	28.85	32.00	34.27
17	5.70	6.41	7.56	8.67	27.59	30.19	33.41	35.72
18	6.26	7.01	8.23	9.39	28.87	31.53	34.81	37.16
19	6.84	7.63	8.91	10.12	30.14	32.85	36.19	38.58
20	7.43	8.26	9.59	10.85	31.41	34.17	37.57	40.00
21	8.03	8.90	10.28	11.59	32.67	35.48	38.93	41.40
22	8.64	9.54	10.98	12.34	33.92	36.78	40.29	42.80
23	9.26	10.20	11.69	13.09	35.17	38.08	41.64	44.18
24	9.89	10.86	12.40	13.85	36.42	39.36	42.98	45.56
25	10.52	11.52	13.12	14.61	37.65	40.65	44.31	46.93
26	11.16	12.20	13.84	15.38	38.89	41.92	45.64	48.29
27	11.81	12.88	14.57	16.15	40.11	43.19	46.96	49.64
28	12.46	13.56	15.31	16.93	41.34	44.46	48.28	50.99
29	13.12	14.26	16.05	17.71	42.56	45.72	49.59	52.34
30	13.79	14.95	16.79	18.49	43.77	46.98	50.89	53.67
40	20.71	22.16	24.43	26.51	55.76	59.34	63.69	66.77
50	27.99	29.71	32.36	34.76	67.50	71.42	76.15	79.49
60	35.53	37.48	40.48	43.19	79.08	83.80	88.38	91.95
70	43.28	45.44	48.76	51.74	90.53	95.02	100.43	104.22
80	51.17	53.54	57.15	60.39	101.88	106.63	112.33	116.32
90	59.20	61.75	65.65	69.13	113.14	118.14	124.12	128.30
100	67.33	70.06	74.22	77.93	124.34	129.56	135.81	140.17

APPENDIX A

A.0 CONCEPTS OF PROBABILITY, RANDOM VARIABLES AND STOCHASTIC PROCESSES

In many physical phenomena the outcome of an experiment may result in fluctuations that are random and cannot be precisely predicted. It is impossible, for example, to determine whether a coin tossed into the air will land with its head side or tail side up. Performing the same experiment over a long period of time would yield data sufficient to indicate that on the average it is equally likely that a head or tail will turn up. Studying this average behavior of events allows one to determine the frequency of occurrence of the outcome (i.e., heads or tails) and is defined as the notion of *probability*.

Associated with the concept of probability are probability density functions and cumulative distribution functions which find their use in determining the outcome of a large number of events. A result of analyzing and studying these functions may indicate regularities enabling certain laws to be determined relating to the experiment and its outcomes; this is essentially known as *statistics*.

A.1 DEFINITIONS OF PROBABILITY

If n_A is the number of times that an event A occurs in N performances of an experiment, the frequency of occurrence of event A is thus the ratio n_A/N. Formally, the probability, P(A), of event A occurring is defined as

$$P(A) = \lim_{N \to \infty} \left[\frac{n_A}{N}\right] \qquad (A.1\text{-}1)$$

Where it is seen that the ratio n_A/N (or fraction of times that an event occurs) asymptotically approaches some mean value (or will show little deviation from the exact probability) as the number of experiments performed, N, increases (more data).

Assigning a number,

$$\frac{n_A}{N},$$

to an event is a measure of how likely or probable the event. Since n_A and N are both positive and real numbers and $0 \leq n_A \leq N$; it follows that the probability of a given event cannot be less than zero or greater than unity. Furthermore, if the occurrence of any one event excludes the occurrence of any others (i.e., a head excludes the occurrence of a tail in a coin toss experiment), the possible events are said to be *mutually exclusive*. If a complete set of possible events A_1 to A_n are included then

$$\frac{n_{A_1}}{N} + \frac{n_{A_2}}{N} + \frac{n_{A_3}}{N} + \ldots + \frac{n_{A_n}}{N} = 1 \qquad (A.1\text{-}2)$$

or

$$P(A_1) + P(A_2) + P(A_3) + \ldots + P(A_n) = 1 \qquad (A.1\text{-}3)$$

Similarly, an event that is absolutely certain to occur has a probability of one and an impossible event has a probability of zero.

In summary:

1. $0 \leq P(A) \leq 1$
2. $P(A_1) + P(A_2) + P(A_3) + \ldots + P(A_n) = 1$, for an entire set of events that are mutually exclusive
3. $P(A) = 0$ represents an impossible event
4. $P(A) = 1$ represents an absolutely certain event

A.2 JOINT PROBABILTY

If more than one event at a time occurs (i.e., events A and B are not mutually excusive) the frequency of occurrence of the two or more events at the same time is called the *joint probability*, P(AB). If n_{AB} is the number of times that event A and B occur together in N performances of an experiment, then

$$P(A,B) = \lim_{N \to \infty} \left[\frac{n_{AB}}{N}\right] \qquad (A.2\text{-}1)$$

A.3 CONDITIONAL PROBABILITY

The probability of event B occurring given that another event A has already occurred is called *conditional probability*. The dependence of the second, B, of the two events on the first, A, will be designated by the symbol P(B)|A) or

$$P(B|A) = \frac{n_{AB}}{n_A} \qquad (A.3\text{-}1)$$

where n_{AB} is the number of joint occurrences of A and B and N_A represents the number of occurrences of A with or without B. By dividing both the numerator and denominator of equation (A.3-1) by N, conditional probability P(B|A) can be related to joint probability, equation (A.2-1), and the probability of a single event, equation (A.1-1)

$$P(B|A) = \left(\frac{n_{AB}}{n_A}\right)\left(\frac{\frac{1}{N}}{\frac{1}{N}}\right) = \frac{P(A,B)}{P(A)} \qquad (A.3\text{-}2)$$

Analogously

$$P(A|B) = \frac{P(A,B)}{P(A)} \qquad (A.3\text{-}3)$$

and combining equations (A.6) and (A.7)

$$P(A|B)\,P(B) = P(A, B) = P(B|A)\,P(A) \qquad (A.3\text{-}4)$$

results in Bayes' theorem

$$P(A|B) = \frac{P(A)\,P(B|A)}{P(B)} \qquad (A.3\text{-}5)$$

Using Bayes' theorem, it is realized that if P(A) and P(B) are *statistically independent* events, implying that the probability of event A does not depend upon whether or not event B has occurred, then $P(A|B) = P(A)$, $P(B|A) = P(B)$ and hence the joint probability of events A and B is the product of their individual probabilities or

$$P(A,B) = P(A)\,P(B) \tag{A.3-6}$$

More precisely, two random events are statistically independent only if equation (A.3-6) is true.

A.4 PROBABILITY DENSITY FUNCTIONS

A formula, table, histogram, or graphical representation of the probability or possible frequency of occurrence of an event associated with variable X, is defined as $f_X(x)$, the *probability density function* (pdf) or *probability distribution function.* As an example, a function corresponding to height histograms of men might have the probability distribution function depicted in *Figure A.4.1.*

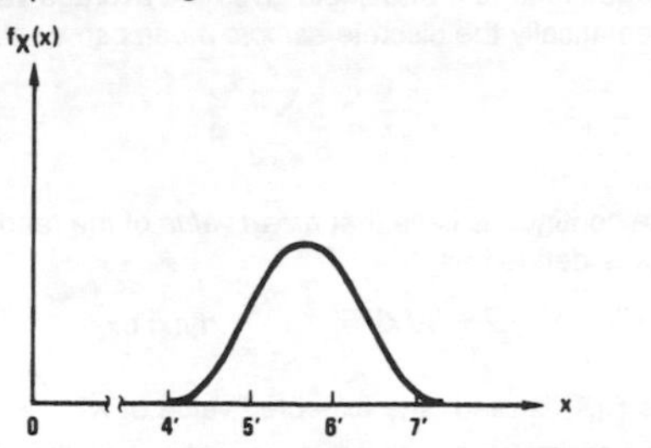

FIGURE A.4.1

The *probability element*, $f_X(x)\,dx$, describes the probability of the event that the random variable X lies within a range of possible values between

$$\left(x - \frac{\Delta x}{2}\right) \text{ and } \left(x + \frac{\Delta x}{2}\right)$$

i.e., the area between the two points 5′5″ and 5′7″ shown in *Figure A.4.2* represents the probability that a man's height will be found in that range. More clearly,

$$\text{Prob}\left[\left(x - \frac{\Delta x}{2}\right) \le X \le \left(x + \frac{\Delta x}{2}\right)\right] = \int_{x - \frac{\Delta x}{2}}^{x + \frac{\Delta x}{2}} f_X(x)\,dx \tag{A.4-1}$$

or

$$\text{Prob}\,[5'5'' \le X \le 5'7''] = \int_{5'5''}^{5'7''} f_X(x)\,dx$$

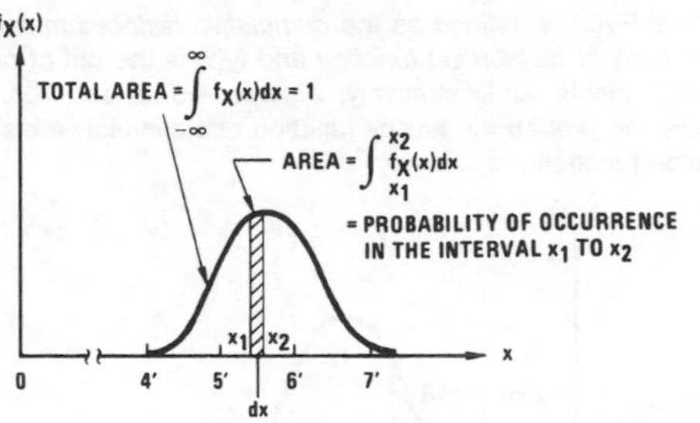

FIGURE A.4.2

Continuing, since the total of all probabilities of the random variable X must equal unity and $f_X(x)\,dx$ is the probability that X lies within a specified interval

$$\left(x - \frac{\Delta x}{2}\right) \text{ and } \left(x - \frac{\Delta x}{2}\right),$$

then,

$$\int_{-\infty}^{\infty} f_X(x)\,dx = 1 \tag{A.4-2}$$

It is important to point out that the density function $f_X(x)$ is in fact a mathematical description of a curve and is not a probability; it is therefore not restricted to values less than unity but can have any non-negative value. Note however, that in practical application, the integral is normalized such that the entire area under the probability density curve equates to unity.

To summarize, a few properties of $f_X(x)$ are listed below.

1. $f_X(x) \ge 0$ for all values of x or $-\infty < x < \infty$
2. $\int_{-\infty}^{\infty} f_X(x)\,dx = 1$
3. $\text{Prob}\left[\left(x - \frac{\Delta x}{2}\right) \le X \le \left(x + \frac{\Delta x}{2}\right)\right]$

$$= \int_{x - \frac{\Delta x}{2}}^{x + \frac{\Delta x}{2}} f_X(x)\,dx$$

A.5 CUMULATIVE DISTRIBUTION FUNCTION

If the entire set of probabilities for a random variable event X are known, then since the probability element, $f_X(x)\,dx$, describes the probability that event X will occur, the accumulation of these probabilities from $x = -\infty$ to $x = \infty$ is unity or an absolutely certain event. Hence,

$$F_X(x) \int_{-\infty}^{\infty} f_X(x)\,dx = 1 \tag{A.5-1}$$

where $F_X(x)$ is defined as the *cumulative distribution function* (cdf) or *distribution function* and $f_X(x)$ is the pdf of random variable X. Illustratively, *Figures A.5.1a* and *A.5.1b* show the probability density function and cumulative distribution function respectively.

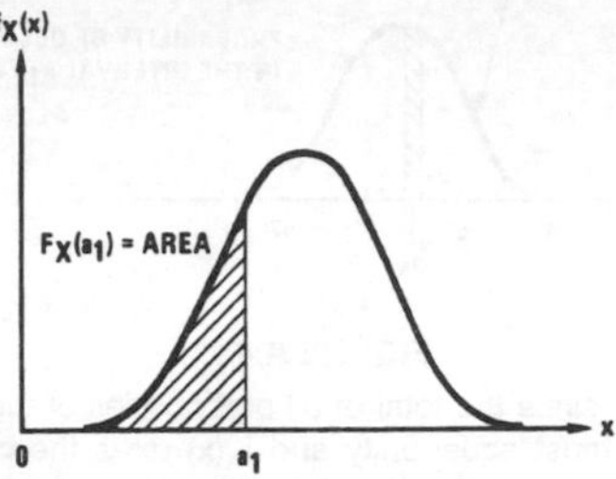

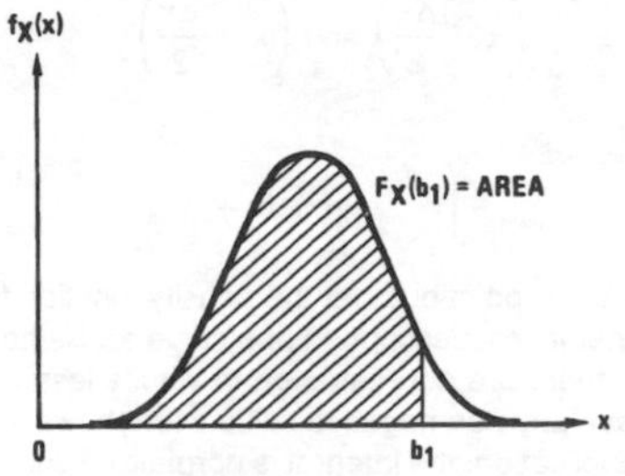

TL/H/8712-14

(a) Probability Density Function

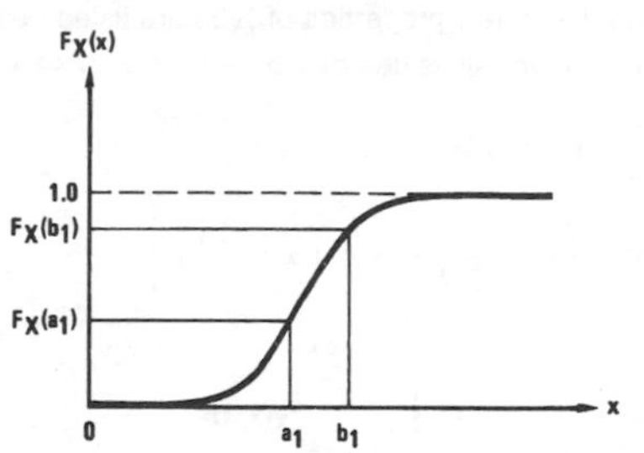

TL/H/8712-15

(b) Cumulative Distribution Function

FIGURE A.5.1

In many texts the notation used in describing the cdf is

$$F_X(x) = \text{Prob}[X \le x] \qquad \text{(A.5-2)}$$

and is defined to be the probability of the event that the observed random variable X is less than or equal to the allowed or conditional value x. This implies

$$F_X(x) = \text{Prob}[X \le x] = \int_{-\infty}^{x} f_X(x)\,dx \qquad \text{(A.5-3)}$$

It can be further noted that

$$\text{Prob}[x_1 \le x \le x_2] = \int_{x_1}^{x_2} f_X(x)dx = F_X(x_2) - F_X(x_1) \qquad \text{(A.5-4)}$$

and that from equation (A.5-1) the pdf can be related to the cdf by the derivative

$$f_X(x) = \frac{d[F_X(x)]}{dx} \qquad \text{(A.5-5)}$$

Re-examining *Figure A.5.1* does indeed show that the pdf, $f_X(x)$, is a plot of the differential change in slope of the cdf, $F_X(x)$.

$F_X(x)$ and a summary of its properties.

1. $0 \le F_X(x) \le 1 \quad -\infty < x < \infty$ (Since $F_X = \text{Prob}\ [X < x]$ is a probability)
2. $F_X(-\infty) = 0 \quad F_X(+\infty) = 1$
3. $F_X(x)$ the probability of occurrence increases as x increases
4. $F_X(x) = \int f_X(x)\,dx$
5. $\text{Prob}\ (x_1 \le x \le x_2) = F_X(x_2) - F_X(x_1)$

A.6 MEAN VALUES, VARIANCES AND STANDARD DEVIATION

The procedure of determining the average weight of a group of objects by summing their individual weights and dividing by the total number of objects gives the average value of x. Mathematically the discrete *sample mean* can be described

$$\bar{x} = \frac{1}{n}\sum_{i-1}^{n} x_i \qquad \text{(A.6-1)}$$

for the continuous case that *mean value* of the random variable X is defined as

$$\bar{x} = E[X] = \int_{-\infty}^{\infty} x f_X(x)\,dx \qquad \text{(A.6-2)}$$

where E[X] is read "the expected value of X".

Other names for the same *mean value* $\bar{x}$ or the *expected value* E[X] are average value and statistical average.

It is seen from equation (A.6-2) that E[X] essentially represents the sum of all possible values of x with each value being weighted by a corresponding value of the probability density function of $f_X(x)$.

Extending this definition to any function of X for example h(x), equation (A.6-2) becomes

$$E[h(x)] = \int_{-\infty}^{\infty} h(x)\, f_X(x)\,dx \qquad \text{(A.6-3)}$$

An example at this point may help to clarify matters. Assume a uniformly dense random variable of density 1/4 between the values 2 and 6, see *Figure A.6.1*. The use of equation (A.6-2) yields the expected value

$$\bar{x} = E[X] = \int_{2}^{6} x\, {}^1\!/_4\, dx = \left.\frac{x^2}{8}\right|_2^6 = 4$$

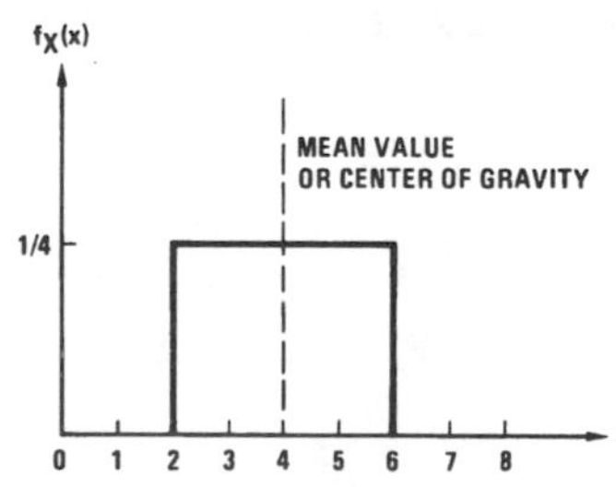

TL/H/8712-16

FIGURE A.6.1

which can also be interpreted as the first moment or center of gravity of the density function $f_X(x)$. The above operation is analogous to the techniques in electrical engineering where the DC component of a time function is found by first integrating and then dividing the resultant area by the interval over which the integration was performed.

Generally speaking, the time averages of random variable functions of time are extremely important since essentially no statistical meaning can be drawn from a single random variable (defined as the value of a time function at a given single instant of time). Thus, the operation of finding the *mean value* by integrating over a specified range of possible values that a random variable may assume is referred to as *ensemble averaging*.

In the above example, $\bar{x}$ was described as the first moment m_1 or DC value. The mean-square value $\overline{x^2}$ or $E[X^2]$ is the second moment m_2 or the total average power, AC plus DC and in general the nth moment can be written

$$m_n = E[X^n] = \int_{-\infty}^{\infty} [h(x)]^2 f_X(x)\,dx \qquad (A.6\text{-}4)$$

Note that the first moment squared m_1^2, $\bar{x}^2$ or $E[X]^2$ is equivalent to the DC power through a 1Ω resistor and is *not* the same as the second moment m_2, $\overline{x^2}$ or $E[X]^2$ which, again, implies the total average power.

A discussion of central moments is next in store and is simply defined as the moments of the difference (deviation) between a random variable and its mean value. Letting $[h(x)]^n = (X - \bar{x})^n$, mathematically

$$\overline{(X - \bar{x})^n} = E[(X - \bar{x})^n] = \int_{-\infty}^{\infty} (X - \bar{x})^n f_X(x)\,dx \qquad (A.6\text{-}5)$$

For n = 1, the first central moment equals zero i.e., the AC voltage (current) minus the mean, average or DC voltage (current) equals zero. This essentially yields little information. The second central moment, however, is so important that it has been named the *variance* and is symbolized by σ^2. Hence,

$$\sigma^2 = E[(X - \bar{x})^2] = \int_{-\infty}^{\infty} (X - \bar{x})^2 f_X(x)\,dx \qquad (A.6\text{-}6)$$

Note that because of the squared term, values of X to either side of the mean $\bar{x}$ are equally significant in measuring variations or deviations away from $\bar{x}$ i.e., if $\bar{x} = 10$, $X_1 = 12$ and $X_2 = 8$ then $(12 - 10)^2 = 4$ and $(8 - 10)^2 = 4$ respectively. The variance therefore is the measure of the variability of $[h(x)]^2$ about its mean value $\bar{x}$ or the expected square deviation of X from its mean value. Since,

$$\sigma^2 = E[(X - \bar{x})^2] = E\left[X^2 - 2\bar{x}X + \bar{x}_1^2\right] \qquad (A.6\text{-}7a)$$

$$= E[X^2] - 2\bar{x}\,E[X] + \bar{x}_1^2 \qquad (A.6\text{-}7b)$$

$$= \overline{x^2} - 2\bar{x}\bar{x} + \bar{x}_1^2 \qquad (A.6\text{-}7c)$$

$$= \overline{x^2} - \bar{x}_1^2 \qquad (A.6\text{-}7d)$$

or

$$m_2 - m_1^2 \qquad (A.6\text{-}7e)$$

The analogy can be made that variance is essentially the average AC power of a function h(x), hence, the total average power, second moment m_2, minus the DC power, first moment squared m_1^2. The positive square root of the variance.

$$\sqrt{\sigma^2} = \sigma$$

is defined as the *standard deviation*. The mean indicates where a density is centered but gives no information about how spread out it is. This spread is measured by the standard deviation σ and is a measure of the spread of the density function about $\bar{x}$, i.e., the smaller σ the closer the values of X to the mean. In relation to electrical engineering the standard deviation is equal to the root-mean-square (rms) value of an AC voltage (current) in circuit.

A summary of the concepts covered in this section are listed in Table A.6.1.

A.7 FUNCTIONS OF TWO JOINTLY DISTRIBUTED RANDOM VARIABLES

The study of jointly distributed random variables can perhaps be clarified by considering the response of linear systems to random inputs. Relating the output of a system to its input is an example of analyzing two random variables from different random processes. If on the other hand an attempt is made to relate present or future values to its past values, this, in effect, is the study of random variables coming from the same process at different instances of time. In either case, specifying the relationship between the two random variables is established by initially developing a probability model for the joint occurrence of two random events. The following sections will be concerned with the development of these models.

A.8 JOINT CUMULATIVE DISTRIBUTION FUNCTION

The joint cumulative distribution function (cdf) is similar to the cdf of Section A.5 except now two random variables are considered.

$$F_{XY}(x,y) = \text{Prob}\,[X \le x, Y \le y] \qquad (A.8\text{-}1)$$

defines the joint cdf, $F_{XY}(x,y)$, of random variables X and Y. Equation (A.8-1) states that $F_{XY}(x, y)$ is the probability associated with the joint occurrence of the event that X is less than or equal to an allowed or conditional value x and the event that Y is less than or equal to an allowed or conditional value y.

TABLE A.6-1

Symbol	Name	Physical Interpretation
$\overline{X}$, E[X], m_1: $\int_{-\infty}^{\infty} x f_X(x)\,dx$	Expected Value, Mean Value, Statistical Average Value	• Finding the mean value of a random voltage (current) is equivalent to finding its DC component. • First moment; e.g., the first moment of a group of masses is just the average location of the masses or their center of gravity. • The range of the most probable values of x.
$E[X]^2$, $\overline{X}^2$, m_1^2		• DC power
$\overline{X^2}$, $E[X^2]$, m_2: $\int_{-\infty}^{\infty} x^2 f_X(x)\,dx$	Mean Square Value	• Interpreted as being equal to the time average of the square of a random voltage (current). In such cases the mean-square value is proportional to the total average power (AC plus DC) through a 1Ω resistor and its square root is equal to the rms or effective value of the random voltage (current). • Second moment; e.g., the moment of inertia of a mass or the turning moment of torque about the origin. • The mean-square value represents the spread of the curve about $\overline{x} = 0$
var[ ], σ^2, $(X - \overline{x})^2$, $E[(X - \overline{x})^2]$,) m_2; $\int_{-\infty}^{\infty} (X - \overline{x})^2 f_X(x)\,dx$	Variance	• Related to the average power (in a 1Ω resistor) of the AC components of a voltage (current) in power units. The square root of the variances is the rms voltage (current) again not reflecting the DC component. • Second movement; for example the moment of inertia of a mass or the turning moment of torque about the value $\overline{x}$. • Represents the spread of the curve about the value $\overline{x}$.
$\sqrt{\sigma^2}, \sigma$	Standard Deviation	• Effective rms AC voltage (current) in a circuit. • A measure of the spread of a distribution corresponding to the amount of uncertainty or error in a physical measurement or experiment. • Standard measure of deviation of X from its mean value $\overline{x}$.

$(\overline{X})^2$ is a result of smoothing the data and then squaring it and $(\overline{X^2})$ results from squaring the data and then smoothing it.

A few properties of the joint cumulative distribution function are listed below.

1. $0 \leq F_{XY}(x,y) \leq 1 \quad -\infty < x < \infty$
   $-\infty < y < \infty$

(since $F_{XY}(x,y) = \text{Prob}[X \leq x, Y \leq y]$ is a probability)

2. $F_{XY}(-\infty,y) = 0$
   $F_{XY}(x,-\infty) = 0$
   $F_{XY}(-\infty,-\infty) = 0$
3. $F_{XY}(+\infty,+\infty) = 0$
4. $F_{XY}(x,y)$ — The probability of occurrence increases as either x or y, or both increase

A.9 JOINT PROBABILITY DENSITY FUNCTION

Similar to the single random variable probability density function (pdf) of sections A.4 and A.5, the joint probability density function $f_{XY}(x, y)$ is defined as the partial derivative of the joint cumulative distribution function $F_{XY}(x, y)$. More clearly,

$$f_{XY}(x,y) = \frac{d^2}{dx\,dy} F_{XY}(x,y) \qquad \text{(A.9-1)}$$

Recall that the pdf is a *density* function and must be integrated to find a probability. As an example, to find the probability that (X, Y) is within a rectangle of dimension ($x_1 \leq X \leq x_2$) and ($y_1 \leq Y \leq y_2$), the pdf of the joint or two-dimensional random variable must be integrated over both ranges as follows,

$$\text{Prob}[x_1 \leq X \leq x_2, y_1 \leq Y \leq y_2] = \qquad \text{(A.9-2)}$$

$$\int_{x_1}^{x_2}\int_{y_1}^{y_2} f_{XY}(x,y)\,dydx = 1$$

It is noted that the double integral of the joint pdf is in fact the cumulative distribution function

$$F_{XY}(x, y) = \int\int_{-\infty}^{\infty} f_{XY}(x, y)\,dxdy = 1 \qquad \text{(A.9-3)}$$

analogous to Section A.5. Again $f_{XY}(x, y)$ dxdy represents the probability that X and Y will jointly be found in the ranges

$$x \pm \frac{dx}{2} \text{ and } y \pm \frac{dy}{2},$$

respectively, where the joint density function $f_{XY}(x, y)$ has been normalized so that the volume under the curve is unity.

A few properties of the joint probability density functions are listed below.

1. $f_{XY}(x,y) > 0$ — For all values of x and y or $-\infty < x < \infty$ and $-\infty < y < \infty$, respectively
2. $\int\int_{-\infty}^{\infty} f_{XY}(x,y)\,dxdy = 1$
3. $F_{XY}(x,y) = \int_{-\infty}^{y}\int_{-\infty}^{x} f_{XY}(x,y)\,dxdy$
4. $\text{Prob}[x_1 \leq X \leq x_2, y_1 \leq Y \leq y_2] = \int_{y_1}^{y2}\int_{x_1}^{x2} f_{XY}(x,y)\,dxdy$

A.10 STATISTICAL INDEPENDENCE

If the knowledge of one variable gives no information about the value of the other, the two random variables are said to be statistically independent. In terms of the joint pdf

$$f_{XY}(x,y) = f_X(x)\,f_Y(y) \qquad \text{(A.10-1)}$$

and

$$f_X(x)\,f_Y(y) = f_{XY}(x,y) \qquad \text{(A.10-2)}$$

imply statistical independence of the random variables X and Y. In the same respect the joint cdf

$$F_{XY}(x, y) = F_X(x)F_Y(y) \qquad \text{(A.10-3)}$$

and

$$F_X(x)F_Y(y) = F_{XY}(x, y) \qquad \text{(A.10-4)}$$

again implies this independence.

It is important to note that for the case of the expected value E[XY], statistical independence of random variables X and Y implies

$$E[XY] = E[X]\,E[Y] \qquad \text{(A.10-5)}$$

but, the converse is *not* true since random variables can be uncorrelated but not necessarily independent.

In summary

1. $F_{XY}(x,y) = F_X(x)\,F_Y(y)$ — reversible
2. $f_{XY}(x,y) = f_X(x)\,f_Y(y)$ — reversible
3. $E[XY] = E[X]\,E[Y]$ — non-reversible

A.11 MARGINAL DISTRIBUTION AND MARGINAL DENSITY FUNCTIONS

When dealing with two or more random variables that are jointly distributed, the distribution of each random variable is called the *marginal distribution.* It can be shown that the marginal distribution defined in terms of a joint distribution can be manipulated to yield the distribution of each random variable considered by itself. Hence, the marginal distribution functions $F_X(x)$ and $F_Y(y)$ in terms of $F_{XY}(x, y)$ are

$$F_X(x) = F_{XY}(x, \infty) \qquad \text{(A.11-1)}$$

and

$$F_Y(y) = F_{XY}(\infty,y) \qquad \text{(A.11-2)}$$

respectively.

The marginal density functions $f_X(x)$ and $f_Y(x)$ in relation to the joint density $f_{XY}(x, y)$ is represented as

$$f_X(x) = \int_{-\infty}^{\infty} f_{XY}(x,y)\,dy \qquad \text{(A.11-3)}$$

and

$$f_Y(x) = \int_{-\infty}^{\infty} f_{XY}(x,y)\,dx \qquad \text{(A.11-4)}$$

respectively.

A.12 TERMINOLOGY

Before continuing into the following sections it is appropriate to provide a few definitions of the terminology used hereafter. Admittedly, the definitions presented are by no means complete but are adequate and essential to the continuity of the discussion.

Deterministic and Nondeterministic Random Processes: A random process for which its future values cannot be exactly predicted from the observed past values is said to be *nondeterministic*. A random process for which the future values of any sample function can be exactly predicted from a knowledge of all past values, however, is said to be a *deterministic* process.

Stationary and Nonstationary Random Processes: If the marginal and joint density functions of an event do not depend upon the choice of i.e., time origin, the process is said to be *stationary*. This implies that the mean values and moments of the process are constants and are not dependent upon the absolute value of time. If on the other hand the probability density functions do change with the choice of time origin, the process is defined *nonstationary*. For this case one or more of the mean values or moments are also time dependent. In the strictest sense, the stochastic process x(f) is stationary if its statistics are not affected by the shift in time origin i.e., the process x(f) and $x(t + \tau)$ have the same statistics for any τ.

Ergodic and Nonergodic Random Processes: If every member of the ensemble in a stationary random process exhibits the same statistical behavior that the entire ensemble has, it is possible to determine the process statistical behavior by examining only one typical sample function. This is defined as an *ergodic* process and its mean value and moments can be determined by time averages as well as by ensemble averages. Further ergodicity implies a stationary process and any process not possessing this property is *nonergodic*.

Mathematically speaking, any random process or, i.e., wave shape x(t) for which

$$\lim_{T \to \infty} \overline{x(t)} = \lim_{T \to \infty} \frac{1}{T} \int_{-T/2}^{T/2} x(t)\,dt = E[x(t)]$$

holds true is said to be an ergodic process. This simply says that as the averaging time, T, is increased to the limit $T \to \infty$, time averages equal ensemble averages (the *expected value* of the function).

A.13 JOINT MOMENTS

In this section, the concept of joint statistics of two continuous dependent variables and a particular measure of this dependency will be discussed.

The joint moments m_{ij} of the two random variables X and Y are defined as

$$m_{ij} = E[X^iY^j] = \int_{-\infty}^{\infty} \int_{-\infty}^{\infty} x^i y^j f_{XY}(x,y)\,dx\,dy \qquad \text{(A.13-1)}$$

where $i + j$ is the order of the moment.

The second moment represented as μ_{11} or σ_{XY} serves as a measure of the dependence of two random variables and is given a special name called the *covariance* of X and Y. Thus

$$\mu_{11} = \sigma_{XY} = E[(X - \overline{x})(Y - \overline{y})] = \qquad \text{(A.13-2)}$$

$$\int \int_{-\infty}^{\infty} (X - \overline{x})(Y - \overline{y})\, f_{XY}(x,y)\,dx\,dy$$

$$= E[XY] - E[X]\,E[Y] \qquad \text{(A.13-3)}$$

or

$$= m_{11} - \overline{x}\overline{y} \qquad \text{(A.13-4)}$$

If the two random variables are independent, their covariance function μ_{11} is equal to zero and m_{11}, the average of the product, becomes the product of the individual averages hence.

$$\mu_{11} = 0 \qquad \text{(A.13-5)}$$

implies

$$m_{11} = E[(X - \overline{x})(Y - \overline{y})] = E[X - \overline{x}]\,E[Y - \overline{y}] \qquad \text{(A.13-6)}$$

Note, however, the converse of this statement in general is not true but does have validity for two random variables possessing a joint (two dimensional) Gaussian distribution.

In some texts the name cross-covariance is used instead of covariance, regardless of the name used they both describe processes of two random variables each of which comes from a separate random source. If, however, the two random variables come from the same source it is instead called the autovariance or auto-covariance.

It is now appropriate to define a normalized quantity called the *correlation coefficient*, ρ, which serves as a numerical measure of the dependence between two random variables. This coefficient is the covariance normalized, namely

$$\rho = \frac{\text{covar}[X,Y]}{\sqrt{\text{var}[X]\,\text{var}[Y]}} = \frac{E\{[X - E[X]]\,[Y - E[Y]]\}}{\sqrt{\sigma_X^2\,\sigma_Y^2}} \qquad \text{(A.13-7)}$$

$$= \frac{\mu_{11}}{\sigma_X\,\sigma_Y} \qquad \text{(A.13-8)}$$

where ρ is a dimensionless quantity

$$-1 \le \rho \le 1$$

Values close to 1 show high correlation of i.e., two random waveforms and those close to −1 show high correlation of the same waveform except with opposite sign. Values near zero indicate low correlation.

A.14 CORRELATION FUNCTIONS

If x(t) is a sample function from a random process and the random variables

$$x_1 = x(t_1)$$

$$x_2 = x(t_2)$$

are from this process, then, the *autocorrelation* function $R(t_1, t_2)$ is the joint moment of the two random variables;

$$R_{xx}(t_1,t_2) = E[x(t_1)x(t_2)] \quad \text{(A.14-1)}$$

$$= \int\int_{-\infty}^{\infty} x_1 x_2 \, f_{x_1 x_2}(x_1,x_2)\, d_{x_1}\, d_{x_2}$$

where the autocorrelation is a function of t_1 and t_2.

The auto-covariance of the process x(t) is the covariance of the random variables $x(t_1)$ $x(t_2)$

$$c_{xx}(t_1,t_2) = E\{[x(t_1) - \overline{x(t_1)}]\,[x(t_2) - \overline{x(t_2)}]\} \quad \text{(A.14-2)}$$

or rearranging equation (A.14-1)

$$c(t_1,t_2) = E\{[x(t_1) - \overline{x(t_1)}]\,[x(t_2) - \overline{x(t_2)}]\}$$

$$= E\{x(t_1)\,x(t_2) - x(t_1)\,\overline{x(t_2)} - \overline{x(t_1)}\,x(t_2) + \overline{x(t_1)x(t_2)}\}$$

$$= E\{x(t_1)\,x(t_2) - x(t_1)\,E[x(t_2)] - E[x(t_1)]\,x(t_2) + E[x(t_1)]\,E[x(t_2)]\}$$

$$= E[x(t_1)\,x(t_2)] - E[x(t_1)]\,E[x(t_2)] - E[x(t_1)]\,E[x(t_2)] + E[x(t_1)]\,E[x(t_2)]$$

$$= E[x(t_1)\,x(t_2)] - E[x(t_1)]\,E[x(t_2)] \quad \text{(A.14-3)}$$

or

$$= R(t_1, t_2) - E[x(t_1)]\,E[x(t_2)] \quad \text{(A.14-4)}$$

The autocorrelation function as defined in equation (A.14-1) is valid for both stationary and nonstationary processes. If x(t) is stationary then all its ensemble averages are independent of the time origin and accordingly

$$R_{xx}(t_1,t_2) = R_{xx}(t_1 + T,\, t_2 + T) \quad \text{(A.14-5a)}$$

$$= E[x(t_1 + T),\, x(t_2 + T)] \quad \text{(A.14-5b)}$$

Due to this time origin independence, T can be set equal to $-t_1$, $T = -t_1$, and substitution into equations (A.14-5a, b)

$$R_{xx}(t_1,t_2) = R_{xx}(0,t_2 - t_1) \quad \text{(A.14-6a)}$$

$$= E[x(0)\,x(t_2 - t_1)] \quad \text{(A.14-6b)}$$

imply that the expression is only dependent upon the time difference $t_2 - t_1$. Replacing the difference with $\tau = t_2 - t_1$ and suppressing the zero in the argument $R_{XX}(0,\, t_2 - t_1)$ yields

$$R_{xx}(\tau) = E[x(t_1)\,x(t_1 - \tau)] \quad \text{(A.14-7)}$$

Again since this is a stationary process it depends only on τ. The lack of dependence on the particular time, t_1, at which the ensemble was taken allows equation (A.14-7) to be written without the subscript, i.e.,

$$R_{xx}(\tau) = E[x(t)\,x(t + \tau)] \quad \text{(A.14-8)}$$

as it is found in many texts. This is the expression for the autocorrelation function of a stationary random process.

For the autocorrelation function of a nonstationary process where there is a dependence upon the particular time at which the ensemble average was taken as well as on the time difference between samples, the expression must be written with identifying subscripts, i.e., $R_{xx}(t_1,\, t_2)$ or $R_{xx}(t_1,\, \tau)$.

The time autocorrelation function can be defined next and has the form

$$\mathcal{R}_{xx}(\tau) = \lim_{t \to \infty} \frac{1}{T} \int_{-T/2}^{T/2} x(t)\, x(t + \tau)\, dt \quad \text{(A.14-9)}$$

For the special case of an ergodic process (Ref. Appendix A.12) the two functions, equations (A.14-8) and (A.14-9), are equal

$$\mathcal{R}_{xx}(\tau) = R_{xx}(\tau) \quad \text{(A.14-10)}$$

It is important to point out that if $\tau = 0$ in equation (A.14-7) the autocorrelation function

$$R_{xx}(0) = E[x(t_1)\,x(t_1)] \quad \text{(A.14-11)}$$

would equal the mean square value or total power (AC plus DC) of the process. Further, for values other than $\tau = 0$, $R_x(\tau)$ represents a measure of the similarity between its waveforms x(t) and $x(t + \tau)$.

In the same respect as the previous discussion, two random variables from two different jointly stationary random processes x(t) and y(t) have for the random variables

$$x_1 = x(t_1)$$
$$y_2 = y(t_1 + \tau)$$

the *crosscorrelation* function

$$R_{xy}(\tau) = E\{x(t_1)\, y(t_1+\tau)] \qquad \text{(A.14-12)}$$
$$= \int\int_{-\infty}^{\infty} x_1 y_2\, f_{x_1 y_2}(x_1, y_2)\, dx_1\, dy_2$$

The crosscorrelation function is simply a measure of how much these two variables depend upon one another.

Since it was assumed that both random processes were jointly stationary, the crosscorrelation is thus only dependent upon the time difference τ and, therefore

$$R_{xy}(\tau) = R_{yx}(\tau) \qquad \text{(A.14-13)}$$

where

$$y_1 = y(t_1)$$
$$x_2 = x(t_1 + \tau)$$

and

$$R_{yx}(\tau) = E\{y(t_1)\, x(t_1+\tau)] \qquad \text{(A.14-14)}$$
$$= \int\int_{-\infty}^{\infty} y_1 x_2\, f_{y_1 x_2}(y_1, x_2)\, dy_1\, dx_2$$

The time crosscorrelation functions are defined as before by

$$\mathcal{R}_{xy}(\tau) = \lim_{t \to \infty} \frac{1}{T} \int_{-T/2}^{T/2} x(t)\, y(t+\tau)\, dt \qquad \text{(A.14-15)}$$

and

$$\mathcal{R}_{yx}(\tau) = \lim_{t \to \infty} \frac{1}{T} \int_{-T/2}^{T/2} y(t)\, x(t+\tau)\, dt \qquad \text{(A.14-16)}$$

and finally

$$\mathcal{R}_{xy}(\tau) = R_{xy}(\tau) \qquad \text{(A.14-17)}$$
$$\mathcal{R}_{yx}(\tau) = R_{yx}(\tau) \qquad \text{(A.14-18)}$$

for the case of jointly ergodic random processes.

APPENDIX B

B.0 INTERCHANGING TIME INTEGRATION AND EXPECTATIONS

If f(t) is a nonrandom time function and a(t) a sample function from a random process then,

$$E\left[\int_{t_1}^{t_2} a(t)\, f(t)\, dt\right] = \int_{t_1}^{t_2} E[a(t)]\, f(t)\, dt \qquad \text{(B.0-1)}$$

This is true under the condition

a) $\int_{t_1}^{t_2} E[|a(t)|]\, |f(t)|\, dt < \infty$ (B.0-2)

b) a(t) is bounded by the interval t_1 to t_2. [t_1 and t_2 may be infinite and a(t) may be either stationary or nonstationary]

APPENDIX C

C.0 CONVOLUTION

This appendix defines convolution and presents a short proof without elaborate explanation. For complete definition of convolution refer to National Semiconductor Application Note AN-237.

For the *time convolution* if

$$f(t) \longleftrightarrow F(\omega) \qquad \text{(C.0-1)}$$
$$x(t) \longleftrightarrow X(\omega) \qquad \text{(C.0-2)}$$

then

$$y(t) = \int_{-\infty}^{\infty} x(\tau)\, f(t-\tau)\, d\tau \longleftrightarrow Y(\omega) = X(\omega) * F(\omega) \qquad \text{(C.0-3)}$$

or

$$y(t) = x(t) * f(t) \longleftrightarrow Y(\omega) = X(\omega) * F(\omega) \qquad \text{(C.0-4)}$$

proof:

Taking the Fourier transform, F[ ], of y(t)

$$F[y(t)] = Y(\omega) = \int_{-\infty}^{\infty}\left[\int_{-\infty}^{\infty} x(\tau)\, f(t-\tau)\, d\tau\right]\epsilon^{-j\omega t}\, dt \qquad \text{(C.0-5)}$$

$$Y(\omega) = \int_{-\infty}^{\infty} x(\tau)\left[\int_{-\infty}^{\infty} f(t-\tau)^{-j\omega t}\, dt\right] d\tau \qquad \text{(C.0-6)}$$

and letting $k = t - \tau$, then, $dk = dt$ and $t = k + \tau$.

Thus,

$$Y(\omega) = \int_{-\infty}^{\infty} x(\tau) \left[\int_{-\infty}^{\infty} f(k)\, \epsilon^{-j\omega(k+\tau)}\, dk \right] d\tau \quad \text{(C.0-7)}$$

$$= \int_{-\infty}^{\infty} x(\tau)\, \epsilon^{-j\omega\tau}\, d\tau$$

$$\int_{-\infty}^{\infty} f(k)\, \epsilon^{-j\omega k}\, dk \quad \text{(C.0-8)}$$

$$Y(\omega) = X(\omega) \bullet F(\omega) \quad \text{(C.0-9)}$$

For the *frequency convolution* of

$$f(t) \longleftrightarrow F(\omega) \quad \text{(C.0-10)}$$

$$x(t) \longleftrightarrow X(\omega) \quad \text{(C.0-11)}$$

then

$$H(\omega) = \frac{1}{2\pi} \int_{-\infty}^{\infty} F(\nu)\, X(\omega - \nu)\, d\nu \longleftrightarrow h(t) = f(t) \bullet x(t) \quad \text{(C.0-12)}$$

or

$$H(\omega) = \frac{1}{2\pi} F(\omega) * X(\omega) \longleftrightarrow h(t) = f(t) \bullet x(t) \quad \text{(C.0-13)}$$

proof:

Taking the inverse Fourier transform $F^{-1}[\ ]$ of equation (C.0-13)

$$h(t) = F^{-1}\left[\frac{x(\omega) * F(\omega)}{2\pi}\right] \quad \text{(C.0-14)}$$

$$= \frac{1}{2\pi} \int_{-\infty}^{\infty}$$

$$\left[\frac{1}{2\pi} \int_{-\infty}^{\infty} F(\nu)(\omega - \nu)\, d\nu\right] \epsilon^{j\omega t}\, d\omega$$

$$= \left(\frac{1}{2\pi}\right)^2 \int_{-\infty}^{\infty} F(\nu) \int_{-\infty}^{\infty} X(\omega - \nu)\, \epsilon^{j\omega t}\, d\omega\, d\nu \quad \text{(C.0-15)}$$

and letting $g = \omega - \nu$, then $dg = d\omega$ and $\omega = g + \nu$.

Thus,

$$F^{-1} \frac{X(\omega) * F(\omega)}{2\pi} \quad \text{(C.0-16)}$$

$$h(t) = \left(\frac{1}{2\pi}\right)^2 \int_{-\infty}^{\infty} F(\nu) \int_{-\infty}^{\infty} X(g)\, \epsilon^{j(g+\nu)t}\, dg\, d\nu$$

$$h(t) = \frac{1}{2\pi} \int_{-\infty}^{\infty} F(\nu)\, \epsilon^{j\nu t}\, d\nu \bullet \int_{-\infty}^{\infty} X(g)\, \epsilon^{jgt}\, dg \quad \text{(C.0-17)}$$

$$h(t) = f(t) \bullet x(t) \quad \text{(C.0-18)}$$

APPENDIX D

D.0 REFERENCES

1. Brigham, E. Oran, *The Fast Fourier Transform*, Prentice-Hall, 1974.
2. Chen, Carson, *An Introduction to the Sampling Theorem*, National Semiconductor Corporation Application Note AN-236, January 1980.
3. Chen, Carson, *Convolution: Digital Signal Processing*, National Semiconductor Corporation Application Note AN-237, January 1980.
4. Conners, F.R., *Noise*.
5. Cooper, George R.; McGillen, Clare D., *Methods of Signal and System Analysis*, Holt, Rinehart and Winston, Incorporated, 1967.
6. Enochson, L., *Digital Techniques in Data Analysis*, Noise Control Engineering, November-December 1977.
7. Gabel, Robert A.; Roberts, Richard A., *Signals and Linear Systems*.
8. Harris, F.J. *Windows, Harmonic Analysis and the Discrete Fourier Transform*, submitted to IEEE Proceedings, August 1976.
9. Hayt, William H., Jr.; Kemmerly, Jack E., *Engineering Circuit Analysis*, McGraw-Hill, 1962.
10. Jenkins, G.M.; Watts, D.G., *Spectral Analysis and Its Applications*, Holden-Day, 1968.
11. Kuo, Franklin F., *Network Analysis and Synthesis*, John Wiley and Sons, Incorporated, 1962.
12. Lathi, B.P., *Signals, Systems and Communications*, John Wiley and Sons, Incorporated, 1965.
13. Liu, C.L.; Liu, Jane W.S., *Linear Systems Analysis*.
14. Meyer, Paul L., *Introductory Probability and Statistical Applications*, Addison-Wesley Publishing Company, 1970.
15. Mix, Dwight F., *Random Signal Analysis*, Addison-Wesley Publishing Company, 1969.
16. Oppenheim, A.V.; Schafer, R.W., *Digital Signal Processing*, Prentice-Hall, 1975.
17. Otnes, Robert K.; Enochson, Loran, *Applied Time Series Analysis*, John Wiley and Sons, Incorporated, 1978.
18. Otnes, Robert K.; Enochson, Loran, *Digital Time Series Analysis*, John Wiley and Sons, Incorporated, 1972.
19. Papoulis, Athanasios, *The Fourier Integral and Its Applications*, McGraw-Hill, 1962.
20. Papoulis, Athanasios, *Probability, Random Variables, and Stochastic Processes*, McGraw-Hill, 1965.
21. Papoulis, Athanasios, *Signal Analysis*, McGraw-Hill, 1977.
22. Rabiner, Lawrence R.; Gold, Bernard, *Theory and Application of Digital Signal Processing*, Prentice-Hall, 1975.

23. Rabiner, L.R.; Schafer, R.W.; Dlugos, D., *Periodogram Method for Power Spectrum Estimation*, Programs for Digital Signal Processing, IEEE Press, 1979.
24. Raemer, Harold R., *Statistical Communications Theory and Applications*, Prentice-Hall EE Series.
25. Roden, Martin S., *Analog and Digital Communications Systems*, Prentice-Hall, 1979.
26. Schwartz, Mischa, *Information Transmission Modulation, and Noise*, McGraw-Hill, 1959, 1970.
27. Schwartz, Mischa; Shaw, Leonard, *Signal Processing: Discrete Spectral Analysis, Detection, and Estimation*, McGraw-Hill, 1975.
28. Silvia, Manuel T.; Robinson, Enders A., *Digital Signal Processing and Time Series Analysis*, Holden-Day Inc., 1978.
29. Sloane, E.A., *Comparison of Linearly and Quadratically Modified Spectral Estimates of Gaussian Signals,* IEEE Translations on Audio and Electroacoustics Vol. Au-17, No. 2, June 1969.
30. Smith, Ralph J., *Circuits, Devices, and Systems*, John Wiley and Sons, Incorporated, 1962.
31. Stanley, William D., *Digital Signal Processing*, Reston Publishing Company, 1975.
32. Stearns, Samuel D., *Digital Signal Analysis*, Hayden Book Company Incorporated, 1975.
33. Taub, Herbert; Schilling, Donald L., *Principles of Communication Systems*, McGraw-Hill, 1971.
34. Tretter, Steven A., *Discrete-Time Signal Processing*, John Wiley and Sons, Incorporated, 1976.
35. Turkey, J.W.; Blackman, R.B., *The Measurement of Power Spectra*, Dover Publications Incorporated, 1959.
36. Welch, P.D., *On the Variance of Time and Frequency Averages Over Modified Periodograms*, IBM Watson Research Center, Yorktown Heights, N.Y. 10598.
37. Welch, P.D., *The Use of Fast Fourier Transforms for the Estimation of Power Spectra: A Method Based on Time Averaging Over Short Periodograms*, IEEE Transactions on Audio and Electroacoustics, June 1967.
38. Programs for *Digital Signal Processing, Digital Signal Processing Committee*, IEEE Press, 1979.
39. Bendat, J.S.; Piersol, A.G., *Random Data: Analysis and Measurement Procedures*, Wiley-Interscience, 1971.

BASIC DIGITAL FILTER THEORY

By John R. Mick

INTRODUCTION

Digital filtering applications are rapidly expanding as new developments in technology provide increased packing density in complex integrated circuits. Until recently digital filter processing algorithms have been used primarily in computer simulations, sampled data analysis and data reduction computations. The variety of complex integrated circuits suitable in size, weight, power and cost for real-time processing of video signals by digital techniques is increasing steadily. With the increasingly extensive application of digital processors to many systems, more and more importance is placed on the development of mathematical tools for the analysis and design of sampled data systems. In particular the classical methods of difference equation solutions are available to the designer as well as the "z-transform" calculus solutions. The latter analytical method results in considerable simplification and understanding of the problems associated with sampled-data systems.

This application note presents a brief review of these concepts. A brief introduction to sampling theory is presented and a review of the difference equation as applicable to digital filtering follows. Several digital filter configurations are outlined and a summary of the most useful transforms for designing digital filters is also presented.

Definition

The term "digital filter" refers to a computational algorithm performed on a sampled input signal resulting in a transformed output signal. The input signal is a sequence of numbers from either an analog-to-digital converter or a direct digital input source. The computational process can correspond to high-pass filtering, low-pass filtering, band-pass filtering, integration, differentiation etc. The output signal is either a direct digital sequence or a regenerated analog signal from a digital-to-analog converter.

Advantages

Several unique advantages are offered by the digital approach to signal processing. These include:

1. Performance from unit to unit is stable and repeatable.
2. Arbitrarily high precision is achieved that is limited only by the number of bits carried in memory and by the input and output resolution capabilities.
3. No impedance matching problems exist in the digital domain.
4. Critical filter break frequencies can be placed without restriction (influences the precision required).
5. Component value variation problems normally associated with capacitors and resistors due to temperature changes or age are nonexistent.
6. Greater flexibility is achieved since filter response can be changed by varying the proper arithmetic coefficients.
7. The intrinsic possibility of time-sharing major implementation sections exists (adders, subtractors, multipliers etc).
8. Small size results from integrated circuit implementation.
9. Periodic calibration as is required with analog circuits is eliminated.
10. Performance limitations of physical analog components are avoided.

THE SAMPLING PROCESS

It's convenient to think of the sampling process as an impulse modulation of a continuous input signal. Accordingly if the input signal v(t) is sampled every T seconds, an output signal results denoted $v^*(t)$. This is shown in Figure 1.

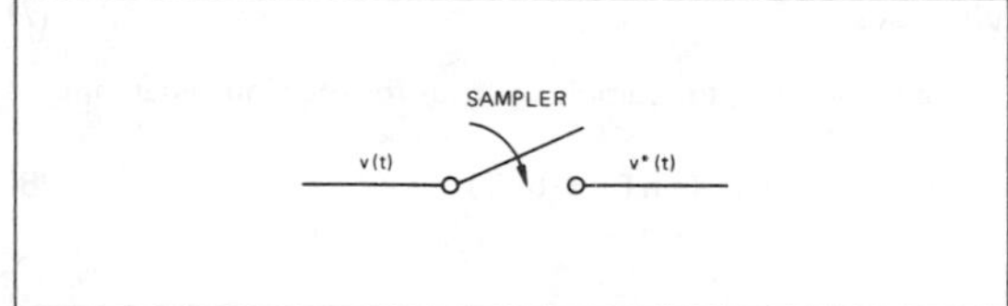

Figure 1. Sampler Representation.

The ideal sampler is represented by using the Dirac delta function to express a unit impulse train $\delta(t-nT)$. This notation represents impulses occurring at each t = nT seconds for n equal to positive integers. The ideal sampler $\delta_T(t)$ for a continuous train of regularly spaced pulses is described for positive time sequences by the following equation:

$$\delta_T(t) = \sum_{n=0}^{\infty} \delta(t-nT) \qquad (1)$$

The sampler output signal is written in terms of a continuous input signal and the ideal sampler unit impulse train as

$$v^*(t) = v(t) \sum_{n=0}^{\infty} \delta(t-nT) \qquad (2)$$

This equation is rewritten to include the input signal as a time function when t = nT as

$$v^*(t) = \sum_{n=0}^{\infty} v(nT)\,\delta(t-nT) \qquad (3)$$

This equation shows that the sampler output is an impulse train with an amplitude equal to the continuous input signal amplitude at the sampling instant.

The Laplace transform of the ideal sampler is the Laplace transform of the Dirac impulse train $\delta_T(t)$ and is given by

$$\mathcal{L}[\delta_T(t)] = \mathcal{L}\left[\sum_{n=0}^{\infty} \delta(t-nT)\right] = \sum_{n=0}^{\infty} e^{-nTs} \qquad (4)$$

5

since the Laplace transform of the unit impulse function $\delta(t-nT)$ is e^{-nTs}.

Using equation 4, the Laplace transform of equation 3, the sampler output becomes

$$V^*(s) = \sum_{n=0}^{\infty} v(nT)e^{-nTs} \quad (5)$$

This is very similar to the definition of the continuous Laplace transform

$$V(s) = \int_0^{\infty} v(t)e^{-st}dt \quad (6)$$

except that the integral is replaced by a summation evaluated at the sampling instants $t = nT$ of the unit impulse train.

Time Domain Sampling

The time domain analysis of the above described sampler is best understood by considering a continuous sinusoidal input signal

$$v(t) = A\sin(wt) \quad (7)$$

Using equation 3, the sampler output for the sinusoidal input is

$$V^*(t) = \sum_{n=0}^{\infty} A\sin(wnT)\ \delta(t-nT) \quad (8)$$

Figure 2 shows the time domain response of a sinusoidal input signal, a sampler, and a sampler output as described by the above equations, where the sampling rate is considerably higher than the continuous input frequency.

Using equation 5, the Laplace transform of equation 8 describing the sampler output for the sinusoidal input is

$$V^*(s) = \sum_{n=0}^{\infty} A\sin(wnT)e^{-nTs} \quad (9)$$

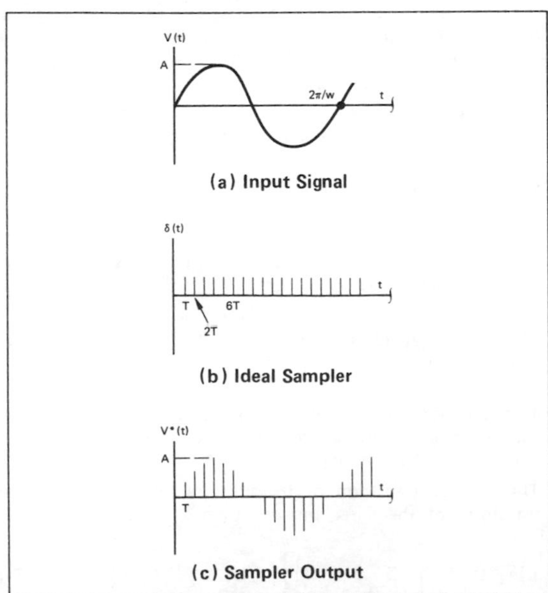

Figure 2. Time Domain Sampling.

The time domain analysis gives a useful picture of the sampler characteristics as a function of time; however, the complete picture also requires an analysis of the sampler in the frequency domain.

Frequency Domain Sampling

To examine the characteristics of the ideal sampler in the frequency domain, the Laplace transform of the ideal sampler as established in equation 4 is expanded as

$$\sum_{n=0}^{\infty} e^{-nTs} = 1 + e^{-st} + e^{-2sT} + e^{-3sT} + \ldots \quad (10)$$

The closed form of this geometric series is

$$\sum_{n=0}^{\infty} e^{-nTs} = \frac{1}{1-e^{-sT}} \quad (11)$$

Thus the Laplace transform of the sampler output is given by the convolution of the ideal sampler and the continuous input signal as

$$V^*(s) = V(s) * \left[\frac{1}{1-e^{-sT}}\right] \quad (12)$$

Since convolution in the s-domain requires contour integration[21], only the result is stated; the sampler output for a continuous input is

$$V^*(s) = \frac{1}{T}\sum_{n=-\infty}^{\infty} V(s + jnw_s) \quad (13)$$

It is important to note that the sampler and its output are periodic with period jw_s. This means $V^*(s)$ is equal to $V^*(s+jw_s)$ and is represented in the s-plane by periodic strips along the jw axis. As a result, the sampler causes a periodic single line spectrum in the frequency domain occurring at each integer multiple of the sampling frequency.

Assuming a continuous sinusoidal input signal

$$v(t) = A\sin(w_a t) \quad (14)$$

and a sampler operating at radian frequency w_s, the output spectrum is periodic with spurious sidebands located at all multiples of w_s. The input spectrum is centered around each of these spurious multiples of the sampling frequency. This is shown in Figure 3.

The Z-Transform

The z-transform is used to describe a sampled data system in much the same way as the Laplace transform is used to describe a continuous time system. The z-transform of a signal describes the signal at the sampling instant and therefore contains information about the corresponding time function at the sampling instants only. The z-transform is obtained by making the substitution

$$z = e^{sT} \quad (15)$$

or

$$s = \frac{1}{T}\ln(z) \quad (16)$$

where z is interpreted as a complex transform variable. Thus, every continuous signal that has a Laplace transform also has a z-transform by a simple substitution.

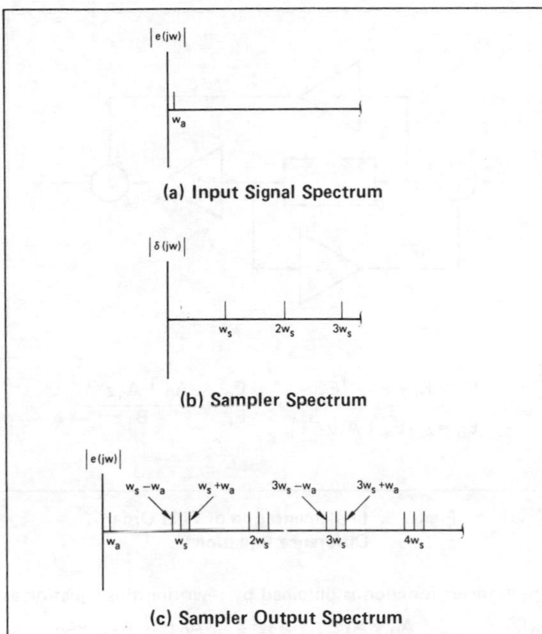

(a) Input Signal Spectrum

(b) Sampler Spectrum

(c) Sampler Output Spectrum

Figure 3. Frequency Domain Sampling.

Since the Laplace transform of a sampler is periodic, the z-transform performs a change of variable which retains the s-plane pole-zero configuration while stripping the function of its repetitive character. Thus, the z-transform allows simple algebraic manipulation of the polynomials in the z-plane just as the Laplace transform does for the polynomials in the s-plane.

The above substitution maps the periodic strip from $-w_s/2$ to $+w_s/2$ of the jw-axis of the s-plane onto the unit circle of the z-plane where w_s is the sampling frequency. The remainder of this strip in the left-hand, s-plane is mapped inside the unit circle in the z-plane. This is shown in Figure 4.

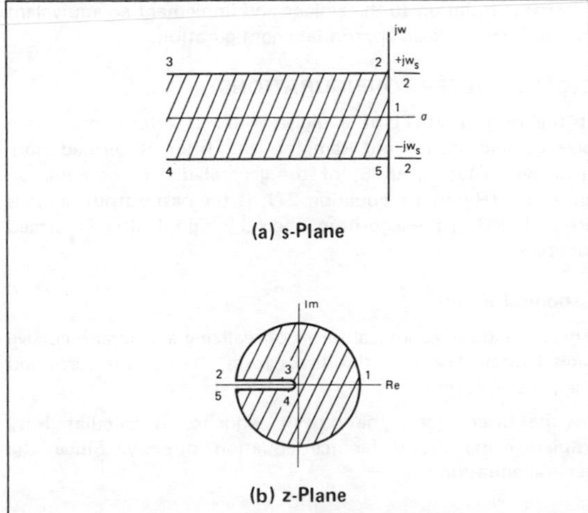

(a) s-Plane

(b) z-Plane

Figure 4. S-Plane to z-Plane Transformation.

Successive $w_s = 2\pi/T$ strips of the left hand side of the s-plane are mapped into the same unit circle of the z-plane. Likewise, the corresponding right-half strip is mapped as the exterior of the unit circle in the z-plane. If a transfer function is to be stable, its poles are in the left half of the s-plane; thus, the poles of the transformed function must lie within the unit circle in the z-plane. It follows that the z-plane poles and zeros occur on the real axis or in complex conjugate pairs.

The interval from $-w_s/2$ to $+w_s/2$ is known as the Nyquist interval. This interval places a bound on the bandwidth of the input signal to the sampler such that if the input signal is not bandlimited to below the radian frequency $w_s/2$, it cannot be recovered exactly at the output. Figure 5a shows the aliasing problems on the input spectrum after sampling if the input signal is not bandlimited while Figure 5b shows the spectrum of a bandlimited signal before and after sampling.

Using the substitution $z = e^{sT}$ on equation 3, the z-transform output for a sampled input signal is found as

$$V^*(z) = \sum_{n=0}^{\infty} v(nT)z^{-n} \tag{17}$$

where z^{-n} is a delay operator and n is an integer representing the number of past unit delays.

THE DIFFERENCE EQUATION

In linear continuous (analog) filter theory, linear differential equations is one mathematical tool available to describe the transfer function. Similarly, in linear digital (sampled) filter theory, the linear difference equation is available as a mathematical tool for analysis and synthesis.

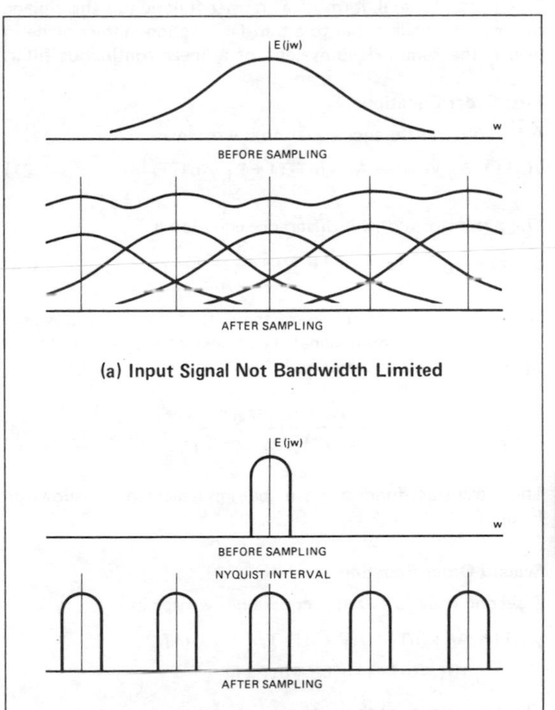

(a) Input Signal Not Bandwidth Limited

(b) Input Signal Bandwidth Limited

Figure 5. Effect of Bandlimiting Before Sampling.

The linear difference equation is used to define the sampled output pulse amplitude y(t) as a function of the present input pulse and any number of past input and output pulses. A general form of the difference equation[13] is

$$y(nT) = \sum_{i=0}^{N} A_i\, x(nT-iT) + \sum_{i=1}^{M} B_i\, y(nT-iT) \qquad (18)$$

where the notation x(nT) represents the present input sample and the x(iT) are past input samples. Similarly, y(nT) is the present output sample and the y(iT) are past output samples. The A_i and B_i coefficients are a set of constants which determine the response of the filter.

The z-transform of the general difference equation 18 is derived by using equation 17 and is given as

$$y(z) = x(z) \sum_{i=0}^{N} A_i z^{-i} + y(z) \sum_{i=1}^{M} B_i z^{-i} \qquad (19)$$

This equation is interpreted as: the present output is equal to the present and past inputs each multiplied by the respective coefficient A_i plus the past outputs each multiplied by the respective coefficient B_i. Equation 19 is rewritten in the normal transfer function form as

$$H(z) = \frac{y(z)}{x(z)} = \frac{\sum_{i=0}^{N} A_i z^{-i}}{1 - \sum_{i=1}^{M} B_i z^{-i}} \qquad (20)$$

This is the general form of a transfer function in the z-plane that can be made equal to a transfer function in the s-plane to realize the sampled equivalent of a linear continuous filter.

First Order Equation

A first order difference equation is written as

$$y(nT) = A_0\, x(nT) + A_1\, x(nT\text{-}T) + B_1\, y(nT\text{-}T) \qquad (21)$$

The z-transform of this difference equation is

$$E_o(z) = A_0 E_i(z) + A_1 z^{-1} E_i(z) + B_1 z^{-1} E_o(z) \qquad (22)$$

where E_o is used to represent the output signal and E_i is used to represent the input signal. The transfer function is obtained by rewriting this equation as

$$\frac{E_o(z)}{E_i(z)} = H(z) = \frac{A_0 + A_1 z^{-1}}{1 - B_1 z^{-1}} = \frac{A_0 z + A_1}{z - B_1} \qquad (23)$$

This transfer function can be implemented as shown in Figure 6.

Second Order Equation

A second order difference equation is written as

$$y(nT) = A_0\, x(nT) + A_1\, x(nT-T) + A_2\, x\,(nT\text{-}2T) + B_1\, y(nT-T) + B_2\, y(nT\text{-}2T)$$

The z-transform of this difference equation is

$$E_o(z) = A_0 E_i(z) + A_1 z^{-1} E_i(z) + A_2 z^{-2} E_i(z) + B_1 z^{-1} E_o(z) + B_2 z^{-2} E_o(z) \qquad (25)$$

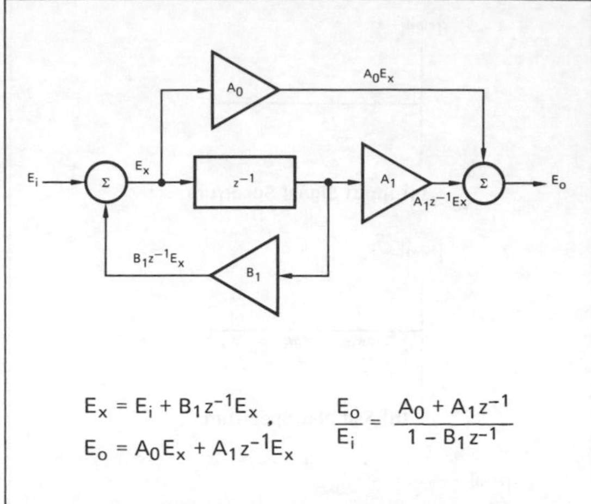

Figure 6. Implementation of First Order Difference Equation.

The transfer function is obtained by rewriting this equation as

$$\frac{E_o(z)}{E_i(z)} = H(z) = \frac{A_0 + A_1 z^{-1} + A_2 z^{-2}}{1 - B_1 z^{-1} - B_2 z^{-2}} = \frac{A_0 z^2 + A_1 z + A_2}{z^2 - B_1 z - B_2} \qquad (26)$$

This transfer function can be implemented as shown in Figure 7.

Difference Equation Summary

The first and second order difference equations, their z-transform functions, and their circuit implementations serve as illustrative examples of the equivalence of the mathematical description and the hardware associated with digital filters. Since the z-transform is equal to the Laplace transform by means of the substitution $z = e^{sT}$, the first and second order implementations are mathematically related to s-plane transfer functions. A great wealth of information for design and synthesis of analog filters using Laplace transforms is available in the literature. It is therefore possible to use these procedures to design an equivalent analog transfer function, then transform this function to the z-plane and implement an equivalent digital filter using an appropriate configuration.

DIGITAL FILTER CONFIGURATIONS

If the output y(nT) of a digital filter is a function of the present and past input samples, the filter is termed nonrecursive. That is, all B_i of the general difference equation and zero. (Reference equation 27). If the past output samples are included in the algorithm, then the digital filter is termed recursive.

Canonical Forms

There are three canonical forms of realizing a general recursive digital filter. These are the direct form, the cascade form and the parallel form.

In the direct form the output sequence is calculated by implementing the difference equation directly. Since the general equation is

$$y(nT) = \sum_{i=0}^{N} A_i\, x(nT-iT) + \sum_{i=1}^{M} B_i\, y(nT-iT) \qquad (27)$$

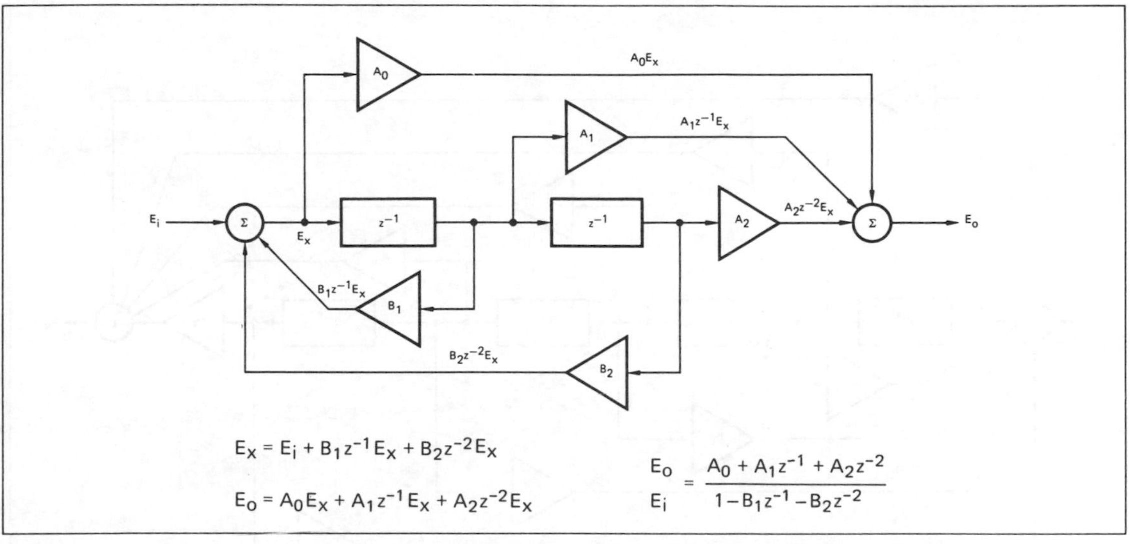

Figure 7. Implementation of Second Order Difference Equation.

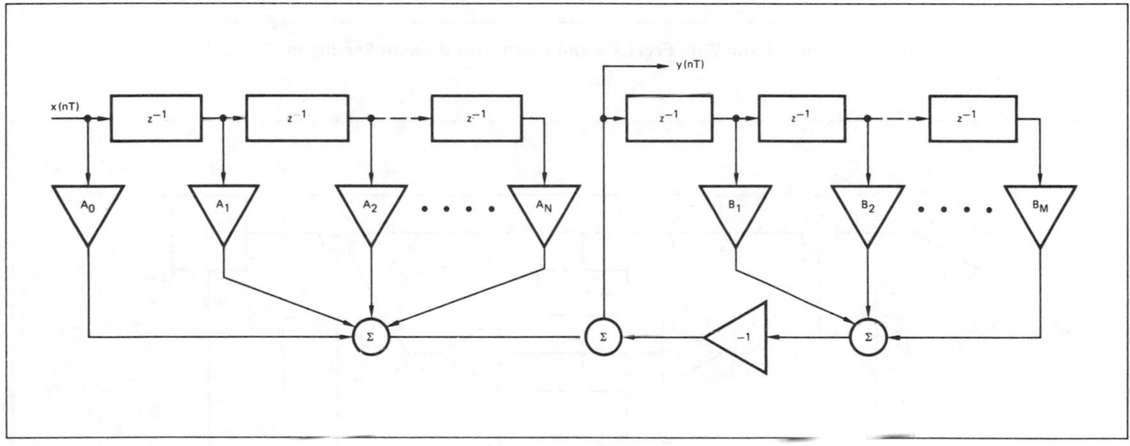

Figure 8. Direct Canonical Form.

the direct form digital filter takes the configuration of Figure 8. Figure 9 depicts another configuration of the direct canonic form in which the memory elements (z^{-1}) are shared by the feedback and feedforward loops. This direct form suffers from the fact that the pole locations are extremely sensitive functions of the coefficients B_i for higher order filters[7]. This directly affects the precision required for the entire digital filter.

In the cascade form the digital filter is implemented from the transfer function written as a product of factors.

$$H(z) = A \frac{\prod_{i=1}^{K_1} (1 + a_iz^{-1} + b_iz^{-2})}{\prod_{i=1}^{K_2} (1 + c_iz^{-1} + d_iz^{-2})} \qquad (28)$$

This configuration consists of a series of lower order filters connected in cascade as shown in Figure 10. The pole coefficients c_i and d_i are not nearly as sensitive as the direct figuration[4]. Therefore, this form is especially practical for higher order filters.

5

The parallel canonical form is implemented by writing the transfer function as a sum of partial fractions.

$$H(z) = A_0 + \sum_{i=1}^{K} \left[B_i \frac{a_i + b_iz^{-1}}{1 + c_iz^{-1} + d_iz^{-2}} \right] \qquad (29)$$

This configuration consists of a group of lower order filters each operating on the input signal with the output of the parallel bank of filters summed together as in Figure 11.

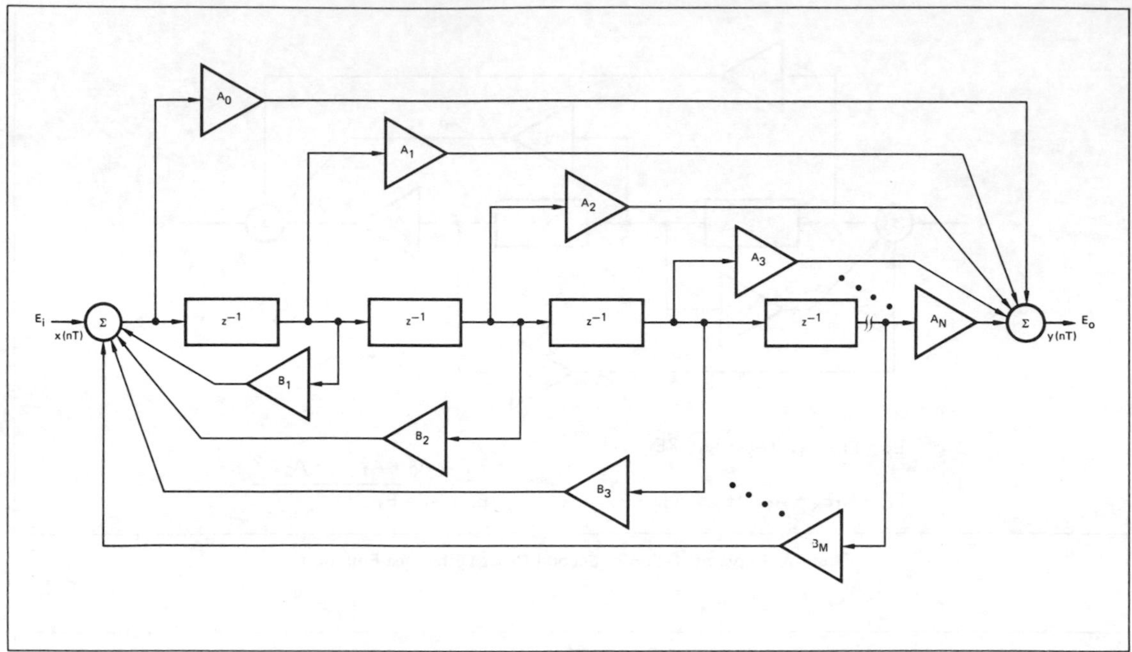

Figure 9. Direct Canonical Form With Feedback and Feedforward Loops Sharing the Same Memory.

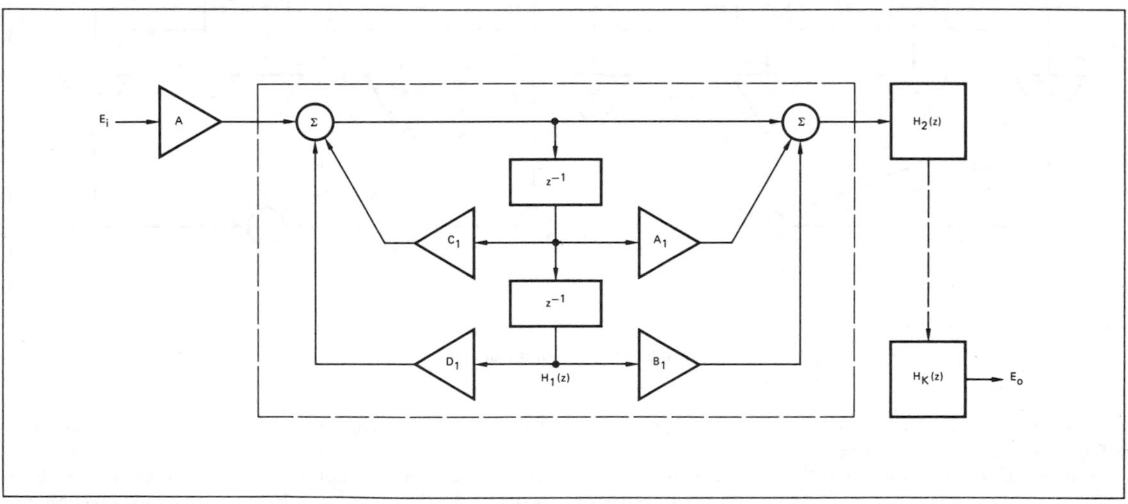

Figure 10. Cascade Canonical Form.

Other Configurations

Figure 12 illustrates a recursive one-pole, one-zero building block while Figure 13 represents a recursive two-pole, two-zero building block. In these structures, the B_i determine the pole locations in the z-plane while the A_i determine the zero locations in the z-plane.

A general recursive two-pole building block with the z-plane transfer function is shown in Figure 14.

It is apparent that many digital filter configurations can be designed. Each configuration has properties that may or may not be desirable depending on the particular application. Thus, each application must be treated individually and it is difficult to generalize that one configuration is always superior.

SYNTHESIS TECHNIQUES

There are three transform techniques that find the greatest application in the design of digital filters from continuous transfer functions. These are the standard z-transform, the bilinear z-transform and the matched z-transform.

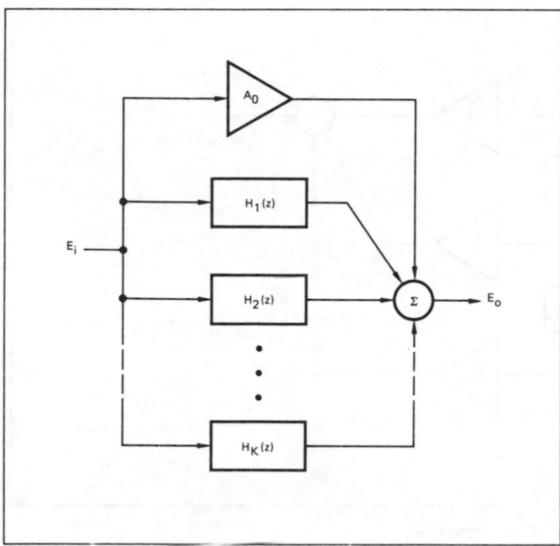

Figure 11. Parallel Canonical Form.

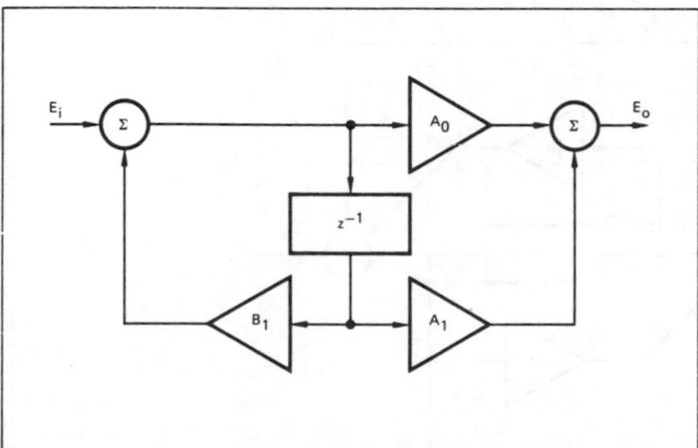

Figure 12. Recursive One-Pole Structure.

Standard Z-Transform Method

The standard z-transform, also known as the impulse invariant technique, utilizes the partial fraction expansion of the continuous filter transfer function. This transformation preserves the impulse response of the sampled continuous filter and is best suited for low-pass and band-pass applications.

In using this technique the Laplace transform partial fraction expansion terms are replaced by the appropriate z-transform terms. The substitutions used are shown in Table 1.

From this it is seen that the standard z-transform technique gives a transfer function of the form

$$H(z) = A_0 + \sum_{i=1}^{K} \frac{a_i}{1-b_i z^{-1}} + \sum_{j=1}^{L} \frac{a_j + b_j z^{-1}}{1+c_j z^{-1} + d_j z^{-2}} \tag{30}$$

where any of the above coefficients may be zero.

Thus, the coefficients are defined uniquely and the digital filter can be implemented directly using one-pole and two-pole building blocks in the parallel canonical form as described in the previous section.

Bilinear Transformation Method

The bilinear z-transform is an algebraic mapping transformation utilizing the substitution

$$s = \frac{2(1-z^{-1})}{T(1+z^{-1})} \tag{31}$$

This transform maps a sampling interval from $-jw_s/2$ to $+jw_s/2$ in the s-plane onto the unit circle in the z-plane. It should be noted however, that this transform does not yield a linear map as does the substitution $z = e^{st}$; that is, a non-linear warping of the frequency scale in the z-plane results. The magnitude of this warping is given by

$$W_A = \frac{2}{T} \, \text{Tan}\left(\frac{w_D T}{2}\right) \tag{32}$$

where w_A = s-plane frequency
w_D = z-plane warped frequency

In searching the literature on the bilinear z-transform,[3,4,5] another substitution is presented that is very similar to equation 31; that is

$$s \rightarrow \frac{z-1}{z+1} \tag{33}$$

with the warping function given as

$$w_A \rightarrow \text{Tan}\left(\frac{w_D T}{2}\right) \tag{34}$$

Occasionally this causes confusion since the units are not the same. In practice, however, both substitutions yield the same pole-zero configuration since the w/T terms will factor and cancel. This is best understood by making the respective substitutions in a general transfer function for a complex pole-pair such as

$$H(s) = \frac{w^2}{s^2 + 2\delta ws + w^2} \tag{35}$$

and comparing the z-plane pole positions that result.

When using the bilinear z-transform care must be taken when the break frequencies are near the half sampling frequency. An illustration of this fact and an appreciation of the warping required is best illustrated by an example. Table 2 shows various digital filter break frequencies and the required warped analog frequencies for a 1000 Hz sampling rate. Figure 15 shows graphically the non-linear frequency scale mapping of the bilinear z-transform.

Thus the bilinear z-transform is a powerful tool in digital filtering and may be utilized with either the partial fraction expansion or the rational fraction form of the Laplace transfer function. It is especially useful in MTI radar filters since the break frequencies of the high-pass and band-pass filters used are normally very low compared with the sampling frequency. This means that the warping is very small and the digital design very closely approximates the analog design.

5

Matched Z-Transform

The matched z-transform is somewhat of a compromise between the standard and bilinear z-transforms. It is an exponential mapping transform which gives a z-plane transfer function with poles and zeros matched to those of the continuous function.

Real poles or zeros are mapped using the substitution

$$s-a \rightarrow 1-z^{-1} e^{aT} \tag{36}$$

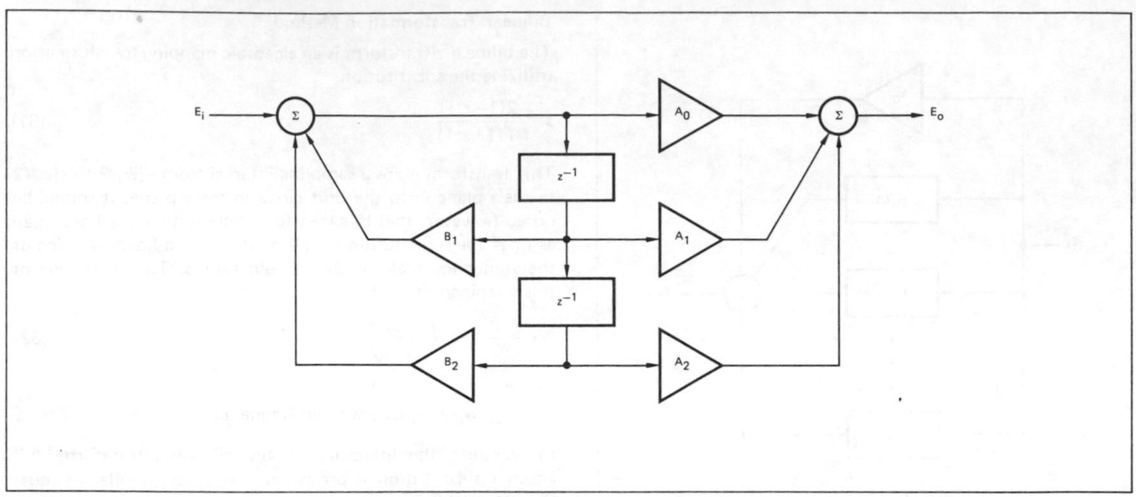

Figure 13. Recursive Two-Pole Structure.

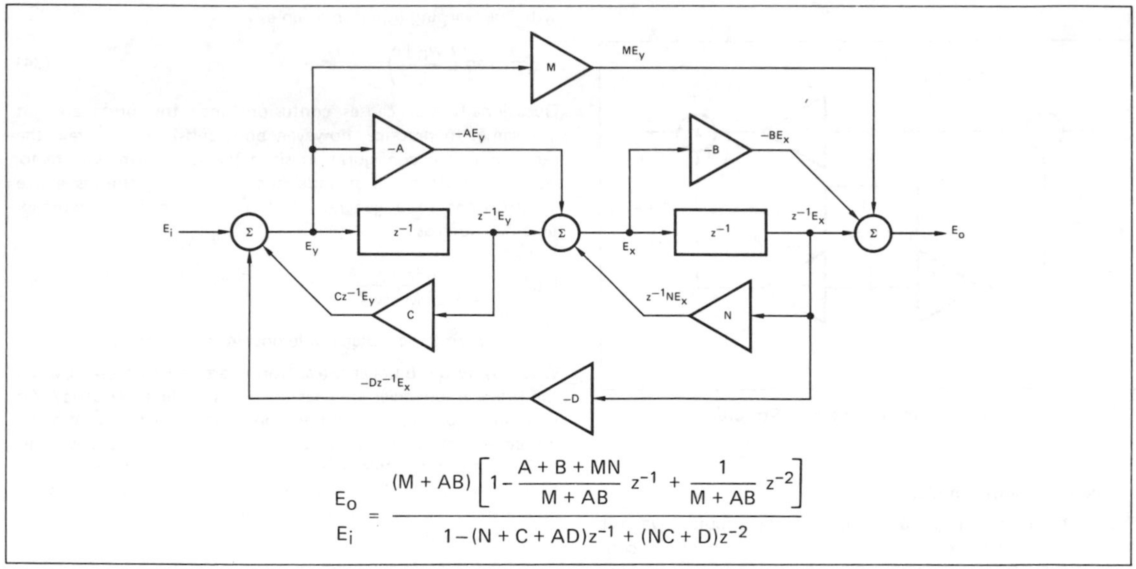

$$\frac{E_o}{E_i} = \frac{(M + AB)\left[1 - \frac{A + B + MN}{M + AB} z^{-1} + \frac{1}{M + AB} z^{-2}\right]}{1 - (N + C + AD)z^{-1} + (NC + D)z^{-2}}$$

Figure 14. General Two-Pole Building Block.

while complex poles or zeros are mapped using the substitution

$$(s - a)^2 + b^2 \rightarrow 1 - 2z^{-1} e^{aT} \operatorname{Cos}(bT) + z^{-2}e^{2aT} \qquad (37)$$

The resulting z-plane transfer function is normally factored into numerator and denominator polynomials that are easily implemented using one and two-pole building blocks in the cascade canonical form.

The matched z-transform yields the same pole configuration as the standard z-transform; however, the zero configuration may require some modification to give satisfactory results. This usually entails the addition of zeros at the half sampling frequency (z= −1) to give the desired result.[8]

TABLE I. Standard z-Transform Substitutions.

f(t)	F(s)	F(z)
e^{-aT}	$\frac{1}{s + a}$	$\frac{z}{z - e^{-aT}}$
$\sin(w_0 t)$	$\frac{w_0}{s^2 + w_0^2}$	$\frac{z \sin(w_0 T)}{z^2 - 2z \cos(w_0 T) + 1}$
$\cos(w_0 t)$	$\frac{s}{s^2 + w_0^2}$	$\frac{z(z - \cos(w_0 T))}{z^2 - 2z \cos(w_0 T) + 1}$

TABLE II. Digital Filter Warping Relation.

$$f_A = \frac{1}{\pi T} \tan(\pi f_D T) \qquad T = 10^{-3} \text{ seconds}$$

Desired Digital Filter Break Frequency f_D Hz	Prewarped Analog Frequency f_A Hz	$\tan(\pi f_D T)$
50	50.5	.15838
100	103.2	.32492
200	231	.72654
250	318	1.0000
300	438	1.3764
400	978	3.0777
450	2010	6.3138
475	4050	12.706
500	?	∞

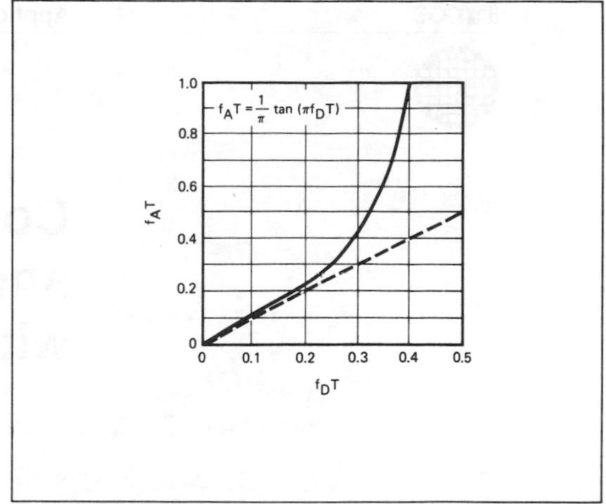

Figure 15. Bilinear z-Transform Frequency Scale Warping.

BIBLIOGRAPHY

1. Burrus, C. S., and T. W. Parks, "Time Domain Design of Recursive Digital Filters," *IEEE Transactions on Audio and Electroacoustics,* Vol. AU-18, No. 2, June, 1970, pp. 137-141.
2. Crystal, T. H., and L. Ehrman, "The Design and Application of Digital Filters with Complex Coefficients," *IEEE Transactions on Audio and Electroacoustics,* Vol. AU-16, September, 1968, pp. 315-320.
3. Gold, B., and C. M. Rader, *Digital Processing of Signals,* McGraw-Hill, Inc., New York, 1969.
4. Golden, R. M., "Digital Filter Synthesis by Sampled-Data Transformation," *IEEE Transactions on Audio and Electroacoustics,* Vol. AU-16, September, 1968, pp. 321-329.
5. Golden, R. M., and J. F. Kaiser, "Design of Wideband Sampled-Data Filters," *Bell System Technical Journal,* July 1964, pp. 1533-1546.
6. Helms, H. D., "Nonrecursive Digital Filters; Design Methods for Achieving Specifications on Frequency Response," *IEEE Transactions on Audio and Electroacoustics,* Vol. AU-16, September, 1968, pp. 336-342.
7. Jackson, L. B., J. F. Kaiser, and H. S. McDonald, "An Approach to the Implementation of Digital Filters," *IEEE Transactions on Audio and Electroacoustics,* Vol. AU-16, September, 1968, pp. 413-421.
8. Jury, E. I., *Sampled-Data Control Systems,,* John Wiley & Sons, New York, 1958, Chapters 1, 8.
9. Knowles, J. B., and E. M. Olcayto, "Coefficient Accuracy and Digital Filter Response," *IEEE Transactions on Circuit Theory,* Vol. CT-15, March, 1968, pp. 31-41.
10. Koivo, A. J., "Quantization Error and Design of Digital Control Systems," *IEEE Transactions on Automatic Control,* Vol. 1, February, 1969, pp. 55-58.
11. Kuo, F. F., and J. F. Kaiser, *System Analysis by Digital Computer,* John Wiley & Sons, New York, 1966, Chapter 7.
12. Monroe, A. J., *Digital Processes for Sampled-Data Systems,* John Wiley & Sons, New York, 1962, Chapters 2, 6, 7, 11.
13. Nowak, D. J., and P. E. Schmid, "Introduction to Digital Filters," *IEEE Transactions on Electromagnetic Compatibility,* Vol. EMC-10, June, 1968, pp. 210-220.
14. Otnes, R. K., "An Elementary Design Procedure for Digital Filters," *IEEE Transactions on Audio and Electroacoustics,* Vol. AU-16, September, 1968, pp. 330-335.
15. Rader, C.M., and B. Gold, "Effects of Parameter Quantization on the Poles of a Digital Filter," *Proceedings of the IEEE,* Vol. 55, May, 1967, pp. 688-689.
16. Rader, C. M., and B. Gold, "Digital Filter Design Techniques in the Frequency Domain," *Proceedings of the IEEE,* Vol. 55, February, 1967, pp. 149-171.
17. Rader, C. M., and B. Gold, "Effects of Quantization Noise in Digital Filters," *1966 Spring Joint Computer Conference, AFIPS Proceedings,* Vol. 28, 1966, pp. 213-219.
18. Rader, C. M., et al., "On Digital Filtering," *IEEE Transactions on Audio and Electroacoustics,* Vol. AU-16, September, 1968, pp. 303-314.
19. Ragazzini, J. R., and G. F. Franklin, *Sampled-Data Control Systems,* McGraw-Hill, New York, 1958, Chapters 4, 5.
20. Steiglitz, K., "Computer-Aided Design of Recursive Digital Filters," *IEEE Transactions on Audio and Electroacoustics,* Vol. AU-18, No. 2, June, 1970, pp. 123-129.
21. Tou, J. T., *Digital and Sampled-Data Control Systems,* McGraw-Hill, New York, 1959.
22. Truxal, J. G., *Control System Synthesis,* McGraw-Hill, New York, 1955, Chapter 9.
23. Tufts, D. W., D. W. Rorabacher, and W. E. Mosier, "Designing Simple, Effective Digital Filters," *IEEE Transactions on Audio and Electroacoustics,* Vol. AU-18, No. 2, June, 1970, pp. 142-158.

inmos

Application note

Correlation and convolution with the IMS A100

Contents

Correlation and convolution with the IMS A100

1 Introduction

The correlation process is widely used in many electronic systems including instrumentation, communication, medical ultrasonics, radar, sonar, control systems and other signal processing environments. The basic reasons for this widespread use can be attributed to the many useful characteristics exhibited by the correlation process. These properties include

- The ability to recover a desired signal masked by noise or other interferences. This is particularly useful in noisy environments that arise in communication, radar, sonar and ultrasonic applications.
- Delay estimation capability which is essential in many applications including range measurement in navigation systems, radar, sonar and also system identification.
- The ability to recognize a given pattern within a signal.
- The auto-correlation of a signal is closely related to the power spectrum which has resulted in the application of the correlation process to spectral analysis.
- The correlation process provides a good characterization of many signals and has therefore been used in many prediction and estimation algorithms.

Convolution is closely related to the correlation process. Mathematically convolution is what happens in the process of filtering. It will be shown in the next section that both these functions involve a large number of multiplications and additions. Up to now, for the time domain implementation of these processes, many systems have used multiply-accumulator devices. Because of their inherent concurrency, the numerical evaluations involved in the convolution and correlation functions can be performed in parallel. But due to the high cost, power consumption requirement, and size restriction many digital systems use only a single (or possibly two) multiply-accumulator(s). This has resulted in a processing bottle-neck in the time domain evaluation of these functions. For example using a 16-bit multiply and accumulator chip available today it is possible, for a 32-point digital correlator, to achieve at best a sampling frequency of around 100 to 300KHz. This is further reduced as the number of correlation points increases. Additional complexities occur as some form of address generator has to be used to sequence the data and the reference coefficients through the multiply-accumulator chip.

The IMS A100 VLSI chip overcomes these problems by incorporating 32 multiply-accumulators on a single chip. The sampling speed of the IMS A100 ranges from 2.5 MHz to 10 MHz depending on the reference-waveform word-size. (4, 8, 12 or 16 bits). It is the true parallelism incorporated in the systolic structure of the IMS A100 that allows such speed increases. The architecture of the IMS A100 has been designed in such a way that large numbers of these chips can be cascaded to perform high precision correlations involving more than 32 points at full speed. Alternatively it is possible to use multidimensional index mapping to decompose a long correlation/convolution into a number of short ones which can then be carried out by usoing a single or a small number of devices.

By suitable allocation of the coefficients, the IMS A100 can be used to perform 3×3, 5×5, or larger two-dimensional image convolutions.

In this application note the concepts of correlation and convolution are first introduced followed by their IMS A100 implementation issues. Partitioning techniques for decomposing a long correlation/convolution into a number of short ones are then described. Next an example of a two-dimensional image convolution is given. Finally some application areas of correlation and convolution are summarised.

2 Correlation concepts

The correlation between two functions is a measure of their similarity. This is illustrated in figure 1 where three extreme cases are depicted. Figure 1a shows two waveforms which are absolutely identical and they have maximum positive correlation. The two waveforms in figure 1b are similar, except for their polarities and as such they have maximum negative correlation. Finally figure 1c shows one of the waveforms of figure 1a and a noise like signal. As these two waveforms are completely dissimilar the correlation between them is expected to be very small or even zero.

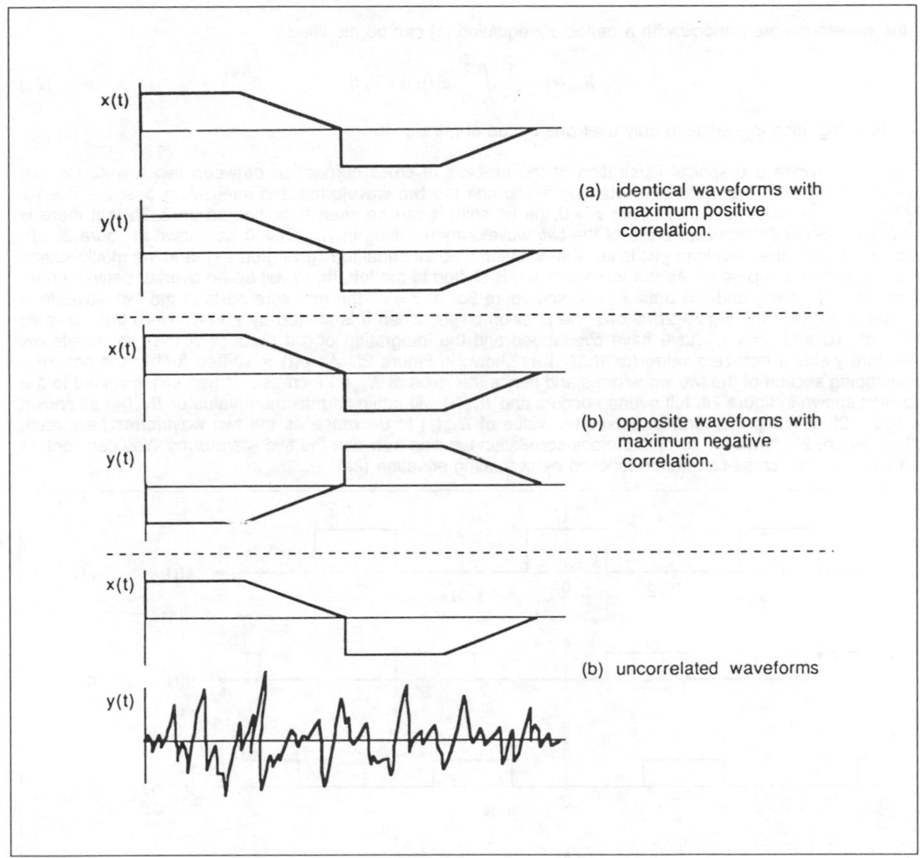
x(t)
y(t)
(a) identical waveforms with maximum positive correlation.
x(t)
y(t)
(b) opposite waveforms with maximum negative correlation.
x(t)
y(t)
(b) uncorrelated waveforms

Figure 1 Illustration of correlation process

Mathematically the correlation function between two waveforms $x(t)$ and $y(t)$ is expressed as

$$R_{xy}(\tau) = \lim_{T \to \infty} \frac{1}{T} \int_{-\frac{T}{2}}^{+\frac{T}{2}} x(t)y(t+\tau)dt \qquad (1)$$

$R_{xy}(\tau)$ is also referred to as cross-correlation between the two waveforms. For identical waveforms (ie correlating a waveform $x(t)$ with itself) the correlation function is denoted by $R_{xx}(\tau)$ and is called the auto-correlation function.

Equation (1) can be interpreted as follows:

The cross-correlation function, $R_{xy}(\tau)$ between the two waveforms $x(t)$ and $y(t)$ is obtained by shifting one of the two signals in time by an amount equal to τ (i.e. modifying $y(t)$ to $y(t+\tau)$), multiplying the shifted waveforms by the other signal and integrating the product.

If the waveforms are periodic with a period T_0, equation (1) can be modified to:

$$R_{xy}(\tau) = \frac{1}{T_0}\int_{\frac{-T_0}{2}}^{\frac{+T_0}{2}} x(t)y(t+\tau)dt \qquad (2a)$$

i.e. the integration is evaluated only over one period of the signal.

Figure 2 provides a graphical illustration of the process of cross correlation between two waveforms $x(t)$ (figure 2a) and $y(t)$ figure 2b. We start by multiplying the two waveforms and integrating over the interval $\frac{-T_0}{2} \leq t \leq \frac{+T_0}{2}$ yielding $R_{xy}(0)$. With $\tau = 0$, (ie no shift) it can be seen from figures 3a & 3b that there is no overlap between non-zero parts of the two waveforms resulting in $R_{xy}(0) = 0$ as shown in figure 2f. To evaluate $R_{xy}(\tau)$, the waveform $y(t)$ is left shifted by an amount equal to τ, giving $y(t+\tau)$, and the multiplication and integration is repeated. As the waveform $y(t)$ is shifted to the left, there will be no overlap between non-zero parts of $y(t+\tau)$ and $x(t)$ untill $\tau > \tau_1$, see figure 2c. At $\tau = \tau_1$, the non-zero parts of the two waveforms just begin to overlap. Figure 2d shows the position of $y(t)$ when it is shifted by $\tau = \tau_2$. Here the non-zero parts of $x(t)$ and $y(t+\tau_2)$ have have overlapped and the integration of the product of the two waveforms therefore yields a non-zero value for $R_{xy}(\tau_2)$ as shown in Figure 2f. As $y(t)$ is shifted further the non-zero overlapping section of the two waveforms and hence the value of $R_{xy}(\tau)$ increase. When $y(t)$ is shifted to the position shown in figure 2e, full overlap occurs and $R_{xy}(\tau)$ will attain its maximum value of $R_{xy}(\tau_3)$ as shown in figure 2f. Shifting $y(t)$ further causes the value of $R_{xy}(\tau)$ to decrease as the two waveforms pass each other. Figure 2f shows the complete cross-correlation function between the two waveforms. You can confirm the shape of this cross-correlation function by evaluating equation (2a).

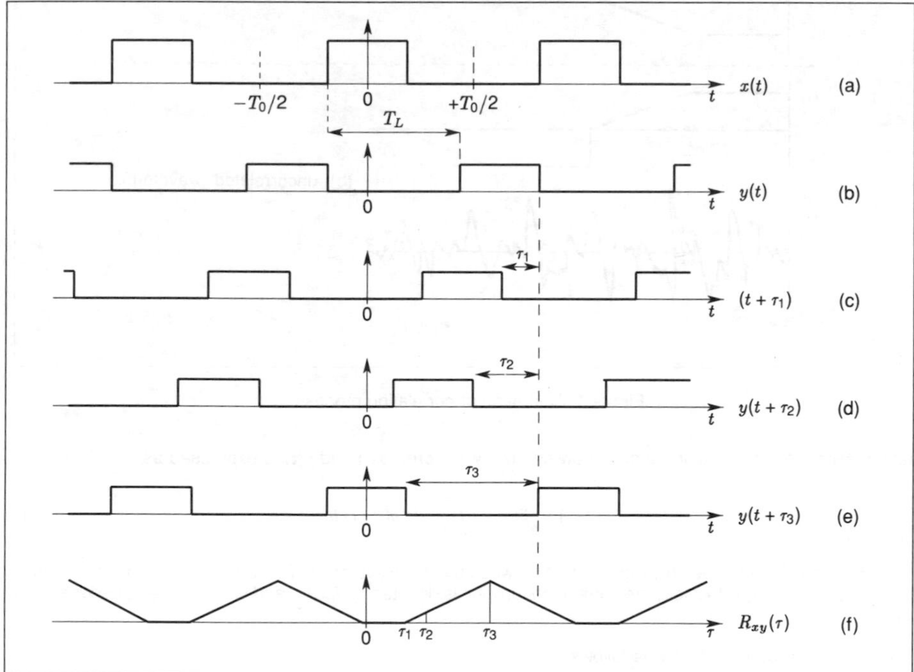

Figure 2 Graphical illustration of the correlation process

One interesting point to note here is that the maximum value of $R_{xy}(\tau)$ occurs at $t = \tau_3$ which is equal to the time-lag, T_L, between the two waveforms $x(t)$ and $y(t)$. This is how the correlation process is used to measure delays.

From figure 2 it can be seen that the cross correlation function could have been evaluated by shifting $x(t)$ to the right instead of left shifting $y(t)$. Mathematically this can be confirmed by defining a new variable $t' = t + \tau$ and substituting in (2a) which gives:

$$R_{xy}(\tau) = \frac{1}{T_0} \int_{\frac{-T_0}{2}}^{\frac{+T_0}{2}} x(t' - \tau) y(t) dt \qquad (2b)$$

So far we have dealt with the correlation of analogue signals. For digital processing both waveforms $x(t)$ and $y(t)$ have to be sampled and digitalized. For discrete-time signals the process of correlation can be expressed as:

$$R_{xy}(mT) = \frac{1}{N} \sum_{k=0}^{N-1} x(kT) y((k+m)T) \qquad (3a)$$

At time $t = kT$ equation (3a) requires future samples of $y(t)$. Similar to the analogue case, (equation 2b), the above equation can be modified so that it only uses past samples of $y(t)$, i.e.

$$R_{xy}(mT) = \frac{1}{N} \sum_{k=0}^{N-1} x((k-m)T) y(kT) \qquad (3b)$$

In equation (3b), T, donates the sampling period and should be chosen to ensure that the sampling rate is greater than twice the signals bandwidth (Nyquist rate). For the sake of simplicity the factor, T, is usually dropped from the indices of equations (3a) & (3b), i.e.

$$R_{xy}(m) = \frac{1}{N} \sum_{k=0}^{N-1} x(k) y(k+m) \qquad (4a)$$

and

$$R_{xy}(m) = \frac{1}{N} \sum_{k=0}^{N-1} x(k-m) y(k) \qquad (4b)$$

Where k and m are used to index the samples and N is the number of correlation points involved. In practice the correlation size N will depend on the duration of the two functions, and on their periodicity if they are periodic.

From equations (4a) or (4b) it can be observed that direct evaluation of M samples of the cross-correlation function, R_{xy}, will involve $M \times N$ multiply-and-accumulate operations.

3 Convolution concepts

The convolution function is closely related to that of correlation. The convolution of two signals $x(t)$ and $y(t)$ is mathematically defined by:

$$C_{xy}(\tau) = \lim_{T \to \infty} \frac{1}{T} \int_{\frac{-T}{2}}^{\frac{+T}{2}} x(t) y(\tau - t) dt \qquad (5)$$

This equation is very much similar to equation (1) defining the correlation process. Their difference is that in convolution the signal $y(t)$ is first time-reversed (i.e. is mirrored around $t = 0$) and then shifted by τ. This time-reversed and shifted signal is then multiplied by $x(t)$ and the product is integrated over all t's. Figure 3 graphically illustrates the process of convolution.

The process of convolution occurs in filters where the output of a filter is in fact the convolution of the input function, $d(t)$, and the impulse response, $h(t)$, of the filter (see the application note entitled 'Digital Filtering with the IMS A100'):

$$f(\tau) = \int_{-\infty}^{+\infty} d(t) h(\tau - t) dt \qquad (6)$$

where $f(\tau)$ is the filter output.

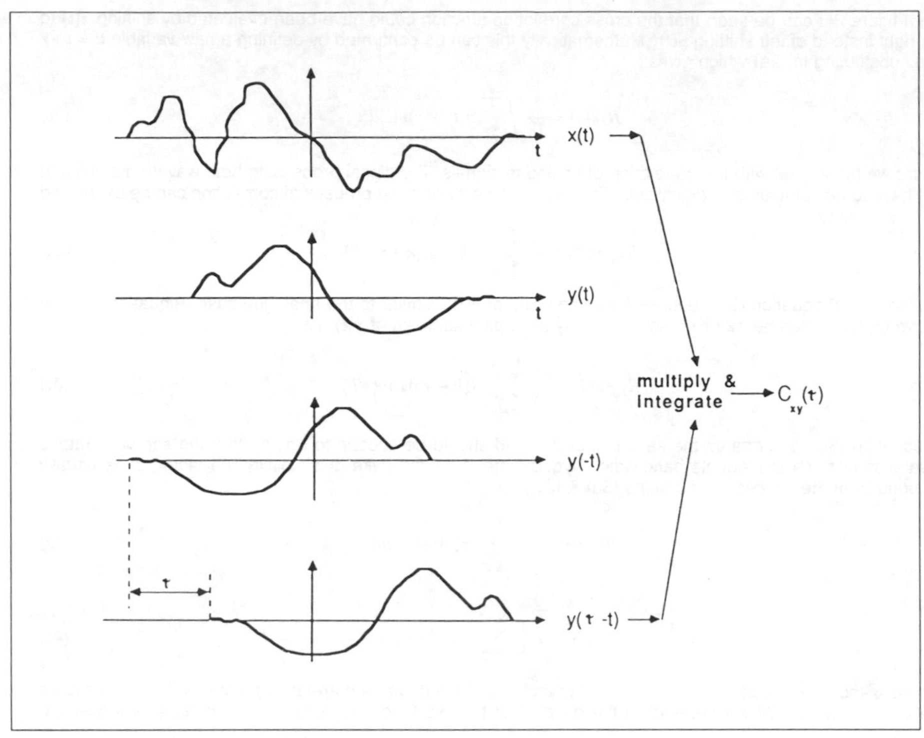

Figure 3 Illustrating the convolution process

For discrete-time signals equation (5) becomes

$$C_{xy}(m) = \frac{1}{N} \sum_{k=0}^{N-1} x(k)y(m-k) \qquad (7)$$

which defines the convolution function for digital signals. Notice that like correlation, convolution involves carrying out N multiply-and-accumulations for each sample of $C_{xy}(m)$. Due to the high degree of similarity between correlation and convolution functions, the same hardware can be used to perform both functions. All that needs to be done is to time-reverse one of the waveforms for the convolution process.

The following two sections deal with hardware implementations for the correlation and convolution functions. The first section deals with the conventional approach involving multiply-and-accumulator chips and points out the processing bottle-necks associated with these solutions. The second section shows how the IMS A100 signal processor can be configured to perform these functions efficiently and simply, at speeds not economically feasible with the conventional approach.

4 Conventional hardware for time-domain evaluation of correlation

As discussed earlier, the processes of correlation and convolution are based on multiplying a delayed version of a sequence of samples by another sequence and summing the products. Conceptually this could be mechanised, as shown in figure 4, by providing two shift registers to hold all the values of x's and y's required for the computation, a further shift register to provide the delay (mT), an array of multipliers for forming the products, and a multi-input adder for the final summation. In the example of figure 4 the output would correspond to $R_{xy}(2)$, as a delay of two stages is incorporated in the path of signal $x(kT)$ giving $x((k-2)T)$ (see equation 3b).

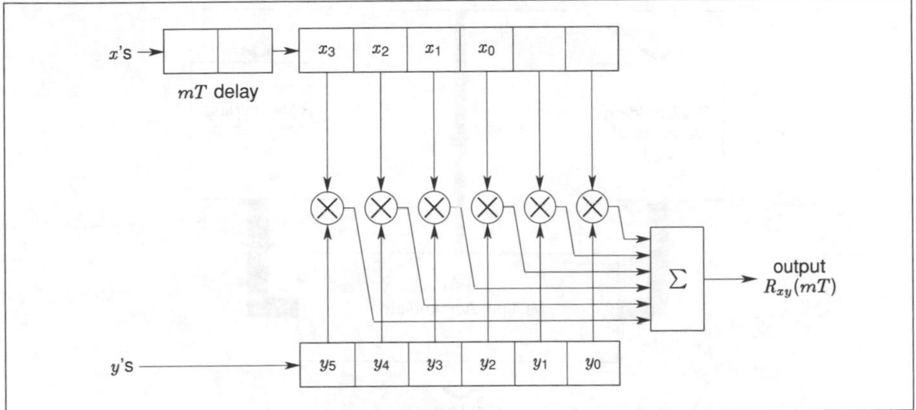

Figure 4 Schematic diagram for an ideal correlator hardware

Up to now, due to the large number of multipliers and adders involved, it has not been possible to economically implement high precision correlators directly in the form given by figure 4. Instead to minimize the size, cost and power consumption, a single multiply-and-accumulator is usually used and time-shared between all the multiplications. Figure 5 shows a schematic block diagram of a conventional correlator implementation. The system consists of memories to hold samples of the two signals to be correlated, a multiply-accumulator and an address generator hardware which is responsible for sequencing the correct order of signal samples through the multiply-and-accumulator. The obvious disadvantage of this arrangement is the processing bottle-neck caused by using a single multiply-and-accumulator to sequentially evaluate what is inherently a concurrent problem. Assuming a multiply-accumulate time of T_{mac}, for an N-point correlator implemented using a single multiply-accumulator, the maximum sampling rate would be

$$f_{max_s} = \frac{1}{NT_{mac}} \tag{8}$$

For example if $T_{mac} = 100ns$ and $N = 100$, then a signal sampling frequency of at most 100kHz would be possible.

Many applications such as radar and communication require faster processing rates than can be achieved using a single multiply-and-accumulator. (Some improvements can be achieved by carrying out the processing in the frequency domain at the cost of introducing some complexity. However here we are only concerned with the time domain approach. A separate application note entitled 'Discrete Fourier Transform with the IMS A100' deals with the time-domain to frequency-domain transformations)

In applications where a fast processing rate is essential, a trade-off is often made between the correlator precision and its speed. For example if one or both of the signals x and y are assumed binary, the multiplications become simple binary AND operations, and it would be possible to implement a high speed low precision correlator. In fact many correlator chips available today are of this type and have very low precision compared to those implemented from multiply-accumulators.

The IMS A100 chip on the other hand is the first high-precision high-speed VLSI implementation of a single-chip correlator. It provides a numerical accuracy in excess of that of the 16-bit multiply and accumulators while allowing sampling rates in the MHz region. The next section illustrates how this chip can be used to perform fast and highly accurate correlation and convolution functions.

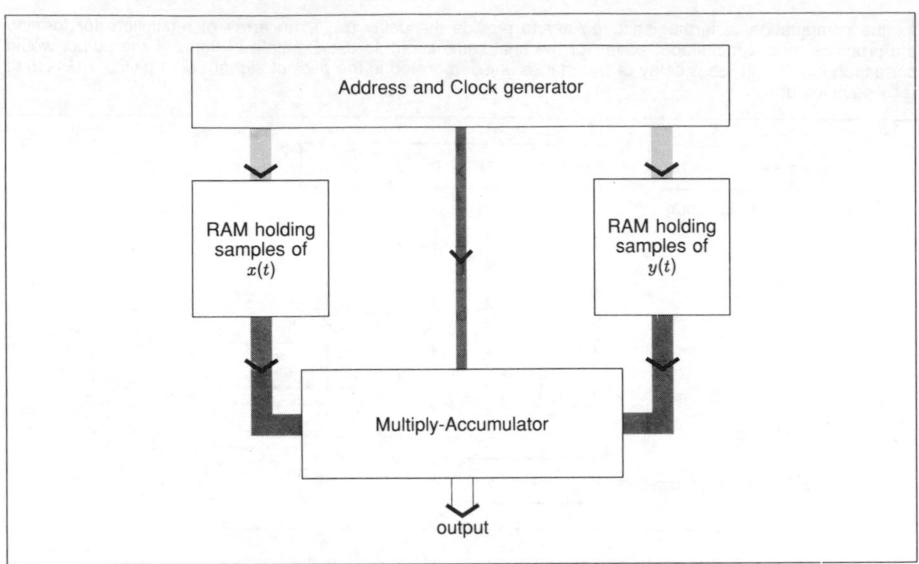

Figure 5 Block diagram of a conventional correlator/convolver

5 The IMS A100 implementation of correlation/convolution

The IMS A100 is a 32-stage correlator (convolver) in which the samples of the two signals to be correlated can be represented as up to 16-bit words. This corresponds to a signal dynamic range of 96 dB's. A number of these devices can also be cascaded, without the need for any external components, to provide much longer correlators while preserving a high degree of accuracy. The IMS A100 chip (or cascaded chips) can be fully memory mapped and used as a peripheral accelerator to a host processor.

To understand the architecture of the IMS A100 let us first consider the basic function of a simple 3-point correlator shown in figure 6a. The three samples of the first signal (reference signal) i.e. x_0, x_1 and x_2 are loaded in three registers feeding an array of three multipliers. The samples of the second signal i.e. y_0, y_1, y_2, ... are fed into a 3-stage shift register whose outputs are also connected to the multipliers. A three input adder combines the products to give the correlation function. As the samples of the signal y are shifted through the shift register, the output of this hypothetical correlator (assuming the shift register is reset at the start) will be:

$$y_0x_2,\ y_0x_1 + y_1x_2,\ y_0x_0 + y_1x_1 + y_2x_2,\ y_1x_0 + y_2x_1 + y_3x_2, ...$$

The correlator structure in figure 6a can be modified to that given in figure 6b without affecting the functionality. In figure 6b the multi-input summation process is avoided and replaced by a chain of delay-and-add units. The input, supplying the signal y, is also made common to all of the multipliers. Note also that the signal samples x_0, x_1, x_2 are stored in the opposite direction to that of figure 6a. Supplying the input sequence of samples $y_0, y_1, y_2, y_3, ...$ to the structure of figure 6b and simultaneously shifting the partial products along the delay-and-add chain, it is straightforward to confirm that the output sequence would be

$$y_0x_2,\ y_0x_1 + y_1x_2,\ y_0x_0 + y_1x_1 + y_2x_2,\ y_1x_0 + y_2x_1 + y_3x_2, ...$$

This sequence is absolutely identical to that obtained from figure 6a. In other words the structure in figure 6a & b have identical functionality and both can be used to perform correlation between two sequences. The IMS A100 architecture is based around this modified structure. The major processing part of the chip incorporates 32 multipliers and a 32-stage delay-and-add chain as shown in figure 7.

At this point the interested reader is advised to consult the data sheet of the IMS A100 for full details.

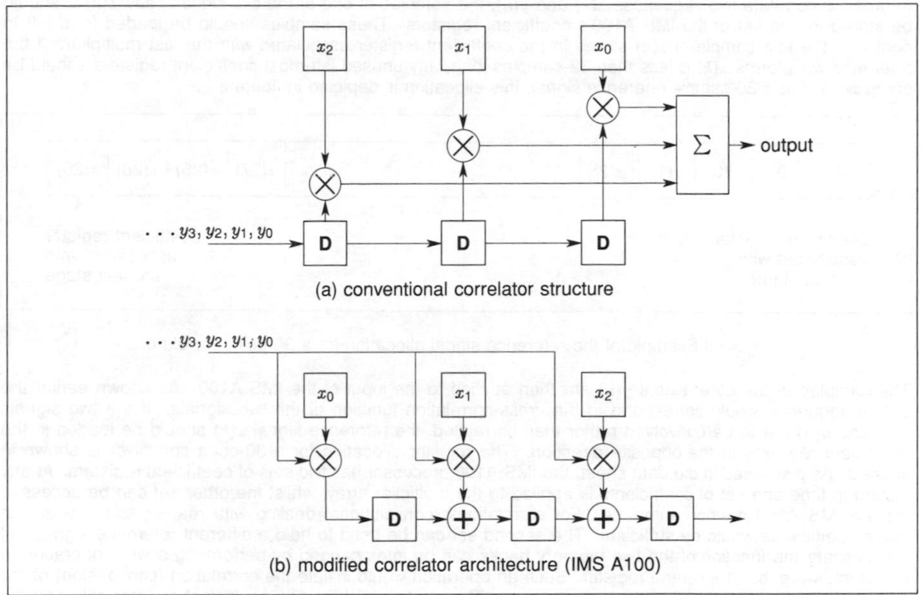

Figure 6 Relating the IMS A100 architecture to that of a correlator

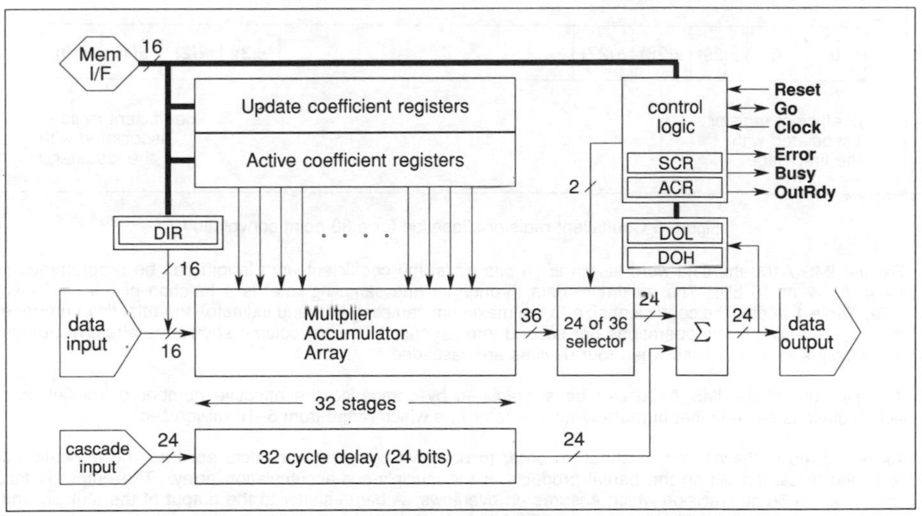

Figure 7 User's model of the IMS A100

In order to correlate two sequences $x(k)$ and $y(n)$, the samples of one of the two signals, say $x(k)$'s, should be stored in one set of the IMS A100's coefficient registers. These samples should be loaded from left to right with the last sample of $x(k)$ stored in the coefficient register associated with the last multiplier. If the reference waveforms $x(k)$ is less than 32-samples long, any unused left-most coefficient registers should be set to zero. For a 30-sample reference signal, this allocation is depicted in figure 8.

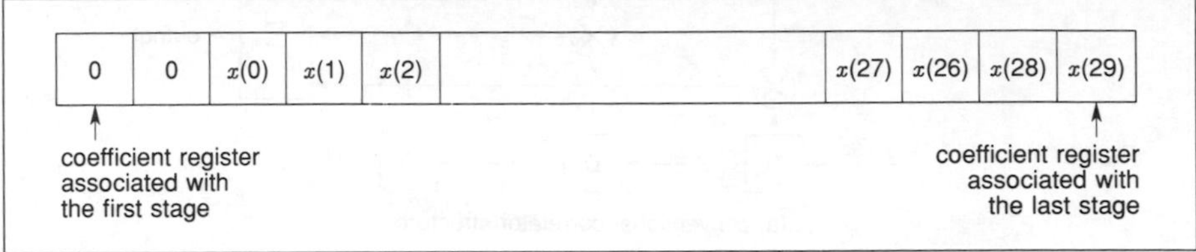

Figure 8 Example of the reference signal allocation for a 30 point correlator

The samples of the other signal $y(n)$ are then applied to the input of the IMS A100. As shown earlier the output sequence would correspond to the cross-correlation function of the two signals. If the two signals $x(k)$ and $y(n)$ are to be convolved rather than correlated, the reference signal $x(k)$ should be loaded in the coefficient registers in the opposite direction. The register allocation for a 30-point convolver is shown in figure 9. As discussed in the data sheet, the IMS A100 processor has two sets of coefficient registers. At any instant in time one set of coefficients is applied to the multiplier array, whilst the other set can be accessed via the IMS A100 memory interface. For correlations (convolutions) dealing with real signals, one set of these coefficients would be sufficient. The second set can be used to hold a different reference signal and if necessary the function of the two memory banks can be interchanged by performing a write operation to the 'Bank swap' bit of a control register. Such an operation would initiate the correlation (convolution) of the input signal with the second reference waveform. The existance of the two coefficient register sets and the continuous bank-swap mode allows the IMS A100 to perform complex (correlation)convolution, where both the reference and the input signal have real and imaginary components. This configuration is discussed in the application note 'Complex Processing with the IMS A100'.

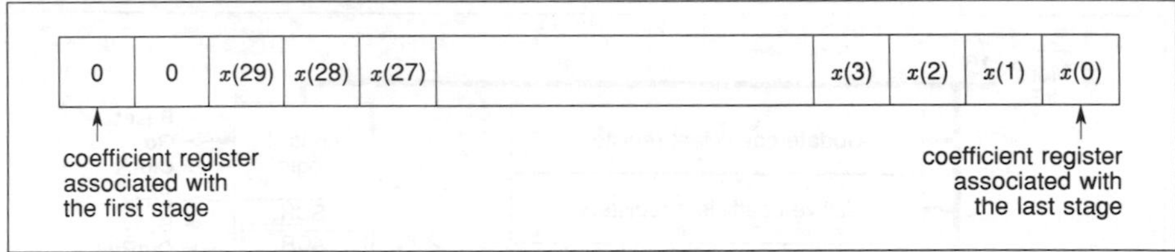

Figure 9 Coefficient register allocation for a 30 point convolution

For the IMS A100 the data-word length is 16 bits whilst the coefficient-word length can be programmed to be 4, 8, 12 or 16 bits. The maximum data throughput (the sampling rate) is a function of the coefficient size. Table 1 relates the coefficient size to the maximum sampling rate and indicates the effective number of multiply-and-accumulate operations per second in each case. The last column shows the effective number of multiply-and-accumulates when four devices are cascaded.

The strength of the IMS A100 can be appreciated by comparing the effective number of multiply-and-accumulations/sec with that of multiply-accumulator IC's which range from 5–10 million/sec.

As discussed in the A100 data sheet, in order to preserve complete numerical accuracy no truncation or rounding is carried out on the partial products in the multiply-and-accumulation array. The output is thus calculated to 36 bit precision which ensures no overflows. A barrel-shifter at the output of the multiply-and-accumulate array allows 24 bits from these 36 bits to be selected (sign-extended if necessary) and rounded for output. This selection can be programmed via a control register. The programming details can be found in the IMS A100 data sheet.

Coefficient word size (bits)	Sampling rate (MHz)	Effective number of multiply-and-accumulates (Millions/second)	
		Single A100	Four A100s
4	10	320	1280
8	5	160	640
12	3.3	106	424
16	2.5	80	320

Table 1

The architecture of the IMS A100 has been designed to allow large numbers of these devices to be cascaded for correlations (convolutions) involving more than 32-stages, the devices can be cascaded while preserving a high degree of accuracy and without the need for any external components. This is made possible by incorporating on chip a 32 stage, 24-bit wide, shift register and a 24-bit adder which combines the output of the barrel-shifter with that of the 32-stage shift register (see figure 7).

The IMS A100 chips can be cascaded by simply connecting the output of the each device to the cascade input of the following device. The input is common to all cascaded devices. The effect of such an arrangement is that the output of the first device is delayed by 32 cycles, before being added to that of the next device. Figure 10 illustrates how, for example, a 64-point correlator can be implemented using two IMS A100 devices. The allocation of the reference signal samples is also indicated in this illustration. In this arrangement the barrel-shifter in each device acts as a data scalar (with rounding). The cascading process can be considered as a block-floating point operation where the common exponent is determined by the extent of the shift carried out by the barrel-shifter. With this cascading technique a very high degree of accuracy is preserved because the output scaling is only performed after every 32-multiply and accumulation stages and not at any intermediate stage.

For convolution purposes the reference signal should be loaded into the coefficient stage in the opposite direction to that shown in figure 10.

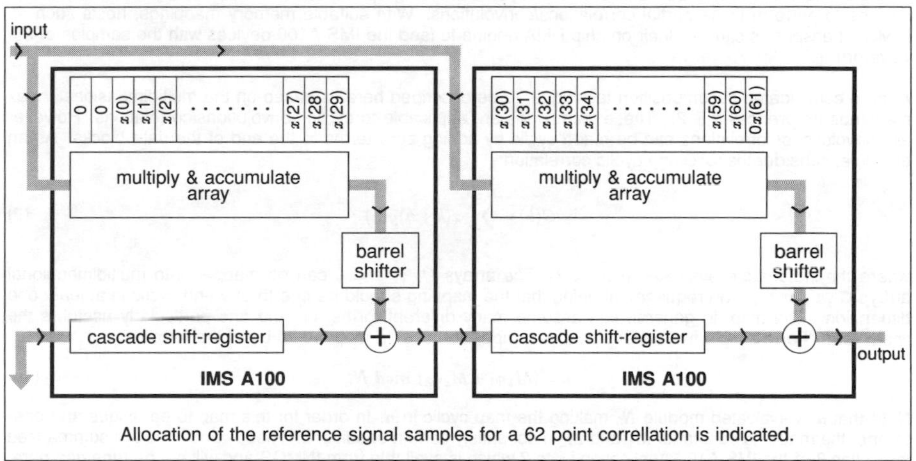

Figure 10 Cascading two IMS A100 devices to obtain a 64 point correlator

A very important feature of the IMS A100 transversal filter is that the part is fully memory mappable. Apart from the two coefficient memory banks, which can be accessed via the IMS A100 standard memory interface, the input and output of the device are also accessible from the same interface. This feature allows the part to be used either with its input and output data communicated through the dedicated ports or through the memory interface. The latter options allows the device to be easily interfaced to a host processor and used as a high speed peripheral. The status and control registers of the IMS A100, accessible via the memory interface, provide full control of the part by the host processor.The memory interface can also be used as a facility for system diagnostics, as the host processor can act as a watch-dog in systems involving arrays of IMS A100's. Full specification of the IMS A100, its status and control registers and its standard memory interface are detailed in the data sheet available from INMOS.

6 Decomposition of long correlations and convolutions

A single IMS A100 device is effectively a 32-tap correlator (convolver) in which the samples of the two signals to be correlated can be expressed in upto 16-bit words. As described earlier, one method to deal with correlations/convolutions involving more than 32 points is to use several cascaded devices to achieve a longer correlator/convolver. For such an arrangement, and with 16-bit coefficients, tha data rate can be as high as 2.5 Million samples/sec.

Alternatively it is possible to use various decomposition techniques to partition a long correlation/convolution into a number of shorter ones, which can then be carried out by a single or a small number of IMS A100 devices. The host machine would merely combine the results from these short correlations/convolutions to obtain the overall result. The advantage of this approach, compared to using single MAC based processors, is a significant reduction in the required memory bandwidth. This is why even a medium-speed general purpose microprocessor can achieve a very high performance when combined with the IMS A100.

A simple way to decompose a long correlation/convolution of length N, between waveforms x and y, is to break up one of the waveforms, say x, into consecutive blocks of 32 sample. Each one of these blocks can then be correlated/convolved with the whole of the waveform y by loading each block into the IMS A100 coefficient registers, and using y as the input sequence. The output from these correlations/convolutions can then be combined by displacing each partial result by 32 samples, with respect to the previous one, and performing an addition operation. Note that the coefficient registers, containing blocks of waveform x, need only be updated once every time the whole of the waveform y is fed through the device, resulting in a significant saving in the memory bandwidth. The block size of 32, suggested above, whould mean that a single IMS A100 would be sufficient. However processing speed can be improved by using cascading devices to perform these partial correlations/convolutions. With suitable memory mappings, hosts such as INMOS transputers can use their on-chip DMA engine to feed the IMS A100 devices with the samples of the waveform y.

A more complicated decomposition technique, to be described here, is based on the multidimensional index mappings (references 1 & 2). These techniques are applicable to cyclic convolutions/correlations. However all convolutions/correlations can be made cyclic by adding zero terms to the end of the data blocks. As an example, consider the following cyclic correlation:

$$C(k) = \sum_{n=0}^{N-1} x(k+n)y(n) \tag{9}$$

where the indices are evaluated modulo N. The arrays C, x, and y can be mapped into multidimensional arrays C', x', and y', the requirement being that the mapping should be one-to-one and cyclic in at least one dimension. The map, in general, can assume many different forms, but the one particularly useful is the linear form. For a simple two-dimensional decomposition such a map would be of the form:

$$n = (M_1 n_1 + M_2 n_2) \ mod \ N. \tag{10}$$

Note that n is evaluated modulo N, making the map cyclic in n. In order for this map to be unique and one-to-one, the mapping constants M_1 and M_2 must satisfy certain conditions. These conditions are summarised in section 6 of the IMS A100 Application Note 2 which is available from INMOS and will not be repeated here.

As an example let us map the arrays in equation (9) into two-dimensional matrices of dimensions N_1 and N_2 where $N = N_1 \times N_2$, we can use the mapping given by equations (10) for n and k giving

$$C(M_1k_1 + M_2k_2) = \sum_{n_1=0}^{N_1-1} \sum_{n_2}^{N_2-1} x(M_1k_1 + M_2k_2 + M_1n_1 + M_2n_2)y(M_1n_1 + M_2n_2) \tag{11}$$

or

$$C'(k_1, k_2) = \sum_{n_1}^{N_1} \sum_{n_2}^{N_2} x'(k_1 + n_1, k_2 + n_2)y'(n_1, n_2) \tag{12}$$

This is now a true two-dimensional convolution which can be made cyclic along n_1 if M_1 is made a multiple of N_2, and/or cyclic along n_2 if M_2 is made a multiple of N_1. With these conditions, inspection of equation (12) shows that the long N-point circular correlation can be performed by N_1^2, N_2-point correlations or N_2^2, N_1-point correlations. This involves correlating each row (or column) of the matrix y' with all the rows (or columns) of the matrix x'. These short circular correlations can be efficiently performed by the IMS A100, with the host merely adding partial results. The approach is particularly efficient as it is possible to load one row (or column) of the matrix y' into the coefficient memory of the device and to feed all the rows (or columns) of the matrix x' successively to the input of the device to obtain partial results, for the elements in the matrix C'. The fact that with this algorithm, the coefficient memories need only be updated occasionally (once every time all the elements of the matrix x' are fed into the device) results in an impressive reduction in the memory bandwidth requirement. This is why, even with a general purpose microprocessor, as the host, very impressive perfomance can be achieved.

In the example given here, we concentrated around a two-dimensional mapping. It is important to realise that the same decomposition concepts can be extended to more dimensions. The easiest way to see this is to start with a two dimensional decomposition and then partition the rows of the two-dimensional matrices further. For example if

$$N = N_1 \times N_2 \times N_3$$

the original N-point correlation can be carried out via N_3^2, $N_1 \times N_2$-point correlations. However, each one of the $N_1 \times N_2$-point correlations can further be decomposed, as before into N_1^2, N_2-point correlations.

7 2-D image convolutions with the IMS A100

Many applications including image processing require 2-D convolutions and correlations. Such operations are needed in image filtering, edge detection, etc. There are many ways that the IMS A100 can be used to speed up these operations. This section gives an example of how the device can be used to perform 3×3, 5×5, or larger convolutions.

Figure 11a shows a 20×20-pixel image which is to be convolved with the 3×3 reference matrix given by figure 11b. One way to achieve this is to load the reference matrix, as shown in figure 11c, in one of the IMS A100 coefficient register banks, and sequence the image data through the device as shown by the arrowed path in figure 11a. In this way every third output sample of the IMS A100 would correspond to a valid filtered pixel for the second row of the image. To proceed, the same sequence, moved down by one row, is then passed through the device which provides the filtered results for the nest row and so on. A single IMS A100 can deal with reference matrices as big as 5×5.

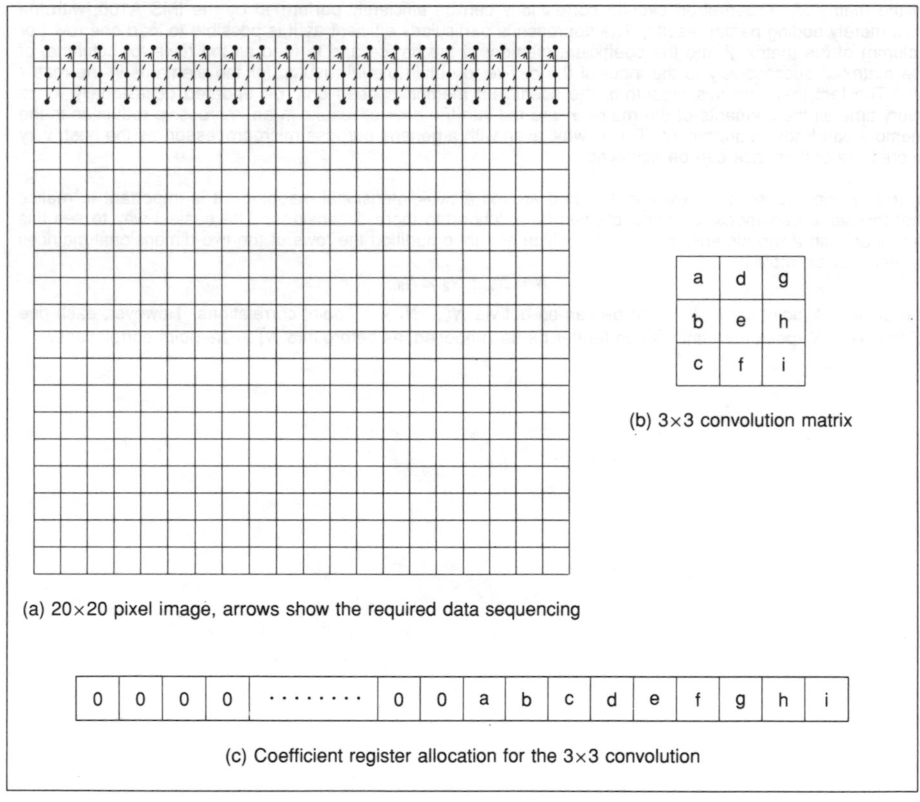

Figure 11 Example of a 3×3 image convolution/correlation with the IMS A100

An alternative arrangement which gives a better throughput is one where, as shown in figure 12a, 7 zeroes are inserted in the IMS A100 coefficient registers (between terms corresponding to the columns of the reference matrix). The data sequencing would be as shown in figure 12c, where ten pixels from a given column are fed through the device before moving to the next column. In this scheme the first nine rows of the image are filtered in one scan, with 80% of the output data samples being valid. (Note that, using a single device, the number of inserted zeroes can be increased from 7 to 11, allowing 13 image rows to be filtered in each scan.)

The examples given here are just a small subset of possible arrangements. Remembering that the IMS A100 devices can be cascaded or used in parallel, numerous other implementations for image processing become possible.

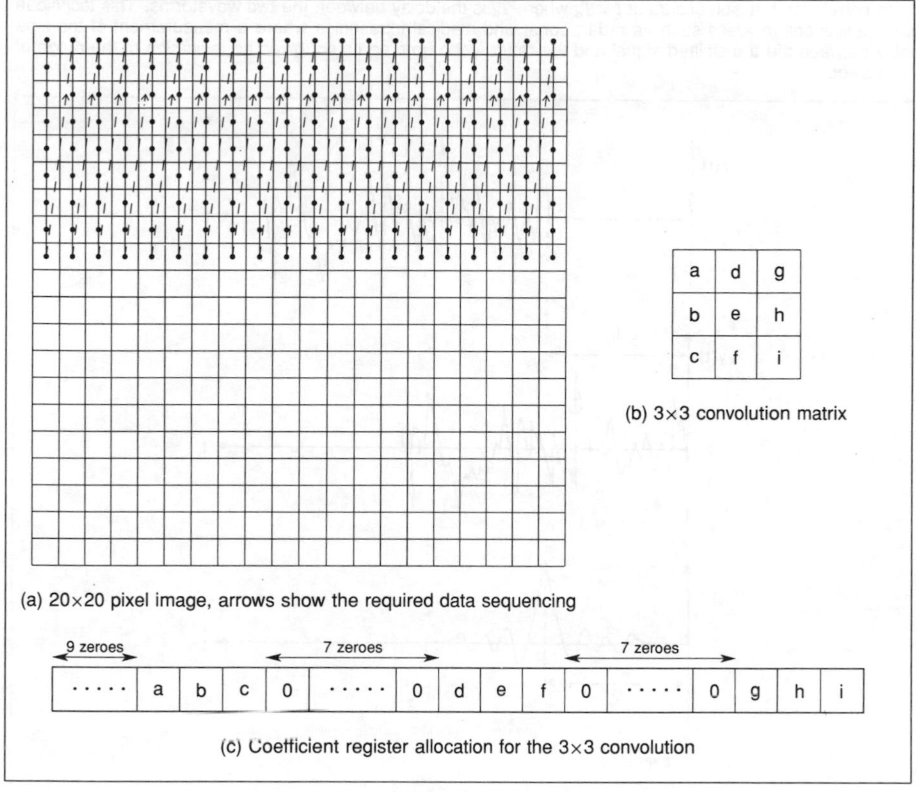

(a) 20×20 pixel image, arrows show the required data sequencing

(b) 3×3 convolution matrix

(c) Coefficient register allocation for the 3×3 convolution

Figure 12 Improved version of the 3×3 image convolution/correlation with the IMS A100

8 Some application examples of correlation and convolution

Correlation and convolution are encountered in numerous applications of digital signal processing, this section summarises some of the application areas where these techniques are used.

8.1 Delay and periodicity estimation

The correlation process can be used to estimate the time delay between two similar signals. Figure 13 shows two signals $x(t)$ and $y(t)$ which are identical in shape but have a time delay between them. If these two signals are correlated the cross-correlation function would attain a maximum when $y(t)$ is delayed by an amount equal to the delay between the two waveforms. This is illustrated in figure 13c where the peak of the cross-correlation function occurs at $t = T_d$ where T_d is the delay between the two waveforms. This technique has applications in areas such as radar, sonar and medical ultrasonics where a measurement of the time delay between the transmitted signal and the return echo from an object gives an indication of the range of that object.

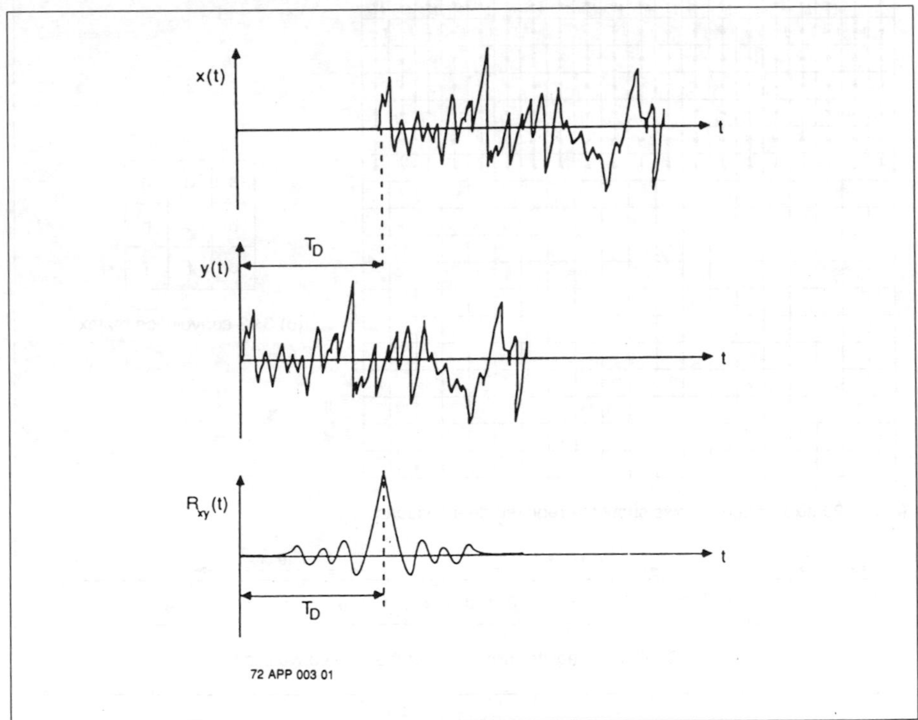

Figure 13 Delay estimation using correlation process

The same technique can also be used to measure the period of a repetitive signal. This can be achieved by correlating the signal with itself i.e. by calculating its auto-correlation function as illustrated in figure 14 the auto-correlation of a periodic signal exhibits peaks, spaced a distance, T_0, apart where T_0 is the period of the signal. One application of this technique is pitch-period measurement in speech signals. The time gap between the peaks in the auto-correlation function of a segment of speech provides an estimate for the pitch period of voiced speech.

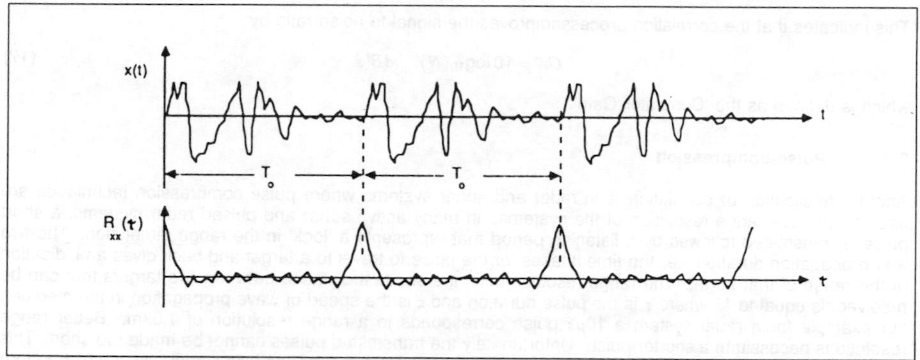

Figure 14 Application of correlation to the periodicity measurement

8.2 Noise reduction using correlation techniques

In many real-world applications the signals to be processed are immersed and possibly masked by noise. Such situations occur in noisy communication channels, long-range radar and sonar systems. In such cases correlation techniques can be used to extract and detect the signal from the background additive noise. This is achieved by correlating the noisy signal with a replica of the expected signal waveform. While the noise is uncorrelated with the replica signal, the signal immersed in noise will strongly correlate with the replica signal giving a large output value, well above the background noise. Mathematically this can be argued as follows (the proof here is not rigorous but does make the point):

Let the signal waveform consist of N samples values $s_0, s_1, s_2, ...s_{N-1}$. Suppose this signal is correlated by samples of a white noise having a standard deviation of σ_n (and variance σ_n^2). The ratio of the signal power to that of the noise prior to any processing is thus:

$$\frac{SignalPower}{NoisePower} = \frac{\sigma_s^2}{\sigma_n^2}$$

where σ_s^2 is the signal variance. Suppose we correlate this noisy signal with a replica of the original signal in an N-point correlator. At the instant when the signal waveform masked by the background noise is aligned with its replica in the correlator, the output attains its maximum. At this instant the amplitude of signal component at the output of the correlator would be

$$S_{out} = s_0^2 + s_1^2 + s_2^2 + s_3^2 + ... + s_{N-1}^2 \simeq N\sigma_s^2 \tag{13}$$

The corresponding output signal power would thus be:

$$Output\ Signal\ Power = N^2\sigma_s^4. \tag{14}$$

The noise would also be modified by the operation of the correlator. In this case each output noise sample is equal to the sum of weighted input noise samples-the weighting coefficients being, of course, the samples of the reference signal. Hence each output noise sample is equal to the summation of N independent random numbers having standard deviations $s_0\sigma_n, s_1\sigma_n, s_2\sigma_n, ..., s_{N-1}\sigma_n$. Since variances are additive in this case, the variance of the output noise samples is therefore

$$\sigma_{n_{OUT}}^2 = s_0^2\sigma^2 + s_1^2\sigma^2 + ...s_{N-1}^2\sigma^2 = N\sigma_n^2\sigma_s^2 \tag{15}$$

The ratio of the output signal power to that of the output noise is thus

$$\left(\frac{S}{N}\right)_{OUT} = \frac{Output\ Signal\ Power}{\sigma_{n_{OUT}}^2} = \frac{N^2\sigma_s^4}{N\sigma_n^2\sigma_s^2} = N\frac{\sigma_s^2}{\sigma_n^2} = N\left(\frac{S}{N}\right)_{INPUT} \tag{16}$$

This indicates that the correlation process improves the signal to noise ratio by

$$C_G = 10\log_{10}(N) \quad dB's \tag{17}$$

which is defined as the 'Correlator Gain'.

8.3 Pulse-compression

Another application of correlation is in radar and sonar systems where pulse compression techniques are used to improve range resolution of the systems. In many active sonar and pulsed radar systems, a short pulse is transmitted followed by a listening period that represents a 'look' in the range dimension. The two way propagation duration, i.e. the time it takes for the pulse to travel to a target and back gives an indication of the range of that target. The range resolution i.e. the shortest distance between two targets that can be resolved, is equal to $\frac{\tau c}{2}$ where τ is the pulse duration and c is the speed of wave propagation in the medium. For example for a radar system a $10\mu s$ pulse corresponds to a range resolution of 1.5km. Better range resolutions necessitate a shorter pulse. Unfortunately the transmitted pulses cannot be made too short. This is because most systems are peak-power limited and a shorter pulse means less signal power which in turn can severely limit operational range of the system.

Pulse compression techniques allow a radar or sonar to utilize a long pulse to achieve large radiated energy, but simultaneously to obtain the range resolution of a short pulse. This is accomplished by using a coded signal instead of a simple CW pulse. At the receiver the returned signal is correlated with a replica of the coded transmit signal. The returned signal would only correlate heavily with the replica for a short time, corresponding to when the echoes are aligned with the replica. This results in a narrow pulse appearing at the output of the correlator, everytime a match occurs. A signal that is commonly used in pulse-compression techniques is the lineal FM signal. An example of such a signal is depicted in figure 15a. The autocorrelation of such a waveform is shown in figure 15b. Note that the autocorrelation function has a narrow peak at the origin, with small side lobes elsewhere, i.e. the initially long FM pulse is 'compressed' into a narrow pulse after the autocorrelation process.

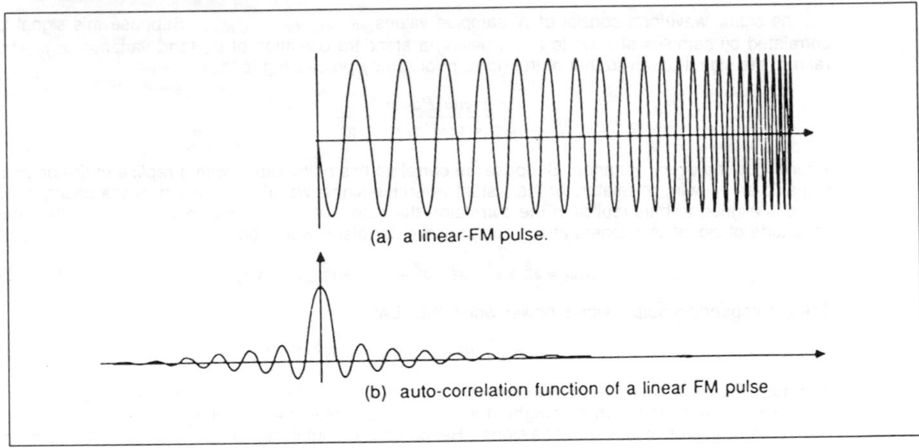

(a) a linear-FM pulse.
(b) auto-correlation function of a linear FM pulse

Figure 15 Pulse compression using a coded signal

It can be shown that the degree of compression is equal to BT where B is the bandwidth of the coded pulse and T is its duration. The effective pulse duration, as far as the range resolution is concerned, will thus be:

$$Effective\ Pulse\ Duration = \frac{T}{BT} = \frac{1}{B} \tag{18}$$

If the $10\mu s$ pulse in the previous example is coded in such a way that its bandwidth becomes 5MHz, the effective pulse duration would be:

$$\frac{1}{5 \times 10^6} = 0.2\mu s$$

This corresponds to a range resolution of 30 metres.

8.4 System identification using correlation

Another important application of cross-correlation is in the use of random-noise test signals to identify the impulse response of a system. For a system with an unknown impulse response $h(t)$, the output $y(t)$ is related to an input $x(t)$ by

$$y(t) = \int_{-\infty}^{+\infty} h(u)x(t-u)dt \tag{19}$$

The cross-correlation between the input $x(t)$ and the output $y(t)$ is defined by

$$\begin{aligned} \Phi_{xy}(\tau) &= \lim_{T\to\infty} \frac{1}{T}\int_0^T x(t)y(t+\tau)dt \\ &= \lim_{T\to\infty} \frac{1}{T}\int_0^T x(t)\int_{-\infty}^{+\infty} h(u)x(t+\tau-u)dudt \end{aligned} \tag{20}$$

Using simple mathematical manipulation it can be shown that

$$\Phi_{xy}(\tau) = \int_{-\infty}^{+\infty} h(u)\Phi_{xx}(\tau-u)du \tag{21}$$

i.e. The cross-correlation between $x(t)$ and $y(t)$ is the convolution of the impulse response $h(t)$ with the auto-correlation of the input signal.

If the input signal consists of broad-band white noise then its auto-correlation function, $\Phi_{xx}(\tau)$, would be an impulse (since a noise signal only correlates with itself at zero delay, $\tau = 0$). Referring to equation (21) it therefore follows that for broad-band noise input, the output $\Phi_{xy}(\tau)$ would be a direct measure of $h(\tau)$ since

$$\Phi_{xy}(\tau) = \int_{-\infty}^{+\infty} h(u)\delta(\tau-u)du = h(\tau) \tag{22}$$

Figure 16 illustrates the technique.

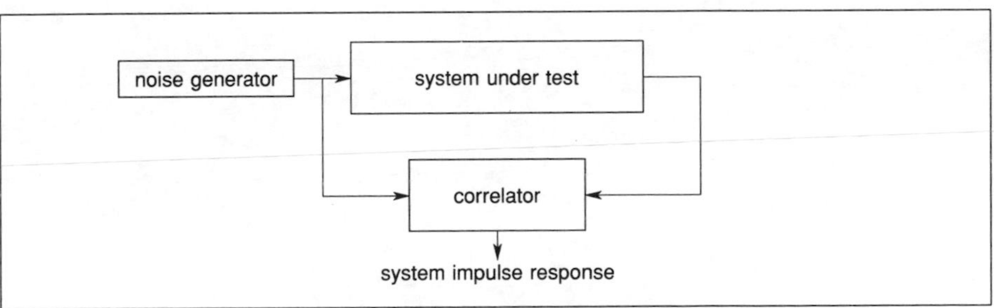

Figure 16 System identification using correlation

8.5 The Discrete Fourier Transform (DFT)

DFT has application in many signal processing areas, including speech processing, radar, sonar, image processing and control. This transform has often been performed using Cooley and Tukey radix-2 FFT algorithm (reference 3). This algorithm reduces the number of multiplications compared to direct evaluation of the DFT at the expense of complicating the required indexing.

Other algorithms have also been developed which allow the evaluation of the DFT via correlation (convolution) techniques (references 1, 2, 4, & 5). The IMS A100 device can be used to perform high speed DFT's based on these convolutional algorithms. Using the IMS A100 as a peripheral to a general-purpose microprocessor converts a slow host into a high-performance DFT processor. A separate application note available from INMOS describes how these algorithms can be implemented using the IMS A100 devices.

9 References

1 Burrus C.S., 'Index mappings for multidimensional formulation of the DFT an convolution', *IEEE Trans. on ASSP*, Vol.25, June l977, pp 239-242.

2 Burrus C.S., Parks T.W., 'DFT/FFT and convolutional algorithms-theory and implementation', Wiley-interscience Publication, New York l985.

3 Cooley J.W., Tukey J.W. 'An algorithm for the machine calculation of complex Fourier series', *Math. Comput.*, Vol.19, pp 297-301, April l965.

4 Rader C.M., 'Discrete Fourier transforms when the number of data samples is prime', *Proc. IEEE,* Vol.56, pp 1107-1109, June l968.

5 Rabiner L.R., *et al*, 'The chirp z-transform algorithm and its application', The Bell System Technical Journal, May-June l969, pp 1249-1292.

HARDWARE DEVELOPMENT FOR A UNIVERSAL DIGITAL FILTER COMPUTING MACHINE

ABSTRACT

The basic concepts of a general purpose microprocessor are expanded to create a high speed digital filter computing machine. Recursive and nonrecursive filters of any structure can be implemented.

The main emphasis here is on the arithmetic logic unit (ALU); It performs the necessary data arithmetic required to implement the desired digital filter. It is different from typical microprocessor ALUs in that its architecture is designed to perform operations which are sequence rather than instruction oriented. This implies a multidata bus configuration with local storage and single cycle multiply-add capability. As a result, a sequence of basic computer-type microinstructions (such as read, multiply, store) can be performed within a single microcycle, rather than taking the large number of instructions common in most small computing machines.

The computer control unit is similar to that of most microprocessors, having a sequencer, microprogram ROM, and some assorted logic. It receives user programmed instructions and directs the ALU to do the prescribed arithmetic. Under the topic of control are the coefficient and memory offset generators. These are separate logic subsystems which provide multiplier coefficients and the correct data delays for the filter. They run in parallel with the general machine sequencer to ensure maximum computational speed. The reader will note throughout the paper that individual tasks are handled by separate dedicated elements rather than by a single processor. This fact, paralleled with the enhanced ALU structure, provides a great improvement in speed and efficiency over general purpose machines.

1. INTRODUCTION

Speed flexibility, word size, and architecture are important considerations when designing a digital signal processor. As with the design of any computing machine, these items are closely interrelated, often making initial machine conception confusing. Fortunately, even the most flexible and complex digital filter computers need not have the complications of general computers. For instance, data and instruction memories should be kept separate. Also, the computer control section of the processor should be isolated from the main arithmetic logic unit (ALU) data ports. Thus, the controller does not borrow the ALU for computing effective addresses for instruction or data acquisition purposes. Speed requirements usually preclude this sort of inner computer activity since controller functions usually waste so much time.

Little emphasis need be placed on elementary microoperations such as load accumulator, add, subtract, etc. These are typical instructions found in most microcomputer instruction sets. They provide the user with the utmost in program flexibility, but they are not efficient for implementing digital filters. The most complex filter configuration can be broken down into a manageable number of high level microinstructions of which loading, adding, multiplying, etc., are implied in a single statement. Normally these instructions might be considered subroutines or branch routines in computer terms, since each involves the execution of a number of elementary computer operations; except that each instruction will take but one microcycle instead of the large number typical of most micros. For example, a common operation in digital filter synthesis involves reading data from memory, multiplying it by another number, adding it to a previous result stored in some working register, and restoring the result in the same register.

An ALU architecture is introduced in Section 3 where this and more complicated sequences may occur under the direction of a single microcode, within a single clock cycle.

Separation of data and control functions provides design simplification. It allows the creation of an efficient ALU with minimum regard for the effects on controller requirements. One finds that whether handling batch or realtime data, control and address functions are best handled by a small amount of specialized hardware. Figure 1 illustrates the difference between a simplified modern day microprocessor and a microprogrammed fast digital sign processor. The obvious isolation of data, address, and instruction busses is an immediate contrast to the basic microstructure.

The fast processor can be a relatively noninteractive machine in that all the computational operations are performed under the unconditional direction of the controller. The controller needs little or no decision making capability on the basis of ALU outcomes. Hence, no feedback is generally required from the ALU to the controller. A general microprocessor would lose much of its capability if this feature were omitted.

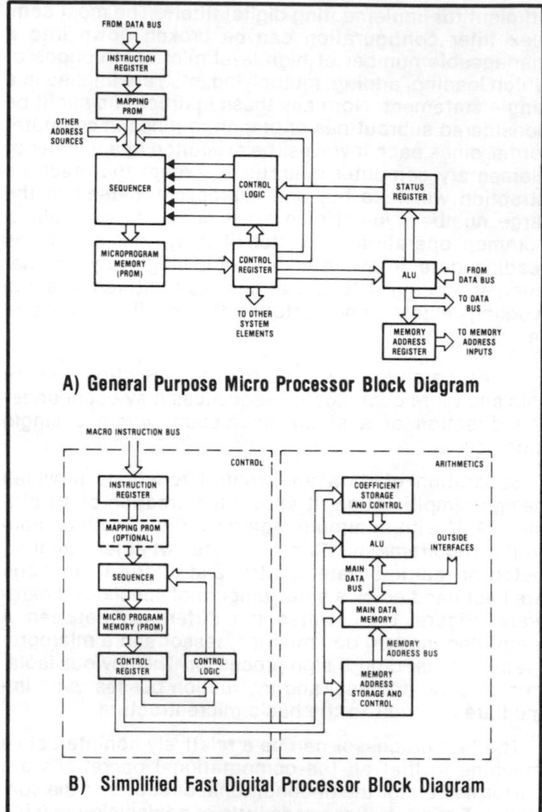

Figure 1. Micro Program Machine Block Diagrams

2. DIGITAL FILTER CONFIGURATIONS

There are two classes of digital filters: recursive and nonrecursive. There are applications requiring a mixture of the two. Recursive filter outputs depend on present input samples and past output values, and therefore they employ feedback in their implementation. In nonrecursive filters the filter output at any given time depends on the input at that time and on a finite number of past input values.

The data input to the digital filter is discrete in time (i.e., data samples rather than continuous data). For so-called batch processing, this data will probably reside in a buffer memory on the outskirts of the arithmetic section of the processor, and be updated in chunks, and processed as such. For realtime data processing where the output data rate is always the same as the input sampling rate, the data will most likely come from one or more A/D converters. The system development discussed in this paper pertains to both situations. The reader may find more emphasis on the realtime case in some instances (e.g., the computer control unit).

2.1 NONRECURSIVE FILTERS

Nonrecursive (finite impulse response or FIR) filters are generally implemented in two ways:

1) Via the summation of N-weighted outputs from a tapped delay line. The delay may be implemented using random access memory or if the order of the filter is not too high, digital shift registers may be used (see Figure 2). The Z^{-1} terms represent data delays of one sample time.

2) A bank of second order "elemental" filters preceded by a comb filter (Figure 3). The resulting outputs from the elemental filters are summed to form output y_n.

These implementations are the natural outgrowths of two mathematical criteria which lead to finite impulse response forms*. The Figure 2 configuration stems from a mathematical development known as the window technique, in which coefficients $a_0 - a_{n-1}$ are the first N terms of the inverse DFT of the desired response.

**A clear interpretation and mathematical development of FIR forms is included in TRW application note TDSP103 on finite impulse response filters.*

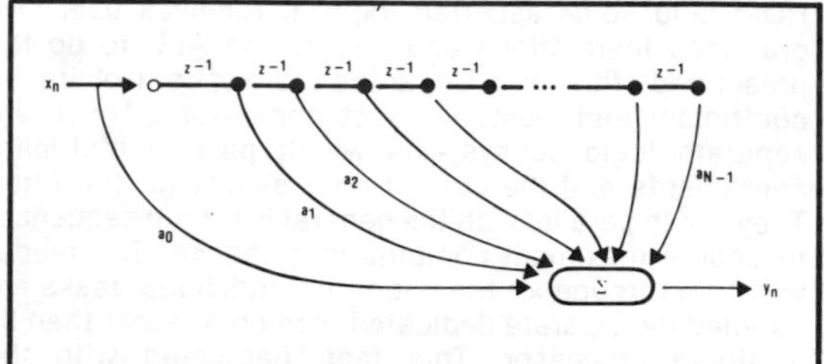

Figure 2. Basic FIR Implementation

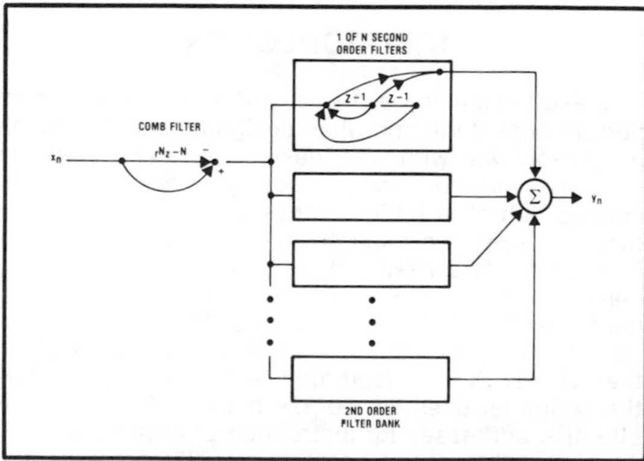

Figure 3. FIR Implementation Using the Frequency Sampling Technique

The frequency sampling technique leads to the configuration in Figure 3. The designer selects a sufficient number of frequency samples (amplitude and phase) to specify the desired spectrum. The frequency sampling filter provides the magnitude and phase at these points exactly.*

**The frequency sampling structure of Figure 3 has an equivalent Figure 2 type implementation but is more conveniently implemented as such one should consult TDSP103 for a comparative analysis of the two forms.*

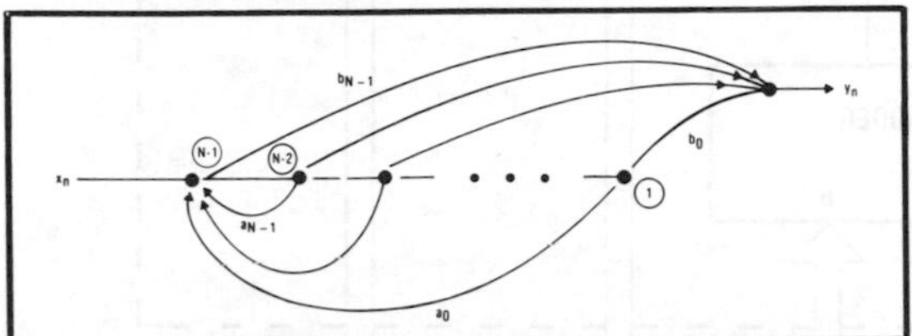

Figure 4. Nth Order Canonical IIR Implementation

2.2 INFINITE IMPULSE RESPONSE FILTERS

Infinite impulse response filters (IIR or recursive filters) have an infinite length impulse response, and unless they are implemented with an infinitely long delay line, feedback must be used. Of the number of well-known equivalent configurations of this type of filter, probably the most common is the canonical** form shown in Figure 4. Although a filter of this type can be of any order, it is recommended that high order filters be broken down into first and second order sections. Filter parameters are less susceptible to coefficient rounding effects for low order filters.

***Several references are included at the end of this paper which provide information of various filter architectures.*

The signal flow graphs of Figures 2, 3, and 4 give the impression that data is available at the signal nodes for immediate and simultaneous processing. In contrast to analog filters where this is the case, digital implementations are done through a precise sequence of steps. The architectures developed for digital filtering are directed toward very efficient program execution, and combine a number of basic computer operations to form higher level micro instructions. Nonetheless, elimination of the sequential nature of the machine is unrealistic. For instance, those signal nodes which represent points on a tapped delay line may be stationary or precessing locations in memory and must be accessed at separate times (i.e., during different machine cycles). Figure 4 indicates data being extracted from memory (notes 1 through N-2), added to input X_n and stored back into memory at node N-1. A typical machine may take up to 2N machine cycles to complete the sequence.

For a realtime system, data samples may be received from an A/D converter at a given rate. Thus, the output sample rate must be effectively the same (ignoring resampling possibilities). Upon arrival of a data sample x_n the computation sequence commences. The sequence must be completed and ready to begin again before the arrival of the next sample x_{n+1}. Signals which contain relatively high frequency components must be sampled at high rates to avoid aliasing effects, and therefore allow limited time between samples for computation.

3. ALU ARCHITECTURE

Confronted with the myriad of possible digital filter configurations, the designer's first compulsion is to create a completely universal ALU bus structure as shown in Figure 5. Each ALU element has direct and independent access to the output of any other ALU element. This allows data to run simultaneously on any number of paths, minimizing sequential steps and computer time, while allowing the synthesis of all filter forms.

The flexibility afforded in the Figure 5 configuration is not necessary, as efficient execution of even the most complicated algorithms involves variations of relatively few routines with the TDC1010J multiplier-accumulator. These routines can be designed to require a maximum of two data buses running at any given execution time with little or no compromise in speed. Arriving at the exact structure for a given dedicated processor may require tracing out data paths on the architecture for some typical program executions. After a number of these the design can eliminate unused paths, minimize his architecture, and arrive at a suitable microinstruction set.

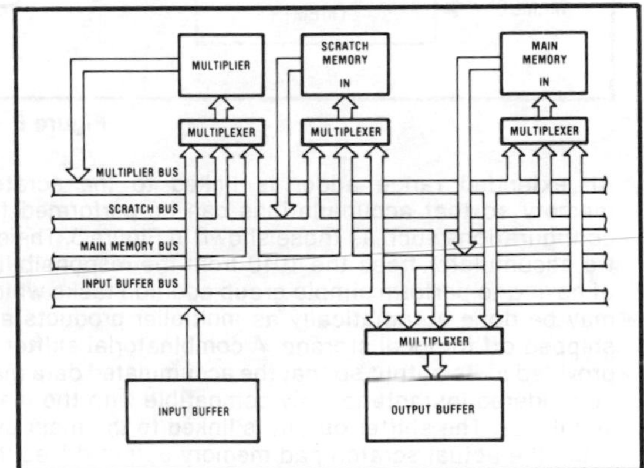

Figure 5. Type I-ALU Architecture

Figure 6 illustrates an ALU section which provides very efficient execution of all digital filters at modest hardware cost. Data is channeled in and out of the ALU via the input and output storage elements (X and OR). These may be A/D and D/A interfaces or additional memory. The heart of the ALU is the TCD1010J multiplier-accumulator. Most of the data computations are done in it. Local rapid access storage is handled in the scratch pad memory (Sc), and the bulk data storage is done in the main memory which provides the necessary data delays for digital filter implementation.

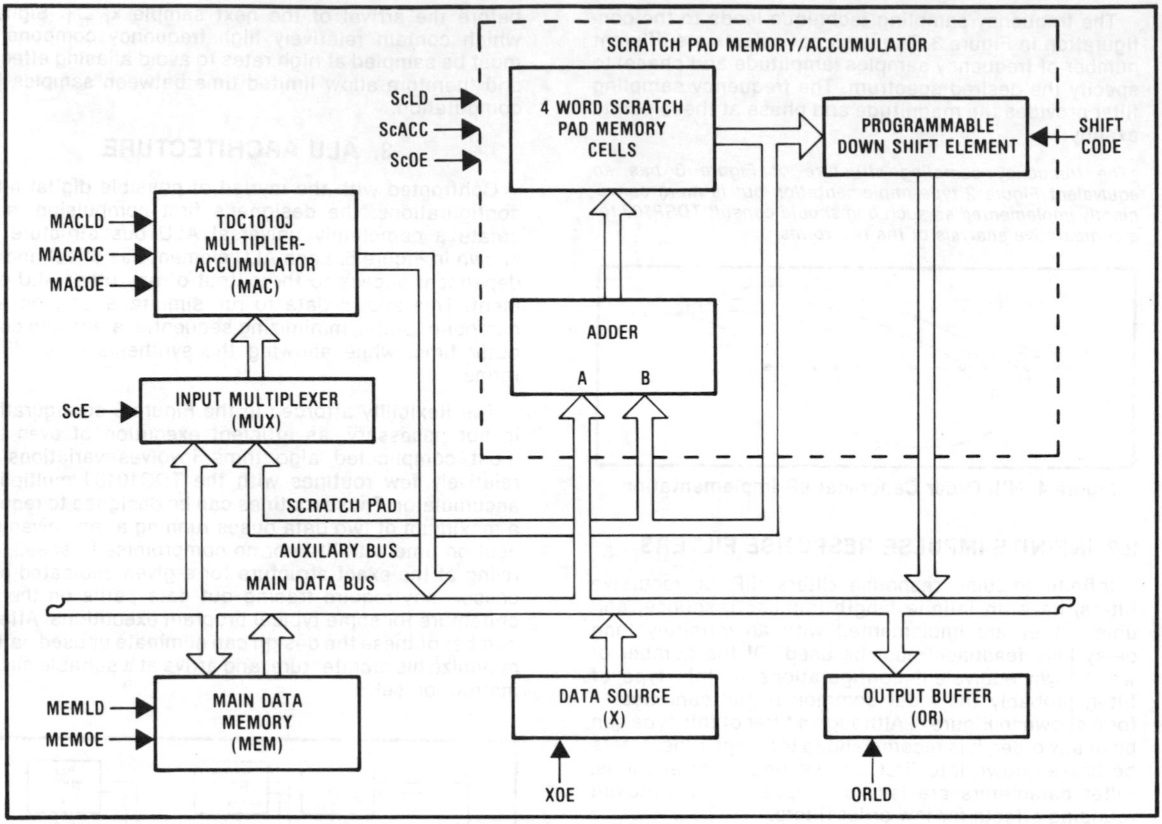

SCRATCH PAD MEMORY/ACCUMULATOR
ScLD
ScACC
ScOE
4 WORD SCRATCH PAD MEMORY CELLS
PROGRAMMABLE DOWN SHIFT ELEMENT
SHIFT CODE
MACLD
MACACC
MACOE
TDC1010J MULTIPLIER-ACCUMULATOR (MAC)
ADDER
A
B
ScE
INPUT MULTIPLEXER (MUX)
SCRATCH PAD
AUXILARY BUS
MAIN DATA BUS
MEMLD
MEMOE
MAIN DATA MEMORY (MEM)
DATA SOURCE (X)
XOE
OUTPUT BUFFER (OR)
ORLD

Figure 6. Type 2 Architecture

An extended range adder is linked to the scratch memory so that accumulations can be performed for configurations such as those shown in Figure 3. The extra accumulator frees the 1010 from the responsibility of having to perform simple group addition tasks which may be done automatically as multiplier products are shipped off for local storage. A combinatorial shifter is provided at its output so that the accumulated data may be rendered instantaneously compatible with the main 16 bit bus. The shifter output is linked to the main bus while the actual scratch pad memory output drives the auxillary bus into the multiplier input multiplexer (MUX).

As an introduction to the efficiency afforded by the auxillary bus the following example is given: a second order canonical filter is to be implemented. The signal flow graph is shown in Figure 7.

Input x_n is added to past (extracted from memory) data products $-b_0X_1$ and $-b_1X_2$. The result is written back into memory as indicated (node 3). During this write cycle the MAC input section is free to do another multiply or multiply-accumulate. Unfortunately, during a write cycle the memory I/O bus is tied up and cannot provide the necessary data (X_1, X_2, or X_3) for the multiplications to form output y_n. If during the recursive operation X_1 (or X_2) is simultaneously stored in Sc while it is being transferred to the MAC, it can then be used during the write cycle when the MAC would otherwise be idle. Data will be running on both buses during this time resulting in the elimination of one microcycle for the implementation of the filter. This constitutes a substantial savings in machine time since four to six cycles are necessary to perform the complete algorithm (see instructions 3 and 5, Table 1, Section 3.2).

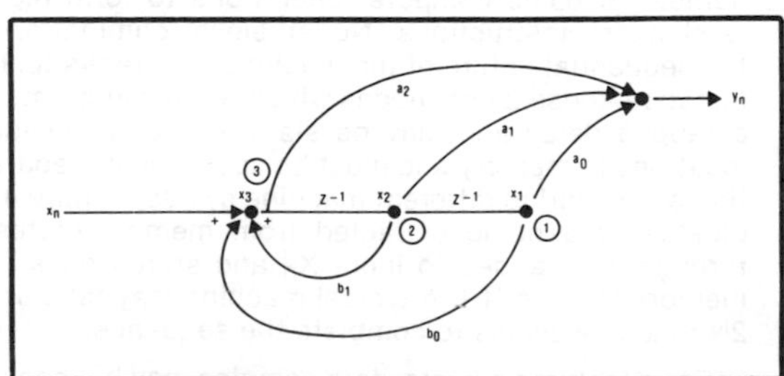

Figure 7. Signal Flow Graph

3.1 CONTROLLING THE ALU

Directing the traffic in and out of each ALU element is a relatively straightforward task and is performed by the controller (Section 4).

The main control signals at each ALU element are defined below:

a) Scratch Pad Memory/Accumulator (Sc) Controls
- Scratch Pad Load (ScLD). Clock enable used to load the memory.
- Scratch Pad Read Address, Scratch Pad Write Address (SRA, SWA)—Memory address directs data in and out of memory; addresses may be tied together or remain separate.
- Scratch Accumulate (ScACC). Logic level activates scratch adder to perform accumulation. A logic 1 will allow data on the bus to be added to data in memory location specified by SRA and stored in memory location specified by SWA. When this signal is inactive a simple data write or read can be performed.
- Scratch Output Enable (ScOE). Logic level enables shifter data onto the main bus.
- Shift Code (S). n bit code specifies the number of places the scratch pad word is to be shifted down; hence, division by 2^n is possible.

b) Main Memory Controls
- Memory Output Enable (MEMOE). Logic level enables memory data onto the main bus.
- Memory Load (MEMLD). Clock enable for loading data into memory.
- Memory Address (MEMA). N bit code specifies the location of the memory word which is accessed.

c) TDC1010J Multiplier-Accumulator (MAC) Controls
- Multiplier-Accumulator Load (MACLD). Clock enable for loading data into X, Y, and accumulator registers.
- Multiply/Accumulate (MACACC). Active logic level directs an accumulate operation within the TDC1010J after multiplication.
- Multiplier Output Enable (MACOE). Active logic level allows the multiplier product onto the main data bus.

d) Multiplexer Input (MUX) Controls
- Scratch Pad Enable (ScE). Active logic level enables data into the MAC from the scratch memory via the auxiliary bus.

e) Input Source (X) Controls
- Source Output Enable (XOE). Active logic level enables X data onto the main bus.

f) Output Buffer (OR) Controls
- Output load (ORLD). Clock enable for loading this element (register on buffer memory).

Data may be sent on the bus connecting one ALU element to another by enabling the output (e.g., MACOE) of the sender and the input clock enable for the receiver (e.g., ScLD).

All control lines are assumed to be a logic 1 for an enabling condition. This simplifies matters in the preliminary stages of the design. It is understood that each ALU element may have its particular requirements for loading and enabling. For example, the TDC1010J requires three input clocks and has five output enables. Also most logic will be tristate, requiring logic zero for an output enabling condition. These particulars may be taken care of with a small amout of random logic in the ALU control interface. The controller outputs during any given microcycle report the state of the machine and may be used as enabling signals when necessary, as well as the actual control signals.

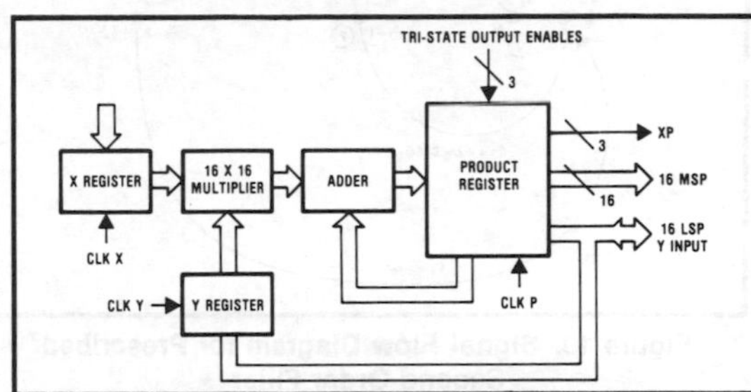

Figure 8. Simplified TDC1010J Multiplier-Accumulator Block Diagram

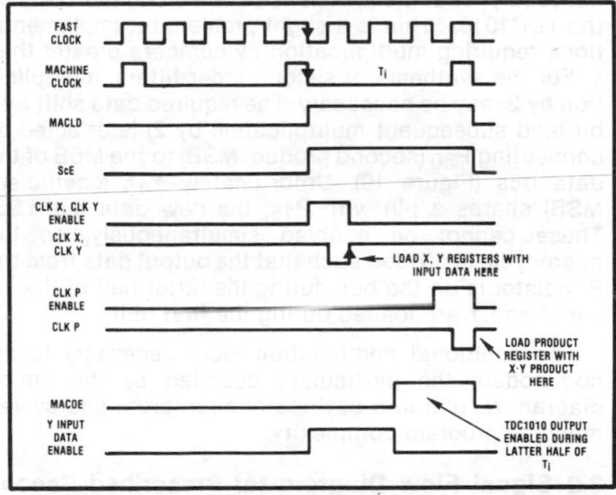

Figure 9. Typical TDC1010J Loading Sequence Timing Diagram

Figure 8 is a simplified block diagram of the TDC1010J MAC, and Figure 9 is a timing diagram illustrating a loading and output enabling sequence of the 1010 during a typical computer cycle. Machine cycle T_i contains an instruction requiring a multiplication or multiply accumulate and that the previous product which is stored in the product register be transferred to the main bus. The 1010 contains input registers for X and Y operands. It is not necessary to waste a complete clock cycle to load X and Y registers since typical multiply-accumulate time for the MAC is under 150 nsec and loading is not useful for anything other than multiplier operations. The clock interval may be subdivided such that loading, multiplying, adding, and final storage are carried out in various portions of a single microcycle (Figure 10).

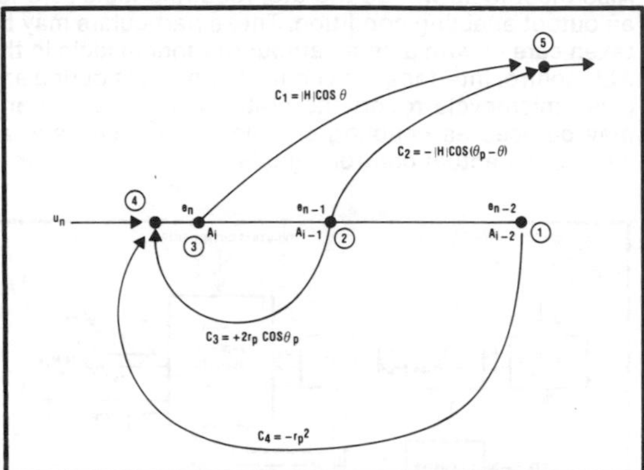

Figure 10. Signal Flow Diagram for Prescribed Second Order Filter

Due to the large number of inputs and outputs required for 16 x 16 double precision multiplication, it is necessary for some functions to share common pins on the TDC1010J. This is a slight problem for implementations requiring multiplication by numbers greater than 1. For the synthesis of second order filters, multiplication by 2 may be necessary. The required data shift by 1 bit (and subsequent mulitplication by 2) is effected by connecting P_{30} (second product MSB) to the MSB of the data bus (Figure 10). Unfortunately Y_{15} (coefficient MSB) shares a pin with P_{15}, the new data bus LSB. These cannot be enabled simultaneously, so the microcycle is divided such that the output data from the P register is on the bus during the latter half of the cycle; X and Y are loaded during the first half.

The additional combination logic necessary to accommodate the particulars depicted by the timing diagram results in a savings of microprom and overall machine program complexity.

3.2 Signal Flow Diagram for Prescribed Second Order Filter

This is probably the hardest filter to implement since it involves many types of instructions and the use of all of the ALU's capabilities. In addition, "fencing" of data memory into many sections is required to form the second order elemental filters. This causes a slight complication which will be ignored for the time being and taken up in Section 5.

The configuration calls for the implementation of many second order recursive(*) filters with transfer functions of the form:

$$H(Z) = \frac{|H(\theta_p)|[\cos\theta - \cos(\theta_p - \theta)\,Z^{-1}]}{1 - 2r_p\cos\theta_p Z^{-1} + r_p^2 Z^{-2}}$$

**A discussion of recursive filter configurations in the synthesis of nonrecursive filters is given in TDSP103.*

The Z^{-1} and Z^{-2} terms denote delay operations of one and two sample times (T and 2T). As sighted in Section 2.1, these delays are usually implemented with RAM but may be envisioned as taps at T seconds and 2T seconds on a delay line. A discussion on the particulars of implementation and control of the delay line is also deferred until Section 5.

The output of each filter is summed to form the composite filter output y_n (See Figure 3). Filter parameters $H(\theta 8_p)$, $\cos\theta$, $\cos(\theta_p - \theta)$, and $2r_p\cos\theta_p$ can be combined and the signal flow graph for the transfer function h(Z) is shown in Figure 10.

The hardware operations indicated by the flow diagram proceed as follows: input signal u_n is a digital word residing in a register (scratch pad, A/D register, etc.) at a time nT. It is summed with words C_3e_{n-1} and C_4e_{n-2} to form $e_n = C_oU_n + C_3e_{n-1} + C_4e_{n-2}$; e_n is then written into memory at effective address A_i (node 3, the input of the delay line). e_{n-1} and e_{n-2} are the results of identical operations having occurred one and two data sample times before. They reside in memory at locations A_{i-1} and A_{i-2}, during sample time period nT. During the next sample time, (n+1)T, when x_{n+1} is to be processed, e_n will advance to node 2 (address A_{i-1}), e_{n-1} to node 1 (A_{i-2}), and e_{n+1} will be generated and written into memory at node 3 (A_i). The filter output is formed in a fashion similiar to that just described for the feedback operation. C_1 and C_2 are the multiplying coefficients and $y_n = C_1e_n + C_2e_{n-1}$.

Including the input scaling multiplication (C_oU_n), five multiplications and four additions are necessary to implement this filter. If we assume a 150 nsec machine cycle time and can achieve one cycle per multiply/accumulate (multiply/add), the elemental filter computation will take 750 nsec. The usefulness of the scratch accumulation capability is realized for this computation sequence. The filter outputs can be accumulated as each is piped from the 1010 output register through the adder into a scratch pad location. Two machine cycles per filter are eliminated. Ignoring initial setup and the small amount of overhead time which is likely to occur, a 100 frequency sample filter, (i.e., a 200th order filter), will take 75 usec to implement.

Table 1 lists a feasible instruction sequence for the frequency sampling filter. The active data paths are indicated in the group of diagrams in Figure 11. Figure 12 is a detailed timing diagram for the sequence.

The comb filter which precedes the filter bank is implemented in instructions 2, 3, and 4. Instruction 1 contains a NOP but may be needed for program initialization. Instructions 6 through 11 implement the first second order section. Twelve through 17, 18 through 23, etc., are repeats of 6 through 11 with the exception that coefficients differ for all the filters. All the filter outputs are automatically accumulated (added) in ScO.

It is appropriate to emphasize the last instruction of each filter (MACACCO). Data taken from the 1010J is added to the value residing in ScO and restored in ScO. This running accumulation of all the elemental outputs forms the final filter output. Instruction number 5 (C1Sc1) could have been eliminated if 11 were simply a load ScO instruction (MACSTOO). There would have been no need for an empty location to prepare for the first filter output data dump. However, from a programming standpoint, the complete sequence (steps 6 through 11) can be stored as a canned routine in the microprogram PROM, thereby simplifying the program. The programmer need only call up "second order filter", specify the coeficients, and forget about the details of the microinstruction sequence. Instead of six program steps for each filter, he would have to program only one each time he wanted to use the filter*. Section 4.2 describes the workings of the computer controller as it alternates between the "canned routine" mode and the single cycle mode.

**More complicated programs might be designed to implement a large number of identical filters with very few programs steps. This could be achieved with a more complex computer control unit than is implied here.*

A point which should be addressed at this time deals with the shifter in the scratch pad memory. The accumulation of the elemental filter otputs in the frequency sampling structure may very well lead to word growth beyond 16 bits. This is not a problem in the accumulation instructions (i.e., MACACCO) since ScO is an extended range register. The problem lies in the extraction of this data for user evaluation since it must be brought to the main data bus before it is shipped off. The nonrecursive coefficients (C_1 and C_2) could be scaled down in anticipation of the situation, but objectionable signal-to-noise degradation could result due to the attenuated signals on each filter output path. The shifter allows the arithmetic to be performed at its basic 16 bit precision. The result is rescaled only after the accumulations are completed, and round off errors may be substantially reduced.*

**The subject of noise generation in digital filters is discussed in detail in TDSP110.*

Table 1. Macro Instruction Sequence for the Frequency Sampling Filter

Instruction Description	Instruction	Data Notation	Multiplier, C
1. Nothing (or program initialization)	NOP	—	—
2. Multiply MEM data by r^N; store in MAC register	MemMult	C•MEM→MAC	$-r^N$
3. Multiply input x data by C_i, add result to value residing in MAC register, restore in MAC register, write input x into memory	XWM/XMAC	X+MAC→MAC X→MEM	C_i(Comb input scaler)
4. Store MAC output register value in Sc address 1	MACSTO1	MAC→Sc1	
	END OF COMB COMPUTATIONS		
5. Clear scratch pad location zero	C1Sc	0→ScO	—
6. Multiply data in Sc address 1 by C_o, place product in MAC output register	Sc1Mult	C•Sc1→MAC	C_o(canonical input scaler)
7. Multiply e_{n-1} by C_3, add result to value residing in MAC output register, place this sum in MAC output register; store e_{n-1} in Sc address 2	MemMAC/STO2	C•MEM+MAC→MAC MEM→Sc2	$C_3 \cdot 2r_p \cos\theta_p$
8. Multiply e_{n-2} by C_4, add this product to value residing in the MAC output register, place this sum in the MAC output register	MemMAC	C•MEM+MAC→MAC	$C_4 = -r_p^2$

Table 1. Macro Instruction Sequence for the Frequency Sampling Filter (Continued)

Instruction Description	Instruction	Data Notation	Multiplier, C
9. Load MAC data into MEM; multiply present data in Sc address 2 by C_2, store in MAC output register	MWM/Sc2Mult	MAC→MEM C•Sc2→MAC	$C_2 = -\|H\| \cos(\theta_p - \theta)$
10. Multiply e_n by C_1, add it to data residing in MAC register, restore result in MAC register	MemMAC	C•Sc2 + MAC→MAC	$C_1 = \|H\| \cos\theta$
11. Add data in MAC register to data in Sc address 0 and restore the result in Sc address 0	MACACC0	MAC + Sc0→Sc0	——
FILTER NO. 1 COMPLETE			
12. Multiply data in Sc address 1 by C_o, place product in MAC output register	Sc1Mult	C•Sc1→MAC	C_o = input scaler
13. Multiply e_{n-1} by C_3, add result to value residing in MAC output register, place this sum in MAC output register; store e_{n-1} in Sc address 2	MemMAC/STO2	C•MEM + MAC MEM→Sc2	$C_3 = 2r_p \cos\theta_p$
14. Multiply e_{n-2} by C_4, add this product to value residing in the MAC output register, place this sum in the MAC output register	MemMAC	C•Mem + MAC→MAC	$C_4 = -r_p^2$
15. Load MAC data into MEM; multiply present data in Sc address 2 by C_2, store in MAC output register	MWM/Sc2Mult	MAC→MEM C•Sc2 + MAC	$C_2 = -\|H\| \cos(\theta_p - \theta)$
16. Multiply e_n by C_1, add it to data residing in MAC register, restore result in MAC register	MemMAC	C•Sc2 + MAC→MAC	$C_1 = \|H\| \cos\theta$
17. Add data in MAC register to data in Sc address 0 and restore the result in Sc address 0	MACACC0	MAC + Sc0→Sc0	——
FILTER NO. 2 COMPLETE			

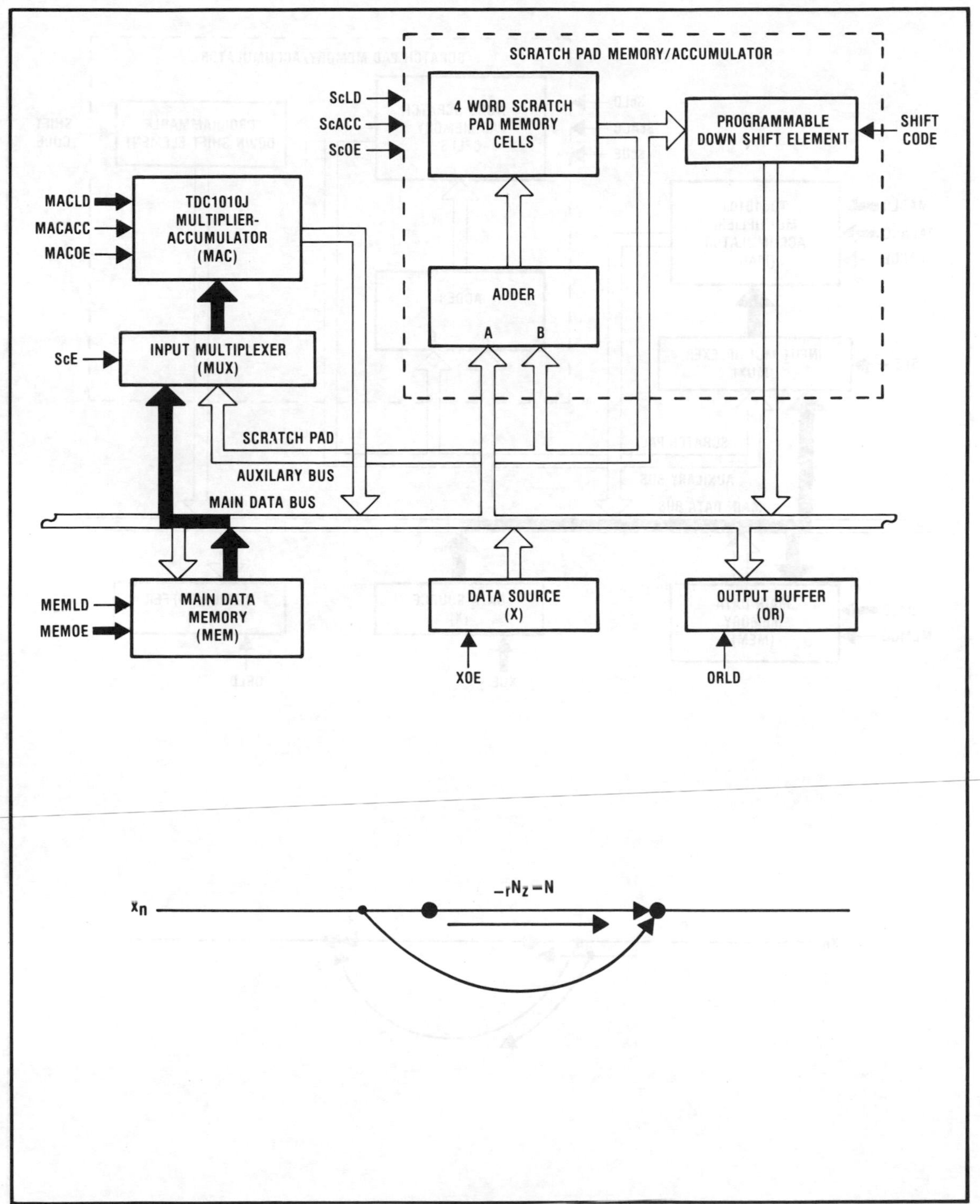

Instruction 2 - MemMult
Figure 11.

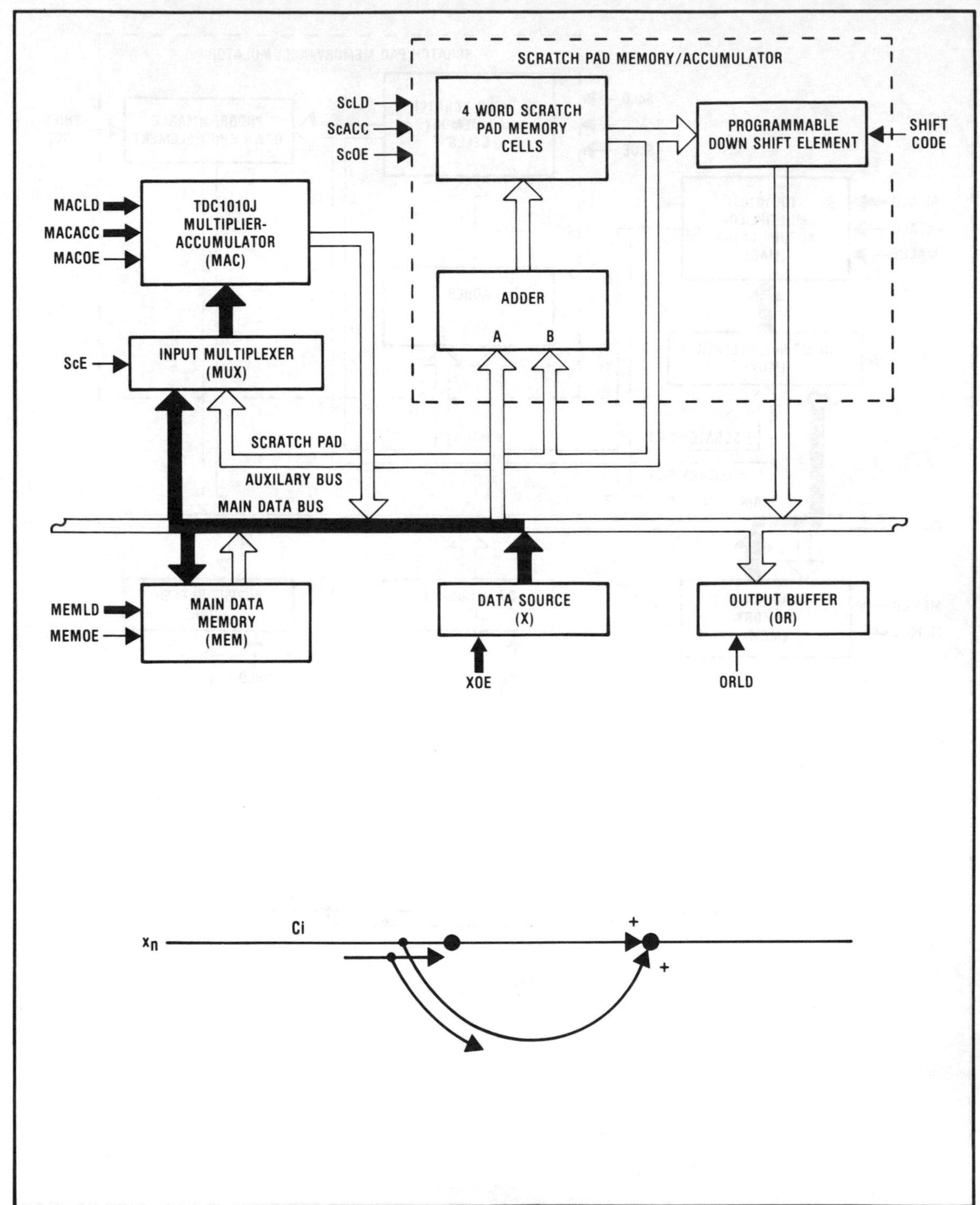

Instruction 3 - XWM/MAC

Figure 11.

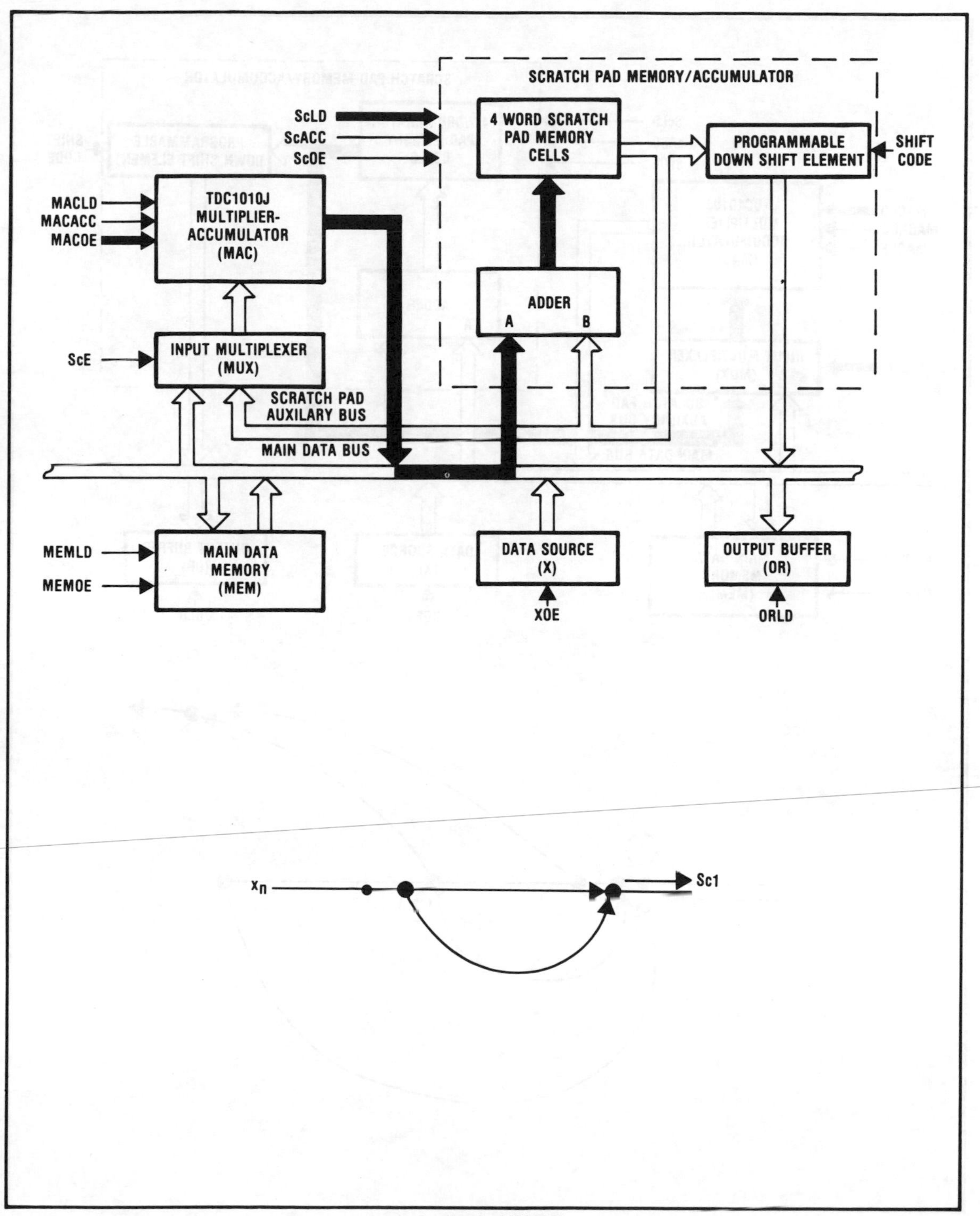

Instruction 4 - MACST01
Figure 11.

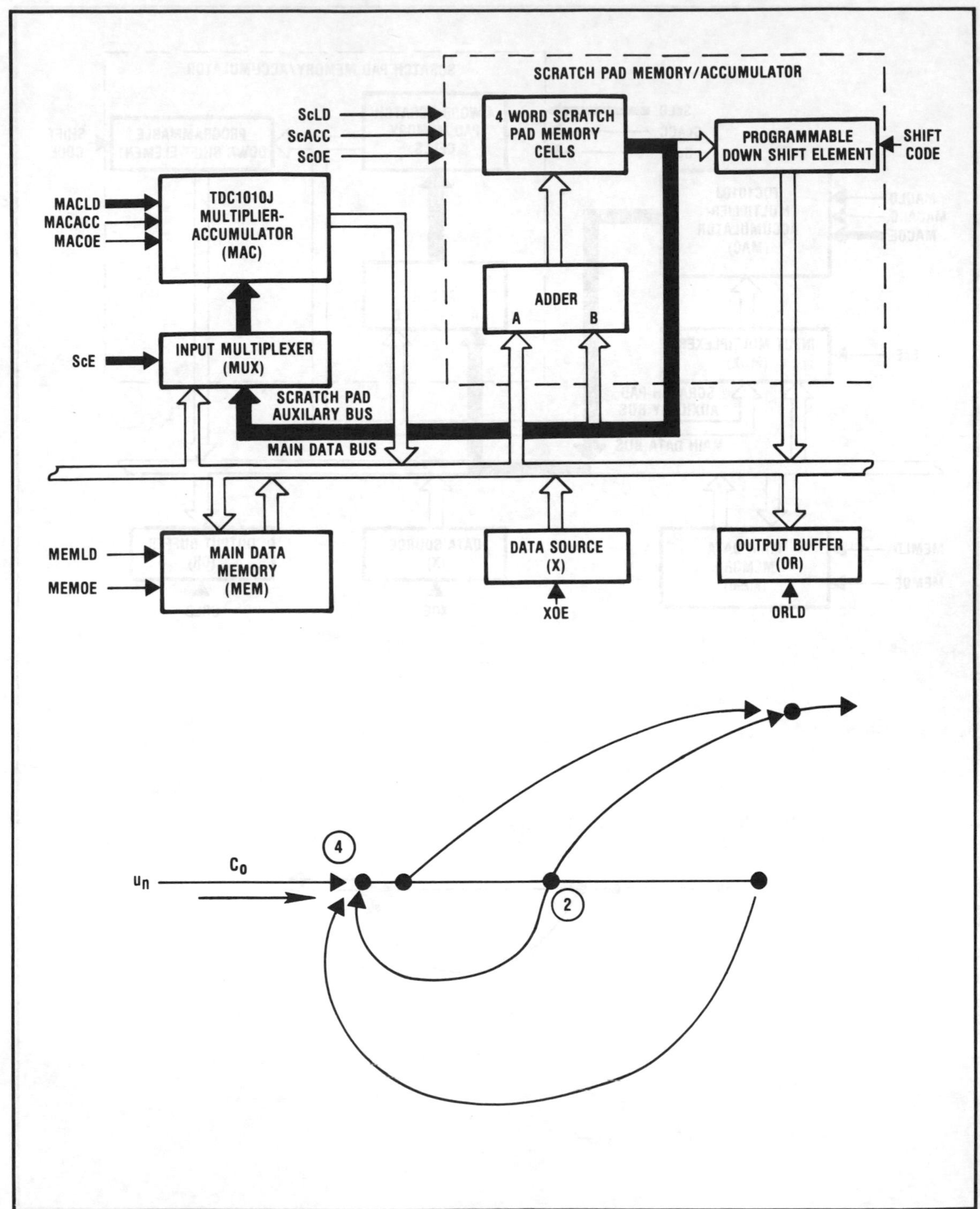

Instruction 6 - Sc1 Mult
Figure 11.

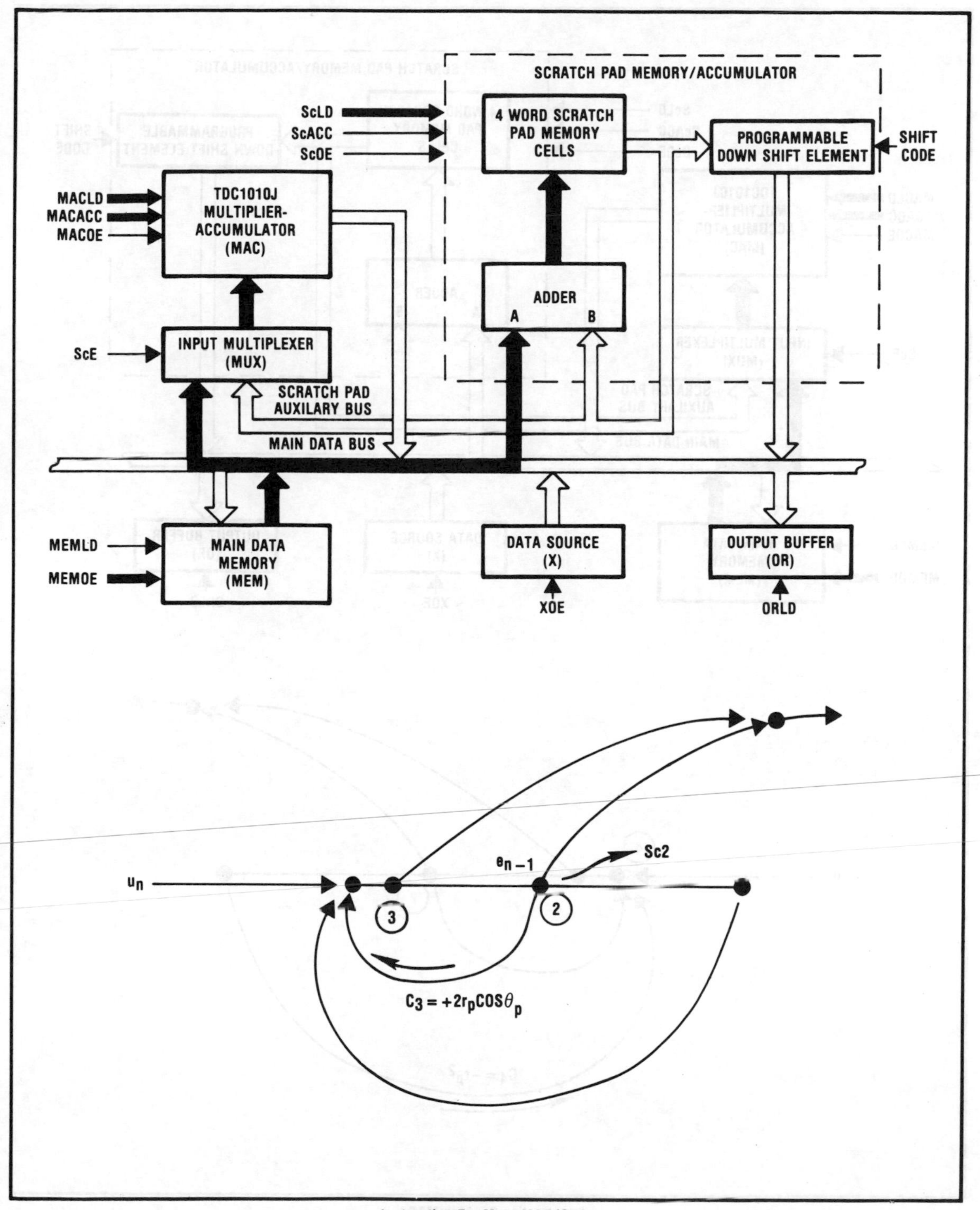

Instruction 7 - Mem MAC/STO2

Figure 11.

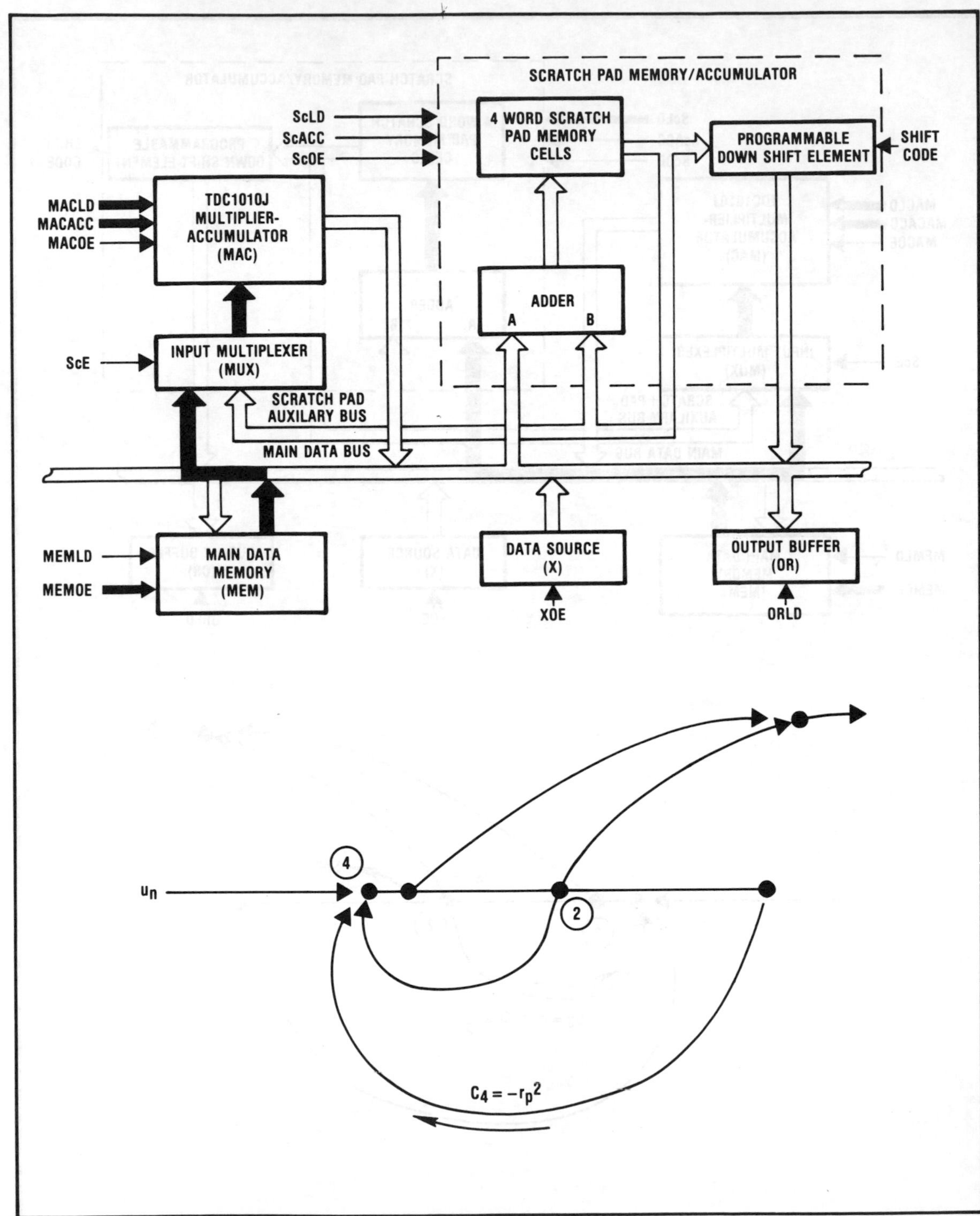

Instruction 8 - MemMAC
Figure 11.

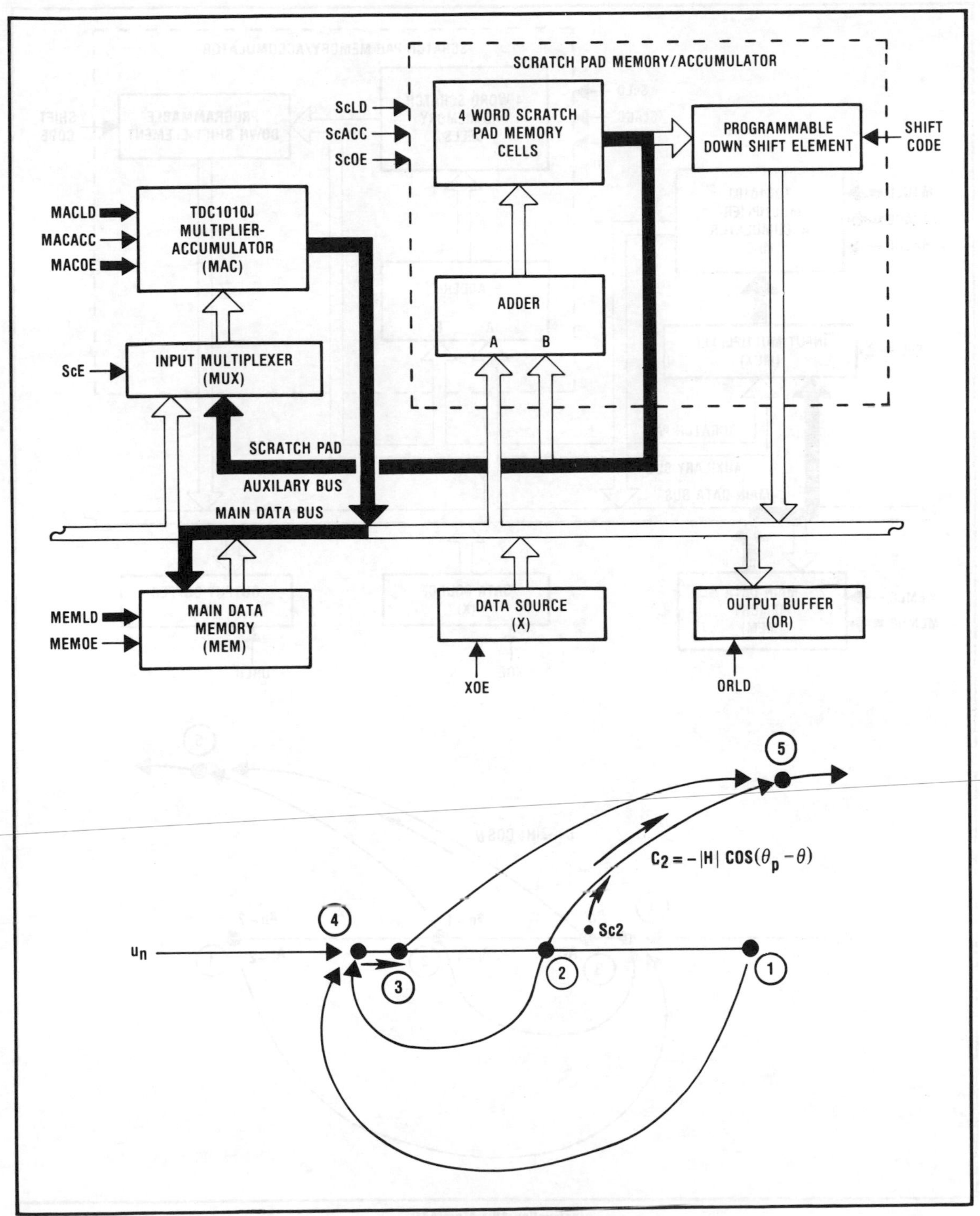

Instruction 9 - MWM/Sc2Mult
Figure 11.

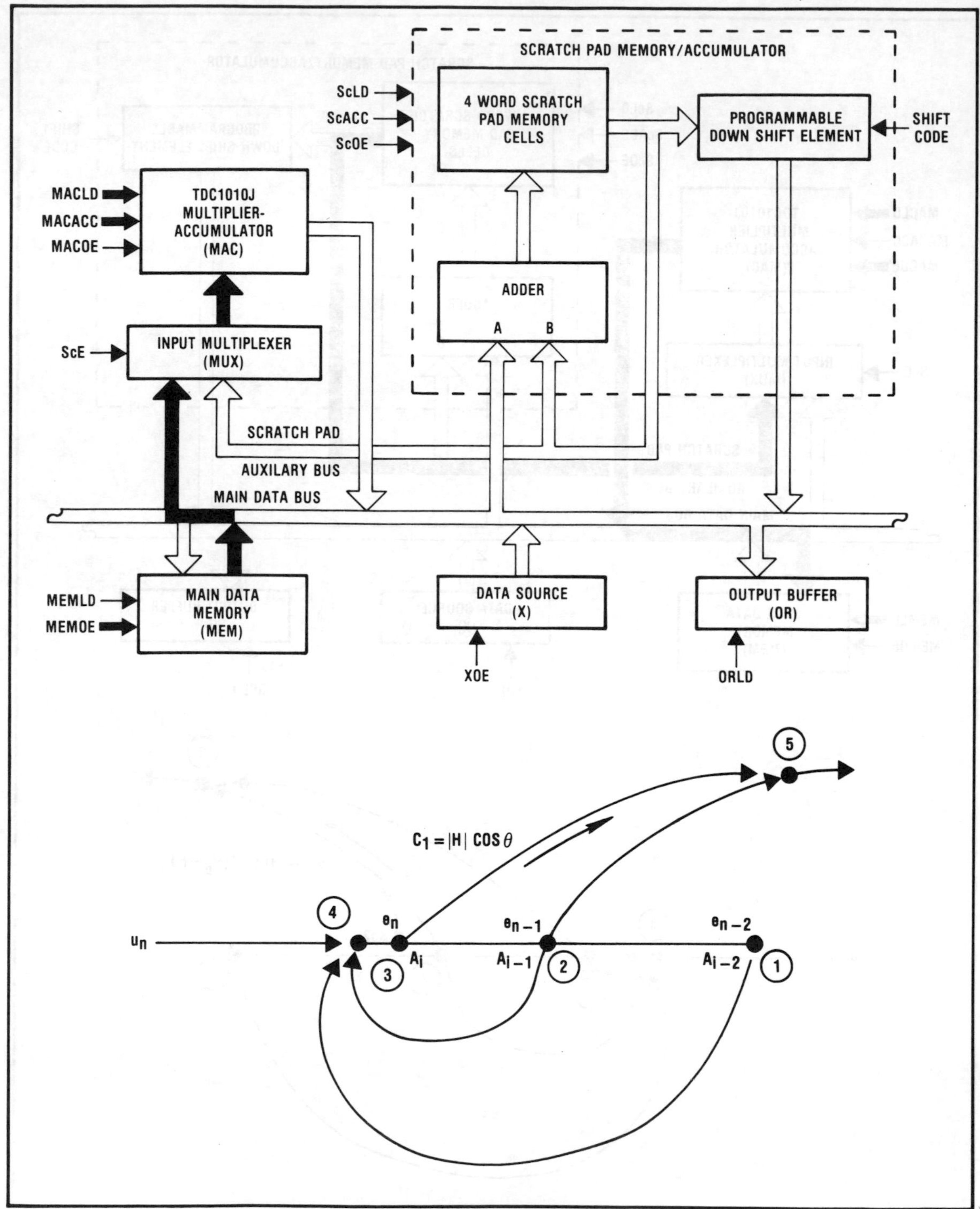

Instruction 10 - MemMAC
Figure 11.

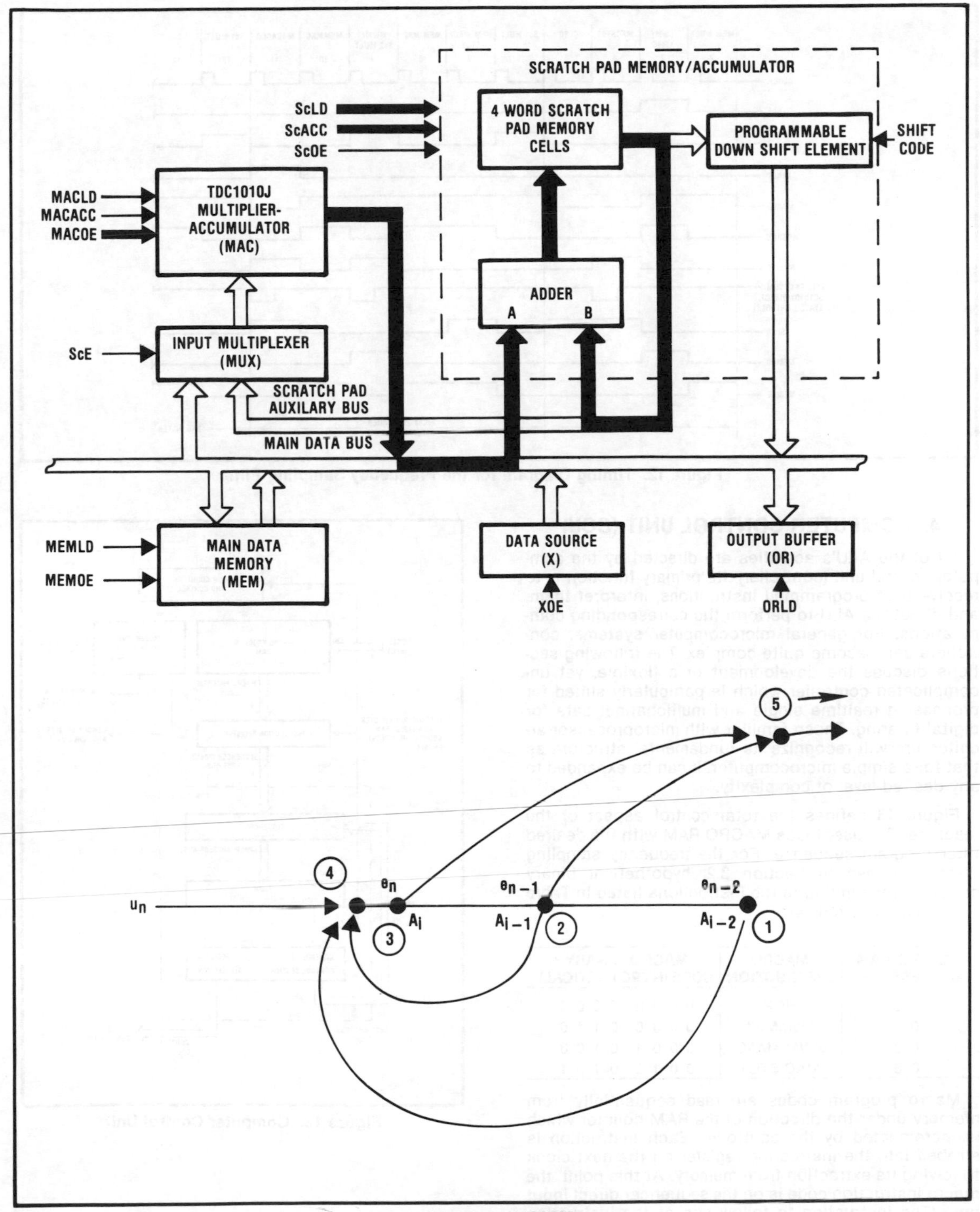

Instruction 11-MACACCO
Figure 11.

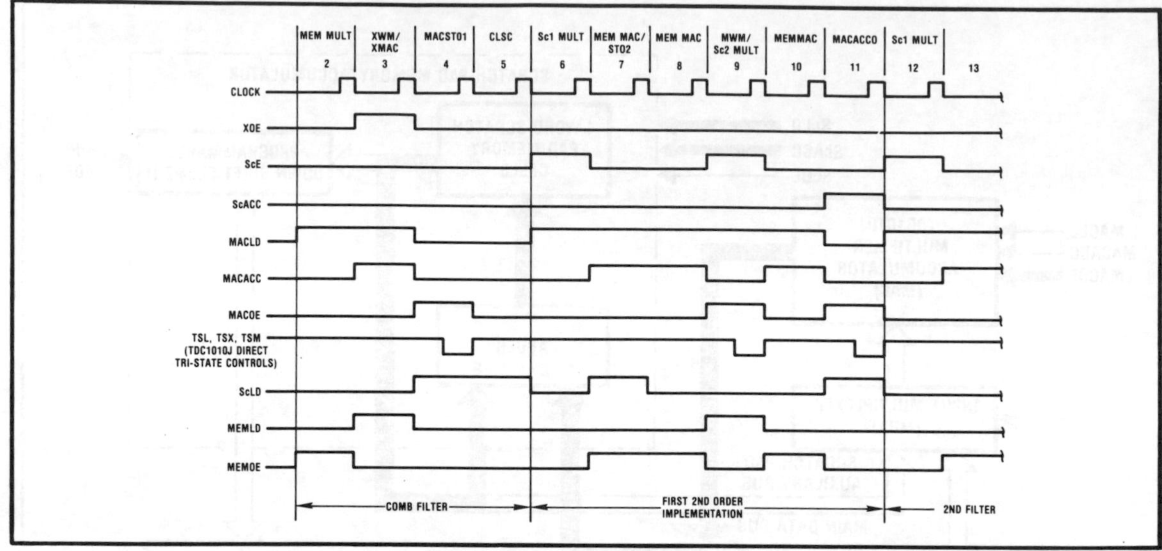

Figure 12. Timing Diagram for the Frequency Sampling Filter

4. COMPUTER CONTROL UNIT (CCU)

All of the ALU's activities are directed by the computer control unit (controller). Its primary function is to receive user programmed instructions, interpret them, and direct the ALU to perform the corresponding computations. For general microcomputer systems, controllers can become quite complex. The following sections discuss the development of a flexible, yet uncomplicated controller which is particularly suited for processing realtime single and multichannel data for digital filtering. Those familiar with microprocessor architecture will recognize its fundamental structure as that for a simple microcomputer. It can be expanded to any desired level of complexity.

Figure 13 defines the total control aspect of the machine. The user loads MACRO RAM with the desired filter program sequence. For the frequency sampling filter described in Section 3.2, hypothetical binary codes corresponding to the instructions listed in Table 1 are loaded as follows:

MACRO RAM ADDRESS (HEX)	MACRO INSTRUCTION	MACRO BINARY CODE (HYPOTHETICAL)
0 0	NOP	0 0 0 0 0 0 0 0
0 1	MemMult	0 0 0 0 0 1 1 0
0 2	XWM/XMAC	0 0 0 1 0 1 0 0
0 3	MAC STO1	0 0 1 0 0 1 0 1

Macro program codes are read sequentially from memory under the direction of the RAM counter which is incremented by the controller. Each instruction is strobed into the instruction register on the next clock following its extraction from memory. At this point, the macro instruction code is on the sequencer direct input bus. (The instruction to follow sits at the instruction

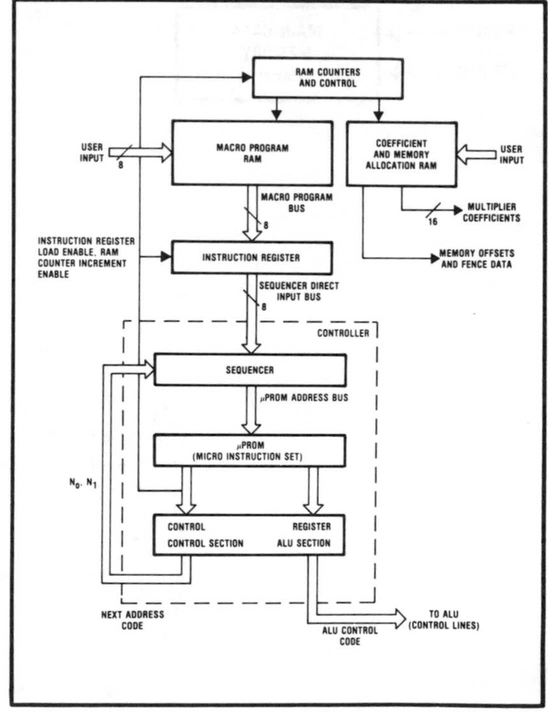

Figure 13. Computer Control Unit

register input awaiting entry into the register.) When the sequencer is not busy, it will allow the code to pass through it to the μprom address bus. This code is the address of the location in prom which contains the desired ALU microcode. For example, locations A_i and A_{i+1} in μprom may contain microinstructions MEM MAC and MEM MULT, respectively (see Figure 14).

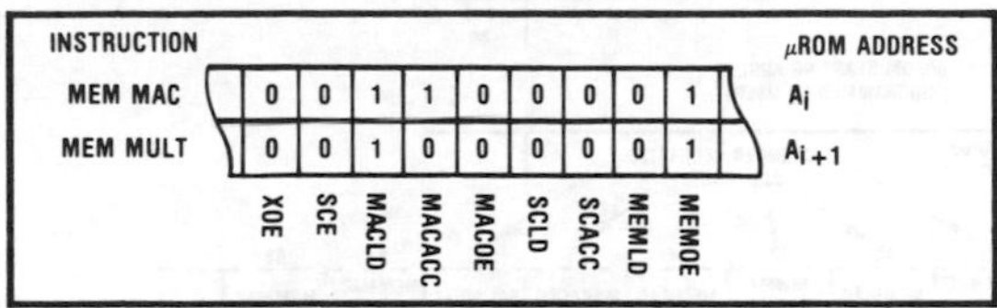

Figure 14. μprom Format

If the macro instruction residing in macro address 04 required a MEM MAC operation for some sequence, the binary equivalent of A_i would be placed in MACRO RAM 04. When A_i eventually arrived on the μprom address bus, the ALU control code corresponding to MemMAC would appear at the output of the prom. On the next clock this code would be loaded into the control register whose outputs would activate the required ALU elements (i.e., MACLD, MACACC, and MEMOE). The next microinstruction to be executed would be present at the μprom output awaiting its entry into the control register.

The sequencer's true capabilities are realized when large filters with many identical sections are to be implemented, or when individual filters are to be timeshared. The frequency sampling filter is a perfect example since each second order canonical filter in the filter bank is identical in form. The microcode for instructions 6 through 11 can be stored in the microprom in the exact sequence that they appear in Table 1. The filter can be executed by placing the μprom address which contains the first microinstruction (Sc1 Mult) into the desired macro RAM location. This μprom address is termed the "starting address". After it is extracted from memory and then stored in the instruction register, the sequencer gates it through to the μprom address bus as previously described. The microprom outputs the correct microcode for SC1Mult which is clocked into the control register. At this point, the sequencer does not take the next macro instruction which is still in the macro memory. Instead, it generates the next μprom addresses internally while the MACRO RAM counter and instruction register states remain frozen. If μprom addresses 80, 81, 82, 83, 84, and 85 contained the filter sequence,the user would program 80 into macro RAM.The sequencer would recognize the control portion of the microcode associated with address 80 as the first of six microaddresses of a canned routine. It would then generate 81 through 85. At the end of the routine, with the microcode corresponding to address 85 in the control register, the sequencer would allow the next macro instruction (which would be residing in the instruction register) through to the μprom bus. The prom output at this time would represent the starting instruction for the next canned sequence or simply another single cycle macro instruction.

The timing diagram (Figure 15) for the Table 1 program gives a clear picture of the events described. The first five macro instructions occupy one machine clock interval each. Each takes two clock times to get from macro memory to control register output. There is a one cycle delay between macro memory and instruction register and another delay from there to the control register depicted in Figure 15.

The sixth and seventh macro instructions, represented by decimal 80, are the starting addresses for the first two second order filters. Instruction 6 is clocked into the instruction register, and the macro counter is incremented to instruction register, and the macro counter is incremented to instruction 7 which appears at the instruction register input. Meanwhile, 6, which is on the direct bus to the sequencer is gated through to the μprom address bus; 5 is in the control register being executed at this time.

On the next clock cycle the Sc1 Mult microinstruction jumps into the control register, but the clock to the macro counter and the instruction register (IRCLK) is inhibited. The instruction register retains the first filter starting address and the macro memory output still shows the second filter starting address.

IRCLK is inhibited for the next four cycles while the rest of the sequence is performed (i.e., MemMAC/STO2, MemMAC, MWM/Sc2Mult...). When the last instruction (MACACO) is on the μprom output, the clock is enabled. The macro counter increments to 8 and the instruction register receives instruction 7 as MACACCO is dumped into the control register. The control portion of the MACACCO μcode instructs the sequencer to gate in instruction 7 (micro prom address 80).

IRCLK is enabled by a μprom control signal which is activated when the last microinstruction of each routine is in μprom. This rule is perfectly general and pertains to routines which are only one microcycle in duration (i.e., single cycle instructions). Hence, all single cycle microinstructions have the IRCLK enable bit in μprom enabled.

Figure 16 emphasizes the control portion of the μprom. N_0 and N_1 are the control bits which decide in which mode the sequencer will operate. In its simplest mode the sequencer allows instruction register outputs through to the μprom. For canned routines the sequencer generates its own μprom addresses.

Given that a certain microinstruction resides in the control register (and is therefore being executed) the control portion of its microcode is used to direct the sequencer into a certain mode. If the microinstruction represents a single cycle macro instruction, then the code N_0, N_1 should instruct the sequencer to allow a new address through to the μprom. If the microinstruction is the starting instruction or some intermediate instruction for a filter sequence, then the code N_0, N_1 should direct the sequencer to continue generating successive μprom addresses. When the final microinstruction is in the control register the corresponding control code will once again gate the instruction register output through to the μprom. N_0, N_1 is therefore termed the

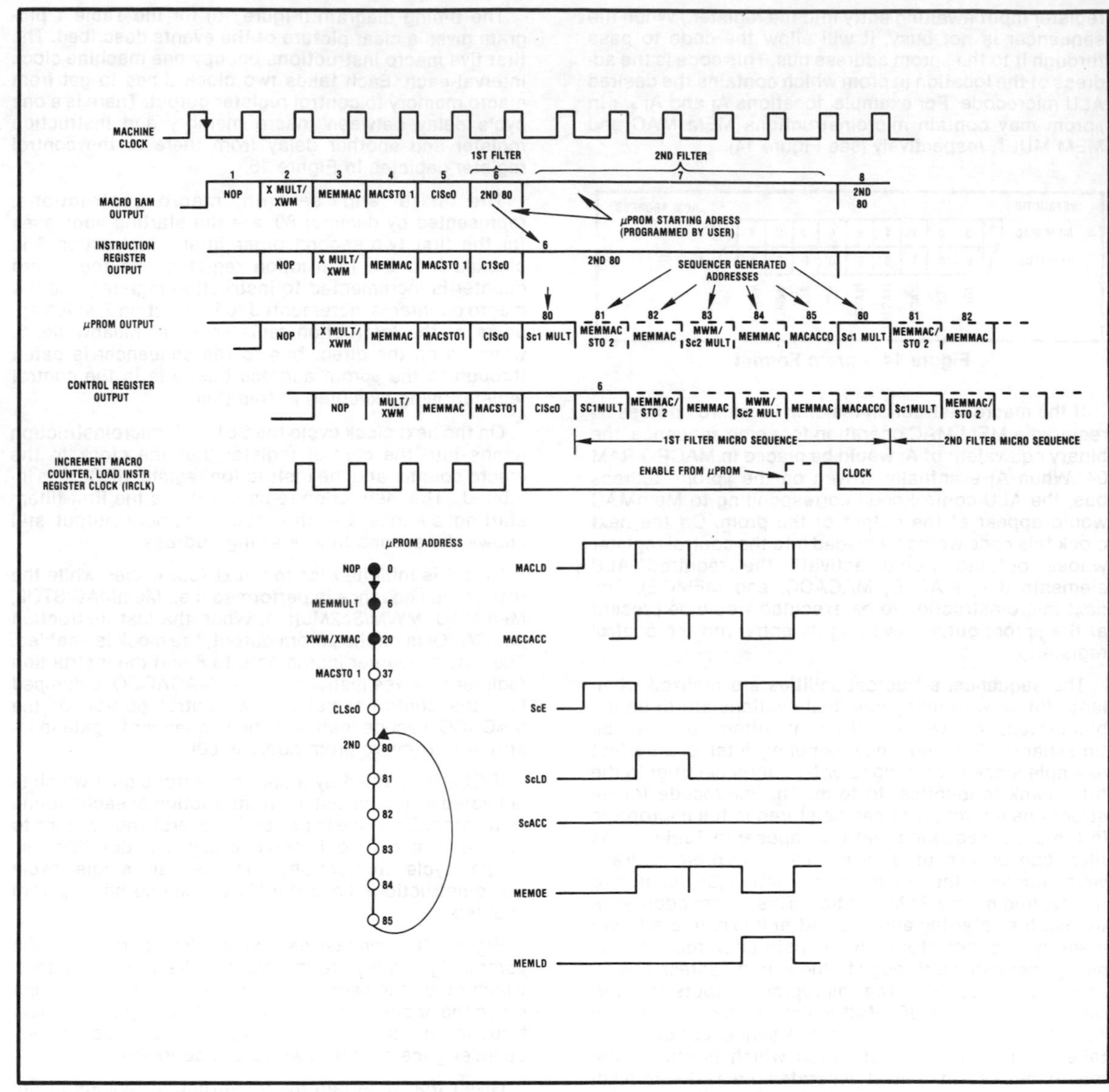

Figure 15. Sequencer Timing for Frequency Sampling Filter

"next address" code, since it indicates from where the μprom address for the next instruction will originate. Exiting from the filter sequence may be done using a third variation of N_0, N_1. A looping routine may be set up whereby a sequence of ALU no operation (NOP) instructions are performed while the sequencer awaits the arrival of a new data sample clock. This may be a requirement for systems with fixed sample rates and relatively short program length. The following section defines the main functions of the sequencer with respect to the "next address" code.

4.1 MICROPROM AND CONTROL REGISTER

As implied in the previous sections, the microprom is divided into two main parts: ALU and control. A typical ALU section is illustrated in Figure 14. The following control signals which have not been mentioned so far require special attention:

Scratch read address (SRA)	2 bits
Scratch write address (SWA)	2 bits
Shift code (S)	n bits

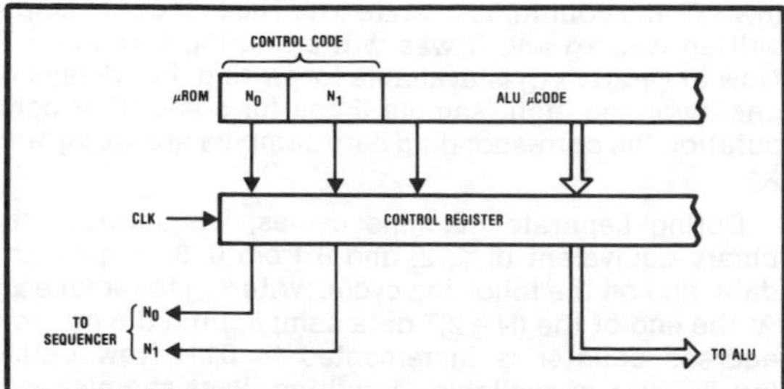

Figure 16. Microprom and Control Register Format

SRA and SWA specify locations in scratch memory. These locations may be addressed for canned routine executions in which they are defined exactly by μprom. On the other hand, many single cycle instructions require the use of scratch pad locations which should be user programmable at the macro instruction level. The designer may wish to instate some parameter bypassing logic such that when the sequencer is not executing a canned routine, the macro memory field which contains these items overrides the corresponding μprom outputs.

This implementation also pertains to the shift code bits which may be used for single cycle or canned arrangements.

The details, though not discussed, should become fairly obvious as the control portion of the microcode is defined in the following discussion.

4.2 SEQUENCER

The sequencer block diagram is shown in Figure 17. The multiplexer selects the source of the next address to the μprom. The direct input (D_1) provides starting addresses for canned routines or single cycle instruction addresses. The microprgram counter (μPC) generates the successive addresses for the canned routines as described earlier.

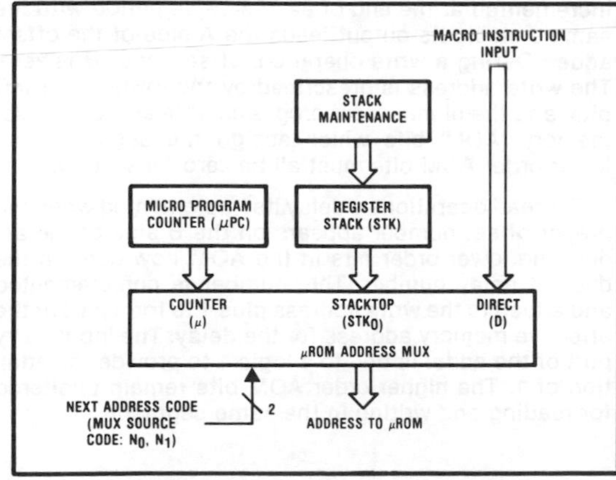

Figure 17. Sequencer Block Diagram

The μPC is configured so that it always generates the last multiplexer output ($Y_i + 1$, and Y_i is the μprom address). When a continuous train of single cycle microaddresses is channeled through the D_i input, the incrementer portion of the μPC instantaneously adds one to each address and on the next clock cycle stores this sum in the μPC register (which is considered as part of the μPC).

When the first microinstruction for a canned sequence is in the control register, the N_0, N_1 code selects the μPC as the next address source. For the frequency sampling filter this instruction is Sc1Mult, and its microcode resides in address 80. The μPC register reads 81 which is gated through the multiplexer to the μprom address bus. Meanwhile, the μPC has generated 82. On the next clock 82 is strobed into the μPC register, and MemMAC/STO2, which corresponds to address 81, is simultaneously transferred into the control register. Its N_0, N_1 code selects the μPC once again. Finally, when instruction MACACCO (address 85) is in the control register, the N_0, N_1 code selects D_i for the next μprom address. Address 86 in the μPC register is ignored.

Generally, microprocessors employ a register stack for address linkage when subroutines are programmed. For digital filter applications, the stack may be used to generate a timing loop for fixed sample period processors. For example, assume that a machine had a 32 μsec sample rate and a microcycle time (machine clock period) of ⅓ μsec. Three instructions per μsec and 96 microinstructions per sample period could be executed. For machine diagnostic tests or simple filter applications it may be desirable to run a very short program, say 20 microcycles worth. This would leave 76 available microcycles which would have to be programmed as NOPs.

A simple exit instruction could be loaded into macro RAM at the end of the main program. This instruction would be the starting address for a three or four cycle microsequence which generated NOPs to the ALU until the sample clock came along. At this point, the first microinstruction of the program would be gated into the sequencer and the program would begin again.

The following discussion describes a typical exit routine which utilizes the register stack. When the exit instruction macro code, say decimal 90, reaches the μprom input, the prom displays the ALU NOP microcode and an N_0, N_1 combination which commands a push operation, and the μPC as the next address. On the following clock cycle the exit microcode is in the control register. The N_0, N_1 push code selects the PC which reads 91, and loads this value plus 1 (92) in the stack register. On the next clock instruction 91 is dumped into the control register and number 92 is loaded into the top of the stack for later selection as a μprom address. Microinstruction 91 should contain an ALU NOP, and its control code should select the μPC which is now at 92. Instruction 92 also contains an ALU NOP but its N_0, N_1 select code should be conditional. That is, when the data sample clock enable (SE) is not present N_0, N_1 should select the stack (STK_0) as the address source; STK_0 contains 92. As long as the sample clock does not show up the sequencer continues to issue NOPs. When

SE is present and instruction 92 is in the instruction register, the N_0, N_1 code is reinterpreted by the sequencer multiplexer. The address source selected is the μPC which reads 93. This instruction may be used for resetting counters and general program initiation.

Present bit slice technology has provided a number of commercially available address controllers and sequencers which can be readily adapted to filter applications. The details for setting up the actual logic designs can be derived from available literature on these devices.

5. USING RANDOM ACCESS MEMORY AS TAPPED DELAY LINE

Discrete delays can be implemented with random access memory by storing data samples in contiguous locations in memory and then reading data from locations which are offset by a prescribed amount from the write locations.

This scheme is easily implemented with binary arithmetic if the number of memory locations is a power of 2 (i.e., 2^N). Figure 18 shows a sequence of data samples being stored in a RAM which is eight samples long.

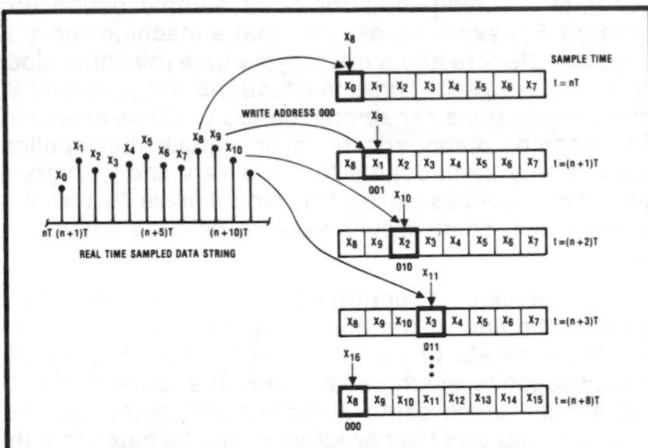

Figure 18. RAM Data Loading Sequence

At arbitrary t = nT the first data sample (x_0) is in location 000 and x_7 is in location 111. Before t = (N + 1)T, X_0 is replaced by X_8, by t = (N + 2)T x_1 is replaced by X_9. By t = (N + 8)T the original eight data samples x_0 through x_7 are erased and replaced by the following samples. This process goes on ad infinitum. Therefore, at the time that a data sample, x_n, is available for memory storage, its past eight values x_{n-2} ... x_{n-8} reside in memory. A delay of between one and eight sample times can be obtained by subtracting that number from the current write address. The write address is contained in the memory address counter which addresses memory and is incremented at the end of each sample time. Referring to Figure 18, at t = (n + 2)T, the counter is in state 010. The last data sample written was x_9 and it was written during t = (n + 1)T. Now at (+2)T, x_{10} is available for writing. For delays of one, two, and eight sample times for some filter computation the corresponding data samples are x_9, x_8, and x_2.

During separate machine cycles, we subtract the binary equivalent of 1, 2, and 8 from 010, acquire the data, and on the following cycle, write x_{10} to replace x_2. At the end of the (N + 2)T data sample time the memory address counter is incremented to 011. Now during (n + 3)T, x_{11} is available for writing. Past samples x_{10}, x_9, and x_3 now correspond to data delays of 1, 2, and 8 from 011. Table 2 lists the steps for the process described above. Two's complement subtraction is used to define the read address. The address is found by complementing the base 2 number representing delay time, adding one, and then adding the result to the present write address. For a delay time of 1, the binary equivalent for the 3 bit case (8 memory locations) is 001. The two's complement value for − 1 is 100 + 001 = 111.

5.1 MEMORY FIELD ALLOCATION FOR MULTIPLE FILTER ARRANGEMENTS (FENCING)

When implementing filters requiring many sections it is necessary to break the main memory down into segments. In the case of the frequency sampling filter which requires N/2 second order filters, each segment contains two words of memory. An additional N words of memory are also needed for the Nth order delay for the comb filter which precedes the second order filter bank.

Implementation of the delays is done in the same manner as in the previous example, except that the arithmetic involves only the necessary lower order bits of the address word. For second order filters, the least significant bit is involved. Overflows beyond the first bit are ignored. The higher order bits are fixed for read and write operations according to which memory segments are active.

Figure 19 is a block diagram of the main memory address controller. The memory address counter (MC) is incremented at the end of each sample period with the sample clock. Its output feeds the A side of the offset adder. During a write operation, offset input B is zero. The write address is prescribed by the lower order MC bits, and the higher order programmable address offset memory (AOM) bits which assign the segment. The lower order AOM bits must all be zero for writing.

For read operations a delay its implemented when the proper offset number appears on the B side of the adder. The lower order bits in the AOM now contain the desired delay number. This number is complemented and added to the write address plus 1 to form part of the effective memory address for the delay. The input carry port of the adder is tied to a logic 1 to provide the addition of 1. The higher order AOM bits remain unaltered for reading and writing in the same segment.

The and/or select collection is used to mask out higher order bits from the base 23 arithmetic address generation. A separate AOM field is programmed with mask number R. This number is decoded by the mask prom which disables the first R most significant bits out of the adder. When a zero is programmed, no address bits are masked and the main memory is not segmented. Thus, AO_1 through AO_n is the address offset word. A delay of N sample times can be attained by programming all zeros into the AOM, reading from main memory, and then writing into the same address (as shown in Section 5).

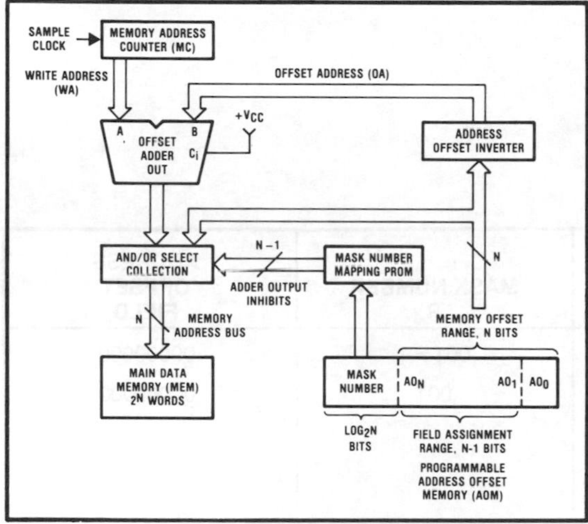

Figure 19. Address Memory Offset Generator Block Diagram

The offset portion of AOM is formatted to contain both field select data as well as memory offset data in complementing amounts in accordance with the mask number. The R most significant bits of main memory address are assigned via the and/or select logic to the R most significant bits of the AOM offset field.

When binary 1 (i.e., 0001, R = 1), is the programmed mask number, the MSB of the memory address bus is disabled from the adder output. AO_1 through AO_{n-1} constitute the address offset word. The memory is thus divided into two halves. Each is selectable according to the programmed state of AO_n.

When binary 2 (i.e., 0010) is programmed into the mask section of the AOM, the mask prom disables the two most significant memory address bits from the adder and assigns them to AO_n and AO_{n-1}. Now A_1 through A_{n-2} is the offset word. The memory is segmented into four fields, each with individual and independent dalay capabilities of 2^{n-2} data sample times. Each field is uniquely selected by the AO_n, AO_{n-1} code assignments (i.e., 00, 01, 10, 11).

Memory segments need not be of equal size. For instance, for the case of R = 2, the programmer is constrained to one field of one-fourth the total memory size. The second field may be also one-fourth size or subdivided again. The third may be one-half the total memory size (R = 1) if desired. In general, each segment may be treated as a separate memory and broken down into smaller segments in powers of one-half. The following example illustrates the flexibility of this scheme without rigorous development.

Cycle	Sample Time	Operation	Input Data Sample	Memory Write Address	Desired Delay	Read Address	Data Extracted
i	nT	Read	x_9	001	T(001)	001 + 111 000	x_8
i + 1		Read	x_9	001	2T(010)	001 + 110 111	x_7
i + 3		Read	x_9	001	8T(000)	001 + 000 001	x_1
i + 4		Write	x_9	001	0(000)	——	——
i	(n + 1)T	Read	x_{10}	010	T(001)	010 + 111 001	x_9
i + 1		Read	x_{10}	010	2T(010)	010 + 110 000	x_8
i + 2		Read	x_{10}	010	8T(000)	010 + 110 010	x_2
i + 3		Write	x_{10}	010	0(000)	——	——

Table 2. Memory Address Sequence for Delays of 1, 2, and 8 Sample Times

5.1.1 Frequency Sampling Filter

For a 128th order filter, 64 second order sections must be implemented. In addition, a 128th order comb filter is required. The canonical sections take 64 independent two word memories: the comb takes a 128 word memory. The net memory requirement is 256 words (2^8) divided as shown in Figure 20.

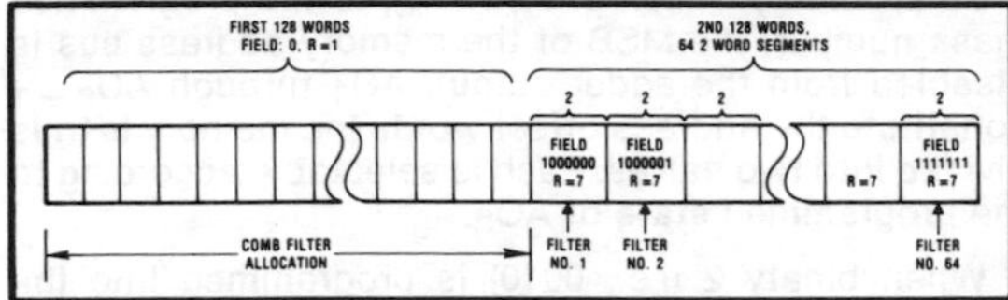

Figure 5-20. Memory Field Allocation for 128th Order Filter

During the comb implementation when a memory read is required, R is set to 1, and $AO_8 = 0$ (signifying address offset generation int he first half of memory) and AO_1 through AO_7 is the offset word. During the 64 canonical implementations R is set to 7, leaving the least significant memory address bit for performing the delays of one and two sample times for each filter. The seven most significant bits are used to assign the correct field for each second order filter. Table 3 shows the AOM states for the frequency sampling filter of Table 1 (Section 3.2).

INSTRUCTION	DATA DELAY	MASK NUMBER R	ADDRESS OFFSET FIELD
2 MemMult	$z-N$	001 = 1	00000000
3 XWM/XMAC	Write Address	001	00000000
4 MAC STO1			
5 C1Sc			
..................Filter Number 1			
6 Sc1Mult			
7 MemMAC	$z-2$	111 = 7	10000000
8 MemMAC	$z-1$	111	10000001
9 MWM/Sc2Mult	Write Address	111	10000000
10 MemMAC	Write Address	111	10000000
11 MACACCO			
..................Filter Number 2			
12 Sc1Mult			
13 MemMAC	$z-2$	111	10000010
14 MemMAC	$z-1$	111	10000011
15 MWM/Sc2Mult	Write Address	111	10000010
16 MemMAC	Write Address	111	10000010
17 MACACCO			
.			
.			
.			
..................Filter Number 31			
191 Sc1Mult			
192 MemMAC	$z-2$	111	30 10111100

Table 3. Memory Address Offset Sequence for the 128th Order Filter

6. CONCLUSIONS

The ALU architecture provides fast implementation for virtually any filter structure. Applications may arise where the flexibility and speed provided by this machine are not required. Such a condition may indicate a machine with less computing capability, but which contains considerably less hardware. For instance, elimination of the scratch pad memory results in considerable simplification in hardware and instruction set. If speed is not a problem, such a configuration will provide the same flexibility as the type 2 architecture.

For batch processing applications, other hardware changes are indicated. The designer would most likely include a buffer memory as the input source. He may also find once again that the services fo the scratch pad are not necessary, or that placement of the auxillary bus in another area provides a more efficient filter execution. Hence, architectural changes may be indicated for fixed applications.

As an example, filters which require a relatively small amount of memory may be implemented with the main memory read and write address buses separtated. Data may be simultaneously read and written into memory. This would give rise to a completely different bus structure. Large memories usually provide a common I/O or at least a single set of address lines, which prohibits this type of configuration.

inmos **Application note**

Digital filtering with the IMS A100

Contents

Digital filtering with the IMS A100

1 Introduction

When an analogue signal is sampled in time, the sampled signal is referred to as a discrete-time signal. If each sample in this discrete-time signal is also quantised in amplitude, (e.g. represented by an arbitrary n-bit number), then it is usually referred to as a digital signal. In the subject of digital filtering it is these types of signals which are processed and operated on. The fact that the digital signals are quantised both in time and amplitude gives one greater control over the processing as compared to analogue signal processing.

In these application notes the concept of the digital filtering is first introduced. This is done by starting from a simple RC analogue filter and deriving a corresponding digital filter. The classification of digital filters is then summarized, followed by giving a summary of techniques applicable to filter design using the IMS A100 device.

2 From analogue to digital

Figure 1a shows a simple first-order RC filter. The simple differential equation describing this circuit in terms of its input and output voltages is:

$$v_0(t) + RC\frac{dv_0(t)}{dt} = v_i(t) \tag{1}$$

where $v_0(t)$ and $v_i(t)$ are analogue output and input voltage waveforms. In the analogue world both input and output voltages are continuous-time waveforms and the complexity of the solution would depend on the input voltage function $v_i(t)$. Given an input waveform $v_i(t)$, the solution can be obtained using:

(i) Standard mathematical techniques which solve the differential equation and obtain the output waveform in closed form.

(ii) Numerical techniques which calculate the approximate output waveform in a digital computer. This would necessitate the sampling of the input and output waveforms.

The second method above provides the basis for digital filtering techniques. Consider that the input and the output voltages are sampled with a sampling interval T such that $v_i(nT)$ and $v_o(nT)$ represent the values of $v_i(t)$ and $v_o(t)$ at time $t = nT$.

If T is sufficiently small then the derivative $\frac{dv_o(t)}{dt}$ at time $t = NT$ can be approximated by:

$$\frac{dv_o(nT)}{dt} \simeq \frac{v_o(nT) - v_o((n-1)T)}{T} \tag{2}$$

substituting this in equation (1) we obtain:

$$v_o(nT) + \frac{RC}{T}v_o(nT) - \frac{RC}{T}v_o((n-1)T) = v_i(nT) \tag{3a}$$

Equation (3a) is a linear difference equation that approximates the differential equation (1). Equation (3a) can be rewritten as:

$$v_o(nT) = \frac{1}{1+(RC/T)}v_i(nT) + \frac{(RC/T)}{1+(RC/T)}v_o((n-1)T) \tag{3b}$$

This is now a recursion formula in which the present input sample and the previous output sample are used to calculate the present output sample. The notation can be simplified to:

$$v_o(n) = b_0 v_i(n) + a_1 v_o(n-1) \tag{4a}$$

where $b_0 = \frac{1}{1+(RC/T)}$ and $a_1 = \frac{(RC/T)}{1+(RC/T)}$.

The signal-flow diagram for this filter is shown in figure 1b. The block labelled 'D' represents a delay equal to one sampling period T. In digital filter notations a delay of n sampling periods is usually denoted by z^{-n}. Therefore a delay of one sampling period can be represented by z^{-1}.

It is important to note that a common element in all filter structures is the concept of storage. In the analogue RC filter (figure 1a) the storage is present in the form of a capacitor and in its digital equivalent (figure 1b) the storage takes the form of a delay stage. In fact the storage element is the essential ingredient for any filter whether analogue or a digital. This is because filters are used to operate on the signal 'changes' and as such they need to have some knowledge of the history of the signal to allow them to perform their function.

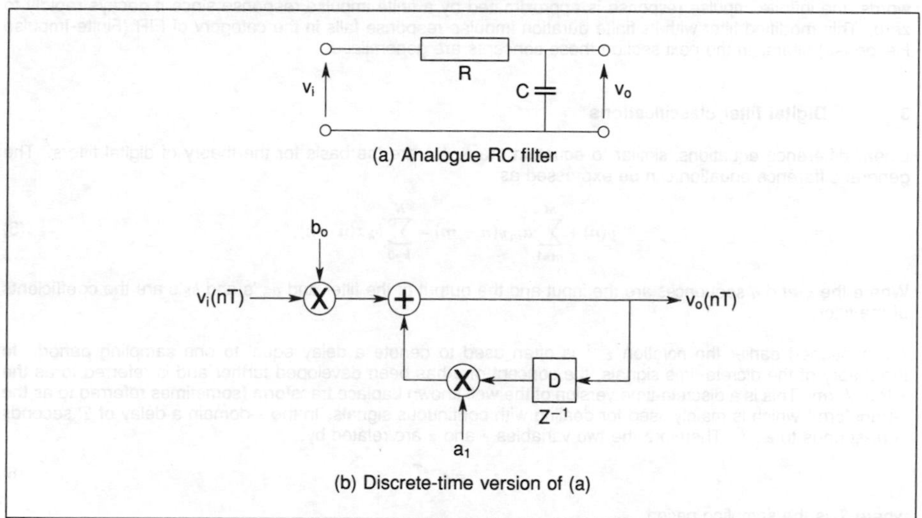

Figure 1 Analogue RC filter and its discrete-time equivalent

An important characteristic feature of any filters is its so called 'impulse response'. This is defined as the output waveform of the filter when a unity impulse is applied to the input. Using equation (4a) and assuming a unity impulse as the input waveform i.e.

$$\begin{aligned} v_i(0) &= 1 \\ v_i(n) &= 0 \quad for \quad n > 0 \end{aligned}$$

then the output sequence would be:

$$b_0,\ a_1 b_0,\ a_1^2 b_0, \ldots\ldots,\ a_1^n b_0, \ldots\ldots\ldots$$

or in short

$$v_0(n) = a_1^n b_0$$

It should be noted that the above impulse response has, in theory, infinite length. This is due to the recursive nature of this particular filter structure. This types of filters are often referred to as infinite-impulse-response (IIR) filters.

An alternative way of looking at the filter in this example is to use equation (4a) in successive substitutions i.e.

$$\begin{aligned} v_o(n) &= b_0 v_i(n) + a_1 v_o(n-1) \\ &= b_0 v_i(n) + a_1[b_0 v_i(n-1) + a_1 v_o(n-2)] \\ &= b_0 v_i(n) + a_1 b_0 v_i(n-1) + a_1^2[b_0 v_i(n-2) + a_1 v_o(n-3)] \\ &= \ldots\ldots \\ &= \ldots\ldots \\ &= b_0 v_i(n) + a_1 b_0 v_i(n-1) + a_1^2 b_0 v_i(n-2) + a_1^3 b_0 v_i(n-3) + \ldots\ldots \end{aligned} \tag{4b}$$

Equation (4b) expresses the output waveform as a linear combination of input samples only, but this involves infinite number of input samples. Notice also that the coefficients b_0 and a_1 have positive values less than unity (R and C are assumed to be finite and non-zero). This means that in equation (4b) the coefficients decrease for older input samples. It may therefore be reasonable to assume that these coefficients approximate to zero beyond a certain point. In this way only a finite number of terms would be involved in equation (4b), or in other words, the infinite impulse response is approximated by a finite impulse response since it decays rapidly to zero. This modified filter with its finite duration impulse response falls in the category of FIR (Finite-Impulse Response) filters. In the next section these concepts are generalized.

3 Digital filter classifications

Linear difference equations, similar to equation (4a & 4b) are the basis for the theory of digital filters. The general difference equation can be expressed as:

$$y(n) + \sum_{m=1}^{M} a_m y(n-m) = \sum_{k=0}^{N} b_k x(n-k) \qquad (5)$$

Where the x and y sequences are the input and the output of the filter and a_m's and b_k's are the coefficients of the filter.

As mentioned earlier the notation z^{-1} is often used to denote a delay equal to one sampling period. In the theory of the dicrete-time signals, the concept of z has been developed further and is referred to as the z-transform. This is a discrete-time version of the well known Laplace transform (sometimes referred to as the s-transform) which is mainly used for dealing with continuous signals. In the s-domain a delay of T seconds corresponds to e^{-sT}. Therefore the two variables s and z are related by:

$$z^{-1} = e^{-sT} \qquad (6)$$

where T is the sampling period.

In the s-domain the spectrum of a signal with a bandwidth B and sampled at a frequency f_s, is periodic with a period equal to f_s. This is depicted in figure 2. This periodicity in the spectrum of a sampled signal is the basic reason behind the Nyquist criterion which requires a minimum sampling frequency of twice the signal bandwidth (i.e. $f_{s_{min}} = 2 \times B$), in order to avoid aliasing effects.

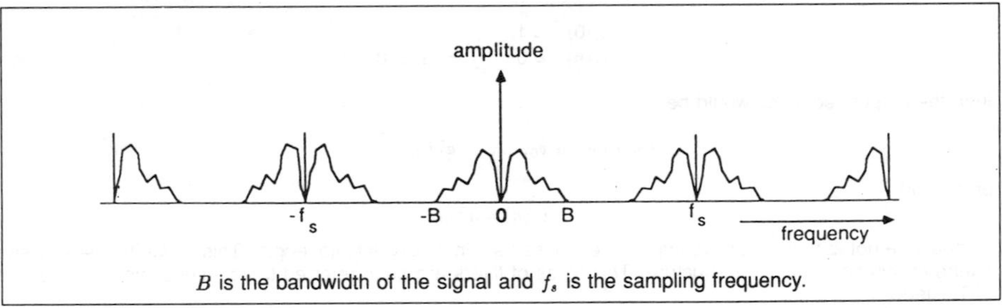

B is the bandwidth of the signal and f_s is the sampling frequency.

Figure 2 Spectrum of a sampled signal

Equation (6) allows a mapping between the two domains. Part of the imaginary axis between $-\frac{f_s}{2}$ to $+\frac{f_s}{2}$, in the s-plane, is mapped into a unit circle in the z-domain as shown in figure 3. The fact that the imaginary axis in the s-plane is mapped onto a circle is a consequence of the periodic nature of the spectrum. As shown in figure 3, the left-hand half of the s-plane (between $-\frac{f_s}{2}$ and $+\frac{f_s}{2}$) is mapped onto the inside of the unit circle, while the right-hand half is mapped onto the outside of the circle.

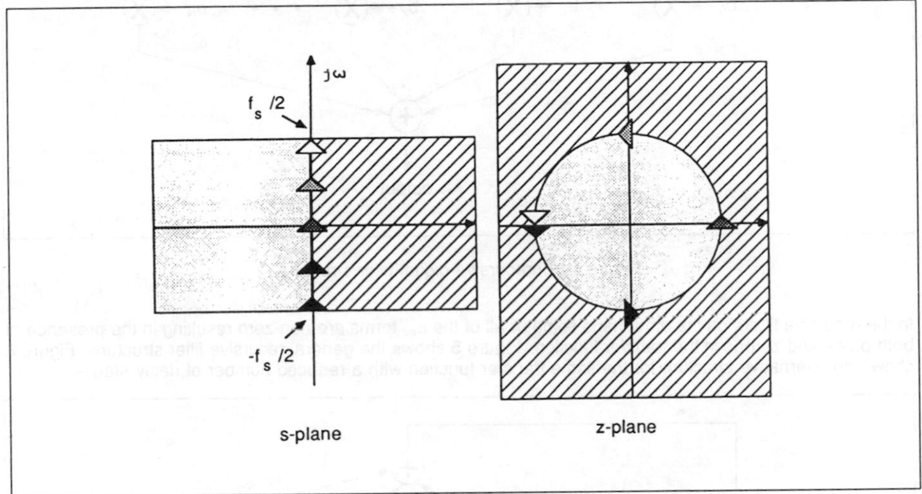

Figure 3 Relationship between the s-domain and the z-domain

As in the analogue design (s-domain) where a pole in the wrong place, i.e. in the right-half plane, indicates instability, in the case of discrete-time signals (z-domain) a pole outside the unit circle causes instabilities. In both cases zeroes can be anywhere.

Using the z-transform notation, the general linear equation (5) can be expressed as:

$$Y(z)(1 + \sum_{m=1}^{M} a_m z^{-m}) = X(z) \sum_{k=o}^{N} b_k z^{-k} \quad (7)$$

Where $X(z)$ and $Y(z)$ are the z-transforms of the input and output waveforms. The discrete-time (or digital) transfer function of the general filter is thus given by:

$$H(z) = \frac{Y(z)}{X(z)} = \frac{\sum_{k=0}^{N} b_k z^{-k}}{1 + \sum_{m=1}^{M} a_m z^{-m}} \quad (8)$$

In terms of realization, digital filters are classified into nonrecursive and recursive types. The nonrecursive structure contains only feed-forward paths and as such all the a_m terms (equation (8)) are zero. This means that for the nonrecursive filters the output is a sum of linearly weighted present and a number of past samples of the input signal as shown in figure 4. Referring to equation (8), for the nonrecursive filters the transfer function has only zeroes and as such is always stable.

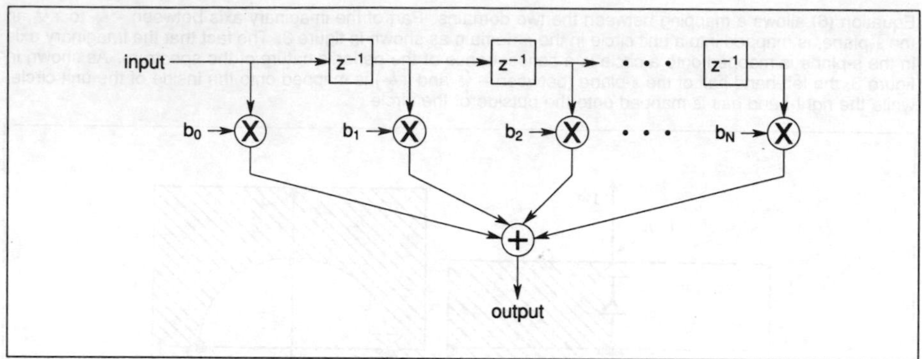

Figure 4 Nonrecursive digital filter structure

In the recursive filters on the other hand some or all of the a_m terms are non-zero resulting in the presence of both poles and zeroes in the transfer function. Figure 5 shows the general recursive filter structure. Figure 6 shows an alternative structure for the same transfer function with a reduced number of delay stages.

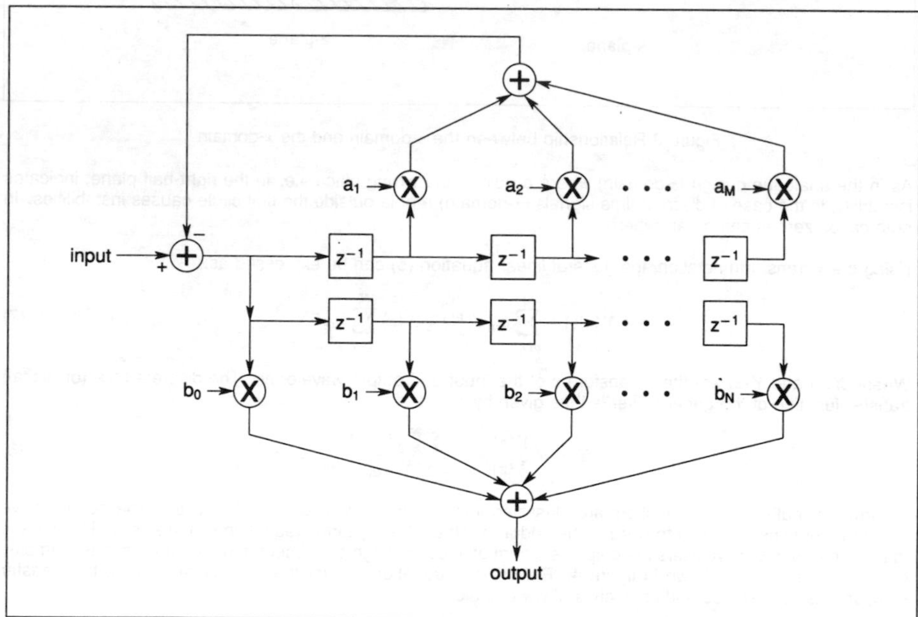

Figure 5 Recursive (IIR) digital filter structure

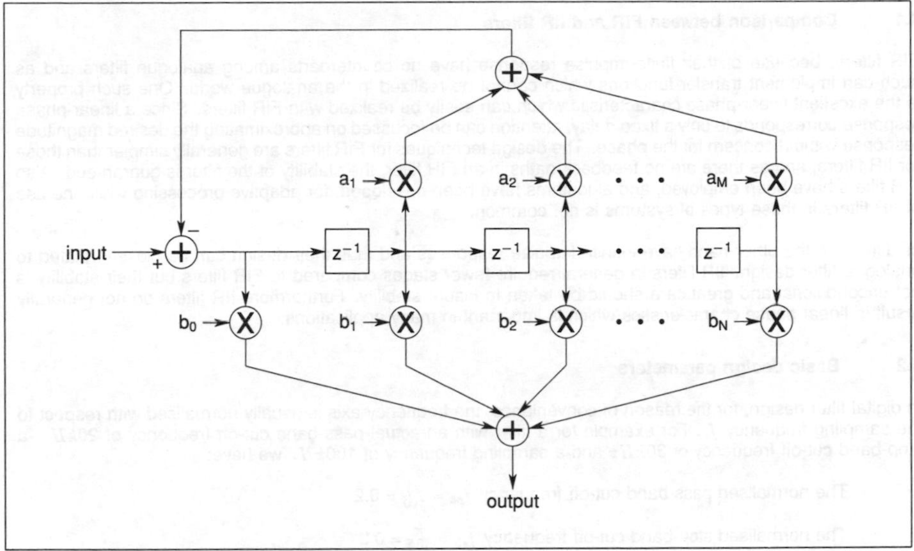

Figure 6 Alternative recursive (IIR) digital filter structure with reduced number of delay stages

Digital filters are also classified in terms of their impulse responses. In this classification those filters with a finite duration impulse response are referred to as FIR filters and those with an infinite duration impulse response are called IIR filters. The simplest FIR filter realization is in the nonrecursive form. For example in figure 4, if a unit impulse is clocked through the filter, the sequence,

$$b_0, b_1, b_2, ...b_N, 0, 0, 0, 0, 0, ...0, 0, 0 \tag{9}$$

will be output. Notice that the response consists of a sequence of samples corresponding to the filter coefficients followed by zeroes, i.e. the nonrecursive structure is an FIR filter. On the other hand the impulse response of the recursive structure (figures 5 & 6), because of the feedback paths, is infinite in duration, making the configuration an IIR filter.

4 Digital filter design

Digital filter design methods can be divided into two categories:

(a) Design techniques suitable for FIR filters.

(b) Design techniques suitable for IIR filters.

In both cases the requirement is simply the choice of filter coefficients in such a way that the specification for the required transfer function is met. The IMS A100 can be used to implement high performance FIR filters directly. It can also be used to implement IIR filters, although the general problems associated with IIR filter design are then introduced. In this section a brief comparison between FIR and IIR filters is given and some of their associated design techniques are summarized. Where necessary the IMS A100 implementation issues are also discussed.

4.1 Comparison between FIR and IIR filters

FIR filters, because of their finite-impulse response have no counterparts among analogue filters and as such can implement transfer functions which cannot be realized in the analogue world. One such property is the excellent linear-phase characteristic which can easily be realized with FIR filters. Since a linear-phase response corresponds to only a fixed delay, attention can be focussed on approximating the desired magnitude response without concern for the phase. The design techniques for FIR filters are generally simpler than those for IIR filters, and as there are no feedback paths in an FIR filter, the stability of the filter is guaranteed. Also FIR filters have been employed, and algorithms have been developed, for adaptive processing while the use of IIR filters in these types of systems is not common.

IIR filters on the other hand have infinite impulse responses and thus their design can be closely related to analogue filter design. IIR filters in general require fewer stages compared to FIR filters but their stability is not unconditional and great care should be taken to insure stability. Furthermore IIR filters do not generally result in linear-phase characteristics which is important in many applications.

4.2 Basic design parameters

In digital filter design, for the reason of convenience, the frequency axis is usually normalised with respect to the sampling frequency f_s. For example for a filter with an actual pass-band cut-off frequency of $20kHz$, a stop-band cut-off frequency of $30kHz$ and a sampling frequency of $100kHz$ we have:

The normalised pass-band cut-off frequency $f_{pb} = \frac{20}{100} = 0.2$

The normalised stop-band cut-off frequency $f_{sb} = \frac{30}{100} = 0.3$

As shown in figure 7 the useful frequency axis (normalised) extends from 0.0 to 0.5, because the Nyquist sampling theorem requires a signal to be sampled at more than twice its highest frequency. This means that the ratio of the frequency of any component in the signal to the sampling frequency must always be less than 0.5.

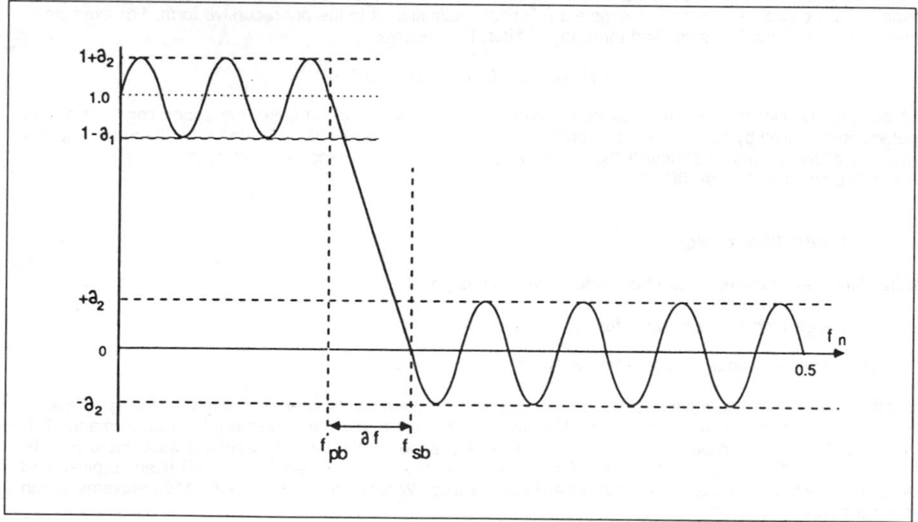

Figure 7 Specification parameters for a low pass filter.
Similar parameters exist for high pass and band pass filters.

Referring to figure 7, the pass-band and the stop-band ripples are usually expressed in dBs i.e:

pass-band ripple (dB)= $20 \log_{10}(1 + \delta_1)$

stop-band ripple (dB)= $-20 \log_{10}(\delta_2)$.

The parameters f_{pb}, f_{sb}, δ_1, δ_2 and the sampling frequency define the basic specification of a filter prior to its design.

4.3 Design techniques suitable for FIR filters

As mentioned earlier one of the major advantages of FIR filters is the ease with which linear-phase behaviour can be obtained from these types of filters. Before summarizing the design techniques for FIR filters let us briefly consider the necessary conditions for linear-phase behaviour. It can readily be shown that in order to obtain an FIR filter with a linear-phase characteristic, the following condition has to be met (references 1 & 2):

$$\begin{aligned} h(i) &= \pm h(N-i) \quad for \quad 0 \le i \le N \\ &= 0 \quad otherwise \end{aligned} \tag{10}$$

This condition requires that the the impulse response of the FIR filter, $h(i)$, to have either positive or negative symmetry.

In the case of positive symmetry the frequency response will be of the form

$$H(e^{j\omega T}) = A(\omega T)e^{-j\omega TN/2} \tag{11a}$$

where $A(\omega T)$ is a real function of ω. Notice that the phase is a linear function of frequency. These types of filters are appropriate for frequency selective filters.

In the case of negative symmetry the filter transfer function will have the following form:

$$H(e^{j\omega T}) = jB(\omega T)e^{-j\omega TN/2} \tag{11b}$$

Again $B(\omega T)$ is a real function of ω. Note that the phase is again linear with frequency, but we also have a j term which indicates an extra phase shift of $\frac{\pi}{2}$. These types of frequency responses are required to realise approximate differentiators and Hilbert transforms which implement a $\frac{\pi}{2}$ phase shift over a specified frequency range.

There are essentially three well-established classes of design methods for (linear phase) FIR filters which are:

(i) window method

(ii) frequency sampling

(iii) optimal design (Remez Exchange Algorithm)

Each one of these techniques has its own merits and the choice of which would depend on the application requirements and the design time involved.

Window method

This is the most straight-forward approach to the design of FIR filters. In this method having defined an ideal frequency-response function, the corresponding ideal impulse response is determined by evaluating the inverse Fourier transform of the ideal frequency response. In the selection of the ideal frequency response, the linear phase condition may or may not be applied depending on the application.

As mentioned earlier because digital filters deal with signals sampled at a frequency f_s,it therefore follows that this frequency response is periodic in frequency with a period equal to f_s (Nyquist theorm). It is therefore possible to relate the impulse response and the frequency response of a digital filter via the following Fourier pairs:

$$H(\omega) = \sum_{n=-\infty}^{+\infty} h(n)e^{-jn\omega T} \quad (12)$$

$$h(n) = \frac{1}{\omega_s}\int_{\frac{-\omega_s}{2}}^{\frac{+\omega_s}{2}} H(\omega)e^{jn\omega T}\,d\omega \quad (13)$$

where ω_s, is the sampling frequency in radians/s and T is the sampling period. Having defined an ideal frequency response, $H(\omega)$, equation (13) can be used to obtain the impulse response, $h(n)$, of the filter. As an example consider the ideal low-pass frequency response characteristics with a cut-off frequency ω_c as shown in figure 8a. Using equation (13), and equating $H(\omega)$ to 1.0 for $-\omega_c \leq \omega \leq +\omega_c$ and to zero elsewhere, we can calculate the impulse response $h(n)$ which is given by:

$$h(n) = \frac{\omega_c T}{\pi}\frac{\sin(n\omega_c T)}{(n\omega_c T)} \quad (14)$$

where $-\infty < n < +\infty$. This impulse response is shown in figure 8b. There are two problems associated with this impulse response obtained in this way:

(i) The filter impulse response is infinite in duration and as such an FIR filter of infinite length is required (remember as discussed earlier for FIR filter the impulse response sample values are effectively the filter coefficients).

(ii) The filter is unrealizable since the impulse response begins at $-\infty$, indicating that no finite amount of delay can make the impulse response realizable.

One way to obtain an FIR filter which approximates the required frequency response is to truncate the infinite impulse response at $n = \pm\frac{N}{2}$, (see figure 8c), and shift the impulse response to the right to avoid negative time (figure 8d). This would result in a realizable FIR filter with $N + 1$ coefficients which are equal to the impulse response samples.

The problem with this direct truncation of the impulse response is that it results in a fixed amount of overshoot (approximately 9%) before and after the discontinuity in the frequency response. In the literature this problem is referred to as the Gibbs phenomenon. For this reason, direct truncation is not often a reasonable way of designing FIR filters.

The frequency response of a truncated time series can be improved considerably by using a window function, $w(n)$, which modifies the impulse response to $w(n) \times h(n)$. In the previous example the window was simply a rectangular window. Figure 9 shows the application of a different window function to the example of the ideal low-pass filter. Figure 9a shows the ideal infinite duration impulse response. Figure 9b shows the window function and figure 9c shows the impulse response after the application of the window function. Figure 9d shows the shifted impulse response which avoids unrealizable negative delays. The filter coefficients (b_k's) correspond to the sample values of this modified impulse response which is now finite and realizable. Several window functions have been suggested in the literature some of which are:

(i) Hamming window

(ii) Hanning window

(iii) Kaiser window

(iv) Dolph-Chebyshev window

(v) Blackman window

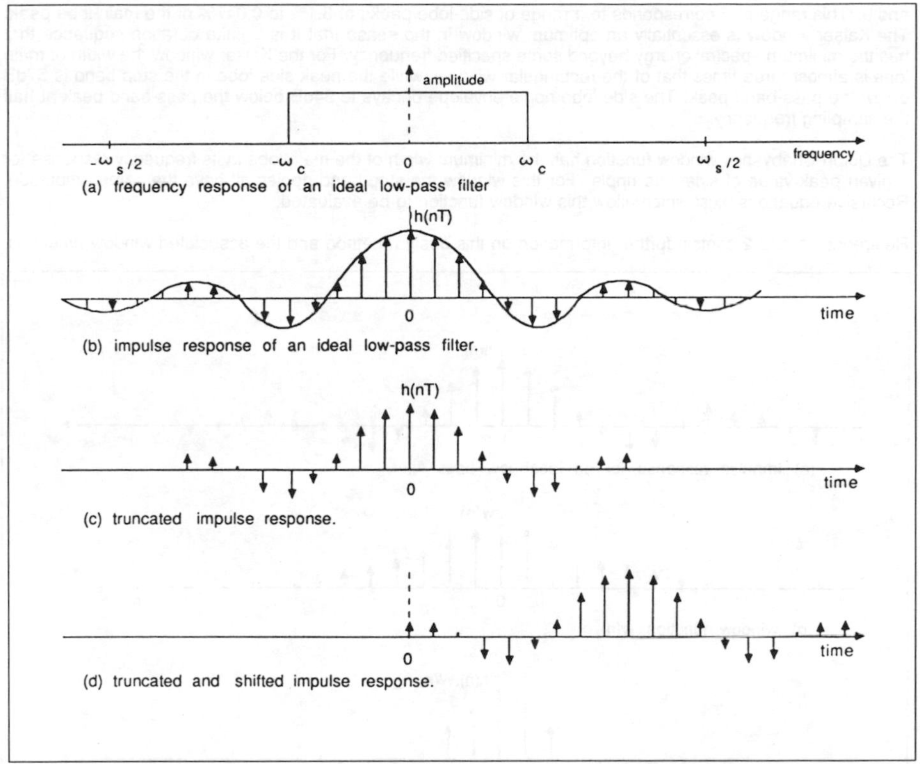

(a) frequency response of an ideal low-pass filter
(b) impulse response of an ideal low-pass filter.
(c) truncated impulse response.
(d) truncated and shifted impulse response.

Figure 8

The generalized Hamming window function is given by:

$$\begin{aligned} w_H(n) &= \alpha + (1-\alpha)\cos(\tfrac{2\pi n}{N}) \quad for \quad -(\tfrac{N-1}{2}) \leq n \leq -(\tfrac{N-1}{2}) \\ &= 0 \quad otherwise \end{aligned} \tag{15}$$

where $0 \leq \alpha \leq 1$. If $\alpha = 0.54$ the window is called a Hamming window, and if $\alpha = 0.50$ it is called a Hanning window.

For the Hamming window the main lobe of the frequency response is twice the width of that of the simple rectangular window. The amplitudes of the ripples of the Hamming window frequency response are considerably smaller than those of the rectangular window. For the rectangular window the peak side lobe (in the stop band) is only 14dB below the main-lobe (pass-band) peak. For the Hamming window the peak side lobe ripple is about 40dB below the pass band peak. Furthermore for the Hamming window 99.96% of the spectral energy is in the main-lobe peak.

Another family of windows are those proposed by Kaiser:

$$\begin{aligned} W_K(n) &= \frac{I_0(\beta\sqrt{1-[2n(N-1)]^2})}{I_0(\beta)} \quad for \quad -(\tfrac{N-1}{2}) \leq n \leq -(\tfrac{N-1}{2}) \\ &= 0 \quad otherwise \end{aligned} \tag{16}$$

Where I_0 is the modified Bessel function of the first kind. The parameters β is used to specify the main-lobe width and the side-lobe level of the frequency response. β is usually specified to have a value between 4 and 9. This range of β corresponds to a range of side-lobe peaks of 3.1% to 0.047% of the main-lobe peak. The Kaiser window is essentially an optimum window in the sense that it is a finite duration sequence that has the minimum spectral energy beyond some specified frequency. For the Kaiser window the width of main lobe is almost three times that of the rectangular window, while the peak side lobe in the stop band is 57dB below the pass-band peak. The side-lobe ripple envelope decays to 94dB below the pass-band peak at half the sampling frequency.

The Dolph-Chebyshev window function has the minimum width of the main lobe in its frequency response for a given peak value of side-lobe ripple. For this window the stop-band ripples all have the same amptitude. Recursive equations exist which allow this window function to be evaluated.

References 1 and 2 contain further information on this design method and the associated window functions.

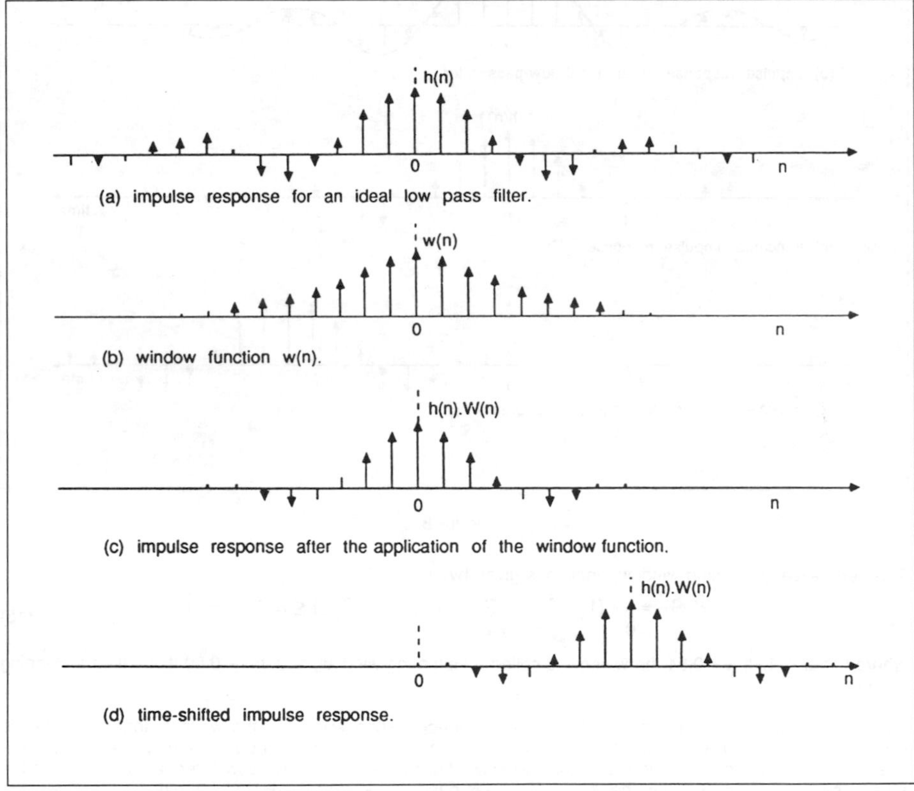
h(n)
0
n
(a) impulse response for an ideal low pass filter.
w(n)
0
n
(b) window function w(n).
h(n).W(n)
0
n
(c) impulse response after the application of the window function.
h(n).W(n)
0
n
(d) time-shifted impulse response.

Figure 9

Frequency sampling technique

This technique is less common than the other two design methods, however for the sake of completeness it is briefly mentioned here.

The basic idea behind this technique is that the given (desired) frequency response is approximated by sampling it at N equally-spaced points along the frequency axis between 0 and f_s (corresponding to N

samples on the unit circle in the z-plane). An N-point inverse DFT is then performed on these N frequency samples to give N samples of the impulse response $h(n)$ which corresponds to the filter coefficients. The z-transform of the filter impulse response is then given by

$$H(z) = \sum_{k=0}^{N-1} h(n)z^{-n}$$

Substituting $e^{j\omega T}$ for z, the resulting frequency response of the filter may be evaluated which would be an approximation of the desired frequency response. The approximation error would be exactly zero at points where the desired frequency response was sampled and would be finite between them. This process is depicted in figure 10.

To reduce these approximation errors a number of frequency samples (particularly those in the transition band between band-pass and band-stop regions, i.e. points T_1, T_2, T_3 and T_4 in figure 10) can be made unconstrained variables. The values of these unconstrained variables are then optimised using computer optimisation techniques involving linear-programming methods. This involves the solution of a set of linear unequalities in the unconstrained frequency samples. In this way, by adjusting the frequency sample values at T_1, T_2, T_3 and T_4, considerable ripple cancellation, both in the pass-band and stop-band, can be achieved resulting in very good filter characteristics. The detail of these techniques are beyond the objectives of this application note, however interested readers can refer to reference 1 for further information.

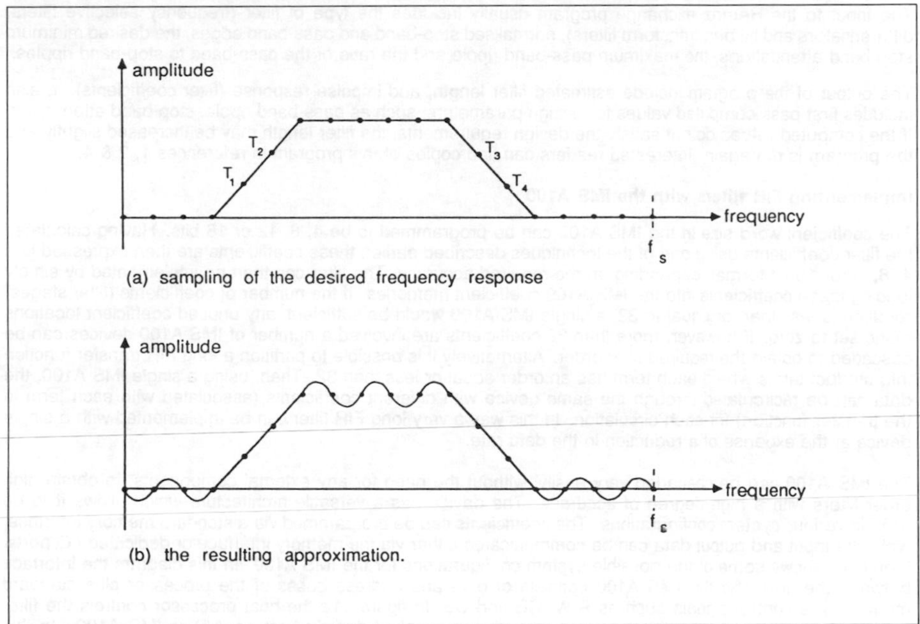

Figure 10

Optimal filter design – (Remez exchange algorithm)

In the frequency sampling technique, discussed in the previous section, some degree of improvement in the filter characteristics is obtained by allowing only a few of the frequency samples to be adjusted via a linear-programming technique.

An even more powerful technique which results in truly optimal filters, in the sense of having the sharpest transition between pass bands and stop bands (for a given filter length and a given approximation error) has been formulated based on the so-called Chebyshev approximations. Computer optimisation techniques based on linear programming have been developed (references 3, 4, 5 & 6) which allowed engineers to design optimal FIR filters with a minimum amount of knowledge about the actual optimisation algorithm. These iterative algorithms are based upon the principles of the Remez exchange algorithm. This algorithm yields optimal filters that satisfy the so-called minimax error criterion (reference 1), where for a given number of coefficients, the filter minimizes the maximum ripple amplitude in the pass band. The implications of this optimal design are:

(a) The Remez exchange algorithm results in an FIR filter with the smallest number of coefficients satisfying the required specification.

(b) The pass-band ripple components all have the same magnitude and need not be equal to the stop-band ripples, but their ratio must be specified.

The input to the Remez exchange program usually includes the type of filter (frequency selective filters, differentiators and Hilbert transform filters), normalised stop-band and pass-band edges, the desired minimum stop-band attenuations, the maximum pass-band ripple and the ratio of the pass-band to stop-band ripples.

The output of the program include estimated filter length, and impulse response (filter coefficients). It also includes first pass computed values for design parameters, such as pass-band ripple, stop-band attenuation. If the computed values do not satisfy the design requirements, the filter length may be increased slightly and the program is run again. Interested readers can find copies of this program in references 1, 2 & 4.

Implementing FIR filters with the IMS A100

The coefficient word size in the IMS A100 can be programmed to be 4, 8, 12 or 16 bits. Having calculated the filter coefficients using one of the techniques described earlier, these coefficients are then expressed in a 4, 8, 12 or 16-bit format, depending on the required accuracy. The filter can then be implemented by simply loading these coefficients into the IMS A100 coefficient memories. If the number of coefficients (filter stages) required is less than or equal to 32, a single IMS A100 would be sufficient, any unused coefficient locations being set to zero. If however, more than 32 coefficients are involved a number of IMS A100 devices can be cascaded to obtain the required filter order. Alternatively it is possible to partition a long FIR transfer function into product terms where each term has an order equal or less than 32. Then, using a single IMS A100, the data can be recirculated through the same device with different coefficients (associated with each term in the transfer function) for each circulation. In this way a very long FIR filter can be implemented with a single device at the expense of a reduction in the data rate.

The IMS A100 can be cascaded very easily, without the need for any external components, to obtain high order filters with a high degree of accuracy. The device has a versatile architecture which allows it to be used in various system configurations. The coefficients can be programmed via a standard memory interface, while the input and output data can be communicated either via the memory interface or dedicated I/O ports. Figure 11 shows some of the possible system configurations for the IMS A100. In this diagram the interface between the host and the IMS A100 consists of data and address buses of the processor plus standard memory-type control signals such as R/W, CE and CS. In figure 11a the host processor controls the filter coefficients, while the actual data to be processed is supplied directly from an A/D to IMS A100. In this example the filtered output is fed directly to a D/A. Using the IMS A100 and a host processor it is possible to supply the input data to the device and also to collect the filtered samples via the memory interface. This allows system configuration such as those shown in figures 11b&c. In figure 11b the host processor receives the input data from a peripheral such as an A/D and writes it (may be after some preprocessing) into the data-input register (DIR) of the IMS A100. The filtered output sample is also collected by the host via the memory interface and output (possibly after post processing) to a peripheral such as a D/A. Figure 11c shows a configuration where the IMS A100 is used purely as a signal processing accelerator to the host. Numerous other configurations are possible including integrating an IMS A100 into existing microprogrammed systems in order to improve the overall system performance.

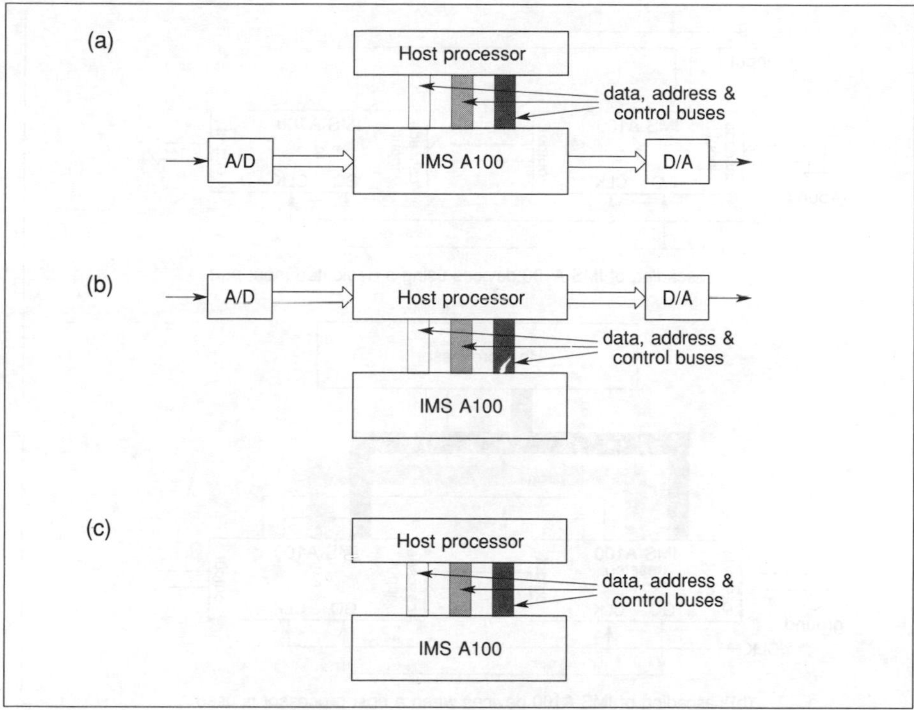

Figure 11 Possible system configurations using the IMS A100 in digital filtering applications

As mentioned earlier large numbers of the IMS A100 devices can be cascaded to construct FIR filters of a high order. The cascading does not involve any external components and is simply a matter of connecting the output of the previous device to the cascade input of the next chip and joining the data input ports together (if they are being used rather than the memory interface). In normal operation the cascade input of the first device should be grounded. Figure 12a shows this cascading arrangement for two IMS A100 devices and figure 12b depicts the block diagram of a system consisting of a host processor and two cascaded devices. In the latter case the data-input register (DIR) of both devices should be associated with the same address in the host's address space; and one of the devices should be selected as a master to generate the GO signal (see product data sheet for further detail).

Another important feature of the IMS A100 is a selector that is incorporated after the multiply-accumulator array. As discussed in the data sheet, the 32 multiply and accumulation in the array are performed to a precision of 36 bits which ensures that no intermediate overflows occur. The output selector can then be used to select and round a 24-bit word from this 36 bit result. This selection and rounding can be programmed to start from bits 7, 11, 15 or 20 and the selected word is sign extended if needed. One particularly useful selection is available when the input data and coefficients are in the form of 16 bit two's complement numbers normalised to between +1 and −1. In this case, if the selection is taken to start from bit 15, the output will have the same format as the input data (i.e. normalised to between +1 and −1).

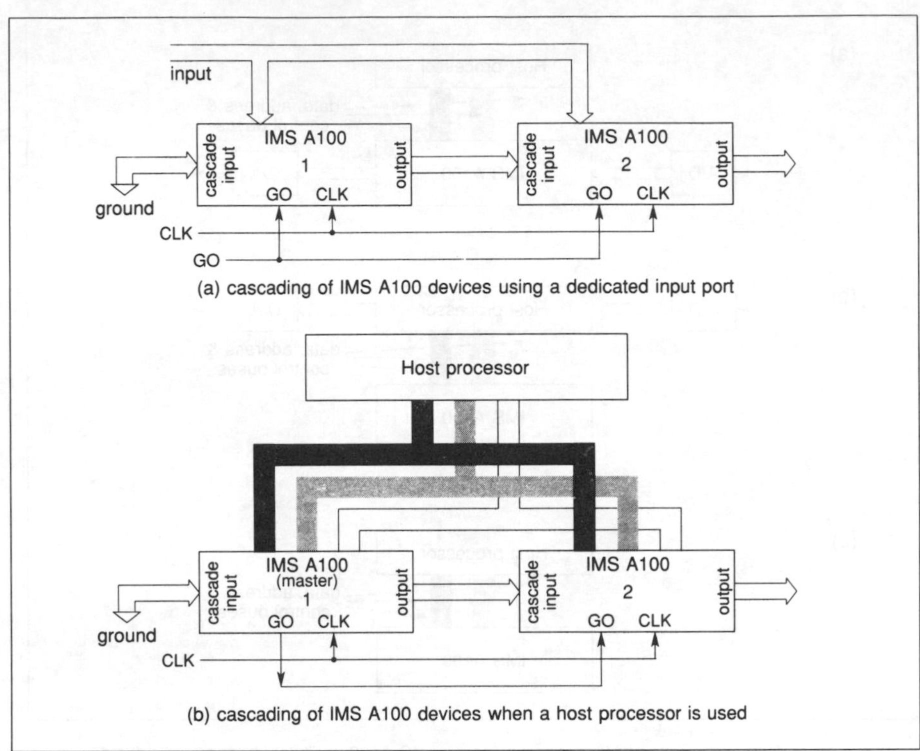

Figure 12 Cascading IMS A100 devices

4.4 The IMS A100 and IIR filters

Although the IMS A100 is designed primarily for FIR type filter implementations, it can also be used in realizing IIR filters. Referring to figure 5 it can be seen that two IMS A100 devices can be used to implement an IIR filter of order 32 or less in the direct form. One chip performing the calculation in the feed-forward path while the other does the feed-back path. Note that in figure 5 the output of the feed-back filter has to be combined with the input sequence in a subtractor and fed into the input of the second chip. This subtraction can be performed either by the host processor controlling the two IMS A100s or by an external adder.

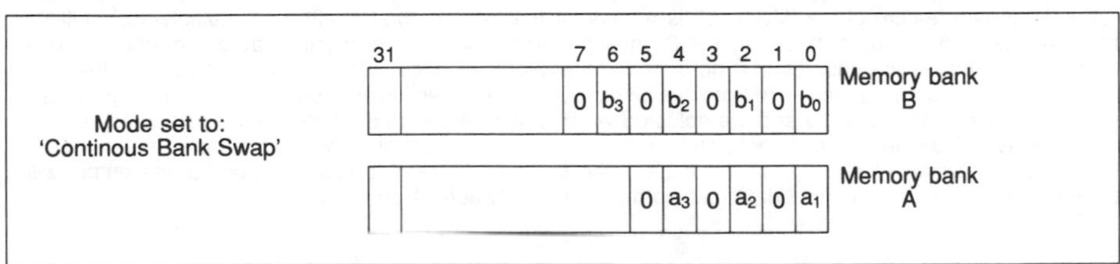

Figure 13 Coefficient memory allocation for IIR filter implementation

A simpler and more elegant technique to implement IIR filters using IMS A100 is to make use of the continuous bank swap feature on the IMS A100 coefficient memories. This allows a single IMS A100 to be sufficient for the implementation of IIR filters whose order is less than or equal to 16. (Before describing how this can be achieved it is worth noting that IIR filters generally require considerably fewer stages than their FIR counterparts, and as such a 16th order IIR filter implementable on a single IMS A100 can be considered as having quite a high order). Figure 13 shows the coefficient memory allocations in this approach, where a's and b's are the feedback and feedforward coefficients of the IIR filter respectively (see figures 5 & 6) and are loaded by the host processor. Note that in figure 13 alternate coefficients are set to zero in the two memory banks. The chip is also set to the continuous bank swap mode so that in one cycle the feedback coefficients (a's) and in the next cycle the feedforward coefficients (b's) are used in the calculation. It will be shown in the following paragraphs that if the difference between data samples and alternate output samples are written to the data input register of the IMS A100, then the remaining output samples would correspond to the correct filter output. The sequence of operations is as follows:

The host starts the filter operation by writing the first data value, x_0, to the data input register of the IMS A100. Remembering that the coefficient allocation is as shown in figure 13, the first output of the device would be a_1x_0. Referring to figure 6, it can readily be seen that this is indeed the feed back contribution needed to be subtracted from the next data sample x_1. The host reads this value (a_1x_0) from the data output registers (DOH and/or DOL) and stores it and then writes x_0, for a second time, to the IMS A100 input. This time the coefficient memory banks would have been swapped and the output would correspond to b_0x_0 which can readily be confirmed to be the first correct filter output (see figure 6). The host then reads this result as the first valid sample of the filtered output.

Next the host subtracts the feedback factor, read in earlier (a_0x_0), from the second data sample x_1, and writes the difference to the input register of the IMS A100. Remembering that the memory banks are automatically swapped every cycle, the corresponding ouput of the IMS A100 will be:

$$a_2x_0 + a_1(x_1 - a_1x_0)$$

Referring to figure 6 you should be able to confirm that this value corresponds to the feedback contribution needed for the third input sample. The host reads this value and stores it and as before writes the input value $(x_1 - a_1x_0)$ to the IMS A100 input register for a second time. This will yield the second valid filtered sample i.e:

$$b_1x_0 + b_0(x_1 - a_1x_0) \tag{17}$$

The process is then continued in the same manner. The output of the IMS A100 will alternate between the feedback contribution and the filtered output samples. It should be emphasized that although the host is performing a single subtraction for every output value, it is the IMS A100 device which is performing the bulk of the processing. Having established how the IMS A100 can be configured to implement IIR filters, the next section deals with some of the design techniques that are used for determining the IIR filter coefficients.

4.5 Summary of the IIR filter design techniques

The problem of designing recursive filters is one of determining the feedforward and feedback coefficients (i.e. b_n's and a_m's in equation (8). The design techniques for IIR filters can be categorised into two basic groups:

(i) Indirect approaches.

(ii) Direct approaches.

Indirect approaches for the design of IIR filters

As mentioned earlier digital recursive filters are closely related to conventional analogue filters. In the indirect method this similarity is exploited and the digital filter coefficients are determined from a suitable analogue filter, using some form of transformation technique. In other words the indirect approach uses the wealth of knowledge already available on analogue filters (such as Butterworth, Chebyshev and Elliptic filters) and develops a corresponding recursive digital filter. This method involves the following two steps:

(1) the determination of a suitable analogue filter transfer function $H(s)$

(2) transformation and digitization of this analogue filter

Some of the most popular design techniques falling into the indirect category are:

(a) Impulse-invariant transformation.

(b) Bilinear z-transform.

(c) Matched z-transform.

These three techniques can be employed to derive recursive digital filters from conventional analogue filter structures. Before discussing these three techniques the basic characteristics of the common analogue filters, from which IIR filters are derived, will be briefly reviewed. The starting point in the indirect IIR design techniques is often one of the following analogue filter types.

1 **Butterworth filters:** These filters are characterised by the property that their magnitude characteristic is maximally flat at the origin of the s-plane. Butterworth filters are specified by their magnitude-square functions i.e:

$$|H(s)|^2 = \frac{1}{1 + (\frac{s}{s_c})^{2n}} \quad (18)$$

The pole locations in the s-plane are equally spaced around a circle of radius ω_c $(s_c = j\omega_c)$. These filters have a monotonically decreasing amplitude function with a roll-off of approximately $6n$ dB/decade. Figure 14 shows the overall amplitude response of this type of filter.

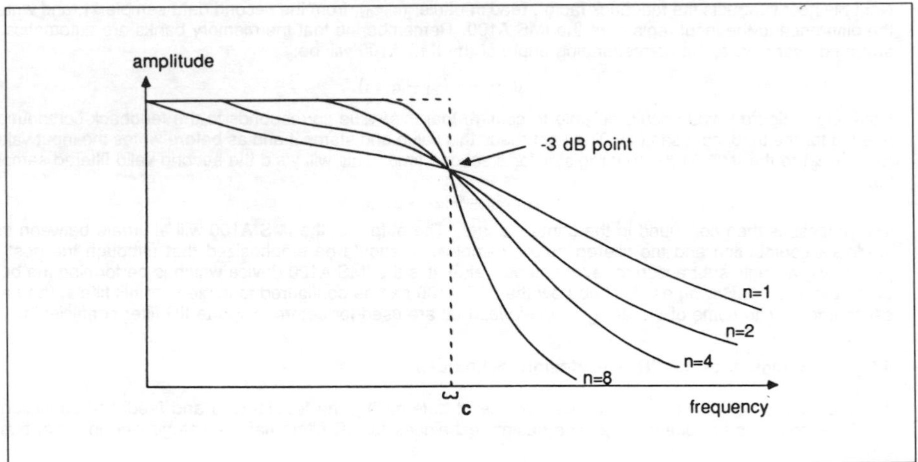

Figure 14 Frequency response of the Butterworth filter

2 **Chebyshev filters:** In these types of filters the peak magnitude of the approximation error is minimized over a prescribed band of frequencies and is also equiripple over the band. Chebyshev filters are specified by the magnitude-square function:

$$|H(s)|^2 = \frac{1}{1 + \epsilon^2 C_N^2(\frac{s}{s_c})} \tag{19}$$

where $C_N(s)$ is a Chebyshev polynomial of order N. The parameter ϵ is used to specify a magnitude function with equal ripple in the pass band and monotonic decay in the stop band. Figure 15 shows the magnitude-square transfer function for the Chebyshev filter (type I) where the amplitude of the ripple is given by:

$$\delta = 1 - \frac{1}{\sqrt{1 + \epsilon^2}} \tag{20}$$

The poles of the Chebyshev filter lie on an ellipse determined from the parameters ϵ, N and s_c. Chebyshev filters of type II on the other hand have monotonic behaviour in the pass band (maximally flat around ω_0) and exhibit equiripple behaviour in the stop band. For further details refer to references 1 & 2.

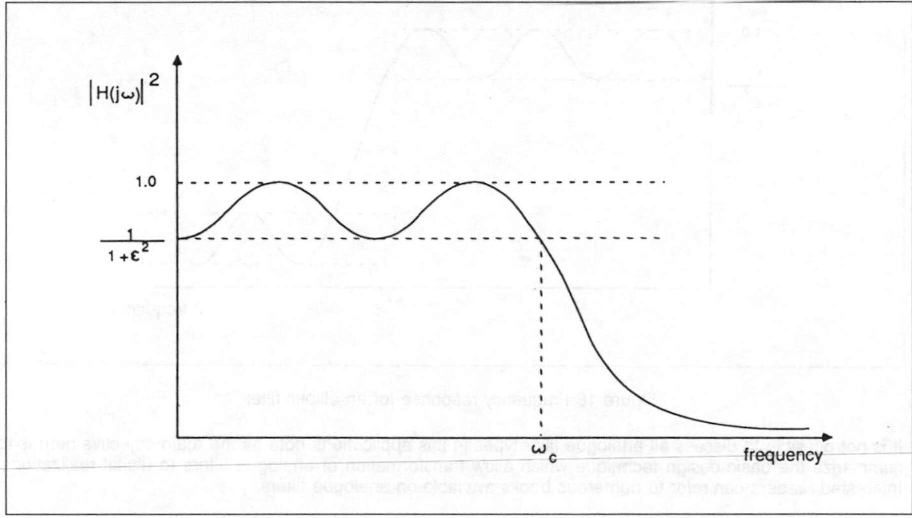

Figure 15 Frequency response of the Chebyshev filter (type I)

3 **Elliptic filters:** These filters exhibit a magnitude response that is equiripple in both the pass band and the stop band. These filters are optimum in the sense that for a given order and for a given ripple specification the transition band is the shortest possible. Elliptic filters are specified by the magnitude-square transfer function:

$$|H(j\omega)|^2 = \frac{1}{1 + \epsilon^2 C_N^2(\omega)} \tag{21}$$

Where $C_N(\omega)$ is a rational Chebyshev function involving elliptical functions. Figure 16 illustrates the magnitude-square response for an elliptic filter.

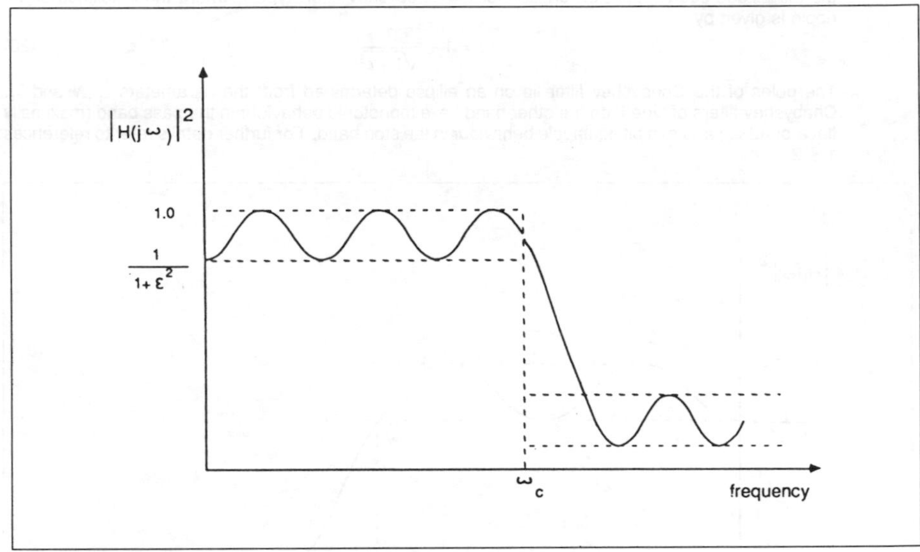

Figure 16 Frequency response for an elliptic filter

It is not possible to discuss all analogue filter types in this applications note as the main objective here is to summarize the basic design technique which allow transformation of analogue filters to digital realizations. Interested readers can refer to numerous books available on analogue filters.

Having decided the type and the specification of the analogue filter that satisfies the requirement, the next step in the indirect design method is to use one of the three following techniques to obtain the corresponding digital filter.

Impulse Invariant Transformation

One of the most common techniques for deriving a digital filter from a given analogue filter is the impulse-invariant transformation. As the name suggests this technique consists of using a sampled version of the impulse response of the analogue filter as the impulse response of the digital filter, i.e. the transformation does not change the impulse response of the analogue filter. Figure 17 illustrates the relationship between the analogue and the resulting digital responses of a typical low-pass filter obtained via the impulse-invariant method. The important point to note here is that sampling the analogue impulse response results in the frequency response of the resulting digital filter being periodic with a period equal to the sampling frequency f_s. This means that the digital filter will have a frequency response similar to a repetitive version of that of the analogue filter. If the frequency response of the analogue filter does not decay to near zero beyond $\frac{f_s}{2}$ then serious aliasing would occur and the digital filter response would be corrupted. This aliasing problem means that this design technique is not suitable for high pass filters. However for low-pass and band-pass filters the problem can be avoided by choosing the sampling frequency high enough to ensure that the magnitude of the analogue filter response is negligible beyond $\frac{f_s}{2}$. (Note that the IMS A100 is capable of a sampling rate of 2.5MHz for 16-bit data and coefficients).

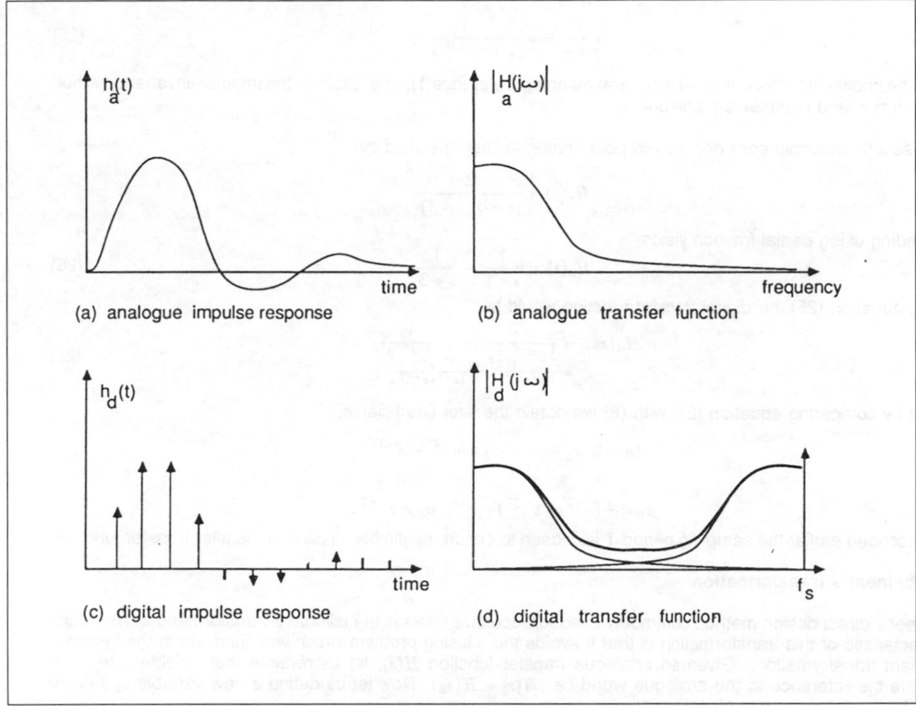

Figure 17 The impulsive invariance transformation relationship between analogue and digital impulse and frequency responses

To demonstrate how the impulse-invariant transformation is used to digitize an analogue filter, consider the simple case of an analogue filter with an impulse response $h_a(t) = Ae^{-\alpha t}$ i.e. a simple RC filter (the s-domain transfer function of this filter is $\frac{A}{s+a}$). We start by sampling the impulse response of this analogue filter with a sampling interval T to obtain the corresponding impulse response for the digital filter, i.e.

$$h_a(kT) = Ae^{-\alpha kT} \tag{22}$$

The z-transform of equation (22) is

$$H_d(z) = \sum_{k=0}^{\infty} Ae^{-\alpha kT} z^{-k} \tag{23}$$

Noting that as equation (23) is a geometric series the result of the summation would be

$$H_d(z) = \frac{A}{1 - z^{-1}e^{-\alpha T}} \tag{24}$$

Equation (24) provides the z-domain transfer function of the resulting digital filter. To determine the filter coefficient (b_k's and a_m's), equation (24) can be compared with equation (8). For this simple example it can be seen that we have

$$a_1 = -e^{-\alpha T} \qquad and \qquad b_0 = A.$$

In this example, for the sake of clarity, the impulse responses were used to arrive at the z-domain transfer function. As analogue filters are often specified in the s-domain, it is more convenient to perform the impulse-invariant transformation directly from the s-domain to the z-domain. It should be obvious to the reader from the previous example that the required mapping is of the form

$$\frac{1}{s+\alpha} \Longrightarrow \frac{1}{1-e^{-kT}\,z^{-1}} \tag{25}$$

It can be shown that this is indeed a general mapping (reference 1), applicable to the impulse-invariant method for both real and complex s-plane poles.

As a second example consider the two-pole analogue filter specified by:

$$H_a(s) = \frac{2}{(s+3)(s+1)}$$

expanding using partial fraction yields

$$H_a(s) = \frac{1}{s+1} - \frac{1}{s+3} \tag{26}$$

Using equation (25) the digital transfer function would be:

$$\begin{aligned} H_d(z) &= \frac{1}{1-e^{-T}\,z^{-1}} - \frac{1}{1-e^{-3T}\,z^{-1}} \\ &= \frac{(e^{-T}-e^{-3T})z^{-1}}{1-(e^{-T}+e^{-3T})z^{-1}+e^{-4T}z^{-2}} \end{aligned} \tag{27}$$

Again by comparing equation (27) with (8) we obtain the filter coefficients,

$$b_0 = 0 \qquad b_1 = e^{-T} - e^{-3T}$$

and

$$a_1 = -(e^{-T} + e^{-3T}) \qquad a_2 = e^{-4T}.$$

As described earlier the sampling period T is chosen to ensure negligible aliasing in the filter transfer function.

The bilinear z-transformation

Another indirect design method commonly used for recursive filters is the bilinear z-transformation. The major characteristic of this transformation is that it avoids the aliasing problem which was inherent in the impulse-invariant transformation. Given an analogue transfer function $H(s)$, let us rename the variable s to s_a to indicate the reference to the analogue world i.e. $H(s) = H(s_a)$. Now let us define a new variable s_d related to s_a by the following mapping:

$$s_a = j\frac{2}{T}\tan\left(\frac{s_d T}{2j}\right) \tag{28}$$

where T is the sampling period.

Since the analogue frequency variable ω_a is related to the s-plane variable by $s_a = j\omega_a$, we can also express the above mapping as:

$$\omega_a = \frac{2}{T}\tan\left(\frac{\omega_d T}{2}\right) \tag{29}$$

where w_d is defined as $s_d = j\omega_d$.

Starting from an analogue transfer function $H(j\omega_a)$, figure 18 illustrates the effect of this mapping on this transfer function. It can be seen from this diagram that the bilinear transformation compresses the entire analogue frequency range ($\omega_a = 0 \rightarrow \infty$) into a finite range equal to half the sampling frequency. This means that the spectral folding problem is completely eliminated and aliasing is therefore avoided. This compression of analogue frequency axis is usually referred to as frequency warping.

The price that is paid for this advantage is a distorted digital frequency scale resulting from this frequency warping. It can be seen from figure 18 that due to the non-linear mapping the specification of the resulting filter, such as the cut-off frequency, would be somewhat different from the starting analogue filter. This distortion can be taken into account in the course of digital filter design. For example the cut-off frequency of the original analogue filters are modified slightly so as after the mapping the resulting filter has the desired cut-off frequencies.

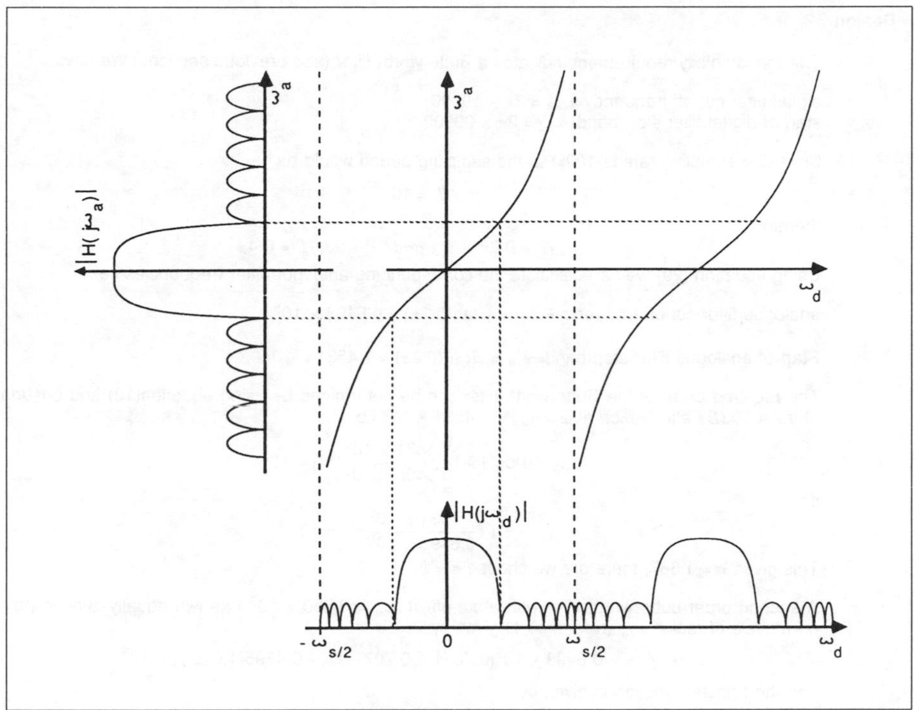

Figure 18 Graphical illustration of the bilinear z-transform

Returning to the transformation equation (28), we can rewrite it as:

$$s_a = \frac{2}{T}\left(\frac{1 - e^{-s_a T}}{1 + e^{-s_d T}}\right) \tag{30}$$

and remembering that $z^{-1} = e^{-s_d T}$ we can write

$$s_a = \frac{2}{T}\left(\frac{1 - z^{-1}}{1 + z^{-1}}\right) \tag{31}$$

Equation (31) provides the means for bilinear transformation directly from the s-domain to the z-domain suitable for digital filter implementation. To illustrate how the bilinear transformation technique is used consider the following example:

Filter specification

Low pass: 0 → 10kHz pass band
Sampling rate: 100kHz
Transition band: 10kHz to 20kHz
Stop-band attenuation: $-10dB$ (starting at 20kHz)
Filter must be monotonic in pass and stop band.

Design

The monotonicity requirement indicates a Butterworth filter (see previous sections).We have:

digital filter cut-off frequency=$\omega_{cd} = 2\pi \times 10000$
start of digital filter stop band=$\omega_{sd} = 2\pi \times 20000$.

Since the sampling rate is 100kHz, the sampling period would be

$$T = 10^{-5}$$

therefore

$$\omega_{cd}T = 0.2\pi \qquad and \qquad \omega_{sd}T = 0.4\pi$$

Using equation (29) we can calculate the corresponding analogue filter frequencies i.e.

analogue filter cut-off frequency=$\omega_{ca} = \frac{2}{T}\tan(0.1\pi) = 0.6498 \times 10^5$

Start of analogue filter stop band=$\omega_{sa} = \frac{2}{T} tan(0.2\pi) = 1.4531 \times 10^5$.

The required order of the Butterworth filter can be determined by using equation(18) and ensuring at least $10dBs$ attenuation at $\omega = \omega_{sa} = 1.4531 \times 10^5$ i.e.

$$10\log[1 + (\frac{1.4531 \times 10^5}{0.6498 \times 10^5})^{2n}] = 10$$

or

$$1 + (\frac{1.4531 \times 10^5}{0.6498 \times 10^5})^{2n} = 10$$

This gives $n = 1.367$, therefore we choose $n = 2$.

A second order butterworth filter with a cut-off at $\omega_{ca} = 0.650 \times 10^5$ has two equally-spaced poles on a circle of radius ω_{ca} (reference 1) given by

$$s_1,\ s_2 = -0.6498 \times 10^5(0.7071 \pm 0.7071j) = -0.4595(1.0 \pm j) \times 10^5$$

and the transfer function is given by:

$$H(s) = \frac{s_1 s_2}{(s - s_1)(s - s_2)} = \frac{4.223 \times 10^9}{s^2 + 0.919 \times 10^5 + 4.223 \times 10^9}$$

Now we apply the bilinear-z transformation by substituting for s in the above transfer function from equation (31). This gives the following digital filter transfer function:

$$H(z) = \frac{0.0675 + 0.1349z^{-1} + 0.0675z^{-2}}{1 - 1.1430z^{-1} + 0.4128z^{-2}} \qquad (32)$$

The digital filter coefficients can be obtained by comparing equation (32) with (8) giving:

$$\begin{array}{ll} b_0 = 0.0675 & --- \\ b_1 = 0.1349 & a_1 = -1.1430 \\ b_2 = 0.0675 & a_2 = 0.4128 \end{array}$$

These coefficient values are then expressed in binary with the number of bits governed by the required accuracy. The factors affecting the necessary accuracy are discussed in section 5 of this application note.

Matched z-transform

This transformation is a direct mapping from the poles and zeroes in the s-plane to the poles and zeroes in the z-plane.

In general the two previous method i.e. the impulse invariant and the bilinear transformations are preferred to the matched z-transform as there are many cases where the matched z-transformation is not applicable. For this reason this technique is not detailed here. It would be sufficient to point out that the mapping is defined by the replacement relationship:

$$s + k = 1 - z^{-1}e^{-kT}. \qquad (33)$$

The direct design techniques for IIR filters

The IIR design techniques described so far were based on transforming a known analogue transfer function into the required digital filter transfer function. It is however possible to design digital IIR filters directly without reference to an analogue filter. Direct design methods fall into two categories namely direct closed form designs and optimisation techniques.

The direct closed form design techniques begin with the desired response of the filter from which one can often decide where to place poles and zeroes to approximate this response. These techniques are not very common and as such will not be discussed here.

The second classes of direct IIR filter design techniques are based on computer optimisation. In these approaches the set of design equations cannot be solved explicitly, instead mathematical optimization techniques are employed to determine the filter coefficients that minimize some error criterion, subject to a set of design equations. The algorithms involved in these optimisation techniques are of an iterative nature and are terminated when the error reaches a minimum or the number of iterations exceeds a specified limit.

Among the most commonly used optimisation technique is one which minimizes the pass-band ripples in filters exhibiting a given stop-band attenuation. This technique is sometimes referred to as the minimax method and the optimization algorithm involved has been developed by Fletcher and Powell (reference 7). The Fletcher-Powell optimization algorithm generates the filter coefficients by using a convergent descent method.

The spectral flatness approach is another optimisation technique and is based on the fact that multiplying the desired frequency response by its inverse should result in unity throughout the frequency spectrum (i.e. a flat spectral line). Any deviations from the ideal response would result in ripples in this flat spectral line. Optimisation techniques have been developed which attempt to minimize these ripples (reference 8). The difficulty with this technique is the modeling of the desired frequency response.

Mean-square-error optimization techniques have also been developed for IIR filter design. One such technique has been described by Steiglitz (reference 9) which involves minimizing the square of the difference between actual filter behaviour and the desired performance. This algorithm searches an error *.vs.* design-parameter curve for a local minimum.

The details of the above optimisation techniques are beyond the objectives of this application note. However the references given should prove adequate for interested readers.

5 Finite word-length considerations and problems

In implementing digital filters both the input samples and the filter coefficients have to be quantised and expressed in a limited number of bits. In the IMS A100 chip both the coefficients and data samples can be quantized up to 16-bits of accuracy, although smaller word-lengths can be used if desired.

The problems of finite word length in digital filters apply to both FIR and IIR filters but their implications are much more severe for the IIR filters, due to their inherent feedback nature. In the fixed-point implementations of digital filters it is usual to normalize the numbers so as to make their absolute values less than one i.e. in the form of

$$d_{N-1}.d_{N-2}d_{N-3} - - - - - - - - - d_5d_4d_3d_2d_1d_0$$

where d_n represents the nth bit in the word and (.) indicates the binary point. Using this format (and two's complement notation) the number

$$0.1111\ldots1111$$

would represent a vlaue very nearly equal to +1, while the number

$$1.0000\ldots0000$$

would represent a value equal to -1.0.

If purely integer numbers were to be used the process of truncation or rounding after multiplications would become meaningless. However using the above fractional-number representation, where the numbers are normalized to be less than one, the problems would not arise as the product of two numbers which are less than one would also be less than one.

In general there are three sources of error arising in the implementation of digital filters these are:

(i) Finite precision of the filter coefficients

(ii) Limited word length of the input data

(iii) Round-off and truncation errors in the multiplication and addition operations.

The finite precision in the representation of the filter coefficients will obviously cause the frequency response of the filter to depart to some extent from that desired for both FIR and IIR filters.

Furthermore in the case of the recursive IIR implementation, because of the existence of feedback paths, this finite precision may cause instabilities in the filter behaviour. This happens because the inaccuracies may move the z-plane poles outside the unit circle hence causing instabilities. The chances of this happening depends on how close the poles are to the unit circle in the first place. If multiplication and addition operations are followed by truncation and rounding (in order to contain word growth) further difficulties may arise. These problems may manifest themselves in undesirable oscillations in the form of 'limit cycle' or 'overflow' oscillations (discussed later). It is therefore absolutely essential for the filter behaviour to be simulated using the precision and roundings involved in the intended implementation. This is particularly relevant to recursive IIR filter where a risk of instability exists.

One of the consequences of rounding and quantisation in the digital recursive(IIR) filters is the limit-cycle phenomenon, which takes the form of a stable periodic non-zero output for zero or constant input. The limit cycle behaviour of a digital filter in general is complex and difficult to analyse. However for simple first order filters, it is possible to illustrate the effect by way of an example. Consider the first order recursive filter with the following equation:

$$y(n) = 0.09x(n) + 0.91y(n-1)$$

Assume that each output y(n) gets rounded to the nearest integer, also assume that the input is constant at 100 and the previous output is 90.

The following table shows the resulting rounded output sequence for each iteration.The last column shows the perfect output (without rounding) for comparison.

n	x(n)	y(n)	rounded y(n)	perfect y(n)
0	100	–	90	–
1	100	90.9	91	90.9
2	100	91.81	92	91.72
3	100	92.72	93	92.46
4	100	93.63	94	93.14
5	100	94.54	95	93.76
6	100	95.45	95	94.32
7	100	95.45	95	94.83
8	100	95.45	95	95.30
9	100	95.45	95	95.72
10	100	95.45	95	96.11
..	...	.....	..	.....
..	...	.....	..	.....
..	...	.....	..	.....
..	100	95.45	95	100.0

It is observed that the output sticks at a value of 95. However if the same filter is implemented with very high precision and no rounding the filter output would closely approach 100 (last column in the table).

If we approach the limit from the opposite side by starting with a value of $y(n)$ of say 110, the output would arrive at a limit of 105. You can see from this example that the system has a dead zone of ±5 units around the ideal output of 100.

In fact it can mathematically be shown that for a first-order recursive filter of the form

$$y(n) = bx(n) + ay(n-1)$$

The dead zone is given by

$$|dead\ zone| \leq \frac{\frac{q}{2}}{1-|a|} \tag{34}$$

where q is the quantisation step. In the above example a quantised step of 1 was used and equation (34) gives a dead zone of ±5 too.

For second-order systems similar results to (34) have been derived in the literature (reference 1).

Overflow oscillation is another problem associated with digital recursive filters. In the IMS A100 chip the full internal precision ensures that no overflow occurs in the multiply-accumulator array. The only source of possible overflow is the external addition which is performed in combining the feedback terms with the input samples (see section 4.4). A simple but effective way to eliminate these oscillations is to perform this addition in a saturating manner (similar to analogue adders). This operation can easily be taken care of by the controlling host processor.

In the IMS A100 device the data and coefficients can be expressed to a precision as high as 16 bits. The 32 multiplications and additions are carried out to 36-bit precision. This ensures that no overflow occurs in the multiply-accumulation array (unless all the coefficients and 32 consecutive data items have values equal to the most negative 16-bit number i.e. 1000000000000000 in binary, which is of course highly unlikely). The selector at the output of the multiply and accumulate array allows the rounding and selection of 24 bits out of this 36 bits. The combination of full internal accuracy, the selector functionality and the fact that the IMS A100 devices can easily be cascaded allows high quality FIR filters to be readily implemented. As described earlier the device can also be used to implement efficient IIR filters only in direct forms. It is well known that for high order filters direct implementations of IIR filters are more prone to instabilities compared to cascade or parallel arrangements. However the full internal precision of the IMS A100 combined with comprehensive filter simulations should minimize these instabilities. It should however be emphasized that it is possible to implement a high order high precision IIR filter in the cascade form on the IMS A100 at the expense of processing speed. In this case the IMS A100 should be used to implement low order (2nd or 4th order) sections of a cascade arrangement in turn by reloading suitable coefficients. The functionality of the whole filter is obtained by recirculating the first output batch through the chip with its coefficients modified to implement the 2nd section in the cascade array and so on.

For the IIR filter implementations figure 11c & 11b can be considered as possible system configurations.

6 Adaptive filters

So far we have discussed digital filters with fixed characteristics. Fixed filters are used in many practical situations to combat noise or interfering signals (e.g. a matched filter) or to select a desired frequency band (e.g. a band-pass filter). In digital signal processing the parameters of such fixed filters are determined once and remain unchanged during processing. Adaptive filters on the other hand automatically adjust their own parameters and seek to optimize their performance according to a specific criterion. The adaptive nature of such filters makes them particularly suitable for situations where signal properties are unknown or variable with time.

Figure 19 illustrates the basic structure of an adaptive filter. The input signal $x(t)$ is filtered or weighted in a programmable filter to yield an output $y(t)$. The filter output $y(t)$ is then compared with a reference (sometimes called a training signal) waveform to yield an error signal $e(t)$. This error is then used to update the filter coefficients in such a way that the error is progressively minimized. Several algorithms for updating the filter coefficients have been developed and can be found in references 10, 11 & 12.

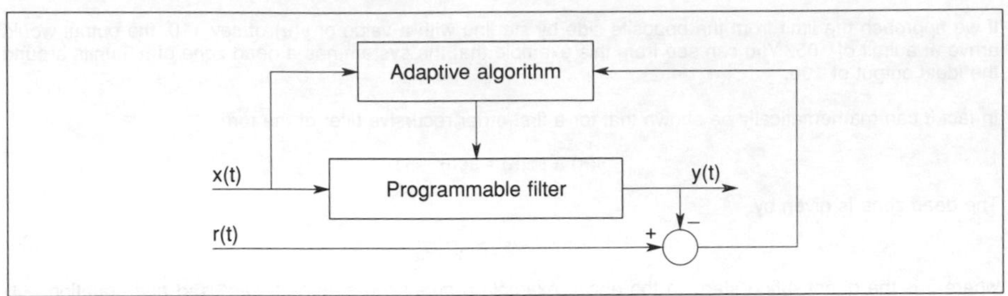

Figure 19 Basic structure of an adaptive filter

One example of adaptive filtering is echo cancellation in telephony. Echoes are the result of impedance mismatches in the communication circuits. The hybrid couplers which are used at the interface between two-wire and four-wire circuits are a major source of echoes. Figure 20 shows how an adaptive filter arrangement can be used to cancel these echoes at the hybrid interface. Notice that in this case the training signal contains the echo, while the input to the adaptive filter is the signal arriving at the hybrid. Effectively the filter adaptively models the echo path and produces a synthetic antiphase echo return which cancels the echo in the 4-wire path returning from the hybrid.

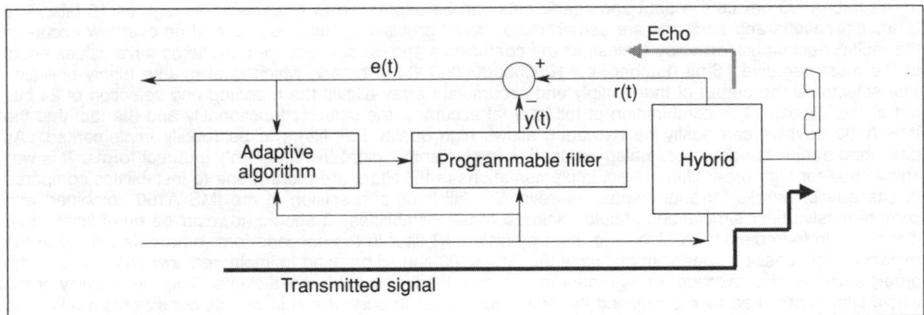

Figure 20 Application of adaptive filtering techniques to echo cancellation

Adaptive filters have application in low-bit rate speech coding based on linear prediction where the filter coefficients, after adaption, are transmitted instead of the speech signal itself.

The programmability of the IMS A100 can be exploited in the implementation of adaptive filters as well as fixed filters discussed earlier.

7 References

1 *Theory and application of digital signal processing*, Rabiner L.R., and Gold B., Prentice-Hall Inc, Englewood Cliffs, NJ, l975.

2 *Digital signal processing*, Oppenheim A.V., and Schafer R.W., Prentice-Hall Inc, Englewood Cliffs, NJ, l975.

3 *A computer program for designing optimum FIR linear-phase digital filters*, McClellan J.H., Parks T.W., and Rabiner L.R., IEEE Transactions on Audio and Electroacoustics, Vol. AU-21, No.8 , December l973.

4 *Digital signal processing*, Peled A., and Liu B., John Wiley and Sons Inc., New York, l976.

5 *Practical design rules for optimum finite impulse response low-pass digital filters*, Rabiner L.R., Bell Systems Technical Journal, Vol. 52, No.6, July-August l973.

6 *Approximate design relationships for low-pass FIR digital filters*, Rabiner L.R., IEEE Transactions on Audio and Electroacoustics, Vol. AU-21, No.5, October 1973.

7 *A rapidly convergent descent method for minimization*, Fletcher R., and Powell M., Computer Journal 6 (No.2) l963, pp 163-168.

8 *Time series analysis: Forcasting and control*, Box G., and Jenkins G., Holden Day, San Francisco, CA, l975

9 *Computer aided design of recursive digital filters*, Steiglitz K., IEEE Transactions on Audio and Electroacoustics 18 (No.2), June l970, pp123-129.

10 *Adaptive noise cancelling: principle and applications*, Widrow B. *et al*, Proc. IEEE, l975, Vol.63, No.12, pp 1692-1716.

11 *Design and application of adaptive filters*, Grant P.M., Cowan C.F.N., Electronics & Power, February l985.

Chapter 8 Digital signal processors

This chapter provides details of four typical digital signal processor chips. As the full data manuals for each of these devices are hundreds of pages long, we have used the manufacturers' abridged technical specifications in this section. These provide an overview of the devices and cover their salient features without going into detail.

MOTOROLA
SEMICONDUCTOR
TECHNICAL DATA

BRE505/D

DSP56001

Technical Summary

56-Bit General Purpose Digital Signal Processor

This document provides a technical summary of the architecture and instruction set for the DSP56001, a fourth generation, low power, HCMOS, user-programmable digital signal processor. The DSP56001 is a member of Motorola's family of general purpose Digital Signal Processors. The DSP56001 features 512 words of full speed on-chip program RAM memory, two preprogrammed data ROMs, and special on-chip bootstrap hardware to permit convenient loading of user programs into the program RAM.

The core of the processor consists of three execution units operating in parallel — the data ALU, the address ALU, and the program controller. The DSP56001 has MCU-style on-chip peripherals, program and data memory, as well as a memory expansion port. The MPU-style programming model and instruction set make the generation of efficient, compact code straightforward. The DSP56001 instruction set is identical to the DSP56000 instruction set.

Core Features

- 10.25 Million Instructions Per Second (MIPS)
- Single Cycle Data ALU
  - 24 x 24 ⧫ 56-Bit Parallel Multiply/Accumulate
  - Two 56-Bit Accumulators
  - Ten Data Registers
  - Two Data Bus Shifter/Limiters
- DSP Oriented Address ALU
  - 24 Address Registers
  - Dual Modulo Arithmetic Units
  - Linear, Modulo, and Bit Reversed Address Generation
- Advanced Program Controller
  - 15 Level Hardware Stack
  - Nested Hardware DO Loops
  - No Overhead Auto-Return (Fast) Interrupts
- Highly Orthogonal Instruction Set
  - 62 MPU-Style Instruction Types
  - Makes Pipeline Invisible
  - Suitable for High Level Language (HLL) Compilers
- Multiple Buses
  - Four Data Buses
  - Three Address Buses

On-Chip MCU-Style Peripherals

- 24 Programmable I/O Port Pins or a Combination of I/O Port Pins and
  - 8-Bit Parallel Host MPU/DMA Interface
  - Serial Communication Interface with Baud Rate Generator
  - Synchronous Serial (Codec) Interface with Clock Generator

On-Chip Memory

- Two Independent 256 x 24-Bit Data RAMs
- Two 256 x 24-Bit Preprogrammed Data ROMs
- 512 x 24-Bit Program RAM

Off-Chip Memory Expansion

- 128K x 24-Bit Data Memory
- 64K x 24-Bit Program Memory
- Programmable Off-Chip Access Times (Wait States)

This document contains information on a new product. Specifications and information herein are subject to change without notice.

PROGRAM RAM (PRAM)

The DSP56001 program memory contains 512 words of high speed, on-chip Program RAM (PRAM). The DSP56001 PRAM can be loaded with the user's program after power-up reset. The bootstrap mode, described in Appendix I, provides a convenient, low cost method to load the DSP56001 PRAM.

The PRAM DSP56001 offers many advantages:

- The DSP56001 is an off-the-shelf product because it contains user programmable PRAM instead of factory programmable ROM.
- PRAM allows quick program development.
- On-chip PRAM operates at the full DSP56001 speed with no bus contention with external Data Memory space accesses.
- Programs can be changed dynamically, allowing efficient overlaying of DSP software algorithms.
- The user can write to PRAM from an application program using the MOVEM instruction.
- The bootstrap mode allows the designer to easily initialize the PRAM. Bootstrap mode can:
  - load PRAM from a single, inexpensive EPROM
  - load PRAM through the Host Interface from a microprocessor.

SIGNAL DESCRIPTION

The DSP56001 is an 88-pin integrated circuit available in surface mount or pin-grid array packaging. Its input and output signals are organized into seven functional groups which are listed below and shown in Figure 1.

Address and Data Buses
Bus Control
Interrupt and Mode Control
Power and Clock
Host Interface or PI/O
SCI Interface or PI/O
SSI Interface or PI/O

Descriptions of the signals in each group are given in the following paragraphs.

ADDRESS AND DATA BUS

The following paragraphs describe the address and data bus signals.

Address Bus (A0-A15)

These three-state output pins specify the address for external program and data memory accesses. A0-A15 do not change state when external memory spaces are not being accessed in order to minimize power dissipation.

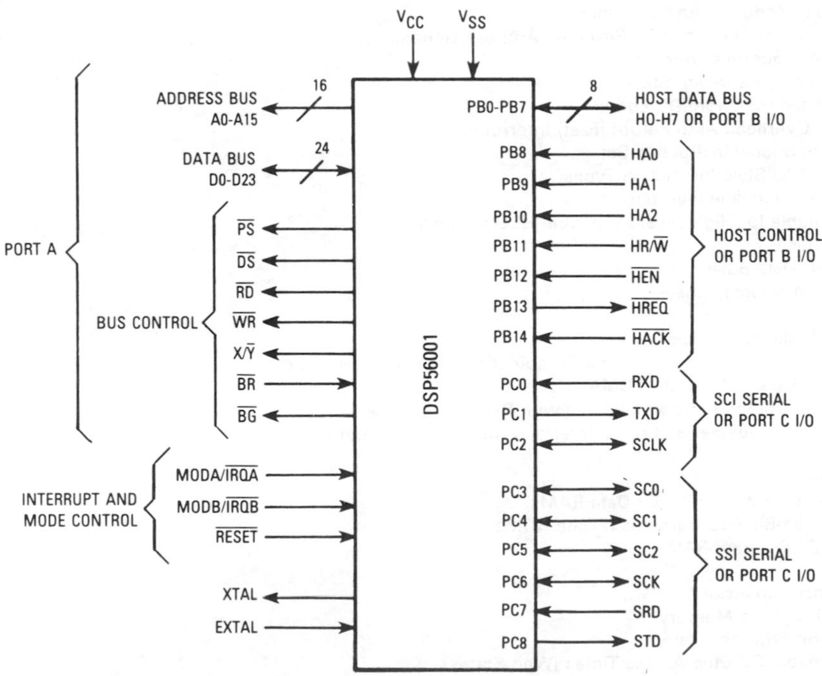

Figure 1. Functional Signal Groups

Data Bus (D0-D23)

These pins provide the bidirectional data bus for external program and data memory accesses. D0-D23 are in the high-impedance state when the bus grant signal is asserted in order to minimize power dissipation.

BUS CONTROL

The following paragraphs describe the bus control signals.

Program Memory Select ($\overline{PS}$)

This three-state output is asserted only when external program memory is referenced.

Data Memory Select ($\overline{DS}$)

This three-state output is asserted only when external data memory is referenced.

X/$\overline{Y}$ Select (X/$\overline{Y}$)

This three-state output selects which external data memory space (X or Y) is referenced by data memory select $\overline{DS}$.

Read Enable ($\overline{RD}$)

This three-state output is asserted to read external memory on the data bus D0-D23.

Write Enable ($\overline{WR}$)

This three-state output is asserted to write external memory on the data bus D0-D23.

Bus Request ($\overline{BR}$)

The bus request input $\overline{BR}$ allows another device such as a processor or DMA controller to become the master of the external data bus D0-D23 and external address bus A0-A15. When $\overline{BR}$ is asserted, the DSP56001 will release control of the external data bus D0-D23, address bus A0-A15 and bus control pins $\overline{PS}$, $\overline{DS}$, X/$\overline{Y}$, $\overline{RD}$, and $\overline{WR}$. These pins will be placed in the high-impedance state and the bus grant, $\overline{BG}$, output will be asserted.

Bus Grant ($\overline{BG}$)

This three-state output is asserted to acknowledge an external bus request.

INTERRUPT AND MODE CONTROL

The following paragraphs describe the interrupt and mode control signals.

Mode Select A, B (MODA, MODB) or External Interrupt Request A, B ($\overline{IRQA}$, $\overline{IRQB}$)

These two inputs have two functions — 1) to select the initial chip operating mode and, 2) to receive an interrupt request from an external source. MODA and MODB are read and internally latched in the DSP when the processor exits the RESET state. Several clock cycles after leaving the RESET state, the MODA and MODB pins automatically change to external interrupt requests $\overline{IRQA}$ and $\overline{IRQB}$. After leaving the RESET state the chip operating mode can be changed by software. $\overline{IRQA}$ and $\overline{IRQB}$ may be programmed to be level sensitive or negative edge triggered.

Reset ($\overline{RESET}$)

This input pin is used to reset the DSP56001. When $\overline{RESET}$ is asserted, the DSP56001 is initialized and placed in the RESET state. When the $\overline{RESET}$ signal is removed, the initial chip operating mode is latched from the MODA and MODB pins.

POWER AND CLOCK

The following paragraphs describe the power and clock signals.

V_{CC} and V_{SS}

V_{CC} is the power input and V_{SS} is ground.

External Clock/Crystal Input (EXTAL)

EXTAL may be used to interface the internal crystal oscillator input to an external crystal or an external clock. The maximum clock rate is 20.5 MHz.

Crystal Output (XTAL)

This output connects the internal crystal oscillator output to an external crystal. If an external clock is used, XTAL should not be connected.

HOST INTERFACE

The following paragraphs describe the Host Interface.

Host Data Bus (H0-H7)

This bidirectional data bus is used to transfer data between the host processor and the DSP56001. This bus is an input unless enabled by a host processor read. H0-H7 may be programmed as general purpose parallel I/O pins called PB0-PB7 when the Host Interface is not being used.

Host Address (HA0, HA1, HA2)

These inputs provide the address selection for each Host Interface register. HA0-HA2 may be programmed as general purpose parallel I/O pins called PB8-PB10 when the Host Interface is not being used.

Host Read/Write (HR/$\overline{W}$)

This input selects the direction of data transfer for each host processor access. HR/$\overline{W}$ may be programmed as a general purpose I/O pin called PB11 when the Host Interface is not being used.

Host Enable ($\overline{HEN}$)

This input enables a data transfer on the host data bus. When $\overline{HEN}$ is asserted and HR/$\overline{W}$ is high, H0-H7 become outputs and DSP56001 data may be read by the host processor. When $\overline{HEN}$ is asserted and HR/$\overline{W}$ is low, H0-H7 become inputs and host data is latched inside the DSP when $\overline{HEN}$ is negated. Normally a chip select signal derived from host address decoding and an enable clock, is used to generate $\overline{HEN}$. $\overline{HEN}$ may be programmed as a general purpose I/O pin called PB12 when the Host Interface is not being used.

Host Request (HREQ)

This open-drain output signal is used by the DSP56001 Host Interface to request service from the host processor, DMA controller or simple external controller. HREQ may be programmed as a general purpose I/O pin (not open-drain) called PB13 when the Host Interface is not being used.

Host Acknowledge (HACK)

This input has two functions — 1) to provide a Host Acknowledge handshake signal for DMA transfers and, 2) to provide a Host Interrupt Acknowledge compatible with MC68000 Family processors. HACK may be programmed as a general purpose I/O pin called PB14 when the Host Interface is not being used.

SERIAL COMMUNICATIONS INTERFACE (SCI)

The following paragraphs describe the serial communications interface.

Receive Data (RXD)

This input receives byte-oriented serial data and transfers the data to the SCI Receive Shift Register. Input data is sampled on the positive edge of the Receive Clock. RXD may be programmed as a general purpose I/O pin called PC0 when the SCI RXD function is not being used.

Transmit Data (TXD)

This output transmits serial data from the SCI Transmit Shift Register. Data changes on the negative edge of the transmit clock. This output is stable on the positive edge of the transmit clock. TXD may be programmed as a general purpose I/O pin called PC1 when the SCI TXD function is not being used.

SCI Serial Clock (SCLK)

This bidirectional pin provides an input or output clock from which the transmit and/or receive baud rate is derived. SCLK may be programmed as a general purpose I/O pin called PC2 when the SCI SCLK function is not being used.

SYNCHRONOUS SERIAL INTERFACE (SSI)

The following paragraphs describe the synchronous serial interface.

Serial Control Zero (SC0)

This bidirectional pin is used for control by the SSI serial interface. SC0 may be programmed as a general purpose I/O pin called PC3 when the SSI SC0 function is not being used.

Serial Control One (SC1)

This bidirectional pin is used for control by the SSI serial interface. SC1 may be programmed as a general purpose I/O pin called PC4 when the SSI SC1 function is not being used.

Serial Control Two (SC2)

This bidirectional pin is used for control by the SSI serial interface. SC2 may be programmed as a general purpose I/O pin called PC5 when the SSI SC2 function is not being used.

SSI Serial Clock (SCK)

This bidirectional pin provides the serial bit rate clock for the SSI interface. SCK may be programmed as a general purpose I/O pin called PC6 when the SSI interface is not being used.

SSI Receive Data (SRD)

This input pin receives serial data and transfers the data to the SSI Receive Shift Register. SRD may be programmed as a general purpose I/O pin called PC7 when the SSI SRD function is not being used.

SSI Transmit Data (STD)

This output pin transmits serial data from the SSI Transmit Shift Register. STD may be programmed as a general purpose I/O pin called PC8 when the SSI STD function is not being used.

BLOCK DIAGRAM DESCRIPTION

DSP56001 architecture has been designed to maximize throughput in data intensive Digital Signal Processing (DSP) applications. This objective has resulted in a dual natured, expandable architecture with sophisticated on-chip peripherals and general purpose I/O. It is dual natured in that there are two independent expandable data memory spaces, two address arithmetic units, and a data ALU which has two accumulators and two shifter/limiters. The duality of the architecture makes it easier to write software for DSP applications. For example, data is naturally partitioned into X and Y spaces for graphics and image processing applications, into coefficient and data spaces for convolution sums, and into real and imaginary spaces for performing complex arithmetic.

The major components of the DSP56001 are:

- Data Buses
- Address Buses
- Data ALU
- Address ALU
- X Data Memory
- Y Data Memory
- Program Controller
- Program Memory
- Bootstrap ROM
- Input/Output
  - Expansion Port
  - General Purpose I/O
  - Host Interface
  - SCI Interface
  - SSI Interface

These components are depicted in Figure 2 and described in the following paragraphs.

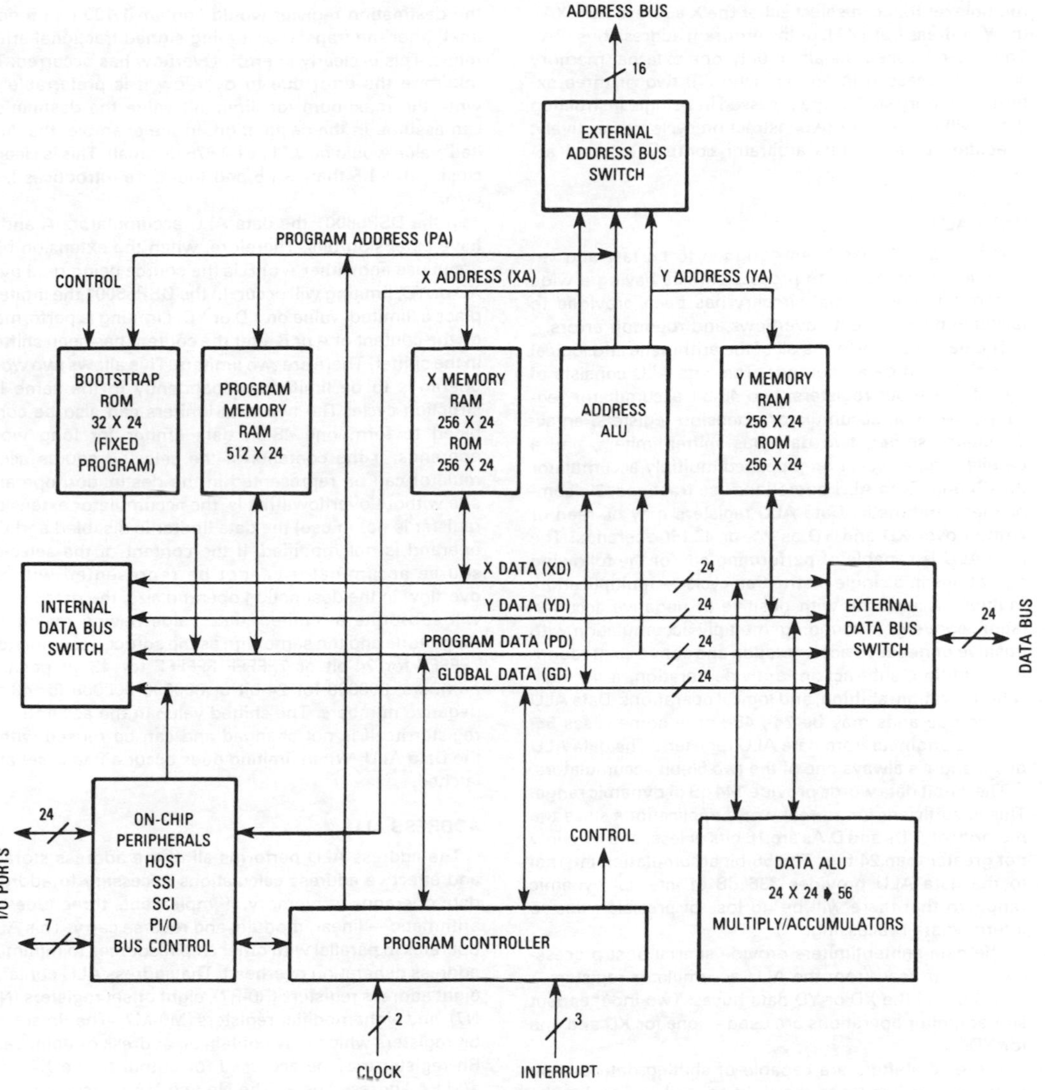

Figure 2. DSP56001 Block Diagram

DATA BUSES

Data movement on the chip occurs over four bidirectional 24-bit buses — the X data bus (XD), the Y data bus (YD), the program data bus (PD), and the global data bus (GD). The X and Y data buses may also be treated by certain instructions as one 48-bit data bus by concatenation of XD and YD. Data transfers between the data ALU and the X data memory and Y data memory occur over XD and YD. XD and YD are kept local on the chip to maximize speed and minimize power dissipation. All other data transfers such as I/O transfers to peripherals occur over GD. Instruction word pre-fetches take place in parallel over PD. Transfers between buses are accomplished in the internal bus switch.

ADDRESS BUSES

Addresses are specified for internal X data memory and Y data memory on two unidirectional 16-bit buses — X address bus (XA) and Y address bus (YA). Program memory addresses are specified on the program address bus (PA). External memory spaces are addressed via a single 16-bit unidirectional address bus driven by a three input multiplexer that can select either the X address bus (XA), the Y address bus (YA), or the program address bus (PA). There is no speed penalty if only one external memory space is accessed in an instruction. If two or three external memory spaces are accessed in a single instruction there will be a one or two instruction cycle, respectively, execution delay. A bus arbitrator controls external access.

DATA ALU

The data ALU has been designed to be fast and yet provide the capability to process signals having a wide dynamic range. Special circuitry has been provided to facilitate handling data overflows and roundoff errors.

The data ALU performs all of the arithmetic and logical operations on data operands. The data ALU consists of four 24-bit input registers, two 48-bit accumulator registers, two 8-bit accumulator extension registers, an accumulator shifter, two data bus shifter/limiters, and a parallel single cycle non-pipelined multiply-accumulator (MAC) unit. Data ALU operations use fractional 2's complement arithmetic. Data ALU registers may be read or written over XD and YD as 24- or 48-bit operands. The data ALU is capable of performing any of the following operations in a single instruction cycle — multiplication, multiply-accumulate with positive or negative accumulation, convergent rounding, multiply-accumulation with positive or negative accumulation and convergent rounding, addition, subtraction, a divide iteration, a normalization iteration, shifting, and logical operations. Data ALU source operands may be 24-, 48-, or in some cases 56-bits and originate from data ALU registers. The data ALU destination is always one of the two 56-bit accumulators.

The 24-bit data words provide 144 dB of dynamic range. This is sufficient for all real world applications since the majority of A/Ds and D/As are 16 bits or less, and certainly not greater than 24 bits. The 56-bit accumulation internal to the data ALU provides 336 dB of internal dynamic range so that there will be no loss of precision due to intermediate processing.

The data shifter/limiters provide special post processing on data read from the ALU accumulator registers A and B out to the XD or YD data buses. Two independent shifter/limiter operations are used — one for XD and one for YD.

The data shifters are capable of shifting data one bit to the left or one bit to the right as well as passing the data unshifted. Each data shifter has a 24-bit output with overflow indication. The data shifters are controlled by the scaling mode bits in the status register. These shifters permit dynamic scaling of fixed point data without modifying the program code. This permits block floating-point algorithms to be implemented in a regular fashion. FFT routines for example, can use this feature to selectively scale each butterfly pass.

Saturation arithmetic is provided to minimize errors due to overflow. Overflow is said to occur when a source operand requires more bits for accurate representation than there are available in the destination.

For example, if the source operand were 01.100 (+1.5 decimal) and the destination register were only four bits, the destination register would contain 1.100 (−1.5 decimal) after the transfer assuming signed fractional arithmetic. This is clearly in error. Overflow has occurred. To minimize the error due to overflow it is preferrable to write the maximum (or 'limited') value the destination can assume in the destination. In the example, the 'limited' value would be 0.111 (+0.875 decimal). This is clearly closer to +1.5 than −1.5 and therefore introduces less error.

In the DSP56001 the data ALU accumulators A and B have extension bits. Therefore, when the extension bits are in use and either A or B is the source being read over XD or YD, limiting will occur. In the DSP56001 the limiters place a 'limited' value on XD or YD. Limiting is performed on the content of A or B after the content has been shifted in the shifter. There are two limiters. This allows two word operands to be limited independently in the same instruction cycle. The two data limiters can also be combined to form one 48-bit data limiter for long word operands. If the contents of the selected source accumulator can be represented in the destination operand size without overflow (that is, the accumulator extension register is not in use) the data limiter is disabled and the operand is not modified. If the content of the selected source accumulator cannot be represented without overflow in the destination operand size, the data limiter will substitute a 'limited' data value having maximum magnitude and the same sign as the source accumulator: 7FFFFF for 24-bit or 7FFFFF FFFFFF for 48-bit positive numbers, 800000 for 24-bit or 800000 000000 for 48-bit negative numbers. The shifted value in the accumulator register itself is not changed and can be reused within the Data ALU. When limiting does occur a flag is set and latched.

ADDRESS ALU

The address ALU performs all of the address storage and effective address calculations necessary to address data operands in memory. It implements three types of arithmetic — linear, modulo, and reverse carry. This ALU operates in parallel with other chip resources to minimize address generation overhead. The address ALU contains eight address registers (R0-R7), eight offset registers (N0-N7), and eight modifier registers (M0-M7). The Rn are 16-bit registers which may contain an address or data. Each Rn register may be accessed for output to the XA, YA, and PA address buses. The Nn and Mn registers are 16-bit registers which are normally used to control updating of the Rn registers.

Address ALU registers may be read or written via the global data bus as 16-bit operands. The Address ALU has two module arithmetic units which can generate two 16-bit addresses every instruction cycle — one for any two of the XA, YA, or PA buses. The address ALU can directly address 65,536 locations on the XA bus, 65,536 locations on the YA bus, and 65,536 locations on the PA bus — a total capability of 196,608 24-bit data words.

MEMORIES

Three independent memory spaces of the DSP56001, X Data, Y Data, and Program, are shown in Figure 3. The memory spaces are configured by control bits in the Operating Mode Register (Figure 8). MA and MB control the Program Memory map and select the reset vector address. DE controls the X and Y Data Memory maps, enabling the internal X and Y data ROMs.

X Data Memory

On-chip X data RAM is a 24-bit wide internal memory which occupies the lowest 256 locations in X memory space. The on-chip X data ROM occupies locations 256 through 511 in X data memory space when enabled by setting DE = 1 in the Operating Mode Register. The X data ROM is factory programmed with positive Mu-law and A-law expansion tables (see **Appendix II**) useful in telecommunication applications. The on-chip peripherals occupy the top 64 locations. Addresses are received from the XA bus and data transfers to the data ALU occur on the XD bus. X memory may be expanded off chip.

Y Data Memory

On-chip Y data RAM is a 24-bit wide internal memory which occupies the lowest 256 locations in Y memory space. The on-chip Y data ROM occupies locations 256 through 511 in Y data memory space when enabled by setting DE = 1. The Y data ROM is factory programmed with a full sine wave table (see **Appendix III**) useful for FFTs, DFTs, and waveform generation. The off-chip peripherals are optimally mapped into the top 64 locations. Addresses are received from the YA bus and data transfers to the data ALU occur on the YD bus. Y memory may be expanded off chip.

Program Memory

On-chip program memory consists of a 512 location by 24-bit RAM which is enabled by the MA and MB bits in the Operating Mode register. Addresses are received from the program control logic (usually the program counter).

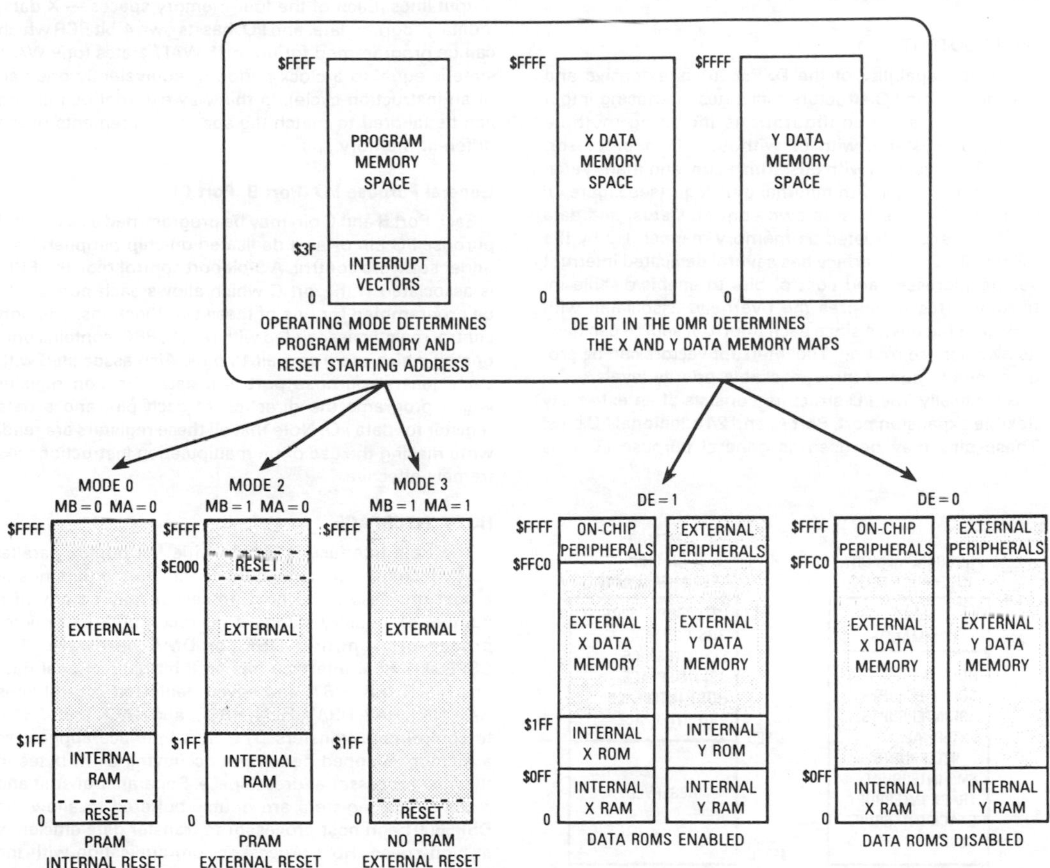

Figure 3. DSP56001 Memory Map

Program memory may be written using MOVEM instructions. The interrupt vectors for the on-chip resources are located in the bottom 64 locations of program memory. Program memory may be expanded off-chip.

Bootstrap ROM

Bootstrap ROM is a 32 location by 24-bit factory programmed ROM which is used only in the bootstrap mode, Operating Mode 1. The Bootstrap ROM is not accessible by the user and is disabled in normal operating modes. See **Appendix I** for a full description of the bootstrap feature of the DSP56001.

PROGRAM CONTROLLER

The program controller performs instruction prefetch, instruction decoding, hardware DO loop control, and exception processing. It contains six directly addressable registers — the program counter (PC), loop address (LA), loop count (LC), status register (SR), operating mode register (OMR), and stack pointer (SP) — and a 15 level by 32-bit system stack memory. The 16-bit PC can address 65,536 locations in program memory space.

INPUT/OUTPUT

The I/O capability of the DSP56001 is extensive and advanced. This I/O structure facilitates interfacing into a variety of system configurations including multiple DSP56001 systems with or without a host processor, global bus systems with bus arbitration, and many serial configurations, all with minimal glue logic (see Figure 4). Each I/O interface has its own control, status, and data registers, and is treated as memory-mapped I/O by the DSP56001. Each interface has several dedicated interrupt vector addresses and control bits to enable/disable interrupts. This minimizes the overhead associated with servicing the device since each interrupt source can have its own service routine. The interrupt vectors can be programmed to one of three maskable priority levels.

Specifically, the I/O structure consists of an extremely flexible expansion port, Port A, and 24 additional I/O pins. These pins may be used as general purpose I/O pins called Port B and Port C, or allocated to an on-chip peripheral under software control. Three on-chip peripherals are provided on the DSP56001 — an 8-bit parallel host MPU/DMA interface, a serial communications interface (SCI), and a synchronous serial interface (SSI). Port B is a 15-bit I/O interface which may be used as general purpose I/O pins or as host MPU/DMA interface pins. Port C is a 9-bit I/O interface which may be used as general purpose I/O pins or as SCI and SSI serial interface pins. These interfaces are described in the following paragraphs.

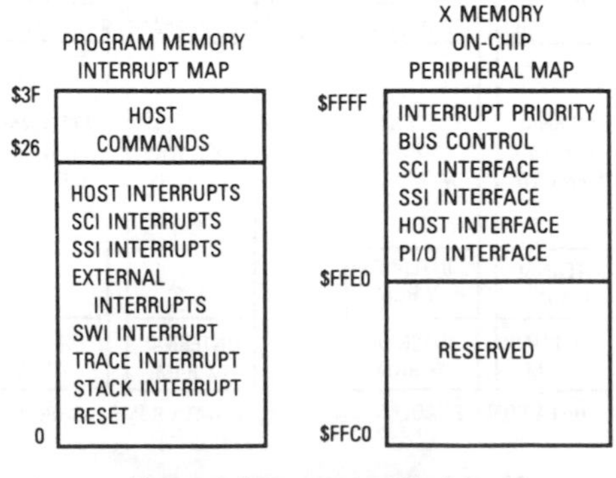

PROGRAM MEMORY INTERRUPT MAP

\$3F – \$26	HOST COMMANDS
\$26 – 0	HOST INTERRUPTS SCI INTERRUPTS SSI INTERRUPTS EXTERNAL INTERRUPTS SWI INTERRUPT TRACE INTERRUPT STACK INTERRUPT RESET

X MEMORY ON-CHIP PERIPHERAL MAP

\$FFFF – \$FFE0	INTERRUPT PRIORITY BUS CONTROL SCI INTERFACE SSI INTERFACE HOST INTERFACE PI/O INTERFACE
\$FFE0 – \$FFC0	RESERVED

Figure 4. Interrupt and Peripheral Register Memory Maps

Expansion Port (Port A)

DSP56001 expansion port is designed to synchronously interface over a common 24-bit data bus with a wide variety of memory and peripheral devices. They include high speed static RAMs, slower memory devices, as well as other DSPs and MPUs in master/slave configurations. This is possible because the expansion bus timing is programmable. The expansion bus timing is controlled by a bus control register (BCR). The BCR controls the timing of the bus interface signals $\overline{RD}$ and $\overline{WR}$, and the data output lines. Each of the four memory spaces — X data, Y data, program data, and I/O, has its own 4-bit BCR which can be programmed for up to 15 WAIT states (one WAIT state is equal to a clock period or equivalently one-half of an instruction cycle). In this way external bus timing can be tailored to match the speed requirements of the different memory spaces.

General Purpose I/O (Port B, Port C)

Each Port B and C pin may be programmed as a general purpose I/O pin or as a dedicated on-chip peripheral pin under software control. A 9-bit port control register, PCC, is associated with port C which allows each port pin to be programmed for one of these two functions. The port control register associated with port B, PBC, contains only one bit which programs all 15 pins. Also associated with each general purpose port is a data direction register which programs the direction of each pin, and a data register for data I/O. Note that all these registers are read/write making the use of bit manipulation instructions extremely effective.

HOST INTERFACE

The Host Interface is a byte-wide full duplex parallel port which may be connected directly to the data bus of a host processor. The host processor may be any of a number of industry standard microcomputers or microprocessors, another DSP, or DMA hardware. The DSP56001 Host Interface has an 8-bit bidirectional data bus H0-H7 (PB0-PB7) and seven dedicated control lines HA0, HA1, HA2, HR/$\overline{W}$, $\overline{HEN}$, $\overline{HREQ}$, and $\overline{HACK}$ (PB9-PB15) to control data transfers. The Host Interface appears as a memory mapped peripheral occupying eight bytes in the *host processor* address space. Separate transmit and receive data registers are double-buffered to allow the DSP56001 and host processor to transfer data efficiently at high speed. Host processor communication with the Host Interface is accomplished using standard host processor data move instructions and addressing modes. Handshake flags are provided for polled or interrupt driven data transfers with the host processor. DMA hardware may be used with the handshake flags to transfer data without host processor intervention.

One of the most innovative features of the Host Interface is the Host Command feature. With this feature the host processor can issue vectored exception requests to the DSP56001. The host may select any one of 32 DSP56001 exception routines to be executed by writing a Vector Address Register in the Host Interface. This flexibility allows the host programmer to execute up to 32 preprogrammed functions inside the DSP56001. For example, host exceptions can allow the host processor to read or write DSP56001 registers, X, Y, or program memory locations, force exception handlers for SSI, SCI, $\overline{IRQA}$, and $\overline{IRQB}$ exception routines, and perform control and debugging operations if the appropriate exception routines are implemented in the DSP56001 to do these tasks.

SERIAL COMMUNICATION INTERFACE

The Serial Communications Interface (SCI) provides a full-duplex port for serial communication to other DSPs, microprocessors, or peripherals such as modems. The communication can be either direct or via RS232C-type lines. This interface uses three dedicated pins — transmit data (TXD), receive data (RXD), and SCI serial clock (SCLK). It supports industry standard asynchronous bit rates and protocols as well as high speed (up to 2.5 Mbits/sec) synchronous data transmission. The asynchronous protocols include a multidrop mode for master/slave operation. The Serial Communication Interface consists of separate transmit and receive sections whose operations can be asynchronous with respect to each other. A programmable baud rate generator is included to generate the transmit and receive clocks. An enable and interrupt vector have been included so that the baud rate generator can function as a general purpose timer when it is not being used by the SCI peripheral.

SYNCHRONOUS SERIAL INTERFACE (SSI)

The Synchronous Serial Interface (SSI) is an extremely flexible full-duplex serial interface which allows the DSP56001 to communicate with a variety of serial devices. These include one or more industry standard codecs, other DSPs, microprocessors, and peripherals. The following characteristics of the SSI interface can be independently defined by the user — the number of bits per word, the protocol or mode, the clock, and the transmit/receive synchronization. There are three modes which can be selected. They are the Normal, On-Demand, and Network modes. The Normal mode is typically used to interface with devices on a regular or periodic basis. In this mode the SSI interface functions with one data word of I/O per frame. The On-Demand mode is a data driven mode. There are no timeslots defined. This mode is intended to be used to interface to devices on a non-periodic basis. The Network mode provides time slots in addition to a bit clock and frame synchronization pulse. The SSI functions with from 2 to 32 words of I/O per frame in the Network mode. This mode is typically used in star or ring Time Division Multiplex (TDM) networks with other DSP56000s and/or codecs. The clock can be programmed to be continuous or gated. Since the transmitter and receiver sections of the SSI interface are independent they may be programmed to be synchronous (use a common clock) or asynchronous with respect to each other. The SSI interface supports a subset of the Motorola SPI interface. The SSI requires three to six pins depending on the operating mode selected. A matrix of SSI operating modes and typical applications is provided in Table 1.

Table 1. SSI Operating Modes

Mode (Protocol)	Serial Clock	Relative Tx-Rx Timing	Typical Applications
Normal	Continuous	Asynchronous	Asynchronous Codec
Normal	Continuous	Synchronous	Synchronous Codecs
Normal	Gated	Asynchronous	DSP-to-DSP
Normal	Gated	Synchronous	DSP-to-A/D and D/A
On-Demand	Continuous	Asynchronous	DSP-to-MCU
On-Demand	Continuous	Synchronous	P-to-S and S-to-P Conversion
On-Demand	Gated	Asynchronous	DSP-to-DSP
On-Demand	Gated	Synchronous	DSP-to-SPI Peripherals
Network	Continuous	Asynchronous	TDM Codec Networks
Network	Continuous	Synchronous	TDM DSP Networks
Network	Gated	Asynchronous	—
Network	Gated	Synchronous	—

PROGRAMMING MODEL DESCRIPTION

The programmer can view the DSP56001 architecture as three execution units operating in parallel. The three execution units are the Data ALU, Address ALU, and Program Controller. It was possible to make the programming model like that of conventional MPUs and eliminate the need to refer to the detailed chip architecture when programming the DSP56001 because the parallel execution units make the pipeline virtually invisible. The programming model is shown in Figure 5 and is described in the following paragraphs.

DATA ALU

The data ALU features 24-bit input/output data registers which can be concatinated to handle 48-bit data, two 56-bit accumulators, programmable scaling, and saturation arithmetic.

DATA ALU INPUT REGISTERS (X1, X0, Y1, Y0)

X1, X0, Y1, and Y0 are four 24-bit general purpose data registers. They may be treated as four independent 24-bit registers or as two 48-bit registers called X and Y developed by the concatenation of X1:X0 and Y1:Y0 respectively. The register with the highest number is the most significant word. The registers serve as input pipeline registers between the XD and YD data buses and the multiply-accumulator unit (MAC). They are used as data ALU source operands and allow new operands to be loaded for the next instruction while the register contents are used by the current instruction. They may also be read back out to the appropriate data bus to implement memory delay operations and save/restore operations for interrupt service routines.

DATA ALU ACCUMULATOR REGISTERS (A2, A1, A0, B2, B1, B0)

The six data ALU registers A2, A1, A0, B2, B1, and B0 form two general purpose 56-bit accumulators, A and B, developed by the concatenation of A2:A1:A0 and B2:B1:B0 respectively. These registers are used for arithmetic calculations and data manipulation. The four registers A1, A0, B1, and B0 are 24 bits wide and the two registers A2 and B2 are 8 bits wide. All of these registers are treated as word operands. The register with the highest number is the most-significant word in the accumulator; the register with the lowest number is the least-significant word.

These accumulators are 48 bits long with 8-bit extensions to accommodate word growth in vector arithmetic. The registers A2 and B2 are called accumulator extension registers. Automatic sign extension is provided when writing to the 56-bit accumulators A or B with a 48- or 24-bit operand. The low-order portion will be automatically zeroed when a 24-bit operand is written to form a valid 56-bit operand. The registers may also be written without sign extension or zero fill by specifying the individual register name.

When the accumulator registers A or B are read, they may be optionally scaled one bit left or one bit right for block floating-point arithmetic. Reading the A or B accumulators over XD and YD is protected against overflow by substituting a limiting constant for the data that is being transferred. The content of A or B is not affected should limiting occur; only the value transfered over XD and YD is limited. This overflow protection is performed after the content of the accumulator has been shifted according to the scaling mode. Note that only when A or B as opposed to A0, A1, A2, B0, B1, or B2, is specified as the source for a parallel data move over XD, YD will shifting and limiting be performed. The accumulator registers serve as pipeline registers between the MAC unit and the XD and YD data buses. They are used as both data ALU source and destination operands.

ADDRESS ALU

The programmer's model for the address ALU consists of three banks of register files — address register files, offset register files, and modifier register files. They provide all the registers necessary to generate address register indirect effective addresses.

Address Register Files (R0-R3 and R4-R7)

The eight address registers, R0-R7, are 16 bits wide and may contain addresses or general purpose data. The 16-bit address in a selected address register is used in the calculation of the effective address of an operand. When supporting parallel X and Y data memory moves the address registers must be thought of as separate two files, R0-R3 and R4-R7. The content of an Rn may point to data directly or may be pre- or post-updated according to the addressing mode selected. Modifier registers, Mn, are always used if an Rn is updated. Offset registers, Nn, are used for the update by offset addressing modes. The address register modification is performed by one of the two modulo arithmetic units.

Offset Register Files (N0-N3 and N4-N7)

The eight offset registers, N0-N7, are 16 bits wide and may contain offset values used to increment and decrement address registers in address register update calculations or they may be used for 16-bit general purpose storage. For example, the contents of an offset register may be used to step through a table at some rate (e.g., five locations per step for waveform generation), or may specify the offset into a table or the base of the table for indexed addressing. Each address register, Rn, has its own offset register, Nn, associated with it.

Modifier Register Files [M0-M3 and M4-M7]

The eight modifier registers, M0-M7, are 16 bits wide. The content of Mn defines the type of address arithmetic to be performed for addressing mode calculations. The address ALU supports linear, modulo, and reverse carry arithmetic types for all address register indirect addressing modes. For the case of modulo arithmetic, the content of Mn also specifies the modulus. Each address register, Rn, has its own modifier register, Mn, associated with it. Each modifier register is set to $FFFF on processor reset which specifies linear arithmetic as the default type for address register update calculations.

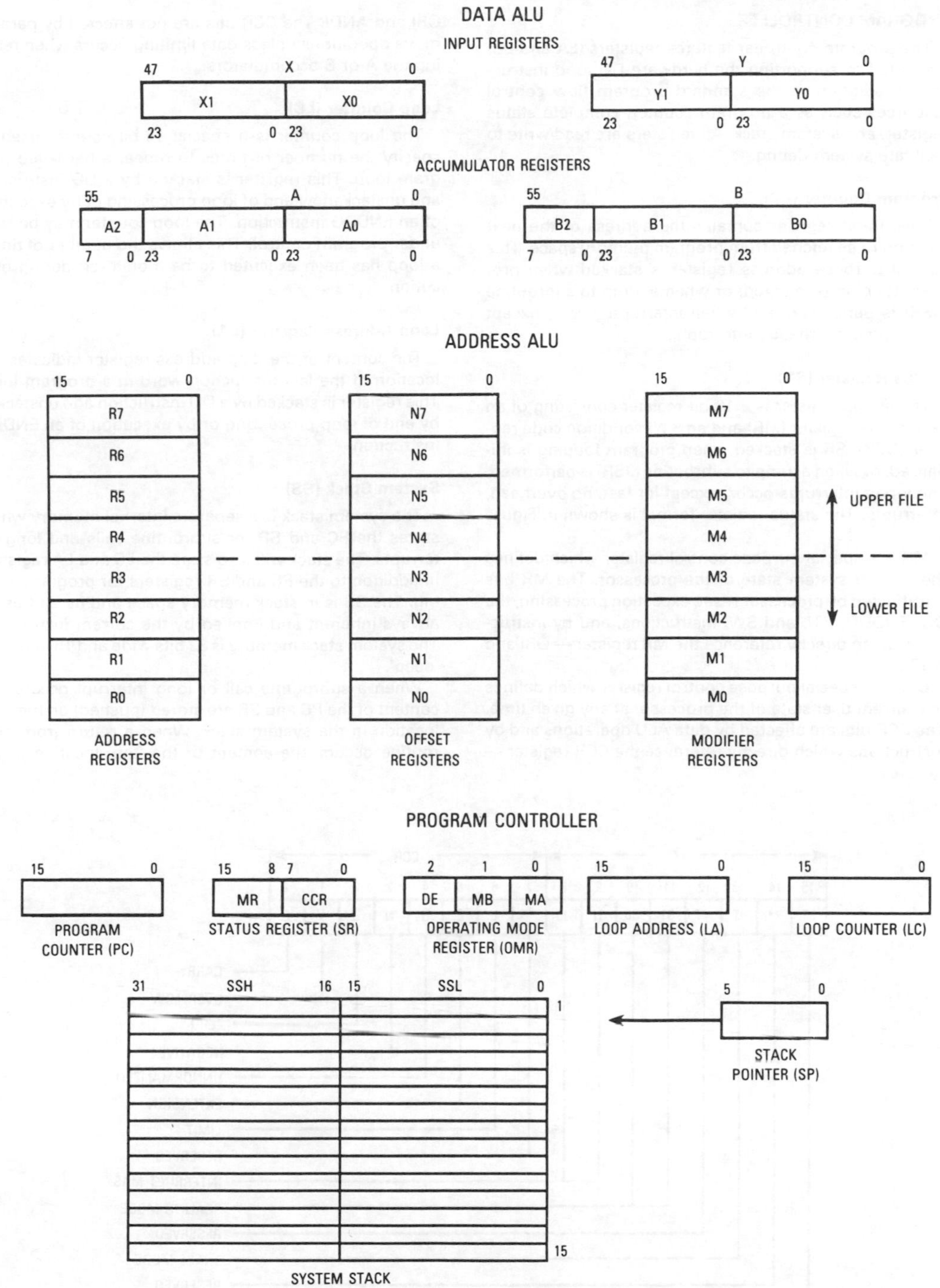

Figure 5. DSP56001 Programming Model

PROGRAM CONTROLLER

The program controller features registers (LA and LC) dedicated to supporting the hardware DO loop instruction in addition to the standard program flow control resources such as a program counter, complete status register, and system stack. All registers are read/write to facilitate system debug.

Program Counter (PC)

This 16-bit register contains the address of the next location to be fetched from program memory space. This special purpose address register is stacked when program looping is initiated, or when a jump to subroutine (JSR) is performed, and when interrupts occur, except for fast, 'no overhead', interrupts.

Status Register (SR)

The status register is a 16-bit register consisting of an 8-bit mode register (MR) and an 8-bit condition code register (CCR). SR is stacked when program looping is initialized, or when a jump to subroutine (JSR) is performed, and when interrupts occur, except for fast, no overhead, interrupts. The status register format is shown in Figure 6.

MR is a special purpose control register which defines the current system state of the processor. The MR bits are affected by processor reset, exception processing, the DO, ENDDO, RTI, and SWI instructions, and by instructions which directly reference the MR register — ORI and ANDI.

CCR is a special purpose control register which defines the current user state of the processor at any given time. The CCR bits are affected by data ALU operations and by instructions which directly reference the CCR register — ORI and ANDI. The CCR bits are not affected by parallel move operations unless data limiting occurs when reading the A or B accumulators.

Loop Counter (LC)

The loop counter is a special 16-bit counter used to specify the number of times to repeat a hardware program loop. This register is stacked by a DO instruction and unstacked by end of loop processing or by execution of an ENDDO instruction. The loop counter may be read under program control. This allows the number of times a loop has been executed to be monitored during execution.

Loop Address Register (LA)

The content of the loop address register indicates the location of the last instruction word in a program loop. This register is stacked by a DO instruction and unstacked by end of loop processing or by execution of an ENDDO instruction.

System Stack (SS)

The system stack is a separate internal memory which stores the PC and SR for subroutine calls and long interrupts. The stack will also store the LC and LA registers in addition to the PC and SR registers for program looping. The SS is in stack memory space and its address is always inherent and implied by the current instruction. The system stack memory is 32 bits wide and 15 locations 'deep'.

When a subroutine call or long interrupt occurs, the content of the PC and SR are stored (pushed) on the 'top' location in the system stack. When a return from subroutine occurs, the content of the 'top' location in the

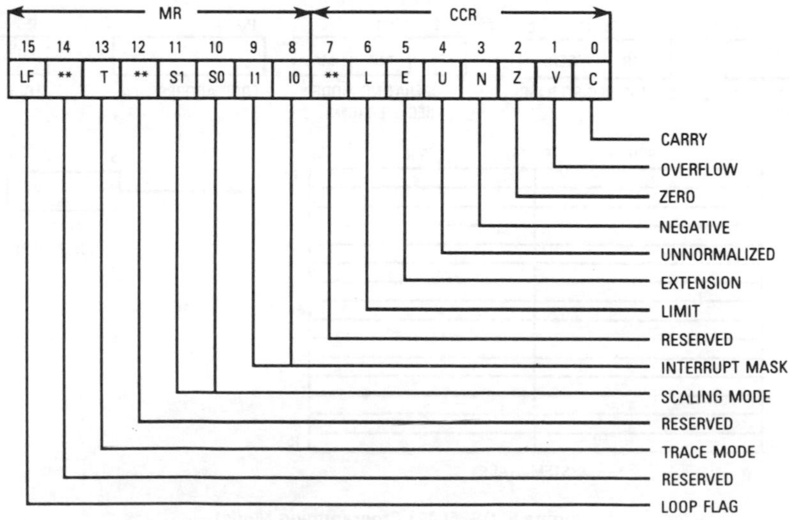

Figure 6. Status Register Format

system stack are transferred (pulled) to the PC only. When a return from interrupt occurs, the content of the 'top' location in the system stack are transferred (pulled) to both the PC and SR.

The interrupt subsystem of the DSP56001 is vector based and prioritized. Interrupt vectors point to two consecutive locations in program memory. If one of the two words fetched by the interrupt controller is a jump to subroutine instruction, a long interrupt routine is formed, and a context switch is performed using the stack. If neither interrupt instruction word causes a change of control flow, then the two interrupt instruction words fetched constitute a fast interrupt routine. The fast interrupt routine provides exception processing with no overhead. This mechanism is commonly used to move data between memory and an I/O device.

The system stack is also used to implement no-overhead nested hardware DO loops. Each stack location can be addressed as two separate 16-bit registers, system stack high (SSH) and system stack low (SSL). This facilitates creating software stacks for unlimited nesting.

Stack Pointer (SP)

The stack pointer register (SP) is a 6-bit register that indicates the location of the top of the system stack and the status of the stack (underflow, empty, full, and overflow conditions). The stack pointer is referenced implicitly by some instructions (DO, REP, JSR, RTI, etc.) or directly by the MOVEC instruction. The stack pointer register format is shown in Figure 7.

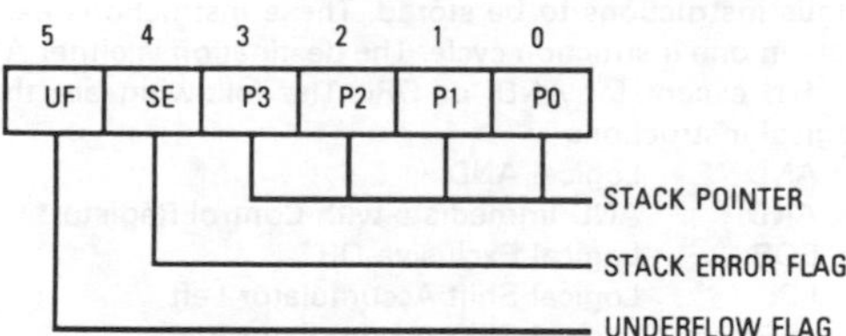

Figure 7. Stack Pointer Format

Operating Mode Register (OMR)

The operating mode register (OMR) is a 3-bit register which defines the current operating mode of the processor, i.e., the memory maps for program and data memories as well as the startup procedure. The OMR bits are only affected by processor reset and by instructions which directly reference the OMR. During processor reset the chip operating mode bits, MB and MA, will be loaded from the external mode select A, B, pins. The Data ROM Enable bit (DE) will be cleared, disabling the X and Y on-chip lookup table ROMs. Figure 3 shows the effect of the OMR on the DSP56001 memory maps. The operating mode register format is shown in Figure 8. Table 2 summarizes the DSP56001 operating modes. Tables 3 and 4 show the program and data memory spaces.

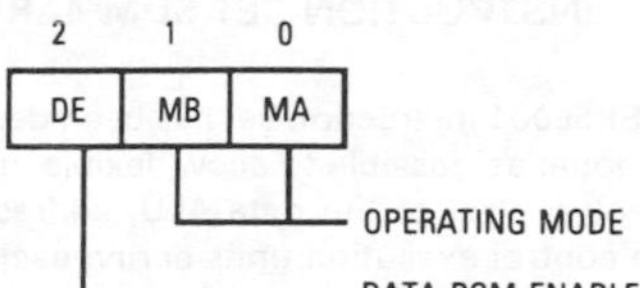

Figure 8. Operating Mode Register Format

Table 2. Operating Mode Summary

Operating Mode	M B	M A	Description
0	0	0	PRAM enabled, Reset at $0000 (internal).
1	0	1	Special Bootstrap mode, after PRAM loading mode 2 is automatically selected.
2	1	0	PRAM enabled, Reset at $E000 (external).
3	1	1	PRAM disabled, Reset at $0000 (external).

Table 3. Program Memory Space

Operating Mode	M B	M A	Program Memory Space Map
0 & 2	X	0	internal RAM: $0000-$01FF external: $0200-$FFFF
3	X	1	external: $0000-$FFFF

Table 4. Data Memory Space

DROM Enable DE	Y Data Memory Space Map	X Data Memory Space Map
0	internal RAM: $0000-$00FF external: $0100-$FFFF	internal RAM: $0000-$00FF external: $0100-$FFBF on-chip peripherals: $FFC0-$FFFF
1	internal RAM: $0000-$00FF internal ROM: $0100-$01FF external: $0200-$FFFF	internal RAM: $0000-$00FF internal ROM: $0100-$01FF external: $0200-$FFBF on-chip peripherals: $FFC0-$FFFF

INSTRUCTION SET SUMMARY

The DSP56001 instruction set has been designed to be as orthogonal as possible to allow flexible, independent, concurrent control of the data ALU, address ALU, and program control execution units during each instruction cycle. This maximizes throughput and minimizes program storage requirements.

The rich instruction set features DSP oriented instructions such as CMPM, NORM, RND, MACR, SUBL, SUBR, ADDL, ADDR, DO, and REP, which are summarized below.

The instruction set is divided into the following groups:

Arithmetic
Logical
Bit Manipulation
Loop
Move
Program Control

ARITHMETIC INSTRUCTIONS

The arithmetic instructions perform all of the arithmetic operations within the data ALU. They may affect all of the condition code register bits. Arithmetic instructions are register-based so that the data ALU operation indicated by the instruction does not use the X data bus, the Y data bus, or the global data bus. This allows for parallel data movement over the X and Y data buses or over the global data bus during a Data ALU operation. This permits new data to be pre-fetched for use in following instructions, and results calculated by previous instructions to be stored. These instructions execute in one instruction cycle. The destination is either of the 56-bit accumulators. The following are the arithmetic instructions.

ABS	Absolute Value
ADC	Add Long with Carry
ADD	Add
ADDL	Shift Left then Add Accumulators
ADDR	Shift Right then Add Accumulators
ASL	Arithmetic Shift Accumulator Left
ASR	Arithmetic Shift Accumulator Right
CLR	Clear Accumulator
CMP	Compare
CMPM	Compare Magnitude
DIV	Divide Iteration*
MAC	Signed Multiply-Accumulate
MACR	Signed Multiply-Accumulate and Round
MPY	Signed Multiply
MPYR	Signed Multiply and Round
NEG	Negate Accumulator
NORM	Normalize Accumulator Iteration*
RND	Round Accumulator
SBC	Subtract Long with Carry
SUB	Subtract
SUBL	Shift Left then Subtract Accumulators
SUBR	Shift Right then Subtract Accumulators
Tcc	Transfer Conditionally*
TFR	Transfer Data ALU Register
TST	Test Accumulator

*These instructions do not allow parallel data moves.

The CMPM affects the condition code bits according to the results of the subtraction of the absolute values of two operands. This instruction, together with TCC, is useful in determining maxima and minima in blocks of data.

The NORM instruction normalizes the content of an accumulator register and updates the content of the specified address register according to the normalization. This is particularly useful in implementing floating-point routines.

The RND instruction performs convergent rounding on the content of an accumulator register in a manner consistent with the scaling mode operation.

The MACR instruction is one of the most powerful instructions in the instruction set. It performs a signed multiply-accumulate with convergent rounding and allows two parallel data moves in one instruction. These rounding instructions facilitate minimizing the effects of roundoff errors.

The ADDL, ADDR, SUBL, and SUBR, instructions multiply or divide the content of the accumulator register by two before the addition or subtraction operation is performed. They are particularly useful for implementing Radix-2 Decimation In Time (DIT) Fast Fourier Transforms (FFT).

LOGICAL INSTRUCTIONS

The logical instructions perform all of the logical operations within the data ALU. They affect all of the condition code register bits. Logical instructions are register-based. Optional data transfers may be specified with most logical instructions. This allows for parallel data movement over XD, YD, or GD during a Data ALU logical operation. This allows new data to be pre-fetched for use in following instructions, and results calculated in previous instructions to be stored. These instructions execute in one instruction cycle. The destination is either A1 or B1, except for ANDI or ORI. The following are the logical instructions.

AND	Logical AND
ANDI	AND Immediate with Control Register*
EOR	Logical Exclusive OR
LSL	Logical Shift Accumulator Left
LSR	Logical Shift Accumulator Right
NOT	Logical Complement on Accumulator
OR	Logical Inclusive OR
ORI	OR Immediate with Control Register*
ROL	Rotate Accumulator Left
ROR	Rotate Accumulator Right

*These instructions do not allow parallel data moves.

BIT MANIPULATION INSTRUCTIONS

There are two basic groups of bit manipulation instructions. One group tests the state of any single bit in a memory location and then optionally sets, clears, or inverts the bit. The other group tests the state of any single bit in a memory location and jumps (or jumps to subroutine) if the bit is set or clear. The carry bit of the condition code register will contain the result of the bit test for the first group. The following are the bit manipulation instructions. Note that parallel data moves are not allowed with any of these instructions.

BCLR	Bit Test and Clear
BSET	Bit Test and Set

BCHG	Bit Test and Change
BTST	Bit Test on Memory
JCLR	Jump if Bit Clear
JSET	Jump if Bit Set
JSCLR	Jump to Subroutine if Bit Clear
JSSET	Jump to Subroutine if Bit Set

LOOP INSTRUCTIONS

The DO and ENDDO instructions make writing straight line code practically unnecessary. The DO instruction sets up a hardware loop by initiating a program loop, setting up looping parameters and then 'cleans up' the system stack when terminating a loop. Initialization includes saving registers LA and LC used by a program loop on the system stack so that program loops can be nested. The address of the first instruction in a program loop is also saved to allow no-overhead looping. Single instruction DO loops can be implemented. The ENDDO instruction is used to terminate a DO loop prematurely. It is used to 'clean up' the stack. These instructions do not allow parallel data moves. The following are the loop instruction definitions.

DO	Start Hardware Loop
ENDDO	Exit from Hardware Loop

MOVE INSTRUCTIONS

The move instructions perform data movement over XD, YD, and GD data buses as well as the program data bus, PD. Move instructions do not affect the condition code register except the limit bit, L, if limiting is performed when reading Data ALU accumulator registers A or B. The MOVE instruction provides all of the parallel data move operations and can be considered to be a data ALU no-op with parallel moves. The following are the move instructions.

LUA	Load Updated Address
MOVE	Move Data
MOVEC	Move Control Register
MOVEM	Move Program Memory
MOVEP	Move Peripheral Data

PROGRAM CONTROL INSTRUCTIONS

The program control instructions include jumps, conditional jumps, and other instructions which affect the PC and system stack. Program control instructions may affect the condition code register bits as specified in the instruction. Optional parallel data transfers over XD, YD, and GD are not allowed in the program control instructions. The REP instruction repeats the next instruction without refetching the instruction to maximize throughput without resorting to using straight line code. Because the REP instruction is not refetched it is not interruptable. An interruptable repeat instruction can be implemented using a single instruction DO loop. All processor activity is suspended and the oscillator is gated off after a STOP instruction has been executed. When the WAIT instruction is executed internal processing is halted and the processor waits for an interrupt. The STOP and WAIT states are low power states. The following are the program control instructions.

Jcc	Jump Conditionally
JMP	Jump
JScc	Jump to Subroutine Conditionally
JSR	Jump to Subroutine
NOP	No Operation
REP	Repeat Next Instruction
RESET	Reset On-Chip Peripheral Devices
RTI	Return from Interrupt
RTS	Return from Subroutine
STOP	Stop Processing (Low Power Standby)
SWI	Software Interrupt
WAIT	Wait for Interrupt (Low Power Standby)

INSTRUCTION FORMATS

Instructions are one or two words in length. The instruction and its length are specified by the first word of the instruction. The second word may contain information about an operand for the instruction. The assembly language source code for a typical one word instruction is shown below. The source code is organized into four columns.

Opcode	Operands	X Bus Data	Y Bus Data
MAC	X0,Y0,A	X:(R0)+,X0	Y:(R4)+,Y0

The Opcode column typically indicates the Data ALU operation to be performed; it may also specify an Address ALU or Program Control operation. The operands column specifies the operands to be used by the opcode. The X bus data column specifies an optional data transfer over the X bus and the addressing mode to be used. The Y bus data column specifies an optional data transfer over the Y bus and the addressing mode to be used. The memory space qualifiers X:, Y:, P: and L: (long memory space) indicate which memory space is being referenced. The opcode column must always be included in the source code.

The DSP56001 allows parallel processing by the data ALU, address ALU, and program controller. For example, in the instruction word above the DSP56001 will perform the designated ALU operation (data ALU), the data transfers specified with address register updates (address ALU), and will also decode the next instruction and fetch an instruction word from program memory (program controller), all in one instruction cycle. In addition, the program controller may be processing an active hardware DO loop. When an instruction is more than one word in length, an additional instruction execution cycle may be required. Most operations involving the data ALU are register-based (that is, all operands are in data ALU registers) and therefore do not utilize the data buses. This allows the programmer to keep each execution unit busy by specifying memory accesses in parallel over the XD, YD, or GD buses. An instruction which is memory-oriented (such as a bit manipulation instruction) or an instruction that causes a control flow change (such as a jump) does not allow parallel data moves during its execution.

ADDRESSING MODES

The addressing modes are grouped into three categories — register direct, address register indirect, and special. These addressing modes are summarized in Table 5. All address calculations are performed in the address ALU to minimize execution time and loop overhead. Addressing modes specify whether the operand(s) is (are) in a register, memory, or encoded in the instruction (as immediate data) and provide the specific addresses of the operand.

The register direct addressing mode can be subclassified according to the specific register addressed. The data registers include X1, X0, Y1, Y0, X, Y, A2, A1, A0, B2, B1, B0, A, and B. The control registers include SR, OMR, SP, SSH, SSL, LA, LC, CCR, and MR.

Address register indirect modes use an address register, Rn, to point to locations in memory. The content of Rn is the effective address (ea) except in the indexed by offset mode where the ea is Rn+Nn. Address register indirect modes use a modifier register, Mn, to specify the type of arithmetic to be used to update Rn. If a mode using an offset is specified an offset register, Nn, is also used for the update. The Nn and Mn registers are assigned to the Rn with the same n. Thus the assigned register sets are R0;N0;M0, R1;N1;M1, R2;N2;M2, R3;N3;M3, R4;N4;M4, R5;N5;M5, R6;N6;M6, and R7;N7;M7. This structure is unique and extremely powerful in general, and particularly powerful in setting up DSP oriented data structures. All address register indirect modes use at least one set of address registers, and the XY memory reference uses two sets of address registers, one set for X memory space and one set for Y memory space.

The special addressing modes include immediate and absolute modes as well as implied references to the PC, system stack, and program memory.

ADDRESS MODIFIERS [Mn]

The address modifiers allow the DSP Address ALU to support linear, reverse-carry, and modulo address arithmetic for all address register indirect modes. These special address arithmetic types allow the creation of data structures in memory for FIFOs (queues), delay lines, circular buffers, stacks and bit-reversed FFT buffers. Data is manipulated by updating address registers rather than moving large blocks of data. The content of the address modifier register, Mn, defines the type of address arithmetic to be performed for addressing mode calculations.

Table 5. Addressing Modes Summary

Addressing Mode	Modifier MMMM	Operand Reference								
		S	C	D	A	P	X	Y	L	XY
Register Direct										
Data or Control Register	No	x	x	x						
Address Register	No				x					
Address Modifier Register	No				x					
Address Offset Register	No				x					
Address Register Indirect										
No Update	Yes					x	x	x	x	x
Postincrement by 1	Yes					x	x	x	x	x
Postdecrement by 1	Yes					x	x	x	x	x
Postincrement by Offset Nn	Yes					x	x	x	x	x
Postdecrement by Offset Nn	Yes					x	x	x	x	
Indexed by Offset Nn	Yes					x	x	x	x	
Predecrement by 1	Yes					x	x	x	x	
Special										
Immediate Data	No					x				
Absolute Address	No					x	x	x	x	
Immediate Short Data	No					x				
Short Jump Address	No					x				
Absolute Short Address	No					x	x	x	x	
I/O Short Address	No						x	x		
Implicit	No	x	x			x				

Where:
MMMM = Address Modifier
S = Stack Reference
C = Program Controller Register Reference
D = Data ALU Register Reference
A = Address ALU Register Reference
P = Program Memory Reference
X = X Memory Reference
Y = Y Memory Reference
L = L Memory Reference
XY = XY Memory Reference

For the case of modulo arithmetic, the content of Mn also specifies the modulus. The three types of arithmetic are discussed below.

Linear Arithmetic [Mn = $FFFF]

The address modification is performed using normal 16-bit (modulo 65,536) linear arithmetic (2's complement). A 16-bit offset, Nn, may be used in the address calculations. The range of values may be considered as signed (Nn from −32,768 to +32,767) or unsigned (Nn from 0 to +65,535).

Reverse-Carry Arithmetic [Mn = $0000]

The address modification is performed by propagating the carry in the reverse direction, that is, from the most-significant bit (MSB) to the least-significant bit (LSB). This is equivalent to bit-reversing (i.e., redefining the MSB as the LSB and the next MSB as bit 1, etc.) the content of Rn and the offset value Nn, adding normally and then bit-reversing the result. If the (Rn) + Nn addressing mode is used with this address modifier type, and Nn contains the value two to the power of k-1 (2^{k-1}), then post-incrementing by +Nn is equivalent to incrementing Rn by 1 and bit-reversing the k LSBs of Rn. This address arithmetic is useful for performing 2^k point Fast Fourier Transforms (FFTs). The range of values for Nn is 0 to +65,535. This allows bit-reversed addressing for FFTs having up to 65,536 points.

As an example, consider a 1,024 point FFT (k = 10) with real data stored in X memory and imaginary data stored in Y memory. Then Nn would contain the value 512 and postincrementing by +Nn would generate the address sequence 0, 512, 256, 768, 128, 640, ... This is the scrambled FFT data order for sequential frequency points from 0 to 2pi. The base address must have at least k zeros so that the reverse-carry modifier also works when the base address of the FFT data buffer is a multiple of 2_k, such as 2048, 3072, in our example. The use of addressing modes other than postincrement by + Nn is possible but may not provide a useful result.

Modulo Arithmetic [Mn = modulus − 1]

The address modification is performed modulo M, where M ranges from 2 to +32,768. Modulo M arithmetic causes the address register value to remain within an address range of size M defined by a lower and upper address boundary. The value m = M-1 is stored in the modifier register Mn. The lower boundary (base address) value must have zeroes in the k LSBs, where $2^k \geq M$, and therefore must be a multiple of 2^k. The upper boundary is the lower boundary plus the modulo size minus one (base address plus M-1). For example, to create a circular buffer of 21 stages, M is 21 and the lower address boundary must have its five least-significant bits equal to zero ($2^k \geq 21$, thus $k \geq 5$). The Mn register is loaded with the value 20. The lower boundary may be chosen as 0, 32, 64, 96, 128, 160, etc. The upper boundary of the buffer is then the lower boundary plus 20. The address pointer is not required to start at the lower address boundary or end on the upper address boundary; it may initially point anywhere within the defined modulo address range. Note that neither the lower nor the upper boundary of the modulo region is stored; only the size of the modulo region is stored in Mn. Assuming the (Rn) + indirect addressing mode, if the address register pointer increments past the upper boundary of the buffer (base address plus M-1) it will wrap around to the base address (lower boundary). Alternatively, assuming the (Rn) − indirect addressing mode, if the address decrements past the lower boundary (base address) it will wrap around to the base address plus M-1 (upper boundary).

If an offset, Nn, is used in the address calculations, the 16-bit value |Nn| must be less than or equal to M. The range of values for Nn is −32,768 to +32,767. The modulo arithmetic unit will automatically wrap the address pointer around by the required amount. This type of address modification is useful in creating circular buffers for FIFOs (queues), delay lines, and sample buffers up to 32,768 words long. It is also useful for decimation, interpolation, and waveform generation.

APPENDIX I
BOOTSTRAP MODE — OPERATING MODE 1

The bootstrap feature of the DSP56001 consists of four special on-chip modules: the 512 words of PRAM, a 32-word bootstrap ROM, the bootstrap control logic and the bootstrap firmware program.

BOOTSTRAP ROM

This 32-word on-chip ROM has been factory programmed to perform the actual bootstrap operation from the memory expansion port (Port A) or from the Host Interface. Users have no access to the bootstrap ROM other than through the bootstrap process. Control logic will disable the bootstrap ROM during normal operations.

BOOTSTRAP CONTROL LOGIC

The bootstrap mode control logic is activated when the DSP56001 is in Operating Mode 1. The control logic maps the bootstrap ROM into program memory space as long as the DSP56001 remains in Operating Mode 1. The bootstrap firmware changes operating modes when the bootstrap load is completed.

When the DSP56001 exits the reset state in Mode 1, the following actions occur.

1. The control logic maps the bootstrap ROM into the internal DSP program memory space starting at location $0000.

2. The control logic forces the user PRAM to be write-only memory during the bootstrap loading process.

3. Program execution begins at location $0000 in the bootstrap ROM.

   The bootstrap ROM program is able to perform the load of PRAM through either the memory expansion port from a byte-wide external memory, or through the Host Interface.

4. The bootstrap ROM program executes the following sequence to end the bootstrap operation and begin user program execution.

   A. Enter Operating Mode 2 by writing to the OMR. This action will be timed to remove the bootstrap ROM from the program memory map, and re-enable read/write access to the PRAM.

   B. The change to Mode 2 is timed exactly to allow the boot program to execute a NOP then a JMP #00 and begin execution of the user's program at location $0000.

The user may also select the bootstrap mode by writing into the OMR. This will cause the bootstrap ROM to be mapped into the program address space after a delay to allow execution of a NOP then a JMP #00. This technique allows the DSP56001 user to reboot the system (with a different program if desired). Steps 1 and 2 below detail this technique.

1. From any operating mode, program the OMR to Operating Mode 1. This begins a timed operation to map the bootstrap ROM into the program address space.

2. This delay is exactly timed to allow the DSP program to execute a NOP then a JMP #00 and begin the bootstrap process as described above in steps 1-4.

BOOTSTRAP FIRMWARE PROGRAM

Bootstrap ROM contains the bootstrap firmware program that performs initial loading of the DSP56001 PRAM.

The program is written in DSP56000/DSP56001 assembly language. It contains two separate methods of initailizing the PRAM: loading from a byte-wide memory starting at location $C000 or loading through the Host Interface. The particular method used is selected by the level of program memory location $C000, bit 23.

If location P:$C000, bit 23 is read as a one, the external bus version of the bootstrap program will be selected. Typically, a byte wide EPROM will be connected to the DSP56001 Address and Data Bus as shown in Figure A-2. The data contents of the EPROM must be organized as shown below.

Address of External Byte Wide P Memory	Contents Loaded to Internal PRAM at:	
P:$C000	P:0000	low byte
P:$C001	P:0000	mid byte
P:$C002	P:0000	high byte
•	•	
•	•	
•	•	
P:$C5FD	P:01FF	low byte
P:$C5FE	P:01FF	mid byte
P:$C5FF	P:01FF	high byte

If location P:$C000, bit 23 is read as a zero, the Host Interface version of the bootstrap program will be selected. Typically a host microprocessor will be connected to the DSP56001 Host Interface. The host microprocessor must write the host interface registers TXH, TXM, and then TXL with the desired contents of PRAM from location P:$0000 up to P:$01FF. If less than 512 words are to be loaded, the host programmer can exit the bootstrap program and force the DSP56001 to begin executing at location P:0000 by setting HF0 = 1 in the host interface. In most systems, the DSP56001 response is so fast that handshaking between the DSP56001 and the host is not necessary.

The bootstrap program is shown in flowchart form in Figure A-1 and in assembler listing format below.

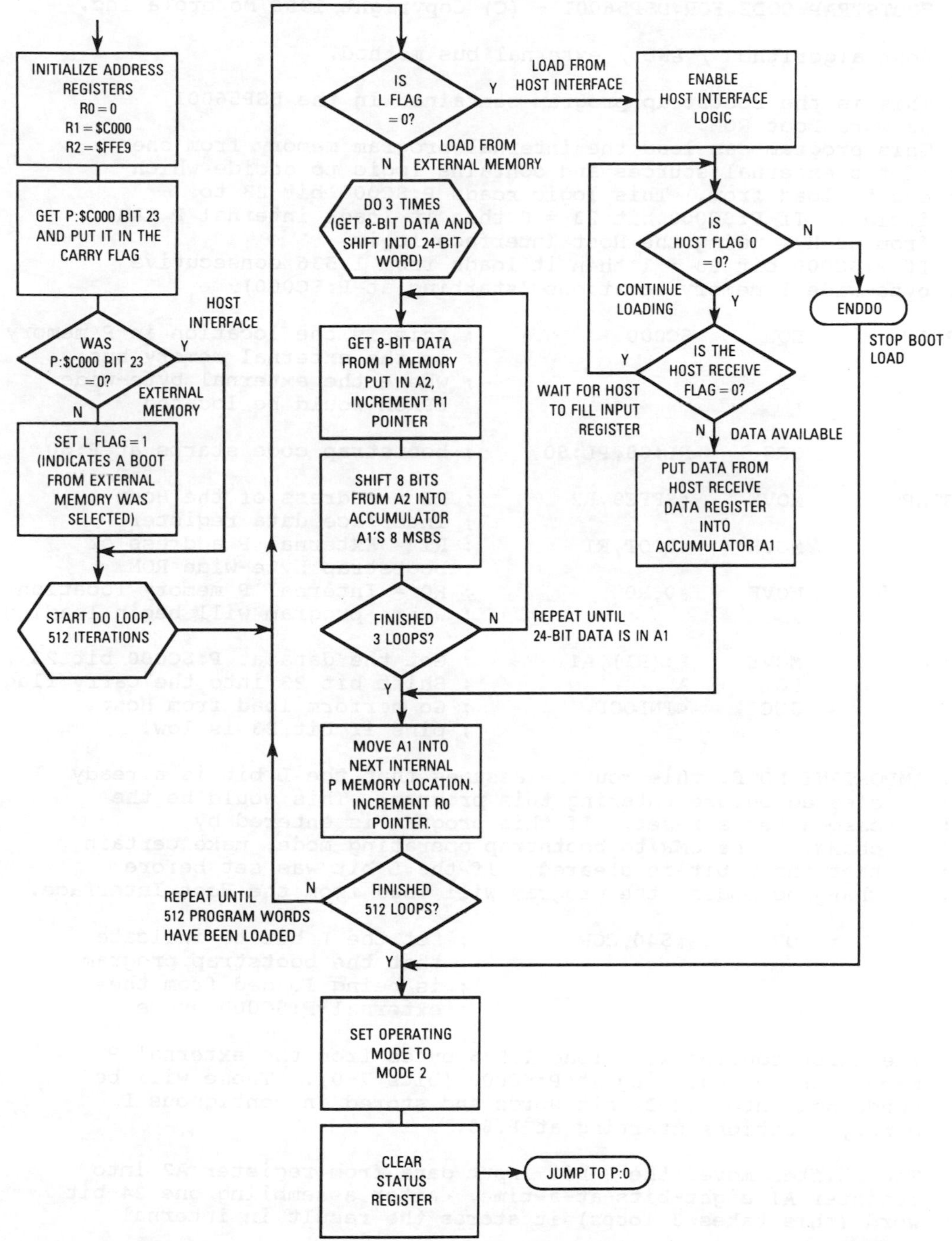

Figure A-1. Bootstrap Program Flowchart

```
; BOOTSTRAP CODE FOR DSP56001 - (C) Copyright 1986 Motorola Inc.
;
; Host algorithm  / AND / external bus method.
;
; This is the Bootstrap program contained in the DSP56001
; 32 word Boot ROM.
; This program can load the internal program memory from one
; of two external sources and contains logic to decide which
; one to load from.  This logic reads P:$C000 bit 23 to
; decide.  If P:$C000 bit 23 = 0 then it loads internal P RAM
; from H0-H7, using the Host Interface logic.
; If P:$C000 bit 23 = 1 then it loads from 1,536 consecutive
; byte-wide P memory locations (starting at P:$C000)

BOOT     EQU    $C000          ; this is the location in P memory
                               ; on the external memory bus
                               ; where the external byte-wide
                               ; EPROM would be located

         ORG    PL:$0,PL:$0    ; bootstrap code starts at P:$0

START    MOVE   #$FFE9,R2      ; R2 = address of the Host
                               ; Interface data register
         MOVE   #BOOT,R1       ; R1 = External P address of
                               ; bootstrap byte-wide ROM
         MOVE   #0,R0          ; R0 = Internal P memory location
                               ; where program will begin loading

         MOVE   P:(R1),A1      ; Get the data at P:$C000 bit 23
         ROL    A              ; Shift bit 23 into the Carry flag
         JCC    <INLOOP        ; Go perform load from Host
                               ; pins if bit 23 is low.

; IMPORTANT NOTE: this routine assumes that the L bit is already
;    cleared before entering this program.  This would be the
;    case after a reset.  If this program is entered by
;    changing the OMR to bootstrap operating mode, make certain
;    that the L bit is cleared.  If the L bit was set before
;    changing modes, the program will load from the Host Interface.

         ORI    #$40,CCR       ; Set the L bit to indicate
                               ; that the bootstrap program
                               ; is being loaded from the
                               ; external P:$C000 space.

; The first routine will load 1,536 bytes from the external P
; memory space beginning at P:$C000 (bits 7-0).  These will be
; condensed into 512 24-bit words and stored in contiguous P
; memory locations starting at P:$0.

; The shifter moves the 8-bit input data from register A2 into
; register A1 eight-bits-at-a-time.  After assembling one 24-bit
; word (this takes 3 loops) it stores the result in internal
```

```
; P memory and continues until internal P memory is filled.
; Note that the first routine loads data starting with the least
; significant byte of P:$0 first.

;       The second routine loads the internal P memory space
; using the Host Interface interface logic.  If the Host only wants
; to load a portion of the P memory, he can kill the Host Interface
; bootstrap load program, and start execution of the loaded
; program, by setting the Host Flag 0 (HF0) = 1.

INLOOP   DO      #512,_LOOP1        ; Load 512 instruction words

; This is the context switch

         JLC     <_HOSTLD           ; Go load from the Host Interface
                                    ; if the Limit flag is clear

; This is the first routine. It loads from external P memory.

         DO      #3,_LOOP2          ; Each instruction has 3 bytes
         MOVE    P:(R1)+,A2         ; Get the 8 LSB from ext. P mem.
         REP     #8                 ; Shift 8 bit data into A1
         ASR     A
_LOOP2                              ; Get another byte.
         JMP     <_STORE            ; Then put the word in P memory

; This is the second routine. It loads from the Host Interface pins.

_HOSTLD  BSET    #0,X:$FFE0         ; Configure Port B as Host
_LBLA    JCLR    #3,X:$FFE9,_LBLB   ; if HF0=1, stop loading data.
         ENDDO                      ; Must terminate the do loop
         JMP     <_BOOTEND

_LBLB    JCLR    #0,X:(R2),_LBLA    ; Wait for HRDF to go high
                                    ; (meaning data is present).
         MOVE    X:$FFEB,A1         ; Put 24-bit host data in A1

_STORE   MOVE    A1,P:(R0)+         ; Store 24-bit result in P mem.

_LOOP1                              ; and go get another 24-bit word.

; This is the exit handler that returns execution to normal
; expanded mode and jumps to the RESET vector.

_BOOTEND MOVEC   #2,OMR             ; Set the operating mode to 2
                                    ; (and trigger an exit from
                                    ; bootstrap mode).
         ANDI    #$0,CCR            ; Clear SR as if RESET to 0.
                                    ; Delay needed for Op. Mode change
         JMP     <$0                ; Then execute the RESET vector.
```

APPLICATION EXAMPLES

The lowest cost DSP56001-based system uses no external memory and requires two chips, the DSP56001 and a low cost EPROM. The EPROM read access time should be less than 780 nanoseconds when the DSP56001 is operating at it's maximum clock rate of 20.5 MHz.

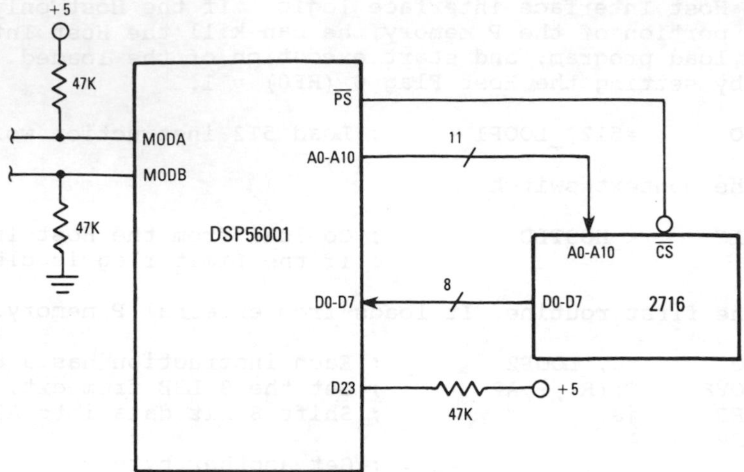

Figure A-2. No Glue, Low Cost Memory Port Bootstrap

A system with external data RAM memory requires no glue logic to select the external EPROM from bootstrap mode. $\overline{PS}$ is used to enable the EPROM and $\overline{DS}$ is used to enable the high speed data memories as shown in Figure A-3.

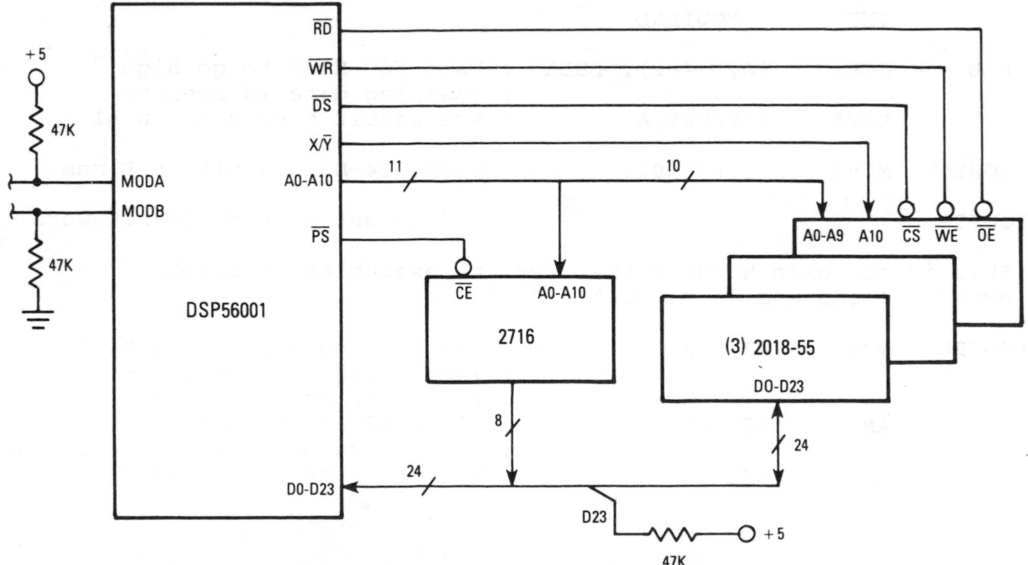

Figure A-3. Bootstrap with External Data RAM

This example (Figure A-4) shows the DSP56001 bootstrapping via the Host Port from a 12.5 MHz MC68000.

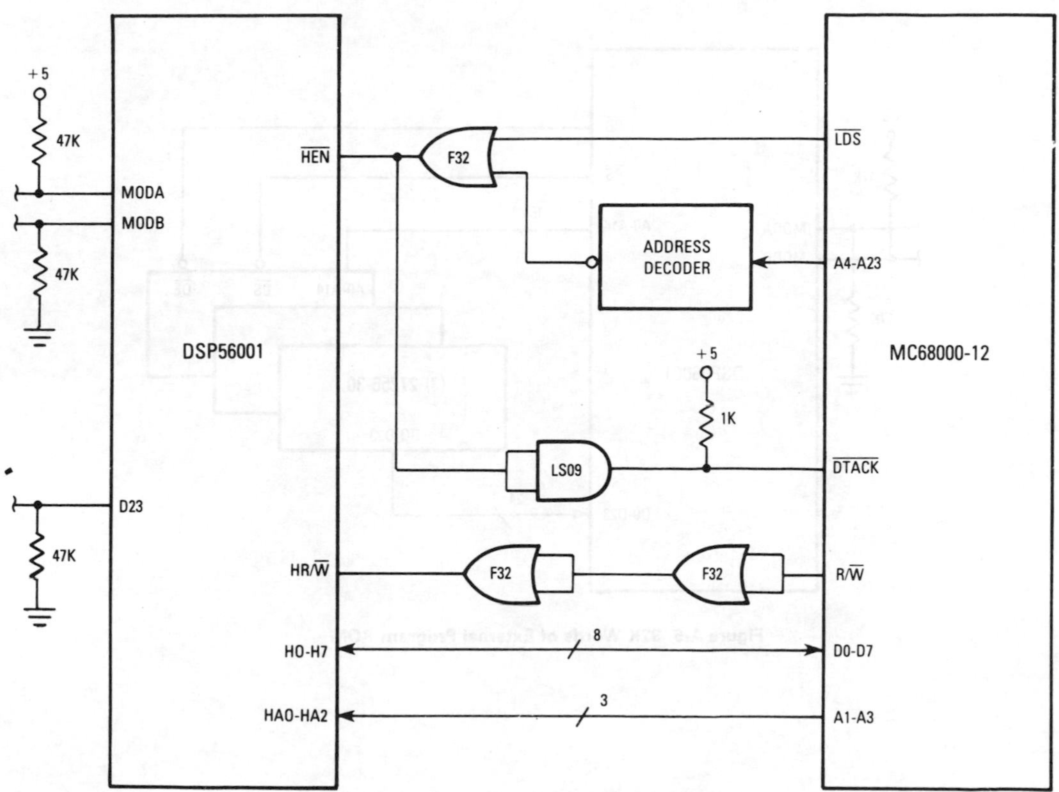

Figure A-4. DSP56001 Bootstrapping Example

Systems with external program memory can load the on-chip PRAM without using the bootstrap mode. In this example (Figure A-5), the DSP56001 is operated in mode 1 with external program memory at location \$E000. The programmer can overlay the high speed on-chip PRAM with DSP algorithms by using the MOVEM instruction.

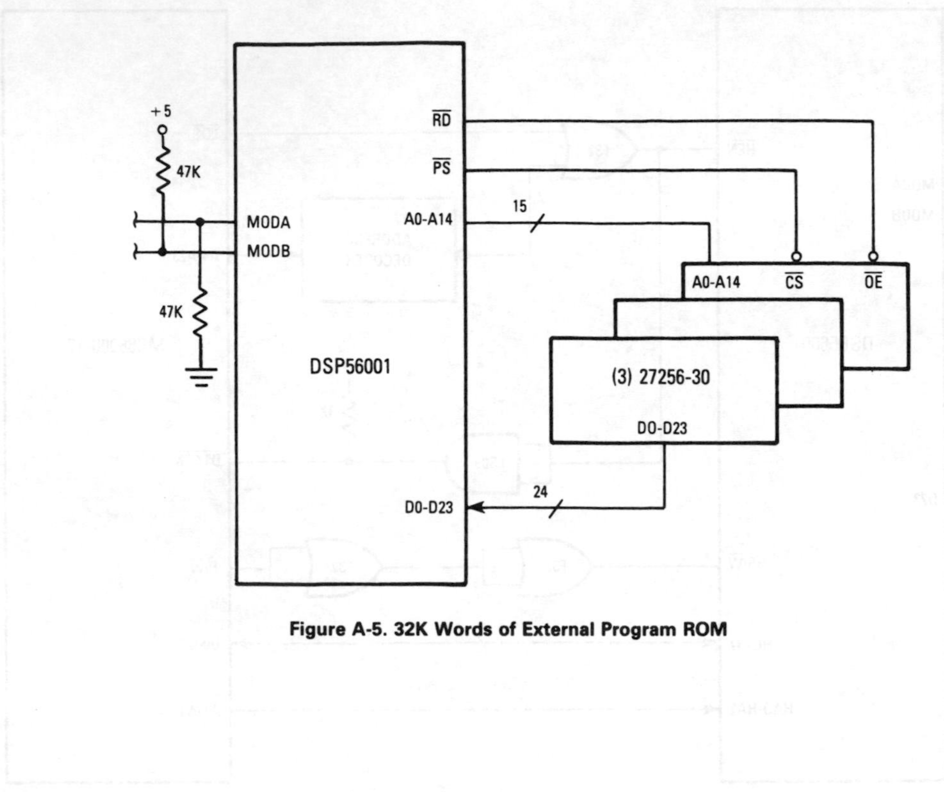

Figure A-5. 32K Words of External Program ROM

The last example (Figure A-6) shows the DSP56001 connected to the bus of an IBM-PC computer.

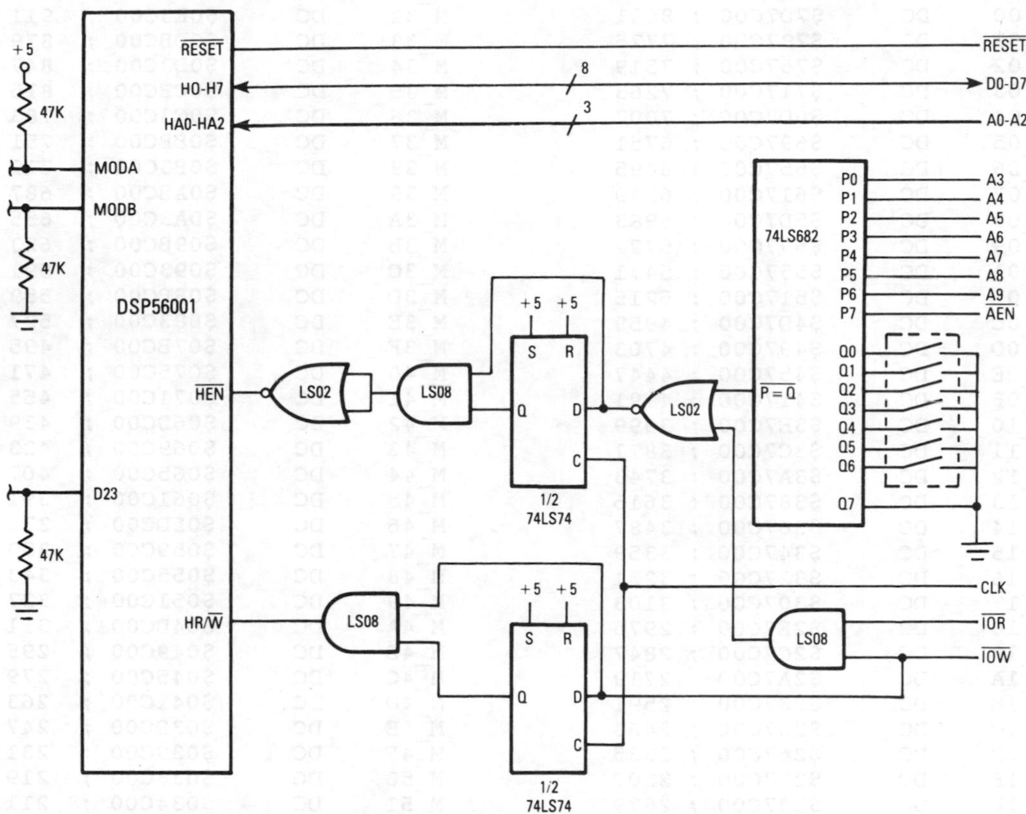

Figure A-6. IBM-PC to DSP56001 Host Port Interface

IBM-PC is a registered trademark of International Business Machines Corp.

APPENDIX II
MU-LAW/A-LAW EXPANSION TABLES

```
        org     x:$100
;
M_00    DC      $7D7C00 ; 8031
M_01    DC      $797C00 ; 7775
M_02    DC      $757C00 ; 7519
M_03    DC      $717C00 ; 7263
M_04    DC      $6D7C00 ; 7007
M_05    DC      $697C00 ; 6751
M_06    DC      $657C00 ; 6495
M_07    DC      $617C00 ; 6239
M_08    DC      $5D7C00 ; 5983
M_09    DC      $597C00 ; 5727
M_0A    DC      $557C00 ; 5471
M_0B    DC      $517C00 ; 5215
M_0C    DC      $4D7C00 ; 4959
M_0D    DC      $497C00 ; 4703
M_0E    DC      $457C00 ; 4447
M_0F    DC      $417C00 ; 4191
M_10    DC      $3E7C00 ; 3999
M_11    DC      $3C7C00 ; 3871
M_12    DC      $3A7C00 ; 3743
M_13    DC      $387C00 ; 3615
M_14    DC      $367C00 ; 3487
M_15    DC      $347C00 ; 3359
M_16    DC      $327C00 ; 3231
M_17    DC      $307C00 ; 3103
M_18    DC      $2E7C00 ; 2975
M_19    DC      $2C7C00 ; 2847
M_1A    DC      $2A7C00 ; 2719
M_1B    DC      $287C00 ; 2591
M_1C    DC      $267C00 ; 2463
M_1D    DC      $247C00 ; 2335
M_1E    DC      $227C00 ; 2207
M_1F    DC      $207C00 ; 2079
M_20    DC      $1EFC00 ; 1983
M_21    DC      $1DFC00 ; 1919
M_22    DC      $1CFC00 ; 1855
M_23    DC      $1BFC00 ; 1791
M_24    DC      $1AFC00 ; 1727
M_25    DC      $19FC00 ; 1663
M_26    DC      $18FC00 ; 1599
M_27    DC      $17FC00 ; 1535
M_28    DC      $16FC00 ; 1471
M_29    DC      $15FC00 ; 1407
M_2A    DC      $14FC00 ; 1343
M_2B    DC      $13FC00 ; 1279
M_2C    DC      $12FC00 ; 1215
M_2D    DC      $11FC00 ; 1151
M_2E    DC      $10FC00 ; 1087
M_2F    DC      $0FFC00 ; 1023
M_30    DC      $0F3C00 ;  975
M_31    DC      $0EBC00 ;  943
M_32    DC      $0E3C00 ;  911
M_33    DC      $0DBC00 ;  879
M_34    DC      $0D3C00 ;  847
M_35    DC      $0CBC00 ;  815
M_36    DC      $0C3C00 ;  783
M_37    DC      $0BBC00 ;  751
M_38    DC      $0B3C00 ;  719
M_39    DC      $0ABC00 ;  687
M_3A    DC      $0A3C00 ;  655
M_3B    DC      $09BC00 ;  623
M_3C    DC      $093C00 ;  591
M_3D    DC      $08BC00 ;  559
M_3E    DC      $083C00 ;  527
M_3F    DC      $07BC00 ;  495
M_40    DC      $075C00 ;  471
M_41    DC      $071C00 ;  455
M_42    DC      $06DC00 ;  439
M_43    DC      $069C00 ;  423
M_44    DC      $065C00 ;  407
M_45    DC      $061C00 ;  391
M_46    DC      $05DC00 ;  375
M_47    DC      $059C00 ;  359
M_48    DC      $055C00 ;  343
M_49    DC      $051C00 ;  327
M_4A    DC      $04DC00 ;  311
M_4B    DC      $049C00 ;  295
M_4C    DC      $045C00 ;  279
M_4D    DC      $041C00 ;  263
M_4E    DC      $03DC00 ;  247
M_4F    DC      $039C00 ;  231
M_50    DC      $036C00 ;  219
M_51    DC      $034C00 ;  211
M_52    DC      $032C00 ;  203
M_53    DC      $030C00 ;  195
M_54    DC      $02EC00 ;  187
M_55    DC      $02CC00 ;  179
M_56    DC      $02AC00 ;  171
M_57    DC      $028C00 ;  163
M_58    DC      $026C00 ;  155
M_59    DC      $024C00 ;  147
M_5A    DC      $022C00 ;  139
M_5B    DC      $020C00 ;  131
M_5C    DC      $01EC00 ;  123
M_5D    DC      $01CC00 ;  115
M_5E    DC      $01AC00 ;  107
M_5F    DC      $018C00 ;   99
M_60    DC      $017400 ;   93
M_61    DC      $016400 ;   89
```

```
M_62    DC    $015400 ;   85
M_63    DC    $014400 ;   81
M_64    DC    $013400 ;   77
M_65    DC    $012400 ;   73
M_66    DC    $011400 ;   69
M_67    DC    $010400 ;   65
M_68    DC    $00F400 ;   61
M_69    DC    $00E400 ;   57
M_6A    DC    $00D400 ;   53
M_6B    DC    $00C400 ;   49
M_6C    DC    $00B400 ;   45
M_6D    DC    $00A400 ;   41
M_6E    DC    $009400 ;   37
M_6F    DC    $008400 ;   33
M_70    DC    $007800 ;   30
M_71    DC    $007000 ;   28
M_72    DC    $006800 ;   26
M_73    DC    $006000 ;   24
M_74    DC    $005800 ;   22
M_75    DC    $005000 ;   20
M_76    DC    $004800 ;   18
M_77    DC    $004000 ;   16
M_78    DC    $003800 ;   14
M_79    DC    $003000 ;   12
M_7A    DC    $002800 ;   10
M_7B    DC    $002000 ;    8
M_7C    DC    $001800 ;    6
M_7D    DC    $001000 ;    4
M_7E    DC    $000800 ;    2
M_7F    DC    $000000 ;    0

A_80    DC    $158000 ;  688
A_81    DC    $148000 ;  656
A_82    DC    $178000 ;  752
A_83    DC    $168000 ;  720
A_84    DC    $118000 ;  560
A_85    DC    $108000 ;  528
A_86    DC    $138000 ;  624
A_87    DC    $128000 ;  592
A_88    DC    $1D8000 ;  944
A_89    DC    $1C8000 ;  912
A_8A    DC    $1F8000 ; 1008
A_8B    DC    $1E8000 ;  976
A_8C    DC    $198000 ;  816
A_8D    DC    $188000 ;  784
A_8E    DC    $1B8000 ;  880
A_8F    DC    $1A8000 ;  848
A_90    DC    $0AC000 ;  344
A_91    DC    $0A4000 ;  328
A_92    DC    $0BC000 ;  376
A_93    DC    $0B4000 ;  360
A_94    DC    $08C000 ;  280
A_95    DC    $084000 ;  264
A_96    DC    $09C000 ;  312
A_97    DC    $094000 ;  296
A_98    DC    $0EC000 ;  472
A_99    DC    $0E4000 ;  456
A_9A    DC    $0FC000 ;  504
A_9B    DC    $0F4000 ;  488
A_9C    DC    $0CC000 ;  408
A_9D    DC    $0C4000 ;  392
A_9E    DC    $0DC000 ;  440
A_9F    DC    $0D4000 ;  424
A_A0    DC    $560000 ; 2752
A_A1    DC    $520000 ; 2624
A_A2    DC    $5E0000 ; 3008
A_A3    DC    $5A0000 ; 2880
A_A4    DC    $460000 ; 2240
A_A5    DC    $420000 ; 2112
A_A6    DC    $4E0000 ; 2496
A_A7    DC    $4A0000 ; 2368
A_A8    DC    $760000 ; 3776
A_A9    DC    $720000 ; 3648
A_AA    DC    $7E0000 ; 4032
A_AB    DC    $7A0000 ; 3904
A_AC    DC    $660000 ; 3264
A_AD    DC    $620000 ; 3136
A_AE    DC    $6E0000 ; 3520
A_AF    DC    $6A0000 ; 3392
A_B0    DC    $2B0000 ; 1376
A_B1    DC    $290000 ; 1312
A_B2    DC    $2F0000 ; 1504
A_B3    DC    $2D0000 ; 1440
A_B4    DC    $230000 ; 1120
A_B5    DC    $210000 ; 1056
A_B6    DC    $270000 ; 1248
A_B7    DC    $250000 ; 1184
A_B8    DC    $3B0000 ; 1888
A_B9    DC    $390000 ; 1824
A_BA    DC    $3F0000 ; 2016
A_BB    DC    $3D0000 ; 1952
A_BC    DC    $330000 ; 1632
A_BD    DC    $310000 ; 1568
A_BE    DC    $370000 ; 1760
A_BF    DC    $350000 ; 1696
A_C0    DC    $015800 ;   43
A_C1    DC    $014800 ;   41
A_C2    DC    $017800 ;   47
A_C3    DC    $016800 ;   45
A_C4    DC    $011800 ;   35
A_C5    DC    $010800 ;   33
A_C6    DC    $013800 ;   39
A_C7    DC    $012800 ;   37
A_C8    DC    $01D800 ;   59
A_C9    DC    $01C800 ;   57
A_CA    DC    $01F800 ;   63
A_CB    DC    $01E800 ;   61
A_CC    DC    $019800 ;   51
A_CD    DC    $018800 ;   49
A_CE    DC    $01B800 ;   55
A_CF    DC    $01A800 ;   53
A_D0    DC    $005800 ;   11
A_D1    DC    $004800 ;    9
A_D2    DC    $007800 ;   15
```

```
A_D3    DC    $006800 ;  13
A_D4    DC    $001800 ;   3
A_D5    DC    $000800 ;   1
A_D6    DC    $003800 ;   7
A_D7    DC    $002800 ;   5
A_D8    DC    $00D800 ;  27
A_D9    DC    $00C800 ;  25
A_DA    DC    $00F800 ;  31
A_DB    DC    $00E800 ;  29
A_DC    DC    $009800 ;  19
A_DD    DC    $008800 ;  17
A_DE    DC    $00B800 ;  23
A_DF    DC    $00A800 ;  21
A_E0    DC    $056000 ; 172
A_E1    DC    $052000 ; 164
A_E2    DC    $05E000 ; 188
A_E3    DC    $05A000 ; 180
A_E4    DC    $046000 ; 140
A_E5    DC    $042000 ; 132
A_E6    DC    $04E000 ; 156
A_E7    DC    $04A000 ; 148
A_E8    DC    $076000 ; 236
A_E9    DC    $072000 ; 228
A_EA    DC    $07E000 ; 252
A_EB    DC    $07A000 ; 244
A_EC    DC    $066000 ; 204
A_ED    DC    $062000 ; 196
A_EE    DC    $06E000 ; 220
A_EF    DC    $06A000 ; 212
A_F0    DC    $02B000 ;  86
A_F1    DC    $029000 ;  82
A_F2    DC    $02F000 ;  94
A_F3    DC    $02D000 ;  90
A_F4    DC    $023000 ;  70
A_F5    DC    $021000 ;  66
A_F6    DC    $027000 ;  78
A_F7    DC    $025000 ;  74
A_F8    DC    $03B000 ; 118
A_F9    DC    $039000 ; 114
A_FA    DC    $03F000 ; 126
A_FB    DC    $03D000 ; 122
A_FC    DC    $033000 ; 102
A_FD    DC    $031000 ;  98
A_FE    DC    $037000 ; 110
A_FF    DC    $035000 ; 106
END
```

```
_START xromdos 0000 0000

_DATA X 0100
7D7C00 797C00 757C00 717C00 6D7C00 697C00 657C00 617C00 5D7C00 597C00 557C00
517C00 4D7C00 497C00 457C00 417C00 3E7C00 3C7C00 3A7C00 387C00 367C00 347C00
327C00 307C00 2E7C00 2C7C00 2A7C00 287C00 267C00 247C00 227C00 207C00 1EFC00
1DFC00 1CFC00 1BFC00 1AFC00 19FC00 18FC00 17FC00 16FC00 15FC00 14FC00 13FC00
12FC00 11FC00 10FC00 0FFC00 0F3C00 0EBC00 0E3C00 0DBC00 0D3C00 0CBC00 0C3C00
0BBC00 0B3C00 0ABC00 0A3C00 09BC00 093C00 08BC00 083C00 07BC00 075C00 071C00
06DC00 069C00 065C00 061C00 05DC00 059C00 055C00 051C00 04DC00 049C00 045C00
041C00 03DC00 039C00 036C00 034C00 032C00 030C00 02EC00 02CC00 02AC00 028C00
026C00 024C00 022C00 020C00 01EC00 01CC00 01AC00 018C00 017400 016400 015400
014400 013400 012400 011400 010400 00F400 00E400 00D400 00C400 00B400 00A400
009400 008400 007800 007000 006800 006000 005800 005000 004800 004000 003800
003000 002800 002000 001800 001000 000800 000000 158000 148000 178000 168000
118000 108000 138000 128000 1D8000 1C8000 1F8000 1E8000 198000 188000 1B8000
1A8000 0AC000 0A4000 0BC000 0B4000 08C000 084000 09C000 094000 0EC000 0E4000
0FC000 0F4000 0CC000 0C4000 0DC000 0D4000 560000 520000 5E0000 5A0000 460000
420000 4E0000 4A0000 760000 720000 7E0000 7A0000 660000 620000 6E0000 6A0000
2B0000 290000 2F0000 2D0000 230000 210000 270000 250000 3B0000 390000 3F0000
3D0000 330000 310000 370000 350000 015800 014800 017800 016800 011800 010800
013800 012800 01D800 01C800 01F800 01E800 019800 018800 01B800 01A800 005800
004800 007800 006800 001800 000800 003800 002800 00D800 00C800 00F800 00E800
009800 008800 00B800 00A800 056000 052000 05E000 05A000 046000 042000 04E000
04A000 076000 072000 07E000 07A000 066000 062000 06E000 06A000 02B000 029000
02F000 02D000 023000 021000 027000 025000 03B000 039000 03F000 03D000 033000
031000 037000 035000
_END
```

APPENDIX III
SINE WAVE TABLE

```
        ORG     Y:$100

S_00    DC      $000000 ;  0.0000000000000000000000
S_01    DC      $03242B ;  0.0245412588119506835937 5
S_02    DC      $0647D9 ;  0.0490676164627075195312 5
S_03    DC      $096A90 ;  0.0735645294189453125000
S_04    DC      $0C8BD3 ;  0.0980170965194702148437 5
S_05    DC      $0FAB27 ;  0.1224106550216674804687 5
S_06    DC      $12C810 ;  0.1467304229736328125000
S_07    DC      $15E214 ;  0.1709618568420410156250 0
S_08    DC      $18F8B8 ;  0.1950902938842773437500 0
S_09    DC      $1C0B82 ;  0.2191011905670166015625 0
S_0A    DC      $1F19F9 ;  0.2429801225662231445312 5
S_0B    DC      $2223A5 ;  0.2667127847671508789062 5
S_0C    DC      $25280C ;  0.2902846336364746093750 0
S_0D    DC      $2826B9 ;  0.3136817216873168945312 5
S_0E    DC      $2B1F35 ;  0.3368898630142211914062 5
S_0F    DC      $2E110A ;  0.3598949909210205078125 0
S_10    DC      $30FBC5 ;  0.3826833963394165039062 5
S_11    DC      $33DEF3 ;  0.4052413702011108398437 5
S_12    DC      $36BA20 ;  0.4275550842285156250000 0
S_13    DC      $398CDD ;  0.4496113061904907226562 5
S_14    DC      $3C56BA ;  0.4713966846466064453125 0
S_15    DC      $3F174A ;  0.4928982257843017578125 0
S_16    DC      $41CE1E ;  0.5141026973724365234375 0
S_17    DC      $447ACD ;  0.5349975824356079101562 5
S_18    DC      $471CED ;  0.5555702447891235351562 5
S_19    DC      $49B415 ;  0.5758081674575805664062 5
S_1A    DC      $4C3FE0 ;  0.5956993103027343750000 0
S_1B    DC      $4EBFE9 ;  0.6152316331863403320312 5
S_1C    DC      $5133CD ;  0.6343933343887329101562 5
S_1D    DC      $539B2B ;  0.6531728506088256835937 5
S_1E    DC      $55F5A5 ;  0.6715589761734008789062 5
S_1F    DC      $5842DD ;  0.6895405054092407226562 5
S_20    DC      $5A827A ;  0.7071068286895751953125 0
S_21    DC      $5CB421 ;  0.7242470979690551757812 5
S_22    DC      $5ED77D ;  0.7409511804580688476562 5
S_23    DC      $60EC38 ;  0.7572088241577148437500 0
S_24    DC      $62F202 ;  0.7730104923248291015625 0
S_25    DC      $64E889 ;  0.7883464097976684570312 5
S_26    DC      $66CF81 ;  0.8032075166702270507812 5
S_27    DC      $68A69F ;  0.8175848722457885742187 5
S_28    DC      $6A6D99 ;  0.8314696550369262695312 5
S_29    DC      $6C2429 ;  0.8448535203933715820312 5
S_2A    DC      $6DCA0D ;  0.8577286005020141601562 5
S_2B    DC      $6F5F03 ;  0.8700870275497436523437 5
S_2C    DC      $70E2CC ;  0.8819212913513183593750 0
S_2D    DC      $72552D ;  0.8932243585586547851562 5
S_2E    DC      $73B5EC ;  0.9039893150329589843750 0
S_2F    DC      $7504D3 ;  0.9142097234725952148437 5
```

```
S_30    DC    $7641AF ;  0.923879504203796386718750
S_31    DC    $776C4F ;  0.932992815971374511718750
S_32    DC    $788484 ;  0.941544055938720703125000
S_33    DC    $798A24 ;  0.949528217315673828125000
S_34    DC    $7A7D05 ;  0.956940293312072753906250
S_35    DC    $7B5D04 ;  0.963776111602783203125000
S_36    DC    $7C29FC ;  0.970031261444091796875000
S_37    DC    $7CE3CF ;  0.975702166557312011718750
S_38    DC    $7D8A5F ;  0.980785250663757324218750
S_39    DC    $7E1D94 ;  0.985277652740478515625000
S_3A    DC    $7E9D56 ;  0.989176511764526367187500
S_3B    DC    $7F0992 ;  0.992479562759399414062500
S_3C    DC    $7F6237 ;  0.995184779167175292968750
S_3D    DC    $7FA737 ;  0.997290492057800292968750
S_3E    DC    $7FD888 ;  0.998795509338378906250000
S_3F    DC    $7FF622 ;  0.999698877334594726562500
S_40    DC    $7FFFFF ;  0.999999880790710449218750
S_41    DC    $7FF622 ;  0.999698877334594726562500
S_42    DC    $7FD888 ;  0.998795509338378906250000
S_43    DC    $7FA737 ;  0.997290492057800292968750
S_44    DC    $7F6237 ;  0.995184779167175292968750
S_45    DC    $7F0992 ;  0.992479562759399414062500
S_46    DC    $7E9D56 ;  0.989176511764526367187500
S_47    DC    $7E1D94 ;  0.985277652740478515625000
S_48    DC    $7D8A5F ;  0.980785250663757324218750
S_49    DC    $7CE3CF ;  0.975702166557312011718750
S_4A    DC    $7C29FC ;  0.970031261444091796875000
S_4B    DC    $7B5D04 ;  0.963776111602783203125000
S_4C    DC    $7A7D05 ;  0.956940293312072753906250
S_4D    DC    $798A24 ;  0.949528217315673828125000
S_4E    DC    $788484 ;  0.941544055938720703125000
S_4F    DC    $776C4F ;  0.932992815971374511718750
S_50    DC    $7641AF ;  0.923879504203796386718750
S_51    DC    $7504D3 ;  0.914209723472595214843750
S_52    DC    $73B5EC ;  0.903989315032958984375000
S_53    DC    $72552D ;  0.893224358558654785156250
S_54    DC    $70E2CC ;  0.881921291351318359375000
S_55    DC    $6F5F03 ;  0.870087027549743652343750
S_56    DC    $6DCA0D ;  0.857728600502014160156250
S_57    DC    $6C2429 ;  0.844853520393371582031250
S_58    DC    $6A6D99 ;  0.831469655036926269531250
S_59    DC    $68A69F ;  0.817584872245788574218750
S_5A    DC    $66CF81 ;  0.803207516670227050781250
S_5B    DC    $64E889 ;  0.788346409797668457031250
S_5C    DC    $62F202 ;  0.773010492324829101562500
S_5D    DC    $60EC38 ;  0.757208824157714843750000
S_5E    DC    $5ED77D ;  0.740951180458068847656250
S_5F    DC    $5CB421 ;  0.724247097969055175781250
S_60    DC    $5A827A ;  0.707106828689575195312500
S_61    DC    $5842DD ;  0.689540505409240722656250
S_62    DC    $55F5A5 ;  0.671558976173400878906250
S_63    DC    $539B2B ;  0.653172850608825683593750
S_64    DC    $5133CD ;  0.634393334388732910156250
S_65    DC    $4EBFE9 ;  0.615231633186340332031250
S_66    DC    $4C3FE0 ;  0.595693103027343750000000
S_67    DC    $49B415 ;  0.575808167457580566406250
```

```
S_68    DC    $471CED ;  0.55557024478912353515625
S_69    DC    $447ACD ;  0.53499758243560791015625
S_6A    DC    $41CE1E ;  0.51410269737243652343750
S_6B    DC    $3F174A ;  0.49289822578430175781250
S_6C    DC    $3C56BA ;  0.47139668464660644531250
S_6D    DC    $398CDD ;  0.44961130619049072265625
S_6E    DC    $36BA20 ;  0.42755508422851562500000
S_6F    DC    $33DEF3 ;  0.40524137020111083984375
S_70    DC    $30FBC5 ;  0.38268339633941650390625
S_71    DC    $2E110A ;  0.35989499092102050781250
S_72    DC    $2B1F35 ;  0.33688986301422119140625
S_73    DC    $2826B9 ;  0.31368172168731689453125
S_74    DC    $25280C ;  0.29028463363647460937500
S_75    DC    $2223A5 ;  0.26671278476715087890625
S_76    DC    $1F19F9 ;  0.24298012256622314453125
S_77    DC    $1C0B82 ;  0.21910119056701660156250
S_78    DC    $18F8B8 ;  0.19509029388427734375000
S_79    DC    $15E214 ;  0.17096185684204101562500
S_7A    DC    $12C810 ;  0.14673042297363281250000
S_7B    DC    $0FAB27 ;  0.12241065502166748046875
S_7C    DC    $0C8BD3 ;  0.09801709651947021484375
S_7D    DC    $096A90 ;  0.07356452941894531250000
S_7E    DC    $0647D9 ;  0.04906761646270751953125
S_7F    DC    $03242B ;  0.02454125881195068359375
S_80    DC    $000000 ;  0.00000000000000000000000
S_81    DC    $FCDBD5 ; -0.02454125881195068359375
S_82    DC    $F9B827 ; -0.04906761646270751953125
S_83    DC    $F69570 ; -0.07356452941894531250000
S_84    DC    $F3742D ; -0.09801709651947021484375
S_85    DC    $F054D9 ; -0.12241065502166748046875
S_86    DC    $ED37F0 ; -0.14673042297363281250000
S_87    DC    $EA1DEC ; -0.17096185684204101562500
S_88    DC    $E70748 ; -0.19509029388427734375000
S_89    DC    $E3F47E ; -0.21910119056701660156250
S_8A    DC    $E0E607 ; -0.24298012256622314453125
S_8B    DC    $DDDC5B ; -0.26671278476715087890625
S_8C    DC    $DAD7F4 ; -0.29028463363647460937500
S_8D    DC    $D7D947 ; -0.31368172168731689453125
S_8E    DC    $D4E0CB ; -0.33688986301422119140625
S_8F    DC    $D1EEF6 ; -0.35989499092102050781250
S_90    DC    $CF043B ; -0.38268339633941650390625
S_91    DC    $CC210D ; -0.40524137020111083984375
S_92    DC    $C945E0 ; -0.42755508422851562500000
S_93    DC    $C67323 ; -0.44961130619049072265625
S_94    DC    $C3A946 ; -0.47139668464660644531250
S_95    DC    $C0E8B6 ; -0.49289822578430175781250
S_96    DC    $BE31E2 ; -0.51410269737243652343750
S_97    DC    $BB8533 ; -0.53499758243560791015625
S_98    DC    $B8E313 ; -0.55557024478912353515625
S_99    DC    $B64BEB ; -0.57580816745758056640625
S_9A    DC    $B3C020 ; -0.59569931030273437500000
S_9B    DC    $B14017 ; -0.61523163318634033203125
S_9C    DC    $AECC33 ; -0.63439333438873291015625
S_9D    DC    $AC64D5 ; -0.65317285060882568359375
S_9E    DC    $AA0A5B ; -0.67155897617340087890625
S_9F    DC    $A7BD23 ; -0.68954050540924072265625
```

```
S_A0    DC    $A57D86 ; -0.7071068286895751953125O
S_A1    DC    $A34BDF ; -0.7242470979690551757812 5
S_A2    DC    $A12883 ; -0.7409511804580688476562 5
S_A3    DC    $9F13C8 ; -0.7572088241577148437500 0
S_A4    DC    $9D0DFE ; -0.7730104923248291015625 0
S_A5    DC    $9B1777 ; -0.7883464097976684570312 5
S_A6    DC    $99307F ; -0.8032075166702270507812 5
S_A7    DC    $975961 ; -0.8175848722457885742187 5
S_A8    DC    $959267 ; -0.8314696550369262695312 5
S_A9    DC    $93DBD7 ; -0.8448535203933715820312 5
S_AA    DC    $9235F3 ; -0.8577286005020141601562 5
S_AB    DC    $90A0FD ; -0.8700870275497436523437 5
S_AC    DC    $8F1D34 ; -0.8819212913513183593750 0
S_AD    DC    $8DAAD3 ; -0.8932243585586547851562 5
S_AE    DC    $8C4A14 ; -0.9039893150329589843750 0
S_AF    DC    $8AFB2D ; -0.9142097234725952148437 5
S_B0    DC    $89BE51 ; -0.9238795042037963867187 5
S_B1    DC    $8893B1 ; -0.9329928159713745117187 5
S_B2    DC    $877B7C ; -0.9415440559387207031250 0
S_B3    DC    $8675DC ; -0.9495282173156738281250 0
S_B4    DC    $8582FB ; -0.9569402933120727539062 5
S_B5    DC    $84A2FC ; -0.9637761116027832031250 0
S_B6    DC    $83D604 ; -0.9700312614440917968750 0
S_B7    DC    $831C31 ; -0.9757021665573120117187 5
S_B8    DC    $8275A1 ; -0.9807852506637573242187 5
S_B9    DC    $81E26C ; -0.9852776527404785156250 0
S_BA    DC    $8162AA ; -0.9891765117645263671875 0
S_BB    DC    $80F66E ; -0.9924795627593994140625 0
S_BC    DC    $809DC9 ; -0.9951847791671752929687 5
S_BD    DC    $8058C9 ; -0.9972904920578002929687 5
S_BE    DC    $802778 ; -0.9987955093383789062500 0
S_BF    DC    $8009DE ; -0.9996988773345947265625 0
S_C0    DC    $800000 ; -1.0000000000000000000000 0
S_C1    DC    $8009DE ; -0.9996988773345947265625 0
S_C2    DC    $802778 ; -0.9987955093383789062500 0
S_C3    DC    $8058C9 ; -0.9972904920578002929687 5
S_C4    DC    $809DC9 ; -0.9951847791671752929687 5
S_C5    DC    $80F66E ; -0.9924795627593994140625 0
S_C6    DC    $8162AA ; -0.9891765117645263671875 0
S_C7    DC    $81E26C ; -0.9852776527404785156250 0
S_C8    DC    $8275A1 ; -0.9807852506637573242187 5
S_C9    DC    $831C31 ; -0.9757021665573120117187 5
S_CA    DC    $83D604 ; -0.9700312614440917968750 0
S_CB    DC    $84A2FC ; -0.9637761116027832031250 0
S_CC    DC    $8582FB ; -0.9569402933120727539062 5
S_CD    DC    $8675DC ; -0.9495282173156738281250 0
S_CE    DC    $877B7C ; -0.9415440559387207031250 0
S_CF    DC    $8893B1 ; -0.9329928159713745117187 5
S_D0    DC    $89BE51 ; -0.9238795042037963867187 5
S_D1    DC    $8AFB2D ; -0.9142097234725952148437 5
S_D2    DC    $8C4A14 ; -0.9039893150329589843750 0
S_D3    DC    $8DAAD3 ; -0.8932243585586547851562 5
S_D4    DC    $8F1D34 ; -0.8819212913513183593750 0
S_D5    DC    $90A0FD ; -0.8700870275497436523437 5
S_D6    DC    $9235F3 ; -0.8577286005020141601562 5
S_D7    DC    $93DBD7 ; -0.8448535203933715820312 5
```

```
S_D8      DC      $959267 ; -0.83146965503692626953125
S_D9      DC      $975961 ; -0.81758487224578857421875
S_DA      DC      $99307F ; -0.80320751667022705078125
S_DB      DC      $9B1777 ; -0.78834640979766845703125
S_DC      DC      $9D0DFE ; -0.77301049232482910156250
S_DD      DC      $9F13C8 ; -0.75720882415771484375000
S_DE      DC      $A12883 ; -0.74095118045806884765625
S_DF      DC      $A34BDF ; -0.72424709796905517578125
S_E0      DC      $A57D86 ; -0.70710682868957519531250
S_E1      DC      $A7BD23 ; -0.68954050540924072265625
S_E2      DC      $AA0A5B ; -0.67155897617340087890625
S_E3      DC      $AC64D5 ; -0.65317285060882568359375
S_E4      DC      $AECC33 ; -0.63439333438873291015625
S_E5      DC      $B14017 ; -0.61523163318634033203125
S_E6      DC      $B3C020 ; -0.59569931030273437500000
S_E7      DC      $B64BEB ; -0.57580816745758056640625
S_E8      DC      $B8E313 ; -0.55557024478912353515625
S_E9      DC      $BB8533 ; -0.53499758243560791015625
S_EA      DC      $BE31E2 ; -0.51410269737243652343750
S_EB      DC      $C0E8B6 ; -0.49289822578430175781250
S_EC      DC      $C3A946 ; -0.47139668464660644531250
S_ED      DC      $C67323 ; -0.44961130619049072265625
S_EE      DC      $C945E0 ; -0.42755508422851562500000
S_EF      DC      $CC210D ; -0.40524137020111083984375
S_F0      DC      $CF043B ; -0.38268339633941650390625
S_F1      DC      $D1EEF6 ; -0.35989499092102050781250
S_F2      DC      $D4E0CB ; -0.33688986301422119140625
S_F3      DC      $D7D947 ; -0.31368172168731689453125
S_F4      DC      $DAD7F4 ; -0.29028463363647460937500
S_F5      DC      $DDDC5B ; -0.26671278476715087890625
S_F6      DC      $E0E607 ; -0.24298012256622314453125
S_F7      DC      $E3F47E ; -0.21910119056701660156250
S_F8      DC      $E70748 ; -0.19509029388427734375000
S_F9      DC      $EA1DEC ; -0.17096185684204101562500
S_FA      DC      $ED37F0 ; -0.14673042297363281250000
S_FB      DC      $F054D9 ; -0.12241065502166748046875
S_FC      DC      $F3742D ; -0.09801709651947021484375
S_FD      DC      $F69570 ; -0.07356452941894531250000
S_FE      DC      $F9B827 ; -0.04906761646270751953125
S_FF      DC      $FCDBD5 ; -0.02454125881195068359375
          END
```

```
_START sinegen 0000 0000

_DATA Y 0100
000000 03242B 0647D9 096A90 0C8BD3 0FAB27 12C810 15E214 18F8B8 1C0B82 1F19F9
2223A5 25280C 2826B9 2B1F35 2E110A 30FBC5 33DEF3 36BA20 398CDD 3C56BA 3F174A
41CE1E 447ACD 471CED 49B415 4C3FE0 4EBFE9 5133CD 539B2B 55F5A5 5842DD 5A827A
5CB421 5ED77D 60EC38 62F202 64E889 66CF81 68A69F 6A6D99 6C2429 6DCA0D 6F5F03
70E2CC 72552D 73B5EC 7504D3 7641AF 776C4F 788484 798A24 7A7D05 7B5D04 7C29FC
7CE3CF 7D8A5F 7E1D94 7E9D56 7F0992 7F6237 7FA737 7FD888 7FF622 7FFFFF 7FF622
7FD888 7FA737 7F6237 7F0992 7E9D56 7E1D94 7D8A5F 7CE3CF 7C29FC 7B5D04 7A7D05
798A24 788484 776C4F 7641AF 7504D3 73B5EC 72552D 70E2CC 6F5F03 6DCA0D 6C2429
6A6D99 68A69F 66CF81 64E889 62F202 60EC38 5ED77D 5CB421 5A827A 5842DD 55F5A5
539B2B 5133CD 4EBFE9 4C3FE0 49B415 471CED 447ACD 41CE1E 3F174A 3C56BA 398CDD
36BA20 33DEF3 30FBC5 2E110A 2B1F35 2826B9 25280C 2223A5 1F19F9 1C0B82 18F8B8
15E214 12C810 0FAB27 0C8BD3 096A90 0647D9 03242B 000000 FCDBD5 F9B827 F69570
F3742D F054D9 ED37F0 EA1DEC E70748 E3F47E E0E607 DDDC5B DAD7F4 D7D947 D4E0CB
D1EEF6 CF043B CC210D C945E0 C67323 C3A946 C0E8B6 BE31E2 BB8533 B8E313 B64BEB
B3C020 B14017 AECC33 AC64D5 AA0A5B A7BD23 A57D86 A34BDF A12883 9F13C8 9D0DFE
9B1777 99307F 975961 959267 93DBD7 9235F3 90A0FD 8F1D34 8DAAD3 8C4A14 8AFB2D
89BE51 8893B1 877B7C 8675DC 8582FB 84A2FC 83D604 831C31 8275A1 81E26C 8162AA
80F66E 809DC9 8058C9 802778 8009DE 800000 8009DE 802778 8058C9 809DC9 80F66E
8162AA 81E26C 8275A1 831C31 83D604 84A2FC 8582FB 8675DC 877B7C 8893B1 89BE51
8AFB2D 8C4A14 8DAAD3 8F1D34 90A0FD 9235F3 93DBD7 959267 975961 99307F 9B177
9D0DFE 9F13C8 A12883 A34BDF A57D86 A7BD23 AA0A5B AC64D5 AECC33 B14017 B3C020
B64BEB B8E313 BB8533 BE31E2 C0E8B6 C3A946 C67323 C945E0 CC210D CF043B D1EEF6
D4E0CB D7D947 DAD7F4 DDDC5B E0E607 E3F47E E70748 EA1DEC ED37F0 F054D9 F3742D
F69570 F9B827 FCDBD5
_END
```

Chapter 1

TMS320C4x Introduction

Texas Instruments TMS320C4x generation of 32-bit processors are designed specifically to meet the needs of parallel-processing and other real-time embedded applications. TMS320C4x products consist of both parallel-processing devices and development tools. With world-class parallel-processing development tools, designers are able to fully utilize the extraordinary performance of 275 MOPS (millions of operations per second) and 320 Mbytes per second throughput made available by the TMS320C4x generation.

This chapter provides a brief overview of the TMS320C4x generation. Major topics covered are as follows:

1.1 The TMS320 Family

The TMS320C4x is one of five generations in the TMS320 family of digital signal processors. The TMS320C1x, TMS320C2x, and TMS320C5x offer designers a complete line of general-purpose and application-specific 16-bit DSPs. The TMS320C3x and TMS320C4x generations round out the TMS320 family, providing an ensemble of 32-bit DSPs. The TMS320 family has grown from a single device introduced in 1982, the TMS32010, to nearly thirty different products across five CPU architectures. On-chip hardware multipliers, register files, barrel shifters, ALUs, ROM, RAM, caches, and I/O peripherals along with massive internal busing (all within a product as programmable as a general-purpose microprocessor), make TI TMS320 devices well-suited for a wide range of computer-intensive applications.

Figure 1–1. TMS320 Family of Devices

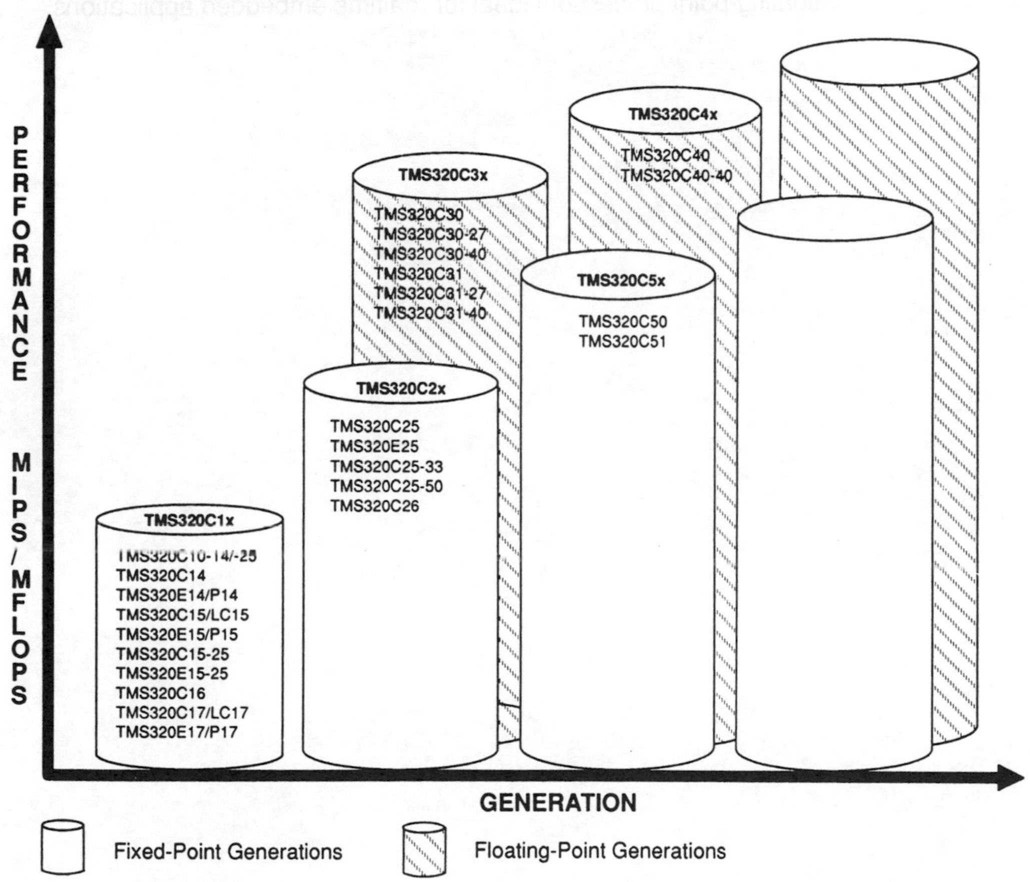

1.2 Parallel Processing

The need for parallel processing is growing rapidly. As floating-point performance requirements grow exponentially, semiconductor manufacturers can no longer meet the need with individual processing elements. Processors not designed for parallel processing are inadequate for the task, as interprocessor communication quickly saturates device I/O and adversely affects computing efficiency. Products in the TMS320C3x generation made the first step in addressing the need for parallel processing by providing designers with two external interface ports, each with a comprehensive memory interface. This yields an immense amount of I/O bandwidth. Devices in the TMS320C4x generation go several steps further by incorporating on-chip hardware to facilitate high-speed interprocessor communication and concurrent I/O without degrading CPU performance. These features, coupled with a host of sophisticated parallel-processing development tools, make the TMS320C4x generation of floating-point processors ideal for realtime embedded applications.

1.3 TMS320C4x Features

The TMS320C4x generation consists of two equally important aspects, parallel-processing devices and parallel-processing development tools.

1.3.1 TMS320C40 Device Key Features

The primary features of the TMS320C40 are:

- Six communication ports for high-speed interprocessor communication. Communication port key features include:
  - 20-Mbytes/sec asynchronous transfer rate at each port for maximum data throughput
  - Direct (glueless) processor-to-processor communication for ease of use
  - Bidirectional transfers for maximum communication flexibility
- Six-channel DMA coprocessor for concurrent I/O and CPU operation, thereby maximizing sustained CPU performance by alleviating the CPU of burdensome I/O. DMA coprocessor key features include:
  - Concurrent data transfers and CPU operation for sustained CPU performance
  - Self-programming (autoinitialize) capability for each channel, thereby not requiring the CPU for initialization, maximizing sustained CPU performance
  - Data transfers to and from anywhere in the processor's memory map for maximum flexibility
- High-performance DSP CPU capable of 275 MOPS and 320 Mbytes/sec. CPU key features include:
  - Eleven operations per cycle throughput, resulting in massive computing parallelism and sustained CPU performance
  - 40-ns and 50-ns instruction cycle times
  - 40/32-bit single-cycle floating-point/integer multiplier for high performance in computationally intensive algorithms
  - Single-cycle IEEE floating-point conversion for efficient interface to IEEE-compatible processors
  - Hardware divide and inverse square root support for high performance
  - Byte and half-word manipulation capabilities for fast data (un)packing
  - Source code compatible with TMS320C3x generation for easy upward and downward mobility

- Support for linear, circular, and bit-reversed addressing for high performance
- Single-cycle branches, calls, and returns for fast program control
- Single-cycle barrel shifter for 0–31 single-cycle right or left shifts for fast bit manipulation
- Relocatable reset and interrupt vectors for easy integration into parallel-processing systems

❑ Two identical external data and address buses supporting shared memory systems and high data rate, single-cycle transfers. Key features include:

- High port data-transfer rate of 100 Mbytes/sec
- 16-Gbyte continuous program/data/peripheral address space for maximum design flexibility
- Status pins that signal type of memory access requested for fast, intelligent bus arbitration in shared memory systems
- Separate address, data, and control-enable pins for high-speed bus arbitration
- Four sets of memory-control signals support different speed memories in hardware, enabling efficient use of low- and high-speed memories

❑ On-chip analysis module supporting efficient, state-of-the-art parallel-processing debug. Key features include:

- Separate breakpoint comparators for program, data, and DMA accesses, providing onchip hardware breakpoint capabilities for fast debug and development
- Discontinuity stack for hardware trace, facilitating fast debug and development
- Event counter for accurate benchmarking and profiling
- JTAG interface for standard system connection

- On-chip program cache and dual-access/single-cycle RAM for increased memory access performance. On-chip memory key features include:
  - 512-byte instruction cache for increased system performance
  - 8K bytes of single-cycle dual access program or data RAM for increased system performance and lower system cost
  - Bootloader (ROM based) supporting program bootup via 8-, 16- or 32-bit memories over any one of the communication ports
- Separate internal program, data, and DMA coprocessor buses for support of massive concurrent I/O of program and data throughput, thereby maximizing sustained CPU performance.

Total device performance is 275 MOPS and 320 Mbytes/sec as noted below.

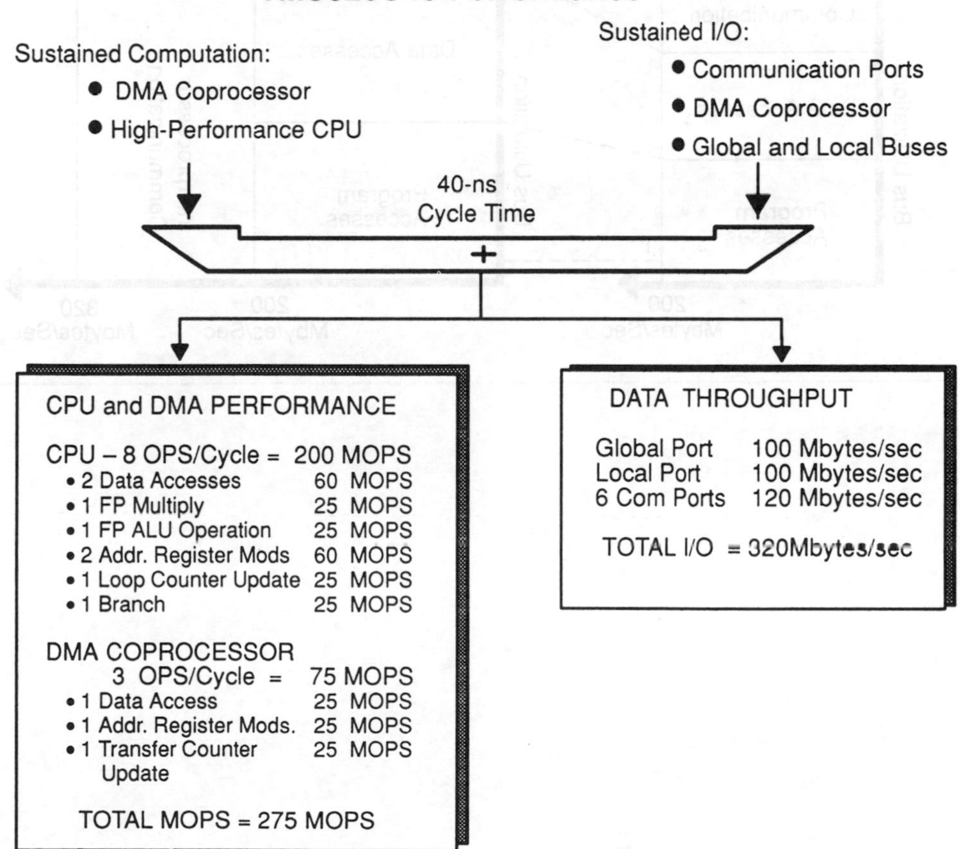

TMS320C40 Performance
Sustained Computation:
● DMA Coprocessor
● High-Performance CPU
Sustained I/O:
● Communication Ports
● DMA Coprocessor
● Global and Local Buses
40-ns
Cycle Time
+
CPU and DMA PERFORMANCE
CPU – 8 OPS/Cycle = 200 MOPS
• 2 Data Accesses 60 MOPS
• 1 FP Multiply 25 MOPS
• 1 FP ALU Operation 25 MOPS
• 2 Addr. Register Mods 60 MOPS
• 1 Loop Counter Update 25 MOPS
• 1 Branch 25 MOPS
DMA COPROCESSOR
3 OPS/Cycle = 75 MOPS
• 1 Data Access 25 MOPS
• 1 Addr. Register Mods. 25 MOPS
• 1 Transfer Counter Update 25 MOPS
TOTAL MOPS = 275 MOPS
DATA THROUGHPUT
Global Port 100 Mbytes/sec
Local Port 100 Mbytes/sec
6 Com Ports 120 Mbytes/sec
TOTAL I/O = 320Mbytes/sec

1.3.2 Communication Port Benefits

Without the six communication ports, 120 Mbytes/sec of processor throughput must be squeezed over one or both of the external memory interfaces, thereby saturating processor throughput and turning the system into a complex shared memory architecture.With the communication ports, bandwidth is expanded (illustrated in Figure 1–2).

Figure 1–2. TMS320C40 Throughput Increases Use of Communication Ports

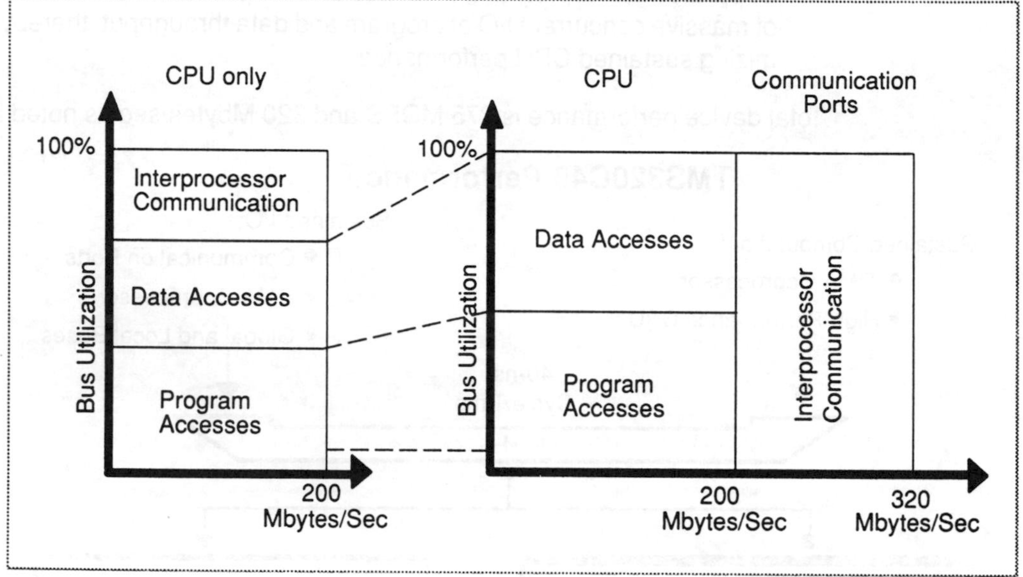

1.3.3 DMA Coprocessor Benefits

Without the DMA coprocessor, the CPU would have to use computational MOPS to transfer data within the processor's memory map.With the DMA coprocessor, the CPU can focus its entire 200 MOPS of performance on quality computational tasks while the DMA coprocessor takes care of the burdensome I/O. This is illustrated in Figure 1–3.

Figure 1–3. TMS320C40 Throughput Increases Use of DMA Coprocessor

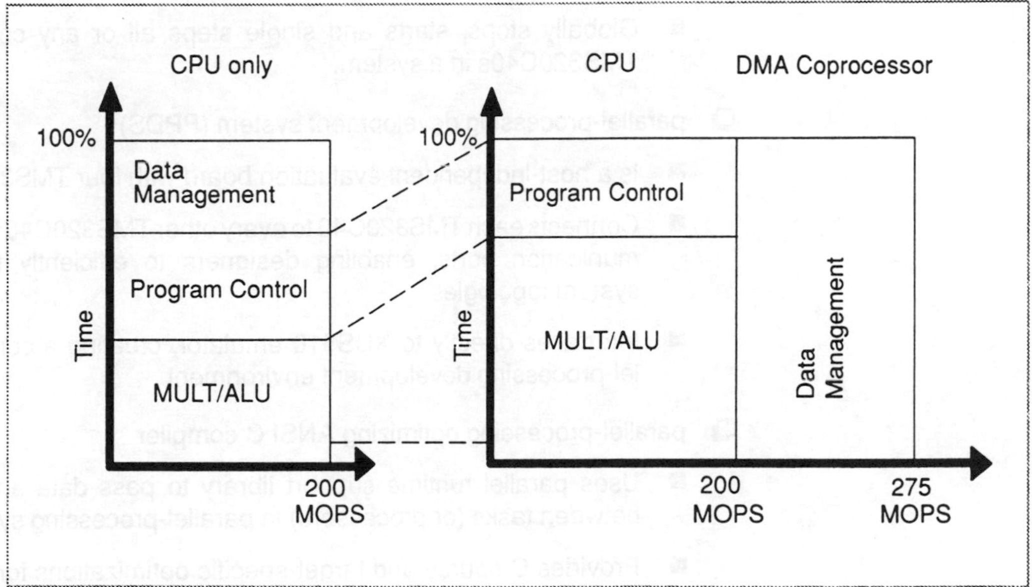

1.3.4 TMS320C40 Parallel-Processing Development Tools Key Features

The primary TMS320C4x development tools are as follow:

- parallel-processing in-system emulator (XDS510)
  - Debugs both C and assembly code simultaneously using the graphical user-friendly source-level debugger
  - Debugs any number of TMS320C4x devices in a system with a single XDS510 controller card
  - Globally stops, starts and single steps all or any combination of TMS320C40s in a system.
- parallel-processing development system (PPDS)
  - Is a host-independent evaluation board with four TMS320C40s
  - Connects each TMS320C40 to every other TMS320C40 via their communication ports, enabling designers to efficiently test different system topologies
  - Interfaces directly to XDS510 emulator, creating a complete parallel-processing development environment.
- parallel-processing optimizing ANSI C compiler
  - Uses parallel runtime support library to pass data and messages between tasks (or processors) in parallel-processing systems
  - Provides C-source and target-specific optimizations for dense, optimal code
  - Is Plum-Hall validated to ANSI standard for maximum code portability
- parallel-processing assembler/linker
  - Provides directives to map program and data code on specific processors for fast integration and debug of parallel-processing code
  - Has relocatable modules for maximum code flexibility
- Hardware verification and full functional models
  - Simulates hardware operation of multiple TMS320C40s and associated logic for accurate development (via software simulation) of parallel-processing systems
  - Facilitates fast development of product hardware through accurate simulation of device bus cycles and function execution
  - Supports various workstation and PC environments

- State accurate simulator
  - Provides cycle-by-cycle simulation of all aspects of the TMS320C4x
  - Offers low-cost way to simulate key software kernels
  - Is supported on a host of workstation and PC platforms
  - Uses standard C and assembly source interface

1.4 Applications

Below is a list of classical DSP applications along with a number of embedded real-time applications which need the computational performance offered by TMS320 devices. The real-time performance, low device costs, and comprehensive development support are the primary reasons that Texas Instruments TMS320 devices are the preferred solution in the following applications:

Figure 1–4. Matrix of TMS320 DSP Applications

General-Purpose DSP	**Graphics/Imaging**	**Instrumentation**
Digital Filtering Convolution Correlation Hilbert Transforms Fast Fourier Transforms Adaptive Filtering Windowing Waveform Generation	3-D Transformations Rendering Robot Vision Image Transmission/Compression Pattern Recognition Image Enhancement Homomorphic Processing Workstations Animation/Digital Map	Spectrum Analysis Function Generation Pattern Matching Seismic Processing Transient Analysis Digital Filtering Phase-Locked Loops
Voice/Speech	**Control**	**Military**
Voice Mail Speech Vocoding Speech Recognition Speaker Verification Speech Enhancement Speech Synthesis Text-to-Speech Neural Networks	Disk Control Servo Control Robot Control Laser Printer Control Engine Control Motor Control Kalman Filtering	Secure Communications Radar Processing Sonar Processing Image Processing Navigation Missile Guidance Radio Frequency Modems Sensor Fusion
Telecommunications		**Automotive**
Echo Cancellation ADPCM Transcoders Digital PBXs Line Repeaters Channel Multiplexing 1200- to 19200-bps Modems Adaptive Equalizers DTMF Encoding/Decoding Data Encryption	FAX Cellular Telephones Speaker Phones Digital Speech Interpolation (DSI) X.25 Packet Switching Video Conferencing Spread Spectrum Communications	Engine Control Vibration Analysis Antiskid Brakes Adaptive Ride Control Global Positioning Navigation Voice Commands Digital Radio Cellular Telephones
Consumer	**Industrial**	**Medical**
Radar Detectors Power Tools Digital Audio/TV Music Synthesizer Toys and Games Solid-State Answering Machines	Robotics Numeric Control Security Access Power Line Monitors Visual Inspection Lathe Control CAM	Hearing Aids Patient Monitoring Ultra Sound Equipment Diagnostic Tools Prosthetics Fetal Monitors MR Imaging

Chapter 2

Architectural Overview

The TMS320C40's high performance is achieved through the precision and wide dynamic range of the floating-point units, large on-chip memory, a high degree of parallelism, and the six-channel DMA coprocessor. Figure 2–1, beginning on the next page, is a block diagram of the TMS320C40.

This chapter gives an architectural overview of the TMS320C40 processor. An in-depth description of the features are available in the *TMS320C4x User's Guide*. Major areas of discussion are listed below.

Section **Page**

Figure 2–1. TMS320C40 Block Diagram

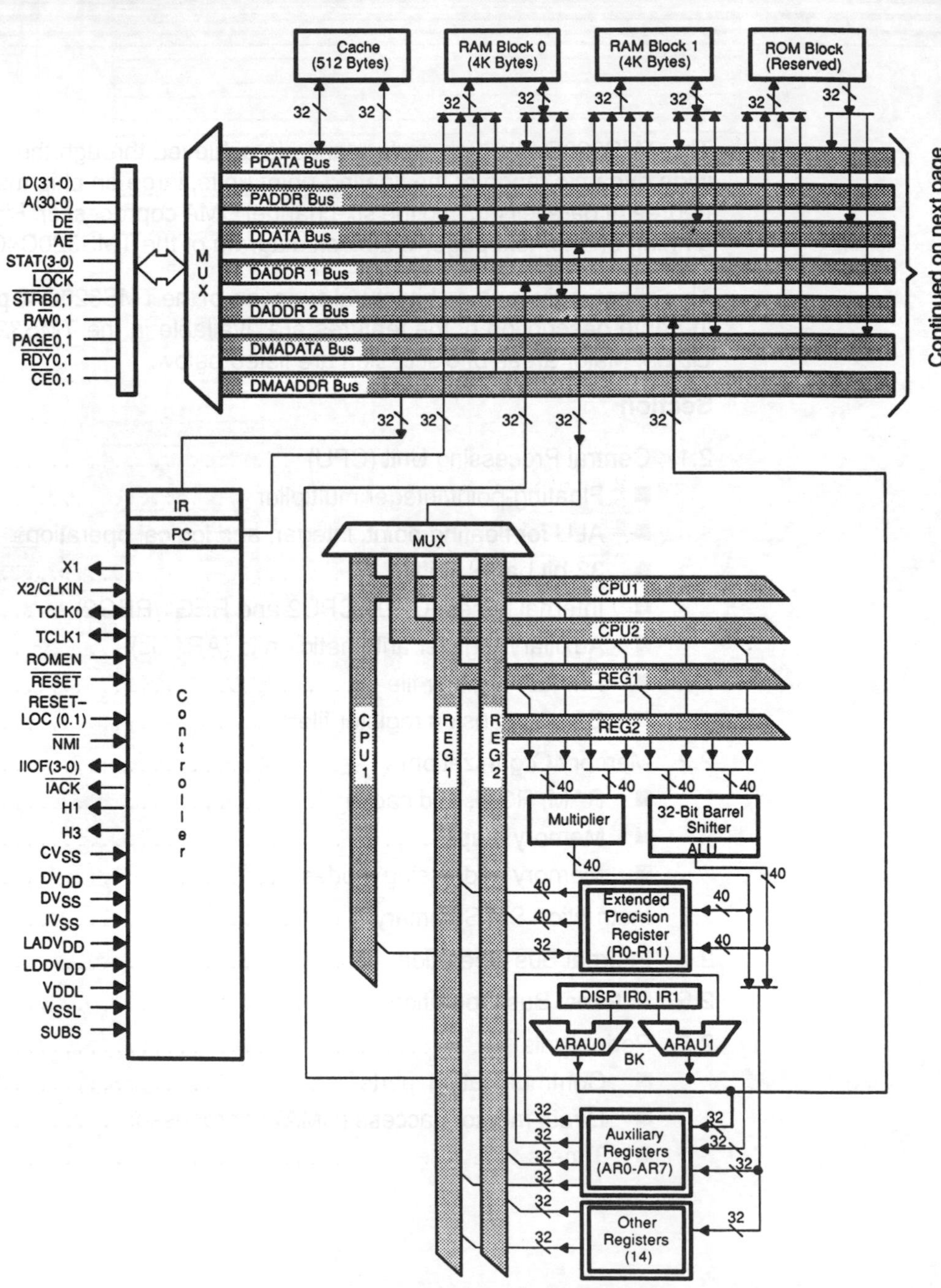

Figure 2–1. TMS320C40 Block Diagram (Concluded)

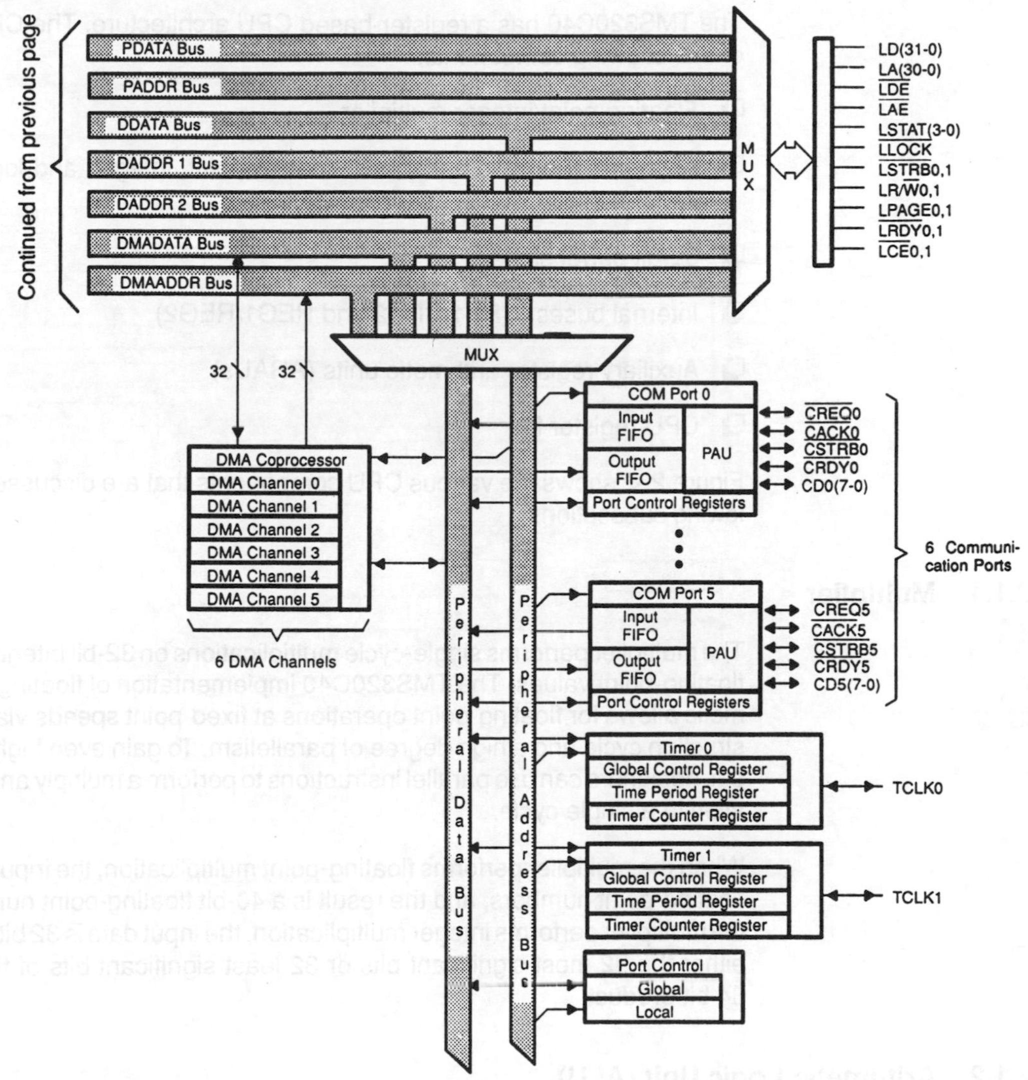

2.1 Central Processing Unit (CPU)

The TMS320C40 has a register-based CPU architecture. The CPU consists of the following components:

- Floating-point/integer multiplier
- ALU for performing arithmetic: floating-point, integer, and logical operations
- 32-bit barrel shifter
- Internal buses (CPU1/CPU2 and REG1/REG2)
- Auxiliary register arithmetic units (ARAUs)
- CPU register file

Figure 2–2 shows the various CPU components that are discussed in the following subsections.

2.1.1 Multiplier

The multiplier performs single-cycle multiplications on 32-bit integer and 40-bit floating-point values. The TMS320C40 implementation of floating-point arithmetic allows for floating-point operations at fixed-point speeds via a 40-ns instruction cycle and a high degree of parallelism. To gain even higher throughput, designers can use parallel instructions to perform a multiply and ALU operation in a single cycle.

When the multiplier performs floating-point multiplication, the inputs are 40-bit floating-point numbers, and the result is a 40-bit floating-point number. When the multiplier performs integer multiplication, the input data is 32 bits and yields either the 32 most-significant bits or 32 least significant bits of the resulting 64-bit product.

2.1.2 Arithmetic Logic Unit (ALU)

The ALU performs single-cycle operations on 32-bit integer, 32-bit logical, and 40-bit floating-point data, including single-cycle integer and floating-point conversions. Results of the ALU are always maintained in 32-bit integer or 40-bit floating-point formats. The barrel shifter is used to shift up to 32 bits left or right in a single cycle.

Internal buses, CPU1/CPU2 and REG1/REG2, carry two operands from memory and two operands from the register file, thus allowing parallel multiplies and adds/subtracts on four integer or floating-point operands in a single cycle.

Figure 2–2. Central Processing Unit (CPU)

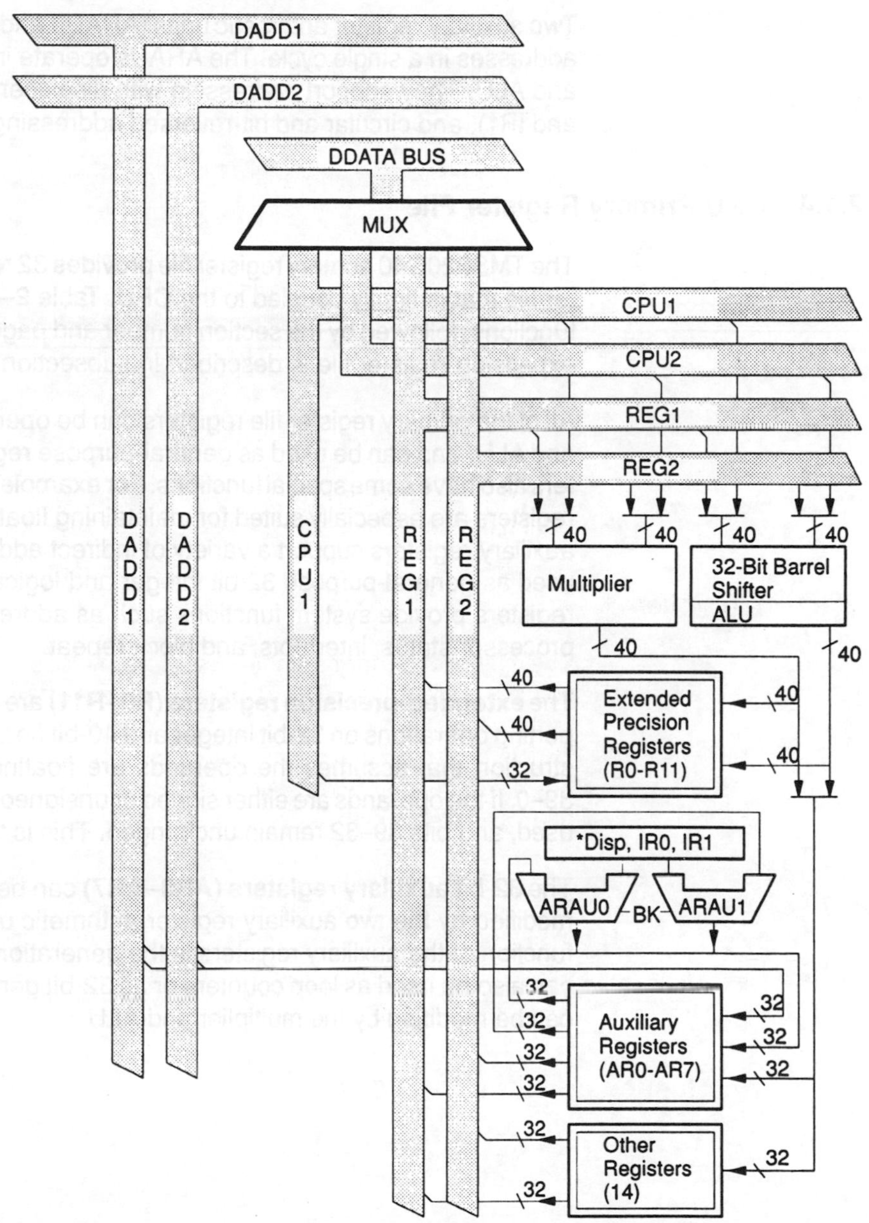

* Disp = an 8-bit integer displacement carried in a program control instruction

2.1.3 Auxiliary Register Arithmetic Units (ARAUs)

Two auxiliary register arithmetic units (ARAU0 and ARAU1) can generate two addresses in a single cycle. The ARAUs operate in parallel with the multiplier and ALU. They support addressing with displacements, index registers (IR0 and IR1), and circular and bit-reversed addressing.

2.1.4 CPU Primary Register File

The TMS320C40 primary register file provides 32 registers in a multiport register file that is tightly coupled to the CPU. Table 2–1 lists register names and functions, followed by the section number and page of each description. (The expansion register file is described in subsection 2.1.5 on page 2-9.)

All of the primary register file registers can be operated upon by the multiplier and ALU, and can be used as general-purpose registers. However, the registers also have some special functions. For example, the 12 extended-precision registers are especially suited for maintaining floating-point results. The eight auxiliary registers support a variety of indirect addressing modes and can be used as general-purpose 32-bit integer and logical registers. The remaining registers provide system functions such as addressing, stack management, processor status, interrupts, and block repeat.

The **extended-precision registers (R0–R11)** are capable of storing and supporting operations on 32-bit integer and 40-bit floating-point numbers. Any instruction that assumes the operands are floating-point numbers uses bits 39–0. If the operands are either signed or unsigned integers, only bits 31–0 are used, and bits 39–32 remain unchanged. This is true for all shift operations.

The 32-bit **auxiliary registers (AR0–AR7)** can be accessed by the CPU and modified by the two auxiliary register arithmetic units (ARAUs). The primary function of the auxiliary registers is the generation of 32-bit addresses. They can also be used as loop counters or as 32-bit general-purpose registers that can be modified by the multiplier and ALU.

Table 2–1. CPU Primary Registers

Assembler Syntax	Assigned Function Name
R0	Extended-precision register 0
R1	Extended-precision register 1
R2	Extended-precision register 2
R3	Extended-precision register 3
R4	Extended-precision register 4
R5	Extended-precision register 5
R6	Extended-precision register 6
R7	Extended-precision register 7
R8	Extended-precision register 8
R9	Extended-precision register 9
R10	Extended-precision register 10
R11	Extended-precision register 11
AR0	Auxiliary register 0
AR1	Auxiliary register 1
AR2	Auxiliary register 2
AR3	Auxiliary register 3
AR4	Auxiliary register 4
AR5	Auxiliary register 5
AR6	Auxiliary register 6
AR7	Auxiliary register 7
DP	Data-page pointer
IR0	Index register 0
IR1	Index register 1
BK	Block-size register
SP	System stack pointer
ST	Status register
DIE	DMA Coprocessor interrupt enable
IIE	Internal-interrupt enable register
IIF	IIOF flag register
RS	Repeat start address
RE	Repeat end address
RC	Repeat counter

The **data page pointer (DP)** is a 32-bit register. The 16 LSBs of the data page pointer are used by the direct addressing mode as a pointer to the page of data being addressed. The TMS320C40 can address up to 64K pages, each page containing 64K words.

The 32-bit **index registers** contain the value used by the auxiliary register arithmetic unit (ARAU) to compute an indexed address.

The ARAU uses the 32-bit **block size register (BK)** in circular addressing to specify the data block size.

The **system stack pointer (SP)** is a 32-bit register that contains the address of the top of the system stack. The SP always points to the last element pushed onto the stack. A push performs a preincrement, and a pop performs a post-decrement of the system stack pointer. The SP is manipulated by interrupts, traps, calls, returns, and the PUSH and POP instructions.

The **status register (ST)** contains global information relating to the state of the CPU. Typically, operations set the condition flags of the status register according to whether the result is zero, negative, etc. This includes register load and store operations as well as arithmetic and logical functions. When the status register is loaded, however, a bit-for-bit replacement is performed with the contents of the source operand, regardless of the state of any bits in the source operand. Therefore, following a load, the contents of the status register are identically equal to the contents of the source operand. This allows the status register to be easily saved and restored.

The **DMA coprocessor interrupt enable register (DIE)** is a 32-bit register containing 2- and 3-bit fields to designate the interrupt synchronization scheme for each of the six DMA channels. It allows each DMA channel to service a corresponding input communication port and output communication port. Also, each DMA channel can be synchronized with external interrupts or the on-chip timers.

The **CPU internal interrupt enable register (IIE)** is also a 32-bit register. This register enables/disables interrupts for the six communication ports, both timers, and the six DMA coprocessor channels.

The **IIOF flag register (IIF)** controls the function (general-purpose I/O or interrupt) of the four external pins (IIOF0 to IIOF3). Interrupts can be level or edge triggered.

The 32-bit **repeat counter (RC)** register specifies the number of times a block of code is to be repeated when performing a block repeat. When the processor is operating in the repeat mode, the 32-bit **repeat start address register (RS)** contains the starting address of the block of program memory to be repeated, and the 32-bit **repeat end address register (RE)** contains the ending address of the block to be repeated.

The **program counter (PC)** is a 32-bit register containing the address of the next instruction to be fetched. Although the PC is not part of the CPU register file, it is a register that can be modified by instructions that modify the program flow.

2.1.5 CPU Expansion Register File

Besides the CPU primary register file (just covered in subsection 2.1.4, starting on page 2-6), the expansion register file contains two special registers that act as pointers:

- IVTP register (points to the interrupt-vector table,
- TVTP register (points to the trap vector table (TVT), which defines vectors for 512 interrupts.

2.2 Memory Organization

The total memory reach of the TMS320C40 is 4G (giga or billion) 32-bit words (4 Gbytes). Program memory (on-chip RAM or ROM and external memory) as well as registers affecting timers, communication ports, and DMA channels are contained within this space. This allows tables, coefficients, program code, and data to be stored in either RAM or ROM. Thus, memory usage is maximized, and memory space is allocated as desired.

By manipulating one external pin (ROMEN, pin AK4), the first one-megaword area of memory (0000 0000h to 000F FFFFh) can be configured to be part of the local address bus *or* configured to address the on–chip ROM when using the boot loader (with remaining space reserved).

2.2.1 RAM, ROM, and Cache

Figure 2–3 shows how the memory is organized on the TMS320C40. RAM blocks 0 and 1 are 4K bytes (1K x 32 bits) each. The ROM block is reserved and contains a boot loader. Each RAM and ROM block is capable of supporting two accesses in a single cycle. The separate program buses, data buses, and DMA buses allow for parallel program fetches, data reads and writes, and DMA operations. For example: the CPU can access two data values in one RAM block and perform an external program fetch in parallel with the DMA coprocessor loading another RAM block, all within a single cycle.

The reserved ROM block (upper right in Figure 2–3) contains a boot loader. This loader supports loading of program and data at reset time. Loading is from 8-, 16-, or 32-bit wide memories or any one of the six communication ports.

A 128 x 32-bit instruction cache is provided to store often-repeated sections of code, thus greatly reducing the number of needed off-chip accesses. This allows for code to be stored off-chip in slower, lower-cost memories. The external buses are also freed for use by the DMA, external memory fetches, or other devices in the system.

Figure 2–3. Memory Organization

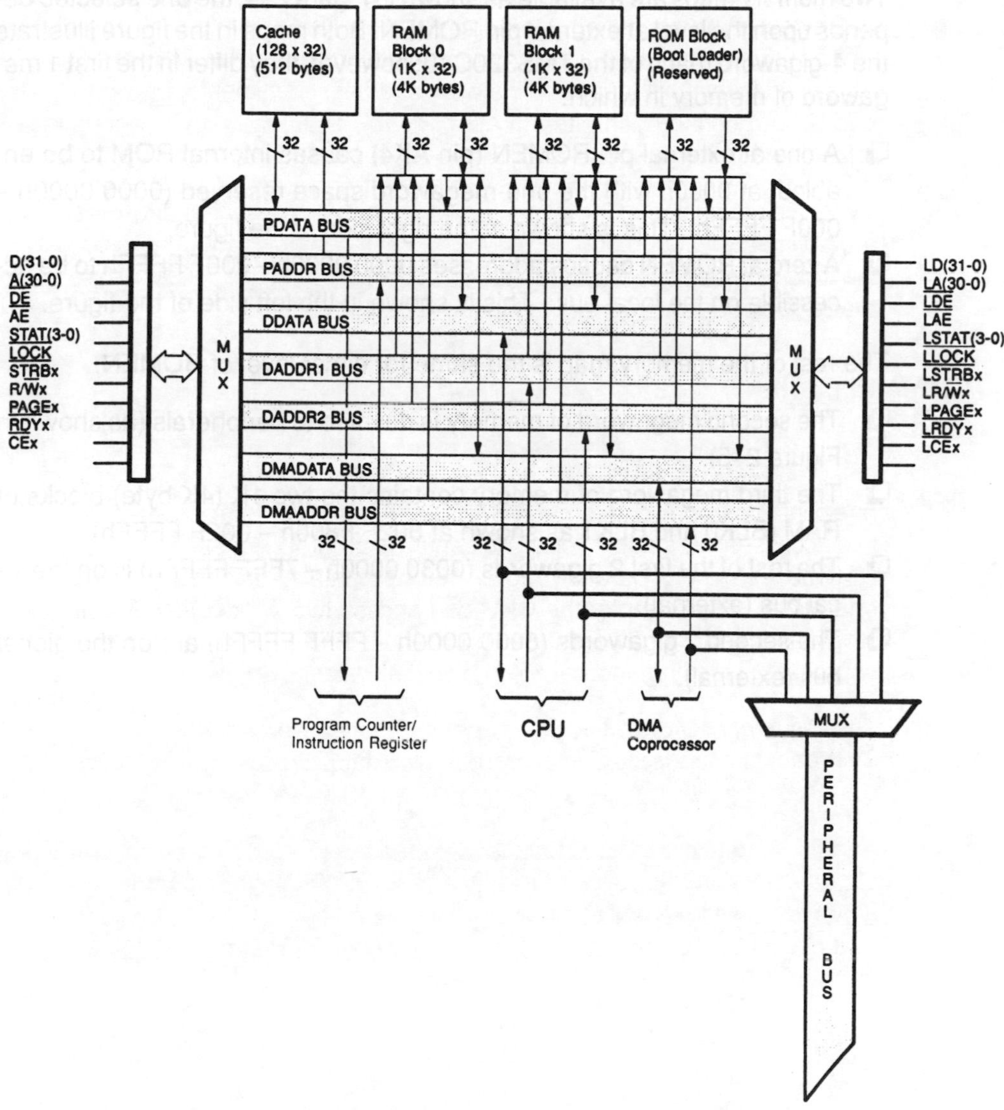

2.2.2 Memory Maps

Two memory maps are available as shown in Figure 2–4; the one selected depends upon the level at external pin ROMEN. Both maps in the figure illustrate the 4-gigaword reach of the TMS320C40; however, they differ in the first 1 megaword of memory in which:

- A one at external pin ROMEN (pin AK4) causes internal ROM to be enabled at 0000h with the one-megaword space reserved (0000 0000h – 000F FFFFh). This is shown in the right side of the figure.
- A zero at ROMEN causes addresses 0000 0000h – 000F FFFFh to be accessible on the local bus. This is shown in the left side of the figure.

The rest of the memory map is the same for either level of ROMEN:

- The second megaword of memory is devoted to peripherals (as shown in Figure 2–5).
- The third megaword of memory contains the two 1K (4K-byte) blocks of RAM (BLK0 and BLK1 as shown at 002F F800h – 002F FFFFh).
- The rest of the first 2 gigawords (0030 0000h – 7FFF FFFFh) is on the local bus (external).
- The second 2 gigawords (8000 0000h – FFFF FFFFh) are on the global bus (external).

Figure 2–4. Memory Maps

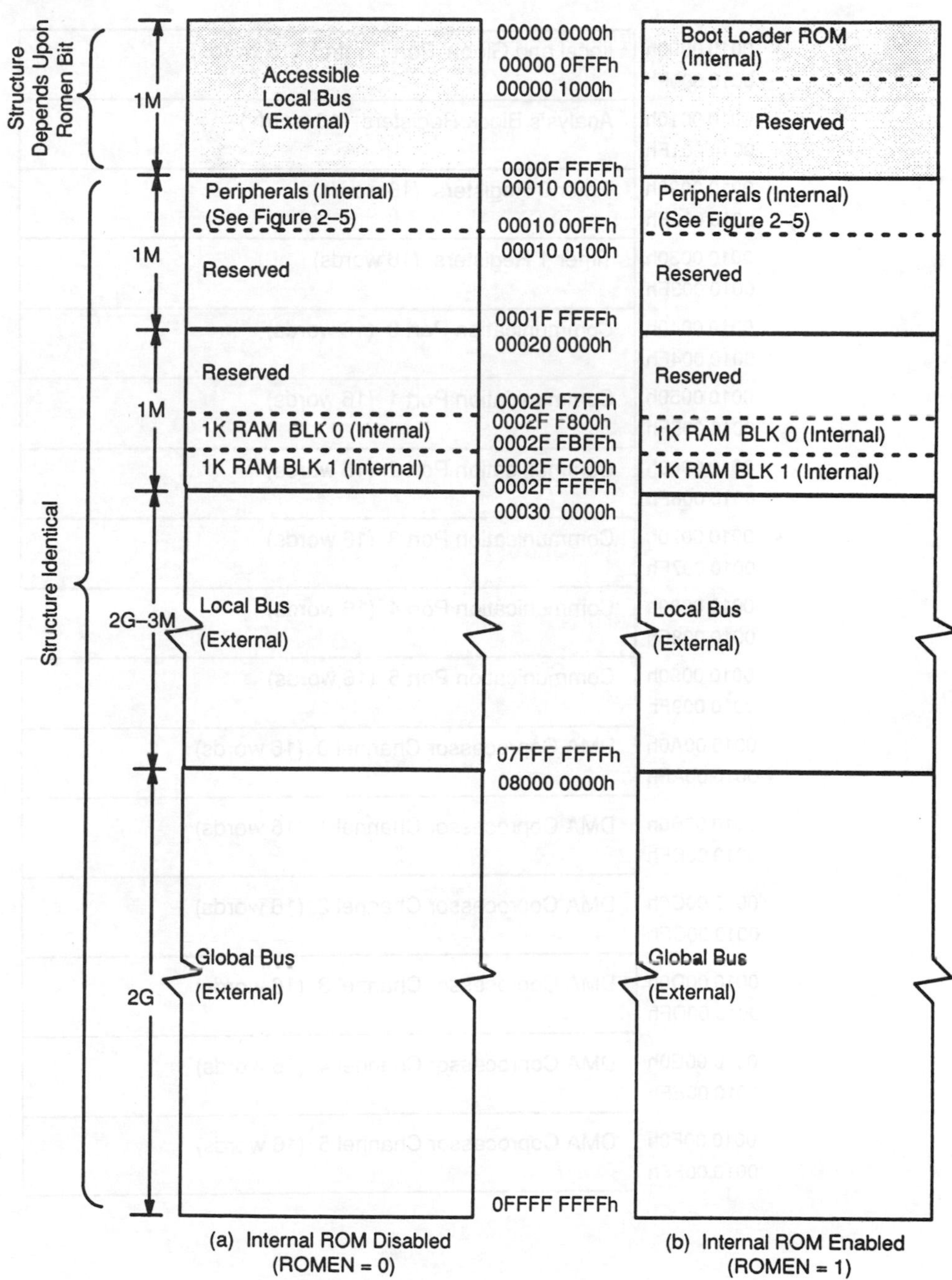

Figure 2–5. Peripheral Memory Map

0010 0000h – 0010 000Fh	Local and Global Port Control (16 words)
0010 0010h – 0010 001Fh	Analysis Block Registers (16 words)
0010 0020h – 0010 002Fh	Timer 0 Registers (16 words)
0010 0030h – 0010 003Fh	Timer 1 Registers (16 words)
0010 0040h – 0010 004Fh	Communication Port 0 (16 words)
0010 0050h – 0010 005Fh	Communication Port 1 (16 words)
0010 0060h – 0010 006Fh	Communication Port 2 (16 words)
0010 0070h – 0010 007Fh	Communication Port 3 (16 words)
0010 0080h – 0010 008Fh	Communication Port 4 (16 words)
0010 0090h – 0010 009Fh	Communication Port 5 (16 words)
0010 00A0h – 0010 00AFh	DMA Coprocessor Channel 0 (16 words)
0010 00B0h – 0010 00BFh	DMA Coprocessor Channel 1 (16 words)
0010 00C0h – 0010 00CFh	DMA Coprocessor Channel 2 (16 words)
0010 00D0h – 0010 00DFh	DMA Coprocessor Channel 3 (16 words)
0010 00E0h – 0010 00EFh	DMA Coprocessor Channel 4 (16 words)
0010 00F0h – 0010 00FFh	DMA Coprocessor Channel 5 (16 words)

2.2.3 Memory Addressing Modes

The TMS320C40 supports a base set of general-purpose instructions as well as arithmetic-intensive instructions that are particularly suited for digital signal processing and other numeric-intensive applications.

Four groups of addressing modes are provided on the TMS320C40 (major headings below). Each group uses two or more of several different addressing types, as shown for each group in the following list:

1) General addressing modes:
   - Register. The operand is a CPU register.
   - Immediate. The operand is a 16-bit immediate value.
   - Direct. The operand is the contents of a 32-bit address (concatenation of 16 bits of the data page pointer and a 16-bit operand).
   - Indirect. A 32-bit auxiliary register indicates the address of the operand.

2) Three-operand addressing modes:
   - Register (same as for general addressing mode).
   - Indirect (same as for general addressing mode).
   - Immediate (same as for general addressing mode).

3) Parallel addressing modes:
   - Register. The operand is an extended-precision register.
   - Indirect (same as for general addressing mode).

4) Branch addressing modes:
   - Register (same as for general addressing mode).
   - PC-relative. A signed 16-bit displacement *or* a 24-bit displacement is added to the PC.

2.3 Instruction Set Summary

Table 2–2 lists the TMS320C40 instruction set in alphabetical order. Each table entry shows the instruction mnemonic, description, and operation.

Table 2–2. Instruction Set Summary

Mnemonic	Description	Operation
ABSF	Absolute value of a floating-point number	|*src*| → Rn
ABSI	Absolute value of an integer	|*src*| → Dreg
ADDC	Add integers with carry	*src* + Dreg + C → Dreg
ADDC3	Add integers with carry (3-operand)	*src1* + *src2* + C → Dreg
ADDF	Add floating-point values	*src* + Rn → Rn
ADDF3	Add floating-point values (3-operand)	*src1* + *src2* → Rn
ADDI	Add integers	*src* + Dreg → Dreg
ADDI3	Add integers (3-operand)	*src1* + *src2* + → Dreg
AND	Bitwise logical-AND	Dreg AND *src* → Dreg
AND3	Bitwise logical-AND (3-operand)	*src1* AND *src2* → Dreg
ANDN	Bitwise logical-AND with complement	Dreg AND $\overline{src}$ → Dreg
ANDN3	Bitwise logical-ANDN (3-operand)	*src1* AND $\overline{src2}$ → Dreg
ASH	Arithmetic shift	If count ≥ 0: (Shifted Dreg left by count) → Dreg Else: (Shifted Dreg right by |count|) → Dreg
ASH3	Arithmetic shift (3-operand)	If count ≥ 0: (Shifted *src* left by count) → Dreg Else: (Shifted *src* right by |count|) → Dreg

LEGEND:

src	general addressing modes	**Dreg**	register address (any register)
src1	three-operand addressing modes	**Rn**	register address (R0 — R11)
src2	three-operand addressing modes	**Daddr**	destination memory address
Csrc	conditional-branch addressing modes	**ARn**	auxiliary register n (AR0 — AR7)
Sreg	register address (any register)	**cond**	condition code
count	shift value (general addressing modes)	**ST**	status register
SP	stack pointer	**RE**	repeat interrupt register
GIE	global interrupt enable register	**RS**	repeat start register
RM	repeat mode bit	**PC**	program counter
TOS	top of stack	**C**	carry bit

Table 2–2. Instruction Set Summary (Continued)

Mnemonic	Description	Operation
B*cond*	Branch conditionally (standard)	If *cond* = true: If C*src* is a register, C*src* → PC If C*src* is a value, C*src* + PC + 1 → PC Else: PC + 1 → PC
B*cond*AF	Branch conditionally delayed and annul if false	If *cond* is true: If *src* is a register: *src* → PC If *src* is a displacement: *src* + PC of branch + 3 → PC Else: If *cond* is false, annul execute phase results of next 3 instructions and continue
B*cond*AT	Branch conditionally delayed and annul if true	If *cond* is true: If *src* is a register: *src* → PC annul execute phase results of next 3 instructions If *src* is a displacement: *src* + PC of branch + 3 → PC annul execute phase results of next 3 instructions Else: continue
B*cond*D	Branch conditionally (delayed)	If *cond* = true: If C*src* is a register, C*src* → PC If C*src* is a value, C*src* + PC + 3 → PC Else: PC + 1 → PC
BR	Branch unconditionally (standard)	C*src* + PC + 1 → PC
BRD	Branch unconditionally (delayed)	C*src* + PC + 3 → PC
CALL	Call subroutine	PC + 1 → TOS C*src* + PC + 1 → PC
CALL*cond*	Call subroutine conditionally	If *cond* = true: PC + 1 → TOS If C*src* is a register, C*src* → PC If C*src* is a value, C*src* + PC → PC Else: PC + 1 → PC
CMPF	Compare floating-point values	Set flags on Rn – *src*
CMPF3	Compare floating-point values (3-operand)	Set flags on *src*1 – *src*2
CMPI	Compare integers	Set flags on Dreg – *src*
CMPI3	Compare integers (3–operand)	Set flags on *src*1 – *src*2
DB*cond*	Decrement and branch conditionally (standard)	ARn – 1 → ARn If *cond* = true and ARn ≥ 0: If C*src* is a register, C*src* → PC If C*src* is a value, C*src* + PC + 1 → PC Else: PC + 1 → PC

Table 2–2. Instruction Set Summary (Continued)

Mnemonic	Description	Operation
DB*cond*D	Decrement and branch conditionally (delayed)	ARn – 1 → ARn If *cond* = true and ARn ≥ 0: If C*src* is a register, C*src* → PC If C*src* is a value, C*src* + PC + 3 → PC Else: PC + 1 → PC
FIX	Convert floating-point value to integer	Fix (*src*) → Dreg
FLOAT	Convert integer to floating-point value	Float(*src*) → Rn
FRIEEE	Convert from IEEE format	Convert *src* from IEEE format → Dreg
IACK	Interrupt acknowledge	Perform a dummy read with $\overline{IACK}$ = 0 At end of dummy read, set $\overline{IACK}$ = 0
IDLE	Idle until interrupt	PC + 1 → PC, then Idle until next interrupt
LAT*cond*	Link and trap conditionally	If *cond* is true: ST(GIE) → ST(PGIE) ST(CF) → ST(PCF) 0 → ST(GIE) 1 → ST(CF) PC of LA*cond* + 4 → R11 trap vector N → PC Else: continue
LAJ	Link and jump	PC + 4 → R11 PC of LAJ + 3 + *src* → PC
LAJ*cond*	Link and jump conditional	If *cond* is true and *src* is a gegister: PC of LAJ*cond* + 4 → R11 & *src* → PC If *cond* is true and *src* is a displacement:: PC of LAJ*cond* + 4 → R11, & *src* + PC of LAJ*cond* + 3 + → PC Else, continue
LBb	Load byte	Sgn extended byte (byte 3,2,1,0) of *src* → Dreg
LBUb	Load byte unsigned	Unsigned byte (byte 3,2,1,0) of *src* → Dreg
LDA	Load address register	*src* → Dreg
LDE	Load floating-point exponent	*src*(exponent) → Rn(exponent)
LDEP	Load integer from exppansion register file to primary register file	*src* → Dreg

LEGEND:

src	general addressing modes	**Dreg**	register address (any register)
src1	three-operand addressing modes	**Rn**	register address (R0 — R11)
src2	three-operand addressing modes	**Daddr**	destination memory address
Csrc	conditional-branch addressing modes	**ARn**	auxiliary register n (AR0 — AR7)
Sreg	register address (any register)	**cond**	condition code
count	shift value (general addressing modes)	**ST**	status register
SP	stack pointer	**RE**	repeat interrupt register
GIE	global interrupt enable register	**RS**	repeat start register
RM	repeat mode bit	**PC**	program counter
TOS	top of stack		

Table 2–2. Instruction Set Summary (Continued)

Mnemonic	Description	Operation
LDF	Load floating-point value	*src* → Rn
LDFcond	Load floating-point value conditionally	If *cond* = true, *src* → Rn Else: Rn is not changed
LDFI	Load floating-point value, interlocked	Signal interlocked operation *src* → Rn
LDHI	Load 16 MSBs with 16–bit immediate	*src* → 16 MSBs of Dreg
LDI	Load integer	*src* → Dreg
LDI*cond*	Load integer conditionally	If *cond* = true, *src* → Dreg Else: Dreg is not changed
LDII	Load integer, interlocked	Signal interlocked operation *src* → Dreg
LDM	Load floating-point mantissa	*src* (mantissa) → Rn (mantissa)
LDP	Load data page pointer	*src* → data page pointer
LDPE	Load integer from primary register file to expansion register file	*src* → Dreg
LDPK	Load data page pointer immediate	*src* → DP
LHw	Load half word	Sign-extended half word of *src* → Dreg
LHUw	Load half word unsigned	Unsigned half word of *src* → Dreg
LSH	Logical shift	If count ≥ 0: (Dreg left-shifted by count) → Dreg Else: (Dreg right-shifted by |count|) → Dreg
LSH3	Logical shift (3-operand)	If count ≥ 0: (*src* left-shifted by count) → Dreg Else: (*src* right-shifted by |count|) → Dreg
LWLct	Load word, left shifted	*src* << (0,1,2,3) bytes and merged with Dreg → Dreg
LWRct	Load word, right shifted	*src* >> (0,1,2,3) bytes and merged with Dreg → Dreg
MBct	Merge byte, left shifted	8 LSBs of *src* << (0,1,2,3) bytes and merged with Dreg → Dreg
MHct	Merge half word, left shifted	16 LSBs of *src* << (0,1) half words and merged with Dreg → Dreg
MPYF	Multiply floating-point values	*src* × Rn → Rn
MPYF3	Multiply floating-point value (3-operand)	*src*1 × *src*2 → Rn
MPYI	Multiply integers	*src* × Dreg → Dreg
MPYI3	Multiply integers (3-operand)	*src*1 × *src*2 → Dreg

Table 2–2. Instruction Set Summary (Continued)

Mnemonic	Description	Operation
MPYSHI	Multiply signed integer and produce 32 MSBs	*dst* × *src* → Dreg
MPYSHI3	Multiply signed integer and produce 32 MSBs, 3 operand	*src1* × *src2* → Dreg
MPYUHI	Multiply unsigned integer and produce 32 MSBs	Dreg × *src* → Dreg
MPYUHI3	Multiply unsigned integer and produce 32 MSBs, 3 operand	*src1* × *src2* → Dreg
NEGB	Negate integer with borrow	0 – *src* – C → Dreg
NEGF	Negate floating-point value	0 – *src* → Rn
NEGI	Negate integer	0 – *src* → Dreg
NOP	No operation	Modify ARn if specified
NORM	Normalize floating-point value	Normalize (*src*) → Rn
NOT	Bitwise logical-complement	$\overline{src}$ → Dreg
OR	Bitwise logical-OR	Dreg OR *src* → Dreg
OR3	Bitwise logical-OR (3-operand)	*src*1 OR *src*2 → Dreg
POP	Pop integer from stack	*SP– – → Dreg
POPF	Pop floating-point value from stack	*SP– – → Rn
PUSH	Push integer on stack	Sreg → *++ SP
PUSHF	Push floating-point value on stack	Rn → *++ SP
RCPF	Reciprocal floating point	16-bit reciprocal of *src* → *dst*
RETS*cond*	Return from subroutine conditionally	If *cond* = true or missing: *SP– – → PC Else: continue
RND	Round floating-point value	Round (*src*) → Rn

LEGEND:

src	general addressing modes	**Dreg**	register address (any register)
src1	three-operand addressing modes	**Rn**	register address (R0 — R11)
src2	three-operand addressing modes	**Daddr**	destination memory address
Csrc	conditional-branch addressing modes	**ARn**	auxiliary register n (AR0 — AR7)
Sreg	register address (any register)	**cond**	condition code
count	shift value (general addressing modes)	**ST**	status register
SP	stack pointer	**RE**	repeat interrupt register
GIE	global interrupt enable register	**RS**	repeat start register
RM	repeat mode bit	**PC**	program counter
TOS	top of stack	**C**	carry bit

Table 2–2. Instruction Set Summary (Continued)

Mnemonic	Description	Operation
ROL	Rotate left	Dreg rotated left 1 bit → Dreg
ROLC	Rotate left through carry	Dreg rotated left 1 bit through carry → Dreg
ROR	Rotate right	Dreg rotated right 1 bit → Dreg
RORC	Rotate right through carry	Dreg rotated right 1 bit through carry → Dreg
RPTB	Repeat block of instructions	*src* → RE 1 → ST (RM) Next PC → RS
RPTBD	Repeat block delayed	If *src* is an immediate value (displacement) *src* + PC +3 → RE Else: *src* → RE 1 → ST (RM) PC of RPTBD + 4 → RS
RPTS	Repeat single instruction	*src* → RC 1 → ST (RM) Next PC → RS Next PC → RE
RSQRF	Reciprocal of square root floating point	16-bit reciprocal of square root of *src* → Dreg
SIGI	Signal, interlocked	Signal interlocked operation Wait for interlock acknowledge Clear interlock
STF	Store floating-point value	Rn → Daddr
STFI	Store floating-point value, interlocked	Rn → Daddr Signal end of interlocked operation
STI	Store integer	Sreg → Daddr
STII	Store integer, interlocked	Sreg → Daddr Signal end of interlocked operation
STIK	Store integer immediate value	*src* → Dreg
SUBB	Subtract integers with borrow	Dreg – *src* – C → Dreg
SUBB3	Subtract integers with borrow (3-operand)	*src*1 – *src*2 – C → Dreg
SUBC	Subtract integers conditionally	If Dreg – *src* ≥ 0: [(Dreg – *src*) << 1] OR 1 → Dreg Else: Dreg << 1 → Dreg

Table 2–2. Instruction Set Summary (Concluded)

Mnemonic	Description	Operation
SUBF	Subtract floating-point values	Rn – *src* → Rn
SUBF3	Subtract floating-point values (3–operand)	*src*1 – *src*2 → Rn
SUBI	Subtract integers	Dreg – *src* → Dreg
SUBI3	Subtract integers (3-operand)	*src*1 – *src*2 → Dreg
SUBRB	Subtract reverse integer with borrow	*src* – Dreg – C → Dreg
SUBRF	Subtract reverse floating-point value	*src* – Rn → Rn
SUBRI	Subtract reverse integer	*src* – Dreg → Dreg
SWI	Software interrupt	Perform emulator interrupt sequence
TOIEEE	Convert to IEEE format	Convert *src* to IEEE format → *dst*
TRAP*cond*	Trap conditionally	If *cond* = true or missing: Next PC → * ++ SP Trap vector N → PC 0 → ST (GIE) Else: continue
TSTB	Test bit fields	Dreg AND *src*
TSTB3	Test bit fields (3-operand)	*src*1 AND *src*2
XOR	Bitwise exclusive-OR	Dreg XOR *src* → Dreg
XOR3	Bitwise exclusive-OR (3-operand)	*src*1 XOR *src*2 → Dreg

LEGEND:

src general addressing modes
src1 three-operand addressing modes
src2 three-operand addressing modes
Csrc conditional-branch addressing modes
Sreg register address (any register)
count shift value (general addressing modes)
SP stack pointer
GIE global interrupt enable register
RM repeat mode bit
TOS top of stack
Dreg register address (any register)
Rn register address (R0 — R11)
Daddr destination memory address
ARn auxiliary register n (AR0 — AR7)
addr 24-bit immediate address (label)
cond condition code
ST status register
RE repeat interrupt register
RS repeat start register
PC program counter
C carry bit

Table 2–3. Parallel Instruction Set Summary

Mnemonic	Description	Operation
	Parallel Arithmetic With Store Instructions	
ABSF || STF	Absolute value of a floating-point	|*src2*| → *dst1* || *src3* → *dst2*
ABSI || STI	Absolute value of an integer	|*src2*| → *dst1* || *src3* → *dst2*
ADDF3 || STF	Add floating-point	*src1* + *src2* → *dst1* || *src3* → *dst2*
ADDI3 || STI	Add integer	*src1* + *src2* → *dst1* || *src3* → *dst2*
AND3 || STI	Bitwise logical-AND	*src1* AND *src2* → *dst1* || *src3* → *dst2*
ASH3	Arithmetic shift	If count ≥ 0: *src2* << count → *dst1* || *src3* → *dst2* Else: *src2* >> |count| → *dst1* || *src3* → *dst2*
FIX || STI	Convert floating-point to integer	Fix(*src2*) → *dst1* || *src3* → *dst2*
FLOAT || STF	Convert integer to floating-point	Float(*src2*) → *dst1* || *src3* → *dst2*
FRIEEE || STF	Parallel FRIEEE and STF	Convert *src2* from IEEE format → *dst1* in parallel with *src3* → *dst2*
LDF || STF	Load floating-point	*src2* → *dst1* || *src3* → *dst2*
LDI || STI	Load integer	*src2* → *dst1* || *src3* → *dst2*
LSH3	Logical shift	If count ≥ 0: *src2* << count → *dst1* || *src3* → *dst2* Else: *src2* >> |count| → *dst1* || *src3* → *dst2*

LEGEND (for parallel instructions):

src1 register addr (R0 — R11)
src3 register addr (R0 — R11)
dst1 register addr (R0 — R11)
op3 register addr (R0 or R1)
src2 indirect addr (disp = 0, 1, IR0, IR1)
src4 indirect addr (disp = 0, 1, IR0, IR1)
dst2 indirect addr (disp = 0, 1, IR0, IR1)
op6 register addr (R2 or R3)

op1,op2,op4,op5 – Two of these operands must be specified using register addr, and two must be specified using indirect.

Table 2–3. Parallel Instruction Set Summary (Continued)

Mnemonic	Description	Operation
MPYF3 || STF	Multiply floating-point and store	*src1* x *src2* → *dst1* || *src3* → *dst2*
MPYI3 || STI	Multiply integer	*src1* x *src2* → *dst1* || *src3* → *dst2*
NEGF || STF	Negate floating-point	0– *src2* → *dst1* || *src3* → *dst2*
TOIEEE || STF	Convert to IEEE floating point format	convert *src2* to IEEE format → *dst1* || *src3* → *dst2*
Parallel Arithmetic With Store Instructions (Concluded)		
NEGI || STI	Negate integer	0 – *src2* → *dst1* || *src3* → *dst2*
NOT || STI	Complement	$\overline{src1}$ → *dst1* || *src3* → *dst2*
OR3 || STI	Bitwise logical-OR	*src1* OR *src2* → *dst1* || *src3* → *dst2*
STF || STF	Store floating-point	*src1* → *dst1* || *src3* → *dst2*
STI || STI	Store integer	*src1* → *dst1* || *src3* → *dst2*
SUBF3 || STF	Subtract floating-point	*src1* – *src2* → *dst1* || *src3* → *dst2*
SUBI3 || STI	Subtract integer	*src1* – *src2* → *dst1* || *src3* → *dst2*
XOR3 || STI	Bitwise exclusive-OR	*src1* XOR *src2* → *dst1* || *src3* → *dst2*

LEGEND (for parallel instructions):

src1 register addr (R0 — R11)
src3 register addr (R0 — R11)
dst1 register addr (R0 — R11)
op3 register addr (R0 or R1)

src2 indirect addr (disp = 0, 1, IR0, IR1)
src4 indirect addr (disp = 0, 1, IR0, IR1)
dst2 indirect addr (disp = 0, 1, IR0, IR1)
op6 register addr (R2 or R3)

op1,op2,op4,op5 – Two of these operands must be specified using register addr, and two must be specified using indirect.

Table 2–3. Parallel Instruction Set Summary (Concluded)

Mnemonic	Description	Operation
	Parallel Load Instructions	
LDF || LDF	Load floating-point	*src2* → *dst1* || *src4* → *dst2*
LDF || STF	Load floating point and store floating point	*src2* → *dst1* || *src3* → *dst2*
LDI || LDI	Load integer	*src2* → *dst1* || *src4* → *dst2*
LSH3 || STI	Logical shift, 3 operand, and store integer	If *count* ≥ 0: *src2* << *count* → *dst1* Else: *src2* >> |*count*| → *dst1* || *src3* → *dst2*
LSH3 || STI	Logical shift 3 and store integer	*src2* → *dst1* || *src3* → *dst2*
	Parallel Multiply And Add/Subtract Instructions	
MPYF3 || ADDF3	Multiply and add floating-point	op1 x op2 → op3 || op4 + op5 → op6
MPYF3 || SUBF3	Multiply and subtract floating-point	op1 x op2 → op3 || op4 – op5 → op6
MPYI3 || ADDI3	Multiply and add integer	op1 x op2 → op3 || op4 + op5 → op6
MPYI3 || SUBI3	Multiply and subtract integer	op1 x op2 → op3 || op4 – op5 → op6

LEGEND (for parallel instructions):

src1 register addr (R0 — R11)
src3 register addr (R0 — R11)
dst1 register addr (R0 — R11)
op3 register addr (R0 or R1)

src2 indirect addr (disp = 0, 1, IR0, IR1)
src4 indirect addr (disp = 0, 1, IR0, IR1)
dst2 indirect addr (disp = 0, 1, IR0, IR1)
op6 register addr (R2 or R3)

op1,op2,op4,op5 – Two of these operands must be specified using register addr, and two must be specified using indirect.

2.4 Internal Bus Operation

A large portion of the TMS320C40's high performance is due to internal busing and parallelism. Separate buses allow for parallel program fetches, data accesses, and DMA accesses:

- **program buses** PADDR and PDATA
- **data buses** DADDR1, DADDR2, and DDATA
- **DMA buses** DMAADDR and DMADATA

These buses connect all of the physical spaces (on-chip memory, off-chip memory, and on-chip peripherals) supported by the TMS320C40. Figure 2–3 shows these internal buses and their connection to on-chip and off-chip memory blocks.

The program counter (PC) is connected to the 32-bit program address bus (PADDR). The instruction register (IR) is connected to the 32-bit program data bus (PDATA). These buses can fetch a single instruction word every machine cycle.

The 32-bit data address buses (DADDR1 and DADDR2) and the 32-bit data data bus (DDATA) support two data memory accesses every machine cycle. The DDATA bus carries data to the CPU over the CPU1 and CPU2 buses. The CPU1 and CPU2 buses can carry two data memory operands to the multiplier, ALU, and register file every machine cycle. Also internal to the CPU are register buses REG1 and REG2, which can carry two data values from the register file to the multiplier and ALU every machine cycle. Figure 2–2 shows the buses internal to the CPU section of the processor.

The DMA controller is supported with a 32-bit address bus (DMAADDR) and a 32-bit data bus (DMADATA). These buses allow the DMA to perform memory accesses in parallel with the memory accesses occurring from the data and program buses.

2.5 External Bus Operation

The TMS320C40 provides two identical external interfaces: the global memory interface and the local memory interface. Each consists of a 32-bit data bus, a 31-bit address bus, and two sets of control signals. Both buses can be used to address external program/data memory or I/O space. The buses also have external $\overline{\text{RDY}}$ signals for wait-state generation with wait states inserted under software control.

2.5.1 Interrupts

The TMS320C40 supports four external interrupts (IIOF3–0), a number of internal interrupts, a nonmaskable, external NMI interrupt, and a nonmaskable external RESET signal, which sets the processor to a known state. The DMA and communication ports have their own internal interrupts. When the CPU responds to the interrupt, the $\overline{\text{IACK}}$ pin can be used to signal an external interrupt acknowledge.

2.5.2 Interlocked Instructions

In order for multiple processors to access global memory and share data in a coherent manner, arbitration is necessary. This arbitration (handshaking) is the purpose of the TMS320C40's interlocked operations, handled through the Interlocked instructions.

2.6 Peripherals

All TMS320C40 peripherals are controlled through memory-mapped registers on a dedicated peripheral bus. This peripheral bus is composed of a 32-bit data bus and a 32-bit address bus. This peripheral bus permits straightforward communication to the peripherals. The TMS320C40 peripherals include two timers and two serial ports. Figure 2–6 shows the peripherals with associated buses and signals.

Figure 2–6. Peripheral Modules

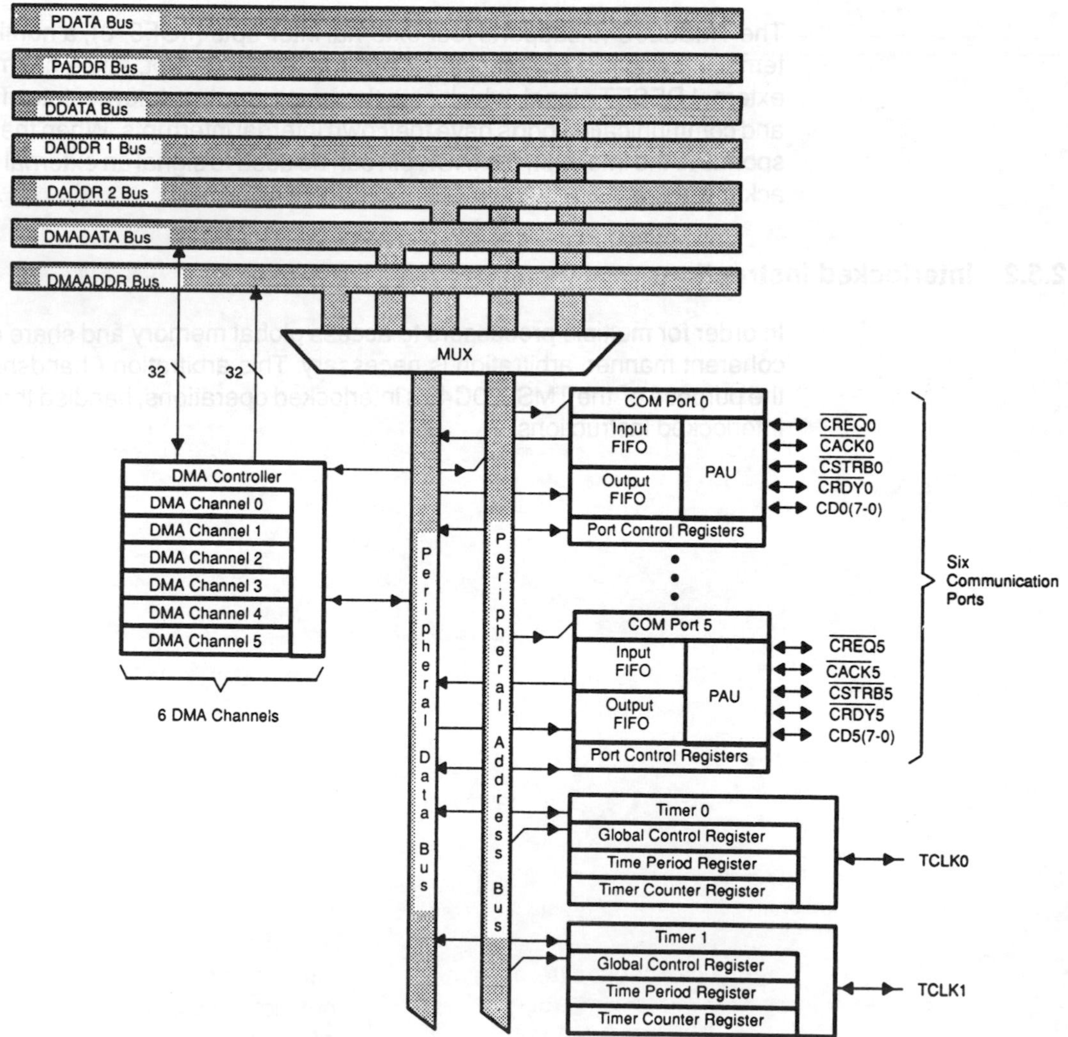

2.6.1 Communication Ports

Six high-speed communication ports provide rapid processor-to-processor communication through each port's dedicated communication interfaces. Coupled with the TMS320C40's two memory interfaces (global and local), this allows you to construct a parallel processor system that attains optimum system performance by the distribuiing of tasks among several processors. Each TMS320C40 can pass the results of its work to another, enabling each TMS320C40 to continue working.

Communication port features:

- 160-megabit per second (20-Mbytes or 5-Mwords per second) bidirectional data transfer operations (at 40-ns cycle time)
- direct (glueless) processor-to-processor communication via eight data lines and four control lines
- buffering of all data transfers, both input and output
- automatic arbitration provided to ensure communication synchronization
- synchronization between the CPU or direct-memory access (DMA) coprocessor and the six communication ports via internal interrupts and internal ready signals.

2.6.2 Direct Memory Access (DMA)

The six channels of the on-chip Direct Memory Access (DMA) coprocessor can read from or write to any location in the memory map without interfering with the operation of the CPU. This allows interfacing to slow external memories and peripherals without reducing throughput to the CPU. The DMA coprocessor contains its own address generators, source and destination registers, and transfer counter. Dedicated DMA address and data buses allow for minimization of conflicts between the CPU and the DMA coprocessor. A DMA operation consists of a block or single-word transfer to or from memory. A key feature of the DMA coprocessor is its ability to automatically reInitialize each channel following a data transfer.

2.6.3 Timers

The two timer modules are general-purpose 32-bit timer/event counters with two signaling modes and internal or external clocking. They can signal internally to the TMS320C40 or externally to the outside world at specified intervals, or they can count external events. Each timer has an I/O pin that can be used as an input clock to the timer, as an output signal driven by the timer, or as a general-purpose I/O pin.

12.5 MIPS DSP Microprocessor

ADSP-2100/ADSP-2100A

FEATURES
Pin- and Code-Compatible DSP Microprocessors
ADSP-2100, 6.144MHz and 8.192MHz
ADSP-2100A, 10.24MHz and 12.5MHz
Separate Program and Data Buses, Extended Off-Chip
Single-Cycle Direct Access to 16K × 16 of Data Memory
Single-Cycle Direct Access to 32K × 24 of Program Memory
Dual Purpose Program Memory for Both Instruction and Data Storage
Three Independent Computational Units: ALU, Multiplier/Accumulator and Barrel Shifter
Two Independent Data Address Generators
Powerful Program Sequencer
Internal Instruction Cache
Provisions for Multiprecision Computation and Saturation Logic
Single-Cycle Instruction Execution
Multifunction Instructions
Four External Interrupts
80ns Cycle Time (ADSP-2100A)
790mW Maximum Power Dissipation (ADSP-2100A, J and K Grades)
100-Pin Grid Array, 100-Lead PQFP (JEDEC Style), 100-Lead CQFP

APPLICATIONS
Optimized for DSP Algorithms Including
Digital Filtering
Fast Fourier Transforms
Applications Include
Image Processing
Radar, Sonar
Speech Processing
Telecommunications

GENERAL DESCRIPTION

The ADSP-2100 and ADSP-2100A are pin- and code-compatible single-chip microprocessors optimized for digital signal processing (DSP) and other high-speed numeric processing applications. The ADSP-2100 and ADSP-2100A are both fabricated in a low-power double-layer metal CMOS process. Together, they offer a span of performance from 6MHz to 12.5MHz. All descriptions of the ADSP-2100 in the text of this data sheet refer to both the ADSP-2100A and the ADSP-2100 versions since they have identical architectures and instruction sets. Timing and electrical specifications differ as shown in those sections of the data sheet.

Both processors integrate computational units, data address generators and a program sequencer in a single device. The ADSP-2100 architecture makes efficient use of external memories for program and data storage, freeing silicon area for increased processor performance. The resulting processor combines the functions and performance of a bit-slice/building block system with the ease of design and development support of a general purpose microprocessor.

The ADSP-2100A (K grade) operates at 12.5MHz. Every instruction executes in a single 80ns cycle. The ADSP-2100A (J and K grades) dissipates less than 790mW while the ADSP-2100 dissipates less than 475mW.

The ADSP-2100's flexible architecture and comprehensive instruction set support a high degree of operational parallelism. Because all instructions execute in a single cycle, MHz = MIPS. In one cycle the ADSP-2100 can:

- generate the next program address
- fetch the next instruction
- perform one or two data moves
- update one or two data address pointers
- perform a computational operation.

DEVELOPMENT SYSTEM

The ADSP-2100 and ADSP-2100A are supported by a complete set of tools for software and hardware system development. The Cross-Software System provides a System Builder for defining the architecture of simulated systems under development, an Assembler, a Linker and a interactive Simulator. An ANSI (draft) Standard C Compiler supports program development in this widely used programming language, producing ADSP-2100 Assembly code which may be assembled, linked and simulated with the other development system tools. A PROM Splitter generates PROM burner compatible files. An In-Circuit Emulator is available for hardware debugging.

An Evaluation Board is available for quick assessment of actual processor performance in a prepackaged hardware environment.

One Technology Way; P. O. Box 9106; Norwood, MA 02062-9106 U.S.A.
Tel: 617/329-4700 Twx: 710/394-6577
Telex: 924491 Cables: ANALOG NORWOODMASS

ADDITIONAL INFORMATION

For additional information on the architecture and instruction set of the processor, refer to the *ADSP-2100 User's Manual*. For more information about programming and the Development System, refer to the *ADSP-2100 Cross-Software Manual* and the *ADSP-2100 Emulator Manual*. For examples of applications routines, refer to the *ADSP-2100 Applications Handbook, Volume 1, 2 or 3*. Manuals are available only from your local Analog Devices sales office. There is also a quarterly newsletter, *DSPatch*™, supporting Analog Devices' digital signal processing customers.

ARCHITECTURE OVERVIEW

Figure 1 is an overall block diagram of the ADSP-2100. The processor contains three independent computational units: the ALU, the multiplier/accumulator (MAC) and the Shifter. The computational units process 16-bit data directly and have provisions to support multiprecision computations. The ALU performs a standard set of arithmetic and logic operations; division primitives are also supported. The MAC performs single-cycle multiply, multiply/add and multiply/subtract operations. The Shifter performs logical and arithmetic shifts, normalization, denormalization and derive exponent operations. The Shifter can be used to efficiently implement any degree of numeric format control, up to and including full floating point representations. The computational units are arranged side-by-side instead of serially for flexible operation sequencing. The internal result (R) bus directly connects the computational units so that the output of any unit may be the input of any unit on the next cycle.

A powerful program sequencer and two dedicated data address generators ensure efficient use of these computational units. The program sequencer generates the next instruction address. To minimize overhead cycles, the sequencer supports conditional jumps, subroutine calls and returns in a single cycle. With internal loop counters and loop stacks, the ADSP-2100 executes looped code with zero overhead; no explicit jump instructions are required to maintain the loop.

The data address generators (DAGs) handle address pointer updates. Each DAG keeps track of up to four address pointers. Whenever the pointer is used to access external data (indirect addressing), it is modified by a prespecified value. A length value may be associated with each pointer to implement automatic modulo addressing for circular buffers. With two independent DAGs, the processor can generate two addresses simultaneously for dual operand fetches.

Efficient data transfer is achieved with the use of five internal buses.

- Program Memory Address (PMA) bus
- Program Memory Data (PMD) bus
- Data Memory Address (DMA) bus
- Data Memory Data (DMD) bus
- Result (R) bus

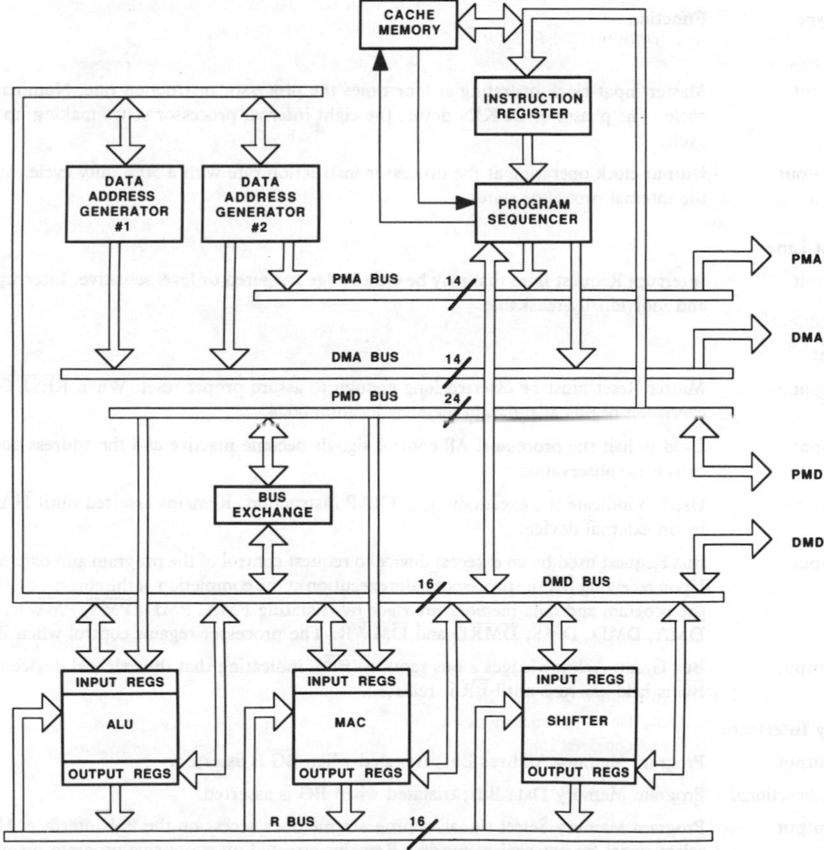

Figure 1. ADSP-2100 Block Diagram

DSPatch is a trademark of Analog Devices, Inc.

The program memory (PMD, PMA) buses and data memory (DMA, DMD) buses extend off-chip to provide direct connections to external memories. The DMD bus is the primary bus for routing data internally and to/from external data memory. The 14-bit DMA bus provides direct addressing of 16K × 16 of external memory. Although the primary function of the program memory is for storing instructions, it can also store data. In this case, the PMD bus provides a path for routing data to/from program memory, permitting dual operand fetches. The 14-bit PMA bus provides direct addressing of 16K × 24 of external memory, expandable to 32K × 24 by using the program memory data access (PMDA) signal as the 15th address line.

When a data fetch from program memory is required, an extra memory cycle is automatically appended to enable the next instruction fetch. To avoid this extra cycle, the ADSP-2100 has an internal instruction cache (16 instructions deep) which serves as an alternate source for the next instruction. The cache monitor circuit transparently determines when the cache contents are valid. When the next instruction is in the cache, no extra cycle is necessary.

The data memory interface supports slower memories and memory-mapped peripherals with wait states. The data memory acknowledge (DMACK) signal provides the necessary handshake. External devices can gain control of program or data buses independently with bus request/ grant signals ($\overline{BR}$, and $\overline{BG}$).

The ADSP-2100 can respond to four external interrupts, which are internally prioritized, maskable and independently programmable as either edge- or level-sensitive. Additional external controls are provided by the $\overline{RESET}$, $\overline{HALT}$ and TRAP signals. With both $\overline{BR}$ and $\overline{RESET}$ recognized, the ADSP-2100 idles, consuming the least possible current.

The ADSP-2100 instruction set provides flexible data moves and multifunction (data moves with a computation) instructions. Every instruction can be executed in a single processor cycle. The ADSP-2100 assembly language uses an algebraic syntax for ease of coding and readability. A comprehensive set of development tools supports program development.

A pin description and detailed discussion of each section of the ADSP-2100 follows.

Pin Description

This section summarizes the pin description of the processor by interface. In this data sheet, when groups of pins are identified with subscripts, as in PMD_{23-0}, the highest numbered pin (PMD_{23}) is the MSB.

Pin Name	Type	Function
Clocks:		
CLKIN	Input	Master input clock operating at four times the processor instruction rate. Nominally 50% duty cycle. The phases of CLKIN define the eight internal processor states making up one instruction cycle.
CLKOUT	Output	Output clock operating at the processor instruction rate with a 50% duty cycle. Synchronized to the internal processor states.
Interrupt Request Lines:		
$\overline{IRQ}_{3-0}$	Input	Interrupt Request lines that may be either edge triggered or level sensitive. Interrupts are prioritized and individually maskable.
Control Interface:		
$\overline{RESET}$	Input	Master Reset must be asserted long enough to assure proper reset. When $\overline{RESET}$ is released, execution begins at program memory location 0004.
$\overline{HALT}$	Input	Used to halt the processor. All control signals become inactive and the address and data buses are driven for observation.
TRAP	Output	Used to indicate the execution of a TRAP instruction. Remains asserted until $\overline{HALT}$ is asserted by an external device.
$\overline{BR}$	Input	Bus Request used by an external device to request control of the program and data memory interface. Upon receiving $\overline{BR}$ the processor halts execution at the completion of the current cycle and relinquishes the program and data memory interface by tristating PMA, PMD, $\overline{PMS}$, $\overline{PMWR}$, $\overline{PMRD}$, PMDA, DMA, DMD, $\overline{DMS}$, $\overline{DMRD}$ and $\overline{DMWR}$. The processor regains control when $\overline{BR}$ is released.
$\overline{BG}$	Output	Bus Grant. Acknowledges a bus request ($\overline{BR}$), indicating that the external device may take control. $\overline{BG}$ is held asserted until $\overline{BR}$ is released.
Program Memory Interface:		
PMA_{13-0}	Output	Program Memory Address Bus; tristated when $\overline{BG}$ is asserted.
PMD_{23-0}	Bidirectional	Program Memory Data Bus; tristated when $\overline{BG}$ is asserted.
$\overline{PMS}$	Output	Program Memory Select signals a program memory access on the PM interface. Usable as a chip select signal for external memories. Remains asserted on successive program memory accesses. HI only when the processor is halted or after execution of a TRAP instruction. Tristated when $\overline{BG}$ is asserted.

Program Memory Interface:

$\overline{\text{PMRD}}$	Output	Program Memory Read indicates a read operation on the PM interface. Also usable as a read strobe or output enable signal. Tristated when $\overline{\text{BG}}$ is asserted.
$\overline{\text{PMWR}}$	Output	Program Memory Write establishes the direction of data transfer on the PM interface. Also usable as a write strobe. Tristated when $\overline{\text{BG}}$ is asserted.
PMDA	Output	Program Memory Data Access used to distinguish instruction and data fetches from PM. Asserted high when data, as opposed to instruction, are accessed. Also usable as a fifteenth PM address bit. Tristated when $\overline{\text{BG}}$ is asserted.

Data Memory Interface:

DMA_{13-0}	Output	Data Memory Address Bus; tristated when $\overline{\text{BG}}$ is asserted.
DMD_{15-0}	Bidirectional	Data Memory Data Bus; tristated when $\overline{\text{BG}}$ is asserted.
$\overline{\text{DMS}}$	Output	Data Memory Select signals a Data Memory Access on the Data Memory interface. Usable as a chip select signal for external memories. Remains asserted on successive data memory accesses. HI only when the processor is halted or after execution of a TRAP instruction. Tristated when $\overline{\text{BG}}$ is asserted.
$\overline{\text{DMRD}}$	Output	Data Memory Read indicates a read operation on the Data Memory interface. Also usable as a read strobe or output enable signal. Tristated when $\overline{\text{BG}}$ is asserted.
$\overline{\text{DMWR}}$	Output	Data Memory Write indicates a write operation on the Data Memory interface. Also usable as a write strobe. Tristated when $\overline{\text{BG}}$ is asserted.
DMACK	Input	Data Memory Acknowledge signal used for asynchronous transfers across the DM interface. Indicates that data memory or memory-mapped peripherals are ready for data transfer. If DMACK is not asserted when checked by the processor, wait states are automatically generated until DMACK is asserted.

Supply Rails:

V_{DD}	Supply	Power supply rail nominally +5VDC. There are four V_{DD} pins.
GND	Ground	Power supply return. There are nine GND pins.

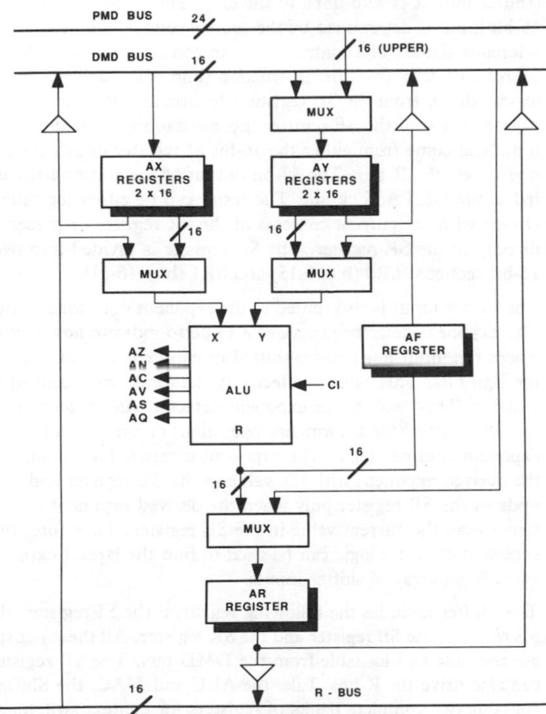

Figure 2. ALU Block Diagram

Arithmetic/Logic Unit

Figure 2 shows a block diagram of the Arithmetic/Logic Unit (ALU).

The ALU provides a standard set of general purpose arithmetic and logic functions: add, subtract, negate, increment, decrement, absolute value, AND, OR, Exclusive OR and NOT. Two divide primitives are also provided to facilitate division. The ALU takes two 16-bit inputs, X and Y, and generates one 16-bit output, R. It accepts the carry (AC) bit in the arithmetic status register (ASTAT) as the carry-in (CI) bit. The carry-in feature enables multiprecision computations. Six arithmetic status bits are generated: AZ (zero), AN (negative), AV (overflow), AC (carry), AS (sign) and AQ (quotient). These status bits are latched in ASTAT.

The X input port can be fed by either the AX register file or any result registers on the R-bus (AR, MR0, MR1, MR2, SR0, or SR1). The AX register file contains two registers, AX0 and AX1. The AX registers can be loaded from the DMD bus. The Y input port can be fed by either the AY register file or the ALU feedback (AF) register. The AY register file contains two registers, AY0 and AY1. The AY registers can be loaded from either the DMD bus or the PMD bus.

The register file outputs are dual ported so that one register can drive the ALU input while either one simultaneously drives the DMD bus. The ALU output can be latched in either the AR register or the AF register.

The AR register has a saturation capability; it can automatically output plus or minus the maximum value if an overflow or underflow occurs. The saturation mode is enabled by a bit in the mode status register (MSTAT). The AR register can drive both the R-bus and the DMD bus and can be loaded from the DMD bus.

The ALU contains a duplicate bank of registers shown in Figure 2 as a "shadow" behind the primary registers. The secondary set contains all the registers described above (AX0, AX1, AY0, AY1, AF, AR). Only one set is accessible at a time. The two sets of registers allow fast context switching for interrupt servicing. The active set is determined by a bit in MSTAT.

Multiplier/Accumulator

The multiplier/accumulator (MAC) implements high-speed multiply, multiply/add and multiply/subtract operations. Figure 3 shows a block diagram of the MAC section.

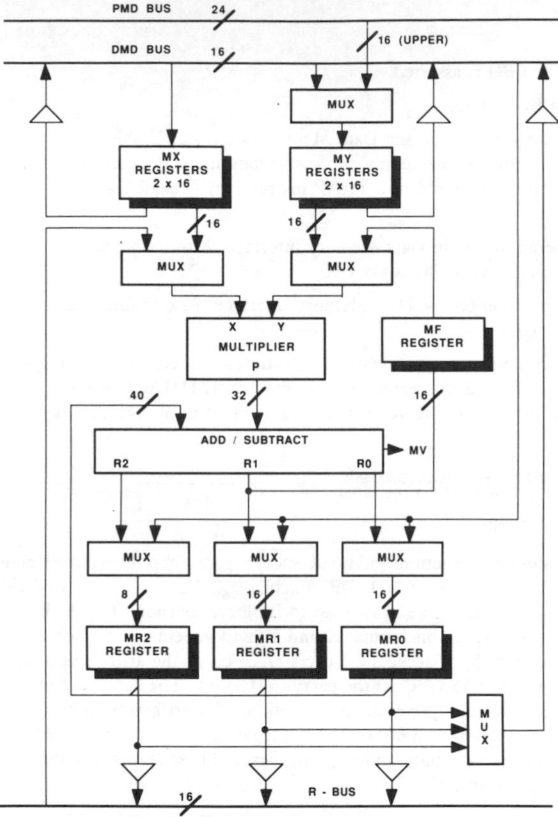

Figure 3. MAC Block Diagram

The multiplier takes two 16-bit inputs, X and Y, and generates one 32-bit output, P. The 32-bit output is routed to a 40-bit accumulator which can add or subtract the P output from the value in MR. MR is a 40-bit register which is divided into three sections: MR0 (bits 0-15), MR1 (bits 16-31), and MR2 (bits 32-39). The result of the accumulator is either loaded into the MR register or into the 16-bit MAC feedback (MF) register. The multiplier accepts the X and Y inputs in either signed or unsigned formats. The result is shifted one bit to the left automatically to remove the redundant sign bit for fractional justification. The accumulator generates one status bit, MV, which is set when the accumulator result overflows the 32-bit boundary. A saturate command is available to change the content of the MR register to the maximum or minimum 32-bit value when MV is set. The accumulator also has the capability for rounding the 40-bit result at the boundary between bit 15 and bit 16.

The MAC and ALU registers are similar. The X input port can be fed by either the MX register file (MX0, MX1) or any result registers on the R-bus (AR, MR0, MR1, MR2, SR0 or SR1). The MX register file is readable and loadable from the DMD bus and has dual-ported outputs.

The Y input port can be fed by either the MY register file (MY0, MY1) or the MF register. The MY register file is readable from the DMD bus and readable and loadable from both the DMD and the PMD bus. Its outputs are dual ported.

The accumulator output can be latched in either the MR register or the MF register. The MR register is connected to both the R-bus and the DMD-bus. Like the ALU section, the MAC section contains two complete banks of registers (MX0, MX1, MY0, MY1, MF, MR0, MR1, MR2) to allow fast context switching.

Shifter

The Shifter gives the ADSP-2100 its unique capability to handle data formatting and numeric scaling. Figure 4 shows a block diagram of the Shifter.

The Shifter can be divided into the following components: the shifter array, the OR/PASS logic, the exponent detector and the exponent compare logic. These components give the Shifter its six basic functions: arithmetic shift, logical shift, normalization, denormalization, derive exponent and derive block exponent.

The shifter array is a 16×32-barrel shifter. It accepts a 16-bit input and can place it anywhere in the 32-bit output field, from off-scale right to off-scale left. The Shifter can perform arithmetic shifts (shifter output is sign-extended to the left) or logical shifts (shifter output is zero-filled to the left). The placement of the 16-bit input is determined by the control code (C) and the HI/LO reference signal. The control code can come from one of three sources: directly from the instruction (immediate arithmetic or logical shift), from the SE register (denormalization) or the negated value of the SE register (normalization). The shifter input can come from either the 16-bit SI register or any result register on the R-bus. The 32-bit output of the shifter array is fed to the OR/PASS circuit. The result can be either logically OR-ed with the current contents of the SR register or passed directly to the SR register. The SR register is divided into two 16-bit sections: SR0 (bits 0-15) and SR1 (bits 16-31).

The shifter input is also routed to the exponent detector circuitry. The exponent detector generates a value to indicate how many places the input must be up-shifted to eliminate all but one of the sign bits. This value is effectively the base 2 exponent of the number. The result of the exponent detector can be latched into the SE register (for a normalize operation) or can be sent to the exponent compare logic. The exponent compare logic compares the derived exponent with the value in the SB register and updates the SB register only when the derived exponent value is larger than the current value in the SB register. Therefore, the exponent compare logic can be used to find the largest exponent value in an array of shifter inputs.

The Shifter includes the following registers: the SI register, the SE register, the SB register and the SR register. All these registers are readable and loadable from the DMD-bus. The SR register can also drive the R-bus. Like the ALU and MAC, the Shifter contains two complete banks of registers for context switching. Each set contains all the registers described above, but only one set is accessible at a time. The active set is determined by a bit in MSTAT.

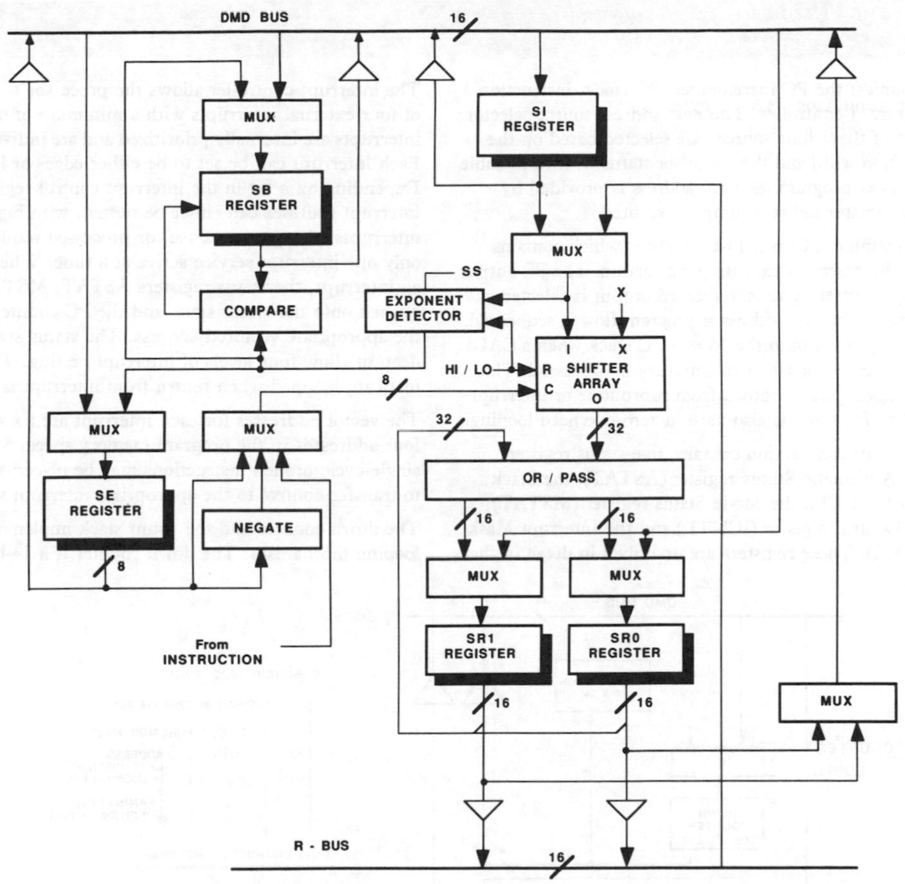

Figure 4. Shifter Block Diagram

Data Address Generators
Figure 5 shows a block diagram of a data address generator.

The data address generators (DAGs) provide indirect addressing for data stored in external memories. The processor contains two independent DAGs so that two data operands (one in program memory and one in data memory) can be addressed simultaneously. The two data address generators are identical except that DAG1 has a bit reversal option on the output and can only generate data memory addresses, while DAG2 can generate both program and data memory addresses but has no bit reversal capability.

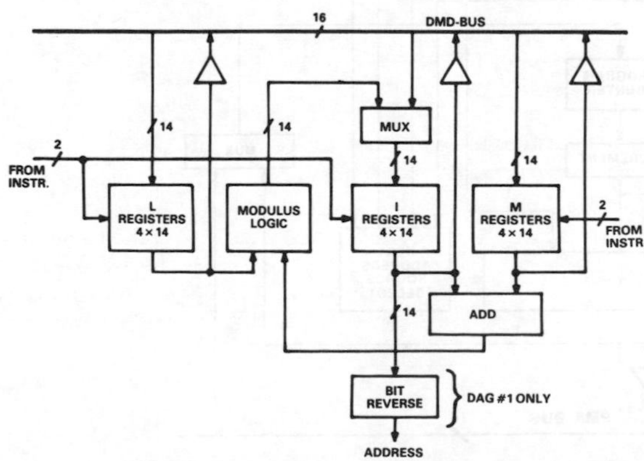

Figure 5. Data Address Generator

There are three register files in each DAG: the modify (M) register file, the indirect (I) register file, and the length (L) register file. Each of these register files contain four 14-bit registers which are readable and loadable from the DMD-bus. The I registers hold the actual addresses used to access external memory. When using the indirect addressing mode, the selected I register content is driven onto either the PMA or DMA bus. This value is post-modified by adding the content of the selected M register. The modified address is passed through the modulus logic. Associated with each I register is an L register which may contain the length of the buffer addressed by the I register. The L register and the modulus logic together enable circular buffer addressing with automatic wrap around at the buffer boundary. The modulus logic is disabled by setting the length of the associated buffer to zero.

Program Sequencer
The program sequencer incorporates powerful and flexible mechanisms for program flow control such as zero-overhead looping, single-cycle branching (both conditional and unconditional), and automatic interrupt processing. Figure 6 shows a block diagram of the program sequencer.

The sequencing logic controls the flow of the program execution. It outputs a program memory address onto the PMA bus from one of four sources: the PC incrementer, PC stack, instruction register or interrupt controller. The next address source selector controls which of these four sources are selected based on the current instruction word and the processor status. A fifth possible source for the next program memory address is provided by DAG2 when a register indirect jump is executed.

The program counter (PC) is a 14-bit register which contains the address of the currently executing instruction. The PC output goes to the incrementer. The incremented output is selected as the next program memory address if program flow is sequential. The PC value is pushed onto the 16 × 14 PC stack when a CALL instruction is executed or when an interrupt is processed. The PC stack is popped when a return from subroutine or interrupt is executed. The PC stack is also used in zero-overhead looping.

The program sequencer section contains five status registers. These are the Arithmetic Status register (ASTAT), the Stack Status register (SSTAT), the Mode Status register (MSTAT), the Interrupt Control register (ICNTL) and the Interrupt Mask register (IMASK). These registers are described in detail in the next section.

The interrupt controller allows the processor to respond to one of four external interrupts with a minimum of overhead. The interrupts are internally prioritized and are individually maskable. Each interrupt can be set to be either edge- or level-sensitive. Depending on a bit in the interrupt control register (ICNTL), interrupt routines can either be nested, with higher priority interrupts taking precedence, or processed sequentially, with only one interrupt service active at a time. When responding to an interrupt, the status registers ASTAT, MSTAT, IMASK are pushed onto the status stack and the PC counter is loaded with the appropriate vectored address. The status stack is four levels deep to allow four levels of interrupt nesting. The stack is automatically popped when return from interrupt is executed.

The vector addresses for each interrupt are fixed at the lowest four addresses in the program memory space. Single-word, single-cycle branch instructions may be placed at these locations to transfer control to the appropriate interrupt service routine.

The down counter and the count stack implement a powerful looping mechanism. The down counter is a 14-bit register with

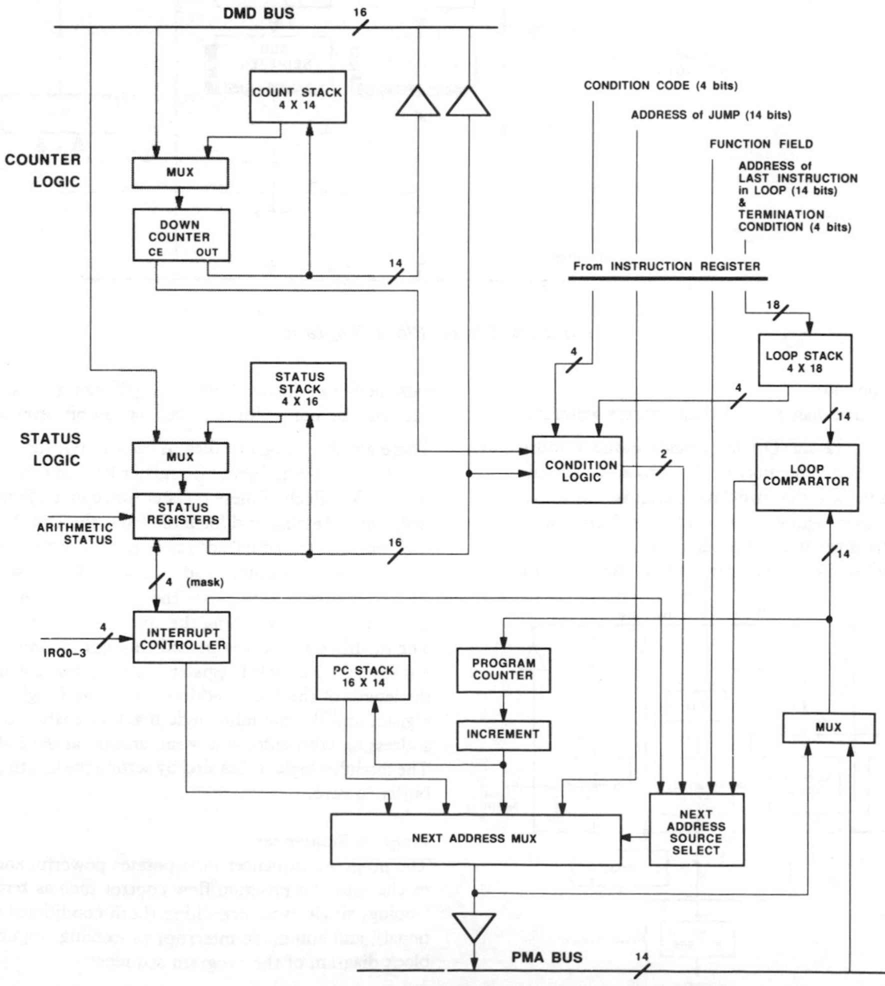

Figure 6. Program Sequencer

auto-decrement capability. It is loaded from the DMD bus with the loop count. The count is decremented every time the counter value is checked; when the count expires, the counter expired (CE) flag is set. The count stack allows the nesting of loops by storing temporarily dormant loop counts. When a new value is loaded into the counter from the DMD bus, the current counter value is automatically pushed onto the count stack as program flow enters a loop. The count stack is automatically popped whenever the CE flag is tested and is true, thereby resuming execution of the code outside the loop.

The DO UNTIL instruction executes a zero-overhead loop using the loop stack and the loop comparator. For a DO UNTIL instruction, a 14-bit termination address and a 4-bit termination condition are pushed onto the 18-bit loop stack. The address of the next instruction (which identifies the top of the loop) is pushed onto the PC stack. The loop comparator continuously compares the current PC value against the termination address on the top of the loop stack. When the termination address is detected, the processor checks if the termination condition is met. If the termination condition is not met, then the top of the PC stack is used as the next PC address, returning program flow to the beginning of the loop. If the termination condition is met, then the PC stack is popped, the current PC is incremented by one, and program flow falls out of the loop. The loop stack is four levels deep, permitting four levels of zero-overhead loop nesting.

Instruction Cache Memory

The instruction cache memory is 16 levels deep and one instruction (24 bits) wide. The cache memory maintains a short history of previously executed instructions so they can be fetched internally if they are needed again.

Every time an instruction is fetched from external memory, it is also written into the cache memory. When the program enters a loop which fits within the cache, all the instructions in the loop are stored in cache during the first pass. On subsequent passes, the instructions can be fetched from the instruction cache when a program memory data access is required. This allows the program memory to be used for data access without penalty. The ADSP-2100 then becomes, in effect, a three-bus system with two data buses and one program bus. For the multiply/accumulate operations typical of digital signal processing algorithms, this gives significant speed advantages.

Instructions are fetched from cache memory *only* when a program memory data fetch is required. The cache monitor circuit automatically keeps track of when the next instruction is contained in the cache. No maintenance or overhead is needed to store externally fetched instructions in the cache or to read previously fetched instructions from cache.

PMD-DMD Bus Exchange

The PMD-DMD bus exchange circuit couples the PMD and DMD buses. The PMD bus is 24 bits wide and the DMD bus is 16 bits wide. The upper 16 bits of PMD are connected to the DMD bus. An 8-bit register (PX) allows transfer of the full width of the PMD bus. When data is read from the PMD bus, the lower 8 bits of the PMD bus are loaded into PX. When writing to the PMD bus, the contents of PX are appended to the upper 16 bits, forming a 24-bit value. The PX register is readable and loadable from the DMD bus.

STATUS REGISTERS

The ADSP-2100 maintains five status registers, each of which can be read over the DMD bus and four of which can be written. These registers are:

ASTAT	Arithmetic Status register
SSTAT	Stack Status register (read-only)
MSTAT	Mode Status register
ICNTL	Interrupt Control register
IMASK	Interrupt Mask register

ASTAT

ASTAT is 8 bits wide and holds the status information generated by the computational sections of the processor. The bits in ASTAT are defined as follows:

0	AZ	(ALU result zero)
1	AN	(ALU result negative)
2	AV	(ALU overflow)
3	AC	(ALU carry)
4	AS	(ALU X input sign)
5	AQ	(ALU quotient flag)
6	MV	(MAC overflow)
7	SS	(Shifter input sign)

The bits which express a particular condition (AZ, AN, AV, AC, MV) are all positive sense (1 = true, 0 = false). Each of the bits are automatically updated whenever a new status is generated by an arithmetic operation. As such, each bit is affected only by a certain subset of arithmetic operations, as defined by the following table:

Status Bit	Updated on:
AZ, AN, AV, AC	Any ALU operation except division
AS	ALU absolute value operation
AQ	ALU divide operations
MV	Any MAC operation except saturate MR
SS	Shifter exponent detect operation

SSTAT

SSTAT is 8 bits wide and holds the status of the four internal stacks. The bits in SSTAT are:

0	PC Stack Empty
1	PC Stack Overflow
2	Count Stack Empty
3	Count Stack Overflow
4	Status Stack Empty
5	Status Stack Overflow
6	Loop Stack Empty
7	Loop Stack Overflow

All of the bits are positive sense (1 = true, 0 = false). The *empty* status bits indicate that the number of pop operations for the stack is greater than or equal to the number of push operations (if no stack overflow has occurred) since the last reset. The *overflow* status bits indicate that the number of push operations for the stack has exceeded the number of pop operations by an amount that is greater than the depth of the stack. When this occurs, the item(s) most recently pushed will be missing from the stack (old data is considered more important than new). The stack overflow status bits "stick" once they are set, so that subsequent pop operations have no effect on them. A processor reset must be executed to clear the stack overflow status.

MSTAT

MSTAT is a 4-bit register that defines various operating modes of the processor. The Mode Control instruction enables or disables the four operating modes. The bits in MSTAT are:

0 Data Register Bank Select
1 Bit Reverse Mode (DAG1 only)
2 ALU Overflow Latch Mode
3 AR Saturation Mode

The data register bank select bit determines which set of data registers is currently active (0 = primary, 1 = secondary). The data registers include all of the result and input registers to the ALU, MAC, and Shifter (AX0, AX1, AY0, AY1, AF, AR, MX0, MX1, MY0, MY1, MF, MR0, MR1, MR2, SB, SE, SI, SR0 and SR1). At initialization, the data register bank select bit is cleared.

The bit reverse mode, when enabled, bit-wise reverses all addresses generated by DAG1. This is most useful for reordering the input or output data in a radix-2 FFT algorithm.

The ALU overflow latch mode causes the AV (ALU overflow) status bit to "stick" once it is set. In this mode, when an ALU overflow occurs, AV will be set and remain set, even if subsequent ALU operations do not generate overflows. AV can then only be cleared by writing a zero into it from the DMD bus.

The AR saturation mode, when set, causes ALU results to be saturated to the maximum positive (H#7FFF) or negative (H#8000) values when an ALU overflow occurs.

IMASK

IMASK is four bits wide and allows the four interrupt inputs to be individually enabled or disabled. The bits in IMASK are:

0 $\overline{\text{IRQ0}}$ Enable
1 $\overline{\text{IRQ1}}$ Enable
2 $\overline{\text{IRQ2}}$ Enable
3 $\overline{\text{IRQ3}}$ Enable

The bits are all positive sense (0 = disabled, 1 = enabled). IMASK is set to zero upon a processor reset so that all interrupts are disabled initially.

ICNTL

ICNTL is a 5-bit register configuring the interrupt modes of the processor. The bits in ICNTL are:

0 $\overline{\text{IRQ0}}$ Sensitivity
1 $\overline{\text{IRQ1}}$ Sensitivity
2 $\overline{\text{IRQ2}}$ Sensitivity
3 $\overline{\text{IRQ3}}$ Sensitivity
4 Interrupt Nesting Mode

The IRQ sensitivity bits determine whether a given interrupt input is edge- or level-sensitive (0 = level-sensitive, 1 = edge-sensitive). These bits are all undefined after a processor reset.

The interrupt nesting mode determines whether nesting of interrupt service routines is allowed. When set to zero, all interrupt levels will be masked automatically when an interrupt service routine is entered. When set to one, IMASK will be set so that only equal and lower priority interrupts will be masked, permitting higher priority interrupts to interrupt the current interrupt service routine. This bit is undefined after a processor reset.

CONDITION CODES

The condition codes are used to determine whether a conditional instruction, such as a jump, trap, call, return, MAC saturation or arithmetic operation, is performed. The sixteen composite status conditions and their derivations are given in Table I. Since arithmetic status is latched into ASTAT at the end of a processor cycle, the condition logic outputs represent conditions generated on a previous cycle.

Code	Status Condition	True If:
EQ	ALU Equal Zero	AZ = 1
NE	ALU Not Equal Zero	AZ = 0
LT	ALU Less Than Zero	AN .XOR. AV = 1
GE	ALU Greater Than or Equal Zero	AN .XOR. AV = 0
LE	ALU Less Than or Equal Zero	(AN .XOR. AV) .OR. AZ = 1
GT	ALU Greater Than Zero	(AN .XOR. AV) .OR. AZ = 0
AC	ALU Carry	AC = 1
NOT AC	Not ALU Carry	AC = 0
AV	ALU Overflow	AV = 1
NOT AV	Not ALU Overflow	AV = 0
MV	MAC Overflow	MV = 1
NOT MV	Not MAC Overflow	MV = 0
NEG	ALU X Input Sign Negative	AS = 1
POS	ALU X Input Sign Positive	AS = 0
NOT CE	Not Counter Expired	CE ≠ 0
TRUE	True	Always True

Table I. Condition Codes

SYSTEM INTERFACE

Figure 7 shows a basic system configuration with the ADSP-2100.

Clock Signals

The ADSP-2100 takes a TTL-compatible clock signal, CLKIN, running at four times the basic processor cycle time as an input. Using this clock input, the processor divides the internal processor cycle into eight states, defined by the edges of the input clock. The active processor cycle consists of states 1 through 7. State 8 is a dead zone to provide a neutral stopping point for halting the processor.

A clock output (CLKOUT) signal is generated by the processor to synchronize external devices to the processor's internal cycles. CLKOUT is high during states 8, 1, 2 and 3, and low during states 4, 5, 6 and 7. Its frequency is one-fourth of that of CLKIN. Except during $\overline{\text{RESET}}$, the CLKOUT signal runs continuously.

Bus Interface

The ADSP-2100 can relinquish control of the memory buses to an external device. When the external device requires access to memory, it asserts the Bus Request ($\overline{\text{BR}}$) signal. After completing the current instruction, the processor halts program execution, tristates the PMA, PMD, $\overline{\text{PMS}}$, $\overline{\text{PMRD}}$, $\overline{\text{PMWR}}$ and PMDA output drivers and the DMA, DMD, $\overline{\text{DMS}}$, $\overline{\text{DMRD}}$ and $\overline{\text{DMWR}}$ output drivers, and asserts the Bus Grant ($\overline{\text{BG}}$) signal. When the $\overline{\text{BR}}$ signal is released, the processor re-enables the output drivers, releases the $\overline{\text{BG}}$ signal, and continues program execution from the point where it stopped.

Program Memory Interface

The Program Memory Interface supports two buses: the program memory address bus (PMA) and the program memory data bus (PMD). The 14-bit PMA bus directly addresses up to 16K words. The PMD bus is bidirectional and 24 bits wide.

Since program memory can be used for both instruction code and data storage, the Program Memory Data Access (PMDA) signal is asserted whenever data, as opposed to an instruction code, is fetched. There is no placement restriction for instruction code and data in program memory area if less than 16K words are used. Since the timing of PMDA is compatible with that of the PMA lines, it may be used as a 15th address line if desired. This effectively doubles the program memory area to 32K, which must be split into 16K dedicated to instruction codes and 16K to data.

The program memory data lines are bidirectional. The Program Memory Select ($\overline{\text{PMS}}$) signal indicates access to the Program Memory and can be used as a chip select signal. The Program Memory Write ($\overline{\text{PMWR}}$) signal indicates a write operation and can be used as a write strobe. The Program Memory Read ($\overline{\text{PMRD}}$) signal indicates a read operation and can be used as a read strobe or output enable signal.

Although the processor internal data bus is only 16 bits, the ADSP-2100 can write to the full 24-bit program memory using the PX register.

Data Memory Interface

The Data Memory Interface supports two buses: the Data Memory Address bus (DMA) and the Data Memory Data bus (DMD). The 14-bit DMA bus directly addresses up to 16K words of data. The DMD bus is bidirectional and 16 bits wide. The Data Memory Select ($\overline{\text{DMS}}$) signal indicates access to the Data Memory and can be used as a chip select signal. The Data Memory Write ($\overline{\text{DMWR}}$) signal indicates a write operation and can be used as a write strobe. The Data Memory Read ($\overline{\text{DMRD}}$) signal indicates a read operation and can be used as a read strobe or output enable signal.

The ADSP-2100 supports memory-mapped I/O, with the peripherals memory mapped into the data memory address space and accessed by the processor in the same manner as data memory.

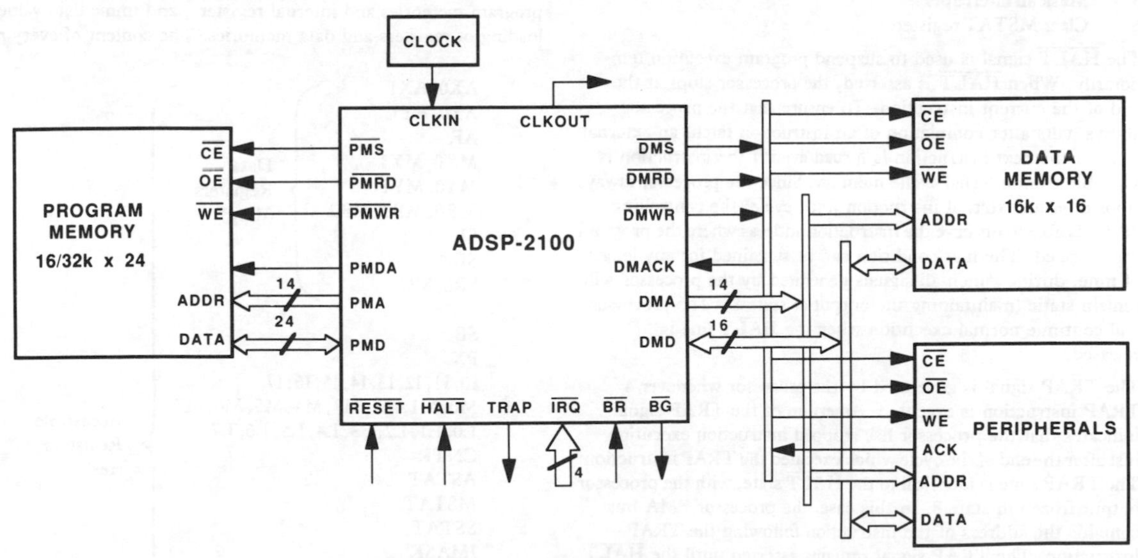

Figure 7. Basic System Configuration

To allow interfacing to slower peripherals, the data memory acknowledge (DMACK) signal is provided. The ADSP-2100 checks the status of the DMACK signal at the end of each processor cycle. If the DMACK signal is not asserted, the processor extends the current cycle by another full cycle. This extension occurs as many times as necessary until the DMACK signal is asserted and the access is completed.

Interrupt Handling

The ADSP-2100 provides four direct interrupt input pins, $\overline{\text{IRQ}_0}$ to $\overline{\text{IRQ}_3}$. Each interrupt pin corresponds to a particular interrupt priority level from 3 (highest) to 0 (lowest). The four interrupt levels are internally prioritized and individually maskable. These input pins can be programmed to be either level- or edge-sensitive.

The ADSP-2100 supports a vectored interrupt scheme: when an external interrupt is acknowledged, the processor switches program control to the interrupt vector address corresponding to the interrupt level (program memory locations 0000 to 0003). Interrupts can optionally be nested so that a higher priority interrupt can preempt the currently executing interrupt service routine.

Processor Control Interface

The processor control interface provides external control over the activity of the processor. The control signals are $\overline{\text{RESET}}$, $\overline{\text{HALT}}$ and TRAP.

The $\overline{\text{RESET}}$ signal initiates a master reset of the ADSP-2100. The $\overline{\text{RESET}}$ signal must be asserted after the chip is powered up to assure proper initialization. The master reset performs the following:

1 Initialize internal clock circuitry
2 Reset all internal stack pointers
3 Clear the cache memory monitor
4 If there is no pending bus request, PMA is driven with 0004
5 Mask all interrupts
6 Clear MSTAT register.

The $\overline{\text{HALT}}$ signal is used to suspend program execution temporarily. When $\overline{\text{HALT}}$ is asserted, the processor stops at the end of the current instruction. To ensure that the processor always halts after completion of an instruction fetch, an external fetch of the next instruction is forced even if the instruction is available from internal cache memory. Since the processor always stops after an external instruction fetch cycle, the controlling device is able to observe the instruction address where the program was stopped. The halt condition can be sustained for any length of time, during which all signals generated by the processor will remain static (maintaining the output at state 8). The processor will continue normal execution when the $\overline{\text{HALT}}$ line is released.

The TRAP signal is generated by the processor whenever a TRAP instruction is executed. Assertion of the TRAP signal indicates that the processor has stopped instruction execution just after the end of the cycle which executed the TRAP instruction. The TRAP state is identical to the $\overline{\text{HALT}}$ state, with the processor output frozen in state 8. In this case, the processor PMA bus contains the address of the instruction following the TRAP instruction. The TRAP signal remains asserted until the $\overline{\text{HALT}}$ signal is asserted externally. When the $\overline{\text{HALT}}$ signal assertion is sensed, the processor releases the TRAP signal. However, the processor remains in the halt condition until the $\overline{\text{HALT}}$ line is released.

Multiprocessor Synchronization

Even when multiple ADSP-2100s are driven from the same CLKIN signal, there is a phase ambiguity between the various processors. This ambiguity can be prevented by using a single master $\overline{\text{RESET}}$ signal synchronized to CLKIN. When the master $\overline{\text{RESET}}$ is released, all the processors begin state 5 on the same edge of CLKIN. Once initialized in this manner, the cycle states of the processors remain synchronized with each other.

INSTRUCTION SET DESCRIPTION

The ADSP-2100 assembly language uses an algebraic syntax for ease of coding and readability. The sources and destinations of computations and data movements are written explicitly in each assembly statement, eliminating cryptic assembler mnemonics. Nevertheless, every instruction assembles into a single 24-bit word and executes in a single cycle. The instructions encompass a wide variety of instruction types along with a high degree of operational parallelism. There are five basic categories of instructions: data move instructions, computational instructions, multifunction instructions, program flow control instructions and miscellaneous instructions. Each of these instruction types is described briefly. The complete instruction set is summarized in Table IV at the end of this section.

Data Move Instructions

Table II gives a list of all registers that are accessible using the data move instructions. (Only the program counter (PC), the instruction register, the arithmetic feedback register (AF) and the multiplier feedback register (MF) are not on this list.) This set of registers is denoted as *reg* in the instruction set summary given in Table IV. A subset of the *reg* group associated with the computational units, which generally hold data as opposed to address or status information, is denoted as *dreg*.

The data move instructions include transfers between internal registers, between data memories and internal registers, between program memories and internal registers, and immediate value loading of registers and data memories. The content of every *reg*

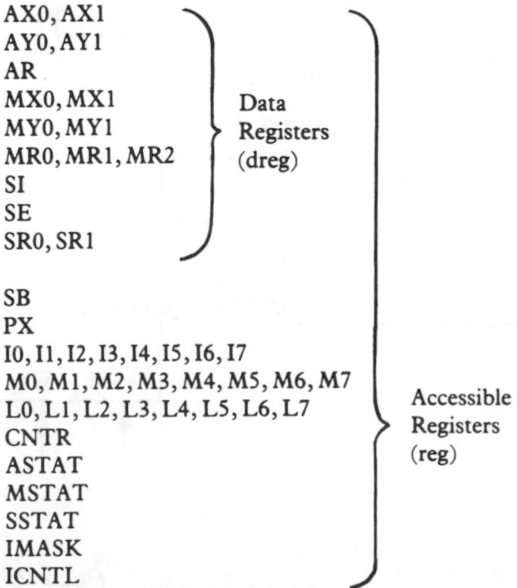

Table II. Register Classification

can also be loaded to any other *reg*. Every *reg* can be loaded with an immediate value which is the full width of the particular register being loaded.

Two addressing modes are supported for data memory transfers: direct addressing and indirect addressing. In direct addressing, the memory address is supplied from the instruction word. In indirect addressing, one of the data address generators provides the address. Using direct addressing, the content of a data memory location can be written and read by any *reg*. Using indirect addressing, the content of a data memory location can only be written and read by a *dreg*. Immediate data load to data memory is permitted with indirect addressing. Only the indirect addressing mode is supported for program memory data transfers, and contents of a program memory location can be read and written to any *dreg*.

Computational Instructions

There are three types of operations associated with the computational units: ALU operations, MAC operations and shifter operations. With few exceptions, all these computational instructions can be made conditional. (The permissible conditions are specified in Table I.) Each computational unit has a set of input registers and output registers. A list of permissible input operands and result registers for each of the units is given in Table III.

Multifunction Instructions

Multifunction instructions execute one computational operation with one or two data moves. All of the multifunction instructions utilize various combinations of the computational and data move operations described above. Since the instruction word is only 24 bits wide, only certain combinations are valid. In general, the following rules are followed.

1 Only one unconditional computational operation can be specified

2 Any memory transfer must use the indirect addressing mode

3 Data move operations can only involve data registers (*dregs*)

4 Only an ALU or a MAC operation can be specified with two operand fetches, one from program memory and one from data memory.

Program Flow Control Instructions

Program flow control instructions include JUMP, CALL, return from subroutine, return from interrupt, DO UNTIL and TRAP. All of these instructions can be made conditional. The JUMP and CALL instructions support both direct addressing, with the destination address specified by the instruction word, and indirect addressing, with the destination address specified by one of the I registers in DAG2.

Miscellaneous Instructions

Miscellaneous instructions include indirect register modify, stack control, mode control and NOP operations.

ALU

Source for X input port (xop)	Source for Y input port (yop)	Destination for output port R
AX0, AX1 AR MR0, MR1, MR2 SR0, SR1	AY0, AY1 AF	AR AF

MAC

Source for X input port (xop)	Source for Y input port (yop)	Destination for output port R
MX0, MX1 AR MR0, MR1, MR2 SR0, SR1	MY0, MY1 MF	MR (MR2, MR1, MR0) MF

Shifter

Source for Shifter input (xop)		Destination for Shifter output
SI AR MR0, MR1, MR2 SR0, SR1		SR (SR1, SR0)

Table III. Computational Input/Output Registers

These conventions are used in Table IV:

1. All keywords are shown in capital letters.
2. Brackets enclose optional parts of the syntax.
3. Vertical lines indicate that one parameter must be chosen from those enclosed.
4. Table I defines the conditions for *condition*.
5. Table II defines the set of registers for *dreg* and *reg*.
6. Table III defines the set of registers for *xop* and *yop*.
7. <data> represents an immediate value.
8. <address> may be an immediate value or label.
9. <comp>, in a multifunction instruction, represents all legal ALU, MAC or Shifter operations with these restrictions:
   - All operations are performed unconditionally
   - Shift Immediate operations are not allowed
   - ALU division (DIVS, DIVQ) is not allowed

DATA MOVE INSTRUCTIONS

Register Move

```
reg = reg;
```

Load Register Immediate

```
reg = <data>;
```

Data Memory Read (direct address)

```
reg = DM(<address>);
```

Data Memory Read (indirect address)

```
dreg = DM( | I0 | , | M0 | );
           | I1 |   | M1 |
           | I2 |   | M2 |
           | I3 |   | M3 |
           |----|   |----|
           | I4 |   | M4 |
           | I5 |   | M5 |
           | I6 |   | M6 |
           | I7 |   | M7 |
```

Program Memory Read (indirect address)

```
dreg = PM ( | I4 | , | M4 | );
            | I5 |   | M5 |
            | I6 |   | M6 |
            | I7 |   | M7 |
```

Data Memory Write (direct address)

```
DM(<address>) = reg;
```

Data Memory Write (indirect address)

```
DM( | I0 | , | M0 | ) = | dreg   | ;
    | I1 |   | M1 |     | <data> |
    | I2 |   | M2 |
    | I3 |   | M3 |
    |----|   |----|
    | I4 |   | M4 |
    | I5 |   | M5 |
    | I6 |   | M6 |
    | I7 |   | M7 |
```

Program Memory Write (indirect address)

```
PM ( | I4 | , | M4 | ) = dreg;
     | I5 |   | M5 |
     | I6 |   | M6 |
     | I7 |   | M7 |
```

COMPUTATIONAL INSTRUCTIONS: ALU

Add/Add with Carry

```
[IF condition] | AR | = xop | +yop     | ;
               | AF |       | +C       |
                            | +yop+C   |
```

Subtract X-Y/Subtract X-Y with Borrow

```
[IF condition] | AR | = xop | -yop       | ;
               | AF |       | -yop+C-1   |
```

Subtract Y-X/Subtract Y-X with Borrow

```
[IF condition] | AR | = yop | -xop       | ;
               | AF |       | -xop+C-1   |
```

AND, OR, Exclusive OR

```
[IF condition] | AR | = xop | AND | yop ;
               | AF |       | OR  |
                            | XOR |
```

Pass/Clear

```
[IF condition] | AR | = PASS | xop | ;
               | AF |        | yop |
```

Negate

```
[IF condition] | AR | = - | xop | ;
               | AF |     | yop |
```

NOT

```
[IF condition] | AR | = NOT | xop | ;
               | AF |       | yop |
```

Absolute Value

```
[IF condition] | AR | = ABS | xop | ;
               | AF |       | yop |
```

Increment

```
[IF condition] | AR | = yop +1 ;
               | AF |
```

Decrement

```
[IF condition] | AR | = yop -1 ;
               | AF |
```

Divide

```
DIVS yop, xop ;
DIVQ xop ;
```

COMPUTATIONAL INSTRUCTIONS: SHIFTER

Arithmetic Shift

```
[IF condition] SR = [SR OR] ASHIFT xop | (HI) | ;
                                       | (LO) |
```

Logical Shift

```
[IF condition] SR = [SR OR] LSHIFT xop | (HI) | ;
                                       | (LO) |
```

Normalize

```
[IF condition] SR = [SR OR] NORM xop | (HI) | ;
                                     | (LO) |
```

Derive Exponent

```
[IF condition] SE = EXP xop | (HI)  | ;
                            | (LO)  |
                            | (HIX) |
```

Block Exponent Adjust

```
[IF condition] SB = EXPADJ xop ;
```

Arithmetic Shift Immediate

SR = [SR OR] ASHIFT xop BY <data> | (HI) / (LO) | ;

Logical Shift Immediate

SR = [SR OR] LSHIFT xop BY <data> | (HI) / (LO) | ;

COMPUTATIONAL INSTRUCTIONS: MAC

Multiply

[IF condition] | MR / MF | = xop*yop (| SS / SU / US / UU / RND |) ;

Multiply Accumulate

[IF condition] | MR / MF | = MR + xop*yop (| SS / SU / US / UU / RND |) ;

Multiply Subtract

[IF condition] | MR / MF | = MR − xop*yop (| SS / SU / US / UU / RND |) ;

Clear

[IF condition] | MR / MF | = 0 ;

Transfer MR

[IF condition] | MR / MF | = MR [(RND)] ;

Conditional MR Saturation

IF MV SAT MR ;

PROGRAM FLOW CONTROL INSTRUCTIONS

Jump

[IF condition] JUMP (| I4 / I5 / I6 / I7 / <address> |) ;

Call

[IF condition] CALL (| I4 / I5 / I6 / I7 / <address> |) ;

Return from Subroutine

[IF condition] RTS ;

Return from Interrupt

[IF condition] RTI ;

Do Until

DO <address> [UNTIL condition] ;

Trap

[IF condition] TRAP ;

MULTIFUNCTION INSTRUCTIONS

Computation with Memory Read

<comp> , dreg = | DM(| I0 / I1 / I2 / I3 / I4 / I5 / I6 / I7 | , | M0 / M1 / M2 / M3 / M4 / M5 / M6 / M7 |) | PM (| I4 / I5 / I6 / I7 | , | M4 / M5 / M6 / M7 |) | ;

Computation with Data Register Move

<comp> , dreg = dreg ;

Computation with Memory Write

| DM(| I0 / I1 / I2 / I3 / I4 / I5 / I6 / I7 | , | M0 / M1 / M2 / M3 / M4 / M5 / M6 / M7 |) | PM (| I4 / I5 / I6 / I7 | , | M4 / M5 / M6 / M7 |) | = dreg, <comp> ;

Data & Program Memory Read

AX0	= DM (	I0	,	M0	) ,	AY0	= PM (	I4	,	M4	) ;
AX1		I1		M1		AY1		I5		M5	
MX0		I2		M2		MY0		I6		M6	
MX1		I3		M3		MY1		I7		M7	

ALU/MAC Operation with Data & Program Memory Read*

<ALU>	,	AX0	= DM (	I0	,	M0	) ,	AY0	= PM (	I4	,	M4	) ;
<MAC>		AX1		I1		M1		AY1		I5		M5	
		MX0		I2		M2		MY0		I6		M6	
		MX1		I3		M3		MY1		I7		M7	

*ALU Division operations not allowed.

MISCELLANEOUS INSTRUCTIONS

Stack Control

[PUSH | POP STS] [, POP CNTR] [, POP PC] [, POP LOOP] ;

[PUSH	STS]	[, POP CNTR]	[, POP PC]	[, POP LOOP]	;
[POP					

Mode Control

[ENA	BIT_REV]	[,	ENA	AV_LATCH]	[,	ENA	AR_SAT]	[,	ENA	SEC_REG]	;
[DIS	]	[	DIS	]	[	DIS	]	[	DIS	]	

Modify Address Register

MODIFY	(	I0	,	M0	) ;
		I1		M1	
		I2		M2	
		I3		M3	
		I4		M4	
		I5		M5	
		I6		M6	
		I7		M7	

No Operation

NOP ;

Table IV. Instruction Set Summary

SPECIFICATIONS

RECOMMENDED OPERATING CONDITIONS

Parameter		ADSP-2100/ADSP-2100A J, K, AJ, AK Grades Min	Max	S, AS, AT, AU Grades Min	Max	Unit
V_{DD}	Supply Voltage	4.75	5.25	4.50	5.50	V
T_{AMB}	Ambient Operating Temperature	0	+70	−55	+125	°C

ELECTRICAL CHARACTERISTICS

Parameter		Test Conditions	ADSP-2100 J & K Grades Min	Max	S Grade Min	Max	Unit
V_{IH}	Hi-Level Input Voltage[1]	@V_{DD} = max	2.0		2.2		
V_{IL}	Lo-Level Input Voltage[1]	@V_{DD} = min		0.8		0.8	V
V_{OH}	Hi-Level Output Voltage[2]	@V_{DD} = min, I_{OH} = −1mA	2.4		2.4		V
V_{OL}	Lo-Level Output Voltage[2]	@V_{DD} = min, I_{OL} = 4mA		0.4		0.6	V
I_{IH}	Hi-Level Input Current[3]	@V_{DD} = max, V_{IN} = max		10		10	μA
I_{IL}	Lo-Level Input Current[3]	@V_{DD} = max, V_{IN} = 0V		10		10	μA
I_{OZH}	Tristate Leakage Current[4]	@V_{DD} = max, V_{IN} = max[7]		10		10	μA
I_{OZL}	Tristate Leakage Current[5]	@V_{DD} = max, V_{IN} = 0V[7]		10		10	μA
I_{OZL}	Tristate Pullup Current[6]	@V_{DD} = max, V_{IN} = 0V[7]		150		150	μA
I_{DD}	Supply Current (Power-Down)[9]	@V_{DD} = max, V_{IN} = 0V[6,7]		10		15	mA
I_{DD}	Supply Current (Dynamic)	@V_{DD} = max, max clock rate[8]		90		100	mA

Parameter		Test Conditions	ADSP-2100A AJ&AK Grades Min	Max	AS Grade Min	Max	AT Grade Min	Max	AU Grade Min	Max	Unit
V_{IH}	Hi-Level Input Voltage[1]	@V_{DD} = max	2.0		2.2		2.2		2.2		V
V_{IH}	Hi-Level Input Voltage at CLKIN	@V_{DD} = max	2.2		2.4		2.4		2.4		V
V_{IL}	Lo-Level Input Voltage[1]	@V_{DD} = min		0.8		0.8		0.8		0.8	V
V_{IL}	Lo-Level Input Voltage at CLKIN	@V_{DD} = min		0.8		0.8		0.8		0.8	V
V_{OH}	Hi-Level Output Voltage[2]	@V_{DD} = min, I_{OH} = −1mA	2.4		2.4		2.4		2.4		V
V_{OL}	Lo-Level Output Voltage[2]	@V_{DD} = min, I_{OL} = 4mA		0.4		0.6		0.6		0.6	V
I_{IH}	Hi-Level Input Current[3]	@V_{DD} = max, V_{IN} = max		10		10		10		10	μA
I_{IL}	Lo-Level Input Current[3]	@V_{DD} = max, V_{IN} = 0V		10		10		10		10	μA
I_{OZH}	Tristate Leakage Current[4]	@V_{DD} = max, V_{IN} = max[7]		10		10		10		10	μA
I_{OZL}	Tristate Leakage Current[5]	@V_{DD} = max, V_{IN} = 0V[7]		10		10		10		10	μA
I_{OZL}	Tristate Pullup Current[6]	@V_{DD} = max, V_{IN} = 0V[7]		180		180		180		180	μA
I_{DD}	Supply Current (Power-Down)[9]	@V_{DD} = max, V_{IN} = 0V[6,7]		10		15		15		15	mA
I_{DD}	Supply Current (Dynamic)	@V_{DD} = max, max clock rate[8]		150		130		180		200	mA

NOTES

[1] Applies to pins: PMD_{0-23}, $\overline{DMD}_{0-15}$, $\overline{BR}$, $\overline{IRQ}_{0-3}$, DMACK, $\overline{RESET}$, $\overline{HALT}$, (48 input pins for ADSP-2100A). Includes CLKIN for ADSP-2100 (49 input pins).

[2] Applies to pins: PMA_{0-13}, $\overline{PMS}$, PMD_{0-23}, $\overline{PMRD}$. $\overline{PMWR}$, PMDA, $\overline{BG}$, DMA_{0-13}, $\overline{DMS}$, DMD_{0-15}, $\overline{DMRD}$. $\overline{DMWR}$, TRAP, CLKOUT (78 output pins).

[3] Applies to pins: $\overline{BR}$, $\overline{IRQ}_{0-3}$, DMACK, $\overline{RESET}$, $\overline{HALT}$, CLKIN (9 input only pins).

[4] Applies to pins: PMA_{0-13}, $\overline{PMS}$, PMD_{0-23} $\overline{PMRD}$. $\overline{PMWR}$, PMDA, DMA_{0-13}, $\overline{DMS}$, DMD_{0-15}, $\overline{DMRD}$. $\overline{DMWR}$ (75 tristateable pins).

[5] Applies to pins: PMA_{0-13}, PMDA, DMA_{0-13} (29 tristateable pins w/o pullup).

[6] Applies to pins: PMD_{0-23}, $\overline{PMS}$, $\overline{PMRD}$. $\overline{PMWR}$, DMD_{0-15}, $\overline{DMS}$, $\overline{DMRD}$. $\overline{DMWR}$ (46 tristateable pins w/pullup).

[7] Additional Test Conditions: V_{IN} = 0V on $\overline{BR}$ and $\overline{RESET}$, CLKIN active, forces tristate condition.

[8] Additional Test Conditions: Outputs loaded TTL loads w/100pF capacitance, V_{IH} = 2.4V, V_{IL} = 0.4V, clock rate = max.

[9] "Power-down" refers to an idle state. While the processor does not have any special standby or low-power mode, these conditions represent the lowest power consumption state.

ABSOLUTE MAXIMUM RATINGS*

Supply Voltage . . . . −0.3V to +7V
Input Voltage . . . . −0.3V to V_{DD} +0.3V
Output Voltage Swing . . . . −0.3V to V_{DD} +0.3V
Operating Temperature Range (Ambient) . . −55°C to +125°C
Storage Temperature Range . . . . −65°C to +150°C
Lead Temperature (10sec) PGA . . . . +300°C
Lead Temperature (5sec) PQFP . . . . +280°C

*Stresses above those listed under "Absolute Maximum Ratings" may cause permanent damage to the device. These are stress rating only and functional operation of the device at these or any other conditions above those indicated in the operational sections of this specification is not implied. Exposure to absolute maximum rating conditions for extended periods may affect device reliability.

ORDERING INFORMATION

Part Number	Speed (MHz)	Temperature Range	Package
ADSP-2100JG	6.144	0 to +70°C	100-Pin Grid Array
ADSP-2100KG	8.192	0 to +70°C	100-Pin Grid Array
ADSP-2100AJG	10.24	0 to +70°C	100-Pin Grid Array
ADSP-2100AKG	12.50	0 to +70°C	100-Pin Grid Array
ADSP-2100JP	6.144	0 to +70°C	100-PQFP
ADSP-2100KP	8.192	0 to +70°C	100-PQFP
ADSP-2100AJP	10.24	0 to +70°C	100-PQFP
ADSP-2100AKP	12.50	0 to +70°C	100-PQFP
ADSP-2100SG	6.144	−55°C to +125°C	100-Pin Grid Array
ADSP-2100ASG	8.192	−55°C to +125°C	100-Pin Grid Array
ADSP-2100ATG	10.24	−55°C to +125°C	100-Pin Grid Array
ADSP-2100AUG	12.50	−55°C to +125'C	100-Pin Grid Array
ADSP-2100SG/883B	6.144	−55°C to +125°C	100-Pin Grid Array
ADSP-2100ASG/883B	8.192	−55°C to +125°C	100-Pin Grid Array
ADSP-2100ATG/883B	10.24	−55°C to +125°C	100-Pin Grid Array
ADSP-2100AUG/883B	12.50	−55°C to +125°C	100-Pin Grid Array
ADSP-2100SZ	6.144	−55°C to +125°C	100-CQFP
ADSP-2100ASZ	8.192	−55°C to +125°C	100-CQFP
ADSP-2100ATZ	10.24	−55°C to +125°C	100-CQFP
ADSP-2100AUZ	12.50	−55°C to +125°C	100-CQFP
ADSP-2100SZ/883B	6.144	−55°C to +125°C	100-CQFP
ADSP-2100ASZ/883B	8.192	−55°C to +125°C	100-CQFP
ADSP-2100ATZ/883B	10.24	−55°C to +125°C	100-CQFP
ADSP-2100AUZ/883B	12.50	−55°C to +125°C	100-CQFP

ADSP-2100/ADSP-2100A Development Tools

Part Number	Description
ADDS-2110	Cross-Software and Simulator (VAX/VMS)
ADDS-2121	Cross-Software (IBM PC/DOS)
ADDS-2122	Simulator (IBM PC/DOS)
ADDS-2123-C	Cross-Software and Simulator (Sun 2/3, Unix BSD 4.2)
ADDS-2130	C Compiler, Cross-Software and Simulator (VAX/VMS)
ADDS-2131	C Compiler, Cross-Software and Simulator (IBM PC/DOS)
ADDS-2133-C	C Compiler, Cross-Software and Simulator (Sun 2/3, Unix BSD 4.2)
ADDS-2150A-8	ADSP-2100A 8MHz In-Circuit Emulator (110V)
ADDS-2150AE-8	ADSP-2100A 8MHz In-Circuit Emulator (220V)
ADDS-2160-8	ADSP-2100A 8MHz Evaluation Board
ADDS-2169	University Package (ADDS-2131 and ADDS-2160)
ADDS-2190	Three Day ADSP-2100 Workshop (U.S.)
ADDS-2190E	Three Day ADSP-2100 Workshop (Europe)

ESD SENSITIVITY

The ADSP-2100 and ADSP-2100A feature proprietary input protection circuitry. Per Method 3015 of MIL-STD-883, the ADSP-2100 has been classified as a Class 1 device and the ADSP-2100A as a Class 2 device.

Proper ESD precautions are strongly recommended to avoid functional damage or performance degradation. Charges as high as 4000 volts readily accumulate on the human body and test equipment and discharge without detection. Unused devices must be stored in conductive foam or shunts, and the foam should be discharged to the destination socket before devices are removed. For further information on ESD precautions, refer to Analog Devices' *ESD Prevention Manual*.

SWITCHING CHARACTERISTICS

GENERAL NOTES

Use the exact timing information given. Do not attempt to derive parameters from the addition or subtraction of others. While this addition or subtraction would yield meaningful results for an individual part, the values given in this data sheet reflect statistical variations and worst cases. Consequently, you cannot meaningfully add up parameters to derive or "verify" longer times.

TIMING NOTES

Switching characteristics specify how the processor is switching its signals. The user has no control over this operation. It is dependent on the internal design. Timing requirements specify the timing of signals that the user has control over such as the placement of data on the DMD bus as input for a read operation.

Timing requirements are used by a designer to guarantee that the processor operates correctly with another device while switching characteristics inform the designer what the device is doing under any given circumstance. Switching characteristics are also referenced to ensure that any timing requirement of a device connected to the processors (such as a memory) is satisfied.

MEMORY REQUIREMENTS

This chart links common memory device specification names and ADSP-2100/ADSP-2100A timing parameters for your convenience.

Parameter Number	Parameter Name	Common Memory Device Specification Name
41	PMA Valid to $\overline{\text{PMWR}}$ Low	Address Set Up to Write Start
79	DMA Valid to $\overline{\text{DMWR}}$ Low	Address Set Up to Write Start
42	$\overline{\text{PMWR}}$ High to PMA Invalid	Address Hold Time
80	$\overline{\text{DMWR}}$ High to DMA Invalid	Address Hold Time
55	PMD Out Valid to $\overline{\text{PMWR}}$ High	Data Set Up Time
91	DMD Out Valid to $\overline{\text{DMWR}}$ High	Data Set Up Time
54	$\overline{\text{PMWR}}$ High to PMD Out Invalid	Data Hold Time
90	$\overline{\text{DMWR}}$ High to DMD Out Invalid	Data Hold Time
58	$\overline{\text{PMRD}}$ Low to PMD Input Valid	$\overline{\text{OE}}$ to Data Valid
94	$\overline{\text{DMRD}}$ Low to DMD Input Valid	$\overline{\text{OE}}$ to Data Valid
59	PMA Valid to PMD Input Valid	Address Access Time
95	DMA Valid to DMD Input Valid	Address Access Time
41 + 40	PMA Valid to $\overline{\text{PMWR}}$ Low + $\overline{\text{PMWR}}$ Width Low	Address Set Up to Write End
79 + 78	DMA Valid to $\overline{\text{DMWR}}$ Low + $\overline{\text{DMWR}}$ Width Low	Address Set Up to Write End

Notes 1 and 2 and information about the Derating Factors and Test Codes appear on page 32.

ADSP-2100 Clock Signals		Test Code	J Grade Min	J Grade Max	K Grade Min	K Grade Max	S Grade Min	S Grade Max	Units	Derating Factor
Timing Requirements										
1	CLKIN Period[1]	A	40.5		30.5		40.5		ns	
2	CLKIN Width Low	A	11		8		11		ns	
3	CLKIN Width High	A	18		12		18		ns	
Switching Characteristics										
4	CLKIN Low (3-4) to CLKOUT Low	B	13	34	13	29	11	34	ns	
5	CLKIN Low (7-8) to CLKOUT High	B	6	24	6	20	5	24	ns	
6	CLKOUT Width Low	A	60		45		60		ns	4

Notes 1 and 2 and information about the Derating Factors and Test Codes appear on page 32.

ADSP-2100A Clock Signals	Test Code	AJ Grade Min	AJ Grade Max	AK Grade Min	AK Grade Max	AS Grade Min	AS Grade Max	AT Grade Min	AT Grade Max	AU Grade Min	AU Grade Max	Units	Derating Factor
Timing Requirements													
1 CLKIN Period[1]	A	24.4		20		30.5		24.4		20		ns	
2 CLKIN Width Low	A	7		4		8		7		4		ns	
3 CLKIN Width High	A	9		8		12		9		8		ns	
Switching Characteristics													
4 CLKIN Low (3-4) to CLKOUT Low	B		24		22		29		24		22	ns	
5 CLKIN Low (7-8) to CLKOUT High	B		20		18		20		20		18	ns	
6 CLKOUT Width Low	A	36		28		45		36		28		ns	4

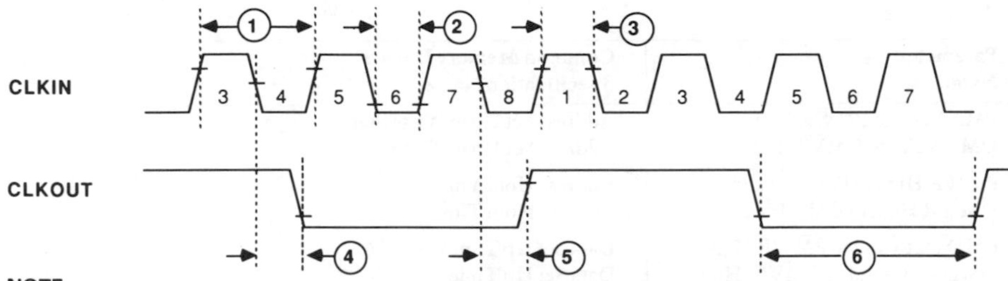

NOTE
The Processor Cycle is Divided into 8 Internal States Determined by the Rising and Falling Edges of CLKIN. CLKOUT is Synchronized to the Processor States as Shown Above.

Figure 8. Clock Signals

Notes 1 and 2 and information about the Derating Factors and Test Codes appear on page 32.

ADSP-2100 Control Signals	Test Code	J Grade Min	J Grade Max	K Grade Min	K Grade Max	S Grade Min	S Grade Max	Units	Derating Factor
Timing Requirements									
7 $\overline{\text{RESET}}$ Low to CLKIN High	B	2		2		2		ns	
8 CLKIN High to $\overline{\text{RESET}}$ High	B	6	36	4	26	6	36	ns	2 (max only)
9 $\overline{\text{RESET}}$ Width Low	A	162		122		170		ns	8

ADSP-2100A Control Signals	Test Code	AJ Grade Min	AJ Grade Max	AK Grade Min	AK Grade Max	AS Grade Min	AS Grade Max	AT Grade Min	AT Grade Max	AU Grade Min	AU Grade Max	Units	Derating Factor
Timing Requirements													
7 $\overline{\text{RESET}}$ Low to CLKIN High	B	2		2		2		2		2		ns	
8 CLKIN High to $\overline{\text{RESET}}$ High	B	4	20	4	16	6	26	4	20	4	16	ns	2 (max only)
9 $\overline{\text{RESET}}$ Width Low	A	98		80		128		98		80		ns	8

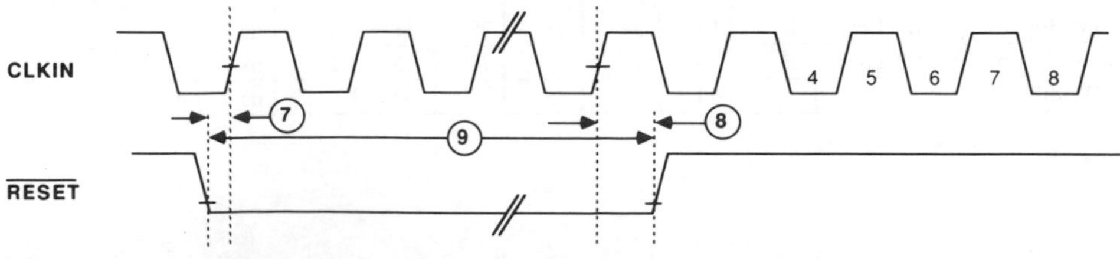

NOTE
The Reset signal determines the phase of the processor cycle.
The processor starts from state 4 after the release of the Reset signal.

Figure 9. $\overline{\text{RESET}}$ Signal

Notes 1 and 2 and information about the Derating Factors and Test Codes appear on page 32.

ADSP-2100 Control Signals	Test Code	J Grade Min	J Grade Max	K Grade Min	K Grade Max	S Grade Min	S Grade Max	Units	Derating Factor
Timing Requirements									
10 $\overline{\text{HALT}}$ Valid to CLKIN Low (3-4)	B	0		0		0		ns	
11 CLKIN Low (3-4) to $\overline{\text{HALT}}$ Invalid	B	12		10		12		ns	
Switching Characteristics									
12 CLKIN Low (7-8) to TRAP Valid	B		25		20		25	ns	
Interrupts									
Timing Requirements									
13 CLKIN Low (7-8) to $\overline{\text{IRQ}}$ Valid	B		2		2		1	ns	
14 CLKIN Low (7-8) to $\overline{\text{IRQ}}$ Invalid	B	21		17		21		ns	

ADSP-2100A Control Signals	Test Code	AJ Grade Min	AJ Grade Max	AK Grade Min	AK Grade Max	AS Grade Min	AS Grade Max	AT Grade Min	AT Grade Max	AU Grade Min	AU Grade Max	Units	Derating Factor
Timing Requirements													
10 $\overline{\text{HALT}}$ Valid to CLKIN Low (3-4)	B	2		2		2		2		2		ns	
11 CLKIN Low (3-4) to $\overline{\text{HALT}}$ Invalid	B	10		8		10		10		8		ns	
Switching Characteristics													
12 CLKIN Low (7-8) to TRAP Valid	B		18		16		20		18		16	ns	
Interrupts													
Timing Requirements													
13 CLKIN Low (7-8) to $\overline{\text{IRQ}}$ Valid	B		1		1		1		1		1	ns	
14 CLKIN Low (7-8) to $\overline{\text{IRQ}}$ Invalid	B	14		14		17		14		14		ns	

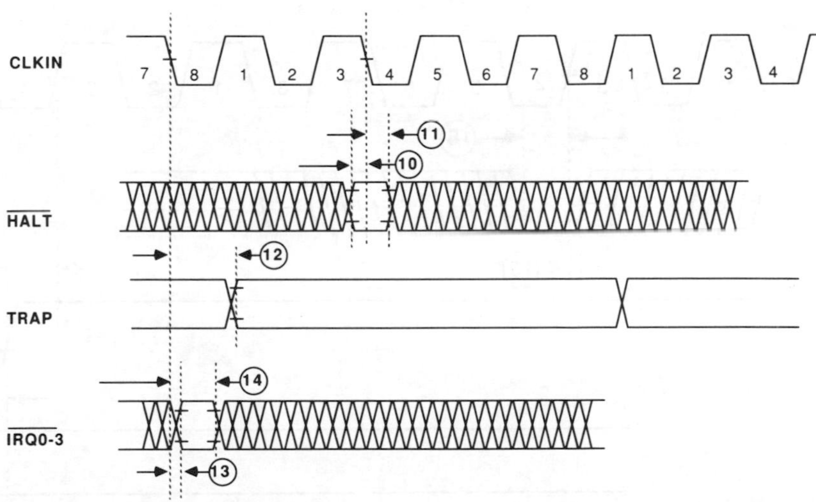

NOTE
The Control Signals are Shown in Relationship to the Processor States in Which They are Recognized or Asserted as Defined by CLKIN. There is No Implied Relationship between HALT, TRAP, and IRQ_{0-3}.

Figure 10. Control Signals

Notes 1 and 2 and information about explaining the Derating Factors and Test Codes appear on page 32.

ADSP-2100 **Bus Request Asserted**	**Test Code**	**J Grade Min**	**J Grade Max**	**K Grade Min**	**K Grade Max**	**S Grade Min**	**S Grade Max**	**Units**	**Derating Factor**
Timing Requirements									
15 $\overline{BR}$ Valid to CLKIN Low (3-4)	B	1		1		1		ns	
16 CLKIN Low (3-4) to $\overline{BR}$ Invalid	B	10		7		10		ns	
Switching Characteristics									
17 CLKIN Low (3-4) to $\overline{BG}$ Low	B		38		30		38	ns	
19 $\overline{BG}$ Low to xMxx Disable[2]	D		22		17		22	ns	

ADSP-2100A **Bus Request Asserted**	**Test Code**	**AJ Grade Min**	**AJ Grade Max**	**AK Grade Min**	**AK Grade Max**	**AS Grade Min**	**AS Grade Max**	**AT Grade Min**	**AT Grade Max**	**AU Grade Min**	**AU Grade Max**	**Units**	**Derating Factor**
Timing Requirements													
15 $\overline{BR}$ Valid to CLKIN Low (3-4)	B	4		4		1		4		4		ns	
16 CLKIN Low (3-4) to $\overline{BR}$ Invalid	B	4		4		7		4		4		ns	
Switching Characteristics													
17 CLKIN Low (3-4) to $\overline{BG}$ Low	B		26		24		30		26		24	ns	
19 $\overline{BG}$ Low to xMxx Disable[2]	D		16		16		17		16		16	ns	

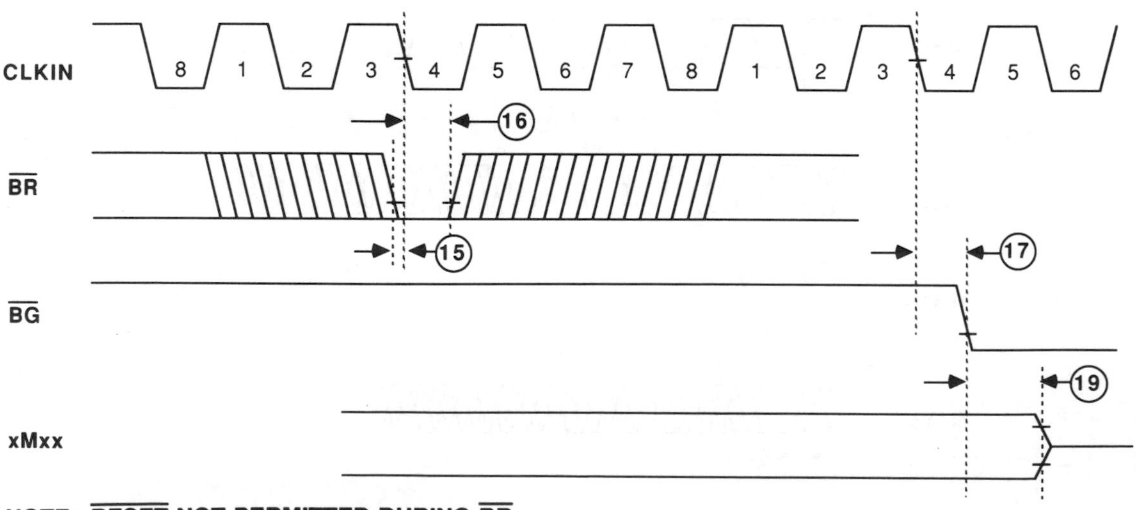

NOTE: $\overline{\text{RESET}}$ NOT PERMITTED DURING $\overline{\text{BR}}$.

Figure 11. Bus Request Asserted

Notes 1 and 2 and information about the Derating Factors and Test Codes appear on page 32.

ADSP-2100 Bus Request Negated		Test Code	J Grade Min	J Grade Max	K Grade Min	K Grade Max	S Grade Min	S Grade Max	Units	Derating Factor
Timing Requirements										
15	$\overline{BR}$ Valid to CLKIN Low (3-4)	B	1		1		1		ns	
16	CLKIN Low (3-4) to $\overline{BR}$ Invalid	B	10		7		10		ns	
Switching Characteristics										
18	CLKIN Low (7-8) to $\overline{BG}$ High	B		31		25		31	ns	
20	xMxx Enable to $\overline{BG}$ High[2]	F		12		10		12	ns	

ADSP-2100A Bus Request Negated		Test Code	AJ Grade Min	AJ Grade Max	AK Grade Min	AK Grade Max	AS Grade Min	AS Grade Max	AT Grade Min	AT Grade Max	AU Grade Min	AU Grade Max	Units	Derating Factor
Timing Requirements														
15	$\overline{BR}$ Valid to CLKIN Low (3-4)	B	4		4		1		4		4		ns	
16	CLKIN Low (3-4) to $\overline{BR}$ Invalid	B	4		4		7		4		4		ns	
Switching Characteristics														
18	CLKIN Low (7-8) to $\overline{BG}$ High	B		24		20		25		24		20	ns	
20	xMxx Enable to $\overline{BG}$ High[2]	F		10		8		10		10		8	ns	

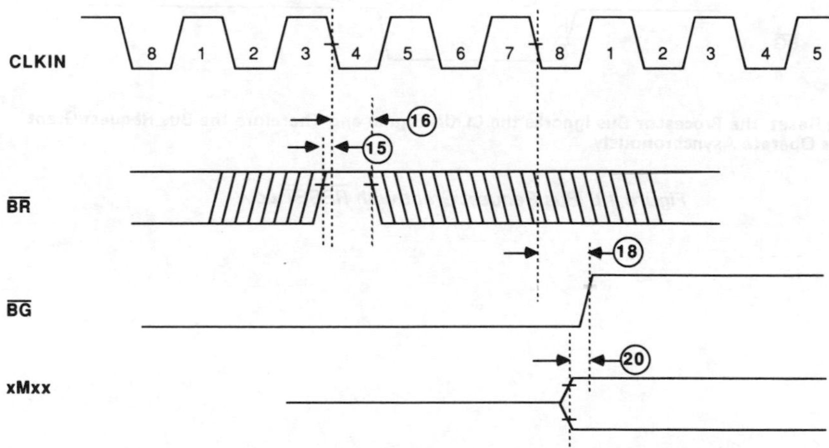

Figure 12. Bus Request Negated

Notes 1 and 2 and information about the Derating Factors and Test Codes appear on page 32.

ADSP-2100 **Bus Request/Grant with $\overline{\text{RESET}}$ Low**	**Test Code**	**J Grade** Min	Max	**K Grade** Min	Max	**S Grade** Min	Max	**Units**	**Derating Factor**
Switching Characteristics									
21 $\overline{\text{BR}}$ Low to $\overline{\text{BG}}$ Low during reset	A		28		23		28	ns	
22 $\overline{\text{BR}}$ High to $\overline{\text{BG}}$ High during reset	A		21		18		21	ns	

ADSP-2100A **Bus Request/Grant with $\overline{\text{RESET}}$ Low**	**Test Code**	**AJ Grade** Min	Max	**AK Grade** Min	Max	**AS Grade** Min	Max	**AT Grade** Min	Max	**AU Grade** Min	Max	**Units**	**Derating Factor**
Switching Characteristics													
21 $\overline{\text{BR}}$ Low to $\overline{\text{BG}}$ Low during reset	A		18		16		23		18		16	ns	
22 $\overline{\text{BR}}$ High to $\overline{\text{BG}}$ High during reset	A		16		14		18		16		14	ns	

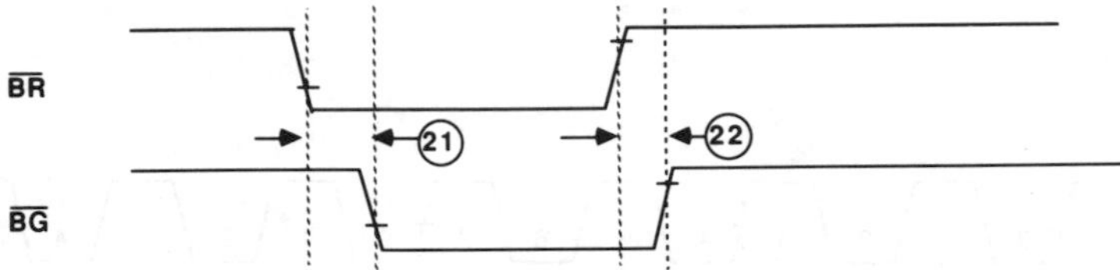

NOTE
During Reset, the Processor Bus Ignores the CLKIN Signal and Therefore the Bus Request/Grant Signals Operate Asynchronously.

Figure 13. Bus Request/Grant with $\overline{\text{RESET}}$ Low

Notes 1 and 2 and information about the Derating Factors and Test Codes appear on page 32.

ADSP-2100 Program Memory Read		Test Code	J Grade Min	J Grade Max	K Grade Min	K Grade Max	S Grade Min	S Grade Max	Units	Derating Factor
Switching Characteristics										
31	$\overline{PMRD}$ Width Low	A	60		45		60		ns	4
32	PMA Valid to $\overline{PMRD}$ Low	A	18		11		18		ns	3
33	$\overline{PMRD}$ High to PMA Invalid	A	20		16		20		ns	1
34	PMDA Valid to $\overline{PMRD}$ Low	A	41		31		41		ns	3
35	$\overline{PMRD}$ High to PMDA Invalid	A	23		18		22		ns	1
36	$\overline{PMS}$ Valid to $\overline{PMRD}$ Low	A	55		40		55		ns	3
37	$\overline{PMRD}$ High to $\overline{PMS}$ Invalid	A	16		12		16		ns	1
Timing Requirements										
58	$\overline{PMRD}$ Low to PMD Input Valid	A		45		37		45	ns	4
59	PMA Valid to PMD Input Valid	A		57		50		57	ns	7
60	$\overline{PMS}$ Valid to PMD Input Valid	A		90		65		90	ns	7
97	$\overline{PMRD}$ High to PMD Input Invalid	A	0		0		0		ns	

ADSP-2100A Program Memory Read		Test Code	AJ Grade Min	AJ Grade Max	AK Grade Min	AK Grade Max	AS Grade Min	AS Grade Max	AT Grade Min	AT Grade Max	AU Grade Min	AU Grade Max	Units	Derating Factor
Switching Characteristics														
31	$\overline{PMRD}$ Width Low	A	36		28		45		36		28		ns	4
32	PMA Valid to $\overline{PMRD}$ Low	A	6		4		14		6		4		ns	3
33	$\overline{PMRD}$ High to PMA Invalid	A	8		6		10		8		6		ns	1
34	PMDA Valid to $\overline{PMRD}$ Low	A	20		18		24		20		15		ns	3
35	$\overline{PMRD}$ High to PMDA Invalid	A	10		10		12		10		10		ns	1
36	$\overline{PMS}$ Valid to $\overline{PMRD}$ Low	A	32		26		40		32		26		ns	3
37	$\overline{PMRD}$ High to $\overline{PMS}$ Invalid	A	8		6		8		8		6		ns	1
Timing Requirements														
58	$\overline{PMRD}$ Low to PMD Input Valid	A		28		20		33		28		18	ns	4
59	PMA Valid to PMD Input Valid	A		46		32		50		46		32	ns	7
60	$\overline{PMS}$ Valid to PMD Input Valid	A		50		45		65		50		35	ns	7
97	$\overline{PMRD}$ High to PMD Input Invalid	A	0		0		0		0		0		ns	

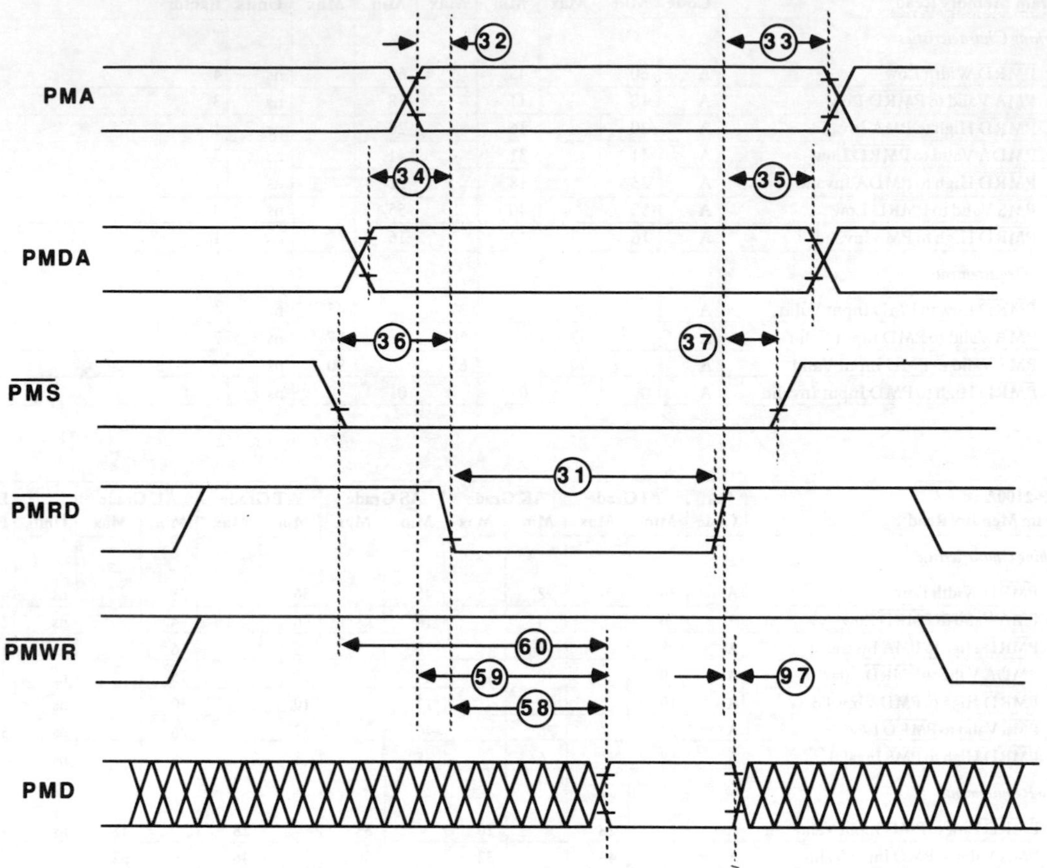

Figure 14. Program Memory Read

Notes 1 and 2 and a table explaining the Derating Factors and Test Codes appear on page 32.

ADSP-2100 Program Memory Write		Test Code	J Grade		K Grade		S Grade		Units	Derating Factor
			Min	Max	Min	Max	Min	Max		
Switching Characteristics										
40	$\overline{PMWR}$ Width Low	A	60		45		60		ns	4
41	PMA Valid to $\overline{PMWR}$ Low	A	16		10		16		ns	3
42	$\overline{PMWR}$ High to PMA Invalid	A	19		15		19		ns	1
43	PMDA Valid to $\overline{PMWR}$ Low	A	39		29		39		ns	3
44	$\overline{PMWR}$ High to PMDA Invalid	A	20		16		21		ns	1
45	$\overline{PMS}$ Valid to $\overline{PMWR}$ Low	A	54		40		54		ns	3
46	$\overline{PMWR}$ High to $\overline{PMS}$ Invalid	A	15		11		14		ns	1
51	$\overline{PMWR}$ Low to PMD Out Enable	F	15		10		15		ns	1
52	$\overline{PMWR}$ High to PMD Out Disable	D		43		37		43	ns	1
53	$\overline{PMWR}$ Low to PMD Out Valid	A		40		32		40	ns	1
54	$\overline{PMWR}$ High to PMD Out Invalid	A	23		18		21		ns	1
55	PMD Out Valid to $\overline{PMWR}$ High	A	33		25		33		ns	3

ADSP-2100A Program Memory Write		Test Code	AJ Grade		AK Grade		AS Grade		AT Grade		AU Grade		Units	Derating Factor
			Min	Max	Min	Max	Min	Max	Min	Max	Min	Max		
Switching Characteristics														
40	$\overline{PMWR}$ Width Low	A	36		28		45		36		28		ns	4
41	PMA Valid to $\overline{PMWR}$ Low	A	8		4		12		8		4		ns	3
42	$\overline{PMWR}$ High to PMA Invalid	A	8		6		10		8		6		ns	1
43	PMDA Valid to $\overline{PMWR}$ Low	A	20		16		28		20		16		ns	3
44	$\overline{PMWR}$ High to PMDA Invalid	A	10		8		12		10		8		ns	1
45	$\overline{PMS}$ Valid to $\overline{PMWR}$ Low	A	32		26		40		32		26		ns	3
46	$\overline{PMWR}$ High to $\overline{PMS}$ Invalid	A	6		4		8		6		4		ns	1
51	$\overline{PMWR}$ Low to PMD Out Enable	F	8		6		8		8		6		ns	1
52	$\overline{PMWR}$ High to PMD Out Disable	D		32		29		38		32		29	ns	1
53	$\overline{PMWR}$ Low to PMD Out Valid	A		29		26		32		29		26	ns	1
54	$\overline{PMWR}$ High to PMD Out Invalid	A	10		8		12		10		8		ns	1
55	PMD Out Valid to $\overline{PMWR}$ High	A	16		13		25		16		13		ns	3

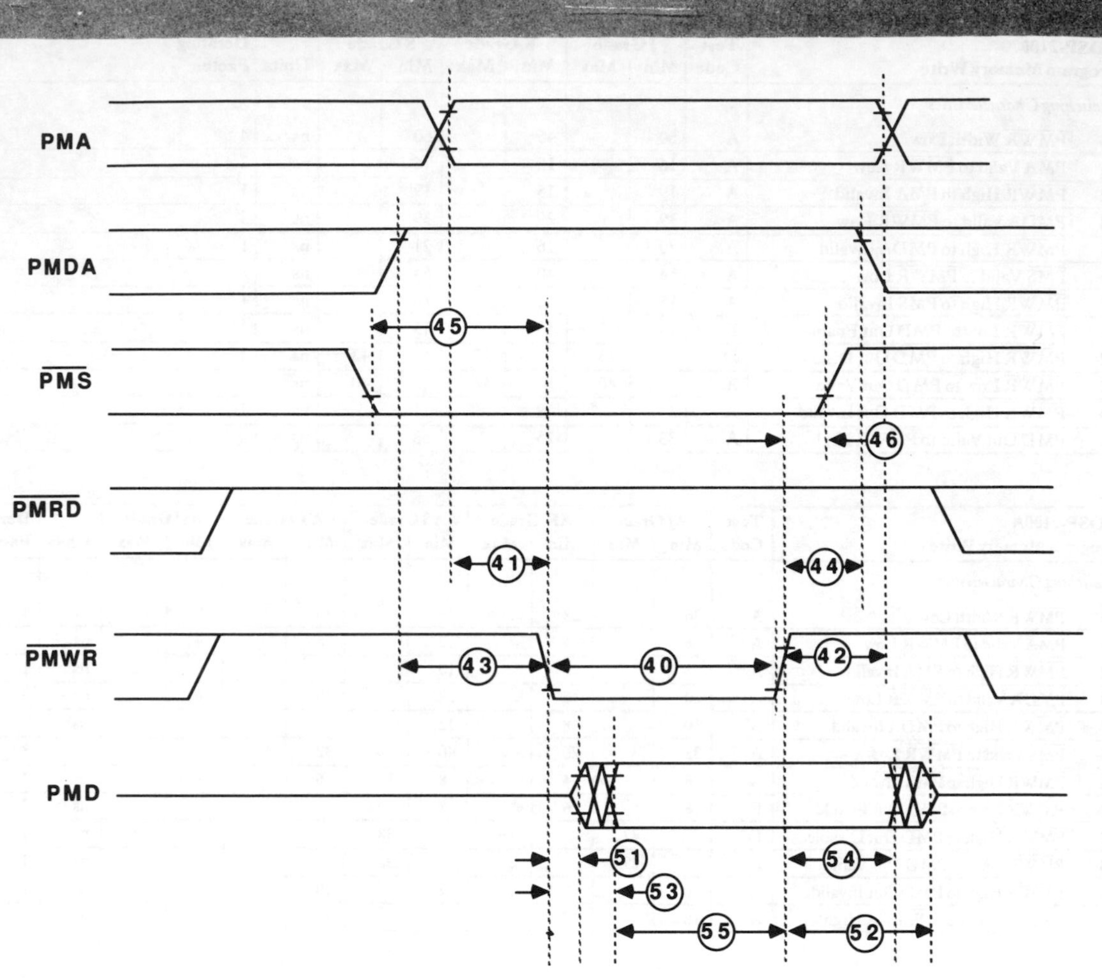

Figure 15. Program Memory Write

Notes 1 and 2 and information about the Derating Factors and Test Codes appear on page 32.

ADSP-2100 Data Memory Read		Test Code	J Grade Min	J Grade Max	K Grade Min	K Grade Max	S Grade Min	S Grade Max	Units	Derating Factor
Switching Characteristics										
67	$\overline{\text{DMRD}}$ Width Low	A	60		45		60		ns	4
68	DMA Valid to $\overline{\text{DMRD}}$ Low	A	21		16		21		ns	3
69	$\overline{\text{DMRD}}$ High to DMA Invalid	A	19		15		19		ns	1
70	$\overline{\text{DMS}}$ Valid to $\overline{\text{DMRD}}$ Low	A	35		27		35		ns	3
71	$\overline{\text{DMRD}}$ High to $\overline{\text{DMS}}$ Invalid	A	22		18		21		ns	1
Timing Requirements										
74	$\overline{\text{DMRD}}$ Low to DMACK Valid	A	0	31	0	21	0	31	ns	3
75	DMA Valid to DMACK Valid	A	0	57	0	42	0	57	ns	6
94	$\overline{\text{DMRD}}$ Low to DMD Input Valid	A		57		41		55	ns	4
95	DMA Valid to DMD Input Valid	A		82		61		79	ns	7
96	$\overline{\text{DMS}}$ Valid to DMD Input Valid	A		96		70		96	ns	7
98	$\overline{\text{DMRD}}$ High to DMD Input Invalid	A	0		0		0		ns	
103	CLKOUT High to DMACK Invalid	A	0	60	0	45	0	60	ns	4

ADSP-2100A Data Memory Read		Test Code	AJ Grade Min	AJ Grade Max	AK Grade Min	AK Grade Max	AS Grade Min	AS Grade Max	AT Grade Min	AT Grade Max	AU Grade Min	AU Grade Max	Units	Derating Factor
Switching Characteristics														
67	$\overline{\text{DMRD}}$ Width Low	A	36		28		45		36		28		ns	4
68	DMA Valid to $\overline{\text{DMRD}}$ Low	A	6		4		14		6		4		ns	3
69	$\overline{\text{DMRD}}$ High to DMA Invalid	A	8		6		10		8		6		ns	1
70	$\overline{\text{DMS}}$ Valid to $\overline{\text{DMRD}}$ Low	A	18		14		27		18		14		ns	3
71	$\overline{\text{DMRD}}$ High to $\overline{\text{DMS}}$ Invalid	A	8		6		10		8		6		ns	1
Timing Requirements														
74	$\overline{\text{DMRD}}$ Low to DMACK Valid	A	0	16	0	10	0	21	0	16	0	10	ns	3
75	DMA Valid to DMACK Valid	A	0	30	0	20	0	42	0	30	0	20	ns	6
94	$\overline{\text{DMRD}}$ Low to DMD Input Valid	A		30		20		37		28		18	ns	4
95	DMA Valid to DMD Input Valid	A		48		32		59		46		32	ns	7
96	$\overline{\text{DMS}}$ Valid to DMD Input Valid	A		52		45		67		50		35	ns	7
98	$\overline{\text{DMRD}}$ High to DMD Input Invalid	A	0		0		0		0		0		ns	
103	CLKOUT High to DMACK Invalid	A	0	36	0	28	0	45	0	36	0	28	ns	4

NOTE ON GENERATING WAIT STATES

See the application note "Wait State Generation on the ADSP-2100/2100A" for information on using DMACK to generate wait states.

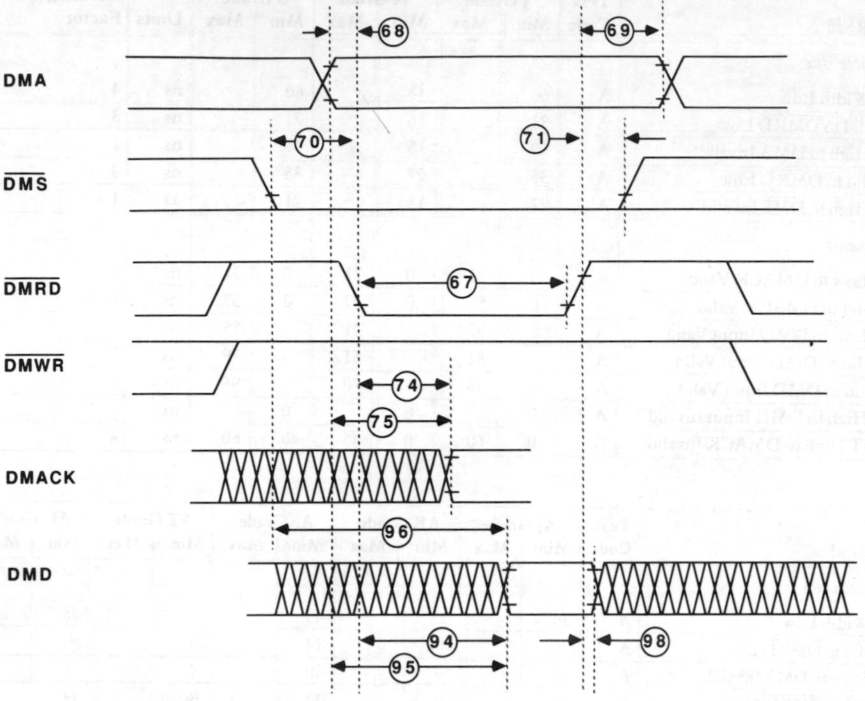

Figure 16a. Data Memory Read

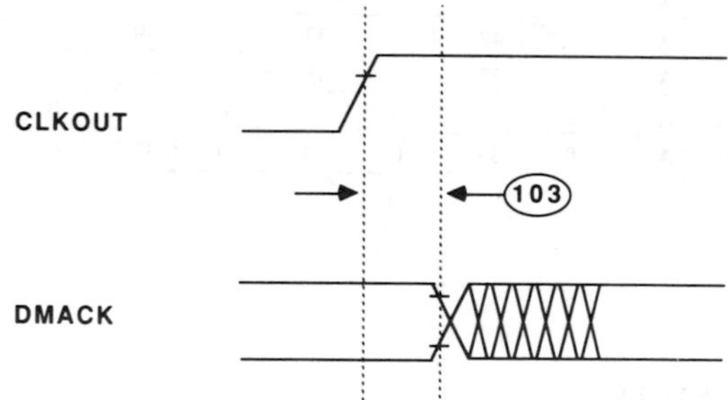

Figure 16b. Data Memory Wait States Extended with DMACK

Notes 1 and 2 and information about the Derating Factors and Test Codes appear on page 32.

ADSP-2100 Data Memory Write	**Test Code**	**J Grade Min**	**J Grade Max**	**K Grade Min**	**K Grade Max**	**S Grade Min**	**S Grade Max**	**Units**	**Derating Factor**
Switching Characteristics									
78 $\overline{DMWR}$ Width Low	A	60		45		60		ns	4
79 DMA Valid to $\overline{DMWR}$ Low	A	24		17		24		ns	3
80 $\overline{DMWR}$ High to DMA Invalid	A	20		15		19		ns	1
81 $\overline{DMS}$ Valid to $\overline{DMWR}$ Low	A	37		28		37		ns	3
82 $\overline{DMWR}$ High to $\overline{DMS}$ Invalid	A	22		19		22		ns	1
87 $\overline{DMWR}$ Low to DMD Out Enable	F	14		9		14		ns	1
88 $\overline{DMWR}$ High to DMD Out Disable	D		40		35		40	ns	1
89 $\overline{DMWR}$ Low to DMD Out Valid	A		38		32		38	ns	1
90 $\overline{DMWR}$ High to DMD Out Invalid	A	21		16		19		ns	1
91 DMD Out Valid to $\overline{DMWR}$ High	A	33		21		33		ns	3
Timing Requirements									
75 DMA Valid to DMACK Valid	A	0	57	0	42	0	57	ns	6
99 $\overline{DMWR}$ Low to DMACK Valid	A	0	31	0	21	0	31	ns	3
103 CLKOUT High to DMACK Invalid	A	0	60	0	45	0	60	ns	4

ADSP-2100A Data Memory Write	**Test Code**	**AJ Grade Min**	**AJ Grade Max**	**AK Grade Min**	**AK Grade Max**	**AS Grade Min**	**AS Grade Max**	**AT Grade Min**	**AT Grade Max**	**AU Grade Min**	**AU Grade Max**	**Units**	**Derating Factor**
Switching Characteristics													
78 $\overline{DMWR}$ Width Low	A	36		28		45		36		28		ns	4
79 DMA Valid to $\overline{DMWR}$ Low	A	8		4		17		8		4		ns	3
80 $\overline{DMWR}$ High to DMA Invalid	A	8		6		10		8		6		ns	1
81 $\overline{DMS}$ Valid to $\overline{DMWR}$ Low	A	20		16		28		20		16		ns	3
82 $\overline{DMWR}$ High to $\overline{DMS}$ Invalid	A	6		4		8		6		4		ns	1
87 $\overline{DMWR}$ Low to DMD Out Enable	F	8		6		8		8		6		ns	1
88 $\overline{DMWR}$ High to DMD Out Disable	D		32		29		38		32		29	ns	1
89 $\overline{DMWR}$ Low to DMD Out Valid	A		29		26		32		29		26	ns	1
90 $\overline{DMWR}$ High to DMD Out Invalid	A	10		8		12		10		8		ns	1
91 DMD Out Valid to $\overline{DMWR}$ High	A	18		13		25		16		13		ns	3
Timing Requirements													
75 DMA Valid to DMACK Valid	A	0	30	0	20	0	42	0	30	0	20	ns	6
99 $\overline{DMWR}$ Low to DMACK Valid	A	0	16	0	10	0	20	0	16	0	10	ns	3
103 CLKOUT High to DMACK Invalid	A	0	36	0	28	0	45	0	36	0	28	ns	4

NOTE ON GENERATING WAIT STATES

See the application note "Wait State Generation on the ADSP-2100/2100A" for information on using DMACK to generate wait states.

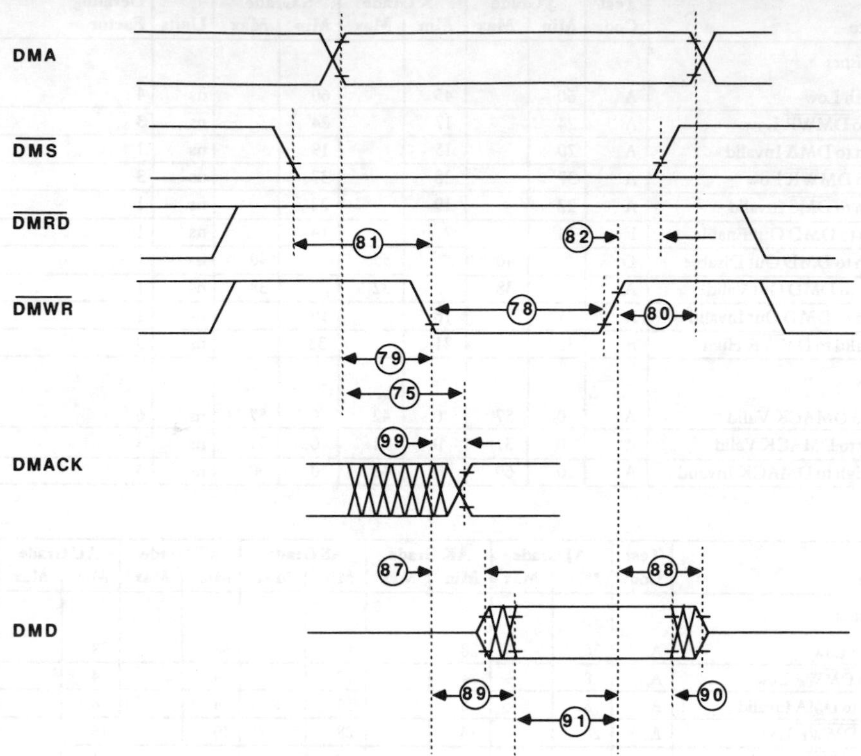

Figure 17a. Data Memory Write

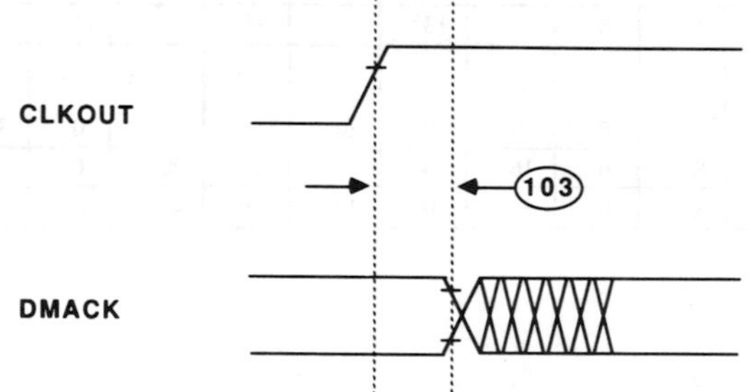

Figure 17b. Data Memory Wait States Extended with DMACK

NOTES

[1]Rise and fall times ≤4ns for ADSP-2100A, 5ns for ADSP-2100.
[2]"xMxx" refers to PMA_{0-13}, $\overline{PMS}$, $\overline{PMRD}$, $\overline{PMWR}$, PMDA, DMA_{0-13}, $\overline{DMS}$, $\overline{DMRD}$ and $\overline{DMWR}$.

TEST CODES

Code	Test Type	Level Reference
A	Inputs, Outputs	Low = 0.8V, High = 2.0V
B	CLKIN *to/from* Inputs, Outputs	1.5V Low = 0.8V, High = 2.0V
D	Output *to* Output Disable	Low = 0.8V, High = 2.0V Low = V_{OL} + 0.5V, High = V_{OH} − 0.5V
F	Output *to/from* Output Enable	Low = 0.8V, High = 2.0V Low = VT − 0.1V, High = VT + 0.1V

VT = 1.5V, the voltage to which tristated outputs are forced.

DERATING FACTOR

The value **N** in the Derating Column shows, for each timing parameter affected, how many of the eight internal clock states are used by this timing parameter; **N**, therefore, ranges between 1 and 8. The formula for changing any individual parameter T uses timing parameter number one, CLKIN Period, shown as P#1:

$$T_{new} = T_{old} + N\,((P\#1_{new} - P\#1_{old})\,/2)$$

You determine the new value of P#1 based on the derating you wish to accomplish. If no **N** value is given for derating, that timing parameter does not change with clock changes.

CAPACITANCE IN PGA PACKAGE

Input capacitance	C_{IN}	10pF typical
Output capacitance	C_{OUT}	10pF typical

Note that output-only pads (PMA_{13-0}, PMDA and DMA_{13-0}) and bidirectional pads (PMD_{23-0} and DMD_{15-0}) have 50kΩ (typical) pull-up resistors between the output and V_{DD} present when the output driver is off.

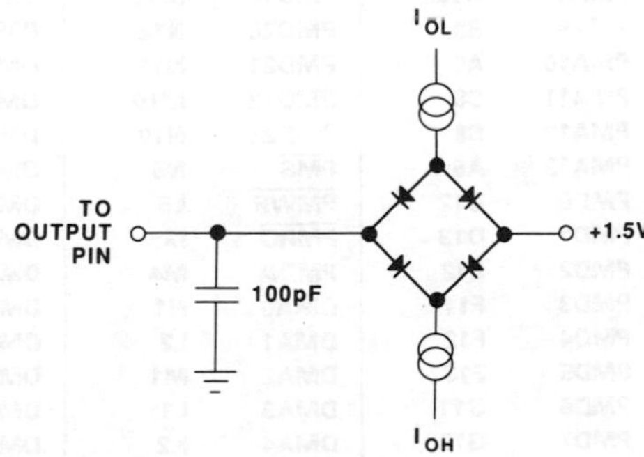

Figure 18. Normal Load for ac Measurements

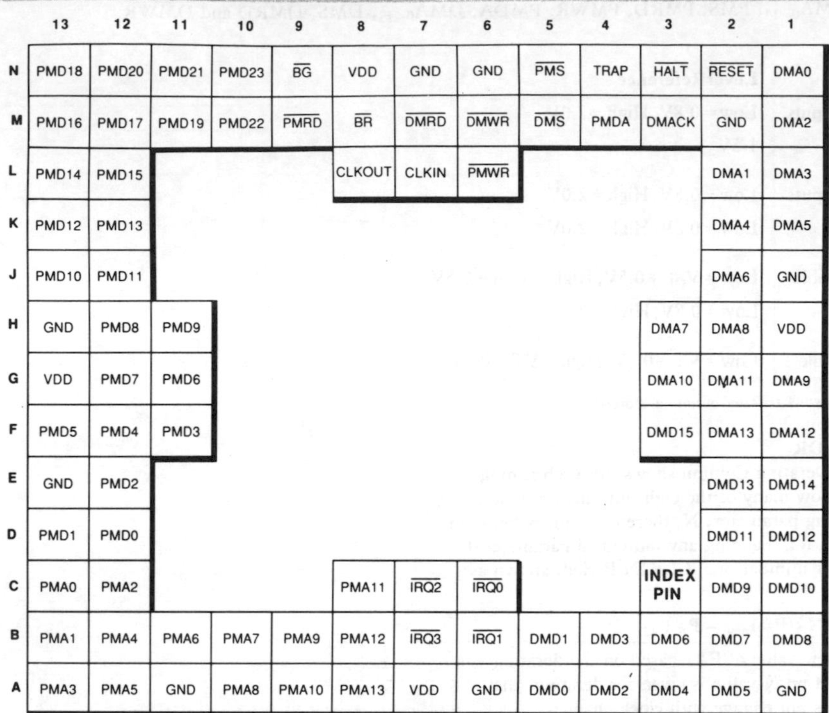

Figure 19. ADSP-2100 Pins, Top View, Pins Down

Function	Location	Function	Location	Function	Location	Function	Location
V_{DD}	A7	PMA1	B13	PMD12	K13	DMA9	G1
V_{DD}	G13	PMA2	C12	PMD13	K12	DMA10	G3
V_{DD}	H1	PMA3	A13	PMD14	L13	DMA11	G2
V_{DD}	N8	PMA4	B12	PMD15	L12	DMA12	F1
GND	A1	PMA5	A12	PMD16	M13	DMA13	F2
GND	A6	PMA6	B11	PMD17	M12	DMD0	A5
GND	A11	PMA7	B10	PMD18	N13	DMD1	B5
GND	E13	PMA8	A10	PMD19	M11	DMD2	A4
GND	H13	PMA9	B9	PMD20	N12	DMD3	B4
GND	J1	PMA10	A9	PMD21	N11	DMD4	A3
GND	M2	PMA11	C8	PMD22	M10	DMD5	A2
GND	N6	PMA12	B8	PMD23	N10	DMD6	B3
GND	N7	PMA13	A8	$\overline{PMS}$	N5	DMD7	B2
CLKIN	L7	PMD0	D12	$\overline{PMWR}$	L6	DMD8	B1
CLKOUT	L8	PMD1	D13	$\overline{PMRD}$	M9	DMD9	C2
$\overline{BR}$	M8	PMD2	E12	PMDA	M4	DMD10	C1
$\overline{BG}$	N9	PMD3	F11	DMA0	N1	DMD11	D2
$\overline{IRQ0}$	C6	PMD4	F12	DMA1	L2	DMD12	D1
$\overline{IRQ1}$	B6	PMD5	F13	DMA2	M1	DMD13	E2
$\overline{IRQ2}$	C7	PMD6	G11	DMA3	L1	DMD14	E1
$\overline{IRQ3}$	B7	PMD7	G12	DMA4	K2	DMD15	F3
$\overline{RESET}$	N2	PMD8	H12	DMA5	K1	$\overline{DMS}$	M5
$\overline{TRAP}$	N4	PMD9	H11	DMA6	J2	$\overline{DMWR}$	M6
$\overline{HALT}$	N3	PMD10	J13	DMA7	H3	$\overline{DMRD}$	M7
INDEX PIN	NC	PMD11	J12	DMA8	H2	DMACK	M3
PMA0	C13						

Table V. ADSP-2100 Pins by Function – G-100A

ADSP-2100 MECHANICAL INFORMATION
100-PIN GRID ARRAY

Dimensions shown in inches and (mm).

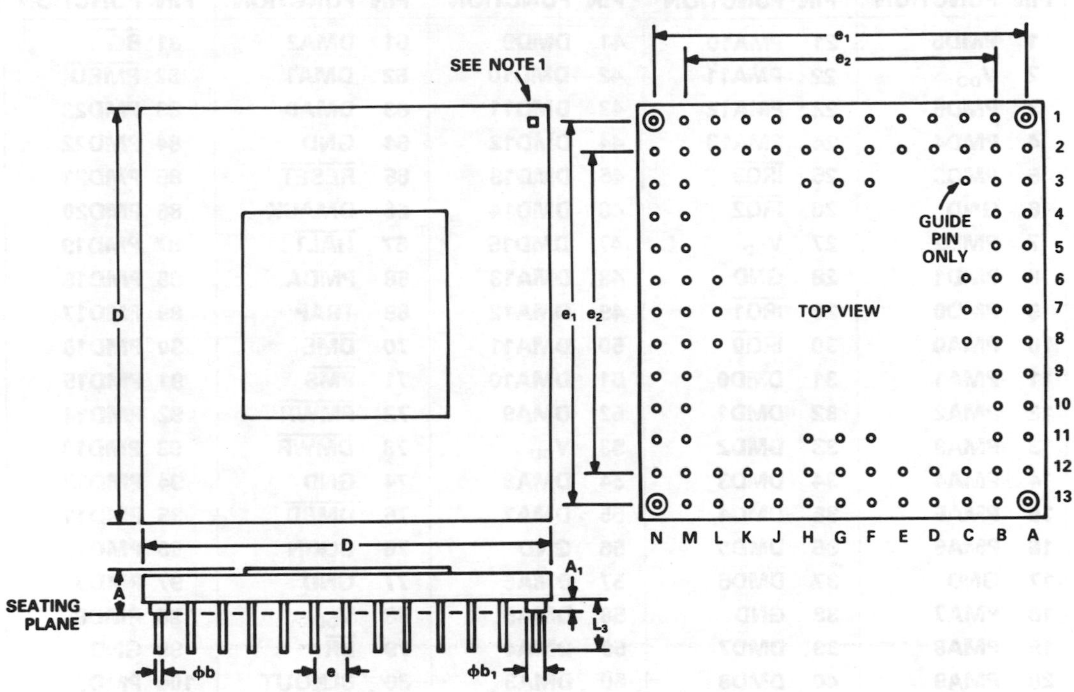

SYMBOL	INCHES MIN	INCHES MAX	MILLIMETERS MIN	MILLIMETERS MAX	NOTES
A		0.169		4.29	3
A_1	0.025	0.055	0.64	1.40	3
φb	0.016	0.020	0.41	0.51	8
$φb_1$	0.040	0.055	1.02	1.40	2, 8
D	1.308	1.332	33.22	33.83	4, 9
e_1	1.188	1.212	30.18	30.78	7
e_2	0.988	1.024	25.10	26.01	7
e	0.095	0.105	2.41	2.67	5
L_3	0.165	0.190	4.19	4.83	

NOTES

1. **Index area; a notch or a lead one identification mark is located adjacent to lead one.**
2. **The minimum limit for dimension $φb_1$ may be 0.023" (0.58mm) for all four corner leads only.**
3. **Dimension shall be measured from the seating plane to the base plane.**
4. **This dimension allows for off-center lid, meniscus and glass overrun.**
5. **The basic pin spacing is 0.100" (2.54mm) between centerlines.**
6. **Applies to all four corners.**
7. **Lead center when α is 0°; e_1 shall be measured at the centerline of the leads.**
8. **All leads – increase maximum limit by 0.003" (0.08mm) measured at the center of the flat, when hot solder dip lead finish is applied.**
9. **All four sides.**
10. **Gold plating 50μ inches over 100μ inches ref. Thickness of nickel.**

PIN	FUNCTION	PIN	FUNCTION	PIN	FUNCTION	PIN	FUNCTION	PIN	FUNCTION
1	PMD6	21	PMA10	41	DMD9	61	DMA2	81	$\overline{\text{BG}}$
2	V_{DD}	22	PMA11	42	DMD10	62	DMA1	82	$\overline{\text{PMRD}}$
3	PMD5	23	PMA12	43	DMD11	63	DMA0	83	PMD23
4	PMD4	24	PMA13	44	DMD12	64	GND	84	PMD22
5	PMD3	25	$\overline{\text{IRQ3}}$	45	DMD13	65	$\overline{\text{RESET}}$	85	PMD21
6	GND	26	$\overline{\text{IRQ2}}$	46	DMD14	66	DMACK	86	PMD20
7	PMD2	27	V_{DD}	47	DMD15	67	$\overline{\text{HALT}}$	87	PMD19
8	PMD1	28	GND	48	DMA13	68	PMDA	88	PMD18
9	PMD0	29	$\overline{\text{IRQ1}}$	49	DMA12	69	TRAP	89	PMD17
10	PMA0	30	$\overline{\text{IRQ0}}$	50	DMA11	70	$\overline{\text{DMS}}$	90	PMD16
11	PMA1	31	DMD0	51	DMA10	71	$\overline{\text{PMS}}$	91	PMD15
12	PMA2	32	DMD1	52	DMA9	72	$\overline{\text{PMWR}}$	92	PMD14
13	PMA3	33	DMD2	53	V_{DD}	73	$\overline{\text{DMWR}}$	93	PMD13
14	PMA4	34	DMD3	54	DMA8	74	GND	94	PMD12
15	PMA5	35	DMD4	55	DMA7	75	$\overline{\text{DMRD}}$	95	PMD11
16	PMA6	36	DMD5	56	GND	76	CLKIN	96	PMD10
17	GND	37	DMD6	57	DMA6	77	GND	97	PMD9
18	PMA7	38	GND	58	DMA5	78	V_{DD}	98	PMD8
19	PMA8	39	DMD7	59	DMA4	79	$\overline{\text{BR}}$	99	GND
20	PMA9	40	DMD8	60	DMA3	80	CLKOUT	100	PMD7

Table VI. ADSP-2100 Pins by Function – P-100 and F-100A

P-100
Plastic Quad Flat Pack
(JEDEC Style)

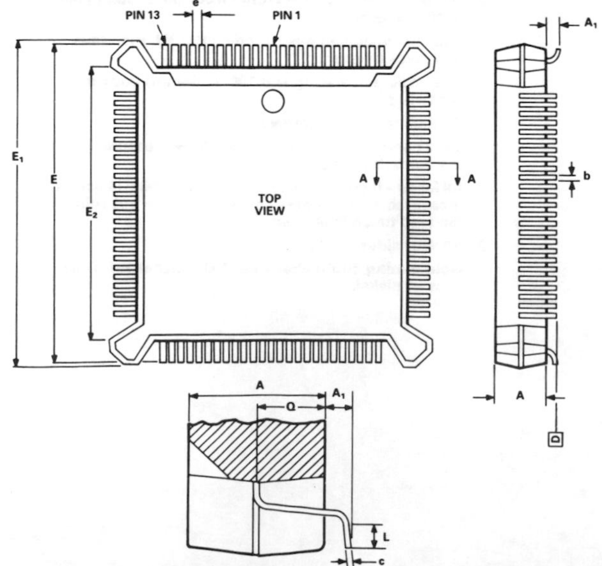

SYMBOL	INCHES MIN	INCHES MAX	MILLIMETERS MIN	MILLIMETERS MAX
A	0.160	0.180	4.06	4.57
A_1	0.020	0.040	0.51	1.02
b	0.010	0.013	0.25	0.33
c	0.006	0.008	0.15	0.20
E	0.875	0.885	22.23	22.48
E_1	0.897	0.903	22.78	22.94
E_2	0.747	0.753	18.97	19.13
e	0.020	0.030	0.51	0.76
L	0.020	0.030	0.51	0.76
Q	0.065	0.075	1.65	1.91
D		0.008		0.20

IMS A100
IMS A100M
Cascadable Signal Processor

Engineering Data

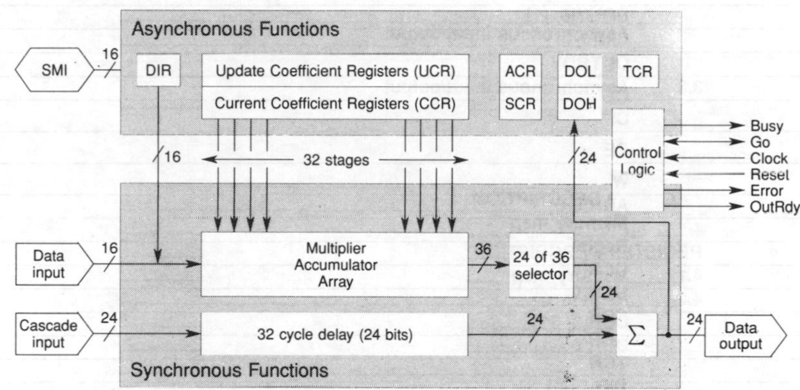

FEATURES

- Variants for full MIL temperature range (−55°C to +125°C)
- MIL-STD-883C processing
- Full 16 bit, 32 stage, transversal filter
- Fully cascadable with no speed degradation or reduction in dynamic range
- Coefficients selectable as 4, 8, 12, or 16 bits wide
- Data throughput to 15.0 MHz
- High speed microprocessor compatible interface
- Data input and output through dedicated ports or via the microprocessor interface
- Fully static high speed CMOS implementation
- Single +5V ±5% or ±10% power supply variants
- TTL and CMOS compatibility
- Less than 2W power dissipation
- Standard 84-pin PGA or flatpack package

APPLICATIONS

- Digital FIR filtering
- High speed adaptive filtering
- Correlation and Convolution
- Discrete Fourier Transform
- Speech processing using Linear Predictive Coding
- Image processing
- Waveform synthesis
- Adaptive and fixed equalizers and echo cancellers
- Spread spectrum communication
- Beamforming and beamscanning in sonar and radar
- Pulse compression
- High speed fixed point matrix multiplication

SGS-THOMSON MICROELECTRONICS

INMOS is a member of the SGS-THOMSON Microelectronics Group

Contents

1 INTRODUCTION

The IMS A100 is a high speed, high accuracy 32 stage transversal filter. Its flexible architecture allows it to be used as a 'building block' in a wide range of Digital Signal Processing (DSP) applications. The part is capable of performing high speed DFTs, convolution and correlation, as well as many filtering functions.

The input data word length is 16 bits, and coefficients are programmable to be 4, 8, 12 or 16 bits wide; two's complement numerical formats are used for both data and coefficients. The coefficients can be updated asynchronously to the system clock during normal operation, allowing the chip to be used in a variety of adaptive systems. The IMS A100 can also be cascaded to construct longer transversal filters with no additional logic or degradation in speed, whilst preserving a high degree of accuracy. The device is controlled through a standard memory interface, allowing use with any general purpose microprocessor. Data communications can be either through the memory interface, or through dedicated data ports.

2 DESCRIPTION

The IMS A100 is a 32 stage, cascadable, digital transversal filter. The general canonical transversal filter is shown in figure 1. An alternative, and functionally equivalent filter is shown in figure 2. It is this second realisation that is used in the IMS A100, where the input signal is supplied in parallel to all 32 multipliers, and the delay and summation operations are performed in a distributed manner.

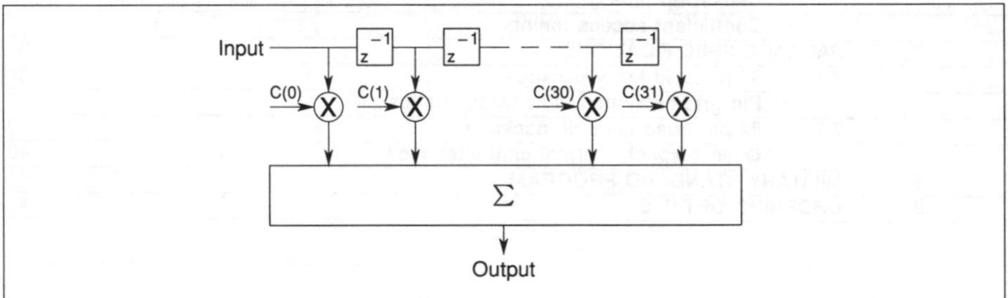

Figure 1 Canonical transversal filter architecture

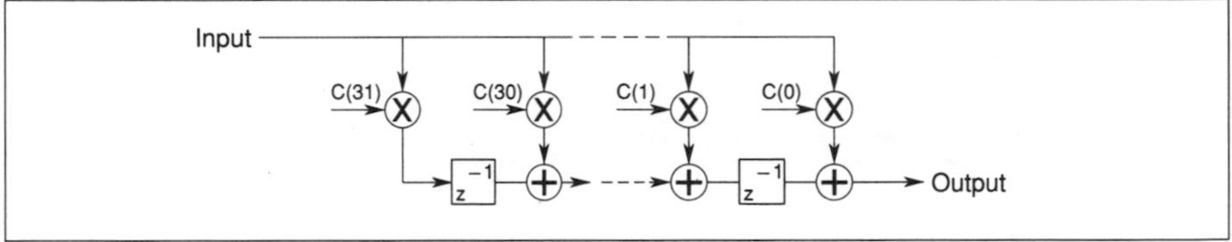

Figure 2 Modified transversal filter architecture

Each data sample loaded into the IMS A100 is fed in parallel to all 32 stages. At each stage the current input sample is multiplied by a coefficient stored in memory, and added to the output of the previous stage delayed by one clock cycle. The filter output at time $t = kT$ is given by:

$$y(kT) = C(0) \times x(kT) + C(1) \times x((k-1)T) + \ldots$$

$$\ldots + C(N-1) \times x((k-N+1)T)$$

where $x(kT)$ represents the kth input data sample, and $C(0)$ to $C(N-1)$ are the coefficients for the N stages.

While the IMS A100 architecture is designed as a transversal filter it contains many features which allow it to be used in a wide range of signal processing applications, e.g. adaptive filtering, matrix multiplication, discrete Fourier transforms, correlation and convolution. Figure 3 shows the users view of the IMS A100.

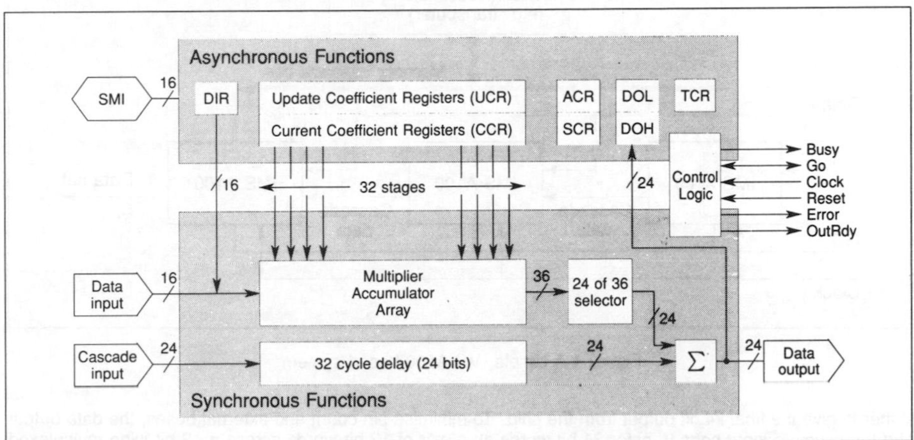

Figure 3 IMS A100 Users Model

The IMS A100 has four interfaces through which data can be transferred. The memory interface port allows access to the coefficent registers, the configuration and status registers and the data input and output registers for the multiplier accumulator array. Three dedicated ports are also provided, allowing high speed data input and output to the IMS A100 and the cascading of several devices.

Typically a microprocessor will configure the IMS A100 via the memory interface, then in a simple system data input and output can be performed through the data input (DIR) and data output (DOL, DOH) registers. Alternatively in a higher performance system data transfer may be performed via the dedicated input and output ports. A typical IMS A100 based system is shown in figure 4. Simple high-throughput fixed-configuration systems can be implemented by clocking the configuration information into the IMS A100 from a ROM.

The IMS A100 input data word width is 16 bits. The coefficient words can be programmed to be 4, 8, 12, or 16 bits wide. There is a trade off between the coefficient size and the speed of operation. If the coefficient word is L_C bits wide and the clock frequency applied to the IMS A100 is F then the maximum data throughput is $\frac{2 \times F}{L_C}$. So, for an IMS A100 operating from a 20.8 MHz clock and using 4-bit coefficients the maximum data throughput is 10.4 MHz, similarly for 16-bit coefficients the throughput is 2.6 MHz.

To preserve complete numerical accuracy, no truncation or rounding is performed on the partial products in the multiplier accumulator array. The output of this array is calculated to full precision (36 bits). A programmable barrel shifter is located at the output of this array, which allows one of five 24 bit fields to be selected from the 36 bit result. The selected 24 bits are always correctly rounded and are sign extended before being output. The selection required can be determined from analysis of the coefficients and input data used in a given application.

Two banks of coefficients are provided. At any instant one set of coefficients is in use within the multiplier accumulator array, the other set being accessible via the memory interface. Once a new set of coefficients has been loaded, the two coefficient banks can be interchanged by performing a write operation to the 'Bank Swap' bit of a control register.

So that devices can be cascaded (eg. to construct longer transversal filters), a 32 stage, 24 bit wide, shift register and 24 bit adder is included on chip. The output of one chip is connected directly to the cascade input of the next. The output of the shift register is added internally to the output of the programmable barrel

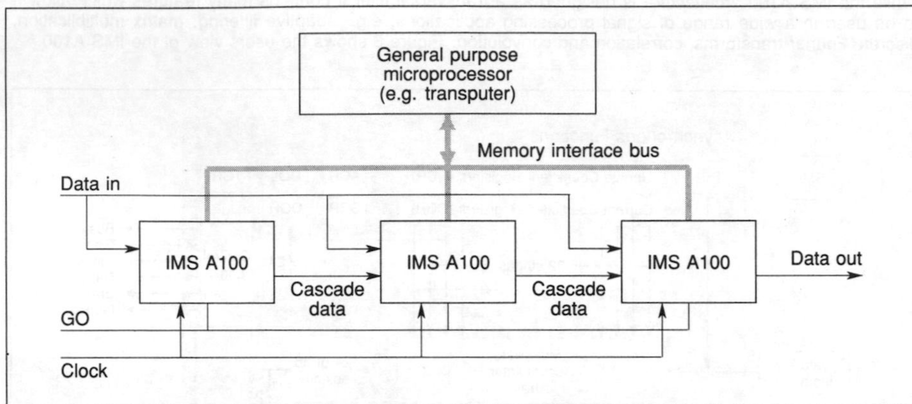

Figure 4 A simple IMS A100 based system

shifter to give the final 24 bit output from the chip. To minimise pin count and external buses, the data output and the cascade input ports transfer 24 bit words as a pair of 12 bit words across a 12 bit wide multiplexed interface.

As IMS A100s can be cascaded there is a price / performance trade off for most IMS A100 systems. For example, a correlation application could achieve high performance by using a cascade of IMS A100s sufficiently long to hold one of the waveforms being correlated in its coefficient registers and sending the other waveform involved in the correlation along the cascade of IMS A100s. A cheaper and slower solution would be to use a smaller number of IMS A100s and to decompose the single long correlation into a sequence of shorter correlations, the results of which are then summed.

3 PIN DESIGNATIONS

System services

Pin	In/out	Function
VCC, GND		Power supply and return
CLK	in	Input clock
$\overline{\text{RESET}}$	in	System reset
$\overline{\text{ERROR}}$	out	Numerical overflow error
BUSY	out	Bank swap in progress

Synchronous input/output

Pin	In/out	Function
GO	in/out	Initiate input/computation/output cycle
DIN[0-15]	in	Data input port
DOUT[0-11]	out	Data output port
CIN[0-11]	in	Cascade input port
OUTRDY	out	Output data ready

Asynchronous input/output

Pin	In/out	Function
D[0-15]	in/out	Memory interface data bus
ADR[0-6]	in	Memory interface address bus
$\overline{\text{CS}}$	in	Memory interface select
$\overline{\text{CE}}$	in	Memory interface enable
$\overline{\text{W}}$	in	Memory interface write enable

Notes

Signal names are shown with an $\overline{\textbf{overbar}}$ if they are active low, otherwise they are active high.
Pinout details are given in section 7

3.1 System services

System services include all the necessary logic to start up and maintain the IMS A100.

Power

Power is supplied to the device via the **VCC** and **GND** pins. Several of each are provided to minimise inductance within the package. All supply pins must be connected. The supply must be decoupled close to the chip by at least one 100nF low inductance (e.g. ceramic) capacitor between **VCC** and **GND**. Four layer boards are recommended; if two layer boards are used, extra care should be taken in decoupling.

Input voltages must not exceed specification with respect to **VCC** and **GND**, even during power-up and power-down ramping, otherwise *latchup* can occur. CMOS devices can be permanently damaged by excessive periods of latchup.

CLK

The clock input signal **CLK** controls the timing of input and output on the three dedicated ports and controls the progress of data through the multiplier accumulator array.

$\overline{\text{RESET}}$

When the IMS A100 is reset the control logic within the IMS A100 will be reset and the ACR, SCR and TCR will be initialised to their default values. Note that neither the internal data path registers nor the coefficient registers are affected by the reset. Resetting the device initialises the SCR to its default setting. So, depending on the setting of SCR before a reset, a reset may also be a device reconfiguration. The sequence of operations required to return the device to a defined state following reconfiguration is described under SCR in the register description.

A reset is initiated automatically when power is first applied to the device. This reset will be completed once four cycles of **CLK** have occured after **VCC** is valid. Alternatively reset can be initiated by taking $\overline{\textbf{RESET}}$ low. This reset will be completed after at least two cycles of **CLK** have occured while $\overline{\textbf{RESET}}$ is held low. $\overline{\textbf{RESET}}$ should be held low for at least 200ns. Normal device operation can then continue after $\overline{\textbf{RESET}}$ is taken high.

The reset should be completed before either the synchronous or asynchronous parts of the device are used.

$\overline{\text{ERROR}}$

If asserted, this pin indicates an error condition has occured, and that the condition has not been cleared. The error condition results from a numerical overflow in either the final adder or in the field selector. To allow this signal to be wire ORed between all the devices in a cascade and hence to be used as an interrupt signal to the host processor, the $\overline{\textbf{ERROR}}$ outputs are open collector.

If suitably armed before the error occured the ACR error bits can be read to discriminate the two error sources. The error bits in the ACR and the error condition can be cleared and then the error bits armed to detect further errors by writing values to the ACR. The sequence of values that should be written to the ACR error bits is 0 followed by 1. An error condition can only be cleared if the error bits were suitably armed before the most recent error occured.

The ACR error bits may not observe an error occuring between clearing and arming the error bits. So, when clearing an error and arming the error bits precautions should be taken to ensure that no new error occurs. For example, first prevent the IMS A100 from initiating computation on new data; second wait for any results pending to be output; then clear and rearm. The ACR error bits will observe any error occuring after they are armed. Thus, if an error occured before the ACR error bits were armed it may be necessary to arm the error and then force an error before proceeding to clear the error (as described above).

Following power up the contents of the multiplier accumulator array and cascade path are indeterminate. As this indeterminate data flushes through a system of one or more IMS A100s errors are likely to occur. Similarly, altering the device configuration defined by the SCR is likely to result in errors. The sequence of operations required to return the device to a defined state following reconfiguration is described under SCR in the register description section of this specification.

BUSY

When high this pin indicates that an exchange of data between the Current and Update Coefficient Registers is in progress. Under certain conditions the duration of **BUSY** may be vanishingly small. **BUSY** will be active if the bank swap is caused by setting ACR[0] to request a single bank swap or when SCR[2] is set selecting Continuous Swap mode. The detailed behaviour is described in the bankswap timing diagrams.

3.2 Synchronous input/output

GO

The **GO** signal initiates a cycle of data input, computation and output. An IMS A100 configured as a slave will monitor the **GO** signal on the rising edge of **CLK** one cycle before it is ready to accept more data and on every rising edge thereafter until **GO** is found to be high. If **GO** is high then data input will occur on the next rising edge of **CLK**. If **GO** is low when it is sampled no new data input will occur.

In a cascade of IMS A100s one IMS A100 may be configured as a master. The master IMS A100 will drive its **GO** pin high after data has been written into its Data Input Register indicating that new data is available and that the slave IMS A100s in the casacade should start an input, computation, output cycle. When the **GO** signal goes low new data can be written to the IMS A100s. Typically a host processor will write simultaneously to the Data Input Registers of all the IMS A100s in the cascade. The host will then monitor the **GO** signal before writing new data to the cascade.

DIN[0-15]

This 16 bit wide data input port allows high speed data input to the IMS A100. The timing of this input is controlled by the **CLK** and **GO** signals. In a cascade of IMS A100s the 16 bit wide input data path and the **CLK** and **GO** signals will be bussed to all devices.

DOUT[0-11]

This 12 bit data port outputs the result from the IMS A100. The 24 bit result is multiplexed through this port as two 12 bit words, the least significant word being output first. The most significant word is output second and remains on the data pins until a new data output sequence is about to start. The **OUTRDY** signal can be used to latch these words into external circuitry. In a cascade of IMS A100s the **DOUT** pins of one device connect to the **CIN** pins of the next device in the cascade.

CIN[0-11]

The Cascade Input allows multiple IMS A100s to be cascaded. A 24 bit word is input as two 12 bit words the least significant word being input first. The 24 bit word is delayed by a shift register and summed with the output of the multiplier accumulator array. The delay from a word being input on the cascade input to that word affecting the data output is 32 data input cycles. In a typical IMS A100 based system the cascade input of each device will be connected to the data output **DOUT[0-11]** of the previous IMS A100 in the cascade. The Cascade Input of the first device in the cascade will normally be connected to ground.

OUTRDY

The output ready signal **OUTRDY** goes low just after the least significant data output word is available on the **DOUT** pins and goes high just after the most significant word is available. The rising edge of **OUTRDY** also indicates that the Data Output registers (DOL, DOH) contain the new result word. Thus the **OUTRDY** signal can either be used to latch the output of the IMS A100 into external logic or to indicate that output of the IMS A100 can be read through the memory interface from the Data Output registers.

3.3 Asynchronous input/output

$\overline{CS}$

This pin selects the chip; if chip select **$\overline{CS}$** is low an access to the memory interface will be enabled. This signal is usually asserted by the host processor's address decoder at the beginning of a memory cycle.

$\overline{CE}$

The chip enable pin. The memory interface on the IMS A100 appears to the system controlling it as 128 words of static RAM. The chip enable **$\overline{CE}$** signal is similar in operation to the chip enable signal found on static RAMs. When **$\overline{CE}$** is high the chip select, write enable and the address inputs are ignored and the memory interface data bus is tri-state. When chip enable is low a single read or write access is made to one of the registers within the IMS A100. Accesses to the memory interface can occur completely asynchronously to operations on the data in, cascade in and data output ports **DIN[0-15]**, **CIN[0-11]** and **DOUT[0-11]**.

$\overline{W}$

The write enable pin indicates whether the access to the IMS A100 memory interface is to be a write or a read. If **$\overline{W}$** is low a write access is indicated.

ADR[0-6]

The seven bit address bus comprises pins **ADR[0-6]**. The seven bit binary value applied to the address inputs of the IMS A100 indicates which register is to be accessed.

D[0-15]

During a write to the memory interface a 16 bit word is applied to data bus pins **D[0-15]**. This word will be latched on the rising edge of chip enable **$\overline{CE}$** at the end of the cycle. During a read cycle the contents of the location accessed are placed on the data pins. When **$\overline{CE}$** is high the data signals are tri-state.

4 REGISTER DESCRIPTION

The memory map shown below indicates the primary addresses for each register. All locations between decimal addresses 64 and 75 inclusive are uniquely decoded. This group of registers is shadowed at other locations up to the 128 word boundary. The effect of reading and writing to areas in the memory map other than those shown in the table is undefined.

If the user wishes to initialise the device from a ROM addressed by a clocked counter, one of the following options applies:

1 Restrict the counter to count only from 0 to 68; this avoids writing to the data registers as well as the shadow locations.

2 Count down from 127 to zero. The initialization at the lower addresses will override spurious ones at the higher shadowed addresses.

4.1 Memory map †

Register	Address decimal	Address hex	Function
CCR[0-31]	32–63	20–3F	Current Coefficient Registers
UCR[0-31]	0–31	00–1F	Update Coefficient Registers
SCR	64	40	Static Control Register
	65	41	Unused location
ACR	66	42	Active Control Register
	67	43	Unused location
TCR	68	44	Test Control Register
DIR	72	48	Data Input Register
DOL	74	4A	Data Output Register (Least Significant Word)
DOH	75	4B	Data Output Register (Most Significant Word)

† All other locations accessible via the memory interface of the IMS A100 are reserved.

4.2 Registers

CCR[0–31]

The Current Coefficient Registers contain the coefficients currently being used by the multiplier accumulator array. CCR[0] (decimal address 32) corresponds to the coefficient register of the multiplier accumlator nearest the output of the IMS A100; i.e. this location is equivalent to C(0) in figure 2.

Similarly CCR[31] (decimal address 63) corresponds to C(31). The Current Coefficient Registers can be read from at any time and can be written to provided that no data processing is taking place. The effect of writing to the Current Coefficient Registers while data is being processed is undefined.

UCR[0–31]

The Update Coefficient Registers are equivalent to the Current Coefficient Registers, with the exception that the values in the Update Coefficient Registers are not currently in use within the multiplier accumlator array and can therefore be written to at any time.

A bank swap operation is equivalent to an exchange of data between the Update Coefficient Registers and the Current Coefficient Registers.

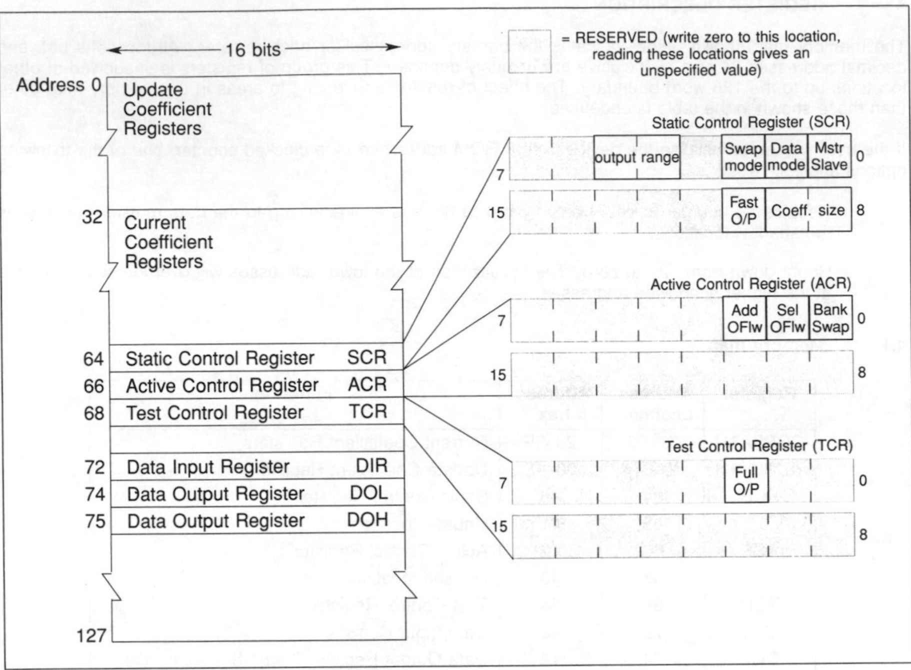

Figure 5 IMS A100 memory map

SCR

The Static Control Register contains the control bits which configure the IMS A100 and are unlikely to need updating after their initial configuration. The contents of the Static Control Register are not affected by the IMS A100 and can be read at any time.

Reconfiguring the SCR may result in indeterminate data values within the IMS A100 system. These values may in turn result in errors. After reconfiguring the SCR the following sequence should be followed to return the IMS A100 system to a defined, error free condition:

1 Arm error bits in ACR.

2 After SCR has been reconfigured **GO** should be held low for 20 cycles of **CLK**.

3 A series of suitable data values should then be flushed through the IMS A100 system.

4 Any errors generated should then be cleared.

5 The IMS A100 system is then ready to commence normal operation.

ACR

The Active Control Register contains status and control bits which are likely to be accessed during normal operation of the IMS A100; i.e. when handling error conditions and when requesting single coefficient bank swaps.

TCR

The Test Control Register is used for test purposes. One of the test modes provides access to the least significant part of the multiplier accumulator array output.

DIR

The Data Input Register. The IMS A100 can be configured to either take its input data from the **DIN** pins or from the Data Input Register. If the IMS A100 is configured as the master of a cascade of IMS A100s the **GO** signal will be driven in response to writing data into the Data Input Register.

In a small IMS A100 based system the Data Input Registers of all the devices in the cascade will normally be mapped into the same location within the address space of the processor controlling the cascade. Thus a single write operation can write data to all devices, the master IMS A100 generating the **GO** signal for the slaves. The Data Input Register is write only.

DOL

The least significant word of the Data Output Register. The output data from the IMS A100 is available from both the **DOUT[0–11]** pins and from the Data Output Registers. The value held in the Data Output Registers is the 24 bit output word, sign extended to 32 bits. DOL contains the least significant 16 bits of the 24 bit result; the register is read only.

DOH

The most significant word of the Data Output Register. The DOH register contains the most significant 8 bits of the 24 bit output word generated by the IMS A100. The most significant 8 bits of DOH are the sign extension of the output word. DOH is read only.

The remainder of this section describes the register details bit by bit. Each section commences with the name of the register with the bit number(s) followed by the default value, in the general format:

Name REGISTER[MSB–LSB] Default: MSB LSB

The least significant bit of a register is bit 0.

† in the tables indicates the default state of the register bit(s).

4.3 Static control register

Fast Output SCR[10] Default: 0

The Fast Output bit controls the way in which the 24 bit output of the IMS A100 is multiplexed across the 12 bit wide **DOUT** port. The interval between data output cycles is the same for both Normal and Fast output modes.

The difference between the modes is the time division between the least and most significant words. In fast output mode the least significant 12 bit word is available for the minimum period possible, thus allowing the most significant word to be output at the earliest possible instant. In normal output mode the least significant word is available for the same length of time as the most significant word (unless the duration of the most significant word is extended by idle cycles).

The timing constraints on data output in Normal mode are significantly simpler than those in Fast mode. Fast mode should be considered a special mode which is only used where the early availability of the output words is important, e.g. an adaptive system where the filter coefficients are being modified in response to the output data.

All devices in a cascade of IMS A100s should be configured for the same output mode. The Fast Output bit should not be altered during data processing. If it is altered the data output of the cascade will be undefined until new input data has flushed through all stages of the cascade. If the coefficient size is 4 bits there is no difference between the fast and normal modes.

SCR[10]	Output mode
0	Normal †
1	Fast

Coefficient Size SCR[9–8] Default: 1 1

Defines size of coefficient used, in terms of word width. This also determines the minimum interval between data input cycles and thus the data throughput of the IMS A100. The Coefficient Size bits should not be altered during data processing. If they are altered the data output of the cascade will be undefined until new input data has flushed through all stages of the cascade.

In each mode the coefficient data is the least signifcant bits of the 16 bit word; e.g. in 4 bit mode, a two's complement number should be programmed into bits 0–3 of the 16 bit register. The remaining bits 4–15 are ignored.

SCR[9–8]	Coefficient size	Data input interval
0 0	4 bits	2 cycles
0 1	8 bits	4 cycles
1 0	12 bits	6 cycles
1 1	16 bits	8 cycles †

Reserved SCR[7–6] Default: 0 0

These locations are reserved. The user should write 0,0 to these locations to maintain compatability with future products. The value read from this location is undefined.

Reserved SCR[3] Default: 0

This location is reserved. The user should write 0 to this location to maintain compatability with future products. The value read from this location is undefined.

Output Word Selection SCR[5–4] Default: 1 0

These bits determine the 24 bit wide field selected from the 36 bit wide output of the multiplier accumulator array (bit positions numbered 0 to 35). The word selected will be rounded and sign extended before being output. Note that ranges '10' and '11' imply sign extension of the result.

The Output Word Selection bits should not be altered during data processing. If they are altered the data output of the cascade will be undefined until new input data has flushed through all stages of the cascade.

SCR[5–4]	Field
0 0	[7–30]
0 1	[11–34]
1 0	[15–38] †
1 1	[20–43]

Continuous Swap SCR[2] Default: 0

The Continuous Swap bit selects whether the two banks of coefficient registers are automatically exchanged after each data input and computation cycle or if individual bank swaps occur under the direction of the Bank Swap bit in the Active Control Register, ACR[0]. SCR[2] should not be set if a bankswap has been requested (by setting ACR[0]) and is still pending.

SCR[2]	Swap Mode
0	Swap on asserting ACR[0] †
1	Swap after end of each input cycle

Input Data Source SCR[1] Default: 0

The data source for the multiplier accumulator array can come from one of two sources, selected by SCR[1]. Data can either be input from the **DIN** port or it can be written into the Data Input Register via the memory interface. See also the following section.

SCR[1]	Data Source
0	From **DIN** port †
1	From DIR

Master not Slave SCR[0] Default: 0

The Master not Slave bit selects whether the IMS A100 samples the **GO** input to determine the start of a data input cycle (slave mode), or drives the **GO** pin when data is written to the DIR (master mode). If input data is supplied through the DIR one IMS A100 in the cascade should be configured as a master. If data is supplied to the **DIN** port by an external data source all the IMS A100s in the cascade should be configured as slaves and **GO** should be driven by an external system. Note that an illegal mode results if SCR[1] is 0 and SCR[0] is 1; i.e. a master cannot obtain data from the **DIN** port.

SCR[0]	Mode
0	Slave †
1	Master

4.4 Active control register

Cascade Adder Overflow ACR[2] Default: 0

If previously armed this status bit will be set if the addition of the 24 bit words output by the 24 from 36 bit selector (on the output of the multiply accumulator array) and the cascade shift register causes an arithmetic overflow. The **ERROR** pin will be driven low while this or any other error condition is active. This error bit and the error condition can be cleared by writing a zero to ACR[2], provided the data in the adder is no longer in error. After clearing this error bit the error bit should be armed (by writing a one to ACR[2]) to ensure that any future error is detected. See **ERROR** section.

Selector Overflow ACR[1] Default: 0

If previously armed this status bit will be set if the 24 bit output range of the selector does not include all the significant binary digits in the 36 bit result generated by the multiply accumulator array. The **ERROR** pin will be driven low while this or any other error condition is active. This error bit and the error condition can be cleared by writing a zero to ACR[1]. After clearing this error bit the error bit should be armed (by writing a one to ACR[1]) to ensure that any future error is detected. See **ERROR** section.

Initiate Bank Swap ACR[0] Default: 0

Writing a one into this control bit requests an exchange of data between the Current and Update Coefficient Registers. The bank swap will occur as soon as the current computation cycle is completed, or on the next clock cycle if the IMS A100 is idle. This control bit is cleared to zero by the IMS A100 when the bank swap is complete. No access should be made to either set of coefficient registers while a bank swap is in progress. ACR[0] should not be set if SCR[2] is already set. For a detailed description of the behaviour see the bankswap and coefficient access timing diagrams.

4.5 Test control register

Examine Full Output Word TCR[2] Default: 0

This bit overrides the output word selection normally made by bits SCR[5–4]. The output word selection determines the 24 bit wide field selected from the 36 bit wide output of the multiply accumulator array (bit positions numbered 0 to 35). When TCR[2] is set to '1' the output word selection is bits '−1' to 22, where bit '−1' is set to zero. The output word selection should not be altered during data processing. If altered the data output of the cascade will be undefined until new input data has flushed through all stages of the cascade.

TCR[2]	Field
0	Set by SCR[5–4] †
1	[−1 to 22]

Reserved TCR[1] Default: 0

This location is reserved for INMOS test purposes. For normal operation the user should write 0 to this location.

Reserved TCR[0] Default: 0

This location is reserved for INMOS test purposes. For normal operation the user should write 0 to this location.

5 DEVICE APPLICATIONS

The IMS A100 can be used in a variety of different applications requiring high performance computation. Some of these are described below, and are covered in detail in the IMS A100 Application Note series, available from INMOS.

5.1 Filtering and adaptive filtering

The IMS A100 device can be used to implement high speed FIR and IIR digital filters. The maximum sampling frequency of the input signal ranges between 2.125MHz and 15MHz, depending on the coefficient word length and speed variant that has been selected.

The continuous bank swap mode allows a single device to filter complex (I & Q) data streams. High speed random access coefficient registers enable high performance adaptive filters and equalisers to be realised with minimal complexity.

The cascadability of the device enables FIRs of greater than 32 stages to be constructed, with no degradation in data throughput.

5.2 Convolution and correlation

The IMS A100 is the first single-chip digital correlator capable of highly accurate computation of correlation and convolution functions (16-bit coefficients, 16-bit data and 36-bit accumulation). These functions have applications in matched filtering, noise reduction and pulse compression in communication, radar and sonar systems.

For correlations and convolutions involving a large number of data points, devices can be cascaded to several thousand stages with careful design. Alternatively, it is possible to use algorithms which allow decomposition of long correlation and convolutions into several smaller ones, which can then be carried out by a single or smaller number of devices.

5.3 Matrix multiplication

The architecture of the IMS A100 allows very high speed fixed point matrix multiplication. In this application the columns of the multiplier matrix are circulated as inputs to the chip while the coefficients are programmed in a suitable manner with the elements of the multiplicant matrix. Larger matrices can be handled by either cascading several chips or by decomposing the matrices into smaller ones.

5.4 Fourier transforms

Two algorithms, namely the Prime Number Transform (PNT) and the Chirp-Z Transform (CZT), can be used to perform high speed Fourier transforms using IMS A100s. The Fourier transform of long data sequences can be evaluated either by using cascaded IMS A100s or by using decomposition algorithms to convert a long transform into a number of short transforms (e.g. <32 points). These short transforms can then be carried out using the IMS A100s and a host processor.

The speed of transform can be traded off against the number of chips employed. Any microprocessor with a standard memory interface could be used to handle intermediate results and to control the overall system. Two IMS A100s can be used to perform a transform of about 1000 points in around 1ms. to 2ms using look-up ROMs for address generation and high speed DSP controllers, or 5ms to 10ms using a microprocessor as the controller. More IMS A100s can be used if higher performance is required.

5.5 Waveform synthesis

The programmability of this digital transversal filter allows the IMS A100 to be used for flexible waveform generation and synthesis, by exploiting the ability to change coefficients randomly, quickly and simply. Such a configuration could be attractive for PC based synthesisers, as the chip can generate very accurate high bandwidth signals.

5.6 General purpose accelerator

By attaching one or more IMS A100s to any computer with DMA capability, a useful accelerator can be constructed, capable of handling all of the above applications without reconfiguration. The cascadability of the device enables users to add IMS A100s as required for extra processing performance, with minimal impact on the driving software.

6 ELECTRICAL SPECIFICATION

The IMS A100 is available in several speed, package and temperature variants (see section 9 – Ordering details) and the electrical characteristics of each are described in this section. When no variant is identified the information refers to all variants.

6.1 DC electrical characteristics

Absolute maximum ratings

Symbol	Parameter	Min.	Max.	Units	Notes (1)
VCC	DC supply voltage	0	7.0	V	2,3
VI, VO	Voltage on input and output pins	−1.0	VCC+0.5	V	2,3
TS	Storage temperature	−65	150	°C	2
TA	Temperature under bias	−55	125	°C	2
PDmax	Power dissipation		2.0	W	2

Notes

1 All voltages are with respect to **GND**.

2 This is a stress rating only and functional operation of the device at these or any other conditions beyond those indicated in the operating sections of this specification is not implied. Stresses greater than those listed may cause permanent damage to the device. Exposure to absolute maximum rating conditions for extended periods may affect reliability.

3 This device contains circuitry to protect the inputs against damage caused by high static voltages or electrical fields. However, it is advised that normal precautions be táken to avoid application of any voltage higher than the absolute maximum rated voltages to this high impedence circuit. Unused inputs should be tied to an appropriate logic level such as **VCC** or **GND**.

DC operating conditions

Symbol	Parameter	Min.	Nom.	Max.	Units	Notes (1)
VCC	DC supply voltage	4.5		5.5	V	4
VCC		4.75		5.25	V	5,6,7
VIH	Input Logic '1' Voltage **CLK**	4.0		VCC+0.5	V	2
	Input Logic '1' Voltage $\overline{\textbf{RESET}}$	2.4		VCC+0.5	V	2
	Input Logic '1' Voltage other pins	2.0		VCC+0.5	V	2
VIL	Input Logic '0' Voltage **CLK**	−0.5		0.5	V	2
	Input Logic '0' Voltage $\overline{\textbf{RESET}}$	−0.5		0.8	V	2
	Input Logic '0' Voltage other pins	−0.5		0.8	V	2
TA	Ambient Operating Temperature	0		70	°C	3,4,7
TA		−55		125	°C	3,5,6

Notes

1 All voltages are with respect to **GND**. All **GND** pins must be connected to **GND**.

2 Input signal transients up to 10 ns wide, are permitted in the voltage ranges (**GND** − 0.5 V) to (**GND** − 1.0 V) and **VCC** + 0.5 V to **VCC** + 1.0 V.

3 400 linear ft/min transverse air flow.

4 IMS A100-G21S, IMS A100-Q21S.

5 IMS A100-G21M, IMS A100-Q21M.

6 IMS A100-G17M.

7 IMS A100-G30S.

DC characteristics

Symbol	Parameter	Min.	Max.	Units	Notes (1,2)
VOH	Output Logic '1' Voltage	2.4	VCC	V	4
VOL	Output Logic '0' Voltage	0	0.4	V	5
II	Input current @ GND<VI<VCC		±10	μA	
IOZ	Tristate output current @ GND<VI<VCC		±10	μA	
ICC	Average power supply current		360	mA	3

Notes

1 All voltages are with respect to **GND**. All **GND** pins must be connected to **GND**.

2 Parameters measured over variants full voltage and temperature operating range.

3 Power dissipation is application dependent and varies with output loading. The maximum given here is for worst case data patterns and activity on all interfaces, with no DC load on outputs.

4 **OUTRDY**, **DOUT**: **IOut** ≤ -4.4 mA; $\overline{\textbf{ERROR}}$ is open collector; other outputs: **IOut** ≤ -5.5 mA.

5 **OUTRDY**, **DOUT**: **IOut** ≤ 4.4 mA; $\overline{\textbf{ERROR}}$: **IOut** ≤ 5.5 mA; other outputs: **IOut** ≤ 5.5 mA.

Capacitance

Pin	Typ.	Units	Notes
CLK	12	pF	1,2
All other pins	5	pF	1,2

1 This parameter is supplied for engineering guidance and is not guaranteed.

2 TA=25°C , F=1 MHz.

6.2 AC timing characteristics

AC test conditions

Output loads (except output turn-off tests)

Pin	Device mode	Load	Unit
GO	Master	20	pF
DOUT, **OUTRDY**	Fast output	15	pF
DOUT, **OUTRDY**	Normal output	30	pF
All other outputs	All modes	30	pF

Output load (output turn-off tests)

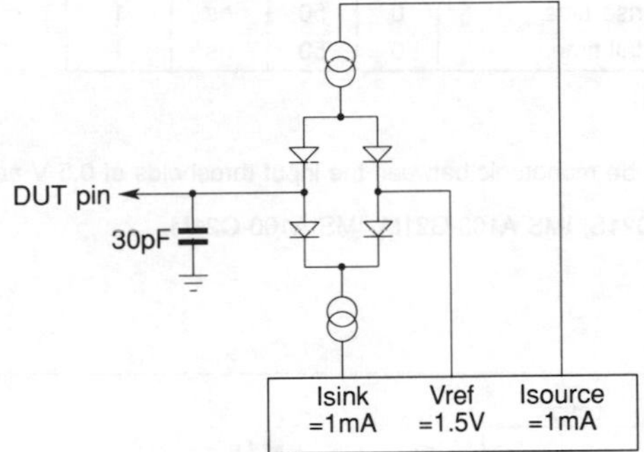

Timing reference levels

Pin	Reference levels	Notes
INPUTS	0.8V, 2.0V	1
CLK	0.5V, 4.0V	
OUTPUTS	0.4V, 2.4V	2,3
OUTPUTS	±100mV change from previous steady output voltage	4

Notes

1 Except **CLK**.

2 Output continuously driven.

3 Timings are tested using VOL=0.8V and with a suitable allowance for the time taken for the output to fall from 0.8V to 0.4V.

4 Output turn-off tests.

Clock

Symbol	Parameter	Min.	Max.	Units	Notes
t_{CHCL}	Clock pulse width high	19		ns	2
t_{CHCL}		24		ns	3
t_{CHCL}		13		ns	4
t_{CLCH}	Clock pulse width low	19		ns	2
t_{CLCH}		24		ns	3
t_{CLCH}		13		ns	4
t_{CHCH}	Clock period	48		ns	2
t_{CHCH}		58		ns	3
t_{CHCH}		33		ns	4
t_R	Clock rise time	0	50	ns	1
t_F	Clock fall time	0	50	ns	1

Notes

1 Clock input transitions should be monotonic between the input thresholds of 0.5 V and 4.0 V.

2 IMS A100-G21S, IMS A100-Q21S, IMS A100-G21M, IMS A100-Q21M.

3 IMS A100-G17M

4 IMS A100-G30S

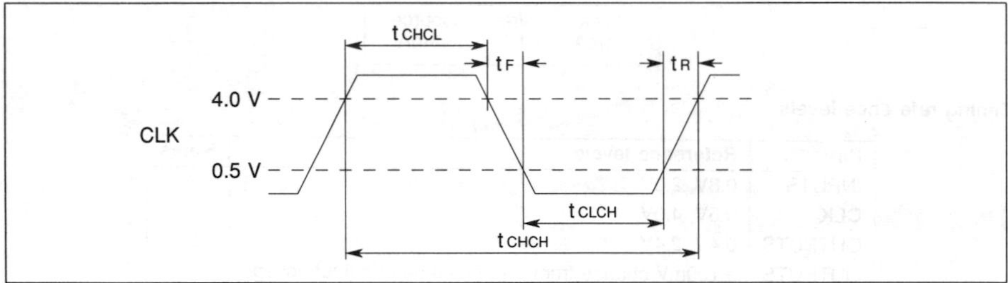

 INMOS is a member of the SGS-THOMSON Microelectronics Group

Memory interface read cycle

Symbol	Parameter	Min.	Max.	Units	Notes
t_{ELEH}	$\overline{CE}$ pulse width low	60		ns	
t_{EHEL}	$\overline{CE}$ pulse width high	50		ns	
t_{SLEL}	$\overline{CS}$ setup time	15		ns	
t_{EHSX}	$\overline{CS}$ hold time	5		ns	
t_{AVEL}	Address setup time	15		ns	
t_{EHAX}	Address hold time	5		ns	
t_{WHEL}	Read Command setup	15		ns	
t_{EHWX}	Read Command hold	5		ns	
t_{ELQX}	Output turn on delay	0		ns	
t_{ELQV}	Read data access		60	ns	
t_{EHQX}	Read data hold	0		ns	
t_{EHQZ}	Output turn off delay		25	ns	

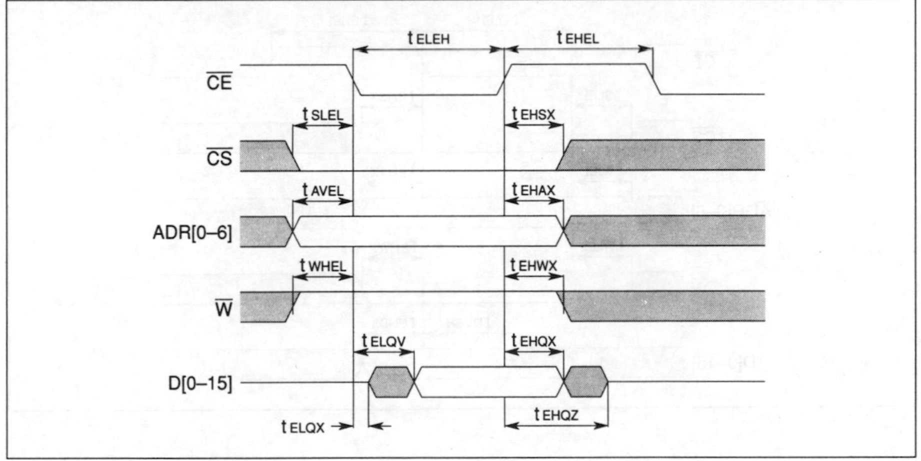

Memory interface write cycle

Symbol	Parameter	Min.	Max.	Units	Notes
t_{ELEH}	$\overline{CE}$ pulse width low	50		ns	
t_{EHEL}	$\overline{CE}$ pulse width high	50		ns	
t_{SLEL}	$\overline{CS}$ setup time	15		ns	
t_{EHSX}	$\overline{CS}$ hold time	5		ns	
t_{AVEL}	Address setup time	15		ns	
t_{EHAX}	Address hold time	5		ns	
t_{WLEL}	Write Command setup	15		ns	
t_{EHWX}	Write Command hold	5		ns	
t_{DVEH}	Write data setup	45		ns	
t_{EHDX}	Write data hold	5		ns	

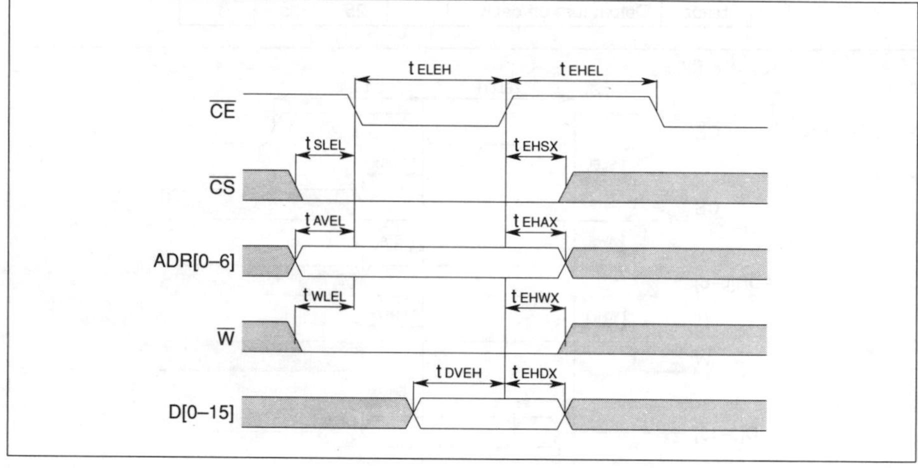

Static read accesses to DOL and DOH registers

Certain applications require to read results from the IMS A100 at high speeds. To ensure full system performance it may be necessary to read results from the DOL and DOH registers using a continuous 'static' access rather than using the normal clocked access.

During static access the $\overline{\textbf{CE}}$ signal is held low continuously. Under this condition it is possible to monitor either DOL or DOH continuously to observe new output words as they become available or alternatively to switch between DOL and DOH without the restriction of having to sequence $\overline{\textbf{CE}}$.

Symbol	Parameter	Min	Max	Units	Notes
t_{AVQV}	Address access time		75	ns	1
t_{CHQV}	Data input access time		$\mathcal{T}$+75	ns	2
t_{ELQV}	$\overline{\textbf{CE}}$ access time		60	ns	3
t_{AXQX}	Data hold after address change	0		ns	
t_{CHQX}	Data hold after new data input	$\mathcal{T}$+0		ns	2
t_{EHQX}	Data hold after end of read	0		ns	

Notes

1 The address access time is specified for address transitions between decimal 74 (DOL register) and decimal 75 (DOH register) only.

2 The parameter $\mathcal{T}$ describes the time taken from the input of a data word to that data word first affecting the most significant word (MSW) output. This is the time at which the DOL and DOH registers are updated.

The duration of $\mathcal{T}$ depends on the coefficient size selected and whether fast or normal output is selected.

Coefficients	Output mode	$\mathcal{T}$ time
4 bit	Fast	8 CLK cycles
8 bit		10 CLK cycles
12 bit		12 CLK cycles
16 bit		14 CLK cycles
4 bit	Normal	Not defined
8 bit		11 CLK cycles
12 bit		14 CLK cycles
16 bit		17 CLK cycles

N.B. The data value read from either DOL or DOH will change as new results are computed by the device.

3 This parameter is the normal read access time for reading any register through the microprocessor interface. In the special case of performing reads from only DOL and DOH any number of reads from these registers can be made with $\overline{\textbf{CE}}$ held low continuously.

It is required that a static access (as described above) should commence like a normal clocked, random, read access to either DOL or DOH. That is **ADDRESS**, $\overline{\textbf{CS}}$ and $\overline{\textbf{W}}$ should be established with setup times to $\overline{\textbf{CE}}$ specified for a normal read access.

During a DOL/DOH static access sequence accesses to locations other than DOL and DOH are undefined.

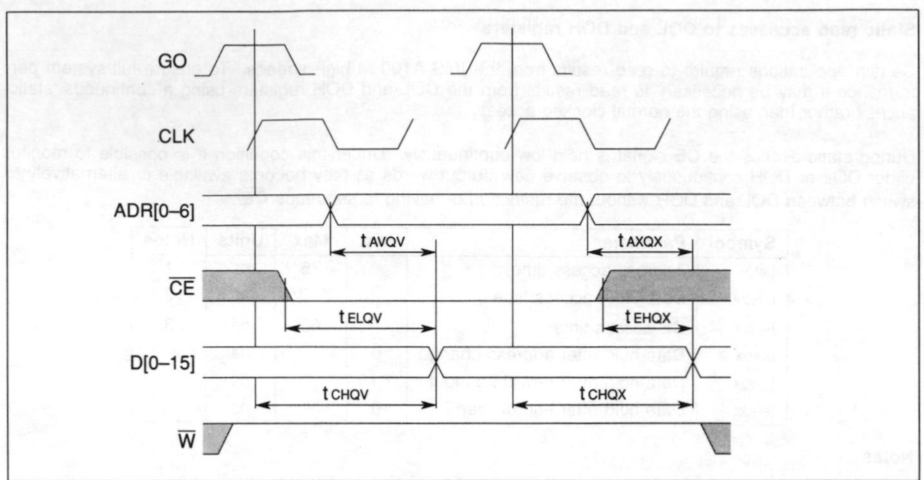
GO
CLK
ADR[0–6]
t AVQV
t AXQX
CE
t ELQV
t EHQX
D[0–15]
t CHQV
t CHQX
W

Typical sequence — 8 bit coefficients, normal output

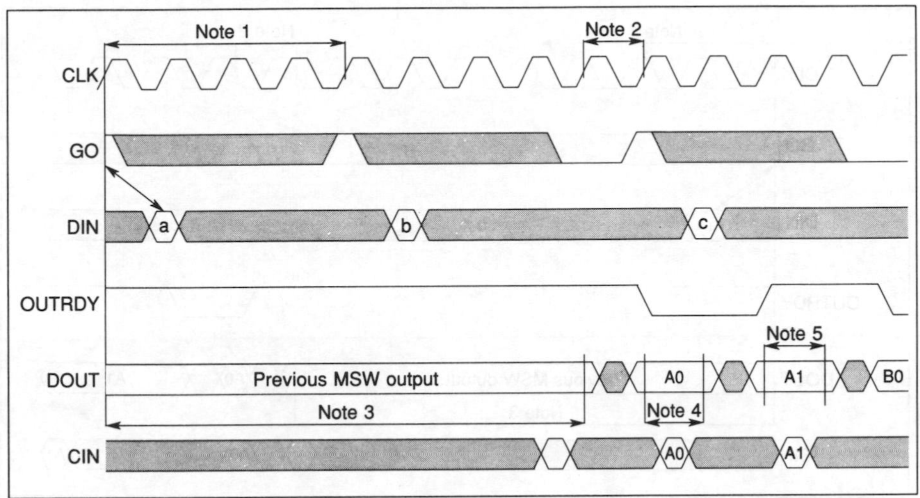

Notes

1 The minimum period between sampling the **GO** input is four clock cycles for 8 bit coefficients, see the table below for the other cases.

2 After the minimum period described in note 1 has elapsed **GO** is sampled on every rising edge of **CLK** until **GO** is high.

3 The delay from an output being initiated by **GO** to the output completing its previous output sequence and starting the new output sequence is 8 clock cycles for 8 bit coefficients, see the table below for the other cases.

4 The least significant word is available at the output across one complete **CLK** cycle for the 8 bit coefficient, normal output case, see the table below for the other cases.

5 The most significant word is available for the minimum period described in note 4, but will be extended by a clock cycle for each additional idle cycle inserted between data inputs.

Coefficients	Min. Output Period note 1	Delay To Output note 3	Min. LSW Output Duration notes 4 and 5
4 bit	2 CLK cycles	6 CLK cycles	Undefined, no normal output
8 bit	4 CLK cycles	8 CLK cycles	1 CLK cycle
12 bit	6 CLK cycles	10 CLK cycles	2 CLK cycles
16 bit	8 CLK cycles	12 CLK cycles	3 CLK cycles

Typical sequence — 8 bit coefficients, fast output

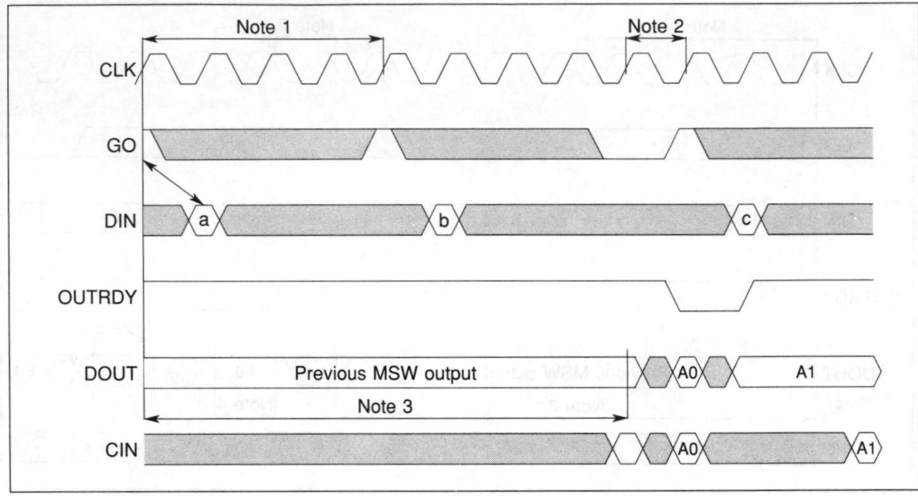

Notes

1 The minimum period between sampling the **GO** input is four clock cycles for 8 bit coefficients, see the table below for the other cases.

2 After the minimum period described in note 1 has elapsed **GO** is sampled on every rising edge of **CLK** until **GO** is high.

3 The delay from an output being initiated by **GO** to the output completing its previous output sequence and starting the new output sequence is 8 clock cycles for 8 bit coefficients, see the table below for the other cases.

Coefficients	Min. Output Period note 1	Delay To Output note 3
4 bit	2 CLK cycles	6 CLK cycles
8 bit	4 CLK cycles	8 CLK cycles
12 bit	6 CLK cycles	10 CLK cycles
16 bit	8 CLK cycles	12 CLK cycles

Typical sequence — 4 bit coefficients

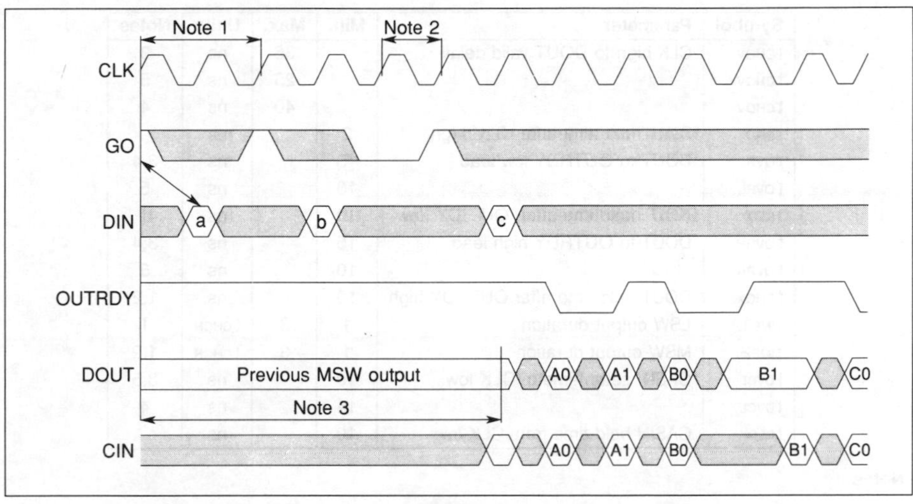

Notes

1 The minimum period between sampling the **GO** input is two clock cycles for 4 bit coefficients, see the table below for the other cases.

2 After the minimum period described in note 1 has elapsed **GO** is sampled on every rising edge of **CLK** until **GO** is high.

3 The delay from an input being initiated by **GO** to the output completing its previous output sequence and starting the new output sequence is 6 clock cycles for 4 bit coefficients, see the table below for the other cases.

Coefficients	**Min. Output Period** note 1	**Delay To Output** note 3
4 bit	2 CLK cycles	6 CLK cycles
8 bit	4 CLK cycles	8 CLK cycles
12 bit	6 CLK cycles	10 CLK cycles
16 bit	8 CLK cycles	12 CLK cycles

Normal output timing — 8 bit coefficient case shown

Symbol	Parameter	Min.	Max.	Units	Notes
t_{CHQV}	CLK high to DOUT valid delay		36	ns	3
t_{CHQV}			25	ns	5
t_{CHQV}			40	ns	4
t_{CHQX}	DOUT hold time after CLK high	2		ns	
t_{QVRL}	DOUT to OUTRDY low lead	15		ns	3,4
t_{QVRL}		10		ns	5
t_{RLQX}	DOUT hold time after OUTRDY low	10		ns	1
t_{QVRH}	DOUT to OUTRDY high lead	15		ns	3,4
t_{QVRH}		10		ns	5
t_{RHQX}	DOUT hold time after OUTRDY high	10		ns	1,2
TIME1	LSW output duration	1	3	t_{CHCH}	1
TIME2	MSW output duration	1	3	t_{CHCH}	1,2
t_{DVCL}	CASIN setup time to CLK low	10		ns	3,5
t_{DVCL}		14		ns	4
t_{CLDX}	CASIN hold time from CLK low	10		ns	

Notes

1 This parameter is determined by the coefficient size in use. The minimum value given is correct for 8 bit coefficients. This parameter is extended by 1 (or 2) periods of **CLK** if 12 (or 16) bit coefficients are used. This mode of operation is not defined if 4 bit coefficients are used.

2 These parameters are extended by one tCHCH for each idle cycle inserted between data input sequences.

3 IMS A100-G21S, IMS A100-Q21S, IMS A100-G21M, IMS A100-Q21M.

4 IMS A100-G17M.

5 IMS A100-G30S.

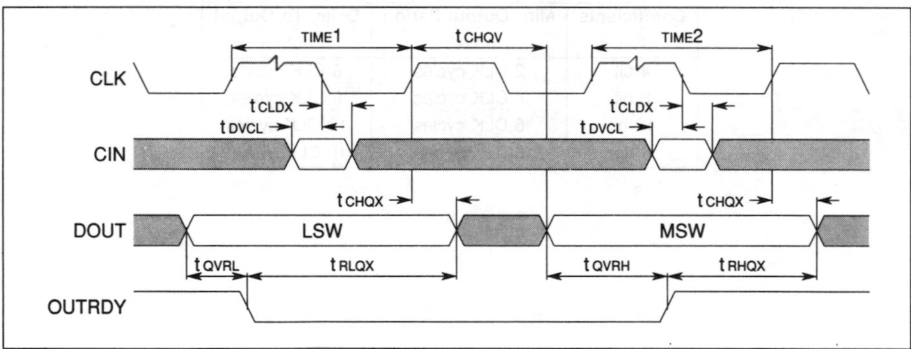

Fast output timing — 4 bit coefficient case shown

Symbol	Parameter	Min.	Max.	Units	Notes
t_{CHQV}	CLK high to DOUT valid delay		36	ns	1,3
t_{CHQV}			22	ns	1,5
t_{CHQV}			40	ns	1,4
t_{CHQX}	DOUT hold time after CLK	2		ns	2
t_{QVRL}	DOUT to OUTRDY low lead	5		ns	1,6
t_{RLQX}	DOUT hold time after OUTRDY low	10		ns	6
t_{QVRH}	DOUT to OUTRDY high lead	5		ns	1,6
t_{RHQX}	DOUT hold time after OUTRDY high	10		ns	2,6
t_{DVCH}	CASIN setup time to CLK high	10		ns	3,5
t_{DVCH}		14		ns	4
t_{CHDX}	CASIN hold time to CLK high	0		ns	7

Notes

1 These parameters assume that each **DOUT** signal is loaded with a maximum of 15 pF.

2 t_{CHQX} and t_{RHQX} for the MSW are shown here for the case where 4 bit coefficients are being used. In the other cases (8, 12 and 16 bit coefficients) the MSW is available for an additional 2, 4 or 6 **CLK** periods. In all cases the MSW will be available for an additional period of **CLK** for each idle cycle inserted between data input sequences.

3 IMS A100-G21S, IMS A100-Q21S, IMS A100-G21M, IMS A100-Q21M.

4 IMS A100-G17M.

5 IMS A100-G30S.

6 The **OUTRDY** signal should not be used in this mode using the IMS A100-G30S variant at clock frequencies above 20.8 MHz.

7 Not tested. Guaranteed by design.

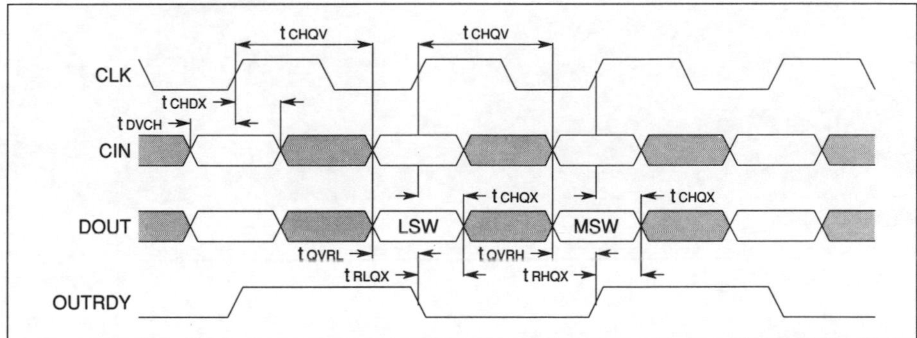

External GO and data input timing

Symbol	Parameter	Min.	Max.	Units	Notes
t_{GHCH}	GO setup time	10		ns	
t_{CHGX}	GO hold time	5		ns	
t_{DVCH}	DIN setup time	30		ns	1
t_{DVCH}		17		ns	2
t_{CHDX}	DIN hold time	5		ns	

Notes

1 IMS A100-G21S, IMS A100-Q21S, IMS A100-G21M, IMS A100-Q21M, IMS A100-G17M.

2 IMS A100-G30S.

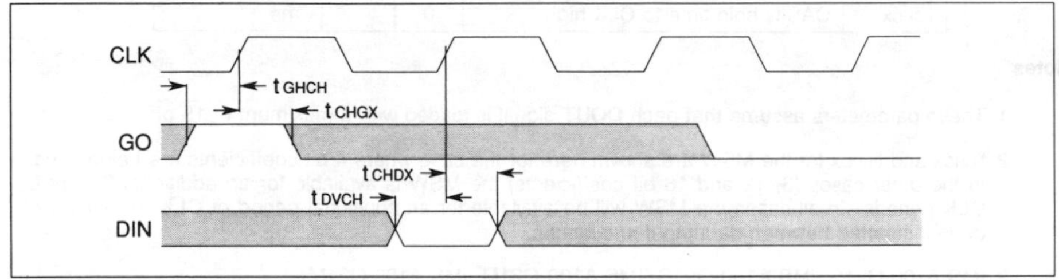

Master generated GO

Symbol	Parameter	Min.	Max.	Units	Notes
t_{EHGH}	Write to DIR to GO high delay	25		ns	1,4
t_{GHCH}	GO high before GO sampled	10		ns	2,4
t_{GLEL}	GO low to write to DIR	0		ns	4
t_{GLCH}	GO low before GO next sampled	10		ns	2,4

Notes

1 The maximum delay from a write to the DIR to **GO** going high is $2 * t_{CHCH} + 50$ ns.

2 This parameter assumes the capacitive load on **GO** is less than 20 pF. **GO** is specified so that one master IMS A100 can drive three slave IMS A100s without buffering.

3 Accesses can be made through the external memory interface to any register other than DIR.

4 This mode should not be used with the IMS A100-G30S variant at clock frequencies above 20.8 MHz.

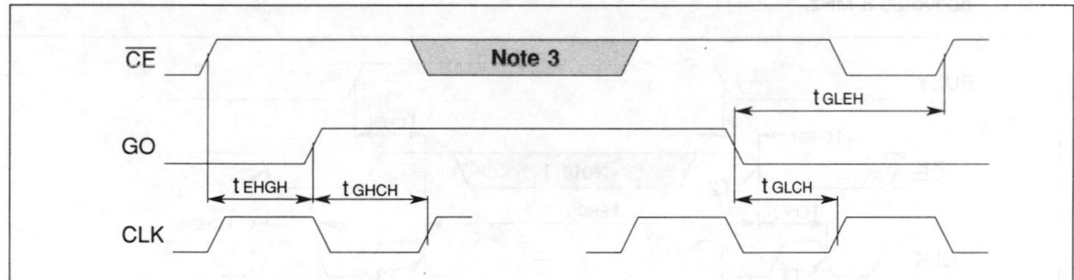

Bankswap timing

Symbol	Parameter	Min.	Max.	Units	Notes
t_{EHBH}	ACR[0] set to BUSY high delay		55	ns	4
t_{CHBL}	BUSY hold after bankswap		50	ns	4
t_{CHEH}	ACR[0]=0 hold after last input	20		ns	3,4
t_{EHCH}	ACR[0]=1 setup to next input	10		ns	3,4

Notes

1 The activity on $\overline{\mathbf{CE}}$ shown is for writing ACR[0]=1. During the period **Note 1** it may be possible to access other registers (subject to their own access constraints).

2 For small t_{EHCH}, **BUSY** may only occur for a short time or not occur at all.

3 If t_{CHEH} or t_{EHCH} is exceeded then bankswap may be synchronised to the previous or next input cycle.

4 This mode should not be used with the IMS A100-G30S variant at clock frequencies above 20.8 MHz.

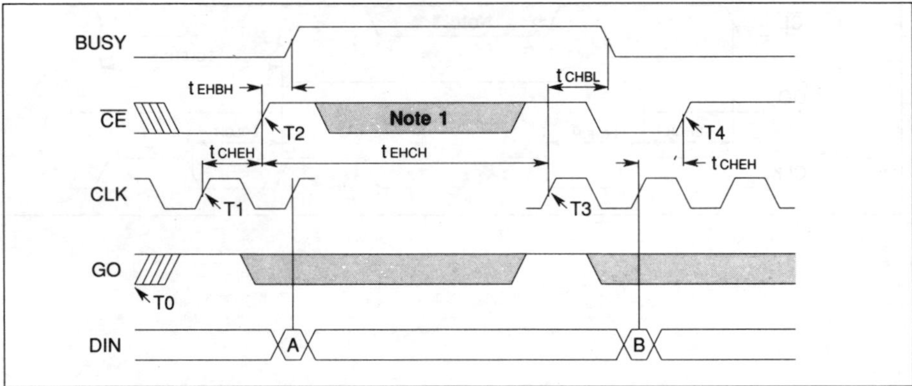

The bankswap timing diagram shows how successive data samples (A and B) can be processed by different sets of coefficients by causing a bankswap to occur between the input of sample A and sample B.

The sequence of events is as follows:

T0 No bankswap pending.

T1 **GO** sampled and found to be high, thus initiating input of data sample A.

T2 Bankswap requested by writing ACR[0]=1. If the minimum timing requirement, t_{CHEH}, from T1 to T2 is not met it is possible (but not guaranteed) that the bankswap requested at T2 will occur immediately and thus affect the processing of data sample A.

T3 Bankswap occurs on the first rising edge of **CLK** upon which **GO** is sampled (without reference to the state of **GO**). If the minimum timing requirement, t_{EHCH}, from T2 to T3 is not met it is possible (but not guaranteed) that the bankswap requested at T2 will not occur at T3 but at the next sampling of **GO**.

T4 This is the earliest time at which another bankswap can be requested.

Coefficient access timing

Symbol	Parameter	Min.	Max.	Units	Notes
t_{EHCH}	End coefficient access before bankswap	0		ns	
t_{CHEL}	Start coefficient access after bankswap	0		ns	

Notes

1 During this period accesses may be made to registers other than the coefficient registers (subject to their own access constraints).

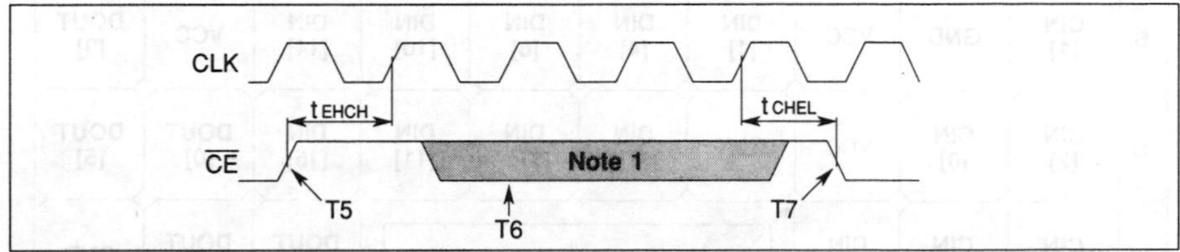

If a bankswap (caused by setting either ACR[0]=1 or SCR[2]=1) occurs at the **GO** sampling point T6, then no access should be made to the coefficient registers between T5 and T7.

7 PACKAGE SPECIFICATIONS

7.1 84 pin grid array package

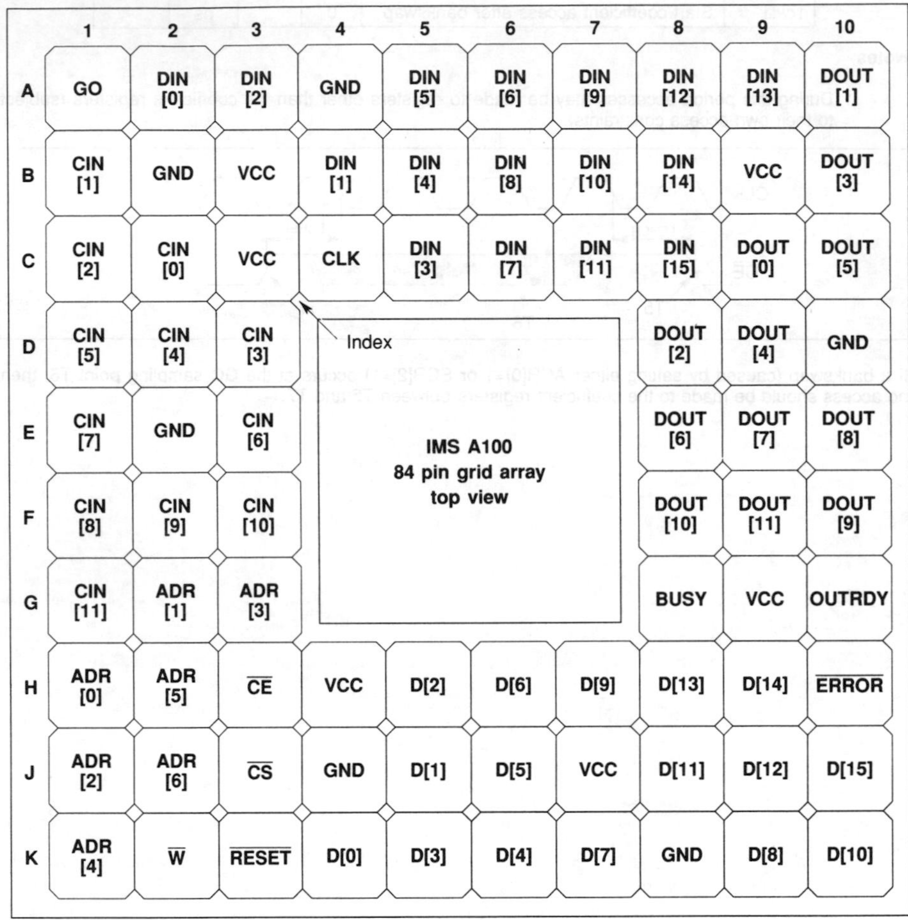

Figure 6 IMS A100 pin configuration

Note

All **VCC** pins **must** be connected to the 5 Volt power supply.
All **GND** pins **must** be connected to ground.

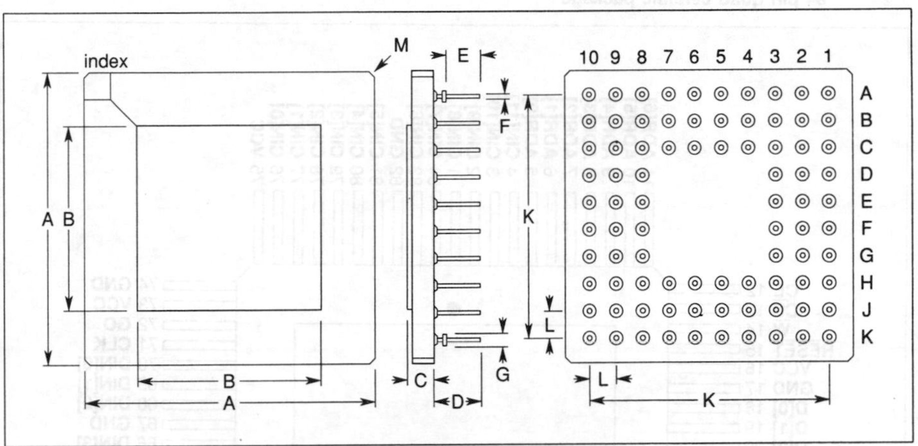

Figure 7 84 pin grid array package dimensions

DIM	Millimetres		Inches		Notes
	NOM	TOL	NOM	TOL	
A	26.924	±0.254	1.060	±0.010	
B	17.019	±0.127	0.670	±0.005	
C	2.456	±0.278	0.097	±0.011	
D	4.572	±0.127	0.180	±0.005	
E	3.302	±0.127	0.130	±0.005	
F	0.457	±0.025	0.018	±0.001	Pin diameter
G	1.143	±0.127	0.045	±0.005	Flange diameter
K	22.860	±0.127	0.900	±0.005	
L	2.540	±0.127	0.100	±0.005	
M	0.508		0.020		Chamfer
Package weight is approximately 7.2 grams					

Table 1 84 pin grid array package dimensions

Pin grid array thermal characteristics

Symbol	Parameter	Min	Nom	Max	Units	Notes
θ JA	Junction to ambient thermal resistance			35	°C /W	1,2

Notes

1 Measured at 400 linear ft/min transverse air flow.

2 This parameter is sampled and not 100% tested.

7.2 84 pin quad ceramic package

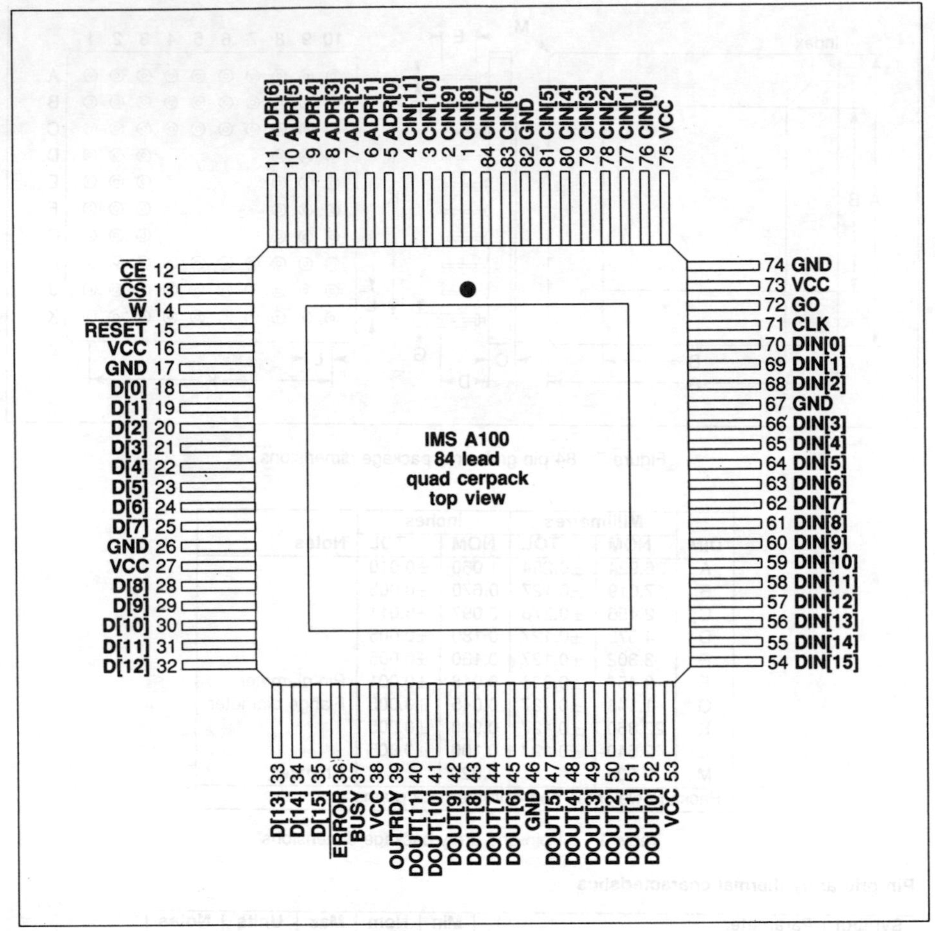

Figure 8 IMS A100 pin configuration

Note

All **VCC** pins **must** be connected to the 5 Volt power supply.
All **GND** pins **must** be connected to ground.

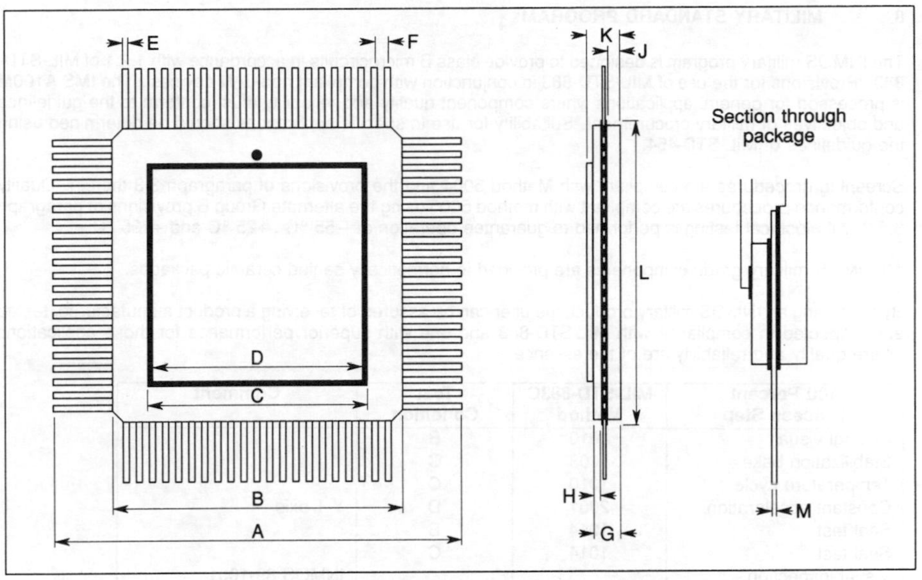

Figure 9 84 lead quad cerpack package dimensions

DIM	Millimetres NOM	Millimetres TOL	Inches NOM	Inches TOL	Notes
A	38.100	±0.508	1.500	±0.020	
B	26.924	±0.305	1.060	±0.012	
C	20.574	±0.203	0.810	±0.008	
D	19.558	±0.254	0.770	±0.010	
E	0.508		0.020		
F	1.270	±0.051	0.050	±0.002	
G	2.489	±0.305	0.098	±0.012	
H	0.635	±0.076	0.025	±0.003	
J	1.143	±0.102	0.045	±0.004	
K	3.099		0.122		Max.
L	27.940		1.100		Max.
M	0.178	±0.025	0.007	±0.001	

Table 2 84 lead quad cerpack package dimensions

Quad cerpack thermal characteristics

Symbol	Parameter	Min	Nom	Max	Units	Notes
θ JA	Junction to ambient thermal resistance			35	°C /W	1,2

Notes

1 Measured at 400 linear ft/min transverse air flow.

2 This parameter is sampled and not 100% tested.

8 MILITARY STANDARD PROGRAM ‡

The INMOS military program is designed to provide class B microcircuits in accordance with 1.2.1 of MIL-STD-883, 'Provisions for the use of MIL-STD-883 in conjunction with compliant non-JAN devices'. The IMS A100M is processed for general applications where component quality and reliability must conform to the guidelines and objectives of military procurement. Suitability for use in specific applications should be determined using the guidelines of MIL-STD-454.

Screening procedures are compliant with Method 5004 and the provisions of paragraph 3.3 therein. Quality conformance procedures are compliant with method 5005 using the alternate Group B provisions of paragraph 3.5.2. All electrical testing is performed to guarantee operation at −55 °C , +25 °C and +125 °C .

All INMOS military grade components are provided in hermetically sealed ceramic packages.

By specifying an INMOS military product, the user can be assured of receiving a product manufactured, tested and inspected in compliance with MIL-STD-883 and one with superior performance for those applications where quality and reliability are of the essence.

100 Percent Process Step	MIL-STD-883C Method	Test Condition	Comment
Internal visual	2010	B	
Stabilization bake	1008	C	
Temperature cycle	1010	C	
Constant acceleration	2001	D	Y-1 axis
Seal test	1014	B	
Seal test	1014	C	
Visual inspection			INMOS 89-1001
Pre burn-in electrical			+25°C data sheet
Burn-in	1015	D	
Post burn-in electrical			+25°C data sheet
PDA			5% max
Final electrical			+125°C data sheet
Final electrical			−55°C data sheet
External visual	2009		
Group A	5005	3.5.1	A1-A11
Group B	5005	3.5.2	
Group C	5005		MIL-STD-883C 1.2.1.b.17
Group D	5005		MIL-STD-883C 1.2.1.b.17

‡ See INMOS document 49-9047 'Military General Processing Specification' for full details.

9 ORDERING DETAILS

The following table indicates the designation of the IMS A100 variants.

INMOS designation	Package	Clock speed	Military/commercial
IMS A100-G21M	Ceramic pin grid array	21 MHz	military
IMS A100-G21S	Ceramic pin grid array	21 MHz	commercial
IMS A100-Q21M	Flatpack	21 MHz	military
IMS A100-Q21S	Flatpack	21 MHz	commercial
IMS A100-G17M	Pin grid array	17 MHz	military
IMS A100-G30S	Pin grid array	30 MHz	commercial

Chapter 9 Digital signal processor applications

This is by far the largest chapter in the book and is composed of a number of extended tutorials on DSP applications. We begin with a detailed tutorial on the implementation of both infinite impulse response (IIR) and finite impulse response (FIR) filters. These are the basic bread and butter applications of DSP.

The second tutorial describes the implementation of fast fourier transforms using DSPs. One of the most common DSP applications is the transformation of signals in the time domain into signals in the frequency domain. The discrete fourier transform carries out the time to frequency conversion, and the fast fourier transform is the practical algorithm employed to implement the conversion.

The introduction of high-speed modems leads to the search for efficient data encoding algorithms. One type of encoding is called convolutional encoding and one of the decoders for this algorithm is called a trellis decoder (because the graph of its decoding states looks like a trellis). The third tutorial describes the convolutional encoder and decoder.

Although DSP was first applied to voice-band signal applications, today's high-speed hardware has made it possible to perform real-time video signal processing. We include the data sheet and application note of an image resampling sequencer. This device can be used for a wide range of real-time video applications, from data compression to filtering.

One of the most common impairments of the public switched telephone network is the echo. The effect of an echo can be particularly irritating on an intercontinental speech link. Unsuppressed echoes increase the error rate of modems and fax transmissions. We include a tutorial on algorithms for echo cancellation.

Another important application of DSP is in sonar. A sonar system, like terrestrial radar, requires a scanning antenna to focus the soundbeam and to pick it up. Unfortunately, an underwater rotating antenna is not entirely practicable. However, by employing a fixed array of transducers and processing the signals applied to them, it is possible to synthesize a beam (an action called beamforming). We provide a tutorial that describes the use of DSP in sonar beamforming.

The final tutorial returns to the subject of the digital signal processing of video signals. The tutorial provides an overview of video DSP and looks at topics ranging from edge detection to thresholding and scaling.

Motorola Digital Signal Processors

Implementing IIR/FIR Filters with Motorola's DSP56000/DSP56001

by
John Lane and Garth Hillman
Digital Signal Processor Operation

MOTOROLA INC., I990 APR7/D

TABLE OF CONTENTS

Paragraph Number	Title	Page Number

Section 1
Introduction

Section 2
Second-Order Direct-Form IIR Digital Filter Sections

Section 3
Single-Section Canonic Form (Direct Form II)

Section 4
Single-Section Transpose Form

TABLE OF CONTENTS (Continued)

Paragraph Number	Title	Page Number

Section 5
Cascaded Direct Form

Section 6
Transmission Lines

Section 7
FIR Filters

References

LIST OF FIGURES

LIST OF FIGURES (Continued)

LIST OF FIGURES (Concluded)

SECTION 1
INTRODUCTION

This application note considers the design of frequency-selective filters, which modify the frequency content and phase of input signals according to some specification. Two classes of frequency-selective digital filters are considered: infinite impulse response (IIR) and finite impulse response (FIR) filters. The design process consists of determining the coefficients of the IIR or FIR filters, which results in the desired magnitude and phase response being closely approximated. Therefore, this application note has a two-fold purpose: 1) to provide some intuitive insight into digital filters, particularly how the coefficients are calculated in the digital domain so that a desired frequency response is obtained, and 2) to show how to implement both classes of digital filters (IIR and FIR) on the DSP56001.

It is assumed that most readers are analog designers learning digital signal processing (DSP). The approach used reflects this assumption in that digital filters are initially presented from an analog point of view. Hopefully, this approach will simplify the transition from the analog s-domain transfer functions to the equivalent functions in the digital z-domain. In keeping with this analog perspective, IIR filters will be discussed first since the equivalent of FIR filters are infrequently encountered in the analog world.

SECTION 1 INTRODUCTION is a brief review of lowpass, highpass, bandpass, and bandstop analog filters. The s-domain formulas governing the key characteristics, magnitude-frequency response, $G(\Omega)$, and phase-frequency response, $\phi(\Omega)$, are derived from first principals. Damping factor, d, cutoff frequency, Ω_C, for lowpass and highpass filters, center frequency, Ω_O, for bandpass and bandstop filters, and quality factor, Q, are defined for the various filter types.

In **SECTION 2 SECOND-ORDER DIRECT-FORM IIR DIGITAL FILTER SECTIONS,** the bilinear transformation is introduced so that analog s-domain designs can be transformed into the digital z-domain and the correct coefficients thereby determined. The form of the formulas for the z-domain filter coefficients thus determined are generalized in terms of the key filter characteristics in the z-domain so that the engineer can design digital filters directly without the necessity of designing the analog equivalent and transforming the design back into the digital domain.

In the analog domain, the performance of the filter depends on the tolerance of the components. Similarly, in the digital domain, the filter performance is limited by the precision of the arithmetic used to implement the filters. In particular, the performance of digital filters is extremely sensitive to overflow, which occurs when the accumulator width is insufficient to represent all the bits resulting from many consecutive additions. This condition is similar to the condition in the analog world in which the signal output is larger than the amplifier power supply so that saturation occurs. A short analysis of the gain at critical nodes in the filters is given in **SECTION 3 SINGLE-SECTION CANONIC FORM (DIRECT FORM II)** and **SECTION 4 SINGLE-SECTION TRANSPOSE FORM** to provide some insight into the scaling requirements for different forms of IIR filters. For this reason, the signal flow graphs developed are centralized about the accumulator nodes.

The analysis of IIR filters focuses on second-order sections. Clearly, higher order filters are often required. Therefore, a brief discussion of how second-order sections can be cascaded to yield higher order filters is given in **SECTION 5 CASCADED DIRECT FORM.** Because the analysis becomes complex quickly, the discussion naturally leads to using commercially available filter design software such as Filter Design and Analysis System (FDAS) from Momentum Data Systems, Inc. **SECTION 6 FILTER DESIGN AND ANALYSIS SYSTEM** concludes by showing how the filter coefficients just discussed can be used in DSP56000 code to implement practical digital filters. Examples of complete filter designs are given, including the code, coefficients, frequency response, and maximum sample frequency.

FIR filters are discussed in **SECTION 7 FIR FILTERS**. Initially, FIR filters are contrasted with IIR filters to show that in many ways they are complementary, each satisfying weaknesses of the other. An intuitive approach is taken to calculating the filter coefficients by starting from a desired arbitrary frequency response. The importance of and constraint imposed by linear phase is emphasized. Having developed an intuitive appreciation of what FIR filter coefficients are, the use of FDAS to accelerate the design process is described. **SECTION 7 FIR FILTERS** concludes by showing how the filter coefficients just determined can be used in DSP56000 code to implement practical digital filters. An example of a passband digital filter using a Kaiser-window design approach is given.

1.1 ANALOG RCL FILTER TYPES

In the following paragraphs, the analog RCL filter network will be analyzed for the four basic filter types: lowpass, highpass, bandpass, and bandstop. Analyzing analog filter types shows that designing digital IIR filters is, in many cases, much simpler than designing analog filters.

In this analysis, as in all of the following cases, the input is assumed to be a steady-state signal containing a linear combination of sinusoidal components whose rms (or peak) amplitudes are constant in time. This assumption allows simple analytic techniques to be used in determining the network response. Even though these results will then be applied to real-world signals that may not satisfy the original steady-state assumption, the deviation of the actual response from the predicted response is small enough to neglect in most cases. General analysis techniques consist of a linear combination of steady-state and transient response solutions to the differential equations describing the network.

1.2 ANALOG LOWPASS FILTER

The passive RCL circuit forming a lowpass filter network is shown in Figure 1-1 where the transfer function, H(s), is derived from a voltage divider analysis of the RCL network. This approach is valid since the effect of C and L can be described as a complex impedance (or reactance, X_C and X_L) under steady-state conditions; s is a complex variable of the complex transfer function, H(s). The filter frequency response is found by evaluating H(s) with $s = j\Omega$, where $\Omega = 2\pi f$ and f is the frequency of a sinusoidal component of the input signal. The output signal is calculated from the product of the input signal and $H(j\Omega)$. To facilitate analysis, the input and output signal components are described by the complex value, $e^{j\Omega t} = \cos \Omega t + j \sin \Omega t$. The actual physical input and output signal components are found by taking the real part of this value. The input is $R\{e^{j\Omega t}\} = \cos \Omega t$; the output is $R\{H(j\Omega)e^{j\Omega t}\} = G(\Omega) \cos [\Omega t + \phi(\Omega)]$. The previous technique is based upon the solution of the differential equations describing the network when the input is steady state. Describing the circuit response by H(s) instead of solving the differential equation is a common simplification used in this type of analysis.

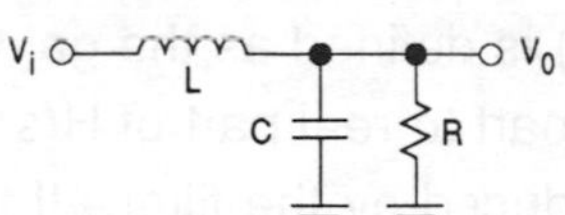

$$\frac{V_0}{V_i} = \frac{X_C \| R}{X_L + X_C \| R}$$

Vi L C R V0

$X_C = 1 / j\Omega C$

$$= \frac{(R / j\Omega C) / (R + 1 / j\Omega C)}{j\Omega L + (R / j\Omega C) / (R + 1 / j\Omega C)}$$

$X_L = j\Omega L$

$$= \frac{1 / LC}{-\Omega^2 + j\Omega / RC + 1 / LC}$$

$\Omega_C = \frac{1}{\sqrt{LC}}$

$$= \frac{\Omega_C^2}{-\Omega^2 + jd\Omega_C \Omega + \Omega_C^2}$$

$d = \sqrt{\frac{L}{R^2 C}}$

Let $s = j\Omega$ and define $H(s) = V_0 / V_i$; then,

$$H(s) = \frac{1}{(s / \Omega_C)^2 + d(s / \Omega_C) + 1}$$

which is the s-domain transfer function. The gain, $G(\Omega)$, of the filter is

$$G(\Omega) \equiv \sqrt{H(s) H^*(s)}\Big|_{s = j\Omega}$$

$$= \frac{1}{\sqrt{(1 - \Omega^2 / \Omega_C^2)^2 + (d\Omega / \Omega_C)^2}}$$

where $*$ denotes complex conjugate. The phase angle, $\phi(\Omega)$, is the angle between the imaginary and real components of $H(s)$:

$$\phi(\Omega) \equiv \tan^{-1} [I \{H(s)\} / R \{H(s)\}]$$

$$= -\tan^{-1}\left[\frac{d(\Omega / \Omega_C)}{1 - (\Omega / \Omega_C)^2}\right] \quad \text{for } \Omega \le \Omega_C$$

$$= -\pi - \tan^{-1}\left[\frac{d(\Omega / \Omega_C)}{1 - (\Omega / \Omega_C)^2}\right] \quad \text{for } \Omega > \Omega_C$$

Figure 1-1. s-Domain Analysis of Second-Order Lowpass Analog Filter

The magnitude of H(s) is defined as the gain, $G(\Omega)$, of the system; whereas, the ratio of the imaginary part to real part of H(s), $I\{H(j\Omega)\}/R\{H(j\Omega)\}$, is the tangent of the phase, $\phi(\Omega)$, introduced by the filter. If the input signal is $A_k \sin(\Omega_k t + \phi_k)$, then the output signal is $A_k G(\Omega_k) \sin[\Omega_k t + \phi_k + \phi(\Omega_k)]$. Figure 1-2 shows the gain, $G(\Omega)$, and phase, $\phi(\Omega)$, plots for the second-order lowpass network of Figure 1-1 for various values of damping factor, d; d also controls the amplitude and position of the peak of the normalized response curve.

The frequency corresponding to the peak amplitude can be easily found by taking the derivative of $G(\Omega)$ (from the equation for $G(\Omega)$ in Figure 1-1) with respect to Ω and setting it equal to zero. Solving the resultant equation for Ω then defines Ω_M as the frequency where the peak amplitude occurs. The peak amplitude is then $G_M = G(\Omega_M)$:

$$\Omega_M = \Omega_c\sqrt{(1-d^2/2)} \qquad (1\text{-}1)$$

$$G_M = \frac{1}{d\sqrt{(1-d^2/4)}} \qquad (1\text{-}2)$$

for $d < \sqrt{2}$. For $d > \sqrt{2}$, $\Omega_M = 0$ is the position of the peak amplitude where $G_M = 1$. When $d = \sqrt{2}$, $G_M = 1$, which gives the maximally flat response curve used in the Butterworth filter design (usually applies only to a set of cascaded sections). Note that Ω_c for a lowpass filter is that frequency where the gain is $G(\Omega c) = 1/d$ and the phase is $\phi(\Omega c) = -\pi/2$.

1.3 ANALOG HIGHPASS FILTER

The passive RCL circuit forming a highpass filter network is shown in Figure 1-3 where the transfer function, H(s), is again derived from a voltage divider analysis of the RCL network. The gain and phase response are plotted in Figure 1-4 for different values of damping coefficient. As evidenced, the high-pass filter response is the mirror image of the lowpass filter response.

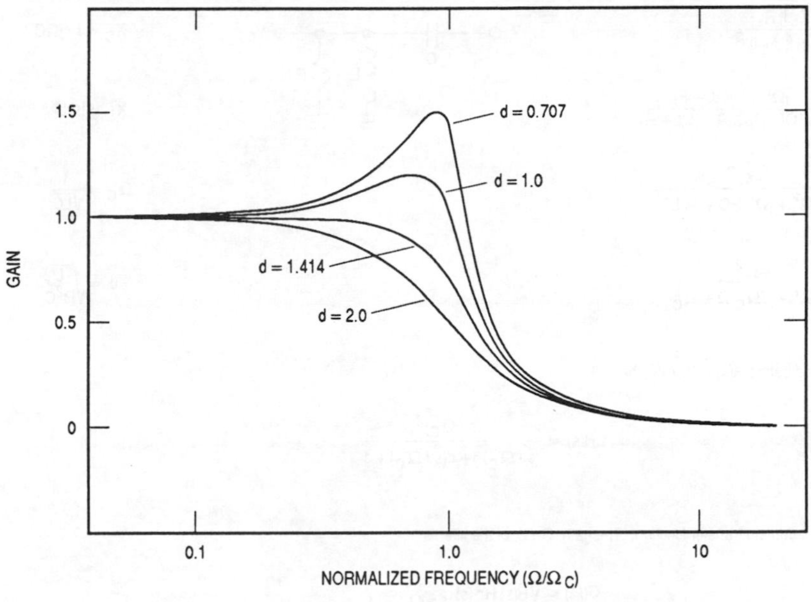

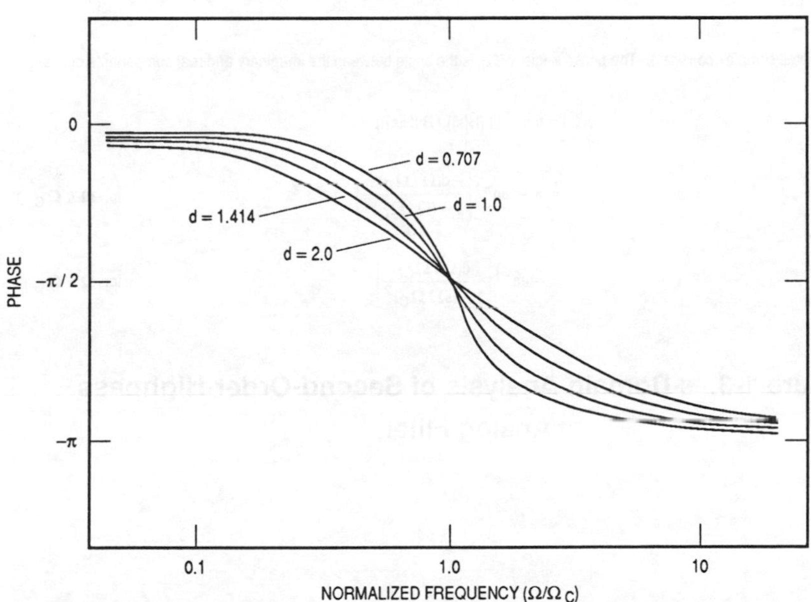

Figure 1-2. Gain and Phase Response of Second-Order Lowpass Analog Filter at Various Values of Damping Factor, d

$$\frac{V_0}{V_i} = \frac{X_L \| R}{X_C + X_L \| R}$$

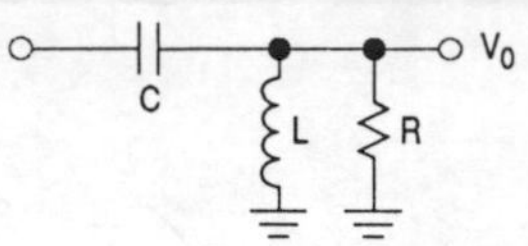

$$= \frac{j\Omega LR / (j\Omega L + R)}{1 / j\Omega C + j\Omega LR / (j\Omega L + R)}$$

$$= \frac{-\Omega^2}{-\Omega^2 + j\Omega / RC + 1 / LC}$$

$$= \frac{-\Omega^2}{-\Omega^2 + jd\Omega_C \Omega + \Omega_C^2}$$

V_i — C — L, R — V_0

$$X_C = 1 / j\Omega C$$

$$X_L = j\Omega L$$

$$\Omega_C = \frac{1}{\sqrt{LC}}$$

$$d = \sqrt{\frac{L}{R^2C}}$$

Let $s = j\Omega$ and define $H(s) = V_0 / V_i$; then,

$$H(s) = \frac{(s / \Omega_C)^2}{(s / \Omega_C)^2 + d(s / \Omega_C) + 1}$$

which is the s-domain transfer function. The gain, $G(\Omega)$, of the filter is

$$G(\Omega) \equiv \sqrt{H(s)\, H^*(s)}\,\Big|_{s = j\Omega}$$

$$= \frac{(\Omega / \Omega_C)^2}{\sqrt{(1 - \Omega^2 / \Omega_C^2)^2 + (d\Omega / \Omega_C)^2}}$$

where $*$ denotes complex conjugate. The phase angle, $\phi(\Omega)$, is the angle between the imaginary and real components of H(s):

$$\phi(\Omega) \equiv \tan^{-1} [\, I\{H(s)\} / R\{H(s)\}]$$

$$= \pi - \tan^{-1}\left[\frac{d(\Omega / \Omega_C)}{1 - (\Omega / \Omega_C)^2}\right] \qquad \text{for } \Omega \le \Omega_C$$

$$= -\tan^{-1}\left[\frac{d(\Omega / \Omega_C)}{1 - (\Omega / \Omega_C)^2}\right] \qquad \text{for } \Omega > \Omega_C$$

Figure 1-3. s-Domain Analysis of Second-Order Highpass Analog Filter

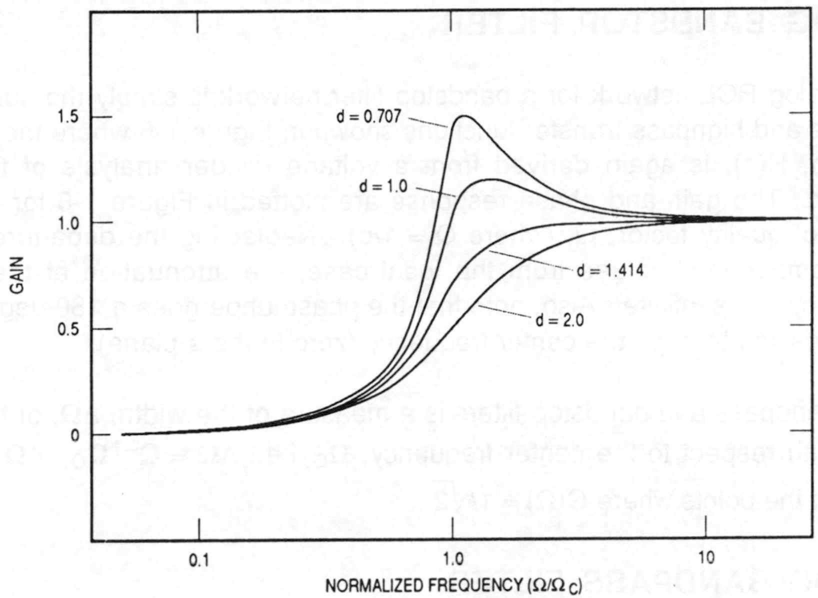

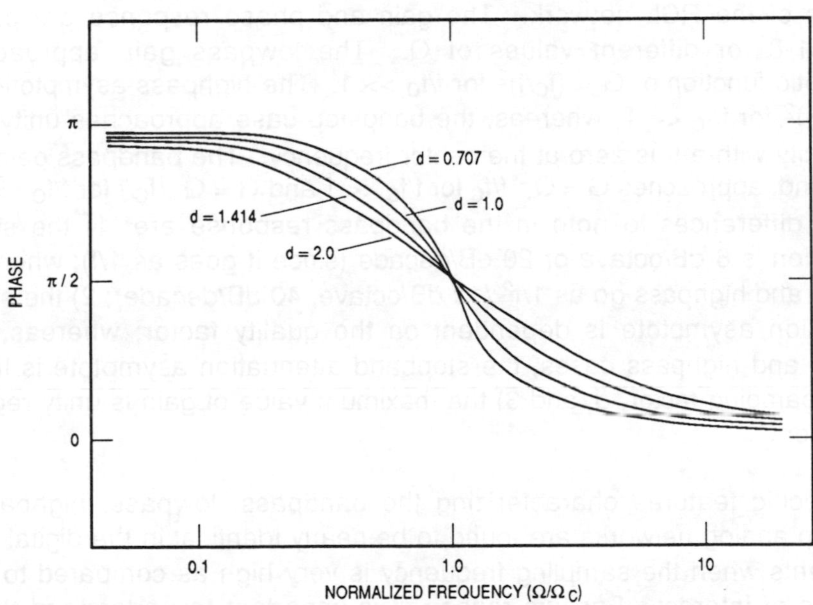

Figure 1-4. Gain and Phase Response of Second-Order Highpass Analog Filter at Various Values of Damping Factor, d

1.4 ANALOG BANDSTOP FILTER

The analog RCL network for a bandstop filter network is simply the sum of the lowpass and highpass transfer functions shown in Figure 1-5 where the transfer function, H(s), is again derived from a voltage divider analysis of the RCL network. The gain and phase response are plotted in Figure 1-6 for different values of quality factor, Q, (where Q = 1/d). Neglecting the departure of real RCL components' values from the ideal case, the attenuation at the center frequency, f_o, is infinite. Also, note that the phase undergoes a 180-degree shift when passing through the center frequency (zero in the s-plane).

Q for bandpass and bandstop filters is a measure of the width, $\Delta\Omega$, of the stopband with respect to the center frequency, Ω_o, i.e., $\Delta\Omega = Q^{-1}\Omega_o$. $\Delta\Omega$ is measured at the points where $G(\Omega) = 1/\sqrt{2}$.

1.5 ANALOG BANDPASS FILTER

The passive RCL circuit forming a bandpass filter network is shown in Figure 1-7 where the transfer function, H(s), is again derived from a voltage divider analysis of the RCL network. The gain and phase response are plotted in Figure 1-8 for different values of Q. The lowpass gain approaches an asymptotic function of $G = (f_c/f)^2$ for $f/f_o >> 1$. The highpass asymptotic gain is $G = (f/f_c)^2$ for $f/f_c << 1$; whereas, the bandstop case approaches unity at zero and infinity with a true zero at the center frequency. The bandpass gain, on the other hand, approaches $G = Q^{-1}f/f_c$ for $f/f_c << 1$ and $G = Q^{-1}f_c/f$ for $f/f_c >> 1$. The primary differences to note in the bandpass response are: 1) the stopband attenuation is 6 dB/octave or 20 dB/decade (since it goes as 1/f); whereas, the lowpass and highpass go as $1/f^2$ (12 dB/octave, 40 dB/decade); 2) the stopband attenuation asymptote is dependent on the quality factor; whereas, for the lowpass and highpass cases, the stopband attenuation asymptote is independent of damping factor, d; and 3) the maximum value of gain is unity regardless of the filter Q.

The specific features characterizing the bandpass, lowpass, highpass, and bandstop analog networks are found to be nearly identical in the digital IIR filter equivalents when the sampling frequency is very high as compared to the frequencies of interest. For this reason, it is important to understand the basic properties of the four filter types before proceeding to the digital domain.

$$\frac{V_0}{V_i} = \frac{R}{X_C \parallel X_L + R}$$

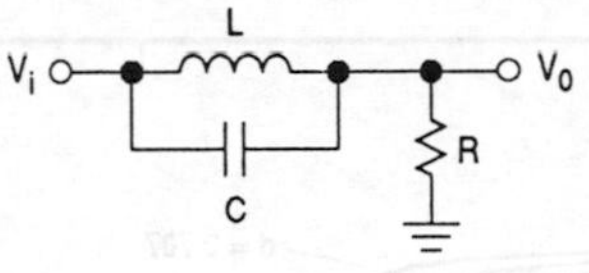

$$= \frac{R}{R + (j\Omega L / j\Omega C) / (j\Omega L + 1 / j\Omega C)}$$

$$= \frac{-\Omega^2 + 1/LC}{-\Omega^2 + j\Omega / RC + 1/LC}$$

$$= \frac{-\Omega^2 + \Omega_0^2}{-\Omega^2 + j\Omega_0\Omega / Q + \Omega_0^2}$$

$$X_C = 1 / j\Omega C$$

$$X_L = j\Omega L$$

$$\Omega_0 = \frac{1}{\sqrt{LC}}$$

$$Q = 1/d = \sqrt{R^2 C / L}$$

Let $s = j\Omega$ and define $H(s) = V_0 / V_i$; then,

$$H(s) = \frac{(s/\Omega_0)^2 + 1}{(s/\Omega_0)^2 + s/\Omega_0 Q + 1}$$

which is the s-domain transfer function. The gain, $G(\Omega)$, of the filter is

$$G(\Omega) \equiv \sqrt{H(s)\,H^*(s)}\,\Big|_{s = j\Omega}$$

$$= \frac{|1 - (\Omega/\Omega_0)^2|}{\sqrt{(1 - \Omega^2/\Omega_0^2)^2 + (\Omega/\Omega_0 Q)^2}}$$

where * denotes complex conjugate. The phase angle, $\phi(\Omega)$, is the angle between the imaginary and real components of H(s):

$$\phi(\Omega) \equiv \tan^{-1}[\,I\{H(s)\} / R\{H(s)\}]$$

$$= -\tan^{-1}\left[\frac{(\Omega/\Omega_0 Q)}{1 - (\Omega/\Omega_0)^2}\right]$$

Figure 1-5. s-Domain Analysis of Second-Order Bandstop Analog Filter

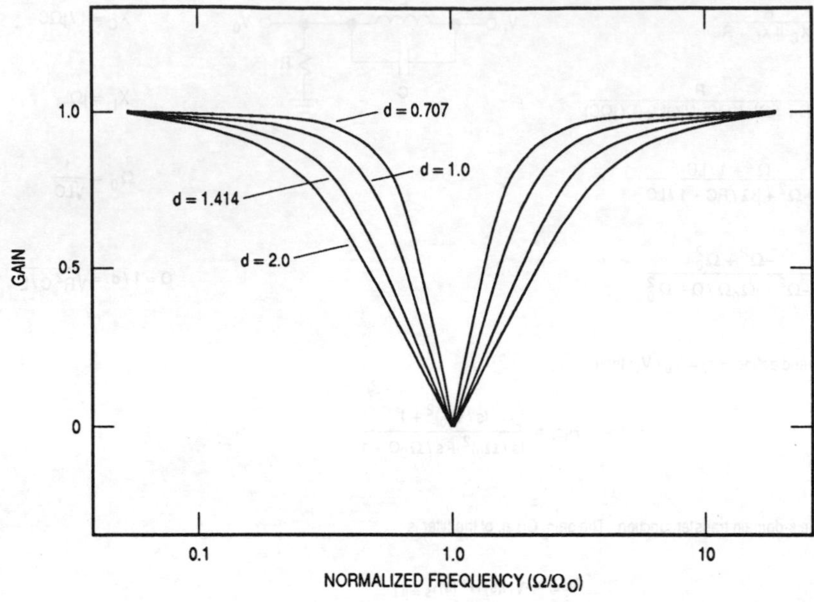

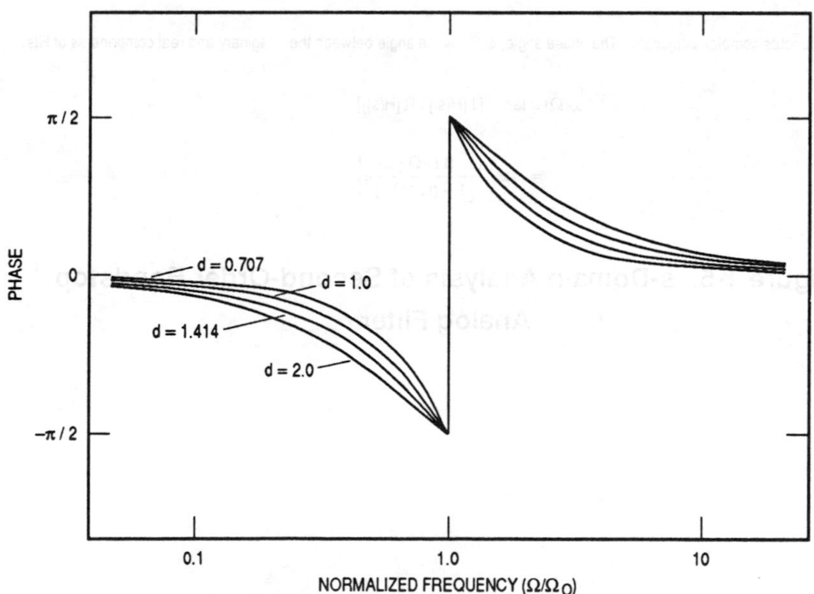

Figure 1-6. Gain and Phase Response of Second-Order Bandstop Analog Filter at Various Values of Damping Factor, d

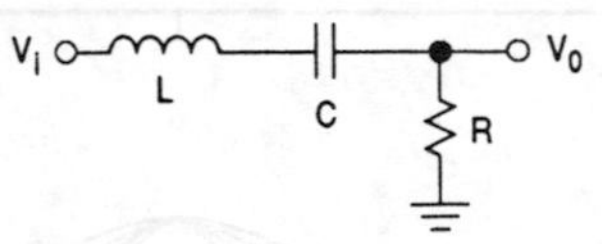

$$\frac{V_0}{V_i} = \frac{R}{X_L + X_C + R}$$

$$= \frac{R}{j\Omega L + 1/j\Omega C + R}$$

$$= \frac{j\Omega R/L}{-\Omega^2 + j\Omega R/L + 1/LC}$$

$$= \frac{j\Omega_0 \Omega/Q}{-\Omega^2 + j\Omega_0\Omega/Q + \Omega_0^2}$$

$X_C = 1/j\Omega C$

$X_L = j\Omega L$

$\Omega_0 = \frac{1}{\sqrt{LC}}$

$Q = 1/d = \sqrt{L/R^2C}$

Let $s = j\Omega$ and define $H(s) = V_0 / V_i$; then,

$$H(s) = \frac{s/\Omega_0 Q}{(s/\Omega_0)^2 + s/\Omega_0 Q + 1}$$

which is the s-domain transfer function. The gain, $G(\Omega)$, of the filter is

$$G(\Omega) \equiv \sqrt{H(s)\,H^*(s)}\,\Big|_{s=j\Omega}$$

$$= \frac{\Omega/\Omega_0 Q}{\sqrt{(1-\Omega^2/\Omega_0^2)^2 + (\Omega/\Omega_0 Q)^2}}$$

where $*$ denotes complex conjugate. The phase angle, $\phi(\Omega)$, is the angle between the imaginary and real components of H(s):

$$\phi(\Omega) \equiv \tan^{-1}[\,I\{H(s)\}/R\{H(s)\}]$$

$$= \tan^{-1}\left[\frac{\Omega/\Omega_0 Q}{1-(\Omega/\Omega_0)^2}\right]$$

Figure 1-7. s-Domain Analysis of Second-Order Bandpass Analog Filter

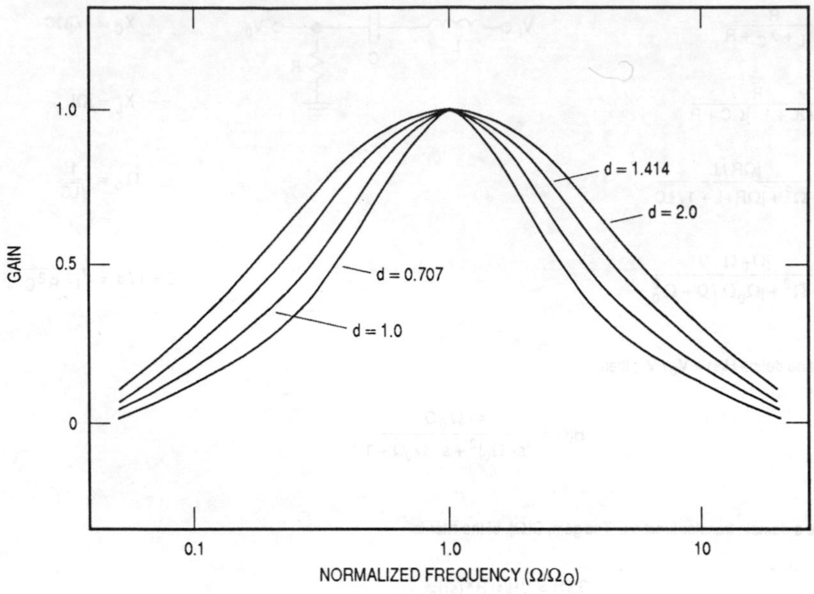

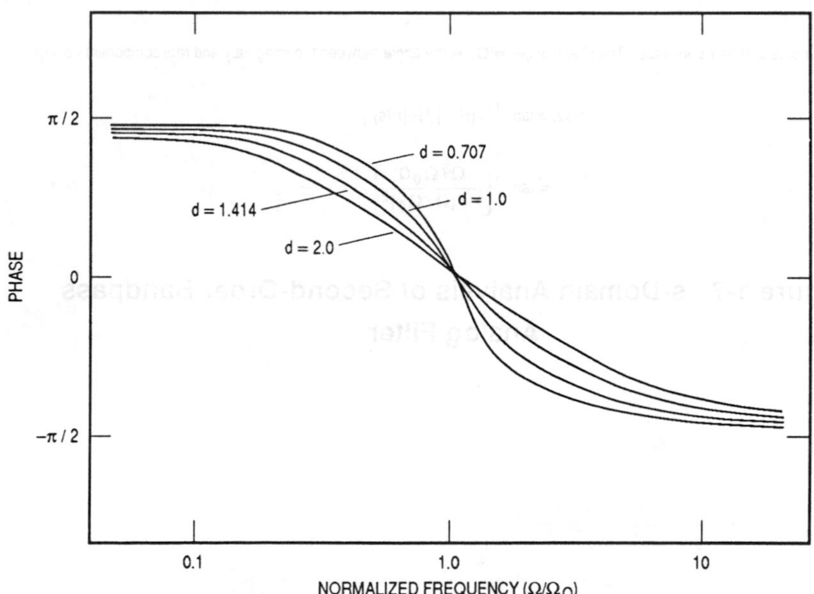

Figure 1-8. Gain and Phase Response of Second-Order Bandpass Analog Filter at Various Values of Damping Factor, d

SECTION 2
SECOND-ORDER DIRECT-FORM IIR DIGITAL FILTER SECTIONS

The traditional approach to deriving the digital filter coefficients has been to start with the digital z-domain description, transform to the analog s-domain to understand how to design filters, then transform back to the digital domain to implement the filter. This approach is not used in this report. Instead, formulas are developed relating the s-domain filter to the z-domain filter so transformations to and from one domain to the other are no longer necessary.

The Laplace or s-transform in the analog domain was developed to facilitate the analysis of continuous time signals and systems. For example, using Laplace transforms the concepts of poles and zeros, making system analysis much faster and more systematic. The Laplace transform of a continuous time signal is

$$X(s) = L\{x(t)\}$$

$$= \int_{0}^{\infty} x(t)e^{-st}\,dt \qquad (2\text{-}1)$$

where L is the Laplace transform operator and implies the operation described in Equation (2-1).

In the digital domain, the continuous signal, x(t), is first sampled and then quantized by an analog-to-digital (A/D) converter before being processed. That is, the signal is only known at discrete points in time, which are at multiples of the sampling interval, $T = 1/f_S$, where f_S is the sampling frequency. Because of the sampled characteristic of a digital signal, its z-transform is given by a summation (as opposed to an integral):

$$X(z) = Z\{x(n)\}$$

$$= \sum_{n=-\infty}^{\infty} x(n)z^{-n} \qquad (2\text{-}2)$$

where Z is the z-transform operator as described by the operation of Equation (2-2), and x(n) is the quantized values from the A/D converter of the continuous time signal, x(t), at discrete times, t = nT.

One property of the z-transform which will be used later in this report is the time shifting property. The time shifting property states:

$$x(n-k) = Z^{-1}\{z^{-k}\, X(z)\} \qquad (2\text{-}3)$$

The proof of this property follows directly from the definition of the z-transform.

An obvious question arises: "If the s-domain of a signal, x(t), is known and that same signal is digitized, what is the relationship between the s-domain transform and the z-transform?" The relationship or mapping is not unique and depends on the viewpoint used. It is obvious that the trivial mapping, s = z, is inappropriate; the signal has been sampled. When a bandlimited signal is sampled (i.e., multiplied by a periodic impulse function), the spectrum of the resulting signal is repetitive as shown in Figure 2-1 (see Reference 10).

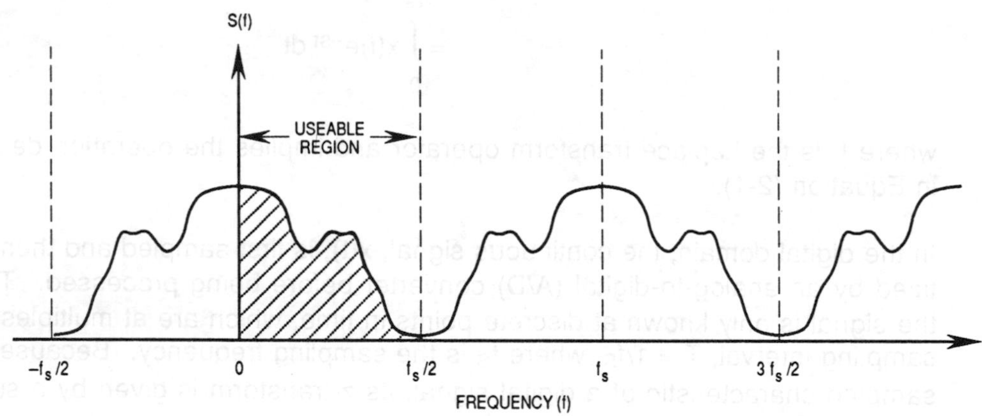

Figure 2-1. Spectrum of Bandlimited Signal Repeated at Multiples of the Sampling Frequency, f_s.

Clearly, the spectrum consists of the spectrum of the bandlimited signal repeated at multiples of the sampling frequency, $\omega_S = 2\pi f_S$. That is, the resulting spectrum is unique only between 0 and $\omega_S/2$ or multiples thereof; whereas, before the signal was sampled, the energy at frequencies greater than those in the signal was independent of the signal. Therefore, acceptable mappings would either reflect the cyclic nature of the spectrum of the sampled test or at least be linear over the frequencies of interest.

One mapping or transformation from the s-domain to the z-domain discussed later in this report is the bilinear transformation. To understand the origin of this transformation, consider the simple first-order linear analog filter with the system function:

$$H(s) = \frac{Y(s)}{X(s)} = \frac{b}{s+a} \tag{2-4}$$

and recall the differentiation property of the s-transform when $x(t) = L^{-1}\{X(s)\}$ (where L^{-1} is the inverse Laplace transform operator); then the time derivative of $x(t)$ is

$$\frac{d}{dt}x(t) = L^{-1}\{s\,X(s)\} \tag{2-5}$$

where $\frac{d}{dt}x(t)$ is the time derivative of $x(t)$.

Using the differentiation property of Equation (2-5), the linear system described by Equation (2-4) can be expressed as follows:

$$\frac{d}{dt}y(t) + ay(t) = bx(t) \tag{2-6}$$

If this differential equation is solved by expressing $y(t)$ with the trapezoidal integration formula,

$$y(t) = \int_{t_0}^{t} \frac{d}{d\tau}y(\tau)\,d\tau + y(t_0) \tag{2-7}$$

where the approximate solution is given by

$$y(t) = \frac{1}{2}\left[\frac{d}{dt}y(t) + \frac{d}{dt}y(t_0)\right](t-t_0) + y(t_0) \tag{2-8}$$

then using $\frac{d}{dt}y(t)$ from Equation (2-6) with $t = nT$ and $t_0 = (n-1)T$, Equation (2-8) can be expressed as follows:

$$(2+aT)\,y(n) - (2-aT)\,y(n-1) = bT\,[x(n) + x(n-1)] \tag{2-9}$$

Taking the z-transform of this difference equation and using the time shifting property of the z-transform, Equation (2-3) results in the z-domain system function:

$$H(z) = \frac{Y(z)}{X(z)} = \frac{b}{\frac{2}{T}\left(\frac{1-z^{-1}}{1+z^{-1}}\right) + a} \tag{2-10}$$

Clearly, the mapping between the s-plane and the z-plane is

$$s = \frac{2}{T}\left(\frac{1-z^{-1}}{1+z^{-1}}\right) \tag{2-11}$$

This mapping is called bilinear transformation.

Although this transformation was developed using a first-order system, it holds, in general, for an Nth-order system (see Reference 14). By letting $s = \sigma + j\Omega$ and $z = re^{j\theta}$, it can be shown that the left-half plane in the s-domain is mapped inside the unit $r = 1$ circle in the z-domain under the bilinear transformation. More importantly, when $r = 1$ and $\sigma = 0$, the frequencies in the s-domain and the z-domain are related by

$$\Omega = \frac{2}{T} \tan \frac{\theta}{2} \tag{2-12}$$

or equivalently:

$$\theta = 2 \tan^{-1} \frac{\Omega T}{2} \tag{2-13}$$

where θ is the digital domain normalized frequency equal to $2\pi f/f_s$, and Ω is the analog domain frequency used in the analysis of the previous section.

On the $j\Omega$ axis or equivalently along the frequency axis, the scale has been changed nonlinearly. The gain and phase values depicted on the vertical axis of Figures 1-2, 1-4, 1-6, and 1-8 remain exactly the same in the digital domain (or z-plane). The horizontal (frequency) axis is modified so that an infinite frequency in the analog domain maps to one-half of the sample frequency, $f_s/2$, in the digital domain; whereas, for frequencies much less than $f_s/2$, the mapping is approximately 1:1 with $\theta = \Omega$ In summary, the bilinear transformation is a one-to-one nonlinear mapping from the s-domain into the z-domain in which high frequencies ($\Omega > 2\pi f_s/4$) in the s-domain are compressed into a small interval in the z-domain. Therefore, the gain and phase expressions of the previous section can be directly transformed into the digital domain by simply substituting Equation (2-12) into the corresponding expressions. This will be done for each filter type in the following paragraphs.

First, it is appropriate to introduce the direct-form implementation of a digital filter by noting that, in general, if the bilinear transformation of Equation (2-11) is substituted into the transfer function, H(s), of the previous section, the resulting H(z) will have the following generalized form:

$$H(z) = \frac{b_0 + b_1 z^{-1} + b_2 z^{-2}}{1 + a_1 z^{-1} + a_2 z^{-2}} \qquad (2\text{-}14)$$

where the digital domain coefficients, a_i and b_i, are exactly related to the s-domain characteristics of the system such as the center frequency, bandwidth, etc. In the direct-form implementation, the a_i and b_i are used directly in the difference equation, which can be easily programmed on a high-speed DSP such as the DSP56001. The time-domain difference equation is derived from the z-domain transfer functions by applying the inverse z-transform in general and the inverse time shifting property in particular as follows:

$$Z^{-1}\{H(z)\} = Z^{-1}\{Y(z)/X(z)\}$$

$$= Z^{-1}\{[b_0 + b_1 z^{-1} + b_2 z^{-2}] / [1 + a_1 z^{-1} + a_2 z^{-2}]\} \qquad (2\text{-}15)$$

so that

$$Z^{-1}\{Y(z)[1 + a_1 z^{-1} + a_2 z^{-2}]\} = Z^{-1}\{X(z)\,[b_0 + b_1 z^{-1} + b_2 z^{-2}]\}$$

therefore, using the inverse time shifting property of Equation (2-3):

$$Z^{-1}\{X(z)z^{-k}\} = \{x(n-k)\}$$

and

$$Z^{-1}\{Y(z)z^{-k}\} = \{y(n-k)\}$$

Equation (2-15) becomes

$$y(n) = b_0 x(n) + b_1 x(n-1) + b_2 x(n-2) - a_1 y(n-1) - a_2 y(n-2) \qquad (2\text{-}16)$$

Equation (2-16) can be directly implemented in software, where x(n) is the sample input and y(n) is the corresponding filtered digital output. When the filter output is calculated using Equation (2-16), y(n) is calculated using the direct-form implementation of the digital filter.

There are other implementations which can be used for the same system (filter) transfer function, H(z). The canonic-form implementation and the transpose-form implementation are discussed in subsequent sections. First, the direct-form implementation will be applied to the transfer function, H(s), developed in **SECTION 1 INTRODUCTION**.

2.1 DIGITAL LOWPASS FILTER

Using the analog transfer function, H(s), from Figure 1-1 and Equations (2-11) and (2-12), the digital transfer function, H(z), becomes that shown in Figure 2-2, where the coefficients α, β, and γ are expressed in terms of the digital cutoff frequency, θ_C, and the damping factor, d. The value of the transfer function at $\theta = \theta_C$ in the digital domain is identical to the value of the s-domain transfer function at $\Omega = \Omega_C$:

$$H_Z(e^{j\theta_C}) = H_S(j\Omega_C) \qquad (2\text{-}17)$$

As shown in Figure 2-3, the digital gain and phase response calculated from the equations of Figure 2-2 are similar to the analog plots shown in Figure 1-2, except for the asymmetry introduced by the zero at $f_S/2$. That is, the frequency axis is modified so that a gain of zero at $f = \infty$ in the s-domain corresponds to a gain of zero at $f = f_S/2$ in the z-domain. The fact that the magnitude of the transfer functions, H(s) and H(z), is identical once the proper frequency transformation is made is very useful for understanding the digital filter and its relationship to the analog equivalent. This fact is also useful for purposes of scaling the gain since the maximum magnitude of $G_S(\Omega_M) = G_Z(\theta_M)$, where Ω_M and θ_M are related by Equation (2-12). In other words, scaling analysis of the digital transfer function, H(z), can be done in the s-domain (the algebra is often easier to manage). Scaling of the gain is an essential part of digital filter implementation since the region of numeric calculations on fixed-point DSPs such as the DSP56001 are usually restricted to a range of –1 to 1.

Using the formulas given in Figure 2-2 with Equation (1-2) guarantees the behavior of the digital filter. Since the gain is scaled to unity at f = 0 (DC), the input data, x(n), in Figure 2-2 must be scaled down by a factor of $1/G_M$ from Equation (1-2) if the entire dynamic range of the digital network is to be utilized. The alternative procedure is automatic gain control to insure that x(n) is smaller than $1/G_M$ before it arrives at the filter input. For $d \geq \sqrt{2}$, the input does not require scaling since the gain of the filter will never exceed unity.

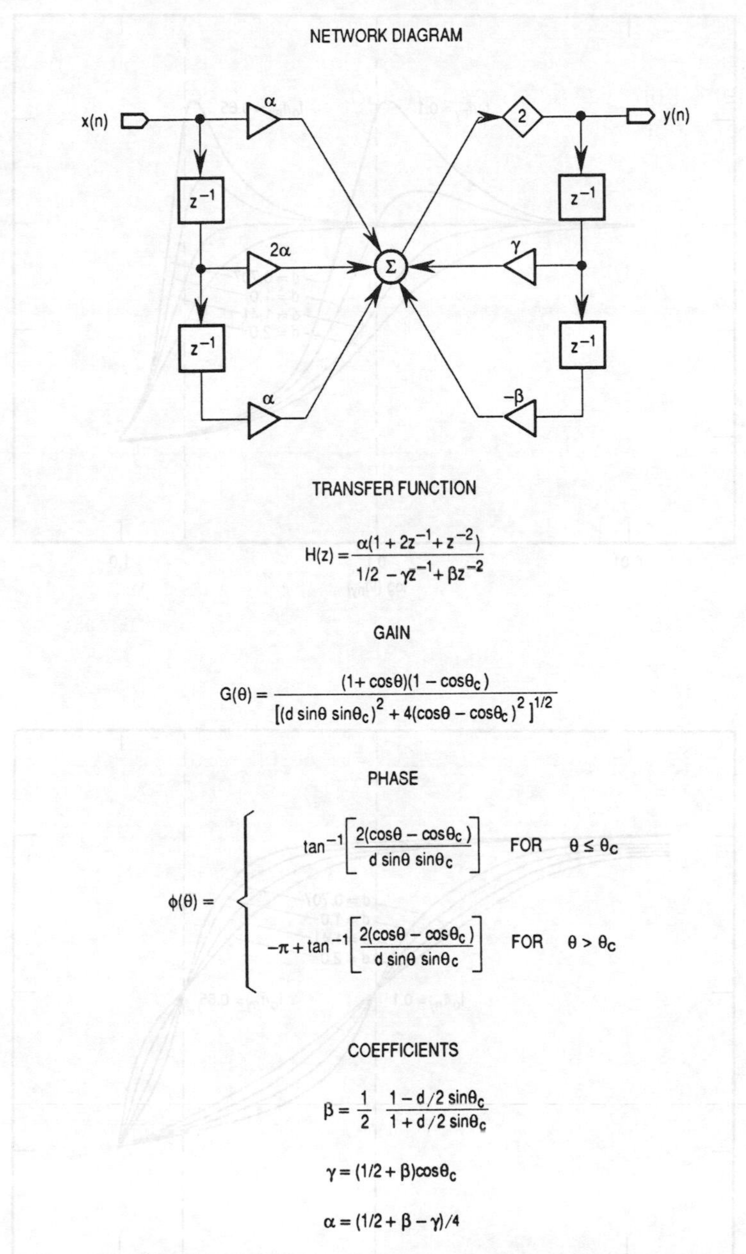

NETWORK DIAGRAM
x(n)
y(n)
α
2α
α
2
γ
−β
z^{-1}
z^{-1}
z^{-1}
z^{-1}
Σ
TRANSFER FUNCTION
$H(z) = \frac{\alpha(1 + 2z^{-1} + z^{-2})}{1/2 - \gamma z^{-1} + \beta z^{-2}}$
GAIN
$G(\theta) = \frac{(1 + \cos\theta)(1 - \cos\theta_c)}{[(d \sin\theta \sin\theta_c)^2 + 4(\cos\theta - \cos\theta_c)^2]^{1/2}}$
PHASE
$\phi(\theta) = \begin{cases} \tan^{-1}\left[\frac{2(\cos\theta - \cos\theta_c)}{d \sin\theta \sin\theta_c}\right] & \text{FOR } \theta \le \theta_c \\ -\pi + \tan^{-1}\left[\frac{2(\cos\theta - \cos\theta_c)}{d \sin\theta \sin\theta_c}\right] & \text{FOR } \theta > \theta_c \end{cases}$
COEFFICIENTS
$\beta = \frac{1}{2} \frac{1 - d/2 \sin\theta_c}{1 + d/2 \sin\theta_c}$
$\gamma = (1/2 + \beta)\cos\theta_c$
$\alpha = (1/2 + \beta - \gamma)/4$

Figure 2-2. Direct-Form Implementation of Second-Order Lowpass IIR Filter and Analytical Formulas Relating Desired Response to Filter Coefficients

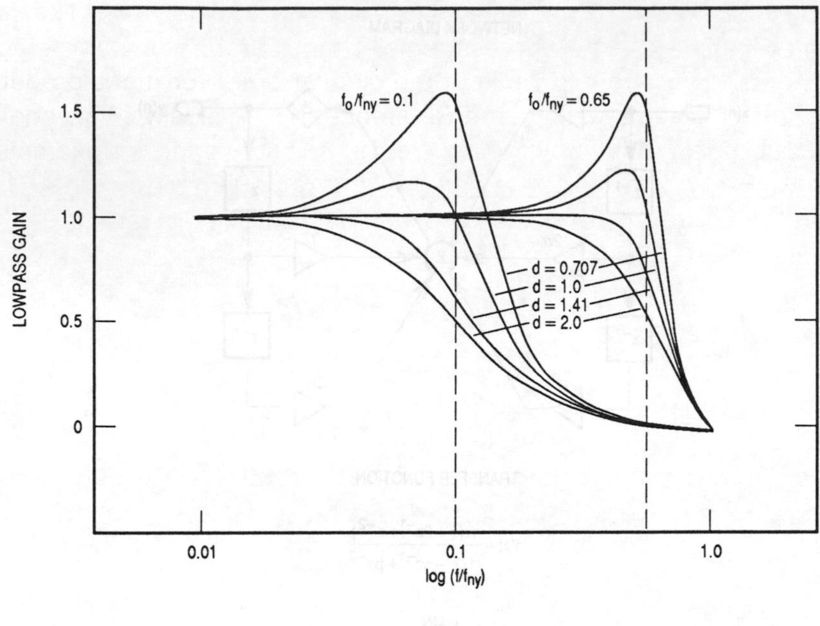

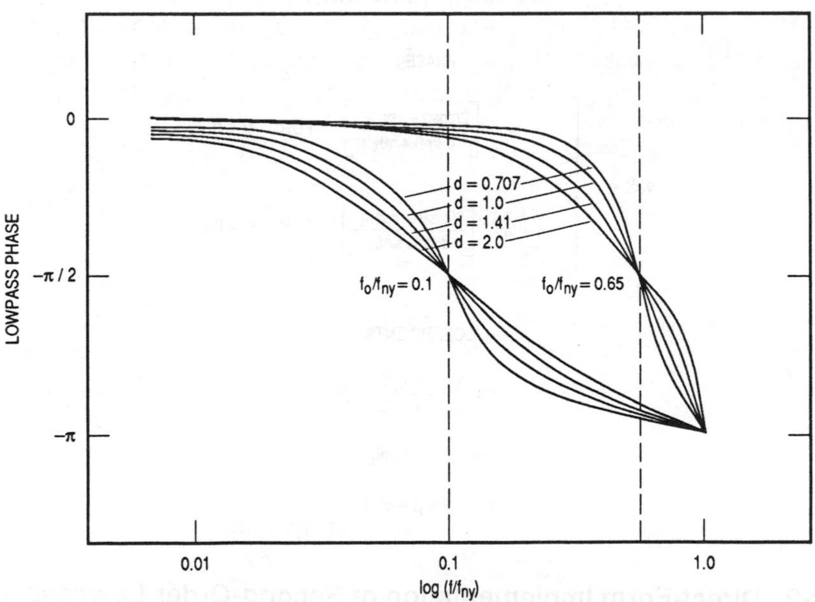

NOTE: Nyquist frequency, f_{ny}, is equal to one-half the sample frequency, f_s.

Figure 2-3. Gain and Phase Response of Second-Order Lowpass IIR Filter

The DSP56001 code to implement the second-order lowpass filter section is shown in Figure 2-4. The address register modifiers are initially set to M0 = 4, M4 = 1, and M5 = 1 to allow use of the circular buffer or modulo addressing in this particular implementation (see Reference 8). Typically, this code would be an interrupt routine driven by the input data (A/D converter, for example) sample rate clock. The basic filter code and the interrupt overhead and data I/O moves for this second-order filter could be executed at a sample rate of nearly 1 MHz on the DSP56001.

2.2 DIGITAL HIGHPASS FILTER

The highpass filter is nearly identical to the lowpass filter as shown by the formulas in Figure 2-5. As with the analog case, the digital highpass filter is just the mirror image of the lowpass filter (see Figure 2-6). The frequency transformation from high to low in the analog case is $\Omega \rightarrow 1/\Omega$; whereas, in the digital case, it is $\theta \rightarrow \pi - \theta$.

The DSP56001 code is shown in Figure 2-7; as seen by comparison to the code shown in Figure 2-4, the same instruction sequence is used. The only difference is the coefficient data, which is calculated by the formulas given in Figure 2-5. The scaling mode is turned on so that a move from the A or B accumulator to the X or Y register or to memory results in an automatic multiply by two. The scaling mode is used not only in the code for the lowpass case but also in the code for the highpass, bandstop, and bandpass cases.

2.3 DIGITAL BANDSTOP FILTER

The formulas and network diagram for the digital bandstop filter are presented in Figure 2-8. The DSP56001 code from Figure 2-9 is identical to that for the lowpass and highpass cases except for the coefficient data calculated from the equations of Figure 2-8. Scaling of this filter is not a problem for the single-section case since the gain from the equation in Figure 2-8 never exceeds unity (as is true in the analog case as seen by the gain equation from Figure 1-5). Figure 2-10 is the calculated gain and phase of the digital filter, which should compare to the response curves of the equivalent analog filter plotted in Figure 1-6.

DIFFERENCE EQUATION:

$$y(n)=2\{\alpha[x(n)+2x(n-1)+x(n-2)]+\gamma y(n-1)-\beta y(n-2)\}$$

DATA STRUCTURES:

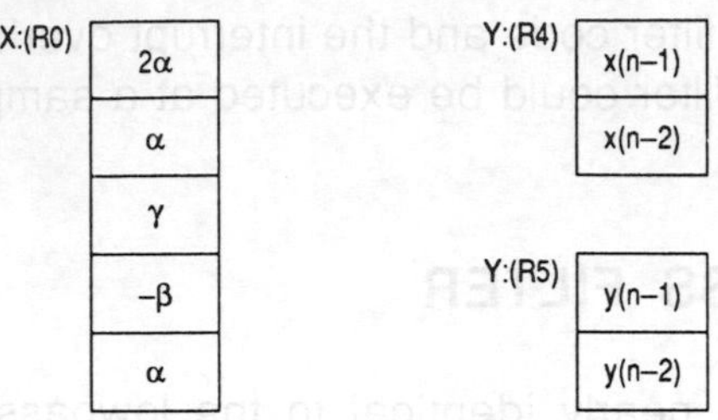

DSP56001 CODE:

```
                                                  ;Y1=x(n) (Input)
                                                  ;X0=α
MPY    X0,Y1,A    X:(R0)+,X0    Y:(R4)+,Y0        ;A=αx(n)
MAC    X0,Y0,A    X:(R0)+,X0    Y:(R4),Y0         ;A=A+2αx(n−1)
MAC    X0,Y0,A    X:(R0)+,X0    Y:(R5)+,Y0        ;A=A+αx(n−2)
MAC    X0,Y0,A    X:(R0)+,Y0    Y:(R5),Y0         ;A=A+γy(n−1)
MAC    X0,Y0,A    X:(R0)+,X0    Y1,Y:(R4)         ;A=A−βy(n−2)
MOVE              A,X1          A,Y:(R5)          ;y(n)=2A (assumes scaling
                                                  ;mode is set). X1 is Output.
```

TOTAL INSTRUCTION CYCLES:

6 Icyc @ 20 MHz = 600 ns

Figure 2-4. DSP56001 Code and Data Structures for Second-Order Direct-Form Implementation of a Lowpass IIR Filter

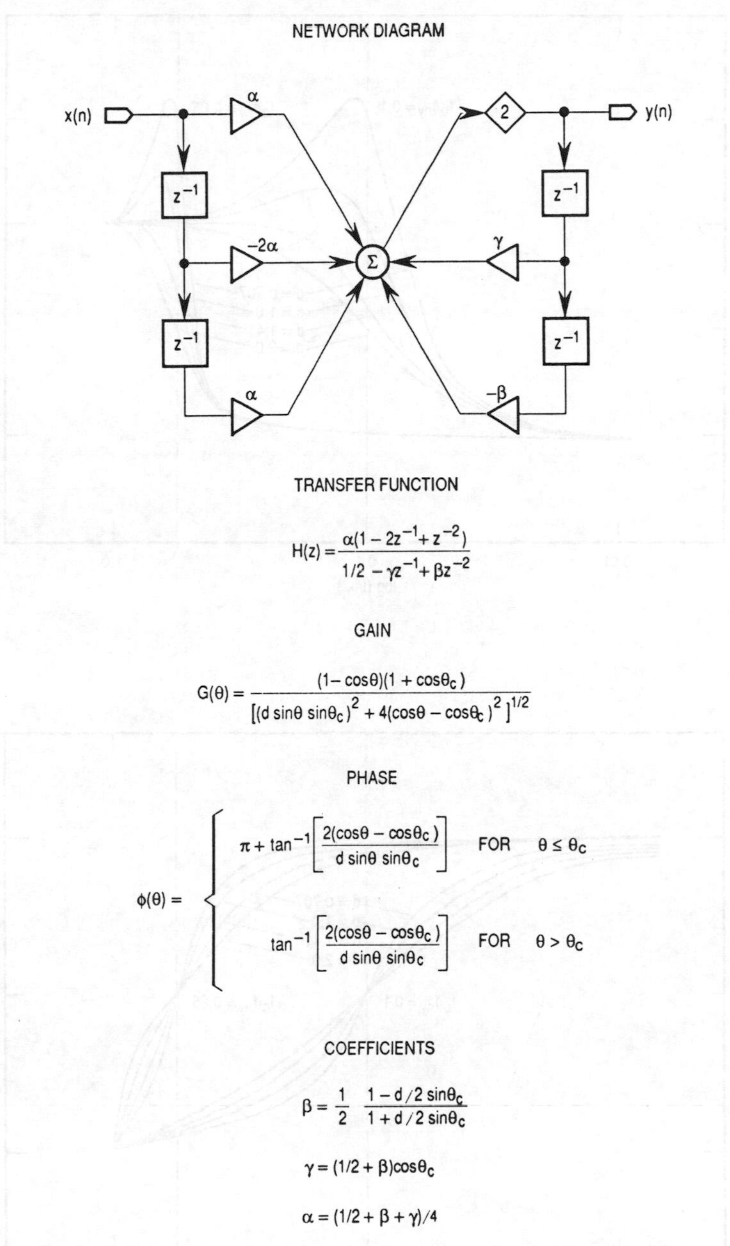

Figure 2-5. Direct-Form Implementation of Second-Order Highpass IIR Filter and Analytical Formulas Relating Desired Response to Filter Coefficients

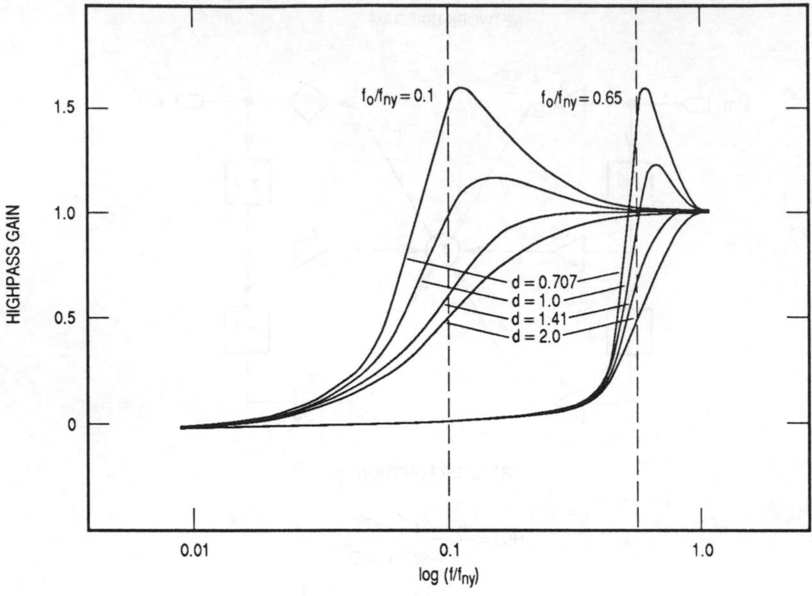

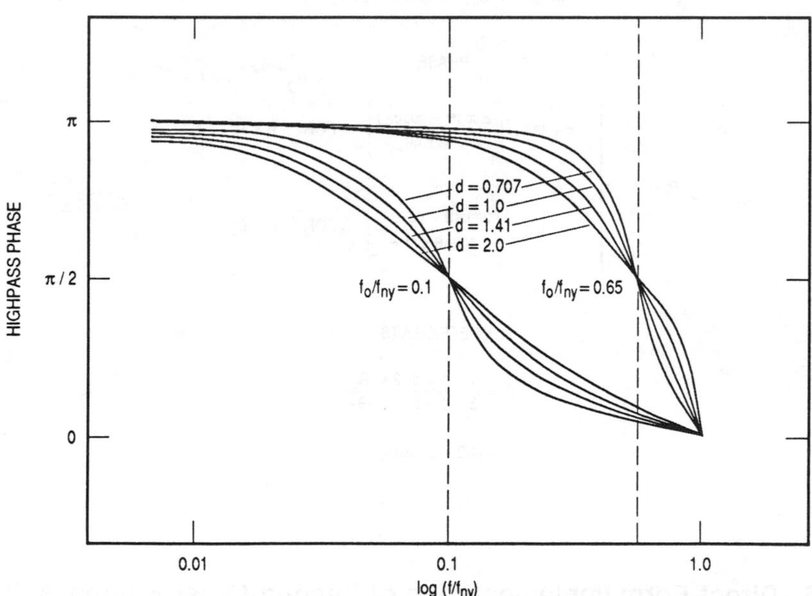

NOTE: Nyquist frequency, f_{ny}, is equal to one-half the sample frequency, f_s.

Figure 2-6. Gain and Phase Response of Second-Order Highpass IIR Filter

DIFFERENCE EQUATION:

$$y(n)=2\{\alpha[x(n)-2x(n-1)+x(n-2)]+\gamma y(n-1)-\beta y(n-2)\}$$

DATA STRUCTURES:

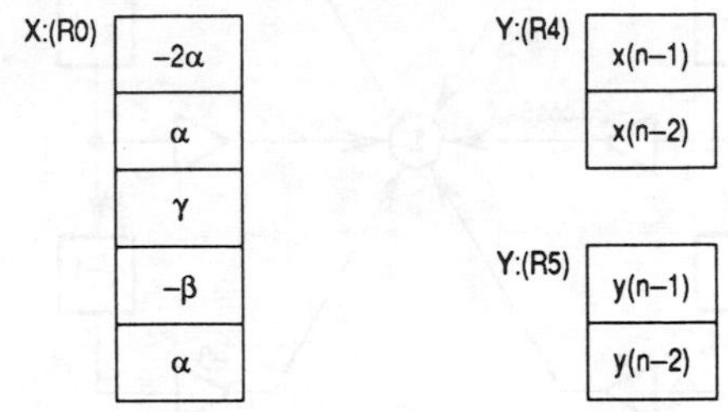

DSP56001 CODE:

```
                                                   ;Y1=x(n) (Input)
                                                   ;X0=α
MPY     X0,Y1,A   X:(R0)+,X0   Y:(R4)+,Y0          ;A=αx(n)
MAC     X0,Y0,A   X:(R0)+,X0   Y:(R4),Y0           ;A=A–2αx(n–1)
MAC     X0,Y0,A   X:(R0)+,X0   Y:(R5)+,Y0          ;A=A+αx(n–2)
MAC     X0,Y0,A   X:(R0)+,Y0   Y:(R5),Y0           ;A=A+γy(n–1)
MAC     X0,Y0,A   X:(R0)+,X0   Y1,Y:(R4)           ;A=A–βy(n–2)
MOVE              A,X1         A,Y:(R5)            ;y(n)=2A (assumes scaling
                                                   ;mode is set). X1 is Output.
```

TOTAL INSTRUCTION CYCLES:

6 Icyc @ 20 MHz = 600 ns

Figure 2-7. DSP56001 Code and Data Structures for Second-Order Direct-Form Implementation of a Highpass IIR Filter

NETWORK DIAGRAM

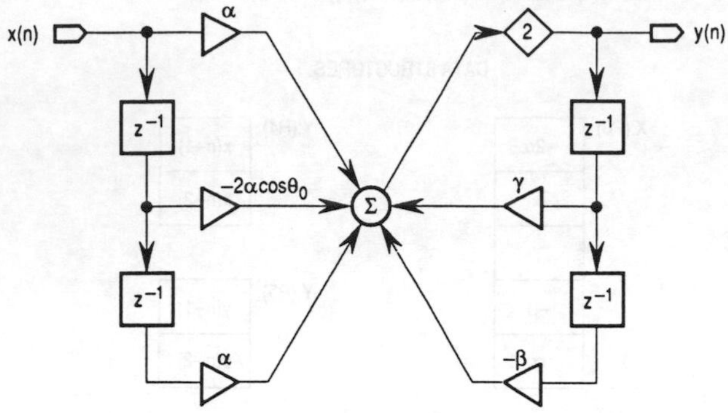

TRANSFER FUNCTION

$$H(z) = \frac{\alpha(1 - 2\cos\theta_0\, z^{-1} + z^{-2})}{1/2 - \gamma z^{-1} + \beta z^{-2}}$$

GAIN

$$G(\theta) = \frac{|\cos\theta - \cos\theta_0|}{[(d\sin\theta\sin\theta_0)^2 + 4(\cos\theta - \cos\theta_0)^2]^{1/2}}$$

PHASE

$$\phi(\theta) = \tan^{-1}\left[\frac{2(\cos\theta - \cos\theta_0)}{d\sin\theta\sin\theta_0}\right]$$

COEFFICIENTS

$$d = \frac{2\tan(\theta_0 / 2Q)}{\sin\theta_0}$$

$$\beta = \frac{1}{2}\;\frac{1 - \tan(\theta_0 / 2Q)}{1 + \tan(\theta_0 / 2Q)}$$

$$\gamma = (1/2 + \beta)\cos\theta_0$$

$$\alpha = (1/2 + \beta) / 2$$

Figure 2-8. Direct-Form Implementation of Second-Order Bandstop IIR Filter and Analytical Formulas Relating Desired Response to Filter Coefficients

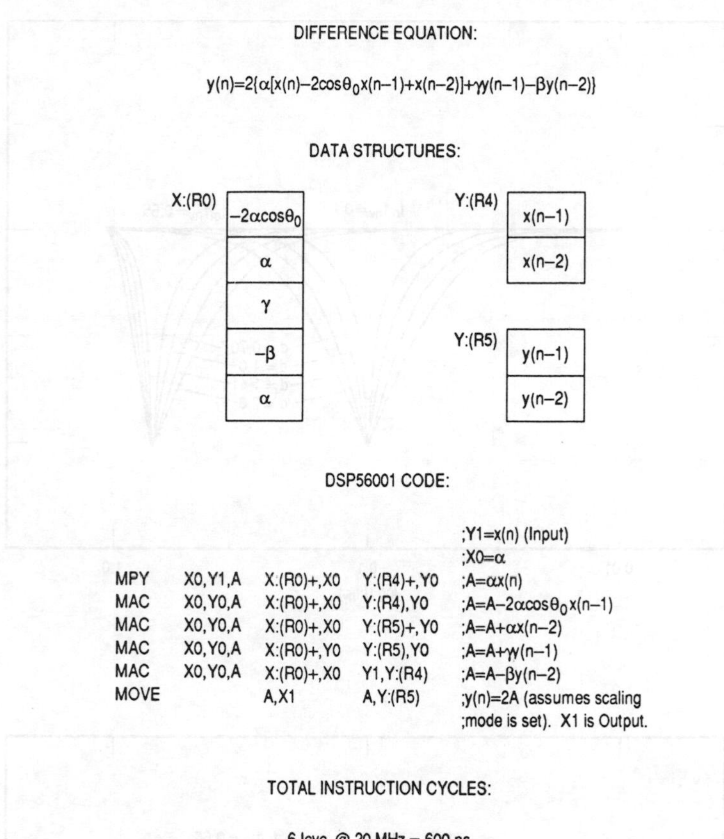
DIFFERENCE EQUATION:
y(n)=2{α[x(n)–2cosθ0x(n–1)+x(n–2)]+γy(n–1)–βy(n–2)}
DATA STRUCTURES:
X:(R0)
–2αcosθ0
α
γ
–β
α
Y:(R4)
x(n–1)
x(n–2)
Y:(R5)
y(n–1)
y(n–2)
DSP56001 CODE:
;Y1=x(n) (Input)
;X0=α
MPY X0,Y1,A X:(R0)+,X0 Y:(R4)+,Y0 ;A=αx(n)
MAC X0,Y0,A X:(R0)+,X0 Y:(R4),Y0 ;A=A–2αcosθ0x(n–1)
MAC X0,Y0,A X:(R0)+,X0 Y:(R5)+,Y0 ;A=A+αx(n–2)
MAC X0,Y0,A X:(R0)+,Y0 Y:(R5),Y0 ;A=A+γy(n–1)
MAC X0,Y0,A X:(R0)+,X0 Y1,Y:(R4) ;A=A–βy(n–2)
MOVE A,X1 A,Y:(R5) ;y(n)=2A (assumes scaling
;mode is set). X1 is Output.
TOTAL INSTRUCTION CYCLES:
6 Icyc @ 20 MHz = 600 ns

Figure 2-9. DSP56001 Code and Data Structures for Second-Order Direct-Form Implementation of a Bandstop IIR Filter

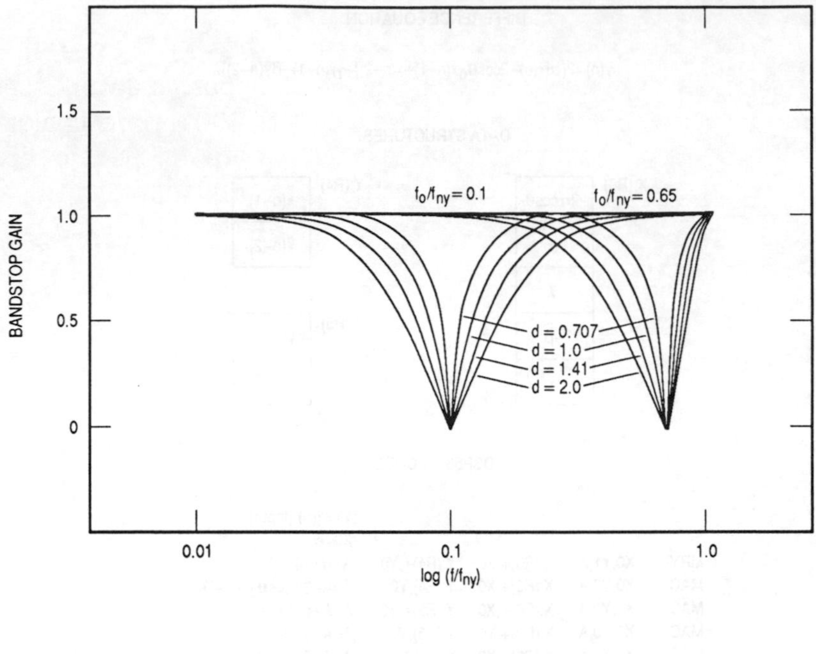

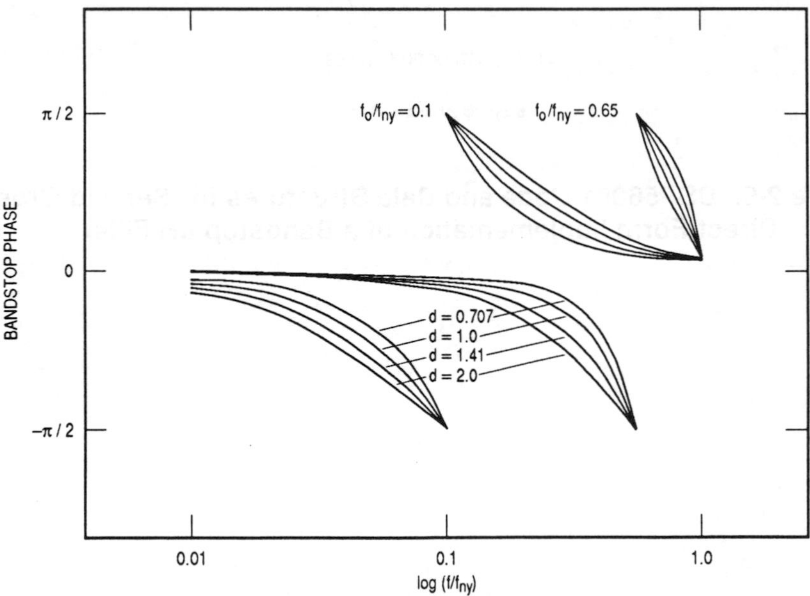

NOTE: Nyquist frequency, f_{ny}, is equal to one-half the sample frequency, f_s.

Figure 2-10. Gain and Phase Response of Second-Order Bandstop IIR Filter

2.4 DIGITAL BANDPASS FILTER

Because there is one less coefficient in the bandpass network (see Figure 2-11), one instruction can be saved in the DSP56001 code implementation shown in Figure 2-12. Otherwise, the instructions are identical to those in the other three filter routines. Like the second-order bandstop network, the maximum response at the center frequency, θ_o, is unity for any value of Q so that scaling need not be considered in the implementation of a single-section bandpass filter. This is true when the formulas for α, β, and γ from Figure 2-11 are used in the direct-form implementation in Figure 2-12. Figure 2-13 is the calculated gain and phase of the digital filter, which should compare to the response curves of the equivalent analog filter plotted in Figure 1-8.

2.5 SUMMARY OF DIGITAL COEFFICIENTS

Figure 2-14 gives a summary of the coefficient values for the four basic filter types. Note that the coefficient β has the same form for all four filter types and that it can only assume values between 0 and 1/2 for practical filters. β is bounded by 1/2 because Q (or d) and θ_o are not independent. For $Q >> 1$, $\beta \rightarrow 1/2$; whereas, for $\theta_o = f_s/4$ and $Q = 1/2$, $\beta \rightarrow 0$. These properties are independent of the form of implementation; they are only dependent on the form of the transfer function. Alternate implementations (difference equations) will be described in the following sections.

Note that the Q described in Figure 2-14 meets the traditional requirements (i.e., Q is the ratio of the bandwidth at the –3 dB points divided by the center frequency). The formula for β can be modified in the case of the bandpass or bandstop filter by replacing the damping coefficient, d, with the formula for Q. When the coefficients are described in this manner, a constant Q filter results. When the bandwidth is any function of center frequency, this relationship between d and Q makes it impossible to implement a bandpass or bandstop filter by replacing Q with the desired function of bandwidth and center frequency.

Figure 2-15 shows the relationship between the pole of the second-order section and the center frequency. Note that the pole is on the real axis for $d > 2$, where d is also constrained by $d < 2/\sin \theta_o$.

NETWORK DIAGRAM

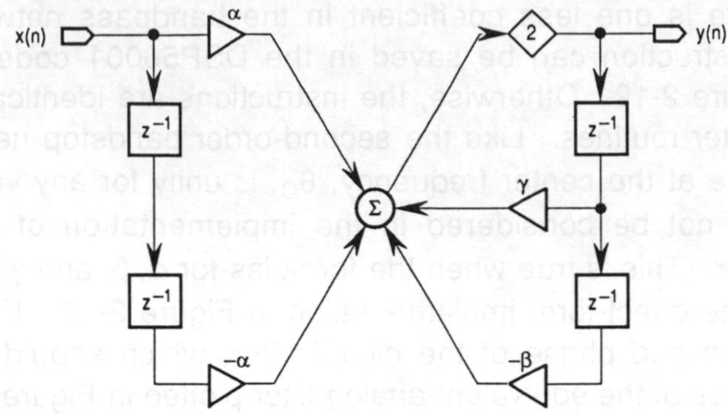

TRANSFER FUNCTION

$$H(z) = \frac{\alpha(1 - z^{-2})}{1/2 - \gamma z^{-1} + \beta z^{-2}}$$

GAIN

$$G(\theta) = \frac{d \sin\theta_0 \sin\theta}{\left[(d \sin\theta \sin\theta_0)^2 + 4(\cos\theta - \cos\theta_0)^2\right]^{1/2}}$$

PHASE

$$\phi(\theta) = -\tan^{-1}\left[\frac{2(\cos\theta - \cos\theta_0)}{d \sin\theta \sin\theta_0}\right]$$

COEFFICIENTS

$$d = \frac{2\tan(\theta_0 / 2Q)}{\sin\theta_0}$$

$$\beta = \frac{1}{2}\,\frac{1 - \tan(\theta_0 / 2Q)}{1 + \tan(\theta_0 / 2Q)}$$

$$\gamma = (1/2 + \beta)\cos\theta_0$$

$$\alpha = (1/2 - \beta) / 2$$

Figure 2-11. Direct-Form Implementation of Second-Order Bandpass IIR Filter and Analytical Formulas Relating Desired Response to Filter Coefficients

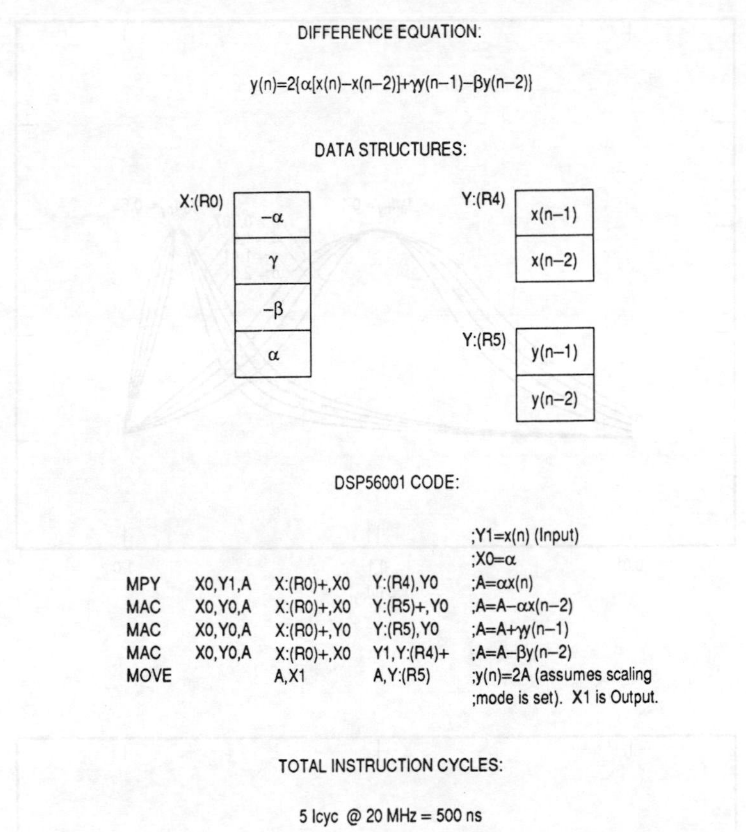
DIFFERENCE EQUATION:
y(n)=2{α[x(n)−x(n−2)]+γy(n−1)−βy(n−2)}
DATA STRUCTURES:
X:(R0)
−α
γ
−β
α
Y:(R4)
x(n−1)
x(n−2)
Y:(R5)
y(n−1)
y(n−2)
DSP56001 CODE:
;Y1=x(n) (Input)
;X0=α
MPY X0,Y1,A X:(R0)+,X0 Y:(R4),Y0 ;A=αx(n)
MAC X0,Y0,A X:(R0)+,X0 Y:(R5)+,Y0 ;A=A−αx(n−2)
MAC X0,Y0,A X:(R0)+,Y0 Y:(R5),Y0 ;A=A+γy(n−1)
MAC X0,Y0,A X:(R0)+,X0 Y1,Y:(R4)+ ;A=A−βy(n−2)
MOVE A,X1 A,Y:(R5) ;y(n)=2A (assumes scaling
;mode is set). X1 is Output.
TOTAL INSTRUCTION CYCLES:
5 Icyc @ 20 MHz = 500 ns

Figure 2-12. DSP56001 Code and Data Structures for Second-Order Direct-Form Implementation of a Bandpass IIR Filter

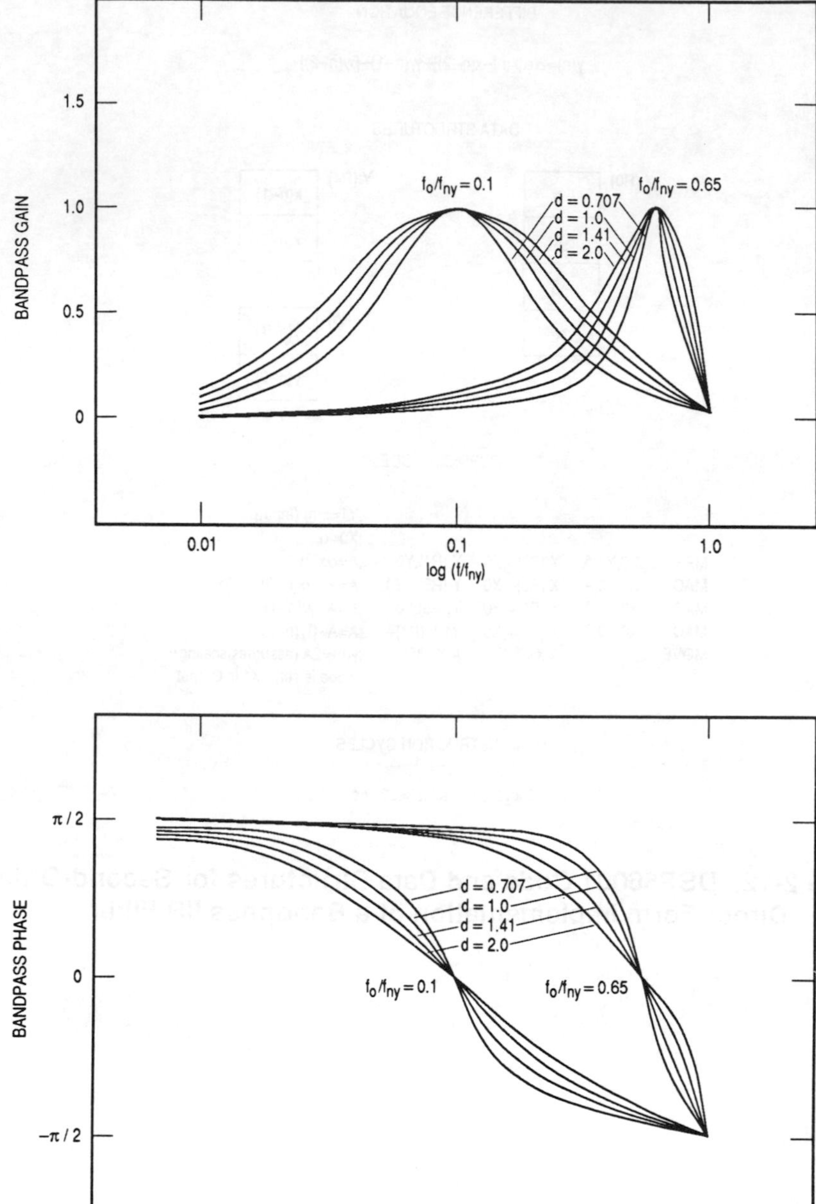

NOTE: Nyquist frequency, f_{ny}, is equal to one-half the sample frequency, f_s.

Figure 2-13. Gain and Phase Response of Second-Order Bandpass IIR Filter

z-DOMAIN TRANSFER FUNCTION

$$H(z) = \frac{\alpha(1 + \mu z^{-1} + \sigma z^{-2})}{1/2 - \gamma z^{-1} + \beta z^{-2}}$$

DIFFERENCE EQUATION (DIRECT FORM)

$$y(n) = 2\{\alpha[x(n) + \mu x(n-1) + \sigma x(n-2)] + \gamma y(n-1) - \beta y(n-2)\}$$

COEFFICIENTS

$$\beta = 1/2\left[\frac{1 - 1/2d\sin\theta_0}{1 + 1/2d\sin\theta_0}\right] \qquad d = \frac{2\tan(\theta_0/2Q)}{\sin\theta_0} \qquad \gamma = (1/2 + \beta)\cos\theta_0$$

where $0 < \beta < 1/2$ and

$$Q = \frac{\theta_0}{\Delta\theta} = \frac{2\pi f_0 / f_s}{2\pi(f_2 - f_1)/f_s} = \frac{f_0}{f_2 - f_1}$$

where f_0 is the center frequency of the bandpass or bandstop filter, f_1 and f_2 are the half-power points (where gain is equal to $1/\sqrt{2}$), and f_s is the sample frequency. Note that f_0 is replaced with f_c in the lowpass and highpass cases.

Numerator Coefficients

Type	α	μ	σ	Unity Gain at
Lowpass	$(1/2 + \beta - \gamma)/4$	2	1	$f = 0$
Highpass	$(1/2 + \beta + \gamma)/4$	−2	1	$f = f_s/2$
Bandpass	$(1/2 - \beta)/2$	0	−1	$f = f_0$
Bandstop	$(1/2 + \beta)/2$	$-2\cos\theta_0$	1	$f = 0$ and $f = f_s/2$

NOTE: $\theta_0 = 2\pi f_0 / f_s$.

Figure 2-14. Summary of Digital Coefficients for the Four Basic Filter Types

POLE EQUATION OF H(z)

$$
\begin{aligned}
Z_p &= r\cos\theta_p + j\,r\sin\theta_p \\
&= \gamma \pm j\sqrt{2\beta - \gamma^2} \\
&= \frac{\cos\theta_0 \pm j\sin\theta_0\sqrt{1-(1/2d)^2}}{1+\frac{1}{2}d\sin\theta_0} \qquad \text{FOR } d < 2
\end{aligned}
$$

WHERE $\beta = \frac{1}{2}(2 - d\sin\theta_0)\,/\,(2 + d\sin\theta_0)$ and $\gamma = (1/2 + \beta)\cos\theta_0$. DISTANCE FROM ORIGIN TO POLE IS $|Z_p| = \sqrt{2\beta}$.

FOR d > 2

$$
Z_p = \gamma - \sqrt{\gamma^2 - 2\beta}
$$

$$
= \frac{\cos\theta_0 - \sin\theta_0\sqrt{(1/2d)^2 - 1}}{1+\frac{1}{2}d\sin\theta_0}
$$

WHERE $\theta_p = 0$. TO SATISFY REQUIREMENT $0 < \beta < 1/2$ RESULTS IN $\frac{1}{2}d\sin\theta_0 < 1$.

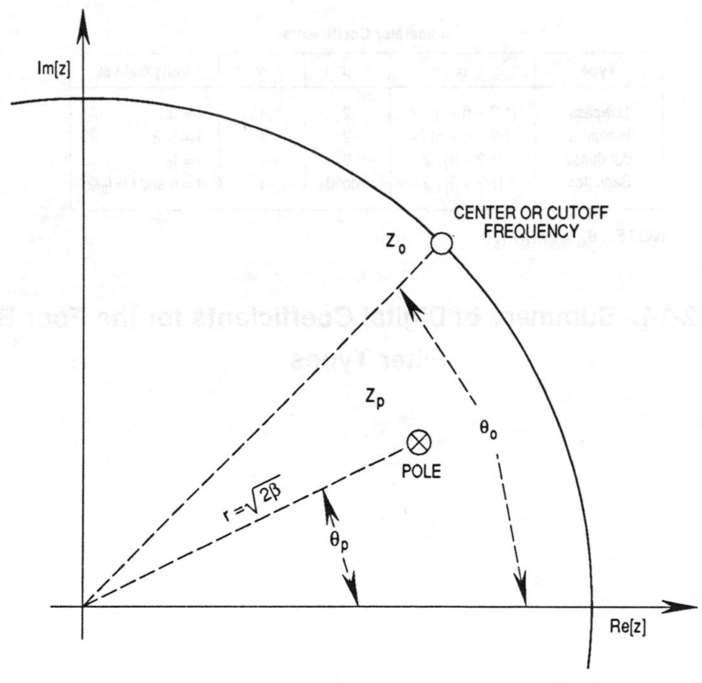

Figure 2-15. Pole Location and Analysis of Second-Order Section

SECTION 3
SINGLE-SECTION CANONIC FORM (DIRECT FORM II)

The single-section canonic form network is discussed in the following paragraphs.

3.1 THE CANONIC-FORM DIFFERENCE EQUATION

The direct-form difference equation, rewritten from Equation (2-16), is

$$y(n) = \sum_{i=0}^{2} b_i\, x(n-i) - \sum_{j=1}^{2} a_j\, y(n-j) \tag{3-1}$$

Equation (3-1) can be represented by the diagram of Figure 3-1(a). This diagram is the same as those shown in Figures 2-2, 2-5, 2-8, and 2-11, except that the summations have been separated to highlight the correspondence with Equation (3-1). From this diagram, it is clear that the direct-form implementation requires four delay elements or, equivalently, four internal memory locations.

The diagram of Figure 3-1(b) represents the same transfer function as that implemented by Equation (3-1), but now the delay variable is w(n). Comparison of this network with that of the direct-form network of Figure 3-1(a) shows that interchanging the order of the left and right halves does not change the overall system response (see Reference 11). The delay elements can then be collapsed to produce the final canonic-form network shown in Figure 3-1(c). As a result, the memory requirements for the system are reduced to the minimum (two locations); therefore, this realization of the IIR filter is often referred to as the canonic form. The system difference equations for the canonic realization are

$$y(n) = \sum_{i=0}^{2} b_i\; w(n-i) \tag{3-2(a)}$$

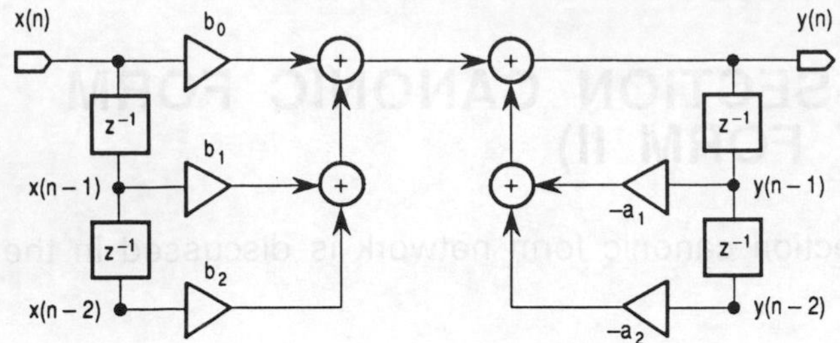

(a) Direct-Form Network Diagram of Equation (3-1)

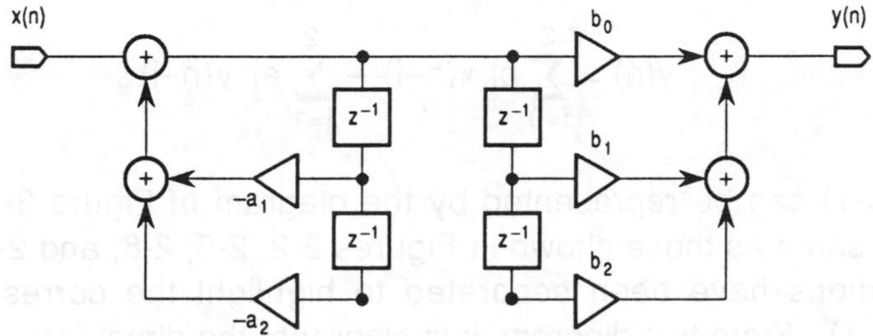

(b) Interchanged Left and Right Halves of Direct-Form Network

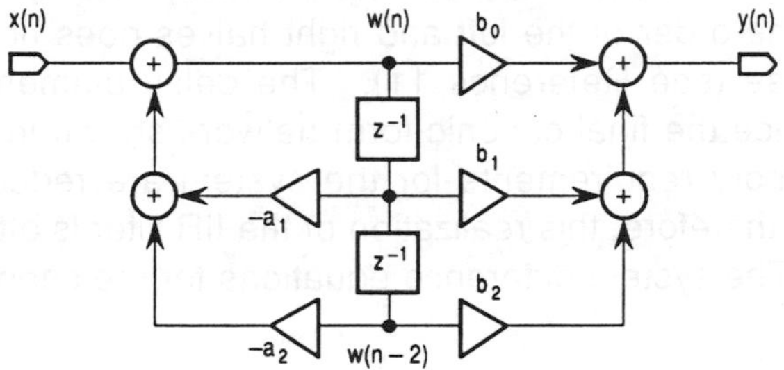

(c) Collapsing Delay Terms from Figure 3-1(a) Resulting in Canonic-Form Network Diagram

Figure 3-1. The Second-Order Canonic (Direct Form II) IIR Filter Network

where

$$w(n) = x(n) - \sum_{j=1}^{2} a_j \, w(n-j) \qquad (3\text{-}2(b))$$

To prove that Equations (3-2(a)) and (3-2(b)) are equivalent to Equation (3-1), the following procedure can be used. First, combine Equations (3-2(a)) and (3-2(b)):

$$y(n) = \sum_{i=0}^{2} b_i \left[x(n-i) - \sum_{j=1}^{2} a_j \, w(n-j-i) \right]$$

$$= \sum_{i=0}^{2} b_i \, x(n-i) - \sum_{j=1}^{2} a_j \sum_{i=0}^{2} b_i \, w(n-i-j)$$

$$= \sum_{i=0}^{2} b_i \, x(n-i) - \sum_{j=1}^{2} a_j \, y(n-j) \qquad (3\text{-}3)$$

The last step uses the definition for y(n) from Equation (3-2(a)). The result is exactly equivalent to Equation (3-1), the direct-form difference equation. To utilize the scaling mode on the DSP56001, it is advantageous to write Equations (3-2(a)) and (3-2(b)) as follows:

$$y(n) = 2\left\{\frac{1}{2}\,w(n) + \frac{\mu}{2}\,w(n-1) + \frac{\alpha}{2}\,w(n-2)\right\} \qquad (3\text{-}4(a))$$

and

$$w(n) = 2\{\alpha x(n) + \gamma w(n-1) - \beta w(n-2)\} \qquad (3\text{-}4(b))$$

where the coefficients have been substituted for the a_j and b_i.

From these equations, it can be seen that both y(n) and w(n) depend on a sum of products and therefore are the output of an accumulator. The accumulators on the DSP56001 have eight extension bits; thus, the sum can exceed unity by 255 without overflowing. However, when the contents of a 56-bit accumulator are transferred to a 24-bit register or memory location (i.e., delay element), an overflow error may occur. The digital filter designer must insure that only values less than unity are stored. Of course, even if memory had the precision of the accumulators, overflow can still occur if the accumulator itself experiences overflow. However, intermediate sums are allowed to exceed the capacity of the accumulator if the final result can be represented in the accumulator (a result of the circular nature of twos-complement arithmetic).

To insure that overflow does not occur, it is necessary to calculate the gain at the internal nodes (accumulator output) of a filter network. Although canonic realization is susceptible to overflow, it is advantageous because of the minimum storage requirements and implementation in fewer instruction cycles.

3.2 ANALYSIS OF INTERNAL NODE GAIN

Calculating the gain at any internal node is no different than calculating the total network gain. In Figure 3-2, the transfer function of node w(n) is $H_W(z)$. $H_W(z)$ represents the transfer function from the input node, x(n), to the internal node, w(n). The gain, $G_W(\theta)$, at w(n) is found in the standard manner by evaluating the magnitude of $H_W(z)$ as shown in Figure 3-3. Note that the flow diagram of Figure 3-2 uses the automatic scaling mode (multiply by a factor of two when transferring data from the accumulator to memory) so that the coefficients are by definition less than one.

The peak gain, g_o, of $G_W(\theta)$ is found by taking the derivative of $G_W(\theta)$ with respect to θ and setting the result equal to zero. A simple expression for g_o is derived as follows:

$$g_o = \frac{\alpha}{\left(\frac{1}{2} - \beta\right)\sin\theta_p} \quad \text{for} \quad \left[\frac{\left(\frac{1}{2} + \beta\right)^2}{2\beta}\right]\cos\theta_o \le 1 \tag{3-5}$$

where θ_p is the angle (see Figure 2-15) to the pole of the filter. Since this result does not depend on the numerator of the filter transfer function, H(z), the result is valid for all four basic filter types. However, as shown by the example given for a bandpass network in Figure 3-3, g_o may exceed unity by a large amount, especially for filters having poles at frequencies much less than $f_s/2$ where $\sin\theta_p << 1$. To compensate, the input must be scaled down by an amount equal to $1/g_o$ or guaranteed not to exceed $1/g_o$ before arriving at the filter. This scaling aspect of the canonic-form network is a disadvantage, but this network has the advantage of being implemented in one less instruction than the other filter realizations for the lowpass, highpass, and bandstop filters; of course, less memory is required since only two intermediate variables are stored.

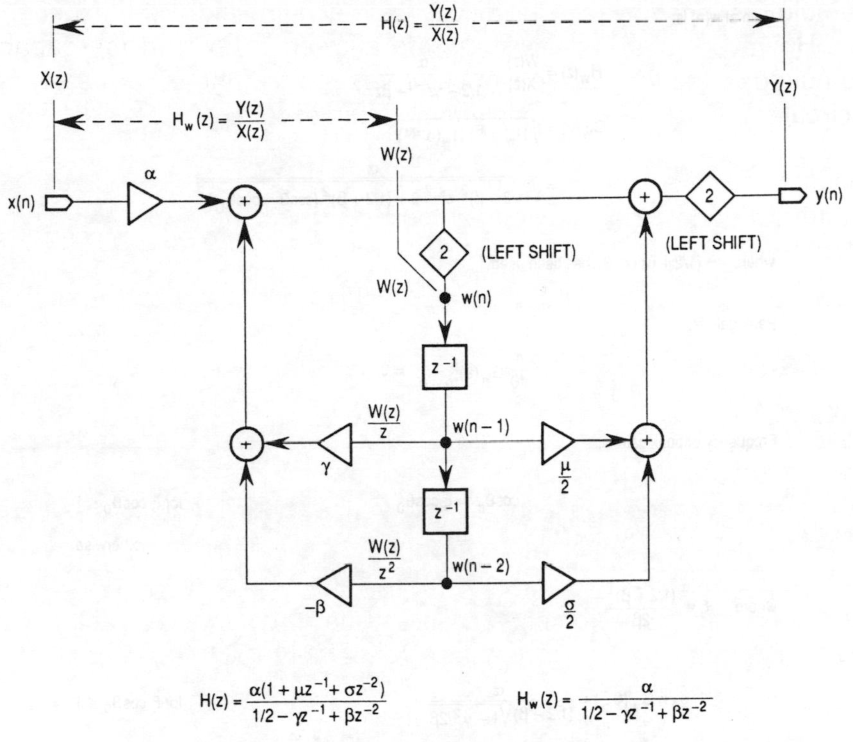

$$H(z) = \frac{\alpha(1 + \mu z^{-1} + \sigma z^{-2})}{1/2 - \gamma z^{-1} + \beta z^{-2}} \qquad H_w(z) = \frac{\alpha}{1/2 - \gamma z^{-1} + \beta z^{-2}}$$

Figure 3-2. Internal Node Transfer Function, $H_W(z)$, of Canonic (Direct Form II) Network

Gain at w(n) is:

$$H_w(z) = \frac{W(z)}{X(z)} = \frac{\alpha}{1/2 - \gamma z^{-1} + \beta z^{-2}}$$

$$G_w(\theta) = \sqrt{H_w(e^{j\theta})\, H_w(e^{-j\theta})}$$

$$= \frac{\alpha}{\sqrt{(1/2-\beta)^2 \sin^2\theta + (1/2+\beta)^2 (\cos\theta - \cos\theta_0)^2}}$$

where $\gamma = (1/2 + \beta)\cos\theta_0$ has been used.

Peak gain is:

$$\frac{d}{d\theta} G_w(\theta)\Big|_{\theta = \theta_m} = 0$$

Frequency of peak gain is:

$$\cos\theta_m = \xi \cos\theta_0 \qquad \text{for } \xi \cos\theta_0 \le 1$$

$$= 1 \qquad \text{otherwise}$$

where $\xi \equiv \frac{(1/2+\beta)^2}{2\beta}$

$$g_0 \equiv G_w(\theta_m) = \frac{\alpha}{(1/2-\beta)\sqrt{1-\gamma^2/2\beta}} \qquad \text{for } \xi \cos\theta_0 \le 1$$

$$= \frac{\alpha}{(1/2+\beta)(1-\cos\theta_0)} \qquad \text{otherwise}$$

EXAMPLE: MAXIMUM INTERNAL NODE GAIN FOR BANDPASS FILTER

$$g_0 = \frac{1}{2\sin\theta_p} \qquad \text{for } \xi \cos\theta_0 \le 1$$

where $\gamma^2 = 2\beta \cos^2\theta_p$ has been used and where $\alpha = (1/2 - \beta)/2$ for a band-pass filter. If $\sin\theta_p > 1/2$, then $g_0 < 1$; otherwise, an overflow (i.e., $g_0 > 1$) may occur at the internal node, w(n), unless the input is scaled down by $1/g_0$.

Figure 3-3. Internal Node Gain Analysis of Second-Order Canonic Form

In general, the behavior of systems at internal nodes can be unexpected. For instance, it is interesting to note that the frequency at which the gain at the internal node of the canonic IIR filter section peaks is not the same as the frequency at which the gain of the filter peaks as given by the poles of the filter. From Figure 2-15, this behavior is expressed by

$$\frac{\sqrt{2\beta}}{\frac{1}{2}+\beta}=\frac{\cos\theta_o}{\cos\theta_p} \tag{3-6}$$

The canonic (direct form II) network has tradeoffs that must be carefully understood and analyzed for the particular application.

3.3 IMPLEMENTATION ON THE DSP56001

Figure 3-4 shows the DSP56001 code and data structures for implementation of the single-section canonic-form network. Note that the modifier register M4 is set equal to 4 to allow circular operation for addressing coefficient data. M0 is set to FFFF to turn off circular addressing for w(n–1) and w(n–2).

DIFFERENCE EQUATION:

$$y(n) = 2\left(\frac{1}{2}w(n) + \frac{\mu}{2}w(n-1) + \frac{\sigma}{2}w(n-2)\right)$$
$$w(n) = 2\left(\alpha\, x(n) + \gamma\, w(n-1) - \beta\, w(n-2)\right)$$

DATA STRUCTURES:

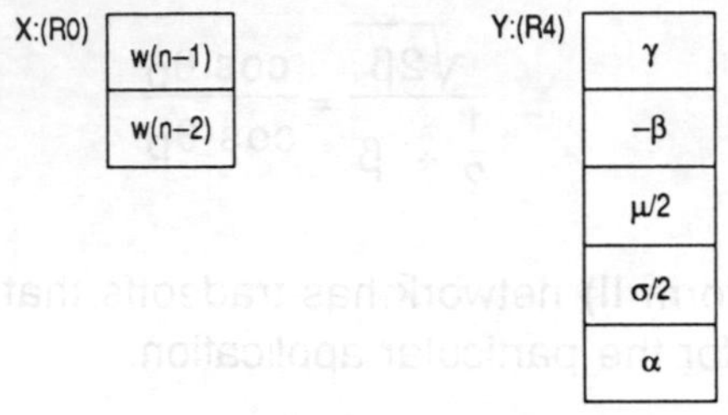

DSP56001 CODE:

```
                                                ;Y1=x(n) (Input)
                                                ;X0=α
MPY    X0,Y1,A   X:(R0)+,X0    Y:(R4)+,Y0       ;A=αx(n)
MAC    X0,Y0,A   X:(R0),X1     Y:(R4)+,Y0       ;A=A+γw(n–1)
MACR   X1,Y0,A   X0,X:(R0)–    Y:(R4)+,Y0       ;A=A–βw(n–2)
MAC    X0,Y0,A   A,X:(R0)      Y:(R4)+,Y0       ;A=1/2w(n)+μ/2w(n–1)
MACR   X1,Y0,A   A,X0          Y:(R4)+,Y0       ;A=A+σ/2w(n–2).  X0=2A
                                                ;(assumes scaling mode is
                                                ;set).  X0 is Output.
```

TOTAL INSTRUCTION CYCLES:

5 Icyc @ 20 MHz = 500 ns

Figure 3-4. DSP56001 Code and Data Structures for Single-Section Second-Order Canonic Form (Direct Form II)

SECTION 4
SINGLE-SECTION TRANSPOSE FORM

A third realization of IIR filters is the transpose form (direct form I) shown in Figure 4-1. This network implementation can be derived directly from the direct-form difference equation (see Figure 4-1) or by taking the transpose of the canonic network (see References 10 and 11).

The transpose realization is characterized by three accumulator operations. One reason for the popularity of this realization is that, like the canonic realization, it only requires two memory locations; however, unlike the canonic realization, it is much less prone to overflow at internal nodes. The disadvantage is that this realization requires more instructions to implement.

4.1 GAIN EVALUATION OF INTERNAL NODES

Using the same techniques used to calculate $H_W(z)$ for the canonic realization, $H_U(z)$ and $H_V(z)$ are found as shown in Figures 4-2 and 4-3. The resulting expressions, unlike the canonic-form results, depend on the numerator of the transfer function; thus, the internal gains, $G_U(\theta)$ and $G_V(\theta)$, have different forms for the different filter types. In the case of the bandpass filter, these results simplify significantly so that a closed-form expression for the maximum gain at the internal nodes can be derived by calculating the maxima of the gain functions. For the bandpass and bandstop networks, the maximum of $G_U(\theta)$ and $G_V(\theta)$ is $g_m = \beta + 1/2$. Since $\beta < 1/2$, then $g_m < 1$ so that no overflow occurs at these nodes in the bandpass or bandstop case.

Figure 4-4 contains example plots of $G_U(\theta)$ and $G_V(\theta)$ for the second-order transpose-form lowpass filter with various values of cutoff frequency, θ_C, and damping factor, d. In most cases, $G_U(\theta)$ and $G_V(\theta)$ never exceed the maximum value of $G(\theta)$, so that, if the total gain does not exceed unity, the internal nodes will not exceed unity (i.e., no overflow).

DIRECT-FORM DIFFERENCE EQUATION

$$
\begin{aligned}
y(n) &= 2\{\alpha[x(n) + \mu x(n-1) + \sigma x(n-2)] + \gamma y(n-1) - \beta y(n-2)\} \\
&= 2\{\alpha x(n) + \alpha\mu x(n-1) + \gamma y(n-1) + [\alpha\sigma x(n-2) - \beta y(n-2)]\} \\
&= 2\{\alpha x(n) + [\alpha\mu x(n-1) + \gamma y(n-1) + \tfrac{1}{2}u(n-2)]\} \\
&= 2\{\alpha x(n) + \tfrac{1}{2}v(n-1)\}
\end{aligned}
$$

TRANSPOSE-FORM DIFFERENCE EQUATION

$$y(n) = 2\{\alpha x(n) + \tfrac{1}{2}v(n-1)\}$$

where

$$v(n) = 2\{\alpha\mu x(n) + \gamma y(n) + \tfrac{1}{2}u(n-1)\}$$

and

$$u(n) = 2\{\alpha\sigma x(n) - \beta y(n)\}$$

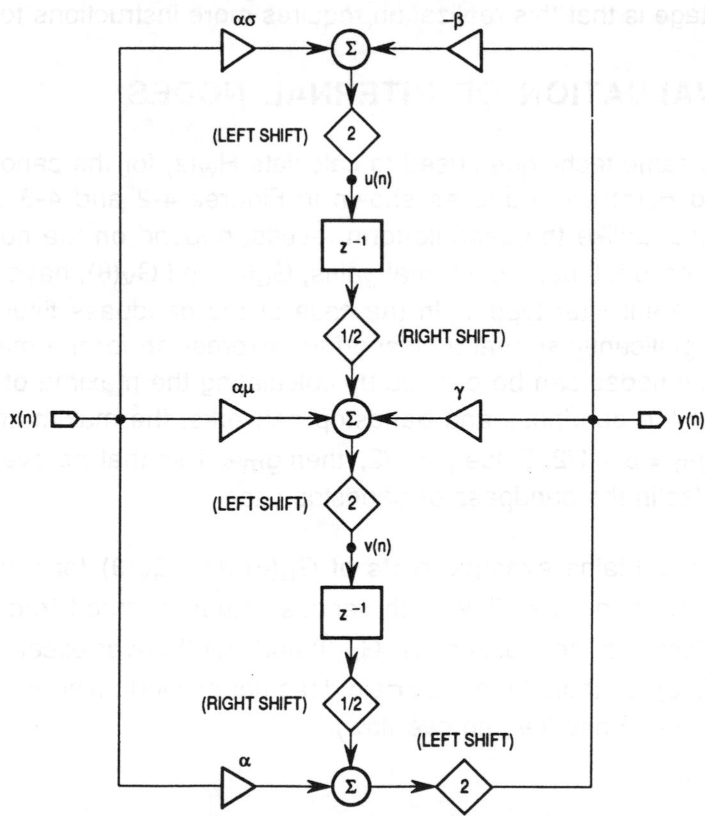

Figure 4-1. Network Diagram for Transpose-Form (Direct Form I) Implementation of a Single-Section Second-Order Filter

GAIN AT INTERNAL NODE, u(n)

Transfer function of u(n) is

$$U(z) = 2[\alpha\sigma X(z) - \beta Y(z)]$$
$$= H_u(z)\,X(z)$$

where

$$H_u(z) = 2[\alpha\sigma - \beta H(z)]$$
$$= 2\left[\frac{\alpha\sigma(1/2 - \gamma z^{-1} + \beta z^{-2}) - \alpha\beta(1 + \mu z^{-1} + \sigma z^{-2})}{1/2 - \gamma z^{-1} + \beta z^{-2}}\right]$$
$$= \frac{2\alpha(A - Bz^{-1})}{1/2 - \gamma z^{-1} + \beta z^{-2}}$$

where $H_u(z)$ is the transfer function from input node, x(n), to internal node, u(n), and $A = \sigma/2 - \beta$ and $B = \sigma\gamma + \mu\beta$.

Gain of u(n) is

$$G_u(\theta) = |H_u(z)|_{z=e^{j\theta}}$$
$$= \frac{2\alpha\sqrt{A^2 + B^2 - 2AB\cos\theta}}{\sqrt{(1/2-\beta)^2\sin^2\theta + (1/2+\beta)^2(\cos\theta - \cos\theta_0)^2}}$$

BANDPASS EXAMPLE

Bandpass coefficients are

$$\sigma = -1 \qquad \mu = 0 \qquad \alpha = (1/2 - \beta)/2$$

so that

$$A = -(1/2 + \beta)$$
$$B = -\gamma$$
$$= -(1/2 + \beta)\cos\theta_0$$

and

$$G_u(\theta) = \frac{(1/2-\beta)(1/2+\beta)\sqrt{\sin^2\theta + (\cos\theta - \cos\theta_0)^2}}{\sqrt{(1/2-\beta)^2\sin^2\theta + (1/2+\beta)^2(\cos\theta - \cos\theta_0)^2}}$$

Peak gain is

$$g_m = G_u(\theta_m)$$

which is found by evaluating

$$\frac{d}{d\theta} G_u(\theta)\Big|_{\theta = \theta_m} = 0$$

after evaluating derivative,

$$\theta_m = \theta_0$$

so that

$$g_m = \frac{(1/2-\beta)(1/2+\beta)\sin\theta_0}{(1/2-\beta)\sin\theta_0} = 1/2 + \beta < 1 \quad (\text{since } \beta < 1/2)$$

Figure 4-2. Gain Evaluation at First Internal Node, u(n), of Transpose Network

GAIN AT INTERNAL NODE, v(n)

Transfer function of v(n) is

$$V(z) = [Y(z) - 2\alpha X(z)] / z^{-1}$$
$$= H_v(z)\, X(z)$$

where

$$Y(z) = 2[\tfrac{1}{2}V(z)z^{-1} + \alpha X(z)]$$
$$H_v(z) = \frac{H(z) - 2\alpha}{z^{-1}}$$

and

$$= \frac{\alpha[(1 + \mu z^{-1} + \sigma z^{-2}) - 2(1/2 + \gamma z^{-1} + \beta z^{-2})]}{z^{-1}(1/2 - \gamma z^{-1} + \beta z^{-2})}$$
$$= \frac{2\alpha(C + Az^{-1})}{1/2 - \gamma z^{-1} + \beta z^{-2}}$$

where $H_v(z)$ is the transfer function from input node, x(n), to internal node, v(n), and $A = \sigma / 2 - \beta$ and $C = \mu / 2 + \gamma$.

Gain of v(n) is

$$G_v(\theta) = |H_v(z)|_{z = e^{j\theta}}$$
$$= \frac{2\alpha\sqrt{A^2 + C^2 + 2AC\cos\theta}}{\sqrt{(1/2 - \beta)^2 \sin^2\theta + (1/2 + \beta)^2(\cos\theta - \cos\theta_0)^2}}$$

BANDPASS EXAMPLE

Bandpass coefficients are

$$\sigma = -1 \qquad \mu = 0 \qquad \alpha = (1/2 - \beta) / 2$$

so that

$$A = -(1/2 + \beta)$$
$$C = \gamma$$
$$= (1/2 + \beta)\cos\theta_0$$

and

$$G_v(\theta) = \frac{(1/2 - \beta)(1/2 + \beta)\sqrt{\sin^2\theta + (\cos\theta - \cos\theta_0)^2}}{\sqrt{(1/2 - \beta)^2 \sin^2\theta + (1/2 + \beta)^2(\cos\theta - \cos\theta_0)^2}}$$

Peak gain is

$$g_m = G_v(\theta_m)$$

which is found by evaluating

$$\frac{d}{d\theta} G_v(\theta)\Big|_{\theta = \theta_m} = 0$$

after evaluating derivative,

$$\theta_m = \theta_0$$

so that

$$g_m = \frac{(1/2 - \beta)(1/2 + \beta)\sin\theta_0}{(1/2 - \beta)\sin\theta_0} = 1/2 + \beta < 1 \quad (\text{since } \beta < 1/2)$$

Figure 4-3. Gain Evaluation at Second Internal Node, v(n), of Transpose Network

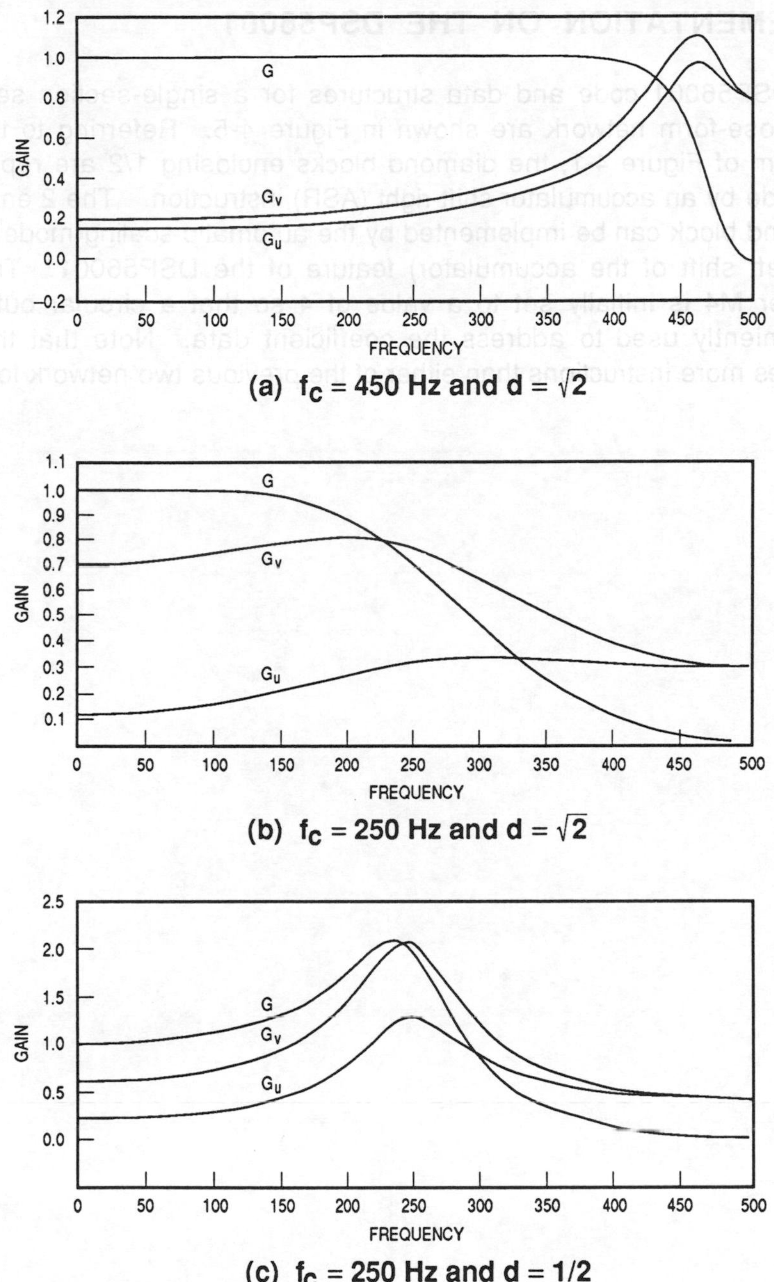

(a) $f_C = 450$ Hz and $d = \sqrt{2}$

(b) $f_C = 250$ Hz and $d = \sqrt{2}$

(c) $f_C = 250$ Hz and $d = 1/2$

Figure 4-4. Total Gain and Gain at Internal Nodes of Lowpass Transpose Filter Network of Figure 4-1 for $f_S = 1000$ Hz

4.2 IMPLEMENTATION ON THE DSP56001

The DSP56001 code and data structures for a single-section second-order transpose-form network are shown in Figure 4-5. Referring to the network diagram of Figure 4-1, the diamond blocks enclosing 1/2 are represented in the code by an accumulator shift right (ASR) instruction. The 2 enclosed by a diamond block can be implemented by the automatic scaling mode (equivalent to a left shift of the accumulator) feature of the DSP56001. The modifier register M4 is initially set to a value of 4 so that a circular buffer can be conveniently used to address the coefficient data. Note that this network requires more instructions than either of the previous two network forms.

DIFFERENCE EQUATION:

$$y(n)=2\{\alpha x(n)+\frac{1}{2}v(n-1)\}$$
$$v(n)=2\{\alpha\mu x(n)+\gamma y(n)+\frac{1}{2}u(n-1)\}$$
$$u(n)=2\{\alpha\sigma x(n)-\beta y(n-1)\}$$

DATA STRUCTURES:

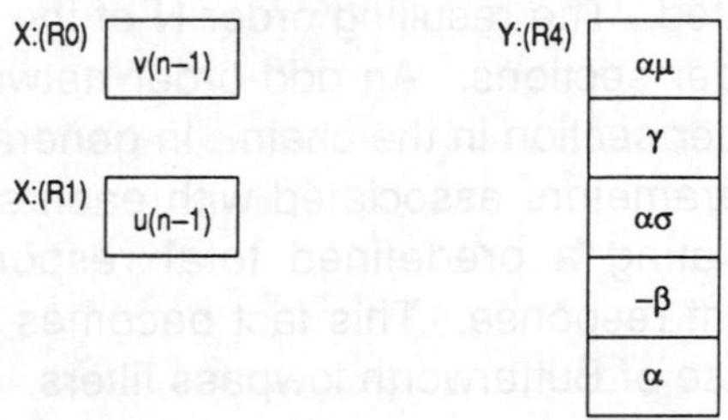

DSP56001 CODE:

```
                                             ;Y1=x(n) (Input)
                                             ;Y0=α.  A=v(n−1)/2
MACR  Y0,Y1,A   X:(R1),B   Y:(R4)+,Y0        ;A=A+αx(n)
ASR   B         A,X0                         ;y(n)=X0=2A (scale mode on),
MAC   Y0,Y1,B              Y:(R4)+,Y0        ;B=u(n−1)/2+αμx(n)
MACR  X0,Y0,B              Y:(R4)+,Y0        ;B=B+γy(n)
MPY   Y0,Y1,B   B,X:(R0)   Y:(R4)+,Y0        ;v(n)=2B.  B=ασx(n)
MACR  X0,Y0,B   X:(R0),A                     ;B=B−βy(n)
ASR   A         B,X:(R1)   Y:(R4)+,Y0        ;A=v(n)/2. u(n)=2B
                                             ;Y0=α.  y(n)=X0 (Output)
```

TOTAL INSTRUCTION CYCLES:

7 Icyc @ 20 MHz = 700 ns

Figure 4-5. DSP56001 Code and Data Structures for Single-Section Second-Order Transpose Form

SECTION 5
CASCADED DIRECT FORM

By placing any of the direct-form second-order filter networks from Figures 2-2, 2-5, 2-8, and 2-11 in series (i.e., connect the y(n) of one to the x(n) of the next), a cascaded filter is created. The resulting order N of the network is two times the number of second-order sections. An odd-order network can be made simply by adding one first-order section in the chain. In general, to achieve a particular response, the filter parameters associated with each second-order section are different since generating a predefined total response requires that each section have a different response. This fact becomes more obvious when discussing the special case of Butterworth lowpass filters.

5.1 BUTTERWORTH LOWPASS FILTER

The Butterworth filter response is maximally flat in the passband at the expense of phase linearity and steepness of attenuation slope in the transition band. For lowpass or highpass cascaded second-order sections, all sections have the same center frequency (not the case for bandpass filters). For this reason, it is easy to design since all that remains to be determined are the damping factors, d_k, of each individual k^{th} section. The damping coefficients, d_k, are calculated from a simple formula for any order N of response.

The s-domain transfer function for the Nth-order lowpass Butterworth filter is

$$H(s) = \frac{1}{(s/\Omega_C)^2 + d_1(s/\Omega_C) + 1} \; \frac{1}{(s/\Omega_C)^2 + d_2(s/\Omega_C) + 1} \cdots$$

$$= \prod_{k=1}^{N/2} \frac{1}{(s/\Omega_C)^2 + d_k(s/\Omega_C) + 1} \qquad (5\text{-}1)$$

where d_k is the kth damping coefficient, $s = j\Omega$, and Ω_C is the common cutoff frequency (see Reference 14). Only filter orders of even N will be considered in this discussion to minimize the complexity of mathematical results. The analysis can be extended to include odd values of N by inserting an additional term of $[(s/\Omega_C) + 1]^{-1}$ in Equation (5-1). Equation (5-1) can be generalized if Ω_C of each second-order section is an arbitrary value (corresponding to a different cutoff frequency, Ω_k, of each section). However, since this discussion is limited to Butterworth polynomials, Equation (5-1) will serve as the basis of all following derivations.

A filter of order N has N/2 second-order sections. The second-order section of a Butterworth filter can be derived from the simple RCL network of Figure 1-1. However, the Butterworth damping factors are predetermined values, which can be shown to yield a maximally flat passband response (see Reference 14). The Butterworth damping coefficients are given by the following equation:

$$d_k = 2 \sin \frac{(2k-1)\pi}{2N} \tag{5-2}$$

Equation (5-2) is the characteristic equation that determines a Butterworth filter response. Note that for a single-section (k = 1) second-order (N = 2) lowpass filter, $d_k = 2 \sin(\pi/4) = \sqrt{2}$ as expected for a maximally flat response. For a fourth-order filter with two second-order sections, $d_1 = 2 \sin(\pi/8)$, where k = 1 and N = 4, and $d_2 = 2 \sin(3\pi/8)$, where k = 2 and N = 4.

Equation (5-1) represents an all-pole response (the only zeros are at plus and minus infinity in the analog s-domain). The poles of a second-order section are the roots of the quadratic denominator as given by

$$p_{k1} = -d_k/2 - j(1-d_k^2/4)^{1/2} \qquad \text{(5-3(a))}$$

and

$$p_{k2} = d_k/2 + j(1-d_k^2/4)^{1/2} \qquad \text{(5-3(b))}$$

Using Equation (5-3), Equation (5-1) becomes:

$$H(s) = \prod_{k=1}^{N/2} \frac{1}{[(s/\Omega_C) - p_k]\left[(s/\Omega_C) - p_k^*\right]} \tag{5-4}$$

where $p_k = p_{k1}$ and $p_k^* = p_{k2}$ (complex conjugate of p_k).

Equation (5-4) is useful in that the response of the system can be analyzed entirely by studying the poles of the polynomial. However, for purposes of transforming to the z-domain, Equation (5-1) can be used as previously shown in Figure 2-2.

To examine the gain and phase response (physically measurable quantities) of the lowpass Butterworth filter, the transfer function, H(s), of Equation (5-1) will be converted into a polar representation. The magnitude of H(s) is the gain, $G(\Omega)$; the angle between the real and imaginary components of H(s), $\phi(\Omega)$, is the arctangent of the phase shift introduced by the filter:

$$G(\Omega) = \sqrt{H(s)H^*(s)}\ \Big|_{s=j\Omega}$$

$$= \prod_{k=1}^{N/2} \frac{1}{\sqrt{[(\Omega/\Omega_c)^2 - 1]^2 + (d_k\Omega/\Omega c)^2}} \tag{5-5}$$

$$\phi(\Omega) = \sum_{k=1}^{N/2} \tan^{-1} \frac{d_k\Omega/\Omega_c}{(\Omega/\Omega_c)^2 - 1} \tag{5-6}$$

Equations (5-2), (5-5), and (5-6) describe the response characteristics of an Nth-order lowpass Butterworth filter in the continuous frequency analog domain. Since the quantity of interest is usually 20 log $G(\Omega)$, Equation (5-5) can be transformed into a sum (over the second-order sections) of 20 log (G_k), where G_k is the gain of the kth section. Similarly, the total phase is just the sum of the phase contribution by each section from Equation (5-6).

The bilinear transformation is used to convert the continuous frequency domain transfer function into the digital domain representation, where θ, the normalized digital domain frequency equal to $2\pi f/f_s$, can be thought of as the ratio of frequency to sampling frequency scaled by 2π. Substituting s from Equation (2-11) and Ω_c from Equation (2-12) into the kth section of Equation (5-1) yields the digital domain form of the Butterworth lowpass filter (for the kth second-order section):

$$H_k(z) = \frac{\alpha_k(1 + 2z^{-1} + z^{-2})}{\frac{1}{2} - \gamma_k z^{-1} + \beta_k z^{-2}} \tag{5-7}$$

where

$$\alpha_k = [\tan^2(\theta_c/2)]/A_k(c) \tag{5-8(a)}$$

$$\beta_k = [1 - d_k \tan(\theta_c/2) + \tan^2(\theta_c/2)]/A_k(c) \tag{5-8(b)}$$

$$\gamma_k = 2[1 - \tan^2(\theta_c/2)]/A_k(\theta_c) \tag{5-8(c)}$$

and

$$A_k(\theta_c) = 2[1 + d_k \tan(\theta_c/2) + \tan^2(\theta_c/2)] \qquad (5\text{-}8(d))$$

Equation (5-8(a))–(5-8(d)) provides a complete description of the digital lowpass N^{th}-order Butterworth filter. Given θ_c and d_k, these formulas allow precise calculation of the digital coefficients, α_k, β_k, and γ_k, used to implement each kth second-order section of the filter.

Equation (5-8(a))–(5-8(d)) can be further simplified into the following set of formulas:

$$\beta_k = \frac{1 - (d_k/2)\sin(\theta_c)}{2[1 + (d_k/2)\sin(\theta_c)]} \qquad (5\text{-}9(a))$$

$$\gamma_k = (1/2 + \beta_k)\cos(\theta_c) \qquad (5\text{-}9(b))$$

$$\alpha_k = (1/2 + \beta_k - \gamma_k)/4 \qquad (5\text{-}9(c))$$

where d_k is given by Equation (5-2) and θ_c is the digital domain cutoff frequency (actual operating cutoff frequency of the digital filter).

Figure 5-1 shows an example of a sixth-order lowpass Butterworth filter (three second-order sections) in both the analog domain and digital domain. Note that the gain of the first section (k = 1) is greater than unity near the cutoff frequency but that the overall composite response never exceeds unity. This fact allows for easy implementation of the Butterworth filter in cascaded direct form (i.e., scaling of sections is not needed as long as the sections are implemented in the order of decreasing k). Overflow at the output of any section is then guaranteed not to occur (the gain of the filter never exceeds unity). Note that the digital response (see Figure 5-1) is identical to the analog response but warped from the right along the frequency axis. Imagine the zero at plus infinity in the analog response mapping into the zero at $f_s/2$ in the digital case. Also note that, because of this mapping, the digital response falls off faster than the –12 dB/octave of the analog filter when the cutoff is near $f_s/2$.

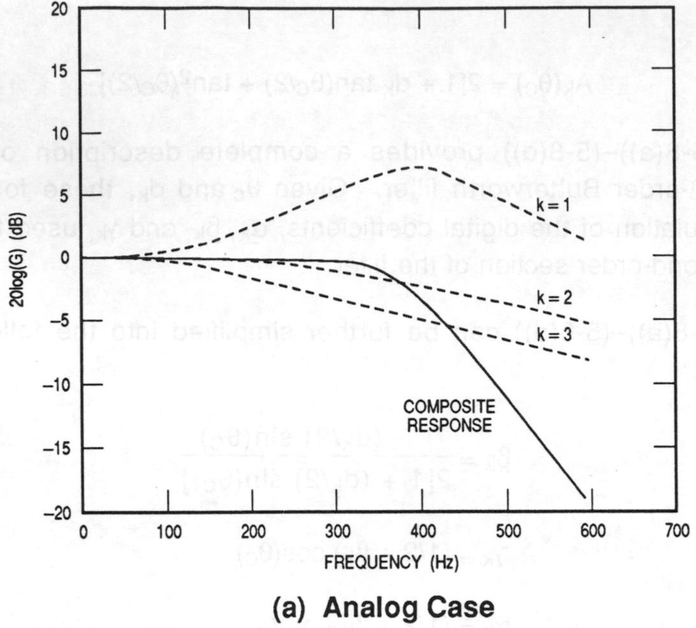

(a) Analog Case

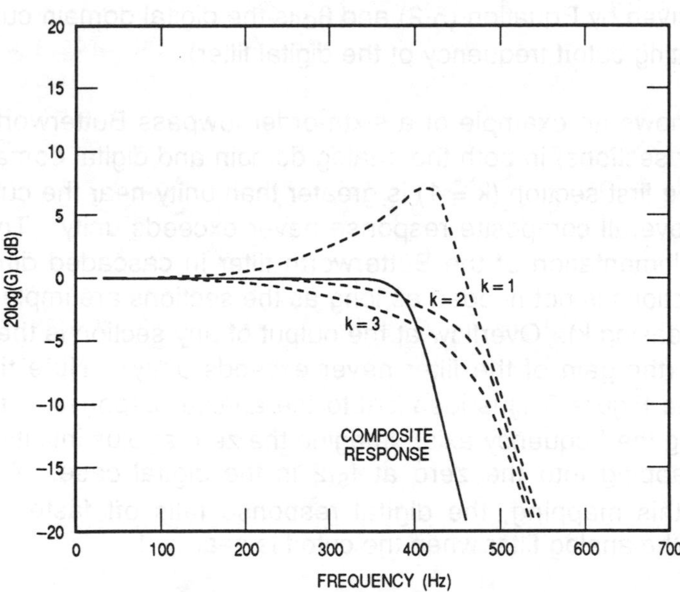

(b) Digital Filter with $f_S = 1200$ Hz and $f_C = 425$ Hz

Figure 5-1. Composite Response of Cascaded Second-Order Sections for Lowpass Butterworth Filter (k^{th} Section Damping Factor According to Equation (5-2)).

The previous analysis is nearly identical to the case of the highpass filter except the coefficients (see Figure 3-1) have slightly different values. Since the bandstop case is just the sum of a lowpass and highpass case, it can be analyzed by these techniques. The bandpass case, however, is more difficult and requires considerably more work (see Reference 14) because the center frequency of each section is now different and the formula for calculating these frequencies is not as simple as the formulas for the previous filter types. In addition, to complicate matters further, scaling between sections becomes more of a problem since the offset of the center of each section reduces the final center response, which must be compensated at some point in the filter network. For cases such as higher order Butterworth bandpass filter designs, commercially available filter design packages such as FDAS are useful. The use of FDAS is discussed in **SECTION 6 FILTER DESIGN AND ANALYSIS SYSTEM (FDAS)** and **SECTION 7 FIR FILTERS**.

5.2 CASCADED DIRECT-FORM NETWORK

Figure 5-2 shows the cascaded direct-form network and data structures for the DSP56001 code implementation of Figure 5-3. By cascading network diagrams presented in **SECTION 2 SECOND-ORDER DIRECT-FORM IIR DIGITAL FILTER SECTIONS**, the set of delays at the output of one section can be combined with the set of delays at the input of the next section, thus reducing the total number of delays by almost a factor of two. For this reason, the cascaded direct-form network becomes canonic as the filter order N increases. The DSP56001 code (see Figure 5-3) shows an example of reading data from a user-supplied memory-mapped analog-to-digital converter (ADC) and writing it to a memory-mapped digital-to-analog converter (DAC). The number of sections (nsec) in these examples is three; thus, the filter order is six. The total instruction time for this filter structure is 600 nsec + 800 ns, including the data I/O moves (but excluding the interrupt overhead).

NETWORK DIAGRAM

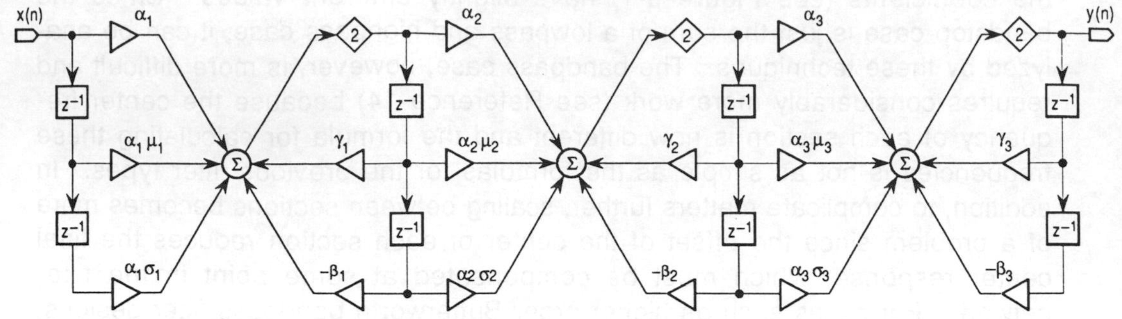

DIFFERENCE EQUATIONS

$$y_i(n) = 2\{\alpha_i [x_i(n) + \mu_i x_i(n+1) + \sigma_i x_i(n-2)] + \gamma_i y_i(n-1) - \beta_i y_i(n-2)\}$$

$$x_{i+1}(n-k) = y_i(n-k)$$

DATA STRUCTURES

Y: (RO)

coeff
$-\beta_1$
α_1
$\alpha_1 \sigma_1$
$\alpha_1 \mu_1$
γ_1
$-\beta_2$
α_2
$\alpha_2 \sigma_2$
$\alpha_2 \mu_2$
γ_2
$-\beta_3$
α_3
$\alpha_3 \sigma_3$
$\alpha_3 \mu_3$
γ_3

Y: (R4)

xbuf		
$x_1(n-2)$		$x(n-2)$
$x_1(n-1)$		$x(n-1)$
$x_2(n-2)$		$y_1(n-2)$
$x_2(n-1)$	=	$y_1(n-1)$
$x_3(n-2)$		$y_2(n-2)$
$x_3(n-1)$		$y_2(n-1)$
$y(n-2)$		$y_3(n-2)$
$y(n-1)$		$y_3(n-1)$

Figure 5-2. Network Diagram for Cascaded Direct-Form Filter and Data Structures Used in Code Implementation (Three-Section Example)

DSP56001 CODE

```
        MOVE    #coeff,R0                                 ;Pointer to coefficients
        MOVE    #1,M4                                     ;Modulo of length 2 for xbuf
        MOVE                X:(R0)+,X1                    ;β1
        MOVE                X:(R0)+,X0                    ;α1
        MOVE                (R4)+                         ;Point to next xbuf entry
                                                          ;Input: Y1=x(n)
        DO      X:nsec,Sectn                              ;Loop on number of sections
        MPY     X0,Y1,A     X:(R0)+,X0   Y:(R4)+,Y0       ;A=αi xi(n)
        MAC     X0,Y0,A     X:(R0)+,X0   Y:(R4)+N4,Y0     ;A=A+αi σi xi(n−2)
        MAC     X0,Y0,A     X:(R0)+,X0   Y:(R4)+,Y0       ;A=A+αi μi xi(n−1)
        MAC     X0,Y0,A                  Y:(R4)−N4,Y0     ;A=A+γi yi (n−1)
        MAC     −X1,Y0,A    X:(R0)+,X1   Y1,Y:(R4)+N4     ;A=A−βi yi(n−2). Save x(n)
        MOVE    A,Y1        X:(R0)+,X0                    ;yi(n)=2A, assumes scale
                                                          ;mode on
Sectn                                                     ;X1=βi+1. X0=αi+1

                                                          ;Output: y(n)=Y1
        MOVE                X:Buflen,M4                   ;Filter Order+1
        NOP
        MOVE                             Y1,Y:(R4)+NA     ;Save y(n)
```

TOTAL INSTRUCTION CYCLES

(6 * nsec + 10)icyc @ 20 MHz = (0.6 * nsec + 1.0) ns

NOTE: nsec is number of sections.

Figure 5-3. DSP56001 Code for Cascaded Direct-Form Filter

SECTION 6
FILTER DESIGN AND ANALYSIS SYSTEM (FDAS)

The design of a cascaded filter in both the direct-form and canonic implementations, using a software package (FDAS) available from Momentum Data Systems, Inc., is discussed in the following paragraphs. The filter example used in the following paragraphs is a sixth-order Butterworth lowpass filter with a cutoff frequency of approximately 225 Hz and a sample frequency of 1000 Hz. Figure 6-1 is the log magnitude (gain) plot from the system output. Figure 6-2 is the phase as a function of frequency in wrapped format ($-\pi$ wraps to $+\pi$). In addition, Figure 6-3 is a zero/pole plot, and Figure 6-4 is the group delay, which is the negative of the derivative of the phase with respect to frequency. FDAS will also generate an impulse response, step response, and a linear magnitude plot. The results of the design are written to a file, FDAS.OUT, which contains much useful information. The coefficient data is written to COEFF.FLT. The DSP56001 code generator (MGEN) reads the COEFF.FLT file and generates a DSP56001 assembly source file, COEFF.ASM, which can be assembled by the DSP56001 assembler or linker software. Examples of these files are shown in Figures 6-5 to 6-10.

6.1 CANONIC IMPLEMENTATION

Figure 6-5 is the output file associated with the FDAS design session, containing information on the analog s-domain equivalent filter as well as the final digital coefficients (listed again in Figure 6-6), which have been properly scaled to prevent overflow at the internal nodes and outputs of each cascaded section. This procedure is done automatically by the program in a matter of seconds. Executable code is generated by MGEN (also from Momentum Data Systems, Inc.), as shown in Figure 6-7. The code internal to each cascaded section is five instructions long; thus, 500 ns (assuming a DSP56001 clock of 20 MHz) is added to the execution time for each additional second-order section.

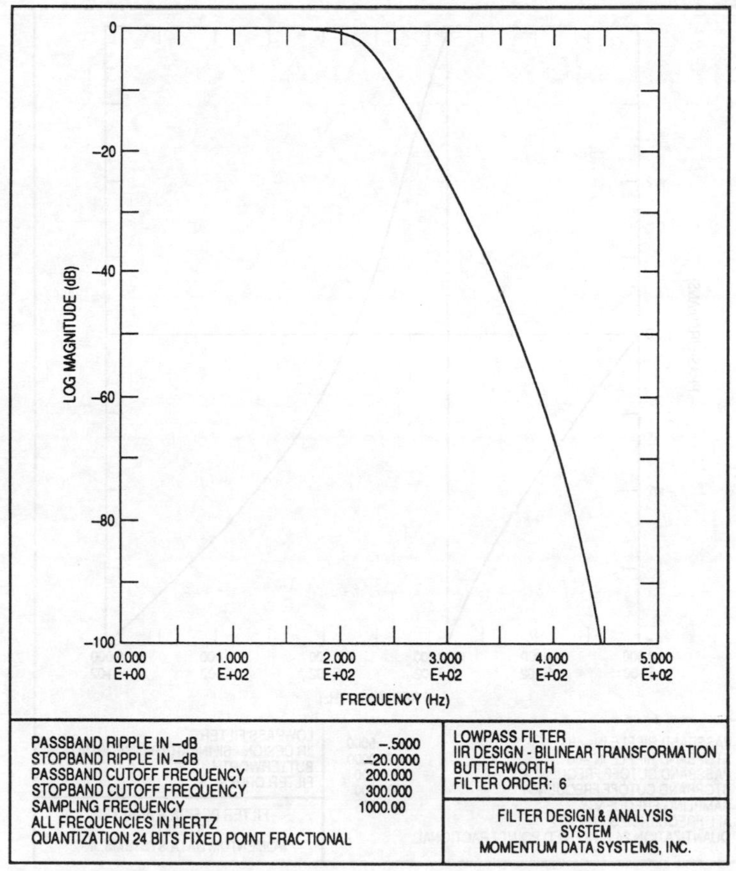

Figure 6-1. Log Magnitude Plot of Example Lowpass Butterworth Filter

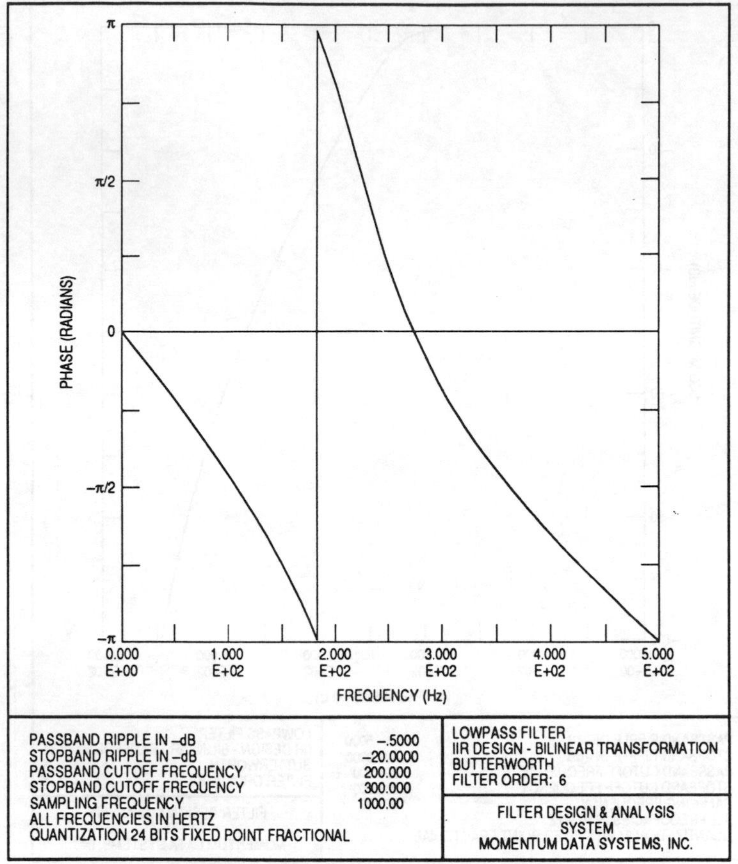

Figure 6-2. Phase versus Frequency Plot for Example Filter

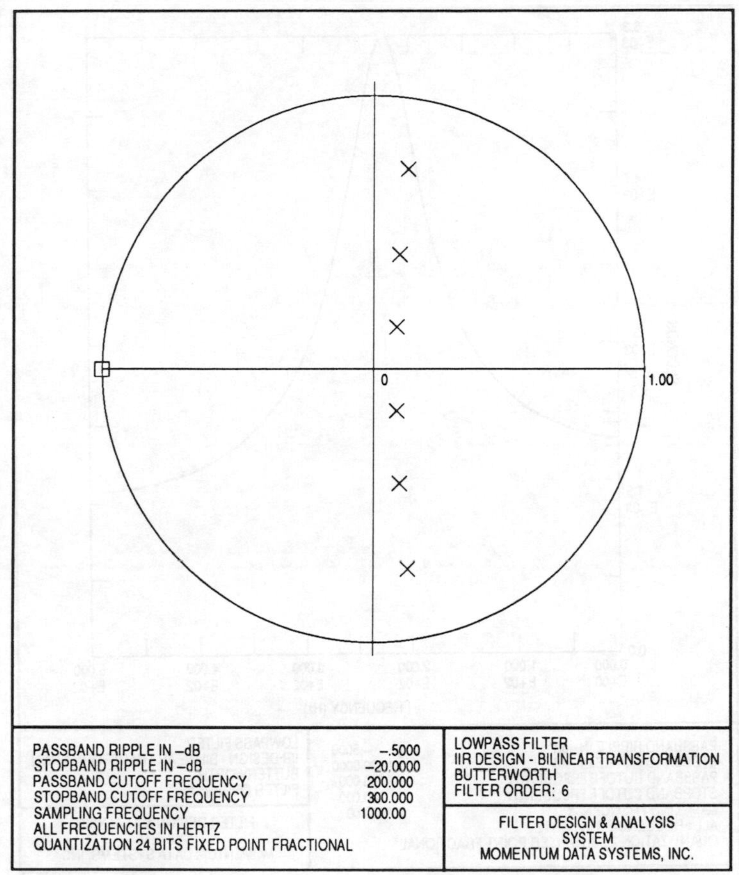

Figure 6-3. Zero/Pole Plot of Sixth-Order Lowpass Example Filter

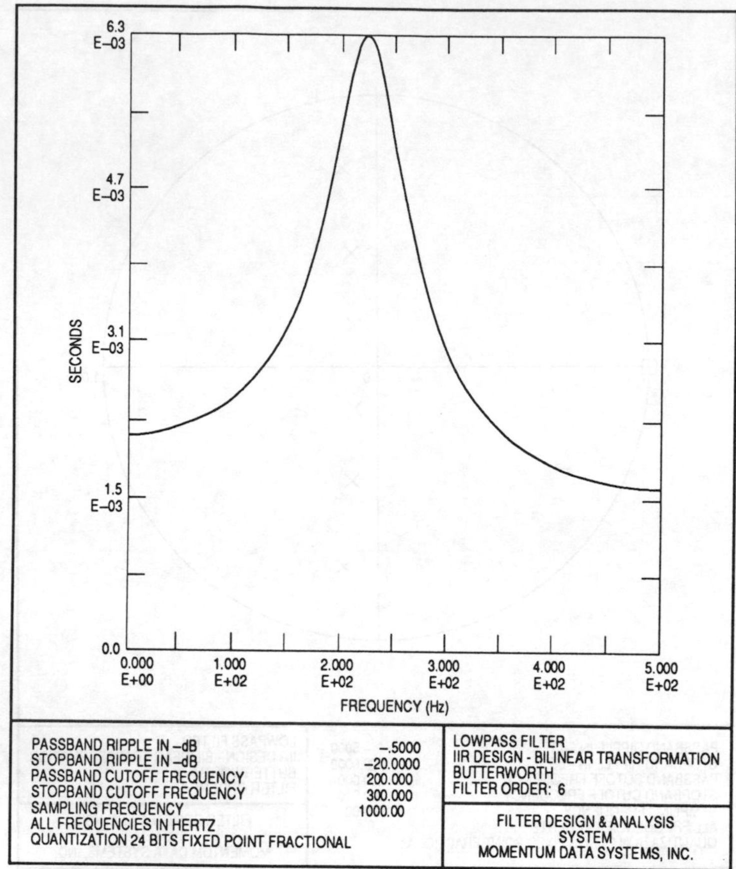

Figure 6-4. Group Delay versus Frequency for FDAS IIR Example

```
FILTER TYPE                      LOW PASS
ANALOG FILTER TYPE               BUTTERWORTH
PASSBAND RIPPLE IN -dB              -.5000
STOPBAND RIPPLE IN -dB             -20.0000
PASSBAND CUTOFF FREQUENCY        200.000      HERTZ
STOPBAND CUTOFF FREQUENCY        300.000      HERTZ
SAMPLING FREQUENCY               1000.00      HERTZ
FILTER ORDER:       6
FILTER DESIGN METHOD: BILINEAR TRANSFORMATION

COEFFICIENTS OF Hd(Z) ARE QUANTIZED TO    24 BITS
QUANTIZATION TYPE: FIXED POINT FRACTIONAL
COEFFICIENTS SCALED FOR CASCADE FORM II

              NORMALIZED ANALOG TRANSFER FUNCTION T(s)

     NUMERATOR COEFFICIENTS                 DENOMINATOR COEFFICIENTS
**************************************  **************************************
 S**2 TERM      S TERM     CONST TERM    S**2 TERM     S TERM      CONST TERM
.000000E+00  .000000E+00  .100000E+01  .100000E+01  .193185E+01  .100000E+01
.000000E+00  .000000E+00  .100000E+01  .100000E+01  .141421E+01  .100000E+01
.000000E+00  .000000E+00  .100000E+01  .100000E+01  .517638E+00  .100000E+01
     INITIAL GAIN  1.00000000

             UNNORMALIZED ANALOG TRANSFER FUNCTION T(s)

     NUMERATOR COEFFICIENTS                 DENOMINATOR COEFFICIENTS
**************************************  **************************************
 S**2 TERM      S TERM     CONST TERM    S**2 TERM     S TERM      CONST TERM
.000000E+00  .000000E+00  .299809E+07  .100000E+01  .334500E+04  .299809E+07
.000000E+00  .000000E+00  .299809E+07  .100000E+01  .244871E+04  .299809E+07
.000000E+00  .000000E+00  .299809E+07  .100000E+01  .896290E+03  .299809E+07
     INITIAL GAIN  1.00000000

                     DIGITAL TRANSFER FUNCTION Hd(z)

          NUMERATOR COEFFICIENTS                    DENOMINATOR COEFFICIENTS
 ****************************************  ****************************************
  Z**2 TERM       Z TERM      CONST TERM    Z**2 TERM      Z TERM       CONST TERM
.2520353794   .5040707588   .2520353794    1.00000   -.1463916302   .0225079060
.2364231348   .4728463888   .2364231348    1.00000   -.1684520245   .1765935421
.3606387377   .7212774754   .3606387377    1.00000   -.2279487848   .5921629667
       INITIAL GAIN  .876116210

                     ZEROES OF TRANSFER FUNCTION Hd(z)

    REAL PART      IMAGINARY PART      REAL PART      IMAGINARY PART      RADIUS
-.100000000E+01  .000000000E+00  -.100000000E+01  .000000000E+00  .100000000E+01
-.999290168E+00  .000000000E+00  -.100071034E+01  .000000000E+00  .999290168E+00
-.100000000E+01  .000000000E+00  -.100000000E+01  .000000000E+00  .100000000E+01

                     POLES OF TRANSFER FUNCTION Hd(z)

    REAL PART      IMAGINARY PART      REAL PART      IMAGINARY PART      RADIUS
.731958151E-01  .130959072E+00  .731958151E-01  -.130959072E+00  .150026351E+00
.842260122E-01  .411703195E+00  .842260122E-01  -.411703195E+00  .420230344E+00
.113974392E+00  .761034036E+00  .113974392E+00  -.761034036E+00  .769521258E+00

          NUMERATOR COEFFICIENTS - HIGHEST ORDER FIRST (in z)

.214893785E-01  .128936282E+00  .322340720E+00  .429787634E+00  .322340720E+00
.128936282E+00  .214893785E-01

          DENOMINATOR COEFFICIENTS - HIGHEST ORDER FIRST (in z)

.100000000E+01  -.542792439E+00  .887692610E+00  -.267088214E+00  .143235132E+00
-.184597152E-01  .235370025E-02

          IMPULSE RESPONSE  MIN = -.179722E+00, MAX =  .452734E+00
```

Figure 6-5. FDAS.OUT File of Example Filter for Cascaded Canonic Implementation

```
FILTER COEFFICIENT FILE
IIR DESIGN
FILTER TYPE                        LOW PASS
ANALOG FILTER TYPE                 BUTTERWORTH
PASSBAND RIPPLE IN -dB                 -.5000
STOPBAND RIPPLE IN -dB               -20.0000
PASSBAND CUTOFF FREQUENCY          .200000E+03 HERTZ
STOPBAND CUTOFF FREQUENCY          .300000E+03 HERTZ
SAMPLING FREQUENCY                 .100000E+04 HERTZ
FILTER DESIGN METHOD: BILINEAR TRANSFORMATION
FILTER ORDER                  6 0006h
NUMBER OF SECTIONS            3 0003h
NO. OF QUANTIZED BITS        24 0018h
QUANTIZATION TYPE - FRACTIONAL FIXED POINT
COEFFICIENTS SCALED FOR CASCADE FORM II
        0 00000000      /* shift count for overall gain */
  7349395 00702493      /* overall gain                 */
        0 00000000      /* shift count for section  1 values                   */
  2114226 002042B2      /* section  1 coefficient B0                           */
  4228452 00408564      /* section  1 coefficient B1                           */
  2114226 002042B2      /* section  1 coefficient B2                           */
  1228022 0012BCF6      /* section  1 coefficient A1                           */
  -188810 FFFD1E76      /* section  1 coefficient A2                           */
        0 00000000      /* shift count for section  2 values                   */
  1983261 001E431D      /* section  2 coefficient B0                           */
  3966523 003C863B      /* section  2 coefficient B1                           */
  1983261 001E431D      /* section  2 coefficient B2                           */
  1413078 00158FD6      /* section  2 coefficient A1                           */
 -1481374 FFE96562      /* section  2 coefficient A2                           */
        0 00000000      /* shift count for section  3 values                   */
  3025257 002E2969      /* section  3 coefficient B0                           */
  6050514 005C52D2      /* section  3 coefficient B1                           */
  3025257 002E2969      /* section  3 coefficient B2                           */
  1912173 001D2D6D      /* section  3 coefficient A1                           */
 -4967423 FFB43401      /* section  3 coefficient A2                           */
 .2520353794097900D+00 3FD0215900000000  .25203538E+00 /* section  1 B0 */
 .5040707588195801D+00 3FE0215900000000  .50407076E+00 /* section  1 B1 */
 .2520353794097900D+00 3FD0215900000000  .25203538E+00 /* section  1 B2 */
 .1463916301727295D+00 3FC2BCF600000000  .14639171E+00 /* section  1 A1 */
-.2250790596008301D-01 BF970C5000000000 -.22507921E-01 /* section  1 A2 */
 .2364231348037720D+00 3FCE431D00000000  .23642325E+00 /* section  2 B0 */
 .4728463888168335D+00 3FDE431D80000000  .47284651E+00 /* section  2 B1 */
 .2364231348037720D+00 3FCE431D00000000  .23642325E+00 /* section  2 B2 */
 .1684520244598389D+00 3FC58FD600000000  .16845204E+00 /* section  2 A1 */
-.1765935420989990D+00 BFC69A9E00000000 -.17659356E+00 /* section  2 A2 */

 .3606387376785278D+00 3FD714B480000000  .36063876E+00 /* section  3 B0 */
 .7212774753570557D+00 3FE714B480000000  .72127753E+00 /* section  3 B1 */
 .3606387376785278D+00 3FD714B480000000  .36063876E+00 /* section  3 B2 */
 .2279487848281860D+00 3FCD2D6D00000000  .22794882E+00 /* section  3 A1 */
-.5921629667282104D+00 BFE2F2FFC0000000 -.59216306E+00 /* section  3 A2 */
```

Figure 6-6. COEFF.OUT File of Example Filter Design—Scaled for Cascaded Canonic Implementation

```
                    COEFF     ident   1,0
                              include 'head2.asm'
                              page    132,66,0,10
                              opt     cex,mex
                    ;
                    ; This program implements an IIR filter in cascaded canonic sections
                    ; The coefficients of each section are scaled for cascaded canonic sections
                    ;
        00FFFF      datin     equ     $ffff                          ;location in Y memory of input file
        00FFFF      datout    equ     $ffff                          ;location in Y memory of output file
        00FFF0      m_scr     equ     $fff0                          ; sci control register
        00FFF1      m_ssr     equ     $fff1                          ; sci status register
        00FFF2      m_sccr    equ     $fff2                          ; sci clock control register
        00FFE1      m_pcc     equ     $ffe1                          ; port c control register
        00FFFF      m_ipr     equ     $ffff                          ; interrupt priority register
        000140      xx        equ     @cvi(20480/(64*1.000))         ;timer interrupt value
        FFD8F1      m_tim     equ     -9999                          ;board timer interrupt value
        000003      nsec      equ     3                              ;number of second order sections
                              include 'cascade2.asm'
                    ;
                    ; This code segment implements cascaded biquad sections in canonic form
                    ;
                    cascade2  macro   nsec
  m                 ;
  m                 ;     assumes each section's coefficients are divided by 2
  m                 ;
  m                           mpy     y0,y1,a    x:(r0)+,x0  y:(r4)+,y0  ;x0=1st section w(n-2),y0=a12/2
  m                           do      #nsec,_ends                       ;do each section
  m                           mac     x0,y0,a    x:(r0)-,x1  y:(r4)+,y0  ;x1=w(n-1) ,y0=ai1/2
  m                           macr    x1,y0,a    x1,x:(r0)+  y:(r4)+,y0  ;push w(n-1) to w(n-2),y0=bi2/2
  m                           mac     x0,y0,a    a,x:(r0)    y:(r4)+,y0  ;push w(n) to w(n-1),y0=bi1/2
  m                           mac     x1,y0,a    x:(r0)+,x0  y:(r4)+,y0  :get this iter w(n),y0=bi0/2
  m                           mac     x0,y0,a    x:(r0)+,x0  y:(r4)+,y0  ;next iter:x0=w(n-2),y0=ai2/2
  m                           _ends
  m                           endm
                    ;
                    ;  multiple shift left macro
                    ;
                    mshl      macro   scount
  m                           if      scount
  m                           rep     #scount
  m                           asl     a
  m                           endif
  m                           endm
        X:0000                org     x:0
        X:0000      states    ds      2*nsec
        Y:0000                org     y:0
```

Figure 6-7. COEFF.ASM File Generated by MGEN for Example Design and Cascaded Canonic Implementation (Sheet 1 of 3)

```
                         coef
d   Y:0000 FE8F3B                dc      $FE8F3B                 ;a(*,2)/2      =-.01125395    section number  1
d   Y:0001 095E7B                dc      $095E7B                 ;a(*,1)/2      = .07319582    section number  1
d   Y:0002 102159                dc      $102159                 ;b(*,2)/2      = .12601769    section number  1
d   Y:0003 2042B2                dc      $2042B2                 ;b(*,1)/2      = .25203538    section number  1
d   Y:0004 D02159                dc      $D02159                 ;b(*,0)/2-0.5=-.37398231      section number  1
d   Y:0005 F4B2B1                dc      $F4B2B1                 ;a(*,2)/2      =-.08829677    section number  2
d   Y:0006 0AC7EB                dc      $0AC7EB                 ;a(*,1)/2      = .08422601    section number  2
d   Y:0007 0F218E                dc      $0F218E                 ;b(*,2)/2      = .11821157    section number  2
d   Y:0008 1E431D                dc      $1E431D                 ;b(*,1)/2      = .23642319    section number  2
d   Y:0009 CF218E                dc      $CF218E                 ;b(*,0)/2-0.5=-.38178843      section number  2
d   Y:000A DA1A01                dc      $DA1A01                 ;a(*,2)/2      =-.29608148    section number  3
d   Y:000B 0E96B6                dc      $0E96B6                 ;a(*,1)/2      = .11397439    section number  3
d   Y:000C 1714B4                dc      $1714B4                 ;b(*,2)/2      = .18031937    section number  3
d   Y:000D 2E2969                dc      $2E2969                 ;b(*,1)/2      = .36063874    section number  3
d   Y:000E D714B4                dc      $D714B4                 ;b(*,0)/2-0.5=-.31968063      section number  3
    Y:00C8                       org     y:200
d   Y:00C8 381249        igain   dc      3674697
    000000               scount  equ     0                       ;final shift count
                                 include 'body2.asm'

    000040               start   equ     $40                                  ;origin for user program

    P:0000                       org     p:$0                                 ;origin for reset vector
    P:0000 0C0040                jmp     start                                ;jump to 'start' on system reset

    P:001C                       org     p:$1c                                ;origin for timer interrupt vector
    P:001C 0BF080                jsr     filter                               ;jump to 'filter' on timer interrupt
           000051

    P:0040                       org     p:start                              ;origin for user program

    P:0040 0003F8                ori     #3,mr                                ;disable all interrupts
    P:0041 08F4B2                movep           #(xx-1),x:m_sccr             ;cd=xx-1 for divide by xx
           00013F
    P:0043 08F4A1                movep           #$7,x:m_pcc                  ;set cc(2;0) to turn on timer
           000007
    P:0045 08F4B0                movep           #$2000,x:m_scr               ;enable timer interrupts
           002000
    P:0047 08F4BF                movep           #$c000,x:m_ipr               ;set interrupt priority for sci
           00C000

    P:0049 300000                move            #states,r0                   ;initialize internal state storage
    P:004A 200013                clr     a                                    ;*   set memory to zero
    P:004B 0606A0                rep     #nsec*2                              ;*
    P:004C 565800                move            a,x:(r0)+                    ;*

    P:004D 4FF000                move                      y:igain,y1         ;y1=initial gain/2
           0000C8
    P:004F 00FCB8                andi    #$fc,mr                              ;allow interrupts
    P:0050 0C0050                jmp     *                                    ;wait for interrupt

```

Figure 6-7. COEFF.ASM File Generated by MGEN for Example Design and Cascaded Canonic Implementation (Sheet 2 of 3)

```
                           filter
    P:0051 0008F8                  ori     #$08,mr                              ;set scaling mode
    P:0052 300000                  move            #states,r0                   ;point to filter states
    P:0053 340000                  move            #coef,r4                     ;point to filter coefficients
    P:0054 09463F                  movep           y:datin,y0                   ;get sample

                                   cascade2 nsec                                ;do cascaded biquads
+                          ;
+                          ;    assumes each section's coefficients are divided by 2
+                          ;
+   P:0055 F09880                  mpy     y0,y1,a    x:(r0)+,x0   y:(r4)+,y0   ;x0=1st section w(n-2),y0=a12/2
+   P:0056 060380                  do      #nsec,_ends                          ;do each section
           00005C
+   P:0058 F49002                  mac     x0,y0,a    x:(r0)-,x1   y:(r4)+,y0   ;x1=w(n-1) ,y0=ai1/2
+   P:0059 F418E3                  macr    x1,y0,a    x1,x:(r0)+   y:(r4)+,y0   ;push w(n-1) to w(n-2),y0=bi2/2
+   P:005A F80002                  mac     x0,y0,a    a,x:(r0)     y:(r4)+,y0   ;push w(n) to w(n-1),y0=bi1/2
+   P:005B F098E2                  mac     x1,y0,a    x:(r0)+,x0   y:(r4)+,y0   ;get this iter w(n),y0=bi0/2
+   P:005C F098D2                  mac     x0,y0,a    x:(r0)+,x0   y:(r4)+,y0   ;next iter:x0=w(n-2),y0=ai2/2
+                          _ends
                                   mshl    scount
+                                  if      scount
+                                  endif
    P:005D 200011                  rnd     a                                    ;round result
    P:005E 09CE3F                  movep           a,y:datout                   ;output sample
    P:005F 000004                  rti


                                   end
0    Errors
0    Warnings
```

Figure 6-7. COEFF.ASM File Generated by MGEN for Example Design and Cascaded Canonic Implementation (Sheet 3 of 3)

6.2 TRANSPOSE IMPLEMENTATION (DIRECT FORM I)

Figure 6-8 is the output file from FDAS for a transpose-form implementation. As before, it contains information on the analog s-domain equivalent filter as well as the final digital coefficients (listed again in Figure 6-9), which have been properly scaled to prevent overflow at the internal nodes and outputs of each cascaded section. Because of the stability of the internal node gain of the transpose form and because of the response of each second-order Butterworth section, scaling was not done by the program because it was not needed. Executable code shown in Figure 6-10 is again generated by MGEN. The code internal to each cascaded section is seven instructions long; thus, 700 ns (assuming a DSP56001 clock of 20 MHz) is added to the execution time for each additional second-order section.

```
FILTER TYPE                      LOW PASS
ANALOG FILTER TYPE               BUTTERWORTH
PASSBAND RIPPLE IN -dB                -.5000
STOPBAND RIPPLE IN -dB              -20.0000
PASSBAND CUTOFF FREQUENCY       200.000      HERTZ
STOPBAND CUTOFF FREQUENCY       300.000      HERTZ
SAMPLING FREQUENCY              1000.00      HERTZ
FILTER ORDER:      6
FILTER DESIGN METHOD: BILINEAR TRANSFORMATION

COEFFICIENTS OF Hd(Z) ARE QUANTIZED TO   24 BITS
QUANTIZATION TYPE: FIXED POINT FRACTIONAL
COEFFICIENTS SCALED FOR CASCADE FORM I (TRANSPOSE FORM)


             NORMALIZED ANALOG TRANSFER FUNCTION T(s)

    NUMERATOR COEFFICIENTS                 DENOMINATOR COEFFICIENTS
**************************************  **************************************
  S**2 TERM     S TERM     CONST TERM   S**2 TERM     S TERM     CONST TERM
.000000E+00  .000000E+00  .100000E+01  .100000E+01  .193185E+01  .100000E+01
.000000E+00  .000000E+00  .100000E+01  .100000E+01  .141421E+01  .100000E+01
.000000E+00  .000000E+00  .100000E+01  .100000E+01  .517638E+00  .100000E+01
     INITIAL GAIN  1.00000000


            UNNORMALIZED ANALOG TRANSFER FUNCTION T(s)

    NUMERATOR COEFFICIENTS                 DENOMINATOR COEFFICIENTS
**************************************  **************************************
  S**2 TERM     S TERM     CONST TERM   S**2 TERM     S TERM     CONST TERM
.000000E+00  .000000E+00  .299809E+07  .100000E+01  .334500E+04  .299809E+07
.000000E+00  .000000E+00  .299809E+07  .100000E+01  .244871E+04  .299809E+07
.000000E+00  .000000E+00  .299809E+07  .100000E+01  .896290E+03  .299809E+07
     INITIAL GAIN  1.00000000
```

```
                         DIGITAL TRANSFER FUNCTION Hd(z)

          NUMERATOR COEFFICIENTS                   DENOMINATOR COEFFICIENTS
  ****************************************  ****************************************
   Z**2 TERM      Z TERM       CONST TERM   Z**2 TERM     Z TERM        CONST TERM
  .2190289497   .4380580187   .2190289497    1.00000   -.1463916302   .0225079060
  .2520353794   .5040707588   .2520353794    1.00000   -.1684520245   .1765935421
  .3410534859   .6821070910   .3410534859    1.00000   -.2279487848   .5921629667
          INITIAL GAIN  1.00000000


                         ZEROES OF TRANSFER FUNCTION Hd(z)

   REAL PART      IMAGINARY PART     REAL PART     IMAGINARY PART      RADIUS
-.999262530E+00  .000000000E+00  -.100073801E+01  .000000000E+00  .999262530E+00
-.100000000E+01  .000000000E+00  -.100000000E+01  .000000000E+00  .100000000E+01
-.999408962E+00  .000000000E+00  -.100059139E+01  .000000000E+00  .999408962E+00


                         POLES OF TRANSFER FUNCTION Hd(z)

   REAL PART      IMAGINARY PART     REAL PART     IMAGINARY PART      RADIUS
 .731958151E-01  .130959072E+00   .731958151E-01 -.130959072E+00  .150026351E+00
 .842260122E-01  .411703195E+00   .842260122E-01 -.411703195E+00  .420230344E+00
 .113974392E+00  .761034036E+00   .113974392E+00 -.761034036E+00  .769521258E+00


              NUMERATOR COEFFICIENTS - HIGHEST ORDER FIRST (in z)

 .188271907E-01  .112963161E+00  .282407928E+00  .376543916E+00  .282407928E+00
 .112963161E+00  .188271907E-01


             DENOMINATOR COEFFICIENTS - HIGHEST ORDER FIRST (in z)

 .100000000E+01 -.542792439E+00  .887692610E+00 -.267088214E+00  .143235132E+00
-.184597152E-01  .235370025E-02


            IMPULSE RESPONSE  MIN = -.179722E+00,  MAX =  .452734E+00
```

Figure 6-8. FDAS.OUT File of Example Filter for Cascaded Transpose-Form Implementation

```
FILTER COEFFICIENT FILE
IIR DESIGN
FILTER TYPE                     LOW PASS
ANALOG FILTER TYPE              BUTTERWORTH
PASSBAND RIPPLE IN -dB               -.5000
STOPBAND RIPPLE IN -dB             -20.0000
PASSBAND CUTOFF FREQUENCY      .200000E+03 HERTZ
STOPBAND CUTOFF FREQUENCY      .300000E+03 HERTZ
SAMPLING FREQUENCY             .100000E+04 HERTZ
FILTER DESIGN METHOD: BILINEAR TRANSFORMATION
FILTER ORDER              6 0006h
NUMBER OF SECTIONS        3 0003h
NO. OF QUANTIZED BITS    24 0018h
QUANTIZATION TYPE - FRACTIONAL FIXED POINT
COEFFICIENTS SCALED FOR CASCADE FORM I
       1 00000001     /* shift count for overall gain */
 4194304 00400000     /* overall gain                 */
       0 00000000     /* shift count for section  1 values           */
 1837348 001C0924     /* section  1 coefficient B0                   */
 3674697 00381249     /* section  1 coefficient B1                   */
 1837348 001C0924     /* section  1 coefficient B2                   */
 1228022 0012BCF6     /* section  1 coefficient A1                   */
 -188810 FFFD1E76     /* section  1 coefficient A2                   */
       0 00000000     /* shift count for section  2 values           */
 2114226 002042B2     /* section  2 coefficient B0                   */
 4228452 00408564     /* section  2 coefficient B1                   */
 2114226 002042B2     /* section  2 coefficient B2                   */
 1413078 00158FD6     /* section  2 coefficient A1                   */
-1481374 FFE96562     /* section  2 coefficient A2                   */
       0 00000000     /* shift count for section  3 values           */
 2860964 002BA7A4     /* section  3 coefficient B0                   */
 5721929 00574F49     /* section  3 coefficient B1                   */
 2860964 002BA7A4     /* section  3 coefficient B2                   */
 1912173 001D2D6D     /* section  3 coefficient A1                   */
-4967423 FFB43401     /* section  3 coefficient A2                   */
 .219028949737548 8D+00 3FCC092400000000  .21902905E+00 /* section  1 B0 */
 .438058018684387 2D+00 3FDC092480000000  .43805810E+00 /* section  1 B1 */
 .219028949737548 8D+00 3FCC092400000000  .21902905E+00 /* section  1 B2 */
 .146391630172729 5D+00 3FC2BCF600000000  .14639171E+00 /* section  1 A1 */
-.225079059600830 1D-01 BF970C5000000000 -.22507921E-01 /* section  1 A2 */
 .252035379409790 0D+00 3FD0215900000000  .25203538E+00 /* section  2 B0 */
 .504070758819580 1D+00 3FE0215900000000  .50407076E+00 /* section  2 B1 */
 .252035379409790 0D+00 3FD0215900000000  .25203538E+00 /* section  2 B2 */
 .168452024459838 9D+00 3FC58FD600000000  .16845204E+00 /* section  2 A1 */
-.176593542098999 0D+00 BFC69A9E00000000 -.17659356E+00 /* section  2 A2 */
 .341053485870361 3D+00 3FD5D3D200000000  .34105356E+00 /* section  3 B0 */
 .682107090950012 2D+00 3FE5D3D240000000  .68210712E+00 /* section  3 B1 */
 .341053485870361 3D+00 3FD5D3D200000000  .34105356E+00 /* section  3 B2 */
 .227948784828186 0D+00 3FCD2D6D00000000  .22794882E+00 /* section  3 A1 */
-.592162966728210 4D+00 BFE2F2FFC0000000 -.59216306E+00 /* section  3 A2 */
```

Figure 6-9. COEFF.FLT File for Example Filter Design—Scaled for Cascaded Transpose Form

```
                 COEFF     ident     1,0
                           include   'head1.asm'

                           page      132,66,0,10
                           opt       cex,mex
                           ;
                           ; This program implements an IIR filter in cascaded transpose sections
                           ; The coefficients of each section are scaled for cascaded transpose sections
                           ;

       00FFFF    datin     equ       $ffff                              ;location in Y memory of input file
       00FFFF    datout    equ       $ffff                              ;location in Y memory of output file
       00FFF0    m_scr     equ       $fff0                              ; sci control register
       00FFF1    m_ssr     equ       $fff1                              ; sci status register
       00FFF2    m_sccr    equ       $fff2                              ; sci clock control register
       00FFE1    m_pcc     equ       $ffe1                              ; port c control register
       00FFFF    m_ipr     equ       $ffff                              ; interrupt priority register
       000140    xx        equ       @cvi(20480/(64*1.000))             ;timer interrupt value
       FFD8F1    m_tim     equ       -9999                              ;board timer interrupt value
       000003    nsec      equ       3                                  ;number of second order sections
                           include   'cascade1.asm'
                           ;
                           ; This code segment implements cascaded biquad sections in transpose form
                           ;

                           cascade1  macro     nsec
 m                         ;
 m                         ;         assumes each section's coefficients are divided by 2
 m                         ;
 m                         do        #nsec,_ends                        ;do each section
 m                         macr      y0,y1,a   x:(r1),b    y:(r4)+,y0   ;a=x(n)*bi0/2+wi1/2,b=wi2,y0=bi1/2
 m                         asr       b         a,x0                     ;b=wi2/2,x0=y(n)
 m                         mac       y0,y1,b   y:(r4)+,y0               ;b=x(n)*bi1/2+wi2/2,y0=ai1/2
 m                         macr      x0,y0,b   y:(r4)+,y0               ;b=b+y(n)*ai1/2,y0=bi2/2
 m                         mpy       y0,y1,b   b,x:(r0)+   y:(r4)+,y0   ;b=x(n)*bi2/2,save wi1,y0=ai2
 m                         macr      x0,y0,b   x:(r0),a    a,y1         ;b=b+y(n)*ai2/2,a=next iter wi1,
 m                                                                      ;y1=output of section i
 m                         asr       a         b,x:(r1)+   y:(r4)+,y0   ;a=next iter wi1/2,save wi2,
 m                                                                      ;y0=next iter bi0
 m                         _ends
 m                         endm


       X:0000              org       x:0
       X:0000    state1    dsm       nsec
       X:0004    state2    dsm       nsec
       Y:0000              org       y:0
                 coef
 d     Y:0000 0E0492       dc        $0E0492                            ;b(*,0)/2  =  .10951447  section number  1
 d     Y:0001 1C0924       dc        $1C0924                            ;b(*,1)/2  =  .21902901  section number  1
 d     Y:0002 095E7B       dc        $095E7B                            ;a(*,1)/2  =  .07319582  section number  1
 d     Y:0003 0E0492       dc        $0E0492                            ;b(*,2)/2  =  .10951447  section number  1
```

Figure 6-10. COEFF.ASM File Generated by MGEN for Example Design and Cascaded Transpose-Form Implementation (Sheet 1 of 3)

```
d    Y:0004 FE8F3B                  dc      $FE8F3B                      ;a(*,2)/2    =-.01125395    section number  1
d    Y:0005 102159                  dc      $102159                      ;b(*,0)/2    = .12601769    section number  2
d    Y:0006 2042B2                  dc      $2042B2                      ;b(*,1)/2    = .25203538    section number  2
d    Y:0007 0AC7EB                  dc      $0AC7EB                      ;a(*,1)/2    = .08422601    section number  2
d    Y:0008 102159                  dc      $102159                      ;b(*,2)/2    = .12601769    section number  2
d    Y:0009 F4B2B1                  dc      $F4B2B1                      ;a(*,2)/2    =-.08829677    section number  2
d    Y:000A 15D3D2                  dc      $15D3D2                      ;b(*,0)/2    = .17052674    section number  3
d    Y:000B 2BA7A4                  dc      $2BA7A4                      ;b(*,1)/2    = .34105355    section number  3
d    Y:000C 0E96B6                  dc      $0E96B6                      ;a(*,1)/2    = .11397439    section number  3
d    Y:000D 15D3D2                  dc      $15D3D2                      ;b(*,2)/2    = .17052674    section number  3
d    Y:000E DA1A01                  dc      $DA1A01                      ;a(*,2)/2    =-.29608148    section number  3
                                    include 'body1.asm'
     000040          start          equ     $40                          ;origin for user program
     P:0000                         org     p:$0                         ;origin for reset vector
     P:0000 0C0040                  jmp     start                        ;jump to 'start' on system reset
     P:001C                         org     p:$1c                        ;origin for timer interrupt vector
     P:001C 0BF080                  jsr     filter                       ;jump to 'filter' on timer interrupt
            000057
     P:0040                         org     p:start                      ;origin for user program
     P:0040 0003F8                  ori     #3,mr                        ;disable all interrupts
     P:0041 08F4B2                  movep           #(xx-1),x:m_sccr     ;cd=xx-1 for divide by xx
            00013F
     P:0043 08F4A1                  movep           #$7,x:m_pcc          ;set cc(2;0) to turn on timer
            000007
     P:0045 08F4B0                  movep           #$2000,x:m_scr       ;enable timer interrupts
            002000
     P:0047 08F4BF                  movep           #$c000,x:m_ipr       ;set interrupt priority for sci
            00C000
     P:0049 300000                  move            #state1,r0           ;point to filter state1
     P:004A 310400                  move            #state2,r1           ;point to filter state2
     P:004B 340000                  move            #coef,r4             ;point to filter coefficients
     P:004C 0502A0                  move            #nsec-1,m0           ;addressing modulo nsec
     P:004D 050EA4                  move            #5*nsec-1,m4         ;addressing modulo 5*nsec
     P:004E 0502A1                  move            #nsec-1,m1           ;addressing modulo nsec
     P:004F 200013                  clr     a                            ;initialize internal state storage
     P:0050 0603A0                  rep     #nsec                        ;*    zero state1
     P:0051 565800                  move            a,x:(r0)+            ;*
     P:0052 0603A0                  rep     #nsec                        ;*    zero state2
     P:0053 565900                  move            a,x:(r1)+            ;*
     P:0054 F88000                  move            x:(r0),a  y:(r4)+,y0 ;a=w1 (initially zero) ,y0=b10/2
     P:0055 00FCB8                  andi    #$fc,mr                      ;allow interrupts
     P:0056 0C0056                  jmp     *                            ;wait for interrupt
```

Figure 6-10. COEFF.ASM File Generated by MGEN for Example Design and Cascaded Transpose-Form Implementation (Sheet 2 of 3)

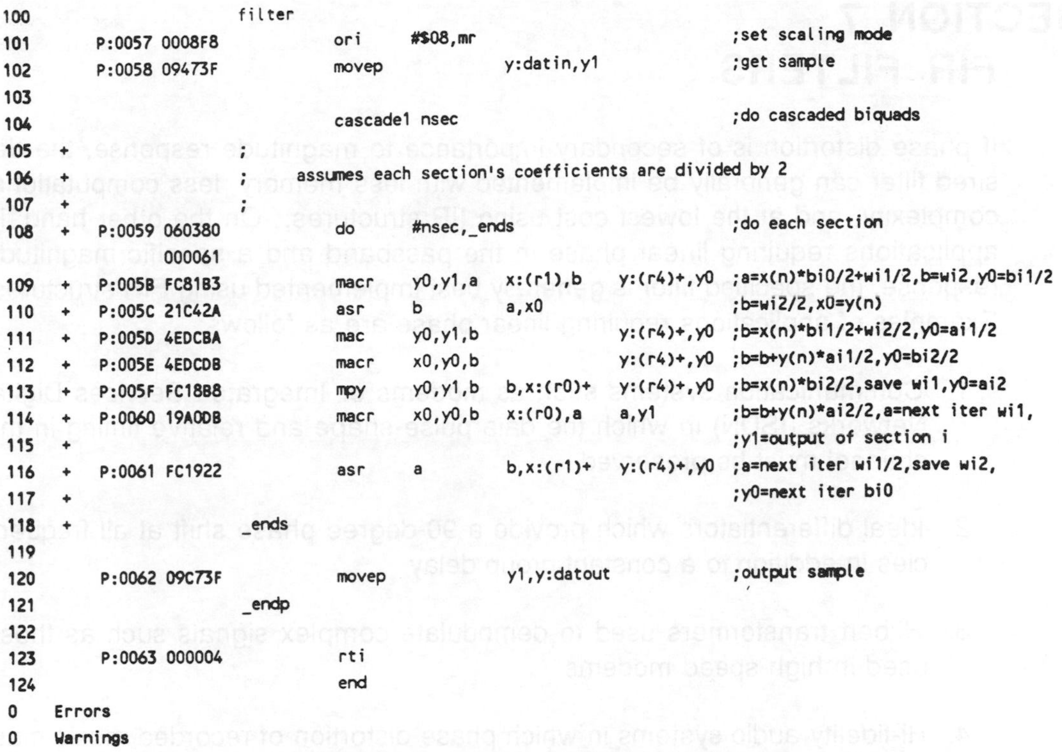

```
                        filter
      P:0057 0008F8                  ori     #$08,mr                                 ;set scaling mode
      P:0058 09473F                  movep                  y:datin,y1                ;get sample

                                     cascade1 nsec                                    ;do cascaded biquads
   +                    ;
   +                    ;  assumes each section's coefficients are divided by 2
   +                    ;
   +  P:0059 060380                  do      #nsec,_ends                              ;do each section
             000061
   +  P:005B FC81B3                  macr    y0,y1,a    x:(r1),b     y:(r4)+,y0     ;a=x(n)*bi0/2+wi1/2,b=wi2,y0=bi1/2
   +  P:005C 21C42A                  asr     b          a,x0                        ;b=wi2/2,x0=y(n)
   +  P:005D 4EDCBA                  mac     y0,y1,b                  y:(r4)+,y0     ;b=x(n)*bi1/2+wi2/2,y0=ai1/2
   +  P:005E 4EDCDB                  macr    x0,y0,b                  y:(r4)+,y0     ;b=b+y(n)*ai1/2,y0=bi2/2
   +  P:005F FC18B8                  mpy     y0,y1,b    b,x:(r0)+    y:(r4)+,y0     ;b=x(n)*bi2/2,save wi1,y0=ai2
   +  P:0060 19A0DB                  macr    x0,y0,b    x:(r0),a     a,y1           ;b=b+y(n)*ai2/2,a=next iter wi1,
   +                                                                                  ;y1=output of section i
   +  P:0061 FC1922                  asr     a          b,x:(r1)+    y:(r4)+,y0     ;a=next iter wi1/2,save wi2,
   +                                                                                  ;y0=next iter bi0
   +                    _ends

      P:0062 09C73F                  movep              y1,y:datout                 ;output sample
                        _endp

      P:0063 000004                  rti
                                     end
0   Errors
0   Warnings
```

Figure 6-10. COEFF.ASM File Generated by MGEN for Example Design and Cascaded Transpose-Form Implementation (Sheet 3 of 3)

SECTION 7
FIR FILTERS

If phase distortion is of secondary importance to magnitude response, the desired filter can generally be implemented with less memory, less computational complexity, and at the lowest cost using IIR structures. On the other hand, in applications requiring linear phase in the passband and a specific magnitude response, the specified filter is generally best implemented using FIR structures. Examples of applications requiring linear phase are as follows:

1. Communication systems such as modems or Integrated Services Digital Networks (ISDN) in which the data pulse shape and relative timing in the channel must be preserved

2. Ideal differentiators which provide a 90-degree phase shift at all frequencies in addition to a constant group delay

3. Hilbert transformers used to demodulate complex signals such as those used in high-speed modems

4. Hi-fidelity audio systems in which phase distortion of recorded music must be minimized to reproduce the original sound with as much fidelity as possible

5. System synthesis in which the system impulse response is known as *a priori*. FIR filters are also important because they are all-zero filters (i.e., no feedback) and are therefore guaranteed to be stable.

7.1 LINEAR-PHASE FIR FILTER STRUCTURE

The basic structure of a FIR filter is simply a tapped delay line in which the output from each tap is summed to generate the filter output. This is shown in Figure 7-1. This structure can be represented mathematically as

$$y(n) = \sum_{i=0}^{N-1} h(i)x(n-i) \qquad (7\text{-}1)$$

where x(n) is the most recent (t = nT) input signal sample, x(n–i) is signal samples delayed by i sample periods (iT), h(i) is the tap weights (or filter coefficients), and y(n) is the filter output at time t = nT.

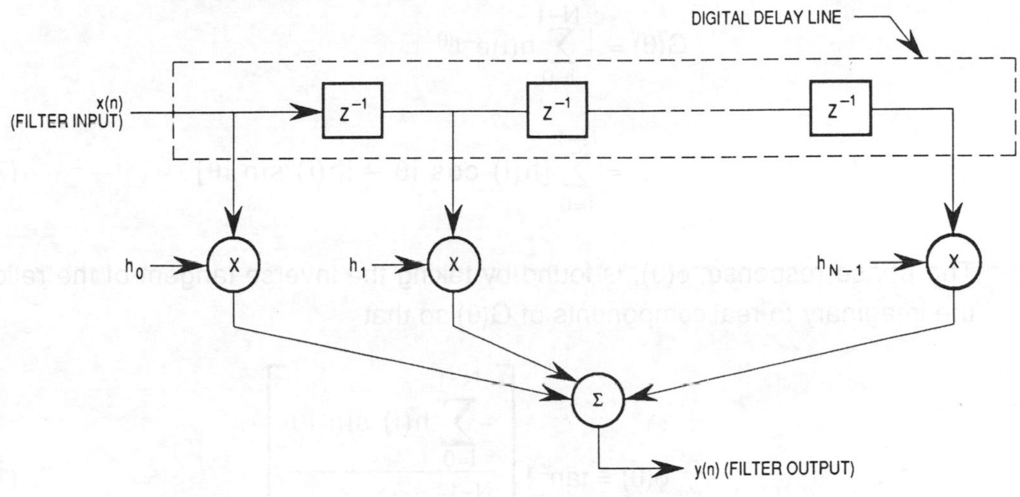

Figure 7-1. FIR Structure

From this structure, it is easy to see why the filter is termed finite—the impulse response of the filter will be identically zero after N sample periods because an impulse input (i.e., x(n) = 1, x(n–i) = 0 for i ≠ 0) will have traversed the entire delay line at time t = NT. That is, the impulse response of FIR filters will only last for a finite period of time. This response is in contrast to IIR filters which will "ring" in response to an impulse for an infinite period of time. The values of the coefficients represent the impulse response of the FIR filter as can be seen by evaluating Equation (7-1) over N sample periods for a single-unit input pulse at time t = 0.

There are no feedback terms in the structure (i.e., Equation (7-1) has no denominator) but rather only N zeros. By taking the z-transform of Equation (7-1), the transfer function of the filter becomes

$$H(z) = \frac{Y(z)}{X(z)} = \sum_{i=0}^{N-1} h(i)z^{-i} \qquad (7\text{-}2)$$

Equation (7-2) is a polynomial in z of order N. The roots of this polynomial are the N zeros of the filter.

The same procedures used to calculate the magnitude and phase response of IIR filters can be applied to FIR filters. Accordingly, the gain, $G(\theta)$, can be obtained by substituting $z = e^{j\theta}$ into Equation (7-2) (where $\theta = 2\pi f/f_s$ is the normalized digital frequency), then taking the absolute value so that

$$G(\theta) = \left| \sum_{i=0}^{N-1} h(i)e^{-ji\theta} \right|$$

$$= \sum_{i=0}^{N-1} [h(i) \cos i\theta - jh(i) \sin i\theta] \qquad (7\text{-}3)$$

The phase response, $\phi(\theta)$, is found by taking the inverse tangent of the ratio of the imaginary to real components of $G(\theta)$ so that

$$\phi(\theta) = \tan^{-1} \left[\frac{-\sum_{i=0}^{N-1} h(i) \sin i\theta}{\sum_{i=0}^{N-1} h(i) \cos i\theta} \right] \qquad (7\text{-}4)$$

where it has been implicitly assumed that h(i) is real.

Intuitively, it can be reasoned that, for any pulse to retain its general shape and relative timing (i.e., pulse width and time delay to its peak value) after passing through a filter, the delay of each frequency component making up the pulse must be the same so that each component recombines in phase to reconstruct the original shape. In terms of phase, this implies that the phase delay must be linearly related to frequency or

$$\phi(\theta) = -\tau\theta \qquad (7\text{-}5)$$

This relationship is illustrated in Figure 7-2 in which a pulse having two components, sin ωt and sin 2ωt, passes through a linear-phase filter having a delay of two cycles per Hz and a constant magnitude response equal to unity. That is, the fundamental is delayed by two cycles (4π) and the first harmonic is delayed by four cycles (8π). The group delay of a system, τ_g, is defined by taking the derivative of $\phi(\theta)$:

$$\tau_g \equiv -\frac{d\phi}{d\theta} \qquad (7\text{-}6)$$

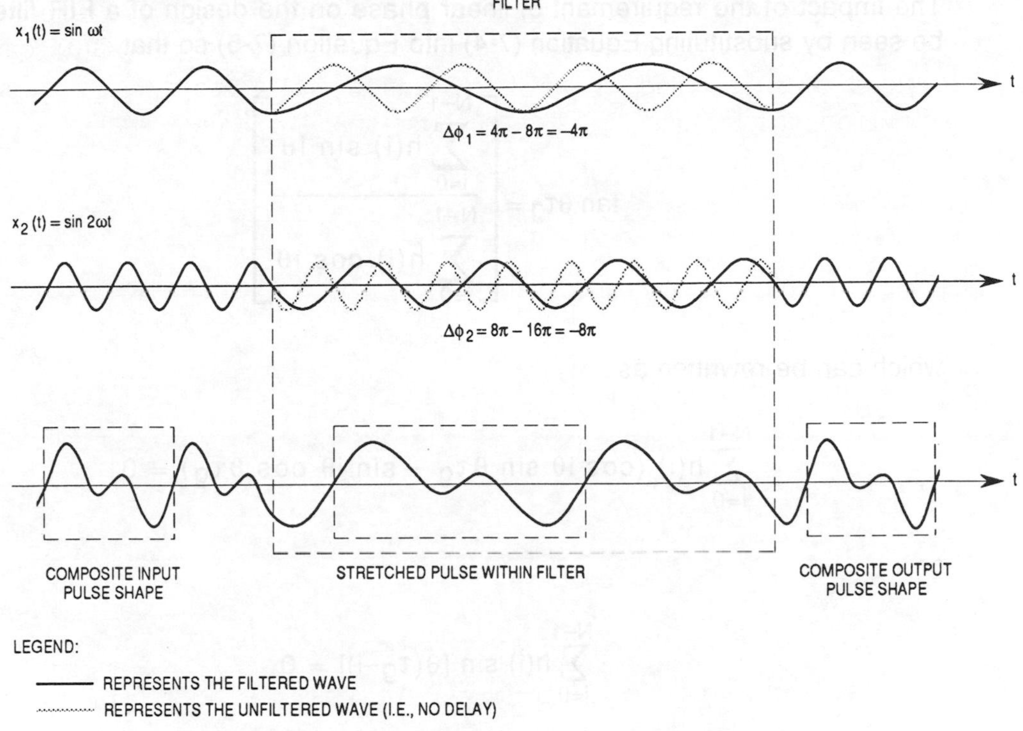

Figure 7-2. Signal Data through a FIR Filter

Since θ is the normalized frequency, τ_g in Equation (7-6) is a dimensionless quantity and can be related to the group delay in seconds by dividing by the sample frequency, f_s. For a linear-phase system, τ_g is independent of θ and is equal to τ. This fact can be seen by substituting Equation (7-5) into Equation (7-6).

In this example, τ_g is two cycles per Hz. Note that the pulse shape within the filter has been retained but is twice the width of the original pulse; that is, the change in the pulse width is equal to the group delay. The negative sign indicates the phase is retarded or delayed (i.e., a causal system).

The impact of the requirement of linear phase on the design of a FIR filter can be seen by substituting Equation (7-4) into Equation (7-5) so that

$$\tan \theta\tau_g = \left[\frac{\sum_{i=0}^{N-1} h(i) \sin i\theta}{\sum_{i=0}^{N-1} h(i) \cos i\theta}\right] \quad (7\text{-}7)$$

which can be rewritten as

$$\sum_{i=0}^{N-1} h(i)\,(\cos i\theta \sin \theta\tau_g - \sin i\theta \cos \theta\tau_g) = 0$$

or

$$\sum_{i=0}^{N-1} h(i) \sin\left[\theta(\tau_g - i)\right] = 0 \quad (7\text{-}8)$$

The solution to Equation (7-8) is the constraint on τ_g for a FIR filter to be linear phase. The solution to Equation (7-8) can be found by expanding the left-hand side (LHS) as follows:

$$\text{LHS} = h_0 \sin \theta\tau_g + h_1 \sin \theta(\tau_g - 1) + h_2 \sin \theta(\tau_g - 2) + \ldots$$

$$+ h_{N-3} \sin \theta(\tau_g - N + 3) + h_{N-2} \sin \theta(\tau_g - N + 2) + h_{N-1} \sin \theta(\tau_g - N + 1)$$

so that if

$$\tau_g = \frac{(N-1)}{2} \quad (7\text{-}9)$$

then for every positive argument there will be a corresponding negative argument. For example, the argument for the h_1 term becomes

$$\theta\left(\frac{N-1}{2} - 1\right) = \theta\left(\frac{N-3}{2}\right)$$

which is the negative of the argument for the h_{N-2} term, i.e.,

$$\theta\left(\frac{N-1}{2} - N + 2\right) = -\theta\left(\frac{N-3}{2}\right)$$

so that if

$$h(i) = h(N-1-i) \quad \text{for } 0 \le i \le N-1 \tag{7-10}$$

then Equations (7-9) and (7-10) represent the solution to Equation (7-8). By substituting Equation (7-9) into Equation(7-5), the phase of a linear-phase FIR filter is given by

$$\phi(\theta) = -\left(\frac{N-1}{2}\right)\theta \tag{7-11}$$

Therefore, a nonrecursive filter (FIR), unlike a recursive filter (IIR), can have a constant time delay for all frequencies over the entire range (from 0 to $f_s/2$). It is only necessary that the coefficients (and therefore impulse response) be symmetrical about the midpoint between samples (N–2)/2 and N/2 for even N or about sample (N–1)/2 for odd N (see Equation (7-10)). When this symmetry exists, τ_g for the filter will be $\left(\frac{N-1}{2}\right)$T seconds.

7.2 LINEAR-PHASE FIR FILTER DESIGN USING THE FREQUENCY SAMPLING METHOD

Implementing a FIR filter in DSP hardware such as the DSP56001 is a relatively simple task. Determining a set of coefficients that describe a given impulse response of a filter is also a straightforward procedure. However, deriving the optimal coefficients necessary to obtain a particular response in the frequency domain is not always as easy. The following paragraphs introduce a simple method to determine a set of coefficients based on a desired arbitrary frequency response (often referred to as the frequency sampling method). This method has the distinct advantage of being done in real time. However, the most efficient determination of coefficients is best done by utilizing a software filter design system such as FDAS to perform numerical curve-fitting and optimization, a procedure that necessitates using a computer. For example, the inverse Fourier transform integral is used to determine the FIR coefficients from a response specification in the frequency domain, which generally requires a numerical integration procedure. The inverse discrete Fourier transform (IDFT), which can be implemented using the inverse fast Fourier transform (IFFT) algorithm, is discussed in the following paragraphs. One reason for choosing this approach is to demonstrate a method that can be used to determine coefficients in real time since the fast Fourier transform (FFT) can be implemented in the same DSP56001 hardware as the FIR filter.

The question is "How must the filter be specified in the frequency domain so that Equations (7-10) and (7-11) are realized?". That is, starting with the definition of the filter in the frequency domain, a method for calculating the filter coefficients is required. Beginning with Equation (7-2), the z-transform of the FIR filter, and evaluating this transfer function on the unit circle at N equally spaced normalized frequencies (i.e., $z = e^{j2\pi k/N}$) produces

$$H(k) = H(e^{j2\pi k/N}) = \sum_{i=0}^{N-1} h(i)\ e^{-j2\pi ki/N} \tag{7-12}$$

where $0 \le k \le N-1$. h(i) can be solved in terms of the frequency response at the discrete frequencies, H(k), by multiplying both sides of Equation (7-12) by $e^{j2\pi km/N}$ and summing over k as follows:

$$\sum_{k=0}^{N-1} H(k)\ e^{j2\pi km/N} = \sum_{k=0}^{N-1}\ \sum_{i=0}^{N-1} h(i)\ e^{-j2\pi ki/N}\ \ e^{j2\pi km/N}$$

$$= \sum_{i=0}^{N-1} h(i) \sum_{k=0}^{N-1} e^{-j2\pi ki/N}\ \ e^{j2\pi km/N}$$

$$= \sum_{i=0}^{N-1} h(i)\ N\delta_{im}$$

$$= N\ h(m) \tag{7-13}$$

where δ_{im} is the Kronecker delta, which is equal to one when i = m but is zero otherwise. Equation (7-13) can be used to find h(i) by simply setting i = m:

$$h(i) = \frac{1}{N} \sum_{k=0}^{N-1} H(k)\ e^{j2\pi ki/N} \tag{7-14}$$

Equation (7-14) is the IDFT of the filter response.

In general, the discrete Fourier coefficients are complex; therefore, H(k) can be represented as

$$H(k) = A(k)\ e^{j\phi(k)}$$

where

$$A(k) \equiv | H(k) |$$

so that Equation (7-14) becomes

$$h(i) = \frac{1}{N} \sum_{k=0}^{N-1} A(k)\, e^{j\phi(k)}\, e^{j2\pi ki/N} \tag{7-15}$$

At this point, the linear-phase constraints, Equations (7-10) and (7-11), can be applied. The constraint that the coefficients be real and symmetrical when h(i) is complex (see Reference 1) implies that

$$h(i) = h^*(N-1-i)$$

where * signifies the complex conjugate; therefore,

$$h^*(N-1-i) = \frac{1}{N} \sum_{k=0}^{N-1} A(k)\, e^{-j\phi(k)}\, e^{-j2\pi k(N-1-i)/N}$$

$$= \frac{1}{N} \sum_{k=0}^{N-1} A(k)\, e^{-j[\phi(k)+2\pi k(N-1)/N]}\, e^{j2\pi ki/N} \tag{7-16}$$

Equation (7-16) will be identical to Equation (7-15) if

$$\phi(k) = -\phi(k) - \frac{2\pi k(N-1)}{N} + 2\pi r \qquad \text{for } r = 0, 1, 2, \ldots$$

or

$$\phi(k) = \pi r - \pi k \frac{(N-1)}{N} \tag{7-17}$$

What are the constraints on r and A(k) which will guarantee a purely real response for N even? Substituting Equation (7-17) into Equation (7-15) yields

$$h(i) = \frac{1}{N} \sum_{k=0}^{N-1} A(k)\, e^{j[\pi r - \pi k(N-1)/N + 2\pi ki/N]} \tag{7-18}$$

Expanding Equation (7-18) for even N yields

$$h(i) = \frac{1}{N} \{A(0)\, e^{j\pi r} + A(1)\, e^{j[\pi r - \pi(N-1)/N + 2\pi i/N]} + \ldots$$

$$+ A(N/2)\, e^{j[\pi r - \pi(N-1)/2 + \pi i]} + \ldots + A(N-1) e^{j[\pi r - \pi(N-1)(N-1)/N + 2\pi(N-1)i/N]}\}$$

Consider the A(k) and A(N–k) terms; if r = 0 for the A(k) term and r = 1 for the A(N–k) term, then the argument for the A(N–k) term is the negative of the argument for the A(k) term, given that N is even. That is,

$$e^{j[\pi - \pi(N-k)(N-1)/N + 2\pi(N-k)i/N]} = e^{j[\pi k(N-1)/N - 2\pi ki/N]}\ e^{j[\pi - \pi(N-1) + 2\pi i]}$$

$$= e^{-j[-\pi k(N-1)/N + 2\pi ki/N]}\ e^{-j[\pi(N-2) + 2\pi i]}$$

$$= e^{-j[-\pi k(N-1)/N - 2\pi ki/N]}$$

Therefore, if A(k) = A(N–k), then all imaginary components in Equation (7-18) will cancel and h(i) will be purely real, which is the desired result. In general, for N even, the formulas for the frequency sampling method can be reduced to

$$h(i) = \frac{1}{N}\left\{A(0) + \sum_{k=1}^{N-1} A(k)\, e^{j[\pi r - \pi k(N-1)/N + 2\pi ki/N]}\right\} \qquad r = 0 \text{ for } 0 \le k < \frac{N}{2}$$

$$r = 1 \text{ for } \frac{N}{2} < k \le N-1$$

$$= \frac{1}{N}\left\{A(0) + \sum_{k=1}^{N/2-1} A(k)\, [e^{j[-\pi k(N-1)/N + 2\pi ki/N]} + e^{j[\pi - \pi(N-k)(N-1)/N + 2\pi(N-k)i/N]}]\right\}$$

$$= \frac{1}{N}\left\{A(0) + \sum_{k=1}^{N/2-1} A(k)\, [e^{j[-\pi k(N-1)/N + 2\pi ki/N]} + e^{-j[-\pi k(N-1)/N + 2\pi ki/N]}]\right\}$$

$$= \frac{1}{N}\left\{A(0) + \sum_{k=1}^{N/2-1} A(k)\, [e^{-j\pi k}\, e^{j[\pi k/N + 2\pi ki/N]} + e^{j\pi k}\, e^{-j[\pi k/N + 2\pi ki/N]}]\right\}$$

$$= \frac{1}{N}\left\{A(0) + 2\sum_{k=1}^{N/2-1} A(k)\, (-1)^k \cos\frac{\pi k}{N}(1+2i)\right\} \qquad (7\text{-}19)$$

given

$$A(k) = A(N-k) \qquad \text{for } 1 \le k \le \frac{N}{2} - 1$$

with

$$A(N/2) = 0$$

and

$$\phi(k) = -\pi k \frac{(N-1)}{N} \qquad \text{for } 0 \le k \le \frac{N}{2} - 1$$

$$\phi(k) = \pi - \pi k \frac{(N-1)}{N} \qquad \text{for } \frac{N}{2} + 1 \le k \le N - 1$$

In summary, if a filter with an even number of symmetrical real coefficients is desired, then the phase must be linear, and the frequency response must be symmetrical about N/2 and zero at N/2 (see Reference 1).

As an example, consider the arbitrary filter specified in Figure 7-3. In this example, N = 32 and an arbitrary lowpass and bandpass combination is specified. Figure 7-4 shows the result of transforming the polar coordinate filter specification into rectangular coordinates, yielding real and imaginary components of the transfer function. This transformation is accomplished by treating the magnitude of H(k) as the length of a vector in polar coordinates and the phase as the angle. The x component (real part) is the product of the length (magnitude of H(k)) and the cosine of the phase. Likewise, the y component (imaginary part) is the product of the length and the sine of the phase angle. This transformation to rectangular coordinates is necessary to perform the IDFT (or IFFT) calculation:

$$h(n) = \frac{1}{N} \sum_{k=0}^{N-1} H(k)\ e^{jk\theta_n} \tag{7-20}$$

where H(k) is a complex number and $\theta_n = 2\pi n/N$.

When the filter's transfer function, H(k), is described as in Figures 7-3 and 7-4 (i.e., the real part symmetric about N/2 and the imaginary part asymmetric about N/2), h(n) is strictly real. If h(n) were complex, the FIR filter would be much more difficult to implement since twice as many terms would be present. Figure 7-5 shows the results of the IDFT applied to the example arbitrary filter specification.

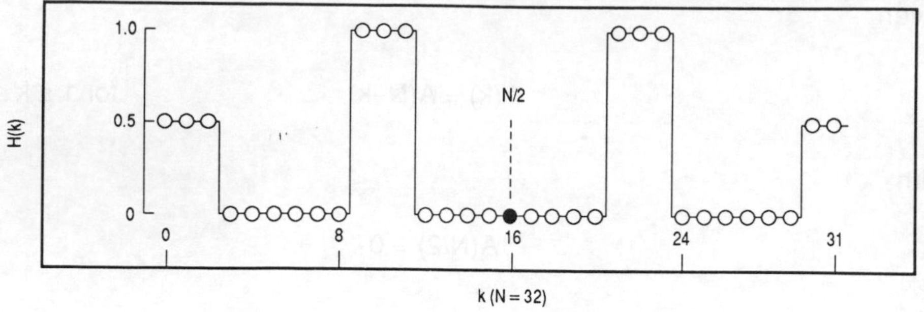

(a) Arbitrary Input Specification for Frequency Response Magnitude

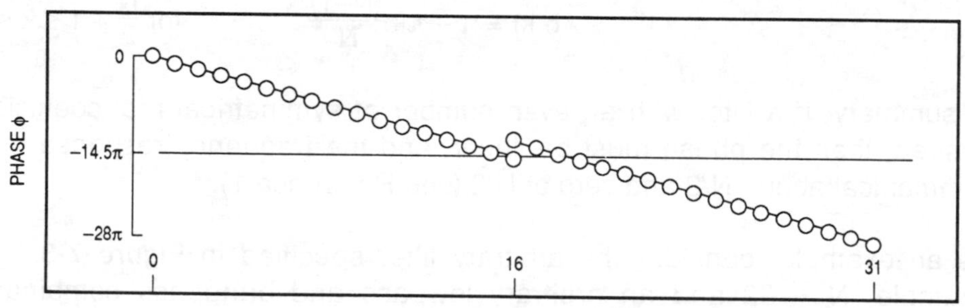

(b) Necessary Phase Specifications for Frequency Response Magnitude

Figure 7-3. Arbitrary Filter Example

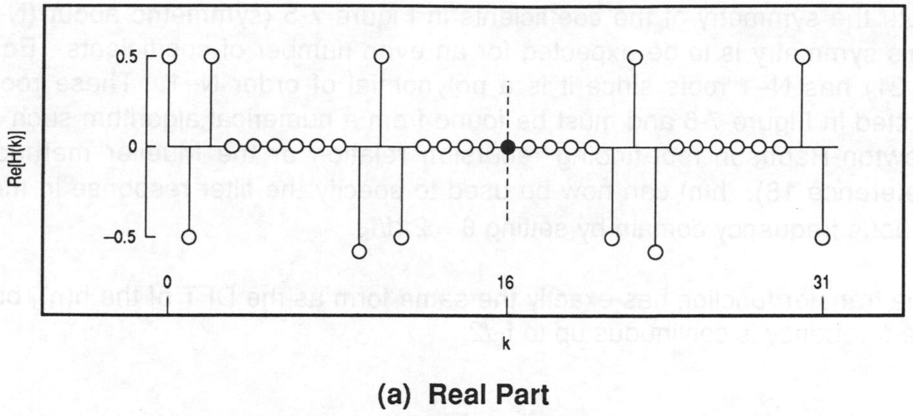

(a) Real Part

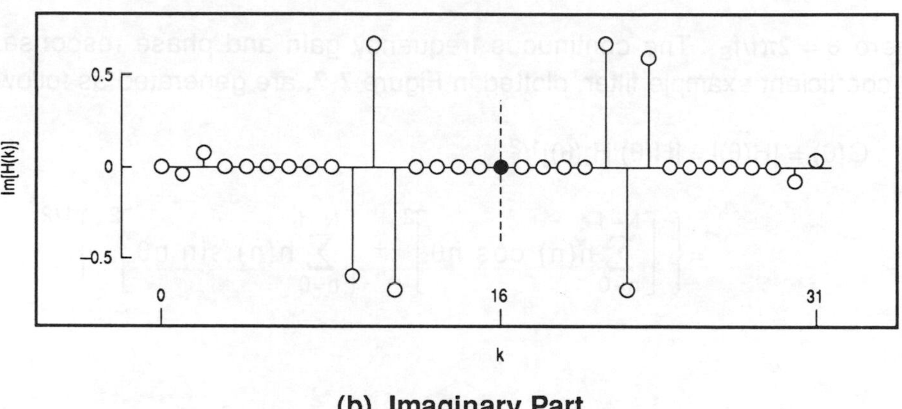

(b) Imaginary Part

Figure 7-4. Response Transformed from Polar Coordinates

Note the symmetry of the coefficients in Figure 7-5 (symmetric about (N–1)/2). This symmetry is to be expected for an even number of coefficients. Equation (7-21) has N–1 roots since it is a polynomial of order N–1. These roots are plotted in Figure 7-6 and must be found from a numerical algorithm such as the Newton-Raphson root-finding recursion relation or the Mueller method (see Reference 18). h(n) can now be used to specify the filter response in the continuous frequency domain by setting $\theta = 2\pi f/f_S$.

The transfer function has exactly the same form as the DFT of the h(n), but now the frequency is continuous up to $f_S/2$:

$$H(\theta) = \sum_{n=0}^{N-1} h(n)\ e^{-jn\theta} \tag{7-21}$$

where $\theta = 2\pi f/f_S$. The continuous frequency gain and phase response of the 32-coefficient example filter, plotted in Figure 7-7, are generated as follows:

$$G(\theta) = |H(\theta)| = [H(\theta)\ H^*(\theta)]^{1/2}$$

$$= \left\{ \left[\sum_{n=0}^{N-1} h(n) \cos n\theta \right]^2 + \left[\sum_{n=0}^{N-1} h(n) \sin n\theta \right]^2 \right\}^{1/2} \tag{7-22}$$

and

$$\phi(\theta) = -\tan^{-1} \left\{ \frac{\sum_{n=0}^{N-1} h(n) \sin n\theta}{\sum_{n=0}^{N-1} h(n) \cos n\theta} \right\} \tag{7-23}$$

where h(n) is the value obtained from Equation (7-20) or equivalently from Equation (7-19). Note that the gain plot (see Figure 7-7) exactly intersects the discrete frequency points originally specified in Figure 7-3. Clearly, $G(\theta)$ and $\phi(\theta)$ have many discontinuities. Due to the symmetry of linear-phase FIR filters, analytic expressions can be found for $H(\omega)$ as shown in Equations (7-27) and (7-28).

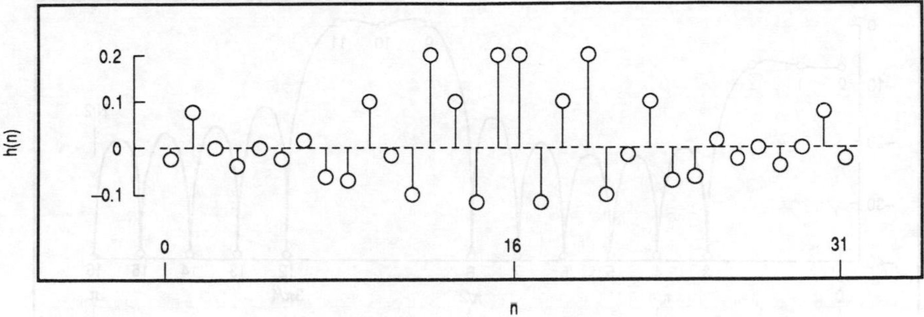

Figure 7-5. FIR Coefficients from Equation (7-2) for Filter Example

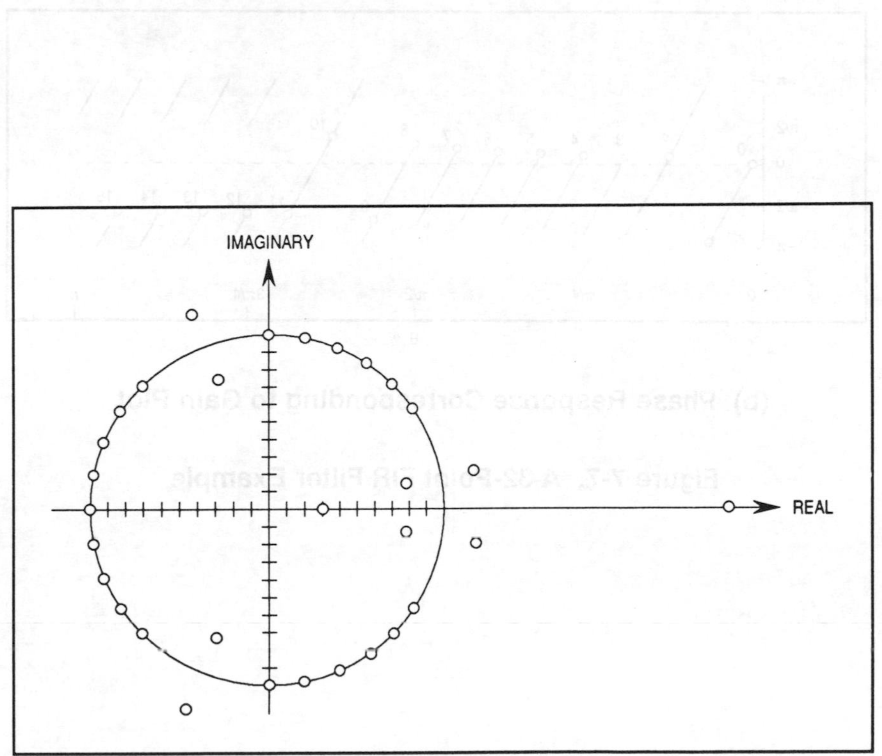

Figure 7-6. Roots (Zeros) of Equation (7-21) for Filter Example

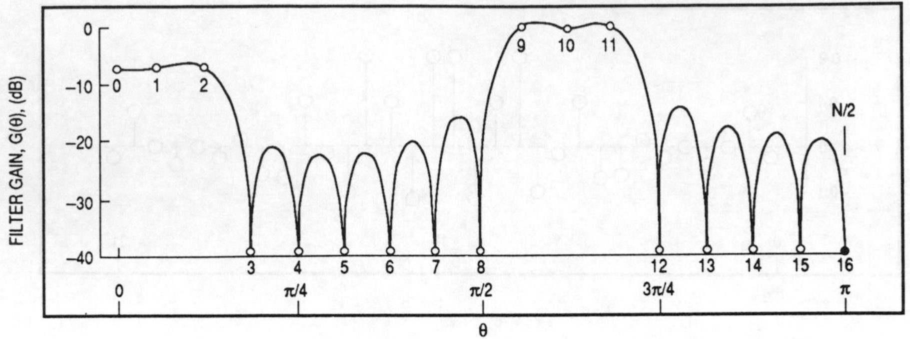

(a) Gain (Magnitude of H(z)) Plot

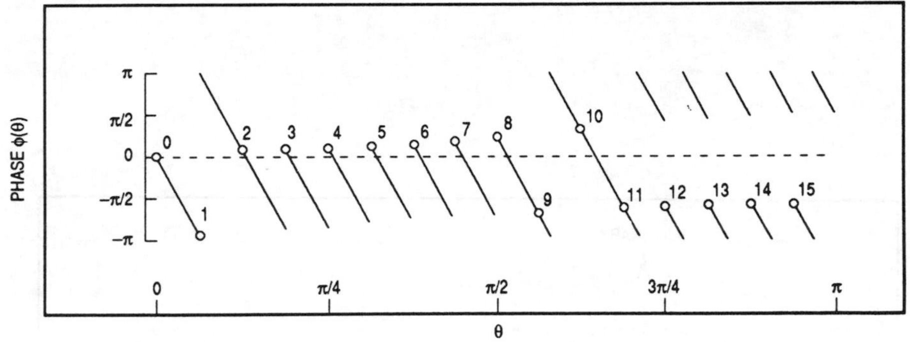

(b) Phase Response Corresponding to Gain Plot

Figure 7-7. A 32-Point FIR Filter Example

For comparison, Figure 7-8 shows the log magnitude response of the example filter with larger values of N. The stopband attenuation is not improved as much as expected when the number of coefficients is increased; however, the sharpness of the transition band is significantly enhanced because the approach discussed has implicitly assumed a rectangular window function. To improve this situation, a smoothing window, w(n), can be used (see Reference 10).

If h(n) is modified by a window function, w(n), $h_w(n) = h(n)w(n)$ for $0 \leq n \leq N-1$, the gain of Equation (7-22) (shown in Figure 7-9(a)) can be greatly improved. The window function, w(n), goes to zero at both ends (n = 0 and n = N) and is unity at the center; w(n) is symmetrical so as not to disrupt the linear-phase characteristics of the filter. Figure 7-9(b) uses a window function described by $w(n) = \sin^2[\pi n/(N-1)]$ (also known as a Hanning window) to demonstrate the sensitivity of the gain to windowing. w(n) is basically an envelope function used to taper the ends of h(n) smoothly. The rounding of the transition-band edges in Figure 7-9(b) is the tradeoff for windowing; however, in most applications, this tradeoff is well worthwhile.

Window functions have a powerful effect on the stopband attenuation as well as the passband transition slope.The best passband transition slope performance is achieved with the rectangular window, but this window results in very poor stopband performance and often severe passband fluctuation (or ripple). All window functions have the effect of increasing the stopband attenuation and reducing the passband ripple at the expense of increasing the width of the transition region. However, for most window functions, the passband ripple is relatively insensitive to N.

Windowing is described in virtually any DSP textbook (see **REFERENCES**). Also, windowing is discussed with practical examples in *Implementation of Fast Fourier Transforms on Motorola's DSP56000/DSP56001 and DSP96002 Digital Signal Processors* (see Reference 19).

7.3 FIR FILTER DESIGN USING FDAS

Figure 7-10 shows an example (log magnitude and impulse response) of a bandpass filter generated with the FDAS software package using the Kaiser window.

A totally different approach to FIR filter design, the equiripple method, is based on finding an optimum approximation to the ideal or desired response, D(θ). An optimum approximation can be found because of the inherent symmetry of the coefficients of linear-phase filters.

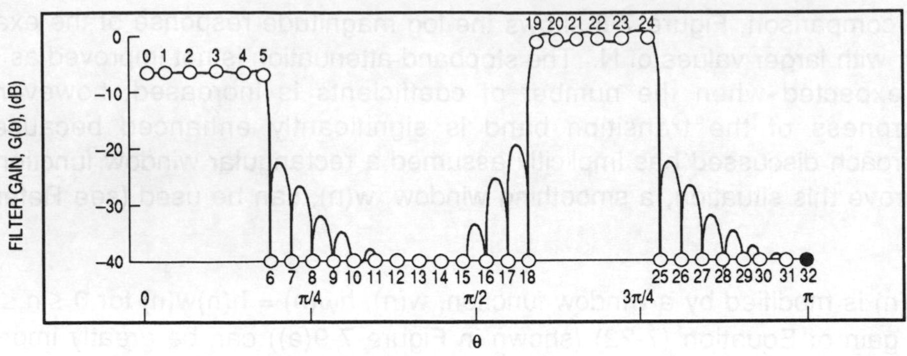

(a) N = 64

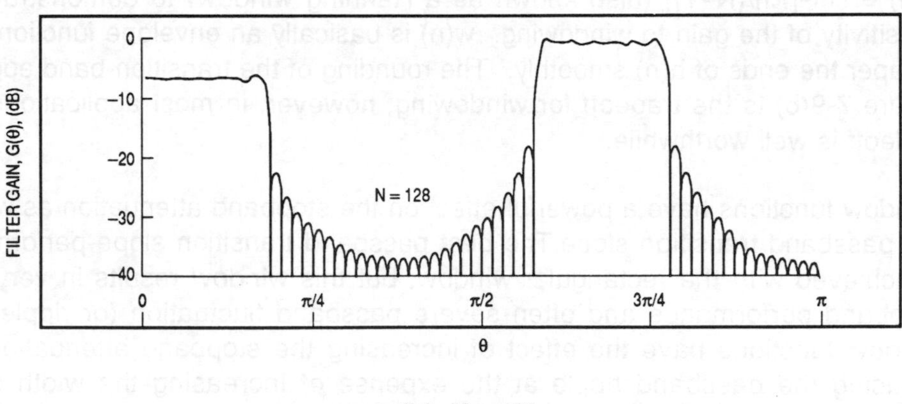

(b) N = 128

Figure 7-8. Log Magnitude Response of Filter Example with Larger N Values

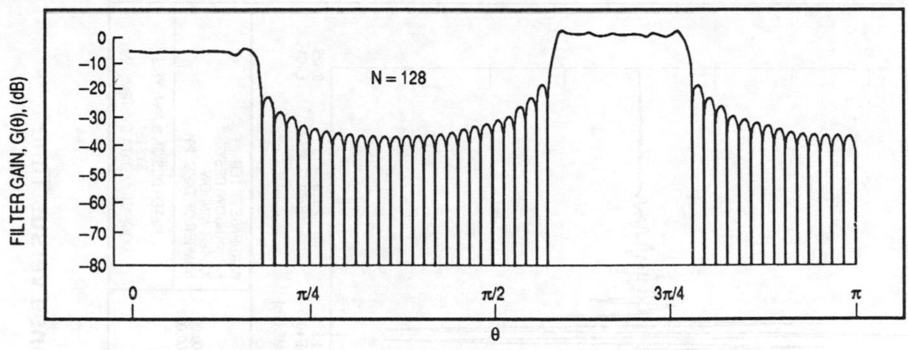

(a) Rectangular Window

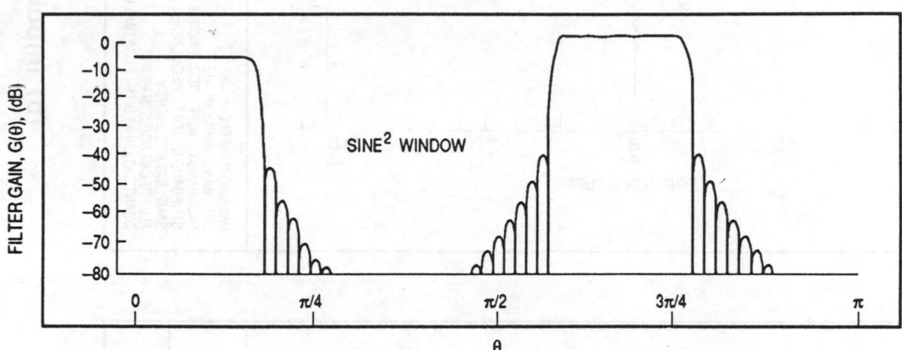

(b) Sine² (Hanning) Window

Figure 7-9. Window Function Effects on Filter Example

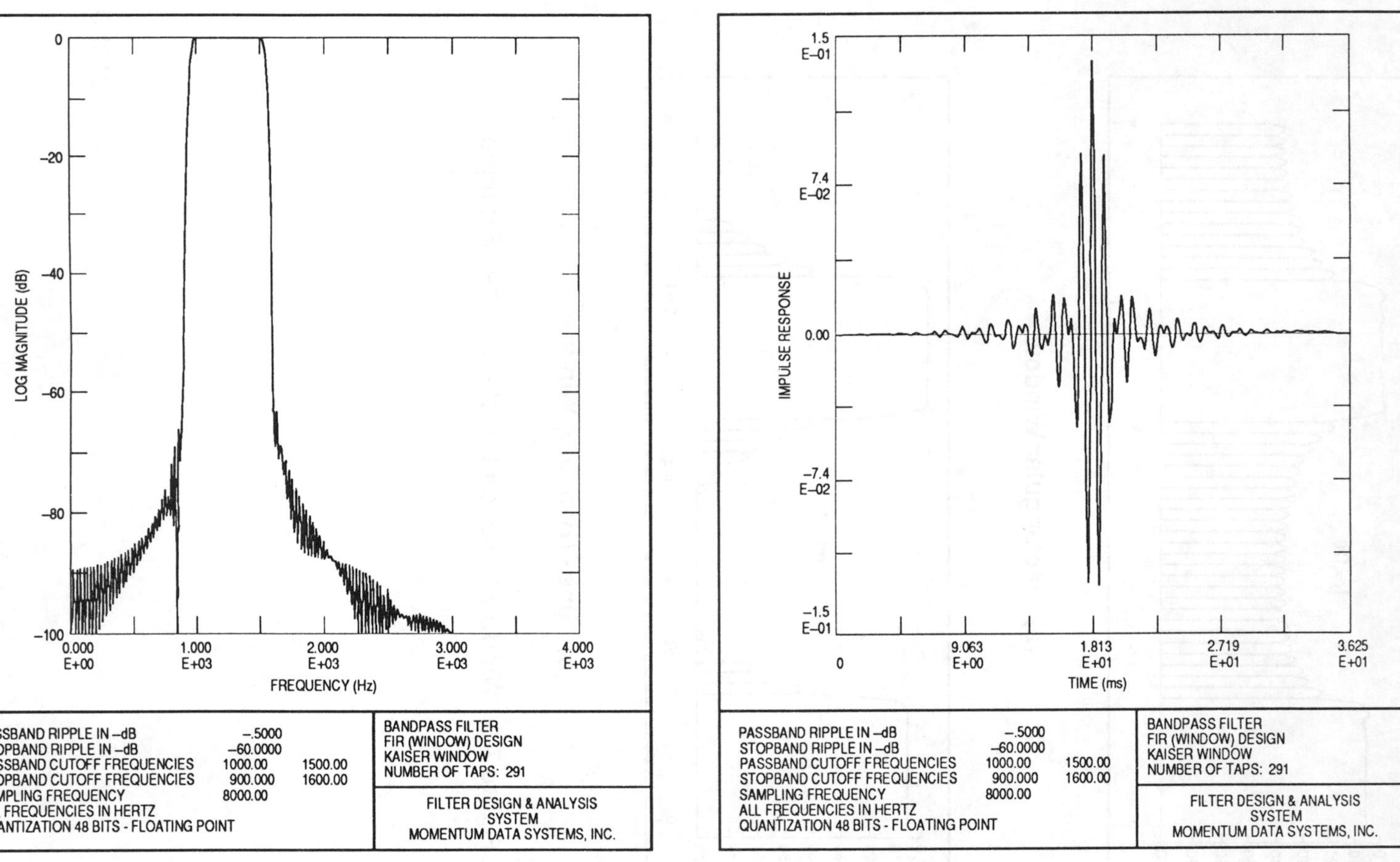

(a) Log Magnitude versus Frequency

(b) Impulse Response versus Time

Figure 7-10. FDAS Output for FIR Bandpass Filter Example with a Kaiser Window

Recall that the continuous frequency response of a FIR filter can be found by setting $z = e^{j\theta}$ in the z-transform so that

$$H(\theta) = \sum_{n=0}^{N-1} h(n)\ e^{-j\theta n} \tag{7-24}$$

If the linear-phase factor is factored out, Equation (7-24) can be written as

$$H(\theta) = e^{-j\theta(N-1)/2} \sum_{n=0}^{N-1} h(n)\ e^{j\theta[(N-1)/2-n]}$$

$$= e^{-j\theta(N-1)/2}\{h(0)\ e^{j\theta(N-1)/2)} + h(1)\ e^{j\theta[((N-1)/2)-1]} + \ldots$$

$$h(N-1))\ e^{-j\theta(N-1)/2)} + h(N-2)\ e^{-j\theta[((N-1)/2)-1]}\} \tag{7-25}$$

Using Euler's identity, Equation (7-25) can be grouped into sine and cosine terms as follows:

$$H(\theta) = e^{-j\theta(N-1)/2}\{[h(0)+h(N-1)]\cos\left[\theta\ \frac{(N-1)}{2}\right] + j[h(0)-h(N-1)]\sin\left[\theta\ \frac{(N-1)}{2}\right]$$

$$+ [h(1)+h(N-2)]\cos\left[\theta\ \frac{(N-1)}{2} - 1\right] + j[h(1)-h(N-2)]\sin\left[\theta\ \frac{(N-1)}{2} - 1\right]$$

$$+ \ldots\} \tag{7-26}$$

If it is assumed that linear phase is achieved by even symmetry (i.e., h(n) = h(N–1–n) and that N is even, Equation (7-26) reduces to

$$H(\theta) = e^{-j\theta(N-1)/2} \sum_{n=0}^{N/2-1} 2h(n)\cos\left[\theta\left(\frac{N-2}{2} - n\right)\right]$$

which, with a change of variable, k = N/2–n–1, can be written as follows:

$$H(\theta) = e^{-j\theta(N-1)/2} \sum_{n=0}^{N/2-1} 2h\left(\frac{N}{2} - k - 1\right)\cos\left[\theta\left(k + \frac{1}{2}\right)\right] \tag{7-27}$$

which is in the following form:

$$H(\theta) = A(\theta)e^{jP(\theta)} \tag{7-28}$$

where $A(\theta)$ is a real value amplitude function and $P(\theta)$ is a linear-phase function. The linear functions, $A(\theta)$ and $P(\theta)$, are to be contrasted with the inherently nonlinear absolute value and arctangent functions in Equations (7-22) and (7-23). For all four types of linear-phase filters (N even or odd and symmetry even or odd), $A(\theta)$ can be expressed as a sum of cosines (see Reference 21).

Since an ideal filter cannot be realized, an approximation must be used. If the desired frequency response, $D(\theta)$, can be specified in terms of a deviation, δ, from the ideal response, then the error function in Equation (7-29) can be minimized

$$||E(\theta)|| = \max_{\theta} |D(\theta) - A(\theta)| \qquad (7\text{-}29)$$

by finding the best $A(\theta)$. Because $A(\theta)$ can be expressed as a finite sum of cosines as shown in Equation (7-27), it can be shown (see Reference 18) that the optimal $A(\theta)$ will be unique and will have at least N/2 + 1 extremal frequencies, where extremal frequencies are points such that for

$$\theta_1 < \theta_2 < \ldots < \theta_{N/2} < \theta_{N/2+1}$$

$$E(\theta_e) = -E(\theta_{e+1}) \qquad \text{for } e = 1, 2, \ldots, \frac{N}{2} + 1$$

and

$$|E(\theta_e)| = \max_{\theta} [E(\theta)]$$

Thus, the best approximation will exhibit an equiripple error function. The problem reduces to finding the extremal frequencies since, once they are found, the coefficients, 2h(N/2–k–1), can be found by solving the set of linear equations:

$$D(\theta) \pm \delta = A(\theta_e) = \sum_{k=0}^{N/2-1} 2h\left(\frac{N}{2} - k - 1\right)\cos\left[\theta_e\left(k + \frac{1}{2}\right)\right] \qquad (7\text{-}30)$$

The Remez exchange algorithm is used to systematically find the extremal frequencies (see Reference 5). Basically, a guess is made for the initial N/2 + 1 extremal frequencies. (Usually, this guess consists of N/2 + 1 equally spaced frequencies in the Nyquist range.) Using this guess, Equation (7-30) is solved for the coefficients and δ. Using these coefficients, $A(\theta)$ is calculated for all frequencies and the extrema, and frequencies at which the extrema are attained are determined. If the extrema are all equal and equal to or less than that specified in the initial filter specification, the problem is solved. However, if this is not the case, the frequencies at which the extrema were attained are used as the next guess. Note that the final extrema frequencies do not have to be equally spaced. Clearly, the equiripple design approach is calculation intensive.

What is the benefit of equiripple designs over window designs? In general, equiripple designs require fewer taps for straightforward requirements. When the specification requires a sharp cutoff and/or a large stopband attenuation or a narrow bandpass, the equiripple approach may fail to converge. In general, when N is decreased, an equiripple design tends to maintain its transition band while sacrificing stopband attenuation; window designs tend to do the opposite. Of the window alternatives, the Kaiser window is preferred for designing filters because the passband ripple and stopband attenuation can be varied relatively independent of the transition width (see Reference 1). For spectral analysis, the Blackman-Harris window is preferred (see Reference 10).

Figure 7-11 shows an example of an equiripple design generated from the FDAS software package. This example is the same as that used for the Kaiser window example of Figure 7-10. The number of coefficients in the equiripple design is far less than that generated by the Kaiser window method (179 versus 291). However, the passband ripple is larger.

7.4 FIR IMPLEMENTATION ON THE DSP56001

The DSP56001 has several architectural features that make it ideally suited for implementing FIR filters:

1. Dual Harvard architecture uses two data memories with dedicated buses and address generation units, allowing two addresses to be generated in a single cycle. If one address is pointing to data and another address is pointing to coefficients, a word of data and a coefficient can be fetched in a single cycle.

2. Modulo addressing makes the shifting of data unnecessary. If an address pointer is incremented (or decremented) with the modulo modifier in effect, data shifting can be accomplished by just "backing up" the address register by one to overwrite the data that would normally be shifted out. This procedure allows very efficient addressing of operands without wasting time shifting the data or reinitializing pointers.

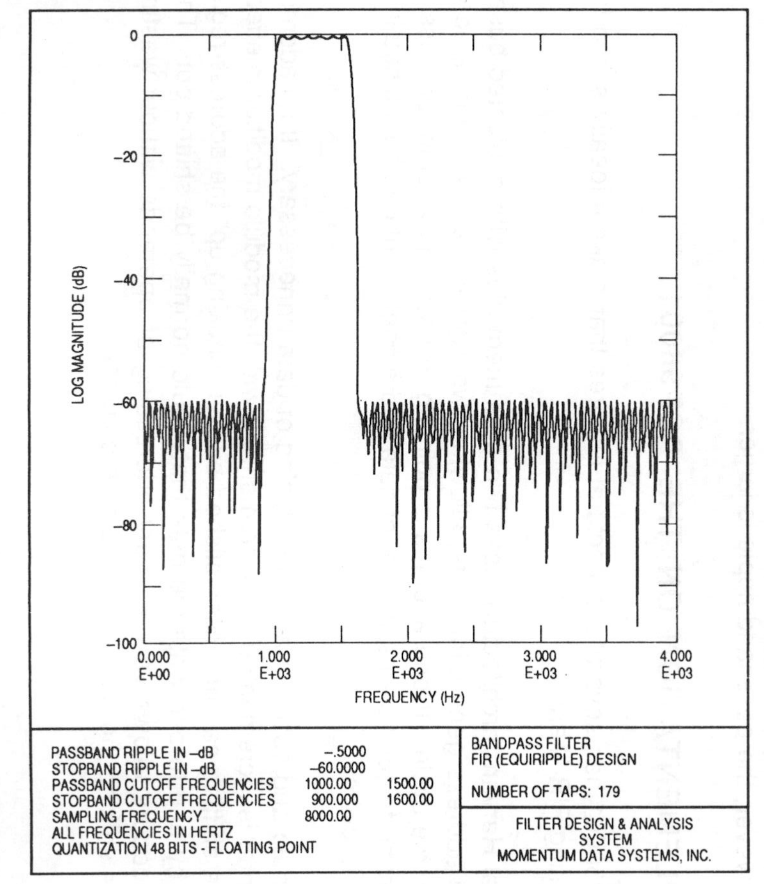

(a) Log Magnitude versus Frequency

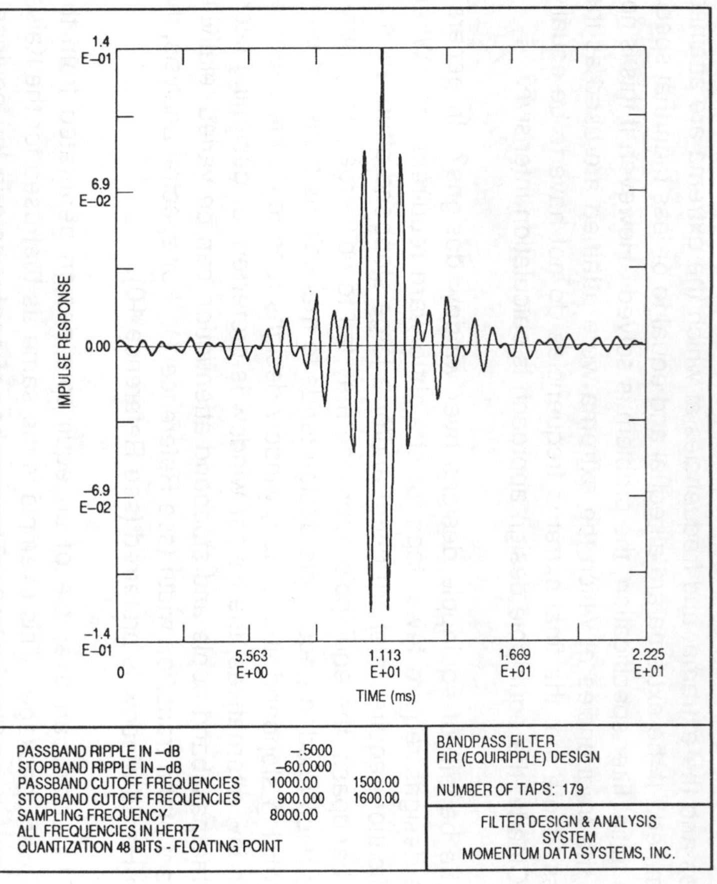

(b) Impulse Response versus Time

Figure 7-11. FDAS Output for FIR Bandpass Filter Example with Equiripple Design

3. Hardware DO loops execute without overhead once the loop is started. After a three-cycle initialization of the DO loop, the body of the loop executes as if it were straight-line code. Since the DO loop does not require any overhead cycles for each pass, the need for straight-line code is eliminated.

For a four-coefficient example of a linear-phase FIR filter, the input-output difference equation can be found by expanding Equation (7-1):

$$y(n) = h_0x(n) + h_1x(n-1) + h_2x(n-2) + h_3x(n-3)$$

This difference equation can be realized with the discrete-time four-tap filter example shown in Figure 7-12. The filter can be efficiently implemented on the DSP56001 by using modulo addressing to implement the shifting and parallel data moves to load the multiplier-accumulator. The filter network is shown in Figure 7-12(a); the memory map for the filter inputs and coefficients is shown in Figure 7-12(b). The following DSP56001 code is used to implement the direct-form FIR filter:

```
CLR    A          X0,X:(R0)+   Y:(R4)+,Y0   ;Save input sample, fetch coef.
REP    #NTAPS–1                             ;Repeat next instruction.
MAC    X0,Y0,A    X:(R0)+,X0   Y:(R4)+,Y0   ;FIR Filters.
MACR   X0,Y0,A    (R0)–                     Round result and adjust R0.
```

Register R0 points to the input variable buffer, M0 is set to three (modulo 4), R4 points to the coefficient buffer, and M4 is set to three. The input sample is in X0.

The CLR instruction clears accumulator A and performs parallel data moves. The data move saves the most recent input value to the filter (assumed to be in X0) into the location occupied by the oldest data in the shift register and moves the first coefficient in the filter (h_0) into the data ALU.

The REP instruction repeats the next instruction NTAPS–1 times. Since there are four taps in this filter, the next instruction is repeated three times.

The MAC instruction multiplies the data in X0 by the coefficient in Y0 and adds the result to accumulator A in a single cycle. The data move in this instruction loads the next input data variable into X0 and the next coefficient into Y0. Both address registers R0 and R4 are incremented.

The MACR instruction calculates the final tap of the filter, rounds the result using convergent rounding, and address register R0 is decremented.

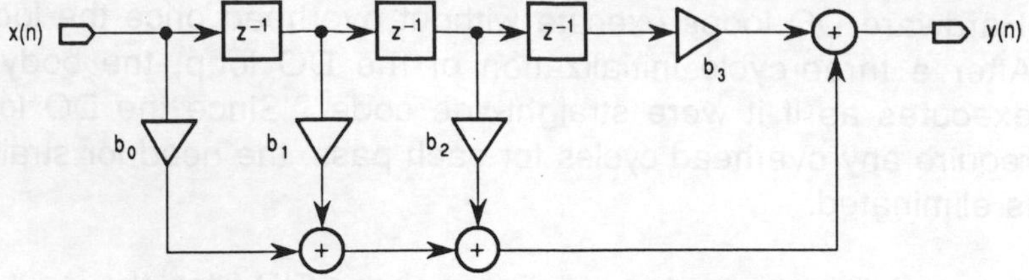

(a) Direct-Form FIR Filter Network with Four Coefficients

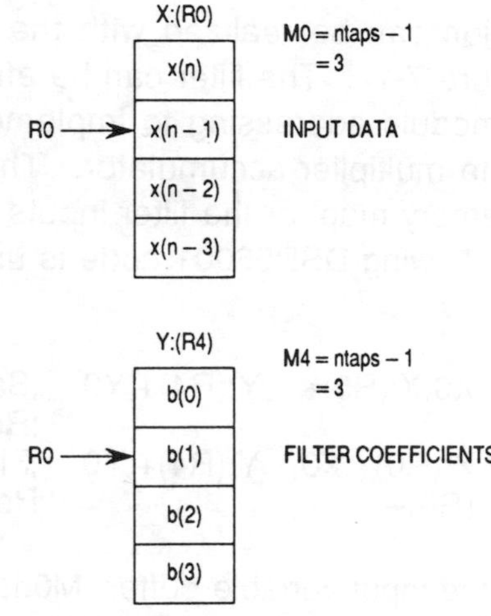

(b) Data Structures for Four-Tap FIR Example

Figure 7-12. FIR Filter Example

Address register R4 is incremented once before the REP instruction and three times due to the REP instruction for a total of four increments. Since the modulus for R4 is four, the value of R4 wraps around pointing back to the first coefficient.

The operation of R0 is similar. The value input to the filter is saved, and then R0 is incremented pointing to the first state. The REP instruction increments R0 three times. Since the modulo on R0 is four and R0 is incremented four times, the value in R0 wraps around pointing to the new input sample. When the MACR instruction is executed, the value of R0 is decremented, pointing to the old value, x(n–3). The next sample time overwrites the value of x(n–3) with the new input sample, x(n). Thus, the shifting of the input data is accomplished by simply adjusting the address pointer, and the modular addressing wraps the pointer around at the ends of the input data buffer. Instruction cycle counts for this filter are NTAPS+3. Thus, for a four-tap filter, seven instructions are required.

REFERENCES

1. Antoniou, Andreas. *Digital Filters: Analysis and Design.* New York, NY: McGraw-Hill, 1974.

2. Berlin, Howard M. *Design of Active Filters with Experiments.* Indianapolis, IN: Howard W. Sams & Co., Inc., 1977.

3. Bogner and Constantinides, eds. *Introduction to Digital Filtering.* New York, NY: John Wiley & Sons, 1975.

4. Brophy, J. J. *Basic Electronics for Scientists.* New York, NY: McGraw-Hill, 1966.

5. Cheney, E. W. *Introduction to Approximation Theory.* New York, NY: McGraw-Hill, I966.

6. Chrysafis, A. *Fractional and Interger Arithmetic Using the DSP56000 Family of General-Purpose Digital Signal Processors* (APR3/D). Motorola, Inc., 1988.

7. *Digital Stereo 10-Band Graphic Equalizer Using the DSP56001* (APR2/D). Motorola, Inc., 1988.

8. "Digital Filters on DSP56000/DSP56001." Motorola Technical Bulletin, 1988.

9. "Filter Design & Analysis System." Version 1.3, Momentum Data Systems, 1985.

10. Harris, Fredric J. "On the Use of Windows for Harmonic Analysis with the Discrete Fourier Transform." *Proc. of IEEE*, vol. 66, no. 1, 1978, pp.51-83.

11. Jackson, Leland B. *Digital Filters and Signal Processing.* Boston, MA: Kluwer Academic Publishers, 1986.

12. Lancaster, D. *Active Filter Cookbook.* Indianapolis, IN: Howard W. Sams & Co., Inc., 1975.

13. McClellan, J. H. and T. W. Parks. "A Unified Approach to the Design of Optimum FIR Linear-Phase Digital Filters." *IEEE Trans. Circuits Systems CT-20*, 1973, pp. 697-701.

14. Moschytzm, G. S. and P. Horn. *Active Filter Design Handbook.* New York, NY: John Wiley & Sons, 1981.

15. Oppenheim, A. V. and R. W. Schafer. *Digital Signal Processing.* Englewood Cliffs, NJ: Prentice-Hall, 1975.

16. Parks, T. W. and J. H. McClellan. "Chebyshev Approximation for Non-recursive Digital Filters with Linear Phase." *IEEE Trans. Circuit Theory CT-19*, 1972, pp.189-194.

17. Proakis, John G. et al. *Introduction to Digital Signal Processing.* New York, NY: MacMillan, 1988.

18. Rabiner, L. R. and B. Gold. *Theory and Application of Digital Signal Processing.* Englewood Cliffs, NJ: Prentice-Hall, 1975.

19. Sohie, Guy R. L. *Implementation of Fast Fourier Transform on Motorola's DSP56000/DSP56001 and DSP96002 Digital Signal Processors* (APR4/D). Motorola, Inc., 1989.

20. Strawn, J., et al. *Digital Audio Signal Processing—An Anthology.* William Kaufmann, 1985.

21. Williams, Arthur B. *Electronic Filter Design Handbook.* New York, NY: McGraw Hill, 1981.

Implementation of Fast Fourier Transforms on Motorola's DSP56000/DSP56001 and DSP96002 Digital Signal Processors

by
Guy R. L. Sohie
Digital Signal Processor Operation

PREFACE

The human body has inherently slow perception mechanisms. This is illustrated, for instance, when listening to music, or speech: we do not hear individual pressure variations of the sound as they occur very fast in time. Instead, we hear a changing pitch, or frequency. Similarly, our eyes do not "see" individual oscillations of electromagnetic fields (light); rather, we see colors. In fact, we do not directly perceive any fluctuations (or oscillations) which change faster than approximately 20 times per second. Any faster changes manifest themselves in terms of the frequency or rate of change, rather than the change itself. Thus, the concept of frequency is as important and fundamental as the concept of time.

TABLE OF CONTENTS

TABLE OF CONTENTS (Continued)

LIST OF ILLUSTRATIONS

LIST OF TABLES

SECTION 1 — INTRODUCTION TO THE FOURIER INTEGRAL

1.1 — DEFINITION AND HISTORY

The scientific and engineering communities have attempted to represent changing signals in two fundamental domains: time and frequency. Temporal changes are easily shown on oscilloscopes, for instance, where change in time is directly proportional to distance on a screen. Representation of signals in terms of frequencies falls under the general category of "spectrum analysis", and has generated a lot of attention in the more recent past, due to increased availability of hardware which makes such representations possible. The first formal approach to spectrum analysis probably dates back to the work of Fourier, who showed how to represent a general class of time-varying phenomena in terms of sine and cosine functions of particular frequencies. His work is best known in terms of the *Fourier integral* (inverse Fourier transform) (see Reference 1):

$$x(t) = \int_{-\infty}^{+\infty} X(f) e^{j2\pi ft} dt \tag{1}$$

where $j = \sqrt{-1}$ and $e^{j2\pi ft} = \cos(2\pi ft) + j\sin(2\pi ft)$. When interpreted as an infinite summation, the previous integral is simply a linear combination of a number of sine and cosine functions (expressed by the complex exponential), each one of which is "weighted" by the (complex) amplitude X(f). Conversely, the complex frequency function ("amplitude") X(f) can be derived from the time-varying signal x(t) by the *Fourier transform*:

$$X(f) = \int_{-\infty}^{+\infty} x(t) e^{-j2\pi ft} dt \tag{2}$$

The two expressions shown in equations (1) and (2) define a *Fourier transform pair* x(t) and X(f). The Fourier transform X(f) determines the frequency content of the signal in question, while x(t) shows the way the signal varies as a function of time. Note that, in general, x(t) can be directly measured (for instance, displayed on an oscilloscope). X(f) remains a mathematical expression which attempts to express our intuitive perception of "frequency". Unfortunately, it is not always true that the theoretical concept of "frequency", as defined by the Fourier transform in (2), and the intuitive concept of frequency as we perceive it, are identical. For instance, music consists of tones (frequencies) which vary over time. Although we can clearly perceive time-varying frequencies, equation (2) does not allow for Fourier's concept of frequency to have any time-varying character: X(f) is a function of frequency only.

1.2 — USE OF THE FOURIER TRANSFORM

Because of the basic nature of the frequency concept, practical applications of the Fourier transform are abundant. As more cost-efficient methods become available to compute the Fourier transform, the number of practical solutions to frequency-based problems will grow even larger. In these frequency-based applications, a digital signal processor can be used to efficiently compute the Fourier transform (as defined in **1.1 DEFINITION AND HISTORY**), and to perform specific frequency-domain tasks such as elimination of certain frequency components, etc.

One can distinguish three general types of Fourier transform applications as shown below:

1. *Number-Based* — Most spectrum analysis applications require the direct evaluation of the Fourier transform as in equation (2). Since the Fourier transform is a mathematical expression, these applications are based on numerical computations, and can be termed "number based". Examples range from spectrum analysis laboratory instrumentation and professional audio equipment to velocity estimation in radar. Note that in number-based applications the accuracy of the computed numbers is of vital importance to the performance of the overall system. For instance, the quality-conscious audio industry requires full 16-bit result precision in order to provide no audible distortion.

2. *Pattern-Based* — Many problems involve the recognition and detection of signals with a specific frequency content (a predefined spectral pattern). For instance, speech consists of segments of sound with very specific frequency characteristics. In this type of application, the conversion to the "frequency domain" is often only a small step in the overall task. It is important that this conversion process be as fast as possible, to allow for enough time to perform computationally intensive pattern matching techniques. In addition to providing fast Fourier transform computations, the processor in question needs to retain a general-purpose nature such that a variety of frequency-based calculations for pattern matching can be done.

3. *Convolution-Based* — The third class of applications of Fourier transforms uses the transform as a simple mathematical tool to perform general filtering in a very efficient manner. This concept is based on the property that the Fourier transform of the *convolution* of two time-signals:

$$y(t) = \int_{-\infty}^{+\infty} x(t-\tau)h(\tau)d\tau \qquad (3)$$

is equal to the *product* of the individual transforms:

$$Y(f) = X(f)H(f) \qquad (4)$$

Equation (3) (better known as the *convolution integral*) represents the output of a linear filter with *impulse response* h(t) and input signal x(t). Clearly, in the frequency domain, the output of a filter can be obtained by a simple multiplication, whereas in the time-domain, a more complicated convolution integral needs to be solved. The amount of computation involved in evaluating the integral in equation (3) becomes particularly large when the impulse response h(t) has a long time duration. This sometimes prevents real-time implementation. Clearly, if the Fourier transform X(f) of the signal can be computed efficiently, the filtering operation itself can be achieved by simple multiplications. The combined number of computations (for computing the Fourier transform, for filtering in the frequency domain, and for obtaining the inverse Fourier Transform) is often less than the total number of calculations required to compute equation (3) directly. This is especially true when the filter in question performs a simple frequency discrimination function (lowpass, bandpass, highpass, bandreject, etc.). In this case, the multiplications in the frequency domain can be replaced by a simple "masking" operation, which deletes the stopbands and leaves the passband(s) unchanged.

Although no direct frequency information is extracted from the signal, the Fourier transform is used as a mathematical tool for fast-filtering applications. Note that again, fast Fourier transform and inverse Fourier transform "engines" are required to provide the real-time filtering operation.

In summary, the basic nature of the concept of frequency indicates that the number of possible frequency domain applications is as large as more conventional time domain applications. In the past, these applications were either impossible to implement or could not be realized in a cost-efficient manner because of the lack of low-cost, high-performance hardware. In this report, it will be demonstrated that the DSP56000/DSP56001 and DSP96002 Families of digital signal processors fulfill the demanding requirements imposed by frequency-domain problems. In addition to providing a *fast* implementation of *high-precision* Fourier transform computations, the *general-purpose nature of the instruction set* allows for a **complete, single-chip, low-cost**, integrated solution to a wide variety of frequency domain problems.

SECTION 2 — THE DISCRETE FOURIER TRANSFORM

2.1 — THE DISCRETE-TIME FOURIER TRANSFORM (DTFT)

In order to compute the Fourier transform using digital hardware, equation (2) needs to be approximated by a form which makes machine computation feasible. The first step in this process consists of eliminating the theoretical integral symbol, and replacing it by a computable sum:

$$X(f) \approx \tilde{X}(f) = T \sum_{n=-\infty}^{+\infty} x(nT)e^{-j2\pi fnT} \qquad (5)$$

The previous expression uses a *sampled signal* x(nT), where the sampling period T is taken as small as possible to reduce approximation errors. $\tilde{X}(f)$ is appropriately called the *discrete-time Fourier transform (DTFT)*. As T (the sampling period) becomes infinitely small, the previous summation approaches the original Fourier transform in equation (2). In order to assess the accuracy of this approximation, it suffices to realize that the resulting expression for $\tilde{X}(f)$ is a *periodic* function of frequency:

$$\tilde{X}(f) = \tilde{X}\left(f + \frac{1}{T}\right) \qquad (6)$$

due to the fact that

$$e^{-\left(j2\pi fnT + j2\pi n \frac{T}{T}\right)} = e^{-j2\pi fnT} e^{-j2\pi n} = e^{-j2\pi fnT} \qquad (7)$$

In general, the original spectrum X(f) is not periodic, and the approximation is only justified for a range of small values of f. In Figure 2-1, the DTFT magnitude and the Fourier transform magnitude of a simple rectangular function are shown for several values of the sample rate $f_s = 1/T$. Note the periodicity of the resulting function, as well as the approximation errors due to the sampling process.

A well-accepted criterion for the sampling rate is given by the *Nyquist sampling theorem*, which says that a signal needs to be sampled faster than twice its highest frequency. In other words, if

$$X(f) = 0 \qquad (8)$$

for $|f| \geqslant B$ (B is referred to as the *bandwidth* of the signal), then the sampling frequency needs to satisfy:

$$f_s \geqslant 2B \qquad (9)$$

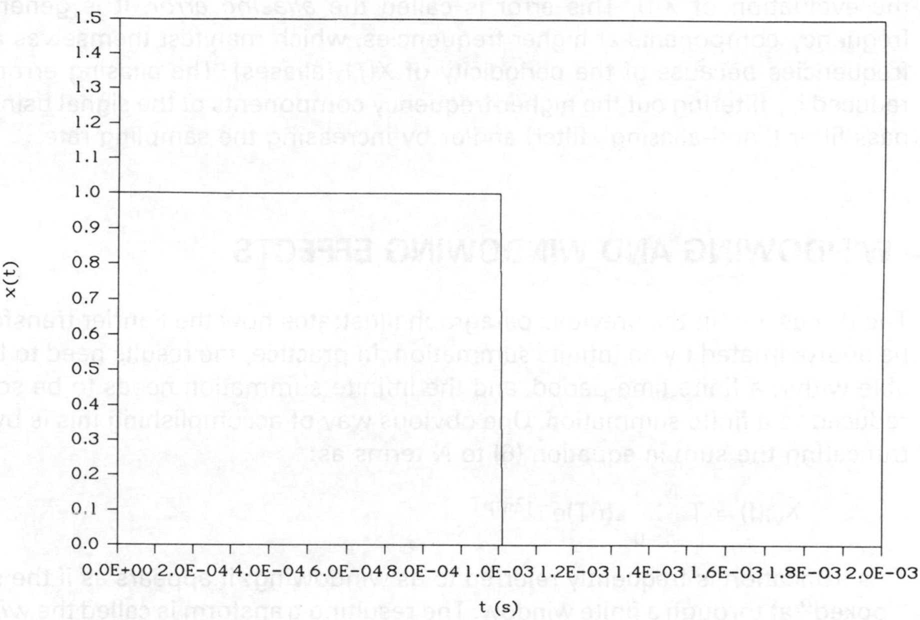

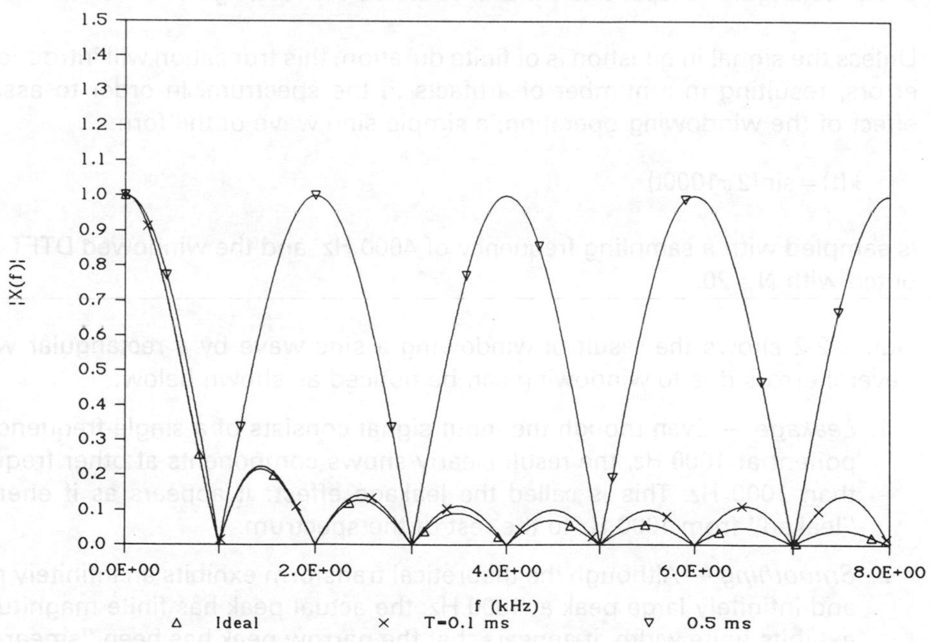

Figure 2-1. Fourier Transform of a Rectangular Function

In practice, signals rarely satisfy equation (9), and some error can be expected in the evaluation of X(f). This error is called the *aliasing error*. It is generated by frequency components at higher frequencies, which manifest themselves at lower frequencies because of the periodicity of $\tilde{X}(f)$ (aliases). The aliasing error can be reduced by filtering out the higher-frequency components of the signal using a low-pass filter ("anti-aliasing" filter) and/or by increasing the sampling rate.

2.2 — WINDOWING AND WINDOWING EFFECTS

The discussion in the previous paragraph illustrates how the Fourier transform can be approximated by an infinite summation. In practice, the results need to be available within a finite time-period, and the infinite summation needs to be somehow reduced to a finite summation. One obvious way of accomplishing this is by simply truncating the sum in equation (6) to N terms as:

$$\tilde{X}_w(f) = T \sum_{n=0}^{N-1} x(nT)e^{-j2\pi fnT} \qquad (10)$$

This truncation is frequently referred to as windowing: it appears as if the signal is "looked" at through a finite window. The resulting transform is called the *windowed discrete-time Fourier transform* (WDTFT). In mathematical terms, windowing is nothing but the multiplication of the signal by a "window" sequence of finite-length, w(n). In the simple case above, $w(n)=1$ for $0 \leq n \leq N-1$; otherwise, $w(n)=0$. Because of its rectangular shape, this window is called the *rectangular* window.

Unless the signal in question is of finite duration, this truncation will introduce other errors, resulting in a number of artifacts in the spectrum. In order to assess the effect of the windowing operation, a simple sine wave of the form:

$$x(t) = \sin(2\pi 1000t) \qquad (11)$$

is sampled with a sampling frequency of 4000 Hz, and the windowed DTFT is computed with $N=20$.

Figure 2-2 shows the result of windowing a sine wave by a rectangular window. Several errors due to windowing can be noticed as shown below:

1. *Leakage* — Even though the input signal consists of a single-frequency component at 1000 Hz, the result clearly shows components at other frequencies than 1000 Hz. This is called the leakage effect: it appears as if energy has "leaked" from 1000 Hz to the rest of the spectrum.
2. *Smoothing* — Although the theoretical transform exhibits an infinitely narrow, and infinitely large peak at 1000 Hz, the actual peak has finite magnitude and exhibits finite width. It appears that the narrow peak has been "smeared" out in the frequency domain as a result of the windowing function in the time domain. This effect is appropriately termed the smoothing effect.

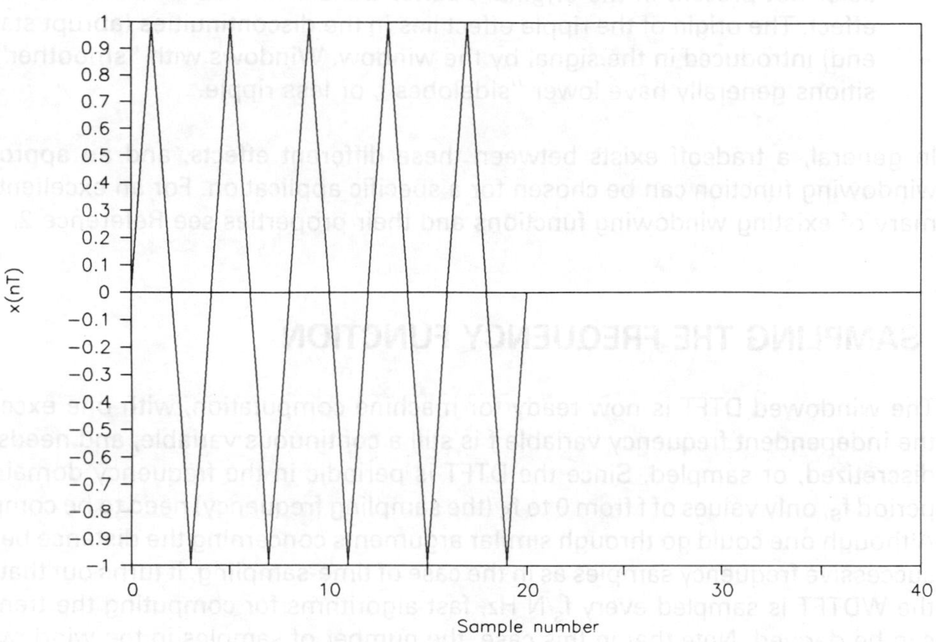

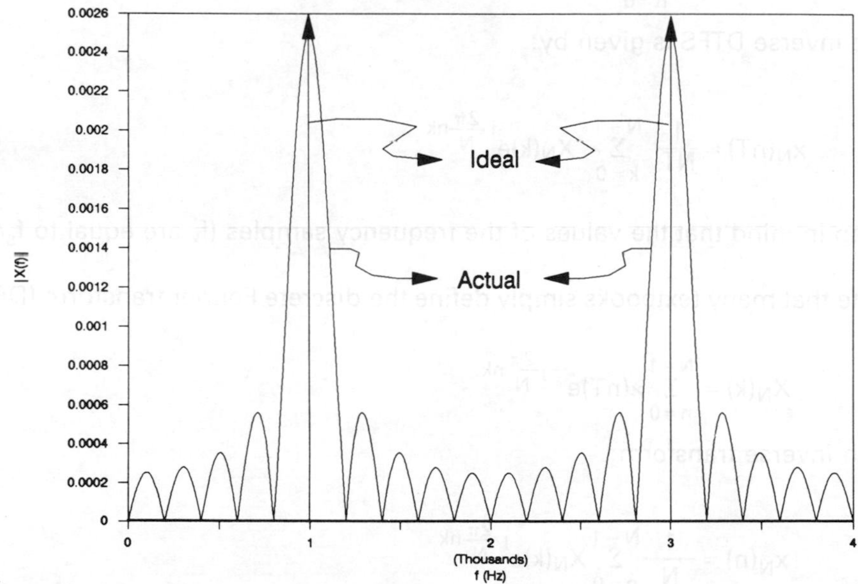

Figure 2-2. Windowing Effects when Windowing a Single Sine Wave

3. *Ripple* — The overall magnitude plot in Figure 2-2 shows an oscillatory character not present in the original Fourier transform: this is called the "ripple" effect. The origin of the ripple effect lies in the discontinuities (abrupt start and end) introduced in the signal by the window. Windows with "smoother" transitions generally have lower "sidelobes", or less ripple.

In general, a tradeoff exists between these different effects, and an appropriate windowing function can be chosen for a specific application. For an excellent summary of existing windowing functions and their properties see Reference 2.

2.3 — SAMPLING THE FREQUENCY FUNCTION

The windowed DTFT is now ready for machine computation, with one exception: the independent frequency variable f is still a continuous variable, and needs to be discretized, or sampled. Since the DTFT is periodic in the frequency domain with period f_S, only values of f from 0 to f_S (the sampling frequency) need to be computed. Although one could go through similar arguments concerning the distance between successive frequency samples as in the case of time-sampling, it turns out that when the WDTFT is sampled every f_s/N Hz, fast algorithms for computing the transform can be derived. Note that in this case, the number of samples in the window (=N) and the number of samples in the frequency domain (=N) are equal. The resulting transform is called the discrete-time Fourier series (DTFS):

$$\tilde{X}_N(k) = T \sum_{n=0}^{N-1} x(nT) e^{-j\frac{2\pi}{N}nk} \qquad (12)$$

The inverse DTFS is given by:

$$x_N(nT) = \frac{1}{NT} \sum_{k=0}^{N-1} \tilde{X}_N(k) e^{j\frac{2\pi}{N}nk} \qquad (13)$$

Keep in mind that the values of the frequency samples (f_k are equal to f_s/N k.

Note that many textbooks simply define the discrete Fourier transform (DFT) $X_N(k)$:

$$X_N(k) = \sum_{n=0}^{N-1} x(nT) e^{-j\frac{2\pi}{N}nk} \qquad (14)$$

with inverse transform:

$$x_N(n) = \frac{1}{N} \sum_{n=0}^{N-1} X_N(k) e^{j\frac{2\pi}{N}nk} \qquad (15)$$

Obviously, the DFT and DTFS differ only by a scaling factor of T, making the spectrum independent of the sampling period. Consequently, explicit T dependence is dropped in the previous expression.

Although the sequence $x_N(n)$ corresponds to the original sampled and windowed sequence $x(nT)$ for sampling instants 0 through $N-1$, the complete sampled sequence $x(nT)$ for any n cannot necessarily be recovered from it. Indeed, $x_N(n)$ appears to be periodic with period N due to the periodicity of $e^{j\frac{2\pi}{N}nk}$, whereas the original sampled signal was not assumed to be periodic.[1] This must be kept in mind in "convolution-based" applications, where the forward as well as inverse transforms are used: the incoming signal stream needs to be segmented, and the computed signal segments need to be "pieced together" to construct the complete output stream. Techniques for accomplishing this are discussed in most basic textbooks on digital signal processing (see, for instance, Reference 3).

[1]The error introduced in the time-domain by sampling a frequency function is termed "aliasing in time". This is completely analogous to the "aliasing in frequency" caused by sampling a time function, as discussed in **2.1 THE DISCRETE-TIME FOURIER TRANSFORM (DTFT)**. That is, if a frequency spectrum is not sampled sufficiently densely, the signal constructed in the time domain through the inverse "discrete-frequency Fourier transform" will show distorion.

SECTION 3 — THE FAST FOURIER TRANSFORM

3.1 — MOTIVATION

Upon closer examination of equation (15), it becomes clear that for every frequency point, $N-1$ complex summations and N complex multiplications need to be evaluated. Since there are N frequency points to be evaluated, this gives a total of $N(N-1)$ complex sums, and N^2 complex multiplications. Counting two real sums for every complex one, and four real multiplications plus two real summations for every complex multiplication, this is a total of $4N^2-2N$ real summations and $4N^2$ real multiplications.

The above numbers grow rapidly for increasing N. In fact, for $N=1024$ (1024-point DFT), 4,194,304 real multiplications are required. If this is computed on a DSP56000/DSP56001 with a 27-MHz clock, it takes 0.31 seconds just to execute that many real multiplications. Since the DFT computation needs to be completed by the time the next 1024 data points are collected for real-time performance, this limits the sampling rate to a maximum of 3.3 kHz. Obviously, faster solutions need to be sought.

3.2 — DIVIDE AND CONQUER

A faster algorithm for computing the DFT can easily be derived. The principle behind this is very basic; in fact, it is one with which we are all very familiar, as shown in Figure 3-1: a square of only half the dimensions of a larger square only has one-fourth the surface area.

This is because the surface area is quadratically proportional to the dimensions of the square. Similarly, the number of multiplications needed to compute the DFT is proportional to the square of the DFT's length (N). Thus, if we could replace the DFT over N points by two DFTs over N/2 points, computations would be reduced in order of magnitude of 0.5 ($=0.25+0.25$).

Since there are two independent variables (time and frequency) in the Fourier transform, dividing (or *decimating*) the DFT into smaller ones can be done in two ways. We can attempt to represent an N-point transform in terms of DFTs over half the number (N/2) of time-samples. This approach is appropriately called the decimation-in-time or DIT approach. Alternatively, the N-point DFT can be represented in terms of DFTs with N/2 frequency samples. This approach is called the decimation-in-frequency or DIF approach.

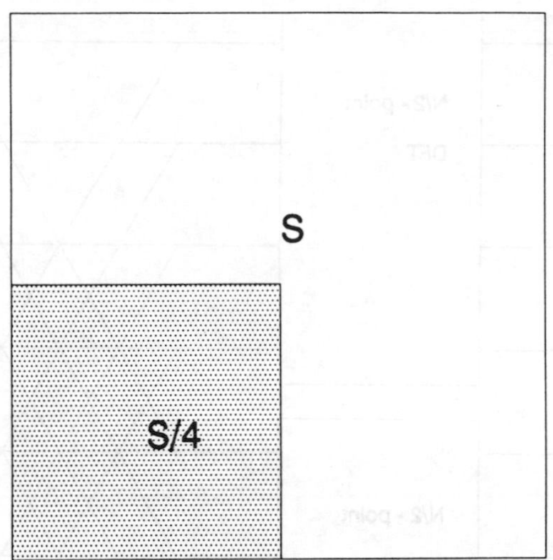

Figure 3-1. The FFT Principle in Layman's Terms

3.3 — THE DECIMATION-IN-TIME AND DECIMATION-IN-FREQUENCY, RADIX-2, FAST FOURIER TRANSFORMS

It is easily shown that equation (15) can be rewritten when N is even as:

$$X_N(k) = \sum_{r=0}^{(N/2)-1} x(2rT)e^{-j\frac{2\pi}{(N/2)}rk} + e^{-j\frac{2\pi}{N}} \sum_{r=0}^{(N/2)-1} X[(2r+1)T]e^{-j\frac{2\pi}{(N/2)}rk} \qquad (16)$$

As illustrated in Figure 3-2, this expression shows how two N/2-point DFTs can be combined to obtain one N-point DFT. If N is an integer power of 2, this process can be repeated, as shown in Figures 3-3 and 3-4, until a simple, two-point DFT is obtained. This gives rise to the flow diagram of a DIT fast Fourier transform (FFT) as shown in Figure 3-5, which represents a complete 8-point FFT computation.

The basic flow diagram of Figure 3-5 can be further simplified by rearranging the terms in the basic building block (the butterfly) as in Figure 3-6. Also, it is seen from Figure 3-5 that input samples no longer occur in normal, sequential order. When the indices are represented in their binary equivalent, however, the input samples appear in "bit-reversed" order. Figure 3-7 shows how the diagram can be rearranged for normally-ordered inputs and bit-reversed outputs.

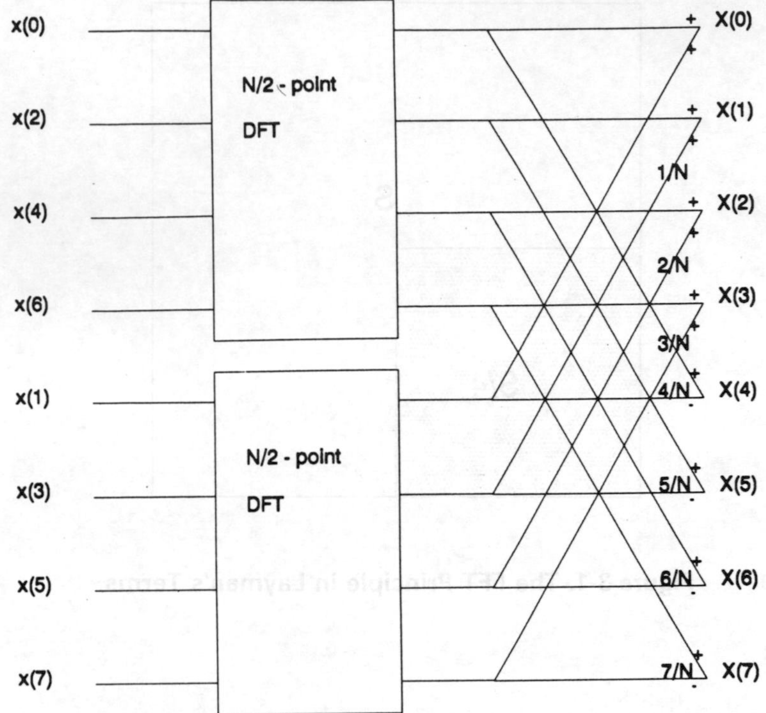

Figure 3-2. Decimation-in-Time of an N-Point FFT

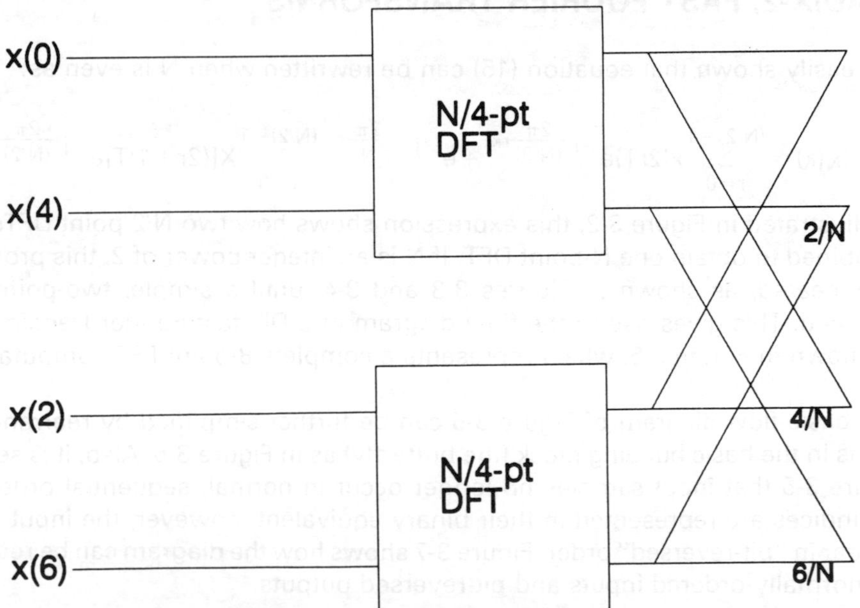

Figure 3-3. Decimation-in-Time FFT: Step Two

k/N denotes multiplication by the "twiddle factors" $e^{-j\frac{2\pi}{N}k}$ throughout this document

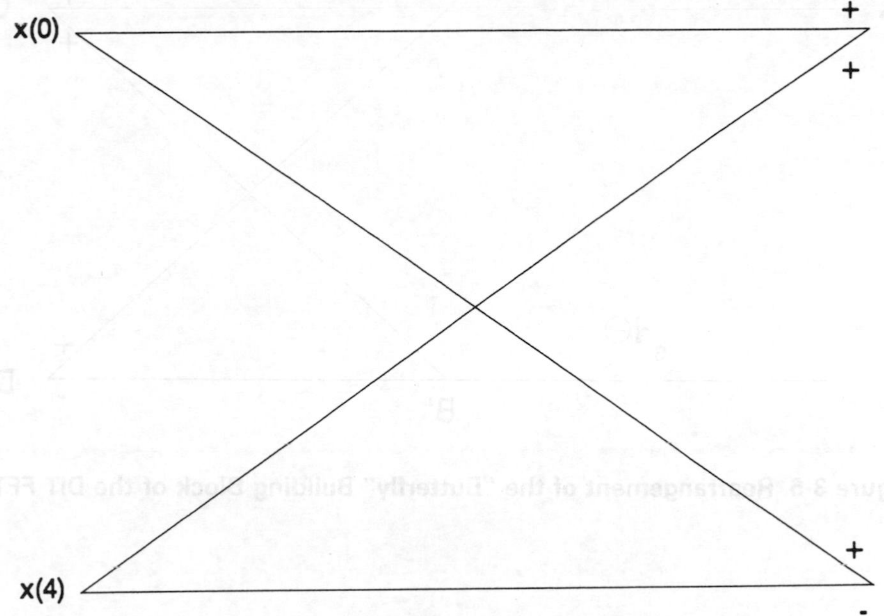

Figure 3-4. Decimation-in-Time FFT: Final Step (2-Point DFT)

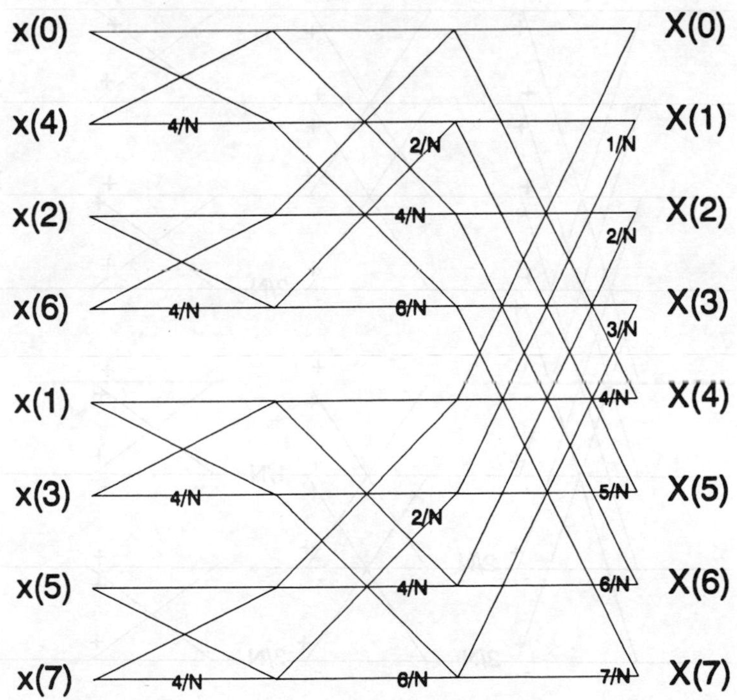

Figure 3-5. An 8-Point, Radix-2, Decimation-in-Time FFT

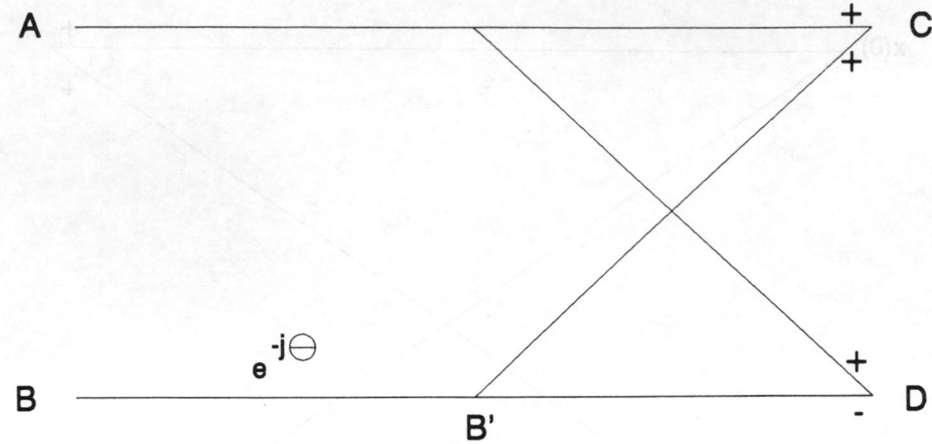

Figure 3-6. Rearrangement of the "Butterfly" Building Block of the DIT FFT

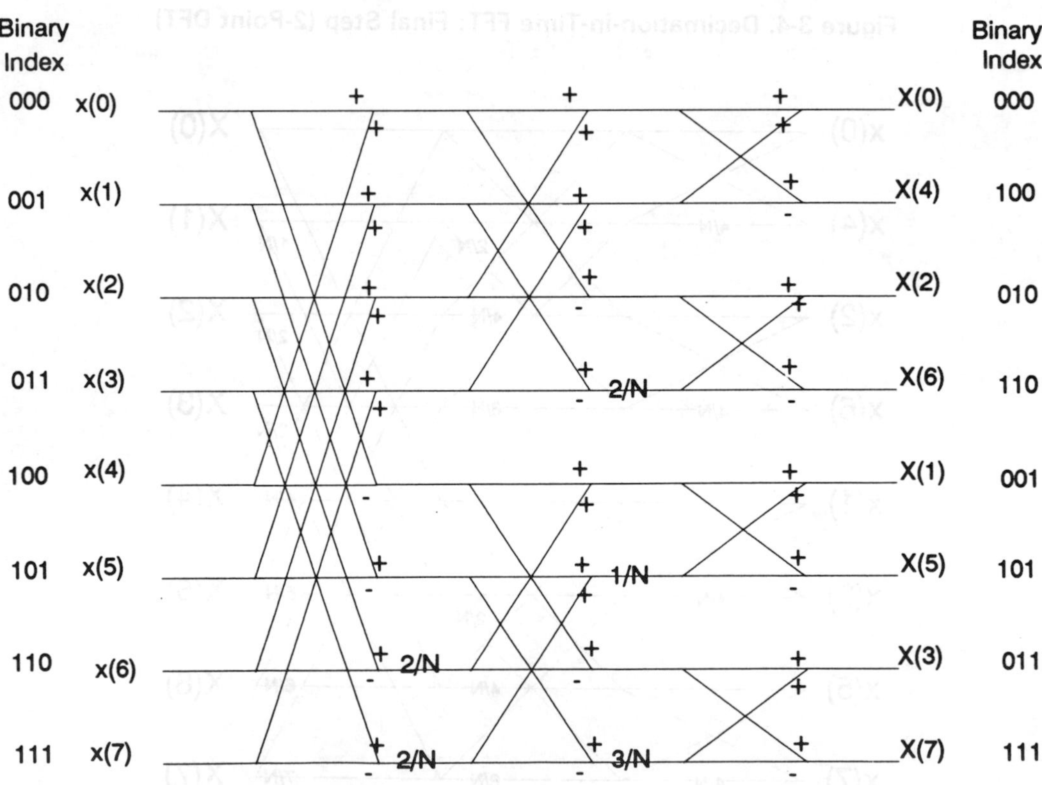

Figure 3-7. Rearrangement of the DIT Computation of Figure 3-6

Figures 3-8 and 3-9 show how the DFT with N frequency points can be obtained in terms of DFTs with a smaller number of frequency samples (decimation-in-frequency FFT). Note that the basic building block (butterfly) is different than for the DIT case (see Figure 3-9).

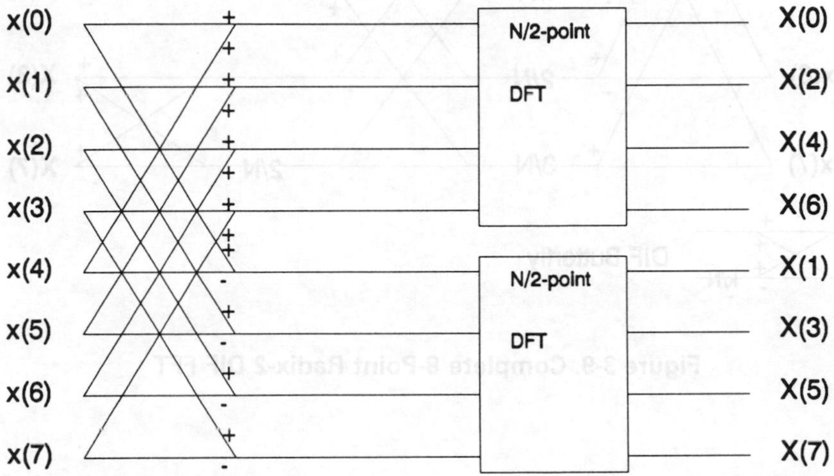

Figure 3-8. Decimation-in-Frequency Concept

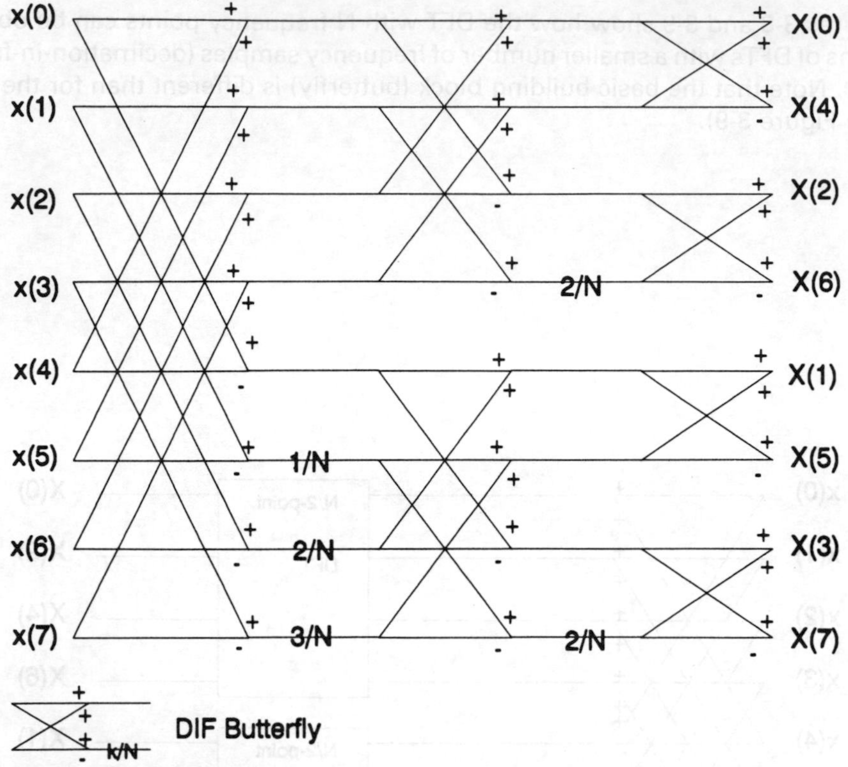

Figure 3-9. Complete 8-Point Radix-2 DIF FFT

SECTION 4 — IMPLEMENTATION OF THE FFT ON THE DSP56000/DSP56001 AND DSP96002

4.1 — REQUIREMENTS FOR IMPLEMENTATION

The basic building block of the DIT[2] FFT routine is the butterfly computation of Figure 3-6. Consequently, the architecture and instruction set of a DSP device should allow efficient computation of this basic butterfly. Since the butterfly consists of additions and multiplications, a hardware adder/subtractor and multiplier are paramount. Since the butterfly data are complex, the architecture must easily support complex arithmetic. The input and output data to the butterflies are moved between the processor's arithmetic unit and memory; consequently, efficient moves are needed, which impact the execution time of the butterfly as little as possible.

The overall DIT FFT algorithm is a collection of many such butterflies, the number of which depends upon the number of points (N) in the FFT. In order to write general FFT routines (for any N), efficient means are required for repetitive execution of the basic butterfly element. Although this could be accomplished in software, a hardware solution is preferred which has minimum impact on execution speed and program length.

In real-life applications, time as well as frequency data is used in normal order, even though the diagram of Figure 3-7 delivers the frequency data in bit-reversed order. Thus, an efficient method for bit-reverse addressing is needed. Again, a time-consuming software solution is to be avoided.

The input data (time samples) of the FFT are in practice obtained from an external source (A/D converter). This data collection must occur in parallel with the FFT computation to make real-time performance possible. Consequently, a DSP device must provide easy interface with a variety of A/D converters, and must support low-overhead interrupt schemes which can load data from an external device with minimal effect on the FFT computation.

[2]The DIT FFT is the main one discussed here, since the DSP56000/DSP56001 instruction set lends itself particularly well for its implementation. DIF FFTs can also be implemented.

4.2 — IMPLEMENTATION ON MOTOROLA'S DSP56000/DSP56001

4.2.1 — Minimum Program Length

The parallel architecture and instruction set of Motorola's DSP56000/DSP56001 (see Reference 4) lends itself particularly well to the radix-2, DIT FFT computation. The architecture is shown in Figure 4-1.

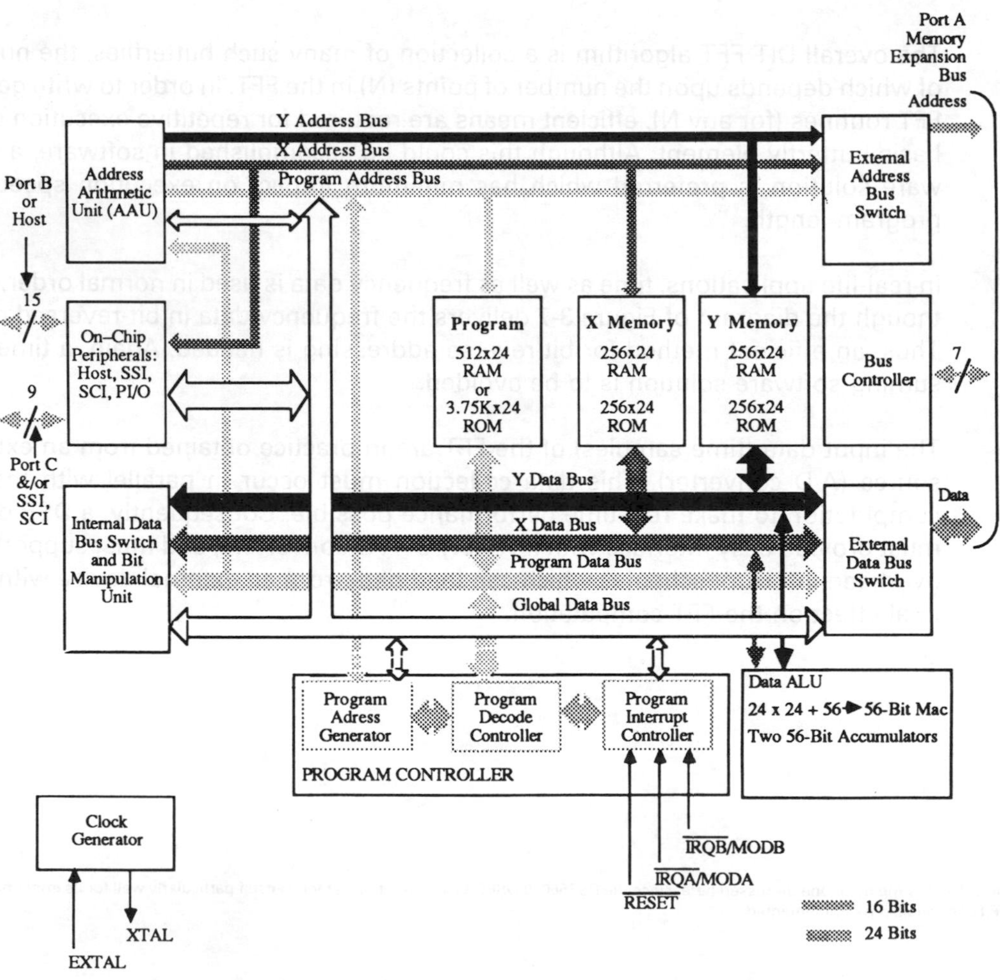

Figure 4-1. DSP56000/DSP56001 Architecture

The DIT butterfly equations are programmed on Motorola's DSP56000/DSP56001 as given below:

$$
\begin{aligned}
C_r &= A_r - B_r(-\cos(\theta)) - B_i(-\sin(\theta)) \\
C_i &= A_i - B_i(-\cos(\theta)) + B_r(-\sin(\theta)) \\
D_r &= 2A_r - C_r \\
D_i &= 2A_i - C_i
\end{aligned}
\tag{17}
$$

where the variables refer back to Figure 3-6. The basic butterfly "kernel" is implemented in assembly language in Figure 4-2.

```
;r0 → A
;r1 → B
;r4 → C
;r5 → D

mac     x1,y0,b     y:(r1)+,y1               ;Ai+Br(-sin) → b, Bi → y1
macr    -x0,y1,b    a,x:(r5)+    y:(r0),a    ;Ai+Br(-sin)-Bi(-cos) → b,Ai → a
subl    b,a         x:(r0),b     b,y:(r4)    ;2Ai-b → a, Ar → b
mac     -x1,x0,b    x:(r0)+,a    a,y:(r5)    ;Ar-Br(-cos) → b, Ar → a
macr    -y1,y0,b    x:(r1),x1                ;Ar-Br(-cos)-Bi(-sin) → b,Br → x1
subl    b,a         b,x:(r4)+    y:(r0),b    ;2Ar-b → a, Ai → b
```

Figure 4-2. The Radix-2, DIT Butterfly Kernel on the DSP56000/DSP56001

Note that the previous equations are written in this particular form such that the SUBL instruction can be used. This instruction allows efficient implementation of the DIT butterfly in a two-accumulator ALU.

The kernel of Figure 4-2 executes in six instruction cycles, or a total of 12 clock cycles[3]. This is made possible because of the parallel architecture of the DSP56000/DSP56001, which allows up to two data ALU operations (multiply/accumulate) in parallel with two data moves to/from memory in a single instruction cycle. The dual data spaces X and Y with the appropriate X and Y buses are ideally suited for complex arithmetic: the real parts are stored in X memory, and imaginary parts are stored in Y memory.

The simplest way of combining all of the butterflies into a complete program is shown in Figure 4-3. The FFT diagram (Figure 3-7) is first divided into FFT passes. On each pass, the data is fetched from memory, the butterfly calculations are done, and the results are moved back out to memory. It is easily shown that there are $\log_2 N$ passes. Within each pass, the butterflies are seen to cluster in groups. From

[3]These figures assume that the input and output data points to the butterfly are stored in internal memory.

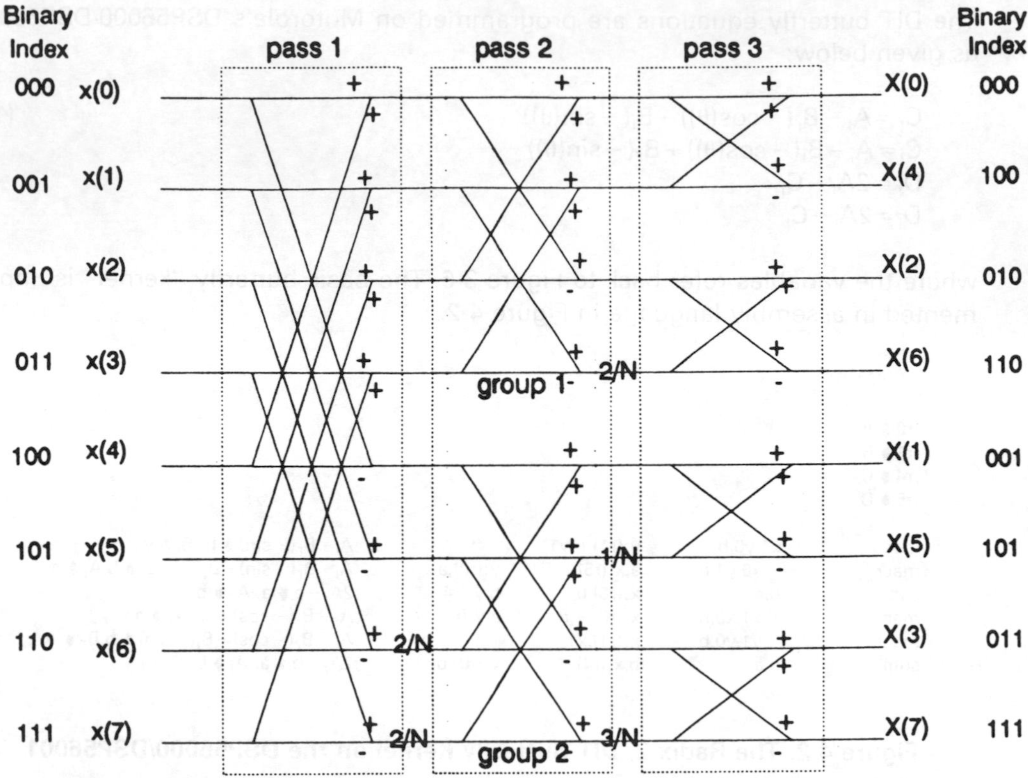

Figure 4-3. Grouping of Butterflies in the FFT Calculation

one pass to the next, the number of groups doubles, while the number of butterflies per group is divided by two. Note that the "twiddle factors" are the same for all butterflies within each group, and that the order of the twiddle factors from one group to the next is bit-reversed. This is easily implemented on the DSP56000/DSP56001 by setting the appropriate modifier register (m6) equal to zero and the offset register (n6) equal to N/4 (= coefficient table size/2), such that the twiddle factors are addressed in bit-reversed manner. THis gives rise to the simple, triple-nested DO-loop program of Figure 4-4: the outer DO loop steps through passes, the middle loop goes through all the groups within a pass, and the inner loop cycles through all the butterflies inside a group. The DSP56000/DSP56001 is particularly well suited for looped program execution, as it has hardware DO loop capability. Once a loop is entered via the DO instruction, this loop is executed without any time penalty. The resulting program takes 40 words in program memory. This is the most compact implementation of the radix-2 DIT FFT. A 1024-point complex FFT using this code executes in 3.19 ms when using a 40-MHz clock.

```
;
;This program originally available on the Motorola DSP bulletin board.
;It is provided under a DISCLAIMER OF WARRANTY available from
;Motorola DSP Operation, 6501 Wm. Cannon Drive W., Austin, Tx., 78735.
;
;Radix 2, In-Place, Decimation-In-Time FFT (smallest code size).
;
;Last Update 30 Sept 86     Version 1.1
;
fftr2a    macro     points, data, coef
fftr2a    ident     1,1
;
;Radix 2 Decimation in Time In-Place Fast Fourier Transform Routine
;
;   Complex input and output data
;      Real data in X memory
;      Imaginary data in Y memory
;   Normally ordered input data
;   Bit reversed output data
;      Coefficient lookup table
;      -Cosine values in X memory
;      -Sine values in Y memory
;
;Macro Call — fftr2a     points,data,coef
;
;      points      number of points (2-32768, power of 2)
;      data        start of data buffer
;      coef        start of sine/cosine table
;
;Alters Data ALU Registers
;      x1     x0     y1     y0
;      a2     a1     a0     a
;      b2     b1     b0     b
;
;Alters Address Registers
;      r0     n0     m0
;      r1     n1     m1
;             n2
;
;      r4     n4     m4
;      r5     n6     m5
;      r6     n6     m6
;
;Alters Program Control Registers
;      pc     sr
;
;Uses 6 locations or System Stack
;
;Latest Revision — September 30, 1986
;
;r0 points to A
;r1 points to B
;r4 points to C
;r5 points to D
;r6 points to twiddle factor
        move      #points/2,n0          ;initialize butterflies per group
        move      #1,n2                 ;initialize groups per pass
        move      #points/4,n6          ;initialize C pointer offset
        move      #-1,m0                ;initialize A and B address modifiers
        move      m0,m1                 ;for linear addressing
        move      m0,m4
        move      m0,m5
        move      #0,m6                 ;initialize C address modifier for
                                        ;reverse carry (bit-reversed) addressing
```

Figure 4-4. A Simple, Triple-Nested DO Loop Radix-2 DIT FFT on DSP56000/DSP56001 (Sheet 1 of 2)

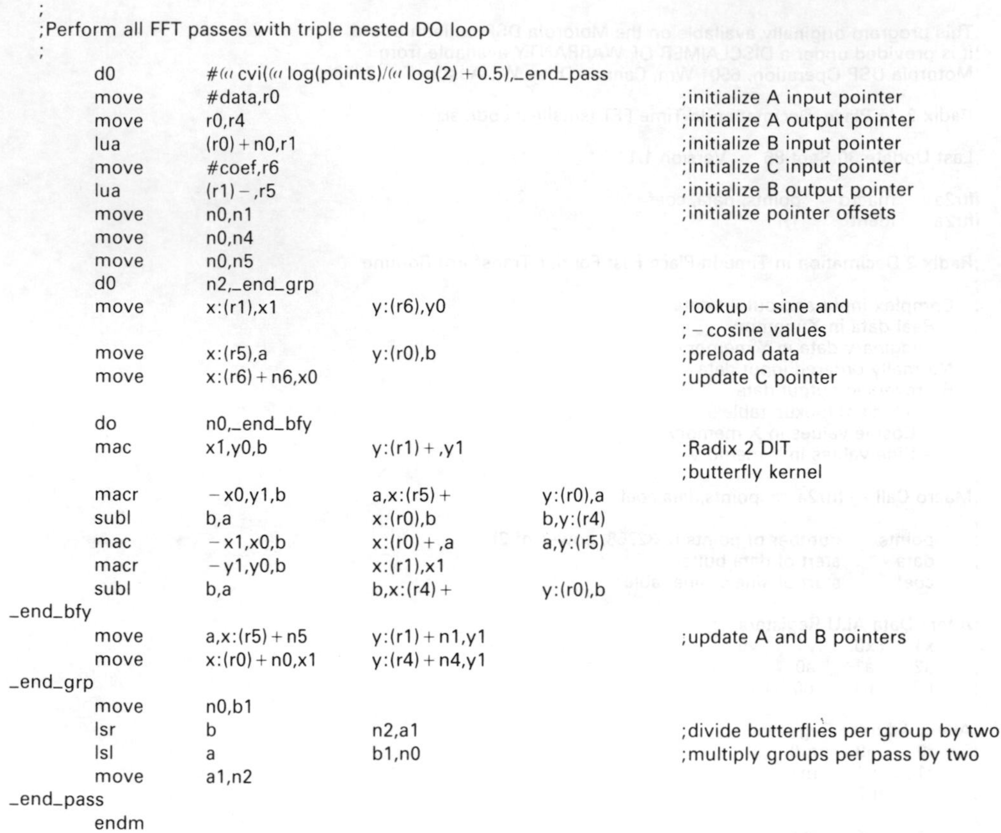

```
;
;Perform all FFT passes with triple nested DO loop
;
        d0      #@cvi(@log(points)/@log(2)+0.5),_end_pass
        move    #data,r0                                 ;initialize A input pointer
        move    r0,r4                                    ;initialize A output pointer
        lua     (r0)+n0,r1                               ;initialize B input pointer
        move    #coef,r6                                 ;initialize C input pointer
        lua     (r1)-,r5                                 ;initialize B output pointer
        move    n0,n1                                    ;initialize pointer offsets
        move    n0,n4
        move    n0,n5
        d0      n2,_end_grp
        move    x:(r1),x1       y:(r6),y0                ;lookup -sine and
                                                         ;-cosine values
        move    x:(r5),a        y:(r0),b                 ;preload data
        move    x:(r6)+n6,x0                             ;update C pointer

        do      n0,_end_bfy
        mac     x1,y0,b         y:(r1)+,y1               ;Radix 2 DIT
                                                         ;butterfly kernel
        macr    -x0,y1,b        a,x:(r5)+     y:(r0),a
        subl    b,a             x:(r0),b      b,y:(r4)
        mac     -x1,x0,b        x:(r0)+,a     a,y:(r5)
        macr    -y1,y0,b        x:(r1),x1
        subl    b,a             b,x:(r4)+     y:(r0),b
_end_bfy
        move    a,x:(r5)+n5     y:(r1)+n1,y1             ;update A and B pointers
        move    x:(r0)+n0,x1    y:(r4)+n4,y1
_end_grp
        move    n0,b1
        lsr     b               n2,a1                    ;divide butterflies per group by two
        lsl     a               b1,n0                    ;multiply groups per pass by two
        move    a1,n2
_end_pass
        endm
```

Figure 4-4. A Simple, Triple-Nested DO Loop Radix-2 DIT FFT on DSP56000/DSP56001 (Sheet 2 of 2)

4.2.2 — Optimization for Faster Execution

Although the previously discussed program executes extremely efficiently, some applications may exist which impose less stringent requirements on program size, but which demand even faster execution. Faster execution can be obtained by further optimizing the above algorithm. Several steps can be taken to achieve this optimization as shown below:

1. The twiddle factors for the first stage are all equal to one. Consequently, the multiplications in pass one are trivial, and need not be calculated. The first two passes can be combined into one, and are actually computed as four-point butterflies.

2. The groups in the last pass all consist of one single butterfly each. A triple nested DO loop is thus no longer required in this pass: it can be "split out" and handled by a single DO loop.

3. For longer FFTs (>256 points), internal memory in the DSP56000/DSP56001 is no longer sufficient to contain the complete data set. Consequently, the butterflies execute more slowly when the processor needs to fetch a data value in external X and in external Y memory in the same instruction cycle. This causes the instruction cycle to be "stretched", resulting in slower execution time. Through intelligent memory usage, however, this effect can be minimized. In a further optimized routine (see Reference 5), the first two passes, combined into one pass as above, are executed. Next, separate 256-point FFTs are computed, whereby the data is moved into internal memory, and the results are not moved to external memory until the final pass. This avoids stretching the instruction cycle on the middle passes, and makes optimal use of the available internal memory.

With these optimizations, a significantly faster routine is obtained. For instance, a 1024-point optimized complex FFT routine is available for the DSP56000/ DSP56001 which executes in 1.68 ms (see Reference 6) (using a 40-MHz clock). Note, however, that more "straight-line" code always results in longer programs: this routine takes up 105 words in program memory.

4.2.3 — FFTs with Real Inputs

In most practical situations, the data to be analyzed by the FFT is real: it is usually obtained from a single analog-to-digital (A/D) converter.[4] This knowledge can be exploited in several ways to increase the speed of the FFT calculation even further:

1. Since the input data is real, there is no need to multiply, add, or subtract the imaginary parts.

2. Use can be made of symmetries within the FFT:

$$XN(k) = X^*N(N - k) \qquad (18)$$

when x(nT) is real, * denotes complex conjugate. Clearly, not all of the frequency points need to be calculated, as many of them can be obtained by taking a simple complex conjugate of other, previously computed points. Taking a complex conjugate can be easily achieved by moving the same values to different memory locations, after taking the negative of the value which goes to Y memory (imaginary part). Figure 4-5 shows the procedure for a 16-point, real

[4]Complex data is obtained when sampling in-phase and quadrature components of a bandpass signal.

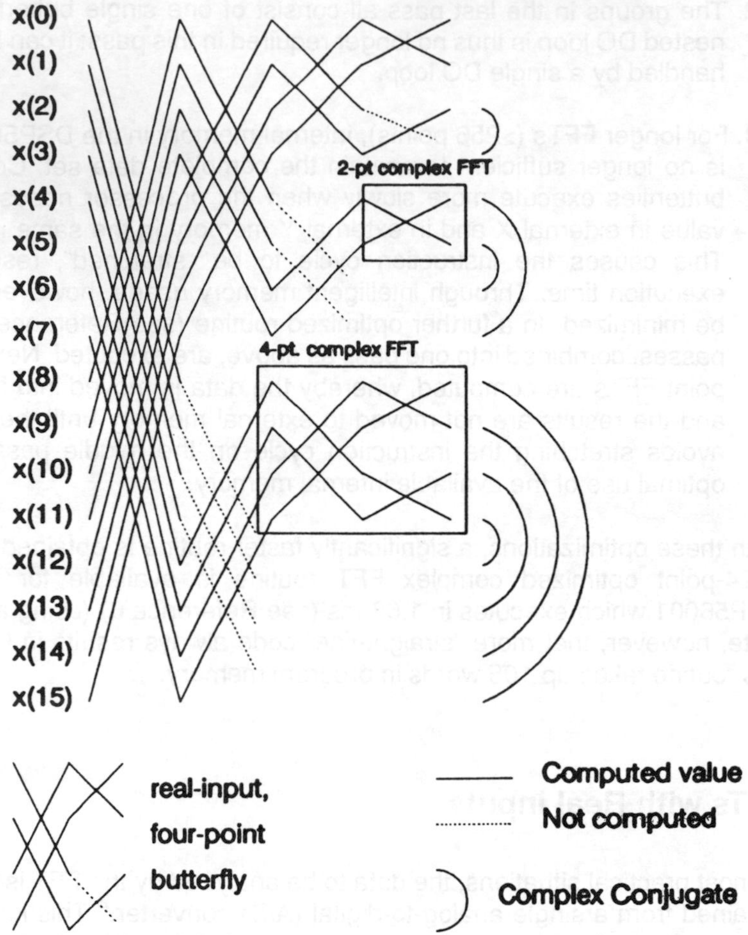

Figure 4-5. Computation of the Real-Input, DIT FFT

FFT in greater detail. A real-input FFT routine is available for DSP56000/ DSP56001, which executes in 1.01 ms using a 40-MHz clock. This also includes the amount of time necessary to bring in 1024 sampled data points from an external A/D converter. Because of the fast interrupt capability of the DSP56000/ DSP56001, data sampling creates very little overhead. As a result, the maximum sampling rate at which a 1024-point real FFT can be executed equals:

$$(fsmax = \frac{1024}{1.01 \times 10^{-3}} = 1.014)\ \text{MHz}$$

Comparing this with the sampling rate of 3.3 kHz mentioned in **3.1 MOTIVATION**, a more than 200-fold improvement is obtained by carefully optimizing the Fourier transform algorithm!

4.3 — IMPLEMENTATION ON MOTOROLA'S DSP96002

4.3.1 — Minimum Program Length

The butterfly equations implemented in the radix-2, DIT FFT on DSP96002 are the following:

$$
\begin{aligned}
B_r' &= B_r\cos\theta + B_j\sin\theta \\
B_j' &= B_j\cos\theta - B_r\sin\theta \\
D_r &= A_r - B_r' \\
D_j &= A_j - B_j' \\
C_r &= A_r + B_r' \\
C_j &= A_j + B_j'
\end{aligned}
\qquad (19)
$$

where the variables again refer back to Figure 3-6. The implementation of this basic butterfly in DSP96002 assembly is shown in Figure 4-6. This kernel executes in only four instruction cycles, or eight clock cycles.[5] Since four real multiplications are needed, and only one real multiplier is available, this is the most efficient implementation possible. In addition to the features available on the DSP56000/DSP56001 (discussed in **4.2.1 Minimum Program Length**), this efficient execution is obtained by the FADDSUB instruction which delivers the sum and the difference of two operands, in parallel with a multiplication and two data moves. With this feature, a total of three floating-point operations can be executed in one instruction cycle, resulting in a peak performance of 60 million floating-point operations per second (MFLOPS) with a 40-MHz clock.

The triple-nested DO loop routine, which computes the radix-2, DIT FFT on the DSP96002 takes only 30 words in program memory. A 1024-point complex FFT is executed in only 1.55 ms, assuming a 40-MHz clock.

[5]Assuming no additional clock cycles are used for treatment of unnormalized numbers, memory wait states, etc.

```
; r0 ➡ A
; r1 ➡ B
; r4 ➡ C
; r5 ➡ D
fmpy     d8,d6,d2    fadd.s     d3,d0  x:(r0),d4.s    d2.s,y:(r5)+   ;B_r*sin ➡ d2
                                                                     ;B_j*sin + B_r*cos ➡ d0
                                                                     ;A_r ➡ d4,D_j ➡ mem.

fmpy     d8,d7,d3    faddsub.s  d4,d0  x:(r1)+,d6.s   d5.s,y:(r4)+   ;B_j*sin ➡ d3
                                                                     ;A_r +B_r^1 ➡d0
                                                                     ;A_r - B_r^1 ➡ d4
                                                                     ;B_r ➡ d6
                                                                     ;C_j ➡ mem.

fmpy     d9,d6,d0    fsub.s     d1,d2  d0.s,x:(r4)    y:(r0) +,d5.s  ;B_r*cos ➡ d0
                                                                     ;B_r*sin - B_j*cos ➡ d2
                                                                     ;C_r ➡ mem.
                                                                     ;A_j ➡ d5

fmpy     d9,d7,d1    faddsub.s  d5,d2  d4.s,x:(r5)    y:(r1),d7.s    ;B_j*cos ➡ d1
                                                                     ;A_j + B_j^1 ➡ d2
                                                                     ;A_j - B_j^1 ➡ d5
                                                                     ;D_r ➡ mem.
                                                                     ;B_j ➡ d7
```

Figure 4-6. The Radix-2, DIT FFT Butterfly Kernel on the DSP96002

4.3.2 — Optimization for Faster Execution

The techniques employed to optimize execution speed of DSP56000/DSP56001 FFTs discussed in **4.2.2 Optimization for Faster Execution** can be invariably applied to the DSP96002 case. Note that the dual external buses available on the DSP96002 avoid the "stretching" of instruction cycles when a parallel move to/from external X and Y space is attempted in one instruction (assuming zero wait state external memory): external X memory and external Y memory can be mapped to different external buses, and FFTs of virtually unlimited length can be executed without any penalty in execution time.

In addition to the methods described in **4.2.2 Optimization for Faster Execution**, the large number of internal registers in the DSP96002's data ALU allow the two butterflies per group in the next to the last pass to be combined into one, four-point butterfly. A routine is available on the DSP96002 which executes a 1024-point complex FFT in 1.05 ms with a 40-MHz clock. This routine takes 137 words in program memory.

4.3.3 — FFTs with Real Inputs

The ideas explained in 4.2.3 FFTs with Real Inputs can also be applied to the DSP96002. In addition to the fast interrupts available on its fixed-point counterpart, the DSP96002 has a two-channel DMA controller, which operates unobtrusively in parallel with the ALU. Consequently, the DSP96002 can collect a block of data from an external location, such as another processor or an A/D converter, while the FFT is being computed and without adding any execution time to this FFT. An FFT for real inputs is available on the DSP96002, which runs in 611 microseconds with a 40-MHz clock, and takes up a total of 317 words in program memory. This limits the sampling rate for real-time performance to a maximum of 1.67 MHz. Table 4-1 gives a summary of execution speeds and program memory requirements for the different routines discussed.

Table 4-1. Required Program Memory and Execution Times for Several FFTs

1024-Point FFT	DSP56001		DSP96002	
	Program Size	Execution Time	Program Size	Execution Time
Triple-Nested DO Loop	40 Words	3.19 ms	30 Words	1.56 ms
Minimum Execution Time	105 Words	1.68 ms	137 Words	1.05 ms
Real Input FFT	254 Words	1.01 ms	317 Words	0.61 ms

SECTION 5 — FIXED POINT, BLOCK FLOATING POINT, AND IEEE FLOATING POINT

Whenever mathematical algorithms are implemented in digital hardware, one must realize that results are only obtained with finite precision. The precision is generally limited by the number of bits used in the number representation, and depends on how the arithmetic limits its results to these bits (truncation, rounding).

When analyzing the effects of finite-precision arithmetic relative to FFT results, it is important to understand how the magnitude of the complex FFT data changes throughout the FFT calculation. The easiest way to characterize behavior of numbers in FFTs is achieved using vector notation. Figure 5-1 shows how the two complex numbers (vectors) at the input of the DIT butterfly of Figure 3-6 are combined to give the two outputs.

First, vector B is multiplied by the twiddle factor. Since all twiddle factors in the FFT have the form $e^{-j\theta}$, which has unit magnitude, the magnitude of B′ is the same as that of B. As shown in Figure 5-1, the multiplication by the twiddle factor can be

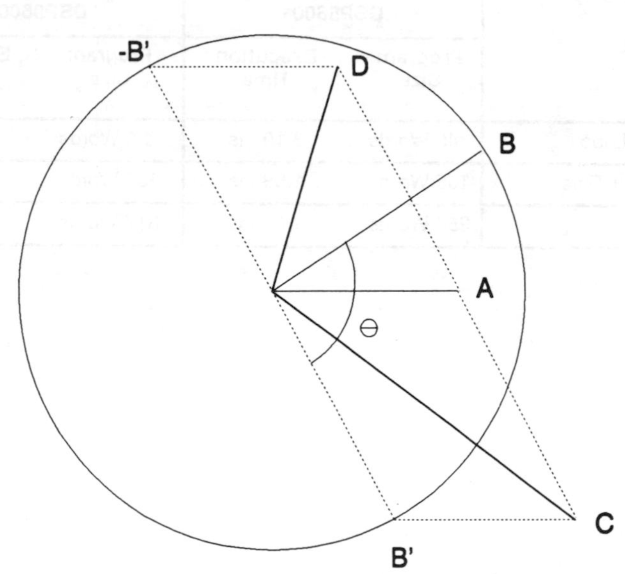

Figure 5-1. Vector Representation of the DIT Butterfly

interpreted as a simple rotation of the vector B over an angle θ. The vectors B′ and A are next added and subtracted to give the butterfly outputs C and D, respectively.

Limits on the output vectors' magnitudes are derived in Figure 5-2. Obviously, the largest magnitude is achieved if the vectors B′ and A line up, such that the total magnitude of either C (B′ and A point in the same direction), or D (B′ and A point in opposite directions) is the sum of the magnitudes of A and B′. It is also clear from Figure 5-2 that either the magnitude of C or the magnitude of D is at least equal to the magnitude of the larger of the two vectors A and B. This leads to the relationships:

$$\max(|A|,|B|) \leq \max(|C|,|D|) \leq 2\max(|A|,|B|) \qquad (20)$$

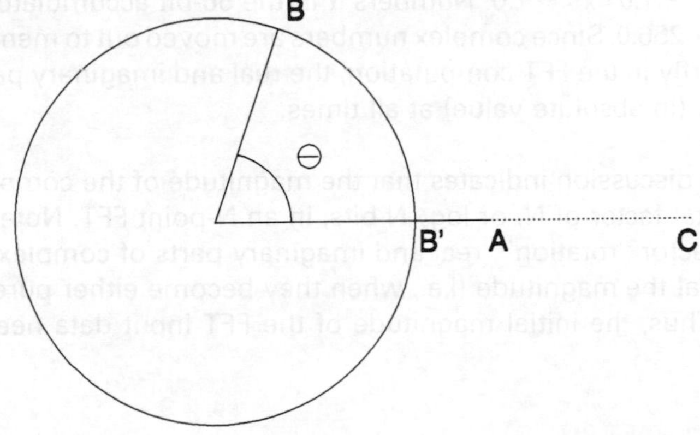

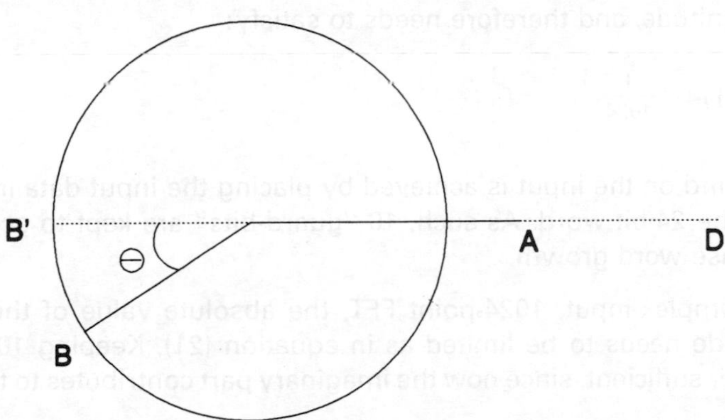

Figure 5-2. Bounds on the Butterfly's Output Magnitude

It is apparent from the previous discussion, that the complex numbers at the output of the butterflies "grow" in magnitude from stage to stage. The maximum growth of the magnitude, as shown by expression (20), is a factor of two per stage. In digital terms, one can conclude that the maximum *word growth* of the magnitude in the FFT that can occur is one bit per pass.

5.1 — FIXED-POINT IMPLEMENTATION

5.1.1 — Input Bounds

The DSP56000/DSP56001 Family implements 24-bit, fixed-point, fractional arithmetic (see Reference 7). This means that in practice, 24-bit numbers $\varkappa$ stored in memory are limited as $-1.0 \leq \varkappa < +1.0$. Numbers α in the 56-bit accumulators are limited by $-256.0 \leq \alpha < +256.0$. Since complex numbers are moved out to memory at the output of each butterfly in the FFT computation, the real and imaginary parts must be kept less than one (in absolute value) at all times.

The previous discussion indicates that the magnitude of the complex numbers can grow by a total factor of N, or $\log_2$ N bits, in an N-point FFT. Note that, because of the twiddle factor "rotation", real and imaginary parts of complex numbers can at any time equal the magnitude (i.e., when they become either purely real or purely imaginary). Thus, the initial magnitude of the FFT input data needs to be limited by:

$$M < \frac{1.0}{N} \tag{21}$$

Examples:

1. For a real-input, 1024-point FFT, the absolute value of the input data equals the magnitude, and therefore needs to satisfy:

   $$|\, x(nT) \,| < \frac{1}{1024} \tag{22}$$

   This bound on the input is achieved by placing the input data in the lower 14 bits of the 24-bit word. As such, 10 "guard-bits" are kept to accomodate the worst-case word growth.

2. For a complex-input, 1024-point FFT, the absolute value of the input data's magnitude needs to be limited as in equation (21). Keeping 10 guard-bits is no longer sufficient, since now the imaginary part contributes to the magnitude

as well. The simplest way to place a bound on the data such that equation (21) is satisfied, is by forcing the inputs to satisfy:

$$-\frac{0.5}{1024} \leq R\{x(nT)\} \leq \frac{0.5}{1024} \quad (23)$$

$$-\frac{0.5}{1024} \leq I\{x(nT)\} \leq \frac{0.5}{1024}$$

This bound forces the real and imaginary parts to the lower 13 bits of the input word, and results in an input magnitude bound of $(\sqrt{2}/2)/1024$, which definitely satisfies equation (21).

The previous examples indicate that the following general rules must be obeyed for an N-point FFT:

$$|x(nT)| < \frac{1}{N} \quad (24)$$

for real-input FFTs, and:

$$|R\{x(nT)\}| \leq \frac{1}{2N} \quad (25)$$

$$|I\{x(nT)\}| \leq \frac{1}{2N}$$

for complex-input FFTs. This can be directly translated into the number of guard-bits needed. For a real-input FFT, a total of

$$G = \log_2 N \quad (26)$$

guard-bits are needed, while for a complex-input:

$$G = \log_2 N + 1 \quad (27)$$

The number of guard-bits, in turn, determines the maximum number of bits in the A/D converter used to digitize the input signal. Clearly, for a fixed-point DSP with P-bit memory storage, the maximum number of A/D bits (A) equals

$$A_{max} = P - \log_2 N \quad (28)$$

for real-input FFTs, and

$$A_{max} = P - \log_2 N - 1 \quad (29)$$

Example:

For a 24-bit processor, like the DSP56000/DSP56001, the theoretical maximum A/D size for a 1024-point real-input FFT is 14 bits. For a 16-bit processor, this is reduced to six bits.

In general, the bounds above are quite strict and are affected by windowing parameters, the type of signal that is analyzed (harmonic signal versus broadband signal), etc. For instance, it is found that a 1024-point FFT of a Blackman-Harris windowed sine wave only requires eight guard-bits. The reason for this apparent contradiction with equation (26) lies in the fact that the windowing operation "smooths" out the peak energy (as discussed previously in **2.2 WINDOWING AND WINDOWING EFFECTS**). Consequently, when using a Blackman-Harris window, a 16-bit A/D converter can be used with the DSP56000/DSP56001, where the A/D bits are placed in the 16 least significant bits of the 24-bit word.

5.1.2 — Roundoff Errors

When mathematical algorithms are implemented using finite-precision arithmetic, roundoff errors or truncation errors occur throughout the computations. These errors become especially important if the number of computations is large, and if the results are not appropriately scaled relative to the errors. This may happen in fixed-point implementations.

Roundoff errors are caused when the results of multiplications are "reduced" to the number of bits used in the processor's data storage. The mantissa size of the result of a multiplication is twice the mantissa size of the operands when the multiplier is implemented with "infinite precision", i.e., all bits of the multiplication result are computed. In the DSP56000/DSP56001 Family, other arithmetic operations (add and subtract) can use a total of 56 bits, and thus accomodate the complete multiplication result. Arithmetic results are "rounded" (using convergent rounding or "round to nearest" (see Reference 4)) only before moving these results out to 24-bit memory locations. Thus, a total error of at most plus or minus one-half of one least significant bit (LSB) of the 24-bit word occurs. In the next FFT pass, results are moved from memory to the arithmetic unit, new calculations are performed, the new results are rounded again and moved back to memory. The successive rounding errors tend to accumulate and appear as roundoff noise in the final results. Because of the accumulative effect, not only the LSB becomes unreliable, but other bits as well. Obviously, one needs to keep as many reliable bits as possible in the final result, and thus the rounding error per operation needs to be kept as small as possible. This is achieved by making the number of bits in the number representation as large as possible, and by making sure the rounding errors do not add coherently. The 24-bit number representation in DSP56000/DSP56001 has proved sufficient for most applications requiring high-precision results. In addition, the convergent rounding scheme used in the DSP56000/DSP56001 assures that rounding errors occur in a truly random fashion, and do not add coherently (hence the name "convergent" rounding: the mean rounding error converges to zero over a large number of operations).

Expressions for the signal-to-roundoff noise ratio (SRNR) are derived in **APPENDIX A**. The SRNR on a digital signal processor (DSP) with infinite-precision internal arithmetic and convergent rounding, such as the DSP56000/DSP56001, is shown to be:

$$SRNR_{dB} = 10\ \log_{10}[3 \cdot 2^{2P-1}] - 10\ \log_{10}(N-1) \tag{30}$$

The following example illustrates how the SRNR shows up in the FFT results.

Example:

For a 1024-point FFT on the DSP56000/DSP56001, $P = 24$ and $N = 1024$. This results in a theoretical SRNR of 116 dB. In Figure 5-3, two Blackman-Harris windowed sine waves of the form:

$$x(nT) = \{\sin(2\pi \cdot 0.1255 \cdot n) + j \cdot 2^{-15} \cdot \sin(2\pi \cdot 0.02505 \cdot n)\} \cdot w(n) \cdot K \tag{31}$$

are digitized using a 16-bit A/D converter, and the magnitude of the FFT is plotted. w(n) denotes the Blackman-Harris window (see Reference 2). The scale factor K is taken such that the output peak magnitude is as close to one as possible ($K \approx 2^{-8}$). The resulting noise floor around −120 dB is clearly visible.

It is important to realize that the signal-to-quantization noise ratio (SQNR) and the SRNR are separate, independent quantities. The quantization noise is due to the finite-precision representation of the input signal, and does not change because of the FFT calculations: it can be adequately represented as (white) noise added to the ideal input signal. The roundoff noise, on the other hand, is independent of how the input signal is quantized, and increases with every butterfly calculation. If the DSP makes full use of the available number of bits by using the maximum-sized A/D converter (LSB of the A/D = LSB of the processor), this LSB is soon affected by the growing roundoff error, and the roundoff error is dominant in the final results. This is illustrated in Figure 5-4.

The quantization noise and the roundoff noise error affect the same bits in the FFT results (become of the same order of magnitude) when (as is derived in **APPENDIX A**, equation A-6):

$$A = P - 0.5 \cdot \log_2(N-1) - G \tag{32}$$

for complex-input FFTs, and (equation A-8)

$$A = P - 0.5 - 0.5 \cdot \log_2(N-1) - G \tag{33}$$

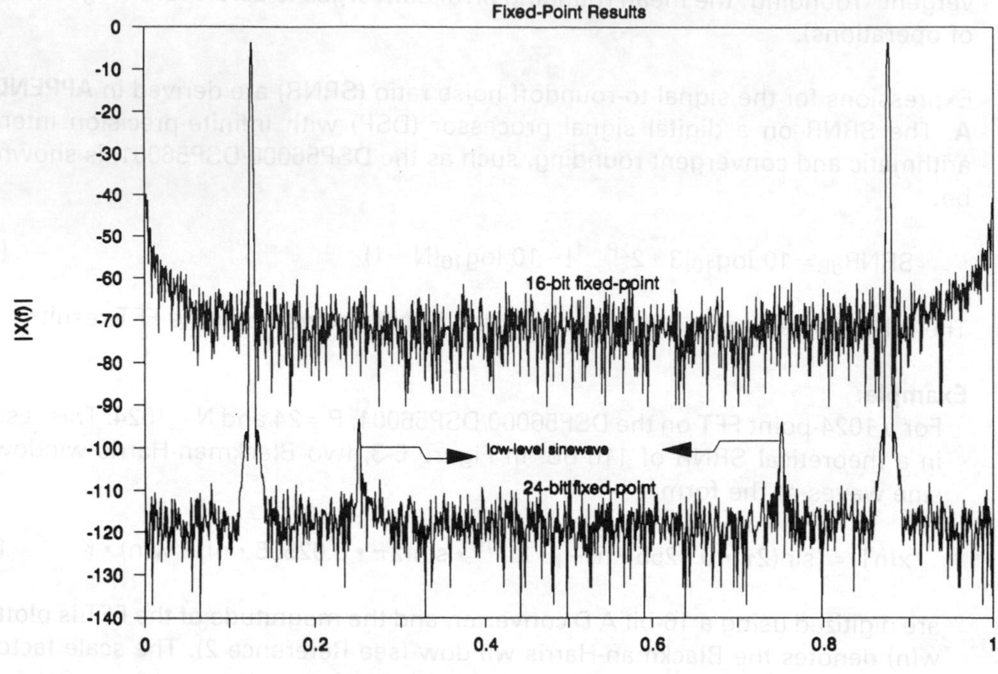

Figure 5-3. Fixed-Point FFT Results: 16-Bit Truncation Arithmetic versus 24-Bit DSP56000/DSP56001 ($r = f/f_s$)

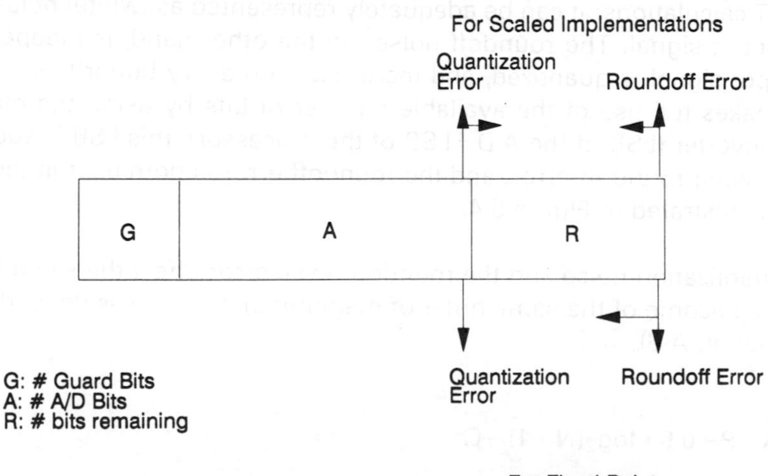

Figure 5-4. Quantization Noise versus Roundoff Noise

for real-input FFTs. A is the number of bits in the A/D converter, and G is the number of guard bits used to avoid overflow or limiting of the FFT output. This A/D size may well be called "optimal": a larger A/D precision does not result in more precise results, as the roundoff noise becomes dominant. A smaller A/D size does not offer the maximum precision obtainable with the processor, since the quantization noise dominates.

Example:

The "cutoff" point at which the roundoff noise and the quantization noise become of the same order of magnitude, assuming complex-input FFTs as in the previous example, is roughly equal to 11, using equation (31). Consequently, a 1024-point FFT which uses more than 11-bit A/D converter inputs, is limited in precision by roundoff noise. Fewer than 11 bits at the input result in more quantization than roundoff noise. Alternatively, if $P=16$ (16-bit DSP), a 1024-point complex FFT is limited by roundoff noise up to an A/D size of only $A=3$.

While the DSP56000/DSP56001 Family implements infinite-precision internal arithmetic, with one final convergent rounding operation, many existing DSPs use a different approach. Often, a multiplication result is simply truncated to the number of bits P used in storage of variables in memory. This results in introduction of errors after every real multiplication in the FFT flow diagram. Clearly, the noise sources are located in different locations than is the case for a DSP56000/DSP56001 type machine. The signal-to-truncation noise ratio (STNR) is derived in **APPENDIX A** (equation A-12).

$$STNR_{dB} = 10\log_{10}[3 \cdot 2^{2P-2}] - 10\log_{10}(N-1) \tag{34}$$

In addition, it is shown in **APPENDIX A** (equation A-9) that truncation arithmetic results in a bias in the FFT magnitude, which is prevalent at DC.

Example:

The signal of equation (31) is digitized and the 1024-point FFT computed on a 16-bit DSP, which implements truncation arithmetic after every multiplication. The same number of guard-bits needs to be kept as in the 24-bit implementation to avoid overflow problems ($G=8$). According to equation (28), the maximum number of A/D bits A_{max} is consequently equal to eight. Figure 5-3 shows the resulting FFT magnitude. Note the bias at DC, which is predicted in **APPENDIX A** (equation A-7) as a result of the truncation arithmetic. The STNR, given in equation (34) for $N=1024$ is predicted to be 65 dB. The actual STNR obtained in Figure 5-3 is seen to be roughly equal to 70 dB. This discrepancy is due to the fact that in the calculations of **APPENDIX A**, it is assumed that every multiplication contributes to truncation noise. In practice, trivial multiplications (such as in pass one) do not contribute to the noise, and the STNR is actually larger. Note that the smaller harmonic component, which was clearly visible in the 16-bit input, DSP56000/DSP56001 computation, is now completely lost in the truncation noise.

5.2 — AUTOMATIC SCALING

5.2.1 — Input Bounds

The main problem in the previous fixed-point implementation lies in the requirement of a number of guard-bits to avoid overflow or limiting when the magnitude of the butterfly results grows every stage. This growth can be avoided, however, by automatically scaling down the outputs of all FFT butterflies by a factor of 0.5. Since this scaling is uniform throughout the FFT (all points are scaled equally), it can be easily taken into account by keeping in mind that a common scale factor of $2^{\log_2 N} = N$ needs to be applied to the FFT results if absolute results are required. This automatic scaling mode is implemented on the DSP56000/DSP56001 in hardware, with no penalty in execution speed.[6]

Since the real and imaginary outputs, and thus the magnitudes, are automatically divided by two at every stage, the bounds on the magnitude at the butterfly outputs are directly obtained from equation (20) as:

$$\frac{\max(|A|, |B|)}{2} \leq \max(|C|, |D|) \leq \max(|A|, |B|) \qquad (35)$$

Thus, with a two's complement, fixed-point notation, it is now sufficient to make sure that the magnitude of the butterfly inputs is less than one at all times. Since there is no more growth of the magnitude of butterfly results, it is sufficient to require that the input magnitude is less than one.

For real-input FFTs, the magnitude of the input is equal to the absolute value of the input signal, and thus:

$$|x(nT)| < 1.0 \qquad (36)$$

Thus, no guard-bits are needed, and the maximum number of bits of the A/D converter A_{max} becomes equal to the number of bits used in the DSP's memory storage, i.e.:

$$A_{max} = P \qquad (37)$$

For complex-input FFTs, the previous limit is no longer sufficient, since both real and imaginary parts affect the magnitude. However, by keeping one guard bit:

$$|R\{x(nT)\}| \leq 0.5 \qquad (38)$$
$$|I\{x(nT)\}| \leq 0.5$$

[6]The user can set the scaling bits S0 and S1 in the status register prior to executing the FFT.

the magnitude of the input is limited to $\sqrt{2}/2 \approx 0.707$, and is definitely less than one. In this case, the maximum number of bits that can be used for digitizing the input signal equals:

$$A_{max} = P - 1 \tag{39}$$

In addition to allowing a larger number of A/D bits at the input, the main advantage of the automatic scaling mode lies in the fact that FFTs of virtually unlimited size can be computed: in the fixed-point (unscaled) version, the maximum number of points in the FFT is obviously limited by the required A/D bits and the number of bits P used for number storage.

5.2.2 — Roundoff Errors

The roundoff errors which occur in the FFT butterfly calculation using automatic scaling are analyzed in **APPENDIX B**. The total SRNR in decibels is given by:

$$SRNR_{dB} = 10\ \log_{10}[3 \cdot 2^{2P-1}] - 10\ \log_{10}[\log_2 N] \tag{40}$$

The SRNR in this case is seen to grow more slowly ($\sim \log_2 N$) than the SRNR for the fixed-point case ($\sim N$) . As is shown in **APPENDIX B**, this is due to the fact that the roundoff errors in early FFT passes are attenuated throughout the calculation because of the division by two at every successive stage. If the maximum-length A/D converter is used, the roundoff noise again exceeds the quantization noise. The cutoff point at which the quantization and roundoff errors are of equal order of magnitude is derived in **APPENDIX B** to be:

$$A = P - 1 - \log_2 N - 0.5 \cdot \log_2(\log_2 N) \tag{41}$$

for complex-input FFTs and

$$A = P - 0.5 - \log_2 N - 0.5 \cdot \log_2(\log_2 N) \tag{42}$$

for real-input FFTs.

Example:

For the complex-input, 1024-point FFT of the previous examples, equation (40) suggests a total SRNR of approximately 145 dB. This is right at the limit of the 24-bit representation. Using this technique, a theoretical input A/D size of 23 bits is possible. The roundoff noise in this case exceeds the quantization noise in the final results. Figure 5-5 shows the magnitude of the complex FFT of the same signal as in the previous examples, but with automatic scaling turned on. The cutoff point at which the quantization noise becomes of the same order of magnitude as the roundoff noise is calculated to occur for an input A/D size of approximately 11 bits.

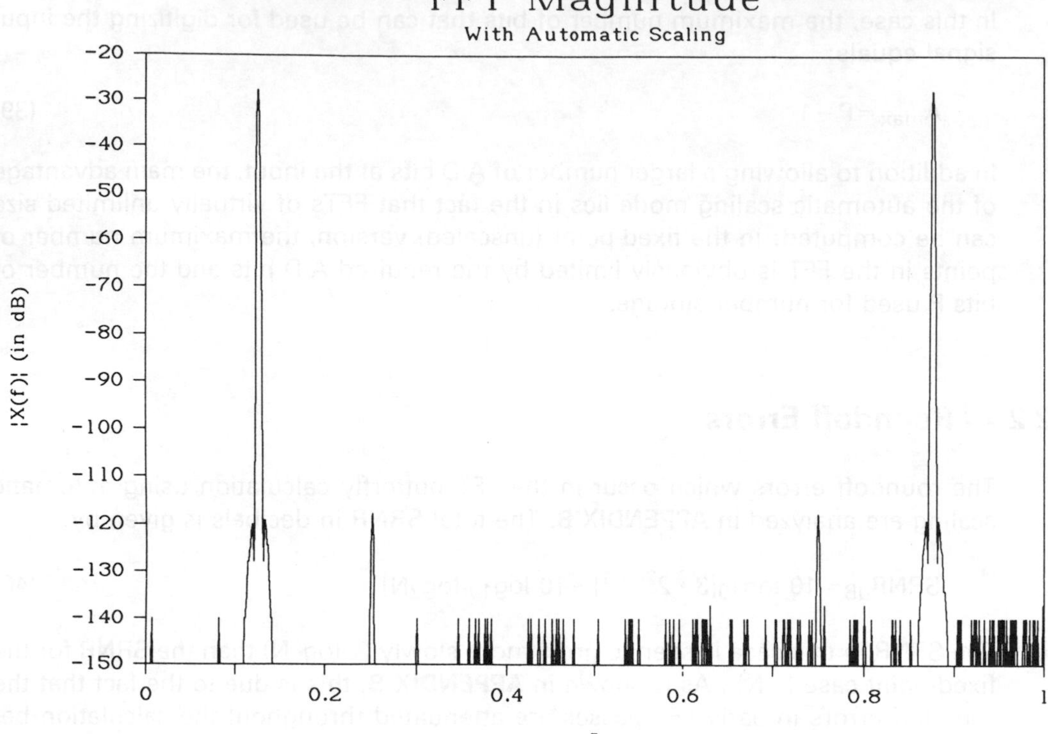

Figure 5-5. Magnitude of the FFT of the Same Signal as in Figure 5-3, with Automatic Scaling Turned On

Note from Figure 5-5 that the noise appears very irregular. The reason for this is that the roundoff error is so small (because of indiscriminate down shifts, − 145 dB) that it only sporadically affects the least significant bit of the result. The resulting FFT has mainly zero values except for at the signal peak. The low-level side-lobes of the Blackman-Harris window are not at all visible, as they "fell off the edge" due to the down shifts. Also note that the unconditional down shifts in the FFT result in a peak magnitude which is less than one. This means that full 1-bit word growth of the magnitude did not occur in every stage. Consequently, the SRNR is in practice worse than predicted: an effective SRNR roughly equal to the fixed-point case is obtained.

Although automatic scaling does offer the ability of implementing FFTs of virtually unlimited size, it is clear from the previous example that a better SRNR is not necessarily obtained in practice. Also, the A/D size at which quantization noise "overtakes" roundoff noise (the "optimal" A/D precision) is roughly equal to that for fixed-point results, unless the number of points N is very large. Consequently, for FFTs of sizes 1024 and less, automatic scaling does not offer any main advantages. However, the advantages do become more pronounced for larger FFTs. For example, when N = 4096, the optimal A/D size for a fixed-point realization is seven, whereas the size for an automatically scaled version is nine.

5.3 — BLOCK FLOATING POINT

5.3.1 — General

Although it becomes clear from the previous discussions that scaling each FFT butterfly improves the maximum SRNR in some cases, this improvement is often not fully used in practical applications. For instance, as is seen in equation (35), automatic scaling also decreases the lower bound on the FFT butterfly output magnitude. Although all of the discussions above are based on a scaled input such that the output peak magnitude is equal to one, practical situations arise where the input is not *a priori* known. For signals without pronounced harmonic components, the maximum output magnitude will be closer to its lower bound. Hence, there is no word growth in the magnitude for unscaled, fixed-point FFTs, and the peak output magnitude is much less than one. As a result, the previously derived expressions are no longer correct, and the SRNR and SQNR are actually found to be much worse: the same noise components now need to be scaled according to the actual peak signal magnitude. In the case of automatic scaling, the peak magnitude of the output may even be much smaller than the maximum magnitude of the input. As becomes evident from Figure 5-5, even with pronounced harmonic components the peak magnitude is considerably smaller than 1 (0 dB). Consequently, the SQNR and SRNR for FFTs of signals without pronounced harmonic components is much worse for the automatically scaled FFT than for the fixed-point FFT. For signals with a combined strong harmonic input as well as an underlying broad-band signal, the harmonic peak is generally reliably computed, while the broadband component is not retained: it "falls off the edge" (is shifted below the least significant bit) of the P-bit representation.

Clearly, some decision mechanism is needed which selectively scales the output of butterflies when needed, and doesn't scale the outputs when there is no appropriate growth in the magnitude of the results.

The simplest way of accomplishing conditional scaling is by a technique termed *block floating point*. With this technique, complete passes of the FFT calculation are selectively scaled, depending on whether sufficient growth in magnitude occurred in the previous pass. This method is especially of interest on fixed-point machines with a hardware scaling mode, such as the DSP56000/DSP56001. When implemented correctly, minimal additional hardware and software is required for implementing block floating-point algorithms with a negligible penalty in speed performance.

5.3.2 — Implementation of Block Floating Point on DSP56002[7]

On the DSP56002, block floating point is easily implemented by using the "scaling bit" S in the status register (bit 7). This bit is set upon moving a result to memory which is larger than 0.25 in absolute value. Upon completion of each FFT pass, this bit is tested. If S is set, the automatic scaling mode is turned on8 before the next FFT pass, indicating that in at least one butterfly the magnitude has increased. If S is not set, no growth has occurred and the scale-down mode is turned off; butterflies on the next pass are computed as in **5.1 FIXED-POINT IMPLEMENTATON.** Testing S also resets it, such that it is ready for the next pass. Since this testing of S only occurs once for every FFT pass, only a few instruction cycles are added. Compared to the total number of instructions required to calculate the FFT, these few cycles per pass are negligible in the overall execution time, as is evident from Table 4-1.

The operation of this algorithm is graphically depicted in Figure 5-6. For complex input FFTs, the input data is limited to be less than or equal to 0.25 in absolute value, while for real-input FFTs, a maximum input magnitude of 0.5 is retained. This should avoid all limiting at the output of the first pass, where scaling is always off. The grey area depicts all real and imaginary values less than 0.25, where scaling down is turned off. Note that any results moved to memory, which fall outside the grey area, will result in scale-down mode being turned on for the next FFT pass. Thus, at all times the results in memory are forced to lie inside the outer circle, and thus the real and imaginary values are always less than or equal to $\sqrt{2}/2 \approx 0.707$, and no overflow or limiting occurs.

Since scaling is implemented for complete "blocks" of memory, a common scaling factor can be used if absolute numbers are required. This scaling factor can be constructed in terms of an "exponent" by adding up the number of passes for which scale-down was turned on. This common exponent is easily obtained by incrementing a counter or memory location each time S is detected to be set. The name *block floating point* indicates this common exponent for a complete memory block.

Example:
The signal of the previous examples is used as input to the block floating-point FFT on the DSP56002. The results are shown in Figure 5-7.

[7]The technique discussed here is available only on the DSP56002.
[8]By setting SO = 1 and S1 = 0 in the status register (scale down mode).

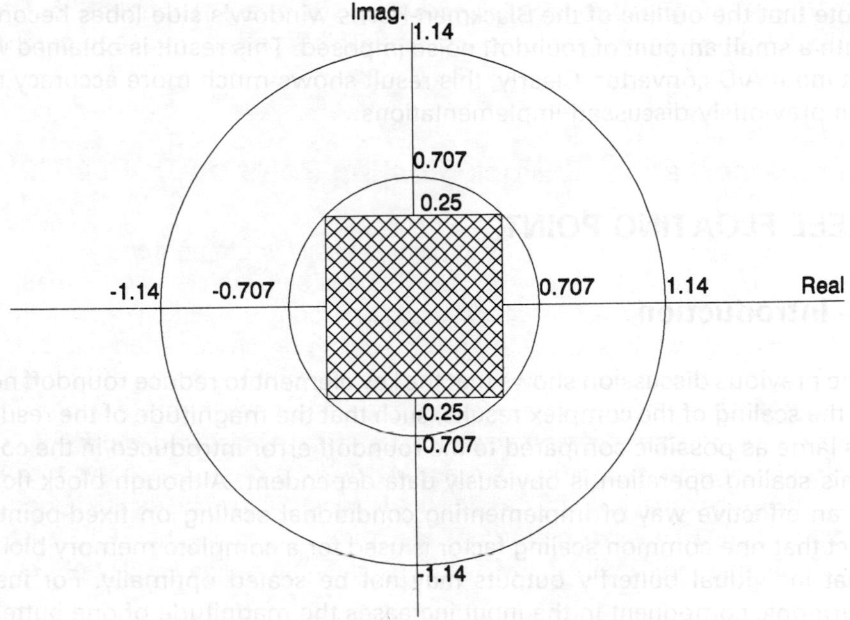

Figure 5-6. Block Floating-Point Bounds

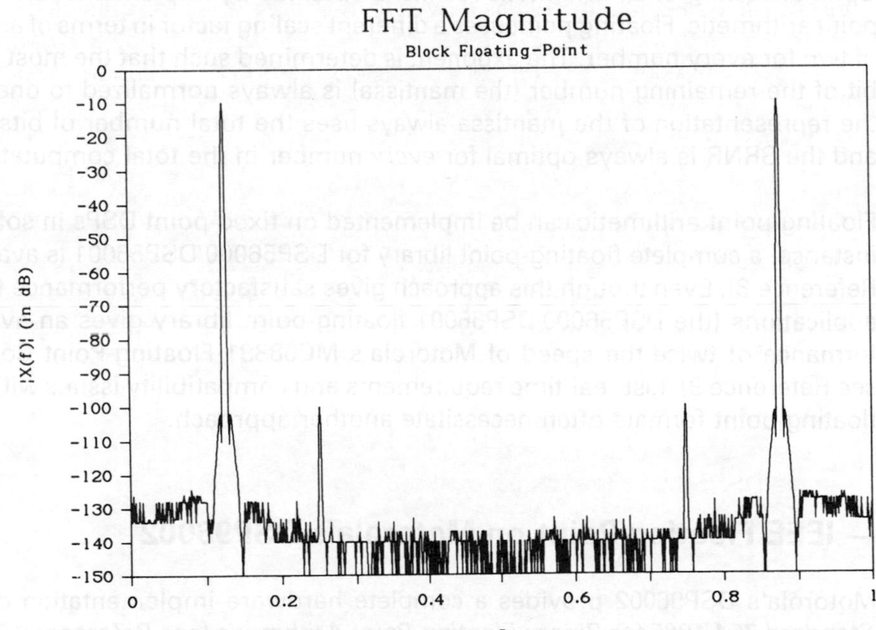

Figure 5-7. Magnitude of the Block Floating-Point FFT of the Same Signal as in Figure 5-3

Note that the outline of the Blackman-Harris window's side lobes becomes visible, with a small amount of roundoff noise imposed. This result is obtained with full 22-bit input A/D converter. Clearly, this result shows much more accuracy than any of the previously discussed implementations.

5.4 — IEEE FLOATING POINT

5.4.1 — Introduction

The previous discussion shows that the key element to reduce roundoff noise effects is the scaling of the complex results, such that the magnitude of the result is always as large as possible compared to the roundoff error introduced in the computation. This scaling operation is obviously data dependent. Although block floating point is an effective way of implementing conditional scaling on fixed-point DSPs, the fact that one common scaling factor is used for a complete memory block indicates that individual butterfly outputs may not be scaled optimally. For instance, if a harmonic component in the input increases the magnitude of one butterfly output, other butterfly outputs may not experience any word growth and thus should not be scaled, since unnecessary scaling invariably reduces the SRNR at the butterfly output.

Optimal scaling of all arithmetic results is obtained by implementation of floating-point arithmetic. Floating point uses a different scaling factor in terms of an exponent of two for every number. The exponent is determined such that the most significant bit of the remaining number (the mantissa) is always normalized to one. As such, the representation of the mantissa always uses the total number of bits available, and the SRNR is always optimal for *every* number in the total computation.

Floating-point arithmetic can be implemented on fixed-point DSPs in software. For instance, a complete floating-point library for DSP56000/DSP56001 is available (see Reference 8). Even though this approach gives satisfactory performance for various applications (the DSP56000/DSP56001 floating-point library gives an average performance of twice the speed of Motorola's MC68881 Floating-Point Coprocessor, see Reference 9), fast real-time requirements and compatibility issues with standard floating-point formats often necessitate another approach.

5.4.2 — IEEE Floating-Point on Motorola's DSP96002

Motorola's DSP96002 provides a complete hardware implementation of the *IEEE Standard 754-1985 for Binary Floating-Point Arithmetic* (see Reference 10). Because of the similarity between the DSP96002's architecture with that of the DSP56000/DSP56001, compatibility between the two processors is maintained. Consequently, implementation of FFTs is very similar on both processors, as illustrated in **4.3 IMPLEMENTATON ON MOTOROLA'S DSP96002**. The result clearly shows both harmonic components, as well as the spectral shape of the Blackman-Harris window. Because of the hardware IEEE floating-point arithmetic, FFTs execute as fast as or even faster than their fixed-point counterpart, as illustrated in Table 4-1.

In addition to the optimal scaling offered by floating-point arithmetic, conformance to the IEEE standard provides several other advantages. These advantages are discussed in greater detail in Reference 11, and include guaranteed error bounds, standard error indication, and portability across many different implementations. This last aspect is of particular importance in spectrum analysis algorithm development: adherence to the IEEE standard guarantees results which are bit-for-bit equal with results obtained in simulations using high-level languages that conform with the standard. Since much time is spent in algorithm development for spectrum-based applications, compatibility of results often saves an additional step in application development, where the use of fixed-point arithmetic or other floating-point formats may require considerable further effort in comparison with simulated results.

Example:

The signal of the previous examples is generated in IEEE standard floating point, and the magnitude of the 1024-point complex FFT is shown in Figure 5-8. Clearly, the optimal scaling of the floating-point numbers has reduced the roundoff noise to a negligeable level, and the figure clearly displays the smooth outline of the Blackman-Harris window. Bit-per-bit compatibility is obtained with high-level language implementations which use the standard. However, one must keep in mind that sine and cosine tables, generated in high-level languages using single-precision sine and cosine calls, are not obtained in full IEEE single precision (the standard only defines addition, multiplication, division, and square root). The sine and cosine tables in the DSP96002 were generated in double precision, and rounded to single precision to obtain correct results for all 32 bits. Consequently, FFTs generated in high-level languages may actually demonstrate additional "noise", created by the lack of precision of the sine and cosine tables. This is illustrated in Figure 5-9. The results do become bit-for-bit identical with the DSP96002 when the sine and cosine tables are generated in the same way. This is illustrated in Figure 5-10.

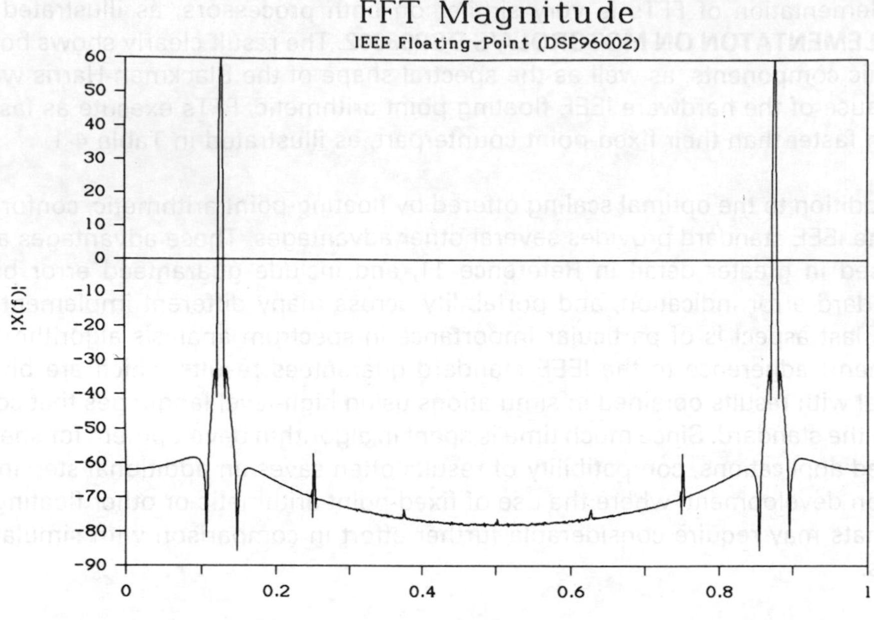

Figure 5-8. Magnitude of the IEEE Standard Floating-Point FFT of the Same Signal as in Figure 5-3

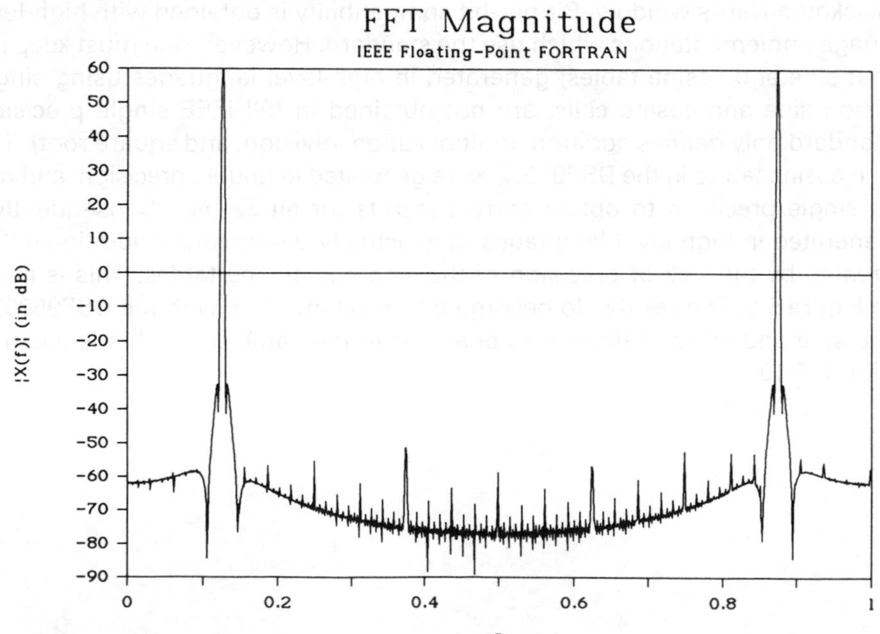

Figure 5-9. Magnitude of an IEEE Floating-Point FORTRAN FFT of the Same Signal as in Figure 5-3

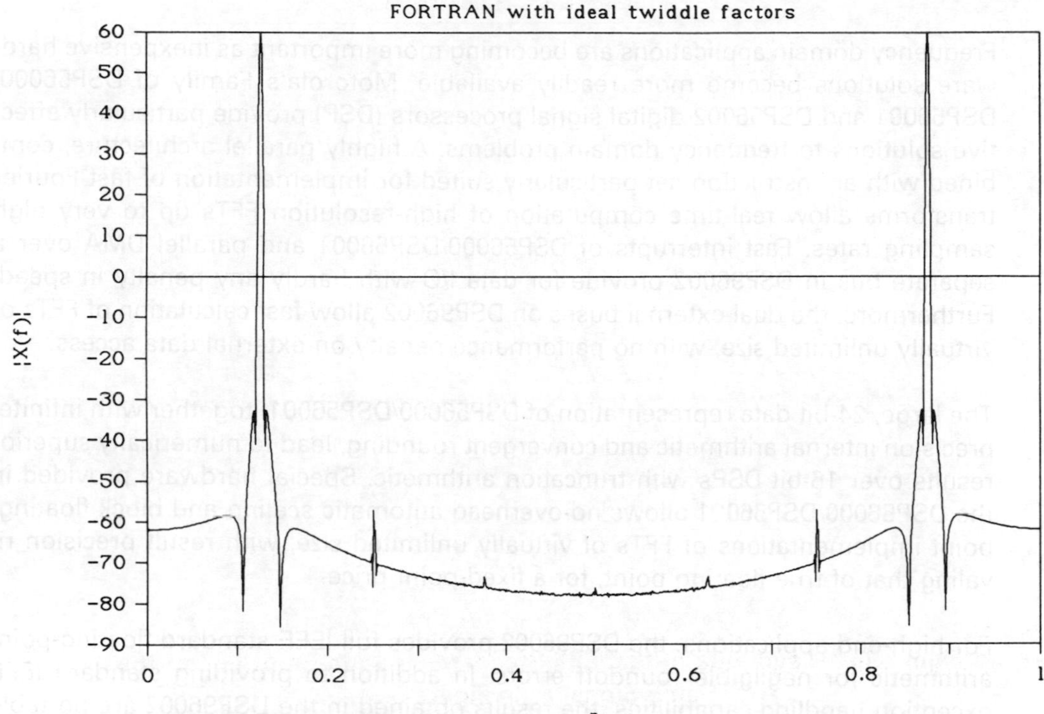

Twiddle factors are generated in double precision and rounded to signal precision.

Figure 5-10. Magnitude of an IEEE Floating-Point FORTRAN FFT of the Same Signal as in Figure 5-3

SECTION 6 — CONCLUSIONS

Frequency domain applications are becoming more important as inexpensive hardware solutions become more readily available. Motorola's Family of DSP56000/DSP56001 and DSP96002 digital signal processors (DSP) provide particularly effective solutions to frequency domain problems. A highly parallel architecture, combined with an instruction set particularly suited for implementation of fast Fourier transforms allow real-time computation of high-resolution FFTs up to very high sampling rates. Fast interrupts of DSP56000/DSP56001 and parallel DMA over a separate bus in DSP96002 provide for data I/O with hardly any penalty in speed. Furthermore, the dual external buses on DSP96002 allow fast calculation of FFTs of virtually unlimited size, with no performance penalty on external data access.

The large, 24-bit data representation of DSP56000/DSP56001, together with infinite-precision internal arithmetic and convergent rounding, lead to numerically superior results over 16-bit DSPs with truncation arithmetic. Special hardware provided in the DSP56000/DSP56001 allows no-overhead automatic scaling and block floating-point implementations of FFTs of virtually unlimited size, with result precision rivaling that of true floating point, for a fixed-point price.

For high-end applications, the DSP96002 provides full IEEE standard floating-point arithmetic for negligible roundoff errors. In addition to providing standard IEEE exception handling capabilities, the results obtained in the DSP96002 are portable across many applications that use the standard, such as high-level language simulations, data bases, etc. Motorola's family of digital signal processors, combined with Motorola's data conversion parts (see Reference 12), provide a complete, cost-efficient solution to frequency domain problems; from low-end small-size FFT applications, to high-end instrumentation and computer workstations for scientific computing.

APPENDIX A — NOISE EXPRESSIONS FOR FIXED-POINT FFTs

A-1 — ROUNDOFF NOISE EXPRESSIONS FOR FIXED-POINT FFTs ON DSP56000/DSP56001

Although a detailed analysis of roundoff errors is beyond the scope of this work (for a more detailed tutorial on roundoff errors in FFTs, see Reference 13[9]), the source of roundoff noise within butterfly calculations is depicted in Figure A-1.

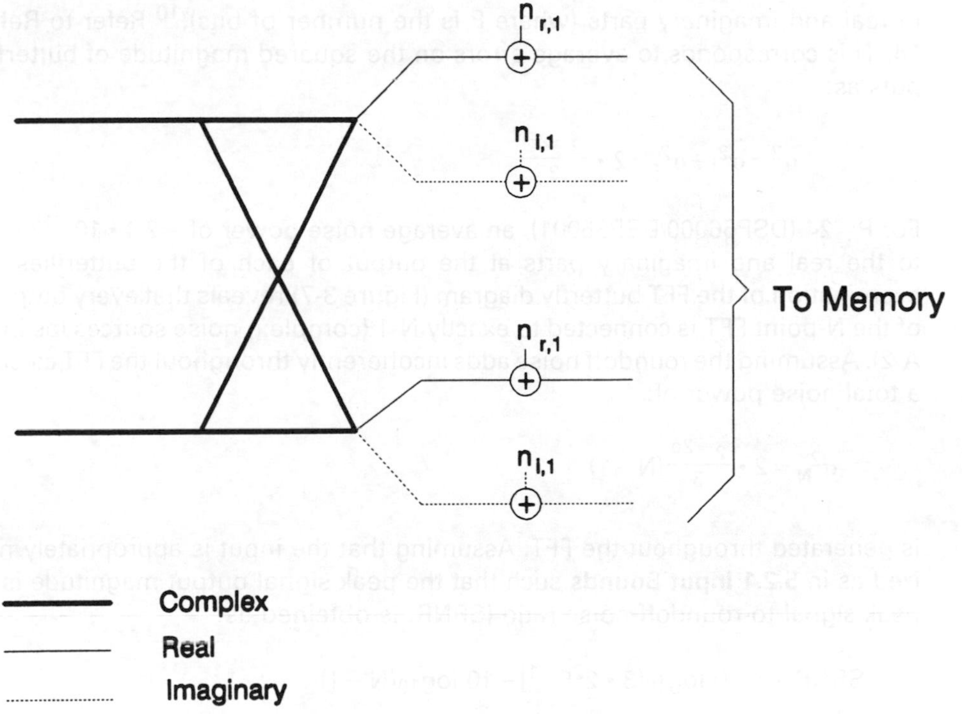

Figure A-1. Roundoff Noise Sources in DSP56000/DSP56001 Butterfly Calculation

[9]The reader should be cautioned that most roundoff noise analysis in the literature holds for "generic" fixed-point problems. Specific features of DSPs such as infinite internal precision, different multiplier and accumulator sizes, number of available registers, etc. are usually not taken into account.

Figure A-1 shows roundoff noise in the FFT butterfly when implemented on the DSP56000/DSP56001. In this particular case, the four sources of roundoff noise occur prior to moving each of the butterfly's four outputs to memory. Note that the errors on the computations for D (bottom butterfly output) are the same as the errors on the computation of C because of the use of the SUBL instruction. These are the only sources of roundoff noise. Internal to the CPU, infinite-precision arithmetic is obtained by the 56-bit accumulators, followed by only one final rounding operation before the result is moved to memory. The noise sources $n_{i,r}$ thereby have zero mean because of the convergent rounding ("round to nearest even") which is implemented in the DSP56000/DSP56001, and have variance (or average "noise power") of:

$$\sigma^2_{i,r} = \frac{2^{-2p}}{3} \qquad \text{(A-1)}$$

in real and imaginary parts (where P is the number of bits).[10] Refer to Reference 14. This corresponds to average errors on the squared magnitude of butterfly outputs as:

$$\sigma^2 = \sigma^2_i + \sigma^2_r = 2 \cdot \frac{2^{-2P}}{3} \qquad \text{(A-2)}$$

For P = 24 (DSP56000/DSP56001), an average noise power of $\approx 2.4 \cdot 10^{-15}$ is added to the real and imaginary parts at the output of each of the butterflies. Closer examination of the FFT butterfly diagram (Figure 3-7) reveals that every output point of the N-point FFT is connected to exactly N-1 (complex) noise sources (as in Figure A-2). Assuming the roundoff noise adds incoherently throughout the FFT calculation, a total noise power of:

$$\sigma^2_N = 2 \cdot \frac{2^{-2p}}{3}(N-1) \qquad \text{(A-3)}$$

is generated throughout the FFT. Assuming that the input is appropriately normalized as in **5.2.1 Input Bounds** such that the peak signal output magnitude is one, a peak signal-to-roundoff-noise ratio (SRNR) is obtained as:

$$SRNR_{dB} = 10\ \log_{10}[3 \cdot 2^{2P-1}] - 10\ \log_{10}(N-1) \qquad \text{(A-4)}$$

The quantization noise in fixed-point FFTs is the error which is caused by the representation of the *input signal* by a finite number of bits. This error affects the least significant bit of the A/D converter. When the A/D converter bits are loaded in the DSP memory, G guard bits are kept. Consequently, the quantization noise and the

[10]Note that here P is the total number of bits and not the number of mantissa bits only, as in Reference 13.

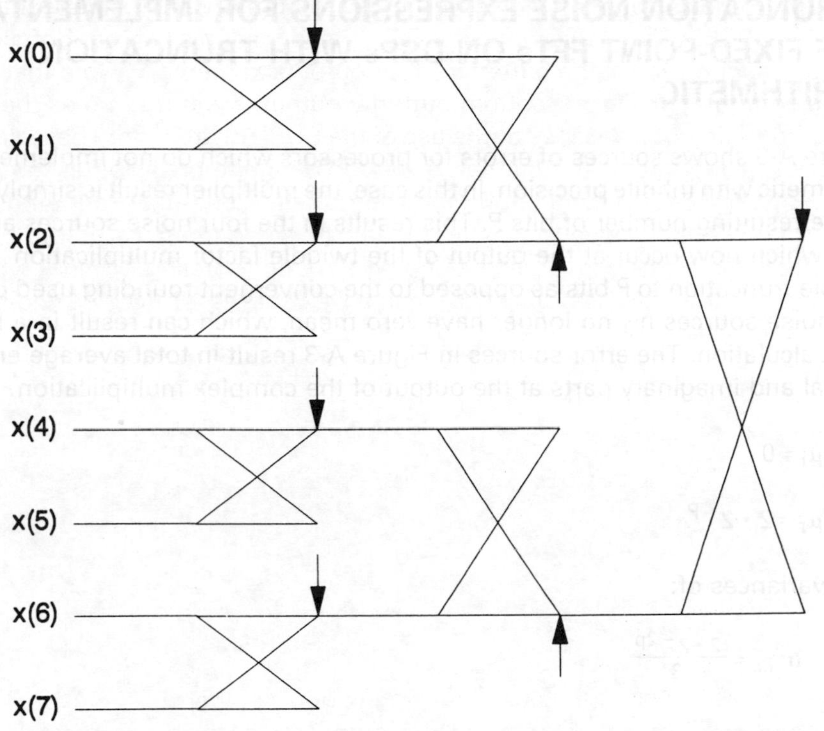

Figure A-2. Combination of Noise Sources to the FFT Result

roundoff noise at the output of the FFT affect the same bits (become of the same order of magnitude) when the associated errors become equal (assuming all noise sources are "white"):

$$\frac{2^{1-2(A+G)}}{3} = \frac{2^{1-2P}}{3}(N-1) \qquad \text{(A-5)}$$

or:

$$A = P - G - 0.5 \cdot \log_2(N-1) \qquad \text{(A-6)}$$

for complex-input FFTs and

$$\frac{2^{-2(A+G)}}{3} = \frac{2^{1-2P}}{3} \cdot (N-1) \qquad \text{(A-7)}$$

or:

$$A = P - G - 0.5 - 0.5 \cdot \log_2(N-1) \qquad \text{(A-8)}$$

for real-input FFTs. A equals the number of bits in the A/D converter and G is the number of guard bits.

A-2 — TRUNCATION NOISE EXPRESSIONS FOR IMPLEMENTATION OF FIXED-POINT FFTs ON DSPs WITH TRUNCATION ARITHMETIC.

Figure A-3 shows sources of errors for processors which do not implement internal arithmetic with infinite precision. In this case, the multiplier result is simply truncated to the resulting number of bits P. This results in the four noise sources as in Figure A-3, which now occur at the output of the twiddle factor multiplication. Assuming simple truncation to P bits as opposed to the convergent rounding used previously, the noise sources $n_{i,r}$ no longer have zero mean, which can result in a bias in the FFT calculation. The error sources in Figure A-3 result in total average error values in real and imaginary parts at the output of the complex multiplication:

$$\mu_i = 0 \qquad \text{(A-9)}$$

$$\mu_r = 2 \cdot 2^{-P}$$

and variances of:

$$\sigma^2_{i,r} = \frac{2 \cdot 2^{-2p}}{3} \qquad \text{(A-10)}$$

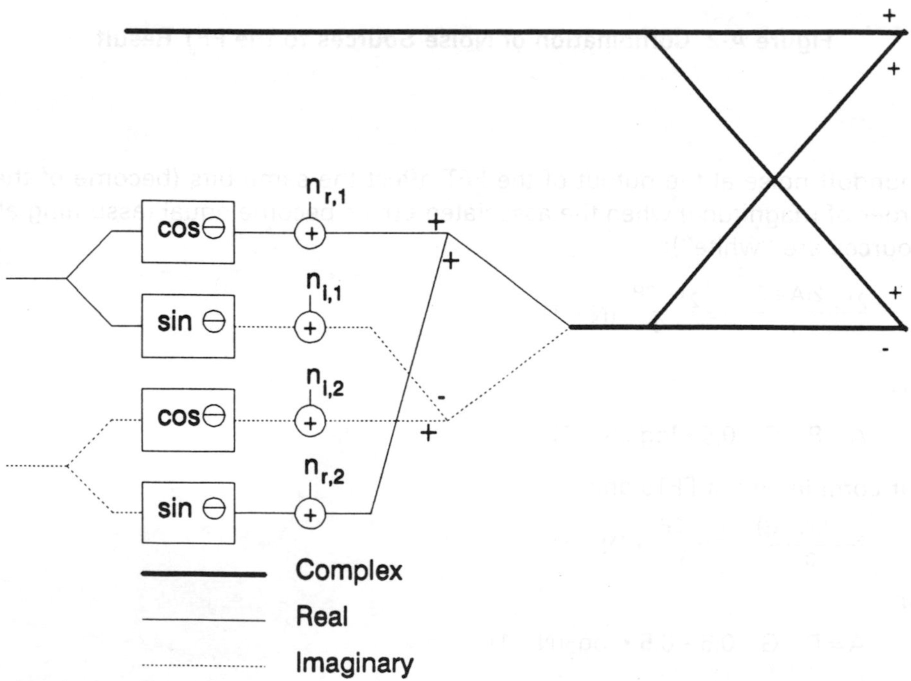

Figure A-3. Truncation Error Sources in the DIT Butterfly

on real and imaginary parts, since the imaginary part is obtained by the difference of two products, while the real part is obtained by the sum. Note that this variance is twice the value obtained when using infinite-precision internal arithmetic as in **A-1 ROUNDOFF NOISE EXPRESSIONS FOR FIXED-POINT FFTs ON DSP56000/ DSP56001**. The nonzero average error in the calculation of the imaginary part of equation (A-7) results in a bias in the FFT calculation. Since this value is constant across the FFT flow diagram, a low-frequency bias will result. The variance, expressed in equation (A-6), propagates as truncation noise variance to the output of the FFT: as before, each output point of the flow-diagram is connected to exactly N-1 butterflies. Assuming that the noises add incoherently, a total variance on the complex outputs is given by:

$$\sigma^2_N = (N-1)\frac{2^{-2(P-1)}}{3} \qquad \text{(A-11)}$$

This past expression can again be used to determine the total Signal-to-Truncation noise ratio (STNR) as:[11]

$$STNR_{dB} = 10\ \log_{10}[3 \cdot 2^{2P-2}] - 10\ \log_{10}(N-1) \qquad \text{(A-12)}$$

[11]It is again assumed that the input is appropriately scaled such that the peak output signal magnitude is one.

APPENDIX B — ROUNDOFF NOISE EXPRESSIONS FOR FFTs ON DSP56000/DSP56001 WITH AUTOMATIC SCALING

Figure B-1 shows the appropriate butterfly when automatic scaling is introduced at every pass. The noise sources introduced have the same characteristics as the ones in the fixed-point representation (equation (A-1)) in **APPENDIX A**. These noise sources still add incoherently throughout the FFT. However, they are no longer of equal strength when they add: noise which is created in the first stage of the FFT shows up scaled down $\log_2 N - 1$ times before contributing to the final output noise, while

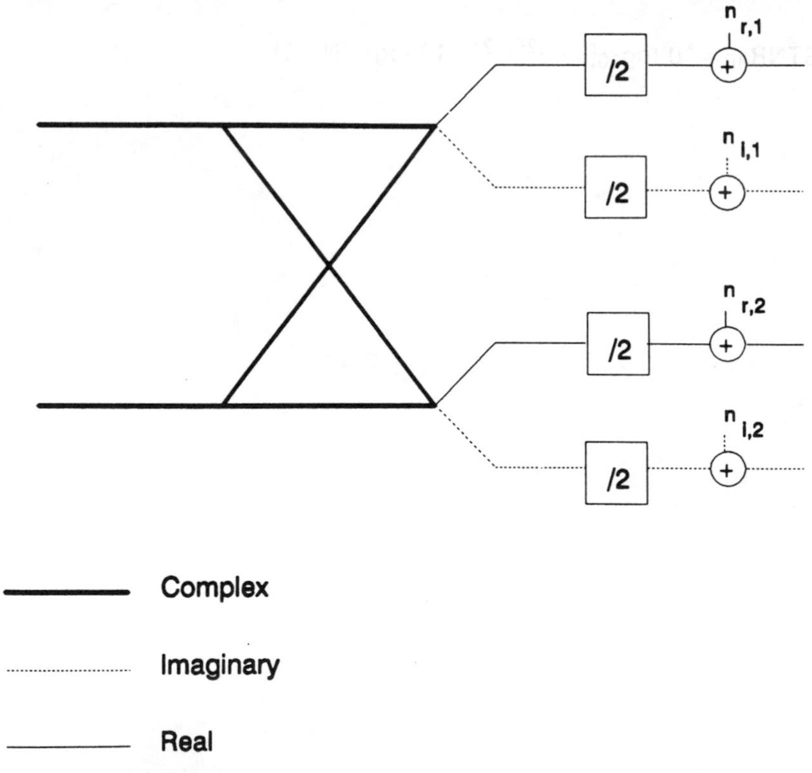

Figure B-1. Roundoff Noise Sources with Automatic Scaling

the noise source in the final stage contributes fully. Thus, an equation for the total noise variance on the complex output is given by:

$$\sigma^2{}_N = \sigma^2 \left\{ \frac{N}{2}\left(\frac{1}{2}\right)^{\log_2 N - 1} + \frac{N}{4}\left(\frac{1}{2}\right)^{\log_2 N - 2} + \cdots + 1 \right\} \qquad \text{(B-1)}$$

$$= \sigma^2 \sum_{k=1}^{\log_2 N} \frac{N}{2^k}\left(\frac{1}{2}\right)^{\log_2 N - k}$$

$$= \sigma^2 \log_2 N$$

where σ^2 is given in equation (A-2). The total SRNR in this case is thus given by:[12]

$$\mathrm{SRNR}_{dB} = 10\ \log_{10}[3 \cdot 2^{2P-1}] - \log_{10}[\log_2 N] \qquad \text{(B-2)}$$

The bit affected by the quantization noise when automatic scaling is turned on, is now shifted down after every FFT butterfly. The point at which the quantization error and the roundoff error affect the same bits can be obtained as in **A-1 ROUND-OFF NOISE EXPRESSION FOR FIXED-POINT FFTs ON DSP56000/DSP56001**. This gives:

$$\frac{2^{1-2(A+G)}}{3 \cdot N^2} = \frac{2^{1-2P}}{3} \cdot \log_2 N \qquad \text{(B-3)}$$

or:

$$A = P - 1 - \log_2 N - 0.5 \cdot \log_2(\log_2 N) \qquad \text{(B-4)}$$

for complex-input FFTs (G = 1). For real-input FFTs, G = 0 as explained in **5.2.1 Input Bounds**, and the result:

$$\frac{2^{-2(A+G)}}{3 \cdot N^2} = \frac{2^{1-2P}}{3} \cdot \log_2 N \qquad \text{(B-5)}$$

$$A = P - 0.5 - \log_2 N - 0.5 \cdot \log_2(\log_2 N) \qquad \text{(B-6)}$$

is obtained.

[12]Assuming that the input is appropriately scaled such that the peak output magnitude is equal to one.

REFERENCES

1. A. Papoulis, *The Fourier Integral and its Applications*

2. F. J. Harris, "On the Use of Windows for Harmonic Analysis with the Discrete Fourier Transform" *Proc. IEEE*, Vol. 66, No. 1, Jan. 1978, pp. 57-84

3. A. V. Oppenheim and R. W. Schafer, *Digital Signal Processing* Englewood Cliffs, NJ: Prentice-Hall, 1975

4. Kevin L. Kloker, "The Motorola DSP56000 Digital Signal Processor" *IEEE Micro*, December 1986

5. Motorola Electronic Bulletin Board (Dr. BuB) FFTR2C Fast Fourier Transform Routine

6. EDN's DSP Benchmarks EDN, September 29, 1988

7. *Fractional and Integer Arithmetic Using the DSP56000 Family of General-Purpose Digital Signal Processors* Motorola, Inc., Application Report APR3/D

8. Motorola's DSP Bulletin Board (Dr. BuB) DSP56000 Floating-Point Library

9. *MC68881/MC68882 Floating-Point Coprocessor User's Manual* Publication No. MC68881 UM/AD, Motorola, Inc., Austin, TX 1987

10. *IEEE Standard for Binary Floating-Point Arithmetic* ANSI/IEEE Std. 754-1985 New York, NY: IEEE, 1985

11. Guy R.L. Sohie and K. Kloker, "A Digital Signal Processor with IEEE Floating-Point Arithmetic" *IEEE Micro*, Vol. 8, No. 6, December 1988

12. Motorola's DSP56ADC16 Data Sheet

13. A. V. Oppenheim and C. J. Weinstein, "Effects of Finite Register Length in Digital Filtering and the Fast Fourier Transform" *Proc. IEEE*, Aug. 1972, pp. 957-976

14. See ref. [3], pp. 404 - 418

Motorola Digital Signal Processors

Convolutional Encoding and Viterbi Decoding Using the DSP56001 with a V.32 Modem Trellis Example

by
Dion D. Messer
Digital Signal Processor Operation

APR6/D

TABLE OF CONTENTS

Paragraph Number	Title	Page Number

Section 1
Introduction

Section 2
Trellis (Convolutional) Encoding

Section 3
Viterbi Decoding

Section 4
Summary

References

Appendix A
DSP56001 Encoding Program Listing

Appendix B
DSP56001 Decoding Program Listing

Appendix C
DSP56001 Bound.d Data File

LIST OF ILLUSTRATIONS

LIST OF TABLES

SECTION 1
INTRODUCTION

Coding techniques have long been used for error correction to decrease the bit error rate (BER) in data transmission systems by adding redundant data bits to the transmitted data bits and, in some cases, scrambling the order of the original data bits. There are many types of coding techniques (Hamming, BCH, and Reed-Solomon) used to correct different error phenomena that occur during data transmission (see Reference 1). This discussion will be limited to convolutional encoding, a good method for correcting burst errors occurring during data transmission.

Convolutional encoders are implemented in the form of a shift register type circuit with particular locations of the shift register exclusive ORed together to produce an output. Figure 1-1 shows one such implementation. The locations that are exclusive ORed together can be referred to as taps. The placement of these taps defines possible state transitions where the number of states for a particular code is defined by $2^{(k-1)}$ (see Reference 2). In this case, K is the constraint length of the code and is also the length of the shift register. The state transitions may also be represented by a trellis diagram. The trellis for the encoder of Figure 1-1 is shown in Figure 1-2 (see Reference 2).

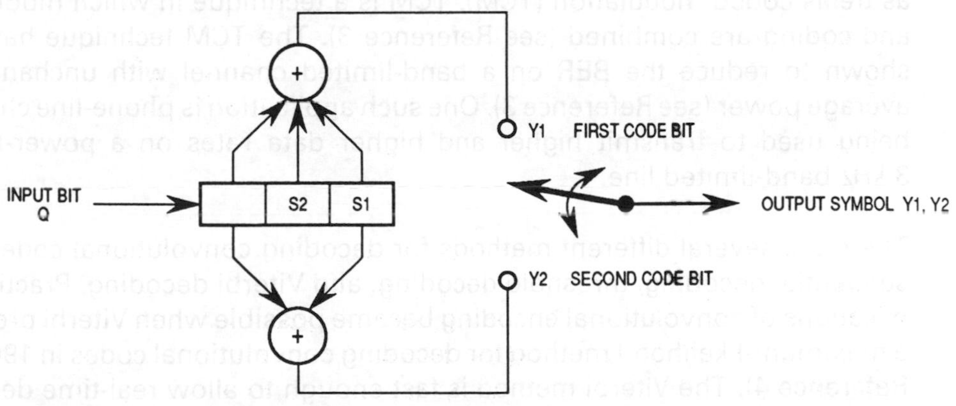

Figure 1-1. Convolutional Encoder Shift Register Implementation

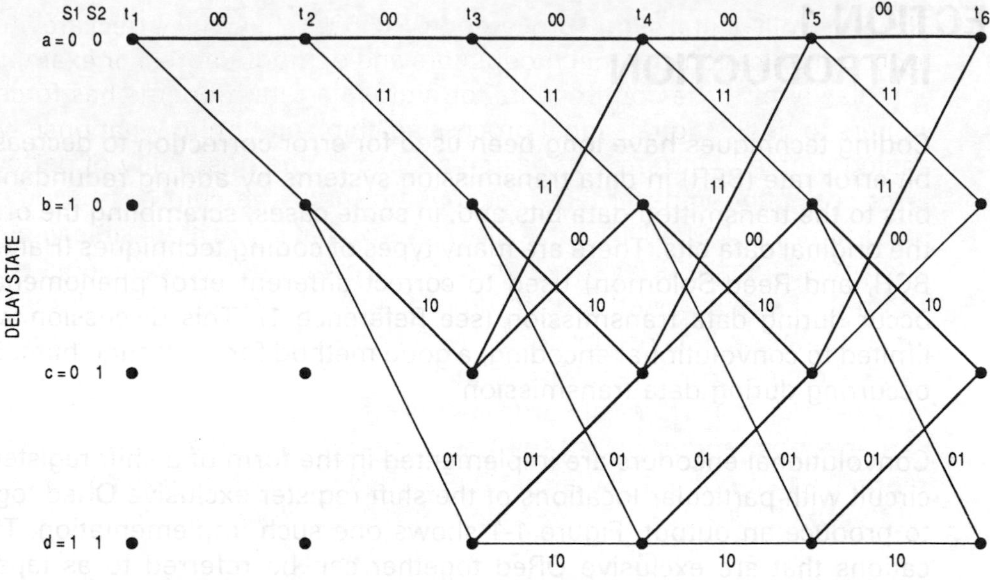

Figure 1-2. Trellis Representation

Many data transmission systems use convolutional encoding when transmitting data; however, some systems use a method of transmission known as trellis coded modulation (TCM). TCM is a technique in which modulation and coding are combined (see Reference 3). The TCM technique has been shown to reduce the BER on a band-limited channel with unchangeable average power (see Reference 3). One such application is phone-line channels being used to transmit higher and higher data rates on a power-limited 3-kHz band-limited line.

There are several different methods for decoding convolutional codes: e.g., sequential decoding, threshold decoding, and Viterbi decoding. Practical applications of convolutional encoding became possible when Viterbi proposed a maximum-likelihood method for decoding convolutional codes in 1967 (see Reference 4). The Viterbi method is fast enough to allow real-time decoding for short constraint length (K) codes with high-speed processors (made possible by recent advances in VLSI technology). Long constraint length codes require so much path memory that it is not practical to use the Viterbi algorithm when decoding. This maximum-likelihood method is equivalent to a dynamic programming solution to the problem of finding the shortest path through a trellis (see Reference 1).

The Motorola digital signal processor (DSP56001) has the perfect architecture for communication channel modulators and demodulators. For example, the complexity of these components for high data rate modems has forced designers to find a digital signal processing chip solution to what once was an analog problem. In terms of design simplicity, the possibility of performing error correcting and modem functions on the same chip has become very important. The DSP56001 has many features that make it possible to perform Viterbi decoding quickly and efficiently, allowing not only a single-chip solution to the high-speed modem application but also a single-chip solution for a higher speed Viterbi decoder. The dual memory architecture allows parallel moves concurrent with arithmetic operations, and the instruction set provides flexibility to easily program using this capability (see Reference 5). Perhaps most important for the Viterbi decoding, however, is the ability to set up circular buffers using modulo addressing (see Reference 5). This feature of the DSP56001 is fully exploited in the included code.

Because the Viterbi algorithm is a dynamic programming model, a trellis has been chosen as the example for this report. Using this trellis, each step in designing the code will be explained, giving enough description to duplicate or modify the existing code for a different trellis. This paper specifically addresses the trellis for the CCITT V.32 modem standard, which uses a TCM consisting of quadrature amplitude modulation (QAM) combined with a differential and convolutional encoder (see Reference 3). The techniques described can be applied to any trellis, and the included DSP56001 code can be modified to work with any trellis, not only the V.32.

SECTION 2
TRELLIS (CONVOLUTIONAL) ENCODING

The theory and implementation of convolutional encoding is discussed in the following paragraphs.

2.1 THEORY

A convolutional encoder is a function of the number of input bits (N) for the number of output bits (M) and the constraint length (K). It can be completely described by N, M, K and the given generator function. The generator function of the encoder is the impulse response of the encoder — that is, the output of the encoder when the input sequence is a one followed by zeros. Thus, the encoder equation can then be expressed as:

$$v = u * g \qquad (2\text{-}1)$$

where ∗ denotes a convolution operation, v is the output, u is the input, and g is the generator polynomial (see Reference 1).

As evidenced, a convolutional encoder adds redundant bits to a signal data stream. Adding more bits generally increases the BER since there are more encoder output bits per input bits to transmit with the same average power. When this happens, a reduction in the signal-to-noise ratio (SNR) occurs since there is less signal power per bit and, therefore, an increase in the BER. However, when a convolutional encoder adds a redundant bit to a data stream, it is done in such a way that the SNR is increased by allowing only certain transitions to occur. The coding gain associated with convolutional encoding is given by:

$$\text{Coding Gain} = 10 \log_{10}[(d^2{}_{min}/P_{av})_{encoded}(d^2{}_{min}/P_{av})_{unencoded}] \qquad (2\text{-}2)$$

where $d^2{}_{min}$ is the minimum distance between possible transitions and P_{av} is the average power of the signal space for the encoded and unencoded cases, respectively (see Reference 6).

The minimum distance must now be examined. Figure 2-1 shows a 16QAM signal constellation where four bits are required to represent one point on the constellation (see Reference 7). Figure 2-2 shows a 32QAM constellation where five bits represent one point on the constellation (see Reference 7). For the 16QAM case, the minimum distance from one point to the next is 2. For the 32QAM case in which the extra bit is a redundant bit obtained from a convolutional encoder so that the actual data rate is the same as the 16QAM case, the points are located such that the minimum distance in this case is now $\sqrt{10}$. Note that transitions between nearest neighbors on an encoded constellation are not allowed in contrast to those on an unencoded constellation. This is how a larger d^2_{min} and, consequently, coding gain is realized with convolutional encoding. Substituting these values into Equation (2-2) results in a coding gain of approximately 4 dB. For the 32QAM case using TCM, this would result in a decrease in the BER over the unencoded 16QAM case for the same average signal power.

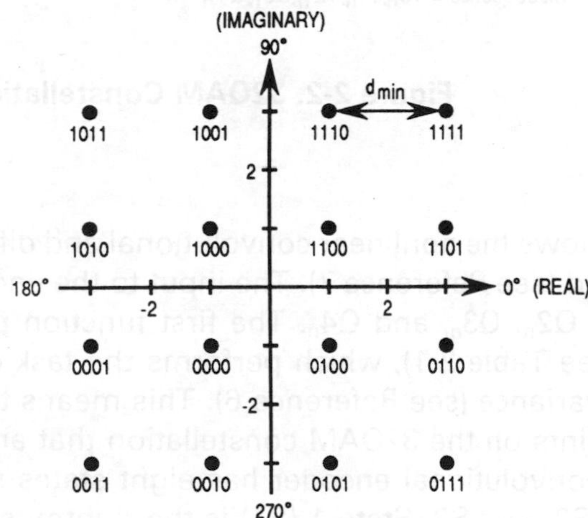

Figure 2-1. 16QAM Constellation

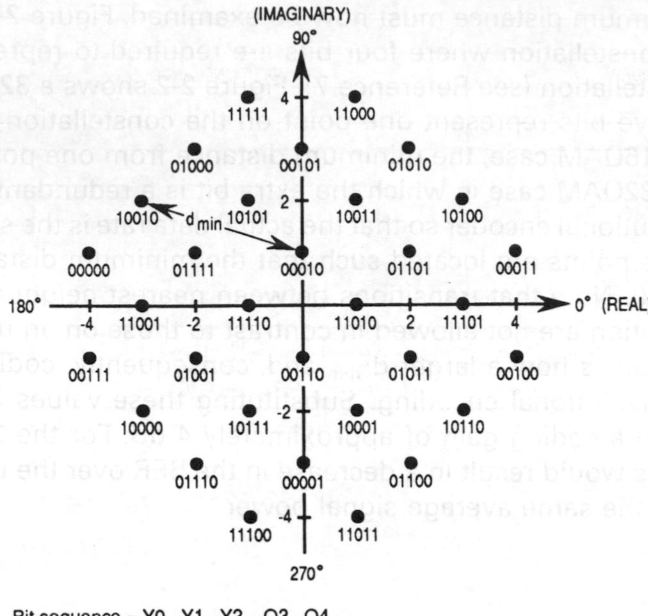

Figure 2-2. 32QAM Constellation

Figure 2-3 shows the nonlinear convolutional and differential encoder for the V.32 standard (see Reference 7). The input to the encoder is the parallel data stream $Q1_n$, $Q2_n$, $Q3_n$, and $Q4_n$. The first function performed is differential encoding (see Table 2-1), which performs the task of providing 90 degrees of phase invariance (see Reference 6). This means that $Q3_n$ and $Q4_n$ are the same for points on the 32QAM constellation that are 90 degrees from each other. The convolutional encoder has eight states resulting from the three delays: S1, S2, and S3. State 1 (S1) is the rightmost state, with state 2 (S2) and state 3 (S3) to the left, respectively. The output of the encoder is $Y0_n$, $Y1_n$, $Y2_n$, $Q3_n$, and $Q4_n$, where $Y0_n$, $Y1_n$, and $Y2_n$ are now considered a path when referring to the trellis. The trellis for the convolutional encoder is shown in Figure 2-4 (see Reference 8).

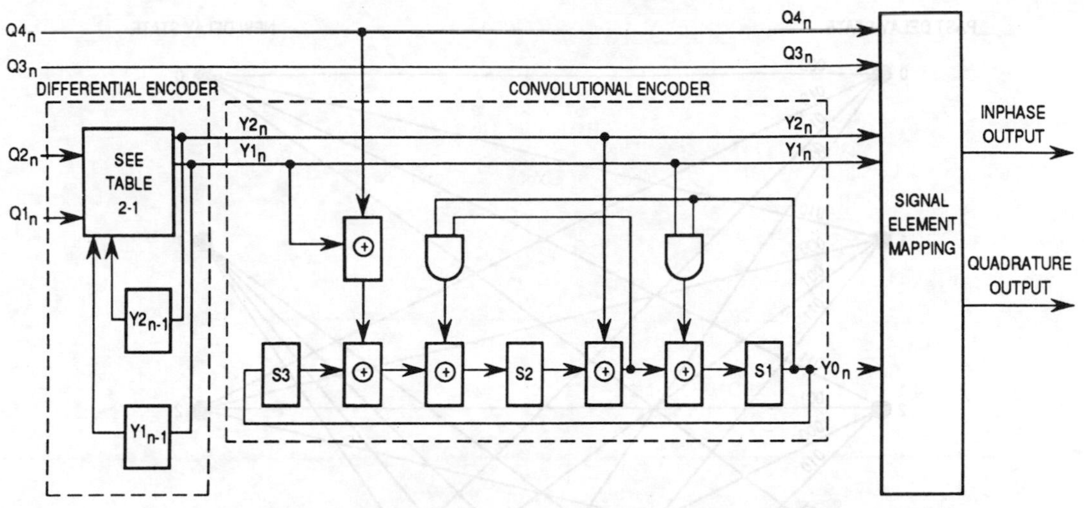

Figure 2-3. V.32 Encoding Diagram

Table 2-1. Differential Encoder

Input Bits		Past Output Bits		Output Bits	
$Q1_n$	$Q2_n$	$Y1_{n-1}$	$Y2_{n-1}$	$Y1_n$	$Y2_n$
0	0	0	0	0	0
0	0	0	1	0	1
0	0	1	0	1	0
0	0	1	1	1	1
0	1	0	0	0	1
0	1	0	1	0	0
0	1	1	0	1	1
0	1	1	1	1	0
1	0	0	0	1	0
1	0	0	1	1	1
1	0	1	0	0	1
1	0	1	1	0	0
1	1	0	0	1	1
1	1	0	1	1	0
1	1	1	0	0	0
1	1	1	1	0	1

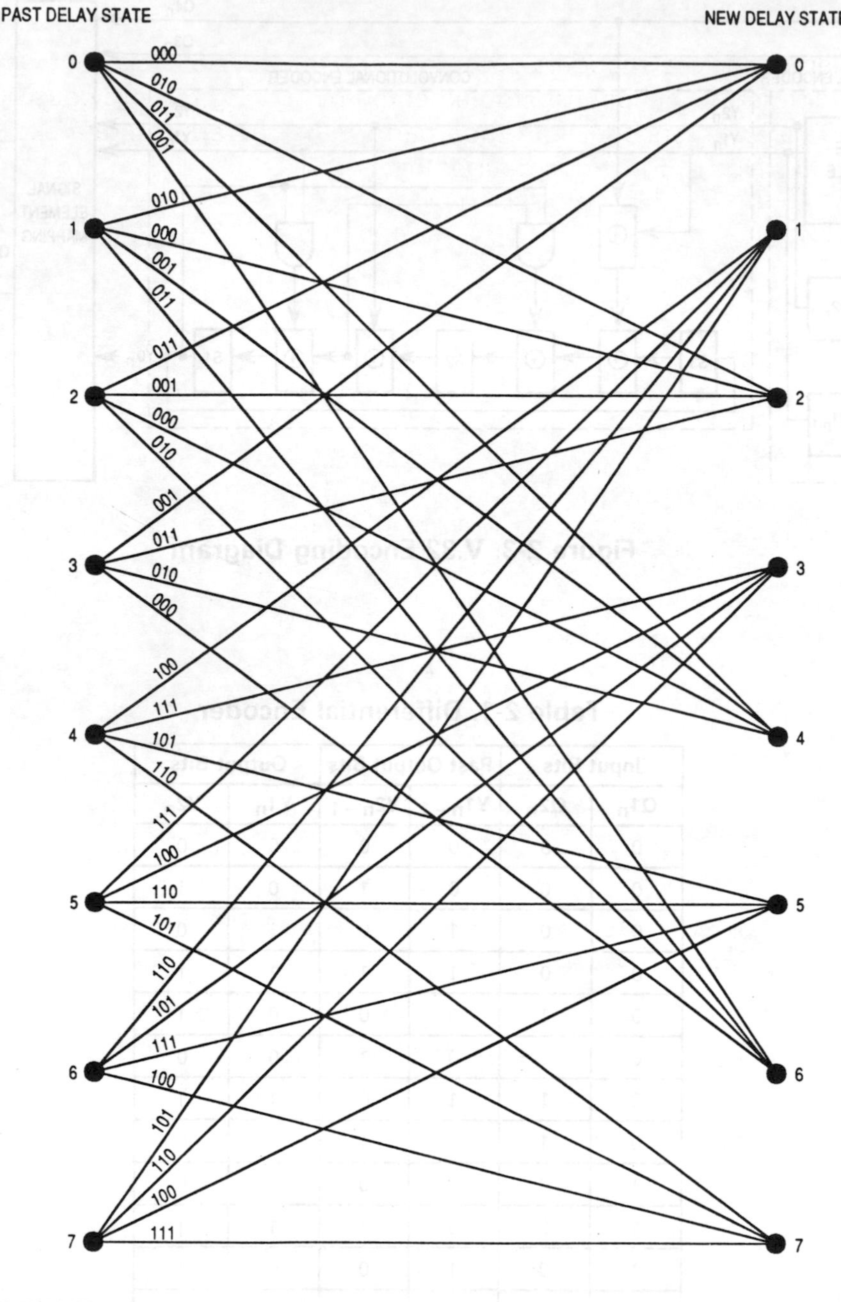

Figure 2-4. V.32 Trellis Diagram

2.2 IMPLEMENTATION

Implementing the encoder of Figure 2-3 on the DSP56001 is relatively simple. The differential encoder is implemented by storing the previous outputs of the differential encoder and then performing the appropriate exclusive OR ($\underline{\vee}$) and AND ($\wedge$) functions defined by:

$$Y1_n = Q1_n \;\underline{\vee}\; Y1_{n-1} \tag{2-3}$$

$$Y2_n = (Q1_n \wedge Y1_{n-1}) \;\underline{\vee}\; Y2_{n-1} \;\underline{\vee}\; Q2_n \tag{2-4}$$

The convolutional encoder is implemented in much the same way. There are three delays (S1, S2, and S3) that each require a memory location. The information from each delay is used at each input and then updated, based on the configuration of Figure 2-3. The output ($Y0_n$) at each time period is the value of delay 1 (S1) before it is updated. Figure 2-5 shows a flowchart of the encoding process. S1, S2, and S3 are referred to as the delay state of the encoder and the decoder; $Y0_n$, $Y1_n$, and $Y2_n$ are referred to as the path state of the encoder and decoder.

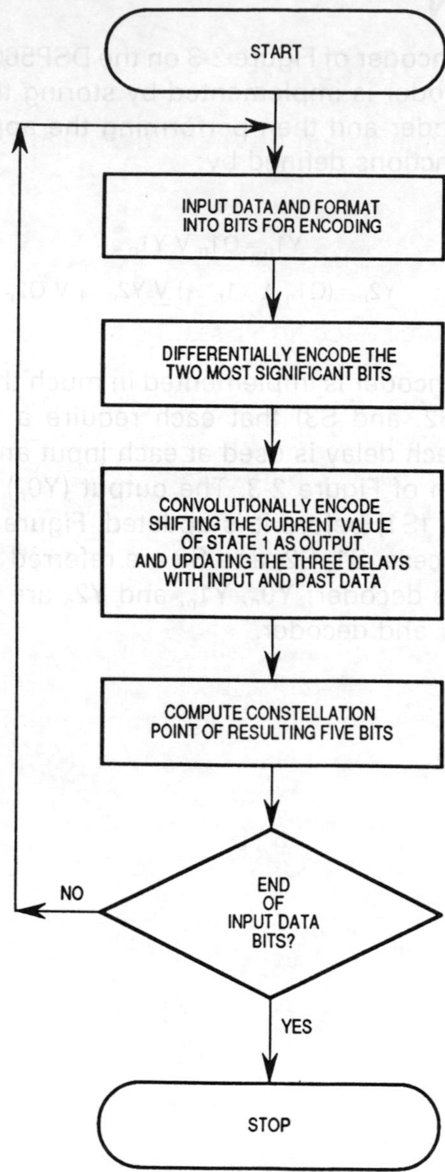

Figure 2-5. Encoding Flowchart

SECTION 3 VITERBI DECODING

The theory and implementation of Viterbi decoding is presented in the following paragraphs.

3.1 THEORY

The Viterbi algorithm for decoding uses the structure of the trellis (i.e., the allowed transitions) and the input data to determine the most likely path through the trellis. The output for time (t_0) reflects a decision made by the decoder on data received up to N time periods in the future. This means that the output for time (t_0) is necessarily delayed by N time periods or that the latency of the decoder is N time periods. N is determined by the constraint length of the code and, for near-optimum decoding, is four or five times the constraint length (see Reference 9).

The most likely path through the trellis is one that is a minimum-distance path for the input data or the path closest to the received data in Euclidean distance. In other words, the Viterbi algorithm minimizes the distance (see Reference 1):

$$d(r, v) = \sum_{i=0}^{N-1} d(r_i, v_i) \tag{3-1}$$

where r_i and v_i are the received and the decoded signal sequence, respectively.

At each time period, every delay state in the trellis can have several paths (defined by each trellis) going into it, but only one will be the minimum distance for that delay state. Thus, the delay state with the smallest accumulated distance is the beginning point, at that time period, to trace the minimum-distance path through the past N − 1 time periods of the trellis. The minimum-distance paths to the next delay state are then determined by evaluating the input to determine which point on the constellation in each path it is closest to, determining the Euclidean distance to each of those points, and then, based on the trellis structure and the minimum-distance paths, determining the minimum accumulated distance to each delay state. After defining the trellis, the steps taken to decode the data are as follows (see Reference 1):

1. At each input, compute the minimum-distance path states and the corresponding Euclidean distances and store them for each path state.
2. Compute the accumulated distance to each delay state by adding the distance for each path state going into a delay state to the distance of the delay state where the path state originated, keeping the smallest of these distances and storing the path state and the delay state from which it came. Eliminate all other path states going into that delay state.
3. Find the delay state with the smallest accumulated distance and trace it back N times to read the path state, which is the output of the decoder for that time period.

Figure 3-1 shows the possible paths to delay state 010 for the V.32 trellis and how the minimum distance to 010 is chosen from the possible paths.

When the minimum-distance path is found at each delay state, the path state taken to get there from the last delay state must also be stored (i.e., 001 in Figure 3-1 assuming $C+\gamma$ was the minimum) so that, in N time periods, the output can be determined from the endpoint of the minimum-distance path at time t_0+N. By storing the minimum-distance path state ($Y0_n$, $Y1_n$, $Y2_n$) to each delay state as well as the delay state (S1, S2, S3) the path originated from, the most likely path can be traced. This tracing is done by starting at the minimum accumulated distance delay state, backtracking to the delay

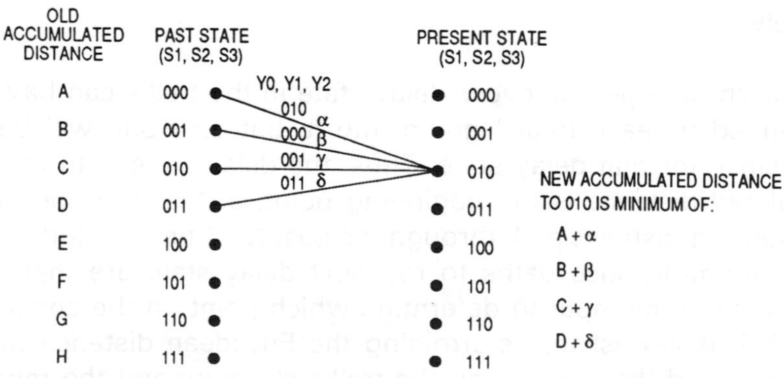

Figure 3-1. Possible Paths to State 010

state it came from, and repeating this process N − 1 times. That is, the minimum accumulated distance for all eight states identifies the state to be used as the starting point from which to trace back N time periods. Once the state for t_0 is found, the path taken to get to that state becomes the output of the decoder for the time period t_0. For instance, in Figure 3-1, if at t_0, the end point of the minimum-distance path was 010, then the output of the Viterbi decoder would be 001 if C + 8 was the minimum-distance path.

At every time period, the accumulated distance to each delay state is calculated and updated, and the minimum-distance path state ($Y0_n$, $Y1_n$, $Y2_n$) to each delay state is stored as well as the delay state it came from (S1, S2, S3). Storing this data creates a history so that it is possible to trace back to get the correct output of the decoder.

A block diagram of the V.32 decoder showing inputs and outputs is shown in Figure 3-2. It can be compared to the block diagram of the encoder shown in Figure 2-3 to keep track of the input and output bit order. Decoding must be done by performing each decoder function in the reverse order in which it was encoded. In this case, the trellis decoding is performed before the differential decoding.

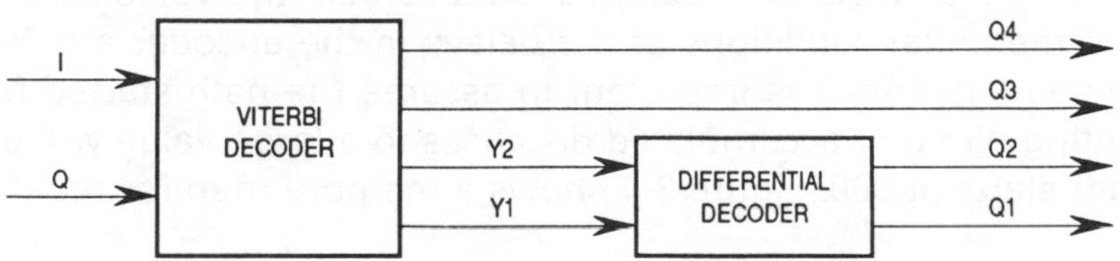

Figure 3-2. V.32 Decoder Block Diagram

3.2 IMPLEMENTATION

Implementing the Viterbi decoder on the DSP56001 involves:

1. Properly segregating memory locations
2. Recognizing boundary conditions of the trellis
3. Utilizing the modulo addressing capability of the DSP56001

First, a brief description of the functions of the decoder must be analyzed to realize the importance of the three previously mentioned ideas. At initialization, the x and y components of the points on the constellation are stored in memory for distance computations. For every input, the Euclidean distance to the closest point in each path state is computed. After computing this distance, the minimum accumulated distance to each delay state can be computed. Then, the minimum-distance delay state is used as the starting point to trace back to find the output for the previous 16 time periods. Recall that the number of time periods needed should be four or five times the constraint length (in this case, K=4). Since four times the constraint length in this case is 16 (4×K), this makes modulo addressing easier than using 20 (5×K), because 16 is a power of two. Once the output delay state is found, the closest point in the path state to the original input at that time period is computed and is the output of the decoding process. Figure 3-3 shows the decoder flowchart. Each of these functions is discussed separately in the following sections. This decoder code is included in **APPENDIX B DSP56001 DECODING PROGRAM LISTING**.

3.2.1 Initialization

During initialization, the x and y components of the constellation points are stored in internal memory since they are accessed frequently. The modulo settings for other parts of the code are also set here. All distance tables must be initialized as well. Since all paths should begin at the 000 delay state, this accumulated distance location should contain the value zero. This assumes that the initial conditions of the delays in the encoder are 000. In practical implementation, it is important to assume the path started from state 000. Setting all other accumulated distances to a large value will ensure that the path starts at 000. Figure 3-4 shows a memory map for the decoder.

3.2.2 Finding the Minimum Distance

This routine should analyze the input data point and determine the Euclidean distance to the closest point in each of the eight path states, where the Euclidean distance is defined to be:

$$d=\sqrt{(x_c-x_i)^2+(y_c-y_i)^2} \tag{3-2}$$

where x_c and y_c are the x and y coordinates of the point on the constellation and x_i and y_i are the coordinates of the input data. This calculation can be done by computing the Euclidean distance to each point in each path state and only keeping the smallest computed distance for that path state; however, there is a more efficient way to calculate this distance (see Reference 8). Looking at Figure 3-5 for the location of points for path state

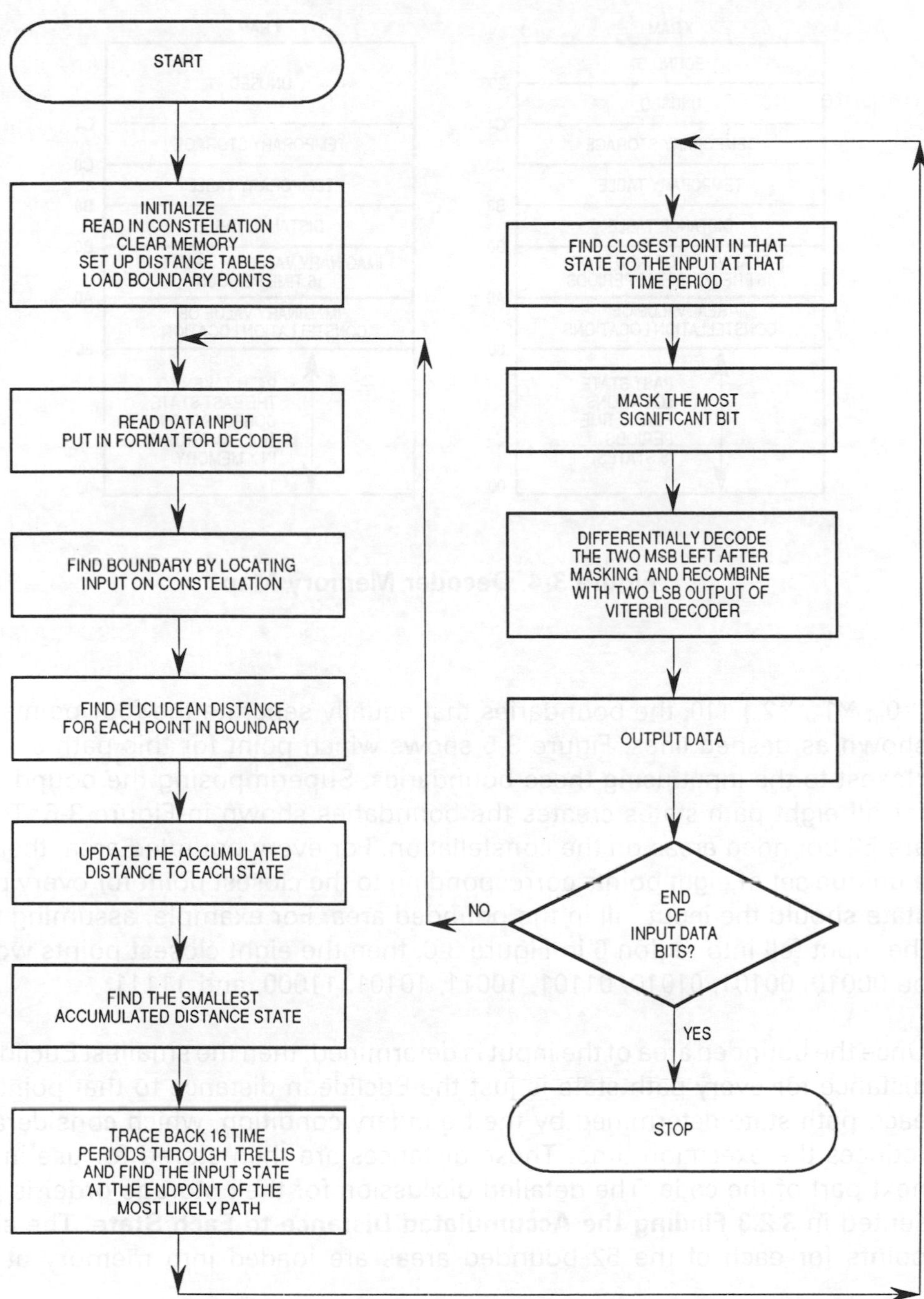

Figure 3-3. Decoder Flowchart

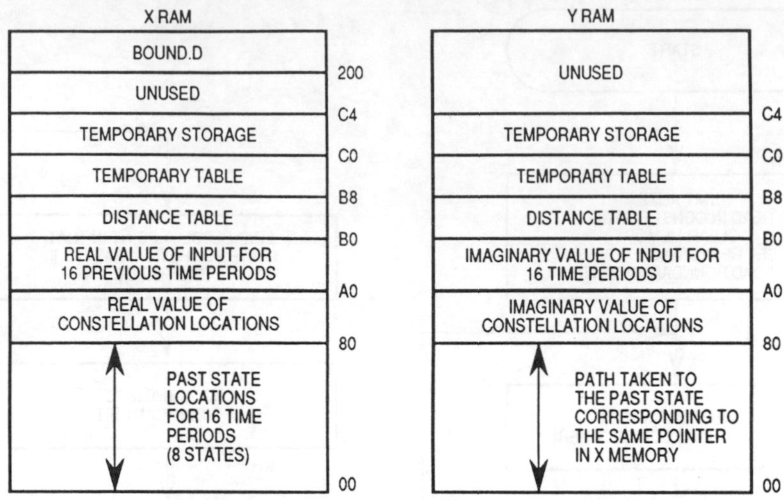

Figure 3-4. Decoder Memory Map

($Y0_n$, $Y1_n$, $Y2_n$) 110, the boundaries that equally separate the four points are shown as dashed lines. Figure 3-5 shows which point for this path state is closest to the input using these boundaries. Superimposing the boundaries for all eight path states creates the boundaries shown in Figure 3-6. There are 52 bounded areas on the constellation. For every bounded area, there is a unique set of eight points corresponding to the closest point for every path state should the input fall in the bounded area. For example, assuming that the input fell into region 6 in Figure 3-6, then the eight closest points would be 00010, 00101, 01010, 01101, 10011, 10101, 11000, and 11111.

Once the bounded area of the input is determined, then the smallest Euclidean distance for every path state is just the Euclidean distance to that point for each path state determined by the boundary condition, which considerably reduces the execution time. These distances are then stored for use in the next part of the code. The detailed discussion for their storage order is presented in **3.2.3 Finding the Accumulated Distance to Each State**. The eight points for each of the 52 bounded areas are loaded into memory at the

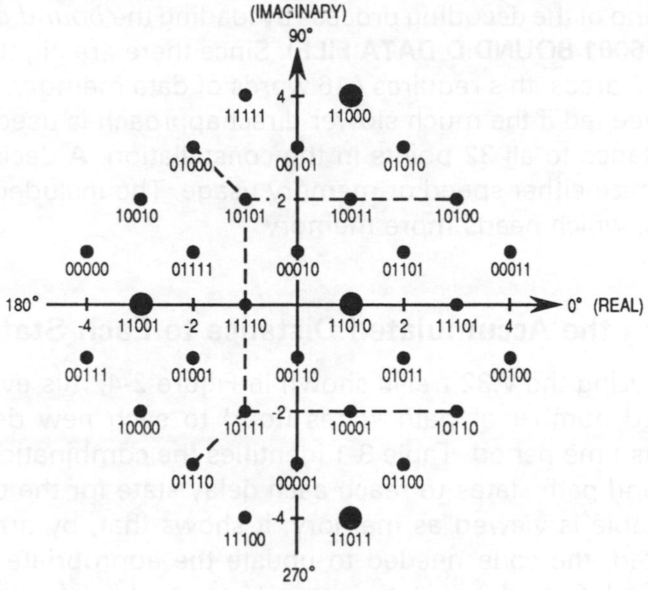

Bit sequence = $Y0_n, Y1_n, Y2_n, Q3_n, Q4_n$.

Figure 3-5. Boundary for State 010

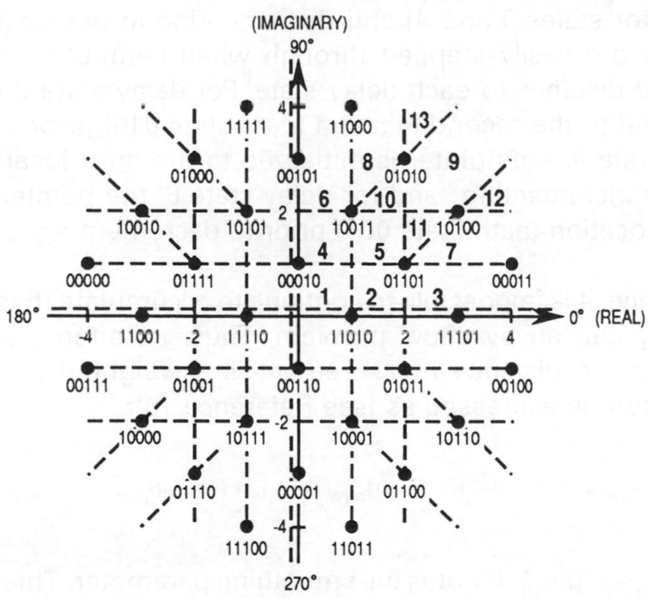

Bit sequence = $Y0_n, Y1_n, Y2_n, Q3_n, Q4_n$.

Figure 3-6. Superimposed Boundaries for All States

beginning of the decoding process by loading the *bound.d* file (see **APPENDIX C DSP56001 BOUND.D DATA FILE**). Since there are eight path states in each of the 52 areas, this requires 416 words of data memory. The larger memory is not needed if the much slower direct approach is used — namely, finding the distance to all 32 points in the constellation. A decision must be made to optimize either speed or memory usage. The included code uses the fast version, which needs more memory.

3.2.3 Finding the Accumulated Distance to Each State

By analyzing the V.32 trellis shown in Figure 2-4, it is evident that there are a limited number of path states (four) to each new delay state from the previous time period. Table 3-1 identifies the combination of previous delay states and path states to reach each delay state for the current time period. If this table is viewed as memory, it shows that, by arranging the data as illustrated, the code needed to update the appropriate delay state can be minimized. Even delay states are only reached by the previous delay states 000, 001, 010, 011, and odd delay states are only reached by the previous delay states 100, 101, 110, and 111. Similarly, the path state to each new even delay state can only be 000, 001, 010, 011, and the path state for each new odd delay state can only be 100, 101, 110, 111. In the even case, if the path states are put in the order 000, 010, 011, 001 by incrementing in two cases (for states 0 and 4) and decrementing in two cases (for states 2 and 6), they are easily stepped through when computing the minimum accumulated distance to each delay state. For delay state 2 (010), the pointer is initialized to the second location (path state 010) prior to decrementing; for delay state 4, the pointer is initialized to the third location (path state 011) prior to incrementing; and for delay state 6, the pointer is initialized to the fourth location (path state 001) prior to decrementing.

In practice, it is impossible to continue to accumulate these distances without running into an overflow problem. Thus, an alternate way to obtain the accumulated distance measurement is a weighted accumulation method, which can be expressed as (see Reference 10):

$$d_{new} = \beta d_{old} + (1 - \beta) d_{path} \tag{3-3}$$

where $0 << \beta < 1$ denotes the smoothing parameter. This method (essentially a low-pass filter) ensures that the new accumulated distance is a bounded arithmetic value. This method has been shown to give unbiased estimates (see Reference 10). Although Equation (3-3) uses all past values to compute

Table 3-1. Minimum Path Table

New State S1, S2, S3	Old State S1, S2, S3	Path Y0, Y1, Y2	New State S1, S2, S3	Old State S1, S2, S3	Path Y0, Y1, Y2
000	000	000	001	100	100
000	001	010	001	101	111
000	010	011	001	110	110
000	011	001	001	111	101
010	000	010	011	100	111
010	001	000	011	101	100
010	010	001	011	110	101
010	011	011	011	111	110
100	000	011	101	100	101
100	001	001	101	101	110
100	010	000	101	110	111
100	011	010	101	111	100
110	000	001	111	100	110
110	001	011	111	101	101
110	010	010	111	110	100
110	011	000	111	111	111

a current accumulated distance, the value of β is directly related to the time constant, τ, which gives the number of recent past values to estimate the accumulated distance as:

$$\tau \sim \frac{2}{1-\beta} \quad (3\text{-}4)$$

Using this equation, 85% of d_{new} comes from the points in the time constant, τ, and the remaining 15% is contributed by points previous to τ. The value of $\beta = 0.9$ was used for this code and simulation; however, it should be considered a performance parameter that can be modified for different applications and performance specifications.

At several points in the code, two paths can have the same minimum distance. In these situations, an arbitrary choice is made to keep one or the other path.

In the included code, the arbitrary choice has been made to keep the last path found; however, it could be changed to pick the first path encountered or the first path in even cases and the last path in odd cases. The performance of the included code is not affected by changing this arbitrary choice (see Reference 1).

Once the minimum distance for each state is accumulated, the path taken to get there and the previous state are stored in memory to be used when tracing back through the trellis to determine an output. Since the memory is 16 time periods deep and there are eight states, the memory is set up to be 128 words in each memory. The previous state information is stored in a circular buffer in x memory, and the path information is stored in a circular buffer in y memory. In this manner, the new states and paths will overwrite the oldest information in the buffer each time period.

3.2.4 Traceback

Tracing back through the 16 time periods along the most likely path is now done by taking advantage of the fact that the memory is set up to be circular around 128 points. Starting at the most current location in which the latest paths for each state were updated as well as the state from which the path came, the memory is decremented by eight to the previous time period. When this is done, the pointer is then updated (a number between zero and seven) to correspond with the previous state from which the most likely path came. The information stored at this pointer location is the state information for the next time period. This procedure is done until the last time period in which the path is read, instead of the state from which it came. This path determines the output of the decoder.

3.2.5 Data Out

By taking the path found at the end of traceback, the output of the Viterbi decoder can be determined. The path corresponds to one of eight paths ($Y0_n$, $Y1_n$, $Y2_n$), each with a point on the constellation represented by five unique bits. The three most significant bits correspond to the path ($Y0_n$, $Y1_n$, $Y2_n$), and the two least significant bits correspond to the unencoded bits ($Q3_n$ and $Q4_n$). The closest of the four points from that path to the input at that time period is the output of the decoder. The most significant bit ($Y0_n$) is stripped off at this point (the redundant bit added in the coding process).

3.2.6 Differential Decoding

The decoding of the differential encoding is now performed by taking the two most significant bits of the Viterbi decoder output to perform the following:

$$Q1_n = Y1_n \veebar Y1_{n-1} \quad (3\text{-}5)$$

$$Q2_n = (Q1_n \wedge Y1_{n-1}) \veebar Y2_{2n-1} \veebar Y2_n \quad (3\text{-}6)$$

where $Y1_n$ and $Y2_n$ are the most significant bit of the Viterbi decoder output. $Q1_n$ and $Q2_n$ are now combined with the two least significant bits of the Viterbi output to complete the decoding process.

SECTION 4
SUMMARY

The included decoder code takes approximately 700 instruction cycles for every four input bits. For the V.32 case, this is about 15% of the processor. It should be fairly straightforward to modify the code in **APPENDIX A DSP56001 ENCODING PROGRAM LISTING** and **APPENDIX B DSP56001 DECODING PROGRAM LISTING** to work with any trellis diagram. This modification can be done by going through each step explained and defining the memory, order of storage, and modulo settings for the new trellis. Boundary conditions will obviously change; therefore, **APPENDIX C DSP56001 BOUND.D DATA FILE** will have to be redone for any new constellation.

It is the hope of the author that enough explanation and description have been given here for the reader to understand how to implement a convolutional encoder with a Viterbi decoder on the DSP56001 for any trellis. Some of the many textbooks and fundamental papers dealing with coding theory are listed in References 11, 12, and 13 for the reader's convenience.

REFERENCES

1. Lin, S. and D. Costello, *Error Control Coding: Fundamentals and Applications*, Englewood Cliffs, NJ: Prentice Hall, 1983.

2. Sklar, B., *Digital Communications Fundamentals and Applications*, Englewood Cliffs, NJ: Prentice Hall, 1988.

3. Ungerboeck, G., "Trellis Coded Modulation with Redundant Signal Sets Part 1: Introduction," *IEEE Communications Magazine*, vol. 25, no. 2, February 1987, pp. 5-11.

4. A. J. Viterbi, "Error Bounds for Convolutional Codes and An Asymptotically Optimum Decoding Algorithm," *IEEE Trans. Inf. Theory*, vol. IT-13, April 1967, pp. 260-269.

5. *DSP56000/DSP56001 Digital Signal Processor User's Manual*, (DSP56000UM/AD), Motorola Inc., 1989.

6. Wei, L. F., "Rotationally Invariant Convolutional Channel Coding with Expanded Signal Space — Part 1:180," *IEEE Journal on Selected Areas in Communications*, vol. SAC-2, no. 5, September 1984, p. 661.

7. The International Telegraph and Telephone Consultative Committee, *CCITT Red Book*, "Data Communication over the Telephone Network," vol. 8, 1985, pp. 221-238.

8. Fagen, A., et al., "Single DSP Implementation of a High Speed Echo Cancelling Modem Employing Trellis Coding," *Proc. of the Intnl. ESA Workshop on DSP Techniques Applied to Space Communications*, November 1988.

9. Heller, J. A. and I. W. Jacobs, "Viterbi Decoding for Satellite and Space Communication," *IEEE Trans. Commun. Tech.*, vol. COM-19, no. 5, October 1971, pp. 835-848.

10. Magotra, N., et al., "A Comparison of Two Parametric Estimation Schemes," *Proc. IEEE*, vol. 74, no. 5, May 1986, pp.760-761.

11. Lee, E. A. and D. G. Messershmitt, *Digital Communication*, Boston, MA: Kluwer Academic Publishers, 1988.

12. Clark, G. C. and J. B. Cain, *Error-Control Coding for Digital Communications*, New York, NY: Plenum Press, 1982.

13. Viterbi, A. J. and J. K. Omura, *Pinciples of Digital Communications and Coding,* New York, NY: McGraw-Hill, 1979.

APPENDIX A
DSP56001 ENCODING PROGRAM LISTING

```
                  ;This is a differential and convolutional encoder for the V.32 Modem standard.
                  ;The diagram for this encoder can be found in the CCITT Red Book on the V
                  ;standard.  Data input is in the form of four bits representing the numbers
                  ;0-15.  This is how the standard defines the input.  This code takes the 4 bit
                  ;input and separates each bit out for coding.  This code may be modified to
                  ;take data which comes one bit at a time (serially) instead of in parallel
                  ;form as this program assumes.


                  ;Set up memory for the 3 delays in the convolutional encoder (statemem)
                  ;Set up memory for the 4 bits of data input (input)
                  ;Alot memory for the past outputs of the differential encoder (y1past)
                  ;Create a location for the redundant bit created by the convolutional encoding
                  ;(output)

                  ;If using a different trellis than V.32, set up your memory corresponding to
                  ;the trellis you are using.  You will also  have to adjust your XOR and
                  ;AND functions to fit your trellis.  Refer to the applications document
                  ;describing this proceedure.


                            page    132,66,3,3,0
X:0000                      org     x:$0
X:0000            statemem  ds      3
X:0003            input     ds      4

Y:0000                      org     y:$0
Y:0000            y1past    ds      2
Y:0002            output    ds      1

000040            start     equ     $40
0000EE            locate    equ     $ee

P:0040                      org     p:start
P:0040 330000               move              #statemem,r3
P:0041 350000               move              #y1past,r5


                  ;Loop here either infinitely, or for the number of inputs you have (104 in this
                  ;case) or set it up for an input to interrupt and perform this routine once for
```

```
                              ;every input.
        P:0042 066880          do      #104,code
               00005C
        P:0044 36EE00          move                    #locate,r6
        P:0045 320600          move                    #input+3,r2
        P:0046 5EE600          move                                  y:(r6),a
        P:0047 44F400          move                    #>$1,x0
               000001

                              ;Mask off each bit of the 4 bit input and put it in its own memory location
                              ;Adjust this mask for your trellis and for your input data bit stream if it
                              ;is different.

        P:0049 060480          do      #4,loop
               00004E
        P:004B 21C546          and     x0,a            a,x1
        P:004C 545200          move                    a1,x:(r2)-
        P:004D 20AE00          move                    x1,a
        P:004E 200022          asr     a
                              loop

                              ;Differential encoding is done in the diff subroutine

        P:004F 0BF080          jsr     diff
               00005D


                              ;Convolutional encoding is done in the encode subroutine



        P:0051 0BF080          jsr     encode
               000069


                              ;Now the redundant bit is put in the msb location and the output bits
                              ;are recombined with it.  This may also be modified to be serial rather
                              ;than parallel.


        P:0053 36EF00          move                    #locate+1,r6
        P:0054 320300          move                    #input,r2
```

```
P:0055 20001B            clr     b
P:0056 59E413            clr     a                          y:(r4),b0
P:0057 200012            addl    b,a

P:0058 060480            do      #4,loop2
       00005B
P:005A 51DA00            move                  x:(r2)+,b0
P:005B 200012            addl    b,a
                 loop2

                 ;Output data here

P:005C 586600            move                               a0,y:(r6)

                 code

                 ;The next subroutine computes Y1n and Y2n as found in the table in the CCITT
                 ;standard for V.32.  Two logic equations express this table. The logic
                 ;expressions can be found in the documemtation for this code.
                 ;Y1n = Q1n EOR Y1n-1, Y2n = (Q1n AND Y1n-1) XOR Y2n-1 EOR Q2n

P:005D 310300    diff    move                  #input,r1
P:005E 4EDD00            move                               y:(r5)+,y0
P:005F E9B900            move                  x:(r1)+,a    y:(r5)-,y1
P:0060 57D100            move                  x:(r1)-,b
P:0061 21C453            eor     y0,a          a,x0
P:0062 21C57B            eor     y1,b          a,x1
P:0063 208E00            move                  x0,a
P:0064 21E756            and     y0,a          b,y1
P:0065 20AF73            eor     y1,a          x1,b
P:0066 BF3900            move                  b,x:(r1)+    b,y:(r5)+
P:0067 AA2100            move                  a,x:(r1)     a,y:(r5)-
P:0068 00000C            rts

                 ;This subroutine does straightforward convolutional encoding as seen in the
                 ;V.32 standard for creating the redundant bit.  This diagram can be found in
                 ;the applications documentation for this code as well as in the CCITT
                 ;Red book for this standard.
```

```
                     encode
     P:0069 300300             move          #input,r0
     P:006A 340200             move          #output,r4
     P:006B 310000             move          #statemem,r1
     P:006C 45D800             move          x:(r0)+,x1
     P:006D 56D900             move          x:(r1)+,a
     P:006E 5E6400             move                        a,y:(r4)
     P:006F 44E066             and    x1,a   x:(r0),x0
     P:0070 57D100             move          x:(r1)-,b
     P:0071 21C64B             eor    x0,b   a,y0
     P:0072 21E75B             eor    y0,b   b,y1
     P:0073 CF1900             move          b,x:(r1)+     y:(r4),b
     P:0074 208E7E             and    y1,b   x0,a
     P:0075 205900             move          (r1)+
     P:0076 44E163             eor    x1,a   x:(r1),x0
     P:0077 4FE443             eor    x0,a                 y:(r4),y1
     P:0078 21E600             move          b,y0
     P:0079 475153             eor    y0,a   y1,x:(r1)-
     P:007A 565900             move          a,x:(r1)+
     P:007B 00000C             rts

0    Errors
0    Warnings
```

APPENDIX B
DSP56001 DECODING PROGRAM LISTING

```
                    ;This is a Viterbi and differential decoder for  the V.32 standard modem
                    ;as defined by the CCITT Red Book for V standards.  This code may be modified
                    ;to use a different trellis than is used by V.32.
                    ;Input for this code is expected in parallel form, one of 32 points on the
                    ;constellation.  It is then perturbed so that the data is not perfect to
                    ;simulate channel noise before decoding and broken down to it's
                    ;X and Y locations.  This is not practical for a real implementation, which
                    ;will expect input in the form of the X and Y outputs of the receiver.
                    ;In this case, the code should be modified to to eliminate the perturbing
                    ;of the data and also to eliminate finding the X and Y cordinates, since
                    ;that will be your input directly.
                    ;If your trellis is different than the V.32  trellis, this code can be used as
                    ;an example in writing code for the different trellis.  Using this code and the
                    ;corresponding applications documentation will allow a step by step implement-
                    ;ion of any trellis.  At each new subroutine adjustments will need to be made
                    ;in the memory storage requirements, the modulo addressing, and order of
                    ;operation.

                             page     132,66,3,3,0
                             opt      cex


                    ;First memory is allocated for the 16 time period path memory
                    ;(8 states x 16 time periods = 128) This must be in long memory because
                    ;both the previous state information and path taken must be stored for
                    ;every state at every time period.
                    ;The constellation points X and Y coordinates are also store in long memory
                    ;X in x memory and Y in y.
                    ;The minimum distance path and minimum accumulated distance tables are also
                    ;allocated.
                    ;Making these modulo n makes buffers processing in the decoder very
                    ;efficient.


L:0000                       org      l:$0000

L:0000              period   dsm      128
L:0080              location dsm      32
L:00A0              input    dsm      16
L:00B0              tables   dsm      8
L:00B8              temp     dsm      8
```

```
0000C0          endlong   equ     *


                ;Now various locations are set aside for temporary storage of values, such as
                ;ynow and ypast which are used in the differential decoder.


X:00C0                    org     x:endlong
X:00C0          storr6    ds      1
X:00C1          ynow      ds      4
Y:00C0                    org     y:endlong
Y:00C0          ypast     ds      2


                ;This memory stores the eight points in each one of 52 bounded areas for finding
                ;the minimum distance point in each of the 8 states to the input.

X:0200                    org     x:512
X:0200          boundry1  ds      32
X:0220          boundry2  ds      32
X:0240          boundry3  ds      32
X:0260          boundry4  ds      32
X:0280          boundry5  ds      32
X:02A0          boundry6  ds      32
X:02C0          boundry7  ds      32
X:02E0          boundry8  ds      32
X:0300          boundry9  ds      32
X:0320          boundry10 ds      32
X:0340          boundry11 ds      32
X:0360          boundry12 ds      32
X:0380          boundry13 ds      32


                ;The following are used when setting up the constellation memory.
                ;The value of four corresponds to four volts input on the V.32 trellis.
                ;Correspondingly, the value of mfour corresponds to -4 volts on the
                ;constellation.  This will allow adjusting the full scale value without
                ;changing the rest of the code, and therefore, the value used for full
                ;scale becomes a performance parameter.  It can be modified to find the
                ;best performance.  The symblol four will correspond to the full four volts
                ;possible on the constellation and three will equal three volts.  The values
                ;picked to equate to the symbols should be scaled realtive to four volts.
```

```
                        ;Example, if fourvolts is equated to .5, then three volts would be (.5)(3/4)
                        ;or .375.

000040                  start     equ     $40
200000                  four      equ     $200000
180000                  three     equ     $180000
100000                  two       equ     $100000
080000                  one       equ     $080000
000000                  zero      equ     $000000
F80000                  mone      equ     $f80000
F00000                  mtwo      equ     $f00000
E80000                  mthree    equ     $e80000
E00000                  mfour     equ     $e00000

                        ;The weighted accumultion is discussed in the applications documentation and is
                        ;implemented with these two parameters which become another performance
                        ;parameter which can be changed for different applications and specifications.

0.900000                large     equ     .9
0.100000                small     equ     .1

                        ;Offset is the value used to perturb the data to simulate channel distortion.
                        ;This value can be eliminated for already distorted data.

010000                  offset    equ     $010000

P:0040                            org     p:start
P:0040 0BF080                     jsr     initialize
       00004E

                        ;loop here for the number of inputs you have, or infinitely, or set it up on
                        ;an interrupt so that it is performed once every interrupt.

P:0042 067380                     do      #115,endrun
       00004D
P:0044 0BF080                     jsr     readdata
       0000A6
P:0046 0BF080                     jsr     findmindist
       0000B6
P:0048 0BF080                     jsr     accumdist
       000141
```

```
P:004A 0BF080          jsr     traceback
       0001D3
P:004C 0BF080          jsr     outputdata
       0001E0
                endrun

                ;this initialization routine initializes register and modifiers
                ;as well as clears the memory.
                ;the constellation is also loaded into memory here.
                ;the accumulated distance array is set so that state zero starts out
                ;at a value of zero and all others start out larger, forcing the paths
                ;to start at the zero state.

                initialize
P:004E 057FA1          move            #127,m1
P:004F 057FA5          move            #127,m5
P:0050 050FA6          move            #15,m6
P:0051 310000          move            #0,r1
P:0052 30001B          clr     b       #$0,r0
P:0053 221513          clr     a       r0,r5

                ;Move zero into every memory location to clear the memory.

P:0054 060081          do      #256,clrmem
       000056
P:0056 BB3800          move            a,x:(r0)+  b,y:(r5)+
                clrmem
P:0057 37B100          move            #tables+1,r7

                ;Now put .5 in all state locations of the accumulted distance table except
                ;state 000.  leaving state 000 at a value of zero will force a start from
                ;that state.

P:0058 54F400          move            #$400000,a1
       400000
P:005A 0607A0          rep     #7
P:005B 545F00          move            a1,x:(r7)+

P:005C 2DA000          move            #input,b1
P:005D 557000          move            b1,x:storr6
       0000C0
```

```
                          ;Now load full scale values of the constellation in the table location.
                          ;Each point on the constellation has cordinates on the x and y axis from
                          ;minus four to plus four volts. The coordinates  of each point are now loaded in
                          ;memory for later use when determining minimum distance paths and paths which
                          ;fall closest to the input.  This method of loading the constellation points is
                          ;particularly good for test and simultaion.  When the final configuration is
                          ;defined, these constellation points can be put into a data file and the code
                          ;removed.  The file can then be loaded with the program to save program words.



P:005F 308000             move           #location,r0
P:0060 221400             move           r0,r4
P:0061 2EE000             move           #mfour,a
P:0062 2F0800             move           #one,b
P:0063 BB1800             move           a,x:(r0)+   b,y:(r4)+
P:0064 2E0000             move           #zero,a
P:0065 2FE800             move           #mthree,b
P:0066 BB1800             move           a,x:(r0)+   b,y:(r4)+
P:0067 2F0800             move           #one,b
P:0068 BB1800             move           a,x:(r0)+   b,y:(r4)+
P:0069 2E2000             move           #four,a
P:006A BB1800             move           a,x:(r0)+   b,y:(r4)+
P:006B 2FF800             move           #mone,b
P:006C BB1800             move           a,x:(r0)+   b,y:(r4)+
P:006D 2E0000             move           #zero,a
P:006E 2F1800             move           #three,b
P:006F BB1800             move           a,x:(r0)+   b,y:(r4)+
P:0070 2FF800             move           #mone,b
P:0071 BB1800             move           a,x:(r0)+   b,y:(r4)+
P:0072 2EE000             move           #mfour,a
P:0073 BB1800             move           a,x:(r0)+   b,y:(r4)+
P:0074 2EF000             move           #mtwo,a
P:0075 2F1800             move           #three,b
P:0076 27F800             move           #mone,y1
P:0077 BB1800             move           a,x:(r0)+   b,y:(r4)+
P:0078 B91800             move           a,x:(r0)+   y1,y:(r4)+
P:0079 2E1000             move           #two,a
P:007A BB1800             move           a,x:(r0)+   b,y:(r4)+
P:007B B91800             move           a,x:(r0)+   y1,y:(r4)+
P:007C 2F0800             move           #one,b
```

```
P:007D 27E800    move    #mthree,y1
P:007E B91800    move    a,x:(r0)+    y1,y:(r4)+
P:007F BB1800    move    a,x:(r0)+    b,y:(r4)+
P:0080 2EF000    move    #mtwo,a
P:0081 B91800    move    a,x:(r0)+    y1,y:(r4)+
P:0082 BB1800    move    a,x:(r0)+    b,y:(r4)+
P:0083 2E0800    move    #one,a
P:0084 21C400    move    a,x0
P:0085 2EE800    move    #mthree,a
P:0086 2F1000    move    #two,b
P:0087 21E600    move    b,y0
P:0088 2FF000    move    #mtwo,b
P:0089 BB1800    move    a,x:(r0)+    b,y:(r4)+
P:008A B31800    move    x0,x:(r0)+   b,y:(r4)+
P:008B B81800    move    a,x:(r0)+    y0,y:(r4)+
P:008C B01800    move    x0,x:(r0)+   y0,y:(r4)+
P:008D 2E1800    move    #three,a
P:008E 21C400    move    a,x0
P:008F 2EF800    move    #mone,a
P:0090 B01800    move    x0,x:(r0)+   y0,y:(r4)+
P:0091 B81800    move    a,x:(r0)+    y0,y:(r4)+
P:0092 B31800    move    x0,x:(r0)+   b,y:(r4)+
P:0093 BB1800    move    a,x:(r0)+    b,y:(r4)+
P:0094 2E0800    move    #one,a
P:0095 2F0000    move    #zero,b
P:0096 21E600    move    b,y0
P:0097 2F2000    move    #four,b
P:0098 BB1800    move    a,x:(r0)+    b,y:(r4)+
P:0099 24E800    move    #mthree,x0
P:009A B01800    move    x0,x:(r0)+   y0,y:(r4)+
P:009B B81800    move    a,x:(r0)+    y0,y:(r4)+
P:009C 2FE000    move    #mfour,b
P:009D BB1800    move    a,x:(r0)+    b,y:(r4)+
P:009E 2EF800    move    #mone,a
P:009F BB1800    move    a,x:(r0)+    b,y:(r4)+
P:00A0 241800    move    #three,x0
P:00A1 B01800    move    x0,x:(r0)+   y0,y:(r4)+
P:00A2 B81800    move    a,x:(r0)+    y0,y:(r4)+
P:00A3 2F2000    move    #four,b
P:00A4 BB1800    move    a,x:(r0)+    b,y:(r4)+
P:00A5 00000C    rts
```

```
                    ;readdata reads in the data from a simulator file.  Since it is read in as
                    ;a point on the constellation, it must be converted to real and imaginary
                    ;components by indexing into a table.
                    ;it is also offset by a value "offset" so it is not considered to be perfect
                    ;data.
                    ;Recall notes at the beginning.  This routine will be much different in the
                    ;case of actual  data where most of this code is not neccessary.


                    readdata
P:00A6 5EF000                 move                              y:$efe,a
       000EFE
P:00A8 21DA00                 move                  a,n2
P:00A9 328000                 move                  #location,r2
P:00AA 66F000                 move                  x:storr6,r6
       0000C0
P:00AC 044A14                 lua       (r2)+n2,r4
P:00AD 44F400                 move                  #>offset,x0
       010000
P:00AF 56E400                 move                  x:(r4),a
P:00B0 5FE440                 add       x0,a                    y:(r4),b
P:00B1 566648                 add       x0,b        a,x:(r6)
P:00B2 5F5E00                 move                              b,y:(r6)+
P:00B3 667000                 move                  r6,x:storr6
       0000C0
P:00B5 00000C                 rts

                    ;First the bounded area which the new data point is located is determined and
                    ;the minimum distance is found to the closest point in every state and stored.
                    ;the values are stored so that indexing is made easier, state 0,2,3,1,4,7,6,5.
                    ;this will greatly reduce the number of cycles needed in the next subroutine.
                    ;a smoothing function is used to accumulate distances in the accumulated
                    ;table so this minimum distance is multiplied by .1.
                    ;Before finding the bounded area the input lies in, it can be noted that
                    ;if the areas are labled correctly, the bounded area can be found in the
                    ;positive quadrant by looking only at the magnitude of the input coordinates.
                    ;Find the bounded area by first seeing if the input is in area 1,4, or 6.
                    ;When a bounded area is determined, testing is terminated.
                    ;Now a test is made to see if the data falls in areas 2 and then 5.
                    ;The final bounded area checked before determining the slope of
                    ;the lines to test those areas is area 3.  If the input
                    ;does not fall in any of these areas, it must then be determined if it is
```

```
                          ;above the top diagonal line (y2), below the bottom diaganal line (y1) or
                          ;if it is in the middle of the two lines.  In this manner, the rest of the
                          ;areas are checked, in the order 7,12,10,11,9,8 and 13.  Refer to the
                          ;document which shows a picture of these bounded areas on the constellation
                          ;as well as explaining how they are derived.

                          findmindist

P:00B6 56FE00                      move              x:-(r6),a
P:00B7 240800                      move              #one,x0
P:00B8 5FE647                      cmpm    x0,a               y:(r6),b

                          ;x>1

P:00B9 0E70C4                      jgt     <bigone
P:00BA 62F44F                      cmpm    x0,b      #boundry1,r2
       000200

                          ;x<1,y<1, load r2 with boundary 1 and continue

P:00BC 0E90EF                      jlt     <continue
P:00BD 251000                      move              #two,x1
P:00BE 62F46F                      cmpm    x1,b      #boundry4,r2
       000260

                          ;x<1,y>1andy<2, load r2 with boundary 4, go on

P:00C0 0E90EF                      jlt     <continue
P:00C1 62F400                      move              #boundry6,r2
       0002A0

                          ;x<1,y>2, load r2 with boundary 6 and continue

P:00C3 0C00EF                      jmp     <continue
P:00C4 251000             bigone   move              #two,x1
P:00C5 200067                      cmpm    x1,a

                          ;x>2, jmp to that case

P:00C6 0E70CD                      jgt     <bigtwo
P:00C7 62F44F                      cmpm    x0,b      #boundry2,r2
       000220
```

```
                                 ;x>1 ans x<2, y<1 load boundary 2 and continue

P:00C9 0E90EF                    jlt     <continue
P:00CA 62F46F                    cmpm    x1,b        #boundry5,r2
       000280

                                 ;x>1,y<2 load boundary 5 and continue

P:00CC 0E90EF                    jlt     <continue

P:00CD 62F44F          bigtwo    cmpm    x0,b        #boundry3,r2
       000240

                                 ;x>2 and y<1 so load boundary 3 and continue

P:00CF 0E90EF                    jlt     <continue

                                 ;now check to see where it falls above, below or inbetween the lines

P:00D0 261026                    abs     a           #two,y0
P:00D1 21C52E                    abs     b           a,x1
P:00D2 21E754                    sub     y0,a        b,y1
P:00D3 20004C                    sub     x0,b
P:00D4 20EF0F                    cmpm    a,b         y1,b
P:00D5 0E70DC                    jgt     <greatery1
P:00D6 62F45D                    cmp     y0,b        #boundry7,r2
       0002C0
P:00D8 0E90EF                    jlt     <continue
P:00D9 62F400                    move                #boundry12,r2
       000360
P:00DB 0C00EF                    jmp     <continue

                                 ;input is above y1

P:00DC 20AE5C          greatery1 sub     y0,b        x1,a
P:00DD 200044                    sub     x0,a
P:00DE 20AE0F                    cmpm    a,b         x1,a
P:00DF 0E70EA                    jgt     <greatery2
P:00E0 62F455                    cmp     y0,a        #boundry10,r2
       000320
P:00E2 0E90EF                    jlt     <continue
```

```
P:00E3 20EF00                  move                    y1,b
P:00E4 62F45D                  cmp       y0,b          #boundry11,r2
       000340
P:00E6 0E90EF                  jlt       <continue
P:00E7 62F400                  move                    #boundry9,r2
       000300
P:00E9 0C00EF                  jmp       <continue

               ;input is above y2

P:00EA 62F455  greatery2 cmp       y0,a          #boundry8,r2
       0002E0
P:00EC 0E90EF                  jlt       <continue
P:00ED 62F400                  move                    #boundry13,r2
       000380

               ;Now find out which quadrant it is in so that the pointer will point to the
               ;correct 8 points.
               ;For each bounded area, the points are stored in such a way that for boundary
               ;one, there are 32 points, 8 points for each quadrant area.  The pointer is
               ;updated by 0, 8, 16 or 24 based on which quadrant the input is in.

P:00EF 45E613  continue  clr       a             x:(r6),x1
P:00F0 4FE665            cmp       x1,a                         y:(r6),y1
P:00F1 0E70F7            jgt       <negx
P:00F2 3A1875            cmp       y1,a          #24,n2
P:00F3 0E70F5            jgt       <posxnegy
P:00F4 0C00FD  posxposy  jmp       <findist
P:00F5 44CA00  posxnegy  move                    x:(r2)+n2,x0              ;update r2 by 24
P:00F6 0C00FD            jmp       <findist
P:00F7 3A0875  negx      cmp       y1,a          #8,n2
P:00F8 0E70FB            jgt       <negxnegy
P:00F9 44CA00  negxposy  move                    x:(r2)+n2,x0              ;update r2 by 8
P:00FA 0C00FD            jmp       <findist
P:00FB 44CA00  negxnegy  move                    x:(r2)+n2,x0              ;update r2 by 16
P:00FC 44CA00            move                    x:(r2)+n2,x0


               ;After locating the boundary, find the Euclidean distance to each
               ;of the eight states defined by the boundary.  The Square Root is not
               ;performed.
               ;The x coordinate  of the input is subtracted from the x coordinate of the
```

```
                        ;constellation point in question and correspondingly the same for y.  These
                        ;values are then squared and added together.  This is done for each of the
                        ;eight points that are pointed to by the values in the boundary table found
                        ;previously.

                        findist
P:00FD 60DA00                 move                x:(r2)+,r0
P:00FE 34B000                 move                #tables,r4
P:00FF 56E000                 move                x:(r0),a
P:0100 5FE064                 sub      x1,a                      y:(r0),b
P:0101 21C47C                 sub      y1,b       a,x0
P:0102 21E680                 mpy      x0,x0,a    b,y0
P:0103 60DA92                 mac      y0,y0,a    x:(r2)+,r0
P:0104 10B400                 move                #small,x0      a,y0
       0CCCCD
P:0106 2000D0                 mpy      x0,y0,a
P:0107 BA8000                 move                x:(r0),a       a,y:(r4)+
P:0108 5FE064                 sub      x1,a                      y:(r0),b
P:0109 10DC7C                 sub      y1,b       a,x0           y:(r4)+,y0
P:010A 21E680                 mpy      x0,x0,a    b,y0
P:010B 60DA92                 mac      y0,y0,a    x:(r2)+,r0
P:010C 10B400                 move                #small,x0      a,y0
       0CCCCD
P:010E 5FDCD0                 mpy      x0,y0,a                   y:(r4)+,b
P:010F AA8000                 move                x:(r0),a       a,y:(r4)-
P:0110 5FE064                 sub      x1,a                      y:(r0),b
P:0111 10D47C                 sub      y1,b       a,x0           y:(r4)-,y0
P:0112 21E680                 mpy      x0,x0,a    b,y0
P:0113 60DA92                 mac      y0,y0,a    x:(r2)+,r0
P:0114 10B400                 move                #small,x0      a,y0
       0CCCCD
P:0116 2000D0                 mpy      x0,y0,a
P:0117 BA8000                 move                x:(r0),a       a,y:(r4)+
P:0118 5FE064                 sub      x1,a                      y:(r0),b
P:0119 21C47C                 sub      y1,b       a,x0
P:011A 21E680                 mpy      x0,x0,a    b,y0
P:011B 60DA92                 mac      y0,y0,a    x:(r2)+,r0
P:011C 10B400                 move                #small,x0      a,y0
       0CCCCD
P:011E 2000D0                 mpy      x0,y0,a
P:011F BA8000                 move                x:(r0),a       a,y:(r4)+
P:0120 5FE064                 sub      x1,a                      y:(r0),b
```

```
P:0121 10DC7C         sub     y1,b        a,x0          y:(r4)+,y0
P:0122 21E680         mpy     x0,x0,a     b,y0
P:0123 60DA92         mac     y0,y0,a     x:(r2)+,r0
P:0124 10B400         move                #small,x0     a,y0
       0CCCCD
P:0126 2000D0         mpy     x0,y0,a
P:0127 BA8000         move                x:(r0),a      a,y:(r4)+
P:0128 5FE064         sub     x1,a                      y:(r0),b
P:0129 10DC7C         sub     y1,b        a,x0          y:(r4)+,y0
P:012A 21E680         mpy     x0,x0,a     b,y0
P:012B 60DA92         mac     y0,y0,a     x:(r2)+,r0
P:012C 10B400         move                #small,x0     a,y0
       0CCCCD
P:012E 5FDCD0         mpy     x0,y0,a                   y:(r4)+,b
P:012F AA8000         move                x:(r0),a      a,y:(r4)-
P:0130 5FE064         sub     x1,a                      y:(r0),b
P:0131 21C47C         sub     y1,b        a,x0
P:0132 21E680         mpy     x0,x0,a     b,y0
P:0133 60DA92         mac     y0,y0,a     x:(r2)+,r0
P:0134 10B400         move                #small,x0     a,y0
       0CCCCD
P:0136 2000D0         mpy     x0,y0,a
P:0137 AA8000         move                x:(r0),a      a,y:(r4)-
P:0138 5FE064         sub     x1,a                      y:(r0),b
P:0139 21C47C         sub     y1,b        a,x0
P:013A 21E680         mpy     x0,x0,a     b,y0
P:013B 60DA92         mac     y0,y0,a     x:(r2)+,r0
P:013C 10B400         move                #small,x0     a,y0
       0CCCCD
P:013E 2000D0         mpy     x0,y0,a
P:013F 5E6400         move                              a,y:(r4)
P:0140 00000C         rts

                ;the accumulated distance routine  adds the smallest distance from the
                ;previously computed table for all paths going into a state and
                ;does this for all eight states.  Refer to the documentation for an extended
                ;explanation of this routine.


                accumdist
P:0141 30B013         clr     a           #tables,r0
```

```
P:0142 54F400              move                #$7fffff,a1
       7FFFFF
P:0144 221400              move                r0,r4
P:0145 32B800              move                #temp,r2
P:0146 0503A0              move                #3,m0
P:0147 0464A0              move                m0,m4
P:0148 390200              move                #2,n1
P:0149 233D00              move                n1,n5
P:014A 223500              move                r1,r5


                   ;In the following routine, distances in the accumulted distance table are
                   ;added to distances in the path table and compared for the four paths.  This
                   ;is done by incrementing through the state table and incrementing, in this
                   ;state through the specially ordered path table.
                   ;Note that the two tge  b,a, test the same condition but move different pointer
                   ;values.  The condition codes are the same from the compare so that this allows
                   ;two conditional pointer moves without extra code.

                   ;find minimum distance to state zero
P:014B 060480              do        #4,statezero
       000152
P:014D C38000              move                x:(r0),x0   y:(r4),b
P:014E 200048              add       x0,b
P:014F 200005              cmp       b,a
P:0150 031003              tge       b,a       r0,r3
P:0151 031407              tge       b,a       r4,r7
P:0152 F39800              move                x:(r0)+,x0  y:(r4)+,b
                   statezero
P:0153 634900              move                r3,x:(r1)+n1
P:0154 FB1A00              move                a,x:(r2)+   y:(r4)+,b
P:0155 6F4D13              clr       a                     r7,y:(r5)+n5
P:0156 54F400              move                #$7fffff,a1
       7FFFFF



                   ;find minimum distance to state two
P:0158 060480              do        #4,statetwo
       00015F
P:015A C38000              move                x:(r0),x0   y:(r4),b
P:015B 200048              add       x0,b
```

```
P:015C 200005            cmp     b,a
P:015D 031003            tge     b,a       r0,r3
P:015E 031407            tge     b,a       r4,r7
P:015F E39800            move              x:(r0)+,x0  y:(r4)-,b
                statetwo
P:0160 634900            move              r3,x:(r1)+n1
P:0161 FB1A00            move              a,x:(r2)+   y:(r4)+,b
P:0162 6F4D13            clr     a                     r7,y:(r5)+n5
P:0163 54F400            move              #$7fffff,a1
       7FFFFF

                ;find minimum distance to state four
P:0165 060480            do      #4,statefour
       00016C
P:0167 C38000            move              x:(r0),x0   y:(r4),b
P:0168 200048            add     x0,b
P:0169 200005            cmp     b,a
P:016A 031003            tge     b,a       r0,r3
P:016B 031407            tge     b,a       r4,r7
P:016C F39800            move              x:(r0)+,x0  y:(r4)+,b
                statefour
P:016D 634900            move              r3,x:(r1)+n1
P:016E FB1A00            move              a,x:(r2)+   y:(r4)+,b
P:016F 6F4D13            clr     a                     r7,y:(r5)+n5
P:0170 54F400            move              #$7fffff,a1
       7FFFFF

                ;find minimum distance to state six
P:0172 060480            do      #4,statezsix
       000179
P:0174 C38000            move              x:(r0),x0   y:(r4),b
P:0175 200048            add     x0,b
P:0176 200005            cmp     b,a
P:0177 031003            tge     b,a       r0,r3
P:0178 031407            tge     b,a       r4,r7
P:0179 E39800            move              x:(r0)+,x0  y:(r4)-,b
                statezsix
P:017A 634100            move              r3,x:(r1)-n1
P:017B 565A00            move              a,x:(r2)+
P:017C 6F6500            move                          r7,y:(r5)
P:017D 34B400            move              #tables+4,r4
P:017E 229000            move              r4,r0
```

```
P:017F 56C100          move            x:(r1)-n1,a
P:0180 57D113          clr      a      x:(r1)-,b
P:0181 54F400          move            #$7fffff,a1
       7FFFFF
P:0183 223500          move            r1,r5

                ;find minimum distance to state one
P:0184 060480          do       #4,stateone
       00018B
P:0186 C38000          move            x:(r0),x0   y:(r4),b
P:0187 200048          add      x0,b
P:0188 200005          cmp      b,a
P:0189 031003          tge      b,a    r0,r3
P:018A 031407          tge      b,a    r4,r7
P:018B F39800          move            x:(r0)+,x0  y:(r4)+,b
                stateone
P:018C 634900          move            r3,x:(r1)+n1
P:018D FB1A00          move            a,x:(r2)+   y:(r4)+,b
P:018E 6F4D13          clr      a                  r7,y:(r5)+n5
P:018F 54F400          move            #$7fffff,a1
       7FFFFF

                ;find minimum distance to state three
P:0191 060480          do       #4,statethree
       000198
P:0193 C38000          move            x:(r0),x0   y:(r4),b
P:0194 200048          add      x0,b
P:0195 200005          cmp      b,a
P:0196 031003          tge      b,a    r0,r3
P:0197 031407          tge      b,a    r4,r7
P:0198 E39800          move            x:(r0)+,x0  y:(r4)-,b
                statethree
P:0199 634900          move            r3,x:(r1)+n1
P:019A FB1A00          move            a,x:(r2)+   y:(r4)+,b
P:019B 6F4D13          clr      a                  r7,y:(r5)+n5
P:019C 54F400          move            #$7fffff,a1
       7FFFFF
P:019E 205C00          move            (r4)+

                ;find minimum distance to state five
P:019F 060480          do       #4,statefive
       0001A6
```

```
P:01A1 C38000           move              x:(r0),x0    y:(r4),b
P:01A2 200048           add     x0,b
P:01A3 200005           cmp     b,a
P:01A4 031003           tge     b,a       r0,r3
P:01A5 031407           tge     b,a       r4,r7
P:01A6 E39800           move              x:(r0)+,x0   y:(r4)-,b
                 statefive
P:01A7 634900           move              r3,x:(r1)+n1
P:01A8 EB1A00           move              a,x:(r2)+    y:(r4)-,b
P:01A9 6F4D13           clr     a                      r7,y:(r5)+n5
P:01AA 54F400           move              #$7fffff,a1
       7FFFFF

                 ;find minimum distance to state seven
P:01AC 060480           do      #4,stateseven
       0001B3
P:01AE C38000           move              x:(r0),x0    y:(r4),b
P:01AF 200048           add     x0,b
P:01B0 200005           cmp     b,a
P:01B1 031003           tge     b,a       r0,r3
P:01B2 031407           tge     b,a       r4,r7
P:01B3 F39800           move              x:(r0)+,x0   y:(r4)+,b
                 stateseven
P:01B4 635900           move              r3,x:(r1)+
P:01B5 FB1A00           move              a,x:(r2)+    y:(r4)+,b
P:01B6 6F5D1B           clr     b                      r7,y:(r5)+
P:01B7 55F400           move              #$7fffff,b1
       7FFFFF

                 ;now move new accumulated distances into the  accumulated distance
                 ;table from the temporary table
                 ;also find the min distance state and store in r4 which is no longer used
P:01B9 05F420           move              #$ffff,m0
       00FFFF
P:01BB 05F424           move              #$ffff,m4
       00FFFF
P:01BD 33B800           move              #temp,r3
P:01BE 30B000           move              #tables,r0
P:01BF 45F400           move              #large,x1
       733333
P:01C1 380200           move              #2,n0
```

```
P:01C2 060480          do      #4,endtable
       0001C7
P:01C4 44DB00          move            x:(r3)+,x0
P:01C5 2000A0          mpy     x1,x0,a
P:01C6 56480D          cmp     a,b     a,x:(r0)+n0
P:01C7 03100C          tge     a,b     r0,r4
                 endtable
P:01C8 30B100          move            #tables+1,r0
P:01C9 060480          do      #4,endtablex
       0001CE
P:01CB 44DB00          move            x:(r3)+,x0
P:01CC 2000A0          mpy     x1,x0,a
P:01CD 56480D          cmp     a,b     a,x:(r0)+n0
P:01CE 03100C          tge     a,b     r0,r4
                 endtablex
                 ;store in r0 instead of r4

P:01CF 229000          move            r4,r0
P:01D0 390800          move            #8,n1
P:01D1 204000          move            (r0)-n0
P:01D2 00000C          rts


                 ;the traceback routine now goes back through every time period starting
                 ;with the current time period and finds the state from which the path
                 ;came from one time period previous.  At the end of this search, the
                 ;last state found will also point to the path at that state, which is the
                 ;output  of the trellis.

                 traceback

                 ;find the displacement from the pointer to table and store value in n5
P:01D3 38B000          move            #tables,n0
P:01D4 204100          move            (r1)-n1
P:01D5 04401D          lua     (r0)-n0,n5
P:01D6 223500          move            r1,r5
P:01D7 060F80          do      #15,endtrace
       0001DC
P:01D9 204100          move            (r1)-n1
P:01DA 60ED00          move            x:(r5+n5),r0
P:01DB 223500          move            r1,r5
```

```
P:01DC 04401D            lua     (r0)-n0,n5
                endtrace
P:01DD 308000            move            #location,r0
P:01DE 5EED00            move                        y:(r5+n5),a
P:01DF 00000C            rts

                ;the output data routine unscrambles the path order and finds one
                ;of the four points on the constellation corresponding to the output state
                ;which is closest to the original input at that time period.
                ;This is neccessary because the paths were stored in a special order so that
                ;it was faster to find the accumulated distance.

                outputdata
P:01E0 21CF00            move            a,b
P:01E1 44F400            move            #>$b1,x0
       0000B1
P:01E3 46F445            cmp     x0,a    #>$b2,y0
       0000B2
P:01E5 02A058            teq     y0,b
P:01E6 44F455            cmp     y0,a    #>$b3,x0
       0000B3
P:01E8 02A048            teq     x0,b
P:01E9 46F445            cmp     x0,a    #>$b1,y0
       0000B1
P:01EB 02A058            teq     y0,b
P:01EC 44F400            move            #>$b5,x0
       0000B5
P:01EE 46F445            cmp     x0,a    #>$b7,y0
       0000B7
P:01F0 02A058            teq     y0,b
P:01F1 200055            cmp     y0,a
P:01F2 02A048            teq     x0,b

                ;Now that the path is unscrambled, mask the 2 lsb's so that the path state
                ;is stored as a pointer to the constellation locations.

P:01F3 21F200            move            b,r2
P:01F4 3AB000            move            #tables,n2
P:01F5 66F000            move            x:storr6,r6
       0000C0
P:01F7 04421B            lua     (r2)-n2,n3
P:01F8 236E00            move            n3,a
```

```
P:01F9 200032        asl    a
P:01FA 200032        asl    a
P:01FB 21D800        move              a,n0
P:01FC 22D300        move              r6,r3
P:01FD 044814        lua    (r0)+n0,r4
P:01FE 45F400        move              #>$7fffff,x1
       7FFFFF
P:0200 229000        move              r4,r0

               ;Now that the path state is known, find out which of the four paths is the
               ;output by comparing it to the input at that time period.  The point
               ;corresponding to the smallest Euclidean distance is the output.
P:0201 060480        do     #4,endout
       00020C
P:0203 CBC300        move              x:(r3),a    y:(r6),b
P:0204 F09800        move              x:(r0)+,x0  y:(r4)+,y0
P:0205 200044        sub    x0,a
P:0206 21C45C        sub    y0,b       a,x0
P:0207 21E680        mpy    x0,x0,a    b,y0
P:0208 200092        mac    y0,y0,a
P:0209 20AE09        tfr    a,b        x1,a
P:020A 20006D        cmp    x1,b
P:020B 039007        tlt    b,a        r0,r7
P:020C 21C500        move              a,x1
               endout

               ;Now mask off msb, which is the redundant bit and differentially decode the
               ;next two MSB's

P:020D 205713        clr    a          (r7)-
P:020E 388000        move              #location,n0
P:020F 22F000        move              r7,r0
P:0210 2C0F00        move              #$f,a1
P:0211 044017        lua    (r0)-n0,r7
P:0212 22E400        move              r7,x0
P:0213 200046        and    x0,a
P:0214 0BF080        jsr    diff
       000219

               ;ouput data, this can be done in the format needed for your system
```

```
P:0216 587000          move                       a0,y:$eff
       000EFF
P:0218 00000C          rts




                ;This subroutine performs the differential decoding on the output
                ;of the Viterbi decoder.  This is done by first making the output of
                ;the decoder serial (each bit stored in its own memory), then using the
                ;two MSB's to implement the equations defining the output.  These equations
                ;Q1n = Y1n EOR Y1n-1, Q2n = (Q1n AND Y1n-1) EOR Y2n-1 EOR Y2n
                ;are discussed in the documentation for this code.  After implementing
                ;this logical equation, the data  is recombined to be parallel.  This
                ;may be eliminated if the output is preferred to be in the serial form.


P:0219 30C400   diff   move              #ynow+3,r0
P:021A 44F400          move              #>$1,x0
       000001
P:021C 37C000          move              #ypast,r7
P:021D 060480          do      #4,diffloop1
       000222
P:021F 21C546          and     x0,a      a,x1
P:0220 545000          move              a1,x:(r0)-
P:0221 20AE00          move              x1,a
P:0222 200022          asr     a
                diffloop1
P:0223 F8F800          move              x:(r0)+,a   y:(r7)+,y0
P:0224 E9F800          move              x:(r0)+,a   y:(r7)-,y1
P:0225 BEF000          move              x:(r0)-,b   a,y:(r7)+
P:0226 5F5700          move                          b,y:(r7)-
P:0227 21C453          eor     y0,a      a,x0
P:0228 21C57B          eor     y1,b      a,x1
P:0229 21E756          and     y0,a      b,y1
P:022A 20AF73          eor     y1,a      x1,b
P:022B 575800          move              b,x:(r0)+
P:022C 565000          move              a,x:(r0)-
P:022D 200013          clr     a
P:022E 20001B          clr     b
P:022F 060480          do      #4,diff2
       000232
```

```
      P:0231 51D800            move           x:(r0)+,b0
      P:0232 200012            addl     b,a
                       diff2
      P:0233 00000C            rts


0   Errors
0   Warnings
```

APPENDIX C
DSP56001 BOUND.D DATA FILE

```
_DATA XE 200 82
 86 8b 8d 93 95 9a 9e 82
 86 89 8f 93 95 9a 9e 82
 86 89 8f 91 97 9a 9e 82
 86 8b 8d 91 97 9a 9e 82
 86 8b 8d 93 94 9a 9d 82
 86 89 8f 92 95 99 9e 82
 86 89 8f 90 97 99 9e 82
 86 8b 8d 91 96 9a 9d 83
 84 8b 8d 93 94 9a 9d 80
 87 89 8f 92 95 99 9e 80
 87 89 8f 90 97 99 9e 83
 84 8b 8d 91 96 9a 9d 82
 85 8a 8d 93 95 9a 9e 82
 85 88 8f 93 95 9a 9e 81
 86 89 8e 91 97 9a 9e 81
 86 8b 8c 91 97 9a 9e 82
 85 8a 8d 93 94 9a 9d 82
 85 88 8f 92 95 99 9e 81
 86 89 8e 90 97 99 9e 81
 86 8b 8c 91 96 9a 9d 82
 85 8a 8d 93 95 98 9f 82
 85 88 8f 93 95 98 9f 81
 86 89 8e 91 97 9b 9c 81
 86 8b 8c 91 97 9b 9c 83
 84 8a 8d 93 94 9a 9d 80
 87 88 8f 92 95 99 9e 80
 87 89 8e 90 97 99 9e 83
 84 8b 8c 91 96 9a 9d 82
 85 8a 8d 93 94 98 9f 82
 85 88 8f 92 95 98 9f 81
 86 89 8e 90 97 9b 9c 81
 86 8b 8c 91 96 9b 9c 83
 85 8a 8d 93 94 98 9d 80
 85 88 8f 92 95 99 9f 81
 87 89 8e 90 97 99 9c 81
 84 8b 8c 91 96 9b 9d 82
 85 8a 8d 93 94 98 9d 82
 85 88 8f 92 95 99 9f 81
 86 89 8e 90 97 99 9c 81
 86 8b 8c 91 96 9b 9d 83
 85 8a 8d 93 94 9a 9d 80
 85 88 8f 92 95 99 9e 81
 87 89 8e 90 97 99 9c 81
 84 8b 8c 91 96 9a 9d 83
 84 8a 8d 93 94 98 9d 80
 87 88 8f 92 95 99 9f 80
 87 89 8e 90 97 99 9c 83
 84 8b 8c 91 96 9b 9d 83
 85 8a 8d 93 94 98 9f 80
 85 88 8f 92 95 98 9f 81
 87 89 8e 90 97 9b 9c 81
 84 8b 8c 91 96 9b 9c
_END
```

TMC2301

CMOS Image Resampling Sequencer
15, 18MHz

The TMC2301 is a VLSI circuit which supports image resampling, rotation, rescaling, and filtering by generating input bit plane, convolution coefficient lookup table, and output bit plane memory addresses along with external multiplier–accumulator control signals. The TMC2301 can process data fields of up to 4096 x 4096 multibit words at a clock rate of up to 18MHz. An IRS–based system can nearest–neighbor resample a 512 x 512 image in 15 milliseconds, translating, zooming, rotating, or warping it, depending on the coefficient set loaded into the lookup table. A complete bilinear interpolation of the same image can be completed in 60 milliseconds. Image resampling speed is independent of the angle of rotation, degree of warp, or amount of zoom specified.

A high performance, TMC2301–based system can execute bilinear and cubic convolution algorithms that rotate images accurately and in real time. Keystone or other perspective correction, image plane distortion, and numerous other second order polynomial transformations can be programmed and executed under direct user control. Direct access to the coefficient lookup table allows dynamic modification of the algorithm.

Following an initialization with the transform coefficient parameters and control bits defining the operation to be executed, the IRS assumes control of the input and output data fields and executes unattended. Data word size is user selectable. All inputs except INTER and all outputs are registered on the rising edge of clock. All outputs are three–state controlled except $\overline{\text{ACC}}$, $\overline{\text{CZERO}}$, END, and DONE.

Fabricated in TRW's OMICRON–C™ one micron CMOS process, the TMC2301 operates at clock rates of up to 18MHz over the full commercial (0 to 70°C) temperature and 15MHz over the extended (–55 to +125°C) temperature and supply voltage ranges. All signals are TTL compatible.

Features

- Rotation, Warping, Panning, Zooming, And Compression Of Images In Real Time
- 18MHz Clock Rate
- 4096 x 4096 Image Field Addressing Capability
- User–Selectable Nearest–Neighbor, Bilinear Interpolation, And Cubic Convolution Resampling Algorithms
- Static Convolutional Filtering Of Up To 16 x 16 Pixel Windows
- Single–Pass Or Two–Pass Convolution Operations
- Low Power–Consumption CMOS Process
- Single 5V Power Supply
- Available In A 68 Lead Pin Grid Array

Applications

- Video Special–Effects Generators
- Image Recognition Systems, Robotics
- Artificial Intelligence
- High–Precision Image Registration (LANDSAT Processing)
- High–Speed Data Encoding/Decoding
- General Purpose Image Processing
- Image Data Compression

Functional Block Diagram

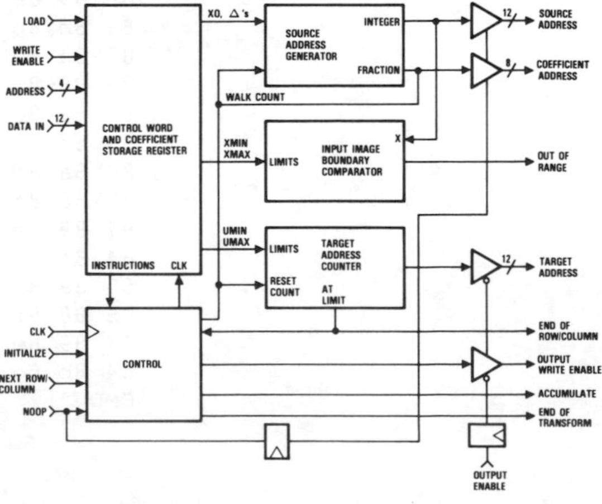

TRW LSI Products
P.O. Box 2472
La Jolla, CA 92038

Phone: (619) 457–1000
Telex: 697–957
TWX: 910–335–1571

40G05061 Rev. C–2/88
Printed in the U.S.A.

Functional Block Diagram

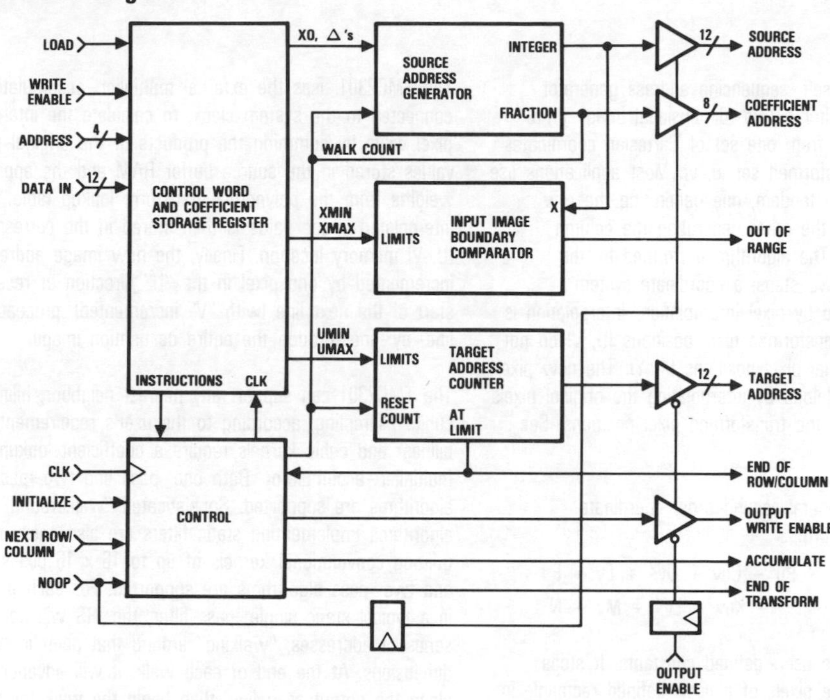

Pin Assignments

68 Pin Grid Array – G8 Package

Pin	Name	Pin	Name	Pin	Name	Pin	Name
B2	INIT	K2	U_{10}	K10	X_1	B10	P_6
B1	$\overline{OETA}$	L2	U_{11}	K11	X_2	A10	P_5
C2	INTER	K3	$\overline{UWRI}$	J10	X_3	B9	P_4
C1	END	L3	$\overline{ACC}$	J11	X_4	A9	P_3
D2	DONE	K4	$\overline{CZERO}$	H10	X_5	B8	P_2
D1	U_0	L4	CA_0	H11	X_6	A8	P_1
E2	U_1	K5	CA_1	G10	X_7	B7	P_0
E1	U_2	L5	V_{DD}	G11	X_8	A7	CLK
F2	GND	K6	GND	F10	GND	B6	GND
F1	U_3	L6	CA_2	F11	X_9	A6	V_{DD}
G2	U_4	K7	CA_3	E10	X_{10}	B5	NOOP
G1	U_5	L7	CA_4	E11	X_{11}	A5	LDR
H2	U_6	K8	CA_5	D10	P_{11}	B4	B_0
H1	U_7	L8	CA_6	D11	P_{10}	A4	B_1
J2	U_8	K9	CA_7	C10	P_9	B3	B_2
J1	U_9	L9	X_0	C11	P_8	A3	B_3
K1	GND	L10	GND	B11	P_7	A2	$\overline{WEN}$

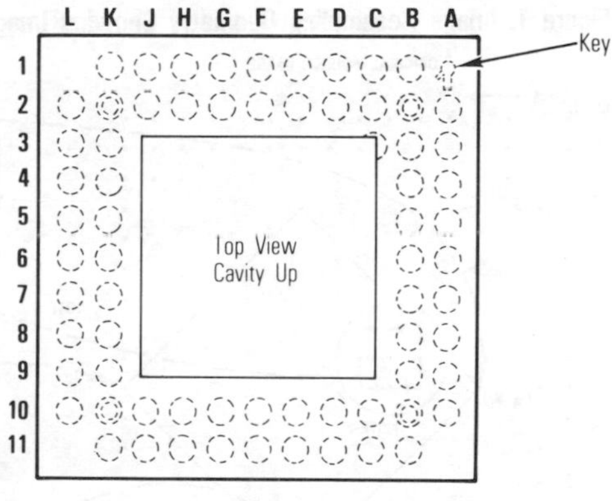

Functional Description

General Information

The IRS is a versatile self-sequencing address generator designed primarily to filter a two-dimensional image or to remap and resample it from one set of Cartesian coordinates (x, y) into a new, transformed set (u, v). Most applications use two identical devices in tandem, one generating the row coordinates (X and U), the other generating the column coordinates (Y and V). The algorithm performed by the TMC2301 consists of two steps: a coordinate system transformation, followed by pixel interpolation. Interpolation is necessary when the transformed pixel positions (U, V) do not coincide with the original pixel positions (X, Y). The new pixel intensity values are obtained by interpolating the original pixels in the neighborhood of the transformed pixel positions. See Figure 1.

The IRS executes a general second order coordinate transformation of the form:

$$X(u, v) = Au^2 + Bu + Cuv + Dv^2 + Ev + F$$
$$Y(u, v) = Gu^2 + Hu + Kuv + Lv^2 + Mv + N$$

where A through N are user-defined constants. It steps sequentially through the pixels of a user-defined rectangle in the new set of coordinates, computing the "old" address (X, Y) corresponding to each "new" location (U, V).

The TMC2301 uses the external multiplier-accumulator, connected to the system clock, to calculate the interpolated pixel value by summing the products of the original pixel values stored in the source buffer RAM and the appropriate weights from the polynomial transform lookup table. The new interpolated image value is then stored in the corresponding (U, V) memory location. Finally, the new image address is incremented by one pixel in the "U" direction or reset to the start of the next line (with "V" incremented), proceeding line-by-line through the entire destination image.

The TMC2301 can support any nearest neighbor, bilinear, or cubic resampling, according to the user's requirements. The bilinear and cubic kernels require a coefficient lookup table and multiplier-accumulator. Both one-pass and two-pass algorithms are supported. Sophisticated "walkaround" algorithms implementing static filters are also easily realized, utilizing convolutional kernels of up to 16 x 16 pixels. Both one and two-pass algorithms are supported. For each output point in a typical static single-pass filter, the IRS will generate a series of addresses, "walking" around that point in two dimensions. At the end of each walk, it will advance one pixel along the output scan line, then begin the walk for the next pixel.

Figure 1. Image Resampling Geometry Showing Image Rotation and Expansion

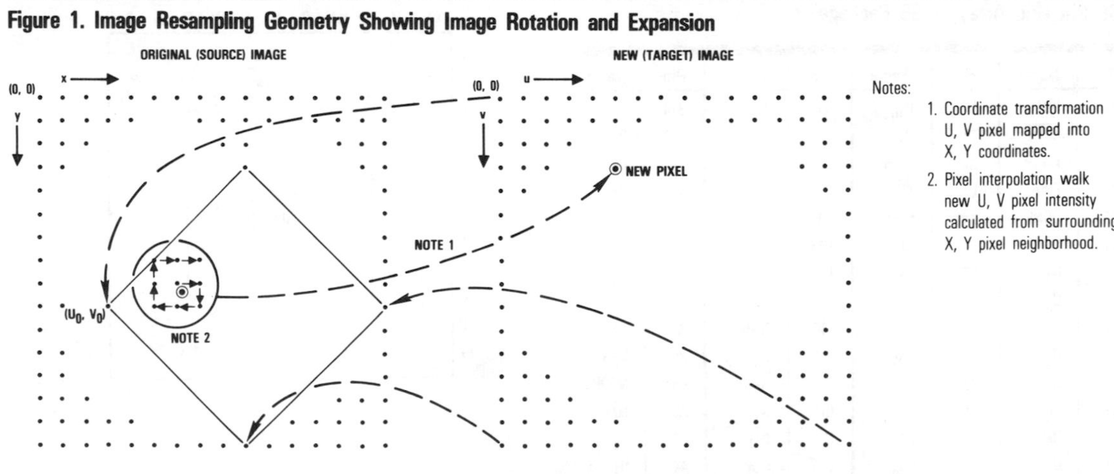
ORIGINAL (SOURCE) IMAGE
NEW (TARGET) IMAGE
(0, 0)
x
y
(0, 0)
u
v
NEW PIXEL
NOTE 1
NOTE 2
(U_0, V_0)
Notes:
1. Coordinate transformation U, V pixel mapped into X, Y coordinates.
2. Pixel interpolation walk new U, V pixel intensity calculated from surrounding X, Y pixel neighborhood.

A basic TMC2301-based system is shown in Figure 2. In this typical system, two Image Resampling Sequencers process the image. The only other external parts needed are a multiplier-accumulator, external interpolation coefficient lookup table RAM, and the user-specified Source and Destination Image Memory.

Figure 2. Basic 2-D Image Convolver Using TMC2301 Image Resampling Sequencer Utilizing Typical 8-Bit Data Path

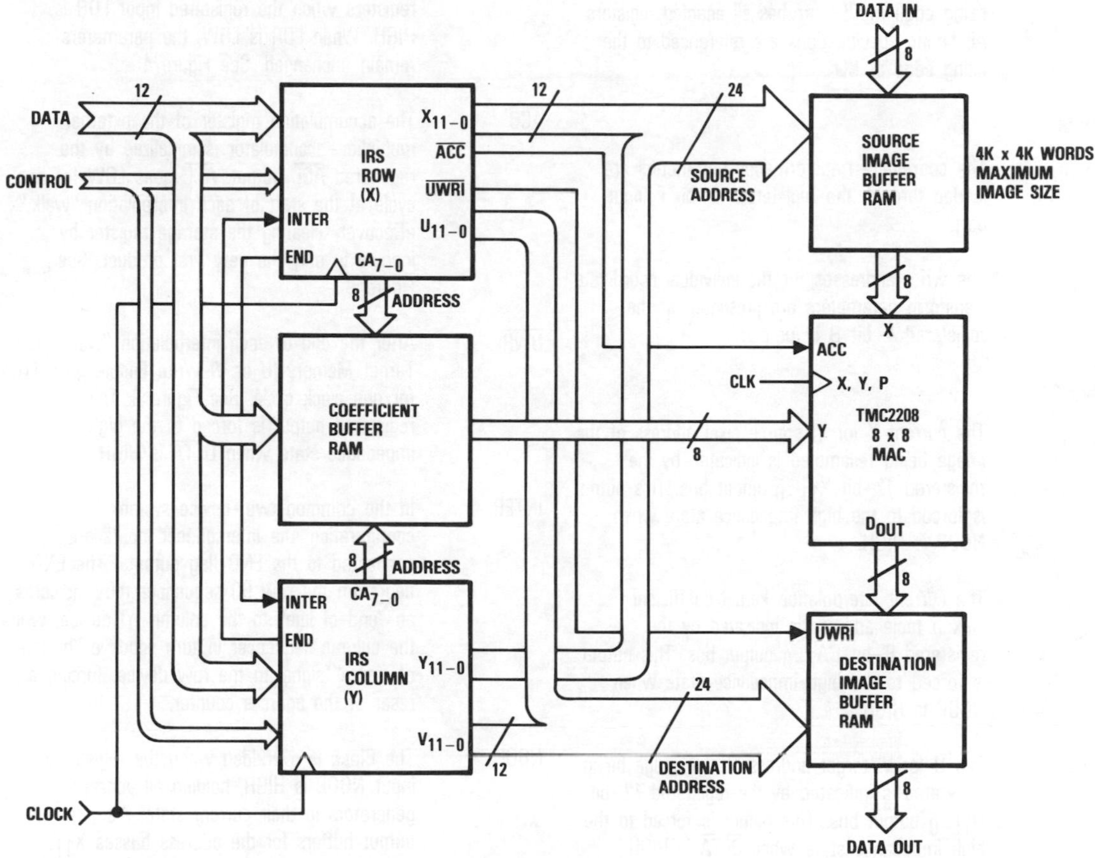

TMC2301

Signal Definitions

Power

V_{DD}, GND — The TMC2301 operates from a single +5V supply. All pins must be connected.

Clock

CLK — The TMC2301 has a single clock input. The rising edge of CLK strobes all enabled registers. All timing specifications are referenced to the rising edge of CLK.

Inputs

P_{11-0} — The coordinate transformation parameters are loaded through the registered 12-bit P input port.

B_{3-0} — The write addresses for the individual coordinate transform parameters are presented at the registered 4-bit B input port.

Outputs

X_{11-0} — The current X (or Y) source pixel address of the image being resampled is indicated by the registered 12-bit X_{11-0} output bus. This output is forced to the high impedance state when NOOP is HIGH.

CA_{7-0} — The current interpolation kernel coefficient lookup table address is indicated by the registered 8-bit CA_{7-0} output bus. This output is forced to the high impedance state when NOOP is HIGH.

U_{11-0} — The U (or V) target address of the image being generated is indicated by the registered 12-bit U_{11-0} output bus. This output is forced to the high impedance state when $\overline{OETA}$ is HIGH.

Controls

INIT — The control logic is cleared and initialized for the start of a new image transformation when the registered INIT input is HIGH for a minimum of two clock cycles. Normal operation begins after INIT goes LOW.

$\overline{WEN}$ — The registered Write Enable input allows the transformation parameters to be written into the preload register indicated by the address at the B input port when LOW. See Figure 4.

LDR — The data held in all transformation parameter preload registers is latched into the storage registers when the registered input LDR is HIGH. When LDR is LOW, the parameters remain unchanged. See Figure 4.

$\overline{ACC}$ — The accumulation register of the external multiplier–accumulator is initialized by the registered $\overline{ACC}$ output. $\overline{ACC}$ goes LOW for one cycle at the start of each interpolation "walk," effectively clearing the storage register by loading in only the new first product. See Figure 9.

$\overline{UWRI}$ — After the end of each interpolation "walk," the Target Memory (U or V) Write Enable goes LOW for one clock cycle. See Figure 9. This registered output is forced to the high impedance state when $\overline{OETA}$ is HIGH.

INTER — In the common two-device system configuration, the Interconnect inputs are connected to the END flag outputs. The END flag from the row (X) sequencer thus indicates an "end of line" to the column (Y) device, while the column sequencer in turn sends a "bottom of frame" signal to the row device, forcing a reset of the address counter.

NOOP — The Clock is overidden when the registered input NOOP is HIGH, holding all address generators in their current state. Also, the output buffers for the address busses X_{11-0} and CA_{7-0} are forced to the high impedance state. This allows the user access to all external memory. When NOOP goes LOW, normal operation resumes on the next clock cycle.

$\overline{OETA}$ — The target memory outputs $\overline{UWRI}$ and address bus U_{11-0} are in the high-impedance state when the registered Output Enable input is HIGH. When $\overline{OETA}$ is LOW, they are enabled on the next clock cycle.

TMC2301

TRW

Flags

$\overline{CZERO}$ When the current X (or Y) source address is outside the defined image space (XMIN or XMAX, or YMIN or YMAX), the registered output $\overline{CZERO}$ goes LOW. This signal may be used to force a zero on the coefficient input to the external multiplier-accumulator. When $\overline{CZERO}$ is HIGH, a valid (within the defined boundaries) address is indicated. In the standard two-device IRS set, the $\overline{CZERO}$ outputs may be ORed together to provide a single active-HIGH signal.

END The registered END flag goes HIGH during the last pixel of the last walk in a row in the case of the row chip, and the last pixel of the last walk in a column in the column chip, in the two-device architecture. This output is used as the end-of-line and end-of-frame indicator in conjunction with the INTER inputs of both TMC2301s.

DONE In the standard two-device system, a row sequencer DONE flag HIGH after the last walk at the end of the last row of an image (during $\overline{UWRI}$ LOW) indicates the end of the transform. This registered output is usually ignored on the column device. See the Control parameter AUTOINIT.

Package Interconnections

Signal Type	Signal Name	Function	G8 Package
Power	V_{DD}	Supply Voltage	L5, A6
	GND	Ground	F2, K1, K6, L10, F10, B6
Clock	CLK	System Clock	A7
Inputs	P_{11-0}	Parameter Register Data	D10, D11, C10, C11, B11, B10, A10, B9, A9, B8, A8, B7
	B_{3-0}	Parameter Register Address	A3, B3, A4, B4
Outputs	X_{11-0}	Source Address	E11, E10, F11, G11, G10, H11, H10, J11, J10, K11, K10, L9
	CA_{7-0}	Coefficient Address	K9, L8, K8, L7, K7, L6, K5, L4
	U_{11-0}	Target Address	L2, K2, J1, J2, H1, H2, G1, G2, F1, E1, E2, D1
Controls	INIT	Initialize	B2
	$\overline{NOOP}$	No Operation	B5
	$\overline{WEN}$	Coefficient Write Enable	A2
	$\overline{LDR}$	Latch Parameter Data Registers	A5
	$\overline{ACC}$	Accumulate	L3
	$\overline{OETA}$	Target Memory Output Enable	B1
	$\overline{UWRI}$	Target Memory Write Enable	K3
	INTER	Interconnect	C2
Flags	$\overline{CZERO}$	Coefficient Zero	K4
	END	End of Row/Page	C1
	DONE	End of Transform	D2

Transformation Control Parameters

The TMC2301 is a self-sequencing device which requires no cycle-to-cycle intervention from the host system. To program the device, the user loads the 16 operating parameters, which define the transformation to be performed, which sections of the original and resampled image spaces are to be utilized, and various control words. Filtering operations are further defined by the values the user loads into the external coefficient memory. The transform parameters are described below. See also Tables 1 through 3.

XMIN, XMAX, YMIN, YMAX These four parameters outline the "source" rectangular region of the original image. Whenever the IRS pair generate an (X, Y) address within this boundary, both $\overline{\text{CZERO}}$ flags remain HIGH, denoting a valid memory read. Addresses outside of this region cause one or both $\overline{\text{CZERO}}$s to go LOW, denoting an invalid memory read. Care must be taken to ensure that XMAX > XMIN and YMAX > YMIN. Each parameter is expressed in 12-bit unsigned binary integer notation. See Figure 12.

UMIN, UMAX, VMIN, VMAX These four parameters outline the "target" region of the (u, v) plane, into which the resampled image will be written. The IRS will generate, line by line, a scan that fills only this portion of the plane, permitting the user to assemble a mosaic of multiple rectangular subimages. Care must be taken to ensure that UMAX > UMIN and VMAX > VMIN. Each parameter is expressed in 12-bit unsigned binary integer notation. See Figure 12.

(X_0, Y_0) These are the coordinates of the first pixel to be read from the original image. In many applications, this point will be one of the four corners of the original image to be resampled. The pixels near (X_0, Y_0) in the original image will be used to compute the upper left pixel of the transformed image. In non-inverting, non-reversing applications (X_0, Y_0) will be the upper left corner of the original subimage. Each coordinate is expressed in 13-bit integer plus 5-bit fraction, two's complement notation.

dX/dU_0 Is the initial horizontal partial first derivative indicating the displacement along the X axis which corresponds to each one-pixel movement along the U axis. Usually, $0 < dX/dU_0 < 1$ corresponds to magnification, whereas $dX/dU_0 > 1$ represents reduction and $dX/dU_0 < 0$ denotes reflection about a vertical axis. The first derivatives are expressed in 8-bit integer, 12-bit fraction two's complement notation.

dX/dV_0 Is the initial horizontal-vertical partial first derivative. It indicates the displacement along the X axis corresponding to each one-pixel movement along the V axis. The coefficients dX/dV_0 and dY/dU_0 define image rotation and shear.

dY/dU_0 Is the initial vertical-horizontal partial first derivative. It indicates the displacement along the Y axis corresponding to each one-pixel movement along the U axis.

dY/dV_0 Is the initial vertical partial first derivative. It indicates the displacement along the Y axis corresponding to each one-pixel step along the V axis. Since dX/dU_0 and dY/dV_0 are separate parameters, vertical magnification and reflection need not match their horizontal counterparts.

NOTE: For each incremental move along the U axis, the starting point of the new "walk around spiral" is indexed to the ENDING point of the previous walk around spiral, rather than to its center. Therefore, the terms dX/dU_0 and dY/dU_0 must be adjusted accordingly. Since each new line is referenced back to the previous line's initial spiral starting point, no similar dX/dV_0 or dY/dV_0 correction is needed.

d^2X/dU^2 Is the second order horizontal derivative. It indicates the rate of change of the horizontal-horizontal first derivative with each step along a line in the output image space. All six second-order derivatives are 4-bit integer, 20-bit fractional two's complement parameters.

d^2X/dV^2 Is the second order horizontal-vertical-vertical derivative. It indicates the rate of change of the horizontal-vertical first derivative with each step down a column in the output image space.

d^2Y/dU^2 Is the second order vertical-horizontal-horizontal derivative. It indicates the rate of change of the the vertical-horizontal first derivative with each step along a line of the output image space.

d^2Y/dV^2 Is the second order vertical derivative. It indicates the rate of change of the vertical-vertical first derivative with each step down a column of the output image space.

$d^2X/dUdV$ Is the mixed second order derivative indicating the rate of change of the first order horizontal derivative as one proceeds downwards through the output image space. This is also the rate of change of the first order horizontal-vertical derivative during horizontal sweeps in the output image space.

$d^2Y/dUdV$ Is the mixed second order derivative indicating the rate of change of the first order vertical derivative as one moves horizontally across the output space, or, equivalently, the rate of change of the first order vertical-horizontal derivative as one moves vertically in the output image space.

Row/Column Select Sets the mode to either Row (1) or Column (0) operation.

Mode This 2-bit control word defines three unique instructions:

Code	Instruction
00, 01	single-pass operation
10	pass 1 of two-pass operation
11	pass 2 of two-pass operation

In single-pass operation, the device walks through the entire (k + 1) x (k + 1) kernel for each output pixel, where k is the value written into the Kernel section (see below) of the parameter register. Two-pass operation, which requires a dimensionally separable kernel, is executed first for a (k + 1) element kernel in one direction, then for a (k + 1) element kernel in the other direction. For kernel sizes exceeding 2 x 2, the two-pass algorithm is obviously beneficial, requiring 2n samples per output point instead of n x n. In this case, the intermediate image data stored in the destination image memory following the first pass is used as the source image data on the second pass. The user may design his system to switch source and destination memory bank addresses in place, or could utilize a second TMC2301 pair in a pipelined architecture. This would require a third image buffer for the final destination image. Both devices of a system pair are usually set to the same mode.

Kernel The effective kernel width (heighth) exceeds this 4-bit unsigned number by 1, thereby providing kernels of 1 x 1 to 16 x 16 source pixels per output, for either resampling or filtering. Simple static filters can be implemented with kernels of up to 16 x 16 pixels (Kernel = 15), while resampling interpolation kernels are limited to 4 x 4 pixels (Kernel = 3), due to the four bits of fractional X (or Y) address generated by the TMC2301. See the Applications Discussion, below. Again, both devices in a pair are generally initialized with equal Kernel values.

Field of View (FOV) As the device walks through its kernel coefficients, each corresponding step in (x, y) space is normally one pixel length or height; this is a field of view of 1. However, the user can subsample the original space before filtering or resampling, by applying the coefficient kernel over a view field of up to 7 units. At a field of view of F, the pixels selected for each kernel operation are F pixels apart. This is useful in oversampled pictures, whose intensity changes only slowly from pixel to pixel.

Autoload (ALR) When set to 1 (HIGH), the LDR control is automatically asserted when INIT is strobed, loading the coefficient set currently stored in the preload registers.

Autoinit (AIN) At the end of an image, if the AIN bit is 1 (HIGH) the DONE flag goes HIGH for one clock cycle and a new transform begins. If 0 (LOW), $\overline{\text{UWRI}}$ and the DONE flag remain HIGH during the sequence until the user strobes the INIT control to begin a new image transformation.

Pipe (PIPE)	Adjusts the timing of the target memory write controls, to compensate for buffered source image RAM. If the PIPE bit is 1 (HIGH), outputs $\overline{ACC}$ and $\overline{UWRI}$ will be delayed one clock cycle relative to the generation of the target address (U or V). See Figure 9.
Test Mode (TM)	This mode is available for user inspection of the coefficient data. The source image and coefficient addresses are calculated by an internal 28-bit accumulator. When TM is 1 (HIGH), the sign bit, normally discarded, and the lower 11 bits of internal data are substituted for the upper 12 bits appearing at the source address port (X) during a standard transform cycle. This allows user verification of algorithm mathematics during debug. Since the TM bit is registered and cannot be changed during a single clock cycle, two distinct clock cycles are required to access both the MSW and LSW of the internal accumulator. See Figure 3.

Figure 3. Test Mode Data Routing

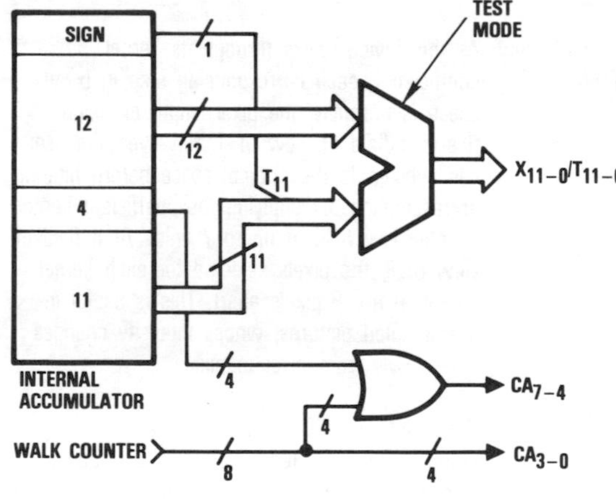

Table 1. Parameter Registers – Row Sequencer

Address	Name	Description
0000	XMIN	Left side of Source Window
0001	XMAX	Right side of Source Window
0010	X_0 (LSW)	Source starting point – X coordinate
0011	X_0 (MSW)	Source starting point – X coordinate
0011	Controls	Mode Select Bits
0100	dX/dU_0 (LSW)	Row/Row first differential
0101	dX/dU_0 (MSW)	Row/Row first differential
0101	TM, FOV	Test Mode, Field of View
0110	dX/dV_0 (LSW)	Row/Column first differential
0111	dX/dV_0 (MSW)	Row/Column first differential
0111	Kernel	Resampling/Filtering Kernel
1000	$d^2X/dUdV$ (LSW)	Mixed second differential
1001	$d^2X/dUdV$ (MSW)	Mixed second differential
1010	d^2X/dU^2 (LSW)	Row second differential
1011	d^2X/dU^2 (MSW)	Row second differential
1100	d^2X/dV^2 (LSW)	Row/Column second differential
1101	d^2X/dV^2 (MSW)	Row/Column second differential
1110	UMIN	Left edge of Final Image
1111	UMAX	Right edge of Final Image

Table 2. Parameter Registers – Column Sequencer

Address	Name	Description
0000	YMIN	Top of Source Window
0001	YMAX	Bottom of Source Window
0010	Y_0 (LSW)	Source starting point – Y coordinate
0011	Y_0 (MSW)	Source starting point – Y coordinate
0011	Controls	Mode Select Bits
0100	dY/dU_0 (LSW)	Column/Row first differential
0101	dY/dU_0 (MSW)	Column/Row first differential
0101	TM, FOV	Test Mode, Field of View
0110	dY/dV_0 (LSW)	Column/Column first differential
0111	dY/dV_0 (MSW)	Column/Column first differential
0111	Kernel	Resampling/Filtering Kernel Size
1000	$d^2Y/dUdV$ (LSW)	Mixed second differential
1001	$d^2Y/dUdV$ (MSW)	Mixed second differential
1010	d^2Y/dU^2 (LSW)	Column/Row second differential
1011	d^2Y/dU^2 (MSW)	Column/Row second differential
1100	d^2Y/dV^2 (LSW)	Column second differential
1101	d^2Y/dV^2 (MSW)	Column second differential
1110	VMIN	Top edge of Final Image
1111	VMAX	Bottom edge of Final Image

Table 3. Parameter Registers Binary Format (Row Or Column Sequencer)

Addr	Format												Limits	
	MSB											LSB	Dec	Hex
0000*	2^{11}	2^{10}	2^{9}	2^{8}	2^{7}	2^{6}	2^{5}	2^{4}	2^{3}	2^{2}	2^{1}	2^{0}	4095 0	FFF 000
0001*	2^{11}	2^{10}	2^{9}	2^{8}	2^{7}	2^{6}	2^{5}	2^{4}	2^{3}	2^{2}	2^{1}	2^{0}	4095 0	FFF 000
0010	2^{6}	2^{5}	2^{4}	2^{3}	2^{2}	2^{1}	2^{0}	2^{-1}	2^{-2}	2^{-3}	2^{-4}	2^{-5}	$4096-2^{-5}$ -4096	0FFF.F8 F000.00
0011							-2^{12}	2^{11}	2^{10}	2^{9}	2^{8}	2^{7}		
0011 (Control)	ALR	AIN	PIPE	R/C	M_1	M_0								
0100	2^{-1}	2^{-2}	2^{-3}	2^{-4}	2^{-5}	2^{-6}	2^{-7}	2^{-8}	2^{-9}	2^{-10}	2^{-11}	2^{-12}	$128-2^{-12}$ -128	007F.FFF FF80.000
0101					-2^{7}	2^{6}	2^{5}	2^{4}	2^{3}	2^{2}	2^{1}	2^{0}		
0101* (TM, FOV)	TM	2^{2}	2^{1}	2^{0}										
0110	2^{-1}	2^{-2}	2^{-3}	2^{-4}	2^{-5}	2^{-6}	2^{-7}	2^{-8}	2^{-9}	2^{-10}	2^{-11}	2^{-12}	$128-2^{-12}$ -128	007F.FFF FF80.000
0111					-2^{7}	2^{6}	2^{5}	2^{4}	2^{3}	2^{2}	2^{1}	2^{0}		
0111* (Kernel)	2^{3}	2^{2}	2^{1}	2^{0}										
1000	2^{-9}	2^{-10}	2^{-11}	2^{-12}	2^{-13}	2^{-14}	2^{-15}	2^{-16}	2^{-17}	2^{-18}	2^{-19}	2^{-20}	$8-2^{-20}$ -8	0007.FFFFF FFF8.00000
1001	-2^{3}	2^{2}	2^{1}	2^{0}	2^{-1}	2^{-2}	2^{-3}	2^{-4}	2^{-5}	2^{-6}	2^{-7}	2^{-8}		
1010	2^{-9}	2^{-10}	2^{-11}	2^{-12}	2^{-13}	2^{-14}	2^{-15}	2^{-16}	2^{-17}	2^{-18}	2^{-19}	2^{-20}	$8-2^{-20}$ -8	0007.FFFFF FFF8.00000
1011	-2^{3}	2^{2}	2^{1}	2^{0}	2^{-1}	2^{-2}	2^{-3}	2^{-4}	2^{-5}	2^{-6}	2^{-7}	2^{-8}		
1100	2^{-9}	2^{-10}	2^{-11}	2^{-12}	2^{-13}	2^{-14}	2^{-15}	2^{-16}	2^{-17}	2^{-18}	2^{-19}	2^{-20}	$8-2^{-20}$ -8	0007.FFFFF FFF8.00000
1101	-2^{3}	2^{2}	2^{1}	2^{0}	2^{-1}	2^{-2}	2^{-3}	2^{-4}	2^{-5}	2^{-6}	2^{-7}	2^{-8}		
1110*	2^{11}	2^{10}	2^{9}	2^{8}	2^{7}	2^{6}	2^{5}	2^{4}	2^{3}	2^{2}	2^{1}	2^{0}	4095 0	FFF 000
1111*	2^{11}	2^{10}	2^{9}	2^{8}	2^{7}	2^{6}	2^{5}	2^{4}	2^{3}	2^{2}	2^{1}	2^{0}	4095 0	FFF 000

* unsigned binary notation

A "–" indicates MSB is sign bit

Operation of the Transformation Parameter Registers

Numerous applications require the ability to update the coordinate transformation parameters "on the fly." Because the parameters are double-buffered, the user can load any or all of them into the preload registers without upsetting the operation in progress. Then LDR (load data registers) will update all transform parameters to the new values simultaneously. This feature is particularly valuable for "pin cushion" and "fish eye" transformations, or polar-to-rectangular conversions, which cannot be performed with constant second derivatives. The Autoload function updates the preload registers at the beginning of a new image automatically. See page 8. Note also that data can be loaded in to the registers while NOOP is active (HIGH).

Figure 4. Operation of LDR Control for Parameter Update

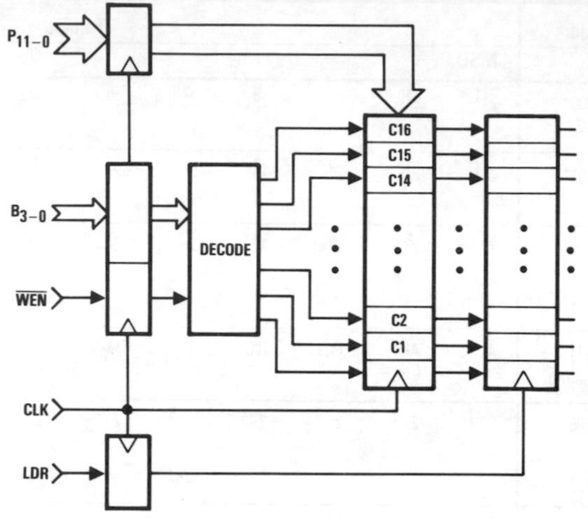

Figure 5. Timing Diagram

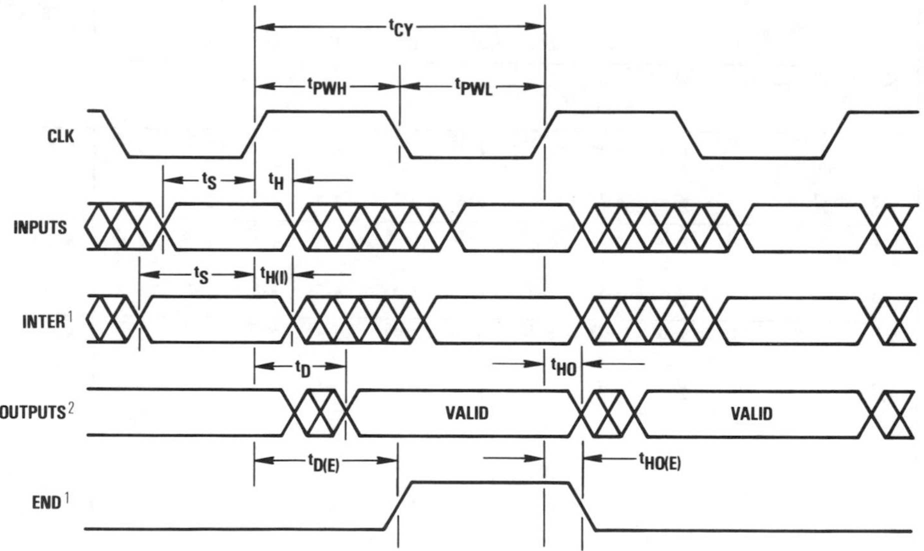

Notes:

1. t_S and $t_{D(E)}$ are guaranteed to allow full speed operation in the standard two-device architecture. See text.
2. All outputs except END. See text.

TMC2301

Figure 6. Equivalent Input Circuit

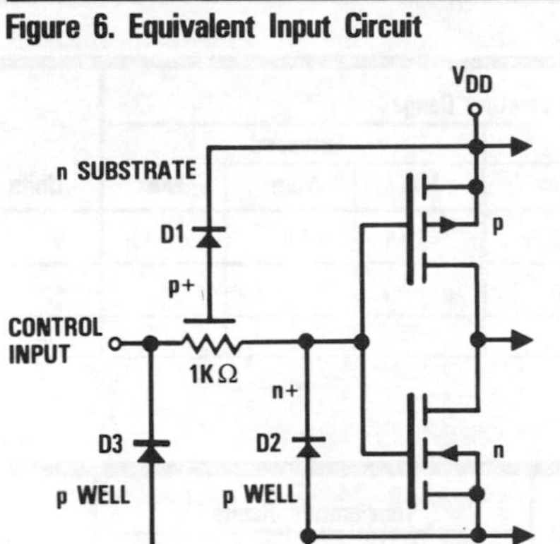

Figure 7. Equivalent Output Circuit

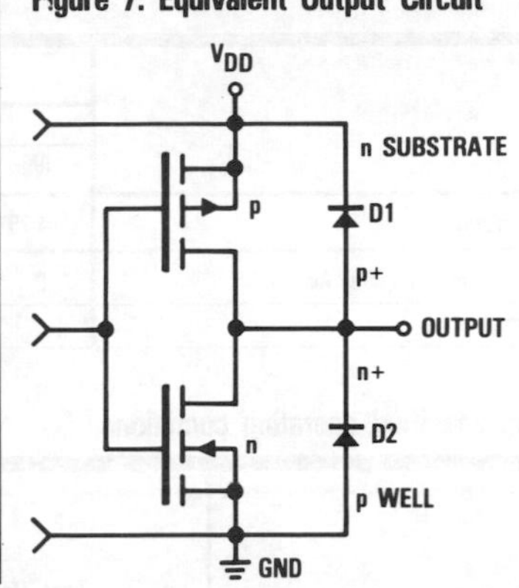

Figure 8. Transition Level for Three-State Measurement

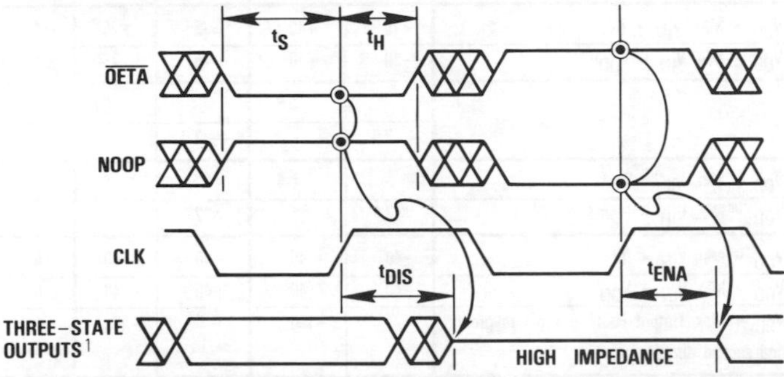

Note:

1. All outputs except $\overline{CZERO}$, $\overline{ACC}$, END and DONE.

Absolute maximum ratings (beyond which the device may be damaged)[1]

Supply Voltage .... -0.5 to +7.0V

Input Voltage .... -0.5 to (V_{DD} +0.5)V

Output

Applied voltage[2] .... -0.5 to (V_{DD} +0.5)V

Forced current[3,4] .... -1.0 to +6.0mA

Short-circuit duration (single output in HIGH state to ground) .... 1 sec

Temperature

Operating, case .... -60 to +130°C

junction .... 175°C

Lead, soldering (10 seconds) .... 300°C

Storage .... -65 to +150°C

Notes:

1. Absolute maximum ratings are limiting values applied individually while all other parameters are within specified operating conditions. Functional operation under any of these conditions is NOT implied.
2. Applied voltage must be current limited to specified range, and measured with respect to GND.
3. Forcing voltage must be limited to specified range.
4. Current is specified as conventional current flowing into the device.

TMC2301

Operating conditions

Parameter		Standard Min	Standard Nom	Standard Max	Extended Min	Extended Nom	Extended Max	Units
		Temperature Range						
V_{DD}	Supply Voltage	4.75	5.0	5.25	4.5	5.0	5.5	V
T_A	Ambient Temperature, Still Air	0		70				°C
T_C	Case Temperature				−55		125	°C

DC characteristics within specified operating conditions [1]

Parameter		Test Conditions	Standard Min	Standard Max	Extended Min	Extended Max	Units
			Temperature Range				
I_{DDQ}	Supply Current, Quiescent	V_{DD} = Max, V_{IN} = 0V		5		5	mA
I_{DDU}	Supply Current, Unloaded	V_{DD} = Max, f = 15MHz		75		75	mA
I_{IL}	Input Current, Logic LOW	V_{DD} = Max, V_{IN} = 0V	−10	+10	−75	+75	μA
I_{IH}	Input Current, Logic HIGH	V_{DD} = Min, V_{IN} = V_{DD}	−10	+10	−75	+75	μA
V_{IL}	Input Voltage, Logic LOW			0.8		0.8	V
V_{IH}	Input Voltage, Logic HIGH		2.0		2.0		V
V_{OL}	Output Voltage, Logic LOW	V_{DD} = Min, I_{OL} = 8mA		0.4		0.4	V
V_{OH}	Output Voltage, Logic HIGH	V_{DD} = Min, I_{OH} = −4mA	2.4		2.4		V
I_{OZL}	Hi-Z Output Leakage Current, Output LOW	V_{DD} = Min, V_{IN} = 0V	−40	+ 40	−40	+ 40	μA
I_{OZH}	Hi-Z Output Leakage Current, Output HIGH	V_{DD} = Min, V_{IN} = V_{DD}	−40	+ 40	−40	+ 40	μA
I_{OS}	Short-Circuit Output Current [2]	V_{DD} = Max, Output HIGH, one pin to ground, one second duration max.		−100		−100	mA
C_I	Input Capacitance	T_A = 25°C, f = 1MHz		10		10	pF
C_O	Output Capacitance	T_A = 25°C, f = 1MHz		10		10	pF

Notes:

1. Actual test conditions may vary from those shown, but guarantee operation as specified.
2. Guaranteed but not tested.

TMC2301

TRW

AC characteristics within specified operating conditions

Parameter		Test Conditions	Standard -1 Min	Standard -1 Max	Standard Min	Standard Max	Extended Min	Extended Max	Units
t_{CY}	Cycle Time	V_{DD} = Min	55		66		66		ns
t_{PWL}	Clock Pulse Width LOW	V_{DD} = Min	25		30		30		ns
t_{PWH}	Clock Pulse Width HIGH	V_{DD} = Min	25		30		30		ns
t_S	Input Setup Time[1]		18		20		20		ns
t_H	Input Hold Time		2		2		2		ns
$t_{H(I)}$	Input Hold Time, INTER		10		10		10		ns
t_D	Output Delay[2]	V_{DD} = Min, C_{LOAD} = 40pF		27		35		35	ns
$t_{D(E)}$	Output Delay, END[1]	V_{DD} = Min, C_{LOAD} = 10pF		37		45		45	ns
t_{HO}	Output Hold Time[2]	V_{DD} = Max, C_{LOAD} = 40pF	5		5		5		ns
$t_{HO(E)}$	Output Hold Time, END	V_{DD} = Max, C_{LOAD} = 10pF	10		10		10		ns
t_{DIS}	Three-State Disable Delay	V_{DD} = Min, C_{LOAD} = 40pF		18		20		20	ns
t_{ENA}	Three-State Enable Delay	V_{DD} = Min, C_{LOAD} = 40pF	27		35		35		ns

Notes:

1. $t_S + t_{D(E)} = t_{CY}$ max.
2. Excluding output pin END.

Applications Discussion

Basic Operation

Each TMC2301 pair contains address controllers which execute patterns much like the following FORTRAN 3-level nested DO loop:

1. The inner loop is a clockwise outgoing spiral "walk" through the N-element coefficient kernel.
2. The middle loop is a left-to-right "scan" along each row of the output image space.
3. Finally, the outer loop is a top-to-bottom "scan" down each column of the output image space.

A typical one-pass image transformation proceeds as follows:

1. The device pair outputs the addresses (X_0, Y_0), which is the first point in the source image, and (CAX, CAY), the interpolation lookup table address for the first pixel in the kernel. The output $\overline{ACC}$ goes LOW, causing the external accumulator to load the first product without summation, clearing the accumulator.
2. For the next N cycles, the IRS walks through an outward clockwise spiral in (x, y) space, accumulating pixel-interpolation coefficient products. The spiral sequence is depicted in Figure 9.
3. After the completion of the first spiral walk, the IRS outputs the target address of the first pixel, (UMIN, VMIN) and the control $\overline{UWRI}$, along with the initial (X, Y) values of the next spiral walk. $\overline{ACC}$ and $\overline{UWRI}$ can be delayed by one clock cycle by setting the control bit PIPE to 1 (HIGH), simplifying the task of interfacing the TMC2301 to buffered source image memory.
4. After the last cycle of the next spiral, $\overline{UWRI}$ again goes LOW for one clock, and the target address outputs are updated, pointing to the location of the pixel calculation just completed, (UMIN + 1, VMIN).

5. The third spiral walk begins with $\overline{ACC}$ going LOW, and ends with (UMIN + 2, VMIN) output and $\overline{UWRI}$ going LOW.
6. The procedure continues until (UMAX, VMIN) is reached, at which point the device resets to U (position within row) and increments V (number of row). Thus, the next (U, V) set after (UMAX, VMIN) will be (UMIN, VMIN + 1), followed by (UMIN + 1, VMIN + 1), etc.
7. Upon completion of the walk corresponding to (UMAX, VMAX), the TMC2301 will generate a DONE flag with the final $\overline{UWRI}$, and begin a new sequence.

Figure 9. Timing Diagram and Pixel Map Showing Outward Clockwise Spiral Walk Generated by TMC2301 (2 x 2 Kernel Shown)

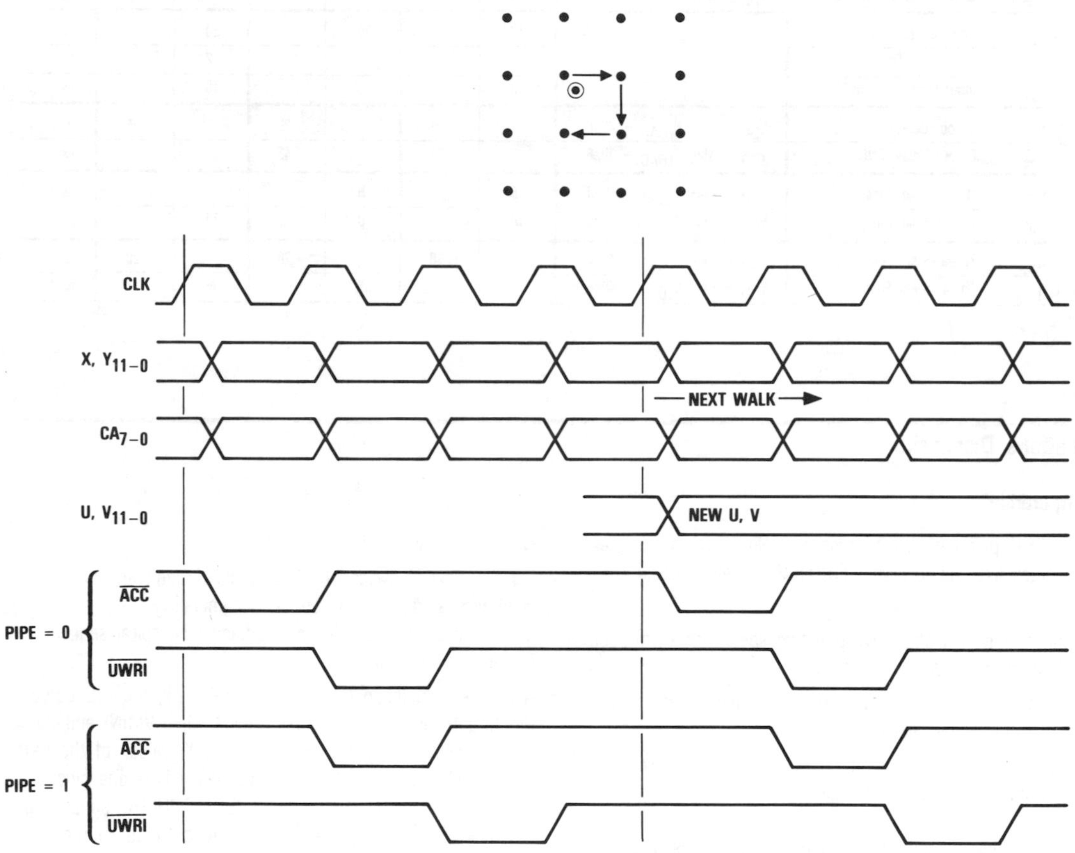

Notes:

1. Assumes that $\overline{OETA}$ is HIGH and NOOP is LOW.
2. Timing Parameters are not shown on this diagram.

On any given clock cycle, the actual (X, Y) and (U, V) outputs of the IRS are given by the following equations:

$$x = X_0 + dX/dU_0*m + dX/dV_0*n + d^2X/dUdV*m*n + d^2X/dU^2*(m^2 - m)/2 + d^2X/dV^2*(n^2 - n)/2 + FOV*CAX(w) + FOV*m*CAX(Ker)$$

$$y = Y_0 + dY/dU_0*m + dX/dV_0*n + d^2Y/dUdV*m*n + d^2Y/dU^2*m(m^2 - m)/2 + d^2Y/dV^2*(n^2 - n)/2 + FOV*CAY(w) + FOV*n*CAY(Ker)$$

$$u = UMIN + m$$

$$v = VMIN + n$$

where FOV is the 4-bit field of view parameter, normally set to 1 so that the spiral walk proceeds in single-pixel steps. Setting FOV to 4 would expand the spiral walk, allowing the user to trade two bits of image size for two bits of additional interpixel positioning resolution. CAX(w) and CAY(w) are the current value of the coefficient address outputs, and CAX(KER) and CAY(KER) are the terminal values of each pixel walk. The CA(KER) terms arise because the IRS computes each new walk's starting point from the previous spiral walk's end point, rather than its starting point.

Interpolation Coefficient Lookup Table Addressing

The external coefficient lookup table RAM stores the interpolation values used to calculate the value of the new pixel. These values are selected by the user, allowing maximum filtering flexibility. In simple filtering applications, all 8 bits of coefficient address are available to access up to 256 interpolation coefficients, for kernels of 16 x 16 pixels. This address is generated by the internal walk counter of the TMC2301. In most applications, the same Kernel parameter value is selected in both IRS devices; thus, the Coefficent Address outputs CA_{7-0} for the X and Y devices are identical, and the user needs only one of the 8-bit buses for memory access.

Applications executing a coordinate transformation, however, will almost always generate non-integer source pixel addresses; that is, the U (or V) locations will not map to the X (or Y) addresses exactly, and fractional address components are generated. The user then must account for this spatial offset in both dimensions by storing the appropriate corrected interpolation kernel values in the lookup table. The 8-bit address bus is broken up into two parts: the fractional portion (upper 4 bits), and the walk counter (lower 4 bits). Thus, in resampling applications, the maximum kernel size is 4 x 4 pixels, or 16 locations. As in the filtering example, assuming that the user has selected the same kernel size for both IRS devices, the 4 bits of least-significant address generated by both devices will be identical, and redundant. The four most significant address bits, however, will reflect the current fractional offsets of the resampled pixel from the nearest X (Y) location, to a spatial resolution of 4 bits, in the X (or Y) directions. Utilization of the 12 bits (total) of lookup table address is left to the user, to be arranged as desired for memory access. See Figure 3.

Application Examples

One of the more common applications for the TMC2301 is simple static filtering. In this case the source and target memories locations are identical and no coordinate transformation is performed. The (X, Y) and (U, V) outputs listed in Table 4 show the address sequencing generated by the TMC2301 to execute the walk of a 5 x 5 pixel interpolation kernel. The normalized coefficients shown implement a first-order Butterworth Low Pass Filter with cutoff radius of $1/\sqrt{2}$. Note that the (U, V) output address is updated following the completion of the walk for that location.

Figure 10. Pixel Map Showing Walk Sequence for 5 x 5 Static Filter

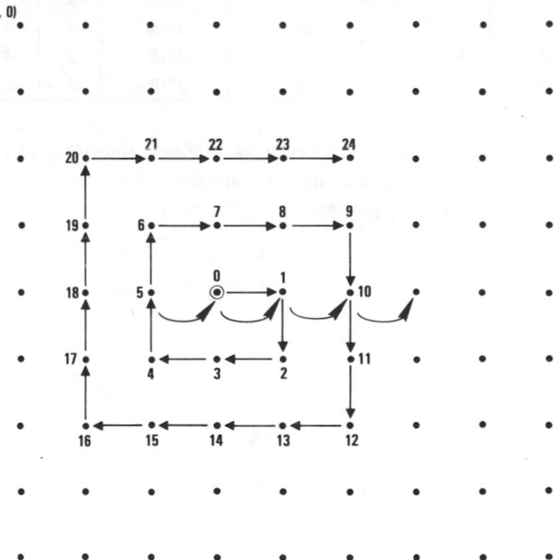

Table 4. IRS Outputs for Static Filter Illustrated in Figure 10

Cycle	X	Y	Index (CA)	Coefficient	U	V
1	3	4	0	0.2176	2	4
2	4	4	1	0.0725	2	4
3	4	5	2	0.0435	2	4
4	3	5	3	0.0725	2	4
5	2	5	4	0.0435	2	4
6	2	4	5	0.0725	2	4
7	2	3	6	0.0435	2	4
8	3	3	7	0.0725	2	4
9	4	3	8	0.0435	2	4
10	5	3	9	0.0198	2	4
11	5	4	10	0.0272	2	4
12	5	5	11	0.0198	2	4
13	5	6	12	0.0128	2	4
14	4	6	13	0.0198	2	4
15	3	6	14	0.0272	2	4
16	2	6	15	0.0198	2	4
17	1	6	16	0.0128	2	4
18	1	5	17	0.0198	2	4
19	1	4	18	0.0272	2	4
20	1	3	19	0.0198	2	4
21	1	2	20	0.0128	2	4
22	2	2	21	0.0198	2	4
23	3	2	22	0.0272	2	4
24	4	2	23	0.0198	2	4
25	5	2	24	0.0128	2	4
26	4	4	0	0.2175	3	4

Figure 11 illustrates the sequence for a bilinear resampling of a 63° rotation. The starting point is translated +1 in the Y-direction. A common rotation matrix might be:

$dX/dU_0 = \cos(a) = .6 \qquad dY/dU_0 = \sin(a) = .8$

$dX/dV_0 = -\sin(a) = -.8 \qquad dY/dV_0 = \cos(a) = .6$

However, we have included a linear compression factor of 5:1, and must accommodate the fact that each time u is incremented, the start of the new walk is referenced to the END of the previous walk. Given these corrections, the rotation matrix becomes:

$dX/dU_0 = 5\cos(a) = 3 \qquad dY/dU_0 = 5\sin(a) - FOV = 3$

$dX/dV_0 = -5\sin(a) = -4 \qquad dY/dV_0 = 5\cos(a) = 3$

Kernel = 1

Figure 11. Pixel Map Showing Parameters for 63° Rotation and 5:1 Compression Listed in Table 5

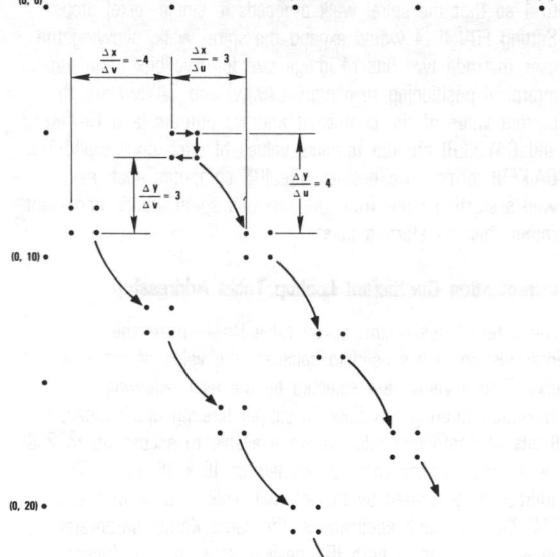

Table 5. IRS Outputs for Operation Illustrated in Figure 11

Cycle	X	Y	Index	U	V
1	5	5	0	4	5
2	6	5	1	4	5
3	6	6	2	4	5
4	5	6	3	4	5
5	8	9	0	5	5
6	9	9	1	5	5
7	9	10	2	5	5
8	8	10	3	5	5
9	11	13	0	6	5
10	12	13	1	6	5
11	12	14	2	6	5
12	11	14	3	6	5
13	14	17	0	7	5
14	15	17	1	7	5
15	15	18	2	7	5
16	14	18	3	7	5
17	1	8	0	8	5
18	2	8	1	8	5
19	2	9	2	8	5
20	1	9	3	8	5
21	4	12	0	5	6
22	5	12	1	5	6
23	5	13	2	5	6
24	4	13	3	5	6
25	7	16	0	6	6
26	8	16	1	6	6
27	8	17	2	6	6
28	7	17	3	6	6
29	10	20	0	7	6
30	11	20	1	7	6
31	11	21	2	7	6
32	10	21	3	7	6
33	0	15	0	8	6

Figure 12 may help clarify the relationships among (X_0, Y_0), (XMIN, YMIN), (XMAX, YMAX), (UMIN, VMIN), and (UMAX, VMAX). With positive first derivatives, (X_0, Y_0) and (UMIN, VMIN) represent the upper left corners of the original image and the new destination field, respectively. The lower right corner of the transformed image is located at (UMAX, VMAX); the location of the corresponding corner of the original image depends on the values of the derivatives. Not to be confused with (X_0, Y_0), the points (XMIN, YMIN) and (XMAX, YMAX) define the "usable" rectangular portion of the original image; points (X, Y) lying outside this region are ignored in most resampling and filtering applications. This feature permits one to construct a mosaic of several abutting subimages in the (x, y) plane, without danger of edge effect interference between adjacent subimages. Note in the figure that the upper left and lower left corners of the original image lie outside the admissible region; in practice, the values fetched at these locations will not be included in the convolutional sums.

Figure 12. Pixel Maps Demonstrating Source and Destination Image Boundaries and Image Clipping (Note Shaded Area)

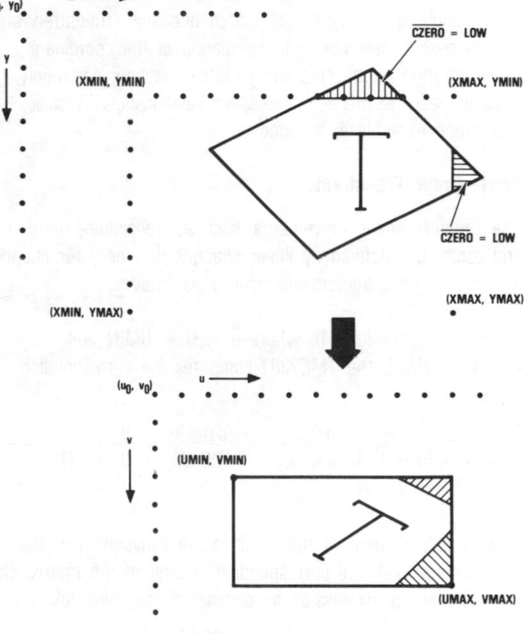

Using Matrix Notation To Build Image Manipulation Algorithms With The TMC2301 Image Resampling Sequencer

John Eldon, Rich Wegner

The TMC2301 Image Resampling Sequencer offers a powerful solution to many image manipulation problems. However, in order to fully utilize the device the user must understand the role of the image transform coefficients. Once the contributions of these user-programmable variables are understood, construction of the desired algorithm becomes straightforward. This discussion describes the calculation of the coordinate transformation coefficients using matrix notation, commonly used to describe image manipulation relationships. A basic application example is included.

First-Order Transforms

The simplest image movements, such as translation, rotation, and zoom, are defined by linear changes in x or y for changes in u or v. These vectors are defined as follows:

Given the vector (u v 1), where u = U − UMIN and v = V − VMIN, the TMC2301 computes the corresponding vector (x y 1);

$$(x\ y\ 1) = (u\ v\ 1) \begin{vmatrix} dX/dU_0 & dY/dU_0 & 0 \\ dX/dV_0 & dY/dV_0 & 0 \\ X_0 & Y_0 & 1 \end{vmatrix} \quad \mathbf{(1)}$$

where each element of the result is the convolution of the (u, v) vector with the corresponding column of the matrix. One can simplify the expression by eliminating the third column:

$$(x\ y) = (u\ v\ 1) \begin{vmatrix} dX/dU_0 & dY/dU_0 \\ dX/dV_0 & dY/dV_0 \\ X_0 & Y_0 \end{vmatrix} \quad \mathbf{(2)}$$

The MIN and MAX parameters, which define the working spaces utilized during the image transformation, must be carefully managed to avoid image clipping and device programming errors. See the Applications Discussion in the TMC2301 Data Sheet.

Pure Second Order Transforms

Interesting warp and motion effects can be realized using the second-order transforms. We will consider their effect seperately from those of the mixed second-order terms. Combined transforms are discussed under Compound Operations, below.

If there is no u-v mixed second-order term, then equation **(2)** becomes:

$$(x\ y) = (u\ v\ 1) \begin{vmatrix} dX/dU & dY/dU \\ dX/dV & dY/dV \\ X_0 & Y_0 \end{vmatrix} \quad \mathbf{(3)}$$

where dX/dU, dX/dV, dY/dU, and dY/dV are no longer constants. In this mode,

$$\begin{aligned} dX/dU &= dX/dU_0 + ((u-1)/2)dX/dU^2 \\ dY/dV &= dY/dV_0 + ((v-1)/2)dY/dV^2 \\ dX/dV &= dX/dV_0 + ((v-1)/2)dX/dV^2 \\ \text{and} \quad dY/dU &= dY/dU_0 + ((u-1)/2)dY/dU^2 \end{aligned}$$

This equation assumes that the cross-terms of dX/dUdV and dY/dUdV are zero. We can expand equation **(3)** as:

$$(x\ y) = (u\ v\ 1) \begin{vmatrix} dX/dU_0 + ((u-1)/2)dX/dU^2 & dY/dU_0 + ((u-1)/2)dY/dU^2 \\ dX/dV_0 + ((v-1)/2)dX/dV^2 & dY/dV_0 + ((v-1)/2)dY/dV^2 \\ X_0 & Y_0 \end{vmatrix} \quad \mathbf{(4)}$$

Alternatively, we can represent the dX/dU etc. terms as:

$$(dX/dU\ \ dY/dU) = \begin{vmatrix} (u-1)/2 & 1 \end{vmatrix} \begin{vmatrix} dX/dU^2 & dY/dU^2 \\ dX/dU_0 & dY/dU_0 \end{vmatrix} \quad \mathbf{(5)}$$

and:

$$(dX/dV\ \ dY/dV) = \begin{vmatrix} (v-1)/2 & 1 \end{vmatrix} \begin{vmatrix} dX/dV^2 & dY/dV^2 \\ dX/dV_0 & dY/dV_0 \end{vmatrix} \quad \mathbf{(6)}$$

Mixed Second-Order Transforms

In applications such as LANDSAT image correction, which requires perspective correction for a type of distortion called "keystoning," the mixed second-order coefficients are used. These transforms involve the cross-terms dX/dUdV and dY/dUdV, multiplied by the product uv. Since they involve both u and v, we can arbitrarily assign each to either of two places in the matrix, yielding any of four equations for dX/dU, etc., including the following:

$$\begin{aligned} dX/dU &= dX/dU_0 + ((u-1)/2)dX/dU^2 + v\,dX/dUdV && \mathbf{(7a)} \\ dY/dV &= dY/dV_0 + ((v-1)/2)dY/dV^2 + u\,dY/dUdV && \mathbf{(7b)} \\ dX/dV &= dX/dV_0 + ((v-1)/2)dX/dV^2 && \mathbf{(7c)} \\ dY/dU &= dY/dU_0 + ((u-1)/2)dY/dU^2 && \mathbf{(7d)} \end{aligned}$$

or:

$$dX/dU = dX/dU_0 + ((u-1)/2)dX/dU^2 + vdX/dUdV \quad \textbf{(8a)}$$
$$dY/dV = dY/dV_0 + ((v-1)/2)dY/dV^2 \quad \textbf{(8b)}$$
$$dX/dV = dX/dV_0 + ((v-1)/2)dX/dV^2 \quad \textbf{(8c)}$$
$$dY/dU = dY/dU_0 + ((u-1)/2)dY/dU^2 + vdY/dUdV \quad \textbf{(8d)}$$

Many applications require only the mixed second-order terms, reducing equation **(7)** to:

$$dX/dU = dX/dU_0 + vdX/dUdV \quad \textbf{(9a)}$$
$$dY/dV = dY/dV_0 + udY/dUdV \quad \textbf{(9b)}$$
$$dX/dV = dX/dV_0 \quad \textbf{(9c)}$$
$$dY/dU = dY/dU_0 \quad \textbf{(9d)}$$

In this case, the governing equation becomes:

$$(x\ y) = (u\ v\ 1) \begin{vmatrix} dX/dU_0 + vdX/dUdV & dY/dU_0 \\ dX/dV_0 & dY/dV_0 + udY/dUdV \\ X_0 & Y_0 \end{vmatrix} \quad \textbf{(10)}$$

Applications

Every first-order image transformation can be decomposed into a product of several matrices, each describing a single operation, such as translation, rotation, or scale. By relating these expressions back to equations **(1)** through **(3)** above, the user can determine how to initialize the IRS device pair to perform a given operation.

Translation

The pure translation matrix is:

$$(x\ y\ 1) = (u\ v\ 1) \begin{vmatrix} 1 & 0 & 0 \\ 0 & 1 & 0 \\ X_0 & Y_0 & 1 \end{vmatrix} \quad \textbf{(11)}$$

Recalling that $u = U-UMIN$ and $v = V-VMIN$, we obtain:

$$x = U - UMIN + X_0 \text{and } y = V - VMIN + Y_0 \quad \textbf{(12)}$$

If $X_0 > UMIN$, then $x > U$, i.e., all points are moved leftward in the remap from the (x, y) space to the (u, v) space. Similarly, if $Y_0 < VMIN$, then $y < V$, and all points are moved upward during the transformation. Thus, horizontal translation is governed by the relative values of UMIN and X_0, whereas vertical translation depends on VMIN and Y_0.

Scaling

The pure scaling matrix is:

$$(x\ y\ 1) = (u\ v\ 1) \begin{vmatrix} CX & 0 & 0 \\ 0 & CY & 0 \\ 0 & 0 & 1 \end{vmatrix} \quad \textbf{(13)}$$

or $x = u(CX)$ and $y = v(CY)$. The remap from (x, y) space to (u, v) space will compress an image horizontally if $CX > 1$, will expand it vertically if $CY < 1$, will invert it about a horizontal axis if $CY < 0$, and will invert it about a vertical axis if $CX < 0$.

Rotation

The pure (flat) rotation matrix is:

$$(x\ y\ 1) = (u\ v\ 1) \begin{vmatrix} \cos R & -\sin R & 0 \\ \sin R & \cos R & 0 \\ 0 & 0 & 1 \end{vmatrix} \quad \textbf{(14)}$$

where $R > 0$ rotates an object counterclockwise about the (common) (x, y)/(u, v) origin.

Compound Operations

Usually, one will need to combine two or more of the above operations to execute a desired transformation. For example, rotating an object about an arbitrary point (XP, YP) entails three steps:

- One – ranslate the object so that (XP, YP) lies at the (u, v) plane origin;
- Two – rotate the object about this new origin (which is actually (XP, YP));
- Three – translate the object in the opposite direction, bringing (XP, YP) back to its original position, but in the (u, v) plane.

This same procedure is used to rescale an object without moving it, since normal scaling is referenced to the origin and will therefore generate translation.

To compute the coefficients for a compound operation, merely multiply the corresponding matrices, in the proper order:

$$(x\ y\ 1) = (u\ v\ 1)\ (M3)\ (M2)\ (M1)\ (15) \quad \textbf{(15)}$$

where M1 is the matrix of the first operation to be performed, M2 is the second, etc. The user computes a new matrix,

$$MC = (M3)\ (M2)\ (M1),$$

whose elements are then used to program the IRS. Since these matrix operations are associative but not commutative, the right-to-left sequence of M1 through M3 must be strictly observed. (It is easy to see that moving an object upward and leftward to the origin, rotating the entire picture about the origin, then moving the object downward and rightward to its original center point is not equivalent to moving it downward and rightward first, then rotating the picture, including the center point, and finally displacing it back upward and leftward.)

Application Example

Problem: Rotate an object 37 degrees counterclockwise and reduce its linear dimensions by 5:1, without moving its center point of (5, 6). This problem requires four steps:

One – translate the object 5 pixels left and 6 pixels up, bringing its center to the upper left origin (0, 0):

$$\begin{vmatrix} 1 & 0 & 0 \\ 0 & 1 & 0 \\ 5 & 6 & 1 \end{vmatrix} = M1 \quad \textbf{(16)}$$

Two – rotate object 37 degrees counterclockwise about this new origin:

$$\begin{vmatrix} .8 & -.6 & 0 \\ .6 & .8 & 0 \\ 0 & 0 & 1 \end{vmatrix} = M2 \quad \textbf{(17)}$$

Three – rescale the object by 5:1:

$$\begin{vmatrix} 5 & 0 & 0 \\ 0 & 5 & 0 \\ 0 & 0 & 1 \end{vmatrix} = M3 \quad \textbf{(18)}$$

Four – move the origin back out to (5, 6):

$$\begin{vmatrix} 1 & 0 & 0 \\ 0 & 1 & 0 \\ -5 & -6 & 1 \end{vmatrix} = M4 \quad \textbf{(19)}$$

The compound matrix is the product of these four matrices, or (M4) (M3) (M2) (M1):

$$MC = \begin{vmatrix} 1 & 0 & 0 \\ 0 & 1 & 0 \\ -5 & -6 & 1 \end{vmatrix} \begin{vmatrix} .8 & -.6 & 0 \\ .6 & .8 & 0 \\ 0 & 0 & 1 \end{vmatrix} \begin{vmatrix} 5 & 0 & 0 \\ 0 & 5 & 0 \\ 0 & 0 & 1 \end{vmatrix} \begin{vmatrix} 1 & 0 & 0 \\ 0 & 1 & 0 \\ 5 & 6 & 1 \end{vmatrix}$$

$$= \begin{vmatrix} 4 & 3 & 0 \\ 3 & 4 & 0 \\ -33 & -3 & 1 \end{vmatrix} \quad \textbf{(20)}$$

Thus, we can see that the coordinate transform coefficients we have calculated are:

$$X_0 = -33 \qquad Y_0 = -3$$
$$dX/dU_0 = 4 \qquad dY/dU_0 = 4$$
$$dX/dV_0 = 3 \qquad dY/dV_0 = 4$$

Notes

Using the TMC2301 Image Resampling Sequencer

John Eldon, Rich Wegner

The TMC2301 Image Resampling Sequencer is a controller/address generator, around which an image filtering and resampling system can be built. Under the limited supervision of a host processor, the TMC2301 will generate the sequence of memory read and write addresses to transform, resample, and/or filter a two-dimensional image. In all cases, it fetches data from one image buffer, governs its convolution with a user-specified kernel of coefficients, and directs the results to another image memory space. With 12-bit address buses, the device can operate between frame sizes of up to 4096 x 4096 pixels, with spatial resolution of 1/16 pixel.

As shown in Figure 2, the basic system comprises data source and target memories, coefficient lookup table, multiplier-accumulator, TMC2301 pair, and host processor. The host loads the original image into source memory, initializes the TMC2301s, and starts the transform. The host system must also provide screen or other refresh, if needed, for the image memory buffers.

Although it is an address generator, the TMC2301 is intended to be a coprocessor building block, to be used with an existing host video processing system. The host system must supervise loading of original image data and transform coefficients, plus any needed video and/or memory refresh. In most systems, the IRS subsystem will be used only when a filter or a coordinate transformation is required. All image storage is assumed to be in the form of bit planes, with the number of bit planes dictated by the user's desired intensity resolution. The examples shown are 8-bit systems. Processing of color images, commonly in RGB format, involves three seperate 8-bit image planes which may all be controlled by the same TMC2301 chip pair. Figure 3 illustrates such a system, with seperate interpolation coefficient storage for each color. This same scheme may be utilized for images stored in composite format, however the user must be sure to compensate for the differences in sampling rates, data word size, intensities. etc. to obtain satisfactory results.

The TMC2301 generates three sets of addresses:

One – the (X, Y) original pixel location;

Two – the (CAX, CAY) "walk index" coefficient address; and

Three – the (U, V) resampled pixel location.

The flow of data is from the source data memory and coefficient lookup table, through the multiplier-accumulator, and into the target data memory. Data Bus width is user-selectable.

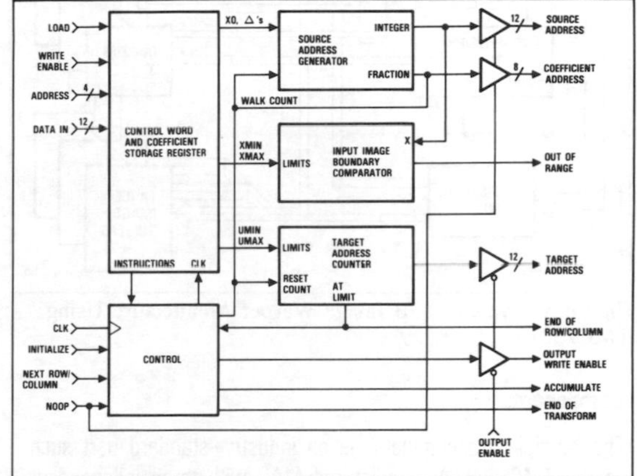

Figure 1. Simplified Block Diagram

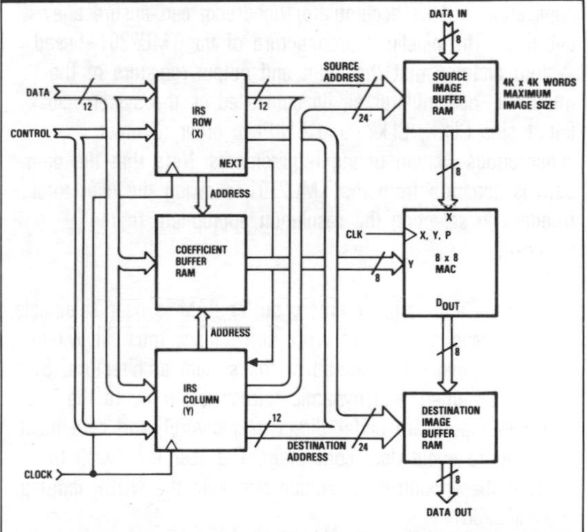

Figure 2. Basic 2-D Image Convolver Using TMC2301 Image Resampling Sequencer

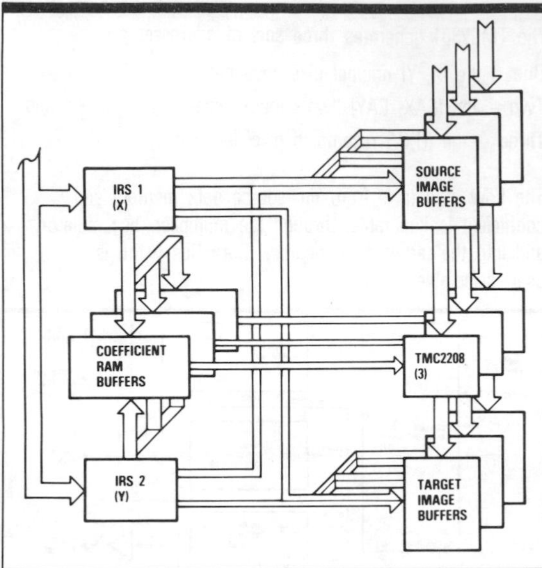

Figure 3. Typical RGB Image Warper Architecture Using TMC2301

Components

The multiplier-accumulator is an industry-standard part such as the TMC2208. Any registered MAC with three independent ports (data input, coefficient input, and data output) should be suitable as long as its speed and resolution satisfy the application and its accumulator headroom can ensure against overflow. The pipelined architecture of the TMC2301-based system dictates that the input and output registers of the multiplier-accumulator all be controlled by the system clock (input pins CLKX, CLKY, and CLKP) in order to meet synchronous system timing requirements. Note that the data path is separate from the TMC2301, allowing the user total freedom in selecting the resolution appropriate to the application.

The source and target memory buffer RAM is user selectable, however read access and write setup times must fit within the pipelined speed requirements of the system architecture. Static RAM is assumed, and dynamic refresh is left up to the user. The R/W inputs are under direct user control, and care must be taken to avoid data contention. The user may wish to operate these controls in conjunction with the NOOP input of the TMC2301.

The coefficient lookup table can be either ROM or RAM, or even several memories in parallel. The user must preprogram this table with an application-specific kernel of interpolation coefficients, which will differ according to kernel size, number of passes (1 or 2), operation type (filtering versus resampling), etc.

Most two-dimensional systems will require two TMC2301 Image Resampling Sequencers, one to compute horizontal ("X" and "U") addresses, the other to compute the vertical ("Y" and "V") addresses. The two devices communicate across a two-pin handshake bus, which permits the horizontal sequencer to announce that it has reached the end of a row, and the vertical that it has reached the end of a column. The TMC2301 pair alerts the system when both of these end flags are high, signifying end of page or transform.

Function of the TMC2301 Image Resampling Sequencer

Address Generation

The TMC2301's primary function is address generation for the source and target data memories and the coefficient lookup table. It always steps across the U-direction scan lines in the rectangular (U, V) target space defined by the parameters (UMIN, VMIN), and (UMAX, VMAX). At each point in the target memory, one or more corresponding source (X, Y) pixel addresses are generated, according to the kernel size programmed by the user.

Controls

The TMC2301's second function is to tell the multiplier-accumulator when to accumulate and when to clear the accumulation and begin a new summation (i.e., at the beginning of each convolutional "walk").

The TMC2301 pair requires fairly minimal supervision from the host. The INIT (start transform), NOOP (halt or freeze operation), and LDR and WEN (update parameters) signals are the only user control inputs. In addition, the host must provide the appropriate transform parameters and their respective addresses.

Internal Operation of the TMC2301

The TMC2301's major blocks are the X (or Y) address generator, the U (V) address generator, the walk counter, and the central controller.

The X (Y) address generator comprises a double-level accumulator designed to compute the necessary second-order difference equations, using the transformation coefficients programmed by the user. The 28-bit width of the accumulator allows single-pixel resolution with second- order coefficients within a 1024 x 1024 image, utilizing the 1/16 pixel spatial resolution of the coefficient addresses.

Below is an example listing the X generator's outputs during the nearest neighbor sampling of a 5 x 3 output image. For a kernel greater than 1, there would also be a walk around each of the data points.

The U (V) address generator is a simple counter. In the horizontal device, it increments at the end of each kernel walk and resets at the end of each row, running from UMIN to UMAX. The vertical device increments at the end of each row and resets at the end of each page, and runs from VMIN to VMAX.

The walk counter is used in all modes except nearest neighbor resampling (where each output value is based on only one input value). In single-pass resampling or single-pass filtering, the device first identifies the (X_0, Y_0) coordinate pair cooresponding in position to the desired (U_0, V_0) output point. The index generator then takes over, stepping the TMC2301 through all (X, Y) in the kernel surrounding (X_0, Y_0), in an outward clockwise spiral. During the spiral walk, it outputs an upward binary count on the CA (Coefficient Address) output port.

Timing

After the INIT input is pulsed, the TMC2301 begins a resampling sequence, according to the address parameters and transformation coefficients programmed by the user. The first address is output two clock cycles later, due to the internal latency of the X address generators two-level accumulator. Subsequent activities are continuous unless the NOOP pin is asserted, in which case the clock is disabled internally and the source image and coefficient lookup table address outputs are disabled. This allows user access to both memory blocks for examination or modification of data, system timing synchronization, or the insertion of delays for slow memory access. Normal operation resumes when NOOP returns LOW, on the next clock.

The resampling parameters' double-buffered storage and dedicated loading port allow them to be updated "on the fly," as often as one parameter per clock cycle, if desired. Typically, the user will preload the desired set of parameters and bring the LDR control HIGH at the start of the transform. During the transform, the user can then preload the next transform's parameters. When the AUTOLOAD (ALR) feature is utilized, the new coefficient load occurs automatically. If the ALR bit is not set, the TMC2301 will continue to use the old parameters until the LDR control is asserted. In applications doing nearest neighbor resampling, the new parameter UMIN is not loaded internally until the sequence begins the second row of the transform. In this situation, potential errors can be avoided by loading the parameters at least one clock prior to the end of the last image, which could be implemented by LDR = NAND (ENDROW, ENDCOL), on both TMC2301s.

At the completion of an image transformation and after the end of the last pixel walk, the address counters have reached the programmed UMAX/VMAX locations in the target image. The DONE flag output on the row (X) device goes HIGH, the final (U, V) address is output, and UWRI is asserted. On the next clock, the counters are reinitialized and a new sequence begins, as defined by the current transform parameters.

Uses of the TMC2301

Filtering

In its basic mode, the TMC2301 governs static filtering, in which the intensity value of each point in a given image is replaced by the convolutional sum of that point's N x N nearest neighbors with a kernel of N x N coefficients. In this mode, the TMC2301 generates N x N read addresses for every write address. Typically, for each convolutional sum, the new pixel address is identical to the address of the center pixel of the source image kernel. If desired, the user may combine this operation with a translation or other remapping operation, such that the kernel center address and target memory write address would differ by the chosen displacement. With its 8-bit coefficient address generator, the TMC2301 will support convolutional kernels of up to 16 x 16 pixels. System throughput, or the rate at which the TMC2301 system completes image transformations, falls geometrically with kernel size due to the number of pixels processed during each summation. Numerous performance improvements are possible, using memory banking/pixel offset schemes. These systems may utilize the TMC2301 pair solely as an address generator, implementing the pixel summation with additional external arithmetic circuitry.

Filtering Using a 3 x 3 Pixel Kernel

For each output point in typical static single-pass filtering, the IRS pair will generate a series of addresses which "walk" in two dimensions around that point. At the end of each walk, it will advance one pixel in the U direction and begin the walk for the next pixel. A portion of a 3 x 3 filtering sequence is shown below:

(X, Y)	(U, V)	Comment
2, 2	1, 2	center; start first walk
3, 2	1, 2	center right
3, 3	1, 2	bottom right
2, 3	1, 2	bottom center
1, 3	1, 2	bottom left
1, 2	1, 2	center left
1, 1	1, 2	top left.
2, 1	1, 2	top center

Table continues on next page.

(X, Y)	(U, V)	Comment
3, 1	1, 2	top right; end first walk
3, 2	2, 2	center; start second walk
4, 2	2, 2	center right
4, 3	2, 2	bottom right
3, 3	2, 2	bottom center
2, 3	2, 2	bottom left
2, 2	2, 2	center left
2, 1	2, 2	top left
3, 1	2, 2	top center
4, 1	2, 2	top right; end second walk
4, 2	3, 2	center; start third walk
...	...	...
5, 1	3, 2	top right; end third walk
5, 2	4, 2	begin fourth walk, write new pixel

Note of course that the U, V outputs are not updated to indicate the address of the pixel being calculated until after the end of the walk.

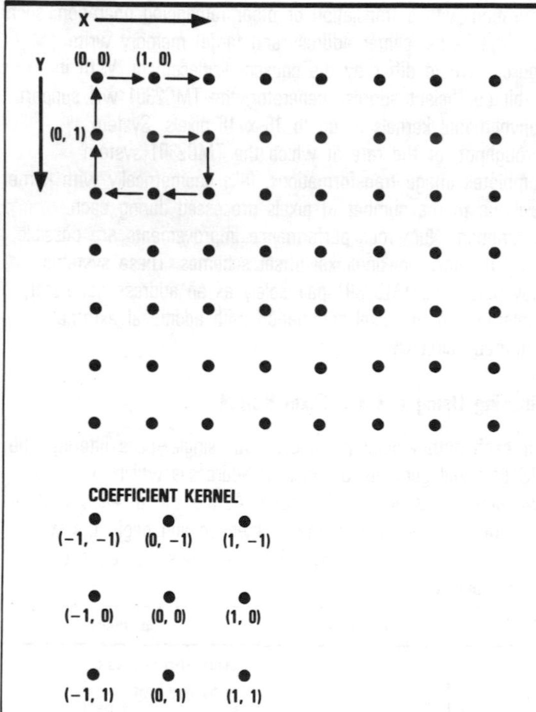

Figure 4. Pixel Map Showing 3 x 3 Filtering, One−Pass

In detail, the coefficient kernel is calculated as follows:

$$z(u_h, v_i) = \sum_{j=-1}^{+1} \sum_{k=-1}^{+1} z(x_{h+j}, y_{i+k})K(j, k)$$

and the first two kernels are calculated as follows:

$$\begin{aligned} z'(1, 1) = {} & z(1, 1)K(0, 0) + z(2, 1)K(1, 0) + z(2, 2)K(1, 1) \\ & + z(1, 2)K(0, 1) + z(0, 2)K(-1, 1) + z(0, 1)K(-1, 0) \\ & + z(0, 0)K(-1, -1) + z(1, 0)K(0, -1) + z(2, 0)K(1, -1) \end{aligned}$$

$$\begin{aligned} z(2, 1) = {} & z(2, 1)K(0, 0) + z(3, 1)K(1, 0) + z(3, 2)K(1, 1) \\ & + z(2, 2)K(0, 1) + z(1, 2)K(-1, 1) + z(1, 1)K(-1, 0) \\ & + z(1, 0)K(-1, -1) + z(2, 0)K(0, -1) + z(3, 0)K(1, -1) \end{aligned}$$

Resampling

In many applications, the new target image pixels do not overlap the source image locations exactly. The other mode of the TMC2301 remaps or transforms each new pixel coordinate pair (U, V) to its corresponding location in the original (X, Y) space. But, since this (X, Y) location is generally not one of the exact pixel locations or integral grid points of the original image, the new pixel value must be obtained by interpolation from the neighboring (X, Y) pixel values. This is calculated by executing a user−defined pixel "walk," as is implemented in simple filtering. See the examples below.

Since the (U, V) locations do not map to the (X, Y) addresses exactly, fractional address components are generated. The user then must account for this spatial offset in both dimensions by storing the appropriate corrected interpolation kernel coefficient values in the lookup table. The 8−bit coefficient address bus is broken up into two parts: the fractional portion (upper 4 bits), and the walk counter (lower 4 bits). Thus, in resampling applications, the maximum kernel size is 4 x 4 pixels, or 16 locations. Assuming that the user has selected the same kernel size for both IRS devices, the 4 bits of least−significant address generated by both devices will be identical, and redundant. The four most significant address bits, however, will reflect the current fractional offsets of the resampled pixel from the nearest X (Y) location, to a spatial resolution of 4 bits, in the X (or Y) directions. Utilization of the 12 bits of lookup table address is left to the user, to be arranged as desired for coefficient memory access.

Resampling Using the Nearest Neighbor Algorithm

Nearest neighbor resampling is the simplest and fastest technique. No interpolation is performed; the TMC2301 simply computes the (X, Y) location corresponding to each (U, V), and outputs only the integer portion of the address. The system then merely calls up the intensity value of that (X, Y) point and copies it into the memory location corresponding to (U, V) without the use of a multiplier−accumulator. Spatial truncation can be avoided by adding one−half pixel spacing to the source image starting points, effecting address rounding. The calculations for X generated by the TMC2301 during a nearest neighbor algorithm utilizing the various transformation coefficients are as follows:

(U, V)	X Address Calculation
(UMIN, VMIN)	X_0
(UMIN + 1, VMIN)	$X_0 + dX/dU_0$
(UMIN + 2, VMIN)	$X_0 + 2dX/dU_0 + d^2X/dU^2$
(UMIN + 3, VMIN)	$X_0 + 3dX/dU_0 + 3d^2X/dU^2$
(UMIN + 4, VMIN)	$X_0 + 4dX/dU_0 + 6d^2X/dU^2$
(UMIN, VMIN + 1)	$X_0 + dX/dV_0$
(UMIN + 1, VMIN + 1)	$X_0 + dX/dV_0 + dX/dU_0 + d^2X/dUdV$
(UMIN + 2, VMIN + 1)	$X_0 + dX/dV_0 + 2dX/dU_0 + d^2X/dU^2 + 2d^2X/dUdV$
(UMIN + 3, VMIN + 1)	$X_0 + dX/dV_0 + 3dX/dU_0 + 3d^2X/dU^2 + 3d^2X/dUdV$
(UMIN + 4, VMIN + 1)	$X_0 + dX/dV_0 + 4dX/dU_0 + 6d^2X/dU^2 + 4d^2X/dUdV$
(UMIN, VMIN + 2)	$X_0 + 2dX/dV_0 + d^2X/dV^2$
(UMIN + 1, VMIN + 2)	$X_0 + 2dX/dV_0 + d^2X/dV^2 + dX/dU_0 + 2d^2X/dUdV$
(UMIN + 2, VMIN + 2)	$X_0 + 2dX/dV_0 + d^2X/dV^2 + 2dX/dU_0 + d^2X/dU^2 + 4d^2X/dUdV$
(UMIN + 3, VMIN + 2)	$X_0 + 2dX/dV_0 + d^2X/dV^2 + 3dX/dU_0 + 3d^2X/dU^2 + 6d^2X/dUdV$
(UMIN + 4, VMIN + 2)	$X_0 + 2dX/dV_0 + d^2X/dV^2 + 4dX/dU_0 + 6d^2X/dU^2 + 8d^2X/dUdV$

Resampling Using Bilinear Interpolation

Using bilinear interpolation, each output pixel intensity is computed from those of its four nearest neighbors in the original image. This requires a local "walk," analogous to that used in filtering. In this algorithm, the device pair first outputs the (x, y) coordinates of the original pixel to the upper left of the desired output point, then those to the upper right, then the lower right, and finally the lower left, completing the cycle. The external multiplier–accumulator multiplies each of these intensities by the corresponding kernel coefficient and computes the sum of the four products.

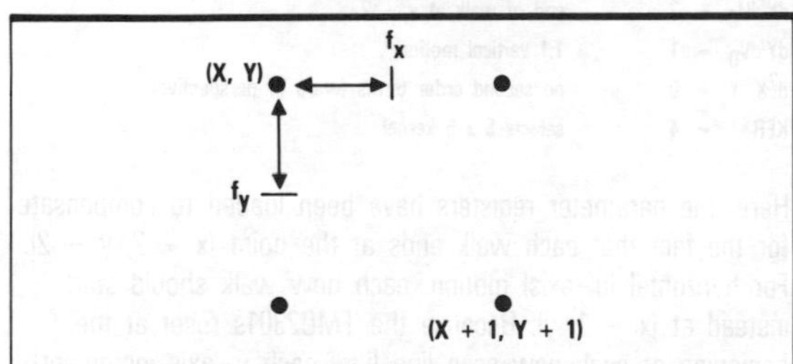

Figure 5. Bilinear Interpolation Kernel Demonstrating Fractional Address Components

$$z'(u, v) = z(x, y)K_1 + z(x+1, y)K_2 + z(x+1, y+1)K_3 + z(x, y+1)K_4$$

Procedure:

One – Compute where the point (u, v) would lie in the (x, y) plane (coordinate transformation).

Two – The points of the square surrounding that location are:

(INT(X), INT(Y)) (1 + INT(X), INT(Y))
(INT(X), 1 + INT(Y)) (1 + INT(X), 1 + INT(Y)),

where INT(A) is the integer portion of A.

Three – The coefficients for each of the data points are:

$$K_1 = (1 - f_x)(1 - f_y)$$
$$K_2 = (f_x)(1 - f_y)$$
$$K_3 = (f_x)(f_y)$$
$$K_4 = (1 - f_x)(f_y)$$

where f_x, f_y are the fractional parts of the address of the new point.

A typical pixel walk sequence for a Bilinear Interpolation implemented by the TMC2301 is:

(X, Y)	(U, V)	Comment
3, 5	1, 2	upper left; start first walk
4, 5	1, 2	upper right
4, 6	1, 2	lower right
3, 6	1, 2	lower left; end first walk
7, 8	2, 2	upper left; start second walk
8, 8	2, 2	upper right
8, 9	2, 2	lower right
7, 9	2, 2	lower left; end second walk
11, 11	3, 2	upper left; start third walk
12, 11	3, 2	upper right
12, 12	3, 2	lower right
11, 12	3, 2	lower left
	4, 2	final target address

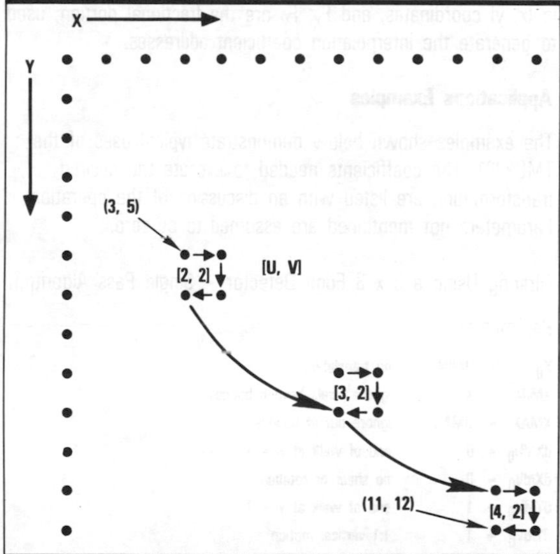

Figure 6. Pixel Map Showing Rotation and Compression

In this example, the image undergoes a 37 degree rotation and a 5:1 linear compression. (Each single pixel motion along the u axis creates a corresponding 4–pixel motion along the x axis and 3 along the y, defining a 3–4–5 right triangle). See the application example on page 9.

Resampling Using Cubic Convolution

Perhaps the most popular resampling algorithm utilizes a group of 16 pixels. In cubic convolution, the nearest 4 x 4 pixel field is resampled and multiplied against the coefficients to generate the new pixel value.

X, Y_{11-0}	CA_{7-0}	Internal U, V	
(I_x0, I_y0)	(F_x0, F_y0)	(u0, v0)	start walk 1
$(I_x0 + 1, I_y0)$	$(F_x0 + 1, F_y0 + 1)$	(u0, v0)	
$(I_x0 + 1, I_y0 + 1)$	$(F_x0 + 2, F_y0 + 2)$	(u0, v0)	
$(I_x0, I_y0 + 1)$	$(F_x0 + 3, F_y0 + 3)$	(u0, v0)	
$(I_x0 - 1, I_y0 + 1)$	$(F_x0 + 4, F_y0 + 4)$	(u0, v0)	
$(I_x0 - 1, I_y0)$	$(F_x0 + 5, F_y0 + 5)$	(u0, v0)	
$(I_x0 - 1, I_y0 - 1)$	$(F_x0 + 6, F_y0 + 6)$	(u0, v0)	
$(I_x0, I_y0 - 1)$	$(F_x0 + 7, F_y0 + 7)$	(u0, v0)	
$(I_x0 + 1, I_y0 - 1)$	$(F_x0 + 8, F_y0 + 8)$	(u0, v0)	
$(I_x0 + 2, I_y0 - 1)$	$(F_x0 + 9, F_y0 + 9)$	(u0, v0)	
$(I_x0 + 2, I_y0)$	$(F_x0 + 10, F_y0 + 10)$	(u0, v0)	
$(I_x0 + 2, I_y0 + 1)$	$(F_x0 + 11, F_y0 + 11)$	(u0, v0)	
$(I_x0 + 2, I_y0 + 2)$	$(F_x0 + 12, F_y0 + 12)$	(u0, v0)	
$(I_x0 + 1, I_y0 + 2)$	$(F_x0 + 13, F_y0 + 13)$	(u0, v0)	
$(I_x0, I_y0 + 2)$	$(F_x0 + 14, F_y0 + 14)$	(u0, v0)	
$(I_x0 - 1, I_y0 + 2)$	$(F_x0 + 15, F_y0 + 15)$	(u0, v0)	end walk 1
(I_x1, I_y1)	(F_x1, F_y1)	(u1, v1)	start walk 2

where I_X, I_Y are the integer portion of the new pixel address in (x, y) coordinates, and F_X, F_Y are the fractional portion, used to generate the interpolation coefficient addresses.

Applications Examples

The examples shown below demonstrate typical uses of the TMC2301. The coefficients needed to execute the desired transformation are listed with an discussion of the operation. Parameters not mentioned are assumed to be zero.

Filtering Using a 3 x 3 Edge Detector – Single Pass Algorithm

Parameters:

X_0	= UMIN	no translation
XMIN	= X_0	ignore pixels beyond bounds
XMAX	= UMAX	ignore out of bounds
dX/dU_0	= 0	end of walk at x + 1
dX/dV_0	= 0	no shear or rotation
dY/dU_0	= 1	end of walk at y + 1
dY/dV_0	= 1	1:1 vertical motion
d^2X, Y	= 0	no second order terms (warp or perspective)
KER	= 2	selects 3 x 3 kernel

A typical vertical edge detection kernel would be:

−1	0	+1
−1	0	+1
−1	0	+1

Convolution of this kernel with a subimage will generate a large negative or positive score if a pronounced vertical edge is encountered. In contrast, a horizontal edge will not contribute to the output. Because of the single pass spiral, this kernel would be stored in the lookup table as follows:

Address		Coefficient
Index CA_{7-4}	Fraction CA_{3-0}	
0000	0000	0
0001	0000	+1
0010	0000	+1
0011	0000	0
0100	0000	−1
0101	0000	−1
0110	0000	−1
0111	0000	0
1000	0000	+1

Filtering Using a 5 x 5 Low Pass Filter

1. One–Pass Algorithm

Parameters:

X_0	= UMIN	no translation
XMIN	= X_0	ignore pixels beyond bounds
XMAX	= UMAX	ignore out of bounds
dX/dU_0	= −1	end of walk at x + 2 instead of x + 1
dX/dV_0	= 0	no shear or rotation
dY/dU_0	= 2	end of walk at y − 2
dY/dV_0	= 1	1:1 vertical motion
d^2X, Y	= 0	no second order terms (warp or perspective)
KER	= 4	selects 5 x 5 kernel

Here, the parameter registers have been loaded to compensate for the fact that each walk ends at the point (x + 2, y − 2). For horizontal (u–axis) motion, each new walk should start instead at (x + 1, y). Because the TMC2301s reset at the beginning of each new scan line (i.e., each v–axis increment), no correction is required for the dX/dV and dY/dV terms.

A typical 5 x 5 kernel low–pass filter kernel might be:

1	2	4	2	1
2	4	8	4	2
4	8	16	8	4
2	4	8	4	2
1	2	4	2	1

Because of the spiral walk, these 25 values are loaded into coefficient memory as follows:

Address		Coefficient
Index CA_{7-4}	Fraction CA_{3-0}	
0000	0000	16
0001	0000	8
0010	0000	4
0011	0000	8
0100	0000	4
0101	0000	8
0110	0000	4
0111	0000	8
1000	0000	4
1001	0000	2
1010	0000	4
1011	0000	2
1100	0000	1
1101	0000	2
1110	0000	4
1111	0000	2
0000	0001	1
0001	0001	2
0010	0001	4
0011	0001	2
0100	0001	1
0101	0001	2
0110	0001	4
0111	0001	2
1000	0001	1

Note that the pixel walk address count continues into the fraction field after the 4-bit index field is exhausted.

2. Two-Pass Algorithm

First pass: horizontal motion along the u axis, replacing the value at (x, y) by the weighted sum of the values at (x − 2, y), (x − 1, y), (x, y), (x + 1, y), and (x + 2, y).

Parameters:

(X_0, Y_0)	= (UMIN − 2, VMIN)	compensate for address resets
(XMIN, YMIN)	= $(X_0 + 2, Y_0)$	ignore pixels beyond bounds
(XMAX, YMAX)	= (UMAX, VMAX)	ignore out of bounds
dX/dU_0	= 3	end of walk at (x + 4, y), instead of (x + 1, y)
dX/dV_0	= 0	no shear or rotation
dY/dU_0	= 0	end of walk in same row
dY/dV_0	= 0	1:1 vertical motion
d^2X, Y	= 0	no second order terms (warp or perspective)
KER	= 4	5 x 5 kernel

Second pass: vertical motion along the v axis, replacing the value at (x, y) by the weighted sum of the values at (x, y − 2), (x, y − 1), (x, y), (x, y + 1), and (x, y + 2).

Parameters:

(X_0, Y_0)	= (UMIN, VMIN − 2)	compensate for address resets
(XMIN, YMIN)	= $(X_0, Y_0 + 2)$	ignore pixels beyond bounds
(XMAX, YMAX)	= (UMAX, VMAX)	ignore out of bounds
dX/dU_0	= 1	end of walk at x instead of x + 1
dX/dV_0	= 0	no shear or rotation
dY/dU_0	= −4	end of walk at y + 4 instead of y
dY/dV_0	= 0	1:1 vertical motion
d^2X, Y	= 0	no second order terms (warp or perspective)
KER	= 4	5 x 5 kernel

Here, a small 5-element kernel can be used twice, since the 25-element kernel of our one pass example is seperable (factorable along the x and y axes).

Address		Coefficient
Index CA_{7-4}	Fraction CA_{3-0}	
0000	0000	1
0001	0000	2
0010	0000	4
0011	0000	2
0100	0000	1

If the kernel is not seperable, a single pass algorithm must be employed.

Resampling Applications with Zeroeth Order Terms

Copy, Using Nearest Neighbor Algorithm

Parameters:

(X_0, Y_0)	= (UMIN, VMIN)	no translation
(XMIN, YMIN)	= (X_0, Y_0)	ignore all pixels outside original image
(XMAX, YMAX)	= (UMAX, VMAX)	ignore anything outside original image
dX/dU_0	= 1	move 1 pixel along x per pixel along u
dX/dV_0	= 0	no x motion for v motion (no rotation or shear)
dY/dU_0	= 0	no y motion for u motion (no rotation or shear)
dY/dV_0	= 1	move 1 pixel along y per pixel along v
d^2X, Y	= 0	no second-order terms (no warp or perspective)
KER	= 0	no pixel walk

To estimate each z'(u, v), simply take the z of the nearest (x, y).

Copy, Using One-Pass Cubic Convolution

Parameters:

(X_0, Y_0)	=	(UMIN, VMIN)	no translation
(XMIN, YMIN)	=	(X_0, Y_0)	ignore all pixels outside original image
(XMAX, YMAX)	=	(UMAX, VMAX)	ignore anything outside original image
dX/dU_0	=	2	end of walk at x + 3 instead of x + 1
dX/dV_0	=	0	no x motion for v motion (no rotation or shear)
dY/dU_0	=	−2	end of walk at y + 2 instead of y
dY/dV_0	=	1	move 1 pixel along y per pixel along v
d^2X, Y	=	0	no second-order terms (no warp or perspective)
KER	=	3	4 x 4 kernel

Since each walk begins at (x_0, y_0) and ends at $(x_0 + 1, y_0 + 2)$,

dX/dU_0 must be increased to 2, and dY/dU_0 must by decreased to −2.

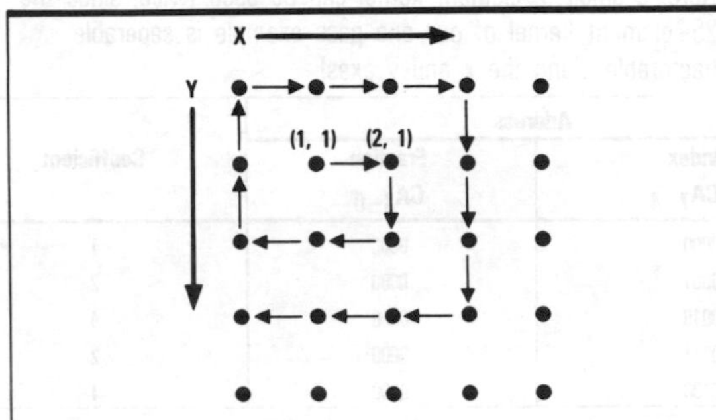

Figure 7. Pixel Walk Showing End−of−Walk Offset

Reflect Around Vertical Axis, Using Nearest Neighbor

Parameters:

(X_0, Y_0)	=	(7, 0)	set vertical axis
(UMIN, VMIN)	=	(0, 0)	start at upper left corner
dX/dU_0	=	−1	move 1 pixel left on x for every pixel right on u
dX/dV_0	=	0	no x motion for v motion (no rotation or shear)
dY/dU_0	=	0	no y motion for u motion (no rotation or shear)
dY/dV_0	=	1	move 1 pixel along y per pixel along v
d^2X, Y	=	0	no second-order terms (no warp or perspective)
KER	=	0	nearest neighbor

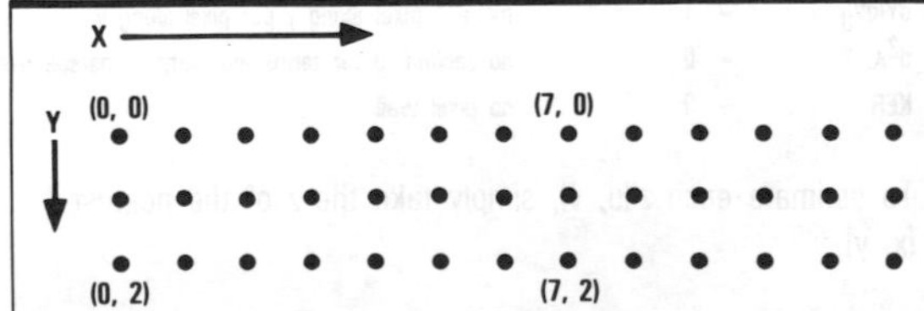

Figure 8. Pixel Map Demonstrating Reflection

$z'(u, v) = z(x, y)$, where $u = (7 - x)$ and $v = y$

Reflect About Vertical Axis, Using Bilinear Interpolation

Parameters:

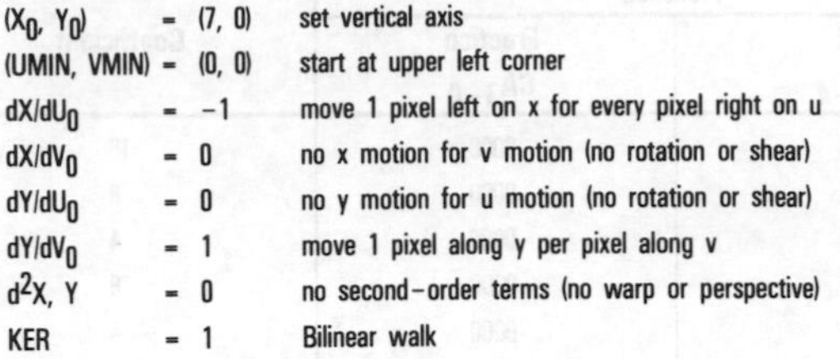

(X_0, Y_0)	=	(7, 0)	set vertical axis
(UMIN, VMIN)	=	(0, 0)	start at upper left corner
dX/dU_0	=	−1	move 1 pixel left on x for every pixel right on u
dX/dV_0	=	0	no x motion for v motion (no rotation or shear)
dY/dU_0	=	0	no y motion for u motion (no rotation or shear)
dY/dV_0	=	1	move 1 pixel along y per pixel along v
d^2X, Y	=	0	no second-order terms (no warp or perspective)
KER	=	1	Bilinear walk

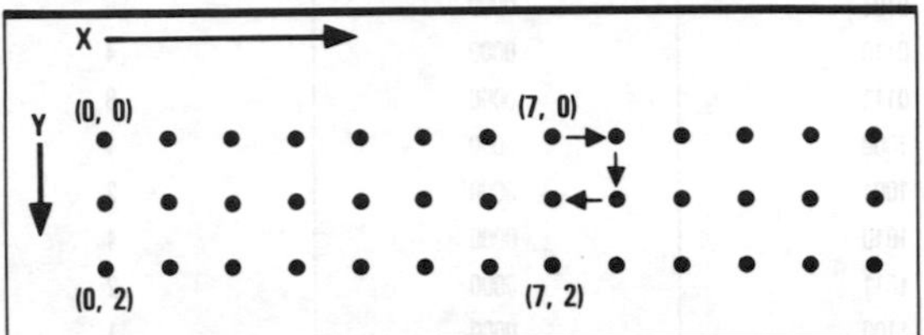

Figure 9. Pixel Map Showing Reflection Using 2 x 2 Kernel

Since each walk begins where the previous one ends, at $(x_0, y_0 + 1)$, compensate in the derivative terms.

Translate (Pan) P Pixels Rightward

Parameters:

$(X_0 - P, Y_0)$	=	(UMIN, VMIN)	puts contents of (x − P, y) at (u, v)
(XMIN, YMIN)	=	(X_0, Y_0)	ignore all pixels outside original image
(XMAX, YMAX)	=	(UMAX, VMAX)	ignore all pixels outside original image
dX/dU_0	=	1	move 1 pixel along x per pixel along u
dX/dV_0	=	0	no x motion for v motion (no rotation or shear)
dY/dU_0	=	0	no y motion for u motion (no rotation or shear)
dY/dV_0	=	1	move 1 pixel along y per pixel along v
d^2X, Y	=	0	no second-order terms (no warp or perspective)

This algorithm uses nearest-neighbor resampling to translate an image. For larger kernels, first differential terms must be adjusted to make up for the change in walk endpoints.

Resampling with First Order Terms

Compression Using Nearest Neighbor

(X_0, Y_0)	=	(0, 0)	
(XMIN, YMIN)	=	(0, 0)	
(XMAX, YMAX)	=	(1023, 1023)	
(UMIN, VMIN)	=	(412, 412)	
(UMAX, VMAX)	=	(611, 611)	
dX/dU_0	=	5.12	horizontal compression
dX/dV_0	=	0	no rotation or shear
dY/dU_0	=	0	no rotation or shear
dY/dV_0	=	5.12	vertical compression
d^2X, Y	=	0	no second-order terms (warp or perspective)
KER	=	0	nearest neighbor

These coefficients will execute a 5.12:1 compression in x and y, from 1024 x 1024 pixels to 200 x 200 pixels, with no change in image center.

Expand (Zoom) Using Nearest Neighbor

(X_0, Y_0)	= (160, 240)	to maintain centering at (200, 300)
(XMIN, YMIN)	= (0, 0)	
(XMAX, YMAX)	= (511, 511)	
(UMIN, VMIN)	= (0, 0)	
(UMAX, VMAX)	= (600, 600)	
dX/dU_0	= 0.2	5:1 horizontal expansion
dX/dV_0	= 0	no rotation or shear
dY/dU_0	= 0	no rotation or shear
dY/dV_0	= 0.2	5:1 vertical expansion
d^2X, Y	= 0	no second-order terms (warp or perspective)
KER	= 0	nearest neighbor

Note that the starting point for this transformation is offset to keep the new image centered at the desired location, in this case at the point (200, 300) on a 600 x 600 pixel image.

Expansion, Using Bilinear Interpolation

(X_0, Y_0)	= (160, 240)	to maintain centering at (200, 300)
(XMIN, YMIN)	= (0, 0)	
(XMAX, YMAX)	= (511, 511)	
(UMIN, VMIN)	= (0, 0)	
(UMAX, VMAX)	= (600, 600)	
dX/dU_0	= 0.2	5:1 horizontal expansion
dX/dV_0	= 0	no rotation or shear
dY/dU_0	= −1	no rotation or shear
dY/dV_0	= 0.2	5:1 vertical expansion
d^2X, Y	= 0	no second-order terms (warp or perspective)
KER	= 1	bilinear interpolation

Changing to a bilinear interpolation, the term dY/dU_0 has been changed to bring the walk endpoint to the correct location.

Expansion, Using Single-Pass Cubic Convolution

(X_0, Y_0)	= (160, 240)	to maintain centering at (200, 300)
(XMIN, YMIN)	= (0, 0)	
(XMAX, YMAX)	= (511, 511)	
(UMIN, VMIN)	= (0, 0)	
(UMAX, VMAX)	= (600, 600)	
dX/dU_0	= 1.2	5:1 horizontal expansion
dX/dV_0	= 0	no rotation or shear
dY/dU_0	= −2	no rotation or shear
dY/dV_0	= 0.2	5:1 vertical expansion
d^2X, Y	= 0	no second-order terms (warp or perspective)
KER	= 1	bilinear interpolation

Larger kernels require careful management of the differentials to ensure that the walks end at the next (U, V) location.

Rotation and Compression, Using Bilinear Interpolation

Parameters:

dX/dU_0	= 4	move 4 horizontally for every horizontal step
dX/dV_0	= −3	move 3 left for every for every vertical step
dY/dU_0	= 2	move 3 down for every horizontal step
dY/dV_0	= 4	move 4 down for every vertical step
KER	= 1	bilinear interpolation

In this operation, the image is compacted 5:1 linearly and rotated 37 degrees ($\tan^{-1}$ 3/4) counterclockwise. The upper left corner is translated 1 pixel left and 3 pixels down. Note that the dY/dU_0 term includes the single step down contributed by the pixel walk.

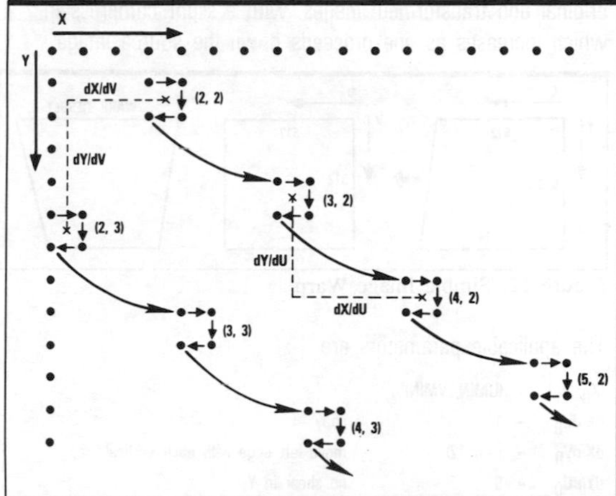

Figure 10. Pixel Map Showing Rotation and Compression

Compensation of the Center Position for Rotation

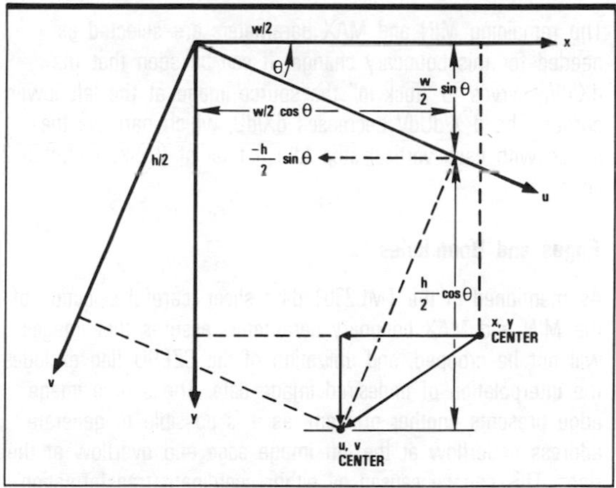

Figure 11. Geometry of Image Rotation Correction

The center of the new image (u, v) moves from (w/2, h/2) to ((w/2)cos θ − (h/2)sin θ, (w/2)sin θ + (h/2)cos θ)).

To align the (x, y) center with the (u, v) center, move (x_0, y_0):

$x_0 = u_{min} + (w/2)(1 - \cos\theta) + (h/2)\sin\theta$, and
$y_0 = v_{min} + (h/2)(1 - \cos\theta) - (w/2)\sin\theta$

This same logic applies for rotation around any desired point in the (x, y) plane.

Resampling with Second Order Terms

Pseudoperspective Warp Using Nearest Neighbor

This simplified example of a common application aligns the original and transformed images, with a slight compression which increases as one proceeds down the source image.

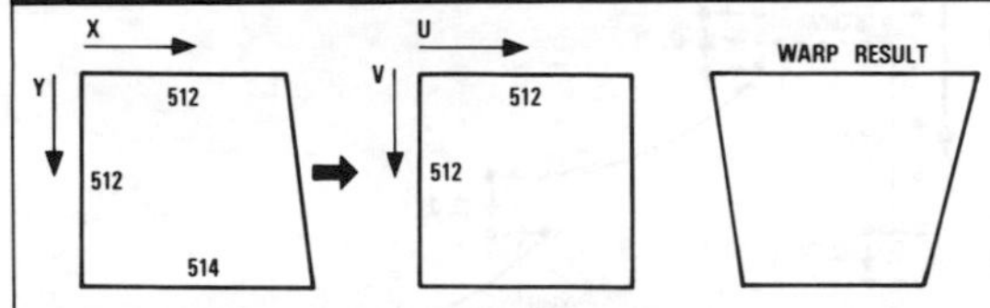

Figure 12. Simple Image Warp

The applicable parameters are:

(X_0, Y_0)	=	(UMIN, VMIN)	
dX/dU_0	=	1	copy
dX/dV_0	=	(−1/512)	move left edge with each vertical step
dY/dU_0	=	0	no shear in Y
dY/dV_0	=	1	copy
$d^2X/dUdV$	=	1/(256*512)	warp X
KER	=	0	nearest neighbor

The remaining MIN and MAX parameters are selected as needed for this boundary change. It can be seen that the dX/dV_0 serves to "tuck in" the source image at the left lower corner. The $d^2X/dUdV$ decreases dX/dU, which narrows the image with each vertical step (dV) a total of 2 pixels after 512 lines.

Edges and Boundaries

As mentioned in the TMC2301 data sheet, careful selection of the MIN and MAX boundary parameters assures that images will not be cropped, and utilization of the CZERO flag excludes the interpolation of undesired image data. The source image edge presents another problem, as it is possible to generate address underflow at the left image edge and overflow at the right. This can be caused by either coordinate transformation calculations or interpolation walks which generate addresses outside the source image extremities. The first solution is careful management of the transformation algorithm, so that this image boundary violation never occurs. Additionally, the user can specify boundary parameters which avoid the problem. For instance, setting $X_0 > XMIN \geqslant 0$, with $X_0 = XMIN + (Kernel + 1)/2$, will assure that an interpolation walk will not step over the left edge of the source image. Including a slight zoom factor in the coordinate transformation algorithm recovers this slight loss in image size, if desired. Again, utilization of the CZERO flag is the most basic tool in managing illegal image address situations.

Appendix 1. General Algorithm Development Procedure

This section details how the user can select and load the parameter registers for any combination of three lower-order operations: translation (pan), magnification (zoom or compress), and rotation.

One – Select the kernel size (0 through 15), number of passes (1 or 2), and pass number (1 or 2, for 2-pass only). K selects a total transform kernel size of (K + 1)(K + 1), per Table 1.

Table 1. Kernel Size: Input Points Per Output Point

K	One-Pass N	Two-Pass (Per Pass)	Typical Application
0	1	--	nearest neighbor
1	4	2	bilinear interp
2	9	3	3 x 3 filter
3	16	4	cubic convolun.
4	25	5	5 x 5 filter
...	...	...	
15	256	16	16 x 16 filter

In single-pass operation, each output pixel's intensity will be a linear combination of the intensities of the N nearest neighbors in input image space. In each pass of a two-pass operation, each output pixel's intensity will be a linear combination of the intensities of the corresponding input point and the next K input pixels (rightward/downward) on the same (horizontal/vertical) scanline (Figure 13). Therefore, most two-pass transformations require a kernel-size-dependent pan correction (step 4 below) to recenter each output pixel onto its input kernel. In addition, most transformations other than nearest neighbor require kernel-size-dependent first difference corrections (step 5 below), because the IRS references the starting point of each new kernel to the ending point of the previous kernel.

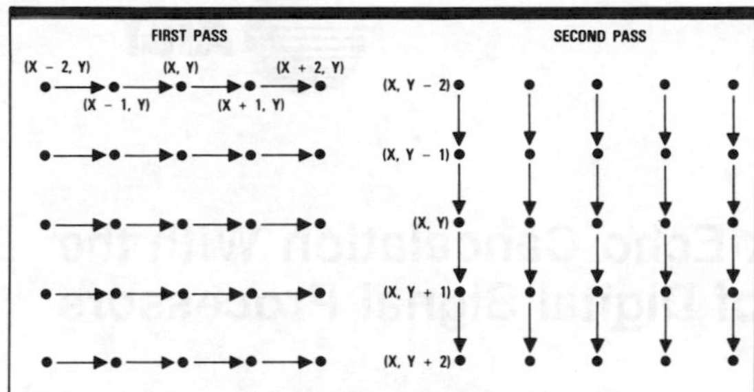

Figure 13. Two−Pass 5 x 5 Static Filter

Two – Set desired output image boundaries. The TMC2301 set will generate write addresses only within these bounds, leaving the rest of the output image memory unaltered.

(UMIN, VMIN) sets the upper left vertex of the (rectangular) output image field.

(UMAX, VMAX) sets the lower right vertex.

Three – Set the input image boundaries. The coefficient zero (CZ) flag of the TMC2301 will go high anytime a pixel outside of these boundaries is addressed, telling the typical system to ignore these values.

(XMIN, YMIN) sets the upper left vertex of the (rectangular) input image field.

(XMAX, YMAX) sets the lower right vertex.

Four – Set the translation of the upper left vertex. The IRS counts upward in two dimensions from this origin, from which all translations and rotations are referred.

(X_0, Y_0) = (UMIN−TX−PX, VMIN−TY−PY) defines a translation of this vertex by TX pixels rightward and TY pixels downward (in source image space). Pan compensation terms PX and PY, which recenter each output pixel on its input pixel ensemble, are set according to Table 2.

Table 2. Input Pan Compensation

	One Pass		Two Pass			
			Horizontal		Vertical	
K	PX	PY	PX	PY	PX	PY
0	0	0	0	0	0	0
1	0	0	0	0	0	0
2	0	0	1	0	0	1
3	0	0	1	0	0	1
4	0	0	2	0	0	2
Odd	0	0	(K − 1)/2	0	0	(K − 1)/2
Even	0	0	K/2	0	0	K/2

The diagrams of 3 x 3 filtering should clarify the need for the pan compensator. In single pass operation, for each output point the TMC2301 correctly selects the corresponding input image point and its eight nearest neighbors (Figure 1). However, in two pass operation, the output point would ordinarily be aligned with the upper left of these 9 input points, rather than the (desired) center one; unlike the single pass case, the two passes are not centered. Subtracting 1 from each X address on the horizontal pass and 1 from each Y address on the vertical pass moves the input sampling area to the proper location.

Five – Set the rotation and magnification about the upper left vertex of the output image.

$dX/dU_0 = M \cos R - KX$ $dY/dU_0 = M \sin R - KY$

$dX/dV_0 = -M \sin R$ $dY/dV_0 = M \cos R$

$0 < $ Magnification factor $M < 1$ denotes magnification (zoom). $1 < M$ denotes compression by a factor of M.

R is the angle of counterclockwise rotation of the image about the upper left vertex of the output image.

Kernel size compensators KX and KY are set according to Table 3.

Table 3. First Difference Kernel Compensators

	One Pass		Two Pass			
			Horizontal		Vertical	
K	KX	KY	KX	KY	KX	KY
0	0	0	0	0	0	0
1	0	1	1	0	0	1
2	1	−1	2	0	0	2
3	−1	2	3	0	0	3
4	2	−2	4	0	0	4
Odd	(1 − K)/2	(1 + K)/2	K	0	0	K
Even	K/2	−K/2	K	0	0	K

Application Note

Network Echo Cancelation With the *WE*® DSP16 Family of Digital Signal Processors

Contents

Contributed by: *S. L. Gay,*
Member of Technical Staff
AT&T Bell Laboratories

1. Introduction

This application note describes the implementation of a network echo canceler using the *WE* DSP16 or DSP16A Digital Signal Processor [1].* The echo canceler is based on the leaky normalized least mean square (NLMS) adaptive filter with double-precision coefficients. When sampling at 8 kHz and using the DSP16 at 36 MHz, echo path lengths of up to 18 ms can be canceled. If single-precision coefficients without tap leakage are used, then echo path lengths of up to 29 ms can be canceled. Using the DSP16A at 60 MHz, echo path lengths of up to 32 ms or 53 ms can be canceled, depending on whether or not double-precision coefficients with tap leakage are used.

The *WE* DSP16 Digital Signal Processor Demonstration Board (DEMO16), described in a companion application note [2], is used as an evaluation platform. DEMO16 provides for the interfacing of both the network- and hybrid-side signals as well as an RS-232C port for the real-time display and control of the coefficients of the adaptive filter.

First, a brief description of the network echo problem and how echo cancelers are used to alleviate it is given [3]. Next, a description of the algorithms that comprise the echo canceler are presented [4,5,6]. Then, the hardware and software implementation of a 16 ms echo canceler is

* [ ] indicates a reference listed at the conclusion of this application note.

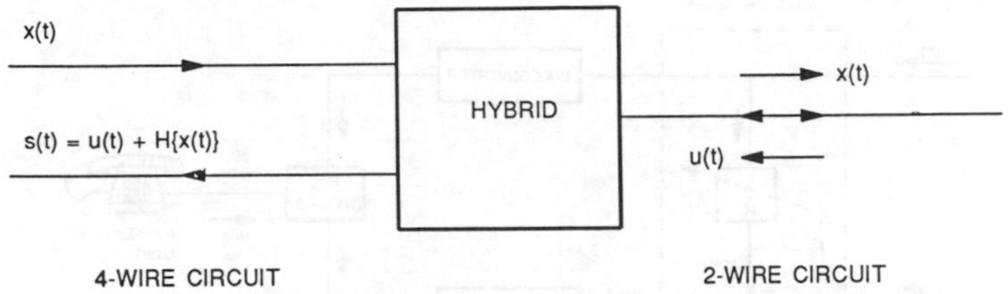

Figure 1. Two- and Four-Wire Interface

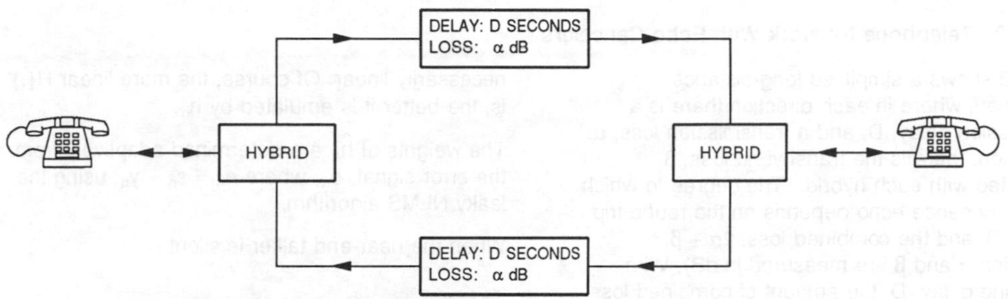

Figure 2. Simplified Long-Distance Connection

described and performance with respect to the CCITT recommendation for echo cancelers is discussed [7]. Finally, a demonstration system that uses a PC and the DEMO16 board is described.

2. The Echo Canceler: Basic Principles

Echoes in telecommunications networks primarily occur at interfaces between two-wire and four-wire circuits. A two-wire circuit carries signals in both directions on the same pair of wires. Four-wire circuits consist of two circuits, each carrying signals in only one direction.

Two wire circuits are found in a telephone company's local loop plant; that is, those circuits that connect the telephone in a home or office to the telephone company's local Central Office facility. The advantage of two-wire circuits is that, at least as far as wire is concerned, their cost is half that of four-wire circuits. The disadvantage is that it is difficult to amplify signals in this type of circuit since amplifiers are inherently unidirectional devices.

Since it is easy to amplify signals in four-wire circuits, they are typically used in long-distance connections between Central Offices.

Two- and four-wire circuits are interfaced to each other by a device called a hybrid circuit. The interface is depicted in Figure 1. On the four-wire side the received signal, x(t), enters the hybrid and the return signal, s(t), departs. On the two-wire side both x(t) and the transmitted signal from the telephone, u(t), are present. The return signal is composed of the transmitted signal as well as a modified version of the received signal, s(t) = u(t) + H{x(t)}. Ideally, the hybrid would attenuate the reflected received signal to the extent that the return signal contains only the transmit signal. Actually, the average transhybrid loss is 11 dB with a standard deviation of 3 dB over the range of frequencies between 500 Hz and 2500 Hz [3]. This is the most critical range of frequencies with respect to echoes.

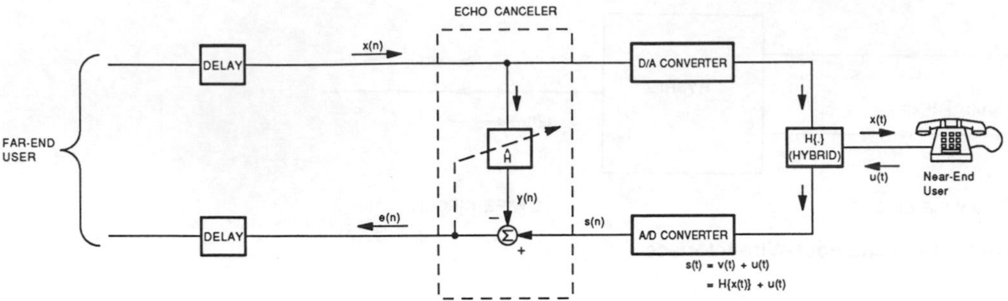

Figure 3. Telephone Network With Echo Cancelers

Figure 2 shows a simplified long-distance connection, where in each direction there is a transmission delay, D, and a transmission loss, α. In addition, there is the transhybrid loss, β, associated with each hybrid. The degree to which users experience echo depends on the round-trip delay, 2D, and the combined loss, $2\alpha + \beta$, (assuming α and β are measured in dB). With increasing delay, D, the amount of combined loss necessary to alleviate the perception of echo increases. Loss plans in ground-based networks regulate the value of α such that connections under 2000 miles in length do not require additional echo control. For longer connections or those that involve satellite hops (270 ms of delay per hop), some type of echo control is required.

Figure 3 shows the addition of an echo canceler, a digital to analog (D/A) converter, and an analog to digital (A/D) converter at the near-end side of the four-wire circuit, where the A/D and D/A converters include the appropriate anti-aliasing and reconstruction filters. The echo canceler must eliminate that part of the return signal, s(t), that is due to the far-end signal, x(t). The eliminated signal is known as the "reflected signal," v(t). Simultaneously, it must preserve that part of s(t) that is due to the near-end signal, u(t).

For the moment assume that u(t)=0, i.e., s(t) = v(t). The echo canceler produces from the far-end signal, x_n, a replica, y_n, of the response of the hybrid from the far-end signal, v_n, and then subtracts y_n from v_n, thus canceling the echo. The canceling is accomplished by modeling the hybrid with a finite impulse response (FIR) filter, $\hat{h}_n$. The response of the hybrid is denoted as v(t) = H{x(t)}, where H{.} is the hybrid operator on x(t) and is not necessarily linear. Of course, the more linear H{.} is, the better it is emulated by $\hat{h}_n$.

The weights of $\hat{h}_n$ are determined adaptively from the error signal, e_n, where $e_n = s_n - y_n$, using the leaky NLMS algorithm.

When the near-end talker is silent,

$$1)\quad e_n = v_n - y_n = H\{x_n\} - \hat{h}(n)*x_n$$

where * denotes convolution. This situation is desired because the NLMS determines the weights of $\hat{h}_n$ by attempting to drive the average power of e_n to zero. To do this, it must drive $\hat{h}_n$ to model H{.}.

In some situations where a particularly long delay and a poor transhybrid loss exist, the residual echo from the adaptive filter can still be strong enough to be perceived by the far-end user. To prevent this, additional echo suppression is usually applied to e_n. It typically consists of a center clipper and is only applied when the near-end signal is not active.

Now consider the situation when the far-end signal is silent and the near-end signal is not silent, i.e., s(t) = u(t). Here,

$$2)\quad e_n = s_n - \hat{y}_n = u_n - \hat{h}(n) * x_n$$

Since x_n is silent, $e_n = u_n$. The error signal provides no information about the relationship between x_n, $\hat{h}_n$, and $\hat{v}_n$. Hence, it is not desirable to use e_n to update $\hat{h}_n$. In fact, if $\hat{h}_n$ is allowed to adapt using this e_n, the weights diverge from an estimate of H{.}.

The situation above is addressed by the addition of a near-end speech detector. When near-end speech is detected, the weights of $\hat{h}_n$ are not updated.

3. Algorithms for Echo Cancelation

Here, the algorithms selected for the various tasks associated with echo cancelation are discussed. In addition, modifications to these algorithms that make them easily implemented on a 16-bit fixed-point processor, such as the DSP16, are presented. First, the NLMS algorithm with tap leakage is stated. Then, the residual echo suppressor and near-end speech detector are discussed.

The NLMS Algorithm With Tap Leakage

Here, the leaky NLMS algorithm is briefly outlined. A detailed development of this algorithm is provided in Appendix A.

For convenience, the adaptive filter, $\hat{h}_n$, is represented as the vector, $\hat{\mathbf{h}}(n)$, where

3) $\hat{\mathbf{h}}^t(n) = [\hat{h}_1(n), \hat{h}_2(n), ..., \hat{h}_N(n)]$

and the superscript, t, denotes transpose. It is also convenient to refer to the state variables in the adaptive filter using the vector, $\mathbf{x}_n$, where

4) $\mathbf{x}_n^t = [x_n, x_{n-1}, ..., x_{n-N}]$

When using this vector notation, the vector inner product, $\mathbf{x}_n^t\hat{\mathbf{h}}(n)$, corresponds to convolution. That is,

5) $y_n = \mathbf{x}_n^t\hat{\mathbf{h}}(n)$

The leaky NLMS algorithm is defined by the following equations:

Step 1)

6a) $\sigma_x^2(n) = (1-\alpha)\sigma_x^2(n-1) + \alpha x_n^2$

Step 2)

6b) if $Cx_n^2 < \sigma_x^2(n)$, goto Step 4

Step 3)

6c) $\sigma_x^2(n) = x_n^2$

Step 4)

6d) $y_n = \mathbf{x}_n^t\hat{\mathbf{h}}(n)$

Step 5)

6e) $e_n = v_n - y_n$

Step 6)

6f) $\hat{\mathbf{h}}(n+1) = \zeta\hat{\mathbf{h}}(n) + \frac{\beta}{\sigma_x^2(n)}(e_n\mathbf{x}_n)$

where $\alpha = .01$, $C = 2^{-8}$, $\zeta = 1-2^{-26}$, and $\beta = 2^{-9}$.

In Step 1, the energy of the far-end signal at sample n, $\sigma_n^2(n)$, is estimated using a "leaky integrator" with forgetting factor, $1-\alpha$. Steps 2 and 3 ensure that, should the power of the far-end signal grow quickly, the estimate of the power responds quickly as well, preventing instability in Step 6 at the onset of speech. In Step 4, the estimate of the return signal is calculated, and the error with respect to the actual return signal (the output of the linear part of the echo canceler) is found in Step 5. Finally, in Step 6, the coefficients of the filter are updated.

The NLMS algorithm, as defined by equations 6a to 6f, needs to be slightly modified for efficient implementation for 16-bit processors. Equation 6a, for example, requires double-precision arithmetic to implement exactly, because the maximum dynamic range of the input signal requires a signed 13-bit representation. If x_n comes from either of these sources, the dynamic range of x_n^2 cannot be represented by a 16-bit signed number. One solution is to discard the lower bits of x_n^2. Another solution, the one used here, is to perform leaky intergration on $|x_n|$ which obviously is representable by a 16-bit number. Thus, equation 6a becomes

Step 1)

7a) $\gamma_x(n) = (1-\alpha)\gamma_x(n-1) + \alpha|x_n|$

The value $\gamma_x(n)$ is a measure of the standard deviation of x_n, σ_x. Similarly, the remainder of the steps in equation 6 are modified as follows:

Step 2)

7b) if $C|x_n| < \gamma_x(n)$, goto Step 4

Step 3)

7c) $\gamma_x(n) = |x(n)|$

Step 4)

7d) $y_n = \mathbf{x}_n^t \hat{\mathbf{h}}(n)$

Step 5)

7e) $e_n = v_n - y_n$

Step 6)

7f) $\hat{\mathbf{h}}(n{+}1) = \zeta\hat{\mathbf{h}}(n) + \dfrac{\beta}{\gamma_x^2(n)}(e_n\mathbf{x}_n)$

In equation 7f, the value $\dfrac{\beta e_n}{\gamma_x^2(n)}$ is required at each sample period, n. For the best use of the available precision of the DSP16, equation 7f is implemented as follows: first, the value $\gamma_x(n)$ is inverted, multiplied by e_n, scaled (i.e., shifted), and once again multiplied by the inverse of $\gamma_x(n)$. The resulting value is scaled again before multiplication with the vector $\mathbf{x}_n$.

Inverting $\gamma_x(n)$ is accomplished by using an inversion algorithm. Prior to this, $\gamma_x(n)$ is first lower-bounded at a level of –34 dBm. The bounding of the signal effectively slows the adaptation speed when the far-end signal is very small.

The choice for ζ is $(1-2^{-26})$. Because this is so close to 1, it is not possible to represent it as a signed 16-bit number. The first term of the right side of equation 7f cannot be calculated by simply using the 16-bit multiplier. Instead, it is calculated by subtracting $2^{-26}\hat{\mathbf{h}}(n)$ from $\hat{\mathbf{h}}(n)$. The factor, $2^{-26}\hat{\mathbf{h}}(n)$, is obtained by first moving the elements of $\hat{\mathbf{h}}(n)$ from one of the accumulators to another, shifting the value to the right by 8 bits, transferring the result to the y register, multiplying it by -2^{-18}, and adding the result to the original accumulator.

The coefficients, $\hat{\mathbf{h}}(n)$, are stored as double-precision values in the DSP16. Although this incurs some cost over single precision in terms of update time and RAM storage, the additional accuracy improves the performance of the echo canceler.

Residual Echo Suppression

A residual echo suppressor is a nonlinear operation on the signal e_n that further increases the echo return loss enhancement (ERLE) of the echo canceler [7]. The nonlinear operation consists of center clipping the residual echo signal. A center clipper forces a signal to zero whenever its magnitude is below a certain level. Of course, it is only desirable to do this when there is no near-end signal. To assure this condition, the residual echo suppression is only enabled when the residual echo's power is sufficiently smaller than that of the far-end signal's power.

As with the near-end signal, the power of the residual echo is estimated using a leaky integrator:

8) $\sigma_e^2(n) = (1-\alpha)\sigma_e^2(n-1) + \alpha e_n^2$

Then the condition

9) $\sigma_e^2(n) < T^2\sigma_x^2(n)$

(where T is a threshold) is tested. If equation 9 is found to be true, e_n is center clipped. The level used for the center clipper is $T\sigma_x$.

The effects of 16-bit arithmetic are felt in the residual echo suppressor both in the estimation of σ_e^2, and in the suppressor application test. The estimation of σ_e^2 is modified to

10) $\gamma_e(n) = (1-\alpha)\gamma_e(n-1) + \sigma|e_n|$

and the test becomes

11) $\gamma_e(n) < (2^{-4})\gamma_x(n)$

where the threshold, T, is set to 2^{-4}. This means that a 24 dB differential between e_n's and x_n's power is necessary to enable the residual echo suppressor.

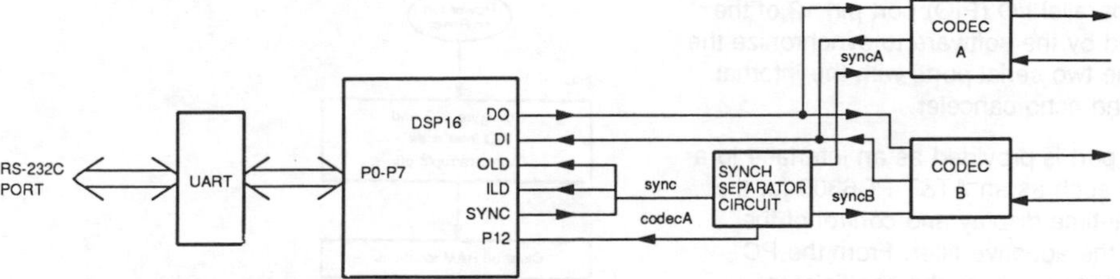

Figure 4. Functional Block Diagram of DEMO16

Near-End Signal Detection

The assumption that the near-end signal, u_n, is zero is central to the proper convergence of the filter $\hat{h}_n$. Since the near-end signal is not always silent, it is necessary to detect its presence and temporarily suspend the canceler adaptation process.

The algorithm by A. A. Geigel [4] examines the far-end and returned signals, x_n and s_n, to find the event in equation 12.

$$12)\quad |s_n| \geq \frac{1}{2}\max\{|x_n|, |x_{n-1}|, \ldots, |x_{n-N}|\}$$

The factor $\frac{1}{2}$ is based on the fact that there is usually at least 6 dB of transhybrid loss. The reason for examining the history of $|x_n|$ is that the echo path's delay is unknown. Equation 12 is used to trigger a "hang-over" counter of 75 ms. As long as the hang-over counter has not timed out, near-end speech is declared to be present. This algorithm is particularly convenient because the values x_n through x_{n-N} are already stored as the state variables of the adaptive filter.

Previously, the absolute value function was used to replace the squaring function because of the ease with which the DSP16 can manipulate 16-bit values rather than 32-bit values. Therefore, it may seem surprising that Geigel's near-end signal detector of equation 12 is modified to replace the absolute value function with squaring. The previous algorithms required the squared values to be stored in RAM, but equation 12 does not. In addition, the DSP16 is more efficient at computing a value's square than its absolute value. Efficiency is important in Geigel's algorithm since N separate squaring operations are performed each sample period.

For these reasons, equation 12 is replaced with the equivalent expression in equation 13.

$$13)\quad s_n^2 \geq \frac{1}{4}\max\{x_n^2, x_{n-1}^2, \ldots, x_{n-N}^2\}$$

4. Hardware

As previously mentioned, the echo canceler is implemented on a standalone demonstration board called DEMO16. A detailed explanation of this board is given in a companion application note [1]. A functional block diagram is shown in Figure 4. DEMO16 provides two analog ports and one RS-232C port to the DSP16.

The analog ports are used for the network and hybrid interfaces of the echo canceler. The network-side signals (the far-end and return signals) are interfaced through port B while the hybrid-side signals are interfaced through port A. The ports are provided via two high-precision codecs (AT&T T7520 Codecs With Filters [8]) that are multiplexed onto the duplex serial port of the DSP16.

The multiplexing is provided by the "sync separator circuit." The DSP16 provides a 16 kHz synchronization signal (sync), which is also used for the input load (ILD) and output load (OLD) signals. The sync separator divides sync into two 8 kHz synchronization signals, syncA and syncB. These signals are used to initiate the serial transfer of data to and from codecs A and B, respectively. When syncA is active, the sync separator forces the signal codecA high, and when syncB is active, the signal codecA is low. This signal, which is connected to parallel I/O (PIO) port pin 12 of the DSP16, is used by the software to synchronize the accesses of the two serial ports with the internal processes of the echo canceler.

The RS-232C port is provided as an interface to a host computer such as an AT&T PC6300 for the purpose of real-time display and control of the coefficients of the adaptive filter. From the PC keyboard, the user can zero the coefficients, change the adaptation rate, or freeze the coefficients.

DEMO16 can be configured in a number of ways. Control switch S2 is used to control the configuration. For proper operation, S2 should be set to the positions described in Table 1.

Table 1. DEMO16 Control Switch S2 Settings

switch	position	switch	position
1	closed	6	open
2	closed	7	closed
3	closed	8	open
4	closed	9	open
5	closed	10	open

5. Software

The overall software architecture is separated into initialization, foreground, and background tasks. During initialization, various control registers, pointers, and counters are set to their initial values. The foreground tasks are primarily dedicated to serial and parallel I/O functions and are interrupt driven. The background tasks consist of the algorithms that perform the echo cancelation.

The source code for a 16 ms echo canceler with double-precision, leaky coefficients is given in Appendix B. The lines marked by the comment "DEMO" have been included to support the PC demonstration described in Section 7. These lines may be deleted if the coefficients of the adaptive filter are not to be displayed.

Initialization Tasks

Figure 5 is a flowchart of the initialization tasks. It consists of two parts. The first part begins at the address label, init. It is executed when the DSP16 is powered-on or reset. The second part begins at the address label, synchro. It is used to synchronize the internal and external serial I/O processes.

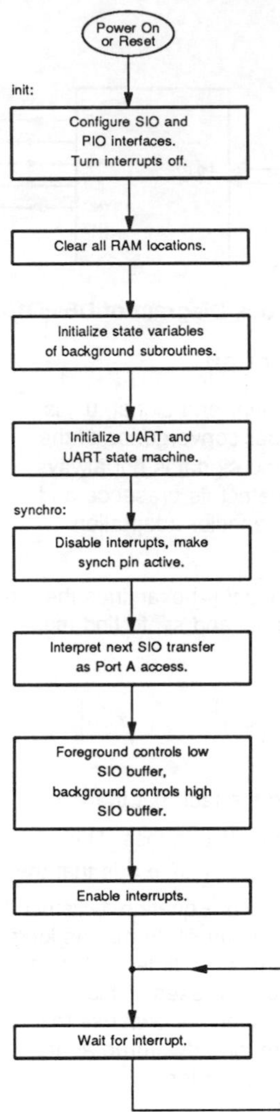

Figure 5. Flowchart of the Initialization Tasks

First, at the label, init, the serial I/O (SIO) is configured such that the input and output clocks, ICK and OCK, and the input and output load signals, ILD and OLD, are to be applied externally. Next, the parallel I/O is configured such that the parallel output and input data strobes, PODS and PIDS, are active. The strobe width of those signals is set to 4T, where T is twice the system clock period.

Network Echo Cancelation

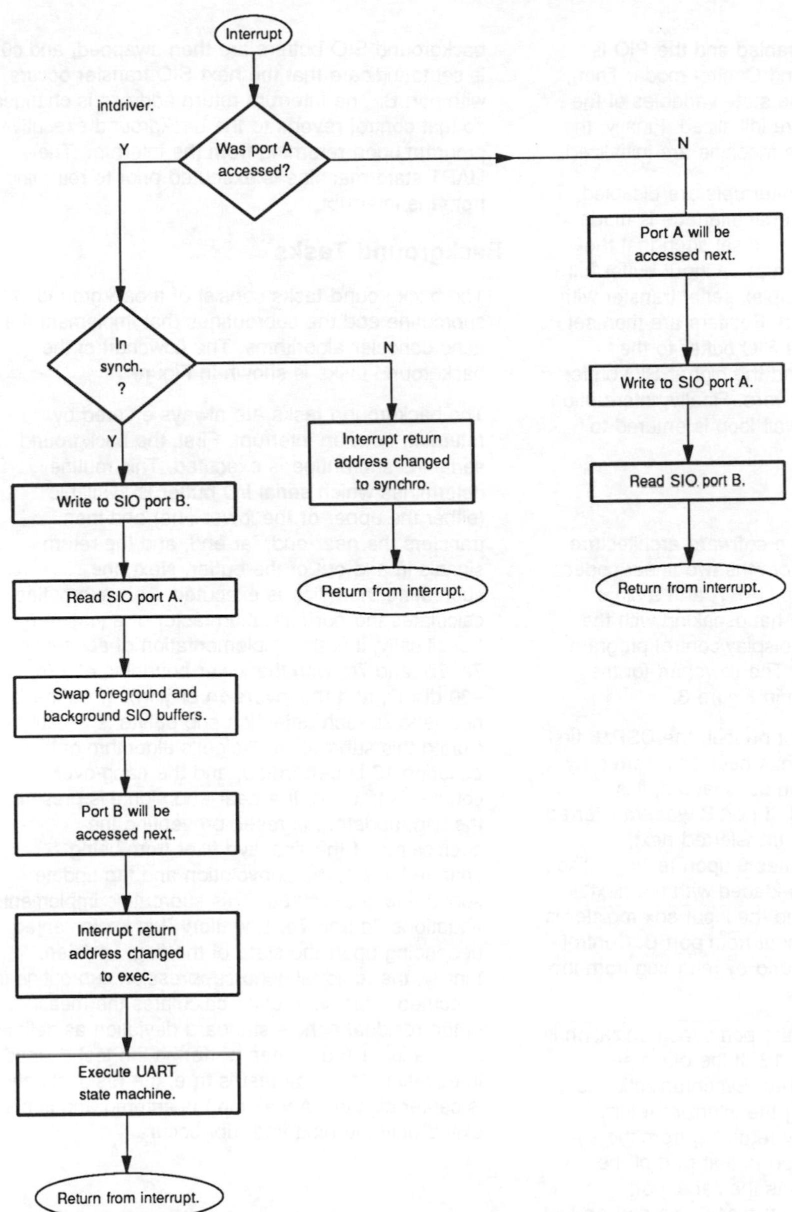

Figure 6. Flowchart of the Foreground Tasks

Also, the interrupts are disabled and the PIO is configured to the Status and Control mode. Then, the RAM is cleared and the state variables of the background subroutines are initialized. Finally, the UART and the UART state machine are initialized.

At the label, synchro, the interrupts are disabled, and the pin, sync, of the serial interface is made active. Then, the c0 counter is set such that the software interprets an interrupt on input buffer full (ibuf) to mean that a full-duplex serial transfer with analog port A has occurred. Pointers are then set to give control of the lower SIO buffer to the foreground I/O software and the higher SIO buffer to the background I/O software. Finally, interruption on ibuf is enabled and a wait loop is entered to await the first interrupt.

Foreground Tasks

The foreground tasks of the software architecture include serial I/O to and from the two linear codecs, parallel I/O to and from the UART, and a state machine that supports the handshaking with the UART and the coefficient display/control program that is running on the PC. The flowchart for the foreground tasks is shown in Figure 6.

Responding to the interrupt on ibuf, the DSP16 first checks to see which port has been transferred by checking the c0 flag. When c0 is tested, it is automatically incremented. If port B was transferred (i.e., c0 was –1), port A is transferred next, because the flag c0 becomes 0 upon testing. The output sdx register is then loaded with the next output value for port A, and the input sdx register is read to receive the new input from port B. Control is returned to the background by returning from the interrupt.

If port A was transferred, the port synchronization is verified by testing PIO pin 12. If the pin is set, resynchronization is required. Synchronization is accomplished by changing the interrupt return address to synchro and by returning from the interrupt. Control is returned to that part of the initialization routine that sets the serial port synchronization to assume the next interrupt on ibuf is due to a port A transfer.

If the serial ports are synchronized, the output sdx register is loaded with the next output value for port B, and the input sdx register is read to receive the new input from port A. The foreground and background SIO buffers are then swapped, and c0 is set to indicate that the next SIO transfer occurs with port B. The interrupt return address is changed so that control reverts to the background executive program upon returning from the interrupt. The UART state machine is executed prior to returning from the interrupt.

Background Tasks

The background tasks consist of a background I/O subroutine and the subroutines that implement the echo canceler algorithms. The flowchart of the background tasks is shown in Figure 7.

The background tasks are always entered by returning from an interrupt. First, the background serial I/O subroutine is executed. This routine determines which serial I/O buffer is available (either the upper or the lower one) and then transfers the near-end, far-end, and the return signals in and out of the buffer. Next, the subroutine, normlizr, is executed. This subroutine calculates the normalization factor of equation 6. Specifically, it is the implementation of equations 7a, 7b, and 7c, with the lower bounding of $\gamma_x(n)$ to –30 dBm0, and the inversion of $\gamma_x(n)$. Then the near-end speech detection subroutine is executed. During this subroutine, Geigel's algorithm of equation 13 is performed, and the hang-over counter is tracked. If a near-end signal is present, the flag, updaten, is reset, preventing the coefficients of the adaptive filter from being changed. Next, the convolution and tap update subroutine is executed. This subroutine implements equations 7d and 7e. Equation 7f is implemented depending upon the state of the flag, updaten. Finally, the residual echo suppression subroutine is executed. This subroutine calculates the measure of the residual echo's standard deviation as defined in equation 10, and then performs the test defined in equation 11. If the test is true, the residual echo is center clipped. A wait loop is entered and is not exited until the next interrupt occurs.

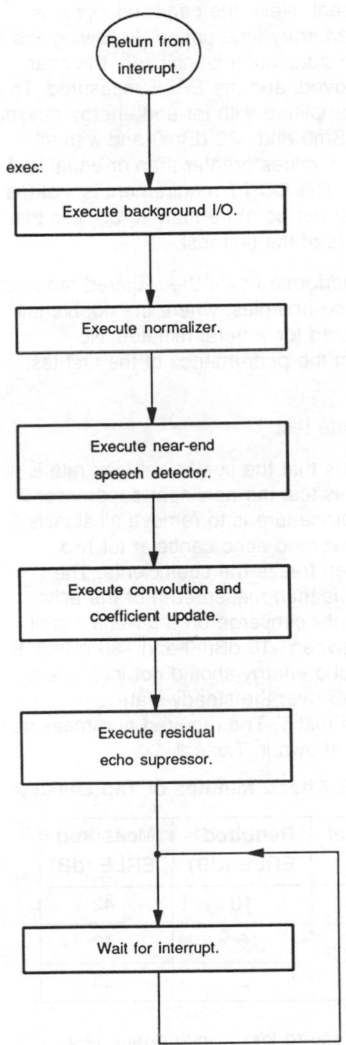

Figure 7. Flowchart of the Background Tasks

6. Performance

The real time required to execute this echo canceler, based on the leaky NLMS algorithm with double-precision taps, is 394 + 13N instruction cycles, where N is the number of taps. The cost of tap leakage is 3 instruction cycles per tap, and the cost of double-precision coefficients is 2 instructions per tap. Thus, an echo canceler with single-precision coefficients without tap leakage would require roughly 394 + 8N instructions.

Table 2 shows the maximum length of an echo path that can be canceled by this program as a function of features and clock speed. The echo path length is defined as the round-trip delay between the echo canceler and the hybrid, and the length of the significant part of the impulse response of the hybrid. The echo path lengths, EPL, shown in Table 2, are related to N by EPL = N/f_s, where the sampling frequency f_s = 8 kHz.

The required RAM for double-precision taps is 18 + 3N 16-bit words per tap, reducing to 18 + 2N for single-precision taps. The required ROM is 380 words without the UART state machine and 553 words with the UART state machine.

Table 2. Maximum Echo Path Length vs. Features and Cycle Time

Features	Cycle Time	
	55 ns	33 ns
Double-Precision Leaky	18 ms	32 ms*
Single-Precision	29 ms	53 ms*

* DSP16A only

The performance of the echo canceler in Appendix B was measured with respect to the six tests described in the CCITT Recommendation G.165 [7]. In these tests, the near- and far-end test signals consist of white noise band-limited to 300 Hz—3400 Hz, and the echo path consists of a delay and an attenuation greater than or equal to 6 dB.

The six tests are as follows:

Test 1: Steady-state residual and returned echo level test.

This test measures the ERLE over the range of far-end signal power between −30 dB and −10 dB with nonlinear echo suppression disabled. Table 3 shows both the CCITT requirement and the echo canceler performance. The requirement is exceeded by at least 28 dB. In fact, the value slightly exceeds the required performance for the echo canceler to operate without nonlinear suppression. However, although CCITT Recommendation G.165 provides a specified performance of echo cancellation without nonlinear suppression, the CCITT has not yet given approval of such cancelers. The topic is under study.

Table 3. Steady-State ERLE

Far-End Signal Level (dBm0)	Required ERLE (dB)	Measured ERLE (dB)
—10	20	48
—20	16	48
—30	12	42

Test 2: Convergence test.

In this test, the nonlinear suppressor is enabled. Initially, the coefficients of the adaptive filter are set to zero, the adaptation is disabled, and the near-end signal is present. Next, the near-end signal is removed and, simultaneously, adaptation of the coefficients is enabled. The coefficients are allowed to adapt for 500 ms when, once again, adaptation is disabled and the resulting ERLE is measured. The CCITT-required ERLE is at least 21 dB. The measured ERLE had a minimum value of 36 dB.

Test 3: Performance during double-talk.

Double-talk occurs when both the near- and far-end signals are active at the same time. This test consists of two parts:

Part A ensures that the double-talk (near-end speech) detector is not overly sensitive. An overly sensitive detector allows echo and low-level near-end speech to cause the disabiling of coefficient adaptation such that sufficient adaptation does not occur.

The coefficients are initially zero, and nonlinear suppression is inhibited. For far-end signal energy ranging between –10 dBm0 and –25 dBm0 and near-end signal strength 15 dB below the far-end signal, the coefficients should converge within one second such that the residual signal energy is at or below that of the near-end signal. The echo canceler of this application note meets this requirement.

Part B ensures that the double-talk (near-end) detector is sufficiently sensitive and operates quickly enough to prevent the adaptive filter from being diverged by the near-end signal.

The nonlinear suppressor is disabled for this test. First, the canceler is fully converged with no near-end signal present. Next, the near-end signal is applied for an arbitrary time period. Following this time period, the adaptation is inhibited, the near-end signal removed, and the ERLE measured. The test must be completed with far-end energy ranging between –10 dBm0 and –30 dBm0 and with all near-end energy values greater than or equal to the far-end energy. The CCITT requirement is that the final ERLE must not be more than 10 dB less than the requirements of the first test.

This test was performed over the required range of near- and far-end energies, where the double-talk condition persisted for several minutes. No divergence from the performance of the first test was detected.

Test 4: Leak rate test

This test ensures that the coefficient leak rate is not excessive. In this test the nonlinear suppressor is disabled. The procedure is to remove all signals from a fully converged echo canceler for two minutes and then freeze the coefficients. The resulting ERLE is then measured. For the echo canceler to initially converge on a far-end signal with energy between –10 dBm0 and –30 dBm0, the final residual echo energy should not increase by more than 10 dB over the steady-state requirements of test 1. The required and measured performance is shown in Table 4.

Table 4. ERLE After 2 Minutes of Tap Leakage

Far-End Signal Level (dBm0)	Required ERLE (dB)	Measured ERLE (dB)
—10	10	42
—20	6	42
—30	2	39

Test 5: Infinite return loss convergence test

The adaptive filter actually constitutes another possible echo path for the far-end signal. This test ensures that when the primary echo path through the hybrid is broken, the secondary echo path through the echo canceler does not produce a significant echo after 500 ms. The echo canceler is initially converged when the echo path is removed. Adaptation is inhibited and the return echo measured 500 ms later. Nonlinear suppression is disabled. The CCITT requirement is that for far-end signal power between –10 dBm0 and –30 dBm0,

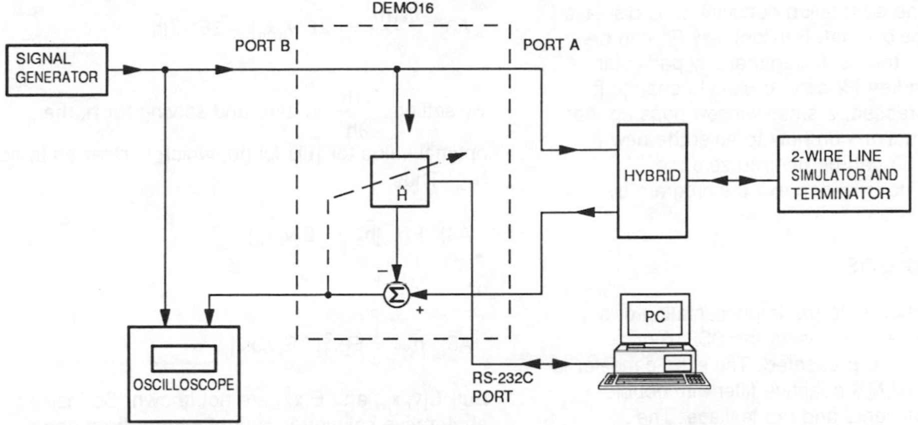

Figure 8. Echo Canceler Test Configuration

the returned echo level should be at or below −37 dBm0. The measured level was at most −43 dB, with the far-end signal at −10 dBm0 and the transhybrid loss at 6 dB.

Test 6: Stability test

This test ensures the stability of the echo canceler for narrow-band signals. Nonlinear suppression is disabled for this test. The coefficients of the adaptive filter are initially cleared and a far-end signal consisting of a sinusoid at 1300 Hz ± 50 Hz with power between −10 dBm0 and −15 dBm0 is applied. According to CCITT requirements, after an initial convergence period of less than one second, the residual echo level should be at least 18 dB below that of the far-end signal for a period of three minutes, with the transhybrid loss of at least 6 dB. The measured value was never greater than 38 dB below the far-end signal.

7. PC Demonstration

The RS-232C port on DEMO16 is used to provide real-time display and control of the coefficients of the adaptive filter in the echo canceler. The port is connected by an RS-232C cable to the serial port of a host computer, e.g., an AT&T PC6300. The program listing in Appendix B includes a state machine in the foreground tasks to control the UART and handle the handshaking and exchange of information between DEMO16 and the PC. In addition, software for the PC to drive this interface and to display and control the tap coefficients can

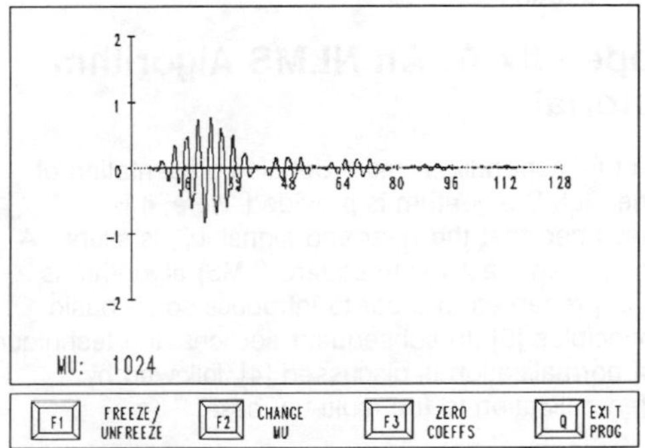

Figure 9. Sample Display Screen

be obtained from your AT&T Account Manager.

Figure 8 shows a typical echo canceler test configuration. A signal generator is used to simulate a far-end signal that is fed into the port B input and passed through the port A output to the hybrid, which is connected to a two-wire line simulator with a termination to simulate a telephone. The output of the hybrid on the four-wire side is passed to the input of port A, and the output of the echo canceler is placed on the output of port B. The far-end signal and echo canceler output are monitored on an oscilloscope, while the coefficients of the echo canceler are controlled and monitored on the PC.

A typical PC display for such a demonstration is shown in Figure 9. The coefficients of the adaptive filter are displayed on the x/y plot at the top of the screen, and the adaptation constant, β, is displayed immediately below that. Function key F1 can be used to freeze the coefficients at any particular time. Function key F2 can be used to change β. When F2 is pressed, a small window pops up that provides instructions on how to enter the new β. Function key F3 can be used to zero the coefficients. The user can exit the program by pressing Q.

8. Conclusions

In this application note the implementation of a network echo canceler using the DSP16 or DSP16A has been presented. The echo canceler is based on the NLMS adaptive filter with double-precision coefficients and tap leakage. The algorithms used in the echo canceler were developed and modified for efficient implementation on a 16-bit digital signal processor. The listings are included in Appendix B.

Appendix A: An NLMS Algorithm Tutorial

In this appendix, a more detailed presentation of the NLMS algorithm is provided. Here, it is assumed that the near-end signal, u_n, is silent. A single-tap least mean square (LMS) algorithm is first presented in order to introduce some basic principles [5]. In subsequent sections, the technique of normalization is discussed [4], followed by generalization to the multi-tap case.

The Single-Tap LMS Filter

The value of the coefficient, $\hat{h}(n)$, of a single-tap adaptive filter that minimizes the mean square of e_n, $\varepsilon(\hat{h})$, is sought. Since it is a measure of the performance of the filter, $\varepsilon(\hat{h})$ is referred to as the *performance index*. From the definition of e_n, it is easy to see that

A1) $\varepsilon(\hat{h}) = E[(e_n)^2]$

A2) $$= E[(v_n - x_n\hat{h})^2]$$
$$= E[v_n^2] - 2E[v_n x_n]\hat{h} + E[x_n^2]\hat{h}^2$$

Differentiating $\varepsilon(\hat{h})$ with respect to $\hat{h}$ yields:

A3) $$\frac{\partial\varepsilon(\hat{h})}{\partial\hat{h}} = -2E[v_n x_n] + 2E[x_n^2]\hat{h}$$

By setting $\frac{\partial\varepsilon(\hat{h})}{\partial\hat{h}}$ to zero and solving for $\hat{h}$, the optimal value for $\hat{h}$ is found, which is referred to as $\hat{h}_{opt}$. Thus,

A4) $$E[x_n^2]\hat{h}_{opt} = E[v_n x_n]$$

or

A5) $$\hat{h}_{opt} = E[x_n^2]^{-1}E[v_n x_n]$$

But, $E[v_n x_n]$ and $E[x_n^2]$ are not known. So instead, an iterative solution is sought where, from one sample period to the next, $\hat{h}$ is updated and, hopefully, improved with each update. That is,

A6) $$\hat{h}(n+1) = \hat{h}(n) + \Delta\hat{h}(n)$$

To find $\Delta\hat{h}(n)$, it is desirable to choose $\Delta\hat{h}$ such that

A7) $$\varepsilon(\hat{h}+\Delta\hat{h}) \le \varepsilon(\hat{h})$$

Expanding $\varepsilon(\hat{h} + \Delta\hat{h})$ using the Taylor Series representation and dropping the higher-order terms results in

A8) $$\varepsilon(\hat{h}) + \Delta\hat{h}\frac{\partial\varepsilon(\hat{h})}{\partial\hat{h}} \le \varepsilon(\hat{h})$$

This inequality is guaranteed if

A9) $$\Delta\hat{h} = -\mu\frac{\partial\varepsilon(\hat{h})}{\partial\hat{h}}$$

is chosen, where μ is a constant chosen small enough to guarantee that dropping the higher terms of the Taylor expansion was valid.

Combining equations A6 and A9, it is found that

A10) $$\hat{h}(n+1) = \hat{h}(n) - \mu\frac{\partial\varepsilon(\hat{h}(n))}{\partial\hat{h}}$$

Alternately, from equation A1 it is seen that $\frac{\partial\varepsilon[\hat{h}(n)]}{\partial\hat{h}}$ can be expressed in the following manner:

A11) $\frac{\partial\varepsilon(\hat{h})}{\partial\hat{h}} = 2E[e_n\frac{\partial e_n}{\partial\hat{h}}]$

and since,

A12) $\frac{\partial e_n}{\partial\hat{h}} = \frac{\partial}{\partial\hat{h}}(v_n - \hat{h}x_n) = -x_n$

it is seen that

A13) $\frac{\partial\varepsilon(\hat{h})}{\partial\hat{h}} = -2E[e_n x_n]$

Combining equations A10 and A13 results in

A14) $\hat{h}(n+1) = \hat{h}(n) + 2\mu E[e_n x_n]$

an update equation for $\hat{h}(n+1)$. However, as mentioned above, $E[e_n x_n]$ is not actually known. Therefore, the *stochastic approximation* is made. That is, the expectation is simply ignored. Thus, $E[e_n x_n]$ becomes $e_n x_n$, and the adaptive algorithm is shown in equation A15.

A15) $\hat{h}(n+1) = \hat{h}(n) + 2\mu e_n x_n$

Equation A15 is the coefficient update part of the LMS algorithm.

Convergence and the Normalized LMS Algorithm

Now consider the convergence properties of the single-tap LMS algorithm. From equations A3 and A10 it is seen that the coefficient update can alternately be expressed as

A16) $\hat{h}(n+1) = (1-2\mu E[x_n^2])\hat{h}(n) + 2\mu E[v_n x_n]$

Solving this difference equation results in

A17) $\hat{h}(n) = \hat{h}_{opt} + (1-2\mu E[x_n^2])^n(\hat{h}(0) - \hat{h}_{opt})$

For this to converge, it is required that μ satisfy

A18) $0 < \mu < E[x_n^2]^{-1}$

Note that when μ is close to zero, n must become very large before there is a significant change in $\hat{h}(n)$. Therefore, it takes a long time before $\hat{h}(n)$ reaches its optimal value. On the other hand, if μ is close to $E[x_n^2]^{-1}$, $\hat{h}(n)$ converges to its optimum value very quickly. So, the adaptation speed is highly dependent on μ and the energy of x_n.

The signal x_n that excites this adaptive filter is speech. The energy in speech varies over quite a large range, roughly 40 dB. So, if the adaptation parameter μ is fixed, it is apparent from equation A17 that the speed of convergence varies by a factor of 100.

This situation can be alleviated [3] by estimating the recent past energy of x_n and replacing μ with $\frac{\beta}{\sigma_x^2(n)}$, where $\sigma_x^2(n)$ is the estimate of x_n's recent energy and β is a positive constant chosen to be significantly less than 1. This technique is knon as normalized LMS or NLMS.

A simple way to estimate $\sigma_x^2(n)$ is the use of what is commonly called the "leaky integrator." That is

A19) $\sigma_x^2(n) = (1-\alpha)\sigma_x^2(n-1) + \alpha x_n^2$

where α is confined to the range $0 < \alpha < 1$ and it controls the "leakyness" or memory of the integration. The closer α is to 0, the less leaky, or alternately, the more memory the estimate $\sigma_x^2(n)$ has in it. A measure of the effective memory of equation A19 is $\frac{1-\alpha}{\alpha}$. A good choice is $\alpha = .01$ for an effective memory of 99 samples (12.375 ms).

One problem with using leaky integration for normalization is that it adapts to a change in energy asymptotically. Adaptation can be a problem when the energy of x_n increases suddenly from a relatively small value to a relatively large value as often happens at the onset of speech. In this situation, $\sigma_x^2(n)$ can react slowly enough that, for a brief period of time, $\mu = \frac{\beta}{\sigma_x^2(n)}$ can be greater than $E[x_n^2]$, violating the stability condition of equation A18. This problem is circumvented by testing each sample's energy and comparing it with the estimated energy, $\sigma_x^2(n)$. If the sample's energy is significantly greater, it replaces the estimated energy. The response of the energy estimate to speech onsets is extremely fast and, if there is an error due to a statistical outlyer, the error is on the side of stability.

This technique, known as "normalized" LMS or NLMS, is now summarized as follows:

Step 1)

A20a) $\sigma_x^2(n) = (1-\alpha)\sigma_x^2(n-1) + \alpha x_n^2$

Step 2)

A20b) if $Cx_n^2 < \sigma_x^2(n)$, goto Step 4

Step 3)

A20c) $\sigma_x^2(n) = x_n^2$

Step 4)

A20d) $y_n = x_n\hat{h}(n)$

Step 5)

A20e) $e_n = v_n - y_n$

Step 6)

A20f) $\hat{h}(n+1) = \hat{h}(n) + \frac{\beta}{\sigma_x^2(n)}(e_n x_n)$

To summarize: the main advantage of NLMS over LMS is that it has a relatively constant speed of convergence over a wide dynamic range of x_n.

Multi-Tap LMS Algorithm

This algorithm is now generalized to the more practical case of an adaptive filter with multiple taps. Here, the scalar $\hat{h}(n)$ generalizes to the vector $\hat{\mathbf{h}}(n)$, where

A21) $\hat{\mathbf{h}}^t(n) = [\hat{h}_1(n), \hat{h}_2(n), \ldots, \hat{h}_N(n)]$

where the superscript t denotes transpose. The scalar x_n generalizes to the vector $\mathbf{x}_n$, where

A22) $\mathbf{x}_n^t = [x_n, x_{n-1}, \ldots, x_{n-N}]$

both e_n and v_n remain scalars. Note that when using this notation, the vector inner product, $\mathbf{x}_n^t\hat{\mathbf{h}}(n)$, corresponds to convolution. So, Step 4 of the NLMS algorithm generalizes to:

A23) $y_n = \mathbf{x}_n^t\hat{\mathbf{h}}(n)$

Similarly, Step 6 becomes

A24) $\hat{\mathbf{h}}(n+1) = \hat{\mathbf{h}}(n) + \frac{\beta}{\sigma_x^2(n)}(e_n\mathbf{x}_n)$

The remaining steps are not affected.

The convergence properties of the multi-tap algorithm are somewhat more involved than the single-tap algorithm. In fact, to ensure convergence for all types of far-end signals, it is necessary to slightly modify equation A24 by incorporating the technique of *tap leakage* [6].

It is easy to show that equation A4 generalizes to

A25) $\mathbf{R}\hat{\mathbf{h}}_{opt} = \mathbf{r}$

where **R** is the auto-correlation matrix of the far-end signal and **r** is the cross-correlation of s_n and x_n. That is, $\mathbf{R} = E[\mathbf{x}_n\mathbf{x}_n^t]$ and $\mathbf{r} = E[s_n\mathbf{x}_n]$. These are the so called "normal equations" whose solution or solutions yield the optimal weights for the filter, $\hat{\mathbf{h}}_{opt}$. Once again, these correlations and cross-correlations are not actually known. So, the relation is not used to find $\hat{\mathbf{h}}_{opt}$, but rather to gain insight into the convergence properties of the adaptive algorithm.

When the far-end signal, x_n, is a narrow-band signal, equation A25 represents a system of underdetermined equations, meaning there are many different possible optimal filters for $\mathbf{h}_{opt}$. As a result, after the coefficients of the multi-tap echo canceler have converged, they tend to wander within the subspace of all possible $\hat{\mathbf{h}}_{opt}$s. Wandering is a problem when any particular coefficient attempts to take on a value with a magnitude greater than the dynamic range of the register in which it is stored. When this happens, the algorithm diverges very quickly from the subspace of optimal solutions.

Heuristically, the idea behind tap leakage is to prevent the coefficients from wandering by forcing them to take on a *particular* optimal solution. The coefficient update algorithm with tap leakage is

A26) $\hat{\mathbf{h}}(n+1) = (\zeta)\hat{\mathbf{h}}(n) + \frac{\beta}{\sigma_x^2(n)}(e_n\mathbf{x}_n)$

with $0 < \zeta < 1$. If the second term of the right-hand side (the coefficient update term) is zero, the coefficients leak asymptotically to zero. The speed with which the coefficients leak to zero is very important. For example, when the far-end speaker is silent, the coefficient update term is very small, and it is desirable that the coefficients change very little. On the other hand, enough tap leakage must be provided to prevent the optimal solutions from wandering. So, the selection of the value ζ involves a design trade-off. A good compromise is $\zeta = 1 - 2^{-26}$.

In order to investigate the stability and speed of convergence of equation A26, it is convenient to assume that $\zeta = 1$. Considering the above choice of ζ, this is a good assumption.

Equation A17, from which the speed of convergence and stability restrictions on μ are determined for the multi-tap case, becomes

A27) $\hat{\mathbf{h}}(n) = \hat{\mathbf{h}}_{opt} + (I-2\mu\mathbf{R})^n(\hat{\mathbf{h}}(0)-\hat{\mathbf{h}}_{opt})$

Now, for $\hat{\mathbf{h}}(n)$ to converge to its optimal value, each eigenvalue of the matrix $(I - 2\mu\mathbf{R})$ must have magnitude less than 1. It can be seen that the stability requirement is then

A28) $0 < \mu < \frac{1}{\lambda_{max}}$

where λ_{max} is the eigenvalue of $\mathbf{R}$ with the largest magnitude. Of course, calculating λ_{max} is not practical in real-time, so a better approach must be found. Note that λ_{max} can be bounded by

A29) $\lambda_{max} < Tr[\mathbf{R}] = \sum_{i=0}^{M}\mathbf{R}[i,i] = \sum_{i=0}^{M}R(0) = (M+1)R(0)$

where Tr[.] denotes the trace of a matrix. So, stability is guaranteed if it is required that

A30) $\mu < [(M+1)R(0)]^{-1}$

But $R(0) = E[x_n^2]$, the energy of the far-end signal. So, the inequality of equation A30 differs from equation A18 only by the factor $\frac{1}{(M+1)}$. This can be embedded into the factor β in Step 6 of equation A20. To summarize, the NLMS algorithm with tap leakage (i.e., the leaky NLMS algorithm) is:

Step 1)

A31a) $\sigma_x^2(n) = (1-\alpha)\sigma_x^2(n-1) + \alpha x_n^2$

Step 2)

A31b) if $Cx_n^2 < \sigma_x^2(n)$, goto Step 4

Step 3)

A31c) $\sigma_x^2(n) = x_n^2$

Step 4)

A31d) $y_n = \mathbf{x}_n^t\hat{\mathbf{h}}(n)$

Step 5)

A31e) $e_n = v_n - y_n$

Step 6)

A31f) $\hat{\mathbf{h}}(n+1) = \zeta\hat{\mathbf{h}}(n) + \frac{\beta}{\sigma_x^2(n)}(e_n\mathbf{x}_n)$

Appendix B: Source Code

Program Listing 1. Source Code

```
/* ekobgone.s*/

/* Copyright © 1988 AT&T                                               */
/* All Rights Reserved                                                 */

/* This is a 16 ms network echo canceler based on a 128-tap double-   */
/* precision NLMS adaptive filter with tap leakage.                   */
/*                                                                     */
/* Includes: near-end speech detector, residual echo suppressor,      */
```

```
/* dual port serial I/O, and coefficient display and control.         */
/*                                                                     */
/* Serial I/O:                                                         */
/*     Port A:  Output: fardata                                        */
/*              Input:  nrdata                                         */
/*     Port B:  Output: ecout                                          */
/*              Input:  fardata                                        */
/*                                                                     */
/* Parallel I/O: Only used for demonstration purposes, i.e., the      */
/* display and control of the adaptive coefficients of the NLMS       */
/* adaptive filter are controlled by a PC. All instructions related   */
/* to this interface are marked by the comment, DEMO, and may be      */
/* removed if PIO is not used.                                         */
/*                                                                     */
/* Note: Register r3 is reserved for serial I/O pointer.              */

#define lngm2 126        /* length of NLMS filter minus 2             */
#define lngm2d2 63       /* (length of NLMS filter minus 2) over 2    */
#define lngm4 124        /* length of NLMS filter minus 4             */
#define lngm4d2 62       /* (length of NLMS filter minus 4) over 2    */
#define lngm8 120        /* length of NLMS filter minus 8             */
#define lngt2 256        /* length of NLMS filter times 2             */
#define lngt2m2 254      /* (length of NLMS filter times 2) minus 2   */

/* Definition of commands sent by PC */                         /*DEMO*/

# define       NEW_COEFF       0x4d    /* 'M' */                  /*DEMO*/
# define       ZERO_COEFFS     0x5a    /* 'Z' */                  /*DEMO*/
# define       SEND_FRAME      0x53    /* 'S' */                  /*DEMO*/

/* UART status register bits */                                 /*DEMO*/

# define       TxEMPTY         1                                  /*DEMO*/
# define       RxREADY         2                                  /*DEMO*/

/*********************************************************************/
/*********************************************************************/
/*                                                                   */
/*                  Beginning of Source Code                         */
/*                                                                   */
/*********************************************************************/
/*********************************************************************/

        goto init                /* Jump to initialization routine.  */

/*********************************************************************/
/*                                                                   */
/*                  Interrupt Service Routine                        */
/*                                                                   */
/*********************************************************************/

intdriver:if c0ge goto synch     /*If c0=0, then check SIO synchroni- */
                                 /*zation, update SIO pointers, and   */
                                 /*execute PIO state machine.         */

        sdx=*r3                  /*Else, output to SIO port A.        */
        *r3++=sdx                /*Then, input from SIO port B.       */

        ireturn                  /*Return from interrupt.             */

/*      Check SIO synchronization:  Since the PIO is double-buffered, */
/*      two reads are required to transfer data from external pins    */
/*      to the accumulator. Because PODS and PIDS strobe width is     */
/*      set to 4T, four instruction cycles are required between       */
```

```
/*     active read and data availability in accumulator.              */

synch:  a0=pdx1              /*codecA signal is on PIO port 1, bit 12. */
        4*;
        a0=pdx1
        2*;
        y=0x1000
        a0=a0&y
        if ne goto insynch        /*Not in sync if bit 12 is set.       */
          pi=synchro              /*If not in synch, reinitialize,      */
          ireturn                 /*return from interrupt.              */

insynch:sdx=*r3                   /*Output to SIO port B.               */
        *r3++=sdx                 /*Input from SIO port A.              */

/*      Next, swap the serial I/O and working buffers.                  */
/*      During one 125 us frame, the forgeround performs serial I/O    */
/*      on one buffer while the background works on the other. In      */
/*      the next frame the roles of the buffers are swapped.           */

        r0=iopnt           /*iopnt contains pointer to working buffer.  */
        y=endsio
        a0=r3
        a0-y
        y=lowsio           /*if r3 < endsio:working buffer addr=lowsio. */
        if le goto inmidst
          r3=lowsio        /*Else set r3 = lowsio,                      */
          y=hisio          /*and set working buffer address = hisio.    */
inmidst:*r0=y              /*Load address of working buffer into iopntr.*/
        c0=-1              /*Next interrupt -- only perform serial I/O. */

        pi=exec            /*Upon return from interrupt -- goto exec.   */

/***********************************************************************/ /*DEMO*/
/*                                                                    */ /*DEMO*/
/*                      UART STATE MACHINE                            */ /*DEMO*/
/*                                                                    */ /*DEMO*/
/*                                                                    */ /*DEMO*/
/*                                                                    */ /*DEMO*/
/* This routine processes all RS-232C related requests. It is         */ /*DEMO*/
/* designed to be entered at a maximum rate of once per sample        */ /*DEMO*/
/* interval. This is substantial overkill due to the fact that at     */ /*DEMO*/
/* 9600 baud, a single byte can only be sent or received once every   */ /*DEMO*/
/* 1/960 s.                                                           */ /*DEMO*/
/*                                                                    */ /*DEMO*/
/* This code is implemented as a state machine. The C language        */ /*DEMO*/
/* description of the state machine is shown below.                   */ /*DEMO*/
/*                                                                    */ /*DEMO*/
/* ------------------------------------------------------------------ */ /*DEMO*/
/*                                                                    */ /*DEMO*/
/*switch(state) {                                                     */ /*DEMO*/
/*      case S_IDLE:                                                  */ /*DEMO*/
/*              if(char_avail()) {                                    */ /*DEMO*/
/*                      switch(getchar()) {                           */ /*DEMO*/
/*                              case CHANGE_UPDATE_COEFF:             */ /*DEMO*/
/*                                      state = S_CHANGE_COEFF_1;     */ /*DEMO*/
/*                                      break;                        */ /*DEMO*/
/*                                                                    */ /*DEMO*/
/*                              case ZERO_COEFFS:                     */ /*DEMO*/
/*                                      zero_coeffs();                */ /*DEMO*/
/*                                      break;                        */ /*DEMO*/
/*                                                                    */ /*DEMO*/
/*                              case SEND_NEXT_FRAME:                 */ /*DEMO*/
/*                                      frame_ptr = frame_buffer;     */ /*DEMO*/
```

```
/*                              state = S_SEND_NEXT;              */ /*DEMO*/
/*                              break;                            */ /*DEMO*/
/*                      default: /* Stay in S_IDLE                */ /*DEMO*/
/*              }                                                 */ /*DEMO*/
/*              break;                                            */ /*DEMO*/
/*                                                                */ /*DEMO*/
/*      case S_CHANGE_COEFF_1:                                    */ /*DEMO*/
/*                      if(char_avail()) {                        */ /*DEMO*/
/*                              update_temp = getchar();          */ /*DEMO*/
/*                              state = S_CHANGE_UPDATE_COEFF_2;  */ /*DEMO*/
/*                      }                                         */ /*DEMO*/
/*                      break;                                    */ /*DEMO*/
/*                                                                */ /*DEMO*/
/*      case S_CHANGE_COEFF_2:                                    */ /*DEMO*/
/*                      if(char_avail()) {                        */ /*DEMO*/
/*                              update_coeff = (getchar() << 8) | */ /*DEMO*/
/*                                              update_temp;      */ /*DEMO*/
/*                              state = S_IDLE;                   */ /*DEMO*/
/*                      }                                         */ /*DEMO*/
/*                      break;                                    */ /*DEMO*/
/*                                                                */ /*DEMO*/
/*      case S_SEND_NEXT:                                         */ /*DEMO*/
/*                      if(transmitter_empty()) {                 */ /*DEMO*/
/*                              putchar(*frame_ptr++ & 0xff);     */ /*DEMO*/
/*                                                                */ /*DEMO*/
/*                              if(frame_ptr == frame_end)        */ /*DEMO*/
/*                                      state = S_IDLE;           */ /*DEMO*/
/*                      }                                         */ /*DEMO*/
/*                      break;                                    */ /*DEMO*/
/*      }                                                         */ /*DEMO*/
/*}                                                               */ /*DEMO*/
/*                                                                */ /*DEMO*/
/* -------------------------------------------------------------- */ /*DEMO*/

uart_io:                                                             /*DEMO*/
        auc  = 0        /* Must clear a0l, a1l, yl on load.       */ /*DEMO*/
        r0 = state      /* Jump to the appropriate state code.    */ /*DEMO*/
        pt = *r0                                                     /*DEMO*/
        goto pt                                                      /*DEMO*/

/* -------------------------------------------------------------- */ /*DEMO*/

idle:                                                                /*DEMO*/
        a0 = pdx1       /* Check if a character has been received. */ /*DEMO*/
        do 4 {          /* Wait for PDX1 access to complete.      */ /*DEMO*/
                ;                                                    /*DEMO*/
                }                                                    /*DEMO*/
        a0 = pdx1                                                    /*DEMO*/
        redo 4          /* Wait for PDX1 access to complete.      */ /*DEMO*/
        y = RxREADY                                                  /*DEMO*/
        a0 & y                                                       /*DEMO*/

        if ne goto idle0                                             /*DEMO*/

        ireturn         /* Return from interrupt.                 */ /*DEMO*/

idle0:                                                               /*DEMO*/
        a0 = pdx0       /* Read received character.               */ /*DEMO*/
        do 4    {       /* Wait for PDX0 access to complete.      */ /*DEMO*/
                ;                                                    /*DEMO*/
                }                                                    /*DEMO*/
        a0 = pdx1                                                    /*DEMO*/
        redo 4          /* Wait for PDX1 access to complete.      */ /*DEMO*/
```

```
        y  = 0x00ff      /* Zero out upper 8 bits of PIO data.       */ /*DEMO*/
        a0 = a0 & y                                                      /*DEMO*/

        y  = NEW_COEFF                                                   /*DEMO*/
        a0 - y                                                           /*DEMO*/
        if ne goto idle1                                                 /*DEMO*/
        pt  = coeff_low                                                  /*DEMO*/
        *r0 = pt                                                         /*DEMO*/

        ireturn          /* Return from interrupt.                   */ /*DEMO*/

idle1:                                                                   /*DEMO*/
        y  = ZERO_COEFFS                                                 /*DEMO*/
        a0 - y                                                           /*DEMO*/
        if ne goto idle2                                                 /*DEMO*/

        /* zero coeffs */                                                /*DEMO*/
        r0 = coeffs                                                      /*DEMO*/
        y  = 0                                                           /*DEMO*/

        do 64 {                                                          /*DEMO*/
                *r0++ = y                                                /*DEMO*/
                *r0++ = y                                                /*DEMO*/
                *r0++ = y                                                /*DEMO*/
                *r0++ = y                                                /*DEMO*/
        }                                                                /*DEMO*/
        ireturn          /* Return from interrupt.                   */ /*DEMO*/

idle2:                                                                   /*DEMO*/
        y  = SEND_FRAME                                                  /*DEMO*/
        a0 - y                                                           /*DEMO*/
        if eq goto idle3                                                 /*DEMO*/
        ireturn          /* Return from interrupt.                   */ /*DEMO*/

idle3:                                                                   /*DEMO*/
        pt     = send_frame                                              /*DEMO*/
        *r0    = pt                                                      /*DEMO*/
        r0     = frame_ptr                                               /*DEMO*/
        pt     = coeffs_end                                              /*DEMO*/
        *r0    = pt                                                      /*DEMO*/

        ireturn          /* Return from interrupt.                   */ /*DEMO*/

/* -------------------------------------------------------------- */ /*DEMO*/

coeff_low:                                                               /*DEMO*/
        a0 = pdx1        /* Check if a character has been received.  */ /*DEMO*/
        do 4     {       /* Wait for PDX1 access to complete.        */ /*DEMO*/
                 ;                                                       /*DEMO*/
                 }                                                       /*DEMO*/
        a0 = pdx1                                                        /*DEMO*/
        redo 4           /* Wait for PDX1 access to complete.        */ /*DEMO*/
        y = RxREADY                                                      /*DEMO*/
        a0 & y                                                           /*DEMO*/

        if ne goto coeff_low0                                            /*DEMO*/
        ireturn          /* Return from interrupt.                   */ /*DEMO*/

coeff_low0:                                                              /*DEMO*/
        a0 = pdx0        /* Read received character.                 */ /*DEMO*/
        do 4     {       /* Wait for PDX0 access to complete.        */ /*DEMO*/
                 ;                                                       /*DEMO*/
                 }                                                       /*DEMO*/
        a0 = pdx1                                                        /*DEMO*/
```

```
        redo 4          /* Wait for PDX1 access to complete.        */ /*DEMO*/

        y  = 0x00ff     /* Zero out upper 8 bits of PIO data.       */ /*DEMO*/
        a0 = a0 & y                                                    /*DEMO*/

        r0  = coeff_tmp /* Save low 8 bits of 16-bit update coeff.  */ /*DEMO*/
        *r0 = a0                                                       /*DEMO*/

        r0  = state                                                    /*DEMO*/
        pt  = coeff_hi                                                 /*DEMO*/
        *r0 = pt                                                       /*DEMO*/

        ireturn         /* Return from interrupt.                   */ /*DEMO*/

/* ------------------------------------------------------------------ */ /*DEMO*/

coeff_hi:                                                              /*DEMO*/
        a0 = pdx1       /* Check if a character has been received.  */ /*DEMO*/
        do 4     {      /* Wait for PDX1 access to complete.        */ /*DEMO*/
                 ;                                                     /*DEMO*/
                 }                                                     /*DEMO*/
        a0 = pdx1                                                      /*DEMO*/
        redo 4          /* Wait for PDX1 access to complete.        */ /*DEMO*/
        y = RxREADY                                                    /*DEMO*/
        a0 & y                                                         /*DEMO*/

        if ne goto coeff_hi0                                           /*DEMO*/
        ireturn         /* Return from interrupt.                   */ /*DEMO*/

coeff_hi0:                                                             /*DEMO*/
        a0 = pdx0       /* Read received character.                 */ /*DEMO*/
        do 4     {      /* Wait for PDX0 access to complete.        */ /*DEMO*/
                 ;                                                     /*DEMO*/
                 }                                                     /*DEMO*/
        a0 = pdx1                                                      /*DEMO*/
        redo 4          /* Wait for PDX1 access to complete.        */ /*DEMO*/

        y  = 0x00ff     /* Zero out upper 8 bits of PIO data.       */ /*DEMO*/
        a0 = a0 & y                                                    /*DEMO*/

        a0 = a0 << 8                                                   /*DEMO*/
        r1 = coeff_tmp                                                 /*DEMO*/
        y  = *r1                                                       /*DEMO*/
        a0 = a0 | y                                                    /*DEMO*/

        r1  = delta                                                    /*DEMO*/
        *r1 = a0                                                       /*DEMO*/

        pt  = idle                                                     /*DEMO*/
        *r0 = pt                                                       /*DEMO*/

        ireturn         /* Return from interrupt.                   */ /*DEMO*/

/* ------------------------------------------------------------------ */ /*DEMO*/

send_frame:                                                            /*DEMO*/
        a0 = pdx1       /* Check if a transmitter is empty.         */ /*DEMO*/
        do 4     {      /* Wait for PDX1 access to complete.        */ /*DEMO*/
                 ;                                                     /*DEMO*/
                 }                                                     /*DEMO*/
        a0 = pdx1                                                      /*DEMO*/
        redo 4          /* Wait for PDX1 access to complete.        */ /*DEMO*/
        y = TxEMPTY                                                    /*DEMO*/
        a0 & y                                                         /*DEMO*/
```

```
        if ne goto send_frame0                                            /*DEMO*/
        ireturn          /* Return from interrupt.                    */ /*DEMO*/

send_frame0:                                                              /*DEMO*/
        r1 = frame_ptr                                                    /*DEMO*/
        r2 = *r1                                                          /*DEMO*/
        j  = -2                                                           /*DEMO*/

        a0   = *r2++j                                                     /*DEMO*/
        a0   = a0 >> 4                                                    /*DEMO*/
        pdx0 = a0                                                         /*DEMO*/

        *r1 = r2                                                          /*DEMO*/

        y = coeffs_strt                                                   /*DEMO*/
        a0 = r2                                                           /*DEMO*/
        a0 - y                                                            /*DEMO*/

        if eq goto send_frame1                                            /*DEMO*/
        ireturn          /* Return from interrupt.                    */ /*DEMO*/

send_frame1:                                                              /*DEMO*/
        pt  = idle                                                        /*DEMO*/
        *r0 = pt                                                          /*DEMO*/
        ireturn                    /*Return from interrupt.           */

/*******************************************************************/ /*DEMO*/
/*                                                                 */ /*DEMO*/
/*                   END OF UART STATE MACHINE                     */ /*DEMO*/
/*                                                                 */ /*DEMO*/
/*******************************************************************/ /*DEMO*/

/*******************************************************************/
/*                                                                 */
/*                   END OF FOREGROUND TASKS                       */
/*                                                                 */
/*******************************************************************/

/*******************************************************************/
/*                                                                 */
/*                     INITIALIZATION                              */
/*                                                                 */
/*******************************************************************/

init:   auc=0x00         /*Clearing of a01, a11, and y1 enabled,       */
                         /*saturation on overflow enabled, p=x*y.      */

        sioc=0x0         /*ICK, OCK, ILD, OLD are passive; LSB first;  */
                         /*and 16-bit serial I/O.                      */

        pioc=0x7c00      /*PODS signal is active, strobe width = 4T,   */
                         /*PIDS is active, in Status & Control mode,   */
                         /*interrupts are disabled.                    */

/****************************ZERO ALL RAM***************************/
        r0=0
        y=0
        do 64    {
        *r0++=y
                 }
        redo 64
        redo 64
        redo 64
        redo 64
        redo 64
```

```
        redo 64
        redo 64
/****Initialize hang-over counter of the near-end speech detector.****/
        pt=maxcnt
        r1=nearcnt
        y=*r0   x=*pt++
        a0=x
        *r1=a0
/****Initialize oldest value state pointer of the circular buffer.****/
        r0=oldest
        y=states
        *r0=y

/*Initialize updaten flag such that NLMS coefficient updates are      */
/*enabled.                                                              */
        r0=updaten
        y=1
        *r0=y

/*Initialize the adjustable factor (from the PC) of the step size in */
/*the adaptive filter.                                                 */
        r0=delta
        y=0x0400
        *r0=y

/*********************************************************************/ /*DEMO*/
/*                                                                   */ /*DEMO*/
/*          Initialization of UART and UART State Machine            */ /*DEMO*/
/*                                                                   */ /*DEMO*/
/*********************************************************************/ /*DEMO*/

        do 127 {                                                        /*DEMO*/
                8 * ;           /* Wait for UART to stabilize.       */ /*DEMO*/
        }                                                               /*DEMO*/

        /* UART mode byte                                            */ /*DEMO*/
        pdx1 = 0x4e                                                     /*DEMO*/

        /* Wait for mode byte write recovery time of 6 Tcyc.         */ /*DEMO*/
        /* At Tcyc = 2.048 MHz, this time is 2.93 us.                */ /*DEMO*/
        /* At DSP16 execution rate of 55 ns, this is equal           */ /*DEMO*/
        /* to approx 55 single-word instructions (nops)              */ /*DEMO*/
        /* plus the 4 cycles for the width of pods.                  */ /*DEMO*/

        do 59 {                                                         /*DEMO*/
                ;                                                       /*DEMO*/
        }                                                               /*DEMO*/

        /* UART command byte                                         */ /*DEMO*/
        pdx1 = 0x27             /*05                                 */ /*DEMO*/

        /* Wait for command byte write recovery time of 8 Tcyc.      */ /*DEMO*/
        /* At Tcyc = 2.048 MHz, this time is 3.91 us                 */ /*DEMO*/
        /* At DSP16 execution rate of 55 ns, this is equal           */ /*DEMO*/
        /* to approx 72 single word instructions (nops)              */ /*DEMO*/
        /* plus the 4 cycles for the width of pods.                  */ /*DEMO*/

        redo 76                                                         /*DEMO*/

        /* Set up initial state for the RS-232C state machine.       */ /*DEMO*/

        r0 = state                                                      /*DEMO*/
        pt = idle                                                       /*DEMO*/
```

```
        *r0 = pt                                                          /*DEMO*/
/************End of UART and UART State Machine Initialization********/ /*DEMO*/

/***************Initialize Synchronization of Serial Interface********/
synchro:pioc=0x7c00     /* Disable interrupts.                        */
        auc=0x0
        sioc=0x0
        tdms=0x1        /* Sync pin is active.                        */

        c0=0
        r3=lowsio/*Initially, foreground controls lower SIO buffer.   */
        r0=iopnt
        y=hisio /*Initially, background controls higher SIO buffer.   */
        *r0=y

        pioc=0x7e00     /* Enable interrupts.                         */
        goto loop       /* Go and wait for first interrupt.           */
/*********************************************************************/
/*                                                                   */
/*                      BACKGROUND TASKS                             */
/*                                                                   */
/*********************************************************************/

exec:   call io
        call normlizr
        call nearspch
        call upncon
        call supress

loop:   3*;       /*Wait for interrupt -- start of next 125 us frame.*/
        goto loop

/*********************************************************************/
/*                                                                   */
/*              Background Serial I/O Subroutine                     */
/*                                                                   */
/*********************************************************************/

io:     r1=iopnt
        r0=*r1       /*  Load r0 with serial I/O pointer.             */

        a0=*r0++     /*  Read input for SIO port B.                   */

        r2=fardata   /*  Port B input to fardata.                     */
        *r2=a0

        a1=*r0--     /*  Read input for SIO port A.                   */

        r2=nrdata
        *r2=a1       /*  SIO port A input to nrdata.                  */

        *r0++=a0     /*  Write output for SIO port A.                 */
                     /*  fardata to A out.                            */

        r2=echout
        a0=*r2       /*  echout to B out.                             */

        *r0++=a0     /*  Write output for SIO port B.                 */

        *r1=r0       /*  Load background SIO buffer pointer with r0.*/

        return
```

```
/************************************************************************/
/*                                                                      */
/*                          NORMLIZR                                    */
/*                                                                      */
/*      Purpose:  Calculate a measure of the standard deviation of     */
/*                the far-end signal and invert it. This process is     */
/*                accomplished by performing leaky integration on       */
/*                the absolute value of the far-end signal and then     */
/*                inverting the value. Should the magnitude of a        */
/*                single far-end sample exceed the sd measure by        */
/*                24 dB, the sd measure is replaced by that value.      */
/*                This prevents instability in the tap update           */
/*                algorithm when the far-end signal strength            */
/*                suddenly increases.                                   */
/*                                                                      */
/*      Inputs:   fardata --  The far-end signal.                       */
/*                                                                      */
/*      State Variable: sdest -- Measure of standard deviation of       */
/*                      fardata, calculated using the leaky             */
/*                      integrator:                                     */
/*                      sdest = a*sdest + (1-a)*|fardata|.              */
/*                      Where a=.99 .                                   */
/*                                                                      */
/*      Output: normal -- The inverted value of the measure of the      */
/*                      standard deviation of the far-end signal.       */
/*                                                                      */
/************************************************************************/

normlizr:auc=0x02       /* Set auc such that register p=4*x*y.          */
        r1=sdest        /* Measure of standard deviation of fardata.    */
        r0=fardata
        r2=normal
        pt=onemina      /* pt now points to (1-a) -- a is stored next.*/
                a1=*r0
                a1=a1
                if mi a1=-a1            /*a1 is |fardata|.              */
                                y=a1    x=*pt++
                        p=x*y   y=*r1   x=*pt++
                a0=p    p=x*y
                a0=a0+p                   /*a0 now contains new sd measure.*/

        a1=a1>>4
        y=a1
        a0-y
        if mi a0=a1<<4          /* If |fardata|/16 > sd measure.        */
                                /* then sd measure = |fardata|.         */

        *r1=a0                 /*sdest and a0 contain new sd measure.*/

                                y=512 /*Defined as min. sdest value. */
                a0-y                  /*This prevents instability at */
                if le a0=y            /*low far-end signal strengths.*/

/*      now, invert the estimated standard deviation.                   */

                                y=a0    x=*pt++ /* dummy load to x.  */
                x=0x1
                a0=x                    /*Here, y is divisor,           */
                                        /*x is dividend.                */
        do 15 {
                a0=a0<<1
                a1=a0-y
                if pl a0=a1+1
                        }
```

Network Echo Cancelation

```
tstpt0:         a0=a0<<16

        *r2=a0                                  /*a0 now contains "normal."  */

        return
onemina:        int     .01
aaaaaaa:        int     .99

/*********************************************************************/
/*                                                                   */
/*                      NEARSPCH                                     */
/*                                                                   */
/* Purpose:  Detect the near-end speech and set the flag, updaten,   */
/*      if none is found. If updaten is not set, the subroutine,     */
/*      upncon, will not update the adaptive filter coefficients.    */
/*      This prevents near-end signals from diverging the            */
/*      coefficients of the echo canceler. If the statement          */
/*                                                                   */
/*       s(n)**2 >= .3125*(max[x(n)**2, x(n-1)**2, ..., x(n-N)**2])  */
/*                                                                   */
/*      is true, where s(n) is nrdata and x(n) is fardata, then we   */
/*      declare the presence of near-end speech. This is Geigel's    */
/*      algorithm.                                                   */
/*                                                                   */
/*   Input:  nrdata:     The near-end sample.                        */
/*           fardata:    The new far-end sample.                     */
/*           states:     The state variables of the adaptive filter. */
/*                                                                   */
/*   State Variables:                                                */
/*           nearcnt:    The detector's hang-over counter.           */
/*                                                                   */
/*   Output: updaten:    Flag to upncon subroutine that enables the  */
/*                       leaky NLMS coefficient update.              */
/*                                                                   */
/*********************************************************************/

nearspch:auc=0x10
        r1=nrdata
        r2=fardata
        r0=states
        c1=1
        c2=-1
                a0=p            y=*r2
halt1:          a0=a0-p         x=*r2++
        do 2 {
                        p=x*y   y=*r0
                a1=a0-p         x=*r0++
        if mi a0=p
                        }
        redo lngm8      /* a0 = max[x(n)**2, ..., x(n-N)**2]         */

        y=a0
        x=0x1000
        a0=a0>>1
        a0=a0>>1

                p=x*y
        a0=a0+p         /* a0 = .3125*(max[x(n)**2, ..., x(n-N)**2]) */

                        y=*r1
                        x=*r1
                p=x*y
        a1=a0-p         /* Perform Giegel's test as described above. */
        ifc mi a1=a1
```

```
        a0=c2
        c1=a0       /* Store result in c1. c1=1 if near speech present,*/
                    /* c1=-1 if near speech not present.               */

        r0=nearcnt
        pt=maxcnt
                                y=*r0    /*y=nearcnt                   */
        a1=p
        a1=a1-p                          /*a1=0                        */
        a0=y                    y=a1  x=*pt++   /*a0=nearcnt,x=maxcnt*/
        if c1ge a0=a1             /*if 4*neardata**2 larger, nearcnt=0*/
        a0h=a0h+1                        /*increment nearcnt.          */
        a1=x
        y=a1                             /*y=maxcnt.                   */
        a1=p
        a1=a1-p                          /*a1=0 again.                 */
        a0-y
        if gt goto stopcnt        /*If nearcnt gt maxcnt, stop supres.*/

                                    /*Else, suppress coeff update.    */
          *r0=a0                    /*Store new nearcnt.              */

          r0=updaten                /*Disable tap leakage.            */
          y=0
          *r0=y
          return

stopcnt:a0=y                    /*Store new nearcnt=maxcnt+1.         */
        a0h=a0h+1
        *r0=a0

        r0=updaten              /*Enable tap leakage.                 */
        y=1
        *r0=y
        return

/*********************************************************************/
/*                                                                   */
/*                          UPNCON                                   */
/*                                                                   */
/*     Purpose: This subroutine performs the convolution, error      */
/*              calculation, and optionally, the coefficient update  */
/*              algorithm of the leaky NLMS adaptive filter.         */
/*                                                                   */
/*     Inputs: fardata -- The far-end signal.                        */
/*             nrdata  -- The near-end signal.                       */
/*             normal  -- The normalization factor.                  */
/*                                                                   */
/*     State Variables:                                              */
/*             states  -- The delay line of the filter.              */
/*             coeffs  -- The filter coefficients.                   */
/*             oldest  -- The address of the oldest sample in filter.*/
/*                                                                   */
/*     Output: echout  -- The "canceled" echo.                       */
/*                                                                   */
/*********************************************************************/

upncon: auc=0x02        /* Set auc such that register p=4*x*y.        */

        r0=coeffs       /* The adaptive filter coefficients.          */

        r1=fardata

        rb=states       /* Set the modulo addressing registers        */
```

```
re=enstates         /* to enclose the state variables.                */

r2=oldest           /* Oldest contains the address of the             */
r2=*r2              /* oldest sample in the filter.                   */

j=2                 /* Only high 16 bits of coeffs used in            */
                    /* convolution.                                   */

      a0=p                x=*r0++j      /* x=last coeff,              */
      a0=a0-p             y=*r2++       /* y=oldest state             */
                                        /*    variable.               */
do 2    {
            p=x*y         x=*r0++j      /* convolve from              */
      a0=a0-p             y=*r2++       /* oldest to newest           */
                                        /* samples in filt.           */
      }
2*;                       /* Stop cache operations periodically*/
redo lngm4d2              /* to allow pending interrupts to be */
2*;                       /* serviced.                         */
redo lngm4d2

            p=x*y         x=*r0++j /* Perform multiply/accum-   */
      a0=a0-p             y=*r1    /* ulate on newest sample    */
            p=x*y         *r2zp:y  /* and place that sample     */
                                   /* into delay line.          */

      a0=a0-p          /*a0 now contains prediction of nrdata.*/

r0=oldest      /* Save address of oldest sample in the filter. */
*r0++=r2

*r0=y     /* Save former oldest state variable in oldold,      */
          /* to be used later in coefficient update algorithm.*/

r2=nrdata
y=*r2               /* Read in nrdata.                        */
a0=a0+y             /* a0 now contains prediction error of filter,*/
r0=echout           /* also known as -- return signal, echout.   */
*r0=a0

/*      Read in adaptive step size.                                */

r1=normal
                          x=*r1++
                          y=a0
            p=x*y
      a0=p                /*a0 = prediction error times normal.*/

            a0=a0<<8/*Shift result to maintain precision.*/
            a0=a0<<4

                          y=a0
            p=x*y
      a0=p                /* a0 = prediction error times normal*/
                                 /* squared.                    */

            a0=a0<<8             /* Once again, shift result  */
            a0=a0<<1             /* to maintain precision.    */

                          y=a0 /* Apply additional modification*/
                          x=*r1/* of stepsize from PC.         */
            p=x*y
      a0=p
      a0=a0<<4                   /* Shift again for precision.*/
```

```
        r0=factor
        *r0=a0                              /* Store result into RAM loc,*/
                                            /* factor.                 */

                                    /* Next, update the coefficients. */

        r0=updaten
        a0=*r0
        a0=a0                   /* If update enable flag is not set, */
        if eq return            /* then do not update coefficients.  */

update: auc=0x11        /* auc set such that register p=x*y/4       */
                        /* and clearing of a0l upon loading of      */
                        /* a0h is inhibited.                        */

        r0=coeffs       /* The adaptive coefficients.               */

        rb=states       /* Set the modulo addressing to surround    */
        re=enstates     /* the state variables of the filter.       */

        r2=oldest       /* Let r1 point to oldest state variable    */
        r1=*r2++        /* in the filter; r2 now points to previous */
                        /* oldest value.                            */

        j=-2            /* Set YAAU auto-incrementers.              */
        k=3

        i=0             /* Set XAAU auto incrementer.               */
        pt=leak         /* pt points to leak factor.                */

                                y=*r2           /*Read in oldold.         */
        r2=factor
                                x=*r2           /*Read in update factor.*/
                        p=x*y   a0=*r0++        /*Read in last double-  */
                                a0l=*r0++       /*precision coefficient.*/

        do 2    {
                a1=a0>>8
                a0=a0+p         y=a1    x=*pt++i /* a0 has new coeff.*/
                                                 /* sans leak        */
                        p=x*y   x=*r2
                a0=a0+p         y=*r1++   /*a0 now has new coeff with*/
                        p=x*y   *r0jk:a0  /*leak store previous coeff*/
                                *r0jk:a0l /*and bring in next one.   */
                }
        2*;                     /* Stop cache operations periodically*/
        redo lngm2d2            /* to allow pending interrupts to be */
        2*;                     /* serviced.                         */
        redo lngm2d2

        return

leak:   int -0x1

/***********************************************************************/
/*                                                                     */
/*              Residual Echo Suppression Subroutine                   */
/*                                                                     */
/*      Purpose:  Detect the state when only the far-end signal is     */
/*                present and NLMS algorithm has reduced echo to a     */
/*                low level. Then further suppress the residual echo   */
/*                by center clipping.                                  */
/*                                                                     */
/*      Inputs:   echout --  Output of NLMS algorithm.                 */
/*                sdest --   Standard deviation measure of fardata.    */
```

Network Echo Cancelation

```
/*                                                                    */
/*      State Variable: sdest2 -- Measure of standard deviation of    */
/*                      echout calculated using the leaky integrator:*/
/*                      sdest2 = a*sdest2 + (1-a)*|echout|.           */
/*                      Where a=.99 .                                 */
/*                                                                    */
/*      Output: echout -- center clipped if sdest2 < sdest*(2**-4).   */
/*                                                                    */
/*                                                                    */
/**********************************************************************/

supress:auc=0x02        /* Set auc such that register p=4*x*y.        */
        r1=sdest2
        pt=onemina      /* pt now points to (1-a) -- a is stored next.*/
        r0=echout
                a0=*r0
                a0=a0
                if mi a0=-a0            /* a0 is |eckout|.            */
                                y=a0     x=*pt++
                        p=x*y   y=*r1    x=*pt++
                a0=p    p=x*y
                a0=a0+p
        *r1=a0                  /* sdest2 and a0 contain new measure of */
                                /* standard deviation of echout.        */

        r1=sdest
        a1=*r1
        a1=a1>>4                /* a1 now contains sdest*(2**-4).       */
        y=a0
        a1-y
        if mi return    /* If sdest2 > sdest*(2**-4), then do nothing.*/

                /* Else center clip echout.                             */

          y=a1                  /* y contains the clipping threshold.*/
          a1=p
          a1=a1-p       a0=*r0  /* a1=0 & a0 contains echout.           */
          a0=a0
          if mi a1=a1+1         /* If echout was negative then a1=1.  */
          a0=a0
          if mi a0=-a0          /* a0 now contains abs(echout).         */
          a0=a0-y
          y=0
          if mi a0=y      /* a0 now contains cntr clipped abs(echout).*/
          a1=a1
          if gt a0=-a0        /* a0 now contains center clipped echout.*/
          *r0=a0        /* echout replaced with center clipped value.*/
        return
endrom:
maxcnt: int     600
/**********************************************************************/

/********************************* RAM ********************************/

        /*****serial I/O buffer***********/

lowsio:         int 0   /* Low SIO buffer.                             */
                int 0
hisio:          int 0   /* High SIO buffer.                            */
endsio:         int 0

        /*****end of serial I/O buffer****/
```

```
iopnt:          int 0     /* Pointer to beginning background        */
                          /* controlled SIO buffer.                  */

sdest2:         int       /* Standard deviation measure of echout.   */
sdest:          int       /* Standard deviation measure of fardata.  */
normal:         int       /* Updated scale factor.                   */
delta:          int       /* Varies scale factor--MUST FOLLOW normal.*/
factor:         int       /*                                         */
updaten:        int       /* Update enable flag.                     */
nrdata:         int       /* Near-end data sample.                   */
echout:         int       /* Output of echo canceler.                */
fardata:        int       /* New data from far-end echo.             */

states:         lngm2*int /* Echo canceler tap delay line.       */
enstates:       int

coeffs_strt:
oldest:         int          /* Pointer to oldest sample in states. */
oldold:         int          /* This must immediately follow oldest.*/
coeffs:         lngt2m2*int  /* The time varying coefficients.      */
                             /* Double-precision 128 coeffs.        */
coeffs_end:     2*int
nearcnt:        int          /* The hang-over near-end signal.      */
                             /* detector counter.                   */
state:          int                                                      /*DEMO*/
frame_ptr:      int                                                      /*DEMO*/
coeff_tmp:      int                                                      /*DEMO*/
.endram
```

References

[1] *WE® DSP16 and DSP16A Digital Signal Processor Information Manual*, November 1988, AT&T Technologies.

[2] *WE® DSP16 Digital Signal Processor Demonstration Board Application Note*, November, 1988, AT&T Technologies.

[3] "Transmission Systems for Communications," Bell Telephone Laboratories, Inc., Holmdel, NJ, 1982.

[4] D.L. Duttweiler, "A Twelve Channel Digital Echo Canceler," IEEE Trans. on Comm., Vol-com-26, No.5, May 1978.

[5] S.J. Orfanidis, *Optimum Signal Processing*, Macmillan, New York, NY, 1985.

[6] B. Widrow and S.D. Stearns, *Adaptive Signal Processing*, Prentice-Hall, Inc., Englewood Cliffs, NJ, 1985.

[7] Recommendation G.165, "Echo Cancelers," CCITT, Geneva (1980).

[8] "T7500 PCM Codec With Filters," *AT&T Communication Devices Data Book*, March 1988.

For additional information contact
your AT&T Account Manager, or call:

□ AT&T Microelectronics
Dept. 50AL330240
555 Union Boulevard
Allentown, PA 18103
1-800-372-2447

In Canada, call:
1-800-553-2448

□ AT&T Microelectronics
AT&T Deutschland GmbH
Bahnhofstr. 27A
D-8043 Unterfoehring
West Germany
Tel. 089/950 86-0
Telefax 089/950 86-111

□ AT&T Microelectronics Asia/Pacific
14 Science Park Drive
#03-02A/04 The Maxwell
Singapore 0511
Tel. (65) 778-8833
FAX (65) 777-7495
Telex RS 42898 ATTM

□ AT&T Microelectronics
AT&T Japan Ltd.
31-11, Yoyogi 1-chome
Shibuya-ku, Tokyo 151
Japan
Tel. (03) 5371-2700
FAX (03) 5371-3556

□ AT&T Microelectronica España
Albacete 5
28027 Madrid
Spain
Tel. (34) 1-404-6012
FAX (34) 1-404-6252

November 1988

AP88-10DMOS

Sonar Beamforming 6

6.1 OVERVIEW

This chapter describes a real-time digital beamforming system for passive sonar. The design of this system is based on several ADSP-2100s that independently perform the beamforming calculations under the supervision of an ADSP-2100 master processor. The modular architecture allows you to tailor the size of the system to your performance needs. Code listings for the master and slave processors are included.

A sonar system can use two different methods to analyze and evaluate possible targets in the water. The first is called *active* sonar. This method involves the transmission of a well defined acoustic signal which can reflect from objects in water. This provides the sonar receiver with a basis for detecting and locating the targets of interest. The limitations of this method are mainly due to the loss of the signal strength during propagation through the water and reverberation caused by the signal reflections. Simplistically, active sonar can be thought of as the underwater equivalent of radar.

The second method is called *passive* sonar. This one bases its detection and localization on sounds which are emitted from the target itself (machine noise, flow noise, transmissions of its active sonar). Its limitations are due to the imprecise knowledge of the characteristics of the target sources and to the dispersion of the target signals by the water and objects in the water. A generic passive sonar system is shown in Figure 6.1 which can be found on the next page.

Sonar systems have a wide variety of military and commercial uses. Some of the military applications include detection, localization, classification, tracking, parameter estimation, weapons guidance, countermeasures and communications. Some of the commercial applications include fish location, bottom mapping, navigation aids, seismic prospecting and acoustic oceanography. More detailed information about sonar technology can be found in Winder, 1975 and Baggeroer, 1978.

6 Sonar Beamforming

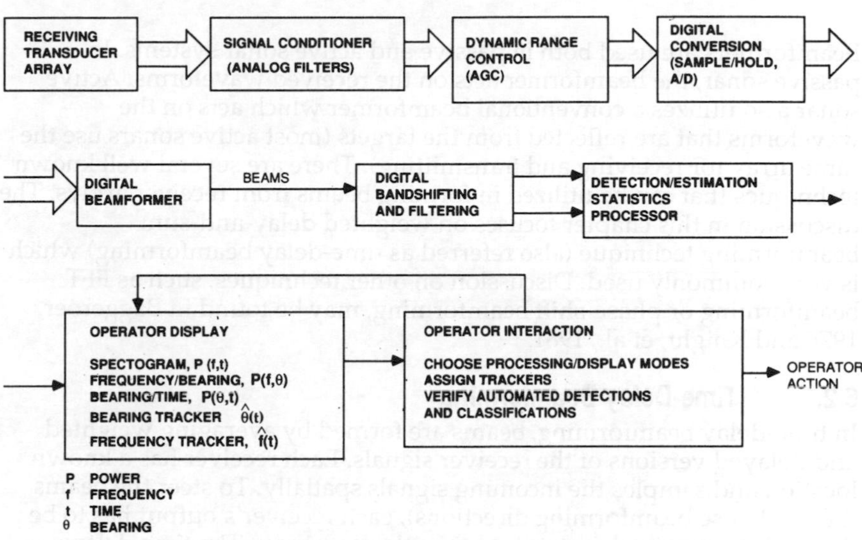

Figure 6.1 Generic Passive Sonar System

6.2 SONAR BEAMFORMING

In its simplest form, sonar beamforming can be defined as "the process of combining the outputs from a number of omnidirectional transducer elements, arranged in an array of arbitrary geometry, so as to enhance signals from some defined spatial location while suppressing those from other sources" (Curtis and Ward, 1980). Thus, a beamformer may be considered to be a spatial filter. It is generally assumed that the waves arriving at the transducers all propagate with the same speed c, so that the signals of interest lie on the surface of the cone defined by $\omega = c|k|$ in (k, ω) space. Ideally the passband of the beamformer lies on the intersection of this cone with the plane containing the desired direction vector.

The beamforming operation is accomplished through a series of operations that involve the weighting, delay and summation of the signals received by the spatial elements. The summed output that contains information about a particular direction is called a beam. This output is then sent to a signal processor and/or a display for frequency and temporal discrimination. A beamforming system can employ analog or digital components and techniques; this chapter focuses on a digital beamforming technique.

Sonar Beamforming 6

Beamformers are used both in passive and active sonar systems. In passive sonar, the beamformer acts on the received waveforms. Active sonar also utilizes a conventional beamformer which acts on the waveforms that are reflected from the targets (most active sonars use the same array for receiving and transmitting). There are several well known techniques that can be utilized in forming beams from receiver arrays. The discussion in this chapter focuses on weighted delay-and-sum beamforming technique (also referred as time-delay beamforming) which is very commonly used. Discussion on other techniques, such as FFT beamforming or phase-shift beamforming, may be found in Baggeroer, 1978 and Knight, et al., 1981.

6.2.1 Time-Delay Beamforming

In time-delay beamforming, beams are formed by averaging weighted and delayed versions of the receiver signals. Each receiver has a known location and samples the incoming signals spatially. To steer the beams (i.e. to choose beamforming directions), each receiver's output has to be delayed appropriately relative to the other receivers. The time delays compensate for the differential travel time between sensors for a signal from the desired beam direction.

In order to describe this operation mathematically, let us assume that the array of receivers is composed of a three dimensional distribution of equally weighted omnidirectional sensors. Their spatial locations are specified in the Cartesian coordinate system of Figure 6.2. The

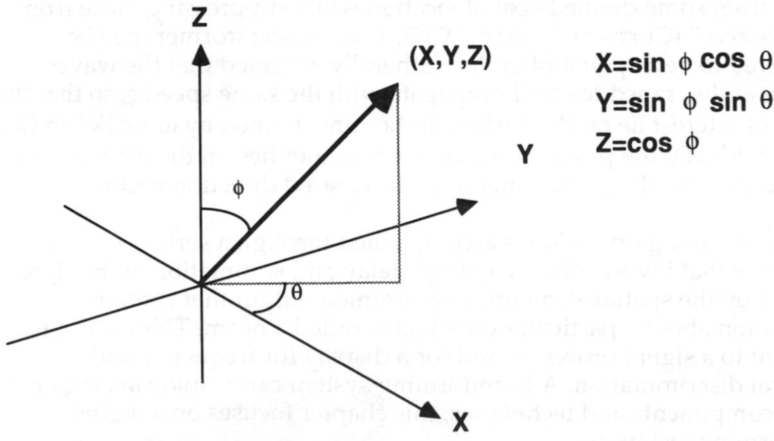

Figure 6.2 Cartesian Coordinate System

6 Sonar Beamforming

beamforming task consists of generating the waveform $b_m(t)$ (or its corresponding sample sequence) for each desired steered beam direction $\mathbf{B}_m$. Each $b_m(t)$ consists of the sum of suitably time delayed replicas of the individual sensor outputs $e_n(t)$.

Let the output of an element located at the origin of coordinates be s(t). Under the assumption of plane wave propagation, a source from direction $\mathbf{S}_l$ (the lth source direction unit vector) produces the following sensor outputs

$$(1) \qquad e_n(t) = s(t + ((\mathbf{E}_n \bullet \mathbf{S}_l)/v))$$

where $\mathbf{E}_n$ is the nth element position vector and v is the speed of propagation for acoustic waves in the ocean ($v \approx 1500$ m/s). Delaying each individual sensor output by an appropriate amount to point a beam in the direction $\mathbf{B}_m$ yields the beamformer output

$$(2) \qquad b_m(t) = \sum_{n=1}^{N} e_n(t - ((\mathbf{E}_n \bullet \mathbf{B}_m)/v))$$

The operation defined in equation (2) is known as *beamforming*. The beamforming operation is computationally demanding because this summation must be calculated in real time for a large number of sensors and a large number of beams.

For simplicity, the rest of this discussion is limited to one-dimensional (line) arrays with regularly spaced hydrophones (underwater omnidirectional acoustic sensors). This discussion can be generalized to multi-dimensional arrays and line arrays with variable spacing.

6.2.2 Digital Beamforming

Assume that the presence of a plane wave signal $s(t - (\mathbf{E} \bullet \mathbf{B})/v)$ needs to be detected. It is propagating with a known direction **B**, and is measured at **E** in a background of spatially white noise (**B** and **E** are vectors). The line array of Figure 6.3 is used. The signal has the same value at each wavefront and the noise is uncorrelated from sensor to sensor. Thus, in order to enhance the signal from the noise, the sensor outputs are delayed and summed. The delays account for the propagation delay of the wavefront to each sensor. This yields

Sonar Beamforming 6

$$(3) \qquad b(n\Delta) = \quad 1/N \sum_{i=0}^{N-1} X_i(n\Delta + ((E_i \bullet B)/v))$$

where $X_i(n\Delta)$ is the sampled output of the *ith* sensor. Note that for the *ith* sensor the following relationship holds:

$$(4) \qquad (E_i \bullet B)/v = -i\,(d/v)\sin\theta$$

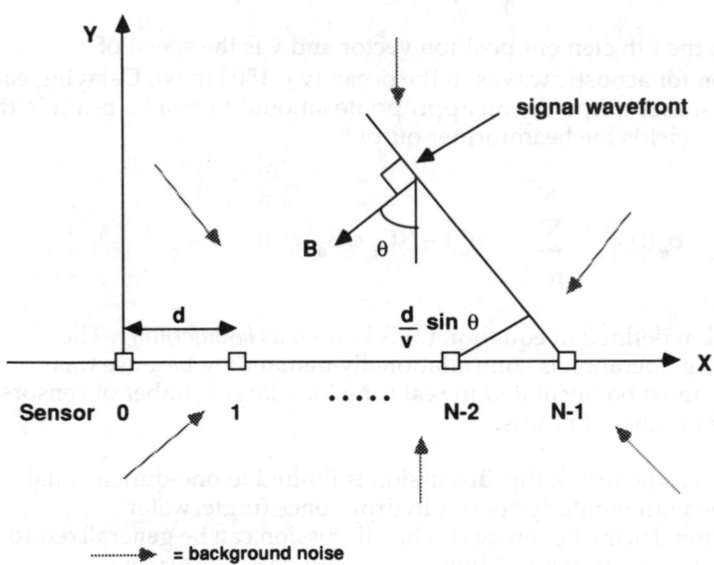

Figure 6.3 Line Array of Equally Spaced Hydrophones

The input to the beamformer is a set of time series. The input is usually one set of time series for each sensor, while the output of the beamformer is another set of time series, generically referred to as beams.

The beamformer is spatially discriminating because for a plane wave with a propagation direction θ, different than θ_0 assumed by the beamformer, the sensor outputs are not coherently combined. This leads to partial

6 Sonar Beamforming

cancellation of the incoming signals with $\theta \neq \theta_0$. Thus, for a plane wave signal

$$X_i(n\Delta) = e^{j(\omega n\Delta + (\omega i d / v \sin\theta))} \quad (5)$$

and

$$b(n\Delta) = 1/N \sum_{i=0}^{N-1} e^{j\omega i d(\sin\theta - \sin\theta_0)/v} \, e^{j\omega n\Delta} \quad (6)$$

where *d* is the spacing between the sensors. Thus the amplitude of the plane wave arriving from a direction θ at the output of a beamformer steered to θ_0 has an attenuation given by

$$|B(\omega, \theta)| = \frac{\sin[N(\omega/2)\ d(\sin\theta - \sin\theta_0)/v]}{N\sin[(\omega/2)\ d(\sin\theta - \sin\theta_0)/v]} \quad (7)$$

This function is known as the *beam pattern* of the array. An example beam pattern for a line array is shown in Figure 6.4. For a given wave frequency and steering direction θ_0, all plane waves with $\theta \neq \theta_0$ are attenuated, leading to the interpretation of a beamformer as a spatial filter. In most applications, it is desirable to have a beam pattern with a very narrow main lobe and very low level sidelobes for maximum noise rejection.

Increasing ω (the operating frequency) and/or N results in narrower main lobes even though the side lobe level does not change. Increasing the sensor spacing also results in narrower main lobes. But this is limited by the fact that spatial aliasing will occur for $\Delta x > \lambda_{min}/2$, where λ_{min} is the signal wavelength for the highest frequency of interest (Knight, et al., 1981). Spatial aliasing exhibits itself in terms of extra main lobes near the endfire region. In order to reduce the sidelobe levels, the sensor outputs must be weighted. This procedure is known as *shading*. Thus, in equation (6), we replace the 1/N factor by w_k. The corresponding beam pattern is

$$|B(\omega, \theta)| = \sum_{i=0}^{N-1} w_i \, e^{j\omega i d(\sin\theta - \sin\theta_0)/v} \quad (8)$$

Sonar Beamforming 6

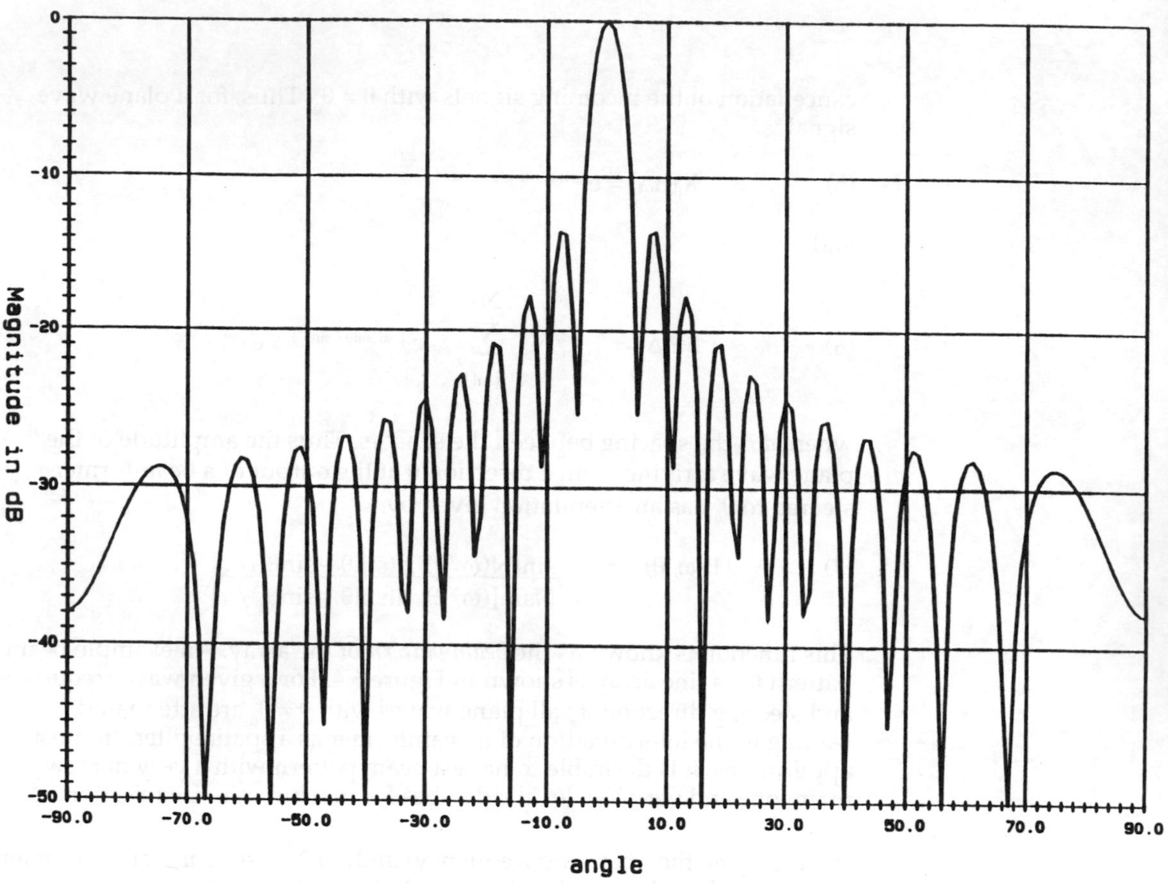

Beam pattern at 0 degrees (perpendicular to line of array) with a sampling rate of 500Hz, for a line of 32 sensors spaced 1m apart

Figure 6.4 Example Beam Pattern

The usual windowing techniques of Fourier transform theory can be used to reduce the sidelobes (Knight, et al., 1981). By employing a Hamming window, for example, the sidelobes may be reduced to –40db at the expense of widening the main lobe.

Another problem that has to be dealt with in a line array with fixed shading is the quantization errors that are introduced from the insertion of the delays. In a digital beamformer, for ideal operation, the beamforming

6 Sonar Beamforming

directions are limited such that the delays *t* are multiples of the sampling interval. Any other choice of directions (and thus delays) introduces errors onto the beam pattern (Gray, 1985). One method that is used to reduce such errors involves the interpolation of the incoming samples. Further discussion on this topic can be found in Pridham and Mucci, 1978.

An overall block diagram for a conventional time delay digital beamformer is shown in Figure 6.5.

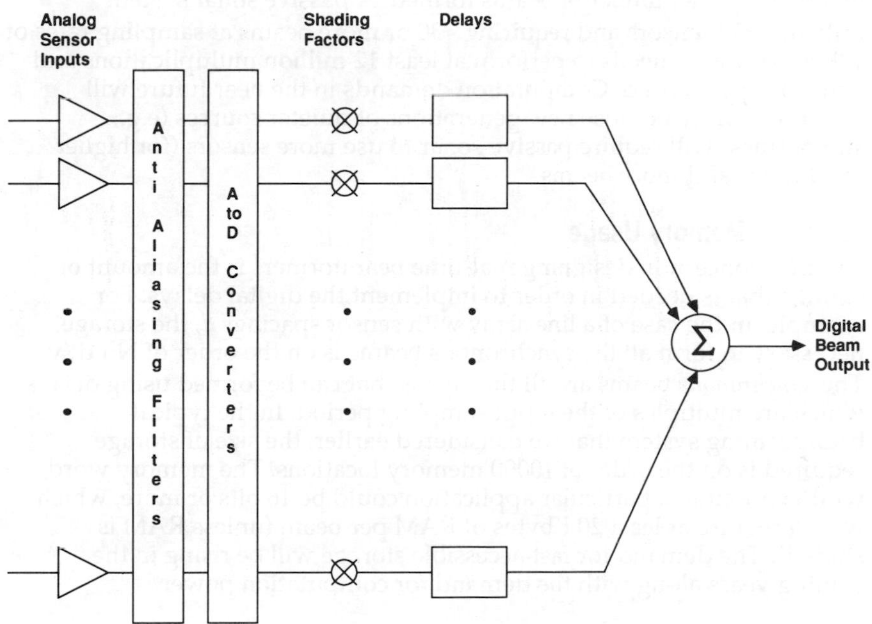

Figure 6.5 Conventional Time-Delay Beamformer

6.3 DIGITAL BEAMFORMER IMPLEMENTATION

The beamforming task can be performed using analog or digital systems. Implementing analog tapped delay lines and forming multiple beams in real time using analog hardware results in big, inflexible and cumbersome systems. A digital beamformer results in a smaller, more accurate, and much more flexible hardware/software unit than an analog technique. In this section, some of the important issues in digital time-delay beamforming systems are discussed.

Sonar Beamforming 6

6.3.1 Computational Power

The computational capacity required by a real-time beamformer can be computed from equation (3) as roughly Nf_s multiplications and additions per second per beam, where N is the number of sensors and f_s is the input sampling frequency. This requirement does not seem by itself very demanding. However, multiple simultaneous beams are usually needed in order to span the space around the sensors and consequently a more realistic requirement is N_bNf_s multiplications and additions per second, where N_b is the number of beams formed. A passive sonar system utilizing ≈30 sensors and requiring ≈30 or more beams at sampling rates of 10kHz or higher needs to perform at least 12 million multiplications and additions per second. Computation demands in the near future will increase rapidly because new generations of quieter sources (e.g., submarines) will require passive sonar to use more sensors (for higher resolution) and more beams.

6.3.2 Memory Usage

Another concern in designing real-time beamformers is the amount of storage that is needed in order to implement the digital delays. For example, in the case of a line array with sensor spacings *d*, the storage necessary to form all the synchronous beams is on the order of N^2f_sd/v. The *synchronous* beams are all the beams that can be formed using delays which are multiples of the input sampling period. In the typical beamforming system that we considered earlier, the size of storage required is on the order of 10000 memory locations. The memory word width chosen in a particular application could be 16 bits or more, which would require at least 20 Kbytes of RAM per beam (unless RAM is shared). The demand for fast accessible storage will be rising in the coming years along with the demand for computation power.

6.3.3 Other Issues

Further discussion on beamforming system issues can be found in Knight, et al., 1981, Janssen, 1987, and Hodgkiss and Anderson, 1981.

6.4 EXAMPLE BEAMFORMER

The example beamformer is able to take inputs from an arbitrary array of up to 32 sensors, form multiple beams in real time, and is user-configurable from an IBM PC personal computer (or compatible). The acoustic frequencies of interest in this example range from 0Hz to 2000Hz. The input sampling rate is 10kHz, which is higher than the Nyquist rate because of the need for a higher resolution in the beamforming directions. The acoustic data is collected by hydrophones and the analog data is

6 Sonar Beamforming

digitized to 12 bits. The gain applied to the analog inputs is manually selectable (1, 10, 100 or 1000). The output beams are sent to another system over a specific parallel interface for further processing. The output beams are also available at analog output points, one at a time, for testing, monitoring and some other signal processing tasks.

6.4.1 System Architecture

Several important issues were considered in selecting an architecture for this beamforming system. One consideration is the need for modularity. A modular system gives a user the ability to start small with the option of expanding the system's capabilities later.

Another consideration is the demand for speed. As discussed earlier, in order to form multiple simultaneous beams, a large number of summations have to be computed in real time. This requirement, along with modularity, led to a distributed processing architecture.

One more consideration is flexibility. A user is able to specify any array configuration and form any beam. Ease of use requires a friendly and interactive user interface, through which the users can specify many different system parameters.

6.4.2 Building Blocks

The beamformer consists of the building blocks that are shown in Figure 6.6. There are several parallel buses in the system for data transfers and communication. A wide common bus is used for the subsystems to communicate and exchange data with each other. Another bus facilitates the communications with the IBM PC. This bus is used to download the user system configuration data from the PC into the beamformer. Finally, a bus is used to send the output beams to a sonar signal processing system which performs further processing on the data.

There are three main types of subsystems in this beamformer; each of these subsystems is implemented as a separate board in the example system:

- The *master* module, which is responsible for controlling the data exchange among all internal modules and the data flow over the I/O buses.
- The *slave* module, which is responsible for the actual beamforming task. Each slave can beamform in multiple directions and more slaves may be added in order to form a larger number of beams.

Sonar Beamforming 6

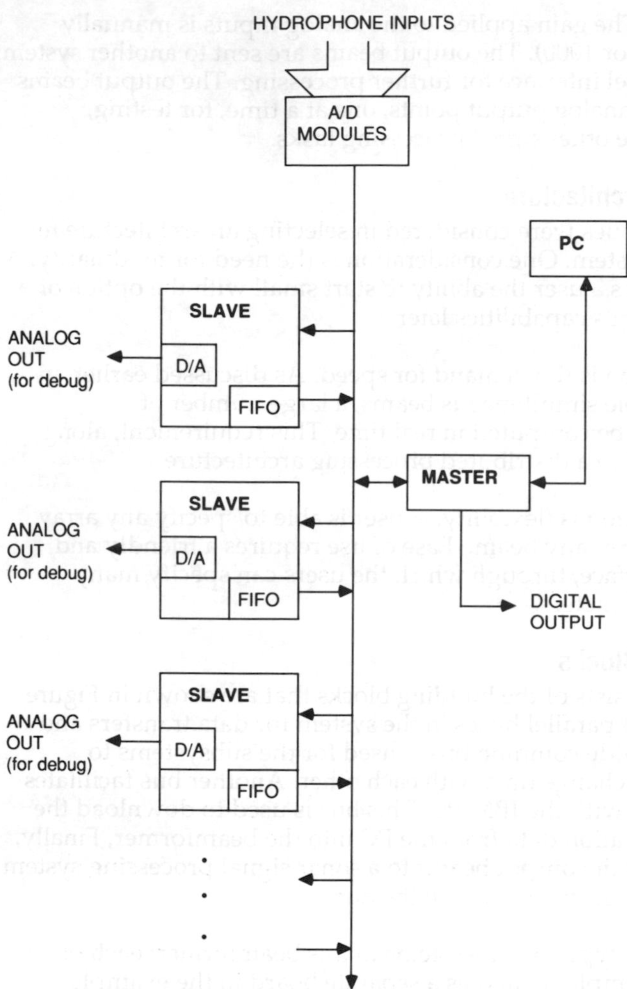

Figure 6.6 Example Beamformer Block Diagram

- The *analog-to-digital conversion* (A/D) module. This module takes the hydrophone outputs as inputs, is responsible for sampling the incoming analog signals and converting them into a digital format. Each A/D module can handle a limited number of hydrophones, but more modules may be attached to the common bus in order to handle a larger number of inputs.

6 Sonar Beamforming

6.4.3 System Operation

The internal operation of the system is controlled by the master. The handling of the incoming samples and the I/O exchanges are also under the master's control. The operation of the system is synchronized to the input sampling clock which has a period of 100 μs (10kHz). This implies that the calculated beam samples have to be sent out every 100μs. Let's call this duration a *system cycle*.

Initially, the master has to accomplish a one-time task of handling of the system configuration data sent from the PC. Thereafter, the master has to go through its duties within one system cycle and be ready to handle the next set of incoming samples. The system events that make up the system configuration (illustrated in Figure 6.7) are as follows:

1. The operator enters system configuration variables (number of sensors, beam directions, shading factors, etc.) through an interactive program on the PC. The same program does some calculations and downloads the data to the master ADSP-2100. Communications between the master and PC are accomplished using a simple protocol over a parallel interface card located in the PC.

2. The master keeps the configuration variables in several of its internal registers. It sends this information to all the slaves so that each of them can identify the beams that they are responsible for.

3. The master waits for a signal from each one of the slaves confirming that they are ready to beamform.

The sampling clock is running during the configuration, but the interrupts initiated by A/D conversions are not recognized until all the slaves are ready. Once each of the slaves has sent the signal that it is ready to beamform, the master starts the cyclic operation of the beamformer.

1. The master responds to the A/D conversion interrupt by reading the results of the A/D conversions, which correspond to a simultaneous snapshot of the incoming waveforms at the hydrophone locations. It reads all the results in sequence and writes them into the memories of all the slaves. Thus, all slaves receive identical copies of the incoming waveform samples.

2. The master initiates an interrupt which orders all the slaves to start beamforming.

MASTER	SLAVE
Power up (or RESET)	Power up (or RESET)
Wait for the user to input the system setup parameters	
PC communications	Idle
Download setup data to slaves	Accept setup data
Idle	Self-prepare using the setup data
Sample all the A/Ds 255 times and send samples to slaves	Receive the first 255 sets of conversion results (fill the sample buffer)
Idle	Beamform
Sample A/Ds, send samples	Receive samples
Read FIFOs, output beams	Beamform
Sample A/Ds, send samples	Receive samples
Read FIFOs, output beams	Beamform
etc.	etc.

Figure 6.7 Sequence of Events

3. Once a slave finalizes a summation (i.e. forms a beam sample), it shifts the result into a FIFO memory for collection by the master. Each slave computes and stores its own beam samples independent of other slaves. Once a slave has completed its assigned set of beams, it waits for the next interrupt (initiated by the master) that orders all the slaves to beamform again.

4. While the slaves are busy forming and storing the current set of beam samples, the master reads the sets of beam samples that were formed during the previous system cycle. After finishing this output duty, the master waits for the next A/D interrupt.

6 Sonar Beamforming

Each slave keeps its input samples in a circular buffer (255 slots) which is located in the slave's data memory space. New incoming samples are put into consecutive locations in this buffer. Once the end of the buffer is reached, the oldest snapshot of samples gets overwritten by the newest samples and this cyclic process goes on.

Each slave performs the beamforming task by reading the appropriate locations in its sample buffer, by multiplying those values with a shading factor and by keeping a running sum of these weighted samples until the summation is finished.

Each slave writes out beam samples to its own dedicated a first-in-first-out memory (FIFO). Only the master can shift the beam samples out of the FIFOs. The master reads and sends each beam sample, one at a time, over its output bus.

The user must realize that the first 255 sets of beam samples produced are invalid. This is due to the fact that the sample buffer is not full until the end of the 255th system cycle. Therefore, during that period, the locations read by the slaves contain meaningless data. Valid system outputs are produced ≈25.5ms after the system starts beamforming.

6.4.4 Timing Issues

There are a number of important operational timing issues due to the length of the system cycles. The number of different beam samples that can be formed by each slave is limited by the system cycle length. This constraint exists because the slaves have to release the control of their individual memory buses in order to allow write operations by the master. Another constraint is that the master needs to read all the incoming samples and also send all the beam samples out within a system cycle. The maximum number of beams that can be formed in this system are directly limited by these timing constraints. The beam allocations per slave must be calculated carefully by the PC during the system configuration phase. Otherwise, incomplete beams and invalid outputs may result because of the master not having enough time to send out all the beam samples or other complications.

6.4.5 Digital Output

The example system provides the ability to send out all the computed beams through a 16-bit parallel output port. This parallel port is located on the master module. It can be used to communicate the beam data to an external signal processor for further processing. The parallel port is

Sonar Beamforming 6

comprised of octal latches, D-flops and address decoding circuitry which allow the master to communicate with the outside world using a simple protocol.

6.4.6 Analog Output

The example system provides the ability to observe some beams through analog outputs. A digital-to-analog (D/A) converter is included on each slave board and the desired analog beam output can be selected using a thumbwheel switch. Each slave sends the desired beam sample to its D/A converter before shifting it into the FIFO. The overhead of this analog port is minimal, and the port is a very useful test point for system debugging.

6.4.7 System Configurations

The minimum system configuration consists of a master, a slave and an A/D module. The maximum possible system configuration is limited by the speed of the internal hardware and the maximum data rate capability of the output port. All buses must be present in any system configuration.

6.5 SYSTEM HARDWARE

The system hardware includes ADSP-2100 DSP processors, high speed hybrid A/D converters and very high speed CMOS and bipolar LSI components. Hardware selection, design, operation and interface issues in the system are discussed in the following sections.

6.5.1 Component Selection

The ADSP-2100 fulfills the high computational requirements of the slave modules. It also fulfills the CPU requirements of the master module by enabling high speed input data transfers between the A/D modules and the slaves as well as the output transfers. It provides easy handling of memory mapped peripherals and can handle four external interrupts. Design time is saved by using the same processor for both master and slave.

The 12-bit fast A/D converters, high precision sample and hold circuits, low noise operational amplifiers used in the input gain section and anti-aliasing filters are also critical for a high performance system. Front-end analog signal conditioning and A/D circuit design and production using discrete components is a difficult task in noisy digital environments such as the one assumed in the example system. A hybrid A/D converter with on-board voltage references along with sample and hold circuits can perform the required duties better than any discrete circuit with similar functionalities.

6 Sonar Beamforming

Analog Devices' AD1332 A/D converter is appropriate and convenient for several reasons:

- The AD1332 integrates several of the necessary components inside. Its central element is a 12-bit 5µs AD7672 A/D converter. A voltage reference for the converter is included. The converter is preceded by an on-board AD585 sample and hold circuit which itself is preceded by an optional 4-pole Butterworth low-pass filter (anti-aliasing filter). The converter output is fed into a 12-bit latch with tristate output buffers (an optional integrated FIFO is also available for the temporary storage of the conversion results).
- The AD1332 is easily addressed from a microprocessor, which makes it ideal for this application.
- Because the front end circuitry must be duplicated for all incoming sensor inputs, the use of a hybrid helps reduce design, prototyping and production times. Multiplexing the sample and hold outputs into fewer converters is not desirable because of system performance considerations.

The selection of the rest of the high speed VLSI and LSI components in the system is not as crucial to system performance. Several levels of address decoding and buffering that are present in the system result in high demands on the memory components. Integrated Device Technologies' (IDT) 2Kx8 CMOS static RAMs with 25ns access times are used as the data memory components on the slaves. The program memories for the master and the slaves also need to be very fast. Cypress Semiconductor's 2Kx8 CMOS EPROMs with 35ns access times are used as the program memory components for all ADSP-2100s.

The slave FIFOs are IDT's 72413L35 64x5 CMOS FIFOs. Analog Devices' AD569 16-bit D/A converters provide analog output ports on the slaves. The rest of the LSI components are off-the-shelf Advanced Schottky (Fairchild's FAST and Texas Instruments' 74AS series), Advanced Low Power Schottky (Texas Instruments' 74ALS series) or very high speed CMOS (IDT's 74FCT series) integrated circuits. More detailed information and specifications for these components are available from their manufacturers.

The interface card between the PC and the master should be a parallel I/O card that can easily be plugged into the PC's backplane and addressed from a high level program. There are a large number of such I/O boards available. A short development time discourages spending effort on a complicated protocol, and thus a simple protocol and a very flexible I/O card is preferable. Analog Devices' RTI-817 parallel I/O board contains three 8-bit bidirectional ports that accept user-configured directions. These ports are memory mapped and addressable from high level programs. More detailed information about the card can be found in the *RTI-817 User's Manual* published by Analog Devices.

6.5.2 Master Board Hardware

A high level block diagram for the master board is shown in Figure 6.8. The circuitry on the master board is centered around an ADSP-2100 processor running at 8MHz. This master CPU takes its instructions from three CY7291-35 2Kx8 EPROMs (program memory) mapped to the CPU's program memory address space (see Figure 6.9 on page 6-18). The master board does not contain any data memory. The ADSP-2100's internal registers are sufficient for most operations except during the configuration phase. The details of the configuration operation are discussed later in the firmware section.

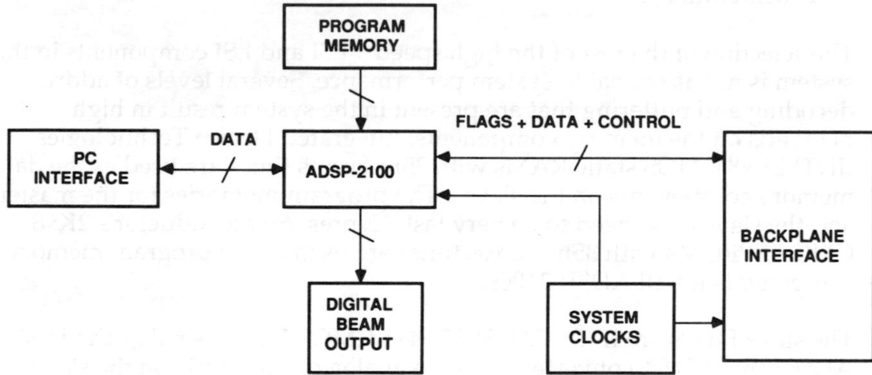

Figure 6.8 Master Module Block Diagram

The master recognizes two interrupts. The first interrupt occurs every 100µs and notifies the master about the availability of new A/D conversion results. This interrupt comes directly from the sampling clock. The second one notifies the master that the external signal processor has received a beam sample and is ready to receive the next one.

6 Sonar Beamforming

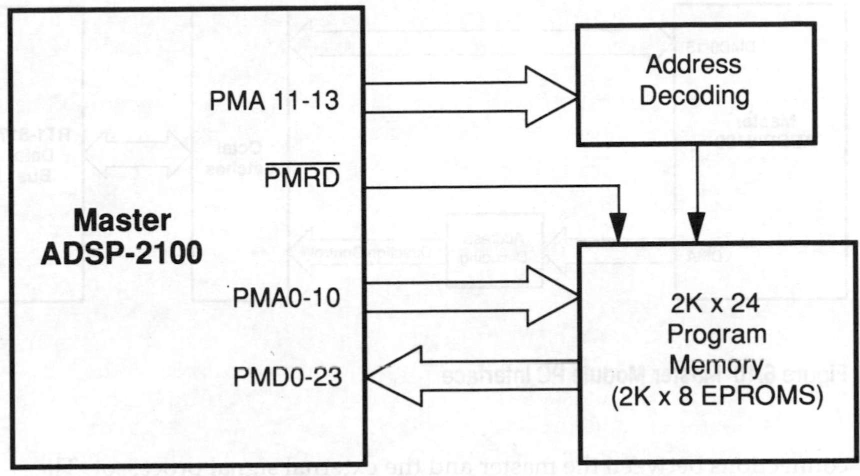

Figure 6.9 Master Module Program Memory Interface

A set of devices generates the sampling clock that is used to sample the analog sensor inputs and to initiate the A/D conversions. These clock signals are sent over the backplane to the A/D boards.

The master board has a number of decoders, latches and buffers that it uses to write to slave data memories, to read slave FIFOs, to send control information to the slaves, to access the PC communication ports and to receive status flags over the backplane bus. See Figure 6.10.

The master board contains a large number of devices dedicated for the CPU's external bus interfaces. These are shown in Figure 6.11. A bank of bus drivers and transceivers provide the necessary buffering for the signals that are traveling over the backplane bus. The communications with the PC are handled through three octal latches that provide a direct interface to the I/O board that is plugged into the PC's backplane. This board also has two inverting octal latches which facilitate the beam sample transfers over the digital output bus.

The interface to the three buses requires four separate connectors: a 96-pin Eurocard connector which is the connection to the backplane, a 50-pin flat cable connector which connects the master and the PC, a 28-pin connector and a 50-pin flat cable connector which are used on the output bus

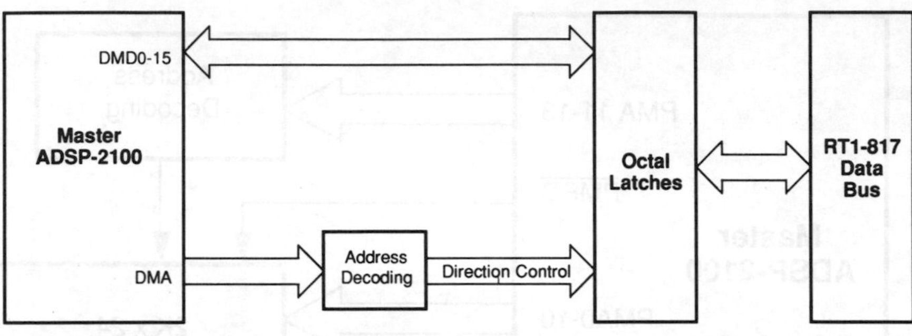

Figure 6.10 Master Module PC Interface

connections between the master and the external signal processor. The master only requires +5V power.

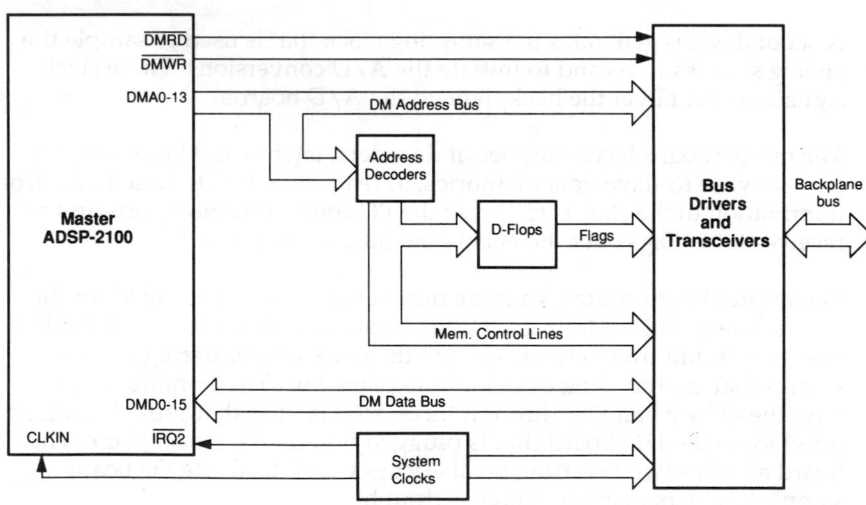

Figure 6.11 Master Module Backplane Interface

6 Sonar Beamforming

6.5.3 Slave Board Hardware

A high level block diagram for the slave board is shown in Figure 6.12. The circuitry on the slave board is centered around an ADSP-2100 processor running at 8MHz. The slave CPU takes its instructions from three CY7291-35 2Kx8 EPROMs (program memory) which are mapped into the program memory address space of the slave CPU (see Figure 6.13). Two 2Kx8, 35ns static RAMs are also mapped into the program memory address space and available for data storage using the upper 16 bits of the ADSP-2100's PMD bus. A decoder provides the necessary address decoding for the PMA lines.

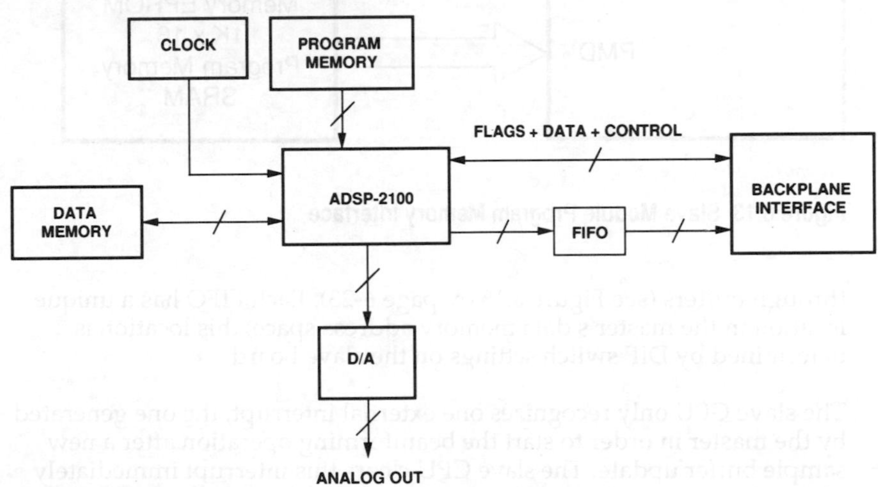

Figure 6.12 Slave Module Block Diagram

There are ten IDT6116LA-25 2Kx8 static RAM chips on the slave board. These devices are mapped onto the data memory address space of the slave CPU (see Figure 6.14 on page 6-22). The first 8K locations of this space are dedicated to the circular sample buffer. The remaining 2K locations are used for additional data storage. A decoder provides the necessary address decoding for the static data RAMs as well as the data memory address mapping for some additional devices and flags.

The output FIFO for the slave module consists of four IDT72413L35 64x5 FIFO components. These can be loaded (written to) from the slave DMD bus. The contents of the FIFO can be read from the backplane data bus

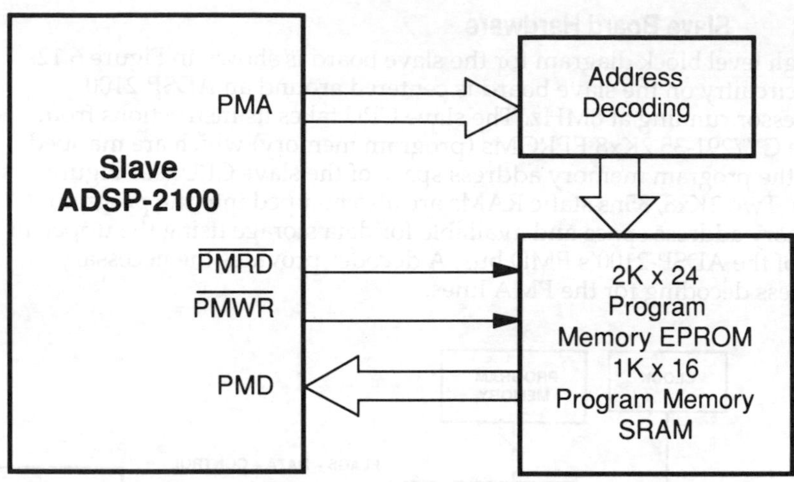

Figure 6.13 Slave Module Program Memory Interface

through buffers (see Figure 6.15 on page 6-23). Each FIFO has a unique location in the master's data memory address space; this location is determined by DIP switch settings on the slave board.

The slave CPU only recognizes one external interrupt, the one generated by the master in order to start the beamforming operation after a new sample buffer update. The slave CPU clears this interrupt immediately after it finishes forming its assigned beams.

The slave board, like the master, contains a large number of bus drivers and transceivers for easy interface with the backplane bus. Several lines on the backplane bring control information from the master. Some of these controls cause the slave CPU to halt its operation and surrender the control of its buses to the master. Some status flags are also sent to the master over the backplane.

The AD569 16-bit D/A converter is mapped into the slave data memory address space (see Figure 6.16 on page 6-23). An AD588 ±5V voltage reference is used with this D/A converter. Because the D/A converter has a slow access time, the slave write cycle must be extended using the DMACK signal (this is an input to the slave CPU). A small circuit is used to generate DMACK during a write cycle to the D/A converter.

6 Sonar Beamforming

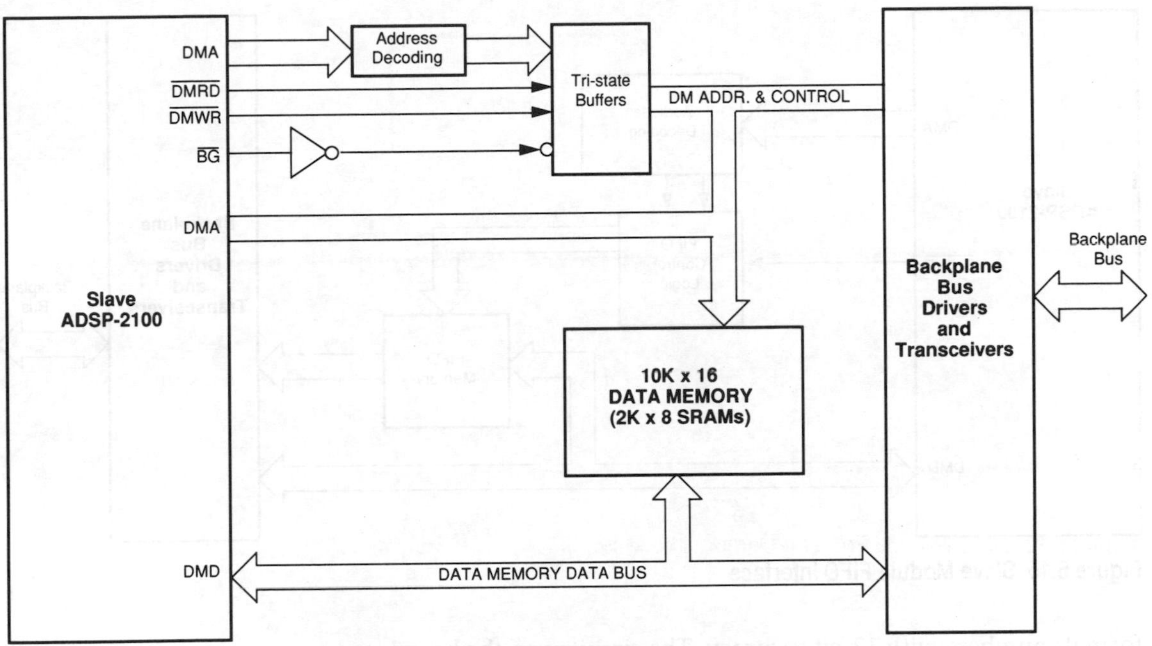

Figure 6.14 Slave Module Data Memory Interface

The slave board has a 16-position thumbwheel switch which selects the beam that is sent out through the D/A converter. A set of four DIP switches give each slave board its own identity. Setting these switches before power-up allows the slave CPU to read them later to determine the beams that are under its responsibility.

There are two connectors on the slave board: a 96-pin Eurocard connector and a male BNC connector. The first is used to interface to the backplane bus, and the second is connected to the analog output of the D/A converter. You can use the BNC connector to send the switch-selected output beam to another device. The slave board requires +5V digital and ±12V analog power supplies.

6.5.4 A/D Board Hardware

A high level block diagram for the A/D board is shown in Figure 6.17, on page 6-24. Each A/D board can receive up to four analog inputs ranging between ±5V and can convert them into 16-bit signed fixed-point (1.15

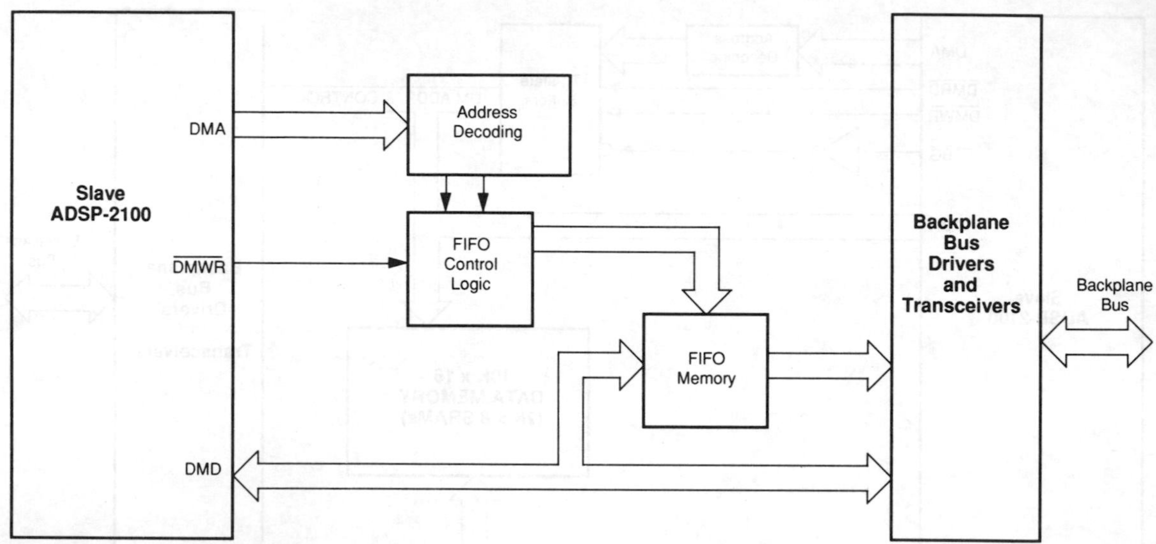

Figure 6.15 Slave Module FIFO Interface

format) numbers with 12-bit accuracy. The circuitry on the board is designed around four AD1332 12-bit hybrid A/D converters.

For each input, a gain stage which uses ADOP07 operational amplifiers is included on the A/D board to provide the necessary amplification of the

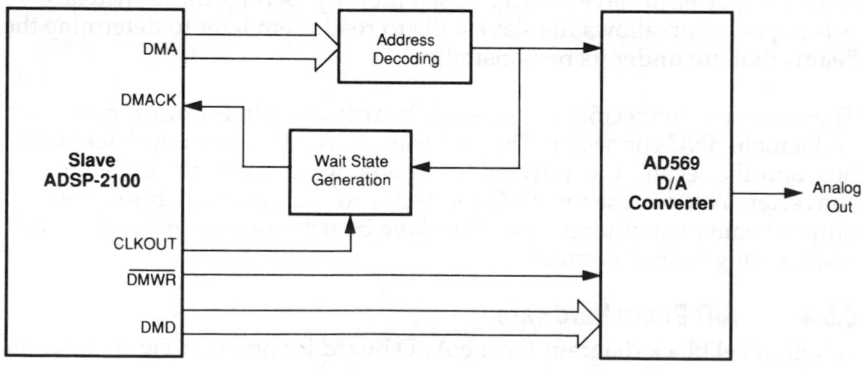

Figure 6.16 Slave Module D/A Converter Interface

6 Sonar Beamforming

hydrophone outputs. The gain is selected (from 1, 10, 100 and 1000) using an external 4-position rotary switch. AD7590 analog CMOS switches select the resistor combination for the amplifier's feedback loop. The CMOS switches are used to allow the future possibility of using CPU-generated signals to make gain selections (i.e., to implement automatic gain control).

The board contains two octal bus drivers to provide adequate buffering for the AD1332 outputs. The outputs of these buffers are tied to the backplane DMD bus (see Figure 6.18). The A/D output, which is in offset binary format, has to have its most significant bit inverted because of the fixed-point format that is used in the system.

A decoder provides a unique location for each AD1332 in the master's data memory address space. The configuration of this decoder must be different in each A/D board for each A/D converter to have a unique location. The A/D board requires +5V digital and ±12V analog power supplies. There are five connectors on the A/D board: a 96-pin male Eurocard connector that is used to interface to the backplane bus and four male BNC connectors that accept the hydrophone outputs.

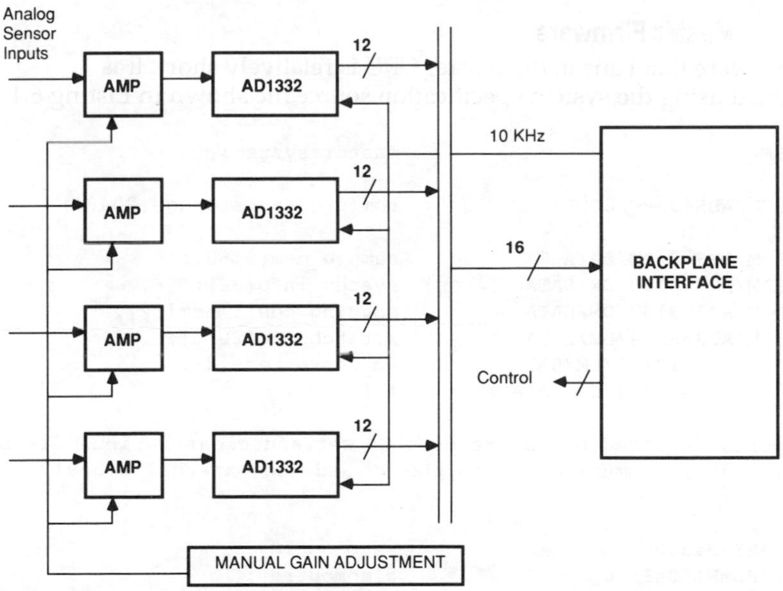

Figure 6.17 A/D Module Block Diagram

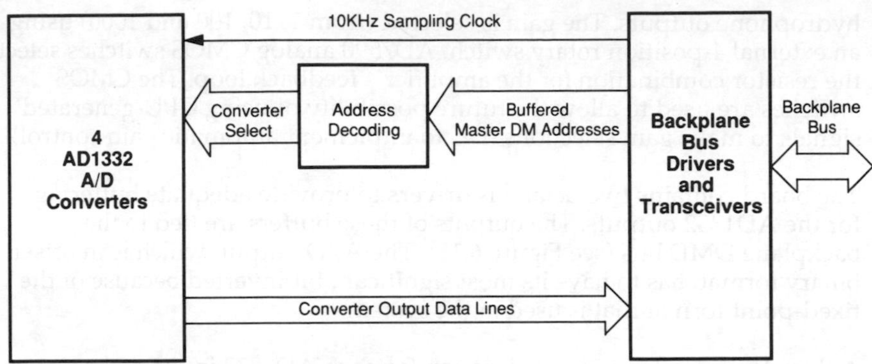

Figure 6.18 A/D Module Backplane Interface

6.6 SYSTEM FIRMWARE

The ADSP-2100 assembly code that is responsible for the master and slave CPUs' operation is discussed in this section.

6.6.1 Master Firmware

The firmware that runs in the master CPU is relatively short. It is assembled using the system specification source file shown in Listing 6.1.

```
.SYSTEM                              master_system;

.SEG/ROM/ABS=0/PM/CODE               rom_program_storage[2048];

.SEG/RAM/ABS=0/DM/DATA               sample_mem[8160];
.SEG/RAM/ABS=8160/DM/DATA            system_info[32];
.SEG/RAM/ABS=8192/DM/DATA            shading_coeff_mem[32];
.SEG/RAM/ABS=8224/DM/DATA            scratch_mem[2016];
.SEG/RAM/ABS=10240/DM/DATA           ad_converters[32];
.SEG/RAM/ABS=14337/DM/DATA           fifos[7];

{The ports declared below are used to set and clear various flags
as well as to communicate with the PC and the external signal
processor}

.PORT/ABS=H#300F                     setbmoutrdy;
.PORT/ABS=H#304F                     clrbmoutrdy;
.PORT/ABS=H#308F                     pcwe;
.PORT/ABS=H#310F                     setsbr;
.PORT/ABS=H#314F                     clrsbr;
```

(listing continues on next page)

6 Sonar Beamforming

```
.PORT/ABS=H#318F                clrslhalt;
.PORT/ABS=H#31CF                setslhalt;
.PORT/ABS=H#30C8                setbpoe;
.PORT/ABS=H#30C9                clrbpoe;
.PORT/ABS=H#30CA                clrbmtaken;
.PORT/ABS=H#30CD                setsmemrd;
.PORT/ABS=H#30CE                clrsmemrd;
.PORT/ABS=H#30CF                setslint;
.PORT/ABS=H#3007                pcrd;
.PORT/ABS=14336                 beamsend;

.ENDSYS;
```

Listing 6.1 System Specification for Master Firmware

The master code, shown in Listing 6.2, only occupies 180 locations in program memory. It can be divided into two sections: the PC communications section and the A/D result handling section.

The PC communications section of the master code starts at the beginning of the file and ends at the *wait* routine. The beginning contains a series of port, variable and interrupt declarations. The program starts by clearing certain flags and then entering the *pc_init_wait* routine which causes the master CPU to wait until the PC is ready to download the system configuration data. Then, the *pc_comm* routine sets some of the internal CPU registers to be used during the communication. At the end of this routine, the PC is notified that the master is ready and the master CPU enters the *pc_wait1* loop.

The first six pieces of data downloaded by the PC are handled differently than the rest. These first six pieces are the following information:

- The number of sensors in the system
- The number of beams to be formed
- The total number of indexes to be used (the indexes are used by the slave CPUs to pick the desired input samples from the circular buffer)
- The number of slaves in the system
- The number of beams assigned to each slave
- The number of beams assigned to the last slave

This data is stored by the *init_sto* routine in the master CPU's data memory (which actually is the same as the slave data memories) in consecutive locations that are declared at the beginning of the program.

Next, the indexes that were calculated by the PC are received and stored in data memory by the *index_store* routine. Then in the *checksum* routine, the master compares the number of pieces of data it received with the number that the PC indicates that it sent (the PC sends a count value to the master in this routine). If these values are equal, that implies that the download was successful.

The *comm_end* routine prepares the master, using the downloaded data, for its next set of tasks (responding to A/Ds, etc.). This routine and the *init_sto* routine are the only times the master CPU reads a value from a slave data memory. This type of read operation is only allowed from a single slave which must be plugged into a designated slot on the backplane.

Another piece of data that is calculated in this routine is the shading coefficient that will be used in the current set of beams. The system currently can only handle a rectangular window with a magnitude of 1. Additional code to handle several different shading windows can be added later.

After configuration, the master CPU issues a signal which commands the slaves to prepare themselves using the recently downloaded data (the slave CPUs were in a TRAP state until now). The master waits for 1000 cycles, more than enough time for the slaves to get prepared. It then enables two nested interrupts and enters the *wait* loop.

The second major section of the code starts at the *wait* loop. The master waits for the "conversion complete" interrupt to occur, then it jumps to the *adcomplete* routine. During this routine, the master reads the output latches (conversion results) of each AD1332 and writes these values into all of the slaves' sample buffers. The master keeps track of a few pointers in order to address the circular buffers and the A/D boards properly.

Next, in the *sendbeam* routine the master reads beam samples from all the slave FIFOs in sequence and writes them to the system output port. Output beam samples are sent one at a time. The handshaking with the external signal processor is executed for each beam sample. The master starts the handshake by asserting the BMOUTRDY signal and then waits for the BMTAKEN interrupt (set by the external processor) that indicates that the external processor is ready to receive the next value. Then the interrupt is cleared by the master and the next beam sample is sent out in the same manner.

6 Sonar Beamforming

For a given number of desired beam directions, it is not always possible to divide the job evenly among the slaves. Thus, the number of beams formed by the last slave may be different than the number formed by each of the rest of the slaves. Consequently, the output FIFO of the last slave is handled by the *oneslave* routine, which serves a similar purpose to the *sendbeam* routine except it only handles the last slave.

Once the beam output tasks are completed, a RTI (return from interrupt) instruction is executed and the program returns to the *wait* loop where the master waits for the next *adcomplete* interrupt.

The master continues its cyclic, double-interrupt-driven operation until the assertion of $\overline{\text{RESET}}$ or a system power-down.

```
.MODULE/ROM/ABS=0       master_code;

{The following are declarations for ports that are used to set and
clear several flags. Data memory mapped D-flipflops are used to
generate flags}

.PORT                   setbmoutrdy; {Beam output ready flag}
.PORT                   clrbmoutrdy;
.PORT                   pcwe;        {PC output port}
.PORT                   setsbr;      {Slave bus request flag}
.PORT                   clrsbr;
.PORT                   setslhalt;  {Slave halt flag}
.PORT                   clrslhalt;
.PORT                   setbpoe;     {Backplane output enable flag}
.PORT                   clrbpoe;
.PORT                   clrbmtaken;
.PORT                   setsmemrd;  {Slave memory read flag}
.PORT                   clrsmemrd;
.PORT                   setslint;    {Slave interrupt}
.PORT                   pcrd;        {PC input port}
.PORT                   beamsend;    {Interrupt the external processor}

{The following are variables that contain the system configuration
info.}

.VAR/DM/ABS=8160        sensor_num;
.VAR/DM/ABS=8161        beam_num;
.VAR/DM/ABS=8162        index_num;
.VAR/DM/ABS=8163        slave_num;
.VAR/DM/ABS=8164        beams_per_slave;
.VAR/DM/ABS=8165        last_slave_beam_num;
```

Sonar Beamforming 6

```
{The following is the main body of the master program}

{Interrupt vectors occupy the first four PM locations}
                    JUMP adcomplete;  {Vectored address for IRQ0}
                    RTI;
                    JUMP beamtaken;   {for IRQ2}
                    RTI;

{Program execution starts here}
                    IMASK=b#0000;
                    ICNTL=b#00000;
                    AY0=h#0FFF;
                    DM(clrbmoutrdy)=MX0;  {Clear all flags by}
                    DM(clrbmtaken)=MX0; {writing a value into}
                    DM(clrsbr)=MX0;     {their respective DM}
                                        {mapped port locations}
                    DM(clrslhalt)=MX0;
                    DM(clrbpoe)=MX0;
                    DM(clrsmemrd)=MX0;
pc_init_wait:       AX0=DM(pcrd);      {Wait for the ready}
                    AR=AX0-AY0;        {message from the PC}
                    IF EQ JUMP pc_comm;
                    JUMP pc_init_wait;

pc_comm:            MX0=h#FF0F ;       {Initial set up before the}
                    AY0=5;             {PC communications.}
                    AF=AY0+1;   {A message is sent to the PC}
                    AY0=h#1FFF; {at the end of this routine}
                    AX1=h#FFFF; {signaling that the master is}
                    MX1=h#FF0F; {ready}
                    M0=0;
                    I1=8160;
                    L1=6;
                    I2=0;
                    M2=1;
                    L2=8160;
                    M3=-1;
                    I4=8166;
                    M4=1;
                    I5=9000;
                    M5=0;
                    L5=1;
                    I6=0;
                    M6=1;
                    L6=2055;
                    DM(setsbr)=AY0; {Set slave Bus Request flag}
```

(listing continues on next page)

6 Sonar Beamforming

```
                     DM(pcwe)=MX0;    {Send Ready mesage to the PC}
                     JUMP pc_wait1;

pc_comm_end_check:   AY1=I3;     {Counter register to check for}
                     AR=AY1-1;    {the end of the index}
                     IF LT JUMP pc_end;      {downloading}

pc_wait1:            AY1=DM(pcrd);     {Wait for FFFF from the PC}
                     AR=AY1-AX1;
                     IF EQ JUMP pc_read;
                     JUMP pc_wait1;

pc_read:             MODIFY(I6,M6);  {Write the initial 6 pieces}
                     AF=AF-1 ;       {of data and then start}
                     IF LT JUMP index_store; {the index storage}

init_read:           AX0=DM(pcrd);    {Read data sent from the PC}
                     AY1=h#E000;
                     AR=AX0 AND AY1; {Mask out bottom 13 bits}
                     AY1=h#A000;
                     AR=AR-AY1;
                     IF EQ JUMP init_sto; {Compare to h#A000}
                     JUMP init_read;
init_sto:            AR=AX0 AND AY0;       {Mask out top 3 bits}
                     DM(setbpoe)=MX0;  {Enable backplane drivers}
                     DM(I1,M1)=AR;
                     DM(setsmemrd)=MX0; {Enable reading from}
                                        {slave DM}
                     L4=DM(beam_num);
                     I3=DM(index_num);
                     DM(clrsmemrd)=MX0; {Disable reading from}
                                        {slave DM}
                     L3=I3;
                     DM(clrbpoe)=MX0;  {Disable backplane drivers}
                     JUMP pc_wait1;

index_store:         AX0=DM(pcrd);     {Read the PC output}
                     AY1=h#E000;
                     AR=AX0 AND AY1;   {Mask out bottom 13 bits}
                     AY1=h#A000;
                     AR=AR-AY1;
                     IF EQ JUMP sto;   {Compare to h#A000}
                     JUMP index_store;

sto:                 AR=AX0 and AY0;
                     DM(setbpoe)=MX0;
```

```
                    DM(I2,M2)=AR ;      {Store index}
                    MODIFY(I3,M3);
                    DM(clrbpoe)=MX0;
                    JUMP pc_comm_end_check;
pc_end:             AX1=h#FF0F;
pc_end_loop:        AY1=DM(pcrd);       {Read PC output and decide}
                    AR=AX1-AY1;     {whether communication should}
                    IF EQ JUMP checksum;     {be completed or not}
                    JUMP pc_end_loop;
checksum:           AX1=I6;             {Perform cheksum operation}
checksum_loop:      AY1=DM(pcrd);
                    AR=AX1-AY1;
                    IF EQ JUMP comm_end;
                    JUMP checksum_loop;

comm_end:           MX0=h#F0;           {This routine ends the PC}
                    DM(pcwe)=MX0;  {communications and does some}
                    DM(setbpoe)=MX1; {reads from the slave DM in}
                    AY0=h#0201;         {order to prepare for the}
                    AF=AY0-1 ;      {beamforming tasks}
                    AY0=0;
                    DM(setsmemrd)=MX0;
                    SI=DM(sensor_num); {Read #of sensors from}
                    DM(clrsmemrd)=MX0; {slave DM and do the}
                    AY0=SI;     {necessary division to obtain a}
                    AR=AY0-1;    {scaled magnitude value for the}
                    IF EQ JUMP onesensor;    {rectangular shading}
                    AY0=0;                   {window}
                    SR=LSHIFT SI BY 9 (LO);
                    AX1=SR0;
                    ASTAT=0;
                    DIVQ AX1;
                    DIVQ AX1;DIVQ AX1;DIVQ AX1;
                    DIVQ AX1;DIVQ AX1;DIVQ AX1;
                    DIVQ AX1;DIVQ AX1;DIVQ AX1;
                    DIVQ AX1;DIVQ AX1;DIVQ AX1;
                    DIVQ AX1;DIVQ AX1;DIVQ AX1;
                    DM(8192)=AY0;       {Shading coefficient}
                    JUMP allsensor;
onesensor:          AY0=h#7FFF; {This is shading coefficient}
                    DM(8192)=AY0; {in case of a single sensor}
allsensor:          DM(setsmemrd)=MX1;
                    AY0=DM(slave_num); {Do the rest of the}
                    AX1=DM(beams_per_slave); {necessary reads}
                    MY1=DM(last_slave_beam_num);
                                {before releasing slave BR flag}
```

(listing continues on next page)

6 Sonar Beamforming

```
                        MX0=DM(sensor_num);
                        DM(clrsmemrd)=MX0; {Clear the slave DM read}
                        MY0=255;           {flag and set up some}
                        MR=MX0*MY0(UU);{pointers to be used later on}
                        SI=MR0 ;           {#of samples to be placed}
                        SR=LSHIFT SI BY -1 (HI);
                                           {into the sample buffer}
                        L2=SR1;
                        AR=AY0-1;
                        DM(clrsbr)=MX1;
                        AX0=AR;
                        I1=10240;
                        L1=MX0;
                        I2=0;
                        M2=1;
                        IF EQ JUMP fix_base;
                        I3=h#3800;         {Base addr. for the FIFOs}
                        AR=AY0+1;          {with multiple slaves}
                        L3=AR;
                        JUMP normal;
fix_base:               I3=h#3801;         {Base addr. for the FIFO}
                        L3=0;                    {with single slave}
normal:                 DM(setslhalt)=MX1; {HALT the slaves, this}
                        DM(clrslhalt)=MX1; {will cause them to get}
                        DM(clrbpoe)=MX1; {out of their TRAP state}
                        CNTR=1000;
                        DO slave_wait UNTIL CE; {Wait until all}
slave_wait:                NOP;            {slaves are ready to go}
                        AR=1;
                        AY1=1;
                        ICNTL-b#00101;
                        IMASK=b#0001;
                                     {Enable sampling interrupt IRQ0}

wait:                   JUMP wait;

adcomplete:             AR=AR-AY1;               {Ignore the first}
                        IF EQ JUMP first_adcomp; {IRQ0 by using this}
                        DM(setsbr)=MX1;
                        AF=AY0-1;
                        DM(setbpoe)=MX1;   {This routine is used}
                        CNTR=MX0;          {to fill up the sample}
                        DO sample_store UNTIL CE; {memory after each}
                           MX1=DM(I1,M1);        {A/D conversion}
sample_store:              DM(I2,M2)=MX1;
                        DM(clrsbr)=MX1;
```

```
                        DM(setslint)=MX1;
                        DM(clrbpoe)=MX1;

sendbeam:               IF EQ JUMP oneslave; {Read out the beams in}
                        MODIFY(I3,M1);       {a sequential manner in}
                        CNTR=AX0;   {each FIFO in order of numbering}
                        DO beamout UNTIL CE; {Send beam outputs via}
                           CNTR=AX1;       {the digital output port}
                           DO fifo_out UNTIL CE;
                              DM(setbpoe)=MX1;
                              MX1=DM(I3,M0);
                              DM(clrbpoe)=MX1;
                              DM(beamsend)=MX1;
                              IMASK=b#0100;{Enable data receive intr}
                              DM(setbmoutrdy)=MX1;     {Set a flag}
                              CNTR=6;  {for the external processor}
                              DO resp_wait UNTIL CE;
resp_wait:                       NOP;
fifo_out:                     NOP;
beamout:                   MODIFY(I3,M1);


oneslave:               CNTR=MY1;
                        DO endfifo UNTIL CE;
                           DM(setbpoe)=MX1; {This routine is for}
                           MX1=DM(I3,M0);   {single slave}
                           DM(clrbpoe)=MX1; {configurations and is}
                                            {also used for handling}
                           DM(beamsend)=MX1; {the last FIFO read out}
                           IMASK=b#0100;        {It handles the}
                           DM(setbmoutrdy)=MX1; {irregularity of the}
                           CNTR=6;              {last set of beams,}
                           DO resp_wt UNTIL CE; {i.e. possibly fewer}
resp_wt:                      NOP;              {beams}
endfifo:                   NOP;
first_adcomp:           AR=0;
                        RTI;

beamtaken:              DM(clrbmoutrdy)=MX1;
                                    {Interrupt routine to handle}
                        DM(clrbmtaken)=MX1;
                                    {the data receive confirmation}
                        RTI;        {from the external processor}

.ENDMOD;
```

Listing 6.2 Master Firmware

6 Sonar Beamforming

6.6.2 Slave Firmware

The slave board requires less firmware than the master. It is assembled using the system specification source file shown in Listing 6.3.

```
.SYSTEM                              slave_system;

.SEG/ROM/ABS=0/PM/CODE               rom_program_storage[2048];
.SEG/RAM/ABS=2048/PM/DATA            index_mem[2048];

.SEG/RAM/ABS=0/DM/DATA               sample_mem[8160];
.SEG/RAM/ABS=8160/DM/DATA            system_info[32];
.SEG/RAM/ABS=8192/DM/DATA            shading_coeff_mem[32];
.SEG/RAM/ABS=8224/DM/DATA            scratch_mem[2016];

{The ports declared below are used.to set and clear some flags as
well as to write to the DAC and to read from some hardware
switches}

.PORT/ABS=H#2800                     beamout;
.PORT/ABS=H#3000                     beamdac;
.PORT/ABS=H#3800                     clrslint;
.PORT/ABS=H#3900                     slave_id;
.PORT/ABS=H#3A00                     dac_beam_sel;

.ENDSYS;
```

Listing 6.3 System Specification for Slave Firmware

The slave firmware code occupies only 100 locations in program memory. The slave firmware, shown in Listing 6.4, can be divided into two sections: the system set-up section and the beamforming section.

The set-up section of the code starts at the beginning of the file and ends at the *wait* routine. The beginning contains a series of port, variable and interrupt declarations. Only one interrupt, SLINT, is recognized by the slave; this is the interrupt initiated by the master to start the beamforming operation.

After system parameters are declared, the slave enters a TRAP state. The slave gets reactivated after the master is finished communicating with the PC. The master asserts the SLHALT signal, which wakes up the slave and causes the program execution to continue from the location following the TRAP instruction.

Sonar Beamforming 6

In the first part of the program, the slave moves the indexes from its data memory into its "index memory" which is located in its program memory space (the indexes are in data memory initially because the master has to store them there temporarily). Then, the slave sets up its address registers using the downloaded beamforming information. Next, the slave enters the *wait* loop to wait for the beamforming interrupts issued by the master.

The slave constantly monitors the D/A beam selection switch while it is in the *wait* loop. Since the slave returns to the *wait* loop every 100µs, it can decide, in real time, which beam to send out through the analog port.

The second major section of the program starts at the *wait* loop. As soon as the SLINT interrupt is received, the slave jumps to the *beam_form* routine. The *beam_form* routine contains very tight loops which allows the slave to form a large number of beams. The routine reads the index memory and picks the indexed samples from the sample buffer. These samples are the delayed samples that are needed for the beam summation.

The frame of the circular buffer is rotated every time a new set of samples comes in. Therefore the indexes that are read must be modified before being used, because they are referenced to the absolute origin of the circular buffer.

The samples that are read are multiplied by the shading factor (which is currently 1) and accumulated in the MR register. The resulting beam sample is written into the FIFO. If the beam sample belongs to the beam that is requested at the analog port, it is then written to the *beamdac* port (the D/A converter). Before returning from the interrupt, the slave clears the SLINT flag. Then the program returns to the *wait* loop to wait for the next SLINT interrupt.

The slave continues its cyclic, single-interrupt-driven operation until the assertion of $\overline{\text{RESET}}$ or a system power-down.

6 Sonar Beamforming

```
.MODULE/ROM/ABS=0     slave_code;

{The following are declarations for data memory mapped ports. One
is used to clear a flag, while others are used to write data to
the DAC, FIFO and read the hardware switches on the slave board}

.PORT               beamout;        {FIFO}
.PORT               beamdac;        {DAC}
.PORT               clrslint;       {Clear the slave interrupt}
.PORT               slave_id;       {Slave identity dipswitch}
.PORT               dac_beam_sel;   {Analog output selection switch}

{The following are variables that contain the system configuration info}

.VAR/DM/ABS=8160        sensor_num;
.VAR/DM/ABS=8161        beam_num;
.VAR/DM/ABS=8162        index_num;
.VAR/DM/ABS=8163        slave_num;
.VAR/DM/ABS=8164        beams_per_slave;
.VAR/DM/ABS=8165        last_slave_beam_num;

{The following is the main body of the slave program}

                        JUMP beam_form;    {Vectored addr. for IRQ0}
                        RTI;
                        RTI;
                        RTI;

   IMASK=b#0000; {Disable interrupts}
   ICNTL=b#00000;
   DM(clrslint)=MX0;
   TRAP;    {TRAP until pc_comm ends}

pc_comm_end:            I1=0;          {This initial routine is used}
                        M1=1;          {to transfer the indexes from}
                        L1=DM(index_num); {sample_mem into the}
                        I4=2048;                    {index_mem in PM}
                        M4=1;
                        L4=L1;
                        CNTR=L1;
                        DO index_store UNTIL CE;
                           MX0=DM(I1,M1);
index_store:               PM(I4,M4)=MX0;

                        MX0=DM(sensor_num);
                        MY0=255;
```

```
                        MR=MX0*MY0(UU);
                        SI=MR0;
                        SR=LSHIFT SI BY -1 (HI);
                                            {SR1 contains the length}
                        I1=MX0;             {of the circular sample}
                        M1=MX0;             {buffer}
                        L1=SR1;
                        SI=DM(slave_id);  {Determine this slave's}
                        SR=LSHIFT SI BY -12 (HI);
                                            {ID# and the starting}
                        AX0=DM(slave_num); {location of the first}
                        AY0=SR1;            {index.Also determine}
                        AF=AX0-AY0;         {the # of indexes}
                        AF=AF-1;            {for this slave}
                        IF EQ JUMP last_slave;
                        SE=DM(beams_per_slave);
                        JUMP all_slave;
last_slave:             SE=DM(last_slave_beam_num);

all_slave:              MX0=DM(sensor_num);
                                            {This routine calculates}
                        MY0=DM(beams_per_slave);
                                            {the starting address of}
                        MR=MX0*MY0(UU);     {this slave's indexes}
                        SI=MR0;
                        SR=LSHIFT SI BY -1 (HI);
                        MX0=SR1;
                        AR=AX0-AY0;
                        AY1=AR;
                        AR=AY1-1;
                        MY0=AR;
                        MR=MX0*MY0(UU);
                        SI=MR0;
                        SR=LSHIFT SI BY -1 (HI);
                        AX1=2048;
                        AY1=SR1;
                        AR=AX1+AY1;
                        I5=AR;              {Starting address of the}
                        M5=1;               {indexes for this slave}
                        MX0=SE;
                        MY0=DM(sensor_num); {# of indexes per beam}
                        MR=MX0*MY0(UU);
                        SI=MR0;
                        SR=LSHIFT SI BY -1 (HI);
                        L5=SR1;             {Total # of indexes to}
                        M3=0;               {be used by this slave}
```

(listing continues on next page)

```
                     L3=L1;
                     I6=8192;
                     M6=0;
                     L6=1;
                     MY1=DM(I6,M6); {Shading coefficient; there}
                     AY1=DM(sensor_num);
                                    {is only one now since it is}
                     AR=AY1-1;      {a rectangular window}
                     AX1=AR;
                     ICNTL=b#00001;
                     IMASK=b#0001; {Enable slave interrupt IRQ0}

wait:                SI=DM(dac_beam_sel); {Setup down counter to}
                     SR=LSHIFT SI BY -12(HI);
                                    {be used in deciding}
                     AY1=SR1;       {which beam to send out}
                     AF=AY1+1;      {to the DAC}
                     JUMP wait;  {Wait for sample buffer update}

beam_form:           CNTR=SE;
                     DO beam_end UNTIL CE;
                        MR=0;
                        AY0=PM(I5,M5); {Read index}
                        M3=AY0;
                        I3=I1;
                        MODIFY(I3,M3); {Modify index}
                        MX0=DM(I3,M3); {Get the desired sample}
                        CNTR=AX1;
                        DO single_beam_sample UNTIL CE;{Beamform}
                           MR=MR+MX0*MY1(SS), MX0=DM(I3,M3),
                                 AY0=PM(I5,M5);
                           M3=AY0;
                           I3=I1;
single_beam_sample:        MODIFY(I3,M3);
                        MR=MR+MX0*MY1(RND);{Last MAC w/rounding}
                        AF=AF-1;       {Check which beam to send}
                        IF EQ JUMP dac_write; {out to the DAC}
beam_end:               DM(beamout)=MR1; {Write result to FIFO}
                     MODIFY(I1,M1);    {Advance circular sample}
                     DM(clrslint)=MX0; {buffer pointer}
                     RTI;

dac_write:           DM(beamdac)=MR1;           {Write result to DAC}
                     JUMP beam_end;

.ENDMOD;
```

Listing 6.4 Slave Firmware

Sonar Beamforming 6

6.7 SYSTEM SOFTWARE

The system software consists of the PC program that is responsible for the user interface and the downloading of the system configuration data. The code is written in the C language and compiled on the Microsoft C Compiler.

The declaration section at the beginning of the program includes certain useful libraries, defines a number of variables and declares the 8-bit parallel I/O port addresses. These ports are located on an Analog Devices RTI-817 parallel I/O card which is plugged into the PC's backplane. Following this section there are a series of function definitions and the execution loop of the program.

The program interactively takes in the system variables from the user. Some questions are displayed on the screen which are answered by the user via the keyboard. The values that have to be entered by the user are: the number of sensors, the number of beams to be formed, the number of slaves in the system, the cartesian coordinates for the sensor locations and the spherical coordinates for the desired beams. The program assigns these values to variables and arrays in order to calculate the necessary tap delays for beamforming.

The program converts the spherical coordinate beam directions to cartesian coordinates. Then it calculates the necessary delays using the equation (2). The propagation speed of sound in water is assumed to be 1470 m/s, which is typical for the ocean water. (This value can be changed easily in the code to conform to the application environment.) The program identifies the beams that the system is unable to produce with the given array configuration. This task is accomplished by checking whether any one of the required delays falls outside of the sample buffer length.

Next, the program determines the maximum number of beams that can be formed with the given system configuration. It also calculates the number of beams to be assigned to each slave and the number of beams to be assigned to the last slave. These values are stored as variables to be downloaded to the master.

The program downloads the system configuration information to the master, beginning with six pieces of system set-up data, as explained earlier in the master firmware section. Then the program sends all the calculated indexes to the master, followed by a checksum, which, as explained earlier, corresponds to the number of pieces of data (number of

6 Sonar Beamforming

indexes + 6) just downloaded. If the master acknowledges that the download was successful, the downloading operation is completed by sending a confirmation message to the PC screen. If the master indicates an unsuccessful download, the downloading operation is terminated by sending a failure message to the the screen. In this case, the user is given the choice of aborting or retrying the download.

The download is easily executable by the user. Once the system parameters are entered, it takes at most a few seconds for the PC to download all the information to the master.

6.8 ENHANCEMENTS

There are several ways to improve the performance, functionality and the user interface of the example beamformer. Possible additional features as well as some architectural and circuit level enhancements are briefly discussed in this section.

6.8.1 Additional Features

A large number of features can be added to this system without great difficulty. One feature is the choice of frequencies for the input sampling clock. It is possible to route an external clock to the A/D boards by incorporating additional hardware on the master board. The external sampling clock option would an additional piece of data collected by comm.*c* during the system parameter configuration. The program would download the information to the master, which would activate the necessary signals for clock selection. The program would also have to modify the beam assignments, because the system cycle may be shorter or longer depending on the choice of sampling frequency.

A software enhancement is the ability to save the current system parameters into a file. This allows the system to be restarted by instructing *comm.c* to boot up the system using the saved parameters instead of getting them from the user. A user could edit this file to restart the system with a new set of configuration parameters in a very short time. This feature would make *comm.c* even more user-friendly.

Another feature is the addition of a filtering and smoothing circuit for the output of the D/A converter. Smoothing the output of the D/A converters would make it possible to feed the analog output beams into a spectrum analyzer or a general purpose data acquisition system for further signal analysis.

The availability of various shading windows for the inputs is another useful enhancement. The modifications would have to be done in *comm.c* and also the system firmware programs. In *comm.c,* the shading window option would be gotten from to the user and the shading coefficients would be calculated on the fly and downloaded to the master. The master would send these coefficients to the slaves instead of the unit rectangular window.

The shading factors would reside in the program memory space of each slave and could ultimately be used by the slaves during beamforming. The downloading overhead would be minimal. The additional program memory accesses during the beamforming loop should not result in a performance degradation since they can be performed in parallel with the data accesses. Careful calculations are needed to determine the exact performance consequences of such a system modification.

6.8.2 Performance Improvements

There are several ways to improve the performance of the beamformer described in this chapter by making relatively minor modifications to the hardware and software. Some modifications, with ascending levels of complexity, are discussed in this section.

An important performance issue for a real-time beamformer is the beam throughput. The main goal of such a system is to form as many simultaneous beams as possible using the existing technology. There are several ways to improve the beam throughput within the existing distributed processing architecture. The most obvious and relatively easy way to achieve this goal is to replace the ADSP-2100s with ADSP-2100As. The ADSP-2100A has an 80ns instruction cycle time as opposed to 125ns cycle time of the ADSP-2100. This upgrade would result in ≈50% increase in system beam throughput.

The ADSP-2100A is pin and source code compatible with the ADSP-2100. This allows the easy upgrade of the system with no firmware changes. The modifications that are needed are mostly in hardware. The timing requirements during the ADSP-2100A's data and program memory access operations must be carefully analyzed and faster devices should be placed on the critical data paths. It is possible to upgrade only the existing memory components to compensate for the new shorter data and program memory access cycles.

A modification to *comm.c* would also be necessary because it would have to be able to assign a larger number of beams per slave. The input

6 Sonar Beamforming

bandwidth of the external signal processor would have to be carefully evaluated, because it is likely that the output bandwidth of the upgraded system, in maximum configuration, would be higher than the input capacity of the external processor. If such an incompatibility resulted, you could use fewer slaves to match the output bandwidth requirements. The overall consequence would be a cost reduction for the less demanding users and higher performance for the more demanding users.

Another improvement is increasing the maximum number of sensor inputs. It is possible to add more input channels to the beamformer. However, each added A/D converter would have to be placed in a unique location in the master CPU's data memory address space, requiring some additional address decoding circuitry on the A/D boards. There is enough room for more digital components on these boards and the changes in the wiring would be relatively simple.

One important effect of adding channels is a reduction of the maximum number of beams that can be formed simultaneously, because the master will have to read more inputs within one system cycle and consequently will have less time to read the results from the slave FIFOs. It is possible to keep the beam throughput at the current level by upgrading the processors while increasing the number of sensor inputs. These performance tradeoffs should be considered carefully before expanding the A/D capabilities of the system.

A major improvement is to redesign the A/D boards with multi-channel A/D converter hybrids replacing the single-channel AD1332s. Analog Devices' AD1334 would be the optimal choice. The system redesign effort that is necessary to implement the substitution of AD1334s is of moderate difficulty. The savings in the number of A/D cards would prove this redesign effort to be very valuable, especially if the need for input channels is expected to rise.

The AD1334 contains four sample and hold circuits, a 4-to-1 analog multiplexer, an AD7672 12-bit, 5μs A/D converter and an output FIFO. The sample and hold circuit (AD585) and the A/D converter are the same as the ones used in the AD1332. The output FIFO is 12 bits wide and 64 locations deep. It is possible to use this hybrid in a mode where all of the sample and hold circuits sample the inputs simultaneously.

Some overhead analog circuitry must be added externally because of the lack of on board low-pass filters in the AD1334. The AD1334 has the same package as the AD1332, so it would be possible to fit as many as three

A/D hybrid packages on the same board even with the additional digital and analog overhead circuitry that is needed. Such a construction strategy would allow each A/D board to handle up to 12 sensor inputs.

Some minor modifications in the master firmware would also be needed in order to properly address the AD1334s.

6.9 REFERENCES

Baggeroer, A. B. 1978. "Sonar Signal Processing." In *Applications of Digital Signal Processing*, A. V. Oppenheim, Ed. Englewood Cliffs, NJ: Prentice-Hall, Inc.

Curtis, T. E. and R. J. Ward. 1980. "Digital Beam Forming for Sonar Systems." *IEEE Proceedings*, Vol. 127, Pt. F, No. 4.

Gray, D. A. 1985. "Effect of Time-Delay Errors on the Beam Pattern of a Linear Array." *IEEE Journal of Oceanic Engineering*, Vol. OE-10, No. 3.

Hodgkiss, W. S. and V. C. Anderson. 1981. "Hardware Dynamic Beamforming." *Jour. Acoust. Soc. Am.* 69(4).

Janssen, R. J. 1987. "Sonar Beamforming and Signal Processing." *Electronic Progress,* Vol. 28, No. 1, Raytheon Co.

Karagozyan, K. 1988. "A Multi-Processor Based Digital Beamforming System." Master's thesis, Massachusetts Institute of Technology, Cambridge, Mass.

Knight, W. C., R. Pridham and S. M. Kay. 1981. "Digital Signal Processing For Sonar." *IEEE Proceedings*, Vol. 69, No. 11.

Pridham, R. G. and R. A. Mucci. 1978. "A Novel Approach to Digital Beamforming." *Jour. Acoust. Soc. Am.* 63(2).

Winder, A. A. 1975. "Sonar System Technology," *IEEE Transactions on Sonics and Ultrasonics*, Vol. SU-22, No. 5.

inmos **Application note**

Image processing with the IMS A100

Contents

List of Figures

Image processing with the IMS A100

1 Introduction

1.1 The aims of this document

The IMS A100 performance makes the real time processing of digital images a practical possibility. This document is a practical guide, which explains how the device is used to process digital images. The processing done by the IMS A100 will be some form of feature extraction, such as line, corner or edge detection. Feature extraction is often the first stage in the analysis of an image. Further analysis of an image, for example, deciding that a group of features in an image is a vehicle number plate, is a higher level function, beyond the scope of this document. This application note describes the following.

- The operations of filtering and edge detection of a picture or image using a technique of 2-dimensional convolution are explained. Some simple filter types including edge detection and contrast enhancement are described.
- The use of the IMS A100, to perform the 2 dimensional convolution, in order to process an image is described. This shows the simplicity of use of the IMS A100 in this particular application.
- The estimation of performance and cost, for processing an image using the IMS A100 is described. Several possible systems consisting of IMS A100 devices are given, to illustrate how easily the cost and performance may be controlled, by using different numbers of devices, and by altering the complexity of the system.
- The processing of images at real time speeds (20 frames per second) is described, and a hardware implementation of this is given. This shows the high performance possible using the device.

1.2 Document structure

The remainder of section 1 gives an introduction to signal processing, and shows the position of the IMS A100 within the field of signal processing, and more specifically its capabilities for the processing of digital images.

Section 2 gives gives a practical explanation of some of the concepts of image processing. Included is an explanation of how filtering and edge detection of a picture operates, and how this may be applied to the IMS A100.

Section 3 gives two possible systems which may be constructed using the IMS A100, from a medium performance system to a very high performance system which will operate at real time speeds. Included in this section is a description of how the performance of a prospective system may be estimated by trading off performance, complexity and cost.

Section 4 concludes and summarises the findings of this application note.

Section 6 gives an implementation of 2-D image processing using the IMS B009, running on an IBM PC. This is included as a practical illustration of the techniques described in section 2 and 3.

1.3 An overview of signal processing

Signal processing is an area of engineering which fills many people with dread. This is not entirely surprising when one considers both the theoretical and practical aspects of the subject. On the one side there are the mathematical algorithms required to solve even the simplest problem. This has long been regarded as the territory of academics and not to be tackled by the average engineer. On the other side there is the circuitry required to implement these algorithms. Historically, systems have often required many complex circuits, with system design requiring a knowledge of analogue design, and also, in the more recent past, digital design.

Not surprisingly, there are very few scientists in the world with the knowledge or experience required to deal with all the aspects of signal processing design. Signal processing design now covers both analogue and digital design from the low end audio spectrum (40 KHz) through the video spectrum (100 MHz) to the top end of the radio spectrum (100 GHz). When signal processing in all these areas began the techniques used were purely analogue. The power of digital signal processing now approaches the top end of the video spectrum. Although it is not yet possible to process pictures the size of a TV screen in real time, it will undoubtedly become possible within the next decade. One of the main applications of the IMS A100, as described in this document, is the processing of pictures in real time.

In the radio frequency (RF) spectrum, specialised devices are used as the first stage processing elements. These devices use components such as wave-guides, to give the necessary processing bandwidth (GHz). The fastest devices use materials such as Gallium Arsenide, often super-cooled to improve its performance. However, these devices are expensive and their use is avoided if possible. The information extracted by these devices from a signal may be used by todays digital devices operating at speeds approaching 100 MHz. In the future, todays digital devices may improve to a level where they encroach on the radio spectrum. However, it is likely that RF devices will always be required as the front-end processing elements at these high frequencies. The reason for this may remain that it is impossible to either sample or synthesise an analogue signal at speeds in excess of 100 MHz, without resort to cost prohibitive technology.

1.4 Analogue and digital conversion

Signal processing techniques in both the Audio and Radio spectrum are advancing both theoretically with the development of new algorithms, and practically with the increase in the level of integration of integrated circuits. Wherever possible, the new levels of integration in conjunction with efficient digital algorithms are used, so that problems which were previously solved using analogue design are now solved using digital design.

Of course, it is nearly always necessary to communicate with the real world using analogue signals, so analogue to digital (A-D) and digital to analogue (D-A) converters are a necessity. This is why so much work is done to increase the speed and accuracy of the conversion which must ultimately limit the speed of the complete system.

The current range of A-D and D-A converters on the market can sample at up to 100 MHz. As might be expected, the limiting speed depends very much on the required accuracy, with slower conversion required to get improved accuracy. Of course, there is little point being able to do digital processing faster than the analogue conversion devices, so that in practice, the performance of conversion devices and digital processing devices proceed together.

So there are fundamentally two problems which hinder DSP development, one is analogue/digital conversion and the other is the digital signal processing itself.

1.5 Techniques for digital signal processing (DSP)

Digital signal processing has advanced rapidly since the major semiconductor manufacturers started to tackle the problem. Since then they have attempted to cram more and more raw processing power onto a single chip. At the same time they have realised that the signal processing devices need to be integrated into an entire system. So, they have devised families of devices which, however, require some considerable expertise to use. This evolution of devices has split into two directions.

The first approach is the more complex and achieves the best performance. It often involves hardware design which is not trivial, and the systems generated will generally only perform one task. Any slight change to the task (algorithm) may require a complete system redesign, which is both lengthy and expensive. However, the performance of these so called bit-slice machines has been and still is very high and has a permanent place in the field. Bit-slice machines use dedicated multipliers, accumulators and address sequencers often with several address and data bus paths to achieve high speed.

The second approach is simpler, and more versatile. However, the performance is considerably lower than the bit-slice engines previously described. Design involves using a general purpose processor (CPU) which has dedicated instructions to perform reasonably fast multiply, divide, add and subtract operations. The CPU does this by having dedicate parallel multipliers and barrel shifters integrated on the chip. The performance limit is not so much the on-chip operation as the time required to get the data off and on chip (memory bandwidth). Possibly the best known example of a signal processing CPU is the TMS 32010[1] and its derivatives the TMS 32020 and TMS 32030.

[1] TMS is a trademark of Texas Instruments

The previous two approaches provide solutions to a large number of signal processing problems. However, one must accept either the performance limitations of the general purpose processor or the complexities of bit-slice design. In both cases the problem is bandwidth into the basic processing element. The fundamental limit is the rate at which memory can be accessed rather than the performance of the processing element itself. If the processing performed by the basic processing element can be increased and the required memory bandwidth can be reduced, an improved performance will be immediate. The IMS A100 uses a novel architecture to achieve these aims.

The IMS A100 is a processing element with considerable processing power, yet having an interface with moderate bandwidth requirement. This is achieved by having data storage on chip, processing the data in parallel, and storing the intermediate results of calculations. The IMS A100 has also been designed to accommodate many of the commonest DSP algorithms; including the discrete Fourier transform [2], correlation and convolution [3], and digital filtering [1].

1.6 Overview of image processing with the IMS A100

The IMS A100 is a digital processing device at the forefront of digital signal processing performance. It is capable of processing video bandwidth signals, as well as many other types of high bandwidth signals. The maximum input sampling rate of the IMS A100 is 10 MHz, which means that it could, for example, process a [512 × 512] image at a rate of 40 frames per second. The device operates on digital data with a width of 16 bits, and will perform 80 million multiply accumulate operations per second (80 MOPS) a performance well in excess of most bit-slice machines.

The IMS A100 will perform calculations on signed 16 bit integers without any loss of accuracy or overflow, perform rounding correctly, and will also perform complex number processing [4] without any additional hardware. This makes it an extremely simple device to use in a wide variety of applications, as it deals with so many of the problems which have historically plagued signal processing design. Immense care has been taken to ensure that the device is simple to use, for example, the microprocessor interface, which can be interfaced very simply with almost any industry standard processor.

Probably the most important aspect of the IMS A100 is that several can be used in parallel, with almost no 'glue' logic. In principle, there is no limit to the number and a system with 30 devices on a single board has been shown to work well. The processing of large images at high speed requires vast processing performance, making the IMS A100 capability of being able to use many devices in parallel absolutely invaluable.

2 Practical methods of 2 dimensional convolution

2.1 2-dimensional convolution

The process of 2-dimensional convolution of an image is the action of comparing a reference template with a group of pixels, at every pixel point on an image. For example, if a [3 × 3] template were compared at every point on an image of size [5 × 5], there would be 9 valid comparison points as shown in figure 1. The first of these valid comparisons surrounds pixel 1, the second pixel 2, and so on. The comparison is done in practice by a number of multiply and add operations. Consider the example with the first row being compared with the template. The result of the [3 × 3] convolution for the first 3 positions will be

1 $a.? + b.? + c.? + d.? + e.1 + f.2 + g.? + h.4 + i.5$

2 $a.? + b.? + c.? + d.1 + e.2 + f.3 + g.4 + h.5 + i.6$

3 $a.? + b.? + c.? + d.2 + e.3 + f.? + g.5 + h.6 + i.?$

which is a total of 9 multiply-accumulate operations for every pixel in the image. The magnitude of the image data and the magnitude and sign of the template elements determine the type of features which will be extracted from the image. Some simple templates are described later in this section.

In a real image, the magnitude of the pixels which is a measure of their blackness, is referred to as grey scale, having typically 8 bit accuracy. The alternative, which uses a single bit for each pixel, was used in the past, before digital grey scale processing was possible. Future picture processing will undoubtedly be capable of processing colour images. This is a complex field, little understood at the present time, outside the scope of this application note.

With grey scale images it is important that the result of any image transformation yields grey scale values within the limits of the original image. This being so, the sign and magnitude of the elements of the template must be chosen with care. It may be necessary to scale and/or invert the results of an image transformation, so that the resultant image can be observed in a normal grey scale.

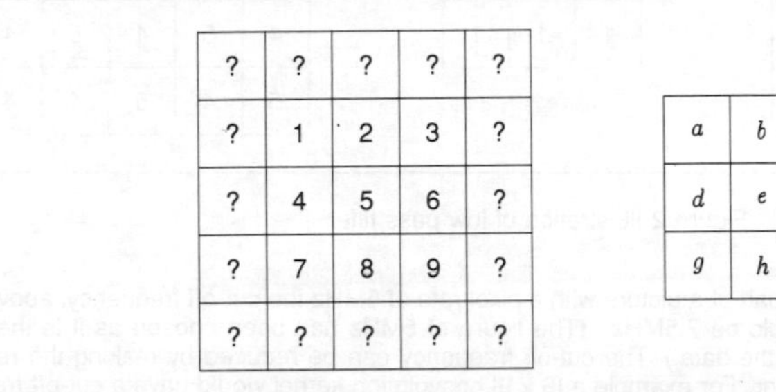

Figure 1 [3 × 3] convolution on a [5 × 5] image

It is usual for the template to be square, although it may be rectangular, and of any size. It is also normal when scanning a real image to traverse the picture as shown in the diagram, i.e. traversing a row, moving down, traversing again and so on until the entire image is scanned.

One point of interest concerns the outermost pixels, which represent invalid data. For a [3 × 3] template a perimeter of one pixel width is invalid, for a [5 × 5] template the outermost 2 pixels are invalid and this redundancy increases as template sizes increase. This does not matter much for large image sizes, but must be borne in mind if large templates, with small images are being used. For the remainder of this section edge effects will, for convenience, be ignored.

2.2 Convolution template types

Low pass filter

The effect shown in figure 2, is of a low pass filter. The numbers have been chosen to show the smoothing effect of the filter. Notice that this is indeed a low pass filter, and that the pixel values are changing at a frequency which is approximately the cut-off frequency of the filter. The filter has effectively changed a black and white image into a blurred grey image.

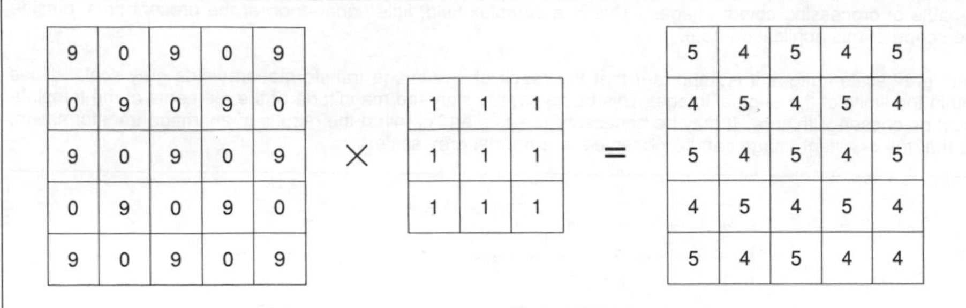

Figure 2 Illustration of low pass filter

If this convolution is regarded as part of a picture with a pixel rate of 5MHz the cut-off frequency, above which all frequencies are removed, would be 2.5MHz. (The figure of 5MHz has been chosen as it is the rate at which an IMS A100 can process the data.) The cut-off frequency can be reduced by making the reference template (convolution kernel) larger. For example a [9 × 9] convolution kernel would have a cut-off frequency of 870KHz.

For the low pass filter kernel no sign modification or scaling of the final image is necessary. Only when the result is outside grey scale limits will any modification be required.

Edge detection

Edge detection is illustrated below with a Sobel operator. This operator combines a vertical and a horizontal edge detector into a single Sobel operator as shown in figure 3.

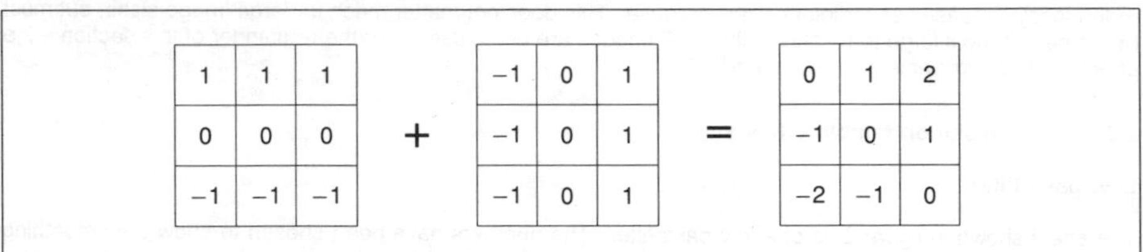

Figure 3 Sobel operator formation

It may be observed that the effect of applying a horizontal edge detection to an image, followed by applying a vertical edge detection to an image, and summing the results, will be exactly the same as directly applying the sobel operator to the image. This same principle of adding operators together, may be applied to many different operators with some interesting results. It is not within the scope of this application note to investigate this subject further.

The following operations, shown in figure 4, on part of an image illustrate the effect of the Sobel operator. It is possible to obtain similar results, by doing a vertical and a horizontal edge detection, squaring and adding the results, and taking the square root to give a result for each final pixel. This is the ideal edge detector, but the cost of squaring twice and a square root is often cost prohibitive, with the Sobel operator a very satisfactory alternative.

Figure 4 shows the result of 2 convolution kernels operating on the different images. This illustrates the requirement for scaling and sign inversion. Before the resultant images can be displayed all negative numbers must be sign inverted and a scaling factor of 4 must also be applied. It is interesting to note that the reason for the sign change is the direction of travel of the convolution kernel across an edge transition. Also, as will be shown later, the steepness of the edge transition is important.

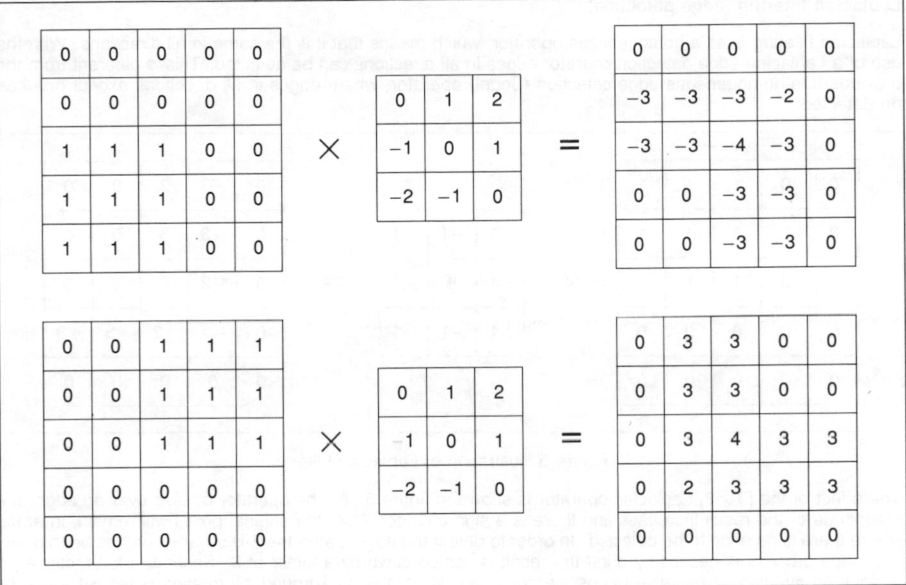

Figure 4 Illustration of edge detection

Laplacian filtering (edge detection)

Laplacian filtering uses a homogeneous operator, which means that it is the same in all directions. With the use of a Laplacian edge detection operator edges in all directions can be detected. This is different from the previous non-homogeneous edge detection (Sobel) operator, where edges in all directions except one can be detected.

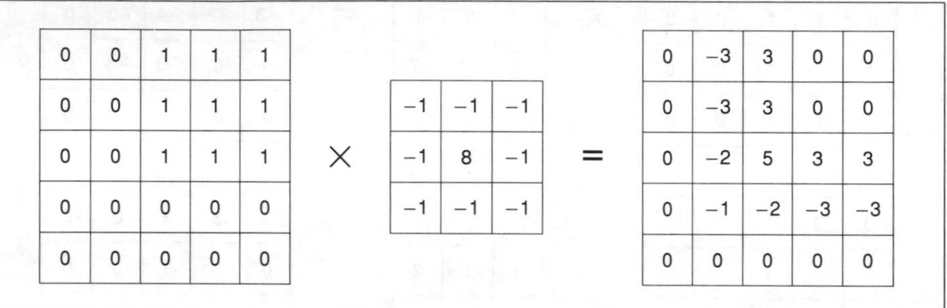

Figure 5 Illustration of Laplacian Filter

The effect of the [3 × 3] Laplacian operator is shown in figure 5. As the operator passes over an edge, the magnitude of the result increases and there is a sign change. Also, the original pixels will remain, in areas where there is no edge to be detected. In order to detect the edges, after the 2-D convolution has been done, a 3 stage process is necessary. First the result is scaled down by a factor of 9. Second, a full rectification is done, converting all negative to positive numbers. Third, the background information is thrown away, by introducing a suitable threshold below which the result is considered to be zero.

2.3 Effect of template size

The previous sections have shown the effect of several [3 × 3] convolution kernels. This kernel size is very effective in many applications, while requiring moderate processing for its calculation. One of the main reasons for not using larger templates is that the processing requirement becomes excessive, mainly due to the large number of multiply operations. The advantages of large kernels are twofold, firstly the larger kernels have a filtering effect which reduces the effect of noise, and secondly the larger kernels are able to detect gradual changes in brightness across a group of pixels.

It must be remembered that single bit pixels are not used, and the pixels are represented in grey scale with range from 0 to 255. This means that in a real image, edges will span several pixels, and to detect these edges will require a larger convolution kernel.

If, for example, a [3 × 3] kernel is used to detect an edge which changes from black to white linearly over 5 pixels, then the maximum and minimum resultant pixel is 1.6 and −0.6 respectively, as shown in figure 6. Each pixel is represented by a single step. This result must be compared with the result in figure 5, where an instantaneous change between 2 pixels gives an output of −3.0 and 3.0. The results of −0.6 and 1.6 are barely enough to detect an edge.

3 Hardware requirements for 2-D convolution

Two possible hardware implementations of 2-D convolution using the IMS A100 are described in this section. Because these two implementaions use exactly the same principle of operation, the IMS A100 device, which is common to both, will first be described. The fundamental difference between the two designs is as follows. In the lower performance system all image data is transferred to the IMS A100 across a comparatively slow memory interface. In the high performance system all image data uses the dedicated input and output ports of the IMS A100. These dedicated ports permit, with the addition of some dedicated hardware, a processing

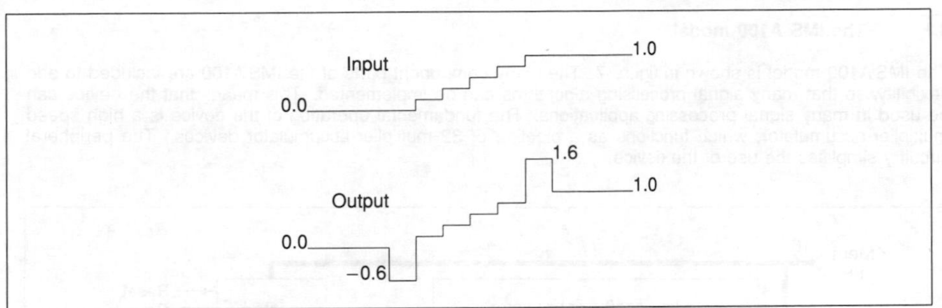

Figure 6 Effect of gradual edges on convolution output

rate of 5 million pixels per second.

Implementation of 2-dimensional convolution on the IMS A100 involves loading the elements of the convolution kernel into the coefficient registers, and passing the entire image through the device while storing the resultant image. To obtain maximum throughput this should be a continuous operation, and will consist of a sequence of alternate read/write operations starting at the first pixel of the first row of the image and finishing with the last pixel of the last row of the image. The two fundamental problems are to arrange that first, the convolution kernel elements are loaded into the appropriate coefficient registers, and second that the pixel data is ordered correctly both before and after processing. The required initialisations of the IMS A100 are also described.

3.1 The IMS A100 model

The IMS A100 model is shown in figure 7. The many component parts of the IMS A100 are included to add flexibility, so that many signal processing algorithms can be implemented. This means that the device can be used in many signal processing applications. The fundamental operation of the device is a high speed multiplier-accumulator, which functions as a pipeline of 32 multiplier-accumulator devices. The peripheral circuitry simplifies the use of the device.

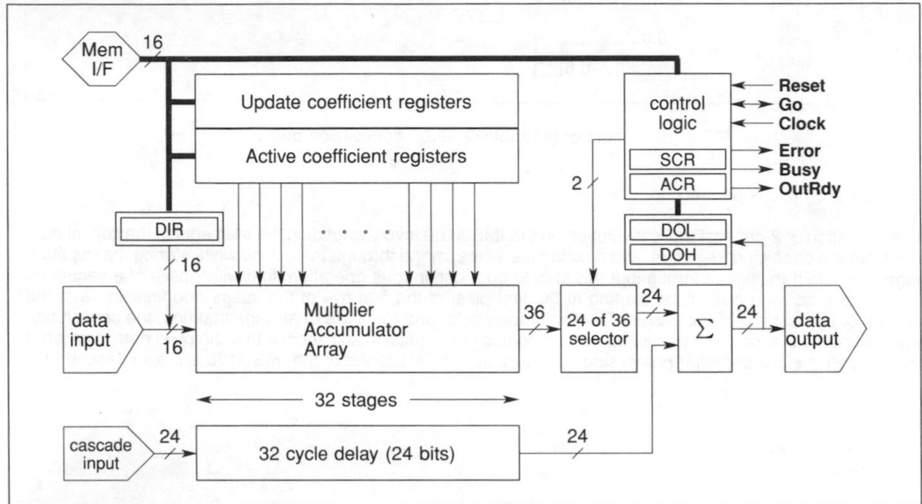

Figure 7 User's model of the IMS A100.

The essential elements of the device, in so far as they are important to this discussion of 2-D convolution, will now be described.

- **The multiplier accumulator array** is the powerhouse of the device. This is a 32 stage pipeline of multipliers, which multiply 32 elements of input data with the contents of the current coefficient registers in between 2 and 8 cycles. The cycle time is between $100ns$ and $400ns$, and is a function of the coefficient width. There is no loss of accuracy in this section because all calculations are at 36 bit accuracy.
- **The data input** is either from the dedicated data input port or via the data input register (DIR). The DIR can be accessed from the memory interface port, which may be connected to a microprocessor. The fastest data access is by the direct input port, with input using the DIR register being usually 2 to 4 times slower.
- **The data output** is taken from the high speed multiplexed data output port or from the data output registers (DOL and DOH), which contain the 24 bits of output data. These registers, like the DIR register, will normally be accessed much slower than the direct data output port.
- **The coefficient registers** are used to store the convolution coefficients, for which only the current coefficient registers are required. Because there is no need to bank-swap coefficient and update registers, neither the update coefficients nor the bank swap capability are ever used in this application.
- **The cascade input** is used for simple connection of devices. The cascade input port is multiplexed in exactly the same way as the data output port, so that direct connection between the two and use

of the GO signal for synchronisation are all that is required to cascade devices. For 2 dimensional convolution requirements, only one IMS A100 is required for doing a convolution with a kernel containing less than 33 elements, although more devices can be used for improved performance as will be shown later. Whenever more than one device is used the cascade input port is required.

- **The 32 cycle delay** element delays the cascade input data by exactly the same time as the multiplier accumulator array. It can be used in conjunction with the data input port to add together 2 streams of pipelined data. This is very useful, particularly for convolution requiring data partitioning. Real time 2-dimensional convolution of images using the IMS A100 requires the use of this delay element.

- **The control registers** are used to initialise the device, and for some of the working operations of the device. These are referred to as the Static control register (SCR) and the Active control register (ACR) respectively. The ACR can be altered during the operation of the device whereas the SCR cannot. The use of these registers as regards 2-dimensional convolution will be described later.

- **The output signals** are described adequately in the IMS A100 data sheet [6] and will not be described further here, except for the GO signal which is relevent to the discussion. The GO pins, of all the IMS A100 devices which are cascaded together, will be joined. GO is used for synchronisation of a cascade of devices, and is not needed in a system with only a single IMS A100, unless the cascade input of that device is used. GO is set up from the SCR to be either a master or slave, and there is never more than one master.

- **GO** is a special signal used to synchronise the cascade and data input ports. If the data input port **Din** or the cascade input port **Cin** is driven by external hardware, then all the IMS A100 devices will be set to slave mode and external hardware will be used to drive the GO pin. If neither the cascade input port or data input port are driven by external hardware, (when all data will use the memory interface) then one of the IMS A100 devices in the cascade will be configured as a master. The master which should be the **last** device in the cascade, drives the GO signal, and all the other devices synchronise their cascade and data inputs from the GO signal they receive. The GO signal master could in theory be driven by any of the devices in a cascade, and this would work for a short cascade. However, operation cannot be guaranteed, whereas an infinite length cascade will work if the master is the last device in the cascade.

3.2 IMS A100 initialisations for convolution

The following description summarised the initialisations of the IMS A100 devices which will be required, prior to the operation of 2-D convolution. A full understanding will require the use of the IMS A100 data sheet [6]. The settings necessary for a 2-dimensional convolution, using 8 bit grey scale data and 8 bit coefficients are described.

The coefficient size is set to 8 bits by setting bits 8 (=1) and 9 (=0) of the SCR. As 8 bit grey scale is used the top 8 data bits from either the data input port or DIR (each 16 bits wide) will be zero.

The result of the 8 by 8 multiplication will require 16 bits, and the 32 stages of accumulation will require a further 5 bits so that the final result, will require 21 bits accuracy. The result required is manipulated internally by a selector so as to be invisible to the user. The significant 8 bits of the result are obtained by setting bits 4 (=0) and 5 (=0) of the SCR, and reading data from the bottom 8 bits of the DOL register.

If there is a cascade of devices the lower 8 bits of output appear on the lowest 8 bits of the multiplexed data output port, which will be connected to the cascade input of the next device in the cascade. By this means scaling is done automatically, and is invisible to the user. The whole purpose of this is that many devices can be cascaded, and appear like a single device with a number of stages which is a multiple of 32.

The remaining SCR register settings are as follows. Bank swap mode will be set to off. Data mode will be set to either input data from the DIR register or data input port depending on the application. Fast output will be set to off for this application.

The ACR will not generally be needed for this application as no bank swapping between the active and update coefficient registers is necessary. It may be necessary to examine the selector overflow and cascade adder overflow bits of the ACR should an error occur (error pin goes low).

3.3 IMS A100 coefficient placement and data flow

Section 2.2 describes some of the convolution kernels which are used to perform feature extraction and filtering on an image. The following discussion describes how these kernel elements are mapped onto the coefficient registers of the IMS A100 so that 2-D convolution is performed.

The IMS A100 can be regarded as a 32 stage multiplier accumulator with 32 constant coefficients, which will be consecutively multiplied with a stream of incoming pixels. The current coefficients are labelled from C(0) to C(31) where C(0) is closest to the output, and C(31) is closest to the input, as shown in figure 8.

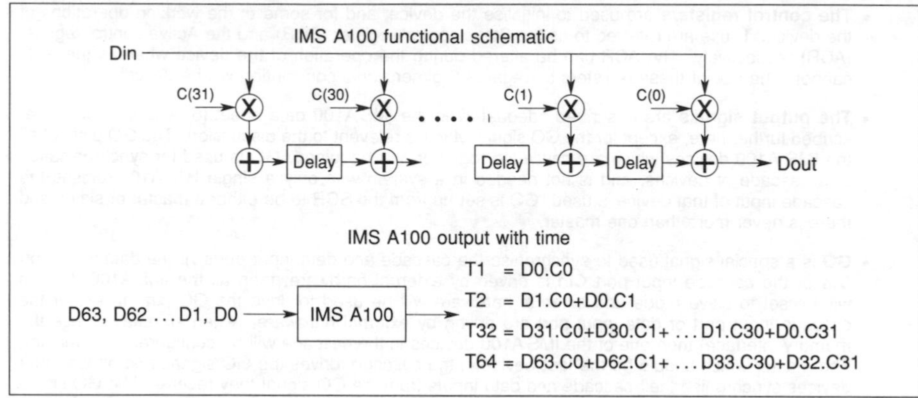

Figure 8 Illustration of IMS A100 pipelined calculation

In figure 8, pixel data presented at the Din port (or DIN register) at time 0 is referred to as D0. Immediately after data is written, at time T1, a result will be read from the Dout port (or DOL/DOH register). For the first 32 cycles (T1 to T32) of the IMS A100, partial results for data D0 to D31, and coefficients C(0) to C(31) will be output from the device. The results at time T1 and T2 are given.

From T32 onwards the device presents full results at its output, and the result at time T32 and T64 are given to illustrate this. The steady state of the device yields the accumulation of 32 multiply operations which have taken place over the previous 32 cycles. Notice also that at any instant the machine contains 32 pieces of information (state), which are the 32 partially accumulated results, as they proceed through the 32 stage pipeline.

If there is a cascade of 2 devices, there are 64 coefficients which can be referred to as C[0] to C[63]. The output from the second device in the cascade is the sum of 64 multiply operations which have accumulated over the previous 64 cycles. This principle can be extended to many IMS A100 devices, so that long multiply-accumulation operations can be done. It is essential to be able to perform long cascades so that large convolutions are possible. For example, a 128 point convolution will require 4 IMS A100 devices in cascade.

This also applies to 2-Dimensional convolution. For instance, an [11 × 11] convolution using 121 stages, will require 4 IMS A100 devices. Of course, 7 stages are not required, which means that 7 of the coefficients (C[127] to C[121]) of the first IMS A100 in the cascade will be set to zero.

As can be recalled from section 2, a [3 × 3] convolution requires the accumulation of 9 multiply operations. Similarly, a [5 × 5] convolution, illustrated in figure 9, will require 25 stages of multiply-accumulation. The only problem is that the coefficients must be loaded in the correct coefficient locations, and the input and output data must be ordered correctly, so that the IMS A100 architecture can be utilised. This is described in the following section.

3.4 Image scanning for a microprocessor based system

The following description will normally only apply to a system using a memory interface, for the transfer of all data to and from the IMS A100. It is perfectly possible to use the following pixel sequencing operations, for transferring data to the IMS A100 devices across the high speed data input and output ports. However, this is not advised as the sequencing operations using normal hardware are complex, but are quite easy with a microprocessor. The additional hardware could be better used for implementing an extremely high performance system, such as described later.

Image scanning for 2-D convolution implementation

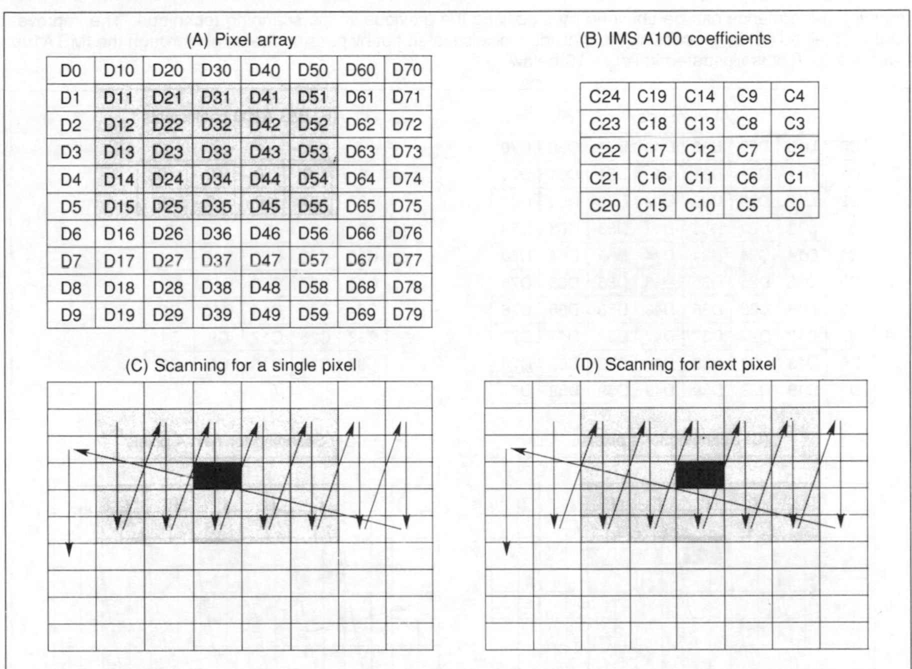

D0	D10	D20	D30	D40	D50	D60	D70
D1	D11	D21	D31	D41	D51	D61	D71
D2	D12	D22	D32	D42	D52	D62	D72
D3	D13	D23	D33	D43	D53	D63	D73
D4	D14	D24	D34	D44	D54	D64	D74
D5	D15	D25	D35	D45	D55	D65	D75
D6	D16	D26	D36	D46	D56	D66	D76
D7	D17	D27	D37	D47	D57	D67	D77
D8	D18	D28	D38	D48	D58	D68	D78
D9	D19	D29	D39	D49	D59	D69	D79

C24	C19	C14	C9	C4
C23	C18	C13	C8	C3
C22	C17	C12	C7	C2
C21	C16	C11	C6	C1
C20	C15	C10	C5	C0

Figure 9 Pixel scanning and coefficient ordering

A pixel array (A) with 10 rows and 8 columns is used purely for convenience. The convolution kernel with 25 coefficients is shown in (B). The order of the coefficients is critical, starting at the bottom right and proceeding one column at a time (Remember that C0 is the coefficient of the last stage of the cascade). The scanning pattern for the image is shown in (C) and (D). The dark black squares are valid output pixels, each of which represent the convolution of 25 pixels with 25 coefficients.

If the light grey area of pixels is written to **Din** as shown in (C) the order will be D11 then D12 and so on in columns until D55 is written. Immediately after D55 is written to **Din** a valid pixel is read from **Dout**. The value of this pixel will be

D33out = C0.D55 + C1.D54 + C2.D53 + ... + C23.D12 + C24.D11

After this D51 is written followed by D52, D53, D54, D55 after which another valid pixel can be read.

D34out = C0.D65 + C1.D64 + C2.D63 + ... + C23.D22 + C24.D21

In other words, for every 5 pixels written one valid pixel is read, from the beginning until the end of the row. At the end of the row go back to the start of the row and move down a row repeating until the entire image is scanned. The net effect is a completely convolved image. This is inefficient as the entire image is effectively written to the IMS A100 **FIVE** times.

There is fortunately an optimisation which can be incorporated. The principle is that the some of the IMS A100 coefficients are set to zero, so that those stages act only to store and delay accumulated results. This is described in the following section.

Improved image scanning for 2-D convolution

Improved performance can be obtained by modifying the previous image scanning technique. The improvement is obtained not by processing the individual pixels faster, but by passing the pixels through the IMS A100 fewer times. This is illustrated in figure 10 below.

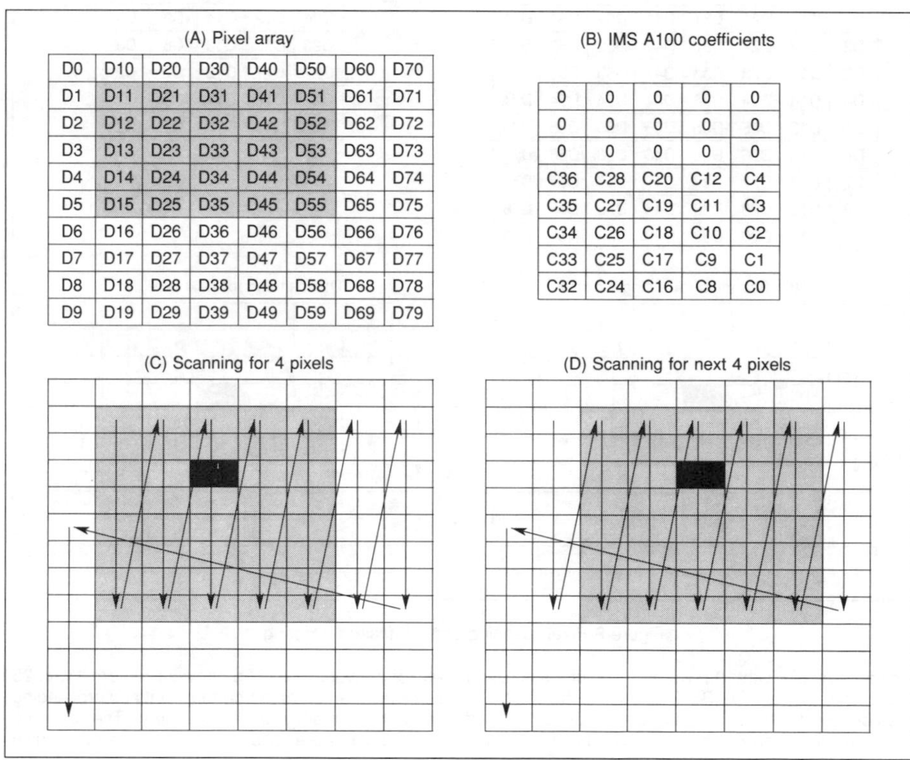

Figure 10 Pixel scanning and coefficient ordering – high performance

Data is written to the IMS A100 devices as before except this time data is scanning over 8 rows at a time, starting at D11 and finishing at D58. Scanning over the fifth column of 8 pixels from D51 to D58 will yield 4 valid pixels, starting at the black square.

D33 out = C0.D55 + C1.D54 + C2.D53 + ... + C23.D12 + C24.D11
D34 out = C0.D56 + C1.D55 + C2.D54 + ... + C23.D13 + C24.D12
D35 out = C0.D57 + C1.D56 + C2.D55 + ... + C23.D14 + C24.D13
D36 out = C0.D58 + C1.D57 + C2.D56 + ... + C23.D15 + C24.D14

Immediately after D33out is read from **Dout**, D56 is written to **Din**. The partially accumulated result for pixel D43out is then stored in the empty slot (The empty slot is the position in the IMS A100 which would accumulate C5 with the data at **Din**. As this coefficient is set to zero no accumulation takes place, and this stage acts as delay and storage of data only) The next pixels D56 and D57 are written, and the partially accumulated result for D44out and D45out are then stored in the IMS A100 pipeline. At this time there will be 3 partially accumulated results D43out, D44out and D45out, which will be required for processing the next column.

At the end of the column scan, after the pixel D58 is written, D61 followed by the remainder of that column of 8 pixels, are written. This yields a further 4 pixels as given below.

D43out = C0.D65 + C1.D64 + C2.D63 + ... + C23.D22 + C24.D21
D44out = C0.D66 + C1.D65 + C2.D64 + ... + C23.D23 + C24.D22
D45out = C0.D67 + C1.D66 + C2.D65 + ... + C23.D24 + C24.D23
D46out = C0.D68 + C1.D67 + C2.D66 + ... + C23.D25 + C24.D24

The scanning technique which scans across 8 rows at a time, while 4 rows of pixels are written, is 2.5 times more efficient than the previous technique, where only 5 rows are written, for a single output row. It is simple to calculate the efficiency for any number of *zeros* inserted into the coefficients of the IMS A100.

Convolution efficiency

For any system, there will be a fundamental pixel processing rate. As shown in section 3.5 the processing rate, writing pixels across a high performance memory interface is unlikely to be better than 4 Mpixels per second. Realistically 2Mpixels per second is a more probable performance. However, as shown above, there will be an efficiency factor, dependent upon the the convolution technique used, which will reduce the performance further. The best that can be done uses an image scanning pattern as shown in figure 10.

To calculate the efficiency, the number of stages and the number of zeros must be known. These calculations assume that the maximum number of zeros will be used, for whatever number of A100s are selected.

$$stages := number.of.A100s \times 32 \qquad (1)$$

$$zeros := \frac{stages}{(filter.size)^2} DIV(filter.size - 1) \qquad (2)$$

$$Efficiency := \frac{zeros + 1}{zeros + filter.size} \qquad (3)$$

As a simple example, it is known to take 500ns to process a single pixel, and the efficiency is calculated at 50%. The expected processing rate will be 1Mpixels per sec.

There is an obvious trade off between the number of A100 devices used and efficiency. A small number of zeros increases efficiency greatly. However, as efficiency approaches 100% the added cost of more IMS A100 devices, to give more stages, will not give a proportionate increase in performance. No figures are given here, as it is simple to calculate the numbers, for any given application.

3.5 Moderate speed image convolution

A moderate image convolution rate can be obtained by using a very simple design incorporating an 8 or 16 bit processor, which controls one or several IMS A100 devices. A typical system is shown in figure 11. The system chosen uses an extremely high performance 16 bit processor, the IMS T212. The limiting speed of this system is either the rate at which data can be transferred across the IMS A100/IMS T212 memory interface, or the rate at which data can be transferred to and from an external system. The external system may consist of camera, frame grabbing hardware and some form of image displaying capability. For the purpose of argument it will be assumed that the IMS A100/IMS T212 memory interface is the limiting factor.

The performance of this system may be easily estimated, for whatever processor is used. This estimation assumes that the image resides in memory before processing starts, and that the data input port is not used. The resultant image will also reside in memory after processing.

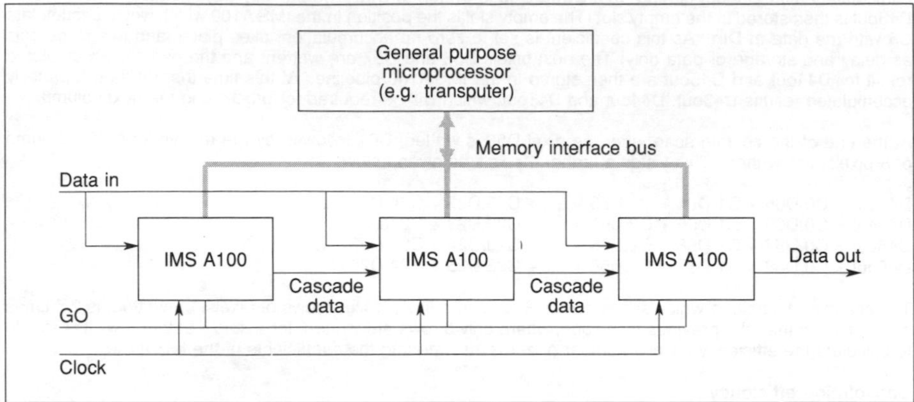

Figure 11 IMS A100 coefficient loading

The processing of a single pixel involves 5 steps which are in order

read from memory	= 100ns
write to IMS A100	= 100ns
IMS A100 processing,	= 200ns
read result from IMS A100	= 100ns
write result to memory	= 100ns

This shows that the time to process a single pixel will be 600ns so that for a complete picture of size [512×512] a processing rate of 6 frames per second may be possible. While this may be achievable in theory there are several problems which lead to a reduction in this performance.

- The data must be transferred both to and from the system, before and after processing. In practice this may take longer than the processing itself.
- The data must be read and then written by the processor, usually into an internal register, which will consume at least 2 extra cycles (100ns minimum)
- The data accessed by the processor will be in the form of a 2 dimensional array of pixels. The processor has to calculate the array subscripts for every pixel in the image, which will consume at least 4 processor cycles (200ns). The order with which pixels are loaded is not simply row by row and is described elsewhere. (section 3.4)
- The nature of the convolution algorithm means that the image may need to be split up into small blocks, which must be overlapped, to give a continuous convolution of the entire screen. This will result in an inefficiency of between 10% and 50% depending on the size of the blocks and the degree of overlap.
- The algorithm implemented on the IMS A100 has a fundamental efficiency as described in section 3.4. Equations are given for the calculation of this efficiency. The algorithm should be arranged so that the efficiency is better than 50%.

The effect of all these factors that a simple microprocessor based system is likely to have a processing rate of between 1 and 4 frames per second. Even this will require an high performance processor with optimised software.

3.6 Very high speed image convolution

In section 3.2 it is shown that complete images can be processed on the IMS B009 at between 1 and 4 frames per second. In appendix A, implementation of the convolution algorithm running on the IMS B009, under an IBM PC environment is less than 1 frame per second. The actual IMS A100 processing time is only 200ns per pixel, which is less than 10% of the total processing time. As the IMS A100 device is such a fast device it seems wasteful to reduce the performance to this extent. In practice, many users of the IMS A100 will wish to extract full performance which requires a hardware design.

Faster speeds require a slightly altered algorithm and dedicated hardware. The following describes a hardware implementation capable of processing speeds up to 20 frames per second. The hardware setup is shown in figure 12 This figure illustrates the hardware required to perform a [3 × 3] convolution on a [512 × 512] image. Larger convolutions on larger images are possible with the addition of extra hardware. For example, a [31 × 31] image convolution could be done using 31 IMS A100's and 30 sets of shift registers. Each shift register has a 512 + 32 stage delay.

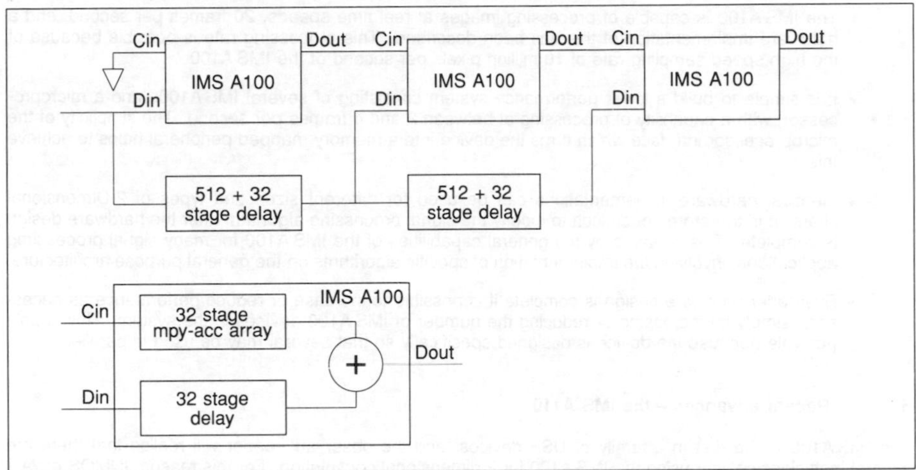

Figure 12 Hardware for real time image convolution

In the example shown, 2 rows of data are stored in long shift registers, while the third row enters the data input port of the first IMS A100 in the cascade. The first IMS A100 in the cascade has its cascade inputs grounded, while all other IMS A100 devices have their cascade inputs connected to the data outputs of the previous IMS A100 in the cascade.

The arrangement ensures that 2 pixels one above the other in an image are fed simultaneously into the cascade input and data input of each IMS A100. The IMS A100 devices will process one line of pixels at a time, and the entire image will eventually pass through every IMS A100. Because the system is fully pipelined this results in no performance degradation.

Each stage of the cascade requires an IMS A100 and a long shift register. The IMS A100 is like a 32 stage delay element, where the cascade input delayed by 32 stages is added to the data input after it has been through a 32 stage multiplier-accumulator array. The 32 stage delay of the IMS A100 means that the shift register delay must be a single line delay (512 pixels in this case) plus 32 stages.

The data throughput rate depends on the coefficient size selected for the IMS A100's, and is unaffected by the number of stages. If 8 bit pixels with 8 bit coefficients are selected, a data rate of 5 KHz is achieved. For a [512 × 512] image this gives a convolution time of one frame in 50 mS. This is a frame rate of 20 Hz with faster speeds achieved by selecting regions of interest or multiplexing frames between several boards.

As a further example application, it is required to pass a [31 × 31] pixel kernel template over a [1024 × 1024] image at 50 frames per second. Processing will require 310 IMS A100 devices, segmented over several boards. One configuration would use 30 [1024 + 32] delay stages, and would require 31 IMS A100's to process the image in 200 ms. Therefore 10 such boards would be required, and a means of multiplexing the individual images at a bandwidth of 50 Mpixels per second. This is mainly a problem of data distribution. The processing problem has been solved by the capability of the IMS A100.

4 Conclusions

This application note has shown how the IMS A100 may be used to perform fast processing of digital images. Some typical image processing functions have been described, including edge detection and filtering of a picture.

The versatility of the IMS A100 has been shown by the following observations.

- The IMS A100 is capable of processing images at real time speeds, 20 frames per second, and a hardware implementation of this has been described. This processing rate is possible because of the high speed sampling rate of 10 million pixels per second of the IMS A100.
- It is simple to build a lower performance system consisting of several IMS A100s and a microprocessor, with a capability of processing at between 2 and 4 frames per second. The simplicity of the microprocessor interface which turns the device into a memory mapped peripheral helps to achieve this.
- Identical hardware implementations can be used for different sizes and types of 2-Dimensional filters. It is therefore not difficult to modify the signal processing algorithm after the hardware design is complete. This is eased by the general capabilities of the IMS A100 for many signal processimg applications, involving the implementation of specific algorithms on the general purpose architecture.
- Even after hardware design is complete it is possible to increase or reduce performance as necessary, simply by increasing or reducing the number of IMS A100 devices in the system. This is only possible because the device is designed specifically so that several may be used in parallel.

5 Recent advances – the IMS A110

The IMS A100 is the first in a family of DSP devices, and the observant reader will realise that there are several inefficiencies with using the IMS A100 for 2 dimensional convolution. For this reason, INMOS have a dedicated device aimed specifically at 2D convolution applications.

The device, known as the IMS A110, has an architecture similar to that shown in figure 12. The device handles 8 bit data and a single device is capable of performing 7×3 image convolution at a rate of 20 M samples per second (4 times as fast as the IMS A100). Further, the line delay elements, as shown in this figure, are integrated onto the chip so that the device can process video signals **directly**, without any frame buffering. The device has several other useful features including

- **Cascadability.** It is possible to perform convolution in any multiple of 7×3, so for example 14×3, or 7×6 or 21×9 convolution is possible. No additional hardware is required for this, only a multiple of IMS A110 devices.
- **An output look up table.** At the output of the device is a look up table which may be modified across a microprocessor interface. This facilitates such useful things as dynamic range enhancement and operates at the full speed of the device.
- **A max/min register and statistics monitor post processing unit.** It can be very useful to monitor the magnitude of signals passing through the device, and this can be done without stopping processing.

The IMS A110 achieves all this by having on-chip programmable delay lines, and an on chip post-processor for data transformation, as well as the the basic 21 stage multiplier-accumulator. It is dedicated to high speed video applications, and minimises system hardware requirements in these applications. The IMS A100 by contrast, is a general purpose device which can for example perform 2 dimensional FFT calculations, 1 dimensional FFT, convolution, correlation and complex processing on 16 bit data.

Despite the advantages of the IMS A110 in video applications, it may be practical to use the IMS A100 for large 2-Dimensional convolution sizes. For example, a 28x30 convolution would require 40 IMS A110 devices. The same could be achieved with 28 IMS A100 devices, plus line delay elements.

6 Implementation of convolution on the IMS B009

One possible application of the IMS A100 cascadable signal processor is the processing of two dimensional images, such as those obtained from a TV camera. The analogue signal from the camera must first undergo analogue to digital conversion, and may then be presented to the IMS A100 for processing. The size of the image used in this application note is by convenience [512 × 512] pixels. The reason why this size was selected was that a camera and frame grabber board were readily available. The IMS A100 is capable of processing any size of image including, for example, larger images such as high resolution satellite pictures of the earth.

This section shows how the IMS A100 may be used to process a [512 × 512] pixel image. The work described involves no hardware design and uses two standard boards supplied for the IBM PC. This shows how the IMS A100 can be used to perform 2 dimensional filtering, convolution or correlation.

The system, shown in figure 13 is composed of a camera, monitor, digitising frame grabber board and IMS B009 board. Both the frame grabber board and the IMS B009 [5] board are standard plug in boards for the IBM PC.

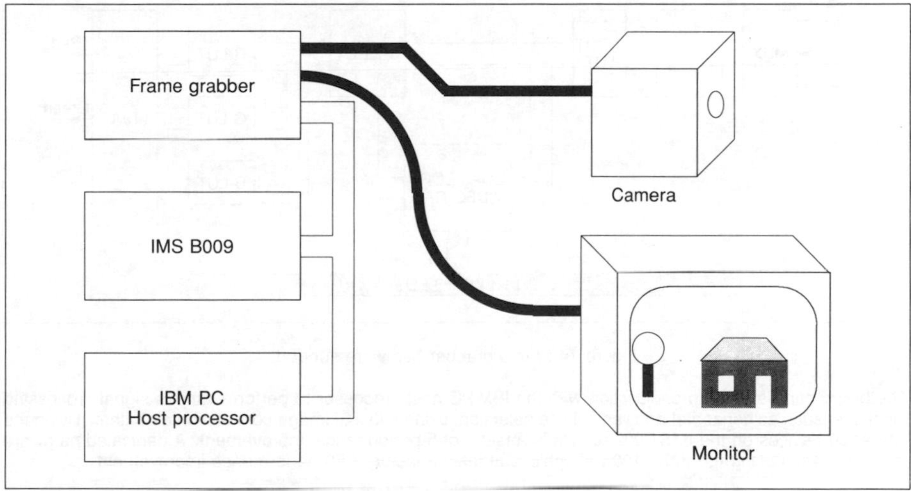

Figure 13 IBM PC and associated hardware

Software which controls the system runs on the IBM PC host processor and is written in Turbo Pascal, Version 3.0. This program commands the frame grabber board to grab images which are then transferred to the IMS B009 via the IBM PC host processor. The image is then transformed, using the convolution algorithm running on the IMS B009, and transferred to the frame grabber board to be displayed on the monitor. It is also possible to dump pre-processed and post-processed images to disc. This can be useful where further processing is required. The software running on the IMS B009 is written in occam, although it could also be written in C or Pascal. A knowledge of occam will be helpful for a full understanding of the implementation of this algorithm. The program is a modified version of the IMS D703 development software [5], which is generally available for DSP development with the IMS A100.

The image transformation technique uses a 2 dimensional image convolution algorithm, and is demonstrated by performing edge detection and filtering on an image. The purpose of the computer program is only to demonstrate the technique, not to investigate the enormous number of image transformations which are possible. The program enables further investigation of 2 dimensional convolution kernels with the minimum of effort. This program is not available as a product but may be obtained by contacting the DSP group based at Bristol.

The performance of the system does not utilise the full performance of the IMS A100. There are two major reasons for this. Firstly, all the data which is processed by the IMS A100 is transferred between the processor and IMS A100, across a comparatively slow memory interface. Secondly, all the data is transferred to the processor across comparatively slow links. The net effect of these factors is to reduce the effective data rate by between 1 and 2 orders of magnitude. The bandwidth possible through the dedicated ports of the IMS A100 is 10 Mbytes/sec. This bandwidth yields a maximum frame rate of 20 frames per second, while the performance shown in the rest of this section is no better than one frame per second. A brief description of how to obtain full performance from the IMS A100 is given in section 3.6.

6.1 Frame Grabber support

The frame grabber board is a card available for the IBM PC which can perform frame grabbing operations on a video signal from an external source. The board used is a matrox PIP-1024 capable of storing a [1024 × 1024] image or 4 individual [512 × 512] images, with a resolution of 8 bits per pixel.

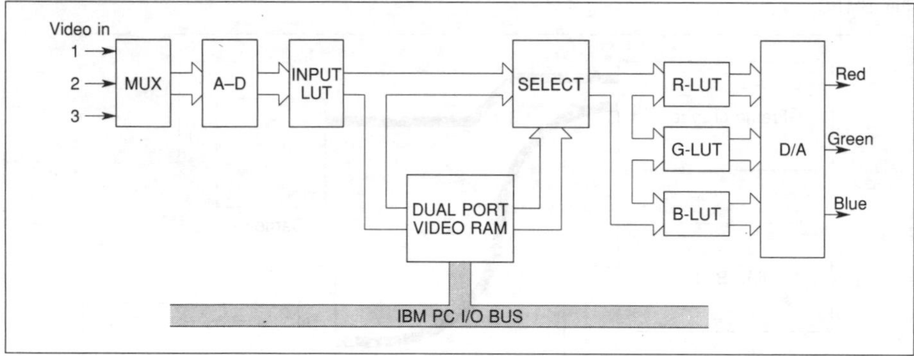

Figure 14 Frame grabber hardware support

The board can be used in conjunction with the IBM PC host processor to perform various signal processing functions, such as horizontal and vertical line detection, using a [3 × 3] image convolution algorithm. Using the IMS A100 devices on the IMS B009 results in a factor of 5 performance improvement. A dedicated hardware implementation using the IMS A100 will achieve at least a factor of 50 performance improvement.

The board is used to continuously grab pictures and display them, to freeze frames, grab single frames and to accept single frames from the IBM PC bus which may then be displayed. The board is controlled directly by the Turbo-pascal program which runs on the IBM PC host processor.

6.2 The IMS B009 hardware

An overview of the IMS B009 is shown in figure 15 , and a more detailed description of the key components is shown in figure 16 The IBM PC host processor communicates with the IMS C011 across the IBM PC data bus. Conversion into the standard inmos link protocol is performed by the IMS C011, and an optional link jumper is used to connect this link to one of the links of the IMS T414. A further fixed serial link communicates with the IMS T212. There are several other links which may optionally connect to other parts of a system. It is possible for example to connect several IMS B009 boards together and form a larger system. Another possibility is to use several links between the IMS T414 and IMS T212 to increase communication bandwidth.

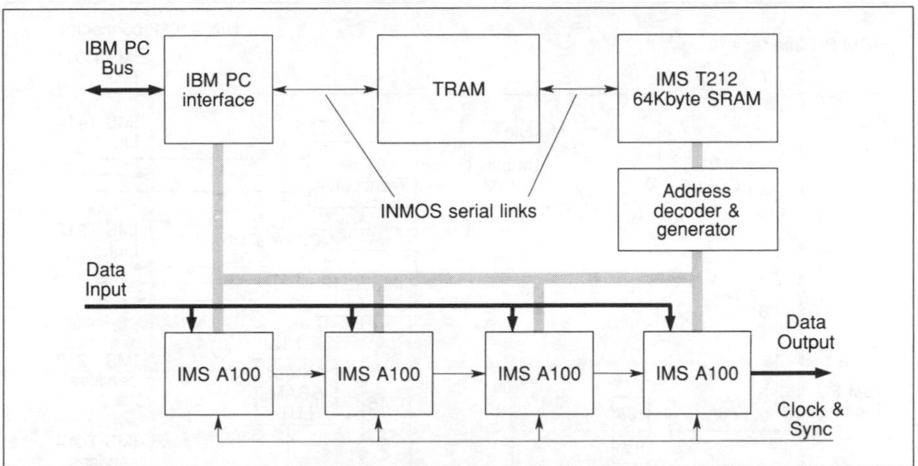

Figure 15 IMS B009 overview

The 64 Kbytes of SRAM act as program and data memory for the IMS T212. The high speed data in and data out interfaces to the IMS A100 cascade are available at an external connector, for maximum speed operation of the IMS A100 cascade. In this application all data input/output is performed by the IMS T212, across the slower microprocessor interface with the IMS A100.

The 4Kx12 SRAM look-up table, multiplexer and address decoder are used to speed the transfer of data during processing. The flow of data during the application of a typical signal processing algorithm will be from the frame grabber, across the IBM PC data bus, through the IMS C011 link adapter and into the memory of the IMS T414. Data is then transferred using the transputer block move engine, across the transputer link, at about 900 Kbytes per second, into the memory of the IMS T212. The data is then processed by the IMS T212 in combination with the IMS A100 cascade and the result transferred to the IMS T414. From the IMS T414 the result may be transferred back to the IBM PC host processor, either to be filed on disc or to be displayed.

The IMS B009 also has a direct interface between the IBM PC bus and the IMS A100's. This is only used when the IMS T212 is disabled. This mode of operation of the IMS B009 is not used in this application and will not be discussed further. The use of this interface is for slow access to the IMS A100 from a program written in any programming language on the IBM PC.

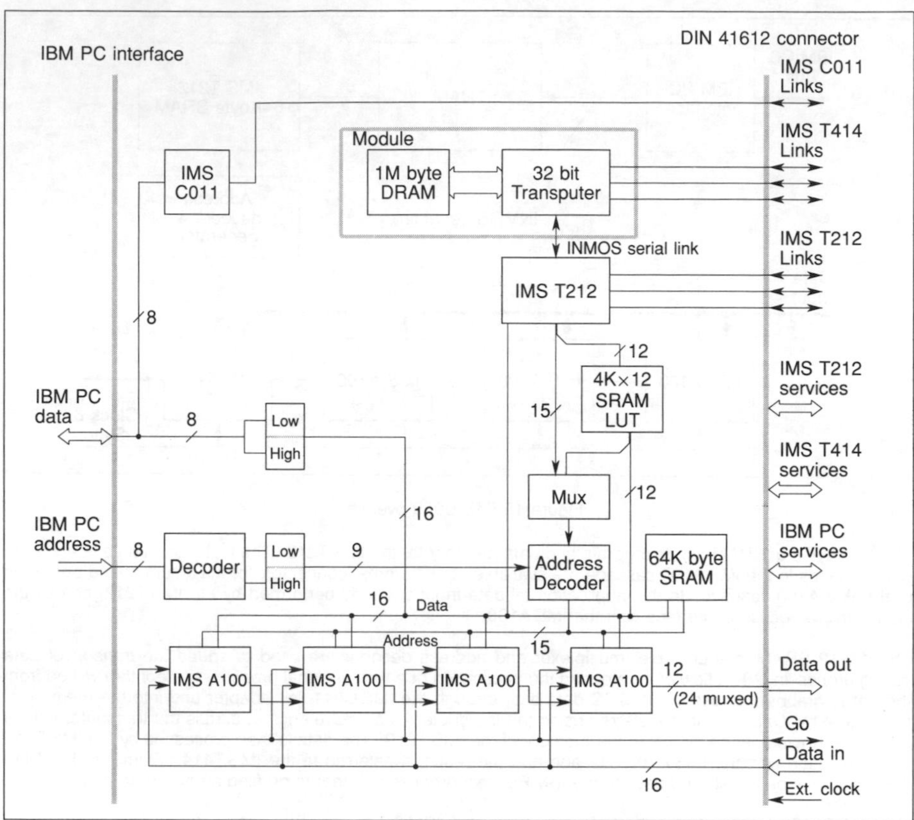

Figure 16 Key components of the IMS B009

6.3 Transputer block move capability

The transputer block move capability is used to transfer data at maximum memory bandwidth, from one position in memory to another. This following describes the mechanics of the block move, and shows how it is modified so that it may be specifically used for this application.

The technique uses hardware external to the transputer to modify memory accesses. Therefore the transputer is not in total control of what is happening during the block move. The transputer is responsible for the initialisation of the external hardware prior to execution of the block move operation. This technique, while useful for improving performance, requires caution for its use.

All transputers have dedicated hardware support for moving blocks of data from one area of memory to another. The IMS T212 doing a block move on the IMS B009 will transfer a 16 bit word from one memory location to another in 300ns, with 150ns required for the read operation and 150 ns for the write operation. One possible occam implementation is given in the code below. In this case 1024 words are transferred from position #4000 to #5000. The compiler deals with setting up the counter and address registers.

```
[1024]INT array1:
[1024]INT array2:
PLACE array1 AT #4000 :
PLACE array1 AT #5000 :
SEQ
  ...  Load array with source data
  Array1 := Array2
```

The block move capability can be used to transfer data into and out of the IMS A100 devices extremely quickly. However, the data must be in the correct order and word aligned. If processing is required to position the data correctly in memory, then the performance of the system will be impaired.

The sequence of operations required to access the IMS A100 devices from the IMS T212 is a read from memory followed by a write to the DIR [2] register, followed by a read from the DOL [3] register followed by a write to memory. There is little similarity between the simple transputer block move operation and the transfer of data into and out of the IMS A100.

The hardware required to modify the simple block move operation is an address decoder and look-up table (LUT), which are included as part of the IMS T212 memory interface. Whenever a memory address is output by the IMS T212 and the LUT is active the address is translated by the LUT. The 4Kbytes of LUT are used to convert a block of sequential addresses into an arbitrary sequence of addresses, with **no processing overhead**.

In addition to the address translation LUT the address output by the IMS T212 is decoded along with the information of a read or a write cycle. The result is used to decide if a write to the DIR register, a read from the DOL register, or a normal read/write cycle is required.

For the 2-D convolution, data is placed in the memory of the IMS T212 one pixel at a time and one line at a time. However, the pixels are loaded into the DIR register of the IMS A100's one column at a time. The address translation LUT is used to map the rows of pixel data into columns of pixel data so that the data enters the IMS A100 cascade in the correct order. In exactly the same way the output data from the IMS A100 is in columns and must be translated into rows, so that it may be redisplayed on a monitor.

The size of the LUT enables 4096 possible translations, half of which are used for the data input, and half of which are used for the data output from the IMS A100 cascade. Therefore it is only possible to have a block of data of 2048 pixels. This means that for example an image of [512 × 512] pixels must be split into, say, 128 blocks each with [16 × 32] pixels. For efficency reasons it is best to keep the blocks as close to square as possible.

The sequence of operations involved with a 2-D convolution is as follows:

1 Read from memory through the address translation LUT so that the correct pixel is accessed.

2 Write pixel to the DIR register of all the IMS A100's in the cascade.

3 Read result of the convolution from the DOL register of the last IMS A100 in the cascade.

4 Write result to memory location given by the address transtation LUT.

The result of doing this operation for every pixel in the block, is stored sequentially in memory and is then transferred across the transputer link to the IMS T414, where all the blocks are recombined into one image. The complete image is then transferred to the matrox board to show the result of the convolution.

[2]The DIR register is the Data Input Register

[3]The DOL register contains the Low byte of Output Data

6.4 Implementation of the 2D convolution algorithm

The convolution algorithm is implemented in occam on the IMS B009 which consists essentially of 2 processors running in parallel. The main purpose of these processors is to prepare the data for processing by the IMS A100 devices. This following description gives the details of this implementation with respect to both the hardware and software.

The IMS T414 and IMS T212

The IMS T414 module with 1 megabyte of memory is connected to the IBM PC host processor via a link adaptor. This processor runs the main program and also stores the complete image buffer and result of the convolution. Both the result of convolution and the image buffer contain 256 Kbytes of image data.

The IMS T212 handles reading and writing of data across the IMS A100/IMS T212 memory interface. All data processed by the IMS A100 passes across this interface.

The IMS T212 which is connected to the IMS T414 by a transputer link, runs several specialised procedures and has limited data space for the storage of images. Therefore, the IMS T212 will at any moment during program execution be processing only a small portion of the image. The following program runs on the IMS T414 processing individual blocks of the image one after the other. The operation is described below in pseudo-occam. Notice that the three dots at the beginning of some lines hide code within them.

```
PROC 2D.convolve(VAL [][]BYTE input.image, [][]BYTE convolved.image)
  ...  calculate block sizes, rows and columns
  ...  set up address mapper
  SEQ block.row = 0 FOR total.block.rows
    SEQ block.col = 0 FOR total.block.cols
      SEQ
        ...  Calculate new pixel coordinates of the block
        ...  Dump block of input image array to IMS T212 from IMS T414
        ...  Convert data into IMS A100 format
        ...  Flush IMS A100 cascade
        ...  Block move data through IMS A100's (DO THE REAL WORK)
        ...  Convert result into bytes
        ...  Send result from IMS T212 to IMS T414
        ...  Place result into convolved image array (final result)
:
```

The block sizes, the number of blocks in each row and the number of rows in each image are first calculated and apply for the duration of the convolution of one entire [512 × 512] pixel image. Each block is made up of a precalculated number of rows and columns of individual pixels. The maximum number of pixels in each block is 2048 pixels. This is a limitation of the address mapper which can perform a maximum of 4096 address translations, 2048 for input and 2048 for output. Without this address mapper the pixel data would need to be reordered by the transputer, which is extremely time consuming. The address mapper makes possible arbitrary address sequences, so that data in the transputers memory space may be in any arbitrary order prior to processing by the IMS A100.

The address mapper is set up once before processing the image. The code to do this is as follows:

```
SEQ
  SEQ i = 0 FOR block.size
    SEQ
      j := 2 * i
      mapper.array[j]   := i
      mapper.array[j+1] := block.size + i
  ...  write array to mapper
```

This address mapping, between the memory and the transputer, enables the use of the transputer block move engine. Under normal operation the block move engine transfers a block of data in contiguous memory to another position in memory one word at a time at maximum speed. This is considerably faster than doing individual read/write operations.

In order to use this block move facility without the address mapper the data would need to be interleaved with the result. This would lead to inelegant and inefficient software, and it is much better to have arrays in contiguous memory. The transputer is reading and writing at consecutive locations, but the external hardware is "cheating" so that the transputer is actually reading or writing at locations defined by the address mapper. The previous piece of code is used to set up this interleaved addressing.

The operation of passing data through the IMS A100's may be considered as 4 distinct actions, which use up two transputer block move cycles.

During the first block move, data is read from the address mapped memory location output by the transputer. The transputer then attemps to write this data to the block move output address. However, the external hardware recognises this cycle and intercepts it, so that data is written to the DIR register of all the IMS A100's and not to memory. During the write to DIR the address mapper is unused.

During the second block move, the transputer attempts to read from the next memory location of the input array. However the external hardware again recognises this and the data is read from the DOL register of the last IMS A100 in the cascade. The transputer then attempts to write data to the next memory location of the output array. However this is again intercepted and the data is written to the address mapped memory location output by the Transputer.

The net effect is that the image block residing in IMS T212 memory is passed through the IMS A100 DIR and DOL registers and the convolution result read from the DOL register now resides in contiguous memory ready to be transferred back across a transputer link to the IMS T414.

Performance

The performance of this setup is easily calculated as the sum of the time for two transputer block moves (2 x 300ns for IMS T212 with 1 wait state) plus the time for a single cycle of the IMS A100's (200 ns for 8 bit coefficients). This gives a total pixel transformation time of 800 ns.

To obtain the time required for the convolution of a complete image it is only necessary to calculate the number of blocks of pixels comprising a complete image. As will later be shown the number of blocks is not just a function of image size, but depends on the size of the convolution kernel, and the number of pixels in each block.

The best possible performance assumes that 256K pixels each require 800 ns of processing. This corresponds to a processing rate of 5 frames per second. In other words the performance degredation caused by using a memory interface, as opposed to using the dedicated cascade and data ports, to get the data into and out of the IMS A100 yields 25% of the available performance of the IMS A100. As will now be explained, the actual frame processing rate will be somewhat less than this, because the blocks of image data comprising the complete image must be overlapped.

Image segmentation

The image is segmented into several blocks of pixels. This is illustrated in figure 17. In this example each block overlaps in both the x and y directions dependent on the size of the convolution kernel. This image convolution kernel with size [5 × 5] requires an overlap of 4 pixels on each edge. If this is not done the result of the convolution of the image will have vertical and horizontal lines of incorrectly convolved data running down and across.

Overlapping of image blocks means that pixels at the edge of a block will be passed through the IMS A100 twice. This does not matter except that the time to process an entire image will be increased. Also pixels at the edge of the image must be ignored. In this example the 2 outermost pixels at the edge of an entire frame do not contain useful information.

To process a single block will require the processing of 144 pixels, only 64 of which will be valid. This represents 44% efficiency for processing the entire image. The larger the kernel or the smaller is each individual processing block, the less efficient is this technique of image segmentation. Unfortunately this technique is the best that can be done using the IMS B009, with data transfer across the relatively slow memory interface.

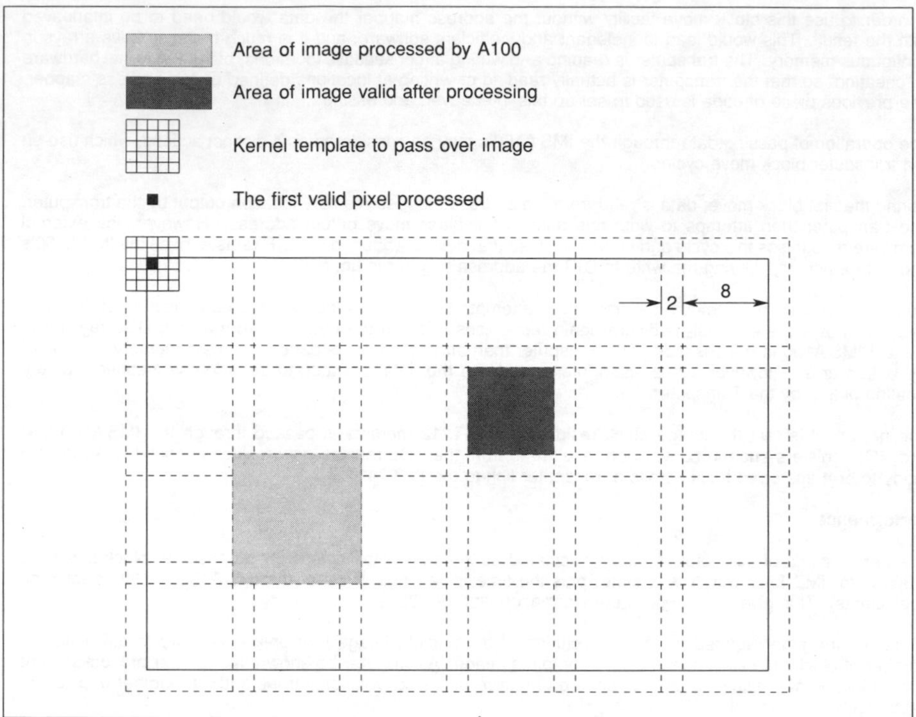

Figure 17 Image segmentation

Thresholding and scaling using software LUT

A software LUT is used to do scaling of the data output from the DOL register of the IMS A100. Thresholding has not been done though it is simple to add.

The data from the camera consists of [512 × 512] pixels each with 8 bit grey scale. Each pixel is operated upon by the convolution kernel, which may have negative components. This operation is done inside the IMS A100 and the result of the convolution for each pixel may be either negative or positive. Also, because the data is only 8 bits, the limits are −128 to +127. As these numbers are inappropriate for output to the monitor, scaling is applied to convert into a grey scale value between 0 and 255. However, if the values output by the IMS A100 are known to be positive, because all the kernel elements are positive, the output does not require scaling. The program enables the optional use of a predefined software LUT in order to create images with the maximum dynamic range.

The LUT facility can be used for other techniques such as non-linear scaling and thresholding. A side effect of the thresholding operation is further deterioration of performance. Each pixel must be read from memory, transformed through the LUT and written back to memory. This takes 800 ms (40-50 cycles) for the IMS T414, with all the data in off-chip memory. The operation in occam is

```
pixel[i][j] := table[pixel[i][j]]
```

Transfer of image across links

Data transfer across links involves the transfer of 512 Kbytes of data which will take approximately 500 ms using a single 20 Mbit/sec link. Also, because of the image segmentation method described earlier, the perimeters of each image block will in effect be transferred twice. This inefficiency is worse for large kernel sizes and small block sizes.

In the examples used in this application note the time taken for data transfer lies between 500 ms and 1000 ms.

6.5 The Demonstration Program

The following information is only relevant to users of the IMS D703 [5] software, which is available from the DSP group, based at Bristol. Readers not using this software can ignore the following.

The demonstration program can be used to execute several functions including the convolution of a single image. Images may be grabbed from the frame grabber board and processed, and the result may be displayed. Also the original images and convolution results may be stored on disc, although a lot of disc space is required at 256 Kbytes per image. An image can also be read from disc instead of from the frame grabber which is useful if images need to be processed several times. Storing away this amount of data is however quite slow. An optional post processing program is also available which transforms these grey scale images on disc into postscript format. The pictures in this document are created in this manner.

IMS D703 Development software

The program operates a modified version of the IMS D703B development software. For more information on this please consult the IMS D703 user guide and reference manual. Briefly the differences are as follows:

- The routines for accessing the matrox board are included in the software but not with the standard IMS D703.
- Routines which enable byte wide data to be transferred across the links between the IMS T212 and IMS T414 have been enabled. With the IMS D703 all data is transferred as 16 bit words both to and from the IMS T212. The 16 bit word format is directly suitable for the IMS A100's assuming the use of 16 bit coefficients. The IMS A100's process data in 400 ns in this mode.
- Because only 8 bit data is used in this application (8 bit grey scale) it would be very wasteful to transfer 16 bit words as half the data would be redundant. However, the 8 bit data must be transformed into 16 bit data with the top 8 bits held low (zero). This is because the IMS A100 does require 16 bit data even when in 8 bit coefficient mode.
- The 5 applications in the IMS D703 system have been removed, and replaced by the single application. It would be possible to add this application to the original 5 but because this application uses a lot of memory, (512 Kbytes for the image buffers alone) it was found that the 6 applications overflowed the 1 Mbyte of memory on the IMS B009-2. This may be aleviated in the future when modules with 2 mega bytes or more are available.

Injection of noise onto images

The program has a facility to add random noise onto an image. The reason for doing this is to show the benifit of large convolution kernel sizes.

Qualitatively, the effect of a large kernel is to locally average pixels surrounding a pixel point. This results in a blurring of the image but does give the effect of reducing noise. This is the analogue equivalent of a low pass filter.

Convolution kernel file

A convolution kernel file is read by the program each time it is executed. This means that many convolution kernels may be investigated without the need for recompilation of the program. An example format of this file is given below. It is only necessary to edit this file to investigate any number of convolution kernels.

The file must start with the number of convolution **KERNELS**. When the program runs, 4 kernels in this example will be sequentially executed, and the resultant image for each is displayed on the monitor. Each **KERNEL** has an optional description which is shown simultaneously while the convolution is being done. The **SIZE** of each kernel must also be given. This is used to check for correctly entered kernel elements. The **SCALE** is multiplied by each element of the kernel to give the resultant convolution kernel. The **SIGN** is used to determine the software LUT which operates on the output of the IMS A100. There are 2 possible LUTs available for the existing program, '+' which assumes that all elements in the array are positive and '–' which accepts either positive or negative integers. These LUTs are used because the IMS A100 is a 2's complement integer machine, and it is therefore necessary to know if the output from the IMS A100 requires conversion or may be assumed positive.

It is possible to add other software LUTs, for example to do nonlinear scaling and saturation control. As different convolution kernels may require different output conversions through the look-up table this attribute is made individual to each kernel.

```
KERNELS 4

KERNEL Simple filter
SIZE 3 SCALE 28 SIGN +
  ROW 1 1 1
  ROW 1 1 1
  ROW 1 1 1
FINISH

KERNEL Simple filter
SIZE 9 SCALE 3 SIGN +
  ROW 1 1 1 1 1 1 1 1 1
  ROW 1 1 1 1 1 1 1 1 1
  ROW 1 1 1 1 1 1 1 1 1
  ROW 1 1 1 1 1 1 1 1 1
  ROW 1 1 1 1 1 1 1 1 1
  ROW 1 1 1 1 1 1 1 1 1
  ROW 1 1 1 1 1 1 1 1 1
  ROW 1 1 1 1 1 1 1 1 1
  ROW 1 1 1 1 1 1 1 1 1
FINISH

KERNEL Sobel Operator Edge Detection
SIZE 3 SCALE 32 SIGN -
  ROW -2 -1  0
  ROW -1  0  1
  ROW  0  1  2
FINISH

KERNEL Sobel Operator Edge Detection
SIZE 9 SCALE 1 SIGN -
  ROW  -7 -7 -7  -4  -3  -4  0  0  0
  ROW  -7 -7 -7  -3  -4  -3  0  0  0
  ROW  -7 -7 -7  -4  -3  -4  0  0  0
  ROW  -4 -3 -4   0   0   0  4  3  4
  ROW  -3 -4 -3   0   0   0  3  4  3
  ROW  -4 -3 -4   0   0   0  4  3  4
  ROW   0  0  0   4   3   4  7  7  7
  ROW   0  0  0   3   4   3  7  7  7
  ROW   0  0  0   4   3   4  7  7  7
FINISH

FINISH kernel file
```

7 References

1 *Digital filtering with the IMS A100*, Application note 1, Hossein Yassaie, INMOS Limited, Bristol.

2 *Discrete Fourier transform with the IMS A100*, Application note 2, Hossein Yassaie, INMOS Limited, Bristol.

3 *Correlation and convolution with the IMS A100*, Application note 3, Hossein Yassaie, INMOS Limited, Bristol.

4 *Complex (I & Q) processing with the IMS A100*, Application note 4, Hossein Yassaie, INMOS Limited, Bristol.

5 *IMS D703 reference manual*, INMOS Limited, Bristol.

6 *IMS A100 data sheet*, INMOS Limited, Bristol.

7 *Index mappings for multidimensional formulation of the DFT and convolution*, Burrus C.S., IEEE Trans. on ASSP, 25:239–242, June 1977.

8 *DFT/FFT and convolution algorithms – theory and implementation*, Burrus C.S. and Parks T.W., IEEE Trans. on ASSP, 25:239-242, June 1977.

Index